Awards

Generally only Gold Medals (GM) or their equivalent are listed for the various awarding agencies throughout the world.

All-American Rose Selection (AARS)
American Rose Society Awards:
- Award of Excellence (AOE) for Miniature Roses
- American Rose Center Trial Ground Gold, Siver and Bronze Certificates
- Daved Fuerstenberg Prize
- Dr W Van Fleet Medal
- Gertrude M. Hubbard GM
- James Alexander Gamble Rose Fragrance Award
- John Cook Medal
- Miniature Rose Hall of Fame

ADR Anerkannte Deutsche Rose (Germany)
Baden-Baden GM
Bagatelle (Paris) GM
Belfast GM
Copenhagen GM
Dublin GM
Geneva GM
Genova (Genoa, Italy) GM
Glasgow (Tollcross, Scotland) GM and Lord Provost Prize
Le Roeulx (Belgium) GM
Lyon GM
Madrid GM
Monza GM
New Zealand GM
NZ Gold Star of the South Pacific
NZ Silver Star of the South Pacific
Orleans GM
Portland (Oregon, USA) GM
RINZ Auckland Rose of the Year
Rome GM
Rose of The Year, Great Britain ROTY
RNRS President's International Trophy PIT
RNRS GM
RNRS James Mason Medal
The Hauge GM and the Golden Rose
Tokyo GM
World Federation of Rose Societies (WFRS) Hall of Fame

AARS	All-American Rose Selections
ADR	Anerkannte Deutsche Rose (Germany)
AOE	Award of Excellence for Miniature Roses
ARS	American Rose Society
ARF	American Rose Foundation
B&A	Bobbink & Aitkins
B.C.	British Columbia
ca.	circa
CIOPORA	Communaute Internationale des Obtenteurs de Plantes Ornamentales et Fruitieres de Reproduction Asexuee
cms	centimeters
C-P	The Conard-Pyle Co.
dble.	double
Dr	Doctor, Docteur
EFR	Edition Francais des Roses
ft	foot, feet
GM	Gold Medal (or equivalent) awards
H&S	Howard & Smith
in	inch, inches
IRAR	International Registration Authority for Roses
J&P	Jackson & Perkins
Mlle	Mademoiselle
Mme	Madame
Mr	Mister
Mons	Monsieur
Mt	Mount
NRS (RNRS)	National Rose Society of Great Britain, (now Royal National Rose Society)
OGR	Old Garden Rose
Prof.	Professor, Professeur
ROTY	Rose of The Year
Souv	Souvenir
URS	Universal Rose Selection
WFRS	World Federation of Rose Societies

MODERN ROSES XI

THE WORLD ENCYCLOPEDIA OF ROSES

MODERN ROSES XI

THE WORLD ENCYCLOPEDIA OF ROSES

The Most Comprehensive Listing of Roses in the World
Over 24,000 roses described
Includes a comprehensive list of roses of historical and botanical importance
Old Garden Roses at Sangerhausen included in listing together with an article on this unique rose repository
Official Modern Rose Registrations filed with the International Registrar for Roses, the American Rose Society
Includes Unregistered Roses in Commerce
Report on Classifications of Roses
Report on Registration Process of Roses
Report on the History of the 'Peace' Rose

EDITOR

Tommy Cairns, PhD, DSc
Vice President, American Rose Society

ASSOCIATE EDITORS
Marily Young
Jolene Adams
Bob Edberg

ACADEMIC PRESS
A Harcourt Science and Technology Company

San Diego San Francisco New York
Boston London Sydney Tokyo

Front Cover
'Peace'
Rose breeder, Francis Meilland (Antibes, France) gave the world of roses a genetic treasure trove when he pollinated an un-named seedling derived from four prize hybrid teas with 'Margaret McGredy' on June 15, 1935 and produced the elegant and classical hybrid tea now recognized throughout the world as 'Peace'. The legacy of the action has lived on through the many cultivars who share that family tree with many progeny becoming award winners. Now 55 years after the formal introduction of 'Peace' into commerce in 1945, this variety is being heralded as holding a special role in rose history as 'The Rose of the 20th Century'.
(photo © Rich Baer)

Parentage: [('George Dickson' x 'Souvenir de Claudius Pernet') x
('Joanna Hill' x Charles P. Kilham')] x 'Margaret McGredy'
Synonyms: 'Gioia' (Spain); 'Gloria Dei' (Germany); 'Mme A. Meilland' (France)

Awards: Portland Gold Medal in 1944; National Rose Society of England Gold Medal in 1947;
All-America Rose Selection in 1946; Golden Rose of The Hague in 1965.

This book represents information obtained from authentic registration files and from past issues of Modern Roses, updated as required by changes approved by the appropriate ARS Committees. Every effort has been made to give reliable data and information, but the publisher cannot assume responsibility for the validity of all materials or for the consequences of their use. The absence of ™ or ® symbols in this publication should not be regarded as an indication that these words, designations, or names are not trademarks.

Direct all enquiries to the American Rose Society, P.O. 30,000, Shreveport, Louisiana 71130-0030

Academic Press
A Harcourt Science and Technology Company
Harcourt Place, 32 Jamestown Road, London NW1 7BY UK
http://www.academicpress.com

Academic Press
A Harcourt Science and Technology Company
525 B Street, Suite 1900, San Diego, California 92101-4495, USA
http://www.academicpress.com

ISBN 0-12-155053-2
CD 0-12-155054-0

A catalogue record for this book is available from the Library of Congress

Typeset by Bibliocraft Ltd, Dundee, Scotland
Printed in Great Britain by MPG Books Ltd, Bodmin, Cornwall

00 01 02 03 04 05 MP 9 8 7 6 5 4 3 2 1

Contents

Preface

The last such cumulative listing, *Modern Roses 10*, was restricted to official registrations as well as those older varieties still in cultivation/commerce or of botanical/historical importance. An appendix of non-registered roses had been added in an attempt to achieve a more comprehensive data base. Recognizing the need for an encyclopedic approach for this present edition, a major effort was undertaken to enlarge the listing of cultivars beyond the previous boundaries to reach international status.

To this end, the help and assistance of the World Federation of Rose Societies (WFRS) was solicited via the Conservation Committee to include those cultivars grown in other countries that were of botanical/historical importance or in danger of becoming extinct. As a result the data in the compilation has been enlarged mainly by incorporation of classical roses grown in various European gardens of historical importance. This effort has been detailed in a following report from the WFRS President, Helga Brichet and amplified by individual reports on the collections at Sangerhausen by Hella Brumme and at Cavriglia, Italy by Prof. Giancarlo Fineschi.

Furthermore, every effort has been made to list unregistered roses in commerce. Sources for this information have been the colorful rose catalogs from breeders/introducers from all over the world as well as the various compilations that often have the title '*Find That Rose*'.

Therefore, the resulting listing of cultivars in this present edition has grown dramatically accounting for the increased size of the book. With the global increase of home computers, the simultaneous issuance of a CD-ROM containing the data base with a search engine became a necessity for the ardent rose researcher who likes to follow genealogy. With this CD-ROM the average rose grower can interrogate the data base using several fields to narrow down the search for a variety with particular attributes.

To aid the reader in understanding the rose registration process, a report on the development of the registration process has been added together with reproduction of the official registration form and guidelines should anyone wish to register their own varieties. Additionally, an explanation of the current classification scheme for roses adopted by the American Rose Society in carrying out its official duties as International Registrar has been provided.

In conclusion, the data base capturing all rose registration information of the 20th century as well as information on non-registered varieties must be viewed as an evolving process rather than ever reaching the end of the project. Therefore, readers are encouraged to submit data for inclusion in the next edition. While infallibility may be a goal, it is rarely reached especially with a volume of this size and word count. Care has been taken to avoid mistakes. Should any mistakes be detected, please communicate these to the editorial staff at ARS.

As always I have been educated by the process of producing such a volume and never fail to find rose growing and its diverse avenues of interest a journey of immense enjoyment and pleasure.

TOMMY CAIRNS

Studio City
California, USA
February 2000

Acknowledgements

There can be no doubt that the assistance received from the Associate Editors contributed greatly to sharing the burden of work and the success of the final volume. These volunteers from the American Rose Society, namely Marily Young, Jolene Adams and Bob Edberg worked diligently with great passion for their hobby to achieve as near a perfect product as any human can. Other unsung heros without whose help and cooperation this volume would not have been published in time are the chairmen of the ARS Classification Committee, Dr. Anthony Liberta and the ARS Old Garden Rose Committee, Bunny Skran and their respective committee members. With the help of all these dedicated individuals this volume has been published in time for the start of the next millennium.

The rose community also owes a great deal of thanks to Helga Brichet, WFRS President who spearheaded the drive to collect information on roses of botanical/historical importance for inclusion in this volume. In this respect, I would like to recognize Hella Brumme from Sangerhausen, and Professor Giancarlo Fineschi from Italy for their willingness to share their data bases for direct incorporation into this volume.

To the many other individuals from across the world who also participated in this gathering process I thank you for your cooperation in this mammoth undertaking. In particular, I would like to thank Dr. Stefan Wagner, President of the Romanian Rose Society; Dr. Josef Thomas of the Czech Republic and Lila Weatherly from Tasmania, Australia for their dedicated interest in assisting the project.

However, I would be remiss not to mention the Editors of the various 'Rose Directories' published throughout the world. These publications served as valuable source documents as well as a verification guide to information gleaned from catalogs. To these hard working Editors I say a very special thanks: '*Rose Directory*' by Dianne Ackland from Australia; '*Find That Rose*' by Dawn Eagle from New Zealand; '*Find That Rose*' by Angela Pawsey from England; and '*Combined Rose List*' by Peter Schneider from America.

Finally, I thank my colleagues within the American Rose Society and various other National Rose Societies for their continued support. This labour of love often took me to the highest point of enjoyment only to be immediately followed by the lowest until problems were resolved successfully. Only with the help of my friends in the rose world was the publication of this volume possible and stands as a testimony to all their volunteer spirit, support and encouragement.

T.C.

The Classification of Roses

by Dr. Anthony Liberta and Marily Young
ARS Classification Committee

Over the past 100+ years, since Darwin's work on the origin of species, Mendel's experiments in genetics, and Linnaeus' development of a common naming system, science has studied the family relationships among all organisms. From the most general categorization as either 'plant' or 'animal', down to the detail of parent and child, zoologists and botanists have attempted to organize all living things into schemes of classification. This overall study of the kinds and diversity of organisms along with their inter-relationships is called Systematics. Taxonomy is the discipline within Systematics which focuses on building systems of classification; these systems are intended to divide organisms into groups (taxa) based on a set of defined characteristics that is shared by all members of the group.

To the average hobbyist rose grower, however, Systematics and Taxonomy, with their resulting classification systems, seem to have little to do with the day-to-day pleasures of gardening or with the requirements for feeding, watering, disease-prevention, or winter protection. In reality, a system of classification has everything to do with the rose garden! Being able to place organisms in groups based on shared characteristics allows both amateurs and professionals alike to apply information that they discover about one organism (i.e., one rose cultivar) to other related organisms. Such utilization of shared information can be applied to the care of the rose cultivar, the monitoring of its health, its use in breeding programs, and its ultimate survival in our gardens.

The study of taxonomy, and the building of appropriate classification schemes for the genus Rosa, are based on the idea of common ancestry among members of the genus. All members of Rosa are part of the family called Rosaceae, which also includes other shrubby, thorny, plants like raspberries. The development and evolution of Rosa, while still subject to much conjecture and debate, is becoming more clear as modern cytologists and molecular geneticists examine the relationships among members of the genus at the most elemental level.

There is, today, general acceptance of division of the genus into 4 sub-genera: Hulthemia, Platyrhodon, Hesperhodos, and Eurosa. All four are native to the northern hemisphere. The first three sub-genera, while historically important, have been involved very little in the heritage of today's garden roses. These groups include such species as Hulthemia persica, Rosa roxburghii, and Rosa stellata. The Eurosa, on the other hand, are the ancestral basis of most roses that gardeners know today.

The Eurosa are divided into 10 sections, many of which have contributed substantially to the genealogy of modern roses. The 10 sections and their general geographic ranges are as follows:

Banksianae – southeast Asia
Bracteatae – southern Northern America
Caninae – Europe and the mid-east; includes R. canina (and the descendant Albas), R. eglanteria, and R. glauca.
Carolinae – native to the south-eastern U.S.; includes R. nitida and R. palustris
Chinensis (Indicae) - native to southeast Asia; includes all precursors of the China and Tea hybrids
Cinnamomeae – widely distributed across northern North America, eastern Europe and most of China; includes R. rugosa.
Gallicanae – native to northern Europe, Mongolia and China.
Laevigatae – southeast Asia
Pimpinellifoliae – native to Europe, Mongolia and China; includes R. foetida and its derivatives, which are primarily responsible for the yellow and orange hues in modern roses.
Synstylae – broad, diverse range across Europe, south-east Asia, and northeast North America; includes arvensis, filipes, moschata, multiflora, sempervirens, setigera and wichurana species and forms.

When developing a classification system that recognizes the above genealogical groupings and is useful to botanists, hybridizers and commercial nurserymen, many different perspectives must be reconciled. Such a system must retain its scientific basis, but must also allow the rose gardener to visit his local retailer and know that, should he purchase a Floribunda, he will get a repeat-blooming rose which bears its flowers in clusters, grows as a moderate, bushy plant, and will require protection from winter in certain climates. Furthermore, it must recognize that the strictly botanical system which is appropriate to species roses and old garden roses is not suitable for complex modern hybrids.

The classification system in use by the American Rose Society attempts to satisfy all of the above requirements. However, ongoing study and modification of this system illustrates the inherent nature of all classification systems.

They are not meant to be static and unchanging, but must remain flexible enough to incorporate the latest developments and available knowledge about the type of plant being studied. As the International Registration Authority for Roses, the American Rose Society bears the additional responsibility for gathering educated opinions from around the world and incorporating them where appropriate. In general, it has been the policy of the ARS to use the class or type designated in the registration forms for a given cultivar.

Within the American Rose Society the Classification Committee, whose chair is appointed by the ARS president, is the reviewing body for the classification system. This committee is obligated to keep itself informed of developments in knowledge about the genus Rosa and to make appropriate recommendations to the ARS Board of Directors. The editors of various ARS publications, including *Modern Roses XI*, have no authority to assign or change rose classifications. In addition, we must always recognize that for each 'expert' who presents an opinion on this subject, there will be one or more equally qualified 'experts' who disagree.

Finally, the rules and considerations which apply to precedence of naming must be taken into account where possible. In past editions of Modern Roses, whose audience consisted primarily of rose enthusiasts in the United States, cultivars were listed according to the name used in the U.S. This does not conform to the International Code for naming plant cultivars, but it has not been possible to make the wholesale changes required to strictly comply with the code. The editors of Modern Roses welcome all information which will assist in correcting this situation in future editions.

Classification Characteristics

Species Roses

Species roses are usually single-petalled (5–12 petals), once-blooming and wide-ranging in size. Various species occur naturally across the entire Northern Hemisphere.

Old Garden Roses

In 1966, the American Rose Society defined an 'old garden rose' as any rose belonging to a class which was in existence prior to 1867 (the year of introduction of the first Hybrid Tea). Within the Old Garden Rose group, a number of subdivisions exist and are based on natural historical developments and physical characteristics. The OGR classes can be grouped most logically as either once-bloomers or repeat bloomers, though exceptions exist in each collection.

Once-Blooming OGR Classes

- Alba
- Ayrshire
- Boursault
- Centifolia
- Damask
- Hybrid Bracteata
- Hybrid Englanteria
- Hybrid Gallica
- Hybrid Sempervirens
- Hybrid Setigera
- Hybrid Spinosissima
- Miscellaneous OGRs

Repeat-Flowering OGR Classes

- Bourbon and Climbing Bourbon
- Hybrid China and Climbing Hybrid China
- Hybrid Foetida
- Hybrid Multiflora
- Hybrid Perpetual and Climbing Hybrid Perpetual
- Moss and Climbing Moss
- Noisette
- Portland
- Tea and Climbing Tea

Modern Roses

The era of modern roses was established in 1867 with the introduction of 'La France', the first Hybrid Tea. Since then, more than 10,000 Hybrid Teas have been introduced, along with entries in the even newer classes of Polyantha, Floribunda, Grandiflora, Miniature, Mini-Flora, Shrub, Large-Flowered Climber, and Hybrid Wichurana. Most, but not all, modern roses provide recurrent bloom.

Other Classification Systems

Although the American Rose Society serves as the International Registration Authority for Roses, other organizations around the world have adopted different classification systems using different terminology. Most widely known among these alternative systems is that used by the World Federation of Rose Societies and accepted by many of its member countries. This system uses a family tree divided first into three major groups: Wild Roses, Old Garden Roses, and Modern Roses. Within these groups are subdivisions based strictly on characteristics of growth and bloom, including climbing or non-climbing habit, recurrent or non-recurrent flowering, and clustered or non-clustered flowering.

Summary

Classification of roses has long been a topic for discussion. The debate over how best to preserve the description of botanical relationships while providing commercial benefit to the gardening public is intense and ongoing. Members of the world-wide rose community, no matter what their particular role, need a common classification to enable data sharing and communication. In other words, we need a system which gives us all a common 'language'. We have not yet achieved a perfect system, and, perhaps we never will. But with the dedicated work of rose enthusiasts and scholars, we can constantly improve our classification systems, improve our communications, and enhance our opportunities to enjoy the world's favorite flower.

Genus Rosa Family Tree

GENUS ROSA

SPECIES ROSES

- Species (Sp)

OLD GARDEN ROSES

- Alba (A)
- Ayrshire (Ayr)
- Bourbon & Climbing Bourbon (B)
- Boursault (Bslt)
- Centifolia (C)
- Damask (D)
- Hybrid Bracteata (HBc)
- Hybrid China & Climbing Hybrid China (HCh)
- Hybrid Eglanteria (HEg)
- Hybrid Foetida (HFt)
- Hybrid Gallica (HGal)
- Hybrid Multiflora (HMult)
- Hybrid Perpetual & Climbing Hybrid Perpetual (HP)
- Hybrid Sempervirens (HSem)
- Hybrid Setigera (HSet)
- Hybrid Spinosissima (HSpn)
- Miscellaneous OGRs (Misc. OGR)
- Moss & Climbing Moss (M)
- Noisette (N)
- Portland (P)
- Tea & Climbing Tea (T & ClT)

MODERN ROSES

- Floribunda & Climbing Floribunda (F & Cl F)
- Grandiflora & Climbing Grandiflora (Gr & Cl Gr)
- Hybrid Kordesii (HKor)
- Hybrid Moyesii (HMoy)
- Hybrid Musk (HMsk)
- Hybrid Rugosa (HRg)
- Hybrid Wichurana (HWich)
- Hybrid Tea & Climbing Hybrid Tea (HT & Cl HT)
- Large-Flowered Climber (LCl)
- Miniature & Climbing Miniature (Min & Cl Min)
- Mini-Flora (MinFl)
- Polyantha & Climbing Polyantha (Pol & Cl Pol)
- Shrub (S)

Rose Registration Process

by Dr. Tommy Cairns

Overview

Today there exists some confusion in the minds of the average rose growers throughout the world why two different rose varieties with exactly the same name can be sold. The reasons are simple. Most breeders prefer to officially register their roses using a unique generic 'variety denomination'. That denomination is often a name, difficult to pronounce, constructed according to the code system supported by CIOPORA (**C**ommunaute **I**nternationale des **O**btenteurs de **P**lantes Ornementales et Fruitieres de **R**eproduction **A**sexuee). At the same time the breeder makes a formal announcement of their intended fancy commercial names under which the variety will be sold. Unless those fancy names are trademarked, there is no legal protection against using such a fancy name over and over again. For instance, Kordes in Germany has introduced 'Valencia' and 'Liebeszauber' several times decades apart as is their right!

As we enter the next millennium, there is comfort in the knowledge that the naming and protecting of 'variety denominations' has quickly evolved by taking a few giant steps in the right direction. While the solution to preventing two roses with the same fancy name from occurring again is by no means completely resolved, the designation of the 'variety denomination' for registration purposes should help proper identification. To understand this tale of nomenclature, the following brief explanation should assist the reader in the complexities of the process of naming to preserve valuable data for future generations of rose growers.

Introduction

In 1913 the American Rose Society pioneered the process of recording new rose varieties from all over the world. By 1916 it had successfully promulgated registration guidelines and began the long tradition of publishing such registrations. In 1921 the *American Rose Annual* contained a feature section called 'New Roses of The World' which has continuously appeared every year into the modern era. Then in 1930 the late Dr. J. Horace McFarland, then Editor of the Society's publications, began to compile lists of roses names and descriptions which he published under the title of *Modern Roses*. Each subsequent edition of *Modern Roses* built on the initial policies and procedures of including only those varieties currently in cultivation. In the 1958 edition, *Modern Roses V*, the goals had been enlarged to also include all the names of species and varieties presently in cultivation or of current interest together with many varieties of historical and botanical importance.

However, all editions up to and including *Modern Roses 10* have been published strictly using the fancy name for intended use by American rose growers. Names were always directly applicable for the United States while the rose might enjoy other names in other countries. For instance, 'Peace' was the fancy name in America while in France it was called 'Mme A. Meilland', in Germany it was sold under the name 'Gloria Dei' and in Spain as 'Gioia'. These other names are often referred to as synonyms. Unless protected by trademark or registered, the names could in theory be used by another grower or introducer. So today many breeders rely heavily on trademarks to protect the assigned fancy names of their roses. Without such protection the names could be re-used in the commercial environment. A good example of this situation is the name 'Lady in Red'. Originally registered as a miniature rose with IRAR, the name has recently been re-used to describe a new red hybrid tea by Jackson & Perkins for sale in New Zealand.

With the growing popularity of roses throughout the world, the 14th International Horticultural Congress (IHC) meeting in Scheveningen, Holland in 1955 appointed the American Rose Society as International Registration Authority for Roses (IRAR). This act by the IHC was a solid declaration for the need for a single central focus for the registration system to prevent duplication of 'variety denominations'. ARS subsequently accepted the role and more importantly the responsibility and implementation. It was clearly stated for 'the good of the industry and botanical and horticultural science that all hybridizers or nurserymen register the names of their roses with one of the many National Registration Centers set up in association with various National Rose Societies. During the years since 1955 the ARS has continued to approve the registration of a 'variety denomination' ensuring that no duplication was permitted. However, as the regulatory authority ARS had no compliance or enforcement policy should an independent grower wish to market a rose under a name already registered but unprotected.

Around 1958 it became apparent that controlling the name of the rose was a much more complex process. In spite of guidelines and controls, there still existed the possibility that different roses with the same name could be sold to the public. On this occasion, the breeders took matters into their own hands and developed a simple system of unique coding to avoid duplication at least in the registration process. Code names were simply a combination of the first three letters of their name or company in capital letters followed by some additional letters usually derived from the slang name given the rose during it's evaluation before introduction. In this way it became easy to identify the roses from a specific breeder, for example KOR for Kordes in Germany, MAC for McGredy, HAR for Harkness in the UK, MEI for Meilland in France, SIM for Nola Simpson from NZ etc. While such code names were included in registration submissions for approval by ARS, they were not included with the descriptions of roses in *Modern Roses* until the edition published in 1986.

In 1978 the Convention of the International Union for the Protection of New Plant Varieties in Geneva, Switzerland voted for the notion that each new rose variety registered be given a unique descriptor, the code name, to clearly identify the plant whatever the final name given the rose for sale to the public. This code name was intended to become the designated 'variety denomination' within the formal registration process. In 1981 the ARS was appointed to serve as the Registrar for Communaute Internationale des Obtenteurs des Plantes Ornementales et Fruitieres de Reproduction Asexuee (CIOPORA). Then in 1985, the US Patent & Trademark Office, proposed rule making in the *Federal Register* to implement the consensus reached by the Convention in Geneva. However, they fully intended to adopt the registered names listed in the latest edition of *Modern Roses* which, in many cases, were popular fancy commercial names and not the unique 'variety denomination' (i.e. usually the code name). To correct this situation, the ARS in combination with American growers and CIOPORA moved to change their registration strategy by declaring the code name as the 'variety denomination'. Within this agreement the breeders agreed that they would indicate on the registration application which cultivars were patented or trademarked under a fancy name. The new registration form was approved for use in January 1998.

These actions caused the American Rose Society (ARS) to redirect its policy with regard to exhibition names for use in their official rose shows. Generally the show name was the registered name, usually a fancy name. Therefore, with the current changes the show name would now become the 'variety denomination', i.e. the majority of show names would be code names. Such a situation would have been unacceptable due to the lack of understanding by the public of almost un-pronounceable names such as 'WEKfrag' instead of 'Ain't She Sweet'. To correct this future catastrophe, ARS quickly moved to identify the fancy name as the ARS Approved Exhibition Name in bold typeface within the published registration file.

For the Future

In the registration of a new variety, the unique code name will automatically become the 'variety denomination'. For most registrations, this guideline will be directly applicable. In most instances, when an amateur breeder registers their rose the lack of a code name places the given fancy name as the 'variety denomination'. Trademarks under which the variety enters commerce will be annotated only at the request of the breeder.

Arrangement of Data

Official IRAR Registration Entries

Listing of main entries will generally be by 'variety denomination' or code name (i.e. the cultivar name appears in single quotes with the first three letters in capitals indicating breeder and the remainder in small caps) with the ARS exhibition name identified in bold typeface carrying a trademark symbol if requested by the registrant (e.g. 'ANTopp'). The ARS approved exhibition name will appear in bold face type.

'WEKPLAPIC', F, rb, 1999; (**Betty Boop**™); bud pointed; flowers rosy edged, yellow at base when fresh, fading to white; bright stamens, single, 6–12 petals, 4 in., borne in small cluster of 3 to 5; floriferous, moderate, fruity fragrance; prickles moderate; foliage medium, dark green, glossy, dark red new growth; rounded, bushy, medium (4 ft) growth; PP10126; AARS, U.S., 1999; [Playboy X Picasso]; Carruth, Tom; Weeks Roses, 1999

In a small number of cases the lack of a code name will automatically default to the fancy name becoming both the registered cultivar name (single quotes, small caps) and the ARS exhibition name (bold), e.g. 'ADRIENNE BERMAN'.

'AGNES WINCHEL', HT, pb, 1989; bud pointed; flowers light pink with deep pink border, medium, dbl., 28 petals, exhibition form, borne singly, slight, fruity fragrance; globular, medium, orange fruit; prickles slightly hooked, medium, green; foliage medium, dark green, semi-glossy; upright, medium growth; PP007826; [Dorothy Anne X Seedling]; Winchel, Joseph F.; Coiner Nursery, 1990

To locate the data using a fancy name or ARS exhibition name, a cross reference will be provided to the main entry under code name or registered cultivar name, e.g. Ain't She Sweet.

Ain't She Sweet™ *see* 'WEKFRAG'

INSTRUCTIONS

FOR

COMPLETING THE IRAR REGISTRATION FORM

To avoid delays, be sure to submit all requested information. Incomplete applications will be returned.

ITEM 1.

CULTIVAR NAME/DENOMINATION OF THE VARIETY WHICH IS PROPOSED FOR REGISTRATION:

Please enter here the generic name of your rose variety, whether in the form of a coded denomination of a fancy denomination. If you wish to use a coded denomination, and have not previously been assigned a three-letter prefix for inclusion the Cultivar Name/ Denomination, contact the IRAR. In all cases the generic cultivar name/denomination in Item 1 will be the IRAR registered name/ denomination for the registered variety.

DO NOT place any trademark information in item.1 (see Item 7).

Please note the following information to facilitate your registration:

If you are proposing to register the rose variety with a generic name/denomination which includes the surname of an individual or the registered name of an organization, please submit a letter of authorization from said individual (or appropriate family member) or organization.

- Cultivar names/denominations must be different from any other rose name/denomination.
- Cultivar names/denomination must not be a translation of the name/denomination of any other rose variety.
- Cultivar names/denominations must not exaggerate the merits of a rose variety.
- Cultivar names/denominations must not be the same as the botanical or common name of another plant species.
- Cultivar names/denominations must not include the words cross, hybrid, grex, group, maintenance, seedling, mutant, selection, sport, or strain, or the plural forms of these words, improved and transformed, (or equivalents in any language).
- Cultivar names/denominations may not contain punctuation marks except for the apostrophe ('), the comma (,), a single exclamation mark (!), the hyphen (-), or the period or full-stop(.).
- Cultivar names/denominations must consist of no more than 10 syllables and no more than 30 letters or characters overall, excluding spaces and the demarcating marks.

It is preferred that the coded denomination be pronounceable.

No other condition than those referred to above are imposed for the formation of rose variety names/denominations identifying rose varieties.

ITEM 2.

SYNONYMS:

Specify all non-proprietary synonyms and the countries where each is used.

DO NOT place any trademark information in Item 2. (see Item 7)

ITEM 5.

DESCRIPTION:

ARS HORTICULTURAL CLASS:

Use the list published by the ARS as a guideline to determine the class. Note: If the classification proposed is not listed on the application, enter it after 'Other'.

COLOR:

Indicate preferred ARS Color Classification (see attached list). Use common color names such as white, yellow, orange, orange-red, pink, orange-pink, red, mauve, etc. For each color, use a prefix of light, medium, or dark. Describe both the color of the upper petal surface (the surface seen when looking down into a typical, fresh, open bloom) and the under petal surface (reverse). Information from RHS Color Charts may also be indicated. Enclose a color photo or slide of the new rose variety. If the actual color of the rose variety differs from the photo/slide, describe the differences(s). Photos and descriptions will enable the IRAR to register and describe your rose variety more accurately.

PARENTAGE:

Specify, if known. If the parents are unnamed seedling, specify the class and color of each parent, if known.

SPORTS (MUTATIONS)

Information must be provided as to the variety of origin and as to how this sport (mutation) is distinct from its variety of origin or other sports/mutations of the same variety of origin which are known to the applicant. For a more accurate description, applicants are invited to supply photos or slides of the variety for which the registration application is being submitted. Applicants may also contact the IRAR for assistance in this determination before submitting the application.

ITEM 7.

TRADEMARK(S):

Breeders or introducers who are using trademarks for the promotion and marketing of the above-identified rose variety are invited to indicate them here, using the proper symbols: ™ for non-registered trademarks and ® for registered trademarks. Any relevant information so that anyone concerned may be informed of their proprietary nature and of the obligation to obtain the trademark owner's permission as to their commercial use may also be supplied.

ITEM 8.

SIGNATURE/DATE

Application must be signed by the Breeder or Introducer and dated.

The IRAR will assist you with any questions you may have.

AMERICAN ROSE SOCIETY

IRAR,

P.O. Box 30,000,

Shreveport,

Louisiana 71130-0030 USA

Voice: (318)938-5402

FAX: (318) 938-5405

E-Mail: ars@ars-hq.org

International Registration Authority for Roses (IRAR)
Application for Registration of a New Rose Variety

Please refer to the instructions on the reverse side of this form before completing each item. Incomplete forms will be returned for additional information. This form may be reproduced locally. All previous editions are obsolete and may not be used.

Send completed forms to: **IRAR, PO Box 30,000, Shreveport, Louisiana 71130-0030, USA**

1. Denomination/Cultivar name of the Variety which is proposed for the registration: [do not list Trademark(s) here]

2A. Synonyms, if any: [do not list Trademark(s) here]	2B. Countries of use:

	3A. Name/Address of Breeder:	3B. Name/Address of Introducer:
NAME		
ADDRESS		
PHONE		
FAX/E-MAIL		

4. Probable First Introduction:		Country:		Year:	

5. Description: For each category, please CHECK, CIRCLE or FILL IN the appropriate information:

ARS Horticultural Class: ☐ Hybrid Tea ☐ Grandiflora ☐ Floribunda ☐ Miniature ☐ Shrub Other: ☐

Flower: (See Instructions: submit color photo or slide, if available)

Colour: Preferred ARS Color Class: ☐ Upper Petal Surface: ☐ Under Petal Surface ☐

Petalage: ☐ Single (4-11) ☐ Semi-dbl. (12-16) ☐ Double (17-25) ☐ Full (26-40) ☐ Very full (41+)

Size: (Relative to other rose varieties in the same class): ☐ Small ☐ Medium ☐ Large Inches/Centimeters: ☐

Fragrance: ☐ None ☐ Slight ☐ Fragrant ☐ Very Fragrant

Blooming Habit: ☐ Mostly Solitary ☐ Small Clusters ☐ Large Clusters

Prickles: Size: ☐ Shape: ☐

Quantity: ☐ None ☐ Few ☐ Moderate ☐ Many

Foliage:

Size: (Relative to other rose varieties in the same class) ☐ Small ☐ Medium ☐ Large

Color: ☐ Light Green ☐ Medium Green ☐ Dark Green Other: ☐

Surface: ☐ Dull/Matte ☐ Semi-glossy ☐ Glossy Other: ☐

Growth Habit:

Ground: ☐ Upright ☐ Compact ☐ Spreading ☐ Bushy Other: ☐

Height: ☐ Low ☐ Medium ☐ Tall Inches/Feet/Cm/Meters: ☐

Use of variety: ☐

Parentage:

Seed (Female): ☐

Pollen (Male): ☐

Sport (Mutation of): ☐

6. Other Important Characteristics, if any:

7. Trademarks used in commerce by applicant with respect to above named variety: *(The providing of this information is optional)*

8. Signature:		Date:	

For Official IRAR Use Only:

Registration Number: ____________ NRC: ____________ Date Submitted: ____________ Date Approved: ____________

Final Decision: ☐ Approved ☐ Denied ☐ Resubmit

Committee Member Decision: ☐ Approved ☐ Denied Reason for Denial: ____________

ARS Exhibition Name: ____________ Hort. Class: ____________ Color Class: ____________

Publications: ARM (Year/Month): ____________ ARA (year): ____________ Supplement (year): ____________ MR (Edition) ____________

The Conservation Role of The World Federation of Rose Societies

by Helga Brichet
WFRS President & Chairman, Conservation Committee

The World Federation of Rose Societies (WFRS) was founded during an International Rose Conference in London, England in 1968. The First World Rose Convention was held in Hamilton, New Zealand in 1971, after which the nine founding Member National Rose Societies released a press statement giving notice of the formal establishment of the Federation, the approval of its Constitution and Rules, and the election of its principal officers. It's main goals focus on the exchange of information and cooperation amongst Members but also the coordination of rose classification, registration and the judging of new seedlings as well as sponsoring research in matters regarding the rose.

In the intervening thirty-odd years World Rose Conventions have been hosted regularly every two to three years by Member Societies, generally alternating between those staged in Europe and elsewhere. Membership has increased rapidly so that today 34 National Rose Societies from all the continents make up the WFRS.

Since its inception the WFRS had rightly considered the Classification and Registration Committees to be of cardinal importance. However, it was not until the 1988 World Rose Convention in Sydney, Australia, that the idea of forming a committee to report on 'Research, Conservation and Environment' was born. At the same meeting it was hoped that the WFRS would collaborate in some way to assist the Editors of *Modern Roses* published by the American Rose Society (ARS) into becoming a truly comprehensive reference work.

During the Roseworld '94 Convention in Christchurch, New Zealand, the Chairman of the Committee called an informal meeting to clarify the aims of what would henceforth be called the 'WFRS Conservation Committee' – to seek out rare and historically important roses and ensure that they are not lost to the world.

At the 1997 Benelux Rose Convention the WFRS Conservation Committee invited the representatives of the various European rose data bases to explain the workings of each system and the progress which had so far been made. The 'Roseto di Cavriglia' and the 'Europa-Rosarium Sangerhausen' had both established data bases to accomodate their famous collections. The 'French Network', based at the research institutes of INGRA-GEVES, visualised cooperation between private and institutional collections, breeders and rosaries in France, and at a later moment aimed at attracting European partners in a joint venture. Unfortunately, due to a lack of EC funding, this project has been forced to a halt. Many National Rose Societies too had records of the varieties to be found in their own gardens. The WFRS Conservation Committee thus agreed that a small group of experts should look into the programme as a whole, in order to coordinate the European initiatives and, hopefully, expand it into an international operation.

In November 1997, the Specialized Conservation Committee met in St. Albans, UK for the first time. WFRS Members represented were the American Rose Society, the Royal National Rose Society, the Norwegian Rose Society and the Italian Rose Society plus INRA (France) and the rose gardens of Cavriglia and Sangerhausen. After a review of the major existing data bases, it was decided to entrust their technical merger to the experts concerned. It was also resolved to ask all WFRS Member Societies to participate in a joint project to gather information in their respective countries. Each National Rose Society was requested to contact all known rose gardens and rosarians and to check the lists of content received against the varieties listed in *Modern Roses*. For each rose not found in this compilation a separate information form was to be returned to the owners/curators concerned soliciting a comprehensive description. The eight Vice Presidents of the WFRS were asked to be responsible for collecting all this information from the countries within their regions. In this manner each National Rose Society became responsible for the construction of its part of the world's rose mosaic giving notice not only of 'endangered varieties' but also of their location. All information would then be combined to constitute an International Rose Data Base, the beneficiaries of which would not only be the Rose Societies but all rose lovers.

Two further meetings of the Specialized Committee have been held; at St. Albans in August 1998 and at Sangerhausen in August 1999. The Scientific Advisor to the Royal National Rose Society and the Presidents of the Czech and German Rose Societies also participated. All incoming data, notably from the comprehensive collections at Cavriglia and Sangerhausen was added to that of *Modern Roses*. The ultimate inventory of the collection at L'Hay-les-Roses is, for the moment, being treated with caution pending the results of an internal commission verifying the garden's present content. At the Sangerhausen meeting the ARS editorial staff could finally predict the publication of all information so far assembled for the spring of 2000 in the form of this present volume.

The very nature of the International Rose Data Base is at the same time urgent and long term; the timetable for future editions is already in preparation. Much depends on the assistance and support, not only of the large collections, institutes and societies, but also on the active participation of individuals, be they professional or amateur, who have the well-being of the rose at heart.

Should anyone with this volume in hand feel able to help trace and preseve those varieties lost or in danger of extinction, the Specialized Conservation Committee would be most grateful for your collaboration.

The Europa-Rosarium at Sangerhausen
A place of pilgrimage for rose lovers from all over the world

by Hella Brumme
Curator, Sangerhausen

The small city of Sangerhausen, located in the approximate center of Germany has, for the last 100 years, had something unprecedented to offer visitors – the Europa-Rosarium with the most important rose collection in the world, a place of pilgrimage for many thousands of garden and rose lovers every year. Even in ancient Rome, roses were planted in a 'Rosaria'. In comparison to the commonly loved rose gardens, Rosariums are living museums, where the development and cultivation of roses can be vividly followed through into the present.

In the Rosarium at Sangerhausen, roses from all countries and periods have been brought together. The current inventory of approximately 6800 different roses mirrors in a near unbroken sequence the development of the modern garden rose. Near the end of the nineteenth century, the initiative of the rose breeder Peter Lambert, German rose lovers passed a resolution to establish a Rose Society. Old roses, threatened by extinction and those forgotten, were to be gathered, reintroduced and maintained. At the instigation of the active rose lovers Albert Hoffmann and Prof Ewald Gnau from Sangerhausen, the city of Sangerhausen provided a piece of land to their disposal. The garden architect Friedrich Dorr developed a design for this area in the formal garden style. Here was the place where the Hybrid Tea rose cultivars of the time were set out.

Even before the completion of this first area, the rose breeder and garden architect Peter Lambert designed a park extension in the natural landscape style. The rose gardener Richard Vogel from Frankfurt/Main was summoned to oversee the practical construction work. In 1903, on the occasion of the German Rose Congress, Sangerhausen's Rosarium opened its gates for the first time. During the first summer visitors could admire 2000 different varieties of roses on an area of land of approximately 1.5 hectares.

As a result of a generous bequest from a citizen of Sangerhausen, who emigrated to America, the area of the Rosarium was extended several times up to the beginning of the first world war. The assortment of roses grew, and conferences and exhibitions were held. After first exhibiting his world class collection of wild roses from Europe, Asia and America at the Rosarium L'Hay-Les-Roses during the World Exhibition in Paris, the botanist George Dieck brought his collection to Sangerhausen in 1909. To further develop the park, a tree nursery from Holstein sent trees and shrubs to the Rosarium. Among them were many rare and unusual species which have made the Rosarium into a dendrologically valuable park with 350 different types of trees and shrubs. Today they provide a structural green background for the typical rose pyramids of the 'once in a lifetime' flowering climbing roses.

It is thanks to Ewald Gnau, one of the founders of the Rosarium, that the Rosarium stayed unscathed during the confusion of the first world war and the following world economic crisis. His convincing publicity campaigns secured the ever so necessary and important financial support. In the year 1935 a Central Office for Rose Research was established as part of the Rosarium which, until 1944, was headed by the scientist Dr. Harald von Rathlef. During the second world war, under the direction of the rose gardener Max Vogel, the rose collection was maintained with great difficulty and only under the most makeshift of conditions. At the end of the war, by order of the Soviet Military Administration, several thousand roses were bred to be sent to Moscow and Kiev as a war indemnity. The valuable library of the German Rose Society was also lost during this time.

After the founding of the German Democratic Republic, the German Rose Society was no longer allowed to be active. The Rosarium was placed under the management of the City of Sangerhausen. In the year 1949 the city administration appointed Hans Vonhold as the head of the Rosarium. He and his assistant, the landscape contractor Paul Tackelburg completed and organised the rose collection. It was not long before the collection could be presented and seen in an organized form on 12.5 hectares of land. Even the August Jager Rose Lexikon published in 1960 mentions the great initiative of these two professionals.

Undaunted by difficult travelling conditions, both rose lovers and professionals world wide, came to Sangerhausen last but not least to see the old rose varieties. In 1962, Paul Tackelburg drew up the first assortment catalogue of the Rosarium. The current available edition is a print of our databank and serves as an important work of reference for rose lovers. From the middle of the 1970's, under the direction of Ingomar Lang, the constructively difficult entrance area was redesigned and many small renovations were carried out throughout the Rosarium. The Rosarium has now developed into a major cultural attraction for the former mining region around Sangerhausen. Since the German unification, its international importance has greatly increased. In the year 1993, on the occasion of it's ninetieth anniversary, the Rosarium received the recognition a 'Europa-Rosarium Sangerhausen'.

For the city of Sangerhausen, the maintenance for the grounds of the Rosarium is an obligation with an extremely high financial expenditure. Investments into a Rosarium which plans for the future, require additional (supplementary) financial resources. Through government aid from the province of Lower Saxony the maintenance of the collection has improved significantly. Special gardens, an information pavilion, a small museum, an exhibition hall and a large number of information panels about roses make a round-tour of the present Rosarium more exciting than ever before. An important obligation of the Rosarium is the preservation of the many species and varieties of roses. For this reason, 8000 roses are propagated every year to compensate for any losses. Associations and contacts to other Rosariums, rose breeders and rose lovers world wide, help keep the rose collection complete. The Rosarium of today can be termed as an unprecedented gene bank for roses.

The most important cultivated forms of 6800 different roses are divided into the following groups:

25	Alba Roses
55	Bengal Roses
65	Bourbon Roses
40	Centifolia Roses
200	Climbing Roses
650	Climbing Roses of various descent
30	Damask Roses
780	Floribunda Roses
160	Gallica Roses
80	Ground Cover Roses
225	Hybrid Polyantha Roses
2380	Hybrid Tea Roses
310	Modern Shrub Roses
130	Moss Roses
55	Noisette Roses
170	Pernetiana Roses
280	Polyantha Roses
465	Rambler Roses
80	Rugosa Roses
115	Tea Roses
500	Wild Roses

The unsurpassed speciality of our Rosarium obliges us to plan for the future. As a result, the city of Sangerhausen, as owner of the Europa-Rosarium, has brought in the landscape architects Dane Landschaftsarchitekten of Weimar, Germany, to draw up a complete Masterplan development concept for the future of the Rosarium. The concept includes a Park Maintenance Program for the existing part of the Park and a new design concept for the 2.5 hectare extension area. The realisation of these projects could fulfil the wishes of rose lovers throughout the world.

In the second century of its founding, the rose gene bank of the world will finally receive the framework it deserves and can look forward to being a place of pilgrimage for rose lovers from near and far, well into the next millennium.

The Botanical Rose Garden 'Carla Fineschi'

by Professor Gianfranco Fineschi
Cavriglia, Italy

This garden is situated in the small municipality of Cavriglia, in the province of Arezzo, central Tuscany, Italy. Its creation dates from 1967, due to the initiative of Prof. Gianfranco Fineschi, don in orthopaedic science, and his wife Carla. The owners' initial desire was to save the approximately 50 rose varieties growing in the garden surrounding the villa inherited from their parents. These roses had been planted during the period 1850 to 1910 and were no longer to be found in nurserymen's catalogues. Thus the idea had come to mind to preserve them after the fashion of museum pieces. From there it was a short step to envisaging a project to safeguard the greatest possible number of those rose species and varieties considered rare and 'endangered' in Europe and, indeed, elsewhere.

Additional impetus was given to the project, so deemed all the more urgent, by the fact that, at the time, rose breeders were turning almost exclusively to modern varieties for their hybridizing programmes in order to keep up with the market's incessant demand for novelties. The result would naturally have been an impoverishment of the bio-diversity within the genus Rosa. Fortunately at Cavriglia two indispensable elements were already at hand – the agricultural land around the Fineschi homestead was sufficient for the planned collection, and the local climate, with four distinct seasons, was considered ideal for the growing of roses.

The realization of the garden, preceded by intense scientific research, proved to be extremely onerous and continues to be so, given the maintenance of almost 7000 plants as well as the annual additions of a large number of the latest varieties in commerce. However the 'Rose Garden Carla Fineschi' has achieved its purpose and has become what many call a living museum in which the genus Rosa is presented from an historical-botanical point of view, displaying the ten sections of the sub-genus Eurosa which are again divided into almost thirty species and their first hybrids. Modern varieties, those created during the last 150 years since man put his hand to selective breeding, are grouped under the breeder's name, and in many cases, generations of family hybridizing can be admired. Others still are assembled according to their country of origin. Thus this museum offers its many hundreds of annual visitors the possibility of discovering and enjoying the Rose from their own personal point of view.

It must be stressed that, although a museum, this garden is a living one. The inevitable circle of life brings with it the risk of losing this or that component of the collection. Fortunately, the close collaboration which exists between the most important European rose gardens manages often, but not always, to avoid catastrophy.

Naturally it is gratifying that the 'Rose Garden Carla Fineschi' should today find itself in the company of the gardens at L'Hay-les-Roses and Sangerhausen, historically of much greater note, in a comradely alliance to ensure the future of the rose. In the same sense the Carla Fineschi Foundation has cooperated with the WFRS Conservation Committee in its efforts to establish a global data base of roses, now included in the present volume. The Foundation has also published (December 1999) a separate catalogue of all the species and varieties at Cavriglia in the hope and belief that this work will be appreciated for its intrinsic value.

The Proud Legacy of 'Peace'
The Rose of the 20th Century'

by Dr. Tommy Cairns

Almost 55 years ago 'Peace' was introduced into the U.S. by the Conard Pyle Company. There was never any doubt that 'Peace' was a great rose and marked a new zenith in the history of Hybrid Teas. Since that date numerous hybridizers have utilized the genetic pool of 'Peace' both as seed parent and pollen parent to produce a staggering 285 off-spring. It is timely to review both the historical background to the production of 'Peace' as well as review it's many offspring.

By far the most popular class of modern roses, easily recognized by the large shapely blooms, usually containing 30–50 petals and borne on long stems either singly or with several side buds, are the Hybrid teas and grandiflora. When the famous Scottish poet, Robert Burns, immortalized the rose in his sonnet composed in the 18th century,

O, my Love is like a red, red rose
That's newly sprung in June

he had no perception of what a modern hybrid tea rose might look like. To him, the personification of the rose was most likely the simple five pealed Ayrshire Rose or perhaps a ruffled damask rose with heavy fragrance. The hybrid tea as an established classification made its debut in France sometime after 1867 with the introduction of the first member cultivar, 'La France'. It was to be some eighty years later when in 1944 a major breakthrough in hybridizing brought the world 'Peace'.

A Special Moment in Time

On June 15, 1935 at Antibes in the south of France, the twenty three year old Francis Meilland made his master pollinating stroke that resulted in the conception of 'Peace'. History has recorded that 55 flowers were pollinated that day from which emerged 52 hips giving 800 seedlings. Of these 800 seedlings, 50 were propagated in 1936 for further trial and observation. In the autumn of 1936 Francis and his father Antoine were able to admire for the first time what a bloom of 'Peace' might look like. It was inscribed under the #3-35-40 indicating that it was the third cross made in 1935 and it was the fortieth of the 50 seedlings considered for further evaluation. Francis Meilland recounted in his article entitled 'My Masterpiece' published in The Rose Annual 1953 (National Rose Society of Great Britain, now Royal National Rose Society):

'It was not very sturdy, this little 3-35 plant, and there was nothing about it to attract attention. It was during the summer of 1936 that a few eyes were budded for the first time. Under the influence of extremely favourable weather conditions during that Autumn of 1936, these few buds produced flowers quite marvellous in shape and size with a greenish tinge, warming to yellow, and progressively impregnated with carmine round the edges of the petals'.

By June 1939 the plant of 'Peace' was receiving great accolades from visitors to Antibes. During that fateful summer bud eyes of 'Peace' were dispatched to the Conrad Pyle Company in the U.S. as well to Germany and Italy. World War II broke out on September 3, 1939 and communications ceased between rose hybridizers. As a direct result of lack of communication the German company introduced 3-35-40 as 'Gloria Dei' while the Italian company chose the name 'Gioia'. Meanwhile in France, Francis Meilland and his father had decided to dedicate this seedling to the memory of his mother, 'Mme A. Meilland' who had died a few years previously.

During the last days before the French armistice, Francis Meilland was visited in his greenhouse by the Comte and Comtesse de Martel (their neighbours) and the Duke of Windsor. On that occasion the Duke of Windsor expressed his enchantment for 3-35-40 – '*I have never seen another rose like it. It is certainly the most beautiful rose in the world*'. In the meantime the Conrad Pyle Company had decided to introduce the rose to the American rose growers under the name of 'Peace' to commemorate the end of the suffering of World War II. Under the auspices of the American Rose Society 'Peace' was formally introduced on April 29, 1945, a date which also marked the fall of Berlin. Shortly after this event, at a meeting of 49 delegations of the United Nations in San Francisco each head of the delegations received in their hotel room a small vase with a single 'Peace' rose accompanied by a card carrying the following message:

'This is the 'Peace' rose wish was christened at the Pacific Rose Society exhibition in Pasadena on the day Berlin fell. We hope that the 'Peace' rose will influence men's thoughts for everlasting world peace'.

In the closing paragraph of his article Francis Meilland commented about the legacy of this seedling 3-35-40 achieving three wonderful names:

> *'If circumstances have decided that it should be known by different name in different countries, this at least is true that each of these names reminds men of good will that the love of flowers, and in particular the admiration of this rose will for ever provide them with the occasion to praise God, with Gloria Dei, to face life with a smile, with Gioia to wish for peace, and, as far as we ourselves are concerned, to perpetuate a loving memory, with Mme A. Meilland'.*

Francis Meilland unfortunately died a young man at the age of 46 in 1958. For most hybridizers, great success is often slow in arriving in their lives and usually after many years of hard research before achieving their masterpieces. 'Peace' arrived right at the very beginning of the career of Francis Meilland and allowed him to enjoy the royalty rewards of his efforts to establish the House of Meilland as the premier rose hybridizing company in France.

Parentage of 'Peace'

There has been considerable discussion as to the exact parentage of 'Peace' but the scholars of the time believe that, in spite of the confusion, the seed parent was from an un-named seedling resulting from a cross of two other un-named seedlings whose seed parents were 'George Dickson' and 'Joanna Hill' respectively. The respective pollen parents of these seedlings were 'Souvenir de Claudius Pernet' and 'Charles P. Kilhan'. The direct pollen parent was 'Margaret McGredy'. In his book '*Roses*', Jack Harkness commented that such a genealogy makes sense since 'Peace' has some of the vigour characteristics portrayed by the Hybrid Perpetual 'George Dickson'; the kind of rose that could result from a cross of 'Souvenir de Claudius Pernet' and 'Joanna Hill'; and the look of 'Margaret McGredy' in its foliage and blooms.

Whatever the final verdict there can be no doubt that the genetic legacy involved in the creation of 'Peace' was indeed a masterpiece that has advanced the future rose and high status of hybrid teas. While there is no additional genealogy data reported on 'George Dickson', 'Charles P. Kilham' and 'Margaret McGredy' it should be stated that all three cultivars were recognized by receiving the Gold medal of the National Rose Society in 1911, 1927, and 1925 respectively. There is no doubt that these three cultivars were deliberately hand picked by young Francis Meilland because of their ability to pass onto the future generations their demonstrated qualities. Any attempt to discuss the parentage of 'Peace' must therefore revolve around the remaining two cultivars, namely the Hybrid Perpetual, 'Souvenir de Claudius Pernet', and the Hybrid tea, 'Joanna Hill'. In the case of 'Souvenire de Claudius Pernet' its ancestors can be traced back to an F2 seedling of Hybrid Perpetual 'Antonio Ducher' and the species 'Rosa foetida persiana' (Persian Yellow Rose). The resulting children from this cross carried the yellow gene all the way to the hybrid tea 'Souvenir de Claudius Pernet'.

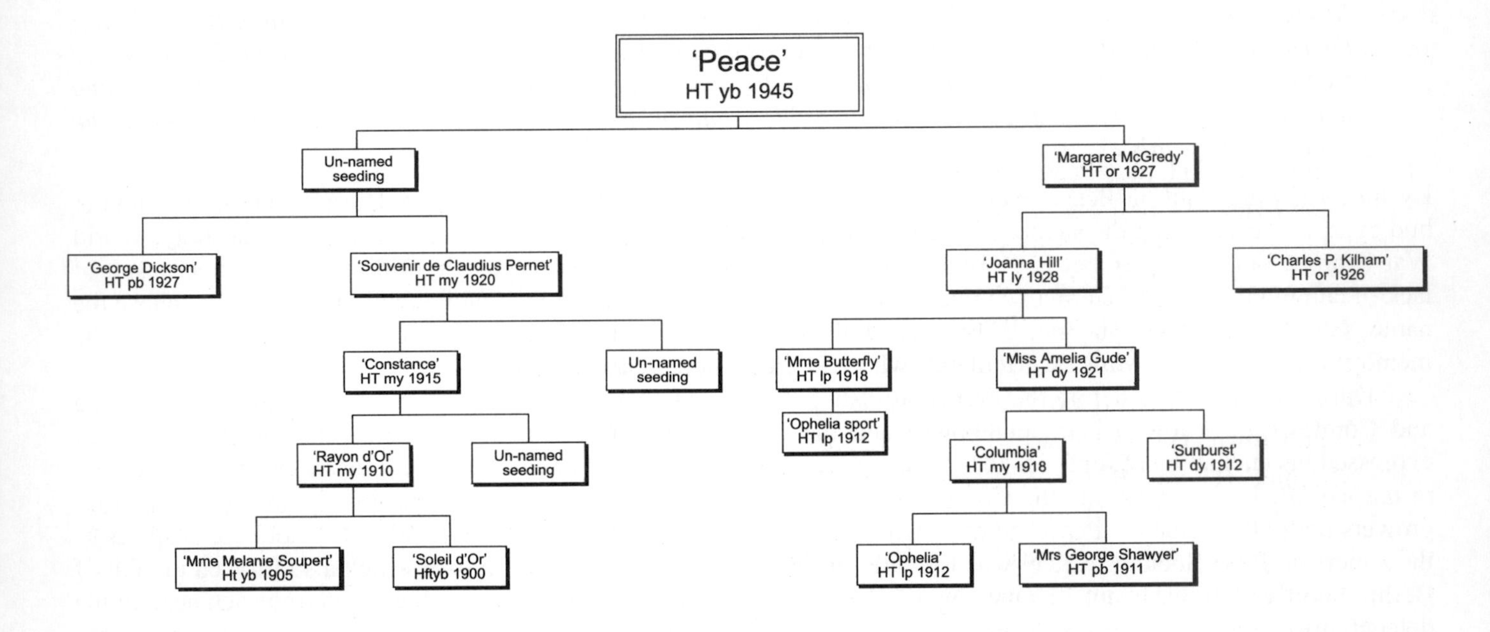

Figure 1. The genealogy of the 'Peace' rose: seed parents are always listed to the left immediately underneath their offspring while pollen parents are listed to the right. The horticultural classification, color class, and year of introduction used are those according to ARS approved nomenclature as listed in *Modern Roses XI*

The genealogy of 'Joanna Hill' is much more complex to understand but deserves some discussion. While many of its early parents were pink ('La France') or white ('Mme Bravy') the introduction of the yellow gene must have resulted from its grandparent, 'Sunburst', which successfully passed it onto its daughter 'Miss Amelia Gude'. What can be decided for certainty is that the rich legacy of 'Joanna Hill' was in great measure responsible for the production of 'Peace'. However, for a proper diagnostic review it would be necessary to examine the characteristics of each rose in the family tree and estimate what genetic characteristics were passed on.

'Peace' has garnered many honors for its beauty and perfection:

1944, Gold Medal, Portland;
1946, All-America Rose Selection;
1947, Gold Medal, NRS of Great Britain;
1947, ARS National Gold Medal Certificate; and
1965, Golden Rose of The Hague.

Hybridized Progeny of 'Peace'

The rich legacy of the genetic information embedded in 'Peace' has been utilized by a host of both amateur and professional hybridizers (Mallerin, Kordes, Weeks, Von Abrams, Delbard, Dickson, Fryer, de Ruiter, Warriner, Armstrong, Gaujard, Tantau, Kriloff, McGredy, Swim, etc) all over the world over the last 50 years. The pool of information contained in both the pollen and seed abilities of 'Peace' have produced a staggering number of off-spring - 382 offspring.

As a seed parent 'Peace' has produced 157 cultivars with an additional 49 cultivars where it participated as an indirect grandfather seed parent. In the case of acting as the direct pollen parent, 'Peace' has been responsible for 132 cultivars with an additional 44 as the indirect grandfather seed parent. Furthermore, 'Peace' has produced an unprecedented number of sports – 19 to be exact. Another angle to observe the rich legacy of 'Peace' is to view the distribution of types of cultivars developed. While the majority of offspring are 338 hybrid teas, there were 35 floribundas, 4 shrubs and 13 grandiflora. Again quite an impressive record of achievement.

'Peace' as a Seed Parent

It is obvious from the registration data of the period 1950–1960 that most amateur and professional hybridizers were cognizant of the potential of 'Peace' and made many crosses to capitalize on worthy characteristics. Many of the already established cultivars such as 'Virgo', 'Independence' and 'Spartan' were often employed as pollen parents. Within the 157 cultivars where 'Peace' was the direct seed parent, there have been a large number of incidences where further hybridizing has been continued. Such cultivars have in themselves been highly successful in achieving popularity and will be the subject of future reports. A short list of cultivars derived from 'Peace' organized by year is given below.

Selected List of Offspring

'Confidence' HT pb ['Peace' X 'Michele Meilland']; Meilland 1954
'Anne Letts' HT pb ['Peace' X 'Charles Gregory']; Letts 1955
'White Knight' HT w [('Virgo' X 'Peace') X 'Virgo']; Meilland 1956
'Grace de Monaco' HT lp ['Peace' X 'Michele Meilland']; Meilland 1956
'Isabelle de France' HT or ['Peace X (Mme Joseph Perraud' X 'Opera')]; Mallerin 1957
'Rose Gaujard' HT rb ['Peace' X 'Opera' seedling]; Gaujard 1959
'Pink Peace' HT mp [('Peace' X 'Monique') X ('Peace' X 'Mrs John Laing')]; Meilland 1960
'Gold Crown' HT dy ['Peace' X 'Golden Scepter']; Kordes 1961
'Mischief' HT op ['Peace' X 'Spartan']; McGredy 1961
'Memoriam' HT lp [('Blanche Mallerin' X 'Peace') X ('Peace' X 'Frau Karl Druschke')]; Von Abrams 1962
'Belle Epoque' HT pb ['Peace' X 'Independence']; Kriloff 1962
'Isabel de Ortiz' HT pb ['Peace' X 'Kordes Perfecta']; Kordes 1965
'Diorama' HT yb ['Peace' X 'Beaut']; de Ruiter 1973
'Lustige' HT rb ['Peace' X 'Brandenburg']; Kordes 1973
'Perfume Delight' HT mp ['Peace' X (('Happiness' X 'Chrysler Imperial') X 'El Capitan')]; Weeks 1984
'O Sole Mio' HT my [('Peace' X 'Marcelle Gret') X 'Velizy' seedling]; Delbard 1984

'Peace' as a Pollen Parent

In many cases in rose hybridizing cultivars are often good seed parents or good pollen parents but rarely play both roles. In the case of 'Peace' it has been equally successful as a pollen parent at producing offspring which have themselves had great influence on future hybridizing. Of the 132 direct descendants of 'Peace' the following lists perhaps the more widely acclaimed as important in the history of the rose. Many of descendents of 'Peace' as the pollen parent (see list below) have been AARS winners (e.g. 'Christian Dior', 'Garden Party', 'Royal Highness'). However, many of the direct offspring have been used in further hybridizing experiments to extend the family tree of 'Peace'. For instance, 'Karl Herbst' is the seed parent of such great cultivars as 'Silver Lining', 'King of Hearts', and 'Redgold'. As a pollen parent it has produced 'Kordes Perfecta', 'Piccadilly', and 'Miss All-American Beauty'.

Selected List of Offspring

'Karl Herbst' HT mr ['Independence' X 'Peace']; Kordes 1954
'Lady Elgin' HT yb ['Mme Kriloff' X ('Peace' X 'Geneve')]; Meilland 1954
'Harmonie' S pb ['R. eglanteria hybrid' X 'Peace']; Kordes 1957
'Sterling Silver' HT m [seedling X 'Peace']; Fisher 1957
'Prima Ballerina' HT dp [Seedling X 'Peace']; Tantau 1958
'Christian Dior' HT mr [('Independence' X 'Happiness') X ('Peace' X 'Happiness')]; Meilland 1959
'Garden Party' HT w ['Charlotte Armstrong; X 'Peace']; Swim & Weeks 1963
'Pink Peace' HT mp [('Peace' X 'Monique') X ('Peace' X 'Mrs John Laing')]; Meilland 1962
'Royal Highness' HT lp ['Virgo' X 'Peace']; Swim & Weeks 1963
'Swarthmore' HT pb [('Independence' X 'Happiness') X 'Peace]; Meilland 1968
'Baronne Edmond de Rothschild' HT rb [('Baccar' X 'Crimson King') X 'Peace']; Meilland 1968
'Princess Margaret of England' HT mp [Queen Elizabeth' X ('Peace' X 'Michele Meilland')]; Meilland 1976
'Better Homes & Gardens' HT pb ['Tropicana' X 'Peace']; Warriner 1976
'Half Time' HT rb [((Fandango' X 'Roundelay') X ('Happiness' X 'Tiffany')) X 'Peace']; Weeks 1976
'Promise' HT lp ['South Seas' X 'Peace']; Warriner 1977
'Allspice' HT my ['Buccaneer' X 'Peace']; Armstrong 1980
'Home & Country' HT yb [Seedling X 'Peace']; Kriloff 1982
'Princess de Monaco' HT w ['Ambassador' X 'Peace']; Meilland 1982

Sports of 'Peace'

There is no other cultivar in the history of the rose with a demonstrated list of sports comparable to that produced from Peace. Since 1954 the number of sports reported from 'Peace' has been 19. This large number of sports reported may well have been because the importance of 'Peace' was recognized and anything out of the ordinary was pre-supposed to be equally spectacular. The range of color of these sports is in keeping with the genealogy of 'Joanna Hill' and 'Souvenir de Claudius Pernet'.

Sports of 'Peace'

'Dorothy Goodwin' HT yb ['Peace' sport]; Goodwin 1955
'Lady Dallas Brooks' HT mp ['Peace' sport]; Downes 1959
'Improved Peace' HT yb ['Peace' sport]; Dean 1959
'Pearly Peace' HT lp ['Peace' sport]; Fryer 1960
'Peaceport' HT op ['Peace' sport]; Rokos 1960
'Narre Peace' HT yb ['Peace' sport]; Brundrett 1960
'Gay Debutante' HT pb ['Peace' sport]; Curtis 1962
'Baby Peace' Min yb ['Peace' sport]; De Mott & Johnson 1962
'Chicago Peace' HT pb ['Peace' sport]; Johnston 1962
'Lucky Piece' HT pb ['Peace' sport]; Gordon 1962
'Speaker Sam' HT yb ['Peace' sport]; Dean 1962
'Jean McGregor Reid' HT w ['Peace' sport]; Sunter 1964
'Tony Peace' HT ab ['Peace' sport]; Brundrett 1964
'Bernadette' HT lp ['Peace' sport]; Kelly 1966
'Flaming Peace' HT yb ['Peace' sport]; McGredy 1966
'New Orleans' HT yb ['Peace' sport]; Tate 1971
'Jay' HT w ['Peace' sport]; Kern Rose Nursery 1985
'Julie Anne Ashmore' HT yb ['Peace' sport]; Owen

Conclusions

There can be no doubt that Francis Meilland gave the world of roses a genetic treasure when he pollinated the unnamed seedling with 'Margaret McGredy' on June 15, 1935 and produced 'Peace'. The legacy of that action has lived on through the many cultivars who share that family tree, many of them becoming All-America Rose Selections (AARS) winners. Now almost 55 years after the formal introduction of 'Peace' into commerce in 1945 in the U.S. there is a renewal and awareness of its important role in history as 'The Rose of the 20th Century'.

Let us hope this tribute will remind the residents of our fragile planet that 'Peace' is as important as it was in 1945. There can be no doubt that 'Peace' deserves the title of 'The Rose of the 20th Century' for the rich legacy it has imparted to the wonderful world of roses, in particular the development of the modern hybrid teas.

Postscript

'Books are Man's memory and his aspirations,
The link between his present and his past,
The tools he builds with'.

Stephen Vincent Benet

A few rosarians indulge in collecting American Rose Annuals – I profess to belong to that category with this strange malady. One of my great pastimes is to read past editions of American Rose Annuals to gain a feeling of rose evolution and tradition. In the American Rose Annual 1965 there appeared a poem about 'Peace' that conveyed the strong feelings of the time expressed by Carolyn Cunningham of West Helena, Arkansas. It is reproduced here to remind us all that progress has been made in both the evolution of the rose and 'Peace' in our times.

'Peace'

There is a rose – we call it 'Peace';
Its beauty pleads that wars shall cease.
A legend grew with each new leaf,
As we prayed for 'Peace' for war is grief.
The rose went round and about the world,
As each new petal was unfurled;
And as it travelled all about,
Within those countries filled with doubt,
How do we know, perhaps this rose
Sent out a message to our foes?
And I should think the message read,
'Please think of our beloved dead,
I know I am but one small rose,
But what man reaps is what he sows;
And if you sow a bitter seed,
Perhaps a gun is what you need;
But if you plant a seed of love,
For all mankind, and God above,
Then here I seek 'Peace' and friendship true
And I should love to be a part
Of both your garden and you heart,
If you should look then you will find,
Though but a rose,
I'm like mankind;
First plant the seed – a kindly word;
Try to be sure that all have heard,
Then fertilize and feed the thought,
And reap the friendship that you sought,
Nourish it and work each day,
Striving for a better way,
Search the garden of your heart,
Let 'Peace' fill all – not just a part,
And as the years go drifting by,
Help the legend live instead of die.
Plant the rose of 'Peace' and watch it there
As you treat the world with loving care.

A

1997 Traditional Home™ *see* 'WAMHOME'

21 Again! *see* 'MEInimo'

A Caen la Paix® *see* ORAkah

'A Capella', HT, mr, 1984; flowers large blooms, dbl., 35 petals, borne singly, slight fragrance; foliage large, medium green, matt; upright growth; Bronze, ARC TG, 1984; [Command Performance X Tiffany]; Stoddard, Louis

'A Feuilles de Chanvre', A, lp, 1800; flowers medium, dbl.; Eurosa, Gallicanae; Braun, H.; (Sangerhausen)

A Fleurs de Rose Tremier de la Chine, C; Unknown; (Hay)

A Fleurs Doubles Violettes, C, m; (Cavriglia)

'A Fleurs Gigantesques', HGal, mp, 1813; flowers rich deep pink, large, very dbl., intense fragrance; bushy growth

A Fleurs Roses de Laffay, HSem; Eurosa, Synstylae; Laffay, M.

'A Longs Pédoncules', M, lp, 1854; bud pale green, mossy, long peduncles; flowers pink, flushed lilac, many petalled, small blooms, nodding in clusters; foliage small, round, soft green; vigorous growth; Eurosa, Gallicanae; Robert

'A Night in June', HT, or, 1935; flowers coral-red, very large, dbl., slight fragrance; foliage leathery; vigorous, bushy growth; [Evening Star sport]; Elmer's Nursery

A Shropshire Lad *see* 'AUSled'

'A. Denis', HT, my, 1935; flowers lemon-yellow, streaked carmine, large, dbl., intense fragrance; [Gorgeous X Marion Cran]; Böhm, J.

'A. Drawiel', HP, dr, 1887; flowers large, dbl., moderate fragrance; Lévêque; (Sangerhausen)

'A. Dvorak', HT, op, 1933; flowers light pinkish orange, open, large; foliage glossy, dark; bushy growth; [Mme Butterfly X Gorgeous]; Böhm, J.

'A. G. A. Rappard', HT, op, 1934; (A.G.A. Ridder van Rappard); flowers salmon, verging on neyron pink, well-shaped, dbl., slight fragrance; foliage glossy, bright green; vigorous, bushy growth; Buisman, G. A. H.

'A. G. Furness', HT, mr, 1941; flowers rich red; semi-climbing growth; [Sensation seedling]; Clark, A.

'A. Geoffroy de St Hilaire', HP, mr, 1878; flowers medium, dbl., intense fragrance; tall growth; Verdier; (Sangerhausen)

A. H. Kirk, HT, op; flowers salmon pink with carmine, large, dbl., moderate fragrance; (Sangerhausen)

A. J. Herwig, F, 1965; Buisman, G. A. H.; (Cavriglia)

'A. MacKenzie', S, rb, 1985; (Alexander MacKenzie); bud ovoid; flowers medium red, reverse lighter, dbl., 45 petals, 3 in., cupped, blooms in clusters of 6-12, repeat bloom, moderate fragrance; purple prickles; foliage yellow-green, glossy, leathery; upright growth; [Queen Elizabeth X (Red Dawn X Suzanne)]; Svedja, Felicitas; Agriculture Canada, 1994

'A. N. W. B. Rose', HT, w, 1933; flowers white, tinted yellow, open, large, dbl., slight fragrance; very vigorous growth; [Frau Karl Druschki X Souv. de Claudius Pernet]; Buisman, G. A. H.

'A. W. Jessep', HT, dp, 1952; flowers rich cerise-pink, large, dbl., 38 petals, moderate fragrance; vigorous growth; Clark, A.

A.G.A. Ridder van Rappard *see* **'A. G. A. Rappard'**

A.R.S. Centennial *see* 'SAVars'

Aachener Dom *see* 'MEIcapinal'

'Aafje Heynis', HT, mr, 1964; flowers bright red, well-shaped, large; foliage glossy, light green; vigorous, upright growth; [Prima Ballerina X Salvo]; Buisman, G. A. H.

Aaland, S, dp

'Aalsmeer Gold'®, HT, dy, 1978; (Bekola); bud long, pointed; flowers deep yellow, dbl., 23 petals, exhibition form, borne singly, slight fragrance; foliage glossy; vigorous, upright, bushy growth; [Berolina X Seedling]; Kordes

Aasmeer, HT, lp; flowers peachish-orange with yellow tones

Aba Saheb, HT, ob, 1993; Chiplunkar

'Abaillard', HGal, pb; (Abalard); flowers rose marbled, dbl.; Eurosa, Gallicanae; Vibert, 1845

Abalard *see* **'Abaillard'**

'Abasanta', HT, dp, 1956; bud pointed, cerise; flowers carmine, occasionally streaked white, semi-dbl., 15–25 petals, 3.5 in., moderate fragrance; foliage dark; compact growth; [Red Columbia X (Red Columbia X Tausendschön sport)]; Motose

'Abba Dabba', Min, mr, 1979; bud long, pointed; flowers, dbl., 38 petals, borne 1-5 per cluster, slight fragrance; tiny, curved prickles; foliage small, medium green; compact, bushy growth; [Red Can Can X Seedling]; Lyon

Abbaye de Cluny™ *see* 'MEIbrinpay'

Abbé Berlèze, HP, dp; Eurosa, Chinensis; Guillot et Fils, 1864; (Roseraie de l'Hay)

'Abbé Bramerel', HP, dr, 1871; flowers large, dbl., slight fragrance; Guillot et Fils; (Sangerhausen)

'Abbé Girardin', B, mp, 1881; flowers carmine-pink, center darker, large, dbl., moderate fragrance; [Louise Odier X Hermosa]; Bernaix, A.

'Abbé Giraudier', HP, mr, 1869; flowers cherry red, very large, dbl.; Eurosa, Chinensis; Levet, 1869; (Sangerhausen)

Abbe Lemire *see* ORAnat

'Abbé Millot', HT, lp, 1900; flowers large, dbl.; Corboeuf; (Sangerhausen)

Abbeyfield Rose *see* 'COCbrose'

'Abbotswood', S, mp, 1954; dbl.; habit similar to R. canina; [Chance hybrid of R. canina X Unknown garden variety]; Hilling

Abdul Hamid, Misc OGR, mr

'Abel Carrière', HP, dr; flowers velvety crimson, brighter center, large, dbl., 45 petals; Eurosa, Chinensis; [Baron de Bonstetten X Seedling]; Verdier, E., 1875

'ABEL GRANT', HP, lp, 1866; flowers large, dbl., intense fragrance; medium growth; Damaizin; (Sangerhausen)

Abelzieds, HRg, mp, 1957; flowers large, semi-dbl., slight fragrance; medium height; Rieksta; (Sangerhausen)

'ABENDRÖTE', HP, op, 1919; flowers light coral-red; compact growth; [Frau Karl Druschki X Juliet]; Ebeling; Teschendorff

Aberdeen Celebration *see* COCmystery

'ABERDONIAN', F, rb, 1974; flowers golden bronze and scarlet, large, dbl., 20 petals, slight fragrance; foliage glossy; [(Evelyn Fison X Manx Queen) X (Sabine X Circus)]; Cocker

'ABHAYA', HT, pb, 1997; (**Magic Medley**); flowers pink with dark red markings, white reverse, dbl., 26–40 petals, 6 in., borne mostly singly, slight fragrance; foliage large, medium green, dull; bushy, medium (3 ft.) growth; [Pristine X Priyatama]; Viraraghavan, M.S.

'ABHISARIKA', HT, rb, 1977; bud long, pointed; flowers red blend, striped, small, dbl., 48 petals, 2.5–3 in., exhibition form, slight fragrance; foliage glossy, yellowish; vigorous growth; [Induced mutant X Kiss of Fire]; IARI

'ABIDING FAITH', HT, dp, 1954; bud semi-ovoid; flowers deep rose-pink, dbl., 20–35 petals, 4–5 in., intense fragrance; vigorous, upright growth; [La France sport X (Senator X Florex)]; Motose

'ABIDING FAITH, CLIMBING', Cl HT, dp, 1957; Motose

Abigaile® *see* 'TANELAIGIB'

Abington Park Northampton, HT, op, 1997; flowers very full, large, very dbl., 41 petals, 7+ in., borne in small clusters, intense fragrance; some prickles; foliage medium, medium green; bushy, medium (3ft.)growth; [Cynthia Brooke X Little Darling]; Jones, L.J.

'ABOL', HT, w, 1927; flowers ivory-white tinted blush, large, dbl., intense fragrance; Evans; Beckwith

'ABONDANT', Pol, mp, 1914; flowers medium-large, dbl.; Turbat; (Sangerhausen)

Abracadabra *see* 'JACBUTE'

Abraham Darby™ *see* 'AUSCOT'

'ABRAHAM ZIMMERMAN', HP, mr, 1879; flowers large, dbl., moderate fragrance; Lévêque; (Sangerhausen)

Abraxas, S, dr, 1973; flowers medium-large, dbl.; VEG; (Sangerhausen)

'ABRICOT', HT, ab, 1929; flowers apricot and coral-salmon, reverse coral-red; [Mrs Aaron Ward X Jean C.N. Forestier]; Barbier

'ABRICOTÉ', T, ab; flowers apricot, margins flesh, dbl., cupped; Dupuis, Prior to 1848

Absolute Hit *see* POUlrougel

Absolutely™ *see* 'SAVALUTE'

Abu, HT, dp; flowers large, dbl., intense fragrance; VEG; (Sangerhausen)

'ABUNDANCE', F, mp, 1974; dbl., 30 petals, 4.5 in., slight fragrance; foliage dark; low, bushy growth; [Seedling X Firecracker]; Gandy, Douglas L.

'AC DE MONTARVILLE', HKor, mp, 1999; (**De Montarville**); flowers medium pink, reverse light pink, dbl. (17–25 petals), medium (7.8 cms) blooms bourne mostly single; slight frangrance; few prickles; foliage small, blue-green, dull; compact, medium growth; Agriculture et Agroalimentaire Canada

'AC MARIE VICTORIN' HKor, pb, 1999; (**Marie-Victorin**); flowers medium pink-yellow, reverse pink-yellow, full (26 petals), medium (7.6 cms) blooms bourne in small clusters; slight fragrance; few prickles; foliage medium. Light green, semi-glossy; arched, medium (1.5 m) growth; [Arthur Bell X (R. kordesii X Max Graf O.P.); Agriculture et Agroalimentaire Canada

'AC WILLIAM-BOOTH', HKor, mr, 1999; (**William-Booth**); flowers medium red, reverse medium pink, single (4–11 petals, small (5 cm) blooms borne in large clusters; fragrant; moderate prickles; foliage medium, dark green, semi-glossy; arching, tall (2 cm) growth; [(R kordesii X Max Graf O.P.) X (Arthur Bell X Apple Jack)]; Agriculture et Agroalimentaire Canada.

Academy® *see* 'MACGUTSY'

'ACADIAN', HCh, dp, 1986; flowers deep, bright pink, medium, single, 5 petals, intense fragrance; fine prickles; foliage small, medium green, matt; upright, bushy, hardy growth; [R. nitida X R. chinensis semperflorens]; James, John

Acapella R *see* TANallepal

'ACAPULCO', F, yb, 1962; bud pointed; flowers yellow and light orange, often flushed pink, medium, dbl., 30–40 petals, cupped, clusters, slight fragrance; foliage dark, glossy; bushy, compact growth; [Seedling X Masquerade]; Von Abrams; Peterson & Dering

Acapulco *see* DICblender

Acaritha, HRg; Eurosa, Cinnamomeae; Wartz; (Hay)

'ACCENT', F, mr, 1977; bud ovoid; flowers cardinal-red, dbl., 25 petals, 2–2.5 in., flat, slight fragrance; foliage small, dark, leathery; bushy, compact growth; [Marlena X Seedling]; Warriner, William A.; J&P

Acclaim *see* 'JACAIM'

'ACCOLADE', HT, rb, 1979; bud ovoid; flowers bright red, shaded darker, dbl., 48 petals, exhibition form, borne mostly singly, slight fragrance; hooked, brown prickles; foliage dark, matt; vigorous growth; [(Daily Sketch X Charles Mallerin) X Peter Frankenfeld]; Dawson, George; Rainbow Roses

'ACCORD', F, ab, 1965; flowers apricot-peach, reverse carmine, well formed, exhibition form, clusters, moderate fragrance; foliage dark; bushy growth; [Circus X Seedling]; Mason, P.G.

Ace of Diamonds *see* 'BRIACE'

Ace of Hearts *see* 'KORRED'

'ACERVATE', Pol, mp, 1934; Miers

Acey Deucy™ *see* 'SAVATHREE'

Achantha, Gr, dr, 1987; Viraraghavan, M.S.

'ACHIEVEMENT', HWich, dp, 1925; foliage variegated; [Dorcas sport]; English

'ACHILLE', Pol, mp, 1936; flowers medium-large, semi-dbl.; Guinoisseau; (Sangerhausen)

'ACHILLE CESBRON', HP, mr, 1894; flowers very large, dbl., moderate fragrance; Rousset; (Sangerhausen)

Achille Gonod, HP; Eurosa, Chinensis; Gonod, 1864; (Roseraie de l'Hay)

'ACHILLE GONOD', HP, mr, 1864; flowers large, dbl., intense fragrance; Gonod; (Sangerhausen)

'ACIDALIE', B, w; flowers white with blush center, large, dbl., globular, recurrent bloom; vigorous growth; Eurosa, Chinensis; Rousseau, 1838

'ACQUA CHETA', HT, rb, 1962; flowers magenta-red, reverse silvery, dbl., 50 petals; foliage dark; compact growth; [Crimson Glory X Peace]; Giacomasso

'ACTRICE', HT, rb, 1966; bud ovoid; flowers red and light pink, large, dbl.; foliage dark; [Tzigane X Kordes' Perfecta]; Verschuren, A.; Van Engelen

'ADA PERRY', Min, op, 1978; bud ovoid; flowers soft coral-orange, medium, dbl., 40 petals, exhibition form, slight fragrance; foliage dark; vigorous, upright growth; [Little Darling X Coral Treasure seedling]; Bennett, Cecilia 'Dee'; Tiny Petals Nursery

'ADABEL', S, rb, 1987; (**Belles and Beaus**); flowers red-white bicolor, floribunda type, medium, dbl., 15–25 petals, exhibition form, no fragrance; foliage medium, dark green, glossy, disease resistant; upright, moderately hardy growth; [Little Darling X Seedling]; Adams, Dr. Neil D.

ADAburi, F, w, 1994; (**Dame de L'Etoile**®); Adam, M.

'ADADISRES', S, mr, 1987; (**Disraeli**); flowers medium, dbl., 26–40 petals, no fragrance; foliage large, dark green, glossy; upright, bushy, tall, broad growth; [Hamburger Phoenix X Seedling]; Adams, Dr. Neil D.

ADAfetap *see* **Rose d'Annecy®**

'ADAGIO', HT, dr, 1971; bud very long, pointed; flowers blood-red, dbl., 28–35 petals, 3.5–4 in., globular; foliage dark; vigorous, bushy growth; [Seedling X Uncle Walter]; Lens

ADAharos, HT, lp, 1995; (**Rose de Rennes®**); Adam, M.

'ADAIR ROCHE', HT, pb, 1968; flowers deep pink, reverse silver, well-formed, large, dbl., 30 petals, slight fragrance; foliage glossy; GM, Belfast, 1971; [Paddy McGredy X Femina seedling]; McGredy, Sam IV; McGredy

'ADAJAN', S, yb, 1993; (**Jan's Wedding**); dbl., 15–25 petals, 2–2.5 in., borne in large and medium clusters; some prickles; foliage medium, dark green, semi-glossy; bushy growth; tall (6 ft.) winter hardy; [Dornroschen X Lichtkonigin Lucia]; Adams, Dr. Neil D.; Rosehaven Nursery, 1993

'ADAM', T, mp; (President); flowers rich rosy-salmon, very large, dbl., globular; Adam, 1833

'ADAM MESSERICH', B, mr; flowers rose-red, semi-dbl., cupped, recurrent bloom; foliage glossy, light; vigorous, bushy growth; Eurosa, Chinensis; [Frau Oberhofgärtner Singer X (Louise Odier seedling X Louis Philippe)]; Lambert, P., 1920

'ADAM RACKLES', T, pb; flowers marbled and spackled pink; Rommel, 1905

ADAmona, F, mr; (**Santiago®**); Adam

ADAmonbu, F, w; (**Montauban de Bretagne®**); Adam

Adam's Smile™ *see* 'SAVASMILE'

ADAnuamn, HT, mp, 1992; (**Parfum d'Armor®**); Adam, M.

'ADASNOW', S, w, 1993; (**Snow Bear**); dbl., 15–25 petals, 3–3.25 in., borne in clusters of 3-5; some prickles; foliage medium, dark green, semi-glossy; upright (4 ft) growth; [Dornroschen X Rosanna]; Adams, Dr. Neil D.; Rosehaven Nursery, 1993

'ADASUN', LCl, my, 1993; (**Summer Sun**); dbl., 15–25 petals, 1.5–2.75 in., borne in large clusters, repeat bloom; some prickles; foliage medium, dark green, semi-glossy; tall (6 ft.), spreading growth; hardy; [Prairie Princess X Lichtkonigin Lucia]; Adams, Dr. Neil D.; Rosehaven Nursery, 1993

ADAtapora, S, mr, 1996; (**Bourgogne No. 2**); Adam, M.

ADAtonysil, F, dy, 1995; (**Rose d'Or de Montreux**); Adam, M.

ADAzombar, F, lp; (**Venise**); Adam, 1997

Added Touch *see* 'LYOAD'

Addo Heritage *see* POUldotage

Adela® *see* FEgama

'ADÉLAIDE D'ORLÉANS', HSem, w, 1826; bud very small, well-formed; flowers pale rose, yellow stamens, semi-dbl., borne in clusters; vigorous (15 ft) growth; Jacques

'ADELAIDE HOODLESS', S, dp, 1975; bud ovoid; flowers light red, medium, semi-dbl., repeat bloom, slight fragrance; foliage glossy; vigorous, bushy growth; [Fire King X (J.W. Fargo X Assiniboine)]; Marshall, H.H.; Canadian Ornamental Plant Foundation, 1972

'ADELAIDE LEE', Min, pb, 1986; flowers white with red petal edges, urn-shaped, dbl., 25 petals, 1 in., borne singly, slight fragrance; few, green, straight prickles; foliage medium green, matt; upright growth; [Gene Boerner X Magic Carrousel]; Stoddard, Louis

'ADÉLAÏDE MOULLÉ', HWich, pb, 1902; flowers lilac-pink, center carmine, clusters, midseason bloom; [R. wichurana X Souv. de Catherine Guillot]; Barbier

Adelaide Tonight *see* 'TOMDEL'

'ADÈLE', HT, dy, 1935; flowers clear gold, base deeper; [Roselandia X Clarice Goodacre]; Lens

'ADÈLE', HGal, pb, 1815; Vibert

'ADÈLE COURTOISÉ', HGal, dp; flowers rosy red, small, very dbl.

'ADÈLE CROFTON', HT, yb, 1928; flowers yellow overlaid scarlet-orange, dbl., moderate fragrance; Dickson, A.

'ADÈLE DESCEMET', HGal; Descemet, M., prior to 1814

'ADÈLE FREY', Cl HT, dp, 1911; flowers large, very dbl., slight fragrance; Walter; (Sangerhausen)

'ADÈLE HEU', HGal, 1816; flowers purple-pink with white shadings, medium large, dbl., moderate fragrance; Vibert, 1816; (Sangerhausen)

'ADÈLE PAVIC', lp, 1850; flowers medium large, dbl., moderate fragrance; Vibert; (Sangerhausen)

Adèle Pradel *see* **'MME BRAVY'**

'ADÈLE PRÉVOST', HGal, lp; flowers blush, center pink, large, dbl.; vigorous, upright growth; Eurosa, Gallicanae

'ADELINE', C, pb, 1830; flowers vivid rose, paler toward edge, dbl.; foliage dark; compact, branching growth; Vibert

'ADELINE', M, dp; bud well mossed; flowers lilac-rose, dbl., 2 in.; foliage light green; compact, well-branched growth

'ADELINE GENÉE', F, my, 1967; bud plump; dbl., 50 petals, 4 in., blooms in clusters, slight fragrance; foliage glossy; low, bushy growth; [Paddy McGredy X Seedling]; Harkness

Adesmano®, HT, pb; Adam

ADharman, HT, dr, 1993; (**Commandant Cousteau**); Adam, M.

'ADIANTIFOLIA', HRg, mp, 1907; Eurosa, Cinnamomeae; Cochet-Cochet; (Hay)

'ADIEU DE BORDIER', HGal, mr; flowers vivid red, very dbl.

Aditya, HT, dy, 1990; K&S

Admirable *see* 'SEARODNEY'

Admiral *see* **'waves'**

'ADMIRAL DEWEY', HT, mr, 1899; flowers large, dbl.; Taylor; (Sangerhausen)

'ADMIRAL RODNEY', HT, pb, 1973; flowers pale rose-pink, reverse deeper, dbl., 45 petals, 4–4.5 in., intense fragrance; foliage large, glossy, dark; vigorous growth; Trew, C.; Warley Rose Gardens

'ADMIRAL SCHLEY', HT, r, 1901; flowers red; [Col. Joffé X Général Jacqueminot]; Cook, J.W.

'ADMIRAL WARD', HT, rb, 1915; flowers crimson-red, shaded fiery red and velvety purple, large, dbl., globular, moderate fragrance; [Seedling X Château de Clos Vougeot]; Pernet-Ducher

'ADMIRATION', HT, rb, 1922; flowers cream, shaded vermilion, pointed, large, dbl., intense fragrance; foliage light green; moderately bushy growth; McGredy

Admired Miranda® *see* 'AUSMIR'

'ADOLF DEEGEN', HT, pb, 1935; flowers rosy pink with fiery streaks, large, moderate fragrance; [Ophelia X Wilhelm Kordes]; Böhm, J.

'ADOLF GRILLE', F, dp, 1940; bud pointed, ovoid; flowers scarlet-carmine, dbl., 25 petals, 4–5 in., cupped, in clusters, slight fragrance; foliage leathery, dark, wrinkled; vigorous, bushy, compact growth; [Dance of Joy X (Cathrine Kordes X E.G. Hill)]; Kordes; J&P

'ADOLF HORSTMANN'®, HT, yb, 1971; (Adolph Horstmann); flowers yellow-orange, dbl., 5 in., exhibition form, slight fragrance; foliage glossy; vigorous, upright growth; [Colour Wonder X Dr. A.J. Verhage]; Kordes, R.; Kordes

'ADOLF KÄRGER', HT, my, 1918; flowers golden yellow, fading, slight fragrance; [Cissie Easlea X Sunburst]; Kordes, H.

'ADOLF KOSCHEL', HT, ab, 1918; flowers orange-yellow, moderate fragrance; [Harry Kirk X Louise Catherine Breslau]; Kordes, H.

Adolf Papeleu, F, w, 1998; RvS-Melle

'ADOLPH GUDE', HT, pb, 1941; flowers rose-pink, reverse darker, dbl., 30 petals, 5.5–6 in., exhibition form, moderate fragrance; very vigorous growth; [Red Radiance sport]; Gude

Adolph Horstmann *see* **'ADOLF HORSTMANN'®**

'ADONIS', HT, ly, 1921; flowers ivory-yellow, shaded lemon, dbl., moderate fragrance; GM, NRS, 1920; [Sunburst X American Beauty]; Bees

'ADORA', Pol, mp, 1936; bud deep flame-pink; flowers rose-pink, dbl., intense fragrance; vigorous, dwarf growth; Beckwith

Adora, HT, op, 1986; Kasturi

'ADORABLE', HT, lp, 1930; flowers flesh-pink; [Columbia sport]; Eichholz

'ADORATION', HT, op, 1940; (Ile de France); bud long, pointed; flowers bright salmon, dbl., 22 petals, 5 in., moderate fragrance; foliage leathery; vigorous, bushy growth; [(Mme Joseph Perraud X Seedling) X Seedling]; Gaujard; J&P

'ADORN', Min, pb, 1986; flowers medium pink blending lighter, reverse medium pink, urn-shape, dbl., 30 petals, exhibition form, borne singly, no fragrance; slender, light prickles; foliage medium, dark, semi-glossy; medium, upright, bushy growth; [Seedling X Seedling]; McDaniel, Earl; McDaniel's Min. Roses

Adrian, F; Camprubi, C.; (Cavriglia)

'ADRIAN BAILEY', F, op, 1972; bud globular; flowers orange-scarlet, dbl., 28 petals, 3.5 in.; foliage glossy; bushy, upright growth; [Fragrant Cloud X Evelyn Fison]; Bailey

'ADRIE STOKMAN', HT, mp, 1948; flowers darker pink than briarcliff; Stokman

Adrien Mercier®, F, rb

'ADRIENNE BERMAN', HT, mr, 1996; flowers very large blooms spiraled, thick-textured, dbl., 26–40 petals, borne mostly singly, moderate fragrance; some prickles; foliage large, dark green, semi-glossy; upright, medium growth; [(Precious Platinum X Silver Jubilee) X Cardiff Bay]; Poole, Lionel; F. Haynes & Partners, 1997

'ADRIENNE DE CARDOVILLE', B, lp, 1859; Guillot Père

'ADRIENNE LEAL', F, dp, 1965; bud ovoid; flowers deep pink to soft mauve, medium, dbl., slight fragrance; foliage dark, leathery; vigorous, upright growth; [Roundelay sport]; Leal

'ADRIENNE MARTIN', HT, dp, 1930; flowers carmine, base yellow, very dbl., cupped, moderate fragrance; foliage glossy, dark; vigorous, bushy, low growth; [Recuerdo de Angel Peluffo X The Queen Alexandra Rose]; Buatois

'ADVANCE', HT, rb, 1940; flowers orange-flame, shaded cerise, reverse cerise, large, dbl., slight fragrance; foliage leathery, dark; long, strong stems; vigorous, bushy growth; [Comtesse Vandal X Mrs Sam McGredy]; LeGrice; C-P

Advance Guard *see* **'WILHELM TEETZMANN'**

Adventure *see* **'AVENTURE'**

'ADVOCATE', HT, mr, 1928; flowers crimson-scarlet, dbl., moderate fragrance; Dickson, A.

Aebleblomst, HMsk, lp, 1955; Petersen

'AËLITA', S, w, 1952; flowers white tinted green, large, dbl., 50 petals, intense fragrance; foliage glossy; vigorous growth; [New Dawn seedling]; Shtanko, E.E.

'AENNE BURDA', HT, mr, 1973; flowers blood-red, large, dbl., exhibition form, slight fragrance; foliage large, glossy; vigorous, upright, bushy growth; Kordes

'AENNE KREIS', HT, ab, 1930; flowers orange-yellow, reverse lighter, well-formed, dbl.; vigorous, branching growth; [Wilhelm Kordes sport]; Kreis

'AFFIRM', HT, mp, 1998; (McMillan's Pink); flowers medium pink full (26–40 petals), high pointed centre, medium (4 in.) Blooms borne singly or small clusters; slight frangance; numerous prickles; foliage medium, medium green, dull; spreading, low (36 in.) Growth; [Seedling X Seedling]; McMillan, Thomas; Hortico, Inc., 1998

'AFLAME', F, op, 1954; flowers orange-strawberry, semi-dbl., 4 in., blooms in clusters, intense fragrance; foliage dark, glossy; vigorous, low, spreading growth; [(Poulsen's Pink X Ellinor LeGrice) X Mrs Pierre S. duPont]; LeGrice

'AFRICA STAR', HT, m, 1965; dbl., 65 petals, 3.5–4 in.; foliage coppery; bushy growth; West; Harkness

African Queen *see* TANeliet

'AFRICAN SUNSET', HT, ab, 1966; flowers apricot-orange, pointed, dbl., 4.5 in., moderate fragrance; free growth; [Sutter's Gold X Chantré]; Herholdt, J.A.

African Sunset *see* JACpik

After Glow *see* JOHilmar

After Midnight™ *see* 'KINNIGHT'

'AFTERGLOW', HT, ab, 1930; flowers apricot-yellow, dbl., intense fragrance; [Seedling X Souv. de Claudius Pernet]; Hill, Joseph H., Co.

'AFTERGLOW', HT, ob, 1938; (Sam Buff); bud long, pointed, orange; flowers golden yellow, reverse golden orange, large, dbl., slight fragrance; foliage bluish green, glossy; long stems; vigorous, bushy growth; [Mrs Sam McGredy sport]; LeGrice

Afterglow, F, dp

Afternoon Delight *see* 'JOLAFT'

'AGAR', HGal, pb, 1843; flowers dark rose, spotted, medium, dbl.; Vibert, 1843

'AGATE POURPRE', HT, mr, 1965; flowers currant-red, dbl., 5–6. in., cupped, intermittent bloom; foliage dark, serrated; [Impeccable X Rome Glory]; Delbard-Chabert; Cuthbert

Agatha *see* **R. GALLICA AGATHA**

'AGATHA CHRISTIE', F, dp, 1966; bud ovoid; flowers pink-red, blooms in clusters, dbl.; foliage dark; compact growth; Buisman, G. A. H.

'AGATHE FATIME', HGal, mp; Vibert, before 1815

'AGATHE INCARNATA', HGal, mp; flowers pink, quartered; habit intermediate between the Gallicas and the Damasks

Agathe Nabonnand *see* **'MME AGATHE NABONNAND'**

'AGE TENDRE', HT, dp, 1966; bud long, pointed; flowers rose, large, dbl., exhibition form; strong stems; vigorous growth; [Queen Elizabeth X Spartan]; Croix, P.

Agemy, HT, ob

Agemy Sport, HT, dy

Agéna® *see* 'DELCUS'

'AGÉNOR', HGal, m; flowers reddish purple, medium, dbl.; Vibert, ca 1832

Agkon, HT, dp; Agel

'AGLAIA', HMult, ly, 1896; (Yellow Rambler); flowers straw-yellow to white, dbl., in clusters, moderate fragrance; foliage glossy; vigorous growth; [R. multiflora X Rêve d'Or]; Schmitt; Lambert, P.

'AGNES', HRg, my, 1900; flowers pale amber, center deeper, open, dbl., profuse, non-recurrent bloom, moderate; foliage light green, glossy, rugose; short stems; vigorous (6 ft.), bushy growth; very hardy; (21); [R. rugosa X R. foetida persiana]; Saunders, 1900; Central Exp. Farm, 1922

'AGNÈS AGERON', HT, mr, 1958; flowers cerise-red, reverse tinted currant-red, well formed; very vigorous growth; [Mme Méha Sabatier X Léonce Colombier]; Arles; Roses-France

'AGNES BARCLAY', HT, yb, 1927; flowers yellow and reddish salmon, moderate fragrance; Clark, A.; NRS Victoria

Agnes Bernauer® *see* KORnauer

'AGNES DE PUY', HT, dp, 1930; flowers geranium-red veined gold, large blooms, dbl., 22 petals, exhibition form, moderate fragrance; long stems; very vigorous growth; hardy.; [Lady Battersea X (Honeymoon X Mme Butterfly)]; De Puy

'AGNES EMILY CARMAN', HRg, mr, 1898; flowers bright crimson, large, dbl., some repeat bloom; foliage large, rugose; vigorous (5 ft.) growth; Carman

'AGNES GLOVER', HT, dr, 1924; flowers deep velvety crimson, moderate fragrance; [Admiral Ward X George Dickson]; Chaplin Bros.

'AGNES KRUSE', F, 1936; flowers velvety red, fiery, large; vigorous growth; [Mme Edouard Herriot X Eblouissant]; Tantau

'AGNES LAURA BLACKBURN', F, yb, 1989; bud pointed; flowers bicolor yellow, medium, single, 5 petals, borne in sprays; [City of Portsmouth X Seedling]; Cants of Colchester, Ltd.

'AGNES LUCKÉ', HT, pb, 1959; bud long, pointed; flowers cerise-pink, reverse whitish pink, large to medium size, dbl., 50 petals, exhibition form, moderate fragrance; foliage leathery; vigorous, upright growth; [Happiness X Peace, Climbing]; Armbrust; Langbecker

'AGNES MARGUERITE', HT, or, 1953; flowers reddish orange, shaded apricot; foliage glossy; Bide

'AGNES ROGGEN', HT, pb, 1926; flowers pale pink, reverse carmine, dbl., moderate fragrance; Leenders, M.

'AGNES UND BERTHA', HMult, lp, 1926; flowers small, single; Bruder Alfons; (Sangerhausen)

'AGNES WINCHEL', HT, pb, 1989; bud pointed; flowers light pink with deep pink border, medium, dbl., 28 petals, exhibition form, borne singly, slight, fruity fragrance; globular, medium, orange fruit; prickles slightly hooked, medium, green; foliage medium, dark green, semi-glossy; upright, medium growth; PP007826; [Dorothy Anne X Seedling]; Winchel, Joseph F.; Coiner Nursery, 1990

Agni, HT, or, 1985; G&L

Agnihotri, HT, dp, 1981; Kasturi

'AGREEMENT', F, pb, 1971; flowers deep glowing pink, base golden, dbl., 30 petals, 3 in., exhibition form; foliage glossy, bright green; tall, very free growth; LeGrice

'AGRIPPINA', HCh, lp; flowers crimson with typical china white line down center of petals

Agrippina, Climbing *see* **'CRAMOISI SUPÉRIEUR, CLIMBING'**

Ah Mow, T, lp; Hay

Ahalya, F, lp, 1969; Kasturi

'AHIMSA', HT, my, 1996; (Orient Silk); dbl., 26–40 petals, 5 in., borne mostly singly, moderate fragrance; no prickles; foliage large, medium green, semi-glossy; upright, medium (3 ft) growth; [Mme Charles Sauvage X Seedling]; Viraraghavan, M.S.; Hortico Roses, 1996

'AHLIN', F, ob, 1979; bud ovoid; flowers orange, 15 petals, exhibition form, borne 4-6 per cluster, slight fragrance; hooked prickles; foliage large, leathery; upright growth; [Seedling X Little Darling]; Fong

'AHOI', F, or, 1964; bud urn shaped; flowers bright orange-red, dbl., blooms in clusters; low, bushy growth; Tantau, Math.

'AÏCHA', HSpn, dy, 1966; bud long, pointed; flowers deep yellow, large, semi-dbl., intense fragrance; foliage light green; vigorous, bushy growth; [Souv. de Jacques Verschuren X Guldtop]; Petersen

'AÏDA', HT, mr, 1956; bud urn shaped; flowers clear rose-red, dbl., 25 petals, 5 in., cupped, intense fragrance; foliage leathery; vigorous, upright, symmetrical growth; [Crimson Glory X Signora]; Mansuino, Dr. Andrea; J&P

'AIGLON', HT, rb, 1961; bud long, pointed; flowers coppery red, reverse yellow, large, dbl., 38 petals, moderate fragrance; foliage glossy, light green; vigorous, upright growth; [Viola X Opera seedling]; Gaujard

'AILEEN', F, ab, 1975; flowers light apricot, peach blended, dbl., 3.5 in., moderate fragrance; upright, free growth; [Elizabeth of Glamis sport]; Wood

'AIMABLE AMIE', HGal, dp; flowers medium, dbl.; Pronville, before 1900

'AIMABLE ROUGE', G, prior to 1845; Godefroy

'AIMABLE ROUGE', HGal; (Amiable Rouge); Vibert, 1819 or 1820

'AIMÉE VIBERT', N, w, 1828; (Bouquet de la Mariée, Nivea); flowers pure white, dbl., blooms in small clusters, recurrent bloom; vigorous growth; [Champneys' Pink Cluster X R. sempervirens hybrid]; Vibert

'AIMÉE VIBERT SCANDENS', N, w, 1841; Curtis

Ain't Misbehavin' *see* 'SEABLA'

Ain't She Sweet℠ *see* 'WEKFRAG'

Air France *see* 'MEIFINARO'

'AIR FRANCE', HT, dr, 1958; flowers red to cherry-red, medium, dbl., cupped, moderate fragrance; vigorous, bushy growth; RULED EXTINCT 9/82 ARM; [Seedling X Poinsettia]; Asseretto, A.; Pin-Blanchon

Air France Meillandina *see* 'MEIFINARO'

Airain *see* **'CHÂTELET'**

'AIRBORNE', F, dp, 1949; flowers deep pink, large, dbl., moderate fragrance; vigorous growth; [Donald Prior X Rosamunde]; Leenders, M.

'AISLING', F, w, 1966; flowers ivory-cream, dbl., 4 in., slight fragrance; foliage small, glossy; very vigorous growth; [Queen Elizabeth X Allgold]; Slattery

Ajaccio, HT, 1968; Croix, P.; (Cavriglia)

'AJANTA', HT, m, 1978; bud tapered; dbl., 35–40 petals, 5–5.5 in., exhibition form, moderate, spicy fragrance; foliage large, glossy; bushy growth; [Lady X seedling X Memoriam]; Thakur; Doon Valley Roses

Aka Tombo, Min, rb

Akashdeep, F, or, 1993; Chiplunkar

Akashi, S, lp

Akashsundari, HT, m, 1982; Pal, Dr. B.P.

'AKEBONO', HT, yb, 1964; flowers light yellow flushed carmine, high pointed, large, dbl., 56 petals; foliage dark, glossy; vigorous, upright growth; [Ethel Sanday X Narzisse]; Kawai

'AKEMI', F, or, 1977; bud ovoid; dbl., 33 petals, 2–2.5 in., cupped, slight fragrance; foliage dark; vigorous, upright growth; [(Sarabande X Hawaii) X (Sarabande X Ruby Lips)]; Keisei Rose Nurseries, Inc.

Akito *see* 'TANITO'

Al Moise, N, lp

'ALABAMA', HT, pb, 1976; bud long, pointed; flowers deep pink, reverse near white, dbl., 25 petals, 3.5–4 in., exhibition form, moderate, tea fragrance; foliage dark, leathery; upright growth; [Mexicana X Tiffany]; Weeks, 1977

'ALABASTER', HT, w, 1961; dbl., 70–80 petals, 5 in., exhibition form, moderate fragrance; moderate, upright growth; [Blanche Mallerin X McGredy's Ivory]; Wyant

'ALADDIN', F, mr, 1965; bud ovoid; flowers red, very dbl., blooms in clusters; many prickles; foliage dark; [Miracle X Edith Piaf]; Verbeek

Aladdin *see* 'SUNALAD'

Aladdins Dream *see* TWOlad

Aladin®, Min, ob

'ALAIN', F, mr, 1948; bud ovoid; flowers bright carmine-red, semi-dbl., blooms in clusters, slight fragrance; foliage glossy, dark; upright, bushy growth; (28); GM, Geneva, 1948; [(Guineé X Skyrocket) X Orange Triumph]; Meilland, F.; C-P

'ALAIN BLANCHARD', HGal, m, 1839; flowers purplish-crimson, becoming mottled giving a spotted look, semi-dbl., cupped, moderate fragrance; foliage medium green; vigorous, medium growth; [Probably R. centifolia X R. gallica]; Vibert

'ALAIN, CLIMBING', Cl F, mr, 1957; (Grimpant Alain); Delforge

'ALAIN, CLIMBING', Cl F, mr, 1957; Roth

'ALAMEIN', F, mr, 1963; flowers scarlet, semi-dbl., 10 petals, 3 in., flat, blooms in clusters, slight fragrance; foliage dark; vigorous, bushy growth; [Spartan X Queen Elizabeth]; McGredy, Sam IV; McGredy

Alamode *see* 'BROALA'

'ALANNA HOLLOWAY', HT, dr, 1974; flowers ovoid, slight fragrance; foliage glossy, leathery; very vigorous growth; [Uncle Walter sport]; Holloway

'ALASKA', HT, w, 1949; flowers ivory-white, well-formed, large, dbl., 50 petals, moderate fragrance; vigorous, upright growth; [Peace X Blanche Mallerin]; Meilland, F.

'ALASKA CENTENNIAL', Gr, dr, 1967; bud long, pointed; flowers dark blood-red-scarlet, dbl., 3.5–5 in., exhibition form, moderate fragrance; foliage dark, leathery, glossy; vigorous, upright growth; [Siren X Avon]; Morey, Dr. Dennison; Country Garden Nursery

Alba, 1965; Mansuino, Q.; (Cavriglia)

'ALBA FOLIACEA', A, w, 1824

Alba Garcia® *see* FEstro

'ALBA MAXIMA', A, w; (Great Double White, Maxima, R. alba maxima, The Jacobite Rose); flowers similar to maiden's blush, but center creamy; height 6-8 ft.

Alba Meidiland™ *see* 'MEIFLOPAN'

Alba Meidiland™ *see* 'MEICOUBLAN'

Alba Meillandécor *see* 'MEIFLOPAN'

Alba Meillandina® *see* 'MEINABRON'

Alba Minima, A, mp, 1997; Scarman

'ALBA ODORATA', HBc, w, 1834; (R. bracteata alba odorata, R. microphylla alba odorata); flowers anthers yellow, large, moderate fragrance; straight prickles; foliage 5-7 oval leaflets; wood greenish-purple; Mariani

Alba Rosea *see* **'MME BRAVY'**

Alba Semi-plena, A, w, prior to 1867; flowers white, golden stamens, semi-double (8–12 petals); large crop of red hips in fall; a very old form obsereved as a sport from Alba Maxiima

Alba Suaveolens, A, w, cultivated prior to 1970; (R. alba suaveolens Dieck); flowers white, semi-double, 2 in. Diameter; source of attar of roses

Albania, Min, 1991; Dot, Simon; (Cavriglia)

'ALBAST', HT, op, 1928; bud ovoid; flowers salmon-pink, open, large, dbl., slight fragrance; foliage bronze, glossy; vigorous growth; [Morgenglans X Mrs Wemyss Quin]; Van Rossem

Albatross, HWich, w, 1988

'ALBÉRIC BARBIER', HWich, ly, 1900; bud small, yellow; flowers creamy white, center yellow, dbl., borne in clusters, non-recurrent, moderate; foliage glossy, dark; vigorous growth; (14); [R. wichurana X Shirley Hibberd]; Barbier

Alberich® *see* **'HAPPY'**

'ALBERT', F, op, 1962; flowers orange salmon, dbl., 50 petals, exhibition form, blooms in clusters; foliage glossy, bronze; vigorous, bushy, compact growth; Jones; Hennessey

'ALBERT EDWARDS', S, ly, 1961; flowers creamy yellow, single, 5 petals, 1.5–2 in., moderate fragrance; vigorous growth; [R. spinosissima altaica X R. hugonis]; Hillier

'ALBERT GILLES', HT, op, 1943; flowers pink tinted coral; [Julien Potin X Mme Joseph Perraud]; Mallerin, C.; A. Meilland

'ALBERT HOFFMAN', HT, ly, 1904; flowers very large, dbl., intense fragrance; Welter; (Sangerhausen)

'ALBERT LA BLOTAIS', HP, dr, 1881; Moreau et Robert

'ALBERT LA BLOTAIS, CLIMBING', Cl HP, mr, 1888; flowers large, very dbl.; Pernet Père; (Sangerhausen)

'ALBERT MAUMENÉ', S, or, 1934; bud large, pointed, ovoid, copper-red; flowers carrot red shaded copper, semi-dbl., cupped, profuse, repeated bloom; foliage glossy, dark; very vigorous, bushy, open habit growth; [Mme Edouard Herriot X R. hugonis]; Sauvageot, H.

'ALBERT PAYÉ', HP, lp, 1873; flowers flesh-pink, large; vigorous growth; Touvais

'ALBERT PIKE', HT, dp, 1926; flowers glowing cerise, flushed peach at times; [Columbia sport]; Vestal

Albert Poyet, LCl, pb, 1979; Eve, A.

Albert Weedall *see* 'SCRIVLEO'

'Alberta', F, dp; flat, 10 petals, slight; medium matt green; [Carefree Beauty X Red Hot]; Fleming, Joyce L., 1995

'ALBERTA HUNTER', F, ly, 1985; flowers red stamens, large, dbl., 25 petals, cupped, borne singly, moderate fragrance; small, globular, dull orange fruit; small, straight, red prickles; foliage medium, dark, semi-glossy; medium, upright, bushy growth; [Eleanor Perenyi X Lillian Gish]; French, Richard

'ALBERTAN', S, mp, 1962; flowers bright pink, semi-dbl., 15–18 petals; (14); [Athabasca seedling]; Erskine

Albert-Georg Pluta Rose, F, mr, 1986; flowers luminous red, large, dbl.; Tantau, Math.; (Sangerhausen)

'ALBERTINE', LCl, op, 1921; bud ovoid; flowers light salmon pink, large, dbl., cupped, blooms in clusters, non-recurrent, moderate fragrance; vigorous growth; [R. wichurana X Mrs Arthur Robert Waddell]; Barbier

Alberton Amor *see* LUDsportiana

Albinia, S, op, 1996; Poulsen

'ALBION', HT, mp, 1969; flowers urn-shaped, large, dbl., borne singly, slight fragrance; foliage semi-glossy; bushy growth; [Frileuse X Seedling]; Poulsen, Niels D.; Vilmorin-Andrieux

'ALBION', S, w; bud cream; dbl., recurrent bloom; foliage leaflets small, firm, dark; height 4 ft.; hardy; Skinner

Albuquerque Enchantment *see* 'MORALBUQUE'

'ALCAZAR', F, or, 1961; bud pointed; flowers coppery red, semi-dbl., 18 petals, 3 in., slight fragrance; foliage glossy, bronze; bushy growth; [Jolie Princesse X (Opera X Miss France)]; Gaujard

Alcha, S, mp

Alchemist *see* **'ALCHYMIST'**

'ALCHYMIST', S, ab, 1956; (Alchemist, Alchymiste); bud ovoid; flowers yellow shaded orange, pink and red, large, very dbl., heavy, non-recurrent bloom, moderate fragrance; foliage glossy, bronze; vigorous, upright (6 ft) growth; [Golden Glow X R. eglanteria hybrid]; Kordes

Alchymiste *see* **'ALCHYMIST'**

'ALCIME', HGal, dp; (Alcine); flowers deep rosy pink, spotted white, edged lilac-blush, large, dbl., cupped; vigorous, upright growth; Vibert, 1834; (Sangerhausen)

Alcine *see* **'ALCIME'**

Alcmaria, F, mr, 1971; flowers blood red, large, dbl.; Verbeek; (Sangerhausen)

Alconbury *see* 'TALALCONBURY'

Alden Biesen, HMsk, 1996; Lens, Louis; (Cavriglia)

'ALEC ROSE', F, or, 1969; flowers scarlet, well-formed, blooms in trusses; free growth; [Hassan X John Church]; McGredy, Sam IV; McGredy

Alec's Red® *see* 'CORED'

'ALEC'S RED, CLIMBING'®, Cl HT, mr, 1975; [Alec's Red sport]; Harkness

'ALEGRIAS', F, rb, 1978; bud globular; flowers medium yellow, reverse orange-red, blooms, 23 petals, cupped, borne 3-6 per cluster, no fragrance; straight, reddish-green prickles; foliage medium, glossy; upright, bushy growth; [Seedling X Charleston seedling]; Rose Barni-Pistoia

Alejandra Conde® *see* FEgosa

'ALENA', Gr, ob, 1970; flowers orange, large, dbl., cupped, moderate fragrance; foliage large, glossy, dark; vigorous, upright, bushy growth; [Seedling X Tropicana]; Raffel; Port Stockton Nursery

Alessa, LCl, mr, 1988; Jensen

Alex C. Collie *see* COCocrust

Alexander® *see* 'HARLEX'

'ALEXANDER EMSLIE', HT, dr, 1918; flowers deep crimson, base slightly white, moderate fragrance; Dickson, A.

'ALEXANDER HILL GRAY', T, dy, 1911; (Yellow Cochet, Yellow Maman Cochet);

flowers deep lemon-yellow, aging deeper, large, intense fragrance; vigorous growth; Dickson, A.

Alexander Laquemont *see* **'ALEXANDRE LAQUEMENT'**

Alexander MacKenzie *see* **'A. MACKENZIE'**

'ALEXANDER MARGHILOMAN', HT, w, 1928; flowers cream-white, center salmon; [Harry Kirk seedling]; Mühle

Alexander Milne *see* 'LAVMILNE'

'ALEXANDER VON HUMBOLDT', HKor, mr, 1960; flowers scarlet, borne in large clusters; foliage glossy; vigorous (9-12 ft) growth; [R. kordesii X Cleopatra]; Kordes, R.; Kordes

Alexandra *see* 'HARLEX'

'ALEXANDRA'®, T, yb, 1900; flowers buff and yellow; Paul, W.

Alexandra *see* KORbaxand

'ALEXANDRA LEEK', F, ob, 1995; flowers medium, soft orange, single (5 petals), medium blooms borne in small clusters; fragrant; some prickles; foliage medium, medium green, semi-glossy; upright, bushy, tall growth; [Masquerade X Mrs John Lang]; Fleming, Joyce; Hortico Roses, 1995

Alexandra Rose *see* 'AUSDAY'

'ALEXANDRE CHOMER', HP, m, 1875; Liabaud

'ALEXANDRE DUMAS', D, lp, 1969; very dbl., cupped, blooms in clusters, abundant, non-recurrent bloom, intense, damask fragrance; foliage small, light green, glossy, leathery; vigorous, upright (6-8 ft.) growth; [Ma Perkins X R. damascena versicolor]; Fankhauser

'ALEXANDRE DUPONT', HP, dr, 1892; flowers dark, velvety red, very large, dbl., moderate fragrance; Liabaud; (Sangerhausen)

'ALEXANDRE GIRAULT', LCl, pb, 1909; flowers carmine-red, base salmon, large, dbl., non-recurrent, moderate fragrance; vigorous growth; [R. wichurana X Papa Gontier]; Barbier

'ALEXANDRE LAQUEMENT', HGal, m; (Alexander Laquemont); flowers violet spotted with red

Alexandre Tremouillet, HWich, dp, 1988

Alexandria Rose *see* 'WILALEX'

'ALEXANDRINE CHAPUIS', HT, my, 1935; bud long, pointed, yellow, shaded carmine; flowers well formed, large, dbl.; foliage bright green; [Feu Joseph Looymans X Seedling]; Vially

Alexia *see* 'CANLOT'

'ALEXIA WILSON', F, ob, 1972; flowers cream veined orange; [Elizabeth of Glamis sport]; Horner

Alexis *see* 'HARLEXIS'

'ALEXIS LEPÈRE', HP, mr, 1875; flowers very large, dbl.; Vigneron; (Sangerhausen)

'ALEZANE', HT, ab, 1935; bud urn shaped, reddish brown; flowers deep apricot, large, dbl., cupped, moderate, fruity fragrance; foliage glossy, bronze, dark; very vigorous growth; [Angèle Pernet X Comtesse de Castilleja]; Pahissa; J&P

Alfi™ *see* 'POULFI'

Alfie™ *see* 'POULFI'

Alfie Luv *see* 'TALALF'

'ALFIERI', HGal, m, 1833; flowers medium large, dbl.; Vibert; (Sangerhausen)

'ALFRED A. BUCKWELL', HT, mr, 1952; flowers red, dbl., 30 petals, 3 in., exhibition form, intense fragrance; foliage dark, leathery; vigorous growth; [Hector Deane X Betty Uprichard]; Buckwell

'ALFRED COLOMB', HP, dp, 1865; flowers strawberry-red, reflexes crimson-carmine, large, dbl., 45 petals, exhibition form, recurrent bloom, intense fragrance; growth dense; [Général Jacqueminot X Seedling]; Lacharme, F.

'ALFRED DE DALMAS', M, lp, 1855; (Mousseline); bud rose-colored; flowers light pink with blush edges, of poor quality, small, dbl., in corymbs, some recurrent bloom, moderate fragrance; very prickly; vigorous, straggling growth; Laffay, M.

'ALFRED DE ROUGEMONT', HP, dr, 1863; flowers crimson-magenta, well formed, very large, dbl.; vigorous, upright growth; [Général Jacqueminot seedling]; Lacharme, F.

Alfred Dietrich, LCl, rb, 1980

'ALFRED K. WILLIAMS', HP, mr, 1877; flowers carmine-red, changing to magenta; [Général Jacqueminot sport]; Schwartz, J.

'ALFRED LEVEAU', HP, 1860; Vigneron; (Cavriglia)

'ALFRED NEWTON', HT, yb, 1959; flowers pale yellow edged crimson, dbl., 35–40 petals, 5 in., exhibition form, intense fragrance; foliage dark; vigorous growth; [Moonbeam X Karl Herbst]; Kemp, M.L.

'ALFRED PÉTOT', HT, dr, 1935; bud elongated; flowers crimson, well formed, large blooms, dbl., intense fragrance; foliage dark; strong stems; very vigorous growth; [Jeanne Excoffier X Yves Druhen]; Buatois

'ALFRED W. MELLERSH', HT, op, 1918; flowers salmon-yellow, shaded rose, center amber; Paul, W.

'ALFREDO MOREIRA DA SILVA', HT, yb, 1946; bud long, pointed; flowers golden yellow tinted coral, large, semi-dbl.; foliage dark; very vigorous, upright growth; [Dr. Kirk X Peace]; Mallerin, C.; A. Meilland

'ALGER', HT, op, 1943; bud pointed; flowers clear pink, reverse salmon-pink, very large, dbl.; foliage leathery; vigorous growth; Gaujard

'ALGONQUIN', S, pb, 1928; flowers purplish rose, center white, large, single, flat, non-recurrent; large, bottle shaped, red fruit; foliage dull, yellow-green; very vigorous (10 ft.) growth; hardy; [R. rubrifolia X R. rugosa hybrid seedling]; Central Exp. Farm

'ALI-BABA', F, or, 1963; flowers large, semi-dbl., 15–20 petals, blooms in clusters; [(Corail X Baccará) X Seedling]; Croix, P.

'ALIBI', F, w, 1960; flowers well formed, dbl., 30 petals, 3–4 in.; vigorous growth; [Kaiserin Auguste Viktoria X Pink Fragrance]; deRuiter

'ALICE', HT, lp, 1978; bud long; flowers light pink, darker center, dbl., 45 petals, exhibition form, borne singly, shy bloom, no fragrance; long, red prickles; foliage large; spreading (to about 4 ft.) growth; [Royal Highness X Christian Dior]; Allender, Robert William

'ALICE', Pol, 1925; flowers pink, fringed, borne in large clusters; RULED EXTINCT 11/80 ARM; [Echo X Orléans Rose]; Spek

'ALICE', 1935; dbl.; RULED EXTINCT 11/80 ARM; Wright, Percy H.

'ALICE ALDRICH', HRg, lp, 1901; flowers clear, bright pink, large, dbl., repeat bloom; [R. rugosa X Caroline de Sansal]; Lovett, J.T.; Conard & Jones

'ALICE AMOS', F, pb, 1922; bud long, pointed; flowers cerise, white eye, single, blooms in clusters on strong stem; very vigorous growth; [Tip-Top X Seedling]; Spek; Prior

Alice Faye *see* 'SEAODD'

'ALICE GRAHAM', HT, w, 1903; flowers large, very dbl., moderate fragrance; Dickson, A.; (Sangerhausen)

'ALICE GRAY', Ayr, w; flowers creamy white, semi-dbl.; moderately vigorous growth; P.T., 1848

'ALICE HAMILTON', Ch, dr, 1903; Nabonnand

'ALICE HARDING', HT, my, 1937; bud ovoid; flowers golden yellow, large, dbl., moderate fragrance; foliage glossy, dark; vigorous growth; [Souv. de Claudius Pernet X Mrs Pierre S. duPont]; Mallerin, C.; J&P

'ALICE HOFFMANN', Ch, rb, 1897; Hoffmann

Alice Jarrett *see* 'EVEJAUNTY'

'ALICE KAEMPFF', HT, pb, 1921; flowers silvery rose-pink, center coppery yellow, dbl., moderate fragrance; [Gen. MacArthur X Radiance]; Felberg-Leclerc

'ALICE KING', Cl F, mr, 1988; flowers medium, luminous red, aging lighter, small, very dbl., 40–50 petals, borne in sprays, no fragrance; prickles small, medium red; foliage light green, glossy; vigorous, slight perpetual growth; [Dublin Bay X Seedling]; Harrison, G.

Alice Lee *see* 'TALALI'

'ALICE LEROI', M, pb, 1842; bud well-mossed; flowers lilac-blush shaded rose, center deep rose, very large, dbl.; vigorous growth; Vibert

'ALICE LINDSELL', HT, lp, 1902; flowers large, dbl.; Dickson, A.; (Sangerhausen)

'ALICE MANLEY', HT, my, 1958; bud long, pointed; flowers mimosa-yellow, dbl., 45–50 petals, 3.5–4 in., exhibition form, slight fragrance; foliage leathery; vigorous, upright, bushy growth; [Seedling X Golden Rapture]; Hill, Joseph H., Co.

'ALICE MARION WHYTE', HT, lp, 1932; flowers soft pink, fading pure white, moderate fragrance; Evans

Alice Mavis, S, lp; Peden, G.H., 1994; (Weatherly, L.)

'ALICE PAT', F, rb, 1981; bud ovoid; flowers red shading to white, reverse white shaded pink, dbl., 38 petals, flat, borne 1-6 per cluster, slight fragrance; slightly hooked, reddish-tan prickles; foliage dark, glossy; upright, short growth; Bronze, ARC TG, 1983; [Seedling X Seedling]; Jerabek, Paul E.

Alice Springs, LCl, lp; cupped, moderate; Peden; (Weatherly, L.)

'ALICE STERN', HT, w, 1926; bud long, pointed; flowers white, center cream, sometimes salmon, large, dbl., 30–40 petals, moderate fragrance; foliage dark, bronze; vigorous, bushy growth; [Grange Colombe X Sunburst]; Gillot, F.

'ALICE VENA', HGal, m; flowers plum-purple, large, blooms in clusters

'ALICE WIEMAN', Pol, dp, 1965; flowers pink, open, small, dbl., continuous bloom, moderate fragrance; growth moderate, bushy; [Seedling X Rita Sammons]; Bodley

'ALICIA', HT, pb, 1972; flowers rose-pink, edged cream, full, dbl., 36 petals, 3.5 in., moderate fragrance; foliage dark; vigorous growth; [Golden Scepter X Ena Harkness]; Lees

Alicja, LCl, lp

'ALIDA', HT, mr, 1938; bud long, pointed; flowers well formed, very dbl.; foliage bright; vigorous, bushy growth; [Charles P. Kilham X E.G. Hill]; Lens

'ALIDA LOVETT', LCl, lp, 1905; bud long, pointed; flowers shell-pink, base shaded sulfur, large, dbl., blooms in clusters, seasonal bloom, slight fragrance; foliage glossy; vigorous, climbing growth; [Souv. du Prés. Carnot X R. wichurana]; Van Fleet, 1905; J.T. Lovett

'ALIETTE', HT, 1958; flowers salmon-pink tinted orange, well formed, slight fragrance; vigorous, bushy growth; [Pres. Herbert Hoover X Signora]; Arles; Roses-France

'ALIJ PACHA CHERIFF', HP, 1886; Lévêque, P.; (Cavriglia)

'ALIKA', HGal, mr, 1906; (Gallica Grandiflora, R. gallica grandiflora); flowers brilliant red with no purple, many stamens, petalage variable, large, semi-dbl., moderate fragrance; vigorous growth; Hansen, N.E., 1906; In commerce, 1930

'ALINE', HT, or, 1950; flowers peach, compact, dbl., 22 petals, 4 in., moderate fragrance; foliage dark; vigorous growth; [Picture sport]; Astolat Nursery

Aline, Min, or, 1994; Laperrière

Alinka, HT, ob, 1985; Kordes

Alisha *see* 'SPODAY'

Alison *see* 'COCLIBEE'

'ALISON BROWN', HT, dp, 1993; flowers deep pink, dbl., 26–40 petals, 3–3.5 in., borne mostly singly, some small clusters, intense fragrance; some prickles; foliage medium, dark green, semi-glossy; medium (125 cms), bushy growth; [Ena Harkness sport]; Brown, Mrs. Ann; Brown, 1993

'ALISON WHEATCROFT', F, ab, 1959; flowers apricot flushed crimson; [Circus sport]; Wheatcroft Bros.

Alistair Sheridan *see* 'PEARICH'

'ALISTER CLARK', F, lp, 1999; dbl., 26–40 petals, 3–3.5 in., borne in small clusters, intense fragrance; prickles moderate; foliage small, dark green, glossy; spreading, medium (3 ft) growth; [Marjory Palmer sport]; Newman, Laurie, 1990; Reliable Roses, 1990

Alister Clark's Pink, HT, lp; Clark, A.; (Weatherly, L.)

'ALISTER STELLA GRAY', N, ly, 1894; (Golden Rambler); bud long, pointed; flowers pale yellow, center orange, fading to white, dbl., blooms in blusters, recurrent bloom, moderate fragrance; vigorous, climbing growth; (14); Gray, A.H.

'ALIX ROUSSEL', T, yb, 1908; Gamon

All Ablaze™ *see* 'WEKSAMSOU'

All Gold *see* **'ALLGOLD'**®

All In One *see* 'MEILIDER'

All That Jazz™ *see* 'TWOADVANCE'

All-American Bride *see* WILBRID

'ALLAMAND-HO', S, pb, 1984; bud ovoid; flowers pink and yellow blend, petals edged ruby red, color intensifying with ag, dbl., 38 petals, cupped, blooms in clusters of 1-6, repeat bloom, moderate, sweet fragrance; awl-like, red-brown prickles; foliage dark, leathery, semi-glossy; erect, bushy growth; hardy; [(Hawkeye Belle X Prairie Star) X Iobelle]; Buck, Dr. Griffith J.; Iowa State University

Allambie, HT, 1973; Dovanan, R.; (Weatherly, L.)

'ALLARD', (form or hybrid of R. xanthina), dy; (R. xanthina Allard); flowers chrome-yellow, dbl.

Allegeo '80 *see* 'MEIFIKALIF'

'ALLÉGRESSE', HT, mr, 1952; flowers red, becoming lighter, well formed, very dbl.; strong stems; vigorous, upright growth; [Fantaisie X Sensation]; Robichon; Vilmorin-Andrieux

'ALLEGRETTO', HT, or, 1975; bud long, pointed; semi-dbl., 14–17 petals, 4 in., slight fragrance; foliage leathery; upright growth; [Fragrant Cloud X Sutter's Gold]; Huber

Allegro® *see* 'MEIARLO'

Alleluia, Cl Pol, 1960; Cazzaniga, F. G.; (Cavriglia)

Allelulia® *see* 'DELATUR'

Allen Box, Pol, lp; Box, S., 1928; (Weatherly, L.)

Allen Brundrett, HT, dr, 1994; Brundrett

'ALLEN CHANDLER', Cl HT, mr, 1923; bud long, pointed; flowers brilliant crimson, large blooms, semi-dbl., borne 3-4 per cluster, recurrent bloom, slight fragrance; foliage dark, leathery, glossy; vigorous, pillar growth; (28); GM, NRS, 1923; [Hugh Dickson X Seedling]; Chandler; Prince

'ALLEN'S FRAGRANT PILLAR', Cl HT, rb, 1931; bud long, pointed; flowers cerise, base flushed yellow, open, large blooms, dbl., recurrent bloom; foliage glossy, bronze; long, strong stems; moderate climbing growth; [Paul's Lemon Pillar X Souv. de Claudius Denoyel]; Allen

'ALLEN'S GOLDEN CLIMBER', LCl, ob, 1933; flowers orange, large, dbl., non-recurrent; foliage glossy; Allen

Allen's Jubilee *see* **'JUBILEE'**

'ALLGOLD'®, F, my, 1958; (All Gold); flowers bright buttercup-yellow, 15–22 petals, 3 in., borne singly and in large trusses, slight fragrance; foliage small, glossy, dark; vigorous growth; GM, NRS, 1956; [Goldilocks X Ellinor LeGrice]; LeGrice, 1956

'ALLGOLD, CLIMBING', Cl F, my, 1961; (Grimpant All Gold, Grimpant Allgold); [Allgold sport]; Gandy, Douglas L.

'ALLIANCE', F, dp, 1966; bud ovoid; flowers light red and deep pink, large blooms, dbl., moderate fragrance; foliage soft;

strong stems; very vigorous, bushy growth; [Rosita X Queen Elizabeth]; Delforge

Alliance *see* 'MEIBLERI'

'ALLIANCE FRANCO-RUSSE', T, my, 1899; Goinard

'ALLISON SWEETIE', Gr, mr, 1975; bud long, pointed, oval; dbl., 21–25 petals, 3–3.5 in., exhibition form, moderate fragrance; foliage large, glossy, dark; tall, vigorous growth; [Tropicana X Mister Lincoln]; Miller, F.

'ALLISWELL', HT, w, 1956; bud ovoid; flowers white overcast pink, center cameo-pink, peony shaped, dbl., 50–60 petals, 5–6 in., moderate fragrance; vigorous, bushy growth; [Neige Parfum X (Charlotte Armstrong X Blanche Mallerin)]; Motose

Allotria® *see* 'TANAL'

Allspice *see* 'AROALL'

Allumbie

'ALLURE', HT, pb, 1950; bud long, pointed, carmine-rose; flowers neyron rose, base yellow, dbl., 28–30 petals, 5–6 in., exhibition form, slight fragrance; foliage leathery, glossy, light green; very vigorous, upright growth; [Mrs Pierre S. duPont X Charlotte Armstrong]; Swim, H.C.; Inter-State Nursery

Allux Symphony *see* 'AUSLETT'

'ALMA', F, dr; flowers dark crimson, dbl., blooms in large clusters; dwarf growth; [Orange Triumph X Eutin]; Riethmuller

Alma Bierbauer, S, op, 1997; flowers salmon pink with yellow eye and pink stamens, single; Williams, J. Benjamin

'ALMA MATER', HT, lp, 1929; flowers lighter in color; [Columbia sport]; Good & Reese

Alma-atinskaj Aromatnaja, HT, m, 1959; flowers large, very dbl., intense fragrance; Besschetnowa; (Sangerhausen)

Almandet, HT, mp, 1966; Croix

'ALMIRANTE AMÉRICO TOMÁS', HT, rb, 1955; (Miramar); flowers geranium-red, reverse old-gold, large, dbl.; very vigorous growth; [Sultane X Peace]; da Silva, Moreira

Almond Glory *see* 'ZIPALM'

Almondeen *see* 'AROFRAP'

'ALOHA', Cl HT, mp, 1949; bud ovoid; flowers rose-pink, reverse deeper, dbl., 58 petals, 3.5 in., cupped, recurrent bloom, moderate fragrance; foliage leathery, dark; vigorous, pillar (8-10 ft.) growth; [Mercedes Gallart X New Dawn]; Boerner; J&P

'ALOIS JIRÁSEK', HT, ob, 1931; bud long, pointed; flowers dark orange, tinted brownish yellow, large, dbl., moderate fragrance; foliage glossy; vigorous growth; [Mme Butterfly X Mrs George Shawyer]; Böhm, J.

'ALOUETTE', Pol, op, 1971; bud long, pointed; flowers salmon-orange, medium, semi-dbl., cupped, profuse, continuous bloom, slight fragrance; foliage leathery; vigorous, bushy growth; [Ambassadeur Baert X Seedling]; Delforge

'ALOYSIA KAISER', S, my, 1937; bud ochre-yellow; flowers reddish bright yellow, round, large, dbl., blooms in clusters, seasonal bloom; vigorous growth; [Miss G. Mesman X Belle Doria]; Lambert, P.

'ALPAIDE DE ROTALIER', HP, lp, 1863; Campy

'ALPENFEE', HSet, m, 1890; flowers medium large, dbl.; Geschwind, R.; (Sangerhausen)

Alpenglühen *see* **'ALPINE GLOW'**

Alpengrüss, F, 1974; Noack, Werner; (Cavriglia)

Alpenkonigin®, F, mr

Alpha *see* 'MEINASTUR'

Alpha Moe *see* 'MOEALPHA'

Alphee, HT, 1972; Croix, P.; (Cavriglia)

Alphonse Daudet *see* MEIrouve

'ALPHONSE DE LAMARTINE', HP, lp, 1853; flowers rosy blush, medium, dbl.; moderate growth; Ducher

'ALPHONSE KARR', T, m, 1878; flowers crimson-purple, center lighter, large, dbl.; [Duchess of Edinburgh seedling]; Nabonnand, G.

'ALPHONSE SOUPERT', HP, mp, 1883; flowers large, dbl., moderate fragrance; Lacharme, F.; (Sangerhausen)

'ALPHONSE TROUSSARD', HT, 1951; Buatois; (Cavriglia)

'ALPIN', LCl, or, 1960; flowers bright orange-red, large, dbl., recurrent bloom, slight fragrance; foliage dark, glossy; vigorous growth; [Spectacular X Seedling]; Combe; Minier

'ALPINE', HT, my, 1954; bud long, pointed; flowers clear yellow, open, dbl., 30 petals, 5.5 in., moderate fragrance; foliage leathery; vigorous, upright growth; [Sunnymount sport]; Grillo

'ALPINE GLOW', F, or, 1954; (Alpenglühen); bud deep orange-red; flowers vermilion-red, dbl., 28 petals, 3–4 in., cupped, blooms in clusters of 4-10, moderate fragrance; foliage glossy; vigorous, bushy growth; [Cinnabar X (Cinnabar X Käthe Duvigneau)]; Tantau, Math.; J&P

Alpine Rock, HT, w

Alpine Rose *see* **R. PENDULINA**

'ALPINE SUNSET'®, HT, ab, 1973; flowers peach-pink, apricot reverse, dbl., 30 petals, 7–8 in., moderate fragrance; foliage glossy, medium green; vigorous, upright growth; [Dr. A.J. Verhage X Irish Gold]; Cants of Colchester, Ltd., 1974

'ALSACE', HT, op, 1946; flowers salmon-pink, base gold, dbl., 6 in., moderate fragrance; foliage dark; free growth; [Peace X Mme Joseph Perraud]; Meilland, F.

Alsace *see* PEKtarampe

'ALSACE-LORRAINE', HP, dr, 1879; flowers dark velvety red, large; Duval

'ALSTERUFER', HT, dr, 1909; flowers medium large, dbl., intense fragrance; Lambert, P.; (Sangerhausen)

'ALT WIEN', HT, pb, 1965; flowers carmine-rose, dbl., 4 in.; foliage dark, glossy; compact growth; [Queen of Bermuda sport]; Prinz; Wohlt

Altaica *see* **R. SPINOSISSIMA ALTAICA**

'ALTALARIS', S, w, 1941; bud large, pointed; flowers white, sometimes flushed pink, open, single, abundant, non-recurrent bloom; apple shaped, bright red fruit; foliage leathery; vigorous, bushy, prickly growth; [R. spinosissima altaica X R. acicularis]; Skinner

Alte Liebe, HT, or, 1974; flowers large, dbl., slight fragrance; Berger, A.; (Sangerhausen)

Alte Liebe, F, op, 1983; flowers salmon-orange and pink, medium large, dbl.; Kordes, W. Söhne; (Sangerhausen)

Altenburg, F, or, 1977; flowers medium large, dbl.; Berger, A.; (Sangerhausen)

'ALTESSE', HT, dp, 1950; bud ovoid; flowers strawberry-red, large, dbl., 35–40 petals, globular; vigorous, upright growth; [Vercors X Léonce Colombier]; Meilland, F.; URS

Altesse *see* 'MEIDINRO'

Altesse 75, HT, 1975; Meilland, L.; (Cavriglia)

'ALTHEA', HT, mr, 1930; flowers glowing crimson, moderate, damask fragrance; Pemberton; Bentall

Altissimo® *see* 'DELMUR'

Alt-marburg, S, lp, 1979; flowers large, dbl., moderate fragrance; Weihrauch; (Sangerhausen)

'ALTMÄRKER', HT, yb, 1908; flowers ochre-yellow tinted garnet, large, dbl., moderate fragrance; vigorous growth; [Kaiserin Auguste Viktoria X Luciole]; Türke

'ALTONIA', (natural variation of R. setigera), mp; flowers brilliant pink, single or semi-double, semi-dbl.; ((found in Illinois, 1835))

'ALT-ROTHENBURG', HT, dp, 1939; flowers large, dbl., moderate fragrance; Burkhardt; (Sangerhausen)

Altus *see* 'DELMUR'

'ALUPKA', N, w, 1800; flowers medium large, dbl., moderate fragrance; (Sangerhausen)

'ALVARES CABRAL', F, mr; flowers bright red; [Pinocchio X Alain]; da Silva, Moreira

'ALWAYS', HT, dr, 1959; bud long, pointed; flowers dark cardinal-red, dbl., 28 petals, 6 in., moderate fragrance; foliage leathery; vigorous growth; [(Charlotte Armstrong X Applause) X Ena Harkness]; Leon, Charles F., Sr.

Always A Lady *see* 'TINLADY'

Always Mine™ *see* 'DEVSIEM'

'AMA', F, or, 1955; bud ovoid; flowers deep orange-scarlet, large, dbl., exhibition form, blooms in clusters (up to 20); foliage dark, glossy, leathery; vigorous, bushy growth; [Obergärtner Wiebicke X Independence]; Kordes

Amadeus® *see* 'BARAMAD'

Amadeus®, HT, ob

Amadine, P, mp

'AMADIS', Bslt, dr, 1829; (Crimson Boursault); flowers deep crimson-purple, large, semi-dbl., cupped; no prickles; young wood whitish green, old wood red-brown; vigorous, upright growth; excellent pillar; [R. chinensis X R. pendulina]; Laffay, M.

'AMAEVELINA', Min, lp, 1992; (**Evelina**); dbl., 26–40 petals, 1.5–2.75 in., borne mostly singly, no fragrance; foliage medium, medium green, semi-glossy; medium (100-150 cms), bushy growth; [Rosa Maria X Seedling (Pink 078)]; Mansuino, Dr. Andrea, 1990

Amaglia, LCl, w, 1980

'AMALFI', Pol, dy, 1971; bud long, pointed; flowers deep yellow to salmon-pink, small, dbl., cupped, abundant, continuous bloom, slight fragrance; foliage small, dark; moderate, bushy growth; Delforge

Amalia™ *see* 'MEICAUF'

'AMALIA JUNG', HT, mr, 1934; flowers crimson-red, large, dbl., moderate fragrance; foliage glossy, cedar-green; vigorous growth; [Mrs Henry Winnett X Lady Helen Maglona]; Leenders, M.

'AMALIE DE GREIFF', HT, pb, 1912; flowers brick-rose, center salmon and orange-yellow, dbl.; [Herrin von Lieser X Mme Mélanie Soupert]; Lambert, P.

Amalinda, HT, mr

'AMAMI', HT, op, 1927; flowers peach-pink, overlarge, semi-dbl., 12–15 petals; foliage light; very vigorous growth; Easlea

Amanda *see* 'BEESIAN'

Amanda™ *see* 'DEVNINA'

Amanda *see* **'AMRUDA'**

Amanda, F, ob, 1979; Noack, Werner

Amanda Kay *see* 'JUSAMANDA'

Amanda Marciel™ *see* 'DEVNINA'

'AMANDA PATENAUDE', P, mp

Amanecer, HT; Dot, Pedro; (Cavriglia)

Amara *see* GAUhari

Amarancha, F, ab

'AMARANTE', Pol, mr, 1916; flowers dark crimson, sometimes striped white, blooms in clusters of 25-70; Barbier

'AMARILLO', HT, dy, 1961; bud pointed; flowers deep yellow, dbl., 30 petals, 5 in., moderate fragrance; foliage leathery, light green; vigorous, upright growth; [Buccaneer X Lowell Thomas]; Von Abrams; Peterson & Dering

Amarillo, F, dy, 1994; Select Roses, B.V.

'AMATEUR E. BIRON', HT, op, 1928; flowers shrimp-pink, center tinted copper, reverse old-rose, camellia-shaped; Biron

'AMATSU-OTOME', HT, yb, 1960; flowers golden yellow edged deep orange, dbl., 48 petals, 4.5–5 in., exhibition form, slight fragrance; foliage semi-glossy; vigorous, compact growth; [Chrysler Imperial X Doreen]; Teranishi, K.; Itami Rose Nursery

'AMAURY FONSECA', Pol, w, 1914; flowers white, suffused light pink in fall, well formed; Soupert & Notting

'AMAZING GRACE', HT, mp, 1973; flowers rich pink, high pointed, dbl., 43 petals, 5 in., slight fragrance; [Carina X Mischief]; Anderson's Rose Nurseries

Amazon *see* JACap

'AMAZONE', LCl, mr, 1961; flowers bright red; [Spectacular sport]; Delforge

'AMAZONE', T, my, 1872; flowers yellow, reverse veined rose, well formed; [Safrano X ?]; Ducher

Amazone™, F, mr; de Ruiter, G.

Ambasciatore Marco Fracisci, T, lp

'AMBASSADEUR BAERT', F, op, 1964; flowers salmon; foliage bronze; low, compact growth; [Sumatra X Seedling]; Delforge

'AMBASSADEUR NEMRY', F, dp, 1949; flowers deep rose-pink, reverse salmon-carmine, large, dbl., moderate fragrance; vigorous growth; Leenders, M.

'AMBASSADOR', HT, ob, 1930; bud long, pointed; flowers bronze-salmon, large, dbl., moderate fragrance; foliage leathery, dark; vigorous, bushy growth; [Mme Butterfly X Souv. de Claudius Pernet]; Premier Rose Gardens

Ambassador® *see* 'MEINUZETEN'

Ambassador *see* MEIbigoud

'AMBER', Cl Pol, 1908; flowers pale amber, single, early; low growth; RULED EXTINCT; [Jersey Beauty seedling]; Paul, W.

'AMBER', HT, 1930; flowers amber; RULED EXTINCT; [Ophelia sport]; Jordan, B.L.; Beckwith

'AMBER BEAUTY', F, pb, 1962; flowers pink tinted brown, dbl., intense fragrance; [Goldilocks X Lavender Pinocchio]; Leenders, J.

Amber Flash™ *see* 'WILDAK'

'AMBER GOLD', HT, dy, 1962; bud ovoid; flowers deep golden yellow, dbl., 40–45 petals, 5–5.5 in., cupped, moderate fragrance; foliage leathery; vigorous, upright growth; [Golden Rapture X Golden Scepter]; Moro; J&P

Amber Light *see* **'AMBERLIGHT'**

Amber Nectar *see* 'MEHAMBER'

Amber Queen® *see* 'HARROONY'

Amber Ribbon *see* 'ZIPAMB'

'AMBER SANDS', Min, ab, 1984; flowers large, dbl., 35 petals, exhibition form, slight fragrance; foliage medium, medium green, semi-glossy; upright, bushy growth; [Fragrant Cloud X Poker Chip]; Hardgrove, Donald & Mary; Rose World Originals

Amber Spire, S, ab, 1995; Kordes

Amber Star, HT, ab

Amber Star *see* 'MANSTAR'

Amber Sunset *see* 'MANAMSUN'

Amber Waves™ *see* 'JACAMQUE'

Amberglo™ *see* 'MINAPCO'

'AMBERLIGHT', F, yb, 1961; (Amber Light); flowers clear amber, dbl., 3.5–4 in., blooms in open clusters, intense, fruity fragrance; vigorous, upright, bushy growth; [(Seedling X Lavender Pinocchio) X Marcel Bourgouin]; LeGrice, 1962

Amberlight *see* 'COCBAMBER'

'AMBIANCE', HT, lp, 1955; bud long, pink, passing to cream; [Comtesse Vandal X Pres. Macia]; Delforge

Ambiance *see* NIRPnufdeu

Ambiance, HMsk, pb, 1994; Lens

'AMBOSSFUNKEN', HT, rb, 1961; (Anvil Sparks); flowers coral-red streaked golden yellow, well-formed, large, dbl., 33 petals, moderate fragrance; bushy growth; Meyer; Kordes

'AMBRA'®, F, ob, 1982; flowers deep, large, dbl., 20 petals, cupped, blooms in clusters of 3-5, slight fragrance; reddish prickles; foliage medium, brownish-green, glossy; upright, bushy growth; [Seedling X Seedling]; Rose Barni-Pistoia

'AMBRE', HT, ob, 1950; bud ovoid; flowers brilliant orange and yellow, very large, dbl., moderate fragrance; foliage glossy, dark; very vigorous, bushy growth; [Peace X Seedling]; Gaujard

'AMBRE SOLAIRE', Pol, pb, 1966; bud ovoid; flowers red, pink and yellow, dbl., blooms in clusters, slight fragrance; foliage dark; [Masquerade X Seedling]; Ebben

Ambridge Rose *see* 'AUSWONDER'

'AMBROGIO MAGGI', HP, mr, 1879; flowers very large, dbl.; Pernet Père; (Sangerhausen)

'AMBROISE PARÉ', HGal, m, 1846; Vibert

'AMBROSIA', F, ob, 1962; flowers amber, single, 7–10 petals, 2.5–3 in., flat, blooms in large clusters; foliage dark; vigorous, bushy growth; [Seedling X Shepherd's Delight]; Dickson, Patrick; A. Dickson

Ambrosia *see* **'SUNAMBRO'**

'AMDO', HRg, mp, 1927; semi-dbl., 16 petals, in clusters of 7-10, non-recurrent; [Tetonkaha X La Mélusine]; Hansen, N.E.

'AMEDÉE DE LANGLOIS', B, 1872; Vigneron; (Cavriglia)

'AMÉDÉE PHILIBERT', HP, m; flowers purple and red; Lévêque, 1879

Amélia, A, mp; flowers bright pink, anthers deep yellow, large, dbl., non-recurrent bloom, moderate fragrance; Vibert, J. P., 1823

Amelia Anderson, T, pb

'AMELIA BARTER', HT, mr, 1963; flowers bright scarlet tinted silvery, dbl., 5.5 in.; foliage dark, glossy; [Queen Elizabeth X Claude]; Barter

'AMELIA EARHART', HT, yb, 1932; (Président Charles Hain); flowers golden yellow, center flushed pink, large, dbl., intense fragrance; vigorous growth; [Souv. de Claudius Pernet X (Louise Catherine Breslau X Paul Neyron)]; Reymond; J&P

'AMELIA FLEMING', F, mp, 1995; flowers medium pink with pink tones, prominent stamens, single, 5 petals, 1.5–2.75 in., borne in clusters, moderate fragrance; foliage medium, medium green, matt; upright (120-140 cms) growth; [Marchenland X Bambula]; Fleming, Joyce L.; Hortico Roses, 1994

'AMELIA LOUISE', Min, mp, 1994; flowers small, dbl., 26–40 petals, 1.5–3 in., borne mostly singly, slight fragrance; few prickles; foliage small, medium green, semi-glossy; low, bushy growth; Haynes, F., 1993; F. Haynes & Partners, 1993

Amelia Louise, HP, mp, 1844; flowers purple-violet, dbl., cupped; [Seedling X Seedling]

Amelia Rey Colaco, F, 1960; Moreira da Silva, A.; (Cavriglia)

'AMÉLIE DE BETHUNE', HT, rb, 1923; flowers coral-red shaded carmine, dbl.; Pernet-Ducher

'AMÉLIE DE MANSFIELD', HGal, mp; flowers vivid pink, medium, dbl.

'AMÉLIE GRAVEREAUX', HRg, mr, 1903; bud ovoid; flowers medium red, fading, dbl., recurrent bloom; foliage dark, rugose; vigorous, spiny growth; [(Général Jacqueminot X Maréchal Niel) X Conrad Ferdinand Meyer]; Gravereaux

'AMERICA', HMult, mp, 1915; (Walsh's Rambler); flowers pink, center white, single, in large clusters (often to 75); [R. wichurana X R. multiflora]; Walsh

America™ *see* 'JACCLAM'

'AMERICAN BANNER', N, pb, 1879; flowers carmine striped white, small, semi-dbl.; [Bon Silène sport]; Cartwright; P. Henderson

'AMERICAN BEAUTY', HP, dp; bud globular; flowers deep pink, large, dbl., 50 petals, cupped, sometimes recurrent bloom, intense fragrance; vigorous growth; Lédéchaux, 1885; Bancroft and Field Bros., 1886

'AMERICAN BEAUTY, CLIMBING', LCl, dp, 1909; flowers deep rose-pink, large, cupped, non-recurrent, moderate fragrance; height 12-15 ft; [(R. wichurana X Marion Dingee) X American Beauty]; Hoopes, Bro. & Thomas

'AMERICAN CLASSIC', HT, dr, 1994 (America's Classic); flowers moderately full, large blooms, dbl., 15–25 petals, 3–3.5 in., borne mostly singly, slight fragrance; some prickles; foliage large, medium green, matt; tall, upright growth; [Seedling X Seedling]; Winchel, Joseph F., 1997; Certified Roses, Inc., 1997

'AMERICAN DAWN', HT, rb, 1976; bud ovoid; flowers rose-red, base white, dbl., 30 petals, 4–5 in., exhibition form, slight fragrance; foliage glossy, dark; upright growth; [Personality X Seedling]; Warriner, William A.; J&P

American Dream™ *see* 'WINBUR'

American Fantasy™ *see* 'TWOFAN'

'AMERICAN FLAGSHIP', HT, mr, 1946; bud urn-shaped; flowers bright scarlet, semi-dbl., 15 petals, 3.5–4.5 in., slight, spicy fragrance; foliage leathery; vigorous, upright, bushy growth; [Crimson Glory X Self]; Lammerts, Dr. Walter; C.R. Burr

'AMERICAN GIRL', HT, mr, 1929; flowers large, dbl., exhibition form, moderate fragrance; foliage leathery, dark; long stems; vigorous growth; [Hollywood sport]; Maton

American Glory *see* 'TWOADMIRE'

American Heritage® *see* 'LAMLAM'

'AMERICAN HERITAGE, CLIMBING', Cl HT, yb, 1971; Arora, Bal Raj; The Rosery, India

'AMERICAN HOME', HT, dr, 1960; bud ovoid; dbl., 30 petals, 4.5 in., cupped, intense fragrance; foliage leathery; vigorous, upright growth; [Chrysler Imperial X New Yorker]; Morey, Dr. Dennison; J&P

American Honor *see* 'TWOHONOR'

American Independence *see* 'MEIFINARO'

American Legion *see* **'LEGION'**

'AMERICAN PILLAR', HWich, pb, 1902; flowers carmine-pink, white eye, golden stamens, single, blooms in large clusters, non-recurrent; red fruit; foliage leathery, glossy; vigorous (15-20 ft.) growth; (21); [(R. wichurana X R. setigera) X Red Hybrid Perpetual]; Van Fleet, 1902; Conard & Jones, 1908

American Pride *see* 'JACARED'

'AMERICAN PRIDE', HT, 1928; bud long, pointed; flowers pure white, outside petals occasionally tinted pink, dbl., 35 petals, 4.5 in., moderate fragrance; foliage leathery; very vigorous growth; RULED EXTINCT 2/79 ARM; [Grillodale sport]; Grillo

American Rose Centennial™ *see* 'SAVARS'

'AMERICAN ROSEATE', HEg, dp; flowers bright rose, semi-dbl.; vigorous growth; Prince Nursery, prior to 1846

American Spirit *see* 'JACTRED'

'AMERICAN WHITE', HEg, w; flowers creamy white, semi-dbl.; vigorous growth; Prince Nursery, about 1840

'AMERICANA', HT, mr, 1961; bud ovoid; flowers bright red, dbl., 28 petals, 5.5 in., exhibition form, moderate fragrance; foliage leathery; vigorous, upright growth; [Poinsettia seedling X New Yorker]; Boerner; J&P

America's Classic *see* **'AMERICAN CLASSIC'**

America's Choice *see* 'POULANDER'

'AMERICA'S JUNIOR MISS', F, lp, 1964; (Junior Miss); bud ovoid; flowers soft coral-pink, medium, dbl., moderate fragrance; foliage glossy; vigorous, bushy growth; [Seventeen X Demure seedling]; Boerner; J&P

Amerlock®, S, op

'AMES 5', S, mp, 1932; (Ames Climber); flowers good size, in clusters, non-recurrent; thornless stems; red in winter; very vigorous growth; very hardy; [R. multiflora X R. blanda]; Maney; Iowa State College

'AMES 6', S, mp; (14)

Ames Climber *see* **'AMES 5'**

Amethyst, HT, m, 1978; flowers violet-pink, large, dbl., moderate fragrance; Urban, J.; (Sangerhausen)

'AMÉTHYSTE', HWich, m, 1911; flowers violet-crimson; [Non Plus Ultra sport]; Nonin

Ametista®, HT, m, 1985; Barni, V.

'AMI CHARMET', HP, dp, 1900; flowers very large, dbl., moderate fragrance; Dubreuil; (Sangerhausen)

Ami Chenault, HT; Moreira da Silva, A.; (Cavriglia)

Ami Clement®, F, dp

Ami des Jardins *see* 'KORAMI'

'AMI DESVIGNES', HT, mr, 1954; flowers base of petals veined coral, petals waved, large, semi-dbl.; foliage glossy; vigorous growth; Privat

Ami Dietrich, HT; Moreira da Silva, A.; (Cavriglia)

'AMI F. MAYERY', HT, mr, 1938; bud long, pointed, vermilion-red; flowers poppy-red, very large blooms, dbl., cupped; foliage dark; stiff stems; vigorous growth; [Huguette Vincent X Seedling]; Denoyel, Mme.; C. Chambard

'AMI L. CRETTÉ', HT, rb, 1931; bud long, pointed, coppery oriental red; flowers crimson-red, reverse light coral-rose and yellow, medium to, semi-dbl., cupped, moderate fragrance; foliage dark; very vigorous, bushy growth; Chambard, C.; C-P

'AMI LÉON CHENAULT', F, mr, 1929; flowers dark garnet, slightly striated white, blooms in clusters; [Lafayette sport]; Nonin

'AMI LÉON PIN', HT, ly, 1947; bud long, pointed; flowers pale yellow, reverse pink, very large, dbl., moderate fragrance; foliage dark; very vigorous growth; Gaujard

'AMI MARTIN', HP, or, 1906; flowers vermilion orange-red, very large, very dbl., intense fragrance; Chédane-Guinoisseau; (Sangerhausen)

'AMI QUINARD', HT, dr, 1927; bud long, pointed; flowers blackish garnet and coppery scarlet, medium, semi-dbl., 17 petals, cupped, moderate fragrance; foliage leathery; vigorous growth; [Mme Méha Sabatier X (Mrs Edward Powell X R. foetida bicolor]; Gaujard, 1927; C-P, 1930

'AMI RENÉ BADEL', F, mr, 1961; bud pointed; flowers carthamus-red, dbl., 36 petals, exhibition form, intense fragrance; foliage bronze; [Belle Créole X (Gloire du Midi X Paul Crampel)]; Arles; Roses-France

Amiable Rouge

'AMICA', HT, 1966; flowers cinnabar-red, large, moderate fragrance; foliage light green; [Coup de Foudre X Lampo]; Cazzaniga, F. G.; F. Giuseppe

'AMIGA MIA', S, mp, 1978; bud ovoid, pointed; flowers empire-rose, dbl., 30 petals, 4–5 in., exhibition form, moderate fragrance; foliage large, dark, leathery; vigorous, upright, bushy growth; [Queen Elizabeth X Prairie Princess]; Buck, Dr. Griffith J.; Iowa State University

'AMIGO', F, mr, 1951; bud short, pointed; flowers currant-red, semi-dbl., 17–20 petals, 3–3.5 in., cupped, moderate, spicy fragrance; foliage leathery; upright, bushy growth; [World's Fair X Adolf Grille]; Whisler; Germain's

Amigo Roger Sucret, F, 1970; Dot, Simon; (Cavriglia)

'AMIRAL GRAVINA', HP, m, 1860; flowers blackish purple, shaded with scarlet; Moreau et Robert

'AMISTAD SINCERA', HT, w, 1963; (Sincera); bud ovoid; flowers pure white, dbl., 55 petals, 4–4.5 in., exhibition form; foliage leathery; vigorous, well-branched growth; [Alaska X Virgo]; Camprubi, C.; C-P

'AMITIÉ', HT, ob, 1951; flowers coppery orange-yellow, well shaped, large, dbl.; vigorous, branching growth; [Mandalay X Schéhérazade]; Mallerin, C.; EFR

Amitié, HGal, pb

'AMLEGER', HT, mr, 1972; flowers blood-red to cardinal-red, imbricated, dbl., 40 petals, 4.5 in., cupped, slight, tea fragrance; foliage dark, leathery; vigorous growth; [Baccará sport]; Molina; URS

Ammerland®, S, my, 1986; Noack, Werner

'AMOR', F, mp, 1957; bud pointed; flowers clear pink, medium, dbl., 22 petals; vigorous, bushy growth; [(Golden Rapture X Floribunda seedling) X Self]; deRuiter; Blaby Rose Gardens

'AMORE', F, mp, 1957; flowers rosy pink, reverse lighter, medium, dbl., moderate fragrance; foliage semi-glossy; moderate growth; [Orange Triumph X Spring Song (F)]; Riethmuller

Amoretta® *see* 'AMORU'

Amorette *see* 'AMORU'

Amorosa, F; Moreira da Silva, A.; (Cavriglia)

Amorosa *see* KORignale

Amorous *see* 'JACARINA'

'AMORU', Min, w, 1980; (Amoretta®, **Amorette**, Snowdrop); bud pointed; flowers white, ivory center, dbl., borne 10-15 per cluster, slight fragrance; long, narrow, red prickles; foliage light, mid-green; short, dense, bushy growth patio; [Rosy Jewel X Zorina]; deRuiter; Fryer's Nursery, Ltd., 1979

Amour Ardent *see* **'BURNING LOVE'**

'AMOURETTE', HCh, mp; flowers rose-pink, edged lighter, large, pointed, petals recurving; foliage dark; stems red

'AMOUREUSE', HT, ab, 1966; flowers coppery yellow, large, dbl., 34 petals, borne singly; foliage large, medium green; [Peace X Rose Gaujard seedling]; Gaujard

Amours de Savernem *see* SAUdime

'AMPÈRE', HT, or, 1937; flowers nasturtium-red, edges lighter, reverse orange-yellow, very large, dbl., intense fragrance; foliage bright green; [Charles P. Kilham X Condesa de Sástago]; Meilland, F.

'AMRUDA', Min, dr, 1979; (Amanda, Red Ace); flowers medium, dbl., 20 petals, slight fragrance; foliage small, medium green, semi-glossy; bushy growth; [Scarletta X Seedling]; deRuiter, 1977; Fryer's Nursery, Ltd., 1982

Amstelveen, S, op, 1993; Meilland

Amsterdam® *see* 'HAVAM'

'AMULETT' *see* TANaluma

Amulett®, Min, pb, 1991; dbl.; foliage medium green; wide, bushy (20 in). growth; Tantau

'AMURENSIS', (form of R. blanda carpohispida), mp; flowers large, semi-dbl.; sparsely armed; wood red; tall growth

'AMY', F, mp, 1954; bud ovoid; flowers carmine-rose, dbl., 38 petals, 2.5–3 in., exhibition form, blooms in loose clusters, moderate fragrance; foliage leathery; compact, dwarf growth; [Show Girl X Fashion]; Von Abrams; Peterson & Dering

Amy Brown® *see* 'HARKUSHI'

Amy Grant™ *see* 'TUCKAMY'

'AMY JOHNSON', LCl, mp, 1931; bud ovoid; flowers large, dbl., cupped, moderate fragrance; foliage wrinkled; vigorous (12-15 ft) growth; [Souv. de Gustave Prat X ?]; Clark, A.; NRS Victoria

Amy Powell, F, op, 1996

'AMY REBECCA', Min, dy, 1986; flowers deep yellow, small, dbl., 38 petals, exhibition form, borne usually singly, slight fragrance; small, brown prickles; foliage small, medium green, semi-glossy; medium, upright growth; [Rise 'n' Shine X Summer Butter]; Jolly, Marie; Rosehill Farm, 1987

'AMY ROBSART', HEg, dp, 1894; flowers deep rose, large, semi-dbl., summer bloom, moderate fragrance; foliage fragrant; vigorous growth; [R. eglanteria X HP or B]; Penzance; Keynes, Williams & Co.

Amy Sis, Min, 1999

'AMY VANDERBILT', F, m, 1956; bud globular; flowers lavender-lilac, dbl., 70

petals, 3 in., cupped, blooms in pyramidal clusters, moderate fragrance; foliage dark, glossy; upright, bushy growth; [Lavender Pinocchio seedling X Lavender Pinocchio]; Boerner; J&P

'AMY'S DELIGHT', Min, mp, 1980; bud ovoid; flowers clear medium pink, opening imbricated, dbl., 60 petals, borne usually singly, no fragrance; straight, tan prickles; foliage small, medium green, very glossy; compact, bushy growth; [Little Darling X Little Chief]; Williams, Ernest D.

Ana de Cuevas *see* FEmiento

Anabell® *see* 'KORBELL'

'ANACRÉON', HGal, mp, 1836; flowers carmine-pink, medium large, very dbl.; Vibert; (Sangerhausen)

'ANAÏS', B, mp

'ANAÏS SÉGALAS', HGal, pb, 1837; flowers rosy crimson, edged rosy lilac, expanded, large, dbl.; branching growth; Vibert

Anaise, B, mp; flowers medium large, dbl.; (Sangerhausen)

Anant, HT, pb, 1991; Chiplunkar

'ANASTASIA', HT, w, 1980; bud ovoid, pointed; flowers, dbl., 30 petals, exhibition form, borne usually singly, no fragrance; bronze prickles, turning brown with age; foliage large, dark; vigorous, dense growth; [John F. Kennedy X Pascali]; Greff, N.P.

'ANATOLE', HGal, mr, 1827; Noisette

'ANATOLE DE MONTESQUIEU', HSem, w, 1860; Van Houtte

Ancestry *see* JAYANC

Anchieta, HT; (Brazil); (Cavriglia)

Anci Böhm *see* **'ANCI BÖHMOVA'**

'ANCI BÖHMOVA', HMult, mp, 1929; (Anci Böhm); [Marietta Silva Tarouca sport]; Böhm, J.

'ANCIENT ART', Min, op, 1985; flowers orange-pink blend, yellow reverse, medium, dbl., slight fragrance; foliage medium, medium green, matt; upright, bushy growth; [Rise 'n' Shine X Picnic]; Hardgrove, Donald L.; Rose World Originals

Anda *see* 'LENDA'

Andalusien® *see* 'KORDALU'

'ANDANTE', S, op, 1962; bud ovoid, pointed; flowers light salmon-pink, medium, dbl., cupped, repeat bloom, slight fragrance; foliage dark, bronze, leathery; vigorous (5-6 ft.), upright, arching growth; [Sea of Fire X (Josef Rothmund X R. laxa)]; Buck, Dr. Griffith J.; Iowa State University

'ANDANTE', HT, mr, 1962; flowers large, dbl., 40 petals; foliage bronze; vigorous, bushy growth; Laperrière; EFR

'ANDELI', HT, rb, 1977; (**Double Delight™**); bud long, pointed to urn-shaped; flowers creamy white becoming strawberry-red, dbl., 30–35 petals, 5.5 in., exhibition form, borne mostly singly, moderate, spicy fragrance; foliage large, deep green; upright, spreading, bushy growth; AARS, US, 1977 GM, Baden-Baden, 1976 GM, Rome, 1976 Hall of Fame, WFRS, 1985 James Alexander Gamble Fragrance Medal, ARS, 1986; [Granada X Garden Party]; Swim, H.C. & Ellis, A.E.; Armstrong Nursery

'ANDENKEN AM J. DIERING', Cl HT, 1902; Hinner, W.; (Cavriglia)

'ANDENKEN AN ALMA DE L'AIGLE', HMsk, lp, 1948; (Andenken an Alma de l'Aigle Ilsabella); Kordes

Andenken an Alma de l'Aigle Ilsabella *see* **'ANDENKEN AN ALMA DE L'AIGLE'**

'ANDENKEN AN BRESLAU', HWich, mr, 1913; flowers carmine-red, medium large, dbl.; Kiese; (Sangerhausen)

'ANDENKEN AN FRANZ HEINSOHN', F, dr, 1938; flowers urn-shaped, medium, dbl., in clusters, slight fragrance; upright, bushy growth; [D.T. Poulsen X Seedling]; Poulsen, S.

'ANDENKEN AN GARTENDIREKTOR SIEBERT', HMult, pb, 1923; flowers carmine-rose, with yellow, blooms in clusters; [Eisenach X Polyantha seedling]; Kiese

Andenken an Gustav Frahm, Pol, mr

Andenken an Hermann Thiess, F, or, 1959; flowers medium large, semi-dbl.; Krause; (Sangerhausen)

'ANDENKEN AND JOHANNES GEHLHAAR', HT, w, 1925; flowers large, dbl., moderate fragrance; Gehlhaar; (Sangerhausen)

'ANDENKEN AND MORITZ VON FRÖHLICH', HT, mr, 1905; flowers carmine red, large, dbl., slight fragrance; Hinner, W.; (Sangerhausen)

'ANDERSEN'S YELLOW', HT, my, 1984; flowers large, dbl., slight fragrance; foliage medium, dark, semi-glossy; [Queen Elizabeth X Seedling]; Walter, J.C.; Kimbrew-Walter Roses

'ANDERSONII', S, mp, 1935; flowers pink, medium, single, 2–2.5 in.; spreading habit growth; [Chance hybrid of R. canina X possibly R. arvensis]

'ANDGEO', F, ob, 1982; (**Georgie Anderson**); flowers shades of orange, medium, dbl., slight fragrance; foliage medium, dark, semi-glossy; upright growth; [Elizabeth of Glamis X Unnamed seedling]; Anderson's Rose Nurseries

'ANDGLO', F, yb, 1985; (**Glowing Embers**); flowers yellow, red reverse, medium, dbl., 35 petals, slight fragrance; foliage medium, medium green, glossy; bushy growth; [Manx Queen X Daily Sketch]; Anderson's Rose Nurseries, 1982

'ANDMAC', Cl HT, yb, 1985; (**Whisky Mac, Climbing**, Whisky, Climbing); Anderson's Rose Nurseries

'ANDORRA', HT, ab, 1973; bud long, pointed; flowers large, dbl., 24 petals, exhibition form, intense fragrance; foliage glossy, dark; vigorous, upright, bushy growth; [Dr. A.J. Verhage X Seedling]; Kordes; Fey

'ANDOUR', HT, yb, 1984; (**Our Love**); flowers yellow-orange; [Doris Tysterman sport]; Anderson's Rose Nurseries

'ANDPAI', F, dr, 1987; (**Paisley Anniversary**); flowers medium, dbl., 15–25 petals, slight fragrance; foliage medium, dark green, glossy; upright growth; [Michele X Smiling Through]; Anderson's Rose Nurseries

'ANDPIN', F, w, 1984; (**Pink Ice**); flowers white with deep pink petal edges, medium, dbl., 20 petals, no fragrance; foliage medium, light green, semi-glossy; bushy growth; [Unnamed seedling X Iceberg]; Anderson's Rose Nurseries

'ANDRÉ GAMON', HT, mp, 1908; flowers carmine; Pernet-Ducher

'ANDRÉ LE TROQUER', HT, ob, 1946; flowers orange shading to apricot, dbl., 30 petals, 5 in., cupped, intense fragrance; foliage very dark; very vigorous, upright growth; GM, Bagatelle, 1946 GM, NRS, 1951; Mallerin, C.; A. Meilland

'ANDRÉ LEROY D'ANGERS', HP, mr, 1862; bud dark violet; flowers crimson, shaded with violet, often ill-formed, large, dbl.; vigorous growth; Trouillard; Standish

'ANDRÉ LOUIS', HMult, w, 1920; flowers white, center flesh-pink; Tanne; Turbat

'ANDRÉ PERNET', HT, mr, 1956; bud long; flowers red, becoming lighter at center and purplish on outer petals, dbl.; foliage dark; very vigorous growth; [Peace X Mme Elie Dupraz seedling]; Gaujard

'ANDRÉ SCHWARTZ', T, mr; flowers crimson, sometimes striped white; Schwartz, J., 1884

'ANDREA', Min, pb, 1971; bud pointed; flowers deep pink, silver reverse, small, dbl., 20 petals, 1.5 in., exhibition form; foliage dark; vigorous, bushy, spreading growth; [Little Darling X ?]; Moore, Ralph S.; Sequoia Nursery, 1978

Andrea, HT, ob, 1968

Andrea Stelzer *see* KORfachrit

'ANDRÉE JOUBERT', HT, op, 1952; bud very long, dark orange-coral; flowers pastel salmon, large; [Soeur Thérèse X Duquesa de Peñaranda]; Mallerin, C.; EFR

'ANDRÉE LENOBLE', Pol, dp, 1915; flowers unfading rose or red; Turbat

'ANDRÉE PALTHEY', HT, mr, 1946; bud long, pointed; flowers bright red, large, very dbl., moderate fragrance; foliage bronze; vigorous growth; [Mme Joseph Perraud X Seedling]; Gaujard

'ANDRÉE PERRIER', HT, ob, 1932; bud long; flowers orange-yellow, shaded carmine, very large, dbl., cupped; [Souv. de F. Bohé X Seedling]; Chambard, C.

'ANDRÉE ROUX', HT, rb, 1927; flowers coral-red, tipped carmine, reverse yellow; Pernet-Ducher; Gaujard

'ANDRÈE SAUVAGER', HT, yb, 1935; flowers light orange-yellow, large, dbl., moderate fragrance; Mallerin, C.; (Sangerhausen)

'ANDRÉE VANDERSCHRICK', HWich, w, 1935; bud greenish; flowers opening well, small, very dbl., blooms in clusters, profuse seasonal bloom, moderate fragrance; foliage dark; very vigorous, climbing growth; Buatois

'ANDRÉE-SOPHIE GIRARD', F, mr, 1958; flowers currant-red, edged silvery, dbl.; [Alain X Independence]; Arles; Roses-France

Andreo Vettori, T, mp

'ANDRES BATLLE', HT, dr, 1951; bud long, pointed; flowers crimson, large blooms, dbl., exhibition form, slight fragrance; strong stems; vigorous growth; [Comtesse Vandal X Sensation]; Camprubi, C.

Andrew Barton Patterson, HT, lp; Allender, Robert William, 1999; (Weatherly, L.)

Andrew's Comfort, S, ly, 1996

'ANDREWSII', M, mp, 1807

'ANDROI', F, rb, 1986; (**Royal Mail**); flowers red with yellow petal edges, small, dbl., 20 petals, no fragrance; foliage small, medium green, glossy; upright, bushy growth; [Seedling X Manx Queen]; Anderson's Rose Nurseries

'ANDROS', Cl HT, rb, 1985; (**Rosy Cheeks, Climbing**); Anderson's Rose Nurseries

'ANDROY', F, rb, 1983; (**Royal Touch**); flowers red, reverse silver, medium, dbl., 20 petals, slight fragrance; foliage medium, medium green, semi-glossy; upright growth; [Orange Sensation X Elizabeth of Glamis]; Anderson's Rose Nurseries

'ANDSHY', F, rb, 1984; (**Shy Maiden**); flowers white with red petal edges, large, dbl., 35 petals, slight fragrance; foliage medium, light green, glossy; bushy growth; [Iceberg seedling X Iceberg]; Anderson's Rose Nurseries

'ANDSUN', F, yb, 1983; (**Sunshine Princess**); flowers medium, dbl., 20 petals, slight fragrance; foliage medium, medium green, semi-glossy; bushy growth; Anderson's Rose Nurseries

'ANDULKA', HT, op, 1935; bud long, pointed; flowers pink to salmon-pink, large, dbl.; vigorous, bushy growth; Brada, Dr.; Böhm

'ANDWEE', Min, mr, 1988; (**Wee Topper**); flowers small, dbl., 15–25 petals, slight fragrance; foliage small, light green, semi-glossy; bushy growth; [Starina sport]; Anderson's Rose Nurseries

'ANDWEL', F, mr, 1984; (**Welcome Home**); dbl., 20 petals; foliage medium, light green, semi-glossy; upright growth; [Orange Sensation X Michelle]; Anderson's Rose Nurseries

'ANDWIT', HT, yb, 1983; (**With Love**); flowers yellow, petal edges pink, moderate fragrance; foliage medium, medium green, semi-glossy; upright growth; [Irish Gold X Daily Sketch]; Anderson's Rose Nurseries

'ANEMONE', S, lp, 1896; (Anemone Rose, Pink Cherokee, R. X anemonoides); flowers silver pink, single, 5 petals, 4 in., spring bloom, then scattered bloom, moderate fragrance; foliage glossy; vigorous, bushy growth; (14); [Possibly R. laevigata X Tea]; Schmidt, J.C.; J.C. Schmidt

Anemone *see* R. X anemonoides

Anemone Rose *see* **'ANEMONE'**

Anemone Rose *see* R. X anemonoides

'ANEMONOIDES', C, mr, 1814; Poilpre

Anette, F, 1978; Noack, Werner; (Cavriglia)

'ANGARA', HT, rb, 1983; flowers dark red with hues of orange; [Montezuma sport]; Gupta, Dr. M.N. & Shukla, R.; National Botanical Research Institute

Angel™ *see* 'DEVITE'

Angel Bells® *see* 'HERMELA'

'ANGEL CREAM', F, w, 1989; bud broadly ovate; flowers creamy white, turning pure white; [Angel Face sport]; Ravi, Professor N., 1988

'ANGEL DARLING', Min, m, 1976; flowers lavender, small, 10 petals, 1.5 in., slight fragrance; foliage leathery; vigorous growth; [Little Chief X Angel Face]; Moore, Ralph S.; Sequoia Nursery

'ANGEL DELIGHT', HT, ab, 1976; flowers peach shaded salmon; [Femina sport]; Fryers Nursery, Ltd.

'ANGEL DUST', Min, w, 1978; bud ovoid; flowers small, 18–20 petals, exhibition form; foliage dark; vigorous, upright, spreading growth; [Magic Carrousel X Magic Carrousel]; Bennett, Cecilia 'Dee'; Tiny Petals Nursery

'ANGEL EYES', Min, dp, 1978; bud ovoid; flowers spinel-red, semi-dbl., 10 petals, 1 in., slight fragrance; foliage small; very compact, bushy growth; Lyon

'ANGEL FACE', F, m, 1968; bud pointed; flowers deep mauve-lavender, petal edges blushed ruby, dbl., 25–30 petals, 3.5–4 in., exhibition form, borne in large clusters, intense, strongly citrus fragrance; foliage dark, leathery, glossy; vigorous, upright, bushy growth; AARS, 1969 John Cook Medal, ARS, 1971; [(Circus X Lavender Pinocchio) X Sterling Silver]; Swim & Weeks; C-P

Angel Face, Climbing, Cl F, 1981; Haight; (Cavriglia)

Angel Girl *see* 'WELAN'

'ANGEL GUIMERA', HT, my, 1926; flowers amber-yellow, dbl.; [Frau Karl Druschki X Souv. de Claudius Pernet]; Dot, Pedro; S. Dot

Angel Pink *see* 'MORGEL'

Angel Rose *see* **R. CHINENSIS MINIMA**

'ANGEL WINGS', HT, yb, 1958; bud ovoid; flowers yellow, shading to white, edged pink, dbl., 23 petals, 3.5–4 in., cupped, moderate fragrance; foliage leathery; upright growth; GM, Portland, 1959; [Golden Rapture X Girona]; Lindquist; Howard Rose Co.

'ANGELA', F, yb, 1957; flowers golden yellow shaded crimson, dbl., 28 petals, 2.5 in., trusses of 15-20, very free bloom, slight fragrance; foliage glossy, dark; vigorous, upright growth; [Masquerade X Golden Scepter]; Kordes; Morse

Angela® *see* 'KORDAY'

Angela *see* JACcream

Angela Daffey, Min, lp; Hannemann, F., 1989; (Weatherly, L.)

Angela Lansbury *see* TWOangel

Angela Merici, HT, w, 1987; Lens

Angela Rippon® *see* 'OCARU'

'ANGELA'S CHOICE', F, pb, 1973; flowers light pink, reverse deep pink, semi-dbl., 15 petals, 3 in.; foliage matt; vigorous, upright growth; [Dainty Maid X Anna Wheatcroft]; Gobbee, W.D.

'ANGÈLE', HT, w, 1933; bud ovoid; flowers creamy white, large, dbl., blooms in clusters, moderate fragrance; foliage leathery; long stems; vigorous, bushy growth; [Seedling X Kaiserin Auguste Viktoria]; Vestal

'ANGÈLE PERNET', HT, ob, 1924; bud ovoid; flowers reddish-orange shaded yellow, large, dbl., globular, moderate, fruity fragrance; foliage dark, bronze, leathery; vigorous, bushy growth; GM, Bagatelle, 1924 GM, NRS, 1925; [Bénédicte Seguin X HT]; Pernet-Ducher

Angelglo™ *see* 'MINACO'

Angelica *see* 'KORDAY'

Angelica Renae™ *see* 'WELANGEL'

Angelika, F, dp

'ANGELINA', S, pb, 1976; flowers rose-pink, white eye and reverse, 11 petals, 3 in., blooms in clusters, moderate fragrance; foliage matt, light green; bushy, spreading growth; [(Tropicana X Carine) X (Cläre Grammerstorf X Frühlingsmorgen)]; Cocker, 1975

Angelina Lauro *see* **'ANGELINE LAURO'**

'ANGELINE LAURO' HT, or, 1967; (Angelina Lauro); flowers medium, dbl.; foliage dark; Lens, 1968; Spek

'ANGELIQUE', F, op, 1961; bud ovoid; flowers coral-pink to salmon-pink, open, dbl., 20–25 petals, 2.5–3 in., in clusters, slight fragrance; vigorous, spreading growth; [World's Fair X Pinocchio]; Swim, H.C.; C.R. Burr

Angelique® *see* 'KORANGELI'

'ANGÉLIQUE QUÉTIER', m, 1839; flowers violet-pink, large, very dbl., slight fragrance; Quétier; (Sangerhausen)

'ANGELIS', HT, w, 1960; bud long; single, 8–10 petals, intense fragrance; upright, bushy growth; [Virgo X Ibiza]; Dot, Pedro

Angelita® *see* 'MACANGELI'

'ANGELITA RUAIX', HT, ob, 1940; flowers orange-yellow, large, dbl., exhibition form; foliage glossy, dark; very vigorous growth; [Duquesa de Peñaranda X Pres. Herbert Hoover]; Dot, Pedro

Angelo Sgaravatti, HT, 1981; (Cavriglia)

Angel's Blush *see* 'MICANGEL'

Angel's Blush

Angel's Camp White, T, w

'ANGELS MATEU', HT, ab, 1934; bud ovoid; flowers salmon, overlaid gold, large, dbl., 40 petals, globular, moderate, blackberry fragrance; foliage dark; vigorous, bushy growth; GM, Bagatelle, 1934 GM, Rome, 1934; [Magdalena de Nubiola X Duquesa de Peñaranda]; Dot, Pedro; C-P

'ANGELUS', HT, w, 1921; flowers white, center cream, large, dbl., 40–45 petals, moderate fragrance; foliage leathery, dark; vigorous growth; [Columbia X Ophelia seedling]; Lemon

Angelus, HT, ob, 1980; Gaujard

'ANGELUS, CLIMBING', Cl HT, w, 1933; Dixie Rose Nursery

'ANGIE HEATWOLE', F, mp, 1996; flowers, dbl., 15–25 petals, 1.5–2.75 in., borne in small clusters, slight fragrance; few prickles; foliage medium, medium green, semi-glossy; medium, upright growth; [Seedling X Dreamer]; Heatwole, Robert E.

Angkor® *see* 'DELTRAC'

'ANGLE', Ayr, lp; ('JESSICA', Rose Angle); flowers flesh tinged with rose, large, semi-dbl.

Anglica Minor, A, w; flowers center muddled, dbl.; foliage dark gray-green; dwarf growth

'ANGOLA', F, dr; [Seedling X Alain]; da Silva, Moreira

Angora, HT; (Cavriglia)

'ANGUS MACNEIL', F, op, 1967; flowers salmon-pink and cream, well formed, in clusters; foliage dark; free growth; [The Optimist X Ma Perkins]; Vincent

'ANIBAL DAVID', HT, my, 1961; da Silva, Moreira

'ANICKA', Pol, dr, 1940; flowers small, dbl.; Valàsek; (Sangerhausen)

Anika®, HT, dy

Anikall, HT, dy, 1987; Haschke

'ANIMO', F, yb, 1962; flowers yellow, becoming copper-red, open, semi-dbl., 12–16 petals, 2.5 in., blooms in clusters; bushy growth; [Masquerade X Beauté]; deRuiter

Anisley Dickson *see* 'DICKIMONO'

'ANITA', F, pb, 1982; bud ovoid, pointed; flowers large, dbl., 43 petals, blooms in cluster of 3-7, slight, tea fragrance; large prickles; foliage large, glossy; medium growth; [Rumba X Marmalade]; Swim, H.C. & Christensen, J.E.; Armstrong Nursery

Anita Charles™ *see* 'MORNITA'

Anita Pereire *see* DRAdiwu

'ANITA STAHMER', F, dp, 1973; bud long, pointed; flowers deep pink, dbl., 27 petals, 2.5 in., exhibition form, slight fragrance; foliage soft; vigorous, upright, bushy growth; [Zorina sport]; Kordes

Anita's Apricot Fantasy, Min, ab

Anita's Briar Rose, S, lp

Anita's Pink Splash, Min, pb

Anita's Quartered Tropicana, HT, mr, 1998

Anja *see* KORkompo

Anjani, Pol, rb, 1970; Kasturi

Anjou® *see* KORNICKEN

Anjou Festival®, F, mr

'ANKARA', HT, op, 1940; flowers salmon-orange, center coppery, very large, dbl., cupped, slight fragrance; foliage leathery; vigorous, upright growth; [Joanna Hill X Mme Joseph Perraud]; Meilland, F.; A. Meilland

Ankori *see* 'KORANGELI'

Ann *see* 'AUSFETE'

'ANN ABERCONWAY', F, ab, 1976; flowers apricot-orange, dbl., 20 petals, 3 in., moderate fragrance; foliage dark, leathery; [Arthur Bell X Seedling]; Mattock

'ANN BARTER', HT, mr, 1962; flowers cerise, semi-dbl., 18 petals, 5 in., moderate fragrance; foliage dark; vigorous growth; [Peace X Ena Harkness]; Barter

Ann Delforge, HT, ob

'ANN ELIZABETH', F, mp, 1962; flowers clear rose-pink, large, semi-dbl., 15 petals, open clusters, slight fragrance; foliage glossy; vigorous, quite tall growth; Norman; Harkness

'ANN ENDT', HRg, dr, 1978; bud long-sepaled; flowers medium, single, 5 petals, moderate, cinnamon fragrance; foliage small, soft; [R. rugosa X R. foliolosa]; Nobbs

'ANN FACTOR', HT, ab, 1975; bud ovoid; flowers pastel apricot, large, very dbl., exhibition form, intense fragrance; foliage large, glossy, bronze, leathery; vigorous, bushy growth; [Duet X Jack O'Lantern]; Ellis & Swim; Armstrong Nursery

'ANN HOLBROOK', Min, yb, 1981; bud globular; flowers yellow-pink blend, small, 15–25 petals, slight fragrance; foliage small, medium green, glossy; spreading growth; [Patricia Scranton X Little Darling]; Dobbs, Annette E.

Ann Kercher, HT, lp; flowers variable form, small

Ann Moore *see* 'MORBERG'

Anna® *see* PEKCOUGEL

Anna Aguilera, F, 1972; Dot, Simon; (Cavriglia)

'ANNA ALEXIEFF', HP, mr, 1858; flowers rose pink, large, dbl., cupped, borne in clusters of 3, moderate fragrance; medium, upright growth; Margottin, 1858

'ANNA CHARTRON', T, w; flowers white tinted pink; [Kaiserin Auguste Viktoria X Luciole]; Schwartz, Vve., 1897

'ANNA DE DIESBACH', HP, dp; (Anna Von Diesbach, Gloire de Paris, Glory of Paris); bud long, pointed; flowers deep pink, center darker, dbl., 40 petals, 5–6 in., cupped, intense fragrance; vigorous, tall growth; [La Reine X Seedling]; Lacharme, F., 1869

'ANNA DE NOAILLES', HT, dr, 1941; flowers crimson-red, medium, semi-dbl., exhibition form, intense fragrance; very vigorous, bushy growth; [Étoile de Hollande X Seedling]; Gaujard

Anna Ford® *see* 'HARPICCOLO'

'ANNA HARTMANNOVÁ', HT, w, 1933; flowers cream-white, very dbl., moderate fragrance; [Frau Luise Kiese sport]; Brada, Dr.; Böhm

Anna Katherine *see* 'ORTANA'

Anna Livia *see* 'KORMETTER'

'ANNA LOUISA', F, lp, 1967; flowers soft pink, dbl., 2.5 in., blooms in large clusters; vigorous, low, bushy growth; [Highlight X Valeta]; deRuiter

'ANNA MARIA', HT, dp, 1948; (Anna Marie); bud ovoid; flowers rosy pink, dbl., 40–70 petals, 5 in., exhibition form, intense fragrance; foliage leathery, dark; vigorous, upright growth; [Soeur Thérèse X (Duquesa de Peñaranda X Mrs Pierre S. duPont)]; Ohlhus; C-P

'ANNA MARIA', HSet, dp, 1843; Feast

Anna Marie *see* **'ANNA MARIA'**

'ANNA MÜLLER-IDSERDA', F, mp, 1966; bud ovoid; flowers pink, medium, dbl., intense fragrance; foliage dark; [Duet X Juliette E. van Beuningen]; Buisman, G. A. H.

'ANNA NEAGLE', HT, rb, 1937; flowers bright currant-red, base sunflower-yellow, dbl.; foliage dark; free, branching growth; McGredy

'ANNA OLIVIER', T, pb, 1872; flowers yellowish flesh, shaded salmon, reverse rose, well-formed, large, dbl.; vigorous growth; (14); Ducher

Anna Pavlova, HT, lp, 1981; Beales, Peter

'ANNA RÜBSAMEN', HWich, mp, 1904; flowers clear pink, large, dbl., blooms in compact clusters; vigorous growth; Weigand, C.

Anna Saheb, HT, pb, 1993; Chiplunkar

'ANNA SCHARSACH', HP, mp, 1890; flowers large, dbl., intense fragrance; Geschwind, R.; (Sangerhausen)

'ANNA SOUPERT', HT, yb, 1934; flowers yellow, center orange, very large, cactus-dahlia form, dbl., slight fragrance; foliage bronze, dark; dwarf growth; [Sunburst X Prince de Bulgarie]; Soupert, G.; C. Soupert

'ANNA STAVE', HT, w, 1973; bud ovoid; flowers white, tipped pink, medium, dbl., exhibition form, intense fragrance; foliage dark, soft; moderate, upright growth; [Pink Parfait X Kordes' Perfecta]; Curtis, E.C.; Kimbrew

Anna Vena, Pol, dr, 1972; flowers medium, dbl.; Zyla; (Sangerhausen)

Anna Von Diesbach *see* **'ANNA DE DIESBACH'**

'ANNA WHEATCROFT', F, or, 1958; flowers light vermilion, gold stamens, single, 4 in., blooms in clusters, slight fragrance; foliage dark, glossy; vigorous growth; [Cinnabar seedling X Seedling]; Tantau, Math.; Wheatcroft Bros.

Anna Zinkeisen *see* 'HARQUHLING'

'ANNABELLA'®, HT, my, 1981; bud ovoid; flowers deeply, medium, dbl., 35 petals, cupped, borne singly, no fragrance; straight, light yellow prickles; foliage medium, dark, matt; upright growth; [Ambassador X Seedling]; Rose Barni-Pistoia

'ANNABELLA', HT, 1940; flowers light buff-gold, dbl., 55 petals, 5 in., moderate fragrance; RULED EXTINCT 5/83 ARM; [Joanna Hill sport]; Grillo

Annabelle *see* 'KORBELL'

Annabelle Kolle, F, dp, 1993; Hetzel

Anna-Maria de Montravel *see* **'ANNE-MARIE DE MONTRAVEL'**

Annan's Orchard, M, dp

'ANNAROY', HT, pb, 1951; bud ovoid; flowers pink with slight salmon undertone, imbricated, very dbl., 110 petals; foliage glossy; [Pink Princess X Los Angeles]; Shepherd; Bosley Nursery

Anna's Red Glory, HT, mr, 1994

Anna's Yellow Glory, HT, my

Ænnchen Müller *see* **'ANNCHEN MÜLLER'**

'ANNCHEN MÜLLER', Pol, dp, 1907; (Ænnchen Müller, Annie Mueller); flowers warm rose, fading, large, dbl., cupped, blooms in clusters, slight fragrance; foliage glossy; vigorous, bushy growth; [Crimson Rambler X Georges Pernet]; Schmidt, J.C.

'ANNCHEN VON THARAU', HMult, w, 1866; flowers medium large, dbl.; Geschwind, R.; (Sangerhausen)

'ANNE', HT, mr, 1925; flowers cherry-red, large, dbl., globular, moderate, damask fragrance; foliage leathery; vigorous, bushy, compact growth; Pemberton

'ANNE COCKER'®, F, op, 1970; flowers vermilion, medium, dbl., 36 petals, 2.5 in.; foliage glossy, light to medium green; vigorous, upright growth; [Highlight X Colour Wonder]; Cocker, A.; Cocker, 1971

Anne Dakin, LCl, pb, 1974; Holmes

Anne de Bretagne® *see* 'MEITURAPHAR'

Anne Diamond *see* **'LANDIA'**

'ANNE D'ORNANO', HT, m, 1967; bud long, pointed; flowers bright purple-crimson, very large, dbl., moderate fragrance; foliage dark, leathery; vigorous, bushy growth; [John S. Armstrong X Rose Gaujard]; Gaujard

Anne Elizabeth, HT, lp; Thomas, Dr. A.S., 1979; (Weatherly, L.)

'ANNE FARNWORTH', HT, rb, 1964; flowers like parent, in sunset shades; [Tzigane sport]; Court

Anne Ford *see* 'HARPICCOLO'

'ANNE GREGG', HT, ab, 1994; flowers moderately full, large blooms, dbl., 15–25 petals, 3–3.5 in., slight fragrance; some prickles; foliage large, dark green, matt; medium (80 cms), upright growth; [Diorama sport]; Rearsby Roses, Ltd., 1994; Rearsby Roses, Ltd., 1994

Anne Harkness® *see* 'HARKARAMEL'

'ANNE JACKSON', LCl, dp, 1973; flowers cerise, medium, dbl., 30–35 petals, cupped, moderate fragrance; foliage glossy, bronze; vigorous growth; [Spectacular X LCl]; Jackson, J.R.

Anne Kercher *see* 'SIMJEZBEL'

'ANNE LAFERRÈRE', HP, dr, 1916; flowers deep velvety blood-red; vigorous growth; Nabonnand, C.

Anne Laure *see* MASlako

'ANNE LETTS', HT, pb, 1954; flowers rose-pink, reverse silvery, pointed, dbl., 28 petals, 4.5 in., moderate fragrance; foliage glossy; bushy growth; [Peace X Charles Gregory]; Letts

'ANNE LEYGUES', T, lp, 1905; flowers flesh-pink; [Gén. Schablikine X Comtesse Bardi]; Nabonnand, P.&C.

'ANNE MARIA', HSet, dp, 1843; flowers pale rose, large, dbl., blooms in clusters; Feast

Anne Marie *see* 'MEIFOUR'

Anne Marie Laing *see* 'JOSPINK'

Anne Marie Trechslin® *see* 'MEIFOUR'

'ANNE MCDONALD', F, pb, 1992; flowers rose pink and creamy yellow, classical, 30 petals, 3.25 in., exhibition form, borne in clusters of 10-30, slight fragrance; foliage medium green, glossy; medium to tall (1-3 m), upright growth; [Granada X Kordes Perfecta]; Spriggs, Ian Raymond, 1984; Treloar Roses Pty. Ltd., 1991

Anne Morrow Lindbergh *see* 'JACYAP'

'ANNE OF GEIERSTEIN', HEg, dr, 1894; flowers deep crimson, single, summer bloom, moderate fragrance; bright scarlet fruit; foliage fragrant; vigorous growth; [R. eglanteria X HP or B]; Penzance; Keynes, Williams & Co.

'ANNE POULSEN', F, mr, 1935; (Anne-Mette Poulsen); bud long, pointed; flowers bright crimson-red, darkening, large, dbl., blooms in clusters, moderate fragrance; vigorous growth; [Ingar Olsson X Seedling]; Poulsen, S.; J&P, 1935; McGredy, 1935

'ANNE SCRANTON', F, lp, 1971; bud ovoid; flowers light pink, center flesh-white, medium, dbl.; foliage leathery; vigorous, upright growth; [Queen Elizabeth X Katherine T. Marshall]; Dobbs, Annette E.

'ANNE VANDERBILT', HT, or, 1941; bud pointed; flowers reddish orange, open, dbl., 28 petals, 4–5 in., intense fragrance; foliage leathery, glossy; very vigorous, bushy growth; [Seedling X Stargold]; Brownell, H.C.

'ANNE WATKINS', HT, ab, 1962; flowers deep cream shaded yellow, reverse apricot, well-shaped, dbl., 30 petals, 5 in.; foliage dark, glossy; vigorous, upright growth; [Ena Harkness X Grand'mère Jenny]; Watkins Roses

Anne-Aymone Giscard d'Estaing® *see* DORalp

Anneka *see* 'HARRONVER'

'ANNEKE DOORENBOS', F, pb, 1956; flowers silver-pink, reverse darker; [Buisman's Triumph sport]; Doorenbos; Boerma

'ANNEKE KOSTER', Pol, dr, 1927; flowers deep red; [Greta Kluis sport]; Koster, D.A.

Anneliese, HT, dr, 1971; Miessler, Herbert; (Sangerhausen)

Anneliese, HT, lp, 1990; flowers light pink to white, large, semi-dbl.; Wänninger, Franz; (Sangerhausen)

Anneliesse Rothenberger *see* 'TANOLG'

'ANNE-MARIE COTTE', N, w, 1875; flowers medium, dbl.; Guillot; (Sangerhausen)

'ANNE-MARIE DE MONTRAVEL', Pol, w, 1879; (Anna-Maria de Montravel); flowers pure white, of irregular form when fully open, sometimes showing stamens, dbl., 1.5 in., moderate, lily-of-the-valley fragrance; very few prickles; foliage dark green above, grayish beneath, glossy, 3-5 leafl; dwarf, compact growth; [Dbl. flowered Multiflora X Mme de Tartas]; Rambaux

'ANNE-MARIE MILLIAT', HT, w, 1939; bud very long; flowers large, very dbl.; very vigorous growth; Gaujard

'ANNEMARIE VAN ONSEM', F, mr, 1971; bud ovoid; flowers vivid red, open, medium, semi-dbl.; foliage large, glossy, dark; very vigorous, upright, bushy growth; [Circus X Korona]; Institute of Ornamental Plant Growing

Anne-Mette Poulsen *see* **'ANNE POULSEN'**

Annerose, F, or, 1976; Croix

'ANNE'S DELIGHT', Min, dp, 1981; bud pointed; flowers, dbl., 40 petals, exhibition form, borne usually singly, slight fragrance; long, thin prickles; foliage small, dark, glossy; upright, bushy growth; [Little Darling X Over the Rainbow]; Williams, Ernest D.; Mini-Roses

'ANNETTE', HT, op, 1952; bud long, pointed; flowers salmon-pink, dbl., 20–25 petals, 4–4.5 in., exhibition form, moderate fragrance; foliage glossy, leathery, dark; vigorous, compact growth; [Charlotte Armstrong X Contrast]; Swim, H.C.; Earl May Seed Co.

Annette Dobbs *see* 'MORNET'

Annette Elizabeth, LCl, lp; [Clair Matin X Wedding Day]; Stewart, L., 1992; (Weatherly, L.)

'ANNETTE GATWARD', HT, ab, 1954; flowers peach, large, dbl., moderate fragrance; foliage light green; upright growth; [Mrs Charles Lamplough X Barbara Richards]; Gatward

'ANNETTE GRAVEREAUX', HT, my, 1929; flowers lemon-yellow, shaded orange, dbl.; [Mev. C. van Marwijk Kooy X Golden Emblem]; Leenders, M

'ANNI JEBENS', HT, dy, 1932; bud large, long, pointed, golden yellow; flowers blood-red, reverse golden yellow, dbl., exhibition form, moderate fragrance; foliage leathery, glossy, bronze; bushy, dwarf growth; [Charles P. Kilham X Mev. G.A. van Rossem]; Kordes

'ANNI WELTER', M, mp, 1906; flowers large, moderate fragrance; [Crested Moss X La France]; Welter

Annie Beaufais, F, or, 1962; deRuiter

'ANNIE BRANDT', HT, op, 1932; bud long, pointed; flowers pink tinted coral, open, large, semi-dbl., moderate fragrance; foliage leathery, glossy; very vigorous, bushy growth; [Mrs Pierre S. duPont X Colette Clément]; Mallerin, C.; C-P

'ANNIE BURGESS', HP, lp, 1926; flowers pale pink, in clusters, early; [Lyon Rose X Frau Karl Druschki]; Burgess, S.W.

'ANNIE COOK', T, rb, 1888; flowers blush-tinted; [Bon Silène sport]; Cook, J.W.

'ANNIE CRAWFORD', HP, mp, 1915; (Miss Annie Crawford); flowers bright pink, very large, dbl., 30–35 petals, exhibition form, recurrent bloom; vigorous growth; GM, NRS, 1914; Hall

'ANNIE DE METZ', HT, or, 1932; bud large; dbl., exhibition form; [Golden Emblem X R. foetida bicolor seedling]; Mallerin, C.; C-P

'ANNIE DREVET', HT, rb, 1939; bud long, yellow; flowers fiery red, reverse yellow, large, semi-dbl., cupped, slight fragrance; foliage leathery, glossy, dark; vigorous growth; [Charles P. Kilham X (K. of K. X Mari Dot)]; Caron, B.; A. Meilland, 1939; Port Stockton Nursery, 1941

'ANNIE DUPEYRAT', HT, op, 1935; bud ovoid; flowers very large, dbl., moderate fragrance; foliage leathery; bushy growth; [Mrs T. Hillas X Elvira Aramayo]; Mallerin, C.; C-P

Annie East *see* 'BOSREXEVER'

Annie Girardot, HT, 1979; Kriloff, Michel; (Cavriglia)

'ANNIE LAURIE', HT, pb, 1918; (Double Mme Butterfly); bud long, pointed; flowers flesh-pink, base yellow, large, dbl., cupped, moderate fragrance; foliage glossy; very vigorous growth; [Ophelia sport]; Stuppy Floral Co.

Annie Mueller *see* **'ANNCHEN MÜLLER'**

Annie R. Mitchell *see* 'MORMITCHELL'

'ANNIE VIBERT', N, w, 1871; Vibert

Annie Wood *see* **'MLLE ANNIE WOOD'**

Annie Wood Mille, HP, mr

Annie's Red, F, dr

Annie's Song, F, lp, 1990; Spriggs, Ian Raymond

'ANNIVERSARY', HT, dr, 1981; bud long, pointed; flowers, dbl., 23 petals, exhibition form, borne 1-3 per cluster, moderate fragrance; prickles hooked down; foliage medium to large; upright growth; [Love Affair X Seedling]; Hoy, Lowel L.; Joseph H. Hill, Co.

'ANNIVERSARY', HT, 1961; bud ovoid; flowers yellow, dbl., 55–60 petals, 4–5 in., exhibition form, moderate fragrance; foliage leathery, dark; strong stems; vigorous, upright growth; RULED EXTINCT 5/81 ARM; [Mary Jo X Lamplighter]; Hill, Joseph H., Co.

Anniversary *see* MATtlace

Ann's Rose *see* 'JUDANN'

'ANN'S WEDDING', HT, my, 1975; flowers very full, dbl., 40 petals, 4 in., moderate fragrance; foliage glossy, dark; free growth; [Whisky Mac sport]; Rosemount Nursery

Annulet, Pol, lp; Miers, A., 1935; (Weatherly, L.)

'ANNY', Min, w, 1949; flowers pale pink fading white, micro-mini, dbl., 30 petals, .5 in.; height 6 in; [Rouletii X Perla de Montserrat]; Dot, Pedro

'ANNY BRANDT', HT, yb, 1951; bud pointed; flowers creamy yellow, edge and reverse tinted lilac, petals waved, dbl.; vigorous growth; Mallerin, C.; EFR

'ANOTHER CHANCE', HT, w, 1994; flowers, dbl., 26–40 petals, 1.5–2.75 in., borne mostly singly, slight fragrance; some prickles; foliage medium to large, medium green, matt; medium-tall, bushy growth; [Mount Shasta X Seedling]; Heyes, Alex, 1996; Rainbow Roses, 1996

Anri, HT, yb, 1997; flowers full, large, very dbl., 26–40 petals, 4.5 in., slight fragrance; few prickles; foliage large, dark green, glossy; bushy, upright, medium (180cms) growth; [Mme. Sachi X Yarna]; Ohtsuki, Hironaka

Ans *see* 'BENBURGUN'

Antares *see* BARtares

'ANTHEA', HT, yb, 1949; flowers pale yellow flushed rose, compact, dbl., 20–25 petals, 4–5 in., slight fragrance; foliage dark; vigorous growth; [McGredy's Yellow X Phyllis Gold]; Bees

Anthea Fortescue *see* 'PEASHINE'

Anthea Turner, HT, yb, 1997; flowers very full, large, very dbl., 41 petals, 4 in., slight fragrance; few prickles; foliage medium, dark green, semi-glossy; compact, medium growth; [Alec's Red X Harry Wheatcroft]; Thomas, D.

'ANTHÉOR', HT, ab, 1948; bud long, furled; flowers reddish apricot, dbl., intense fragrance; [(Joanna Hill X Duquesa de Peñaranda) X (Charles P. Kilham X Mme Joseph Perraud)]; Meilland, F.

Anthony Meilland *see* 'MEITALBAZ'

Anticipation *see* 'MEIDINRO'

Antico Amore® *see* 'BARANAM'

Antigone® *see* 'GAUHTI'

Antigua *see* 'JACTIG'

Antike, F, yb, 1966; flowers reddish-yellow, medium, dbl.; Kordes, W. Söhne; (Sangerhausen)

Antike 89® *see* KORdalen

Antique Gold *see* 'LAVTIQUE'

'ANTINEA', HT, op, 1934; flowers salmon-orange, base yellow, very large, dbl., moderate fragrance; foliage glossy; very vigorous, bushy growth; [Julien Potin X Seedling]; Gaujard; Dreer; H&S

'ANTIQUE', F, rb, 1967; flowers crimson and gold, blooms in clusters; bushy growth; [Honeymoon X Circus]; Kordes; A. Dickson

Antique 89, LCl, rb; Kordes, 1988

'ANTIQUE LACE', F, mp, 1991; bud ovoid; flowers medium pink, light pink center, urn-shaped, medium, semi-dbl., 25–35 petals; foliage medium, dark green, semi-glossy; bushy, medium growth; [Seedling X Little Cameo]; Strahle, B. Glen, 1985; Coyier's Roses, 1990

Antique Rose™ *see* 'MORCARA'

Antique Silk *see* 'KORAMPA'

Antique Tapestry *see* 'CLETAPE'

'ANTIQUE VELVET', Min, dr, 1993; flowers velvet dark red, moderately full (15–25 petals), medium (4–7 cms) blooms borne in large clusters; patio; fragrant; few prickles; foliage large, medium green, semi-glossy; tall, upright, bushy growth; [Valerie Jeanne X (Party Girl X Pillow Talk)]; Jobson, Dan; Jobson. 1993

'ANTOINE DUCHER', HP rb, 1866; flowers violet-red, large; (28); Ducher

Antoine Mouton, HP, mp; Levet, 1874

'ANTOINE NOAILLY', HT, mr, 1958; bud long; flowers clear red, scalloped petals; [Seedling X Mme G. Forest-Colcombet]; Croix, P.

'ANTOINE RIVOIRE', HT, lp, 1895; bud ovoid; flowers light pink shaded darker, imbricated, dbl., moderate fragrance; foliage dark; vigorous growth; [Dr. Grill X Lady Mary Fitzwilliam]; Pernet-Ducher

'ANTOINE SCHURZ', HP, 1890; Geschwind, R.; (Cavriglia)

'ANTOINE VERDIER', HP, pb, 1871; flowers pink, shaded muddy lilac, dbl.; Jamain, H.

'ANTOINETTE', HT, ab, 1968; bud long, pointed; flowers open, medium size, dbl., moderate fragrance; foliage large, glossy, leathery; vigorous, upright growth; [Queen Elizabeth X Peace]; Patterson; Patterson Roses

Antoinette *see* KORiganta

'ANTONELLA', HT, mp, 1964; bud globular; flowers camellia-pink, dbl., 5 in., exhibition form, moderate fragrance; strong stems; vigorous growth; Mondial Roses

Antonella Fineschi, F, 1985; Fineschi, G.; (Cavriglia)

'ANTONELLIANA', HT, ob, 1952; flowers orange and deep yellow tipped, well-formed, dbl.; foliage glossy; vigorous growth; [Gaiezza X Margaret McGredy]; Giacomasso

'ANTONIA', F, w, 1979; bud long, pointed; flowers blushed white, palest pink flush, well-formed, dbl., blooms in large clusters, slight fragrance; vigorous, bushy, low growth; [Tantau's Tip Top sport]; Bazeley, B.L.; Highfield Nursery

'ANTONIA D'ORMOIS', HGal, lp; flowers blush, fading at edge, medium, dbl., cupped; Vibert, 1835

'ANTONIA PAHISSA', HT, ob, 1935; bud long, pointed; flowers rich orange, large blooms, dbl., cupped, moderate fragrance; foliage glossy, dark; long stems; very vigorous, bushy growth; Pahissa

Antonia Ridge® *see* 'MEIPARADON'

'ANTONIA ROLLERI DE PELUFFO', HT, dr, 1926; flowers brilliant red, center darker, dbl., moderate fragrance; [Gen. MacArthur X Mme Edouard Herriot]; Soupert & Notting

'ANTONIETTA INGEGNOLI', Pol, pb, 1923; flowers golden pink, opening in two distinct tones on same plant, dbl., moderate fragrance; [R. wichurana X R. chinensis]; Ingegnoli

'ANTONIN DVORÁK', HT, op, 1933; flowers light salmon-pink, very large, very dbl., intense fragrance; Böhm, J.; (Sangerhausen)

Antony Meilland *see* 'MEITALBAZ'

'ANTOPP', Gr, dy, 1996; (**Golden Opportunity™**); bud urn-shaped; flowers golden yellow, light golden yellow, blooms borne in small clusters, dbl., 15–25 petals, slight fragrance; some prickles; foliage medium, medium green, semi-glossy; medium growth; [Broadway X Delta Gold]; Qerry, Anthony; Certified Roses, Inc., 1995

'ANURAG', HT, pb, 1980; bud long, pointed; flowers tyrian rose, dbl., 54 petals, exhibition form, borne singly, intense fragrance; hooked, brown prickles; foliage large, smooth, light green; upright, bushy growth; [Sweet Afton X Gulzar]; Division of Vegetable Crops and Floriculture

Anurupa, HT, mp, 1997; Friends Rosery

Anuschka® *see* 'TANKANUSCH'

Anusheh *see* 'PAYABLE'

Anvil Sparks *see* **'AMBOSSFUNKEN'**

'ANYTIME', Min, op, 1973; flowers salmon-orange, purplish eye, semi-dbl., 12 petals, .5–1 in., moderate fragrance; foliage dark; [New Penny X Elizabeth of Glamis]; McGredy, Sam IV; McGredy

'ANZAC', HT, op, 1943; flowers azalea-pink with coppery scarlet sheen, camellia form, dbl., 42–50 petals, 5 in., moderate, fruity fragrance; foliage leathery; long stems; very vigorous, upright, compact growth; [Miss Rowena Thom X Seedling]; Howard, F.H.; H&S

'AORANGI', F, w, 1979; bud pointed; flowers cream, dbl., 41 petals, 3 in., exhibition form, slight fragrance; foliage large; upright growth; [Arthur Bell X Red Devil]; Murray, Nola

Aorangi *see* SANaran

Aotearoa-New Zealand *see* 'MACGENEV'

'AOZORA', HT, m, 1973; (Blue Sky); flowers deep lilac-blue, large, dbl., exhibition form, moderate fragrance; foliage large, leathery; vigorous, upright growth; [Sterling Silver seedling X Seedling]; Suzuki, Seizo; Keisei Rose Nursery, 1972

'APACHE', S, yb, 1961; bud ovoid, flushed red; flowers medium to buff-yellow, dbl., 60 petals, 5–6. in., intense fragrance; foliage leathery; vigorous (5-6 ft), spreading, open growth; [Fred Howard X Buccaneer]; Von Abrams; Peterson & Dering

'APACHE BELLE', HT, rb, 1969; bud ovoid; flowers orange-red, very dbl., 5 in., slight fragrance; foliage glossy; vigorous, upright, compact growth; [The Alamo sport]; Sitton; Co-Operative Rose Growers

Apache Princess *see* 'TWOMIN'

'APACHE TEARS', F, rb, 1971; flowers cream to creamy pink, petals edged red, medium, dbl., exhibition form, slight fragrance; foliage large, light; vigorous, bushy growth; [Karl Herbst X China Doll]; Pikiewicz; Edmunds Roses

'APACHE WELLS', F, yb, 1971; bud ovoid; flowers canary-yellow, washed pink, medium-small, dbl., exhibition form, slight fragrance; foliage leathery; vigorous, bushy growth; [Circus X The Optimist]; Williams, J. Benjamin

Apachi *see* **'TAI-GONG'**

Apart, HRg, m, 1981; flowers mauve blend, dbl.; prolific fruit; 2.5 ft growth; Uhl, J.

Aparte *see* **'SPARTAN'**

'APELES MESTRES', LCl, dy, 1931; flowers sunflower-yellow, large, dbl., globular, moderate fragrance; foliage dark, glossy; vigorous, climbing growth; [Frau Karl Druschki X Souv. de Claudius Pernet]; Dot, Pedro; C-P

Aperitif®, HT, lp, 1998; bud pointed, yellow; dbl., slight fragrance; foliage medium, dark green, semi-glossy, disease-resistant; vigorous, upright (3 ft) growth; McGredy, Sam IV

'APÉRITIF', F, pb, 1972; bud ovoid; flowers ivory, petals edged rose-pink, medium, dbl., exhibition form, slight fragrance; foliage leathery; vigorous, upright, bushy growth; [Seedling X Starbright]; Boerner; J&P

Apertif *see* 'MACWAIRAR'

Apfelblute, S, w, 1991; flowers light pink-white, medium, semi-dbl., no fragrance; Noack, Werner; (Sangerhausen)

'APFELBLÜTE', Pol, w, 1907; flowers light pink-white, small, dbl., moderate fragrance; Wirtz & Eicke; (Sangerhausen)

'APHRODITE', HT, or, 1928; bud long, pointed; flowers coral-red, shaded gold, large, semi-dbl., intense fragrance; foliage dark, glossy; vigorous, bushy growth; [Hortulanus Budde X Toison d'Or]; Easlea

'APHRODITE, CLIMBING', Cl HT, or, 1933; Hillock

Apogée® *see* 'DELBAT'

'APOLLINE', B, mp, 1848; flowers bright rose-pink, cupped; [Pierre de St. Cyr seedling]; Verdier, V.

Apollo® *see* 'ARMOLO'

'APOLLO', HT, dy, 1941; flowers golden yellow; [Mme Joseph Perraud sport]; Armstrong, J.A.; Armstrong Nursery

Apollo, HT, pb, 1998

Apollo Tribute, HT, ab

Apoman®, HT, yb; Adam

Apothecary's Rose *see* **R. GALLICA OFFICINALIS**

(The) Apothecary's Rose of Provins *see* **R. GALLICA OFFICINALIS**

'APOTHEKER FRANZ HAHNE', S, op, 1919; flowers salmon-rose on orange-yellow ground; Müller, Dr. F.

'APOTHEKER GEORG HÖFER', HT, or, 1900; flowers coppery-red, very large, very dbl., moderate fragrance; Welter; (Sangerhausen)

'APOTHEKER GEORGE HÖFER, CLIMBING', Cl HT, or, 1941; flowers very large, dbl., intense fragrance; Vogel, M.; (Sangerhausen)

'APOTHEOSE', F, yb, 1963; flowers indian yellow edged red, becoming garnet-red; vigorous growth; [Arc-en-Ciel X Seedling]; Delforge

'APPEAL', HT, mp, 1957; flowers clean pink, dbl., intense fragrance; long stems; vigorous, bushy growth; [Ena Harkness X Treasure]; Fletcher; Tucker

'APPLAUSE', HT, dp, 1949; bud long, pointed; flowers light red, dbl., 50 petals, 4–4.5 in., exhibition form, slight fragrance; foliage leathery, dark; vigorous, upright, bushy growth; GM, Bagatelle, 1947; [Contrast X Charlotte Armstrong]; Swim, H.C.; Armstrong Nursery

'APPLAUSE' *see* 'SAVAPPLE'

'APPLE BLOSSOM', HMult, lp, 1932; flowers light pink, center lighter, petals crinkled, dbl., borne in huge clusters; vigorous growth; [Dawson X R. multiflora]; Burbank; Stark Bros.

Apple rose *see* **R. POMIFERA**

'APPLEBLOSSOM', S, lp, 1963; flowers apple blossom-pink, dbl., cupped, recurrent bloom; bushy (2 ft.) growth; Skinner

Appleblossom, S, mp, 1997; Noack, Werner

Appleblossom Festival, S, pb, 1999; Williams, J. Benjamin

'APPLEDORE', F, lp, 1962; flowers medium, dbl., 24 petals, moderate fragrance; foliage glossy; vigorous growth; [Karl Herbst X Pink Charming]; Allen, E.M.

'APPLEJACK', S, pb, 1973; bud small, long, pointed, ovoid; flowers neyron rose, stippled crimson, large, semi-dbl., repeat bloom, intense fragrance; foliage leathery; vigorous, upright, bushy growth; [Goldbusch X (Josef Rothmund X R. laxa Retzius)]; Buck, Dr. Griffith J.; Iowa State University

'APPLETON'S LIMELIGHT', HT, dy, 1934; bud long, pointed; flowers deep golden yellow, open, large, semi-dbl.; foliage leathery, glossy; vigorous, bushy growth; [Lady Forteviot sport]; Appleton

'APPRECIATION', HT, mr, 1971; flowers light red shading crimson, pointed, dbl., 27 petals, 4 in., slight fragrance; foliage glossy; vigorous growth; [Queen Elizabeth X Seedling]; Gregory

Apps Rose *see* KORalogen

Apricot Angel, HT, ab, 1999

Apricot Beauty, S, ab, 1980; Rendu

'APRICOT BRANDY', F, ab, 1972; flowers apricot, base yellow, dbl., 22 petals, 4 in., slight fragrance; foliage bronze-green; Fryers Nursery, Ltd.

Apricot Charm™ *see* 'MINAECO'

Apricot China, F, mr, 1996

'APRICOT CRÈME', Min, ab, 1988; bud pointed; flowers light apricot, edges cream, reverse apricot to cream; [Yellow Doll, Climbing sport]; Bell, Douglas & Judy; Michigan Mini Roses, 1989

'APRICOT DAWN', HT, ab, 1938; flowers apricot, base yellow, moderate fragrance; [Golden Dawn sport]; Wyant

Apricot Delicious, Min, ab, 1996; Welsh

Apricot Delight *see* ATTlight

Apricot Doll *see* 'LAVDOLL'

Apricot Gem, F, ab, 1978; Delbard-Chabert

'APRICOT GLOW', LCl, ab, 1936; flowers apricot, turning apricot-pink, dbl., large trusses, intense fragrance; foliage very glossy; long stems; very vigorous (20 ft) growth; [(Emily Gray X Dr. W. Van Fleet) X Jacotte]; Brownell, H.C.; B&A, 1936; C-P, 1937; Dreer, 1937

Apricot Hit®, Min, ab

Apricot Kisses *see* 'GEACOT'

Apricot Medinette *see* 'POULCOT'

Apricot Midinette *see* 'POULCOT'

Apricot Mist™ *see* 'SAVAMIST'

Apricot Mist *see* 'SOCSPO'

Apricot Moon, F, ab, 1995; Lowery/Eisen

'APRICOT NECTAR', F, ab, 1965; bud ovoid; flowers pink-apricot, base golden, dbl., 4–4.5 in., cupped, intense, fruity fragrance; foliage glossy, dark; vigorous, bushy growth; AARS, 1966; [Seedling X Spartan]; Boerner; J&P

'APRICOT PARFAIT', HT, ab, 1978; bud ovoid; flowers apricot-pink blend, dbl., 53 petals, 4 in., exhibition form, slight fragrance; foliage large, dark; upright growth; [Seedling X South Seas]; Warriner, William A.; J&P

Apricot Perfection *see* 'CLEPERF'

Apricot Prince *see* 'AROSNAP'

'APRICOT QUEEN', HT, ab, 1940; bud pointed; flowers salmon-pink, base apricot-orange, large, dbl., 45 petals; foliage leathery; very vigorous, bushy growth; AARS, 1941; [Mrs J.D. Eisele X Glowing Sunset]; Howard, F.H.; H&S

Apricot Queen Elizabeth, Gr, ab, 1980; Verschuren

'APRICOT QUEEN, CLIMBING', Cl HT, ab, 1950; Maranda

'APRICOT SILK', HT, ab, 1965; flowers apricot, large, dbl., exhibition form, moderate fragrance; foliage dark, glossy; vigorous, upright growth; [Souv. de Jacques Verschuren X ?]; Gregory

Apricot Spice *see* 'SANSPIC'

Apricot Summer *see* KORpapiro

Apricot Sunblaze *see* 'SAVAMARK'

Apricot Sunblaze® *see* 'MEIFRUIJE'

Apricot Surprise, S, ab

Apricot Twist *see* 'MORBROWN'

'APRICOT WINE', F, ab, 1978; flowers burnt apricot, semi-dbl., 12 petals; foliage dark, glossy; low, compact growth; [Allgold X Seedling]; Slack

'APRIHEART', Min, ab, 1983; bud small; flowers light apricot, center deeper, small, dbl., borne singly, intense fragrance; very few prickles; foliage small, medium green, semi-glossy; bushy growth; [Picnic X Rise 'n' Shine]; Hardgrove, Donald L.

April Fools Day *see* 'GREGSIL'

'APRIL HAMER', HT, pb, 1983; flowers shell pink with bright pink edges, large, dbl., 40 petals, exhibition form, moderate fragrance; foliage dark; vigorous, upright growth; [Mount Shasta X Prima Ballerina]; Bell, Ronald J.; Treloar Roses Pty. Ltd.; Wisconsin Roses, 1998

'APRIL MOON', S, my, 1984; bud small; flowers lemon yellow, dbl., 28 petals, cupped, borne 5-10 per cluster, repeat bloom, moderate, sweet fragrance; awl-like, tan prickles; foliage dark, leathery; erect, short, bushy growth; hardy; [Serendipity X (Tickled Pink X Maytime)]; Buck, Dr. Griffith J.; Iowa State University

'APRILIA', HT, dp, 1937; bud ovoid; flowers old-rose, open, very large, dbl., slight fragrance; foliage leathery; vigorous, upright growth; GM, Rome, 1937; Cazzaniga, F. G.

Aprutina, F, 1969; Borgatti, G.; (Cavriglia)

'APSARA', HT, op, 1983; flowers light pink shaded salmon, medium, dbl., 75 petals, flat, intense fragrance; hooked, brown prickles; foliage medium, roundish, leathery; upright growth; [Sonia X Sabine]; Pal, Dr. B.P.; K.S.G. Son's Roses, 1982

'APSARA', F, 1966; bud ovoid; flowers salmon-pink, open, medium, semi-dbl.; foliage glossy; vigorous, upright, open growth; RULED EXTINCT 3/84 ARM; Pal, Dr. B.P.; Indian Agric. Research Inst.

'AQUARELLE', F, dy, 1969; bud ovoid; flowers open, dbl., 18–25 petals, 2 in., flat, moderate, fruity fragrance; foliage dark; very vigorous growth; [Gold Strike X Golden Garnette]; Lens

Aquarelle, HT, r; Croix

Aquarius *see* 'ARMAQ'

'AQUILLA BRIGHT', HT, ob, 1988; flowers medium, dbl., 26–40 petals, slight fragrance; foliage medium, medium green, semi-glossy; upright growth; [Sunblest X Matador]; Lea, R.F.G.

'ARA PACIS', HT, w, 1955; bud tubular, well formed; flowers ivory-white edged reddish purple, very large, dbl., 50 petals; foliage glossy, bright green; long stems; vigorous growth; [Peace X Marguerite Chambard]; Giacomasso

'ARABELLA', HT, dp, 1918; flowers crimson-pink, pointed, large, moderate fragrance; vigorous growth; [Mme Caroline Testout sport]; Tantau

'ARABESQUE', F, lp, 1978; bud pointed; flowers soft pink, 10 petals, 3 in., slight fragrance; vigorous growth; [(Gavotte X Tropicana) X Tropicana]; Sanday, John

Arabesque *see* INTerstreep

Arabia® *see* 'TANIBARA'

'ARABIAN NIGHTS', F, op, 1963; flowers light salmon-orange, well-formed, dbl., 25 petals, 4.5 in., moderate fragrance; vigorous growth; [Spartan X Beauté]; McGredy, Sam IV; McGredy

'ARABY', Gr, lp, 1973; bud long, pointed; flowers light orchid-pink, center white, open, large, dbl., intense fragrance; foliage large, glossy, dark, leathery; vigorous, upright growth; [Honey Chile X Rose Merk]; Thomson

'ARACELI LEYVA', HT, op, 1940; bud long, pointed; flowers rose-salmon, large blooms, dbl., cupped, moderate fragrance; foliage leathery; strong stems; vigorous, upright growth; [Mme Butterfly X Comtesse Vandal]; Dot, Pedro

'ARAKAN', F, lp, 1968; flowers, dbl., blooms in trusses, slight fragrance; [Pink Parfait X Ivory Fashion]; Harkness

'ARAMIS', F, dr, 1964; flowers bright scarlet, semi-dbl., 3 in., blooms in clusters of 7-8; foliage dark; very bushy, compact growth; [Bel Ami X (Java X Alain)]; Laperrière; EFR

'ARASHIYAMA', HT, w, 1997; (Kyoto 1200; YKH 501); flowers ivory white with pink shade, very full (41+ petals), large blooms, frangant; moderate prickles; foliage medium, medium green, semi-glossy; medium (1–2 m) growth; [Garden Party X Pristine]; Kameyama, Yasushi

'ARATAMA', HT, yb, 1976; bud pointed; flowers yellow and red, dbl., 25 petals, 6 in., exhibition form, slight fragrance; foliage glossy, light green; vigorous growth; [Kordes' Perfecta X (Garden Party X Christian Dior)]; Takahashi, Takeshi

Aravali Princess, HT, op, 1987; Pal, Dr. B.P.

Arbelle *see* GAUrama

ARC Angel *see* FRYorst

'ARC DE TRIOMPHE', HT, yb, 1955; bud globular; flowers yellow-copper, large, dbl., moderate fragrance; foliage glossy, olive-green; upright growth; RULED EXTINCT 1/86; Frères, Buyl

Arc de Triomphe *see* 'JACALE'

Arcade, Min, mr, 1996; Spooner, Raymond A.

'ARCADIA', HWich, dr, 1913; flowers crimson-scarlet, dbl., rosette, blooms in clusters; very vigorous growth; Walsh

'ARCADIA', HT, rb, 1938; bud ovoid; flowers reddish copper, very large, dbl.; foliage glossy; long stems; vigorous growth; Gaujard

Arcadia Louisiana Tea, T, pb

Arcadian *see* 'MACNEWYE'

Arcanto *see* BARarc

Arcata Light Yellow Wichurana, HWich, ly

Arcata Pink Globs, N, lp

'ARC-EN-CIEL', F, pb, 1961; flowers rich yellow to salmon-pink, carmine and crimson, dbl., 65–70 petals, borne in tight clusters; foliage dark; growth moderate; [Masquerade X Maria Delforge]; Delforge

'ARCH. REVENTÓS', HT, ab, 1935; bud ovoid, apricot; flowers cream-yellow, large, dbl.; foliage glossy, dark; vigorous, bushy growth; Leenders, M.

Arch. Reventós, Climbing, Cl HT; Leenders, M.; (Cavriglia)

'ARCHANGEL', S, lp, 1979; flowers delicate pink, frilled, single, 9–10 petals, 3.5 in., slight fragrance; foliage light green; tall growth; [Little Darling X Gypsy Moth]; Hawker, U.

'ARCHDUCHESS CHARLOTTE', HCh, dp, 1975; bud pointed; flowers intense deep solid pink, very dbl., 76 petals, 2.5–3 in., cupped, profuse bloom early summer, moderate fragrance; foliage glossy, smooth; climbing growth; [Archduke Charles sport]; Earing, F.E.; Kern Rose Nursery

'ARCHDUKE CHARLES', HCh, rb; flowers rose with paler edges, aging to rich crimson; moderate growth; Laffay, M., prior to 1837

'ARCHIDUC JOSEPH', T, pb, 1892; flowers purplish pink, center flesh-pink; vigorous growth; (14); [Mme Lombard seedling]; Nabonnand, G.

'ARCHIDUCHESSE ELISABETH D'AUTRICHE', HP, mp, 1881; flowers rose-pink, dbl.; Moreau et Robert

'ARCHIDUCHESSE ELISABETH-MARIE', Pol, ly, 1898; flowers canary-yellow fading white, imbricated, medium, dbl., intense fragrance; vigorous growth; [Mignonette X Luciole]; Soupert & Notting

'ARCHIE GRAY', HT, dr, 1920; flowers deep crimson, shaded scarlet, dbl.; Dickson, H.

'ARCTIC EMERALD', S, w, 1982; bud globular, pointed; flowers white, with yellow-green center, small to medium, semi-dbl., 12 petals, borne 1-5 per cluster, repeat bloom, slight fragrance; foliage small, light green; low, compact growth; [Thérèse Bugnet X Europeana]; James, John

'ARCTIC FLAME', HT, mr, 1955; bud pointed; flowers bright red, dbl., 55 petals, 5 in., moderate fragrance; vigorous, bushy growth; [(Queen o' the Lakes X Pink Princess) X Mirandy]; Brownell, H.C.; Stern's Nursery

'ARCTIC GLOW', S, rb, 1982; bud globular, pointed; flowers scarlet shading to white center, large, dbl., 28 petals, borne singly, repeat bloom, slight fragrance; foliage dark, rough; compact growth; [Pike's Peak X Show Girl]; James, John

'ARCTIC PINK', F, lp, 1966; flowers pink fading lighter, 3 in., cupped, blooms in trusses, intense fragrance; vigorous growth; [Dearest sport]; Smith, E.

Arctic Rose *see* **R. ACICULARIS**

Arctic Snow™ *see* 'MINLCO'

Arctic Sunrise *see* 'BARARCSUN'

'ARDELLE', HT, w, 1957; bud long, pointed; flowers creamy white, dbl., 72 petals, 5 in., exhibition form, moderate fragrance; foliage glossy; very vigorous, compact growth; [Mrs Charles Lamplough X Peace]; Eddie; Harkness; Wyant

Ardennes *see* 'INTERPEEL'

'ARDENTE', Cl F, op; flowers salmon and orange; [Seedling X Alain]; da Silva, Moreira

'ARDOISÉE DE LYON', HP, m, 1858; flowers violet-rose; vigorous growth; Damaizin

'ARDON', HWich, mp, 1925; flowers bright neyron rose, stained white, borne in pyramidal clusters of 30-40; Turbat

'ARDORE', HT, mr, 1973; bud ovoid, globular; flowers orient red, large, dbl., cupped, moderate fragrance; foliage large, dark, leathery; very vigorous, upright, bushy growth; [Seedling X Ninfa]; Calvino

'ARDOUR', S, ly, 1999; (**Joyce Barden**); dbl., 26–40 petals, 4 in., borne in small clusters, intense fragrance; prickles moderate, outer shoots mostly thornless; foliage medium, dark green, dull, light green when new; upright, spreading, medium (5-6 ft) growth; [Sweet Juliet X Souv de la Malmaison]; Barden, Paul

Ards Beauty *see* 'DICJOY'

'ARDS PILLAR', HT, mr, 1903; flowers large, dbl.; Dickson, A.; (Sangerhausen)

'ARDS RAMBLER', Cl HT, or, 1908; flowers large, dbl., intense fragrance; Dickson, A.; (Sangerhausen)

'ARDS ROVER', Cl HP, dr, 1898; flowers crimson, shaded maroon, large, dbl., sometimes recurrent bloom, moderate fragrance; pillar growth; Dickson, A.

Arejay *see* 'KIRRAD'

Arena's Dream *see* KORdreweer

'AREND HERWIG', F, ob, 1966; bud ovoid; flowers orange-red, medium, dbl.; foliage dark; [Korona X Heureux Anniversaire]; Buisman, G. A. H.

'ARETHUSA', HCh, yb, 1903; flowers yellow, tinted apricot; vigorous growth; Paul, W.

Argental, HT, w; Croix

'ARGENTINA', F, ab, 1941; flowers reddish apricot, semi-dbl., moderate fragrance; [Mev. Nathalie Nypels X Orange Glory]; Leenders, M.

'ARGENTINE CRAMON', HT, w, 1915; flowers white, center tinted salmon-rose; Chambard, C.

'ARGOSY', HT, op, 1938; flowers salmon, flushed pink, dbl.; long stems; [Souv. de Gustave Prat X Seedling]; Clark, A.; NRS New South Wales

Argovia, HT, pb, 1998; Huber

'ARGYLE', HT, w, 1921; flowers pure white; vigorous growth; [Mme Caroline Testout X Marquise de Sinéty]; Dobbie

'ARIA', F, op, 1957; flowers salmon shaded pink, large, dbl.; bushy growth; [Duchess of Rutland X Fashion]; deRuiter

'ARIADNE', HCh, rb, 1918; flowers bright crimson, center shaded yellow; Paul, W.

'ARIAKE', HT, w, 1976; bud globular; flowers ivory, dbl., 47–50 petals, 5 in., cupped, moderate fragrance; foliage light green; vigorous, upright growth; [(Lady X X Garden Party) X Seedling]; Teranishi, K.; Itami Rose Nursery

Arianna® *see* 'MEIDALI'

Arianna *see* MEIkruza

'ARIBAU', HT, mr, 1936; bud long, pointed; flowers brilliant red, large blooms, dbl.; foliage glossy; long stems; very vigorous growth; [K. of K. X Director Rubió]; Dot, Pedro; H. Guillot

'ARIEL', HT, my, 1921; flowers golden yellow, streaked crimson, large blooms, dbl., globular, moderate fragrance; foliage dark; long stems; vigorous, bushy growth; GM, NRS, 1920; [Mme Edouard Herriot X Natalie Boettner]; Bees

Arielle Dombasie® *see* 'MEIHOURAG'

Arioso, HT, or, 1970; flowers light vermilion, full, dbl., 25 petals, 5 in., slight, fruity fragrance; foliage glossy, dark; very vigorous, upright growth; [(Paris-Match X Baccará) X Marella]; Paolino; URS

Arioso *see* MEImucas

'ARISTE', HT, my, 1960; bud very pointed, deep yellow; flowers light yellow, medium, single, 5 petals, moderate fragrance; foliage dark, leathery; vigorous, bushy growth; [Joanna Hill X Souv. de Mme Boullet]; Jones; Hennessey

'ARISTIDE BRIAND', HWich, m, 1928; flowers mauve-pink, borne in clusters of 10-29, moderate fragrance; [Yseult Guillot X Seedling]; Penny

'ARISTIDE DUPUY', HP, 1866; Trouillard; (Cavriglia)

'ARISTOBULE', M, dp, 1849; flowers dark rose with touches of clear rose, dbl.; growth good; Foulard

'ARISTOCRAT', HT, pb, 1949; bud long, pointed; flowers clear light pink, reverse darker, dbl., 28–35 petals, 4.5–5 in., exhibition form, moderate fragrance; foliage leathery, dark; very vigorous, upright growth; [Pink Delight sport]; Holmes, M.A.; Avansino; Mortensen

Arizona *see* 'WERINA'

'ARIZONA SUNSET', Min, yb, 1985; flowers light yellow, flushed orange-red, medium, cupped, 20 petals, cupped, slight; prickles slanted downward; foliage small, medium green, semi-glossy; bushy, spreading growth; PP006559; [(Orange Sweetheart X Zinger) X Party Girl]; Jolly, Nelson F.; Rosehill Farm

'ARJUN', HT, or, 1980; bud long, pointed; flowers, dbl., 35 petals, cupped, borne singly or 8 per cluster, slight fragrance; hooked prickles; foliage large, smooth; tall, upright growth; [Blithe Spirit X Montezuma]; Division of Vegetable Crops and Floriculture

'ARKANSAS', HT, or, 1980; bud ovoid, pointed; flowers urn-shaped, dbl., 48 petals, borne singly or 2-4 per cluster, slight, spicy fragrance; long, oval-based prickles, hooked downward; foliage leathery; upright, vigorous growth; [Seedling X Seedling]; Weeks

Arkansas Rose *see* **R. ARKANSANA**

'ARKANSAS SUNSHINE', F, my, 1962; flowers golden yellow, dbl., 30 petals, cupped, blooms in clusters; foliage dark, leathery; vigorous, bushy growth; [Goldilocks X Seedling]; Jones; Hennessey

Arkavathi, F, dp, 1971; Kasturi

'ARKLE', HT, or, 1977; bud cupped; flowers dark tangerine, dbl., 45–50 petals, 4 in., cupped, intense fragrance; foliage glossy; vigorous growth; [Whisky Mac sport]; Hughes Roses

'ARLENE FRANCIS', HT, my, 1957; bud long, pointed; flowers golden buttery yellow, dbl., 25–30 petals, 5 in., exhibition form, borne mostly singly, intense, sweet licorice fragrance; foliage dark, glossy; vigorous growth, upright, medium sized; [Eclipse seedling X Golden Scepter]; Boerner; J&P

'ARLEQUIN', HT, ob, 1945; flowers orange-yellow and coppery red, very large, dbl., globular, moderate fragrance; foliage dark, glossy; bushy growth; Gaujard

'ARLEQUIN', HGal, m, 1837; flowers purple-violet, dbl.; medium growth; Paillard

Arles *see* **'FERNAND ARLES'**

Arlette *see* POUlske

Armada® *see* 'HARUSEFUL'

'ARMAGH', HT, ab, 1950; flowers creamy pink, apricot and buff, pointed, large, dbl., 49 petals, slight fragrance; foliage dark; free growth; [Sam McGredy X Admiration]; McGredy, Sam IV

'ARMAQ', Gr, pb, 1971; (**Aquarius**); bud ovoid; flowers medium pink blend, medium, dbl., exhibition form, slight fragrance; foliage large, leathery; vigorous, upright, bushy growth; AARS, 1971 GM, Geneva, 1970; [(Charlotte Armstrong X Contrast) X (Fandango X (World's Fair xFloradora))]; Armstrong, D.L.; Armstrong Nursery

'ARMÉNIE', HT, ab, 1936; bud purple-garnet; flowers blood-red, shaded, large, very dbl., moderate fragrance; very vigorous, bushy growth; [Rhea Reid X Yves Druhen]; Buatois

'ARMIDE', A, w, 1817; flowers medium, moderate fragrance; Vibert

'ARMILLA', Gr, pb, 1973; (**Cherry-Vanilla**); bud pointed; flowers pink, center creamy yellow, medium, dbl., cupped, moderate fragrance; foliage dark, leathery; vigorous, upright, bushy growth; [Buccaneer X El Capitan]; Armstrong, D.L.; Armstrong Nursery

'ARMINDA', HT, op, 1956; bud ovoid; flowers bright pink tinted coral, large blooms, very dbl., globular, moderate fragrance; foliage glossy; strong stems; vigorous growth; [Peace X Symphonie]; Camprubi, C.

'ARMLU', Gr, lp, 1966; (**Lucky Lady**); bud long, pointed; flowers light pink, reverse darker, large, dbl., 28 petals, exhibition form, slight fragrance; foliage dark, glossy; vigorous, upright growth; AARS, 1967; [Charlotte Armstrong X Cherry Glow]; Armstrong, D.L. & Swim; Armstrong Nursery

'ARMMA', HT, w, 1965; (**Matterhorn**®); flowers medium to large, dbl., exhibition form; foliage leathery; very tall, upright growth; AARS, 1966 GM, Portland, 1964; [Buccaneer X Cherry Glow]; Armstrong, D.L. & Swim; Armstrong Nursery

'ARMOLO', HT, my, 1971; (**Apollo**®); bud long, pointed; flowers large, dbl., moderate fragrance; foliage large, glossy, dark, leathery; vigorous, upright, bushy growth; AARS, 1972; [High Time X Imperial Gold]; Armstrong, D.L.; Armstrong Nursery

'ARMONIA', HT, mr, 1951; bud long, pointed; flowers bright red, blooms, dbl., exhibition form, moderate fragrance; foliage dark, leathery; long stems; very vigorous, upright growth; Cazzaniga, F. G.

Armorique Nirpaysage *see* 'POULAPS'

Armosa *see* **'HERMOSA'**

'ARM-ROY BEAUTY', HT, mr, 1945; very vigorous, tall growth; [Better Times sport]; Armacost & Royston

Arnaud Delbard® *see* 'DELTEP'

'ARNDT', LCl, lp, 1913; bud yellowish red; flowers pale pink, dbl., in large clusters, recurrent bloom; foliage dark; half climbing growth; [Hélène X Gustav Grünerwald]; Lambert, P.

'ARNELDA MAE', Min, lp, 1983; bud small; flowers small blooms, dbl., 30 petals, exhibition form, borne singly and in clusters up to 7, slight fragrance; foliage small, light green, matt; bushy growth; [Sheri Anne X Seedling]; Pencil, Paul S.

'ARNHEM GLORY', HT, dr, 1959; flowers deep velvety red, large, intense fragrance; vigorous growth; Verschuren

'ARNOLD', HRg, mr, 1893; (Arnoldiana); flowers scarlet, single, some recurrent bloom; vigorous growth; [R. rugosa X Génèral Jacqueminot]; Dawson; Eastern Nursery

Arnold Greensitt *see* 'NOSTARN'

Arnoldiana *see* **'ARNOLD'**

'AROALL', HT, my, 1977; (**Allspice**); bud ovoid, pointed, deep yellow; dbl., 35 petals, 4.5 in., moderate, honey and tea rose fragrance; foliage large, olive-green; vigorous, upright, bushy growth; [Buccaneer X Peace]; Armstrong, D.L.; Armstrong Nursery

'AROART', HT, dr, 1981; (**Jennifer Hart**); bud ovoid, pointed; flowers, dbl., 45 petals, exhibition form, borne singly, slight, tea fragrance; foliage medium, medium green, semi-glossy; medium, upright, bushy growth; PP005219; [Pink Parfait X Yuletide]; Swim, H.C. & Christensen, J.E.; Armstrong Nursery, 1982

'AROBIPY', HT, w, 1987; (**Crystalline**™, Valerie Swane); flowers large, pure white, dbl., 30–35 petals, exhibition form, borne usually singly, moderate, spicy or sweet tea fragrance; globose, large, orange fruit; normal, light green-tan prickles; foliage medium, medium green, semi-glossy; upright, bushy, tall growth; PP006714; [Bridal Pink X (Blue Nile X (Ivory Tower X Angel Face))]; Christensen, Jack & Carruth, Tom; Armstrong Nursery, 1986

'AROBLAVEET', HT, dr, 1982; (**Centenary College**); flowers deep red, large, dbl., 35 petals, intense fragrance; foliage large, medium green, matt; vigorous, upright, tall growth; [Angel Face X Typhoo Tea]; Christensen, Jack E.; Armstrong Nursery

'AROBRI', Min, ab, 1978; (**Holy Toledo**™); bud ovoid, pointed; flowers brilliant apricot-orange, reverse yellow-orange, imbricated, dbl., 28 petals, 1.5–2 in.; foliage small, glossy, dark; vigorous, bushy growth; AOE, 1980; [Gingersnap X Magic Carrousel]; Christensen, Jack E.; Armstrong Nursery

'AROBRISP', HT, dr, 1988; (**Tuxedo**); bud ovoid; flowers urn-shaped, large, borne singly, dbl., 45 petals; prickles broad, hooked downward, red-brown; foliage medium, medium green, semi-glossy; upright growth; [Portland Trailblazer X Olympiad]; Christensen, Jack E.; Bear Creek Gardens, 1988

'AROCAD', HT, ab, 1981; (**Brandy**™); bud long, pointed; flowers deep apricot, dbl., 25–30 petals, 5 in., exhibition form, borne mostly singly, slight, tea fragrance; straight prickles; foliage large, dark green; vigorous, medium growth; PP005168; AARS, 1982; [First Prize X Dr. A.J. Verhage]; Swim, H.C. & Christensen, J.E.; Armstrong Nursery

'AROCANT', HT, ob, 1985; (**Spring Fever**); flowers apricot, pink and orange blend, large, dbl., 35 petals, slight fragrance; foliage large, dark, semi-glossy; [Gingersnap X Brandy]; Christensen, Jack E.; Armstrong Nursery

'AROCHARM', F, pb, 1987; (**Origami**™); flowers clear, soft pink, outstanding form, medium, borne usually singly, dbl., 25 petals, exhibition form, moderate, spicy fragrance; normal, light green-tan prickles; foliage medium, medium green, semi-glossy; upright, bushy, medium growth; [Coquette X Zorina]; Christensen, Jack & Carruth, Tom; Armstrong Nursery, 1991

'AROCHER', HT, rb, 1981; (**Mon Cheri**™); bud ovoid, pointed; flowers medium pink, suffusing to near yellow at base, aging to dark, dbl., 38 petals, slight, spicy fragrance; short prickles; foliage semi-glossy, medium green; upright, medium growth; PP005156; AARS, 1982; [(White Satin X Bewitched) X Double Delight]; Christensen, Jack E.; Armstrong Nursery

'AROCLIDD', Cl HT, rb, 1982; (**Double Delight**, **Climbing**, Grimpant Double Delight); PP005155; [Double Delight sport]; Christensen, Jack E.; Armstrong Nursery, 1985

'AROCOMU', HT, m, 1983; ('AROCUMU', **Patsy Cline**™); flowers light lavender, petals edged ruby lavender, large, dbl., 35 petals, exhibition form, intense fragrance; foliage medium, dark, matt; upright, bushy growth; PP005556; [Angel Face X Double Delight]; Christensen, Jack E.; Armstrong Nursery

'AROCORE', HT, mr, 1986; (**Rodeo Drive**™, Sunset Strip); flowers bright deep red, large, dbl., 32 petals, exhibition form, borne usually singly, slight fragrance; fruit not observed; many attenuated, medium, reddish prickles aging light brown; foliage medium to large, medium green, semi-glossy; bushy, medium growth; PP006813; [Merci X Pharaoh]; Christensen, Jack E.; Armstrong Nursery

AROcruby, F, rb, 1987; (**Park Place**); Christensen, Jack E.

'ARODER', HT, dr, 1972; (**Kentucky Derby**); flowers large, dbl., exhibition form, slight fragrance; foliage large, glossy, leathery; vigorous, upright, bushy growth; [John S. Armstrong X Grand Slam]; Armstrong, D.L.; Armstrong Nursery

'ARODI', Min, mp, 1978; (**Heidi**™); bud mossy; flowers clear medium pink, mini-moss, small, dbl., 35 petals, 1.5 in., intense fragrance; foliage glossy; very vigorous, bushy growth; [Fairy Moss X Iceberg]; Christensen, Jack E.; Armstrong Nursery

'ARODOUSNA', HT, rb, 1986; (**Givenchy**™, Paris Pink); flowers pink, blushed red, reverse pink, yellow base, blooms in sprays, dbl., 30 petals, exhibition form, intense, spicy fragrance; no fruit; medium, brown, hooked prickles; foliage medium, dark; medium, upright, bushy growth; [Gingersnap X Double Delight]; Christensen, Jack E.; Armstrong Nursery, 1985

'AROFEIGEL', HT, lp, 1982; (**Peggy Lee**); flowers pale pink; PP005467; [Century Two sport]; Feigel, John R.; Armstrong Nursery

'AROFIRIC', F, rb, 1987; (**Fire 'n' Ice**™); flowers red, reverse white, fading purplish-red, medium, borne usually singly, dbl., 40 petals, exhibition form, slight fragrance; no fruit; foliage medium, dark green, glossy, very attractive, pointed; upright, bushy, tall growth; [Bluhwunder X Love]; Christensen, Jack & Carruth, Tom, 1981; Armstrong Nursery, 1985

'AROFRAP', HT, pb, 1982; (**Almondeen**); flowers almond and pink blend, good form, large, dbl., slight fragrance; foliage large, medium green, semi-glossy; upright, bushy growth; PP005704; [Angel Face X First Prize]; Christensen, Jack E.; Armstrong Nursery, 1984

'AROFRICHEE', HT, yb, 1986; (**Pink Lemonade**); flowers yellow turning bright pink, well-formed, large, dbl., 35 petals, moderate fragrance; foliage large, medium green, semi-glossy; upright, bushy growth; [Friendship X Rosy Cheeks]; Christensen, Jack E.; Armstrong Nursery

'AROFUTO', HT, mr, 1989; (City of Warwick, **Grand Marshall**™); bud ovoid, pointed; flowers large, borne usually singly, dbl., 35 petals, exhibition form, slight fragrance; prickles hooked, medium, red to brown; foliage medium, medium green, semi-glossy; upright, bushy, medium growth; [Futura X Olympiad]; Christensen, Jack E.; Michigan Bulb Co., 1989

'AROGLOFY', F, dy, 1988; (**Bugle Boy**™); bud ovoid, pointed; flowers deep yellow, good substance, medium, borne in sprays of 5-6, dbl., 35 petals, cupped, slight, tea fragrance; fruit unknown; prickles hooked slightly downward, medium, red to tan; foliage medium, medium green, very glossy; bushy, medium growth; [Sunsprite X (Katherine Loker seedling X Gingersnap)]; Christensen, Jack E.; Bear Creek Gardens, 1988

'AROGLOR', HT, m, 1983; (**Governor's Lady Gloria**); flowers pastel mauve, well-formed, large, dbl., 35 petals, intense fragrance; foliage large, medium green, matt; upright, bushy growth; [Sweet Afton X Blue Nile]; Christensen, Jack E.; Armstrong Nursery

'AROGOBI', Min, ob, 1984; (**Fool's Gold**); flowers gold, reverse bronze, well-formed, small, dbl., 20 petals, slight fragrance; foliage medium, dark, semi-glossy; upright growth; [Cricket X Dr. A.J. Verhage]; Christensen, Jack E.; Armstrong Nursery

'AROGRAJU', HT, m, 1985; (**Plum Crazy**); flowers deep lavender, well-formed, large, dbl., 35 petals, moderate fragrance; foliage medium, dark, matt; bushy growth; [(Ivory Tower X Angel Face) X Blue Nile]; Christensen, Jack E.; Armstrong Nursery

'AROGRESH', HT, w, 1983; (**Mint Julep**); flowers pale green and pink blend, large, dbl., 35 petals, slight fragrance; foliage medium, medium green, semi-glossy; upright growth; [White Masterpiece X Queen Elizabeth]; Christensen, Jack E.; Armstrong Nursery

'AROGREWOD', HT, m, 1987; (**Silverado**); flowers soft silver blushed ruby, reverse white, large blooms, dbl., 28 petals, exhibition form, borne singly, slight fragrance; large prickles, hooked slightly downward; foliage medium, dark, matt; medium, bushy growth; PP006861; [(Ivory Tower X Angel Face) X Paradise]; Christensen, Jack E.; Armstrong Nursery

Aroha *see* 'MURHA'

'AROHAICLO', Cl Gr, or, 1982; (**Olé, Climbing**™); Haight, George S. & Swim, H.C.; Armstrong Nursery

'AROJECHS', F, rb, 1980; (**Confetti**); bud ovoid; flowers deep yellow, aging orange-red, 18–25 petals, exhibition form, borne 3-7 per cluster, slight, tea fragrance; prickles hooked downward; foliage medium green; upright growth; PP005399; [Jack O'Lantern X Zorina]; Swim, H.C. & Christensen, J.E.; Armstrong Nursery, 1983

'AROKET', Min, ob, 1978; (**Cricket**™); bud ovoid; flowers light orange to yellow, dbl., 25 petals, 1–1.5 in., globular, slight fragrance; foliage dark; upright, bushy growth; [Anytime X Katherine Loker]; Christensen, Jack E.; Armstrong Nursery

'AROKR', F, my, 1978; (**Katherine Loker**); bud pointed; flowers medium golden yellow, classic form, imbricated, dbl., 28 petals, 3.5 in., slight fragrance; upright, spreading growth; [Zorina X Dr. A.J. Verhage]; Swim, H.C. & Christensen, J.E.; Armstrong Nursery, 1979

AROkris, F, dy, 1990; (**Golden Wedding**); Bear Creek Gardens

'AROKUNCE', HT, ob, 1982; (**Inferno**); flowers well-formed, large, slight fragrance; foliage medium, dark, semi-glossy; upright, bushy growth; PP005558; [Zorina X Yankee Doodle]; Christensen, Jack E.; Armstrong Nursery

'AROLALA', HT, m, 1983; (**Azure Sea**); flowers silvery lavender, petals edged ruby, deeper reverse, well-formed, large, dbl., 30 petals, slight fragrance; foliage large, dark, matt; upright, bushy growth; PP005693; [(Angel Face X First Prize) X Lady X]; Christensen, Jack E.; Armstrong Nursery

'AROLAQUELI', Gr, m, 1986; (**Lagerfeld**™, Starlight); flowers silvery lavender, blooms in sprays of 5-15, dbl., 30 petals, 4–5 in., exhibition form, intense fragrance; large, globular, orange fruit; medium, light brown prickles, hooked downward; foliage medium, medium green, matt; tall, upright, bushy growth; [Blue Nile X (Ivory Tower X Angel Face)]; Christensen, Jack E.; Armstrong Nursery, 1985

'AROLEMO', HT, ly, 1986; (**Lanvin**™); dbl., 30 petals, exhibition form, blooms in sprays of 3-5, moderate fragrance; no fruit; medium, straight, light brown to red prickles; foliage medium, dark green tinted red, semi-glossy; medium, upright bushy growth; [Unnamed seedling X Katherine Loker]; Christensen, Jack E.; Armstrong Nursery, 1985

'AROLICAL', HT, m, 1984; (**Blue Ribbon**); bud ovoid; flowers lilac-blue, well-formed, dbl., 35–40 petals, 5–6 in., borne mostly singly, intense, sweet rose fragrance; foliage very large, dark, semi-glossy; upright growth; PP006000; [(Angel Face X First Prize) X Blue Nile]; Christensen, Jack E.; Armstrong Nursery, 1986

'AROLYME', HT, my, 1987; (**Sundial**™); flowers medium, bright yellow, medium, borne usually singly, dbl., 30 petals, exhibition form, slight, spicy fragrance; no fruit; many, normal, small and large, light green-tan prickles; foliage medium, medium green, glossy; upright, bushy, medium growth; [(Golden Wave X (American

Heritage X First Prize)) X ((Camelotx First Prize) X Yankee Doodle)]; Christensen, Jack E.; Armstrong Nursery, 1986

'AROMA', HT, mr, 1931; bud ovoid; flowers crimson, large, dbl.; vigorous, bushy growth; Cant, B. R.; J&P

Aromatic, HT, pb, 1970

'AROMICLEA', HT, ob, 1984; (**Voodoo**®); flowers salmon, yellow, orange and pink blend, large, dbl., 30–35 petals, 5–6 in., exhibition form, intense, rich, sweet fragrance; foliage medium, dark, very glossy; upright, bushy, tall growth; PP006121; AARS, 1986; [((Camelot X First Prize) X Typhoo Tea) X Lolita]; Christensen, Jack E.; Armstrong Nursery, 1986

'AROMIKEH', Min, mr, 1981; (**Hotline**®); bud ovoid, pointed, lightly mossed; flowers bright medium red, mini-moss, dbl., 22 petals, exhibition form, borne usually singly, moderate, moss fragrance; straight, thin prickles; foliage medium green; compact (near 12 in) growth; PP005672; [Honest Abe X Trumpeter]; Christensen, Jack E.; Armstrong Nursery

'AROMONTELIB', Cl HT, yb, 1980; (Caribia, Climbing, **Harry Wheatcroft, Climbing**); Mungia, Fred A., Sr.; Montebello Rose Co., Inc.

'ARON', Min, dr, 1978; (**Honest Abe**®); bud mossy; flowers deep velvety crimson-red, mini-moss, small, dbl., 33 petals, 1.5 in., slight, tea fragrance; foliage glossy; vigorous, bushy growth; [Fairy Moss X Rubinette]; Christensen, Jack E.; Armstrong Nursery

'ARONEMO', Min, ly, 1977; (**Honey Moss**); bud mossy; flowers near white, toward honey, mini-moss, dbl., 52 petals, 1 in., flat, moderate fragrance; foliage dark, leathery; spreading growth; [Fairy Moss X ?]; Sudol, Julia

'ARONESUF', F, dy, 1988; (**South Pacific**®); flowers color holds well, medium, borne in sprays of 4-6, dbl., 26 petals, cupped, slight fragrance; no fruit; prickles hooked slightly downward, medium, few, green to tan; foliage medium, medium green, glossy; bushy, medium growth; [Sunsprite X Unnamed seedling]; Christensen, Jack E.; Bear Creek Gardens, 1988

'ARONEWP', HT, yb, 1988; (**New Peace**®); flowers yellow-cream with bright red margins aging larger red margin, dbl., 48–52 petals, exhibition form, slight fragrance; prickles pointed, small, dark tan; foliage medium, medium green, matt; upright, medium growth; [Gingersnap X Young Quinn]; Christensen, Jack E.; Michigan Bulb Co., 1988

'AROPICLU', S, pb, 1984; (**Dream Cloud**); flowers light to dark salmon-pink, dbl., 20 petals, blooms in large pyramidal clusters, slight fragrance; foliage medium, long, narrow, medium green, matt; spreading, bushy growth semi-pendulous habit; PP005998; [Zorina X Gartendirektor Otto Linne]; Christensen, Jack E.; Armstrong Nursery, 1985

'AROPLUMI', HT, m, 1990; (**Fragrant Plum**); bud long, ovoid; flowers light lavender blushing purple, smoky edges, dbl., 20–25 petals, 4–4.5 in., exhibition form, borne singly and in large clusters, intense, fruity fragrance; tall, upright growth; [Shocking Blue X (Blue Nile X (Ivory Tower X Angel Face))]; Christensen, Jack E.; Armstrong Nurseries, 1990

'AROPRAWN', Min, lp, 1982; (**Helen Boehm**); flowers soft pink, small, dbl., 20 petals, exhibition form, slight fragrance; foliage small, medium green, semi-glossy; upright, bushy growth; PP005397; [Foxy Lady X Deep Purple]; Christensen, Jack E.; Armstrong Nursery, 1983

'ARORAJU', F, ob, 1986; (Lady Glencora, **Orange Juice**); flowers clear orange, blooms in sprays of 3-5, dbl., 33 petals, exhibition form, slight fragrance; ovoid, medium, orange-red fruit; long, red prickles; medium, upright, bushy growth; [Katherine Loker X Gingersnap]; Christensen, Jack E.; Michigan Bulb Co.

'ARORASP', Min, dy, 1981; (**Sunspray**®); bud ovoid, long, pointed; flowers bright deep yellow, semi-dbl., 16 petals, borne singly or several per cluster, slight, tea fragrance; straight prickles; foliage semi-glossy, dark; vigorous, upright growth; PP005035; [Gingersnap X Magic Carrousel]; Christensen, Jack E.; Armstrong Nursery

'AROREROY', HT, mr, 1987; (Royal Delight, **Royal Success**®); flowers large, dbl., 30 petals, exhibition form, borne usually singly; fruit not observed; normal, light green to tan prickles; foliage medium, medium green, semi-glossy; upright, bushy, tall growth; PP006910; [Red Success X Royalty]; Christensen, Jack & Carruth, Tom; Armstrong Nursery, 1986

'ARORESAS', Gr, mr, 1987; (**Scarlett O'Hara**®); flowers medium, dbl., 35 petals, exhibition form, borne usually singly, slight fragrance; nearly thornless; foliage medium, medium green, semi-glossy; upright, tall growth; PP006670; [Red Success X Mary DeVor]; Christensen, Jack & Carruth, Tom; Armstrong Nursery, 1986

AROshrel, HT, op, 1991; (**Michelle Joy**); Bear Creek Gardens

'AROSHRIM', Min, op, 1980; (**Foxy Lady**®); bud ovoid, pointed; flowers salmon and creamy blend, imbricated, dbl., 25 petals, 1.5 in.; foliage small; tall, vigorous, bushy growth; [Gingersnap X Magic Carrousel]; Christensen, Jack E.; Armstrong Nursery

Arosia, HT, mp, 1998; Noack, Werner

'AROSILHA', HT, m, 1986; ('AROSILMA', **Silk Hat**®); flowers red purple, cream reverse, large blooms borne usually singly, dbl., 45 petals, exhibition form, moderate fragrance; no fruit; foliage large, medium green, matt; medium, upright, bushy growth; [Ivory Tower X (Night 'n' Day X Plain Talk)]; Christensen, Jack E.; Armstrong Nursery, 1985

'AROSNAP', F, ob, 1978; (Apricot Prince, **Gingersnap**, Prince Abricot); bud long, pointed; flowers pure orange, imbricated to ruffled, dbl., 30–35 petals, 4 in., borne in small clusters, slight fragrance; foliage dark green; vigorous, upright, bushy growth; PP4330; [(Zambra X (Orange Triumph X Floradora)) X (Jean de la Lune x(Spartan X Mandrina))]; Delbard-Chabert; Armstrong Nursery

'AROSTAL', HT, m, 1985; (**Quicksilver**); flowers pale lavender gray, large blooms borne singly, dbl., 25 petals, exhibition form, moderate fragrance; medium, yellow-gray prickles; foliage large, dark, matt; tall, upright, bushy growth; [Blue Nile X Brandy]; Christensen, Jack E.; Michigan Bulb Co.

AROsumo, HT, my, 1983; (**Morning Sun**); Christensen, Jack E.

'AROTIGY', HT, yb, 1986; (**Polo Club**®); flowers yellow bordered red, fading cream with pink edges, dbl., slight fragrance; fruit not observed; straight, small, few, greenish to light brown prickles; foliage medium, dark green, semi-glossy; upright, bushy, tall growth; PP006758; [Gingersnap X Young Quinn]; Christensen, Jack E.; Armstrong Nursery, 1986

'AROTRUSIM', S, mr, 1988; (**Bloomin' Easy**®, Blooming Easy); flowers clear, bright red, urn-shaped, medium, semi-dbl., 22–25 petals, borne singly, repeat bloom, moderate, tea fragrance; round, medium, bright red-orange fruit; prickles hooked slightly downward, medium, red; foliage large, medium green, glossy; upright, bushy, medium growth; PP007157; [Trumpeter X Simplicity]; Christensen, Jack E.; Armstrong Nursery, 1987

'AROVIDIL', Min, mr, 1980; (**Little Red Devil**); bud ovoid, pointed; flowers imbricated blooms borne1-8 per cluster, dbl., 44 petals, 1.5 in., slight fragrance; small, narrow prickles; foliage small, semi-glossy, irregularly serrated; vigorous, bushy, fairly tall growth; [Gingersnap X Magic Carrousel]; Christensen, Jack E.; Armstrong Nursery

'AROVULC', HT, ob, 1984; ('AROVULE', Harlequin, **Miss Liberté**, Miss Liberty); flowers coral orange to dusty deep red, well-formed, large, dbl., 35 petals, slight fragrance; foliage large, dark, semi-glossy; upright, bushy growth; [(Camelot X First Prize) X Gingersnap]; Christensen, Jack E.; Armstrong Nursery

'AROVULE', HT, ob, 1984; ('AROVULC', Harlequin, **Miss Liberté**, Miss Liberty); flowers coral orange to dusty deep red, well-formed, large, dbl., 35 petals, slight fragrance; foliage large, dark, semi-glossy; upright, bushy growth; [(Camelot X First Prize) X Gingersnap]; Christensen, Jack E.; Armstrong Nursery

'AROWAGO', HT, rb, 1986; (Show Off, **Showoff**); flowers velvety brilliant red, reverse silvery blend, large, dbl., 35 petals, slight fragrance; foliage large, dark, semi-glossy; upright, bushy growth; [Typhoo Tea X Snowfire]; Christensen, Jack E.; Armstrong Nursery

'AROWEDYE', HT, dy, 1982; (**Candlelight**™); bud ovoid, pointed; flowers deep yellow, spiraled, dbl., 30 petals, borne mostly singly, slight fragrance; large-based prickles; foliage large, semi-glossy; medium-tall, upright, branching growth; PP005398; [Shirley Laugharn X (Bewitched X King's Ransom)]; Christensen, J.E. & Swim, H.C.; Armstrong Nursery

'AROWHIF', Gr, w, 1980; (**White Lightnin'**™); bud ovoid, pointed; flowers clear white, ruffled, dbl., 30 petals, 3.5–4 in., exhibition form, borne in small clusters, intense, citrus fragrance; foliage glossy, medium green; upright, bushy growth; PP4670; AARS, 1981; [Angel Face X Misty]; Swim, H.C. & Christensen, J.E.; Armstrong Nursery

AROwillip, MinFl, lp; (**Strawberry Fayre**); Bear Creek Gardens, 1991

AROyefel

'AROYEFELT', F, my, 1987; (AROyefel, **Haiku**™); flowers medium, borne usually singly or in sprays of 2-3, dbl., 38 petals, exhibition form, slight fragrance; no fruit; normal, light green to tan prickles; foliage medium, dark green, glossy; upright, bushy, medium growth; [Bridal Pink X Sunspray]; Christensen, Jack & Carruth, Tom; Armstrong Nursery, 1986

'AROYOL', Min, my, 1979; (**Hopscotch**); bud ovoid, pointed; flowers golden yellow, imbricated, small, dbl., 28 petals, 1.5 in.; foliage small; vigorous, bushy growth; [Gingersnap X Magic Carrousel]; Christensen, Jack E.; Armstrong Nursery

'AROYQUELI', Gr, my, 1982; (**Gold Medal**®); bud ovoid, long, pointed; flowers deep golden yellow sometimes flushed orange, classic shape, dbl., 30–35 petals, 4.5–5 in., exhibition form, mostly borne singly, slight, fruity fragrance; foliage large, dark; tall, upright, bushy growth; PP005177; Gold Star of the South Pacific, Palmerston North, NZ, 1983; [Yellow Pages X (Granada X Garden Party)]; Christensen, Jack E.; Armstrong Nursery

'AROYUMI', Min, yb, 1984; (**Ferris Wheel**); flowers yellow, turning pink, orange and red (striped), medium, dbl., 20 petals, slight fragrance; foliage small, dark, semi-glossy; bushy growth; PP005703; [Golden Angel X Cricket]; Christensen, Jack E.; Armstrong Nursery

Arpége, HT, 1978; Dorieux, Francois; (Cavriglia)

'ARPEGGIO', F, dp, 1961; bud pointed; flowers light red, semi-dbl., 12–18 petals, 3 in., blooms in clusters, slight fragrance; foliage dark, glossy; vigorous, compact growth; Von Abrams; Peterson & Dering

'ARRAS', Pol, mr, 1924; flowers crimson-red; [Triomphe Orléanais sport]; Turbat

'ARRILLAGA', HP, lp, 1929; flowers light pink, base golden, large, dbl., 50 petals, moderate fragrance; vigorous growth; [(R. centifolia X Mrs John Laing) X Frau Karl Druschki]; Schoener; B&A

'ARROGANCE', HT, pb, 1970; flowers coral-pink, dbl., 25 petals, 4 in., slight fragrance; foliage dark, leathery; compact growth; [Mischief X John S. Armstrong]; Poulsen, Niels D.; Poulsen

Arromanches, F, op, 1975; Eve, A.

Arrowtown, LCl, dy, 1982; Martin

Art Deco, HMsk, mr, 1993; Lens

'ARTAMA', HT, yb, 1978; bud pointed; 25 petals, exhibition form, slight fragrance; few prickles; foliage light green, pointed, glossy; sturdy growth; [Kordes' Perfecta X (Garden Party X Christian Dior)]; Takahashi, Takeshi

'ARTEK', HT, dr, 1939; flowers dark, velvety red, medium, dbl.; Kosteckij; (Sangerhausen)

Artful Dodger *see* 'SABBELIEF'

'ARTHUR BELL', F, my, 1965; flowers yellow to creamy yellow, large, semi-dbl., 15 petals, intense fragrance; foliage heavily veined; vigorous growth; [Cläre Grammerstorf X Piccadilly]; McGredy, Sam IV; McGredy

'ARTHUR BELL, CLIMBING', Cl F, my, 1979; Pearce, C.A.; Limes Rose Nursery

'ARTHUR COOK', HT, dr, 1924; bud long, pointed; flowers deep crimson, large blooms, dbl., moderate fragrance; foliage light, glossy; long, strong stems; vigorous, bushy, compact growth; GM, NRS, 1925; McGredy

Arthur Cox *see* JAYART

'ARTHUR DE SANSAL', P, m, 1855; flowers rich crimson-purple, fully damask-like, dbl.; Cochet, Sc.

'ARTHUR HILLIER', HMoy, dp, 1961; flowers rose-crimson, single, 5 petals, 2.5–3 in., repeat bloom, slight fragrance; vigorous growth; [R. macrophylla X R. moyesii]; Hillier

'ARTHUR J. TAYLOR', HT, dr, 1947; flowers large, dbl.; Wheatcroft Bros.; (Sangerhausen)

'ARTHUR OGER', HP, m, 1875; flowers purple/pink, very large, dbl.; Oger; (Sangerhausen)

'ARTHUR R. GOODWIN', HT, or, 1909; flowers coppery orange-red, passing to salmon-pink; [Seedling X Soleil d'Or]; Pernet-Ducher

'ARTHUR R. GOODWIN, CLIMBING', Cl HT, or

'ARTHUR SCARGILL', Min, mr, 1985; [Amruda sport]; Thompson, M.L.

'ARTHUR SCHULTE', HT, rb, 1986; flowers cherry to blood red with ivory white at base of petals, large, dbl., 43 petals, exhibition form, borne singly and in sprays of 1-3, moderate, damask fragrance; rounded, medium, pumpkin-orange fruit; medium-large, light green-bronze prickles; foliage large, dark green, semi-glossy, disease resistant; upright, bushy, medium, very vigorous, some interbranching growth; [Colorama seedling X Chrysler Imperial]; Williams, J. Benjamin; Krider Nursery, 1989

'ARTHUR YOUNG', M, m, 1863; flowers dark purple, large, cupped, recurrent bloom; vigorous growth; Portemer fils

Artiste® *see* 'DORISTEL'

'ARTISTIC', F, ob, 1971; flowers orange, fading red, pointed, semi-dbl., 10–15 petals, 2 in., moderate fragrance; foliage small; LeGrice

Artistry™ *see* 'JACIRST'

Arturo Toscanini *see* 'MEILUCRE'

Aruba *see* SPEcawijk

'ARUBA-CARIBE', HT, pb, 1967; bud ovoid; flowers rose-pink and ivory, dbl., 38 petals, 5 in., exhibition form, intense, fruity fragrance; foliage leathery; vigorous, upright growth; [Diamond Jubilee seedling X Fashion seedling]; Boerner; J&P

'ARUNA', HT, or, 1968; flowers bright orange-scarlet, medium, dbl., cupped; foliage glossy; bushy, compact growth; [Independence X Seedling]; Pal, Dr. B.P.; Indian Agric. Research Inst.

'ARUNIMA', F, dp, 1975; bud ovoid; flowers deep pink, dbl., 50 petals, 2 in.; foliage glossy; vigorous, bushy, compact growth; [Frolic X Seedling]; IARI

As de Coeur *see* **'HERZ AS'**®

'ASABORAKE', HT, dy, 1979; flowers deep yellow, sometimes tipped pink, large, dbl., 30 petals, exhibition form, slight fragrance; foliage medium green, semi-glossy; bushy, upright growth; [(Golden Scepter X Narzisse) X Kordes' Perfecta]; Ota, Kaichiro

'ASAGUMO', HT, yb, 1973; (Oriental Dawn); bud ovoid; flowers large, dbl., cupped, moderate fragrance; foliage glossy, dark; vigorous, upright growth; [Peace seedling X Charleston seedling]; Suzuki, Seizo; Keisei Rose Nursery

ASAlav, Min, m, 1985; (**Lavender Crystal**); (ASAlav)

'ASBACH', F, or, 1960; [Ambassadeur Nemry X Cinnabar]; Leenders, M.

'ASCHENBRÖDEL', Pol, 1903; flowers rose with salmon, small, dbl., moderate fragrance; bushy, dwarf growth; [Petite Léonie X R. foetida bicolor]; Lambert, P.

'ASCHERMITTWOCH', LCl, w, 1955; (Ash Wednesday); bud silvery gray; flowers snow-white, large, dbl., blooms in large trusses; vigorous growth; Kordes

'ASCHERSONIANA', S, m; (R. X aschersoniana); flowers bright, light purple, very numerous, small; growth to 6 ft.; (14); [R. blanda X R. chinensis]

'ASCOT', F, ab, 1962; flowers salmon-coral, semi-dbl., 18 petals, 4 in., blooms in clusters; low growth; [Brownie X Seedling]; Dickson, Patrick; Dickson, A.

Ascot Jubilee, HT, lp; Welsh, Eric; (Weatherly, L.)

'ASEPALA', M, w; flowers white, shaded flesh, sometimes edged rose, petal edges curled, small, dbl.; compact, erect growth

Ash Wednesday *see* **'ASCHERMITTWOCH'**

Asha, HT, pb, 1968; Agarwal

Ashgrove Jubilee *see* 'WELCO'

Ashley Marie *see* 'JUDMARIE'

Ashley's Surprise, Min, m, 1997; King

Ashwini '89, HT, dr, 1989; K&S

Asja, HT, or, 1998; McGredy, Sam IV

Askari *see* JACkari

'ASMODÉE', HGal, mr, 1849; flowers rosy crimson, large, dbl.; Vibert

Aspen *see* 'POULURT'

'ASPIRANT MARCEL ROUYER', HT, ab, 1919; bud long, pointed; flowers apricot, tinted salmon-flesh, veined yellow, large blooms, dbl., moderate fragrance; foliage glossy, bronze; long, strong stems; very vigorous growth; [Sunburst X Seedling]; Pernet-Ducher

'ASPIRANT MARCEL ROUYER, CLIMBING', Cl HT, ab, 1934; Brenier, E.C.

Aspirin *see* TANiripsa

'ASSEMBLAGE DES BEAUTÉS', HGal, dr, 1823; flowers brilliant crimson, medium, dbl.; erect growth; Delaage

'ASSINIBOINE', S, dp, 1962; flowers purplish red, large to medium blooms, semi-dbl., intermittent bloom, slight fragrance; foliage glossy; weak stems; vigorous growth; [Donald Prior X R. arkansana]; Marshall, H.H.; Canada Dept. of Agric.

Asso di Cuori® *see* 'KORRED'

Asso di Cuori, **Climbing**, Cl HT, dr

'ASSO FRANCESCO BARACCA', HT, pb, 1936; bud long; flowers golden salmon, center deeper, well formed, dbl., intense fragrance; long stems; vigorous growth; [Julien Potin X Seedling]; Giacomasso

'ASTA VON PARPAT', HMult, m, 1909; flowers medium, dbl.; Geschwind, R.; (Sangerhausen)

Asterix® *see* 'LENPON'

'ASTOLAT CHARM', HT, pb, 1951; flowers flesh-pink, base apricot, dbl., 30 petals, 4 in., intense fragrance; foliage dark; vigorous growth; [McGredy's Salmon sport]; Astolat Nursery

'ASTORIA', HT, ob, 1963; bud pointed; flowers bright orange shaded coral, moderate fragrance; foliage bronze, glossy; [Opera X Demoiselle]; Delforge

Astra™ *see* 'WILSMA'

Astra *see* 'BENSTAR'

Astra *see* RUIzesac

Astra Desmond, LCl, yb

'ASTRAL', HT, dp, 1976; flowers deep rose-pink, low-centered, dbl., 24 petals, 4 in., intense fragrance; foliage glossy, dark; vigorous growth; [Tropicana X Pink Favorite]; Bees

Astral, LCl, dr; Croix

Astrea® *see* 'BARORP'

'ASTRÉE', HT, pb, 1956; flowers medium pink, reverse shaded orange, large, dbl., intense fragrance; [Peace X Blanche Mallerin]; Croix, P.

Astreé, **Climbing**, Cl HT, op; Croix

Astrée, Climbing, Cl HT, 1969; Croix, P.; (Cavriglia)

Astrid Lindgren®, S, lp, 1991; Poulsen

Astrid Späth *see* **'FRAU ASTRID SPÄTH'**

'ASTRID SPATH STRIPED', F, rb, 1933

'ASTROROSE', HT, mr, 1971; bud long, pointed; flowers cardinal-red to crimson, medium, dbl., moderate fragrance; foliage large, dark, bronze, leathery; vigorous growth; [Helene Schoen X Chrysler Imperial]; Porter; General Bionomics

'ASTURIAS', F, rb, 1956; flowers red, reverse carmine, large, dbl., 40 petals, borne in corymbs; foliage glossy; strong stems; very vigorous growth; [Méphisto X Coralín]; Dot, M.

'ASUN GALINDEZ DE CHAPA', HT, pb, 1923; flowers salmon-pink, reverse darker, base yellow; [Mons. Paul Lédé X Jacques Vincent]; Ketten Bros.

Atago, HT, or; Keisei Rose Nurseries, Inc.

'ATALANTA', LCl, 1927; bud coppery pink; flowers flesh-pink, dbl., moderate fragrance; [Paul Ploton X William Allen Richardson]; Williams, A.

'ATALANTE', F, mr, 1958; flowers geranium-red, large; very vigorous growth; Frères, Buyl

'ATARA', HT, rb, 1975; flowers medium red, flecked and striped near white, reverse near white with red; [Suspense sport]; Nevo, Motke; Maoz Haim Rose Nursery, 1974

Atco Royal *see* FRYwinner

Atena®, HT, dy

'ATHABASCA', (R. macounii variant), dp, 1930; flowers deep pink, semi-dbl., non-recurrent; very vigorous growth; hardy; ((found growing wild at Vilna, Alberta, Canada, about 1930))

'ATHALIN', B, mr, 1830; flowers cherry-red; (28); Jacques

Athena *see* **'RÜHKOR'**

'ATHENE', HT, lp, 1975; bud green; flowers porcelain-pink, base yolk-yellow, dbl., 37 petals, 5.5 in., intense, apple fragrance; foliage large; moderate, bushy growth; [Peace X Diamond Jubilee]; Murray & Hawken; Rasmussen's

Athene *see* BENcreberg

'ATHERTON', HT, my, 1981; bud long, pointed; flowers urn-shaped, 35 petals, borne 1-3 per cluster, moderate, fruity fragrance; small, recurved prickles; foliage medium, matt; medium growth; [Seedling X Sunblest]; Perry, Astor

'ATHLETE', HT, rb, 1964; flowers nasturtium-red shaded gold, medium, semi-dbl., 16 petals, slight fragrance; foliage light green; bushy growth; [Tzigane X Claude]; Barter

'ATHLONE', F, yb, 1965; flowers cream edged orange-scarlet, open, blooms in clusters, slight fragrance; foliage small, dark; free growth; [Circus X Cinnabar]; McGredy, Sam IV

Athos® *see* 'LAPWON'

'ATIDA', HT, dr, 1941; flowers velvety dark scarlet, well formed, quite large, semi-dbl., moderate fragrance; very vigorous growth; [Mme Van de Voorde X Dance of Joy]; Sauvageot, H.; Sauvageot

Atida *see* SAUblim

Atida 93, HT, 1993; Sauvageot, H.; (Cavriglia)

Atkins Beauty *see* 'SAVACENT'

Atlanta *see* 'WEKDOCLEM'

Atlanta, Min, mp, 1992; deRuiter

Atlantic® *see* 'GAVAL'

Atlantic City *see* 'JOLLMEA'

Atlantic Star *see* FRYworld

'ATLANTIDA', HT, op, 1939; flowers coppery salmon, shaded peach; vigorous growth; Pahissa

'ATLANTIS', F, m, 1970; flowers deep mauve, single, 5 petals, 3 in., moderate fragrance; foliage glossy, purple tinted; GM, Rome, 1969; [Orangeade X Lilac Charm]; Harkness

Atlantis Palace *see* POUlsiana

Atlas® *see* 'DELKORT'

Atoll *see* 'MEIBYSTER'

Atom Bomb *see* **'ATOMBOMBE'**

'ATOMBOMBE', F, mr, 1953; (Atom Bomb, Atomflash, Velvet Robe); bud pointed; flowers deep scarlet-orange, dbl., 28 petals, 2.5 in., clusters; very vigorous growth; [Obergärtner Wiebicke X Independence]; Kordes

Atomflash *see* **'ATOMBOMBE'**

'ATOMIC WHITE', HT, ly, 1948; bud long, pointed; flowers white, center tinted yellow, large, dbl., exhibition form, moderate fragrance; vigorous, bushy growth; [Pink Princess X Shades of Autumn]; Brownell, H.C.

'ATROPURPUREA', HRg, mr, 1899; flowers carmine-crimson, single, blooms in clusters; vigorous (3-5 ft) growth; [R. rugosa X R. damascena]; Paul

'ATROPURPUREA', Pol, m, 1910; flowers purple-red; [Mme Norbert Levavasseur X Perle des Rouges]; Levavasseur

'ATTAR OF ROSES', HT, w, 1936; bud globular; flowers creamy white, edged pink, dbl., cupped, moderate fragrance; foliage glossy, bronze; long, strong stems; vigorous, bushy, compact growth; Cant, B. R.

Attila, A, dp; flowers deep pink, large, semi-dbl., cupped

ATTlight, HT, ab, 1999; (**Apricot Delight**); Attfield

'ATTRACTION', HT, yb, 1931; flowers yellow and orange, dbl., globular, slight fragrance; foliage glossy, bronze; dwarf, bushy growth; Dickson, A.

'ATTRAKTION'®, F, op, 1963; bud long, pointed; flowers salmon-copper, reverse golden yellow, dbl., borne in large clusters, intense fragrance; foliage glossy, dark; vigorous, bushy growth; Tantau, Math.

'AUBADE', HT, ob, 1963; flowers yellow-orange, medium, dbl.; foliage dark; [Docteur Valois X Seedling]; Verbeek

'AUBREY COBDEN', HT, my, 1948; flowers clear yellow, medium, dbl., 30 petals, intense fragrance; foliage glossy, stems red; free growth; [Oswald Sieper X Seedling]; Mee; Fryer's Nursery, Ltd.

Auckland Metro *see* 'MACBUCPAL'

Audace® *see* 'BUCBI'

'AUDIE MURPHY', HT, mr, 1957; bud long, pointed, crimson; dbl., 20 petals, 4.5–5.5 in., exhibition form, moderate, spicy fragrance; foliage semi-glossy, dark, bronze; vigorous, upright, bushy growth; GM, Portland, 1957; [Charlotte Armstrong X Grande Duchesse Charlotte]; Lammerts, Dr. Walter; Roseway Nursery

'AUDINE', HT, lp, 1950; flowers shell-pink, reflexed, dbl., 22 petals, exhibition form, moderate fragrance; foliage dark; vigorous, upright growth; [Percy Izzard X William Moore]; Gatward

'AUDREY', HT, dr, 1922; flowers deep crimson, well formed, exhibition form, moderate fragrance; Paul, W.

Audrey Gardner *see* 'PEASPECIAL'

'AUDREY HARRISON', Cl HT, ab, 1969; flowers pale apricot, semi-dbl., 4 in., free bloom, early, intense fragrance; foliage glossy, dark; [Shot Silk sport]; Harrison; Harkness

Audrey Hepburn *see* 'TWOADORE'

'AUDREY MARIE', F, mr, 1994; flowers medium red, dbl., 26–40 petals, 1.5–2.75 in., borne mostly singly, slight fragrance; some prickles; foliage medium, medium green, semi-glossy; low (2 1/2 ft), bushy growth; [Tamanga X Anne Scranton]; Dobbs, Annette E., 1994; Michael's Roses, 1994

Audrey McCormack *see* 'KENOBSESS'

Audrey Mieklejohn, HT, my

'AUDREY STELL', HT, pb, 1937; flowers soft strawberry-pink, reverse sulfur-yellow, semi-dbl.; foliage light; long, strong stems; vigorous, bushy growth; [Soeur Thérèse sport]; Stell; Stell Rose Nursery

Audrey Wilcox® *see* 'FRYWILREY'

Augie Boy *see* 'RENBOY'

August Kordes *see* **'LAFAYETTE'**

'AUGUST NOACK', HT, dr, 1928; bud long, pointed; flowers deep scarlet, open, large, dbl., exhibition form, moderate fragrance; foliage bronze, dark; vigorous, bushy growth; [Columbia sport]; Kordes

'AUGUST NOACK, CLIMBING', Cl HT, dr, 1935; Lens

'AUGUST SEEBAUER', F, dp, 1944; (The Queen Mother); bud long, pointed; flowers deep rose-pink, large, dbl., exhibition form, blooms in clusters; foliage glossy; vigorous growth; (28); [Break o' Day (HT) X Else Poulsen]; Kordes

Augusta *see* 'POULDAVA'

Augusta, HRg, lp; flat, intense; Wright, Percy H., 1953

'AUGUSTE CHAPLAIN', HP, mr, 1921; flowers large, dbl.; Tanne; (Sangerhausen)

'AUGUSTE COMTE', T, mp, 1896; flowers large, dbl.; Soupert & Notting; (Sangerhausen)

'AUGUSTE DELOBEL', LCl, mr, 1924; flowers carmine, large white eye, yellow stamens, borne in clusters; Turbat

'AUGUSTE FINON', HMult, yb, 1923; flowers golden yellow passing to coppery and salmon, borne in clusters, moderate fragrance; [Goldfinch X Seedling]; Turbat

'AUGUSTE GERVAIS', LCl, ab, 1918; flowers coppery yellow and salmon, fading white, semi-dbl., 4–5 in., blooms in clusters, seasonal bloom, moderate fragrance; very vigorous, climbing growth; [R. wichurana X Le Progrès]; Barbier

'AUGUSTE HALEM', HT, 1891; Guillot, J. B.; (Cavriglia)

'AUGUSTE KORDES', Cl F, mr, 1928; (Lafayette, Climbing); flowers light scarlet, borne in clusters, fine seasonal bloom, later intermittent; vigorous, climbing growth; [Lafayette sport]; Kordes

Auguste Renoir® *see* 'MEITOIFAR'

'AUGUSTE ROUSSEL', LCl, pb, 1913; flowers salmon-pink to flesh-pink, well-formed, petals undulated, large, semi-dbl., seasonal bloom; very vigorous, climbing growth; [R. macrophylla X Papa Gontier]; Barbier

'AUGUSTE VERMARE', HT, or, 1958; bud pointed; flowers coral-red; vigorous, upright growth; [Comtesse Vandal X Seedling]; Arles; Roses-France

'AUGUSTINE GUINOISEAU', HT, w, 1889; (White La France); flowers white, tinted light flesh; [La France sport]; Guinoiseau, B.

Augustine Halem, HT, op

'AUGUSTUS HARTMANN', HT, dp, 1914; flowers tyrian rose, large, dbl., slight fragrance; foliage rich green, leathery; vigorous growth; GM, NRS, 1914; Cant, B. R.

Augustus Hartmann, Climbing, Cl HT, 1980; Fineschi, G.; (Cavriglia)

'AUGUSTUS STONE', HT, ab, 1999; flowers light apricot/amber, resist bad weather, dbl., 26–40 petals, 4–4.5 in., borne in small clusters, moderate fragrance; some prickles; foliage medium, dark green, dull; bushy, medium (3 ft) growth; [Gavotte X Pot of Gold]; Poole, Lionel; David Lister, Ltd., 2000

'AULD LANG SYNE', HT, ob, 1956; bud ovoid; flowers orange edged pale orange-yellow, dbl., 35–45 petals, 4 in., moderate, fruity fragrance; bushy growth; [Tawny Gold X (Talisman X Tawny Gold)]; Motose

Aunt Belle's Tea, T, pb

'AUNT GERRY', HT, ly, 1992; flowers golden yellow, more golden yellow than lanvin, slight petal, dbl., 35 petals, 3–3.5 in., slight, spicy fragrance; prickles prickly peduncle; foliage medium, medium green, semi-glossy; medium (90-120 cms), upright, very vigorous growth; [Lanvin sport]; Sheldon, John & Jennifer; Trophy Roses, Ltd., 1992

'AUNT HARRIET', HMult, rb, 1918; flowers scarlet-crimson, white eye, yellow stamens; vigorous growth; [Apolline X R. wichurana]; Van Fleet; Phila. Farm Journal

'AUNT HONEY', S, mp, 1984; bud large, ovoid; flowers, dbl., 38 petals, exhibition form, borne 5-10 per cluster, repeat bloom, moderate, damask fragrance; awl-like, tan prickles; foliage medium, dark olive-green; erect, short, bushy growth; hardy; [Music Maker X Habanera]; Buck, Dr. Griffith J.; Iowa State University

'AUNT RUTH', LCl; pb; 1995; Flowers streaked medium pink, white eye, reverse white streaked pink, semi-dbl, (6–14 petals), medium (3 in.) blooms borne in small and large clusters, some singly; slight frangance; few prickles; foliage medium, medium green, semi-glossy, medium, spreading, bushy growth; [Unknown X Unknown]; Jerabek, Paul E.

Auntie's Favorite, HT, dr

'AUNTY DORA', F, m, 1970; flowers magenta, semi-dbl., blooms in trusses, slight fragrance; low growth; [Dearest X Lilac Charm]; Deamer; Warley Rose Gardens

Aunty Lil *see* 'KENMASNIA'

'AUREA', HT, yb, 1948; flowers yellow, reverse red, well formed, moderate, fruity fragrance; foliage glossy; vigorous growth; Dot

Aurea *see* MEIbosat

'AUREATE', HT, ob, 1932; flowers orange and scarlet to yellow, large, dbl., moderate fragrance; foliage bronze; vigorous, bushy growth; GM, NRS, 1929; Dickson, A.

'AURELIA CAPDEVILA', HT, pb, 1933; flowers pink, base salmon, moderate fragrance; foliage dark; vigorous growth; Dot, Pedro

'AURELIA LIFFA', HSet, mr, 1896; Geschwind, R.

'AUREOLA', HT, my, 1934; flowers golden yellow; [Mev. G.A. van Rossem sport]; Böhm, J.

'AURÉOLE', HT, ob, 1951; bud large; flowers orange-yellow, semi-dbl., moderate fragrance; foliage glossy, dark; very vigorous growth; Gaujard

'AURIOL', Pol, rb; Alderton, 1930

Auriu de Cluj, F, lp; flat, 15 petals, slight; medium, light green, glossy; [Orangeade X Western Sun]; St. Wagner, 1984; Res. Stn. f. Horticulture, Cluj, 1991, Romania

'AURORA', LCl, ob, 1956; flowers orange-yellow, large, dbl.; vigorous growth; Kordes

'AURORA', HMsk, ly, 1928; flowers yellow, passing to creamy white, single, moderate fragrance; [Danaë X Miriam]; Pemberton

'AURORA', F, op, 1941; flowers salmon-pink tinted golden yellow and orange, dbl., moderate fragrance; GM, Bagatelle, 1940; [Mev. Nathalie Nypels X Seedling]; Leenders, M.

'AURORA', (variety of R. acicularis), dp; flowers have wide, thick petals, single, 5 petals; foliage red and purple in fall; ((found by Erskine in Alberta, Canada, about 1950))

Aurora, F, my

Aurora, HT, lp; Strnad; (Czech Rosa Club)

'AURORA BOREAL', HT, rb, 1935; flowers geranium, shaded fiery red, semi-dbl.; foliage dark; very vigorous growth; [Étoile de Hollande X (Ville de Paris X Sensation)]; Munné, B.

'AURORE', HCh, yb; flowers yellow passing to salmon-pink, loose form, recurrent bloom; (14); Touvais

'AURORE', HT, pb, 1936; flowers pink, base yellow, dbl., moderate fragrance; vigorous growth; Capiago

Aurore, F, or

'AURORE', HP, mp, 1861; Laffay, M.

'AURORE D'ESPAGNE', HT, pb, 1966; bud long, pointed; flowers salmon-pink to nankeen yellow, medium, semi-dbl., cupped, slight fragrance; vigorous growth; [Zambra X Queen Elizabeth]; Dot; Minier

'AURORE DU MATIN', HP, mr, 1857; flowers light red, large, dbl., intense fragrance; Rolland; (Sangerhausen)

'AURORE SAND', LCl, pb, 1964; flowers two-

AUSapple, S, lp, 1983; (**Dapple Dawn**); Austin, David

'AUSBAKER', S, yb, 1998; (**Teasing Georgia**); very dbl., 110 petals, 3.5 in., borne mostly singly, moderate fragrance; few prickles; foliage medium, semi-glossy, disease-resistant; branching, strong, graceful, medium (3.5 ft) growth; [Charles Austin X seedling]; Austin, David; David Austin Roses, Ltd., 1998

'AUSBAND', S, dy, 1998; (**Buttercup**™); flowers loosely cupped, large heads of semi-double golden yellow florets, 21 petals; bright green semi-glossy foliage; almost thornless; light airy growth 3–4 ft; excellent repeat, disease resistant; Austin, David; David Austin Roses, Inc., 1998

'AUSBEAM', S, ab, 1992; (**Moonbeam**); flowers apricot yellow, blooms borne in small clusters, semi-dbl., 6–14 petals, 1.5–2.75 in., intense fragrance; some prickles; foliage medium, light green, semi-glossy; medium (100 cms), bushy growth; Austin, David; David Austin Roses, Ltd., 1983

'AUSBELLS', S, dp, 1994; (**Bow Bells**™); flowers deep pink, dbl., 15–25 petals, 1.5–2.75 in., borne in large clusters, moderate fragrance; some prickles; foliage medium, medium green, semi-glossy; upright, bushy (120 cms) growth; [Seedling X Graham Thomas]; Austin, David, 1991; David Austin Roses, Ltd., 1991

'AUSBLOOM', S, dr, 1994; (**The Dark Lady**); flowers dark red, very full, dbl., 140 petals, 2.75–3 in., borne in small clusters, intense fragrance; some prickles; foliage medium, dark green, semi-glossy; medium (3 ft), upright, bushy growth; [Mary Rose X Prospero]; PP8677 Austin, David, 1991

'AUSBLOSSOM', S, lp, 1994; (**Peach Blossom**); flowers blush pink, semi-dbl., 6–14 petals, 1.5–2.75 in., borne in large clusters, slight fragrance; some prickles; foliage medium, medium green, semi-glossy; bushy, spreading (120 cms) growth; [The Prioress X Mary Rose]; Austin, David, 1990; David Austin Roses, Ltd., 1990

'AUSBLUSH', S, lp, 1985; (**Heritage**®); flowers medium, dbl., cupped, recurrent bloom, intense fragrance; foliage small, dark, semi-glossy; upright, bushy growth; [Unnamed seedling X Iceberg seedling]; Austin, David; David Austin Roses, Ltd., 1984

AUSboard, S, mp, 1986; (**Gertrude Jekyll**®); PP7220 Austin, David

X Tamora]; Austin, David; David Austin Roses, Ltd., 1990

'AUSBRED', S, ab, 1985; (**Bredon®**); flowers medium, dbl., repeat bloom, moderate fragrance; foliage small, light green, matt; upright growth; [Wife of Bath X Lilian Austin]; Austin, David; David Austin Roses, Ltd., 1984

AUSbrid, S, lp, 1997; (**Mayor of Casterbridge**); flowers very full, old rose form, medium, very dbl., 91 petals, 5.5–6.5 in., borne in small clusters, moderate fragrance; some prickles; foliage medium, light green, leathery; upright, medium (3.5x2.5ft.)growth; Austin, David

'AUSBUFF', S, ab, 1990; (**English Garden™**); flowers soft apricot yellow, very large, very dbl., cupped, intense fragrance; foliage clear green; bushy, vigorous growth; PP007214; [Lilian Austin X (Seedling X Iceberg)]; Austin, David; Delbard Roses, 1988

'AUSBURN', S, lp, 1986; (**Robbie Burns**); flowers light pink, white center, small, single, 5 petals, moderate fragrance; foliage small, medium green, matt; bushy, strong growth; [Wife of Bath X R. pimpinellifolia]; Austin, David, 1985

'AUSBURTON', S, lp, 1994; (**Emily**); flowers pale pink, very dbl., 3–3.5 in., borne in small clusters, moderate fragrance; many prickles; foliage medium, medium green, semi-glossy; upright, bushy (75 cms) growth; [The Prioress X Mary Rose]; PP88388 Austin, David; David Austin Roses, Ltd., 1992

'AUSCA', S, w, 1983; (**Fair Bianca®**); flowers light yellow to white, very dbl., quartered, moderate fragrance; foliage medium, light green, semi-glossy; upright growth; Hardy, medium; Austin, David; David Austin Roses, Ltd., 1982

AUScam, S, mp, 1995; (**Marinette**); flowers double, large, semi-dbl., 15–25 petals, borne singly or in small clusters, slight fragrance; prickles few to some; foliage medium, medium green, semi-glossy; bushy, medium growth; [Lucetta X Red Coat]; Austin, David

'AUSCANTERBURY', S, mp, 1969; (**Canterbury**); flowers rose-pink, medium, semi-dbl., 12 petals, repeat bloom, intense fragrance; [(Monique X Constance Spry) X Seedling]; Austin, David

'AUSCAT', S, w, 1995; (**Winchester Cathedral**); flowers white, with a tendency to revert to pink, very dbl., 2–2.75 in., cupped, borne in small clusters, moderate fragrance; some prickles; foliage medium, medium green, semi-glossy; medium (120 cms), upright, bushy growth; PP008141; [Mary Rose sport]; Austin, David; David Austin Roses, Ltd., 1988

'AUSCENT', S, dp, 1994; (**John Clare**); flowers very full, medium, very dbl., 110–130 petals, borne in small clusters, slight fragrance; few prickles; foliage medium, dark green, semi-glossy; upright, medium (90cms) growth; [Wife of Bath X Seedling]; Austin, David

'AUSCHAR', S, yb, 1994; (**Charity™**); flowers full, 90 + petals, soft apricot yellow; strong myrrh frangance; medium upright growth, few thorns; Austin, David; David Austin Roses, Inc., 1994

'AUSCHILD', S, dr, 1987; (**Fisherman's Friend®**); flowers deep crimson, reverse lighter, fading crimson-purple, large, very dbl., cupped, repeat bloom, intense, damask fragrance; no fruit; broad, straight, large, red-brown prickles; foliage medium, dark green, semi-glossy; bushy growth; [Lilian Austin X The Squire]; Austin, David

AUSclough, S, dp, 1987; (**Sir Clough**); Austin, David

'AUSCLUB', S, lp, 1995; (**Kathryn Morley**); flowers pale pink, very dbl., 3–3.5 in., cupped, borne in small clusters, moderate fragrance; many prickles; foliage medium, dark green, glosssy; tall (5-6 ft), bushy growth; [Mary Rose X Chaucer]; Austin, David; David Austin Roses, Ltd., 1990

'AUSCOMP', S, my, 1994; (**Happy Child**); flowers bright yellow, very dbl., 3–3.5 in., borne in small clusters, intense fragrance; some prickles; foliage medium, medium green, glossy; medium (100 cms), bushy growth; [Seedling X Hero]; PP9007 Austin, David; David Austin Roses, Ltd., 1993

'AUSCON', S, mp, 1981; (**Chaucer®**); bud globular; cupped, borne 1-5 per cluster, intense fragrance; slightly hooked, red prickles; foliage medium green; vigorous, upright, bushy growth; [Seedling X Constance Spry]; PP10618 Austin, David; David Austin Roses, Ltd., 1970

'AUSCOOK', S, rb, 1997; (**Heather Austin**); flowers very full, medium, very dbl., 50 petals, 1.5–2.5 in., borne in small clusters, moderate fragrance; some prickles; foliage medium, medium green, semi-glossy; upright to bushy, medium (3x2.5ft.) growth; Austin, David

'AUSCOT', S, op, 1990; (**Abraham Darby™**); bud rounded, dark pink base with yellow; flowers pink peach-apricot, cup form, very large, very dbl., 3–4 in., quartered, intense fragrance; foliage dark green, shiny; vigorous, bushy, angular growth; PP007215; [Yellow Cushion X Aloha]; Austin, David; 1985

'AUSCOUNTRY', S, lp, 1994; (**Country Living**); flowers pale pink, very dbl., 3–3.5 in., borne in small clusters, moderate fragrance; some prickles; foliage medium, medium green, semi-glossy; upright, bushy (105 cms) growth; [Wife of Bath X Graham Thomas]; Austin, David; David Austin Roses, Ltd., 1991

'AUSCRESS', S, ab, 1992; (**Cressida**); flowers apricot peach, blooms borne in small clusters, very dbl., 3–3.5 in., intense fragrance; some prickles; foliage small, light green, semi-glossy; tall (180 cms), upright growth; Austin, David; David Austin Roses, Ltd., 1983

'AUSCRIM', S, dr, 1993; (L. D. Braithwaite, **Leonard Dudley Braithwaite**, Braithwaite); flowers bright red, informal, large, very dbl., 3–3.5 in., borne in small clusters, intense fragrance; some prickles; foliage medium, dark green, semi-glossy; medium (110 cms), bushy growth; [Mary Rose X The Squire]; Austin, David; David Austin Roses, Ltd., 1983

'AUSCROSS', S, lp, 1997; (**Windflower**); flowers light pink, full (26–40 petals), medium blooms; slight fragrance; foliage medium, medium green, semi-glossy, very disease resistant; bushy, medium (120 cms) growth; [Shropshire Lass X Seedling]; Austin, David; David Austin Roses, Ltd.

'AUSCUP', S, ab, 1985; (**Ellen®**); flowers old rose form, large, dbl., intense fragrance; foliage large, medium green, semi-glossy; bushy growth; Austin, David, 1984

'AUSDAY', S, pb, 1994; (**The Alexandra Rose**); flowers coppery pink, single, 5 petals, 1.5–2.75 in., borne in large clusters, slight fragrance; few prickles; foliage medium, medium green, semi-glossy; bushy, compact (130 cms) growth; [Shropshire Lass X Heritage]; Austin, David; David Austin Roses, Ltd., 1992

'AUSDIMINDO', S, dp, 1995 (**Bibi Maizoon**; Bibi Mezoon); flowers rich pink, very full (41 + petals), large (7 + cms) blooms borne in small clusters; very fragrant; some prickles; foliage medium, dark green, semi-glossy; medium (100 cms), arching, bushy growth; [The Reeve X Chaucer]; Austin, David; David Austin Roses, Ltd. 1989

'AUSDIR', S, dr, 1994; (**Tradescant**); flowers very dark red, tends to burn in hot weather, very dbl., 1.5–2.75 in., rosette, borne in small clusters, moderate fragrance; some prickles; foliage medium, dark green, semi-glossy; bushy, spreading (75 cms) growth; PP9009; [Prospero X Seedling]; Austin, David; David Austin Roses, Ltd., 1993

'AUSDOCTOR', S, mr, 1992; (**Doctor Jackson**); flowers scarlet, golden stamens, blooms borne mostly singly, single, 5 petals, 1.5–2.75 in., no fragrance; few prickles; foliage medium, medium green, semi-glossy; medium (120 cms), spreading growth; Austin, David; David Austin Roses, Ltd., 1987

'AUSDOR', HRg, mp, 1994; (**Mrs Doreen Pike**); flowers medium, very dbl., 1.5–2.75 in., borne in large clusters, moderate fragrance; many prickles; foliage small, light green, semi-glossy; medium (90 cms), bushy, spreading, compact growth; [Martin Frobisher X Roseraie de l'Hay]; Austin, David; David Austin Roses, Ltd., 1993

'AUSDOVE', S, lp, 1986; (**Dove**); flowers medium, dbl., slight fragrance; foliage medium, dark, semi-glossy; spreading growth; [Wife of Bath X Iceberg seedling]; Austin, David, 1984

AUSecret, S, lp; flowers soft warm pink, very dbl., 110 petals, 3 in., rosette, light fragrance; almost thornless; foliage shiny, disease-resistant; arching growth; Austin, David, 1999

'AUSELLE', S, lp, 1985; (**Belle Story**®); flowers light pink, yellow stamens, large, dbl., 35 petals, cupped, intense fragrance; foliage medium, medium green, semi-glossy; bushy growth; PP007213; [(Chaucer X Parade) X (The Prioress X Iceberg)]; Austin, David; David Austin Roses, Ltd., 1984

'AUSEMI', S, ab, 1992; (**Lucetta**®); flowers pale peach, blooms borne in small clusters, semi-dbl., 6–14 petals, 5.25 in., moderate fragrance; some prickles; foliage medium, medium green, semi-glossy; medium (120 cms), spreading growth; Austin, David; David Austin Roses, Ltd., 1983

'AUSFAR', S, lp, 1999; (**Dr Herbert Gray**); flowers clear light pink, reverse medium pink, very dbl., 41 petals, 3.5 in., cupped, borne in small clusters, moderate fragrance; few prickles; foliage small, dark green, glossy; spreading, medium (3 ft) growth; [Heritage X seedling]; Austin, David; David Austin Roses, Ltd., 1998

'AUSFATHER', S, ab, 1981; (**Charles Austin**®); bud globular; flowers apricot tinged pink, fading to light pink, rosette, borne 1-3 per cluster, dbl., 70 petals, moderate fragrance; hooked, red prickles; foliage medium green, dense; vigorous, upright, bushy growth; [Chaucer X Aloha]; Austin, David; David Austin Roses, Ltd., 1973

'AUSFETE', S, mp, 1999; (**Ann**); single, 5–11 petals, 2.5 in., borne in clusters, moderate fragrance; prickles moderate; foliage medium, medium green, semi-glossy; bushy, medium (3 ft) growth; Austin, David; David Austin Roses, Ltd., 1997

'AUSFIN', S, dp, 1994; (**Financial Times Centenary**); flowers deep pink, very dbl., 3–3.5 in., borne in small clusters, intense fragrance; some prickles; foliage medium, dark green, semi-glossy; medium (110 cms), upright, bushy growth; PP8142; Austin, David; David Austin Roses, Ltd., 1988

AUSfire, S, pb, 1997; (**Morning Mist**); flowers single, medium, single, 5 petals, 4 in., borne in small clusters, slight fragrance; some prickles; foliage large, light green, leathery; attractive, shrubby, spreading, tall (4.5x4ft.) growth; Austin, David, 1996

'AUSFIRST', S, lp, 1961; (**Constance Spry**®); flowers rose pink, blooms in clusters, dbl., 5 in., cupped, moderate, myrrh fragrance; foliage dark; vigorous (5-6 ft) growth; [Belle Isis X Dainty Maid]; Austin, David

'AUSGLISTEN', S, mp, 1994; (**Cottage Rose**, The Cottage Rose); flowers warm pink, very dbl., 1.5–2.75 in., borne in large clusters, moderate fragrance; many prickles; foliage medium, medium green, semi-glossy; medium (90-105 cms), upright, bushy growth; [Mary Rose X Wife of Bath]; Austin, David; David Austin Roses, Ltd., 1991

'AUSGLOBE', S, mp, 1994; (**Brother Cadfael**™); flowers rich pink, dbl., 45 petals, cupped, intense, old rose fragrance; few thorns; foliage dark green; straight stems; strong, bushy growth; PP008681; Austin, David, 1990

'AUSGOLD', S, dy, 1993; (**Golden Celebration**); flowers old fashioned, blooms borne in small clusters), very dbl., 55–75 petals, 3–3.5 in., borne in small clusters, intense fragrance; some prickles; foliage large, dark green, semi-glossy; medium (120 cms), bushy growth; [Charles Austin X Abraham Darby]; Austin, David; David Austin Roses, Ltd., 1992

AUSguard, S, lp; rich deep pink, very dbl., 110 petals, 3 in., shallow-cupped, intense, old rose fragrance; foliage dark, slighly glossy; broad, bushy growth; Austin, David, 1999

'AUSHAM', S, mp, 1999; (**Geoff Hamilton**); flowers rounded, very dbl., 108 petals, 2.5 in., borne in small clusters, moderate fragrance; prickles moderate; foliage medium, medium green, semi-glossy, good disease resi; sturdy, compact, upright, medium (4 ft) growth; [Heritage X seedling]; Austin, David; David Austin Roses, Ltd., 1997

'AUSHERO', S, mp, 1983; (**Hero**®); flowers glistening medium pink, deeply, large, dbl., 20 petals, cupped, intense fragrance; foliage medium, medium green, semi-glossy; spreading growth; [The Prioress X Unnamed seedling]; Austin, David; David Austin Roses, Ltd., 1982

'AUSIRE', S, dr, 1976; (**The Squire**®); bud globular; flowers very dark red, rosette-shaped blooms borne 1-3 per cluster, very dbl., 120 petals, intense fragrance; straight prickles; foliage dark; open, bushy growth; [The Knight X Château de Clos Vougeot]; Austin, David, 1977

'AUSJESS', S, dp, 1992; (**Pretty Jessica**); flowers deep pink, blooms borne in small clusters, very dbl., 1.5–2.75 in., intense fragrance; some prickles; foliage medium, dark green, semi-glossy; low (65 cms), bushy growth; [Wife of Bath X Unnamed seedling]; Austin, David; David Austin Roses, Ltd., 1983

AUSjo, S, my, 1997; (**Jude the Obscure**); flowers very full, large, globular, very dbl., 55–70 petals, globular, borne singly or in small clusters, intense fragrance; some prickles; foliage medium, medium green, semi-glossy; bushy, medium growth; [Abraham Darby X Windrush]; Austin, David

'AUSJOLLY', S, pb, 1999; (**Mary Magdalene**); very dbl., 41 petals, 4 in., borne in small clusters, intense, myrrh fragrance; prickles moderate; foliage medium, medium green, dull; spreading, medium (3 ft) growth; [seedling X seedling]; Austin, David, 1998

'AUSJUNO', S, dp, 1992; (**Immortal Juno**); flowers deep pink, blooms borne in small clusters, very dbl., 3–3.5 in., intense fragrance; some prickles; foliage medium, medium green, semi-glossy; tall (150 cms), upright growth; Austin, David; David Austin Roses, Ltd., 1983

AUSky, S, mp, 1997; (**Mistress Quickly**); flowers very full, informally arranged, small, very dbl., 41 petals, borne in very large clusters, slight fragrance; few prickles; foliage small to medium, medium green, semi-glossy; bushy, medium growth; disease resistant; [Blush Noisette X Martin Frobisher]; PP10617; Austin, David, 1998

'AUSLAND', S, lp, 1997; (**Scepter'd Isle**); flowers double, cupped, medium, dbl., 45 petals, 4–5 in., borne in small clusters, intense fragrance; some prickles; foliage medium, dark green, semi-glossy; narrow, bushy, low (2.5x2ft.) growth; PP10969; Austin, David

'AUSLEA', S, ab, 1983; (**Leander**®); flowers small blooms in clusters, very dbl., flat, moderate fragrance; foliage medium, medium green, semi-glossy; spreading growth; [Charles Austin X Unnamed seedling]; Austin, David; David Austin Roses, Ltd., 1982

'AUSLEAF', S, ob, 1985; (Autumn Leaves, **English Elegance**®); flowers large, dbl., moderate fragrance; foliage medium, medium green, semi-glossy; upright growth; PP007557; Austin, David

'AUSLEAN', S, lp, 1983; (**Cymbeline**); flowers medium, dbl., intense fragrance; foliage medium, medium green, semi-glossy; spreading growth; [Seedling X Lilian Austin]; Austin, David; David Austin Roses, Ltd., 1982

'AUSLEAP', S, ab, 1994; (**Sweet Juliet**); very dbl., 1.5–2.75 in., cupped, borne in small clusters, moderate fragrance; some prickles; foliage medium, medium green, semi-glossy; upright, bushy (100 cms) growth; PP8153; Belfast 1992; [Graham Thomas X Admired

Miranda]; Austin, David; David Austin Roses, Ltd., 1989

'AUSLED', S, yb, 1997; (**A Shropshire Lad**); flowers very full, rosette form, large, very dbl., 100 petals, 3 in., rosette, borne in small clusters, intense fragrance; prickles moderate; foliage medium, medium green, semi-glossy; bushy, medium (4x3ft.)growth; Austin, David, 1996

'AUSLETT', S, ly, 1994; (Allux Symphony, **Symphony**); flowers soft yellow, very dbl., 3–3.5 in., borne in small clusters, slight fragrance; foliage medium, light green, glossy; medium (120 cms), upright, bushy growth; [The Friar X Yellow Cushion]; Austin, David; 1986; David Austin Roses, Ltd., 1986

'AUSLEVEL', S, w, 1994; (**Glamis Castle**); very dbl., 1.5–2.75 in., cupped, borne in small clusters, intense, myrrh fragrance; many prickles; foliage medium, medium green, semi-glossy; medium (100 cms), bushy growth; [Graham Thomas X Mary Rose]; PP8765; Austin, David; David Austin Roses, Ltd., 1992

'AUSLIAN', S, dp, 1992; (**Warwick Castle**®); very dbl., 3–3.5 in., borne in small clusters, moderate fragrance; some prickles; foliage small, medium green, matt; low (75 cms), spreading growth; [The Reeve X Lilian Austin]; Austin, David; David Austin Roses, Ltd., 1986

'AUSLIGHT', S, mp, 1990; (**Claire Rose**); flowers very large, cupped, intense fragrance; foliage clear green; bushy, vigorous growth; [Charles Austin X (Seedling X Iceberg)]; Austin, David; 1988

'AUSLILAC', S, pb, 1994; (**Lilac Rose**); flowers lilac pink, very dbl., 3–3.5 in., borne in small clusters, intense fragrance; some prickles; foliage medium, medium green, semi-glossy; upright (90 cms) growth; PPAF; [Seedling X Hero]; PP8837; Austin, David; David Austin Roses, Ltd., 1990

'AUSLO', S, mr, 1990; (**Othello**); flowers clear red, fading to purple-red, very large, very dbl., 4–5 in., cupped, intense fragrance; many large thorns; foliage dark green; vigorous, bushy growth; PP007212; [Lilian Austin X The Squire]; Austin, David; 1989

'AUSLOT', S, rb, 1999; (**Sophy's Rose**); flowers red-purple, dome-shaped, very dbl., 82 petals, 3.5 in., borne in small clusters, moderate fragrance; prickles moderate; foliage medium, medium green, semi-glossy; strong, bushy, vigorous, medium (3.5 ft) growth; [Prospero X seedling]; Austin, David; David Austin Roses, Ltd., 1997

'AUSMAK', S, lp, 1994; (**Eglantyne**, Eglantyne Jebb); flowers light, delicate pink, very dbl., 3–3.5 in., cupped, borne in small clusters, intense fragrance; some prickles; foliage medium, medium green, matt; medium (115-130 cms), bushy growth; [Seedling X Mary Rose]; PP9526; Austin, David; David Austin Roses, Ltd., 1994

'AUSMAN', S, mp, 1987; (**The Countryman**®); flowers rosette, medium, borne usually singly and in sprays of 3-5, dbl., 40 petals, repeat bloom, intense, damask fragrance; ovoid, medium, red fruit; hooked, small, pale prickles; foliage medium, medium green, matt; spreading, medium growth; PP007556; [Unnamed seedling X Comte de Chambord]; Austin, David

'AUSMARY', S, mp, 1983; (**Mary Rose**®); flowers large, very dbl., cupped, recurrent bloom, intense fragrance; foliage medium, medium-green, matt; upright, bushy growth; [Unnamed seedling X The Friar]; Austin, David; David Austin Roses, Ltd.

'AUSMAS', S, dy, 1983; (**Graham Thomas**®); flowers rich deep yellow, medium, dbl., 35 petals, cupped, recurrent bloom, intense fragrance; foliage small, dark, glossy; bushy growth; [Unnamed seedling X (Charles Austin X Iceberg seedling)]; Austin, David; David Austin Roses, Ltd.

'AUSMASH', S, pb, 1997; (**Heavenly Rosalind**); flowers soft pink center, deepening to medium pink on outer edges, wild-rose effect, single (5 petals), medium blooms borne in small clusters; slight fragrance; some prickles; foliage medium, dark green, dull, leathery; upright, bushy, medium growth; Austin, David

'AUSMIAN', S, mp, 1983; (**Charmian**®); flowers large, dbl., intense fragrance; foliage medium, medium green, semi-glossy; spreading growth; [Seedling X Lilian Austin]; Austin, David; David Austin Roses, Ltd., 1982

'AUSMIR', S, lp, 1983; (**Admired Miranda**®); flowers opening flat, then reflexing, large, dbl., flat, intense fragrance; foliage medium, medium green, semi-glossy; upright growth; [The Friar X The Friar]; Austin, David, 1982

'AUSMIT', S, my, 1987; (**St Cecilia**); flowers pale buff yellow, fading cream, medium, borne usually singly, dbl., 40 petals, cupped, repeat bloom, moderate, myrrh fragrance; no fruit; small, brown prickles; foliage small, medium green, matt; bushy, low, medium growth; [Unnamed seedling X Unnamed seedling]; PP8157; Austin, David

'AUSMOL', S, dy, 1994; (**Molineux**); very full rosettes of flowers; strong tea frangance; 110–120 petals borne in small-medium clusters; bushy upright growth; few thorns; 3–4 ft tall; very good repeat flowering; excellent disease resistance; foliage dark-green, semi-glossy; PP9624; Austin, David; David Austin Roses, Inc., 1994

'AUSMOON', S, 1997; (**Pegasus**); flowers very full, large, camelia-like, very dbl., 110 petals, borne in small clusters, intense, rich tea-rose fragrance; few prickles; foliage large, dark green, semi-glossy, leathery; bushy, branching, medium growth; Austin, David

'AUSMOUND', S, op, 1981; (**Lilian Austin**®); bud globular; dbl., 33 petals, flat, borne 1-5 per cluster, moderate fragrance; hooked, brown prickles; foliage glossy, dark; spreading growth; [Aloha X The Yeoman]; Austin, David, 1973

'AUSMOVE', S, dr, 1999; (**Tess of the D'Urbervilles**); flowers dark, crimson red, very dbl., 41 petals, 3.5 in., cupped, borne in small clusters, moderate fragrance; many prickles; foliage large, dark green, semi-glossy; branching, medium (3 ft) growth; [The Squire X seedling]; Austin, David; David Austin Roses, Ltd., 1998

'AUSMUM', S, 1997; (**Pat Austin**™); bud pointed, ovoid; flowers very full, large, bright copper inside, paler outside, very dbl., 50 petals, 1–2 in., borne in small clusters, some prickles; foliage large, dark green, glossy; branching, rounded, medium growth; PP9527; Austin, David

'AUSMURR', S, mp, 1985; (**Hilda Murrell**®); flowers old rose form, large, dbl., intense fragrance; foliage large, medium green, matt; bushy growth; [Unnamed seedling X (Parade X Chaucer)]; Austin, David, 1984

'AUSNUN', S, w; (**The Nun**); Austin, David, 1987

'AUSOIL', S, ab, 1992; (**Troilus**); flowers creamy apricot, blooms borne in large clusters, very dbl., 3–3.5 in., intense fragrance; some prickles; foliage large, dark green, semi-glossy; medium (110 cms), upright growth; [(Duchesse de Montebello X Chaucer) X Charles Austin]; Austin, David; David Austin Roses, Ltd., 1983

'AUSOLD', S, dp, 1997; (**Trevor Griffiths**™); flowers very full, old rose form, large, very dbl., 41 petals, borne in small clusters, intense fragrance; some prickles; foliage medium, dark green, dull, rough texture; upright, medium (105cms) growth; Austin, David

'AUSONIUS', HMsk, pb, 1932; bud oval, yellow-red; flowers yellowish pink, center white, semi-dbl., borne in pyramidal trusses of 20-50, free, recurrent boom, moderate fragrance; foliage leathery; semi-climbing, bushy growth; [(Chamisso X Léonie Lamesch) X (Geheimrat Dr. Mittweg X Tip-Top)]; Lambert, P.

'AUSPALE', S, lp, 1994; (**Redoute**); pale pink sport of Mary Rose; more petals, otherwise identical; Austin, David; David Austin Roses, Inc., 1994

'AUSPERD', S, ab, 1992; (**Perdita**®); flowers blush apricot, blooms borne in small clusters, very dbl., 3–3.5 in., intense fragrance; some prickles; foliage medium, medium green, semi-glossy; medium (100 cms), bushy growth; Edland Fragrance Medal, ARS, 1984; [The Friar X (Seedling X Iceberg)]; Austin, David; David Austin Roses, Ltd., 1983

'AUSPERO', S, dr, 1983; (**Prospero**®); flowers large, dbl., flat, intense fragrance; foliage medium, dark, matt; weak, upright growth; [The Knight X Unnamed seedling]; PP9008; Austin, David, 1982

'AUSPOLY', S, ly, 1994; (**Charlotte**, Elgin Festival); flowers soft yellow, very dbl., 3–3.5 in., borne in small clusters, moderate fragrance; few prickles; foliage medium, medium green, semi-glossy; medium (100 cms), upright, bushy growth; [Seedling X Graham Thomas]; Austin, David; David Austin Roses, Ltd., 1993

'AUSPOM', S, w, 1997; (**Snow Goose**); flowers full, small, very dbl., 26–40 petals, 1–2 in., borne in small clusters, slight fragrance; few prickles; foliage small, dark green, smooth; bushy, (8ft.); Austin, David

'AUSPORT', S, m, 1983; (**Wise Portia**); flowers large, dbl., intense fragrance; foliage medium, dark, semi-glossy; bushy growth; [The Knight X Seedling]; Austin, David, 1982

AUSpot, S, mp, 1988; (**Potter & Moore**); Austin, David

'AUSPRIMA', S, mr, 1995; (**Sir Edward Elgar**); flowers cerise-crimson, very dbl., 1.5–2.75 in., borne mostly singly, intense fragrance; many prickles; foliage medium, medium green, semi-glossy; medium (75 cms), upright, bushy, compact growth; [Mary Rose X The Squire]; Austin, David; David Austin Roses, Ltd., 1992

AUSrace, S, lp; flowers pale apricot in center, whole bloom fading to cream, rounded, very dbl., 118 petals, 3 in., rounded, moderate, tea fragrance; foliage slighly glossy, dark green; bushy growth; Austin, David, 1999

'AUSRAM', S, w, 1994; (**Francine Austin**); very dbl., 1.5 in., borne in large clusters, slight fragrance; some prickles; foliage small, medium green, semi-glossy; groundcover; bushy, spreading (90 cms) growth; PPAF; [Alister Stella Gray X Ballerina]; PP8156a; Austin, David; David Austin Roses, Ltd., 1988

'AUSREEF', S, lp, 1995; (**Sharifa Asthma**, Sharifa); flowers pale pink, fading to pinkinsh white, very full (41+ petals), large (7+ cms) blooms borne in small clusters; very fragrant; many prickles; foliage medium, dark green, semi-glossy; medium (90 cms), bushy strong growth; [Mary rose X Admired Miranda]; Austin, David; David Austin Roses Ltd., 1989

'AUSREEVE', S, dp, 1981; (**The Reeve**®); bud globular; flowers deep pink, blooms borne 1-5 per cluster, dbl., 58 petals, cupped, repeat bloom, intense fragrance; hooked, red prickles; foliage red, turning to green; spreading, shrubby growth; [Lilian Austin X Chaucer]; Austin, David, 1979

'AUSREN', S, pb, 1994; (**Charles Rennie Macintosh**); flowers lilac pink, very dbl., 1.5–2.75 in., borne in small clusters; many prickles; foliage medium, dark green, matt; medium (110 cms), bushy growth; PP8155; [Seedling X Mary Rose]; Austin, David; David Austin Roses, Ltd., 1988

'AUSRON', S, lp, 1983; (**Lordly Oberon**); flowers large, very dbl., cupped, intense fragrance; foliage large, medium green, matt; upright growth; [Chaucer X Seedling]; Austin, David; David Austin Roses, Ltd., 1982

'AUSROYAL', S, dr, 1987; (**William Shakespeare**®); flowers deep crimson-purple, fading rich purple, rosette, small, bor, dbl., repeat bloom, intense, damask fragrance; no fruit; broad based, straight, medium, red prickles; foliage large, dark green, semi-glossy; upright, tall growth; [The Squire X Mary Rose]; Austin, David

'AUSRUSH', S, ly, 1985; (**Windrush**®); flowers large, semi-dbl., intense fragrance; foliage medium, light green, matt; vigorous, branching growth; [Unnamed seedling X (Canterbury X Golden Wings)]; Austin, David, 1984

'AUSSAL', S, mp, 1997; (**Radio Times**); flowers very full, old fashioned form, medium, very dbl., 95–110 petals, borne in small clusters, intense fragrance; many prickles; foliage medium, medium green, semi-glossy; bushy, low (75cms) growth; PP9525; Austin, David

'AUSSAUCER', S, ab, 1992; (**Evelyn**); flowers old fashioned rosette form, blooms borne in small clusters, very dbl., 4–5 in., intense fragrance; some prickles; foliage medium, medium green, semi-glossy; medium (110 cms), upright, bushy growth; [Graham Thomas X AUStamora]; PP8680 Austin, David; David Austin Roses, Ltd., 1991

AUSschool, S, lp; flowers soft yellow, dbl., 51 petals, 2.25 in., loose, produced in large sprays, moderate, musk fragrance; foliage bright green, matte; bushy growth; Austin, David, 1999

'AUSSEMI', S, dp, 1994; (Herbalist, **The Herbalist**™); flowers deep pink, semi-dbl., 6–14 petals, 1.5–2.75 in., borne in small clusters, slight fragrance; some prickles; foliage medium, medium green, semi-glossy; medium (90 cms), bushy, spreading growth; [Seedling X Louise Odier]; Austin, David; David Austin Roses, Ltd., 1991

'AUSSNOW', HWich, w, 1985; (**Mountain Snow**); flowers small blooms in large sprays, dbl., 20 petals, slight fragrance; foliage large, dark, semi-glossy; vigorous (to 20 ft) growth; Austin, David

'AUSSPRY', S, mp, 1994; (**Sir Walter Raleigh**™); flowers warm medium pink, very dbl., 3–3.5 in., borne mostly singly, intense fragrance; few prickles; foliage medium, medium green, semi-glossy; medium (120 cms), upright, bushy growth; PP7213; [Lilian Austin X Chaucer]; Austin, David; David Austin Roses, Ltd., 1985

'AUSTAMORA', S, ab, 1992; (**Tamora**); flowers apricot yellow, deeply, old rose form, blooms borne in small, very dbl., 3–4 in., cupped, intense, unusual myrrh fragrance; some prickles; foliage small, dark green, semi-glossy; medium (90 cms), bushy growth; [Chaucer X Conrad Ferdinand Meyer]; Austin, David; David Austin Roses, Ltd., 1983

'AUSTANIA', S, w, 1983; (**Proud Titania**); dbl., 35 petals, flat, intense fragrance; foliage small, medium green, semi-glossy; upright growth; [Unnamed seedling X Unnamed seedling]; Austin, David, 1982

'AUSTOP', S, lp, 1999; (**Barbara Austin**); very dbl., 73 petals, 3 in., borne mostly singly, moderate fragrance; prickles moderate; foliage medium, medium green, semi-glossy, good disease resi; upright, medium (4 ft) growth; [Fair Bianca X seedling]; Austin, David, 1997; David Austin Roses Ltd., 1997

Austragold® *see* 'MACKUNG'

Austral, HT, 1976; Bees; (Cavriglia)

'AUSTRALIA FELIX', HT, pb, 1919; bud small, globular; flowers pink and silver shaded lavender, semi-dbl., cupped, intense fragrance; foliage dark, glossy; vigorous, bushy growth; [Jersey Beauty X La France]; Clark, A.

'AUSTRALIAN BEAUTY', Cl HT, dr, 1912; Kerslake

'AUSTRALIAN BEAUTY', Cl HP, lp; [President X Lord Macaulay]; Kerslake, 1907; (Weatherly, L.)

Australian Bicentennial, HT, lp; [Daily Sketch X Red Planet]; Bell, Ronald J., 1987; (Weatherly, L.)

Australian Centre Gold, Min, my

Australian Gold® *see* 'KORMAT'

'AUSTRALIE', HT, dp, 1907; flowers large, dbl.; Kerslake; (Sangerhausen)

'AUSTREAM', S, w, 1997; (**Rushing Stream**); flowers single, small, single,

4–7 petals, 1 in., borne in large clusters, clove fragrance; few prickles; foliage large, light green, glossy; broad, bushy, medium (1.5ft.) growth; Austin, David

Austrian Briar *see* **R. FOETIDA**

Austrian Brier Rose *see* **R. FOETIDA**

Austrian Copper Rose *see* **R. FOETIDA BICOLOR**

Austrian Yellow Rose *see* **R. FOETIDA**

Austriana *see* TANanairan

'AUSUEL', S, ab, 1992; (**Emanuel®**, Emmanuelle); flowers apricot pink, opening rosette, blooms borne in small clusters, dbl., 3–3.5 in., flat, intense fragrance; some prickles; foliage small, medium green, semi-glossy; medium (110 cms), bushy growth; [(Chaucer X Parade) X (Seedling X Iceberg)]; Austin, David; David Austin Roses, Ltd., 1985

'AUSVELVET', S, dr, 1993; (**The Prince™**); flowers large, very dbl., 125 petals, deeply cupped, intense fragrance, strong fragrance; many thorns; foliage dark green; short, upright, bushy growth; PP008813; Austin, David, 1990

AUSverse, S, lp; flowers dark crimson turning to purple, very dbl., 105 petals, 3.25 in., cupped; dark, thorny stems; foliage tough, dark green, slighly glossy; vigorous upright bush growth; Austin, David, 1999

'AUSWALKER', S, my, 1993; (**The Pilgrim™**); very dbl., 170 petals, cupped, intense, strong fragrance; few large thorns; foliage bright green leaves; strong, upright growth; PP008678; Austin, David, 1991

AUSway, S, mr, 1997; (**Noble Antony**); flowers very full, deeply domed, outer petals recurve, large, 85–90 petals, borne singly or in small clusters, moderate fragrance; some prickles; foliage medium, dark green, semi-glossy; bushy, medium, growth; [Seedling X Seedling]; PP10779; Austin, David

'AUSWEBB', S, ab, 1985; (**Mary Webb®**); flowers very large, dbl., cupped, intense fragrance; foliage large, light green, matt; bushy growth; [Unnamed seedling X Chinatown]; Austin, David, 1984

'AUSWEN', S, mr, 1985; (**Wenlock®**); flowers large, dbl., intense fragrance; foliage large, dark, semi-glossy; vigorous growth; [The Knight X Glastonbury]; Austin, David, 1984

'AUSWHITE', S, w, 1987; (**Swan**); flowers white, tinged buff, reverse white, rosette, large, very dbl., borne usually singly, repeat bloom, moderate, fruity fragrance; no fruit; hooked, medium, red prickles; foliage large, light green, semi-glossy; upright, tall growth; PP007564; [Charles Austin X (Seedling X Iceberg)]; Austin, David

'AUSWIFE', S, pb, 1969; (The Wife of Bath, **Wife of Bath**); flowers deep rose-pink, reverse blush, medium, semi-dbl., cupped, repeat bloom, intense fragrance; foliage small; [Mme Caroline Testout X (Ma Perkins X Constance Spry)]; Austin, David

'AUSWILL', S, ab, 1999; (**William Morris**); flowers apricot blend, reverse light pink, very dbl., 120 petals, 3.5 in., borne in small clusters, intense fragrance; prickles moderate; foliage medium, dark green, glossy; branching, medium (4 ft) growth; [Abraham Darby X seedling]; Austin, David; David Austin Roses, Ltd, 1998

'AUSWINE', S, m, 1967; (**Chianti**); flowers purplish maroon, borne in small clusters, semi-dbl., repeat bloom, slight fragrance; foliage dark, glossy; vigorous growth; [Dusky Maiden X Tuscany]; Austin, David; Sunningdale Nursery, 1965

'AUSWING', S, ly, 1994; (**Wildflower**); flowers pale yellow, single, 5 petals, 1.5 in., borne mostly singly, slight fragrance; some prickles; foliage small, light green, semi-glossy; low (60 cms), bushy, spreading growth; [Canterbury X Seedling]; Austin, David; David Austin Roses, Ltd., 1986

AUSwinter, S, lp; flowers bright apricot orange, very dbl., 120 petals, 3 in., cupped, intense, fruity fragrance; almost thornless; dark glossy foliage; strong, arching growth; Austin, David, 1999

'AUSWITH', S, lp, 1994; (**St Swithun™**); flowers pale pink, very dbl., 3–3.5 in., borne in small clusters, moderate fragrance; some prickles; foliage medium, medium green, semi-glossy; medium (90 cms), bushy growth; [Mary Rose X Seedling]; PP9010; Austin, David; David Austin Roses, Ltd., 1993

'AUSWONDER', S, ab, 1994; (**Ambridge Rose**); flowers apricot pink, very dbl., 1.5–2.75 in., cupped, borne in small clusters, intense fragrance; some prickles; foliage medium, dark green, semi-glossy; small (2.5 ft), bushy growth; PPAF; [Charles Austin X Seedling]; PP8679; Austin, David; David Austin Roses, Ltd., 1990

'AUTOCRAT', HT, pb, 1925; flowers ochre to flesh-pink, reverse prawn-red, large; Beckwith

'AUTUMN', HT, ob, 1928; flowers burnt-orange, streaked red, dbl., 70 petals, cupped, moderate fragrance; foliage dark, glossy; [Sensation X Souv. de Claudius Pernet]; Coddington

'AUTUMN BOUQUET', S, mp, 1948; bud long, pointed; flowers carmine-rose-pink, large, dbl., recurrent bloom, intense fragrance; foliage leathery; vigorous, upright, compact growth; [New Dawn X Crimson Glory]; Jacobus; B&A

'AUTUMN DAMASK', D, mp; (Castilian, Four Seasons, Quatre Saisons, Rose of Castille, Old Castilian, R. bifera, R. bifera semperflorens, R. damascena bifera, R. damascena semperflorens, R. semperflorens, Rose des Quatre Saisons); tending to bloom in autumn as well as early summer; (28)

Autumn Dawn *see* 'BRIAUTUMN'

'AUTUMN DELIGHT', HMsk, w, 1933; flowers stamens red, single, borne in large clusters; Bentall

'AUTUMN DUSK', Gr, pb, 1976; bud ovoid, pointed; flowers pale tyrian rose and white, dbl., 33 petals, 3.5–4.5 in., exhibition form, moderate fragrance; foliage leathery; upright, bushy growth; [Music Maker X (Dornröschen X Peace)]; Buck, Dr. Griffith J.; Iowa State University

Autumn Fire *see* 'MORANIUM'

Autumn Fire *see* **'HERBSTFEUER'**

'AUTUMN FLAME', HT, ob, 1953; bud long, pointed; flowers orange-yellow tinted red, dbl., 35–40 petals, 4 in., cupped; foliage dark, leathery; upright, bushy, compact growth; [Ednah Thomas X Autumn]; Thomson

Autumn Frost *see* 'LITLIN'

'AUTUMN GLOW', HT, ob, 1978; bud pear shaped; flowers yellow-orange, dbl., 55 petals, 5 in., exhibition form, slight fragrance; foliage small; bushy growth; [Pascali X Bayadère]; Anderson's Rose Nurseries

'AUTUMN GOLD', HT, yb, 1969; bud pointed; flowers brown-butterscotch-yellow, dbl., 42 petals, 3.5–4 in., globular, moderate fragrance; foliage glossy, dark, leathery; tall, upright growth; [Seedling X Seedling]; Weeks

'AUTUMN HUES', F, ob, 1962; bud pointed; flowers orange, yellow and scarlet, dbl., 35 petals, 3–3.5 in., exhibition form, blooms in clusters, slight fragrance; foliage glossy; upright growth; [Pinocchio X Fred Edmunds]; Von Abrams; Peterson & Dering

Autumn Kiss *see* 'WALKISS'

Autumn Leaves *see* 'AUSLEAF'

Autumn Magic *see* 'FOUTUM'

'AUTUMN QUEEN', HT, ob, 1933; (Vice-President Curtis); bud long, pointed; flowers burnt-orange, pink, gold, open, semi-dbl., slight fragrance; foliage leathery, bronze; vigorous, bushy growth; Vestal

Autumn Shades, Min, lp; [Silver Jubilee X Oz Gold]; Hannemann, F., 1997; The Rose Paradise; (Weatherly, L.)

Autumn Song, F, pb

Autumn Splendor *see* 'MICAUTUMN'

'AUTUMN SPRAY', F, yb, 1964; flowers gold edged red, dbl., 40 petals, 3 in., flat;

foliage glossy; [Masquerade X Isobel Harkness]; Norman; Harkness

Autumn Sunblaze™ *see* 'MEIFERJAC'

'AUTUMN SUNLIGHT', LCl, or, 1965; flowers orange-vermilion, semi-globular, dbl., 30 petals, blooms in clusters, moderate fragrance; foliage glossy, bright green; [Spectacular X Goldilocks, Climbing]; Gregory

'AUTUMN SUNSET'™, S, ab, 1986; flowers medium apricot, with touches of orange and golden yellow, fading lighter, cupped, 20 petals, cupped, repeat bloom, intense, fruity fragrance; round, medium, orange fruit; curved, medium, red prickles; foliage medium, medium green, glossy, disease resistant; bushy, tall, climbing growth; [Westerland sport]; Lowe, Malcolm; Lowe*s Own Root Roses, 1988, U.S.A.

'AUTUMN TINTS', yb, 1914; flowers coppery-pink with yellow, medium, dbl.; Cant, B. R.; (Sangerhausen)

'AUTUMN, CLIMBING', Cl HT, ob, 1951; deVor, W.L.; Amling-DeVor Nursery

Avalanch II, S, w, 1998; Williams, J. Benjamin

'AVALANCHE', HT, w, 1922; flowers creamy white, center deeper, dbl.; Lippiatt

'AVALANCHE', HT, w, 1936; bud long; flowers large; foliage slightly bronze; vigorous, bushy, compact growth; Chambard, C.

Avalanche *see* 'JACAY'

Avalanche Rose® *see* 'DELAVAL'

'AVALON', HT, ab, 1935; bud ovoid; flowers apricot-yellow, center deeper, very large, dbl., globular, slight fragrance; foliage glossy, bronze; vigorous, bushy growth; [Duchess of Atholl sport]; Western Rose Co.; Germain's

Avance *see* BARvan

Avandel *see* 'MORVANDEL'

Avanti *see* 'JACSAY'

'AVE MARIA', HT, w, 1957; flowers pure white, well-formed, dbl., 4–6 in.; vigorous, bushy growth; RULED EXTINCT 12/85; [Seedling X Break o' Day, Climbing]; Brownell, H.C.; Stern's Nursery

Ave Maria® *see* 'KORAV'

Avebury, LCl, lp; McLeod, J., 1995; Honeysuckle Cottage; (Weatherly, L.)

'AVENANT', HGal, lp; flowers deep flesh-pink, fading, large, very dbl.; erect growth

'AVENIR', HWich, lp, 1910; flowers small, semi-dbl.; Corboeuf; (Sangerhausen)

'AVENTURE', HT, or, 1964; (Adventure); flowers large, dbl., 55 petals, exhibition form, slight fragrance; foliage leathery, glossy; vigorous, upright, bushy growth; [(Corail X Baccará) X Seedling]; Croix, P., 1964; C-P, 1965

Aventure No. 2 *see* CROpal

Avenue's Red *see* 'KORJET'

Aveu, HT, dp; Croix

'AVIATEUR BLÉRIOT', HWich, yb, 1910; flowers pale orange-yellow, fading white, dbl., 34 petals, borne in clusters, non-recurrent, moderate, magnolia fragrance; foliage glossy, dark; vigorous, climbing growth; [R. wichurana X William Allen Richardson]; Fauque

'AVIATEUR MICHEL MAHIEU', HT, mp, 1912; flowers large, dbl., intense fragrance; Soupert & Notting; (Sangerhausen)

'AVIATOR PARMENTIER', HT, op, 1941; bud long, pointed; flowers peach-pink to orange, dbl., 30–35 petals, 4–4.5 in., exhibition form; foliage glossy; very vigorous growth; [Seedling X Briarcliff]; Verschuren-Pechtold; J. Parmentier

'AVIGNON', F, my, 1974; flowers medium, dbl., 23 petals, 2 in., slight fragrance; foliage light, glossy; vigorous growth; [Zambra X Allgold]; Cants of Colchester, Ltd.

'AVIORA', HT, mr, 1960; flowers clear red, dbl., 40 petals; foliage glossy, dark; very large growth; [Happiness X Seedling]; Verbeek

'AVÔ ALBINA', HT, rb, 1956; flowers crimson, reverse silvery, large, exhibition form; very vigorous growth; [Peace X Crimson Glory]; da Silva, Moreira

'AVÔ ALFREDO', HT, rb, 1956; flowers spectrum-red, reverse carmine; [Seedling X Independence]; da Silva, Moreira

'AVOCA', HT, mr, 1907; flowers crimson-scarlet, large, dbl., exhibition form, moderate fragrance; very vigorous growth; Dickson, A.

'AVOCAT DUVIVIER', HP, m, 1875; dbl., moderate fragrance; Lévêque; (Sangerhausen)

Avocet *see* 'HARPLUTO'

'AVON'®, HT, dr, 1961; bud ovoid; dbl., 23 petals, 4.5–5.5 in., exhibition form, intense fragrance; foliage leathery; vigorous, upright growth; [Nocturne X Chrysler Imperial]; Morey, Dr. Dennison; J&P

Avon *see* POUlmulti

'AVON, CLIMBING', Cl HT, dr, 1975; Kumar

'AVRIL SHERWOOD', F, my, 1975; bud ovoid; flowers golden yellow fading to buttercup-yellow, semi-dbl., 18 petals, 3–3.5 in., slight fragrance; foliage glossy, dark, leathery; upright growth; [Pink Parfait X Allgold]; Sherwood; F. Mason

'AWAKENING', LCl, lp, 1992; dbl., 26–40 petals, 1.5–2.75 in., quartered, moderate fragrance; foliage light green, glossy; 10' x 8' growth; [New Dawn sport]; Hortico, Inc., 1992; Peter Beales Roses, 1992

'AWARD', HT, mr, 1953; flowers velvety red, intense fragrance; foliage very glossy, dark; short stems; vigorous growth; [Will Rogers X Mme Henri Guillot]; Taylor, C.A.

Awareness *see* FRYbingo

Awayuki, S, w

Axeline, LCl, mp

Axelle, HT, 1964; Gaujard; (Cavriglia)

Ayako™ *see* 'MALMIYA'

Ayaori, F, rb

'AYLSHAM', S, dp, 1948; bud ovoid; flowers deep pink, large, dbl., non-recurrent, slight fragrance; foliage light green, glossy; vigorous (to 5 ft) growth; [Hansa X R. nitida]; Wright, Percy H.

'AYRSHIRE QUEEN', Ayr, dr, 1835; (Ayrshire Splendens); flowers purplish crimson, semi-dbl.; [Blush Ayrshire X Tuscany]; Rivers

Ayrshire Rose; (**R. ARVENSIS AYRESHIREA**, R. arvensis capreolata, R. capreolata)

Ayrshire Splendens *see* **'AYRSHIRE QUEEN'**

'AZALEA ROSE', Pol, mp, 1940; flowers bright rose-pink, resemble azaleas; [Ellen Poulsen sport]; Griffing Nursery

'AZEEZ', F, op, 1965; bud pointed; flowers coral-pink, reverse lighter, medium, dbl., exhibition form; foliage leathery; upright, compact growth; Pal, Dr. B.P.; Indian Agric. Research Inst.

'AZELDA', HT, op, 1963; bud long, pointed; flowers coral-rose, dbl., cupped, intense fragrance; foliage glossy, light green; long stems; vigorous, upright growth; [Queen Elizabeth X Rosenelfe]; Matthews

'AZTEC', HT, or, 1957; bud long, pointed; flowers scarlet-orange, dbl., 25 petals, 4–5 in., exhibition form, moderate fragrance; foliage glossy, leathery; vigorous, spreading growth; [Charlotte Armstrong X Seedling]; Swim, H.C.; Armstrong Nursery

Azulabria *see* 'RUIBLUN'

'AZUR', F, m, 1967; bud ovoid; flowers deep lavender-mauve, pointed, semi-dbl., 15 petals, 2.5–3 in., intense fragrance; foliage dark reddish-green; vigorous, compact, bushy growth; [Sterling Silver X (Gold Strike X Golden Garnette)]; Lens

Azure Sea *see* 'AROLALA'

B

B. C.® *see* 'KINBEE'

'B. S. BHATCHARJI', HT, my, 1933; flowers buttercup-yellow, well formed, large; foliage glossy, dark; vigorous growth; Dickson, A.

'B. W. PRICE', HT, mp, 1943; bud long, pointed; flowers cerise-pink, open, large, single, 6–8 petals; foliage soft; vigorous, upright, bushy growth; [Night X Mme Butterfly]; McGredy; J&P

Babe, Pol, pb, 1958; Hazelwood

'BABE RUTH', HT, op, 1950; bud ovoid; flowers coral, reverse rose-coral, dbl., 35–40 petals, 4–4.5 in., cupped, moderate fragrance; foliage leathery, glossy, bronze; vigorous, upright growth; [Los Angeles X Seedling]; Howard, F.H.; H&S

'BABETTE', F, m, 1969; flowers lavender-pink, small, single, cupped, slight fragrance; foliage leathery, small; vigorous, bushy growth; [Seedling X Eminence]; Gaujard

'BABETTE RAMBLER', HWich, mr, 1906; flowers carmine-red, medium, dbl.; Walsh; (Sangerhausen)

'BABs', HT, dy, 1984; (**Coton Gold**); flowers glowing yellow; [Whisky Mac sport]; Babb, J.T.

'BABT DEITZ', Pol, or, 1924; flowers oriental red, edged salmon, dbl.; Opdebeeck

'BABY ALAN', Pol, mp, 1930; flowers shining pink, resembling English daisy in form, very dbl., borne in clusters; Kessler; J.T. Lovett

'BABY ALBERIC', Pol, ly, 1932; bud yellow; flowers creamy white, small, dbl., recurrent bloom; vigorous growth; Chaplin Bros.

Baby Ashley® *see* 'KINASH'

Baby Baccará® *see* 'MEIBYBA'

Baby Ballerina, Min, pb, 1997; Villegas

Baby Bath, Pol, mp, 1989; Robinson

'BABY BETSY MCCALL', Min, lp, 1960; flowers micro-mini, dbl., 20 petals, 1 in., cupped, moderate fragrance; foliage leathery, light green; vigorous, dwarf, compact (8 in) growth; [Cécile Brünner X Rosy Jewel]; Morey, Dr. Dennison; J&P

Baby Bettina *see* 'MEIDACINU'

'BABY BETTY', Pol, mp, 1929; bud ovoid, yellow, tinged red; flowers pink, base lighter, small, dbl., cupped, abundant, recurrent bloom, moderate fragrance; foliage leathery, dark, bronze; vigorous, compact, bushy growth; [Eblouissant X Comtesse du Cayla]; Burbage Nursery

'BABY BIO', F, dy, 1977; flowers deep golden yellow, patio, dbl., 28 petals, 3 in., slight fragrance; foliage glossy; GM, Rome, 1976; [Golden Treasure X Seedling]; Smith, E.; Rosemont Nursery Co.

Baby Blanket *see* 'KORFULLWIND'

'BABY BLAZE', F, mr, 1954; (Lund's Jubiläum); bud ovoid; flowers cherry-red, white eye, dbl., 33 petals, 3 in., cupped, blooms in clusters of 10-25, moderate fragrance; foliage light, glossy; vigorous, bushy, compact growth; [World's Fair X Hamburg]; Kordes, W.; J&P

'BABY BUNTING', Min, dp, 1953; flowers deep pink, small, dbl., 20 petals, 1.5 in., moderate fragrance; [Ellen Poulsen X Tom Thumb]; deVink, J.; T. Robinson, Porchester Nursery

Baby Cakes, *see* 'TALBABY'

Baby Carnaval *see* 'TANBAKEDE'

Baby Carnival *see* 'TANBAKEDE'

'BABY CAROLINE LOUISE', Min, w, 1998; bud tall; flowers white, slight lemon center, white reverse, very dbl., 26–40 petals, 2 in., exhibition form, borne in small, well-formed clusters, opens flat, slight fragrance; prickles moderate; foliage medium, medium green, semi-glossy; tall, bushy growth; [Sexy Rexy X Laura Ford]; Barker, S.J.L.

Baby Cécile Brunner *see* 'MORCEBRU'

'BABY CHÂTEAU', F, mr, 1936; (Baby Vougeot, Château); bud ovoid, crimson; flowers red shaded garnet, large, dbl., moderate fragrance; foliage glossy, bronze; very vigorous, bushy growth; (28); [Aroma X (Eva X Ami Quinard)]; Kordes

'BABY CHERYL', Min, lp, 1965; bud pointed; flowers light pink, reverse lighter, micro-mini, small, dbl., moderate, spicy fragrance; foliage leathery; vigorous, dwarf growth; [Spring Song X Seedling]; Williams, Ernest D.; Mini-Roses

Baby Chicks, Min, my

Baby Crimson *see* **'PERLA DE ALCAÑADA'**

Baby Crimson, Min, mr

'BABY DARLING', Min, ab, 1964; bud pointed; flowers apricot-orange, small, dbl., 20 petals; dwarf, bushy (12 in.) growth; [Little Darling X Magic Wand]; Moore, Ralph S.; Sequoia Nursery

'BABY DARLING, CLIMBING', Cl Min, ab, 1972; Trauger, F.; Sequoia Nursery

Baby Diana® *see* 'SAVADI'

Baby Doll *see* **'TIP-TOP'**®

Baby Dominic *see* 'JUDNIC'

Baby Donnie, Min, mr, 1972

Baby Dorothy *see* **'MAMAN LEVAVASSEUR'**

Baby Eclipse® *see* 'MORECLI'

'BABY ELEGANCE', Pol, ab, 1912; flowers pale yellow-orange, single; Hobbies

Baby Face *see* 'LAVABY'

'BABY FAURAX', Pol, m, 1924; flowers violet, small, dbl., blooms in large clusters, moderate fragrance; dwarf growth; Lille, L.

'BABY GARNETTE', Min, mr, 1962; flowers blood-red, small, dbl.; foliage dark; vigorous (10-12 in), compact growth; [Red Imp X Sparkler]; Morey, Dr. Dennison; J&P

'BABY GLORIA', Pol, or, 1936; flowers salmon-red, larger than parent; very dwarf (6-8 in) growth; [Gloria Mundi sport]; Böhm, J.

'BABY GOLD STAR', Min, dy, 1940; (Estrellita de Oro); bud pointed; flowers golden yellow, semi-dbl., 14 petals, slight fragrance; foliage small, soft; [Eduardo Toda X Rouletii]; Dot, Pedro; C-P

'BABY GOLD STAR, CLIMBING', Cl Min, dy, 1964; Williams, Ernest D.; Mini-Roses

Baby Grand® *see* 'POULIT'

Baby Herriot *see* **'ÉTOILE LUISANTE'**

'BABY JAYNE', Cl Min, mp, 1957; (Fairy Hedge, Pixie Hedge); flowers soft pink, very small, dbl., 45 petals, borne in clusters; foliage small, glossy; height 3-4 ft; [Violette X Zee]; Moore, Ralph S.; Germain's

'**BABY KATIE**'™, Min, pb, 1978; bud ovoid, pointed; flowers cream and pink blend, small, dbl., 28 petals, exhibition form, slight fragrance; foliage matt, green; vigorous, compact, bushy growth; PP4471; [Sheri Anne X Watercolor]; Saville, F. Harmon; Nor'East Min. Roses

'**BABY LILIAN**', Min; flowers begonia-rose, base tinted orange-yellow, rather large for the class, dbl., slight fragrance; foliage small, light, glossy; vigorous (12-15 in) growth

Baby Love™ *see* 'SCRIVLUV'

Baby Love®, Min, lp

'**BABY LYON ROSE**', Pol, rb, 1916; flowers coral-rose, shaded chrome-yellow or shrimp-red, dbl.; Turbat

Baby Maria, Min, dp

Baby Mascarade *see* 'TANBAKEDE'

Baby Maskarade *see* 'TANBAKEDE'

Baby Maskerade *see* 'TANBAKEDE'

Baby Masquerade® *see* 'TANBAKEDE'

'**BABY MASQUERADE, CLIMBING**', Cl Min, rb, 1974; Sykes, R.O.; Mini-Roses

Baby Michael *see* 'JUSMICHAEL'

'**BABY MINE**', Pol, my, 1929; flowers sulfur to butter-yellow; [Cécile Brunner, Climbing seedling]; Moore, Ralph S.

'**BABY OPHELIA**', Min, lp, 1961; bud pointed; flowers soft pink, dbl., 33 petals, 1 in., moderate fragrance; foliage glossy; vigorous, bushy (8-12 in) growth; [(R. wichurana X Floradora) X Little Buckaroo]; Moore, Ralph S.; Sequoia Nursery

Baby Orange Triumph, Min, or; flowers small, dbl.; (Sangerhausen)

'**BABY PEACE**', Min, yb, 1962; bud urn shaped; flowers ivory-yellow tipped pink, dbl., 50–55 petals, .5–1 in., slight fragrance; very vigorous growth; [Peace sport]; De Mott & Johnson, G.E.

'**BABY PINOCCHIO**', Min, pb, 1967; bud ovoid; flowers salmon-pink blend, small, dbl., moderate fragrance; foliage glossy, leathery; vigorous, bushy growth; [Golden Glow X Little Buckaroo]; Moore, Ralph S.; Mini-Roses;, Sequoia Nursery

Baby Rambler *see* 'CHEWRAMB'

Baby Rambler, Climbing *see* '**MISS G. MESMAN**'

'**BABY ROSAMUNDE**', Pol, mp, 1930; flowers rose-pink, semi-dbl.; Kessler; J.T. Lovett

Baby Secret *see* BRIbaby

'**BABY SUNBEAM**', LCl, ab, 1934; flowers light apricot, passing to cream-yellow, large yellow center, borne in large clusters; foliage bronze; very vigorous growth; Burbank; Stark Bros.

Baby Sunrise *see* 'MACPARLEZ'

'**BABY SYLVIA**', F, op, 1959; flowers flesh-salmon-pink, dbl., 25–30 petals, 3 in.; very vigorous growth; [Lady Sylvia X Seedling]; Fryers Nursery, Ltd.

Baby Talisman *see* '**PRESUMIDA**'

'**BABY TALK**', F, m, 1980; bud small, ovoid, pointed; flowers dusty mauve-pink, patio, dbl., 26 petals, borne mostly singly, moderate, tea fragrance; foliage small to medium, moderately thin; low to medium, compact, dense growth; [Plain Talk X Angel Face]; Weeks

Baby Tausendschön *see* '**ECHO**'

Baby Tom *see* 'BOSEYEBALL'

'**BABY TYPHOON**', Min, ab, 1988; flowers bright, clear apricot, center golden-yellow, medium, dbl., 35–40 petals, exhibition form, borne singly or in sprays of 3-5, moderate, damask fragrance; prickles pointed, large, straw; foliage medium, dark green, semi-glossy; bushy, spreading, tall growth; [Gold Coin sport]; Halevi, A.M., 1987

Baby Vougeot *see* '**BABY CHÂTEAU**'

Babyflor® *see* TANrolfy

'**BABYLON**', HT, op, 1976; flowers deep coral-pink, low-centered, dbl., 36 petals, 5 in., intense fragrance; foliage glossy, dark; moderately vigorous growth; [Tropicana X Pink Favorite]; Bees; Sealand Nursery

Baby's Blush, HCh, mp, 1995; Scarman

'**BACARDI**', HT, ob, 1987; flowers uniform iridescent coral-orange, fading lighter, small, dbl., 30 petals, exhibition form, borne singly, slight, fruity fragrance; straight, small, reddish prickles; foliage medium, medium green, semi-glossy; upright, medium growth; PP006862; [Seedling self pollination sport]; Weeks, O.L.; Weeks Wholesale Rose Growers, 1988

Baccará® *see* 'MEIGER'

Baccará, Climbing® *see* 'MEGERSAR'

'**BACCHANTE**', HGal, dr, 1848

'**BACCHUS**', HT, dp, 1951; flowers cherry-pink, medium, dbl., 25 petals, moderate fragrance; vigorous growth; GM, NRS, 1952; Dickson, A.

Back Home *see* 'GELBACK'

Bad Bergzabern, HT, mr, 1993; Hetzel

Bad Birnbach, F, dp, 1999; Kordes

Bad Ems, S, mp, 1990; Schultheis

Bad Füssing® *see* 'KORBAD'

Bad Homburg, HT, ab, 1995; Eichelmann

'**BAD LANGENSALZA**', F, pb, 1954; flowers carmine-pink, large, dbl., moderate fragrance; Berger, W.; (Sangerhausen)

Bad Nauheim *see* '**NATIONAL TRUST**'

Bad Naukeim *see* '**NATIONAL TRUST**'

'**BAD NEUENAHR**', HKor, mr, 1958; bud ovoid; flowers scarlet, dbl., 50 petals, 4 in., cupped, blooms in clusters (to 15), moderate fragrance; foliage dark, leathery; vigorous (6 ft) growth; Kordes, W.

'**BAD PYRMONT**', F, or, 1976; bud pointed; dbl., 40 petals, 4 in., exhibition form, abundant bloom, slight fragrance; foliage dark, soft; vigorous, upright, bushy growth; [Duftwolke X Seedling]; Kordes; Horstmann

Bad Salzuflen®, F, mp

'**BAD WÖRISHOFEN**'®, F, mr, 1972; (Gruss an Worishofen); bud ovoid; semi-dbl., 18 petals, 2.5 in., cupped, slight fragrance; foliage dark, soft; vigorous, bushy growth; [Sarabande X Marlena]; Kordes, W.

'**BADEN-BADEN**', HT, dr, 1952; flowers deep crimson, large, dbl., intense fragrance; foliage dark, leathery; vigorous, upright growth; [Poinsettia X Crimson Glory]; Kordes

Badener Gold®, F, ob, 1974; McGredy, Sam IV

Badia, HT; Mansuino, Dr. Andrea; (Cavriglia)

'**BADINAGE**', HT, op, 1951; flowers salmon-pink flushed coppery, dbl., 30 petals, 4–4.5 in., moderate fragrance; foliage leathery, dark; vigorous growth; [Peace X Seedling]; Gaujard

Badner Traberchampion, HT, lp

'**BAGATELLE**', HT, mr, 1943; bud large, oval; flowers bright red, very dbl., moderate fragrance; foliage leathery; very vigorous, bushy growth; Gaujard

'**BAGDAD**', HT, or, 1953; bud ovoid to urn shape; flowers nasturtium-red to orange, becoming cupped, dbl., 48–55 petals, 4.5 in., exhibition form, intense fragrance; foliage glossy; very vigorous, bushy growth; [Charlotte Armstrong X Signora]; Swim, H.C.

Bagheera *see* 'KORGERA'

'**BAGLIORE**', HT, mr, 1955; flowers crimson-red, intense fragrance; stiff stems; very vigorous growth; Aicardi, D.

'**BAHAMA**', F, dr, 1968; semi-dbl., intense fragrance; vigorous, low growth; [Fidélio X Hanne]; Soenderhousen

'**BAHIA**', F, ob, 1974; bud ovoid; flowers orange, medium, dbl., cupped, moderate, spicy fragrance; foliage glossy, dark, leathery; vigorous, upright, bushy growth; AARS, 1974; [Rumba X Tropicana]; Lammerts, Dr. Walter; Armstrong Nursery

'BAHRS LIEVELING', F, mr, 1950; flowers velvety crimson-red; [Donald Prior X Orange Triumph]; Leenders, M.

Bailey Red, F, mr

'BAISER', HT, w, 1953; flowers pearly white, edges tinted pink, dbl., 40 petals, 3 in., exhibition form, moderate fragrance; vigorous growth; [Mme Joseph Perraud X Independence]; Mallerin, C.; EFR

'BAJAZZO', HT, rb, 1961; flowers velvety blood-red, reverse white, well-formed, large, intense fragrance; vigorous, upright growth; Kordes, R.; Kordes

Bakels, F, 1980; Gaujard; (Cavriglia)

Bakewell Scots Briar, HSpn, w

Bakker's Newcomer®, HT, m

Bakker's Orange, HT, ob

'BAKTAF', Min, yb, 1988; (**Taffy**); flowers soft, pastel butter yellow, tipped light pink, reverse pure, dbl., 33 petals, exhibition form, slight fragrance; prickles nearly straight, small, light brown; foliage small, medium green, matt; bushy, low, compact, full growth; [Unnamed seedling X Unnamed seedling]; Baker, Larry

Baladin, F; Combe, M.; (Cavriglia)

Balaji, HT, ob, 1998; Patil, B.K.

'BALALAIKA', F, dr, 1978; bud ovoid; flowers blood-red, dbl., moderate fragrance; foliage glossy; low, bushy growth; Hubner; O. Baum

'BALCON', LCl, or, 1960; bud ovoid; flowers geranium-red, open, medium, dbl., slight fragrance; foliage dark, glossy; very vigorous growth; [Spectacular X Seedling]; Combe; Minier

'BALDUIN', HT, pb, 1896; (Crimson Maman Cochet, Helen Gould, Red Maman Cochet); flowers pink, edged darker, large, dbl., moderate fragrance; vigorous growth; [Charles Darwin X Marie van Houtte]; Lambert, P., 1896; Dingee & Conard, 1901

Baldwin *see* **'SLEEPY'**

'BALÉARES', HT, w, 1957; bud pointed; flowers base and reverse of petals white suffused carmine, deeper at edges, dbl., 35 petals, intense fragrance; foliage dark, glossy; stiff stems; upright growth; [Peace X Flambee]; Dot, Simon

'BALI', Pol, 1960; flowers orange-yellow, open, large, single, 6 petals; foliage glossy, light green; bushy growth; [Masquerade X Golden Rain]; Leenders, J.

'BALI-HI', HT, mp, 1959; flowers peach-pink to shell-pink, medium, exhibition form, moderate fragrance; foliage glossy; vigorous, upright growth; Lowe

'BALINESE', HT, m, 1963; bud ovoid; flowers brown and lavender tones, dbl., 35–40 petals, 4.5–5 in., cupped, intense, fruity fragrance; foliage glossy; vigorous, upright growth; [Grey Pearl X Brownie]; Boerner; J&P

'BALJIT', HT, dp, 1979; bud pointed; flowers spirea-red, open but full, dbl., 30–40 petals, 5 in., long-lasting, profuse bloom; foliage dark, leathery; free growth; [Velsheda sport]; Lucknow

Balkan Star, HT, m, 1997; Tapanchev

'BALL OF SNOW', N, w, 1887; Henderson; (Sangerhausen)

'BALLADE', F, op, 1960; flowers deep orange-salmon, well formed, dbl., 25 petals, borne in clusters; bushy growth; [Signal Red X Polyantha seedling]; deRuiter

Ballade *see* TANedallab

'BALLADY', C, lp, 1934; flowers large, dbl.; Perrot; (Sangerhausen)

'BALLERINA', HMsk, mp, 1937; flowers bright soft pink, white eye, small, single, blooms in very large clusters; vigorous (3 ft) growth; Bentall

'BALLERINA', F, my, 1941; flowers naples yellow, large, dbl., moderate fragrance; Leenders, M.

'BALLERINE', HT, w, 1955; bud long; flowers satiny snow-white, open, medium, dbl., intense fragrance; foliage glossy; vigorous, bushy growth; Buyl Frères; Delforge

Ballet® *see* 'KORFLOT'

'BALLET, CLIMBING', Cl HT, w, 1962; Kordes

BALlgance, HT, dr, 1996; (**Red Elegance ™**); Twomey, Jerry

'BALLILA', HT, mr, 1935; flowers medium, dbl.; Bräuer; (Sangerhausen)

Ballindalloch Castle *see* 'COCNEEL'

Balmain Climber, LCl, lp

'BALPARTY', HT, w, 1999; (**Rose Odyssey 2000**); flowers cream with pink edge, dbl., 26–40 petals, 4–5 in., exhibition form, borne mostly singly, moderate fragrance; prickles moderate; foliage medium to large, medium to dark green, semi-glossy; upright, bushy, medium (3.5 to 4.5 ft) growth; [Garden Party sport]; Ballin, Don & Paula, 2000; Arena Rose Co., 2000

Baltik, F, m; flowers violet, large, dbl., slight fragrance; VEG; (Sangerhausen)

'BALTIMORE', HT, lp, 1898; flowers blush tinted; [Mme Antoine Rivoire X Lady Mary Fitzwilliam]; Cook, J.W.

'BALTIMORE BEAUTY', LCl, ly, 1927; flowers buff-yellow, fading white, semi-dbl., borne in clusters, moderate fragrance; long, strong stems; Schluter

'BALTIMORE BELLE', HSet, lp, 1843; (Belle de Baltimore); flowers pale blush to rose-white, very dbl., borne in clusters, non-recurrent, moderate fragrance; vigorous growth; [Probably R. setigera X Noisette]; Feast

Balwant, HT, ab; [Christian Dior sport]; Phadtare

'BAMBEY', HT, mr, 1979; bud long, pointed; dbl., 60 petals, exhibition form, intense, fruity fragrance; small, recurved prickles; foliage matt; medium growth; [(Fragrant Cloud X Peace) X Alec's Red]; Perry, Astor, 1981

'BAMBI', F, mp, 1962; flowers bright pink, dbl., 24 petals, 2 in., blooms in clusters, slight fragrance; foliage dark; vigorous, bushy, low growth; [The Optimist X Korona]; Watkins Roses

'BAMBI', F, mp, 1962; bud pointed; flowers light apricot-pink, semi-dbl., 20 petals, 3 in., cupped, blooms in clusters, slight fragrance; foliage glossy; vigorous, bushy, compact growth; Von Abrams; Peterson & Dering

Bambi *see* DELmistri

'BAMBINA', F, w, 1962; bud yellow; flowers large, semi-dbl., borne in clusters; [Seedling X Virgo]; da Silva, Moreira

'BAMBINO', Min, mp, 1953; [Perla de Alcañada pink sport]; Dot, Pedro

Bambino *see* 'SAVABINO'

Bambolina®, Min, pb

Bamboo Rose *see* **R. MULTIFLORA WATSONIANA**

'BAMBULA'®, F, op, 1970; dbl., 28 petals, 5 in., slight fragrance; foliage glossy, dark; Tantau, Math.; Wheatcroft & Sons, 1969

Banana Split *see* 'ZIPBAN'

'BANARAS DAWN', HT, ab, 1977; bud tapered; flowers apricot-buff, dbl., 30 petals, 4.5 in., exhibition form, intense, fruity fragrance; foliage glossy, light green; vigorous growth; Saxena; Doon Valley Roses

'BANATER ROSE', HT, w, 1927; flowers cream-white, center orange-yellow, dbl., slight fragrance; [Harry Kirk seedling]; Mühle

'BANBRIDGE', F, pb, 1967; flowers rose-red and yellow, well shaped, 3 in., blooms in clusters, slight fragrance; [Mme Léon Cuny X Cläre Grammerstorf]; McGredy, Sam IV

'BANCO', HT, op, 1956; flowers salmon-pink becoming gold tinted, well-formed, large, dbl., 50 petals; foliage bright green; vigorous, upright growth; [Peace X Seedling]; Laperrière; EFR

Banco 86® *see* 'LAPDEF'

'BANGOR', F, or, 1972; flowers geranium-lake, ovate, dbl., 24 petals, 3.5 in., slight fragrance; foliage leathery; free growth; [Jubilant X Marlena]; Dickson, A.

Bangor Cathedral *see* KIRmelody

Bangsbo *see* POUlrine

'BANJARAN', F, yb, 1969; flowers gold and orange-red, small, dbl., cupped, blooms in clusters, slight fragrance; foliage leathery; vigorous, upright, compact growth; Pal, Dr. B.P.; K.S.G. Son's Roses

Banks' Rose *see* **R. BANKSIAE**

Banksiae Alba *see* **R. BANKSIAE BANKSIAE**

'BANKSIAEFLORA', HSem, w; flowers white, center cream, small, very dbl.; [Possibly R. arvensis X Noisette hybrid]

Banksian Rose *see* **R. BANKSIAE**

'BANNER', HT, pb, 1951; flowers deep pink, striped white; [Charlotte Armstrong sport]; Raffel; Port Stockton Nursery

Banquise®, HT, w

Banquise *see* LAPruni

'BANSHEE', S, mp; flowers pink, troubled by balling, of poor texture, very dbl., non-recurrent, moderate fragrance; Origin unknown, 1928

'BANTRY BAY'®, LCl, mp, 1967; flowers soft pink, reverse bright pink, 4 in., blooms in clusters, slight fragrance; [New Dawn X Korona]; McGredy, Sam IV; McGredy

Banzai *see* 'MEILIMONA'

Banzai, T, or

Banzai '76 *see* 'MEILIMONA'

Banzai 83®, HT, or

'BAPTISTE LAFAYE', Pol, dp, 1910; flowers carmine-pink, medium, dbl.; Puyravaud; (Sangerhausen)

BARafne, S, pb, 1995; (**Dafne**); Barni, V.

BARafne, S, op, 1998; (**Meriggio**); Barni, V.

Barakura, S, lp, 1999

'BARAMAD', S, rb, 1991; (**Amadeus®**); Barni, V.

'BARANAM', HT, mp, 1988; (**Antico Amore®**); Barni, V.

BARarc, S, mp, 1995; (**Arcanto**); Barni, V.

'BARARCSUN', Min, w, 1989; (**Arctic Sunrise**); bud pointed; flowers small, dbl., 30 petals, flat, borne in sprays of 40-60, no fragrance; no fruit; prickles long, thin, small, pale tan; foliage small, medium green, glossy; spreading, low growth; [Snow Carpet X Tranquillity]; Barrett, F.H., 1991

Barba Blue, HT, m

'BARBARA', HT, rb, 1923; flowers bright red, base yellow, reverse pale yellow; Paul, W.

'BARBARA', HT, yb, 1962; flowers amber-yellow lightly flushed pink, semi-dbl., 20 petals, 5 in., exhibition form; foliage dark; very vigorous growth; Gaujard; Gandy Roses, Ltd.

Barbara® *see* GAUbor

Barbara Allen, S, w, 1999; flowers ivory with pink washing on petal edges, semi-dbl.; spreading growth; Williams, J. Benjamin

Barbara Austin *see* 'AUSTOP'

Barbara Bush™ *see* 'JACBUSH'

Barbara Carrera, F, op, 1994; Beales, Peter

Barbara Dawson, HT, lp, 1987; [Mount Shasta X Saffron]; Dawson

'BARBARA FRIETCHIE', HT, dp, 1959; bud long, pointed; flowers rose-red, open, large, dbl.; vigorous, upright growth; [Heart's Desire X The Chief]; Silva; Plant Hybridizers of Calif.

Barbara Hauenstein, HT, mp, 1957; flowers large, dbl., moderate fragrance; Poulsen; (Sangerhausen)

Barbara Hendricks *see* DORgran

'BARBARA JOYCE', Min, mr, 1999; very dbl., 41 petals, 1.5 in., borne in small clusters, no fragrance; few prickles; foliage medium, dark green, resistant, semi-glossy; upright, medium (16 in) growth; [Miss Dovey X Kristin]; Jolly, Betty J.

Barbara Mandrell™ *see* 'KINBARB'

'BARBARA MASON', HT, my, 1947; bud long, pointed; flowers medium, semi-dbl., exhibition form, moderate fragrance; foliage glossy; vigorous, upright growth; [Eclipse X Luis Brinas]; Moss; F. Mason

'BARBARA MEYER', Gr, or, 1971; bud ovoid; flowers medium, very dbl., exhibition form, slight fragrance; foliage glossy, dark; vigorous, upright growth; [Queen Elizabeth X Tropicana]; Meyer, H.M.; Aloe Vera Nursery

'BARBARA RICHARDS', HT, yb, 1930; flowers yellow, reverse flushed pink, very large, dbl., moderate fragrance; weak necks; vigorous, bushy growth; Dickson, A.

'BARBARA ROBINSON', HT, w, 1925; bud long, pointed; flowers creamy white, large, dbl., exhibition form, slight fragrance; Dickson, A.

'BARBARA STRAUS', HT, dr, 1978; bud long, pointed; flowers , dbl., 22 petals, borne singly, slight fragrance; foliage glossy; spreading, upright growth; [Mister Lincoln X Unknown]; Schwartz, Ernest W.; Flora World, Inc.

'BARBARA WARD', HT, mr, 1931; flowers crimson-scarlet, large, dbl., 42 petals, moderate fragrance; vigorous growth; [Royal Red X Columbia]; Ward, F.B.

Barbara Worl, B, mp, 1991

Barbara's Rose, HBc, yb

'BARBARELLA', Min, pb, 1981; flowers deep pink, reverse creamy yellow, small, dbl., 20 petals, slight fragrance; foliage small, dark, matt; bushy growth; [Seedling X Seedling]; Barni-Pistoia, Rose

Barbarina, F, yb, 1965; flowers large, dbl., moderate fragrance; Berger, A.; (Sangerhausen)

Barbarosa, S, lp, 1978

'BARBAROSSA', HP, m, 1906; flowers carmine-purple, large, dbl., 55 petals, moderate fragrance; vigorous growth; [(Frau Karl Druschki X Captain Hayward) X Princesse de Béarn]; Welter

BARbea, HT, w, 1995; (**Beatrice**); Barni, V.

'BARBECUE', F, mr, 1961; flowers rich red, dbl., 30 petals, 4 in., flat, blooms in large clusters, moderate fragrance; foliage dark; vigorous, upright, bushy growth; [Seedling X Lilli Marleen]; Dickson, Patrick; A. Dickson

Barbeuce, F, mr, 1961; flowers large, dbl., moderate fragrance; Dickson, Patrick; (Sangerhausen)

'BARBIE', F, mp, 1977; bud pointed, ovoid; flowers sweetheart, dbl., 35–40 petals, 2.5 in., slight fragrance; foliage small, glossy; upright growth; [Escort X Jazz Fest]; Swim, H.C. & Ellis, A.E.; Armstrong Nursery

Barbie® *see* 'JACMOBLI'

Barbra Streisand *see* 'WEKQUANEZE'

BARbrio, F, yb; (**Briosa**); Barni, V.

Barby *see* KORaby

'BARCAROLLE', HT, mr, 1959; flowers crimson tinted geranium-red, large, dbl., 50 petals; vigorous, bushy growth; [Paulette X Tonnerre]; Laperrière; EFR

BARcast, S, lp, 1987; (**Castore®**); Barni, V.

'BARCELONA', HT, dr, 1932; bud long, pointed; flowers crimson, large, dbl., 75 petals, moderate, spicy fragrance; foliage dark; vigorous growth; [(Sensation X Templar) X Lord Charlemont]; Kordes

Barcelona 95® *see* FEcasa

BARciak, LCl, rb, 1994; (**Ciak®**); Barni, V.

BARcora, MinFl, ob, 1993; (**Corallina**); Barni, V.

BARdem, F, lp; (**Tempi Moderni**); Barni, V., 1996

BARdon, HT, lp, 1988; (**Donna Marella Agnelli®**); Barni, V.

BARdord, HT, ab, 1998; (**Dorada**); Barni, V.

'BARDORG', Min, yb, 1993; (**Newbury Angel**); flowers open yellow turning to pale orange, blooms borne in large clusters, dbl., 26–40 petals, 1.5 in., no fragrance; some prickles; foliage small, medium green, matt; low (40 cms), bushy growth; [Freegold X Orange Honey]; Barrett, F.H.; Hills Nurseries, 1993

'BARDOU JOB', B, dr, 1887; flowers crimson, shaded blackish, semi-dbl.; vigorous, semi-climbing growth; [Gloire des Rosomanes X Général Jacqueminot]; Nabonnand, G.

BARemoz, HT, ab, 1999; (**Emozione**); Barni, V.

BARenic, S, lp, 1993; (**Berenice**); Barni, V.

BARfeel, F, lp, 1994; (**Feeling®**); Barni, V.

BARfiob, F, w, 1988; (**Fiocco Bianco**); Barni, V.

BARflam, S, dr, 1993; (**Flaming Star**); Barni, V.

BARflor, MinFl, dp, 1994; (**Florita®**); Barni, V.

BARfort, Min, op, 1999; (**Fortunella**); Barni, V.

BARfri, S, mp, 1992; (**Frine®**); Barni, V.

BARfunn, F, ob, 1998; (**Funny Girl**); Barni, V.

BARgai, MinFl, pb, 1993; (**Gaia®**); Barni, V.

BARger, S, mr, 1996; (**Ranger**); Barni, V.

BARgold, F, my, 1991; (**Golden Dance®**); Barni, V.

BARhop, HT, mr, 1999; (**Hot Point**); Barni, V.

'BARILLET', M, mr, 1850; flowers dark carmine, large, dbl., cupped; Verdier, V.

Barillet, HP, lp; flowers brilliant rose; Deschamps, 1867

BARIMP, HT, mr, 1990; (**Impulse®**); Barni, V.

BARinas, S, pb, 1989; (**Rinascimento®**); Barni, V.

BARitzia, HT, w, 1996; (**Letizia**); Barni, V.

BARjou, S, dy, 1999; (**Bijou**); Barni, V.

Barkarole *see* 'TANELORAK'

BARkel, HT, lp; (**Helen Keller**); Barni, V., 1991

'BARKHATNAIA KRASAVITSA', HT, dr, 1938; (Velvet Beauty); flowers velvety red, medium, semi-dbl., cupped, moderate fragrance; foliage dark; low growth; Gubonen

BARlar, F, mr, 1999; (**La Rossa**); Barni, V.

BARleo, LCl, dy; (**Eleonora®**); Barni, V.

BARlev, F, ab, 1991; (**Rita Levi Montalcini®**); Barni, V.

Barley Sugar, Cl Min, lp; Peden, R., 1997; (Weatherly, L.)

BARliz, MinFl, mr, 1993; (**Liza®**); Barni, V.

BARluc, Min, ob, 1992; (**Lucy**); Barni, V.

BARmar, HT, dr, 1992; (**Marina Marini®**); Barni, V.

BARmeri, HT, op, 1994; (**Meridiana®**); Barni, V.

BARmiss, HT, yb, 1999; (**Rosita Missoni**); Barni, V.

BARmiz, S, dy, 1999; (**Mizar**); Barni, V.

'BARN DANCE', S, op, 1975; bud ovoid, long, pointed; flowers light salmon-pink, dbl., 23 petals, 2.5–3.5 in., cupped, slight fragrance; foliage light to dark, leathery; [Tickled Pink X Prairie Princess]; Buck, Dr. Griffith J.; Iowa State University

BARnec, HT, pb, 1986; (**Mirella®**); Barni, V.

Barni, HT, ob

Barock *see* TANkorab

BARoly, S, rb, 1998; (**Olympus**); Barni, V.

'BARON DE BONSTETTEN', HP, dr, 1871; flowers dark velvety crimson, large, dbl., 80 petals, sometimes recurrent bloom, moderate fragrance; vigorous, compact growth; [Général Jacqueminot X Géant des Batailles]; Liabaud

'BARON DE WASSENAER', M, dp, 1854; flowers light crimson, cupped, blooms in clusters; vigorous growth; Verdier, V.

'BARON DE WOLSELEY', HP, 1882; Verdier, E.; (Cavriglia)

'BARON GIROD DE L'AIN', HP, rb, 1897; flowers bright red, petals edged white; [Eugène Furst sport]; Reverchon

'BARON HAUSSMANN', HP, mr, 1867; flowers light red, large, very dbl.; Lévêque; (Sangerhausen)

'BARON J.B. GONELLA', B, pb, 1859; Guillot Père

'BARON JACQUES RISTON', HT, op, 1936; bud long, pointed; flowers salmon-pink, well shaped, large, dbl., 45–50 petals, moderate fragrance; very vigorous growth; [Mme Butterfly X (Rev. David R. Williamson X Gorgeous)]; Ketten Bros.

Baron Meillandina® *see* 'MEITIFRAN'

'BARON PALM', HT, mr, 1913; flowers large, dbl., moderate fragrance; Lambert, P.; (Sangerhausen)

Baron Sunblaze *see* 'MEITIFRAN'

'BARON TAYLOR', HP, mr, 1880; flowers light red, large, dbl.; Dugat; (Sangerhausen)

'BARON T'KIND DE ROODENBECKE', HP, m, 1897; flowers shaded purple; Lévêque

'BARONESA DE OVILAR', HT, 1935; flowers carmine tinted yellow, very large, dbl., intense fragrance; foliage dark; strong stems; very vigorous growth; [Sensation X Souv. de Claudius Pernet]; Munné, B.

Baroness Henrietta Snoy *see* **'BARONNE HENRIETTE DE SNOY'**

'BARONESS KRAYENHOFF', HT, pb, 1931; flowers peach-pink, center orange, very large, dbl., moderate fragrance; foliage light green; long stems; vigorous, compact growth; [Mrs Henry Bowles X Lady Roundway]; Buisman, G. A. H.

'BARONESS ROTHSCHILD', HP, lp, 1868; (Baronne Adolphe de Rothschild); flowers very soft rose, tinted white, large, dbl., 40 petals, cupped, some recurrent bloom; vigorous, erect growth; [Souv. de la Reine d'Angleterre sport]; Pernet Père

Baronesse® *see* 'TANESSENOR'

'BARONESSE A. VAN HÖVELL TÔT WESTERFLIER', HT, m, 1933; flowers carmine-purple, base yellow, large, dbl., moderate fragrance; foliage bronze; very vigorous, bushy growth; Leenders, M.

'BARONESSE H. VON GEYR', HT, pb, 1928; flowers pale flesh and vermilion-red, dbl., moderate fragrance; [(Farbenkonigin X Juliet) X Sunburst]; Leenders, M.

'BARONESSE M. VAN TUYLL VAN SEROOSKERKEN', HT, pb, 1922; flowers rose and lilac-white, base apricot, semi-dbl.; [Jonkheer J.L. Mock X Mme Mélanie Soupert]; Leenders, M.

'BARONESSE MANON', F, mp, 1938; flowers clear pink, firm petals, 3 in., moderate fragrance; foliage glossy, holly-like; vigorous, bushy growth; [Else Poulsen X Dame Edith Helen]; Poulsen, S.; C-P, 1952

'BARONESSE S. H. W. VAN DEDEM', HT, yb, 1923; flowers yellow and coppery, open, large, semi-dbl., moderate fragrance; foliage dark, glossy; vigorous, bushy growth; Leenders, M.

'BARONESSE VAN ITTERSUM', HMult, rb, 1910; flowers light crimson, shaded deeper, open, semi-dbl., borne in clusters, free, non-recurrent bloom, slight fragrance; foliage dark, glossy; vigorous, climbing (15 ft.) growth; [Crimson Rambler X Mme Laurette Messimy]; Leenders, M.

'BARONIN ANNA VON LÜTTWITZ', HMult, lp, 1909; flowers cream-pink, medium, dbl.; Walter; (Sangerhausen)

'BARONIN VON ADELEBSEN', S, mr, 1938; flowers medium, single; Vogel, M.; (Sangerhausen)

Baronne Adolphe de Rothschild *see* **'BARONESS ROTHSCHILD'**

'BARONNE CHARLES D'HUART', HT, pb, 1910; flowers lilac-rose suffused white, reflexed, dbl.; [Pharisaer X Seedling]; Ketten Bros.

'BARONNE DE MAYNARD', N, w, 1865; flowers white, edges tinged pink, cupped; moderately vigorous growth; [Blanche Lafitte X Sappho (an unrecorded variety)]; Lacharme, F.

'BARONNE DE MEDEM', HP, 1876; Verdier, E.; (Cavriglia)

Baronne de Nervo, S, 1969; Delbard-Chabert; (Cavriglia)

Baronne de Noirmont, B, mp; Granger

'BARONNE DE PRAILLY', HP, 1871; Liabaud, J.; (Cavriglia)

Baronne de Rothschild *see* 'MEIGRISO'

Baronne de Rothschild, Climbing *see* 'MEIGRISOSAR'

'BARONNE DE SAINT DIDIER', HP, 1886; Leveque, P.; (Cavriglia)

Baronne de Schorlemmer, HT, mp, 1985; flowers large, dbl.; Lens, Louis; (Sangerhausen)

'BARONNE DE STAEL', HGal, ob, 1820; flowers salmon-pink, open, large, dbl.; vigorous, branching growth; Vibert

'BARONNE DE VIVARIO', Pol, w, 1925; flowers , dbl., in clusters; [Orléans Rose X Jeanny Soupert]; Soupert & Notting

Baronne Edmond de Rothschild® *see* 'MEIGRISO'

Baronne Edmond de Rothschild, Climbing® *see* 'MEIGRISOSAR'

'BARONNE FINAZ', HT, mp, 1961; flowers bright pink, large blooms, dbl., 35 petals, moderate fragrance; foliage glossy, dark; long stems; very vigorous, upright growth; [Peace seedling X Opera]; Gaujard

'BARONNE G. DE ST PAUL', HP, lp, 1894; flowers large, dbl.; Glatenet; (Sangerhausen)

'BARONNE HENRIETTE DE LOEW', T, w, 1888; flowers pinkish white, center yellow, dbl., moderate fragrance; Nabonnand, G.

'BARONNE HENRIETTE DE SNOY', T, pb, 1897; (Baroness Henrietta Snoy, Baroness Henriette Snoy); flowers flesh, reverse carmine-pink, well-formed, large, dbl.; vigorous growth; [Gloire de Dijon X Mme Lombard]; Bernaix, A.

'BARONNE MAURICE DE GRAVIERS', HP, 1866; Verdier, E.; (Cavriglia)

'BARONNE NATHALIE DE ROTHSCHILD', HP, lp, 1885; flowers very large, very dbl., moderate fragrance; Pernet Père; (Sangerhausen)

Baronne Presval, HP, dp

'BARONNE PRÉVOST', HP, mp, 1842; flowers rose-pink, shading lighter, large, dbl., flat, recurrent bloom, moderate fragrance; vigorous, erect growth; Desprez

Baronne Prévost, Climbing, Cl HP, mp

Baronne Surcouf *see* 'BRIANCOUF'

Baroque *see* HARbaroque

'BARORP', Min, m, 1985; (**Astrea®**); Barni, V.

Barossa Dream *see* 'SUNAUCK'

'BAROUT', HT, ab, 1982; (Handout, Tranquility, **Tranquillity**); flowers pale apricot, peach and yellow blend, blooms borne singly, dbl., 35 petals, slight fragrance; foliage medium, dark, glossy; upright growth; [Whisky Mac X Pink Favorite]; Barrett, F.H.; John Mattock, Ltd.

BARpipp, MinFl, pb, 1995; (**Pippy**); Barni, V.

BARpoll, S, mr, 1987; (**Polluce®**); Barni, V.

BARprett, LCl, dp, 1992; (**Pretty Pink**); Barni, V.

BARprimo, S, ly, 1995; (**Primo Passo**); Barni, V.

BARpris, HT, my, 1987; (**Primo Sole®**); Barni, V.

BARred, F, mr, 1990; (**Red Sea®**); Barni

BARrez, HT, op, 1996; (**Carezza**); Barni, V.

BARrian, MinFl, mr, 1995; (**Brian**); Barni, V.

Barricade, LCl, 1971; Combe, C.; (Cavriglia)

'BARRIE', F, pb, 1973; bud ovoid; flowers salmon-pink, medium, dbl., cupped, free bloom, slight fragrance; foliage glossy, light; [Sumatra X Fashion]; Schloen, J.

BARros, F, mp, 1990; (**Rosellana®**); Barni, V.

Barry Fearn *see* KORschwama

Barry Stephens *see* 'HORCABELLERO'

BARsan, F, w, 1995; (**Sans Souci**); Barni, V.

BARsav, F, op, 1999; (**Mafalda di Savoia**); Barni, V.

BARser, HT, op, 1998; (**Serenata**); Barni, V.

'BARSHIFLO', F, w, 1986; (**Snowgoose**); flowers small blooms in clusters, dbl., 20 petals, no fragrance; foliage medium, medium green, semi-glossy; upright growth; [Seaspray X Iceberg]; Barrett, F.H.

BARsib, HT, dr, 1987; (**Sibilla®**); Barni, V.

BARsor, HT, ob, 1990; (**Eros**); Barni, V.

BARstel, F, lp; (**Stelvio Coggiatti**); Barni, V., 1999

BARsus, F, pb; (**Suni®**); Golden Rose, Geneva, 1993; Barni, V., 1993

BARtares, S, pb, 1996; (**Antares**); Barni, V.

'BARTER'S PINK', HT, ob, 1963; flowers bright coral-pink, deeply veined, dbl., 4 in., slight fragrance; foliage light green; vigorous, bushy growth; [Queen Elizabeth X Claude]; Barter

'BARTHELEMY JOUBERT', HP, 1877; Moreau et Robert; (Cavriglia)

BARtusc, HT, mr; (**Tuscia®**); Barni, V., 1993

Barunka, F, op, 1983; flowers salmon-pink, large, very dbl., slight fragrance; Urban, J.; (Sangerhausen)

BARvan, S, mp, 1999; (**Avance**); Barni, V.

BARvelv, LCl, lp; (**Velvet Star**); Barni, V., 1998

BARwast, HT, ob; (**Wall Street®**); Barni, V., 1988

BARwrock, LCl, lp; (**White Rock**); Barni, V., 1996

'BASHFUL', Pol, pb, 1955; (Giesebrecht); flowers reddish pink, white eye, small, single, blooms in trusses; bushy, compact growth; deRuiter; Gregory & Willicher Baumschulen

'BASILDON BELLE', F, or, 1964; flowers vermilion; [Anna Wheatcroft sport]; Maarse, G.; Basildon Rose Gardens

Basildon Bond *see* 'HARJOSINE'

Basilika, HT, op; flowers salmon-pink, large, dbl., slight fragrance; VEG; (Sangerhausen)

Bassino® *see* 'KORMIXEL'

Bastei, S, or, 1973; Schmadlak, Dr.

'BASTOGNE', HT, or; flowers vermilion, shaded darker, well formed, moderate fragrance; foliage dark; very vigorous growth; Grandes Roseraies

Basye's Blueberry, S, mp, 1982; Basye, Robert

Basye's Myrhh-Scented, S; Basye, Robert

Basye's Purple Rose, S, m, 1968; Basye, Robert

'BATMERCURY', HT, my, 1994; (**Freddie Mercury**); dbl., 26–40 petals, 1.5–2.75 in., borne mostly singly, moderate fragrance; some prickles; foliage large, dark green, glossy; medium, bushy growth; [Tina Turner sport]; Stainthorpe, Eric; Battersby Roses, 1994

'BATSYBIL', HT, dr, 1993; (**Dr Sybil Johnson**); dbl., 26–40 petals, 1.5–2.75 in., borne mostly singly, moderate fragrance; some prickles; foliage medium, medium green, semi-glossy; medium (85 cms), upright growth; [Tropicana X Prima Ballerina]; Stainthorpe, Eric; Battersby Roses, 1994

Battersby Beauty *see* 'HOBATBEAUT'

'BATTLE OF BRITAIN', HT, yb, 1970; flowers yellow-orange, medium, dbl., 30 petals, moderate fragrance; [Miss Ireland X Summer Sunshine]; Gandy, Douglas L.; Wheatcroft & Sons

'BATTOO', HT, pb, 1992; (**Myra**); dbl., 15–25 petals, 1.5–2.75 in., borne in small clusters, slight fragrance; many prickles; foliage medium, dark green, glossy; medium (75 cms), bushy growth; [Matangi X Mood Music]; Stainthorpe, Avril E.; Battersby Roses, 1990

Baty's Pink Pillar, LCl, mp; (found rose)

Bavarian Girl *see* 'KORLEEN'

Baveria *see* 'KORMUN'

Baxter Beauty, HT, ab, 1938; light yellow to sulphur overlaid outside with light salmon pink; [Lorraine Lee sport]; Clark, A.

Bay Glow, Min, yb

'BAYADÈRE', HT, ab, 1954; flowers salmon-pink to canary yellow, tinted pink, large, dbl., 52 petals, exhibition form, slight fragrance; foliage dark, bronze; vigorous, bushy growth; GM, NRS, 1954; [R.M.S. Queen Mary X Seedling]; Mallerin, C.; EFR

Bayerngold® *see* 'TANYAB'

Bayreuth, S, rb, 1965; flowers medium, dbl., moderate fragrance; Kordes, W. Söhne; (Sangerhausen)

'BE GLAD', HT, rb, 1988; flowers white changing to red, reverse white, aging deep pink, medium, dbl., 25–30 petals, exhibition form, moderate, fruity fragrance; no fruit; prickles normal; foliage medium, dark green, glossy; upright, medium growth; Silver, ARC TG, 1988; [Paradise X Color Magic]; McMillan, Thomas G.

BEAcath, HT, my, 1997; **Norwich Cathedral**; flowers full, medium, very dbl., 26–40 petals, 4 in., borne mostly singly, intense fragrance; some prickles; foliage medium, medium green, leathery, semi-glossy; upright, low (3ft.)growth; [Diamond Jubilee sport]; Beales, Peter

'BEACH BOY', Min, my, 1981; bud pointed; flowers golden buff, blooms borne usually singly, dbl., 48 petals, exhibition form, slight fragrance; short, tan prickles; foliage small, dark, slightly glossy; upright, bushy growth; [Tom Brown X Seedling]; Williams, Ernest D.; Mini-Roses

Beachcomber *see* 'MACLAPAZ'

'BEACON', HT, 1952; Swim, H.C.; (Cavriglia)

'BEACON BELLE', Ayr, lp, 1919; flowers flesh passing to white, borne in clusters, dbl., moderate fragrance; [Orléans Rose X (Katharina Zeimet X R. arvensis hybrid)]; Farquhar

'BEACON LODGE', HT, dr, 1976; flowers deep red, full, dbl., 40–45 petals, 4–5 in., moderate fragrance; foliage glossy, dark; very vigorous growth; [Chopin X Ena Harkness]; Ellick

BEAdix, Cl HT, ab, 1997; (**Dixieland Linda**); flowers very full, large, very dbl., 41 petals, 4.5 in., borne in small clusters, intense fragrance; foliage medium, dark green, glossy; upright, medium (8-10ft.)growth; [Aloha sport]; Beales, Peter

'BEAHOR', S, mp, 1998; (**Horatio Nelson**); flowers clear pink, very full, opens flat, very dbl., 41–50 petals, rosette, borne in small clusters; prickles moderate; foliage medium, medium green, glossy; compact (3 ft) growth; [Centenaire de Lourdes X Aloha]; Beales, Peter; Peter Beales Roses, 1997

Beales' Mons Tillier, T, mr

Beales' The Bride, T, w

'BEAMAC', S, w, 1998; (**Macmillan Nurse**); flowers white, very full, very dbl., 41–50 petals, borne in large clusters, moderate fragrance; prickles moderate; foliage medium, dark green, glossy; compact, 3 ft growth; [Bonica X Maigold]; Beales, Peter; Peter Beales Roses, 1998

'BEAPAW', S, mp, 1999; (**Paws**); very dbl., 41 petals, 4 in., borne mostly singly, intense fragrance; many prickles; foliage medium, medium green, dull; bushy, medium (2.5 x 2 ft) growth; [Silver Jubilee X Constance Spry]; Beales, Peter; Peter Beales Roses, 1999

'BEATRICE', F, dp, 1968; flowers deep rose pink, in large sprays, dbl., slight fragrance; [Paddy McGredy X (Kordes' Perfecta X Montezuma)]; McGredy, Sam IV

Beatrice *see* BARbea

Beatrice Berkeley, HT, lp; Fitzhardinge, before 1932; (Weatherly, L.)

'BEATRICE BOEKE', HT, dp, 1966; bud ovoid; flowers pink-red, large, dbl.; [Montezuma X Detroiter]; Buisman, G. A. H.

'BEATRICE MCGREGOR', HT, dr, 1938; flowers large, dbl., intense fragrance; [Sensation X Seedling]; Clark, A.; NRS Victoria

Beatrice Samzin, F, mp, 1970; flowers large, very dbl., slight fragrance; Scholle, E.; (Sangerhausen)

'BEATRIX', HP, 1855; Cherpin; (Cavriglia)

'BEATWE', F, dr, 1997; (**Twenty-Fifth**); flowers semi double, medium, semi-dbl., 8–14 petals, 2.5 in., borne in large clusters, slight fragrance; some prickles; foliage medium, dark green, semi-glossy; compact, low (2ft.)growth; [Redbreast X Pearl Drift]; Beales, Peter

'BEAU NARCISSE', HGal, m; flowers purple, striped; Miellez, before 1828

Beau Rose, Misc OGR, mp

Beau Rose Primaplant, Pol, dp, 1963; flowers large, dbl.; Vlaeminck; (Sangerhausen)

'BEAUCAIRE', Pol, rb, 1937; flowers edges chamois-pink, center coppery yellow, reverse coppery-red, dbl.; vigorous, dwarf growth; Grandes Roseraies

'BEAUJOLAIS', HT, mr, 1932; flowers crimson-carmine, very large, dbl., globular, intense fragrance; very vigorous growth; [Hadley X Laurent Carle]; Croibier

Beaulieu *see* 'DICOBEY'

'BEAULIEU', HP, 1883; Moreau et Robert; (Cavriglia)

Beaulieu *see* 'TANZAHDE'

'BEAULIEU ABBEY', F, ab, 1964; flowers creamy yellow suffused pink, well formed, dbl., 4 in., moderate fragrance; foliage dark, glossy, leathery; vigorous growth; [Masquerade X Docteur Valois]; Cobley; Blaby Rose Gardens

Beauregard, LCl, 1972; Croix, P.; (Cavriglia)

'BEAUTÉ', HT, ab, 1953; bud long; flowers light apricot, well-formed, large, dbl., moderate fragrance; vigorous growth; [Mme Joseph Perraud X Seedling]; Mallerin, C.; EFR

'BEAUTÉ D'AUTOMNE', Pol, dp, 1918; flowers bright rose-pink, borne in clusters of 50-70, dbl.; [Phyllis X Seedling]; Turbat

'BEAUTÉ DE FRANCE', HT, w, 1920; flowers creamy white to pure white, inside yellow, dbl., moderate fragrance; [Mme Mélanie Soupert X Kaiserin Auguste Viktoria]; Toussaint Mille Fils

'BEAUTÉ DE FRANCE', HT, 1952; bud long, pointed; flowers brick flushed coppery, medium, dbl., moderate fragrance; foliage leathery, dark; upright growth; [(Comtesse Vandal X Seedling) X Seedling]; Gaujard

'BEAUTÉ DE L'EUROPE', T, op, 1881; flowers light orange-pink, yellow base, large, dbl., intense fragrance; Gonod; (Sangerhausen)

'BEAUTÉ DE LYON', HP, or, 1910; flowers coral-red, tinted yellow, dbl., moderate fragrance; GM, Bagatelle, 1911; [Seedling X Soliel d'Or]; Pernet-Ducher

'BEAUTE FRANCAISE', HP, mr, 1862; Lartay

'BEAUTÉ INCONSTANTE', T, ob, 1892; flowers orange to coppery red, shaded carmine and yellow, moderate fragrance; vigorous growth; Pernet-Ducher

'BEAUTÉ LYONNAISE', HT, w, 1895; flowers white tinted pale yellow, large; Pernet-Ducher

'BEAUTE ORLÉANAISE', HWich, w, 1919; flowers white to flesh-pink, borne in clusters of 20-25, dbl., slight fragrance; Turbat

Beauté Spatiale® *see* 'DELDROP'

Beauté Virginale, A, w

Beautiful Black, HP, dr

Beautiful Britain *see* 'DICFIRE'

'BEAUTIFUL DOLL', Min, mp, 1982; dbl., 25 petals, exhibition form, borne singly and in small clusters, moderate fragrance; foliage small, dark, semi-glossy; bushy, spreading growth; [Seedling X Zinger]; Jolly, Betty J.; Rosehill Farm

'BEAUTIFUL DREAMER', F, ob, 1977; flowers orange to sunset-gold, full, pointed, dbl., 35 petals, 2–2.5 in., slight fragrance; foliage glossy; [Seedling X Seedling]; Herholdt, J.A.

'BEAUTIFUL SUNDAY', Min, mr, 1977; bud ovoid; flowers bright geranium-red, small, very dbl., 60 petals, cupped, moderate fragrance; foliage glossy, dark, leathery; compact, bushy growth; [Camelot X Camelot seedling]; Takatori, Yoshiho; Japan Rose Nursery

'BEAUTY', HT, mr, 1931; flowers american beauty red, but darker, very large, dbl., moderate fragrance; long stems; vigorous growth; [(Crusader X Premier) X American Beauty]; Ward, F.B.

'BEAUTY CREAM', F, w, 1956; flowers cream, large, moderate fragrance; foliage dark; vigorous, upright growth; Verschuren; Gandy Roses, Ltd.

Beauty of Badgen, HT, lp; French, A., 1924; (Weatherly, L.)

'BEAUTY OF BEESTON', HP, 1882; Frettingham; (Cavriglia)

Beauty of Brisbane, Pol, lp; [Goldlachs sport]; Perrot, 1932; (Weatherly, L.)

'BEAUTY OF DROPMORE', S, w, 1956; dbl., non-recurrent; bushy, erect growth; [R. spinosissima altaica X R. spinosissima cultivar]; Skinner

Beauty of England, HP, dp

Beauty of Festival, HT, 1955; Klimenko, V. N.; (Cavriglia)

Beauty of Glazenwood *see* **'FORTUNE'S DOUBLE YELLOW'**

Beauty of Glenhurst, HCh, dp, 1983; Morley, Dr B.

'BEAUTY OF HURST', LCl, w, 1926; flowers creamy buff, dbl., moderate fragrance; foliage dark; very vigorous growth; Hicks

Beauty of Leafland, HSpn, pb; dbl., intense; [Butterball X Haidie]; Erskine

'BEAUTY OF NEW SOUTH WALES', Pol, rb, 1931; flowers bright crimson, center white, small blooms in clusters of 20, single, moderate fragrance; bushy, dwarf growth; [Orléans Rose X Alice Amos]; Knight, G.

'BEAUTY OF ROSEMAWR', T, pb, 1903; flowers carmine rose veined vermilion and white, medium, dbl.; Conard & Jones

'BEAUTY OF STAPLEFORD', HT, rb, 1879; flowers red and violet, well formed, large; moderate growth; [Mme Bravy X Comtesse d'Oxford]; Bennett

Beauty of the Prairies *see* **'QUEEN OF THE PRAIRIES'**

'BEAUTY OF WALTHAM', HP, mr, 1862; flowers rosy crimson, large, dbl., recurrent bloom, moderate fragrance; vigorous growth; Paul, W.

Beauty Queen *see* 'CANMISS'

'BEAUTY SECRET', Min, mr, 1965; bud pointed; flowers cardinal-red, small, dbl., 1.5 in., exhibition form, intense fragrance; foliage small, glossy, leathery; vigorous, bushy growth; AOE, ARS, 1975 Miniature Rose Hall of Fame, ARS, 1999; [Little Darling X Magic Wand]; Moore, Ralph S.; Sequoia Nursery

Beauty Star *see* 'FRYSTAR'

Beautyglo™ *see* 'MINWCO'

'BEAUTY'S BLUSH', HRg, dp, 1955; flowers deep pink, becoming lighter, dbl., non-recurrent, moderate fragrance; vigorous (6 ft) growth; hardy on the Canadian prairies.; [Tetonkaha X Pink Pearl]; Univ. of Saskatchewan

'BEAYAR', HT, ly, 1999; (**Yardley Baroque**); dbl., 26–40 petals, 4.5 in., borne mostly singly, moderate fragrance; many prickles; foliage medium, dark green, semi-glossy; upright, medium (2 ft) growth; [Alpine Sunset sport]; Beales, Peter; Peter Beales Roses, 1997

'BÉBÉ BLANC', Pol, w, 1922; dbl., borne in large clusters; dwarf growth; Turbat

'BÉBÉ FLEURI', HCh, mp, 1906; flowers small, semi-dbl.; Dubreuil; (Sangerhausen)

'BÉBÉ LEROUX', Pol, w, 1901; flowers medium, in trusses of 20-40, moderate fragrance; compact growth; [Mignonette X Archiduchesse Elisabeth-Marie]; Soupert

Bebe Lune® *see* 'DELFIB'

Becca Godman, F, pb, 1998; Wells

Becker Street, HT, op

Beckett's Single, HGal, mp; Quest-Ritson, before 1995

'BECKY', HT, dp, 1925; flowers glowing rose-pink, single; vigorous growth; Beckwith

Bedazzled™ *see* 'JACHOTTA'

BEDchild, HT, dy, 1996; (**Pudsey Bear**); Chessum, Paul

'BEDFORD CRIMSON', HT, dr, 1926; flowers velvety crimson, well formed, dbl., 40 petals, moderate fragrance; [Richmond X Château de Clos Vougeot]; Laxton Bros.

'BEDFORDIA', HT, pb, 1931; flowers pink, outer petals lighter, center salmon, dbl., 52 petals, moderate fragrance; vigorous growth; Laxton Bros.

'BEDRICH SMETANA', HT, w, 1933; flowers pearly white, open, very large, semi-dbl.; vigorous, bushy growth; [Modesty X Ophelia]; Böhm, J.

BEDswap, MinFl, lp; (**Rest in Peace**); Chessum, Paul

'BEEBOP', HT, lp, 1981; (**Cleo**); flowers soft light pink, blooms borne 1-2 per cluster, dbl., 37 petals, exhibition form, slight fragrance; red prickles; foliage light green, semi-matt; strong, bushy growth; [Kordes' Perfecta X Prima Ballerina]; Bees

'BEEJES', HT, yb, 1982; (**Jessie Mathews**); flowers light yellow, petals edged pink, medium, dbl., 35 petals, slight fragrance; foliage medium, light green, semi-glossy; bushy growth; [Ernest H. Morse X Rosenella]; Bees

'BEELAH', HT, yb, 1983; (**Tallulah**); flowers gold with deep pink petal edges, large, dbl., 20 petals, slight fragrance; foliage medium, dark, semi-glossy; bushy growth; [Astral X Piccadilly]; Bees

'BEERIL', HT, w, 1983; (**Cyril Fletcher**); flowers creamy white, cabbage rose form, large, dbl., 35 petals, intense fragrance; foliage medium, dark, semi-glossy; upright, bushy growth; [Fragrant Cloud X Whisky Mac]; Bees

'BEE'S FROLIC', S, lp, 1976; bud small, pointed; semi-dbl., 14 petals, borne 2-3 per cluster, repeat bloom, intense fragrance; small prickles; foliage medium, matt; upright, strong growth; [Schneezwerg X Clair Matin]; Dawnay, Mrs. E.

Bees Knees™ *see* 'JACKEE'

'BEESIAN', F, my, 1979; (**Amanda**); bud globular; dbl., 25 petals, 3.5 in., exhibition form, slight fragrance; foliage small, light green; upright growth; [Arthur Bell X Zambra]; Bees

'BEEVAL', F, pb, 1983; (**Vale of Clwyd**); flowers yellow, pink petal edges, medium, dbl., no fragrance; foliage large, medium green, glossy; upright growth; [Handel X Arthur Bell]; Bees

Begonia, HT, op

Behold™ *see* 'SAVAHOLD'

Bekola *see* **'AALSMEER GOLD'**®

Bel Ami® *see* 'LAPCI'

Bel Ami *see* **'PHILIPPE'**

'BEL ANGE', HT, mp, 1962; (Bella Epoca, Belle Ange, Belle Epoque); flowers soft pink, reverse darker, dbl., 35 petals, 4.5 in., moderate fragrance; foliage dark; vigorous growth; GM, Kortrijk, 1965; [(Independence X Papillon Rose) X (Charlotte Armstrong X Floradora)]; Lens; J&P, 1965

Bel Ange, Climbing, Cl HT, lp; [Bel Ange sport]; Ruston, D., 1970; (Weatherly, L.)

'BEL CANTO', HT, or, 1964; flowers bright geranium-red, well formed; foliage coppery; upright growth; Mondial Roses

Bel Esprit, HMsk, pb, 1992; Lens

Bel-Air, HT, dr, 1986; Swane

Belami *see* 'KORPRILL'

'BELCHARM', Min, ob, 1992; (**Bell Charmer**); flowers distinctive soft orange with light yellow petal base and reverse, dbl., 26–40 petals, 1.5–2.75 in., slight fragrance; some prickles; foliage medium, medium green, semi-glossy; medium (45-55 cms), upright, bushy growth; [(Cherish X Avandel) X Seedling]; Bell, Charles E., Jr.; Kimbrew Walter Roses, 1993

Belfast Belle *see* 'DICROBOT'

Belfield, HCh, mr

Belgian Lace®, S, lp, 1994; Lens, Louis; (Sangerhausen)

'BELGICA', HT, dr, 1929; flowers crimson-red, shaded garnet, large, very dbl.; vigorous growth; Buyl Frères

'BELINDA', HMsk, mp, 1936; semi-dbl., borne in very large, erect trusses, moderate fragrance; vigorous (4-6 ft) growth; a good hedge or pillar rose; Bentall

Belinda *see* 'TANBEEDEE'

'BELINDA'S DREAM', S, mp, 1992; (Belinda's Rose); very dbl., 3.25–4.25 in., borne in small clusters, repeat bloom, moderate, fruity, raspberry fragrance; foliage medium, medium green, matt; medium (5x4'), bushy growth; [Jersey Beauty X Tiffany]; Basye, Robert, 1988

Belinda's Rose *see* **'BELINDA'S DREAM'**

Bell Charmer *see* 'BELCHARM'

'BELL RINGER', Min, dp, 1989; (Bellringer); flowers deep pink, silver at base, small, dbl., 34 petals, slight, fruity fragrance; foliage medium, medium green, semi-glossy; upright, bushy growth; [(Fragrant Cloud X Avandel) X Bonny]; Bell, Charles E., Jr.; Kimbrew Walter Roses, 1989

Bella *see* DEVsmooth

Bella *see* POUlijill

'BELLA DONNA', D, lp; flowers soft lilac-pink, large, dbl.

'BELLA DONNA', F, mr, 1964; bud ovoid; flowers bright red medium, in clusters, dbl.; foliage dark; [Baccará seedling X Miracle]; Verbeek

Bella Epoca *see* **'BEL ANGE'**

Bella Multiflora, HMult, mp, 1994; flowers large, borne in sprays; few prickles; arching growth; Uhl, J., 1994

Bella Nitida, S, mp, 1994; flowers large; medium growth; very hardy; Uhl, J.

Bella Renaissance *see* POUljill

Bella Rosa® *see* 'KORWONDIS'

Bella Via *see* 'ZIPVIA'

Bella Weiss® *see* 'KORTUEL'

'BELLARD', HGal; Roseraie de l'Hay, Prior to 1857

Belle *see* 'MARBEL'

'BELLE ADÉLAIDE', HGal, dp; flowers cerise-red, very dbl., flat; Miellez

'BELLE AMOUR', A, lp; flowers soft pink with salmon tones, prominent yellow stamens, semi-dbl., moderate, myrrh fragrance; height 5-6ft; Found at a convent at Elboeuf in the 1940's

Belle Ange *see* **'BEL ANGE'**

Belle Anglaise, HT, dy, 1995

Belle au Bois Dormant, S, pb, 1960; Kordes

'BELLE AURORE', A, lp; flowers flesh shaded purple, large, very dbl.; Descemet, about 1815

'BELLE BIBLIS', HGal, m, 1815; flowers medium, dbl., moderate fragrance; Descemet, M.; (Sangerhausen)

'BELLE BLANCA', LCl, w; [Probably a Belle Portugaise white sport]

Belle Blonde *see* 'MENAP'

Belle Champenoise, HT, my, 1997; Orard

'BELLE CLEMENTINE', A, lp; flowers mottled flesh-color

Belle Courtisanne *see* **'KÖNIGIN VON DÄNEMARK'**

'BELLE CRÉOLE', F, rb, 1958; flowers variegated red and brick-red, borne in clusters; foliage dark; vigorous, upright growth; [(Gruss an Teplitz X Independence) X (Floradora X Independence)]; Arles; Roses-France

'BELLE CUIVRÉE', HT, or, 1924; flowers coral-red, shaded coppery yellow, semi-dbl.; Pernet-Ducher

Belle d'Automne, HT, 1969; Ducher, Ch.; (Cavriglia)

Belle de Baltimore *see* **'BALTIMORE BELLE'**

'BELLE DE BORDEAUX', T, pb, 1861; (Gloire de Bordeau, Gloire de Bordeaux); flowers pink with crimson center, large, dbl.; vigorous growth; Lartay

'BELLE DE CRÉCY', HGal, m; flowers cerise and purple, becoming lavender-gray, center green, dbl., flat, moderate fragrance; lax growth; prior to 1829

Belle de Dom *see* MASdomo

Belle de Juin, F, 1974; Croix, P.; (Cavriglia)

Belle de Londres *see* **'COMPASSION'**®

'BELLE DE MARLY', HGal, dp; flowers bright rose, shaded violet, large, dbl.

'BELLE DE PARME', F, mp, 1962; flowers lilac-mauve; low, spreading growth; [(Lafayette X (Gruss an Teplitz X Independence)) X R. rugosa rubra]; Arles; Roses-France

'BELLE DE PROVINS', HT, dr, 1954; flowers velvety dark red, well formed, large, dbl., intense fragrance; vigorous growth; [Crimson Glory X E.G. Hill]; Robichon

Belle de Remalard, HMult, pb, 1998; d'Andlau

'BELLE DE SÉGUR', A, lp; (Joséphine Beauharnais, Joséphine de Beauharnais); flowers soft rosy flesh, edges blush, dbl., cupped; foliage dark; vigorous, upright growth; Vibert, prior to 1848

'BELLE DE YÈBLES', HGal, mr; flowers bright red; Desprez, before 1835

'BELLE DES JARDINS', HGal, m, 1872; flowers purplish violet-red, variegated carmine, striped white, dbl.; vigorous growth; [Village Maid X Seedling]; Guillot et Fils

Belle des Jardins *see* **'VILLAGE MAID'**

'BELLE DORIA', HGal, pb; flowers pink with red flecks and stripes, cupped; Parmentier, before 1847

'BELLE D'ORLÉANS', LCl, or, 1958; flowers large blooms in clusters, dbl., recurrent bloom; foliage glossy; vigorous growth; [Seedling X Independence]; Robichon

Belle Epoque *see* **'BEL ANGE'**

'BELLE ÉPOQUE', HT, pb, 1962; (Royale); flowers fuchsia-pink, reverse creamy, dbl., 40 petals, 6 in., moderate fragrance; foliage glossy; vigorous, upright growth; [Peace X Independence]; Kriloff, Michel; Cramphorn's Nursery

'BELLE ÉTOILE', Gr, my, 1961; bud long, pointed; flowers golden yellow, well formed, blooms in clusters of 5-7, dbl., 25 petals; long stems; vigorous, upright growth; [Joanna Hill X Tawny Gold]; Lens

'BELLE HÉLÈNE', HGal, m; Vibert, before 1815

'BELLE HERMINIE', HGal, m; Coquerel (?), before 1838

'BELLE ISIS', HGal, lp, 1845; flowers pale flesh-pink, dbl.; Parmentier

'BELLE IVRYENNE', HP, dp, 1891; flowers large, dbl.; Lévêque; (Sangerhausen)

'BELLE LILETTE', HT, mr, 1926; flowers carmine-red, very large, dbl., moderate fragrance; Gemen & Bourg; (Sangerhausen)

Belle Loire, S; Croix, P.; (Cavriglia)

'BELLE LYONNAISE', Cl T, ly, 1870; flowers canary yellow, fading white, large, dbl., moderate fragrance; vigorous, climbing growth; [Gloire de Dijon seedling]; Levet, F.

Belle Meillandina® *see* 'MEIDANEGO'

'BELLE NANON', B, dp, 1872; Lartay

'BELLE NORMANDE', HP, w; flowers silvery rose; [La Reine sport]; Oger, 1864

Belle of Berlin *see* TANireb

Belle of Portugal *see* **'BELLE PORTUGAISE'**

'BELLE OF PUNJAB', F, mp, 1965; bud ovoid; flowers pink, dbl., exhibition form, slight fragrance; foliage dark, glossy; very vigorous, compact growth; [Montezuma X Flamenco]; Pal, Dr. B.P.; Indian Agric. Research Inst.

'BELLE OF TASMANIA', F, or, 1968; bud pointed; flowers velvety scarlet, center darker, medium, dbl., exhibition form; foliage glossy; vigorous, tall, compact growth; [Korona X Étoile de Hollande]; Holloway

'BELLE POITEVINE', HRg, mp, 1894; bud long, pointed; flowers rose-pink to magenta-pink, large, semi-dbl., recurrent bloom; foliage dark, rugose; vigorous (3 1/2-4 ft.), bushy growth; (14); Bruant

'BELLE PORTUGAISE', LCl, lp, 1903; (Belle of Portugal); bud very long, pointed (to 4 in.); flowers light flesh-pink, semi-dbl., 4–6 in., long spring bloom; foliage glossy; very vigorous (20 ft.) growth; not hardy north; (21); [R. gigantea X Reine Marie Henriette]; Cayeux, H.

'BELLE ROSINE', HGal, dp; flowers deep pink, edged lighter, open, large, dbl.; erect growth; Vibert, 1829 or 1830

'BELLE ROUGE'™, HT, dr, 1956; flowers dark velvety red becoming carmine-purple, large; [Happiness seedling X Impeccable]; Delbard-Chabert

Belle Rubine *see* **'VILLAGE MAID'**

'BELLE SANS FLATTERIE', HGal, m, 1820; Hardy, before 1806

Belle Siebrecht *see* **'MRS W.J. GRANT'**

Belle Siebrecht, Climbing *see* **'MRS W.J. GRANT, CLIMBING'**

Belle Story® *see* 'AUSELLE'

'BELLE SUISSE', HT, mr, 1936; flowers large, dbl.; Heizmann, E.; (Sangerhausen)

'BELLE SULTANE', m, 1800; flowers purple-red, medium, dbl.; (Sangerhausen)

Belle Sunblaze *see* 'MEIDANEGO'

Belle Symphonie *see* MEIrivoui

'BELLE VICHYSOISE', N, lp, 1897; flowers pink or pinkish white, small blooms in clusters of 20-50; very vigorous growth; Lévêque; Lévêque

Belle Villageoise *see* **'VILLAGE MAID'**

Belle Yvrienne, HP, rb

Belles and Beaus *see* 'ADABEL'

Bellevue® *see* 'POULENA'

'BELLINA', F, op, 1959; flowers shrimp-pink, blooms in large clusters, dbl., 40 petals, 2–2.5 in., exhibition form, moderate fragrance; foliage glossy; low, compact growth; [Pinocchio X (Fashion X Orange Triumph)]; Von Abrams; Peterson & Dering

Bellisima *see* 'SUNLAMPO'

Bellissima® *see* 'ZIPBELL'

Bellissima® *see* 'LAPBAU'

Bellona *see* 'KORILONA'

Bellringer *see* **'BELL RINGER'**

Belmont, HCh, lp

'BELOVED', HT, w, 1978; bud tinted pink; flowers reflexing, dbl., 33–35 petals, 6–7 in., moderate fragrance; foliage glossy; [Memoriam sport]; Hill, E.H.; Shropshire Roses

'BELVÉDÈRE', F, dr, 1928; flowers velvety dark red, large, dbl.; vigorous, dwarf growth; [Eblouissant X Château de Clos Vougeot]; Kiese

'BELVÉDÈRE', HT, dr, 1955; [Reine Elisabeth X Christopher Stone]; Delforge

Belvedere Park *see* 'LECBELPA'

'BEN ARTHUR DAVIS', HT, yb, 1935; flowers yellow, reverse pinkish gold, dbl., cupped; foliage leathery, glossy, dark; bushy growth; [Edith Nellie Perkins sport]; Bostick

'BEN CANT', HP, dr; flowers crimson, center darker, very large, dbl., 25 petals, exhibition form, intense fragrance; vigorous growth; [Suzanne-Marie Rodocanachi X Victor Hugo]; Cant, B. R., 1901

'BEN STAD', LCl, pb, 1925; (Rev. Floris Ferwerda); flowers pink, center yellow, edged white, reverse flesh-pink; [Silver Moon X Mme Jules Grolez]; Undritz; B&A

'BENALAV', Min, m, 1987; (**Lavender Jade**™); flowers lavender-white bicolor-mauve blend, large, borne usually singly, dbl., 32–35 petals, exhibition form, intense, damask fragrance; fruit not observed; short, straight prickles; foliage medium, dark green, semi-glossy; upright, tall growth for a miniature.; [Rise 'n' Shine X Laguna]; Benardella, Frank A.; Nor'East Min. Roses, 1987

'BENBLACK', Min, dr, 1985; (**Black Jade**™); bud near black; flowers deep red, medium, dbl., 35 petals, 1–1.5 in., exhibition form, borne singly, no fragrance; foliage medium, dark, semi-glossy; upright growth; PP005925; AOE, 1985; [Sheri Anne X Laguna]; Benardella, Frank A.; Nor'East Min. Roses

'BENBURGUN', Min, dr, 1990; (**Ans**); Benardella, Frank A.

BENcamelia, HT, pb, 1993; (**Roedean**); Benardella, Frank A.

BENcreberg, HT, w, 1996; (**Athene**); Benardella, Frank A.

'BENDAY', Min, mr, 1988; (**Old Glory**™); flowers bright post office red, aging blood-red to crimson, large, dbl., 23–25 petals, 1.5–2 in., exhibition form, borne singly and in small clusters, no fragrance; prickles long, thin, curved downward, gray-red; foliage medium, medium green, semi-glossy; mini-flora upright, tall, vigorous growth; PP005658; AOE, 1988; [Rise 'n' Shine X Harmonie]; Benardella, Frank A.; Nor'East Min. Roses, 1988

'BENDIGOLD', F, or, 1979; bud globular; dbl., 3 in., moderate fragrance; foliage glossy, bronze; vigorous, upright growth; [Rumba X Redgold]; Murley, J.J.; Brundrett

'BÉNÉDICTE SEGUIN', HT, ob, 1918; flowers ochre, shaded coppery orange, dbl.; Pernet-Ducher

'BENEDICTUS XV', HT, w, 1917; flowers rosy white, dbl., moderate fragrance; [Jonkheer J.L. Mock X Marquise de Sinéty]; Leenders, M.

'BENEDIKT ROEZL', HRg, lp, 1925; flowers light carmine-rose, large, very dbl., moderate fragrance; foliage rugosa-like; vigorous, very bushy growth; [R. rugosa seedling X La France]; Berger, V.; Faist

'BENELUX', F, dr, 1949; flowers crimson-red, semi-dbl., moderate fragrance; [Donald Prior X Rosamunde]; Leenders, M.

Benelux Star, HT, 1997; 37 petals, 4 in., cupped, moderate fragrance; foliage matt; strong grower growth; RvS-Melle; (Cavriglia)

Benevolence *see* 'SANOLENCE'

'BENFIG', Min, w, 1991; (**Figurine**™); bud delicate coloration, well-formed; flowers ivory white tinged pink, moderately full, semi-dbl., slight fragrance; few prickles; foliage medium, dark green, matt; very long stems suitable for cutting; medium (40-50 cms), upright, bushy growth; AOE, 1992; [Rise 'n' Shine X Laguna]; Benardella, Frank A.; Weeks Roses, 1992

'BENGAL CENTIFOLIA', HCh, 1804; Noisette

Bengal Cramoisi Double *see* **'SANGUINEA'**

Bengal Ordinaire *see* **'PALLIDA'**

Bengal Rose *see* **R. CHINENSIS**

'BENGALE ANIMÉE', HCh, 1832; Laffay, M.; (Cavriglia)

Bengale Cerise, HCh, mr

'BENGALE D'AUTOMNE', HCh, dp, 1825; Laffay, M.

'BENGALE ROUGE', HCh, mr, 1955; bud ovoid; flowers bright carmine-red, open, very large, recurrent bloom; foliage abundant; very vigorous growth; [Gruss an Teplitz X Seedling]; Gaujard

Bengali® *see* 'KORAL'

'BENGEE', Min, ly, 1987; (**Gee Gee**™); flowers medium yellow, fading lighter, loose, small, dbl., 20–25 petals, cupped, borne usually singly, slight, fruity fragrance; no fruit; pointed, beige prickles; foliage medium, light green, matt, edges toothed; upright, bushy, medium growth; PP006783; [Rise 'n' Shine X Patricia]; Benardella, Frank A., 1981; Kimbrew Walter Roses, 1987

'BENGT M. SCHALIN', S, dp, 1956; flowers rose-red, blooms in clusters (up to 10), semi-dbl., non-recurrent, slight fragrance; foliage light green, leathery, glossy; very vigorous growth; [R. kordesii X Eos]; Kordes

'BENHILE', Min, rb, 1999; (**Hilde**); flowers white with red washing, reverse ivory, dbl., 17–25 petals, 1.5–2 in., borne mostly singly, moderate fragrance; few prickles; foliage medium, dark green, glossy; bushy, medium (12-18 in) growth; [Figurine X Kristin]; Benardella, Frank A.

'BEN-HUR', Gr, mr, 1960; bud long, pointed; flowers crimson, dbl., 23 petals, 4–5 in., exhibition form, moderate fragrance; foliage leathery, glossy; vigorous growth; [Charlotte Armstrong X (Charlotte Armstrong X Floradora)]; Lammerts, Dr. Walter; Germain's

Benihime, Min, mr

BENimbro, Min, yb, 1990; (**Imbroglio**); Benardella, Frank A.

Benita *see* 'DICQUARREL'

'BENJAMIN DROUET', HP, 1878; Verdier, E.; (Cavriglia)

'BENJAMIN FRANKLIN', HT, lp, 1969; bud ovoid; flowers dawn-pink, large, dbl., 55 petals, exhibition form, slight fragrance; foliage dark, leathery; upright growth; Von Abrams

'BENJEN', Min, pb, 1985; (**Jennifer**™); flowers light pink, white reverse, small, dbl., 35 petals, exhibition form, intense fragrance; foliage medium, dark, semi-glossy; bushy, spreading growth; PP005857; AOE, 1985; [Party Girl X Laguna]; Benardella, Frank A.; Nor'East Min. Roses

'BENJIM', Min, rb, 1988; (**Jim Dandy**™); bud pointed; flowers medium red, reverse yellow flushed red, aging lighter, medium, exhibition form, slight, spicy fragrance; no fruit; no prickles; foliage medium, medium green, semi-glossy; upright, bushy, medium growth; PP007166; AOE, 1989; [Rise 'n' Shine X Marina]; Benardella, Frank A.; Nor'East Min. Roses, 1989

Benkey, HT, 1969; Itami, B.; (Cavriglia)

BENlavscent, Min, m, 1997; (**Moon River**); Benardella, Frank A.

'BENMAGIC', Min, rb, 1992; (**Kristin**™); flowers white/red bicolor, do not open beyond 1/2 open stage, dbl., 27–30 petals, 1.5 in., exhibition form, borne singly, no fragrance; foliage large, dark green, semi-glossy; upright, bushy, medium growth; PP8603; AOE, 1993; [DICmickey X Tinseltown]; Benardella, Frank A.; Nor'East Min. Roses, 1993

BENmfig, Min, w, 1996; (**Jilly Jewel**); Benardella, Frank A.

'BENNETT'S SEEDLING', Ayr, w, 1840; (Thoresbyana); flowers pure white, expanded form, medium, dbl., moderate fragrance; Lord Manners at Thoresby, England

'BENOIST PERNIN', HP, 1889; Kiard; (Cavriglia)

'BENOIT FRIART', HT, m, 1978; flowers rosy lilac, medium blooms borne 1-7 per cluster, dbl., 47 petals, flat, intense fragrance; foliage matt; upright growth; [Fragrant Cloud X Astée]; Rijksstation Voor Sierplantenteelt

Benoni '75 *see* 'POULONI'

BENorchid, Min, m, 1995; (**Orchid Lace**); Benardella, Frank A.

BENraar, HT, rb; (**Zebra**); Benardella, Frank A., 1995

'BENRAD', Min, or, 1987; (**Radiant**™); flowers brilliant orange-red, urn-shaped, medium, dbl., 23–27 petals, exhibition form, borne singly, moderate, spicy fragrance; no fruit; prickles long, straight, pointed slightly downward, gray-red; foliage dark green, semi-glossy; mini-flora long, straight stems, upright, tall growth; PP006569; [Sheri Anne X Sheri Anne]; Benardella, Frank A.; Nor'East Min. Roses, 1988

'BENROS', Min, pb, 1987; (**Rosie**™); flowers cream with pink edges, excellent form, medium, borne in sprays, dbl., 30–33 petals, exhibition form, slight fragrance; fruit not observed; long, thin, straight prickles; foliage medium, medium green, exceptional substance; upright, bushy, medium growth; [Rise 'n' Shine X (Sheri Anne X Laguna)]; Benardella, Frank A.; Nor'East Min. Roses, 1987

Ben's Gold, HT, dy, 1996; flowers star-shaped, semi-dbl.; Williams, J. Benjamin

Benson & Hedges Gold® *see* 'MACGEM'

Benson & Hedges Special *see* 'MACSHANA'

'BENSTAR', Min, pb, 1995; (Astra, **Soroptimist International**); flowers shrimp pink and ivory, opening to star shape, very dbl., 1.5 in., slight fragrance; no prickles; foliage large, dark green, glossy; tall (24in), upright, bushy growth; [Party Girl X Rosie]; Benardella, Frank A.; Bell Roses, Ltd., 1994

'BENTALL'S SCARLET', HT, or, 1935; flowers bright scarlet, slight fragrance; very vigorous growth; Bentall

'BENTEM', Min, dr, 1989; (**Rosetime**); flowers medium red, brushed with dark red, upright, medium, dbl., 26–40 petals, slight fragrance; foliage medium, medium green, matt; upright growth; [Rise 'n' Shine X Black Jade]; Benardella, Frank A.; Kimbrew Walter Roses, 1989

BENtintot, Min, lp; (**Tiny Tot**); Benardella, Frank A., 1990

'BENTVELD', HT, op, 1932; flowers carmine-orange, edges lighter than parent; vigorous growth; [Charles P. Kilham sport]; Posthuma; Low

Benvenuto® *see* 'MEIELPA'

'BÉRANGÈRE', M, lp, 1849; flowers delicate pink, large, dbl.; Vibert

'BERbow'™, Min, or, 1997; (**Oxbow**); flowers double, medium, dbl., 15–25 petals, borne mostly singly, no fragrance; foliage medium, medium green, dull; compact, medium (12-15in.) growth; [Luis Desamero X Rainbow's End]; Berg, David H.

'BERBUT', Min, w, 1994; (**Butter 'n' Sugar**); flowers vary from white to yellow shading to white, very dbl., 1.5–2.75 in., borne mostly singly, slight fragrance; few prickles; foliage medium, medium green, semi-glossy; low (15-18 in), spreading growth; [Klima X (Intrigue X Poker Chip)]; Berg, David H.

'BERCEUSE', HT, my, 1950; bud globular, yellow; flowers chamois, very large, very dbl., intense fragrance; foliage leathery; very vigorous, upright growth; [Signora X Mrs Pierre S. duPont]; Robichon

Berenice, F, 1979; Croix, P.; (Cavriglia)

Berenice *see* BARenic

'BÉRÉNICE', HGal, rb, 1818; flowers rose and crimson, shaded with slate, large, dbl., globular; pendulous growth habit; Vibert

Berenice Neville *see* 'KIRBELL'

'BERGERS ERFOLG', HRg, mr, 1925; flowers fire-red, stamens yellow, blooms in clusters, single, 3–4 in., occasionally recurrent bloom; foliage dark; very vigorous, bushy growth; [R. rugosa seedling X Richmond]; Berger, V.; Pfitzer

Bergers Koralle, F, op, 1956; flowers dark salmon-pink, large, dbl.; Berger, W.; (Sangerhausen)

Bergers Morgenröte, S, lp, 1959; flowers creamy pink, medium, dbl.; Berger, W.; (Sangerhausen)

Bergers Roma, HT, rb, 1965; flowers large, dbl.; Berger, W.; (Sangerhausen)

'BERGESLOH', Pol, mr, 1929; flowers carmine-red, small, dbl.; Vogel, M.; (Sangerhausen)

Bergfeuer, F, 1959; Leenders, J.; (Cavriglia)

Bergfeuer Superior, F, 1968; Leenders, J.; (Cavriglia)

'BERGME', F, mr, 1977; (**Gabriella**®, Gabrielle); bud ovoid; dbl., 33 petals, 3 in., cupped, slight fragrance; foliage glossy; vigorous, bushy growth; PP004452; Berggren; W. Kordes Sons

'BERGRAT OTTO BERGER', HT, w, 1924; flowers creamy white to sulfur, center deeper, dbl., moderate fragrance; [Pharisaer X Prince de Bulgarie]; Berger, V.; Faist

Bering Renaissance *see* POUlberin

Berkeley *see* 'JACIENT'

Berkeley Beauty™ *see* 'MORBERK'

Berleburg *see* POUlbella

'BERLEI', Min, ob, 1995; (**Leila**); flowers light orange with yellow reverse, blooms borne mostly single, dbl., 26–40 petals, 1.5 in., slight fragrance; some prickles; foliage medium, medium green, semi-glossy; medium (15 in), upright growth; [Arizona Sunset X June Laver]; Berg, David H.

Berlengas, HT; Moreira da Silva, A.; (Cavriglia)

'BERLIN', S, ob, 1949; bud long, pointed; flowers orange-scarlet, center golden, large, single, borne in large clusters, repeat bloom, moderate fragrance; prickles large; foliage leathery, dark; very vigorous, upright growth; (28); [Eva X Peace]; Kordes

Berlin Beauty, HT, lp

Berliner Luft®, F, ob, 1985; (Jura); Hauser

'BERMUDA PINK', HT, lp, 1974; bud ovoid; flowers flesh-pink, dbl., 35–40 petals, 4–5 in., globular, slight, rose fragrance; foliage glossy, light; compact, moderate growth; [Queen of Bermuda X Montezuma]; Golik; Dynarose

Bermuda's Catherine Mermet, T, w

Bermuda's Emmie Gray, Ch, rb

Bermuda's Kathleen, S, mr

'BERMUDIANA', HT, mp, 1966; bud ovoid; dbl., 35–60 petals, 5–6 in., exhibition form, moderate fragrance; foliage leathery; vigorous, upright growth; [Golden Masterpiece seedling X Golden Masterpiece seedling]; Boerner; J&P

'BERN'®, HT, dp, 1975; bud long, pointed; flowers deep pink, shallow, dbl., 24 petals, 4 in., intense, spicy fragrance; foliage dark, leathery; spreading growth; [Crimson Glory X Lilac Charm]; Huber

'BERNADETTE', HT, lp, 1964; flowers large, dbl., exhibition form, slight fragrance; foliage dark, leathery; vigorous, bushy growth; [Peace sport]; Kelly

Bernadette, HT, w, 1976; Dorieux

Bernadette Chirac® *see* 'DELBÉCHIR'

'BERNAIX, CLIMBING', Cl HT, dr, 1935; [Souv. d'Alexandre Bernaix sport]; Shamburger, C.S.

Bernalene *see* KORcountry

'BERNALIA', S, mp, 1961; flowers pink, center white, small blooms in clusters, semi-dbl., recurrent bloom, moderate fragrance; foliage bright green; vigorous growth; [Mosqueta X Cecilia]; Bernal

'BERNARD', P, op; (Pompon Perpetual); flowers salmon-pink; [Rose du Roi sport]; (France), 1846

'BERNARD VERLOT', HP, rb; flowers red-scarlet, center violet; Baron-Veillard, 1874

'BERND CLÜVER', F, mp, 1974; bud pointed; 2.5–3 in., moderate, fruity fragrance; moderate, upright growth; [Nordia X Sans Souci]; Reinold

Bernensis, S, mr, 1994; Meilland

Bernhard Daneke Rose *see* 'TANWEIEKE'

'BERNICE', HT, mp, 1927; flowers carmine-pink on yellow base, semi-dbl., moderate fragrance; Pemberton

'BERNICE', Pol, mp, 1937; bud ovoid; flowers brilliant cerise-pink, small blooms in clusters, dbl., globular, recurrent bloom, intense fragrance; foliage glossy, light; dwarf growth; [Baby Tausendschon (probably Echo) X Gloria Mundi]; Nicolas; J&P

Bernina®, F, w, 1979; deRuiter

'BERnora'™, Min, yb, 1997 (**Nora**); flowers double, medium, dbl., 15–25 petals, borne mostly singly, slight fragrance; foliage medium, medium green, dull; upright, tall (15-18in.) growth; [Rainbow's End X Leila]; Berg, David H.

Bernstein, F, dy, 1987; flowers golden yellow with copper tints, large, very dbl.; Tantau, Math.; (Sangerhausen)

Bernstein, Pol, dy; flowers dark yellow and copper, medium, dbl.; VEG; (Sangerhausen)

Bernstein-Rose® *see* TANeitber

'BEROLINA', F, my, 1977; bud long, pointed; dbl., 24 petals, 3.5 in., exhibition form, slight fragrance; foliage dark, soft; vigorous, upright growth; [Mabella X Seedling]; Kordes

Berolina *see* 'KORPRIWA'

'BERPAR', Min, w, 1988; (**Wintonbury Parish**); flowers white with top third of petal light red, medium, borne usually singly, semi-dbl., 14 petals, exhibition form, slight fragrance; prickles curved, large, light greenish-white; foliage medium, medium green, semi-glossy; upright, medium growth; PP006843; [Poker Chip X Lady X]; Berg, David H.

Berries 'n' Cream™ *see* 'POULCLIMB'

Berry Berry Grape, Min, m, 1997; King

Berry Berry Red *see* 'TALBER'

'BERSAGLIERA', HT, dp, 1960; bud globular; flowers crimson, phlox-pink and fuchsia-pink, medium to large blooms, dbl.; strong stems; vigorous growth; Luigi

Bert Hinkler, Pol, lp; Harrison, A., 1928; (Weatherly, L.)

'BERTHA', S, lp, 1946; delicate pink, very large, borne like hollyhock flowers, single, non-recurrent; erect stems; height 8 ft; [(R. rugosa X Hybrid Perpetual) X (R. multiflora X R. blanda)]; Wright, Percy H.

'BERTHA AIKMAN', HT, op, 1977; bud high pointed; flowers two-toned salmon-pink, dbl., 45 petals, 4 in., moderate fragrance; foliage matt; moderate, slightly spreading growth; [Gypsy Moth X Percy Thrower]; Simpson, J.W.; Manawatu Rose Soc.

'BERTHA GORST', HT, mr, 1933; flowers crimson-cerise, base gold, veined bronze, very large, dbl., intense fragrance; foliage bronze; [Autumn sport]; Beckwith

'BERTHA KIESE', HT, my, 1913; flowers medium, semi-dbl., slight fragrance; Jacobs; (Sangerhausen)

'BERTHA TURNER', HT, op, 1925; flowers salmon-peach, moderate fragrance; Pemberton

'BERTHA VON SUTTNER', HT, ly, 1918; flowers light yellow with copper tints, medium, dbl.; Verschuren; (Sangerhausen)

'BERTHE BARON', HP, lp, 1869; flowers delicate rose shaded with white, large, dbl.; [Jules Margottin X Seedling]; Baron-Veillard

'BERTHE DUMESNIL DE MONTCHAUVAU', HP, lp, 1876; flowers medium, dbl.; Jamain, H.; (Sangerhausen)

'BERTHE LÉVÊQUE', HP, 1866; Cochet, A.; (Cavriglia)

'BERTHE MALLERIN', HT, or, 1960; flowers red tinted orange, dbl.; strong stems; vigorous growth; Mallerin, C.

'BERTI GIMPEL', HP, mp, 1913; flowers large, semi-dbl.; Altmüller; (Sangerhausen)

Bertram *see* **'SNEEZY'**

'BERTRAM PARK', HT, mr, 1929; (Coquette); flowers rosy crimson, base yellow, single, slight fragrance;

[Eblouissant X Mme Edouard Herriot]; Burbage Nursery

'BERWICK', HSpn, pb; flowers rose shading to white at edges, large, semi-dbl.; dwarf growth

Beryl, HT, ly

'BERYL AINGER', HT, w, 1955; flowers cream, base golden yellow, moderate fragrance; [The Doctor sport]; F. Cant

Beryl Bach *see* 'HARTESIA'

'BERYL FORMBY', HT, yb, 1948; flowers golden yellow shaded crimson, dbl., 36–40 petals, intense fragrance; foliage glossy; bushy growth; [McGredy's Sunset sport]; Fryers Nursery, Ltd.

'BERYL FORMBY, CLIMBING', Cl HT, yb, 1956; Letts

'BERYL WEARMOUTH', F, mp, 1973; semi-dbl., 19 petals, 4.5 in.; foliage light green, matt; [(Ann Elizabeth X Orange Sensation) X Sea Pearl]; Harkness

Besançon, HT, 1968; Sauvageot; (Cavriglia)

'BESS LOVETT', LCl, dp, 1915; flowers light red, well-formed blooms in clusters, dbl., cupped, moderate fragrance; foliage glossy, dark; vigorous, climbing growth; Van Fleet; J.T. Lovett

'BESSIE BROWN', HT, ly, 1899; flowers yellowish white, large, very dbl., moderate fragrance; foliage light, leathery, glossy; Dickson, A.

'BESSIE CHAPLIN', HT, mp, 1921; flowers bright pink, center deeper, very dbl., slight fragrance; GM, NRS, 1923; [Lady Pirrie X Gorgeous]; Chaplin Bros.

'BESSIE JOHNSON', HP, 1873; Curtis; (Cavriglia)

Bessie Lee, Pol; (Cavriglia)

Bessy, S, ob, 1998; Interplant

'BEST REGARDS', HT, pb, 1944; flowers pink bicolor, dahlia form, dbl., 50–60 petals, 6–7 in., moderate fragrance; foliage leathery, dark; very vigorous, compact growth; [Soeur Thérèse X Signora]; Morris; Germain's

'BEST REGARDS, CLIMBING', Cl HT, pb, 1940; Elmer, C.A.; Germain's

'BEST WISHES', HT, dp, 1959; bud long, pointed; flowers currant-red, dbl., 25–35 petals, 5.5–6 in., exhibition form, moderate fragrance; foliage leathery; strong stems; very vigorous growth; Fisher, G.; Arnold-Fisher Co.

Best Wishes *see* CHESnut

'BETAHAT', HT, rb, 1987; (**Malahat**); flowers scarlet, reverse white, fading slightly, medium, borne usually singly, dbl., 40 petals, exhibition form, intense fragrance; round, medium, brown fruit; foliage medium, red-brown to dark green, glossy; bushy, vigorous growth; [Pristine X Shockling Blue]; Betts, John; Wisbech Plant Co., 1988

'BETANO BEACH', F, rb, 1966; bud long, pointed; flowers light salmon-pink, reverse scarlet and burnt crimson, medium, dbl., exhibition form; foliage glossy, dark, leathery; vigorous, upright growth; [Ma Perkins X Detroiter]; Fankhauser; A. Ross & Son

'BETH', F, ly, 1966; flowers buff-yellow; [Elizabeth of Glamis sport]; O'Connell; Rumsey

Beth, HSpn, lp, 1973; Mertens

'BETHANY HELENA', HT, pb, 1997; flowers full, large, very dbl., 26–40 petals, borne mostly singly, moderate fragrance; foliage large, dark green, semi-glossy; upright, medium growth; [Tom Foster X Gavotte]; Poole, Lionel

'BETINHO', HT, rb, 1958; flowers velvety red and brown; [Charles Mallerin X Monte Carlo]; da Silva, Moreira

'BETSAT', HT, dr, 1987; (**Black Satin**); dbl., 30 petals, exhibition form, borne singly, slight fragrance; rounded, medium, red fruit; prickles few, pointed slightly down, small, light brown; foliage medium, dark green, semi-glossy; upright, medium growth; [Folklore X Loving Memory]; Betts, John; Wisbech Plant Co., 1987

'BETSIE JANE', HT, ab, 1976; bud long, pointed; flowers soft apricot-pink, full, large, dbl., moderate fragrance; foliage light green; tall, vigorous growth; [Bewitched sport]; Tresise

'BETSUE', F, or, 1987; (**Sue Betts**); flowers long lasting color, large blooms borne in sprays of 65-120, dbl., cupped, moderate fragrance; round, red fruit; prickles sharp, thin, brown; foliage medium, medium green, semi-glossy; bushy, low growth; [Europeana sport]; Betts, John; Wisbech Plant Co., 1987

'BETSY MCCALL', F, op, 1956; bud ovoid; flowers shrimp-pink, open, blooms in large clusters, dbl., 25–30 petals, 3–3.5 in., moderate fragrance; foliage glossy; vigorous, bushy growth; [Seedling X Fashion]; Boerner; J&P

Betsy Murchison *see* DESbet

Betsy Ross *see* 'DELUP'

'BETSY ROSS', HT, ob, 1931; flowers like parent but marked russet-orange; [Talisman sport]; Samtmann Bros.

Betsy Taaffe, S, lp; [Abraham Darby sport]; Taaffe, 1996; (Weatherly, L.)

'BETSY VAN NES', Pol, mr, 1914; flowers small, semi-dbl.; van Ryn; (Sangerhausen)

'BETTE IRENE', S, ab, 1986; flowers apricot, reverse deeper, fading light pink, imbricated, medium, semi-dbl., 9 petals, repeat bloom, moderate fragrance; rounded, medium, ornamental, bright red fruit; awl-like, medium, brown prickles; foliage medium, dark green, matt; upright, medium-tall growth; [Dairy Maid X Seedling]; Schneider, Peter

'BETTER HOMES & GARDENS', HT, pb, 1976; bud ovoid; flowers rose, ivory reverse, dbl., 38 petals, 3–3.5 in., exhibition form, slight fragrance; foliage glossy, dark; medium-tall, upright growth; [Tropicana X Peace]; Warriner, William A.; J&P

Better Homes & Gardens Diamond Jubilee™ *see* 'WINBILEE'

'BETTER TIMES', HT, mr, 1934; flowers cerise, large, dbl., exhibition form, slight fragrance; foliage dark, leathery; very vigorous, compact growth; [Briarcliff sport]; Hill, Joseph H., Co.

'BETTER TIMES, CLIMBING', Cl HT, mr, 1937; Parmentier, J.

Bettie Herholdt *see* 'HERTIE'

Bettina® *see* MEpal

Bettina '78 *see* 'MEIBRICO'

Bettina, Climbing® *see* 'MEPALSAR'

'BETT'S CARDINAL SPIRIT', F, dr, 1999; flowers cardinal red, dbl., 15–25 petals, 4.5 in., borne in small clusters, slight fragrance; few prickles; foliage medium, medium green, semi-glossy; upright, medium (3 ft) growth; [High Spirit X (Moody Blues X Melina)]; Walters, Betty & Richard

'BETT'S LITTLE GEM', Min, mp, 1996; flowers medium pink with white ring around stamens, open, cup-shaped, dbl., 26–40 petals, slight fragrance; few prickles; foliage medium, light green, semi-glossy; upright, medium growth; [Pink Petticoat X Unknown]; Walters, Betty & Richard

Bett's Little Rhapsody *see* 'WALTRAP'

'BETT'S PINK LACE', Min, pb, 1996; flowers white with pink edge, changing to salmon pink, blooms borne, dbl., 26–40 petals, intense fragrance; few prickles; foliage medium, dark green, glossy; compact, low growth; [Pink Petticoat X Seedling]; Walters, Betty & Richard

'BETT'S SNOW DANCER', F, w, 1999; bud hint of green, opening to white; flowers outer edge white with cream center, numerous golden stamens, dbl., 26–40 petals, 4.5 in., borne in large clusters, slight fragrance; some prickles; foliage medium, medium green, semi-glossy; compact, medium (4 ft) growth; [Mt Hood X Aorangi]; Walters, Betty & Richard

'BETT'S WHITE DELIGHT', Min, w, 1992; flowers creamy white opening to pure white, rosette, very dbl., 58–86 petals, 2.5 in., borne in clusters of 5, slight fragrance; foliage medium to dark green, matt; tall, compact growth;

[Pink Petticoat X Seedling]; Walters, Betty & Richard, 1987

'BETTY', HT, pb, 1905; flowers coppery rose, shaded yellow, large, dbl., moderate fragrance; very vigorous growth; Dickson, A.

'BETTY ALDEN', Cl Pol, lp, 1919; flowers appleblossom-pink passing to white, single; [Orléans Rose X (Katharina Zeimet X R. arvensis hybrid)]; Farquhar

'BETTY BAUM', HT, lp, 1927; flowers delicate pink, base yellow, dbl., moderate fragrance; [Premier sport]; Baum

Betty Bee™ *see* 'BLABEE'

'BETTY BERKELEY', T, mr, 1904; flowers bright red; Bernaix, A.

'BETTY BLAND', S, dp, 1925; flowers deep rose, fading pink, center deeper, dbl., non-recurrent bloom, moderate fragrance; foliage rich green, soft; twigs ruby-red; vigorous (6 ft.), bushy growth; very hardy; (14); [R. blanda X HP]; Skinner

'BETTY BLOSSOM', HWich, mp, 1923; flowers clear rose-pink, loose cluster, semi-dbl.; vigorous pillar or bush growth; [R. wichurana X Mrs W.J. Grant]; Dawson; Eastern Nursery

Betty Boop™ *see* 'WEKPLAPIC'

'BETTY CUTHBERT', HT, or, 1964; bud long, pointed; flowers medium, very dbl., intense fragrance; foliage soft, glossy; very vigorous, upright growth; [Roundelay sport]; Palmer; Palmer & Engall

Betty Driver *see* 'GANDRI'

'BETTY FREE', F, mp, 1950; bud pointed; flowers neyron rose, borne in clusters, dbl., 25 petals, slight fragrance; vigorous growth; [Fortschritt sport]; LeGrice

'BETTY GRACE CLARK', HT, ob, 1933; flowers orange-yellow, reverse streaked red, exhibition form, moderate fragrance; vigorous, bushy growth; [Marie Adélaide sport]; Clarke Bros.

Betty Harkness *see* HARette

Betty Herholdt *see* 'HERTIE'

'BETTY HULTON', HT, dy, 1923; flowers deep saffron-yellow, dbl., moderate fragrance; GM, NRS, 1923; Dickson, A.

'BETTY MAY WOOD', HT, ab, 1968; flowers apricot to buff, reverse coral-salmon, well formed, dbl., 30 petals; foliage light green; free growth; [Mischief sport]; Wood

'BETTY MORSE', HT, mr, 1950; bud long, pointed; dbl., 25 petals, 4 in., slight fragrance; foliage olive-green; vigorous growth; [Crimson Glory X (Crimson Glory X Cathrine Kordes)]; Kordes; Morse

'BETTY NEUSS', HT, mp, 1973; bud small, long, pointed; flowers pure pink, medium, dbl., slight fragrance; foliage small; very vigorous, upright growth; Dawson, George; C. Brundrett

'BETTY PAUL', HT, mp, 1988; flowers medium pink blend, medium, borne usually singly or in sprays, dbl., 20–25 petals, exhibition form, slight fragrance; ovoid, small, yellow-orange fruit; prickles straight, medium, light brown; foliage medium, medium green, semi-glossy, disease resistant; upright, medium, tall growth; Bronze, ARC TG, 1987; [Queen Elizabeth X (Tiffany X Tropicana)]; Warner, A.J.

'BETTY PEARSON', HT, w, 1929; flowers cream, center apricot, petals shell shaped, large; Burbage Nursery

'BETTY PRIOR', F, mp, 1935; bud ovoid, dark carmine; flowers carmine-pink, borne in clusters, single, 5 petals, cupped, moderate fragrance; vigorous, bushy growth; GM, NRS, 1933; [Kirsten Poulsen X Seedling]; Prior; J&P, 1938

Betty Prior, Climbing *see* 'COOPRIOR'

Betty Sheriff, LCl, w; Sheriff

'BETTY STIELOW', HT, dp, 1928; flowers dark pink, almost red at times, dbl., moderate fragrance; [Premier sport]; Stielow Bros.

'BETTY SUTOR', HT, pb, 1929; flowers pale pink, veined rose, reverse rosy, large, dbl., moderate fragrance; foliage light, glossy; vigorous, bushy growth; McGredy

'BETTY UPRICHARD', HT, ab, 1922; flowers delicate salmon-pink, reverse carmine with coppery sheen, large, dbl., 20 petals, intense fragrance; foliage light, leathery, glossy; very vigorous, tall growth; GM, NRS, 1921; Dickson, A.

'BETTY UPRICHARD, CLIMBING', Cl HT, ab, 1936; Krause

'BETTY WILL', S, pb, 1963; flowers bright pink, lighter reverse, dbl., 35 petals; some large prickles; foliage dark, leathery; red canes; tall growth; [George Will X Betty Bland]; Erskine

Betty Wilson *see* 'SHERISCENT'

Betty Wright™ *see* 'BURBET'

'BETTY, CLIMBING', Cl HT, pb, 1926; Hohman

Betty's Baby *see* 'JUDBABY'

Betty's Pride®, Min, rb

'BETZEL'S PINK', HT, op, 1955; bud long; flowers coral-pink, blooms, dbl., 55–60 petals, 5 in., exhibition form, moderate fragrance; foliage leathery, glossy; strong stems; vigorous, bushy growth; [Pres. Herbert Hoover sport]; Betzel; Edmunds Roses

'BEVERLEY ANNE', HCh, mr, 1987; flowers red-purple, open, borne in sprays of 3-7, dbl., 21 petals, flat, slight fragrance; no prickles; foliage pivoted, serrated; semi-dwarf growth; [Seedling X Seedling]; Nobbs, Kenneth J., 1986

Beverley Stoop *see* 'JAYDON'

Beverley Watson *see* KORdreischi

Beverly Hills *see* 'DELMATOR'

'BEVERLY JAYNE', S, lp, 1999; flowers center deep pink, outer light pink, very dbl., 41 petals, 1.3in., borne in large clusters, slight fragrance; few prickles; foliage small, light green, dull; upright, bushy, medium (3 ft) growth; [Angela Rippon X New Dawn]; Jones, L.J.

'BEVERLY NICOLS', HT, op, 1939; flowers cream, reverse salmon, well formed, large, exhibition form, moderate fragrance; vigorous growth; Burbage Nursery

'BEWITCHED', HT, mp, 1967; bud urn shaped; flowers cotton candy pink, large, dbl., 27–30 petals, 5 in., exhibition form, borne singly, moderate, damask fragrance; foliage large, apple green, glossy; new stems and foliage is red; vigorous, medium, rounded growth; AARS, 1967 GM, Portland, 1967; [Queen Elizabeth X Tawny Gold]; Lammerts, Dr. Walter; Germain's

Bewitched *see* POUlbella

Bewitched, Climbing, Cl HT, mp, 1989

Beyreuth, S, rb

'BEZRUC', HT, mr, 1938; flowers large, dbl.; Böhm, J.; (Sangerhausen)

'BHAGMATI', F, mr, 1977; bud ovoid; 15–20 petals, 2.5–3 in., cupped, slight fragrance; foliage glossy, light green; dwarf, vigorous, bushy growth; [Charleston X ((Roman Holiday X Flamenco) X Goldgleam)]; Viraraghavan, M.S.; Gopalsinamiengar

Bhanu, HT, yb, 1990; K&S

Bharami *see* **'BHARANI'**

'BHARANI', Min, m, 1979; (Bharami); flowers small, semi-dbl., slight fragrance; foliage small, light green, matt; upright growth; [Seedling X Seedling]; Goralaswamiengar, K.S.; Gandy Roses, Ltd., 1973

Bhargav, HT, op, 1993; K&S

Bhavani, HT, op, 1986; Kasturi

'BHIM', HT, dr, 1970; bud long, pointed; flowers scarlet-red, open, large, very dbl., abundant, intermittent bloom, moderate fragrance; vigorous, upright growth; [Charles Mallerin X Delhi Princess]; IARI; Div. of Vegetable Crops & Flori.

'BIANCA', HT, 1913; bud long, pointed; flowers pale peach tinted pinkand violet, center shaded rose, dbl.; RULED EXTINCT 3/83 ARM; Paul, W.

'BIANCA', HT, 1927; flowers white, sometimes lightly flushed cream or pink, well-formed, dbl., moderate fragrance; RULED EXTINCT 3/83 ARM; Pemberton

Bianca®, HT, w, 1987; Kuhn

'Bianca Camelia', HT, w, 1933; bud ovoid, pointed; flowers snow-white, center light blush-yellow, very large, dbl., 23–25 petals, intense fragrance; foliage light green; very vigorous, upright growth; [Nuntius Pacelli X Sachsengruss]; San Remo Exp. Sta.

Bianco *see* 'COCblanco'

Bibi Maizoon™ *see* AUSdimindo

Bibiché® *see* DORflo

Bibiché, Climbing, Cl F, 1980; Dorieux, Francois; (Cavriglia)

'Bicentennial', F, rb, 1975; flowers deep pink and red blend, medium, dbl., exhibition form, slight fragrance; foliage leathery; vigorous, upright, bushy growth; Meyer, C.; C-P

'Bichette', HT, mp, 1968; flowers persian rose, large, dbl., 20–25 petals; foliage dark, leathery; vigorous, upright growth; [Diamond Jubilee X Seedling]; Verschuren, A.; Stassen

Bicolette, HT, rb, 1980; Tschanz, E.

'Bicolore', HP, 1877; Oger; (Cavriglia)

'Bicolore Incomparable', HGal, 1861; Touvais; (Cavriglia)

Biddulph Grange *see* FRYdarkeye

Biddy, Min, w

Bienkie *see* KORsisten

'Bienvenu', Gr, ob, 1969; bud long, pointed; flowers reddish orange, large, dbl., 70 petals, exhibition form, intense fragrance; foliage leathery, matt; vigorous, upright growth; [Camelot X (Montezuma X War Dance)]; Swim & Weeks; Weeks Wholesale Rose Growers

Bifera Italica, HGal, lp; flowers medium, dbl.; (Sangerhausen)

Big and Beautiful *see* 'TINbab'

'Big Apple', HT, mr, 1983; flowers large, dbl., moderate fragrance; foliage large, medium green, matt to semi-glossy; upright, spreading growth; [Mister Lincoln X (Suspense X King's Ransom)]; Weeks, O.L.; Weeks Wholesale Rose Growers, 1984

Big Apricot, HT, ab; Wells

'Big Bang'®, F, or, 1980; bud pointed; flowers deep orange-red, shallow-blooms borne 5-10 per cluster, semi-dbl., 13 petals, cupped, no fragrance; reddish-green prickles; foliage matt, light green; bushy growth; [Sarabande X Sarabande seedling]; Barni-Pistoia, Rose

'Big Ben', HT, dr, 1964; flowers well-formed, 5–6 in., intense fragrance; foliage dark; tall growth; [Ena Harkness X Charles Mallerin]; Gandy, Douglas L.

Big Cabbage of Holland *see* **'Gros Choux d'Hollande'**

Big Chief *see* **'Portland Trailblazer'**

Big Daddy® *see* 'MACchome'

'Big Duke', HT, dp, 1991; (Duke Wayne); bud pointed; flowers deep pink, silvery pink reverse, large blooms borne usually, dbl., exhibition form, moderate, fruity fragrance; foliage large, medium green, semi-glossy; upright, tall growth; [The Duke X Seedling]; Weddle, Von C., 1984

'Big Jack Charlton', HT, w, 1994; flowers ivory, pale pink edge, blooms borne mostly single, dbl., 26–40 petals, 3–3.5 in., slight fragrance; some prickles; foliage large, dark green, glossy; medium (85 cms), upright growth; [Gavotte X Queen Esther]; Poole, Lionel; Battersby Roses, 1995

Big Jim Larkin *see* 'SEAlark'

'Big John', Min, mr, 1979; bud pointed; flowers deep medium red, base yellow, dbl., 42 petals, 1–1.5 in., exhibition form; foliage small, glossy, bronze; upright, bushy growth; [Starburst X Over the Rainbow]; Williams, Ernest D.; Mini-Roses

Big John *see* **'Galleria'**

Big John, LCl

Big Purple *see* 'STEbigpu'

'Big Red', HT, dr, 1967; bud pointed; dbl., 52 petals, 4.5–6 in., exhibition form, slight fragrance; foliage leathery; vigorous, upright, bushy growth; [Chrysler Imperial X Seedling]; Meilland, Mrs. Marie-Louise; C-P

'Big Splash', Cl HT, rb, 1969; bud pointed; flowers flame-red, reverse lighter, base yellow, large, dbl., exhibition form, moderate fragrance; foliage glossy, leathery; vigorous, climbing (8-10 ft) growth; [Buccaneer X Bravo]; Armstrong, D.L.; Armstrong Nursery

'Biggi', HT, yb, 1975; bud ovoid; dbl., 27 petals, 4.5 in., exhibition form, intermittent bloom, moderate fragrance; foliage glossy, dark; vigorous, upright growth; [Dr. A.J. Verhage X Seedling]; Kordes; Fey

'Bijou', Pol, lp, 1932; flowers old rose; deRuiter; Sliedrecht & Co.

Bijou *see* BARjou

Bijou, MinFl, pb, 1991; Lens

'Bijou de Royat-les-Bains', HCh, mp, 1891; flowers medium, dbl.; Veysset; (Sangerhausen)

'Bijou des Amateurs', HGal, dr, 1848

Bijou des Prairies *see* **'Gem of the Prairies'**

Bijou d'Or *see* TANledolg

'Bijou Superior', Pol, lp; flowers have more lasting color; [Bijou sport]; deRuiter

'Bikini Red', HT, mr, 1974; bud ovoid; flowers rose-red, tinged white, ruffled, dbl., 40 petals, 5 in., moderate, fruity fragrance; foliage glossy, dark; [Queen of Bermuda X Peace]; Golik; Dynarose

Bila Junior Miss, HT, lp; Strnad; (Czech Rosa Club)

'BILfan', Min, yb, 1986; (**Jack's Fantasy**); flowers yellow blushed with orange-red from edge, reverse medium yellow, dbl., 21 petals, exhibition form, slight fragrance; no fruit; few, beige prickles slightly sloped downward; foliage medium, medium green, semi-glossy; upright, bushy, medium growth; [Little Darling X Over the Rainbow]; Bilson, Jack M., Jr. & Bilson, Jack M. III

'BILice', Min, pb, 1987; (**Phoebe's Choice**™); flowers pink, edges yellow, reverse same, pink softens, white edges, dbl., exhibition form, no fragrance; round, light orange & green fruit; reddish-green prickles, sloped downwards; foliage medium, medium green, glossy; upright, bushy growth; [Little Darling X Over the Rainbow]; Bilson, Jack M., Jr. & Bilson, Jack M. III

'Bilitis', F, or, 1969; flowers vermilion-red, center yellow; [Tabarin X Golden Slippers]; Gaujard

'Bill Beaumont', F, mr, 1982; flowers crimson, dbl., 20 petals, slight fragrance; foliage medium, medium green, matt; bushy growth; [Evelyn Fison X Redgold]; Fryer, Gareth; Fryer's Nursery, Ltd.

Bill Cone *see* 'MICone'

'Bill Hunt', HT, op, 1976; flowers deep coral, full, medium, dbl., 30 petals, moderate fragrance; foliage glossy; fairly vigorous growth; [Mischief X Serenade]; Blakemore

Bill Slim *see* 'HARquito'

'Bill Temple', HT, w, 1975; flowers cream, dbl., 30 petals, 5–6 in., slight fragrance; foliage glossy, dark; [Crimson Halo X Piccadilly]; Harkness

Bill Warriner™ *see* 'JACsur'

'Billard et Barré', Cl T, my, 1898; flowers golden yellow, dbl., intense fragrance; [Mlle Alice Furon X Duchesse d'Auerstadt]; Pernet-Ducher

Billie and Lew, HWich, m, 1995; Nobbs

Billie Teas *see* 'HOOtea'

'Billionaire', HT, mr, 1972; bud ovoid, long, pointed; flowers large, dbl., exhibition form, slight fragrance; foliage leathery; vigorous, upright growth; [Fragrant Cloud X Proud Land]; Warriner, William A.; J&P

Billy, S, rb, 1993; Hauser

'BILLY BOILER', Cl HT, mr, 1927; vigorous, tall growth; Clark, A.; NRS Victoria

'BILLY BOY', HT, my, 1926; flowers sunflower-yellow, semi-dbl.; [Golden Emblem X Christine]; McGredy; Beckwith

Billy Boy *see* 'MORBOY'

Billy Graham(TM) *see* 'JACGRAY'

'BILPAT', Min, pb, 1989; (**Naughty Patricia**); flowers medium pink, edges blush, outer petals quill, large, borne u, dbl., 29 petals, exhibition form, slight, fruity fragrance; globular, medium, green with orange-red blotch fruit; prickles slight downward slope, reddish-tan; foliage large, medium green, matt; upright, tall growth; [Rise 'n' Shine X Redgold]; Bilson, Jack M., Jr. & Bilson, Jack M. III

'BIMBORO'®, HT, dr, 1978; bud globular; dbl., 46 petals, 3–3.5 in., exhibition form, moderate fragrance; foliage glossy; vigorous, bushy, upright growth; [Seedling X Kardinal]; Kordes, W. Söhne

'BING CROSBY', HT, ob, 1980; bud ovoid; flowers strong clear orange, very dbl., 40–45 petals, 5 in., cupped, borne mostly singly, slight, light spice fragrance; long prickles hooked downward; foliage medium, heavy, leathery, wrinkled, dark; long stems; vigorous, upright, bushy growth; PP4695; AARS, 1981; [Seedling X First Prize]; Weeks

'BINGO', HT, dr, 1955; (Dyna); bud pointed; flowers cardinal-red, dbl., 55 petals, 4–5 in., exhibition form, intense fragrance; foliage glossy; upright growth; [(Hadley seedling X Ami Quinard) X Crimson Glory]; Robichon; Ilgenfritz Nursery

Bingo Meidiland® *see* 'MEIPOTAL'

Bingo Meillandecor® *see* 'MEIPOTAL'

Bingo Queen *see* 'TINQUEEN'

Bipontina, S, mr

Bird of Fire *see* 'TAYBIRD'

'BIRDIE BLYE', S, mp, 1904; bud long, pointed, carmine; flowers rose-pink, large blooms in clusters, dbl., cupped, recurrent bloom, slight fragrance; foliage light; vigorous (4-5 ft.) growth; [Helene X Bon Silene]; Van Fleet; Conard & Jones

'BIRDSONG', LCl, mp, 1991; semi-dbl., 6–14 petals, 1.5 in., borne in very large clusters, moderate fragrance; few prickles; foliage small to medium, medium green, semi-glossy; tall (300+ cms), spreading, large, rambling growth; [Seedling X Seedling]; Seward, Grace

'BIRGITTA', HT, dr, 1961; bud ovoid; flowers large, dbl.; de Boer

Birgitte de Villenfagne, HMsk, pb

'BIR-HACKEIM', HT, mr, 1946; bud long, pointed; flowers fiery red, open, semi-dbl., cupped; very vigorous, upright, bushy growth; Mallerin, C.; A. Meilland

Birichina, F; (Italy); (Cavriglia)

'BIRMINGHAM BOERNER', F, w, 1975; flowers light flesh-pink turning white; [Gene Boerner sport]; Schoepfle

'BIRMINGHAM POST', F, dp, 1968; flowers deep pink, large, dbl., intense fragrance; foliage leathery; vigorous growth; [Queen Elizabeth X Wendy Cussons]; Watkins Roses

'BIRTHDAY PARTY', Min, mp, 1979; bud ovoid; flowers pink, medium, dbl., 28 petals, 1.5–2 in., exhibition form, moderate fragrance; foliage dark; upright, spreading growth; [Attraktion X Sheri Anne]; Strawn; Pixie Treasures Min. Roses

'BIRTHDAY PRESENT', Cl HT, dr, 1950; bud ovoid; dbl., 20 petals, exhibition form, non-recurrent, intense fragrance; foliage dark, leathery; vigorous, climbing growth; [Guinee X Rouge Mallerin]; Toogood

Birthday Wishes *see* GUESDELAY

'BISCAY', Min, mp, 1988; bud ovoid; flowers medium pink, reverse slightly darker, medium, borne usually, dbl., 26 petals, exhibition form, slight fragrance; prickles slightly downward pointed, medium, deep pink; foliage medium, medium green, semi-glossy; upright, medium growth; [Summer Spice X Seedling]; Bridges, Dennis A.; Bridges Roses, 1989

'BISCHOF DR KORUM', HP, pb, 1921; flowers yellowish rose, dbl., moderate fragrance; [Frau Karl Druschki X Laurent Carle]; Lambert, P.

'BISCHOFSSTADT PADERBORN'®, S, or, 1964; (Fire Pillar); flowers cinnabar-scarlet, saucer shaped, semi-dbl.; vigorous (3-4 ft), bushy growth; ADR, 1968; Kordes, R.

'BISCOF', Min, op, 1983; (**Sadler**); flowers orange pink, small, dbl., 43 petals, exhibition form, no fragrance; foliage medium green, matt; upright growth; [Faberge X Darling Flame]; Bischoff, Francis J.; Kimbrew-Walter Roses

'BISFRA', Min, lp, 1983; (**Penny Annie**); flowers small, dbl., 35 petals, exhibition form, no fragrance; foliage medium, medium green, matt; bushy growth; [Little Darling X Unnamed seedling]; Bischoff, Francis J.

'BISHOP DARLINGTON', HMsk, ab, 1926; bud ovoid; flowers cream to flesh-pink, with yellow glow, large, semi-dbl., 17 petals, cupped, recurrent bloom, moderate, fruity fragrance; foliage bronze, soft; semi-climbing growth; [Aviateur Bleriot X Moonlight]; Thomas; Dreer, 1928;, H&S

Bishop Elphinstone *see* 'COCJOLLY'

'BISHOP OF SHERWOOD', HT, op, 1976; flowers pale salmon flushed pink, full, dbl., 34 petals, 5 in., intense fragrance; vigorous, upright growth; [Mischief X (Wendy Cussons X Peace)]; Bracegirdle

Bishop's Rambler, HWich, dr, 1996; Scarman

'BISHOP'S ROSE', C, pb; vigorous (to 5 ft.) growth; [Apparently R. gallica X R. centifolia]

'BISJEN', Min, rb, 1981; (**Ginny**); bud ovoid; flowers white edged medium red, yellow at hinge, yellow stamens, small, dbl., 45 petals, exhibition form, no fragrance; straight red prickles; foliage dark, leathery, reddish tinge on new growth; upright, compact growth; PP005275; [Little Darling X Toy Clown]; Bischoff, Francis J.; Kimbrew-Walter Roses

'BISMAR', Min, op, 1984; (**Marty's Triumph**(TM)); flowers bright coral pink, white reverse, small, dbl., 28 petals, exhibition form, borne singly, slight fragrance; pale green, straight prickles; foliage medium, dark, semi-glossy; upright, bushy growth; PP006099; [Little Darling X Unnamed seedling]; Bischoff, Francis J., 1985

'BISRED', Min, rb, 1996; (**JuJu**(TM)); flowers dark red with a little white at base, good color, very good, dbl., 26–40 petals, no fragrance; few prickles; foliage medium, dark green, glossy; upright, bushy, medium growth; [Little Darling X Black Jade]; Bischoff, Francis J.; Kimbrew Walter Roses, 1997

'BISyel', Min, my, 1997; (**Lida O**) flowers full, small, very dbl., 26–40 petals, 1.5 in., borne in small clusters, no fragrance; foliage medium, medium green, semi-glossy; medium (23 cms) compact growth; [Party Girl X Miss Dovie]; Bischoff, Francis J.

'BIT O' GOLD', Min, dy, 1981; bud ovoid; flowers deep yellow, imbricated form, micro-mini, dbl., 40 petals, moderate fragrance; foliage smll, dark, semi-glossy; compact, bushy growth; PP005306; [Seedling X Golden Angel]; Williams, Ernest D.; Mini-Roses

'BIT O' MAGIC', Min, pb, 1979; bud pointed; flowers deep pink, reverse nearly white, micro-mini, dbl., 50 petals, 1 in., exhibition form; foliage small, dark, glossy; compact, spreading growth; [Over the Rainbow X Over the Rainbow]; Williams, Ernest D.; Mini-Roses

'BIT O' SPRING', Min, pb, 1980; bud long, pointed; flowers medium buffy pink, reverse lighter yellow-pink, borne usually singly, dbl., 45 petals, exhibition form, moderate fragrance; thin, tan prickles curved down; foliage deep green, matt; upright, bushy growth;

[Tom Brown X Golden Angel]; Williams, Ernest D.

'BIT O' SUNSHINE', Min, dy, 1956; (Little Bit o' Sunshine); flowers bright buttercup-yellow, small, 18–20 petals, 1.5 in., moderate fragrance; bushy (12-14 in), compact growth; [Copper Glow X Zee]; Moore, Ralph S.; Sequoia Nursery

'BIT OF HONEY', Min, dy, 1980; dbl., 48 petals; [Sunnydew sport]; Vastine, Gilbert; Gulf Stream Nursery

Bizarre Triomphante *see* **'CHARLES DE MILLS'**

'BLABEE', Min, pb, 1983; (**Betty Bee**™); flowers pink, white reverse, small, dbl., exhibition form, no fragrance; foliage small, medium green, semi-glossy; compact, bushy growth; PP005448; [Little Darling X Toy Clown]; Blazey, Daniel; Nor'East Min. Roses

'BLABY COURIER', HT, mr, 1956; bud deep crimson; flowers vivid scarlet, moderate fragrance; strong stems; Verschuren; Blaby Rose Gardens

Blaby Jubilee *see* **'DRIES VERSCHUREN'**

'BLABY MONARCH', HT, lp, 1960; flowers rose-pink, dbl., 40 petals, 5–6. in., exhibition form, intense fragrance; foliage light green; vigorous growth; [Briarcliff X Seedling]; Verschuren; Blaby Rose Gardens

'BLACK BEAUTY', HT, dr, 1973; flowers garnet-red, large, dbl., slight fragrance; bushy growth; [(Gloire de Rome X Impeccable) X Papa Meilland]; Delbard; Bees

'BLACK BESS', F, dr, 1939; bud long, pointed; flowers blackish crimson, borne in clusters, semi-dbl., exhibition form, slight fragrance; foliage dark, bronze; vigorous, bushy growth; [Dance of Joy X Crimson Glory]; Kordes; Morse

Black Bourbon, B, dr

'BLACK BOY', LCl, dr, 1919; (Blackboy); flowers very dark red, large, semi-dbl., moderate fragrance; foliage sparse, wrinkled, light; vigorous growth; [Étoile de France X Bardou Job]; Clark, A.; NRS South Australia

'BLACK BOY', M, dr, 1958; bud ovoid, lightly mossed; flowers deep crimson, large, very dbl., non-recurrent, intense fragrance; foliage light green, leathery; vigorous, upright, bushy growth; [World's Fair X Nuits de Young]; Kordes

Black Delight, HT, dr, 1985; G&L

'BLACK FIRE', Pol, dr, 1969; bud ovoid; flowers open, medium, dbl., slight fragrance; foliage dark, soft; vigorous, bushy growth; [Red Favorite X Seedling]; Delforge

'BLACK GARNET', HT, dr, 1980; bud ovoid, pointed; flowers dark, black-red, velvety, dbl., 50–55 petals, 5.5 in., exhibition form, borne singly or 2-3 per cluster, slight, tea fragrance; short to medium prickles, hooked downward; foliage medium to large, moderately leathery, dark grayish-g; bushy, upright, branching growth; PP4738; [Mister Lincoln X Mexicana]; Weeks

Black Gold *see* CLEblack

'BLACK ICE', F, dr, 1971; dbl., 24 petals, 4 in., slight fragrance; foliage glossy, dark; [(Iceberg X Europeana) X Megiddo]; Gandy, Douglas L.

Black Jack™ *see* 'MINKCO'

Black Jade™ *see* 'BENBLACK'

'BLACK KNIGHT', HT, dr, 1934; flowers crimson shaded blackish, dbl., 30–35 petals, moderate fragrance; foliage glossy, dark; vigorous growth; [Ami Quinard X Château de Clos Vougeot]; Hillock

Black Lady® *see* 'TANBLADY'

'BLACK MAGIC', LCl, dr, 1953; flowers blackish crimson, medium, occasional repeat bloom, intense fragrance; [Guinee seedling]; Hamilton

Black Magic *see* TANkalgic

Black Night® *see* 'HUBAR'

'BLACK OPAL', HT, dr, 1957; flowers dark velvety red; [Mirandy X Tassin]; Ulrick, L.W.

Black Pearl *see* 'DELURT'

'BLACK PRINCE', HP, dr, 1866; flowers dark crimson shaded black, large, dbl., cupped, recurrent bloom, intense fragrance; vigorous growth; Paul, W.

'BLACK RUBY', HT, dr, 1965; flowers center crimson, large, dbl., 40 petals, cupped; tall growth; [Rome Glory X Impeccable]; Delbard-Chabert; Cuthbert

Black Ruby, LCl, mr

Black Sapphire *see* 'LAVSAPH'

Black Satin *see* 'BETSAT'

'BLACK TAQUIN', LCl, dr, 1955; flowers dark crimson-maroon; vigorous growth; [Honour Bright X Guinee]; Eacott

'BLACK TEA', HT, r, 1973; flowers brown, urn-shaped, blooms borne usually singly, dbl., 32 petals, slight fragrance; deep brown, hooked prickles; foliage medium, dark, semi-glossy; medium, bushy growth; [Hawaii X (Aztec X (Goldilocks X Fashion))]; Okamoto, K.; K. Hirakata Nursery

'BLACK VELVET', HT, dr, 1960; bud ovoid; dbl., 28 petals, 5–5.5 in., exhibition form, intense fragrance; foliage leathery, dark; vigorous, upright growth; [New Yorker X Happiness]; Morey, Dr. Dennison; J&P

Blackberry Blossom, Pol, w, 1980; Semple

Blackberry Nip *see* 'SOMNIP'

Blackberry Rose *see* **R. RUBUS**

Blackboy *see* **'BLACK BOY'**

'BLAIRII NO. 1', HCh, mp; flowers bright rose, sometimes tinged red, large, semi-dbl., cupped, intense fragrance; branching growth; liable to injury from severe cold (less hardy than Blairii No.2); [Parks' Yellow Tea-scented China X Hardy rose]; Blair, 1845

'BLAIRII NO. 2', HCh, lp, 1845; flowers rosy blush, large, dbl., moderate fragrance; ; vigorous (up to 15 ft. hardy.; [Parks' Yellow Tea-scented China X Hardy rose]; Blair

'BLAKENEY'S RED', HT, mr, 1962; bud long, pointed; flowers currant-red, base yellow, dbl., 45 petals, 5 in., exhibition form, moderate fragrance; foliage leathery; moderate, bushy growth; [Karl Herbst X Peace]; Blakeney; Eddie

'BLANC DE VIBERT', P, w, 1847; dbl., sometimes recurrent bloom; foliage light green; Vibert

Blanc Dot *see* **'BLANCHE DOT'**

'BLANC DOUBLE DE COUBERT', HRg, w, 1892; bud pleasing; flowers half-open white, fairly large, dbl., repeat bloom, intense fragrance, even at night; foliage very rugose; vigorous (5-7 ft.) growth; [R. rugosa X Sombreuil]; Cochet-Cochet

Blanc Lafayette *see* **'DAGMAR SPÄTH'**

Blanc Meillandécor *see* 'MEICOUBLAN'

Blanc Parfait *see* **'POMPON BLANC PARFAIT'**

'BLANC PUR', Misc OGR, w, 1827; Mauget

Blanc Queen Elizabeth *see* **'WHITE QUEEN ELIZABETH'**

'BLANCA', F, w, 1965; flowers pure white, large blooms in clusters, dbl., slight fragrance; vigorous growth; [Purpurine X (Papillon Rose X Sterling Silver)]; Lens

'BLANCHE AMIET', HT, op, 1921; flowers coppery salmon, passing to clear rose; Turbat

Blanche Colombell *see* DELgribla

Blanche Comète *see* 'DELOBLAN'

'BLANCHE DE BELGIQUE', A, w; (Blanche Superbe); flowers pure white, very large, dbl.

'BLANCHE DE BELGIQUE', A, lp; (Blush Belgique); flowers blush, large, semi-dbl.; [R. alba X R. canina]; Vibert, 1817

'BLANCHE DOT', HT, w, 1962; (Blanc Dot); flowers snow-white, well-formed, large, dbl., 33 petals, slight fragrance; [White Knight X Virgo]; Dot, Pedro

'BLANCHE DURANTHON', N, w, 1800; flowers medium, dbl.; Nabonnand; (Sangerhausen)

'BLANCHE FROWEIN', LCl, yb, 1916; flowers coppery yellow, moderate fragrance; Leenders, M.

'BLANCHE MALLERIN', HT, w, 1941; bud long, pointed; flowers pure white, dbl., 33 petals, 4 in., exhibition form; foliage leathery, glossy; vigorous growth; [Edith Krause X White Briarcliff]; Mallerin, C.; A. Meilland;, C-P

'BLANCHE MESSIGNY', HT, ly, 1923; flowers creamy yellow, large, dbl., 45 petals, moderate fragrance; vigorous, bushy growth; Gillot, F.

'BLANCHE MOREAU', M, w, 1880; bud well mossed; flowers pure white, large blooms in clusters, dbl., repeat bloom, moderate fragrance; lax growth; [Comtesse de Murinais X Perpetual White Moss]; Moreau et Robert

Blanche Mousseuse *see* **'SHAILER'S WHITE MOSS'**

'BLANCHE NABONNAND', T, w, 1883; flowers creamy white tinged lemon or flesh, dbl., globular, moderate fragrance; Nabonnand, G.

Blanche Neige *see* 'MACCARPE'

'BLANCHE NEIGE', Pol, w, 1929; Koster

Blanche Nouvelle, M, w

'BLANCHE ODORANTE', HT, w, 1952; bud very long, pointed; flowers purest white, large, very dbl., intense fragrance; vigorous growth; [Pole Nord X Neige Parfum]; Caron, B.; EFR

Blanche Pasca® *see* 'LENIP'

'BLANCHE REBATEL', Pol, pb, 1889; flowers carmine and white; Bernaix, A.

'BLANCHE SIMON', w, 1862; flowers large, dbl.; Moreau et Robert; (Sangerhausen)

Blanche Simple, HGal, w

Blanche Superbe *see* **'BLANCHE DE BELGIQUE'**

Blanche Unique *see* **'UNIQUE BLANCHE'**

'BLANCHE WIMER', Min, pb, 1986; flowers light yellow and pink blend, blooms borne usually singly, dbl., 66 petals, cupped, slight, fruity fragrance; no fruit; slightly hooked, reddish-green prickles; foliage medium, medium green, semi-glossy; tall, upright growth; [Pink Petticoat X Seedling]; Shaw, Dr. John, 1985

'BLANCHEFLEUR', C, w, 1835; flowers white tinted blush, dbl.; vigorous growth; Vibert

'BLANDA EGRETA', HMult, mp, 1926; flowers small, dbl.; Bruder Alfons; (Sangerhausen)

Blanik, F, dr, 1964; flowers medium, dbl.; Vecera, L.; (Sangerhausen)

Blank, HP, mp

'BLARNEY', Cl HT, mp, 1934; flowers pink, base apricot, dbl., recurrent bloom, moderate fragrance; foliage leathery, dark; vigorous, climbing growth; [Irish Charm sport]; Howard Rose Co.

Blastoff™ *see* 'MORFLASH'

'BLATENSKÀ KRÀLOVNA', HWich, mp, 1937; flowers medium, dbl.; Böhm, J.; (Sangerhausen)

'BLATNÁ', HT, dr, 1927; flowers velvety dark red, dbl., slight fragrance; dwarf growth; [Lieut. Chaure X Oskar Cordel]; Böhm, J.

'BLAUWE DONAU', F, m, 1971; (Blue Danube); bud ovoid; flowers full, open, semi-dbl., 10–12 petals, 3 in., intense fragrance; foliage glossy; bushy growth; [Orangeade X Sterling Silver]; Verschuren, Ted; Verschuren

'BLAYDON RACES', F, yb, 1976; flowers scarlet shading to yellow, 2.5 in., slight fragrance; foliage glossy, dark, leathery; vigorous, tall, upright growth; [Bobby Shafto X (Arthur Bell X Piccadilly)]; Wood

'BLAZE', LCl, mr, 1932; flowers bright scarlet, dbl., 20–25 petals, 2–3 in., cupped, blooms in large clusters, recurrent bloom, slight fragrance; foliage leathery, dark; very vigorous, climbing growth; [Paul's Scarlet Climber X Gruss an Teplitz]; Kallay; J&P

'BLAZE AWAY', F, or, 1979; bud pointed; flowers scarlet-vermilion, single, 7 petals, 3.5 in., slight fragrance; vigorous, bushy growth; [(Karl Herbst X Crimson Glory) X Sarabande]; Sanday, John

Blaze Superier *see* **'DEMOKRACIE'**

Blazing Lights, HT, ob, 1995; K&S

Bleak House, HRg, lp; Bleak House, 1995; (Weatherly, L.)

'BLEBAR', F, ob, 1981; (**Wagbi**); bud pointed; flowers orange-pink, blooms borne 3-5 per cluster, dbl., 20 petals, flat, moderate fragrance; large, red prickles; foliage large, dark; vigorous, upright growth; Barrett, F.H.; John Mattock, Ltd.

Blenheim *see* TANmurse

Bles Bridges *see* KORampli

'BLESMA SOUL', HT, lp, 1982; bud long, pointed; dbl., 36 petals, borne singly or in small clusters, intense fragrance; large based, reddish-brown prickles; foliage light green; upright growth; [Pascali X Fragrant Cloud]; Anderson's Rose Nurseries

'BLESS MY TIME', Gr, lp, 1983; flowers clear light pink, veins sometimes darker; [Queen Elizabeth sport]; Orr, Rudolph F.

Blessed Event *see* 'LAVFUN'

'BLESSINGS'®, HT, op, 1967; flowers medium coral-salmon, large, dbl., 30 petals, exhibition form, moderate fragrance; [Queen Elizabeth X Seedling]; Gregory, 1968

Blessings, Climbing, Cl HT, 1975; (Cavriglia)

'BLESTOGIL', S, pb, 1982; (**Festival Fanfare**); flowers dark pink with paler pink stripes, which become nearly white; [Fred Loads sport]; Ogilvie, W.D.

'BLEU MAGENTA', HMult, m; flowers dark crimson-purple, borne in clusters, dbl., slight fragrance; foliage dark

Blickfang, F, dr; flowers reddish-violet, large, very dbl., no fragrance; VEG; (Sangerhausen)

'BLITHE SPIRIT', HT, lp, 1964; flowers medium, dbl., slight fragrance; foliage leathery; vigorous, spreading, upright growth; [Fandango X Seedling]; Armstrong, D.L. & Swim; Armstrong Nursery

Blizzard *see* 'JACDRIFT'

Bloemfontein *see* KORtrolle

'BLÖHM & VOSS', F, or; flowers bright orange-scarlet good trusses

'BLONDE BOMBSHELL', F, my, 1995; flowers medium yellow fading to pale yellow with pink edging, full (26–40 petals), medium blooms borne in small clusters; fragrant, numerous prickles; foliage medium, dark green, glossy, upright, medium growth; [Seedling X Seedling]; Bees of Chester; L. W. van Geest Farms Ltd., 1995

Blondie *see* 'JACLIY'

Blondie *see* LAPdil

'BLONDINE', HT, lp, 1954; bud pointed, globular; flowers blush-pink, dbl., 60 petals, 4 in., moderate fragrance; foliage leathery; upright growth; [Catalina sport]; Grillo

'BLONDINE', Cl HT, w; flowers pearl-white, well formed, large, intense fragrance; foliage clear green; vigorous growth; [Comtesse Vandal, Climbing X Michèle Meilland, Climbing]; Arles; Roses-France

Blood-red ChinaRose *see* **'SANGUINEA'**

'BLOODSTONE', HT, or, 1951; flowers large, dbl., 24 petals, exhibition form, slight fragrance; foliage dark, coppery green; vigorous growth; [The Queen Alexandra Rose X Lord Charlemont]; McGredy, Sam IV

Bloomer Girl *see* 'TINGIRL'

Bloomfest™ *see* 'JUDFEST'

'BLOOMFIELD ABUNDANCE', F, lp, 1920; flowers light salmon-pink, dbl.; foliage glossy, dark; bushy growth; GM, Portland, 1919; [Sylvia (probably R.) X Dorothy Page-Roberts]; Thomas; B&A

'BLOOMFIELD BEVERLY', HT, or, 1924; flowers orange-crimson, dbl., moderate fragrance; vigorous growth; [Mary, Countess of Ilchester X Mme Edouard Herriot]; Thomas

'BLOOMFIELD BRILLIANT', LCl, op, 1931; flowers light salmon with orange glow, large, semi-dbl., 18 petals, moderate fragrance; vigorous growth; [Mme Abel Chatenay X Kitty Kininmonth]; Thomas; H&S

Bloomfield Cemetary, T, ab

'BLOOMFIELD COMET', HMsk, op, 1924; bud long, pointed, reddish orange; flowers orange, base yellow, single, 5 petals, 3.5 in., moderate fragrance; foliage sparse, light bronze, soft; [Duchess of Wellington X Danaë]; Thomas; B&A

'BLOOMFIELD COMPLETENESS', HMsk, ab, 1931; flowers deep orange-yellow, dbl., moderate fragrance; [Bloomfield Perfection X Mme Butterfly]; Thomas

'BLOOMFIELD COURAGE', HWich, rb, 1925; flowers dark velvety red, center white, prominent yellow stamens, small, single, non-recurrent; foliage dark; vigorous, climbing or pillar (20 ft.) growth; Thomas; B&A;, H&S

'BLOOMFIELD CULMINATION', HMsk, pb, 1924; bud long, pointed; flowers rose-pink, center white, single, 3 in., recurrent bloom, slight fragrance; foliage leathery; moderately vigorous growth; [Sheila Wilson X Danaë]; Thomas; B&A

'BLOOMFIELD DAINTY', HMsk, my, 1924; bud long, pointed, deep orange; flowers clear canary-yellow, single, 2 in., moderate fragrance; foliage glossy; moderately vigorous growth; [Danaë X Mme Edouard Herriot]; Thomas; B&A

'BLOOMFIELD DAWN', HMsk, pb, 1931; bud long, slender, rose-pink; flowers light pink, base yellow, reverse deep pink, large blooms, semi-dbl., moderate fragrance; long, strong stems; [Seedling, Climbing X Bloomfield Progress]; Thomas; Armstrong Nursery

'BLOOMFIELD DECORATION', HMsk, pb, 1925; flowers cerise-pink, center white, prominent golden stamens, open, single, moderate fragrance; foliage glossy; [Sylvia X Arndt]; Thomas; B&A, 1927;, H&S, 1927

'BLOOMFIELD DISCOVERY', HMsk, pb, 1925; flowers pink, reverse darker, single; foliage dark; moderately vigorous growth; [Danaë X (Frau Karl Druschki X Mme Caroline Testout)]; Thomas; B&A

Bloomfield Endurance *see* **'W. FREELAND KENDRICK'**

'BLOOMFIELD EXQUISITE', Cl HT, mp, 1924; flowers clear pink, dbl., recurrent bloom, moderate fragrance; vigorous growth; [Gloire de Dijon X Gruss an Teplitz]; Thomas; H&S

'BLOOMFIELD FASCINATION', HMsk, ly, 1924; flowers light canary yellow, small, dbl., slight fragrance; foliage rich bronze green, soft; [Danaë X Mme Lautette Messimy]; Thomas; B&A

'BLOOMFIELD FAVORITE', HMsk, w, 1924; flowers pinkish cream, dbl., slight fragrance; [Debutante X Moonlight]; Thomas

'BLOOMFIELD FLAME', HT, rb, 1930; bud long, pointed, flame-red; flowers crimson-flame, center orange-yellow, large, dbl., 22 petals, moderate, spicy fragrance; foliage leathery, glossy, dark bronze; vigorous, bushy growth; [Louise Crette X Mme Charles Lutaud]; Thomas; H&S

Bloomfield Improvement *see* **'EDNAH THOMAS'**

Bloomfield Loveliness *see* **'SOPHIE THOMAS'**

'BLOOMFIELD LUSTRE', LCl, op, 1931; flowers salmon-pink, base yellow, dbl., moderate fragrance; [Hortulanus Budde X Souv. de Mme Léonie Viennot]; Thomas; H&S

'BLOOMFIELD MAGIC', LCl, op, 1924; flowers light salmon to cream, flat, moderate fragrance; moderately vigorous growth; [Gloire de Dijon X Frau Berta Gurtler]; Thomas

'BLOOMFIELD MYSTERY', LCl, lp, 1924; flowers silver-pink, tinged yellow, 2 in., flat, moderate fragrance; moderately vigorous growth; [Blanche Frowein X Bloomfield Abundance]; Thomas; B&A

'BLOOMFIELD PERFECTION', HMsk, w, 1925; flowers cream-yellow suffused lilac, dbl., recurrent bloom, slight fragrance; [Danaë X Bloomfield Abundance]; Thomas; B&A, 1927

'BLOOMFIELD PERPETUAL', HP, w, 1920; flowers resembling the cherokee rose, single; [Iceberg X Frau Karl Druschki]; Thomas; B&A

'BLOOMFIELD PROGRESS', HT, mr, 1920; flowers glowing red, dbl., intense fragrance; vigorous growth; [Mary, Countess of Ilchester X Gen. MacArthur]; Thomas; B&A

'BLOOMFIELD QUAKERESS', Cl T, ly, 1931; flowers small blooms, semi-dbl., free recurrent bloom; foliage glossy, light; long stems; vigorous growth; [Chance Safrano seedling]; Thomas; Armstrong Nursery

'BLOOMFIELD ROCKET', LCl, dp, 1925; flowers dark pink, center lighter, very large, single, slight fragrance; very vigorous (6-8 ft) growth; [Mme Caroline Testout X Ulrich Brunner Fils]; Thomas; B&A

Bloomin' Easy™ *see* 'AROtrusim'

Bloomin' Pretty, LCl, w, 1999; flowers white with pale pink paint, semi-dbl.; Williams, J. Benjamin

Blooming Easy *see* 'AROtrusim'

'BLOOMSDAY', F, ob, 1981; flowers orange, marked brown, reverse deep gold, loose, 4-5 per cluster, dbl., 28 petals, moderate fragrance; straight, red-brown prickles; foliage matt, green; vigorous, upright growth; [Belinda X (Maxi X Joyfullness)]; McCann, Sean; Hughes Roses

Bloomtown *see* 'TALblo'

'BLOSSOM', HT, pb, 1925; flowers red over peach-pink, reverse pinkish yellow; Beckwith

'BLOSSOM HILL', F, ob, 1957; flowers orange, medium in clusters, single, 10 petals, 2.5 in., cupped, slight fragrance; foliage dark, glossy; vigorous, upright, bushy growth; Kordes; Morse

Blossom Magic *see* MEImanoir

'BLOSSOMTIME', LCl, mp, 1951; bud pointed; flowers pink, reverse deeper, blooms in clusters of 3-8, dbl., 38 petals, 4 in., exhibition form, repeat bloom, intense fragrance; tall shrub or moderate climbing (6-7 ft.) growth; [New Dawn X Seedling]; O'Neal; Bosley Nursery

Blue Angel *see* 'RENangel'

Blue Bajou® *see* KORkultop

Blue Bell *see* MEInalpir

Blue Bird, HT; Poulsen; (Cavriglia)

'BLUE BOY', M, m, 1958; bud ovoid; flowers deep reddish violet, large, dbl., exhibition form, non-recurrent, intense fragrance; foliage light green, glossy; vigorous (3 ft.), upright, bushy growth; [Louis Grimmard X Independence]; Kordes

Blue Boy, HT, m

Blue Carpet *see* 'HARquillypond'

'BLUE CHATEAU', HT, m, 1999; flowers pale violet, very dbl., 50 petals, exhibition form, intense fragrance; to 4 ft growth; [Madame Violet X seedling]; Teranishi, K.; Itami Rose Nursery, 1994, Japan

Blue Chip *see* 'JACave'

Blue Danube *see* **'BLAUWE DONAU'**

Blue Delight, HT, m, 1980; Kasturi

'BLUE DIAMOND', HT, m, 1963; flowers lavender, dbl., 35 petals, 4 in., moderate fragrance; foliage dark, coppery; vigorous, compact, bushy growth; [Purpurine X (Purpurine X Royal Tan)]; Lens

Blue Friendship, S, m, 1984; Verschuren

Blue Girl *see* 'KORgi'

'BLUE GLOW', HT, m, 1982; bud pointed; flowers pinkish-mauve, large blooms borne singly and 3-5 per cluster, dbl., 45 petals, exhibition form, slight fragrance; light brown, triangular prickles; foliage light green; upright, branching growth; [Silent Night X Blue Moon]; Cattermole, R.F.

'BLUE HEAVEN', HT, m, 1971; bud ovoid; flowers large, dbl., exhibition form, intense fragrance; foliage large, glossy; vigorous, upright growth; [(Sterling Silver X Simone) X Song of Paris]; Whisler, D.; Gro-Plant Industries

Blue Ice *see* 'LAVCHIP'

Blue Jay, F, m

Blue Magic, HT; Dot, Pedro; (Cavriglia)

Blue Magic, Min, m, 1986

Blue Mikey, HP, m

'BLUE MIST', Min, m, 1970; bud short, rounded; flowers soft pink to lavender, micro-mini, small, dbl., 23 petals, intense fragrance; foliage soft; vigorous, bushy, rounded growth; [Seedling X Seedling]; Moore, Ralph S.; Sequoia Nursery

Blue Monday *see* 'TANSI'

Blue Monday, Climbing *see* **'BLUE MOON, CLIMBING'**

Blue Moon® *see* 'TANSI'

'BLUE MOON, CLIMBING', Cl HT, m, 1981; (Blue Monday, Climbing, Mainzer Fastnacht, Climbing, Sissi, Climbing); [Blue Moon sport]; Mungia, Fred A., Sr.; Montebello Rose Co., Inc.

Blue Nile® *see* 'DELNIBLE'

Blue Ocean, HT, m, 1983; Hande

Blue Parfum® *see* 'TANTIFUM'

Blue Perfume *see* 'TANTIFUM'

Blue Peter *see* 'RUIBLUN'

Blue Rambler *see* **'VEILCHENBLAU'**

Blue Ribbon *see* 'AROLICAL'

Blue River® *see* 'KORSICHT'

'BLUE RIVER', HT, m, 1973; bud long, pointed; flowers mauve-magenta, large, dbl., cupped, intense fragrance; foliage glossy; vigorous, upright growth; RULED EXTINCT 6/84 ARM; [Mainzer Fastnacht X Silver Star]; Kordes; Horstmann

Blue Rosalie *see* **'VEILCHENBLAU'**

Blue Skies *see* 'BUCBLU'

Blue Sky *see* **'AOZORA'**

Blue Sky, HT, m, 1982; Dot

'BLUE STAR, CLIMBING', Cl HT, m, 1963; flowers lavender-blue, small blooms in cluster, dbl., flat, profuse, non-recurrent bloom, slight fragrance; foliage glossy; vigorous (8 ft.) growth; Thompson, M.L.

Blue Violet, HT, m, 1981; Perry

Blue Work, HT, m

Blueberry Hill® *see* 'WEKCRYPLAG'

Blueblood® *see* 'LAVBLU'

Bluenette *see* 'RUIBLUN'

Bluesette® *see* 'LENMAU'

Bluewunder *see* KORedan

Bluhendes Barock *see* NOAbell

Bluhwunder® *see* KORedan

'BLUMEN DANKERT', HMult, mp, 1904; flowers carmine-pink, medium, semi-dbl.; Kiese; (Sangerhausen)

'BLUMENSCHMIDT', T, yb, 1906; flowers primrose-yellow, outer petals rose-pink; [Mlle Franziska Kruger sport]; Schmidt, J.C.

'BLUMENSCHMIDT'S ELFENKÖNIGIN', HT, w, 1939; flowers ivory-white, center orange, very large, dbl., exhibition form, moderate, lily-of-the-valley fragrance; long, strong stems; bushy, dwarf growth; [Ophelia X Julien Potin]; Weigand, C.; Schmidt, J.C.

'BLUMENSCHMIDTS SONNTAGSKIND', F, mp, 1945; flowers medium, dbl.; Vonholdt; (Sangerhausen)

Blush Baby, Min, 1986; Pearce, C.A.; (Cavriglia)

Blush Belgique *see* **'BLANCHE DE BELGIQUE'**

'BLUSH BOURSAULT', Bslt, pb; (Calypso); flowers blush, center deep flesh, very large blooms, tending to ball, very dbl., globular; foliage remains longer than other boursaults; pendulous growth habit

'BLUSH CHINA', HCh, lp, 1903; flowers medium, semi-dbl., slight fragrance; Cant, B. R.; (Sangerhausen)

'BLUSH DAMASK', D, lp, 1759; flowers center rose, shading to pale blush on outside petals, small, dbl.

'BLUSH HIP', A, lp; flowers soft pink, green eye, dbl.; vigorous growth

Blush Maman Cochet *see* **'WILLIAM R. SMITH'**

'BLUSH MOSS', M, lp; flowers blush, center pinkish when first open, well-mossed, large, dbl., cupped; moderate branching growth

Blush Musk *see* **'FRASER'S PINK MUSK'**

'BLUSH NOISETTE', N, w; flowers pinkish white; (14); [Champneys' Pink Cluster seedling]; Noisette, before 1817

'BLUSH QUEEN', HT, lp, 1924; flowers blush-pink, dbl., moderate fragrance; Cant, F.

'BLUSH RAMBLER', HMult, lp, 1903; flowers blush-pink, blooms in clusters, semi-dbl., cupped, non-recurrent bloom; vigorous, climbing (10-12 ft.) growth; [Crimson Rambler X The Garland]; Cant, B. R.

'BLUSHING BEAUTY', LCl, lp, 1934; flowers three-toned shell-pink, very large; very vigorous growth; Burbank; Stark Bros.

Blushing Blue *see* 'RENBLUE'

'BLUSHING BRIDE', HT, w, 1918; flowers white, center blush, dbl.; RULED EXTINCT 5/90; Dickson, H.

'BLUSHING BRIDE', HT, W, 1930; flowers white tinged pink, large, dbl., intense fragrance; vigorous growth; RULED EXTINCT 5/90; [Mme Butterfly X Premier]; Hill, Joseph H., Co.

Blushing Bride *see* 'GREBLUB'

Blushing Bride *see* HARfling

Blushing Dawn *see* 'WEEBLUSH'

Blushing Groom *see* 'SEAGRU'

'BLUSHING JEWEL', Min, lp, 1958; bud ovoid; flowers blush-pink, overcast rose-pink, open, dbl., 45–50 petals, .5–.75 in., moderate fragrance; low, compact growth; [Dick Koster sport X Tom Thumb]; Morey, Dr. Dennison; J&P

Blushing June *see* 'LECBLU'

'BLUSHING LUCY', LCl, lp, 1938; flowers pale pink, white eye, large cluster, semi-dbl., profuse, late bloom, intense fragrance; foliage glossy; vigorous growth; Williams, A.

Blushing Maid *see* 'MACKEPA'

'BLUSHING QUEEN', Gr, w, 1976; bud pointed; flowers near white, center blush-pink, dbl., 38–40 petals, exhibition form, moderate fragrance; vigorous, upright growth; [Queen Elizabeth sport]; Baker, Larry

Blushing Rose *see* **'COY COLLEEN'**

Bo *see* PEKoubo

Bo, HT, lp; flowers small

'BOB COLLARD', F, mr, 1986; flowers very luminous, brilliant red, reverse deeper red, medium, single, 11 petals, cupped, borne singly or in sprays of 5-10, slight, fruity fragrance; round, cupped, average, red fruit; foliage medium, medium green, glossy; upright, bushy, medium growth; [Happiness X Copenhagen]; Turley, V.G., 1987

Bob Greaves *see* FRYzippy

'BOB HOPE', HT, mr, 1966; bud urn-shaped; flowers scarlet red, dbl., 35–40 petals, 6 in., exhibition form, borne mostly singly, intense, damask fragrance; foliage dark, leathery; vigorous, tall, upright growth; [Friedrich Schwartz X Kordes' Perfecta]; Kordes, R.; J&P

Bob Kennedy, HT; Ingegnoli; (Cavriglia)

'BOB WOOLLEY', HT, ab, 1970; flowers peach-pink, reverse lemon, dbl., 60 petals, 5 in., slight fragrance; foliage matt, green; [Gavotte X Golden Scepter]; Sanday, John

'BOBBIE JAMES', HWich, w, 1961; flowers creamy white, blooms in trusses, single, 7–9 petals, 2 in., cupped, intense fragrance; foliage glossy; vigorous growth; Sunningdale Nursery

'BOBBIE LUCAS', F, op, 1967; flowers deep salmon-orange, well-formed, blooms in clusters, 3.5 in., slight fragrance; foliage dark; [Margot Fonteyn X Elizabeth of Glamis]; McGredy, Sam IV

Bobbie Robbie *see* **'SOUR DE J. CHABERT'**

Bobbie Vesely *see* 'RESBOBIE'

'BOBBINK WHITE CLIMBER', LCl, w, 1951; flowers creamy yellow to pure white, dbl., 45–50 petals, 3 in., recurrent bloom, moderate fragrance; foliage dark, glossy; vigorous, climbing growth; [Dream Girl X Seedling]; Jacobus; B&A

'BOBBY CHARLTON', HT, pb, 1974; flowers deep pink, reverse silver, well-formed, dbl., 38 petals, 6 in., moderate, spicy fragrance; foliage dark, leathery; GM, Baden-Baden, 1976 GM, Portland, 1980; [Royal Highness X Prima Ballerina]; Fryer, Gareth; Fryer's Nursery, Ltd.

Bobby Dazzler *see* 'ROSELLA'

'BOBBY SHAFTO', HT, my, 1967; [Piccadilly sport]; Wood

Bobino, F, mr

'BOBO', HT, mr, 1979; bud long; flowers urn-shaped blooms borne singly, dbl., 30 petals, intense, fruity fragrance; Perry, Astor, 1981

'BOBOLINK', Min, dp, 1959; flowers rose-pink, base near white, dbl., 50 petals, 1–1.5 in., slight, fruity fragrance; foliage leathery, glossy; vigorous, bushy (18 in) growth; [(R. wichurana X Floradora) X (Oakington Ruby X Floradora)]; Moore, Ralph S.; Sequoia Nursery

Bobravka, S, op, 1975; flowers dark orange-pink with red tones, medium, dbl., moderate fragrance; Urban, J.; (Sangerhausen)

Bob's Koster, Pol, dp

Bob's Peach *see* 'WEBBREAD'

'BOCCA NEGRA', HMult, dr, 1910; flowers purple-crimson, center white, borne in clusters of 15-20, cupped; vigorous growth; Dubreuil

'BOCCACCIO', F, or, 1963; flowers bright scarlet-red, large blooms in clusters, dbl., 30–40 petals; foliage glossy; vigorous, upright growth; [Atombombe X Seedling]; Verschuren, A.; Stassen

'BOCCACE', HP, 1859; Moreau et Robert; (Cavriglia)

'BOCHER', HT, pb, 1990; very dbl., slight fragrance; foliage medium, medium green, matt; upright growth; [Prima Ballerina X Alec's Red]; Rodgers, Shafner R.

Bodhisatva *see* 'VIRSPLASH'

'BOHEMIA', HT, mp, 1928; flowers pure rose-pink, semi-dbl., moderate fragrance; [Mrs Franklin Dennison X Mrs Henry Morse seedling]; Böhm, J.

'BOHÉMIENNE', HT, or, 1954; dbl., 45 petals; bushy growth; Mallerin, C.; EFR

'BÖHM JUNIOR', HT, dp, 1935; flowers carmine-red, dbl., intense fragrance; foliage glossy; bushy growth; [Laurent Carle X (Paul's Scarlet Climber X Ethel Somerset)]; Böhm, J.

'BÖHM SENIOR', HT, mr, 1938; flowers large, dbl.; Böhm, J.; (Sangerhausen)

'BÖHMOROSE', HT, mp, 1935; flowers rosy carmine to light rose-pink, very large, dbl.; foliage glossy, dark; very vigorous growth; [Gen. MacArthur X Laurent Carle]; Böhm, J.

'BÖHMOVA AZUROVÀ', HP, m, 1934; flowers dark violet-pink, large, dbl.; Böhm, J.; (Sangerhausen)

'BÖHMOVA POPELKA', Pol, dr, 1934; flowers dark blood-red; foliage curiously variegated; Böhm, J.

'BÖHM'S CLIMBER', Cl HT, mr, 1935; flowers medium, dbl., moderate fragrance; Böhm, J.; (Sangerhausen)

'BÖHM'S TRIUMPH', HT, dr, 1934; flowers very large, semi-dbl., exhibition form; foliage dark; vigorous, bushy growth; [Vaterland X Lord Charlemont]; Böhm, J.

Bojangles *see* 'JACSUN'

Boksburg Fantasia *see* 'KORNITZEL'

Bolchoi *see* MEIruchka

Bolcholi *see* MEIzuzes

Bolero *see* POUlbo

Bolero, HT, 1963; Pironti, N.; (Cavriglia)

'BOLÉRO', HT, rb, 1937; flowers nasturtium-red and gold, dbl.; foliage leathery; long stems; very vigorous, bushy growth; Gaujard

'BOLÉRO', F, ob, 1958; bud pointed; flowers orange, large, dbl.; very vigorous growth; Buyl Frères

Bombon, Min, 1967; Dot, Simon; (Cavriglia)

'BON ACCORD', HT, pb, 1967; flowers pink shaded silver, high-pointed, 4.5 in., moderate fragrance; foliage glossy; [Prima Ballerina X Percy Thrower]; Anderson's Rose Nurseries

'BON SILÈNE', T, dp; bud well-formed; flowers deep rose, large, dbl., recurrent bloom, moderate fragrance; vigorous growth; Hardy, before 1837

'BON SILÈNE BLANC', T, ly, 1885; (White Bon Silène); flowers pale yellow to creamy white, large; Morat

'BON VOYAGE', HT, 1960; (Voeux de Bonheur); Delbard-Chabert; Stark Bros., 1969

'BONA WEILSCHOTT', HT, mp, 1889; flowers large, dbl., moderate fragrance; Soupert & Notting; (Sangerhausen)

Bonanza® *see* 'KORMARIE'

'BONANZA', HT, 1957; flowers yellow shaded red, slight fragrance; foliage dark; vigorous growth; RULED EXTINCT 5/82 ARM; Verbeek

'BONAVISTA', HRg, lp, 1977; bud ovoid; flowers open, dbl., 20 petals, 2 in., abundant, repeat bloom, intense fragrance; foliage yellow-green; upright, bushy growth; [Schneezwerg X Nemesis]; Svedja, Felicitas; Canada Dept. of Agric.

'BON-BON', F, pb, 1974; bud ovoid; flowers deep rose-pink, reverse white, large, semi-dbl., slight fragrance; foliage dark; vigorous, bushy growth; AARS, 1974; [Bridal Pink X Seedling]; Warriner, William A.; J&P

Bonbon Hit® *see* POUlbon

'BOND STREET', HT, op, 1965; flowers deep salmon-pink, dbl., 75 petals, 4.5 in., intense fragrance; [Radar X Queen Elizabeth]; McGredy, Sam IV; McGredy

Boneyard Yellow, F, my, 1996

'BONFIRE', HMult, mr, 1928; flowers scarlet, in clusters of 20-25, dbl.; foliage light; very vigorous growth; [Crimson Rambler X R. wichurana]; Turbat

'BONFIRE NIGHT'®, F, rb, 1971; flowers red, shaded yellow-orange, 19 petals, 3.5 in., globular, slight fragrance; foliage matt; [Tiki X Variety Club]; McGredy, Sam IV; McGredy

Bonfire of Artec, HT; (Cavriglia)

Bonhomme *see* 'LAVHOMME'

'BONICA'®, F, or, 1958; flowers scarlet, dbl., 70 petals, exhibition form, blooms in clusters, slight fragrance; foliage dark, leathery; vigorous, bushy growth; [(Alain X Independence) X Moulin Rouge]; Meilland, F.; C-P;, URS

Bonica™ *see* 'MEIDOMONAC'

Bonica '82 *see* 'MEIDOMONAC'

'BONITA', LCl, dr, 1958; flowers crimson, medium, in clusters of 3-8, semi-dbl., 15 petals, moderate fragrance; foliage dark, glossy; vigorous (10-15 ft) growth; Knight, A.T.

'BONJOUR', Gr, mr, 1965; flowers bright red, medium, dbl., slight fragrance; foliage leathery; very vigorous, bushy growth; [Mignonne X Miss Universe]; Gaujard

'BONN', HMsk, or, 1950; flowers orange-scarlet, blooms in trusses of 10, dbl., 25 petals, 4 in., moderate, musk fragrance; foliage glossy; upright, bushy (4-5 ft) growth; [Hamburg X Independence]; Kordes

'BONNE CHERE', S, rb, 1985; flowers bright velvet red, creamy eye, blooms in large clusters, single, 5 petals, 3.5 in., repeat blooming, slight fragrance; foliage medium, dark, leathery; vigorous, arching (to 4 ft.) growth; [(R. nutkana X Baronne Prevost) X Alika]; James, John

'BONNE FÊTE', HT, or, 1960; bud long, pointed; flowers orange-coral, large blooms, dbl., 25–30 petals, slight fragrance; foliage leathery, glossy; strong stems; vigorous, bushy growth; [Independence X Barcelona]; Delbard-Chabert

Bonne Nouvelle *see* **'GOOD NEWS'**

'BONNE NUIT', HT, dr, 1966; bud pointed; flowers blackish red, medium, dbl., exhibition form, moderate fragrance; foliage sparse, glossy; bushy growth; Combe; Wyant, 1956

'BONNE NUIT, CLIMBING', Cl HT, dr, 1964; Kashimoto; Itami Rose Nursery

'BONNIE', S, dp, 1956; similar to Aylsham but taller and more rapid and vigorous in growth; [Hansa X R. nitida]; Wright, Percy H.

'BONNIE ANNE', HT, pb, 1975; flowers pink, reverse yellow and pink, dbl., 30 petals, 3–3.5 in., moderate fragrance; foliage large, dark, glossy; [Prima Ballerina X Wendy Cussons]; MacLeod

'BONNIE BELLE', HWich, mp, 1911; flowers rose, large, single, slight fragrance; height 8-10 ft; Walsh

'BONNIE BESS', HT, op, 1929; flowers deep coral-pink, suffused copper, dbl., slight fragrance; [Wilhelm Kordes X (Crusader X Sunburst)]; Dale

'BONNIE DOONE', Cl HT, lp; [City of Little Rock seedling X Seedling]; Clark, A., 1941

'BONNIE HAMILTON', F, or, 1976; flowers vermilion-red, well-formed, dbl., 26 petals, 2.5 in., slight fragrance; foliage dark; [Anne Cocker X Allgold]; Cocker

'BONNIE JEAN', HT, rb, 1933; flowers carmine-cerise, base white, large, single; vigorous growth; Archer

'BONNIE MAID', F, pb, 1951; flowers silvery pink reverse deep pink, blooms in clusters of 5-8, semi-dbl., 17 petals, 3 in.; foliage leathery, dark; vigorous, bushy growth; GM, NRS, 1955; LeGrice

'BONNIE PINK', F, lp, 1964; bud ovoid; flowers geranium-pink overcast begonia-rose, with upright center petals, dbl., 35–40 petals, 4.5 in., exhibition form, moderate fragrance; foliage glossy; vigorous, upright growth; [White Garnette seedling X Hawaii]; Boerner; J&P

'BONNIE PRINCE', HMult, w, 1916; flowers white, center tinged yellow, dbl., moderate fragrance; excellent pillar growth; [Tausendschön X Seedling]; Cook, T.N.; Portland Rose Soc., 1924

'BONNIE PRINCE CHARLIE', HT, mr, 1960; bud pointed; flowers bright red, well formed, dbl., moderate fragrance; Cuthbert

Bonnie Prince Charlie's Rose *see* White Rose of York

'BONNIE SCOTLAND', HT, dp, 1976; flowers light red, dbl., 43 petals, 5 in., exhibition form, intense, damask fragrance; foliage glossy; [Wendy Cussons X Percy Thrower]; Anderson's Rose Nurseries

'BONNY', Min, mp, 1974; flowers deep pink, lighter reverse, dbl., globular, slight fragrance; foliage small, light, wrinkled; dwarf growth; [Zorina X Seedling]; Kordes

Bonny *see* KORniebon

Bonsoir *see* 'DICBO'

'BOOKER T. WASHINGTON', HMult, dr, 1930; flowers deep maroon, borne in clusters; Turbat; C-P

Boomerang *see* 'SPOBOOM'

'BOOYOL', S, mp, 1977; (**Rachel**); flowers medium, dbl., 35 petals, slight, delicate fragrance; foliage medium, medium green, matt; upright growth; Booth, Mrs. Rachel Y.

'BO-PEEP', Min, mp, 1950; bud ovoid, pointed; flowers rose-pink, micro-mini, very small, dbl., 28 petals, cupped, slight fragrance; foliage small, glossy; bushy, dwarf (5-8 in) growth; [Cécile Brunner X Tom Thumb]; deVink; C-P

'BORDEAUX', HMult, mr, 1908; flowers wine-red, small blooms in large cluster, dbl.; very vigorous growth; [Crimson Rambler X Blanche Rebatel]; Soupert & Notting

'BORDER BEAUTY', F, mr, 1957; flowers bright scarlet, medium in very large trusses (10-12 in), semi-dbl., 2–2.5 in.; foliage dark, glossy; vigorous, upright growth; [Floribunda seedling X Signal Red]; deRuiter; Gregory

'BORDER CORAL', F, op, 1957; flowers coral-salmon, blooms in trusses, semi-dbl., 2–2.5 in.; foliage dark, glossy; vigorous, spreading growth; [Signal Red X Fashion]; deRuiter; Gregory

'BORDER CORAL, CLIMBING', Cl F, op, 1966; Sanday; Gregory

'BORDER GEM', F, pb, 1961; bud ovoid; flowers geranium-pink, center light salmon-orange, blooms in clusters, dbl., 20 petals, 2.5–3 in., moderate fragrance; foliage leathery; compact, low growth; [(Navajo X Golden Dawn) X Pinocchio]; Morey, Dr. Dennison; J&P

'BORDER GOLD', F, dy, 1966; bud ovoid; flowers medium to deep yellow, very dbl., slight fragrance; foliage dark, leathery; vigorous, low, bushy growth; [Allgold X Pigmy Gold]; Morey, Dr. Dennison; Country Garden Nursery

'BORDER KING', Pol, mr, 1952; (Roi des Bordures); flowers bright strawberry-red, small large truss, semi-dbl., 16 petals; foliage dark, glossy; very vigorous growth; deRuiter; Gregory

'BORDER KING, CLIMBING', Cl Pol, mr, 1960; Gregory

'BORDER PRINCESS', F, op, 1951; flowers coral-pink shaded orange, reflexed, dbl., 22 petals, 3 in., exhibition form; foliage light, glossy; vigorous growth; Verschuren-Pechtold; Bentley

'BORDER QUEEN', F, op, 1951; (Reine des Bordures); flowers salmon-pink, center paler, reverse darker, large truss, single, 9 petals, 2.5 in.; foliage leathery, olive-green; vigorous, compact growth; (28); GM, NRS, 1950; deRuiter; Gregory;, Harkness

'BORDERER', Pol, pb, 1918; flowers salmon, fawn and pink, semi-dbl., slight fragrance; dwarf, spreading growth; [Jersey Beauty seedling]; Clark, A.; NRS Victoria

Bordura de Nea, F, lp; short; flowers small to medium sized, white-pink, flat, 8 petals, borne in clusters, slight; small, semi-glossy, medium green; [Bonica '82 X Incandescent]; St. Wagner, 1994; Res. Stn. f. Fruit Growing, Cluj, 1995, Romania

'BORDURE', Pol, mp, 1911; flowers carmine-pink, small, dbl.; Barbier; (Sangerhausen)

Bordure Blanche *see* DELbobla

Bordure de Nacrée *see* 'DELCROUF'

Bordure d'Or® *see* DELbojaune

Bordure Magenta, F, 1994; Delbard, Georges; (Cavriglia)

Bordure Nacrée® *see* 'DELCROUF'

Bordure Rose® *see* 'DELBARA'

Bordure Rose #2 *see* DELcoussi

Bordure Vermillon® *see* DELbover

Bordure Vive® *see* DELboviv

Bordurella®, F, pb

'BOREALIS', HKor, w, 1980; bud globular, pointed; dbl., 27 petals, exhibition form, borne singly, repeat bloom, moderate, sweet fragrance; gray prickles; foliage shiny; bushy growth; [Blanche Mallerin X Leverkusen]; James, John

'BORN FREE', Min, or, 1978; bud long, pointed; flowers brilliant orange-red, small, dbl., 20 petals, 1.5 in.; foliage dark; bushy, upright growth; [Red Pinocchio X Little Chief]; Moore, Ralph S.; Sequoia Nursery

Borsalino® *see* INTerhyro

Borussia®, S, ly, 1992; Poulsen

'BORYANA', HT, dr, 1977; flowers large, dbl., 55 petals; foliage dark, glossy; vigorous, upright growth; [Tallyho X Spartan]; Staikov, Prof. Dr. V.; Kalaydjiev and Chorbadjiiski

'BOSADSON', LCl, dr, 1997; (**Louisa Jane Morris**); flowers double, frilly, medium, 15–25 petals, slight fragrance; some prickles; foliage medium, medium green, semi-glossy; climbing (15ft.) growth; [Admiral Rodney X Evelyn Fison]; Bossom, W.E.

'BOSANNCOAT', Cl F, dp, 1996; (**Country Garden**); flowers bright pink, white center, blooms borne in large clusters, semi-dbl., 8–14 petals, 4 in., slight fragrance; few prickles; foliage medium, light green, glossy; spreading, climbing, tall (15 ft) growth; [Anne Harkness X Seedling]; Bossom, W.E.

'BOSANNE', F, op, 1990; (**Middlesex County**); bud pointed; flowers golden peach, medium, borne in sprays of 20-36, semi-dbl., 10 petals, flat, slight fragrance; foliage medium, medium green, glossy; tall, upright, very vigorous growth; [Anne Harkness X Greensleeves]; Bossom, W.E., 1985

'BOSANNEVES', F, my, 1996; (**Enfield in Bloom**); flowers deep yellow, reddish tinge to petal edge, reverse lighter yellow, dbl., 26–40 petals, 3.5 in., slight fragrance; moderate prickles; foliage large, medium green, semi-glossy; upright, tall (115 cms) growth; [Anne Harkness X Greensleeves]; Bossom, W.E.

'BOSARTHRIC', Cl F, pb, 1997; (**Caring For You**); flowers very full, large, very dbl., 41 petals, borne in small clusters, moderate fragrance; some prickles; foliage medium, medium green, semi-glossy; climbing (15ft.) growth; [English Miss X Summer Wine]; Bossom, W.E.

'BOSBIGSOUTH', F, lp, 1999; (**Lady Yvonne**); dbl., 17–25 petals, 3.5 in., borne in small clusters, moderate fragrance; prickles moderate; foliage medium, medium green, semi-glossy; medium (3 ft) growth; [Southampton X seedling]; Bossom, W.E.

'BOSCHERRY', F, pb, 1995; (**Darling Diane**); flowers lavender pink blend, blooms, dbl., 26–40 petals, 3–3.5 in., slight fragrance; foliage large, dark green, semi-glossy; tall, upright growth; [Champagne Cocktail X Seedling]; Bossom, W.E.

'BOSCHERRYDRIFT', F, pb, 1997; (**Deborah Devonshire**); flowers old fashioned form, very full, medium, very dbl., 41 petals, 2.75 in., borne in small clusters, moderate fragrance; few prickles; foliage medium, dark green, glossy; bushy, medium (90cms.) growth; [Pearl Drift X CherryAde]; Bossom, W.E.

Boscobel® *see* 'KINBOSCO'

'BOSCONPEA', F, pb, 1997; (Fab, **Lady of Hertford**); flowers semi double, medium, semi-dbl., 8–14 petals, 2.5 in., borne in small clusters, slight; few prickles; foliage small, medium green, glossy; upright, tall (120cms.) growth; [Conversation X Pearl Drift]; Bossom, W.E.

'BOSELADNEE', F, dr, 1996; (**Linda Mary**); flowers dark red, silver reverse, blooms borne in small clusters, dbl., 15–25 petals, 3.25 in., slight fragrance; few prickles; foliage medium, dark green, semi-glossy; bushy, medium (70 cms) growth; [Guinee X Glad Tidings]; Bossom, W.E.

'BOSELFTAY', F, mr, 1997; (**Connie Crook**); flowers full, large, very dbl., 26–40 petals, 3.5 in., borne in large clusters, slight fragrance; few prickles; foliage large, dark green, glossy; upright, tall (120cms.)growth; [Seedling X Selfridges]; Bossom, W.E.

'BOSEMWINE', Cl F, dp, 1995; (**Mrs Cecily McMullen**); flowers salmon pink, blooms borne in small clusters, very dbl., 2.75 in., moderate fragrance; few prickles; foliage large, medium green, semi-glossy; tall, spreading (10 ft x 10 ft), climbing growth; [English Miss X Summer Wine]; Bossom, W.E.

'BOSEYEBALL', S, mr, 1997; (**Baby Tom**); flowers single, single, 4–7c petals, borne in small clusters, slight fragrance; few prickles; foliage small, light green, glossy; spreading, medium (60cms.) growth; [Eyepaint X Ballerina]; Bossom, W.E.

'BOSGREEN', F, w, 1992; (**Peppermint Ice**); flowers creamy green, blooms borne singly or in sprays of 3-5, semi-dbl., 10 petals, 3.25 in., cupped, slight fragrance; foliage medium, medium green, semi-glossy; upright, medium growth; [Anne Harkness X Greensleeves]; Bossom, W.E.; E.B. LeGrice Roses, Ltd., 1991

'BOSHIPEACON', F, op, 1994; (**Emily Victoria**); flowers salmon pink, blooms borne in small clusters, very dbl., 1.5–2.75 in., moderate fragrance; some prickles; foliage medium, medium green, semi-glossy; low to medium (60 cms), bushy, compact growth; [Conservation X (Pearl Drift X Highfield)]; Bossom, W.E.

'BOSHOTEARL', F, lp, 1997; (**Lady Barbara Blossom**); flowers full, frilly, large, very dbl., 26–40 petals, 4 in., borne in small clusters, slight fragrance; some prickles; foliage medium, light green, glossy; upright, tall (90cms.) growth; [Savoy Hotel X Pearl Drift]; Bossom, W.E.

'BOSILJURIKA', F, mr, 1995; (**Preservation**); flowers bright red, blooms borne in small clusters, very dbl., 2.75 in., slight fragrance; some prickles; foliage medium, medium green, glossy; medium, upright growth; [Silver Jubilee X Paprika]; Bossom, W.E.

'BOSLAURAMBER', Min, dy, 1998; (**Good Morning Sunshine**); flowers small, golden yellow, very dbl., 26–40 petals, 2.5 in., borne mostly singly, moderate fragrance; prickles numerous; foliage small, dark green, semi-glossy; low, compact growth; [Laura Ford X Forever Amber]; Bossom, W.E.

'BOSLORVET', F, ob, 1993; (**Sharon Anne**); flowers orange salmon, blooms borne in large clusters, dbl., 15–25 petals, 1.5–2.75 in., slight fragrance; few prickles; foliage large, medium green, semi-glossy; tall to medium (120 cms), upright growth; [Sharon Lorraine X Brown Velvet]; Bossom, W.E.; Bossom, 1993

'BOSNEECH', S, dr, 1996; (**William Stubbs**); dbl., 15–25 petals, 3.25 in., borne in small clusters, early flowering (May), slight fragrance; moderate prickles; foliage small, medium green, glossy; spreading (10x12 ft.) growth; [Seedling X Guinee]; Bossom, W.E.

Bosom Buddy, HT, or, 1992; Wells

'BOSPARDON', F, m, 1998; (**Nancy Bennett**); flowers lavender pink, dbl., 15–25 petals, 2.75 in., borne in small clusters, slight fragrance; prickles moderate; foliage large, dark green, semi-glossy; tall, spreading growth; [City of London X Paprika]; Bossom, W.E.

'BOSPEABAY', F, mp, 1993; (**Tender Loving Care**); flowers medium pink, moderately full, dbl., 20 petals, 4 in., borne in large clusters, moderate fragrance; some prickles; foliage medium, dark green, semi-glossy; medium to tall (90 cms), upright growth; [Pearl Drift X (Dublin Bay X Seedling)]; Bossom, W.E.; F. Haynes & Partners, 1995

'BOSREXCITY', F, w, 1995; (**Frances Perry**); dbl., 26–40 petals, 3–3.5 in., borne in small clusters, moderate fragrance; some prickles; foliage medium, medium green, semi-glossy; low, bushy growth; [Sexy Rexy X City of London]; Bossom, W.E.

'BOSREXEVER', S, ly, 1998; (**Annie East**); flowers cream yellow, semi-dbl., 8–14 petals, 3 in., large clusters, slight fragrance; prickles moderate; foliage large, dark green, semi-glossy; spreading, medium growth; [Sexy Rexy X Forever Amber]; Bossom, W.E.

'BOSREXEYE', F, or, 1998; (**Irene Smith**); flowers coral with lighter reverse, very dbl., 26–40 petals, 2.75 in., borne in large clusters, slight fragrance; prickles moderate; foliage medium, dark green, glossy; bushy, medium growth; [Sexy Rexy X Eyepaint]; Bossom, W.E.

'BOSSA NOVA', HT, dy, 1964; flowers deep golden yellow, dbl., 28 petals, 4 in., exhibition form; foliage dark; [Leverkusen X Buccaneer]; McGredy, Sam IV; McGredy

Bossa Nova *see* POUloma

'BOSSEXEYE', F, mr, 1999; (**Olive Elsie**); flowers medium red, lighter reverse, 3 in., borne in small clusters, slight fragrance; prickles moderate; foliage medium, light green, semi-glossy; medium (2.5 ft) growth; [Sexy Rexy X Eyepaint]; Bossom, W.E.

'BOSSUET', HGal, mr; flowers scarlet, edges darker

'BOSTON', HT, mp, 1917; dbl., intense fragrance; [Mrs George Shawyer X Seedling]; Montgomery Co.

'BOSTON BEAUTY', Pol, mp, 1919; flowers clear pink, dbl., moderate fragrance; [Orléans Rose X (Katharina Zeimet X Old Ayrshire rose)]; Farquhar

Botanica *see* 'TOMBOT'

Botaniste Abrial *see* **'LOWELL THOMAS'**

'BOTANISTE HENRI GRIMM', F, my, 1958; flowers golden to straw-yellow, becoming pink tinted, large, blooms in clusters; vigorous growth; [Goldilocks X Fashion]; Gaujard, R.; G. Truffaut

Botany Bay, HT, rb

'BOTZARIS', D, w, 1856; flowers creamy white, dbl., flat; foliage light green

'BOUDOIR', HT, pb, 1942; (Paul Fromont); flowers tyrian rose, reverse white, large, dbl., 50 petals, exhibition form; foliage leathery; vigorous, upright, bushy growth; [Ampere X (Charles P. Kilham X Margaret McGredy)]; Meilland, F.; C-P

Boudoir *see* CLEbou

'BOUGAINVILLE', N, pb, 1822; bud red; flowers pink in center, becoming paler and tinged with lilac at the base, very dbl., cupped; branches very prickly; foliage narrow, glossy; Vibert

Boule de Nanteuil *see* **'COMTE DE NANTEUIL'**

'BOULE DE NEIGE', B, w, 1867; flowers pure white, compact, dbl., occasional recurrent bloom, moderate fragrance; foliage dark; [Blanche Lafitte X Sappho (Vibert's)]; Lacharme, F.

Boulie's Dream™ *see* 'WILFOLK'

'BOUNTIFUL', F, pb, 1972; flowers strawberry-salmon, reverse deeper, dbl., 33 petals, 3 in., exhibition form, moderate fragrance; foliage small; tall, erect growth; [Vesper X Seedling]; LeGrice

'BOUQUET', F, dp, 1940; (Lied, Siegeslied); flowers deep pink, dbl., 33 petals, cupped, slight fragrance; foliage dark, leathery; vigorous, bushy, compact growth; [Ingar Olsson X Heidekind]; Tantau; C-P

'BOUQUET BLANC', Pol, w, 1914; flowers small, dbl.; Conard; (Sangerhausen)

'BOUQUET BLANC', HP, w, 1856; flowers white to light pink, medium, dbl.; Robert; (Sangerhausen)

'BOUQUET CHARMANT', HGal, mp; flowers rosy purple, large, dbl.; Hardy, before 1811

Bouquet de la Mariée *see* **'AIMÉE VIBERT'**

Bouquet de Marie, HP, w; Damaizin, 1858

'BOUQUET DE NEIGE', Pol, w, 1900; flowers medium, dbl.; Vilin; (Sangerhausen)

Bouquet de Venus, HGal, lp; Lerouge

'BOUQUET D'OR', N, yb, 1872; flowers yellow, center coppery salmon, large, dbl.; [Gloire de Dijon X Seedling]; Ducher

'BOUQUET D'OR', HT, ly, 1922; flowers light golden yellow, well formed; vigorous growth; Lippiatt

Bouquet Fait *see* 'LEN 1'

Bouquet Parfait®, HMsk, pb, 1989; Lens

'BOUQUET ROSE', Pol, lp, 1928; flowers flesh-pink and peach-blossom, clusters of 30-40; Granger-Gaucher; Turbat

'BOUQUET ROUGE', F, mr, 1963; flowers large blooms in clusters of 14-16, dbl., 45 petals; vigorous growth; [(Gruss an Teplitz X Independence) X (Independence X Floradora)]; Arles; Roses-France

'BOUQUET TOUT FAIT', N, w; Laffay, M., before 1836

Bouquet Vanille® *see* DELblatine

Bouquetterie, Min, my

Bourbon Queen *see* **'QUEEN OF BOURBONS'**

'BOURBON ROSE', 1817; (R. canina borboniana, R. X borboniana); flowers pink, red or purple solitary or in few-flowered corymbs, semi-dbl., 3 in., some are recurrent

Bourgogne®, S, mr, 1983; Interplant

Bourgogne No. 2 *see* ADAtapora

'BOURSAULT ROSE'; (R. boursaultii, R. l'heritierana, R. reclinata); flowers pink to purple, nodding, in corymbs, semi-dbl., non-recurrent; subglobose, smooth fruit; climbing to 12 ft. growth; (14 and probably 21)

Bouton Red Bourbon, B, dr

'BOUTONNIERE', HT, op, 1940; flowers salmon-pink, dbl., 40–50 petals, 3 in.; foliage dark, glossy, bronze; vigorous, bushy, compact growth; [Lulu X Mrs Sam McGredy]; Lammerts, Dr. Walter; Armstrong Nursery

'BOUZLOUDJA', HT, mr, 1974; flowers large, dbl., 40 petals, slight fragrance; foliage dark, glossy; vigorous, bushy growth; [Sarah Arnot X Rina Herholdt]; Staikov, Prof. Dr. V.; Kalaydjiev and Chorbadjiiski

Bow Bells™ *see* 'AUSBELLS'

Bowie Pink, S, mp

Bowie Pink Lady, HT, lp, 1995; bud dark pink; flowers soft pink with hint of lemon at heart, exhibition form; Williams, J. Benjamin

Bowie White Patio, Min, w; Williams, J. Benjamin

Bowie Yellow Patio, MinFl, my, 1995;, exhibition form; Williams, J. Benjamin

Boy Crazy *see* 'DICREVIVAL'

Boy O Boy™ *see* 'DICUNIFORM'

'BOY SCOUT', HT, or, 1946; (Scout); bud long, pointed; flowers flame, large, dbl., exhibition form, moderate fragrance; foliage dark, glossy; very vigorous, upright, bushy growth; [Joanna Hill X Olympiad]; Duehrsen; California Roses

Boys' Brigade® *see* 'COCDINKUM'

'BOZENA NEMCOVÁ', HT, dp, 1931; flowers pure dark pink, very large, dbl., cupped, intense fragrance; foliage bronze, thick; strong stems; vigorous growth; [Sylvia X Priscilla]; Böhm, J.

'BRAD', HT, ab, 1999; dbl., 26–40 petals, 5.5 in., borne mostly singly, moderate fragrance; few prickles; foliage medium, medium green, dull; bushy, medium (3 ft) growth; [Gavotte X Ravenswood]; Poole, Lionel

'BRADGATE', HT, yb, 1970; flowers yellow and deep red bicolor, pointed, dbl., 27 petals, 4–5 in., slight fragrance; foliage dark; free growth; [Piccadilly sport]; Worth; Lowe

Bradley Craig® *see* 'MACSTEWAR'

Bradley Graig *see* 'MACSTEWAR'

'BRADOVA GERMANIA', HFt, op, 1932; flowers coppery-pink, large, semi-dbl.; Brada, Dr.; (Sangerhausen)

'BRADOVA LOSOSOVA DRUSCHKI', HP, op, 1937; flowers salmon-pink, large; Brada, Dr.; (Sangerhausen)

Braine l'Alleud, HT, 1984; Delforge, H.; (Cavriglia)

'BRAISWICK CHARM', HWich, ob, 1914; flowers orange-yellow, edges almost white clusters, moderate fragrance; foliage dark, glossy, leathery; long, strong stems; very vigorous growth; Cant, F.

Braithwaite *see* 'AUSCRIM'

'BRANDENBURG', HT, or, 1965; flowers deep salmon, reverse darker, dbl., 40 petals, 5 in., exhibition form; vigorous, upright growth; [(Spartan X Prima Ballerina) X Karl Herbst]; Kordes, R.; McGredy

Brandenburg Gate *see* 'JACGATE'

'BRANDON', LCl, mr, 1964; bud long, pointed; flowers dark carmine-red, large, dbl., very free, recurrent bloom; vigorous growth; Combe

Brandy® *see* 'AROCAD'

'BRANDY BUTTER', HT, ob, 1982; bud long, pointed; flowers pale gold, blooms borne 4-5 per cluster, dbl., 28 petals, exhibition form, moderate fragrance; thick, triangular-shaped prickles; foliage mid-green; tall, upright growth; [Fred Gibson X Royal Gold]; Northfield, G.

Brandy Snap® *see* 'DICQUEUE'

Brandy, Climbing, Cl HT, lp; [Brandy sport]; Swane, 1994; (Weatherly, L.)

'BRANDYGLOW', F, yb, 1993; (**Vanessa Belinda**); flowers yellow blend, dbl., 15–25 petals, 3–3.5 in., borne in small clusters, moderate fragrance; some prickles; foliage medium, dark green, glossy; medium (100 cms), bushy growth; [Champagne Cocktail X Seedling (Glenfiddich X Priscilla Burton)]; Bracegirdle, A.J.; Bracegirdle, 1993

'BRANDYPINK', F, mp, 1992; (**Ronald George Kent**); dbl., 15–25 petals, 3–3.5 in., borne in small clusters, slight fragrance; few prickles; foliage large, dark green, semi-glossy; medium (86 cms), bushy growth; [Pink Favorite X Piccasso]; Bracegirdle, A.J.

Brandysnap, F, ab, 1999

Brandysnap *see* 'DICQUEUE'

'BRANDYWINE', HT, ly, 1941; bud long, pointed; flowers buff-yellow, dbl., 25–30 petals, 4–5 in., slight fragrance; foliage olive-green, leathery; very vigorous, upright growth; John Cook Medal, ARS, 1945; [Seedling X Souvenir]; Thompson's, J.H., Sons; C-P;, J&P

Brasero, F, mr

Brasier *see* **'BRAZIER'**

'BRASILIA', HT, rb, 1968; flowers scarlet, reverse gold, dbl., 4 in., slight fragrance; [Kordes' Perfecta X Piccadilly]; McGredy, Sam IV; McGredy

Brass Band® *see* 'JACCOFL'

Brass Monkey *see* HARxaglen

Brass Ring *see* 'DICGROW'

'BRATOWIN', HT, dp, 1993; (**George Armer**); flowers deep pink, blooms borne mostly single, dbl., 15–25 petals, 1.5–2.75 in., exhibition form, slight fragrance; some prickles; foliage medium, bronze-red, matt; medium (12 cms), upright growth; [Gavotte X First Prize]; Bracegirdle, A.J.; Bracegirdle, 1993

Braunwald®, S, w

Bravado *see* 'JACRO'

Brave Heart *see* 'HORBONDSMILE'

Braveheart *see* CLEbravo

'BRAVO', HT, mr, 1951; bud ovoid; flowers cardinal-red, borne in clusters, dbl., 35 petals, 4–5 in., exhibition form, slight fragrance; foliage leathery; vigorous, upright, moderately bushy growth; [World's Fair X Mirandy]; Swim, H.C.; Armstrong Nursery

Braz Ornelas, HT; Moreira da Silva, A.; (Cavriglia)

'BRAZIER', HT, mr, 1937; (Brasier); flowers flame-scarlet, large, dbl., slight fragrance; foliage glossy; very vigorous, bushy growth; [Charles P. Kilham X Seedling]; Mallerin, C.; A. Meilland;, C-P

'BRAZIL', HT, rb, 1947; bud long, pointed; flowers saturn-red, reverse saffron-yellow, well shaped, dbl.; foliage light green; Caron, B.; URS

Bread 'n' Butter *see* 'TINBUTT'

Break o' Dawn *see* 'MINFCO'

'BREAK O' DAY', Pol, pb, 1937; flowers copper-pink, center yellow, fading to shell-pink, blooms in, semi-dbl., cupped, recurrent bloom; vigorous growth; Archer

'BREAK O' DAY', HT, ob, 1939; (Delta); flowers orange shades, large, dbl., 50 petals, intense fragrance; vigorous growth; [Seedling X Glenn Dale]; Brownell, H.C.

'BREAK O'DAY, CLIMBING', Cl HT, ob, 1944; Brownell, H.C.

'BREAKAWAY', Min, mr, 1980; bud ovoid, pointed; dbl., 23 petals, borne singly, slight fragrance; straight, tiny, brownish-green prickles; foliage small, medium green; compact, upright, bushy growth; [Dandy Lyon X Seedling]; Lyon, Lyndon

Breath of Life *see* 'HARQUANNE'

Breathless *see* 'JACCHRY'

Breathtaking *see* 'LEOBRETAK'

Bredon® *see* 'AUSBRED'

'BREEZE HILL', LCl, ab, 1926; flowers flesh tinted apricot, center rose, paling, blooms in clusters, dbl., 55 petals, 3 in., cupped, non-recurrent, moderate fragrance; bushy growth, heavy canes; [R. wichurana X Beauté de Lyon]; Van Fleet; American Rose Society

Breezy® *see* 'SAVABREZ'

'BREGINA', F, or, 1965; flowers vermilion-red; [Mandrina sport]; deRuiter

'BRENDA', HEg, lp, 1894; flowers peach-blossom-pink; foliage fragrant; very vigorous growth; Penzance; Keynes, Williams & Co.

'BRENDA ANN', HT, dy, 1974; flowers amber-yellow, full, dbl., 37 petals, 5 in., slight fragrance; foliage glossy, tinted bronze; vigorous, tall, upright growth; [Piccadilly sport]; Watts

Brenda Burg *see* CLEswan

'BRENDA COLVIN', LCl, lp, 1970; (R. filipes 'Brenda Colvin'); single, 5 petals, 1 in., blooms in trusses, intense fragrance; foliage dark, glossy; very vigorous growth; Colvin; Sunningdale Nursery

Brenda Lee *see* 'MICLEE'

'BRENDA OF TASMANIA', F, w, 1971; bud ovoid; flowers white, center pink, large, dbl., cupped, moderate fragrance; foliage large, glossy; upright, bushy growth; [Queen Elizabeth sport]; Holloway

'BRENDA'S FRAGRANCE', HT, pb, 1967; flowers cerise edged white, exhibition form, intense fragrance; foliage leathery; vigorous growth; [Lieut. Chaure X Hector Deane]; Smith, W.H.

Brennende Liebe *see* **'BURNING LOVE'**

Brennpunkt, F, 1972; Haenchen, E.; (Cavriglia)

'BRENNUS', HCh, dr, 1830; flowers deep red, shaded with violet; Laffay, M.

'BRESILIENNE', Pol, dr, 1971; bud long, pointed; flowers deep red, medium, dbl., cupped, slight fragrance; foliage bronze, leathery; vigorous, dwarf, bushy growth; [Red Favorite X Ena Harkness]; Delforge

Brewood Belle, LCl, mr, 1996; Scarman

'BRIACE', Min, mr, 1998; (**Ace of Diamonds**); flowers bright medium red, good substance, very dbl., 26–40 petals, 1–1.5 in., exhibition form, borne mostly singly and small clusters, intense fragrance; few prickles; foliage small, dark green, semi-glossy; strong; bushy, spreading, medium growth; Disease resistant; Bridges, Dennis A.; Bridges Roses, 1998

Brian *see* BARrian

Brian Donn *see* 'TINDONN'

Brian Lee *see* 'TINLEE'

'BRIANCOUF', S, mp, 1989; (**Baronne Surcouf**); Briant

'BRIAND-PANEUROPA', HT, mp, 1931; flowers carmine-rose, base yellow, very large, dbl., slight fragrance; vigorous growth; [Franklin X Willowmere]; Böhm, J.

Brianna *see* 'DEVCARLOS'

'BRIAN'S SONG', S, op, 1977; dbl., 35 petals, 4 in., exhibition form, intermittent bloom, moderate fragrance; foliage large, glossy, dark; [Independence X Pike's Peak]; Smith, R.L.; Smith's Greenhouse & Nursery

Briant Hill®, Min, ob, 1994; Poulsen

'BRIARCLIFF', HT, pb, 1926; flowers center deep rose-pink, outer petals lighter; [Columbia sport]; Pierson, P.M.

'BRIARCLIFF BRILLIANCE', HT, dp, 1932; flowers rose to rose-red, large, dbl.;

vigorous growth; [Briarcliff sport]; Pierson, P.M.

'BRIARCLIFF SUPREME', HT, pb, 1947; bud long, pointed; flowers briarcliff pink, very large, dbl., intense fragrance; foliage dark; very vigorous, upright growth; [Briarcliff sport]; Hinner, P.; Bauské Bros. & Hinner

'BRIARCLIFF, CLIMBING', Cl HT, pb, 1929; Parmentier, J.

'BRIAUTUMN', Min, op, 1999; (**Autumn Dawn**); flowers orange-pink, reverse slightly lighter, dbl., 17–25 petals, 1.3in., slight fragrance; prickles moderate; foliage medium, dark green, semi-glossy; spreading, medium (2 ft) growth; [Carrot Top X Unknown]; Bridges, Dennis A.

BRIbaby, Min, w, 1997; (**Baby Secret**); Bridges, Dennis A.

'BRIBEN', HT, pb, 1993; (**Dr Jack Bender**); flowers pink and white, blooms borne mostly single, dbl., 26–40 petals, 3–3.5 in., slight fragrance; many prickles; foliage medium, dark green, semi-glossy; bushy, spreading (100 cms) growth; [Lady X X Flaming Beauty]; Bridges, Dennis A.; Bridges Roses, 1993

'BRIBOLT', HT, pb, 1990; (**Thunderbolt**); bud ovoid; flowers pink blend, white reverse, color intensifies with age, dbl., 45 petals, slight fragrance; foliage medium, dark green, matt; bushy, medium growth; [Lady X X Flaming Beauty]; Bridges, Dennis A., 1980; Bridges Roses, 1991

'BRICHARM', Min, yb, 1992; (**Southern Charm**); flowers shades of light yellow and pink, pink intensifying with sun, dbl., 25–40 petals, 1.5–2.75 in., slight fragrance; no prickles; foliage large, medium green, semi-glossy; medium (40-45 cms), upright, bushy, slightly spreading growth; [Baby Katie X Unknown (select pollen)]; Bridges, Dennis A.; Bridges Roses

'BRICHAT', Min, my, 1999; (**Chit Chat**); dbl., 26–40 petals, .75 in., borne mostly singly, no fragrance; few prickles; foliage medium, dark green, semi-glossy; compact, bushy, medium (18-20 in.) growth; [Cal Poly X Unknown]; Bridges, Dennis A.

'BRICLASS', HT, pb, 1992; (**Carolina Classic**); flowers medium pink edged deeper pink, blooms borne mostly single, very dbl., 3–3.5 in., slight fragrance; few prickles; foliage medium, medium green, matt; medium (90-100 cms), upright growth; [Just Lucky X Flaming Beauty]; Bridges, Dennis A.; Bridges Roses

'BRICOS', Min, Ob, 1997; (**Cosmic**); flowers light to medium orange with light yellow reverse; full (26–40 petals), medium (4–7 cms.) blooms borne in small clusters; slight fragrance; foliage large, medium green, semi-glossy; upright (24–26 in.), bushy growth; [Suzy X Seedling]; Bridges, Dennis A.; Bridges Roses

Bridal Blush™, HT, w; DeVor Nurseries, Inc, 1989

'BRIDAL BOUQUET', HT, w, 1976; dbl., 23 petals, 3–4 in., slight fragrance; foliage matt, green; vigorous, compact growth; [Lady Sylvia sport]; Ormerod

Bridal Delight, HT, lp; [Bridal White sport]; Day, 1992; (Weatherly, L.)

Bridal Pink™ *see* 'JACBRI'

'BRIDAL ROBE', HT, w, 1955; flowers ivory-white, high-pointed, dbl., 54 petals, 4 in., moderate fragrance; foliage glossy, olive-green; vigorous growth; GM, NRS, 1953; [McGredy's Pink X Mrs Charles Lamplough]; McGredy, Sam IV

Bridal Shower™ *see* 'JACTAFL'

Bridal Shower *see* DEVblush

Bridal Sonia, Gr, mp

Bridal Sunblaze® *see* 'MEILMERA'

'BRIDAL VEIL', F, ly, 1954; bud ovoid, cream; flowers white overcast sea-foam-yellow, open, medium in clusters, dbl., 75–80 petals, 2.5 in., intense fragrance; vigorous, bushy growth; [Pinocchio seedling X Pigmy Gold]; Boerner; Stark Bros.

Bridal White™ *see* 'JACWHY'

'BRIDAWN', Min, m, 1991; (**Purple Dawn**); flowers velvety mauve, long-lasting, urn-shaped, blooms borne usuall, dbl., 25 petals, 1.5 in., exhibition form, slight fragrance; foliage medium, dark green, semi-glossy; upright, tall growth; [Party Girl X Unknown]; Bridges, Dennis A.; Bridges Roses

Bride *see* FRYyearn

'BRIDE'S BLUSH', HT, w, 1923; flowers creamy white, at times blush-pink, single, 6 petals, moderate fragrance; [Columbia sport]; Amling Co.

Bride's Dream *see* 'KOROYNESS'

'BRIDE'S WHITE', F, w, 1968; flowers pure white, small, dbl., cupped; vigorous, upright, bushy growth; Mansuino, Q.; Carlton Rose Nurseries

'BRIDESMAID', T, lp, 1893; [Catherine Mermet sport]; Moore

'BRIDESMAID, CLIMBING', Cl T, lp, 1893; Moore

'BRIDGET', HT, op, 1947; bud long, pointed; flowers brilliant orange-scarlet, base bright golden yellow, dbl., 35–40 petals, 4–5 in., flat; foliage glossy, bright green; [Mrs Henry Bowles X Phyllis Gold]; Fletcher; Tucker

Bridget *see* CLEbridge

Bridget *see* KOReibei

Bridgets Joy, HT, lp, 1997; flowers full, large, very dbl., 26–40 petals, 4 in., borne mostly singly, slight fragrance; foliage medium, medium green, semi-glossy; compact, medium (140cms.) growth; [Coral Fiesta sport]; Johnstone, Leonard E.

'BRIDGWATER PRIDE', F, op, 1982; flowers rich salmon, medium, dbl., 20 petals, moderate fragrance; foliage medium, dark, semi-glossy; bushy, compact growth; [Vera Dalton X Allgold]; Sanday, John; Sanday Roses, Ltd.

'BRIDIXIE', Min, rb, 1992; (**Dixieland**); flowers in shades of vibrant pink, red and white intensifying with sun, dbl., 15–25 petals, 1.5–2.75 in., exhibition form, slight fragrance; no prickles; foliage medium, medium green, semi-glossy; tall (50-60 cms), upright growth; [Fancy Pants X Seedling]; Bridges, Dennis A.; Bridges Roses

Brier Bush *see* **R. CANINA**

'BRIFIRE', Min, rb, 1991; (**Gabriel's Fire**); bud pointed; flowers creamy light yellow turning red, urn-shaped, blooms borne usually singly, semi-dbl., 20–22 petals, 2 in., exhibition form, intense fragrance; foliage medium, medium green, semi-glossy; bushy, spreading, medium growth; [Sachet X Unknown]; Bridges, Dennis A., 1988; Bridges Roses, 1992

'BRIFREE', HT, w, 1994; (**Freemont**); flowers near white blushed with salmon, blooms borne mostly singly, very dbl., 3–3.5 in., moderate fragrance; some prickles; foliage large, dark green, semi-glossy; tall, upright growth; [Thriller X Tiki]; Bridges, Dennis A.; Bridges Roses, 1994

'BRIGADEIRO FRANÇA BORGES', F, or, 1960; flowers bright orange-red; [Independence X Seedling]; da Silva, Moreira

Brigadoon™ *see* 'JACPAL'

BRIgarold, HT, mr, 1989; (**Roland Garros**®); Briant

'BRIGAUNT', Min, dr, 1991; (**Gauntlet**); bud ovoid; flowers dark red, lighter reverse, lightens slightly with age, medium, semi-dbl., 20–22 petals, exhibition form, slight fragrance; foliage large, medium green, semi-glossy; spreading, medium growth; [Kitty Hawk X Unnamed seedling]; Bridges, Dennis A., 1986; Bridges Roses, 1990

'BRIGENTLE', Min, mr, 1998; (**Gentleman's Agreement**); bud long, slender; flowers medium red, excellent substance, very dbl., 26–40 petals, 1–1.5 in., exhibition form, borne mostly singly, no fragrance; few prickles; foliage large, medium green, semi-glossy; tall, upright, bushy growth; Disease resistant; [Purple Dawn X Select pollen]; Bridges, Dennis A.

'BRIGHT ANGEL', S, yb, 1977; bud ovoid; flowers light yellow, edged pink, dbl., 48 petals, 4 in., exhibition form, intermittent bloom, intense fragrance; foliage large, glossy, dark; vigorous, bushy growth; [Dornroschen X Golden Wings]; Smith, R.L.; Smith's Greenhouse & Nursery

'BRIGHT BEAM', Gr, pb, 1972; bud ovoid; flowers cream, edged pink, large, dbl., cupped, profuse, intermittent bloom, moderate fragrance; foliage glossy, dark; very vigorous, upright growth; [Peace X Little Darling]; Fuller; Wyant

Bright Beauty, HT, or

'BRIGHT BOY', HT, lp; brilliant red; Clark, A., 1948

Bright Eyes *see* 'SANMAR'

'BRIGHT EYES', F, 1948; bud ovoid; flowers light yellow to primrose, medium, large truss, dbl., 25 petals, slight fragrance; foliage leathery, glossy, dark; dwarf growth; RULED EXTINCT 6/83 ARM; [Joanna Hill X (Heidekind X Betty Uprichard)]; Duehrsen; H&S

'BRIGHT FIRE' *see* PEAxi

'BRIGHT GARBS', F, or, 1977; bud pointed; semi-dbl., 12 petals, 2.5 in., moderate fragrance; foliage large; dwarf growth; [Orangeade X Seedling]; Hardikar, Dr. M.N.

'BRIGHT JEWEL', Min, pb; bud pointed; flowers rose-pink, white center, very small blooms in clusters, semi-dbl.; low, compact growth

Bright Lights *see* 'KSEELITE'

Bright Meadow *see* **'YASNAYA POLIANA'**

'BRIGHT MELODY', S, mr, 1984; bud ovoid, pointed; flowers large, dbl., 30 petals, cupped, borne 1-10 per cluster, repeat bloom, slight fragrance; awl-like, tan prickles; foliage medium-large, dark olive green, leathery; erect, bushy growth; [Carefree Beauty X (Herz As X Cuthbert Grant)]; Buck, Dr. Griffith J.; Iowa State University

'BRIGHT MORNING', HT, my, 1958; flowers golden yellow, moderate fragrance; Ratcliffe

'BRIGHT RED', Pol, dr, 1938; flowers velvety dark red, large cluster, dbl.; bushy growth; deRuiter

Bright Sight *see* 'JUDBRIGHT'

Bright Smile® *see* 'DICDANCE'

Bright Spark *see* RUbrispa

Bright Star, MinFl, 1999; single; Williams, J. Benjamin

'BRIGHT WINGS', HT, op, 1942; (Gitane); bud long, pointed, rosy orange; flowers centers orange shading to pink, large, dbl., 22 petals, cupped, moderate, fruity fragrance; foliage bronze; strong stems; vigorous, bushy growth; [Mme Arthaud X Annie Drevet]; Mallerin, C.; A. Meilland;, C-P

'BRIGHTNESS', HT, rb, 1958; bud long, pointed; flowers scarlet, reverse golden yellow, merging to orange, dbl., 28–35 petals, 4.5–5.5 in., exhibition form, intense fragrance; foliage leathery, glossy; bushy, moderately vigorous growth; [Doreen sport]; Fryers Nursery, Ltd.

'BRIGHTSIDE', Min, or, 1974; dbl., 25 petals, 1 in., exhibition form, moderate fragrance; foliage small, matt; upright, bushy growth; [Persian Princess X Persian Princess]; Moore, Ralph S.; Park Seed Co.

'BRIGHTSIDE CREAM', N, w; flowers creamy white, yellow stamens blooms in clusters, semi-dbl., 18 petals, 3 in.; foliage dark; short stems; vigorous growth, long canes

Brigitte de Landsvreugd, LCl, w

Brigitte de Villenfagne, HMsk, lp, 1993; flowers small to medium, single, moderate fragrance; Lens, Louis; (Sangerhausen)

'BRIGITTE JOURDAN', F, or, 1960; bud ovoid; flowers pomegranate-red, large, dbl., globular, slight fragrance; foliage dark, glossy; vigorous, bushy growth; [Belle Créole X Independence]; Arles; Roses-France

'BRIGRAN', HT, mp, 1990; (**Grand Romance**); bud ovoid; flowers reverse slightly lighter, aging slightly lighter, urn-shaped, very dbl., 50 petals, moderate fragrance; foliage medium, dark green, semi-glossy; upright, medium growth; [Lady X X Wini Edmunds]; Bridges, Dennis A., 1987; Bridges Roses

'BRIHEART', Min, dr, 1996; (**Hearts A'Fire**); flowers velvety, blooms borne mostly singly, single, 15–25 petals, 1.25 in., slight fragrance; moderate prickles; foliage medium, dark green, glossy, disease-resistant; strong, upright growth; [Merrimac X Seedling]; Bridges, Dennis A.; Bridges Roses, 1996

'BRIICE', HT, lp, 1994; (**Ice Breaker**); dbl., 15–25 petals, 1.5–2.75 in., borne mostly singly, intense fragrance; some prickles; foliage medium, medium green, semi-glossy; tall, upright growth; [Thriller X Just Lucky]; Bridges, Dennis A.; Bridges Roses, 1994

'BRIINCOG', Min, m, 1995; (**Incognito**™); flowers mauve blend with yellow reverse, dbl., 15–25 petals, 1.5 in., borne mostly singly, slight fragrance; few prickles; foliage medium, dark green, semi-glossy; tall (30-34 in), upright, bushy growth; [Jean Kenneally X Twilight Trail]; Bridges, Dennis A.; Bridges Roses, 1995

BRIletjui, S, rb, 1989; (**Juillet**®); Briant

'BRILLANT', HT, mr, 1952; flowers somewhat large, semi-dbl., moderate fragrance; Kordes, W. Söhne; (Sangerhausen)

Brillant, S, or, 1983; flowers medium, dbl., slight fragrance; Kordes, W. Söhne; (Sangerhausen)

'BRILLIANCE', F, or, 1958; bud ovoid; flowers coral, dbl., 50 petals, 2 in., cupped, moderate fragrance; foliage glossy; compact growth; [Seedling X Independence]; Boerner; J&P

'BRILLIANCY', HT, or, 1936; flowers brilliant scarlet, dbl., intense fragrance; foliage leathery; short stems; bushy growth; [Étoile de Hollande X Daily Mail Scented Rose]; LeGrice

Brilliant *see* **'DETROITER'**

Brilliant Cover *see* POUlterp

'BRILLIANT ECHO', Pol, mp, 1927; flowers rosy pink; [Echo sport]; Western Rose Co.

'BRILLIANT KING', F, mr, 1961; flowers bright red, shallow, dbl., 32 petals, 4 in., cupped, slight fragrance; [Cocorico X Orange Delight]; Leenders, J.

Brilliant Light *see* **'KAGAYAKI'**

Brilliant Meillandina *see* 'MEIRANOGA'

Brilliant Pink Iceberg™ *see* 'PROBRIL'

'BRILLIANT RED', HT, mr, 1938; bud very long, pointed; flowers brilliant red, well formed, very large, dbl.; foliage bright green; vigorous, bushy growth; [Charles P. Kilham X Étoile de Hollande]; Lens

'BRILLIANT STAR', F, mr, 1965; flowers bright red, center shaded yellow, semi-dbl., 12–20 petals, 2.5–3 in., borne in clusters; foliage glossy; vigorous growth; [Masquerade X Dicksons Flame]; Watkins Roses

'BRIMAN', Min, mp, 1993; (**Manteo**); dbl., 15–25 petals, 1.5 in., borne mostly singly, slight fragrance; some prickles; foliage small, medium green, semi-glossy; tall (55-60 cms), upright, bushy growth; [Rise 'n' Shine X Seedling]; Bridges, Dennis A.; Bridges Roses, 1993

'BRINDIS', LCl, rb, 1962; flowers geranium-red, center yellow, single; foliage dark; vigorous growth; [Orange Triumph, Climbing X (Phyllis Bide X Baccará)]; Dot, Simon

Brinessa®, HT, pb, 1985; Delforge

'BRINEY', Min, w, 1994; (**Whitney**); flowers near white with a hint of pink, blooms bornemostly single or, dbl., 15–25 petals, 1.5 in., moderate fragrance; few prickles; foliage small, medium green, semi-glossy; medium (38 cms), upright, spreading growth; [Jennifer X Unknown]; Bridges, Dennis A.; Bridges Roses, 1994

Briosa *see* BARbrio

'BRIROSE', Min, m, 1998; (**Moonlight and Roses**); flowers light lavenderwith rosy edges, with darker lavender and lavender overlay, very dbl., 26–40 petals, 2–2.5 in., exhibition form, borne mostly singly, slight fragrance; some prickles; foliage medium, dark green, glossy; bushy, upright growth; [Seedling X Select pollen]; Bridges, Dennis A.; Bridges Roses, 1998

Brisbane Blush *see* 'PALOK'

'BRISE PARFUMÉE', F, mr, 1950; bud pointed; dbl., borne in clusters, intense fragrance; many spines; foliage glossy, bronze; bark and twigs reddish brown; very vigorous growth; [Böhm's Triumph X Baby Chateau]; Truffaut, G.

BRIseflu, S, w; (**White Fleurette®**); Briant, 1989

BRIsiness, Min, w, 1990; (**Riviere de Diamant**); Briant

'BRISIS', Min, w, 1998; (**Sis**); flowers delicate white with a hint of pink, firm substance, dbl., 15–25 petals, 1.5 in., exhibition form, borne mostly singly, slight fragrance; few prickles; foliage small, medium green, semi-glossy; strong, bushy, slightly spreading growth; Disease resistant; [Party Girl X Cape Hatteras]; Bridges, Dennis A.; Bridges Roses, 1998

'BRISTEP', HT, mp, 1995; (**Steppin' Out**); flowers medium pink, full (26–40 petals), large (7 cms) blooms borne mostly single; fragrant; few prickles; foliage medium, dark green, semi-glossy; medium (4½ ft.), bushy growth; [Kardinal X Thriller]; Bridges, Dennis; Bridges Roses, 1995

'BRISTOL', HT, rb, 1968; flowers bright crimson, reverse lighter, large, dbl., moderate fragrance; foliage dark; compact growth; [Gavotte X Tropicana]; Sanday, John

'BRISTOL POST', HT, op, 1972; flowers pale salmon-pink, base orange, pointed, dbl., 29 petals, 4.5 in., slight fragrance; foliage slightly glossy, dark; upright growth; [Vera Dalton X Parasol]; Sanday, John

'BRISUN', Min, yb, 1993; (**Summer Sunset**); flowers bright yellow with salmon pink shading, blooms borne mostly, dbl., 15–25 petals, 1.5 in., slight fragrance; some prickles; foliage small, medium green, semi-glossy; medium (45 cms), bushy, compact growth; [Fancy Pants X Seedling]; Bridges, Dennis A.; Bridges Roses, 1993

'BRISUZY', Min, mp, 1990; (**Suzy**); bud pointed; flowers medium pink, near white at base blending to light pink, ligh, dbl., 38–40 petals, exhibition form, slight fragrance; foliage medium, dark green, semi-glossy; bushy, medium growth; PP007738; AOE, 1991; [Party Girl X Unknown seedling]; Bridges, Dennis A., 1986; Bridges Roses, 1991

'BRITANNIA', Pol, rb, 1929; flowers crimson, center white, small, borne in clusters of 30-40, single, recurrent bloom; foliage small, leathery, light; compact, bushy growth; [Coral Cluster X Eblouissant]; Burbage Nursery

Britannia *see* FRYcalm

Britannica, Pol, rb

Brite Blue, HT, m, 1984

Brite Lites® *see* 'HARTANNA'

Britestripe *see* 'CLEBRITE'

Britestripes *see* 'CLEBRITE'

'BRITIDE', Min, w, 1991; (**Tidewater**); bud pointed; flowers white with slight pink tints, blooms borne usually singly, dbl., 30–32 petals, 2.5 in., exhibition form, moderate fragrance; foliage medium, medium green, matt; long stems for cutting; bushy, spreading, medium growth; [Jennifer X Unknown]; Bridges, Dennis A., 1988; Bridges Roses

'BRITISH COLUMBIA CENTENNIAL', HT, pb, 1971; bud ovoid; flowers light rose-pink, reverse white, large, dbl., exhibition form, slight fragrance; foliage leathery; vigorous growth; [Pink Masterpiece X Seedling]; Boerner; Pan American Bulb Co.

'BRITISH QUEEN', HT, w, 1912; flowers creamy white, center flushed, open, dbl., slight fragrance; foliage light green, soft; long, weak stems; bushy growth; GM, NRS, 1912; McGredy

'BRITRICK', Min, rb, 1995; (**Trickster**); flowers red with white reverse, blooms borne mostly singly, dbl., 28–30 petals, 1.5 in., intense fragrance; foliage small, medium green, semi-glossy; medium (45 cms) growth; [Jennifer X Red Beauty]; Bridges, Dennis A.; Bridges Roses, 1995

Brittany's Glowing Star *see* 'MANSTAR'

'BRITY', HT, mr, 1997; (**Tycoon**); flowers very full, large, very dbl., 41 petals, 2.75 in., borne mostly singly, slight fragrance; foliage medium, dark green, semi-glossy; medium (4ft.) upright, bushy growth; [Kardinal X Thriller]; Bridges, Dennis A.

'BRIYOUNG', Min, pb, 1997; (**Forever Young**); flowers full, small, very dbl., 26–40 petals, 1.5 in., borne mostly singly, slight fragrance; foliage medium, dark green, semi-glossy; low (12-14in.)compact growth; [Trickster X Select Pollen]; Bridges, Dennis A.

'BRNO', HT, ob, 1933; flowers maroon, orange, gold, open, large, semi-dbl.; foliage glossy; vigorous, bushy growth; [Souv. de George Beckwith X Rosemary]; Böhm, J.

'BROADCASTER', Gr, or, 1969; bud ovoid; flowers medium, dbl., slight fragrance; foliage dark; vigorous growth; [Queen Elizabeth X Circus]; Perry; Conklin

Broadlands *see* TANmirson

Broadway™ *see* 'BURWAY'

Broadway, LCl, lp; large single apricot flowers in spring, flat; healthy evergreen foliage; Clark, A.

Broadway (F), F, dp

'BROADWAY (S)', S, yb, 1933; Clark

'BROALA', Min, w, 1997; (**Alamode**); flowers single, medium, single, 4–7 petals, borne in small clusters, slight fragrance; foliage medium, dark green, glossy; bushy, medium (2.5ft.) growth; [Esprit X Party Girl]; Brown, Ted

'BROCADE', HT, pb, 1960; (Jeune Fille); bud ovoid; flowers soft rose, base cream-white, dbl., 54 petals, 4–5 in., cupped, moderate fragrance; foliage leathery; vigorous, upright growth; [Charlotte Armstrong X Baiser]; Combe; Hémeray-Aubert

Brocade, S, lp, 1998; flowers ivory with mauve tint, dbl.; Williams, J. Benjamin

'BROCAKE', Min, w, 1997; (**Cheese Cake**); flowers double, medium, dbl., 15–25 petals, borne mostly singly, slight fragrance; foliage medium, dark green, glossy; bushy, medium (5ft.) growth; [Esprit X Seedling]; Brown, Ted

'BROCARP', Min, dp, 1997; (**Fragrant Carpet**); flowers double, medium, dbl., 15–25 petals, borne in small clusters, intense fragrance; foliage medium, dark green, semi-glossy; spreading (2'x5')growth; [Seedling X Seedling]; Brown, Ted

'BROCOR', F, ab, 1997; (**Corinne**); flowers double, medium, dbl., 15–25 petals, borne in small clusters, moderate fragrance; foliage medium, dark green, glossy; bushy, medium (30in.)growth; [Esprit X Indian Summer]; Brown, Ted

'BRODAN', Min, mp, 1997; (**Danielle**); flower medium pink fading to light pink as bloom opens, reverse light pink, full (26–40 petals), medium (1 ½ in.) blooms borne in small clusters; very fragrant, foliage medium, medium green, semi-glossy; medium (18 in.) growth; [Esprit X Seedling]; Brown, Ted.

'BRODREAM', F, m, 1997; (**Canadian Dream**); flowers single, medium, single, 4–7 petals, 3 in.., borne in small clusters, moderate fragrance; foliage medium, dark green, glossy; upright, tall (4.5x6ft.)growth; [Nymphenburg X Seedling]; Brown, Ted

'BROEMP', F, dr, 1997; (**Red Emperor**); flowers semi double, large, semi-dbl., 15–25 petals, borne in small clusters, moderate fragrance; foliage medium, dark green, glossy; spreading, low (2ft.)growth; [Esprit X Seedling]; Brown, Ted

'BROFOX', F, ob, 1997; (**Foxtrot**); flowers single, medium, single, 4–7 petals, borne mostly singly, slight fragrance; foliage medium, burgundy turning dark green, glossy; spreading, low (1ft.)growth; [Esprit X Stretch Johnson]; Brown, Ted

'BROGAR', F, rb, 1997; (**Red Garter**); flowers single, medium, single, 4–7 petals, borne in small clusters, slight fragrance; foliage medium, dark red turning dark green, glossy; upright, medium (4ft.)growth; [Esprit X Stretch Johnson]; Brown, Ted

'BRÖG'S CANINA', (strain of R. canina), lp; almost thornless; vigorous growth; in use as an understock; Brög

'BROHEART', Min, mr, 1997; (**Heartstrings**); flowers very full, large, opening to OGR-type bloom, very dbl., 41 petals, borne in small clusters; foliage medium, dark green, dull; bushy, medium (16in.)growth; [Esprit X Seedling]; Brown, Ted

'BROMAN', Min, ob, 1997; (**Music Man**); flowers double, medium, dbl., 15–25 petals, borne mostly singly, slight fragrance; foliage medium, medium green, semi-glossy; upright, tall (20-24in.)growth; [Esprit X Seedling]; Brown, Ted

Bronce Masterpiece *see* **'BRONZE MASTERPIECE'**

'BRONEW', Min, lp, 1997; (**New Antique**); flowers full, large, very dbl., 26–40 petals, borne in small clusters, no fragrance; foliage medium, dark green, semi-glossy; bushy, tall (30in.)growth; Brown, Ted

Bronte, LCl, dy

Bronze Baby *see* PEAxanthous

'BRONZE BEAUTY', F, ob, 1974; bud ovoid-pointed; flowers golden yellow to orange-yellow, open, dbl., 27 petals, 4 in., slight fragrance; foliage large, leathery; vigorous growth; [Electra X Woburn Abbey]; Warriner, William A.; J&P

'BRONZE BEDDER', HT, yb, 1920; flowers bronzy yellow, large, single; Paul, W.

'BRONZE MASTERPIECE', HT, ab, 1960; (Bronce Masterpiece); bud long; flowers bronze-apricot, becoming orange-yellow, dbl., 48 petals, 5.5–6. in., exhibition form, moderate fragrance; foliage leathery, glossy; vigorous, upright growth; GM, Geneva, 1958; [Golden Masterpiece X Kate Smith]; Boerner; J&P

'BROOK SONG', S, my, 1985; flowers imbricated form, large blooms borne 1-8 per cluster, dbl., 40 petals, repeat bloom, moderate fragrance; awl-like, tan prickles; foliage leathery, dark; erect growth; hardy.; [Prairie Star X Tom Brown]; Buck, Dr. Griffith J.; Iowa State University, 1984

Brookdale Giant White *see* **'JACKMAN'S WHITE'**

Brooks Red, HT, lp

'BROOKVILLE', HT, ly, 1942; flowers cream-yellow; [Leonard Barron sport]; Brookville Nursery

Broomfield Novelty *see* **'MARGARET ANDERSON'**

'BROROY', Gr, dp, 1997; (**Viceroy**); flowers double, medium, dbl., 15–25 petals, borne in large clusters, intense fragrance; foliage medium, dark green, glossy; upright, medium (4.5ft.)growth; [Esprit X Harmonie]; Brown, Ted

'BROSUN', F, ob, 1997; (**Harvest Sun**); flowers semi double, large, semi-dbl., 8–14 petals, borne in large clusters (8-10), moderate fragrance; foliage medium, dark green, glossy; upright, tall (4.5ft.)growth; [Esprit X Mountbatten]; Brown, Ted

Brother Cadfael™ *see* 'AUSGLOBE'

'BROTHER SUN', Gr, my, 1998; flowers medium yellow, deeper color at base, flattened top, long lasting, very dbl., 41–50 petals, 2.5–3 in., borne mostly singly, moderate fragrance; prickles moderate; foliage medium, medium green, dull; compact, upright growth; [Goldener Olymp X Australian Gold]; Fleming, Joyce L.; Hortico, Inc., 1997

'BROTHER WILFRID', HT, mp, 1976; [Alec's Red sport]; Wood

Brown County Splendor™ *see* 'WILKBSP'

'BROWN STUDY', F, r, 1999; flowers russet, turning brown, darker at edges, dbl., 15–25 petals, 2.75 in., borne singly and in small clusters, slight fragrance; prickles moderate; foliage medium, dark green, semi-glossy; spreading, medium (3 ft) growth; Jerabek, Paul E.; Freedom Gardens, 1998, U.S.

Brown Sugar *see* 'TALBRO'

Brown Velvet *see* 'MACULTRA'

'BROWNELL YELLOW RAMBLER', HMult, my, 1942; (Yellow Rambler); flowers petals recurved, dbl., borne in more open clusters than Dorothy Perkins, slight fragrance; [(Emily Gray X Ghislaine de Feligonde) X Golden Glow]; Brownell, H.C.

'BROWNIE', F, r, 1959; bud ovoid, tan shades, edged pinkish; flowers brownish tan, reverse yellow, blooms in small clusters, dbl., 38 petals, 3.5–4 in., cupped, moderate fragrance; foliage leathery; vigorous, upright, bushy growth; [Lavender Pinocchio seedling X Grey Pearl]; Boerner; J&P

'BROWSHOLME ROSE', Ayr, w, 1900

Brundrette Centenary, S, dp

Brunella®, Min, mr

'BRUNETTE', F, yb, 1970; bud ovoid; flowers amber-yellow to orange, dbl., 25 petals, 1.5–2.5 in., cupped; foliage glossy; [(Purpurine X Lavender Pinocchio X Fillette) X (Gold Strike X Golden Garnette)]; Lens; Spek

Bruocsella® *see* 'LENBRU'

Brushstrokes *see* 'GUESCOLOUR'

Bruun, F, 1995; Delforge, H.; (Cavriglia)

Bryan, LCl, my, 1993; Mekdeci

Bryan Freidel Pink Tea, T, w

Bryte White, HT, w

Bubble Bath, HMsk, lp, 1980; Matson

Bubble Gum *see* 'BUSBUGUM'

Bubbles *see* 'ZIPBUB'

Bubbles *see* FRYbubbly

Bubikopf® *see* TANkobi

'BUCBI', S, mp, 1977; (Audace®, **Carefree Beauty**™); bud ovoid, long, pointed; flowers light rose, 15–20 petals, 4.5 in., repeat bloom, moderate fragrance; foliage olive-green, smooth; vigorous, upright, spreading growth; PP004225; [Seedling X Prairie Princess]; Buck, Dr. Griffith J.; C-P

'BUCBLU', HT, m, 1983; (**Blue Skies**); flowers large, dbl., 35 petals, moderate fragrance; foliage large, semi-glossy; upright, bushy growth; PP005756; [((Sterling Silver X Intermezzo) X (Sterling Silver X Simone))x (Music Maker X (Blue Moon X Tom Brown))]; Buck, Dr. Griffith J.; J.B. Roses, Inc., 1988

'BUCCANEER', Gr, my, 1952; bud urn-shaped; flowers buttercup-yellow, dbl., 30 petals, 3–3.5 in., cupped, moderate fragrance; foliage dark, leathery; vigorous, upright, tall growth; GM, Geneva, 1952; [Golden Rapture X (Max Krause X Capt. Thomas)]; Swim, H.C.; Armstrong Nursery

'BUCKEYE BELLE', S, pb, 1956; bud globular; flowers pale to deep pink, open, small, borne in compact clusters, semi-dbl., 15 petals, 1.5 in., abundant, recurrent bloom, moderate fragrance; foliage dark; vigorous, upright, bushy growth; [R. hugonis X Seedling]; Garwood

Bucks Fizz *see* 'POULGAV'

Budapest, HT, ab

'BUENOS AIRES', F, r, 1957; flowers burnt brick to dark red-ochre, dbl.; low growth; [Mme Henri Guillot X Pinocchio]; Silva

'BUFF BEAUTY', HMsk, ab, 1939; flowers apricot-yellow blooms in clusters of 12, dbl., 50 petals, 4 in., moderate fragrance; foliage large, medium green, semi-glossy; vigorous (to 6 ft (21)) growth; [William Allen Richardson X Seedling]; Bentall, Ann

'BUFF KING', LCl, ab, 1939; bud ovoid, deep amber; flowers amber and buff, large, cupped; foliage glaucous green; long, strong stems; very vigorous (10-12 ft) growth; Horvath; Wayside Gardens Co.

Buffalo, HRg, dp, 1989; Uhl, J.

Buffalo Bill *see* 'MACYOUMIS'

Buffalo Gal *see* 'UHLATER'

Buffy® *see* 'KINBUFF'

Buffy Sainte-Marie *see* MANCLASSIC

Bugle Boy® *see* 'AROGLOFY'

'BUISMAN'S GLORY', F, mr, 1952; flowers currant-red, open, medium, single; foliage light green; [Karen Poulsen X Sangerhausen]; Buisman, G. A. H.

Buisman's Gold, F, dy

'BUISMAN'S TRIUMPH', F, mp, 1952; flowers bright pink, becoming lighter, large, semi-dbl., 13 petals; foliage dark; vigorous growth; [Käthe Duvigneau X Cinnabar]; Buisman, G. A. H.

'BUISSON ARDENT', Gr, mr, 1956; flowers bright red, medium, moderate fragrance; foliage dark; [Peace X Seedling]; Gaujard

'BUISSON D'OR', HFt, my, 1928; flowers canary-yellow, dbl., good seasonal bloom, moderate fragrance; height 3-5 ft; [Mme Edouard Herriot X R. X harisonii]; Barbier

Bukala, LCl, ob

Bukavu, F, rb, 1998; Lens

'BULLATA', C, mp; (R. X centifolia bullata, Rose à Feuilles de Laitue); flowers over-lapping petals, very dbl., 3 in., globular, solitary or in clusters on long, slender peduncles, summer bloom, intense fragrance; foliage very large and crinkled like lettuce; (28)

Bulls Red *see* 'MACRERO'

'BULMEG', F, mr,1995; (**Meghan's Arrival**); flowers medium red, full (26–40 petals), small blooms borne in large clusters; slight fragrance; some prickles; foliage medium, light green, glossy; upright, bushy, medium growth; [Doris Tysterman X Seedling]; Bull, Derek Gordon; Bull, 1986

'BUNDESRAT HÄBERLIN', HFt, yb, 1940; flowers medium, dbl.; Soupert & Notting; (Sangerhausen)

'BUNKER HILL', HT, dp, 1949; bud pointed; flowers rose-red, large, dbl., 25–40 petals, exhibition form, intense fragrance; foliage leathery, dark; very vigorous, bushy growth; [Rome Glory X Better Times]; Fisher, G.; Arnold-Fisher Co.

Bunny Hop *see* 'JACCLIP'

'BUNTE FRAU ASTRID SPÄTH', Pol, lp, 1940; flowers light pink with red, medium, semi-dbl.; Vogel, M.; (Sangerhausen)

Bunter Kobold, MinFl, yb, 1995; Dickson, Patrick

'BURALP', HT, dr, 1988; (**National Velvet**®); flowers dark, deep velvet red, urn-shaped, medium, semi-dbl., 35 petals, exhibition form, borne usually singly, slight fragrance; globular, average, orange fruit; prickles recurved, average, dark; foliage large, medium green, semi-glossy; upright, tall growth; PP007236; [Poinsettia X National Beauty]; Burks, Larry; Co-Operative Rose Growers, 1990

'BURAPWARD', HT, yb, 1996; (**Dreamward**); flowers light yellow with pink blush, blooms borne mostly single, dbl., 15–25 petals, 4.5 in., slight fragrance; moderate prickles; [Peace X Granada]; Perry, Astor; Certified Roses, Inc., 1997

'BURBANK', T, mp, 1900; (Santa Rosa); flowers bright rose-pink shading lighter, dbl., moderate fragrance; [Hermosa X Bon Silene]; Burbank; Burpee

Burbank White Rambler, HWich, w; Burbank

'BURBET', Gr, m, 1988; (**Betty Wright**®); flowers lilac-mauve, darker at tips, aging lighter lilac, urn-shaped, semi-dbl., cupped, slight, fruity fragrance; globular, average, orange-yellow fruit; prickles slight recurve, average, brown; foliage medium, medium green, matt; bushy, medium growth; [Angel Face X Seedling]; Burks, Larry; Co-Operative Rose Growers

'BURBRAHMVIDYA', HT, w, 1996; (**Simply Irresistible**, gust Simon); flowers white with pink tones, reverse cream with pink tones, blooms, dbl., 15–25 petals, moderate fragrance; few prickles; foliage medium, medium green, semi-glossy; upright, meidum (4 1/2 ft) growth; [First Prize X Garden Party]; Datt, Braham; Certified Roses, Inc.

BURbrindley, F, or, 1995; (**Katie Crocker**); Burrows, Steven

'BURCEN', HT, or, 1988; (**Redlands Century**®); flowers medium orange-red, aging slightly lighter, urn-shaped, medium, semi-dbl., slight fragrance; globular, small, orange fruit; prickles average, yellow-green; foliage medium, medium green, semi-glossy; upright, medium growth; [World Peace X Command Performance]; Perry, Anthony; Co-Operative Rose Growers, 1988

'BURDANCER', Gr, w, 1995; (**Cloud Dancer**®); flowers white blend, double, blooms borne in small clusters, 15–25 petals, slight fragrance; few prickles; foliage medium, medium green, semi-glossy; upright, medium growth; [White Queen Elizabeth X Unknown]; Burks, Larry; Certified Roses, Inc., 1995

'BURDEL', HT, yb, 1988; (**Delta Gold**®); flowers red and yellow blend, reverse red-yellow, urn-shaped, medium, semi-dbl., 35 petals, slight fragrance; round, average, orange fruit; prickles slight recurve, average, dark; foliage medium, dark green, glossy; strong stems; bushy, medium growth; [Arizona X World Peace]; Perry, Anthony; Co-Operative Rose Growers, 1989

'BUREURO', Cl F, dr, 1987; (**Europeana, Climbing**); [Europeana sport]; Burks, Joe J.; Cooperative Rose Growers

Burg Baden, LCl, mp, 1955; Kordes

'BURGEMEESTER BERGER', HT, lp, 1934; flowers soft pink, marked white; vigorous growth; [Dame Edith Helen sport]; Leenders Bros.

'BURGEMEESTER SANDBERG', HT, op, 1920; flowers silvery pink, shaded coral-rose-pink, dbl., moderate fragrance; [Pharisaer X Lady Alice Stanley]; Van Rossem

'BURGEMEESTER VAN OPPEN', HT, my, 1939; flowers golden yellow, moderate fragrance; [Golden Ophelia X Pardinas Bonet]; Leenders, M.

'BÜRGERMEISTER CHRISTEN', HT, dp, 1911; flowers large, dbl., intense fragrance; Bernaix, A.; (Sangerhausen)

Burghausen® *see* KORonto

Burgund® *see* 'KORGUND'

Burgund '81 *see* 'KORGUND '81'

'BURGUNDIAN ROSE', C, pb, 1664; (Burgundy Rose, Parvifolia, Pompon de Bourgogne, Pompon de Burgogne, R. burgundensis, R. burgundica, R. centifolia parvifolia, R. ehrrhartiana, R. gallicaremensis, R. parvifolia, R. pomponia, R. remensis); flowers deep pink suffused purple, center paler, rosette form, dbl., 1 in.; few prickles; foliage dark gray-green; 3-5 ft. growth; (14)

'BURGUNDY', HT, mr, 1939; flowers wine-red broad, evenly arranged petals, dbl.; very vigorous growth; [Vaterland X Seedling]; H&S

Burgundy Queen *see* 'WILJAME'

Burgundy Rose *see* **'BURGUNDIAN ROSE'**

'BURJES', HT, mp, 1996; (**Jesse's Jewels**); semi-dbl., 6–14 petals, borne mostly singly, intense fragrance; few prickles; foliage medium, medium green, dull; upright, medium growth; [(Unknown X Pink Apache Belle) X Unknown]; Burks, Larry; Certified Roses, Inc.

Burkhard *see* **'GRUMPY'**

Burkhardt *see* **'GRUMPY'**

'BURLOV', HT, w, 1988; (**Pure Love**®); flowers urn-shaped, large, semi-dbl., borne singly, slight fragrance; globular, average, gray-red fruit; prickles slight recurve, average, brown-green; foliage medium, dark green, semi-glossy; upright, medium growth; PP006781;

[Queen Elizabeth X World Peace]; Perry, Anthony; Co-Operative Rose Growers, 1987

'BURMA STAR', F, ab, 1974; flowers light apricot yellow, dbl., 22 petals, 3.5 in., moderate fragrance; foliage large, glossy; [Arthur Bell X Manx Queen]; Cocker

'BURNABY', HT, w, 1954; (Gold Heart, Golden Heart); flowers creamy white, dbl., 56 petals, 4–6 in., exhibition form, slight fragrance; foliage dark, glossy; vigorous, bushy growth; GM, NRS, 1954 GM, Portland, 1957; [Phyllis Gold X Pres. Herbert Hoover]; Eddie; Peterson & Dering

Burnet Rose *see* R. spinosissima

Burning Gold, F, ob, 1990; Poulsen

'BURNING LOVE', Gr, mr, 1956; (Amour Ardent, Brennende Liebe); flowers scarlet, blooms in trusses of 3-5, dbl., 22 petals, 4 in., moderate fragrance; foliage dark, glossy; vigorous, bushy growth; GM, Baden-Baden, 1954; [Fanal X Crimson Glory]; Tantau, Math.

Burning Sky *see* 'WEZIP'

'BURNT ORANGE', F, ob, 1973; flowers deep orange; [Woburn Abbey sport]; Hamilton

Burr Rose *see* **R. ROXBURGHII**

'BURRED', HT, dr, 1987; (**Theodore Roosevelt**); flowers deep red, fading slightly lighter, medium, borne singly, semi-dbl., 6–14 petals, exhibition form, slight fragrance; globular, large, dark red fruit; average, light yellow-green prickles; foliage large, medium green, semi-glossy; upright, tall growth; [Alamo sport]; Burks, Larry; Co-Operative Rose Growers, 1986

'BURR'S MULTIFLORA', (clone of R. multiflora); vigorous; used as understock

BURspec, F, w, 1996; (**Good Luck**); Burston

'BURSRTPARA', HT, ob, 1996; (**Tropical Paradise**); flowers orange blend, reverse orange blend with yellow gold, blooms, dbl., 15–25 petals, 4.5 in., moderate fragrance; few prickles; foliage large, dark green, glossy; upright, tall (5 ft) growth; [South Seas X First Prize]; Rodgers, Shafner R.; Certified Roses, Inc.

'BURSTEIN', HT, op, 1982; (**Easter Bonnet**, Super Derby); [Queen Elizabeth sport]; Burks, Joe J.; Co-Operative Rose Growers

'BURWAH', Cl HT, dp, 1953; flowers deep rose-pink, large, dbl., cupped; very vigorous climbing growth; [Editor McFarland X Black Boy (Cl HT)]; Ulrick, L.W.

'BURWAY', HT, yb, 1985; (**Broadway**™); bud pointed; flowers golden yellow, blended orange-pink, well-formed, medium, dbl., 30–35 petals, 4–4.5 in., exhibition form, borne singly, moderate, spice and damask fragrance; foliage medium to large, dark, semi-glossy; upright growth, moderately tall; PP5827; AARS, 1986; [(First Prize X Gold Glow) X Sutter's Gold]; Perry, Anthony; Cooperative Rose Growers, 1986

'BURWIN', HT, rb, 1988; (**Double Perfection**™); flowers red, reverse white, aging darker red, reverse cream, medium, semi-dbl., exhibition form, slight fragrance; round, average, orange fruit; prickles slight recurve, average, brown; foliage medium, dark green, semi-glossy; bushy, medium growth; [(My Dream X First Prize) X Seedling]; Winchel, Joseph F.; Co-Operative Rose Growers, 1987

'BURWINLADHOM', HT, m, 1998; (**Ladies Home Journal**®); flowers pinkish mauve, medium, semi-double, semi-dbl., 8–14 petals, borne mostly singly, slight fragrance; few prickles; foliage medium, medium green, semi-glossy; upright, medium, 4 1/2 - 5 ft growth; [Unknown X Unknown]; Winchel, Joseph F.; Certified Roses, 1998

'BURWINTRADHOM', HT, m, 1998; (**Traditional Home**®); flowers pinkish, double, exhibition, dbl., 15–25 petals, 4 in., exhibition form, borne mostly singly, slight fragrance; prickles moderate; foliage medium, dark green, semi-glossy; upright, medium, 5 ft growth; [Unknown X Unknown]; Winchel, Joseph F.; Certified Roses Inc., 1998

'BURWORPE', HT, pb, 1987; (**World Peace**™); flowers pink blend suffused with cream, tips of petals tinged dark pink, dbl., 30–35 petals, exhibition form, moderate, fruity fragrance; globular, orange fruit; average prickles; foliage large, medium green, semi-glossy, disease resistant; upright, tall growth; [First Prize X Gold Glow]; Perry, Anthony, 1972; Co-Operative Rose Growers, 1989

'BURYEAR', Cl, Gr, ob, 1995; (**New Year, Climbing**); flowers spanish orange, semi-dbl. (6–14 petals), medium blooms borne in small clusters; slight fragrance; some prickles; foliage medium, medium green, glossy; spreading, tall growth; [New Year sport]; Burks, Joe J.; Certified Roses, Inc., 1995

'BURYELLOW', Cl F, my, 1987; (**Sun Flare, Climbing**™, Yellow Blaze); flowers medium yellow, fading lighter, medium, semi-dbl., 15–25 petals, 3.5 in., cupped, borne in sprays of 3-5, slight, licorice fragrance; globular, medium, orange-red fruit; short, yellow-green prickles, slightly curved; foliage medium, medium green, semi-glossy; spreading, tall growth, 14 ft.; PP006509; [Sun Flare sport]; Burks, Joe J.; Co-Operative Rose Growers, 1987

'BUSBUGUM', Min, pb, 1999; (**Bubble Gum**); flowers pink, reverse ivory, dbl., 17–25 petals, 1.2in., borne mostly singly, slight fragrance; prickles moderate; foliage medium, medium green, semi-glossy; upright, tall (2 ft) growth; [Garden State X seedling]; Buster, Larry S.; Kimbrew-Walter Roses, 1998

Bush Baby *see* 'PEANOB'

Bush Garden Climber, LCl, op; Thomas

'BUSHFIRE', HWich, mr, 1917; flowers bright crimson, yellow zone around center, small, very large, dbl.; Clark, A.

Bushfire *see* POUlbufi

'BUSHU', HT, ob, 1984; bud ovoid; flowers orange-red, reverse lighter, large, dbl., 35 petals, exhibition form, no fragrance; prickles slanted downward; foliage medium, dark, semi-glossy; tall, vigorous, upright growth; [Dolce Vita X Roklea]; Yasuda, Yuji

Bushveld Dawn *see* 'MEIMAINGER'

'BUSIMEL', Min, m, 1993; (**Simply Elegant**); flowers light lavender inside with darker reverse, blooms borne most, dbl., 26–40 petals, 1.5–2.75 in., slight fragrance; some prickles; foliage medium, medium green, matt; upright, compact (50 cms) growth; [Lady X X Winsome]; Buster, Larry S.; Kimbrew Walter Roses, 1993

'BUSMALOU', MinFl, m, 1996; (**Mary Louise**); flowers red lavender, lighter lavender reverse, blooms borne mostly, dbl., 26–40 petals, moderate fragrance; moderate prickles; foliage medium green, dull; upright, tall growth; RR; [Lady X X Winsome]; Buster, Larry S.; Kimbrew Walter Roses, 1996

'BUSTEV', Min, w, 1992; (**Stevie**); flowers well formed, blooms borne mostly singly opens to 3+ cms, very dbl., 53 petals, 1.5 in., slight fragrance; some prickles; foliage small, medium green, matt; low (42 cms), upright, compact growth; [Frau Karl Druschki X Miniature seedling]; Buster, Larry S.

Busy, Min, dp

Busy Lizzie *see* 'HARBUSY'

'BUSYBODY', HT, dy, 1929; flowers rich chrome-yellow, small; [Georges Schwartz X Lena]; Clark, A.; Hazlewood Bros.

'BUTA', HT, w; flowers pure white, well formed, intense fragrance; Cant, B. R.

Butter 'n' Sugar *see* 'BERBUT'

'BUTTERBALL', S, ly, 1950; flowers creamy yellow, single, non-recurrent; large, rounddish, reddish fruit; prickles prickly, arching branches; foliage small; spinosissima type; height 6 ft.; Skinner

'BUTTERCUP', HT, my, 1930; flowers buttercup-yellow, well-formed, semi-dbl., moderate fragrance; vigorous growth; Dobbie

'BUTTERCUP', HT, ab, 1929; flowers apricot-yellow, dbl., cupped; Towill

Buttercup™ *see* 'AUSBAND'

'BUTTERCUP', HWich, my, 1909; Paul

Butterflies, LCl, rb

Butterflies *see* 'MEKKADA'

Butterflies *see* INTermug

Butterflies *see* POUlbut

'BUTTERFLIES OF GOLD', HT, my, 1939; [Mrs Arthur Curtiss James seedling]; Brownell, H.C.

Butterfly®, HT, rb; Herholdt

'BUTTERFLY GLOW', HT, or, 1969; flowers vermilion, full, dbl., 20 petals, 3.5 in., slight fragrance; foliage dark; free growth; [Centre Court X Carla]; Barter

Butterfly Kisses *see* 'GELFLY'

Butterfly Papilio *see* **'MOTYLEK'**

'BUTTERFLY WINGS', F, pb, 1976; flowers ivory, petals edged pink, blooms in clusters, semi-dbl., 12 petals, 4–4.5 in., flat, moderate fragrance; foliage large; [Dainty Maid X Peace]; Gobbee, W.D.; Harkness

'BUTTERMERE', HWich, yb, 1932; flowers creamy yellow flushed pink, large truss; foliage glossy; vigorous, erect growth; Chaplin Bros.

Buttermint *see* 'MORSNOP'

'BUTTERSCOTCH', HT, yb, 1942; bud long, pointed; flowers lemon-chrome, reverse pale orange-yellow, dbl., 28 petals, 4.5–5.5 in., slight fragrance; foliage leathery, glossy, dark; upright, compact growth; [Souv. de Claudius Pernet X R.M.S. Queen Mary]; Hill, Joseph H., Co.; H&S, 1946;, Wayside Gardens Co., 1946

Butterscotch *see* 'JACTAN'

Buttons *see* 'LEMBUT'

Buttons *see* 'DICMICKEY'

'BUTTONS 'N' BOWS', Min, dp, 1981; (Felicity II, Teeny-Weeny); bud small; flowers deep pink, and, reflexing at maturity, borne singly or in sprays, dbl., 28 petals, exhibition form, moderate, fruity fragrance; straight prickles; compact, upright growth; [Mini-Poul X Harriet Poulsen]; Poulsen Roser APS; Windy Hill Nursery, 1982

'BUZBY', F, or, 1976; flowers light vermilion, conical, small, semi-dbl., 11 petals, 1.5 in., slight fragrance; dwarf, compact, upright growth; [Irish Mist X Topsi]; Plumpton, E.

Buzzy's White Seedling, N, w

By Appointment *see* 'HARVOLUTE'

By Design *see* 'LAVSIGN'

'BY JOE', Min, w, 1990; bud pointed; flowers ivory white, blooms borne singly, dbl., 25 petals, exhibition form, no fragrance; foliage medium, dark green, matt; upright, bushy, tall growth; [Pink Petticoat X Pink Petticoat]; Gatty, Joseph, 1990; Keith Keppel, 1990

'BYALA VALENTINA', Gr, w, 1974; flowers creamy white, large, dbl., 25 petals, slight, delicate fragrance; foliage dark, glossy; vigorous, upright growth; [Queen Elizabeth X Seedling]; Staikov, Prof. Dr. V.; Kalaydjiev and Chorbadjiiski

'BYRBRADBY', HT, op, 1999; (**Mary Bradby**™); flowers unusual orange and russet, reverse salmon, ruffled petals, dbl., 17–25 petals, 3 in., borne mostly singly, slight fragrance; few prickles; foliage medium, dark green, semi-glossy, disease-resistant; upright, medium (5 ft) growth; hardy; [Queen Elizabeth X Taboo]; Byrnes, Robert; Overbrooke Gardens, 1999

'BYRLYNNE', S, pb, 1999; (**Lynne Elizabeth**™); flowers light pink, reverse medium pink, dbl., 15–25 petals, 4 in., borne mostly singly, slight fragrance; few prickles; foliage medium, medium green, semi-glossy, disease-resistant; spreading, bushy, medium (4.5 ft) growth; winter hardy; [Carefree Beauty X Carefree Beauty]; Byrnes, Robert; Overbrooke Gardens, 1999

'BYRSHERRILL', S, pb, 1999; (**Sherrill Anne**™); flowers blush pink, reverse medium pink, very dbl., 41 petals, 4 in., slight fragrance; few prickles; foliage medium, medium green, semi-glossy; upright, medium (4-5 ft) growth; [Country Dancer X Country Dancer]; Byrnes, Robert; Overbrooke Gardens, 1999

C

'C. A. FLETCHER', HT, dp, 1947; flowers clear rose-crimson, well formed, dbl., 35–40 petals, 5–6 in., slight fragrance; vigorous growth; [May Wettern X Mrs Henry Bowles]; Fletcher; Tucker

'C. CHAMBARD', HT, 1934; flowers deep yellow, reverse tinted red, dbl., cupped; foliage dark; vigorous growth; Bel

'C. H. MIDDLETON', HT, dr, 1939; flowers dark crimson, large, very dbl., exhibition form, intense fragrance; foliage glossy; long stems; vigorous, bushy growth; Cant, B. R.

'C. V. HAWORTH', HT, dr, 1917; flowers intense black-scarlet with rich crimson bloom, massive shell-shaped petals, moderate fragrance; vigorous growth; GM, NRS, 1919; Dickson, A.

'C. V. HAWORTH, CLIMBING', Cl HT, dr, 1932; Cant, F.

'C. W. COWAN', HT, mr, 1912; flowers warm carmine-cerise, dbl., moderate fragrance; Dickson, A.

C.F. Worth *see* **'MME CHARLES FREDERIC WORTH'**

'C.S.R.', HP, 1934; flowers white with pink striping, large, dbl.; Böhm, J.; (Sangerhausen)

C.W.S. *see* **'CANADIAN WHITE STAR'**®

Cabaret *see* 'JACARET'

'CABARET', F, or, 1963; flowers vermilion-salmon, camellia shaped, dbl., 35–40 petals, 2.5 in., borne in clusters; vigorous, upright growth; RULED EXTINCT 6/83 ARM; [Dacapo X Floribunda seedling]; deRuiter; Blaby Rose Gardens

Cabaret® *see* LAPed

'CABBAGE ROSE', C, mp; (Provence Rose, R. centifolia, R. gallica centifolia); flowers overlapping petals, blooms borne singly or in clusters, very dbl., 3 in., summer blooming, moderate fragrance; (28, 21)

'CACAPHONY', HT, op, 1974; flowers pink tinged orange, dbl., 28 petals, 4 in., moderate, fruity fragrance; foliage glossy, dark; [Baccará X Golden Showers]; Golik; Dynarose

Cachet™ *see* 'TUCKACH'

Cacilda Backer, HT; (Brazil)

Cactus Blanc® *see* 'DELTROB'

'CADDY', HT, my, 1943; bud large, well formed, yellow; vigorous growth; [Soeur Thérèse X Prof. Deaux]; Meilland, F.

'CADENZA', LCl, dr, 1967; bud ovoid; dbl., 2.5–3 in., blooms in clusters, recurrent bloom, slight fragrance; foliage glossy, dark, leathery; compact, moderate growth; [New Dawn X Embers, Climbing]; Armstrong, D.L.; Armstrong Nursery

'CADETTE', F, ab, 1971; bud pointed; flowers pastel pink-apricot, dbl., 35 petals, 2 in., cupped, moderate, fruity fragrance; foliage glossy, dark; vigorous, upright growth; [Poupee X Fillette]; Lens

Cadillac *see* KORveril

Cadillac DeVille *see* 'WEKCRYLAND'

'CAECILIE SCHARSACH', HP, lp, 1887; flowers large, very dbl., intense fragrance; Geschwind, R.; (Sangerhausen)

Caesar *see* 'VARBOLE'

Caesar's Rose *see* 'JUDSAR'

'CAFÉ', F, r, 1956; flowers coffee-with-cream color, blooms in clusters, very dbl., flat, moderate fragrance; foliage olive-green; vigorous growth; [(Golden Glow X R. kordesii) X Lavender Pinocchio]; Kordes; McGredy

Café Olé *see* 'MOROLÉ'

'CAFOUGNETTE', HT, op, 1956; flowers soft orange-salmon, reverse carmine-red, open, well shaped, ; foliage dark; strong stems; [Happiness X Peace]; Dorieux; Pin

Cagul, HT, op, 1971; flowers salmon-orange and pink, large, very dbl., moderate fragrance; Moreira da Silva, A.; (Sangerhausen)

Cahto Maid, S, lp

Caid *see* 'DELSIRP'

'CAIRNGORM', F, ob, 1973; flowers tangerine and gold, dbl., 25 petals, 2.5 in., slight fragrance; foliage glossy, dark; upright growth; [Anne Cocker X Arthur Bell]; Cocker

Caitlin *see* 'TRACAIT'

Cajun Dancer *see* 'TALCAJ'

Cajun Spice *see* 'TALSPI'

Cajun Spice, HT, or, 1996; McMillan, Thomas G.

Cal Poly *see* 'MORPOLY'

Calay *see* 'TROBELLE'

Calcutta 300, HT, pb, 1994; V&B

Caldwell Pink, Pol, mp

Caldwill Pink Cl., Cl Pol, mp

'CALEDONIA', HT, w, 1928; bud long, pointed; flowers large, dbl., 25 petals, exhibition form, slight fragrance; foliage leathery, dark; vigorous growth; RULED EXTINCT 7/83 ARM; Dobbie

'CALEDONIA, CLIMBING', Cl HT, w, 1936; Bel

'CALEDONIAN', HT, ly, 1983; flowers creamy yellow, large, dbl., slight fragrance; foliage medium, medium green, semi-glossy; bushy growth; [Kordes' Perfecta X Irish Gold]; Mayle, W.J.

Calgary, HT, mr, 1997; Twomey, Jerry

'CALGOLD', Min, dy, 1977; bud pointed; flowers deep clear yellow, small, dbl., 23 petals, 1.5 in., slight fragrance; foliage small to medium, glossy; bushy growth; [Golden Glow (Brownell) X Peachy White]; Moore, Ralph S.; Sequoia Nursery

Calibra *see* KORcrisett

'CALICO', HT, pb, 1976; flowers pink, yellow reverse, dbl., 3.5–4 in., globular, slight, tea fragrance; foliage dark; vigorous, upright to spreading growth; [Seedling X Granada]; Weeks; Weeks Wholesale Rose Growers

Calico Doll *see* 'SAVADOLL'

'CALICO STAR', F, yb, 1977; bud ovoid; flowers golden yellow, edged red, dbl., 25–30 petals, 4 in., exhibition form, slight fragrance; upright growth; [Circus X Lavender Girl]; Fong; United Rose Growers

'CALIENTE', F, dr, 1974; bud ovoid-pointed; flowers deep pure red, medium, dbl., exhibition form, slight fragrance; foliage large, leathery; very vigorous, bushy growth; [Seedling X Seedling]; Warriner, William A.; J&P

'CALIFORNIA', HT, ob, 1940; bud long, pointed; flowers ruddy orange, reverse overlaid pink, dbl., 30 petals, 5–6 in., moderate, fruity fragrance; foliage leathery, glossy; vigorous, bushy, spreading growth; AARS, 1941; [Miss Rowena Thom X Lady Forteviot]; Howard, F.H.; H&S

'CALIFORNIA', HT, ob, 1916; flowers deep orange, base golden yellow; H&S

'CALIFORNIA BEAUTY', HT, my, 1926; Pacific Rose Co.

'CALIFORNIA BEAUTY', HT, dp, 1935; bud long, pointed to ovoid; flowers deep bright pink, very large, dbl.; foliage leathery, dark; vigorous growth; [Dame Edith Helen X Hollywood]; Proietti

California Blonde *see* 'RENBLONDE'

'CALIFORNIA CENTENNIAL', HT, dr, 1949; bud long, pointed; dbl., 28 petals, 3.5–4 in., exhibition form, intense fragrance; foliage leathery, bronze; vigorous, upright growth; [Tango X Mauna Loa]; Howard, F.H.; H&S

California Dreaming *see* 'RENOVE'

California Girl *see* 'RENIRL'

California Glory *see* 'ORTCAL'

'CALIFORNIA GOLD', Pol, ab, 1934; flowers orange-yellow, clusters, dbl., globular, slight fragrance; foliage leathery, light; long stems; very vigorous, bushy growth; [Gloria Mundi sport]; Smith, J.

California Ground Rose *see* **R. SPITHAMEA**

California Sun *see* 'RENSUN'

California Surf *see* 'RENURF'

California Wild Rose *see* **R. CALIFORNICA**

'CALIFORNIA, CLIMBING', Cl HT, ob, 1953; Howard, A.P.; H&S

'CALIFORNIA'S FAVORITE', HT, op, 1949; bud long, pointed; flowers light salmon-pink, base yellow, large, moderate fragrance; foliage soft; vigorous, bushy growth; [Stockton Beauty sport]; Raffel; Port Stockton Nursery

'CALINE', HT, op, 1958; flowers soft geranium-red, medium, dbl., 33 petals; foliage clear green; vigorous, upright growth; Ducher, Ch.; EFR

Calliope®, S, mp, 1985; Barni, V.

Calliope *see* HARfracas

'CALLISTO', HMsk, my, 1920; flowers golden yellow, rosette form, borne in clusters, recurrent bloom; height 3-4 ft; [William Allen Richardson X Self]; Pemberton

'CALOCARPA', HRg, mp; (R. rugosa calocarpa, R. X calocarpa); flowers rose-colored, single; fruit handsome, in abundance; (14); [R. rugosa X Form of R. chinensis]

Calumet *see* 'SOCAPAN'

Calypso *see* **'BLUSH BOURSAULT'**

'CALYPSO', D, lp; flowers rosy blush with paler edges, large, dbl., cupped; Vibert, prior to 1848

'CALYPSO', F, or, 1957; bud globular; flowers orange-red, reverse red, loose, blooms in large pyramidal, semi-dbl., 18 petals, 3–3.5 in., cupped, intense, damask fragrance; foliage dark, leathery; vigorous, bushy growth; [Geranium Red X Fashion]; Boerner; Stuart

Calypso® *see* 'POULCLIMB'

'CALYPSO', N, lp

'CAMAIEUX', HGal, m; flowers white and pale rosy purple, striped, well-shaped, dbl., intense fragrance; vigorous, rather dwarf growth; 1830

Camaieux Fimbriata, HGal, m, 1980; Bell

Camaieux Reversion, HGal, m, 1985; Robinson

Camara® *see* 'DELCAMA'

'CAMAY', HT, dp, 1959; flowers deep rose-pink to carmine, dbl., 40 petals, 5 in., moderate fragrance; foliage light green; vigorous growth; [Ena Harkness X C.A. Fletcher]; Fletcher; Tucker

'CAMBRAI', HFt, yb, 1920; flowers light orange-yellow, medium, dbl.; Smith; (Sangerhausen)

Cambria, S, lp

Cambridgeshire *see* KORhungen

'CAMELEON', F, yb, 1961; flowers yellow, pink and orange, large clusters, dbl., 22–27 petals, intense fragrance; foliage dark, glossy; upright, compact, bushy, symmetrical growth; [Masquerade X Seedling]; Verschuren, A.; van Engelen

'CAMELIA', F, dp, 1953; flowers cerise-pink, medium, clusters; strong stems; moderate growth; [Pinocchio X Seedling]; Klyn

'CAMÉLIA', HT, or, 1948; bud pointed; flowers fiery vermilion-red, medium, semi-dbl., cupped, moderate fragrance; foliage glossy, dark; vigorous, bushy growth; GM, Geneva, 1948; [Vainqueur X (Charles P. Kilham X Katharine Pechtold)]; Heizmann & Co.

'CAMELIA ROSE', HCh, 1830; Prévost, prior to 1830

'CAMELIAROSE', F, lp, 1960; flowers soft pink, camellia form, very free bloom, especially in autumn; upright growth; [Mme Joseph Perraud X Incendie]; Croix, P.; Minier

'CAMELOT', Gr, op, 1964; bud ovoid; flowers shrimp-pink, blooms in clusters, dbl., 48 petals, 3.5–4 in., cupped, moderate, spicy fragrance; foliage leathery, glossy, dark; vigorous, tall growth; AARS, 1965; [Circus X Queen Elizabeth]; Swim & Weeks; C-P

'CAMEO', Pol, op, 1932; flowers salmon-pink, turning soft orange-pink; [Orléans Rose sport]; deRuiter; J&P

Cameo Cream *see* 'MEIPIERAR'

Cameo Cream, HP, lp; flowers purple-violet, dbl., cupped

'CAMEO QUEEN', Min, pb, 1986; flowers light pink, reverse blends of pink, mini-flora, dbl., 34 petals, exhibition form, borne usually singly, slight fragrance; medium, long, light prickles; foliage large, medium green, semi-glossy; medium, bushy growth; [Heartland X Seedling]; Bridges, Dennis A.

'CAMEO SUPERIOR', Pol, mp; flowers have more lasting color; [Cameo sport]; deRuiter

Cameo, Climbing *see* **'PINK CAMEO'**

'CAMILLA', HT, dp, 1954; bud ovoid; flowers strawberry-red, loosely formed, large, dbl., 30–36 petals, moderate fragrance; very vigorous growth; [Fiamma X Talisman]; Aicardi, D.; Olivieri

Camilla Sunsation *see* KORfibi

Camille *see* 'JACFRUIT'

'CAMILLE BERNARDIN', HP, rb, 1866; flowers red with white, very large, dbl., intense fragrance; Gautreau; (Sangerhausen)

Camille Pisarro *see* DELstricol

'CAMILLO SCHNEIDER', HT, mr, 1922; flowers clear blood-red, dbl., slight fragrance; [Lieut Chaure X Comte G. de Rochemur]; Kordes

'CAMOËNS', HT, mp, 1881; flowers bright rose, center shaded yellow, moderate fragrance; moderately vigorous growth; [Antoine Verdier X Seedling]; Schwartz, J.

Camp David, HT, dr, 1984; Tantau

Campanela *see* 'DELTRUT'

Campanile® *see* 'DELTRUT'

'CAMPFIRE', Cl Pol, or, 1956; flowers orange-scarlet, rosette shape, small, borne in clusters; vigorous (6-8 ft) growth; [Cameo sport]; Fryers Nursery, Ltd.

Campfire *see* 'WILCAMP'

Campfire Arteka *see* **'KOSTIOR ARTEKA'**

'CAMPFIRE GIRL', HT, op, 1946; bud long, pointed; flowers deep salmon, large, dbl., exhibition form, moderate fragrance; foliage dark, leathery; vigorous, upright, bushy growth; [Joanna Hill X Gruss an Aachen]; Duehrsen; California Roses

Camphill Glory *see* 'HARKREME'

'CAMPINA', HT, lp, 1937; bud long; flowers flesh-pink; very vigorous growth; [Comtesse Vandal X White Briarcliff]; Lens

'CAMPING', Pol, pb, 1967; flowers deep lavender-pink with white eye, small, borne in large clusters, single, globular; foliage small, olive-green; low, bushy growth; [Paul Crampel sport]; Grabczewski

Camrose, S, mp, 1998; Twomey, Jerry

Can Can *see* 'LEGGLOW'

'CANADIAN CENTENNIAL', F, or, 1965; bud ovoid; flowers coral-red, medium, dbl., cupped, moderate fragrance; foliage glossy; vigorous, upright, compact growth; [Pinocchio seedling X Spartan]; Boerner; J&P

Canadian Dream *see* BRODREAM

'CANADIAN JUBILEE', HT, op, 1927; flowers indian red to pink, base orange, dbl., slight fragrance; [Priscilla X Commonwealth]; Dunlop

'CANADIAN NORTHLIGHT', HT, dp, 1984; flowers deep pink, large blooms in clusters of 6-9, dbl., 28 petals, exhibition form, slight fragrance; dark red prickles; foliage medium green, leathery; spreading growth; [(Fragrant Cloud X Diamond Jubilee) X Super Sun]; Mander, George

Canadian Sunset, HT, rb, 1996; flowers red with yellow reverse; foliage disease-resistant; Williams, J. Benjamin

'CANADIAN WHITE STAR'®, HT, w, 1980; (C.W.S., Dr Wolfgang Pöschl); flowers opening to multi-pointed star, dbl., 43 petals, exhibition form, borne singly, slight fragrance; slightly hooked prickles; foliage dark, leathery, glossy; vigorous, upright growth; PP005852; [Blanche Mallerin X Pascali]; Mander, George, 1980; Hortico Roses, 1985

Canadiana *see* **'IMPERIAL GOLD'**

Canadiana, HT, 1962; Swim, H.C.; (Cavriglia)

'CANANA', HT, mp, 1983; (**Mary Donaldson**); flowers medium salmon-pink, medium, dbl., exhibition form, intense fragrance; foliage large, dark, glossy; upright growth; [Kathleen O'Rourke X Seedling]; Cants of Colchester, Ltd., 1984

'CANARIAS', HT, mr, 1964; flowers bright red, large; Dot, Pedro; Minier

'CANARIENVOGEL', Pol, 1903; (Sunset Glow); flowers saffron-yellow and amber-yellow stained pink and purple, semi-dbl.; strong stems; vigorous growth; [Étoile de Mai X Souv. de Catherine Guillot]; Welter

Canarina, 1965; Mansuino, Q.; (Cavriglia)

'CANARY', T, my, 1852; bud small, well formed; flowers canary-yellow; growth rather weak; Guillot Père

'CANARY', HT, ly, 1929; bud golden yellow, edges flushed; flowers light yellow, deepening, spiral, exhibition form, moderate fragrance; branching growth; Dickson, A.

'CANARY', Min, my; flowers primuline-yellow, star shaped, semi-dbl.

Canary *see* 'TANCARY'

Canary, HSpn, ly

'CANARY BIRD', S, dy, 1907; (R. xanthina 'Canary Bird'); flowers yellow stamens, single, 5 petals; blackish purple fruit; [Probably R. hugonis X R. xanthina]

'CANARY CHARM', HT, my, 1969; flowers very large, dbl., exhibition form; foliage dark, leathery; free growth; [Wiener Charme sport]; Knight, G.

Canary Island, HGal, m

'CANASTA', HT, mr, 1966; bud long, pointed; flowers bright red, large, dbl., exhibition form, moderate fragrance; very vigorous, upright growth; [Karl Herbst X Miss Universe]; Gaujard

'CANBERRA', Pol, op, 1935; flowers salmon-coral-pink; [Gloria Mundi sport]; Knight, G.

'CANBERRA', HT, op, 1927; flowers carmine, reverse buff at base, shading to salmon-pink; [Donald MacDonald X The Queen Alexandra Rose]; Burbage Nursery

'CANBERRA', HT, mp, 1928; very dbl.; Harrison

Canberra Rose *see* 'TOMFED'

'CANCAN', F, ob, 1969; (Diorette); bud short, pointed; flowers mandarin-red, small, dbl., moderate, spicy fragrance; foliage leathery; Jelly; E.G. Hill Co.

Can-Can *see* 'LEGGLOW'

Cancun™ *see* 'WEKIBERTAZ'

Candelabra™ *see* 'JACCINQO'

Candella® *see* 'MACSPEEGO'

Candeur® *see* 'DELCANDE'

'CANDEUR LYONNAISE', HP, w, 1914; bud long, pointed; flowers white, sometimes tinted pale yellow, large, dbl.; very vigorous growth; [Frau Karl Druschki seedling]; Croibier

Candia™ *see* 'MEIBIRANDA'

Candice *see* 'ZIPCAN'

Candid, HT, w; Delbard

'CANDIDA', HT, yb, 1964; flowers creamy yellow, center salmon, well formed; [Tawny Gold X Golden Scepter]; Leenders, J.

'CANDIDE', HT, dy, 1983; (Gold Star, Goldina, **Goldstar**, Point du Jour®); flowers medium, dbl., 35 petals, slight fragrance; foliage small, light green, matt; upright growth; GM, The Hague, 1984; [Yellow Pages X Dr. A.J. Verhage]; Cants of Colchester, Ltd., 1984

Candide, HT, 1970; Gaujard; (Cavriglia)

Candle Light, Cl Min, yb, 1995; Warner, Chris

'CANDLEFLAME', Min, yb, 1956; bud slender; flowers red, yellow and orange, single, 5 petals; foliage leathery; vigorous (10 in), bushy growth; [(Soeur Thérèse X Julien Potin) X (Eblouissant X Zee)]; Moore, Ralph S.; Sequoia Nursery

'CANDLEGLOW', HT, yb, 1951; bud long, pointed; flowers yellow washed shrimp-pink, dbl., 38 petals, 5–6 in., cupped, slight fragrance; foliage glossy, dark; vigorous, upright growth; [Golden Rapture X (Seedling X Joanna Hill)]; Whisler; Germain's

Candlelight™ *see* 'AROWEDYE'

'CANDLELIGHT', HT, 1932; flowers yellow, deeper in hot weather, large, dbl., exhibition form, moderate fragrance; foliage glossy; bushy growth; RULED EXTINCT 9/81 ARM; [Souv. de Claudius Pernet X Mme Butterfly]; Horvath; Bosley Nursery

CANdoodle, F, w, 1990; (**Lady Rachel**); Cants of Colchester, Ltd.

'CANDY', HT, ob, 1950; bud long, pointed; flowers apricot-orange, medium, dbl., exhibition form, moderate fragrance; vigorous, upright, compact, bushy growth; [Pink Princess X Shades of Autumn]; Brownell, H.C.

'CANDY APPLE', Gr, mr, 1966; bud ovoid; flowers bright apple red, very full, very dbl., 40–45 petals, 5 in., cupped, blooms in small clusters, slight, tea fragrance; foliage matt, olive-green; upright, medium, bushy growth; [Jack O'Lantern X (Seedling X El Capitan)]; Weeks, O.L.; Weeks Wholesale Rose Growers

'CANDY CANE', Cl Min, pb, 1958; flowers deep pink, striped white, blooms in loose clusters, semi-dbl., 13 petals, 1.5 in.; vigorous, upright (to 4 ft) growth; [Seedling X Zee]; Moore, Ralph S.; Sequoia Nursery

Candy Corn *see* 'TALCAN'

'CANDY FAVORITE', HT, pb, 1971; flowers carmine, striped pale rose, full, dbl., 25 petals, 4.5–5 in., slight fragrance; foliage very glossy; vigorous growth; [Pink Favorite sport]; Heath, W.L.

Candy Flo, HT, rb

'CANDY FLOSS', F, lp, 1959; flowers bright pink, rosette shape, borne in large clusters, dbl., 40 petals, 3 in., slight fragrance; very vigorous growth; [Lilibet X Seedling]; Fryers Nursery, Ltd.

Candy Mountain™ *see* 'JACCHARI'

'CANDY PINK', Min, lp, 1969; bud ovoid; flowers small, dbl.; foliage small, leathery; vigorous, dwarf, bushy growth; [(R. wichurana X Floradora) X (Oakington Ruby X Floradora)]; Moore, Ralph S.; Sequoia Nursery

Candy Rose® *see* 'MEIRANOVI'

Candy Stick *see* **'CANDYSTICK'**

'CANDY STRIPE', HT, pb, 1963; flowers dusty pink streaked (striped) lighter; [Pink Peace sport]; McCummings; C-P

Candy Sunblaze® *see* 'MEIDANCLAR'

'CANDYSTICK', HT, pb, 1978; (Candy Stick, Red 'n' White Glory); flowers deep pink striped white; [Better Times sport]; Williams, J. Benjamin; Lakeland Nursery Sales

'CANIGÓ', HT, w, 1927; dbl., slight fragrance; [Antoine Rivoire X Mme Ravary]; Dot, Pedro

'CANLISH', F, mp, 1983; (**St Helena**, Union-Rose st Helena®); flowers medium lilac pink, medium, dbl., 20 petals, slight fragrance; foliage medium, medium green, semi-glossy; upright growth; [Jubilant X Prima Ballerina]; Cants of Colchester, Ltd.

CANlloyd, F, ob, 1991; (**Lloyds of London**); Cants of Colchester, Ltd.

'CANLOT', F, pb, 1984; (**Alexia**); flowers cream blended pink, medium, dbl., 20 petals, moderate fragrance; foliage medium, medium green, semi-glossy; upright growth; [Jubilant X Seedling]; Cants of Colchester, Ltd.

'CANMISS', F, mp, 1984; (**Beauty Queen**); flowers medium, dbl., intense fragrance; foliage large, dark, glossy; [English Miss X Seedling]; Cants of Colchester, Ltd.

'CANNES FESTIVAL', HT, yb, 1951; flowers indian yellow veined amber, pointed, dbl., 35 petals, 4 in., moderate fragrance; foliage dark; vigorous, upright, branching growth; [Peace X Prinses Beatrix]; Meilland, F.; URS

Cannes Festival 83 *see* MEIlicafal

'CANREM', HT, pb, 1994; (**Sally's Rose**); dbl., 15–25 petals, 1.5–2.75 in.; many prickles; foliage dark green, glossy; medium (60 cms), bushy growth; [Amber Queen X Remember Me]; Cants of Colchester, Ltd.; Cants of Colchester Ltd., 1994

'CANSEND', F, dp, 1992; (**Colchester Beauty**); flowers candy pink, blooms borne mostly singly, semi-dbl., 6–14 petals, 1.5–2.75 in., intense fragrance; some prickles; foliage medium, dark green, semi-glossy; medium (60-80 cms), bushy growth; [English Miss X Seedling]; Pawsey, P.R.; Cants of Colchester Ltd., 1989

CANsit, F, 1996; (**Jenny's Rose**); Cants of Colchester, Ltd.; (Cavriglia)

CANson, F, mp, 1991; (**Dame Wendy**); Cants of Colchester, Ltd.

'CANTAB', S, dp, 1927; bud long, pointed; flowers deep pink, base white, stamens yellow, saucer-shaped, single, 3.5 in., non-recurrent; foliage dark, 7-9 leaflets; height 6-8 ft; [R. nutkana X Red-Letter Day]; Hurst

'CANTABILE', S, lp, 1962; bud ovoid; flowers light camellia-rose shaded darker, medium, dbl., 25 petals, repeat bloom, moderate fragrance; foliage leathery, bronze; vigorous, upright (5 ft.) growth; [Harmonie X (Josef Rothmund X R. laxa)]; Buck, Dr. Griffith J.; Iowa State University

'CANTABRIGIENSIS', S, ly; (R. X cantabrigiensis, R. X pteragonis cantabrigiensis, The Cambridge Rose); flowers pale yellow, single, 2.25 in.; (14); [R. hugonis X R. sericea hookeri]

'CANTATE', F, ob, 1959; flowers orange; [Red Favorite sport]; van de Water; Spek

Canterbury *see* 'AUSCANTERBURY'

Canterbury *see* **'SCOOP JACKSON'**

Canterbury Pride *see* 'REYCANTPRI'

Cantilena Bohemica, HT, dp, 1977; flowers carmine-pink, large, very dbl.; Havel; (Sangerhausen)

Cantilena Moravica, HT, yb, 1981; Havel

'CANZONETTA', F, or, 1953; bud ovoid, cardinal-red; flowers orange-red, reverse yellow suffused red, open, large, borne, semi-dbl., 10–12 petals, moderate fragrance; foliage dark, glossy, leathery; vigorous, bushy growth; [Lawrence Johnston X Fashion]; San Remo Exp. Sta.; Sgaravatti

Cape Cod *see* 'POULFAN'

'CAPE CORAL', HT, ab, 1964; bud ovoid; flowers orange-coral, dbl., 50–55 wavy petals, 5 in., cupped, moderate fragrance; foliage leathery, glossy, veined red; long, strong stems; vigorous, bushy growth; [Spartan X Golden Masterpiece]; Boerner; J&P

'CAPE HATTERAS', Min, w, 1988; flowers medium, borne singly, dbl., 40 petals, exhibition form, slight, spicy fragrance; prickles long, pointed, medium pink-tan; foliage medium, dark green, glossy; upright, medium, vigorous growth; [Rise 'n' Shine X Seedling]; Bridges, Dennis A.; Bridges Roses

'CAPELINE', LCl, mp, 1965; flowers tyrian rose, semi-dbl., exhibition form, recurrent bloom; foliage leathery; vigorous growth; [Etendard X Diane d'Urfe]; Hémeray-Aubert

Capella *see* 'MEIRILOCRA'

'CAPER', F, dr, 1976; bud long, pointed; flowers open, dbl., 25 petals, 2–2.5 in., slight fragrance; foliage dark; upright growth; [Seedling X Mary DeVor]; Warriner, William A.; J&P

'CAPFLARE', HT, or, 1996; (**Solarflash**, Solar Flare, Solarflare); flowers bright vermilion, vibrant dark orange, with dark petal edge, dbl., 26–40 petals, 3.5 in.; some to numerous prickles; foliage medium, dark green, semi-glossy; medium to low (2 1/2-3 ft), compact growth; [Unknown Seedling X Hot Pewter]; Pallek, Otto; Pallek, 1995

'CAPISTRANO', HT, mp, 1949; bud ovoid; flowers bright pink, dbl., 36 petals, 6 in., globular, moderate fragrance; foliage leathery; vigorous, upright growth; AARS, 1950; Morris; Germain's

'CAPISTRANO, CLIMBING', Cl HT, mp, 1952; Germain's

'CAPITAINE BASROGER', M, rb, 1890; flowers bright carmine-red, shaded purple, large, dbl.; very vigorous, almost climbing growth; Moreau et Robert

'CAPITAINE DYEL DE GRAVILLE', B, pb, 1843; Boutigny

'CAPITAINE GEORGES DESSIRIER', HT, dr, 1919; flowers dark velvety red, shaded crimson, dbl., moderate fragrance; [Seedling X Château de Clos Vougeot]; Pernet-Ducher

'CAPITAINE JOHN INGRAM', M, m; bud well mossed; flowers variously described as dark purple, velvety crimson and reddish purple, dbl.; vigorous growth; Laffay, M., 1855

'CAPITAINE JOUEN', HP, mr, 1900; flowers very large, dbl.; Boutigny; (Sangerhausen)

'CAPITAINE MILLET', T, mr, 1901; flowers bright red, reverse purplish, base golden, large, dbl., moderate fragrance; [Gén. Schablikine X Mme Etienne]; Ketten Bros.

'CAPITAINE PEILLON', HP, m, 1893; flowers purple-pink, large, dbl.; Liabaud; (Sangerhausen)

'CAPITAINE SISOLET', HP, mp; flowers rose, large, very dbl., cupped; vigorous, branching growth; (France), Prior to 1848

'CAPITAINE SOUPA', HT, dp, 1902; flowers carmine-pink, very large, dbl.; Laperrière; (Sangerhausen)

'CAPITAINE SOUPA, CLIMBING', Cl HT, dp, 1938; flowers carmine-pink, large, dbl.; Vogel, M.; (Sangerhausen)

'CAPITAINE WILLIAMS', HGal, dr; flowers medium, very dbl.

Capitalia, HT; Ansaloni; (Cavriglia)

Capitole® *see* LAPovi

'CAPMINA', F, mp, 1996; (**Conservency Rose**); flowers vibrant pink, some variegation, not solid pink, long lasting, dbl., 30 petals, 1.5 in., slight fragrance; some prickles; foliage medium, medium green, glossy, disease resistant; medium (20 in), bushy growth; Pallek, Ruth; Carl Pallek & Son Nurseries, 1995

Caporosso, HRg, mr, 1975; Mansuino, Q.

Cappa Magna® *see* 'DELSAP'

'CAPREOLATA RUGA', Misc OGR, mp, 1820

'CAPRI', F, op, 1956; bud conical; flowers bright coral, reverse lighter, to, blooms in small clusters, dbl., 38 petals, 3–3.5 in., exhibition form, slight fragrance; foliage bright green, glossy, leathery; vigorous growth; [Fashion X Floradora]; Fisher, G.; Arnold-Fisher Co.

Capri, HT, ab

Capri Sun *see* TANnus

'CAPRICE', HT, pb, 1948; (Lady Eve Price); bud ovoid; flowers deep pink, reverse cream, large, dbl., 24 petals, slight fragrance; foliage dark, leathery; vigorous, upright, bushy growth; [Peace X Fantastique]; Meilland, F.; C-P

'CAPRICE', HT, or, 1934; bud long, pointed; flowers orient red and peach-red, large, dbl., slight fragrance; foliage glossy, light; vigorous growth; [Seedling X Gwyneth Jones]; Leenders, M.

Caprice 86, S, 1986; Meilland, L.; (Cavriglia)

Caprice de Mielland *see* MEIsionver

'CAPRICE, CLIMBING', Cl HT, or, 1951; Lens

'CAPRICORN', F, rb, 1961; flowers scarlet, reverse silver, borne in large clusters, semi-dbl., 16 petals, 2 in., slight fragrance; foliage dark, dull, leathery; very vigorous growth; Verschuren; Blaby Rose Gardens

Capriolata, Misc OGR, pb, 1970

'CAPRIOLE', F, mp, 1956; bud cherry-red; flowers vivid pink, stamens prominent, medium, in large clusters, semi-dbl.; foliage dark, semi-glossy; vigorous, dwarf, bushy growth; [Red Favorite X Fanal]; Tantau, Math.

'CAPT. F. S. HARVEY-CANT', HT, pb, 1923; bud long, pointed; flowers peach-pink, reverse deep pink, very large, dbl., moderate fragrance; foliage dark, leathery; vigorous, bushy growth; GM, NRS, 1922; Cant, F.

'CAPT. ROBINSON', Sp, dp; semi-dbl., 13 petals, non-recurrent, moderate fragrance; few prickles; foliage small, dark; height 1 ft; ((discovered in the sand dunes of Porter Co., In 1924))

'CAPT. RONALD CLERK', HT, or, 1923; flowers vermilion-scarlet, semi-dbl., slight fragrance; McGredy

'CAPT. RONALD CLERK, CLIMBING', Cl HT, or, 1935; Austin & McAslan

'CAPTAIN BLIGH', HT, mp, 1939; bud long, pointed; flowers silvery rose, large, very dbl.; foliage leathery, dark; very vigorous growth; [Gustav Grünerwald X Betty Uprichard]; Fitzhardinge; Hazlewood Bros.

'CAPTAIN BLOOD', HT, dr, 1938; bud ovoid; flowers scarlet and crimson, dbl., cupped, intense fragrance; foliage leathery; vigorous, bushy growth; [Gen. MacArthur X E.G. Hill]; Melville Bros.; R. Murrell

'CAPTAIN CHRISTY', HT, lp, 1873; flowers soft flesh-pink, center darker, large, dbl., 40 petals, globular, slight fragrance; wide, compact growth; [Victor Verdier X Safrano]; Lacharme, F.

'CAPTAIN CHRISTY, CLIMBING', Cl HT, lp, 1881; Ducher

Captain Cook *see* 'MACAL'

'CAPTAIN F. BALD', HT, dr, 1919; flowers scarlet-crimson, velvety black sheen, dbl., slight fragrance; Dickson, A.

'CAPTAIN GLISSON', HT, dy, 1935; bud long, pointed; flowers dark yellow, edged lighter, large, dbl., 28–30 petals; very vigorous, compact growth; [Joanna Hill X Sweet Adeline]; Hill, Joseph H., Co.

'CAPTAIN HARRY STEBBINGS', HT, dp, 1980; bud long, pointed; flowers deep pink, dbl., 43 petals, 5–6.5 in., exhibition form, intense, fruity fragrance; foliage large, leathery; upright, bushy growth; [Seedling sport]; Stebbings; Country Garden Nursery

'CAPTAIN HAYWARD', HP, dp, 1893; flowers light crimson, edged lighter, large, dbl., 25 petals, exhibition form, sparsely recurrent, moderate fragrance; large, orange fruit; vigorous growth; (28); [Triomphe de l'Exposition seedling]; Bennett

'CAPTAIN HAYWARD, CLIMBING', Cl HP, dp, 1906; Paul

'CAPTAIN KIDD', HSet, mr, 1934; flowers blood-red, open, large, borne, dbl., cupped, moderate fragrance; foliage leathery, dark; long, strong stems; very vigorous, climbing or tall pillar growth; [R. setigera seedling X Hoosier Beauty]; Horvath

'CAPTAIN KILBEE STUART', HT, dr, 1922; bud long, pointed; flowers scarlet-crimson, very large, dbl., intense fragrance; moderate growth; GM, NRS, 1922; Dickson, A.

'CAPTAIN KILBY', F, mr, 1955; flowers blood-red, medium, large truss, semi-dbl., moderate fragrance; foliage glossy, parsley-green; very vigorous growth; deRuiter; Gandy Roses, Ltd.

'CAPTAIN SAMUEL HOLLAND', S, mr, 1991; flowers moderately medium, dbl., 15–25 petals, slight fragrance; foliage medium, medium green, glossy; spreading, medium growth; Ogilvie, Ian S.; Agriculture Canada, 1992

'CAPTAIN SASSOON', HT, dr, 1938; flowers dark crimson, moderate fragrance; vigorous growth; Gaujard

'CAPTAIN THOMAS', Cl HT, w, 1935; bud long, pointed; flowers lemon to cream, single blooms in clusters, recurrent bloom, moderate fragrance; foliage glossy, light; climbing or pillar (10 ft.) growth; [Bloomfield Completeness X Attraction]; Thomas; Armstrong Nursery, 1938

Captain Watkins, Pol, mp; Heyde, C.W., before 1960

'CAPTAIN WOODWARD', HP, dp; flowers light red

Captivation™ *see* 'KINCAP'

'CAPTIVATOR', HT, mp, 1942; [Better Times sport]; Hill, Joseph H., Co.

'CAPUCINE CHAMBARD', HFt

'CARA MIA', HT, mr, 1969; (Danina, Dearest One, Maja Mauser, Natacha); bud ovoid; flowers large, dbl., moderate fragrance; foliage dark; vigorous, upright growth; McDaniel, G.K.; Carlton Rose Nurseries

'CARABELLA', F, yb, 1960; bud pointed, apricot; flowers cream edged pink, open, borne in clusters, single, moderate fragrance; foliage glossy, light green; moderate, bushy growth; [Gartendirektor Otto Linne X Seedling]; Riethmuller

Caraibes, HT, 1970; Combe, M.; (Cavriglia)

'CARALE', F, dy, 1993; (**Cisco**™); dbl., 26–40 petals, 1.5–2.75 in., borne in small clusters, slight fragrance; foliage medium, medium green, glossy; medium (60 cms), upright growth; [Escort X Golden Nugget]; Strahle, Robert; Carlton Rose Nurseries, 1989

Caramba® *see* 'TANCA'

'CARAMEL CREME', HT, my, 1980; bud ovoid; flowers caramel yellow, blooms borne singly, dbl., 30 petals, cupped, slight, spicy fragrance; long prickles, hooked downward; foliage finely serrated; bushy, moderate growth; [(Sunbonnet X Mister Lincoln) X Oldtimer]; Weeks, O.L.

Caramel Sunset, HT, lp; flowers butterscotch, good form

Caramella® *see* MEInitper

CARange, HT, lp; (**Tropical Passion**); Strahle, B. Glen, 1999

'CARAVELLE', HT, mr, 1964; flowers bright cherry-red, large, dbl., exhibition form; strong stems; vigorous, upright growth; [Better Times X Seedling]; Mondial Roses

'CARBANE', HT, dp, 1993; (**Hot Spot**™); flowers red pink, blooms borne mostly single, dbl., 26–40 petals, 3–3.5 in., intense fragrance; some prickles; foliage large, dark green, semi-glossy; upright (90-100 cms) growth; [Duchess sport]; Strahle, B. Glen; Carlton Rose Nurseries, 1990

'CARCADE', HT, w, 1993; (**White Cascade**™); flowers large, white, full, very dbl., 26–40 petals, 3–3.5 in., borne mostly singly, slight fragrance; some prickles; foliage medium, medium green, semi-glossy; medium, upright growth; [Coquette X Jack Frost]; Strahle, B. Glen; Carlton Rose Nurseries, 1984

'CARDEAL DE ROHAN', F, dr, 1957; flowers deep red, semi-dbl.; da Silva, Moreira

'CARDIFF BAY', HT, mr, 1993; dbl., 26–40 petals, 3–3.5 in., borne mostly singly, intense fragrance; some prickles; foliage medium, red changing to dark green, matt; medium (105 cms), upright growth; [Seedling X Loving Memory]; Poole, Lionel; F. Haynes & Partners, 1994

'CARDINAL', HT, 1904; flowers red; RULED EXTINCT 12/85 ARM; [Liberty X Seedling]; Cook, J.W.

'CARDINAL DE RICHELIEU', HGal, m; (Cardinal Richelieu); flowers dark purple, large, dbl.; bushy growth; Laffay, M., before 1847

Cardinal Hume® *see* 'HARREGALE'

'CARDINAL MERCIER', HT, op, 1930; flowers salmon-pink tinted orange, overlarge, globular; foliage bronze; vigorous growth; Lens

Cardinal Richelieu *see* **'CARDINAL DE RICHELIEU'**

Cardinal Song™ *see* 'MEIMOUSLIN'

'CARDINALE DE LA PUMA', HT, dp, 1938; flowers carmine-red, large, semi-dbl., slight fragrance; foliage glossy; vigorous, bushy growth; GM, Rome, 1937; [Mgr. Lemmens X Lord Charlemont]; Leenders, M.

'CARDINALS HAT', HT, mr, 1960; [Souv. de Jacques Verschuren X Charles Mallerin]; Leenders, J.

Cardinal's Robe *see* 'MACRENO'

Cardinal's Rose *see* 'MACRENO'

Care Deeply *see* 'LYOCA'

'CAREFREE', F, dp, 1959; flowers rose bengal, camellia form, medium, clusters, dbl., 75 petals; foliage dark; free growth; [Alain X Pinocchio]; Fletcher

Carefree Beauty™ *see* 'BUCBI'

Carefree Delight *see* 'MEIPOTAL'

Carefree Wonder™ *see* 'MEIPITAC'

'CARELESS LOVE', HT, pb, 1955; flowers deep pink, streaked (striped) and splashed white; [Red Radiance sport]; Conklin; Golden State Nursery

'CARELESS MOMENT', Min, pb, 1977; bud long, pointed; flowers white, lightly edged pink, dbl., 45 petals, 1–1.5 in., exhibition form, moderate fragrance; foliage small; bushy, spreading growth; [Little Darling X Over the Rainbow]; Williams, Ernest D.; Mini-Roses

'CARESS', HT, yb, 1935; flowers pale buttercup yellow, tinted rose, edged salmon-carmine, dbl., moderate fragrance; foliage bright green, leathery; vigorous growth; Dickson, A.

Carezza *see* BARrez

'CARFAIT', HT, op, 1993; (**Peach Parfait**™); flowers pink coral, blooms borne mostly single, dbl., 26–40 petals, 1.5–2.75 in.; few prickles; foliage medium, medium green, semi-glossy; medium (70-100 cms), upright growth; [Indian Pink X Melissa]; Strahle, B. Glen; Carlton Rose Nurseries, 1992

Caribbean™ *see* 'KORBIRAC'

Caribbean Queen *see* 'RENEEN'

Caribe *see* 'JACIBE'

Caribia *see* **'HARRY WHEATCROFT'**

Caribia, Climbing *see* 'AROMONTELIB'

'CARIBOU', HRg, w, 1946; bud pointed; flowers large, single, 5 petals, flat, non-recurrent bloom, slight fragrance; ornamental fruit; foliage glossy, leathery, dark, rugose, scented like sweetbr; bushy, vigorous growth; very hardy.; [Ross Rambler X (R. rugosa X R. eglanteria) X Seedling)]; Preston; Central Exp. Farm

'CARILLON', HT, or, 1935; bud long, pointed, scarlet-orange; flowers brilliant flame, paling, large, semi-dbl.; foliage glossy, light; bushy growth; [Charles P. Kilham X Mrs Pierre S. duPont]; Nicolas; J&P

'CARILLON', HT, dy, 1953; bud long, pointed, deep yellow spotted carmine; flowers medium, dbl., 30–35 petals; vigorous growth; [(Soeur Thérèse X Orange Nassau) X Orange Nassau]; Moulin-Epinay; Vilmorin-Andrieux

Carina® *see* 'MEICHIM'

'CARINA SUPERIOR', HT, w, 1978; bud pointed; flowers white, reverse light pink, dbl., 35–40 petals, 5 in., exhibition form, moderate fragrance; foliage leathery; upright, bushy growth; [Carina sport]; Takatori, Yoshiho; Japan Rose Nursery

Carina, Climbing® *see* 'MEICHIMSAR'

'CARINE', HT, op, 1911; bud long, pointed; flowers orange-carmine tinted buff, well formed, intense fragrance; strong stems; vigorous growth; Dickson, A.

Caring For You *see* 'BOSARTHIC'

Carinella, HT, w

Carinita *see* MEIvostro

'CARIOCA', HT, ab, 1942; bud long; dbl., 25–38 petals, 4.5 in., exhibition form, slight fragrance; foliage dark, glossy, leathery; vigorous, upright growth; [Talisman sport]; Chase; J&P

'CARIOCA', HT, ob, 1951; bud semi-ovoid; flowers tangerine-orange, large, dbl., globular, moderate fragrance; foliage bright bronze green; long, strong stems; very vigorous growth; [Rubin X Mme Henri Guillot]; Lens

'CARISSIMA', HWich, lp, 1905; flowers delicate flesh, carnation-like, dbl., moderate fragrance; Walsh

Caritas, HT, mp, 1970; flowers large, dbl., slight fragrance; Urban, J.; (Sangerhausen)

'CARITO MACMAHON', HT, my, 1934; bud large, ovoid; dbl., cupped, moderate fragrance; foliage glossy, dark; strong stems; vigorous growth; [Mrs Pierre S. duPont X Cayetana Stuart]; Dot, Pedro

'CARL KEMPKES', F, dr, 1937; bud long, pointed; flowers crimson, open, large, borne in clusters, semi-dbl., slight fragrance; foliage glossy, dark; strong stems; vigorous, bushy growth; [Dance of Joy X Mary Hart]; Kordes; Späth

Carl Luhn's Autumn Taffeta, HT, ob

Carl PhilipKristian IV, The Times Rose *see* 'KORPEAHN'

Carl Red, HT, mr

'CARLA', HT, op, 1963; flowers soft salmon-pink, dbl., 26 petals, 3.5–5 in., moderate fragrance; foliage dark; vigorous growth; [Queen Elizabeth X The Optimist]; deRuiter; Ball Seed Co., 1968

'CARLA', S, mr, 1963; (14); [Will Alderman X Hansa]; Erskine

Carla, Cl HT, lp; Ross, A., & Son, 1969; (Weatherly, L.)

'CARLA, CLIMBING', Cl HT, mr, 1969; Ross, A., & Son

'CARLEA', S, mp, 1965; flowers similar to victory year but darker and more free blooming, non-recurrent; [Betty Bland seedling]; Wright, Percy H.

Carlita, HT, mp, 1975; flowers large, dbl., slight fragrance; Kamrad; (Sangerhausen)

'CARLOS BEAUTY', (variety of R. acicularis), mp; flowers bright pink, semi-dbl., 12–15 petals, intense fragrance; ((found in Alberta, Canada))

'CARLOS REIS', Cl HT, mr; flowers purplish red; [Étoile de Hollande, Climbing X Pres. Herbert Hoover]; da Silva, Moreira

Carl's Rose *see* 'LEMCAR'

Carl's Rose, Min, m

'CARLSHAM', S, mp, 1964; flowers rose-pink, large, dbl., 25 petals, recurrent bloom; foliage glossy; (14); [Hansa X R. nitida]; Erskine

Carmagnole® *see* 'DELROBLA'

Carmargue *see* 'CHEWOZ'

'CARMEL BICE', F, pb, 1959; bud ovoid; flowers pink, reverse lighter, blooms in clusters, semi-dbl., intense fragrance;

foliage leathery, glossy; vigorous, upright growth; [Gartendirektor Otto Linne X Seedling]; Riethmuller

Carmel Sunset *see* 'MACCLOSEUP'

'CARMELA', Min, ob, 1980; bud lightly mossed; flowers orange, yellow center, small, semi-dbl., 15 petals, borne in sprays of 3-5, slight fragrance; small, ovoid to globular, orange fruit; small, brownish prickles; foliage small, light green, matt; medium, bushy, spreading growth; [Fairy Moss X Yellow Jewel]; Moore, Ralph S.; Moore Min. Roses

Carmela® *see* 'INTERGEORGE'

'CARMELITA', HT, mr, 1933; bud ovoid; flowers vivid red, large, dbl., exhibition form; foliage dark; very vigorous, bushy growth; [Matchless X Milady]; Spanbauer

Carmelita, F, dy

'CARMEN', HRg, mr, 1907; flowers crimson, stamens yellow, large blooms in clusters, single; foliage dark; vigorous growth; [R. rugosa rosea X Princesse de Bearn]; Lambert, P.

'CARMEN', HT, dr, 1956; flowers deep red, well-formed, large, dbl., intense fragrance; foliage dark; moderate growth; [Crimson Glory X Seedling]; Delforge

Carmen *see* 'JACLAM'

Carmen *see* TANemrac

Carmen de Bencomo, HT; Camprubi, C.; (Cavriglia)

Carmen Miranda, HT; (Brazil); (Cavriglia)

'CARMEN PAPANDREA', Min, dp, 1989; flowers deep pink; [Magic Carrousel sport]; Papandrea, John T.

'CARMEN SISTACHS', Pol, 1936; flowers rose-pink, base old-gold, cluster, dbl.; foliage sparse, small, soft, light; small vigorous, bushy growth; Dot, Pedro

'CARMEN SYLVA', HT, yb, 1891; flowers yellowish white with pink, large, dbl.; Heydecker; (Sangerhausen)

'CARMEN TALÓN', HT, dr, 1953; bud oval; flowers velvety dark red, large, dbl., 35 petals, moderate fragrance; strong stems; vigorous growth; [Charles Mallerin X Satan]; Dot, Pedro

'CARMEN TESSIER', HT, dr, 1964; bud long, pointed; flowers crimson-red, open, dbl., slight fragrance; foliage leathery; vigorous, bushy growth; [Seedling X Independence]; Mondial Roses

'CARMENCITA', Min, w, 1954; bud ovoid; flowers pure white, dbl., 55 petals; foliage clear green; vigorous growth; [Lady Sylvia X Perla de Alcanada]; Camprubi, C.

'CARMENETTA', S, lp, 1923; flowers pale pink, borne in clusters, single, slight fragrance; foliage leathery, reddish; ; vigorous(7 ft.), spreading (11 ft. very hardy; (28); [R. rubrifolia X R. rugosa]; Central Exp. Farm

Carmine Button, Min, dp

Carmine Pillar *see* **'PAUL'S CARMINE PILLAR'**

Carmosine *see* LAPmiravi

Carnation, HCh, mp

Carnaval® *see* 'KORFRILLA'

Carnaval de Rio® *see* 'DELORFEU'

'CARNE', M, lp; flowers flesh-pink, large, dbl.; Robert

Carnet de Bal, HT, 1972; Dorieux, Francois; (Cavriglia)

Carnival *see* 'KORFRILLA'

'CARNIVAL', HT, op, 1939; flowers glowing orange to soft cerise, dbl.; foliage glossy, light; vigorous growth; RULED EXTINCT, 1/88; Archer

'CARNIVAL GLASS', Min, ob, 1979; bud pointed; flowers yellow-orange blend, dbl., 38 petals, 1–1.5 in., slight fragrance; foliage small, glossy, bronze; bushy, spreading growth; [Seedling X Over the Rainbow]; Williams, Ernest D.; Mini-Roses

'CARNIVAL PARADE', Min, yb, 1978; bud long, pointed; flowers golden yellow, edged red, dbl., 45 petals, 1 in., exhibition form, slight fragrance; foliage small, dark, glossy; upright, bushy growth; [Starburst X Over the Rainbow]; Williams, Ernest D.; Mini-Roses

'CARNIVAL QUEEN', HT, mr, 1965; bud long, pointed; flowers luminous red, large, semi-dbl., moderate fragrance; vigorous growth; [(Tassin X Priscilla) X Charlotte Armstrong]; Armbrust; Langbecker

'CAROL', HT, pb, 1964; flowers cyclamen-rose, becoming orchid at edge, center apricot, well-formed, dbl., 45 petals, slight fragrance; no prickles; moderate growth; [Queen Elizabeth X Confidence]; Herholdt, J.A.; Herholdt's Nursery

Carol *see* **'CAROL AMLING'**

'CAROL AMLING', F, mp, 1953; (Carol, Garnette Carol, Garnette Pink); flowers deep rose-pink, edged lighter; [Garnette sport]; Amling, C.M. & Beltran; Amling Bros.

'CAROL ANN', Pol, op, 1940; flowers orange-salmon, small, tight clusters, dbl., 35–45 petals, 1–1.5 in., cupped, recurrent bloom; dwarf (12 in. or less) growth; [Marianne Kluis Superior sport]; Kluis; Klyn

Carol Ann *see* 'PEAPOST'

Carol Ann, Min, lp; Welsh; (Weatherly, L.)

'CAROL HOWARD', HT, mp, 1935; bud long, pointed; flowers rose-pink, peony shape, very large, moderate fragrance; foliage dark; vigorous growth; Pfitzer; P.J. Howard

Carola®, S, dp, 1988; Noack, Werner

Carole Anne, HT, my

Carolin *see* **'CORALIN'**

Carolin Reiberl *see* HELreib

Carolina Classic *see* 'BRICLASS'

Carolina Daza® *see* FEpuma

'CAROLINA MOON', HT, my, 1989; bud pointed; flowers medium yellow, fading slightly, borne singly, dbl., 28 petals, exhibition form, intense, fruity fragrance; prickles pointed slightly downward, medium, deep pink; foliage medium, medium green, semi-glossy; upright, tall, vigorous growth; [Just Lucky X Thriller]; Bridges, Dennis A.; Bridges Roses, 1990

Carolina Morning *see* 'MICAR'

Carolina Rose *see* **R. CAROLINA**

'CAROLINA SUNSET', LCl, rb, 1990; bud ovoid; flowers crimson, gold reverse, blooms borne in sprays of 3-7, dbl., cupped, slight fragrance; foliage average, dark green, matt; upright, tall growth with 10-12 ft.; [Seedling X Seedling]; Jeremias, Lephon L.

'CAROLINE', HT, op, 1955; flowers cinnabar-salmon, very large, dbl., 70 petals, moderate fragrance; foliage dark, glossy, leathery; very vigorous, bushy growth; [Peace X Seedling]; Gaujard

Caroline Anne, HT, my, 1986; Thomas

'CAROLINE BANK', HMult, lp, 1890; flowers pink with white, medium, dbl.; Geschwind, R.; (Sangerhausen)

Caroline Brian, MinFl, ob, 1997

'CAROLINE BUDDE', HMult, dr, 1913; (Lien Budde); flowers crimson-red, large, dbl.; foliage dark; vigorous growth; [Crimson Rambler X Léonie Lamesch]; Leenders, M.

Caroline Clarke *see* 'HORBONDARC'

'CAROLINE D'ARDEN', HP, mp, 1888; flowers very large, dbl., intense fragrance; Dickson, A.; (Sangerhausen)

Caroline Davison *see* 'HARHESTER'

Caroline de Berry *see* **'FOLIACÉE'**

Caroline de Monaco *see* 'MEIPIERAR'

'CAROLINE DE SANSAL', HP, mp, 1849; flowers pink, center darker, large, dbl., flat, recurrent bloom; vigorous growth; Desprez

'CAROLINE EMMONS', F, rb, 1962; bud ovoid; flowers scarlet-red, dbl., 55–60 petals, 3.5 in., cupped, intense, geranium fragrance; foliage leathery; vigorous, upright, bushy growth; [(Geranium Red X Fashion) X (Diamond Jubilee X Fashion)]; Boerner; Home Nursery Products Corp.

'CAROLINE ESBERG', LCl, mp, 1926; flowers dull rose, cluster, dbl.; Diener

'CAROLINE KAART', HT, op, 1964; flowers salmon-pink veined dark red, large; foliage glossy; upright growth; [Bayadère X Ballet]; Buisman, G. A. H.

'CAROLINE KUSTER', N, ly; Pernet, 1872

'CAROLINE MARNIESSE', N, w, 1848; flowers creamy white, dbl.; Roeser

'CAROLINE OLDREY', HT, lp, 1998; flowers pale pink, very full, exhibition, very dbl., 41–45 petals, 2.5–3 in., borne mostly singly, moderate fragrance; prickles moderate; foliage medium, medium green, semi-glossy; medium, compact, upright growth; [Solitaire X Prima Ballerina]; Jones, L.J.; Haynes Roses, 1997

'CAROLINE PLUMPTON', F, dp, 1975; flowers deep neyron rose, edged lighter, moderately full, dbl., 20–25 petals, 2.5 in., slight fragrance; foliage matt, green; very free growth; [Red Lion X Seedling]; Plumpton, E.

Caroline Schmitt *see* **'MME CAROLINE SCHMITT'**

'CAROL-JEAN', Min, dp, 1977; (Indian Meillandina, Indian Sunblaze); bud pointed; flowers deep pink, dbl., 22 petals, 1 in., slight fragrance; foliage small to medium, dark; upright, very bushy growth; [Pinocchio X Little Chief]; Moore, Ralph S.; Sequoia Nursery

Carol-Joy, HT, mr, 1991; Allender, Robert William

Carolyn *see* **'CORALIN'**

Carolyn *see* 'MACCOLUMB'

'CAROLYN ANN', Min, mp, 1985; flowers small, dbl., 20 petals, moderate fragrance; foliage medium, medium green, semi-glossy; bushy growth; [Gene Boerner X Baby Katie]; Hooper, Clint; Kimbrew Walter Roses

'CAROLYN DEAN', LCl, mp, 1941; bud long, pointed; flowers bright pink, small, borne in clusters, single, 1.75 in., recurrent bloom; foliage glossy; height 5 ft; [Étoile Luisante X Sierra Snowstorm]; Moore, Ralph S.; Sequoia Nursery

'CAROLYN DIANNE', F, lp, 1964; flowers medium, dbl., moderate fragrance; foliage glossy; vigorous, compact growth; [Ma Perkins X Pinocchio]; Patterson; Patterson Roses

'CARON', F, pb, 1972; flowers white and pink, dbl., 38 petals, 3 in., slight fragrance; foliage semi-glossy; upright growth; [Kordes' Perfecta X Saratoga]; Langdale, G.W.T.

Carosi, HT, 1969; Wituszynski, B.; (Cavriglia)

'CAROUBIER', HWich, mr, 1912; flowers very bright light crimson-scarlet, single, early bloom; Nonin

'CAROUGE', HT, dr, 1976; flowers crimson-red, dbl.; fall growth; [Marylene X Credo]; Gaujard

Carpet of Color *see* KORhaugen

'CARPET OF GOLD', LCl, my, 1939; flowers, dbl., 2–3 in.; trailing growth; [(Emily Gray X Yellow Rambler) X Golden Glow]; Brownell, H.C.

'CARRIE CORL', Gr, mr, 1969; bud ovoid; flowers large, dbl., moderate fragrance; foliage dark, leathery; vigorous growth; [Queen Elizabeth X (Queen Elizabeth seedling X Happiness)]; Germain's; Flower of the Month

'CARRIE JACOBS BOND', HT, mp, 1935; bud ovoid; flowers rose-pink, center flushed crimson, large, very dbl., intense fragrance; foliage leathery, dark; long stems; vigorous growth; [Premier Supreme X Lady Leslie]; Howard, F.H.; Dreer

'CARRIE JACOBS BOND, CLIMBING', Cl HT, mp, 1940; H&S

Carrot Top *see* 'POULTOP'

'CARROUSEL', Gr, mr, 1950; flowers medium, dbl., 20 petals, moderate fragrance; foliage leathery, dark, glossy; vigorous, upright, bushy growth; GM, ARS, 1956 GM, Portland, 1955; [Seedling X Margy]; Duehrsen; Elmer Roses Co.

'CARROUSEL, CLIMBING', Cl Gr, mr, 1958; Weeks; Elmer Roses Co.

'CARRY NATION', F, w, 1959; bud ovoid; flowers white, center cream, small, borne in clusters, semi-dbl., globular, slight fragrance; foliage soft, glossy; vigorous, low growth; [Pinocchio X Katharina Zeimet]; Silva; Plant Hybridizers of Calif.

Carte Blanche™ *see* 'MEIRINGA'

Carte d'Or *see* 'MEIROBIDOR'

Carte Rose, F, lp

Cartwheel *see* 'JACCART'

Cary Grant™ *see* 'MEIMAINGER'

'CARYATIDE', HT, dr, 1955; bud long, pointed; semi-dbl.; strong stems; upright growth; [Hens Verschuren X Poinsettia]; Buyl Frères

'CASA BLANCA', LCl, w, 1968; bud tinged carmine-pink; semi-dbl., blooms in clusters, intermittent bloom, slight fragrance; foliage dark, glossy; vigorous, climbing growth; [New Dawn X Fashion]; Sima

'CASANOVA', HT, ly, 1964; flowers straw-yellow, dbl., 38 petals, 6 in., exhibition form, moderate fragrance; [Queen Elizabeth X Kordes' Perfecta]; McGredy, Sam IV; Fisons Horticulture

'CASCABEL', F, rb, 1957; flowers red, reverse pearly, passing to carmine, borne in clusters, dbl., 45–50 petals, globular; foliage glossy; very vigorous, upright, compact growth; [Méphisto X Perla de Alcañada]; Dot, Pedro

'CASCADE', LCl, mr, 1951; flowers bright crimson, almost single, large, in clusters, very dbl.; very vigorous growth; [Holstein X American Pillar]; Mallerin, C.; EFR

Cascade, HT, w, 1990; Strahle, B. Glen

'CASCADIA', HMsk, lp, 1927; flowers blush-pink paling to white, borne in clusters, semi-dbl., 15 petals, 1 in., recurrent bloom, slight fragrance; foliage glossy, dark; tall growth; GM, Portland, 1922; [Mme d'Arblay X Bloomfield Abundance]; Thomas; B&A

'CASHMERE', HT, ab, 1980; bud pointed; flowers soft apricot, borne singly and 3-5 per cluster, dbl., 30 petals, globular, slight, spicy fragrance; long, hooked prickles; foliage leathery, dark; vigorous, bushy growth; [Tanya X Jack O'Lantern]; Weeks, O.L.

'CASIMIR MOULLÉ', HWich, pb, 1910; flowers purplish pink, reverse silvery pink; [R. wichurana X Mme Norbert Levavasseur]; Barbier

Casino® *see* 'MACCA'

'CASLON', HT, rb, 1996; (**Long Island**®); flowers creamy white with red edge, blooms borne mostly single, dbl., 26–40 petals, 4 in., moderate fragrance; foliage medium, light green, semi-glossy; upright, tall (3-4 ft) growth; [Tiki X Color Magic]; Castillo, Angel; Z&B Own Root Roses, 1998

Casper *see* **'MAGNA CHARTA'**

Casque d'Or® *see* 'DELCASCOR'

'CASSANDRA', HT, mr, 1966; flowers cherry, 4.5 in., exhibition form, slight fragrance; foliage dull, serrated; free growth; [(Karl Herbst X Ena Harkness) X (Christian Dior X Peace)]; Dorieux; Bees

Cassandra, F, or, 1998; Noack, Werner

Cassandre® *see* MEIdenji

'CASTA DIVA'®, HT, w, 1982; flowers large, dbl., 35 petals, cupped, no fragrance; flat, yellow prickles; foliage large, dark, glossy; upright growth; [Pascali X Seedling]; Barni-Pistoia, Rose

'CASTANET', F, op, 1960; bud ovoid; flowers orange-pink, reverse lighter, dbl., 45 petals, 3.5 in., moderate fragrance; upright, bushy growth; [Chic X Garnette seedling]; Boerner; J&P

Castel *see* 'DELSET'

Castella® *see* 'TANALLET'

Castellana, F; (Italy); (Cavriglia)

Castilian *see* **'AUTUMN DAMASK'**

Castilian, HT, ob, 1996; Williams, J. Benjamin

Castle Hill *see* 'ZIPHILL'

Castle Howard Tercentenary *see* 'TANTASCH'

Castle of Mey *see* 'COCLUCID'

Castore® *see* BARcast

'CATALINA', HT, pb, 1939; bud long, pointed, carmine, reverse salmon; flowers carmine suffused yellow and rose, very large, dbl., 23 petals, intense fragrance; foliage dark, glossy; very vigorous growth; [Pres. Herbert Hoover X Katharine Pechtold]; San Remo Exp. Sta.

'CATALINA', HT, op, 1940; bud very long, coral-pink edged old-rose; flowers salmon to shrimp-pink, dbl., 25–30 petals, 4.5 in., exhibition form, moderate fragrance; very vigorous, upright growth; [Joanna Hill sport]; Grillo

Catalina Frau® *see* FErama

'CATALONIA', HT, or, 1933; flowers bright orange-crimson, shaded gold, large, very dbl., globular; foliage dark; bushy growth; [(Shot Silk X Mari Dot) X Jean C.N. Forestier]; Dot, Pedro; C-P

'CATALUNYA', Cl HT, mr, 1918; [Gruss an Teplitz sport]; Nonin

Caterpillar® *see* 'POULCAT'

'CATHARINA KLEIN', HT, pb, 1930; bud ovoid; flowers bright pink, yellow background, large, very dbl., moderate fragrance; foliage rich green; very vigorous growth; [Mrs Franklin Dennison X Hadley]; Berger, V.; Münch & Haufe

'CATHAY', F, ly, 1957; dbl., 35–40 petals, 3.5–4.5 in., exhibition form, intense fragrance; foliage glossy, leathery; vigorous, compact growth; [Fandango X Pinocchio]; Swim, H.C.

'CATHCART BEDDER', HT, op, 1939; flowers salmon-shrimp-pink; foliage bronze; vigorous, bushy growth; Austin & McAslan

'CATHEDRAL', F, ab, 1975; (Coventry Cathedral, Houston); flowers apricot shading salmon, dbl., 22–25 petals, 4–5 in., borne in small clusters, slight, anise fragrance; foliage glossy, olive-green; bushy growth, medium, compact; AARS, 1976 GM, Portland, 1974 Gold Star of the South Pacific, Palmerston North, NZ, 1974; [Little Darling X (Goldilocks X Irish Mist)]; McGredy, Sam IV

Cathedral Peak *see* JACwiht

Cathedral Splendour *see* HARbell

'CATHERINE ANNE', Min, w, 1999; flowers medium, dbl., 17–25 petals, borne in small clusters, slight fragrance; many prickles; foliage medium, medium green, semi-glossy; bushy, tall (3 ft) growth; [Rise 'n' Shine X City of London]; Goodall, G. T.

'CATHERINE BLACKBURN', Pol, lp; Matthews, W.J., 1921; (Weatherly, L.)

Catherine Bonnard, B, dp; Verdier, E., 1871

Catherine Cookson *see* 'NOSCOOK'

'CATHERINE DE WÜRTEMBERG', M, lp, 1843; bud well mossed; flowers soft pink, large, very dbl., globular; vigorous growth; Robert

Catherine Deneuve® *see* MEIpraserpi

'CATHERINE GHISLAINE', D, m, 1885; flowers violet-pink, fading to white, small, semi-dbl.; Roseraie de l'Hay; (Sangerhausen)

'CATHERINE GUILLOT', B, dp, 1860; flowers carmine-rose, large, dbl.; vigorous growth; [Louise Odier seedling]; Guillot Père

'CATHERINE II', HCh, lp, 1832; Laffay, M.

'CATHERINE LANGEAIS', HT, mr, 1965; flowers carmine-red, well formed, dbl., 46 petals, exhibition form; vigorous, upright growth; [Michele Meilland X Berthe Mallerin]; Hémeray-Aubert

'CATHERINE MARIE', HT, mp, 1991; bud pointed; flowers medium to shell pink, dbl., 26 petals, exhibition form, intense fragrance; foliage medium, medium green, semi-glossy; [Pristine X Captain Harry Stebbings]; Wambach, Alex A., 1990

Catherine McAuley *see* JACibras

'CATHERINE MERMET', T, lp, 1869; bud well-shaped; flowers flesh-pink, edges tinted lilac-pink, large, dbl., moderate fragrance; vigorous growth; Guillot et Fils

Catherine Pericard, F, 1981; Pineau; (Cavriglia)

'CATHERINE SEYTON', HEg, lp, 1894; flowers soft pink, yellow stamens, single, non-recurrent, moderate fragrance; foliage fragrant (apple); vigorous growth; (35); Penzance; Keynes, Williams & Co.

'CATHERINE SOUPERT', HP, 1879; Lacharme, F.; (Cavriglia)

'CATHIE', HT, m, 1999; flowers mauve, outer petals edges shaded light red, dbl., 26–40 petals, 3.5 in., slight fragrance; foliage medium, dark green, dull; upright, tall (4 ft) growth; [Paradise X Unknown]; Priestly, James L.

Cathie Irwin *see* KORlowi

'CATHRINE KORDES', HT, dr, 1930; bud long, pointed, blood-red, shaded black; flowers dark scarlet, large, dbl., moderate fragrance; foliage dark (blood-red when young), leathery; very vigorous growth; [(Mme Caroline Testout X Willowmere) X Sensation]; Kordes; Dreer;, H&S

'CATHRINE KORDES, CLIMBING', Cl HT, dr, 1938; Krohn; Kordes

'CATHY ANNE', HT, r, 1990; bud pointed; semi-dbl., 25 petals, exhibition form, borne singly, no fragrance; prickles long, slender, red; foliage small, dark green; long, straight stems; medium growth; [Judith Morton X Sylvia]; Wilson, George D., 1986

'CATINAT', HGal, m; flowers violet, spotted; Robert, ca. 1850

'CATINAT', HP, mr, 1874; Oger

'CATO', HMult, lp, 1904; flowers medium, dbl.; Gratama; (Sangerhausen)

'CATORCE DE ABRIL', HT, yb, 1932; flowers yellow streaked red; Padrosa

Cato's Pink Cluster, N, lp

Cat's Meow™ *see* 'DESLING'

'CAULDRON', S, mp, 1984; flowers small, single, 5 petals, slight fragrance; foliage medium, mid-green, matt; upright growth; [R. rubrifolia X R. nutkana]; Holliger, Franc

'CAULDRON', HT, pb, 1971; flowers rose-pink to yellow, dbl., 23 petals, 4.5 in., moderate fragrance; foliage glossy, dark; RULED EXTINCT 6/83 ARM; Waterhouse Nursery

Cauvery, HT, dr, 1973; Kasturi

'CAVALCADE', F, rb, 1950; bud ovoid; flowers oxblood-red and yellow, changing daily to crimson, carmine and silvery pink, dbl., 32 petals, intense, fruity fragrance; foliage glossy, dark; vigorous, bushy growth; Verschuren-Pechtold; Stuart

'CAVALCADE, CLIMBING', Cl F, rb, 1957; Gandy, Douglas L.

'CAVALIER', HT, ob, 1939; flowers burnt-orange to cream-buff, dbl., moderate fragrance; [Mrs Franklin D. Roosevelt sport]; Samtmann Bros.

Cavriglia, HT, 1989; Fineschi, G.; (Cavriglia)

'CAYENNE', HT, ob, 1966; bud short, pointed; flowers deep orange, dbl., 38 petals, 3–4 in., slight fragrance; upright growth; [South Seas X Seedling]; Warriner, William A.; J&P

'CAYETANA STUART', HT, my, 1931; bud long, pointed; flowers very large, dbl., cupped, moderate fragrance; foliage dark, glossy; very vigorous, bushy growth; GM, Bagatelle, 1930; [Isabel Llorach X (Constance X Sunburst)]; Dot, Pedro; C-P

'CECHOSLAVIA', HT, w, 1921; flowers milky white suffused salmon-carmine, center golden yellow, dbl., moderate fragrance; [Pharisaer X Mme Antoine Mari]; Berger, V.; A. Berger

'CECIL', HT, my, 1926; flowers golden yellow, blooms in large clusters, single, 5 petals, 4 in.; bushy growth; Cant, B. R.

'CECIL, CLIMBING', Cl HT, my, 1940; Chaffin; Armstrong Nursery

Cécile Brünner *see* **'MLLE CÉCILE BRÜNNER'**

Cécile Brünner, Climbing *see* **'MLLE CÉCILE BRÜNNER, CLIMBING'**

'CECILE CUSTERS', HT, pb, 1914; flowers lilac-rose, reverse deep rose-pink, dbl.; [Mme Abel Chatenay X Violet Liddell]; Leenders, M.

Cecile Lens, Min, mp

'CECILE MANN', HT, mr, 1939; vigorous growth; [Mrs Albert Nash X Seedling]; Clark, A.; Brundrett

'CÉCILE RATINCKX', HT, my, 1924; flowers coppery yellow; [Louise Catherine Breslau X Mme Edouard Herriot]; Vandevelde

'CÉCILE VERLET', HT, yb, 1926; flowers yellow, passing to rose-pink; [Marianna Rolfs X Nordlicht (HT)]; Walter, L.

'CÉCILE WALTER', HT, op, 1926; bud long, pointed; flowers coral-pink to coppery pink, base gold, very large, dbl., 28 petals, cupped, slight fragrance; foliage rich green, leathery; vigorous growth; [(Mme Mélanie Soupert X Mme Edouard Herriot) X Seedling]; Mallerin, C.; C-P

'CECILIA', S, pb, 1980; bud globular, pointed; flowers light pink blended darker pink, carnation-like form, dbl., 25 petals, borne singly, moderate, spicy fragrance; hooked prickles; foliage small, pointed, glossy; vigorous, compact, bushy growth; [(R. wichurana X Baronne Prevost) X (R. X odorata)]; James, John

'CECILIO RODRIGUEZ', HT, dr, 1950; bud long, pointed; flowers dark velvety red, large, dbl., exhibition form, intense fragrance; foliage dark; strong stems; very vigorous growth; [Tassin X Eugenio d'Ors]; Camprubi, C.

Cecilitas, Min, lp

Cecil's Bright Apricot, HT, ab, 1996; Godman

Cecily Gibson *see* 'EVERBRIGHT'

Cedar Crest College *see* 'WILCREST'

'CEDRIC ADAMS', HT, dr, 1949; bud ovoid; flowers scarlet to carmine, large, dbl., exhibition form, moderate fragrance; foliage dark, bronze; vigorous growth; [Pink Princess X Crimson Glory]; Brownell, H.C.

Cee Dee Moss *see* 'MORCEEDEE'

Ceearcee, HT, ob, 1995; Gokhale

Celebrate America(TM) *see* 'TANCRESSOR'

'CELEBRATION', F, op, 1961; flowers salmon-pink, reverse lighter, blooms in clusters, dbl., 30 petals, 3 in., cupped, slight fragrance; foliage light green; vigorous, bushy growth; [Dickson's Flame X Circus]; Dickson, Patrick; A. Dickson

Celebration 2000 *see* 'HORCOFFITUP'

Celebration Day, HT, mp, 1987; Thames Valley Rose Growers

'CELEBRITY', HT, pb, 1945; (Lustrous); bud pointed; flowers pink, base shaded yellow, dbl., 30–35 petals, 6–7 in., exhibition form, moderate, spicy fragrance; foliage leathery, dark; vigorous, bushy growth; RULED EXTINCT 1/88; [Golden Rapture X Carmelita]; Hill, E.G., Co.

'CELEBRITY', HT, dy, 1988; bud pointed, large; flowers deep yellow, aging clear yellow, large, dbl., 30–35 petals, 4.5–5 in., exhibition form, borne usually singly, moderate, fruity fragrance; prickles pointed slightly downward, small, yellow-brown; foliage large, dark green, glossy; upright, bushy, medium growth; PP7264; [(Sunbonnet X Mister Lincoln) X Yello Yo Yo]; Weeks, O.L.; Weeks Roses, 1989

Céleste *see* **'CELESTIAL'**

'CÉLESTE', A, lp; flowers medium to large, moderate fragrance

'CELESTE MAHLEY', F, pb, 1972; bud long, pointed; flowers medium, dbl., slight fragrance; foliage leathery; vigorous, upright, bushy growth; [Seventeen X Gemini]; Byrum; J.H. Hill Co.

'CELESTIAL', A, lp; (Céleste); flowers light blush, golden stamens, large, dbl., non-recurrent, intense fragrance; foliage bluish; vigorous (to 6 ft.) growth; Kew, Prior to 1797

'CELESTIAL', HSpn, lp; flowers pale flesh, small, dbl., intense fragrance; low growth; [R. eglanteria X R. spinosissima]

'CELESTIAL', HT, lp, 1924; flowers light pink, edged paler, dbl., moderate fragrance; [Premier sport]; Myers & Samtmann

'CELESTIAL STAR', F, or, 1965; bud ovoid; dbl., borne in clusters; foliage glossy; vigorous, compact growth; Pal, Dr. B.P.; Indian Agric. Research Inst.

'CELIA WALKER', F, rb, 1962; flowers cherry-red, reverse silver, clusters, semi-dbl., 15–20 petals, 3 in., slight fragrance; foliage dark, glossy; vigorous, upright growth; [Alain X Golden Scepter]; Fletcher; Tucker

Celica *see* 'MEIFOTA'

Celientje, HT, ly, 1997

Celie's Slidell Pink, Cl T, lp

Celina *see* NOAsun

'CÉLINA', M, m, 1855; bud heavily mossed; flowers reddish-purple, center occasionally streaked white, large, dbl.; Hardy

'CELINA DUBOS', D, lp, 1849; flowers pale pink, sometimes recurrent bloom; Dubos

'CÉLINE', B, mp, 1824; flowers pale rose, large large clusters, dbl., cupped, does not repeat; very vigorous growth; Laffay, M.; Laffay, about 1825

Céline Delbard® *see* 'DELCET'

'CÉLINE FORESTIER', N, ly, 1858; flowers pale yellow, large blooms in clusters of 3-4, dbl., repeat blooming, moderate fragrance; foliage dark, glossy; vigorous (to 6 ft.) growth; (14); Leroy, A.

'CELS MULTIFLORA', HMult, lp; Hardy/Cels, 1836

'CELSIANA', D, lp; flowers pale pink, petals crinkled, blooms in cluster of 3-4, semi-dbl., 4 in., moderate fragrance; foliage smooth, grayish, fragrant; vigorous, upright (4-5 ft) growth

'CENDRILLON', HT, mp, 1951; bud very large; flowers salmon, dbl., intense fragrance; foliage leathery; very vigorous, upright growth; Gaujard

Centenaire de Lourdes® *see* 'DELGE'

Centenaire de Lourdes Rouge® *see* DELfloro

Centenaire du Vesinet *see* 'PIRONIA'

Centenary *see* KOReledas

Centenary College *see* 'AROBLAVEET'

Centennaire de Lourdes *see* 'DELGE'

'CENTENNIAL', HT, ob, 1953; bud ovoid; flowers peach-red, reverse orange-buff, dbl., 40–45 petals, 4–4.5 in., exhibition form, moderate fragrance; foliage rich green; vigorous, bushy growth; [Seedling X Orange Nassau]; Mallerin, C.; J&P

Centennial Gold, HT, my

'CENTENNIAL MISS', Min, dr, 1952; flowers deep wine-red, base tinged white, dbl., 60 petals, 1 in., moderate fragrance; no prickles; foliage small, dark, leathery, glossy; dwarf (10-12 in), bushy growth; [Oakington Ruby X Self]; Moore, Ralph S.; Sequoia Nursery

Centennial Star(TM) *see* 'MEINEREAU'

'CENTENNIAL SWEETHEART', Cl F, mr, 1959; bud globular; flowers bright red, borne in large clusters, dbl., 25 petals, 3.5 in., intense fragrance; foliage glossy, light green; very vigorous growth; [Alain sport]; Greene

Center Gold(TM) *see* 'SAVACENT'

Centerpiece(TM) *see* 'SAVAPIECE'

Centifolia a Fleurs Doubles Violettes, C, m

'CENTIFOLIA FOLIACEA', C, mp, 1810

Centifolia Major *see* **'ROSE DES PEINTRES'**

'CENTIFOLIA MINIMA', C, lp; flowers soft pink, very dbl., 80–100 petals, non-recurrent, moderate fragrance; height 8-15 in, branching

Centifolia Minima *see* **'ROULETII'**

'CENTIFOLIA MUSCOSA', M, mp; (Common Moss, Communis, Moss Rose, Mousseux Ancien, R. centifolia muscosa, R. muscosa); flowers large, peduncles and calyx glandular, mossy, dbl.; (28)

Centifolia Muscosa *see* **COMMUNIS**

'CENTIFOLIA ROSEA', HP, mp, 1863; flowers bright rose, large, cupped; foliage crinkled, light green wood with many red prickles; vigorous growth; Touvais

Centifolia Variegata *see* **'VILLAGE MAID'**

Central Park *see* 'POULPYG'

'CENTRE COURT', HT, w, 1964; flowers white, center tan, well shaped, dbl., 38 petals, 4 in., moderate fragrance; foliage dark; vigorous growth; [Eden Rose X Ena Harkness]; Barter

Centre Piece *see* 'SAVAPIECE'

'CENTREX GOLD', F, my, 1975; dbl., 25 petals, 3 in., slight fragrance; foliage matt, green; [Alison Wheatcroft X Chinatown]; Smith, E.; Wheatcroft

'CENTRO DE LECTURA', HT, mr, 1959; flowers crimson, dbl., 40 petals, moderate fragrance; foliage glossy, bright green; long strong stems; vigorous growth; [Texas Centennial X Peace]; Dot, M.

Centurio® *see* 'HAVOP'

'CENTURION', F, dr, 1975; flowers blood-red, shaded crimson, dbl., 30 petals, 3 in., slight fragrance; foliage glossy, dark; [Evelyn Fison X Seedling]; Mattock

'CENTURY 21', HT, lp, 1962; bud ovoid; flowers shell-pink, reverse slightly darker, large, dbl., exhibition form, moderate fragrance; foliage glossy; vigorous growth; [Condessa de Sástago X Soeur Therese]; Morey, Dr. Dennison; J&P

Century Sunset *see* 'TANSARAS'

'CENTURY TWO', HT, mp, 1971; bud long, pointed; flowers large, hot pink, dbl., 30–35 petals, 5 in., exhibition form, floriferous, borne singly, moderate, damask fragrance; foliage leathery; vigorous, upright, bushy growth; [Charlotte Armstrong X Duet]; Armstrong, D.L.; Armstrong Nursery

'CEREMONY', F, w, 1970; bud cupped; flowers pure white, dbl., 40–45 petals, 2.5 in.; foliage leathery; very bushy growth; [Tiara X Pascali]; Lens

'CERES', HMsk, lp, 1914; flowers pale blush, tinted light yellow, stamens bright yellow, semi-dbl., profuse seasonal bloom; shrub growth; Pemberton

'CERES', HT, ob, 1922; flowers deep orange, center salmon; [Sunburst X Mme Edmond Rostand]; Spek

Ceres *see* TANegnaro

'CERISE', HT, dp, 1945; flowers deep pink, large, dbl., 25 petals, intense fragrance; foliage dark, leathery; vigorous, upright growth; [Crimson Glory X Sterling]; Tantau

Cerise, S, dp; Delbard

'CERISE BOUQUET', S, dp, 1958; flowers cerise-crimson, semi-dbl., flat, moderate fragrance; foliage small, grayish; open, arching growth; [R. multibracteata X Crimson Glory]; Kordes

Cerise Dawn™ *see* 'DEVRISE'

'CERISE D'ORLIN', HGal, dp; flowers deep pink, reverse silver, loose, semi-dbl.

Cerise Margo, Pol, dp

Cerise Nights, S, dp

Cerise Rouge, S, mr

'CERISE TALISMAN', HT, dp, 1933; flowers cerise; [Talisman sport]; Clarke Bros.

Cérisette la Jolie *see* **'SURPASSE TOUT'**

Cernousek, HT, lp; Strnad; (Czech Rosa Club)

Cervanek, HT, lp; Strnad; (Czech Rosa Club)

'CERVANKY', Cl HT, dy, 1935; flowers large, dbl.; Böhm, J.; (Sangerhausen)

Cervena Gloria Dei, HT, lp; Strnad; (Czech Rosa Club)

Cervena Super Star, HT, 1976; Strnad; (Cavriglia)

Cerys Ann *see* 'GUESCAN'

Cesar ™ *see* MEIsardan

'CESARE BECCARIA', HGal, 1870; flowers large, dbl.; Moreau et Robert; (Sangerhausen)

'CESKÁ POHADKA', Pol, rb, 1933; flowers red, salmon, rose and white, borne in clusters; very vigorous, dwarf growth; [Golden Salmon sport]; Böhm, J.

Ceské 'Praci cest', LCl, 1970; flowers small, dbl.; Vecera, L.; (Sangerhausen)

'CÉSONIE', D, dp, 1859; flowers deep rose, large, dbl.; compact growth; Moreau et Robert, Prior to 1848

'CESTIFLORA', HSpn, ly; flowers sulfur-yellow, early bloom; glossy, black fruit; foliage finely divided; height 3-4 ft

Cevennes *see* 'KORLIRUS'

Cha Cha® *see* 'COCARUM'

'CHABLIS', HT, w, 1983; flowers creamy white, medium-large, dbl., exhibition form, slight fragrance; foliage large, medium green, matt; upright growth; [Seedling X Louisiana]; Weeks, O.L.; Weeks Wholesale Rose Growers, 1984

'CHACITA', F, dr, 1947; bud ovoid, globular; flowers deep red, medium, borne in clusters, dbl., cupped, moderate fragrance; foliage dark, leathery; vigorous, bushy growth; [Pinocchio X Crimson Glory]; Boerner

Chacok *see* 'MEICLOUX'

'CHAGIP', F, mr, 1956; (**Embrasement**); flowers fiery red, in clusters of 8-12, dbl.; foliage bronze; vigorous growth; Delbard-Chabert

'CHALICE', HT, ab, 1959; bud ovoid, pointed; flowers apricot-yellow, dbl., 55–60 petals, 5.5–6 in., exhibition form, intense fragrance; foliage leathery, glossy; vigorous, upright growth; [Orange Delight X Golden Rapture]; Verschuren-Pechtold; J&P

Chalice Well, HMult, lp; [R. multiflora X Unknown]; McLeod, J., 1995; Honeysuckle Cottage; (Weatherly, L.)

'CHALLENGE', F, dr, 1962; flowers deep blood-red, camellia shape, borne in clusters, semi-dbl., 20 petals, 2 in.; foliage dark; bushy, low growth; [Alain X Pinocchio]; Fletcher; Tucker

'CHALLENGER', HT, dr, 1938; flowers dark crimson, very large, dbl., exhibition form, intense fragrance; foliage leathery, dark; long stems; vigorous, bushy growth; RULED EXTINCT, 6/89; Cant, B. R.

Challenger™ *see* 'HILRED'

Challis Gold *see* 'NOSCHAL'

Chalom, HT; Moreira da Silva, A.; (Cavriglia)

'CHAMBA PRINCESS', F, op, 1969; flowers salmon-pink, open, medium, semi-dbl., intermittent bloom, slight fragrance; foliage leathery; moderate, bushy growth; IARI

'CHAMBE DI KALI', HT, mp, 1983; flowers large blooms borne singly, dbl., 22 petals, exhibition form, slight fragrance; green to brown prickles; foliage large, dark, glossy; medium, compact growth; [Bewitched X Seedling]; Pal, Dr. B.P.; K.S.G. Son's Roses

'CHAMBORD', F, mp, 1960; bud pointed; flowers pink, becoming darker, open, large, borne in clusters, dbl., slight fragrance; foliage bronze; bushy growth; [Roquebrune X Queen Elizabeth]; Delforge

'CHAMELEON', HT, or, 1918; flowers flame, edged cerise; [Lyon Rose seedling]; Dickson, A.

Chameleon, lp; [Wee Beth X (Starina X Unknown)]; Welsh, Eric, 1995; Biotech Plants Pty, Ltd.; (Weatherly, L.)

'CHAMI', HMsk, mp, 1929; flowers bright rose-pink, stamens yellow, single, recurrent bloom, intense fragrance; bushy growth; Pemberton

'CHAMISSO', HMult, pb, 1922; flowers flesh-pink, center yellowish white, borne in clusters, semi-dbl., profuse, recurrent bloom., moderate fragrance; foliage bronze; long stems; vigorous, trailing growth; [Geheimrat Dr. Mittweg X Tip-Top]; Lambert, P.

Chamois Dore®, F, ab, 1990; Lens

'CHAMP WEILAND', HT, mp, 1915; flowers clear pink; foliage glowing, reddish; [Killarney sport]; Weiland & Risch

'CHAMPAGNE', HT, yb, 1961; bud pointed, ovoid; flowers buff shaded apricot, dbl., 28 petals, 4–5 in., exhibition form, moderate fragrance; foliage leathery, dark; vigorous, upright, bushy growth; [Charlotte Armstrong X Duquesa de Peñaranda]; Lindquist; Howard Rose Co.

Champagne Arches, LCl, lp

Champagne Cocktail *see* 'HORFLASH'

Champagne Pearl, F, ab, 1983; (Champagnerperle); Kordes (Sangerhausen)

Champagner® *see* 'KORAMPA'

Champagnerperle *see* **Champagne Pearl**

'CHAMPION', HT, yb, 1976; flowers yellow-cream, flushed red and pink, dbl., 50–55 petals, 7–8 in., moderate fragrance; foliage large, light; [Irish Gold X Whisky Mac]; Fryer, Gareth; Fryer's Nursery, Ltd.

Champion®, F, or, 1988; Noack, Werner

'CHAMPION OF THE WORLD', HP, mp, 1894; (Mrs de Graw, Mrs DeGraw); flowers rose-pink, large, dbl., seasonal bloom, moderate fragrance; vigorous growth; [Hermosa X Magna Charta]; Woodhouse

'CHAMPION, CLIMBING', Cl HT, yb, 1999; flowers yellow-cream, flushed red and pink, very dbl., 50–55 petals, 3–3.5 in., exhibition form, moderate fragrance; foliage large, medium green; climbing (12 ft) growth; GM, Japan (JRC), 1997; [Champion sport]; Kameyama, Yasushi, 1990; La Vie en Rose Co., Ltd., 1998

'CHAMPLAIN', HKor, dr, 1982; flowers large, bright medium red with darker petal tips, dbl., 30 petals, 2.5 in., borne in small clusters, repeat bloom, slight fragrance; straight, yellow-green prickles; foliage small, dark yellow-green; bushy growth, 3 ft.; hardy; [(R. kordesii X Seedling) X (Red Dawn X Suzanne)]; Svedja, Felicitas; Agriculture Canada

'CHAMPNEYS' PINK CLUSTER', N, lp; (Champneys' Rose); dbl., borne in large clusters, recurrent bloom; moderately vigorous growth; moderately hardy; (14); [R. chinensis X R. moschata]; Champneys, 1811

Champneys' Rose *see* **'CHAMPNEYS' PINK CLUSTER'**

Champs-Elysées® *see* 'MEICARL'

Champs-Elysées, Climbing® *see* 'MEICARLSAR'

'CHANDELLE', F, lp, 1958; flowers light pink, reverse darker, well formed, medium, borne in clusters, dbl., intense fragrance; foliage bronze; vigorous, compact growth; [Gretel Greul X (Lady Sylvia X Fashion)]; Lens

'CHANDERI', HT, lp, 1968; bud long, pointed; flowers light pink, edged deeper, medium, dbl., exhibition form, slight fragrance; foliage soft; moderate, upright, compact growth; [Peace X Seedling]; Singh

'CHANDRAMA', F, w, 1980; bud pointed; dbl., 25 petals, borne in clusters of 3-6, slight fragrance; straight, brown prickles; foliage dark; spreading growth; [White Bouquet X Virgo]; Division of Vegetable Crops and Floriculture

Chandrika, Min, w, 1978; Kasturi

Chanel, HT, dp, 1995; Juneida

'CHANELLE', F, op, 1959; flowers peach-pink shaded rose-pink, well-formed, blooms in clusters, dbl., 20 petals, 3 in., moderate fragrance; foliage dark, glossy, pointed; vigorous, bushy growth; GM, Madrid, 1959; [Ma Perkins X (Fashion X Mrs William Sprott)]; McGredy, Sam IV; McGredy

'CHANOINE TUAILLON', HT, dp, 1931; bud persian red; flowers carmine, very large, dbl., globular, moderate fragrance; very vigorous growth; [Betty Uprichard X Lucie Nicolas Meyer]; Gillot, F.

Chanson d'Été *see* **'SUMMER SONG'**

'CHANTAL', HT, w, 1958; bud long, ovoid; flowers ivory, center yellow, edged carmine, very large, dbl., 60 petals, moderate fragrance; vigorous, bushy, upright growth; [Mme Charles Sauvage X Carillon]; Moulin-Epinay

Chantebrise, LCl, 1969; Croix, P.; (Cavriglia)

'CHANTECLERC', F, mr, 1956; flowers bright red, large; foliage bright green; bushy growth; [Peace X Seedling]; Gaujard

Chantefleur *see* LAParan

Chanterelle *see* 'WARDROSA'

'CHANTILLY', HT, or, 1964; flowers orange to red-lead, dbl., 40 petals; foliage glossy, dark; strong stems; vigorous growth; [Baccará X Seedling]; Verschuren, A.; van Engelen

'CHANTILLY LACE', HT, m, 1978; bud long; flowers red-purple, dbl., 35 petals, 4–5 in., exhibition form, intense fragrance; foliage glossy; vigorous growth; PP004665; [Blue Moon X Angel Face]; deVor, Paul F.; DeVor Nurseries, Inc.

Chantoli *see* 'FRYXOTIC'

'CHANTRÉ', HT, ob, 1958; bud long, pointed; flowers orange and golden yellow, dbl., 20–25 petals, 5 in., exhibition form, moderate fragrance; foliage dark, leathery; very vigorous, upright, bushy growth; [Fred Streeter X Antheor]; Kordes, R.

Chapeau de Napoléon *see* **'CRESTED MOSS'**

'CHAPELAIN D'ARENBERG', HGal, mp; flowers bright pink

'CHAPERON ROUGE', F, mr, 1951; bud globular; flowers velvety crimson, center darker, open, medium, very dbl., moderate fragrance; foliage glossy, bronze, dark; vigorous, upright growth; [Crimson Glory X (Baby Chateau X Seedling)]; Vilmorin-Andrieux

'CHAPLIN'S CRIMSON GLOW', LCl, rb, 1930; flowers deep crimson, base white, large, dbl.; Chaplin Bros.

'CHAPLIN'S PINK CLIMBER', LCl, mp, 1928; flowers bright pink, stamens golden yellow, large blooms in large clusters, semi-dbl., flat, non-recurrent; very vigorous growth; (28); GM, NRS, 1928; [Paul's Scarlet Climber X American Pillar]; Chaplin Bros.

'CHAPLIN'S PINK COMPANION', LCl, lp, 1961; flowers silvery pink, blooms in clusters of up to 30, dbl., 22 petals, 2 in., moderate fragrance; foliage glossy; vigorous growth; [Chaplin's Pink Climber X Opera]; Chaplin, H.J.; Chaplin & Sons

'CHAPLIN'S TRIUMPH', HT, dr, 1936; bud long, pointed; flowers deep velvety crimson, intense fragrance; foliage dark; vigorous growth; Chaplin Bros.

'CHARADE', F, op, 1965; flowers coral-cerise tinted salmon, pointed, borne in clusters, dbl., 3 in.; foliage glossy; free growth; [Queen Elizabeth X Seedling]; Herholdt, J.A.

'CHARADE'®, F, dr, 1988; flowers cardinal-red, small, dbl., 20–25 petals, exhibition form, borne in sprays of 2-5, slight fragrance; prickles straight, medium, lilac; foliage medium, dark green, semi-glossy; mini flora bushy, profuse growth; PP006669; [Cindy X Sassy]; Hill, Joseph H., Co.; DeVor Nurseries, Inc., 1988

Chardonnay *see* 'MACRELEASE'

Chardony *see* 'MACRELEASE'

'CHARIOT OF ROSES', Min, dr, 1986; dbl., 25 petals, borne singly, no fragrance; small, globular fruit; small, nearly straight, reddish prickles; foliage small, dark, matt; long, pendulous growth; [Seedling (S) X Fairy Moss]; Fischer, C.&H.; Alpenflora Gardens

Chariots of Fire, LCl, dr, 1996; free-standing climber growth; Williams, J. Benjamin

Charisma *see* 'JELROGANOR'

'CHARISMA', HT, or, 1973; RULED EXTINCT 3/77 ARM; Meilland

Charisma *see* PEAtrophy

'CHARITY', HT, mr, 1953; flowers bright velvety red, dbl., 40–50 petals, 5 in., exhibition form, moderate fragrance; foliage glossy; vigorous, upright, bushy growth; [Will Rogers X Mme Henri Guillot]; Taylor, C.A.

Charity(TM) *see* 'AUSCHAR'

Charivari *see* 'KORUB'

'CHARLEMAGNE', HGal, dp

Charlemagne, HP, lp

'CHARLES ALBANEL', HRg, mr, 1982; flowers medium, dbl., 20 petals, repeat bloom, moderate fragrance; straight, gray-green prickles; foliage yellow-green, rugose; groundcover; low, spreading growth; [Souv. de Philemon Cochet X Seedling]; Svedja, Felicitas; Agriculture Canada

Charles Austin® *see* 'AUSFATHER'

Charles Aznavour *see* 'MEIBEAUSAI'

'CHARLES BONNET', HP, dp, 1884; flowers dark rose, medium, dbl., repeat bloom; Bonnet

'CHARLES CRETTÉ', HT, mp, 1917; flowers velvety rose, large; Chambard, C.

'CHARLES DARWIN', HP, mp, 1879; flowers brownish crimson, large; [Mme Julian Daran seedling]; Laxton Bros.; W. Paul

Charles de Gaulle® *see* 'MEILANEIN'

'CHARLES DE LAPISSE', HT, lp, 1910; flowers pale blush-pink; [Mme Caroline Testout sport]; Laroulandie

'CHARLES DE LEGRADY', T, 1884; Pernet-Ducher; (Cavriglia)

'CHARLES DE MILLS', HGal, m; (Bizarre Triomphante); flowers dark crimson and purple, many petaled, quartered, 4.5 in., quartered; very few prickles; height to 5 ft; Hardy, before 1746

'CHARLES DICKENS', F, op, 1970; flowers rosy salmon, semi-dbl., 16 petals, 3 in., slight fragrance; [Paddy McGredy X Elizabeth of Glamis]; McGredy, Sam IV; McGredy

'CHARLES DILLON', F, mp, 1970; flowers soft pink, large, dbl., 24 petals, slight fragrance; foliage dark; low, bushy growth; [Orangeade X Piccadilly]; Wood; Homedale Nursery

Charles Dingee *see* **'WILLIAM R. SMITH'**

'CHARLES DUVAL', HP, mr, 1847; flowers scarlet, dbl., cupped; vigorous growth; Laffay, M.

'CHARLES E. SHEA', HT, mp, 1917; flowers rich pink; [Mrs George Shawyer sport]; Hicks

Charles Eyck, F, yb

'CHARLES F. WARREN', HT, mp, 1957; flowers rose-pink, well-formed, large, dbl., 40 petals, moderate fragrance; vigorous growth; [Wilfred Pickles X Karl Herbst]; Mee

'CHARLES FARGAS', HT, pb, 1935; flowers large, dbl.; Dot, Pedro; (Sangerhausen)

'CHARLES GATER', HP, mr, 1893; flowers red, dbl., 40 petals, globular, moderate fragrance; vigorous growth; Paul

'CHARLES GREGORY', HT, ob, 1947; flowers vermilion, shaded gold, well-formed, dbl., 22 petals, moderate fragrance; foliage dark, glossy; vigorous growth; Verschuren; Gregory

'CHARLES GREGORY, CLIMBING', Cl HT, ob, 1960; Gregory

'CHARLES H. RIGG', HT, mr, 1931; flowers bright red fading to pink, dbl., moderate fragrance; large stout, erect stems; vigorous growth; Chaplin Bros.

'CHARLES HENRY', HT, mr, 1967; flowers crimson, moderate fragrance; foliage dark; vigorous, upright growth; [Ena Harkness X Lady Sylvia]; Hooney

'CHARLES J. GRAHAME', HT, mr, 1905; flowers dazzling scarlet, well formed, large, intense fragrance; vigorous growth; Dickson, A.

'CHARLES K. DOUGLAS', HT, mr, 1919; bud long, pointed; flowers large, dbl., 28 petals, moderate fragrance; foliage dark; vigorous growth; Dickson, H.

'CHARLES K. DOUGLAS, CLIMBING', Cl HT, mr, 1934; Leenders Bros.

Charles Kuralt *see* CLEtraveler

'CHARLES LAMB', HP, 1884; Paul, W.; (Cavriglia)

'CHARLES LAWSON', B, dp, 1853; flowers vivid rose; vigorous, compact growth; Lawson

'CHARLES LEFÈBVRE', HP, dr, 1861; flowers reddish crimson shaded purple, large, dbl., 70 petals, cupped, bloom often recurrent, moderate fragrance; vigorous, tall growth; [Général Jacqueminot X Victor Verdier]; Lacharme, F.

'CHARLES LOUIS', HP, mr, 1800; flowers large, very dbl., moderate fragrance; Guinoiseau, B.; (Sangerhausen)

'CHARLES MALLERIN', HT, dr, 1951; flowers blackish crimson, dbl., 38 petals, 6 in., flat, intense fragrance; foliage leathery, dark; vigorous, irregular growth; [(Rome Glory X Congo) X Tassin]; Meilland, F.; C-P

'CHARLES MALLERIN, CLIMBING', Cl HT, dr, 1960; Balducci & Figli

'CHARLES MARGOTTIN', HP, mr, 1864; flowers very large, dbl., moderate fragrance; Margottin; (Sangerhausen)

Charles Metroz, Pol, dp

Charles Notcutt *see* KORhassi

'CHARLES P. KILHAM', HT, or, 1926; flowers red-orange, fading to lincoln red, well formed, large, dbl., 32 petals, slight fragrance; vigorous, bushy growth; GM, NRS, 1927; McGredy; Beckwith

'CHARLES P. KILHAM, CLIMBING', Cl HT, or, 1931; Howard Rose Co.

'CHARLES P. KILHAM, CLIMBING', Cl HT, or, 1934; Morse

'CHARLES QUINT', HGal, m; flowers lilac-rose and white; Robert, 1856

'CHARLES RAVOLLI', T, dp; flowers carmine-rose; moderate growth; Pernet Père, 1875

Charles Rennie Macintosh *see* 'AUSREN'

Charles Turner, HP, mr, 1968; flowers large, dbl.; Margottin; (Sangerhausen)

Charles Walker's Mignonette, Pol, w

'CHARLES WILLIAM', HT, pb, 1989; bud urn-shaped; flowers carmine rose, reverse lighter, yellow at base of petals, pointed, reflexing, very dbl., 56 petals, no fragrance; prickles light brown; foliage matt, medium green, large; upright, tall growth; [Bradenburg X Command Performance]; Cattermole, R.F.

Charles Wood, HP, rb; Laffay, M., 1864

Charleston *see* 'MEIRIDGE'

Charleston, HP, mp

Charleston 88 *see* 'MEIRESTIF'

Charleston Graveyard, HCh, rb

'CHARLESTON, CLIMBING', Cl F, yb, 1966; Rumsey, R.H.

Charlie(TM) *see* 'KINCHA'

Charlie Brown *see* 'MORCHARLIE'

Charlie Chaplin *see* TSCHaka

'CHARLIE MCCARTHY', F, w, 1955; bud creamy white; flowers pure white, blooms in clusters, dbl., 28 petals, 1.5–2 in., moderate fragrance; foliage glossy, leathery, dark; dwarf, compact growth; [Mrs Dudley Fulton X Mermaid]; Wiseman; H&S

'CHARLIE PERKINS', HT, dr, 1970; bud long, pointed; flowers overlaid black, full, large, semi-dbl., moderate fragrance; foliage large, glossy, dark, bronze, leathery; very vigorous, upright, bushy growth; [Carrousel X Circus]; Zombory; General Bionomics

'CHARLIE'S AUNT', HT, pb, 1965; flowers cream, heavily suffused rose, high-pointed, dbl., 65 petals, 5 in., moderate fragrance; foliage dark; [Golden Masterpiece X Karl Herbst]; McGredy, Sam IV; Geest Industries

'CHARLIE'S UNCLE', HT, rb, 1975; flowers cream, suffused carmine, full, dbl., 40 petals, 5–6 in., exhibition form, slight fragrance; foliage large, leathery; [Charlie's Aunt sport]; Haynes

Charlotte *see* 'AUSPOLY'

'CHARLOTTE', HT, op, 1941; bud long, pointed; flowers salmon-pink and coral, base gold, dbl., exhibition form, intense fragrance; foliage glossy; vigorous, bushy growth; [Joanna Hill X

Golden Dawn]; Duehrsen; California Roses

Charlotte *see* TANettola

'CHARLOTTE ANNE', F, m, 1993; flowers lavender, single, 5 petals, 1.5–2.75 in., borne in small clusters, slight fragrance; some prickles; foliage medium, dark green, glossy; medium, spreading growth; [Playgirl sport]; Blankenship, Paul; Roses Unlimited, 1994

'CHARLOTTE ARMSTRONG', HT, dp, 1940; bud long, pointed, blood-red; flowers deep pink, dbl., 35 petals, 3–4 in., moderate fragrance; foliage dark, leathery; vigorous, compact growth; AARS, 1941 David Fuerstenberg Prize, ARS, 1941 Gertrude M. Hubbard, ARS, 1945 GM, Portland, 1941 John Cook Medal, ARS, 1941 NRS, NRS, 1950; [Soeur Thérèse X Crimson Glory]; Lammerts, Dr. Walter; Armstrong Nursery

'CHARLOTTE ARMSTRONG, CLIMBING', Cl HT, dp, 1942; Morris; Armstrong Nursery, 1942;, Mon Reve Nursery, 1950

Charlotte Brownell, HT, yb; Brownell, H.C.

'CHARLOTTE CHEVALIER', HT, dy, 1916; flowers dark canary-yellow; [Arthur R. Goodwin sport]; Chambard, C.

'CHARLOTTE E. VAN DEDEM', HT, my, 1937; bud long, pointed; flowers large, semi-dbl.; foliage glossy; vigorous growth; GM, Portland, 1938; [Roselandia X Ville de Paris]; Buisman, G. A. H.

'CHARLOTTE ELIZABETH', Gr, dp, 1965; flowers deep rose-pink, blooms in clusters, dbl., 26 petals, 3–4 in., exhibition form; foliage glossy; Norman; Harkness

'CHARLOTTE IVES', F, pb, 1965; bud globular; flowers rose-pink, center light yellow, open, medium, borne in clusters, single, slight fragrance; compact, upright, bushy growth; [Ma Perkins X Rose Gaujard]; Warren

Charlotte Kemp, HWich, m, 1997; Nobbs

'CHARLOTTE KLEMM', HCh, or, 1905; flowers red shaded orange, medium, semi-dbl.; Türke

'CHARLOTTE MACKENSEN', HMult, mr, 1938; flowers carmine red with silvery reverse, medium, dbl.; Vogel, M.; (Sangerhausen)

'CHARLOTTE MARIE', S, pb, 1998; flowers pale pink, light at the base, pale reverse, single, 4–7 petals, 3 in.., borne in small clusters, slight fragrance; prickles moderate; foliage medium, narrow, dark green, semi-glossy; low, compact, upright growth; [Marchenland X Rosa virginiana]; Fleming, Joyce L.; Hortico, Inc., 1997

'CHARLOTTE PATE', HT, yb, 1971; bud globular; flowers pink, yellow reverse, dbl., 25 petals, 3.5 in., intense fragrance; foliage medium, medium green, glossy; [Wendy Cussons X Golden Sun]; MacLeod

Charlotte Rampling® *see* MEIhirvin

Charlotte Searle *see* KORnends

'CHARLOTTE VON RATHLEF', HWich, lp, 1936; very dbl., abundant seasonal bloom; foliage leathery, dark; vigorous, climbing growth; [Fragezeichen X American Pillar]; Vogel, M.; Heinemann

'CHARLOTTE WHEATCROFT', F, mr, 1957; flowers bright scarlet, large, large truss, single; foliage dark, glossy; vigorous, tall growth; Wheatcroft Bros.

'CHARLOTTE WIEREL', Pol, w, 1926; flowers cream-white, center bright rose-pink; vigorous growth; [Bebe Leroux X Helene Videnz]; Walter, L.

'CHARLOTTENHOF', S, mr, 1938; flowers medium, dbl.; Vogel, M.; (Sangerhausen)

Charlye Rivel® *see* FEkusa

'CHARM', HT, op, 1920; bud reddish orange, shaded pink and copper; flowers coppery yellow; Paul, W.

'CHARM' *see* **'CHARME'**

Charm Bracelet *see* 'JACFOG'

'CHARM OF PARIS', HT, mp, 1965; flowers pink, medium, dbl., 48 petals, 3.5 in., moderate fragrance; vigorous growth; Edland Fragrance Medal, ARS, 1966; [Prima Ballerina X Montezuma]; Tantau, Math.

'CHARMAINE', Pol, mp, 1929; flowers pink tinged salmon, open, sprays, dbl., recurrent bloom, moderate fragrance; foliage bright, glossy; long stems; very vigorous, bushy growth; [Evelyn Thornton seedling]; Burbage Nursery

Charmant® *see* 'KORZIMKO'

Charmante *see* 'KORZIMKO'

'CHARME', HT, dp, 1930; ('CHARM', 'GERMANIA'); bud long, pointed; flowers cherry-red, large, dbl., moderate fragrance; foliage glossy; vigorous growth; Rice Bros. Co.

Charme d'Amour *see* **'LIEBESZAUBER'**

Charme de Paris, F, 1979; Delbard, Georges; (Cavriglia)

Charme de Vienne *see* 'KORSCHAPRAT'

'CHARMENTE', HT, lp, 1975; bud globular; dbl., 42 petals, 4 in.; foliage bright green, leathery; upright, spreading growth; [Fragrant Cloud X Ena Harkness]; Huber

'CHARMER', HT, lp, 1923; flowers silvery pink; [Pharisaer seedling X Joseph Hill]; Schoener; Doyle

'CHARMER', HT, lp, 1934; flowers light pink, center shaded salmon, large, dbl., exhibition form, moderate fragrance; foliage leathery; vigorous, free branching growth; GM, NRS, 1932; Dickson, A.

'CHARMGLO', Min, pb, 1980; bud long, pointed; flowers creamy white painted deep pink, reverse lighter, borne usually singly, dbl., 35 petals, exhibition form, slight fragrance; long, thin, brown prickles, curved down; foliage small, medium to dark green, slightly matt; bushy, compact growth; [Seedling X Over the Rainbow]; Williams, Ernest D.; Mini-Roses

'CHARMI', HMsk, mp, 1929; flowers light to dark pink, medium, semi-dbl., intense fragrance; Pemberton; (Sangerhausen)

Charmian® *see* 'AUSMIAN'

'CHARMING', HT, mp, 1922; flowers salmon-pink, reverse coral-pink, semi-dbl., slight fragrance; [Alexander Hill Gray X Mme Edouard Herriot]; Van Rossem

Charming Cover *see* POUlharm

Charming Diana™ *see* 'TWODI'

'CHARMING MAID', F, ob, 1953; bud orange-salmon; flowers salmon, base golden, blooms in trusses, single, 5–6 petals, 4–4.5 in., moderate fragrance; foliage dark, glossy; GM, NRS, 1953; [Dainty Maid X Mrs Sam McGredy]; LeGrice

'CHARMING PRINCESS', HT, yb, 1926; flowers deep yellow, edged vermilion large, dbl.; foliage dark; vigorous growth; [The Queen Alexandra Rose sport]; Hancock

Charming Rosamini *see* RUIcharo

Charming Rose, HT, lp

Charming Vienne *see* 'KORSCHAPRAT'

Charter 700, F, my, 1993; Fryer, Gareth

'CHARTREUSE', HT, my, 1940; bud long; flowers canary-yellow, dbl., moderate fragrance; foliage glossy; bushy growth; [Soeur Thérèse X Angels Mateu]; Mallerin, C.; A. Meilland

Chartreuse *see* 'MACYEFRE'

Chartreuse de Parme *see* DELviola

'CHASE BEAUTY', HT, dr, 1947; bud long; flowers rich dark red, dbl., 35–45 petals, 5.5 in., exhibition form, slight fragrance; foliage dark; very vigorous, tall growth; [Better Times sport]; Chase; Chase Gardens

Chasin' Rainbows™ *see* 'SAVACHASE'

Chasing Rainbows *see* 'SAVACHASE'

'CHASTITY', Cl HT, w, 1924; flowers pure white, base lemon, dbl., exhibition form, non-recurrent bloom; foliage light, glossy; vigorous growth; Cant, F.

'CHASTLETON', HP, lp, 1800; flowers large, very dbl., slight fragrance; (Sangerhausen)

Château *see* **'BABY CHÂTEAU'**

Chateau Canon *see* MACghovie

Château d'Amboise® *see* 'DELROUVEL'

Chateau de Bagnols *see* ORAreg

'CHÂTEAU DE CHENONCEAUX', HT, mp, 1973; flowers brilliant pink, large, dbl., 45 petals; tall growth; [Americana X Queen Elizabeth]; Gaujard

'CHÂTEAU DE CLOS VOUGEOT', HT, dr, 1908; flowers deep velvety red, dbl., 75 petals, intense, damask fragrance; foliage dark, leathery; sprawling growth; Pernet-Ducher

'CHÂTEAU DE CLOS VOUGEOT, CLIMBING', Cl HT, dr, 1920; Morse

Chateau de Filain *see* SAUbima

'CHATEAU DE LA JUVENIE', S, lp, 1901

'CHÂTEAU DE NAMUR', HGal, pb; Quétier, before 1842

'CHÂTEAU DE VAIRE', S, dr, 1934; (Vaire); flowers deep red, dbl., cupped, non-recurrent; foliage bronze, dark; height 3 1/2-6 1/2 ft.bushy; [Charles K. Douglas X R. macrophylla]; Sauvageot, H.

Chateau de Versailles® *see* 'DELRICOS'

Chateau Frontenac, S, dp

Chateau la Croix *see* DORcroix

'CHÂTEAU LA SALLE', HT, yb, 1966; bud long, pointed; flowers buff-yellow, large, dbl., exhibition form, intense fragrance; foliage dark, leathery; vigorous, bushy, compact growth; [Joanna Hill X Ellinor LeGrice]; Morey, Dr. Dennison; Country Garden Nursery

'CHÂTEAU PELLES', HT, lp, 1927; bud cream-white; flowers soft pink, shaded salmon, dbl., slight fragrance; [Harry Kirk seedling]; Mühle

Chateauroux, F; Croix, P.; (Cavriglia)

'CHÂTELAINE', F, op, 1957; bud pointed; flowers coral overcast salmon, well formed, borne in small clusters, dbl., 32 petals, 3 in., moderate fragrance; foliage glossy, coppery; vigorous growth; [(Peace X Seedling) X Fashion]; Lens

Châtelaine de Lullier, HT, mr, 1987; flowers carmine-red, large, dbl., moderate fragrance; Meilland; (Sangerhausen)

'CHÂTELET', HT, mp, 1952; (Airain); bud globular, coral; flowers pink heavily tinted salmon, medium, very dbl.; [Yvonne Plassat X Seedling]; Moulin-Epinay; Vilmorin-Andrieux

'CHATILLON RAMBLER', HWich, mp, 1913; flowers salmon-pink, small, semi-dbl., slight fragrance; height 15-20 ft; [Dorothy Perkins X Crimson Rambler]; Nonin

'CHATILLON ROSE', Pol, mp, 1923; flowers bright pink, blooms in large clusters, semi-dbl., cupped, moderate fragrance; foliage glossy; bushy (1-2 ft. growth; [Orléans Rose X Seedling]; Nonin

Chatillon White *see* **'WHITE CHATILLON'**

Chatsworth *see* TAnotax

'CHATTEM CENTENNIAL'™, Min, or, 1979; bud ovoid; flowers medium, dbl., 38 petals, cupped, slight, fruity fragrance; upright, bushy growth; [Orange Sensation X Zinger]; Jolly, Betty J.; Rosehill Farm

'CHATTEM CENTENNIAL, CLIMBING', Cl Min, or, 1990; bud ovoid; flowers orange-red, orange-red reverse, aging light orange, loose, medium, dbl., 38 petals, cupped, borne usually singly or in sprays of 3-5, slight, fruity fragrance; foliage medium, light green, matt; tall (4-6 ft) growth; [Chattem Centennial sport]; Jolly, Marie; Rosehill Farm, 1991

'CHATTER', F, mr, 1947; flowers velvety bright crimson, blooms in large clusters, semi-dbl., 14 petals, cupped, moderate fragrance; bushy, compact growth; (28); [World's Fair X Betty Prior]; Boerner; J&P

'CHATTER, CLIMBING', Cl F, mr, 1960; Schmidt, K.

'CHATTERBOX', F, ob, 1973; flowers bright orange-vermilion, rosette, semi-dbl., 16 petals, 2 in.; foliage glossy; dwarf growth; [Sarabande X Circus]; Sanday, John

Chaucer® *see* 'AUSCON'

Chaumant *see* 'MACLOCKER'

Chaumont *see* 'MACLOCKER'

Chaumont™, F, ob

Checkers *see* 'PIXCHEK'

'CHEER', F, mp, 1941; flowers deep rose-pink, open, clusters, semi-dbl., 4 in., moderate fragrance; foliage leathery; vigorous, upright growth; RULED EXTINCT 1/84 ARM; [Dance of Joy X Golden Rapture]; Kordes; J&P

Cheer Up *see* 'TINCHEER'

'CHEERFUL', HT, ob, 1915; flowers orange-flame, base yellow, very large, dbl., moderate fragrance; foliage rich green, glossy; McGredy

'CHEERFULNESS', F, op, 1981; flowers orange, pink and yellow blend, borne 10 per cluster, 8 petals, slight fragrance; small, brown prickles; foliage small; compact growth; [Seedling X Seedling]; Everitt, Derrick; Gandy Roses, Ltd.

Cheerie *see* **'CHÉRIE'**

Cheerio *see* **'CHÉRIE'**

Cheerio *see* **'PLAYBOY'**®

Cheerleader™ *see* 'MORCHEER'

Cheers™ *see* 'SAVALOT'

Cheery Chatter *see* 'LYOTER'

Chelsea *see* 'MORSEA'

Cheese Cake *see* 'BROCAKE'

'CHELSEA', HT, mr, 1950; flowers carmine shaded orient red, dbl., 4 in., moderate fragrance; vigorous, compact growth; RULED EXTINCT 6/86; LeGrice

Chelsea™ *see* 'ORTSEA'

Chelsea Belle *see* 'TALCHELSEA'

Chelsea Gold *see* 'LANWOOL'

Chelsea Pensioner *see* 'MATTCHE'

'CHÉNÉDOLÉ', HCh, or; flowers crimson, small, dbl., cupped, moderate fragrance; prickles shoot very prickly; vigorous, upright growth; Thierry, prior to 1848

'CHERÉ MICHELLE', Min, op, 1986; (Cherie Michelle); flowers white with coral pink petal edges, reverse same, small, dbl., 30 petals, exhibition form, slight fragrance; medium, globular, orange-green fruit; small, cream to light brown prickles, hooked downward; foliage medium, medium green, semi-glossy; medium, upright growth; [Sheri Anne X Anita Charles]; Jolly, Marie; Rosehill Farm, 1987

'CHÉRIE', F, mp, 1931; (Cheerie, Cheerio); flowers bright rose-pink, small, borne in clusters, dbl., cupped, moderate fragrance; foliage leathery; vigorous, bushy growth; [Else Poulsen sport]; Morse

'CHÉRIE', F, ob, 1964; bud long; flowers bright orange, reverse coppery, well formed, dbl.; Gaujard

Cherie Michelle *see* **'CHERÉ MICHELLE'**

Cherish *see* 'JACSAL'

Cherokee Fire *see* 'LYOCH'

Cherokee Rose *see* **R. LAEVIGATA**

Cherries Jubilee *see* CLEcherry

Cherries 'n' Cream, Min, rb, 1997; King

'CHERRIO', F, dp, 1937; flowers carmine-cerise, large, borne in clusters, semi-dbl., cupped, slight fragrance; vigorous, compact growth; Archer

'CHERRIO', F, mp, 1948; (Planten un Blomen); flowers light pink, reverse darker, semi-dbl.; vigorous, bushy growth; [Holstein X Sapho]; Kordes

'CHERRY', HT, pb, 1928; flowers brilliant carmine-pink flushed yellow, lower half yellow, large, dbl., exhibition form, moderate fragrance; vigorous, bushy growth; McGredy

'CHERRY BLOSSOM', F, mp, 1964; bud orient red; flowers rose-pink to camellia-pink, borne in clusters, dbl., 26 petals; foliage glossy, dark; compact growth; [Fashion X Seedling]; Verschuren, A.; van Engelen

Cherry Bomb *see* 'JOLCHER'

'CHERRY BRANDY'®, HT, ob, 1965; flowers orange, dbl., 30 petals, 5 in., moderate fragrance; foliage dark, glossy, leathery; very vigorous, upright growth; GM, Belfast, 1989; Tantau, Math.

Cherry Brandy '85 *see* 'TANRYRANDY'

'CHERRY CHARM', LCl, pb, 1976; flowers deep pink, reverse silver pink, dbl., 25 petals, 5 in., non-recurrent, slight fragrance; foliage dark; [Norwich Salmon X (Seedling sport X Peeping Tom)]; MacLeod

'CHERRY CHEERFUL', Pol, rb, 1995; flowers cherry red with white shading, semi-dbl. (6–14 petals), small blooms borne in large clusters; slight fragrance; few prickles; foliage medium, medium green, glossy; compact, spreading, bushy, medium (3 ft. X 5ft.) growth; [Anytime X Mountain Mist]; Jobson, Daniel, J.; Jobson, 1995

Cherry Cola, HT, dr

Cherry Cordial *see* 'WELCORD'

Cherry Cream, Gr, w, 1995

'CHERRY GLOW', Gr, mr, 1959; flowers cherry-red, dbl., 23 petals, 3–4 in., cupped, moderate, spicy fragrance; foliage leathery, glossy; vigorous, upright growth; [Floradora X First Love]; Swim, H.C.; C.R. Burr

Cherry Glow *see* **'SWEET CHERRY'**

Cherry Hi *see* 'MORSHODOT'

Cherry Jubilee *see* 'JACSOS'

Cherry Lips, F, dp, 1997; Williams, J. Benjamin

Cherry Magic™ *see* 'MORCHERMAG'

Cherry Meidiland *see* 'MEIRUMOR'

Cherry Meillandecor® *see* 'MEIRUMOR'

'CHERRY PAGE', HT, pb, 1914; flowers carmine-pink, base yellow; [Duchess of Bedford X Le Progres]; Easlea

Cherry Pastel™ *see* 'WILCHER'

'CHERRY PIE', HT, dp, 1965; flowers deep rose-pink, dbl., 4.5 in., exhibition form, slight fragrance; Gaujard; Gandy Roses, Ltd.

'CHERRY RIPE', F, mr, 1949; flowers scarlet, small, dbl., 70 petals, intermittent bloom; vigorous growth; [Orange Triumph sport]; Heers; Pacific Nursery

Cherry Sunblaze™ *see* 'MEIBEKARB'

Cherry Sunblaze® *see* 'MEIBOKARB'

Cherry Velvet® *see* 'POULTRESS'

Cherry Wine *see* 'JALWINE'

'CHERRY, CLIMBING', Cl HT, pb, 1934; Savage Nursery

'CHERRYADE', S, dp, 1961; flowers deep pink, well-formed, dbl., 40 petals, 4 in., moderate fragrance; foliage dark; vigorous, tall growth; [New Dawn X Red Wonder]; deRuiter

'CHERRY-ROSE', HT, mp, 1946; bud long, pointed; flowers cherry-rose, very large, dbl., exhibition form, moderate fragrance; foliage glossy; very vigorous, upright, compact growth; [Pink Princess X Crimson Glory]; Brownell, H.C.

Cherry-Vanilla *see* 'ARMILLA'

Cherry-Vanilla, Climbing, Cl Gr; (Cavriglia)

'CHERUB', HMult, pb, 1923; flowers pink and salmon, small, semi-dbl., cupped, profuse, non-recurrent bloom; foliage rich green, glossy, wrinkled; very vigorous, climbing growth; [Claire Jacquier seedling]; Clark, A.; Brundrett

'CHERVENA GHITA', HT, mr, 1974; flowers bright cerise, large blooms in clusters, dbl., 75 petals; foliage dark, glossy; vigorous, upright growth; [General Stefanik X Peace]; Staikov, Prof. Dr. V.; Kalaydjiev and Chorbadjiiski

Cheryl's Delight™ *see* 'MINRCO'

'CHESAPEAKE', Min, lp, 1984; flowers small, dbl., 50 petals, slight fragrance; foliage medium, medium green, semi-glossy; bushy growth; [Rise 'n' Shine X (Helen Traubel X First Prize)]; Jolly, Nelson F.; Rosehill Farm

Chesapeake Sunset, F, ob, 1999; dbl.; Williams, J. Benjamin

CHEsdeep, MinFl, lp; (**Thank You**); Chessum, Paul

Cheshire *see* 'FRYELISE'

'CHESHIRE CREAM', F, w, 1975; flowers soft buff, becoming cream, dbl., 50 petals, 2.5 in., moderate, spicy fragrance; foliage small, glossy; low growth; [(Anna Wheatcroft X Ivory Fashion) X (Buff Beauty X Masquerade)]; Holmes, R.; Fryer's Nursery, Ltd.

'CHESHIRE LADY', HT, pb, 1968; flowers bright pink to scarlet, large, dbl., 30–40 petals; free growth; [Fragrant Cloud X Gavotte]; Dale, F.

'CHESHIRE LIFE', HT, or, 1972; flowers vermilion, spiral, dbl., 36 petals, 5 in., slight fragrance; foliage dark, leathery; [Prima Ballerina X Princess Michiko]; Fryer, Gareth; Fryer's Nursery, Ltd.

Cheshire Regiment *see* FRYzebedee

'CHESHUNT HYBRID', HT, mr, 1872; flowers red shaded violet, large, dbl.; vigorous growth; [Believed to be Mme de Tartas X Prince Camille de Rohan]; Paul

CHESnut, Cl HT, rb, 1996; (**Best Wishes**); Chessum, Paul

Chess® *see* 'INTERTRA'

Chess Man *see* 'GELMAN'

Chessum's Choice *see* 'CHESTOCK'

CHEssupremo, MinFl, my, 1995; (**Golden Hands**); Chessum, Paul

'CHESTER', F, my, 1976; flowers golden yellow, semi-dbl., 15 petals, 3 in., slight fragrance; foliage glossy, dark; vigorous growth; [Arthur Bell X Zambra]; Bees

Chester Cathedral *see* 'FRANSHINE'

Chestnut rose *see* **R. ROXBURGHII**

'CHESTOCK', (strain of R. canina), w, 1988; (**Chessum's Choice**); used as a stem of standard (tree) roses; [Pfander's Canina sport]; Chessum, Paul; Alan Thompson

'CHEVREUL', M, mp, 1887; flowers salmon-pink, well mossed; large fruit, colorful in fall; Moreau et Robert

Chevreuse *see* 'KORPLAVI'

'CHEVY CHASE', HMult, dr, 1939; flowers dark crimson, small blooms in clusters of 10-20, dbl., 65 petals, non-recurrent, moderate fragrance; foliage soft, light green, wrinkled; vigorous, climbing (to 15 ft.) growth; Dr. W. Van Fleet Medal, ARS, 1941; [R. soulieana X Eblouissant]; Hansen, N.J.; B&A

'CHEWALLOP', Cl Min, ob, 1992; (Rocketeer, **Rosalie Coral**); flowers bright coral-orange, opens flat, dbl., 15–25 petals, 1.5 in., flat, borne in small clusters, slight fragrance; few prickles; foliage small, medium green, glossy; medium (175-190 cms), bushy growth; PP9013; [[Elizabeth of Glamis X (Galway Bay X Sutter's Gold)] X Anna Ford]; Warner, Chris, 1991

CHEwapri, Cl Min, ab, 1994; (**Patio Charm**); Warner, Chris

'CHEWARVEL', Cl Min, my, 1989; (King Tut, **Laura Ford**®, Normandie); bud pointed; flowers medium yellow, reverse lighter yellow, aging pink flushes, dbl., 22 petals, 2 in., exhibition form, borne mostly singly, slight, fruity fragrance; round, large, average fruit; prickles straight, small, infrequent, light brown; foliage small, light green, glossy; upright, bushy, tall growth, 7 ft.; PP9012; [Anna Ford X [Elizabeth of Glamis X (Galway Bay X Sutter's Gold)]]; Warner, Chris, 1990

'CHEWAZE', F, or, 1986; (**Pillar Box**, Pillar BoxWardlip); flowers vivid vermillion, moderately medium, dbl., 15–25 petals, slight fragrance; foliage medium, medium green, semi-glossy; upright growth; [Alexander X (Galway Bay X Elizabeth of Glamis)]; Warner, Chris

'CHEWBEAUT', F, or, 1992; (**Donald Davies**); flowers vermilion, blooms borne in small clusters, dbl., 15–25 petals, 1.5–2.75 in., slight fragrance; few prickles; foliage medium, medium green, semi-glossy; medium (90 cms), upright growth; [Anne Harkness X Beautiful Britain]; Warner, Chris

CHEwdainty, Cl Min, yb, 1997; (**Gilt Edged**); Warner, Chris

'CHEWDOR', LCl, op, 1998; (**Lady Penelope**); flowers salmon pink, pink reverse, looks like a pom-pom dahlia, repeats well, very dbl., 26–40 petals, 2.5–3 in., borne in small clusters, slight fragrance; few prickles; foliage medium, medium green, glossy; spreading, medium (8 ft.) growth; [Laura Ford X Royal Baby]; Warner, Chris; Bransford Garden Plants, 1998

CHEWECREST, Cl Min, lp; (**Taffeta**); Warner, Chris, 1997

'CHEWELL', F, r, 1988; (**Iris Webb**); flowers tan, fading to slate gray, medium, dbl., 15–25 petals, moderate fragrance; foliage medium, dark green, semi-glossy; bushy growth; [Southampton X (Belinda X (Elizabeth of Glamis X (Galway Bay xSutters Gold)))]; Warner, Chris; LeGrice Roses, 1990

'CHEWGLORIOUS', Cl Min, mr, 1999; (**Love Knot**); flowers crimson, reverse light red, dbl., 17–25 petals, 2.5 in., borne in small clusters, slight fragrance; few prickles; foliage small, medium green, glossy; upright, tall (6 ft) growth; [Laura Ford X Ingrid Bergman]; Warner, Chris; Warner's Roses, 2000

'CHEWHARLA', Cl Min, m, 1989; (**Laura Ashley**); bud pointed; flowers lilac-mauve pink, reverse pink, aging same, loose, small, single, moderate, fruity fragrance; oval, small, red fruit; prickles hooked, small, brown; foliage small, medium green, semi-glossy; spreading, low growth; [Marjorie Fair X Nozomi]; Warner, Chris; R. Harkness & Co., Ltd., 1991

'CHEWILY', S, mp, 1999; (**Mix 'n' Match®**); bud short, furled; flowers pastel pearl pink with golden boss of stamens, semi-dbl., 10–15 petals, 1.5–2 in., borne in large clusters, slight fragrance; foliage medium, dark green, quilted; medium, rounded, slight spreading growth; [Seaspray X (Rosa sinowilsonii X Marjorie Fair)]; Warner, Chris

CHEwily, S, lp, 1998; (**Pink Cottage**); Warner, Chris

'CHEWIZZ', Cl Min, or, 1992; (**Warm Welcome**); flowers orange vermilion, blooms borne in small clusters, semi-dbl., 6–14 petals, 1.5 in., moderate fragrance; few prickles; foliage small, dark green, semi-glossy; tall (200+ cms), upright growth; [[Elizabeth of Glamis X (Galway Bay X Sutter's Gold)] X Anna Ford]; Warner, Chris, 1991

'CHEWLEGACY', F, r, 1988; (**Edith Holden**, Edwardian Lady, The Edwardian Lady); flowers russet-brown with yellow center, reverse slightly paler, aging slate grey, semi-dbl., 15 petals, slight fragrance; rounded, small, orange fruit; prickles very few; foliage medium, medium green, glossy; upright, tall, robust growth; [Belinda X (Elizabeth of Glamis X (Galway Bay X Sutters Gold))]; Warner, Chris; E.B. LeGrice Roses, Ltd., 1988

'CHEWOZ', S, ob, 1992; (Carmargue, **Telford's Promise**); flowers coppery salmon, blooms borne in large clusters, semi-dbl., 6–14 petals, 1.5 in., slight fragrance; few prickles; foliage small, medium green, glossy; spreading (60 cms) growth; [Mary Sumner X Nozomi]; Warner, Chris, 1990

'CHEWPAN', Min, mr, 1997; (**Peter Pan**); foliage small, medium green, glossy; bushy, low (10-12cms.)growth; [Eyeopener X Seedling]; Warner, Chris

'CHEWPEARL', Cl Min, mp, 1999; (**Star Performer**); flowers satin pink, reverse medium pink, small, dbl., 17–25 petals, borne in small clusters, slight fragrance; few prickles; foliage small, medium green, glossy; upright, tall (7 ft) growth; for walls, fences, pillars; [Laura Ford X Congratulations]; Warner, Chris; Warner's Roses, 1999

CHEwpiwitel, Cl Min, ly, 1995; (**Patio Honey**); Gold Star of the South Pacific, Palmerston North, NZ, 1995; Warner, Chris

'CHEWPIXCEL'R, lp, 1995; (**Open Arms**); flowers shell pink, semi-dbl. (6–14 petals), small blooms borne in large clusters; fragrant; few prickles; foliage small, dark green, glossy; spreading, medium growth; [Mary Sumner X Laura Ashley]; Warner, Chris; Warner's Roses

'CHEWPOBEY', S, or, 1995; (**Pathfinder**); flowers vermillion, yellow eye, small blooms borne in large clusters, semi-dbl., 6–14 petals, slight fragrance; few prickles; foliage small, medium green, glossy; patio/groundcover; spreading, low growth; Breeders Choice, 1996 TGC, St. Albans, 1994; [(Little Darling X Anna Ford) X Eyeopener]; Warner, Chris & Barbara; Warner's Roses, 1996

CHEwpope, Cl Min, m, 1997; (**Gloriana 97**); flowers double, medium, dbl., 15–25 petals, borne in small clusters, slight fragrance; foliage medium, medium green, semi-glossy; upright, tall (7 ft.) growth; [Laura Ford X Big Purple]; Warner, Chris

'CHEWRAMB', Cl Min, lp, 1994; (**Little Rambler**); flowers pale pink, blooms borne in large clusters, dbl., 26–40 petals, 1.5 in., intense fragrance; few prickles; foliage small, dark green, semi-glossy; tall (7 ft x 7 ft), bushy, spreading growth; [(Cecile Brunner X Baby Faurax) X (Marjorie Fair X Nozomi)]; Warner, Chris; Warner's Roses, 1995

'CHEWSEA', Cl Min, op, 1992; (**Nice Day**, Patio Queen); flowers salmon pink, blooms borne in large clusters, dbl., 15–25 petals, 1.5 in., moderate fragrance; few prickles; foliage small, bronze turning medium green, glossy; upright (200 cms), bushy, climbing growth; [Seaspray X Warm Welcome]; Warner, Chris; Warner's Roses, 1993

'CHEWSOS', Min, ob, 1996; (**Society Special**); flowers yellow with red reverse, small blooms borne in small clusters, dbl., 26–40 petals, slight fragrance; few prickles; foliage small, medium green, semi-glossy; upright, bushy, tall growth; [Laura Ford X Anne Harkness]; Warner, Chris; Warner's Roses, 1997

'CHEWSUNBEAM', Cl Min, dy, 1994; (**Good as Gold**); dbl., 25–26 petals, 1.5 in., borne in small clusters, moderate fragrance; some prickles; foliage small, light green, semi-glossy; tall (7 ft), upright growth; Gold Star of the South Pacific, Palmerston North, NZ, 1996; [Anne Harkness X Laura Ford]; Warner, Chris; Warner's Roses, 1995

'CHEWSUNFORD', Min, dy, 1996; (**Golden Handshake**); flowers clear bright yellow, small blooms borne in small clusters, dbl., 15–25 petals, slight fragrance; few prickles; foliage small, light green, glossy; upright, bushy, tall growth; TGC, St. Albans, 1994; [Pam Ayres X Laura Ford]; Warner, Chris; Warner's Roses, 1997

CHEwwily, S, lp, 1995; (**Cottage Garden**); Warner, Chris

Cheyenne *see* 'SPOCHEY'

'CHEYENNE', Cl HT, lp, 1962; bud long, pointed; flowers base coral pink, dbl., 30–40 petals, 4–5 in., exhibition form, slight fragrance; foliage leathery; vigorous (6-7 ft) growth; RULED EXTINCT 1/85 ARM; [Queen Elizabeth X Seedling]; Von Abrams; Peterson & Dering

'CHEYENNE FRONTIER', HT, w, 1971; bud large; flowers white, slowly changing to red, medium, dbl., exhibition form, slight fragrance; foliage large, glossy, dark; moderate, upright growth; [(Charlotte Armstrong X Vogue) X Peace]; Adams, M.R.

'CHEZ VITO', HT, mr, 1972; bud ovoid; flowers medium red, reverse lighter, medium, dbl., exhibition form, moderate fragrance; foliage large, leathery; vigorous, upright, bushy growth; [Paris-Match X (Baccará X Happiness)]; Meilland

'CHI LO SÀ?', HT, yb, 1965; flowers yellow suffused red; [(Peace X Fiaba) X Seedling]; Giacomasso, Fratelli

Chianti *see* 'AUSWINE'

'CHIARASTELLA', HT, rb, 1948; flowers rose-red and yellow bicolor, well formed, large, dbl.; foliage glossy; strong stems; vigorous growth; [Julien Potin X Mme G. Forest-Colcombet]; Giacomasso, Fratelli

'CHIC', F, pb, 1953; bud ovoid; flowers geranium-pink, blooms in clusters, dbl., 68 petals, 2.5 in., cupped, moderate fragrance; vigorous, branching growth;

[Pinocchio seedling X Fashion]; Boerner; J&P

'CHIC PARISIEN', F, op, 1956; flowers coral-pink, center darker, well-formed blooms in clusters of 4-8, dbl., slight fragrance; foliage dark; vigorous growth; Delbard-Chabert

Chica, HT, lp, 1998; Kordes

'CHICAGO', HT, m, 1928; flowers soft mauve-pink, dbl., intense fragrance; [Premier sport]; Aldous

Chicago Peace® *see* 'JOHNAGO'

Chicago Peace, Climbing, Cl HT, lp; [Chicago Peace sport]; Brundrett, 1978; (Weatherly, L.)

Chick-a-dee *see* 'MORCHICK'

Chickasaw Rose *see* **R. BRACTEATA**

Chidori *see* **'RYOKKOH'**

Chief *see* **'THE CHIEF'**

'CHIEF JUSTICE HOLMES', Cl HT, dr, 1935; flowers very dark red; GM, Portland, 1936; [Jules Margottin X Château de Clos Vougeot]; Schoener

'CHIEF SEATTLE', HT, yb, 1951; bud conical; flowers buff and old-gold, center shrimp-red, dbl., 55 petals, 4–5 in., exhibition form, moderate fragrance; foliage glossy; tall growth; [Charlotte Armstrong X Signora]; Swim, H.C.; Armstrong Nursery

'CHIEFTAIN', HT, rb, 1936; bud ovoid; flowers brilliant red, base yellow, large, dbl., exhibition form; foliage leathery, dark; vigorous growth; [Hadley X Talisman]; Montgomery Co.

'CHIFFON', HT, lp, 1940; flowers blush-pink, tinted light lavender, dbl., 30 petals, 5.5 in., moderate fragrance; [Regina Elena sport]; Grillo

Childhood Memories *see* FERho

Child's Play™ *see* 'SAVACHILD'

Chili Pepper *see* 'GELPEP'

Chill Out *see* 'GELOUT'

Chilterns *see* 'KORTEMMA'

Chimene *see* HARazz

Chimo® *see* 'INTERCHER'

Chin Chin *see* **Promise**

'CHINA BELLE', HCh, pb, 1980; bud ovoid; flowers light and medium pink blend with yellow, blooms borne 5-7 per cluster, dbl., cupped, slight, fruity fragrance; curved, red prickles; foliage glossy; short stems; compact, bushy, upright growth; [(Doubloons X Holiday) X Slater's Crimson China]; James, John

'CHINA DOLL', Pol, mp, 1946; bud pointed; flowers china-rose, base mimosa-yellow, blooms in large trusses, dbl., 24 petals, 1–2 in., cupped, slight, tea fragrance; foliage leathery, with mostly 5 leaflets (similar to pinkie); dwarf (18 in), bushy growth; [Mrs Dudley Fulton X Tom Thumb]; Lammerts, Dr. Walter; Armstrong Nursery

'CHINA DOLL, CLIMBING', Cl Pol, mp, 1977; (Weeping China Doll); Weeks; Weeks Wholesale Rose Growers

China Rose *see* **R. CHINENSIS**

'CHINATOWN'®, F, dy, 1963; (Ville de Chine); flowers yellow, sometimes edged pink, blooms in clusters, dbl., 4 in., intense fragrance; foliage dark; vigorous, tall, bushy growth; GM, NRS, 1962; [Columbine X Clare Grammerstorf]; Poulsen, Niels D.; A. Dickson;, McGredy

Chinatown Moss, M, rb

Chinese Lantern *see* 'FOUCHIN'

Chinese Monthly Rose *see* **'SLATER'S CRIMSON CHINA'**

Chinese Puzzle *see* 'GELPUZZLE'

'CHINGARI', F, yb, 1976; bud pointed; flowers aureolin to currant-red, open, semi-dbl., 17 petals, 3 in., slight fragrance; foliage glossy; vigorous, bushy, compact growth; [Charleston X Seedling]; Pal, Dr. B.P.; Laveena Roses

Chinquapin Rose *see* **R. ROXBURGHII**

Chipie *see* 'POUBICARBE'

Chipmonk *see* 'PIXICHIP'

Chipmunk *see* 'PIXICHIP'

'CHIPPER', Min, ab, 1966; bud ovoid; flowers salmon-pink, small, dbl., slight fragrance; foliage glossy, leathery; vigorous, dwarf growth; [(Dany Robin seedling X Fire King) X Perla de Montserrat]; Meilland, Alain A.; C-P

'CHIPPEWA', S, mp; flowers rose-pink, borne in clusters, semi-dbl.; foliage leathery, bronze; Central Exp. Farm

'CHIQUITA', LCl, op, 1938; flowers orange-yellow to coppery orange and salmon-pink, base yellow, intense fragrance; RULED EXTINCT 1/88; [Sierra Snowstorm X Étoile Luisante]; Moore, Ralph S.; Brooks & Son

Chiquita™ *see* 'MORKITA'

'CHIRAZ'®, HT, pb, 1986; flowers creamy white, flushed pink, petals edged carmine red, large; foliage dark; dense growth; [Kordes' Perfecta X Peace]; Kriloff, Michel

'CHIRIPA', F, mr, 1957; bud pointed; flowers red, reverse carmine, borne in clusters of 3-5, dbl., 26 petals; foliage bright green; upright, compact growth; [Radar X (Rosalia Riviera X Independence)]; de Dot, G.F.

Chit Chat *see* 'BRICHAT'

Chitchor, F, pb, 1996; Pal, Dr. B.P.

Chitra, HT, ob, 1995; IARI

Chitrarajini, HT, pb, 1985; Kasturi

Chivalry® *see* 'MACPOW'

'CHIYO', HT, dp, 1975; bud long, pointed; flowers deep pink, dbl., 25 petals, 4–4.5 in., exhibition form, slight, fruity fragrance; foliage glossy, medium to dark green; vigorous, upright growth; [Karl Herbst X Chrysler Imperial]; Ota, Kaichiro; Eastern Roses, 1970

'CHLORIS', A, lp; (Rosée du Matin); flowers soft pink, many petaled, reflexing with a button eye; few prickles; foliage dark, leathery; vigorous (to 4-5 ft) growth; (France), Prior to 1835

'CHOBEE', F, mr, 1993; dbl., 26–40 petals, 1.5–2.75 in., borne mostly singly, slight fragrance; few prickles; foliage medium, medium green, matt; medium, bushy growth; [Seedling X Congratulations]; Giles, Diann; Giles Rose Nursery, 1993

'CHOO-CHOO CENTENNIAL', Min, lp, 1980; bud ovoid; flowers light pink, edged darker, reverse white, small blooms borne, dbl., 68 petals, exhibition form, slight fragrance; straight prickles; foliage matt, light green; compact, bushy growth; [Rise 'n' Shine X Grand Opera]; Jolly, Betty J.; Rosehill Farm

'CHOO-CHOO'S BABY', Min, rb, 1980; bud urn-shaped; flowers red, shaded to yellow at base, borne usually 1-2 per cluster, dbl., 26 petals, flat, slight fragrance; no prickles; foliage tiny, light green; low, branching, dense growth; [Watercolor X Watercolor seedling]; Jolly, Betty J.; Rosehill Farm

'CHOPIN', HT, mr, 1968; dbl., 38 petals, 4–6 in., moderate fragrance; foliage medium to light green; vigorous growth; [Montezuma X Christian Dior]; Ellick

'CHORALE', S, lp, 1978; bud ovoid, pointed; flowers pale pink, dbl., 48 petals, 3.5 in., exhibition form, moderate fragrance; foliage dark, leathery; vigorous, upright, bushy growth; [(Ruth Hewitt X Queen Elizabeth) X (Morning Stars X Suzanne)]; Buck, Dr. Griffith J.; Iowa State University

Chorus® *see* 'MEIMORE'

'CHORUS GIRL', F, or, 1970; flowers vermilion, semi-dbl., 16 petals, 3 in., exhibition form, moderate fragrance; foliage dark, coppery; [Highlight X Seedling]; Robinson, H.; Victoria Nursery

Chorus Line *see* 'JACDAZ'

Chorus, Climbing *see* MEIjulitasar

'CHOT PESTITELE', HP, mp, 1932; bud oblong; flowers rose-pink, dbl., 22 petals, 6 in., flat, moderate fragrance; foliage glossy; [Frau Karl Druschki sport]; Böhm, J.

'CHOTA', F, or, 1975; flowers light vermilion, very full, dbl., 50 petals, 3.5 in., slight fragrance; foliage small, semi-glossy; moderately low growth; [Violet Carson X Korona]; Sheen

'CHOUETTE', Pol, dr, 1969; bud ovoid; flowers dark fire-red, full, large, dbl.; foliage dark, soft; vigorous, bushy growth; [Atlantic X Seedling]; Delforge

'CHOWAN', HT, ob, 1995; dbl., 26–40 petals, 4.75 in., borne mostly singly, moderate fragrance; some prickles; foliage large, medium green, dull; tall (160 cms), upright growth; [Folklore X Hot Pewter]; Perry, Astor; Hortico Roses, 1995

Chris *see* KIRsan

Chris Evert® *see* 'WEKJUVOO'

'CHRIS JOLLY', MinFl, or, 1985; flowers mini-flora, medium, dbl., 40 petals, exhibition form, slight fragrance; foliage medium, medium green, semi-glossy; upright, bushy growth; [(Orange Sweetheart X Zinger) X Rise 'n' Shine]; Jolly, Nelson F.; Rosehill Farm

'CHRISGOBRO', HT, yb, 1998; (**Joshua Bradley**); flowers deep gold, bronze reverse, strong substance, dbl., 17–25 petals, 3–3.5 in., borne mostly singly, slight fragrance; prickles moderate, medium, hooked; foliage medium, dark green, dull; upright, medium growth; [(Gingersnap X Brandy) X Caramel Creme]; Christensen, Jack E.; Armstrong Garden Centers, 1999

CHRisgood, F, yb; (**Rainbow Warrior**); (CHRisgood), 1996

'CHRISJEVANS', HT, or, 1989; (**Vivacious Dianne**); flowers medium, dbl., 33 petals, slight fragrance; foliage medium, medium green, semi-glossy; upright, bushy growth; [Voodoo X Hello Dolly]; Christensen, Jack E.

'CHRISS AND DIANNI', (clone of R. multiflora), w; used as understock

Chris's Sunshine, HT, my

'CHRISSIE MACKELLAR', HT, op, 1913; bud crimson-carmine on deep madder; flowers orange-pink, reverse deeply zoned orange, semi-dbl., moderate fragrance; Dickson, A.

Christa *see* 'PIXSTA'

Christel von der Post® *see* KORpora

'CHRISTIAN CURLE', HWich, lp, 1910; flowers flesh-pink; [Dorothy Perkins sport]; Cocker

Christian Dior *see* 'MEILIE'

'CHRISTIAN DIOR, CLIMBING', Cl HT, mr, 1966; Chang, Chi-Shiang

'CHRISTIANA WOOD', HT, mr, 1975; bud ovoid; dbl., 31 petals, 3.5 in., moderate fragrance; upright, bushy growth; [Tropicana sport]; Wood, J.

Christiane Horbiger, F, w

'CHRISTINA', HT, pb, 1977; flowers deep pink, reverse light pink, large, dbl., 45 petals; foliage leathery, glossy; vigorous, upright growth; [Rina Herholdt X Seedling]; Staikov, Prof. Dr. V., Kalaydjieve & Chorbadjiiski

'CHRISTINA', HT, pb, 1959; flowers pink, reverse currant red, dbl.; very vigorous growth; RULED EXTINCT 12/85 ARM; [Granat X Radiance]; Crouch; Roseglen Nursery

Christina *see* 'SELSTAR'

Christina Atherton *see* 'MACSEV'

'CHRISTINA NILSSON', HP, 1867; Lévêque, P.; (Cavriglia)

'CHRISTINE', HT, dy, 1918; flowers deep golden yellow, well-shaped, small, moderate fragrance; foliage dark, glossy; GM, NRS, 1916; McGredy

Christine *see* CLEalta

'CHRISTINE CL.', Cl HT, dy, 1936; Willink

'CHRISTINE GANDY', F, dp, 1958; flowers deep pink, blooms in small clusters, semi-dbl., 3 in., moderate fragrance; foliage dark; vigorous growth; [Polyantha seedling X Fashion]; deRuiter; Gandy Roses, Ltd.

Christine Horbiger *see* 'MACMOUHOO'

'CHRISTINE HORBIGER' *see* 'MACMOUHOO'

Christine Lanson, HT, ob

'CHRISTINE PRIOR', HT, rb, 1924; flowers deep rosy red, flushed yellow and peach, base yellow, semi-dbl., moderate fragrance; McGredy

'CHRISTINE WEINERT', Min, or, 1976; flowers brilliant scarlet, shaded deeper, flat to rounded, dbl., 25 petals, 1 in., moderate fragrance; foliage small, leathery; upright, bushy growth; [(Little Darling X Eleanor) X (Little Darling X Eleanor)]; Moore, Ralph S.; Sequoia Nursery

'CHRISTINE WRIGHT', LCl, mp, 1909; flowers wild-rose-pink, very large, semi-dbl., cupped, moderate fragrance; foliage glossy; long stems; height 12-15 ft; [Seedling X Mme Caroline Testout]; Hoopes, Bro. & Thomas

'CHRISTINE WUNDERLICH', HT, op, 1934; flowers yellowish orange-pink, large, moderate fragrance; very vigorous growth; [Golden Ophelia sport]; Wunderlich

'CHRISTINE, CLIMBING', Cl HT, dy, 1936; Willink

Christingle *see* 'HARVALEX'

'CHRISTMAS BEAUTY', HT, dr, 1942; flowers darker red than parent, dbl., 25–30 petals, 4.5–5 in., moderate fragrance; foliage blue-green, leathery; vigorous growth; [Better Times sport]; Krowka

Christmas Card *see* 'RESCARD'

'CHRISTMAS CHEER', HT, mr, 1957; bud long, pointed; flowers cherry-red, dbl., 45–50 petals, 4.5–5.5 in., exhibition form, slight fragrance; foliage dark, leathery; vigorous, upright growth; [Sister Kenny X Happiness]; Hill, Joseph H., Co.

'CHRISTMAS RED', HT, mr, 1948; bud large, long, pointed; flowers spectrum-red, open, dbl., moderate fragrance; foliage glossy; bushy, dwarf growth; [Pink Princess X Crimson Glory]; Brownell, H.C.

'CHRISTOBEL', HT, ab, 1937; flowers apricot-yellow shaded salmon, very large, dbl.; foliage glossy; very vigorous, bushy growth; [Frau Karl Druschki X Mme Butterfly]; Croibier

Christoph Colombus *see* 'MEIRONSSE'

'CHRISTOPH WEIGAND', HT, lp, 1928; flowers very large, dbl., exhibition form, slight fragrance; foliage rich green, wrinkled; vigorous growth; [Frau Karl Druschki X Souv. de Claudius Pernet]; Weigand, C.

Christophe Colomb *see* 'MEIRONSSE'

Christopher *see* 'TINCHRIS'

Christopher *see* COCopher

Christopher Columbus *see* 'MEIRONSSE'

Christopher Columbus *see* POUlstripe

'CHRISTOPHER MILTON', HT, dp, 1965; bud long, pointed, light red; flowers medium pink, edged lighter, large, dbl., exhibition form; foliage dark, glossy; very vigorous, upright growth; [Christian Dior sport]; Martin, W.A.

'CHRISTOPHER STONE', HT, mr, 1935; bud long, pointed; flowers large, dbl., 30 petals, moderate, damask fragrance; foliage bright green; vigorous growth; GM, NRS, 1934 GM, Portland, 1937; Robinson, H.; C-P, 1936;, Wheatcroft Bros., 1935

'CHRISTOPHER STONE, CLIMBING', Cl HT, mr, 1942; Marsh's Nursery

'CHROMATELLA', N, ly, 1843; (Cloth of Gold); flowers creamy white, center yellow, large, very dbl., globular, moderate fragrance; vigorous, climbing growth; [Lamarque seedling]; Coquereau

Chrysandra, S, dp, 1983; flowers carmine-pink, large, semi-dbl., intense fragrance; Weihrauch; (Sangerhausen)

'CHRYSLER IMPERIAL', HT, dr, 1952; bud long, pointed; flowers deep red, velvety, very dbl., 45–50 petals, 4.5–5 in., exhibition form, borne singly, intense fragrance; foliage dark, semi-glossy; vigorous, compact growth; AARS, 1953 GM, ARS, 1956 GM, Portland, 1951 James Alexander Gamble Fragrance Medal, ARS, 1965 John Cook Medal, ARS, 1964; [Charlotte Armstrong X Mirandy]; Lammerts, Dr. Walter; Germain's

'CHRYSLER IMPERIAL, CLIMBING', Cl HT, dr, 1957; (Grimpant Chrysler Imperial); Begonia, P.B.; Germain's

Chrystelle *see* 'GODIALING'

'**CHRYZIA**', HT, dp, 1970; bud long, pointed; flowers light red, large, dbl., moderate fragrance; vigorous, upright, bushy growth; [Chrysler Imperial X Lady Zia]; Wyant

'**CHUCKLES**', F, dp, 1958; bud long, pointed; flowers deep pink, white eye, blooms in large clusters, semi-dbl., 11 petals, 3.5 in., moderate fragrance; foliage dark, leathery; vigorous, bushy growth; [(Jean Lafitte X New Dawn) X Orange Triumph]; Shepherd; Bosley Nursery

Chula Vista *see* 'TINCHULA'

Church Mouse *see* 'FOUMOUSE'

Ciak® *see* BARciak

'**CIBLES**', HRg, mr, 1893; flowers bright red, base yellow; vigorous, upright growth; [R. rugosa rubra X Perle de Lyon]; Kaufmann

Cicely Lascelles, Cl HT, op; flowers pink shaded salmon, semi-dbl.; very vigorous pillar type; [Frau Oberhofgartner Singer X Scorcher]; Clark, A., 1937

'**CICELY O'RORKE**', Cl HT, mp, 1937; flowers pink, shaded salmon, large, semi-dbl., cupped, recurrent bloom; long stems; very vigorous, climbing or pillar growth; [Souv. de Gustave Prat X Seedling]; Clark, A.; NRS Victoria

'**CIDADE DE LISBOA**', LCl, pb, 1939; bud long, pointed; flowers salmon-pink, edged yellow, large, semi-dbl., exhibition form, slight fragrance; long stems; vigorous growth; [Belle Portugaise X Mme Edouard Herriot]; da Silva, Moreira

Cider Cup *see* 'DICLADIDA'

Ciel Bleu®, HT, m, 1982; Dot, Simon

'**CILLY MICHEL**', HT, or, 1928; flowers nasturtium-red, large, dbl., moderate fragrance; [Mme Mélanie Soupert X Felbergs Rosa Druschki]; Felberg-Leclerc

Cimarosa®, F, ob, 1989; Lens

'**CIMARRON**', HT, pb, 1938; bud spiraled, almost red; flowers ruffled, salmon, reverse deep pink; vigorous, compact growth; [Nellie E. Hillock X Golden Dawn]; Hillock

'**CINA**', HT, ob, 1973; bud full; flowers brilliant coral, flushed salmon; foliage large; [Premiere Ballerine X Femina]; Gaujard

'**CINDERELLA**', Min, w, 1953; flowers satiny white tinged pale flesh, micro-mini, dbl., 55 petals, 1 in., moderate, spicy fragrance; no prickles; upright growth; [Cécile Brunner X Tom Thumb]; deVink; C-P

'**CINDERELLA**', HWich, dp, 1909; flowers deep pink, petal tips quilled, small, borne in large clusters, dbl., free, non-recurrent bloom; vigorous growth; Walsh

Cinderella Gold *see* 'MORCINGOLD'

'**CINDERELLA, CLIMBING**', Cl Min, w, 1975; Sequoia Nursery

'**CINDERELLA'S MIDNIGHT ROSE**', HT, mp, 1976; flowers rich pink, 4–5 in., slight fragrance; foliage glossy; vigorous growth; England

Cindy™ *see* 'MINAACO'

Cindy *see* MEItinor

'**CINERAIRE**', HCh, rb; flowers rich red, center white, very small blooms in trusses, single, 5 petals, cupped; E. Murrell, prior to 1964

'**CINERAMA**', HT, pb, 1966; flowers salmon, reverse buff-yellow, pointed, 4.5–5 in., slight fragrance; vigorous growth; [Seedling X Tzigane]; Herholdt, J.A.

'**CINERARIA**', Pol, m, 1934; flowers carmine-purple, center white, open, large, semi-dbl., slight fragrance; foliage soft; vigorous, bushy growth; [Miss Edith Cavell X Tip-Top]; Leenders, M.

'**CINGALLEGRA**', HT, yb, 1958; bud well shaped; flowers lemon-yellow, edged pinkish, dbl.; long, strong stems; [Golden Scepter X Crimson Glory]; Cazzaniga, F. G.

'**CINNABAR**', F, or, 1945; (Tantau's Triumph); bud small, globular; flowers scarlet, semi-dbl., cupped, blooms in clusters, slight fragrance; foliage leathery; bushy, upright growth; [Baby Chateau X R. roxburghii]; Tantau

'**CINNABAR IMPROVED**', F, or, 1951; (Verbesserte Tantau's Triumph); flowers orange-scarlet, borne in trusses, semi-dbl.; (28); [(Cinnabar X Kathe Duvigneau) X Cinnabar]; Tantau

Cinnamon Delight™ *see* 'MINICIN'

Cinnamon Rose *see* **R. CINNAMOMEA**

Cinnamon Toast™ *see* 'SAVACIN'

Cinzia, 1996; Embriaco, B.; (Cavriglia)

'**CIRCE**', HT, lp, 1916; flowers whitish-pink, large, dbl.; Paul, W.; (Sangerhausen)

Circé *see* 'GAUMOVA'

'**CIRCUS**', F, yb, 1956; bud urn-shaped; flowers yellow marked pink, salmon and scarlet, very dbl., 45–58 petals, 2.5–3 in., exhibition form, blooms in large, ruffled clusters, moderate, tea to spicy fragrance; foliage semi-glossy, leathery; bushy growth; AARS, 1956 GM, Geneva, 1955 GM, NRS, 1955; [Fandango X Pinocchio]; Swim, H.C.; Armstrong Nursery

Circus Clown *see* 'MORPICO'

'**CIRCUS KNIE**'®, HT, yb, 1975; bud long, pointed; dbl., 32–38 petals, 3.5–4 in., cupped, slight fragrance; foliage dark, leathery; [Moulin Rouge X Peace]; Huber

'**CIRCUS PARADE**', F, yb, 1963; flowers multicolor, redder than circus; [Circus sport]; Begonia, F.B. & DeVor, P.; Armstrong Nursery

'**CIRCUS, CLIMBING**', Cl F, yb, 1961; House; Armstrong Nursery

Cisco™ *see* 'CARALE'

'**CISSIE**', Min, mp, 1979; bud long, pointed; 18–20 petals, 1 in., exhibition form, borne singly, slight fragrance; small, thin prickles; foliage dark green with touches of red and bronze on edges; bushy, upright growth; [Gene Boerner X Elfinesque]; Bennett, Cecilia 'Dee'; Tiny Petals Nursery

'**CISSIE EASLEA**', HT, my, 1913; [Melanie Soupert X Rayon d'Or]; Pernet-Ducher

'**CITATION**', Gr, dr, 1982; flowers dark red, reverse lighter, blooms in sprays of 2-3, dbl., 35 petals, exhibition form; globular, light orange fruit; short, lilac prickles, hooked downward; foliage large, dark, semi-glossy; medium, bushy growth; [Seedling X Seedling]; Hoy, Lowel L.; Joseph H. Hill, Co.

Citrina, HT, dy

'**CITRON**', HT, ly, 1942; flowers buff, shaded copper, dbl., 28 petals, cupped, moderate fragrance; foliage reddish; vigorous growth; [Julien Potin X Seedling]; Gaujard; J&P

'**CITRONELLA**', F, ly, 1946; flowers lemon-yellow, borne in clusters, semi-dbl., 15 petals, globular, moderate fragrance; foliage glossy; branching growth; [Mev. Nathalie Nypels X Donald Prior]; Leenders, M.; Longley

Citronella® *see* 'SUNSIEN'

Citron-Fraise *see* DELcifra

City Girl *see* 'HARZORBA'

City Lights *see* POUlgan

City of Alexandria *see* 'WILALEX'

City of Auckland *see* 'MACTANE'

'**CITY OF BATH**', HT, pb, 1969; flowers deep candy-pink, reverse lighter, dbl., 55 petals, 4 in., moderate fragrance; foliage matt, green; [Gavotte X Buccaneer]; Sanday, John

City of Belfast® *see* 'MACCI'

'**CITY OF BENALLA**', HT, op, 1983; bud globular, pointed; flowers carmine, opening with outer petals paling, inner petals coral, dbl., exhibition form, slight fragrance; brown prickles, hooked down; foliage dark, dense, glossy; vigorous, tall growth; [My Choice X Extravaganza]; Dawson, George; Rainbow Roses

City of Birmingham *see* 'KORHOLST'

City of Bradford *see* 'HARROTANG'

'**CITY OF CARDIFF**', HT, rb, 1992; very dbl., 3–3.5 in., borne mostly singly, moderate fragrance; some prickles; foliage medium, medium green, matt; medium, upright growth; [Lady Sylvia X Chicago Peace]; Poole, Lionel, 1993

City of Christchurch *see* MACoffer

City of Dunedin *see* 'REYCIDUN'

'CITY OF GISBORNE', HT, op, 1968; bud long, pointed; flowers pink shaded orange-yellow, large, semi-dbl., exhibition form, intense fragrance; foliage glossy; vigorous, upright growth; [Prima Ballerina X Prima Ballerina]; Appleyard

'CITY OF GLASGOW', HT, ab, 1970; flowers apricot, suffused pink; [Femina sport]; Haynes

'CITY OF GLOUCESTER', HT, dy, 1969; flowers saffron-yellow shaded gold, large, dbl., exhibition form; vigorous growth; [Gavotte X Buccaneer]; Sanday, John, 1970

'CITY OF HAMILTON', F, ob, 1972; flowers orange and gold, dbl., 3.5 in., slight fragrance; foliage dull; free growth; [Innisfree X Elizabeth of Glamis]; Dickson, A.

'CITY OF HARVEY', HWich, op, 1944; flowers pink tinted orange, small, borne in clusters, semi-dbl., cupped, profuse bloom, not repeated, slight fragrance; foliage dark, glossy; height 12 ft; [R. wichurana X Orléans Rose]; Wiseman

'CITY OF HEREFORD', HT, mp, 1967; flowers carmine-pink, pointed, 6 in., intense fragrance; foliage dark; [Wellworth X Spartan]; LeGrice

City of Ichalkaranji, HT, pb, 1989; Chiplunkar

City of Invercargill *see* 'REYINVER'

'CITY OF KINGSTON', F, or, 1973; bud ovoid; flowers medium, very dbl., globular, slight fragrance; foliage dark, leathery; moderate, bushy growth; [Malibu X Independence]; Schloen, J.

'CITY OF LEEDS', F, op, 1966; flowers salmon, blooms in clusters, semi-dbl., 19 petals, 4.5 in., slight fragrance; foliage dark; GM, RNRS, 1965; [Evelyn Fison X (Spartan X Red Favorite)]; McGredy, Sam IV; McGredy

'CITY OF LITTLE ROCK', HT, mp, 1924; flowers hydrangea-pink, open, semi-dbl., moderate fragrance; vigorous growth; Hill, E.G., Co.; Vestal

City of London® *see* 'HARUKFORE'

City of Manchester *see* 'NOSMAN'

'CITY OF NEWCASTLE', HT, yb, 1976; flowers yellow, tinged salmon-orange, dbl., 35 petals, 5 in., intense fragrance; foliage semi-glossy; vigorous, tall, upright growth; [Arthur Bell X Mischief]; Wood

City of Newcastle Bicentennary *see* 'JACOPPER'

'CITY OF NORWICH', HT, mr, 1949; bud ovoid; flowers scarlet-crimson, well formed, dbl., 35 petals, 6 in., moderate fragrance; foliage leathery; [Crimson Glory X (Crimson Glory X Cathrine Kordes)]; Kordes; Morse

'CITY OF NOTTINGHAM', F, or, 1962; flowers orange-scarlet, rosette shape, borne in clusters, dbl., 30–40 petals, 2.5 in., slight fragrance; foliage dark; vigorous, bushy, compact growth; [Seedling (F) X Moulin Rouge]; deRuiter

City of Panjim, HT, pb, 1972; Kasturi

City of Pilsen *see* **'PLZEN'**

'CITY OF PORTLAND', F, mr, 1977; bud pointed; flowers geranium-red, base primrose-yellow, single, 5 petals, 2.5–3.5 in., slight fragrance; foliage glossy; very free growth; [Seedling X Cocktail seedling]; Takatori, Yoshiho; Japan Rose Nursery

'CITY OF PORTSMOUTH', F, ob, 1975; flowers copper, dbl., 25 petals, 3–4 in., moderate fragrance; foliage bronze; tall growth; Cants of Colchester, Ltd.

City of Pretoria *see* KORseubel

City of Sheffield *see* 'NOSSHEF'

'CITY OF SPRINGFIELD', F, rb, 1988; flowers non-fading, medium, borne in sprays of 6-10, dbl., 34 petals; bushy, hardy growth; [Pink Parfait X Roman Holiday]; Pencil, Paul S., 1989

City of Wangaratta, HT, mr, 1985; Dawson

City of Warwick *see* 'AROFUTO'

City of Welland, HT, yb, 1996; Kordes

City of Windsor, HT, ob

'CITY OF WORCESTER', HT, mr, 1983; flowers large, dbl., 35 petals, exhibition form, moderate fragrance; foliage medium, medium green, matt; [Red Planet X (Ena Harkness X Fragrant Cloud)]; Crivens, L.

'CITY OF YORK', LCl, w, 1945; (Direktör Benschop); flowers creamy white, large, in clusters of 7-15, semi-dbl., 15 petals, cupped, moderate fragrance; foliage glossy, leathery; vigorous, climbing growth; GM, ARS, 1950; [Prof. Gnau X Dorothy Perkins]; Tantau; C-P

'CIUDAD DE OVIEDO', C, mp; (R. centifolia simplex); single; (28)

Cl. All Ablaze™ *see* 'WEKSAMSOU'

Cl. Berries 'n' Cream *see* 'POULCLIMB'

Cl. Shadow Dancer *see* 'MORSTRORT'

'CLAIR DE LUNE', HT, m, 1967; bud pointed; flowers large, dbl., moderate fragrance; foliage leathery; vigorous, upright growth; [Eminence X Viola]; Gaujard

Clair Matin® *see* 'MEIMONT'

Clair Renaissance *see* POUlsyng

Claire, HT, pb, 1987; Dawson

'CLAIRE DESMET', HT, my, 1932; flowers golden yellow, very dbl., cupped, moderate fragrance; foliage leathery, bronze; long, strong stems; vigorous, bushy growth; [Margaret Dickson Hamill X Souv. de Claudius Pernet]; Buatois

'CLAIRE JACQUIER', N, ly, 1888; flowers yellow, fading to creamy yellow; very vigorous growth; [Possibly R. multiflora X Tea rose]; Bernaix, A.

Claire Jolly, HMsk, 1991; Lens, Louis; (Cavriglia)

Claire Rayner *see* 'MACPANDEM'

Claire Rose *see* 'AUSLIGHT'

Claire Scotland *see* 'COCDIMITY'

'CLAIRE-FRANCE', HT, mp, 1964; bud pointed; flowers clear pink, large, dbl., 40 petals, exhibition form, slight fragrance; foliage glossy; vigorous, upright growth; Mondial Roses

Clairette, HT, 1974; Croix, P.; (Cavriglia)

Clanwilliam, HT, op

Clara *see* 'RESCLARA'

Clara, HT, rb

Clara® *see* FEsato

'CLARA BOW', Cl HT, yb, 1927; (Clara Bow, Climbing); flowers yellow stained crimson, dbl., slight fragrance; height 12-15 ft; [Golden Emblem sport]; Padella Rose Co.; Germain's

Clara Bow, Climbing *see* **'CLARA BOW'**

'CLARA COCHET', HP, lp, 1886; flowers very large, dbl.; Lacharme, F.; (Sangerhausen)

'CLARA CURTIS', HT, my, 1922; flowers rich golden yellow, very dbl., intense fragrance; GM, NRS, 1919; Dickson, A.

Clara d'Arcis *see* **'MME CLARA D'ARCIS'**

Clara Granato, F, mr

'CLARA MUNGER', LCl, m; flowers lavender; Munger

'CLARA WATSON', HT, w, 1894; flowers mother-of-pearl white, center peach, dbl., moderate fragrance; Prince

Clara's Surprise *see* 'KIRWIM'

'CLARE', HT, pb, 1972; flowers cream, edged rose-pink, dbl., 35 petals, 5.5 in., exhibition form, slight fragrance; foliage large, medium green, semi-glossy; vigorous, tall growth; [Ethel Sanday X Rose Gaujard]; MacLeod

'CLARE DE ESCOFET', HT, w, 1920; flowers delicate flesh-white, dbl.; Easlea

'CLÄRE GRAMMSERSTORF', F, dy, 1957; bud ovoid; flowers large, dbl., exhibition form; foliage leathery, glossy; vigorous, bushy growth; [Harmonie X R. eglanteria seedling]; Kordes

'CLARE HELEN', HT, lp, 1985; [Gail Borden sport]; Owen, Fred

'CLARET', Min, m, 1977; bud short, pointed; dbl., 48 petals, 1 in., cupped; very compact, spreading. growth; [Little Chief X Little Chief]; Saville, F. Harmon; Nor'East Min. Roses

'CLARET CUP', Min, rb, 1962; bud globular; flowers dark red, white eye, small blooms in clusters, dbl., moderate fragrance; foliage leathery, dark; vigorous, bushy, compact growth; [Spring Song (Min) X Eutin]; Riethmuller; Hazlewood Bros.

'CLARICE GOODACRE', HT, w, 1916; bud pointed; flowers ivory-white, shaded chrome, dbl., exhibition form, moderate fragrance; foliage dark, soft; vigorous, bushy growth; Dickson, A.

Claridge *see* 'PEKLIPINK'

'CLARION CALL', F, op, 1995; flowers orange pink, red salmon hue, blooms borne 1-5 per cluster, semi-dbl., 6–14 petals, 1.5–2.75 in., slight fragrance; foliage medium, medium green, matt; bushy (up to 75 cms) growth; 1st prize, Unnamed Seedling, Norfolk, VA, National Show, ARS, 1991; [Canadian Centennial X Traumerei]; Fleming, Joyce L.; Hortico Roses, 1994

Clarissa® *see* 'HARPROCRUSTES'

'CLARISSA DANA', HT, pb, 1933; flowers brilliant pink with amber glow, large, dbl., exhibition form, moderate fragrance; vigorous growth; [(HT X La France) X Marechal Niel]; Nicolas; J&P

Clarita *see* 'MEIBYSTER'

'CLARITY', F, w, 1993; flowers white with cream center, dbl., 15–25 petals, 1.5–2.75 in., borne in small clusters, slight fragrance; some large prickles; foliage large, dark green, glossy; medium, upright, bushy growth; [Pristine X (Party Girl X Laureate)]; Jobson, Daniel J.; Jobson, 1993

Clark Ochre Seedling, Cl HT, dy; Clark

Clark Pink, Cl HT, mp; Clark

'CLARKE'S MULTIFLORA', (clone of R. multiflora), w; thornless; used as understock

Class Act *see* 'JACARE'

Classic, HT, dp, 1997; K&S

'CLASSIC BEAUTY', HT, pb, 1994; dbl., 15–25 petals, 3–3.5 in., borne mostly singly, slight fragrance; some prickles; foliage medium, medium green, matt; medium, bushy growth; Winchel, Joseph F.; Certified Roses, Inc., 1997

'CLASSIC CHICK', Min, ob, 1986; flowers orange, blooms borne usually singly, dbl., 35 petals, exhibition form, no fragrance; few, slender prickles; foliage medium, dark, semi-glossy; medium, upright growth; [Seedling X Seedling]; McDaniel, Earl; McDaniel's Min. Roses

Classic Love *see* 'LYOCL'

Classic Orange, Min, ob, 1997

Classic Sunblaze® *see* 'MEIPINJID'

'CLASSIC TOUCH', HT, lp, 1991; flowers large blooms borne mostly singly, dbl., 26–40 petals, slight fragrance; foliage large, medium green, semi-glossy; tall, upright growth; [Touch of Class sport]; Hefner, John; Co-Operative Rose Growers, 1993

Classical Velvet *see* 'JOHILLSTAR'

'CLASSIE LASSIE'™, Gr, pb, 1990; bud pointed; flowers ivory pink, with salmon pink edges, aging to salmon, oval-shaped, semi-dbl., 25–30 petals, exhibition form, moderate, fruity fragrance; prickles average, brown green, slight recurve; foliage medium, medium green, glossy, disease resistant; bushy, medium growth; [Touch of Class X Seedling]; Winchel, Joseph F.; Co-Operative Rose Growers, 1991;, Specialty Roses, 1991

Classy *see* 'TRACLA'

Classy™ *see* HILrap

'CLAUDE', HT, mr, 1950; flowers bright orient red, dbl., 35 petals, 6–7 in., moderate fragrance; foliage glossy, dark; vigorous, upright growth; [(Comtesse Vandal X Brazier) X Seedling]; Mallerin, C.; EFR;, Wheatcroft Bros.

'CLAUDE JACQUET', HP, dr, 1892; flowers dark red/purple, very large, dbl., moderate fragrance; Liabaud; (Sangerhausen)

Claude Monet® *see* JACdesa

'CLAUDE PETIT', HT, mp, 1936; flowers soft salmon-pink, stamens yellow, very dbl., exhibition form, intense fragrance; vigorous growth; [Mlle Marie Mascuraud X Beauté de Lyon]; Buatois

'CLAUDE RABBE', LCl, dp, 1941; flowers carmine-pink, medium, borne in clusters of 6-10; Buatois

'CLAUDIA', F, pb, 1959; flowers deep cherry-coral, dbl.; vigorous growth; [Ma Perkins X Geranium Red]; Broadley; Roseglen Nursery

Claudia *see* MElyacom

'CLAUDIA AUGUSTA', N, pb; Damaizin, 1858

Claudia Cardinale *see* MAScatna

'CLAUDY CHAPEL', HT, yb, 1930; flowers deep yellow and coppery, to salmon-pink, large, dbl., moderate fragrance; vigorous growth; Beaumez; Delhaye

Claus Groth *see* **'KLAUS GROTH'**

Claysnow *see* **'SNOW BALLET'**®

CLEalta, Cl HT, ob, 1998; (**Christine**); Clements, John K.

CLEamour, MinFl, rb, 1992; (**Glamour Girl**); Clements, John K.

'CLEBEAU', Min, yb, 1992; (**Golden Beauty**, Golden Girls (CLEgirl)); flowers gold, edged copper, blooms borne mostly singly, dbl., 26–40 petals, 1.5 in., exhibition form, slight fragrance; few prickles; foliage small, dark green, glossy; medium (30 cms), bushy, spreading growth; [Seedling X Seedling]; Clements, John K.; Heirloom Old Garden Roses, 1990

CLEblack, Min, dr, 1996; (**Black Gold**); Clements, John K.

CLEbou, S, pb, 1995; (**Boudoir**); Clements, John K.

CLEbravo, S, dr, 1998; (**Braveheart**); Clements, John K.

CLEbridge, HMsk, ab, 1996; (**Bridget**); Clements, John K.

'CLEBRITE', Min, rb, 1991; (**Britestripe**, Britestripes); flowers soft pink, striped bright red, blooms borne in small clusters, dbl., 26–40 petals, 1.5 in., no fragrance; some prickles; foliage small, medium green, semi-glossy; medium (30 cms), bushy, compact growth; [Pinstripe X Seedling]; Clements, John K.; Heirloom Old Garden Roses, 1990

CLEcham, S, mp, 1995; (**Pink Champagne**); Clements, John K.

CLEcherry, HWich, mr, 1999; (**Cherries Jubilee**); Clements, John K.

CLEconcert, S, lp, 1999; (**Regina Louise**); Clements, John K.

CLEdan, HMsk, pb, 1995; (**Dancing in the Wind**); Clements, John K.

CLEdare, MinFl, lp; (**Virginia Dare**); Clements, John K., 1998

'CLEDRAG', HCh, dr, 1992; (**Dragon's Eye**, Eye of the Dragon, The Dragon's Eye); very dbl., 1.5–2.75 in., exhibition form, borne mostly singly, moderate fragrance; some prickles; foliage small, dark green, semi-glossy; medium (70 cms), bushy, compact growth; [Seedling X Seedling]; Clements, John K.; Heirloom Old Garden Roses, 1991

'CLEFIRE', Min, or, 1991; (**Firestorm**); flowers fiery oriental lacquer orange-red, blooms borne mostly singly, very dbl., 1.5 in., exhibition form, no fragrance; few prickles; foliage small, medium green, semi-glossy; medium (35 cms), bushy, spreading growth; [Seedling X Seedling]; Clements, John K.; Heirloom Old Garden Roses

'CLEGRAN', S, lp, 1994; (**Grandma's Lace**); flowers pale pink, blooms borne in clusters, dbl., 26–40 petals, 1.5–2.75 in., moderate fragrance; foliage medium, medium green, semi-glossy; medium, spreading growth; [Sexy Rexy X Trier]; Clements, John K.; Heirloom Old Garden Roses

CLEheat, S, ly, 1997; (**Good Old Summertime**); Clements, John K.

CLEhelp, S, mr, 1997; (**Helping Hands**); Clements, John K.

'CLEHON', Min, w, 1991; (**Honey Mini-Delite**); flowers honey cream, blooms borne in small clusters, dbl., 26–40 petals, 1.5 in., no fragrance; few prickles; foliage small, medium green, matt; low (25 cms), bushy, compact growth; [Seedling X Seedling]; Clements, John K.; Heirloom Old Garden Roses

CLEhonor, S, dr, 1999; (**Ora Kingsley**); Clements, John K.

'CLEICE', Min, w, 1991; (**Ice Crystal**); flowers crystal white, blooms borne in small clusters, very dbl., 1.5 in., exhibition form, no fragrance; some prickles; foliage small, medium green, semi-glossy; micro-mini; low (20 cms), bushy, compact growth; [Seedling X Baby Betsy McCall]; Clements, John K.; Heirloom Old Garden Roses, 1989

CLEjoan, S, w, 1996; (**Joan Fontaine**(TM)); Clements, John K.

CLEjoy, S, mp, 1997; (**Memories**); Clements, John K.

CLEkate, S, mp, 1996; (**Kateryna**); Clements, John K.

'CLEKISS', Min, rb, 1991; (**Strawberry Kiss**); flowers white edged red, blooms borne mostly singly, dbl., 15–25 petals, 1.5 in., exhibition form, no fragrance; some prickles; foliage small, light green, semi-glossy; medium (35 cms), upright, bushy growth; [Seedling X Seedling]; Clements, John K.; Heirloom Old Garden Roses, 1990

'CLELADY', Min, pb, 1991; (**Fancy Lady**); flowers medium pink, white reverse, high pointed, blooms borne mostly singly, very dbl., 1.5 in., exhibition form, no fragrance; few prickles; foliage small, dark green, glossy; tall (50 cms), upright growth; [Seedling X Seedling]; Clements, John K.; Heirloom Old Garden Roses, 1990

CLEland, HWich, w, 1996; (**Landscape Splendor**); rambler growth; Clements, John K.

'CLELIGHT', Min, rb, 1991; (**Little Lighthouse**); flowers bright red with yellow to white eye, blooms borne in small clusters, single, 5 petals, 1.5 in., exhibition form, no fragrance; some prickles; foliage small, medium green, semi-glossy; low (18 cms), bushy, spreading, compact, growth; groundcover; [Robin Red Breast X Little Artist]; Clements, John K.; Heirloom Old Garden Roses, 1992

'CLELIPS', Min, rb, 1991; (**Lipstick 'n' Lace**); flowers cream, shaded lipstick red, blooms borne in small clusters, dbl., 26–40 petals, 1.5–2.75 in., exhibition form, moderate fragrance; no prickles; foliage medium, medium green, matt; medium (30 cms), upright, compact growth; [Seedling X Seedling]; Clements, John K.; Heirloom Old Garden Roses, 1992

'CLELOCK', Min, my, 1991; (**Turlock High**); flowers bright yellow, blooms borne mostly singly, dbl., 26–40 petals, 1.5 in., intense fragrance; some prickles; foliage small, dark green, semi-glossy; medium (40 cms), upright, bushy growth; [Rise 'n' Shine X Seedling]; Clements, John K.; Heirloom Old Garden Roses, 1992

CLElou, S, ob, 1996; (**Louise Clements**); Clements, John K.

'CLEMATIS', HWich, rb, 1924; flowers dark red, prominent white eye, small, borne in clusters, single; vigorous, climbing growth; Turbat

Clémence Beaugrand, mp; flowers medium, dbl.; Roseraie de l'Hay; (Sangerhausen)

'CLÉMENCE RAOUX', HP, mp; Granger, 1869

'CLEMENCE ROBERT', M, mp, 1863; flowers heavily mossed, sometimes recurrent bloom; Robert et Moreau

CLEment, LCl, dr, 1995; (**Commitment**); Clements, John K.

'CLÉMENT PACAUD', HT, dp, 1916; flowers brilliant carmine; Chambard, C.

'CLEMENTINA', HT, pb, 1961; flowers carmine and rosy white bicolor; [Grand Gala X Vicky Marfá]; Dot, Pedro

'CLEMENTINA CARBONIERI', T, yb, 1913; Bonfiglio, A.

'CLEMENTINE', HEg; flowers rosy blush

Clementine *see* **'JANET'S PRIDE'**

Clementine *see* TANogrew

'CLEMENTINE DUVAL', B, mp, 1847; flowers bright rose, cupped; Laffay, M.

'CLÉMENTINE SÉRINGE', HP, mp, 1840; flowers rose-pink, dbl.; Wood

CLEmist, MinFl, lp; (**Will-o'-the-Wisp**); Clements, John K., 1998

Cleo *see* 'BEEBOP'

'CLEOPATRA' *see* KORverpea

'CLEORA', Min, mp, 1977; bud globular; flowers medium pink, reverse darker, small, dbl., 50 petals, 1.5 in., flat; foliage small, light, leathery; vigorous growth; [Fairy Moss X Fairy Moss]; Dobbs; Port Stockton Nursery

'CLEPEACH', Min, ab, 1991; (**Peach Silks**); flowers rich peach, blooms borne in small clusters, dbl., 15–25 petals, 1.5 in., no fragrance; few prickles; foliage small, dark green, glossy; tall (50 cms), upright growth; [Seedling X Seedling]; Clements, John K.; Heirloom Old Garden Roses

'CLEPERF', Min, ab, 1992; (**Apricot Perfection**); flowers soft apricot, moderately, dbl., 15–25 petals, 1.5 in., exhibition form, borne mostly singly, slight fragrance; some prickles; foliage small, medium green, matt; tall (50 cms), upright, bushy growth; [My Louisa X Seedling]; Clements, John K.; Heirloom Old Garden Roses, 1990

'CLEPURP', Min, m, 1991; (**Color Purple**); flowers rich deep purple, small, blooms borne mostly singly, very dbl., 1.5 in., moderate fragrance; few prickles; foliage medium, dark green, holly-like, glossy; low (25 cms), spreading, compact growth; [Angel Face X Seedling]; Clements, John K.; Heirloom Old Garden Roses, 1990

'CLERAIN', Min, yb, 1991; (**Oregon Rainbow**); bud high-pointed, exhibition; flowers golden yellow, shaded and edged red, blooms, dbl., 26–40 petals, 1.5 in., no fragrance; foliage small, dark green, semi-glossy; medium (40 cms), bushy growth; [Seedling X Seedling]; Clements, John K.; Heirloom Old Garden Roses

'CLERUFF', S, rb, 1994; (**Ruffles 'n' Flourishes**); flowers violet red and gold, blooms borne in small clusters, dbl., 15–25 petals, 1.5–2.75 in.; few prickles; foliage medium, reddish green, glossy; low (18 in), upright, bushy growth; [Sexy Rexy X Whistle Stop]; Clements, John K.; Heirloom Old Garden Roses, 1994

'CLESCRUB', Min, or, 1991; (Patriot Flame, **Scudbuster**); flowers intense rocket flame orange, blooms borne mostly singly, very dbl., 1.5 in., exhibition form, slight fragrance; some prickles; foliage small, dark green, semi-glossy; medium (35 cms), upright growth; [Seedling X Seedling]; Clements, John K.; Heirloom Old Garden Roses, 1992

'CLESHIR', Min, mp, 1991; (**Sweet Shirley**); flowers medium pink, lighter reverse, blooms borne mostly singly, dbl., 15–25 petals, 1.5 in., exhibition form, slight fragrance; some prickles; foliage small, medium green, semi-glossy; bushy (40 cms), spreading growth; [Tweedle Dee X Seedling]; Clements, John K.; Heirloom Old Garden Roses

'CLESNOB', Min, rb, 1991; ('CLESNOW', **Snowblush**); flowers purest white, edges red, blooms borne mostly singly, very dbl., 1.5–2.75 in., exhibition form, no fragrance; some prickles; foliage medium green, matt; tall (50 cms), upright growth; [Minuette X Seedling]; Clements, John K.; Heirloom Old Garden Roses, 1990

'CLESNOW', Min, rb, 1991; ('CLESNOB', **Snowblush**); flowers purest white, edges red, blooms borne mostly singly, very dbl., 1.5–2.75 in., exhibition form, no fragrance; some prickles; foliage medium green, matt; tall (50 cms), upright growth; [Minuette X Seedling]; Clements, John K.; Heirloom Old Garden Roses, 1990

CLEspirit, S, dp, 1999; (**Laura Clements**); Clements, John K.

CLEswan, S, ob, 1999; (**Brenda Burg**); Clements, John K.

'CLETAPE', Min, rb, 1991; (**Antique Tapestry**); flowers burgundy and gold, dbl., 15–25 petals, 1.5–2.75 in., exhibition form, borne mostly singly, slight fragrance; few prickles; foliage medium, dark green, semi-glossy; medium (40 cms), upright growth; [Redgold X Seedling]; Clements, John K.; Heirloom Old Garden Roses, 1990

'CLETESS', Min, lp, 1991; ('CLETEST', **Tess**); flowers flesh pink, blooms borne mostly singly, semi-dbl., 6–14 petals, 1.5 in., exhibition form, no fragrance; few prickles; foliage small, medium green, semi-glossy; medium (30 cms), bushy growth; [Cupcake X Seedling]; Clements, John K.; Heirloom Old Garden Roses, 1992

'CLETIG', Min, yb, 1992; (**Tiger Stripes**); flowers unusual striped color combination, yellow, striped orange, b, dbl., 26–40 petals, 1.5 in., no fragrance; few prickles; foliage small, medium green, semi-glossy; low (25 cms), bushy, compact growth; [Seedling X Seedling]; Clements, John K.; Heirloom Old Garden Roses, 1991

CLEtraveler, S, mp, 1999; (**Charles Kuralt**); Clements, John K.

'CLETWIN', Min, pb, 1991; (**Twin Pinks**); flowers two-tone pink, blooms borne in small clusters, dbl., 15–25 petals, 1.5 in., exhibition form, slight fragrance; some prickles; foliage small, medium green, semi-glossy; low (30 cms), bushy, compact growth; [Seedling X Seedling]; Clements, John K.; Heirloom Old Garden Roses

'CLEVELAND', HT, pb, 1916; (H.P. Pinkerton); flowers reddish copper on old-rose, base coppery yellow, dbl., moderate fragrance; Dickson, H.

'CLEVELAND BOUQUET', HT, op, 1940; flowers shrimp-pink with salmon-pink undertone, open, semi-dbl.; Horvath

Cleveland II *see* **'MRS DUNLOP BEST'**

CLEwedding, S, my, 1996; (**Morning Has Broken**); Clements, John K.

CLEwonder, S, lp; (**Wonderstripe**); Clements, John K., 1997

Cliff Richard, F, pb, 1993; Van Geest

Cliffs of Dover *see* 'POULEMB'

Clifton Moss *see* **'SHAILER'S WHITE MOSS'**

Climbing Bride *see* **'RUTH VESTAL'**

Climbing Geheimrat Duisberg *see* **'GOLDEN RAPTURE, CLIMBING'**

Climbing Rainbow's End™ *see* 'SAVACLEND'

'CLIMBING ROSES'

Climbing Spek's Yellow *see* **'GOLDEN SCEPTER, CLIMBING'**

'CLIMENTINA', HT, mp, 1955; flowers rosy pink, large, slight fragrance; [Independence X Peace]; Klimenko, V. N.

'CLINORA', Cl HT, dy, 1978; (**Landora, Climbing**); bud pointed; flowers well shaped, 3.5 in., intense fragrance; [Sunblest sport]; Orard, Joseph; Pekmez

'CLIO', HP, lp, 1894; flowers flesh, large blooms in clusters, very dbl., globular, seasonal bloom, moderate fragrance; foliage rich green; vigorous growth; Paul, W.

Clio, HT, or, 1972; Gaujard

Clio®, S, mr

Clive Lloyd, HT, dp

Clivia® *see* 'KORTAG'

Clochermerle® *see* 'DELPÉTRI'

Clo-Clo *see* 'PEKATAN'

Cloris Adriana, HT, dp, 1997; flowers very full, high centered, large, very dbl., 41 petals, borne mostly singly, intense fragrance; some prickles; foliage large, dark green, dull; spreading, bushy, medium growth; [Blue Moon X Gavotte]; Poole, Lionel

Clos de la Pellerie *see* 'DELCLOPEL'

Clos Fleuri, S, 1970; Delbard-Chabert; (Cavriglia)

Clos Fleuri Bicolore, F, 1992; Delbard, Georges; (Cavriglia)

Clos Fleuri Blanc® *see* 'DELBLAN'

Clos Fleuri Champagne, F, w, 1990; Delbard

Clos Fleuri d'Or, F, dy, 1990; Delbard

Clos Fleuri Jaune® *see* 'DELJAUNE'

Clos Fleuri Rose® *see* 'DELPOMP'

Clos Fleuri Rose No. 2 *see* DELodive

Clos Fleuri Rouge® *see* 'DELECLA'

Clos Vougeot® *see* 'DELIFIC'

'CLOTARIA', HT, mr, 1936; (Red Gruss an Coburg); flowers bright fuchsia-red, well formed, medium, dbl., 26–28 petals, intense fragrance; foliage dark, glossy; vigorous, upright, bushy growth; [Gruss an Coburg X J.C. Thornton]; San Remo Exp. Sta.

Cloth of Gold *see* **'CHROMATELLA'**

'CLOTILDE SOUPERT', Pol, w, 1890; flowers pearly white, center soft rose-pink, large blooms in clusters, very dbl., moderate fragrance; foliage rich green, soft; bushy (10-20 in) growth; [Said to be Mignonette X Mme Damaizin]; Soupert & Notting

'CLOTILDE SOUPERT, CLIMBING', Cl Pol, w, 1902; Dingee & Conard

'CLOTILDE SOUPERT, CLIMBING', Cl Pol, w, 1896; Berckmans, P.J., Co.

Cloud Dancer™ *see* 'BURDANCER'

Cloud Nine *see* 'JACLITE'

Cloud Nine, S, w, 1999; flowers white touched with apricot; Williams, J. Benjamin

'CLOUDLAND', HT, m, 1994; dbl., 26–40 petals, 3–3.5 in., moderate fragrance; foliage large, medium green, matt; upright (150 cms) growth; [Blue Moon X Paradise]; Perry, Astor; Perry Roses, 1995

Clouseau *see* 'TALCLOUSEAU'

Clove Scented Musk, Misc OGR, w

'CLOVELLY', HT, dp, 1924; flowers carmine-pink, well formed, moderate fragrance; vigorous growth; GM, NRS, 1924; Hicks

Cloverdene *see* HERclov

'CLUB', F, mr, 1957; flowers bright red, open, medium, single, slight fragrance; foliage glossy, bronze; very vigorous, bushy growth; [Peace X Opera seedling]; Gaujard

Clubrose Lydia *see* **'LYDIA'**

'CLUBROSE SCALA', F, rb, 1973; (Scala); flowers blood-red and orange, medium, dbl., globular, slight fragrance; foliage glossy, dark, leathery, bronze; vigorous, upright, bushy growth; [Marlena X Seedling]; Kordes

Clydebank Centenary *see* 'COCDAZZLE'

'CLYTEMNESTRA', HMsk, op, 1915; bud copper; flowers salmon-chamois, ruffled, small blooms in clusters, recurrent bloom, moderate fragrance; foliage leathery, dark; bushy (3-4 ft.), spreading growth; GM, NRS, 1914; [Trier X Liberty]; Pemberton

'CNOS X', HT, my, 1957; flowers golden yellow tinged copper, moderate fragrance; vigorous growth; [Bettina seedling]; Bronisze (Poland) State Nursery

'COALITE FLAME', HT, or, 1974; very dbl., 60 petals, 5 in., moderate fragrance; foliage large, matt; [Fragrant Cloud X Red Planet]; Dickson, A.

'COBY FANKHAUSER', HT, my, 1971; flowers buttercup-yellow, large, dbl., exhibition form, moderate fragrance; foliage leathery; vigorous, upright, bushy growth; [John S. Bloomfield X Elizabeth Fankhauser]; Fankhauser

'COCABEL', Min, dp, 1980; (**Little Jewel**); flowers deep pink, patio, blooms borne 6-12 per cluster, dbl., 34 petals, slight fragrance; straight prickles; foliage small, glossy, dark; low, compact growth; [Wee Man X Belinda]; Cocker, James

'COCABEST', Min, mr, 1980; (**Wee Jock**®); bud pointed; flowers patio, blooms borne 9-15 per cluster, dbl., 50 petals, slight fragrance; red prickles; foliage small, fairly glossy, medium green; low, compact growth; [National Trust X Wee Man]; Cocker, James

'COCACERT', F, ab, 1982; (**Vital Spark**); flowers gold, flushed coral, medium, dbl., 35 petals, slight fragrance; foliage medium, medium green, semi-glossy; bushy growth; [(Anne Cocker X (Sabine X Circus)) X Yellow Pages]; Cocker, James; Cocker & Sons

'COCADILLY', HT, pb, 1985; (**The Coxswain**); flowers cream and pink blend, large, dbl., 35 petals, intense fragrance; foliage medium, medium green, semi-glossy; bushy growth; [(Tropicana X Ballet) X Silver Jubilee]; Cocker, Alexander M.; Cocker & Sons

'COCAGOLD', HT, my, 1981; (**Golden Jubilee**); dbl., 29 petals, exhibition form, moderate, tea fragrance; narrow, red-brownprickles; foliage large, matt green, glossy; [Peer Gynt X Gay Gordons]; Cocker, James

'COCAMOND', Min, op, 1981; (**Dainty Dinah**); flowers medium salmon-pink, patio, small, semi-dbl., slight fragrance; foliage small, medium green, semi-glossy; bushy growth; [Anne Cocker X Wee Man]; Cocker, James; Cocker & Sons

'COCANEILA', Min, m, 1982; ('COCANELIA', **Pink Posy**); flowers lilac, small, dbl., moderate fragrance; foliage dark, matt; bushy growth; [Trier X New Penny]; Cocker, James; Cocker & Sons

'COCANELIA', Min, m, 1982; ('COCANEILA', **Pink Posy**); flowers lilac, small, dbl., moderate fragrance; foliage dark, matt; bushy growth; [Trier X New Penny]; Cocker, James; Cocker & Sons

'COCAPEER', HT, mp, 1980; (**Sweetheart**); bud ovoid; flowers medium pink, yellow base, blooms borne singly, dbl., 52 petals, exhibition form, moderate fragrance; foliage large, medium green; vigorous, upright growth; Fragrance Award, Belfast, 1982; [Peer Gynt X (Fragrant Cloud X Gay Gordons)]; Cocker, James

COCaquil, F, dr, 1984; (**Cockadoo**); Cocker

'COCARDE JAUNE', HT, pb, 1933; bud large, long, pointed, reddish salmon; flowers yellowish salmon, base yellow, reverse coral-red, dbl., cupped, moderate fragrance; foliage dark; vigorous, bushy growth; [Diana X Marie Adélaide]; Ketten Bros.

'COCARDE PALE', HGal, 1820; Pradel, H.; (Cavriglia)

'COCARED', HT, ab, 1982; (**Roddy MacMillan**); flowers large, dbl., 35 petals, moderate fragrance; foliage medium, medium green, semi-glossy; bushy growth; [(Fragrant Cloud X Postillion) X Wisbech Gold]; Cocker, James; Cocker & Sons

'COCARUM', Min, ob, 1982; (**Cha Cha**®); flowers orange-red with yellow eye, dbl., slight fragrance; foliage small, medium green, semi-glossy; bushy growth; [(Wee Man X Manx Queen) X Darling Flame]; Cocker, James; Cocker & Sons, 1983

'COCASUN', HT, ab, 1981; (**Sunset Song**); bud pointed; flowers golden amber, borne 1-5 per cluster, dbl., 46 petals, exhibition form, slight fragrance; beak-shaped, red-brown prickles; foliage large, glossy, light olive green; upright growth; [(Sabine X Circus) X Sunblest]; Cocker, James

'COCATHES', F, mp, 1980; (**Crathes Castle**); 18 petals, borne 12-15 per cluster, moderate fragrance; triangular prickles; foliage large, dark, glossy; rounded, bushy growth; [Dreamland X Topsi]; Cocker, James

'COCAVOTER', F, ob, 1983; (**Sweet Nell**); flowers orange, large, dbl., 35 petals, slight fragrance; foliage medium, medium green, semi-glossy; upright growth; [Anne Cocker X ((Mischief X ((Sabine X Circus) X (Tropicana X Circus)))]; Cocker, Alexander M.; Cocker & Sons, 1984

'COCBADEN', HT, op, 1985; (**Doctor Dick**, Dr Dick); flowers orange-coral, large, dbl., exhibition form, slight fragrance; foliage large, medium green, matt; upright growth; [Fragrant Cloud X Corso]; Cocker, James & Sons, 1986

'COCBAMBER', HT, pb, 1985; (Amberlight, **Fyvie Castle**); flowers light apricot, amber and pink blend, well-formed, large, dbl., 35 petals, moderate fragrance; foliage large, medium green, semi-glossy; upright growth; Gold Star of the South Pacific, Palmerston North, NZ, 1985; [(Sunblest X (Sabine X Dr. A.J. Verhage)) X Silver Jubilee]; Cocker, Alexander M.; Cocker & Sons

'COCBAY', HT, mp, 1982; (**May Lyon**); flowers large, dbl., 35 petals, slight fragrance; foliage large, medium green, glossy; bushy growth; [((Anne Cocker X Arthur Bell) X National Trust) X Silver Jubilee]; Cocker, James; Cocker & Sons, 1983

'COCBLANCO', Min, w, 1983; (**Bianco**); flowers patio, small, dbl., 35 petals, no fragrance; foliage medium, medium green, semi-glossy; bushy growth; [Darling Flame X Jack Frost]; Cocker, James; Cocker & Sons

'COCBONNE', HT, or, 1989; (**Ena Baxter**); bud pointed; flowers salmon pink, reverse salmon red, medium, borne in sprays of 5, dbl., 26 petals, exhibition form, slight fragrance; urn-shaped, large, brown fruit; prickles triangular, average, green; foliage large, medium green, glossy; bushy, medium growth; GM, Glasgow, 1994; [HARkrispin X Silver Jubilee]; Cocker, James & Sons

'COCBROSE', HT, dp, 1983; (**Abbeyfield Rose**); flowers rose red, large, dbl., 35 petals, slight fragrance; foliage medium, medium green, semi-glossy; bushy growth; Golden Prize, Glasgow, 1990; [National Trust X Silver Jubilee]; Cocker, A.; Cocker & Sons, 1985

'COCCAGES', HT, ob, 1988; (Country Heritage, **Our Jubilee**); flowers medium, dbl., 20 petals, slight fragrance; foliage medium, medium green, semi-glossy; upright growth; [Yellow Pages X Silver Jubilee]; Cocker, James & Sons, 1986

'COCCELESTE', F, mp, 1988; (**Rosabell**); flowers medium, dbl., slight fragrance; foliage medium, medium green, semi-glossy; patio; bushy growth; [(National Trust X Wee Man) X Darling Flame]; Cocker, James & Sons, 1986

'COCCINELLE', LCl, or, 1956; flowers bright geranium-red; Buyl Frères

'COCCLARE', S, my, 1989; (**Ray of Sunshine**); bud pointed; flowers small, borne in sprays of 3-9, semi-dbl., 15 petals, cupped, slight, spicy fragrance; round, small, green fruit; prickles small, green; foliage small, dark green, glossy; bushy, low growth; [Sunsprite X (Clare Grammerstorf X Fruhlingsmorgen)]; Cocker, James & Sons, 1988

'COCCLARION', HT, op, 1992; (**Laura Anne**); flowers pink flushed orange, blooms borne in large clusters, dbl., 26–40 petals, 3–3.5 in., moderate fragrance; some prickles; foliage large, medium green, glossy; bushy (76 cms) growth; [((Sabine X Circus) X Maxi) X Harrinyl]; Cocker, James; James Cocker & Sons, 1990

COCcopna, F, op, 1982; (**Coppa Nob**); Cocker

'COCCORD', Min, or, 1982; (**Little Prince**); flowers orange-red, yellow eye, patio, small, semi-dbl., slight fragrance; foliage medium, medium green, semi-glossy; upright, bushy growth; [Darling Flame X (National Trust X Wee Man)]; Cocker, James; Cocker & Sons, 1983

'COCCRAZY', F, r, 1988; (**Gingernut**); flowers medium, dbl., 43 petals, moderate fragrance; foliage small, medium green, semi-glossy; patio; bushy growth; [(Sabine X Circus) X Darling Flame]; Cocker, James & Sons, 1989

'COCDANA', HT, yb, 1988; (**Fulton MacKay**, Maribel); flowers large, dbl., 20 petals, moderate fragrance; foliage large, medium green, glossy; bushy growth; Golden Prize, Glasgow, 1992; [Silver Jubilee X Jana]; Cocker, James, 1989

'COCDANDY', HT, ob, 1988; (**Royal Volunteer**); flowers orange-red and yellow blended, medium, dbl., 26–40 petals, slight fragrance; foliage medium, light green, semi-glossy; upright growth; [Yellow Pages X Alexander]; Cocker, James & Sons

'COCDAPPLE', Min, lp, 1985; (**Scottish Special**); flowers patio, large, semi-dbl., slight fragrance; foliage small, medium green, semi-glossy; bushy growth; [Wee Man X Darling Flame]; Cocker, Ann G., 1987

'COCDARLEE', Min, op, 1986; (**Coral Reef**); flowers orange, patio, medium, semi-dbl., slight fragrance; foliage small, medium green, glossy; bushy growth; [(Darling Flame X St. Albans) X Silver Jubilee]; Cocker, Ann G.; Cocker & Sons

'COCDAZZLE', F, or, 1988; (**Clydebank Centenary**); flowers orange-vermilion red, medium, dbl., 15–25 fimbriated petals, slight fragrance; foliage medium, medium green, matt; upright growth; [((Highlight X Colour Wonder) X (Parkdirector Riggers X Piccadilly)) X Darling Flame]; Cocker, James & Sons, 1987

'COCDESTIN', HT, ob, 1984; (**Remember Me**®); flowers orange and yellow blend, large, dbl., 20 petals, slight fragrance; foliage small, dark, glossy; bushy, spreading growth; [Ann Letts X (Dainty Maid X Pink Favorite)]; Cocker

'COCDIMITY', Min, ab, 1992; (**Claire Scotland**); flowers light apricot pink, blooms borne in small clusters, dbl., 15–25 petals, 1.5–2.75 in., slight fragrance; some prickles; foliage medium, medium green, semi-glossy; patio; low (50-80 cms), bushy growth; [(National Trust X Wee Man) X Darling Flame]; Cocker, James; James Cocker & Sons, 1990

'COCDIMPLE', Min, pb, 1986; (**Conservation**); flowers patio, medium, semi-dbl., slight fragrance; foliage small, light green, semi-glossy; bushy growth; GM, Dublin, 1986; [((Sabine X Circus) X Maxi) X Darling Flame]; Cocker, James & Sons, 1988

'COCDINKUM', Min, mr, 1983; (**Boys' Brigade**®); flowers medium, patio, blooms in clusters, single, 5 petals, no fragrance; foliage small, medium green, semi-glossy; bushy growth; [(Darling Flame X Saint Alban) X (Little Flirt X Marlena)]; Cocker, Ann G.; Cocker & Sons, 1984

COCember, HT, or, 1991; (**Roxburghe Rose**); Cocker

'COCFLAG', Min, mr, 1989; (**Highland Laddie**); bud pointed; flowers scarlet red, medium, borne in sprays of 5-11, dbl., 19 petals, cupped, slight fragrance; urn-shaped, medium, brown fruit; prickles medium, green; foliage medium, medium green, glossy; upright growth; [National Trust X Dainty Dinah]; Cocker, James & Sons

COCfoster, S, mr, 1988; (**Regal Red**); Cocker

'COCGLEN', F, yb, 1989; (**Honey Bunch**®, Honeybunch); bud ovoid; flowers yellow with salmon-red, reverseyellow, aging honey-yellow, dbl., 45 petals, cupped, moderate fragrance; round, small, green fruit; prickles small, green; foliage small, dark green, glossy; patio; bushy, low growth; [[(Sabine X Circus) X Maxi] X Bright Smile]; Cocker, James & Sons

'COCGOLD', F, dy, 1992; (Dania, **Toprose**); flowers bright yellow, blooms borne in large clusters, dbl., 26–40 petals, 3–3.5 in., slight fragrance; some prickles; foliage large, medium green, glossy; medium (75 cms), upright growth; GM, Baden-Baden, 1987; [((Chinatown X Golden Masterpiece) X Adolph Horstmann] X YellowPages]; Cocker, James; James Cocker & Sons, 1988

'COCGRAND', HT, lp, 1992; (**Myriam**®); very dbl., 3–3.5 in., borne mostly singly, intense fragrance; some prickles; foliage large, medium green, semi-glossy; medium (75 cms), upright growth; [Typhoo Tea X Grandpa Dickson]; Cocker, James; James Cocker & Sons, 1991

'COCHAROD', F, mr, 1993; (**Shirley Spain**); flowers glowing russet red, blooms borne in small clusters, dbl., 15–25 petals, 1.5–2.75 in., slight fragrance; some prickles; foliage medium, dark green, glossy; medium, compact growth; [Seedling X Roddy McMillan]; Cocker; James Cocker & Sons, 1992

'COCHELLO', Min, mr, 1992; (**Hello**); flowers crimson with white eye, blooms borne in large clusters, semi-dbl., 6–14 petals, 1.5–2.75 in., slight fragrance; some prickles; foliage medium, medium green, semi-glossy; patio; medium (50-80 cms), bushy growth; Golden Prize, Glasgow, 1993; [Darling Flame X Seedling]; Cocker, James; James Cocker & Sons, 1991

'COCHINEAL GLORY', HT, mr, 1937; bud pointed; flowers open, large, semi-dbl.; vigorous, bushy growth; Leenders, M.

'COCHINEAL GLORY, CLIMBING', Cl HT, mr, 1945; flowers large, semi-dbl.; Vogel, M.; (Sangerhausen)

'COCHUNTER', HT, or, 1992; ('COCHUSTER', **Ohshima Rose**); dbl., 26–40 petals, 1.5–2.75 in., borne mostly singly, moderate fragrance; some prickles; foliage medium, medium green, matt; tall (91.44 cms), upright growth; [(National Trust X Alexander) X Red Planet]; Cocker, James; James Cocker & Sons, 1991

'COCHUSTER', HT, or, 1992; ('COCHUNTER', **Ohshima Rose**); dbl., 26–40 petals, 1.5–2.75 in., borne mostly singly, moderate fragrance; some prickles; foliage medium, medium green, matt; tall (91.44 cms), upright growth; [(National Trust X Alexander) X Red Planet]; Cocker, James; James Cocker & Sons, 1991

'COCJABBY', F, op, 1992; (**Gordon's College**); flowers coral salmon, blooms borne in large clusters, dbl., 26–40 petals, 3–3.5 in., moderate fragrance; some prickles; foliage large, dark green, glossy, purplish when young; medium, upright growth; [Abbeyfield Rose X Roddy McMillan]; James Cocker & Sons, Ltd.

COCjojo, F, ob; (**UNICEF**); Cocker, 1993

'COCJOLLY', F, dr, 1993; (**Bishop Elphinstone**); flowers crimson, blooms borne in large clusters, dbl., 26–40 petals, 1.5–2.75 in., slight fragrance; some prickles; foliage large, dark green, semi-glossy; tall (3 ft), upright growth; [Seedling X (Tropicana X Baccara)]; Cocker, James; James Cocker & Sons, 1994

Cockadoo *see* COCaquil

Cockle Shells *see* 'LEOCOK'

Cocktail® *see* 'MEIMICK'

Cocktail '80 *see* 'MEITABIFOB'

'COCLAGER', F, ab, 1997; (**William Quarrier**); flowers full, medium, very dbl., 26–40 petals, borne in large clusters, slight fragrance; some prickles; foliage medium, light green, dull; compact, tall (2.5-3ft.) growth; [Silver Jubilee X Geraldine]; Cocker, Ann G.

'COCLAMENT', F, mr, 1998; (**Hope 98**); flowers sparkling geranium red, lighter reverse, frilly petals, dbl., 15–25 petals, 1–1.5 in., borne in large clusters, slight fragrance; prickles moderate, small, slightly hooked; foliage small, medium, glossy; compact, low, bushy growth; [Memento X Evelyn Fison]; Cocker, Ann G.; James Cocker & Sons, 1998

'COCLANDS', HT, lp, 1992; (**Scotland's Trust**); flowers light pink, silver reverse, moderately well shaped lg. blooms, dbl., 15–25 petals, 1.5–2.75 in., intense fragrance; some prickles; foliage medium, medium green, matt; medium (16-20 cms), upright growth; [Sunblest X Prima Ballerina]; Cocker, James & Sons

'COCLENT', F, ab, 1993; (**Lady MacRobert**); flowers light apricot, blooms borne in large clusters, dbl., 15–25 petals, 1.5–2.75 in., slight fragrance; some prickles; foliage medium, light green, semi-glossy; medium (2 1/2 ft), upright, compact growth; [Clydebank Centenary X Seedling]; Cocker; James Cocker & Sons, 1993

'COCLIBEE', F, op, 1996; (**Alison**); flowers peach salmon, 3–3.5 in., borne in large clusters, moderate fragrance; some prickles; foliage medium, medium green, glossy; medium, upright, bushy growth; [Silver Jubilee X (Sabine

X Circus)]; Cocker, James & Sons; James Cocker & Sons, 1995

COClion, HT, op, 1997; (**Press and Journal**); Cocker

'COCLUCID', F, ob, 1992; (**Castle of Mey**); flowers orange gold, blooms borne in small clusters, dbl., 15–25 petals, 1.5–2.75 in., moderate fragrance; some prickles; foliage medium, dark green, semi-glossy; medium, bushy growth; [Anne Cocker X (Yellow Pages X Silver Jubilee)]; Cocker, James; James Cocker & Sons

'COCMARRIS', F, or, 1996; (**Wee Cracker**); flowers orange vermilion, blooms borne in large clusters, dbl., 15–25 petals, 1.5–2.75 in., slight fragrance; some prickles; foliage small, medium green, semi-glossy; low (46-60 cms), bushy, compact growth; [Len Turner X Jean Thomson Harris]; Cocker, James & Sons; James Cocker & Sons, 1995

COCmystery, F, ob, 1994; (**Aberdeen Celebration**); Cocker

'COCNANNE', F, pb, 1993; (**Friend for Life**); flowers pink, semi-double, blooms borne in large clusters, 6–14 petals, 1.5–2.75 in., slight fragrance; some prickles; foliage medium, dark green, glossy; medium (2 1/2 ft), bushy, compact growth; [Seedling X (Anne Cocker X Silver Jubilee)]; Cocker; James Cocker & Sons, 1994

'COCNEEL, F, dp, 1997; (**Ballindalloch Castle**); flowers full, medium, very dbl., 26–40 petals, borne in large clusters, slight fragrance; some prickles; foliage medium, medium green, glossy; bushy, medium (2.5ft.)growth; [(Anne Cocker X Maxi) X Silver Jubilee]; Cocker, Ann G.

'COCNEST', F, ab, 1993; (**Constance Fettes**); dbl., 26–40 petals, 1.5–2.75 in., borne in small clusters, moderate fragrance; some prickles; foliage medium, medium green, glossy; medium (2 1/2 ft), upright, bushy growth; [(Fragrant Cloud X Alexander) X Sunblest]; Cocker; James Cocker & Sons, 1993

COCnilly, F, mr, 1996; (**Ray of Hope**); Cocker

'Coco', F, yb, 1975; flowers deep golden yellow to orange-pink, dbl., 20–25 petals, 3 in., moderate fragrance; foliage glossy, bright; vigorous, free growth; [Pernille Poulsen X Redgold]; Fryer, Gareth; Fryer's Nursery, Ltd.

Coco *see* KORferse

COCocrust, F, mr, 1996; (**Alex C. Collie**); Cocker

'COCONUT ICE', HT, pb, 1991; bud large, pointed; flowers pink blend with mauve shadings, blooms borne 1-3 per stem, dbl., exhibition form, slight fragrance; foliage medium green, semi-glossy; tall, vigorous growth; [Alexander X Vol de Nuit]; Walker, D.R., 1981

COCopher, HT, mr, 1993; (**Christopher**); Cocker

'COCORAY', F, ab, 1997; (**Innocence 96**); flowers semi double, small, semi-dbl., 8–14 petals, borne in small clusters, slight fragrance; few prickles; foliage small, medium green, glossy; bushy, low (2.5ft.) growth; [Clydebank Centenary X Ray of Sunshine]; Cocker, Ann G.

'COCORICO', F, or, 1951; bud pointed; flowers geranium-red, blooms in clusters, single, 8 petals, 3 in., moderate, spicy fragrance; foliage glossy, bright; vigorous, upright, bushy growth; (28); GM, Geneva, 1951 GM, NRS, 1951; [Alain X Orange Triumph]; Meilland, F.; C-P, 1953;, URS, 1951

Cocorico® *see* MEIlasso

'COCORICO, CLIMBING', Cl F, or, 1964; Ruston, D.

'COCORONA', F, ab, 1997; (**Heartbeat 96**); flowers double, medium, dbl., 15–25 petals, slight fragrance; some prickles; foliage medium, dark green, glossy; upright, medium (2.5ft.) growth; Cocker, Ann G.

COCorust, F, lp; (**Alex C Collie**); Cocker

'COCOSIMBER', HT, rb, 1998; (**President Heldar Aliyev**); flowers vibrant salmon red, lighter reverse, very full, very dbl., 41 petals, 2–2.5 in., borne singly or in large clusters, moderate fragrance; prickles moderate, medium, slightly hooked; foliage large, dark green, glossy; upright, medium growth; [Silver Jubilee X Remember Me]; Cocker, Ann G.; James Cocker & Sons, 1999

'COCOTTE', HT, mp, 1958; flowers bright salmon, large, dbl., moderate fragrance; foliage bronze; upright growth; [Peace X (Fashion X Vogue)]; Gaujard

COCover, HT, ab, 1997; (**Gwen Major**); flowers full, large, very dbl., 26–40 petals, borne mostly singly or in large clusters, moderate fragrance; some prickles; foliage medium, dark green, glossy; bushy, upright, medium (2.5ft.)growth; [Silver Jubilee X Remember Me]; Cocker, Ann G.

'COCQUAMBER', HT, r, 1998; (**Home of Time**); flowers cinnamon red with bronze shading, lighter reverse, very dbl., 26–40 petals, 2 in., mostly borne singly, slight fragrance; prickles moderate, medium, slightly hooked; foliage medium, dark green, glossy; upright, medium (2 1/2 ft) growth; [(Sabine X Circus) X Amber Queen]; Cocker, Ann G.; James Cocker & Sons, 1998

'COCQUATION', F, lp, 1999; (**Hi Society**); flowers lilac pink, reverse lighter, dbl., 17–25 petals, 1 in., borne in large clusters, slight fragrance; prickles moderate; foliage small, light green, glossy; upright, bushy, medium (2-2.5 ft) growth; [Conservation X (Chanelle X (Golden Masterpiece X Adolf Horstmann))]; Cocker, A.; James Cocker & Sons, 1999

'COCQUETRUM', F, op, 1998; (**Rose 2000**, Rose2000, Rose Two Thousand); flowers coral vermilion, lighter reverse, semi-dbl., 8–14 petals, .5–1.5 in., borne in large clusters, slight fragrance; prickles moderate amount, small, slightly hooked; foliage medium, dark green, glossy; bushy, low, rounded growth; [Trumpeter X Clydebank Centenary]; Cocker, Ann G.; James Cocker & Sons, 1998

'COCQUIRIAM', F, w, 1998; (**White Gold**, White Gold 98, Whitegold); flowers white with golden yellow center, reverse same, opens wide and flat, very dbl., 26–40 petals, 3 in.., borne in large clusters, moderate fragrance; prickles moderate, medium, slightly hooked; foliage large, medium green, semi-glossy; tall, bushy growth; [Morning Jewel X Myriam]; Cocker, Ann G.; James Cocker & Sons, 1998

'COCREDWARD', F, lp, 1997; (**Marguerite Anne**); flowers full, medium, very dbl., 26–40 petals, borne in large clusters, moderate fragrance; foliage medium, dark green, glossy; bushy, broad, medium (2.5ft.)growth; [Anisley Dickson X Roddy McMillan]; Cocker, Ann G.

'COCTY', HT, rb, 1971; (**Curiosity**); flowers scarlet, reverse gold, dbl., 35 petals, 4 in., cupped, slight fragrance; foliage variegated (only rose to date with this property); [Cleopatra sport]; Cocker

Coed™ *see* 'JELLO'

Coeur d'Alene, HRg, mp, 1996

Coeur d'Amour *see* 'DICAM'

Coeur Farouche® *see* DELfrei

Coffee Country *see* 'VIRBROWN'

'COGAMO', LCl, pb, 1996; (Mon Amour, **My Love**™); flowers phlox pink, reverse coral pink, cream pink at base, blooms b, dbl., 26–40 petals, 4.5 in., moderate fragrance; few prickles; foliage medium, dark green, semi-glossy to glossy; spreading, tall (10 ft.) growth; [Caprice X Peace, Cl.]; Coggiatti, Stelvio

'COGNAC', F, ab, 1956; flowers apricot, reverse darker, stamens dark amber, large, dbl., borne in small clusters; foliage glossy, dark olive-green; moderate growth; [Alpine Glow X Mrs Pierre S. duPont]; Tantau, Math.

'COGNAC, CLIMBING', Cl F, ab, 1962; Kordes

'COIMBRA', HT, dp, 1953; flowers cerise-pink, large, dbl., exhibition form, intense fragrance; [Heinrich Wendland X Crimson Glory]; da Silva, Moreira

'COISTOR', S, w, 1999; (**Hailstorm**™); semi-dbl., 12–16 petals, 1 in., borne in large clusters, slight fragrance; prickles moderate; foliage small, medium green, dull; upright, tall (3 ft) growth;

[seedling X seedling]; Coiner, Jim; Coiner Nursery, 2000

Colbert® *see* 'DELCOLB'

'COLCESTRIA', Cl HT, pb, 1916; flowers rose to silver-pink, petals reflexed, large, dbl., intense fragrance; foliage light; Cant, B. R.

Colchester Beauty *see* 'CANSEND'

'COLCHESTER GAZETTE', F, mr, 1972; flowers bright red, rosette form, dbl., 40–50 petals, 2 in., slight fragrance; foliage light; very free growth; [Evelyn Fison X Etendard]; Cants of Colchester, Ltd.

'COLERAINE', HT, lp, 1970; flowers pale pink, classic form, dbl., 49 petals, 3.5 in., slight fragrance; foliage light; free growth; [Paddy McGredy X (Mme Léon Cuny X Columbine)]; McGredy, Sam IV; McGredy

'COLE'S PINK LAFAYETTE', F, mp, 1930; flowers rose-pink, semi-dbl.; [Lafayette sport]; Cole Nursery Co.

'COLETTA MONTANELLI', HT, dy, 1975; bud globular; flowers deep yellow, dbl., 32 petals, 4 in., exhibition form, slight fragrance; foliage dark, soft; vigorous, upright, bushy growth; [Seedling X Peer Gynt]; Kordes; Barni

'COLETTE', N, pb, 1932; flowers center nankeen yellow, shaded salmon-pink, large; foliage glossy, light; very vigorous growth; RULED EXTINCT 7/90; [William Allen Richardson X Mme Laurette Messimy]; Schwartz, A.

Colette *see* 'JACCOL'

Colette™ *see* 'MEIROUPIS'

'COLETTE BERGES', HT, mr, 1940; flowers crimson, medium, dbl., 45 petals, exhibition form, moderate, damask fragrance; vigorous growth; [Red Columbia X Ami Quinard]; Dot, Pedro

'COLETTE CLÉMENTE', HT, ob, 1932; flowers reddish orange, medium, semi-dbl., slight fragrance; foliage glossy, dark; vigorous growth; [(Mme Mélanie Soupert X Mme Edouard Herriot) X (Mrs Edward Powell X R. foetida bicolor)]; Mallerin, C.; C-P

'COLETTE JELOT', F, yb, 1942; flowers amber-yellow shaded currant-red, open, medium, semi-dbl., borne in clusters, slight fragrance; foliage soft; dwarf growth; [Ampere X ((Charles P. Kilham X (Charles P. Kilham X Capucine Chambard))]; Meilland, F.; A. Meilland

Colette John Keats™ *see* 'MEIROUPIS'

'COLFRAGRASAR', Cl HT, or, 1973; (**Fragrant Cloud, Climbing**, Nuage Parfume, Climbing); Collin, W.C.; W.H. Collin & Sons

Colibre *see* 'MEIMAL'

Colibre 79® *see* 'MEIDANOVER'

Colibri *see* 'MEIMAL'

'COLIBRI', Pol, ly, 1898; flowers soft yellow, fading; Lille

Colibri 80, Min, 1980; Meilland, Alain A.; (Cavriglia)

'COLIN KELLY', HT, dp, 1945; bud pointed; flowers cerise-red, very large, dbl., intense fragrance; foliage leathery; upright, bushy growth; [E.G. Hill X The Queen Alexandra Rose]; Krebs; Marsh's Nursery

'COLIN KELLY, CLIMBING', Cl HT, dp; Marsh's Nursery

Colinda *see* JACredi

Colin's Salmon, HT, lp; [Bel Ange sport]; Bell, Ronald J., 1970; (Weatherly, L.)

'COLISÉE', F, mr, 1965; (Colysée); bud pointed; flowers coppery pink, open, medium, semi-dbl.; foliage dark, glossy; very vigorous, bushy growth; [Atlantic X Circus]; Gaujard; Ilgenfritz Nursery

'COLLEEN', HT, mp, 1914; bud high pointed; flowers bright rose shaded rose-pink, large, dbl., moderate fragrance; vigorous growth; GM, NRS, 1913; McGredy

Colleen Little *see* 'LEECAL'

'COLLEEN MOORE', HT, dr, 1944; flowers velvety carmine, open, semi-dbl., 15–25 petals, 4–4.5 in., intense fragrance; foliage leathery, dark; strong stems; vigorous, upright growth; [((De Luxe X Senior) X Premier) X Chieftain]; Hill, Joseph H., Co.; Bosley Nursery

College Avenue #10, HT, yb

College Avenue Columbia, HT, dp

Collegiate 110 *see* DICknowall

Collegiate Pride, HT, mr, 1989; Bell

Cologne *see* **'KÖLN AM RHEIN'**

Cologne *see* 'MACSUPBOW'

Cologne Carnival *see* 'KORGI'

Coloma's Gold, Min, yb, 1998; [Work of Art sport]; Fischer

Colombe, HT, 1976; Croix, P.; (Cavriglia)

Colombina®, HT, pb, 1985; Barni, V.

Colombine *see* **'COLUMBINE'**

'COLONEL CAMPBELL WATSON', HT, mp, 1936; bud pointed; flowers salmon-pink, dbl., exhibition form; foliage soft; long stems; vigorous growth; GM, NRS, 1935; [Joan Howarth X Portadown]; Bees

'COLONEL DAZIER', HT, w, 1927; flowers rosy white, reverse bright rose, base golden yellow, dbl., intense fragrance; [Le Progrés X Jonkheer J.L. Mock]; Ketten Bros.

'COLONEL DE SANSAL', HP, 1875; Jamain, H.; (Cavriglia)

'COLONEL FOISSY', HP, mp, 1849; flowers medium, dbl.; Margottin; (Sangerhausen)

'COLONEL GRAVEREAUX', HT, op, 1940; flowers salmon-coral, reverse yellow; strong stems; vigorous, upright growth; Mallerin, C.; A. Meilland

'COLONEL JOFFÉ', T, mr, 1893; flowers purplish red, petals wrinkled, flat; Liabaud

'COLONEL LECLERC', HT, mp, 1909; flowers tyrian pink, dbl., moderate fragrance; [Mme Caroline Testout X Horace Vernet]; Pernet-Ducher

'COLONEL LINDBERGH', Pol, op, 1928; flowers salmon-orange; [Juliana Rose sport]; Ouden, Den

'COLONEL NICOLAS MEYER', HT, mr, 1934; bud pointed; flowers brilliant velvety red, open, dbl., moderate fragrance; foliage leathery; strong stems; vigorous, bushy growth; [La Maréchale Petain X Edouard Mignot]; Sauvageot, H.

'COLONEL OSWALD FITZGERALD', HT, dr, 1917; flowers dark velvety crimson, well formed, dbl.; vigorous, branching growth; Dickson, A.

'COLONEL R. S. WILLIAMSON', HT, lp, 1907; flowers white, center deep blush, large, dbl., exhibition form, moderate fragrance; foliage glossy, dark; vigorous, open growth; Dickson, A.

'COLONEL ROBERT LEFORT', M, mr; flowers purple-red; Verdier, E., 1884

'COLONEL SHARMAN-CRAWFORD', HT, mr, 1933; flowers rich velvety crimson, large, dbl., exhibition form, intense fragrance; foliage leathery; long, strong stems; vigorous, bushy growth; GM, NRS, 1931; Dickson, A.

Colonel Svec *see* **'PLUKOVNIK SVEC'**

Colonia, LCl, mr, 1988; flowers large, semi-dbl.; Meilland; (Sangerhausen)

'COLONIAL WHITE', LCl, w, 1959; bud ovoid; dbl., 3 in., flat, recurrent bloom, moderate fragrance; foliage light green; vigorous growth; [New Dawn X Mme Hardy]; Wyant

'COLONIAL WHITE', T, lp

'COLOR GIRL', F, pb, 1966; bud ovoid; flowers whitish, edged red, medium, dbl., exhibition form, intense fragrance; foliage dark, leathery; bushy, low growth; [Little Darling X Cocorico]; Fuller; Wyant

Color Guard *see* 'JOLCOL'

Color Magic *see* 'JACMAG'

Color Purple *see* 'CLEPURP'

Colorado, HT, 1970; Combe, M.; (Cavriglia)

Colorado, F, mp; Select Roses, B.V.

Colorama *see* 'MEIRIGALU'

'COLORANJA', HT, or, 1963; (Mustang); bud ovoid; flowers large to medium, dbl., exhibition form; foliage dark, leathery; vigorous, bushy growth; [Fandango X (Independence X Papillon Rose)]; Lens

Colorbreak *see* 'MACULTRA'

Colorburst *see* TWOcherish

'COLOSO', HT, mp, 1962; dbl., 50 petals, moderate fragrance; very vigorous growth; [Chrysler Imperial X (Peace X Queen Elizabeth)]; Dot, Simon

Colossus® *see* 'PERSUS'

'COLOUR CARNIVAL', F, yb, 1962; flowers primrose-yellow edged pink, well formed, borne in clusters, dbl., 50 petals, 3 in., slight fragrance; vigorous, low, bushy growth; LeGrice; Wayside Gardens Co.

'COLOUR GLOW', F, ob, 1969; flowers orange, reddening to flame, dbl., 36 petals, 3.5 in., globular, slight fragrance; vigorous growth; [Tropicana X Masquerade]; Butter; Wood End Gardens

Colour Magic, MinFl, pb, 1999

Colour Parade *see* 'LAVCOAT'

'COLOUR SERGEANT', F, or, 1972; dbl., 20 petals, 4 in., slight fragrance; foliage glossy; [Queen Elizabeth X (Ann Elizabeth X Circus)]; Harkness

Colour Wonder *see* 'KORBICO'

Colourama *see* 'MEIRIGALU'

'COLUMBIA', HT, mp, 1916; bud long, pointed; flowers glistening rose-pink, large, dbl., 65 petals, intense fragrance; foliage dark; vigorous growth; Gertrude M. Hubbard, ARS, 1919 GM, Portland, 1919; [Ophelia X Mrs George Shawyer]; Hill, E.G., Co.

'COLUMBIA', HWich, mp, 1903;, non-recurrent; moderately vigorous growth; [Seedling X Mme Caroline Testout]; Hooper, Bro. & Thomas

'COLUMBIA, CLIMBING', Cl HT, mp, 1923; Vestal, 1923 & Lens, 1929

'COLUMBINE', F, yb, 1956; (Colombine); flowers creamy yellow tinged pink, well-formed blooms in open clusters, intense fragrance; foliage glossy; vigorous growth; [Danish Gold X Frensham]; Poulsen, S.; McGredy

Columbus® *see* 'WEKUZ'

'COLUMBUS QUEEN', HT, pb, 1962; bud ovoid, pointed; flowers light pink, reverse darker, dbl., 24–30 petals, 4 in., exhibition form, slight fragrance; foliage leathery, dark; vigorous, upright growth; GM, Geneva, 1961; [La Jolla X Seedling]; Armstrong, D.L.; Armstrong Nursery

'COLUMELLE', HGal, 1860; Moreau et Robert; (Cavriglia)

Colysée *see* **'COLISÉE'**

'COMANCHE', Gr, or, 1968; bud pointed; flowers medium, dbl., exhibition form, slight fragrance; foliage leathery; vigorous, upright, bushy growth; AARS, 1969; [Spartan X (Carrousel X Happiness)]; Swim & Weeks; C-P

Come Lei, HT, 1968; Giacomasso; (Cavriglia)

Comedie, F, 1977; Combe, M.; (Cavriglia)

'COMENDADOR NOGUEIRA DA SILVA', HT, rb, 1961; flowers dark red, reverse silvery; [Confidence X Independence]; da Silva, Moreira

'COMET', Cl F, pb, 1934; flowers flesh pink, shades salmon-yellow, large, very dbl., slight fragrance; foliage leathery; short, strong stems; [Gruss an Aachen sport]; Mesman; Bosley Nursery

Comice de Tarn-et-Garonne, B, mr; flowers carmine red, intense fragrance; Pradel, 1852

'COMMAND PERFORMANCE', HT, or, 1970; bud ovoid; flowers medium, dbl., exhibition form, intense fragrance; foliage leathery; vigorous, tall, bushy growth; AARS, 1971; [Tropicana X Hawaii]; Lindquist; Howard Rose Co.

'COMMANDANT BEAUREPAIRE', B, pb, 1874; flowers bright rose-pink streaked (striped) purple-violet and marbled white, dbl., occassional repeats, moderate fragrance; foliage light green; vigorous growth; Moreau et Robert

Commandant Cousteau *see* ADharman

'COMMANDANT FÉLIX FAURE', HP, mr, 1901; flowers light crimson-red, tinted vermilion, dbl., 25 petals, cupped, moderate fragrance; vigorous, upright growth; Boutigny

'COMMANDANT L. BARTRE', HT, rb, 1920; flowers dark carmine-red, tinted brilliant pink, dbl.; vigorous growth; [Lady Ashtown X Louis van Houtte]; Schwartz, A.

'COMMANDANT LETOURNEUX', HT, dp, 1903; flowers carmine-pink, very large, very dbl., intense fragrance; Bahaud; (Sangerhausen)

'COMMANDATORE FRANCESCO INGEGNOLI', Cl HT, or, 1923; flowers geranium-red; Ingegnoli

'COMMANDEUR JULES GRAVEREAUX', HP, mr, 1908; bud pointed; flowers dazzling red, center shaded maroon, large, peony-like, dbl., intense fragrance; vigorous growth; [Frau Karl Druschki X Liberty]; Croibier

'COMMANDEUR JULES GRAVEREAUX, CLIMBING', Cl HP, mr, 1925; Belouet; Grandes Roseraies

'COMMANDO', HT, op, 1945; bud long, pointed; flowers orange-buff, suffused pink, camellia form, dbl., 30–35 petals, 5 in., moderate fragrance; foliage leathery, glossy; upright, bushy growth; [Mrs J.D. Eisele X Glowing Sunset]; Howard, F.H.; H&S

Commitment *see* CLEment

Common Blush China *see* **'OLD BLUSH'**

Common Monthly *see* **'OLD BLUSH'**

Common Moss *see* **'CENTIFOLIA MUSCOSA'**

Common Moss *see* **'COMMUNIS'**

'COMMONWEALTH', HT, dp, 1923; bud pointed; flowers deep pink, large, dbl., moderate fragrance; foliage leathery, rich green; [Ophelia X Seedling]; Montgomery Co.; A.N. Pierson

'COMMONWEALTH', F, pb, 1948; (Herzblut); flowers crimson with white eye, borne in clusters, semi-dbl., 10 petals, slight fragrance; foliage leathery; (28); [Col. Nicolas Meyer X Holstein]; Kordes; Morse

Commonwealth Glory *see* HARclue

'COMMUNIS' *see* Centifolia Muscosa

'COMMUNIS', M, mp; (Centifolia Muscosa, Common Moss, Mousseau Ancien, Old Pink Moss, Pink Moss, R. centifolia muscosa); bud mossed; flowers pale rose, very dbl., globular; vigorous growth

Compactilla, HSpn, ly

'COMPASSION'®, LCl, op, 1972; (Belle de Londres); flowers salmon pink shaded apricot, large blooms borne singly or in, dbl., 36 petals, moderate, sweet fragrance; large, reddish prickles; foliage large, dark; medium, bushy growth; Geneva, 1979, Orleans, 1979 ADR, 1976 Edland Fragrance Medal, ARS, 1973 GM, Baden-Baden, 1973; [White Cockade X Prima Ballerina]; Harkness, 1973

Compassionate Friend *see* 'HARZODIAC'

'COMPLICATA', HGal, pb; flowers deep pink, white eye, yellow stamens, large, single; rampant (to 6 ft) growth

Comptoire des Bourbons, B, mp

'COMRADE', HT, mr, 1948; flowers scarlet-crimson, well formed, medium, moderate fragrance; Dickson, A.

COMsan, HT, w, 1993; (**Cosmos**®); Combe

'COMSOR', HT, dr, 1971; (**Norita**®, Norita-Schwarze Rose®); flowers very deep red, large, dbl., exhibition form, moderate fragrance; foliage dark, leathery; vigorous, bushy growth; [Charles Mallerin X Seedling]; Combe; Kern Rose Nursery, 1966

'COMTE ADRIEN DE GERMINY', HP, mp, 1881; flowers large, dbl.; Lévêque; (Sangerhausen)

Comte Boula de Nanteuil *see* **'COMTE DE NANTEUIL'**

'COMTE DE BOBRINSKY', HCh, dp, 1849; flowers carmine-pink, medium, dbl.; Marest; (Sangerhausen)

'COMTE DE CHAMBORD', P, pb, 1860; flowers pink tinted lilac, very dbl., flat, intense fragrance; vigorous, erect growth; Robert et Moreau

'COMTE DE FALLOUX', HP, dp; Trouillard, 1860

'COMTE DE FLANDRES', HP, 1881; Lévêque, P.; (Cavriglia)

'COMTE DE MORTEMART', HP, mp, 1880; (Comtesse de Mortemart); flowers clear pink, intense fragrance; Margottin

'COMTE DE NANTEUIL', HGal, m; (Boule de Nanteuil, Comte Boula de Nanteuil); flowers crimson purple, center sometimes fiery crimson, compact, large, dbl.; branching growth; Roeser, ca. 1834

'COMTE DE NANTEUIL', HP, pb, 1852; flowers light rose with darker edges, sometimes with a green center, dbl., cupped; vigorous growth; Quétier

Comte de Paris, HP, m; Laffay, M., 1839

Comte de Paris, HP, lp; Verdier, E., 1864

'COMTE DE TORRES', HT, op, 1906; flowers coppery salmon-pink, large, dbl.; Schwartz, A.; (Sangerhausen)

'COMTE D'EPREMESNIL', HRg, m, 1882; flowers violet-pink, large, semi-dbl., intense fragrance; Nabonnand; (Sangerhausen)

'COMTE F. DE CHAVANAC', HT, mp; flowers peach-blossom-pink, center rosy carmine; [Antoine Rivoire X Zephirine Drouhin]

'COMTE F. DE CHAVANAC, CLIMBING', Cl HT, mp, 1929; Siret-Pernet

Comte Florimond de Bergeyck, HP, pb

'COMTE FOY', HGal, lp; (Comte Foy de Rouen); flowers pale rose, very large, dbl., cupped; Savoureux, 1827

Comte Foy de Rouen *see* **'COMTE FOY'**

'COMTE FRÈDÈRIC DE THUN-HOHENST', HP, dr, 1880; flowers large, dbl., moderate fragrance; Lévêque; (Sangerhausen)

'COMTE G. DE ROCHEMUR', HT, 1911; flowers bright scarlet, large, dbl., moderate fragrance; [Xavier Olibo X Gruss an Teplitz]; Schwartz, A.

'COMTE RAIMBAUD', HP, mr, 1867; flowers crimson, large; Rolland

COMteri, LCl, 1976; (**Cortege**); Combe, M.; (Cavriglia)

'COMTESSE ANNE DE BRUCE', HT, pb, 1937; bud pointed; flowers coppery pink to nasturtium-red, very large, semi-dbl., cupped, moderate fragrance; foliage glossy; very vigorous growth; [Charles P. Kilham X (Mrs Pierre S. duPont X R. foetida bicolor seedling)]; Mallerin, C.; H. Guillot

'COMTESSE BARDI', T, pb, 1896; flowers reddish fawn; Soupert & Notting

'COMTESSE BEATRIX DE BUISSERET', HT, 1900; flowers very large, dbl., moderate fragrance; Soupert & Notting; (Sangerhausen)

'COMTESSE BRANICKA', HP, lp, 1888; flowers large, dbl.; Lévêque; (Sangerhausen)

Comtesse Brigitte Chandon-Moet *see* DORfaut

'COMTESSE CAHEN D'ANVERS', HP, mp, 1885; flowers large, dbl., moderate fragrance; Lédéchaux; (Sangerhausen)

'COMTESSE CÉCILE DE CHABRILLANT', HP, pb, 1858; flowers satiny pink, silvery reverse, medium, dbl., globular, moderate fragrance; numerous small, dark prickles; foliage dark, leathery; Marest

'COMTESSE CÉCILE DE FORTON', Cl T, pb, 1916; flowers rose-peach, very large, dbl., moderate fragrance; Nabonnand, G.

Comtesse d'Alcantara *see* 'KRIBATIS'

'COMTESSE D'ANSEMBOURG', HT, ly, 1918; flowers yellowish white, dbl., intense fragrance; [Étoile de France X Marquise de Sinéty]; Leenders, M.

'COMTESSE DE BARBANTANE', B, lp, 1858; Guillot Père, 1858

Comtesse de Barry, F, 1994; Meilland, Alain A.; (Cavriglia)

Comtesse de Bouchard *see* **'COMTESSE DE BOUCHAUD'**

'COMTESSE DE BOUCHAUD', N, my, 1890; (Comtesse de Bouchard); flowers saffron yellow, very large; Guillot et Fils

'COMTESSE DE BRESSON', HP, lp, 1873; flowers large, dbl.; Guinoiseau, B.; (Sangerhausen)

'COMTESSE DE CASERTA', T, mr, 1877; flowers coppery-red, large; Nabonnand, G.

'COMTESSE DE CASSAGNE', HT, mp, 1919; flowers coppery rose, shaded bright rose, sometimes entirely yellow, dbl., moderate fragrance; Guillot, M.

'COMTESSE DE CASTILLEJA', HT, op, 1926; bud orange; flowers coral, cupped, intense fragrance; foliage dark; strong stems; very vigorous growth; [(Mme Edouard Herriot X Juliet) X Seedling]; Chambard, M.

'COMTESSE DE CHAPONAY', LCl, lp, 1924; flowers cream-rose to salmon, dbl., intense fragrance; very vigorous growth; [R. gigantea X Mme Hoste]; Nabonnand, G.

'COMTESSE DE FLANDRE', HP, lp, 1878; flowers large, dbl.; Verdier, E.; (Sangerhausen)

'COMTESSE DE FRIGNEUSE', T, my, 1885; flowers large, dbl., moderate fragrance; [Mme Damaizin X Seedling]; Guillot et Fils

'COMTESSE DE GALARD-BÈARN', N, ly, 1894; flowers light yellow with pink tones, large, dbl.; Bernaix, A.; (Sangerhausen)

'COMTESSE DE LA MORANDIÈRE', HT, op, 1929; flowers shrimp-pink, reverse coral-red, dbl., moderate fragrance; Chambard, C.

Comtesse de Labarathe *see* **'DUCHESSE DE BRABANT'**

'COMTESSE DE LACÉPÈDE', HGal, lp; flowers silvery blush, center sometimes rosy, large, dbl.; moderate growth

'COMTESSE DE LEUSSE', HT, dy, 1878; Nabonnand

'COMTESSE DE MARTEL', HT, mp, 1939; flowers carnation-pink, center coppery, very large, moderate fragrance; very vigorous growth; [Charles P. Kilham X Margaret McGredy]; Meilland, F.

Comtesse de Mortemart *see* **'COMTE DE MORTEMART'**

'COMTESSE DE MURINAIS', M, w, 1843; (White Moss); flowers flesh, opening white, large, dbl., non-recurrent, moderate fragrance; height 4-5 ft; Vibert

'COMTESSE DE NOGHERA', T, op, 1902; flowers light salmon-pink, very large, very dbl., moderate fragrance; Nabonnand; (Sangerhausen)

Comtesse de Paris, HT, dy, 1993; Briant

'COMTESSE DE POLINGAC', HP, dr, 1862; flowers medium, dbl.; Granger; (Sangerhausen)

'COMTESSE DE ROCQUIGNY', B, w, 1874; flowers salmon-red; Vaurin

Comtesse de Segur *see* DELtendre

'COMTESSE DE SERENYE', HP, lp, 1874; flowers soft pink; Lacharme, F.

Comtesse de Vezins, Misc OGR, mp

Comtesse Diana *see* HELlux

'COMTESSE DORIA', M, m, 1854; flowers purple-pink, shaded salmon, heavily mossed; Portemer fils

'COMTESSE D'OXFORD', HP, dp, 1869; (Countess of Oxford); flowers deep pink/red with violet stripes, dbl., globular, recurrent bloom, moderate fragrance; Guillot Père

Comtesse d'Oxford, HP, dp

'COMTESSE DU CAYLA', HCh, ob, 1902; flowers nasturtium-red, tinted orange, semi-dbl., flat, recurrent bloom, moderate fragrance; foliage dark, glossy; vigorous growth; Guillot, P.

'COMTESSE EMMELINE DE GUIGNE', T, pb, 1903; Nabonnand

'COMTESSE FESTECTICS HAMILTON', T, pb, 1897; Nabonnand

'COMTESSE FRESSINET DE BELANGER', HP, mp, 1886; flowers rose; Lévêque

'COMTESSE G. DE ROQUETTE-BUISSON', HP, mp, 1899; flowers medium, dbl.; Lévêque; (Sangerhausen)

'COMTESSE HENRIETTA COMBES', HP, pb, 1881; flowers satin-rose; Schwartz, J.

'COMTESSE ICY HARDEGG', HT, dp, 1911; flowers carmine, large, dbl.; [Mrs W.J. Grant X Liberty]; Soupert & Notting

'COMTESSE MÉLANIE DE POURTALES', HT, w, 1914; flowers creamy white, outer petals shaded red; [Frau Karl Druschki X Mme Ravary]; Walter

'COMTESSE MOENS DE FERNIG', HT, dr, 1961; flowers wine-red, dbl., 25 petals, moderate fragrance; vigorous growth; [Poinsettia X Seedling]; Verbeek

'COMTESSE O'GORMAN', HP, rb, 1888; flowers red and violet; Lévêque

'COMTESSE O'GORMAN', T, lp; Nabonnand, 1892

Comtesse Ouwaroff *see* **'DUCHESSE DE BRABANT'**

'COMTESSE PROZOR', LCl, pb, 1922; flowers salmon-rose, reverse coral red; [R. gigantea X Comtesse de Bouchaud]; Nabonnand, P.

'COMTESSE RENÉ DE BEARN', HP, 1896; Lévêque, P.; (Cavriglia)

Comtesse Risa du Parc *see* **'COMTESSE RIZA DU PARC'**

'COMTESSE RIZA DU PARC', T, mp, 1876; (Comtesse Risa du Parc); flowers rose to carmine, large, dbl., globular, moderate fragrance; [Duchesse de Brabant seedling]; Schwartz, J.

Comtesse Vally de Serenye, HP, lp, 1875; Fontaine

'COMTESSE VANDAL', HT, pb, 1932; (Comtesse Vandale, Countess Vandal); flowers salmon-pink, reverse coppery pink, large, dbl., 30 petals, exhibition form, moderate fragrance; foliage leathery; bushy growth; (28); GM, Bagatelle, 1931; [(Ophelia X Mrs Aaron Ward) X Souv. de Claudius Pernet]; Leenders, M.; J&P

'COMTESSE VANDAL, CLIMBING', Cl HT, op, 1936; (Comtesse Vandale, Climbing, Countess Vandal, Climbing, Grimpant Comtesse Vandal); J&P

Comtesse Vandale *see* **'COMTESSE VANDAL'**

Comtesse Vandale, Climbing *see* **'COMTESSE VANDAL, CLIMBING'**

Concertino® *see* 'MEIBINOSOR'

Concertino™ *see* 'MEIGLUSOR'

'CONCERTO', F, mr, 1953; bud ovoid, pointed; semi-dbl., 12–15 petals, 2–2.5 in., cupped, slight fragrance; foliage dark; upright, bushy growth; GM, NRS, 1953 PIT, RNRS, 1953; [Alain X Floradora]; Meilland, F.; URS

Concerto™ *see* 'MEIROLOUR'

Concerto *see* MEIhailoil

Concerto 94, S, 1994; Meilland, Alain A.; (Cavriglia)

'CONCERTO, CLIMBING', Cl F, mr, 1968; Truffant, G.; URS

'CONCHITA', Pol, mp, 1935; flowers clear salmon, borne in clusters, dbl., cupped, slight fragrance; foliage glossy; vigorous growth; Jordan, H.; Low

Concorde *see* **'FOREVER YOURS'**

Concorde *see* MEIreibat

Concorde *see* MEIdorsun

Concorde, HT, lp

'CONCORDIA', HT, mp, 1924; flowers glowing pink, edged silver-pink, semi-dbl., moderate fragrance; Brix; Teschendorff

'CONCORDIA', HT, rb, 1946; flowers red, reverse deep yellow, well formed, dbl., moderate fragrance; vigorous growth; [Charles P. Kilham X Crimson Glory]; Giacomasso

Condesa de Barcelona® *see* FEodase

'CONDESA DE BENAHAVIS', HT, mp, 1949; flowers salmon-pink, well formed, intense fragrance; [Étoile de Hollande X Sensation]; La Florida

'CONDESA DE GLIMES', F, pb; flowers begonia-pink, center ochre-yellow, single; Bofill; Torre Blanca

'CONDESA DE MAYALDE', HT, rb, 1956; bud pointed; flowers white edged carmine, dbl., 30 petals, exhibition form; foliage glossy; compact, upright growth; [Peace X Flambee]; Dot, Pedro

'CONDESA DE MAYALDE, CLIMBING', Cl HT, w, 1964; Samuels

'CONDESA DE MUNTER', HT, or, 1932; flowers geranium-red, tinted orange-yellow, large, semi-dbl., cupped; long, strong stems; vigorous growth; [Souv. de Josefina Plà X Souv. de Claudius Pernet]; Munné, B.

'CONDESA DE SALDANHA', HT, ob, 1961; flowers reddish orange, mottled yellow, large, dbl., 45 petals, moderate fragrance; vigorous growth; Munné, M.

'CONDESA DE SÁSTAGO', HT, pb, 1932; bud ovoid; flowers deep pink, reverse yellow, large, dbl., 55 petals, cupped, moderate fragrance; foliage glossy, dark; vigorous, tall growth; (28); GM, Rome, 1933; [(Souv. de Claudius Pernet X Maréchal Foch) X Margaret McGredy]; Dot, Pedro; C-P

'CONDESA DE SÁSTAGO, CLIMBING', Cl HT, pb, 1936; Vestal

'CONDESA DE VILLARREA', HT, mr, 1960; flowers crimson, reverse cardinal-red, large, dbl., 35 petals; long, stiff stems; vigorous growth; [Chrysler Imperial X Texas Centennial]; Dot, M.

Conditorum *see* **R. GALLICA CONDITORUM**

'CONESTOGA', S, w, 1946; bud ovoid; flowers open, borne in clusters, dbl., 30 petals, 2 in., non-recurrent, slight fragrance; foliage soft, sparse, small; upright, vigorous growth; hardy.; [Betty Bland X Seedling]; Preston; Central Exp. Farm

Confection *see* 'JACUTE'

'CONFEDERATION', HT, mp, 1964; flowers compact, large, dbl., 70 petals, intense fragrance; foliage dark, glossy; moderate, vigorous growth; [Queen o' the Lakes X Serenade]; Golik; Ellesmere Nursery

'CONFERENCE 63', HT, dr, 1965; flowers deep crimson, dbl., 4.5 in., slight fragrance; foliage light green; tall growth; Quentin

Confetti *see* 'AROJECHS'

'CONFIDENCE', HT, pb, 1951; bud ovoid; flowers pearly light pink to yellow blend, large, dbl., 28–38 petals, exhibition form, moderate fragrance; foliage dark, leathery; vigorous, upright, bushy growth; GM, Bagatelle, 1951; [Peace X Michèle Meilland]; Meilland, F.; C-P, 1953;, URS, 1951

Confidence™ *see* 'MEIGAFOR'

'CONFIDENCE, CLIMBING', Cl HT, pb, 1961; Hendrickx; URS

'CONGO', HT, dr, 1943; bud long, pointed; flowers velvety maroon, open, medium, dbl., intense fragrance; foliage leathery, bronze; bushy, dwarf growth; [Admiral Ward X Lemania]; Meilland, F.; A. Meilland

'CONGOLAISE', HT, dr; flowers velvety dark red, well formed, large, dbl., moderate fragrance; vigorous growth; Tantau

Congratulations *see* 'KORLIFT'

Connie *see* 'BOSELFTAY'

'CONNIE', Min, dy, 1991; bud pointed; flowers dark yellow fading to light yellow, semi-dbl., 11 petals, flat, borne singly or in small clusters, slight fragrance; foliage medium, dark green, glossy; bushy, medium (36 cms), very dense growth; Jerabek, Paul E.

Connie Crook *see* 'BOSELFTAY'

'CONNIE MACK', F, dr, 1952; flowers dark velvety crimson, medium, borne in clusters, dbl., 25 petals, slight fragrance; foliage glossy, dark; vigorous growth; [Seedling X Margy]; Duehrsen; H&S

'CONQUEROR', HT, ly, 1929; flowers saffron-yellow, fading pale yellow, semi-dbl., 15 petals, moderate fragrance; vigorous, bushy growth; Chaplin Bros.

Conqueror's Gold *see* 'HARTWIZ'

Conquest *see* 'HARBRILL'

Conquete, HT, 1967; Combe, M.; (Cavriglia)

Conquistador, LCl, 1983; Pineau; (Cavriglia)

'CONRAD FERDINAND MEYER', HRg, lp, 1899; flowers silver pink, large blooms in clusters, dbl., cupped, repeat bloom, intense fragrance; foliage leathery; vigorous (8-10 ft.), good pillar rose, bushy growth; (28); [R. rugosa hybrid X Gloire de Dijon]; Müller, Dr. F.

'CONRAD HILTON', F, my, 1962; bud ovoid; flowers golden yellow, outer petals sometimes white, dbl., 30–45 petals, 2.5 in., flat, moderate fragrance; foliage leathery, dark, glossy, crinkled; very vigorous, upright, bushy growth; [(R. X dupontii X Pinocchio) X (Goldilocks X Feu Pernet-Ducher)]; Shepherd; Bosley Nursery

'CONRAD O'NEAL', S, dp, 1966; bud ovoid; flowers deep pink, medium, very dbl., intense fragrance; foliage dark, glossy; vigorous, upright growth; [Blossomtime seedling X Don Juan]; O'Neal; Wyant

'CONRAD'S CRIMSON', S, rb, 1972; flowers crimson, shaded purple, dbl., 30 petals, 3 in., flat, early bloom, moderate fragrance; foliage light green to bronze; [Sweet Sultan X Conrad F. Meyer]; Eacott

Conservation *see* 'COCDIMPLE'

Conservency Rose *see* 'CAPMINA'

'CONSOLATA', HT, yb, 1936; bud pointed; flowers yellow and coppery nasturtium-red; foliage bronze; long, strong stems; Capiago

'CONSPICUOUS', HT, mr, 1930; flowers glowing scarlet, very large, dbl.; vigorous growth; Dickson, A.; B&A, 1932

'CONSTANCE', HT, my, 1915; flowers yellow to golden yellow, dbl., exhibition form, slight fragrance; foliage rich green, glossy; bushy growth; GM, Bagatelle, 1916; [Rayon d'Or X Seedling]; Pernet-Ducher

'CONSTANCE CASSON', HT, pb, 1920; flowers carmine, flushed apricot, dbl., moderate fragrance; [Queen Mary X Gorgeous]; Cant, B. R.

Constance Fettes *see* 'COCNEST'

Constance Finn *see* HAReden

'CONSTANCE MORLEY', HT, ob, 1981; flowers orange-gold tinged red, diffused center, borne singly, dbl., 39 petals, slight fragrance; elongated, orange prickles; foliage dark, glossy; spreading, bushy growth; [Piccadilly X Seedling]; Gregory, C. & Sons, Ltd.

Constance Spry® *see* 'AUSFIRST'

'CONSTANCE, CLIMBING', Cl HT, my, 1927; Pacific Rose Co.

'CONSTANTIA', HT, mp, 1960; bud pointed; flowers neyron rose, well formed, dbl., 40 petals, 4–4.5 in., moderate fragrance; moderate growth; [Baccará X Grace de Monaco]; Herholdt, J.A.; Herholdt's Nursery

'CONSTANZE', HT, ob, 1966; bud pointed; flowers orange, well-formed, large, dbl., 25–30 petals; foliage dark, glossy; upright, bushy growth; Tantau, Math.

Constanze *see* TANzecon

'CONSTELLATION', HT, ob, 1949; bud long, pointed; flowers coppery orange, well formed, very large; very vigorous, erect growth; [Peace X Seedling]; Gaujard

Constellation, LCl, ab, 1999

Constellation, Min, lp

Consuelo, HT, mr

Consul M. Mezin, HT; Verbeek; (Cavriglia)

'CONTEMPO', F, ob, 1971; bud ovoid; flowers orange blending to gold, medium, dbl., exhibition form, moderate fragrance; foliage light, leathery; vigorous, bushy growth; [Spartan X (Goldilocks X (Fandango X Pinocchio))]; Armstrong, D.L.; Armstrong Nursery

'CONTENTMENT', HT, pb, 1956; bud globular; flowers soft pink suffused yellow, dbl., 65–70 petals, 5.5 in., exhibition form, intense fragrance; foliage glossy, leathery; vigorous, upright growth; [(Seedling X Lilette Mallerin) X Orange Delight]; Boerner; J&P

Contessa *see* 'JACRIS'

Contessa Mona Bismark, Gr; (Cavriglia)

'CONTINENTAL', HT, mr, 1966; bud ovoid; flowers cardinal-red, large, dbl., moderate, fruity fragrance; foliage leathery; vigorous, upright growth; [Baccará X Yuletide]; Lammerts, Dr. Walter; Amling-DeVor Nursery

'CONTRAST', HT, pb, 1937; flowers china-pink and bronze, reverse white and bronze, dbl., exhibition form, moderate fragrance; foliage leathery, glossy; very vigorous, bushy, compact growth; [Seedling X Talisman]; H&S

Contribute, HT, my, 1977; Dawson

Cool Dude *see* 'RENDUDE'

Cool Wave *see* 'ZIPCOOL'

'COOLNESS', F, w, 1958; bud ovoid, cream; dbl., 55–60 petals, 2.5–3 in., moderate fragrance; foliage leathery, glossy; vigorous, bushy, compact growth; [Glacier seedling X Starlite seedling]; Boerner; J&P

Coon Carnival® *see* 'KORCOON'

'COOPERI', S, 1927; (Cavriglia)

'COOPEROO EMBLEM', Pol, lp; Harrison, A., 1927; (Weatherly, L.)

'COOPER'S BURMESE', S, w, 1927; (Gigantea Cooperi, R. X cooperi); flowers near white, large, single; foliage glossy; vigorous (to 20 ft) growth; (14); [Possibly a natural hybrid of R. gigantea X R. laevigata]

'COOPRIOR' Cl F, mp, 1995; (**Betty Prior, Climbing**™); flowers medium carmine pink, single (5 petals), medium blooms borne in small clusters, slight fragrance; some prickles; foliage large, medium green, dull; upright, spreading, medium growth; Betty Prior sport; Cooper, Donell; Certified Roses, 1997

'COORAN', F, dp, 1953; bud long, pointed; flowers deep rose-pink, very large, borne in clusters, dbl., intense fragrance; very vigorous growth; [Mrs Tom Henderson X Ming Toy]; Ulrick, L.W.

Coorg *see* 'VIRBROWN'

'COOROY', F, mp, 1953; flowers rose-pink, bornein clusters, very dbl., exhibition form; [Mrs Tom Henderson X Self]; Ulrick, L.W.

'COOSYL', Min, rb, 1977; (**My Own**); flowers red, yellow center, small, single, 5 petals, moderate fragrance; foliage medium, medium green, semi-glossy; [(Scarlet Knight X Soeur Therese) X Willie Winkie]; Cook, Sylven S.

'COOSYN', LCl, yb, 1981; (**Golden Penny**); flowers gold, reverse creamy, dbl., intense, lemon fragrance; foliage medium, medium green, semi-glossy; upright growth (7-8 ft.); [Queen Elizabeth X Scarlet Knight]; Cook, Sylven S.

'COPACABANA', LCl, or, 1966; dbl., 40 petals, 3.5 in., globular, slight fragrance; foliage dark; [Coup de Foudre X Seedling]; Dorieux; Bees

Copacabana *see* DORneye

'COPENHAGEN', Cl HT, mr, 1964; flowers scarlet, dbl., 5 in., moderate fragrance; foliage coppery; vigorous growth; [Seedling X Ena Harkness]; Poulsen, Niels D.; McGredy

'COPIA', F, mr, 1964; flowers bright cardinal-red, large, semi-dbl.; vigorous, bushy growth; [(Independence X Seedling) X Tour de France]; Mondial Roses

Coppa Nob *see* COCcopna

'COPPÉLIA', HT, ob, 1952; flowers rosy shades, deepening to orange, medium, dbl., 28 petals, cupped, slight fragrance; foliage leathery; vigorous, upright growth; [Peace X Europa]; Meilland, F.

Coppélia 76 *see* MEIgurami

Copper Arch *see* KORhurtlen

'COPPER CLIMBER', LCl, op, 1938; bud pointed, coppery; flowers glowing coppery salmon, edged pink, large; Burbank; Stark Bros.

'COPPER CORONET', HT, ab, 1986; flowers coppery amber, reverse blush pink, fading amber and cream, large, very dbl., 60 petals, exhibition form, slight fragrance; foliage medium, medium green, semi-glossy; upright, medium growth; [Ginger Rogers X Royal Highness]; Strange, J.F.

Copper Crown *see* 'WILCROWN'

'COPPER DELIGHT', F, ob, 1956; flowers clear orange, large blooms in large clusters, semi-dbl., 14 petals, moderate fragrance; foliage olive-green; vigorous, upright, bushy growth; [Goldilocks X Ellinor LeGrice]; LeGrice

Copper Gem, HT, ab, 1984; Cocker

'COPPER GLOW', LCl, ob, 1940; flowers copper, dbl., 27 petals, 4 in., seasonal bloom, intense fragrance; foliage glossy; vigorous, climbing (20 ft), open habit growth; [Golden Glow X Break o' Day]; Brownell, H.C.

'COPPER KETTLE', F, ob, 1978; bud elongated; flowers brilliant copper-orange and yellow, dbl., 28 petals, 2.5–3.5 in., exhibition form, moderate fragrance; foliage glossy, bronze, dark; upright growth; [Queen Elizabeth X Golden Slippers]; Williams, J. Benjamin; J.B. Williams & Associates

Copper Kettle, S, ob, 1997; bud tangerine; flowers brilliant copper orange with yellow, moderate fragrance; foliage disease-resistant; hardy to zone 5; Williams, J. Benjamin

Copper King *see* 'HERCOP'

'COPPER LUSTER', HT, op, 1945; flowers coppery pink, loosely, large, dbl., 23 petals, cupped, slight fragrance; foliage glossy, bronze; vigorous, upright growth; [Better Times X Orange Nassau]; Roberts; Totty

'COPPER NUGGET', HT, op, 1942; bud ovoid to urn shaped; flowers orange-salmon, small, dbl., 50–60 petals, exhibition form, slight fragrance; foliage leathery, glossy, dark; strong stems; dwarf, bushy growth; [Charles P. Kilham X Capt. Thomas]; Lammerts, Dr. Walter; Armstrong Nursery

Copper Pot *see* 'DICPE'

'COPPER RUFFLES', Min, or, 1982; flowers small, dbl., exhibition form, slight fragrance; foliage small, medium-green, semi-glossy; upright growth; [Anytime X Sheri Anne]; Dobbs, Annette E.

Copper Sunset™ *see* 'SAVACOP'

'COPPERKINS', HT, ob, 1957; bud long, pointed, dark orange-flame; flowers orange, well formed, dbl., 25 petals, intense, fruity fragrance; foliage glossy; vigorous growth; [Mme Henri Guillot X Golden Scepter]; Ratcliffe

Coppers Grand, S, mr, 1997

Coppertone *see* 'KOROL'

'COPPERY HEART', S, yb, 1958; flowers coppery yellow shaded red, large, dbl., repeat bloom, moderate fragrance; foliage dark, glossy; long stems; very vigorous growth; [Peace X Conrad Ferdinand Meyer]; Gaujard

Copy Cat *see* 'MORCAT'

'COQ DE ROCHE', HT, dr, 1945; flowers blood-red, large, very dbl.; very vigorous growth; [Duquesa de Peñaranda X J. B. Meilland]; Meilland, F.

'COQUELLICOT', HT, or, 1942; flowers orange-red with gold, medium, semi-dbl., slight fragrance; Meilland; (Sangerhausen)

Coquette *see* **'BERTRAM PARK'**

Coquette *see* 'JACCO'

'COQUETTE', HT, lp, 1929; flowers pale flesh-pink, well formed, slight fragrance; vigorous growth; Dobbie

'COQUETTE BORDELAISE', HP, pb, 1896; flowers dark pink with white stripes, very large, dbl., moderate fragrance; Duprat; (Sangerhausen)

'COQUETTE DE LYON', HP, lp, 1859; flowers flesh-pink; Lacharme, F.

'COQUETTE DE LYON', T, ly, 1872; (La Coquette de Lyon); flowers canary-yellow, medium size; Ducher

'COQUETTE DE LYON', T, lp, 1870; Ducher

'COQUETTE DES ALPES', B, w, 1867; flowers white tinged blush, semi-, medium to large, cupped; vigorous growth; [Blanche Lafitte X Sappho (Vibert's)]; Lacharme, F.

'COQUETTE DES BLANCHES', B, w, 1871; flowers white, lightly washed pink, dbl., cupped, moderate fragrance; vigorous growth; [Blanche Lafitte X Sappho (Vibert's)]; Lacharme, F.

'COQUINA', HWich, pb, 1909; flowers rose-pink fading lighter, base creamy white, borne in large, single, cupped, moderate fragrance; foliage dark, almost evergreen; long strong stems; very vigorous (20-24 ft) growth; Walsh

'CORA', HGal, m, 1885; flowers small, dbl.; Savoureux; (Sangerhausen)

Cora Marie *see* 'KORLIMIT'

'CORAIL', LCl, pb, 1931; flowers light peach-blossom-pink, reverse coral-pink and carmine, opening well, dbl., recurrent bloom; foliage bright, glossy; very vigorous growth; [William Allen Richardson X Orléans Rose]; Schwartz, A.

'CORAL', LCl, mp; foliage massive; vigorous, climbing growth; half-hardy.; [R. sinowilsonii X Seedling]; F.C. Stern

'CORAL', HT, pb, 1931; flowers bright coral, base buttercup-yellow, dbl., globular, moderate fragrance; wiry erect stems; vigorous growth; Dickson, A.

Coral® *see* FEloma

'CORAL ANNE GRIFFITHS', HT, mp, 1978; bud very tight; very dbl., 72 petals, moderate fragrance; foliage glossy, dark; very large (6 in) very vigorous growth; [Red Devil sport]; Henson

'CORAL BAY', HT, op, 1971; bud ovoid; flowers silvery coral-orange, medium, dbl., cupped, moderate fragrance; foliage glossy, leathery; vigorous growth; [Seedling X Seedling]; Swim & Weeks; Weeks Wholesale Rose Growers

'CORAL BEAUTY', Pol, mp, 1941; flowers spinel-pink, small borne in clusters, dbl., 1.5 in., flat; vigorous, branching growth; [Sport of Orléans Rose sport]; deRuiter; J&P

'CORAL BELLE', F, ob, 1962; bud pointed, ovoid; flowers vermilion, medium, dbl., 45 petals, 1.5–2.5 in., exhibition form, moderate fragrance; foliage leathery; strong stems; vigorous, upright growth; [Stoplite X Orange Sweetheart]; Jelly; E.G. Hill Co.

Coral Bells, Pol, op, 1962; flowers coral pink, medium, dbl.; J&P; (Sangerhausen)

Coral Border *see* POUlalo

Coral Button, Min, op

Coral Cameo *see* 'MORCALYN'

Coral Carpet, MinFl, mp, 1997; flowers coral pink to light pink; groundcover growth; Williams, J. Benjamin

'CORAL CASCADE', F, or, 1980; flowers coral-red, blooms borne singly or 3-5 per cluster, dbl., 50–75 petals, globular, slight, carnation fragrance; reddish-gray prickles; foliage opens russet, turning darkgreen, glossy; vigorous, compact, bushy growth; [Van Bergen X Pink Hat sport]; James, John

'CORAL CLUSTER', Pol, op, 1920; flowers coral-pink; GM, NRS, 1921; [Orléans Rose sport]; Murrell, R.

'CORAL CREEPER', LCl, dp, 1938; bud deep red; flowers coral to light pink, semi-dbl., 4 in., intense fragrance; foliage leathery; upright stems; very vigorous growth; [(Dr. W. Van Fleet X Emily Gray) X Jacotte]; Brownell, H.C.

'CORAL CROWN', F, or, 1960; flowers coral-red, blooms in clusters, dbl., 35 petals, 3 in., exhibition form, moderate fragrance; foliage glossy; low, compact growth; [Else Poulsen X (Fashion X Orange Triumph)]; Von Abrams; Peterson & Dering

'CORAL CUP', Pol, op, 1936; flowers soft coral, borne in clusters, very dbl., cupped, profuse, repeated bloom, slight fragrance; very vigorous, bushy growth; [Gloria Mundi sport]; B&A

'CORAL DAWN'®, LCl, mp, 1952; bud ovoid; flowers rose-pink, blooms in clusters, dbl., 30–35 petals, 5 in., cupped, moderate fragrance; foliage leathery; vigorous (8-12 ft) growth; [(New Dawn seedling X Seedling) X Seedling]; Boerner; J&P

'CORAL DESTINY', HT, op, 1984; flowers medium coral pink, large, dbl., 35 petals, slight fragrance; foliage medium green, semi-glossy; upright growth; [Joanna Hill X Queen Elizabeth]; Perry, Anthony; Ball Seed Co.

'CORAL DROPS', HMoy, op; flowers pale coral-pink, single, non-recurrent

Coral Fairy *see* 'MORCOFAIR'

Coral Fantasy *see* 'LYOCO'

Coral Fiesta® *see* 'DOTRAMES'

Coral Floc, F, or; Dot

'CORAL GEM', F, pb, 1958; bud ovoid; flowers light coral-pink, open, borne in clusters, dbl., 40–45 petals, 2.5 in., cupped, moderate, fruity fragrance; foliage leathery, dark; vigorous, bushy growth; [((Pinocchio X Mrs Sam McGredy, Climbing) X (Pinocchio X MrsSam McGredy, Climbing)) X Fashion]; Boerner; J&P

'CORAL GLOW', LCl, op, 1964; flowers salmon-pink blooms in clusters; [Spectacular X Seedling]; Croix, P.; Minier

Coral Ice *see* INTerice

Coral Island, Min, op

Coral Meidiland *see* 'MEIPOPAL'

Coral Midinette, Min, op, 1991; Moore

'CORAL MIST', HT, op, 1966; bud ovoid; flowers coral-pink, dbl., exhibition form, moderate fragrance; foliage glossy, light green; vigorous, bushy, open growth; [Spartan X Good News]; Patterson; Patterson Roses

Coral 'n' Gold *see* 'MANCORAL'

Coral Palace *see* 'POULDRON'

'CORAL PILLAR', Cl HT, mr, 1945; flowers geranium-pink, large, dbl., exhibition form, intense fragrance; foliage glossy, dark; very vigorous, upright growth; [Crimson Glory X Capt. Thomas]; Lammerts, Dr. Walter; Univ. of Calif.

'CORAL PRINCESS', F, ob, 1966; bud ovoid; flowers coral-orange, large, dbl., 25 petals, cupped, moderate fragrance; foliage leathery; vigorous, bushy growth; [(Fashion seedling X Garnette seedling) X Spartan]; Boerner; J&P

'CORAL QUEEN', HT, or, 1928; flowers coral-red, large, dbl., cupped, slight fragrance; foliage dark, glossy; long, strong stems; vigorous, bushy growth; [The Queen Alexandra Rose sport]; Reeves

'CORAL QUEEN ELIZABETH', F, op, 1966; flowers coral-salmon, borne in clusters, dbl., 3 in., moderate fragrance; foliage glossy; very vigorous growth; [Queen Elizabeth X Seedling]; Gregory

Coral Reef *see* 'COCDARLEE'

'CORAL REEF', HT, mp, 1948; bud long, pointed; flowers pink, medium, dbl., 25–30 petals, moderate fragrance; foliage leathery, wrinkled, dark; vigorous, upright, bushy growth; RULED EXTINCT 4/86; [Joanna Hill X R.M.S. Queen Mary]; Hill, Joseph H., Co.

Coral Rosamini, Min, op

Coral Sand *see* 'RENSAND'

'CORAL SATIN', LCl, op, 1960; bud ovoid; flowers coral, dbl., 25 petals, 3.5–4 in., exhibition form, moderate fragrance; foliage leathery, glossy; vigorous (6-8 ft) growth; [New Dawn X Fashion]; Zombory; J&P

'CORAL SEA', HT, dp, 1942; bud globular, old-rose; flowers light red, large, dbl., 40–45 petals, exhibition form, slight fragrance; foliage leathery, dark; long, strong stems; vigorous, upright, much branched growth; [Katharine Pechtold X R.M.S. Queen Mary]; Hill, Joseph H., Co.

Coral Sea, HT, op, 1999

'CORAL SILK', F, op, 1972; flowers coral and peach, semi-dbl., 18 petals, 3 in., flat, moderate fragrance; foliage glossy, dark; very free growth; Gregory

Coral Spire *see* KORkragor

Coral Sprite™ *see* 'JACORAL'

'CORAL STAR', HT, op, 1967; flowers coral-pink, well-formed, medium, intense fragrance; vigorous, upright growth; [Tropicana X Stella]; Robinson, H.

'CORAL SUNSET', HT, or, 1966; (Coucher de Soleil); bud ovoid; flowers coral-red, large, dbl., moderate fragrance; foliage dark, leathery; vigorous growth; [Garnette seedling X Hawaii]; Boerner; J&P

'CORAL TREASURE', Min, ob, 1971; bud ovoid; flowers coral-orange, medium, dbl.; foliage glossy, leathery; dwarf, bushy growth; [Seedling X Little Buckaroo]; Moore, Ralph S.; Sequoia Nursery

'CORALGLO', F, op, 1960; bud ovoid; flowers orange-rose, large, dbl., 42 petals, cupped, slight fragrance; foliage bronze; bushy growth; [Independence X Fashion]; Boerner; Stark Bros.

'CORALI', LCl, rb, 1931; flowers coral-red to rosy salmon, profuse seasonal bloom; foliage bright green; vigorous growth; Dot, Pedro

'CORALIE', M, lp; flowers flesh, well-formed, medium, dbl.; Miellez, ca 1860

'CORALIE', D, lp; flowers soft pink, cupped; foliage grayish; vigorous growth

'CORALIE', LCl, pb, 1919; flowers coral-red to deep pink, dbl., non-recurrent; foliage glossy; vigorous growth; [Hiawatha X Lyon Rose]; Paul, W.

'CORALIN', Min, or, 1955; (Carolin, Carolyn, Karolyn); flowers coral-red, dbl., 40 petals; low, compact growth; [Méphisto X Perla de Alcañada]; Dot, M.

'CORALÍN SUPERB', Min, dr, 1958; [Coralín sport]; Will; Kordes

'CORALITA', LCl, or, 1964; bud ovoid, deep red; flowers orange-coral, dbl., 40–45 petals, 4 in., moderate fragrance; foliage dark, leathery; vigorous (6-8 ft) growth; David Fuerstenberg Prize, ARS, 1968; [(New Dawn X Geranium Red) X Fashion]; Zombory; J&P

Coralitos *see* 'SUNCOR'

'CORALLINA', T, rb, 1900; flowers coppery red, large, dbl.; Paul, W.

Corallina *see* BARcora

'CORALLINE', Pol, or, 1938; flowers very large clusters, profuse, repeated bloom; very rampant growth; [Gloria Mundi X Golden Salmon Superieur]; Smith, J.; Eddie

Corallovy Surprise, HT, 1982; Klimenko, V. N.; (Cavriglia)

'CORBEILLE ROYALE', HT, mp, 1956; bud globular; flowers salmon-pink, dbl., moderate fragrance; vigorous, upright growth; Buyl Frères

'CORDELIA', HT, ab, 1975; flowers peach shaded deeper, pointed, dbl., 20 petals, 4–4.5 in., moderate fragrance; foliage dark; tall growth; LeGrice

'CORDELIA DE GREY', S, dp, 1995; flowers deep, bright pink, small white eye, golden stamens, single, 5 petals, 2 in., borne in small clusters, slight fragrance; numerous prickles; foliage small, dark green, dull; medium, upright growth, arching canes; de Grey, Cordelia

'CORDIAL', HT, mp, 1965; bud ovoid; flowers pink, medium, in clusters, dbl.; foliage dark; [Satisfaction X Seedling]; Verbeek

Cordon Bleu *see* 'HARUBASIL'

Cordon Rouge, LCl, or, 1970; flowers large, dbl.; Combe; (Sangerhausen)

Cordula® *see* 'KORTRI'

'CORED', HT, mr, 1973; (**Alec's Red®**); dbl., 45 petals, 6 in., moderate fragrance; foliage matt, green; vigorous, upright growth; ADR, 1973 Edland Fragrance Medal, ARS, 1969 GM, RNRS, 1970 PIT, RNRS, 1970; [Fragrant Cloud X Dame de Coeur]; Cocker, 1970

'COREHEAD', HT, mr, 1960; flowers bright crimson, intense fragrance; [Ena Harkness X Hazel Alexander]; Dicksons of Hawlmark

Corina *see* 'JACVEP'

'CORINIUM', F, m, 1970; flowers mauvish pink, full rosette form, dbl., 48–50 petals, 3 in., intense fragrance; foliage semi-glossy; [Alamein sport]; Cooper

Corrine *see* 'BROCOR'

Corky *see* 'ZIPCORK'

'CORLIA', HT, or, 1921; flowers terra-cotta, moderate fragrance; moderate growth; Bees

Corne Herholdt, F, pb

'CORNELIA', HMsk, pb, 1925; flowers strawberry flushed yellow, rosette form, small, dbl., borne in flattish sprays, recurrent bloom, moderate fragrance; foliage dark bronze, leathery, glossy; very vigorous growth; (14); Pemberton

'CORNELIA', HT, pb, 1919; flowers light pink, base orange, dbl., moderate fragrance; [Ophelia X Mrs Aaron Ward]; Scott, R.

Cornelia Cook *see* **'CORNÉLIE KOCH'**

'CORNÉLIE KOCH', T, w, 1855; (Cornelia Cook); flowers creamy white, tinged lemon-yellow and flesh, well formed, dbl., moderate fragrance; vigorous growth; [Devoniensis seedling]; Koch

'CORNELIS TIMMERMANS', HT, pb, 1919; flowers clear pink, edged deep yellow, dbl., moderate fragrance; [Pharisaer X Le Progres]; Timmermans

'CORNER', HT, pb, 1945; bud pointed; flowers cerise, base yellow, reverse yellow washed pink, semi-dbl., 20 petals, 4.5 in., moderate fragrance; foliage leathery, glossy, dark; vigorous, bushy growth; [Soeur Thérèse X Seedling]; Wyant

'CORNET', HP, m, 1845; flowers rose tinted with purple, very large, very dbl., cupped, moderate, centifolia fragrance; vigorous, branching growth; Lacharme, F.

'CORNICE DE TARN-ET-GARONNE', B, mr, 1852; Pradel

Cornsilk(TM) *see* 'SAVASILK'

'COROB', F, dr, 1970; (**Rob Roy**®); dbl., 30 petals, 4.5 in., exhibition form, slight fragrance; foliage glossy; [Evelyn Fison X Wendy Cussons]; Cocker, 1971

'COROLLE', LCl, mr, 1962; flowers medium, semi-dbl., 12 petals; vigorous growth; [Spectacular X Cocktail]; Dot, Simon

Corona de Oro *see* **'GOLD CROWN'**

'CORONADO', HT, rb, 1961; bud long, pointed; flowers red, reverse yellow, dbl., 40 petals, 5–6 in., exhibition form, moderate fragrance; foliage glossy, dark; vigorous, upright growth; [(Multnomah X Peace) X (Multnomah X Peace)]; Abrams, Von; Peterson & Dering

Coronado, Climbing, Cl HT; Ansaloni; (Cavriglia)

'CORONATION', HWich, rb, 1911; flowers red, lightly striped white, small, large clusters; Turner

'CORONATION', HP, mp, 1913; flowers flesh, shaded bright shrimp-pink, well formed, large, dbl., 50 petals, recurrent bloom, moderate fragrance; smooth wood; vigorous growth; GM, NRS, 1912; Dickson, H.

'CORONATION GOLD', F, ab, 1978; (Maja Oetker); bud globular; flowers golden yellow to apricot, dbl., 27 petals, 4 in., slight fragrance; foliage glossy; vigorous, upright growth; [(Sabine X Circus) X (Anne Cocker X Arthur Bell)]; Cocker, A.; Cocker

'CORONATION GOLD', HT, yb, 1954; flowers golden yellow flushed crimson, large, intense, damask fragrance; foliage glossy; vigorous growth; RULED EXTINCT 1980; [Signora X Peace]; Cox

'CORONET', F, dr, 1957; flowers deep crimson, blooms in clusters, semi-dbl., 17 petals, 3 in.; foliage dark, glossy; vigorous, upright growth; [Independence X Red Wonder]; deRuiter; Blaby Rose Gardens

'CORONET SUPREME', HT, mp, 1955; bud short, pointed; flowers rose-pink, dbl., 45–65 petals, 4–5 in., moderate fragrance; foliage dark, glossy; very vigorous, upright growth; [Seedling X Golden Rapture]; Jelly; E.G. Hill Co.

'CORPORAL JOHANN NAGY', HSet, dp, 1890; flowers carmine pink, medium, dbl.; Geschwind, R.; (Sangerhausen)

'CORPUS CHRISTI', HT, dr, 1985; flowers large, dbl., 35 petals, moderate fragrance; foliage large, medium green, semi-glossy; upright, bushy, spreading growth; PP005854; [Seedling X Night Time]; Weeks, O.L.; Weeks Wholesale Rose Growers

Corrida, Gr, 1969; Delbard-Chabert; (Cavriglia)

'CORRIE KOSTER', Pol, dp, 1923; flowers light coral-red, opening to deep pink, borne in clusters, semi-dbl., slight fragrance; [Juliana Rose sport]; Koster, M.; Royer

'CORROBOREE', HT, m, 1962; bud long, pointed; flowers lilac-mauve, open, very large, dbl., intense, damask fragrance; foliage glossy; vigorous, upright growth; [Baccará X My Choice]; Fankhauser

Corry, HT, or, 1978; Kordes

'CORSAGE', F, w, 1965; flowers small blooms in clusters, semi-dbl., globular, slight fragrance; foliage soft; vigorous, bushy growth; [Blanche Mallerin X White Swan]; Belden; Wyant

Corsair *see* 'SANROC'

Corsaire, HT, 1974; Combe, M.; (Cavriglia)

'CORSICA', F, pb, 1964; flowers salmon-pink edged dark pink to red, semi-dbl., 20 petals, intense fragrance; foliage dark, glossy; vigorous, bushy growth; Verschuren, A.; van Engelen

'CORSO'®, HT, ob, 1976; flowers coppery orange, dbl., 33 petals, 4.5 in., slight fragrance; foliage glossy, dark; [Anne Cocker X Dr. A.J. Verhage]; Cocker

'CORSO FLEURI', F, dr, 1956; flowers scarlet-red, petals waved, semi-dbl.; dwarf, compact growth; [Red Favorite sport]; Mondial Roses

Cortege *see* COMteri

Corvette *see* KORveco

'CORYANA', S, mp; (R. X coryana); flowers rich pink, single, 2.5 in.; sparingly prickly; tall, shrub growth; [R. roxburghii X R. macrophylla]; Hurst, C.C., 1926

'CORYLUS', S, mp, 1988; flowers open, medium, borne usually singly or in sprays of 1-4, single, slight fragrance; round, medium, scarlet fruit; prickles small, light brown; foliage medium, dark green, deep veined; bushy growth; [R. nitida X R. rugosa rubra]; LeRougetel, Hazel; Peter Beales, 1988

'COSETTA'®, F, ob, 1984; flowers light orange, reverse deeper with reddish shadings, medium, dbl., 30 petals, no fragrance; foliage medium, dark, matt; [Zorina X Sole di San Remo]; Bartolomeo, Embriaco

Cosette *see* 'HARQUILLYPOND'

Cosima® *see* 'TANCOFEUMA'

'COSIMO RIDOLFI', HGal, m, 1842; flowers old-rose to lilac, spotted crimson, medium, cupped; foliage soft green; compact growth; Vibert

Cosmic *see* 'BRICOS'

Cosmopoliet *see* **'COSMOPOLITAN'**

Cosmopolit *see* **'COSMOPOLITAN'**

'COSMOPOLITAN', F, mp, 1955; (Cosmopoliet, Cosmopolit);, large clusters; bushy growth; [Silberlachs X Seedling]; Buisman, G. A. H.

Cosmos® *see* COMsan

Costa Dorada, HT; Dot; (Cavriglia)

Cote d'Azur, HT, m, 1998

Cote Rotie *see* **'GRAND AMOUR'**

'COTILLION', F, yb, 1967; bud ovoid; flowers yellow, edged pink, medium, exhibition form, slight fragrance; foliage leathery; vigorous, upright growth; [Rumba X Golden Garnette]; Hill, Joseph H., Co.

Cotillion(TM) *see* 'JACSHOK'

Coton Gold *see* 'BABS'

'COTORRITA REAL', Pol, pb, 1931; flowers white, rose and yellow, small, borne in clusters, very dbl., globular, profuse, repeated bloom, slight fragrance; vigorous growth; Padrosa

'COTSWOLD CHARM', HT, dp, 1967; flowers magenta, exhibition form; foliage dark, leathery; moderate growth; [Ophelia seedling X William Moore]; Bennett, V.G.T.

Cotswold Gold *see* 'JAYCOT'

Cotswold Sunset *see* 'JAYSUN'

Cottage Dream, S, mp

Cottage Garden *see* 'HARYAMBER'

Cottage Garden *see* 'CHEWWILY'

Cottage Maid *see* **'VILLAGE MAID'**

Cottage Maid *see* POUlspan

Cottage Pink, HMult, mp

Cottage Rose *see* 'AUSGLISTEN'

Cottage White, HMult, w

'COTTON CANDY', HWich, mp, 1952; bud well-formed; very dbl., 2.5 in., blooms in clusters; foliage very glossy, turning to autumn colors; vigorous (10-15 ft) growth; [R. wichurana X Seedling (F with R. multibracteata ancestry)]; Moore, Ralph S.; Sequoia Nursery

'COTTON TOP', F, w, 1962; bud short, pointed; dbl., 45–50 petals, 2–3 in., flat, slight fragrance; foliage leathery; vigorous, upright, well-branched growth; [Seedling X White Butterfly]; Hill, Joseph H., Co.

Cottontail *see* 'TINTAIL'

Coucher de Soleil *see* **'CORAL SUNSET'**

Coucou *see* DORgold

'COULEUR DE BRENNUS', HGal, mr; flowers medium, dbl.; slender shrub growth

'COUNTESS CADOGAN', HT, lp, 1978; bud long, pointed; dbl., 40–45 petals, 4.5 in., exhibition form, slight fragrance; foliage light green, leathery; very vigorous, upright growth; [Carlita sport]; Buss; H. Buss Nursery

Countess Celeste(TM) *see* 'POULDRON'

'COUNTESS CLANWILLIAM', HT, mr, 1915; flowers pinkish cherry-red, dbl., exhibition form, moderate fragrance; foliage rich green; bushy growth; GM, NRS, 1913; Dickson, H.

'COUNTESS MARY', Cl HT, mp, 1933; [Mary, Countess of Ilchester Climbing sport]; Dixie Rose Nursery

Countess Mary of Ilchester *see* **'MARY, COUNTESS OF ILCHESTER'**

'COUNTESS OF DALKEITH', F, rb, 1957; flowers vermilion flushed orange, very dbl., intense fragrance; vigorous growth; [Fashion sport]; Dobbie

'COUNTESS OF DERBY', HT, w, 1906; flowers creamy white with pink tints, large, dbl., intense fragrance; Dickson, A.; (Sangerhausen)

'COUNTESS OF ELGIN', HT, op, 1925; flowers salmon-pink, reverse deep rose-pink; vigorous growth; [Mme Edouard Herriot sport]; Ferguson

'COUNTESS OF GOSFORD', HT, op, 1906; flowers salmon-pink, long pointed, large, dbl.; GM, NRS, 1905; McGredy

Countess of Ilchester *see* **'MARY, COUNTESS OF ILCHESTER'**

'COUNTESS OF LIEVEN', Ayr, w; flowers creamy white, medium, semi-dbl., cupped

'COUNTESS OF LONSDALE', HT, dy, 1919; dbl., moderate fragrance; Dickson, H.

Countess of Oxford *see* **'COMTESSE D'OXFORD'**

'COUNTESS OF PEMBROKE', HT, pb; flowers pink shaded darker, large, very dbl.; vigorous growth

'COUNTESS OF STRADBROKE', Cl HT, dr, 1928; bud ovoid; flowers dark glowing crimson, well shaped, very large, dbl., globular, free, recurrent bloom, intense fragrance; foliage rich green, wrinkled; vigorous growth; [Walter C. Clark X Seedling]; Clark, A.; Hazlewood Bros.

'COUNTESS OF WARWICK', HT, my, 1919; flowers lemon-yellow, edged pink, large; Easlea

Countess Vandal *see* **'COMTESSE VANDAL'**

Countess Vandal, Climbing *see* **'COMTESSE VANDAL, CLIMBING'**

'COUNTRY DANCER', S, dp, 1973; bud ovoid; flowers rose-red, large, dbl., repeat bloom, moderate fragrance; foliage large, glossy, dark, leathery; vigorous, dwarf, upright, bushy growth; [Prairie Princess X Johannes Boettner]; Buck, Dr. Griffith J.; Iowa State University

'COUNTRY DOCTOR', HT, lp, 1952; bud long, pointed to ovoid; flowers silvery pink, large, dbl., exhibition form, moderate fragrance; foliage glossy; vigorous, bushy growth; [Pink Princess X Crimson Glory]; Brownell, H.C.

Country Fair *see* HARbanner

Country Garden *see* 'BOSANNCOAT'

'COUNTRY GIRL', F, mr, 1958; bud ovoid; flowers geranium-red, medium, dbl.; vigorous growth; [Independence X Salmon Perfection]; Temmerman; Schraven

Country Girl *see* 'HORALCAMSTRIP'

Country Girl *see* 'WILCGIR'

'COUNTRY GIRL, CLIMBING', Cl F, mr, 1962; Buyl Frères

Country Heritage *see* 'COCCAGES'

Country Joy *see* 'MORCOJO'

Country Lady *see* 'HARTSAM'

Country Life *see* HARzap

Country Living *see* 'AUSCOUNTRY'

'COUNTRY MAID', F, lp, 1971;, slight fragrance; low, compact growth; [Tip Top sport]; Whartons Roses; Deamer

Country Morning *see* 'FLOMOR'

'COUNTRY MUSIC', S, dp, 1973; bud ovoid; flowers neyron rose, large, dbl., intermittent bloom, moderate fragrance; foliage large, leathery; vigorous, dwarf, upright, bushy growth; [Paddy McGredy X ((World's Fair X Floradora) X Applejack)]; Buck, Dr. Griffith J.; Iowa State University

Country Music *see* HARcheer

'COUNTRY SONG', S, lp, 1984; bud ovoid; flowers large, dbl., 28 petals, cupped, borne 1-5 per cluster, repeat bloom, moderate, myrrh fragrance; awl-like, brown prickles; foliage leathery, dark; erect, bushy growth; hardy.; [Carefree Beauty X The Yeoman]; Buck, Dr. Griffith J.; Iowa State University

'COUNTRYMAN', S, pb, 1978; bud ovoid, pointed; flowers light rose-bengal, dbl., 25–30 petals, 4–5 in., cupped, moderate fragrance; foliage large, dark, leathery; vigorous, upright, spreading, bushy. growth; [(Improved Lafayette X Independence) X Maytime]; Buck, Dr. Griffith J.; Iowa State University

'COUNTRYWOMAN', HT, my, 1978; bud globular; flowers lemon-yellow, medium, dbl., moderate fragrance; foliage leathery; bushy growth; [Seedling X Peace]; Dawson, George; Australian Roses

'COUNTY FAIR', F, mp, 1960; bud ovoid, pointed; flowers medium to dark pink, fading much lighter, blooms in clusters, 8–10 petals, 2.5–3 in., flat, slight fragrance; foliage leathery, dark, semi-glossy; vigorous, bushy growth; [Frolic X Pink Bountiful]; Swim, H.C.; Armstrong Nursery

County Girl *see* 'HORALCAMSTRIP'

Coup de Coeur *see* MALloeur

'COUP DE FOUDRE', F, or, 1956; bud well-formed; flowers fiery red, cupped; foliage glossy, bronze; vigorous growth; [(Peace X Independence) X Oiseau de Feu]; Hémeray-Aubert

Coup de Foudre, Climbing, Cl F, 1976; Fineschi, G.; (Cavriglia)

'COUPE D'HÉBÉ', B, dp, 1840; flowers deep pink, waxy texture, large, very dbl., cupped, intense fragrance; foliage glossy; vigorous, erect growth; [Bourbon X R. chinensis hybrid]; Laffay, M.

'COUPE D'OR', LCl, my, 1930; flowers canary-yellow, open, dbl., cupped, moderate fragrance; foliage rich green, leathery, glossy; vigorous, climbing or trailing growth; [Jacotte seedling]; Barbier

Coupe d'Or *see* **'GOLD CUP'**

'COURAGE', HT, dr, 1923; flowers deep brilliant maroon-crimson, very large, dbl., exhibition form, intense fragrance; foliage richgreen, leathery; bushy, dwarf growth; McGredy

'COURAGE', HT, rb, 1941; flowers red tinted yellow, open, very large, dbl., slight fragrance; vigorous, bushy growth; [Seedling X Brazier]; Mallerin, C.; A. Meilland

Courage *see* POUlduff

Courageous Indira, HT, ob, 1991; Chandrakant

'COURIER', LCl, pb, 1930; flowers pink on white ground, borne in clusters; vigorous, climbing growth; Clark, A.; Brundrett

Couronne d'Or *see* **'GOLD CROWN'**

'COURT JESTER', F, ob, 1980; flowers orange, reverse yellow, borne 5-7 per cluster, exhibition form, slight fragrance; large hooked, red-brown prickles; foliage mid-green, glossy; tall, upright growth; Cants of Colchester, Ltd.

Courtisane, HT, or, 1965; flowers large, dbl.; Gaujard; (Sangerhausen)

Courtney *see* 'GELCOURT'

'COURTNEY PAGE', HT, dr, 1922; flowers velvety dark scarlet-crimson, dbl., intense fragrance; GM, NRS, 1920; McGredy

Courtoisie® *see* 'DELCOURT'

'COURTSHIP', HT, mp, 1955; bud conical; flowers cerise-pink, reverse lighter, dbl., 28 petals, 4–5 in., exhibition form; foliage dark; vigorous, bushy growth; [Mme Henri Guillot X Peace]; Shepherd; Bosley Nursery

Courvoisier® *see* 'MACSEE'

'COUSIN ESSIE', S, w, 1996; semi-dbl., 8–14 petals, 1.5 in., moderate fragrance; few prickles; foliage medium, medium green, semi-glossy; spreading, busy, tall, (6 ft) growth; [Seedling of Honeyflow sport]; Robertson, Myrtle; Honeysuckle Cottage Nursery, 1988

'COVA DA IRIA', HT, rb, 1963; flowers light red, reverse gold; [Seedling X Crimson Glory]; da Silva, Moreira

'COVENT GARDEN', HT, rb, 1919; flowers rich deep crimson flushed plum-black on reverse, intense fragrance; foliage leathery, glossy; vigorous growth; GM, NRS, 1918; Cant, B. R.

'COVENTRIAN', F, dp, 1962; flowers ruby-cerise, well formed, borne in large clusters, dbl., 30 petals, 3–3.5 in.; foliage dark, glossy; vigorous, bushy growth; [Highlight X Seedling]; Robinson, H.

Coventry Cathedral *see* **'CATHEDRAL'**

'COVER GIRL', HT, ob, 1960; bud long, pointed; flowers orange, copper and gold, dbl., 28–35 petals, 5 in., exhibition form, slight fragrance; foliage glossy, dark; upright, bushy growth; [Sutter's Gold X (Mme Henri Guillot X Seedling)]; Von Abrams; Peterson & Dering

'COWICHAN SUPER', Cl HT, op, 1975; flowers orange-salmon, dbl., 30–35 petals, 5 in., bloom repeats, intense fragrance; foliage glossy; tall, very vigorous growth; Tamarack Roses

Cox's Pink Polyantha, Pol, mp

Cox's Red Polyantha, Pol, mr

'COY COLLEEN', HT, w, 1953; (Blushing Rose); bud pointed, rosy white; flowers milky white, well-formed, borne in clusters, slight fragrance; foliage glossy; vigorous growth; [(Modesty X Portadown Glory) X Phyllis Gold]; McGredy, Sam IV

'CRACKER', HT, lp; flowers bright red with prominent stamens; Clark, A., 1920

'CRACKER, CLIMBING', S, mr, 1927; flowers red with white, large, single, intense fragrance; Clark, A.; (Sangerhausen)

'CRACKERJACK', F, rb, 1959; flowers scarlet flushed yellow, borne in clusters, semi-dbl., 20 petals, moderate fragrance; free growth; [Fashion X Masquerade]; Fryers Nursery, Ltd.

Crackerjack *see* POUlcrack

Crackling Fire™ *see* 'JACORG'

'CRAIGWEIL', HT, pb, 1929; flowers silvery cerise-pink, reverse deeper, dbl., moderate fragrance; strong stems; vigorous growth; [Mme Abel Chatenay seedling]; Hicks

'CRAMOISI DES ALPES', HGal, mr; flowers bright purplish red, dbl., heavy, nonrecurrent bloom; Hardy (?), before 1829

'CRAMOISI ÉBLOUISSANT', HCh, dr, 1839; flowers medium, dbl.; (Sangerhausen)

'CRAMOISI FONCÉ VELOUTÉ', M, dr; flowers deep velvety crimson

'CRAMOISI PICOTÉ', HGal, rb, 1834; flowers crimson, streaked and mottled (striped & spotted) darker, medium, very dbl., 2 in.; Vibert

'CRAMOISI SUPÉRIEUR', HCh, mr, 1832; (Lady Brisbane); flowers crimson-red, small blooms in large clusters, dbl., cupped, recurrent bloom; vigorous growth; (21); Coquereau

'CRAMOISI SUPÉRIEUR, CLIMBING', Cl HCh, mr, 1885; (Agrippina, Climbing, Lady Brisbane, Climbing); (21); Couturier

'CRANBROOK', Pol, lp; Matthews, W.J., 1921; (Weatherly, L.)

'CRARAE', HT, yb, 1981; flowers deep yellow marked with scarlet, large, dbl., 35 petals, slight fragrance; foliage medium, dark, glossy; bushy growth; [Piccadilly X Fred Gibson]; McKirdy, J.M.; John Sanday Roses

Crathes Castle *see* 'COCATHES'

Crazy Dottie *see* 'SEADOT'

Crazy For You *see* 'WEKROALT'

Crazy Horse, F; de Ruiter, G.; (Cavriglia)

Crazy Quilt *see* 'MORTRIP'

Crazy Spire, HT, ob, 1995

'CREAM CRACKER', HT, w, 1979; bud pointed; flowers cream, classical pointed form, blooms borne singly, dbl., 38 petals, 4.5 in., intense, apple fragrance; often without prickles; foliage large, matt; bushy growth; [Columbine X Iceberg]; Murray, Nola; Rasmussen's

'CREAM CRACKER', HT, lp, 1933; flowers creamy buff, reverse shaded salmon, large, dbl.; vigorous growth; RULED EXTINCT 10/78 ARM; Dickson, A.

Cream Delight *see* 'SUNCREDEL'

Cream Dream *see* KORomtar

'CREAM GOLD', Min, my, 1978; bud long, pointed; flowers small, dbl., 38 petals, 1.5 in., exhibition form, moderate fragrance; compact, spreading growth; [Golden Glow X Seedling]; Moore, Ralph S.; Sequoia Nursery

'CREAM PEACH', F, w, 1975; flowers cream, edged pink, full, semi-dbl., 15–20 petals, 3–4 in., slight fragrance; foliage large, glossy; very free growth; [Paddy McGredy X Seedling]; Sheridan

'CREAM PUFF', MinFl, pb, 1981; bud ovoid; flowers cream blushed pink, borne singly or in clusters of 3, semi-dbl., 18 petals, moderate fragrance; long prickles; foliage dark, semi-glossy; spreading, bushy growth; [Little Darling X Elfinesque]; Bennett, Cecilia 'Dee'; Tiny Petals Nursery

Cream Puff, Climbing, Cl Min, lp; [Cream Puff sport]; Trimper, K., 1996; (Weatherly, L.)

Cream Sunsation *see* KORgatine

'CREDO', HT, mr, 1965; flowers purplish red, medium, dbl., exhibition form, moderate fragrance; foliage leathery; vigorous, upright growth; [Eminence X John S. Armstrong]; Gaujard

'CREE', S, lp; flowers pale pink fading to white, large, single, early bloom, nonrecurrent; foliage glossy, bright green; vigorous growth; [R. rugosa albo-plena X R. spinosissima hispida]; Central Exp. Farm

'CREEPING EVERBLOOM', LCl, mr, 1939; dbl., 30 petals, 4 in., borne in clusters, recurrent bloom, moderate fragrance; canes 3 ft long growth; [Frederick S. Peck X (Général Jacqueminot X Dr. W. Van Fleet)]; Brownell, H.C.

Creepy *see* 'MORPAPPLAY'

'CREINA MURLAND', HT, dy, 1934; flowers sunflower-yellow, deepening, dbl., slight fragrance; foliage glossy; very vigorous growth; Dickson, A.

'CREME', HWich, lp, 1895; flowers small, semi-dbl., intense fragrance; Geschwind, R.; (Sangerhausen)

Crème de la Crème *see* GANcre

Crème Glacée *see* 'LAVCREME'

'CRÉOLE', HT, dr, 1962; bud long, pointed; flowers purplish, red base coppery, large, dbl., slight fragrance; foliage dark, glossy; very vigorous, bushy growth; [Peace X Josephine Bruce]; Gaujard

Crêpe de Chine® *see* 'DELTOP'

'CREPE MYRTLE', F, mr, 1937; (Harold Ickes); height 5 ft; [Permanent Wave sport]; Dixie Rose Nursery

Crepe Suzette *see* 'TRASUZ'

'CRÉPUSCOLO', HT, ob, 1955; flowers copper, pointed, dbl.; strong stems; upright growth; [Julien Potin X Sensation]; Aicardi, D.; Giacomasso

'CRÉPUSCULE', N, ab, 1904; flowers orange, fading to apricot-yellow; makes a tall hedge growth; Dubreuil

'CRESCENDO', HT, rb; flowers bright red, base touched with white, well formed, large, dbl., moderate fragrance; tall growth

'CRESSET', Gr, mr, 1961; bud long, pointed; flowers scarlet, semi-dbl., exhibition form, blooms in clusters, moderate fragrance; foliage leathery, glossy, dark; vigorous, bushy growth; [Queen Elizabeth X Cocorico]; Francis; F. Mason

Cressida *see* 'AUSCRESS'

'CRESTED JEWEL', M, mp, 1971; bud long, pointed, "mossed" similar to crested moss; flowers bright rose-pink, medium, semi-dbl., exhibition form; foliage leathery; vigorous growth; [Little Darling X Crested Moss]; Moore, Ralph S.; Sequoia Nursery

'CRESTED MOSS', C, mp, 1827; (Chapeau de Napoléon, Crested Provence Rose, Cristata, R. centifolia cristata, R. centifolia muscosa cristata); flowers similar to centifolia muscosa, but the mossy excrescences confined to sepal edges; Vibert

Crested Provence Rose *see* **'CRESTED MOSS'**

Crested Sweetheart *see* 'MORSWEET'

Cricket™ *see* 'AROKET'

Cricket Cl., Cl Min, ob

Cricri® *see* 'MEICRI'

Cri-Cri *see* 'MEICRI'

Crimean Night, HT, dr; USSR

'CRIMSON BEAUTY', HT, dr, 1930; flowers crimson, dbl.; very vigorous growth; [Hoosier Beauty X Crimson Champion]; Dingee & Conard

'CRIMSON BEAUTY', HT, dr, 1935; flowers red, shaded scarlet and maroon, large, dbl., intense fragrance; foliage leathery; vigorous growth; [Daily Mail Scented Rose X Étoile de Hollande]; LeGrice

'CRIMSON BEDDER', HP, mr, 1874; flowers medium, dbl.; Cranston; (Sangerhausen)

Crimson Blush *see* SIEson

Crimson Bouquet™ *see* 'KORBETEILICH'

Crimson Boursault *see* **'AMADIS'**

'CRIMSON BROCADE', HT, dr, 1962; flowers bright scarlet-crimson, large, exhibition form, moderate fragrance; strong stems; vigorous growth; Robinson, H.

Crimson Cascade *see* FRYclimbdown

'CRIMSON CHAMPION', HT, dr, 1916; flowers velvety crimson-red, dbl., moderate fragrance; dwarf growth; [Étoile de France X Seedling]; Cook, J.W.

'CRIMSON CHATENAY', HT, dr, 1915;, moderate fragrance; [Mme Abel Chatenay X Leuchtfeuer (HCh)]; Merryweather

Crimson China Rose *see* **'SLATER'S CRIMSON CHINA'**

'CRIMSON CONQUEST', Cl HT, dr, 1931; flowers scarlet crimson, small, semi-dbl., 2.5 in.; foliage light green; vigorous growth; [Red-Letter Day sport]; Chaplin Bros.

'CRIMSON CROWN', HT, mr, 1905; flowers medium, semi-dbl.; Dickson, A.; (Sangerhausen)

'CRIMSON DAMASK', D, dr, 1901; Turner

'CRIMSON DAWN', F, rb, 1970; flowers crimson to carmine, full, dbl., 20 petals, 3.5–4.5 in., moderate fragrance; foliage dark; vigorous growth; [(Anne Poulsen X Dainty Maid) X (Bonn X Opera)]; Ellick

Crimson Delight *see* 'JOHILLGOLF'

'CRIMSON DESCANT', LCl, mr, 1972; flowers crimson, dbl., 30 petals, 5 in., slight fragrance; [Dortmund X Etendard]; Cants of Colchester, Ltd.

'CRIMSON DIAMOND', HT, dr, 1947; bud long, pointed; flowers crimson-red, open, large, dbl., 35–40 petals, moderate fragrance; bushy growth; [Crimson Glory X Charlotte Armstrong]; Lammerts, Dr. Walter; L.C. Lovett

'CRIMSON DUKE', HT, mr, 1963; bud ovoid; flowers crimson, dbl., 45–55 petals, 4–5 in., exhibition form, moderate fragrance; foliage leathery, dark; vigorous, upright, bushy growth; [(Happiness X Independence) X Peace]; Meilland, Alain A.; C-P

Crimson Elegance *see* 'LEOCREL'

'CRIMSON EMBLEM', HT, dr, 1916; flowers brilliant crimson-scarlet, large, dbl., cupped; vigorous growth; McGredy

'CRIMSON ERECTA', HT, 1954; Cazzaniga, F. G.; (Cavriglia)

'CRIMSON FRAGRANCE', HT, dr, 1979; bud long, slender; dbl., 38 petals, borne singly, intense fragrance; reddish-bronze prickles; foliage slender, mid-green; branching, upright growth; [Fragrant Cloud X Seedling]; Wright, R. & Sons

Crimson Gallica, HGal, mr

'CRIMSON GEM', Min, dr, 1974; (Flammette); bud ovoid; flowers deep red, medium, very dbl., cupped, slight fragrance; foliage bronze, soft; vigorous, bushy growth; [Lillan X Polyantha seedling]; deRuiter; C-P

'CRIMSON GLOBE', M, dr, 1890; flowers deep crimson, large, dbl., globular; vigorous growth; Paul, W.

'CRIMSON GLORY', HT, dr, 1935; bud long, pointed; flowers deep velvety crimson, large, dbl., 30–35 petals, 5 in., cupped, borne mostly singly, intense, damask fragrance; foliage leathery; vigorous, bushy, spreading growth; (28); GM, NRS, 1936 James Alexander Gamble Fragrance Medal, ARS, 1961; [Cathrine Kordes seedling X W.E. Chaplin]; Kordes; Dreer;, J&P

'CRIMSON GLORY, CLIMBING', Cl HT, dr, 1946; (Grimpant Crimson Glory); J&P

'CRIMSON GLOW', Pol, dr, 1945; flowers oxblood-red, semi-dbl., cupped, intense fragrance; foliage glossy, dark; vigorous, upright growth; [Night X Mrs Dudley Fulton]; Lammetts; Univ. of Calif.

Crimson Glow *see* **'OUR PRINCESS'**

'CRIMSON HALO', HT, dr, 1964; flowers deep rose-red, dbl., 30 petals, 5 in., globular, intense fragrance; vigorous growth; [Karl Herbst X Crimson Glory]; Park; Harkness

'CRIMSON KING', HT, dr, 1943; (Liebesglut); bud ovoid, long, pointed; flowers deep velvety crimson, large, dbl., exhibition form, intense, damask fragrance; foliage leathery; vigorous, bushy growth; [Crimson Glory X Kardinal]; Kordes; C-P

Crimson Lace™ *see* 'JACEZZ'

Crimson Maman Cochet *see* **'BALDUIN'**

Crimson Masse *see* **'LIBERTÉ'**

Crimson Medinette *see* 'POULCRIM'

Crimson Meillandecor *see* MElouscki

Crimson Midinette *see* 'POULCRIM'

'CRIMSON MME DESPREZ', B, dr; flowers crimson, large, cupped

'CRIMSON MOSS', M, dr; flowers crimson, dbl.; vigorous growth; Lee, before 1846; Lee, prior to 1846

'CRIMSON ORLÉANS', Pol, dr, 1922; [Orléans Rose sport]; Koster, M.

Crimson Promise *see* 'SPRINGPROMISE'

'CRIMSON QUEEN', HT, dr, 1912; flowers rich crimson, turning blue with age, very large, dbl., globular, intense fragrance; vigorous growth; [(Liberty X Richmond) X Gen. MacArthur]; Montgomery, A.

'CRIMSON QUEEN', HP, dr, 1890; Paul

Crimson Rambler *see* **'TURNER'S CRIMSON RAMBLER'**

Crimson Rosamini, Min, mr

'CRIMSON ROSETTE', F, dr, 1948; bud small, ovoid; flowers dark crimson, rosette, blooms in clusters, dbl., 30 petals, 1–1.5 in., slight fragrance; foliage leathery, dark; vigorous, bushy, dwarf growth; Krebs; H&S

'CRIMSON SHOWER', HWich, mr, 1951; flowers clear crimson, pompon, dbl., 20 petals, 1.25 in., slight fragrance; foliage glossy, light; vigorous (10 ft) growth; [Excelsa seedling]; Norman; Harkness

Crimson Tide *see* 'MACMOTA'

Crimson Velvet, F, dr; flowers medium, single, slight fragrance; (Sangerhausen)

Crimson Wave *see* 'MEIPERATOR'

Crinkles *see* 'FROCRIN'

'CRINOLINE', LCl, or, 1964; semi-dbl., cupped, recurrent bloom; foliage bronze; vigorous, climbing growth; [Diane d'Urfe X Étendard]; Hémeray-Aubert

Crispin-Morwenna *see* 'HARKITTEN'

'CRISSY', Min, ob, 1979; bud pointed; flowers deep bright coral, dbl., 20–25 petals, 1.5–2 in., slight fragrance; foliage dark; [Liverpool Echo X Sheri Anne]; Strawn; Pixie Treasures Min. Roses

Crissy, HT, ob

'CRISTAL', HT, ob, 1937; bud long, pointed; flowers orange-yellow, large, dbl.; foliage leathery, light; vigorous, bushy growth; [Julien Potin X Seedling]; Gaujard

Cristata *see* **'CRESTED MOSS'**

Cristian *see* METset

'CRISTIN CIRA', Gr, op, 1996; flowers orange, yellow, salmon blend, blooms borne in small clusters, dbl., 15–25 petals, slight fragrance; some prickles; foliage large, medium green, semi-glossy; upright, medium growth; Vanderkruk, William (Discoverer); Hortico Roses, 1995

Cristina, HT, 1972; Zandri, R.; (Cavriglia)

Cristobal Colon *see* 'MEIRONSSE'

Cristoforo Colombo *see* 'MEIRONSSE'

'CRISTOFORO COLOMBO', HT, yb, 1953; (Christophe Colomb); bud long; flowers reddish yellow, reverse tinged pink, large, dbl., 25–35 petals, slight fragrance; foliage glossy; long stems; very vigorous growth; RULED EXTINCT 4/92; [Julien Potin X Frau Karl Druschki]; Aicardi, D.; V. Asseretto

'CRITERION', HT, dp, 1966; flowers rose-red, dbl., 5 in., moderate fragrance; foliage dark; vigorous, tall growth; [(Independence X Signal Red) X Peace]; deRuiter

Criterion Cl., Cl HT, mr

'CROCK O' GOLD', HT, my, 1970; flowers clear golden yellow, classic form, dbl., 30–35 petals, 4.5 in.; [Beauté sport]; Anderson's Rose Nurseries

'CROFT ORIGINAL', HT, RB, 1971; flowers old-gold and red, full, dbl., 30 petals, 5 in., moderate fragrance; foliage light; Cocker; Wheatcroft Bros.

Croix Blanc, HT, w, 1982; Croix, P.

Croix d'Or, Cl HT; Croix, P.; (Cavriglia)

Cromwell School, Misc OGR, rb

CROpal, HT, ob; (**Aventure No. 2**); Croix

'CROWN JEWEL', F, or, 1964; flowers bright orange-red, medium, in clusters, dbl., moderate fragrance; foliage dark, leathery; moderate growth; [Pink Bountiful X Spartan]; Boerner; J&P

'CROWN OF GOLD', HT, yb, 1937; bud pointed; flowers deep gold, edged lemon-yellow, large, dbl., exhibition form; foliage light green, leathery, vigorous growth; [Seedling X Joanna Hill]; Duehrsen; H&S

Crown of Jewels *see* **'LITTLE BEAUTY'**

'CROWN PRINCE', HP, rb, 1880; flowers reddish crimson tinged with purple; Paul & Son

Crucencia *see* **'HOT PEWTER'**

Crucenia *see* **'HOT PEWTER'**

Crumble Bar, F, dy, 1982; LeGrice

'CRUSADER', HT, dr, 1920; flowers crimson-red, center brighter, large, dbl., 65 petals, slight fragrance; foliage leathery, rich green; vigorous growth; Montgomery Co.; A.N. Pierson

Cruzenia *see* **'HOT PEWTER'**

'CRYSTAL', Min, pb, 1985; flowers light pink, deep pink reverse, small, dbl., 35 petals, exhibition form, moderate fragrance; foliage large, dark, glossy; bushy growth; [Zinger X Seedling]; Bridges, Dennis A.; Bridges Roses

Crystal Brook, HWich, pb, 1994; Sutherland

Crystal Palace *see* POUlrek

'CRYSTAL STAR', Min, w, 1994; dbl., 26–40 petals, 1.5 in., borne mostly single and in small clusters, slight fragrance; some prickles; foliage small, medium green, semi-glossy; medium growth; [Pink Petticoat X Tooth of Time]; Muha, Julius; Mori Miniature Roses, 1994

'CRYSTAL WHITE', HT, w, 1965; flowers clear white, large, dbl., exhibition form, moderate fragrance; foliage leathery; moderate, bushy growth; [Princess White X White Queen]; Boerner; J&P

Crystalline™ *see* 'AROBIPY'

Crystal's Double Dark Red, HT, dr

Csárdás, F, op, 1965; flowers luminous orange-pink, large, dbl.; Berger, W.; (Sangerhausen)

'CSL CERVENY KRIZ', Pol, mr, 1928; flowers small, dbl.; Böhm, J.; (Sangerhausen)

'CSL LEGIE', Cl HT, dr, 1933; flowers large, dbl.; Böhm, J.; (Sangerhausen)

'CUBA', HT, rb, 1926; flowers cardinal-red, tinted yellow, fading quickly, large, semi-dbl., globular, intense fragrance; foliage dark, bronze; very vigorous growth; Pernet-Ducher

Cuddle Up *see* 'TINCUDDLE'

'CUDDLES', Min, op, 1978; bud ovoid; flowers deep coral-pink, dbl., 55–60 petals, 1–1.5 in., exhibition form, slight fragrance; compact growth; AOE, 1979; [Zorina X Seedling]; Schwartz, Ernest W.; Nor'East Min. Roses

Cuidad de Oviedo, C, mp

'CUISSE DE EMUE NYMPHE', A, lp; flowers whitish rose; Dumont de Courset, 1802

Cuisse de Nymphe *see* **'GREAT MAIDEN'S BLUSH'**

'CULVERBRAE', S, dr, 1973; flowers crimson-purple, dbl., 58 petals, 3.5–4 in., some repeat bloom, intense fragrance; foliage light; vigorous growth; [Scabrosa X Francine]; Gobbee, W.D.

Cumba Meillandina® *see* 'MEINEYTA'

'CUMBERLAND BELLE' *see* Le Poilu

'CUNNINGHAM LADY BANKS', HBank, ly, 1900

'CUNNINGHAM WEST CLIMBER', N, ly, 1900

Cunosa *see* **'MY GIRL'**

Cup Final *see* 'MACSINGAP'

Cupcake™ *see* 'SPICUP'

'CUPID', Cl HT, lp, 1915; flowers glowing flesh tinted peach, large, single; large, orange fruit; vigorous, pillar growth; Cant, B. R.

'CUPIDO', Min, lp; flowers shell-pink; height 6 in; Maarse, G.

'CUPIDON', F, dr, 1966; flowers brilliant crimson, small, semi-dbl., cupped, slight fragrance; foliage leathery; vigorous, bushy growth; [Chanteclerc X Red Favorite]; Gaujard

'CUPID'S BEAUTY', Min, ob, 1978; bud long, pointed; flowers light orange and cream, small, dbl., 40–45 petals, 1.5 in., exhibition form, moderate fragrance; foliage small, dark; compact, spreading growth; [Seedling X Over the Rainbow]; Williams, Ernest D.; Mini-Roses

'CUPID'S CHARM', F, pb, 1964; bud pointed; flowers salmon-pink, medium, dbl., 22 petals, moderate fragrance; vigorous, bushy growth; [Little Darling X First Love.]; Fuller; Wyant

Cupid's Heart, HT, rb, 1997

Cupie Doll *see* 'TINCUPIE'

'CURBAL', Min, pb, 1988; (**Little Ballerina**); flowers white flushed pink, urn-shaped, medium, borne singly, very dbl., 34–40 petals, exhibition form, slight fragrance; no fruit; no prickles; foliage medium, medium green, semi-glossy; bushy, medium growth; [Little Darling X Little Pioneer]; Curtis, Thad; Oregon Miniature Roses, 1988

Curiosa, F, op, 1964; flowers salmon-orange, large, dbl.; deRuiter; (Sangerhausen)

Curiosity *see* 'COCTY'

'CURLEM', Min, my, 1985; (**Lemon Fluff**); flowers small blooms borne singly and in clusters, dbl., 20 petals, exhibition form, no fragrance; foliage small, medium green, matt; upright, bushy growth; [(Unnamed seedling X Rise 'n' Shine) X Summer Butter]; Curtis, Thad; Hortico Roses

'CURLY LOCKS', Min, lp, 1954; flowers soft pink; dwarf, compact growth; Robinson, T.

'CURLY PINK', HT, mp, 1948; bud long, pointed, rose-red; dbl., 3.5–5 in., moderate fragrance; foliage glossy, dark; vigorous, compact growth; [Pink Princess X Crimson Glory]; Brownell, H.C.

'CURMIST', Min, mr, 1986; (**Mister Otis**); flowers small blooms borne singly, dbl., 35 petals, exhibition form, no fragrance; foliage small, medium green, matt; upright, bushy growth; [Rise 'n' Shine X Fire Princess]; Curtis, Thad; Hortico Roses

'CURNEER', Min, mr, 1986; (**Little Pioneer**); flowers gold stamens, small blooms borne singly and in clusters, dbl., 35 petals, exhibition form, moderate fragrance; foliage small to medium, medium green, matt; upright, bushy growth; [(Rise 'n' Shine X Summer Butter) X Sheri Anne]; Curtis, Thad; Hortico Roses

'CURRENT AFFAIR', Min, dr, 1991; flowers red, reverse red with yellow, ages dark pink, medium, very dbl., 80 petals, cupped, no fragrance; foliage medium, dark green, semi-glossy; upright, medium growth; [Red Ace X Seedling]; Gruenbauer, Richard, 1984; Flowers 'n' Friends Miniature Roses, 1993

'CURTAIN CALL', HT, dp, 1977; bud pointed; flowers cherry-red, dbl., 32–35 petals, 4–5 in., exhibition form, slight fragrance; foliage dark, leathery; vigorous, growth; [First Prize X Seedling]; Weeks; Weeks Roses

'CURTIS YELLOW', HT, my, 1973; flowers clear yellow, large, dbl., exhibition form, moderate fragrance; foliage light, leathery; vigorous, upright growth; [Golden Scepter X Miss Hillcrest]; Curtis, E.C.; Kimbrew

'CURTOY', Min, rb, 1985; (**Toy Soldier**); flowers red-white blend, medium, dbl., 20 petals, no fragrance; foliage medium, medium green, semi-glossy; bushy growth; [Pink Parfait X Over the Rainbow]; Curtis, Thad; Hortico Roses

'CUTHBERT GRANT', S, dr, 1967; bud ovoid; flowers deep purplish red, large, semi-dbl., cupped, intermittent bloom, slight fragrance; foliage glossy; vigorous, bushy growth; [(Crimson Glory X Assiniboine) X Assiniboine]; Marshall, H.H.; Canada Dept. of Agriculture

'CUTIE', Min, mp, 1952; bud pointed; flowers clear pink, base white, semi-dbl., 16 petals, 1 in., flat, slight fragrance; very few prickles; foliage small, glossy, bright green; dwarf (10 in), bushy growth; [Dancing Doll X Oakington Ruby]; Moore, Ralph S.; Sequoia Nursery

Cutie Pie *see* 'RENPIE'

Cuwaert, HT, 1992; Delforge, H.; (Cavriglia)

Cybele, HT, 1962; Hauser, V.; (Cavriglia)

Cyclamen *see* 'DELBRE'

Cyclamen La Sevillana *see* ORAuna

Cyclamen Meillandecor *see* 'MEIPELTA'

Cygne Blanc *see* **'WHITE SWAN'**

Cymbeline *see* 'AUSLEAN'

'CYNOSURE', HT, rb, 1973; bud ovoid; flowers red, striped pink, dbl., 50 petals, 4.5 in., intense fragrance; foliage glossy, soft; vigorous, upright growth; [Scarlet Knight X Festival Beauty]; Hardikar, Dr. M.N.

'CYNTHIA', HT, mr, 1934; flowers rich oriental red, dbl.; Verschuren-Pechtold; Dreer;, H&S

Cynthia *see* 'WARDROSA'

'CYNTHIA ANN PARKER', HT, w, 1929; flowers white tinged yellow or cream, dbl., moderate fragrance; Vestal

'CYNTHIA BROOKE', HT, yb, 1943; flowers empire yellow, reverse light salmon, dbl., 45 petals, 4 in., globular, moderate, fruity fragrance; foliage leathery, dark; moderate, compact, bushy growth; [Le Progres X (Mme Mélanie Soupert X Le Progres)]; McGredy; J&P

'CYNTHIA E. HOLLIS', HMult, lp; flowers pale pink, dbl.; foliage small, glossy; Dawson; Eastern Nursery

Cynthia Westcott, S, mp, 1997; Williams, J. Benjamin

'CYNTHIE', HGal, lp; flowers pale rose, circumference almost blush, large, dbl., cupped; erect, moderate growth; Descemet, M. (?), prior to 1833

'CYPRIENNE', Pol, mr, 1969; bud ovoid; flowers light brilliant red, full, large, dbl., abundant, continuous bloom, slight fragrance; foliage glossy, light; moderate, bushy growth; [Sumatra X Fashion]; Delforge

'CYRANO', HT, rb, 1954; flowers bright red shaded purple, large, dbl., moderate fragrance; long stems; very vigorous, upright growth; [(Opera X Seedling) X Seedling]; Gaujard

Cyril Fletcher *see* 'BEERIL'

'CZARDAS', HT, rb, 1956; flowers red, becoming pink; vigorous growth; [Tango X Seedling]; Delforge

D

'D. Ana Guedes', HT, ob, 1938; bud ovoid; flowers orange and salmon-pink, veined yellow, large, dbl., cupped, slight fragrance; foliage soft; vigorous, bushy growth; [Angèle Pernet X Mme Méha Sabatier]; da Silva, Moreira

'D. Angelica Pereira da Rosa', HT, pb, 1936; flowers pink, shaded golden orange, reverse red, open, very large, dbl.; foliage light; [Angèle Pernet X Edith Nellie Perkins]; da Silva, Moreira

D. D. Ruaux *see* DORyeco

D. H. Lawrence *see* 'ROSLAW'

'D. Laura Pinto d'Azevedo', HT, pb, 1936; flowers shrimp-pink, center coral-red, base orange-yellow, very large, dbl., exhibition form; foliage light, soft; [Pink Pearl X Constance]; da Silva, Moreira

D. Malvina Loureiro, F; Moreira da Silva, A.; (Cavriglia)

'D. Maria Antonia Pacheco', HT, mp, 1935; flowers deep carmine-pink, well formed, large, dbl.; foliage rich green, glossy; vigorous growth; [Mme Butterfly X Johanniszauber]; da Silva, Moreira

'D. Maria do Carmo de Fragoso Carmona', HT, pb, 1939; bud long, pointed; flowers flesh-pink, edged yellow, large, dbl., exhibition form; foliage glossy; long stems; vigorous growth; [Charles P. Kilham X Souv. de Claudius Pernet]; da Silva, Moreira

'D. Maria José de Melo', HT, op; flowers salmon-pink; da Silva, Moreira

'D. Maria Navarro', HT, rb, 1962; flowers carmine; [First Love X Paramount]; da Silva, Moreira

'D. Silvia Ferreira', HT, ob; flowers salmon and red, reverse old-gold; da Silva, Moreira

'D. T. Poulsen', F, mr, 1930; flowers bright blood-red, open, semi-dbl., slight fragrance; foliage dark, leathery; bushy growth; (21); [Orléans Rose X Vesuvius]; Poulsen, S.

'D. T. Poulsen Improved', F, mr, 1940; bud small, globular, blood-red; flowers solid cherry-red, open, dbl., 30–35 petals; foliage leathery, wrinkled, dark; short, strong stems; vigorous, compact, bushy growth; Van der Vis; C-P

'Dab', HT, ab, 1984; flowers apricot center, pink reverse, petals tipped deep pink, large, dbl., 35 petals, exhibition form, no fragrance; foliage large, dark, glossy; bushy growth; [Lady X X Flaming Beauty]; Bridges, Dennis A.; Bridges Roses, 1985

'Dacapo', F, op, 1960; flowers deep salmon-pink, blooms in clusters, dbl., 28 petals, 3–4 in., slight fragrance; vigorous, compact, bushy growth; [Fashion X Floribunda seedling]; deRuiter; Horstmann

'Daddies Girl', HT, yb, 1963; flowers milky white flushed pink, base golden, dbl., 40 petals, 5–6 in., moderate fragrance; foliage dark; vigorous, spreading growth; [McGredy's Ivory X Peace]; McTeer, Gilbert

Daddy's Pink, A, lp

Dad's Pink Hill, S, mp, 1994

Dady® *see* 'GAUZINE'

Dafne *see* BARafne

'Dagenham Show', F, ob, 1976; flowers salmon-orange, dbl., 25 petals, 3.5 in., intense fragrance; foliage matt; bushy growth; [Elizabeth of Glamis X Seedling]; Warley Rose Gardens

Dagmar, F, lp; Urban, J., 1991; (Czech Rosa Club)

'Dagmar Späth', F, w, 1936; (Blanc Lafayette, White Lafayette); flowers white, edge flushed pink, fading pure white; [Lafayette sport]; Wirtz & Eicke; Spath

'Dagmar Späth, Climbing', Cl F, w, 1940; Buisman, G. A. H.

'Dagmar Späth, Climbing', Cl F, w, 1943; Howard Rose Co.

'Dagmar Späth, Climbing', Cl F, w, 1961; Huber

'D'Aguesseau', HGal, mr; flowers fiery crimson, occasionally shaded dark purple, large, dbl.; compact, erect growth; Vibert (?), 1836

'Dahlila', F, yb, 1962; flowers yellow, becoming red; [Golden Perfume X Peace]; Leenders, J.

'Daidala'®, F, m, 1975; bud ovoid; dbl., 35 petals, 4 in., exhibition form, slight fragrance; foliage dark, soft; vigorous, upright, bushy growth; [Seedling X Silver Star]; Kordes; Willemse

Daily Express *see* FRYchambi

'Daily Herald', HT, yb, 1942; bud pointed; flowers yellowish orange, dbl., 4–5 in., cupped; foliage glossy, dark; vigorous, upright growth; Robinson, T.; J&P

Daily Mail Rose *see* **'Mme Edouard Herriot'**

Daily Mail Rose, Climbing *see* **'Mme Edouard Herriot, Climbing'**

'Daily Mail Scented Rose', HT, rb, 1927; flowers crimson, shaded maroon and vermilion, reverse dark crimson, , intense, damask fragrance; Cup for Best New Scented Seedling, Daily Mail, 1927; [Château de Clos Vougeot X K. of K.]; Archer

'Daily Mail Scented Rose, Climbing', Cl HT, rb, 1930; Archer

Daily Post *see* 'FRYTROOPER'

Daily Sketch *see* 'MACAI'

Daily Telegraph *see* 'PEAHIGH'

'Daimonji', HT, or, 1981; flowers large blooms borne usually singly, dbl., 48 petals, exhibition form, slight fragrance; foliage medium, dark; medium, bushy growth; [Seedling X (Miss Ireland X Polynesian Sunset)]; Shibata, T.; K. Hirakata Nursery

'Dainty', HT, pb, 1921; flowers rosy apricot, tinted cherry-pink, edges and reverse deeper pink, dbl., moderate fragrance; Dickson, H.

'Dainty', Pol, op, 1931; flowers salmon-pink, cupped; deRuiter

'Dainty Bess', HT, lp, 1925; flowers soft rose-pink, very distinct maroon stamens, medium, single, 4–11 petals, 5 in., borne in clusters of 5 to 9 single blooms, moderate, tea fragrance; foliage leathery, dark green; vigorous growth; GM, NRS, 1925; [Ophelia X K. of K.]; Archer

'Dainty Bess, Climbing', Cl HT, lp, 1935; van Barneveld; California Roses

Dainty Bouquet *see* 'LAVFUHR'

'Dainty Dawn', Pol, mp, 1931; flowers cerise-pink to mauve, blooms in clusters, semi-dbl., cupped, moderate fragrance; foliage bronze; [Amaury Fonseca X Annchen Müller]; Knight, G.

'DAINTY DELIGHT', Cl HT, mp, 1949; bud ovoid; flowers darker pink than Dainty Bess, medium, borne in clusters, semi-dbl., globular, slight fragrance; foliage glossy; vigorous (6-8 ft) growth; [Ednah Thomas X Dainty Bess]; Duehrsen; California Roses

Dainty Dinah *see* 'COCAMOND'

'DAINTY LADY', HT, mp, 1959; bud long, pointed; flowers pink, medium, dbl., exhibition form, moderate fragrance; foliage leathery; vigorous, upright, compact growth; [(Girona X Pres. Herbert Hoover) X Michèle Meilland]; Peden, G.H.

'DAINTY LADY', F, op, 1963; flowers coppery salmon-pink, well formed, borne in clusters; moderate growth; [Baby Sylvia X Seedling]; Fryers Nursery, Ltd.

'DAINTY MAID', F, pb, 1940; bud pointed, cerise; flowers silvery pink, reverse carmine, blooms in clusters, single; foliage leathery, dark; vigorous, compact, bushy growth; (28); GM, Portland, 1941; [D.T. Poulsen X Seedling]; LeGrice; C-P

'DAINTY SUPERIOR', Pol, op; flowers have more lasting color; [Dainty sport]; deRuiter

'DAIRY MAID', F, ly, 1957; bud yellow, splashed carmine; flowers cream, fading white, blooms in large clusters, single, 5 petals, 3.5 in.; foliage glossy; vigorous growth; [(Poulsen's Pink X Ellinor LeGrice) X Mrs Pierre S. duPont]; LeGrice

'DAISY', HT, ob, 1923; flowers orange-flamed; [Mme Edouard Herriot sport]; Hicks

'DAISY BRASILEIR', HMult, rb, 1918; flowers bright red and purple-red, anthers yellow, borne in clusters, single, slight fragrance; Turbat

'DAISY BUD', HT, pb, 1933; bud well shaped; flowers rosy pink, shaded carmine and silver, large, moderate fragrance; Dickson, A.

'DAISY DOLL', Min, pb, 1977; bud long, pointed; flowers rose-pink, open, small, dbl., 20–25 petals, 1.5 in., slight fragrance; foliage tiny; compact, upright growth; [Little Amy X Seedling]; Lyon; L. Lyon Greenhouses

'DAISY DUMAS', HT, rb, 1962; flowers bright red, reverse gold edged red, medium, dbl., slight fragrance; foliage dark; bushy growth; [Seedling X (Tzigane X Seedling)]; White; A. Ross & Son

'DAISY HILL', S, mp, 1906; flowers large, single, moderate fragrance; abundant fruit; height 8 ft; [R. X waitziana macrantha hybrid]; Kordes

'DAISY HILLARY', F, op, 1963; bud pointed; flowers salmon-pink, large, borne in clusters, dbl., exhibition form; foliage leathery; vigorous, bushy growth; [Spartan sport]; Mell

'DAISY MAE', HT, dy, 1987; flowers deep yellow, reverse medium yellow, large, borne singly, single, 6 petals, flat; round, ovoid, russet-yellow fruit; large, red down-curved prickles, fading gray-tan; foliage large, dark green, semi-glossy; upright, tall growth; [Golden Showers X Golden Sun]; Stoddard, Louis

'DAISY MAY ROGERS', HMsk, w, 1996; single, 4–7 petals, 1 in., borne in large clusters, slight fragrance; small, moderate prickles; foliage small, medium green, semi-glossy; compact, spreading, medium (4 ft x 5 ft) growth; [Mozart X Unknown]; Muia, Charlotte R.; My Green Thumb, 1992

Daisy Rose *see* 'KORDAISY'

'DAKAR', HT, pb, 1932; flowers silvery pink, striped rose-pink, large, intense fragrance; long, strong stems; [Julien Potin X Seedling]; Gaujard

Dakota, HT, dr; flowers exhibition form, dbl., borne one per stems; foliage medium, dark green, matte

'DALE FARM', F, or, 1973; flowers vermilion, dbl., 25 petals, 3 in., moderate fragrance; foliage dark; vigorous growth; Smith, E.; Wheatcroft & Sons

Dale's Pink Glow, HMsk, mp, 1994

Dale's Pink Mini Climber, Cl Min, mp

Dale's Ragged Pink, Misc OGR, mp

Dale's Sunrise® *see* 'KINSUN'

'DALILA', HT, rb, 1943; bud long, pointed; flowers brilliant coppery red, large, semi-dbl., moderate fragrance; foliage leathery; vigorous, erect growth; [Souv. de Claudius Pernet X Seedling]; Gaujard

'DALILA', F, ob, 1958; flowers orange, dbl.; Buyl Frères

Dallas *see* 'KORLIMIT'

'DALLAS', HT, rb, 1963; flowers crimson-carmine, base primrose-yellow, very large, dbl., 40 petals; foliage dark, glossy; vigorous growth; [Peace sport]; Hunter; Waterhouse Nursery, 1963

'DALLAS GOLD', HT, yb, 1987; flowers large, heavy blooms, dbl., 25 petals, slight fragrance; foliage medium, dark green, glossy; bushy growth; Silver, ARC TG, 1984; [Seedling X Flaming Beauty]; Winchel, Joseph F.; Kimbrew Walter Roses, 1987

Dalli Dalli® *see* 'TANLILIDA'

'DALVEY', HT, dp, 1971; flowers deep pink, 5.5 in., exhibition form, intense fragrance; foliage medium, light green, matt; vigorous growth; [Peeping Tom X Seedling]; MacLeod

Dama diCuori *see* **'DAME DE COEUR'**

Damas de Yuste *see* **'FREIHEITSGLOCKE'**

'DAMAS FRANKLIN', D, pb, 1853; flowers flesh-pink, shading to silver; Robert, ca. 1856

Damask Rose *see* **'SUMMER DAMASK'**

'DAME BLANCHE', HWich, w, 1923; flowers greenish white, stamens yellow, borne in clusters, single, non-recurrent bloom; vigorous growth; Turbat

'DAME BLANCHE', HT, w, 1927; (White Lady); flowers white tinged green, dbl., moderate fragrance; [Stadtrat Glaser seedling]; Mühle

Dame Cath *see* 'MACHILVER'

'DAME CATHERINE', HT, dy, 1937; bud ovoid; flowers golden yellow, large, dbl., exhibition form; foliage glossy; long stems; vigorous, bushy growth; Cant, B. R.

'DAME DE COEUR', HT, mr, 1958; (Dama diCuori, Herz-Dame, Queen of Hearts); flowers cherry-red, large, dbl., moderate fragrance; foliage dark, glossy; vigorous growth; [Peace X Independence]; Lens

'DAME DE COEUR, CLIMBING', Cl HT, mr, 1984; [Dame de Coeur sport]; Mungia, Fred A., Sr.

Dame de L'Etoile® *see* ADAburi

'DAME EDITH HELEN', HT, mp, 1926; flowers glowing pink, very large, dbl., cupped, not very free bloom, moderate fragrance; foliage leathery; long, strong stems; vigorous, bushy growth; GM, NRS, 1926; Dickson, A.

'DAME EDITH HELEN, CLIMBING', Cl HT, mp, 1930; H&S

Dame Joyce Frankland *see* 'HORNEWGRAM'

'DAME OF SARK', F, ob, 1976; flowers orange flushed red, reverse yellow, dbl., 33 petals, 4.5 in., slight fragrance; foliage large, dark; [(Pink Parfait X Masquerade) X Tablers' Choice]; Harkness

'DAME PRUDENCE', S, lp, 1969; flowers soft pink, reverse lighter, medium, dbl., 65 petals, flat, intense fragrance; [Ivory Fashion X (Constance Spry X Ma Perkins)]; Austin, David

Dame Vera *see* 'PEAMAX'

Dame Vera Lynn *see* 'PEAMAX'

Dame Wendy *see* CANson

Dames Patronesses d'Orleans, HP, dr; flowers crimson red; Vigneron, 1877

Damien's Amulet, T, lp

'DAMON RUNYON', HT, mr, 1955; bud ovoid; flowers crimson, dbl., 50 petals, 4–5.5 in., exhibition form, moderate fragrance; foliage glossy, coppery green; vigorous, upright, bushy growth;

[Major Shelley X Heart's Desire]; Duehrsen; H&S

'DANA', HT, mp, 1982; bud long, pointed; flowers spiraled, formal blooms borne singly, sometimes 3 per cluster, dbl., 38 petals, slight, carnation fragrance; long, straight prickles; foliage large, semi-glossy; tall, upright, bushy growth; [White Satin X Bewitched]; Swim, H.C. & Ellis, A.E.; Armstrong Nursery

'DANAË', HMsk, ly, 1913; flowers pale buff-yellow, fading white, in clusters, recurrent bloom; height 6 ft; (14); [Reputedly Trier X Gloire de Chedane-Guinoiseau]; Pemberton

'DANCE OF JOY', F, mr, 1931; flowers vivid scarlet-crimson, large, dbl., moderate fragrance; foliage dark; vigorous growth; (28); GM, Bagatelle, 1931; [Paul's Scarlet Climber X Seedling]; Sauvageot, H.

Dance of Joy 95 *see* SAUdero

'DANCING DOLL', Cl F, dp, 1952; bud small, pointed; flowers deep rose-pink, borne in clusters, semi-dbl., 10–14 petals, cupped, profuse, repeated bloom, intense fragrance; foliage leathery, glossy; vigorous, climbing or spreading (10 ft.) growth; [Étoile Luisante X Seedling]; Moore, Ralph S.; Marsh's Nursery

Dancing in the Wind *see* CLEdan

Dancing Pink *see* 'HENDAN'

'DANCING SILK', HT, op, 1966; flowers coral-pink, reflexed, 5 in., slight fragrance; foliage light green; vigorous growth; [Ena Harkness X McGredy's Yellow]; Barter

'DANDEE', Min, dr, 1983; bud globular; dbl., 35 petals, flat, borne singly, moderate fragrance; foliage medium, dark, semi-glossy; upright growth; [Seedling X Libby]; Meredith, E.A. & Rovinski, M.E.; Casa de Rosa Domingo

Dandenong, MinFl, mr

'DANDY', Pol, op, 1945; flowers light orange-pink, small, semi-dbl., cupped, profuse, non-recurrent bloom, slight fragrance; foliage wrinkled, glossy, dark; very vigorous growth; [Gloria Mundi X Unidentified Sp rose]; Wiseman

'DANDY DICK', F, mp, 1967; flowers well-formed, large blooms in clusters, dbl., 25 petals, moderate, spicy fragrance; foliage light; [Pink Parfait X Red Dandy]; Harkness

'DANDY LYON', Min, dy, 1978; bud long, pointed; flowers buttercup-yellow, dbl., 30 petals, 2 in., moderate fragrance; foliage small, dark, glossy; compact, bushy growth; [Seedling X Sunspot]; Lyon

Dania *see* 'COCGOLD'

Danica *see* NOArda

'DANIEL', HT, rb, 1943; bud long, pointed; flowers capucine-red on golden yellow base, dbl., cupped; foliage dark, glossy; vigorous, bushy growth; Mallerin, C.; A. Meilland

Daniel, HT, dr, 1995; flowers large, exhibition form, intense fragrance; Williams, J. Benjamin

Daniel Boone, HT, dr, 1967; Morey, Dr. Dennison

Daniel Gelin *see* 'HARQUINCE'

'DANIEL LACOMBE', HMult, yb, 1885; flowers yellow washed pink, medium, dbl., flat; vigorous growth; [R. multiflora X Général Jacqueminot]; Allard

'DANIEL PHILIP', F, mr, 1999; semi-dbl., 12–16 petals, 1.2–1.5 in., borne in small clusters, slight fragrance; prickles moderate; foliage small, medium green, semi-glossy, disease resistant; upright, medium (6 ft) growth; [(Morden Fireglow X Unknown) X seedling]; Fleming, Joyce L.; Hortico, Inc., 1999

Daniela® *see* KORlunta

Daniele, Min, 1991; Kordes, R.; (Cavriglia)

Danielle®, HT, w, 1992; Cocker

Danielle *see* 'BRODAN'

'DANIELLE DARRIEUX', HT, pb, 1948; bud long, pointed; flowers salmon suffused yellow, reverse salmon-orange-pink, very large, dbl., moderate fragrance; very vigorous, bushy growth; Gaujard

'DANIELLE ROBYN', HT, rb, 1971; bud globular; flowers pinky red and creamy white bicolor, large, dbl., exhibition form, intermittent bloom, slight fragrance; foliage glossy, dark; bushy growth; [Grand Gala X Western Sun]; Hastie

Danina *see* **'CARA MIA'**

'DANIPHYL', HT, mr, 1978; flowers mandarin-red, full, dbl., 45–50 petals, 4–5 in., moderate fragrance; foliage semi-glossy, light; vigorous, upright growth; [Puccini X Chopin]; Ellick; Excelsior Roses

'DANISH GOLD', F, yb, 1949; flowers yellow fading to creamy white, single, 5–9 petals, 2.5–3 in., moderate, spicy fragrance; foliage glossy; vigorous, compact growth; GM, NRS, 1949; [(Golden Salmon X Souv. de Claudius Pernet) X Julien Potin]; Poulsen, S.; McGredy

'DANISH PINK', F, dp, 1965; flowers deep pink, almost medium borne in large clusters, single, 2–2.5 in., moderate, fruity fragrance; vigorous, tall growth; Soenderhousen; Hoersholm Nursery

'DANMARK', HT, mp, 1891; flowers large, dbl., intense fragrance; Zeiner-Larsen; (Sangerhausen)

'DANNENBERG', HT, op, 1916; flowers coral pink, large, dbl., moderate fragrance; Kiese; (Sangerhausen)

'DANNIE'S SMILE', HT, lp, 1994; dbl., 26–40 petals, 1.5–2.75 in., borne mostly singly, slight fragrance; some prickles; foliage medium, dark green, matt; medium, spreading growth; [Elizabeth Taylor sport]; Maxheimer, Joanne; Giles Rose Nursery, 1993

'DANNY BOY', LCl, or, 1969; flowers well-formed, recurrent bloom, intense fragrance; foliage dark; [Uncle Walter X Milord (HT)]; McGredy, Sam IV; McGredy

'DANNY THOMAS', Min, mr, 1980; flowers yellow stamens, borne usually singly, dbl., 35 petals, moderate fragrance; straight prickles; foliage dark; compact growth; [Rose Hills Red X Self]; Wells, V.W., Jr.; Lou McGuire

'DANORANG', HT, or, 1972; (**Orange Tango**); flowers large, very dbl., exhibition form, slight fragrance; foliage bronze; vigorous growth; [From unnamed seedlings sport]; McDaniel, Earl; Carlton Rose Nurseries

'DANSE AZTEQUE', Min, yb, 1985; (Rigobec 3); flowers yellow-pink blend, medium, semi-dbl., flat, slight fragrance; foliage medium, medium green, semi-glossy; vigorous, upright growth; [Baby Masquerade seedling]; Gailloux, Gilles

Danse des Étoiles *see* 'GODSENSOR'

Danse des Sylphes® *see* 'MALCAIR'

Danse du Feu *see* **'SPECTACULAR'**

'DANUBIO AZUL', HT, m, 1957; flowers lilac, medium, dbl., cupped, slight fragrance; upright growth; [Tristesse X Independence]; Camprubi, C.

'DANY ROBIN', F, op, 1958; bud pointed; flowers salmon-pink, open, medium, borne in clusters, dbl., slight fragrance; foliage leathery; vigorous, compact growth; [Goldilocks X Fashion]; Meilland, F.; URS

Danyrose, HT, 1975; Croix, P.; (Cavriglia)

'DANZIG', HT, dr, 1940; flowers shining dark red, medium, borne in clusters, dbl.; upright growth; [Hadley X Kardinal]; Tantau

Danzille *see* **'MME BRAVY'**

'DAPHNE', HMsk, lp, 1912; flowers blush-pink, blooms in clusters, semi-dbl., moderate fragrance; vigorous growth; Pemberton

'DAPHNE', HT, lp, 1925; flowers soft pink, flushed rose, well-formed, dbl., moderate fragrance; Dobbie

'DAPHNE', HGal, mr; flowers bright red, cupped; Vibert, 1819

'DAPHNÉ', HGal, dp, 1819; flowers medium, dbl.; Vibert; (Sangerhausen)

Daphne Claire Jones *see* 'KIRJACK'

'DAPHNE GANDY', F, mr, 1952; flowers blood-red, blooms in large trusses, 3

in., slight fragrance; foliage dark; vigorous growth; [Farida X Crimson Glory]; Leenders, M.; Gandy Roses, Ltd.

'DAPHNIS', HT, mp, 1974; flowers brilliant pink, very large; upright growth; [Marylène X Mignonne]; Gaujard

Dapple Dawn *see* AUSapple

'DARA', HT, ob, 1991; dbl., 30 petals, exhibition form, borne usually singly, intense fragrance; foliage medium, medium green, semi-glossy; upright growth; [Olympiad X Just Joey]; Wambach, Alex A.

Darby *see* 'DEVcico'

Darby O'Gill *see* 'SEAdarby'

Darcelle *see* 'SPOdarc'

'd'ARCET', M, dr, 1851; flowers large, dbl.; Robert; (Sangerhausen)

'DARDANELLE', HT, mr, 1926; flowers cherry-rose, dbl., moderate fragrance; foliage wrinkled, dark; bushy growth; [Premier X Ophelia]; Vestal

'DARIUS', HGal, mr; flowers vivid red, large, dbl.

'DARK BOY', HT, dr, 1965; bud ovoid; flowers velvety dark maroon-red, medium, dbl., slight fragrance; foliage soft; moderate, upright, open growth; [Nigrette X Seedling]; Pal, Dr. B.P.; Indian Agric. Research Inst.

Dark Mirage *see* 'SEAdark'

'DARK SECRET', HT, dr, 1937; flowers large, dbl., globular, intense fragrance; foliage leathery; short stems; dwarf growth; [Radiance X Hollywood]; Amling Co.

Darling *see* 'SUNcredel'

'DARLING', HT, lp, 1958; bud ovoid; flowers large, dbl., moderate, fruity fragrance; foliage dark, glossy; vigorous, bushy growth; [Pink Princess X Charlotte Armstrong]; Taylor, C.A.; California Nursery Co.

Darling *see* KORspatax

Darling, HT, op, 1985; Noack, Werner

'DARLING ANNABELLE', HT, w, 1992; flowers white with faint pink center, excellent blooms, dbl., 26–40 petals, 3 in., exhibition form, moderate fragrance; foliage large, medium green, matt; upright (170 cms) growth; [South Seas X Peace]; Perry, Astor; Hortico Roses, 1993

Darling Diane *see* 'BOScherry'

Darling Flame *see* 'MEIlucca'

'DARLING JENNY', HT, pb, 1996; flowers deep pink shaded peach, large blooms borne mostly single, dbl., 26–40 petals, slight fragrance; some prickles; foliage medium, dark green, semi-glossy; bushy, medium growth; [Solitaire X Gavotte]; Poole, Lionel; Poole, 1997

Darlow's Enigma, HMsk, w; Darlow

'D'ARTAGNAN', D, dr, 1969; flowers wine-red, blooms in clusters, dbl., cupped, non-recurrent, intense, damask fragrance; foliage dark, leathery, wrinkled; vigorous, upright (6 ft.) growth; [Ma Perkins X York and Lancaster]; Fankhauser

D'Artagnan *see* LAPniro

Darthuizer Orange Fire *see* 'INTerfire'

Dart's Dash, HRg, dr

Dart's Defender, S, m, 1971; Darthuis

Dart's Red Dot® *see* INTermunder

'Das Goldene Prag' *see* **'Zlatá Praha'**

Dasher *see* 'SPOdash'

'D'ASSAS', HP, pb, 1850; flowers dark pink, tinged crimson, petals somewhat fringed, medium, dbl.; vigorous, straggling growth; Vibert

Daughter Margaret *see* 'NOBam'

'DAUNTLESS', HT, pb, 1949; bud long, pointed; flowers pink, reverse yellow, large, dbl., 60 petals, exhibition form, moderate fragrance; foliage leathery, dark; upright growth; [Crimson Glory X Feu Pernet-Ducher]; Davis

'DAUPHINE', F, op, 1955; bud ovoid; flowers pink shaded salmon, open, large, borne in clusters, dbl., moderate fragrance; very vigorous growth; [Seedling X Opera seedling]; Gaujard

'DAUPHINE, CLIMBING', Cl F, op, 1959; Gaujard

'DAVE DAVIS', HT, dr, 1964; bud long, pointed; flowers dark velvety red, large, dbl., 60 petals, exhibition form, intense fragrance; foliage leathery; moderate growth; [Seedling X Charles Mallerin]; Davis; Wyant

Dave Hessayon *see* 'DRIscobruce'

'DAVID ARNOT', HT, rb; flowers bright scarlet, reverse old-gold; vigorous growth

David Charles Armstrong *see* 'DRIscogeorge'

David Dot, Min, 1978; Dot, Simon; (Cavriglia)

'DAVID GILMORE', HT, mr, 1923; flowers brilliant scarlet; Dickson, H.

'DAVID GOLD', HT, rb, 1957; flowers cherry-cerise tinted golden yellow, 6 in., exhibition form, intense fragrance; foliage dark, glossy; very vigorous growth; [Shot Silk X Peace]; Robinson, H.

David Leek, F, dp, 1996; Fleming, Joyce L.

'DAVID MCKEE', HT, mr, 1933; flowers carmine-red, large, dbl., exhibition form, intense fragrance; foliage leathery; vigorous, bushy growth; Dickson, A.

'DAVID O. DODD', HT, mr, 1926; flowers rich crimson flushed scarlet, large, dbl., intense fragrance; foliage glossy; bushy growth; Vestal

'DAVID O. DODD, CLIMBING', Cl HT, mr, 1937; Howard Rose Co.

David Ruston, HT, pb

'DAVID THOMPSON', HRg, mr, 1979; bud ovoid; flowers yellow stamens, dbl., 25 petals, 2.5 in., intense fragrance; upright growth; [(Schneezwerg X Frau Dagmar Hartopp) X Seedling]; Svedja, Felicitas; Canada Dept. of Agric.

David Whitfield *see* GANa

Davidoff, HT, op

Davina Jane, HT, ob, 1997; flowers full, medium, very dbl., 26–40 petals, borne mostly singly; foliage medium, medium green, semi-glossy; upright, medium (3ft.)growth; [Sexy Rexy X Corso]; Rawlins, R.

Davit Dot, S, mr

'DAVona', F, dr, 1982; (**Red Splendour**); flowers deeper red; [Europeana sport]; Davies, Gareth, 1979

'DAVY CROCKETT', F, dr, 1956; flowers large, dbl.; vigorous growth; [Étoile de Hollande X Floribunda seedling]; deRuiter; Gandy Roses, Ltd.

'DAWN', HT, op, 1953; bud long, pointed; flowers salmon-pink, base yellow, dbl., 28–36 petals, 5–6. in., exhibition form, moderate, spicy fragrance; vigorous, upright growth; Jelly; E.G. Hill Co

Dawn Chorus *see* 'DICquasar'

Dawn Creeper™ *see* 'WILdawn'

'DAWN FRAGRANCE', LCl, op, 1969; flowers salmon-flesh, flushed rose-pink, well formed, cupped, intermittent bloom, intense fragrance; foliage dark, leathery; vigorous growth; [Blossomtime X Blossomtime seedling]; Mason, P.G.

Dawn Haggle *see* KORboden

'DAWN MIST', F, pb, 1962; bud ovoid; flowers reddish pink, base yellow, dbl., 30–35 petals, 3 in., cupped, intense, spicy fragrance; foliage leathery, glossy; vigorous, bushy, compact growth; [(Goldilocks X Pinocchio) X Vogue]; Boerner; Home Nursery Products Corp.

'DAWN PINK', F, mp, 1962; bud apricot; flowers pink, rosette form, borne in clusters, dbl., 45–50 petals, 3.5 in., moderate fragrance; foliage light green; vigorous, upright growth; [Magenta X Ma Perkins]; Anderson's Rose Nurseries

Dawn Sunsation *see* KORfeining

Dawn Weller *see* POUlsum

'DAWNGLOW', HT, pb, 1937; flowers flesh-pink, dbl., 3.5 in., exhibition form, moderate fragrance; Burbank

'DAWNING, CLIMBING', LCl, op, 1956; flowers salmon-pink, moderate fragrance; foliage glossy; vigorous growth; [New Dawn X Margaret McGredy]; Bennett, H.; Pedigree Nursery

'DAWNLIGHT', F, pb, 1958; bud ovoid; flowers soft pink, dbl., 30–35 petals, 2–2.5 in., flat, slight fragrance; thornless; foliage leathery, light green; vigorous, bushy growth; [Summer Snow, Climbing X Summer Snow]; Motose; G.B. Hart, Inc.

'DAWNS EARLY LIGHT', HT, w, 1972; bud long, palest pink; flowers white, faintly edged pale pink, pointed, bloom lasts well, single, 6 petals, 3.5–4 in., intense fragrance; foliage light; tall growth; [Vesuvius X Vesuvius]; Linscott

'DAWSON', HMult, mp, 1888; flowers bright rose-pink, small blooms in clusters of 10-20, dbl.; vigorous (10-25 ft) growth; [R. multiflora X Général Jacqueminot]; Dawson; Strong

Dawson's Climber, Cl HT, lp; Dawson, 1960; (Weatherly, L.)

Dawson's Delight, HT, pb, 1978; Dawson

Dawson's Yellow Climber, LCl, my

Day Break *see* **'DAYBREAK'**

'DAY DREAM', HT, dp, 1969; bud long, pointed; flowers deep pink, large, dbl.; foliage glossy, leathery; vigorous, upright, bushy growth; [Helen Traubel X Tiffany]; Armstrong, D.L.; Armstrong Nursery

Day Glow *see* 'JACRINK'

'DAY IS DONE', HT, or, 1985; flowers large, dbl., 35 petals, exhibition form, slight fragrance; foliage large, dark, semi-glossy; [Die Welt X Rosalynn Carter]; Schneider, Peter

Day Light® *see* 'INTERLIGHT'

'DAY OF TRIUMPH', HT, mp, 1953; (Rendezvous); bud ovoid; flowers pink edged lighter, dbl., 50–65 petals, 5.5 in., cupped, moderate fragrance; foliage leathery; vigorous, upright, bushy growth; [Peace X Europa]; Meilland, F.; Breedlove Nursery, 1955;, URS, 1953

'DAYBREAK', HMult, op, 1909; flowers deep salmon-pink, borne in clusters, late; [R. wichurana X R. chinensis]; Dawson; Eastern Nursery

'DAYBREAK', HMsk, my, 1918; (Day Break); flowers golden yellow, single, recurrent bloom; foliage dark; vigorous, bushy growth; [Trier X Liberty]; Pemberton

'DAYBREAK', HT, pb, 1935; bud large, long, pointed; flowers salmon-pink and yellow, very dbl., exhibition form, slight fragrance; foliage leathery; long stems; very vigorous growth; [Violet Simpson X Ivy May]; Laxton Bros.

'DAYBREAK', S, dp, 1960; flowers deep pink, dbl., 25–30 petals; [Hansa X R. macounii]; Erskine

Daybreak *see* **'RASSVET'**

Daydream *see* 'DICJEEP'

'DAYDREAM', Cl HT, lp; waterlily form, blush pink, semi-dbl.; vigorousd bush or pillar; [Souvenir de Gustave Prat X Rosy Morn or Gwen Nash]; Clark, A., 1925; Hazelwood Bros., 1925, AUS

Daylight *see* 'INTERLIGHT'

'DAYLIGHT', HT, pb, 1939; flowers creamy blush-pink, base yellow, very dbl., 4 in., exhibition form, moderate fragrance; foliage soft; vigorous growth; RULED EXTINCT 4/92 ARM; [Grange Colombe X Los Angeles]; Hansen, N. J.; B&A

'DAYLIGHT KATY', F, my, 1987; flowers bright yellow, fading slightly paler, urn-shaped, medium, dbl., 22 petals, borne in sprays of 3-15, slight fragrance; round, small, red fruit; pointed, medium, reddish prickles; foliage medium, bronze-green, semi-glossy; bushy, low growth; [Bright Smile X (Princess Michael of Kent X Party Girl)]; Schneider, Peter

Daytona *see* OLIjzouc

'DAZLA', HT, rb, 1930; bud long, pointed; flowers orange-scarlet, base and reverse golden yellow, wavy petals, semi-dbl., 6 in.; foliage dark; vigorous growth; Cant, B. R.

'DAZLA, CLIMBING', Cl HT, rb, 1950; Cant, B. R.

Dazzler *see* **'LEONIE'**

Dazzler™ *see* 'KELDAZ'

Dazzler *see* GENpat

Dazzler, Min, pb, 1992; Kasturi

'DE CANDOLLE', M, lp, 1857; flowers soft pink or rose-tinted, large, dbl.; Portemer fils

'DE CANDOLLE', HWich, op, 1913; Robichon

'DE GREEFF'S JUBILEE', F, dr, 1980; (Fellowship); flowers large, semi-dbl., slight fragrance; foliage medium, dark, semi-glossy; bushy growth; [Diablotin X Gisselfeld]; Verschuren, Ted; Verschuren & Sons

De Kat, HT, w, 1995

'DE LA GRIFFERAIE', HMult, dp, 1845; flowers carmine to pink, medium, dbl.; robust growth; (28); Vibert

'DE LUXE', HT, rb, 1931; flowers bright velvety scarlet, reverse red, semi-dbl., slight fragrance; [Premier sport]; White Bros.; Liggit

De Meaux *see* **'ROSE DE MEAUX'**

De Meaux White *see* **'ROSE DE MEAUX WHITE'**

De Montarville, S, mp, 1998; flat, 19 petals, slight; L'Assomption, 1982

De Montarville *see* 'AC DE MONTARVILLE'

'DE SCHELFHOUT', HGal, lp, 1840; flowers medium, dbl.; Parmentier; (Sangerhausen)

'DEAN COLLINS', Gr, dp, 1953; bud ovoid, red; flowers deep pink, dbl., 53 petals, 4.5–5 in., slight fragrance; foliage dark, glossy, leathery; vigorous growth; [Charlotte Armstrong X Floradora]; Lammerts, Dr. Walter; Roseway Nursery

'DEAN HOLE', HT, pb, 1904; flowers silvery carmine, shaded salmon, dbl., moderate fragrance; Dickson, A.

Deantoni, HT, rb

Dear Eleanor *see* 'GELELE'

Dear One *see* **'KARA'**

Dear Prudence *see* 'TALDEARPRU'

'DEAREST', F, pb, 1960; flowers rosy salmon-pink, gold stamens, well-formed, blooms in clusters, dbl., 30 petals, 3.5 in., moderate fragrance; foliage dark, glossy; vigorous, bushy growth; GM, NRS, 1961; [Seedling X Spartan]; Dickson, A.

Dearest, Cl F, op, 1970; Ruston, D.

Dearest One *see* **'CARA MIA'**

Dearest, Climbing, Cl F, lp; Ruston, D., 1970; (Weatherly, L.)

'DEBBIE', Min, yb, 1966; flowers yellow, edges becoming pink, small, dbl., moderate fragrance; foliage small, leathery; bushy, low, sometimes semi-climbing growth; [Little Darling X Zee]; Moore, Ralph S.; Sequoia Nursery

Debbie, S, dp; Poulsen

Debbie Dawn, F, lp, 1997; flowers double, medium, dbl., 15–25 petals, borne in small clusters, no fragrance; foliage medium, dark green, semi-glossy; upright (2ft.6in.) growth; [Margaret Merril X Melrose]; Rawlins, R.

'DEBBIE LYNN', HT, dr, 1992; flowers urn-shaped, blooms borne usually singly, dbl., 36 petals, 4.75 in., exhibition form, moderate fragrance; foliage medium, dark green, semi-glossy; upright, bushy, medium growth; Jerabek, Paul E., 1986

'DEBBIE THOMAS', HT, rb, 1996; flowers crimson, lighter reverse, large, very dbl., exhibition form, slight fragrance; numerous prickles; foliage medium, medium green, semi-glossy; upright, tall growth; [City of Gloucester X My Joy]; Thomas, D.; F. Haynes & Partners, 1992

'DEBBIE-KAREN', HT, mp, 1983; dbl., 4.5–5.5 in., exhibition form, borne singly, slight fragrance; foliage large, light green, semi-glossy; vigorous, upright growth; [Seedling X Seedling]; Burdett, H.J., 1984

Debidue® *see* 'MICDEB'

'DEBONAIR', HT, my, 1946; bud ovoid; flowers primrose-yellow, dbl., 28–35 petals, 3–4.5 in., exhibition form, moderate fragrance; foliage leathery, glossy, dark; very vigorous, upright, bushy growth; [Golden Rapture X Seedling]; Lammerts, Dr. Walter; Armstrong Nursery

Deborah® *see* MEInoiral

Deborah Devonshire *see* 'BOSCHERRYDRIFT'

Deborah Beggs Moncrief

Deborah Moncrief™ *see* 'WILDEB'

Deborah Moncrief™

Debra Gaye *see* 'TINDEB'

'DEBRAD', F, mr, 1982; (**Royal Baby**); flowers medium, dbl., 20 petals, slight fragrance; foliage small, medium green, semi-glossy; upright growth; [Generosa X Baby Darling]; Bracegirdle, Derek T.; Arthur Higgs Roses

'DEBRAF', HT, yb, 1986; (**Wendy Pease**); flowers medium, dbl., 35 petals, slight fragrance; foliage medium, medium green, semi-glossy; upright growth; [John Waterer X Tenerife]; Bracegirdle, Derek T.

'DEBRAH', F, mr, 1986; (**Mary Sheffield**); flowers moderately medium, dbl., 15–25 petals, moderate fragrance; foliage small, medium green, glossy; spreading growth; [Doris Tysterman X Admiral Rodney]; Bracegirdle, Derek T., 1987

'DEBRARO', HT, dy, 1984; (**William Walker**); flowers medium, dbl., 20 petals, no fragrance; foliage medium, medium green, semi-glossy; upright growth; [Western Sun X Circus]; Bracegirdle, Derek T.

Deb's Delight *see* 'LEGSWEET'

Debut™ *see* 'MEIBARKE'

'DEBUTANTE', HMult, lp, 1902; flowers rose-pink, fading to cameo-pink, borne in clusters of 4-6, dbl., non-recurrent, moderate, sweetbriar fragrance; foliage dark, glossy; short stems; height 6-8 ft; [R. wichurana X Baroness Rothschild]; Walsh

Debutante *see* 'JACINAL'

Deccan Delight, F, yb, 1985; Chiplunkar

Deccan Deluxe, HT, m, 1988; Chiplunkar

'DECEA ANN', F, ab, 1975; flowers flushed peach-pink, full, dbl., 26 petals, 3.5–4 in., intense fragrance; foliage dark, leathery; upright growth; [Queen Elizabeth X Elizabeth of Glamis]; Horsfield

'DECEPTION', HT, dp, 1923; flowers deep rose, large, moderate fragrance; GM, NRS, 1923; Beckwith

Declic, S, m, 1988; Croix

'DÉCOR', LCl, or, 1951; (Record); flowers bright scarlet, blooms in clusters, semi-dbl.; foliage leathery; vigorous growth; [(Love X Paul's Scarlet Climber) X Demain]; Mallerin, C.; URS

Decor Arlequin *see* 'MEIZOURAYOR'

Decor Rose *see* 'MEITURAPHAR'

Decora, C, mp; flowers small, very dbl., slight fragrance; Roseraie de l'Hay; (Sangerhausen)

Decorat *see* 'DEKORAT'

'DECORATOR', HT, rb, 1936; flowers brilliant carmine, base orange-yellow; foliage glossy; very vigorous growth; Dickson, A.

Decosilrose®, HT, pb

Decosilthe®, HT, my

'DEDICATION', F, ly, 1968; flowers creamy ivory, borne in trusses, dbl., slight fragrance; foliage glossy; [Pink Parfait X Circus]; Harkness

Dee Bennett™ *see* 'SAVADEE'

'DEEP PURPLE', F, m, 1980; bud ovoid, pointed; flowers mauve-pink, imbricated, dbl., 30–45 petals, 3–4 in., moderate fragrance; foliage glossy, dark; vigorous, upright, bushy growth; [Zorina X Silver Star]; Kordes, R.; Armstrong Nursery

'DEEP SECRET', HT, dr, 1977; (Mildred Scheel); flowers deep crimson, dbl., 40 petals, 4 in., intense fragrance; foliage glossy, dark; vigorous, upright growth; ADR, 1978; Tantau, Math.; Wheatcroft, 1979

'DEEP SECRET', HT, 1946; flowers rich red becoming darker, compact, intense fragrance; RULED EXTINCT 4/77 ARM; [Matchless X Seedling]; Hildebrandt

'DEEP VELVET'™, Min, dr, 1981; flowers urn-shaped blooms borne singly and 2-3 per cluster, dbl., 33 petals, exhibition form, slight fragrance; straight prickles; foliage tiny, medium green; bushy, compact growth; [(Grand Opera X Jimmy Greaves) X Baby Katie]; Jolly, Betty J.; Rosehill Farm

'DEEPIKA', F, rb, 1975; bud pointed; semi-dbl., 15 petals, 2 in., free bloom, fairly lasting; foliage glossy, dark; vigorous, upright, open growth; [Shepherd's Delight X Seedling]; IARI

'DEEPSHIKHA', F, mr, 1975; bud pointed; flowers open, dbl., 35 petals, 2.5 in., slight, tea fragrance; foliage glossy, light; vigorous, compact, bushy growth; [Sea Pearl X Shola]; IARI

'DÉESSE', HT, rb, 1957; (Goddess); bud long, pointed, white; flowers red spreading from outside to center petals, ending crimson, dbl.; vigorous, upright growth; [Peace X Seedling]; Gaujard

'DEFIANCE', HT, dr, 1914; flowers dark velvety red, dbl., 5–6. in., intense fragrance; foliage dark; vigorous growth; [Gruss an Teplitz X Étoile de France]; Kress

Degenhard *see* **'Doc'**

DeGrazia's Pink *see* 'MINHCO'

'DÉJÀ VOUS', Min, ab, 1987; bud pointed; flowers light apricot, aging lighter, large blooms borne usually singly, dbl., 50 petals, exhibition form, slight fragrance; no fruit; no prickles; foliage medium, medium green, semi-glossy; medium, upright, spreading growth; [Anita Charles X Sheri Anne]; Jolly, Nelson & Marie; Rosehill Farm

'DEKORAT', HT, or, 1977; (Decorat, **Freude**®); flowers vermilion and gold blend, moderate fragrance; foliage dark; vigorous, bushy growth; ADR, 1975; [Fragrant Cloud X Peer Gynt]; Kordes, W. Söhne; Mattock, 1975

Del Mar Fair *see* 'TINDEL'

DELaby, LCl, 1995; (**Papi Delbard**); Delbard; (Cavriglia)

'DELADEL', HT, dr, 1982; (Madame Delbard®, **Mme Georges Delbard**®); flowers large, dbl., 40 petals, exhibition form, no fragrance; foliage large, medium green, semi-glossy; upright growth; PP004391; [(Tropicana X Samourai) X (Tropicana X (Rome Glory X Impeccable))]; Delbard, Georges; Roseraies Delbard, 1980

'DELAMBRE', P, dp, 1863; flowers carmine; Moreau et Robert

'DELANCHE', HT, w, 1981; (**Perle Blanche**); flowers large, dbl., 35 petals, no fragrance; foliage medium, medium green, semi-glossy; upright, bushy growth; [(Virgo X Peace) X (Goldilocks X Virgo)]; Delbard, Georges

Delany Sisters™ *see* 'WILDELS'

DELapo, HT, rb, 1997; (**La Passionate**); Delbard

DELarle, S, pb, 1998; (**Rose des Cisterciens**); Delbard

'DELATUR', HT, rb, 1982; (**Allelulia**®, Hallelujah); flowers velvety deep red, silver reverse, large blooms of heavy substance, dbl., 30 petals; foliage deep green, glossy; [((Impeccable X Papa Meilland) X (Gloire de Rome X Impeccable))x Corrida]; Delbard, Georges, 1980

'DELAVAL', F, mr, 1986; (**Avalanche Rose**®); flowers large, dbl., no fragrance; vigorous, bushy growth; [(François et Joseph Guy X Sultane seedling) X (Alain X Étoile de Hollande)]; Delbard, 1977

'DELBAF', HT, ob, 1966; (**Apogée**®, 'DELBAL', 'DELBAT'); bud ovoid; flowers coppery, large, dbl., cupped, slight fragrance; foliage bronze, glossy; vigorous, upright growth; [(Queen Elizabeth X Provence) X (Sultane seedling X Mme JosephPerraud)]; Delbard-Chabert

'DELBARA', F, pb, 1986; (**Bordure Rose**®, Roslyne, Strawberry Ice); flowers medium pink blending to white at base, dbl., 25 petals, 3–4 in., cupped, no fragrance; foliage thick and bright; low, compact growth; GM, Baden-Baden, 1973 Gold Star of the South Pacific, Palmerston North, NZ, 1978; [((Goldilocks X Virgo) X (Orange Triumph X Yvonne Rabier)) X Fashion]; Delbard, 1975

Delbard, F, or; flowers light orange-red, large, dbl., slight fragrance; Delbard-Chabert; (Sangerhausen)

Delbard's Orange Climber *see* 'DELPAR'

'DELBÉCHIR', HRg, ab, 1979; (**Bernadette Chirac**®); flowers apricot, yellow and orange blend, dbl., 23 petals, 3.5 in., cupped, slight fragrance; foliage rugose; vigorous, bushy growth; [R. rugosa X (First Edition X Floradora)]; Delbard-Chabert

'DELBIR', F, mp, 1965; (**Milrose**®); flowers rose pink, blooms in clusters of 5-15, semi-dbl., cupped, slight fragrance; foliage light green, glossy; vigorous, bushy growth; GM, Baden-Baden, 1964; [Orléans Rose X (Francais X Lafayette)]; Delbard-Chabert

'DELBLAN', F, w, 1988; (**Clos Fleuri Blanc**®); flowers large, dbl., 40 petals, slight fragrance; foliage bright; semi-climbing, vigorous growth; [(Milrose X Legion d'Honneur) X Candeur]; Delbard & Chabert, 1990

DELblatine, S, w, 1993; (**Bouquet Vanille**®); Delbard

'DELBLUE', HT, m, 1984; (**Mamy Blue**®); flowers large, dbl., 35 petals, exhibition form; foliage medium, dark; bushy growth; [((Holstein X Bayadere) X (Prelude X St. Exupery)) X Unnamed seedling.]; Delbard, Georges, 1991

DELbobla, F, w, 1997; (**Bordure Blanche**); Delbard

'DELBOIP', HT, dp, 1986; (**Lancéme**®); flowers deep pink, large, dbl., 28 petals, exhibition form, no fragrance; vigorous, upright, bushy growth; [(Dr. Albert Schweitzer X (Michele Meilland X Bayadere)) X (MElmet X Present Filial)]; Delbard, 1973

DELbojaune, F, my; (**Bordure d'Or**®); Delbard

'DELBOURBO', Pol, rb, 1987; ('DELBOURDO', **Gloire du Bourdonnais**®); flowers center cream, margin carmine, opening turns purple, large, dbl., 35–40 petals, no fragrance; foliage bright; good, dwarf growth; [(Milrose X Legion d'Honneur) X (Zambra X Sensation)]; Delbard-Chabert, 1988

DELbover, F, rb, 1990; (**Bordure Vermillon**®); Delbard

DELboviv, T, mp, 1985; (**Bordure Vive**®); Delbard

'DELBRAT', HT, dp, 1969; (**Mondovision**); flowers cyclamen-pink, dbl., 35–45 petals, 4–5 in., moderate fragrance; [Dr. Albert Schweitzer X (Bayadere X Mme Rene Cassin)]; Schweitzer; Trioreau

'DELBRE', F, mp, 1959; (**Cyclamen**); bud long; semi-dbl., 3 in., blooms in sprays of 4-6; vigorous, bushy growth; [(Frau Karl Druschki seedling X Orange Triumph seedling) X (Orange Triumph seedling X Tonnerre)]; Delbard-Chabert

'DELBRO', HT, pb, 1967; (**Parthenon**®); flowers carmine-pink, reverse soft yellow, large, dbl., cupped; foliage bronze, glossy; vigorous, upright, bushy growth; [Chic Parisien X (Bayadere X Rome Glory)]; Delbard-Chabert

'DELBUT', F, dy, 1965; ('DELCRO', **Jean de la Lune**, Moon Magic, YelloGlo); dbl., cupped, blooms in clusters, slight fragrance; foliage matt; low growth; [(Orléans Rose X Goldilocks) X (Fashion X Henri Mallerin seedling)]; Delbard-Chabert; Cuthbert

'DELCAMA', HT, or, 1978; (**Camara**®); flowers orange-vermilion, recurved, dbl., 33 petals, 4–5 in., slight fragrance; PP004332; [((Chic Parisien X Tropicana) X (Gloire de Rome X Impeccable))x (Tropicana X Samourai)]; Delbard

'DELCANDE', F, w, 1978; (**Candeur**®); dbl., 30 petals, 3.5 in., exhibition form, no fragrance; low, upright growth; GM, Baden-Baden, 1978; [((Robin Hood X Virgo) X (Frau Karl Druschki X (Queen Elizabethx Provence)) X (Virgo X Peace)]; Delbard

'DELCAP', LCl, mr, 1963; (**Ténor**); flowers velvety red, blooms in clustersof 6-20, semi-dbl., early; very vigorous growth; Delbard-Chabert

DELcart, HT, dr, 1973; (**Le Rouge et Le Noir**); Delbard

'DELCASCOR', HT, my, 1979; (**Casque d'Or**®); flowers well-shaped, dbl., 30 petals, 4 in., no fragrance; vigorous, upright, bushy growth; [(Zambra X Jean de la Lune) X (Michèle Meilland X Tahiti]; Delbard

'DELCELI', F, ob, 1986; (**Céline Delbard**®, 'DELCÉLIT', 'DELCET'); flowers salmon, silver reverse, large, dbl., 23 petals, cupped, no fragrance; bushy growth; GM, Monza; [Seedling X (Milrose X Legion d'Honneur)]; Delbard-Chabert, 1983

'DELCHINE', F, pb, 1988; (**Shantung**®); flowers mottled pink, cream and red, large, dbl., 22 petals, flat, slight fragrance; foliage matt; vigorous growth; [(Orléans Rose X Goldilocks) X Unnamed Bordure Rose seedling]; Delbard & Chabert

DELcifra, S, w, 1998; (**Citron-Fraise**); Delbard

DELciste, S, w, 1996; (**Libre Ingenue**); Delbard

'DELCLOPEL', HT, or, 1987; (**Clos de la Pellerie**); flowers medium vermilion red, long, large, dbl., 25–35 petals, slight fragrance; foliage large; good, vigorous, bushy growth; [(Spartan X Baccará) X Seedling]; Delbard-Chabert, 1988

'DELCOLB', HT, pb, 1991; (**Colbert**®); flowers white cream, delicately shaded porcelain pink, cup shaped, large, dbl., 27–35 petals, slight fragrance; foliage dark green, flat; bushy (80-100 cms) growth; [((Peace X Bettina) X (President Herbert Hoover X Tropicana)) xChateau de Versailles]; Delbard-Chabert, 1990

DELcouro, LCl, 1995; (**Rose Cascade**); Delbard; (Cavriglia)

'DELCOURT', F, ob, 1984; (**Courtoisie**®); flowers orange, reverse orange blended with yellow, large, dbl., 20 petals, moderate fragrance; foliage medium, medium green; bushy growth; [Avalanche Rose X Fashion seedling]; Delbard, Georges; Delbard-Chabert

DELcoussi, F, mp, 1993; (**Bordure Rose #2**); Delbard

'DELCRAF', HT, ab, 1967; ('DELCRAFT', Louksor, **Louqsor**®); flowers medium, dbl., globular, slight fragrance; foliage glossy; vigorous, bushy growth; [Dr. Albert Schweitzer X Provence]; Delbard-Chabert

DELcreme, F, pb, 1990; (**Regine Crespin**); Delbard

'DELCRIP', Pol, mp, 1967; (**Mme Dimitriu**®); bud globular; flowers pink tinted lighter, large, borne in clusters of 5-10, dbl.; foliage bronze, glossy; vigorous, bushy growth; GM, Geneva, 1967 GM, Rome, 1967; [Chic Parisien X Provence]; Delbard-Chabert

'DELCRO', F, dy, 1965; ('DELBUT', **Jean de la Lune**, Moon Magic, YelloGlo); dbl., cupped, blooms in clusters, slight fragrance; foliage matt; low growth; [(Orléans Rose X Goldilocks) X (Fashion X Henri Mallerin seedling)]; Delbard-Chabert; Cuthbert

'DELCROUF', Min, ly, 1973; (Bordure de Nacrée, **Bordure Nacrée**®); flowers creamy light yellow, small, dbl., 38 petals, no fragrance; low, bushy growth; GM, Baden-Baden, 1972; [(Orléans Rose X Francois et Joseph

Guy) X (Goldilocks X DELtorche)]; Delbard

'DELcus', HT, op, 1966; (**Agéna**®); bud long, pointed; flowers salmon-pink, large, dbl., moderate fragrance; foliage glossy, leathery; vigorous, bushy growth; [Chic Parisien X (Michèle Meilland X Mme Joseph Perraud)]; Delbard-Chabert

'DELdal', HT, op, 1986; (**Dolce Vita**®, Niagara Pride); flowers rosy salmon, large, dbl., 37 petals, exhibition form, slight fragrance; vigorous, upright, bushy growth; [Voeux de Bonheur X (Chic Parisien X (Michele Meilland X Mme Joseph Perraud))]; Delbard, 1971

'DELde', F, dr, 1957; (**Walko**); flowers dark crimson, blooms in clusters of 6-8, dbl., 23 petals, moderate fragrance; [(Incendie X Holstein) X Rouge Chabert]; Delbard-Chabert

DELdido, HT, pb, 1994; (Fetes Galantes®, **Fêtes Galantes**); Delbard

'DELdiore', Gr, m, 1984; (**Dioressence**®); flowers lavender, well-formed, large, dbl., 35 petals, intense fragrance; foliage large, medium green, semi-glossy; bushy growth; [((Holstein X Bayadère) X (Prelude) X Seedling]; Delbard, Georges; Delbard-Chabert

DELdog, S, dy, 1997; (**Pimpernelle**); Delbard

'DELdra', HT, my, 1966; (**Venusic**®); flowers saffron-yellow, medium, dbl., cupped, moderate fragrance; foliage dark, glossy; vigorous, upright, bushy growth; [(Queen Elizabeth X Provence) X (Mme Joseph Perraud X Bayadere)]; Delbard-Chabert

'DELdrop', HT, mr, 1966; (**Beauté Spatiale**®); flowers velvety red, medium, semi-dbl., exhibition form, slight fragrance; foliage bronze, soft; moderate, bushy growth; [Walko X Impeccable]; Delbard-Chabert

'DELecla', F, mr, 1988; (**Clos Fleuri Rouge**®, Sparkler); flowers medium, dbl., 20 petals, flat, slight fragrance; foliage bright; bushy, vigorous growth; [Orléans Rose X Queen Elizabeth]; Delbard & Chabert, 1990

'DELegran', HT, pb, 1986; (**Grand Siècle**®, Great Century); flowers creamy pink blend, well-formed, large, dbl., 33 petals, cupped, slight fragrance; foliage large; vigorous, bushy, branching growth; [((Queen Elizabeth X Provence) X (Michele Meilland X Bayadere))x ((Voeux de Bonheur X MEImet) X (Peace X Dr. Debat))]; Delbard, 1987

'DELépi', HT, dy, 1981; (**Epidor**®); flowers large, dbl., 35 petals, slight fragrance; foliage large, medium green, matt; bushy growth; GM, Rome; [(Peace X Marcelle Gret) X (Velizy X Jean de la Lune)]; Delbard, Georges

'DELfat', HT, ab, 1967; (**Louvre**®); bud long, pointed; flowers rosy apricot, medium, dbl., slight fragrance; foliage bronze, glossy; moderate, upright growth; [Souv. de J. Chabert X (Walko X Souv. de J. Chabert)]; Delbard-Chabert

'DELfesrou', F, dr, 1980; (**Festival Rouge**®, Red Festival); flowers well-formed, blooms in clusters, semi-dbl.; foliage dark; [Walko X (Happiness X Sonia)]; Delbard, Georges; Delbard Roses

'DELfib', F, my, 1965; (**Bebe Lune**®); Delbard

'DELflip', HT, mr, 1967; (**Escurial**®); flowers velvety cardinal-red, medium, semi-dbl., exhibition form, slight fragrance; foliage dark, glossy; vigorous, bushy growth; [Gay Paris X Impeccable]; Delbard-Chabert

'DELflori', F, mr, 1977; (**Fluorescent**®); flowers medium, dbl., 33 petals, cupped, no fragrance; vigorous, bushy, branching growth; [Zambra X ((DELtorche X Tropicana) X (Alain X Souv. de J. Chabert))]; Delbard

DELfloro, F, mr, 1992; (**Centenaire de Lourdes Rouge**®); Delbard

DELfrei, S, m, 1993; (**Coeur Farouche**®); Delbard

'DELfri', HT, pb, 1965; (**Tourmaline**®); flowers creamy white, widely edged carmine, large, dbl., 28 petals, exhibition form; foliage light green; moderate growth; GM, Madrid, 1965; [Michele Meilland X Chic Parisien]; Delbard-Chabert; Cuthbert

'DELfror', HT, pb, 1966; (**Rose du Ciel**®); flowers cream-white broadly edged carmine-pink, large, dbl., globular; foliage dark, glossy; vigorous, bushy growth; [Chic Parisien X (Michele Meilland X Bayadere)]; Delbard-Chabert

'DELgap', HT, yb, 1968; (**Lido di Roma**); bud long, pointed; flowers deep yellow, shaded red, large, dbl., exhibition form, slight fragrance; foliage glossy, leathery; vigorous, upright, bushy growth; GM, Japan, 1968; [(Chic Parisien X Michele Meilland) X (Sultane X Mme Joseph Perraud)]; Delbard-Chabert

'DELge', F, mp, 1958; (**Centenaire de Lourdes**®, Centennaire de Lourdes, Mrs Jones); flowers soft rose, blooms in clusters of 5-10, semi-dbl., 3.5–4 in.; vigorous, bushy growth; [(Frau Karl Druschki X Seedling) X Seedling]; Delbard-Chabert

'DELgeot', HT, dr, 1986; (Isobel Champion, **La Marseillaise**®); bud large, ovoid; flowers well-formed, large, dbl., 40 petals, moderate fragrance; vigorous, bushy growth; [((Gloire de Rome X Impeccable) X (Rouge Meilland X Soraya)) x(MEIsar X Walko)]; Delbard, 1976

'DELglap', Cl HT, my, 1971; (**Puerta del Sol**®); flowers medium golden yellow, large, dbl., 28 petals, slight fragrance; vigorous, climbing (to 9 ft.) growth; [(Queen Elizabeth X Provence) X (Michele Meilland X Bayadere)]; Delbard

'DELgo', LCl, or, 1961; (**Phare**®); flowers bright orange-red, dbl., 33 petals, 3 in.; foliage glossy, dark; vigorous (to 10 ft) growth; [Spectacular X (Floradora X Unnamed seedling)]; Delbard-Chabert

DELgold, HT, dy, 1980; (**Lord Gold**®); Delbard

'DELgorg', HT, rb, 1986; (Chateau de Versailles®, 'DELricos', Gorgeous George, **Guy Laroche**, La Tour d'Argent™); flowers brilliant red, silver reverse, large blooms borne usually singly, dbl., 30 petals, exhibition form, slight fragrance; foliage medium, medium green, matt; upright, bushy growth; [Unnamed seedling X (Michele Meilland X Carla)]; Delbard, Georges; Armstrong Nursery, 1985

DELgribla, LCl, w, 1995; (**Blanche Colombell**); Delbard

DELgrim, LCl, my, 1993; (**Dune**®); Delbard

'DELgrord', HT, w, 1986; (**Grand Nord**®, Great Nord®, Great North); flowers large, dbl., 28 petals, exhibition form, slight fragrance; vigorous, bushy growth; GM, Paris, 1970 GM, Rome, 1973; [((Queen Elizabeth X Provence) X (Virgo X Carina)) X ((Voeux deBonheur X Virgo) X (Virgo X Peace))]; Delbard, 1975

'DELgus', Pol, or, 1968; (**Prince Tango**®); bud pointed; flowers mandarin color, open, small, semi-dbl.; foliage bronze, glossy; moderate, bushy growth; GM, Madrid, 1968; [(Orléans Rose X Goldilocks) X (Orange Triumph seedling X Floradora)]; Delbard-Chabert

'Delhi Apricot', HT, ab, 1964; bud pointed; flowers apricot-yellow, medium, dbl., slight fragrance; foliage light green, soft; moderate, bushy growth; Pal, Dr. B.P.; Indian Agric. Research Inst.

'Delhi Brightness', F, op, 1963; flowers open, medium, semi-dbl.; foliage glossy; vigorous, upright growth; Pal, Dr. B.P.; Indian Agric. Research Inst.

'Delhi Daintiness', F, lp, 1963; bud pointed; flowers light pink, reverse darker, medium, semi-dbl.; foliage glossy; vigorous, upright, compact growth; Pal, Dr. B.P.; Indian Agric. Research Inst.

'Delhi Maid', F, ob, 1963; bud pointed; flowers flame-orange, base gold, open, medium, single; foliage dark, glossy;

vigorous, upright, compact growth; Pal, Dr. B.P.; Indian Agric. Research Inst.

'DELHI PINK PEARL', HMult, lp, 1962; bud ovoid; flowers pearly pink, open, medium, semi-dbl.; foliage glossy, light green; very vigorous, compact growth; [Echo sport]; Pal, Dr. B.P.; Indian Agric. Research Inst.

'DELHI PRINCE', F, dp, 1963; bud pointed; flowers glowing deep pink, open, medium, semi-dbl., slight fragrance; foliage glossy; vigorous, bushy growth; Pal, Dr. B.P.; Indian Agric. Research Inst.

'DELHI PRINCESS', F, mp, 1963; bud ovoid, cerise-red; flowers deep pink, open, large, borne in clusters, semi-dbl., slight fragrance; foliage glossy, bronze; very vigorous, compact growth; Pal, Dr. B.P.; Indian Agric. Research Inst.

'DELHI ROSETTE', F, ob, 1965; bud ovoid; flowers bright orange-scarlet, open, medium, dbl., 28 petals; foliage dark, glossy; vigorous, compact growth; Pal, Dr. B.P.; Indian Agric. Research Inst.

'DELHI SHERBET', F, mp, 1963; bud ovoid; flowers deep rose-pink, medium, dbl., intense fragrance; foliage glossy; vigorous, bushy, compact growth; [Gruss an Teplitz X Seedling]; Pal, Dr. B.P.; Indian Agric. Research Inst.

'DELHI STARLET', Min, ly, 1963; bud pointed; flowers open, small, semi-dbl., slight, musk fragrance; foliage small, glossy; dwarf, compact growth; [Goudvlinder X Seedling]; Pal, Dr. B.P.; Indian Agric. Research Inst.

'DELHI SUNSHINE', HT, yb, 1963; bud pointed; flowers deep cream, reverse flushed pink, medium, dbl.; foliage light green; bushy, open growth; [Mme Charles Sauvage X Seedling]; Pal, Dr. B.P.; Indian Agric. Research Inst.

'DELHI WHITE PEARL', LCl, w, 1963; bud ovoid; flowers pearly white, blooms in clusters, dbl., repeat bloom, slight fragrance; foliage glossy; vigorous growth; [Prosperity X Seedling]; Pal, Dr. B.P.; Indian Agric. Research Inst.

'DELIC', HT, lp, 1978; (**Éterna®**); bud long; flowers light carmine pink, dbl., 28–32 petals, 4–5 in., slight fragrance; vigorous, upright growth; [((Michèle Meilland X Carla) X (Dr. Schweitzer X Tropicana)) x(Queen Elizabeth X Provence)]; Delbard-Chabert

'DELICADO', HT, op, 1954; flowers shell-pink shaded peach, well shaped, 6 in.; moderate growth; Lowe

'DELICATA', HRg, lp, 1898; flowers soft lilac-pink, large, semi-dbl., recurrent bloom; vigorous growth; Cooling

Delicate Beauty *see* KORlarkon

Delicia *see* 'KORGATUM'

'DELICIOUS', Min, mp, 1995; flowers small blooms borne in small clusters, dbl., 26–40 petals, intense fragrance; few prickles; foliage small, medium green, semi-glossy; tall (65-70 cms), upright, compact growth; [Avandel X [(Seedling X Friesia) X Seedling]]; Welsh, Eric; Rose Hill Roses, 1994

Delicious Weldel, Min, lp; [Avandel X Freesia]; Welsh, Eric, 1991; Rose & Fruit Tree; (Weatherly, L.)

'DÉLIE COMMUNAUDAT', HT, yb, 1933; flowers naples yellow, shaded and edged carmine, dbl., cupped, moderate fragrance; foliage leathery; very vigorous, bushy growth; [Mme Charles Detreaux X Mme Edouard Herriot]; Buatois

'DELIFIC', F, mr, 1983; (**Clos Vougeot®**, Red Prolific, Rouge Prolific); flowers medium, dbl., 28 petals, no fragrance; bushy growth; [(Alain X Charles Mallerin) X (Lafayette X Walko)]; Delbard-Chabert

'DELIGHT', HMult, rb, 1904; flowers bright carmine, base white, stamens yellow, medium to large, semi-dbl., cupped; foliage glossy; long stems; very vigorous, climbing (15-20 ft) growth; Walsh

'DELIGHTFUL', HT, pb, 1931; flowers rose, base yellow, reverse amber-yellow, large, dbl., exhibition form, moderate fragrance; foliage glossy; vigorous growth; [George Dickson seedling]; McGredy

'DELIGHTFUL', HT, yb, 1956; bud pointed; flowers straw-yellow, base shaded red, dbl., 35–50 petals, 4–5 in., exhibition form, moderate fragrance; upright, compact growth; [Curly Pink X Shades of Autumn]; Brownell, H.C.

'DELIGHTFUL KIWI', HT, lp, 1989; bud tapering; flowers blush pink aging to creamy pink, urn-shaped, medium, dbl., 45 petals, borne singly, slight fragrance; prickles light brown, very few; foliage light green, large, shiny; upright, branching growth; [Silent Night X (Prima Ballerina X Irish Mist)]; Cattermole, R.F.; South Pacific Rose Nursery, 1988

'DELIGHTFUL LADY', HT, lp, 1982; flowers large, borne singly in spring, 3-5 per cluster in autumn, dbl., 30–35 petals, exhibition form, slight fragrance; brown prickles; foliage semi-glossy; medium-tall growth; [Pascali X Merry Widow]; Attfield, B.B.

'DELIGHTFUL PINK', F, mp, 1958; bud ovoid; flowers pink, medium, borne in clusters, dbl., 40–45 petals, cupped, moderate fragrance; vigorous, upright growth; [Chic X Demure]; Boerner; J&P

DELilac, S, lp; (**Souv de Louis Amade**); Delbard, 1998

Delilah® *see* 'SUNDEL'

'DELILLE', M, w, 1852; flowers blush-white, nicely mossed, semi-dbl., may repeat; Robert

'DELJABER', HT, dr, 1990; (**Professeur Jean Bernard®**); bud cupped; flowers very large bloom, 25–30 petals, slight fragrance; foliage dark green, abundant; bushy growth; [(Charles Mallerin X Divine) X [Tropicana X (Rome Glory X Impeccable)]]; Delbard & Chabert, 1989

'DELJACQ', HT, or, 1988; (Joconde, **La Joconde**); flowers well-shaped, large, dbl., 40 petals, no fragrance; upright, vigorous growth; [(Tropicana X (Rome Glory X Impeccable)) X (Spartan X MEIger)]; Delbard & Chabert, 1989

DELjapal, HT, lp; (**Folio Courtisane**); Delbard, 1996

'DELJAUNE', F, my, 1988; (**Clos Fleuri Jaune®**); flowers yellow, shaded ochre, opening to yellow-amber, large, dbl., 18 petals, slight fragrance; foliage bright; raised, semi-climbing, vigorous growth; [(Orléans Rose X Goldilocks) X Parure d'Or]; Delbard & Chabert, 1990

'DELJAUNOR', HT, ob, 1990; (**France Libre®**); flowers nasturtium orange and coppery, yellow and gold reverse, large, dbl., 25–30 petals, cupped, slight fragrance; foliage dark green, glossy; upright growth; [(Zambra X Orange Sensation) X (Unnamed seedling X Unnamed seedling)]; Delbard & Chabert, 1981; Delbard & Chabert

DELjavert, S, w, 1997; (**Pur Caprice**); Delbard

'DELJIS', LCl, or, 1956; (**Tropique**); dbl.; foliage dark; Delbard-Chabert

'DELJOFEM', HT, pb, 1982; (**Tendresse**); flowers light pink, reverse apricot-pink, large, dbl., exhibition form, moderate fragrance; foliage dense; [(Michèle Meilland X Bayadère) X (Grace de Monaco X Present Filial)]; Delbard, Georges; Delbard Roses, 1980

'DELJONQ', F, dy, 1982; (**Jonquille**); flowers large, dbl., 35 petals, no fragrance; foliage medium, medium green, matt; bushy growth; [(Peace X Marcelle Gret) X (Velizy X Jean de la Lune)]; Delbard, Georges

'DELJULI', F, ab, 1986; (**Julie Delbard®**); flowers apricot with yellow and orange hues, ht-form, large, dbl., 28 petals, no fragrance; vigorous, bushy growth; GM, Madrid, 1976; [(Zambra X (Orange Triumph X Floradora)) X ((Orléans Rose X Goldilocks) X (Bettina X Henri Mallerin))]; Delbard, 1976

DELki, Min, ob, 1998; (**Pinocchio**); Delbard

'DELKORT', HT, dr, 1966; (**Atlas®**); bud long, pointed; flowers magenta-red edged darker, large, dbl., cupped; foliage dark, glossy; vigorous, upright, bushy growth; [Chic Parisien X Provence]; Delbard-Chabert

'DELKRI', HT, mr, 1969; (**France Inter**); flowers magenta-red, ovoid, dbl., 35–45 petals, 4–5 in., slight fragrance; [(Rome Glory X La Vaudoise) X Divine]; Delbard; Trioreau

Della Balfour *see* HARblend

DELmagsa, F, lp; (**The Prince Imperial**); Delbard, 1995

'DELMANCHE', HT, pb, 1980; (**Maman Chérie**); dbl., blooms in clusters; [Gay Paris X (Baccará X Impeccable)]; Delbard, Georges

'DELMAT', HT, yb, 1970; ('DELNAT', **Mitsouko**®); flowers yellow, petals edged red, large, dbl., 50 petals, moderate, fruity fragrance; bronze-red prickles; foliage medium, clear green; dense, bushy growth; [(Michele Meilland X Chic Parisien) X Peace]; Delbard, Georges

'DELMATOR', HT, ob, 1986; (**Beverly Hills**, Malicorne); flowers dark orange, greenhouse variety, blooms in clusters; PP005673; [(Zambra X Orange Sensation) X (Zambra X (Orange Triumph X Floradora))]; Delbard, Georges, 1982

'DELMIR', LCl, yb, 1968; (**Parure d'Or**®); flowers golden yellow edged orange, medium, semi-dbl., repeat bloom; foliage dark, glossy; vigorous, climbing growth; GM, Bagatelle, 1968; [(Queen Elizabeth X Provence) X (Sultane seedling X Mme JosephPerraud)]; Delbard-Chabert, 1970

DELmistri, Min, rb, 1998; (**Bambi**); Delbard

'DELMOT', LCl, op, 1967; (**Obélisque**®); flowers coppery orange-pink, medium, semi-dbl., globular, abundant, intermittent bloom; foliage bronze, glossy; vigorous, climbing growth; GM, Geneva, 1967; [Spectacular X (Orange Triumph seedling X Floradora)]; Delbard-Chabert

'DELMOUN', Cl HT, mr, 1986; (Sensass, **Sensass Delbard**®); flowers bright velvety red, large, dbl., 28 petals, cupped, no fragrance; vigorous, climbing (to 10 ft) growth; [(Danse du Feu X (Orange Triumph X Floradora)) X Tenor seedling, unnamed]; Delbard, 1973

'DELMUR', LCl, mr, 1966; (**Altissimo**®, Altus); flowers blood-red, single, 7 petals, 4–5 in., cupped, repeat bloom, slight fragrance; foliage dark, serrated; tall growth; [Ténor X Seedling]; Delbard-Chabert; Cuthbert

'DELNAT', HT, yb, 1970; ('DELMAT', **Mitsouko**®); flowers yellow, petals edged red, large, dbl., 50 petals, moderate, fruity fragrance; bronze-red prickles; foliage medium, clear green; dense, bushy growth; [(Michele Meilland X Chic Parisien) X Peace]; Delbard, Georges

'DELNIBLE', HT, m, 1981; (**Blue Nile**®, Nil Bleu); bud ovoid, pointed; flowers deep lavender, dbl., 28–30 petals, 5 in., exhibition form, borne singly or 2-3 per cluster, intense, fruity fragrance; short prickles, hooked downward; foliage large, olive green; tall, upright, spreading growth; PP4671; GM, Bagatelle; [(Holstein X Bayadere) X (Prelude X Saint-Exupery)]; Delbard, Georges; Armstrong Nursery, 1976

'DELNOLLI', F, pb, 1981; (**Tourbillon**); flowers deep pink, yellow reverse, large, dbl., 20 petals, moderate fragrance; foliage small, medium green, semi-glossy; upright growth; [Zambra X ((Orléans Rose X Goldilocks) X (Spartan X Fashion))]; Delbard, Georges

'DELOBLAN', HT, w, 1980; (**Blanche Comète**, White Comet); dbl., blooms in clusters; [(Virgo X Peace) X (Goldilocks X Virgo)]; Delbard, Georges; Delbard Roses

DELodive, F, mp, 1994; (**Clos Fleuri Rose No. 2**); Delbard

Deloitte & Touche *see* KORaucher

'DELOP', HT, ab, 1967; (**Vatican**®); bud ovoid; flowers apricot-yellow shaded carmine, medium, dbl., exhibition form, slight fragrance; foliage glossy; moderate, bushy growth; [Grande Premiere X (Sultane X Mme Joseph Perraud)]; Delbard-Chabert

Delore, HT, ly, 1982

'DELORFEU', F, ob, 1982; (**Carnaval de Rio**®); flowers orange, blooms in clusters, dbl.; [(Zambra X Orange Sensation) X (Zambra X Seedling (Orange Triumph X Floradora))]; Delbard, Georges; Delbard Roses

'DELOSOL', HT, my, 1984; (**O Sole Mio**®); flowers well-formed, large, dbl., 35 petals, no fragrance; foliage medium, medium green, glossy; [(Peace X Marcelle Gret) X Velizy seedling]; Delbard, Georges

DELpapy, S, my; (**Souvenirs de Marcel Proust**); Delbard, 1993

'DELPAR', LCl, or, 1966; (**Delbard's Orange Climber**, Grimpant Delbard); bud ovoid; flowers medium, dbl., exhibition form, repeat bloom, slight fragrance; foliage dark, glossy, leathery; vigorous, climbing, well branched growth; [Spectacular X (Rome Glory X La Vaudoise)]; Delbard-Chabert, 1963; Armstrong Nursery, 1963

DELperl, LCl, dr, 1994; (**Salammbo**®); Delbard

'DELPÉTRI', HT, mr, 1988; (**Clochermerle**®); flowers large, dbl., 38 petals, cupped, slight fragrance; foliage matt; vigorous, bushy growth; [Seedling X [(Michele Meilland X Karla) X Seedling)]]; Delbard & Chabert

Delphin, HT, m; flowers violet-pink, large, dbl., intense fragrance; VEG; (Sangerhausen)

'DELPLI', Pol, ob, 1969; (**Rosier d'Or**); flowers orange-yellow, shaded apricot-yellow, dbl., 20–28 petals, 2.5–3.5 in., slight fragrance; [Zambra X (Orléans Rose X Goldilocks)]; Delbard; Trioreau

'DELPO', F, mr, 1961; (**Diablotin**®, Little Devil); semi-dbl., 17 petals, 2–3 in., blooms in small clusters, no fragrance; bushy, compact growth; [Orléans Rose X Fashion]; Delbard-Chabert

'DELPOC', HT, mp, 1966; (**Diapason**®); flowers porcelain pink, medium, dbl., 40 petals, globular, moderate fragrance; foliage bronze, glossy; very vigorous, bushy growth; [Chic Parisien X (Sultane seedling X Mme Joseph Perraud)]; Delbard-Chabert

'DELPOMP', F, mp, 1988; (**Clos Fleuri Rose**®); flowers medium, dbl., 30 petals, slight fragrance; foliage bright; bushy, vigorous growth; [(Zambra X Orange Sensation) X (Robin Hood X Virgo)]; Delbard & Chabert, 1990

'DELPOSAR', Cl F, mr, 1970; (**Diablotin, Climbing**®, Grimpant Diablotin, Little Devil, Climbing); GM, Geneva, 1970 GM, Paris, 1970 GM, Rome, 1970; [Diablotin sport]; Delbard

'DELPOUS', HT, dp, 1967; (Fontaine Blue, **Fontainebleau**®); bud long, pointed; flowers magenta-pink, large, dbl., globular, slight fragrance; foliage dark, glossy, leathery; vigorous, bushy growth; [Dr. Albert Schweitzer X (Bayadere X Rome Glory)]; Delbard-Chabert

'DELPRAT', HT, lp, 1970; (**Présence**); flowers pink, lighter reverse, large, dbl., 38 petals, slight, fruity fragrance; foliage medium, medium green, matt; upright, bushy growth; [Dr. Albert Schweitzer X (Michele Meilland X Bayadere)]; Delbard, Georges; G. Delbard;, Society of Nurseries

'DELPRE', Gr, op, 1960; (Happy Anniversary, **Heureux Anniversaire**®); bud urn-shaped; flowers salmon-orange, dbl., 28 petals, 3 in., slight, spicy fragrance; foliage glossy; very vigorous, bushy growth; [(Incendie X Chic Parisien) X (Floradora X Independence)]; Delbard-Chabert; Stark Bros., 1963

'DELPRIMA'®, HT, op, 1968; ('DELTUF', **Grand Prix**®); flowers coral-pink shaded ochre, large, semi-dbl., slight fragrance; foliage glossy, leathery; vigorous, upright, bushy growth; GM, Belgium, 1968; [Chic Parisien X (Grande Premiere X (Sultane X Mme Joseph-Perraud))]; Delbard-Chabert

DELredi, HT, ob; (**Renata Tebaldi**); Delbard

'DELRENO', HT, dy, 1987; (**Mme Georges Renoard**); flowers long, large, dbl., 35–40 petals, moderate fragrance; foliage clear, matt; upright, half-divirgent, vigorous growth; [(Peace X Marcelle Gret)

X Unnamed Legion d'Honneur seedling]; Delbard-Chabert, 1988

'DELRICOS', HT, rb, 1986; (Chateau de Versailles®, 'DELGORG', Gorgeous George, **Guy Laroche**, La Tour d'Argent™); flowers brilliant red, silver reverse, large blooms borne usually singly, dbl., 30 petals, exhibition form, slight fragrance; foliage medium, medium green, matt; upright, bushy growth; [Unnamed seedling X (Michele Meilland X Carla)]; Delbard, Georges; Armstrong Nursery, 1985

DELrima, F, rb, 1995; (**Sasa**); Delbard

'DELRIO', HT, m, 1983; (Night Flight, **Vol de Nuit**); flowers deep lilac, large, dbl., 33 petals, intense fragrance; bronze-red prickles; foliage medium, light green, matt; bushy growth; GM, Rome, 1970; [(Holstein X (Bayadere X Prelude)) X Saint-Exupery]; Delbard, Georges, 1970

'DELROBLA', F, w, 1991; (**Carmagnole®**); flowers white cream with soft pink, moderately cup shaped, blooms, dbl., 20–25 petals, slight fragrance; foliage dark green; bushy (80-100 cms) growth; [(Milrose X Legion d'Honneur) X (Zambra X Orange Sensation)]; Delbard & Chabert, 1990

'DELROCELES', Cl HT, lp, 1979; (**Rose Céleste®**); flowers well-formed, large, dbl., 33 petals, cupped, moderate fragrance; vigorous, climbing (to 9 ft) growth; [(Queen Elizabeth X Provence) X (Sultane X Mme J. Perraud)]; Delbard

DELroujan, HT, 1988; Delbard, Georges; (Cavriglia)

'DELROUVEL', HT, rb, 1988; (**Chàteau d'Amboise®**); flowers dark red, opening bright, long, dbl., 23–30 petals, slight fragrance; foliage bright; vigorous, semi-raised growth; [Tropicana seedling X [(Rome Glory X Impeccable) X (Rouge Meilland X Soraya)]]; Delbard & Chabert

DELrovrai, HT, 1995; (**Jardins de Villandry**); Delbard, Georges; (Cavriglia)

DELrugro, LCl, lp; (**Tour Eiffel 2000**); Delbard, 1998

'DELRULA', F, pb, 1988; (**Ville du Perreux**); flowers pink with white and cream, long, large, dbl., 28 petals, slight fragrance; foliage bright; bushy, vigorous growth; [Unnamed seedling X (Milrose X Legion d'Honneur)]; Delbard & Chabert

'DELSAB', HT, pb, 1968; (**Royal Ascot®**); flowers pink, reverse shaded crimson, large, semi-dbl., exhibition form; foliage glossy, leathery; vigorous, bushy growth; [Chic Parisien X (Grande Premiere X (Sultane X Mme Joseph Perraud))]; Delbard-Chabert

'DELSAMAR', F, mr, 1974; (**Légion d'Honneur®**); flowers well-formed, medium, dbl., 55 petals, cupped, slight fragrance; low, bushy growth; GM, Geneva, 1974; [(Souv. de J. Chabert X (Walko X Souv. de J. Chabert)) X ((Tamango X Gay Paris) X (Zambra X Jean de la Lune))]; Delbard

'DELSAMO', HT, dp, 1974; (**Paris 2000**); flowers deep pink, large, dbl., 28 petals, cupped, no fragrance; vigorous, upright, bushy growth; [(DELtorche X (Sultane X Mme Joseph Perraud)) X (Queen Elizabeth X Provence)]; Delbard

'DELSAMOUR', HT, ab, 1973; (**Velizy®**); flowers well-formed, large, dbl., 30 petals, slight fragrance; bushy, branching growth; [((Peace X Marcelle Gret) X (Michele Meilland X Tahiti)) X (Peace X Grand Premiere)]; Delbard

'DELSAP', F, mr, 1965; (**Cappa Magna®**); flowers yellow stamens, wavy petals, blooms in clusters of 20-30, 8–10 petals, 4 in., cupped; foliage dark, glossy, large; upright growth; [Tenor seedling]; Delbard-Chabert; Cuthbert, 1967

'DELSATEL', HT, or, 1982; (**Satellite**); flowers medium, dbl., 28 petals, moderate fragrance; foliage medium, medium green, semi-glossy; [((Tropicana X Samourai) X (Tropicana X (Rome Glory X Impeccable))) X Granada]; Delbard, Georges

'DELSET', HT, lp, 1967; (Castel, **Versailles®**); bud ovoid; flowers soft pink, medium, dbl., cupped, slight fragrance; foliage dark, glossy, leathery; vigorous, upright, bushy growth; GM, Baden-Baden, 1965 GM, Bagatelle, 1966 GM, Geneva, 1966; [(Queen Elizabeth X Provence) X (Michele Meilland X Bayadere)]; Delbard-Chabert

'DELSIRE', Cl HT, dr, 1986; (Grandessa, **Messire Delbard®**); flowers deep crimson red, well-formed, large, dbl., 38 petals, slight fragrance; foliage large; vigorous (to 9 ft) growth; [(Danse du Feu X Guinee) X ((Tenor X Fugue) X (Delbard's OrangeClimber X Gloire de Dijon))]; Delbard, 1976

'DELSIRP', Pol, ob, 1971; (**Caid**); bud ovoid; flowers orange, medium, dbl., repeat bloom, slight fragrance; foliage leathery; vigorous, bushy growth; [Orangeade X Seedling]; Delforge

'DELSOB', F, ob, 1975; (**Puerto Rico**, Sable Chaud, Sable Cloud); flowers orange, reverse blended with yellow, medium, dbl., cupped, slight, spicy fragrance; foliage bronze, leathery; vigorous, bushy growth; [Zambra X (Orange Triumph X Floradora)]; Delbard; Armstrong Nursery

DELsorb, S, pb; (**Sorbet Framboise®**); Delbard, 1994

'DELSOUCHE', Gr, dr, 1980; (**Souv de Maurice Chevalier**); semi-dbl., blooms in clusters; [(Walko X Impeccable) X (Papa Meilland X (Baccará X Michele Meilland))]; Delbard, Georges

DELstrabla, HT, pb, 1995; (**Hei Matisse**); Delbard

DELstricol, F, 1996; (**Camille Pisarro**); Delbard, Georges; (Cavriglia)

DELstror, F, pb, 1997; (**Grimaldi**); Delbard

'DELSULAN', HT, mp, 1988; (**Europe 92**); flowers pink magenta, long, large, dbl., 11 petals, slight fragrance; foliage bright; vigorous, bushy growth; [Seedling X (Michèle Meilland X Karla)]; Delbard & Chabert

Delta *see* **'BREAK O' DAY'**

Delta Dawn *see* 'IVEMAX'

'DELTA GAMMA', HT, ly, 1977; bud pointed; flowers creamy white, base pale yellow, dbl., 38 petals, 4–4.5 in., exhibition form, moderate, spicy fragrance; foliage dark; vigorous, upright, bushy growth; [Queen Elizabeth X Mount Shasta]; Kimbrew-Walter Roses

Delta Gold™ *see* 'BURDEL'

Delta Queen, HT, mp, 1998

'DELTAF', HT, pb, 1962; (**Mme René Cassin**); flowers cyclamen-pink, reverse silvery, dbl., 24 petals, 5 in., exhibition form, moderate fragrance; foliage dark, glossy; vigorous, upright growth; GM, Bagatelle, 1962 GM, Lyon, 1962; [Mme Armand Souzy X Impeccable]; Delbard-Chabert

'DELTANGA', HT, ob, 1978; (**Tango Rose**); bud pointed; flowers salmon-orange, large, dbl., 35–40 petals; foliage matt green; bushy growth; [(Belle Rouge X (Gloire de Rome X Gratitude)) X ((Dr. Schweitzer X Tropicana) X (Ena Harkness X Quebec))]; Delbard

DELtendre, S, mp, 1994; (**Comtesse de Segur**, Tendresse); Delbard

'DELTEP', F, op, 1976; (Arnaud Delbard®, **First Edition**); bud ovoid, pointed; flowers luminous coral, shaded orange, dbl., 28 petals, 2–2.5 in., slight, tea fragrance; foliage glossy, light; upright growth; AARS, 1977; [(Zambra X (Orléans Rose X Goldilocks)) X (Orange Triumph seedling X Floradora)]; Delbard; C-P

'DELTOGO', F, ab, 1961; (**Tobago®**); flowers yellow apricot, outer petals aging pink, large, dbl., 35 petals, moderate fragrance; foliage medium, dark, semi-glossy; upright growth; Golden Rose, Geneva, 1981; [Avalanche X (Zambra X Orange Sensation)]; Delbard, Georges

'DELTOP', HT, mr, 1983; (**Crêpe de Chine®**); flowers large, dbl., 20 petals, slight fragrance; bronze-red prickles; foliage glossy, clear green, dense; vigorous, upright, bushy growth; GM, Madrid, 1970; [Joyeux Noel X (Gloire de Rome X Impeccable)]; Delbard, Georges; Delbard-Chabert, 1970

'DELTOS', HT, m, 1961; ('DELVOR', **Saint-Exupéry®**, Waltz Time); flowers mauve tinted silvery, dbl., 5 in., exhibition form, slight fragrance; vigorous, bushy growth; [(Christopher Stone X Marcelle Gret) X (Holstein X Bayadere)]; Delbard-Chabert

DELtou, F, 1986; Delbard, Georges; (Cavriglia)

'DELTRAC', HT, mr, 1967; (**Angkor®**); flowers carmine-red, large, dbl., globular, slight fragrance; foliage bronze, glossy, leathery; vigorous, bushy growth; [Belle Rouge X (Rome Glory X Gratitude)]; Delbard-Chabert

DELtrap, F, mr, 1994; (**Saint-Vincent®**); Delbard, Georges

'DELTRE', HT, mr, 1959; (**Maurice Chevalier®**); bud long; flowers rich red shaded garnet, large, dbl., 25 petals, moderate fragrance; foliage glossy; vigorous growth; [Incendie X (Floradora seedling X Independence)]; Delbard-Chabert

'DELTROB', Pol, w, 1967; (**Cactus Blanc®**); bud globular; flowers creamy white, small, borne in clusters, dbl., cupped; foliage light green, glossy; moderate, bushy growth; [(Orléans Rose X Orléans Rose) X ((Francais X Lafayette) X (Orléans Rose X Goldilocks))]; Delbard-Chabert

'DELTRUT', LCl, dp, 1967; (Campanela, **Campanile®**); bud globular; flowers deep magenta-pink, large, dbl., repeat bloom, moderate fragrance; foliage glossy, leathery, bronze; vigorous, climbing growth; [(Queen Elizabeth X Provence) X (Sultane seedling X Mme JosephPerraud)]; Delbard-Chabert

'DELTUF', HT, op, 1968; ('DELPRIMA'®, **Grand Prix®**); flowers coral-pink shaded ochre, large, semi-dbl., slight fragrance; foliage glossy, leathery; vigorous, upright, bushy growth; GM, Belgium, 1968; [Chic Parisien X (Grande Premiere X (Sultane X Mme Joseph-Perraud))]; Delbard-Chabert

'DELUP', HT, dr, 1969; (**Betsy Ross**, La Passionata); flowers deep red, large, dbl., exhibition form, slight fragrance; vigorous, upright growth; [(Gloire de Rome X La Vaudoise) X Divine]; Delbard; Trioreau & Stark Bros.

'DELURT', HT, dr, 1986; (Black Pearl, **Perle Noire®**); flowers velvety dark red, well-formed, large, dbl., 38 petals, slight fragrance; vigorous, bushy growth; [((Impeccable X Papa Meilland) X (Gloire de Rome X Impeccable))x ((Charles Mallerin X Gay Paris) X (Rouge Meilland X Soraya)]; Delbard, 1976

'DELVEEN', P, lp; Alderton, 1930; (Weatherly, L.)

DELverjaune, F, dy, 1997; (**Nedbank Rose**); Delbard

DELviola, S, 1996; (**Chartreuse de Parme**); Delbard, Georges; (Cavriglia)

DELvirge, S, lp; (**Vierge Folle**); Delbard, 1998

'DELVOR', HT, yb, 1970; (**Femme**); flowers ivory yellow, tinted pink, large, dbl., 28 petals, slight fragrance; bronze-red prickles; foliage medium, dark, glossy; upright growth; [(Gloire de Rome X Bayadere) X (Queen Elizabeth X Provence)]; Delbard, Georges; Delbard Roses

'DELVOR', HT, m, 1961; ('DELTOS', **Saint-Exupéry®**, Waltz Time); flowers mauve tinted silvery, dbl., 5 in., exhibition form, slight fragrance; vigorous, bushy growth; [(Christopher Stone X Marcelle Gret) X (Holstein X Bayadere)]; Delbard-Chabert

'DELZEN', LCl, or, 1982; (**Zenith**); semi-dbl., blooms in clusters; climbing (to 8 ft) growth; [(Spectacular X Tenor seedling) X (Floradora X Incendie)]; Delbard, Georges; Delbard Roses

DELzinsch, HT, dy, 1998; (**President Armand Zinsch**); Delbard

'DEMAIN', HT, mr, 1945; bud pointed; flowers brilliant cardinal-red, reverse saffron-yellow, semi-dbl., cupped; very vigorous, upright growth; [Mrs Pierre S. duPont X Dr. Kirk]; Mallerin, C.; A. Meilland

Dembrosky *see* **'DEMBROWSKI'**

'DEMBROWSKI', HP, dr, 1849; (Dembrosky); flowers deep crimson-violet; Vibert

'DEMOISELLE', HT, pb, 1960; flowers soft pink becoming darker, open, medium, dbl., 25–30 petals, slight fragrance; foliage bronze; vigorous, bushy growth; [Peace X Opera]; Delforge

'DEMOKRACIE', LCl, dr, 1935; (Blaze Superier);, said to be recurrent; Böhm, J.

Demon *see* 'MEIDOMONAC'

'DÉMONE', HT, dr, 1965; bud ovoid; flowers medium, dbl., intense fragrance; foliage dark; Tantau, Math.

'DEMURE', F, dp, 1952; bud ovoid; flowers rose-pink, borne in clusters, dbl., 52 petals, 2 in., flat, moderate fragrance; foliage leathery; vigorous, compact growth; [Garnette X Seedling]; Boerner; J&P

Deneb, LCl, dy

'DENIS HÉLIE', HP, mr, 1864; (Denis Helye); flowers rosy crimson; vigorous growth; Gautreau

Denis Helye *see* **'DENIS HÉLIE'**

'DENISE', HT, my, 1955; bud oval; flowers citron-yellow, open, large, dbl., slight fragrance; foliage glossy, clear green; vigorous, bushy growth; [Peace X Brandywine]; Buyl Frères; Delforge

'DENISE CASSEGRAIN', Pol, w, 1922; flowers snow-white, clusters of 30-40, very dbl., intense fragrance; Grandes Roseraies

'DENISE CHAMBARD', HT, yb, 1940; bud long, carmine-yellow; flowers sulfur-yellow, shaded carmine, large, cupped; foliage bright green; strong stems; vigorous, upright growth; Chambard, C.; Orard

'DENISE DEWAR', HT, dr, 1968; flowers crimson, pointed; foliage bronze-green; free growth; [Isabelle de France X Karl Herbst]; Trew, C.

Denise Grey® *see* MEIxetal

Denise Hale *see* 'WAMHALE'

Denise Hilling, HMoy, dp

'DENISE LEFEUVRE', HT, rb, 1930; flowers nasturtium-red, center yellow, reverse bright red, large, dbl., cupped; strong stems; very vigorous growth; Chambard, C.

'DENISE MCCLELLAND', HT, dr, 1964; flowers claret-red, large, dbl., 30 petals, moderate fragrance; foliage dark, leathery, glossy; vigorous, upright, bushy growth; [Amy Johnson X New Yorker]; Riethmuller; Akhurst

'DENISE-ANNE', F, op, 1973; flowers blush-pink to orange-apricot, full, dbl., 30–35 petals, 4 in., intense fragrance; foliage small, glossy, dark; vigorous growth; [(Memoriam X Orange Sensation) X (Peace X Memoriam)]; Ellick; Radway Roses

Denk an Mich, F, ab, 1995; Huber

Denman *see* 'LANDEN'

'DENNY BOY', F, yb, 1949; bud small, ovoid; flowers orange-yellow, becoming redder, reverse sulfur-yellow, very dbl., cupped, slight fragrance; very vigorous, bushy, dwarf growth; [Pinocchio X Mrs Erskine Pembroke Thom]; Marsh; Marsh's Nursery

Denoyel *see* 'SOUV DE CLAUDIUS DENOYEL'

Dentelle de Bruges®, S, w, 1991; Lens

Dentelle de Bruxelles®, S, mr, 1988; Lens

Dentelle De Malines®, S, mp, 1986; Lens

Denver's Dream™ *see* 'SAVADEN'

'DER KRAD', HT, dr, 1962; bud long, pointed; flowers maroon-red, striped darker, dbl., 3.5 in., exhibition form, intense fragrance; foliage glossy; vigorous, bushy growth; [Ami Quinard X Crimson Glory]; Wyant

'DERBY', F, or, 1963; bud ovoid; flowers medium, semi-dbl., moderate fragrance; foliage light green, soft; very vigorous, bushy growth; [Miss France X Seedling]; Gaujard

Derdinger Sommer *see* HELstrau

'DEREHAM PRIDE', Pol, dr, 1932; flowers darkest crimson; vigorous, dwarf growth; [Éblouissant X Orange King]; Norfolk Nursery

Derek Nimmo *see* 'MACWHENU'

'DERNBURG', HT, pb, 1916; flowers bright rose, shaded coral-red and yellow, large, dbl.; Krüger

'DERUITER'S HERALD', Pol, dr, 1949; (Herald); flowers blood-red, prominent yellow stamens, small, borne in large trusses, single, 6 petals; foliage glossy, dark; vigorous, bushy growth; (21); GM, NRS, 1948; [Orange Triumph X Seedling]; deRuiter; Gregory

DESbet, HT, dp; (**Betsy Murchison**); dbl., 34 petals, exhibition form; [Southern Lady X Seedling]; Desamero, Luis, 1998

Descamet, C, mp

Descanso Dream *see* 'WEKDESC'

'DESCANSO PILLAR', LCl, pb, 1952; bud urn-shaped, carmine to scarlet-red; flowers begonia-rose to deep rose-pink, inside varying to scarlet, dbl., 33 petals, 4.5–5 in., exhibition form, recurrent bloom, moderate fragrance; foliage dark, glossy; height 6-7 ft; [Crimson Glory X Capt. Thomas]; Lammerts, Dr. Walter; Germain's

'DESCHAMPS', N, mr, 1877; flowers cherry-red, large, cupped; vigorous growth; Deschamps

'DESDECHARDO', HT, dp, 1980; flowers deep pink; [Red Lion sport]; Taylor, W.J.

'DESDEMONA', HT, lp, 1912; flowers large, dbl., intense fragrance; Paul, W.; (Sangerhausen)

'DESERT CHARM', Min, dr, 1973; flowers deep red, medium, dbl., exhibition form, slight fragrance; foliage dark, leathery; vigorous, dwarf, bushy growth; [Baccará X Magic Wand]; Moore, Ralph S.; Sequoia Nursery

'DESERT DANCE', F, ob, 1975; bud pointed; flowers orange, reverse gold, semi-dbl., 15–18 petals, 3 in., cupped, slight fragrance; foliage glossy; moderaterly tall growth; [Impala X Seedling]; Herholdt, J.A.

Desert Dawn, Gr, yb, 1998

'DESERT DREAM', HT, lp, 1955; flowers buff-pink, high pointed, 5 in., slight fragrance; foliage light green; very vigorous growth; [R.M.S. Queen Mary X Mrs Sam McGredy]; McGredy, Sam IV; Kordes

Desert Peace *see* 'MEINOMAD'

'DESERT SANDS', F, ab, 1976; flowers deep apricot, full, dbl., 30 petals, 4.5 in., intense fragrance; vigorous growth; [Arthur Bell X Elizabeth of Glamis]; Bees

'DESERT SONG', HT, rb, 1948; flowers glowing reddish copper, peony form, dbl., 40–45 petals, 4–5 in., intense fragrance; foliage glossy, bronze; [Mrs Sam McGredy X Golden Dawn]; Fletcher; Tucker

'DESERT STORM', Min, op, 1991; bud ovoid; flowers orange in summer, pink in cold weather, urn-shaped, dbl., borne singly; foliage medium, dark green, semi-glossy; upright, medium growth; [Libby X Rise 'n' Shine]; Gruenbauer, Richard, 1984; Flowers 'n' Friends Miniature Roses, 1991

Desert Storm, S, or, 1997; Williams, J. Benjamin

Desert Sun *see* DEVsleek

'DESERT SUNSET', Gr, or, 1962; bud ovoid; single, 10 petals, 3–4 in., cupped; foliage glossy; upright growth; [Floradora X Chrysler Imperial]; Booy, H.; Booy Rose Nursery

'DESGACHES', B, lp, 1850; Plantier

Desi, HT, yb, 1964; flowers golden yellow with red stripes, large, dbl.; Rupprecht-Radke; (Sangerhausen)

'DESIGNER'S CHOICE'™, Gr, yb, 1989; (Hi Teen); bud pointed; flowers yellow marked orange, medium, borne usually singly and in sprays, dbl., 30 petals, exhibition form, slight fragrance; ovoid, small, orange fruit; prickles slight recurve, small, brown; foliage medium, dark green, glossy; upright, medium growth; [Prominent X Bengali]; deVor Nursery; Co-Operative Rose Growers, 1989

'DÉSIR', HT, rb, 1945; flowers purplish red, large, dbl., intense fragrance; foliage glossy; very vigorous, bushy growth; Gaujard

'DESIRE', HT, mr, 1953; bud ovoid; flowers cardinal-red, dbl., 35–45 petals, 5 in., exhibition form, moderate fragrance; foliage dark; upright, compact growth; [Pink Delight sport]; Obertello; Amling-DeVor Nursery

'DÉSIRÉ BERGERA', HWich, pb, 1910; flowers coppery rose, center brighter, dbl., seasonal bloom; vigorous growth; [R. wichurana X Aurore]; Barbier

Desirée® *see* 'TANERISED'

'DÉSIRÉE PARMENTIER', HGal, lp; flowers vivid pink, large, dbl., flat; bushy growth; Parmentier, Prior to 1841

'DESLING', F, ob, 1996; (**Cat's Meow**™); flowers medium orange, petal base medium yellow, bright orange, buff, dbl., 29–31 petals, 3.5 in., slight fragrance; numerous prickles; foliage medium, dark green, semi-glossy; upright, tall (40") growth; [Colour Wonder X Unknown]; Desmet, Paul; Desling Farms

'DESMOND GATWARD', HT, mr, 1966; flowers cerise-crimson, well formed, 4.5 in., slight fragrance; foliage red-bronze; free growth; [Karl Herbst X Dicksons Red]; Kemp, M.L.

'DESMOND JOHNSTON', HT, mr, 1927; flowers brilliant scarlet, base orange, reverse veined orange, large, dbl., exhibition form, moderate fragrance; foliage rich green, leathery, glossy; short stems; bushy growth; GM, NRS, 1927; McGredy

DESmother, Min, w; (**Sadie**); very dbl., exhibition form; [Irresistible sport]; Desamero, Luis, 1998

'DESPERADO', F, yb, 1968; flowers yellow shaded pink, borne in trusses, semi-dbl., slight fragrance; [Pink Parfait X Masquerade]; Harkness

Desprez à Fleur Jaunes *see* **'JAUNE DESPREZ'**

Dessert, F, rb; flowers red with white, medium, semi-dbl.; VEG; (Sangerhausen)

Dessy, F, lp; Williamson, 1959; (Weatherly, L.)

Destin, HT, 1961; Croix, P.; (Cavriglia)

Destin Cl., Cl HT, mr

'DESTINO', HT, pb; flowers salmon-pink, becoming lilac-pink, well formed, dbl., 45 petals; foliage dark; upright growth; Camprubi, C.

'DESTINY', HT, dr, 1935; bud long, pointed, well shaped; flowers rich crimson-scarlet shaded blackish, dbl.; foliage dark, leathery; vigorous, bushy growth; Beckwith

Destiny *see* MEIkiji

'DETROITER', HT, dr, 1952; (Brilliant, Schlosser's Brilliant); bud long, pointed; dbl., 23 petals, 5.5 in., exhibition form, moderate fragrance; vigorous, upright, bushy growth; GM, NRS, 1952; [Poinsettia X Crimson Glory]; Kordes; J&P

'DETROITER, CLIMBING', Cl HT, dr, 1960; (Schlösser Brilliant, Climbing); J&P

Detty, HT, or; Adam

'DEUIL DE DR REYNAUD', B, mp, 1862; flowers deep crimson pink, intense fragrance; Pradel

'DEUIL DE DUC D'ORLEANS', B, m, 1845; Lacharme, F.

'DEUIL DE DUNOIS', HP, dp, 1864; flowers carmine-pink, medium, dbl.; Verdier, E.; (Sangerhausen)

'DEUIL DE PAUL FONTAINE', M, m, 1873; (Paul de Fontainne); bud somewhat mossy; flowers purple-red, reverse mahogany, cupped, repeat bloom; very prickly; vigorous growth; Fontaine

Deuil du Colonel Denfert, HP, dr; Margottin Père, 1878

'DEUTSCHE HOFFNUNG', HT, yb, 1920; flowers salmon-yellow to apricot-yellow; [Mme Caroline Testout X Grossherzogin Feodora von Sachsen]; Kiese

Deutsche Welle®, F, m, 1983; Leenders

'DEUTSCHES DANZIG', Pol, dp, 1935; flowers carmine-pink with white, small, single; Lambert, P.; (Sangerhausen)

'DEVABE', HT, w, 1982; (**White Mystery**); [Paul's Pink sport]; deVor, Paul F.; DeVor Nurseries, Inc.

'DEVAMARILLO', Min, ob, 1989; (**Rainbow Yellow**™); bud pointed, urn-shaped; flowers tangerine orange, aging lighter, small, borne singly, dbl., intense, fruity fragrance; prickles declining, reddish-brown; foliage medium, medium green, matt; bushy, medium growth; PP007638; [Unnamed seedling X Amber Flash]; Marciel, Stanley G.; DeVor Nurseries, Inc.

'DEVAURORA', Min, ob, 1989; (**Rainbow Sunrise**™); bud pointed; flowers orange with tinge of red, reverse same, dbl., slight, spicy fragrance; prickles sparse, declining, brown; foliage small, medium green, glossy; bushy, medium growth; PP006583; [Amber Flash X Rumba]; Marciel, Stanley G.; DeVor Nurseries, Inc.

'DEVBILL', Cl HT, op, 1995; (**Sheer Elegance, Climbing**); flowers pink blend, blooms, dbl., 26–40 petals, 3–3.5 in., slight fragrance; few prickles; foliage large, dark green, semi-glossy; tall (315-345cms), upright, branching growth; [Sheer Elegance sport]; deVor, Bill; DeVor Nursery, 1994;, Youngs' American Rose Nursery, 1994

DEVblush, HT, w, 1996; (**Bridal Shower**); deVor

'DEVCAL', HT, w, 1995; (**Victorian Lace**); dbl., 26–40 petals, 1.5–2.5 in., borne mostly singly, no fragrance; some prickles; foliage medium, medium green, semi-glossy; tall (60 in), upright growth; Marciel, Stanley G.; DeVor Nurseries, Inc., 1995;, Young's American Rose Nursery, 1995

'DEVCALI', HT, dp, 1989; (**Sizzle Pink**™); bud slender, tapering; flowers deep pink, sweetheart, large, dbl., 25 petals, cupped, borne singly, slight fragrance; prickles declining, olive green; foliage large, dark green, matt; upright, tall growth; PP007562; [Unnamed seedling X Unnamed seedling]; Marciel, Stanley G.; DeVor Nurseries, Inc.

'DEVCARLOS', HT, ly, 1992; (**Brianna**); flowers canary yellow, blooms borne mostly singly, dbl., 26–40 petals, 3–3.5 in., moderate fragrance; few prickles; foliage large, medium green, semi-glossy; upright (152 cms) growth; [Seedling 82249-1 (75062-1 X Excitement) X Seedling (75062-1 xCocktail)]; Marciel, Stanley G.; DeVor Nurseries, Inc.

'DEVCICO', HT, mp, 1993; (**Darby**); flowers moderately medium blooms borne mostly single, dbl., 15–25 petals, 1.5–2.75 in., no fragrance; some prickles; foliage large, dark green, semi-glossy; tall, upright growth; [Dolores X Cerise Dawn]; Marciel, Stanley G. & Jeanne A.; DeVor Nurseries, Inc., 1993;, Matsui Nursery, 1993

'DEVCLAVEL', Min, dp, 1989; (**Rainbow Cerise**™); bud tapering, slender; flowers deep pink, medium, borne singly, dbl., 45 petals, cupped, slight, fruity fragrance; foliage medium, dark green, glossy; bushy, medium growth; PP007560; [Scarlet Sunblaze X Unnamed seedling]; Marciel, Stanley G.; DeVor Nurseries, Inc.

'DEVDICHA', Min, rb, 1989; (**Rainbow Bliss**™); bud slender, tapering; flowers white with cream inside with red edges, reverse same, aging, dbl., cupped, slight, damask fragrance; prickles declining, brown-orange; foliage small, dark green, glossy; upright, low growth; [Unnamed seedling X Scarlet Sunblaze]; Marciel, Stanley G.; DeVor Nurseries, Inc.

'DEVDORADO', HT, ly, 1992; (**Mariko**); dbl., 26–40 petals, 3–3.5 in., borne mostly singly, moderate fragrance; some prickles; foliage large, dark green, semi-glossy; tall (210 cms), upright growth; [Unnamed seedling X Unnamed seedling]; Marciel, Stanley G. & Jeanne A.; DeVor Nurseries, Inc.

'DEVECLIPSAR', Min, rb, 1989; (**Rainbow Eclipse**™); bud pointed; flowers crimson-pink edges, center very light whitish-pink, fading t, dbl., slight fragrance; prickles declining, rusty-brown; foliage small, dark green, glossy; bushy, medium growth; PP007637; [Scarlet Sunblaze X Unnamed seedling]; Marciel, Stanley G.; DeVor Nurseries, Inc.

'DEVFRAGO', HT, ab, 1992; (**Fragrant Fantasy**); dbl., 15–25 petals, 3–3.5 in., borne mostly singly, intense fragrance; few prickles; foliage medium, dark green, matt; medium (3-4 ft), upright growth; [Unnamed seedling 80227-20 X Unnamed seedling 82249-1]; Marciel, Stanley G. & Jeanne A.; DeVor Nurseries, Inc.

'DEVHAUTE', HT, dp, 1994; (**Matty**); flowers deep pink, blooms borne mostly single, dbl., 15–25 petals, 3–3.5 in., moderate fragrance; some prickles; foliage large, medium green, matt; tall (97 cms), upright growth; [(Happiness X Seedling) X (Emily Post X Visa)]; Marciel, Stanley G. & Jeanne A.; DeVor Nurseries, Inc., 1993;, Sakai Brothers Rose Co., 1993

Devi Gayatri, F, w, 1992; Sen

'DEVICIO', HT, dp, 1989; (**Touch of Raspberry**™); bud slender, tapering; flowers deep pink, large, dbl., 30 petals, cupped, borne singly, slight, fruity fragrance; prickles declining, light lime green with mauve tinges; foliage large, dark green, semi-glossy; upright, tall growth; PP007548; [Love Affair X Paul's Pink]; Marciel, Stanley G.; DeVor Nurseries, Inc.

'DEVIENNE LAMY', HP, mr, 1868; flowers medium, dbl.; Lévêque; (Sangerhausen)

'DEVIENTE', Min, dp, 1989; (**Rainbow Hot Pink**™); bud long, slightly urn-shaped; flowers deep pink, small, borne singly, dbl.; prickles declining, mauve; foliage small, dark green, semi-glossy; bushy, medium growth; PP007689; [Orange Sunblaze X Unnamed seedling]; Marciel, Stanley G.; DeVor Nurseries, Inc.

'DEVIL DANCER', HT, mr, 1979; bud ovoid; flowers brilliant brick-red, single, 5 petals, 4 in., slight fragrance; foliage dark; bushy growth; [Sonora X Matangi]; Hawken, Una

'DEVILK', HT, ob, 1989; (**Sparkling Orange**™); bud tapering; flowers vermillion, reverse scarlet, aging to rose, sweetheart, large, dbl., borne singly, intense, musk fragrance; prickles variation, reddish-brown; foliage large, dark green, glossy; upright, tall growth; [Sonia X Prominent]; Marciel, Stanley G.; DeVor Nurseries, Inc.

Devil's Dance, F, mr

DEVine, HT, pb; (**Dolores**™); deVor

'DEVITE', S, w, 1982; (**Angel**™); flowers flora-tea, small, dbl., 35 petals, slight fragrance; foliage medium, light green, semi-glossy; upright growth; [(Queen Elizabeth X Seedling) X Jack Frost]; deVor, Paul F.

'DEVLASS', HT, w, 1989; (Spun Glass®, **Spunglass**); bud pointed; flowers very large, borne singly, dbl., 50 petals, cupped, slight, spicy fragrance; prickles declining, pea green; foliage large, dark green, glossy; upright, tall growth; PP007613; [Unnamed seedling X Angel]; Marciel, Stanley G.; DeVor Nurseries, Inc.

DEVlicor, HT, dy, 1995; (**Licorice Twist**); deVor

'DEVLOREN', HT, op, 1989; (**Sophia's Song**™); bud slender, tapering; flowers coral, large, dbl., 32 petals, exhibition form, borne singly, slight, fruity fragrance; prickles declining, copper brown with pea green tinges; foliage medium, dark green, semi-glossy; upright, tall growth; PP007455; [Emily Post X Prominent]; Marciel, Stanley G.; DeVor Nurseries, Inc.

DEVmauve, HT, m, 1995; (**Sonora Sunset**); deVor

'DEVMENTA', HT, rb, 1989; (**Peppermint Swirl**™); bud slender, tapering; flowers currant red, reverse same, aging discolors slightly, large, dbl., 30 petals, cupped, slight, spicy fragrance; prickles declining, copper brown; foliage medium, dark green, semi-glossy; upright, tall growth; PP007612; [Unnamed seedling X Unnamed seedling]; Marciel, Stanley G.; DeVor Nurseries, Inc.

'DEVMESI', Min, dr, 1989; (**Rainbow Crimson**®); bud slender; flowers deep red, aging discolors slightly, small, borne singly, dbl., cupped, slight, damask fragrance; prickles declining, slightly dark mauve; foliage medium, dark green, semi-glossy; upright, low growth; [Unnamed seedling X Unnamed seedling]; Marciel, Stanley G.; DeVor Nurseries, Inc.

'DEVMOMENTO', HT, dy, 1993; (**Glory**); dbl., 26–40 petals, 3–3.5 in., borne mostly singly, no fragrance; some prickles; foliage large, dark green, semi-glossy; tall (205 cms), upright growth; [Capella X Seedling]; Marciel, Stanley G. & Jeanne A.; DeVor Nurseries, Inc., 1992

'DEVMORADA', HT, m, 1989; (**Royal Amethyst**®); bud pointed; flowers lavender, large, borne singly, dbl., 32 petals, intense, fruity fragrance; globular, tangerine orange fruit; prickles declining, henna; foliage medium, medium green, glossy; upright, tall growth; [Angel Face X Blue Moon]; deVor, Paul F.; DeVor Nurseries, Inc.

'DEVNINA', HT, lp, 1989; (Amanda®, **Amanda Marciel**®); bud slender and tapering; flowers very delicate pink, large, borne singly, dbl., 26 petals, cupped, slight, spicy fragrance; no prickles; foliage medium, dark green, glossy; upright, tall growth; [Seedling X Pink Puff]; Marciel, Stanley G.; DeVor Nurseries, Inc.

'DEVNOVIA', HT, mp, 1989; (Megan, **Megan Dolan**); bud urn-shaped; flowers small, sweetheart, 18 petals, cupped, borne singly, slight, spicy fragrance; prickles declining, copper brown with olive green tinges; foliage medium, dark green; upright, tall growth; PP007520; [Angel X Independence '76]; Marciel, Stanley G.; DeVor Nurseries, Inc.

Devon® *see* POUlrijk

'DEVON MAID', LCl, pb, 1982; flowers light pink, reverse medium pink, large blooms in clusters of 3-4, dbl., 22 petals, moderate, fruity fragrance; ovoid, orange fruit; large, curved, orange prickles; foliage large, medium green, glossy; needs support; vigorous, spreading, tall growth; [Casino X Elizabeth of Glamis]; Warner, Chris; Warner's Roses

'DEVONIENSIS', T, w, 1838; (Magnolia Rose); flowers creamy white, center sometimes tinged blush, very large, dbl., recurrent bloom, moderate fragrance; very vigorous growth; [Parentage uncertain, perhaps Elinthii X Yellow China]; Foster; Lucombe, Prince & Co., 1841

'DEVONIENSIS, CLIMBING', Cl T, w, 1858; [Devoniensis sport]; Pavitt; Curtis

Devonshire Maid, LCl, mp, 1993

DEVoran, HT, yb, 1989; (**Karen**); deVor

'DEVORO', Min, yb, 1991; (**Rainbow Gold**®); flowers yellow-orange blend, full petals, blooms borne in small clusters, slight fragrance; foliage small, dark green, semi-glossy; upright, medium (35 cms) growth; [Amber Flash X Rhumba]; Marciel, Stanley G.; DeVor Nurseries, Inc.

'DEVOTION', F, pb, 1971; flowers light pink, flushed deeper, dbl., 32 petals, 4.5 in., moderate fragrance; foliage light; [Orange Sensation X Peace]; Harkness

'DEVPAJARO', HT, mr, 1994; (**Red Satin**); dbl., 26–40 petals, 3–3.5 in., borne mostly singly, slight fragrance; few prickles; foliage large, medium green, matt; low (55-60 cms), upright growth; [Jacqueline X Prominent]; Marciel, Stanley G. & Jeanne A.; DeVor Nurseries, Inc., 1993;, Pajaro Valley Greenhouses, 1993

'DEVPRESA', Min, ab, 1989; (**Rainbow Surprise**®); bud slender, tapering; flowers medium coral, reverse light coral, aging pink, small, borne, dbl., slight, fruity fragrance; prickles declining, mauve; foliage small, dark green, glossy; bushy, medium growth; PP007558; [Orange Sunblaze X Unnamed seedling]; Marciel, Stanley G.; DeVor Nurseries, Inc.

'DEVRICO', Min, rb, 1989; (**Rainbow Stanford**®); bud pointed, slender; flowers bright red, reverse same, aging discolors to a bright red-or, semi-dbl., slight, fruity fragrance; prickles declining, reddish-brown; foliage medium, dark green, glossy; bushy, medium growth; PP007559; [Candia X Unnamed seedling]; Marciel, Stanley G.; DeVor Nurseries, Inc.

'DEVRISE', HT, dr, 1989; (**Cerise Dawn**®); bud urn-shaped; flowers magenta, reverse tyrian purple, aging no discoloration, large, dbl., 30 petals, globular, slight, damask fragrance; prickles declining and well spaced apart, pea green; foliage large, dark green, semi-glossy; upright, tall growth; [Carina X Angel Face]; Marciel, Stanley G.; DeVor Nurseries, Inc.

'DEVROJO', Min, mr, 1989; (**Rainbow Red**®); bud pointed; flowers small, dbl., borne singly, slight, spicy fragrance; prickles declining, brown with orange; foliage small, dark green, glossy; bushy, medium growth; PP006584; [Scarlet Sunblaze X Rumba]; Marciel, Stanley G.; DeVor Nurseries, Inc.

'DEVROSADO', Min, dp, 1989; (**Rainbow Pink**®); bud ovoid, pointed; flowers deep pink, small, dbl., borne singly, slight, spicy fragrance; prickles declining, purple; foliage medium, dark green, semi-glossy; bushy, medium growth; PP006875; [Unnamed seedling X Orange Sunblaze]; Marciel, Stanley G.; DeVor Nurseries, Inc.

'DEVRUDI', S, lp, 1998; (**First Light**); bud pointed, dark candy pink; flowers single, light pink with purple stamens, single, 5–7 petals, 3.5–4 in., borne in clusters, moderate, spicy fragrance; foliage dark green; compact, low growth; AARS, 1998; [Bonica X Ballerina]; Stanley/Marciel

'DEVSIEM', HT, dr, 1989; (**Always Mine**®); bud pointed, tapering; flowers deep red, large, dbl., 39 petals, cupped, borne singly, intense, spicy fragrance; prickles declining, pea green with cinnamon tinges; foliage large, dark green, glossy; upright, tall growth; [Visa X Sassy]; Marciel, Stanley G.; DeVor Nurseries, Inc.

DEVsleek, HT, dy, 1996; (**Desert Sun**); deVor

DEVsmooth, F, mr, 1995; (**Bella**); deVor

'DEVSOLEAR', Gr, dy, 1989; (**Goldlite**®); bud urn-shaped; flowers canary yellow, reverse buttercup yellow, aging no discoloration, dbl., slight, musk fragrance; round, average, tangerine orange fruit; prickles declining, red; foliage medium, dark green, semi-glossy; upright, tall growth; PP007462; [Unnamed seedling X Excitement]; Marciel, Stanley G.; DeVor Nurseries, Inc., 1987

'DEVSPILIO', HT, w, 1992; (**Silver Fox**); dbl., 26–40 petals, 3–3.5 in., borne mostly singly, slight fragrance; some prickles; foliage large, dark green, semi-glossy; tall (210 cms), upright growth; [Unnamed seedling X Unnamed seedling]; Marciel, Stanley G. & Jeanne A.; DeVor Nurseries, Inc.

'DEVSTAR', HT, w, 1992; (**Stardust**); dbl., 26–40 petals, 3–3.5 in., borne mostly singly, no fragrance; some prickles; foliage large, medium green, semi-glossy; tall (230 cms), upright growth; [Coquette X Unnamed seedling 64022-39]; Marciel, Stanley G. & Jeanne A.; DeVor Nurseries, Inc.

'DEVSTICA', HT, mr, 1989; (**Mystique**®); bud pointed, slender, tapering; flowers bright red, sweetheart, large, dbl., 28 petals, cupped, borne singly, intense, musk fragrance; prickles declining, pale red with tinges of green; foliage large, medium green, semi-glossy; upright, tall growth; [Samantha X Royalty]; Marciel, Stanley G.; DeVor Nurseries, Inc.

DEVsunset, HT, lp; (**Sundancer**); deVor, 1995

'DEVTINTA', HT, dr, 1990; (**Obsession**); bud high-centered; flowers imbricated, large blooms borne singly, dbl., 42 petals, cupped, moderate fragrance; prickles wing-shaped, reddish tinge; foliage medium, medium green, semi-glossy; upright, tall growth; [Unnamed seedling X Unnamed seedling]; Marciel, Stanley G.; DeVor Nurseries, Inc.

'DEVUNICAN', F, dy, 1992; (**Golden Sprite**); dbl., 15–25 petals, 3–3.5 in., borne mostly singly, slight fragrance; some prickles; foliage medium, dark green, matt; medium, upright growth; [Golden Fantasie X Excitement]; Marciel, Stanley G. & Jeanne A.; DeVor Nurseries, Inc.

Dew Drop *see* 'ZIPDEW'

'DEWDROP', HT, lp, 1921; flowers pale pink to pale rose, moderate fragrance; RULED EXTINCT 11/91 ARM; McGredy

Dezent, HT, ly; flowers yellowish-white, large, dbl., slight fragrance; VEG; (Sangerhausen)

Diablotin® *see* 'DELPO'

Diablotin, Climbing® *see* 'DELPOSAR'

'DIABOLO', F, op, 1958; flowers bright salmon, medium, semi-dbl., cupped, slight fragrance; foliage bronze; short stems; very vigorous, bushy growth; [Jolie Princess X (Alain X Miss France)]; Gaujard

'DIADEM', HT, ob, 1922; flowers orange-crimson suffused salmon and yellow, very large, dbl., exhibition form, moderate fragrance; foliage rich green, leathery, glossy; vigorous, bushy growth; McGredy

Diadem *see* TANmeda

Diamant® *see* 'KOREB'

Diamant Rose, HMsk, pb, 1995; Lens

'DIAMANTINA', HT, op, 1949; flowers salmon and rose, large; strong stems; [Julien Potin X Ophelia]; Giacomasso

Diamond Anniversary *see* 'MORSIXTY'

Diamond Border *see* POUldiam

Diamond Doll *see* 'JUDDOLL'

'DIAMOND JEWEL', Min, w, 1958; bud globular; flowers white, overcast blush-pink, dbl., 45–50 petals, .5–.75 in., cupped; compact, low, open growth; [Dick Koster sport X Tom Thumb]; Morey, Dr. Dennison; J&P

'DIAMOND JUBILEE', HT, ly, 1947; bud ovoid; flowers buff-yellow, dbl., 28 petals, 5–6 in., cupped, moderate fragrance; foliage leathery; upright, compact growth; AARS, 1948; [Marèchal Niel X Feu Pernet-Ducher]; Boerner; J&P

'DIAN', Min, dp, 1957; flowers soft red, dbl., 45 petals, 1 in., moderate, apple fragrance; foliage small, dark, glossy; vigorous (15 in), bushy growth; [(R. wichurana X Floradora) X (Oakington Ruby X Floradora)]; Moore, Ralph S.; Sequoia Nursery

Diana® *see* 'TANDINADI'

'DIANA', HT, mp, 1921; flowers malmaison pink, very large, dbl., globular, moderate fragrance; vigorous growth; [Mrs Frank Workman X Sunburst]; Bees

'DIANA', Pol, op, 1922; flowers bright orange shaded pink, large, semi-dbl.; Spek

'DIANA ALLEN', HT, op, 1939; flowers salmon-pink, small, dbl.; short stems; bushy, compact growth; [Mrs Aaron Ward X Seedling]; Clark, A.; NRS New South Wales

'DIANA ARMSTRONG', HT, dy, 1992; dbl., 26–40 petals, 1.5–2.75 in., borne mostly singly, intense fragrance; some prickles; foliage medium, medium green, semi-glossy; upright (80 cms) growth; [Seedling X Prima Ballerina]; Thompson, Robert; Battersby Roses, 1993

'DIANA CANT', HT, rb, 1928; flowers carmine-red, base flushed orange, dbl., moderate fragrance; [Isobel X Seedling]; Cant, B. R.

Diana Festival, Min, lp, 1996; Laver, Keith G.

Diana H. Gupta *see* 'JAYGUP'

'DIANA MAXWELL', HT, ob, 1957; flowers orange-cerise, high pointed, dbl., 45 petals, 5 in., intense fragrance; foliage bronze; vigorous growth; [Ena Harkness X Sam McGredy]; Kemp, M.L.

'DIANA MENUHIN', HT, dy, 1963; flowers deep buttercup-yellow, dbl., 30 petals, 5.5 in., globular; foliage dark, glossy; vigorous, upright growth; [(Golden Masterpiece X Ellinor LeGrice) X Forward]; LeGrice

'DIANA ROWDEN', LCl, op, 1976; flowers deep copper-salmon to rose-pink, dbl., 30 petals, 5–6. in., profuse, continuous bloom, intense fragrance; foliage large, copper to green; [Mrs Sam McGredy, Climbing X Red Dandy]; Hawker; Harkness

Diana, Princess of Wales™ *see* 'JACSHAQ'

'DIANE', HT, yb, 1958; flowers clear yellow, center orange-yellow, well formed, large; vigorous growth; [Peace X (Seedling X Opera)]; Gaujard

Diane, HT, lp; Dawson, George, 1976; (Weatherly, L.)

'DIANE DE BROGLIE', HT, ob, 1929; flowers coral-orange, very large, dbl., cupped, moderate fragrance; foliage dark; strong stems; very vigorous growth; Chambard, C.

Diane de Poltiers, HT, dp

'DIANE D'URFÉ', HT, w, 1958; flowers white, becoming red-edged; vigorous growth; [Peace X Incendie]; Croix, A.; Minier

'DIANNA KAY', Min, or, 1981; flowers small, dbl., 20 petals, no fragrance; foliage small, medium green, semi-glossy; upright growth; [Anytime X Sheri Anne]; Dobbs, Annette E.

'DIANNE FEINSTEIN', HT, yb, 1979; bud ovoid; dbl., 23 petals, exhibition form, borne singly, moderate fragrance; triangular prickles; foliage dark; vigorous, spreading growth; [McGredy's Yellow X Sutter's Gold]; Fong, William P.

Diantheflora *see* **'FIMBRIATA'**

Dianthiflora *see* **'FIMBRIATA'**

'DIANY BINNY', LCl, w, 1976; flowers small, single, 5 petals, 2 in., intense fragrance; foliage purplish; [Kiftsgate X R. rubrifolia (?)]; Binny

Diapason® *see* 'DELPOC'

Diavoletta, 1966; Mansuino, Q.; (Cavriglia)

Dic Della, HT, dp, 1996; Chiplunkar

'DICALOW', F, dy, 1977; (**Yellow Ribbon**); flowers deep golden yellow, medium, dbl., 23 petals, cupped, slight fragrance; foliage medium green, semi-glossy; vigorous, compact, bushy growth; [Illumination X Stroller]; Dickson, Patrick; A. Dickson

'DICAM', HT, mr, 1970; (Coeur d'Amour, **Red Devil**); bud ovoid; flowers medium red, reverse lighter, large, dbl., 72 petals, exhibition form, moderate fragrance; foliage glossy; vigorous growth; GM, Belfast, 1969 GM, Japan, 1967 GM, Portland, 1970; [Silver Lining X Prima Ballerina]; Dickson, A.; J&P

'DICBAND', F, ob, 1978; (**Springfields**); flowers orange, red, gold, dbl., 48 petals, 3 in., cupped, slight fragrance; [Eurorose X Anabell]; Dickson, A.

'DICBAR', F, rb, 1978; (**Memento**®); bud globular; flowers salmon-red, dbl., 22 petals, 3 in., cupped; foliage sage green; bushy growth; GM, Belfast, 1980; [Bangor X Anabell]; Dickson, Patrick; Dickson Nurseries, Ltd.

'DICBEE', F, or, 1978; (**High Summer**); bud ovoid; flowers vermilion, dbl., 26 petals, 3 in., cupped; foliage large; bushy growth; [Zorina X Ernest H. Morse]; Dickson, Patrick; Dickson Nurseries, Ltd.

DICblender, HT, rb, 1997; (**Acapulco**); Dickson, Patrick

'DICBO', HT, mp, 1968; (**Bonsoir**); bud ovoid; flowers peach-pink, dbl., 6 in., intense fragrance; foliage glossy; Dickson, A.

'DICDANCE', F, my, 1980; (**Bright Smile**®); bud pointed; flowers empire yellow, patio, convex blooms borne 15 per cluster, semi-dbl., 15 petals, flat, slight fragrance; concave, purple prickles; foliage dense, mid-green; bushy, medium growth; GM, Belfast, 1982; [Eurorose X Seedling]; Dickson, Patrick, 1981

'DICDIP', HT, pb, 1974; flowers large, dbl., 32 petals, moderate fragrance; foliage medium, purple when young; upright, bushy growth; [Eurorose X Typhoon]; Dickson, Patrick

'DICDIVINE', HT, my, 1980; (**Pot o' Gold**); bud pointed; dbl., 32 petals, flat, borne 7 per cluster, intense fragrance; brown prickles; foliage mid-green with strong purple veins; bushy, medium growth; [Eurorose X Whisky Mac]; Dickson, Patrick

'DICDRUM', F, op, 1982; (**Shona**); flowers medium coral pink, blooms in clusters, dbl., 23 petals, slight fragrance; foliage medium, medium green, semi-glossy; bushy growth; [Bangor X Anabell]; Dickson, Patrick

'DICEL', F, or, 1963; (**Scarlet Queen Elizabeth**®); flowers flame-scarlet, medium, dbl., globular, slight fragrance; foliage dark; vigorous, tall growth; GM, The Hague, 1973 Golden Rose, The Hague, 1973; [(Korona X Seedling) X Queen Elizabeth]; Dickson, Patrick; A. Dickson

'DICFATE', F, mp, 1975; flowers medium, 18 petals, slight fragrance; foliage medium, medium green; upright, bushy growth; [Futura X (Pye Colour X Prominent)]; Dickson, Patrick

'DICFIRE', F, or, 1983; (**Beautiful Britain**); flowers orange-red, reverse deeper, medium, in clusters, dbl., 20 petals, slight fragrance; foliage medium, medium green, semi-glossy; upright, bushy growth; ROTY, 1983; [Red Planet X Eurorose]; Dickson, Patrick; Dickson Nurseries, Ltd.

'DICGROW', Min, ob, 1981; (**Brass Ring**, Peek a Boo); bud pointed; flowers coppery orange, fading to rose pink, patio, blooms in large clusters, dbl., flat; foliage small, pointed, glossy; upright, arching growth; [Memento X Nozomi]; Dickson, Patrick; J&P

'DICINFRA', F, or, 1984; (**Disco Dancer**®); flowers orange scarlet, medium, semi-dbl., slight fragrance; foliage medium, medium green, glossy; bushy growth; GM, The Hague, 1982; [Cathedral X Memento]; Dickson, Patrick, 1983

'DICJANA', HT, ly, 1984; (**Elina**®, Peaudouce); flowers pale yellow to ivory, large, luminous, very dbl., 30–35 petals, 5–5.5 in., exhibition form, borne singly, slight fragrance; foliage large, dark, glossy; long stems; vigorous, tall growth; ADR, 1987 Gold Star of the South Pacific, Palmerston North, NZ, 1987; [Nana Mouskouri X Lolita]; Dickson, Patrick, 1983

'DICJEEP', F, rb, 1984; (Daydream, **Len Turner**); flowers ivory petals flushed and edged carmine, large, dbl., 35 petals, slight fragrance; foliage medium, medium green, glossy; bushy, very compact growth; [Electron X Eyepaint]; Dickson, Patrick

'DICJEM', HT, dy, 1984; (**Freedom**®); flowers chrome yellow, large, dbl., 35 petals, exhibition form, moderate fragrance; foliage medium, medium green, glossy; bushy growth; GM, RNRS, 1983; [(Eurorose X Typhoon) X Bright Smile]; Dickson, Patrick; Dickson Nurseries, Ltd.

'DICJOON', HT, rb, 1984; (**Leslie's Dream**); flowers medium red, yellow reverse, large, borne singly and in truss, dbl., 20 petals, no fragrance; foliage medium, dark, glossy; bushy growth; [Bonfire X Typhoon]; Dickson, Patrick

'DICJOY', F, my, 1984; (**Ards Beauty**); bud large; flowers large, dbl., 20 petals, exhibition form, moderate fragrance; foliage medium, medium green, glossy; bushy growth; GM, RNRS, 1983; [(Eurorose X Whisky Mac) X Bright Smile]; Dickson, Patrick, 1986

'DICJUBELL', HT, mp, 1986; (Dickson's Jubilee, **Lovely Lady**); flowers large, dbl., 35 petals, moderate fragrance; foliage medium, mid-green, glossy; bushy growth; GM, Belfast, 1988; [Silver Jubilee X (Eurorose X Anabell)]; Dickson, Patrick

'DICK KOSTER', Pol, dp, 1929; flowers deep pink; [Anneke Koster sport]; Koster, D.A.

'DICK KOSTER FULGENS', Pol, dp, 1940; flowers light red, borne in clusters, semi-dbl.; low, compact growth; Koster, M.

'DICK KOSTER SUPERIOR', Pol, mr, 1955; flowers rosy red; [Dick Koster sport]; Koster, D.A.

Dick Lindner, HT, mp, 1996

'DICK WILCOX', HT, dr, 1949; bud long, pointed to ovoid; flowers rose-red, dbl., 50–60 petals, 4–5.5 in., exhibition form, moderate fragrance; foliage dark; vigorous growth; [Pink Princess X Crimson Glory]; Brownell, H.C.

'DICKERFUFFLE', F, mp, 1985; (Georgie Girl, **Wishing**); flowers medium peachy pink, large, dbl., 35 petals, slight fragrance; foliage medium, medium green, semi-glossy; bushy growth; [Silver Jubilee X Bright Smile]; Dickson, Patrick; Dickson Nurseries, Ltd., 1984

'DICKERRY', F, pb, 1986; (**Laughter Lines**); flowers large, semi-dbl., 6–14 petals, slight fragrance; foliage small, medium green, semi-glossy; bushy growth; GM, RNRS, 1984; [(Pyecolour X Sunday Times) X Eyepaint]; Dickson, Patrick

'DICKIMONO', F, op, 1984; (Anisley Dickson, **Dicky**®, München Kindl); flowers reddish salmon-pink, reverse lighter, large, dbl., 35 petals, slight fragrance; foliage medium, medium green, glossy; bushy growth; GM, RNRS, 1984 PIT, RNRS, 1984; [Cathedral X Memento]; Dickson, Patrick, 1983

DICkindlel, F, ly, 1986; (**Lemon Honey**); Dickson, Patrick

DICkisser, F, or, 1995; (**Hi Doll**); Dickson, Patrick

DICkitty, F, ob, 1987; (**Old Flame**); Dickson, Patrick

DICknowall, F, my; (**Collegiate 110**); Dickson, Patrick

'DICKOOKY', F, my, 1984; (**Tall Story**®); flowers medium, semi-dbl., slight fragrance; foliage medium, light green, glossy; spreading growth; [Sunsprite X Yesterday]; Dickson, Patrick; Dickson Nurseries, Ltd.

Dick's Apricot Rambler, HWich, ab

Dick's Delight™ *see* 'DICWHISTLE'

'DICKSON'S BOUQUET', HT, ab, 1938; flowers salmon, carmine and apricot, blended saffron, dbl., moderate fragrance; long, wiry stems; vigorous growth; Dickson, A.

'DICKSON'S CENTENNIAL', HT, mr, 1936; bud pointed; flowers crimson to scarlet, loosely formed, very large, moderate fragrance; foliage bronze; long, strong stems; vigorous, bushy growth; Dickson, A.; Dreer, 1937;, J&P

'DICKSON'S DELIGHT', HT, ob, 1938; flowers vivid orange, heavily shaded scarlet-orange, moderate fragrance; foliage bronze-green; vigorous growth; Dickson, A.

'DICKSON'S FLAME', F, or, 1958; flowers scarlet-flame, blooms in trusses, dbl., 3.5 in., slight fragrance; vigorous growth; GM, NRS, 1958 PIT, NRS, 1958; [Independence seedling X Nymph]; Dickson, A.

Dickson's Jubilee *see* 'DICJUBELL'

'DICKSON'S PERFECTION', HT, pb, 1937; flowers shrimp-pink, base orange-yellow, large, dbl., intense fragrance; very vigorous growth; Dickson, A.; Port Stockton Nursery

'DICKSON'S RED', HT, dr, 1938; (Dr F.G. Chandler); flowers velvety crimson-scarlet, large, semi-dbl., 18 petals, cupped, intense, spicy fragrance; foliage leathery, dark; vigorous, bushy growth; AARS, 1940 GM, NRS, 1939 GM, Portland, 1941; Dickson, A.

Dickson's Wonder, HT, ob

Dicky® *see* 'DICKIMONO'

'DICLADIDA', Min, ob, 1987; (**Cider Cup**); flowers medium, dbl., 15–25 petals, slight fragrance; foliage medium, medium green, glossy; patio growth; [Memento X (Liverpool Echo X Woman's Own)]; Dickson, Patrick, 1988

DIClady, F, or, 1989; (**Feu Follet®**); Dickson, Patrick

'DICLITTLE', Min, pb, 1986; (**Little Woman**); dbl., 15–25 petals, moderate fragrance; foliage small, medium green, semi-glossy; patio; bushy growth; [Memento X (Liverpool Echo X Woman's Own)]; Dickson, Patrick

'DICLULU', Min, lp, 1986; (**Gentle Touch**); flowers moderately small, dbl., 15–25 petals, slight fragrance; foliage small, medium green, semi-glossy; patio; bushy growth; ROTY, 1986; [(Liverpool Echo X Woman's Own) X Memento]; Dickson, Patrick

'DICMADDER', F, rb, 1987; (**Star Child®**, Tangeglow); flowers small, semi-dbl., 6–14 petals, slight fragrance; foliage small, medium green, glossy; strong, straight stems; bushy, prolific growth; PP007312; [Eyepaint X (Liverpool Echo X Woman's Own)]; Dickson, Patrick, 1988

'DICMAGIC', Min, ob, 1986; (**Sweet Magic**); flowers moderately small, dbl., 15–25 petals, no fragrance; foliage small, medium green, glossy; patio; bushy growth; ROTY, 1987; [Peek A Boo X Bright Smile]; Dickson, Patrick

'DICMICKEY', Min, or, 1986; (**Buttons**); flowers medium, dbl., 15–25 petals, slight fragrance; foliage small, medium green, glossy; patio; bushy growth; [(Liverpool Echo X Woman's Own) X Memento]; Dickson, Patrick, 1987

'DICMOPPET', S, my, 1987; (Goldfächer, Mini Lights, **Minilights**); flowers small, semi-dbl., 6–14 petals, slight fragrance; foliage small, dark green, glossy; spreading, compact growth; [White Spray X Bright Smile]; Dickson, Patrick, 1988

'DICNAME', HT, yb, 1986; (**Peter Goldman**); flowers medium, dbl., 26–40 petals, slight fragrance; foliage medium, medium green, glossy; bushy growth; [Silver Jubilee X Bright Smile]; Dickson, Patrick

DICnifty, HT, lp, 1992; (**Empress Michiko**); Dickson, Patrick

'DICNORTH', F, ob, 1989; (**Harvest Fayre**); flowers medium, dbl., 15–24 petals, slight fragrance; foliage medium, medium green, glossy; bushy growth; ROTY, 1990; [Unnamed seedling X Bright Smile]; Dickson, Patrick; Dickson Nurseries, Ltd., 1990

'DICOBEY', HT, rb, 1988; (Beaulieu, **Tequila Sunrise**); dbl., 40 petals, slight fragrance; foliage medium, medium green, glossy; bushy growth; GM, Belfast, 1991 GM, RNRS, 1988; [Bonfire Night X Freedom]; Dickson, Patrick; Dickson Nurseries, Ltd., 1989

'DICODOUR', HT, ab, 1988; (**Fragrant Dream**); flowers apricot blended orange, large, dbl., 20 petals, intense fragrance; foliage large, medium green, glossy; upright growth; [(Eurorose X Typhoon) X Bonfire]; Dickson, Patrick; Dickson Nurseries, Ltd., 1989

'DICOGLE', F, mp, 1989; (**Valentine Heart**); flowers large, dbl., 15–25 petals, intense fragrance; foliage large, medium green, glossy, purple when young; bushy growth; [Shona X Pot 'o Gold]; Dickson, Patrick; Dickson Nurseries, Ltd., 1990

'DICOMO', Min, w, 1988; (**Tear Drop**, Teardrop); flowers small, semi-dbl., 6–14 petals, flat, slight fragrance; foliage small, medium green, glossy; patio; bushy growth; [Pink Spray X Bright Smile]; Dickson, Patrick; Dickson Nurseries, Ltd., 1989

'DICOR', F, yb, 1971; (**Redgold**, Rouge et Or); bud ovoid; flowers gold edged deep pink, blooms in large clusters, dbl., slight fragrance; vigorous, upright growth; AARS, 1971 GM, Portland, 1969; [((Karl Herbst X Masquerade) X Faust) X Piccadilly]; Dickson, A.; J&P

'DICORSAR', Cl F, yb, 1984; (Grimpant Rouge etOr, **Redgold, Climbing**, Rouge et Or, Climbing); Pekmez, Paul

'DICPAINT', HT, rb, 1989; (**Painted Moon**); flowers large, dbl., 40 petals, cupped, slight fragrance; foliage medium, medium green, semi-glossy; upright, bushy, stocky growth; GM, Belfast, 1992; [Bonfire X Silver Jubilee]; Dickson, Patrick; Dickson Nurseries, Ltd., 1990

DICparty, F, dp; (**Party Trick**); Dickson, Patrick, 1994

'DICPARTY', F, dp, 1999; (**Party Trick™**); flowers cerise pink, single, 5–11 petals, 2 in., borne in small clusters, no fragrance; prickles moderate; foliage small, medium green, glossy; compact, low (18 in.) growth; [Robin Redbreast X seedling]; Dickson, Colin; Dickson Nurseries, Ltd., 1999

'DICPE', F, ob, 1968; (**Copper Pot**); flowers orange-yellow, deeper reverse, large blooms in trusses, semi-dbl., 15 petals, moderate fragrance; foliage glossy, bronze; tall growth; [Seedling X Golden Scepter]; Dickson, A.

'DICPERHAPS', Gr, op, 1991; (**Quaker Star**); flowers orange with silver reverse, aging to salmon with orange petals, very dbl., 3 in., no fragrance; few prickles, straight, small; foliage medium, dark green, glossy; upright, medium (120 cms) growth; [Anisley Dickson X Unnamed seedling]; Dickson, Colin; Roses by Fred Edmunds, 1992

DICplay, F, ob, 1991; (**New Horizon**); GM, Belfast, 1993; Dickson, Patrick

'DICPLEASANT', F, or, 1992; (**Ruth Woodward**); dbl., 26–40 petals, 3–3.5 in., borne in small clusters; some prickles; foliage medium, medium green, semi-glossy; medium, bushy growth; [Wishing X Unknown seedling]; Dickson Nurseries

'DICQUARREL', Gr, my, 1994; (**Benita®**); *flowers saffron yellow*, blooms, dbl., 26–40 petals, 3 in., moderate fragrance; foliage medium, medium green, semi-glossy; upright (90 cms) to bushy growth; Dickson, Colin; Dickson Nurseries, Ltd., 1994

'DICQUASAR', HT, ob, 1994; (**Dawn Chorus**); flowers orpiment orange with buttercup yellow at base, blooms borne, dbl., 15–25 petals, 1.5–2.75 in., slight fragrance; some prickles; foliage medium, medium to dark green, glossy; medium (74 cms), upright to bushy growth; GM, Dublin, 1991 ROTY, 1993; [Wishing X Peer Gynt]; Dickson, Colin; Dickson Nurseries, Ltd., 1993

'DICQUEEN', F, or, 1990; (**Melody Maker**); flowers very full petals, large blooms, slight fragrance; foliage medium, dark green, semi-glossy; bushy growth; ROTY, 1991; [Anisley Dickson X Wishing]; Dickson, Patrick; Dickson Nurseries, Ltd., 1991

'DICQUEUE', HT, ob, 1999; (**Brandy Snap™**, Brandysnap); flowers orange/bold, dbl., 26–40 petals, 3.5 in., borne mostly singly, intense fragrance; prickles moderate; foliage medium, medium green, glossy; upright, medium (30 in.) growth; [seedling X seedling]; Dickson, Colin; Dickson Nurseries Ltd., 1999

'DICQUIET', S, yb, 1994; (**Gypsy Dancer**); bud small, pointed; flowers handpainted light yellow with orange and light yellow reverse, semi-dbl., 6–14 petals, 1.5–2.75 in., borne in clusters, slight, citrus fragrance; few prickles; foliage medium, dark green, glossy; medium (100-110 cms), bushy growth; PP8900; [Sweet Magic X Little Artist]; Dickson, Patrick; Bear Creek Gardens, 1994

'DICRACER', F, or, 1994; (**Duchess of York**, Sarah, Duchess of York, Sunseeker); flowers mandarin red suffused with sulfur yellow, blooms borne in small clusters, dbl., 26–40 petals, 1.5 in., slight fragrance; some prickles; foliage medium, medium green, semi-glossy; patio; low (56 cms), bushy growth; [Little Prince X Gentle Touch]; Dickson, Colin; Dickson Nurseries, Ltd., 1992

'DICREASON', S, mr, 1994; (**Our Molly**); flowers currant red with a silvery white eye, blooms borne in large, single, 5 petals, 1.5–2.75 in., no fragrance; orange fruit; some prickles; foliage medium, medium green; tall (90 cms), spreading growth; GM, Glasgow, 1996; Dickson, Colin; Dickson Nurseries, Ltd., 1994

DICrelax, F, yb, 1993; (**Flair**); GM, Glasgow, 1995; Dickson, Patrick

'DICREVIVAL', S, dp, 1992; (**Boy Crazy**); flowers deep pink, petal base has cream colored "half moon", heavy petal substance, dbl., 15–25 petals, 1.5–2.75 in.; some prickles; foliage medium, dark green, glossy; patio; medium (90 cms), upright, bushy growth; [Sweet Magic X DICmerlin]; Dickson, Patrick; Bear Creek Gardens, 1992

'DICROBOT', HT, dy, 1990; (**Belfast Belle**); flowers large, dbl., 26–40 petals, slight fragrance; foliage medium, medium green, semi-glossy; upright, bushy growth; [Seedling X Pot o'Gold]; Dickson, Patrick; Dickson Nurseries, Ltd., 1991

'DICROCKY', S, lp, 1990; (**Space Invader**); flowers large blooms, dbl., 26–40 petals, moderate fragrance; foliage medium, medium green, semi-glossy; spreading growth; [Seedling X Temple Bells]; Dickson, Patrick; Dickson Nurseries, Ltd.

DICrocky, S, lp, 1992; (**Explorer**); Dickson, Patrick

'DICROYAL', HT, ab, 1992; (**Princess Royal**); dbl., 26–40 petals, 3–3.5 in., borne mostly singly, slight fragrance; many prickles; foliage large, medium green, semi-glossy; very stiff, thorny stems; medium (88 cms), bushy growth; [Tequila Sunrise X Unnamed seedling]; Dickson, Colin; Dickson Nurseries, Ltd.

'DICSILVING', Cl HT, pb, 1997; (**Silver Lining, Climbing**); flowers full, large, very dbl., 26–40 petals, intense fragrance; many prickles; foliage medium, medium green, glossy; tall, spreading growth; has 10-12 canes ranging in size of 10-20ft; [Silver Lining sport]; Dickson, Alex

DICsun, MinFl, lp; (**Mr J C B**); Dickson, Patrick

'DICTALENT', Min, op, 1994; (Just Happy, **Shine On**™); flowers nasturtium red, reverse azalea pink, 2.5 in., borne in small clusters, slight fragrance; prickles moderate; foliage small, medium green, semi-glossy; compact, low (22 in.) growth; [Sweet Magic X seedling]; Dickson, Colin; Dickson Nurseries, Ltd., 1999

'DICTATOR', HT, 1999; (**Pure Bliss**™); flowers pale pink, reverse mid-pink, dbl., 26–40 petals, 2.25–2.5 in., borne mostly singly, moderate fragrance; prickles moderate; foliage medium, medium green, semi-glossy; patio; upright, medium growth; GM, Belfast, 1997; [Elina X seedling]; Dickson, Alex, 1994; Dickson Nurseries Ltd., 1994; (Cavriglia)

'DICUMPTEEN', S, 1999; (**Pretty in Pink**™); flowers pale pink, very dbl., 41 petals, 2 in., borne in small clusters, moderate fragrance; few prickles; foliage medium, medium green, glossy; groundcover; spreading, low (24 in.) growth; [seedling X Grouse]; Dickson, Colin, 1994; Dickson Nurseries Ltd., 1994; (Cavriglia)

'DICUNCLE', S, dp, 1999; (**Wine and Dine**™); flowers rose red, single, 5–11 petals, 2.25 in., moderate fragrance; many prickles; foliage small, medium green, glossy; groundcover; spreading, medium (3 ft) growth; [seedling X seedling]; Dickson, Colin; Dickson Nurseries, Ltd., 1999

'DICUNIFORM', S, mr, 1999; (**Boy O Boy**™); semi-dbl., 12–16 petals, 2.75–3 in., slight fragrance; prickles moderate; foliage small, medium green, glossy; flori-shrub; bushy to spreading (30 in.) growth; [Little Prince X Eye Opener]; Dickson, Colin, 1996; Dickson Nurseries, Ltd., 1996

'DICUPTIGHT', HT, or, 1999; (**Tintinara**™); flowers poppy red, reverse geranium lake, dbl., 26–40 petals, 5 in., borne in small clusters, slight fragrance; prickles moderate; foliage large, medium green, glossy; upright, tall (40 in.) growth; [Melody Maker X seedling]; Dickson, Colin; Dickson Nurseries, Ltd., 1999

'DICVANILLA', F, yb, 1999; (**Happy Ever After**™); flowers pale pink/lemon, revers pale pink, semi-dbl., 12–16 petals, 2 in., borne in large clusters, moderate fragrance; prickles moderate; foliage medium, light green, semi-glossy; bushy (33 in.) growth; [The Fairy X seedling]; Dickson, Colin, 1997; Dickson Nurseries, Ltd., 1999

DICvintage, MinFl, 1993; (**Roche Centenary**); Dickson, Patrick; (Cavriglia)

'DICVOOD', S, mr, 1999; (**Glenshane**™); semi-dbl., 12–16 petals, 1.5 in., borne in large clusters, no fragrance; few prickles; foliage medium, medium green, semi-glossy; flori-shrub; spreading to bushy, medium (30 in.) growth; [seedling X Star Child]; Dickson, Colin, 1997

'DICWAFFLE', HT, w, 1999; (**Racy Lady**™); flowers creamy white, reverse cream, dbl., 26–40 petals, 5 in., borne mostly singly, moderate fragrance; many prickles; foliage medium, dark green, glossy; upright, medium (3 ft) growth; [Solitaire X Elina]; Dickson, Colin

'DICWHISTLE', Min, dp, 1999; (**Dick's Delight**™); dbl., 26–40 petals, 1.2in., borne in large clusters, slight fragrance; prickles moderate; foliage small, dark green, glossy; patio carpet; compact to spreading, low (24 in.) growth; [seedling X The Fairy]; Dickson, Colin; Dickson Nurseries, 1999

DICwillynilly, F, ob, 1999; (**Old John**™); flowers mid-orange, reverse orange-red, dbl., 17–25 petals, 2.5 in., borne in large clusters, moderate fragrance; prickles moderate; foliage medium, dark green, glossy; upright, medium (40 in.) growth; [Sunseeker X New Horizon]; Dickson, Colin; Dickson Nurseries, Ltd, 1999

'DICWITNESS', F, yb, 1999; (**Irish Eyes**™); flowers mid-yellow/red, reverse mid-red/yellow, dbl., 26–40 petals, 2.5 in., borne in large clusters, slight fragrance; prickles moderate; foliage medium, medium green, semi-glossy; bushy (30 in.) growth; [Mr J. C. B. X Gypsy Dancer]; Dickson, Patrick; Dickson Nurseries, Ltd., 2000

'DICWONDER', Min, mp, 1999; (**Marry Me**™); dbl., 26–40 petals, 2 in., borne in small clusters, slight fragrance; prickles moderate; foliage medium, dark green, glossy; patio; upright, medium (28 in.) growth; [seedling X Cider Cup]; Dickson, Patrick; Dickson Nurseries, Ltd., 1998

'DICXPLOSION', Min, rb, 1999; (**Rainbow Magic**™); flowers cerise red, reverse yellow, semi-dbl., 12–16 petals, 1.75 in., borne in large clusters, slight fragrance; prickles moderate; foliage small, medium green, glossy; patio; spreading, low (2 ft x 3 ft) growth; [Sunseeker X seedling]; Dickson, Colin; Dickson Nurseries, Ltd., 1999

DICxtol, S, rb, 1998; (**Red Cottage**); Dickson, Patrick

Die Krone, HT, dy

'DIE MUTTER VON ROSA', HT, lp, 1906; flowers medium, dbl., moderate fragrance; Verschuren; (Sangerhausen)

'DIE PRÄSIDENTIN', HT, w, 1928; flowers marble white, center soft yellow, semi-dbl., slight fragrance; [Harry Kirk seedling]; Mühle

Die Rheinpfalz *see* HELrobu

Die Schonste, HT, w, 1998

'DIE SPREE', HT, lp, 1907; flowers large, dbl., moderate fragrance; Nauke; (Sangerhausen)

Die Welt® *see* 'DIEKOR'

'DIEKOR', HT, ob, 1976; (**Die Welt**®, The World); bud long, pointed; flowers orange, red and yellow blend, dbl., 25 petals, 4.5 in., exhibition form, slight fragrance; foliage glossy; vigorous, upright, very tall, bushy growth; [Seedling X Peer Gynt]; Kordes

'DIENER'S BLUE', HMult, m, 1926; flowers violet, large, heavy clusters, dbl., 2 in.; almost thornless; vigorous growth; Diener

'DIENER'S ROSE UNDERSTOCK', HMult, rb, 1932; bud long; flowers rose-red shaded purple, stamens yellowish, borne in clusters, single, 10 petals, 2 in.; foliage small, notched; climbing or trailing growth; producing much new wood in a season; [Veilchenblau X Veilchenblau sport]; Diener

'DIETER WOLF', Pol, or, 1969; flowers salmon-orange, semi-dbl.; foliage glossy, dark; vigorous, compact growth; [Tropicana X Jiminy Cricket]; Buisman, G. A. H.

'DIGNITY', HT, w, 1940; bud long, pointed; flowers creamy white, dbl., intense fragrance; foliage leathery; vigorous, bushy, compact growth; LeGrice

Diletta, HT, op, 1985; Barni, V.

Dil-Ki-Rani, HT, lp, 1985; Pal, Dr. B.P.

Dilly Dilly *see* 'TINDILLY'

'DILLY'S WIEDERKEHR', HFt, 1925; flowers large, dbl.; Schwartzbach; (Sangerhausen)

'DILYS ALLEN', HT, ob, 1952; bud long, pointed, ovoid; flowers orange-red, base saffron, 4 in., slight fragrance; foliage glossy, dark bluish green; vigorous, bushy growth; [Mrs Sam McGredy X Seedling]; Norman; Harkness

'DIMITY', HT, pb, 1956; bud ovoid, pointed; flowers ivory to pure white, edged pink, large, dbl., cupped, moderate fragrance; foliage dark, leathery; upright, bushy growth; [Peace X Seedling]; Taylor, C.A.

'DIMPLES', F, ly, 1968; flowers canary-yellow to ivory, borne in trusses, semi-dbl., slight fragrance; foliage glossy; LeGrice

'DINAH', HT, dr, 1920; flowers deep crimson, shaded darker; Paul, W.

'DINAH SHORE', HT, dp, 1942; flowers cerise-pink, dbl., 65 petals, 5 in., globular, intense fragrance; foliage glossy, dark; vigorous, upright growth; [Jewel sport]; Grillo

'DINKY', Min, or, 1986; flowers orange-red, reverse orange, urn-shaped, blooms borne usually singly, dbl., 20 petals, slight fragrance; long, light red prickles; foliage medium, medium green, semi-glossy; upright growth; [Sheri Anne X Seedling]; Bridges, Dennis A.

Dinsmore *see* **'MME CHARLES WOOD'**

'DINY HAGE', HT, dr, 1956; flowers crimson-red, large, dbl., intense fragrance; vigorous growth; [Ambassadeur Nemry X Crimson Glory]; Leenders, M.

'DIORAMA', HT, yb, 1965; flowers apricot-yellow, dbl., 4.5 in., exhibition form, moderate fragrance; vigorous, upright growth; [Peace X Beauté]; deRuiter

Dioressence® *see* 'DELDIORE'

Diorette *see* **'CANCAN'**

'DIPLOMAT', HT, dr, 1962; bud ovoid; flowers current red edged blood-red, dbl., 50–55 petals, 3.5–4 in., cupped, moderate fragrance; foliage leathery, dark; vigorous, upright growth; [(Poinsettia X Tawny Gold) X Detroiter]; Boerner; Home Nursery Products Corp.

Diplomatka, F, mr, 1967; flowers blood red, large, dbl.; Klimenko, V. N.; (Sangerhausen)

'DIPUTACION DE TARRAGONA', HT, ob, 1967; (Tarragona); bud pointed; flowers orange-coral, large, dbl., 35 petals, exhibition form, moderate fragrance; foliage glossy, bronze; upright, compact growth; [Baccará X (Chrysler Imperial X Soraya)]; Dot, Pedro; Rosas Dot

'DIRECTEUR ALPHAND', HP, m, 1883; flowers blackish purple, large, dbl.; Lévêque

'DIRECTEUR CONSTANT BERNARD', HT, m, 1886; Soupert & Notting

'DIRECTEUR DONATIEN LELIEVRE', F, ob, 1959; flowers coppery orange, medium, dbl.; Privat

'DIRECTEUR GUÉRIN', HT, ob, 1935; flowers orange-yellow, center coppery, overlarge, dbl.; foliage light; long stems; very vigorous growth; Gaujard

'DIRECTEUR N. JANSEN', HP, m, 1883; flowers purple/pink, large, dbl.; Verdier, E.; (Sangerhausen)

Director Plumecock *see* **'PRÉSIDENT PLUMECOCQ'**

'DIRECTOR RUBIÓ', HT, rb, 1929; flowers magenta-red, very large, semi-dbl., moderate fragrance; stiff stems; dwarf, bushy growth; [O. Junyent X Jean C.N. Forestier]; Dot, Pedro; C-P

Direktör Benschop *see* **'CITY OF YORK'**

'DIREKTOR HJELM', Pol, rb, 1927; flowers red; [Prasident Hindenburg sport]; Koster, D.A.

'DIREKTOR REBHUHN', HT, ob, 1929; flowers orange, center reddish, dbl., intense fragrance; [Mme Butterfly X Angèle Pernet]; Kordes

'DIREKTOR RIKALA', F, mp, 1934; [Lafayette sport]; Koster, D.A.

Direktör Rikala *see* **'FRAU ASTRID SPÄTH'**

'DIREKTOR STRUVE', Pol, w, 1924; [Echo sport]; van Nes

'DIRIGENT'®, S, mr, 1956; (The Conductor); bud pointed; flowers blood-red, blooms in clusters of up to 28, semi-dbl., recurrent bloom, slight fragrance; foliage leathery; vigorous (4 ft.) growth; ADR, 1958; [Fanal X Karl Weinhausen]; Tantau, Math.

'DISCO', HT, rb, 1980; bud medium to long, pointed; flowers medium red, reverse cream, blooms borne singly and 2-4 per cluster, dbl., 30 petals, exhibition form, slight, spicy fragrance; prickles hooked downward; foliage leathery, dark; tall, upright growth; GM, Baden-Baden; [Sunrise-Sunset X Seedling]; Weeks, O.L.

Disco®, F, pb, 1980; Harkness

Disco Dancer® *see* 'DICINFRA'

'DISCOVERY', Gr, pb, 1958; flowers soft pink shaded apricot, dbl., 5–6 in., intense fragrance; vigorous growth; [(Peace X Christopher Stone) X Floribunda seedling]; deRuiter; Blaby Rose Gardens

'DISCRETION', HT, pb, 1952; flowers salmon-pink shaded copper, large, dbl., 28 petals, moderate fragrance; foliage glossy; [Peace X Seedling]; Gaujard

'DISPLAY', F, pb, 1956; flowers salmon-pink, becoming cherry-pink, well formed, borne in large clusters, semi-dbl., 13 petals, 2.5 in., slight fragrance; foliage glossy, bronze-green; very vigorous growth; [Orange Triumph X Golden Scepter]; Arnot; Croll

Disraeli *see* 'ADADISRES'

'DISTANT DRUMS', S, m, 1985; bud ovoid, pointed; flowers rose-purple, imbricated, large blooms borne 1-10 per cluster, dbl., 40 petals, repeat bloom, intense, myrry fragrance; awl-like, brown prickles; foliage medium-large, dark, leathery; vigorous, erect, bushy growth; [September Song X The Yeoman]; Buck, Dr. Griffith J.; Iowa State University

Distant Sounds *see* WEBpriace

'DISTINCT', F, rb, 1953; flowers spectrum-red with white eye, imbricated, small, rounded clusters, dbl., 25–30 petals, cupped, moderate fragrance; foliage glossy; vigorous, compact growth; [Triomphe Orléanais X Mrs Pierre S. duPont]; Boerner; J&P

'DISTINCTION', F, dp, 1927; flowers deep rose-pink, center brighter; [Lafayette sport]; Turbat

'DISTINCTION, CLIMBING', Cl F, dp, 1935; Lens

Ditto *see* 'LYODIT'

'DIVA', HT, dr, 1976; flowers dark velvety red, dbl., 30 petals, 6 in., slight fragrance; foliage dark; spreading growth; [Sonia X Gisselfeld]; Poulsen

Diva, HT, dy, 1995; Cocker

Diversity, HT, rb, 1995; K&S

'DIVIDEND', HT, dy, 1931; flowers rich yellow, dbl., globular, slight fragrance; foliage dark; dwarf growth; [Franz Deegen X Seedling]; Clark, A.; NRS Victoria

'DIVINE', HT, mr, 1964; bud dark purplish; flowers cardinal-red, well formed, large, dbl., 45 petals; foliage bright green; strong stems; vigorous, upright growth; GM, Geneva, 1964; Delbard-Chabert

'DIVINE LADY', F, op, 1965; flowers salmon-pink suffused brownish, dbl.; very vigorous, dense growth; [Circus X Queen Elizabeth]; Lens

'DIXIE', HT, op, 1925; flowers salmon-pink, more large, dbl., cupped, intense fragrance; [Radiance sport]; Gray, W.R.

'DIXIE BELLE', HT, lp, 1963; bud ovoid, rose-pink; dbl., 38 petals, 5–5.5 in., cupped, moderate fragrance; foliage leathery; vigorous, upright growth; [Golden Masterpiece X Seedling]; Boerner; J&P

'DIXIE CLIMBER', Cl HT, ob, 1935; flowers salmon and gold; [Gov. Alfred E. Smith sport]; Watkins, A.F.; Dixie Rose Nursery;, J&P

Dixie Dazzle™ *see* 'KINDIXIE'

Dixie Dream *see* **'FESTIVAL'**

'DIXIE HOLIDAY', HT, mr, 1968; bud long, pointed; flowers medium, dbl., exhibition form, slight fragrance; foliage bronze, leathery; very vigorous growth; [Étoile de Hollande sport]; Garrison; Kimbrew

Dixieland *see* 'BRIDIXIE'

Dixieland Linda *see* 'BEADIX'

Dizzy, F, 1974; Delforge, H.; (Cavriglia)

Dizzy Heights *see* 'FRYBLISSFUL'

'DOC', Pol, mp, 1954; (Degenhard); flowers phlox-pink, small blooms in large trusses, semi-dbl., 15 petals; compact growth; [Robin Hood X Polyantha seedling]; deRuiter; Gregory & Willicher Baumschulen, 1954

Docile, F, 1995; Eve, A.; (Cavriglia)

Docter Dorothy, F, pb, 1997; flowers very full, medium, very dbl., 41 petals, borne in small clusters, intense fragrance; some prickles; foliage medium, medium green, semi-glossy; upright, medium (5ft.)growth; [Jubilee X Little Darling]; Jones, L.J.

Docteur F. Debat *see* **'DR DEBAT'**

'DOCTEUR LOUIS ESCARRAS', HT, rb, 1922; flowers dark salmon-red shaded carmine-pink, very dbl., 120 petals; [Constance X Seedling]; Nabonnand, C.

'DOCTEUR MARJOLIN', M, mp, 1860; flowers shell-pink; Robert et Moreau

'DOCTEUR MOREL', HT, m, 1946; flowers carmine with chrome-yellow reflections, large, dbl., 40–45 petals; foliage dark; vigorous, upright growth; [Edith Nellie Perkins X Pres. Herbert Hoover]; Laperrière

'DOCTEUR REYMONT', HMult, w, 1907; flowers pure white on pale green base, pyramidal cluster, dbl.; Mermet

'DOCTEUR ROBERT SALMONT', HT, yb, 1946; bud pointed; flowers capucine and yellow, reverse tinted chrome, base coppery; foliage dark; vigorous growth; Gaujard

'DOCTEUR VALOIS', HT, rb, 1950; flowers geranium shaded vermilion, reverse yellow, semi-dbl., 4 in., moderate fragrance; foliage dark, glossy; vigorous, bushy growth; [(Annie Drevet X Condesa de Sástago) X Vive la France]; Mallerin, C.; URS

Doctor Abrahams, HT, ab, 1991; Hallows

Doctor Behring *see* 'DOTEMIBE'

Doctor Dick *see* 'COCBADEN'

'DOCTOR ELDON LYLE', Gr, dr, 1968; bud pointed; flowers medium, dbl., exhibition form, moderate fragrance; foliage soft, bronze; vigorous, compact growth; [Pres. Eisenhower X Suspense]; Mackay; Texas Rose Research Foundation

Doctor Faust, F, yb, 1957; flowers dark yellow with orange and pink tints, medium, dbl., slight fragrance; Kordes, W. Söhne; (Sangerhausen)

Doctor Goldberg *see* 'GANGO'

Doctor Jackson *see* 'AUSDOCTOR'

Doctor Robert Korns *see* LETrob

'DOCTOR'S WIFE', HT, op, 1967; bud long, pointed; flowers salmon-pink, large, dbl., exhibition form, slight fragrance; foliage dark, glossy, leathery; vigorous, upright growth; Von Abrams; Edmunds Roses

Dog Rose *see* **R. CANINA**

'DOHHAWK', Min, mr, 1988; (Night Hawk, **Nighthawk**™); bud globular, pointed; flowers medium, dbl., 22 petals, exhibition form, borne singly and in sprays of 3-5, intense, damask fragrance; prickles straight, slanted down, medium, reddish-brown; foliage medium, medium green; upright, bushy, medium growth; PP007417; AOE, 1989; [Quinella X Poker Chip]; Hardgrove, Donald L.; Nor'East Min. Roses, 1989

Dolce Vita® *see* 'DELDAL'

Dolcezza®, S, mp

'DOLLAR-ROSE', HT, mr, 1936; flowers carmine-red, medium, dbl.; Tantau, Math.; (Sangerhausen)

Dollie B *see* 'TROBEE'

Dolly® *see* 'POULVISION'

'DOLLY BROWNELL', F, lp, 1926; flowers color same as Dr. W. van Fleet; [Dr. W. Van Fleet seedling]; Brownell, H.C.

'DOLLY DARLING', HT, mp, 1949; bud long, pointed, red; flowers lustrous pink, open, semi-dbl., 20 petals, 4–5 in., moderate fragrance; foliage glossy; vigorous, compact growth; [Pink Princess X Crimson Glory]; Brownell, H.C.

Dolly Dot, MinFl, dy, 1998; J&P

'DOLLY MADISON', HT, dy, 1935; (Super-Dupont); flowers golden yellow; vigorous growth; [Mrs Pierre S. duPont climbing sport seedling sport]; Hillock

'DOLLY PARTON', HT, or, 1984; flowers luminous orange-red, large, dbl., 35 petals, borne mostly singly, intense fragrance; foliage large, medium green, semi-glossy; upright growth; PP005608; Bronze, ARC TG, 1982; [Fragrant Cloud X Oklahoma]; Winchel, Joseph F.; C-P, 1983

'DOLLY VARDEN', HRg, ab, 1914; flowers light apricot-pink, base yellow, large, recurrent bloom; vigorous growth; Paul

'DOLLY VARDEN', Pol, mp, 1930; flowers clear pink, dbl.; deRuiter

'DOLLY'S SISTER', HT, op, 1989; flowers medium coral-pink; [Dolly Parton sport]; Taylor, Thomas E.; Michigan Mini Roses, 1989

Dolly's Sister, Gr, or, 1999; Williams, J. Benjamin

'DOLOMITI', HT, ab, 1933; bud pointed; flowers flesh, with yellow reflex, very large, dbl., moderate fragrance; foliage dark; strong stems; vigorous growth; Ingegnoli

Dolores™ *see* DEVine

'DOMAINE DE CAPUIS', S, m, 1901; Roseraie de l'Hay

Domaine de Courso *see* MEIdrimy

Dometille Beccard, HGal; (Sangerhausen)

Domila® *see* LAPnat

'DOMINA', HT, op, 1943; flowers salmon-pink, large, dbl., slight fragrance; Heizmann, E.; (Sangerhausen)

'DOMINANT', HT, op, 1964; bud ovoid; flowers salmon-pink, medium, borne in clusters, dbl., slight fragrance; foliage dark; [Golden Masterpiece X Spartan]; Boerner; Spek

'DOMINATOR', F, mp, 1961; semi-dbl., 3 in., blooms in clusters, moderate fragrance; vigorous, upright growth; [New Yorker X The Optimist]; deRuiter

'DOMINIE SAMPSON', HSpn, lp; flowers soft pink, semi-dbl., non-recurrent; glossy, black fruit; foliage finely divided; dense, shrubby (3-4 ft.) growth

'DOMINIQUE', Min, lp, 1981; bud ovoid; flowers light peachy pink, dbl., 30 petals, exhibition form, intense, apple fragrance; curved prickles; foliage medium green, arrow-shaped; upright growth; [Electron X Little Chief]; Bennett, Cecilia 'Dee'; Tiny Petals Nursery

'DOMINO', HT, dr, 1956; bud long, pointed; flowers dark crimson, medium, slight fragrance; foliage dark; [Peace X Seedling]; Gaujard

Domino *see* TANdomo

'DOMKAPITULAR DR LAGER', HT, pb, 1903; flowers rose and carmine, moderate fragrance; [Mme Caroline Testout X Princesse de Bassaraba de Brancovan]; Lambert, P.

Domstadt Fulda® *see* KORtanken

'DOMUS AUREA', HT, my, 1940; flowers pure yellow; foliage dark, glossy; strong stems; very vigorous growth; [Julien Potin X Yellow seedling]; Aicardi, D.; Giacomasso

Don Bosco, HT, 1976; Dorieux, Francois; (Cavriglia)

Don Bosco *see* LAPduf

'DON BRADMAN', HT, rb, 1938; bud long, shapely; flowers coppery claret, fading to silvery pink, dbl., 40–50 petals; Wheatcroft Bros.

'DON CHARLTON', HT, pb, 1991; flowers deep rose pink with silver reverse, large blooms, very dbl., moderate fragrance; foliage large, dark green, glossy; upright growth; [Silver Jubilee X (Chicago Peace X Doris Tysterman seedling)]; Thompson, Robert; Battersby Roses, 1992

'DON DON', Min, dr, 1976; flowers red, reverse blending near white at base, dbl., 60 petals, 1–1.5 in., moderate fragrance; foliage small, glossy, bronze; upright, bushy growth; [Seedling X Over the Rainbow]; Williams, Ernest D.; Mini-Roses

'DON JOSÉ', HT, op, 1922; flowers salmon-pink, semi-dbl., slight fragrance; [Archiduc Joseph X Seedling]; Clark, A.; NRS Victoria

'DON JUAN', LCl, dr, 1958; bud ovoid; flowers velvety crimson red, very dbl., 30–35 petals, 5 in., cupped, recurrent bloom, intense fragrance; foliage dark green, glossy, leathery; climbing, height 12-14 ft; [New Dawn seedling X New Yorker]; Malandrone; J&P

Don Marshall *see* 'MORBLACK'

Don Pedro, lp; flowers medium, dbl.; Roseraie de l'Hay; (Sangerhausen)

'DON QUICHOTTE', F, rb, 1964; flowers cherry-red, base yellow, well formed, large, borne in clusters, dbl., slight fragrance; foliage glossy, leathery; vigorous, upright growth; [Charles Gregory X Marcelle Auclair]; Robichon; Ilgenfritz Nursery

'DON ROSE', HT, pb, 1943; bud long, pointed, carmine-red; flowers coppery pink, open, large, dbl., 40 petals, cupped, slight fragrance; foliage leathery, bluish green; vigorous, upright, bushy, rather compact growth; [Souer Thérèse X Seedling]; Mallerin, C.; C-P

'DOÑA CLARA', HT, m, 1965; bud ovoid; flowers purplish pink, large, dbl., 50 petals, exhibition form; vigorous growth; Camprubi, C.

Donald Davies *see* 'CHEWBEAUT'

'DONALD MACDONALD', HT, ob, 1916; flowers orange-carmine, borne in clusters, semi-dbl., moderate fragrance; dwarf growth; GM, NRS, 1916; Dickson, A.

'DONALD PRIOR', F, mr, 1938; bud ovoid; flowers bright scarlet flushed crimson, blooms in large clusters, semi-dbl., 11 petals, 3 in., cupped, moderate fragrance; foliage leathery, dark; vigorous, bushy growth; (28); [Seedling X D.T. Poulsen]; Prior; J&P

'DONALD PRIOR, CLIMBING', Cl F, mr; Farr

Donald Thomas Heald *see* 'KIRSMILE'

'DONALDO', HT, rb, 1979; bud ovoid; flowers red to pink, shapely, dbl., 35 petals, 4 in.; foliage large, glossy, dark; tall growth; [Honey Favorite X Rose Gaujard]; Murray, Nola

Donatella *see* **'GRANADA'**

'DONAU', LCl, m, 1913; Praskac

Donauprinzessin®, F, mp, 1994; Noack, Werner

Donauwelle, S, m, 1991; Weihrauch

Donella *see* JAYdon

'DONNA CLARA', HT, pb; flowers buff, reverse strawberry-pink; Leenders, M.

Donna Darlin' *see* 'WINDONNA'

'DONNA FANNY CAVALIERI', HT, 1953; San Remo Exp. Sta.; (Cavriglia)

'DONNA FAYE', Min, lp, 1976; bud pointed; dbl., 27 petals, 1 in., exhibition form, moderate fragrance; upright growth; [Ma Perkins X Baby Betsy McCall]; Schwartz, Ernest W.; Nor'East Min. Roses

Donna Jean *see* 'TALDON'

Donna Marella Agnelli® *see* BARdon

Donna Maria *see* **'DONNA MARIE'**

'DONNA MARIE', HSem, w, 1830; (Donna Maria); flowers pure white, small, very dbl.; Vibert

Donna Silva Carmine, F, 1974; Cazzaniga, F. G.; (Cavriglia)

Doorenbos Selection, HSpn, dr; Doorenbos

'DOORYARD DELIGHT', HT, pb, 1940; bud short, pointed, spiraled; flowers light pink, reverse rose-pink, petals sharply pointed, large, dbl., 2.5 in., recurrent bloom, slight fragrance; foliage leathery; vigorous, bushy growth; [R. setigera X Lady Alice Stanley]; Horvath; Wyant

'DOPEY', Pol, mr, 1954; (Eberwein); flowers crimson-red, small blooms in trusses, semi-dbl.; compact growth; [Robin Hood X Polyantha seedling]; deRuiter; Gregory & Willicher Baumschulen, 1954

'DORA', HT, or, 1975; bud long; flowers brilliant orange-red, dbl.; foliage bronze; [Tanagra X Rubens]; Gaujard

Dora Delle *see* 'TALDOR'

'DORA HANSEN', HT, mp, 1908; bud long, pointed; flowers thulite-pink, open, large, dbl., slight fragrance; Jacobs

'DORA STOBER', HT, w, 1925; flowers white shaded yellow, dbl., moderate fragrance; Leenders, M.

Dorabella, HT, 1986; Barni, V.; (Cavriglia)

Dorada *see* BARdord

DORalp, F, w, 1993; (**Anne-Aymone Giscard d'Estaing**®); Dorieux

Doralta, HT, 1979; Dorieux, Francois; (Cavriglia)

'DORAPRI', HT, ob, 1988; (**Mango**®); Dorieux

DORastri, S, pb, 1996; (**Hugues Aufray**); Dorieux

'DORCAS', S, pb, 1984; bud ovoid, pointed; flowers light pink, pale yellow blend, flecked deeper pink, dbl., 40 petals, cupped, repeat bloom, slight fragrance; awl-like, tan prickles; foliage dark, leathery; erect, bushy growth; hardy.; [Minigold X Freckle Face]; Buck, Dr. Griffith J.; Iowa State University

'DORCAS', HWich, pb, 1922; flowers deep rose-pink to coral-pink, base yellow, borne in large clusters, dbl.; vigorous growth; RULED EXTINCT 1/85; English

DORcast, F, dy, 1997; (**Soleillade**); Dorieux

DORcroix, HT, ab, 1995; (**Chateau la Croix**); Dorieux

DORdeli, HT, pb; (**Starion**®); Dorieux, 1987

'DOREEN', HT, ob, 1951; flowers deep golden orange flushed scarlet, well-formed, moderate fragrance; foliage dark; vigorous growth; [Lydia X McGredy's Sunset]; Robinson, H.; Baker's Nursery

'DOREEN JOHNSON', HT, lp, 1977; bud long, pointed; flowers pale pink, dbl., slight fragrance; foliage large, light; vigorous, bushy growth; [(Great Venture X Fort Vancouver) X Memoriam]; Dawson, George; Australian Roses

'DOREEN THORN', HT, pb, 1934; flowers deep pink, base yellow, well shaped, large, dbl., moderate fragrance; vigorous growth; Cant, F.

'DOREEN WELLS', F, or, 1970; flowers orange-scarlet, dbl., 25 petals, 3 in., flat, moderate fragrance; foliage glossy, dark; low, bushy growth; [Soraya X Circus]; Watkins Roses

'DORELL', HMult, dp, 1887; Geschwind, R.

'DOREN', HT, 1951; Robinson, H.; (Cavriglia)

DORfaut, HT, ab, 1997; (**Comtesse Brigitte Chandon-Moet**); Dorieux

DORflo, F, ob, 1974; (**Bibiché**®); Dorieux, Francois; (Cavriglia)

DORfuri, HT, 1991; (**Roseraie de Blois**); Dorieux; (Cavriglia)

DORgold, F, dy, 1982; (**Coucou**); Dorieux

DORgran, ab, 1998; (**Barbara Hendricks**); Dorieux

'DORIC', F, yb, 1963; flowers golden salmon, large, borne in well-spaced clusters, dbl., 40 petals; foliage glossy; vigorous, compact growth; [Masquerade X Korona]; LeGrice

'DORIENNE', HT, mr, 1958; bud short; flowers large, dbl., 34 petals, cupped; bushy, spreading growth; [Mrs Nieminen X Seedling]; Buyl Frères

'DORIENT', Gr, m, 1998; (Melodie Parfumee, **Melody Parfumee**™); flowers dark lavender-plum, lighter reverse, very dbl., 26–40 petals, 4.5 in., borne in large clusters, intense, damask fragrance; prickles moderate, straight; foliage medium, dark green, semi-glossy; bushy, upright, 5 ft growth; [Dioressence X Stephens' Big Purple]; Dorieux, Francois; Bear Creek Gardens, Inc., 1995

'DORINA NEAVE', HT, pb, 1926; flowers silvery pink, large, dbl., globular, intense fragrance; stiff stems; compact growth; Pemberton

Dorinda, S, lp; Peden, R., 1998; (Weatherly, L.)

'DORIS', HT, rb, 1939; flowers cerise striped white; [Briarcliff sport]; Spandikow

'DORIS ANN', Min, dr, 1986; flowers small, dbl., 26–40 petals; foliage small, medium green, matt; [Black Jade X Tiki]; Wambach, Alex A.

'DORIS ARCHER', F, rb, 1962; flowers yellow, bronze and red, well formed, borne in clusters, dbl., 30–35 petals, 4 in., moderate fragrance; foliage glossy; vigorous, compact growth; [Circus X Seedling]; Fryers Nursery, Ltd.

'DORIS DICKSON', HT, ob, 1924; flowers orange-cream, veined cherry-red, moderate fragrance; foliage very dark; stiff, wiry stems; vigorous growth; Dickson, S.

Doris Dowman *see* 'HORSILKARE'

'DORIS DOWNES', LCl, pb, 1932; flowers pink, shaded red, overlarge, semi-dbl., cupped, early, intense fragrance; climbing growth; Clark, A.; NRS Victoria

'DORIS FINDLATER', HT, ab, 1936; flowers light apricot, reverse flushed reddish salmon and carmine, dbl.; vigorous growth; Dickson, A.

'DORIS GRACE ROBINSON', HT, w, 1943; bud pointed; flowers creamy white, well shaped, large; foliage olive-green; vigorous, upright growth; Bees

'DORIS HOWARD', F, mr, 1957; flowers blood-red, borne in large clusters; vigorous, bushy growth; Wheatcroft Bros.

Doris J. Robertson, HT, lp; flowers whitish-pink, large, dbl., slight fragrance; (Sangerhausen)

'DORIS NORMAN', F, or, 1958; flowers bright orange, to open, borne in small clusters, dbl., 30 petals, 2 in., exhibition form; foliage purplish to dull green; vigorous, bushy growth; [Paul's Scarlet Climber X Mary]; Norman; Harkness

'DORIS OSBORNE', HT, mr, 1937; bud pointed; flowers ruby-cerise, semi-dbl.; bushy growth; [Mme Abel Chatenay X Seedling]; Clark, A.; NRS Victoria

'DORIS PLEASANCE', HT, w, 1978; flowers blush-pink to white; [Queen Elizabeth sport]; Brewer

Doris Reese *see* 'RESGOLD'

'DORIS RYKER', Pol, ob, 1942; (Dorus Rijkers); flowers salmon-pink, blooms in clusters, dbl., recurrent bloom, moderate fragrance; foliage light green; vigorous, upright growth; Leenders, M.; Klyn

'DORIS TRAYLER', HT, yb, 1924; bud pointed, orange; flowers yellow, reverse flushed crimson and orange, large, dbl., exhibition form, moderate fragrance; foliage light green, leathery, glossy; bushy, dwarf, compact growth; McGredy

'DORIS TYSTERMAN'®, HT, ob, 1975; flowers tangerine and gold, dbl., 28 petals, 4–5 in., slight fragrance; foliage glossy; upright growth; [Peer Gynt X Seedling]; Wisbech Plant Co.

'DORISTEL', F, w, 1986; (**Artiste**®); Dorieux

'DORJURE', HT, rb, 1992; (**Portland Rose Festival**); flowers strawberry red, white reverse, exceptional form with clear s, dbl., 26–40 petals, 3–3.5 in., moderate fragrance; some prickles; foliage large, dark green, glossy; tall (150 cms), upright growth; [Osiria X Pharaon]; Dorieux, Francois; Roses by Fred Edmunds, 1992

DORkade, HT, dr, 1996; (**Georges Truffaut**); Dorieux

DORlain, HT, w, 1997; (**Marie Laforet**); Dorieux

DORlina, HT, w, 1996; (**Provence**); Dorieux

'DORMA', HT, op, 1966; (**Majesté**®); bud oval; flowers salmon, large, dbl., moderate fragrance; foliage glossy, leathery; very vigorous, upright growth; [Radar X Eclipse]; Dorieux; Vilmorin

DORmagi, HT, rb, 1992; (**Imagine**); Dorieux

'DORMAL', HT, mr, 1967; (**Rouge Dorieux**®); bud pointed; flowers cherry-red, open, over large, dbl., slight fragrance; foliage dark, glossy, leathery; vigorous, upright growth; [Seedling X Ena Harkness]; Dorieux; Vilmorin

DORnapa, HT, rb, 1988; (**Flushing Meadow**®); Dorieux

DORneye, F, ob, 1995; (**Copacabana**); Dorieux

DORnice, F, rb, 1997; (**Orange d'Éte**); Dorieux

'DORNRÖSCHEN', S, pb, 1960; bud well-shaped; flowers salmon to deep pink, reverse yellow, large blooms in clusters, dbl., recurrent bloom, moderate fragrance; upright, well-branched growth; [Pike's Peak X Ballet]; Kordes, R.

Dornroschenschloss Sababurg® *see* KORtensei

Dorola® *see* 'MACSHANA'

Dorothe® *see* 'LEGGA'

Dorothea Furrer®, HT, dr

'DOROTHEA HOWARD', HT, pb, 1978; flowers light pink, deeper pink reverse, well-formed, large, dbl., 30 petals, borne singly, moderate, tea fragrance; many hooked, brown prickles; foliage medium green, glossy; [First Prize X Roundelay]; Barclay, Hilary M.

Dorothee Heidorn, Bslt, mp, 1995

'DOROTHY A. GOLIK', HT, ob, 1973; bud ovoid; flowers orange to flaming red, high pointed, dbl., 35 petals, 4 in., moderate, spicy fragrance; foliage glossy; moderate growth; [Tropicana X Peace]; Golik; Dynarose

'DOROTHY ANDERSON', HT, lp, 1949; flowers large, dbl., 33 petals, exhibition form, slight fragrance; free growth; [Sam McGredy X George Dickson]; McGredy

'DOROTHY ANNE', HT, pb, 1985; flowers white blending to deep pink at edges, large, dbl., 35 petals, no fragrance; foliage medium, dark, semi-glossy; upright growth; PP006100; [First Prize X Lady X]; Winchel, Joseph F.; Kimbrew-Walter Roses

'DOROTHY BROSTER', HT, mp, 1978; flowers azalea-pink, full, dbl., 45 petals, 5 in., moderate fragrance; foliage dark; very vigorous growth; [Blue Moon X Karl Herbst]; Ellick; Excelsior Roses

'DOROTHY DENNISON', HWich, lp, 1909; flowers pale pink; [Dorothy Perkins sport]; Dennison

'DOROTHY DIX', Pol, mp, 1923; flowers rose-pink, borne in clusters; Hicks

Dorothy Dix *see* **'MRS E.M. GILMER'**

Dorothy Donnelly, HT, mr, 1997; flowers full, large, very dbl., 26–40 petals, borne mostly singly, intense fragrance; foliage medium, dark green, semi-glossy; upright, bushy, medium growth; [Adrienne Berman X (Royal William X Gabi)]; Poole, Lionel

'DOROTHY DOUGLAS', HT, rb, 1924; flowers vivid cerise-pink; Dobbie

'DOROTHY DROWNE', HWich, pb, 1924; flowers white to pink, center crimson and scarlet; [Sodenia seedling]; Brownell, H.C.

'DOROTHY FOWLER', HRg, mp, 1938; flowers clear pink, well formed, semi-dbl., 3–3.5 in., non-recurrent, intense fragrance; height 3 ft; [R. rugosa X (R. acicularis X R. spinosissima)]; Skinner

'DOROTHY GOODWIN', HT, yb, 1954; (Perfect Peace); flowers yellow tipped cerise-pink, well formed, dbl., 32 petals, 4 in., moderate fragrance; foliage dark holly-green; vigorous growth; [Peace sport]; Goodwin; Gregory

'DOROTHY GRACE', Cl Min, pb, 1986; flowers yellow with pink petal edges, small blooms in sprays of 3-5, exhibition form, 25 petals, exhibition form, none; very few, brown prickles, hooked downward; foliage small, medium green, semi-glossy; upright, climbing (to 6 ft.) growth; [Little Darling X Rise 'n' Shine sport]; Dobbs, Annette E.; Port Stockton Nursery

'DOROTHY HODGSON', HT, ob, 1930; flowers orange-cerise, veined darker, well formed, large, slight fragrance; vigorous growth; Cant, F.

'DOROTHY HOWARTH', Pol, op, 1921; flowers coral-pink, tinted salmon, open, borne in clusters, dbl., moderate fragrance; foliage dark; bushy growth; [Léonie Lamesch X Annchen Müller]; Bees

'DOROTHY JAMES', HT, pb, 1939; flowers peach-pink reverse deep rose; good habit growth; [Golden Dawn sport]; C-P

'DOROTHY KING', HT, rb, 1924; flowers scarlet-crimson and maroon, semi-dbl.; King

'DOROTHY LEE', HT, pb, 1929; flowers silvery shell-pink, base golden yellow, dbl.; Morse

Dorothy Lloyd *see* 'SHERIGREY'

'DOROTHY MARIE', HT, dp, 1935; [Talisman sport]; Scittine; Lainson

'DOROTHY MAY COOPER', Min, w, 1975; flowers pure white, full, dbl., 25–30 petals, 1–2 in.; foliage dark; very vigorous growth; [R. roulettii X Memoriam]; Ellick

'DOROTHY MCGREDY', HT, rb, 1936; flowers deep vermilion, base and reverse yellow, well shaped, moderate fragrance; foliage cedar-green; strong stems; vigorous growth; McGredy

'DOROTHY MOLLISON', HT, dr, 1930; flowers dark crimson; [Mrs R.C. Bell X Seedling]; Clark, A.; NRS Victoria

'DOROTHY PAGE-ROBERTS', HT, pb, 1907; flowers coppery pink, suffused yellow, open, very large, dbl., slight fragrance; vigorous growth; Dickson, A.

'DOROTHY PEACH', HT, yb, 1957; flowers deep yellow flushed pink, dbl., 37 petals, 5 in., exhibition form, moderate fragrance; foliage dark, glossy; vigorous growth; GM, NRS, 1959; [Lydia X Peace]; Robinson, H.

'DOROTHY PEACH, CLIMBING', Cl HT, yb, 1963; Watkins Roses

'DOROTHY PERKINS', HWich, lp, 1901; flowers rose-pink, dbl., moderate fragrance; foliage dark, glossy; very vigorous (10-20 ft) growth; [R. wichurana X Mme Gabriel Luizet]; J&P

'DOROTHY RATCLIFFE', HT, rb, 1910; flowers coral-red shaded fawn-yellow; vigorous growth; McGredy

Dorothy Rose *see* 'JONDOROSE'

Dorothy Virginia *see* 'JUDVIRG'

'DOROTHY WHEATCROFT', F, mr, 1960; flowers oriental red shaded darker, blooms in clusters of 13, semi-dbl., 18 petals, 3.5 in., slight fragrance; foliage bright green; vigorous, bushy growth; GM, NRS, 1961; Tantau, Math.; Wheatcroft Bros.

Dorothy Whitney Wood, HT, op, 1992; Fryer, Gareth

Dorothy Wilson, F, ob, 1995; Beales, Peter

'DOROTHY, CLIMBING', Cl HT, pb, 1935; [Dorothy Page-Roberts sport]; Bostick

Dorothy's Front Porch, Misc OGR, r

'DOROTHY'S REGAL RED', HT, mr, 1984; flowers large blooms in small clusters, dbl., 58 petals, intense fragrance; foliage medium dark, glossy; bushy growth; Jerabek, Paul E.

'DOROTKA DARLING', Min, pb, 1991; flowers pink and cream blend, well-formed, blooms borne mostly singly, dbl., 26–40 petals, 1.5 in., slight fragrance; few prickles; foliage medium, dark green, semi-glossy, clean and abundant; tall (60-80 cms), upright growth; [Orange Darling X Seedling]; Sudol, Julia, 1992

DORouvi, HT, 1992; (Perle du Lac, **Perle du Lac Annecy**); Dorieux; (Cavriglia)

DORpark, F, rb, 1996; (**Le Cid**); Dorieux

DORpurp, HT, rb, 1990; (**Parc des Princes**®); Dorieux

Dorris Lee *see* 'WELLEE'

'DORRIT', F, ob, 1968; flowers orange-yellow, large blooms in trusses, dbl., 30–40 petals, flat; [Seedling X Folie d'Espagne]; Soenderhousen

Dorrit *see* **'SUNSONG'**

DORsafr, HT, yb, 1990; (**Open d'Australie**®); Dorieux

DORsand, HT, ab, 1994; (**Gaby Morlay**); Dorieux

'D'ORSAY ROSE', Misc OGR, dp; bud deep pink; flowers deep pink, outer petals fading to pale pink receptacle wide, dbl., summer bloom, moderate fragrance; prickles paired below each leaf; foliage leaflets 5-7, leaden green; erect growth to 5 ft.

DORster, HT, pb, 1996; (**Ressins Etienne Gautier**); Dorieux

DORtan, F, lp; (**Vieux Chateau Certan**); Dorieux, 1994

DORtiche, HT, dr; (**Marguerite d'Autricho**); Dorieux

DORtive, HT, op, 1991; (**Pierre Troisgros**®); Dorieux

'DORTMUND'®, HKor, mr, 1955; bud long, pointed; flowers red, white eye, large blooms in large clusters, single, recurrent bloom, moderate fragrance; foliage dark, very glossy; vigorous, climbing growth; ADR, 1954 GM, Portland, 1971; [Seedling X R. kordesii]; Kordes

Dortmunder Kaiserhain, S, lp, 1994; flowers medium-large, dbl.; Noack, Werner; (Sangerhausen)

DORto, HT, lp; (**Sika**); Dorieux, 1974

DORtomyl, Pol, rb, 1992; (**Pleins Feux 92**); Dorieux

Dorus Rijkers *see* **'DORIS RYKER'**

DORvizo, HT, dp, 1993; (**Rina Hugo**); Dorieux, 1993

DORwest, S, dy; (**Ellen Hamlyn**); Dorieux

DORyeco, HT, ob, 1998; (**D. D. Ruaux**); Dorieux

'DO-SI-DO', S, mp, 1984; flowers medium lavender-pink, large blooms borne 3-10 per cluster, cupped, repeat bloom, moderate fragrance; awl-like, tan prickles; foliage dark olive green, leathery, glossy; vigorous, erect, bushy growth; hardy.; [(Autumn Dusk X Solitude) X Wanderin' Wind]; Buck, Dr. Griffith J.; Iowa State University

'DOSPEARL', Min, op, 1991; (**Our Coral Pearl**); flowers coral-salmon-pink, blooms borne mostly singly, dbl., 26–40 quilling petals, 1.5–2.75 in., slight fragrance; few prickles, tan, straight and sharp; foliage medium, medium green, semi-glossy; medium growth; [Minnie Pearl sport]; Osburn, Dr. William, 1992

'DOTEMIBE', HT, m, 1978; (**Doctor Behring**); bud pointed; flowers red-purple, dbl., 35 petals, 4 in., cupped, moderate fragrance; foliage dark; tall, upright growth; [Amanecer X Tanya]; Dot, Simon

'DOTHAN', HT, mp, 1983; flowers large, dbl., 35 petals, exhibition form, slight fragrance; foliage large, medium green, matt; upright growth; [Koppies X King of Hearts]; Perry, Astor; Perry Roses, 1984

'DOTRAMES', HT, or, 1983; (**Coral Fiesta**®, Maria Teresa de Esteban, Mme Teresa Estaban); flowers large, dbl., 45 petals, cupped, no fragrance; light yellow prickles; foliage large, dark, matt; bushy growth; [Seedling X Seedling]; Dot, Simon; Rose Barni-Pistoia

DOTraner, HT, rb, 1969; (**Jardinero Ortiz**®); Dot

'DOTSUBEBE', HT, or, 1978; (**Galaty**); dbl., 32 petals, 4 in., cupped, intense fragrance; foliage dark; [Tropicana X Lola Montes]; Dot, Simon

'DOTSURODO', HT, or, 1978; (**Simon Dot**); dbl., 35 petals, 5 in., cupped, intense fragrance; foliage dark, upright growth; [Pharaoh X Rose Dot]; Dot, Simon

'DOTTY', HT, dy, 1931; flowers bronze-yellow, large, semi-dbl., globular, moderate fragrance; foliage glossy; long stems; very vigorous growth; [Souv. de Claudius Pernet X R. foetida bicolor seedling]; Towill

'DOTTY BASS', HT, dr, 1968; bud long, pointed; flowers medium, dbl.; foliage dark, leathery; vigorous, upright growth; Bass; DeVor Nurseries, Inc.

Double Blush, S, lp; flowers medium, dbl.; (Sangerhausen)

'DOUBLE BLUSH BURNET', HSpn, pb; flowers center blush, fading at edges, reverse white

'DOUBLE BRIQUE', HGal, pb; flowers rosy pink, shading silver toward outside, dbl.

Double Cherokee *see* **'FORTUNIANA'**

Double Cinnamon *see* **R. CINNAMOMEA PLENA**

'DOUBLE DARK MARBLED', HSpn, rb; (Petite Red Scotch); flowers red mottled purple, small, semi-dbl., early

Double Date *see* 'LAVFLUSH'

Double Delight™ *see* 'ANDELI'

Double Delight Supreme, HT, rb, 1993; Chiplunkar

Double Delight, Climbing *see* 'AROCLIDD'

Double Else Poulsen *see* **'ELSE'S RIVAL'**

'DOUBLE FEATURE', Gr, m, 1976; (Lakeland's Pride); bud pointed; flowers reddish purple, reverse yellow, dbl., 28 petals, 4 in., exhibition form, slight, damask fragrance; foliage large, dark; vigorous, upright growth; [Angel Face X Granada]; Williams, J. Benjamin; Lakeland Nursery Sales

Double French Rose *see* **R. GALLICA OFFICINALIS**

Double Happy *see* 'SUNDOHA'

'DOUBLE JOY', Min, op, 1979; bud long, pointed; dbl., 35 petals, 1.5 in., moderate fragrance; foliage small, matt, green; bushy growth; [Little Darling X New Penny]; Moore, Ralph S.; Sequoia Nursery

Double Mme Butterfly *see* **'ANNIE LAURIE'**

'DOUBLE OPHELIA', HT, lp, 1916; [Ophelia X Seedling]; Hill, E.G., Co.

'DOUBLE ORLÉANS', Pol, mr, 1924; flowers rosy crimson, center white; [Orléans Rose sport]; Hicks

Double Perfection™ *see* 'BURWIN'

'DOUBLE PINK KILLARNEY', HT, mp, 1910; flowers large, very dbl., moderate fragrance; Scott; (Sangerhausen)

Double Plum, HCh, m

'DOUBLE SCARLET', HEg, mr; flowers bright rosy red, dbl., intense fragrance; weak growth

'DOUBLE STAR', Min, w, 1978; bud ovoid; 10 petals, borne 2-5 per cluster, slight fragrance; no prickles; foliage small, firm, disease-resistant; vigorous growth; [Fairy Moss X Fairy Moss]; Dobbs, Annette E.; Small World Min. Roses

'DOUBLE TALK', F, rb, 1980; bud ovoid, pointed; flowers medium red, creamy white reverse, petals rolled loosely outward, dbl., 48 petals, cupped, slight, spicy fragrance; long prickles, hooked downard; foliage glossy, slightly wrinkled, dark; compact growth; [Plain Talk X Suspense]; Weeks, O.L.

Double Time, HT, rb

Double Treat *see* 'MORTREAT'

'DOUBLE WHITE', HEg, w; flowers flesh-white, dbl.; vigorous growth

Double White *see* **'ELEGANS'**

'DOUBLE WHITE BURNET', HSpn, w;, intense fragrance; vigorous growth

Double White Noisette *see* **'PLENA'**

Double Yellow Scots Rose *see* **'WILLIAM'S DOUBLE YELLOW'**

'DOUBLOONS', LCl, my, 1934; bud ovoid, deep saffron-yellow; flowers rich gold, large blooms in clusters, dbl., cupped, intermittent bloom, moderate fragrance; foliage glossy; vigorous growth; David Fuerstenberg Prize, ARS, 1936; [R. setigera hybrid X R. foetida bicolor hybrid]; Horvath; J&P

Douce Symphonie *see* 'MEIBARKE'

Douceur Normande® *see* 'MEIPOPUL'

Douchka *see* **'MARY DEVOR'**

'DOUÉ RAMBLER', HWich, mp, 1921; flowers bright pink, borne in well-filled clusters; Begault-Pigné

'DOUGLAS MACARTHUR', HT, pb, 1943; bud long, pointed; flowers delft rose, base slightly bronze, dbl., 24–30 petals, 4–4.5 in., exhibition form; foliage leathery; long stems; vigorous, upright, bushy, compact growth; [Mrs J.D. Eisele X Glowing Sunset]; Howard, F.H.; H&S

'DOUGLAS MACARTHUR, CLIMBING', Cl HT, pb, 1949; Howard, F.H.; H&S

'DOUGLASS', Ch, dr, 1848; flowers crimson, medium; vigorous growth; Verdier, V.

'DOULCE FRANCE', F, pb, 1964; bud round; flowers clear pink flushed apricot, large, dbl.; vigorous, upright growth; [(Peace X Seedling) X Lady Sylvia]; Mondial Roses

'DOURADA', HT, dy, 1957; flowers well formed, dbl.; very vigorous growth; [Mme Marie Curie X Julien Potin]; da Silva, Moreira

Doutz, S, lp

Doux Parfum *see* **'TYPHOO TEA'**

Dove *see* 'AUSDOVE'

Dove Dale *see* 'MATDOVE'

'DOVEDALE', HT, rb, 1975; flowers cream, petals edged carmine, dbl., 42 petals, 5 in.; foliage dark; low, bushy growth; [Fragrant Cloud X Stella]; Moorhouse & Thornley

'DOWNLAND CHERRY', HT, mr, 1954; flowers light cerise shaded scarlet, intense, spicy fragrance; foliage dark, leathery, dull green; vigorous growth; [Vanessa X Shot Silk]; Ratcliffe

'DOWNLAND LUSTRE', HT, yb, 1955; bud bronzy gold; flowers maize-yellow, reverse orange, medium, dbl., intense, spicy fragrance; compact, bushy growth; [Vanessa X Shot Silk]; Ratcliffe

Downunder, S, lp; Peden, R., 2000; (Weatherly, L.)

'DR A. HERMANS', HT, ly, 1906; flowers yellowish white; Verschuren

'DR A. I. PETYT', HT, rb, 1924; flowers maroon-crimson shaded scarlet, large, dbl., exhibition form; vigorous, bushy growth; [George Dickson seedling X Edward Mawley]; Burrell

'DR A. J. VERHAGE', HT, dy, 1963; (Golden Wave); flowers large, dbl., 22–30 wavy petals, intense fragrance; foliage dark, glossy; vigorous, bushy growth; [Tawny Gold X (Baccará X Seedling)]; Verbeek; Carlton Rose Nurseries

'DR A. J. VERHAGE, CLIMBING', Cl HT, dy, 1968; Blaby Rose Gardens

'DR A. S. THOMAS', HT, rb, 1951; bud long, pointed; flowers dark crimson shaded darker, large, dbl., 60 petals, exhibition form, intense fragrance; foliage leathery, dark; vigorous, fairly compact growth; Clark, A.; NRS Victoria

'Dr A. Svehla', HT, rb, 1935; flowers dark carmine, very large, dbl.; bushy growth; [Col. Leclerc X Gen. MacArthur]; Böhm, J.

'Dr A. von Erlach', HT, pb, 1932; bud nankeen yellow and salmon; flowers pink and straw-yellow, stamens yellow, semi-dbl., cupped, intense fragrance; stiff stems; vigorous growth; [Prince de Bulgarie X Mrs S.K. Rindge]; Soupert & Notting

Dr Adam Christman® *see* 'WILdac'

'Dr Albert Schweitzer', HT, pb, 1961; flowers opal-pink, reverse rose-red, well-formed, dbl., 30–35 petals, 5–6 in., moderate fragrance; foliage leathery, glossy; vigorous, upright bushy growth; [Chic Parisien X Michele Meilland]; Delbard-Chabert

'Dr Andrew Carnegie', HT, pb, 1927; flowers light silvery pink, base yellowish; [Mrs Henry Morse sport]; Ferguson, R.C.; Dreer, 1930

'Dr Andry', HP, mr, 1864; flowers rosy crimson, semi-, medium, dbl., 45 petals, cupped; foliage glossy; vigorous, upright growth; Verdier, E.

'Dr Antonin Joly', HP, op, 1886; flowers salmon-pink, large, very dbl.; Besson; (Sangerhausen)

'Dr Auguste Krell', HP, 1877; Verdier, E.; (Cavriglia)

'Dr Augustin Wibbelt', HT, yb, 1928; flowers golden yellow, shaded orange, semi-dbl., moderate fragrance; [Los Angeles sport]; Leenders, M.

'Dr Augusto de Castro', HT, rb, 1954; flowers bright red, reverse yellow; [Sultane X Peace]; da Silva, Moreira

Dr B. Benacerraf *see* 'MILben'

'Dr B. P. Pal', HT, m, 1980; bud long, pointed; flowers solferino purple, blooms borne singly, dbl., 70 petals, exhibition form, slight fragrance; straight, brown prickles; foliage dark, leathery; upright growth; [Seedling X Seedling]; Division of Vegetable Crops and Floriculture

'Dr Barnardo', F, dr, 1968; flowers crimson, large blooms in trusses, dbl., 30 petals, slight fragrance; upright, bushy growth; [Vera Dalton X Red Dandy]; Harkness

'Dr Belville', Cl HT, ob, 1931; flowers orange-crimson, base yellow, open, large, semi-dbl., profuse spring bloom, then scattering, moderate fragrance; very vigorous (12 ft.) growth; [Barbara X Sunstar]; Thomas; H&S

Dr Benjamin Pal, HT, mp, 1993; IARI

Dr Bob Harvey *see* 'WINbob'

'Dr Bradas Rosa Druschki', HP, mp, 1934; flowers very large, dbl.; Brada, Dr.; (Sangerhausen)

'Dr Branscom', HT, ab, 1947; bud pointed; flowers peach-blossom flushed apricot-pink, large, dbl., moderate, spicy fragrance; upright growth; [Pink Dawn sport]; Danegger; J.T. Lovett

'Dr Bretonneau', HP, m, 1858; flowers violet-red, medium, dbl.; Trouillard; (Sangerhausen)

'Dr Briere', B, pb, 1860; Vigneron

'Dr Brownell', HT, yb, 1964; bud long, pointed; flowers buff, center chrome-yellow, dbl., 34 petals, 5.5 in., exhibition form, intense fragrance; foliage glossy, dark; vigorous, upright growth; [Helen Hayes X Peace]; Brownell, H.C.; Brownell

'Dr Burt', LCl, rb, 1942; bud long, pointed; flowers deep red to pink flushed orange, large, dbl., 45 petals, non-recurrent; foliage glossy, light; very vigorous, climbing (to 20 ft.), branching growth; [Coral Creeper X Seedling]; Brownell, H.C.

Dr Carabare, HT, m; flowers purple/pink, large, dbl., intense fragrance; (Sangerhausen)

'Dr Carbonaro', HT, pb, 1958; flowers rose, reverse silver; [Happiness X Grand'mere Jenny]; da Silva, Moreira

'Dr Carneiro Pacheco', HT, rb, 1938; flowers carmine, open, large, dbl.; foliage glossy, light; vigorous, bushy growth; [Mev. G.A. van Rossem X Sir David Davis]; da Silva, Moreira

'Dr Cathrall', HT, pb, 1966; flowers deep pink, reverse lighter, 4.5–5 in., intense fragrance; foliage dark, leathery; vigorous growth; [Hector Deane sport]; Hills

'Dr Cazeneuve', HT, dr, 1899; flowers medium, dbl.; Dubreuil; (Sangerhausen)

Dr Charles T. Beaird *see* 'WAMbeaird'

'Dr D. F. Malan', HT, dr, 1960; bud pointed; flowers very dark maroon, large, dbl., 45–50 petals, moderate fragrance; upright growth; [Happiness X Mirandy]; Herholdt, J.A.; Herholdt's Nursery

Dr Darley *see* 'HARposter'

'Dr Debat', HT, pb, 1952; (Docteur F. Debat, Dr F. Debat, La Rosée); bud ovoid, pointed; flowers bright pink tinted coral, dbl., 25–30 petals, 5–6 in., exhibition form, moderate fragrance; foliage leathery, dark; vigorous, upright growth; GM, NRS, 1950; [Peace X Mrs John Laing]; Meilland, F.; C-P

'Dr Debat, Climbing', Cl HT, pb, 1955; Barni, V.; URS

Dr Dick *see* 'COCbaden'

'Dr Domingos Pereira', Cl T, pb, 1925; flowers lilac-rose, center yellow, dbl., moderate fragrance; de Magalhaes

'Dr E. M. Mills', S, yb, 1926; flowers primrose suffused pink, becoming darker, semi-dbl., 2–2.5 in., globular, early; foliage small, dark; vigorous growth; [R. hugonis X Radiance]; Van Fleet; American Rose Society

'Dr Eckener', HRg, pb, 1930; flowers coppery rose on yellow ground, aging soft pink, large, semi-dbl., cupped, repeat bloom, moderate fragrance; vigorous (5-6 ft.) growth; [Golden Emblem X Hybrid Rugosa]; Berger, V.; C-P, 1931;, Teschendorff, 1930

'Dr Edvard Benes', HT, rb, 1935; flowers red with many white streaks, very large, dbl., moderate fragrance; bushy growth; [Étoile de France sport]; Böhm, J.

'Dr Edward Deacon', HT, ob, 1926; flowers deep salmon-orange to shrimp-pink, large, dbl., globular, moderate fragrance; vigorous, bushy growth; [Mme Edouard Herriot X Gladys Holland]; Morse

'Dr Ernst Mühle', HT, pb, 1928; flowers rose-pink, with salmon-white reflex, very large, very dbl.; [Mme Edmée Metz seedling]; Mühle

Dr F. Debat *see* **'Dr Debat'**

Dr F. L. Skinner, HSpn, lp; Simonet

'Dr F. Weigand', HT, mr, 1930; flowers cherry-red, dbl., slight fragrance; [Mme Caroline Testout X Hadley]; Weigand, C.

Dr F.G. Chandler *see* **'Dickson's Red'**

Dr Faust *see* **'Faust'**

'Dr Felix Guyou', T, dy, 1901; Mari

'Dr Ferrandiz', HT, ob; flowers deep orange-red; vigorous growth; Camprubi, C.

'Dr Fleming', HT, pb, 1960; flowers soft pink, flushed crimson, well-formed, dbl., 40 petals; vigorous growth; [Queen Elizabeth X Baleares]; Dot, M.

Dr Franco Nogueira, HT; Moreira da Silva, A.; (Cavriglia)

'Dr G. Krüger', HT, dr, 1913; flowers crimson; [Mme Victor Verdier X Seedling]; Ulbrich; Kiese

'Dr Gallwey', LCl, w, 1937; flowers snow-white, borne in large clusters, single, 2 in., profuse bloom; very vigorous growth; Reiter

Dr Gentil, HT; Moreira da Silva, A.; (Cavriglia)

'Dr Georges Leger', HT, mr, 1935; flowers blood red, large, dbl., slight fragrance; Ketten, Gebrüder; (Sangerhausen)

'Dr Georges Martin', HP, pb, 1908; flowers carmine-pink, very large, very dbl.; Vilin; (Sangerhausen)

Dr Goldberg, HT, 1988; Gandy, Douglas L.; (Cavriglia)

'Dr Grandvilliers', T, pb, 1893; flowers yellowish-pink with darker pink, medium, dbl.; Perny; (Sangerhausen)

'Dr Grill', T, op, 1886; flowers rose shaded coppery, moderate fragrance; [Ophirie X Souv. de Victor Hugo]; Bonnaire

'Dr Guarnero', HT, dp, 1958; flowers deep rose, well formed, large, dbl., intense fragrance; vigorous, bushy growth; [Happiness X Grand'mere Jenny]; da Silva, Moreira

'Dr Guilherme Pereira da Rosa', HT, mr, 1955; flowers cherry-red, well formed; moderate growth; [Charles Mallerin X Lisboa]; da Silva, Moreira

'Dr H. E. Rumble', Min, mr, 1981; flowers scarlet red, small, dbl., 30 petals, cupped, no fragrance; small, brown prickles; foliage large, light green; vigorous, upright growth; [Born Free X Westmont]; Hooper, John C.

'Dr H. I. Gallagher', HT, yb, 1989; bud ovoid; flowers yellow with bright pink on petal tips, reverse same, dbl., 35–50 petals, exhibition form, intense, spicy fragrance; prickles medium, bronze-green; foliage medium, bronze-medium green, semi-glossy heavy, dise; bushy, medium to tall growth; Bronze, ARC TG, 1989; [Spellbinder X Irish Gold]; Anderson, Mrs. Etta S.

'Dr Harry Upshall', S, pb, 1993; dbl., 26–40 petals, 1.5–2.75 in., moderate fragrance; foliage small, medium green, matt; bushy, spreading (2 m) growth; [Liverpool Echo X R. foetida persiana]; Fleming, Joyce L.; Hortico Roses, 1993

'Dr Heinrich Lumpe', HT, pb, 1928; flowers light rose-pink, base yellow, large, dbl., exhibition form, moderate fragrance; strong stems; very vigorous, bushy growth; [Constance X Admiral Ward]; Berger, V.; A. Berger

'Dr Helfferich', HT, pb, 1919; flowers rose, center yellowish orange, edged silvery, dbl., moderate fragrance; [Gustav Grunerwald X Mrs Aaron Ward]; Lambert, P.

'Dr Henri Neuprez', HWich, ly, 1913; flowers canary-yellow to sulfur-white; [R. wichurana X Mme Barthelémy Levet]; Tanne

Dr Herbert Gray *see* 'AUSfar'

'Dr Herbert Hawkesworth', HT, dr, 1927; flowers deep crimson, center almost black, moderate fragrance; Bees

'Dr Hess von Wichdorf', HT, rb, 1936; flowers red, shaded rose-lilac, large, dbl., exhibition form, intense fragrance; vigorous, bushy growth; [Frank W. Dunlop sport]; Vogel, M.; Heinemann

'Dr Homi Bhabha', HT, w, 1968; bud long, pointed; flowers white, center somteimes tinted cream, large, very dbl., exhibition form, slight fragrance; foliage leathery; vigorous, upright growth; [Virgo X Seedling]; Pal, Dr. B.P.; Indian Agric. Research Inst.

'Dr Homi Bhabha, Climbing', Cl HT, w, 1976; IARI

'Dr Hoog', HP, 1880; Laxton; (Cavriglia)

'Dr Huey', LCl, dr, 1914; (Shafter); flowers crimson-maroon, anthers light yellow, blooms in clusters of, semi-dbl., 15 petals, 2 in., slight fragrance; foliage rich green; Gertrude M. Hubbard, ARS, 1924; [Ethel X Gruss an Teplitz]; Thomas; A.N. Pierson, 1920;, B&A, 1920

'Dr Ing. H. Blohm', HP, dr, 1919; flowers dark carmine-red, large, dbl., intense fragrance; Lambert, P.; (Sangerhausen)

'Dr J. G. Fraser', HT, ab, 1926; flowers salmon-apricot, suffused vermilion-pink; vigorous growth; [St. Helena X Muriel Dickson]; Easlea

'Dr J. H. Nicolas', LCl, mp, 1940; flowers rose-pink, blooms in clusters of 3 or 4, dbl., 50 petals, 5 in., globular, recurrent bloom, moderate fragrance; foliage dark, leathery; vigorous, pillar (8 ft.) growth; [Charles P. Kilham X Georg Arends]; Nicolas; J&P

Dr Jack Bender *see* 'BRIben'

'Dr Jaime Lopes Dias', HT, mp, 1961; flowers large; [Confidence X Juno]; da Silva, Moreira

'Dr John Snow', HT, w, 1979; flowers creamy white, dbl., 35 petals, 5 in., exhibition form, moderate fragrance; foliage light green; tall growth; [Helen Traubel X Seedling]; Gandy, Douglas L.

'Dr Joseph Drew', HT, yb, 1918; flowers salmon-yellow, suffused pink, dbl., moderate fragrance; [Mme Mélanie Soupert seedling X Comtesse Icy Hardegg]; Page; Easlea

Dr Jules Bouche, LCl, w

Dr K. C. Chan *see* 'TINchan'

'Dr Karel Kramár', HT, dr, 1937; flowers large, dbl., moderate fragrance; Böhm, J.; (Sangerhausen)

'Dr Kater', Pol, dr

'Dr Kidwai', HT, pb, 1999; flowers light pink, edges blended magenta, reverse light pink, dbl., 26–40 petals, 3–4 in., borne mostly singly, slight fragrance; prickles moderate; foliage medium, medium green, semi-glossy; upright, tall (4-5 ft) growth; [Paradise X (Paradise X Oklahoma)]; Chiplunkar, C. R.; KSG Roses, 1998

'Dr Kirk', HT, op, 1940; bud long, pointed; flowers coral, shaded nasturtium-yellow, very large, dbl., 35 petals, exhibition form, slight fragrance; vigorous growth; [Charles P. Kilham X R. foetida bicolor hybrid]; Mallerin, C.; A. Meilland;, C-P

'Dr Lande', T, 1902; flowers large, dbl., moderate fragrance; Berger, V.; (Sangerhausen)

Dr Lopez Diaz, HT; Moreira da Silva, A.; (Cavriglia)

'Dr M. Euwe', HT, pb, 1936; bud pointed; flowers salmon tinted yellow and pink, dbl., intense fragrance; foliage leathery, bronze; bushy growth; Buisman, G. A. H.

Dr M.S. Randhawa, HT, pb, 1989; Pal, Dr. B.P.

'Dr Manuel Alves de Castro', HT, rb; flowers red, reverse golden yellow; da Silva, Moreira

'Dr Margaretha', F, dr, 1960; flowers velvety dark red, medium, in large clusters, dbl.; vigorous, bushy growth; [Red Pinocchio seedling X Alain]; Maarse, G.

'Dr Marx', HP, pb, 1842; flowers crimson red; Laffay, M.

'Dr Maximo de Carvalho', HT, dr, 1960; flowers crimson-red; [Crimson Glory X Charles Mallerin]; da Silva, Moreira

'Dr Mazaryk', HP, lp, 1930; flowers large, dbl., slight fragrance; Böhm, J.; (Sangerhausen)

Dr McAlpine *see* 'PEAfirst'

'Dr Mendes Correia', HT, mr, 1938; bud pointed; flowers bright red, very large, dbl., intense fragrance; foliage soft; vigorous, bushy growth; [Frau Margarete Oppenheim X Hortulanus Budde]; da Silva, Moreira

'Dr Mengelberg', F, dr, 1952; flowers deep blood-red, large, semi-dbl.; very vigorous growth; Leenders, M.

'Dr Merkeley', (similar to R. spinosissima), dp, 1924; flowers deep pink, dbl., non-recurrent, moderate fragrance; low to medium growth; (discovered in eastern Siberia; named in honor of Dr. Merkeley, who first grew it in Canada)

'Dr Miroslav Tyrs', HP, rb, 1932; flowers crimson, shaded darker, very large, moderate fragrance; [Anna de Diesbach sport]; Böhm, J.

Dr Morse, HT, m

'Dr Müller', HFt, op, 1905; flowers salmon-pink with touches of red, medium, semi-dbl.; Müller, Dr. F.; (Sangerhausen)

'Dr Müller's Rote', HT, m, 1920; flowers purple/pink, medium, dbl., intense fragrance; Müller, Dr. F.; (Sangerhausen)

'Dr Nicolas Welter', HT, op, 1912; flowers large, dbl., moderate fragrance; Soupert & Notting; (Sangerhausen)

'Dr Noshir Wadia', HT, rb, 1999; flowers bright red, white stripes, reverse light red with stripes, dbl., 26–40 petals, 3–4 in., borne mostly singly, slight fragrance; many prickles; foliage large, dark green, glossy; compact, medium (4-5 ft) growth; [Norma sport]; Chiplunkar; KSG Roses, 1992

'DR O'DONEL BROWNE', HT, pb, 1908; flowers carmine-rose, large, well formed, dbl., intense fragrance; vigorous growth; Dickson, A.

'DR OLIVEIRA SALAZAR', HT, ob, 1955; flowers salmon and yellow shaded carmine, large, very dbl., intense fragrance; [Mme Marie Curie X Peace]; da Silva, Moreira

Dr Paul Menzel, HT, mp, 1980; flowers carmine-pink, large, dbl., moderate fragrance; Lucke, G.; (Sangerhausen)

Dr R. Maag *see* 'MEIRIGALU'

'DR RAFAEL DUQUE', HT, rb, 1938; flowers velvety purplish red, large, moderate fragrance; [Frau Margarete Oppenheim X Hortulanus Budde]; da Silva, Moreira

Dr Reiner Klimke, F, lp, 1988; Noack, Werner

'DR RENATA TYRSOVÁ', LCl, op, 1937; flowers salmon-pink, medium, semi-dbl.; Böhm, J.; (Sangerhausen)

'DR REYMONT', HMult, w, 1907; flowers large, dbl.; Mermet; (Sangerhausen)

'DR RICAUD', Pol, lp, 1907; flowers small, dbl., moderate fragrance; Corboeuf; (Sangerhausen)

'DR RICHARD LEGLER', HT, pb; flowers shrimp-pink changing to old-rose and orange, moderate, fruity fragrance; moderate growth

'DR ROUGES', Cl T, rb; flowers large, dbl., moderate fragrance; Schwartz, 1893; (Sangerhausen)

Dr S. S. Bhatnagar, F, dr, 1994; IARI

'DR SCHEINER', HT, dr, 1929; flowers large, dbl., moderate fragrance; Böhm, J.; (Sangerhausen)

'DR SCOTT', (strain of R. multiflora), w; mildew-resistant; used as understock

Dr Selma Lagerlof, HRg, mp

Dr Sybil Johnson *see* 'BATSYBIL'

Dr Tomin, HT, 1990; Tagashira, Kazuso; (Sangerhausen)

'DR TRIGO DE NEGREIROS', HT, dr, 1954; bud long, pointed; flowers deep red, large; very vigorous growth; [Charles Mallerin X Lisboa]; da Silva, Moreira

'DR VALOIS', HT, 1950; Mallerin, C.; (Cavriglia)

'DR VAN DE PLASSCHE', Pol, mp, 1968; bud ovoid; flowers pink, medium, semi-dbl.; foliage dark; [Heureux Anniversaire X Allotria]; Buisman, G. A. H.

'DR VAN RIJN', HT, my, 1952; bud ovoid; flowers lemon-yellow, large, dbl., exhibition form, moderate fragrance; foliage light green, glossy; vigorous, bushy growth; Leenders, M.

'DR VAZQUEZ', HT, op, 1935; flowers salmon, open, medium, semi-dbl.; foliage glossy; upright growth; [Duchess of Atholl X Margaret McGredy]; Camprubi, C.

'DR W. E. HADDEN', HT, rb, 1934; flowers raspberry-red, flushed yellow, deepening at base, well formed; foliage dark; long, strong stems; vigorous growth; McGredy

'DR W. VAN FLEET', LCl, lp, 1910; bud pointed; flowers cameo-pink fading flesh-white, large, dbl., non-recurrent, moderate fragrance; foliage dark, glossy; vigorous, climbing (15-20 ft.) growth; (21); [(R. wichurana X Safrano) X Souv. du Prés. Carnot]; Van Fleet; P. Henderson

Dr Wolfgang Pöschl *see* **'CANADIAN WHITE STAR'**®

'DR ZAMENHOF', HWich, rb, 1935; flowers crimson-red, base yellow, very large, borne in clusters, intense fragrance; very vigorous growth; [R. wichurana X Seedling]; Brada, Dr.; Böhm

Dr Zumel, HT; Kordes, R.; (Cavriglia)

Dr. Robert Korns

DRAdiwu, S, w, 1996; (**Anita Pereire**); Orard

Dragon's Eye *see* 'CLEDRAG'

Dragon's Fire™ *see* 'MINIFIRE'

'DRAMBUIE', HT, rb, 1973; flowers orange-red, reverse red, full, high pointed, dbl., 28–30 petals, 5 in., intense fragrance; foliage glossy; vigorous, bushy growth; [Whisky Mac sport]; Anderson's Rose Nurseries

'DREAM', HT, op, 1938; flowers geranium-pink, dbl., 50–60 petals, 4.5 in., moderate fragrance; strong stems; [Better Times sport]; Dramm

Dream *see* KORmiller

Dream Baby *see* 'RENBABY'

Dream Blush *see* 'TWOAT'

Dream Boat *see* **'DREAMBOAT'**

Dream Cloud *see* 'AROPICLU'

Dream Dolly *see* JACpin

'DREAM DUST', F, mp, 1969; bud pointed; flowers small, borne in clusters, dbl., slight fragrance; foliage leathery; very vigorous, spreading growth; [Lavender Girl seedling X Little Darling seedling]; Gardner, B.C.

'DREAM GIRL', LCl, pb, 1944; (Dreamgirl); flowers salmon-pink overlaid apricot, dbl., 55–65 petals, 3.5 in., recurrent bloom, intense fragrance; foliage glossy; good pillar rose growth; (28); [Dr. W. Van Fleet X Senora Gari]; Jacobus; B&A

Dream Lover *see* 'RENLOVER'

Dream Lover *see* PEAyetti

Dream Orange *see* 'TWOAEBI'

'DREAM PARADE', HT, op, 1938; flowers amber in spring, seashell-pink in hot weather, burnt-orange, dbl., moderate fragrance; vigorous growth; [Condesa de Sástago sport]; Hillock

Dream Pink *see* 'TWOJOAN'

Dream Pink, HT, mp; Twomey, Jerry, 1999

Dream Red *see* 'TWOPAUL'

'DREAM TIME', HT, mp, 1977; (Dreamtime); dbl., 38 petals, 5 in., exhibition form, intense fragrance; foliage light green; moderately vigorous growth; [Kordes' Perfecta X Prima Ballerina]; Bees

'DREAM WALTZ', F, dr, 1969; flowers large blooms in trusses, dbl.; foliage glossy; Tantau, Math.

Dream Weaver™ *see* 'JACPICL'

Dream Yellow *see* 'TWOYEL'

'DREAMBOAT', Min, my, 1981; (Dream Boat); flowers medium, dbl., 60–70 petals, exhibition form, slight fragrance; bushy, spreading growth; [Rise 'n' Shine X Grand Opera]; Jolly, Betty J.; Rosehill Farm, 1982

Dreamboat, HT, lp, 1995; Dawson

Dreamcatcher *see* 'SEACATCH'

Dreamcoat *see* 'SEACOAT'

Dreamer™ *see* 'SAVADREAM'

Dreamgirl *see* **'DREAM GIRL'**

'DREAMGLO', Min, rb, 1978; bud long, pointed; flowers white, tipped and blended red, dbl., 50 petals, 1 in., exhibition form, slight fragrance; foliage small, dark; upright growth; [Little Darling X Little Chief]; Williams, Ernest D.; Mini-Roses

Dreaming *see* 'REIKOR'

Dreaming Parade, Min, op, 1995; Poulsen

'DREAMING SPIRES', LCl, dy, 1973; flowers bright golden yellow, dbl., 25 petals, 3 in., exhibition form, repeat bloom, intense fragrance; foliage dark; GM, Belfast, 1977; [Buccaneer X Arthur Bell]; Mattock

Dreamland *see* **'RESLAND'**

Dreamland *see* **'TRAUMLAND'**

Dreamrider *see* 'MINRID'

Dreams Come True *see* 'MEIVESTAL'

Dreamsicle *see* 'TALDRE'

Dreamtime *see* **'DREAM TIME'**

Dreamward *see* 'BURAPWARD'

Dreamy *see* 'TINDREAM'

Drei Gleichen, HFt, ly; flowers medium, semi-dbl., slight fragrance; (Sangerhausen)

Dreienbrunnen, Pol, mr, 1957; flowers small, dbl.; Berger, W.; (Sangerhausen)

'DRESDEN', HT, w, 1961; (Mathé Altéry); bud ovoid; flowers white lightly suffused pink, dbl., 60 petals, 4–5 in., exhibition form, intense fragrance; foliage leathery, dark; very vigorous, upright growth; [Ophelia X Cathrine Kordes]; Robichon; Ilgenfritz Nursery

'DRESDEN DOLL', Min, lp, 1975; bud mossy; flowers soft pink, mini-moss, small, semi-dbl., 18 petals, 1.5 in., cupped, moderate fragrance; foliage glossy, leathery; low, bushy, compact growth; [Fairy Moss X Moss seedling]; Moore, Ralph S.; Sequoia Nursery

Dresdner Gelbe No. 79311, F, dy; flowers medium-large, semi-dbl., slight fragrance; VEG; (Sangerhausen)

Dresselhuys *see* 'MACJOCEL'

'DRIES VERSCHUREN', HT, my, 1961; (Blaby Jubilee); bud pointed; flowers buttercup-yellow, well formed, large, dbl., 25 petals, moderate fragrance; foliage glossy, bronze; vigorous, upright, bushy growth; [Golden Rapture X Seedling]; Verschuren, A.; Blaby Rose Gardens

'DRIFTER'S ESCAPE', F, or, 1971; flowers vermilion, semi-dbl., 10 petals, 3–3.5 in.; foliage glossy, reddish when young; [Orangeade X Orange Sensation]; Greenway

'DRISCOBRUCE', HT, mp, 1990; (**Dave Hessayon**); dbl., 26–40 petals, moderate fragrance; foliage medium, medium green, glossy; bushy growth; [Silver Jubilee X Pink Favorite]; Driscoll, W.E.; Rosemary Roses, 1989

'DRISCOGEORGE', HT, mr, 1992; (**David Charles Armstrong**); flowers coral red, blooms borne mostly singly, dbl., 26–40 petals, 3–3.5 in., slight fragrance; some prickles; foliage medium, medium green, semi-glossy; tall (106 cms), upright growth; [Silver Jubilee X Fragrant Cloud]; Driscoll, W.E., 1993

'DRISCONUN', F, ly, 1988; (**Maud Nunn**); bud pointed; flowers creamy yellow, aging paler lemon yellow, loose, small, semi-dbl., 16 petals, exhibition form, slight fragrance; prickles green; foliage medium green, semi-glossy; bushy, medium growth; [Rise 'n' Shine X Rise 'n' Shine]; Driscoll, W.E.

'DRISCORED', Min, op, 1999; (**Just William**); flowers salmon pink, reverse pale rose, dbl., 26–40 petals, 2.5 in., no fragrance; few prickles; foliage medium, light green, semi-glossy; upright, tall (40 in.) growth; [Perestroika X Pink Petticoat]; Driscoll, W.E.

'DRISCOROY', HT, dr, 1996; (**Menaka Durga Roy**); flowers deep red, blooms borne mostly single, dbl., 26–40 petals, moderate fragrance; few prickles; foliage medium, medium green, lighter underside, semi-glossy; upright, tall (3 ft) growth; [Silver Jubilee X [Red Planet X Blessings) X (Parkdirector Riggers X Honey Favorite)]]; Driscoll, W.E.

'DRONNING ALEXANDRINE', HT, dp, 1926; flowers deep pink, urn-shaped, medium, dbl., no fragrance; foliage medium green, semi-glossy; medium, bushy growth; Poulsen, S.; Poulsen

Dronning Margrethe *see* 'POULSKOV'

'DROPMORE YELLOW', HFt, my; [R. foetida X R. spinosissima altaica]; Skinner

'DROUJBA', F, yb, 1975; flowers yellow, shaded red, small blooms in clusters of 5-30 per stem, dbl., 42 petals, cupped, moderate, tea fragrance; foliage dark, glossy; bushy growth; [Masquerade X Rumba]; Staikov, Prof. Dr. V.; Kalaydjiev and Chorbadjiiski

'DRU', Min, dy, 1985; flowers small, dbl., 35 petals, exhibition form, slight fragrance; foliage medium, medium green, semi-glossy; upright, bushy growth; Hunt, W. Henry, 1986

'DRUMMER BOY', F, mp, 1965; bud ovoid; flowers soft carmine-rose, small, borne in large clusters, dbl., slight fragrance; foliage leathery; vigorous, tall growth; [Pinocchio X Queen Elizabeth]; Lammerts, Dr. Walter; Germain's

Drummer Boy *see* 'HARVACITY'

'DRUSCHKA', HP, dp, 1932; flowers carmine-pink, large, dbl., slight fragrance; Kordes, W.; (Sangerhausen)

'DRUSCHKI RUBRA', HP, mr; Lambert, P., 1929

'DUARTE DE OLIVERA', N, op, 1879; Brassac

'DUBLIN', HT, mr, 1982; flowers large, smoky red, dbl., 35–40 petals, 5 in., exhibition form, borne mostly singly, intense, raspberry fragrance; foliage large, medium green, matt; long, straight stems; upright growth; [(Seedling X Mister Lincoln) X Ann Letts]; Perry, Astor; Perry Roses, 1983

Dublin Bay® *see* 'MACDUB'

'DUBONNET', Pol, dr, 1958; bud small; flowers cardinal-red, dbl., slight fragrance; foliage leathery; vigorous growth; [Stoplite sport]; Jelly; E.G. Hill Co.

'DUC D'ANJOU', HP, dr, 1862; flowers deep red; Boyau

Duc d'Aremberg, HGal, m

'DUC D'AUDIFFRET-PAQUIER', HP, 1887; Verdier, E.; (Cavriglia)

Duc de Angoulême *see* **'DUCHESSE D'ANGOULÊME'**

'DUC DE BORDEAUX', HGal, pb, 1820; flowers rosy lilac, large, dbl.; Vibert

Duc de Brabant, C, mr; flowers medium, dbl.; (Sangerhausen)

'DUC DE CAMBRIDGE', D, m; flowers deep purplish rose, large, dbl.; foliage dark, edged reddish brown when young

'DUC DE CAZES', HP, m, 1861; flowers velvety purple, dbl., cupped, moderate fragrance; [Général Jacqueminot X Seedling]; Touvais

'DUC DE CONSTANTINE', Ayr, mp, 1857; Soupert & Notting

'DUC DE CRILLON', B, mr, 1858; Moreau et Robert

'DUC DE FITZJAMES', HGal, dr; flowers very dark crimson, shaded purple

'DUC DE GUICHE', HGal, m, 1835; flowers light reddish-violet, large, dbl.; Prévost, before 1810

'DUC DE MAGENTA', T, ab; flowers flesh shaded fawn, large, dbl.; Margottin, 1859

'DUC DE MARLBOROUGH', HP, dr; Lévêque, 1884

'DUC DE MONTPENSIER', HP, dr, 1875; flowers large, dbl., moderate fragrance; Lévêque; (Sangerhausen)

Duc de Rohan *see* **'DUCHESSE DE ROHAN'**

'DUC DE SUSSEX', D, lp, 1856; flowers yellowish-white with pink, large, dbl.; Roseraie de l'Hay; (Sangerhausen)

'DUC DE VALMY', HGal, m; flowers light purplish rose, marbled purple, large, dbl., cupped

Duc de Wellington *see* **'DUKE OF WELLINGTON'**

'DUC D'HARCOURT', HP, mr, 1863; flowers large, dbl.; Moreau et Robert; (Sangerhausen)

'DUC D'ORLÉANS', HGal, m, 1830; flowers purple/red, medium, dbl.; Vibert, before 1848; (Sangerhausen)

Duc Meillandina *see* 'MEIPINJID'

'DUCHER', Ch, w, 1869; flowers pure white, small to medium, dbl., flat; vigorous growth; Ducher

'DUCHESS', HT, mp, 1976; bud ovoid; flowers cameo-pink shaded deeper, dbl., 4–4.5 in., exhibition form, moderate fragrance; foliage glossy, leathery; bushy, upright growth; [White Satin X Seedling]; Van Veen; Carlton Rose Nurseries

Duchess, HT, lp; greenhouse variety

'DUCHESS OF ABERCORN', HT, pb, 1919; flowers creamy white edged bright rose, dbl.; Dickson, H.

'DUCHESS OF ALBANY', HT, dp, 1888; (Red La France); flowers deep pink; [La France sport]; Paul, W.

'DUCHESS OF ATHOLL', HT, ob, 1928; flowers vivid orange, flushed old-rose, large, dbl., cupped, intense fragrance; foliage bronze, leathery; vigorous growth; Dobbie

'DUCHESS OF ATHOLL, CLIMBING', Cl HT, ob, 1933; Howard Rose Co.

'DUCHESS OF BEDFORD', HP, mr, 1879; flowers bright medium red, large, dbl., globular; Postans, R.B.; W. Paul & Sons

'DUCHESS OF CONNAUGHT', HT, pb, 1879; flowers deep silvery pink, large, globular, intense fragrance; dwarf growth; [Adam X Duchesse de Vallombrosa]; Bennett

'DUCHESS OF EDINBURGH', T, dr, 1874; (Prince Wasiltchikoff); flowers crimson, becoming lighter, large, dbl.; moderate growth; [Souv. de David d'Angers seedling]; Nabonnand, G.; Veitch

'DUCHESS OF FIFE', HP, lp, 1893; flowers large, dbl., intense fragrance; Cocker; (Sangerhausen)

'DUCHESS OF KENT', F, dp, 1968; flowers rose-neyron-red, blooms in trusses, cupped; low, bushy growth; [Katharine Worsley sport]; Waterhouse, W.P.; Waterhouse Nursery

Duchess of Kent *see* **'HOMÈRE'**

'DUCHESS OF MARLBOROUGH', HT, pb, 1922; flowers brilliant lilac-rose, reverse carmine-crimson, dbl.; [Jonkheer J.L. Mock seedling X Beauté de Lyon]; Nabonnand, P.

'DUCHESS OF MONTROSE', HT, rb, 1929; flowers vermilion-crimson, large; Dobbie

'DUCHESS OF NORMANDY', HT, pb, 1912; flowers soft salmon-flesh, overlaid yellow, large, dbl., exhibition form; vigorous, branching growth; [Dean Hole sport]; Le Cornu

'DUCHESS OF PADUCAH', HT, w, 1971; bud ovoid; flowers white, edge flushed red, large, dbl., exhibition form, intense fragrance; foliage large, glossy, dark, leathery; vigorous, upright growth; [Kordes' Perfecta X Peace]; Williams, J. Benjamin

'DUCHESS OF PORTLAND', P, mr; (Duchesse de Portland, Portland Rose); flowers bright scarlet

'DUCHESS OF RUTLAND', F, dp, 1956; flowers rich carmine-pink, small, borne on large truss, semi-dbl.; vigorous growth; deRuiter; Gandy Roses, Ltd.

'DUCHESS OF SUTHERLAND', HP, lp, 1839; flowers rosy pink, large, dbl., moderate fragrance; vigorous growth; Laffay, M.

'DUCHESS OF SUTHERLAND', HT, pb, 1912; flowers rose-pink shaded lemon on white base, large, dbl., exhibition form, moderate, sweetbriar fragrance; foliage glossy, olive-green; vigorous growth; Dickson, A.

'DUCHESS OF WELLINGTON', HT, ly, 1909; bud pointed; flowers buff-yellow, deeper toward center, open, large, semi-dbl., 17 petals, moderate fragrance; foliage leathery; long, strong stems; bushy growth; Dickson, A.

'DUCHESS OF WELLINGTON, CLIMBING', Cl HT, ly, 1924; Howard Rose Co.

'DUCHESS OF WESTMINSTER', HT, pb, 1879; flowers pink shaded carmine, large, dbl., slight fragrance; moderate growth; [Adam X Marquise de Castellane]; Bennett

'DUCHESS OF WESTMINSTER', HT, mp, 1911; flowers clear rose-pink, dbl., moderate fragrance; Dickson, A.

Duchess of Windsor *see* **'PERMANENT WAVE'**

Duchess of York *see* 'DICRACER'

'DUCHESS OF YORK', HT, yb, 1925; flowers deep golden yellow, center tangerine, very well formed, large, dbl.; vigorous growth; Dickson, S.

'DUCHESSE D'ABRANTES', M, lp; flowers light pink with dark pink shading, large, dbl.; Robert, 1851; (Sangerhausen)

'DUCHESSE D'ALBE', T, yb, 1903; flowers yellowish salmon, shaded coppery purple rose, base golden yellow, dbl., globular; Lévêque

'DUCHESSE D'ANGOULÊME', HGal, lp; (Duc de Angoulême, Wax Rose); flowers blush, center pink, medium, dbl., cupped; moderate, upright growth; [Probably Gallica X Centifolia hybrid]; Vibert, 1821

Duchesse d'Anjou *see* 'GODZOTY'

'DUCHESSE D'AOSTE', HP, mr, 1867; flowers large, very dbl., moderate fragrance; Margottin; (Sangerhausen)

'DUCHESSE D'ASSUMA', HP, 1877; Jamain, H.; (Cavriglia)

'DUCHESSE D'AUERSTÄDT', N, my, 1888; flowers golden yellow; [Reve d'Or sport]; Bernaix, A.

'DUCHESSE DE BERRY', HGal, pb; Vibert, 1820

'DUCHESSE DE BRABANT', T, lp, 1857; (Comtesse de Labarathe, Comtesse Ouwaroff); flowers soft rosy pink, large, dbl., 45 petals, cupped, intense fragrance; vigorous, spreading growth; Bernède

'DUCHESSE DE BRAGANCE', HP, mp, 1886; flowers very large, dbl.; Verdier; (Sangerhausen)

'DUCHESSE DE BUCCLEUGH', HGal, rb; flowers lively crimson, edges tinged lavender, large, dbl., cupped; vigorous growth; Vibert, 1837

'DUCHESSE DE CAMBACÉRÈS', HP, m, 1854; flowers lilac-rose, dbl.; vigorous growth; Fontaine

'DUCHESSE DE CAYLUS', HP, dp, 1864; flowers brilliant carmine-pink, well formed, large, globular, moderate fragrance; moderate growth; [Alfred Colomb seedling]; Verdier, C.

'DUCHESSE DE DINO', HP, dr, 1889; flowers very large, dbl.; Lévêque; (Sangerhausen)

'DUCHESSE DE GALLIERA', HP, pb; flowers bright rose shaded flesh, large, dbl., cupped; Laffay, M., 1847; 1847

'DUCHESSE DE GRAMMONT', N, w; flowers flesh

Duchesse de Guermantes, M, lp; Morley, Dr B., 1988; (Weatherly, L.)

'DUCHESSE DE LA MOTHE-HOUDANCOURT', HT, dp, 1907; flowers large, dbl., moderate fragrance; Toussaint Mille Fils; (Sangerhausen)

Duchesse de la Tremoille, M, lp; Morley, Dr B., 1988; (Weatherly, L.)

'DUCHESSE DE MONTEBELLO', HGal, lp; flowers rosy pink, changing to flesh-pink, medium, dbl., moderate fragrance; erect, compact growth; Laffay, M., 1824 or 1825

'DUCHESSE DE MORNY', HP, mp, 1863; flowers large, dbl.; Verdier, E.; (Sangerhausen)

Duchesse de Portland *see* **'DUCHESS OF PORTLAND'**

'DUCHESSE DE ROHAN', HP, pb; (Duc de Rohan); flowers rosy crimson, margined with lilac, compact, large, dbl.; (France; possibly Lévêque), before 1880

Duchesse de Savoie® *see* LAPbel

'DUCHESSE DE TALLEYRAND', HT, yb, 1944; bud pointed; flowers egg-yolk-yellow to chrome-yellow, dbl., moderate fragrance; well branched growth; [Mme Joseph Perraud X Fred Edmunds]; Meilland, F.

'DUCHESSE DE THURINGE', B, w, 1847; flowers white tinted lilac; Guillot Père

'DUCHESSE DE VALLOMBROSA', HP, ab; flowers flesh shaded rose, large, dbl.; [Jules Margottin seedling]; Schwartz, J., 1875

'DUCHESSE DE VENDOME', HT, rb, 1924; flowers crimson, coppery reflexes, reverse yellow, dbl., moderate fragrance; [Souv. de Gilbert Nabonnand X Juliet]; Nabonnand, P.

'DUCHESSE DE VERNEUIL', M, pb, 1856; bud heavily mossed; flowers flesh-pink deepening to salmon-pink, camellia form; Portemer fils

'DUCHESSE D'ISTRIE', M, lp; Laffay, M., 1855

'DUCHESSE D'ORLEANS', HP, 1851; Quétier; (Cavriglia)

'DUCHESSE D'ORLEANS', HGal, mp, 1821; Laffay, M.

'DUCHESSE MARIE SALVIATI', T, ob; flowers orange-yellow tinted pink; [Mme Lombard X Mme Maurice Kuppenheim]; Soupert & Notting, 1889

DUCmar, Min, op; (**Marquisette®**); Ducher

Dudley Cross *see* **'MRS DUDLEY CROSS'**

'DUET', HT, mp, 1960; bud ovoid; flowers light pink, reverse dark pink, ruffled, dbl., 25–30 petals, 4 in., exhibition form, borne singly and in small clusters, slight, tea fragrance; foliage leathery, dark green, glossy; vigorous, upright growth; AARS, 1961 GM, Baden-Baden, 1959; [Fandango X Roundelay]; Swim, H.C.; Armstrong Nursery

'DUET SUPREME', HT, pb, 1990; bud rounded; flowers light to medium pink blend, medium pink reverse, urn-shaped, dbl., 35 petals, slight fragrance; foliage medium, dark green, semi-glossy; [Duet sport]; Patterson, William, 1988; Roses Unlimited, 1990

Duett®, F, or, 1982; Noack, Werner

'DUFFEY'S DELIGHT', HT, dr, 1976; bud long, pointed; flowers velvety dark red, very dbl., 60 petals, 5–6 in., exhibition form, slight fragrance; foliage dark; very vigorous, upright growth; [Norman Hartnell sport]; Duffey

'DUFTBELLA'®, F, dr, 1973; bud ovoid; flowers dark velvet red, center lighter, large, dbl., intense fragrance; vigorous, bushy growth; [Fragrant Cloud X (Monique X Mardi Gras)]; Hetzel; GAWA

Duftes Berlin®, HT, ob, 1988; Cocker

Duftgold *see* 'TANDUGOFT'

Duftparadies, F, dr, 1966

Duftrausch® *see* 'TANSCHAUBUD'

Duftrausch *see* 'TANRAUSCH'

'DUFTSTAR', HT, dr, 1974; bud long, pointed; dbl., 24 petals, 4 in., exhibition form, intense fragrance; foliage dark, soft; vigorous, upright growth; [Seedling X Papa Meilland]; Kordes; Dehner & Co.

Duftstern®, HT, op, 1973; Noack, Werner

Duftwolke *see* 'TANELLIS'

'DUFTWUNDER', F, ob, 1972; flowers yellowish orange, large, very dbl., intense fragrance; moderate, upright growth; [Fragrant Cloud X Goldmarie]; Hetzel

Duftzauber *see* 'KORDU'

Duftzauber '84 *see* 'KORZAUN'

'DUHAMEL DUMONCEAU', HP, mr, 1872; flowers large, dbl., moderate fragrance; Villin-Hugues; (Sangerhausen)

'DUILIU ZAMFIRESA', HT, mr, 1938; flowers large, dbl., intense fragrance; Palocsay, R.; (Sangerhausen)

'DUISBURG', HT, dp, 1908; flowers carmine-pink, large, dbl., moderate fragrance; Hinner, W.; (Sangerhausen)

'DUKAT', LCl, my, 1955; flowers golden yellow, large, dbl., moderate fragrance; foliage glossy, leathery; vigorous (10-16 ft), upright growth; [Mrs Pierre S. duPont X Golden Glow]; Tantau, Math.

Dukat, HT, lp; Brabec, 1980; (Czech Rosa Club)

Duke Meillandina *see* 'MEIPINJID'

Duke of Argyll, HSpn, pb

'DUKE OF CONNAUGHT', HP, rb, 1875; flowers dark velvety crimson flushed brighter, large, dbl., intense fragrance; vigorous growth; Paul

'DUKE OF CONNAUGHT', HT, mr, 1879; flowers rose-crimson, very large, dbl.; moderate growth; [Adam X Louis van Houtte]; Bennett

'DUKE OF EDINBURGH', HP, dr, 1868; flowers deep red, large, dbl., moderate fragrance; vigorous, erect growth; [Général Jacqueminot X Seedling]; Paul

'DUKE OF NORMANDY', HT, lp, 1921; flowers silvery pink; [St. Helena X George Dickson]; Jersey Nursery

'DUKE OF PADUCAH', HT, rb, 1970; bud ovoid; flowers dark velvety crimson, full, large, dbl., slight fragrance; foliage large, glossy, dark, leathery; vigorous, bushy growth; [Grand Gala X Josephine Bruce]; Williams, J. Benjamin

'DUKE OF TECK', HP, dp, 1880; flowers deep pink, dbl., 40 petals, globular, moderate fragrance; vigorous growth; [Duke of Edinburgh X Seedling]; Paul

'DUKE OF WELLINGTON', HP, dr, 1864; (Duc de Wellington); flowers velvety crimson-red, large, dbl., cupped, intense fragrance; vigorous growth; Granger

'DUKE OF WINDSOR', HT, ob, 1969; (Herzog von Windsor); bud pointed; flowers orange, well-formed, large, dbl., 27 petals, intense fragrance; foliage dark, glossy; very vigorous, upright growth; ADR, 1970 Edland Fragrance Medal, ARS, 1968; Tantau, Math.

'DUKE OF YORK', HCh, pb, 1894; flowers rosy-pink and white to crimson, variable; Paul & Son

Duke Wayne *see* **'BIG DUKE'**

'DULCINEA', HT, mr, 1963; flowers oriental red, edged darker, reverse yellow, large, dbl., 50–55 petals; foliage dark, glossy; upright, bushy growth; [Condesa de Sástago X Seedling]; Verschuren, A.; Stassen

'DUMBO', F, mr, 1956; flowers bright cherry-red, very large, semi-dbl.; foliage dark; very vigorous growth; [Mme G. Forest-Colcombet X Independence]; Combe

'DUMORTIER', HGal, pb; flowers light red with silvery reflex, medium, very dbl., flat

'DUNDEE', HSpn, w; flowers white blotched pink, reverse pure white, dbl., cupped; moderate growth; Austin, R., prior to 1832

'DUNDEE RAMBLER', Ayr, w; flowers compact, blooms in large clusters; [Thought to be R. arvensis X Noisette]; Martin, prior to 1837

Dune® *see* DELgrim

Dunkelrote Ellen Poulsen *see* **'RED ELLEN POULSEN'**

'DUNKELROTE TAUSENDSCHÖN', HMult, dr, 1942; flowers medium, dbl., slight fragrance; Vogel, M.; (Sangerhausen)

'DUNKERQUE', HT, op, 1940; bud pointed, ovoid; flowers bright pink, slightly coppery, very large, dbl., 30 petals, moderate fragrance; foliage clear olive-green; long, strong stems; vigorus, upright, branching growth; GM, Bagatelle, 1940; [Charles P. Kilham X Seedling]; Laperrière

'DUNKIRK', HT, mr, 1947; flowers rose-red, dbl., 36 petals, 4–5 in., moderate fragrance; foliage glossy; Dickson, A.

'DUNTON GOLD', HT, ob, 1966; flowers deep golden yellow-orange and pink tipped, 6 in., exhibition form, moderate fragrance; foliage glossy; vigorous growth; [Tzigane sport]; Dunton Nursery

Dunwich Rose, HSpn, w

'DUO', F, or, 1955; flowers coppery orange, open, petals fringed, medium, single, moderate fragrance; foliage dark; very vigorous growth; [Peace X Seedling]; Gaujard

'DUPLEX', Misc OGR, mp; (R. pomifera duplex, Wolley-Dod's Rose); flowers clear pink, semi-dbl.; heavy prickles; foliage downy, gray-green; [Chance garden hybrid of R. pomifera X Unidentified garden rose]; Vibert, prior to 1838

'DUPONTII', Misc OGR, w, 1817; (R. freudiana, R. X dupontii); (28); [Perhaps descended from R. gallica X R. moschata hybrid]

'DUPUY JAMAIN', HP, mr, 1868; flowers cerise-red, well-formed, dbl., 30 petals, moderate fragrance; vigorous growth; Jamain, H.

'DUQUESA DE PEÑARANDA', HT, ob, 1931; (Morning Blush); bud pointed; flowers shades of orange, large, dbl., 35 petals, cupped, moderate fragrance; foliage rich green, glossy; vigorous growth; GM, Portland, 1933; [Souv. de Claudius Pernet X Rosella (P. Dot)]; Dot, Pedro; C-P

'DUQUESA DE PEÑARANDA, CLIMBING', Cl HT, ob, 1940; Germain's

Dura, S, w

Durban July *see* 'KORDURBAN'

Durbankor *see* 'KORDURBAN'

Durgapur Delight, HT, lp, 1980; Gupta

'DURHAM PILLAR', HMult, mr, 1958; bud globular; flowers rose-red, small, borne in clusters, single, cupped, free, recurrent boom, slight fragrance; foliage dark, leathery, glossy; moderate climbing or trailing growth; [Chevy Chase X Seedling]; Risley

'DURHAM PRINCE BISHOPS', HT, op, 1990; flowers orange, flushed pink, medium, dbl., 26–40 petals, moderate fragrance; foliage large, dark green, glossy; bushy growth; [Silver Jubilee X Doris Tysterman]; Thompson, Robert; Battersby Roses, 1989

Dusky Dancer *see* FRYtrooper

'DUSKY MAIDEN', F, dr, 1947; flowers deep crimson scarlet, blooms in trusses, single, 3 in., moderate fragrance; foliage dark; vigorous growth; GM, NRS, 1948; [(Daily Mail Scented Rose X Étoile de Hollande) X Else Poulsen]; LeGrice

'DUSKY RED', HT, dr, 1972; flowers medium red, veined, large, dbl., exhibition form, intermittent bloom, intense fragrance; foliage dark, leathery; vigorous, upright, bushy growth; [Karl Herbst X Big Red]; Wyant

'DÜSTERLOHE', Ayr, dp, 1931; flowers rose-red, single, 3 in., non-recurrent bloom, slight fragrance; vigorous, climbing growth; [Venusta Pendula X Miss C.E. van Rossem]; Kordes

'DUSTY PINK', F, lp, 1961; bud ovoid, long, pointed; dbl., 50–60 petals, 3–3.5 in., exhibition form, moderate fragrance; vigorous, upright growth; [Garnette X Garnette seedling]; Jelly; E.G. Hill Co.

'DUSTY ROSE', Min, m, 1974; flowers reddish purple, small, dbl., 40–50 petals, 1.5 in., exhibition form, moderate, spicy fragrance; foliage dark; upright growth; [Amy Vanderbilt X Cécile Brunner]; Morey, Dr. Dennison; Pixie Treasures Min. Roses

Dutch Fork China, Ch, rb; (found rose)

'DUTCH GOLD'®, HT, my, 1978; flowers golden yellow, dbl., 32–34 petals, 6 in., moderate fragrance; foliage glossy, dark; vigorous growth; [Peer Gynt X Whisky Mac]; Wisbech Plant Co.

'DUTCH HEDGE', HRg, lp, 1958; flowers small, single; orange-red fruit; compact, upright growth; [(R. rugosa rubra X R. cinnamomea) X R. nitida]; Nyveldt

'DUTCH MISS', Min, pb, 1986; flowers light pink veining to white base, fading slightly, medium, dbl., 22 petals, exhibition form, borne singly, slight fragrance; straight, pointed, medium, tan prickles; foliage medium, dark green, glossy; upright, medium growth; [Summer Spice X Seedling]; Bridges, Dennis A., 1987

'DUTCH PROVENCE', C, mp; flowers rose, large, resembling 'cabbage rose', but blooms are larger, globular

Dwarf Fairy *see* 'KORFEE'

Dwarf King *see* **'DWARFKING'**

Dwarf Pavement, S, lp; (Rosazwerg)

Dwarf Queen *see* **'QUEEN OF THE DWARFS'**

Dwarf Queen '82 *see* **'KORWERK'**

'DWARFKING', Min, mr, 1957; (Dwarf King, Zwergkönig); flowers carmine, small blooms borne singly and in clusters, dbl., 25 petals, cupped, slight fragrance; foliage glossy; compact (8-10 in) growth; [World's Fair X Tom Thumb]; Kordes; J&P

'DYKNINI', Min, rb, 1995; (**Hanini**); flowers burgundy red stripes serrated by reddish pink, white throat, single, 5 petals, 2 in., moderate, fruity fragrance; foliage medium, medium green, semi-glossy; medium (2 ft), spreading (30in) growth; RR; [Sarabande X Hurdy Gurdy]; Dykstra, Dr. A. Michael; Justice Miniature Roses, 1995

'DYKTICK', F, pb, 1999; (**Tickles**™); flowers pink and dream stripe, reverse same, semi-dbl., 8–14 petals, 2.5 in., borne in small clusters, slight fragrance; few prickles; [(Pristine X Typhoo Tea) X Hurdy Gurdy]; Dykstra, Dr. A. Michael; Certified Roses, 2000

Dyna *see* **'BINGO'**

Dynamite™ *see* 'JACSAT'

Dynastie, S, 1993; Meilland, Alain A.; (Cavriglia)

Dynasty™ *see* 'JACYO'

Dynasty, HT, rb, 1992; Keisel

'DZAMBUL', HP, mr, 1938; flowers medium, dbl., moderate fragrance; Kosteckij; (Sangerhausen)

E

'E. E. Saskavá', HT, dy, 1933; flowers large, dbl.; Böhm, J.

'E. G. Hill', HT, mr, 1929; bud ovoid; flowers dazzling scarlet, well formed, very large, dbl., intense, damask fragrance; vigorous growth; Hill, E.G., Co.

'E. G. Hill, Climbing', Cl HT, mr, 1942; Marlin

'E. Godfrey Brown', HT, dr, 1919; flowers deep reddish crimson, dbl., moderate fragrance; Dickson, H.

'E. I. Farrington', F, mr, 1953; flowers cardinal-red to blood-red, turning almost crimson, dbl., 50–60 petals, 3.5–4 in., exhibition form, moderate fragrance; vigorous, spreading growth; [Queen o' the Lakes X Seedling]; Brownell, H.C.

'E. J. Baldwin', HT, my, 1952; flowers rich golden yellow, high pointed, well formed, large, dbl., 30–40 petals; foliage dark; vigorous, upright, branching growth; [Phyllis Gold X Seedling]; Robinson, H.; Baker's Nursery

'E. J. Ludding', HT, pb, 1931; flowers carmine-pink shaded coral-red and salmon, open, large, dbl., slight fragrance; bushy growth; [Ophelia X Hill's America]; Van Rossem; C-P, Prior

'E. J. Moller', HT, dr, 1924; flowers intense red, deepening toward black, dbl.; [George Dickson seedling]; Moller

'E. P. H. Kingma', HT, ab, 1919; flowers apricot and orange-yellow, dbl.; [Mme Edouard Herriot X Duchess of Wellington]; Verschuren

'E. Pemberton Barnes', HT, pb, 1928; flowers light pink, shaded cerise; Pemberton

'E. V. Lucas', HT, dr, 1934; flowers dark velvety crimson, large, semi-dbl., borne in sprays, slight fragrance; foliage dark; vigorous, upright, branching growth; McGredy

'E. Veyrat Hermanos', Cl T, pb, 1895; (E. Veyrath Hermanos, Pillar of Gold); flowers apricot and carmine-pink, reflexes violet-rose, dbl., intense fragrance; vigorous growth; Bernaix, A.

E. Veyrath Hermanos *see* **'E. Veyrat Hermanos'**

'E. Y. Teas', HP, mp, 1874; (Mons. E.Y. Teas); flowers bright red, large, globular, intense fragrance; [Alfred Colomb seedling]; Verdier, E.

'E.N. Ward', F, lp; Kershaw, 1919

'Eads', F, pb, 1991; flowers pink and yellow, blooms borne in small clusters, semi-dbl., slight fragrance; foliage medium, medium green, semi-glossy; low, compact growth; [(Seedling X Pinocchio) X Seedling]; Burks, Larry; Specialty Roses, 1991

Eagle® *see* 'HAVeal'

Eagle Wings® *see* 'LENsim'

'Earl Beatty', HT, dr, 1923; flowers deep crimson, large, dbl., cupped; bushy growth; [Hoosier Beauty X George Dickson]; Chaplin Bros.

'Earl Godard Bentinck', HT, mr, 1931; flowers red, base orange, large, dbl., slight fragrance; very vigorous growth; [Pharisaer X Covent Garden]; Buisman, G. A. H.

'Earl Haig', HT, mr, 1921; flowers brick-red, large, dbl., moderate fragrance; few thorns; vigorous, bushy growth; GM, NRS, 1920; Dickson, A.

'Earl of Dufferin', HP, dr, 1887; flowers velvety crimson, shaded chestnut-red, very large, dbl., 53 petals, globular, moderate fragrance; vigorous growth; Dickson, A.

'Earl of Eldon', N, ob, 1872; flowers coppery orange; Eldon-Coppin

'Earl of Gosford', HT, dr, 1912; flowers large, dbl., intense fragrance; McGredy

'Earl of Pembroke', HP, mr, 1882; flowers carmine-red, large, dbl., moderate fragrance; Bennett

'Earl of Warwick', HT, lp, 1904; flowers pale pinkish buff, reverse livid pink, large, dbl., moderate fragrance; [Souv. de S.A. Prince X Mrs W.J. Grant]; Paul, W.

'Earldomensis', S, my, 1934; flowers bright yellow; height 6 ft; [R. hugonis X R. sericea pteracantha]; Page

'Early Bird', F, mp, 1965; flowers rose-opal, well formed, dbl., 4 in., moderate fragrance; free growth; [Circus X Fritz Thiedemann]; Dickson, Patrick; A. Dickson

'Early Blush', F, lp; [R. rugosa X Unknown]; Wheen, G., 1992; Gretchen's

'Early Mist', Min, w, 1971; flowers cream, dbl., 25 petals, 2 in., globular, slight fragrance; foliage dull, light; vigorous growth; de Yssel, Van; Warmerdam

'Early Morn', HT, lp, 1944; flowers shell-pink, large, dbl., exhibition form, moderate fragrance; foliage glossy; long stems; vigorous growth; [(Dr. W. Van Fleet X Général Jacqueminot) X Break o'Day]; Brownell, H.C.

Early Peace *see* **'Molodost Mira'**

'Early Red', HSpn, rb; flowers carmine-red with gold, medium-large, semi-dbl.; [R. pimpinellifolia X Claudius Denoyel]

'Earth Song', Gr, dp, 1975; bud long, pointed to urn-shaped; flowers tyrian red to tyrian rose, blooms, dbl., 25–30 petals, 4–4.5 in., cupped, moderate fragrance; foliage glossy, dark, leathery; upright, bushy growth; [Music Maker X Prairie Star]; Buck, Dr. Griffith J.; Iowa State University

Earthquake™ *see* 'MORquake'

Earthquake, Climbing *see* 'MORshook'

'Easlea's Golden Rambler', LCl, yb, 1932; (Golden Rambler); flowers rich buff-yellow marked crimson, blooms in clusters, dbl., 35 petals, 4 in., non-recurrent, moderate fragrance; foliage leathery, rich olive-green; vigorous, climbing growth; GM, NRS, 1932; Easlea; Totty

'East Anglia', HT, mp, 1939; flowers aurora-pink; [Golden Dawn sport]; Morse

'East Europe', F, lp; Urban, J.

Easter Bonnet *see* 'BURstein'

Easter Bunny *see* 'RENbun'

Easter Morn *see* **'Easter Morning'**

'Easter Morning', Min, w, 1960; (Easter Morn); bud pointed; flowers ivory-white, small, dbl., 60–70 petals, 1.5 in.; foliage leathery, glossy; vigorous, dwarf (12-16 in) growth; [Golden Glow (Brownell) X

Zee]; Moore, Ralph S.; Sequoia Nursery

'EASTER PARADE', F, yb, 1951; bud ovoid, golden yellow; flowers salmon-pink and cerise, reverse yellow, becoming light carmine, dbl., 50–55 petals, 2.5–3.5 in., slight fragrance; foliage dark, glossy, bronze; vigorous, bushy growth; [Sunshine X Herrenhausen]; Whisler; Germain's

'EASTER ROSE'; (Rubus coronarius, Rubus rosaefolius coronarius)

Easy *see* 'ZIPEASY'

Easy Cover *see* 'POUleas'

Easy Going™ *see* 'HARFLOW'

'EASY ORANGE', F, ob, 1995; flowers bright orange, full (26–40 petals), medium blooms borne in small clusters, fragrant; numerous prickles; foliage medium, dark green, glossy; upright, medium growth; [(Valerie Jeanne X Eyepaint) X Laureate]; Jobson, Daniel, J.; Jobson, 1995

'Easy to Cut', HMsk, 1997; Lens, Louis

Easy Vibes *see* 'GELEASY'

'EBB TIDE', HT, ab, 1961; bud long, pointed; flowers light yellowish pink, dbl., 28 petals, 5 in., exhibition form, slight fragrance; foliage glossy; vigorous, upright, compact growth; [(Sutter's Gold X Seedling) X Peace]; Von Abrams; Peterson & Dering

'EBERHARD JUNG', Pol, mr, 1930; flowers small, semi-dbl.; Schmitt, L.

Eberwein *see* **'DOPEY'**

'ÉBLOUISSANT', Pol, dr, 1918; flowers dazzling deep red, very dbl., globular, blooms in clusters, slight fragrance; foliage bronze, glossy; bushy growth; [Seedling X Cramoisi Superieur]; Turbat

'EBONY', Gr, dr, 1960; bud ovoid; flowers velvety dark red, dbl., 20–30 petals, 3–4 in., exhibition form, slight fragrance; foliage glossy; vigorous, upright growth; [Carrousel X Charles Mallerin]; Von Abrams; Peterson & Dering

'ÉBORACUM', LCl, w, 1977; flowers creamy white, base yellow, full, small, dbl., 30–35 petals, 2.5–3 in., repeat bloom, moderate fragrance; foliage glossy; strong growth; [Casino X Ice White]; Powell

'ÉCARLATE', HT, mr, 1907; flowers brilliant scarlet, somewhat like Gruss Anteplitz, open, small, semi-dbl., no fragrance; foliage rich green, glossy; vigorous, bushy growth; [Camoens X Seedling]; Boytard

'Echizo', HT; Camprubi, C.

'ECHO', HMult, pb, 1914; (Baby Tausendschön); flowers varying (like tausendschon) from dark pink to almost white, semi-dbl., cupped; bushy growth; [Tausendschön sport]; Lambert, P.

'Echo', S, op, 1970; Lens

'ÉCLAIR', HP, dr, 1883; flowers very dark red shaded blackish, well-shaped, small, dbl., moderate fragrance; tall growth; [Général Jacqueminot X Seedling]; Lacharme, F.

'Eclaireur', F, 1971; Godin, M.

'ÉCLAIREUR', HP, dr, 1895; flowers large, dbl., moderate fragrance; Vigneron

'ECLIPSE', HT, ly, 1935; bud remarkably long, pointed, deep gold, with long, narrow; flowers golden yellow, loose blooms, dbl., 28 petals, moderate fragrance; foliage leathery, dark; vigorous, bushy growth; David Fuerstenberg Prize, ARS, 1938 GM, Bagatelle, 1936 GM, Portland, 1935 GM, Rome, 1935; [Joanna Hill X Federico Casas]; Nicolas; J&P

Eclipse *see* 'SAUdora'

'Eclipse, Climbing', Cl HT, ly

Ecole d'Ecully *see* 'LAPbu'

'ECSTASY', HT, yb, 1935; flowers pale yellow shaded bronze and cerise, dbl., intense fragrance; erect, branching growth; Dickson, A.

Ecstasy *see* 'KORazerka'

'ED STEER', HT, mr, 1999; dbl., 26–40 petals, 5 in., borne mostly singly, slight fragrance; prickles moderate; foliage medium, dark green, semi-glossy; upright, medium (30 in.) growth; [(Royal William X Gabi) X Adrienne Berman]; Poole, Lionel

'EDDA', S, mr, 1969; flowers clear rose-red, open, medium, borne in clusters, semi-dbl., 16 petals, profuse, repeated bloom; foliage dark, glossy, leathery; vigorous growth; [Lichterloh X Scharlachglut]; Lundstad

'EDDIE'S ADVENT', HT, lp, 1938; flowers pale buff, tipped pink, fading almost white, large, dbl., exhibition form, slight fragrance; foliage leathery; vigorous growth; [Mrs Sam McGredy X Edith Krause]; Eddie

'EDDIE'S CREAM', F, w, 1956; flowers cream, large, borne in clusters, dbl., moderate, apricot fragrance; vigorous growth; [Golden Rapture X Lavender Pinocchio]; Eddie; Harkness

'EDDIE'S CRIMSON', HMoy (S), mr, 1956; flowers blood-red, semi-dbl., 4–5 in., non-recurrent; large, globular fruit; vigorous (9-10 ft.) growth; [Donald Prior X R. moyesii hybrid]; Eddie

'EDDIE'S JEWEL', HMoy (S), mr, 1962; flowers fiery red, recurrent bloom; few prickles; bark red; vigorous (8-9 ft.) growth; [Donald Prior X R. moyesii hybrid]; Eddie

'EDEL', HT, w, 1919; flowers ivory-white, passing to pure white, well formed, very large, dbl., moderate fragrance; vigorous growth; [Frau Karl Druschki X Niphetos]; McGredy

Edelweiss *see* **'SNOWLINE'**®

Eden *see* 'MEIVIOLIN'

Eden Climber™ *see* 'MEIVIOLIN'

'EDEN ELLEN', F, ab, 1984; flowers medium, dbl., 35 petals, slight fragrance; foliage medium, medium green, matt; bushy growth; [Seedling X Seedling]; Schneider, Peter

'EDEN ROSE'®, HT, dp, 1950; bud ovoid; flowers tyrian rose, dbl., 50–60 petals, 4.5 in., cupped, intense fragrance; foliage glossy, bright dark green; vigorous, upright growth; GM, NRS, 1950; [Peace X Signora]; Meilland, F.; C-P, 1953, URS, 1950

Eden Rose 88 *see* 'MEIVIOLIN'

'EDEN ROSE, CLIMBING'®, Cl HT, dp, 1962; Meilland, Alain A.; URS

'EDEN SUNGOLD', F, my, 1976; flowers pure yellow, dbl., 30–35 petals, 3 in., slight fragrance; semi-dwarf growth; [Seedling X Seedling]; Herholdt, J.A.

'EDGAR ANDREU', LCl, mr, 1912; flowers bright blood-red, large, borne in clusters of 7-15, dbl.; foliage dark, glossy; [R. wichurana X Cramoisi Supérieur]; Barbier

'EDGAR BLANCHARD', HT, lp, 1912; flowers pink shading to white, large, dbl.; Duron

'EDGAR M. BURNETT', HT, pb, 1914; flowers flesh-pink, center dark pink, dbl., moderate fragrance; GM, NRS, 1913; McGredy

'Edie Anne', HT, mp, 1997

'EDINA', HT, w, 1934; flowers white, occasionally flushed pink, well formed, intense fragrance; foliage bronze red passing to green; vigorous growth; Dobbie

'Edinger White HP', HP, w

'EDITH BELLENDEN', E, mp, 1895; flowers pale rose, single; foliage fragrant; vigorous growth; very hardy.; Penzance

'EDITH CAVELL', HT, ly, 1918; flowers pale lemon-white, dbl.; Chaplin Bros.

Edith Cavell *see* **'MISS EDITH CAVELL'**

'EDITH CLARK', HT, mr, 1928; flowers fiery red, dbl., globular, slight fragrance; foliage rich green; dwarf growth; [Mme Abel Chatenay X Seedling]; Clark, A.; Hackett

'EDITH DE MARTINELLI'®, F, op, 1958; flowers salmon-pink, well-formed; vigorous growth; GM, Geneva, 1958; [(Gruss an Teplitz X Independence) X Floradora]; Arles; Roses-France

Edith de Martinelli, Climbing® *see* 'ORADIT'

'EDITH DE MURAT', B, w, 1858; Ducher

'EDITH DENNETT', F, op, 1973; flowers salmon-pink, rosette form, dbl., 24–30 petals, 2–3 in., slight fragrance; foliage glossy; free growth; Holmes, R.; Fryer's Nursery, Ltd.

'EDITH FELBERG', HT, w, 1931; flowers cream, center slightly darker, rather, dbl., cupped; foliage leathery; [Seedling X Souv. de H.A. Verschuren]; Felberg-Leclerc

'EDITH HAYWARD', Cl HT, lp, 1967; flowers pastel pink, large, dbl., exhibition form, recurrent bloom, moderate fragrance; foliage glossy, dark; vigorous growth; [Fontanelle X Gen. MacArthur]; Hayward

'EDITH HAZELRIGG', HT, ob, 1953; flowers orange-cerise, pointed, dbl., 25 petals, moderate fragrance; foliage dark; vigorous growth; Cant, F.

Edith Holden *see* R. filipes 'Kiftsgate'

'EDITH KRAUSE', HT, w, 1930; flowers greenish white, large, dbl., 30 petals, exhibition form, moderate fragrance; very vigorous growth; [Mrs Charles Lamplough X Souv. de H.A. Verschuren]; Krause; J&P

'EDITH MARY MEE', HT, or, 1936; flowers vivid orient red, flushed orange, base yellow, dbl., moderate fragrance; foliage dark, leathery; vigorous, bushy, compact growth; Mee; Beckwith

'EDITH NELLIE PERKINS', HT, op, 1928; flowers salmon-pink, flushed orange, reverse orange-red, shaded orange, dbl., 35–40 petals, moderate fragrance; few prickles; vigorous, bushy growth; Dickson, A.

'EDITH NELLIE PERKINS, CLIMBING', Cl HT, op, 1936; H&S; Howard Rose Co.

'EDITH OLIVER', HT, lp, 1980; flowers soft light pink, large blooms in clusters of 4, dbl., 35 petals, moderate fragrance; brownish-red prickles; foliage medium, medium green, semi-glossy; bushy growth; [Pink Parfait X Seedling]; Singleton, C.H.

'EDITH PART', HT, pb, 1913; flowers rich red, suffused deep salmon and coppery yellow, moderate fragrance; McGredy

'EDITH PIAF', HT, m, 1964; bud ovoid; flowers purple-red, large, dbl., slight fragrance; foliage dark; [Poinsettia X (Baccará X Seedling)]; Verbeek

Edith Piaf *see* 'MEInical'

'EDITH ROBERTS', HT, ab, 1969; 5 in., exhibition form, moderate fragrance; foliage glossy, dark; [Dorothy Peach sport]; Roberts, P.D.

'EDITH SCHURR', S, yb, 1976; bud globular, sulfur-yellow; flowers light yellow, center pink, dbl., 60 petals, 5 in., recurrent bloom, intense, damask fragrance; foliage glossy; spreading growth; [(Wendy Cussons X Gavotte) X Leverkusen]; Stanard; Edmunds Roses

'Edith Southgate', Cl Min, pb, 1997; flowers double, medium, dbl., 15–25 petals, borne mostly singly, moderate fragrance; some prickles; foliage medium, medium green, semi-glossy; upright, medium (5ft.)growth; [Laura Ford X Admiral Rodney]; Barker, S.J.L.

'EDITH WILLKIE', HT, pb, 1943; bud long, pointed; flowers livid pink, base lemon-chrome, dbl., 25–30 petals, 4.5–5 in., slight fragrance; foliage leathery, dark; vigorous, upright, compact growth; [Joanna Hill X R.M.S. Queen Mary]; Hill, Joseph H., Co., 1943; H&S, 1946, Wayside Gardens Co., 1946

'EDITH YORKE', HMult, lp, 1953; flowers light almond-pink, rosette form, borne in trusses, semi-dbl., intense fragrance; foliage light green; very vigorous growth; [Havering Rambler X Seedling]; Miller, A.I.; Jackman

'EDITOR MCFARLAND', HT, mp, 1931; flowers glowing pink, slightly suffused yellow, large, dbl., 30 petals, intense fragrance; vigorous, bushy growth; [Pharisaer X Lallita]; Mallerin, C.; C-P

'EDITOR MCFARLAND, CLIMBING', Cl HT, mp, 1948; Roseglen Nursery

'EDITOR STEWART', HT, mr, 1939; flowers deep cherry-red, open, large, semi-dbl.; foliage bronze; long stems; vigorous pillar or large bush growth; Clark, A.; NRS Victoria

'EDITOR TOMMY CAIRNS', HT, pb, 1991; flowers bright pink, light pink reverse, blooms borne mostly singly, dbl., 32–40 petals, slight; foliage medium, medium green, semi-glossy; upright, medium growth; [Seedling X Seedling]; Winchel, Joseph F., 1992

'EDMÉ ET ROGER', HT, 1903; Ketten Bros.

'Edmond Charles-Roux', HT, 1990; Dorieux, Francois

'EDMOND DESCHAYES', HT, ly, 1901; flowers yellowish-white, large, dbl.; Bernaix, A.

'EDMOND PROUST', LCl, mr, 1903; flowers pale rose and carmine, large, borne in clusters of 3-6, very dbl., sparse seasonal bloom, moderate fragrance; foliage glossy; short stems; height 5-8 ft; [R. wichurana X Souv. de Catherine Guillot]; Barbier

'EDMUND M. MILLS', HT, ob, 1927; flowers rosy flame, base deep gold, open, large, semi-dbl., intense fragrance; foliage dark, leathery; very vigorous, upright growth; [Red Radiance X Padre]; Hieatt

Edmund Rice *see* 'WELPIN'

'EDNA KAYE', HT, lp, 1959; flowers light pink tinted buff, dbl., 56 petals, 5 in.; foliage bronze; free growth; [Directeur Guerin X Mirandy]; Kemp, M.L.

Edna Marie *see* 'MORED'

'EDNA WALLING', HMult, w, 1940; Clark

Edna Wilson *see* 'GRIFED'

'EDNA-CHRIS', F, w, 1977; bud pointed; flowers off-white, dbl., 30 petals, 3.5 in., slight fragrance; foliage matt, green; tall, upright growth; [Gene Boerner sport]; Ogden

'EDNAH THOMAS', Cl HT, op, 1931; (Bloomfield Improvement); flowers salmon-rose, large, dbl., recurrent bloom, moderate fragrance; strong stems; vigorous, climbing growth; [Seedling climber X Bloomfield Progress]; Thomas; H&S

'EDO BERGSMA', HT, ly, 1932; bud pointed; flowers bright flesh and peach, large, intense fragrance; [Capt. F.S. Harvey-Cant X Étoile de Hollande]; Buisman, G. A. H.

'EDOUARD ANDRÉ', HP, 1880; Verdier, E.

'EDOUARD HERVE', HP, 1884; Verdier, E.

'EDOUARD MIGNOT', HT, m, 1927; flowers purplish garnet-red, reverse amaranth, dbl., moderate fragrance; bushy growth; Sauvageot, H.; F. Gillot

'EDOUARD RENARD', HT, mr, 1933; flowers carmine, base yellow; long, stiff stems; Dot, Pedro

'EDUARD SCHILL', HT, or, 1931; flowers brick-red shaded nasturtium-yellow, very large, semi-dbl., cupped, slight fragrance; foliage glossy; vigorous growth; [Charles P. Kilham X Mev. G.A. van Rossem]; Kordes

'EDUARDO TODA', HT, my, 1947; bud pointed; flowers sunflower-yellow, slight fragrance; [Ophelia X Julien Potin]; Dot, Pedro

'EDWARD BEHRENS', HT, dr, 1921; flowers very dark velvety crimson, very large, dbl., moderate fragrance; [Richmond X Admiral Ward]; Kordes

Edward Colston *see* 'SANCOL'

'EDWARD MAWLEY', HT, dr, 1911; bud almost black; flowers dark crimson, large, semi-dbl., 18 petals, exhibition form, moderate fragrance; bushy growth; GM, NRS, 1910; McGredy

'EDWARD MORREN', HP, mr, 1868; flowers deep cherry-rose, large, dbl., flat; vigorous growth; [Jules Margottin seedling]; Granger

'EDWARD VII', Pol, mp, 1911; flowers clear pink, small; [Mme Norbert Levavasseur sport]; Low

Edwardian Lady *see* R. filipes 'Kiftsgate'

'EDWIN MARKHAM', HT, mp, 1923; flowers bright rose-pink suffused silvery, dbl.; [Ophelia X Hoosier Beauty]; Clarke Bros.

'EDWIN T. MEREDITH', HT, ob, 1979; bud ovoid, pointed; flowers coral-pink, dbl., 30 petals, 5 in., flat, slight fragrance; bushy, upright growth; [Futura X First Prize]; Warriner, William A.; J&P

'EFFECTIVE', LCl, mr, 1913; bud long; flowers crimson, cupped, moderate fragrance; very vigorous growth; [Gen. MacArthur seedling X Paul's Carmine Pillar]; Hobbies

'EFFEKT', HT, rb, 1935; flowers scarlet-red, reverse flushed golden yellow, large, dbl., cupped, slight fragrance; vigorous, bushy growth; [I Zingari X Seedling]; Krause

'Effekt', F, dr, 1975; Berger

'EGALITÉ', F, m, 1946; bud pointed; flowers pale lilac-rose, borne in trusses, dbl., 25 petals, 4 in., moderate fragrance; foliage bronze; vigorous, branching growth; [Irene X Seedling]; Leenders, M.; Longley

'Egas Monitz', HT; Moreira da Silva, A.

'EGERIA', HP, 1877; Schwartz, J.

'EGESKOV'®, F, mp, 1982; flowers bright medium pink, blooms in large clusters, dbl., 20 petals, cupped, slight fragrance; foliage medium, light green, glossy; bushy growth; [Tornado X Matangi]; Olesen, Pernille & Mogens N.; D.T. Poulsen

'EGLANTINE', Pol, dp, 1930; flowers carmine, center white, many yellow stamens, single, moderate fragrance; small vigorous, dwarf, bushy growth; [Amaury Fonseca X Rodhatte]; Soupert & Notting

Eglantine *see* **'R. EGLANTERIA'**

Eglantyne *see* 'AUSMAK'

Eglantyne Jebb *see* 'AUSMAK'

Egoli *see* 'KORameget'

'Egon Schiele'®, S, lp

'EGYPTIAN TREASURE', S, ob, 1973; flowers orange, semi-dbl., 16 petals, 4 in., slight fragrance; foliage dark, glossy; [(Coup de Foudre X S'Agaro) X Vagabonde]; Gandy, Douglas L.

'EICHSFELDIA', HMult, ly, 1925; flowers yellowish-white, small, single; Bruder Alfons

'Eidth Piaf', HT, dr, 1999; Meilland

'EIFFEL TOWER', HT, mp, 1963; (Eiffelturm, Tour Eiffel); bud long, urn-shaped; dbl., 35 petals, 3.5–5 in., exhibition form, intense fragrance; foliage leathery, semi-glossy; vigorous, upright growth; GM, Geneva, 1963 GM, Rome, 1963; [First Love X Seedling]; Armstrong, D.L. & Swim; Armstrong Nursery

'EIFFEL TOWER, CLIMBING', Cl HT, mp, 1967; Laveena Roses

Eiffelturm *see* **'EIFFEL TOWER'**

'EIKO', HT, yb, 1978; bud pointed; flowers yellow and scarlet, pointed, dbl., 30–35 petals, 5–6. in., slight fragrance; foliage large, glossy, light green; vigorous growth; [(Peace X Charleston) X Kagayaki]; Suzuki, Seizo; Keisei Rose Nursery

'EILEEN BOXALL', HT, dr, 1948; flowers cerise, semi-dbl., 18 petals, 5 in., moderate fragrance; foliage dark; vigorous growth; [Betty Uprichard sport]; Boxall

'EILEEN DOROTHEA', HT, mr, 1931; bud pointed; flowers crimson-scarlet, edged darker, base yellow, dbl., exhibition form, intense fragrance; foliage deeply serrated; vigorous growth; Dickson, A.

'EILEEN LOUISE', HT, mp, 1985; [Admiral Rodney sport]; Brown, Harry G.L.S.

'EILEEN LOW', Pol, mp, 1911; flowers small, dbl.; Levavasseur

'EISENACH', HWich, mr, 1910; flowers bright red, borne in clusters, single; very vigorous growth; Kiese

'Eisprinzessin', HT, pb

'EKTA', G, mp, 1927; single, non-recurrent; [Alika X American Beauty]; Hansen, N.E.

'EL CAPITAN', Gr, mr, 1959; flowers cherry to rose-red, blooms in small clusters, dbl., 30 petals, 3.5–4.5 in., exhibition form, slight fragrance; foliage dark, glossy; vigorous, upright, bushy growth; GM, Portland, 1959; [Charlotte Armstrong X Floradora]; Swim, H.C.; Armstrong Nursery

'EL CAPITAN, CLIMBING', Cl Gr, mr, 1963; Armstrong, D.L.

'EL CATALÁ', Gr, rb, 1981; bud ovoid, pointed; flowers medium red, reverse light pink, slightly blooms borne 1-8 per cluster, dbl., 35 petals, cupped, slight fragrance; awl-like prickles; foliage large, glossy; erect, slightly bushy growth; [Wanderin' Wind X ((Dornroschen X Peace) X Brasilia)]; Buck, Dr. Griffith J.; Iowa State University

'EL CID', HT, or, 1969; bud ovoid; flowers large, dbl., exhibition form; foliage soft; vigorous, upright, bushy growth; GM, Rome, 1969; [Fandango X Roundelay]; Armstrong, D.L.; Armstrong Nursery

'EL DORADO', HT, yb, 1972; flowers golden yellow, edged reddish, open, large, dbl., exhibition form, intense fragrance; foliage large, glossy, leathery; vigorous, upright growth; [Manitou X Summer Sunshine]; Armstrong, D.L.; Armstrong Nursery

'El Dorado', HT, my

'El Dorado Cemetary Short Moss', M, mp

'EL PASO', HT, yb, 1988; flowers light yellow on outer petals, deep yellow on inner petals, reverse light yellow, dbl., exhibition form, borne usually singly, slight, spicy fragrance; prickles straight, medium, greenish-brown; foliage medium, medium green, semi-glossy; bushy, medium, floriferous growth; Bronze, ARC TG, 1988; [First Prize X Arlene Francis]; Ohlson, John

El Toro *see* **'UNCLE JOE'**

Elaina *see* 'ZIPELA'

'ELAINE', HT, mp, 1950; flowers rose-pink, large, very dbl., exhibition form, moderate fragrance; [Mrs A.R. Barraclough X Lady Sylvia]; Robinson, H.; Baker's Nursery

'ELAINE', HT, ab, 1951; bud pointed; dbl., 35–45 petals, 4.5 in., exhibition form, slight fragrance; foliage leathery; very vigorous, upright growth; [Eclipse X R.M.S. Queen Mary]; Boerner; J&P

'Elaine Frawley', F, lp; [Cornelia X Unknown]; Weatherly, Lila, 1996; Prophyl Pty, Ltd.

'ELAINE HOLMAN', HT, pb, 1975; bud long, pointed; dbl., 50 petals, 3.5–4 in., exhibition form, moderate fragrance; foliage soft; vigorous growth; [Red Devil X Avon]; Watson

'ELAINE STUART', HT, w, 1932; flowers cream, center yellow; long stems; [Antoine Rivoire X Lillian Moore]; Edwards

'ELAINE WHITE', F, w, 1959; bud pointed; flowers cream, white and pink, open, small, borne in clusters, semi-dbl., moderate fragrance; foliage leathery; very vigorous, upright growth; [Gartendirektor Otto Linne X Seedling]; Riethmuller

'ELAINE'S CHOICE', F, pb, 1972; flowers pale pink-apricot, yellow blend, full, dbl., 40 petals, 5 in., intense fragrance; foliage semi-glossy; vigorous growth; [Orange Sensation X Peace]; Ellick

'Élan', F, 1960; Croix, P.

'ELATION', HT, dy, 1974; bud long, pointed; flowers large, dbl., slight fragrance; foliage large, glossy, dark, leathery; vigorous, upright growth; [Buccaneer X Seedling]; Warriner, William A.; Spek

'ELBA', HT, dr, 1963; flowers deep red, reverse gold; [Confidence X Crimson Glory]; da Silva, Moreira

'Elbefreude', F, mr, 1983; flowers vermilion, large, dbl.; Schmadlak, Dr.

'Elbeglut', Pol, dr, 1978; flowers large, semi-dbl.; Schmadlak, Dr.

'Elbegold', Pol, dy, 1973; flowers golden yellow, medium, dbl.; Schmadlak, Dr.

'ELDORA HARVEY', HT, pb, 1930; flowers pink, center tinted lavender, reverse dark pink, dbl., moderate fragrance; [Red Radiance X Maman Cochet]; Harvey

'ELDORADO', HT, ob, 1923; flowers copper, suffused orange and salmon, very large, dbl., moderate fragrance; bushy growth; [Seedling X Mme Edouard Herriot]; H&S

Eldorado *see* 'JACspri'

'ELEANOR', Min, op, 1960; bud long, pointed; flowers coral-pink, aging darker, dbl., 20–30 petals, 1 in.; foliage leathery, glossy; upright, bushy (12 in.), growth; [(R. wichurana X Floradora) X (Seedling X Zee)]; Moore, Ralph S.; Sequoia Nursery

'ELEANOR FRANCES', HT, dy, 1971; flowers deep aureolin-yellow, full, dbl., 28 petals, 5 in., moderate fragrance; foliage glossy, leathery; vigorous, tall, upright growth; [Vienna Charm X Seedling]; Green Acres Rose Nursery

'ELEANOR HENNING', HT, op, 1920; bud pointed; flowers salmon-pink; Easlea

'ELEANOR PERENYI', Gr, yb, 1985; flowers yellow flushed apricot, reverse yellow flushed salmon, loose, dbl., 25 petals, moderate, fruity fragrance; small, globular, orange-red fruit; very few, hooked, small, red prickles; foliage medium, dark, semi-glossy; medium, upright growth; [America X Sunsong]; French, Richard

'ELECTRA', HT, my, 1970; flowers open, large, dbl., slight fragrance; foliage large, glossy, leathery; vigorous, upright growth; [Eclipse X Seedling]; Boerner; J&P

'ELECTRA', HMult, ly, 1900; flowers yellow, fading white, dbl., globular, slight fragrance; foliage rich green, glossy; very vigorous, climbing growth; [R. multiflora X William Allen Richardson]; Veitch

'ELECTRON'®, HT, dp, 1970; (Mullard Jubilee); flowers shocking pink, classic form, dbl., 32–40 petals, 5 in., exhibition form, borne mostly singly, intense fragrance; foliage large, dark green, glossy; fully branched; upright, medium, bushy growth; AARS, 1973 GM, Belfast, 1972 GM, Portland, 1973 GM, RNRS, 1969 GM, The Hague, 1970; [Paddy McGredy X Prima Ballerina]; McGredy, Sam IV; McGredy

'ELEGANCE', LCl, my, 1937; flowers yellow, fading white at edges, blooms, dbl., 48 petals, 6 in., moderate fragrance; foliage dark, glossy; vigorous growth; [Glenn Dale X (Mary Wallace X Miss Lolita Armour)]; Brownell, H.C.

Elegance *see* 'MEIbicmarj'

'Elegance', Min, m

'ELÉGANCE', HT, pb, 1955; bud globular; flowers rose-copper, large, dbl., moderate fragrance; foliage dark, leathery; moderate growth; Frères, Buyl

'ELEGANS', Ayr, w; (Double White); semi-dbl., borne in large clusters; very vigorous growth

'Elegant', G, mp

Elegant Beauty *see* 'KORGATUM'

Elegant Design *see* 'MORDARCREST'

Elegant Pearl® *see* 'INTERGANT'

'Elegant Touch', HT, lp

'Eleganta', Min, mr

'ÉLÉGANTE', HT, ly, 1918; bud pointed; flowers creamy yellow, large, dbl., globular; branching growth; Pernet-Ducher

'ÉLÉGANTE, CLIMBING', LCl, w; bud long, slender; flowers white, center creamy yellow; foliage dull; LeGrice Roses

'ELEGHYA', Gr, dr, 1975; flowers deep blackish-red, blooms in clusters of 2-5, very dbl., 80 petals, cupped; foliage dark; bushy growth; [Spectacular X Seedling]; Staikov, Prof. Dr. V.; Kalaydjiev and Chorbadjiiski

Elegy *see* 'MEILUCRE'

'Elektra', HT, ly, 1964; flowers yellowish-white, large, dbl., moderate fragrance; Rupprecht-Radke

'ELENA CASTELLO', HT, pb, 1932; flowers apricot-yellow to rose, semi-dbl.; vigorous growth; [Mme Butterfly X Angèle Pernet]; Munné, B.

Eleonora® *see* 'BARleo'

'Eleonore', F, 1973; Noack, Werner

'ELEONORE BERKELEY', HMult, m, 1900

'ELETA', HT, mp, 1934; bud pointed; flowers clear rose-pink, large, dbl., 65 petals, exhibition form; vigorous growth; [Sensation X Seedling]; Dahlgren; Kemble-Smith Co.

'ELETTRA', HT, op, 1940; flowers copper-pink suffused reddish yellow, large, dbl., exhibition form, moderate fragrance; foliage leathery; very vigorous growth; [Julien Potin X Sensation]; Aicardi, D.; Giacomasso

'Eleusine', HT, m

'Elf', Min, dr

Elfe® *see* 'TANELFE'

'ELFE', F, w, 1951; bud pointed; flowers white tinted rose, shell shaped, large, borne in clusters, intense fragrance; foliage glossy, dark; vigorous, upright, bushy growth; RULED EXTINCT 4/85; [Swantje X Hamburg]; Tantau

Elfe *see* 'TANcreif'

'ELFE SUPREME', F, mp; flowers rose-pink; [Rosenelfe sport]

'ELFENREIGEN', HMacrantha (S), dp, 1939; flowers deep rose-pink, center brighter, petals shell shaped large, single, profuse, non-recurrent bloom; foliage reddish orange, later gray-green; very vigorous (5 ft) growth; [Daisy Hill X Seedling]; Krause

'ELFIE'E JOY', HT, mp, 1995; flowers medium hot pink, darker edges, lighter reverse, full (26–40 petals), medium (5") blooms borne mostly single; fragrant; numerous prickles;foliage large, medium green, semi-glossy; upright, medium growth; [Sheer Elegance sport]; Dry, Elfie; Consolidated Nurseries, Inc.

Elfin *see* 'POULFI'

'ELFIN', F, or, 1939; flowers cherry-rose shaded orange-salmon, borne in clusters, dbl., 4.5 in.; low growing growth; RULED EXTINCT 12/85; Archer

'ELFIN CHARM', Min, pb, 1974; bud short, pointed; flowers phlox-pink, dbl., 65 petals, 1 in., moderate fragrance; foliage small, glossy, leathery; bushy, compact growth; [(R. wichurana X Floradora) X Fiesta Gold]; Moore, Ralph S.; Sequoia Nursery

'ELFINESQUE', Min, op, 1974; bud pointed; flowers coral-orange to bright pink, small, semi-dbl., slight fragrance; foliage small, glossy, leathery; vigorous, dwarf, upright, bushy growth; [Little Darling seedling X Yellow Bantam]; Morey, Dr. Dennison; Pixie Treasures Min. Roses, 1973

'ELFINGLO', Min, m, 1977; bud ovoid; flowers red-purple, micro-mini, dbl., 33 petals, .5 in., cupped, moderate fragrance; foliage small, glossy; compact growth; [Little Chief X Little Chief]; Williams, Ernest D.; Mini-Roses

Elgin Festival *see* 'AUSPOLY'

'ELIANE', HT, ab, 1954; flowers bright salmon, well formed, large, moderate fragrance; foliage dark; vigorous growth; [Mme Joseph Perraud X R. foetida bicolor hybrid]; Gaujard; Gandy Roses, Ltd., 1959

Eliane Gillet *see* 'MASelgi'

'ELIDA', HT, or, 1966; flowers vermilion, large, dbl., 30 petals, exhibition form, moderate fragrance; foliage dark, glossy; vigorous, branching growth; Tantau, Math.

'ELIE BEAUVILLAIN', Cl T, mp, 1887; flowers coppery pink, large, dbl.; vigorous growth; [Gloire de Dijon X Ophirie]; Beauvillain

Elina® *see* 'DICJANA'

'ELISA'®, HT, mp, 1981; flowers large, dbl., 40 petals, cupped, no fragrance; foliage large, light green, matt; upright growth; [Seedling X Blessings]; Rose Barni-Pistoia

'ELISA BOËLLE', HP, w, 1869; (Elise Boelle); flowers white, tinted with rose; vigorous growth; Guillot Père

'ELISA ROBICHON', HWich, mp, 1901; flowers salmon-pink, fading pinkish buff, open, large, borne in clusters, semi-dbl., free seasonal bloom, slight fragrance; short, strong stems; vigorous, climbing (10 ft), or trailing growth; [R. wichurana X L'Ideal]; Barbier

'ELISABETH', HMult, lp, 1926; flowers light, creamy pink, small, dbl., moderate fragrance; Bruder Alfons

'ELISABETH DIDDEN', HT, mr, 1918; flowers glowing carmine-red and scarlet, semi-dbl., moderate fragrance; [Mme Caroline Testout X Gen. MacArthur]; Leenders, M.

'ELISABETH FAURAX', HT, w, 1937; bud pointed; flowers white, lightly shaded ivory, large, very dbl.; long stems; upright growth; [Caledonia X Mme Jules Bouche]; Meilland, F.

'ELISABETH FAURAX, CLIMBING', Cl HT, 1937; Meilland, F.

'Elisabeth Tschudin', F, mp, 1981; Hetzel

'ELISE', HT, w, 1969; [Prima Ballerina sport]; Edmunds

Elise Boelle *see* **'ELISA BOËLLE'**

'Elise Flory', Ch, mr

'Elise Lemaire', HP; Roseraie de l'Hay

'ELISE NOELLE', LCl, mr, 1991; dbl., 15–25 petals, 1.5–2.75 in., borne in small clusters, slight fragrance; foliage medium, medium green, semi-glossy; climbing (300 cms) growth; [Dublin Bay X Burgund]; Alde, Robert O.

'ELISE ROVELLA', G, mp; flowers rosy pink, medium, dbl.; tall growth; Roseraie de l'Hay, before 1842

'ELISE SAUVAGE', T, yb; flowers orange to yellow, sometimes yellow to white; Miellez, 1818

'ELISKA KRÁSNOHORSKÁ', HP, mp, 1932; flowers brilliant pink, large, semi-dbl., exhibition form, moderate fragrance; foliage soft, bronze; bushy growth; [Capt. Hayward X Una Wallace]; Böhm, J.

'ELITE', HT, yb, 1936; bud pointed, red; flowers salmon-pink and yellow blend, very large, dbl., exhibition form; foliage leathery, light; upright growth; [Charles P. Kilham X Pres. Herbert Hoover]; Tantau; J&P, 1941

'ELIZA', S, mp, 1961; flowers clear pale rose, borne in clusters, non-recurrent; foliage dark, glossy; bushy, erect (3 ft.) growth; Skinner

Eliza *see* 'KORlis'

'ELIZA BALCOMBE', HP, pb, 1842; flowers pale flesh; foliage gray-green; Laffay, M.

'Eliza Dorothy', HT, my, 1984; Allender, Robert William

'Eliza Wren', Pol, lp, 1997; Williamson

'ELIZABETH', Pol, mp, 1937; flowers rich salmon, borne in large clusters, semi-dbl.; Letts

Elizabeth Abler *see* 'TINABLER'

Elizabeth Ann *see* 'KIRBRONZE'

'ELIZABETH ARDEN', HT, w, 1929; flowers pure white, dbl., intense fragrance; GM, NRS, 1929; [Edith Part X Mrs Herbert Stevens]; Prince

'ELIZABETH BROW', M, 1883; Stamler

'ELIZABETH CONE', HT, lp, 1954; flowers flesh-pink, well shaped; vigorous, low growth; [Picture sport]; Cone; Roger

'ELIZABETH CULLEN', HT, mr, 1921; flowers rich scarlet-crimson, semi-dbl., intense fragrance; GM, NRS, 1917; Dickson, A.

'Elizabeth Fankhauser', HT, op, 1965

'ELIZABETH HAMLIN', LCl, pb, 1987; flowers blush pink, fading to white, small, single, 5 petals, cupped, borne in mass clusters, slight fragrance; few prickles; foliage pennate, 5-7leaflets, medium; rampant growth; [Seedling X Seedling]; Nobbs, Kenneth J., 1986

Elizabeth Harbour *see* 'HABONE'

'ELIZABETH HARKNESS'®, HT, ly, 1969; flowers off-white to creamy buff, often with pastel yellow and pink, dbl., 28 petals, moderate fragrance; foliage dark; upright, bushy growth; [Red Dandy X Piccadilly]; Harkness

'ELIZABETH HARKNESS, CLIMBING'®, Cl HT, w, 1975; Harkness

'Elizabeth Harwood', LCl, lp

'ELIZABETH HASSEFRAS', Pol, mp, 1951; flowers buttercup form, glistening rose-pink with many stamens; B&A

Elizabeth Heather Grierson *see* 'MATTNOT'

'ELIZABETH LEE', HT, dr, 1935; flowers dark velvety red, well shaped, moderate fragrance; Chaplin Bros.

Elizabeth Munn *see* 'SEABET'

Elizabeth of Glamis® *see* 'MACEL'

'ELIZABETH OF YORK', HT, dp, 1928; bud large, pointed, cerise; flowers cerise-pink, dbl., 27 petals, exhibition form, intense fragrance; few thorns; foliage dark, glossy; Dobbie

'ELIZABETH PHILP', F, ab, 1976; flowers creamy peach; [Liverpool Echo sport]; Philp, J.B. & Son

'ELIZABETH ROWE', M, dp; bud large, well mossed, deep pink.

Elizabeth Scholtz™ *see* 'WILSCSO'

'ELIZABETH TAYLOR', HT, dp, 1985; flowers deep pink with smoky edges, dbl., 30–35 petals, 4.5–5 in., exhibition form, borne usually singly, moderate, spicy fragrance; foliage large, dark, semi-glossy; long stems; upright growth; PP6492; [First Prize X Swarthmore]; Weddle, Von C.

'ELIZABETH W. ADAM', HT, dp, 1926; flowers pink veined crimson, base yellow, dbl., 50 petals; Adam & Craigmile

'ELIZABETH ZEIGLER', HWich, dp, 1917; flowers deep rose-pink; [Dorothy Perkins sport]; Pierson, A.N.

'ELKA GAARLANDT', Pol, dp, 1966; bud ovoid; flowers medium, borne in large clusters, dbl.; foliage dark; Gold Star of the South Pacific, Palmerston North, NZ, 1970; [Hobby X Kathleen Ferrier]; Buisman, G. A. H.

'ELLA GUTHRIE', HT, mp, 1937; flowers large, dbl., intense fragrance; vigorous growth; [Premier X Seedling]; Clark, A.; NRS Victoria

'ELLA MCCLATCHY', HMult, mp, 1926; flowers rose, borne in clusters, single, sometimes recurrent bloom; thornless; Diener

'ELLA SCOTT', HWich, dp, 1925; flowers deep rose-pink, approaching red, borne in clusters of 15-20, dbl.; few thorns; vigorous, climbing growth; [Orléans Rose X Chance seedling]; Scott, G.J.; Brundrett (NRS Victoria)

Ellamae™ *see* 'SAVAMAE'

'ELLE', S, dp, 1980; bud pointed; flowers deep pink, borne 3-5 per cluster, semi-dbl., 16 petals, non-recurrent, intense fragrance; curved gray prickles; foliage light green, 5-7 leaflets; vigorous, upright dense growth; [Schneezwerg X Splendens]; Lundstad, Arne; Agricultural University of Norway

Elle *see* **'NOBILITY'**

Ellen® *see* 'AUSCUP'

'ELLEN', HT, dp, 1929; bud pointed; flowers unvarying dark pink, large, very dbl., intense fragrance; bushy growth; RULED EXTINCT 12/85; [Premier sport]; Hinner, P.; Gould

'ELLEN GRIFFIN', Min, lp, 1988; flowers medium, dbl., 15–25 petals, slight fragrance; foliage medium, medium green, semi-glossy; upright, bushy growth; [Uwe Seeler X Party Girl]; Hefner, John; Kimbrew Walter Roses, 1988

Ellen Hamlyn *see* 'DORwest'

'ELLEN MARY', HT, dr, 1963; flowers well-formed, dbl., 34 petals, 5 in., moderate fragrance; vigorous, upright growth; [Wellworth X Independence]; LeGrice

'ELLEN POULSEN', Pol, mp, 1911; flowers bright cherry-pink, large blooms in clusters, dbl., recurrent bloom, moderate fragrance; foliage glossy, dark; bushy growth; (14); [Mme Norbert Levavasseur X Dorothy Perkins]; Poulsen, D.T.; Teschendorff

'ELLEN POULSEN LYSROSA', Pol, op, 1938; flowers light salmon-pink, medium, dbl., moderate fragrance; Poulsen

'ELLEN POULSEN MORK', Pol, dr, 1928; [Ellen Poulsen sport]; Poulsen, S.

'ELLEN TERRY', HT, ly, 1925; flowers soft sulfur-cream, outer petals soft peach, well shaped, moderate fragrance; upright, vigorous growth; Chaplin Bros.

'ELLEN WILLMOTT', HT, yb, 1936; flowers creamy lemon, flushed rosy pink, large, single; foliage leathery, dark; vigorous, upright growth; [Dainty Bess X Lady Hillingdon]; Archer

'ELLEN WILLMOTT', HT, lp, 1898; flowers silvery flesh to shell-pink, dbl., cupped; very vigorous growth; Bernaix, A.

'ELLEN ZINNOW', HT, my, 1930; flowers yellow, shaded coppery orange and pink, dbl., moderate fragrance; [Souv. de H.A. Verschuren X Sunstar]; Krause

'ELLEN'S JOY', S, lp, 1989; bud ovoid; flowers light shell pink, aging lighter, medium, borne singly and in small clusters, dbl., 23 petals, cupped, repeat bloom, moderate, fruity fragrance; globular, orange-red fruit; prickles awl-like, rusty-green; foliage medium, medium green, semi-glossy; upright, bushy, spreading, medium, winter hardy growth; [Vera Dalton X [Dornroschen X (Tickled Pink X Applejack)]]; Buck, Dr. Griffith J.; Historical Roses, 1991, Kimbrew-Walter Roses, 1991

'ELLESMERE', HT, w, 1927; flowers ivory-white to pure white, moderate fragrance; few thorns; [Ophelia seedling]; Allen

'ELLI HARTMANN', HT, my, 1913; flowers yellowish old-gold, dbl., moderate fragrance; [(Souv. du Pres. Carnot X Mme Mélanie Soupert) X Marechal Niel]; Welter

'ELLI KNAB', HT, pb, 1934; flowers flesh-cream flushed bright rose, veined vermilion, very large, dbl., exhibition form, slight fragrance; foliage leathery; upright, very vigorous growth; [Cathrine Kordes X W.E. Chaplin]; Kordes

'ELLI KNAB, CLIMBING', Cl HT, pb, 1953; Tantau

'ELLINOR LEGRICE', HT, my, 1949; bud ovoid; flowers, dbl., 50 petals, 5–5.5 in., cupped, moderate, fruity fragrance; foliage leathery, glossy, dark; vigorous, upright growth; [Mrs Beatty X Yellowcrest]; LeGrice

'ELLINOR LEGRICE, CLIMBING', Cl HT, my, 1959; LeGrice

'ELLIS WOOD', F, yb, 1983; [Arthur Bell sport]; Gateshead Metro. Borough Council

'ELMHURST', HT, pb, 1985; flowers large, peachy pastel pink with yellow, dbl., 30–35 petals, 5 in., exhibition form, borne mostly singly, moderate, fruity fragrance; foliage medium, medium green, matt; upright growth, moderately tall; [Granada X Helmut Schmidt]; Perry, Astor; Perry Roses

'ELMIRA', HRg (S), mr, 1977; bud ovoid; flowers bright red, open, dbl., 25 petals, 1.5–2 in., intense fragrance; foliage yellow-green; upright, bushy growth; [Schneezwerg X Old Blush]; Svedja, Felicitas; Canada Dept. of Agric.

'ELMSHORN', S, dp, 1951; flowers deep pink, pompom type, blooms in large trusses (to 40), dbl., 20 petals, 1 in., cupped, recurrent bloom, slight fragrance; foliage glossy, wrinkled, light green; (21); ADR, 1950; [Hamburg X Verdun]; Kordes; Morse

Elna, HMult, lp, 1964; Petersen

Elna Noack, Pol, or, 1960; flowers orange-red, medium, semi-dbl.; Schmid, P.

'ELNAR TONNING', HT, dp, 1926; flowers fuller and darker; [Ophelia sport]; Gyllin

Elodea® *see* FEgima

Eloira *see* 'VARIORA'

Eloquence *see* 'JACSIL'

'ELSA ARNOT', HT, pb, 1960; flowers golden yellow shaded pink and cerise, dbl., 32 petals, 4 in., intense fragrance; foliage glossy; vigorous, upright growth; GM, NRS, 1959; [Ena Harkness X Peace]; Arnot; Croll

'ELSA KNOLL', HT, op, 1966; flowers shrimp-pink, large, dbl., 30 petals, exhibition form, intense fragrance; foliage dark, glossy, leathery; vigorous, upright growth; [First Love X Castanet]; Morey, Dr. Dennison; Country Garden Nursery

'ELSBETH', F, lp, 1962; flowers soft pink, borne in clusters, dbl.; foliage dark; [Valeta sport]; deRuiter

'ELSBETH MEYER', Pol, dr, 1940; flowers medium, semi-dbl.; Vogel, M.

'ELSE', HT, mr, 1929; flowers medium, dbl.; Vogel, M.

'ELSE CHAPLIN', Pol, dp, 1937; flowers deep rich pink, borne in large trusses, semi-dbl.; vigorous growth; Chaplin Bros.

'ELSE KREIS', Pol, dp, 1913; flowers shining deep pink; Kreis

'ELSE POULSEN', F, mp, 1924; (Joan Anderson); flowers bright rose-pink, blooms in clusters, 10 petals, 2 in., slight fragrance; foliage dark, bronze, glossy; vigorous, bushy growth; (21); [Orléans Rose X Red Star]; Poulsen, S.

'ELSE POULSEN MELDUGSFRI', F, mp, 1937; [Else Poulsen sport]; Poulsen, S.; Poulsen's Roses

'ELSE POULSEN MORKROD', F, dr, 1934; [Else Poulsen sport]; Poulsen, S.

'ELSE POULSEN, CLIMBING', Cl F, mp, 1932; Ley

'ELSE'S RIVAL', F, mr, 1938; (Double Else Poulsen); flowers carmine red, more, dbl.; healthier growth than parent; (21); [Else Poulsen sport]; Boer Bros.

'ELSIE', LCl, lp, 1934; flowers soft pink, single; foliage dark, glossy; very vigorous growth; Chaplin Bros.

'ELSIE ALLEN', HT, lp, 1971; flowers pale pink, medium, dbl., exhibition form, slight fragrance; vigorous, bushy growth; [Montezuma sport]; Allen, L.C.; E.T. Welsh

'ELSIE BECKWITH', HT, mp, 1922; flowers rich rosy pink, center deeper, large, dbl., exhibition form, intense fragrance; foliage dark, shaded red, leathery; upright growth; [Ophelia seedling X Mev. Dora van Tets]; Beckwith

'ELSIE BOLDICK', Min, mr, 1978; bud ovoid, mossy; single, borne 1-5 per cluster, no fragrance; foliage small, soft; height 15 in; [Fairy Moss X Fairy Moss]; Dobbs, Annette E.; Small World Min. Roses

'ELSIE DEVY', F, m, 1967; bud ovoid; flowers soft lavender-pink, reverse mauve-pink, dbl., exhibition form, slight fragrance; foliage light green, leathery; very vigorous, upright, bushy growth; [Ma Perkins X Detroiter]; Fankhauser

'Elsie May', F, lp; Hamilton, 1992

'ELSIE MELTON', HT, pb, 1991; (Picture Perfect); bud pointed; flowers large blooms borne usually singly, dbl., exhibition form, moderate, fruity fragrance; foliage large, dark green, semi-glossy; upright, tall growth; [Pristine X King of Hearts]; Wambach, Alex A., 1986; Alex A. Wambach, 1990

Elsie Warren *see* 'MILSWEET'

'ELSIE WRIGHT', F, dr, 1982; bud long, pointed; flowers large blooms borne usually singly, sometimes 3 per cluster, 15 petals, intense, damask fragrance; light brown prickles; foliage medium, medium green, glossy; upright growth; [Crimson Glory X Crimson Glory]; Cattermole, R.F.

'ELSIEMAE', Min, op, 1986; flowers light coral pink, blooms in sprays of 2-4, dbl., 25 petals, exhibition form, slight fragrance; very few straight, light brown prickles; foliage medium, medium green, semi-glossy; tall, bushy growth; [Anne Scranton X Patricia Scranton]; Dobbs, Annette E.; Port Stockton Nursery

'ELSINORE', F, mr, 1957; (Helsingör); flowers bright scarlet, borne in large, open clusters, semi-dbl.; GM, NRS, 1957; [Floradora X Pinocchio]; Lindquist; Poulsen

'ELUSIVE', F, pb, 1993; flowers soft pink/cream blend, blooms borne in small clusters, dbl., 15–25 petals, 1.5–2.75 in., moderate fragrance; some prickles; foliage medium, dark green, semi-glossy; medium, upright, bushy growth; [(Party Girl X Laureate) X Ivory Fashion]; Jobson, Daniel J.; Jobson, 1993

Elveshörn® *see* 'KORBOTAF'

Elvira *see* 'VARELVI'

'ELVIRA', E, lp; flowers flesh, medium, semi-dbl.; vigorous growth

'ELVIRA ARAMAYO', HT, mr, 1922; flowers indian red, dbl.; bushy growth; GM, Bagatelle, 1922; [Feu Joseph Looymans X (Leslie Holland X Rayon d'Or)]; Looymans

'ELVIRA ARAMAYO, CLIMBING', Cl HT, mr, 1933; Ingegnoli

'ELVIRE POPESCO', HT, my, 1949; bud long, pointed; flowers golden yellow, large, dbl., 25 petals, moderate fragrance; foliage bronze; very vigorous, upright growth; [Comtesse Vandal X Seedling]; Gaujard

'ELVIS', Min, mr, 1978; bud pointed; flowers medium red, base white, dbl., 65 petals, 1 in., exhibition form, slight fragrance; foliage dark; vigorous growth; [Judy Fischer X Seedling]; Wells, V.W., Jr.

'Elvis', F, lp; flowers pinkish orange with darker edges; Edwards, Eddie; Johnny Becnel Show Roses, 1999

Elwina®, Min, or

'ELYSE FLORY', B, lp, 1852; Guillot

Elysium® *see* 'KORUMELST'

'EMALINE ROUGE', HT, dr, 1938; flowers deep red; [Better Times sport]; Hofmann

Emanuel® *see* 'AUSUEL'

'EMBAJADOR LEQUERICA', HT, pb, 1962; bud pointed; flowers strawberry-pink, reverse indian yellow at base passing to brick red at edge, dbl., 30 petals; vigorous growth; La Florida

'EMBASSY', HT, yb, 1967; flowers light gold veined and edged carmine, pointed, large, dbl., moderate fragrance; foliage glossy; [Gavotte X (Magenta X Golden Scepter)]; Sanday, John, 1969

'EMBASSY REGAL', HT, pb, 1976; flowers cream overlaid peach-pink, dbl., 30 petals, 5 in., moderate fragrance; [(Gavotte X Ethel Sanday) X (Crimson Glory X Seedling)]; Sanday, John

Ember *see* 'SAVEMBER'

'EMBERGLOW', HT, mp, 1935; flowers rich salmon-pink, dbl., 50 petals, 5 in.; foliage leathery; long stems; vigorous growth; [Souvenir sport]; Grillo

'EMBERS', F, mr, 1953; bud ovoid; flowers scarlet, dbl., 23 petals, 2.5–3 in., exhibition form, blooms in clusters, moderate, spicy fragrance; foliage dark, semi-glossy; vigorous, bushy, compact growth; [World's Fair X Floradora]; Swim, H.C.; Armstrong Nursery

Embers, S, yb; Delbard

Emblem™ *see* 'JACBLEM'

'EMBRACE', Pol, lp, 1972; flowers medium, very dbl., exhibition form, slight fragrance; foliage leathery; vigorous, upright, bushy growth; [Seventeen X Jack Frost]; Byrum; J.H. Hill Co.

Embrasement *see* 'CHAGIP'

Embruixada *see* **'VIOLETERA'**

'Emely', HT, w, 1998; Kordes

'EMERALD DREAM', F, w, 1976; bud pointed, light to apple green; flowers white, center green, to ivory, loose cup shape, semi-dbl., 12 petals, 1.5–2 in., flat, slight, fruity fragrance; foliage dull, very dark, leathery; low, compact growth; [Pinafore X Ivory Fashion]; Williams, J. Benjamin; Lakeland Nursery Sales

'EMERAUDE D'OR', HT, yb, 1965; flowers yellow suffused carmine-pink, petals serrated, 5 in.; vigorous growth; [Sultane X Queen Elizabeth]; Delbard-Chabert; Cuthbert

'EMERICKROSE', HMult, lp, 1922; flowers small to medium, dbl., moderate fragrance; Bruder Alfons

'EMIL KRUISIUS', F, my, 1943; bud long, pointed; flowers large, dbl., 25–30 petals, borne in clusters, slight fragrance; foliage glossy, light green; vigorous, bushy growth; [Golden Rapture X (Johanna Tantau X Eugenie Lamesch)]; Tantau

'EMILE BARDIAUX', HP, 1889; Lévêque, P.

'EMILE CHARLES', HT, or, 1922; flowers coral-red; [Mme Edouard Herriot sport]; Bernaix, P.

'EMILE CRAMON', HT, or, 1937; bud pointed; flowers coppery carmine, stamens chrome-yellow, very large; foliage dull green; very vigorous growth; Chambard, C.

'EMILE FORTÉPAULE', HWich, w, 1902; flowers white, flushed salmon, borne in large clusters, dbl.; foliage dark; vigorous growth; [R. wichurana X Souv. de Catherine Guillot]; Barbier

'EMILE HAUSBURG', HP, m, 1868; flowers lilac-rose, large, dbl.; Lévêque

'EMILE J. LE DUC', HT, mr, 1931; flowers scarlet-crimson, larger and stronger than the parent; [Scott's Columbia sport]; Le Duc

'EMILE NÉRINI', HWich, dp, 1911; flowers carmine-pink with white, medium, semi-dbl.; Nonin

'EMILIA PLANTIER', N, 1878; flowers medium large to large, dbl., moderate fragrance; Schwartz

'EMILIE', M, w; Roseraie de l'Hay, Prior to 1906

'EMILIE DUPUY', Cl T, op, 1870; flowers light coppery-pink, large, dbl., moderate fragrance; Levet

Emilo Feliu® *see* 'FEsodios'

Emily *see* 'AUSBURTON'

'EMILY', HT, lp, 1949; flowers soft pink, dbl., 40 petals, 5–6 in., moderate fragrance; foliage dark; vigorous, upright growth; [Mme Butterfly X Mrs Henry Bowles]; Baines; F. Cant

'EMILY DODD', HT, w, 1927; flowers milk-white, center cream, large, dbl., slight fragrance; Dickson, A.

'EMILY GRAY', LCl, dy, 1918; flowers deep golden buff, stamens yellow, blooms in clusters, dbl., 25 petals, moderate fragrance; foliage very glossy, dark, bronze; vigorous, climbing growth; (21); GM, NRS, 1916; [Jersey Beauty X Comtesse du Cayla]; Williams, A.; B.R. Cant

Emily Hough *see* 'HOUEMILY'

'EMILY LAXTON', HP, mr, 1878; flowers large, dbl.; Laxton

Emily Louise *see* 'HARwilla'

Emily Post *see* 'OMEGA'

'EMILY RHODES', LCl, mp, 1937; flowers large, dbl., cupped, moderate fragrance; vigorous, climbing or pillar growth; [Golden Ophelia X Zephirine Drouhin]; Clark, A.; NRS Victoria

Emily Victoria *see* 'BOSHIPEACON'

'EMIN PASCHA', HT, dp, 1894; flowers large, dbl.; Drögemüller

Éminence® *see* 'GAXENCE'

'EMIR', HT, rb, 1960; flowers yellow with orange-red, dbl., 45 petals, 6 in.; foliage glossy; free growth; [Seedling X Peace]; Verbeek

'EMJAY SKIBA', Min, mp, 1989; bud pointed; flowers medium pink, outer petals lighter, large, borne usually singly, dbl., 45 petals, exhibition form, moderate, fruity fragrance; round, dark green orange fruit; prickles sharp, pointed slightly downward, light green; foliage large, dark green, edged red, semi-glossy; bushy, tall growth; [Sonia X Pink Petticoat]; Skiba, Norman A., 1990

'EMMA', F, dp, 1980; flowers deep pink, blooms borne 6-10 per cluster, dbl., 70 petals, intense fragrance; large prickles; foliage dark, glossy; upright, branching growth; [Chanelle X Prima Ballerina]; Pearce, C.A.; Limes Rose Nursery

Emma, HT, lp, 1996

Emma Clare, HMult, pb, 1994

'EMMA JANE', F, op, 1970; flowers salmon-pink, base orange, semi-dbl., 16 petals, 3 in.; [Vera Dalton X (Masquerade X Independence seedling)]; Sanday, John

Emma Kate *see* 'JAYEMM'

Emma May, HT, mp

Emma May *see* 'SHERISILVER'

Emma Mitchell *see* 'HORHARPDOS'

'EMMA WRIGHT', HT, op, 1918; flowers orange shaded salmon, semi-dbl., moderate fragrance; foliage rich green, glossy; dwarf growth; McGredy

'EMMA WRIGHT, CLIMBING', Cl HT, op, 1932; Cant, F.

'EMMANUELLA DE MOUCHY', LCl, mp, 1922; flowers delicate transparent rose-pink, semi-dbl., intense fragrance; [R. gigantea X Lady Waterlow.]; Nabonnand, P.

Emmanuelle *see* 'AUSUEL'

'EMMELINE', HT, my, 1921; bud pure deep yellow; flowers lemon-yellow; Paul, W.

Emmeline *see* **'MADELINE'**

'EMMELOORD', Pol, or, 1973; bud cupped; flowers round, semi-dbl.; foliage glossy, dark; [Olala X Finale]; Buisman, G. A. H.

'EMMERDALE', F, mp, 1983; flowers medium, dbl., 35 petals, moderate fragrance; foliage medium, medium green, semi-glossy; bushy growth; [Seedling X Pink Parfait]; Greensitt, J.A.; Nostell Priory Rose Gardens

'Emmie Gray', Ch, mr

'EMMIE KOSTER', Pol, dr, 1956; flowers deep red; [Dick Koster sport]; Koster, D.A.

Emmy®, Min, ab, 1993; Barni, V.

'EMOTION', B, pb, 1879; Fontaine

Emotion®, F, or, 1981; deRuiter

Emozione *see* 'BARemoz'

'EMPEREUR DU MAROC', HP, dr, 1858; flowers crimson, tinged purple, very distinct, small, dbl., 40 petals, moderate fragrance; low, compact growth; [Geant des Batailles seedling]; Guinoiseau, B.; E. Verdier

'EMPEROR', HT, dr, 1958; bud urn shaped; flowers rose-red becoming darker, dbl., 35–45 petals, 3–3.5 in., exhibition form, intense fragrance; foliage glossy; vigorous, compact growth; [Pink Delight sport]; Kuramoto, H.

Emperor *see* 'JACrette'

'EMPIRE GRANGER', HT, mr, 1970; bud long, pointed; flowers velvety blood-red, full, very large, dbl., moderate fragrance; foliage large, glossy, dark, bronze, leathery; very vigorous, upright, bushy growth; [Rose Bowl X Hallmark]; Morey, Dr. Dennison

'EMPIRE QUEEN', HT, mr, 1925; flowers brilliant cerise, large, dbl., slight fragrance; upright growth; [Cherry Page X Vanessa]; Easlea

'EMPIRE STATE', HT, mr, 1934; bud pointed; flowers velvety scarlet, base golden yellow, large, dbl., exhibition form, intense fragrance; foliage leathery; vigorous growth; Nicolas; J&P

'EMPRESS', HT, dr, 1933; flowers dark cerise and red, well formed, moderate fragrance; vigorous, upright growth; [Ophelia seedling X Seedling]; Chaplin Bros.

'EMPRESS JOSEPHINE', Misc. OGR, mp; (Francofurtana, Imperatrice Josephine, R. francofurtana, Souv. de l'Imperatrice Josephine); flowers rich pink veined deeper, loosely shaped, large, wavy petals, semi-dbl.; height 4 ft, well-branched growth

'EMPRESS MARIE OF RUSSIA', T, my; flowers canary-yellow, fading, large, moderate fragrance

Empress Michiko *see* 'DICnifty'

'EMPRESS OF CHINA', Cl Ch, mp, 1896; Jackson

'EMPRESS OF INDIA', HP, 1876; Laxton

Emsie Girl *see* 'WILCGIR'

Ena Baxter *see* 'COCBONNE'

'ENA GLADSTONE', HT, dp, 1936; flowers carmine-pink, base yellow, well shaped, large, moderate fragrance; Chaplin Bros.

'ENA HARKNESS', HT, mr, 1946; flowers large, dbl., exhibition form, intense fragrance; foliage leathery; vigorous, upright growth; GM, NRS, 1945 GM, Portland, 1955; [Crimson Glory X Southport]; Norman; C-P, 1949, Harkness, 1946

'ENA HARKNESS, CLIMBING', Cl HT, mr, 1954; (Grimpant Ena Harkness); Gurteen & Ritson & R. Murrell

Enchanted *see* 'JAChon'

'ENCHANTED AUTUMN', Gr, ob, 1976; bud ovoid, pointed; flowers orange, dbl., 33 petals, 4–4.5 in., cupped, intense fragrance; foliage glossy, dark, coppery; upright, bushy growth; [(Queen Elizabeth X Ruth Hewitt) X Whisky]; Buck, Dr. Griffith J.; Iowa State University

'ENCHANTER', HT, dp, 1903; flowers deep pink; [Mme Caroline Testout X Mlle Alice Furon]; Cook, J.W.

'ENCHANTMENT', HT, pb, 1946; bud long, pointed; flowers shell-pink, base yellow, dbl., 35 petals, 6 in., slight fragrance; foliage leathery; vigorous, upright growth; [R.M.S. Queen Mary X Eternal Youth]; Hill, E.G., Co.; J&P

Enchantment *see* 'POULSKOV'

Enchantment *see* **'Sanka'**

Enchantment *see* 'POUlspan'

'ENCHANTRESS', HRg (S), dr; flowers velvety blood-red, very dbl.; extra strongand hardy.

'ENCHANTRESS', T, mp, 1904; flowers rose-pink; Cook, J.W.

Enchantress, F, lp; flowers cream white; Paul & Son, 1896

Encore *see* 'JACORE'

'ENCORE', F, mp, 1958; bud pointed; flowers creamy pink, reverse rose-pink, large cluster, semi-dbl., 10–14 petals, 3 in., cupped, slight, spicy fragrance; foliage glossy; vigorous, upright, bushy growth; RULED EXTINCT 2/84 ARM; [Else Poulsen X Capt. Thomas]; Von Abrams; Peterson & Dering

'ENDEARMENT', HT, op, 1988; bud pointed; flowers creamy pink, reverse coral pink, large, borne usually singly, semi-dbl., 10 petals, slight, sweet fragrance; no fruit; prickles straight, medium, light brown; foliage large, medium green, matt; upright, tall growth; [Gladiator X First Prize]; Taylor, Thomas E.; Michigan Mini Roses, 1989

Endeavour *see* 'TALEND'

Endless Dream™ *see* 'TWODREAM'

Endless Love *see* 'LYOSS'

Endless Summer *see* 'RENMER'

'ENDLESS TALE', Cl HT, my, 1956; bud ovoid, deep yellow; flowers amber-yellow, outer petals creamy, dbl., 30–35 petals, 6–7 in., abundant, intermittent bloom, moderate fragrance; foliage leathery; vigorous (20 ft.or more) growth; [Lestra Hibberd, Climbing X Lestra Hibbard sport]; Motose

Endora *see* 'ZIPEND'

Enduring Love, HT, yb, 1998; Pallek

'ENEMY OF WAR', HT, pb, 1986; flowers open, dbl., 50–60 petals, slight fragrance; beak-shaped, light green to deep brown prickles; foliage large, dark green, glossy, leathery; very vigorous, profuse growth; [Festival Beauty X Gynosure]; Hardikar, Dr. M.N.

'ENFANT D'ORLÉANS', Pol, m, 1929; flowers neyron rose, tinted purple, fading lighter, borne in clusters; Turbat

'ENFANT DE FRANCE', HP, lp, 1860; flowers silvery pink, very dbl.; Lartay

Enfant de lyon *see* **'NARCISSE'**

Enfield in Bloom *see* 'BOSANNEVES'

'ENG. D. JOSÉ DE MENDIA', HT, mp; flowers rosy salmon; da Silva, Moreira

'ENG. DUARTE PACHECO', HT, dr, 1938; flowers blackish crimson, large, very dbl., cupped, intense fragrance; dwarf growth; [Hadley X Presidente Carmona]; da Silva, Moreira

'ENG. PEREIRA CALDAS', HT, mp, 1954; flowers salmon-pink, base yellow; da Silva, Moreira

'ENG. PULIDE GARCIA', HT, yb, 1961; flowers yellow stained pink; [Grand'mere Jenny X Michele Meilland]; da Silva, Moreira

'ENG. VITÓRIA PIRES', HT, dr, 1954; flowers velvety dark red, moderate fragrance; da Silva, Moreira

'ENGAGEMENT', Gr, op, 1969; bud globular; flowers coral-pink, large, dbl., exhibition form, slight fragrance; foliage dark, leathery; vigorous, bushy growth; [Ma

Perkins X Montezuma]; Patterson; Patterson Roses

Enghien, HT, ob, 1991; RvS-Melle

'ENGLAND'S GLORY', HT, lp, 1904; flowers large, dbl.; Wood

'England's Rose', S, my, 1995; Austin, David

'English Elegance'®, S, pb, 1986; Austin, David

'ENGLISH ESTATES', HT, yb, 1992; flowers deep yellow edged red, blooms borne mostly singly, dbl., 26–40 petals, 1.5–2.75 in., intense fragrance; some prickles; foliage medium, dark green, glossy; upright (80 cms) growth; [Whisky Mac X Catherine Cookson]; Thompson, Robert; Battersby Roses, 1991

English Garden™ *see* 'AUSBUFF'

'ENGLISH HEDGE', HRg (S), mp, 1959; flowers pink, small, single; red fruit; compact, upright growth; [(R. rugosa rubra X R. cinnamomea) X R. nitida]; Nyveldt

'ENGLISH HOLIDAY', F, yb, 1977; flowers yellow, blended with salmon, dbl., 33 petals, 4 in., moderate fragrance; foliage large, glossy; [Bobby Dazzler X Goldbonnet]; Harkness

'ENGLISH MISS', F, lp, 1977; flowers pale pink, blooms in clusters, dbl., 60 petals, 2.5 in., intense fragrance; foliage dark purple to dark green; [Dearest X The Optimist]; Cants of Colchester, Ltd., 1978

English Perfume™ *see* 'JACLEWT'

English Porcelain *see* 'MORPORC'

English Sachet™ *see* 'JACOLFA'

English Sonnet™ *see* 'HARVERAG'

English Violet, S, dp; Austin, David

Enhance *see* 'SANDAYA'

'ENID', Pol, lp, 1936; flowers pale pink, borne in clusters; foliage light; upright growth; Prior

Enjoy *see* 'LAVJOY'

'ENNIO MORLOTTI', F, mp, 1973; bud globular; flowers clear pink, dbl., 35 petals, 2.5–3 in., exhibition form, intense fragrance; foliage glossy; vigorous, upright growth; [Fashion X Queen Elizabeth]; Cazzaniga, F. G.

Enric Palau, HT, m; Dot

Ensa de Rennes *see* 'JADecosill'

'ENTENTE CORDIALE', HT, w, 1908; flowers nasturtium-red, base yellow; [Mme Caroline Testout X Soleil d'Or]; Guillot, P.

'ENTENTE CORDIALE', HT, mr, 1909; flowers creamy white, tinged carmine at edges; [Mme Abel Chatenay X Kaiserin Auguste Viktoria]; Pernet-Ducher

'ENTERPRISE', F, pb, 1957; flowers deep pink edged peach, medium borne in large clusters, semi-dbl., 20 petals, 2 in., moderate fragrance; foliage dark, glossy, vigorous; upright, bushy growth; [Masquerade X Seedling]; Kordes; Morse

'ENVER PASCHA', HT, lp, 1916; flowers large, dbl.; Kiese

'ENZO FUMAGALLI', F, mp, 1966; bud globular; flowers salmon-pink, full, medium, very dbl., abundant, intermittent bloom; foliage glossy; vigorous, bushy growth; [Mount Shasta X Papillon Rose]; Cazzaniga, F. G.

'EOS', S, rb, 1950; bud ovoid; flowers sunset-red becoming brighter, center white, semi-dbl., cupped, borne several together, non-recurrent, slight fragrance; foliage leathery, glossy; shrub or pillar (to 6 ft.) growth; [R. moyesii X Magnifica]; Ruys

'EPIC', F, mp, 1989; bud tapering; flowers medium, borne in sprays of 3-6, dbl., flat, moderate, spicy fragrance; prickles brown; foliage bronze to dark green, glossy; upright, bushy growth; [Silent Night X Irish Mist]; Cattermole, R.F.; South Pacific Rose Nursery

Epidor® *see* 'DELÉPI'

Epinal, F, 1994; Croix, P.

Epoca *see* 'LEN 2'

Epoca Mondadori, Gr, 1966; Lens, Louis

'EPONINE', HMsk (S), w; dbl., cupped, moderate fragrance

'EPOS', Pol, mr, 1971; bud ovoid; flowers medium, semi-dbl., cupped, slight fragrance; foliage bronze, leathery; vigorous, upright growth; [Tommy Bright X Seedling]; Delforge

'ERATO', LCl, w, 1937; bud pointed; flowers open, bornein clusters, semi-dbl., slight fragrance; foliage glossy; long stems; very vigorous climbing growth; [(Ophelia X R. multiflora) X Florex]; Tantau

'ERBPRINZESSIN LEOPOLD VON ANHALT', HT, ly, 1933; flowers whitish-yellow, large, dbl., moderate fragrance; Behrens

'ERFURT', HMsk (S), pb, 1939; bud long, pointed; flowers medium pink, yellow toward base, large blooms in clusters, semi-dbl., recurrent bloom, intense, musk fragrance; foliage leathery, wrinkled, bronze; vigorous (5-6 ft.), trailing, bushy growth; [Eva X Reveil Dijonnais]; Kordes, 1931

'ERIC', F, mr, 1965; bud ovoid; flowers medium, semi-dbl., cupped; foliage dark, glossy, leathery; vigorous growth; [Alain X Coup de Foudre]; Hémeray-Aubert

'ERIC B. MEE', HT, mr, 1937; flowers vivid cerise, well shaped, small; Mee; Beckwith

'ERIC HOBBIS', HT, pb, 1966; flowers pink, reverse peach, 4.5 in., exhibition form; low growth; [Gavotte X Peace]; Sanday, John

'ERIC HOLROYD', HT, mr, 1925; flowers bright scarlet, base shaded gold, moderate fragrance; Chaplin Bros.

'ERIC LOUW', HT, mr, 1964; bud pointed; flowers cyclamen-red, well formed, dbl., 35–40 petals; foliage leathery, glossy; strong stems; vigorous, bushy growth; [Queen Elizabeth X Confidence]; Herholdt, J.A.; Herholdt's Nursery

Eric The Red, F, lp; Welsh, 1997

Eric von Melnibonée, S, op, 1983; flowers yellowish salmon-pink, large, dbl., slight fragrance; Weihrauch

'ERICA', F, or, 1964; flowers orange-scarlet, frilled, borne in large clusters, semi-dbl.; [Seedling X Montezuma]; Herholdt, J.A.; Herholdt's Nursery

Erica *see* 'INTEROP'

Erica Herholdt, F, or, 1964; flowers large, semi-dbl.; Herholdt, J.A.

'ERICH FRAHM', F, mr, 1939; bud long, pointed, yellowish red; flowers carmine-scarlet, center yellow, shell shaped, open, borne in umbels of up to 20; foliage dark, glossy, leathery; vigorous growth, very branching.; [Dance of Joy X Mary Hart]; Kordes; Timm

'ERIE', S, lp, 1946; flowers pale pink, borne in clusters, single, 5 petals, very free, non-recurrent bloom; bottle shaped, bright red fruit; foliage dark, fragrant (sweetbriar); vigorous, spreading growth; hardy.; Preston; Central Exp. Farm

'ERIE TREASURE', HRg (S), w; flowers blush to white, dbl., recurrent bloom, intense fragrance; foliage wrinkled; vigorous (6 ft.), bushy growth; [Souv. de Pierre Leperdrieux X Nova Zembla]; Wedrick

'ERIK HJELM', HT, op, 1929; flowers pure salmon-pink, very dbl., moderate fragrance; [Lieut. Chaure X Sachsengruss]; Kordes

Erika, HT, mp

'ERIKA MY LOVE', Min, my, 1993; flowers yellow with white tips, medium, dbl., 25–30 petals, 2–3 in., borne mostly singly, no fragrance; some prickles; foliage medium green, semi-glossy; tall (72 cms), upright growth; [Rise 'n' Shine X Seedling]; Armstrong, James L.

Erika My Love, Min, my; flowers purple-violet, dbl., cupped; [Rise'n'Shine X Seedling]

'Erika Pluhar', HT, dr

'ERIKA TESCHENDORFF', HT, mr, 1949; bud long, pointed; flowers fiery scarlet, open, very large, dbl., globular, slight fragrance; foliage glossy, dark; very vigorous, upright growth; Berger, V.; Teschendorff

'ERIKONIG', HMult, m, 1886; Geschwind, R.

Erin Fleming, LCl, mp, 1997; flowers very full, medium, very dbl., 41 petals, borne singly and in large clusters, up to 15 buds per cluster, moderate fragrance; foliage medium, dark green, glossy; upright, tall growth; [Sunsation X Henry Kelsey]; Fleming, Joyce L.

'ERINNERUNG AN BROD', HSet, rb; (Souv. de Brod); flowers cerise through crimson to purple, large, dbl., quartered, moderate fragrance; Geschwind, R., 1886

'ERINNERUNG AN SCHLOSS SCHARFENST', HT, m, 1892; flowers purple/pink, large, dbl., intense fragrance; Geschwind, R.

Erlkönig *see* **'ROI DES AUNES'**

'ERNA', HWich, lp, 1929; flowers small, dbl.; Vogel, M.

'ERNA BALTZER', HT, my, 1954; flowers golden yellow, medium, moderate fragrance; vigorous growth; [Tawny Gold X Gaudia]; Leenders, M.

Erna Doris *see* 'LENDORI'

'ERNA GROOTENDORST', F, dr, 1938; flowers deep velvety crimson, large, semi-dbl.; foliage glossy, dark; bushy growth; [Bergers Erfolg X Gloria Mundi]; Grootendorst, R.

'ERNA TESCHENDORFF', Pol, mr, 1911; flowers strawberry-red, open, small, semi-dbl., slight fragrance; foliage rich green, soft; bushy growth; [Mme Norbert Levavasseur sport]; Teschendorff

'ERNEST H. MORSE', HT, mr, 1964; flowers turkey-red, dbl., 30 petals, 4 in., intense fragrance; foliage leathery; vigorous growth; GM, RNRS, 1965; Kordes; Morse

'ERNEST LAURENT', HT, lp, 1914; flowers large, dbl., moderate fragrance; Viaud-Bruant

'ERNEST METZ', T, mp, 1888; flowers rose-pink, center darker, large, dbl.; Guillot, J. B.

'ERNEST MOREL', HP, dr, 1898; flowers bright garnet-red; Cochet, P.

'Ernest's Blue', HT, m, 1992

'ERNESTINE COSME', HWich, rb, 1926; flowers brilliant red, with large white eye, borne in clusters of 75, single; many thorns; very vigorous, climbing growth; Turbat

Ernie *see* 'TINERNIE'

'ERNIE PYLE', HT, mp, 1946; bud long, pointed; flowers deep rose-pink, reverse deeper, dbl., 35–40 petals, 4.5–5 in., cupped, intense fragrance; foliage leathery; vigorous, upright, bushy growth; [((Royal Red X Talisman) X Seedling (red)) X (Talisman X Nutneyron]; Boerner; J&P

'ERNST DECHANT', HWich, w, 1928; flowers white with yellow tints, small, semi-dbl.; Vogel, M.

'Ernst G. Dörell'®, HMult, dp; Geschwind, R., 1897

'ERNST GRANDPIERRE', HWich, w, 1902; flowers pale cream, base yellow, open, borne in clusters, dbl., sparse bloom, slight fragrance; foliage light, glossy; height 8-10 ft; [R. wichurana X Perle des Jardins]; Weigand, C.

Eroica *see* **'EROTIKA'**®

Eroika *see* **'EROTIKA'**®

'EROS', F, dp, 1955; flowers deep rosy pink shaded brick-red, base yellow; dwarf, compact growth; [Pinocchio seedling]; Maarse, G.

Eros *see* 'BARsor'

Erotica *see* **'EROTIKA'**®

'EROTIKA'®, HT, dr, 1968; (Eroica, Eroika, Erotica); bud ovoid; flowers velvety dark red, well-formed, large, dbl., 33 petals, intense fragrance; foliage dark, glossy; vigorous, upright growth; ADR, 1969; Tantau, Math.

Mothers Day *see* **'MOTHERSDAY'**

'ERSKINE', Sp, mp; Hansen, N.E.

'ERUPTION', HT, mr, 1934; flowers fiery scarlet-red, large, semi-dbl.; foliage sea-green; bushy growth; [Red-Letter Day X Columbia]; Van Rossem

'ERWIN HÜTTMANN', HWich, dp, 1941; flowers medium, dbl., intense fragrance; Krause

Escada *see* 'TANadac'

'ESCALADE', Cl HT, mr, 1962; bud pointed; flowers carmine, large, exhibition form; vigorous growth; [Spectacular X Charlotte Armstrong]; Combe; Vilmorin-Andrieux

Escapade® *see* 'HARPADE'

Esco Rose, F; Gregory, C.

'ESCORT', F, dr, 1963; bud pointed to urn-shaped; flowers small to medium, dbl., 30 petals, exhibition form; foliage dark, leathery; vigorous, bushy growth; [Spartan X Garnette]; Swim & Weeks

'ESCULTOR CLARÁ', HT, m, 1956; bud pointed; flowers purple-garnet, reverse magenta, large, dbl., 30 petals, exhibition form, moderate fragrance; foliage dark, glossy; very vigorous, upright, compact growth; [Lilette Mallerin X Floradora]; Dot, Simon

Escurial® *see* 'DELFLIP'

'ESKIL', HT, yb, 1939; flowers light yellow overlaid red and orange; [Mrs Franklin D. Roosevelt sport]; Ringdahl

Eskimo® *see* 'KORcilmo'

'ESMÉ', HT, w, 1920; flowers cream-white, edged rosy carmine, dbl.; [Mme Edouard Herriot X Seedling]; Cant, B. R.

Esme Euvrard *see* 'KORelgas'

'ESMERALDA', F, dp, 1957; flowers deep rose-pink, reverse lighter, small, borne in very large, dbl., moderate fragrance; vigorous growth; [Gartendirector Otto Linne X Seedling]; Riethmuller

Esmeralda® *see* 'KORMALDA'

Especially for You *see* 'FRYworthy'

'ESPERANTO', HT, my, 1932; flowers pure yellow, very large, dbl., 60 petals, globular, intense fragrance; [Miss Lolita Armour sport]; Böhm, J.

Esperanto Jubileo *see* 'SANROZO'

'ESPÉRANZA', F, mr, 1966; bud ovoid; flowers bright red, large blooms in clusters, dbl.; foliage dark, bronze, leathery, glossy; upright growth; GM, Baden-Baden, 1968 GM, The Hague, 1968; [Donald Prior X Reverence]; Delforge

'ESPLANADE', HT, dr, 1961; dbl., 40 petals, moderate fragrance; foliage glossy; vigorous growth; [Soraya X Seedling]; Verbeek

'ESPOIR', HCh, w, 1984; (Interbec); flowers small, single, 5 petals, slight fragrance; foliage very small, medium green, semi-glossy; upright, spreading growth; [HCh seedling X HCh seedling]; Gailloux, Gilles

'ESPOIR', F, 1958; flowers rich salmon, rosette shape, dbl., 50 petals, 4.5 in., moderate fragrance; vigorous, low, bushy growth; RULED EXTINCT 6/83 ARM; [(Oiseau de Feu X Fashion) X (Independence X Seedling)]; Combe; Japan Rose Society

'ESPOIR', HT, 1947; bud long, pointed; flowers pink, center light salmon-pink, large, dbl., 35 petals, abundant bloom, slight fragrance; foliage soft; bushy growth; RULED EXTINCT 6/83?; [Charles P. Kilham X Neville Chamberlain]; Lens

Esprit® *see* 'KORHOLST'

'ESSENCE', HT, dr, 1930; bud pointed; flowers rich scarlet-crimson, becoming bluish, outer petals slightly fimbriated, cupped, moderate, damask fragrance; Cant, B. R.

'ESSENCE, CLIMBING', Cl HT, dr, 1938; Western Rose Co.

Essex *see* 'POUlnoz'

Essie Lee *see* 'JUDLEE'

'ESTAFETTE', F, dr, 1962; flowers open, borne in clusters, semi-dbl., 2.5–3 in.; foliage dark, glossy; vigorous growth; [Alain X Elmshorn]; Delforge

Estelle, HGal, mp; flowers medium, very dbl., intense fragrance; Roseraie de l'Hay

Estelle *see* 'OLIjbrau'

Estelle, HT, op; Croix

Estelle Pradel *see* **'ZÉLIA PRADEL'**

Esterel *see* 'KORIVER'

'ESTHER', F, ob, 1954; bud pointed, turkey-red; flowers golden orange, reverse lighter, open, large, borne in clusters, single, 7–8 petals; foliage glossy, bright green; long stems; very vigorous, bushy growth; [Cocorico X Canzonetta]; San Remo Exp. Sta.

'ESTHER', G, pb, 1849; Vibert

Esther Geldenhuys *see* 'KORSKIPEI'

Esther Jasik *see* 'ZIPEST'

'ESTHER JERABEK', F, mp, 1978; semi-dbl., 18 petals; spreading growth; [The Fairy seedling]; Jerabek, Paul E.

Esther O'Farim *see* 'KORFARIM'

Esther Ofarim *see* 'KORFARIM'

Esther Peiro® *see* 'FEbasa'

'ESTHER RANTZEN', F, ob, 1982; flowers medium, semi-dbl., moderate, fruity fragrance; foliage medium, medium green, semi-glossy; upright, bushy growth; [Spartan X Orangeade]; Dwight, Robert & Sons

Esther's Baby® *see* 'HARKINDER'

Estima, Min, op

Estralia *see* **'ESTRELLITA'**

'ESTRELLITA', Min, w; (Estralia, Pixie Pearl); flowers large blooms in clusters, dbl.; Said to have been grown in US since about 1910

Estrellita de Oro *see* **'BABY GOLD STAR'**

'ESTRU', F, lp, 1975; flowers patio, small, dbl., 50 petals, 1.5 in.; foliage small, glossy, dark; low, compact growth; [Rosy Jewel X Floribunda seedling]; deRuiter

'ETAIN', HWich, op, 1953; flowers salmon-pink, borne in trusses, slight fragrance; foliage glossy, dark, almost evergreen; very vigorous growth; Cant, F.

Été Parfumé *see* **'TYPHOO TEA'**

'ÉTENDARD', LCl, mr, 1956; (New Dawn Rouge, Red New Dawn); flowers bright red, dbl., blooms in large clusters, recurrent bloom, moderate fragrance; foliage very glossy, leathery; vigorous growth; [New Dawn X Seedling]; Robichon

'ETENDARD DE JEANNE D'ARC', N, w, 1882; Margottin

Éterna® *see* 'DELIC'

Eterna Giovanezza *see* **'ETERNAL YOUTH'**

'ETERNAL FLAME', LCl, ob, 1955; (Everblooming Pillar No.12); flowers light orange, semi-dbl., 12–19 petals, 3–4 in., moderate fragrance; upright, climbing growth; [Seedling X Queen o' the Lakes]; Brownell, H.C.

Eternal Flame, F, ob, 1992; Interplant

'ETERNAL SUN', HT, or, 1966; bud ovoid; flowers vermilion, large, dbl., exhibition form, slight fragrance; foliage dark, leathery; vigorous, upright, bushy growth; [Seedling X Jacqueline (HT)]; Hill, Joseph H., Co.

'ETERNAL YOUTH', HT, lp, 1937; (Eterna Giovanezza, Jeunesse Éternelle); bud long, pointed; flowers suffused orange-salmon, dbl., 50 petals, 4–5 in., cupped, intense fragrance; foliage leathery; vigorous, upright, bushy growth; [Dame Edith Helen X Julien Potin]; Aicardi, D.; J&P

Eternally Yours *see* 'MACSPEEGO'

'ETERNITÉ', F, mr, 1947; flowers scarlet, dbl., 25 petals, 4–5 in., moderate fragrance; foliage dark; very free growth; [Mme Joseph Perraud X Holstein]; Gaujard

Eternity *see* 'TWOETERN'

'ETHEL', HWich, lp, 1912; flowers flesh-pink, borne in clusters, semi-dbl.; vigorous growth; [Dorothy Perkins seedling]; Turner

Ethel Austin *see* 'FRYMESTIN'

'ETHEL CHAPLIN', HT, my, 1926; flowers soft lemon-yellow, dbl., moderate fragrance; Chaplin Bros.

'Ethel Dawson', F, lp; [Daily Sketch X (Stella X Charles Mallerin)]; Dawson, George

'Ethel Dawson', F, lp; [Red Meillandina X seedling]; Dawson, George, before 1987

'ETHEL DICKSON', HT, dp, 1917; flowers deep salmon-rose with silvery flesh reflexes; Dickson, H.

'ETHEL JAMES', HT, dp, 1921; flowers softer carmine-red than Isobel, flushed orange-scarlet, center yellow, single, 5 petals; bushy growth; GM, NRS, 1920; McGredy

'ETHEL MALCOLM', HT, w, 1909; flowers ivory-white, very large; vigorous growth; GM, NRS, 1909; McGredy

Ethel Orr™ *see* 'MINAMCO'

'ETHEL SANDAY', HT, yb, 1954; flowers yellow flushed apricot, well-formed, dbl., 34 petals, 4–5 in., slight fragrance; foliage dark; vigorous, upright growth; GM, NRS, 1953; [Rex Anderson X Audrey Cobden]; Mee; Sanday

'ETHEL SLOMAN', HT, mr, 1966; bud ovoid; flowers crimson, very dbl., slight fragrance; foliage leathery; compact, bushy growth; [Baccará X My Choice]; Fankhauser

'ETHEL SOMERSET', HT, mp, 1921; bud pointed; flowers shrimp-pink large, dbl., exhibition form, intense fragrance; vigorous, branching growth; Dickson, A.

'ETHEL UTTER', LCl, my; bud cherry-red; flowers, dbl.; Wilber

Ethical, F, rb, 1995; Dawson

'ETIENNE DUBOIS', HP, dr, 1873; flowers large, dbl.; Damaizin

'ÉTIENNE LEVET', HP, mr, 1871; flowers carmine-red, large, dbl., 70 petals, sometimes recurrent bloom; vigorous, erect growth; [Victor Verdier seedling]; Levet Père

'ÉTIENNE REBEILLARD', HT, pb, 1924; flowers flesh-pink, suffused golden, semi-dbl., slight fragrance; Pernet-Ducher

'ÉTINCELANTE', HT, dr, 1941; flowers brilliant red, tinted purple, dbl., moderate fragrance; [Gruss an Teplitz X Étoile de France]; Chambard, C.

'ÉTINCELLE', HT, or, 1958; flowers bright red tinted orange, medium, dbl., 40–45 petals, intense fragrance; low, bushy growth; [Crimson Glory X Seedling]; Moulin; Vilmorin-Andrieux

'ETNA', M, dr, 1845; flowers crimson shaded purple, large, very mossy, moderate fragrance; Laffay, M.

'ETNA', HT, dr, 1924; flowers deep crimson-maroon, semi-dbl.; [Red-Letter Day X H.V. Machin]; Looymans; Prior

'ÉTOILE D'ALAÏ', HMsk (S), mr, 1946; flowers brilliant red, prominent golden stamens, medium, very dbl., repeat bloom; bushy growth; [Skyrocket seedling]; Meilland, F.

'ÉTOILE D'OR', HT, yb, 1931; flowers golden yellow, reverse shaded orange, large, semi-dbl., intense fragrance; upright, bushy growth; Pernet-Ducher; Gaujard

'ÉTOILE DE BELGIQUE', HT, mr, 1946; flowers brilliant red, very large, dbl., slight fragrance; foliage bronze; vigorous, bushy growth; [Charles P. Kilham X Étoile de Hollande]; Lens

'ÉTOILE DE BELGIQUE', HT, mr, 1956; bud ovoid; flowers geranium-red, large, dbl.; bushy, spreading growth; [Independence X Happiness]; Buyl Frères

'ÉTOILE DE FEU', HT, op, 1921; flowers salmon-pink and coral-red, large, dbl., globular; foliage bronze; vigorous, bushy, branching growth; Pernet-Ducher

'ÉTOILE DE FEU, CLIMBING', Cl HT, ob, 1930; H&S

'ÉTOILE DE FRANCE', HT, dr, 1904; bud pointed; flowers dark rose-red, center cerise, medium to small, dbl., cupped, intense fragrance; bushy growth; [Mme Abel Chatenay X Fisher Holmes]; Pernet-Ducher

'ÉTOILE DE FRANCE, CLIMBING', Cl HT, dr, 1915; Howard Rose Co.

'ÉTOILE DE HOLLANDE', HT, mr, 1919; flowers bright red, large, dbl., 35–40 petals, cupped, intense, damask fragrance; foliage soft; moderate, open growth; (28); [Gen. MacArthur X Hadley]; Verschuren

'ÉTOILE DE HOLLANDE, CLIMBING', Cl HT, mr, 1931; (Grimpant Étoile de Hollande); Leenders, M.

'ÉTOILE DE LYON', T, my, 1881; flowers golden yellow, dbl., sparse intermittent bloom, moderate fragrance; foliage soft; short, weak stems; bushy growth; Guillot, P.

'ÉTOILE DE MAI', Pol, ly, 1893; flowers sulfur-white, small, dbl., intense fragrance; vigorous growth; Gamon

'ÉTOILE DE PORTUGAL', LCl, dp, 1898; flowers rose-red, dbl.; Cayeux, H.

'ÉTOILE LUISANTE', Pol, mr, 1918; (Baby Herriot); bud pointed; flowers cerise-red, shaded coppery, semi-dbl., exhibition form, borne in clusters, slight fragrance; few thorns; foliage bronze, glossy; long stems; bushy growth; Turbat; Michell

Etoile Rouge, Min, rb

'ETTY VAN BEST', HT, ly, 1934; flowers white, shaded yellow, dbl.; foliage leathery; vigorous growth; [Pharisaer X Souv. de H.A. Verschuren]; Buisman, G. A. H.

'ÉTUDE', LCl, dp, 1965; flowers deep rose-pink, semi-dbl., blooms in clusters, recurrent bloom, moderate fragrance; foliage glossy, light green; [Spectacular X New Dawn]; Gregory

'EUCHARIS', G, dp; flowers bright rose, edged lighter, large, dbl.; Descemet, M. (?), before 1815

'EUGÈNE APPERT', HP, dr, 1860; flowers medium, dbl.; Trouillard

'EUGÈNE BARBIER', HP, dy, 1920; flowers brilliant canary-yellow, shaded coppery golden yellow, dbl., globular, moderate fragrance; few thorns; upright growth; [Frau Karl Drushchki X Rayon d'Or]; Barbier

'EUGÈNE BOULLET', HT, dr, 1909; flowers crimson-red, dbl.; vigorous growth; Pernet-Ducher

'EUGÈNE DE BEAUHARNAIS', Ch, m, 1838; (Prince Eugène); flowers purple, large, dbl., moderate fragrance; Hardy

'EUGÈNE DE SAVOIE', M, mr; flowers bright red, shaded, dbl.; Moreau et Robert, 1861

'EUGÈNE E. MARLITT', B, mr, 1900; (Mme Eugène Marlitt); flowers bright carmine shaded scarlet, large, dbl.; few prickles; vigorous growth; Geschwind, R.

'EUGÈNE FÜRST', HP, dr, 1875; flowers crimson-red, shaded purple, large, dbl., globular, recurrent bloom, moderate fragrance; [Baron de Bonstetten X Seedling]; Soupert & Notting

'EUGÈNE JACQUET', HWich, mr, 1916; flowers cherry-red, borne in clusters of 25-30, dbl., very early bloom, moderate fragrance; foliage bright green; vigorous, symmetrical growth; [wichurana hybrid (red) X Multiflora hybrid (pink)]; Turbat

'EUGÈNE JANVIER', G, dp; flowers dark pink, paling to lilac, medium, dbl.

'EUGÈNE PICARD', HT, op, 1938; flowers light coppery-pink with red tints, large, very dbl., moderate fragrance; Sauvageot, H.

'EUGÈNE TRANSON', LCl, or, 1926; flowers orange and copper, reverse orange-red, shaded, borne in clusters; vigorous, climbing growth; [Mme Berard X Constance]; Barbier

'EUGÈNE VERDIER', M, dp, 1872; flowers crimson or light red, center deeper, well-formed, very dbl., moderate fragrance; Verdier, E.

'EUGENIA', HT, rb, 1920; flowers coral-red to prawn-red, flecked or striped yellow, dbl., moderate fragrance; [Mme Edouard Herriot sport]; Collier

Eugénie de Guinoiseau *see* **'EUGÉNIE GUINOISEAU'**

'EUGÉNIE GUINOISEAU', M, mr, 1864; (Eugénie de Guinoiseau); flowers reddish cerise, changing to reddish violet, large, dbl.; vigorous growth; Guinoiseau, B.

'EUGÉNIE LAMESCH', Pol, yb, 1899; flowers ochre-yellow and bright yellow, shaded pink, blooms in clusters, dbl., moderate fragrance; foliage glossy; dwarf, compact growth; [Aglaia X William Allen Richardson]; Lambert, P.

'EUGENIE VERDIER', HP, pb, 1869; Guillot

'EUGENIO D'ORS', HT, dr, 1946; flowers oxblood-red, large, dbl., intense fragrance; [Sensation X Margaret McGredy]; Camprubi, C.

'EUGENIO FOJO', HT, mr, 1953; bud pointed; flowers vermilion-red, well formed, large, dbl., 35 petals, moderate fragrance; vigorous, bushy growth; [Texas Centennial X Carlos Fargas]; Dot, Pedro

'EUGhien', HT, ab; (**Pearly Shores**); (EUGhien), 1995

'EULALIA', HT, mp, 1934; flowers pink, lighter toward base, large, dbl., intense fragrance; vigorous growth; Verschuren-Pechtold; Dreer, H&S

'Eulalia Berridge', Pol, lp; flowers medium, dbl., slight fragrance

'EULALIA LEBRUN', HGal, w, 1844; flowers white with pink stripes, medium, dbl.; Vibert

'EUMUNDI', F, w, 1953; flowers pure white, very dbl., cupped; foliage light green; vigorous, bushy growth; [Yvonne Rabier X Baby Alberic]; Ulrick, L.W.

'EUMUNDI', F, lp; [Yvonne Rabier X Baby Alberic]; Ulrick, L.W., 1954

Euphoria *see* 'INTereup'

Euphrates *see* 'HARUNIQUE'

'EUPHROSYNE', HMult, mp, 1895; flowers pure pink, small, dbl., moderate fragrance; very vigorous growth; [R. multiflora X Mignonette]; Schmitt

'EUREKA', HT, mp, 1914; flowers bright rose; Hobbies

'EUREKA', S, w, 1956; flowers pure white, small, borne in clusters, semi-dbl.; intended for trial as a hardy understock.; [Probably Betty Bland X Ames 5]; Wright, Percy H.

Euro 92, F, mr, 1989; RvS-Melle

'EuroGROOT', F, lp; (**Souvenir**); Kordes/Grootendorst, 1991

'EUROPA', HT, mp, 1928; flowers bright pink, better; [Columbia sport]; Keessen; Nieuwesteeg

Europa *see* 'KORTEXUNG'®

Europas Rosengarten *see* 'HELeuro'

Europawelle Saar, HT, op, 1985; flowers salmon-pink, large, dbl.; Meilland

Europe 92 *see* 'DELSULAN'

Europe Sensation, F, or

European Touch, HT, ab, 1997; bud elongated, edged with pink; flowers large, dbl., 4–5 in., exhibition form, borne singly, no fragrance; foliage large, dark green, matte; long, straight stems; very tall (5 to 6 ft), upright, vigorous growth; Von Koss

'EUROPEANA'®, F, dr, 1968; flowers dark velvety crimson, rosette shape, dbl., 25–30 petals, 3 in., rosette, blooms in large, heavy clusters, slight, tea fragrance; foliage bronze-green to dark green; vigorous growth; AARS, 1968 GM, Portland, 1970 GM, The Hague, 1962; [Ruth Leuwerik X Rosemary Rose]; deRuiter, 1963; C-P, 1968

Europeana, Climbing *see* 'BUREURO'

'EUROROSE', F, yb, 1973; flowers yellow-ochre, flushed fire-red, dbl., 25 petals, 3.5 in., globular; [Zorina X Redgold]; Dickson, A.

Eurosong® *see* 'LENORAN'

Eurostar *see* 'POUlreb'

'EUROVISION', HT, dr, 1961; dbl., 30 petals, 4 in., moderate fragrance; foliage dark; vigorous, bushy growth; [Miss France X Rosita]; Delforge

Euroway *see* 'HELway'

Euryanthe®, S, mp

'EURYDICE', HSet, 1886; Geschwind, R.

Eustace, HT, w, 1989; Robinson

Eustacia *see* 'SEAEUS'

'EUTERPE', LCl, ly, 1937; bud pointed; flowers open, borne in clusters, semi-dbl., slight fragrance; foliage glossy; long stems; very vigorous, climbing growth; [(Ophelia X R. multiflora) X Florex]; Tantau

'EUTIN', F, dr, 1940; (Hoosier Glory); bud globular, pointed; flowers glowing carmine-red, blooms in clusters, dbl., cupped, slight fragrance; foliage leathery, glossy, dark; vigorous growth; [Eva X Solarium]; Kordes

'EUTIN, CLIMBING', Cl F, dr, 1957; Lindquist; Howard Rose Co.

'EVA', HMsk (S), rb, 1933; bud pointed; flowers carmine-red, center white, large blooms in clusters to 75, semi-dbl., intermittent bloom, moderate fragrance; very vigorous growth; (28); [Robin Hood (HMsk) X J.C. Thornton]; Kordes

'EVA DE GROSSOUVRE', HT, mp, 1908; Guillot

'EVA EAKINS', HT, or, 1926; flowers scarlet-carmine, flushed orange, base bright yellow, small, dbl., exhibition form, slight fragrance; foliage leathery; bushy growth; McGredy

Eva Gabor *see* 'POULTAL'

'EVA KNOTT', HT, ob, 1957; flowers coppery orange, well formed, dbl., 35 petals; vigorous growth; [Ethel Sanday X Mrs Sam McGredy]; Mee

'Eva Simone', F, op, 1992; dbl., intense fragrance; Michler, K. H.

'EVA TESCHENDORFF', HMult, w, 1923; flowers greenish-white; [Echo sport]; Grunewald; Teschendorff

'EVA TESCHENDORFF, CLIMBING', Cl Pol, w, 1926; Opdebeeck

'EVALINE', Pol, lp, 1920; flowers light pink, edged brighter, petals quilled, small, borne in, dbl., moderate fragrance; bushy growth; [Orléans Rose X Rayon d'Or]; Prosser

'EVANGELINE', HWich, pb, 1906; flowers rosy white, veined cameo-pink, blooms in clusters, single, 2 in., late seasonal bloom, moderate fragrance; foliage dark, leathery; long stems; very vigorous, climbimg (12-15 ft) growth; [R. wichurana X Crimson Rambler]; Walsh

'EVANGELINE', T, w, 1951; bud deep pink; flowers creamy white, edged blush-pink, medium, dbl., moderate fragrance; almost thornless; vigorous, spreading growth; [Mrs Dudley Cross sport]; Krider Nursery

'EVANGELINE BRUCE', F, yb, 1971; flowers yellow, flushed pink, well-formed, dbl., 24 petals, 4.5 in., moderate fragrance; foliage light; [Colour Wonder X Sea Pearl]; Dickson, A.

Evasion, F, 1972; Hendricks

'EVE', HT, ob, 1954; bud long, pointed; flowers coral-red shaded yellow, very large, dbl., moderate fragrance; foliage glossy; vigorous growth; Gaujard

'EVE ALLEN', HT, rb, 1964; flowers crimson, reverse and base saffron-yellow, dbl., 26 petals, 5 in., moderate fragrance; foliage dark, glossy; vigorous growth; [Karl Herbst X Gay Crusader]; Allen, E.M.; Sanday

'EVEBRIGHT', F, rb, 1991; (**Cecily Gibson**, 'EVERBRIGHT'); bud ovoid; flowers current red, yellow base, blooms borne in sprays of 3-6, dbl., 35 petals, cupped, slight fragrance; foliage medium to large, dark green, glossy; bushy, medium to tall growth; [Southampton X ((Arthur Bell X Maigold) X Glenfiddich)]; Everitt, Derrick, 1990

'EVEGAYTIME', F, ob, 1999; (**Kantha Selvon**); flowers light orange, reverse cream yellow, dbl., 26–40 petals, 3–3.5 in., borne singly and in small clusters, slight fragrance; prickles moderate; foliage medium, medium green, glossy; upright, medium (2.5 ft) growth; TG, RNRS, 1997; [Pot O'Gold X (Mary Sumner X (Glenfiddich X (Arthur Bell X Maigold)))]; Everitt, Derrick

'EVEJAUNTY', F, ob, 1999; (**Alice Jarrett**); flowers orange, reverse cream yellow, dbl., 26–40 petals, 3–3.5 in., borne in small clusters, slight fragrance; prickles moderate; foliage medium, medium green, semi-glossy; upright, tall (3-3.5 ft) growth; [(Mary Sumner X (L'Oreal Trophy X Edith Holden)) X Friendship]; Everitt, Derrick

'EVELEINE', F, ob, 1993; (**Lady Madeleine**); flowers tangerine, paler reverse, blooms borne singly and in small clusters, dbl., 15–25 petals, 1.5–2.75 in., slight fragrance; some prickles; foliage medium, medium green, semi-glossy; medium (30 cms), upright growth; [(Pristine X Edith Holden) X (L'Oreal Trophy X Edith Holden)]; Everitt, Derrick

Evelien *see* 'INTERLIEN'

Evelina *see* 'AMAEVELINA'

'EVELYN', HT, lp, 1918; flowers soft pink, base yellow, dbl., 45–50 petals, moderate fragrance; RULED EXTINCT 1/92; [Ophelia sport]; Pierson, A.N.

'EVELYN', HT, MP, 1918; flowers salmon, shaded and edged rose, base yellow, imbricated, large, dbl.; RULED EXTINCT 1/92; Paul, W.

Evelyn *see* 'AUSSAUCER'

Evelyn, Min, dp; Agel

'EVELYN BUCHAN', HT, pb, 1959; flowers pink tinted yellow, dbl., 23 petals, 3–4 in., exhibition form, moderate fragrance; foliage leathery, dark; vigorous, upright growth; [Luis Brinas X Crimson Glory]; Riethmuller

'EVELYN DAUNTESSEY', HT, PB, 1909; flowers salmon stained carmine-rose; moderately vigorous growth; McGredy

'EVELYN ELLICE', F, LP, 1966; flowers light pink, becoming white, pointed, borne in clusters, 3 in.; foliage light green; vigorous growth; [Queen Elizabeth sport]; Ellice

Evelyn Fison *see* 'MACEV'

Evelyn Grace *see* 'HORAVME'

Evelyn Hough, HT, lp

'EVELYN MAY', HT, dp, 1932; flowers vermilion-pink, dbl., 65 petals, moderate fragrance; free growth; [Lady Alice Stanley X Edith Part]; Edward

'EVELYN MURLAND', HT, pb, 1923; flowers salmon-pink and carmine, veined yellow, reverse veined pink, dbl., moderate fragrance; Dickson, A.

Evelyn Rogers™ *see* 'MINARCO'

Evelyn Taylor, F, ob

'EVELYN THORNTON', Pol, mp, 1919; flowers shell-pink deepening to salmon and lemon shaded orange, open, dbl., moderate fragrance; foliage leathery, glossy, dark bronze; bushy growth; [Léonie Lamesch X Mrs W.H. Cutbush]; Bees

Evening Fire *see* 'FROFIRE'

'EVENING GLOW', HT, r, 1959; bud long, pointed; flowers buff, large, dbl., 35 petals, moderate fragrance; foliage leathery; moderate growth; [Charlotte Armstrong X Narzisse]; Armbrust; Langbecker

Evening Light *see* **'TARDE GRIS'**

'EVENING NEWS', HT, my, 1927; flowers apricot-yellow veined rose, base deep buttercup-yellow, open, dbl., moderate fragrance; foliage glossy, rich green; vigorous, bushy growth; [Mme Edouard Herriot sport]; Letts

Evening Queen *see* 'HADEVENING'

Evening Sentinel *see* 'FRYevenest'

Evening Shadows *see* 'MICEVEN'

'EVENING SKY', HT, ob, 1939; flowers orange, tipped scarlet, base yellow, single, 6–8 ruffled petals, 2.5–3 in., moderate fragrance; foliage bluish green; vigorous growth; [Talisman X Unknown]; Moore, Ralph S.

Evening Star® *see* 'JACVEN'

'EVENING STAR', HT, my, 1919; flowers golden yellow, shaded apricot, large, dbl., slight fragrance; bushy growth; [Mme Edouard Herriot sport]; Morse

'EVENING TELEGRAPH', HT, dy, 1976; [Whisky Mac sport]; Haynes

'EVENSONG', HT, op, 1963; flowers rosy salmon, well-formed, dbl., 25 petals, 5

in.; foliage dark; vigorous growth; [Ena Harkness X Sutter's Gold]; Arnot; Croll

'EVENTAIL', HT, yb, 1989; bud ovoid; flowers light yellow to pink, medium, borne singly, dbl., 50 petals, slight fragrance; prickles downward-pointed, reddish-purple; foliage medium, semi-glossy, slightly denticulated; upright, tall growth; [Sonia X Miyabi]; Kono, Yoshito

'EVENTIDE', HT, dr, 1948; bud ovoid; flowers dark velvety red, open, medium, dbl., moderate fragrance; foliage wrinkled, soft; moderate growth; [Crimson Glory seedling X Rouge Mallerin]; Toogood

'EVEPILK', F, pb, 1993; (**Mary Pilkington**); flowers peach, quickly paling to light pink, creamy yellow reverse, dbl., 15–25 petals, 1.5–2.75 in., moderate fragrance; some prickles; foliage medium, dark green, semi-glossy; tall (80-90 cms), bushy growth; [Mary Sumner X Remember Me]; Everitt, Derrick

'EVER READY', LCl, mr, 1976; flowers bright crimson, dbl., 32 petals, 3 in., recurrent bloom, moderate fragrance; foliage large, medium green, matt; [Aloha X Étoile de Hollande]; MacLeod; Christie Nursery, Ltd.

'EVERBLOOM', Cl Pol, dp, 1939; flowers deep pink, borne in clusters, single, recurrent bloom; foliage glossy; height 3-6 ft first year, 8-10 ft in about 3 years; [Phyllis Bide seedling]; Archer

Everblooming Climbing Cecile Brunner, Cl Pol, lp, 1996; Heirloom

Everblooming Dr W. Van Fleet *see* **'NEW DAWN'**

Everblooming Jack Rose *see* **'RICHMOND'**

'EVERBLOOMING PILLAR NO. 122', LCl, yb, 1954; flowers light yellow and orange, dbl., 90 petals, 3.5–4.5 in., moderate fragrance; growth like a hybrid tea, followed by 4-5 ft. canes; [Seedling X Break o' Day, Climbing]; Brownell, H.C.

'EVERBLOOMING PILLAR NO. 126', LCl, mp, 1955; flowers pink, base yellow, dbl., 35–50 petals, 3.5–4.5 in., moderate fragrance; bushy, upright growth; [Seedling X Queen o' the Lakes]; Brownell, H.C.

'EVERBLOOMING PILLAR NO. 214', LCl, my, 1954; flowers amber-yellow, dbl., 75 petals, 3.5 in.; growth like a hybrid tea, followed by 4-5 ft. canes; [Seedling X Break o' Day, Climbing]; Brownell, H.C.

'EVERBLOOMING PILLAR NO. 340', LCl, pb, 1957; flowers pink and yellow, dbl., 35–40 petals, 4.5–5 in., exhibition form, moderate fragrance; growth like a hybrid tea, followed by longer canes.; [Queen o' the Lakes X Scarlet Sensation]; Brownell, H.C.

Everblooming Pillar No. 83 *see* **'SALMON ARCTIC'**

Everblooming Pillar No.12 *see* **'ETERNAL FLAME'**

Everblooming Pillar No.73 *see* **'SCARLET SENSATION'**

'EVERBRIGHT', F, rb, 1991; (**Cecily Gibson**, 'EVEBRIGHT'); bud ovoid; flowers current red, yellow base, blooms borne in sprays of 3-6, dbl., 35 petals, cupped, slight fragrance; foliage medium to large, dark green, glossy; bushy, medium to tall growth; [Southampton X ((Arthur Bell X Maigold) X Glenfiddich)]; Everitt, Derrick, 1990

'EVERDREAM', HT, my, 1956; bud ovoid; flowers canary-yellow, dbl., 35–40 petals, 4–5 in., moderate fragrance; bushy growth; [Souv. de Claudius Pernet X Kaiserin Auguste Viktoria]; Motose

'EVEREST', HP, w, 1927; flowers cream-white, center tinted green-lemon, very large, dbl., 38 petals, exhibition form, moderate fragrance; foliage light; low, spreading growth; GM, NRS, 1927; [Candeur Lyonnaise X Mme Caristie Martel]; Easlea

'EVEREST DOUBLE FRAGRANCE', F, lp, 1980; bud pointed; dbl., 25 petals, borne 3-7 per cluster, intense fragrance; large prickles; foliage dark, heavily veined; tall, upright growth; [Dearest X Elizabeth of Glamis]; Beales, Peter, 1979

Everglades *see* 'POUledge'

Everglo, Min, or, 1998; Laver, Keith G.

Everglow, F, ob

Evergold, LCl, my

'EVERGREEN GEM', HWich, w, 1899; bud buff; dbl., 2–3 in., borne in clusters, moderate, sweetbriar fragrance; foliage almost evergreen; vigorous, climbing or trailing growth; [R. wichurana X Marechal Niel]; Horvath; W.A. Manda

Everlasting Love *see* 'GRAlove'

'EVERT REGTERSCHOT', F, mr, 1965; bud ovoid; flowers bright red, medium, semi-dbl.; foliage dark; [Korona X Seedling]; Buisman, G. A. H.

'EVERT VAN DYK', HT, mp, 1931; flowers rose-pink tinted salmon, large, dbl., exhibition form, slight fragrance; foliage dark; long stems; bushy growth; [Ophelia X Hill's America]; Van Rossem; H&S

'EVGHENYA', F, op, 1975; flowers coral-orange, darker petal edges, base cream, large, very dbl., 75 petals, borne in clusters of 3-12; foliage dark, glossy; vigorous growth; [Highlight X Masquerade]; Staikov, Prof. Dr. V.; Kalaydjiev and Chorbadjiiski

'EVIAN CACHAT', HT, mp, 1939; flowers bright pink, center copper salmon, very large, dbl., cupped; vigorous growth; Chambard, C.

Evita *see* 'POULVITA'

Evita Bezuidenhout *see* 'JACblar'

'EVODIA', HMult, lp, 1925; flowers small, dbl., moderate fragrance; Bruder Alfons

Evolution *see* 'OLIjfaon'

Evona's Gallica, G, pb

Evona's Yellow, HT, yb

'EVRARD KETTEN', HT, m, 1920; flowers bright unshaded carmine-purple, dbl., intense fragrance; [Farbenkonigin X Ruhm de Gartenwelt]; Ketten Bros.

'EXCALIBUR', F, mr, 1967; flowers scarlet, semi-dbl., 14 petals, 2.5 in., blooms in clusters, slight fragrance; foliage dark, glossy; bushy growth; [Vera Dalton X Woburn Abbey]; Harkness

'EXCELLENZ KUNTZE', S, ly, 1909; flowers creamy yellow, small, borne in large clusters, dbl., moderate fragrance; foliage dark, glossy; vigorous, upright growth; [Aglaia X Souv. de Catherine Guillot]; Lambert, P.

'EXCELLENZ M. SCHMIDT-METZLER', HT, w, 1910; flowers large, dbl., slight fragrance; Lambert, P.

'EXCELLENZ VON SCHUBERT', Pol, dp, 1909; flowers dark carmine-rose, small blooms in clusters, dbl., late; foliage dark; vigorous growth; [Mme Norbert Levavasseur X Frau Karl Druschki]; Lambert, P.

'EXCELSA', HWich, mr, 1909; (Red Dorothy Perkins); flowers irregular blooms in large clusters, dbl., cupped, non-recurrent; foliage rich green, glossy; vigorous, climbing (12-18 ft.) growth; (14); Gertrude M. Hubbard, ARS, 1914; Walsh

'EXCELSIOR', F, mp, 1959; flowers salmon, borne in clusters; foliage dark; vigorous, upright growth; [Pinocchio X Mrs Henri Daendels]; Buisman, G. A. H.

Exception *see* **'MÄRCHENLAND'**

Excitement™ *see* 'HILco'

Exodus *see* 'GODUSEX'

Exotic® *see* 'FRYXOTIC'

Exotic, HT, 1995; McGredy, Sam IV

Exotic Beauty *see* 'LEOEXBEAU'

Exotica *see* 'MEIrecrom'

Exploit® *see* 'MEILIDER'

Explorer *see* 'DICROCKY'

Explorer's Dream *see* 'MICEXPLORE'

Expo 64, F, 1964; Tschanz, E.

'EXPOSITION DE BRIE', F, lp; flowers bright red; Granger, 1865

Express, LCl, dp; flowers carmine-pink, large, dbl.; VEG

'EXQUISITE', HT, lp, 1979; bud long, pointed; dbl., 30 petals, 5.5–6. in., exhibition form, moderate fragrance; vigorous, upright growth; [Memoriam X ((Blanche Mallerin X Peace) X (Peace X Virgo))]; Leon, Charles F., Sr.

'EXQUISITE', HT, ly, 1918; flowers creamy yellow, dbl., slight fragrance; Therkildsen

Exquisite, HT, lp

'EXTASE', HT, dr, 1956; bud long, dark red; vigorous growth; [E.G. Hill X Seedling]; Delforge

'EXTASIS', LCl, dr, 1963; flowers red, shaded darker, blooms in clusters, single, 5 petals; vigorous growth; [Spectacular X Cocktail]; Dot, Simon

'EXTRAVAGANZA', F, pb, 1974; bud ovoid; flowers pink, base cream, medium, dbl.; foliage leathery; vigorous, bushy growth; [Stella X (Sabrina X Golden Giant)]; Dawson, George; Neil

'EYDIE', Min, w, 1981; flowers white with pink blush, fading to white, small blooms in clusters, dbl., 50 petals, moderate, tea fragrance; beige prickles; foliage medium green, purple, when young; vigorous, bushy growth; [Janna X Seedling]; Hooper, John C.; E.M. Brown

Eye Appeal® *see* 'INTERPEEL'

'EYE LINER', HT, dr, 1966; bud long, pointed; flowers blood red, medium, semi-dbl., exhibition form, slight fragrance; foliage dark, leathery; vigorous, upright growth; [Queen Elizabeth X Montezuma]; Armbrust; Langbecker

Eye of the Dragon *see* 'CLEDRAG'

Eye Opener® *see* 'INTEROP'

Eye Paint *see* 'MACEYE'

'EYECATCHER', F, pb, 1976; flowers pink flushed apricot, reverse silvery cream, dbl., 22 petals, 2.5 in., moderate fragrance; foliage glossy, light; [Arthur Bell X Pernille Poulsen]; Cants of Colchester, Ltd., 1977

Eyeopener *see* 'INTEROP'

Eyepaint® *see* 'MACEYE'

F

'F. Cambó', HT, mr, 1933; flowers carmine, large, dbl., cupped, moderate fragrance; foliage glossy; dwarf growth; [Li Bures X Florence L. Izzard]; Dot, Pedro

F. Cuixart *see* **'Francesca de Cuixart'**

'F. Ferrer', LCl, dr, 1940; bud long, pointed; flowers dark velvety red, open, large, somewhat recurrent bloom, slight fragrance; foliage leathery; vigorous, climbing growth; Pahissa; J&P

'F. J. Grootendorst', HRg, mr, 1918; (Grootendorst, Grootendorst Red, Nelkenrose); flowers bright red, edges serrated like a carnation, small, dbl., blooms in clusters of up to 20, recurrent bloom, slight fragrance; foliage small, leathery, wrinkled, dark; vigorous, bushy growth; [R. rugosa rubra X Unknown Polyantha]; de Goey; F.J. Grootendorst

F. Katom®, Gr, op, 1982; McGredy, Sam IV

'F. L. de Voogd', HT, yb, 1920; flowers clear reddish yellow, semi-dbl.; [Mme Mélanie Soupert X Mme Jenny Gillemot]; Timmermans

'F. M. Vokes', HT, my, 1927; flowers yellow, passing to cream, semi-dbl., moderate fragrance; [Ophelia seedling]; Hicks

'F. P. Merritt', Cl HT, dr, 1951; flowers bright fiery crimson, dbl., 50–60 petals, 5 in., exhibition form, intense fragrance; foliage glossy; very long stems; very vigorous, climbing (12 ft) growth; [Hoosier Beauty sport]; Merritt

'F. R. Patzer', HT, pb, 1909; flowers creamy buff, reverse warm pink, large, dbl.; branching growth; Dickson, A.

'F. W. Alesworth', HT, dr, 1954; (Fred W. Alesworth); flowers deep crimson, well shaped, large, intense fragrance; foliage dark; vigorous growth; [Poinsettia X Crimson Glory]; Robinson, H.

'F. W. Lowe', HT, ab, 1936; flowers rich orange-yellow, well shaped; foliage glossy; vigorous growth; Lowe

F.K. Druschkii *see* **'Frau Karl Druschki'**

F.R.M. Undritz *see* **'Gen. John Pershing'**

F.W. Mee *see* **'Fred W. Mee'**

Fab *see* 'BOSconpea'

'Fabergé', F, pb, 1969; bud ovoid; flowers light peach-pink, reverse tinted yellow, large, dbl., exhibition form, slight fragrance; foliage dark, leathery; vigorous, dense, bushy growth; [Seedling X Zorina]; Boerner; J&P

'Fabienne', F, mp, 1958; flowers reddish salmon, small, dbl., 35 petals; foliage clear green; bushy, low growth; [Orange Triumph X (Independence X Floradora)]; Arles; Roses-France

Fabiola *see* **'Queen Fabiola'**

'Fabvier', Ch, mr, 1832; flowers crimson-scarlet, very showy, medium, semi-dbl., recurrent bloom; Laffay, M.

'Facade', HT, mp, 1970; flowers apricot-pink, very large, very dbl., 60 petals, exhibition form, intense fragrance; foliage glossy, dark, leathery; vigorous, upright growth; [Elizabeth Fankhauser X Royal Highness]; Fankhauser

'Fackel', HT, dr, 1937; flowers crimson-red, shaded blackish, very dbl., cupped, slight fragrance; compact growth; [Vaterland X Barcelona]; Krause

'Faïence', HT, op, 1935; flowers peach and apricot, reverse pure yellow, large, dbl., 45 petals, cupped, slight fragrance; foliage leathery; vigorous, bushy growth; [Charles P. Kilham X Julien Potin]; Van Rossem; C-P, 1937

'Faint Heart', HT, w, 1980; flowers cream edged light pink, blooms 1-3 per cluster, dbl., 25 petals, exhibition form, slight fragrance; straight prickles; foliage mid to dark, leathery, semi-glossy; medium, branching growth; [Hawaii X Unnamed seedling]; Pavlick, Mike

Fair Bianca® *see* 'AUSca'

Fair Dinkum *see* 'TINdink'

'Fair Dinkum', F, 1966; flowers peach-salmon, 3–4 in., clusters, moderate fragrance; foliage glossy; RULED EXTINCT 6/83 ARM; [Queen Elizabeth X Circus]; Small

Fair Genie *see* 'LAVmoth'

'Fair Lady', HT, ab, 1959; bud ovoid to pointed; flowers buff overcast pink, dbl., 50 petals, 4.5 in., exhibition form, intense fragrance; foliage glossy; vigorous, upright growth; [Golden Masterpiece X Tawny Gold]; Boerner; J&P

'Fair Maid', HT, mp, 1940; flowers bright rose to strawberry-pink, changing to deep pink, dbl., 60–75 petals, 4–5 in., intense fragrance; foliage light green; vigorous, upright growth; [Talisman sport]; Peirce

'Fair Marjorie', Pol, lp, 1952; bud ovoid, bright pink; flowers blush-pink, lighter in sun, medium, semi-dbl., cupped, intense fragrance; vigorous, bushy growth; [Katharina Zeimet seedling]; Armstrong, P.M.

Fair Molly *see* 'MORfairpol'

Fair Opal *see* **'Fire Opal'**

Fair Play® *see* 'INTerfair'

Fair Pol, Min, 1985; Moore, R. E.; (Cavriglia)

Fair Princess, S, mr, 1999; Williams, J. Benjamin

'Fairest of Fair', Min, ly, 1982; flowers medium yellow, borne singly or 3-5 per cluster, 15–18 petals, exhibition form, slight, tea fragrance; foliage small, medium green; very compact, low growth; [Sunbonnet X Rise 'n' Shine]; Bennett, Cecilia 'Dee'; Tiny Petals Nursery

Fairfield Blaze, LCl, mr

Fairhope *see* 'TALfairhope'

'Fairlane', Min, ly, 1980; flowers near white flushed pink and yellow, urn-shaped, dbl., 20 petals, exhibition form, borne 1-5 per cluster, slight fragrance; prickles slanted downward; foliage glossy, medium green, deeply serrated; compact, bushy growth; [Charlie McCarthy X Unnamed seedling]; Schwartz, Ernest W.; Nor'East Min. Roses

'Fairlie Rede', HT, op, 1937; flowers salmon, flushed fawn, large, dbl., moderate fragrance; vigorous growth; [Mrs E. Willis X Seedling]; Clark, A.; NRS Victoria

'Fairlight', F, op, 1964; flowers coppery salmon to flame, well-formed, in clusters, dbl., 3.5 in., moderate fragrance;

foliage coppery bronze; [Joybells X Seedling]; Robinson, H.

Fairy *see* **'THE FAIRY'**

Fairy Changeling *see* 'HARNUMEROUS'

'FAIRY CLUSTER', F, mp, 1935; flowers rose-pink, borne in clusters, single, slight fragrance; foliage glossy; long stems; very vigorous growth; [Dainty Bess X Ideal]; Archer

Fairy Damsel® *see* 'HARNEATLY'

Fairy Dance® *see* HARward

'FAIRY DANCERS', HT, ab, 1969; flowers buff-pink, small, dbl., moderate fragrance; low, spreading growth; [Wendy Cussons X Diamond Jubilee]; Cocker

Fairy Floss, Min, lp; [Sweet Chariot X seedling]; Hannemann, F., 1992; The Rose Paradise; (Weatherly, L.)

Fairy Gold *see* FRYgoldie

Fairy Hedge *see* **'BABY JAYNE'**

Fairy Like *see* 'HARNIMBLE'

'FAIRY MAGIC', Min, mp, 1979; bud mossy, long, pointed; flowers medium, semi-dbl., 10–15 petals, 1.5 in., moderate fragrance; foliage small, glossy; bushy, upright growth; [Fairy Moss X Unnamed Min. Moss seedling]; Moore, Ralph S.; Sequoia Nursery

Fairy Maid *see* 'HARLASSIE'

Fairy Moon *see* POUlwarb

'FAIRY MOSS', Min, mp, 1969; bud mossy; semi-dbl.; foliage mini-moss, small, light green, leathery; vigorous, bushy, dwarf growth; [(Pinocchio X William Lobb) X New Penny]; Moore, Ralph S.; Sequoia Nursery

Fairy Pompons *see* 'TRAPOM'

Fairy Prince® *see* 'HARNOUGETTE'

'FAIRY PRINCESS', Cl Min, lp, 1955; bud pointed, salmon-apricot; very dbl., 1 in., blooms in clusters; foliage small, fern-like; height to 2 1/2 ft; [Eblouissant X Zee]; Moore, Ralph S.; Sequoia Nursery

Fairy Princess *see* **'LILIBET'**

'FAIRY QUEEN', F, lp, 1971; bud ovoid; flowers bluish pink, center coral-pink, small, dbl., exhibition form, moderate fragrance; foliage small, glossy, bronze; very vigorous, bushy growth; [The Fairy X Queen Elizabeth]; Williams, J. Benjamin

Fairy Queen, Pol, mr; Vurens

Fairy Queen *see* SPErien

Fairy Red 92, S, mr, 1992; Liebig

Fairy Ring *see* 'HARNICELY'

Fairy Snow *see* 'HARLITTLE'

Fairy Snow *see* HOLfairy

'FAIRY TALE', S, lp, 1960; bud ovoid; flowers open, small, dbl., slight fragrance; foliage dark, glossy; very vigorous, upright (6 ft) growth; RULED EXTINCT 3/93 ARM; [The Fairy X Goldilocks]; Thomson

Fairy Tale *see* JACchar

Fairy Tale Queen *see* 'KOROYNESS'

Fairyland® *see* 'HARLAYALONG'

Fairy-Tale *see* KEIren

Faith *see* 'HORFAIWIL'

Faith, HT, lp; [Amatsu Otome sport]; Grayco, 2000; (Weatherly, L.)

'FAITHFUL', F, rb, 1964; flowers crimson edged white, becoming velvety crimson, well shaped, semi-dbl.; foliage small, light green; vigorous growth; [Dusky Maiden X Tabarin]; Latham

Faithful *see* HAResay

'FAJA LOBBI', HT, mr, 1963; flowers bright red; vigorous growth; [Queen Elizabeth X Florence Mary Morse]; Leenders, J.

Fakir *see* 'MEICLOUX'

'FALBALA', HT, mp, 1948; bud long; flowers brilliant salmon, moderate fragrance; foliage glossy, dark; erect growth; Gaujard

Falcon, F, 1989; Dot, Simon; (Cavriglia)

Falcon, LCl, dr, 1998; Kordes

'FALKLAND', HSpn, w; flowers pale pink to nearly white, dbl.; low, bushy growth

Fall Festival™ *see* 'LAVFAL'

Fall Splendor *see* 'MINIDOR'

'FAMA', HT, rb, 1942; flowers amber and red, large, dbl., 25–30 petals, cupped; foliage soft, light green; short stems; vigorous, upright, bushy growth; Dot, Pedro; C-P

Fama, F, dy, 1986; Cocker

Fame!™ *see* 'JACZOR'

'FAMOSA', HT, mp, 1964; flowers pink, well formed; [Tallyho X Flamingo]; Leenders, J.

Famous Cliff *see* **'YAROSLAVNA'**

'FAN FARE 81', HT, ob, 1981; bud short, pointed; dbl., 23 petals, exhibition form, borne singly, slight, rose fragrance; straight, short, broad-based prickles; foliage medium to large; vigorous growth; [Cotillion X Hoosier Gold]; Byrum, Roy L.; Joseph H. Hill, Co.

'FAN MAIL', F, dy, 1973; bud long, pointed; flowers open, medium, dbl., moderate fragrance; foliage large, leathery; very vigorous, bushy growth; [Spanish Sun X Unnamed]; Boerner; Spek

'FANAL', F, mr, 1946; flowers large, bornein clusters of 10-15, dbl., 20 petals, moderate fragrance; foliage dark, glossy; upright growth; [(Johanna Tantau X Heidekind) X Hamburg]; Tantau

'FANCY', HT, mr, 1928; flowers peach shaded cherry-red, base yellow, open, semi-dbl., moderate fragrance; [Souv. de Claudius Pernet X Gen. Smuts]; Van Rossem

Fancy Beauty, F, ob, 1999

'FANCY FREE', HT, pb, 1922; flowers pink, center white, semi-dbl., moderate fragrance; dwarf growth; [Gustav Grunerwald X Seedling]; Clark, A.; NRS New South Wales

Fancy Hit *see* POUlfancy

'FANCY LACE', HT, mp, 1967; bud ovoid; flowers pink tipped silver, large, dbl., exhibition form, moderate fragrance; foliage glossy; very vigorous, upright growth; [Unnamed seedling X Queen Elizabeth]; Patterson; Patterson Roses

Fancy Lady *see* 'CLELADY'

Fancy Pants™ *see* 'KINFANCY'

Fancy Potluck *see* 'LAVFAN'

Fancy Princess *see* **'OTOHIME'**

'FANCY TALK', F, pb, 1965; bud urn-shaped; flowers pink tinted orange, small blooms in clusters, dbl., exhibition form, slight fragrance; foliage leathery; vigorous, bushy, low growth; [Spartan X Garnette]; Swim & Weeks; Weeks Wholesale Rose Growers

'FANCY THAT', Min, pb, 1989; bud pointed; flowers different shades of pink, reverse pink-yellow blend, medium, dbl., 68 petals, exhibition form, slight, spicy fragrance; globular, green-brown fruit; no prickles; foliage medium, medium green, semi-glossy; upright, spreading, medium, vigorous growth; [Rise 'n' Shine X Rainbow's End]; Jolly, Marie; Rosehill Farm

'FANDANGO', HT, mr, 1950; bud ovoid, turkey-red, base yellow; flowers orange-red, open, semi-dbl., 16–25 petals, 3.5–4.5 in., moderate fragrance; foliage leathery, glossy, dark; vigorous, upright, bushy growth; [Charlotte Armstrong X Unnamed seedling]; Swim, H.C.; Armstrong Nursery

Fandango *see* MEIjadel

'FANELY REVOIL', HT, or, 1962; flowers cerise-red, reverse tinted orange, well formed; long stems; [Michele Meilland X Unnamed seedling]; Orard, Joseph

Fanette® *see* 'LAPCAL'

'FANFARE', F, pb, 1956; bud urn shaped; flowers orange and salmon to pink, open, borne in large clusters, dbl., 20–30 petals, 3–4 in., cupped, moderate, spicy fragrance; foliage glossy, leathery; very vigorous, spreading growth; GM, Rome, 1955; [Fandango X Pinocchio]; Swim, H.C.; Armstrong Nursery

'FANION', HT, dy, 1961; (Puregold); bud long, pointed; flowers rich yellow, dbl., 30 petals, 3–3.5 in., exhibition form, intense fragrance; foliage leathery, dark, glossy; vigorous, upright growth; [Helen Fox X (Mrs Pierre S. duPont X Joanna Hill)]; Robichon; Ilgenfritz Nursery

'FANNY', HT, mr, 1935; flowers bright red; vigorous growth; [Hadley X Mrs Henry Winnett]; Lens

'FANNY BIAS', HGal, mp; flowers blush, center rosy, dbl., 3 in., moderate fragrance; erect, bushy growth; Vibert, before 1811

'FANNY BLANKERS-KOEN', HT, ob, 1949; (Luxembourg); bud long, pointed; flowers orange-yellow, flushed and veined red, large, semi-dbl., 16 petals, intense fragrance; foliage glossy; very vigorous, upright, bushy growth; [Talisman X Unnamed seedling]; Verschuren-Pechtold

'FANNY BULLIVANT', HT, lp; [The Rajah sport]; Clark, A., 1941; (Weatherly, L.)

'FANNY ESSLER', HGal, mp; Vibert, 1835

'FANNY OPPENHEIMER', HT, mr, 1923; flowers brilliant cardinal, shaded gold, moderate fragrance; McGredy

Fanny Pavetot, HGal, mp; flowers large, very dbl., moderate fragrance; (Sangerhausen)

Fanny's La France, HT, dp

Fano, S, mp

'FANTAISIE', HT, r, 1948; flowers coppery-salmon, base yellow, slight fragrance; foliage dark; vigorous growth; Gaujard

Fantan *see* 'MEIMEX'

'FANTAN', F, or, 1956; flowers coppery orange, small; foliage small; [Eternite X Seedling]; Gaujard

Fantasi® *see* 'KINFANTA'

Fantasia *see* 'KORFAN'

'FANTASIA', HT, my, 1943; bud long, pointed; flowers golden to lighter yellow, open, medium, dbl., 30–35 petals, intense fragrance; foliage glossy; strong stems; vigorous, bushy, compact growth; RULED EXTINCT; [Seedling X Lord Lonsdale]; Dickson, A.; J&P

'FANTASIA, CLIMBING', Cl HT, my, 1956; Mell

Fantasque *see* **'FANTASTIQUE'**

'FANTASTIQUE', HT, yb, 1943; (Fantasque); flowers yellow, heavily edged carmine, dbl., 38 petals, 2.5–3 in., cupped, moderate, spicy fragrance; foliage dark, leathery, glossy; vigorous, compact growth; [Ampere X (Charles P. Kilham X (Charles P. Kilham X Capucine-Chambard))]; Meilland, F.; C-P

'FANTASY', HT, rb, 1945; bud long, pointed; flowers cerise, reverse yellow ashed pink, open, blooms, semi-dbl., 9–11 petals, 4.5 in., slight fragrance; foliage leathery, glossy; slender stems; very vigorous, upright growth; RULED EXTINCT 1/86 ARM; [Soeur Thérèse X Unnamed seedling]; Wyant

Fantasy *see* JACpow

'FANTIN-LATOUR', C, lp; flowers blush, dbl., flat, moderate fragrance; foliage dark, broad; vigorous, bushy growth

'FAR SIDE', F, ob, 1988; bud ovoid, pointed; flowers orange-vermillion, yellow center, small, borne in sprays of, single, 5 petals, moderate, spicy fragrance; globular, medium, vermillion fruit; prickles straight, medium, stout, pink to brown; foliage large, medium green, glossy, smooth; upright, spreading, tall growth; [[Sunsprite X (Many Moons X Maigold)] X Eyepaint]; Stoddard, Louis, 1990

'FARAH', HT, yb, 1961; flowers coppery yellow, large, dbl., moderate fragrance; foliage glossy; long stems; very vigorous, bushy growth; [Peace X Georges Chesnel]; Gaujard

'FARANDOLE', F, mr, 1959; bud oval; flowers vermilion, open, medium, borne in large clusters, dbl., 25 petals; foliage leathery; vigorous, well branched growth; GM, Rome, 1959; [(Goldilocks X Moulin Rouge) X (Goldilocks X Fashion)]; Meilland, Mrs. Marie-Louise; URS

'FARBENKÖNIGEN', HT, mr, 1902; (Queen of Colors, Reine des Couleurs); flowers rosy carmine, dbl., moderate fragrance; [Grand-Duc Adolphe de Luxembourg X La France]; Hinner, W.

'FARBENSPIEL', F, rb, 1960; flowers pink edged red, dbl., intense fragrance; foliage dark; bushy, compact growth; [Pinocchio X Masquerade]; Verschuren, A.; van Engelen

'FARFADET', F, or, 1955; semi-dbl.; very upright growth; [Méphisto X Incendie]; Combe

Faria *see* OLIjkroet

'FARIDA', F, dp, 1941; flowers deep rose-pink, semi-dbl., moderate fragrance; [Seedling X Permanent Wave]; Leenders, M.

'FARNY WURLITZER', HT, mr, 1968; bud long, pointed; flowers large, dbl., intense fragrance; foliage dark, glossy; vigorous, upright growth; [Poinsettia X Charlotte Armstrong]; Cadey; Ty-Tex Rose Nursery

'FARQUHAR', HWich, mp, 1903; (The Farquhar Rose); flowers bright clear pink, carnation-like, resembling Lady Gay, dbl., late; vigorous, climbing growth; [R. wichurana X Crimson Rambler]; Dawson; Farquhar

'FASCINATING', HT, yb, 1961; bud long, pointed; flowers rose-opal suffused yellow, dbl., 25 petals, 4.5 in., exhibition form, moderate fragrance; foliage leathery, glossy, dark; upright, bushy growth; [Peace X Orange Nassau]; Fisher, G.; C-P

Fascination *see* 'JACOYEL'

'FASCINATION', HT, pb, 1927; flowers rosy cerise, shaded yellow, moderate fragrance; foliage dark, glossy; vigorous growth; RULED EXTINCT 2/81; Chaplin Bros.

'FASHION', F, pb, 1949; bud ovoid, deep peach; flowers lively coral-peach, blooms in clusters, dbl., 23 petals, 3–3.5 in., moderate fragrance; vigorous, bushy growth; (28); AARS, 1950 David Fuerstenberg Prize, ARS, 1950 GM, ARS, 1954 GM, Bagatelle, 1949 GM, NRS, 1948 GM, Portland, 1949; [Pinocchio X Crimson Glory]; Boerner; J&P

'FASHION FLAME', Min, op, 1977; bud ovoid, pointed; flowers coral-orange, dbl., 35 petals, 1–1.5 in., exhibition form, slight fragrance; foliage large, leathery; bushy growth; [Little Darling X Fire Princess]; Moore, Ralph S.; Sequoia Nursery

Fashion Flame, Min, op

Fashion Parade, Min, mp, 1994; Poulsen

'FASHION, CLIMBING', Cl F, pb, 1951; Boerner

'FASHION, CLIMBING', Cl F, op, 1955; Mattock

'FASHIONETTE', F, pb, 1958; bud pointed to ovoid, coral; flowers pinkish coral, borne in irregular clusters, dbl., 35–40 petals, 3 in., cupped, moderate fragrance; foliage glossy; vigorous, upright growth; [Goldilocks X Fashion]; Boerner; J&P

'FASHIONETTE, CLIMBING', Cl F, pb, 1962; Noack, Werner

Fassadenzauber, LCl, lp, 1997; Noack, Werner

Fat 'n' Sassy *see* 'TINSASSY'

Fat Tuesday *see* 'TALFAT'

'FATA MORGANA', F, ob, 1957; flowers orange-yellow, open, borne in large trusses, dbl., 28 petals, 2.5 in., slight fragrance; foliage leathery, glossy; vigorous, upright growth; [Masquerade X Seedling]; Kordes, R.

Father Christmas *see* 'TALCHRIS'

Father David's Rose *see* **R. DAVIDII**

Father Hugo Rose *see* **R. HUGONIS**

Father Hugo's Rose *see* **R. HUGONIS**

Father's Day *see* **'VATERTAG'**®

'FATIMA', HT, ob, 1955; bud long, pointed; flowers orange, reverse bright golden yellow, very large, moderate fragrance; vigorous growth; [Opera seedling X Unnamed seedling]; Gaujard

Fatima '67, HT, 1967; Moreira da Silva, A.; (Cavriglia)

'FATIME', HGal, mp, 1820; flowers pink, dotted and spotted both lighter and darker, medium; Descemet, M.

Fatinitza, HT, op, 1956; flowers yellowish salmon-pink, medium, dbl.; Berger, W.; (Sangerhausen)

'FAUST', F, yb, 1957; (Dr Faust); flowers golden yellow shaded orange-pink, blooms in large clusters, dbl., 25 petals, 2 in., moderate fragrance; foliage dark, glossy; vigorous, bushy growth; GM, NRS, 1956; [Masquerade X Golden Scepter]; Kordes, R.; A. Dickson;, McGredy

Faust *see* **'SCARLANO'**

'FAUST, CLIMBING', Cl F, yb, 1963; deRuiter

Favori *see* 'LEN 3'

'FAVORITA'™, HT, op, 1954; bud ovoid, burnt-orange; flowers salmon overcast orange, dbl., 48 petals, 5.5–6. in., moderate fragrance; foliage dark; vigorous growth; GM, Rome, 1952; [Unnamed HT seedling X Serenade]; Boerner; Stark Bros.

Favorite *see* 'LEN 3'

Favorite *see* LEBbet

Favorite Dream *see* TWOfavor

Favourite Rosamini *see* RUIfarol

'FAY', S, pb, 1997; flowers double, medium, dbl., 15–25 petals, borne mostly singly, moderate fragrance; few prickles; foliage medium, light green, semi-glossy; upright, medium (36in.)growth; [Cecile Brunner X Zepherin Drouhin]; Robert, David

'FAYANNE', F, pb, 1953; flowers deep rose-pink, reverse lighter, small, dbl., 55 petals, 2 in., globular, slight fragrance; foliage leathery; dwarf, bushy growth; [Garnette sport]; Pinchbeck

'FAYBELL', HT, ob, 1974; (**Rebell**); flowers dark orange, large, dbl., exhibition form, intense fragrance; foliage leathery; vigorous, upright growth; [Brandenburg X Seedling]; Kordes; Fey

Faye Reynolds *see* 'REYFAYE'

FAZcanne, HT, mr, 1999; (**Red France**)

'FEAR NAUGHT', F, dp, 1968; flowers deep pink, dbl., slight fragrance; [Queen Elizabeth X Ena Harkness]; Harkness

FEbama, HT, lp; (**Iberflora 95**®); exhibition form, 28 petals, none; [FE-85138 X Dallas]; Roses Noves Ferrer, S L; (Asociacion Espanola de la Rose)

FEbasa, F, lp; (**Esther Peiro**®); exhibition form, 26 petals, none; [Lamada X K-881543-02]; Roses Noves Ferrer, S L, 1996; (Asociacion Espanola de la Rose)

FEbesa, Gr, lp; (**Vanessa Campello**®); exhibition form, 32 petals, none; [Lambada X K-881543-02]; Roses Noves Ferrer, S L, 1996; (Asociacion Espanola de la Rose)

FEcasa, HT, lp; (**Barcelona 95**®); exhibition form, 26 petals, slight; [FE-903021 X FE-89081]; Roses Noves Ferrer, S L, 1995; (Asociacion Espanola de la Rose)

Fecirco, HT, 1987; Ferrer, F.; (Cavriglia)

'FEDERATION', LCl, op, 1938; bud pointed; flowers rosy pink, orange undertone, dbl., 24–36 wavy petals, 3.5 in., cupped, intense fragrance; foliage leathery, glossy, dark; long, strong stems; very vigorous climbing (12-14ft.) growth; [(R. setigera X Mrs F.F. Prentiss) X Director Rubió]; Horvath; Wayside Gardens Co.

'FEDERICO CASAS', HT, op, 1931; flowers coppery pink and orange, open, very large, semi-dbl., intense fragrance; foliage dark; vigorous, bushy growth; [Unnamed variety X Eugene Barbier]; Dot, Pedro; C-P

'FEDERICO CASAS, CLIMBING', Cl HT, op, 1937; Stell; Stell Rose Nursery

Federico Garcia Lorca® *see* FEticos

'FEDRA', F, mr, 1959; flowers brick-red, dbl.; vigorous growth; [(Fiamma X Independence) X Unnamed seedling]; Giacomasso

Fedugia, HT, 1988; Ferrer, F.; (Cavriglia)

FEdugia, HT, lp; (**Marta Salvador**®); globular, 32 petals, moderate; [Zambra X White Satin]; Viveros Fco. Ferrer, S L, 1987; (Asociacion Espanola de la Rose)

'FEE', F, or, 1963; flowers orange, large, dbl.; strong, wiry stems; moderate growth; Kordes, R.

Fee *see* KORmasyl

'FÉE DES CHAMPS', Pol, op, 1965; flowers salmon-orange, dbl., 20–25 petals; foliage bronze; low growth; GM, Bagatelle, 1965; [Queen Elizabeth X Zambra]; Dot; Minier

Fée des Neiges *see* 'KORBIN'

Feeling® *see* BARfeel

Feeling®, HMsk, yb, 1994; Lens

Feerie *see* **'THE FAIRY'**

'FÉERIE', HT, ob, 1938; flowers coppery red, reverse orange-yellow, open, semi-dbl., moderate fragrance; long stems; vigorous, erect growth; Gaujard

FEgama, HT, lp; (**Adela**®); dbl., 32 petals, slight; [Shocking Blue X Kardinal]; Viveros Fco. Ferrer, S L, 1991; (Asociacion Espanola de la Rose)

FEgima, HT, lp; (**Elodea**®); exhibition form, 22 petals, none; [FE-87518 X K-861242-1]; Roses Noves Ferrer, S L, 1994; (Asociacion Espanola de la Rose)

FEgosa, Min, lp; (**Alejandra Conde**®); rosette, 22 petals, none; [M Litigan X Scarlet Meilladina]; Viveros Fco. Ferrer, S L, 1991; (Asociacion Espanola de la Rose)

FEkusa, HT, lp; (**Charlye Rivel**®); exhibition form, 32 petals, none; [Jack-8301365 X Inedita]; Roses Noves Ferrer, S L, 1995; (Asociacion Espanola de la Rose)

'FELBERG'S ROSA DRUSCHKI', HP, mp, 1929; flowers bright rose-pink, large, dbl., 25 petals; [Frau Karl Druschki X Farbenkonigin]; Felberg-Leclerc

'FELICIA', HMsk, pb, 1928; flowers pink fading to blush and partly white, large, branching panicles, semi-dbl., moderate, musk fragrance; pillar or shrub growth; (21); [Trier X Ophelia]; Pemberton

Felicia Teichmann *see* KORciate

Felicitas *see* KORberis

'FÉLICITÉ BOHAIN', M, dp; (Félicité Bohan); flowers vivid pink or bright rose, large, dbl.; (France), before 1866

Félicité Bohan *see* **'FÉLICITÉ BOHAIN'**

'FÉLICITÉ ET PERPÉTUE', HSem, w, 1828; flowers pale flesh changing to white, fairly large blooms in clusters, very dbl., flat; foliage almost evergreen; very vigorous growth; [Thought to be R. sempervirens X Noisette]; Jacques, 1827

'FÉLICITÉ PARMENTIER', A, lp; flowers soft flesh-pink, and then reflexing, very dbl., flat, intense fragrance; foliage gray-green; vigorous, compact growth

'FELICITY', HT, mp, 1919; flowers rose-pink suffused silvery, large, dbl., 50–60 petals, moderate fragrance; foliage dark; vigorous, branching growth; [Ophelia X Hoosier Beauty]; Clarke Bros.

Felicity II *see* **'BUTTONS 'N' BOWS'**

Felicity Kendal *see* 'LANKEN'

'FELIX BRIX', HT, pb, 1921; flowers soft rose, suffused yellow, passing to salmon-rose, semi-dbl.; [Natalie Boettner X Old Gold]; Brix; Teschendorff

'FELIX LAPORTE', HT, m, 1928; flowers blackish velvety purple tinged garnet, dbl., cupped, intense fragrance; foliage dark, leathery, glossy; vigorous, bushy growth; [Yves Druhen X Mme Edouard Herriot]; Buatois

'FELIX MOUSSET', HP, 1878; Verdier, E.; (Cavriglia)

'FELLEMBERG', HCh, mr; flowers bright crimson, cupped, 36 petals, cupped; foliage dark; vigorous, spreading growth; Fellemberg, 1857; Introduced prior to 1835

Fellini *see* INTerdain

Fellowship *see* **'DE GREEFF'S JUBILEE'**

Fellowship *see* 'HARWELCOME'

FEloma, HT, lp; (**Coral®**); exhibition form, 22 petals, none; [FE-87518 X K-861242-1]; Roses Noves Ferrer, S L, 1994; (Asociacion Espanola de la Rose)

FEmental, Gr, lp; (**Juli de Sala®**); exhibition form, 30 petals, none; [FE-85143 X MT-86173]; Roses Noves Ferrer, S L, 1994; (Asociacion Espanola de la Rose)

FEmiento, HT, lp; (**Ana de Cuevas**); exhibition form, 24 petals, slight; [Bucaneer X Cocktail]; Viveros Fco. Ferrer, S L, 1987; (Asociacion Espanola de la Rose)

'FEMINA', HT, op, 1963; bud long, pointed; flowers salmon-pink, large, dbl., moderate fragrance; foliage leathery; vigorous, upright growth; [Fernand Arles X Mignonne]; Gaujard; Ilgenfritz Nursery, 1966

'FEMINA', HT, lp, 1957; flowers soft pink, large, dbl.; vigorous growth; Poulsen, S.

Femme *see* 'DELVOR'

Femnet *see* PEKaledon

'FEN QUEEN', F, lp, 1963; flowers pale flesh, large, dbl., 25 petals, slight fragrance; foliage light green; vigorous, upright growth; [Queen Elizabeth X Unknown Hybrid Tea]; Sharman

Fenja, S, mp, 1965; Petersen

Fennet, HT, rb; Laperrière

Fennica *see* 'RUNTRU'

FEodase, HT, lp; (**Condesa de Barcelona®**); exhibition form, 26 petals, slight; [Mount Shasta X Helmut Schmidt]; Viveros Fco. Ferrer, S L, 1988; (Asociacion Espanola de la Rose)

FEprogo, HT, lp; (**Lourdes Arroyo®**); exhibition form, 26 petals, slight; [Zambra X Inedita]; Viveros Fco. Ferrer, S L, 1990; (Asociacion Espanola de la Rose)

FEpuma, HT, lp; (**Carolina Daza®**); exhibition form, 28 petals, slight; [Zambra X Kolner Karneval]; Viveros Fco. Ferrer, S L, 1991; (Asociacion Espanola de la Rose)

FErama, HT, lp; (**Catalina Frau®**); exhibition form, 28 petals, none; [FE-85146 X Osiana]; Viveros Fco. Ferrer, S L, 1991; (Asociacion Espanola de la Rose)

Ferdi *see* 'KEITOLI'

'FERDINAND CHAFFOLTE', HP, mr, 1879; flowers large, very dbl., moderate fragrance; Pernet Père; (Sangerhausen)

'FERDINAND DE BUCK', HGal, mp; (Feu de Buck); flowers brilliant pink, medium, dbl.

'FERDINAND DE LESSEPS', HP, lp; flowers deep crimson; Verdier, E., 1869

'FERDINAND JAMIN', HP, mr, 1888; flowers vermilion-red; Lévêque

'FERDINAND JAMIN', HT, lp; Pernet-Ducher, 1896

'FERDINAND PICHARD', HP, rb, 1921; flowers streaked (striped) pink and scarlet, yellow stamens, dbl., 25 petals, recurrent bloom; vigorous, tall growth; Tanne

'FERDINAND ROUSSEL', HWich, m, 1902; flowers purple/pink to light red, small, dbl.; Barbier; (Sangerhausen)

Ferdinando Dukei, HT, yb; flowers orange-yellow with red, large, semi-dbl., moderate fragrance; (Sangerhausen)

Ferdy™ *see* 'KEITOLI'

FErecha®, HT, lp; (**Iris**); exhibition form, 27 petals, none; [Zambra X Jelcanodir]; Viveros Fco. Ferrer, S L, 1991; (Asociacion Espanola de la Rose)

Fergie *see* 'GANFER'

Fergus Games, S, op, 1999; flowers light salmon pink, dbl.; Williams, J. Benjamin

FERho, LCl, mp, 1997; (**Childhood Memories**); Fergusson

Feria *see* 'MEIFRISON'

Ferline® *see* LAPdade

'FERN KEMP', HRg, lp, 1918; flowers delicate pink, semi-dbl., 4 in., intense fragrance; vigorous growth; hardy.; [Conrad Ferdinand Meyer X Frau Karl Druschki]; Kemp, J.A.

'FERN ROEHRS', LCl, mp, 1943; flowers same as parent but much more, dbl.; [Paul's Scarlet Climber sport]; Graf; Roehrs

'FERNAND ARLES', HT, op, 1949; (Arles); bud long, pointed; flowers orange-salmon shaded red, very large, dbl., moderate fragrance; foliage bronze; very vigorous, bushy growth; [Mme Joseph Perraud X Unnamed seedling]; Gaujard

Fernand Majorel, F, 1975; Gaujard; (Cavriglia)

'FERNAND POINT', HT, rb, 1964; flowers crimson-red, reverse flesh-pink, base yellow; foliage glossy; [Peace X Unnamed seedling]; Orard, Joseph

'FERNAND RABIER', HWich, dr, 1918; flowers pure deep scarlet, borne in clustersof 40-50, dbl., slight fragrance; vigorous, climbing or trailer growth; [Delight X Unnamed seedling]; Turbat

'FERNAND TANNE', LCl, dy, 1920; flowers deep yellow to cream-yellow, large, dbl., intense fragrance; vigorous growth; Tanne; Turbat

Fernanda, HT; (Brazil); (Cavriglia)

'FERNANDE KRIER', HWich, pb, 1925; flowers peach-pink, occasionally margined red or sometimes entirely; [Excelsa sport]; Walter, L.

'FERNANDE LUMAY', HT, ab, 1922; flowers apricot-nankeen-yellow, edged milk-white, dbl., moderate fragrance; [Mrs Aaron Ward X Unnamed seedling]; Buatois

Ferndale Red China, Ch, dp; (found rose)

'FERNIELEA', HT, ab, 1926; flowers apricot, center deeper, moderate fragrance; Adam & Craigmile

Ferris Wheel *see* 'AROYUMI'

'FERRY PORSCHE', HT, mr, 1971; bud long, pointed; dbl., 34 petals, exhibition form, borne singly and in small clusters, slight fragrance; foliage large, dark, soft; vigorous, upright growth; [Tropicana X Americana]; Kordes, W. Söhne

'FERTRY', S, my, 1998; (**The Didgemere Rose**); flowers medium yellow, full, very dbl., 26–40 petals, 2.5–3 in., borne mostly singly, moderate fragrance; prickles small, slightly hooked; foliage medium, medium green, dull; upright, low (3-4ft) growth; [Sunsprite X Trier]; Ferguson, B.; Burston Nurseries, Inc., 1998

'FERVID', F, or, 1960; flowers scarlet-orange, in clusters, single, 3 in., slight fragrance; foliage glossy, dark; vigorous, upright growth; [Pimpernell X Korona]; LeGrice

FEsaru, HT, lp; (**Neus®**); exhibition form, 25 petals, moderate; [Shocking Blue X Carta Blanca]; Viveros Fco. Ferrer, S L, 1991; (Asociacion Espanola de la Rose)

FEsato, Min, lp; (**Clara®**); rosette, 20 petals, none; [M Litigan X Inedita]; Viveros Fco. Ferrer, S L, 1986; (Asociacion Espanola de la Rose)

FEsodios, HT, lp; (**Emilo Feliu®**); exhibition form, 28 petals, slight; [Zambra X Osiana]; Viveros Fco. Ferrer, S L, 1992; (Asociacion Espanola de la Rose)

'FESTIVAL', HT, mr, 1943; (Dixie Dream); flowers rich red, very dbl., moderate fragrance; entirely prickle free; RULED EXTINCT, 8/93 ARM; [E.G. Hill seedling]; Dixie Rose Nursery; Krider Nursery, 1945

Festival *see* KORdialo

Festival Beauty *see* **'KRASAVITZA FESTIVALIA'**

Festival Fanfare *see* 'BLESTOGIL'

Festival Pink™ *see* 'RUPFESPIN'

'FESTIVAL QUEEN', HT, lp, 1968; bud long, pointed; flowers light pink, edged deeper, well-formed, exhibition form, slight fragrance; foliage dark, leathery; upright, compact growth; Lindquist; Edmunds Roses

Festival Rouge® *see* 'DELFESROU'

'FESTIVAL, CLIMBING', Cl HT, mr, 1945; flowers large, globular, free, intermittent bloom, intense fragrance; thornless; foliage glossy; very vigorous, climbing growth; hardy in South.; [E.G. Hill sport]; Watkins, A.F.; Dixie Rose Nursery

'FESTIVE', Pol, dr, 1975; bud short, pointed; dbl., 30 petals, 3–3.5 in., exhibition form, repeat bloom, slight, sweetbriar fragrance; foliage parsley-green; vigorous, upright growth; [Bacarra X Seedling]; Jelly; E.G. Hill Co.

Festivity *see* 'LENOR'

FEstro, HT, lp; (**Alba Garcia**®); exhibition form, 26 petals, intense; [Shocking Blue X Carta Blanca]; Viveros Fco. Ferrer, S L, 1991; (Asociacion Espanola de la Rose)

FEsuma, HT, lp; (**Gloria Ferrer**®); exhibition form, 30 petals, slight; [Zambra X Osiana]; Viveros Fco. Ferrer, S L, 1992; (Asociacion Espanola de la Rose)

Fête des Mères *see* **'MOTHERSDAY'**

Fête des Pères *see* **'VATERTAG'**®

Fêtes Galantes *see* DELdido

'FÉTICHE', F, mr, 1962; flowers coral-red, center white, open, borne in clusters, semi-dbl., 12 petals, 2 in.; foliage dark; moderate, bushy growth; [Philippe X Tabarin]; Delforge

FEticos, HT, lp; (**Federico Garcia Lorca** R); globular, 30 petals, slight; [Zambra X Susan Hamshire]; Viveros Fco. Ferrer, S L, 1989; (Asociacion Espanola de la Rose)

FEtrone, HT, lp; (**Rocio Elias**®); flat, 32 petals, moderate; [Carinella X Kardinal]; Viveros Fco. Ferrer, S L, 1988; (Asociacion Espanola de la Rose)

FEtural, Min, lp; (**Lucia Cotarelo**®); rosette, 15 petals, none; [M Litigan X Scarlet Meilladina]; Viveros Fco. Ferrer, S L, 1991; (Asociacion Espanola de la Rose)

'FEU D'ARTIFICE', LCl, yb, 1935; (Fireworks, Climbing); bud long, pointed, nasturtium-red; flowers yellow, tinted nasturtium-red, open, borne in clusters, semi-dbl.; long stems; vigorous, climbing (over 8 ft) growth; [R. foetida hybrid X Colette Clément]; Mallerin, C.; B&A, 1939

Feu d'Artifice *see* **'FEUERWERK'**®

'FEU DE BENGALE', F, lp, 1951; flowers pearl-pink, large, dbl., cupped, moderate fragrance; foliage glossy; vigorous growth; [Orange Triumph X Unnamed seedling]; Gaujard, R.; G. Truffaut

Feu de Buck *see* **'FERDINAND DE BUCK'**

Feu de Camp *see* **'LAGERFEUER'**®

'FEU DE JOIE', F, mr, 1951; flowers carmine-red, large, borne in clusters, dbl.; foliage dark, glossy; vigorous growth; [Orange Triumph X Unnamed seedling]; Gaujard, R.; G. Truffaut

'FEU DE SAINT-JEAN', Pol, dr, 1951; flowers blackish red, small, borne in clusters, semi-dbl.; foliage dark, glossy; vigorous, bushy growth; [Orange Triumph X Unnamed seedling]; Gaujard, R.; G. Truffaut

'FEU D'ENFER', F, or, 1958; flowers large, dbl.; upright, well branched growth; [Orange Triumph X Unnamed seedling]; Gaujard, R.; G. Truffaut

'FEU DU CIEL', Pol, ob, 1951; flowers clear orange, small, borne in clusters, semi-dbl.; foliage glossy; vigorous, bushy growth; [Orange Triumph X Unnamed seedling]; Gaujard, R.; G. Truffaut

'FEU FOLLET', F, mp, 1953; flowers salmon, cupped, moderate fragrance; vigorous growth; Gaujard, R.

Feu Follet® *see* DIClady

'FEU JOSEPH LOOYMANS', HT, ob, 1921; flowers orange-yellow, large blooms, dbl., cupped; foliage leathery; weak stems; vigorous growth; [Sunburst X Rayon d'Or]; Looymans

'FEU JOSEPH LOOYMANS, CLIMBING', Cl HT, ob, 1935; Western Rose Co.

'FEU MAGIQUE', F, mr, 1956; flowers cherry-red, semi-dbl., 18 petals; vigorous growth; [Independence X Signal Red]; Frères, Buyl

'FEU PERNET-DUCHER', HT, my, 1935; flowers bright yellow, center apricot, large, dbl., moderate, fruity fragrance; foliage leathery, dark; vigorous, branching growth; GM, Portland, 1936; [Julien Potin X Margaret McGredy]; Mallerin, C.; A. Meilland;, C-P

'FEU ROUGE', F, mr, 1956; bud ovoid; flowers open, large, semi-dbl., slight fragrance; foliage glossy; vigorous growth; [Red Favorite X Fanal]; Tantau, Math.

'FEUDOR', HT, yb, 1963; flowers golden yellow shaded vermilion; foliage bright green; vigorous growth; [Peace X Baccará]; Croix, P.; Minier

'FEUERBALL', HT, mr, 1965; bud globular; flowers red lead color, well formed, large, dbl., 25–30 petals; foliage glossy; very vigorous, bushy growth; Tantau, Math.

Feuerfunken, S, or, 1979; flowers large, dbl.; GPG Bad Langensalza; (Sangerhausen)

'FEUERLAND', F, or, 1977; bud ovoid; dbl., 25 petals, 2.5 in., cupped, slight fragrance; very vigorous, upright growth; [Kathe Duvigneau X Topsi]; Kordes, W. Söhne

Feuermeer *see* **'SEA OF FIRE'**

'FEUERREITER', F, mr, 1968; (Fire-Rider); bud long, pointed; flowers open, large, borne in clusters, semi-dbl., slight fragrance; foliage dark, leathery; very vigorous, upright growth; [Alain X Oskar Scheerer]; Haenchen, E.; Teschendorff

'FEUERSCHEIN', F, mr, 1930; (Krause's Rote Joseph Guy); flowers brilliant red, not turning blue, dbl.; foliage dark; bushy growth; [Lafayette sport]; Krause

'FEUERSCHEIN, CLIMBING', Cl F, mr, 1936; Krause

Feuersturm, F, or; flowers medium, slight fragrance; Verschuren, after 1900; (Sangerhausen)

Feuertaufe, Pol, 1980; dbl.; GPG Bad Langensalza; (Sangerhausen)

'FEUERWERK'®, S, ob, 1962; (Feu d'Artifice, Magneet); flowers bright orange, blooms in clusters, semi-dbl.; foliage glossy; upright, bushy (to 5 ft) growth; Tantau, Math.

Feuerzauber® *see* 'KORFEU'

'FEUERZAUBER', Ch, mr, 1913

'FEURIO', F, or, 1956; bud ovoid; flowers scarlet-red, borne in clusters, dbl., 30 petals, 2.5 in., cupped, moderate fragrance; foliage glossy, light green; vigorous, low, bushy growth; [Rudolph Timm X Independence]; Kordes, R.; J&P, 1957

'FEURIO, CLIMBING', Cl F., or, 1963; Kordes

Fever *see* 'PEAFEVER'

Fewell's Noisette, N, lp

'FIABA', HT, rb, 1963; flowers bright red tipped yellow; foliage dark, glossy; very vigorous growth; [(Fiamma X Sovrana) X Unnamed seedling]; Giacomasso

'FIALAWN', HT, mp, 1985; (**Wood Lawn**); flowers clear medium pink; [Fontainebleau sport]; Fiamingo, Joe

'FIALOPI', F, mp, 1991; (**Loads of Pink**); semi-dbl., 6–14 petals, 1.5–2.75 in., borne in large clusters, moderate fragrance; some prickles; foliage medium green, semi-glossy, disease resistant; tall, upright, very vigorous growth; [Fred Loads sport]; Fiamingo, Joe, 1991

'FIAMETTA', F, dr, 1962; flowers velvety red, center yellow; moderate growth; [Karl Weinhausen X Goldilocks]; Leenders, J.

'FIAMMA', F, or, 1948; flowers vermilion, borne in clusters; vigorous growth; GM, Rome, 1951; [Paul's Scarlet Climber X Talisman seedling]; Aicardi, D.; Giacomasso

Fiamma Nera, HMult, w; flowers medium, semi-dbl., slight fragrance; (Sangerhausen)

'FIAMMETTA', LCl, ab, 1922; flowers warm amber-yellow, streaked yellow, single, intense fragrance; vigorous, climbing

Scenes from THE BOTANICAL ROSE GARDEN 'CARLA FINESCHI'

Il Roseto Botanico 'Carla Fineschi' (ITALY)

Professor Gianfranco Fineschi started collecting roses 30 years ago and today he has one of the world's largest private collections – more than 6500 varieties, each represented by single plant. In this living museum the genus Rosa is presented from a historical and a botanical point of view. Species, sub-species and later hybrids are systematically presented, while climbers and ramblers form walls between the scientific divisions. Many of the beds of modern varieties are grouped according to their hybridizers. In this garden overlooking the hills of Tuscany one may trace the history of the rose from antiquity to the present day while marveling at the beauty of the garden itself.

Scenes from THE EUROPA-ROSARIUM AT SANGERHAUSEN

View of the Pavillion

Floribunda beds

Frau Ida Munch
1919

View of climbers

The Europa-Rosarium at Sangerhausen (GERMANY)

In the Rosarium at Sangerhausen, roses from all countries and periods have been brought together. The current inventory of approximately 6800 different roses mirrors in a near unbroken sequence the development of the modern garden rose. Near the end of the nineteenth century, on the initiative of the rose breeder Peter Lambert, German rose lovers passed a resolution to establish a Rose Society. Old roses, threatened by extinction and those forgotten, were to be gathered,

Scenes from THE EUROPA-ROSARIUM AT SANGERHAUSEN

Princesse ed Lamballe
1850

Dr. Eckener
1929

Malesherbes
1834

Harlickeva Narodni
1935

reintroduced and maintained. At the instigation of the active rose lovers Albert Hoffmann and Prof. Ewald Gnau from Sangerhausen, the city of Sangerhausen provided a piece of land to their disposal. The garden architect Friedrich Dorr developed a design for this area in the formal garden style. Here was the place where the Hybrid Tea rose cultivars of the time were set out.

Sonia Rykiel
(MASdougi)
Shrub, orange pink, 1995
Hybridized by Guillot-Massad
Parentage not released
Photo courtesy Arena Roses

Claudia Cardinale
(MAScanta)
Shrub, deep yellow, 1997
Hybridized by Guillot-Massad
Parentage not released
Photo courtesy Arena Roses

Martine Guillot
(MASmabay)
Shrub, white, 1997
Hybridized by Guillot-Massad
Parentage not released
Photo courtesy Arena Roses

Dream™ Orange
(TWOaebi)
Hybrid Tea, orange red, 2000
Hybridized by Jerry Twomey
[Cherish x (Evening Star x Trumpeter)]
Photo courtesy Anthony Tesselaar International

Dream™ Yellow
(TWOyel)
Hybrid Tea, deep yellow, 2000
Hybridized by Jerry Twomey
[Whiskey Mac x (Sonia x Prominent)]
Photo courtesy Anthony Tesselaar International

Dream™ Pink
(TWOjoan)
Hybrid Tea, medium pink, 2000
Hybridized by Jerry Twomey
[White Masterpiece x Silver Jubilee]
Photo courtesy Anthony Tesselaar International

Dream™ Red
(TWOpaul)
Hybrid Tea, dark red, 2000
Hybridized by Jerry Twomey
[Esmeralda x (Evening Star x Trumpeter)]
Photo courtesy Anthony Tesselaar International

Knock Out
(RADraz)
Shrub, red blend, 2000
Hybridized by William J. Radler
[Carefree Beauty seedling x Razzle Dazzle seedling]
AARS 2000
Photo courtesy Conard-Pyle Co.

Traviata™
(MEILAVIO)
Hybrid Tea, dark red, 1999
Hybridized by Meilland
[(Porta Nigra x Paola) x William Shakespeare]
Photo courtesy Conard-Pyle Co./E. Ulzega

growth; [R. gigantea X Margaret Molyneux]; Nabonnand, P.

Ficksburg, F, ob, 1994; GM, Durbanville, 1992; Delbard

Fiddler's Gold™ *see* 'MINZCO'

Fiddletown Pink Noisette, N, mp; (found rose)

Fidélio® *see* 'MEICHEST'

'FIDÉLIO', F, ab, 1961; bud long, pointed, rosy red; dbl., borne in large clusters; vigorous growth; Horstmann

'FIDELITY', HT, mr, 1962; bud long, pointed; flowers large, dbl., exhibition form, moderate fragrance; foliage glossy; tall growth; RULED EXTINCT 7/80 ARM; [Crimson Glory X Peace]; Abrams, Von

Fidelity *see* **'SCOOP JACKSON'**

Fides, HT, ly, 1970; flowers large, dbl.; Urban, J.; (Sangerhausen)

Field Rose *see* **R. ARVENSIS**

Fieldfare, S, rb, 1989; Delbard

Fiery Hit *see* POUlfiry

Fiery Sunblaze *see* 'MEINEYTA'

Fiery Sunsation *see* 'KORTEMMA'

'FIESTA', HT, rb, 1940; bud ovoid; flowers vermilion, splashed bright yellow, large, dbl., moderate fragrance; foliage glossy, dark; vigorous, bushy, compact growth; [The Queen Alexandra Rose sport]; Hansen, C.B.; Armstrong Nursery

Fiesta *see* 'MACFIRINLIN'

'FIESTA BRAVA', Min, or, 1959; flowers geranium-red, blooms in clusters, semi-dbl., 14 petals; foliage glossy; vigorous, upright, bushy growth; [Méphisto X Perla de Alcañada]; Dot, M.; Combe

Fiesta Charm, F; Schloen, P.; (Cavriglia)

Fiesta Clown *see* 'LAVCLO'

'FIESTA FLAME', F, mr, 1978; bud pointed; flowers intense scarlet, semi-dbl., 15 petals, 3 in., slight fragrance; low, bushy growth; [Sarabande X Ena Harkness]; Sanday, John

'FIESTA GOLD', Min, yb, 1970; bud long, pointed; flowers yellow orange, small, dbl., cupped, slight fragrance; foliage small, glossy, light, leathery; vigorous, dwarf, upright, bushy growth; [Golden Glow X Magic Wand]; Moore, Ralph S.; Mini-Roses

'FIESTA RUBY', Min, mr, 1977; bud ovoid, pointed; dbl., 40 petals, 1 in., exhibition form, slight fragrance; foliage dark; bushy, compact growth; [Red Pinocchio X Little Chief]; Moore, Ralph S.; Sequoia Nursery

Fiesta Time™ *see* 'MINGCO'

Fièvre d'Or *see* 'KORBI'

Fifi *see* 'HENFIF'

'FIFTH AVENUE', HT, dp, 1948; bud large, ovoid; flowers grenadine-pink, dbl., exhibition form, moderate fragrance; vigorous growth; [Orange Nassau sport]; Johnson, W.E.

'FIGARO', HT, dr, 1954; flowers velvety scarlet, dbl., 26 petals, 4.5 in., intense fragrance; vigorous growth; [(Crimson Glory X Grande Duchesse Charlotte) X New Yorker]; Lens

'FIGMENT', F, m, 1998; flowers purple, overlaid with bright magenta in fresh flower, dark magenta reverse, very dbl., 26–40 petals, 4 in., borne in small clusters, moderate fragrance; prickles moderate; foliage medium, medium green, semi-glossy; upright, medium (2 1/2 ft) growth; [Len Turner X Remember Me]; Rawlins, R.

'FIGURINE', HT, pb, 1952; flowers china pink, dbl., 25 petals, moderate fragrance; foliage bluish green; vigorous growth; RULED EXTINCT 11/91 ARM; [Soeur Thérèse X Unnamed seedling]; Lens

Figurine™ *see* 'BENFIG'

'FIJI', Gr, or, 1965; bud ovoid; flowers bright orange-red, medium, dbl., cupped, moderate fragrance; vigorous, upright growth; [Queen Elizabeth X Unnamed seedling]; Schwartz, Ernest W.; Wyant

Fil d'Ariadne® *see* LENfil

'FILAGREE PILLAR', LCl, dp, 1962; bud ovoid; flowers tyrian rose, heavily veined, reverse lighter, dbl., 50 petals, moderate fragrance; foliage glossy, bronze; strong stems; vigorous, upright (to 5 ft) growth; [Titian X Sterling]; Riethmuller; Hazelwood Bros.

Filipes Kiftsgate *see* **'KIFTSGATE'**

'FILLETTE', F, mp, 1965; bud ovoid; dbl., blooms in clusters, intense fragrance; foliage dark; [Circus X Papillon Rose]; Lens; Spek

'FIMBRIATA', HRg, lp, 1891; (Diantheflora, Dianthiflora, Phoebe's Frilled Pink); flowers petals carnation-like, intense fragrance; [R. rugosa X Mme Alfred Carrière]; Morlet

Finale *see* 'KORAMI'

Financial Times Centenary *see* 'AUSFIN'

'FINE FLARE', HT, or, 1978; bud pointed; flowers vermilion, rounded, dbl., 40 petals, 4 in., moderate fragrance; foliage dark; moderately vigorous, upright growth; [Fragrant Cloud X Mildred Reynolds]; Bees

Fine Gold *see* 'WEEGOLD'

Fine Touch *see* 'LYOFIN'

'FINESSE', Min, my, 1983; flowers small blooms borne 1-3 per stem, dbl., exhibition form, no fragrance; foliage small, medium green, semi-glossy; bushy, spreading growth; [Picnic X Rise 'n' Shine]; Hardgrove, Donald L.

Finest Hour *see* 'MICFINEST'

Fingerpaint *see* 'MORFING'

Finkenrech, F, mp, 1989; dbl., intense fragrance; Michler, K. H.; (Sangerhausen)

'FINLANDIA', Pol, op, 1969; flowers orange-salmon; [Greta Kluis sport]; Kraats; Longley

Finnstar® *see* 'RUFIN'

Finstar *see* 'RUFIN'

Fintona, HT, w, 1997; Brundrett

Fiocco Bianco *see* BARfiob

Fiona® *see* 'MEIBELUXEN'

Fiona *see* 'KORFI'

Fiona Ravenscroft, HT, lp; Tebbin, N.H., 1972; (Weatherly, L.)

'FIONIA', HWich, lp, 1914; flowers light pink to rose, small blooms borne in clusters; foliage dark, glossy; vigorous growth; [Mme Norbert Levavasseur X Dorothy Perkins]; Poulsen, D.T.; Poulsen

Fiord, *see* 'MEICAUF'

Fiorella *see* 'MEICAPULA'

Fiorella '82 *see* 'MEICAPULA'

Fiorona, F; Mati; (Cavriglia)

'FIRE BIRD', F, rb, 1992; flowers vermillion, pale yellow center, blooms borne in large clusters, semi-dbl., 6–14 petals, 1.5–2.75 in., no fragrance; some prickles; foliage medium, dark green, leathery; medium (70 cms), upright growth; [Avocet X Evelyn Fison]; Strange, J.F.

'FIRE CHIEF', HT, mr, 1942; bud long, pointed; flowers flame-red, large, dbl., 24–28 petals, exhibition form, moderate fragrance; foliage glossy; very vigorous, bushy growth; [Crimson Glory X Ami Quinard]; Jacobus; B&A

'FIRE DANCE', HT, mr, 1951; bud very long, ovoid; flowers brilliant velvety scarlet, dbl., 25–35 petals, 4–6 in., globular, slight fragrance; foliage glossy, light green; upright, moderately vigorous growth; [(Ulrich Brunner Fils X Westfield Star) X (Chieftain X Better Times)]; Verschuren; Totty

'FIRE FLAME', F, or, 1958; flowers scarlet, center orange, single, 4 in.; moderate growth; Morse

Fire King *see* 'MEIKANS'

Fire King *see* MEIrolyz

Fire Magic *see* 'KORFEU'

Fire Meidiland™ *see* 'MEIPSIDUE'

Fire 'n' Ice™ *see* 'AROFIRIC'

'FIRE OPAL', F, or, 1955; (Fair Opal); bud ovoid; flowers reddish orange-scarlet, reverse lighter, medium, borne in clusters, dbl., 20–25 petals, cupped, moderate fragrance; foliage glossy; vigorous, open habit growth; [Goldilocks X Unnamed orange Polyantha]; Boerner; McGredy

Fire Pillar *see* **'BISCHOFSSTADT PADERBORN'**®

'FIRE PRINCESS', Min, or, 1969; flowers small, dbl.; foliage small, glossy, leathery; vigorous, bushy growth; [Baccará X Eleanor]; Moore, Ralph S.; Sequoia Nursery

'FIRE QUEEN', F, dr, 1963; bud ovoid; flowers bright red, medium, borne in clusters, dbl.; foliage dark, glossy; vigorous, bushy growth; [Fusilier X (Carrousel X Queen o' the Lakes)]; Von Abrams; Peterson & Dering

Fire Robe, Gr, rb

Fire Signal *see* **'SIGNALFEUER'**®

'FIRE SKY', HT, mr, 1952; bud long, pointed; flowers large, cupped; foliage dark, glossy; vigorous growth; [Gen. MacArthur X Étoile de Hollande]; Silva

'FIREBALL', Pol, or, 1931; flowers glowing reddish orange; foliage dark, glossy; deRuiter

'FIREBEAM', F, rb, 1960; flowers yellow, flame, orange and crimson, borne in clusters, semi-dbl., 14 petals, 2.5 in., intense fragrance; foliage glossy; vigorous growth; [Masquerade X Unnamed seedling]; Fryers Nursery, Ltd.

'FIREBIRD', F, ob, 1960; flowers bright orange, base golden yellow, borne in clusters, semi-dbl., 12–18 petals, 2.5–3 in.; foliage glossy, light green; vigorous, upright growth; RULED EXTINCT 4/92; [Masquerade X Mme Henri Guillot]; Watkins Roses

'FIREBRAND', HT, or, 1938; flowers bright scarlet; [Flamingo sport]; Cant, B. R.

'FIREBRAND, CLIMBING', Cl HT, or, 1953; Raffel; Port Stockton Nursery

Fireburst *see* 'TWORIGHT'

'FIRECRACKER', F, mr, 1956; flowers scarlet, base yellow, blooms in clusters, semi-dbl., 14 petals, 4.5 in., moderate fragrance; foliage leathery, light green; dwarf, bushy growth; [Pinocchio seedling X Numa Fay seedling]; Boerner; A. Dickson, 1956;, J&P, 1959;, McGredy, 1956

'FIRECREST', F, mr, 1964; dbl., 35 petals, 3 in., in clusters; vigorous, low growth; [(Cinnabar X Marjorie LeGrice) X Pimpernell]; LeGrice

'FIRECREST, CLIMBING', Cl F, mr, 1969; LeGrice

'FIREDRAGON', HT, mr, 1923; flowers fiery red, dbl., moderate fragrance; Clark, A.

'FIREFALL', Cl Min, dr, 1980; (Roy Rumsey); bud short; dbl., 43 petals, 1.5 in., flat, blooms in clusters; foliage small, glossy; trailing, arching growth; [Dortmund X Little Chief]; Moore, Ralph S.; Sequoia Nursery

'FIREFLAME', F, mr, 1954; bud ovoid; flowers carmine, borne in pyramidal clusters, dbl., 65–70 petals, 2.5–3 in., moderate fragrance; vigorous growth; [Chatter X Red Pinocchio]; Boerner; Stark Bros.

'FIREFLASH', HT, yb, 1960; flowers golden yellow splashed scarlet, well formed, dbl., 25–30 petals, 4.5 in., moderate fragrance; foliage glossy; upright growth; [Marjorie LeGrice X Seedling]; LeGrice; Wayside Gardens Co.

'FIREFLY', F, yb, 1975; bud slightly pointed; flowers medium yellow, dbl., 22 petals, 3.5 in., cupped, slight fragrance; foliage dark; upright growth; RULED EXTINCT 4/78 ARM; [Contempo sport]; Joliffe

Firefly *see* **'HOLSTEIN'**

Firefly *see* 'MACFRABRO'

'FIREGLOW', Pol, or, 1929; flowers brilliant vermilion-red, shaded orange, blooms in clusters, single, slight fragrance; dwarf, compact growth; [Orange King sport]; Wezelenburg

'FIREGLOW, CLIMBING', Cl Pol, or, 1950; Guillot, M.

'FIRELIGHT', HT, or, 1971; flowers large, dbl., exhibition form, moderate fragrance; foliage large, light, leathery; vigorous, upright growth; [Detroiter X Orange Delbard]; Kordes, R.; J&P

Fire-Rider *see* **'FEUERREITER'**

'FIRESIDE', HT, yb, 1977; bud ovoid; flowers yellow, white, red, imbricated, dbl., 35 petals, 5.5–6. in., slight fragrance; foliage large, glossy, dark; vigorous, bushy growth; [Kordes' Perfecta X Belle Blonde]; Lindquist

Firestar®, LCl, mr

Firestorm *see* 'CLEFIRE'

Firetail, Cl Min, lp; Peden, G.H., 1994; (Weatherly, L.)

Fireworks™ *see* 'SAVAFIRE'

Fireworks, Climbing *see* **'FEU D'ARTIFICE'**

Firlefanz, S, yb, 1968; flowers golden yellow plus carmine-red, medium, dbl.; GPG Bad Langensalza; (Sangerhausen)

'FIRMAMENT', HT, dr, 1971; flowers vermilion-crimson, large, dbl.; foliage bronze; [Chrysler Imperial X Credo]; Gaujard

First Affair *see* JACvoo

'FIRST BLUSH', Pol, mp, 1965; flowers phlox-pink, borne in clusters, 3 in., cupped, intermittent bloom; foliage dull, light green; moderate growth; [Francais X Orléans Rose]; Delbard-Chabert; Cuthbert

First Born *see* 'GILFLEUR'

'FIRST CHOICE', F, or, 1958; flowers fiery orange-scarlet, center yellow, in trusses, single, 7 petals, 5 in., moderate fragrance; tall, spreading growth; [Masquerade X Sultane]; Morse

First Class *see* 'JACARE'

First Edition *see* 'DELTEP'

'FIRST FEDERAL', HT, pb, 1965; bud ovoid; flowers geranium-pink tinted scarlet, dbl., 35–40 petals, 5–5.5 in., exhibition form, intense, rose geranium fragrance; strong stems; vigorous, upright growth; [(Radiance X Pageant) X Diamond Jubilee seedling]; Boerner; J&P

'FIRST FEDERAL GOLD', HT, dy, 1967; bud ovoid; flowers gold and yellow, large, dbl., exhibition form, moderate fragrance; foliage glossy, leathery; vigorous, upright growth; [Golden Masterpiece seedling X Golden Masterpiece seedling]; Boerner; J&P

'FIRST FEDERAL'S RENAISSANCE', HT, mp, 1980; bud long, pointed; flowers medium pink, tinted lighter, dbl., 23 petals, 5–7 in., very early bloom, slight fragrance; foliage large; compact growth; [Unnamed variety X First Prize]; Warriner, William A.; J&P

First Hit *see* POUlrang

First Kiss *see* 'JACLING'

'FIRST LADY', HT, dp, 1961; bud ovoid; flowers rose madder to phlox-pink, semi-dbl., 18–22 petals, 3.5–4.5 in., cupped, slight fragrance; foliage leathery, dark, semi-glossy; very vigorous, upright growth; [First Love X Roundelay]; Swim, H.C.; C.R. Burr

First Lady *see* TANrif

'FIRST LADY NANCY', HT, yb, 1981; bud ovoid, long, pointed; flowers light yellow tinged light pink, formal, spiraled, dbl., 36 petals, borne singly, slight, tea fragrance; medium prickles; foliage semi-glossy, medium; medium, upright, bushy growth; [American Heritage X First Prize]; Swim, H.C. & Christensen, J.E.; Armstrong Nursery

'FIRST LADY, CLIMBING', Cl HT, dp, 1964; Burr, C.R.

First Light *see* 'DEVRUDI'

'FIRST LOVE', HT, lp, 1951; (Premier Amour); bud long, pointed; flowers medium, dbl., 25 petals, 2.5–3.5 in., slight fragrance; foliage leathery, light green; moderately bushy growth; [Charlotte Armstrong X Show Girl]; Swim, H.C.; Armstrong Nursery

'FIRST NATIONAL GOLD', Min, dy, 1976; bud pointed; dbl., 38–42 petals, 1–1.5 in., exhibition form, slight fragrance; foliage small; compact, upright growth; [Rise 'n' Shine X Yellow Jewel]; Saville, F. Harmon; Flora World

'FIRST NATIONAL SILVER', Min, w, 1976; bud ovate; flowers small, very dbl., 50–55 petals, 1.5 in., flat, slight fragrance; foliage small, very glossy; compact, spreading growth; [Charlie McCarthy X Little Chief]; Schwartz, Ernest W.; Flora World

'FIRST OFFERING', F, dr, 1975; bud ovoid; flowers open, semi-dbl., 15 petals, 2.5 in., globular, profuse, very lasting bloom, moderate fragrance; foliage large, glossy, bronze, reddish brown when young; vigorous, dwarf, bushy growth; [(Seedling X ?) X Samba]; Viraraghavan, M.S.

'FIRST PRIZE', HT, pb, 1970; bud long, pointed; flowers rose-pink, center blended with old ivory, very large, dbl., 25–30 petals, 5.5–6. in., exhibition form, borne singly, moderate, tea fragrance; foliage large, dark green, leathery; long, stout stems; vigorous, upright growth; AARS, ARS, 1970 Gertrude M. Hubbard, ARS, 1971 GM, ARS, 1971; [Enchantment seedling X Golden Masterpiece seedling]; Boerner; J&P

First Prize, Climbing *see* 'JACCLIST'

First Red® *see* PEKcoujenny

'FIRST ROSE CONVENTION', HT, dr, 1971; bud globular; flowers dark crimson, borne singly, dbl., 48 petals, cupped, no fragrance; hooked, pale cream prickles; foliage red when young, turning dark green; vigorous, upright growth; [Flaming Peace X Helen Traubel]; Hardikar, Dr. M.N.; The Bombay Rose Society

Firstar *see* 'JACARY'

Firstburst *see* 'TWORIGHT'

Fisher & Holmes *see* **'FISHER HOLMES'**

'FISHER HOLMES', HP, dr, 1865; (Fisher & Holmes); bud long, pointed; flowers deep red, well-formed, large, dbl., 30 petals, recurrent bloom; upright growth; (28); [Probably Maurice Bernardin self-seedling sport]; Verdier, E.

Fisherman's Friend® *see* 'AUSCHILD'

FISmelody, Min, lp; (**Sweet Melody**); Fischer, 1998

Five-Colored Rose *see* **'FORTUNE'S FIVE-COLORED ROSE'**

Five-Roses Rose *see* 'JACOPPER'

'FLAIR', HT, op, 1951; bud ovoid; flowers coral-blush, dbl., 30–35 petals, 3.5–4.5 in., exhibition form; vigorous growth; [Lady Sylvia X Unnamed seedling]; Verschuren-Pechtold; J&P

'FLAIR', HT, rb, 1989; bud pointed; flowers medium red to dark red, reverse lighter, medium, borne singly, dbl., exhibition form, slight fragrance; obovate, one inch, green with red splotch fruit; prickles curved, hooked, reddish-brown, sparse; foliage medium, dark green, semi-glossy; upright, medium growth; [Pristine X Ink Spots]; Cummings, Peter E., 1991

Flair *see* DICrelax

'FLAMBEAU', HT, mr, 1940; bud pointed; flowers crimson shaded scarlet, open, large, dbl.; foliage glossy; strong stems; vigorous, bushy, open growth; [Royal Red X Johanniszauber]; Nicolas; J&P

'FLAMBÉE', F, or, 1954; flowers medium, borne in clusters of 5-6, dbl., 25–30 petals; foliage reddish to bronze; upright, bushy growth; GM, Bagatelle, 1952; Mallerin, C.; EFR

Flambo, HT, or, 1993; Kordes

Flamboyance, HT, or

'FLAMBOYANT', Pol, mr, 1931; flowers bright scarlet, aging to crimson-carmine, large blooms in clusters, dbl.; foliage glossy; dwarf growth; Turbat

'FLAMBOYANT', F, mr; flowers bright red; vigorous growth; [Holstein X Incendie]; Croix, P.

'FLAME', HMult, op, 1912; flowers bright salmon-pink, borne in large clusters, semi-dbl.; foliage dark, glossy; very vigorous, compact growth; [Crimson Rambler X ?]; Turner

Flame Bouquet™ *see* 'LAVFIRE'

Flame Dance *see* 'KORFLATA'

Flame Dancer *see* 'HARDANCER'

'FLAME OF FIRE', HT, ob, 1917; flowers orange-flame, open, large, dbl., moderate fragrance; bushy growth; GM, NRS, 1916; McGredy

'FLAME OF LOVE', HT, mr, 1950; Hieatt

Flame of the East *see* **'PLAMYA VOSTOKA'**

'FLAMEBURST', F, dr, 1961; bud pointed; flowers crimson, center yellow, borne in clusters, single, 4–5 petals, 2 in., slight fragrance; foliage dark, glossy; very vigorous growth; [Nearly Wild X Unnamed Hybrid Tea seedling]; Brownell, H.C.

Flameche, F, or

'FLAMEGLO', Min, yb, 1981; flowers deep yellow to orange-red, reverse deep yellow, small, dbl., slight fragrance; foliage small, dark, glossy; upright, bushy growth; PP005176; [Starburst X Over the Rainbow]; Williams, Ernest D.; Mini-Roses

'FLAMENCO', F, lp, 1960; flowers light salmon-pink, blooms in clusters, dbl., 21 petals, 3.5 in., slight fragrance; foliage dark; [Cinnabar X Spartan]; McGredy, Sam IV; McGredy

Flamenco *see* POUltika

Flamendr, HT, dp, 1969; flowers large, dbl.; Urban, J.; (Sangerhausen)

Flames 'n' Sparks *see* LUDsporcoma

Flametta, Min, mp

'FLAMINA', F, my, 1964; bud pointed, flushed red; flowers golden yellow, large, dbl.; foliage clear green; very vigorous, upright, bushy growth; [(Faust X Peace) X Unnamed seedling]; Mondial Roses

'FLAMINAIRE', HT, ob, 1960; (Montagny); bud globular; flowers orange-flame, medium, semi-dbl., slight fragrance; foliage dark, glossy; vigorous, bushy growth; [Eclipse X Independence]; Dorieux; Pin

'FLAMING ARROW', F, or, 1965; bud long, pointed; flowers bright orange-red, medium, dbl., moderate fragrance; foliage glossy; vigorous, upright growth; [Montezuma X Nadine]; Schwartz, Ernest W.; Wyant

'FLAMING BEAUTY', HT, rb, 1978; flowers yellow and red-orange, dbl., 35 petals, 4 in., exhibition form, slight fragrance; foliage matt, green; bushy growth; Silver, ARC TG, 1979; [First Prize X Piccadilly]; Winchel, Joseph F.; Kimbrew-Walter Roses

'FLAMING JUNE', Pol, or, 1931; (Mrs A. Hudig); flowers bright orange-scarlet; vigorous growth; Cutbush

Flaming Peace *see* 'MACBO'

Flaming Potluck *see* 'LAVFLAME'

Flaming Rosamini *see* RUIflami

'FLAMING RUBY', F, dr, 1963; flowers deep ruby-red, borne in clusters, dbl., 60 petals, 3 in.; foliage dark, reddish green, small; vigorous growth; [(Eva self X Guinee) X Guinee]; Hennessey

Flaming Star *see* BARflam

'FLAMING SUNSET', HT, ob, 1948; flowers deep orange, reverse lighter; foliage light bronze; [McGredy's Sunset sport]; Eddie

'FLAMING SUNSET, CLIMBING', Cl HT, ob, 1954; Mattock

Flaming Torch, F, op, 1999

'FLAMINGO', HRg, mp, 1956; bud pointed; flowers rich pink, large blooms in clusters, single, 5 petals, cupped, recurrent bloom; foliage glossy, gray-green; vigorous (3 ft.) growth; [R. rugosa X White Wings]; Howard, F.H.; Wayside Gardens Co.

Flamingo® *see* 'KORFLÜG'

'FLAMINGO', HT, dp, 1929; flowers bright geranium-red to rosy cerise, spiral shape, dbl., exhibition form, intense fragrance; vigorous growth; GM, NRS, 1927; Dickson, A.

'FLAMINGO', F, mp, 1958; flowers rose; long, strong stems; very vigorous growth; Frères, Buyl

'FLAMINGO', F, op, 1961; bud long, pointed; flowers deep scarlet-pink, open, borne in clusters, dbl., 30 petals; vigorous, upright growth; Horstmann

Flamingo *see* 'HERFLA'

Flamingo Garden Tea, T, op

Flamingo Meidiland *see* MEIsolroz

'**FLAMINGO QUEEN**', Gr, dp, 1972; flowers deep pink; [Queen Elizabeth sport]; Chan; Canadian Ornamental Plant Foundation

'**FLAMMÈCHE**', F, ob, 1959; flowers orange, large; vigorous growth; Combe

Flammenmeer *see* **'SHALOM'**®

'**FLAMMENROSE**', HT, ob, 1921; flowers bright orange-yellow, semi-dbl., 16 petals; [Mrs Joseph Hill X Mme Edouard Herriot]; Türke; Kiese

'**FLAMMENSPIEL**', S, mp, 1974; bud ovoid; flowers salmon-pink, dbl.; foliage large, leathery, dark; vigorous, upright growth; [Peer Gynt X Unnamed seedling]; Kordes; Horstmann

Flammentanz® *see* 'KORFLATA'

Flammette *see* **'CRIMSON GEM'**

Flanders Field *see* 'HORFLAN'

'**FLANDRIA**', Pol, mr, 1966; bud ovoid; flowers open, semi-dbl., moderate fragrance; foliage dark, glossy; bushy growth; Delforge

'**FLAPPER**', F, lp, 1998; flowers light pink, slightly curled petals, single, 5–7 petals, 3 in.., borne in small clusters, no fragrance; prickles few, medium sized, straight; foliage medium, dark green, semi-glossy; compact, medium (5 ft.) growth; [Permanent Wave sport]; Bennett, Frank David

'**FLASH**', LCl, rb, 1938; bud ovoid, yellow suffused scarlet; flowers orange-scarlet, reverse and center yellow, dbl., cupped, long blooming season, slight fragrance; foliage leathery, glossy, bronze; pillar (6-8 ft), compact growth; AARS, 1940 GM, Rome, 1939; [Rosella (Cl HT) X Margaret McGredy]; Hatton; C-P

Flash *see* 'LENANA'

Flash *see* **'ROSEMARY GANDY'**

Flash Meidiland *see* MEIstocko

Flashdance *see* 'POULFLASH'

Flashfire *see* 'LEEFIR'

'**FLASHLIGHT**', F, or, 1974; flowers orange-scarlet, dbl., 28 petals, 3–3.5 in., moderate fragrance; foliage large, bronze; tall growth; [Vesper X Unnamed seedling]; LeGrice

Flauce, HT, rb

'**FLAVESCENS**', HSpn, ly; flowers pale lemon, rounded; very hardy.; Prior to 1824

Flavia *see* 'INTERETTE'

'**FLAVIEN BUDILLON**', T, lp; flowers pale flesh, large, globular, intense fragrance

'**FLEET STREET**', HT, dp, 1972; flowers deep rose pink, well-formed, dbl., 40 petals, 6 in., intense fragrance; foliage large, dark, leathery; [Flaming Peace X Prima Ballerina]; McGredy, Sam IV; McGredy

'**FLEETWOOD**', F, ob, 1989; bud pointed; flowers bright orange, reverse lighter, aging darker on outer petals, dbl., 32 petals, exhibition form, slight, fruity fragrance; prickles straight, medium pink; foliage medium, dark green, glossy; bushy, medium, vigorous growth; [Little Darling X Orangeade]; Bridges, Dennis A.; Bridges Roses, 1990

'**FLESH TAITO**', F, ob, 1988; flowers yellow at base to vermilion, reverse orange flushed with pink, dbl., 25–30 petals, cupped, moderate fragrance; prickles ordinary, green; foliage medium, dark green, undulated; bushy, medium growth; [Masquerade X Matador]; Kikuchi, Rikichi, 1989

'**FLEUR COWLES**', F, ly, 1972; flowers cream, center buff, dbl., 35 petals, 3 in., moderate, spicy fragrance; foliage glossy, dark; [Pink Parfait X ?]; Gregory

Fleur d'Amour, S, ab, 1998; Austin, David

'**FLEUR DE FRANCE**', HT, my, 1944; flowers capucine and yellow, medium, semi-dbl., globular, slight fragrance; foliage glossy; vigorous, dwarf growth; Gaujard

'**FLEUR DE PELTIER**', HGal, pb, 1824; Roseraie de l'Hay

Fleurette® *see* 'INTERETTE'

Fleurop® *see* 'KORTEXUNG'®

Fleurs de Pelletier, HGal, dp; flowers medium, very dbl.; Roseraie de l'Hay; (Sangerhausen)

Flicker, F, ab

'**FLIEGERHELD BOELCKE**', HT, my, 1920; flowers nankeen yellow, shaded reddish yellow; [Mme Caroline Testout X Sunburst]; Schmidt, J.C.

'**FLIEGERHELD ÖHRING**', HT, or, 1919; flowers medium, dbl.; Kiese; (Sangerhausen)

'**FLIGHTY**', F, m, 1969; bud cherry-red; flowers light mauve, medium, borne in trusses, semi-dbl.; foliage dark; free growth; [Orangeade X Sterling Silver]; Trew, C.; Basildon Rose Gardens

Flimo, F, lp, 1990; Poulsen

'**FLIPPER**', HT, or, 1973; bud long, pointed; flowers large; foliage large; [Tanagra X John S. Armstrong]; Gaujard

'**FLIRT**', F, mr, 1952; bud pointed; flowers bright cherry-red, reverse yellow, medium, borne in clusters, dbl., 35–40 petals, moderate fragrance; foliage glossy, dark; vigorous, upright growth; [Pink Princess X Shades of Autumn]; Brownell, H.C.

Flirt *see* KORvondra

'**FLIRTATION**', HT, pb, 1953; bud pointed; flowers begonia-rose, reverse lemon to deep yellow, dbl., 30 petals, 5–6 in., exhibition form, moderate fragrance; foliage dark, leathery; strong stems; vigorous, bushy, compact growth; [Fiesta X Peace]; Shepherd; Bosley Nursery

'**FLOCON DE NEIGE**', HCh, w, 1898; Lille

'FLOELLO', Min, w, 1986; (**Hello There**); flowers white, light yellow tints in center, good petal retention, small, very dbl., 108 petals, cupped, borne usually singly, no fragrance; tan, very few prickles; foliage small, medium green, matt; bushy, low growth; [Care Deeply X Red Can Can]; Florac, Marilyn, 1987

'FLOHIH', Min, or, 1985; (**High Hope**); flowers bright orange-red, small, borne singly and in clusters, dbl., 45 petals, cupped, no fragrance; foliage medium, light, glossy; vigorous, upright growth; [Young Love X Little Chief]; Florac, Marilyn; M.B. Farm Min. Roses, Inc.

'FLOMOR', Min, yb, 1986; (**Country Morning**); flowers light yellow, blended pink, blooms borne singly, dbl., exhibition form, moderate, spicy fragrance; small, reddish prickles; foliage medium, medium green, semi-glossy; medium, upright growth; [Avandel X Young Love]; Florac, Marilyn; MB Farm

'FLOMYST', Min, ob, 1986; (**Sweet Mystery**); flowers deep orange fading to pink, medium, borne usually singly, dbl., 90 petals, cupped, moderate, damask fragrance; few, green prickles; foliage medium, dark green, semi-glossy; bushy, medium growth; (21); [Care Deeply X Red Can Can]; Florac, Marilyn, 1987

'**FLON**', D, mr, 1845; (Gloire des Perpetuelles, La Mienne); flowers bright red, very free bloom, occasionally repeated; Vibert

'FLOPER', Min, op, 1985; (**Per Chance**); flowers orange-red fading bright pink, small blooms borne singly and, single, 5 petals, moderate fragrance; foliage small, dark, matt; bushy growth; [Red Can Can X Care Deeply]; Florac, Marilyn; M.B. Farm Min. Roses, Inc.

'**FLOR DE TORINO 61**', HT, w, 1961; flowers white edged violet-pink; [Monte Carlo X Michele Meilland]; da Silva, Moreira

'**FLORA**', HSem, m; (Flore); flowers lilac-pink, center deeper, dbl., seasonal bloom, moderate fragrance; Jacques, 1829

'**FLORA**', HT, op, 1957; flowers pink tinted salmon, large, dbl., intense fragrance;

vigorous, upright growth; [Independence X Charlotte Armstrong]; Maarse, G.

Flora Bama *see* 'TALFLORA'

Flora Danica *see* POUlrim

'FLORA MACLEOD', LCl, lp, 1971; flowers pale rose-pink, dbl., 50 petals, 4.5 in., non-recurrent, intense fragrance; foliage large, medium green, matt; vigorous growth; [New Dawn X Shot Silk, Climbing]; MacLeod

'FLORA MCIVOR', HEg, pb, 1894; flowers rosy pink, white center, yellow stamens, nearly small, single, summer bloom; foliage very fragrant (and flowers); vigorous growth; [R. eglanteria X HP or B]; Penzance; Keynes, Williams & Co.

'FLORABELLE', F, op, 1964; flowers soft salmon-pink, large, dbl., 33 petals, globular, moderate fragrance; foliage soft; vigorous, bushy growth; [Ma Perkins X Unnamed seedling]; Schwartz, Ernest W.; Wyant

'FLORADORA', F, or, 1944; bud globular; flowers cinnabar-red, blooms in sprays of 6-12, dbl., 25 petals, 2 in., cupped, slight fragrance; foliage leathery, glossy; upright, bushy growth; AARS, 1945; [Baby Chateau X R. roxburghii]; Tantau; C-P

'FLORADORA, CLIMBING', Cl F, or, 1951; Shamburger, P.; Shamburger Rose Nursery

Floraece Delattre *see* MASflodel

Floral Choice, F, ab, 1982; Fryer, Gareth

'FLORAL DANCE', Cl HT, or, 1955; flowers orange-cerise; very vigorous, climbing growth; [Souv de Mme Boullet, Climbing X Crimson Glory]; Homan; Roseglen Nursery

'FLORALIES VALENCIENNOISES', LCl, mr, 1955; bud long, pointed; flowers currant-red, large, semi-dbl., intense fragrance; vigorous growth; [Matador X Soliel d'Orient]; Dorieux; Pin

Florange, F, mr, 1990; RvS-Melle

Floranje, F, or; dbl., 23 petals, 3 in., flat; foliage dense dark green; strong grower growth; RvS-Melle, 1985

'FLORANNE', Min, my, 1996; flowers medium yellow, fades in sun to white, color better in partial shade, single, 1.5–2.75 in., slight fragrance; some prickles; foliage medium, medium green, semi-glossy; upright (30 cms) growth; [Crazy Dottie sport]; Pratt, Florence

Flore *see* **'FLORA'**

'FLORE BERTHELOT', Pol, my, 1921; flowers clear lemon-yellow, passing to white, borne in clusters; Turbat

'FLORÉAL', Pol, pb, 1923; flowers flesh and rose-pink, dbl., slight fragrance; [Orléans Rose X Yvonne Rabier]; Turbat

'FLORÉAL', HT, op, 1944; bud pointed; flowers coral-pink tinted yellow, medium, dbl., cupped, moderate fragrance; foliage dark, glossy; vigorous growth; Gaujard

'FLORENCE', HT, lp, 1921; flowers silvery pink; Paul, W.

'FLORENCE', HT, w, 1961; bud long; flowers pure white, large; Dorieux; Pin

Florence *see* KORnagent

'FLORENCE CHENOWETH', HT, yb, 1918; (Yellow Herriot); flowers yellow, shaded coral-red; [Mme Edouard Herriot sport]; Chenoweth

Florence Delattre *see* MASflodel

'FLORENCE EDITH COULTHWAITE', HT, ly, 1908; flowers medium, dbl., moderate fragrance; Dickson, A.; (Sangerhausen)

'FLORENCE EDNA', HT, lp, 1985; [Princesse sport]; Owen, Fred

'FLORENCE FORRESTER', HT, w, 1914; flowers white tinged lemon, large, dbl.; GM, NRS, 1913; McGredy

'FLORENCE HASWELL VEITCH', Cl HT, dr, 1911; flowers bright scarlet, shaded black, large, dbl., intense fragrance; vigorous growth; [Mme Edmee Metz X Victor Hugo]; Paul, W.

'FLORENCE L. IZZARD', HT, dy, 1923; bud pointed; flowers bright deep golden yellow, large, dbl., exhibition form, intense fragrance; foliage dark, bronze, leathery, glossy; very vigorous, bushy growth; McGredy

'FLORENCE LORRAINE', F, ab, 1973; bud ovoid; flowers dark apricot, large, very dbl., exhibition form, moderate fragrance; foliage glossy, dark, leathery, wrinkled; very vigorous, upright, bushy growth; [Royal Highness X (Hawaii X Helen Traubel)]; Middlebrooks; Ellesmere Nursery

'FLORENCE LYDIA', S, ab, 1999; flowers apricot with tint of pink, reverse same, fading to cream, 2 in., borne in large clusters, intense fragrance; prickles moderate; foliage medium, medium green, glossy; upright, tall (6 ft) growth; [Burma Star X Admiral Rodney]; Jones, L.J.

'FLORENCE MARY', HT, dy, 1964; flowers deep yellow-ochre, well formed, large, moderate fragrance; bushy growth; [Doreen sport]; Morse

'FLORENCE MARY MORSE', S, mr, 1951; flowers copper-scarlet, blooms in trusses to 30, semi-dbl., 15 petals, 3 in.; foliage dark, glossy; (28); [Baby Chateau X Magnifica]; Kordes

'FLORENCE MAYER', HT, w, 1998; flowers white and pink blend, reverse white, exhibition, very dbl., 26–40 petals, 4–6 in., exhibition form, borne mostly singly, slight fragrance; prickles moderate amount, large, hooked downward; foliage medium, medium green, semi-glossy; upright, tall growth; [Great Scott X Headliner]; Singer, Steven; Wisconsin Roses, 1998

Florence Nightingale *see* 'GANFLO'

'FLORENCE PAUL', HP, 1886; Paul, W.; (Cavriglia)

'FLORENCE PEMBERTON', HT, w, 1903; flowers creamy white, suffused pink, dbl., exhibition form; foliage rich green, leathery; vigorous growth; Dickson, A.

Florence Rambler, HWich, lp

'FLORENCE RUSS', HT, lp; Russ, 1923; (Weatherly, L.)

'FLORENTIA', HT, ob, 1941; flowers deep orange, well shaped, large; vigorous, bushy, compact growth; GM, Rome, 1940; [Julien Potin X Unnamed variety]; Giacomasso

'FLORENTINA', HT, dr, 1973; (Kordes' Rose Florentine®); bud large, long, pointed;, exhibition form, moderate fragrance; foliage leathery; vigorous growth; ADR, 1974; [Liebeszauber X Brandenberg]; Kordes; Horstmann, 1974

'FLORENTINA', F, pb, 1938; flowers hydrangea-pink, reverse rose-red, large, single; foliage leathery, dark; long stems; very vigorous growth; [Unnamed seedling X Permanent Wave]; Leenders, M.

Florescence, F, ob

Florett, F, op, 1983; flowers large, dbl., slight fragrance; GPG Bad Langensalza; (Sangerhausen)

'FLOREX', HT, op, 1927; bud pointed; flowers deep coral-salmon, suffused orange-carmine, large, dbl., exhibition form, intense fragrance; foliage leathery, glossy, dark; long stems; very vigorous growth; [Mme Butterfly X Premier]; Geiger; A.N. Pierson;, Florex Gardens

Flori®, F, ob

'FLORIADE', Gr, ob, 1963; bud urn shaped; flowers bright orange-scarlet, to open, dbl., 4 in., exhibition form; long, strong stems; vigorous growth; [Montezuma sport]; van der Schilden; Armstrong Nursery

Florian *see* 'MEILAUR'

Florian, Climbing *see* 'MEILAURSAR'

'FLORICEL', HT, or, 1957; bud pointed; flowers red and salmon, large, dbl., 25 petals, exhibition form; foliage dark, glossy; strong stems; upright, compact growth; [Carito MacMahon X Luis Brinas]; Dot, Pedro

Florida International, *see* 'FOXFLORIN'

'FLORIDA RED', HT, mr, 1964; flowers rich red, dbl., 60 petals, 5 in.; vigorous growth; Hennessey

Florida Sun, *see* 'GELFLA'

'FLORIDA VON SCHARBEUTZ', F, ob, 1957; flowers orange-yellow shaded coppery, very large blooms in large clusters, dbl., exhibition form, moderate fragrance; foliage dark, glossy; vigorous, bushy growth; [Golden Scepter X (Munchen X Peace)]; Kordes, R.

'FLORIMEL', F, mp, 1958; flowers silvery pink to deep rose-pink, well formed, borne in clusters; foliage glossy; vigorous growth; [Pinocchio X Unnamed seedling]; Fryers Nursery, Ltd.

'FLORINDA NORMAN THOMPSON', HT, pb, 1920; flowers delicate rose on lemon, base deeper, dbl., moderate fragrance; Dickson, A.

Florine, HT, 1981; Gaujard; (Cavriglia)

Florita® *see* BARflor

Florizel, HT, ab; 36 petals, flat, slight fragrance; strong grower growth; RvS-Melle, 1987

'FLORON', HT, ab, 1948; flowers indian yellow-orange, inside lighter, very large, dbl., exhibition form, slight fragrance; upright growth; [Comtesse Vandal X Pilar Landecho]; Camprubi, C.

'FLORRIE JOYCE', F, dp; flowers cherry-pink, blooms in large clusters, dbl., 43 petals; bushy growth; [Gartendirektor Otto Linne X Borderer]; Riethmuller

'FLOSAR', Min, mp, 1986; (**Fresh Start**); flowers bright medium pink, blooms borne usually singly, dbl., globular, no fragrance; small, red prickles; foliage small, medium green, semi-glossy; low, bushy growth; [Avandel X Little Chief]; Florac, Marilyn; M.B. Farm Min. Roses, Inc.

'FLOSOF', Min, pb, 1986; (**Soft Steps**); flowers creamy, petal edges pink, medium, borne singly, dbl., 33 petals, cupped, no fragrance; small, reddish prickles; foliage medium, medium green, semi-glossy; tall, upright growth; [Avandel X ?]; Florac, Marilyn; M.B. Farm Min. Roses, Inc.

Flossie *see* 'HORFLOSS'

Flower Basket *see* 'SEABASKET'

Flower Carpet™ *see* 'NOATRAUM'

Flower Child *see* 'HORFLOWER'

Flower Girl *see* **'SEA PEARL'**

Flower Girl™ *see* 'FRYYEOMAN'

'FLOWER OF FAIRFIELD', HMult, mr, 1909;, sometimes recurrent bloom; [Crimson Rambler sport]; Ludorf; H. Schultheis

Flower Power *see* FRYcassia

Flower Power *see* KORedan

Flower Power, F, mp, 1994; Kordes

Flower Power *see* FRYcassia

'FLOWER SHOW', HT, rb, 1980; bud ovoid, pointed; flowers scarlet, yellow reverse, borne 3-4 per cluster, dbl., 25 petals, moderate fragrance; dark red prickles; foliage mid-green, semi-matt; vigorous, upright growth; [Fragrant Cloud X Tropicana]; Bees

'FLOWER WORLD', HT, or, 1980; bud pointed, oval; dbl., 20–30 petals, 4 in., exhibition form; foliage dark, leathery; tall growth; [Baccará X South Seas]; Warriner, William A.; Flower World of America

'FLOWIN', Min, mp, 1986; (**Pink Winks**); flowers urn-shaped, small, borne usually singly, semi-dbl., 13 petals, no fragrance; foliage small, medium green, matt; bushy, low growth; (14); [Baby Betsy McCall X Red Can Can]; Florac, Marilyn, 1987

'FLUFFLY RUFFLES', F, pb, 1935; flowers silver-pink, reverse deeper rose, borne in clusters, semi-dbl., cupped, slight fragrance; foliage leathery; vigorous growth; [Miss Rowena Thom X Seedling]; H&S; Dreer

Fluffy *see* 'INTERFLU'

Fluorescent® *see* 'DELFLORI'

'FLUORETTE', F, or, 1971; bud pointed; flowers salmon-orange, dbl., 22 petals, 3–3.5 in., cupped, slight fragrance; vigorous, upright, bushy growth; [(Panache X Soprano) X Coloranja]; Lens

'FLUSH O' DAWN', HT, lp, 1900; flowers light pink changing to white, large, dbl., moderate fragrance; vigorous, upright growth; [Margaret Dickson X Sombreuil]; Walsh

Flushing Creek Poly, Pol, w; (found rose)

Flushing Meadow® *see* DORnapa

Flutterbye™ *see* 'WEKPLASOL'

Flying Colors™ *see* 'SAVAPAINT'

'FLYING COLOURS', LCl, dp, 1922; flowers light red, large, single, slight fragrance; foliage light, leathery, glossy; very vigorous, compact growth; RULED EXTINCT 8/82 ARM; Clark, A.; Hazlewood Bros.

'FLYING TATA', HT, dr, 1983; bud ovoid; dbl., 45 petals, exhibition form, borne singly, moderate fragrance; foliage medium, dark, glossy; vigorous, upright growth; [Scarlet Knight X Cynosure]; Hardikar, Dr. M.N.

Foc de Tabara™, F, lp, 1969; small, ovoid; flowers velvety rede, cupped, 33 petals, slight; medium green, leathery, glossy; [Paprika X Coup de Fondre]; St. Wagner, 1969; Res. Stn. f. Horticulture, Cluj, 1970, Romania

Focus *see* 'LENPAC'

Focus *see* NOAgut

'FOLIACÉE', C, mp, 1810; (Caroline de Berry); flowers light rose, very large, dbl., globular; Descemet, M.

'FOLIE D'ESPAGNE', F, yb, 1965; flowers yellow, orange and scarlet, blooms in clusters, dbl., 20 petals, 2 in., flat, slight fragrance; foliage dark, glossy; Soenderhousen

'FOLIES-BERGÈRE', HT, yb, 1948; flowers yellow shaded coppery, large, dbl., moderate fragrance; foliage leathery, light green; very vigorous, erect growth; [Souv. de Claudius Pernet X Unnamed seedling]; Gaujard

Folio Courtisane *see* DELjapal

Folk Dance *see* 'JUDDANCE'

'FOLK SONG', S, pb, 1964; flowers light pink, reverse darker, medium, dbl., recurrent bloom; foliage glossy; vigorous (3-4 ft.), compact growth; Von Abrams; Edmunds Roses

'FOLKESTONE', F, dr, 1936; semi-dbl., blooms in clusters, slight fragrance; foliage dark; bushy, spreading growth; (28); Archer

Folklore® *see* 'KORLORE'

'FOLKSINGER', S, yb, 1985; flowers yellow flushed with dark peach, slightly, large blooms, dbl., 28 petals, cupped, repeat bloom, moderate fragrance; awl-like, tan prickles; foliage leathery, glossy, coppery mid-green; upright, bushy growth; hardy; [Carefree Beauty X Sunsprite]; Buck, Dr. Griffith J.; Iowa State University, 1984

Folle Courtisane, HT, 1996; Delbard, Georges; (Cavriglia)

Folletto, F, 1961; Borgatti, G.; (Cavriglia)

Folletto, Min, 1962; Mansuino, Q.; (Cavriglia)

'FONDANT CASCADE', F, mp, 1996; flowers mid pink fading to pale pink, bright yellow stamens, blooms, dbl., 15–26 petals, 1.5 in., slight fragrance; some prickles; foliage medium, medium green, glossy; spreading, medium (20 in) growth; patio; Bees of Chester; L W Van Geest Farms, Ltd., 1995

Fondly™ *see* 'HILSET'

Fontaine *see* **'FOUNTAIN'**

Fontaine Blue *see* 'DELPOUS'

Fontainebleau® *see* 'DELPOUS'

'FONTANELLE', HT, my, 1927; flowers lemon-yellow, center gold, very large, dbl., moderate fragrance; foliage leathery; vigorous growth; [Souv. de Claudius Pernet X Columbia]; Hill, E.G., Co.

'FONTANELLE, CLIMBING', Cl HT, my, 1935; Johns

'FONTENELLE', HP, mr, 1877; flowers bright red, very large, dbl., intense fragrance; vigorous, growth; Moreau et Robert

Fool's Gold *see* 'AROGOBI'

Footlights, Pol, w, 1983

Footloose® *see* 'TANOTAX'

For You *see* **'PARA TI'**

'FOR YOU', F, w, 1999; flowers white, deep pink in petal edge, dbl., 30 petals, 2.5 in., slight fragrance; foliage medium green; compact, low (.7 m) growth; [Majorca X Sweet Memory]; Yasuda, Yuji

For You Dad *see* 'JONDAD'

Forbidden® *see* 'MINIBID'

Fordham Rose® *see* 'WILFORD'

Forest Fire *see* 'LEOFIRE'

Forest Queen, F, dp, 1997

Forever, HT, dr, 1978; Armstrong

'FOREVER AMBER', F, ob, 1975; flowers golden amber, suffused fiery orange, semi-dbl., 15 petals, 4 in., flat, intense fragrance; foliage dark, leathery; vigorous growth; [Arthur Bell X Elizabeth of Glamis]; Bees

Forever Free *see* 'HORBRIZESEA'

Forever Friends *see* 'MEIOFFIC'

Forever Mine *see* 'RENMINE'

'FOREVER SCARLET', HT, pb, 1976; flowers deep pink, reverse lighter, large blooms borne singly, very dbl., 80 petals, exhibition form, moderate, spicy fragrance; globular fruit; bright red prickles; tall, upright growth; [Wini Edmunds X Mister Lincoln]; Epperson, Richard G.

Forever Young *see* 'BRIYOUNG'

'FOREVER YOURS', HT, dr, 1964; (Concorde); bud long, pointed; flowers cardinal-red, dbl., 38 petals, 4–5 in., exhibition form, moderate, spicy fragrance; vigorous, upright growth; John Cook Medal, ARS, 1969; [Yuletide X Unnamed seedling]; Jelly; E.G. Hill Co., 1964;, URS, 1967

Forevermore *see* 'LYOFOR'

'FOREZ ROSE', F, mr, 1964; flowers geranium-red, becoming old-rose, borne in clusters, semi-dbl.; vigorous growth; [Sumatra X Antoine Noailly]; Croix, P.

'FORGOTTEN DREAMS', HT, mr, 1981; bud pointed; flowers cardinal red, borne singly and in trusses of 3-5, dbl., 24 petals, intense fragrance; straight, red-brown prickles; foliage medium green, semi-glossy; vigorous, bushy growth; [Fragrant Cloud X Teneriffe]; Bracegirdle, Derek T.; Arthur Higgs Roses

'FORMBY FAVOURITE', HT, or; flowers orange-scarlet on golden yellow; [McGredy's Sunset sport]; Wright, R. & Sons

'FORMBY SHOW', HT, mr, 1986; flowers medium, dbl., 26–40 petals, slight fragrance; foliage medium, dark green, semi-glossy; upright growth; [Fragrant Cloud X Elida]; Dwight, Robert & Sons

'FORNARINA', HGal, pb; flowers deep rose, marbled white; Verdier, 1847

'FORNARINA', M, mp, 1862; flowers deep rose, small; dwarf growth; Moreau et Robert

'FORST', HT, or, 1937; flowers fiery scarlet red, well formed, very large; vigorous growth; [Essence X Fritz Schrodter]; Krause

'FORSTMAISTER HEIM', HSet, 1887; Geschwind, R.; (Cavriglia)

Forsythe *see* 'VERJO'

'FORT KNOX', HT, dy, 1956; bud ovoid; flowers clear yellow, dbl., 20 petals, 3.5–4.5 in., intense fragrance; foliage dark, leathery; vigorous, upright, open growth; [Unnamed seedling X Ville de Paris]; Howard, A.P.; H&S

'FORT VANCOUVER', HT, mp, 1956; bud long, pointed; flowers well-formed, dbl., 42 petals, 5–6 in., intense, damask fragrance; foliage leathery; vigorous growth; [Charlotte Armstrong X Times Square]; Swim, H.C.; Peterson & Dering

Fortissima®, HMsk, pb, 1994; Lens

Fortissimo, S, mr, 1974; flowers large, dbl.; GPG Bad Langensalza; (Sangerhausen)

'FORTSCHRITT', F, yb, 1933; (Progress); flowers yellow-pink, open, large bornein clusters, semi-dbl., slight fragrance; foliage glossy, light; vigorous, bushy growth; [Mrs Pierre S. duPont X Gloria Mundi]; Kordes

Fortuna® *see* 'KORTUNA'

'FORTUNA', HT, mp, 1927; flowers rose-pink becoming lighter, many golden anthers, semi-dbl., 25 petals, moderate, fruity fragrance; dwarf, bushy growth; RULED EXTINCT 12/85; GM, NRS, 1927; [Lady Pirrie X Nur Mahal]; Pemberton

'FORTUNE', HT, yb, 1951; flowers gold shaded peach, dbl., 60 petals, 4 in., moderate fragrance; foliage glossy; vigorous growth; RULED EXTINCT 12/85; [Phyllis Gold sport]; Watkins Roses

Fortuné Besson *see* **'GEORG ARENDS'**

Fortune Cookie *see* SAVACOOK'

Fortune Teller *see* 'JACHEIR'

Fortuneana *see* **'FORTUNIANA'**

Fortunella *see* BARfort

'FORTUNE'S DOUBLE YELLOW', Misc OGR, yb; (Beauty of Glazenwood, Gold of Ophir, R. chinensis pseudindica, R. fortuniana, R. pseudindica, R. X odorata pseudindica, San Rafael Rose); flowers salmon-yellow, outside tinged red, blooms in clusters of 3-4, dbl., moderate, sweet fragrance; (14); 1845

'FORTUNE'S FIVE-COLORED ROSE', T, w, 1844; (Five-Colored Rose, Smith's Parish); bud red-tinged; flowers creamy white tinged with pale blush, fading to white, dbl.; foliage light green; vigorous growth; Fortune

'FORTUNIANA', Misc OGR, w; (Double Cherokee, Fortuneana, R. fortuneana, R. X fortuniana); flowers large, dbl.; climbing growth; (14); [Supposedly R. banksiae X R. laevigata]; 1850

'FORTY-NINER', HT, rb, 1949; bud long, pointed; flowers medium red, reverse yellow, dbl., 33 petals, 3.5–4 in., slight fragrance; foliage leathery, glossy, dark; vigorous, upright, compact growth; AARS, 1949 GM, Portland, 1947; [Contrast X Charlotte Armstrong]; Swim, H.C.; Armstrong Nursery

'FORTY-NINER, CLIMBING', Cl HT, rb, 1952; Moffet; Armstrong Nursery

'FORUM', Pol, mr, 1969; flowers bright red, open, large, dbl.; foliage light green, glossy; vigorous, upright growth; [Veronique X Independence seedling]; Delforge

'FORWARD', HT, my, 1962; flowers clear primrose-yellow, moderately large, dbl.; upright growth; [Ethel Sanday X Peace]; LeGrice

'FORWARD MARCH', HT, mp, 1934; bud pointed, bright old-rose; flowers bright rose-pink, becoming lighter, large, dbl.; foliage dark, bronze, leathery, glossy; long stems; very vigorous, bushy growth; [Better Times sport]; Wolfe

'FOSSE WAY', HT, mp, 1980; bud long; flowers rose pink, paler reverse, dbl., 38 petals, exhibition form, borne singly and several to a cluster, slight fragrance; hooked prickles; foliage matt, light green; vigorous, tall, upright growth; [Colour Wonder X Prima Ballerina]; Langdale, G.W.T.

Foster's Melbourne Cup *see* 'MACMOUHOO'

Foster's Wellington Cup *see* 'MACMOUHOO'

'FOUCHIN', Min, rb, 1986; (**Chinese Lantern**); flowers red, yellow and white hand painting, reverse light pink, fading red and white, semi-dbl., 15 petals, slight, spicy fragrance; round, medium, red-orange fruit; foliage medium, medium green, red when young, glossy; bushy, spreading, medium growth; [Avandel X Old Master]; Jacobs, Betty A.; Four Seasons Rose Nursery

'FOUMAD', Min, my, 1987; (**Summer Madness**®); flowers medium golden yellow, gold stamens, fading white, deeper in, dbl., 20–25 reflexed petals, exhibition form, slight, fruity tea fragrance; round, small, orange-red fruit; long, red to light brown prickles; foliage medium medium green matt; bushy, medium growth; [Party Girl X Sun Flare]; Jacobs, Betty A.; Four Seasons Rose Nursery

'FOUMAGIC', Min, m, 1986; (**Winter Magic**®); flowers light lavender-gray, golden stamens, large blooms, dbl., 30 petals, cupped, moderate, citrus tea fragrance; medium, globular, orange-red fruit; red, slender, straight, medium prickles; foliage medium, medium green, semi-glossy; medium, upright, bushy growth; [Rise 'n' Shine X Blue Nile]; Jacobs, Betty A.; Four Seasons Rose Nursery

'FOUMANDE', Min, ob, 1989; (**Mandarin Delight**); bud pointed; flowers soft, light mandarin orange, reverse lighter, aging peach in cooler weather, dbl., exhibition form, slight fragrance; no fruit; prickles needle, medium, red to tan; foliage medium, medium green, semi-glossy; bushy, medium growth; [Party Girl X Gingersnap]; Jacobs, Betty A.; Four Seasons Rose Nursery

'FOUMOUSE', Min, r, 1989; (**Church Mouse**); bud pointed; flowers tan-brown, with yellow at base, aging light lavender-brown, dbl., 20 petals, moderate, sweet fragrance; no fruit; prickles slightly declining, small, red to tan; foliage large, medium green, matt; low, bushy, compact growth; [Angel Face X Plum Duffy]; Jacobs, Betty A.; Sequoia Nursery, 1989

Found Fancy, HWich, mp; (found rose)

Founder's Pride *see* 'MICPRIDE'

'FOUNTAIN', HT, mr, 1970; (Fontaine, Red Prince®); flowers crimson, dbl., 35 petals, 5 in., cupped, intense fragrance; foliage dark, glossy; ADR, 1971 GM, RNRS, 1971 PIT, RNRS, 1971; Tantau, Math.; Wheatcroft Bros.

'FOUNTAIN OF BEAUTY', HT, ab, 1974; bud ovoid; flowers creamy to salmon, very dbl., 100 petals, 3–4 in.; foliage leathery; moderate growth; [Unnamed seedling X Colour Wonder]; Golik; Dynarose

Fountain Square® *see* 'JACMUR'

'FOUR CHEERS', HT, ob, 1988; bud ovoid; flowers soft orange with yellow center, urn-shaped, medium, borne usually singly, dbl., 30 petals, exhibition form, slight fragrance; globular, medium, orange fruit; prickles falcate, moderate, stout, maroon to ivory; foliage medium, medium green, glossy, smooth, leathery; spreading, medium growth; [Daisy Mae X First Prize]; Stoddard, Louis, 1989

Four Seasons, Quatre Saisons, Rose of Castille *see* **'AUTUMN DAMASK'**

Fourth of July® *see* 'WEKROALT'

'FOUSUN', Min, dy, 1991; (**San Jose Sunshine**); bud pointed; flowers deep golden yellow, aging light yellow with orange highlight, dbl., 25–30 petals, exhibition form, slight, fruity to tea fragrance; foliage medium, medium green, matt; bushy, spreading, medium to tall growth; [(Rise 'n' Shine X Redgold) X Summer Madness]; Jacobs, Betty A., 1987; Sequoia Nursery, 1991

'FOUTELL', Min, ob, 1988; (**Show 'n' Tell**); flowers bright, velvety orange-red, white border, reverse white, dbl., 30 petals, exhibition form, slight, spicy fragrance; foliage medium, medium green, semi-glossy, disease-resistant; spreading, tall growth; PP007375; [Rocky X (Matangi X Honey Hill)]; Jacobs, Betty A.; Four Seasons Rose Nursery

'FOUTOUCH', Min, pb, 1989; (**Touch-Up**); bud pointed; flowers medium pink with cream border, reverse cream, hand-painted, dbl., 30 petals, exhibition form, moderate, tea fragrance; round, medium, russet-orange fruit; prickles straight, large, red to tan; foliage medium, medium green, matt; upright, bushy, tall growth; [Scarlet Knight X (Matangi X Honey Hill)]; Jacobs, Betty A.; Four Seasons Rose Nursery

'FOUTUM', Min, ob, 1987; (**Autumn Magic**); flowers bright golden-orange, with red at petal tips, reverse yellow, fading to red, dbl., 35 petals, exhibition form, no fragrance; no fruit; very few prickles; foliage medium, dark green, semi-glossy; bushy, medium growth; [Confetti X Anita Charles]; Jacobs, Betty A.; Four Seasons Rose Nursery

'FOUTWIST', Min, dy, 1988; (**Lemon Twist**); bud pointed; flowers medium, dbl., 25 petals, exhibition form, borne usually singly and in sprays of up to 3, slight, tea fragrance; prickles slightly declining, long, red to brown; foliage medium, medium green, glossy; bushy, medium growth; PP007450; [Gold Badge X Great Day]; Jacobs, Betty A.; Four Seasons Rose Nursery

'FOXFIRE', HT, mr, 1990; bud pointed; dbl., 30 petals, exhibition form, borne singly, no fragrance; prickles straight, green; foliage medium, medium green, glossy; upright, tall growth; [Unnamed seedling X First Prize]; Stoddard, Louis, 1991

'FOXFLORIN', Min, ab, 1997 (**Florida International**); flowers full, high centered, exhibition, medium, very dbl., 30–35 petals, borne mostly singly, slight fragrance; foliage medium, medium green, semi-glossy; upright, spreading, medium to tall (1.5 to 2ft.) growth; [Pierrine X (Loving Touch X Seedling)]; Fox, Tillie

Foxi *see* 'UHLATER'

Foxi Pavement *see* 'UHLATER'

Foxtrot *see* 'BROFOX'

Foxy, S, lp; Peden, R., 2000; (Weatherly, L.)

Foxy Lady® *see* 'AROSHRIM'

'FR. LAD. RIEGER', HT, mr, 1939; flowers large, dbl.; Böhm, J.; (Sangerhausen)

'FRAGEZEICHEN', HWich, mp, 1910; flowers shining pink, large, in few-flowered trusses, dbl., 25 petals, globular; foliage glossy; vigorous growth; very hardy.; [Dorothy Perkins X Marie Baumann]; Boettner

Fragola, HT, 1980; Croix, P.; (Cavriglia)

'FRAGRANCE', HT, dp, 1965; bud long, pointed; flowers carmine to rose-madder, large, dbl., exhibition form, intense fragrance; foliage bronze, leathery; vigorous, tall, compact growth; [Charlotte Armstrong X Merry Widow]; Lammerts, Dr. Walter; Germain's

'FRAGRANCE', HT, dr, 1924; flowers deep crimson, high pointed, large, dbl., moderate fragrance; vigorous growth; [Hoosier Beauty X George Dickson]; Chaplin Bros.

'FRAGRANT AIR', F, rb, 1977; flowers red changing to magenta-pink, dbl., 20 petals, 2.5 in., intense fragrance; foliage dark; Pearce, C.A.; Limes Rose Nursery

Fragrant Apricot® *see* 'JACGRANT'

'FRAGRANT BEAUTY', S, dp, 1950; bud ovoid; flowers carmine, large, dbl., 22 petals, cupped, profuse, repeated bloom, intense, spicy fragrance; foliage glossy; very vigorous (4-5 ft.), upright, compact growth; [(Pharisaer X Conrad Ferdinand Meyer) X Crimson Glory]; Jacobus; B&A

Fragrant Beauty, HT, mr, 1998; Ghosh

'FRAGRANT BOUQUET', HT, lp, 1922; flowers shell-pink, base yellow, dbl., 30–35 petals, intense fragrance; H&S

Fragrant Carpet *see* 'BROCADE'

Fragrant Charm *see* 'KORDU'

Fragrant Cloud *see* 'TANELLIS'

Fragrant Cloud, Climbing *see* 'COLFRAGRASAR'

'FRAGRANT DELIGHT'®, F, op, 1978; (Wisbech Rose Fragrant Delight®); flowers light orange-salmon, reverse deeper, dbl., 22 petals, 3 in., intense fragrance; foliage glossy, reddish; Edland Fragrance Medal, ARS, 1976 GM, ARS, 1988 James Mason, ARS, 1988; [Chanelle X Whisky Mac]; Wisbech Plant Co.

Fragrant Dream *see* 'DICODOUR'

Fragrant Fantasy *see* 'DEVFRAGO'

'FRAGRANT GLORY', HT, dp, 1950; bud dark red; flowers deep cyclamen-pink, dbl., 36 petals, 6 in., exhibition form, intense fragrance; strong stems; vigorous growth; [Phyllis Gold X Crimson Glory]; Cobley

Fragrant Gold *see* 'TANDUGOFT'

'FRAGRANT HOUR', HT, op, 1973; flowers bronze-pink, high-pointed, dbl., 35

petals, 4.5 in., intense fragrance; foliage light; GM, Belfast, 1975; [Arthur Bell X (Spartan X Grand Gala)]; McGredy, Sam IV

Fragrant Lace™ *see* 'JACSEE'

'FRAGRANT LADY', HT, mp, 1991; bud ovoid; flowers urn-shaped, blooms borne usually singly, semi-dbl., cupped, intense, fruity fragrance; foliage medium, dark green, semi-glossy; upright, medium growth; [Queen Elizabeth X Broadway]; Perry, Anthony; Co-Operative Rose Growers, 1991

Fragrant Lady *see* 'MEINIACIN'

Fragrant Lavendar *see* **'VELVET MIST'**™

'FRAGRANT LOVE'®, HT, mr, 1979; bud globular, pointed; flowers medium purplish red, blooms borne 1-3 per cluster, dbl., 45 petals, cupped, intense fragrance; curved reddish prickles; foliage large, deep green, matt; upright growth; [Chrysler Imperial X Unnamed seedling]; Rose Barni-Pistoia

Fragrant Memory *see* 'JACDIS'

Fragrant Mist *see* 'SMITSBLANC'

Fragrant Morning *see* 'RENMORNING'

Fragrant Obsession *see* 'RENFRAGOBSES'

'FRAGRANT PINK TALISMAN', HT, pb, 1938; flowers pink shades, slightly larger than talisman, semi-dbl., 18 petals, intense fragrance; [Talisman seedling]; Moore, Ralph S.

Fragrant Plum *see* 'AROPLUMI'

Fragrant Queen, HT, op

Fragrant Rhapsody™ *see* 'JACTRIG'

Fragrant Surprise *see* 'HARVERAG'

'FRAÎCHEUR', HWich, lp, 1921; flowers soft pink, borne in pyramidal clusters; foliage glossy, dark; very vigorous growth; Turbat

'FRAÎCHEUR', HT, lp, 1942; bud long, pointed; flowers soft pink tinted pearl-white, medium, semi-dbl., cupped, slight fragrance; foliage leathery, light green; vigorous, bushy growth; [Joanna Hill X ?]; Meilland, F.; A. Meilland

'FRANBURST', HT, dy, 1999; (**Uppingham School**); flowers golden yellow, reverse yellow, dbl., 26–40 petals, 4 in., borne in large clusters, slight fragrance; prickles moderate; foliage medium, dark green, glossy; upright, medium (to 5 ft) growth; [Ann Harkness X (Diamond Jubilee X Picadilly)]; Cowlishaw, Frank

'FRANCA', F, pb, 1961; flowers pink and yellow; [Confidence X Unnamed seedling]; da Silva, Moreira

'FRANÇAIS', F, op, 1951; flowers bright pink tinted orange, borne in clusters, semi-dbl.; vigorous growth; [Holstein X Orange Triumph]; Mallerin, C.; EFR

France Inter *see* 'DELKRI'

France Libre® *see* 'DELJAUNOR'

'FRANCES ASHTON', HT, mp, 1937; bud pointed; flowers carmine, stamens wine-colored, large, single, 5 petals, slight fragrance; foliage leathery; vigorous growth; [Lady Battersea X Hawlmark Crimson]; DePuy; Stocking

'FRANCES BLOXAM', HP, op, 1892; flowers medium, dbl.; Paul, G.; (Sangerhausen)

'FRANCES GAUNT', HT, ab, 1918; flowers apricot to salmon-yellow, large, semi-dbl., cupped, intense fragrance; foliage glossy; vigorous, branching growth; GM, Bagatelle, 1920; Dickson, A.

'FRANCES GAUNT, CLIMBING', Cl HT, ab, 1934; Cazzaniga, F. G.

Frances Neale *see* 'KIRPICE'

Frances Perry *see* 'BOSREXCITY'

Frances Phoebe, HT, w, 1979; LeGrice

'FRANCESCA', HMsk, ab, 1922; single, blooms in large sprays, recurrent bloom, slight fragrance; foliage leathery; long stems; vigorous (5-6 ft.) growth; (21); [Danaë X Sunburst]; Pemberton

'FRANCESCA DE CUIXART', HT, rb, 1960; (F. Cuixart); bud ovoid; flowers lincoln red, reverse fuchsine-red, open, large, dbl., 40 petals, intense fragrance; foliage glossy, bronze; long stems; vigorous, compact growth; [Baccará X Golden Masterpiece]; Dot, Simon; P. Dot

'FRANCESCH MATHEU', HT, ob, 1940; bud long, pointed; flowers rich golden orange, large, dbl., cupped; foliage glossy, dark; strong stems; very vigorous, upright growth; [Luis Brinas X Catalonia]; Dot, Pedro

'FRANCESCO LA SCOLA', HT, or, 1934; flowers bright orange-red, edged nilsson pink, large, semi-dbl., early, slight fragrance; foliage holly-green; vigorous growth; [Hortulanus Budde X Cuba]; Ketten Bros.

'FRANCETTE GIRAUD', F, op, 1961; flowers bright salmon-pink, borne in clusters, dbl.; foliage dark; vigorous growth; [Aloha X (Gloire du Midi X Edith de Martinelli)]; Arles

Francia, HT, 1982; Dorieux, Francois; (Cavriglia)

'FRANCIBEL', HT, m, 1992; (**Tonia**); flowers light lavender/mauve with deeper tones (silvery lavender) at, dbl., 26–40 petals, 3–3.5 in., intense fragrance; some prickles; foliage medium, medium green, matt; upright (175-180 cms), vigorous growth; [Lagerfeld X Remember Me]; Franklin-Smith, Roger, 1995

'FRANCIE SIMMS', HT, mp, 1926; flowers rose-pink, marked carmine, base buttercup yellow, dbl., moderate fragrance; Dickson, A.

'FRANCINE', HT, rb, 1961; bud pointed; flowers crimson, reverse silvery, dbl., 30–35 petals, 6 in., moderate fragrance; foliage glossy; vigorous, bushy growth; Kriloff, Michel; Cramphorn's Nursery

Francine, HMsk, w

Francine Austin *see* 'AUSRAM'

'FRANCINE CONTIER'®, HT, my, 1977; bud long, pointed; flowers canary-yellow, dbl., 40–45 petals, 3.5–4.5 in., exhibition form, slight fragrance; foliage glossy, dark; very vigorous growth; [Peer Gynt X Thalia]; Lens

'FRANCIS', HWich, mp, 1933; flowers rose-pink, dbl.; vigorous growth; [R. wichurana hybrid X Crimson Rambler]; Hauser

'FRANCIS DUBREUIL', T, dr, 1894; bud long; flowers velvety crimson, medium; Dubreuil

'FRANCIS E. LESTER', HMsk, w, 1946; flowers white edged pink, blooms in clusters of 25-30, single, 2 in., recurrent bloom; very vigorous (8-10 ft. in a season) growth; very hardy.; [Kathleen X Unnamed variety]; Lester Rose Gardens

Francis King, HT, dy

Francis Moreau, HT, ob

'FRANCIS SCOTT KEY', HT, pb, 1913; flowers deep pink, reverse lighter, very large, dbl., exhibition form, slight fragrance; foliage dark, leathery, glossy; long, strong stems; very vigorous growth; [Radiance X Unnamed seedling]; Cook, J.W.

'FRANCIS SCOTT KEY, CLIMBING', Cl HT, pb

'FRANCISCO CURBERA', HT, pb, 1923; flowers salmon-pink and yellow, well formed, very dbl.; vigorous growth; Dot, Pedro

'FRANCITA', HT, dp, 1965; flowers light red, medium, very dbl., slight fragrance; foliage dark; [((Dame Edith Helen X Baccará) X Baccará) X Comtesse Vandal]; Mondial Roses

Francofurtana *see* **'EMPRESS JOSEPHINE'**

'FRANÇOIS ALLARD', HT, yb, 1927; flowers creamy yellow, reverse salmon-pink, dbl.; [Mme Mélanie Soupert X Mme Segond Weber]; Felberg-Leclerc

'FRANÇOIS ARAGO', HP, dr, 1858; flowers medium, dbl.; Trouillard; (Sangerhausen)

'FRANÇOIS BOLLEZ', HT, op, 1935; bud shrimp-pink; flowers salmon, base orange, large, dbl.; Gillot, F.

'FRANÇOIS COPPÉE', HP, dr, 1895; flowers dark crimson, large, dbl., moderate fragrance; Lédéchaux

'FRANÇOIS CROUSSE', Cl T, mr, 1900; flowers cerise-crimson shaded darker, dbl., globular; vigorous growth; Guillot, P.

'FRANÇOIS CROUSSE', Cl HT, mr, 1900; flowers large, dbl., moderate fragrance; Guillot; (Sangerhausen)

'FRANÇOIS DE SALIGNAC', M, 1854; Moreau et Robert; (Cavriglia)

Francois Drion, HMsk, mp, 1995; Lens

'FRANÇOIS FOUCARD', HWich, my, 1900; flowers lemon-yellow, borne in clusters, semi-dbl.; very vigorous growth; [R. wichurana X L'Ideal]; Barbier

'FRANÇOIS GUILLOT', HWich, w, 1907; flowers milk-white, dbl., slight fragrance; foliage light green, glossy; vigorous (15-18 ft) growth; [R. wichurana X Mme Laurette Messimy]; Barbier

'FRANÇOIS JURANVILLE', HWich, op, 1906; flowers bright salmon-pink, base yellow, quite distinct, large; very vigorous growth; [R. wichurana X Mme Laurette Messimy]; Barbier

Francois Krige *see* KORharment

'FRANÇOIS LAPLANCHE', HT, w, 1934; flowers flesh-white on yellow ground, veined and edged carmine, large, dbl., cupped, intense fragrance; foliage leathery; bushy growth; [Mme Charles Detreaux X Mme Edouard Herriot]; Buatois

'FRANÇOIS LEVET', HP, dp, 1880; flowers cherry-rose, well formed; Levet, A.

'FRANÇOIS MICHELON', HP, m, 1871; flowers deep rose tinged lilac, large, dbl., globular, moderate fragrance; foliage somewhat wrinkled; upright growth; [La Reine seedling]; Levet, A.

'FRANÇOIS OLIN', HP, dp, 1881; flowers large, dbl.; Ducher; (Sangerhausen)

'FRANÇOIS POISSON', LCl, ly, 1902; flowers pale sulfur-yellow, center shaded orange, passing to white, dbl.; very vigorous growth; [R. wichurana X William Allen Richardson]; Barbier

'FRANÇOIS PREMIER', HP, 1859; Trouillard; (Cavriglia)

Francois Rabelais™ *see* 'MEINUSIAN'

'FRANÇOISE BLONDEAU', HT, op, 1938; flowers coral, large, dbl.; foliage dark; very vigorous growth; [Charles P. Kilham X Colette Clément]; Mallerin, C.; H. Guillot

'FRANÇOISE DE SALIGNAC', M, mp, 1854; flowers rose-pink; vigorous growth; Robert

'FRANK CHAPMAN', HT, yb, 1937; flowers yellowish-orange, medium, dbl., moderate fragrance; Cant, F.; (Sangerhausen)

Frank Leddy *see* **'FRANS LEDDY'**

'FRANK MACMILLAN', HT, dr, 1978; flowers crimson-red, full, dbl., 26 petals, 4 in., intense fragrance; foliage semi-glossy; moderately vigorous, spreading growth; [Uncle Walter X (Ena Harkness X Fragrant Cloud)]; Scrivens

'FRANK NAYLOR', S, rb, 1978; flowers dark red, yellowish eye, small, single, 5 petals, 1.5 in., moderate, musk fragrance; foliage small, plum shaded; Harkness, 1977

'FRANK NEAVE', HT, ly, 1928; flowers pale mustard-yellow, dbl., moderate fragrance; Morse

'FRANK PENN', HT, pb, 1970; bud long, pointed; flowers cerise, reverse pink, full, large, dbl.; foliage dark, leathery; very vigorous, upright growth; [Wendy Cussons X Lys Assia]; Clayworth; F. Mason

'FRANK READER', HT, my, 1927; flowers lemon-yellow, center apricot, fairly, large, dbl., exhibition form, moderate fragrance; strong stems; vigorous growth; [Golden Ophelia X Souv. de H.A. Verschuren]; Verschuren; Dreer;, H&S

'FRANK SERPA', Gr, op, 1960; bud pointed; flowers pink tinted salmon, large, dbl., cupped, moderate fragrance; foliage leathery, glossy, dark; very vigorous growth; [Pres. Macia X ?]; Serpa

'FRANK W. DUNLOP', HT, dp, 1920; flowers deep bright rose-pink, large, dbl., 45 petals, exhibition form, intense fragrance; [Mrs Charles E. Russell X Mrs George Shawyer]; Dunlop; Totty

'FRANK W. DUNLOP, CLIMBING', Cl HT, dp, 1933; Dixie Rose Nursery

Frankenland® *see* 'TANKENFRAM'

'FRANKFORT AGATHÉ', HGal, dp; flowers cerise

Frankfurt *see* **'SPLENDENS'**

'FRANKFURT AM MAIN', F, mr, 1960; flowers blood-red shaded scarlet, well formed, borne in clusters, dbl., 25 petals, 2.5–3 in., slight fragrance; foliage dark; bushy, upright growth; Boerner; Kordes;, Kordes & Tantau

'FRANKLIN', HT, op, 1918; flowers salmon, shaded yellowish salmon, dbl.; Pernet-Ducher

'FRANKLIN D. ROOSEVELT', HT, mr, 1939; flowers bright red fading to rose-purple, medium, dbl., 26 petals, cupped, intense fragrance; foliage glossy, dark; bush and semi-climbing growth; [Betty Uprichard sport]; McClung

'FRANKLIN ENGELMANN', F, dr, 1970; flowers bright scarlet, pointed, very large blooms in trusses, dbl., 36 petals; vigorous growth; [Heidelberg X (Detroiter X Seedling)]; Dickson, A.

Frank's Climber, LCl, yb

'FRANLAC', HT, m, 1995; (**Road to Freedom**); flowers lilac, gold stamens, blooms, dbl., 15–25 petals, 1.5–2.75 in., moderate fragrance; foliage medium, medium green, glossy; low (15-20 in), bushy, compact growth; [Seedling X Lilac Charm]; Cowlishaw, Frank; Rearsby Roses, Ltd., 1996

'FRANLUV', F, ab, 1986; (**Summer Love**); flowers light apricot, large blooms in clusters of 5-10, dbl., 27 petals, exhibition form, moderate, spicy fragrance; foliage large, dark, semi-glossy; upright growth; [Pink Parfait X Cynthia Brooke]; Cowlishaw, Frank

FRAnmilro, MinFl, mp, 1998; (**Lovebird**); Cowlishaw, Frank

Frans Hals, S, mp, 1997; Williams, J. Benjamin

'FRANS LEDDY', Pol, op, 1927; (Frank Leddy); flowers light orange-red, turning pink, small, dbl.; [Kersbergen sport]; Kersbergen; Van Nes

'FRANSHINE', HT, ab, 1990; (**Chester Cathedral**); bud pointed; flowers light apricot with cream/very light gold reverse, medium, dbl., exhibition form, borne usually singly, slight fragrance; foliage medium, dark green, glossy; low, bushy growth; [Honey Favorite X Piccadilly]; Cowlishaw, Frank, 1989

'FRANSMOOV', HT, yb, 1995; (**With Thanks**); flowers gold/pink bicolor, blooms borne in clusters, dbl., 26–40 petals, 1.5–2.75 in., moderate fragrance; foliage medium, medium green, glossy; medium (24in), upright growth; [(Southampton X Prominent) X Summer Love]; Cowlishaw, Frank; Rearsby Roses, Ltd., 1996

'FRANTASIA', S, m, 1999; (**Rhapsody in Blue**); flowers dark purple/blue, reverse lighter, semi-dbl., 12–16 petals, 2.5 in., borne in large clusters, intense fragrance; prickles moderate; foliage medium, light green, glossy; upright, semi-climbing, tall (6-7 ft) growth; [Summer Wine X seedling]; Cowlishaw, Frank; Warner's Roses, 2000

FRAntier, Min, lp; (**Maureen Lipman**); Cowlishaw, Frank

'FRANTISEK VALÀSEK', Pol, dr, 1930; flowers small, dbl.; Valàsek; (Sangerhausen)

'FRANZ DEEGEN', HT, my, 1901; (Yellow Kaiserin Auguste Viktoria); flowers soft yellow to golden yellow, dbl.; [Kaiserin Auguste Viktoria X ?]; Hinner, W.

'FRANZ GRÜMMER', HT, or, 1927; flowers coral-red, dbl.; [Mme Abel Chatenay X Château de Clos Vougeot]; Maass

'FRASER MCLAY', HT, mr, 1974; bud ovoid; flowers glossy medium red, medium, dbl.; foliage glossy, leathery; bushy growth; [Grand Gala X Suspense]; Dawson, George; Neil

'FRASER'S PINK MUSK', N, lp, 1818; (Blush Musk); flowers blush, medium, borne in large clusters, semi-dbl., intense fragrance; Fraser

Frasquita®, F, op

'FRATERNITÉ', F, pb, 1946; flowers neyron rose, reverse pale lilac-rose, borne in clusters, dbl., 25 petals, 3 in., globular, slight fragrance; foliage bronze; vigorous growth; [Florentina X World's Fair]; Leenders, M.; Longley

'FRAU A. VON BRAUER', HWich, lp, 1913; flowers small, very dbl., moderate fragrance; Lambert, P.; (Sangerhausen)

'FRAU A. WEIDLING', S, mp, 1930; flowers small, semi-dbl.; Vogel, M.; (Sangerhausen)

'FRAU ADOLF ANDERS', HT, 1937; Anders, Adolf; (Sangerhausen)

'FRAU ALBERT HOCHSTRASSER', HWich, lp, 1908; flowers small, dbl., intense fragrance; Weigand, C.; (Sangerhausen)

'FRAU ALEXANDER WEISS', Pol, yb, 1909; flowers light yellow with pink, small, dbl.; Lambert, P.; (Sangerhausen)

'FRAU ANNA LAUTZ', HT, mr, 1911; flowers large, dbl., moderate fragrance; Kiese; (Sangerhausen)

'FRAU ANNA PASQUAY', Pol, dp, 1909; flowers small, dbl.; Walter; (Sangerhausen)

'FRAU ANNY BEAUFAYS', F, mr, 1962; (Mrs Annie Beaufays); bud ovoid; flowers salmon-red, borne in clusters, semi-dbl., slight fragrance; low growth; deRuiter; Beaufays

'FRAU ASTRID SPÄTH', F, dp, 1930; (Astrid Späth, Direktör Rikala); flowers clear carmine-rose; [Lafayette sport]; Spath

'FRAU ASTRID SPÄTH, CLIMBING', Cl F, dp, 1935; Lens

'FRAU BERTA GÜRTLER', HMult, lp, 1913; flowers light silky pink, dbl.; Gurtler; P. Lambert

'FRAU BERTHA KIESE', HT, my, 1914; flowers pure golden yellow, dbl.; [Kaiserin Auguste Viktoria X Undine]; Kiese

'FRAU BÜRGERMEISTER KIRSCHSTEIN', HT, dp, 1906; flowers medium, semi-dbl., intense fragrance; Jacobs; (Sangerhausen)

'FRAU CECILIE WALTER', Pol, ly, 1904; flowers small, dbl., intense fragrance; Lambert, P.; (Sangerhausen)

'FRAU CHARLOTTE GIESELER', Pol, dp, 1939; flowers medium, dbl.; Vogel, M.; (Sangerhausen)

'FRAU DAGMAR HARTOPP', HRg, mp; (Frau Dagmar Hastrup, Fru Dagmar Hastrup); flowers silvery pink, single, 5 petals; foliage crinkled, rich green; low growing growth; (14); ca. 1914

Frau Dagmar Hastrup *see* **'FRAU DAGMAR HARTOPP'**

'FRAU DIREKTOR ANNI HARTMANN', HT, ly, 1933; flowers medium, very dbl., moderate fragrance; Brada, Dr.; (Sangerhausen)

'FRAU DR ERRETH', F, dy, 1915; flowers deep golden yellow, passing to white, well shaped, borne in, dbl.; branching growth; [Gruss an Aachen X Mrs Aaron Ward]; Geduldig

'FRAU DR HOOFTMANN', Cl Pol, lp, 1935; flowers large, dbl.; Buisman, G. A. H.; (Sangerhausen)

'FRAU DR KRÜGER', HT, op, 1919; flowers cream-salmon on golden ground; [Baronne Henriette de Loew X Mme Caroline Testout]; Kiese

'FRAU DR SCHRICKER', HCh, rb, 1927; flowers fiery carmine and coppery-red, large, dbl., intense fragrance; dwarf growth; [Gruss an Teplitz X Souv. de Mme Eugene Verdier]; Felberg-Leclerc

'FRAU E. WEIGAND', HT, yb, 1928; (Frau Elisabeth Weigand); flowers canary-yellow, large, dbl., 70 petals, exhibition form, intense fragrance; foliage dark, leathery; strong stems; [Mme Caroline Testout X Souv. de Claudius Pernet]; Weigand, C.; Weigand & H. Schultheis

'FRAU EDUARD BETHGE', HT, dr, 1930; flowers dark crimson and velvety blood-red, large, dbl., intense fragrance; foliage light; [Hadley X Admiral Ward]; Felberg-Leclerc

'FRAU ELISABETH BALZER', HT, w, 1933; flowers white, base orange-yellow, reverse bright flesh-pink, pointed; vigorous growth; [Mrs Henry Morse sport]; Balzer

'FRAU ELISABETH FISHER', HFt, dy, 1930; flowers medium, dbl., moderate fragrance; Fisher & Schulz; (Sangerhausen)

'FRAU ELISABETH MÜNCH', Pol, mr, 1921; flowers scarlet-cherry-red, with deeper reflexes; [Orléans Rose sport. sport]; Münch & Haufe

'FRAU ELISABETH SPRENGER', HT, dy, 1937; flowers medium, dbl.; Wirtz & Eicke; (Sangerhausen)

Frau Elisabeth Weigand *see* **'FRAU E. WEIGAND'**

'FRAU EMMA SASSE', HT, lp, 1908; flowers large, dbl., moderate fragrance; Plog; (Sangerhausen)

'FRAU EMMY HAMMANN', HT, yb, 1923; flowers reddish lemon-yellow shaded sunflower-yellow, dbl.; [Mme Caroline Testout X Mme Hoste]; Weigand, C.; Hammann

'FRAU EVA SCHUBERT', HWich, mp, 1937; Tepelmann

'FRAU FELBERG-LECLERC', HT, my, 1921; flowers pure golden yellow, dbl., slight fragrance; [Louise Catherine Breslau sport]; Felberg-Leclerc

'FRAU FELIX TONNAR', HT, mp, 1924; flowers bright rose, base coppery orange, semi-dbl.; [Mme Mélanie Soupert X Mme Annette Aynard]; Leenders, M.

'FRAU FRITZ PELZER', HT, or, 1927; flowers reddish crimson-orange, dbl.; [Mme Edouard Herriot X Edward Mawley]; Leenders, M.

'FRAU GEHEIMRAT DR STAUB', Cl HT, dr, 1908; flowers large, dbl., intense fragrance; Lambert, P.; (Sangerhausen)

'FRAU GEHEIMRAT SPÄTH', S, lp, 1941; flowers large, dbl., slight fragrance; Tepelmann; (Sangerhausen)

'FRAU GEORG VON SIMSON', HMult, mp, 1909; flowers rose; vigorous growth; [Helene X Rosel Dach]; Walter

'FRAU HEDWIG KOSCHEL', Pol, w, 1921; flowers white, slightly shaded yellow, edges tinted rose-pink, dbl., moderate fragrance; [Ellen Poulsen sport]; Münch & Haufe

'FRAU HEDWIG WAGNER', HT, mp, 1919; dbl., moderate fragrance; [Enchantress X Mrs W.J. Grant]; Krüger; Kiese

'FRAU HELENE KÜHN', HT, w, 1938; flowers medium, dbl.; Vogel, M.; (Sangerhausen)

'FRAU HELENE VIDENZ', HMult, lp, 1904; flowers medium, dbl.; Lambert, P.; (Sangerhausen)

'FRAU HUGO LAUSTER', HT, dy, 1932; flowers deep canary-yellow, edged lighter, well formed, dbl., moderate fragrance; vigorous growth; Lauster; Pfitzer & Dreer

'FRAU IDA MÜNCH', HT, ly, 1919; flowers light golden yellow, center deeper, dbl.; [Frau Karl Druschki X Billard et Barre]; Beschnidt

Frau Jenny Wienke, HT, mr, 1958; flowers large, dbl., intense fragrance; Berger, W.; (Sangerhausen)

'FRAU KARL DRUSCHKI', HP, w, 1901; (F.K. Druschkii, Reine des Neiges, Snow Queen, White American Beauty); bud pointed, tinged carmine-pink; flowers snow-white, center sometimes blush-pink, large, dbl., 35 petals, no fragrance; foliage dark; vigorous growth; (28); [Merveille de Lyon X Mme Caroline Testout]; Lambert, P.

'FRAU KARL DRUSCHKI, CLIMBING', Cl HP, w, 1906; (Grimpant Reine des Neiges); Lawrenson

'FRAU KÄTE SCHMID', HMult, mp, 1931; flowers large, semi-dbl., slight fragrance; Vogel, M.; (Sangerhausen)

'FRAU LIESEL BRAUER', HWich, mp, 1938; flowers medium, semi-dbl., slight fragrance; Thönges; (Sangerhausen)

'Frau Lieselotte Weber', Pol, 1938; Vogel, M.; (Sangerhausen)

'Frau Lina Strassheim', HMult, op, 1907; flowers reddish salmon-pink; [Crimson Rambler sport]; Strassheim

'Frau Luise Kiese', HT, ly, 1921; (Luise Kiese); flowers ivory-yellow, sometimes clear yellow, very dbl.; Kiese

'Frau Luise Lindecke', HT, dr, 1928; flowers deep claret-red, sometimes crimson, dbl., intense fragrance; [Columbia sport]; Lindecke; Lindecke & Kordes

'Frau Margarete Oppenheim', HT, rb, 1928; flowers intense carmine-red shaded brick-red and yellow, semi-dbl.; [Hortulanus Budde X Souv. de Claudius Pernet]; Felberg-Leclerc

Frau Maria Rüdt, S, lp, 1994; flowers medium-large, dbl.; Tagashira, Kazuso; (Sangerhausen)

'Frau Marie Bromme', Pol, dr, 1928; flowers bright dark red; [Dr. Kater sport]; Wirtz & Eicke

'Frau Marie Weinbach', HWich, w, 1906; flowers small, dbl., moderate fragrance; Weigand, C.; (Sangerhausen)

'Frau Martha Schmidt', HT, mr, 1923; flowers carmine-red, dbl., intense fragrance; [Paula Clegg X Edward Mawley]; Kiese

'Frau Math. Noehl', HT, my, 1913; flowers lemon-yellow, dbl., moderate fragrance; [Kaiserin Auguste Viktoria X Mme Ravary]; Welter

'Frau Mathilde Bätz', HT, w, 1929; flowers pure white, stamens yellow, dbl., intense fragrance; [Unnamed seedling X Ophelia]; Felberg-Leclerc

'Frau Mélanie Niedieck', HT, dy, 1916; flowers vivid lemon-yellow, dbl.; [Mme Jenny Gillemot X Prince de Bulgarie]; Leenders, M.

Frau Minka Rödiger, HT, yb, 1959; flowers yellow to light yellow with orange highlights, large, dbl.; Berger, W.; (Sangerhausen)

'Frau O. Plegg', B, dr, 1909; flowers medium, dbl., intense fragrance; Nabonnand; (Sangerhausen)

'Frau Oberhofgärtner Schulze', Pol, mp, 1909; flowers small, dbl.; Lambert, P.; (Sangerhausen)

'Frau Oberhofgärtner Singer', HT, pb, 1908; flowers soft pink edged white, moderate fragrance; [Jules Margottin X Mme Eugenie Boullet]; Lambert, P.

'Frau Oberpräsident von Grothe', HT, op, 1920; flowers rose-orange streaked carmine; [Richmond X Farbenkonigin]; Löbner; P. Lambert

'Frau Peter Lambert', HT, op, 1902; flowers pink shading to salmon; [(Kaiserin Auguste Viktoria X Mme Caroline Testout) X Mme Abel Chatenay]; Welter

'Frau Professor Baranov', Pol, rb, 1947; flowers salmon-red, medium, semi-dbl.; Vogel, M.; (Sangerhausen)

'Frau Professor Gnau', HT, w, 1925; flowers creamy white, large, semi-dbl.; Kiese; (Sangerhausen)

'Frau Professor Grischko', S, mr, 1947; flowers medium, semi-dbl.; Vogel, M.; (Sangerhausen)

'Frau Robert Türke', HT, dr, 1928; flowers dark crimson, very dbl., moderate fragrance; foliage dark, glossy; long stems; vigorous growth; [Hadley X Hugh Dickson]; Türke; Teschendorff

'Frau Rudolf Schmidt', Pol, dr, 1919; flowers dark ruby-red without objectionable blue shades; very dwarf growth; [Jessie sport]; Schmidt, R.

'Frau Sophie Meyerholz', S, mp, 1942; flowers medium, dbl.; Vogel, M.; (Sangerhausen)

'Frau Therese Lang', HT, dp, 1910; flowers large, very dbl., intense fragrance; Welter; (Sangerhausen)

'Fräulein Octavia Hesse', HWich, ly, 1910; flowers yellowish white, center deeper, small, dbl., recurrent bloom, moderate fragrance; vigorous, climbing growth; [R. wichurana X Kaiserin Auguste Viktoria]; Hesse

Frazer's Pink Musk, N, lp

'Frazier Annesley', HT, dp, 1935; bud pointed; flowers carmine, base golden yellow, dbl., exhibition form, slight fragrance; foliage glossy, bronze; very vigorous growth; McGredy

Frechdachs *see* MEHnan

'Freckle Face', S, pb, 1976; bud ovoid; flowers light spirea-red, striped dark spirea, dbl., 23 petals, 3.5–4 in., moderate, clove fragrance; foliage coppery, leathery; bushy, spreading growth; [(Vera Dalton X Dornroschen) X ((World's Fair X Floradora) xApplejack)]; Buck, Dr. Griffith J.; Iowa State University

'Freckles', S, pb, 1976; bud ovoid, pointed to urn-shaped; flowers light scarlet, flushed yellow, dbl., 28 petals, 4–4.5 in., cupped, slight fragrance; foliage dark, coppery, leathery; upright, bushy growth; [Tickled Pink X Country Music]; Buck, Dr. Griffith J.; Iowa State University

'Fred Cramphorn', HT, or, 1961; (Manola, Samoa); flowers geranium-red, dbl., 5–6 in., cupped; foliage dark, glossy; vigorous, upright, bushy growth; [Peace X Baccará]; Kriloff, Michel; Cramphorn's Nursery, 1961;, Ilgenfritz Nursery, 1964

'Fred Edmunds', HT, ob, 1943; (L'Arlésienne); bud long, pointed; flowers coppery orange, dbl., 25 petals, 5–5.5 in., cupped, intense, spicy fragrance; foliage leathery, glossy; bushy, open habit growth; AARS, 1944 GM, Portland, 1942; [Duquesa de Peñaranda X Marie-Claire]; Meilland, F.; C-P

'Fred Edmunds, Climbing', Cl HT, ob, 1989; [Fred Edmunds sport]; Weeks, O.L.; Weeks Wholesale Rose Growers, 1977

'Fred Fairbrother', HT, dp, 1974; flowers bright cerise, dbl., 40 petals, 4–5 in., moderate fragrance; foliage semi-glossy; [(Gavotte X Tropicana) X Fragrant Cloud]; Sanday, John

'Fred Gibson', HT, ab, 1966; flowers apricot suffused gold, dbl., 30 petals, 5 in., slight fragrance; foliage dark; tall, vigorous growth; [Gavotte X Buccaneer]; Sanday, John, 1968

Fred Hollows Vision, F, w, 1998; [Mary McKillop sport]; Stratford's Roses, 1996

'Fred Howard', HT, yb, 1952; bud long; flowers golden orange shaded pink, dbl., 55 petals, 4 in., exhibition form, slight fragrance; vigorous, upright growth; AARS, 1952; [Pearl Harbor X Unnamed seedling]; Howard, F.H.; H&S

'Fred Howard, Climbing', Cl HT, yb, 1954; Howard, A.P.; H&S

'Fred J. Harrison', HT, mr, 1924; flowers cardinal-red shaded crimson, dbl., moderate fragrance; GM, NRS, 1923; Dickson, A.

'Fred Loads', S, or, 1968; flowers vermilion-orange, blooms in clusters, single, 3 in., moderate; foliage glossy; vigorous, tall growth; GM, RNRS, 1967; [Dorothy Wheatcroft X Orange Sensation]; Holmes, R.A.; Fryer's Nursery, Ltd., 1967

'Fred Owen', HT, dy, 1985; [Jan Guest sport]; Owen, Fred

'Fred Streeter', HT, my, 1955; bud pointed; flowers clear yellow, well-formed, dbl., 48 petals, 4 in., moderate fragrance; foliage dark; vigorous growth; [Luis Brinas X Golden Scepter]; Kordes; Wheatcroft Bros.

'Fred Streeter', HMoy, dp, 1951; flowers deep cerise-pink, medium, borne in clusters (up to 3), single; upright, branching growth; Jackman

Fred W. Alesworth *see* **'F. W. Alesworth'**

'Fred W. Mee', HT, mr, 1957; (F.W. Mee); flowers scarlet-cerise, dbl., 30 petals, exhibition form, intense fragrance; vigorous, upright growth; [Karl Herbst X The Doctor]; Mee

'Fred Walker', HT, mp, 1935; flowers glowing pink, base coppery orange, large, dbl., exhibition form, slight fragrance; foliage soft, light; vigorous growth; McGredy

Fredagh of Bellinchamp *see* **'SWANY RIVER'**

Freddie Mercury *see* 'BATMERCURY'

Freddy *see* 'PEAPROOF'

Fredensborg *see* POUlmax

'FRÉDÉRIC II DE PRUSSE', HCh, m, 1847; (Frederick the Second); flowers rich crimson-purple, medium, dbl., moderate fragrance; vigorous growth; Verdier, V.; 1847; (Sangerhausen)

'FRÉDÉRIC LERR', HT, pb, 1950; flowers carmine-red, reverse lighter, large, dbl., exhibition form; foliage bronze; very vigorous growth; [Crimson Glory X (Mrs Pierre S. duPont X Signora)]; Sauvageot, H.; Sauvageot

Frederic Mistral® *see* 'MEITEBROS'

'FRÉDÉRIC SCHNEIDER II', HP, dp, 1885; flowers large, dbl.; Ludovic; (Sangerhausen)

'FRÉDÉRIC SOULIÉ', rb, 1835; flowers carmine-red with purple striping, large, dbl.; Laffay, M.; (Sangerhausen)

Frederic Worth, HRg, dp

'FREDERICA', HT, mr; flowers deep crimson, dbl., 30 petals, 5 in., moderate fragrance; foliage dark, glossy; vigorous growth; About 1953 in England

'FREDERICK S. PECK', LCl, dp, 1938; flowers deep grenadine-pink, center more yellow, semi-dbl., 4 in.; slightly arched stems; [Hybrid creeper X Mrs Arthur Curtiss James]; Brownell, H.C.

Frederick the Second *see* **'FRÉDÉRIC II DE PRUSSE'**

'FREDERICKSBERGROSEN', F, dp, 1942; flowers deep pink, medium, single, 5 petals, cupped, no fragrance; foliage medium, medium green, semi-glossy; vigorous, bushy growth; [Orléans Rose X Unnamed seedling]; Poulsen, S.; Poulsen's Roses

Frederik Mey, F, op, 1980; Eve, A.

Frederiksborg *see* POUldrik

Frederyk Chopin, HT, ly, 1995; flowers large, dbl.; Zyla; (Sangerhausen)

'FREDICA', S, w, 1974; bud oval; flowers medium, single, 5 petals, cupped; thornless; very vigorous, bushy, upright growth; PP004463; [Indica Major X Multiflora Inermis]; INRA

Free As Air *see* 'MEHBRONZE'

'FREE GOLD', HT, 1948; bud long, pointed; flowers yellow, open, large, dbl., exhibition form, moderate fragrance; foliage glossy; bushy, dwarf growth; RULED EXTINCT 5/83 ARM; [Pink Princess X Shades of Autumn]; Brownell, H.C.

Free Gold® *see* 'MACFREEGO'

Free Spirit *see* 'PIXIREE'

Freedom® *see* 'DICJEM'

'FREEDOM', LCl, w, 1918; (White American Beauty, Climbing); flowers white, center yellow, open, dbl., 75 petals, 4 in., exhibition form, slight fragrance; foliage dark, bronze, glossy; long, strong stems; very vigorous, climbing growth; RULED EXTINCT 3/84 ARM; [Silver Moon X Kaiserin Auguste Viktoria]; Undritz

'FREEDOM'S RING', LCl, rb, 1994; flowers red and white stripes, blooms borne in small clusters, semi-dbl., 6–14 petals, 1.5 in., slight fragrance; some prickles; foliage medium, medium green, matt; medium, bushy, spreading growth; [Stars 'n' Stripes X Paradise]; Dykstra, Dr. A. Michael; Certified Roses, Inc., 1994

Freegold® *see* 'MACFREEGO'

'FREELEIGH', HT, my, 1956; flowers buttercup-yellow, well shaped, small, dbl., slight fragrance; foliage glossy, light green; vigorous growth; [Kingcup X Golden Scepter]; LeGrice

Freemont *see* 'BRIFREE'

Fregate, HT, lp, 1979; flowers large, dbl., slight fragrance; Delbard; (Sangerhausen)

'FREIA', HT, ab, 1936; flowers sun-yellow tinted orange, base orange, open, large, dbl.; foliage leathery, glossy; vigorous, bushy growth; [Ville de Paris X Rev. F. Page-Roberts]; Tantau

Freiamt, HT, mr, 1995; Huber

'FREIBURG II', HT, pb, 1917; flowers silver-rose, reverse bright apricot-pink, dbl.; [Dr. G. Kruger X Frau Karl Drucshki]; Krüger

'FREIBURG II, CLIMBING', Cl HT, pb, 1953; Lindecke

'FREIFRAU IDA VON SCHUBERT', HT, dr, 1912; flowers dark crimson-red, dbl., moderate fragrance; [Oskar Cordel X Frau Peter Lambert]; Lambert, P.

'FREIFRAU VON MARSCHALL', HWich, mp, 1913; flowers fresh pink, small, borne in immense, loose clusters, dbl., mid-season bloom; vigorous (8-12 ft.) growth; [Farquhar X Schneewittchen]; Lambert, P.

'FREIHEITSGLOCKE', HT, rb, 1963; (Damas de Yuste, Liberty Bell); flowers claret-rose, reverse light cream, dbl., 50 petals, 5 in., globular, moderate fragrance; foliage leathery; vigorous growth; [Detroiter X Kordes' Perfecta]; Kordes, R.; A. Dickson, 1963;, McGredy, 1963;, Wyant, 1966

Freiheitsglocke *see* Liberty Bell

'FREIHERR VON MARSCHALL', T, mr, 1903; bud pointed; flowers large, dbl.; foliage blood-red when young; vigorous growth; [Princesse Alice de Monaco X Rose d'Evian]; Lambert, P.

Freisinger Morgenrote® *see* KORmarter

'FRENCH CAN CAN', HT, pb, 1956; bud well formed; flowers pink, reverse yellow, large, dbl.; vigorous, bushy growth; Frères, Buyl

French Cancan *see* **Tourbillon**

French Lace *see* 'JACLACE'

French Liberty *see* 'MEIJETTE'

French Perfume™ *see* 'KEIBIAN'

French Rose *see* **R. GALLICA**

'FRENCH VANILLA', Gr, w, 1986; flowers white and faint pink, reverse same, fading white, medium, dbl., exhibition form, borne usually singly, moderate fragrance; no fruit; few, moderate, brown prickles; foliage medium, dark green, matt; spreading growth; [Araby X Royal Highness]; Thomson, R.

French Vanilla, HT, w, 1999

Frenesie, LCl, 1965; Combe, M.; (Cavriglia)

'FRENSHAM', F, dr, 1946; flowers deep scarlet, blooms in large trusses, semi-dbl., 15 petals, slight fragrance; vigorous growth; (21); GM, ARS, 1955 GM, NRS, 1943; [Floribunda seedling X Crimson Glory]; Norman; C-P, 1949;, Harkness, 1946

'FRENSHAM, CLIMBING', Cl F, dr, 1958; Bennett, J.A.; Pedigree Nursery

'FRENSHAM'S COMPANION', F, dp, 1952; flowers cerise, loosely formed, medium, borne in trusses, semi-dbl., 18 petals, slight fragrance; very free growth; [Frensham sport]; Morse

Frenzy *see* 'MEIHIGOR'

Fresco® *see* 'RUICO'

Fresh Cream, HT, w, 1991; Kordes

Fresh Hit *see* POUlfre

'FRESH PINK', Min, lp, 1964; bud ovoid; flowers light pink tipped salmon, blooms in clusters, dbl., 25 petals, cupped, slight fragrance; foliage leathery, glossy; vigorous, bushy growth; [(R. wichurana X Floradora) X Little Buckaroo]; Moore, Ralph S.; Sequoia Nursery

Fresh Pink *see* MACpinderal

Fresh Start *see* 'FLOSAR'

Freshie *see* 'MACSWEETWA'

Freude® *see* 'DEKORAT'

'FREUDENFEUER', Pol, mr, 1918; flowers bright red; moderate growth; Kiese

'FREUDENTANZ', F, mr, 1973; bud pointed; flowers bright red, reverse flamed red, medium, intense fragrance; foliage glossy; vigorous, upright, bushy growth; [Fragrant Cloud X Goldmarie]; Hetzel; GAWA

'FREUND PILZ', HT, mr, 1930; flowers carmine-red, medium, dbl.; Schildmann; (Sangerhausen)

'FREYA', HT, dp, 1956; flowers carmine-red, large, dbl.; vigorous growth; [Seedling X Étoile de Hollande]; Leenders, M.

'FREYA', HCan, pb, 1910; Geschwind, R.

Friction Lights *see* 'HORLIGHTS'

Friday's Child *see* 'HORABI'

Fridolin, Min, yb

'FRIEDA KRAUSE', HT, or, 1935; flowers orange-scarlet-red, large, dbl., exhibition form; foliage leathery, dark; very vigorous, bushy growth; [I Zingari X Unnamed seedling]; Krause

'FRIEDLANDERIANA', HGal, mp; flowers bright rose-pink, single, non-recurrent; [R. gallica hybrid X R. canina]

'FRIEDRICH ALFRED KRUPP', HT, op, 1903; flowers yellowish salmon-pink, large, very dbl., moderate fragrance; Welter; (Sangerhausen)

'FRIEDRICH HEYER', S, ob, 1956; flowers bright orange, in large clusters, 10 petals, 3.5 in., moderate fragrance; foliage dark, glossy, leathery; vigorous, upright growth; Tantau, Math.

'FRIEDRICH SCHWARZ', HT, dr, 1952; flowers crimson, large, dbl., 30 petals, cupped, intense fragrance; foliage dark; very tall, branching growth; [Poinsettia X (Crimson Glory X Lord Charlemont)]; Kordes

'FRIEDRICH WÖRLEIN', F, dy, 1963; bud globular; flowers golden yellow, large, dbl., moderate fragrance; foliage dark; vigorous, upright, bushy growth; [Clare Grammerstorf X Golden Masterpiece]; Kordes; Wörlein

'FRIEDRICHSRUH', HT, dr, 1908; flowers dark crimson, shaded black, turning blue, open, large, dbl., intense fragrance; foliage dark, glossy; bushy, open growth; [Princesse de Bearn X Francis Dubreuil]; Türke

Friend for Life *see* 'COCNANNE'

'FRIEND OF HEART', HT, pb, 1985; flowers large blooms borne singly, dbl., 30 petals, exhibition form, slight fragrance; deep brown prickles; foliage medium, dark, glossy; upright growth; [(Festival Beauty X Scarlet Knight) X Self]; Hardikar, Dr. M.N.

'FRIEND OF PEACE', HT, yb, 1986; dbl., 20–22 petals, globular, intense fragrance; crescent, light brown prickles; foliage large, dark green, glossy, leathery; upright, open growth; [(Scarlet Knight X Festival Beauty) X Festival Beauty]; Hardikar, Dr. M.N.

Friends of Benalla Gardens, Pol, w, 1996; Sutherland

Friendship™ *see* 'LINRICK'

'FRIENDSHIP', HT, dp, 1937; bud pointed; flowers dark red, semi-dbl., exhibition form, intense fragrance; foliage glossy; vigorous growth; RULED EXTINCT 9/77; [Templar X Talisman]; Amling Co.

'FRIENDSHIP', HT, dr, 1938; flowers bright strawberry-red, with bright scarlet undertone, very large, dbl.; very vigorous growth; RULED EXTINCT 9/77; GM, NRS, 1938; Dickson, A.

Friendship *see* 'HAVIPIP'

Friesensöhne® *see* 'TANSENFRIE'

Friesia *see* 'KORRESIA'

'FRIGG', S, m, 1969; flowers small, borne in clusters, semi-dbl., 15 petals, flat, free, recurrent boom; foliage small, rich green; low growth; [Schneezwerg X R. nitida]; Lundstad

'FRILEUSE', Gr, mp, 1966; flowers urn-shaped, large, dbl., slight fragrance; foliage large, light green, glossy; tall, upright, bushy growth; [Queen Elizabeth X Baronesse Manon]; Poulsen, Niels D.; Vilmorin-Andrieux

'FRILLS', HT, op, 1950; flowers deep salmon-pink becoming lighter, petals scalloped, large; compact, bushy growth; Moss

Frilly Dilly *see* 'MURFRI'

Frilly Dilly, MinFl, lp, 1986; Cocker

'FRIMOUSSE', F, op, 1959; flowers orange-pink, base yellow, medium, borne in clusters, single, slight fragrance; vigorous, very bushy growth; [Masquerade X ?]; Vilmorin-Andrieux

Frine® *see* BARfri

'FRINÉ', HT, dr, 1961; flowers crimson suffused strawberry-red, well formed, dbl., 35 petals, moderate fragrance; vigorous growth; [Lila Vidri X (Soraya X Vigoro)]; Dot, Simon

'FRINGETTE', Min, dp, 1964; flowers deep pink, white center, small, dbl., 25 petals; low (8 in), compact growth; [Unnamed seedling X Magic Wand]; Moore, Ralph S.; Sequoia Nursery

Frisco® *see* 'KORFLAPEI'

'FRISETTE', F, or, 1964; flowers bright scarlet, open, borne in clusters, semi-dbl.; vigorous, low to medium height; [Unnamed seeding X Concerto]; Mondial Roses

'FRISKY', HT, mr, 1959; flowers velvety red, open, large, dbl., 50 petals, moderate fragrance; foliage dark, glossy; vigorous, bushy growth; [Charlotte Armstrong X Chrysler Imperial]; Wyant

Frisson Frais®, HMsk, w, 1990; Lens

'FRITZ HEGAR', Pol, dr, 1930; flowers medium, semi-dbl.; Schmitt-Eltville; (Sangerhausen)

'FRITZ HÖGER', HT, dr, 1934; bud pointed; flowers pure crimson, large, dbl., exhibition form; foliage leathery, dark; very vigorous growth; [(Hadley X Comte G. de Rochemur) X Cathrine Kordes]; Kordes

'FRITZ MAYDT', HT, w, 1925; flowers coppery flesh-white, moderate fragrance; [Mev. C. van Marwijk Kooy X Marquise de Sinéty]; Leenders, M.

'FRITZ NOBIS', S, pb, 1940; bud long, pointed, light red; flowers white, reverse reddish salmon-pink, large blooms in clusters, dbl., exhibition form, non-recurrent, intense fragrance; foliage glossy, leathery; vigorous growth; [Joanna Hill X Magnifica]; Kordes

'FRITZ SCHRÖDTER', HT, dr, 1928; flowers brilliant dark scarlet, large, dbl.; foliage bronze, soft; vigorous, bushy growth; [Hortulanus Budde seedling]; Mühle

'FRITZ THIEDEMANN', HT, or, 1959; bud pointed; flowers brick-red, well-shaped, dbl., 36 petals, 4 in., moderate fragrance; foliage dark; bushy growth; [Horstmann's Jubilaumsrose seedling X Alpine Glow seedling]; Tantau, Math.

'FRITZ THIEDEMANN, CLIMBING', Cl HT, or, 1961; Kordes

Fritz Walter *see* HELfriwa

Fritzi *see* HELtzi

'FRIULI', HT, ab, 1933; flowers amber-yellow, large, cupped; vigorous growth; Ingegnoli

'FRIVOLE', F, mr, 1958; flowers geranium-red, very dbl.; vigorous, low growth; [Independence X Country Girl]; Frères, Buyl

'FRIVOLITÉ', HT, or, 1956; flowers scarlet shaded orange and salmon, large, dbl., 50 petals, intense fragrance; vigorous, compact growth; [Peace X Catalonia]; Dot, Pedro

Frivolous *see* 'MINFRIV'

'FROCRIN', Min, w, 1989; (**Crinkles**); bud ovoid; flowers white, aging with flecks of red, slow to fade, crinkled center, very dbl., 60 petals, exhibition form, slight fragrance; fruit not observed; prickles straight, short, red; foliage medium, dark green, semi-glossy; upright, bushy, medium growth; [Rise 'n' Shine X Seedling]; Frock, Marshall J., 1990

'FROFIRE', Min, rb, 1995; (**Evening Fire**); flowers red, pink reverse, blooms borne mostly single, dbl., 15–25 petals, 1.5 in., moderate fragrance; some prickles; foliage medium, medium green, semi-glossy; bushy growth; [Black Jade X Pierrine]; Frock, Marshall J.

'FROHSINN'®, F, ab, 1961; (Joyfulness); bud pointed; flowers apricot, cream and pink blend, large blooms in large clusters, dbl., 20 petals, cupped, slight fragrance; foliage glossy; vigorous, bushy growth; [Horstmann's Jubilaumsrose X Circus]; Tantau, Math.

Frohsinn '82 *see* 'TANSINNROH'

'FROJEAN', Min, dy, 1989; (**Jeannine Michelle**); bud pointed; flowers deep gold-yellow, aging to pale yellow, medium, borne usually singly, dbl., 30 petals, exhibition form, moderate, fruity fragrance; fruit not observed; prickles straight, tan; foliage medium, medium green, semi-glossy; upright, bushy, medium, hardy growth; [Rise 'n' Shine X Unnamed seedling]; Frock, Marshall J.

'FROLIC', F, mp, 1953; flowers bright pink, in large sprays, dbl., 21 petals, 2.5 in., slight fragrance; vigorous, bushy growth; [World's Fair X Pinocchio]; Swim, H.C.; Armstrong Nursery

Frolic, F, pb, 1995; Dawson

'FROMITE', Min, w, 1991; (**White Mite**); flowers small, micro-mini blooms, dbl., 26–40 petals, slight fragrance; foliage small, medium green, semi-glossy; upright growth; [Baby Katie X Unnamed seedling]; Frock, Marshall J.

Frondeuse®, S, pb

'FRONEEN', Min, lp, 1992; (**Jeannine**); flowers light pink aging to white with a few red spots, similar to Royal Highness, dbl., 30 petals, 1.5–1.75 in., exhibition form, intense fragrance; foliage medium, medium green, semi-glossy; medium (40-50 cms), upright growth; [Baby Katie X Unnamed seedling]; Frock, Marshall J.

'FRONORM', Min, pb, 1989; (**Norma Margaret**); bud pointed; flowers pink, blending with copper shading, reverse white with pink, dbl., 30 petals, exhibition form, moderate fragrance; fruit not observed; prickles straight, medium, brown; foliage medium, medium green, semi-glossy; upright, medium growth; [Baby Katie X Unnamed seedling]; Frock, Marshall J., 1990

'FRONTENAC', S, dp, 1992; dbl., 15–25 petals, 3–3.5 in., borne in small clusters, slight fragrance; some prickles; foliage medium, dark green, glossy; medium (100 cms), upright growth; very winter hardy; Ogilvie, Ian S.; Agriculture Canada, 1992

'FRONTIER TWIRL', S, pb, 1984; flowers pink-yellow blend, large blooms borne 1-8 per cluster, dbl., 25 petals, cupped, repeat bloom, moderate fragrance; awl-like, tan prickles; foliage leathery, medium, bronze green; erect, bushy growth; hardy; [Sevilliana X Just Joey]; Buck, Dr. Griffith J.; Iowa State University

'FROSTFIRE', Min, mr, 1963; flowers red, sometimes flecked white, dbl., 30 petals, 1 in.; foliage dark, glossy; bushy, compact (12-14 in) growth; [((R. wichurana X Floradora) X Seedling) X Little Buckaroo]; Moore, Ralph S.; Sequoia Nursery

'FROSTY', Min, w, 1953; bud ovoid, pale pink; flowers clear white, very small blooms in clusters of 3-10 or more, dbl., 45 petals, moderate, honeysuckle fragrance; foliage glossy; vigorous (12-14 in), compact, spreading growth; [(R. wichurana X Unnamed seedling) X Self]; Moore, Ralph S.; Sequoia Nursery

Frosty Morning *see* HORFROST

Frothy *see* MACfrothy

'FROUFROU', HT, yb, 1955; flowers chamois-yellow, center carmine, large, very dbl.; vigorous, bushy growth; [Mme Joseph Perraud X Yvonne Plassat]; Moulin-Epinay

'FROU-FROU', HT, rb, 1957; flowers crimson, reverse carmine-pink, large, dbl., 25 petals; moderate growth; [Comtesse Vandal X Unnamed seedling]; Laperrière; EFR

'FROY', F, or, 1972; bud ovoid; flowers open, medium, semi-dbl., slight fragrance; foliage glossy, dark; bushy growth; [Traumland X Poulsen's Pink]; Lundstad; Norges Landbrukshogskole

Fru Dagmar Hastrup *see* **'FRAU DAGMAR HARTOPP'**

'FRU GERDA HELMUUS', HT, op, 1935; flowers light orange-pink, large, semi-dbl., slight fragrance; Poulsen, D.T.; (Sangerhausen)

Fru Inge Poulsen *see* **'MRS INGE POULSEN'**

'FRU JOHANNE POULSEN', HT, mp, 1924; flowers bright pink, well formed; vigorous growth; [Margrethe Moller seedling]; Poulsen, S.

Fru Julie Poulsen *see* **'POULSEN'S DELIGHT'**

'FRU XENIA JACOBSEN', HT, dr, 1925; flowers deep red, well shaped, dbl., moderate fragrance; vigorous growth; [Étoile de France X Richmond]; Poulsen, S.

'FRÜHLINGSANFANG', HSpn, w, 1950; bud long, pointed; flowers ivory-white, single, 4 in., intermittent bloom, moderate fragrance; foliage leathery; very vigorous, bushy (9 ft. in 6 years) growth; [Joanna Hill X R. spinosissima altaica]; Kordes

'FRÜHLINGSDUFT', HSpn, pb, 1949; bud ovoid, golden yellow; flowers lemon-yellow with light pink, very large, dbl., exhibition form, non-recurrent, intense fragrance; foliage large, leathery; very vigorous, upright, bushy growth; [Joanna Hill X R. spinosissima altaica]; Kordes

'FRÜHLINGSGOLD'®, HSpn, my, 1937; (Spring Gold); bud pointed, nasturtium-red; flowers creamy yellow, single, 3 in., non-recurrent, intense fragrance; foliage large, light, soft, wrinkled; very vigorous, bushy growth; (28); [Joanna Hill X R. spinosissima hispida]; Kordes; B&A, 1951

'FRÜHLINGSMORGEN', HSpn, pb, 1942; (Spring Morning); flowers cherry-pink, center soft yellow, stamens maroon, medium, single, occasionally slightly recurrent bloom; large, red fruit; foliage dark; free growth (6 ft.); (28); [(E.G. Hill X Cathrine Kordes) X R. spinosissima altaica]; Kordes

'FRÜHLINGSSCHNEE', HSpn, w, 1954; bud ovoid; flowers snow-white, very large, single, slight fragrance; foliage leathery, wrinkled, light green; very vigorous, upright growth; [Golden Glow X R. spinosissima altaica]; Kordes

'FRUHLINGSTAG', HSpn, my, 1949; bud ovoid; flowers golden yellow, large, borne in small clusters, semi-dbl., profuse non-recurrent bloom, moderate fragrance; many thorns; foliage leathery; [McGredy's Wonder X Fruhlingsgold]; Kordes

'FRÜHLINGSZAUBER', HSpn, mp, 1942; flowers pink, medium, semi-dbl., non-recurrent; large, dark red fruit; foliage dark; vigorous (7 ft.) growth; [(E.G. Hill X Cathrine Kordes) X R. spinosissima altaica]; Kordes

Fruit Buffet, F, yb

Fruité *see* MEIfructoz

'FRÜLINGSSTUNDE', S, lp, 1942; flowers medium, semi-dbl., slight fragrance; Kordes, W. Söhne; (Sangerhausen)

FRYaffair, HT, lp, 1997; (**Sight Saver**); Fryer, Gareth

FRYbingo, HT, dp, 1997; (**Awareness**); Fryer, Gareth

'FRYBLISSFUL', LCl, mr, 1999; (**Dizzy Heights**); flowers bright red, reverse same, dbl., 17–25 petals, 5–5.5 in., borne in large clusters, slight fragrance; prickles moderate; foliage medium, dark green, semi-glossy; upright, tall (7-10 ft) growth; Fryer, Gareth; Fryer's Nurseries, Ltd., 1999

FRYbountiful, F, dy, 1998; (**Phab Gold**); Fryer, Gareth

FRYbright, F, ob, 1997; (**Razzle Dazzle**); Fryer, Gareth

FRYbubbly, Min, mp, 1998; (**Bubbles**); Fryer, Gareth

FRYcalm, HT, ob, 1998; (**Britannia**); Fryer, Gareth

FRYcassia, MinFl, 1998; (**Flower Power**); Fryer, Gareth

FRYchambi, HT, ab, 1986; (**Daily Express**); Fryer, Gareth

FRYcharm, HT, dy, 1998; (**Lions International**); Fryer, Gareth

FRYclimbdown, LCl, dr, 1991; (**Crimson Cascade**); Fryer, Gareth

FRYdarkeye, S, rb, 1988; (**Biddulph Grange**); GM, Glasgow, 1991; Fryer, Gareth

'FRYELISE', HT, my, 1999; (**Cheshire**); flowers golden honey yellow, large, dbl., 17–25 petals, borne in small clusters, moderate fragrance; prickles moderate; foliage large, dark green, semi-glossy; upright medium (3 ft) growth; Fryer, Gareth; Fryer's Nurseries, Ltd., 1999

'FRYER'S ORANGE', HT, ab, 1934; flowers orange-yellow; [Mrs Sam McGredy sport]; Fryers Nursery, Ltd.

FRYevenest, F, ob, 1986; (**Evening Sentinel**); Fryer, Gareth

FRYgoldie, Min, dy, 1992; (**Fairy Gold**); Fryer, Gareth

'FRYGRAN', HT, ab, 1983; (**Johnnie Walker**); flowers buff apricot, well-formed, large, dbl., 20 petals, intense fragrance; foliage medium, medium green, matt; vigorous, bushy growth; [Sunblest X (Arthur Bell X Belle Blonde)]; Fryers Nursery, Ltd., 1982

'FRYJAM', F, ab, 1984; (**The Flower Arranger**); bud large; flowers peach, well-formed, semi-dbl., slight fragrance; foliage large, dark, glossy; upright growth; [Seedling X Seedling]; Fryers Nursery, Ltd.

'FRYJASSO', F, pb, 1984; (**Inner Wheel**); flowers carmine edged rose pink, large, dbl., 22 petals, slight fragrance; foliage medium, dark red, matt; bushy growth; [Pink Parfait X Picasso]; Fryers Nursery, Ltd.

'FRYJINGO', HT, yb, 1985; (**The Lady**); flowers honey yellow, petals edged salmon, well-formed, large blooms, dbl., 35 petals, exhibition form, slight fragrance; foliage medium, medium green, semi-glossy; upright growth; GM, Baden-Baden, 1987; [Pink Parfait X Redgold]; Fryer, Gareth; Fryer's Nursery, Ltd.

FRYmartor, HT, mr, 1989; (**Marianne Tudor**); Fryer, Gareth

FRYmaxicot, F, ab; (**Summer Dream**); Fryer, Gareth, 1990

'FRYMESTIN', F, dp, 1984; (**Ethel Austin**); flowers deep pink, dbl., 20 petals, moderate fragrance; foliage large, medium green, semi-glossy; upright growth; [Pink Parfait X Redgold]; Fryers Nursery, Ltd.

'FRYMINICOT', F, ab, 1988; (**Sweet Dream**); flowers peach-apricot, medium, dbl., moderate fragrance; foliage medium, medium green, semi-glossy; mini-flora; bushy growth; ROTY, 1988; [Unnamed seedling X Unnamed seedling]; Fryers Nursery, Ltd.

'FRYMINILES', Min, op, 1987; (**Lesley Anne**); flowers pale peach-pink, pale yellow base, fading slightly paler, large, dbl., 45 petals, cupped, no fragrance; oval, orange-red fruit; thin, small, pointed, light brown prickles; foliage medium, light green, matt; upright growth; [Unnamed seedling X Unnamed seedling]; Fryer, Gareth

FRYministar, Min, or; (**Top Marks**); Fryer, Gareth, 1992

FRYnblue, F, m; (**The Colwyn Rose**); Cowlishaw, Frank; Fryer, 1992

FRYorst, HT, ob, 1996; (**ARC Angel**); Fryer, Gareth

'FRYPERDEE', HT, dr, 1988; (**Velvet Fragrance**); flowers deep crimson, large, dbl., intense fragrance; foliage large, dark green, semi-glossy; upright growth; Edland Fragrance Medal, ARS, 1987 Fragrance Prize, Baden-Baden, 1990; [Unnamed seedling X Unnamed seedling]; Fryers Nursery, Ltd.

FRYprincess, F, ob, 1988; (**Julie Cussons**); Fryer, Gareth

FRYrelax, F, ab, 1989; (**Pensioners Voice**); Fryer, Gareth

FRYrhapsody, F, ab, 1990; (**Langdale Chase**); Fryer, Gareth

FRYromeo, HT, ob, 1998; (**Scent-Sation**); Fryer, Gareth

FRYshrewby, HT, dr; (**Shrewsbury Show**); Fryer, Gareth, 1988

'FRYSTAR', HT, or, 1992; (Beauty Star, **Liverpool Remembers**); flowers vermillion, blooms borne mostly singly, dbl., 26–40 petals, 3–3.5 in.; many prickles; foliage medium, medium green, glossy; tall (240 cms), upright growth; [Corso X Unnamed seedling]; Fryer, Gareth; Roses by Fred Edmunds, 1990

FRYstassi, Min, pb, 1988; (**Rosaletta**); Fryer, Gareth

FRYsweetie, Min, op, 1993; (**Mary Gammon**); Fryer, Gareth

FRYtango, HT, ab; (**The Observer**); Fryer, Gareth, 1991

FRYtranquil, HT, dy, 1991; (**Golden Moments**); Fryer, Gareth

'FRYTROOPER', S, dr, 1999; (Daily Post, Dusky Dancer, Karla, Liverpool Daily Post, **Raven**™); flowers deep velvety red, golden stamens, dbl., 20–25 petals, 1.2–1.5 in., borne in very large clusters, slight fragrance; prickles moderate; foliage small, dark green, glossy; upright, bushy, medium (3 - 4 ft) growth; PP009211; [Lavaglut X [(Anytime X Liverpool Echo) X (New Penny X Unknown)]]; Fryer, Gareth; Weeks Roses, 1992, U.S.

FRYvivacious, F, op, 1992; (**Julie Andrews**); Fryer, Gareth

'FRYWILREY', HT, rb, 1987; (**Audrey Wilcox**®); flowers cerise red and silver cream, large blooms, dbl., 26–40 petals, intense fragrance; foliage large, dark green, glossy; [Alpine Sunset X Whisky Mac]; Fryers Nursery, Ltd., 1985

FRYwinner, F, my, 1994; (**Atco Royal**); Fryer, Gareth

FRYworld, F, op, 1993; (**Atlantic Star**); Fryer, Gareth

FRYworthy, HT, my, 1996; (**Especially for You**); Fryer, Gareth

'FRYXOTIC', HT, ab, 1999; (Chantoli, Exotic®, **Sunset Celebration**™, Warm Wishes); flowers creamy apricot/ amber blend, dbl., 35–40 petals, 4.5–5 in., exhibition form, borne mostly singly, moderate, fruity fragrance; prickles moderate; foliage large, medium green, semi-glossy; upright, bushy, medium, (3.5-4 ft) growth; PP009718; AARS, U.S., 1998 GM, Belfast, 1996 Golden Rose, The Hague, 1997; [Pot O'Gold X seedling]; Fryer, Gareth, 1994; Weeks Roses, 1998

FRYxquisite, Min, mp; (**Sweet Petite**); Fryer, Gareth, 1994

FRYyaboo, HT, r, 1994; (**Belle Époque**); Fryer, Gareth

FRYyat, F, ob, 1995; (**Good Morning**); Fryer, Gareth

FRYyat, HT, lp; (**The Cheshire Regiment**); Fryer, Gareth, 1996

FRYyearn, HT, lp, 1995; (**Bride**); Fryer, Gareth

FRYyeh, F, ob; (**Pomona**); Fryer, Gareth

'FRYYEOMAN', S, lp, 1999; (**Flower Girl**™); bud short, ovoid; flowers soft pink fading to light pink centers, semi-dbl., 8–15 petals, 1–1.5 in., borne in very large clusters, slightly pendulous, slight, apple fragrance; few prickles; foliage medium, light green, matte; long, slender, arching stems; spreadying, bushy, medium (4-5 ft) growth; [Amruda X Fairy Snow]; Fryer, Gareth; Weeks Roses, 2000

FRYyippee, S, m, 1993; (**Rosie Larkin**); Fryer, Gareth

FRYyoung, HT, lp; (**Special Occasion**); Fryer, Gareth, 1995

FRYzippy, F, ob, 1998; (**Bob Greaves**); Fryer, Gareth

Fuchsia, HGal, m

Fuchsia Meidiland™ *see* 'MEIPELTA'

Fuchsia Meillandecor® *see* 'MEIPELTA'

Fuchsia Sunblaze *see* MEIrulex

'FUCHSINE GUY', F, m, 1930; flowers lilac-purple, open, large, borne in clusters, semi-dbl.; foliage rich green; bushy growth; [Lafayette sport]; Leenders, M.

'FUËGO', F, or, 1964; bud pointed; flowers chinese vermilion, open, medium, borne in clusters, dbl., 30 petals, slight fragrance; [Aloha X Gabychette]; Arles; Roses-France

Fuggerstadt Augsburg® *see* 'KORTREU'

'FUGITIVE', F, yb, 1965; bud pointed; flowers apricot-yellow, becoming lighter, open, borne in clusters, semi-dbl.; foliage glossy, bronze; vigorous, bushy growth; [Mrs Oakley Fisher X ?]; Pal, Dr. B.P.; Indian Agric. Research Inst.

Fugue® *see* 'MEITAM'

'FUJIZAKURA', Min, mp, 1997; flowers single, small, single, 4–7 petals, no fragrance; some prickles; foliage small, medium green, semi-glossy; spreading, low (20cms) growth; [Azumino X Azumino]; Ohtsuki, Hironaka

'FUKUYAMA', HT, op, 1988; bud pointed; flowers salmon-pink, aging dark, large, dbl., 35 petals, exhibition form, slight fragrance; round, yellow blend fruit; prickles red to dark green; foliage red aging dark green, matt, ovoid; upright, tall growth; [Pristine X Takao]; Tagashira, Kazuso; Hiroshima Rose Nursery, 1988

'FULGENS', HSpn, m; flowers lilac-pink, semi-dbl., early bloom; glossy, black fruit; height 3-4 ft

Fulgens *see* **'MALTON'**

Fulgens, HGal, lp; Vibert, before 1830

Fulgurante® *see* WARoujo

Full Moon *see* 'SPRINGMOON'

Full Sail *see* 'MACLANOFLON'

'FULLCREAM', HT, ly, 1959; flowers creamy yellow, well formed, dbl., 28 petals, 5–7 in., intense, honey fragrance; foliage glossy; vigorous, low growth; [Wellworth X Diamond Jubilee]; LeGrice

Fullerton Centennial *see* 'PIXIFULL'

'FULSTEVE', HT, pb, 1991; (**Steve Silverthorne**); flowers white with pink blushing, large blooms borne mostly singly, dbl., slight fragrance; foliage large, medium green, semi-glossy; medium, upright growth; [Dorothy Anne X Headliner]; Fulgham, Mary, 1992

Fulton MacKay *see* 'COCDANA'

'FULVIA', HT, dp, 1950; flowers pink tinted carmine, medium, dbl., slight fragrance; foliage leathery, light green; vigorous, bushy growth; [Mme Joseph Perraud X Mme Elie Dupraz]; Gaujard

'FUN JWAN LO', S, w; (Indica Major, Odorata 22449); flowers white, center pale pink, small, dbl.; very vigorous growth; not hardy.; ca. 1924

Funchal, HT; Moreira da Silva, A.; (Cavriglia)

'FUNKENMARIECHEN', F, yb, 1973; flowers yellow, red, dbl., globular, slight fragrance; foliage glossy; upright, bushy growth; [Seedling X Samba]; Kordes; Horstmann

Funkuhr® *see* 'KORPORT'

'FUNNY FACE', Min, rb, 1981; flowers white, petals edged red, aging red, medium, dbl., 35 petals, slight fragrance; foliage small, medium green, semi-glossy; upright, bushy growth; [Avandel X Zinger]; Jolly, Betty J.; Rosehill Farm

Funny Girl *see* 'JACFUN'

Funny Girl *see* BARfunn

Furedaiko *see* **'FURE-DAIKO'**

'FURE-DAIKO', Cl F, op, 1974; (Furedaiko); bud ovoid; flowers yellow to light orange-red, medium, dbl., cupped, moderate fragrance; foliage large, glossy, dark; vigorous, climbing growth; [(Goldilocks seedling X Sarabande) X Golden Giant seedling]; Suzuki, Seizo; Keisei Rose Nursery

Furia, HT, 1968; Combe, M.; (Cavriglia)

'FURORE', HT, or, 1965; flowers scarlet, medium, very dbl., slight fragrance; foliage dark; [(Baccará X Unnamed seedling) X Miracle]; Verbeek

'FÜRST BISMARCK', T, dy, 1886; flowers large, very dbl.; Drögemüller; (Sangerhausen)

'FÜRST LEOPOLD IV ZU SCHAUMBURG-LIPPE', HP, dr, 1918; flowers large, dbl., moderate fragrance; Kiese; (Sangerhausen)

'FÜRSTIN BISMARCK', T, mr, 1887; flowers large, very dbl.; Drögemüller; (Sangerhausen)

Fürstin Infantin von Hohenzollern *see* **'FÜRSTIN VON HOHENZOLLERN INFANTIN'**

'FÜRSTIN MARIA HATZFELDT', HT, dr, 1927; flowers bright dark red, dbl., moderate fragrance; [Gen. MacArthur X Unnamed seedling]; Boden

'FÜRSTIN VON HOHENZOLLERN INFANTIN', T, m, 1898; (Fürstin Infantin von Hohenzollern); flowers purple rose, center yellowish salmon, medium, dbl.; Brauer, P.

'FURSTIN VON PLESS', HRg, w, 1911; Lambert, P.

Fushimi, HT, pb, 1987; Keihan

'FUSHINO', HT, lp, 1980; flowers large, dbl., 45 petals, exhibition form, no fragrance; foliage medium green; vigorous, spreading growth; [Utage X Ann Letts]; Ota, Kaichiro

'FUSILIER', F, or, 1957; (Grenadier, Red Soldier); bud globular; flowers orange-scarlet, blooms in heavy clusters, dbl., 40 petals, 3–3.5 in., slight fragrance; foliage dark, glossy, leathery; vigorous growth; AARS, 1958; [Red Pinocchio X Floradora]; Morey, Dr. Dennison; J&P

'FUTTAKER SCHLINGROSE', HMult, dr; Geschwind, R., before 1900

'FUTURA', HT, or, 1975; bud long, pointed; flowers vermilion, large, dbl., cupped, slight fragrance; foliage glossy, light; vigorous, upright, bushy growth; [Unnamed seedling X Unnamed seedling]; Warriner, William A.; J&P

Futura *see* KORoketto

Future Award *see* JACsto

Fuxiana, HT, 1979; Dot, Simon; (Cavriglia)

'FYFIELD PRINCESS', F, my, 1994; dbl., 15–25 petals, 1.5–2.75 in., borne in large clusters, moderate fragrance; many prickles; foliage medium, medium green, semi-glossy; medium (4 ft), upright growth; [Southampton sport]; Shuttleworth, F.I.; F. Haynes & Partners, 1994

Fyvie Castle *see* 'COCBAMBER'

G

'G. Amédée Hammond', HT, ab, 1915; flowers apricot-yellow on ivory yellow, dbl., intense fragrance; GM, NRS, 1913; Dickson, A.

'G. F. Veronica', HMult, ly, 1900; flowers creamy yellow, small, semi-dbl., moderate fragrance; Demitrovisi; (Sangerhausen)

'G. H. Davison', HT, mr, 1988; flowers medium, well-defined red, large, borne in sprays of 1-3, dbl., 45 petals, exhibition form, slight fragrance; oval, large, green fruit; prickles normal, large, red; foliage large, dark green, semi-glossy; upright growth; [Unnamed seedling X Unnamed seedling]; Davison, G.H.; The Central Nursery, 1990

'G. I. Joe', HT, dp, 1943; bud long, pointed; flowers rose-red to deep rose-pink, large, very dbl., moderate fragrance; foliage leathery, dark; vigorous growth; [Red Better Times sport]; Parmentier, J.

G K Rose, HT, dr, 1989; Chiplunkar

G. Nabonnand *see* **'Gilbert Nabonnand'**

G. P. & J. Baker *see* 'HARrango'

'G. W. Peart', HT, mr, 1948; bud ovoid; flowers medium, very dbl.; foliage leathery, dark; moderate, bushy growth; [Guinee seedling X Rouge Mallerin]; Toogood

'G.W. Watkins', HT, lp; Williams, A., 1890; (Weatherly, L.)

'Gabi', Gr, or, 1987; (Gabrielle); flowers very bright orange-red, fading to lighter orange, classic form, dbl., 24 petals, exhibition form, moderate fragrance; rounded, small, light brown-green fruit; prickles fairly flat, small, dark brown; foliage medium, dark green, glossy; bushy, medium growth; [Pink Favorite X Red Dandy]; Poole, Lionel, 1988

'Gabriel Lombart', HT, w, 1932; flowers flesh-white to cream-white, very large, dbl., cupped, moderate fragrance; vigorous, bushy growth; [Dr. A. Hermans X Rayon d'Or]; Buatois

'Gabriel Noyelle', M, ab, 1933; (Gabrielle Noyelle); bud ovoid; flowers apricot, dbl., cupped, recurrent bloom, moderate fragrance; foliage leathery; very vigorous growth; [Salet X Souv de Mme Kreuger]; Buatois

Gabriela Sabatini *see* 'WILsab'

Gabriella® *see* 'BERgme'

Gabrielle *see* **'Gabi'**

Gabrielle *see* 'BERgme'

Gabrielle Noyelle *see* **'Gabriel Noyelle'**

'Gabrielle Privat', Pol, mp, 1931; flowers brilliant carmine-pink, blooms in pyramidal corymbs of 30-50, semi-dbl.; bushy growth; Barthelemy-Privat; Turbat

Gabriel's Fire *see* 'BRIfire'

Gaby Morlay *see* DORsand

'Gabychette', F, rb, 1960; bud ovoid; flowers reddish salmon, base sulfur-yellow, reverse white to rosy white, dbl., exhibition form, slight fragrance; foliage glossy, light green; low growth; [Floradora X Pioupiou]; Arles; Roses-France

'GActa', HT, or, 1969; (**Tanagra®**); bud long, pointed; flowers large, dbl., moderate fragrance; foliage soft; vigorous, upright growth; [Queen Elizabeth X Tropicana]; Gaujard

'GAegui', F, ob, 1963; ('GAegul', **Guitare®**); bud ovoid; flowers gold and orange-red blend, medium, dbl., intense fragrance; foliage light green, leathery; vigorous, bushy growth; GM, Bagatelle, 1966; [Vendome X Golden Slippers]; Gaujard

'GAegul', F, ob, 1963; ('GAegui', **Guitare®**); bud ovoid; flowers gold and orange-red blend, medium, dbl., intense fragrance; foliage light green, leathery; vigorous, bushy growth; GM, Bagatelle, 1966; [Vendome X Golden Slippers]; Gaujard

Gaia® *see* BARgai

'Gaiata', HT, yb, 1956; flowers indian yellow shaded pink; [Boudoir X Peace]; da Silva, Moreira

'Gaiety', HT, ob, 1926; bud pointed; flowers orange, indian red and silver, large, dbl., cupped; foliage light, glossy; vigorous, branching growth; [Mme Butterfly X Souv. de Claudius Pernet]; Hill, E.G., Co.; Hill Floral Products Co.

'Gaiety', HT, op; flowers salmon-pink flushed yellow, moderate fragrance; vigorous growth; Archer

'Gaiezza', HT, ob, 1940; flowers orange touched red, center yellow, intense fragrance; [Julien Potin X Mme G. Forest-Colcombet]; Giacomasso

Gail *see* TINgail

'Gail Borden', HT, pb, 1957; bud ovoid; flowers deep rose-pink, reverse overcast cream, dbl., 53 petals, 5.5 in., exhibition form, moderate fragrance; foliage dark, glossy, leathery; vigorous, upright growth; GM, NRS, 1957; [R.M.S. Queen Mary X Viktoria Adelheid]; Kordes; J&P

'Gail Borden, Climbing', Cl HT, pb, 1960; J&P, 1960

'Gainsborough', Cl HT, lp, 1903; flowers flesh-pink, almost white, large, dbl.; long stems; vigorous growth; [Viscountess Folkestone sport]; Good & Reese

'GAisu', F, dr, 1968; (**Sundra®**); bud pointed; flowers large, semi-dbl., cupped, intense fragrance; foliage glossy; vigorous, bushy growth; [Club X Lilli Marlene]; Gaujard

'Gala', F, pb, 1973; bud short, pointed; flowers light pink, dbl., 28–34 petals, 3–3.5 in., exhibition form, slight spicy fragrance; vigorous, upright, free growth; [Seedling No. 19-64 ps X Seventeen]; Jelly; E.G. Hill Co.

Gala *see* 'SAVagala'

Gala Charles Aznavour *see* MEIsazy

'Gala Day', F, or, 1966; flowers vermilion-scarlet, pointed, 4 in., slight fragrance; foliage light green; free, upright growth; [Queen Elizabeth X Dicksons Flame]; Watkins Roses

Gala Gold *see* 'LAVchro'

Gala Sunrise *see* 'GELrise'

'Galah', HT, mp, 1956; flowers carmine-pink, base lighter, large, borne in clusters, semi-dbl., moderate fragrance; Riethmuller

'Galahad', HT, w, 1986; flowers very large, dbl., moderate, anise fragrance; foliage medium green, semi-glossy; [Micaela X Lara]; Kriloff, Michel

'GALATEA', HMsk, yb, 1914; flowers stone-color, edged pink, small rosette, borne in clusters, recurrent bloom; Pemberton

Galaty *see* 'DOTSUBEBE'

Galaxy® *see* 'MORGAL'

'GALAXY', HWich, dp, 1906; flowers bright carmine; vigorous growth; Walsh

Galaxy *see* MEIhuterb

Galejade, S, op, 1997; Reuter

Galia® *see* 'MEITINIROL'

'GALIA', HT, lp, 1966; flowers light pink, center darker, large, very dbl., slight fragrance; foliage dark, leathery; very vigorous, upright growth; RULED EXTINCT 4/81 ARM; Betzel

Galileo *see* 'MEIGALIL'

Galina, HT, mr, 1979; flowers large, dbl., slight fragrance; Cocker; (Sangerhausen)

Galina, F, ab; flowers apricot and yellow, medium, dbl., slight fragrance; VEG; (Sangerhausen)

'GALLAGHER', F, rb, 1979; bud pointed; flowers cream, edged crimson, shapely, dbl., 41 petals, 2.5 in., moderate, fruity fragrance; foliage leathery; bushy growth; Murray, Nola

'GALLANT', F, mr, 1968; flowers scarlet, dbl., 3–3.5 in., blooms in clusters, moderate fragrance; foliage glossy; [Tropicana X Barbecue]; Dickson, A.

Gallantry, HT, pb; Delbard

'GALLERIA', HRg, mr, 1990; (Big John); bud ovoid; flowers medium watermelon pink, reverse silvery pink, large, semi-dbl., 13 petals, exhibition form, borne in sprays of 3-5, moderate, spicy, fruity fragrance; round, small, green yellow fruit; prickles straight, medium, light green to pink; foliage large, dark green, glossy; bushy, tall (8 ft) growth; [The Duke X Hansa]; Weddle, Von C.; Hortico Roses, 1990

'GALLERIA BORGHESE', HT, w, 1954; flowers flesh streaked coral, very large; foliage glossy; strong stems; [Peace X Crimson Glory]; Giacomasso

Gallica Grandiflora *see* **'ALIKA'**

'GALLICA MACRANTHA', Misc OGR, w; (R. gallica macrantha, R. macrantha, R. X waitziana macrantha); flowers flushed rose at first, changing to nearly white

Gallica Maheca *see* **'LA BELLE SULTANE'**

'GALLI-CURCI', HT, my, 1924; flowers golden yellow; [Columbia sport]; Kinsman

'GALLIVARDA', HT, rb, 1977; (Galsar); bud long, pointed; flowers red, yellow reverse, dbl., 34 petals, 4.5 in., exhibition form, slight fragrance; foliage glossy; vigorous, upright growth; [Colour Wonder X Wiener Charme]; Kordes, W. Söhne; Willemse

Galsar *see* **'GALLIVARDA'**

Galway Bay® *see* 'MACBA'

Gambler *see* 'GELLOR'

'GAMIN DE PARIS', F, dr, 1964; flowers dark blood-red, large, borne in large clusters, dbl.; vigorous, upright growth; [(Orange Triumph X Paprika) X Unnamed seedling]; Mondial Roses

'GAMINE', F, op, 1961; flowers salmon-pink, medium, dbl., moderate fragrance; vigorous growth; [Eclipse X Baccará]; Kriloff, Michel; Verbeek

'GAMMA', HT, pb, 1972; bud pointed; flowers pink, suffused vermilion, intense fragrance; foliage large; [Jouvencelle X American Heritage]; Gaujard

'GAMUSIN', HT, r, 1960; flowers cinnamon to pale pink, dbl., 25 petals, intense fragrance; vigorous, spreading growth; [Grey Pearl X (Lila Vidri X Prelude)]; Dot, Pedro

GANa, F, mp, 1991; (**David Whitfield**); Gandy

GANcre, LCl, w, 1998; (**Crème de la Crème**); Gandy

'GANDOL', HT, my, 1989; (**Doctor Goldberg**, 'GANGO'); flowers large, dbl., 26–40 petals, intense fragrance; foliage large, dark green, matt; upright growth; [Royal Dane X Dutch Gold]; Gandy, Douglas L.; Gandy Roses, Ltd., 1988

'GANDRI', F, pb, 1982; (**Betty Driver**); flowers pale peach and gold blend, yellow stamens, patio, large blooms, semi-dbl., slight fragrance; foliage medium, light green, semi-glossy; low, bushy growth; [Seedling X Topsi]; Gandy, Douglas L.; Gandy Roses, Ltd.

'GANFER', F, ob, 1987; (**Fergie**); flowers large, dbl., slight fragrance; foliage medium, medium green, semi-glossy, disease-resistant; patio; bushy growth; frost-proof; [Unnamed seedling X Copper Pot]; Gandy, Douglas L., 1988

'GANFLO', S, w, 1989; (**Florence Nightingale**); bud pointed; flowers glowing white, flushed buff, reverse tinged pink, aging to white, dbl., 32 petals, moderate, spicy fragrance; rounded, green fruit; prickles very pointed, fawn; foliage medium, medium green, semi-glossy; spreading, medium growth; [Morgengruss X Unnamed seedling]; Gandy, Douglas L.; Gandy Roses, Ltd.

'GANGA', HT, dy, 1970; flowers deep golden yellow, medium, dbl., exhibition form, profuse, intermittent bloom, moderate, tea fragrance; vigorous, upright growth; [Sabina X Unknown]; Division of Vegetable Crops and Floriculture; Indian Agric. Research Inst.

'GANGO', HT, my, 1989; (**Doctor Goldberg**, 'GANDOL'); flowers large, dbl., 26–40 petals, intense fragrance; foliage large, dark green, matt; upright growth; [Royal Dane X Dutch Gold]; Gandy, Douglas L.; Gandy Roses, Ltd., 1988

'GANHOL', HT, ob, 1984; (**Moriah**); flowers creamy yellow, orange reverse, large, dbl., 35 petals, intense fragrance; foliage large, dark, matt; bushy growth; [Fragrant Cloud X Seedling]; Holtzman, Arnold; Gandy Roses, Ltd.

GANjil, F, op, 1998; (**Jill's Rose**); Gandy

'GANSPA', F, lp, 1994; (**Spangles**); flowers pale pink speckled, blooms borne in large clusters, dbl., 15–25 petals, 1.5–2.75 in., moderate fragrance; many prickles; foliage medium, light green, semi-glossy; medium, upright, compact growth; [Florence Nightingale X Silver Jubilee]; Gandy, Douglas L.; Gandy Roses, Ltd., 1995

'GANYMED', F, dr, 1975; bud pointed; dbl., 32 petals, 2.5 in., cupped, slight fragrance; foliage glossy; vigorous, upright, bushy growth; [Europeana X Seedling]; Kordes

'GARDEJÄGER GRATZFELD', Cl Pol, dp, 1940; flowers carmine-red; vigorous growth; [Rodhatte sport]; Gratzfeld

Garden Blanket™ *see* 'JACGRON'

Garden City *see* 'HARKOVER'

Garden Club, HT; Mansuino; (Cavriglia)

'GARDEN DELIGHT', F, dp, 1956; flowers deep rose-pink, rosette form, borne in clusters, dbl., 34 petals, 3.5–4 in., moderate fragrance; vigorous, branching growth; Norman; Harkness

'GARDEN GEM', HT, lp, 1930; flowers satiny pink, dbl.; very vigorous growth; [Mrs E.T. Stotesbury X Hill's America]; Dingee & Conard

'GARDEN GLOW', HT, op, 1937; flowers scarlet, base copper, dbl.; foliage glossy, bronze; very vigorous, bushy growth; Cant, B. R.

Garden Magic *see* **Gartenzauber**®

'GARDEN NEWS', HT, dr, 1962; bud pointed; flowers dark crimson-scarlet, large, dbl., 32 petals, intense fragrance; foliage dull, dark, leathery; strong stems; moderate growth; [New Yorker X Étoile de Hollande]; Verschuren; Blaby Rose Gardens

Garden News *see* POUlrim

'GARDEN PARTY'®, HT, w, 1959; bud urn-shaped; flowers pale yellow to white, often tinged light pink, dbl., 28 petals, 4–5 in., exhibition form, slight fragrance; foliage large, medium green, semi-glossy; vigorous, bushy, well-branched growth; AARS, 1960 GM, Bagatelle, 1959; [Charlotte Armstrong X Peace]; Swim, H.C.; Armstrong Nursery

Garden Party *see* KORmollis

'GARDEN PARTY, CLIMBING', Cl HT, w, 1964; Itami Rose Nursery

Garden Pavilion Roos, HT, op, 1998

Garden Pearl, HT, w, 1992; Tejganga

Garden Perfume *see* KORsenter

'GARDEN PRINCESS', F, my, 1961; flowers yellow, becoming lighter, semi-dbl.; growth moderate; [Goldilocks X Lavender Pinocchio]; Leenders, J.

'GARDEN QUEEN', HT, op, 1960; flowers pink to salmon, dbl., 4 in., intense fragrance; vigorous growth; [Ambassadeur Nemry X Tawny Gold]; Leenders, J.

Garden State *see* 'MEIGENE'

Garden State II, Gr, or, 1997; Williams, J. Benjamin

'GARDEN SUPREME', F, ob, 1959; flowers orange blend shaded reddish, medium, borne in clusters, dbl., 32 petals; foliage leathery, bronze; low, vigorous growth; Jones; Hennessey

'GARDENER'S SUNDAY', F, my, 1975; flowers, dbl., 20 petals, 3 in., moderate fragrance; foliage bright green; [(Pink Parfait X Masquerade) X Arthur Bell]; Harkness

'GARDENIA', HWich, w, 1899; bud pointed, yellow; flowers creamy white, center yellow, well-formed blooms in small sprays; foliage small, dark, glossy; short, strong stems; very vigorous growth; [R. wichurana X Perle des Jardins]; Manda, W.A.

'GARDENIAEFLORA', HMult, w, 1901; flowers pure white, semi-dbl., early bloom, slight fragrance; Benary

Gardens of the World *see* 'JACCOEUR'

'GARETH DAVIES', HT, mr, 1997; flowers full, large, very dbl., 26–40 petals, borne mostly singly, moderate fragrance; foliage medium, dark green, semi-glossy; upright, bushy, tall growth; [Crimson Glory X Loving Memory]; Poole, Lionel

Garibaldi, B, dp; flowers rosy lilac; Damaizin, 1859

'GARISENDA', HWich, pb, 1911; flowers clear rose-pink, tinted silvery, blooms in clusters, dbl.; [R. wichurana X Souv. de la Malmaison]; Bonfiglio, A.

'GARNETTE', F, dr, 1951; (Garnette Red, Red Garnette); flowers garnet-red, base light lemon-yellow, small, dbl., 50 petals, slight fragrance; foliage leathery, dark; bushy growth; [(Rosenelfe X Eva) X Heros]; Tantau; J&P

Garnette Apricot, Pol, ab

Garnette Carol *see* **'CAROL AMLING'**

Garnette Pink *see* **'CAROL AMLING'**

Garnette Red *see* **'GARNETTE'**

Garnette Rose, Pol, dp

'GARNETTE SUPREME', F, dp, 1954; bud ovoid; flowers carmine, dbl., 35–40 petals, 2.5–3 in., cupped, moderate fragrance; foliage glossy, bronze; vigorous, upright, compact growth; [Yellow Pinocchio seedling X Garnette]; Boerner; J&P

Garnette White, Pol, w

Garnette Yellow, Pol, my

'GARNETTE, CLIMBING', Cl F, dr, 1954; Soria; Amling-DeVor Nursery

'GARNIA', F, dp, 1970; flowers deep pink, full, dbl., 48 petals, 4.5 in., intense fragrance; foliage matt, green; moderately vigorous growth; [Lady Sylvia X Garnette]; Butter; Wood End Gardens

'GARO', LCl, dr, 1986; flowers blackish-red, large, borne 1-5 per cluster, very dbl., 48 petals, exhibition form, slight fragrance; large, brown prickles; foliage red turning dark green, semi-glossy; upright growth; [Uncle Walter X Unnamed seedling]; Garelja, Anita, 1978

Garry Brown *see* 'MEHBROWN'

Gartenarchitekt Gunther Schulze®, S, ly

Gartenblut®, HT, dr, 1986; Noack, Werner

'GARTENDIREKTOR GLOCKER', F, or, 1957; bud ovoid; flowers cinnabar-red, large, borne in clusters, very dbl., moderate fragrance; foliage glossy, leathery; vigorous, bushy growth; [Obergärtner Wiebicke X Independence]; Kordes

'GARTENDIREKTOR JULIUS SCHUTZE', HT, pb, 1920; flowers pale rosy pink and peach-blossom-pink; [Mme Jules Gravereaux X Pharisaer]; Kiese

'GARTENDIREKTOR NOSE', HT, dr, 1930; bud pointed; flowers dark crimson, large, dbl., exhibition form, moderate fragrance; foliage dark, glossy; vigorous, bushy growth; [Royal Red X Templar]; Kordes; Dreer;, H&S

'GARTENDIREKTOR OTTO LINNE', S, dp, 1934; flowers dark carmine-pink, edged darker, base yellowish-white, dbl., borne in clusters of up to 30; foliage leathery, light green; long strong stems; vigorous, bushy growth; [Robin Hood (HMsk) X Rudolph Kluis]; Lambert, P.

Gartengold, HT, dy, 1984; Noack, Werner

'GARTENSTADT LIEGNITZ', HMult, m, 1910; flowers violet/red, medium, semi-dbl.; Lambert, P.; (Sangerhausen)

'GARTENSTOLZ', F, op, 1945; flowers rose tinted salmon, large, borne in clusters of 12-15, single, 8–10 petals, moderate fragrance; foliage leathery, light green; vigorous, upright, bushy growth; [Swantje X Hamburg]; Tantau

Gartenstolz, HT, op, 1974; Noack, Werner

Gartenzauber® *see* 'KORNACHO'

'GARTENZAUBER', F, 1961; (Garden Magic, Magie des Jardins); flowers blood-red, tinted cinnabar-red, well formed, large; low growth; RULED EXTINCT 6/81 ARM; Kordes, R.

Gartenzauber '84 *see* 'KORNACHO'

Gärtnerfreude, Pol, or, 1965; flowers medium, dbl.; Kordes, W. Söhne; (Sangerhausen)

'GARVEY', HT, pb, 1961; flowers light geranium, reverse pale red, dbl., 30 petals, 6 in., globular, moderate fragrance; foliage dark, leathery; strong stems; vigorous, upright growth; [McGredy's Yellow X Karl Herbst]; McGredy, Sam IV; McGredy

Gary Lineker *see* 'PEAROBIN'

'GARY PLAYER', HT, ob, 1968; (Goliath); flowers orange-vermilion, dbl., 35 petals, 4–4.5 in., exhibition form; foliage glossy, dark; vigorous growth; [Jolie Madame X Unnamed seedling]; Herholdt, J.A.

'GARY WERNETT', HT, op, 1985; flowers medium coral pink, large blooms borne usually singly, dbl., exhibition form, moderate fragrance; rare to full term fruit; medium, triangular, light red prickles; foliage medium, medium green, matt; bushy growth; [Helen Traubel X Helen Traubel]; French, Richard

'GASPARD MONGE', C, m, 1854; flowers light violet/red, medium, dbl.; Robert; (Sangerhausen)

'GASTON CHANDON', HT, pb, 1884; flowers light pink with coppery highlights, medium, dbl.; Schwartz; (Sangerhausen)

'GASTON LESIEUR', HWich, dp, 1915; flowers carmine-pink, medium, dbl.; Turbat; (Sangerhausen)

'GATESHEAD FESTIVAL', HT, op, 1989; flowers glowing orange flushed salmon with gold at base of petals, dbl., 26–40 petals, intense fragrance; foliage large, dark green, glossy; bushy growth; [Doris Tysterman X Silver Jubilee]; Thompson, Robert; Battersby Roses, 1989

'GAUBIROC', HT, op, 1986; (**Pénélope**®); flowers medium salmon pink, well-formed, large, dbl., 35 petals, moderate fragrance; foliage large, medium green, semi-glossy; upright growth; [(Americana X ?) X Chenonceaux]; Gaujard, Jean; Roseraies Gaujard, 1980

GAUbor, HT, ab; (**Barbara**®); Gaujard

'GAUCOVA', HT, or, 1978; (**Junon**®); bud full; flowers medium, dbl., 45 petals, 3 in.; foliage large, brownish; bushy growth; [Tanagra X Dora]; Gaujard

'GAUDENGI', HT, m, 1981; ('GAUTARA', **Méduse**®); flowers lavender red, dbl., 35 petals, moderate fragrance;

foliage large, dark, semi-glossy; upright growth; [Chenonceaux X Tropicana]; Gaujard, Jean; Roseraies Gaujard, 1980

Gaudi, HT; Dot; (Cavriglia)

'GAUDIA', F, mp, 1946; flowers rose-pink, base gold and salmon, borne in clusters, semi-dbl., 15 petals, 4 in., moderate fragrance; foliage bright green edged red; vigorous, tall growth; [Florentina X Talisman]; Leenders, M.; Longley

'GAUDINO', HT, rb, 1945; (**Renaissance**®); bud pointed; flowers red and gold, medium, dbl., slight fragrance; foliage glossy; low growth; RULED EXTINCT 10/86; Gaujard

GAUesca, HT, dp; (**Scala**); Gaujard

'GAUFRARNER', HT, op, 1977; (**Odeon**); bud globular; flowers coral-pink, dbl., 50 petals, 3 in., globular, moderate fragrance; foliage large, dark; vigorous growth; [(Chateau de Chenonceaux X Mignonne) X Americana]; Gaujard

GAUhari, HT, pb, 1980; (**Amara**); Gaujard

'GAUHTI', HT, yb, 1969; (**Antigone**®); bud pointed; flowers yellow shaded red, large, dbl., moderate fragrance; foliage light green, soft; vigorous, upright growth; GM, Bagatelle, 1967; [Rose Gaujard X Guitare]; Gaujard

Gaujard 985, HT, mr, 1974; flowers large, dbl.; Gaujard; (Sangerhausen)

'GAULIMOR', HT, op, 1981; (**Hermione**); flowers deep salmon, large, dbl., 35 petals, moderate fragrance; foliage large, dark, glossy; upright, bushy growth; [Rose Gaujard X Colour Wonder]; Gaujard, Jean; Roseraies Gaujard

'GAUMO', HT, rb, 1957; (**Rose Gaujard**®); flowers cherry-red, reverse pale pink and silvery white, dbl., 80 petals, 3–4 in., exhibition form, slight fragrance; foliage leathery, glossy; vigorous, bushy growth; GM, NRS, 1958; [Peace X Opera seedling]; Gaujard; Armstrong Nursery, 1964

'GAUMOVA', F, or, 1978; (**Circé**); bud long; flowers brilliant orange-red, semi-dbl., 16 petals, 1.5–2 in.; foliage large, dark; vigorous, compact growth; [Guitare X Prominent]; Gaujard

Gauntlet *see* 'BRIGAUNT'

'GAURA', F, ob, 1957; (**Vendéme**); bud long; flowers bright salmon, medium, dbl., moderate fragrance; foliage dark, glossy; bushy growth; [Comtesse Vandal X (Fashion X Vogue)]; Gaujard

GAUrama, F, or; (**Arbelle**); Gaujard

'GAUSECA', HT, pb, 1980; ('GAUZECA', **Montreal**®); bud long; flowers cream and pink blend, large, dbl., 45 petals, moderate fragrance; green prickles; foliage large, dark; [Americana X Dora]; Gaujard, Jean

'GAUTARA', HT, m, 1981; ('GAUDENGI', **Méduse**®); flowers lavender red, dbl., 35 petals, moderate fragrance; foliage large, dark, semi-glossy; upright growth; [Chenonceaux X Tropicana]; Gaujard, Jean; Roseraies Gaujard, 1980

GAUtara, HT, ob, 1981; (**Simone Merieux**®); Gaujard

'GAUTIRA', HT, or, 1981; (**Molitor**®); flowers deep orange-red, large, dbl., 35 petals, moderate fragrance; foliage medium, dark, semi-glossy; [Junon X Tanagra]; Gaujard, Jean

GAUvera, HT, mr; (**Serenité**); Gaujard

GAUvila, HT, 1979; (**Renee Columb**); Gaujard; (Cavriglia)

'GAUVITOR', HT, mr, 1977; (**Jean Gaujard**®); flowers brilliant red, dbl., 40 petals; vigorous growth; [Canasta X Rose Gaujard]; Gaujard

'GAUZECA', HT, pb, 1980; ('GAUSECA', **Montreal**®); bud long; flowers cream and pink blend, large, dbl., 45 petals, moderate fragrance; green prickles; foliage large, dark; [Americana X Dora]; Gaujard, Jean

GAUzimi, HT, mr; (**Lisima**); Gaujard

'GAUZINE', F, mp, 1975; (**Dady**®);, intense fragrance; [Mignonne X Seedling]; Gaujard

'GAUZOMI', F, or, 1979; (**Odyssée**); flowers well-formed blooms borne 3-5 per cluster, dbl., 25 petals, moderate fragrance; small, brown prickles; foliage dark; [Pampa X Unnamed seedling]; Gaujard, Jean

'GAVÁ', Cl HT, or, 1934; flowers oriental red shaded rose-pink, base yellow, very large, dbl., cupped, intense fragrance; foliage leathery; very vigorous, climbing growth; [Souv. de Claudius Denoyel X Souv. de Claudius Pernet]; Munné, B.; Camprubi

'GAVAL', F, or, 1959; (**Atlantic**®); flowers medium, semi-dbl., slight fragrance; foliage glossy, dark; [Peace X Seedling]; Gaujard

Gavina, HT, w; Dot

Gavno *see* 'POULGAV'

Gavolda *see* 'MEICRI'

'GAVOTTE', HT, pb, 1963; flowers pink, reverse light yellow, dbl., 45 petals, 5 in., moderate fragrance; foliage dark; vigorous, upright growth; [Ethel Sanday X Lady Sylvia]; Sanday, John

'GAVROCHE', F, ob, 1963; flowers orange, center yellow, large; vigorous, bushy growth; Robichon

Gavroche™ *see* 'MEIRIENTAL'

'GAXENCE', HT, m, 1962; (**Éminence**®); flowers lavender, large, dbl., 40 petals, intense fragrance; foliage leathery, light green; vigorous, upright growth; [Peace X (Viola X Seedling)]; Gaujard; Ilgenfritz Nursery, 1965

'GAY CRUSADER', HT, rb, 1948; flowers red fading to pink, reverse deep yellow, large, exhibition form, moderate fragrance; foliage dark; [Phyllis Gold X Catalonia]; Robinson, H.; Baker's Nursery

'GAY DAWN', HT, op, 1958; bud pointed; flowers large, dbl., cupped, moderate, spicy fragrance; foliage dark, semi-glossy; very vigorous, upright growth; [Eclipse X Mme Henri Guillot]; Taylor, C.A.; California Nursery Co.

'GAY DEBUTANTE', HT, pb, 1960; flowers light pink, base yellow, very large, dbl., 40–45 petals, cupped, slight fragrance; foliage leathery, glossy; vigorous, upright growth; [Peace sport]; Curtis, R.F.; C.R. Burr

'GAY DICKY', F, ob, 1956; flowers tangerine-orange; moderate growth; Verschuren-Pechtold; Gandy Roses, Ltd.

'GAY GOLD', HT, my, 1973; flowers, dbl., 30 petals, 5.5 in., intense fragrance; foliage glossy; [King's Ransom X Piccadilly]; Lowe

'GAY GORDONS'®, HT, yb, 1969; flowers orange-yellow and red, dbl., slight fragrance; foliage dark, glossy; bushy, rather low growth; [Belle Blonde X Karl Herbst]; Cocker

'GAY GYPSY', HT, dr, 1949; bud long, pointed; flowers oxblood-red shades maroon, open, semi-dbl., 15–20 petals, 4.5–5 in., cupped, slight fragrance; foliage leathery; vigorous, upright, bushy growth; [Charles K. Douglas sport]; Crane; Bosley Nursery

'GAY HEART', F, mp, 1951; bud ovoid; flowers bright pink, large, borne in large clusters, dbl., 25 petals, exhibition form, moderate fragrance; foliage leathery; vigorous, upright growth; [Joanna Hill X World's Fair]; Boerner; J&P

'GAY JEWEL', Min, lp, 1958; bud globular; flowers light rose-pink, dbl., 35–40 petals, .5 in., cupped, moderate fragrance; foliage glossy; compact (6-8 in) growth; [Dick Koster sport X Tom Thumb]; Morey, Dr. Dennison; J&P

'GAY LADY', HT, mr, 1953; bud ovoid; flowers currant-red, open, dbl., 20–28 petals, 3.5–4.5 in., moderate, spicy fragrance; foliage dark, leathery, glossy; very vigorous, upright growth; [Charlotte Armstrong X Piccadilly]; Swim, H.C.; Breedlove Nursery

'GAY LYRIC', HT, mp, 1971; flowers rose-pink, large, very dbl., exhibition form, moderate fragrance; foliage glossy, dark, leathery; vigorous, upright growth; [Royal Highness X Elizabeth Fankhauser]; Fankhauser

'GAY MAID', F, or, 1969; flowers red suffused orange-pink, borne in trusses, dbl., 26 petals, globular; foliage light green; very vigorous growth; [Masquerade X ?]; Gregory

'GAY MOOD', LCl, dp, 1940; bud large, ovoid to urn shaped, rose-red; flowers deep rose-pink, open, semi-dbl., 15–25 petals, profuse, repeated bloom; foliage glossy, dark; very vigorous, climbing growth; [Joanna Hill X Sanguinaire]; Lammerts, Dr. Walter; Armstrong Nursery

'GAY NINETIES', Cl F, mr, 1955; bud ovoid; flowers rose-red, blooms in clusters of 5-8, dbl., 65 petals, 2–2.5 in., intense fragrance; foliage leathery, glossy; vigorous, pillar (8 ft) growth; [(New Dawn X Red Ripples) X Red Ripples]; Sima

'GAY PARIS', HT, mr, 1960; bud long; flowers bright crimson, well-formed, large, dbl., moderate fragrance; foliage bright green; vigorous growth; [(Floradora X Barcelona) X (Charles Mallerin X Tonnerre)]; Delbard-Chabert

'GAY PRINCESS', F, lp, 1967; bud ovoid; flowers blush-pink, large blooms in clusters, dbl., cupped, moderate fragrance; foliage leathery; vigorous, upright, bushy growth; AARS, 1967; [Spartan X The Farmer's Wife]; Boerner; J&P

'GAY VISTA', S, lp, 1957; flowers very large blooms in very large clusters, single, repeat bloom; height 3 1/2 ft; Riethmuller

Gayness, HT, lp; Reithmuller, 1955; (Weatherly, L.)

'GAYTIME', F, rb, 1966; bud ovoid, pointed; flowers red and yellow, large, dbl., cupped, slight fragrance; foliage dark, glossy, leathery; vigorous, bushy, compact growth; [Unnamed seedling X Circus]; Armstrong, D.L.; Armstrong Nursery

'GAZELLE', HGal, lp; flowers delicate rose, large

G'DayGidday *see* 'TINHAT'

Ge Korsten *see* 'HERGEKO'

'GEAAURA', S, my, 1999; (**Precious Child**); flowers medium, dbl., 17–25 petals, borne in small clusters, slight fragrance; prickles moderate; foliage medium, dark green, semi-glossy; spreading (4 ft) growth; [Hokey Pokey X Patio Princess]; Gear, Ian Robert

'GEAAWARD', S, w, 1999; (**The Wedding Rose**); flowers medium, inner petals tinged lemon to white, opens somewhat quartered, very dbl., 41 petals, borne in small clusters, slight fragrance; prickles moderate; foliage dark blue/green, semi-glossy; upright, compact, bushy, medium (3 ft) growth; [Enchantment X English Miss]; Gear, Ian Robert

'GEACOT', S, ab, 1999; (**Apricot Kisses**); flowers dark apricot, aging to pale apricot, dbl., 26–40 petals, 2.5 in., borne in small clusters, slight, fruity fragrance; prickles moderate; foliage medium, medium green, new growth purple-red, semi-gl; spreading, bushy, medium (4-5 ft) growth; patio climber; [Laura Ford X seedling]; Gear, Ian Robert

'GÉANT DES BATAILLES', HP, mr, 1846; (Giant of Battles); flowers deep fiery crimson, dbl., 85 petals, intense fragrance; moderately vigorous growth; Nerard; Guillot Pere

'GEDGE'S GLORY' *see* WEBheart

Gee Dee, HT, mr, 1979; Dawson

Gee Gee™ *see* 'BENGEE'

'GEE WHIZ', S, yb, 1984; flowers yellow tinted orange-red, blooms borne 1-10 per cluster, dbl., 23 petals, cupped, repeat bloom, moderate, sweet fragrance; needle-like, brown prickles; foliage medium, leathery, dark olive green; low, bushy, free-branching growth; hardy; [Gingersnap X Sevilliana]; Buck, Dr. Griffith J.; Iowa State University

'GEELOVE', HT, mr, 1996; (**I Love You**); flowers bright red, large blooms borne mostly single, dbl., 26–40 petals, slight fragrance; some prickles; foliage medium, dark green, semi-glossy; upright, medium growth; Bees of Chester; L W Van Geest Farms, Ltd., 1995

'GEEPEOP', F, mp, 1997; (**The People's Princess**); flowers full, 28in., very dbl., 26–40 petals, borne in small clusters, slight fragrance; foliage medium, medium green, semi-glossy; upright, medium growth; [Seedling X Seedling]; Bees; L.W. Van Geest Farms Ltd., 1997

'GEESTRAW', Min, yb, 1996; (**Strawberries and Cream**); flowers light yellow, pink and red striped, small blooms borne in small clusters, dbl., 26–40 petals, slight fragrance; some prickles; foliage small, medium green, semi-glossy; compact, medium growth; Bees of Chester; L W Van Geest Farms, Ltd., 1995

'GEHEIMRAT DR METTWEG', S, pb, 1909; flowers rose-red, center yellowish white, large, borne in large clusters, recurrent bloom; foliage dark; vigorous, bushy growth; [(Mme Norbert Levavasseur X Trier) X R. foetida bicolor]; Lambert, P.

Geheimrat Duisberg *see* **'GOLDEN RAPTURE'**

'GEHEIMRAT DUISBERG, CLIMBING', Cl HT, dy, 1954; flowers large, dbl., moderate fragrance; Knackfuss; (Sangerhausen)

'GEHEIMRAT RICHARD WILLSTÄTTER', HT, ab, 1931; flowers apricot-yellow, veined carmine-red, stamens yellow, large, semi-dbl., slight fragrance; foliage bright, thick; vigorous growth; [Constance X Admiral Ward]; Felberg-Leclerc

'GEISHA'®, F, mp, 1964; (Pink Elizabeth Arden); bud long; semi-dbl., 2.5–3 in., blooms in clusters of 1-3; foliage dark; bushy, medium height.; Tantau, Math.

'GEISHA', HT, ob, 1920; bud orange, marked coral-red; flowers golden yellow; [Mme Edouard Herriot sport]; Van Rossem

'GEISHA', HMult, m, 1913; Geschwind, R.

'GEISHA GIRL', F, my, 1964; dbl., 25 petals, 3.5 in., blooms in clusters; foliage long, pointed; tall growth; [Gold Cup X McGredy's Yellow]; McGredy, Sam IV; McGredy

'GELA GNAU', HT, ab, 1926; flowers amber-yellow, reverse apricot, dbl., moderate fragrance; Leenders, M.

'GELA TEPELMANN', HMult, mp, 1950; flowers medium, dbl.; Tepelmann; (Sangerhausen)

'GELANCE', F, rb, 1999; (**Turbulance**); flowers large, single, 5–11 petals, borne in small clusters, slight fragrance; prickles moderate; foliage medium, light green, glossy; bushy, medium growth; [Pink Favorite X Scentimental]; Giles, Diann; Giles Rose Nursery, 1999

'GELANGEL', F. m, 1998; (**Silver Angel**); flowers light silver mauve, much lighter than Angel Face, double, (15–25 petals), medium blooms borne in small clusters; fragrant; foliage medium green, semi-glossy; bushy, medium (3ft) growth; [Angel Face sport]; Giles, Diann; Giles Rose Nursery, 1966

'GELBACK', HT, ly, 1995; (**Back Home**); flowers light yellow, full (26–40 petals), medium (4–7 cms) blooms borne mostly single; very fragrant; few prickles; foliage medium, light green, semi-glossy; medium, spreading growth; [Dr A. J. Verhage X Midas Touch]; Giles, Diann; Giles Rose Nursery, 1995

Gelbe Dagmar Hastrup *see* 'MORYELRUG'

Gelbe Florida von Scharbeutz, F; (Cavriglia)

'GELBE HOLSTEIN', F, ly, 1951; (Yellow Holstein); bud long, pointed; flowers yellow paling to lemon, blooms in large clusters, dbl., 20 petals, 3 in., slight fragrance; foliage glossy, light green; vigorous, upright, bushy growth; [(Eva X Viscountess Charlemont) X Sunmist]; Kordes; Wheatcroft Bros.

'GELBE PHARISÄER', HT, my, 1927; flowers clear yellow, center deeper, moderate fragrance; [Pharisaer X Mrs Aaron Ward]; Hinner, W.

Gelber Kobold, MinFl, dy, 1996

'GELBIRD', LCl, mr, 1997; (**Red Bird**); flowers full, medium, very dbl., 26–40 petals, borne mostly singly, slight fragrance; foliage medium, medium green, semi-glossy; spreading, medium (6-8ft.) growth; [Crepe de Chine X Kardinal]; Giles, Diann

'GELBURN', F, ob, 1997; (**Sunburnt**) flowers double, small, dbl., 15–25 petals, borne in small clusters, no fragrance; foliage medium, medium green, semi-glossy; spreading, medium (3.5ft.) growth; [Dalton X Select Pollen]; Giles, Diann

'GELCOURT', Min, rb, 1999; (**Courtney**); flowers red, reverse gold, dbl., 17–25 petals, 1.5 in., borne in small clusters, no fragrance; many prickles; foliage medium, dark green, glossy; spreading, low (1.5 ft) growth; [Princess Celest X Rainbow's End]; Giles, Diann; Giles Rose Nursery, 1999

'GELDAVIS', HT, pb, 1998; (**Ken Davis**); flowers light pink with darker edges, medium size, dbl., 15–25 petals, borne mostly singly, no fragrance; prickles medium; foliage medium, light green, dull; upright, medium (4 ft.) growth; [Bride's Dream sport]; Davis, Ken; Giles Rose Nursery, 1995

'GELDAY', S, mp, 1997; (**Sunday China**); flowers small, 1 in., borne in large clusters, no fragrance; foliage small, medium green, dull; compact, low (2ft.6in.)growth; [Champney's Pink Cluster X Seedling]; Giles, Kevin

'GELEASY', F, op, 1999; (**Easy Vibes**); flowers medium, dbl., 17–25 petals, borne in small clusters, slight fragrance; few prickles; foliage medium, dark green, glossy; compact, medium growth; [Sun Flare X Easy Living]; Giles, Diann

'GELELE', Min, lp, 1999; (**Dear Eleanor**); dbl., 17–25 petals, borne mostly singly, slight fragrance; few prickles; foliage medium, dark green, dull; upright, medium growth; [Party Girl X Snow Bride]; Giles, Diann; Giles Rose Nursery, 1999

'GELFLA', F, my, 1997; (**Florida Sun**); flowers full, small, very dbl., 26–40 petals, borne in small clusters, slight fragrance; foliage small, medium green, glossy; compost (2.5-3ft.) growth; [Sun Flare X Select Pollen]; Giles, Diann

'GELFLY', F, my, 1999; (**Butterfly Kisses**); flowers medium, single, 5–11 petals, borne in small clusters, moderate fragrance; few prickles; foliage medium, medium green, semi-glossy; upright, spreading, medium growth; [Sun Flare X Summer Snow]; Giles, Diann; Giles Rose Nursery, 1999

'GELGOOD', F, mp, 1999; (**Good Vibes**); flowers medium, dbl., 17–25 petals, borne in large clusters, slight fragrance; few prickles; foliage medium, medium green, glossy; compact, medium growth; [Sun Flare X Simplicity]; Giles, Diann; Giles Rose Nursery, 1999

'GELGRAPE', Min, m, 1999; (**Grape Delight**); flowers small, semi-dbl., 12–16 petals, borne in small clusters, slight fragrance; many prickles; foliage medium, medium green, semi-glossy; compact, low (1.5 ft) growth; [Lavender Sweetheart X Herbie]; Giles, Diann; Giles Rose Nursery, 1999

'GELJAM', HT, mr, 1999; (**Jamie's Love**); flowers medium, dbl., 26–40 petals, borne mostly singly, slight fragrance; few prickles; foliage medium, dark green, dull; upright, medium growth; [Vera Dalton X Special Merit]; Giles, Diann; Giles Rose Nursery, 1999

'GELKEN', HT, mr, 1996; (**Kenny's Rose**); dbl., 26–40 petals, 3–3.5 in., borne mostly singly, moderate fragrance; some prickles; foliage medium, dark green, semi-glossy; medium, spreading, sprawling growth; [Swarthmore X Mister Lincoln]; Giles, Kenneth; Giles Rose Nursery, 1995

'GELLANDO', Min, rb, 1999; (**Rose Magic**); flowers white and red edge, white reverse, medium, dbl., 26–40 petals, borne in small clusters, slight fragrance; few prickles; foliage medium, dark green, semi-glossy; upright, medium growth; [Little Darling X Kristen]; Giles, Diann; Giles Rose Nursery, 1999

'GELLITE', Min, w, 1998; (**Tiffany Lite**); flowers large, white, double, dbl., 15–25 petals, borne in small clusters, slight fragrance; foliage medium, medium green, semi-glossy; upright, bushy, 3-4ft growth; [Tiffany Lynn sport]; Giles, Diann; Giles Rose Nursery, 1997

'GELLOV', HT, w, 1995; (**Gambler**); flowers white, full (26–40 petals), large (7 cms) blooms borne mostly single; slight fragrance; few prickles; foliage large medium green, semi-glossy; tall (5 ft), upright growth; [Grace de Monaco X Seeding]; Gimer, Louis; Giles Nursery, 1995

'GELLYNN', HT, rb, 1999; (**Sierra Lynn**); flowers medium, dbl., 26–40 petals, borne mostly singly, slight fragrance; few prickles; foliage medium, dark green, semi-glossy; upright, medium (4-5 ft) growth; [Vera Dalton X Paradise]; Giles, Diann; Giles Rose Nursery, 1999

'GELMAGIC', F, pb, 1997; (**Magic Baby**); flowers full, medium, very dbl., 26–40 petals, borne mostly singly, no fragrance; foliage medium, medium green, dull; upright, medium (3.5ft.)growth; Giles, Diann

'GELMAN', Min, mr, 1997; (**Chess Man**); flowers full, medium, very dbl., 26–40 petals, borne mostly singly, no fragrance; foliage medium, dark green, dull; upright, medium (3ft.)growth; [Chrysler Imperail X Select Pollen]; Giles, Kevin

'GELMAR', Min, mr, 1995 (**Marvie**); flowers medium red, full (26–40 petals), small (0–4 cms) blooms borne mostly single; slight fragrance; few prickles; foliage small, medium green, semi-glossy; medium upright growth; [Winsome X Seedling]; Giles, Diann; Giles Rose Nursery, 1995

'GELMAUD', Min, mp, 1999; (**Mary Maud**); flowers medium, dbl., 17–25 petals, borne in small clusters, slight fragrance; prickles moderate; foliage medium, medium green, semi-glossy; upright, medium growth; [Little Darling X Unknown]; Giles, Diann; Giles Rose Nursery, 1999

'GELMISS', HT, dp, 1996; (**Miss Mary**); flowers deep pink, blooms, dbl., 26–40 petals, 3–3.5 in., no fragrance; few prickles; foliage large, medium green, matt; tall (5 ft), upright growth; [Big Ben sport]; Miller, Carol; Giles Rose Nursery, 1995

'GELMOON', S, w, 1998; (**South Moon**); flowers pure white, yellow stamens, full, very dbl., 26–40 petals, borne in large clusters, intense fragrance; foliage medium, medium green, semi-glossy; upright, medium (4 ft.) growth; [Belle Story sport]; Giles, Diann; Giles Rose Nursery, 1997

'GELOUT', Min, m, 1999; (**Chill Out**); flowers large, dbl., 17–25 petals, borne mostly singly, no fragrance; few prickles; foliage large, medium green, semi-glossy; upright, tall growth; [Lavender Sweetheart X Herbie]; Giles, Diann; Giles Rose Nursery, 1999

'GELPAM', HT, ob, 1999; (**Pam's Passion**); flowers medium, dbl., 26–40 petals, borne mostly singly, slight fragrance; few prickles; foliage dark green, semi-glossy; upright, medium growth; [Vera Dalton X Unknown]; Giles, Diann; Giles Rose Nursery, 1999

'GELPEACH', HT, op, 1997; (**Just Peachy**); flowers double, small, dbl., 15–25 petals, borne mostly singly, no fragrance; foliage medium, dark green, glossy; upright, medium growth; [Vera Dalton X Select Pollen]; Giles, Diann

'GELPEP', Min, or, 1997; (**Chilli Pepper**); flowers full, medium, very dbl., 26–40 petals, borne in small clusters, no fragrance; foliage large, dark green, glossy; bushy, medium (1.5-2ft.)growth; [Vera Dalton X Select Pollen]; Giles, Diann

'GELPUZZLE', LCl, rb, 1999; (**Chinese Puzzle**); flowers small, dbl., 17–25 petals, borne in large clusters, slight fragrance; many prickles; foliage small, dark green, glossy; climbing, tall growth; [Vera Dalton X Roller Coaster]; Giles, Diann; Giles Rose Nursery, 1999

'GELRED', Min, op, 1995; (**Redneck Girl**); flowers orange pink with white base, full (26–40 petals); few prickles; foliage small, dark green, semi-glossy; spreading growth; [Seedling X Seedling]; Giles, Diann; Giles Rose Nursery, 1995

Gelria, F, 1974; Verbeek; (Cavriglia)

'GELRISE', F, rb, 1997; (**Gala Sunrise**); flowers full, medium, very dbl., 26–40 petals, borne in small clusters, no fragrance; foliage medium, dark green, dull; upright, medium (3.5ft.)growth; [Vera Dalton X Rainbow's end]; Giles, Diann

'GELRITE', F, my, 1996; (**Rite Brite**, Right Bright); flowers medium yellow, full (26–40 petals), small (0–4 cms) blooms; slight fragrance; few prickles; foliage small, medium green, matt; low (3½ ft.), spreading, compact growth; [Sun Flare X Rise 'n' Shine]; Giles, Diann; Giles Nursery, 1995

'GELSHAN', Min, pb, 1997; (**Shannie**); flowers full, medium, very dbl., 26–40 petals, borne mostly singly, no fragrance; foliage medium, medium green, semi-glossy; upright, medium (2.5ft.)growth; [Little Darling X Magic Carrousel]; Giles, Diann

'GELSKIES', Min, ab, 1997; (**Morning Skies**); flowers full, apricot blend, very dbl., 26–40 petals, borne mostly singly, no fragrance; foliage medium, medium green, dull; upright, low (1-1/2 ft) growth; [Little Darling X Rainbow's End]; Giles, Diann; Giles Rose Nursery, 1997

'GELSWEET', F, lp, 1999; (**Sweet Vibes**); flowers medium, dbl., 17–25 petals, borne in small clusters, slight fragrance; few prickles; foliage medium, medium green, glossy; compact, medium growth; [Sun Flare X Simplicity]; Giles, Diann; Giles Rose Nursery, 1999

Geltendorf, HT, 1971; Croix; (Cavriglia)

'GELTWO', HT, ab, 1999; (**Two Thumbs Up**); dbl., 26–40 petals, 5 in., borne mostly singly, slight fragrance; few prickles; foliage medium, medium green, dull; upright, medium (5 ft) growth; [South Seas X St Patrick]; Giles, Diann

'GELWIN', Min, w, 1997; (**Winter Princess**); flowers double, medium, dbl., 15–25 petals, borne in small clusters, no fragrance; foliage medium, medium green, semi-glossy; upright, medium (2.5ft.)growth; [Little Darling X Select Pollen]; Giles, Diann

'GEM', HT, mp, 1960; bud long, pointed; flowers deep soft pink, medium, semi-dbl., exhibition form, moderate fragrance; foliage soft; vigorous, upright growth; [Ena Harkness X Mme Butterfly]; Walker

'GEM OF THE PRAIRIES', HSet, dp, 1865; (Bijou des Prairies); flowers rosy red, occasionally blotched white, large blooms in large clusters, flat, non-recurrent, slight fragrance; vigorous growth; [Believed to be Queen of the Prairies X Mme Laffay]; Burgess, A.

'GEMINI', Pol, ob, 1967; bud pointed; flowers orange, small, dbl., exhibition form, slight fragrance; foliage dark, glossy; vigorous, upright, bushy growth; [Unnamed seedling X Rumba]; Hill, Joseph H., Co.

Gemini® *see* 'JACNEPAL'

Gemma, S, mp

'GEMSTONE', HT, mp, 1978; bud high-centered; dbl., 28 petals, 4.5–5 in., exhibition form, slight fragrance; foliage matt; vigorous, upright growth; [Helen Traubel X Swarthmore]; J&B Roses; Eastern Roses

Gén. Jacqueminot *see* **'GÉNÉRAL JACQUEMINOT'**

'GEN. JOHN PERSHING', LCl, dp, 1917; (F.R.M. Undritz); flowers large, dbl., 53 petals, moderate fragrance; vigorous, climbing growth; [Dr. W. Van Fleet X Mrs W.J. Grant]; Undritz

GENdee, S, lp; (**Pearly King**); Genesis

'GENE BOERNER', F, mp, 1968; bud ovoid; flowers deep pink, medium, dbl., 35 petals, exhibition form; foliage glossy; vigorous, upright growth; AARS, 1969; [Ginger X (Ma Perkins X Garnette Supreme)]; Boerner; J&P

Gene Jones F, W; (Angel's Blush)

Gene Sandberg *see* 'RESAND'

'GENERAAL SMUTS', HT, mr, 1922; flowers cherry-red, shaded deep coral-red, dbl., moderate fragrance; [Gen. MacArthur X Mme Edouard Herriot]; Van Rossem

'GENERAAL SNIJDERS', HT, dp, 1917; flowers deep carmine shaded coral-red, dbl., intense fragrance; [Mme Mélanie Soupert X George C. Waud]; Leenders, M.

'GENERAL ALLARD', B, dp, 1835; Laffay, M.

'GÉNÉRAL BARON BERGE', HP, mr, 1892; flowers red, center occasionally striped white, large, dbl., 50 petals; erect, vigorous growth; Pernet Père

'GENERAL BARRAL', HP, 1867; Damaizin; (Cavriglia)

'GÉNÉRAL BEDEAU', HP, pb, 1851; flowers pink and red, large, dbl.; Margottin; (Sangerhausen)

'GÉNÉRAL BERTHELOT', HT, dp, 1926; flowers dark pink, slightly streaked white, dbl.; [J.B. Clark X Farbenkonigin]; Walter, L.

'GENERAL BROWNE', HSpn, w; flowers blush-white, fading pure white, quilled, dbl., moderate fragrance; prickles twiggy, prickly growth; foliage dark

General Cavaignac, Pol, mr

'GÉNÉRAL CLERC', M, 1845; Laffay, M.; (Sangerhausen)

'GÉNÉRAL DE VAULGRENANT', HT, dp, 1926; flowers rose-pink, very dbl.; [Mme Henriette Schissele X Mme Adele Gance]; Walter, L.

'GENERAL DESAIX', HP, 1867; Moreau et Robert; (Cavriglia)

'GENERAL DOMINGOS DE OLIVEIRA', HT, ab, 1939; flowers yellow-apricot tinted flesh-pink, large, dbl., cupped, slight fragrance; foliage glossy; dwarf growth; [Frank Reader X Golden Gleam]; da Silva, Moreira

'GENERAL DON', HT, pb, 1919; flowers strawberry tinted coppery, base golden yellow, dbl., moderate fragrance; [Mme Mélanie Soupert X Louise Catherine Breslau]; Le Cornu

'GÉNÉRAL DONADIEU', HGal, mr; flowers purplish red, compact, very dbl.

'GÉNÉRAL DROUOT', mr, 1854; flowers carmine-red, medium, semi-dbl.; Robert; (Sangerhausen)

'GÉNÉRAL DROUST', M, m, 1847; flowers purplish crimson, medium, very dbl., recurrent bloom; vigorous growth; Vibert

'GÉNÉRAL DUC D'AUMALE', HP, dr, 1875; flowers large, dbl.; Verdier, E.; (Sangerhausen)

'GÉNÉRAL FETTER', HT, m, 1922; flowers carmine-purple, glossy, very dbl.; [Jonkheer J.L. Mock X Luise Lilia]; Walter, L.

'GÉNÉRAL GALLIÉNI', T, rb, 1899; flowers coppery red, cupped; vigorous growth; [Souv. de Therese Levet X Reine Emma des Pays-Bas]; Nabonnand, G.

'GÉNÉRAL GUISAN', HT, dr, 1945; flowers large, dbl.; Heizmann, E.; (Sangerhausen)

General Jack *see* **'GÉNÉRAL JACQUEMINOT'**

'GÉNÉRAL JACQUEMINOT', HP, rb, 1853; (Gén. Jacqueminot, General Jack, Jack Rose); bud scarlet-crimson; flowers dark red, whitish reverse, dbl., 27 petals, recurrent bloom, intense fragrance; foliage rich green; long, strong stems; vigorous, bushy growth; (28); [Probable seedling of Gloire des Rosomanes sport]; Roussel

'GÉNÉRAL KLÉBER', M, mp, 1856; bud well mossed; flowers pink tinted lilac; Robert

General Labutere, Ch, dp

'GENERAL MACARTHUR', HT, dp, 1905; flowers rose-red, dbl., 20 petals, intense, damask fragrance; foliage leathery; (28); E.G. Hill Co.

'GENERAL MACARTHUR, CLIMBING', Cl HT, dp, 1923; Dickson, H.

'GENERAL ROBERT E. LEE', T, my, 1896; bud deep orange-yellow; flowers canary-yellow; Good & Reese

'GÉNÉRAL SCHABLIKINE', T, op, 1878; flowers coppery; vigorous growth; Nabonnand, G.

'GENERÁL STEFÁNIK', HP, m, 1933; (Krásná Azurea); [La Brillante X Unnamed seedling]; Böhm, J.; J&P

'GÉNÉRAL TARTAS', T, dp; flowers deep rose, large, dbl.; Bernède

'GENERAL TESTARD', HWich, rb, 1918; flowers red, center white, small, bornein large clusters, semi-dbl.; Pajotin-Chédane

'GENERAL TH. PESCHKOFF', HT, lp, 1909; flowers very large, dbl., moderate fragrance; Ketten, Gebrüder; (Sangerhausen)

'GENERAL VON BOTHNIA-ANDREAE', HP, mr, 1900; flowers carmine-red, very large, dbl.; Verschuren; (Sangerhausen)

'GENERAL WASHINGTON', HP, dr, 1861; flowers deep crimson, reflexes maroon, large, very dbl., flat, occasionally recurrent bloom, moderate fragrance; moderate growth; [Triomphe de l'Exposition sport]; Granger

'GÉNÉRALE MARIE RAIEWSKY', HP, pb, 1911; flowers pink with dark yellow, large, dbl.; Ketten, Gebrüder; (Sangerhausen)

'GENERALIN ISENBART', HT, pb, 1915; flowers coppery-pink with dark yellow, large, dbl., moderate fragrance; Lambert, P.; (Sangerhausen)

'GENERALOBERST VON KLUCK', HT, dp, 1917; flowers carmine-pink, large, dbl., intense fragrance; Lambert, P.; (Sangerhausen)

'GENERAL-SUPERIOR ARNOLD JANSSEN', HT, dp, 1912; bud pointed; flowers deep rose-pink, veined darker, reverse much darker, large, dbl., moderate fragrance; [Farbenkonigin X Gen. MacArthur]; Leenders, M.

'GENERAL-SUPERIOR ARNOLD JANSSEN, CLIMBING', Cl HT, dp, 1931; Böhm, J.

Generosa *see* **'MANSUINO ROSE'**

'GENEROSITY', HT, ob, 1982; flowers cream with orange center, large, dbl., 35 petals, moderate fragrance; foliage large, dark, matt; bushy growth; [Fred Gibson X Lady Elgin]; Northfield, G.

'GENESIS', Min, m, 1991; bud ovoid; flowers lavender, reverse white, blooms borne singly, dbl., 45 petals, 2 in., moderate fragrance; foliage medium, medium green, semi-glossy; upright growth; [Lavender Jade X Angel Face]; Jolly, Marie, 1986; Rosehill Farm, 1992

'GENÈVE', HT, or, 1944; bud long; flowers salmon-carmine and capucine-red, high pointed, dbl., moderate fragrance; vigorous growth; [Charles P. Kilham X Mme Joseph Perraud]; Meilland, F.

Genevieve® *see* 'SAVAGEN'

Geneviève Genest *see* **'MRS SAM MCGREDY, CLIMBING'**

'GENEVIÈVE LE GOASTER', HT, w, 1923; flowers white, center pale rose, base salmon-rose; Carrette; Richardier

Genevieve Ros, S, lp; Peden, R., 1998; (Weatherly, L.)

'GENIUS MENDEL', HT, mr, 1935; bud pointed; flowers light fiery red to pure red, large, dbl., exhibition form; foliage glossy; bushy growth; [Mrs Henry Winnett X Sir David Davis]; Böhm, J.

GENpat, MinFl, op, 1995; (**Dazzler**); Genesis

GENpen, S, lp; (**Prince Regent**); Genesis

GENpink, S, dp; (**King William**); Genesis

GENruby, S, lp; (**King Richard**); Genesis

Gentiliana *see* **'POLYANTHA GRANDIFLORA'**

'GENTLE', F, op, 1960; flowers salmon-pink, well formed, borne in clusters, dbl., 26 petals, 2.5–3 in.; vigorous, compact, bushy growth; [Independence X (Lady Sylvia X Fashion)]; Lens

Gentle Clown *see* HARtopper

Gentle Cover *see* 'POULLEN'

Gentle Kiss *see* HARtbrad

'GENTLE LADY', HT, lp, 1971; bud slender, long, pointed; dbl., 35 petals, 3.5 in., cupped, intense fragrance; foliage matt, dark, leathery; upright, bushy growth; [Tiffany X Michele Meilland]; Fuller; Wyant

Gentle Maid *see* HARvilac

'GENTLE PERSUASION', S, yb, 1984; flowers yellow tinted orange, medium-large blooms borne 1-5 per cluster, dbl., 28 petals, cupped, repeat bloom, slight fragrance; awl-like, tan prickles; foliage large, leathery, semi-glossy, dark olive green; vigorous, bushy, erect growth; hardy.; [Carefree Beauty X Oregold]; Buck, Dr. Griffith J.; Iowa State University

Gentle Touch *see* 'DICLULU'

Gentleman's Agreement *see* 'BRIGENTLE'

'GENVAL', HT, mp, 1963; flowers cyclamen-pink, moderate fragrance; foliage bronze, dull; vigorous growth; [Rosita X Margaret]; Delforge

GENwine, S, lp; (**King Henry**); Genesis

'GEOFF BOYCOTT', F, w, 1974; dbl., 35 petals, 3.5 in., slight fragrance; foliage dark; [Ice White X Tip-Top]; McGredy, Sam IV

Geoff Hamilton *see* 'AUSHAM'

'GEORG ARENDS', HP, mp, 1910; (Fortuné Besson); flowers soft pink, large, dbl., 25 petals, intense fragrance; vigorous growth; [Frau Karl Druschki X La France]; Hinner, W.

'GEORG GEUDER', HT, op, 1931; flowers dark salmon-pink, medium, dbl., moderate fragrance; Schmidt, I. C.; (Sangerhausen)

George Armer *see* 'BRATOWIN'

'GEORGE BAKER', HP, dp, 1881; flowers cerise, very dbl.; Paul & Son

George Burns™ *see* 'WEKCALROC'

George Burns Centennial *see* 'WEKCALROC'

'GEORGE C. WAUD', HT, mp, 1908; flowers rose, veined darker, dbl., 4.5 in., intense fragrance; bushy growth; Dickson, A.

George Cuvier, B, dp

'GEORGE DAKIN', HT, pb, 1927; flowers silvery pink, flushed apricot, reverse orange to apricot, dbl., exhibition form, moderate fragrance; foliage glossy, bronze; vigorous, bushy growth; [Ophelia X Mrs Henry Morse]; Burbage Nursery

'GEORGE DICKSON', HT, mr, 1912; flowers large blooms with weak stems, dbl., 36 petals, moderate fragrance; weak stems; GM, NRS, 1911; Dickson, A.

'GEORGE DICKSON, CLIMBING', Cl HT, mr, 1949; Woodward

'GEORGE ELGER', Pol, my, 1912; (Yellow Baby Rambler); bud small, golden yellow; flowers coppery yellow to clear yellow, borne in large clusters, very dbl., slight fragrance; foliage small, dark, soft; bushy, dwarf growth; Turbat

'GEORGE ELLIOT', HT, op, 1970; flowers shrimp-pink, dbl., 35 petals, 4–5 in., intense fragrance; foliage bronze; [Highlight X Dorothy Peach]; Wills

'GEORGE FOX', HT, or, 1939; flowers orange-vermilion, medium, globular; foliage glossy; compact growth; [Charles P. Kilham X Lady Forteviot]; Savage Nursery

'GEORGE GEARY', HT, yb, 1953; flowers golden yellow flushed vermilion, high pointed, 4 in., moderate fragrance; foliage dark, bronze; vigorous growth; [Gwyneth Jones X Unnamed seedling]; Geary; Burbage Nursery

'GEORGE GEUDER', HT, yb, 1931; flowers salmon-pink and bright carmine on yellow ground, moderate fragrance; vigorous growth; Schmidt, J.C.

'GEORGE H. MACKERETH', HT, dr, 1924; flowers crimson shaded velvety maroon, dbl., intense fragrance; Dickson, A.

'GEORGE HEERS', HT, pb, 1961; flowers rich pink, touched apricot and yellow; Langbecker

'GEORGE HOWARTH', HT, dp, 1928; flowers bright carmine, dbl., moderate fragrance; [Gorgeous X The Queen Alexandra Rose]; Bees

George IV *see* **'RIVERS' GEORGE IV'**

'GEORGE R. HILL', HT, w, 1990; dbl., 45 petals; foliage large, dark green; vigorous growth; [Admiral Rodney sport]; Varney, Eric; Battersby Roses, 1991

George Sand *see* **'LUNELLE'**

'GEORGE THOMAS', HT, w, 1975; flowers pure white, tinged pink, full, dbl., 40 petals, 6–8 in., moderate fragrance; foliage dark; vigorous growth; [Ena Harkness X Memoriam]; Ellick

George Vancouver, S, mr, 1994; flat, 24 petals, slight; Ogilvie, Ian S.

'GEORGE WILL', HRg, dp, 1939; flowers deep pink, blooms in clusters, dbl., 3 in., flat, all-summer bloom, moderate, clove fragrance; foliage rugose; slender branches; height 3-4 ft; [(R. rugosa X R. acicularis) X ?]; Skinner

Georgeous® *see* 'INTERGEORGE'

'GEORGES CAIN', HRg, dr, 1909; flowers crimson with purple, large, dbl.; very vigorous growth; [Souv. de Pierre Notting X R. rugosa]; Müller, Dr. F.

Georges Cassagne, Gr, 1980; Croix; (Cavriglia)

'GEORGES CHESNEL', HT, dy, 1935; bud pointed; flowers deep golden yellow, veined copper, dbl.; foliage glossy; [Julien Potin seedling X Étoile d'Or]; Pernet-Ducher; Gaujard

'GEORGE'S CHOICE', F, mr, 1979; bud small, ovoid; flowers currant-red, moderately full, dbl., 35–40 petals, moderate fragrance; compact, bushy growth; [Evelyn Fison X Tabarin]; Ellick; Excelsior Roses

'GEORGES CLEMENCEAU', HT, ob, 1919; flowers bright orange, shaded umber and carmine; [Mme Edouard Herriot sport]; Lévêque

'GEORGES DE CADONEL', B, lp, 1904; Schwartz

'GEORGES HAMONIÈRE', HT, dr, 1937; flowers medium, dbl.; Moulin; (Sangerhausen)

'GEORGES PAQUEL', HT, my, 1934; flowers saffron-yellow, large, dbl., slight fragrance; vigorous, bushy growth; [Unnamed seedling X Souv. de Claudius Pernet]; Leenders, M.

'GEORGES PERDOUX', HT, pb, 1927; flowers reddish pink tinted coppery red, dbl., moderate fragrance; Barbier

'GEORGES PERNET', Pol, mp, 1887; flowers bright peach-pink; [Mignonette X ?]; Pernet-Ducher

'GEORGES ROUSSET', HP, 1893; Rousset; (Cavriglia)

'GEORGES SCHWARTZ', HT, my, 1899; flowers canary-yellow; [Kaiserin Auguste Viktoria X Souv. de Mme Levet]; Schwartz, Vve.

'GEORGES SCHWARTZ, CLIMBING', Cl HT, my, 1917; Knight, G.

Georges Truffaut *see* DORkade

'GEORGES VIBERT', HGal, rb, 1853; flowers purplish red, streaked (striped) white, large, dbl., flat, moderate fragrance; Robert

Georgetown Gateside Yellow, N, ly; (found rose)

Georgetown Tea, T, pb; (found rose)

'GEORGETTE'®, Min, mp, 1981; bud ovoid; flowers medium pink, veined darker, dbl., 30 petals, exhibition form, slight fragrance; straight prickles; foliage medium green, dense; upright, compact, bushy growth; [Electron X Little Chief]; Bennett, Cecilia 'Dee'; Tiny Petals Nursery

Georgette *see* 'INTERORGE'

Georgette *see* KORacona

Georgeus, Min, mp, 1998

'GEORGIA', HT, ab, 1980; bud short, pointed; flowers peach-apricot blend, borne singly, very dbl., 53–55 petals, 5.5–6. in., moderate, tea fragrance; long prickles, hooked downward; foliage large, glossy, leathery; tall, upright growth; PP4712; [Arizona X Unnamed seedling]; Weeks, O.L.

'GEORGIANNA DOAN', HT, pb, 1942; bud long, pointed; flowers two-tone pink, medium, dbl., 25–30 petals, exhibition form, intense fragrance; foliage leathery, wrinkled, dark; vigorous, upright, much branched growth; [Ophelia X Unnamed seedling]; Hill, Joseph H., Co.

Georgie Anderson *see* 'ANDGEO'

Georgie Girl *see* 'DICKERFUFFLE'

Georgie Lad *see* 'HORKORBLUSH'

'GERALD HARDY', HT, mr, 1936; bud pointed; flowers bright scarlet-red, spiral, large, dbl., intense fragrance; strong, erect stems; bushy growth; Dickson, A.

Geraldine *see* 'PEAHAZE'

'GERALDINE', HT, 1924; flowers buff, shaded pink, dbl., moderate fragrance; RULED EXTINCT 11/82 ARM; [Antoine Rivoire X Marie Adélaide]; Chaplin Bros.

'GERALDINE HICKS', HT, ab, 1950; flowers bronze-yellow; [William Moore sport]; Hicks

'GERANIUM', HMoy, mr, 1938; flowers almost scarlet, single, 2 in., blooms in clusters (up to 5); crimson, crimson fruit; upright (8-10 ft), compact growth; Royal Hort. Soc.

Geranium *see* **'INDEPENDENCE'**

Geranium Primaplant, Pol, mr, 1964; flowers medium, semi-dbl.; Vlaeminck; (Sangerhausen)

'GERANIUM RED', F, or, 1947; flowers bright geranium-red, blooms in clusters, dbl., 50 petals, 4 in., globular, intense, geranium fragrance; foliage dark, glossy; bushy growth; [Crimson Glory X Seedling]; Boerner; J&P

Gerard ter Borch, S, dp, 1997; Williams, J. Benjamin

Gerbe d'Or *see* 'MACCA'

'GERBE ROSE', LCl, lp, 1904; flowers delicate pink, large, dbl., slight fragrance; foliage glossy; vigorous growth; [R. wichurana X Baroness Rothschild]; Fauque; Langue

'GERDA HENKEL', HT, dr, 1964; flowers deep blood-red, large, dbl.; foliage dark, leathery; strong stems; vigorous, upright growth; [New Yorker X Prima Ballerina]; Tantau, Math.

Gerdo, HT, ab, 1986

'GERMAINE', HT, w, 1926; flowers creamy white, center salmon; [Unnamed seedling X Sunburst]; Chambard, C.

'GERMAINE CHENAULT', HT, lp, 1910; flowers cream-pink with darker edges, large, dbl., moderate fragrance; Guillot; (Sangerhausen)

'GERMANEA', HT, dp, 1929; flowers deep shining rose-pink, well formed, very large, dbl.; [Columbia sport]; Ravenberg

'GERMANIA' *see* **'CHARME'**

Germania-Africana *see* KORtechna

'GERMANICA', HRg, 1900; Mueller, F.; (Cavriglia)

Germiston Gold *see* 'KORTAKE'

Germiston Gold, Climbing, Cl HT, dy, 1992; Malanseuns

Gero, F, rb, 1984; flowers pink and red, medium, dbl., slight fragrance; Scholle, E.; (Sangerhausen)

'GERT POTGIETER', HT, 1968; foliage light green; vigorous growth; Gowie

Gertrud®, F, dr

'GERTRUD HUCK', HT, dp, 1932; bud pointed; flowers flamingo-red, large, dbl., cupped, moderate fragrance; foliage leathery, bronze; vigorous growth; [Wilhelm Kordes sport]; Huck; C-P

'GERTRUD SCHWEITZER', HT, ob, 1973; bud long, pointed; flowers apricot-orange, large, dbl., cupped, moderate fragrance; foliage glossy, dark; [Colour Wonder X Seedling]; Kordes; Horstmann

'GERTRUD WESTPHAL', F, or, 1951; flowers orange-scarlet, single, 5–7 petals, 3 in., slight fragrance; foliage glossy, dark reddish green; dwarf, bushy, much branched growth; [Baby Chateau X Obergärtner Wiebicke]; Kordes

'GERTRUD WESTPHAL, CLIMBING', Cl F, or, 1961; Buisman, G. A. H.

'GERTRUDE GREGORY', HT, my, 1957; flowers bright golden yellow; [Lady Belper sport]; Gregory

Gertrude Jekyll® *see* AUSbord

'GERTRUDE RAFFEL', F, dp, 1956; flowers pink, center rosy, well formed, borne in large clusters, semi-dbl., 15–20 petals, 2–3 in., slight fragrance; foliage dark; vigorous, bushy growth; Raffel; Port Stockton Nursery

'GERTRUDE REUTENER', F, dp, 1954; flowers crimson-pink; vigorous growth; Leenders, M.

'GERTRUDE SHILLING', HT, dy, 1988; flowers bright, deep yellow, aging paler, urn-shaped, decorative, large, very dbl., 52 petals, moderate, fruity fragrance; prickles broad, fairly flat, large, dark brown; foliage large, medium green, matt; upright, tall, vigorous, good basal growth; [Golden Splendour X Peer Gynt]; Poole, Lionel; Rearsby Roses, Ltd., 1989

'GESCHWIND'S GILDA', HMult, dr, 1887; Geschwind, R.

'GESCHWIND'S GORGEOUS', Cl HT, mr, 1916; flowers medium, semi-dbl., slight fragrance; Geschwind, R.; (Sangerhausen)

'GESCHWIND'S NORDLANDROSE II', HSet, mr, 1928; Geschwind, R.

'GESCHWIND'S NORDLANDSROSE', HSet, mp, 1884; Geschwind, R.

'GESCHWIND'S ORDEN', HMult, m, 1886; Geschwind, R.

'GESCHWIND'S SCHONSTE', HMult, dr, 1900; Geschwind, R.

Geschwister Scholl, Pol, w, 1974; flowers medium, dbl.; GPG Bad Langensalza; (Sangerhausen)

Gessel, HT; Pironti, N.; (Cavriglia)

'GHERGANA', Gr, dr, 1974; flowers deep blackish-red, large blooms in clusters of 2-5, dbl., 55 petals, cupped, moderate, tea fragrance; foliage dark; vigorous, upright growth; [Spectacular X Unnamed seedling]; Staikov, Prof. Dr. V.; Kalaydjiev and Chorbadjiiski

'GHISLAINE DE FÉLIGONDE', HMult, ly, 1916; bud bright yellow; flowers yellowish white tinted flesh, borne in clusters of 10-20; vigorous, climbing (8-10 ft) growth; [Goldfinch X ?]; Turbat

Giana®, F, dr

Gianlauro, 1978; Zandri, R.; (Cavriglia)

Giant of Battles *see* **'GÉANT DES BATAILLES'**

Giant Pink®, S, pb, 1990; Lens

'GIBBY', HT, dp, 1977; bud ovoid, pointed; flowers carmine-pink, dbl., 50 petals, 4.5 in., exhibition form; foliage dull, dark, leathery; vigorous, upright, bushy growth; [Christian Dior sport]; Prof. F. Roses; Ludwigs Roses Pty. Ltd.

Gidday *see* 'TINHAT'

'GIDGET', Min, op, 1975; bud pointed; flowers coral-pink to coral-red, informal shape, 1 in., slight fragrance; foliage small, glossy; vigorous, bushy growth; [(R. wichurana X Floradora) X Fire Princess]; Moore, Ralph S.; Sequoia Nursery

Giesebrecht *see* **'BASHFUL'**

Gift of Life *see* 'HARELAN'

Gigantea Cooperi *see* **'COOPER'S BURMESE'**

'GIGANTÈSQUE', T, dp, 1845; flowers deep pink; Odier

Giggles *see* 'LYOGI'

Giggles *see* 'KINGIG'

'GIGI', HT, mp, 1959; flowers rose-pink, reverse brighter, large, dbl., moderate fragrance; foliage light green; long, strong stems; vigorous growth; [The Doctor X Unnamed seedling]; Verschuren; Blaby Rose Gardens

Gigliola, 1969; Mansuino; (Cavriglia)

'GIGOLETTE', F, rb, 1953; bud ovoid; flowers yellow and red bicolor, medium, borne in clusters, semi-dbl., moderate fragrance; foliage leathery, light green; vigorous growth; Gaujard

'GIL BLAS', HGal, pb, 1843; flowers light pink with white striping, large, dbl.; Roseraie de l'Hay; (Sangerhausen)

Gilbert Bécaud *see* 'MEIRIDORIO'

'GILBERT F. LEVY', F, mr, 1958; flowers currant-red, dbl.; very vigorous growth; [Moulin Rouge X Oiseau de Feu]; Combe

'GILBERT NABONNAND', T, op; (G. Nabonnand); flowers large, semi-dbl., moderate fragrance; Nabonnand, 1883; (Sangerhausen)

'GILDA', HT, ab, 1936; bud long, pointed; flowers pure orange-yellow, large, dbl.; foliage leathery, dark; vigorous growth; RULED EXTINCT 3/87; [Souv. de Claudius Pernet X (Lady Hillingdon X Harry Kirk)]; Towill

Gilda *see* 'PEAHIGH'

'GILFLEUR', HT, pb, 1993; (**First Born**); flowers pink/peach, holds well for cut and arranging, blooms borne mostly singly, very dbl., 47 petals, moderate fragrance; some prickles; foliage medium, dark green, glossy; tall (80-90 cms), upright growth; [Silver Jubilee X Tahiti]; Gilmore, T.O.; Gilmore, 1993

'GILLIAN', HT, op, 1958; bud long, pointed; flowers soft coral-pink; foliage bronze; [Michele Meilland X Mme Butterfly]; Verschuren; Gandy Roses, Ltd.

'GILLIAN DAWN', HT, mr, 1997; flowers full, medium, very dbl., 26–40 petals, borne mostly singly, slight fragrance; foliage medium, medium green, semi-glossy; spreading, medium (24in.) growth; Robinson, Kenneth G.

Gilmore, HT, ly; Select Roses, B.V.

Gilt Edged *see* CHEwdainty

Gin Fizz *see* MEIromar

Gin no suzu *see* **'SILVER BELL'**

'GINA', F, dr, 1960; flowers velvety dark crimson, borne in large clusters, single, 6 petals, slight fragrance; foliage glossy; vigorous, upright growth; [Alain X Independence]; Kriloff, Michel; Cramphorn's Nursery

Gina Lollobrigida® *see* 'MEILIVAR'

Gina Louise *see* 'TROBGINA'

'GINETTE', HT, op, 1924; flowers salmony maize-yellow, dbl., moderate fragrance; [Paul Monnier X Souv. de Claudius Pernet]; Buatois

'GINGER', F, or, 1962; bud ovoid; flowers orange-vermilion, blooms in irregular clusters, dbl., 28 petals, 4 in., cupped, moderate fragrance; foliage leathery; vigorous, compact, bushy growth; [Garnette seedling X Spartan]; Boerner; J&P

Ginger Hill *see* HILcap

Ginger Meggs, F, ob, 1962; Tantau

'GINGER ROGERS', HT, op, 1969; (Salmon Charm); flowers salmon, loosely formed, large, dbl., 30 petals, moderate fragrance; foliage light green; very tall growth; [Tropicana X Miss Ireland]; McGredy, Sam IV; McGredy

Ginger Toddler *see* PEAvesta

Gingerbread Man™ *see* 'POULXAS'

Gingernut *see* 'COCCRAZY'

Gingersnap *see* 'AROSNAP'

'GINGIA', HT, w, 1983; flowers large, dbl., 35 petals, intense fragrance; foliage large, light green, glossy; bushy growth; [Unnamed seedling X Unnamed seedling]; Fumagalli, Niso

Ginny *see* 'BISJEN'

Ginny-Lou *see* 'TROBINKA'

'GINSKY', F, pb, 1983; flowers light salmon pink opening to pale pink to cream; [Liverpool Echo sport]; Barké; L.E.J. Wood

'GINZA KOMACHI', Cl Min, pb, 1980; bud globular; flowers deep pink, white eye, yellow stamens, small blooms in clusters, single, 5 petals, slight fragrance; many, hooked prickles; foliage medium green, glossy; vigorous growth; [Nozomi X Unnamed seedling]; Kono, Yoshito

Gioia *see* **'PEACE'**

Gioia, Climbing *see* **'PEACE, CLIMBING'**

'GIOIELLO', Min, my, 1984; flowers small, dbl., 20 petals, no fragrance; foliage small, dark, matt; [Zorina X Sole Di San Remo]; Bartolomeo, Embriaco

'GION', Cl Min, pb, 1979; bud rounded; single, 5 petals, 1 in., flat, non-recurrent, moderate fragrance; foliage tiny, leathery; bushy, climbing growth; [Nozomi X Unnamed seedling]; Onodera, Toru F.; S. Onodera

'GIOVANE', HT, op, 1965; bud pointed; flowers salmon-orange, large, dbl., 28 petals, exhibition form, moderate fragrance; foliage glossy, bronze; dense growth; [Queen Elizabeth X Orient]; Dot, Simon; Rosas Dot

'GIOVANEZZA', HT, rb, 1933; flowers geranium-red, reverse cream-white, edged lighter; vigorous growth; Ingegnoli

Giovanni Paolo II, HT, 1984; McEntire, J.; (Cavriglia)

'GIPSY', HT, dr, 1931; flowers medium, dbl.; Van Rossem; (Sangerhausen)

'GIPSY BOY', B, dr, 1909; (Zigeunerknabe); flowers dark crimson-red, medium; vigorous (3-5 ft) growth; Lambert, P.

Gipsy Jewel, Min, dp, Moore, 1975

'GIPSY LASS', HT, mr, 1932; (Gypsy Lass); flowers scarlet-crimson shaded blackish, dbl., globular, intense fragrance; long, willowy stems; bushy growth; Dickson, A.

'GIPSY LOVE', HT, or, 1964; flowers orange-vermilion, dbl., 25 petals, 4 in., moderate fragrance; vigorous growth; [Chic Parisien X Fashion]; Delbard-Chabert; Cuthbert

'GIPSY MAID', F, dp, 1955; flowers carmine-scarlet, base golden, borne in small clusters of 3 or more, single, moderate, sweetbriar fragrance; foliage olive-green; LeGrice

Giranu Czecheti, Pol, op, 1955; flowers medium, semi-dbl.; Madarsko; (Sangerhausen)

'GIRASOL', HT, my, 1945; bud oval; flowers sunflower-yellow, dbl., 25–30 petals, slight fragrance; foliage dark, glossy; upright, compact growth; [Joanna Hill X Carito MacMahon]; Dot, Pedro

Giriji, HT, dr, 1988; Friends Rosery

Girl Friend *see* **'PODRUGA'**

Girl Guide, F, op

'GIRL SCOUT', F, my, 1961; bud ovoid; flowers golden yellow, dbl., 50 petals, 3.5–4 in., cupped, moderate fragrance; foliage leathery, glossy; vigorous, medium tall growth; [Gold Cup X Pigmy Gold]; Boerner; J&P

'GIRLIE', Pol, dr, 1923; flowers bright scarlet-crimson; bushy growth; [Orléans Rose sport]; Wezelenburg

Girls' Brigade *see* 'HARBANJO'

'GIRONA', HT, pb, 1936; flowers soft red and yellow, well-formed, large, dbl., 30 petals, exhibition form, intense, damask fragrance; foliage bright green; vigorous, spreading growth; [Li Bures X Talisman]; Dot, Pedro; C-P, 1939;, H. Guillot, 1939

Girtnerfreude *see* KORdisard

'GISELA', F, op, 1961; flowers salmon-pink, base straw-yellow, borne in clusters, dbl., 56 petals; foliage dark, glossy, bronze; upright, bushy, compact growth; [Masquerade X Pinocchio]; Verschuren, A.; van Engelen

'GISÈLE ALDAY', HT, mp, 1933; bud pointed; flowers bright rose-pink tinted flesh, large, semi-dbl., cupped, slight fragrance; foliage glossy, dark; vigorous growth; [Mrs Pierre S. duPont X Lallita]; Mallerin, C.; H. Guillot

Giselle *see* 'JUSELLE'

'GISSELFELD'®, HT, dr, 1972; 17–20 petals, 4–4.5 in., moderate fragrance; foliage dark, leathery; upright growth; [(Tropicana X Champs-Elysees) X Furore]; Poulsen, Niels D.; Poulsen

Gitane *see* **'BRIGHT WINGS'**

Gites de France® *see* MEIwaton

Gitta Grummer *see* 'KORSEE'

Gitte *see* 'KORITA'

'GIULETTA', B, lp, 1859; Laurentius

'GIULIANA BORGATTI', HT, w, 1936; flowers white, center shaded rose and salmon, very well formed, large, dbl.; foliage dark ivy-green; vigorous growth; [Ophelia X Ville de Paris]; Borgatti, G.

'GIUSEPPE MOTTA', HT, lp, 1936; flowers large, semi-dbl., intense fragrance; Heizmann, E.; (Sangerhausen)

'GIUSEPPINA PAPANDREA', Min, pb, 1989; flowers cerise, reverse lighter; [Petite Folie sport]; Papandrea, John T.

Givenchy™ *see* 'ARODOUSNA'

Gizmo™ *see* 'WEKCATLART'

'GLACIER', F, w, 1952; bud ovoid; flowers white, slightly overcast yellow, dbl., 28 petals, 4.5 in., cupped, moderate fragrance; foliage glossy, dark; vigorous, upright growth; [Unnamed white HT X Summer Snow]; Boerner; J&P

Glad Tidings *see* 'TANTIDE'

'GLADIADOR', Cl HT, dp, 1954; bud pointed; flowers carmine, very large, dbl., intense fragrance; vigorous growth; [Texas Centennial X Guinee]; Dot, Pedro

'GLADIATOR', LCl, mr, 1955; bud ovoid; flowers rose-red, dbl., 35 petals, 4.5–5 in., exhibition form, moderate fragrance; foliage dark, leathery; vigorous (10-12 ft) growth; [Charlotte Armstrong X (Pink Delight X New Dawn seedling)]; Malandrone; J&P

'GLADNESS', F, pb, 1959; bud pointed; flowers light pink edged darker, large, borne in clusters, dbl., 25 petals; foliage dark, glossy; upright growth; [Sunny Maid X Cinnabar]; Fletcher

'GLADSOME', HMult, mp, 1937; flowers delicate light pink, single; tall hedge rose; Clark, 1940

'GLADYS BENSKIN', HT, op, 1929; flowers rose-cerise, shaded orange, base deeper orange, large, dbl., exhibition form, moderate fragrance; vigorous growth; GM, NRS, 1929; Dickson, A.; Dreer

'GLADYS HARKNESS', HT, op, 1900; flowers very large, dbl., intense fragrance; Dickson, A.; (Sangerhausen)

'GLADYS HOLLAND', HT, lp, 1917; flowers light pink, shaded buff, dbl., moderate fragrance; GM, NRS, 1916; McGredy

'GLADYS MONCRIEF', HT, yb, 1981; flowers golden yellow to apricot, flushed rose red at petal tips; [Granada sport]; Jack, J.; Girraween Nursery

'GLADYS SAAVEDRA', HT, mp, 1922; flowers rosy peach-blossom-pink, dbl.; [Mme Abel Chatenay X Jonkheer J.L. Mock]; Nabonnand, P.

'GLADYS TWEEDIE', HT, dr, 1950; bud long, pointed; flowers crimson, dbl., 30–35 petals, 5 in., exhibition form, moderate fragrance; foliage wrinkled; very vigorous, bushy growth; [Crimson Glory X William Orr]; Toogood

'GLAIVE', HT, pb, 1951; bud long, pointed; flowers cream, center tipped pink, small, dbl., 25 petals, exhibition form, slight fragrance; foliage glossy; vigorous, bushy, compact growth; Clark, A.; NRS Victoria

Glamis Castle *see* 'AUSLEVEL'

Glamorgan *see* 'THOOM'

'GLAMOROUS', Min, or, 1979; bud pointed; flowers orange-red, base yellow, dbl., 35 petals, 1–1.5 in., exhibition form, slight fragrance; foliage small, glossy, bronze-green; bushy, spreading growth; [Starburst X Over the Rainbow]; Williams, Ernest D.; Mini-Roses

'GLAMOUR', HT, op, 1939; bud long, ovoid; flowers salmon-pink, large, dbl., intense fragrance; vigorous growth; [Comtesse Vandal X Pres. Macia]; Leenders, M.; T. Robinson, Ltd.

'GLAMOUR GIRL', HT, op, 1942; bud pointed, light jasper-red; flowers light salmon, open, dbl., 45–50 petals, 4–5 in.; foliage dark, leathery; strong stems; very vigorous, upright, much branched growth; [Captain Glisson X Justine]; Hill, Joseph H., Co.

Glamour Girl *see* CLEamour

'GLANLIN', HT, ob, 1985; (**Lincoln Cathedral**, Sarong); flowers outer petals pink, inner ones orange, yellow reverse, large, dbl., 28 petals, slight fragrance; many, reddish prickles; foliage medium, medium green glossy; bushy growth; GM, RNRS, 1985; [Silver Jubilee X Royal Dane]; Langdale, G.W.T.

'GLANMUSIC', F, pb, 1997; (**Sir Neville Marriner**); flowers very full, medium, very dbl., 41 petals, borne mostly singly or in small clusters, slight fragrance; few prickles; foliage medium, medium green, semi-glossy; upright, medium (3x2ft.)growth; [Seedling X Painted Doll]; Langdale, G.W.T.

'GLAREABIT', HT, yb, 1992; (**Lincolnshire Poacher**); flowers yellow flushed apricot and pink, blooms borne mostly singly, dbl., 26–40 petals, 3–3.5 in., slight fragrance; some prickles; foliage medium, medium green, semi-glossy; medium (80 cms-1 m), bushy growth; [Silver Jubilee X Woman and Home]; Langdale, G.W.T., 1993

'GLARONA', HT, w, 1922; flowers creamy flesh, center rose; Krüger; Kiese

'GLASTONBURY', S, rb, 1981; bud globular; flowers dark crimson to deep purple, borne 1-5 per cluster, dbl., 55 petals, repeat bloom, intense fragrance; hooked, red prickles; foliage medium green, sparse, spreading growth; [The Knight X Unnamed seedling]; Austin, David; David Austin Roses, Ltd.

'GLEAMING', F, dy, 1958; flowers deep lemon-yellow, borne in trusses, single, 6–8 petals, 4 in., intense fragrance; foliage dark; very free growth; [Goldilocks X Golden Scepter]; LeGrice

'GLEN ALMOND', HT, pb, 1973; flowers pale orient pink, dbl., 30 petals, 3.5–4 in.; foliage glossy, leathery; free growth; [Pascali X Happy Event]; Wallace

'GLEN ARTNEY', HT, dr, 1973; flowers beetroot-purple, dbl., 40 petals, 3.5–4 in., slight fragrance; vigorous growth; [Baccará X Sterling Silver]; Wallace

'GLENARA', HT, dp, 1951; bud long, pointed; flowers deep rosy pink, very large, semi-dbl., 18 petals, slight fragrance; foliage leathery; vigorous, upright bush or pillar growth; Clark, A.; NRS Victoria

Glendora *see* **'JOASINE HANET'**

Glendora *see* KORhuba

'GLENFIDDICH', F, dy, 1976; flowers amber-gold, dbl., 25 petals, 4 in., moderate fragrance; foliage glossy, dark; [Arthur Bell X (Sabine X Circus)]; Cocker

'GLENGARRY', F, or, 1969; flowers vermilion, large, dbl., 32 petals, slight fragrance; foliage semi-glossy; compact, bushy growth; [Evelyn Fison X Wendy Cussons]; Cocker

'GLENITI GOLD', HT, dy, 1973; flowers very deep yellow; [Lady Mandeville sport]; Bone, John, & Son; Trevor Griffiths Ltd.

'GLENN DALE', LCl, ly, 1927; flowers lemon, fading to white, large blooms in clusters (to 20), dbl., 40 petals, slight fragrance; foliage dark, leathery; vigorous (10 ft) growth; GM, Portland, 1920; [Believed to be R. wichurana X Isabella Sprunt]; Van Fleet; American Rose Society

Glenshane® *see* 'DICVOOD'

'GLENYS STEWART', HT, dp, 1968; flowers deep rose-pink, dbl., moderate bloom, slight fragrance; moderate growth; [Montezuma X Pink Favorite]; Kemp, M.L.; G. Stewart

'GLETSCHER', F, m, 1955; bud ovoid; flowers pale lilac, large blooms in large trusses, dbl., exhibition form, intense fragrance; foliage glossy; vigorous, upright, bushy growth; [Seedling X Lavender Pinocchio]; Kordes

Gletscherfee® *see* KORdomal

'GLIMMER', Min, rb, 1988; bud pointed; flowers bright, medium red, yellow at base, reverse slightly darker, dbl., 24 petals, exhibition form, slight fragrance; prickles slightly downward pointed, medium, red; foliage medium, medium green, semi-glossy; upright, medium growth; [Party Girl X Unnamed seedling]; Bridges, Dennis A.; Bridges Roses, 1989

'GLITTERS', Cl HT, mp, 1934; bud pointed; flowers brilliant pink, base orange, open, very large, dbl., globular, recurrent bloom, moderate fragrance; foliage leathery; long stems; very vigorous growth; [Mrs W.J. Grant X Mrs Sam McGredy]; Smith, J.

'GLOAMING', HT, pb, 1935; bud pointed; flowers luminous pink suffused salmon, reverse lighter, open, very large, dbl., 36 petals, intense fragrance; foliage leathery, dark; vigorous, bushy growth; [Charles P. Kilham X Mrs Pierre S. duPont]; Nicolas; J&P

'GLOBE', F, mr, 1956; flowers blood-red, borne in clusters, semi-dbl., 20 petals, 2 in., cupped, moderate fragrance; foliage dark; vigorous, bushy, compact growth; [Fanal X Red Favorite]; Tantau, Math.

'GLOBE HIP', C, w

Gloira Dei-mutace, HT, lp; Lorenc; (Czech Rosa Club)

'GLOIRE D'ANTIBES', HT, dr, 1938; flowers large, dbl.; Mallerin, C.; (Sangerhausen)

Gloire de Bordeau *see* **'BELLE DE BORDEAUX'**

Gloire de Bordeaux *see* **'BELLE DE BORDEAUX'**

'GLOIRE DE BOURG-LA-REINE', HP, mr, 1879; Margottin

'GLOIRE DE BRUXELLES', HP, m, 1889; (Gloire de l'Exposition de Bruxelles); flowers very dark, velvety crimson-purple, large, dbl., 60 petals, moderate fragrance; vigorous, upright growth; [Souv. de William Wood X Lord Macaulay]; Soupert & Notting

'GLOIRE DE CHÉDANE-GUINOISEAU', HP, mr, 1907; flowers bright crimson-red, well-formed, large, dbl., 40 petals, cupped, occasionally recurrent bloom, moderate fragrance; foliage dark, soft; vigorous growth; [Gloire de Ducher X Unnamed variety]; Chedane-Pajotin

'GLOIRE DE CIBEINS', HT, or, 1958; flowers deep vermilion-red, well formed, dbl., 30 petals; long stems; vigorous growth; [Mme Méha Sabatier X Léonce Colombier]; Arles; Roses-France

'GLOIRE DE DEVENTER', T, ly, 1897; flowers light yellow with pink tints, large, dbl., intense fragrance; Soupert & Notting; (Sangerhausen)

'GLOIRE DE DIJON', Cl T, op, 1853; flowers rich buff-pink shaded orange toward center, very large, dbl., moderate fragrance; very vigorous, climbing growth; (28); [Thought to be an unknown Tea X Souv. de la Malmaison]; Jacotot

'GLOIRE DE DUCHER', HP, dr, 1865; flowers very large, dbl., occasional recurrent bloom, moderate fragrance; Ducher

'GLOIRE DE FRANCE', HGal, lp; flowers pale pink, dbl.; low growth; Bizard, 1828

'GLOIRE DE FRANCE', HT, or, 1946; flowers orange-red variegated copper, well formed, large, dbl.; vigorous growth; GM, Bagatelle, 1945; Gaujard

'GLOIRE DE GUILAN', D, lp; flowers clear pink, center incurved, dbl., quartered, spring bloom, intense fragrance; foliage light green; sprawling shrub (4-5 ft.) growth; Nancy Lindsay/Hilling, 1949

'GLOIRE DE HOLLANDE', HT, dr, 1918; Verschuren

Gloire de Hollande, Climbing, Cl HT, dr; flowers large, dbl., moderate fragrance; (Sangerhausen)

'GLOIRE DE LA BRIE', HT, mr; bud long; flowers bright red, large; Grandes Roseraies

Gloire de l'Exposition de Bruxelles *see* **'GLOIRE DE BRUXELLES'**

'GLOIRE DE MARGOTTIN', HP, dr, 1887; flowers large, dbl., 60 petals, globular, occasional recurrent bloom, intense fragrance; Margottin

'GLOIRE DE MEZEL', M, mp; flowers pale rose, very large

Gloire de Paris *see* **'ANNA DE DIESBACH'**

Gloire de Rome *see* **'ROME GLORY'**

'GLOIRE DE VITRY', HP, mp, 1854; flowers bright pink, large, globular; [La Reine seedling]; Masson

Gloire des Anciens *see* INTerglo

'GLOIRE DES BELGES', HT, dp, 1916; flowers vivid carmine; Chambard, C.

'GLOIRE DES CHARPENNES', Pol, mr, 1898; flowers carmine-red, small, dbl.; Lille; (Sangerhausen)

'GLOIRE DES JARDINS', HGal, m; Vibert, before 1815

'GLOIRE DES LAWRANCEANAS', Min, dr, 1837; flowers dark crimson; dwarf growth

'GLOIRE DES MOUSSEUSES', M, mp, 1852; (Gloire des Mousseux); bud heavily mossed; flowers clear bright pink, center deeper, petals imbricated, large, dbl., borne in clusters; foliage light green; vigorous growth; Robert

Gloire des Mousseux *see* **'GLOIRE DES MOUSSEUSES'**

Gloire des Perpetuelles *see* **'FLON'**

'GLOIRE DES POLYANTHA', Pol, mp, 1887; flowers bright pink, well-shaped, small blooms in large clusters, dbl.; dwarf growth; [Mignonette seedling]; Guillot et Fils

'GLOIRE DES ROSOMANES', HCh, mr, 1825; (Ragged Robin, Red Robin); flowers glowing crimson, very large blooms in large clusters, semi-dbl., repeat bloom, moderate fragrance; vigorous growth; (21); Vibert

'GLOIRE D'ORIENT', M, dr, 1856; flowers deep red; Béluze

'GLOIRE D'ORLÈANS', Pol, mr, 1912; flowers small, dbl.; Levavasseur; (Sangerhausen)

Gloire du Beaujolais, HT, 1990; Delbard, Georges; (Cavriglia)

Gloire du Bourbonnais® *see* 'DELBOURDO'

'GLOIRE DU MIDI', Pol, or, 1932; flowers brilliant orange-scarlet; [Gloria Mundi sport]; deRuiter; J&P;, Sliedrecht & Co.

'GLOIRE DU MIDI SUPERIOR', Pol, or; flowers like parent with more lasting color; [Gloire du Midi sport]; deRuiter

'GLOIRE D'UN ENFANT D'HIRAM', HP, mr, 1899; flowers bright red; Vilin

'GLOIRE LYONNAISE', HP, w, 1885; flowers white with trace of yellow at center, very large, very dbl., 84 petals, cupped, moderate fragrance; foliage leathery; very vigorous, bushy growth; notvery hardy.; [Baroness Rothschild X Mme Falcot]; Guillot et Fils

'GLORIA', HT, dr, 1922; flowers brilliant scarlet-crimson, dbl.; Paul, W.

'GLORIA D'AUTUNNO', HT, 1952; Cazzaniga, F. G.; (Cavriglia)

'GLORIA DE GRADO', HT, mp, 1950; flowers pink, tinted carmine, globular; foliage bright green; [Mari Dot X Comtesse Vandal]; La Florida

Gloria Dei *see* **'PEACE'**

Gloria Dei, Climbing *see* **'PEACE, CLIMBING'**

'GLORIA DEL LLOBREGAT', HT, or, 1940; flowers strawberry-red to vermilion, large, dbl., cupped, moderate fragrance; foliage glossy; very vigorous growth; [Sensation X Margaret McGredy]; Camprubi, C.

Gloria di Milano, HT; Ingegnoli; (Cavriglia)

Gloria di Roma *see* **'ROME GLORY'**

Gloria di Venezia, HT; Kordes, R.; (Cavriglia)

Gloria Dot, Min, 1986; Dot, Simon; (Cavriglia)

Gloria Ferrer® *see* FEsuma

'GLORIA MUNDI', Pol, or, 1929; flowers striking orange-scarlet, borne in clusters, dbl.; foliage light, glossy; vigorous, bushy growth; [Superb sport]; deRuiter; Sliedrecht & Co.;, Teschendorff

'GLORIA MUNDI SUPERIOR', Pol, or; flowers like parent with more lasting color; [Gloria Mundi sport]; deRuiter

'GLORIA MUNDI, CLIMBING', Cl Pol, or, 1934; Lens

'GLORIA MUNDI, CLIMBING', Cl Pol, or, 1943; Howard Rose Co.

'GLORIA SOLIS', HT, my, 1949; flowers well formed, 4–5 in.; foliage dark, glossy; vigorous growth; [Ville de Paris X Max Krause]; Giacomasso

'GLORIANA', HT, my, 1936; flowers intense lemon-yellow in heat, deep gold with cerise markings, dbl., 35 petals, cupped; foliage leathery, glossy, dark; vigorous, compact growth; [Condesa de Sástago (probably self seed)]; Hillock

Gloriana 97 *see* 'CHEWPOPE'

Gloriette, Min, rb

Gloriette, S, op, 1979; Cocker

'GLORIFIED LA FRANCE', HT, lp, 1916; flowers silvery pink, deeper than la france, very dbl., 92 petals, slight fragrance; [Frau Karl Druschki X Mrs Charles E. Russell]; Cook, J.W.

'GLORIGLO', Min, ob, 1976; bud pointed; flowers orange, yellow reverse, dbl., 45 petals, 1 in., exhibition form, slight fragrance; foliage small, glossy, bronze; upright, bushy growth; AOE, 1978; [Seedling X Over the Rainbow]; Williams, Ernest D.; Mini-Roses

Glorimontana, F, or, 1974; flowers medium, dbl.; deRuiter; (Sangerhausen)

'GLORIO', HT, mr, 1923; flowers scarlet-cerise, dbl., moderate fragrance; [Premier X Primrose]; Hill, E.G., Co.; Vestal

'GLORIOSA', HT, w, 1920; flowers ivory-white, base yellow, intense fragrance; [Kaiserin Auguste Viktoria X Pharisaer]; Kiese

Glorious *see* 'LEOGLO'

'GLORIOUS', F, dp, 1947; bud pointed; flowers salmon scarlet, shaded orange, open, medium, large trusses, semi-dbl., 15–17 petals, slight fragrance; foliage leathery, dark; vigorous, upright growth; RULED EXTINCT 7/84 ARM; [(Betty Uprichard X Heidekind) X Heidekind]; Duehrsen; H&S

'GLORIOUS EASTER', HT, op, 1965; bud ovoid; flowers salmon, medium, dbl., exhibition form; foliage leathery; moderate, bushy growth; [Unnamed seedling X Penelope]; Howard, P.J.

'GLORIOUS PERNET', HT, r, 1928; flowers copper, center orange; [Souv. de Claudius Pernet sport]; Myers & Samtmann

'GLORIOUS SUNSET', Pol, rb, 1931; flowers bronze, suffused red, small, borne in clusters, semi-dbl., slight fragrance; foliage small, thick; vigorous growth; [Mariposa sport]; Allen

Glory *see* 'DEVMOMENTO'

Glory Be *see* 'SAVABE'

Glory Days *see* 'JACCOR'

'GLORY OF CALIFORNIA', LCl, lp, 1935; flowers large, dbl., moderate fragrance; Clark, A.; (Sangerhausen)

'GLORY OF CEYLON', F, op, 1967; flowers orange-yellow blended pink, blooms in clusters, semi-dbl., 14 petals, moderate fragrance; foliage dark, glossy; [Vera Dalton X Masquerade]; Harkness

'GLORY OF CHESHUNT', HP, dr, 1880; flowers rich crimson, cupped; vigorous growth; [Charles Lefebvre seedling]; Paul & Son

'GLORY OF EDSELL', HSpn, lp

'GLORY OF HURST', Pol, mr, 1921; flowers cherry-red, borne in clusters, semi-dbl.; foliage small, leathery, glossy, rich green; dwarf growth; [Orléans Rose X Jessie]; Hicks

Glory of Paris *see* **'ANNA DE DIESBACH'**

Glory of Rome *see* **'ROME GLORY'**

'GLORY OF SURREY', HT, my, 1935; flowers golden yellow, semi-dbl., moderate fragrance; fairly vigorous growth; Ley

'GLORY OF WALTHAM', HP, dr, 1865; flowers crimson, very large, very dbl., moderate fragrance; vigorous, climbing or pillar growth; Vigneron; W. Paul

Glow Achievement *see* TANelliv

'GLOW WORM', HT, or, 1919; flowers scarlet, suffused coppery orange, semi-dbl., moderate fragrance; Easlea

Glowing Amber *see* 'MANGLOW'

'GLOWING CARMINE', HT, dp, 1936; flowers carmine, large, dbl., globular, slight fragrance; foliage leathery; vigorous, open habit growth; [Miss Rowena Thom X Unnamed seedling]; H&S; Dreer

Glowing Carpet *see* 'MORPAPPLAY'

Glowing Cushion *see* INTerreflet

Glowing Embers *see* 'ANDGLO'

Glowing Peace ™ *see* 'MEIZOELO'

Glowing Petals *see* 'MANPETALS'

'GLOWING SUNSET', HT, ob, 1933; (Wilhelm Breder); bud long, pointed; flowers orange shaded yellow and pink, very large, dbl., exhibition form, intense fragrance; foliage leathery, glossy, dark; vigorous growth; [Fontanelle X Julien Potin]; Kordes; Dreer

'GLOWING VELVET', HT, dr, 1975; flowers deep crimson to scarlet, dbl., 25 petals, 4 in., bloom repeats quickly, intense fragrance; foliage dark; Pasley

Glowry® *see* 'KINGLOW'

Glucksburg, S, pb, 1988; Jensen

'GLÜCKSKIND', HT, dp, 1935; flowers deep pink, well formed, large, moderate fragrance; Berger

'GLÜCKSKIND', F, dr, 1952; flowers dark crimson, medium, semi-dbl.; very vigorous growth; Leenders, M.

'GLYNDYFRDWY', HT, mp, 1978; flowers neyron rose, full, dbl., 35 petals, 4–5 in., moderate fragrance; foliage large, light; very vigorous growth; [Gavotte X George Thomas]; Ellick; Excelsior Roses

'GLYNIS BRYAN', Min, pb, 1997; flowers double, medium, dbl., 15–25 petals, borne in large clusters, no fragrance; some prickles; foliage small, medium green, semi-glossy; spreading, tall (9ft.) growth; [Party Girl X Sheri Anne]; Jones, L.J.

'GNEISENAU', S, w, 1924; flowers snow-white, stamens yellow, borne in clusters, dbl., non-recurrent; height 5-6 ft; [(Schneelicht X Killarney) X Crimson Rambler]; Lambert, P.

Gnom, F, op, 1957; flowers salmon-orange and pink, medium, dbl.; Berger, W.; (Sangerhausen)

'GNOME', Pol, ly, 1936; flowers cream-yellow, large, dbl., intense fragrance; foliage leathery, light; short stems; bushy, dwarf growth; [Unnamed seedling X Mev. Nathalie Nypels]; Leenders, M.

Gnome World *see* JACpat

Godavari, HT, or, 1987; Kasturi

Goddess *see* **'DÉESSE'**

Godescalcus Vulf de Sapprothe, S, lp, 1984; flowers small, semi-dbl.; Scholle, E.; (Sangerhausen)

'GODEURPAN', F, ob, 1973; (**Ibis**); bud ovoid; flowers orange, semi-dbl., 14–16 petals; foliage dark; [Europeana X Orangeade]; Godin, M.

Godewind® *see* KORbraufa

'GODFREY WINN', HT, m, 1968; flowers purplish, dbl., globular, intense fragrance; Dot; Wheatcroft & Sons

'GODFREY'S RED PETITE', HT, dr, 1966; flowers deep red, medium, semi-dbl., 15 petals, slight fragrance; foliage dark; free growth; [Baccará X Audie Murphy]; Godfrey

'GODIALING', HT, or, 1975; (**Chrystelle**); bud oval; flowers deep, dbl., 30–35 petals, 3.5 in., cupped; foliage dark; [(Lady Zia X Wizo) X Silver Lining]; Godin, M.

'GODILOFTER', HT, dr, 1976; (**Maman Pineau**); bud pointed; flowers dark red-purple, dbl., 26–28 petals, 3–3.5 in., cupped, slight fragrance; foliage bronze; vigorous growth; [Maryse Kriloff X Uncle Walter]; Godin, M.

'GODRACHE', HT, m, 1973; (**Grand Duche**); bud ovoid; dbl., 25 petals, 3–3.5 in., cupped, slight fragrance; foliage dark; vigorous growth; [American Heritage X (Seedling X Blue Girl)]; Godin, M.

'GODSENSOR', F, or, 1973; (**Danse des Étoiles**); bud ovoid; flowers red-orange, deep, semi-dbl., 15–18 petals, 2.5 in., cupped; foliage light green; [Orangeade X Orange Sensation]; Godin, M.

'GODUSEX', Gr, mp, 1974; (**Exodus**); bud ovoid; very dbl., 50–52 petals, 3 in., cupped; foliage glossy, dark; [Kordes' Perfecta X Kalinka]; Godin, M.

'GODZOTY', HT, mr, 1976; (**Duchesse d'Anjou**); bud pointed; flowers crimson-red, 3–3.5 in., cupped, slight fragrance; foliage dark; vigorous growth; [Wizo X Soraya]; Godin, M.

Goedele, HT, lp, 1998; RvS-Melle

'GOETHE', M, m, 1911; flowers magenta, small, single; foliage blue-green, rough; new wood bright red; very vigorous growth; Lambert, P.

'GOLCONDA', HT, ly, 1968; bud ovoid; flowers pale yellow, center deep apricot, large, dbl., cupped, intense fragrance; foliage leathery; moderate, bushy, compact growth; [Mme Charles Sauvage X ?]; Pal, Dr. B.P.; Indian Agric. Research Inst.

Gold Badge™ *see* 'MEIGRONURI'

Gold Badge, Climbing *see* MEIgro

'GOLD BLAZE', Min, yb, 1980; bud ovoid, pointed; flowers yellow, dipped red, blooms borne singly or several together, dbl., 23 petals, intense fragrance; tiny, straight prickles; foliage small, glossy, deepgreen; compact, upright growth; [Unnamed seedling X Unnamed seedling]; Lyon

Gold Britannia, HT, my

Gold Bunny® *see* 'MEIGRONURI'

'GOLD COAST', Gr, my, 1958; (Golden Pride); bud ovoid; flowers clear yellow, overcast buff-yellow, dbl., 25–30 petals, 4 in., cupped; foliage leathery, glossy; vigorous, upright growth; [Pinocchio X Peace]; Robinson, H.; J&P

'GOLD COIN', Min, dy, 1967; flowers buttercup-yellow, small, dbl., moderate fragrance; vigorous, bushy growth; [Golden Glow (LCl) X Magic Wand]; Moore, Ralph S.; Sequoia Nursery

Gold Country *see* 'SEAGOLD'

Gold Crest *see* **'GOLDEN CREST'**

'GOLD CROWN', HT, dy, 1960; (Corona de Oro, Couronne d'Or, Gold Krone, Goldkrone); flowers golden yellow, well-formed, dbl., 35 petals, 5 in., moderate fragrance; foliage leathery, dark; vigorous, upright growth; [Peace X Golden Scepter]; Kordes, R.; McGredy

'GOLD CUP', F, dy, 1957; (Coupe d'Or); bud pointed; flowers golden yellow, blooms in clusters, dbl., 28 petals, 4 in., moderate fragrance; foliage dark, glossy; bushy growth; AARS, 1958; [Goldilocks seedling X King Midas seedling]; Boerner; J&P

'GOLD DAME', HT, dy, 1929; flowers deep golden yellow, semi-dbl., moderate fragrance; foliage dark, glossy; vigorous, bushy growth; Dobbie

Gold Dollar *see* 'HERDIO'

'GOLD DOT', HT, my, 1963; flowers large, dbl., 25 petals; vigorous, upright growth; [Queen Elizabeth X Peace]; Dot, Simon

Gold Fantasy, Min, dy, 1998

Gold Fever *see* 'MACPARLEZ'

Gold Fever *see* 'MORFEVER'

'GOLD GLOW', HT, dy, 1959; flowers bright yellow, very dbl., 100 petals, 3.5–4 in., moderate fragrance; foliage leathery, dark, glossy; vigorous, upright growth; [Fred Howard X Sutter's Gold]; Perry, Anthony; C.R. Burr

Gold Glow Bronze Sport, HT, yb

'GOLD GLOW, CLIMBING', Cl HT, dy, 1964; Burr, C.R.

Gold Heart *see* **'BURNABY'**

Gold Heart *see* 'MACYELKIL'

Gold Krone *see* **'GOLD CROWN'**

'GOLD MAGIC', F, dy, 1990; flowers golden yellow, medium, borne in large clusters, dbl., 15–25 petals, slight fragrance; foliage medium, dark green, glossy;

bushy, medium growth; [Gold Badge X Friesensohne]; Christensen, Jack E.; Vaughan's Seed Co., 1991

Gold Magic Carpet *see* 'POULURT'

Gold Medal® *see* 'AROYQUELI'

'GOLD MINE', HT, my, 1925; flowers indian yellow paling toward edges, base deep orange, dbl., intense fragrance; RULED EXTINCT 9/84; [Golden Rule X Mrs Aaron Ward]; Hill, Joseph H., Co.; A.N. Pierson;, J.H. Hill Co.

Gold Moon® *see* 'HAVOON'

'GOLD 'N' FLAME', Min, rb, 1980; bud long, pointed; flowers medium red, deep golden yellow reverse, borne singly, dbl., 33 petals, cupped, slight fragrance; very thin, long, tan prickles, curved down; foliage dark, glossy; upright, bushy growth; [Unnamed seedling X Over the Rainbow]; Williams, Ernest D.; Mini-Roses

'GOLD 'N' HONEY', HT, yb, 1976; bud long, pointed; flowers yellow and peach, edged rose, dbl., 28 petals, 5–6. in., exhibition form, intense fragrance; vigorous, upright, bushy growth; [Helen Traubel X (Unnamed seedling X Ulster Monarch)]; Leon, Charles F., Sr.; Edmunds Roses

'GOLD NUGGET', F, dy, 1972; flowers bright yellow, medium, dbl., exhibition form, moderate fragrance; foliage glossy, abundant; vigorous, upright growth; Patterson; Patterson Roses

Gold of Ophir *see* **'FORTUNE'S DOUBLE YELLOW'**

Gold Patio, MinFl, dy, 1997

'GOLD PIN', Min, dy, 1974; flowers bright golden yellow, semi-dbl., 18 petals, 1 in., slight fragrance; foliage bronze; Mattock

'GOLD PIQUE', Min, my, 1977; bud pointed; flowers small, dbl., 36 petals, 1.5 in., moderate fragrance; foliage small, dark; compact, bushy growth; [Unnamed seedling X Yellow Jewel]; Lyon

Gold Reef *see* POUIdom

'GOLD RUSH', LCl, yb, 1941; flowers gold, dbl., 24 petals, exhibition form, not dependably recurrent, moderate fragrance; foliage glossy, ivy-green; vigorous, climbing growth; Duehrsen; H&S

Gold Rush *see* 'KIRSUN'

'GOLD SPRAY', F, my, 1971; flowers full, medium, dbl.; foliage soft; moderate, bushy growth; [Philippe X Spek's Yellow]; Delforge

'GOLD STAR', HT, yb, 1933; bud pointed, orange; flowers golden yellow shaded orange, large, dbl.; foliage glossy, bronze, leathery; very vigorous growth; RULED EXTINCT 6/83 ARM; [Souv. de Claudius Pernet X Talisman]; Vestal

Gold Star *see* 'CANDIDE'

Gold Star *see* 'TANTERN'

'GOLD STRIKE', F, my, 1955; bud urn shaped; flowers lemon-yellow, borne in rounded clusters, dbl., 30–35 petals, 2–2.5 in., exhibition form, moderate fragrance; foliage leathery; vigorous, bushy, compact growth; [Goldilocks X Pinocchio]; Swim, H.C.; Armstrong Nursery

Gold Strike® *see* 'SUNLUCK'

Gold Sweetheart™ *see* 'WILGOSH'

Gold Symphonie® *see* 'MEITOLEIL'

'GOLD TOP', LCl, dy, 1978; flowers golden yellow, full, dbl., 25 petals, 5.5 in., moderate fragrance; foliage large, light matt green; free growth; Pearce; Limes Rose Nursery

Gold Topaz *see* 'KORTOSSGO'

Goldbeet®, F, dy, 1974; Noack, Werner

'GOLDBONNET', S, my, 1973; flowers, semi-dbl., 13 petals, 4 in., slight fragrance; foliage large, glossy; [(Ann Elizabeth X Allgold) X Golden Showers]; Harkness

'GOLDBUSCH', S, my, 1954; bud long, pointed; flowers yellow, becoming lighter, large blooms in clusters (up to 20), semi-dbl.; foliage leathery, glossy, light green; very vigorous, upright, bushy growth; Kordes

'GOLDDIGGER', HT, dy, 1963; flowers dark saffron-yellow, large, dbl., 50–55 petals, slight, spicy fragrance; foliage glossy, bronze; upright growth; [Marcelle Gret X Dries Verschuren]; Verschuren, A.; Stassen

Golddorf Seppenrade, S, dy, 1969; flowers large, dbl.; Scholle, E.; (Sangerhausen)

'GOLDDUST', HT, my, 1963; flowers golden yellow; foliage clear green; vigorous growth; [Brandywine X Seedling]; Delforge

'GOLDELSE', HT, ob, 1900; flowers golden orange; [Kaiserin Auguste Viktoria seedling]; Hinner, W.

Golden Afternoon, HT, ab, 1984; Pal, Dr. B.P.

'GOLDEN ALTAI', HSpn, ly, 1943; flowers cream to pale yellow, single, non-recurrent; very hardy.; [R. spinosissima altaica X Harison's Yellow]; Wright, Percy H.

'GOLDEN ANGEL', Min, dy, 1975; bud short, pointed; dbl., 65 petals, 1 in., moderate fragrance; foliage matt; bushy, compact growth; [Golden Glow (LCl) X (Little Darling X Seedling)]; Moore, Ralph S.; Sequoia Nursery

Golden Anniversary *see* 'JACARY'

'GOLDEN ANNIVERSARY', HT, dy, 1948; bud ovoid; dbl., 50–60 petals, 4.5–5 in., exhibition form, intense fragrance; foliage leathery; vigorous, upright, bushy growth; RULED EXTINCT 9/82 ARM; [Good News sport]; Mordigan Evergreen Nursery; C-P

Golden Anniversary, MinFl, my, 1997

Golden Arches®, Min, dy

'GOLDEN ARCTIC', LCl, yb, 1954; flowers yellow to orange, dbl., 38 petals, 3.5–4 in., moderate fragrance; growth like a hybrid tea, followed by 4-5 ft. canes; [Unnamed seedling X Free Gold]; Brownell, H.C.

Golden Autumn, HT, dy, 1955; flowers large, dbl., slight fragrance; Klimenko, V. N.; (Sangerhausen)

'GOLDEN BAY', S, yb, 1979; bud ovoid; flowers deep buff-yellow, shapely, dbl., 40 petals, 3.5 in., slight fragrance; foliage large; spreading, bushy growth; [(Tropicana X Sabine) X Zitronenfalter]; Murray, Nola

'GOLDEN BEAUTY', HT, ab, 1937; bud very long; flowers orange buff-yellow, stamens golden, large, semi-dbl.; foliage clear green, glossy; vigorous, bushy growth; Van Rossem

Golden Beauty *see* 'CLEBEAU'

Golden Beryl *see* 'MANBERYL'

Golden Bettina, HT, yb; [Bettina sport]; Ruston, D., 1970

Golden Blush, A, ab, 1988; Sievers

Golden Border, S, my, 1993; Verschuren

Golden Bounty™ *see* 'JACSUYEL'

Golden Bouquet, F, dy

'GOLDEN BOY', HT, dy, 1964; flowers deep yellow, very, dbl., 35 petals, 5.5 in., exhibition form; foliage long, pointed; moderate growth; [Golden Masterpiece X Belle Blonde]; McGredy, Sam IV; Spek

'GOLDEN BUTTERFLY', HT, ab, 1920; flowers apricot-yellow, shaded carmine, dbl., moderate fragrance; [Old Gold seedling]; Therkildsen

Golden Butterfly *see* **'GOUDVLINDER'**

'GOLDEN CALIFORNIA', HT, dy, 1966; flowers golden, large, dbl., cupped, moderate fragrance; foliage bronze, leathery, glossy; tall, bushy growth; [California sport]; Howard, P.J.

'GOLDEN CASCADE', LCl, my, 1962; bud ovoid; flowers chrome-yellow, dbl., 25–30 petals, 4.5–5 in., cupped, moderate, fruity fragrance; foliage leathery; vigorous (10-12 ft) growth; [(Capt. Thomas X Joanna Hill) X Lydia]; Morey, Dr. Dennison; J&P

Golden Celebration *see* 'AUSGOLD'

'GOLDEN CENTURY', Cl Min, ob, 1978; bud pointed; flowers cadmium-orange to nasturtium-red, dbl., 35 petals, 1.5–2 in., intense fragrance; foliage glossy, leathery; moderate climber growth; [(R.

wichurana X Floradora) X (Sister Therese X Unnamed Miniature)]; Moore, Ralph S.; Sequoia Nursery

'GOLDEN CHALICE', HT, dy, 1960; bud ovoid; flowers clear yellow, open, dbl., 40–45 petals, 4 in., moderate fragrance; foliage glossy; vigorous, upright growth; [(Starlite X Snow White) X Golden Masterpiece]; Boerner; J&P

'GOLDEN CHARM', HT, dy, 1933; [Talisman sport]; Groshens & Morrison

'GOLDEN CHARM, CLIMBING', Cl HT, dy, 1948; Krider Nursery

Golden Chersonese *see* 'HILGOLD'

'GOLDEN CHOICE', HT, my, 1967; flowers lemon-yellow; [My Choice sport]; Bardill Nursery; LeGrice Roses

'GOLDEN CITY', HT, my, 1922; bud golden yellow; flowers light buff; [Rayon d'Or X Frau Karl Druschki]; Lippiatt

Golden Climber *see* **'MRS ARTHUR CURTISS JAMES'**

Golden Coach *see* 'ZIPGOLD'

Golden Colonel, F, ob

'GOLDEN COMET', HT, yb, 1937; bud long, pointed; flowers yellow and pink, open, large, semi-dbl., intense fragrance; foliage dark, leathery; vigorous growth; Burbank; Stark Bros.

'GOLDEN CORONET', F, my, 1967; flowers medium, dbl., exhibition form, slight fragrance; foliage glossy, leathery; vigorous, compact growth; [(Lydia X Golden Scepter) X Isobel Harkness]; Morey, Dr. Dennison; Country Garden Nursery

'GOLDEN CREST', Cl HT, my, 1948; (Gold Crest); flowers pure yellow, dbl., 3–4 in., moderate fragrance; foliage glossy, dark; good pillar (6 ft) growth; Archer

Golden Dance® *see* BARgold

'GOLDEN DAWN', HT, my, 1929; bud yellow, flushed pink; flowers well-formed, dbl., 45 petals, intense fragrance; low, spreading growth; [Elegante X Ethel Somerset]; Grant; Hazelwood Bros;, Prior

'GOLDEN DAWN, CLIMBING', Cl HT, my, 1937; Knight, G.

'GOLDEN DAWN, CLIMBING', Cl HT, my, 1935; Armstrong, J.A.

'GOLDEN DAWN, CLIMBING', Cl HT, my, 1947; LeGrice

'GOLDEN DAY', HT, my, 1931; flowers bright golden yellow, center deeper, larger and fuller, intense fragrance; vigorous growth; [Independence Day sport]; Bentley; Harkness

Golden Days *see* 'RUGOLDA'

'GOLDEN DELIGHT', F, my, 1956; flowers canary-yellow, dbl., 58 petals, 3 in., moderate fragrance; foliage dark, glossy; dwarf growth; [Goldilocks X Ellinor LeGrice]; LeGrice

'GOLDEN DIAMOND', HT, dy, 1943; flowers large, dbl., 30 petals; foliage leathery, dark; strong stems; vigorous, upright, compact growth; Verschuren; L.C. Lovett

'GOLDEN DREAM', HRg, my, 1932; (Goldener Traum); bud pointed, streaked red; flowers pure yellow, large, dbl., recurrent bloom, intense fragrance; very vigorous (6 1/2 ft.) growth; [Turkes Rugosa Samling X Constance]; Türke; J.C. Schmidt

'GOLDEN DROP', HT, dy, 1939; flowers rich yellow, small blooms in clusters, semi-dbl.; tall growth; [Mme Mascuraud X Seedling]; Clark, A.

Golden Eagle, Gr, dy, 1997

Golden Earing™ *see* 'WILGOLD'

Golden Elegance, HT, lp; long; flowers large, exhibition form, 30 petals, single or in clusters of 3-7 florets, good cut flower, large, dark green, glossy foliage; [Ambassador sport]; St. Wagner, 1993; Res. Stn. f. Fruit Growing, Cluj, 1995, Romania

Golden Emblem *see* 'JACGOLD'

'GOLDEN EMBLEM', HT, my, 1917; bud yellow, splashed and shaded red; flowers canary-yellow, well formed, large, dbl., moderate fragrance; foliage dark, glossy, leathery; vigorous growth; GM, NRS, 1915; [Mme Mélanie Soupert X Constance]; McGredy

'GOLDEN EMBLEM, CLIMBING', Cl HT, my, 1927; Armstrong Nursery

'GOLDEN EMPIRE', HT, ob, 1957; flowers orange, dbl.; foliage leathery, glossy; [Orange Everglow X Golden Emblem]; Silva

'GOLDEN FAIRY', Pol, ly, 1889; flowers clear buff, yellow and white; dwarf growth; Bennett

Golden Fantasie *see* 'HILGOFAN'

'GOLDEN FICTION', F, my, 1958; free growth; [Yellow Pinocchio X Moonbeam]; Spek

Golden Flame® *see* LAPam

'GOLDEN FLEECE', F, my, 1955; (Toison d'Or); bud ovoid; flowers buff-yellow, in clusters (to 20), dbl., 38 petals, 4.5 in., cupped, intense fragrance; foliage leathery; vigorous, bushy growth; GM, Bagatelle, 1955; [Diamond Jubilee X Yellow Sweetheart]; Boerner; J&P

Golden Flipper, F, dy

Golden Fox, MinFl, dy, 1997; Warner, Chris

Golden Friendship *see* HARtellody

'GOLDEN FRILLS', HT, my, 1936; flowers rich golden yellow, dbl., cupped; foliage glossy, wrinkled; [Feu Joseph Looymans sport]; B&A

Golden Future *see* 'HORANYMOLL'

Golden Galaxy *see* JACyesp

Golden Gardens *see* 'MORGOGARD'

'GOLDEN GARNETTE', F, dy, 1960; bud ovoid; flowers golden yellow, edged lighter, blooms in clusters, dbl., 33 petals, 3–4 in., cupped, intense, fruity fragrance; foliage leathery, dark, glossy; vigorous, upright, bushy growth; [(Goldilocks seedling X Seedling) X Tawny Gold]; Boerner; J&P

'GOLDEN GATE', HT, my, 1972; bud ovoid; flowers large, dbl., exhibition form, slight fragrance; foliage large, glossy; vigorous, upright, bushy growth; [South Seas X King's Ransom]; Warriner, William A.; J&P

'GOLDEN GATE', T, w, 1891; bud pointed; flowers cream-white, anthers golden yellow, very large, dbl., cupped, moderate fragrance; foliage bright green; vigorous growth; [Safrano X Cornelie Koch]; Dingee & Conard

'GOLDEN GEM', HT, my, 1916; flowers golden yellow; [Lady Hillingdon X Harry Kirk]; Towill

Golden Giant® *see* 'KORBI'

'GOLDEN GIANT, CLIMBING', Cl HT, dy, 1967; Laveena Roses

Golden Girl *see* 'MEIVIRGI'

Golden Girls *see* 'CLEBEAU'

'GOLDEN GLAMOUR', F, my, 1951; bud pointed; flowers large, dbl., 25 petals, exhibition form, moderate fragrance; foliage glossy; vigorous, upright, bushy growth; [Joanna Hill X (Mrs Pierre S. duPont X Amelia Earhart)]; Boerner; J&P

'GOLDEN GLEAM', HT, my, 1926; flowers buttercup-yellow, outer petals streaked, dbl., 25 petals, slight, fruity fragrance; foliage dark; McGredy; Beckwith

Golden Globe, Min, dy

'GOLDEN GLORY', HT, dy, 1931; flowers deep golden yellow large, dbl., intense fragrance; Dobbie

Golden Gloves, F, dy, 1991; Bear Creek Gardens

'GOLDEN GLOW', LCl, my, 1937; dbl., 3.5–5 in., exhibition form, blooms in clusters, non-recurrent, moderate fragrance; foliage leathery, glossy, dark; very vigorous, climbing (20 ft.) growth; [Glenn Dale X (Mary Wallace X HT)]; Brownell, H.C.

'GOLDEN GLOW', HT, ab, 1918; flowers apricot, shaded bronzy orange; [Mme Edouard Herriot sport]; Chaplin Bros.

'GOLDEN GRUSS AN AACHEN', F, ob, 1935; bud pointed, red; flowers golden orange, sometimes shaded reddish, very large, dbl., exhibition form, moderate fragrance; foliage glossy; bushy growth; [Mme Butterfly X Gloria Mundi]; Kordes

Golden Halo® *see* 'SAVAHALO'

Golden Hands *see* CHEssupremo

Golden Handshake *see* 'CHEWSUNFORD'

'GOLDEN HARVEST', HT, dy, 1943; flowers clear yellow, dbl., 35 petals, 4.5 in., exhibition form, moderate fragrance; foliage leathery, glossy, bronze; vigorous growth; [McGredy's Ivory X Unnamed seedling]; Mallerin, C.; C-P

'GOLDEN HAZE', HT, yb, 1965; bud ovoid; flowers light golden yellow, large, dbl., moderate fragrance; foliage glossy; vigorous growth; [Peace X Golden Rapture]; Verschuren, H.A.M.; J&P

Golden Heart *see* **'BURNABY'**

'GOLDEN HERITAGE', HT, my, 1974; flowers canary-yellow, pointed, dbl., 25 petals, 3.5 in., moderate, tea fragrance; bushy growth; [Golden Masterpiece X Unnamed seedling]; Herholdt, J.A.

Golden Hit®, Min, dy

Golden Holstein® *see* KORtikel

Golden Hope *see* 'MEHPIC'

'GOLDEN HOUR', HT, ob, 1952; bud ovoid; flowers golden yellow, reverse orange, dbl., 45–55 petals, 4–5 in., exhibition form, intense fragrance; foliage leathery, glossy; very vigorous, upright growth; [Los Angeles X California]; Howard, P.J.

'GOLDEN IDEAL', HT, dy, 1939; bud long, pointed; flowers brilliant chrome-yellow, large, dbl.; vigorous growth; [Roselandia X Joanna Hill]; Lens

Golden Jet *see* 'TANTASCH'

Golden Jet *see* 'TANTASCH'

'GOLDEN JEWEL' *see* TANledolg

Golden Jubilee *see* 'COCAGOLD'

'GOLDEN JUBILEE', F, my, 1948; bud ovoid; flowers golden yellow, becoming buff and chrome, in clusters, dbl., 3 in., slight fragrance; foliage glossy, dark; vigorous, bushy, compact growth; [(Mary Wallace X Talisman) X Mrs Pierre S. duPont]; Jacobus; B&A

Golden Julia® *see* 'RUPGOLJUL'

'GOLDEN KING', HRg, ly, 1935; flowers pale yellow; [Dr. Eckener sport]; Beckwith

'GOLDEN LACE', F, dy, 1962; bud pointed; dbl., 28 petals, 3 in., exhibition form, blooms in clusters, moderate fragrance; foliage glossy; vigorous, upright growth; [Goldilocks X (Golden Scepter X Encore)]; Von Abrams; Peterson & Dering

'GOLDEN LEADER', HT, ab, 1961; flowers apricot, moderate fragrance; [Tawny Gold X Seedling]; Leenders, J.

'GOLDEN LEOPARD', F, my, 1976; bud pointed; flowers daffodil-yellow, small, dbl., 18–25 petals, 2 in., exhibition form, moderate fragrance; foliage glossy, leathery; low, compact growth; [Golden Slippers seedling X Golden Slippers seedling]; Takatori, Yoshiho; Japan Rose Nursery

'GOLDEN LIGHT', LCl, ob, 1939; bud pointed; flowers orange-apricot to buff, edged pink, open, large, dbl., slight fragrance; foliage glossy, dark; strong stems; vigorous, climbing growth; Nicolas

'GOLDEN LION', HSet, my, 1944; flowers clear golden yellow, open, borne in clusters, cupped, non-recurrent; foliage light, glossy; height 8-10 ft; Horvath

'GOLDEN LUSTRE', HT, ab, 1964; bud ovoid; flowers bronze-apricot overcast yellow, dbl., 60–65 petals, 5 in., cupped, moderate fragrance; foliage leathery; moderate growth; [Kate Smith X Tanya]; Boerner; J&P

'GOLDEN MAIN', HT, dy, 1933; (Golden Romance, Goldenes Mainz, Golmain); bud golden yellow striped red; flowers golden yellow, large, dbl., cupped, intense fragrance; foliage glossy; very vigorous, bushy growth; [Fontanelle X Julien Potin]; Kordes; J&P

'GOLDEN MASTERPIECE', HT, my, 1954; bud long, pointed; flowers golden yellow, very large, dbl., 30–35 petals, exhibition form, borne singly, moderate, licorice fragrance; foliage very glossy; vigorous, upright growth; [Mandalay X Golden Scepter]; Boerner; J&P

'GOLDEN MASTERPIECE, CLIMBING', Cl HT, my, 1957; Valdrez

Golden Medaillon *see* 'KORIKON'

'GOLDEN MEDALLION', HT, my, 1981; bud long, pointed; flowers medium golden yellow, loosely imbricated, dbl., 40–50 petals, 7–8 in., slight fragrance; foliage large, light; upright growth; [Medallion sport]; Permenter; J&P

Golden Medallion *see* KORnanze

Golden Meillandina *see* **'RISE 'N' SHINE'**

Golden Melody *see* **'IRENE CHURRUCA'**

'GOLDEN MEMORIES', HT, dy, 1965; bud long, pointed, canary-yellow tinted coppery; flowers deep canary-yellow, to open, dbl., 28–32 petals, 5–6 in., exhibition form, moderate, spicy fragrance; foliage glossy; vigorous, upright growth; [Golden Rapture sport]; Ravine; Endres Floral Co.

'GOLDEN MME SEGOND WEBER', HT, op, 1923; flowers salmon, center yellow, dbl.; [Mme Segond Weber X Primerose]; Soupert & Notting

Golden Moments *see* FRYtranquil

Golden Monica® *see* TANgolca

'GOLDEN MOSS', M, my, 1932; bud globular, peach-yellow, sepals well mossed; flowers tawny yellow, large blooms in clusters of 3-5, dbl., 37 petals, scanty bloom; no repeat, moderate fragrance; foliage almost rugose; vigorous growth; [Frau Karl Druschki X (Souv. de Claudius Pernet X Blanche Moreau)]; Dot, Pedro; C-P

Golden Mozart, S, my, 1986; Verschuren

Golden Mrs Sam McGredy *see* **'GOLDEN SAM MCGREDY'**

Golden Nugget® *see* MEIlsinplox

Golden Oldie *see* 'MACGOLOLD'

'GOLDEN OPHELIA', HT, my, 1918; flowers golden yellow in center, paling slightly on outer petals; foliage glossy; vigorous growth; GM, NRS, 1918; [Ophelia seedling]; Cant, B. R.

'GOLDEN OPHELIA, CLIMBING', Cl HT, my, 1924; Hage; Prior

Golden Opportunity® *see* 'ANTOPP'

'GOLDEN ORANGE CLIMBER', LCl, ob, 1937; flowers orange to orange-scarlet, often overlaid golden yellow, semi-dbl., 5 in.; vigorous, climbing growth; [Mrs Arthur Curtiss James sport]; Brownell, H.C.

'GOLDEN PAMELA', HT, yb, 1969; bud ovoid; flowers yellow-apricot and pink, large, dbl., cupped, slight fragrance; foliage large, glossy; vigorous, upright growth; [Wellworth sport]; Wheatley

'GOLDEN PEACE', HT, my, 1961; flowers canary-yellow, large, dbl., 45 petals, exhibition form, moderate fragrance; foliage dark, dull; very vigorous, tall growth; LeGrice

'GOLDEN PEARL', HT, my, 1967; flowers large, single, slight fragrance; foliage dark; free growth; Warmerdam

Golden Penny *see* 'COOSYN'

'GOLDEN PERFECTION', Pol, my, 1937; flowers golden yellow, small, dbl., intense fragrance; bushy, dwarf growth; Leenders, M.

'GOLDEN PERFUME', F, my, 1959; flowers orange-yellow becoming golden yellow, large, borne in clusters, dbl., 50 petals, moderate fragrance; foliage dark, glossy, leathery; vigorous, bushy growth; [Goldilocks X Fashion]; Leenders, J.

Golden Pernet *see* **'JULIEN POTIN'**

'GOLDEN PERRAUD', HT, my, 1946; bud long, pointed, well formed; flowers brilliant golden yellow, large, dbl., slight fragrance; foliage glossy; vigorous growth; [Mme Joseph Perraud sport]; Lens

'GOLDEN PHEASANT', F, ob, 1951; (Goldfasan); flowers orange and gold, imbricated, dbl., 40 petals, 4 in., moderate fragrance; foliage glossy; low, compact growth; [Pres. Ferier X Dr. Debat]; Kordes; Wheatcroft Bros.

Golden Phoenix *see* 'NAKBET'

Golden Piccolo *see* 'POULTEX'

'GOLDEN PICTURE', HT, yb, 1967; flowers light yellow tinged pink, medium, slight fragrance; foliage leathery; upright growth; [Picture seedling X Marcelle Gret]; Handover, P.&R.

Golden Pillar®, LCl, my, 1999; flowers large, full, loose, golden yellow, nonfading, very dbl., 30–35 petals, 4 in., slight fragrance; foliage medium, deep green, glossy; Meilland

'GOLDEN PIRRIE', HT, ly, 1921; flowers yellowish white; [Lady Pirrie sport]; Dobbie

'GOLDEN PIXIE', Min, my, 1985; flowers medium golden yellow, medium, dbl., 60 petals, slight fragrance; foliage small, medium green, semi-glossy; vigorous, upright, bushy growth; [Yellow Pages X Rise 'n' Shine]; Hardgrove, Donald L.; Rose World Originals

Golden Plover, S, dy, 1991; Poulsen; (Cavriglia)

'GOLDEN POLY', Pol, yb, 1931; flowers pure yellow, edged carmine, dbl., globular; dwarf growth; [Angèle Pernet X Orange King]; Pahissa

'GOLDEN POLY', Pol, my, 1935; bud yellow with red lines; flowers golden yellow to yellowish white, open, large, semi-dbl., slight fragrance; foliage light, glossy; bushy, dwarf growth; Leenders, M.

Golden Pride *see* **'GOLD COAST'**

Golden Prince *see* 'MEIGOLD'

'GOLDEN PRINCESS', S, yb, 1984; bud ovoid, pointed; flowers yellow, petals edged deep pink, large blooms borne 1-5 per cluster, dbl., 33 petals, cupped, repeat bloom, moderate fragrance; large, hooked, tan prickles; foliage medium, olive green, semi-glossy; upright, bushy growth; hardy; [Hawkeye Belle X (Roundelay X Country Music)]; Buck, Dr. Griffith J.; Iowa State University

Golden Promise *see* 'LAVLUV'

Golden Promise, F, dy, 1972; deRuiter

'GOLDEN PYRAMID', LCl, my, 1939; flowers large, semi-dbl., free seasonal bloom; vigorous pyramid to 5-6 ft growth; Brownell, H.C.

'GOLDEN QUEEN', HT, my, 1937; bud gold, slightly marked carmine; flowers golden yellow, large, dbl.; foliage glossy; very vigorous, bushy growth; Chambard, C.

Golden Queen *see* KORgitte

'GOLDEN RAIN', F, my, 1951; (Goldregen); flowers golden yellow, well-formed, large, borne in clusters of 10-15, semi-dbl., 20 petals; foliage glossy; upright, bushy growth; [Swantje X G. Bentheim]; Tantau

Golden Rambler *see* **'ALISTER STELLA GRAY'**

Golden Rambler *see* **'EASLEA'S GOLDEN RAMBLER'**

'GOLDEN RAPTURE', HT, dy, 1933; (Geheimrat Duisberg); bud pointed; flowers golden yellow, very large, dbl., 40 petals, moderate fragrance; foliage glossy; vigorous growth; [Rapture X Julien Potin]; Kordes; Dreer, 1934;, H&S, 1934

Golden Rapture No. 5, HT, my; bud ovoid; flowers clear yellow, dbl., 25 petals, 5 in., exhibition form, intense fragrance; foliage dark, leathery; vigorous growth; [Golden Rapture sport]; Krieter; A.N. Pierson

'GOLDEN RAPTURE, CLIMBING', Cl HT, dy, 1941; Swim, H.C.; Armstrong Nursery

'GOLDEN RAPTURE, CLIMBING', Cl HT, dy, 1954; (Climbing Geheimrat Duisberg); Knackfuss

'GOLDEN REVELRY', HT, my, 1952; flowers golden yellow, dbl., 21 petals, exhibition form, slight fragrance; foliage glossy, bright green; [Phyllis Gold X Blossom]; McGredy, Sam IV

Golden Rider *see* **'GOLDENER REITER'**

Golden Romance *see* **'GOLDEN MAIN'**

Golden Rosamini *see* INTergol

Golden Rose of China *see* **R. HUGONIS**

'GOLDEN RUFFELS', HT, my, 1954; bud long, pointed; flowers golden yellow, medium, dbl., moderate fragrance; upright growth; [Orange Ruffels sport]; Brownell, H.C.

'GOLDEN RULE', HT, my, 1918; flowers clear yellow, dbl.; [Ophelia seedling X Sunburst]; Hill, E.G., Co.

'GOLDEN SALMON', Pol, ob, 1926; (Goldlachs); flowers pure orange, large blooms in huge trusses; vigorous, bushy growth; [Superb sport]; deRuiter

Golden Salmon Improved *see* **'GOLDEN SALMON SUPÉRIEUR'**

'GOLDEN SALMON SUPÉRIEUR', Pol, ob, 1929; (Golden Salmon Improved); flowers different shade of orange; [Golden Salmon sport]; deRuiter; Sliedrecht & Co.

'GOLDEN SALUTE', HT, dy, 1963; bud ovoid; flowers golden yellow, dbl., 33 petals, 5–5.5 in., cupped, moderate fragrance; foliage leathery, glossy; vigorous, moderately tall growth; [Diamond Jubilee seedling X Golden Masterpiece]; Boerner; J&P

'GOLDEN SAM MCGREDY', HT, yb, 1935; (Golden Mrs Sam McGredy); flowers chrome-yellow, reverse salmon with chrome; [Mrs Sam McGredy sport]; Lens

'GOLDEN SÁSTAGO', HT, dy, 1938; flowers clear yellow, large, dbl., globular, moderate fragrance; foliage soft; vigorous growth; [Condesa de Sástago sport]; Dot, Pedro; C-P

'GOLDEN SCEPTER', HT, dy, 1950; (Spek's Yellow); bud pointed; flowers deep yellow, dbl., 35 petals, 4.5 in., exhibition form, moderate fragrance; foliage leathery, glossy; vigorous, upright growth; [Golden Rapture X Unnamed seedling]; Verschuren-Pechtold; J&P;, Spek

'GOLDEN SCEPTER, CLIMBING', Cl HT, dy, 1956; (Climbing Spek's Yellow); Walters

'GOLDEN SÉVERINE', HT, dy, 1929; flowers deep golden yellow; Morse

'GOLDEN SHEEN', F, my, 1966; bud urn shaped; flowers yellow, edged lighter, borne in clusters, dbl., exhibition form; foliage leathery; moderate, upright, bushy growth; [Ophelia X Circus]; Swim & Weeks; Carlton Rose Nurseries

'GOLDEN SHOT', F, dy, 1973; flowers golden yellow, dbl., 24 petals, 4 in., slight fragrance; foliage dark; [Seedling X Allgold]; Martin, J.; Gandy Roses, Ltd., 1976

'GOLDEN SHOWERS'®, LCl, my, 1956; bud long, pointed; flowers daffodil-yellow, open quickly, loose, dbl., 25–28 petals, 4 in., exhibition form, borne singly and in clusters, recurrent bloom, moderate, sweet fragrance; foliage medium, medium green, glossy; vigorous, pillar or climbing (6-10 ft.) growth; AARS, 1956 GM, Portland, 1957; [Charlotte Armstrong X Capt. Thomas]; Lammerts, Dr. Walter; Germain's

'GOLDEN SIGNORA', HT, my, 1954; flowers golden veined orange, large, dbl., 30 petals, moderate fragrance; foliage glossy; vigorous growth; [Signora sport]; Lowe

Golden Silence *see* 'RENSILEN'

'GOLDEN SLIPPERS', F, yb, 1961; bud pointed; flowers yellow flushed vermilion, center golden yellow, blooms in clusters, dbl., 23 petals, 3 in., exhibition form, moderate fragrance; foliage leathery, glossy; vigorous, compact, low growth; AARS, 1962 GM, Portland, 1960; [Goldilocks X Unnamed seedling]; Von Abrams; Peterson & Dering

'GOLDEN SONG', Cl Min, yb, 1980; bud long, pointed; flowers golden yellow, petals edged pink, blooms borne usually singly, dbl., 35 petals, exhibition form, moderate fragrance; long, thin, tan prickles; foliage small, medium to dark, glossy; upright (to about 5 ft) growth; [Little Darling X Golden Angel]; Williams, Ernest D.; Mini-Roses

Golden Spire *see* KORamgis

'GOLDEN SPLENDOR', HT, dy, 1960; bud long, pointed; flowers golden yellow, dbl., 30 petals, 5 in., exhibition form; foliage glossy; vigorous, tall growth; Jones; Hennessey

'GOLDEN SPLENDOUR', HT, my, 1962; flowers clear light yellow, large, dbl., 40 petals, moderate fragrance; [Buccaneer X Golden Sun]; Kordes; Wheatcroft Bros.

'GOLDEN SPRAY', HT, my, 1917; flowers clear lemon-yellow, huge mass of prominent anthers, semi-dbl., borne in long, arching sprays; GM, NRS, 1915; Dickson, H.

Golden Sprite *see* 'DEVUNICAN'

'GOLDEN STAR', HT, dy, 1974; flowers full, dbl., 32 petals, 4–4.5 in., slight fragrance; foliage dark; free growth; [Whisky Mac sport]; Lowe

'GOLDEN STATE', HT, dy, 1937; flowers golden yellow, large, dbl., cupped; foliage leathery, glossy; vigorous growth; GM, Bagatelle, 1937 GM, Portland, 1937; [Souv. de Claudius Pernet X (Charles P. Kilham X Unnamed seedling)]; Meilland, F.; C-P, 1938

Golden Summers *see* 'KORPORT'

'GOLDEN SUN', HT, my, 1957; (Goldene Sonne); bud long, pointed; flowers golden yellow, dbl., 5 in., exhibition form, moderate fragrance; foliage glossy; upright, bushy growth; [(Walter Bentley X Condesa de Sástago) X Golden Scepter]; Kordes, R.

'GOLDEN SUNBLAZE'™ *see* 'MEICUPAG'

Golden Sunblaze *see* **'RISE 'N' SHINE'**

'GOLDEN SUNBURST', HT, yb, 1967; bud ovoid; flowers cadmium-orange and saffron-yellow, large, dbl., intense fragrance; foliage dark, glossy; vigorous, bushy growth; [Golden Wave sport]; Schneeberg; Carlton Rose Nurseries

'GOLDEN SUNSET', LCl, my, 1934; flowers golden yellow, often tipped orange-red, large, dbl., cupped; foliage glossy; long stems; vigorous growth; Burbank; Stark Bros.

'GOLDEN SUNSHINE', HT, my, 1964; bud pointed, ovoid, chrome-yellow splashed red; flowers canary-yellow, dbl., 50 petals, 5 in., exhibition form, intense fragrance; vigorous, upright growth; [Helen Hayes X Golden Masterpiece]; Brownell, H.C.; Brownell

'GOLDEN SURPRISE', F, my, 1978; [Woburn Abbey sport]; Hamilton

'GOLDEN TALISMAN', HT, my, 1931; [Talisman sport]; Hill, E.G., Co.

'GOLDEN TALISMAN, CLIMBING', Cl HT, my, 1935; [Talisman sport]; Elmer's Nursery

Golden Thoughts *see* **'MCGREDY'S ORANGE'**

'GOLDEN TIMES', HT, my, 1970; flowers lemon-yellow, dbl., 40 petals, 4–5 in., slight fragrance; foliage glossy; [Fragrant Cloud X Golden Splendour]; Cocker; Wheatcroft & Sons

Golden Times *see* 'KORTIME'

Golden Token™ *see* 'MINITOK'

Golden Topas *see* 'MANTOPAS'

Golden Touch, S, my

Golden Treasure *see* **'GOLDSCHATZ'**

'GOLDEN TREASURE, CLIMBING', Cl F, my, 1976; Pearson; Burston Nursery

'GOLDEN TZIGANE', HT, ob, 1961; flowers orange; [Tzigane sport]; Gregory

'GOLDEN UNICORN', S, yb, 1985; flowers yellow, petals edged orange-red, large, dbl., 28 petals, cupped, borne 1-8 per cluster, repeat bloom, moderate fragrance; awl-like, tan prickles; foliage dark olive green, leathery; vigorous, upright, bushy, spreading growth; hardy; [Paloma Blanca X (Carefree Beauty X Antike)]; Buck, Dr. Griffith J.; Iowa State University, 1984

Golden Vale, S, lp; [Primula X Unknown]; Sutherland, P, 2000; Golden Vale Nursery; (Weatherly, L.)

'GOLDEN VAN ROSSEM', HT, dy, 1937; flowers chrome yellow; [Mev. G.A. van Rossem sport]; Lens

'GOLDEN VANDAL', HT, dy, 1935; flowers chrome yellow; [Comtesse Vandal sport]; Lens

'GOLDEN VISION', LCl, my, 1922; flowers Marechal Niel yellow, fading nearly white, semi-dbl., moderate fragrance; [Marechal Niel X R. gigantea]; Clark, A.; NRS Victoria

Golden Wave *see* **'DR A. J. VERHAGE'**

'GOLDEN WEDDING', HT, my, 1938; bud pointed, yellow tinted crimson; flowers clear yellow, dbl., 5 in.; foliage leathery, dark; vigorous growth; [Souv. de H.A. Verschuren X Yellow seedling]; Krebs

Golden Wedding *see* AROkris

'GOLDEN WEST', HT, my, 1936; bud pointed; flowers golden yellow, open, large, semi-dbl., moderate fragrance; foliage dark, leathery, glossy; vigorous, bushy, spreading growth; [Duchess of York sport]; Stocking

'GOLDEN WINGS', S, ly, 1956; bud long, pointed; flowers sulfur-yellow, stamens prominent, single, 4–5 in., recurrent bloom, slight fragrance; vigorous, bushy growth; very hardy.; GM, ARS, 1958; [Soeur Thérèse X (R. spinosissima altaica X Ormiston Roy)]; Shepherd; Bosley Nursery

'GOLDEN WONDER', HT, yb, 1973; flowers lemon-yellow, edged red, dbl., 36–40 petals, 4–5 in., exhibition form, moderate fragrance; foliage dark; [Miss Ireland X Princess]; Gandy, Douglas L.; Morse Roses

'GOLDEN WONDER', HT, ab, 1936; flowers golden apricot suffused pink, globular, moderate fragrance; foliage olive-green; vigorous growth; Gunn

Golden Wonders 99 *see* 'RAWANA'

Golden Years® *see* 'HARWEEN'

'GOLDENDALE', HT, my, 1956; bud long, pointed; flowers golden yellow, dbl., 50 petals, 5 in., moderate fragrance; [Annabella sport]; Grillo

Goldene Aue, F, dy, 1964; flowers large, dbl.; GPG Bad Langensalza; (Sangerhausen)

'GOLDENE DRUSCHKI', HP, my, 1936; bud pointed; flowers golden yellow, edged lighter turning creamy yellow in hot weather, cupped, slight fragrance; foliage leathery, dark; very vigorous growth; [Frau Karl Druschki X Friedrich Harms]; Lambert, P.

'GOLDENE GRUSS AN AACHEN', F, ob, 1935; bud pointed, red; flowers golden orange, sometimes shaded reddish, dbl., high centered, very large; fragrant, foliage glossy; bushy growth; [Mme Butterfly X Gloria Mundi]; Kordes

'GOLDENE JOHANNA TANTAU', F, my, 1945; (Wheatcroft's Golden Polyantha); bud ovoid; flowers clear golden yellow, borne in clusters, single, 3.5 in., cupped, moderate fragrance; foliage glossy, dark; compact, bushy growth; (28); [Golden Rapture X (Johanna Tantau X Eugenie Lamesch)]; Tantau

Goldene Sonne *see* **'GOLDEN SUN'**

'GOLDENER ADLER', F, my, 1965; bud ovoid; flowers golden yellow, medium, borne in clusters, semi-dbl.; foliage light green; [Allgold X Seedling]; Verschuren, A.; van Engelen

Goldener Olymp® *see* 'KORSCHNUPPE'

'GOLDENER REITER', F, my, 1969; (Golden Rider); flowers golden yellow, large, dbl., 25 petals, cupped, intense fragrance; foliage dark; vigorous, upright growth; [Circus X Golden Giant]; Haenchen, E.; Teschendorff

Goldener Sommer 83®, F, my, 1983; flowers large, dbl.; Noack, Werner

Goldener Traum *see* **'GOLDEN DREAM'**

'GOLDENES HERZ', HT, dy, 1975; bud ovoid; flowers medium, dbl., cupped, moderate fragrance; foliage glossy, dark, leathery; vigorous, upright, bushy growth; [Dr. A.J. Verhage X Seedling]; Kordes

Goldenes Mainz *see* **'GOLDEN MAIN'**

Goldenes Prag, HT, 1966; dbl., intense fragrance; GPG Bad Langensalza; (Sangerhausen)

Goldfächer *see* 'DICMOPPET'

Goldfasan *see* **'GOLDEN PHEASANT'**

Goldfassade, LCl, my, 1967; Baum

'GOLDFINCH', HMult, ly, 1907; flowers aging white, small blooms in clusters, semi-dbl., non-recurrent, slight fragrance; foliage small, wrinkled; vigorous, climbing growth; [Helene X ?]; Paul

Goldfinger® *see* 'PEAROYAL'

'GOLDGLEAM', F, my, 1966; semi-dbl., 18 petals, 3.5–4 in., blooms in small clusters, intense fragrance; foliage dark, glossy; growth moderate; [Gleaming X Allgold]; LeGrice

Goldglow, HT, 1959; Perry, Anthony; (Cavriglia)

'GOLDIE', F, my, 1958; bud ovoid; flowers golden yellow, edged lighter, dbl., 25–30 petals, 4 in., moderate fragrance; foliage leathery; vigorous, upright, bushy growth; [Goldilocks X Pigmy Gold]; Boerner; J&P

Goldie *see* 'MACAROM'

Goldika®, HT, ob, 1986; Cocker

'GOLDILOCKS', F, my, 1945; flowers deep yellow, fading to cream, blooms in clusters, dbl., 45 petals, 3.5 in., globular, moderate fragrance; foliage leathery, glossy; vigorous, bushy growth; (28); John Cook Medal, ARS, 1947; [Unnamed seedling X Doubloons]; Boerner; J&P

Goldilocks *see* 'SUNLOCK'

'GOLDILOCKS, CLIMBING', Cl F, my, 1951; Caluya; J&P

Goldina *see* 'CANDIDE'

Goldjuwel *see* TANledolg

Goldjuwel *see* TANlewu

Goldkrone *see* **'GOLD CROWN'**

Goldlachs *see* **'GOLDEN SALMON'**

Goldlite™ *see* 'DEVSOLEAR'

Goldmarie *see* 'KORFALT'

'GOLDMARIE', F, dy, 1958; flowers orange-gold, very large, cluster, semi-dbl., intense fragrance; foliage glossy; very vigorous, upright, bushy growth; RULED EXTINCT 6/84 ARM; [Masquerade X Golden Main]; Kordes, R.

Goldmarie 82® *see* 'KORFALT'

Goldmarie Nirp® *see* 'KORFALT'

Goldmarie, Climbing *see* KORkuma

'GOLDMOSS', F, my, 1972; bud long, pointed; flowers clear yellow, medium, dbl., intense fragrance; foliage light, leathery; vigorous, dwarf, bushy growth; [Rumba X Moss hybrid]; Moore, Ralph S.; Sequoia Nursery

Goldpin, Min, dy, 1998; Ilsink

Goldpoint *see* 'JACPO'

'GOLDQUELLE', S, my, 1965; bud ovoid; flowers pure golden yellow, large, borne in clusters, dbl., recurrent bloom, slight fragrance; foliage leathery; strong stems; upright (4 ft.) growth; Tantau, Math.

Goldquelle 88, F, 1988; Tantau, Math.; (Cavriglia)

Goldrausch *see* 'KORBI'

Goldregen *see* **'GOLDEN RAIN'**

'GOLDRUSH', dy, 1941; flowers medium, dbl., moderate fragrance; Howard & Smith; (Sangerhausen)

'GOLDSCHATZ', F, my, 1964; (Golden Treasure); bud pointed; flowers golden yellow, large blooms in clusters (up to 30), dbl.; foliage dark, glossy; bushy, upright growth; Tantau, Math.

Goldschatz *see* 'TANTASCH'

Goldschmied, F, dy; dbl., slight fragrance; VEG; (Sangerhausen)

Goldsmith *see* 'KORBELMA'

Goldstadt Pforzheim *see* HELpforz

Goldstar *see* 'CANDIDE'

Goldstein®, F, ob, 1983; Cocker

Goldstern® *see* 'TANTERN'

Goldstrike, HT, yb, 1997; Perry, Astor

'GOLDSTÜCK', F, my, 1961; flowers lemon-yellow, borne in clusters, dbl., 40–55 petals; foliage glossy, dark; vigorous, upright growth; [Goldilocks X Unnamed seedling]; Verschuren, A.; van Engelen

'GOLDSTÜCK', S, my, 1963; bud long, pointed; flowers golden yellow, large, borne in clusters, abundant, non-recurrent bloom, intense fragrance; vigorous, upright, bushy growth; Tantau, Math.

Goldtopas® *see* 'KORTOSSGO'

Goldy *see* 'KORBEEN'

Golestan *see* 'MEISADINA'

Golf, S, w, 1993; Wageningen

Goliath *see* **'GARY PLAYER'**

Golmain *see* **'GOLDEN MAIN'**

Gomathi, HT, mp, 1987; Kasturi

Gomery, HT, dy, 1996; 28 petals, 4.5 in.; foliage matt; strong grower growth; RvS-Melle, 1999

'GONDUL', F, mr, 1969; flowers cardinal-red, open, medium, borne in clusters, semi-dbl., 18 petals; foliage dark, glossy; vigorous growth; [Lichterloh X Lumina]; Lundstad

Gone Fishin'™ *see* 'SAVAFISH'

'GONSOLI GAELANO', HP, mp, 1874; flowers satiny rose; Pernet

Good as Gold *see* 'CHEWSUNBEAM'

'GOOD CHEER', HT, dp, 1937; bud pointed; flowers cerise, open, dbl.; long, strong stems; very vigorous growth; [Talisman X Templar]; Amling Co.

'GOOD COMPANION', F, mr, 1961; flowers rich red, large, borne in trusses, dbl., 30 petals, flat, moderate fragrance; foliage dark; vigorous growth; Dickson, A.

Good Day Sunshine *see* 'TALGOO'

'GOOD LIFE', F, or, 1970; dbl., 30 petals, 2 in., exhibition form, slight fragrance; free growth; [Elizabeth of Glamis X John Church]; McGredy, Sam IV; McGredy

Good Luck *see* BURspec

'GOOD MORNING', HT, w, 1935; flowers white, reverse faintly tinged pink, very large, dbl.; foliage leathery; long, strong stems; very vigorous, bushy growth; [Premier Supreme sport]; Kaucher; Hill Crest Greenhouses

Good Morning *see* FRYyat

Good Morning America™ *see* 'SAVAGOOD'

Good Morning Sunshine *see* 'BOSLAURAMBER'

'GOOD NEIGHBOR', HT, ob, 1958; flowers burnt-orange, reverse golden, large, dbl., intense fragrance; foliage dark; vigorous, upright growth; [Fred Howard X Unnamed seedling]; Warriner, William A.; H&S

'GOOD NEWS', HT, pb, 1940; (Bonne Nouvelle); flowers silvery pink, center tinged apricot, dbl., 50 petals, 5–6 in., globular, moderate fragrance; vigorous, bushy growth; [(Radiance X Souv. de Claudius Pernet) X (Joanna Hill X Comtesse Vandal)]; Meilland, F.; C-P

Good News, F, my, 1992; Warley Rose Gardens

Good Old Summertime *see* CLEheat

Good Show *see* JACmeen

'GOOD TIMES', HT, pb, 1977; bud ovoid, pointed; flowers ivory to light pink with deep pink edging, dbl., 56 petals, 5–5.5 in., cupped, slight fragrance; foliage large, glossy; compact, upright growth; [Pink Peace X Peace]; Williams, J. Benjamin; Hershey Nursery

Good Vibes *see* 'GELGOOD'

'GOOILAND', HT, mp, 1922; flowers clear rose-pink, reverse dark coral-rose, dbl., slight fragrance; [Sunburst seedling X Red-Letter Day]; Van Rossem

'GOOILAND BEAUTY', HT, ab, 1924; flowers clear golden orange, open, over large, semi-dbl., slight fragrance; foliage dark, leathery; vigorous, bushy growth; GM,

Bagatelle, 1925; [Sunburst X Golden Emblem]; Van Rossem; Van Rossem & Prior

'GOOILAND GLORY', HT, or, 1925; flowers cherry-red shaded coral-red, semi-dbl., slight fragrance; [Mme Edouard Herriot X Gen. MacArthur]; Van Rossem; Van Rossem & Prior

Goose Fair *see* 'MEIPOPUL'

'GOPIKA', F, op, 1969; flowers light salmon-pink, medium, dbl., globular, slight fragrance; foliage glossy; vigorous, bushy growth; [Marlena X Open pollination]; Singh; Gopalsinamiengar

'GORDON DRAKE', F, dp, 1956; flowers cerise-pink, borne in large clusters, dbl.; very vigorous growth; [Eutin sport]; Williams, G.A.

'GORDON EDDIE', HT, ab, 1949; flowers deep apricot, edged lighter, very large, dbl., 40 petals, exhibition form, moderate fragrance; foliage leathery, glossy; very vigorous, bushy growth; GM, NRS, 1950; [Royal Visit X Cynthia Brooke]; Eddie

Gordon's College *see* 'COCJABBY'

'GORGEOUS', HT, dy, 1915; flowers deep orange-yellow, veined copper, well-formed, large, dbl., moderate fragrance; foliage rich, green, soft; bushy, open growth; GM, NRS, 1913; Dickson, H.

'GORGEOUS', HT, mp, 1956; flowers rose-pink, semi-dbl., 18–20 petals, 5 in., exhibition form, moderate fragrance; foliage leathery; vigorous growth; [Pink Delight sport]; Franc; Carlton Rose Nurseries

Gorgeous George *see* 'DELRICOS'

'GOTENHAFEN', F, mp, 1940; flowers pure bright rose, medium, semi-dbl., 12–15 petals; vigorous, well branched growth; [Mev. Nathalie Nypels X Kardinal]; Tantau

'GOTHA', HT, ab, 1932; (Gotha IV); flowers brownish yellow passing to apricot, moderate fragrance; [(Souv. de H.A. Verschuren X Sunset) X Mev. G.A. van Rossem]; Krause

Gotha IV *see* **'GOTHA'**

'GOTTFRIED KELLER', HFt, ab, 1894; flowers apricot-yellow; [((Mme Berard X R. foetida persiana) X (Pierre Notting X MmeBerard)) X R. foetida persiana]; Müller, Dr. F.

'GOUDVLINDER', HT, ab, 1926; (Golden Butterfly); flowers orange-yellow, small, semi-dbl., 12 petals; foliage glossy, brownish red; vigorous, bushy growth; [Lady Hillingdon X Souv. de Claudius Pernet]; Van Rossem

'GOURDAULT', B, m, 1859; Guillot Père

Gourmet Pheasant, Min, dp, 1995

Gourmet Popcorn *see* 'WEOPOP'

'GOVERNADOR BRAGA DA CRUZ', HT, ly, 1954; very vigorous growth; [Peace X Unnamed seedling]; da Silva, Moreira

'GOVERNOR ALFRED E. SMITH', HT, yb, 1933; flowers blend of buff, terra-cotta, gold and salmon, large, dbl., exhibition form; foliage glossy; vigorous growth; [Souv. de F. Bohé seedling X Unnamed seedling]; Denoyel, Vve.; J&P

'GOVERNOR MARK HATFIELD', Gr, dr, 1962; bud pointed; flowers rich red, large, dbl., 40 petals, exhibition form, slight fragrance; foliage leathery; vigorous, upright growth; [Carrousel X Charles Mallerin]; Von Abrams; Peterson & Dering

'GOVERNOR PHILLIP', Cl HT, mr, 1939; flowers ruby-red, flushed darker, open, large, very dbl., moderate fragrance; foliage leathery, glossy, dark, bronze; long stems; vigorous, climbing growth; [Ophelia seedling X Black Boy]; Fitzhardinge; Hazlewood Bros.

'GOVERNOR ROSELLINI', Gr, mr, 1958; flowers rose-red, dbl., 30 petals, 3–4 in., exhibition form, moderate, raspberry fragrance; foliage dark, leathery; vigorous, upright growth; [Baby Chateau X Tiffany]; Lindquist; Howard Rose Co.

Governor's Lady Gloria *see* 'AROGLOR'

Gowirichs Traum, Pol, dp

'GOYA', F, w, 1976; flowers cream, dbl., 30 petals, 5 in., exhibition form, slight fragrance; foliage dark; vigorous growth; [Mildred Reynolds X Arthur Bell]; Bees

Graaff-Reinet *see* 'KORMATE'

'GRACE', HRg, 1923; flowers amber, center apricot, open, very dbl., moderate fragrance; foliage wrinkled; bushy (5-6 ft) growth; RULED EXTINCT ?; [R. rugosa X R. Xharisonii]; Saunders; Central Exp. Farm

'GRACE ABOUNDING', F, w, 1968; flowers ivory, blooms in trusses, semi-dbl., moderate, musk fragrance; foliage glossy; [Pink Parfait X Penelope]; Harkness

'GRACE DARLING', T, w; flowers cream-white shaded golden, large, dbl., globular; vigorous growth; Bennett, 1885

Grace de Monaco® *see* 'MEIMIT'

Grace Donnelly *see* 'HORLEXSTRIP'

'GRACE HASLAM', HT, ob; flowers orange to carmine; [Scarlet Glory X Mrs Sam McGredy]; Fryers Nursery, Ltd.

Grace Kelly *see* 'MEIMAGARMIC'

'GRACE KIMMINS', F, mr, 1973; flowers crimson, dbl., 28 petals, 3 in.; foliage glossy; vigorous, bushy growth; [Dainty Maid X Red Dandy]; Gobbee, W.D.

'GRACE MOLYNEUX', HT, ab, 1909; flowers creamy apricot, center flesh, dbl., slight fragrance; Dickson, A.

'GRACE MOORE', HT, mr, 1948; (Paul Holtge); bud ovoid; flowers crimson-red, large, dbl., cupped; foliage leathery, dark olive-green; vigorous, bushy growth; [Kardinal X Crimson Glory]; Kordes; C-P

Grace Moore's Noisette, N, w

'GRACE NOLL CROWELL', HT, pb, 1929; flowers rose-pink, base slightly shaded cream, large, dbl., exhibition form, moderate fragrance; foliage soft, light; vigorous, bushy growth; Vestal

'GRACE NOTE', S, pb, 1984; bud ovoid, pointed; flowers medium pink, freckled red, imbricated, large blooms borne 3-8 per cluster, dbl., 38 petals, repeat bloom, moderate fragrance; awl-like, tan prickles; foliage large, leathery, dark; vigorous, erect, bushy growth; hardy; [(Tiki X Marigold) X Freckle Face]; Buck, Dr. Griffith J.; Iowa State University

Grace Seward *see* 'TINGRACE'

'GRACE WAYMAN', Cl HT, mp, 1936; flowers pink, very large, dbl., intense fragrance; foliage leathery; vigorous, climbing (10 ft) growth; Wayman

'GRACE WOOD', Min, my, 1993; dbl., 26–40 petals, 1.5 in., borne in small clusters, no fragrance; few prickles; foliage small, medium green, matt; medium, compact growth; [Sun Flare X Rise 'n' Shine]; Giles, Diann; Giles Rose Nursery, 1992

'GRACEFUL', S, ly, 1966; flowers yellow fading to cream, well formed, dbl., 4 in., slight fragrance; foliage glossy; vigorous, bushy growth; [Paul's Lemon Pillar X Marcelle Gret]; Smith, W.H.

Graceland *see* 'JACEL'

Graceland *see* KIRscot

Gracie Allen *see* 'WEKCRYREG'

'GRACIE FIELDS', HT, dy, 1937; flowers vivid buttercup-yellow, moderate, sweetbriar fragrance; foliage glossy; vigorous growth; Letts

'GRACILIS', M, dp; flowers deep pink, well mossed, large, dbl., globular, seasonal bloom; foliage large; vigorous growth; good as a standard; Prévost, prior to 1846

'GRACILIS', Bslt, mr, 1830; flowers cherry shaded lilac-blush, semi-dbl., cupped; large, long prickles; foliage dark; vigorous, branching growth; Wood

'GRACILIS', C; flowers pale rosy pink, edged lilac-blush, compact, dbl.; branching, low-growing growth; Shailer, 1796

'GRACILIS', HSet, mp, 1841; flowers pink to rose, borne in clusters, very dbl.; Prince Nursery

Gracilis *see* **'SHAILER'S PROVENCE'**

'GRACIOSA', HT, w, 1957; flowers white edged rose, well formed; very vigorous growth; [Branca X Peace]; da Silva, Moreira

'GRACIOUS LADY', HT, pb, 1965; bud ovoid; flowers peach-pink, base apricot, large, dbl., exhibition form, moderate fragrance; foliage glossy, dark; vigorous, bushy growth; [(Peace seedling X Gail Borden) X Dorothy Peach seedling]; Robinson, H.; J&P

Graeme Douglas *see* 'MACPIOPI'

'GRAF FRITZ METTERNICH', HP, dr, 1896; flowers large, dbl., intense fragrance; Soupert & Notting; (Sangerhausen)

'GRAF FRITZ VON HOCHBERG', HT, op, 1905; flowers large, dbl.; Lambert, P.; (Sangerhausen)

Graf Lennart, HT, mr, 1991; Meilland

'GRAF SILVA TAROUCA', HT, dp, 1916; flowers carmine-red, very large, dbl., intense fragrance; very vigorous growth; [Étoile de France X Lady Mary Fitzwilliam]; Lambert, P.

'GRAF ZEPPELIN', HMult, dp, 1909; flowers light red to bright pink, medium, borne in large clusters, semi-dbl., very free bloom; vigorous (6-8 ft.) growth; [Non Plus Ultra sport]; Boehm

'GRÄFIN ADA BREDOW', HMult, lp, 1909; flowers small, semi-dbl.; Walter; (Sangerhausen)

Grafin Esterhazy, HCh; Geschwind, R.; (Cavriglia)

'GRÄFIN HARDENBERG', S, dp, 1938; flowers carmine-pink, medium, semi-dbl.; Vogel, M.; (Sangerhausen)

'GRÄFIN MARIE HENRIETTE CHOTEK', HMult, mr, 1911; flowers medium, dbl., moderate fragrance; Lambert, P.; (Sangerhausen)

'GRÄFIN MINNIE SCHAFFGOTSCH', HT, w, 1928; flowers cream-white, center pink, very dbl., slight fragrance; [Clio X Unnamed Hybrid Tea]; Mühle

Grafin Sonja *see* KORfeimot

'GRAFTON PILLAR', LCl, mr, 1958; bud globular; flowers bright red, small, dbl., borne in clusters, free, recurrent boom, slight fragrance; foliage wrinkled; moderate growth; [Second generation Skinner's Rambler X Gruss an Aachen]; Risley

'GRAHAM', HT, dr, 1961; flowers deep crimson-scarlet, large, dbl., 35–40 petals, exhibition form, moderate fragrance; foliage glossy; long stems; vigorous, upright growth; [Eclipse X Unnamed seedling]; Kriloff, Michel; Cramphorn's Nursery

Graham Thomas® *see* 'AUSMAS'

Graham Thomas Musk, HMsk, w

'GRALOVE', F, mr, 1998; (**Everlasting Love**, Millennium®); flowers large, scarlet, dbl., 15–25 petals, 2.75 in., borne in small clusters, slight fragrance; some prickles; foliage medium, dark green, glossy, new growth is red; medium (80 - 100 cm), bushy growth; [Molly McGredy X Satchmo]; Grant, Doug

'GRAN PARADA', F, my, 1967; (Grande Parade); flowers yellow, becoming reddish, open, medium, dbl., 25 petals; foliage glossy; vigorous, upright, compact growth; [Gold Dot X (Queen Elizabeth X Zambra)]; Dot, Simon; Rosas Dot

'GRANADA', HT, 1955; flowers deep red, borne in clusters; [Opera X The Doctor]; Delforge

'GRANADA', HT, rb, 1963; (Donatella); bud urn shaped; flowers blend of rose, nasturtium-red and lemon-yellow, to open, dbl., 25–30 petals, 4–5 in., exhibition form, borne singly, intense, damask and spice fragrance; foliage leathery, crinkled; vigorous, upright growth; AARS, 1964 James Alexander Gamble Fragrance Medal, ARS, 1968; [Tiffany X Cavalcade]; Lindquist; Howard Rose Co.

'GRANADA, CLIMBING', Cl HT, rb, 1964; Swim & Weeks; Comley

'GRANADINA', Min, mr, 1956; (Grenadine); flowers oxblood-red, small, dbl., 30 petals, globular; foliage dark; low, upright growth; [Granate X Coralín]; Dot, Pedro; Kordes

'GRANAT', HT, dr, 1937; flowers blackish red, well formed; very vigorous growth; [Barcelona X Château de Clos Vougeot]; Krause

'GRANATE', Min, dr, 1947; flowers velvety oxblood-red, often streaked white, small; almost no prickles; height 6-8 in; [Merveille des Rouges X Pompon de Paris]; Dot, Pedro

Granatina, HT, 1964; Cazzaniga, F. G.; (Cavriglia)

'GRAND AMOUR', HT, mr, 1956; (Cote Rotie); flowers bright red, large, moderate fragrance; Delbard-Chabert

'GRAND CANARY', HT, my, 1934; dbl., exhibition form; foliage glossy; vigorous, compact growth; [Token sport]; Lowman; U.S. Cut Flower Co.

'GRAND CANYON', S, dy, 1951; bud long, pointed, chrome-yellow; flowers yellow, salmon and copper, turning to crimson, open, medium, recurrent bloom, slight fragrance; foliage glossy, bronze; very vigorous (6 ft.), arching growth; [Herrenhausen X Golden Rapture]; Whisler; Germain's

Grand Chateau *see* 'TANELORAK'

'GRAND CRAMOISI', HGal, dr; flowers crimson, profuse bloom; Vibert (?), ca 1832

Grand Duche *see* 'GODRACHE'

Grand Finale™ *see* 'JACPIHI'

'GRAND GALA'™, HT, rb, 1954; bud globular; flowers rose-red, reverse white suffused pink, dbl., 45–60 petals, 4.5–5 in., exhibition form, slight fragrance; foliage leathery; vigorous, bushy growth; [Peace X Independence]; Meilland, F.; C-P, 1956;, URS, 1954

Grand Gala *see* MEIqualis

'GRAND GALA, CLIMBING', Cl HT, rb, 1961; Yamate; Kakujitsuen

Grand Hotel® *see* 'MACTEL'

Grand Impression™ *see* 'JACLION'

Grand Ivory, HT, lp

'GRAND LADY', HT, mp, 1968; bud ovoid; flowers pink, center lighter, large, dbl., exhibition form, moderate fragrance; foliage leathery; vigorous, upright, bushy growth; [Ma Perkins X Peace]; Patterson; Patterson Roses

Grand Marshall™ *see* 'AROFUTO'

Grand Masterpiece *see* 'JACPIE'

'GRAND MOGUL', HT, w, 1965; flowers creamy white, large, dbl., 33 petals, exhibition form, moderate fragrance; growth moderate; [Sultane X Chic Parisien]; Delbard-Chabert; Cuthbert

Grand Nord® *see* 'DELGRORD'

'GRAND OCCASION', HT, pb, 1970; flowers rosy coral shaded yellow, edged carmine, full, dbl., 30 petals, 3 in., intense fragrance; foliage glossy, light; vigorous growth; [Comtesse Vandal X Mme Henri Guillot]; Delbard; Laxton & Bunyard Nursery

'GRAND OPENING', Min, rb, 1991; bud ovoid; flowers orange-red showing yellow eye, red reverse, ages pale red, dbl., 35 petals, exhibition form, moderate, fruity fragrance; foliage medium, dark green, semi-glossy; upright, medium growth; [Poker Chip X Zinger]; Gruenbauer, Richard, 1984; Flowers 'n' Friends Miniature Roses, 1993

'GRAND OPERA', HT, pb, 1964; bud long, pointed; flowers cream edged pink, becoming pink, dbl., 40 petals, 4–5 in., exhibition form, moderate fragrance; foliage leathery; vigorous, bushy growth; [Masquerade X Peace]; Schwartz, Ernest W.; Wyant

Grand Palace *see* POUlgrad

Grand Prix® *see* 'DELTUF'

'GRAND PRIZE', HT, rb, 1935; flowers petals red, white and spotted, very large, dbl., cupped, intense fragrance; foliage leathery; very vigorous growth; [Red Radiance sport]; Kistler

Grand Romance *see* 'BRIGRAN'

Grand Siècle® *see* 'DELEGRAN'

'GRAND SLAM', HT, mr, 1963; bud urn-shaped; flowers cherry to rose-red, dbl., 28 petals, 4 in., exhibition form, slight fragrance; foliage leathery, dark, semi-glossy; vigorous, upright, spreading growth; [Charlotte Armstrong X Montezuma]; Armstrong, D.L. & Swim; Armstrong Nursery

Grand Trianon, HT, 1967; Truffaut, G.; (Cavriglia)

'GRANDCHILD', Min, mp, 1986; flowers clear, medium pink, small, borne 3-15 per cluster, dbl., 40 petals, exhibition form, moderate, sweet fragrance; hooked, gray-brown prickles; foliage long, narrow, dense, reddish to medium green; bushy, shrub-like, tall growth; [Unnamed Cécile Brunner seedling X Self]; Garelja, Anita, 1987

'GRAND-DUC ADOLPHE DE LUXEMBOURG', HT, rb, 1892; flowers brick-red, reverse carmine, large, dbl., moderate fragrance; weak stems; moderate growth; [Triomphe de la Terre des Roses X Mme Loeben Sels]; Soupert & Notting

'GRAND-DUC ALEXIS', HP, mr, 1892; flowers large, dbl.; Lévêque; (Sangerhausen)

Grande Amore, HT, dr

'GRANDE CENTFEUILLE DE HOLLANDE', C, lp, 1840; flowers large, very dbl., moderate fragrance; Prévost; (Sangerhausen)

Grande Cuisse de Nymphe *see* **'GREAT MAIDEN'S BLUSH'**

'GRANDE DUCHESSE CHARLOTTE', HT, mr, 1942; bud long, pointed; flowers tomato-red, shaded geranium-red, cactus form, dbl., 25 petals, 5–5.5 in., slight fragrance; foliage glossy, dark; vigorous, bushy growth; AARS, 1943 GM, Portland, 1941 GM, Rome, 1938; Ketten Bros.; C-P

Grande Duchesse de Luxembourg *see* **'MARIE ADÉLAÏDE'**

Grande Henriette *see* **'L'ENCHANTRESSE'**

Grande Parade *see* **'GRAN PARADA'**

'GRANDE PREMIÈRE', HT, yb, 1959; bud long, pointed; flowers yellow edged pink, large, dbl., 38 petals, moderate fragrance; foliage bright green, glossy; upright, bushy growth; [Comtesse Vandal X Mme Henri Guillot]; Delbard-Chabert

'GRANDE RENONCULE', C, mp; flowers dull pink, shading to violet, medium, dbl.

'GRANDE ROUGE', F, dr, 1973; bud ovoid; flowers medium, dbl., slight fragrance; foliage soft; upright, bushy growth; [Unknown X Unknown]; Tantau, Math.; Ahrens & Sieberz

'GRANDE WALZER'®, HT, rb, 1980; bud ovoid; flowers deep orange-red, reverse deep yellow, blooms borne singly, dbl., 35 petals, cupped, slight fragrance; curved, light yellow prickles; foliage rather small, light green, glossy; upright growth; Kordes, W.; Rose Barni-Pistoia

Grandee *see* 'MACHOMAI'

Grandessa *see* 'DELSIRE'

'GRANDEUR', HT, dp, 1954; bud long, pointed; flowers cerise-red, dbl., 70 petals, 4 in., exhibition form, moderate fragrance; foliage leathery; very vigorous, upright growth; [Joyance sport]; Grillo

'GRANDEUR', HT; flowers vermilion-red, medium, dbl., 25 petals; moderate, bushy growth; Mallerin, C.

'GRANDEZZA', HT, mp, 1962; bud spiral, pointed; flowers peach-blossom-pink, well formed, large, dbl.; moderate, bushy growth; [Monique X Radar]; Herholdt, J.A.; Herholdt's Nursery

Grandhotel® *see* 'MACTEL'

'GRANDIOSO', HT, mr, 1961; flowers cherry-red, dbl., 30–40 petals, 5 in., moderate fragrance; foliage dark; [(Happiness X Satisfaction) X (Poinsettia X Happiness)]; Verbeek

'GRANDISSIMA', HGal, m, 1824; (Louis Philippe); flowers rosy crimson, sometimes purplish, very large and compact, dbl.; moderate, branching growth; Hardy, Prior to 1848

Grandma's Baby, Min, lp

Grandma's Lace *see* 'CLEGRAN'

Grandma's Pink *see* 'MORBOUQUET'

Grandma's Violet, HT, m

'GRANDMASTER', HMsk, ab, 1954; bud long, pointed; flowers apricot shaded lemon and pink, large blooms in clusters, semi-dbl., 10 petals, recurrent bloom, moderate fragrance; foliage light green; bushy growth; (28); GM, NRS, 1951; [Sangerhausen X Sunmist]; Kordes; Morse

'GRAND'MÈRE JENNY', HT, yb, 1950; (Grem); flowers apricot-yellow, edged and suffused pink, dbl., 30 petals, 4–4.5 in., exhibition form, moderate fragrance; foliage dark, glossy; vigorous growth; GM, NRS, 1950 GM, Rome, 1955; [Peace X (Julien Potin X Sensation)]; Meilland, F.; C-P, 1955;, URS, 1950

'GRAND'MÈRE JENNY, CLIMBING', Cl HT, yb, 1958; (Gremsar, Grimpant Grand'mère Jenny); Meilland, F.; URS

Grandmother's Hat, HP, mp

Grandpa Alex *see* 'RENAL'

Grandpa Dan *see* 'RESPAD'

Grandpa Dickson *see* **'IRISH GOLD'**

Grandpa Ray *see* 'RESPAR'

Grandpa Toni *see* 'RESPAT'

'GRANDPA'S DELIGHT', F, ob, 1984; flowers orange-red, yellow center and stamens, medium, semi-dbl., slight fragrance; foliage medium, medium green, glossy; bushy growth; [Living Fire X Seedling]; Pawsey, Roger; Rearsby Roses, Ltd.

'GRAND-PÈRE LOTTIN', HWich, op, 1918; flowers salmony flesh-pink, center brighter, borne in clusters, very dbl.; [Lady Godiva X Mrs W.H. Cutbush]; Lottin

'GRANGE BRIAR', (strain of R. canina), lp; vigorous; once used as understock

'GRANGE COLOMBE', HT, w, 1912; bud pointed; flowers cream-white, center yellow, large, dbl., cupped, moderate fragrance; vigorous growth; [Mme Caroline Testout X Lady Ashtown]; Guillot, P.

'GRANNIE'S ROSE', S, my; flowers clear pink, borne in clusters of 5-7, semi-dbl., non-recurrent, intense fragrance

Granny Grimmetts, HP, dr; flowers dark purple-red, medium, dbl., moderate fragrance; Hilling, 1955

Granny Savoie, HGal, dr

Granny's Favourite, MinFl, lp

Grape Delight *see* 'GELGRAPE'

Grapeade, Cl Min, m, 1979

Grasten *see* POUlfeld

'GRATIA', HT, w, 1934; bud pointed; flowers creamy white, large, semi-dbl., exhibition form, moderate fragrance; foliage leathery, dark; vigorous growth; [Seedling X Pius XI]; Leenders, M.

'GRATITUDE', HT, rb, 1960; bud ovoid; flowers reddish orange, dbl., 25–35 petals, 3–3.5 in., slight fragrance; foliage leathery, dark, glossy; vigorous, upright, bushy growth; [Impeccable X Incendie]; Delbard-Chabert

Graton Noisette, N, lp

Gratulation, F, 1985; GPG Bad Langensalza; (Sangerhausen)

'GRAVIN D'ALCANTARA', HT, dr, 1985; flowers well-formed, large blooms borne 1-7 per cluster, dbl., 28 petals, cupped, slight fragrance; red prickles; foliage matt, dark; upright growth; [Montezuma X Forever Yours]; Rijksstation Voor Sierplantenteelt, 1982

Gravin Michel d'Ursel®, HMsk, ab, 1994; Lens

'GRAZIA', HT, op, 1941; flowers salmon-pink, center darker, very large; foliage bright green; long stems; [Julien Potin X Mme G. Forest-Colcombet]; Giacomasso

'GRAZIELLA', F, ob, 1960; bud globular; flowers orange, open, medium, borne in clusters, dbl., slight fragrance; foliage glossy, light green; vigorous, bushy growth; [Feu Follet X Unnamed seedling]; Gaujard, R.; G. Truffaut

'GRAZIELLA', T, lp, 1893; Dubreuil

Great Century *see* 'DELEGRAN'

Great Day *see* 'MINBCO'

Great Double White *see* **'ALBA MAXIMA'**

Great Expectations *see* 'LANICAN'

Great Expectations *see* 'JACDAL'

'GREAT MAIDEN'S BLUSH', A, w; (Cuisse de Nymphe, Grande Cuisse de Nymphe, La Royale, La Séduisante, La Virginale, Maiden's Blush, R. alba incarnata, R. alba rubicanda plena, R. alba rubicunda, R. carnea, R. incarnata, R. rubicans); flowers white, tinged pink, dbl., intense fragrance; (42); Roseraie de l'Hay, Cultivated prior to 1738

'GREAT NEWS', F, m, 1973; flowers plum-purple, reverse silver, dbl., 33 petals, 4 in., intense fragrance; foliage large, olive-green; moderate growth; [Rose Gaujard X City of Hereford]; LeGrice

Great Nord® *see* 'DELGRORD'

Great North *see* 'DELGRORD'

Great Ormond Street, F, my, 1991; Beales, Peter

Great Phoebus *see* 'HINGREAT'

'GREAT SCOTT', HT, mp, 1991; flowers large blooms, very dbl.; prickles slightly pubescent peduncles, light green with some; foliage large, medium green, matt; upright (120 cms), bushy growth; [Cleo sport]; Ballin, Don & Paula

'GREAT VENTURE', HT, yb, 1970; bud long, pointed; flowers orange-yellow, flushed pink, medium, dbl., intense fragrance; foliage large, leathery; vigorous, upright growth; [Daily Sketch X Suspense]; Dawson, George; Brundrett

'GREAT WESTERN', B, m, 1840; flowers purplish maroon, large, dbl., blooms mostly in early summer; vigorous growth; Laffay, M.

'GREATER HASTINGS', HT, mp, 1957; Francis

'GREATHEART', HT, op, 1921; flowers pale flesh, shaded salmon, center deeper, dbl., moderate fragrance; [Mrs Walter Easlea sport]; Rosenbluth

'GREBLUB', HT, w, 1990; (**Blushing Bride**); flowers moderately large blooms, dbl., 15–25 petals, intense fragrance; foliage large, dark green, semi-glossy; upright, bushy growth; [Silverado sport]; Greenwood, Chris, 1991

'GREELIZ', F, lp, 1997; (**Turn of the Century**); flowers double, medium, dbl., 15–25 petals, borne in small clusters, slight fragrance; some prickles; foliage medium, semi-glossy; upright, medium (2-2.5) growth; [Elizabeth of Glamis X Seedling]; Sealand Nurseries, Ltd.

'GREEN BUBBLES', Min, w, 1978; bud ovoid; flowers light green, semi-dbl., 12 petals, 1 in., slight fragrance; foliage tiny; very compact, bushy growth; Lyon

'GREEN DIAMOND', Min, w, 1975; bud pointed, dusty pink; flowers soft green, dbl., 25 petals, .5 in., cupped; foliage small, leathery; upright, bushy growth; [Unnamed Polyantha X Sheri Anne]; Moore, Ralph S.; Sequoia Nursery

'GREEN FIRE', F, dy, 1958; bud ovoid, pointed; semi-dbl., 13 petals, 3 in., flat, blooms in clusters, slight fragrance; foliage semi-glossy; vigorous, bushy growth; [Goldilocks X Unnamed seedling]; Swim, H.C.; Armstrong Nursery

'GREEN ICE', Min, w, 1971; bud pointed; flowers white to soft green, small, dbl.; foliage small, glossy, leathery; vigorous, dwarf, bushy growth; [(R. wichurana X Floradora) X Jet Trail]; Moore, Ralph S.; Sequoia Nursery

'GREEN ROSE', Ch, w; (R. chinensis viridiflora, R. viridiflora); flowers green, often touched with bronze, with narrow leaf-like petals, dbl., 1.5–2 in., recurrent bloom; medium, upright growth; (14); Bambridge & Harrison, Cultivated prior to 1856

Green Snake® *see* 'LENWIGA'

Greenalls Glory *see* KIRmac

'GREENMANTLE', HEg, rb, 1895; flowers bright rosy red, white eye, golden stamens, single; foliage richly fragrant; very vigorous, tall growth; Penzance

Greensleeves *see* 'HARLENTEN'

'GREENTAE', F, pb, 1987; (**Hightae**); flowers clear, medium pink outer petals, deeper in color at the heart, dbl., 45–50 petals, exhibition form, intense fragrance; prickles long, curved slightly downward, red; foliage small, bronze aging to medium green, glossy; low, bushy, free-flowering growth; [Duet X Regensberg]; Greenfield, Mrs. P.L.; Hightae Plant Nurs., 1988

'GREER GARSON', HT, pb, 1943; bud pointed; flowers begonia-rose, dbl., 35 petals, 5 in., exhibition form, slight fragrance; foliage leathery, dark; vigorous, tall, bushy growth; Denoyel, Vve.; J&P

'GREET KOSTER', Pol, op, 1933; flowers deep pink, shaded salmon; [Margo Koster sport]; Koster, D.A.

Greetings *see* **'GRÜSS AN BERLIN'**

Greetings *see* JACdreco

Greetings from Alma-Aty *see* **'PRIVET IZ ALMA-ATY'**

Greetje Hennekens, Pol, mr, 1957; flowers carmine-red, medium, semi-dbl.; Loose; (Sangerhausen)

Greg Chappell, HT, ab, 1984

Greg Moore *see* 'HORGREG'

'GREGART', HT, mr, 1997; (**Never Forgotten**); flowers full, medium, very dbl., 26–40 petals, borne in small clusters, moderate fragrance; some prickles; foliage medium, dark green; compact, medium (2-21ft.)growth; Sealand Nurseries, Ltd.

'GRÉGOR MENDEL', F, op, 1955; flowers coral-pink shaded yellow and carmine, well shaped, borne in, dbl.; vigorous growth; [Pinocchio seedling]; Maarse, G.

'GREGSIL', HT, lp, 1997; (**April Fools Day**); flowers full, medium, very dbl., 26–40 petals, borne in small clusters, moderate fragrance; some prickles; foliage medium, medium green, semi-glossy; bushy, medium (2.5-3ft.)growth; [Silver Jubilee sport]; Gregory, C.

Grem *see* **'GRAND'MÈRE JENNY'**

Gremsar *see* **'GRAND'MÈRE JENNY, CLIMBING'**

Grenadier *see* **'FUSILIER'**

'GRENADIER', HT, mr, 1930; flowers brilliant currant-red shaded scarlet, dbl., cupped, slight fragrance; foliage rich green, leathery, glossy; vigorous, bushy growth; Dickson, A.

Grenadine *see* **'GRANADINA'**

'GRENOBLE', HT, mr, 1927; (Ville de Grenoble); flowers clear scarlet, dbl., 30–40 petals, 3.5 in., exhibition form, slight, spicy fragrance; foliage thick; very vigorous growth; [Capt. F. Bald X Mme Van de Voorde]; Mallerin, C.; C-P, 1931

'GRENOBLE, CLIMBING', Cl HT, mr, 1939; Western Rose Co.

Greta, F, w, 1995; Briant

'GRETA FEY', HWich, lp, 1909; flowers small, semi-dbl.; Strassheim; (Sangerhausen)

'GRETA KLUIS', Pol, mr, 1916; flowers carmine-red; [Echo sport]; Kluis & Koning

'GRETA KLUIS SUPERIOR', Pol, mr, 1928; flowers deep carmine-red; [Tausendschön sport]; Kluis

'GRETAG', F, lp, 1991; (**Terrell Anne**); bud ovoid; dbl., 26–40 petals, 1.5–2.75 in., exhibition form, borne in small clusters, slight fragrance; some prickles; foliage medium, medium green, glossy; medium (90 cms), upright growth; [Gene Boerner sport]; Greenwood, Chris

'GRETE BERMBACH', HT, pb, 1925; flowers silvery flesh, center rose, sometimes shaded yellow orange, dbl., moderate fragrance; [Mrs Aaron Ward X Pharisaer]; Leenders Bros.

Grete Schickendanz, HT, 1987; Kordes, R.; (Cavriglia)

'GRETEL GREUL', HT, mr, 1939; [Rote Rapture sport]; Greul

'GRETELEIN', HFt, lp, 1933; flowers large, semi-dbl., slight fragrance; Schmitt, K.; (Sangerhausen)

'GRETHE POULSEN', Pol, dp, 1928; flowers light cherry-red, base yellow, semi-dbl., early; dwarf, well branched growth; [Ellen Poulsen X Mme Laurette Messimy]; Poulsen, S.

'GRETOOTS', F, pb, 1990; (**Tootsie**); flowers deep pink painted white, reverse mostly white, aging with a, semi-dbl., 20 petals, slight fragrance; foliage medium, dark green, glossy, disease resistant; upright, bushy, rounded, medium growth; [Angel Face X Old Master]; Greenwood, Chris, 1985; Armstrong Garden Centers, 1991

Gretta *see* 'SPOGRET'

'GREVINDE ROSE DANNESKJOLD SAMSÖE', HT, dr, 1914; bud dark velvety red; flowers scarlet; Poulsen, D.T.

'GREVINDE SYLVIA KNUTH', LCl, w, 1913; bud yellow; flowers white, center yellow, small, borne in large clusters; Poulsen, D.T.

'GREY DAWN', F, m, 1975; flowers gray, reverse flushed pink and gold, dbl., 45 petals, 3–4 in., moderate fragrance; foliage glossy; bushy growth; [Brownie X News]; LeGrice

Grey Lady Pink, HT, mp

'GREY PEARL', HT, m, 1945; (The Mouse); bud ovoid; flowers lavender-gray, shaded olive and tan, dbl., 43 petals, 4–4.5 in., exhibition form, moderate fragrance; foliage glossy; vigorous growth; [(Mrs Charles Lamplough X Seedling) X (Sir David Davis X Southport)]; McGredy; J&P

'GREY PEARL, CLIMBING', Cl HT, m, 1951; Caluya

'GRIBALDO NICOLA', T, w, 1891; flowers white with touches of yellow, very large, dbl., moderate fragrance; Soupert & Notting; (Sangerhausen)

Grido, F, 1972; Pironti, N.; (Cavriglia)

'GRIFED', HT, my, 1973; (**Edna Wilson**); flowers large, dbl., moderate fragrance; foliage glossy; upright, vigorous growth; [Beauté sport]; Griffiths, Trevor; T. Griffiths, Ltd.

'GRILLODALE', HT, lp, 1926; flowers light pink, center deeper, dbl., 50 petals, 4.5 in., moderate fragrance; foliage dark; [Mme Butterfly sport]; Grillo

Grimaldi *see* DELstror

Grimbeert, HT, 1995; Delforge, H.; (Cavriglia)

'GRIMM', LCl, pb, 1932; flowers apple-blossom-pink, center white, stamens golden, edges fluted, single, non-recurrent; vigorous growth; [(Hiawatha X Altmarker) X (Mme Leon Pain X Marquise de Sinéty)]; Lambert, P.

Grimpant Alain *see* **'ALAIN, CLIMBING'**

Grimpant All Gold® *see* **'ALLGOLD, CLIMBING'**

Grimpant Allgold® *see* **'ALLGOLD, CLIMBING'**

Grimpant Baronne de Rothschild *see* 'MEIGRISOSAR'

Grimpant Bettina® *see* 'MEPALSAR'

Grimpant Carina® *see* 'MEICHIMSAR'

Grimpant Chrysler Imperial *see* **'CHRYSLER IMPERIAL, CLIMBING'**

Grimpant Clair Matin® *see* 'MEIMONT'

Grimpant Comtesse Vandal *see* **'COMTESSE VANDAL, CLIMBING'**

Grimpant Crimson Glory *see* **'CRIMSON GLORY, CLIMBING'**

Grimpant Danse des Sylphes® *see* 'MALCAIR'

Grimpant Delbard® *see* 'DELPAR'

Grimpant Diablotin® *see* 'DELPOSAR'

Grimpant Double Delight® *see* 'AROCLIDD'

Grimpant Ena Harkness *see* **'ENA HARKNESS, CLIMBING'**

Grimpant Étoile de Hollande *see* **'ÉTOILE DE HOLLANDE, CLIMBING'**

Grimpant Exploit® *see* 'MEILIDER'

Grimpant Grand'mère Jenny *see* **'GRAND'MÈRE JENNY, CLIMBING'**

Grimpant Lilli Marleen® *see* 'PEKLIMASAR'

Grimpant Marie-Claire *see* **'MARIE-CLAIRE, CLIMBING'**

Grimpant Michèle Meilland *see* **'MICHÈLE MEILLAND, CLIMBING'**

Grimpant Mrs Herbert Stevens *see* **'MRS HERBERT STEVENS, CLIMBING'**

Grimpant Opera *see* **'OPERA, CLIMBING'**

Grimpant Orange Meillandina® *see* 'MEIJIKATARSAR'

Grimpant Papa Meilland® *see* 'MEISARSAR'

Grimpant Pierre de Ronsard® *see* 'MEIVIOLIN'

Grimpant Queen Elizabeth *see* **'QUEEN ELIZABETH, CLIMBING'**

Grimpant Reine des Neiges *see* **'FRAU KARL DRUSCHKI, CLIMBING'**

Grimpant Rose Gaujard® *see* **'ROSE GAUJARD, CLIMBING'**

Grimpant Rouge et Or *see* 'DICORSAR'

Grimpant Sonia Meilland® *see* 'MEIHELVETSAR'

Grimpant Soraya® *see* 'MEJENORSAR'

Grimpant Spartan *see* **'SPARTAN, CLIMBING'**

Grimpant Sutter's Gold *see* **'SUTTER'S GOLD, CLIMBING'**

Grimpant Tiffany *see* **'TIFFANY, CLIMBING'**

Grisbi *see* **'SUNLIGHT'**

Grisbi *see* MEIgrelou

Grisbi, Climbing *see* **'SUNLIGHT, CLIMBING'**

'GRISELDIS', B, mp, 1895; dbl., flat, borne in clusters of 1-6; Geschwind, R.

Griseldis, F, w; Select Roses, B.V.

Grootendorst *see* **'F. J. GROOTENDORST'**

Grootendorst Red *see* **'F. J. GROOTENDORST'**

'GROOTENDORST SUPREME', HRg, dr, 1936; flowers deeper crimson-red; [F.J. Grootendorst sport]; Grootendorst, F.J.

'GROS CHOUX D'HOLLANDE', C, lp; (Big Cabbage of Holland); flowers soft rose-pink, dbl., intense fragrance; vigorous growth

'GROS PROVINS PANACHE', HGal, m; flowers violet-purple streaked white, dbl.

'GROSSHERZOG ERNST LUDWIG', Cl HT, mp, 1888; (Red Marechal Niel); flowers silvery carmine, very dbl., intense fragrance; Müller, Dr. F.

'GROSSHERZOG FRIEDRICH VON BADEN', HT, mp, 1908; flowers medium, dbl., intense fragrance; Lambert, P.; (Sangerhausen)

'GROSSHERZOG WILHELM ERNST VON SACHSEN', HT, dp, 1915; flowers large, dbl., intense fragrance; Welter; (Sangerhausen)

'GROSSHERZOGIN ELEONORE VON HESSEN', HMult, m, 1907; flowers dark violet-red, dbl.; Strassheim; (Sangerhausen)

'GROSSHERZOGIN FEODORA VON SACHSEN', HT, w, 1914; flowers creamy white, base deep yellow; [Frau Karl Druschki X Kaiserin Auguste Viktoria]; Kiese

Grossherzogin Josefine Schararolle, HT, lp, 1989; flowers medium-large, dbl.; Lens, Louis; (Sangerhausen)

Grossmütterchen, HCh, mp, 1983; flowers large, very dbl., intense fragrance; Weihrauch; (Sangerhausen)

Grouse *see* 'KORIMRO'

Grugakind, Pol, op; flowers small, dbl.; (Sangerhausen)

'GRULIT', Min, rb, 1992; (**Littlest Spartan**); flowers medium red and yellow, blooms borne mostly singly, dbl., 22–30 petals, 1 in., no fragrance; some prickles; foliage medium, medium green, semi-glossy; low (20-30 cms), upright, bushy growth; [Red Ace X Unnamed seedling]; Gruenbauer, Richard; Flowers 'n' Friends Miniature Roses

'GRUMPY', Pol, mp, 1956; (Burkhard, Burkhardt); flowers pink, small blooms in long trusses, dbl.; deRuiter; Gregory & Willicher Baumschulen

'GRUPPENKÖNIGIN', F, pb, 1935; flowers deep bicolor pink, very large, dbl.; foliage light, leathery; vigorous, bushy growth; [Gruss an Aachen X Mme Edouard Herriot]; Kordes, H.

'GRÜSS AN AACHEN', F, lp, 1909; bud orange-red and yellow; flowers flesh-pink fading to creamy white, very dbl., 40–45 petals, 3–3.5 in., borne in clusters, slight, sweet fragrance; foliage rich green, leathery; slender stems, will nod; dwarf growth; (21); [Frau Karl Druschki X Franz Deegen]; Geduldig

'GRÜSS AN AACHEN, CLIMBING', Cl F, lp, 1937; Kordes

Gruss an Aachen, White, F, w

Gruss an Angeln®, HT, mr, 1986; Clausen

Grüss an Bayern® *see* 'KORMUN'

'GRÜSS AN BERLIN', HT, mr, 1963; (Greetings); bud ovoid; flowers pure red, dbl., 40 petals, 5.5 in., exhibition form, slight fragrance; foliage dark, glossy; vigorous, upright, bushy growth; Kordes, R.

'GRÜSS AN BREINEGG', HMult, dp, 1925; flowers small, single; Bruder Alfons; (Sangerhausen)

'GRÜSS AN COBURG', HT, ab, 1927; flowers apricot-yellow, reverse coppery pink, globular, intense fragrance; foliage bronze; vigorous growth; [Alice Kaempff X Souv. de Claudius Pernet]; Felberg-Leclerc

'GRÜSS AN DRESDEN', HT, mr, 1913; flowers fiery red; [Princesse de Bearn seedling]; Türke; Hoyer & Klemm

'GRÜSS AN FÖHR', HFt, yb, 1930; flowers medium, semi-dbl.; Riewers; (Sangerhausen)

'GRÜSS AN FREUNDORF', HWich, dr, 1913; flowers dark velvety crimson, center whitish, stamens bright yellow, semi-dbl.; [R. wichurana hybrid X Crimson Rambler]; Praskac; Teschendorff

'GRÜSS AN FRIEDBERG', N, dy, 1902; flowers medium, dbl., moderate fragrance; Rogmanns; (Sangerhausen)

'GRÜSS AN GERMERSHAUSEN', HMult, mr, 1926; flowers small, single; Bruder Alfons; (Sangerhausen)

'GRÜSS AN HANNOVER', LCl, op, 1938; flowers large, dbl.; vigorous, upright (10-13 ft) growth; Lahmann

Grüss an Heidelberg *see* 'KORBE'

'GRÜSS AN KOBLENZ', LCl, mr, 1963; flowers bright scarlet, blooms in clusters (up to 10), 20 petals, 3 in., recurrent bloom, slight fragrance; vigorous growth; Kordes, R.

Grüss an Lorrach, F, 1983; Hetzel, K.; (Cavriglia)

Gruss an Maiengrun, F, or

Grüss an Munchen, F, 1962; Tantau, Math.; (Cavriglia)

'GRÜSS AN NAUMBURG', HT, rb, 1928; flowers medium, dbl., intense fragrance; Muller, J.F.; (Sangerhausen)

Gruss an Oldenburg, S, yb, 1995; Weihrauch

'GRÜSS AN RENGSDORF', S, mp, 1920; flowers medium; Boden; (Sangerhausen)

'GRÜSS AN SANGERHAUSEN', HT, mr, 1904; flowers large, dbl., moderate fragrance; Müller, Dr. F.; (Sangerhausen)

'GRÜSS AN STEINFURTH', F, my, 1961; flowers open, dbl., 21 petals, cupped, borne in large clusters (to 15); moderate growth; [Goldilocks X Masquerade]; Leenders, J.

'GRÜSS AN STUTTGART', F, mr, 1976; bud ovoid; flowers velvety red, medium, dbl., slight fragrance; vigorous, bushy growth; [(Carina X Seedling) X Sans Souci]; Hetzel

'GRÜSS AN TEPLITZ', HCh, mr, 1897; (Virginia R. Coxe); bud small, ovoid; flowers light crimson, edges sometimes marked dark garnet, dbl., 33 petals, recurrent bloom, intense, spicy fragrance; foliage dark, young growth bronze-red; short, weak stems; vigorous (6 ft.), bushy growth; good for hedges; (28); [((Sir Joseph Paxton X Fellenberg) X Papa Gontier) X Gloire desRosomanes]; Geschwind, R.; P. Lambert

'GRÜSS AN TEPLITZ, CLIMBING', HCh, mr, 1911; (Virginia R. Coxe, Climbing); Storrs & Harrison Co.

'GRÜSS AN WEIMAR', HP, pb, 1919; flowers pink on yellowish ground; [Frau Karl Druschki X Lyon Rose]; Kiese

Gruss an Worishofen *see* **'BAD WÖRISHOFEN'**®

'GRÜSS AN ZABERN', HMult, w; [Euphrosine X Mme Ocker Ferencz]; Lambert, P., 1903

'GRÜSS AN ZWEIBRÜCKEN', HT, mr, 1915; flowers large, dbl., moderate fragrance; Lambert, P.; (Sangerhausen)

Grüss aus Alma-aty, HT, 1958; Sushkov, K. L.; (Cavriglia)

'GRÜSS VOM WESTERWALD', HT, op, 1914; flowers medium, dbl., moderate fragrance; Kettenbeil; (Sangerhausen)

Guadalajara *see* 'MACDEEPO'

Guadalupe's Love, HT, w, 1997; flowers full, ruffled, medium, very dbl., 26–40 petals, borne mostly singly, moderate fragrance; foliage medium, medium green, semi-glossy; upright, medium (2.5 to 3ft.) growth; [Bewitched sport]; Price, Kathleen M.

'GUARDSMAN', HT, or, 1937; flowers bright scarlet, base yellow, large, dbl., slight fragrance; foliage glossy; vigorous, compact growth; [Unnamed seedling X Shot Silk]; Archer

Gudhemsrosen, A, w

Guernsey Gold *see* 'TROBGUERN'

Guernsey Love *see* 'TROBLOVE'

'GUERREIRO', Cl F, dr; [Unnamed seedling X Alain]; da Silva, Moreira

'GUESCAN', HT, yb, 1997; (**Cerys Ann**); flowers full, medium, very dbl., 26–40 petals, borne mostly singly, slight fragrance; some prickles; foliage large, dark green, glossy; bushy, medium (30in.) growth; [Fulton Mackay X Freedom]; Guest, M.M.

'GUESCOLOUR', F, yb, 1999; (**Brushstrokes**); flowers yellow and red striped, reverse paler, dbl., 26–40 petals, 3.5 in., borne in small clusters, slight fragrance; few prickles; foliage medium, dark green, glossy; upright, medium (28 in.) growth; [seedling X Solitaire]; Guest, M.M.; A. J. Palmer & Son, 1999

'GUESDELAY', HT, rb, 1997; (**Birthday Wishes**); flowers full, large, very dbl., 26–40 petals, borne mostly singly, intense fragrance; some prickles; foliage large, dark green, glossy; upright, medium (30in.) growth; [Fragrant Cloud X Honey Favourite]; Guest, M.M.

'GUESGENUS', HT, ly, 1997; (**Ted Allen**); flowers double, medium, dbl., 15–25 petals, borne mostly singly, slight fragrance; some prickles; foliage medium, medium green, dull; upright, medium (28in.) growth; [Manx Queen X (Manx Queen X Rosa Bella)]; Guest, M.M.

'GUESTALL', Cl HT, ob, 1994; (**Just Reward**); flowers orange, blooms borne mostly single, dbl., 15–25 petals, 3–3.5 in., moderate fragrance; some prickles; foliage medium, medium green, glossy; tall (6-8 ft), upright growth; [Basildon Bond X Alexander]; Guest, M.M.; F. Haynes & Partners, 1994

'GUGLIELMO MARCONI', HT, w, 1934; flowers almost white, tinted flesh; [Ophelia X Elisabeth Faurax]; Giacomasso

Guglielmo Marconi *see* HARbleep

Guiding Spirit *see* 'HARWOLAVE'

Guildfordian, HT, w

'GUILLAUME KAEMPFF', HT, dr, 1931; flowers dark crimson-red, edged blackish, large, dbl., intense fragrance; foliage thick; vigorous growth; [Hadley X Admiral Ward]; Felberg-Leclerc

'GUINEA GOLD', HT, ab, 1945; bud long, pointed, buff-yellow; flowers apricot-yellow, open, dbl.; foliage dark, leathery; strong stems; vigorous, upright, much branched growth; [Joanna Hill X Golden Rapture]; Hill, Joseph H., Co.

'GUINÉE', Cl HT, dr, 1938; bud pointed; flowers blackish garnet, sometimes mottled scarlet, large, dbl., intense fragrance; foliage leathery; height 6 1/2-9

ft; [Souv. de Claudius Denoyel X Ami Quinard]; Mallerin, C.; A. Meilland;, C-P

'GUINEVERE', HT, mp, 1967; dbl., 40 petals, 4.5 in., slight fragrance; foliage glossy; GM, Baden-Baden; [Red Dandy X Peace]; Harkness

'GUINGUETTE', F, pb, 1958; flowers pink edged darker, well formed, cupped; vigorous growth; [Alain X Feu de Joie]; Gaujard, R.; Hémeray-Aubert

Guirlande D'Amour, LCl, w, 1993; Lens

'GUIRLANDE FLEURIE', LCl, mr, 1968; bud ovoid; flowers bright red, large, borne in clusters, semi-dbl., cupped, slight fragrance; foliage leathery; very vigorous, climbing growth; [Valenciennes X Paul's Scarlet Climber]; Robichon; Ilgenfritz Nursery

'GUISEPPE MOTTA', HT, 1936; flowers flesh-pink, reverse red and yellow, large, semi-dbl., intense fragrance; vigorous growth; Heizmann, E.

Guitare® *see* 'GAEGUL'

Guitare Cl., Cl F, ob, 1974; Kasturi

'GULAB-E-PAL', HT, m, 1985; flowers mauve, blended with yellow, blooms borne singly, dbl., 70 petals, exhibition form, moderate fragrance; brown prickles; foliage medium, light green; upright growth; [Festival Beauty X (Scarlet Knight X Festival Beauty)]; Hardikar, Dr. M.N.

'GULDTOP', HSpn, dy; [R. Spinosissima seedling]

Gulf Breeze *see* 'TALGUL'

Gulgong Gold, HT, yb

Gulletta *see* 'RUGUL'

'GULLIVER'S GLOW', S, mr, 1954; flowers bright red, small, borne in very large clusters, dbl.; thornless; bushy growth; very hardy.; [Hiawatha X (R. maximowicziana pilosa X Tausendschon)]; Gulliver; Shenandoah Nursery

'GULNARE', HT, my, 1918; flowers golden yellow; Poulsen, D.T.

Gulzar, HT, dr, 1971; IARI

Gumdrop *see* 'JACGUM'

'GUNDY', F, dp, 1966; bud ovoid; flowers deep rose-pink, borne in clusters, semi-dbl., cupped, slight fragrance; foliage dark; vigorous, upright growth; Schloen, P.; Ellesmere Nursery

'GUNSEI', Cl F, pb, 1986; flowers white flushed pink on fringe, borne large number in clusters, dbl., 13–15 petals, cupped, slight fragrance; no prickles; foliage 7 leaflet, green; vigorous, upright growth; [Unnamed seedling X Summer Snow]; Kikuchi, Rikichi

'GUNSTON HALL', HT, mr, 1929; flowers scarlet-crimson; [Seedling X Hoosier Beauty]; U.S. Dept. of Agric.; C-P

'GURNEY BENHAM', HT, my, 1935; flowers buttercup-yellow, large, dbl., cupped; foliage glossy, bronze; vigorous, bushy growth; [Lady Forteviot sport]; Cant, B. R.

'GURNEY HILL', HT, mr, 1924; flowers pure red, dbl., moderate fragrance; Hill, E.G., Co.

'GUSSIE', Min, pb, 1979; bud ovoid; flowers medium red, reverse pale pink and silver, blooms borne singly, very dbl., 90 petals, exhibition form, slight fragrance; few prickles; foliage green, leathery; bushy, dwarf growth; [Unnamed seedling X Unnamed seedling]; Lorenzen, Frederick

'GUSTAV FRAHM', F, mr, 1959; flowers crimson-scarlet, blooms in large clusters, dbl., 25 petals, 3 in., flat, slight fragrance; foliage light, glossy; vigorous, upright growth; [Fanal X Ama]; Kordes; Timm

'GUSTAV GRÜNERWALD', HT, pb, 1903; flowers carmine-pink, center yellow; (28); [Safrano X Mme Caroline Testout]; Lambert, P.

Gustave Courbet® *see* SAUbord

'GUSTAVE PIGANEAU', HP, mr, 1889; flowers bright carmine, very large, dbl., cupped; growth moderate; Pernet-Ducher

'GUSTAVE RÈGIS', HT, ly, 1890; flowers creamy yellow, large, semi-dbl., moderate fragrance; Pernet-Ducher; (Sangerhausen)

'GUSTAVE THIERRY', HP, 1881; Oger; (Cavriglia)

'GUSTEL LÖBNER', HT, w, 1927; flowers large, dbl.; Löbner; (Sangerhausen)

'GUSTEL MAYER', Pol, mr, 1909; flowers small, dbl.; Lambert, P.; (Sangerhausen)

Gutersloh 85®, S, mr

Guy de Maupassant™ *see* 'MEISOCRAT'

'GUY FAWKES', F, yb, 1975; flowers yellow center, reverse shading scarlet, full, semi-dbl., 15 petals, 3.5 in., intense fragrance; foliage glossy; [My Choice X Masquerade]; Cadle's Roses

Guy Laroche *see* 'DELRICOS'

'GUY-GUY', F, ab, 1966; flowers apricot and pink, edged crimson, small, dbl., moderate fragrance; foliage dark, glossy, leathery; low, compact growth; [(Circus X Circus) X (Circus X Circus)]; Fankhauser

Guyscliffe *see* 'LINDLIFFE'

Gwen Fagan *see* POUlgewfa

'GWEN MARIE', F, rb, 1964; flowers dark red, center light cream, open, large, semi-dbl., intense fragrance; foliage soft; tall growth; [Dainty Bess seedling]; Robins

Gwen Mayor *see* 'COCOVER'

'GWEN NASH', Cl HT, pb, 1920; flowers rich pink, center white, large, semi-dbl., cupped, slight fragrance; foliage glaucous, wrinkled; vigorous, climbing growth; [Rosy Morn X ?]; Clark, A.; NRS New South Wales

Gwen Swane *see* 'MACWHAKA'

'GWENDOLINE COLLINS', HT, mr, 1937; flowers cerise shaded cherry, large, dbl., globular; vigorous, bushy growth; Clark, A.; NRS Victoria

Gwent *see* 'POULURT'

'GWYNETH', Pol, ly, 1923; flowers pale yellow, tinted lemon, changing to nearly white, open, semi-dbl., moderate, musk fragrance; foliage light, leathery; bushy growth; (21); [(Trier X Rayon d'Or) sport X (Gottfried Keller X Entente Cordiale) sport]; Woosman; Easlea

'GWYNETH', HT, ly, 1928; flowers canary-yellow, without shading; [Willowmere X Mrs Wemyss Quin]; Chaplin Bros.

'GWYNETH JONES', HT, op, 1925; bud pointed; flowers brilliant carmine-orange, open, semi-dbl., slight fragrance; foliage light, leathery; vigorous, bushy growth; GM, NRS, 1925; McGredy

'GWYNNE CARR', HT, lp, 1924; flowers silvery pink shaded lilac-rose, dbl.; Dickson, A.

'GWYNNE CARR, CLIMBING', Cl HT, lp, 1934; Easlea

'GYLDENORANGE', F, ob, 1952; flowers golden orange, fades to light yellow, large, dbl., no fragrance; foliage medium, medium green, semi-glossy; medium growth; [Poulsen's Yellow X Unnamed seedling]; Poulsen, S.; Poulsen

'GYMPIE', F, pb, 1953; flowers white and pink, borne in clusters, very dbl., moderate fragrance; foliage light green; vigorous, bushy growth; [Yvonne Rabier X Tip-Top]; Ulrick, L.W.

'GYMPIE BEAUTY', F, dr, 1962; flowers deep red, borne in clusters; good growth; Dunstan; Langbecker

'GYPSY', HT, or, 1972; bud ovoid; flowers fiery orange-red, large, dbl., slight fragrance; foliage large, glossy, leathery; vigorous, upright, bushy growth; AARS, 1973; [((Happiness X Chrysler Imperial) X (El Capitan) X Comanche]; Swim & Weeks; C-P

Gypsy *see* **'KIBOH'**

Gypsy Bride, Gr, w

Gypsy Carnival *see* **'KIBOH'**

Gypsy Dancer *see* 'DICQUIET'

Gypsy Fire *see* 'MORGLO'

'GYPSY JEWEL', Min, dp, 1975; flowers deep rose-pink, small, dbl., 50 petals,

1.5 in., exhibition form; foliage dark, leathery; vigorous growth; [Little Darling X Little Buckaroo]; Moore, Ralph S.; Park Seed Co.

Gypsy Jubilee, Gr, yb

Gypsy Lady, Gr, mp, 1995

Gypsy Lass *see* **'GIPSY LASS'**

'GYPSY MOTH', F, op, 1968; flowers salmon, blooms in clusters, dbl., 35 petals, exhibition form; foliage glossy; Tantau, Math.

'GYPSY QUEEN', HMult, mr, 1929; flowers crimson, small, borne in clusters, dbl.; vigorous, climbing growth; [Crimson Rambler X ?]; Moore, Ralph S.

Gypsy Song, HT, pb

Gypsy Sunblaze®™ *see* 'MEIMAGUL'

'GYPSY'S WINE CUP', F, mr, 1968; flowers deep crimson, borne in trusses, slight fragrance; foliage dark; low, bushy growth; [Highlight X ?]; Austin, David

'GYRENE', S, mr, 1987; flowers bright, medium red, large, dbl., 15–25 petals, repeat bloom, moderate fragrance; foliage medium, medium green, matt, disease resistant; upright, bushy, branching, vigorous, hardy growth; [Arctic Glow X ((Pink Hat X R. arkansana) X R. arkansana)]; James, John; Historical Roses, 1987

'GZLATY DECH', HT, ob, 1936; bud long, pointed; flowers orange-yellow, brown, red and gold shadings, large, semi-dbl., cupped, moderate fragrance; foliage glossy, light; vigorous growth; [Admiration X Talisman]; Böhm, J.

H

H. C. Andersen® *see* 'POULANDER'

'H. C. VALETON', HT, yb, 1926; flowers golden yellow overspread with rose, large, moderate fragrance; strong stems; vigorous growth; [Golden Ophelia X Aspirant Marcel Rouyer]; Verschuren

'H. C. YOUNG', HT, op, 1934; bud pointed; flowers shrimp-pink, deepening to salmon, base yellow; strong stems; vigorous growth; Austin & McAslan

'H. CHAUBERT', HT, op, 1928; flowers coppery salmon, open, borne in clusters, semi-dbl., slight fragrance; foliage rich green, glossy; bushy growth; [Mrs Aaron Ward X Unnamed seedling]; Barbier

'H. D. M. BARTON', HT, dr, 1917; flowers deep velvety crimson, large, dbl., moderate fragrance; bushy growth; Dickson, A.

'H. E. RICHARDSON', HT, mr, 1913; flowers dazzling crimson, dbl., exhibition form; vigorous growth; GM, NRS, 1912; Dickson, A.

'H. F. ALEXANDER', LCl, ab, 1952; flowers very large, dbl., 30–40 petals; foliage glossy; very long stems; tall, climbing growth; [Duquesa de Peñaranda X Ruth Alexander]; Wilber; Buckley Nursery Co.

'H. F. EILERS', HT, mr, 1914; flowers large, dbl., moderate fragrance; Lambert, P.; (Sangerhausen)

'H. V. MACHIN', HT, dr, 1914; flowers very dark scarlet-crimson, very large, dbl., globular, slight fragrance; foliage glaucous beech-green; GM, NRS, 1912; Dickson, A.

'H. V. MACHIN, CLIMBING', Cl HT, dr, 1919; Dickson, H.

'H. V. MACHIN, CLIMBING', Cl HT, dr, 1922; H&S

H.G. Hastings *see* **'HARRY G. HASTINGS'**

H.P. Pinkerton *see* **'CLEVELAND'**

H.R.G. Unknown Pink, HP, dp

'HAAKSBERGEN', F, my, 1961; flowers bright yellow, medium, borne in clusters, semi-dbl.; foliage dark; moderate growth; [Mrs Pierre S. duPont X King Boreas]; Buisman, G. A. H.

'HABANERA', S, dr, 1976; bud ovoid, pointed; flowers dark cardinal-red edged lighter, shallow-cupped, dbl., 33 petals, 4–4.5 in., cupped, repeat bloom, slight fragrance; foliage leathery; upright, bushy growth; [(Vera Dalton X Dornroschen) X ((World's Fair X Floradora) X Applejack)]; Buck, Dr. Griffith J.; Iowa State University

'HABONE', HT, pb, 1985; (**Elizabeth Harbour**); bud pointed; flowers light pink, dark pink reverse, well-formed, blooms borne singly, dbl., 30 petals, 3 in., slight fragrance; foliage medium, medium green, semi-glossy; upright (3 ft), bushy growth; [Elizabeth of Glamis X HT seedling (dark red)]; Harbour, E.R.

'HACKEBURG', HMult, pb, 1912; flowers soft lilac-pink, center white, borne in clusters; vigorous, climbing growth; Kiese

'HADANGEL', HT, ab, 1986; (**Smooth Angel**); flowers petals apricot blending to cream at edges, large blooms, dbl., 36 petals, cupped; no fruit; no prickles; foliage medium, medium green, matt; medium, bushy, spreading growth; PP006146; [Smooth Sailing X Royal Flush]; Davidson, Harvey D.; Gurney Seed

'HADANGO', HT, ob, 1996; (**Olde Tango**™); flowers orange, old fashioned, blooms borne mostly singly, very dbl., 3.25–3.5 in., flat, moderate, pine fragrance; moderate prickles; foliage medium, medium green, glossy; bushy (80-90 cms) growth; [Shining Ruby X The World]; Davidson, Harvey D.; L E Cooke Co., 1996

'HADCORAL', HT, pb, 1992; (**Shining Coral**); flowers coral pink, blooms borne in sprays of 1-3, dbl., 22–24 petals, 4 in., cupped, moderate, fruity fragrance; foliage large, dark green, very glossy; medium, bushy growth; [Shining Ruby X [Honey Favorite X (Little Darling X Traviata)]]; Davidson, Harvey D.; C & L Valley Rose Co., 1992

'HADDEN'S VARIETY', S, m, 1948; flowers rosy purple, borne several together, single, non-recurrent; foliage small, gray-green; Hilling

Haddington, HGal, m; flowers dark purple/pink, small to medium, semi-dbl.; (Sangerhausen)

'HADEMP', HT, mp, 1991; (**Pink Empress**); flowers clean medium pink, urn-shaped, large blooms borne usually singly, dbl., 32–36 petals, exhibition form, moderate, fruity fragrance; foliage medium, medium green, glossy; bushy, medium growth; [Smooth Sailing X Medallion]; Davidson, Harvey D., 1984; Hortico Roses, 1991

'HADEVENING', HT, m, 1995; (**Evening Queen**); flowers slight red on edge of petals with aging, dbl., 26–40 petals, 3–3.5 in., borne in small clusters, intense fragrance; foliage medium, medium green, matt; medium (3 ft), bushy growth; [Blue Ribbon X Great News]; Davidson, Harvey D.; Hortico Roses, 1994

'HADFLARE', HT, or, 1992; (**Shining Flare**); dbl., 24–26 petals, 4.5 in., cupped, borne in sprays, slight, fruity fragrance; foliage medium, dark green, very glossy; bushy, medium growth; [Shining Ruby X (Smooth Sailing X Futura)]; Davidson, Harvey D.; C & L Valley Rose Co., 1992

'HADLACE', HT, ly, 1993; (**Olde Lace**); very dbl., 3–3.5 in., moderate fragrance; some prickles; foliage large, dark green, glossy; medium (3-4 ft), bushy growth; [Pink Favorite X (Polly X Peace)]; Davidson, Harvey D.; C & L Valley Rose Co., 1994

'HADLADY', HT, mp, 1986; (**Smooth Lady**); flowers urn-shaped to loose, dbl., 21 petals, blooms in clusters of 1-3, moderate, spicy fragrance; medium, globular, orange fruit; no prickles; foliage large, medium green, very glossy; tall, upright, bushy growth; PP006147; [Smooth Sailing X ((Polly X Peace) X Circus)]; Davidson, Harvey D.; Gurney Seed

'HADLEY', HT, mr, 1914; flowers rich crimson, well-formed, very large, dbl., intense fragrance; foliage rich green; vigorous growth; [(Liberty X Richmond) X Gen. MacArthur]; Montgomery Co.; A.N. Pierson

'HADLEY ELATIOR', HT, mr, 1927; [Hadley sport]; Teschendorff

'HADLEY, CLIMBING', Cl HT, mr, 1927; Teschendorff

'HADMELODY', F, rb, 1990; (**Smooth Melody**); bud ovoid; flowers red to white center, white reverse with red on outer edge, a, dbl., 26 petals, intense, fruity fragrance; round, rarely sets seed; thornless; foliage medium, dark green, semi-glossy; PP007729; [Royal Flush X Smooth Lady]; Davidson, Harvey D., 1979; Hortico & C&L Valley Rose Co., 1990

'HADPERFUME', HT, lp, 1990; (**Smooth Perfume**); bud pointed; flowers light pink to very light mauve on edge, urn-shaped, large, b, dbl., 28–30 petals, exhibition form, intense fragrance; thornless; foliage medium, medium green, semi-glossy; bushy, medium growth; PP007728; [(Smooth Sailing X Medallion) X Blue Moon]; Davidson, Harvey D., 1979; Hortico & C&L Valley Rose Co., 1990

'HADPLEASURE', HT, mp, 1995; (**Pink Treasure**); dbl., 26–40 petals, 4.5 in., borne mostly singly, moderate fragrance; some prickles; foliage large, medium green, semi-glossy; upright, medium (110 cms) growth; [Smooth Sailing X Red Planet]; Davidson, Harvey D.; Hortico Roses, 1995

'HADPRETTIE', HT, dr, 1995; (**Velvet Lady**); flowers very dark, velvet red, single (5 petals), borne in small clusters; fragrant; numerous prickles; foliage medium, dark green, very glossy; spreading, medium growth; [Shining Ruby X Precious Platinum]; Davidson, Harvey; Coiner Nursery, 1995-96

'HADPRINCE', HT, mr, 1990; (**Smooth Prince**); bud ovoid; flowers urn-shaped, large, borne usually singly, dbl., 26–28 petals, exhibition form, slight, fruity fragrance; oblong, rarely sets seed. fruit; thornless; foliage medium, medium green, semi-glossy; [Smooth Sailing X Old Smoothie]; Davidson, Harvey D., 1979; Hortico & C&L Valley Rose Co., 1990

'HADROMANCE', HT, w, 1992; (**Smooth Romance** TM); flowers cream with tinge of pink in center, urn shaped, blooms borne, dbl., 42–45 petals, 4.75 in., exhibition form, fast repeat, moderate fragrance; will not set seed; thornless; foliage medium, medium green, glossy; straight, upright stems; upright, bushy, tall growth; [Smooth Sailing X Portrait]; Davidson, Harvey D., 1984; Hortico Roses, 1991

'HADROMEO', HT, dr, 1993; (**Olde Romeo**); dbl., 26–40 petals, 3–3.5 in., borne mostly singly, intense fragrance; many prickles; foliage medium, medium green, matt; medium (4 ft), upright growth; [Smooth Sailing X (Old Smoothie X ((Polly X Peace) X Simon Bolivar))]; Davidson, Harvey D.; C & L Valley Rose Co., 1994

'HADRUBY', HT, mr, 1992; (**Shining Ruby**); flowers medium red, fading to blue, blooms borne in sprays, dbl., 24–26 petals, 5.5 in., cupped, moderate, spicy fragrance; foliage large, dark green, very glossy; bushy, medium growth; [Pink Favorite X Simon Bolivar]; Davidson, Harvey D.; C & L Valley Rose Co., 1992

'HADSATIN', HT, mp, 1993; (**Smooth Satin**); dbl., 26–40 petals, 3–3.5 in., borne mostly singly, moderate fragrance; 95-100% thornless; foliage large, medium green, glossy; medium (1 m), upright, bushy growth; [Smooth Lady X Smooth Sailing]; Davidson, Harvey D.; C & L Valley Rose Co.

'HADSUN', HT, ab, 1989; (**Western Sunlight**); bud pointed; flowers apricot-orange, aging to yellow, large, dbl., 31 petals, exhibition form, borne usually singly, slight, fruity fragrance; inverted, does not set seed readily, orange fruit; prickles hooked downward, light brown; foliage medium, dark green, glossy, serrated; upright, medium growth; PP007442; [[(Honey Favorite X Irish Mist) X (San Francisco X Prima Ballerina)] X Just Joey]; Davidson, Harvey D.; Hortico Roses, 1990

'HADSWEE', HT, lp, 1996; (**Olde Sweetheart**®); flowers light pink, with large, raspberry colored stamen, single, 4–7 petals, 3.5 in., slight fragrance; moderate prickles; foliage medium, medium green, glossy; upright, medium (100 cms) growth; [Smooth Sailing X Red Planet]; Davidson, Harvey D.; L E Cooke Co., 1997

'HADTREASURE', HT, w, 1995; (**White Treasure**); flowers white, slight cast of light yellow in center, dbl., 26–40 petals, 4.75 in., borne mostly singly, moderate fragrance; few prickles; foliage large, dark green, semi-glossy; upright, medium (100 cms) growth; [Smooth Sailing X Sunset Jubilee]; Davidson, Harvey D.; Hortico Roses, 1995

'HADVELVET', HT, dr, 1986; (**Smooth Velvet**); flowers large blooms, dbl., 42 petals, cupped, borne usually singly, slight, damask fragrance; medium, globular, orange fruit; no prickles; foliage light green, matt; tall, upright growth; PP006152; [(Smooth Sailing X ((Polly X Peace) X Circus)) X Red Devil]; Davidson, Harvey D.; Gurney Seed

Haendel® *see* 'MACHA'

Hafiz®, HT, mp

'HAGOROMO', LCl, op, 1970; bud ovoid; flowers silvery coral-pink, large, dbl., exhibition form, free, intermittent bloom, moderate fragrance; foliage dark, leathery; vigorous, climbing growth; [Aztec seedling X New Dawn seedling]; Suzuki, Seizo; Keisei Rose Nursery

'HAIDEE', S, pb, 1953; flowers clear pink, center cream, large, dbl., cupped, non-recurrent; large, dark red fruit; sometimes very prickly; foliage small, dark; wood red; height 6 ft; [R. laxa X R. spinosissima seedling]; Skinner

Haiku™ *see* 'AROYEFELT'

'HAILEYBURY', HP, mr, 1896; flowers large, very dbl., moderate fragrance; Paul, G.; (Sangerhausen)

Hailstorm™ *see* 'COISTOR'

Haircutter's Pink Climber, LCl, mp; (found rose)

'HAÏSHA', HT, yb, 1947; flowers gold, edges suffused carmine, dbl., 60 petals, 6 in., intense fragrance; foliage leathery, glossy, dark; upright growth; [Peace X Fantastique]; Meilland, F.

Haitian Belle, LCl, mp

'HAKATA KANOKO', Min, rb, 1999; flowers medium red with white eye, 5 petals, 3.75 in., borne in large clusters, no fragrance; foliage small, medium green, disease-resistant; bushy, very compact (6-10 in.) growth; Yamazaki, Kazuko, 1979; Takii & Co., Ltd, 1995

'HAKKODA', HT, pb, 1983; flowers white, pink petal edges, large, dbl., 40 petals, exhibition form, intense fragrance; broad prickles, curved downward; foliage medium green, semi-glossy; bushy growth; [Lady X X Izayoi]; Kodoya, Y.; Kogura Rose Nursery, 1986

'HAKUHOH', HT, w, 1999; bud pale cream, turning white; dbl., 40 petals, 6 in., exhibition form, moderate fragrance; foliage medium green; 4.5 ft growth; [White Prince X Bridal Robe]; Hayashi, Shunzo, 1989

'HAKUSYU', HT, w, 1999; flowers white, center coral pink, dbl., 26–40 petals, 5.5–6 in.; some prickles; foliage medium, medium green; upright, bushy, medium (4.5-5 ft) growth; Bronze Medal, Japan Rose Concours, 1998; [Sizunomai X Hoshizukuyo]; Ohtsuki, Hironaka

'HAKUUN', F, w, 1962; (White Cloud); bud small; flowers creamy white, patio, semi-dbl., 15 petals, 2 in., slight fragrance; foliage light green; low, compact, bushy growth; [Seedling X (Pinocchio X Pinocchio)]; Poulsen, Niels D.; Poulsen

'HALALI', F, dp, 1956; flowers deep pink, large blooms in clusters, semi-dbl.; foliage leathery, dense; vigorous (5-6 ft), spreading growth; [Marchenland X Peace]; Tantau, Math.

Halarious, HT, lp; Clark, A., 1935

Haleakala *see* 'MANHALE'

'HALF TIME', HT, rb, 1976; bud pointed; flowers cherry-red, reverse yellow, dbl., 40 petals, 3.5–4 in., moderate, tea fragrance; foliage dark; upright growth;

[((Fandango X Roundelay) X (Happiness X Tiffany)) X Peace]; Weeks; Weeks Wholesale Rose Growers

'HALKA', HT, lp, 1987; flowers white blush pink, reverse white-silver, large, borne in sprays, dbl., 30 petals, exhibition form, moderate, damask fragrance; straight, medium, red prickles; foliage medium, medium green, glossy; upright, medium growth; [Red Queen X Peace]; Bracegirdle, Derek T., 1988

Hall of Flowers *see* 'MORMINT'

Hallandsasen, HRg

Hallelujah *see* 'DELATUR'

'HALLEY'S COMET', F, mr, 1986; flowers medium, dbl., 20 petals, slight fragrance; foliage medium, medium green, semi-glossy; [Tip Top sport]; Rearsby Roses, Ltd.

'HALLMARK', HT, mr, 1966; bud ovoid; flowers large, dbl., 28 petals, cupped, moderate fragrance; foliage glossy; [Independence X Chrysler Imperial]; Morey, Dr. Dennison; J&P

'HALLOWEEN', HT, yb, 1962; flowers deep yellow, tipped scarlet, large, dbl., 65 petals, intense fragrance; foliage glossy, dark, leathery; vigorous, upright growth; [(Peace X Fred Howard) X Unnamed seedling]; Howard, A.P.; Great Western Rose Co.

'HALO', HT, w, 1956; bud ovoid, seafoam-green; dbl., 25 petals, 4.5–5 in., exhibition form, moderate fragrance; foliage leathery; vigorous, upright growth; [Lady Sylvia X (Virgo X White Briarcliff)]; Lens; J&P

Halo Dolly *see* 'MORWATEYE'

Halo® Fire 'MORHALFIRE'

Halo Karol *see* 'MOREYES'

Halo® Rainbow *see* 'MORRAINBOW'

Halo Star *see* 'MORANYFACE'

Halo Suise *see* MORsuise

Halo Sunrise 'MORSUNRISE'

Halo® Today *see* 'MORTODAY'

'HAMBURG', S, dr, 1935; bud pointed; flowers glowing crimson, very large blooms in clusters, semi-dbl., recurrent bloom, slight fragrance; foliage large, leathery, glossy; vigorous growth; (28); [Eva X Daily Mail Scented Rose]; Kordes

Hamburger Deern *see* KORpalud

'HAMBURGER PHOENIX', HKor, mr, 1954; (Hamburger Phonix®); bud long, pointed; flowers rich red, large blooms in clusters, repeat bloom, slight fragrance; large, orange-red fruit; foliage dark, glossy; vigorous, climbing or trailer growth; [R. kordesii X Seedling]; Kordes

Hamburger Phonix® *see* **'HAMBURGER PHOENIX'**

'HAMBURG'S LOVE', F, dy, 1974; dbl., 28 petals, 3 in., intense fragrance; foliage glossy; compact growth; [Fragrant Cloud X Manx Queen]; Timmerman's Roses

Hamish *see* 'SIMHAHA'

Hammerberg, S, m

Hampshire *see* KORhamp

Hampton Palace *see* POUlgret

Hanabusa *see* **'HANA-BUSA'**

'HANA-BUSA', F, or, 1981; (Hanabusa); bud ovoid; semi-dbl., 18 petals, flat, borne 6-10 per cluster; straight prickles; bushy growth; [Sarabande X (Rumba X Olympic Torch)]; Suzuki, Seizo; Keisei Rose Nursery

Hanae Mory, HT, 1990; Delbard, Georges; (Cavriglia)

'HANAGASA', F, or, 1978; bud globular; flowers vermilion, dbl., 23 petals, 4–4.5 in., cupped, moderate fragrance; foliage large, light green; vigorous growth; [(Hawaii X Unnamed seedling) X Miss Ireland]; Suzuki, Seizo; Keisei Rose Nursery, 1979

Hanagasumi *see* **'HANA-GASUMI'**

'HANA-GASUMI', F, w, 1985; (Hanagasumi); flowers soft white, aging pink, blooms borne 6-12 per cluster, semi-dbl., 13 petals, flat, moderate fragrance; small, hooked prickles, slanted downward; foliage dark, semi-glossy; bushy growth; [Europeana X (Myo-joh X Fidélio)]; Suzuki, Seizo; Keisei Rose Nursery, 1984

'HANAGURUMA', HT, yb, 1977; bud globular; dbl., 58 petals, 6.5–7 in., exhibition form, slight fragrance; foliage light green; upright growth; [Kordes' Perfecta X (Kordes' Perfecta X American Heritage)]; Teranishi, K.; Itami Bara-en

'HANAKAGO', F, or, 1972; bud ovoid; flowers deep salmon-vermilion, medium, dbl., cupped, moderate fragrance; foliage glossy, dark; vigorous, bushy growth; [Sarabande seedling X Rondo seedling]; Suzuki, Seizo; Keisei Rose Nursery

'HANA-KURENAI', HT, pb, 1980; bud ovoid; flowers light pink, flushed yellow, reverse deeper, blooms borne 1-3 per cluster, dbl., 33 petals, exhibition form, slight fragrance; few, sickle-shaped prickles; foliage medium, medium green, glossy; vigorous, upright growth; [Big Red X Star Queen]; Ohata, Hatsuo

Hanamigawa *see* **'HANAMI-GAWA'**

'HANAMI-GAWA', Cl Min, op, 1986; (Hanamigawa); flowers soft salmon-pink, shaded orange, small blooms borne 6-10 per cluster, dbl., 23 petals, moderate fragrance; small, curved prickles, slanted downward; foliage dark, semi-glossy; vigorous, very bushy growth; [Unnamed seedling X Petite Folie]; Suzuki, Seizo; Keisei Rose Nursery, 1985

'HANAMORI', F, rb, 1977; bud circular; dbl., 20–25 petals, 2.5 in., exhibition form, slight fragrance; foliage glossy, dark; bushy growth; [(Tropicana X Karl Herbst) X Lydia]; Teranishi, K.; Itami Bara-en

'HANATIRUSATO', Min, dp, 1999; flowers brilliant rose, very dbl., 70 petals, 1.4in., flat, borne 2-5 per cluster, slight fragrance; foliage medium, medium green; vigorous, bushy, compact (8 in.) growth; [seedling X Red Minimo]; Yamazaki, Kazuko, 1991; Takii & Co., Ltd., 1996

'HANAYUZEN', Min, dy, 1999; flowers deep golden yellow, sometimes flushed orange, very dbl., 45 petals, 1.5 in., exhibition form, moderate fragrance; foliage medium, dark green; very vigorous, upright, bush, compact (12 in.) growth; [Himetatibana X Hanahotaru]; Yamazaki, Kzauko, 1993; Takii & Co., Ltd., 1996

Hand in Hand *see* HARaztec

Handel® *see* 'MACHA'

Handel's Largo *see* **'LARGO D'HAENDEL'**

Handout *see* 'BAROUT'

'HANDSOM RED', HT, mr, 1954; flowers spectrum-red, dbl., 45 petals, 4–5 in., exhibition form, moderate fragrance; upright, bushy growth; [(Pink Princess X Mirandy) X Queen o' the Lakes]; Brownell, H.C.

'HANDY ANDY', HT, ab, 1965; flowers apricot edged pink, 4 in., slight fragrance; free growth; [Kordes' Perfecta X Piccadilly]; McGredy, Sam IV; Geest Industries

Hanib *see* **Cecilai 89**

Hanini *see* 'DYKNINI'

Hanka, HT, ly; flowers creamy yellow, very large, dbl.; (Sangerhausen)

Hanky Panky *see* 'TINPANKY'

Hannah Gordon *see* 'KORWEISO'

Hannah Hansen, HWich, m, 1997; Nobbs

'HANNAH HAUXWELL', F, op, 1991; flowers deep salmon, small, dbl., 26–40 petals, slight fragrance; foliage small, medium green matt; patio; bushy growth; [Unnamed seedling X Unnamed seedling]; Battersby Roses, 1990

'HANNE', HT, mr, 1959; flowers scarlet-crimson, medium to large, dbl., exhibition form, intense fragrance; foliage leathery; upright growth; [Ena Harkness X Peace]; Soenderhousen; Hoersholm Nursery

Hanne Dänomik, HT, mr; flowers large, dbl.; (Sangerhausen)

Hanneli Rupert *see* KORsebue

Hannes *see* **'ROSE HANNES'**

Hannover *see* **'MESSESTADT HANNOVER'**

Hannover's Weisse *see* NOAhan

'HANS', HT, w, 1970; flowers open, large, semi-dbl.; foliage glossy, light; vigorous, upright growth; [Message X Virgo]; IARI; Div. of Vegetable Crops & Flori.

Hans Berger, HT, mr, 1958; flowers carmine-red, large, dbl., intense fragrance; Berger, W.; (Sangerhausen)

'HANS BILLERT', HT, mr, 1928; flowers brilliant red, very dbl., moderate fragrance; [Laurent Carle X Richmond]; Billert; Teschendorff

Hans Christian Andersen *see* 'POULANDER'

Hans Erni, F, op, 1992; Meilland

'HANS HAUBOLD', HT, mp, 1942; flowers large, dbl., slight fragrance; Vogel, M.; (Sangerhausen)

'HANS MACKART', HP, rb, 1884; flowers bright deep geranium red, outer petals tinted carmine purple, dbl., moderate fragrance; Verdier, E.

'HANS SCHMID', HWich, dp, 1934; flowers deep pink, rather well formed, small to medium, borne in clusters, dbl., globular; foliage large; vigorous, climbing growth; [Fragezeichen X American Pillar]; Vogel, M.; P. Lambert & Heinemann

'HANSA', HRg, mr, 1905; flowers mauvy-red, large blooms, dbl., recurrent bloom, intense, clove-rose fragrance; large, red fruit; short, weak stems; vigorous growth; hardy.; Schaum & Van Tol

Hansaland® *see* KORhassi

Hansa-Park® *see* KORfischer

'HANSEAT'®, S, mp, 1961; flowers rose-pink, center lighter, medium, single, 5 petals, cupped, slight fragrance; vigorous (6 ft) growth; Tantau, Math.

Hansen's Red Hedge, S, mr

'HANSESTADT BREMEN', F, op, 1958; bud ovoid, crimson; flowers deep salmon and reddish pink, large, borne in clusters (up to 10), dbl., 47 petals, moderate fragrance; foliage leathery; very vigorous, bushy growth; [Ama X Fanal]; Kordes, R.

'HANSESTADT LÜBECK', F, mr, 1962; (Lübeck); dbl., 3.5 in., slight fragrance; foliage dark; vigorous, tall growth; Kordes, R.; McGredy

'HANSETTE', S, mr, 1938; flowers red, semi-dbl., non-recurrent; [Hansa X R. rubrifolia]; Wright, Percy H.

Hap Renshaw *see* 'RENHAP'

'HAPPENSTANCE', HBc, ly, 1950; single; thorns large, hooked; low, prostrate ground cover; up to 15 ft.; Buss

'HAPPINESS', HT, mr, 1954; (Rim, Rouge Meilland); bud long, pointed; dbl., 38 petals, 5–6 in., exhibition form, slight fragrance; upright, vigorous growth; [(Rome Glory X Tassin) X (Charles P. Kilham X (Charles P. Kilham X Capucine Chambard))]; Meilland, F.; C-P, 1951;, URS, 1949

'HAPPINESS, CLIMBING', Cl HT, mr, 1954; (Rouge Meilland, Climbing); Meilland, F.; URS

'HAPPY', Pol, mr, 1954; (Alberich®); flowers currant-red, very small blooms in large trusses, semi-dbl.; foliage dark, glossy; vigorous, compact (12-15 in) growth; [Robin Hood X Katharina Zeimet seedling]; deRuiter; Gregory & Willicher Baumschulen

Happy Anniversary *see* 'DELPRE'

Happy Anniversary, HT, lp; [Kardinal sport]; Catt, Graeme Charles, 1997; (Weatherly, L.)

'HAPPY BIRTHDAY', HT, dp, 1964; bud ovoid; flowers deep rose, large, dbl., 25 petals, exhibition form, intense fragrance; foliage leathery; vigorous, upright growth; [Peace X The Doctor]; Howard, P.J.

Happy Birthday, MinFl, w, 1997

'HAPPY BUTT', HT, ab, 1995; flowers open apricot, petals deepen to pink as bloom matures, very full (41+ petals), large (7+cms) blooms borne mostly single; fragrant; some prickles; foliage medium, medium green, semi-glossy; upright, tall growth; [Nantucket X Medallion]; Carlson, William

Happy Child *see* 'AUSCOMP'

Happy Day *see* 'SIMPALNO'

'HAPPY DAYS', HT, or, 1932; bud pointed; flowers geranium-red, open, large, dbl.; foliage dark; long stems; very vigorous growth; RULED EXTINCT 4/87; [Briarcliff sport]; Amling, M.C.; Amling Bros.

'HAPPY DAYS', HT, dr, 1962; bud pointed; flowers oxblood-red, dbl., 3–3.5 in., exhibition form; long stems; vigorous growth; RULED EXTINCT 4/87; [Exciting X Grand Gala]; Herholdt, J.A.; Herholdt's Nursery

Happy Days *see* 'MACSEATRI'

'HAPPY EVENT', F, pb, 1964; flowers light chrome-yellow, flushed rose-opal, blooms in clusters, semi-dbl., 12 petals, 3 in.; foliage glossy; growth moderate; [(Karl Herbst X Masquerade) X Rose Gaujard]; Dickson, Patrick; A. Dickson

Happy Ever After™ *see* 'DICVANILLA'

Happy Face™ *see* 'SAVAFACE'

Happy Go Lucky™ *see* 'SAVALUCK'

Happy Go Lucky *see* 'SUNHAP'

Happy Hour™ *see* 'SAVANHOUR'

'HAPPY RED', F, or, 1960; flowers bright brick-red, borne in clusters, single; foliage glossy; moderate growth; [Red Favorite X Cocorico]; Leenders, J.

'HAPPY TALK', F, mr, 1973; flowers cherry-red, small, dbl., slight fragrance; foliage glossy, dark; vigorous, upright, bushy growth; [Escort X Orange Garnet]; Weeks

'HAPPY THOUGHT', Min, op, 1978; bud pointed; flowers pink blended with coral and yellow, dbl., 40 petals; foliage small, glossy; vigorous, bushy growth; [(R. wichurana X Floradora) X Sheri Anne]; Moore, Ralph S.; Sequoia Nursery

'HAPPY TIME', Cl Min, rb, 1974; bud short, pointed; flowers yellow overlaid red, dbl., 35 petals, 1 in., slight fragrance; foliage small, glossy, leathery; climbing growth; [(R. wichurana X Floradora) X (Golden Glow X Zee)]; Moore, Ralph S.; Sequoia Nursery

Happy Times, MinFl, mp, 1995; Chessum, Paul

Happy Trails *see* 'JACCASP'

'HAPPY WANDERER'®, F, mr, 1972; flowers scarlet, slight fragrance; [Seedling X Marlene]; McGredy, Sam IV; McGredy, 1974

'HAPPY WEDDING BELLS', HT, w, 1966; bud long, pointed; flowers large, dbl., 52 petals, exhibition form, moderate, spicy fragrance; foliage leathery; vigorous, upright growth; [White Swan X Virgo]; Morey, Dr. Dennison; Country Garden Nursery

Har Tabor, HT, pb, 1972; Fischel

HARavis, HT, 1995; (**Tower Bridge**); Harkness; (Cavriglia)

HARaztec, MinFl, or, 1994; (**Hand in Hand**); Harkness

HARazz, HT, ab, 1998; (**Chimene**); Harkness

HARbabble, F; (**Sunset Boulevard**); Golden Prize, City of Glasgow, 1998; Harkness, 1997; (Cavriglia)

HARbadge, S, ab, 1997; (**Louisa Stone**); Harkness

'HARBANJO', F, or, 1993; (**Girls' Brigade**); flowers vermilion, orange-red reverse, aging orange-red, dbl., 28 petals, 2 in., no fragrance; foliage small, dark green, glossy; bushy, low growth; [Sexy Rexy X Anna Ford]; Harkness, 1987; Harkness New Roses, Ltd., 1993

HARbanner, S, lp, 1998; (**Country Fair**); Harkness

HARbar, F, 1995; (**Sporting Duo**); Harkness, R.; (Cavriglia)

HARbaroque, S, m, 1995; (**Baroque**); Harkness

HARbell, F, 1995; (**Cathedral Splendour**); Harkness, R.; (Cavriglia)

'HARBELLA', F, pb, 1994; (Peacekeeper, **United Nations Rose**); flowers orange/pink blend, yellowing with age, double, blooms borne, 30 petals, 3.5 in., slight, spicy fragrance; foliage medium, light green, glossy; bushy, medium growth; [Dame of Sark X Bright Smile]; Harkness; Harkness New Roses, Ltd., 1996

'HARBILBO', F, w, 1994; (**St. John**); dbl., 15–25 petals, 1.5–2.75 in., borne in small clusters, slight fragrance; few prickles; foliage medium, medium green, glossy; low (60 cms), bushy, spreading growth; [Prima X Grace Abounding]; Harkness; Harkness New Roses, Ltd., 1994

'HARBINGER', LCl, lp, 1923; bud pointed; flowers soft pink, large, single, slight fragrance; foliage light; vigorous, climbing growth; Clark, A.; Hackett

HARbinger, HT, or, 1994; (**Julie Y**); Harkness

HARbingo, MinFl, my, 1995; (**House Beautiful**); Harkness

HARbleep, F, op, 1996; (**Guglielmo Marconi**); Harkness

HARblend, S, ab, 1994; (**Della Balfour**); Harkness

'HARBONNY', F, mr, 1993; (**Royal Anniversary**); flowers crimson, yellow base, reverse mid-pink, aging dark red, small, dbl., 24 petals, 2.5 in., slight, fruity fragrance; foliage small, dark green, glossy; bushy, low growth; [Intrigue X Anna Ford]; Harkness, 1987; Harkness New Roses, Ltd., 1992

HARboul, MinFl, mp, 1998; (**Pink Pirouette**); Harkness

'HARBRILL', F, my, 1994; (**Conquest**); dbl., 30 petals, 3.5 in., borne in sprays of 7-10, slight, spicy fragrance; foliage medium, light green, glossy; bushy, medium growth; [Dame of Sark X Bright Smile]; Harkness; Harkness New Roses, Ltd., 1994

'HARBRITE', F, lp, 1993; (**Mrs Iris Clow**); flowers blush pink, light pink reverse, paling to near white, loose, dbl., 28 petals, cupped, moderate, spicy fragrance; foliage large, dark green, glossy; upright, medium growth; [Memento X Princess Alice]; Harkness, 1987; Harkness New Roses, Ltd., 1994

'HARBUSY', F, mp, 1970; (**Busy Lizzie**); flowers pink, medium, semi-dbl., 12 petals, 2 in., slight fragrance; foliage glossy, dark; free growth; [(Pink Parfait X Masquerade) X Dearest]; Harkness

HARcheer, S, pb, 1998; (**Country Music**); Harkness

HARchutzpah, HT, 1995; (**June Whitfield**); Harkness, R.; (Cavriglia)

HARclue, HT, ab, 1998; (**Commonwealth Glory**); Harkness

HARcogent, HT, 1996; (**St Christopher**); GM, Belfast, 1998; Harkness, R.; (Cavriglia)

'HARCOMP', LCl, ly, 1980; (**Highfield**®); [Compassion sport]; Harkness, 1981

HARcross, MinFl, dr, 1995; (**Humanity**); Harkness

'HARDANCER', F, ob, 1976; (**Flame Dancer**); flowers orange, red reverse, medium, dbl., cupped, moderate fragrance; large prickles; foliage dense; medium, bushy growth; [Orange Sensation X Alison Wheatcroft]; Harkness; Mason

HARdeed, HT, lp; (**Royal Philharmonic**); Harkness

HARdeluxe, F, 1996; (**Lilian Baylis**); Harkness, R.; (Cavriglia)

HARdenier, S, 1996; (**Marjorie Marshall**); Harkness; (Cavriglia)

HARdimple, F, 1996; (**World Class**); Harkness; (Cavriglia)

HARdinkum, F, w, 1997; (**Princess of Wales**); Harkness

HARdolly, F, lp; (**Tambourine**); Harkness, 1998

HARdwell, LCl, ab, 1998; (**Penny Lane**); Harkness

HAReast, F, 1995; (**Rising Star**); Harkness, R.; (Cavriglia)

HAReco, S, 1997; (**Nipper**); Harkness, R.; (Cavriglia)

HAReden, F, lp, 1997; (**Constance Finn**); Harkness

'HARELAN', HT, yb, 1999; (**Gift of Life**, Poetry in Motion); flowers mid yellow-pink, reverse yellow, dbl., 26–40 petals, 4 in., borne mostly singly, moderate fragrance; many prickles; foliage large, medium green, glossy; bushy, medium (40 in.) growth; GM, Belfast, 1999; [Dr Darley X Elina]; Harkness; Edmunds' Roses, 1999;, Harkness New Roses, 1996

HARelite, F, 1996; (**Wheel Horse Classic**); Harkness; (Cavriglia)

HARencens, S, ab, 1998; (**Miss Dior**); Harkness

HARencore, HT, mr, 1998; (**Pride of England**); Harkness

HARentrap, F, lp; (**St Piers**); Harkness, 1998

HARessay, F, mr, 1998; (**Faithful**); Harkness

HARette, F, ob, 1998; (**Betty Harkness**); Harkness

HAReverso, F, lp; (**Spirit of Tollcross**); Harkness, 1998

Harewood *see* TANinaso

HARexclaim, F, ly, 1998; (**Irish Hope**); Harkness

HARextra, S, dr, 1998; (**Madrigal**); Harkness

HARfab, MinFl, lp; (**Suncharm**); Harkness, 1998

HARfling, F, lp, 1998; (**Blushing Bride**); Harkness

'HARFLOW', F, yb, 1999; (**Easy Going**™); flowers deep gold apricot, reverse same, peach overlay on fresh blooms, dbl., 26–30 petals, 3.5–4 in., borne in small clusters, moderate, fruity fragrance; many prickles; foliage large, very bright, light green, glossy; upright, bushy, medium (3-3.5 ft) growth; PP10478; [Sport of Livin' Easy sport]; Harkness; Harkness New Roses, 1999;, Weeks Roses, 1999

HARfracas, F, ab, 1998; (**Calliope**); Harkness

'HARHERO', S, rb, 1978; (**Marjorie Fair**®, Red Ballerina, Red Yesterday); flowers medium red, white eye, blooms in very large clusters, single, 5 petals, 1 in., slight fragrance; foliage small, light green, semi-glossy; dense, bushy growth; GM, Baden-Baden, 1979 GM, Rome, 1977; [Ballerina X Baby Faurax]; Harkness, 1977

'HARHESTER', F, op, 1980; (**Caroline Davison**); flowers medium salmon-pink, mini-flora, small blooms in clusters of, semi-dbl., 16 petals, slight fragrance; straight, dark green prickles; foliage small, dark reddish-green; low, bushy growth; [Tip Top X Kim]; Harkness, R., & Co., Ltd.

'HARISON LEMON', HSpn, ly, 1929; flowers clear lemon-yellow, semi-dbl., non-recurrent, moderate fragrance; bushy (5 ft.) growth; [Harison's Yellow seedling]; Hamblin

Harison Salmon *see* **'HARISON'S SALMON'**

Harisonii *see* **'HARISON'S YELLOW'**

'HARISON'S HARDY', HSpn, ly, 1943; flowers cream, center tinted yellow, semi-dbl., non-recurrent; very hardy.; [R. spinosissima altaica X Harison's Yellow]; Wright, Percy H.

'HARISON'S SALMON', HSpn, op, 1929; (Harison Salmon); flowers salmon, semi-dbl., non-recurrent, moderate fragrance; [Harison's Yellow seedling]; Hamblin

'HARISON'S YELLOW', HFt, dy; (Harisonii, R. foetida harisonii, R. lutea hoggii, R. X harisonii); flowers bright yellow, yellow stamens, semi-dbl.; Eurosa, Pimpinellifolia (28); [Probably Persian Yellow X R. spinosissima]; Harison, 1830; ca. 1830

'HARJAMES', S, yb, 1989; (**Xerxes**); bud pointed; flowers rich yellow with scarlet redeye at base, reverse yellow, single, 5 petals, cupped, slight fragrance; fruit not observed; prickles narrow, small, reddish; foliage small, grayish-green, matt; upright, medium to tall growth; [H. persica X Canary Bird]; Harkness, R., & Co., Ltd.

'HARJOOBILY', F, op, 1978; (**Mr E.E. Greenwell**); flowers rosy salmon, semi-dbl., 18 petals, 3 in., flat; vigorous, bushy, spreading growth; [Jove X City of Leeds]; Harkness

'HARJOSINE', HT, ab, 1980; (**Basildon Bond**); flowers loose form, dbl., 27 petals, borne 1-3 per cluster, moderate fragrance; large, straight prickles; foliage large, medium green, very glossy; medium, upright growth; GM, Belfast, 1982; [(Sabine X Circus) X (Yellow Cushion X Glory of Ceylon)]; Harkness, R., & Co., Ltd.

'HARKANTABIL', HT, yb, 1979; (**Marion Harkness®**); flowers canary-yellow, flushed orange-red, dbl., 24 petals, 3.5 in., slight fragrance; bushy growth; [((Manx Queen X Prima Ballerina) X (Chanelle X Piccadilly)) X Piccadilly]; Harkness

'HARKARAMEL', F, ab, 1979; (**Anne Harkness®**); bud globular; flowers deep apricot, dbl., cupped, blooms in trusses, slight fragrance; foliage medium green, semi-glossy; vigorous, upright, tall growth; [Bobby Dazzler X ((Manx Queen X Prima Ballerina) X (Chanelle xPiccadilly))]; Harkness, 1980

'HARKINDER', Min, mp, 1979; (**Esther's Baby®**); bud pointed; flowers persian rose, patio, medium, flat; foliage small, glossy; low, spreading growth; [(Vera Dalton X (Chanelle X Piccadilly)) X Little Buckaroo]; Harkness

'HARKING', F, yb, 1977; (**Judy Garland**); flowers yellow, petals edged orange-red, medium-large, borne singly, dbl., 35 petals, slight fragrance; foliage semi-glossy; medium, bushy growth; [((Tropicana X Circus) X (Sabine X Circus)) X Pineapple Poll]; Harkness, 1978

'HARKITTEN', Min, or, 1980; (**Crispin-Morwenna**); flowers salmon-red, small blooms borne 3-7 per cluster, dbl., 25 petals, cupped, slight, spicy fragrance; reddish prickles; foliage small, dark, glossy; low, spreading growth; [(Vera Dalton X (Chanelle X Piccadilly)) X Little Buckaroo]; Harkness, R., & Co., Ltd.

Harkness Marigold *see* 'HARTOFLAX'

'HARKOTUR', F, pb, 1977; (**Softly Softly**); flowers pink and creamy pink, dbl., 35 petals, 5 in., slight fragrance; foliage leathery; [White Cockade X ((Highlight X Colour Wonder) X (Parkdirektor Riggers X Piccadilly))]; Harkness, 1980

'HARKOVER', F, op, 1972; (Garden City, **Letchworth Garden City®**); flowers medium salmon-pink, medium, dbl., 20 petals, 2.5 in., moderate, spicy fragrance; foliage medium green, semi-glossy; vigorous, bushy growth; GM, Monza, 1978; [(Sabine X Pineapple Poll) X (Circus X Mischief)]; Harkness, 1979

'HARKREME', HT, pb, 1980; (**Camphill Glory**); flowers creamy pink, large blooms borne singly, dbl., 54 petals, exhibition form, slight fragrance; many prickles; foliage medium green, matt; vigorous, branching growth; [Elizabeth Harkness X Kordes' Perfecta]; Harkness, R., & Co., Ltd., 1982

'HARKULY', F, w, 1977; (**Margaret Merril®**); flowers blush white, full, dbl., 28 petals, 4 in., exhibition form, intense fragrance; Edland Fragrance Medal, ARS, 1978 GM, Geneva, 1978 GM, Monza, 1978 GM, Rome, 1978 Gold Star of the South Pacific, Palmerston North, NZ, 1982 James Mason Medal, RNRS, 1990; [(Rudolph Timm X Dedication) X Pascali]; Harkness, 1978

'HARKUSHI', F, ob, 1979; (**Amy Brown®**); bud ovoid; flowers burnt-orange to fire-red, rounded, dbl., 28 petals, 2.5–3 in., moderate, fruity fragrance; foliage large, dark; low, bushy growth; [Orange Sensation X ((Highlight X Colour Wonder) X (Parkdirektor Riggers X Piccadilly))]; Harkness

'HARLACAL', S, yb, 1981; (**Rachel Bowes Lyon**); flowers peach pink, reverse yellow, blooms in large clusters, semi-dbl., 14 petals, flat, moderate fragrance; small prickles; foliage small to medium, medium green, low, bushy growth; [Kim X ((Orange Sensation X Allgold) X R. californica)]; Harkness, R., & Co., Ltd.

'HARLASSIE', Pol, lp, 1980; (**Fairy Maid**); bud short, plump; flowers light rose-pink, medium, dbl., 20 petals, cupped; foliage glossy; low, bushy growth; [The Fairy X Yesterday]; Harkness

'HARLAYALONG', Pol, lp, 1980; (**Fairyland®**); bud short, fat; flowers medium, dbl., 24 petals, cupped, moderate fragrance; foliage glossy; spreading growth; [The Fairy X Yesterday]; Harkness

HARlecho, LCl, pb, 1994; (**Morning Light**); Harkness

Harlekijn, F, yb

Harlekin® *see* KORlupo

'HARLENTEN', F, w, 1980; (**Greensleeves**); bud pointed; flowers chartreuse-green, large, semi-dbl., 15 petals, flat; foliage dark; vigorous, upright growth; [(Rudolph Timm X Arthur Bell) X ((Pascali X Elizabeth of Glamis) X (Sabine X Violette Dot))]; Harkness

'HARLEQUIN', HWich, pb, 1935; flowers half pale pink and half dark red, borne in clusters; very vigorous, climbing growth; RULED EXTINCT 12/83; [Excelsa sport]; Cant, F.

Harlequin *see* 'AROVULE'

Harlequin™ *see* 'KORLETTE'

Harlew, F, 1973; Petersen, V.; (Cavriglia)

'HARLEX', HT, or, 1972; (**Alexander®**, Alexandra); bud pointed; flowers bright vermilion-red, dbl., 25 petals, 5 in., exhibition form, slight fragrance; foliage glossy; tall, vigorous growth; ADR, 1974 GM, Belfast, 1974 GM, Hambourg, 1973 James Mason Medal, RNRS, 1987; [Tropicana X (Ann Elizabeth X Allgold)]; Harkness

'HARLEXIS', HT, ob, 1981; (Alexis, **L'Oréal Trophy**); flowers orange; GM, Bagatelle, 1984 GM, Belfast, 1984 Golden Rose, Courtrai, 1986; [Alexander sport]; Harkness, R., & Co., Ltd., 1982

Harley® *see* LENpaga

'HARLIGHTLY', F, my, 1981; (**Princess Michael of Kent®**); flowers large blooms borne 1-3 per cluster, dbl., 38 petals, exhibition form, moderate fragrance; short, thick, red prickles; foliage medium large, mid-green, glossy; low, bushy growth; [Manx Queen X Alexander]; Harkness, R., & Co., Ltd.

'HARLINO', F, op, 1980; (**Sue Ryder®**); flowers salmon orange, reverse shaded yellow, blooms in large clusters, dbl., 20 petals, cupped, slight fragrance; small prickles; foliage medium, mid-green, semi-glossy; vigorous, medium, bushy growth; [Southampton X ((Highlight X Colour Wonder) X (Parkdirektor Riggers X Piccadilly))]; Harkness, 1983

'HARLITTLE', Pol, w, 1980; (**Fairy Snow®**); bud squat; dbl., 22 petals, cupped, borne several per cluster, slight fragrance; slender, dark prickles; foliage small, dark, glossy; short, bushy growth; [The Fairy X Yesterday]; Harkness, R., & Co., Ltd.

'HARLOW', HT, op, 1968; flowers salmon, large, dbl., 29 petals; foliage glossy; [Fragrant Cloud X Melrose]; Cocker

Harlow Carr *see* KIRlyl

Harman, F, 1991; Adam, M.; (Cavriglia)

Harman Inermis®, F, dp; Adam

'HARMANTELLE', F, my, 1982; (**Mountbatten®**); flowers large blooms borne singly or several together, dbl., 45 petals, cupped, moderate fragrance; many, large prickles; foliage large, leathery, glossy; upright, dense growth; GM, Belfast, 1982 GM, Courtrai, 1986 GM, Orleans, 1982 Golden Rose, The Hague, 1986 ROTY, 1982; [Peer Gynt X ((Anne Cocker X Arthur Bell) X Southampton)]; Harkness, R., & Co., Ltd.

'HARMARK', F, yb, 1985; (**Hiroshima's Children**); flowers light yellow, petals edged pink, large, dbl., 35 petals, exhibition form, slight fragrance; foliage medium, medium green, matt; bushy growth; Harkness, R., & Co., Ltd.

Harmonie® *see* 'KORTEMBER'

'HARMONIE', S, pb, 1954; (Kordes' Harmonie); bud ovoid, light red; flowers pink bicolor, large, borne in clusters, moderate fragrance; foliage leathery; very vigorous, upright growth; RULED EXTINCT 5/80; [R. eglanteria hybrid X Peace]; Kordes

Harmonie *see* KRIlamy

'HARMONY', Cl HP, ab, 1933; flowers apricot-pink, very large, semi-dbl., exhibition form, intense fragrance; foliage leathery, dark; strong stems; very vigorous, climbing growth; [Rosella (self)]; Nicolas; C-P

Harmony Parade *see* POUlming

'HARMUSKY', F, mp, 1980; (**Radox Bouquet**, Rosika); flowers soft medium pink, blooms borne 1-3 per cluster, dbl., 30 petals, cupped, moderate fragrance; large, dark prickles; foliage large, glossy, medium green; upright, rather open growth; [(Alec's Red X Piccadilly) X (Southampton X (Clare Grammerstorfx Fruhlingsmorgen))]; Harkness, R., & Co., Ltd., 1981

'HARNEATLY', Pol, dr, 1979; (**Fairy Damsel**®*)*; bud short, plump; flowers oxblood-red, medium, dbl., 24 petals, cupped; foliage glossy; low, spreading growth; [The Fairy X Yesterday]; Harkness

'HARNICELY', Pol, mp, 1979; (**Fairy Ring**); bud short, plump; flowers rose-pink, dbl., 20 petals, 2 in., cupped; foliage glossy; low, bushy growth; Harkness

'HARNIMBLE', Pol, lp, 1979; (**Fairy Like**); bud short, plump; flowers light rose-pink, small, dbl., 20 petals, cupped; foliage small, glossy; low, spreading growth; [The Fairy X Yesterday]; Harkness

'HARNOBLE', HT, dp, 1981; (**Pacemaker**); bud pointed; flowers deep pink, blooms borne singly, sometimes 3 per cluster, dbl., 40 petals, 4 in., intense fragrance; large, broad, dark prickles; foliage large, semi-glossy; vigorous, upright, bushy growth; GM, Belfast, 1983; [Red Planet X Wendy Cussons]; Harkness

'HARNOUGETTE', Pol, mr, 1981; (**Fairy Prince**®); bud short; flowers medium, dbl., 25 petals, cupped; foliage glossy; spreading growth; First Prize, Copenhagen; [The Fairy X Yesterday]; Harkness

'HARNUMEROUS', Pol, mp, 1979; (**Fairy Changeling**); bud short, plump; flowers pink, pompon, medium, dbl., 22 petals, cupped, slight fragrance; foliage small, dark; recumbent, spreading growth; [The Fairy X Yesterday]; Harkness, 1981

'HAROFFEN', S, mp, 1985; ('HARROFFEN', **Rochester Cathedral**); flowers medium to large blooms in clusters, dbl., 58 petals, cupped, repeat bloom, moderate fragrance; foliage medium, dark, matt; medium, dense, spreading growth; [(Unnamed seedling X ((Orange Sensation X Allgold) X R. californica)) X Frank Naylor]; Harkness, R., & Co., Ltd., 1987

Harold Ickes *see* **'CREPE MYRTLE'**

Harold Macmillan *see* 'HARWESTSUN'

'HARPADE', F, m, 1967; (**Escapade**®); flowers magenta-rose, center white, blooms in clusters, semi-dbl., 12 petals, 3 in., slight fragrance; foliage glossy, light green; ADR, 1973 GM, Baden-Baden, 1969 GM, Belfast, 1969; [Pink Parfait X Baby Faurax]; Harkness

'HARPICCOLO', Min, ob, 1980; (**Anna Ford**®, Anne Ford); flowers deep salmon-orange, yellow eye, patio, small, semi-dbl., 18 petals, blooms in large clusters, slight fragrance; small prickles; foliage small, glossy; low, bushy growth; GM, Genoa, 1987 GM, Glasgow, 1989 GM, RNRS, 1981 PIT, RNRS, 1981; [Southampton X Darling Flame]; Harkness

'HARPILLAR', F, mr, 1982; (**Olive**); flowers large blooms in clusters, dbl., 36 petals, exhibition form, moderate, spicy fragrance; dark prickles; foliage large, dark, glossy; branching growth; [((Vera Dalton X Highlight) X Unnamed seedling) X Dublin Bay]; Harkness, R., & Co., Ltd.

'HARPIPPIN', LCl, ob, 1984; flowers pale salmon-red, yellow reverse, blooms borne singly, dbl., 22 petals, slight fragrance; foliage medium, semi-glossy; upright (to 7 ft.) growth; [Royal Dane sport]; Harkness

'HARPLUTO', F, ob, 1983; (**Avocet**); bud pointed; flowers orange edged vermilion, large, semi-dbl., blooms in large clusters, moderate fragrance; many, dark prickles; foliage dark, glossy; medium, bushy growth; [Dame of Sark X Seedling]; Harkness, R., & Co., Ltd., 1984

'HARPOOH', HT, mp, 1983; (**Ruth Harker**); flowers large blooms borne 1-4 per cluster, dbl., 46 petals, intense fragrance; slightly curved, reddish prickles; foliage large, medium green, matt to semi-glossy; upright, bushy growth; [Fragrant Cloud X Compassion]; Harkness, R., & Co., Ltd.

'HARPORTLY', HT, mp, 1982; (**Neville Gibson**); flowers large, dbl., 40 petals, exhibition form, slight fragrance; foliage large, medium green, semi-glossy; medium, upright growth; Golden Rose, Geneva, 1980; [Red Planet X (Carina X Pascali)]; Harkness, R., & Co., Ltd., 1983

'HARPOSTER', HT, mp, 1980; (**Dr Darley**); flowers rose bengal, blooms borne usually singly, dbl., 45 petals, globular, slight fragrance; narrow, reddish prickles; foliage mid-green, semi-glossy; upright, bushy growth; GM, Munich, 1983; [Red Planet X (Carina X Pascali)]; Harkness, R., & Co., Ltd., 1982

'HARPRIER', S, yb, 1986; (**Tigris**); flowers yellow, dark red eye, dbl., 1 in., slight fragrance; gooseberry-like prickles; foliage small, light green, variable in shape; compact, rounded growth; [Hulthemia persica X Trier]; Harkness, 1985

'HARPRINCELY', HT, dr, 1983; (**Innoxa Femille**); dbl., 50 petals, borne singly, sometimes 3 per cluster, slight fragrance; dark prickles; foliage large, semi-glossy; medium, bushy growth; [Red Planet X Eroica]; Harkness, R., & Co., Ltd.

'HARPROCRUSTES', Min, ab, 1982; (**Clarissa**®); flowers small blooms in large clusters, dbl., 43 petals, exhibition form, slight fragrance; small prickles; foliage small, dark, glossy; tall, upright growth; Gold Star of the South Pacific, Palmerston North, NZ, 1982; [Southampton X Darling Flame]; Harkness, R., & Co., Ltd.; Harkness New Roses, Ltd., 1983

'HARPURL', HT, mp, 1979; (**Leigh-Lo**); bud pointed; flowers rose-bengal, urn-shaped, large, dbl., 42 petals, slight fragrance; foliage large; vigorous, upright growth; [Elizabeth Harkness X Red Devil]; Harkness, 1981

'HARQUAKER', F, yb, 1985; (**Volunteer**); flowers large, dbl., 35 petals, slight fragrance; foliage medium, light, glossy; bushy growth; [Dame of Sark X Silver Jubilee]; Harkness, R., & Co., Ltd., 1986

'HARQUANNE', LCl, ab, 1980; (**Breath of Life**); bud plump; flowers apricot to apricot-pink, borne 1-3 per cluster, dbl., 33 petals, moderate fragrance; large, straight, reddish prickles; foliage semi-glossy; upright (to 8 ft) growth; [Red Dandy X Alexander]; Harkness, R., & Co., Ltd., 1982

'HARQUANTUM', F, m, 1984; (**International Herald Tribune**®, Violetta, Viorita®); flowers violet-purple, mini-flora, blooms in trusses, dbl., 20 petals, cupped, moderate fragrance; foliage medium, medium green, semi-glossy; low, bushy growth; GM, Monza, 1984 GM, Tokyo, 1983 Golden Rose, Geneva, 1983; [Unnamed seedling X ((Orange Sensation X Allgold) X R. californica)]; Harkness, R., & Co., Ltd., 1985

'HARQUETERWIFE', HT, op, 1981; (Heart Throb, **Paul Shirville**); flowers light salmon-pink, medium-large blooms borne singly and in3', dbl., 30 petals, exhibition form, moderate, sweet fragrance; large, reddish prickles; foliage large, dark, semi-glossy; medium, bushy growth; Edland Fragrance Medal, ARS, 1982; [Compassion X Mischief]; Harkness, R., & Co., Ltd., 1983

'HARQUHLING', S, ly, 1982; (**Anna Zinkeisen**); bud plump; flowers , dbl., 30 petals, blooms in clusters of 3-7, moderate, musk fragrance; small prickles; foliage small, light green, semi-glossy; medium, spreading, dense growth; [Seedling X Frank Naylor]; Harkness, 1983

'HARQUIBBLER', S, pb, 1989; (**Nigel Hawthorne**); bud pointed; flowers pale salmon-rose, deep scarlet eye at base, reverse same, single, 5 petals, cupped, slight, spicy fragrance; plump, small, infrequent, green fruit; prickles thin, narrow, variable, dark to light; foliage medium, medium green, semi-glossy; spreading, low growth; [H. persica X Harvest Home]; Harkness, R., & Co., Ltd.

'HARQUILLYPOND', F, mp, 1986; (Blue Carpet, **Cosette**); dbl., slight fragrance; small, dark prickles; foliage small, medium green, matt; patio; low, spreading growth; [Seedling X Esther's Baby]; Harkness, R., & Co., Ltd., 1983

'HARQUINCE', F, op, 1986; (Daniel Gelin, **Wandering Minstrel**); flowers pink shaded orange, large blooms in clusters, dbl., 28 petals, slight fragrance; foliage dark green, glossy; medium, bushy growth; [Dame of Sark X Silver Jubilee]; Harkness

'HARQUISP', Min, dp, 1985; (**John Bradshaw**); flowers light rose red, rosette (starry-shaped) blooms in clusters, dbl., 24 petals, flat, slight fragrance; foliage small, semi-glossy; low, bushy growth; [Unnamed seedling X Esther's Baby]; Harkness; White Rose Nurseries, Ltd.

'HARQUITO', F, op, 1987; (**Bill Slim**); bud globular; flowers salmon, deep pink reverse, large blooms borne singly or in clusters, dbl., 30 petals, slight fragrance; foliage glossy; bushy, spreading growth; [Seedling X Silver Jubilee]; Harkness, 1989

'HARQUORGOLD', F, dy, 1984; (**Rediffusion Gold**); flowers deep golden yellow, blooms in clusters of 3-7, dbl., 70 petals, slight fragrance; small prickles; foliage small, light green, matt; medium, bushy growth; [(Orange Sensation X Allgold) X Sunsprite]; Harkness, R., & Co., Ltd.; R. Harkness & Co., Ltd.

'HARRAMIN', Min, mp, 1985; (**Hollie Roffey**); flowers rosette form, small blooms in clusters, dbl., 35 petals, slight fragrance; foliage small, pointed, medium green, semi-glossy; spreading growth; [(Tip Top X (Manx Queen X Golden Masterpiece)) X Darling Flame]; Harkness, R., & Co., Ltd., 1986

'HARRANGO', F, op, 1984; (**G. P. & J. Baker**); flowers salmon orange, reverse lighter, blooms in clusters of 3-11, dbl., 36 petals, flat, slight fragrance; many prickles; foliage dark, glossy; medium, bushy growth; [(Bobby Dazzler X Unnamed seedling) X Marion Harkness]; Harkness, R., & Co., Ltd.

'HARREGALE', S, m, 1984; (**Cardinal Hume®**); bud pointed; flowers violet-purple, blooms in clusters of 3 to many, dbl., 31 petals, cupped, repeat bloom, moderate, musk fragrance; small prickles; foliage variable shades of green, matt; medium, spreading growth; [((Seedling X (Orange Sensation X Allgold) X R. californica)) xFrank Naylor]; Harkness, R., & Co., Ltd.

'HARRIET', Cl HT, my, 1931; flowers golden yellow, edged paler, dbl., recurrent bloom; foliage bronze; vigorous, climbing (12-15 ft.) growth; [Golden Ophelia sport]; Moore, Ralph S.

'HARRIET A. EASLEA', HT, rb, 1922; flowers bright carmine, reverse golden yellow, dbl., moderate fragrance; McGredy

'HARRIET ELIZABETH', S, mp, 1987; flowers pure medium pink, large, borne singly and in clusters, dbl., 26–40 petals, repeat bloom, intense fragrance; foliage medium, red aging dark green, leathery, disease-resi; upright, bushy, vigorous, fully hardy growth; [Paula X (Micki X Northlander)]; James, John; Historical Roses, 1987

'HARRIET MILLER', HT, pb, 1972; bud long, pointed; flowers pink, large, very dbl., globular, moderate fragrance; foliage large, glossy; very vigorous, bushy growth; [Helen Hayes X Traviata]; Brownell, H.C.; Stern's Nursery

'HARRIET NEESE', S, op, 1928; flowers coral blended with yellow, base golden yellow, semi-dbl., abundant non-recurrent bloom, slight fragrance; bushy growth; [Ophelia X R. X harisonii]; Conyers

'HARRIET POULSEN', F, mp, 1912; flowers apple-blossom-pink, single; vigorous growth; [Mme Norbert Levavasseur X Dorothy Perkins]; Poulsen, D.T.

'HARRIET SHEPHERD', Gr, mp, 1982; flowers medium, dbl., 35 petals, no fragrance; foliage medium, medium green, holly-like; upright growth; [Queen Elizabeth X Queen Elizabeth]; Shepherd, David

Harriet Sheppard, HWich, m, 1997; Nobbs

'HARRINY', HT, mp, 1967; flowers clear pink, pointed, large, dbl., 40 petals, intense fragrance; foliage dark; [Pink Favorite X Lively]; LeGrice

'HARROFFEN', S, mp, 1985; ('HAROFFEN', **Rochester Cathedral**); flowers medium to large blooms in clusters, dbl., 58 petals, cupped, repeat bloom, moderate fragrance; foliage medium, dark, matt; medium, dense, spreading growth; [(Unnamed seedling X ((Orange Sensation X Allgold) X R. californica)) X Frank Naylor]; Harkness, R., & Co., Ltd., 1987

'HARRONVER', HT, yb, 1990; (**Anneka**); bud ovoid; flowers high-centered to cupped to reflexed, medium, very dbl., 45 petals, exhibition form, borne usually singly or in sprays of 3-5, moderate, fruity fragrance; ovoid, medium to large, medium green fruit; prickles slightly curved, medium, dark reddish; foliage medium, medium green, glossy; upright, medium growth; [Goldbonnet X Silver Jubilee]; Harkness, R., & Co., Ltd.; R. Harkness & Co., Ltd.

'HARROONY', F, ab, 1983; (**Amber Queen®**); bud plump; flowers apricot gold, large, dbl., 25–30 petals, cupped, blooms in clusters of 3-7, moderate, sweet and spicy fragrance; reddish prickles; foliage large, copper red to medium green, semi-glossy; low, compact, bushy growth; disease resistant; PP5582; AARS, 1988 GM, Genoa, 1986 GM, Orléans, 1987 Gold Star of the South Pacific, Palmerston North, NZ, 1988 ROTY, St. Albans, UK, 1984; [Southampton X Typhoon]; Harkness, 1984

'HARROTANG', F, or, 1986; (**City of Bradford**); flowers small to blooms in large clusters, semi-dbl., cupped, slight fragrance; foliage dark, semi-glossy; medium, upright growth; [(Manx Queen X Whisky Mac) X ((Highlight X Colour Wonder) X (Parkdirektor Riggers X Piccadilly))]; Harkness

'HARROWBOND', HT, op, 1985; (**Rosemary Harkness**); flowers orange-salmon, orange-yellow reverse, medium-large, dbl., 35 petals, intense fragrance; foliage large, dark, semi-glossy; bushy growth; Fragrance Award, Belfast, 1987 GM, Belfast, 1987; [Compassion X (Basildon Bond X Irish Gold)]; Harkness, R., & Co., Ltd.

Harry® *see* **'HARRY WHEATCROFT'**

'HARRY CAMPBELL', F, dp, 1977; bud small, pointed; flowers cerise-pink shading to white, dbl., 45 petals, 2.5 in., intense, wild apple fragrance; foliage dark, leathery; vigorous, bushy growth; Burnet; Benefield's Nursery

'HARRY EDLAND', F, m, 1975; flowers lilac-pink, dbl., 26 petals, 4 in., intense fragrance; foliage dark, glossy; Edland Fragrance Medal, ARS, 1975; [(Lilac Charm X Sterling Silver) X (Blue Moon X (Sterling Silver X Africa Star))]; Harkness, 1978

'HARRY G. HASTINGS', HT, dr, 1965; (H.G. Hastings); flowers large, dbl., slight fragrance; foliage leathery; vigorous growth; [Gov. Mark Hatfield X Helene Schoen]; Von Abrams

'HARRY KIRK', T, ly, 1907; bud pointed; flowers light sulfur-yellow, open, dbl., moderate fragrance; foliage leathery; strong stems; vigorous, bushy growth; Dickson, A.

'HARRY MAASZ', LCl, rb, 1939; bud long, pointed; flowers crimson, center white, very large, single, cupped, slight fragrance; foliage large, leathery, wrinkled, dark; very vigorous, climbing growth; [Barcelona X Daisy Hill]; Kordes

Harry Oppenheimer *see* KORabmask

'HARRY WHEATCROFT', HT, yb, 1972; (Caribia, Harry®); flowers yellow striped red, reverse yellow; [Piccadilly sport]; Wheatcroft & Sons, 1973

Harry Wheatcroft, **Climbing** *see* 'AROMONTELIB'

'HARSAMY', HT, op, 1989; (**Shire County**); bud ovoid; flowers peach on primrose yellow base, reverse salmon rose on primrose, dbl., 33 petals, cupped, moderate fragrance; prickles decurved, medium, reddish; foliage medium, medium green, semi-glossy; bushy, medium growth; [Amy Brown X Bonfire Night]; Harkness, R., & Co., Ltd., 1990

'HARSHERRY', F, yb, 1982; (**Sheila's Perfume**); flowers yellow, petals edged red, large blooms borne singly, dbl., 20 petals, intense fragrance; foliage medium, dark, semi-glossy; medium, bushy growth; Edland Fragrance Medal, ARS, 1991 Gold Star of the South Pacific, Palmerston North, NZ, 1993; [Peer Gynt X (Daily Sketch X (Paddy McGredy X Prima Ballerina))]; Sheridan, John; Harkness New Roses, Ltd., 1985

'HARSPRICE', HT, rb, 1986; (**Potton Heritage**); flowers plum red, reverse straw-yellow, large, dbl., 32 petals, exhibition form, moderate fragrance; foliage large, dark, glossy; bushy, branching growth; [Precious Platinum X Dr. A.J. Verhage]; Harkness, 1987

HARsuma, S, mr; (**Suma**); Onodera, Toru F.; Harkness, 1989

'HARTANNA', F, my, 1985; (Brite Lites®, **Princess Alice**, Zonta Rose); flowers butter yellow, ruffled, dbl., 28–32 petals, 4–4.5 in., exhibition form, blooms in large clusters, slight, fruity fragrance; foliage medium, medium green, semi-glossy; upright growth; disease resistant; PP6953; GM, Dublin, 1984; [Judy Garland X Anne Harkness]; Harkness, R., & Co., Ltd.

HARtbrad, Min, ab, 1989; (**Gentle Kiss**); Harkness

HARtellody, HT, my, 1992; (**Golden Friendship**); Harkness

'HARTESIA', HT, yb, 1985; (**Beryl Bach**); flowers yellow, blended pink, large, dbl., 40 petals, moderate fragrance; foliage large, light green, matt; tall, upright growth; [Sunsprite X Silver Jubilee]; Harkness, R., & Co., Ltd.; R. Harkness & Co., Ltd.

'HARTILION', F, or, 1986; (**Seafarer**); flowers large blooms in clusters of 3-5, dbl., 30 petals, cupped, slight fragrance; medium, green fruit; medium, red prickles; foliage medium, dark, glossy; medium, bushy growth; [Amy Brown X Judy Garland]; Harkness

HARtillery, HT, 1995; (**Reconciliation**); Harkness, J.; (Cavriglia)

Hartina *see* **'KAMMERSANGER TERKAL'**

'HARTOFLAX', F, ob, 1986; (**Harkness Marigold**); flowers well-formed, blooms in clusters, dbl., 35 petals, slight fragrance; foliage medium, medium green, semi-glossy; upright growth; [Judy Garland X Anne Harkness]; Harkness, R., & Co., Ltd.

HARtopper, MinFl, pb, 1994; (**Gentle Clown**); Harkness

'HARTRED', F, mr, 1986; (**Save the Children**); flowers bright red, patio, blooms in clusters, dbl., cupped, slight fragrance; narrow, straight prickles; foliage dark, semi-glossy; low, bushy, compact growth; [Amy Brown X Red Sprite]; Harkness

'HARTSAM', HT, ob, 1987; (**Country Lady**); flowers burnt range, reverse suffused pale scarlet, fading orange-salmon, dbl., 25 petals, slight, spicy fragrance; ovoid, medium, green fruit; decurved, medium, reddish prickles; foliage medium, medium green, semi-glossy; bushy, medium, high-shouldered growth; [Alexander X Bright Smile]; Harkness, R., & Co., Ltd.

'HARTUBOND', HT, op, 1989; (**Lord Houghton of Sowerby**); bud pointed; flowers warm, reddish, salmon-pink, reverse deeper, large, borne singly, exhibition form, slight, spicy fragrance; fruit not observed; prickles broad based, large, reddish; foliage large, dark green, glossy; upright, medium growth; [Silver Jubilee X Basildon Bond]; Harkness, R., & Co., Ltd., 1990

'HARTWIZ', F, yb, 1986; (**Conqueror's Gold**); flowers yellow, petals edged orange-red, blooms in clusters of up to 7, 18 petals, cupped, slight fragrance; foliage medium, dark semi-glossy; medium, bushy growth; [Amy Brown X Judy Garland]; Harkness, R., & Co., Ltd.

'HARUBASIL', HT, ab, 1990; (**Cordon Bleu**); bud pointed; flowers apricot with begonia pink reverse, deepening of apricot with aging, dbl., 20 petals, cupped, moderate, fruity fragrance; prickles recurving, average, reddish; foliage medium to large, dark green, glossy; upright, medium growth; [Basildon Bond X Silver Jubilee]; Harkness, R., & Co., Ltd.; R. Harkness & Co., Ltd., 1992

'HARUBONDEE', HT, op, 1988; (**Queen Charlotte**); flowers deep salmon-red, yellow base, reverse pink-red, aging paler, dbl., exhibition form, slight fragrance; ovoid, large, green fruit; prickles recurved, medium, reddish-green; foliage large, dark green, semi-glossy; upright, tall growth; [Basildon Bond X Silver Jubilee]; Harkness, R., & Co., Ltd.; R. Harkness & Co., Ltd., 1989

Harukaze *see* **'HARU-KAZE'**

'HARU-KAZE', LCl, op, 1986; (Harukaze); bud ovoid; flowers salmon yellow to orange-red, blooms borne 6-8 per cluster, dbl., 33 petals; small, curved prickles, slanted downward; foliage dark, glossy; bushy, creeping growth; [Charleston X Dorothy Perkins]; Suzuki, Seizo; Keisei Rose Nursery, 1985

'HARUKFORE', F, lp, 1986; (**City of London®**); flowers light pink, fading to blush, large, dbl., 15–25 petals, intense fragrance; small, reddish, sparse prickles; foliage medium, medium green, glossy, ovate to pointed; bushy growth; GM, LeRoeulx, 1985; [Radox Bouquet X Margaret Merril]; Harkness, R., & Co., Ltd.; The Rose Garden & Mini Rose Nursery, 1988

'HARUNIQUE', S, pb, 1986; (**Euphrates**); flowers pale salmon red, deep pink eye, small blooms in clusters of, single, 5 petals, slight fragrance; prickly; foliage small, variable form (usually long and narrow), ligh; low, spreading growth; [Hulthemia persica X Seedling]; Harkness

Harunomai, F, mp

'HARUSEFUL', S, mp, 1988; (**Armada®**); flowers medium pink, aging slightly paler, medium, dbl., 17 petals, cupped, borne in sprays of up to 12, repeat bloom, moderate fragrance; rounded, medium green fruit; prickles slightly recurved, medium green; foliage medium, medium green, glossy; spreading growth; [New Dawn X Silver Jubilee]; Harkness, R., & Co., Ltd.; R. Harkness & Co., Ltd.

Haruyo, HT, w

'HARVACITY', F, dr, 1987; (**Drummer Boy**); flowers deep, vivid, bright scarlet, fading slightly paler, loose, semi-dbl., 15 petals, cupped, slight, spicy fragrance; ovoid, small, greenish fruit; fairly straight, small, purplish-red prickles; foliage small, medium green, semi-glossy, oval-pointed; patio; spreading, low growth; [(Wee Man X (Southampton X Darling Flame)) X Red Sprite]; Harkness, R., & Co., Ltd.

'HARVALEX', F, or, 1985; (**Christingle**); flowers large, dbl., 35 petals, slight fragrance; foliage medium, dark, semi-glossy; bushy growth; [Bobby Dazzler X Alexander]; Harkness, R., & Co., Ltd., 1987

'HARVANDER', Min, lp, 1989; (**Phoebe**); bud pointed; flowers pale rose pink, reverse same, aging very little, rosette, medium, very dbl., slight fragrance; fruit not a noticeable feature; prickles needle-like, long, decurved, small, dark green; foliage small, medium green, semi-glossy; bushy, low growth; [Clarissa X (Unnamed seedling X Mozart)]; Harkness, R., & Co., Ltd.

'HARVARD', HT, dr, 1926; bud pointed; flowers deep crimson, open, very large, dbl., intense fragrance; foliage soft, bronze; long stems; [Hoosier Beauty X Unnamed seedling]; Vestal

'HARVEE', Min, rb, 1989; (**Phoenix**); bud urn-shaped; flowers blood red, with yellow base, reverse same, aging orange-carm, dbl., slight fragrance; fruit not a noticeable feature; prickles narrow, recurved, small, green; foliage small, medium green, semi-glossy, pointed; bushy, low growth; [Clarissa X [Wee Man X (Southampton X Darling Flame)]]; Harkness, R., & Co., Ltd.

'HARVERAG', F, ab, 1999; (**English Sonnet**™, Fragrant Surprise, Samaritan®); flowers apricot pink/yellow, very dbl., 41 petals, 4 in., intense fragrance; prickles moderate; foliage medium, dark green, glossy; upright, spreading, medium (3 ft) growth; Harkness; Harkness New Roses, 1999;, J&P, 1999

Harvest Fayre *see* 'DICNORTH'

'HARVEST FESTIVAL', HT, ab, 1980; flowers light apricot-orange, reverse apricot flushed pink, urn-shaped, dbl., 28 petals, slight fragrance; globular, large, orange-yellow fruit; medium, reddish-brown prickles; foliage medium, medium green, semi-glossy; tall, bushy growth; [Blessings X Sunblest]; Law, M.J.

'HARVEST GLOW', LCl, rb, 1941; bud long, pointed, ovoid; flowers red to pink, reverse yellow, large, dbl., 60 petals, exhibition form; foliage light green; long stems; vigorous, climbing growth; [Golden Glow X Mercedes Gallart]; Brownell, H.C.

Harvest Home® *see* 'HARWESI'

'HARVEST MOON', HT, my, 1976; bud long, pointed; dbl., 35 petals, 4–5 in., slight fragrance; vigorous growth; [Whisky sport]; Mason, A.L.; F. Mason

'HARVEST MOON', HT, ly, 1938; flowers cream, open, large, borne in clusters, single; foliage leathery, dark; long stems; vigorous, bushy, compact growth; Cant, B. R.

Harvest Song, S, pb

Harvest Sun *see* 'BROSUN'

'HARVEST TIME', Cl HT, ab, 1939; flowers apricot, reverse sometimes pinkish, open, very large, semi-dbl., free, recurrent boom, moderate fragrance; foliage leathery, dark; very vigorous, climbing (15-20ft.) long stems growth; [Sophie Thomas X Souv. de Claudius Pernet]; Thomas; Armstrong Nursery

'HARVESTAL', Min, lp, 1989; (**Pallas**); bud ovoid; flowers light buff pink, paling to buff white, rosette, medium, born, very dbl., no fragrance; fruit not a noticeable feature; prickles narrow, small, dark green; foliage small, medium green, semi-glossy, pointed, plentiful; bushy, spreading, low growth; [Clarissa X New Penny]; Harkness, R., & Co., Ltd.

'HARVESTER', HT, pb, 1975; flowers carmine, reverse silver, center lilac, dbl., 35 petals, 5 in., intense fragrance; foliage large, matt; [Wendy Cussons X Kordes' Perfecta]; Mayhew

HARvilac, Min, m, 1998; (**Gentle Maid**); Harkness

'HARVINTAGE', HT, lp, 1987; (Integrity, **Savoy Hotel**, Vercors, Violette Niestlé); flowers light phlox pink, reverse shaded deeper, well-formed, large, dbl., 40 petals, exhibition form, slight fragrance; rounded, average, green fruit; small, reddish-green, fairly straight, narrow prickles; foliage medium, dark green, semi-glossy; bushy, medium growth; GM, Dublin, 1988; [Silver Jubilee X Amber Queen]; Harkness, R., & Co., Ltd., 1989

HARvissa, Min, ly, 1990; (**Juliet Ann**); Harkness

'HARVOLUTE', F, ab, 1989; (**By Appointment**); bud ovoid; flowers pale buff apricot, aging paler, urn-shaped becoming, medium, dbl., 22 petals, cupped, slight fragrance; prickles rather narrow, medium, dark reddish-green; foliage medium, dark green, semi-glossy; upright, medium growth; [Anne Harkness X Letchworth Garden City]; Harkness, R., & Co., Ltd., 1990

'HARVOOL', F, mp, 1992; (**Muriel**); dbl., 15–25 petals, 1.5 in., borne in large clusters, slight fragrance; few prickles; foliage small, medium green, semi-glossy; patio; low (30 cms), bushy growth; [Liverpool Echo X Unnamed seedling]; Harkness, R., & Co., Ltd.; Harkness New Roses, Ltd., 1991

'HARWADEROX', F, dp, 1992; (**Rosy Future**); flowers carmine, blooms borne in large clusters, dbl., 15–25 petals, 1.5 in., moderate fragrance; few prickles; foliage small, dark green, semi-glossy; patio; medium (60 cms), upright growth; [Radox Bouquet X Anna Ford]; Harkness, R., & Co., Ltd.; Harkness New Roses, Ltd., 1991

'HARWANNA', S, w, 1988; (**Jacqueline du Pré**®); flowers creamy blush to white, loose, large, borne singly or in sprays, semi-dbl., 15 petals, cupped, repeat bloom, moderate, musk fragrance; oval, medium, green orange fruit; prickles small, dark; foliage medium, dark green, glossy; tall, spreading growth; GM, LeRoeulx, 1988; [Radox Bouquet X Maigold]; Harkness, R., & Co., Ltd.; R. Harkness & Co., Ltd., 1989

'HARWANTED', S, pb, 1988; (Many Happy Returns, **Prima**); flowers blush white, reverse blush pink, aging blush to white, large, semi-dbl., 18 petals, cupped, repeat bloom, moderate, fruity fragrance; round, medium, red fruit; prickles straight, small, red; foliage medium, medium green, semi-glossy; spreading, medium growth; GM, Geneva, 1987; [Herbstfeuer X Pearl Drift]; Harkness, R., & Co., Ltd.; R. Harkness & Co., Ltd., 1990

HARward, Pol, mr, 1980; (**Fairy Dance**®); Harkness

'HARWAZZLE', F, or, 1992; (**Sheer Delight**); dbl., 15–25 petals, 1.5 in., borne in large clusters, no fragrance; few prickles; foliage small, light green, semi-glossy; patio; low (40 cms), bushy growth; [Bobby Dazzler X Little Prince]; Harkness, R., & Co., Ltd.; Harkness New Roses, Ltd., 1991

'HARWEEN', F, my, 1988; (**Golden Years**®); bud ovoid; flowers golden yellow, reverse some bronze tint, large, borne in sprays, dbl., 46 petals, cupped, slight, fruity fragrance; fruit not observed; prickles slightly curved, long, thin, greenish-red; foliage medium, dark green, semi-glossy; bushy, medium growth; GM, Orleans, 1990 Hradec Golden Rose, 1989; [Sunblest X Amber Queen]; Harkness, R., & Co., Ltd.; R. Harkness & Co., Ltd., 1990

'HARWELCOME', F, ob, 1992; (Fellowship, **Livin' Easy**™); bud ovoid, short; flowers frilly, orange-apricot, medium sized, very dbl., 25–30 petals, 4–4.5 in., borne in clusters, moderate, sweet and citrus fragrance; foliage medium, medium green, semi-glossy; rounded, medium growth; PP9161; AARS, 1996 GM, RNRS, 1990; [Southampton X Remember Me]; Harkness

'HARWELLINGTON', HT, or, 1992; (**High Sheriff**); dbl., 15–25 petals, 1.5–2.75 in., borne in small clusters, slight fragrance; few prickles; foliage large, dark green, glossy; tall (95 cms), upright growth; [Seedling X Silver Jubilee]; Harkness, R., & Co., Ltd.; Harkness New Roses, Ltd.

'HARWESI', HRg, mp, 1979; (**Harvest Home**®, Harwest Home); bud pointed; flowers mauve-pink, semi-dbl., 14 petals, 4.5 in., cupped, abundant early bloom, then sporadic, slight fragrance; foliage light green, wrinkled; bushy growth; [R. rugosa scabrosa X ?]; Spicer, Mrs. W.E.; Harkness

Harwest Home *see* 'HARWESI'

'HARWESTSUN', F, or, 1988; (**Harold Macmillan**); flowers medium, borne in sprays of 3-7, dbl., 18 petals, cupped, slight fragrance; rounded, medium, green fruit; prickles broad, medium, green; foliage medium, medium green, glossy, abundant; bushy, medium growth; [Avocet X Remember Me]; Harkness, R., & Co., Ltd.; R. Harkness & Co., Ltd., 1989

'HARWEX', F, or, 1990; (**Spirit of Pentax**); bud pointed; flowers bright red, bright orange-red reverse, aging deeper, to roun, dbl., 21 petals, exhibition form, slight fragrance; prickles straight, narrow, small to medium, medium green; foliage small to medium, dark green, glossy; upright, medium growth; [Alexander X Remember Me]; Harkness, R., & Co., Ltd.; R. Harkness & Co., Ltd.

'HARWHARRY', HT, mr, 1987; (**Malcolm Sargent**, Natascha); flowers shining, bright crimson, urn shaped, loose, medium, borne usually singly, dbl., 25 petals, slight, spicy fragrance; rounded, medium, green fruit; glossy, pointed, narrow, small, reddish-green prickles; foliage medium, dark green; bushy, medium growth; GM, Belfast, 1990; [Herbstfeuer X Trumpeter]; Harkness, R., & Co., Ltd., 1988

'HARWICKLOW', F, op, 1990; (**The Fisherman's Cot**); bud pointed; flowers light salmon pink, aging to deeper pink, medium, borne in sprays, dbl., 28 petals, cupped, moderate, sweet, slightly pungent fragrance; foliage small to medium, dark green, semi-glossy; [Radox Bouquet X Anna Ford]; Harkness, R., & Co., Ltd.; R. Harkness & Co., Ltd., 1991

HARWIGWAM, Cl Min, op, 1991; (**Indian Summer**); Harkness

HARwilla, MinFl, dy, 1990; (**Emily Louise**); Harkness

'HARWINNER', Min, w, 1989; (**Pandora**); bud ovoid; flowers ivory, reverse same, rosette, medium, borne in sprays of 3-15, very dbl., 100 petals, slight fragrance; fruit not a noticeable feature; prickles thin, small, reddish; foliage small, medium green, semi-glossy; bushy, spreading, low, compact growth; [Clarissa X Darling Flame]; Harkness, R., & Co., Ltd.

HARwoey, HT, ab; (**Yesteryear**); Harkness, 1994

'HARWOLAVE', Min, dp, 1988; (**Guiding Spirit**); bud ovoid; flowers deep pink, reverse lighter, medium, borne in sprays of 3-9, dbl., flat, slight fragrance; prickles needle-like, very small; foliage small, dark green, semi-glossy; low, bushy growth; [(Blue Moon X Unnamed seedling) X Little Prince]; Harkness, R., & Co., Ltd.; R. Harkness & Co., Ltd., 1989

'HARWONDER', F, ob, 1990; (**Pride of Maldon**); bud pointed; flowers bright reddish-orange, light orange-yellow reverse, darkening, semi-dbl., 10 petals, cupped, slight fragrance; prickles straight or slightly curved, medium to small, green; foliage medium, dark green, glossy; bushy, medium growth; [Southampton X Wandering Minstrel]; Harkness, R., & Co., Ltd.; R. Harkness & Co., Ltd., 1991

HARwotnext, F, w, 1988; (**Sheila MacQueen**); flowers chartreuse green with apricot tint at certain seasons, mediu, dbl., 24 petals, cupped, slight, peppery fragrance; rounded, medium, green fruit; prickles broad, straight, green; foliage medium, medium green, semi-glossy; upright, medium growth; [Greensleeves X Letchworth Garden City]; Harkness, R., & Co., Ltd.

HARxaglen, LCl, ab, 1998; (**Brass Monkey**); Harkness

'HARXAMPLE', F, mr, 1992; (**Velvia**); dbl., 26–40 petals, 3–3.5 in., borne in small clusters, slight fragrance; some prickles; foliage medium, dark green, semi-glossy; medium, bushy growth; [Dr. Darley X Trumpeter]; Harkness, R., & Co., Ltd.; Harkness New Roses, Ltd., 1991

'HARXAMPTON', F, mr, 1992; (**Remembrance**); dbl., 32 petals, 3.5 in., cupped, borne in sprays of 5-7, slight fragrance; foliage dark green, glossy; bushy growth; Golden Prize, Glasgow, 1995; [Trumpeter X Southampton]; Harkness, R., & Co., Ltd., 1983; R. Harkness & Co., Ltd.

HARxever, F, mp, 1996; (**Joy of Health**); Harkness

'HARYAMBER', Min, ob, 1992; (**Cottage Garden**); flowers deep rich orange, blooms borne in large clusters, dbl., 26–40 petals, 1.5–2.75 in.; some prickles; foliage small, dark green, glossy; low (60 cms), upright growth; [Clarissa X Amber Queen]; Harkness, R., & Co., Ltd.; Harkness New Roses, Ltd.

'HARYEARN', HT, mr, 1990; (**Lady Mitchell**); bud pointed; flowers deep rose-red, reverse to rose-red, paling with age, very dbl., 50 petals, cupped, moderate fragrance; prickles slightly declining, medium, green; foliage medium, medium green, semi-glossy; bushy, low to medium growth; [Dr. Darley X Silver Jubilee]; Harkness, R., & Co., Ltd.; R. Harkness & Co., Ltd., 1991

'HARYEN', HT, ab, 1990; (**Mary Jean**, Mary-Jean); bud ovoid to pointed; flowers large blooms borne usually singly, dbl., 37 petals, cupped, moderate, sweet fragrance; prickles slightly decurved, medium, green; foliage medium to large, oval, medium green, semi-glossy; bushy, medium growth; [Dr. Darley X Amber Queen]; Harkness, R., & Co., Ltd.; R. Harkness & Co., Ltd., 1991

HARyoricks, F, ob, 1991; (**Pat James**); Harkness

'HARYUP', LCl, mp, 1994; (**High Hopes**); dbl., 32 petals, 4 in., borne in sprays of 1-3, moderate, spicy fragrance; foliage medium, medium green, semi-glossy; upright, tall, climbing growth; [Compassion X Congratulations]; Harkness; Harkness New Roses, Ltd., 1992

HARzap, F, op, 1998; (**Country Life**); Harkness

HARzart, HT, w, 1994; (**Renaissance**); Harkness

HARzazz, HT; (**Sue Hipkin**, Sweet Revelation); Harkness, 1997; (Cavriglia)

'HARZEAL', F, mp, 1993; (**Octavia Hill**); flowers very double, blooms borne in sprays of 3-5, 75 petals, 3.25 in., moderate, damask fragrance; foliage medium, dark green, semi-glossy; medium, bushy growth; [Armada X Compassion]; Harkness; Harkness New Roses, Ltd., 1994

HARzelt, HT, or, 1998; (**Hector Berlioz**); Harkness

'HARZEST', F, ob, 1993; (**Saga Holiday**); flowers orange, orange-pink reverse, loose, blooms borne in sprays, dbl., 28 petals, 4 in., exhibition form, slight, spicy fragrance; foliage medium, purplish, glossy; bushy, medium growth; [Seedling X Amber Queen]; Harkness, 1985; Harkness New Roses, Ltd., 1993

'HARZIPPEE', HT, pb, 1999; (**LeAnn Rimes**, Perception); flowers rose pink edged ivory yellow, reverse cream, dbl., 26–40 petals, 5 in., exhibition form, borne mostly singly, intense, citrus and rose fragrance; prickles moderate; foliage large, dark green, glossy; upright, tall (4.5 ft) growth; [Dr Darley X Sweetheart]; Harkness; Edmunds' Roses, 1998;, Harkness New Roses, 1997

'HARZODIAC', F, lp, 1993; (**Compassionate Friend**); flowers medium pink, lighter at center, reverse light pink, dbl., 24 petals, 4 in., borne in sprays of 5-7, moderate, fruity fragrance; foliage medium, medium green, semi-glossy; spreading, low growth; [Seedling X Memento]; Harkness, 1985; Harkness New Roses, Ltd., 1993

'HARZOLA', F, lp, 1999; (L'Aimant, **Victorian Spice**®); flowers medium, soft peach pink, old-fashioned, very dbl., 41–50+ petals, 3.5–4 in., cupped, borne in large clusters, intense, damask fragrance; prickles moderate; foliage medium, dark green, glossy; spreading, mounding, bushy, medium (3.5 ft) growth; Harkness, Robert & Philip; J&P, 1994

'HARZORBA', LCl, pb, 1993; (**Victorian Spice**); flowers light rose pink, primrose base, reverse salmon pink stained, semi-dbl., 12 petals, 4.75 in., moderate, fruity fragrance; foliage large, dark green, glossy; upright, tall, climbing growth; [Armada X Compassion]; Harkness, 1985; Harkness New Roses, Ltd., 1993

HARzumber, HT; (**Welwyn Garden Glory**); Harkness, 1996; (Cavriglia)

'HASSAN', F, mr, 1963; flowers scarlet, dbl., 28 petals, 4 in., moderate fragrance; foliage glossy, light green; vigorous, upright growth; [Tivoli X Independence]; McGredy, Sam IV; Fisons Horticulture

'HASSI-MESSAOUD', LCl, or, 1961; flowers garnet-red shaded orange, borne in clusters of 12-15, abundant, recurrent bloom; Hémeray-Aubert

Hat Pin *see* 'TINPIN'

'HAT TRICK', HT, pb, 1992; flowers pink, darker pink petal edges and reverse, natural recurve, very dbl., 3–3.5 in., slight fragrance; few prickles on upper half of stems; foliage medium, dark green, semi-glossy; medium (90-120 cms), upright growth; [First Prize X Unnamed seedling]; Lienau, David W.; Trophy Roses, Ltd., 1993

'HATAKEYAMA', HT, dp, 1995; flowers deep pink, full (26–40 petals), large (7+ cms) blooms; slight fragrance; some prickles; foliage medium, dark green, matt; medium, upright growth; [Garden Party X Kolner Karneval]; Shimizu, Junji

Hatchell Brown Tea, T, lp; Clark

Hatsukoi, HT, w

'HATUZAKURA', HT, ab, 1999; (Orange Chateau); flowers apricot-orange, very dbl., 80 petals, 5.5 in., exhibition form, moderate fragrance; 4 ft. growth; Teranishi, K.; Itami Rose Nursery, 1997

'HAUFF', HMult, m, 1911; flowers reddish violet, borne in clusters, dbl., recurrent when established; foliage dark, broad; vigorous, climbing growth; [Aimee Vibert X Crimson Rambler]; Lambert, P.

'HAUMI', HT, mp, 1981; (**Sweet Mimi**®); flowers large, dbl., 20 petals, moderate fragrance; foliage medium, reddish green, semi-glossy; bushy growth; [Tropicana X Elizabeth Harkness]; Hauser, Victor; Roseraies Hauser

Hauraki *see* 'MACWYNSCAR'

HAUric, Min, op, 1989; (**Ricky**®); Hauser

'HAUSER', HT, my, 1975; flowers yellow-cream, dbl.; vigorous growth; [Barbara X Guitare]; Gaujard

Haute Pink *see* 'JACHOP'

'HAVAM', F, or, 1972; (**Amsterdam**®); bud ovoid; flowers clear orange-red, semi-dbl., 12–15 petals, 2.5 in.; foliage glossy, brown-red; vigorous growth; GM, The Hague, 1972; [Europeana X Parkdirektor Riggers]; Verschuren, Ted; Verschuren

'HAVANA', HT, pb, 1950; flowers salmon-rose, reverse orange-yellow, dbl., 40 petals, 5–5.5 in., slight fragrance; foliage soft; vigorous, compact growth; [Peace X Orange Nassau]; Fisher, G.; Arnold-Fisher Co.

'HAVAPS', HT, dp, 1985; (**Poker**®); flowers deep pink, large blooms in sprays of 6-8, semi-dbl., 11 petals, cupped, intense fragrance; rounded fruit; foliage medium, medium green, semi-glossy; bushy, medium growth; [Red Planet X Sonia]; Verschuren, Ted; H.A. Verschuren

'HAVEAL', HT, dr, 1984; (**Eagle**®); flowers large blooms in sprays of 3-5, dbl., 28 petals, cupped, moderate fragrance; foliage large, dark, semi-glossy; bushy, tall growth; [Centurio X Red Planet]; Verschuren, Ted; H.A. Verschuren

'HAVERING', HMsk, mp, 1937; flowers china-pink, large, borne in clusters of 4 or 5, moderate fragrance; vigorous growth; Bentall

'HAVERING RAMBLER', HMult, mp, 1920; flowers almond-blossom-pink, rosette form, small, dbl., borne in large clusters; long stems; very vigorous growth; Pemberton

'HAVINK', Cl HT, lp, 1980; (**Pink Ocean**®); flowers large, dbl., 20 petals, intense fragrance; foliage medium, medium green semi-glossy; upright (to 7 ft.) growth; [Pink Showers X Alexander]; Verschuren; H.A. Verschuren

'HAVIPIP', Min, pb, 1981; (**Friendship**); bud ovoid; flowers pink and white blend, small blooms in sprays of 30-50, single, 5–6 petals, no fragrance; very small fruit; foliage medium, light green, semi-glossy; groundcover; spreading, medium growth; [(Swany X Mozart) X Mozart]; Verschuren, Ted; H.A. Verschuren

'HAVLICKOVA NÁRODNI', HP, rb, 1935; flowers dark red with white, medium, semi-dbl.; Böhm, J.; (Sangerhausen)

'HAVOON', Min, dy, 1984; (**Gold Moon**®); bud ovoid; flowers small blooms borne singly, semi-dbl., 15 petals, slight fragrance; no prickles; foliage medium, medium green, glossy; spreading, low growth; (28); [(Aalsmeer Gold X Unnamed seedling) X (Motrea X Golden Times)]; Verschuren, Ted; H.A. Verschuren

'HAVOP', HT, dr, 1974; (**Centurio**®); flowers lasting well; moderate, bushy growth; [Orangeade X Baccará]; Verschuren, Ted; Verschuren

'HAVYCHI', Min, dy, 1984; (**Sunny Child**®); flowers small, single, 8 petals, slight fragrance; no prickles; foliage medium, light green; spreading, low growth; [Gold Bunny X Unnamed seedling]; Verschuren, Ted; H.A. Verschuren

'HAWA MAHAL', HT, op, 1976; flowers salmon-pink, dbl., 25 petals, 5–6. in., moderate fragrance; foliage dark; [Fragrant Cloud X Kordes' Perfecta]; Harkness; Anand Roses

'HAWAII', HT, or, 1960; bud long, pointed; flowers orange-coral, dbl., 33 petals, 6 in., exhibition form, intense fragrance; foliage leathery; vigorous, upright growth; [Golden Masterpiece X Unnamed seedling]; Boerner; J&P

'HAWAIIAN BEAUTY', HT, pb, 1993; flowers have salmon pink outer petals, light pink inner petals, very dbl., 5–5.5 in., moderate fragrance; few prickles; foliage medium, medium green, semi-glossy; medium (4 1/2-5 ft), bushy growth; [Thriller X Dothan]; Wyckoff, Gilbert R.; Wyckoff, 1994

'HAWAIIAN BELLE', Min, pb, 1982; flowers medium pink, aging to pink blend, small, dbl., 35 petals, slight fragrance; foliage small, medium green, matt; bushy growth; [Pink Ribbon X Pink Ribbon]; Dobbs, Annette E.

'HAWAIIAN DELIGHT', F, op, 1968; flowers burnt-orange to pink, small, dbl., cupped; foliage dark, leathery; vigorous, bushy growth; [Orange Sensation X Circus]; deRuiter; Carlton Rose Nurseries

'HAWAIIAN FRAGRANCE', HT, or, 1994; flowers coral, orange-red, similar to fragrant cloud, dbl., 26–40 petals, 5–6. in., intense fragrance; some prickles; foliage large, medium green, semi-glossy; tall (6+ft), upright growth; [Fragrant Cloud X Captain Harry Stebbings]; Wyckoff, Gilbert R.; Wyckoff, 1994

'HAWAIIAN KING', HT, lp, 1993; very dbl., 3–3.5 in., borne mostly singly, moderate fragrance; some prickles; foliage medium, medium to dark green, semi-glossy; tall (5-6 ft), upright growth; [Peggy Lee X Captain Harry Stebbings]; Wyckoff, Gilbert R.; Wyckoff, 1994

'HAWAIIAN LADY', HT, dp, 1993; flowers deep pink, blooms borne mostly single, very dbl., 3–3.5 in., moderate fragrance; many prickles; foliage medium, medium green, semi-glossy; medium (5 ft), upright, bushy growth; [Peggy Lee X Captain Harry Stebbings]; Wyckoff, Gilbert R.; Wyckoff, 1994

'HAWAIIAN QUEEN MARTHA', HT, op, 1994; flowers salmon, interspered with light pink on both petal surfaces, very dbl., 5.5–6. in., intense fragrance; some prickles; foliage medium, medium green, semi-glossy; medium (4 ft), upright growth; [Kordes Perfecta X Dothan]; Wyckoff, Gilbert R.; Wyckoff, 1995

'HAWAIIAN SUNRISE', Min, rb, 1981; bud pointed; flowers red and yellow blend, blooms borne usually singly, dbl., 40 petals, exhibition form, slight fragrance; thin, reddish prickles; foliage small, dense, glossy, bronze; upright growth; [Unnamed seedling X Over the Rainbow]; Williams, Ernest D.; Mini-Roses

'HAWAIIAN SUNSET', HT, ob, 1962; bud ovoid; flowers orange edged yellow, open, dbl., 45–50 petals, 4–5.5 in., moderate fragrance; foliage leathery, glossy; vigorous, upright, well branched growth; [Charlotte Armstrong X Signora]; Swim & Weeks; C.R. Burr

'HAWAIIAN THRILL', HT, pb, 1993; flowers pink, white center, moderately blooms borne mostly singly, dbl., 23–25 petals, 5.5 in., intense fragrance; some prickles; foliage medium, dark green, semi-glossy; medium (4 1/2-5 ft), upright growth; [Fragrant Cloud X Thriller]; Wyckoff, Gilbert R.; Wyckoff, 1994

Hawkesbury Wonder, S, lp; [Hebe's Lip X Unknown]; (Weatherly, L.)

'HAWKEYE BELLE', S, w, 1975; bud ovoid, pointed; flowers white, tinted azalea-pink, dbl., 38 petals, 4–4.5 in., exhibition form, intense fragrance; foliage large, dark, leathery; vigorous, erect, bushy growth; [(Queen Elizabeth X Pizzicato) X Prairie Princess]; Buck, Dr. Griffith J.; Iowa State University

'HAWLMARK CRIMSON', HT, mr, 1920; bud pointed; flowers crimson-scarlet, semi-dbl., moderate fragrance; bushy growth; Dickson, A.

'HAWLMARK SCARLET', HT, mr, 1923; flowers brilliant velvety scarlet-crimson, moderate fragrance; GM, NRS, 1920; Dickson, A.

Haynesville Pink Cluster, N, pb

'HAZEL ALEXANDER', HT, dr, 1933; flowers deep red, intense fragrance; [Ophelia seedling]; Dicksons of Hawlmark

'HAZEL ROSE', HT, lp, 1992; dbl., 26–40 petals, 3–3.5 in., borne singly, slight fragrance; few prickles; foliage medium, dark green, semi-glossy; medium (120 cms), upright growth; [Queen Esther X Selfridges]; Poole, Lionel, 1993

'HAZELDEAN', HSpn, my, 1948; flowers more open, of better form than Persian Yellow, moderate fragrance; ; very hardy (to -60); [R. spinosissima altaica X Persian Yellow]; Wright, Percy H.

Head of Rivers, D, mp

'HEADLEYENSIS', S, ly; (R. headleyensis); flowers creamy yellow, single; vigorous growth; [R. hugonis X R. spinosissima altaica]

Headline, HT, yb, 1970; Dawson

Headliner *see* 'JACTU'

Healdsburg Glory, HWich, m

'HEALING HANDS', HT, yb, 1998; flowers pale yellow, edged pink, exhibition form, very dbl., 26–40 petals, 5.5 in., exhibition form, borne mostly singly, slight fragrance; some prickles; foliage large, dark green, glossy; upright, bushy, medium growth; [Gavotte X (Peer Gynt X Golden Splendour)]; Poole, Lionel; David Lister, Ltd., 1999

Heart O' Gold™ *see* 'WEKDYKSTRA'

'HEART OF ENGLAND', Gr, dp, 1978; flowers carmine-rose, dbl., 42 petals, 4 in., exhibition form, moderate fragrance; moderately vigorous growth; [Pink Parfait X ?]; Gregory

'HEART OF GOLD', HMoy, rb, 1926; flowers crimson, center white, stamens yellow, open, borne in clusters, single; foliage rich green; vigorous (10 ft) growth; [R. wichurana X R. moyesii]; Van Fleet; American Rose Society

Heart O'Gold *see* 'WEKDYKSTRA'

Heart of Gold *see* 'MACYELKIL'

'HEART OF T.D.K.', HT, op, 1983; bud ovoid; flowers salmon-pink, large blooms borne singly, dbl., 50 petals, slight fragrance; heavy hooked prickles; foliage small, light green; vigorous, upright, tall growth; [Sunblest X Red Devil]; Ogawa, Isamu

Heart Throb *see* 'LEONHART'

'HEART THROB', F, 1966; bud urn shaped; flowers deep yellow, edges flushed pink, medium, cluster, very dbl., intense fragrance; foliage dark, glossy; vigorous, bushy growth; RULED EXTINCT 4/82 ARM; [(Circus X Circus) X (Circus X Circus)]; Fankhauser

Heart Throb *see* 'HARQUETERWIFE'

Heartache *see* KOReledas

'HEARTBEAT', F, op, 1970; flowers deep salmon-orange, dbl., 26 petals, 4.5 in., globular, slight fragrance; foliage small, dull; very free growth; [(Castanet X Castanet) X (Cornelia X Seedling)]; Dickson, Patrick; A. Dickson

'HEARTBEAT', HT, mr, 1989; bud pointed; flowers medium red, darker outer petals, reverse lighter, medium, dbl., 45 petals, exhibition form, borne usually singly, slight, fruity fragrance; prickles straight, large, pink to yellow; foliage medium, medium green, semi-glossy; upright growth; [Thriller X Wild Cherry]; Bridges, Dennis A.; Bridges Roses, 1990

Heartbeat 96 *see* 'COCORONA'

Heartbreaker *see* 'WEKSYBIL'

'HEARTH GLOW', F, or, 1963; bud ovoid; flowers brick-red, medium, borne in clusters, dbl.; foliage soft, light green; vigorous, upright growth; [Red Pinocchio X (Carrousel X Queen o' the Lakes)]; Von Abrams; Peterson & Dering

Heartland™ *see* 'SAVSAY'

Heartlight™ *see* 'KINHEART'

Hearts A'Fire *see* 'BRIHEART'

'HEART'S DELIGHT', HT, op, 1933; flowers apricot-coral-orange, veined red, large, very dbl., exhibition form, intense fragrance; foliage soft; vigorous, open habit growth; [Mrs Beckwith sport]; Hart, L.P.

'HEART'S DESIRE', HT, dr, 1942; bud long, pointed; flowers crimson, dbl., 30 petals, 4.5 in., exhibition form, intense, damask fragrance; foliage leathery, dark; vigorous, upright growth; AARS, 1942 GM, Portland, 1941; [Unnamed variety X Crimson Glory]; Howard, F.H.; H&S

Heart's Desire Pink, HWich, pb

'HEART'S DESIRE, CLIMBING', Cl HT, dr, 1945; Howard, F.H.; H&S

Heartsounds *see* 'MINSOUNDS'

Heartstrings *see* 'BROHEART'

'HEAT WAVE', F, or, 1958; (Mme Paula Guisez); bud urn-shaped; flowers orange-scarlet, blooms in clusters, dbl., 30 petals, 3.5–4.5 in., cupped, slight fragrance; foliage dark, semi-glossy, rounded; vigorous, upright, bushy growth; [Unnamed seedling X Roundelay]; Swim, H.C.; Armstrong Nursery

Heather *see* 'JACLAV'

Heather Austin *see* 'AUSCOOK'

'HEATHER CLAIRE', HT, ab, 1982; bud long; flowers apricot pink, blooms borne singly, dbl., 30 petals, no fragrance; slightly hooked prickles; foliage dark, red reverse; medium growth; [Diamond Jubilee X Bonsoir]; Allender, Robert William

Heather Honey *see* 'HORSILBEE'

'HEATHER JENKINS', HT, mp, 1968; bud globular; flowers pink, reverse darker, medium, dbl., exhibition form, slight fragrance; foliage light green, wrinkled; moderate, upright, open growth; [Charlotte Armstrong X Ballet]; Watson

Heather Leigh *see* 'TALHEATHER'

'HEATHER MUIR', sp, (variety of R. sericea), w, 1957; flowers pure white, single, 3 in., blooms over a long period, moderate fragrance; orange fruit; foliage ferny; Sunningdale Nursery

'HEATHER PATON', HT, mr, 1934; flowers carmine, center darker; vigorous growth; Austin & McAslan

Heather Pudney, HT, mp, 1995; Dawson

Heaven *see* 'JACFON'

Heaven Bound, F, mp, 1999; flowers bright pink, old-fashioned, ruffled, dbl.; compact (2.5 ft) growth; Williams, J. Benjamin

'HEAVEN SCENT', F, op, 1968; flowers salmon, large blooms in trusses, dbl., 30 petals, intense fragrance; [Pernille Poulsen X Isabel de Ortiz]; Poulsen; McGredy

Heaven Scent Pink *see* 'JALPINK'

Heavenly Days™ *see* 'SAVAHE'

'HEAVENLY FRAGRANCE', HT, lp, 1963; flowers light pink, reverse darker, well formed, 5 in., intense fragrance; moderate growth; [Tiffany X Mme Gregoire Staechelin]; Hennessey

Heavenly Pink *see* LENneel

Heavenly Rosalind *see* 'AUSMASH'

'HEAVENLY SCENT', S, mp, 1999; flowers cerise pink, lighter reverse, large, very dbl., 58 petals, borne mostly singly in and small clusters, moderate fragrance; prickles moderate; foliage medium, dark green, glossy; upright, compact, medium (5 ft) growth; Hamilton, Noel

Heavenly Vision *see* 'MINIVISION'

'HEBE', F, dp, 1941; flowers deep pink, reverse lighter; Leenders, M.

'HEBE', HT, op, 1949; flowers rosy salmon toned orange and apricot-yellow, high pointed, large, dbl., 27 petals, moderate fragrance; foliage glossy, bronze green; vigorous growth; GM, NRS, 1949; Dickson, A.

Hebe Camargo, HT; (Brazil); (Cavriglia)

'HEBE'S LIP', HEg, w; (R. damascena rubrotincta, Reine Blanche, Rubrotincta); flowers creamy white, petals edged pink, vigorous growth, semi-dbl., cupped, non-recurrent; [Probably R. damascena X R. eglanteria hybrid.]; Lee, before 1846; Re-int. W. Paul, 1912

Heckenfeuer® *see* KORrohe

Heckengold, F, dy, 1986; flowers large, dbl., slight fragrance; Kordes, W. Söhne; (Sangerhausen)

Heckenzauber *see* 'MACREXY'

'HECTOR', HGal, m; flowers purple, faintly striped with white, pompon, small, dbl.; Parmentier, before 1847

Hector Berlioz *see* HARzelt

'HECTOR DEANE', HT, rb, 1938; bud pointed; flowers orange, carmine and salmon-pink, dbl., exhibition form, intense, fruity fragrance; foliage glossy, dark; vigorous, compact growth; [McGredy's Scarlet X Lesley Dudley]; McGredy; J&P

'HEDDA HOPPER', HT, mp, 1952; bud ovoid; flowers light peach passing to pearly pink, dbl., 40 petals, 3.5–4 in., globular, moderate fragrance; foliage coppery; very vigorous growth; [Radiance X Unnamed seedling]; Howard, A.P.; H&S

'HEDE', HT, my, 1934; flowers pure sunflower-yellow, large, dbl., slight fragrance; foliage dark, leathery; strong stems; vigorous growth; [Prof. Gnau X Mev. G.A. van Rossem]; Tantau

Hedgefire, MinFl, mr, 1983; Kordes

Hedgehog Rose *see* **R. RUGOSA**

'HEDWIG FULDA', Cl Pol, mr, 1934; flowers clear vermilion-red, well formed, large, borne in large clusters, dbl.; foliage bright, dark; long strong stems; vigorous growth; [Orléans Rose X Farbenkonigin]; Leenders Bros.

'HEDWIG REICHER', HT, ly, 1912; flowers large, very dbl., moderate fragrance; Hinner, W.; (Sangerhausen)

'HEER', HT, lp, 1969; flowers rose-pink, medium, dbl., exhibition form, slight fragrance; vigorous, upright growth; [Picture X Open pollination]; Singh; Gopalsinamiengar

'HEFQUEEN', Min, pb, 1989; (**Regine**™); flowers soft, light pink, silvery-pink reverse, small, dbl., 30 petals, slight fragrance; foliage medium, medium green, semi-glossy; bushy growth; AOE, 1990; [Little Darling X Party Girl]; Hefner, John; Kimbrew Walter Roses, 1990

Hei Matisse *see* DELstrabla

Hei W. Perron, HT, or

Heidefee®, Pol, dp, 1990; Noack, Werner

Heidefeuer *see* NOAfeuer

'HEIDEGRUSS', F, pb, 1937; flowers salmon-flesh, base light yellow, large, borne in clusters, very dbl.; foliage leathery; vigorous, bushy growth; [Heidekind X Ophelia]; Tantau; Münch & Haufe

'HEIDEKIND', HRg, dp, 1931; flowers brilliant pink shaded copper-red, large blooms in clusters, dbl., slight fragrance; foliage thick, rugose; [Mev. Nathalie Nypels X R. rugosa hybrid]; Berger, V.; Münch & Haufe

Heidekönigin® *see* 'KORDAPT'

Heidelberg *see* 'KORBE'

Heidelinde® *see* KORdehei

'HEIDEMARIE', F, mr, 1945; flowers carmine-red, large, borne in clusters of 4-6, single, 5–7 petals, slight fragrance; foliage dark, leathery; vigorous, upright, bushy growth; [Hamburg X (Heros X Heidekind)]; Tantau

'HEIDEMARIE PLÜCKER', HP, mp, 1940; flowers large, very dbl., intense fragrance; Westphal; (Sangerhausen)

'HEIDERÖSLEIN', LCl, yb, 1932; bud pointed, orange-red; flowers bright yellowish salmon-pink, base sulfur-yellow, large blooms, single, flat, recurrent bloom, moderate fragrance; broad, bushy growth; [Chamisso X Amalie de Greiff]; Lambert, P.

Heideroslein Nozomi *see* **'NOZOMI'**

Heideschnee *see* 'KORCONTA'

Heidesommer® *see* 'KORLIRUS'

Heidetraum™ *see* 'NOATRAUM'

'HEIDEZAUBER', F, dr, 1936; flowers large, borne in clusters, very dbl., slight fragrance; foliage dark, leathery; bushy growth; [Heidekind X Johanniszauber]; Tantau; Münch & Haufe

Heidi™ *see* 'ARODI'

Heidi, S, mp, 1987; Noack, Werner

'HEIDI JAYNE', HT, dp, 1986; flowers bright deep pink, large, dbl., 32 petals, moderate fragrance; numerous prickles; foliage large, light green, glossy; medium, upright growth; [(Piccadilly X Queen Elizabeth) X (Fragrant Cloud X Unnamedseedling)]; Esser; Harkness

Heidi Kabel *see* 'KORDIAM'

Heidi Rossin, S, ly

Heidikind *see* 'KORIVER'

Heidi's Gold, Cl HT, dy, 1994

Heietta, T, m

Heiich Blanc, LCl, w, 1994; Hetzel

Heike *see* KORrundum

'HEIMATLOS', HCan, mp, 1931; bud pointed; flowers rose-pink, borne in clusters, single, non-recurrent, intense fragrance; short stems; vigorous (5-7 ft.), open habit growth; has endured -27; [(R. canina X R. roxburghii) X R. canina]; Lohrberg

'HEIN EVERS', F, mr, 1957; bud pointed; flowers bright blood-red, open, borne in clusters, semi-dbl., slight fragrance; foliage leathery; vigorous, upright growth; [Red Favorite X Fanal]; Tantau, Math.

'HEIN EVERS, CLIMBING', Cl F, mr, 1963; Kordes

'HEIN MÜCK'®, S, dr, 1961; flowers velvety blood-red, blooms in clusters, single, cupped; vigorous (6 ft) growth; Tantau, Math.

'HEINE', S, w, 1912; flowers small, dbl.; Lambert, P.; (Sangerhausen)

Heinfels *see* 'HELFELS'

'HEINRICH CONRAD SOTH', S, pb, 1919; flowers light rosy red, with white eye, small, borne in pyramidal clusters, single, recurrent bloom, moderate fragrance; foliage large, glossy, dark; long, strong stems; very vigorous, bushy growth; [Geheimrat Dr. Mittweg X R. foetida bicolor]; Lambert, P.

'HEINRICH EGGERS', HT, op, 1928; flowers orange-copper, often with lighter outer petals, dbl., moderate fragrance; upright growth; [Mrs Charles E. Russell X Mrs Wemyss Quin]; Kordes

'HEINRICH KARSCH', Pol, m, 1927; flowers violet-rose, semi-dbl.; [Orléans Rose X Joan]; Leenders, M.

'HEINRICH MÜNCH', HP, mp, 1911; flowers soft pink, very large, dbl., 50 petals, occasionally recurrent bloom, moderate fragrance; very vigorous growth; [Frau Karl Druschki X (Mme Caroline Testout X Mrs W.J. Grant)]; Hinner, W.; Münch & Haufe

'HEINRICH SCHULTHEIS', HP, lp, 1882; flowers soft pink, well-formed, very large, dbl., occasionally recurrent bloom, intense fragrance; vigorous growth; [Mabel Morrison X E.Y. Teas]; Bennett

'HEINRICH WENDLAND', HT, mr, 1930; flowers nasturtium-red, reverse deep golden yellow, very large, dbl., exhibition form, intense, fruity fragrance; foliage bronze, leathery, glossy; vigorous growth; [Charles P. Kilham X Mev. G.A. van Rossem]; Kordes; Dreer;, H&S

'HEINRICH WENDLAND, CLIMBING', Cl HT, mr, 1937; Stell; Stell Rose Nursery

'HEINSOHN'S RECORD', HCan, lp; Heinsohn-Wedel

'HEINZ ERHARDT'®, F, mr, 1962; dbl., 25 petals, 3 in., blooms in clusters (up to 8), slight fragrance; foliage coppery; vigorous, bushy growth; GM, Baden-Baden, 1961; Kordes, R.

Heinz Treffinger *see* HELtreff

Heinzelmännchen® *see* 'KORNUMA'

'HEIRESS', HT, mp, 1959; bud pointed; flowers clear rose-pink, well shaped, medium, intense fragrance; Longsdon

Heirloom *see* 'JACLOOM'

'HELDENGRUSS', HT, dr, 1920; flowers pure deep blood-red, dbl., moderate fragrance; [Étoile de France X Baron Girod de l'Ain]; Kiese

HELdoro, HWich, mp; (**Super Dorothy**®); Hetzel, 1986

'HELEN', HT, pb, 1930; flowers salmon-pink, base shaded yellow, semi-dbl., moderate fragrance; vigorous growth; RULED EXTINCT 11/91; Ferguson, W.

Helen *see* 'TINHELEN'

'HELEN ALLEN', HT, dr, 1975; flowers velvety red, dbl., 20 petals, 3.5–4 in., moderate fragrance; upright growth; [Evelyn Fison X Vagabonde]; Clayworth

'HELEN BLAND', S, mp, 1950; flowers rose-pink, center deeper, open, medium, borne several together, semi-dbl., profuse non-recurrent bloom, slight fragrance; thornless; foliage soft; stems red-brown; vigorous (7-8 ft.), upright growth; [Betty Bland X R. blanda (St. Hilaire clone)]; Wright, Percy H.

Helen Boehm *see* 'AROPRAWN'

'HELEN CHAMBERLAIN', HT, yb, 1918; flowers creamy yellow to orange-gold, paling on outer petals; Easlea

'HELEN DAVIS', HT, dr, 1925; flowers large, dbl.; (Sangerhausen)

'HELEN FOX', HT, my, 1928; bud pointed, indian yellow; flowers golden yellow, dbl., cupped, slight fragrance; foliage bronze; vigorous, bushy growth; GM, Bagatelle, 1926; [Mme Mélanie Soupert X Souv. de Claudius Pernet]; Buatois

Helen Gambier *see* **'MLLE HÉLÈNE GAMBIER'**

'HELEN GOOD', T, yb, 1907; bud pointed; flowers delicate yellow suffused pink, edged deeper, dbl.; vigorous growth; [Maman Cochet sport]; Good & Reese

Helen Gould *see* **'BALDUIN'**

'HELEN GOULD, CLIMBING', Cl HT, pb, 1912; [Balduin sport]; Good & Reese

'HELEN HAYES', HT, yb, 1956; bud long, pointed; flowers yellow splashed orange and pink, dbl., 43 petals, 4–5 in., exhibition form, moderate fragrance; foliage glossy; very vigorous growth; [Unnamed R. wichurana hybrid X Sutter's Gold]; Brownell, H.C.

'HELEN JANE BURN', HT, dp, 1999; flowers high-pointed, classic form, dbl., 26–40 petals, 5.5–6 in., borne mostly singly, slight fragrance; some prickles; foliage medium, dark green, semi-glossy; long stems; upright, bushy, medium (3 ft) growth; [Blue Moon X seedling]; Poole, Lionel

'HELEN KELLER', HP, dp, 1895; flowers rosy cerise, petals large, shell shaped, very free bloom; GM, NRS; Dickson, A.

Helen Keller *see* BARkel

Helen Knight, S, my, 1970; Knight

'HELEN LEENDERS', S, mp, 1924; flowers hydrangea-pink, open, large, borne in clusters, semi-dbl., sometimes recurrent bloom, moderate fragrance; foliage large, rich green; very vigorous, (5 ft.), bushy growth; [Orléans Rose X R. foetida bicolor]; Leenders, M.

Helen Margaret *see* 'HORBLUSH'

'HELEN M. GREIG', HT, lp; flowers pastel pink; [Mrs A.R. Barraclough X Marmion]; Dobbie

Helen Naude *see* 'KORDIENA'

'HELEN OF TROY', HT, mp, 1956; flowers rose-pink, very dbl., slight, spicy fragrance; vigorous, bushy growth; [Dame Edith Helen X Mrs Henry Morse]; Stevenson; Waterer

Helen Suzman Rose, F, lp

'HELEN TAYLOR', HT, pb, 1924; flowers rosy salmon, dbl., moderate fragrance; Pemberton

'HELEN TRAUBEL', HT, pb, 1951; bud long, pointed; flowers pink to apricot, dbl., 23–30 petals, 5–6 in., exhibition form, borne singly, moderate, fruity fragrance; foliage leathery, matt, green; weak necks; tall, vigorous growth; AARS, 1952 GM, Rome, 1951; [Charlotte Armstrong X Glowing Sunset]; Swim, H.C.; Armstrong Nursery

Helen Traubel, Climbing, Cl HT, ab, 1970; Treloar

'HELEN WILD', HT, op, 1959; flowers orange-pink veined rose-red, dbl., 35 petals, 5–6. in., exhibition form, moderate fragrance; foliage light green; free growth; [Show Girl X Charlotte Armstrong]; Kemp, M.L.

Helena *see* POUlna

Helena, HT, lp; Strnad; (Czech Rosa Club)

Helena Renaissance *see* POUlna

'HELENA VAN VLIET', Pol, lp, 1931; flowers soft pink tinted salmon, borne in large trusses; vigorous growth; [Salmonea sport]; Kersbergen

Helene, HT, mp; flowers large, dbl., intense fragrance; Vecera, L.; (Sangerhausen)

'HÉLÈNE', HMult, pb, 1897; flowers soft violet-rose, base yellowish white, borne in clusters, semi-dbl., slight fragrance; vigorous, climbing (12-15 ft) growth; [Hybrid Tea seedling X (Aglaia X Crimson Rambler)]; Lambert, P.

'HÉLÈNE DAPPLES', HT, mr, 1932; bud pointed, dark; flowers glowing crimson-red, intense fragrance; vigorous growth; [Mrs Henry Winnett X Lady Maureen Stewart]; Heizmann, E.

Helene de Gerlache, HT, w, 1982; Lens

'HELÈNE DE MONTBRIAND', HT, mr, 1933; flowers deep carmine-red, shaded vermilion, well formed, large, globular; foliage glossy, dark; vigorous growth; [Reine Marie Henriette X Laurent Carle]; Schwartz, A.

'HÉLÈNE DE ROUMANIE', HT, rb, 1949; flowers red to pink, urn shaped, dbl., 35 petals, 5 in., slight fragrance; upright growth; GM, NRS, 1950; [(Mme Joseph Perraud X Unnamed seedling) X (Unnamed seedling xPres. Herbert Hoover)]; Meilland, F.

Helene de Savoie *see* KRInirosy

'HÉLÈNE DUCHÉ', HT, lp, 1921; flowers very large, dbl., moderate fragrance; Buatois; (Sangerhausen)

'HÉLÈNE FRANCÇOIS', HT, pb, 1923; flowers salmon-pink shaded coppery red, center salmon-orange tinted, dbl.; [Mme Edouard Herriot X Viscountess Enfield]; Schwartz, A.

'HÉLÈNE GRANGER', HMult, pb, 1910; flowers pink, center copper-yellow, borne in clusters of 15-20; [Tea Rambler X Aglaia]; Granger

Helene Leenders, Pol, mp; flowers medium large, very dbl., moderate fragrance; medium growth; Leenders, M., 1925; (Sangerhausen)

Helene Marechal, S, 1995; Lens, Louis; (Cavriglia)

'Hélène Robinet', HT, lp, 1928; flowers salmon-white, shaded rose, base yellow, dbl., slight fragrance; [Unnamed variety X Pres. Parmentier]; Sauvageot, H.; F. Gillot

'Helene Schoen', HT, mr, 1963; bud long, pointed; dbl., 60 petals, 6 in., exhibition form, slight fragrance; foliage leathery, glossy; vigorous, upright growth; [Multnomah X Charles Mallerin]; Von Abrams; Peterson & Dering

'Hélène Vacaresco', HT, pb, 1939; flowers salmon, shaded copper-carmine, large, cupped; foliage dark; Chambard, C.

'Hélène Valabrègue', HT, lp, 1953; bud pointed; flowers pale rose, overlarge, dbl., cupped, slight fragrance; foliage leathery; strong stems; vigorous, bushy growth; [Lorraine X Michele Meilland]; Meilland, F.; URS

'Hélène Videnz', Pol, pb, 1905; flowers salmon-pink, borne in clusters to 75, dbl.; vigorous growth; [Euphrosyne X Louis Philippe]; Lambert, P.

Helenka, HT, lp; Tesar, 1989; (Czech Rosa Club)

HELeuro, F, mp, 1989; (**Europas Rosengarten**); flowers medium, dbl.; Hetzel; (Sangerhausen)

HELexa, LCl, mr; (**Super Excelsa**®); Hetzel, 1986

'HELfels', LCl, dr, 1996; (**Super Sparkle**, Heinfels); flowers crimson scarlet, blooms borne in large clusters, dbl., 15–25 petals, 2 in., slight fragrance; foliage medium, dark green, glossy; rambling, climbing (300 cms) growth; Helzel, Karl; Eurosa, 1997

HELfriwa, HT, or, 1981; (**Fritz Walter**); Hetzel

Helga® *see* HelgRUI

'Helga', HT, yb, 1926; Weigand, C.

'Helgoland', F, mr, 1936; bud pointed, dark; flowers crimson to carmine, open, very large, borne in clusters, semi-dbl., slight fragrance; foliage leathery, wrinkled, dark; vigorous, bushy growth; [Else Poulsen X Hybrid Tea seedling (dark crimson)]; Kordes

'Helgoland', F, mr, 1973; bud pointed; flowers copper-red, medium, semi-dbl., slight fragrance; foliage glossy, light; moderate, upright, bushy growth; Tantau, Math.; Horstmann

HELgraf, HT, mp; (**Steffi Graf**); flowers large, dbl., intense fragrance; Hetzel, 1993; (Sangerhausen)

HelgRUI, F, w, 1975; (**Helga**®); deRuiter

HELilast, S, lp; (**Werner von Braun**); Hetzel, 1993

'Héliodore Dober', HGal, mr; flowers deep red edged crimson, ball shaped, quite large

'Helios', HT, dy, 1935; flowers deep sunflower-yellow, open, semi-dbl.; foliage leathery, light; vigorous, bushy growth; Leenders, M.

'HELkleger', LCl, or, 1996; (**Super Elfin**, Strombergzauber); dbl., 15–25 petals, 2 in., borne in large clusters, slight fragrance; foliage medium, dark green, glossy; rambling, medium (300 cms) growth; Helzel, Karl; Eurosa, 1997

Hellen Ann, HT, 1989; Tantau, Math.; (Cavriglia)

Hello *see* 'COChello'

Hello *see* KRImony

Hello There *see* 'FLOello'

HELlux, HT, or, 1989; (**Comtesse Diana**); Hetzel

Helma, Min, 1986; Delforge, H.; (Cavriglia)

HELmedia, HT, mr, 1993; (**Medialis**); Hetzel

Helmut Kohl *see* TANtumleh

Helmut Schmidt® *see* 'KORbelma'

'Help the Aged', HT, mp, 1986; flowers clear pink, reverse slightly darker, satin two-tone effect, dbl., 23 petals, exhibition form, intense, damask fragrance; no fruit; straight, brown, very few prickles; foliage medium, light green, semi-glossy; bushy, medium growth; [Mischief X Fragrant Cloud]; Bracegirdle, A.J.; Rosemary Roses, 1987

HELpforz, HT, dy, 1998; (**Goldstadt Pforzheim**); Hetzel

Helping Hands *see* CLEhelp

Helpmekaar Roos *see* KORogesa

HELprinzess, Pol, dp, 1993; (**Rosenprinzessin**); Hetzel

HELreg, Min, dr, 1983; (**Kleine Regina**®); Hetzel

HELreib, F, dp, 1992; (**Carolin Reiberl**); Hetzel

HELrobu, F, ob, 1993; (**Die Rheinpfalz**); Hetzel

HELsabin, F, mp, 1988; (**Sabine Ruf**); Hetzel

Helsingör *see* **'Elsinore'**

HELstrau, F, mp, 1991; (**Derdinger Sommer**); Hetzel

'HELsufair', LCl, lp, 1996; (**Super Fairy**®); dbl., 15–25 petals, 1.25 in., borne in large clusters, repeat bloom, moderate fragrance; foliage medium, medium green, glossy; rambling (300-500 cms) growth; Helzel, Karl; Eurosa, 1997

HELtoria, HT, 1986; (**Kronprinzessin Victoria**); Hetzel, K.; (Cavriglia)

HELtreff, HT, mp, 1998; (**Heinz Treffinger**); Hetzel

HELtzi, Pol, mp, 1994; (**Fritzi**); Hetzel

'Helvetia', HT, dp, 1911; flowers large, dbl., moderate fragrance; Heizmann, E.; (Sangerhausen)

Helvétia *see* **'Mandalay'**

HELway, HT, mr, 1993; (**Euroway**); Hetzel

HELzwerg, Min, m, 1998; (**Lila Zwerg**); Hetzel

Hemavathy, F, op, 1975; Kasturi

'Hen Kauffmann', F, mp, 1954; flowers rosy pink, dbl., moderate fragrance; vigorous growth; Leenders, M.

'HENdan', F, dp, 1993; (**Dancing Pink**); flowers deep pink, blooms borne in large clusters, semi-dbl., 6–14 petals, 3–3.5 in., moderate fragrance; some prickles; foliage medium, medium green, semi-glossy; medium (75 cms), bushy growth; [Southampton X Dortmund]; Henson, R.W.; Henson, 1993

'HENfif', F, ab, 1993; (**Fifi**); flowers cream pale peach, blooms borne in large clusters, semi-dbl., 6–14 petals, 3–3.5 in., intense fragrance; many prickles; foliage large, dark green, glossy; tall (100 cms), upright growth; [Lichterloh X Liverpool Echo]; Henson, R.W.; Henson, 1993

'Henkell Royal'®, HT, mr, 1964; bud long; flowers blood-red, well-formed, large, intense fragrance; vigorous, bushy growth; GM, Baden-Baden, 1964; Kordes, R.

'HENnev', F, rb, 1993; (**Nevertheless**); flowers orange/red/yellow blend, blooms borne in small clusters, dbl., 26–40 petals, 1.5–2.75 in., moderate fragrance; few prickles; foliage large, medium green, matt; tall (100 cms), upright growth; [Seedling (Silver Jubilee X Trumpeter) X Gold Medal]; Henson, R.W.; Henson, 1993

'Henri Barruet', LCl, pb, 1918; flowers coppery yellow, opening to pink and tinted white, borne in clusters; vigorous, climbing (8 ft) growth; Barbier

'Henri Brichard', HT, w, 1891; flowers large, very dbl.; Bonnaire; (Sangerhausen)

Henri Caillaud, LCl; Moreira da Silva, A.; (Cavriglia)

'HENRI COUPÉ', HP, mp, 1916; dbl., moderate fragrance; [Frau Karl Druschki X Gruss an Teplitz]; Barbier

'HENRI DECLINAND', LCl, mr, 1934; flowers bright magenta-red, quite large, dbl.; foliage dark; Mermet

'HENRI FOUCQUIER', HGal, mp; (Henri Fouquier); flowers pure rose-pink, large, dbl., moderate fragrance

Henri Fouquier *see* **'HENRI FOUCQUIER'**

Henri IV, HP, m; Verdier, 1862

'HENRI LINGER', LCl, yb, 1928; flowers clear yellow-orange, open, semi-dbl., slight fragrance; foliage light, glossy; very vigorous, climbing growth; [R. wichurana X Benedicte Seguin]; Barbier

'HENRI MALLERIN', HT, yb, 1953; bud ovoid; flowers empire-yellow suffused pink, large, dbl., 55–70 petals, slight fragrance; foliage leathery, glossy; bushy growth; [Soeur Thérèse X Duquesa de Peñaranda]; Mallerin, C.; EFR

Henri Mallerin *see* **'ROUGE MALLERIN'**

'HENRI MARTIN', M, mr; (Red Moss); bud sparsely mossed; flowers shining crimson, blooms in clusters of 3-8, semi-dbl.; Laffay, M., 1862

Henri Matisse®, F, 1996; Delbard, Georges; (Cavriglia)

'HENRI PAUTHIER', HT, mr, 1933; flowers bright red, open, large, semi-dbl., moderate fragrance; foliage glossy; bushy growth; [Unnamed seedling X Edouard Mignot]; Sauvageot, H.; C-P

Henrietta *see* 'POULETTA'

'HENRIETTA', HT, op, 1917; bud pointed, orange-crimson; flowers soft coral-salmon, open, semi-dbl., moderate fragrance; foliage dark; vigorous growth; RULED EXTINCT 12/85; [Alister Stella Gray X Andre Gamon]; Merryweather

'HENRIETTE', HT, or, 1916; flowers coppery orange-red, large, dbl.; Dickson, A.; (Sangerhausen)

'HENRIETTE CHANDET', HT, op, 1942; bud oval; flowers orange-coral, large, dbl., slight fragrance; foliage glossy; vigorous, bushy growth; [Rochefort X La Parisienne]; Mallerin, C.; A. Meilland

'HENRIETTE KOSTER', Pol, mr, 1939; [Dick Koster sport]; Koster, D.A.

'HENRIETTE PECHTOLD', HT, mr, 1946; bud long, pointed; flowers red, reverse salmon-red, large, dbl., moderate fragrance; foliage soft; vigorous, bushy growth; [Briarcliff X Katharine Pechtold]; Verschuren-Pechtold

'HENROS', Cl F, dp, 1995; flowers rose pink, blooms borne in small clusters, dbl., 15–25 petals, 1.5–2.75 in., moderate fragrance; some prickles; foliage medium, medium green, glossy; tall (250 cms), climbing growth; [Glenfiddich X Lichterloh]; Henson, R.W.

'HENRY FIELD', HT, mr, 1948; bud ovoid, long, pointed; flowers crimson-red, dbl., 60 petals, 5 in., exhibition form, moderate fragrance; foliage glossy; vigorous, bushy growth; [Pink Princess X Crimson Glory]; Brownell, H.C.; H. Field

Henry Fonda® *see* 'JACYES'

'HENRY FORD', HT, mp, 1954; bud long; flowers silvery pink, dbl., 30 petals, 4–5 in., exhibition form, moderate fragrance; vigorous, upright growth; [Pink Dawn, Climbing X The Doctor]; Howard, A.P.; H&S

'HENRY FORD', HT, my, 1927; flowers yellow edged salmon-orange, opening to lemon-yellow, semi-dbl., slight fragrance; [Mme Edouard Herriot X Golden Emblem]; Deverman; B&A

'HENRY HUDSON', HRg, w, 1976; bud ovoid; flowers yellow stamens, dbl., 25 petals, 2.5–3 in., recurrent bloom, intense fragrance; low, bushy growth; Svedja, Felicitas; Canada Dept. of Agric.

'HENRY IRVING', Cl HP, dp, 1907; flowers medium, semi-dbl.; Conard & Jones; (Sangerhausen)

'HENRY IV', HP, lp; flowers purple-violet, dbl., cupped

'HENRY KELSEY', HKor, mr, 1984; dbl., 28 petals, blooms in clusters of 9-18, moderate, spicy fragrance; foliage glossy; trailing growth; remontant, very winter hardy.; [R. kordesii hybrid X Unnamed seedling]; Svedja, Felicitas

'HENRY KING STANFORD', F, mr, 1973; flowers small, semi-dbl., 15 petals, 1.5 in.; spreading growth; [Red Pinocchio X Open pollination]; Sheridan, V.V.

Henry Lawson, HT, lp; Allender, Robert William, 1999; (Weatherly, L.)

'HENRY MORSE', F, dr, 1958; flowers deep blood-red shaded scarlet, borne in large trusses, semi-dbl., 3 in.; free growth; Kordes; Morse

'HENRY NEVARD', HP, dr, 1924; flowers crimson-scarlet, very large, dbl., 30 petals, cupped, recurrent bloom, intense fragrance; foliage dark, leathery; vigorous, bushy growth; Cant, F.

'HENRY S. BADGERY', HT, lp; Johnson, 1885; (Weatherly, L.)

'HENRY V', HCh, rb; flowers crimson, center white, cupped

'HENS VERSCHUREN', HT, mr, 1948; bud long; flowers bright red, very large; [Mary Hart X Seedling]; Verschuren

'HER MAJESTY', HP, mp, 1885; flowers clear rose, with carmine reflexes toward center, very large, dbl., occasionally recurrent bloom, moderate fragrance; very vigorous growth; [Mabel Morrison X Canary (T)]; Bennett

'HERA', HT, mr, 1924; flowers brilliant carmine shaded blood-red, dbl., 40 petals, intense fragrance; [Gen. MacArthur X Luise Lilia]; Van Rossem

Herald *see* **'DERUITER'S HERALD'**

'HERALDO', Cl HT, m, 1949; bud long, pointed; flowers purple-pink, large, very dbl., intense fragrance; foliage leathery, dark; [Guinee X Texas Centennial]; Dot, M.; P. Dot

'HERANI', HT, pb, 1962; (**Rina Herholdt**); bud long, pointed; flowers milky white, flushed deep pink at edges, darkening with age, semi-dbl., 60 petals, 3.5–4 in., cupped, moderate fragrance; foliage leathery, glossy; vigorous, bushy growth; [Peace X Seedling]; Herholdt, J.A.; Herholdt's Nursery

'HERARO', HT, w, 1977; (**Kilimanjaro**); bud pointed; flowers pure white, dbl., 35–38 petals, 4.5–5 in., moderate fragrance; foliage rich green; vigorous growth; [Unnamed seedling X Pascali]; Herholdt, J.A.

Herb Garden China, Ch, lp; (found rose)

Herbalist *see* 'AUSSEMI'

HERbeau, HT, ob; (**Southern Sun**)

'HERBEMONT'S MUSK CLUSTER', N, w; flowers pure white, very large blooms in large clusters, dbl., recurrent bloom; Herbemont, prior to 1836

'HERBERT BRUNNING', HT, mr, 1940; flowers brilliant red; Clark, A.

'HERBERT WILSON', F, w, 1967; flowers well-formed, slight fragrance; foliage light green; [White Knight X The Optimist]; Latham

'HERBIC', Gr, rb, 1981; (**Herero**); flowers yellow, reverse yellow with red overlay, spreading with age, semi-dbl., no fragrance; foliage large, dark, glossy; upright growth; [Angel Bells X Southern Sun]; Herholdt, J.A.

Herbie *see* 'TINHERB'

Herbiz, Min, op, 1987

'HERBSTFEUER', HEg, dr, 1961; (Autumn Fire); flowers large blooms in clusters (up to 5), semi-dbl., repeat bloom, moderate fragrance; large, pear-shaped, reddish yellow fruit; vigorous (6 ft.) growth; Kordes; Kern Rose Nursery

HERclov, HT, pb; (**Cloverdene**)

'HERCOP', HT, dy, 1982; (**Copper King**); flowers copper-gold, large, dbl., 35 petals, no fragrance; foliage dark, glossy; upright growth; [Vienna Charm X Seedling]; Herholdt, J.A.

'HERCRES', HT, dp, 1982; (**Summer Crest**); flowers deep pink, large, dbl., slight fragrance; foliage medium, medium green, semi-glossy; [Miss All-American Beauty X Unnamed seedling]; Herholdt, J.A.

'HERCULES', LCl, mp, 1938; flowers Dame Edith Helen pink, over large, dbl., cupped, slight fragrance; foliage large, glossy, dark; long stems; very vigorous, climbing growth; [Doubloons X Charles P. Kilham]; Horvath; Wayside Gardens Co.

'HERDIO', HT, dy, 1971; (**Gold Dollar**); flowers large, dbl., 35 petals, slight fragrance; foliage glossy; vigorous growth; [Seedling X Weiner Charme]; Herholdt, J.A.; Herholdt's Nursery

Herero *see* 'HERBIC'

'HERE'S COLETTE', S, pb, 1999; flowers dark pink, white eye, reverse dark pink, single, 5–11 petals, 1.5 in., borne in large clusters, slight fragrance; few prickles; foliage small, light green, dull; spreading, low (3 ft.) growth; [seedling X Ballerina]; Watson, Thomas L. & Glenda, 1994

Here's Ian *see* 'JUDIAN'

'HERFLA', HT, pb, 1981; (**Flamingo**); flowers light pink, silvery reverse, large, dbl., 35 petals, cupped, moderate fragrance; foliage medium green, semi-glossy; upright growth; [Unnamed seedling X Unnamed seedling]; Herholdt, J.A.

Herfordia, F, dp, 1993; flowers medium, dbl., slight fragrance; Hempelmann; (Sangerhausen)

'HERFSTTOOI', HT, dr, 1919; flowers dark crimson, dbl., moderate fragrance; [Gen. MacArthur seedling X Leuchtfeuer (HCh)]; Van Rossem

'HERGALE', HT, pb, 1970; (**Nightingale**); flowers rich rose-red, blended lighter, large, dbl., 25 petals, exhibition form, slight fragrance; growth moderate; [Rina Herholdt X Tiffany]; Herholdt, J.A.; Herholdt's Nursery

HERGEKO, HT, mr; (Ge Korsten)

Heriflor *see* SAUbaflor

Heritage® *see* 'AUSBLUSH'

Herman Steyn *see* KORfolklori

Hermann Berger, HT, or, 1982; flowers large, dbl., slight fragrance; GPG Bad Langensalza; (Sangerhausen)

'HERMANN EGGERS', HT, or, 1930; flowers deep orange-scarlet, very large, dbl., exhibition form, intense fragrance; foliage dark, leathery; very vigorous growth; [(Pink Pearl X Templar) X Florex]; Kordes; Dreer;, H&S

'HERMANN KEGEL', M, m; flowers reddish violet, sometimes streaked crimson, medium, dbl.; vigorous growth; Portemer fils, 1848 (Cavriglia)

'HERMANN KIESE', HT, yb, 1906; flowers yellow and pink, large, semi-dbl., intense fragrance; Geduldig; (Sangerhausen)

'HERMANN LINDECKE', HT, lp, 1929; flowers whitish pink, reverse salmon-pink large, dbl., exhibition form; [General-Superior Arnold Janssen sport]; Lindecke

'HERMANN LÖNS', HT, dp, 1931; flowers shining light red, large blooms in clusters, single, cupped, moderate fragrance; foliage glossy; vigorous growth; [Ulrich Brunner Fils X Red-Letter Day]; Tantau

'HERMANN NEUHOFF', HT, mr, 1923; flowers uniform blood-red, dbl., moderate fragrance; [General-Superior Arnold Janssen sport]; Neuhoff; Kordes

'HERMANN ROBINOW', HT, op, 1918; flowers salmon-orange shaded salmon-rose and dark yellow, large, dbl.; vigorous growth; [Frau Karl Druschki X Lyon Rose]; Lambert, P.

'HERMANN ROBINOW, CLIMBING', Cl HT, mp, 1934; flowers large, dbl., moderate fragrance; Lambert, P.; (Sangerhausen)

Hermann Schmidt, LCl, dp, 1986; Hetzel

'HERMANN SCHÖNFELD', HT, lp, 1925; flowers large, dbl.; Dechan; (Sangerhausen)

'HERMANN TESCHENDORFF', HT, rb, 1949; bud ovoid; flowers copper-red, reverse old-gold, open, very large, dbl., cupped, moderate fragrance; foliage glossy, dark, bronze; very vigorous, upright growth; Berger, V.; Teschendorff

'HERMAWI', HT, rb, 1982; (**Mary Wise**); flowers medium red and gold with red reverse, large, dbl., 40 petals, no fragrance; foliage medium green, semi-glossy; bushy growth; [Madelaine X (Unnamed seedling X Apogee)]; Herholdt, J.A.

'HERMELA', HT, rb, 1964; (**Angel Bells**®); flowers ivory, flushed orange and red, well-formed, large, dbl., borne singly, slight fragrance; foliage dark, glossy; bushy growth; [Peace X Rina Herholdt]; Herholdt, J.A.; Herholdt's Nursery

'HERMELIA CASAS', F, mp, 1956; flowers pearly, reverse carmine, medium, borne in clusters of 3-6, dbl., 30 petals; moderate growth; [Méphisto X Perla de Alcañada]; Dot, Pedro

'HERMEN ANGLADA', HT, w, 1933; bud very large; flowers white tinted pink, single, intense fragrance; very vigorous growth; Dot, Pedro

'HERMES', HT, dy, 1935; flowers large, dbl., moderate fragrance; Teschendorff; (Sangerhausen)

'HERMESY', HT, rb, 1983; (**Mystery**); flowers light orange-red, edged amber, golden reverse, large, dbl., 35 petals, slight fragrance; foliage large, medium green semi-glossy; bushy growth; [(Unnamed seedling X Southern Sun) X Southern Sun]; Herholdt, J.A.

'HERMINA'®, S, pb, 1996; flowers red with white reverse aging neyron rose with white reverse, dbl., 15–25 petals, good repeat, slight fragrance; some prickles; foliage medium, dark green, semi-glossy; upright, bushy, medium growth; hardy to 18-24", resistant to blackspot; [[Tickled Pink X Prairie Princess) X Autumn Dusk] X [{TickledPink X {Carrousel X (Morning Stars X Suzanne)}} X Maytime]]; Buck, Dr. Griffith J.; Kimbrew Walter Roses, 1997

Hermione *see* 'GAULIMOR'

'HERMOSA', HCh, lp; (Armosa, Melanie Lemaire, Mme Neumann); bud pointed; flowers light blush-pink, dbl., 35 petals, exhibition form, recurrent bloom, moderate fragrance; foliage bluish green; small vigorous growth; (21); Marcheseau, before 1837; Prior to 1837

Hero® *see* 'AUSHERO'

'HEROÏCA', HT, dr, 1960; flowers deep velvety red, becoming lighter; vigorous growth; [Rome Glory X Independence]; Lens

'HEROINE', HT, op, 1935; bud pointed; flowers salmon-shrimp-pink, large, dbl., exhibition form; foliage leathery, dark; very vigorous, bushy growth; [Wilhelm Kordes X Mrs Atlee]; Krause

'HEROINE DE VAUCLUSE', B, mp, 1863; Moreau et Robert

'HEROS', HT, mr, 1933; flowers very large, dbl., cupped, intense fragrance; foliage leathery, dark; bushy growth; [Johanniszauber X Étoile de Hollande]; Tantau

'HERPIM', F, op, 1972; (**Impala**); bud ovoid, pointed; flowers coppery, reverse orange, dbl., 30 petals, 2.5–3 in., cupped; foliage bright green; bushy growth; [Zambra X Unnamed seedling]; Herholdt, J.A.

HERpot, HT, yb, 1990; (**Kristo Pienaar**); Herholdt

'HERRENHAUSEN', HMsk, ly, 1938; bud ovoid, greenish yellow; flowers light yellow, fading white, red tints in sun, large, dbl., cupped, borne in clusters, profuse, intermittent bloom, moderate, pansy fragrance; foliage leathery, glossy, light; long stems; vigorous, bushy growth; [Eva X Golden Rapture]; Kordes

'HERRIN VON LIESER', HT, ly, 1907; flowers cream-yellow, center reddish yellow, intense fragrance; [Frau Karl Druschki X G. Schwartz]; Lambert, P.

'HERSELF', F, lp, 1966; semi-dbl., 18 petals, 4 in., borne in clusters; free growth; [The Optimist X Moulin Rouge]; Vincent; Harkness

Hershey's Red, HT, dr, 1996; Williams, J. Benjamin

Hertfordshire *see* KORtenay

'HERTFORDSHIRE GLORY', F, yb, 1970; flowers yellow, tinted red, large, dbl., 20 petals, slight fragrance; foliage glossy; [Isobel Harkness X Circus]; Harkness

'HERTIE', HT, w, 1976; (**Bettie Herholdt**, Betty Herholdt, Messagere); bud pointed; flowers ivory-white, dbl., 50 petals, 5.5 in., moderate fragrance; vigorous growth; [(White Swan X Seedling) X Pascali]; Herholdt, J.A.

HERtroci, HT, pb; (**Potch Pearl**); (HERtroci)

'HERZ AS'®, HT, mr, 1963; (As de Coeur); bud long, pointed; flowers pure blood-red, well-formed, large, dbl., slight fragrance; vigorous, upright growth; Tantau, Math.

Herz As® *see* 'TANSARAS'

'HERZBLÄTTCHEN', Pol, dp, 1889; flowers carmine-pink, small, dbl.; Geschwind, R.; (Sangerhausen)

Herzblut *see* **'COMMONWEALTH'**

Herz-Dame *see* **'DAME DE COEUR'**

'HERZENSGRUSS', HT, dr, 1973; bud pointed; flowers dark velvety red, medium, dbl., intense fragrance; foliage glossy; vigorous, upright, bushy growth; [(Fragrant Cloud X Goldmarie) X Red American Beauty]; Hetzel; GAWA

'HERZOG FRIEDRICH II VON ANHALT', HT, lp, 1906; flowers soft pink, moderate fragrance; [Souv. du President Carnot X Mme Jules Grolez]; Welter

Herzog von Windsor *see* **'DUKE OF WINDSOR'**

'HERZOGIN VIKTORIA ADELHEID VON COBURG-GOTHA', HT, op, 1905; flowers coppery carmine-pink, large, dbl., moderate fragrance; Welter; (Sangerhausen)

'HERZOGIN VON CALABRIEN', HT, ly, 1914; flowers large, dbl., moderate fragrance; Lambert, P.; (Sangerhausen)

'HESSENGRUSS', HT, dp, 1928; flowers deep pink, reverse carmine-rose, dbl., moderate fragrance; [Laurent Carle sport]; Thönges

'HESSENSTAR', HT, or, 1973; bud ovoid; flowers orange-red to geranium-red, medium, dbl., 30–40 petals, moderate fragrance; moderate growth; [Baccará X Prima Ballerina]; Hetzel; GAWA

'HESSIE LOWE', HT, op, 1956; flowers peach-pink, dbl., exhibition form, intense fragrance; foliage glossy; very vigorous growth; Lowe

'HESSOISE', HEg, mp; flowers bright rose, semi-dbl.

Hester Prynne, F, dr, 1993; Robinson

Hestia, F; Arles, F.; (Cavriglia)

'HETKORA', HT, ob, 1983; (**Königin Beatrix**®, Queen Beatrix); flowers orange, large, dbl., 35 petals, exhibition form, intense fragrance; foliage medium, medium green, semi-glossy; upright, bushy growth; [Unnamed seedling X Patricia]; Kordes, W.

'HEURE MAUVE', HT, m, 1962; flowers lilac-mauve tinted blush, well-formed, dbl., 35 petals, 5 in.; foliage glossy, bright green; vigorous growth; [Simone X Prelude]; Laperrière; EFR

Heureka, HT, lp; Urban, J.; (Czech Rosa Club)

Heureux Anniversaire® *see* 'DELPRE'

'HEXHAM ABBEY', F, op, 1975; flowers salmon-pink, base yellow turning copper, 3.5 in., intense fragrance; foliage leathery; vigorous, low to medium height, upright growth; [Fairlight X Arthur Bell]; Wood

Heywood *see* 'KIRTYN'

Hi Doll *see* DICkisser

'HI HO', Cl Min, op, 1964; flowers deep pink, small, in clusters, dbl.; foliage glossy; vigorous, climbimg growth; [Little Darling X Magic Wand]; Moore, Ralph S.; Sequoia Nursery

Hi Society *see* 'COCQUATION'

Hi Teen *see* **'DESIGNER'S CHOICE'**™

'HI, NEIGHBOR', Gr, mr, 1981; bud ovoid, pointed; flowers slightly blooms borne 1-6 per cluster, dbl., 43 petals, cupped, moderate fragrance; awl-like prickles; foliage leathery, dark, matt; bushy, erect growth; [(Queen Elizabeth X Prairie Princess) X Portrait]; Buck, Dr. Griffith J.; Iowa State University

'HIAWATHA', HMult, rb, 1904; flowers deep crimson, center white, anthers golden, blooms in large, single, cupped; foliage rich green, leathery, glossy; very vigorous, climbing (15-20 ft late bloom) growth; [Crimson Rambler X Paul's Carmine Pillar]; Walsh

'HIAWATHA RECURRENT', Cl Pol, op, 1931; flowers carmine suffused orange, white eye, small, borne in clusters, intermittent bloom; foliage small, glossy; long stems; very vigorous, climbing growth; [Hiawatha X Maman Levavasseur]; Sauvageot, H.; C-P

'HIBERNICA', Misc OGR, lp; (R. X hibernica); flowers pale pink, single, 1 in.; foliage glaucous green; low growth; (42); [R. canina X R. spinosissima]; Before 1800 in Great Britain

Hidalgo® *see* 'MEITULANDI'

'HIDCOTE GOLD', sp, (variety of R. sericea omeiensis), my, 1948; flowers canary-yellow, single blooms borne several together, non-recurrent; foliage ferny; Hilling

Hidcote Yellow *see* **'LAWRENCE JOHNSTON'**

Hi-de-hi *see* 'MACANAT'

'HI-FI', F, or, 1958; flowers bright orange-scarlet, borne in clusters, semi-dbl.; foliage glossy; [Independence X ?]; Gregory

High Cloud *see* 'TINDICK'

'HIGH ESTEEM', HT, pb, 1961; bud pointed; flowers phlox-pink, reverse silvery, dbl., 43 petals, 6 in., exhibition form, intense, fruity fragrance; foliage leathery, light green; vigorous, upright, compact growth; [(Charlotte Armstrong X Mme Henri Guillot) X (Multnomah X Charles Mallerin)]; Von Abrams; Peterson & Dering

'HIGH FASHION', HT, dp, 1972; bud ovoid; flowers deep pink, medium, dbl., exhibition form, intense fragrance; foliage glossy, soft; vigorous, upright growth; [Queen Elizabeth X Peace]; Patterson; Patterson Roses

High Five *see* 'SUNHIFI'

High Hope *see* 'FLOHIH'

High Hopes *see* 'HARYUP'

High Jinks™ *see* 'SAVAJINKS'

High Life *see* 'SEALIFE'

'HIGH NOON', Cl HT, my, 1946; flowers lemon-yellow, loosely cupped, dbl., 28 petals, 3–4 in., cupped, moderate, spicy fragrance; foliage leathery, glossy; upright, vigorous, climbing (8 ft) growth; AARS, 1948; [Soeur Thérèse X Capt. Thomas]; Lammerts, Dr. Walter; Armstrong Nursery

High Point *see* 'MOLHIP'

High Sheriff *see* 'HARWELLINGTON'

'HIGH SOCIETY', HT, mr, 1961; flowers bright red, dbl., 30 petals, 4 in., exhibition form, slight fragrance; bushy growth; Kordes, R.; A. Dickson

High Spirits™ *see* 'SAVASPIR'

High Stepper *see* 'MORCLIM'

High Style *see* 'LYOHI'

High Summer *see* 'DICBEE'

'HIGH TIDE', Min, ob, 1986; flowers orange, yellow reverse, medium, borne usually singly, dbl., 25 petals, cupped, slight fragrance; few prickles; foliage small, dark, semi-glossy; upright, bushy growth; [Unnamed seedling X Unnamed seedling]; McDaniel, Earl; McDaniel's Min. Roses

'HIGH TIME', HT, pb, 1959; bud urn-shaped; flowers claret-rose, reverse gold and pink, dbl., 24 petals, 4–5 in., exhibition form, intense, spicy fragrance; foliage dark, glossy; vigorous, upright growth; [Charlotte Armstrong X Signora]; Swim, H.C.; Roseway Nursery

'HIGHDOWNENSIS', S, mr, 1928; (R. X highdownensis); flowers bright medium red, single blooms in clusters, slight fragrance; orange-scarlet fruit; colorful prickles; foliage dark, coppery; vigorous, bushy (10x10 ft) growth; [R. moyesii seedling]; Stern, Sir Frederich

Highfield® *see* 'HARCOMP'

'HIGHLAND BEAUTY', F, mr, 1956; flowers rich red, borne in clusters, semi-dbl., 3 in., slight fragrance; foliage dark, glossy; vigorous growth; [Signal Red X Red Wonder]; deRuiter; Anderson

'HIGHLAND CHARM', F, op, 1956; flowers coral-salmon, borne in large clusters, semi-dbl., 3.5 in., slight fragrance; foliage leathery; vigorous, bushy growth; [Duchess of Rutland X Fashion]; deRuiter; Anderson

Highland Dancer, F, rb

'HIGHLAND FLING', F, or, 1971; flowers orange-scarlet, veined black, dbl., 22 petals, 4–4.5 in., moderate fragrance; foliage glossy; [Dearest X Elizabeth of Glamis]; Anderson's Rose Nurseries

'HIGHLAND GLORY', F, mr, 1956; flowers crimson-red, borne in large clusters, semi-dbl., 3 in., slight fragrance; foliage coppery; vigorous growth; [Sidney Peabody X Floribunda seedling]; deRuiter; Anderson

Highland Laddie *see* 'COCFLAG'

Highland Lass *see* 'ZIPHIGH'

'HIGHLAND PARK', HT, op, 1942; bud tawny salmon; flowers salmon-pink, open, large, dbl., 35 petals, slight fragrance; foliage leathery; vigorous, upright, bushy growth; [E.G. Hill X Mme Henri Guillot]; Mallerin, C.; C-P

'HIGHLAND WEDDING', HT, w, 1971; flowers white, suffused blush, center light gold, dbl., 30 petals, 4.5 in., slight fragrance; foliage large, dark, semi-glossy; vigorous growth; [Virgo X Rose Gaujard]; MacLeod

'HIGHLIGHT', F, ob, 1957; flowers orange-scarlet, blooms in large clusters, dbl., 24 petals, 2.5 in., moderate fragrance; vigorous growth; GM, NRS, 1957; [Unnamed seedling X Independence]; Robinson, H.

Hightae *see* 'GREENTAE'

Highveld Sun *see* 'MORPAPPLAY'

Highway 290 Pink Buttons, Min, mp; (found rose)

Highway Rose, S, dp

'HIGOROMO', HT, dr, 1999; dbl., 30–35 petals, 4–5 in., flat, slight fragrance; foliage dark green, semi-glossy; vigorous (4-5 ft) growth; GM, Japan Rose Concours, 1990; [Duftzauber X (Josephine Bruce X seedling)]; Suzuki, Seizo, 1988; Keisei Rose Nurseries, 1990

Hikari, F, 1969; Suzuki, Seizo; (Cavriglia)

'HILAROMA', HT, pb, 1992; (**Secret**™); flowers light creamy pink edged with deep pink, dbl., 30–40 petals, 4–4.5 in., exhibition form, borne mostly singly, intense, sweet and spicy fragrance; some prickles; foliage large, medium green, semi-glossy; strong stems; tall (120-130 cms), bushy growth; PP8494; AARS, 1994; [Pristine X Friendship]; Tracy, Daniel; The Conard-Pyle Co., 1994

HILcap, HT, w; (**Ginger Hill**); flowers blush pink, petals crêped; (HILcap), 1999

'HILCO', Gr, dy, 1985; (**Excitement**™); dbl., 20 petals, exhibition form, borne usually singly, moderate fragrance; medium, slightly pear-shaped, slightly orange fruit; no prickles; foliage medium, dark, semi-glossy; medium, upright, bushy growth; [Golden Fantasie X Coed]; Jelly, Robert G.; E.G. Hill Co.

'HILDA', HT, pb, 1928; flowers salmon-pink, reverse orange-carmine, very large, dbl., globular, slight fragrance; foliage leathery; long stems; vigorous growth; Cant, B. R.

Hilda Heinemann, HT, mr, 1971; flowers large, dbl.; Meilland; (Sangerhausen)

Hilda Murrell® *see* 'AUSMURR'

'HILDA PHILLIPS', HT, dy, 1948; flowers deep golden yellow, well-shaped, medium, moderate fragrance; foliage glossy, bronze; [Aureate seedling X Mrs Sam McGredy]; Bees

'HILDA SCOTT', HT, my, 1955; flowers butter-yellow, small; [Lady Hillingdon seedling]; Glassford; Morse

Hilde *see* 'BENHILE'

'HILDE APELT', HT, ob, 1927; flowers saffron-yellow, dbl., moderate fragrance; [Unnamed seedling X Souv. de Claudius Pernet]; Leenders, M.

'HILDE STEINERT', HT, rb, 1926; flowers coral-red, reverse reddish salmon and old-gold, semi-dbl., moderate fragrance; Leenders, M.

'HILDEGARDE', HT, ob, 1946; bud long, pointed; flowers saffron-rose-pink, dbl., 30–35 petals, 5 in., exhibition form, intense fragrance; foliage leathery; very vigorous, upright growth; [Briarcliff sport]; Boerner; J&P

'HILDENBRANDSECK', HRg, mp, 1909; flowers shining clear pink, in clusters, single, recurrent bloom; vigorous growth; [Atropurpurea (HRg) X Frau Karl Druschki]; Lambert, P.

'HILDEPUTCHEN', HMult, m, 1922; flowers violet-pink with white, small, semi-dbl., moderate fragrance; Bruder Alfons; (Sangerhausen)

'HILGOFAN', HT, my, 1971; (**Golden Fantasie**, Joan Brickhill®); flowers large, semi-dbl., exhibition form, intense fragrance; foliage large, dark, leathery; vigorous, upright, bushy growth; [Dr. A.J. Verhage X Anniversary]; Byrum; J.H. Hill Co.

'HILGOLD', S, my, 1967; (**Golden Chersonese**); single, 1.5–2 in., borne singly at each node, early, moderate fragrance; foliage leaflets 7-9; vigorous growth; [R. ecae X Canary Bird]; Allen, E.F.

'HILL CREST', HT, rb, 1948; bud short, pointed, oxblood-red; flowers carmine, dbl., 50–55 petals, 3.5–4 in., exhibition form, moderate fragrance; foliage leathery; very vigorous, upright, compact, tall growth; Hill, Joseph H., Co.

'HILL TOP', HT, ab, 1942; bud long, pointed, light coral-red; flowers buff, dbl., 28 petals, 4–5 in., globular, moderate fragrance; foliage leathery, dark, wrinkled; vigorous, upright growth; [Joanna Hill X R.M.S. Queen Mary]; Hill, Joseph H., Co.; H&S, 1946;, Wayside Gardens Co., 1946

Hillary First Lady *see* LAPcli

'HILLCREST PILLAR', S, my, 1930; flowers bright yellow, medium, semi-dbl.; pillar (6 ft) growth; [R. X harisonii seedling]; Hillcrest Gardens

'HILLIER ROSE', S, dr; (R. hillieri, R. X pruhoniciana); flowers deep red, 2 in.; (35); [R. moyesii X R. multibracteata(?)]

'HILL'S AMERICA', HT, mp, 1921; flowers rose-pink, dbl., 44 petals, moderate fragrance; GM, NRS, 1924; [Premier X Hoosier Beauty]; Hill, E.G., Co.

'HILL'S HILLBILLY', F, mp, 1947; bud ovoid, red; flowers pink, open, borne in clusters, single, 5–6 petals, 1–2 in., slight fragrance; foliage leathery, dark; vigorous, upright, bushy growth; [Juanita X Mrs R.M. Finch]; H&S;, Wayside Gardens Co.

'HILL'S VICTORY', HT, rb, 1942; bud red; flowers rose-red to rose-pink, dbl., 50–55 petals, 5–6 in., slight fragrance; foliage leathery, dark, wrinkled; long, strong stems; vigorous, upright, much branched growth; [Chieftain X Sweet Adeline]; Hill, Joseph H., Co.

Hilltop, LCl, pb; Huxley, Ian, 1989

HILrap, HT, mr; (**Classy**™); Hill, E.G., Co., 1992

HILraz, HT, dp; (**Lovely Dream**); (HILraz), 1995

'HILRED', HT, dr, 1989; (**Challenger**™); bud ovoid; flowers reverse medium red, urn-shaped, medium, borne in sprays of 2-3, dbl., 25 petals, exhibition form, moderate fragrance; pear-shaped, small, orange blend fruit; prickles angle downward, small, orange blend; foliage medium, semi-glossy; upright, medium growth; [Jacqueline X Seedling]; Tracy, Daniel; DeVor Nurseries, Inc., 1987

'HILSET', F, lp, 1985; (**Fondly**™); flowers sweetheart, medium, dbl., 20 petals, exhibition form, borne singly and in clusters of 2-4, moderate, spicy fragrance; medium, ovoid, orange-red fruit; few prickles on peduncles; foliage medium, dark, matt; medium, upright growth; PP004983; [Unnamed seedling X Unnamed seedling]; Jelly, Robert G.; E.G. Hill Co.

'HILTACO', F, dr, 1985; (**Sassy**®); flowers sweetheart, small, dbl., 20 petals, exhibition form, borne usually singly, slight fragrance; small, globular, grayed red fruit; prickles only on peduncle; foliage small, medium green, matt; upright growth; PP004559; [Little Leaguer X Mary DeVor]; Jelly, Robert G.; E.G. Hill Co.

Hilton Edward *see* 'STIEDWARD'

Himalayan Musk Rose *see* **R. MOSCHATA NEPALENSIS**

'HIMANGINI', F, w, 1968; bud ovoid; flowers ivory-white, center light buff, open, medium, dbl.; vigorous, bushy, compact growth; [Saratoga X ?]; IARI

'HIMATSURI', HT, dr, 1977; bud circular; 3.5–4 in., exhibition form, slight fragrance; foliage small; upright growth; [(Tropicana X Karl Herbst) X Mainauperle]; Teranishi, K.

Hime *see* 'KEIFUPIE'

'HIMMELSAUGE', HSet, m; Geschwind, R., 1894

Himmelsstürmer, LCl, pb, 1990; flowers carmine-pink with white, small, single; Wänninger, Franz; (Sangerhausen)

'HINEMOA', F, rb, 1963; flowers buttercup-yellow shading to vermilion, borne in clusters, semi-dbl., 17–20 petals, 3–3.5 in., cupped, moderate fragrance; foliage bronze, leathery; upright, bushy growth; [Circus X ?]; Mason, P.G.

'HINGREAT', LCl, ab, 1999; (**Great Phoebus**); bud white, spherical; flowers opening to orange and yellow, with slight pink edge, very dbl., 41 petals, 5.5 in., pompon, borne mostly singly, moderate fragrance; prickles moderate; foliage large, dark green, semi-glossy; upright, vigorous, medium (8 ft) growth; [Abraham Darby X Breath of Life]; Hintlian, Nancy Sears, 1999

'HINRICH GAEDE', HT, ob, 1931; bud pointed, nasturtium-red; flowers orange-yellow tinted nasturtium-yellow, very large, dbl., exhibition form, moderate, fruity fragrance; foliage glossy, bronze; vigorous growth; [Lady Margaret Stewart X Charles P. Kilham]; Kordes

'HINRICH GAEDE, CLIMBING', Cl HT, ob, 1935; Armstrong, J.A.

'HINRUBY', F, dr, 1999; (**Ruby Velvet**); flowers dark red, turning purple with age, lasts well, very dbl., 41 petals, 5 in., borne in small clusters, intense fragrance; prickles many, very large; foliage medium, medium green, semi-glossy; spreading, medium (4 ft) growth; [The Dark Lady X Oklahoma]; Hintlian, Nancy Sears, 1999

'HINSWEET', S, m, 1999; (**Sweet Ecstasy**); flowers purple-crimson, turning more purple with age, very dbl., 41 petals, 5 in., borne mostly singly, intense, lasting fragrance; prickles moderate; foliage large, dark green, semi-glossy; upright, vigorous, medium (5 ft) growth; [The Dark Lady X Oklahoma]; Hintlian, Nancy Sears, 1999

'HINTUSCAN', HGal, m, 1999; (**Tuscan Beauty**); flowers red-lavender, aging to near purple, very dbl., 41 petals, 4–4.5 in., borne mostly singly, intense fragrance; prickles moderate; foliage medium, medium green, semi-glossy; spreading, tall (6 ft) growth; [Superb Tuscan X Big Purple]; Hintlian, Nancy Sears, 1999

'HI-OHGI', HT, or, 1981; flowers deep orange-red, large, dbl., 28 petals, moderate fragrance; prickles slanted downward; foliage dark, semi-glossy; tall, upright growth; [San Francisco X (Montezuma X Peace)]; Suzuki, Seizo; Keisei Rose Nursery

Hipango *see* 'MURGO'

'HIPÓLITO LÁZARO', LCl, dr; flowers carmine, large, intense fragrance; Pahissa

'HIPPOLYTE', HGal, m; flowers vivid carmine shaded violet, small

'HIPPOLYTE JAMAIN', HP, dp, 1874; flowers carmine-red, well-formed, semi-globular, dbl., 38 petals, moderate fragrance; foliage red, when young; vigorous, erect growth; Lacharme, F.

Hippy, HT, 1971; Delbard-Chabert; (Cavriglia)

Hiroshima, HT, lp

Hiroshima Spirit, HT, yb, 1991; flowers dark yellow with red, large, dbl.; Harada, Toshiyuki; (Sangerhausen)

Hiroshima's Children *see* 'HARMARK'

'HIS MAJESTY', HT, dr, 1909; flowers dark crimson shaded deeper, very large, dbl., exhibition form, intense fragrance; long, strong stems; very vigorous growth; McGredy

'HISAMI', HT, lp, 1988; flowers light cream, flushed crimson at tip, large, borne usually singly, very dbl., 50 petals, exhibition form, moderate fragrance; prickles downward curved, red to light green; foliage medium, dark green, matt; upright, medium growth; [Kordes' Perfecta X Christian Dior]; Harada, Toshiyuki

'HISPANIA', HT, dr, 1938; bud pointed; flowers velvety red, open, overlarge, dbl., intense fragrance; foliage leathery; long, strong stems; very vigorous growth; Pahissa

'HIT PARADE', F, ob, 1962; flowers red, orange and gold, borne in clusters, semi-dbl., 16 petals, 4 in.; foliage dark; low growth; [(Independence X Unnamed seedling) X Brownie]; Dickson, Patrick; A. Dickson

'HO NO-O-NO-NAMI', LCl, or, 1968; (Wave of Flame, Waves of Flame); bud pointed; flowers orange-red, reverse lighter, to open, medium, dbl., exhibition form, slight fragrance; foliage glossy, leathery; vigorous, climbing growth; [Spectacular X Aztec]; Suzuki, Seizo; Country Garden Nursery, 1967

Hoag House Cream, HT, w; (found rose)

Hoagy Carmichael *see* 'MACTITIR'

Hobby *see* 'TANOB'

'HOCHSOMMER', F, dp, 1960; dbl., borne in clusters; foliage dark; low growth; [Queen Elizabeth X Unnamed seedling]; Verschuren, A.; van Engelen

'HOCUS-POCUS', Gr, or, 1975; dbl., 30 petals, 4–4.5 in., cupped, slight fragrance; foliage large, dark; vigorous, upright growth; [Fandango X Simon Bolivar]; Armstrong, D.L.; Armstrong Nursery

Hoddy Toddy® *see* 'KINTODDY'

'HOFFMANN VON FALLERSLEBEN', LCl, rb, 1917; flowers salmon-red, shaded yellow and ochre, borne in clusters of 5-2, sometimes repeats; vigorous, climbing growth; [Geheimrat Dr. Mittweg X Tip-Top]; Lambert, P.

Hoffnung, HT, ob, 1995; Huber

'HOFGARTENDIREKTOR GRAEBENER', HT, lp, 1899; flowers creamy pink, medium, dbl., moderate fragrance; Lambert, P.; (Sangerhausen)

'HOFGÄRTNER KALB', HCh, pb; flowers bright carmine-rose, center yellow, outer petals shaded red, dbl., 35 petals, moderate fragrance; vigorous, bushy growth; [Souv. de Mme Eugene Verdier X Gruss an Teplitz]; Felberg-Leclerc, 1913

Hoggar, S; Lens, Louis; (Cavriglia)

'HOH-JUN', HT, pb, 1981; flowers pink flushed rose-red, blooms in clusters of 2-5, dbl., 28 petals, cupped, intense fragrance; large prickles; foliage dark, semi-glossy; compact growth; [Granada X Flaming Peace]; Suzuki, Seizo; Keisei Rose Nursery

'HOHSHUN', HT, dr, 1999; flowers velvet deep red, dbl., 25–27 petals, exhibition form, intense fragrance; foliage dark green, half leathery; 5 ft growth; [Charles Mallerin X Mrs Nieminen]; Ohkawara, Kiyoshi, 1970; Nagashima Rose Nursery, 1971

'HOKEY POKEY', Min, ab, 1980; bud long, pointed; flowers deep apricot, borne singly, dbl., 28 petals, exhibition form, slight, spicy fragrance; straight prickles; foliage finely serrated; compact, bushy growth; [Rise 'n' Shine X Sheri Anne];

Saville, F. Harmon; Nor'East Min. Roses

Hokkaido, HRg, mp, 1988

'HOKU-TO', HT, ly, 1979; flowers soft buff yellow, dbl., 42 petals, exhibition form, moderate fragrance; foliage large, light green; vigorous, upright, spreading growth; [(Myoo-Jo X Chicago Peace) X King's Ransom]; Suzuki, Seizo; Keisei Rose Nursery

Hold Slunci, HP, ly, 1956; flowers medium, dbl., slight fragrance; Blatnà; (Sangerhausen)

Hole in One *see* 'HOReagle'

HOLfairy, Pol, w, 1991; (**Fairy Snow**); Holmes

'HOLIDAY', HT, mr, 1981; flowers small, dbl., 35 petals, slight fragrance; foliage medium green; upright growth; [Cara Mia X Volare]; Strahle, B. Glen; Carlton Rose Nurseries

'HOLIDAY', F, 1948; bud ovoid, orange-yellow flushed pink; flowers flame-pink, reverse clear yellow, clusters of 3-10, semi-dbl., 3–3.5 in., cupped, moderate, clove fragrance; foliage glossy; vigorous, bushy, compact growth; RULED EXTINCT 4/83 ARM; [McGredy's Pillar X Pinocchio]; Boerner; J&P

Holiday Cheer *see* 'MORLIDAY'

'HOLJE', F, w, 1983; (**Jerusalem**); flowers cream, large blooms in sprays of 5-8, dbl., 40 petals, exhibition form, moderate fragrance; ovoid, small, green fruit; foliage medium, light green, matt; upright, bushy growth; [(Queen Elizabeth X Unnamed seedling) X Moriah]; Holtzman, Arnold; Gandy Roses, Ltd.

'HOLLAND DOUBLE WHITE ALTAI', sp, (form of R. spinosissima altaica), w; dbl., non-recurrent; P.H. Wright

Holländerin *see* 'TANSCHWEIGRU'

'HOLLANDIA', Pol, dr; flowers deep red, rosette form, small, dbl.

'HOLLANDIA', HT, mr, 1930; flowers brick-red, shaded copper; [Aspirant Marcel Rouyer sport]; Zijverden

'HOLLANDICA', HRg; thought to have been raised by J. Spek, ca. 1888; selected Dutch clone of R. Rugosa, or perhaps a hybrid with Manetti, widely used as an understock, especially for tree roses

Hollands Rugosa, HRg, mp; flowers medium, single; (Sangerhausen)

Hollie Roffey *see* 'HARRAMIN'

'HOLLY ROCHELLE', HT, dr, 1974; bud pointed; flowers velvety red, very dbl., 60–70 petals, 4 in., exhibition form, moderate fragrance; very vigorous, upright growth; [Charlotte Armstrong X Scarlet Knight]; Graham; South Forrest Rose Nursery

'HOLLYBANK', F, mr, 1966; flowers vermilion, 3–3.5 in., globular, slight fragrance; foliage coppery; upright growth; [Independence X United Nations]; Hooney

'HOLLYWOOD', HT, rb, 1930; bud pointed; flowers dark rose-red, veined darker, dbl., exhibition form, intense fragrance; foliage dark; long, strong stems; bushy growth; [Premier sport]; Scittine

Hollywood, HT, w, 1998

'HOLLYWOOD BEAUTY', HT, dp, 1929; flowers camellia-red, dbl., moderate fragrance; [Rose Marie sport]; Pacific Rose Co.

'HOLSTEIN', F, mr, 1939; (Firefly); bud pointed, dark crimson; flowers clear crimson, blooms in immense clusters, single, 6 petals, 4 in., cupped, slight fragrance; foliage leathery, dark, bronze; very vigorous, bushy growth; GM, Portland, 1939; [Else Poulsen X Dance of Joy seedling]; Kordes; J&P

Holstein 87 *see* 'KORHOLST'

'HOLSTEIN, CLIMBING', Cl F, mr, 1947; Kordes, P.

Holsteinperle® *see* 'KORDIAM'

'HOLSTENROSE', HT, mr, 1937; flowers scarlet-red, large, dbl., cupped, moderate fragrance; foliage glossy; vigorous, bushy growth; [Gen. MacArthur X Amulett]; Tantau

Holstentor *see* MEImuide

'HOLT HEWITT', HT, rb, 1925; flowers rich velvety crimson, flushed and edged scarlet, well shaped, intense fragrance; vigorous, bushy growth; Beckwith

Holtermann's Gold, HT, my, 1989; Swane

Holy Toledo™ *see* 'AROBRI'

Homage, HT, pb, 1986; Pal, Dr. B.P.

'HOMBRE', Min, pb, 1983; flowers light apricot pink, reverse light pink, small, dbl., exhibition form, slight fragrance; foliage small, medium green, semi-glossy; bushy growth; PP005552; AOE, 1983; [Humdinger X Rise 'n' Shine]; Jolly, Nelson F.; Rosehill Farm, 1982

Home & Country *see* 'KRIBATIS'

Home of Time *see* 'COCQUAMBER'

'HOME RUN', HT, pb, 1956; bud ovoid; flowers rose-bengal, dbl., 40 petals, 5 in., exhibition form, moderate fragrance; foliage leathery; vigorous growth; [Pink Delight sport]; Motose

'HOME SWEET HOME', HT, mp, 1941; flowers rich velvety pink, dbl., moderate, damask fragrance; foliage glossy, dark; vigorous growth; Wood & Ingram; C-P

Homecoming *see* 'MICHOME'

Home-Coming *see* 'MICHOME'

'HOMELAND', HT, pb, 1951; flowers neyron rose, base tinted orange, well formed, dbl., 50–60 petals, 5.5–6 in., intense fragrance; foliage leathery, dark; very free growth; [Hybrid Tea (red) X Guinee]; LeGrice

'HOMENAGEM EGAS MONIZ', HT, mp, 1959; [Walter X Juno]; da Silva, Moreira

'HOMENAGEM GAGO COUTINHO', HT, rb, 1959; flowers crimson-red, reverse yellow; [Confidence X Unnamed seedling]; da Silva, Moreira

'HOMENAGEM PINTO D'AZEVEDO', HT, op, 1959; flowers bright salmon-pink, dbl.; [Super-Congo X Independence]; da Silva, Moreira

'HOMÈRE', T, pb, 1858; (Duchess of Kent); flowers pink, center flesh-white, dbl., cupped, moderate fragrance; vigorous, bushy growth; (21); Robert et Moreau

Home's Beauty *see* 'JACED'

Home's Choice *see* 'JACADE'

Home's Pride *see* 'JACARY'

Homestead, HCh, dr

Homestead, HP, mp

Hommage d'Anjou, HT, or, 1965; flowers coppery-red, large, dbl., slight fragrance; Minier; (Sangerhausen)

'HON A. NORTON', HT, lp; [Mme Lombard sport]; Williams, A., 1910; (Weatherly, L.)

'HON. CHARLOTTE KNOLLYS', HT, pb, 1926; flowers rose, edged lighter, center creamy yellow, dbl., slight fragrance; GM, NRS, 1926; [Antoine Rivoire X Willowmere]; Bees

'HON. EDITH GIFFORD', T, w, 1882; flowers flesh-white tinted rose, large, dbl.; [Mme Falcot X Perle des Jardins]; Guillot et Fils

'HON. GEORGE BANCROFT', HT, rb, 1879; flowers red shaded violet-crimson, intense fragrance; moderate growth; [Mme de St. Joseph X Lord Macaulay]; Bennett

'HON. INA BINGHAM', HP, mp, 1905; flowers pink, stamens golden yellow, large, dbl., 23 petals, cupped, intense fragrance; upright growth; Dickson, A.

'HON. JOAN ACTON', HT, pb, 1950; flowers cream edged pink, pointed, very large, very dbl., 90 petals, moderate fragrance; foliage bronze-green; ; hardy.; [Mrs Sam McGredy X Golden Dawn]; Marshall, J.

'HON. LADY LINDSAY', S, pb, 1939; (Honorine Lady Lindsay); flowers pink, reverse darker, dbl., 35 petals, recurrent bloom; foliage dark; bushy (3x3 ft.) growth; not dependably hardy.; [New Dawn X Rev. F. Page-Roberts]; Hansen, N.J.; B&A, 1938

'HON. VIOLET DOUGLAS PENNANT', HT, pb, 1927; flowers blend of cream and rose, dbl., moderate fragrance; Bees

'HONDO', HT, yb, 1989; flowers medium yellow with red-purple on tips, aging red-purple, large, dbl., 33 petals, moderate fragrance; foliage large, dark green, matt; upright growth; [Irish Gold X Las Vegas]; Perry, Astor, 1990

Hondo, HRg, dp, 1988

Honest Abe™ *see* 'ARON'

'HONEST RED', HT, mr, 1991; flowers large blooms borne mostly singly, dbl., 26–40 petals, intense fragrance; foliage medium, dark green, semi-glossy; tall, bushy growth; [Unnamed seedling X Sea Pearl]; Wambach, Alex A.

'HONEY', F, my, 1955; flowers coppery yellow, becoming lemon-yellow, small, semi-dbl.; [Smiles sport]; Marsh; Marsh's Nursery

Honey Bear *see* 'PIXIHON'

Honey Bouquet™ *see* 'JACYIMP'

'HONEY BUN', F, my, 1973; bud ovoid; flowers medium, dbl., cupped, intense fragrance; foliage dark, leathery; very vigorous, upright, bushy growth; [Gold Strike X Golden Garnette]; Ellis; Armstrong Nursery

'HONEY BUNCH', F, dy, 1971; flowers deep gold, shaded peach, long, pointed, dbl., 20 petals, 3 in., slight fragrance; foliage light; moderate, upright growth; [Circus X Soraya]; Watkins Roses

Honey Bunch® *see* 'COCGLEN'

Honey Butter *see* 'PIXHONEY'

Honey Child, F, ab, 1999; Ilsink

'HONEY CHILE', F, lp, 1964; bud pointed; dbl., cupped, blooms in clusters, slight fragrance; foliage leathery; vigorous, upright, bushy growth; [Fashion X Queen Elizabeth]; Thomson; Tillotson

'HONEY FAVORITE', HT, lp, 1962; (Honey Favourite); flowers light yellowish pink, base yellow; [Pink Favorite sport]; Von Abrams; Peterson & Dering

Honey Favourite *see* **'HONEY FAVORITE'**

'HONEY GOLD', F, my, 1956; bud ovoid; flowers maize-yellow overcast buff-yellow, dbl., 43 petals, 3–3.5 in., moderate fragrance; foliage dark, glossy; vigorous, bushy growth; [Yellow Pinocchio X Fashion]; Boerner; J&P

'HONEY HILL', Min, ob, 1981; bud ovoid, pointed; flowers orange, borne singly or several together, dbl., 48 petals, exhibition form, moderate fragrance; curved, light brown prickles; foliage medium green, semi-glossy; vigorous, bushy, upright growth; [Unnamed seedling X Unnamed seedling]; Lyon

Honey Mini-Delite *see* 'CLEHON'

Honey Moss *see* 'ARONEMO'

Honey 'n' Spice™ *see* 'MINAGCO'

'HONEY REA', HT, pb, 1973; bud ovoid; flowers pink, very large, semi-dbl., exhibition form, slight fragrance; foliage large, dark, leathery; very vigorous, bushy growth; Concord Floral Co.

Honey Sweet *see* **'HONEYSWEET'**

Honeybunch *see* 'COCGLEN'

'HONEYCOMB', Min, ly, 1974; flowers soft yellow to near white, small, dbl., 30 petals, 1.5 in., exhibition form, moderate fragrance; foliage small, glossy, light, leathery; dwarf, bushy growth; [(R. wichurana X Floradora) X Debbie]; Moore, Ralph S.; Sequoia Nursery

Honeycup®, HT, ab, 1990

'HONEYFLOW', F, pb, 1957; flowers white edged pink, borne in very large clusters, single, moderate fragrance; foliage glossy; vigorous growth; [Spring Song (F) X Gartendirektor Otto Linne]; Riethmuller

'HONEYGLOW', F, yb, 1956; flowers lemon-yellow, reverse shaded orange, pointed, borne in clusters, semi-dbl., 2 in.; foliage glossy; vigorous growth; [Goldilocks X Ellinor LeGrice]; LeGrice

Honeymoon *see* **'HONIGMOND'**

'HONEYPOT', S, my, 1969; flowers sulfur-yellow, dbl., 40–50 petals, 4 in., cupped, intense fragrance; foliage dark, semi-glossy; vigorous growth; [Honigmond X Constance Spry]; Austin, David

'HONEYSWEET', S, op, 1984; (Honey Sweet); bud medium-large, ovoid, pointed; flowers yellow-red-orange blend, large blooms borne 1-8 per cluster, dbl., 28 petals, cupped, repeat bloom; awl-like, brown-tan prickles; foliage leathery, dark with copper tints; bushy, erect growth; hardy; [Serendipity X Wiener Charme]; Buck, Dr. Griffith J.; Iowa State University

Honeysweet, F, ab

'HONG-KONG', HT, rb, 1962; flowers citron-yellow edged currant-red, becoming red, dbl.; strong stems; bushy growth; [Soraya X (Henri Mallerin X Peace)]; Dot, Pedro

'HONIGMOND', F, my, 1960; (Honeymoon); flowers canary-yellow, rosette shape, blooms in clusters (up to 5), dbl., 40 petals, slight fragrance; foliage dark, veined; vigorous, upright, bushy growth; [Clare Grammerstorf X Golden Scepter]; Kordes, R.; A. Dickson;, McGredy

Honky Tonk *see* 'TALHON'

Honor™ *see* 'JACOLITE'

'HONORABLE KEN', F, ob, 1993; flowers orange/yellow reverse, blooms borne in small clusters, dbl., 26–40 petals, 1.5–2.75 in., moderate fragrance; some prickles; foliage medium, medium green, semi-glossy; compact (100 cms) growth; [John Lawrence X Zorina]; Bracegirdle, Derek T.; Bracegirdle, 1994

Honore de Balzac® *see* 'MEIPARNIN'

'HONORINE DE BRABANT', B, pb; flowers pale lilac-pink, spotted and striped mauve and crimson, dbl., cupped, recurrent bloom; foliage light green; vigorous (to 6 ft) growth

Honorine Lady Lindsay *see* **'HON. LADY LINDSAY'**

Honour *see* 'JACOLITE'

'HONOUR BRIGHT', LCl, dr, 1950; flowers brilliant crimson, medium, semi-dbl., recurrent bloom, moderate fragrance; foliage bright green; very vigorous (4-6 ft) growth; [(New Dawn X Allen Chandler) X (Mrs W.J. Grant, Climbing X Richmond, Climbing)]; Eacott

'HOOPEARL', Min, lp, 1991; (**Miss Pearl**™); dbl., 15–25 petals, 1.5–2.75 in., borne mostly singly, intense, spicy fragrance; few prickles; foliage medium, medium green, matt; tall (80 cms), upright growth; [Gene Boerner X Pacesetter]; Hooper, John C.; AGM Miniature Roses

'HOOSIER BEAUTY', HT, rb, 1915; bud pointed; flowers glowing crimson shaded darker, large, dbl., intense fragrance; foliage sparse, rich green, glossy; bushy growth; GM, NRS, 1915; [Richmond X Château de Clos Vougeot]; Dorner

'HOOSIER BEAUTY, CLIMBING', Cl HT, rb, 1925; Gray, W.R.

Hoosier Glory *see* **'EUTIN'**

'HOOSIER GOLD', F, dy, 1974; bud ovoid; dbl., 30–35 petals, 2.5–3.5 in., slight fragrance; vigorous growth; [Lydia X Golden Wave]; Byrum; J.H. Hill Co.

'HOOSIER HONEY', HT, dy, 1973; bud long, pointed; flowers mimosa-yellow, full, dbl., 25–30 petals, 4–5 in., exhibition form, intense fragrance; foliage dark, leathery; vigorous growth; [Unnamed seedling No. 63-704 X Golden Fantasie]; Byrum; J.H. Hill Co.

'HOOSIER HONEY', HT, my, 1955; bud ovoid; flowers amber-yellow, open, dbl., 45–50 petals, 4.5–5 in., moderate fragrance; vigorous, upright, bushy growth; [Anzac X Golden Rapture]; Hill, Joseph H., Co.

'HOOSIER HYSTERIA', S, dr, 1981; bud ovoid; dbl., 45 petals, exhibition form, borne singly or in sprays of 5-7, slight fragrance; very few, curved, dark green prickles; foliage dark, leathery; vigorous, tall, upright growth; [Karl Herbst X Simone]; Schwartz, Ernest W.; Krider Nursery

Hoot 'n' Holler *see* 'MORPETTI'

Hoot Owl *see* 'MORHOOT'

'HOOTEA', Min, dr, 1992; (**Billie Teas**); dbl., 26–40 petals, 1.5 in., borne mostly singly, no fragrance; some prickles; foliage medium, medium green, matt; medium (38-45 cms), upright growth; AOE, 1993; [Seedling X Merrimac]; Hooper, John C.; Kimbrew Walter Roses, 1992

'HOPE', Min, w, 1985; flowers well-formed, small, dbl., 35 petals, slight fragrance; foliage large, dark, glossy; bushy growth; [Rise 'n' Shine X Party Girl]; Bridges, Dennis A.; Bridges Roses

Hope *see* 'ORAcreps'

Hope 98 *see* 'COCLAMENT'

'HOPE FOR HUMANITY', S, dr, 1996; flowers deep dark red, blooms borne in small clusters, dbl., 15–25 petals, 1.75 in., slight fragrance; moderate prickles; foliage medium, semi-glossy; upright, low (56 cms) growth; [(Prairie Princess X Morden Amorette) X (Morden Cardinette X K1)]; Collicutt, L.M. & Davidson, C.G.; Morden Exp. Farm, 1995

Hopeful 'RESHOPE'

Hopscotch *see* 'AROYOL'

'HORAARDVARK', HT, yb, 1999; (**Mario Lanza**); flowers medium yellow, edged red, reverse paler yellow, dbl., 26–40 petals, 5 in., borne mostly singly, moderate fragrance; prickles moderate; foliage medium, dark green, glossy; bushy, medium (3 ft) growth; [Elina X Remember Me]; Horner Colin P.; Warley Rose Gardens, 2000

'HORABI', HT, mr, 1995; (**Friday's Child**); dbl., 26–40 petals, 5 in., borne mostly singly, intense fragrance; foliage medium, medium green, semi-glossy; tall (120 cms), upright growth; [Spirit of Youth X Seedling]; Horner, Colin P.; Battersby Roses, 1997

'HORACE MCFARLAND', HT, op, 1944 (Président Nomblot); bud mahogany-red; flowers coppery pink, dbl., 43 petals, 4.5–5.5 in., exhibition form, moderate, fruity fragrance; foliage leathery, dark; vigorous, bushy growth; AARS, 1945; [Mme Arthaud X Unnamed seedling]; Mallerin, C.; A. Meilland, 1944; C-P, 1944

'HORACE VERNET', HP, dr, 1866; flowers deep red, large, dbl., 40 petals, exhibition form, repeat bloom, intense fragrance; moderate, erect growth; [Général Jacqueminot X ?]; Guillot et Fils

'HORALCAMSTRIP', F, op, 1999; (**County Girl**); flowers orange, pink cream striped, reverse paler striped, ages more salmon, dbl., 17–25 petals, 4 in., borne in small clusters, slight fragrance; prickles moderate; foliage medium, medium green semi-glossy; bushy, medium (3 ft) growth; [Alexander X (Southampton X ((New Penny X White Pet) X Stars 'n' Stripes))]; Horner, Colin P.; Grange Farm Nursery, 2001

'HORANYME', Gr, mr, 1999; (**Kieran Ross Clark**); flowers bright scarlet red, reverse paler, dbl., 17–25 petals, 3 in., borne in small clusters, slight fragrance; prickles moderate; foliage medium, dark green, glossy; upright, tall (4 ft) growth; [(Anytime X (Liverpool Echo X (Flamenco X Rosa Bella))) X Remember Me]; Horner, Colin P.

'HORANYMOLL', LCl, my, 1997; (**Golden Future**); flowers full, large, very dbl., 26–40 petals, borne in small clusters, moderate fragrance; foliage large, dark green, glossy; upright, medium (9-12ft.) growth; [((Anytime X Liverpool Echo) X (Flamenco X Rosa Bella)) X (Korressia X Kiskadee)]; Horner, Colin P.

'HORAPSUNMOLBABE', S, dy, 1999; (**The Care Rose**); flowers buttercup yellow, reverse medium yellow, small, semi-dbl., 12–16 petals, borne in small clusters, moderate fragrance; few prickles; foliage small, medium green, glossy; compact, low (2 ft); patio growth; [Apricot Sunrise X (Golden Future X Baby Love)]; Horner, Heather M.

Horatio Nelson *see* 'BEAHOR'

'HORAVME', F, ob, 1993; (**Evelyn Grace**); flowers orange/yellow bicolor, blooms borne in small clusters, dbl., 15–25 petals, 1.5–2.75 in., slight fragrance; some prickles; foliage medium, light green, glossy; medium (90 cms), compact growth; [Avocet X Remember Me]; Horner, Heather M.; Horner, 1995

'HORBARJEFF', HT, my, 1989; (**Taylors Gold**); bud ovoid; flowers does not fade, medium, borne singly and in sprays of 5, dbl., 48 petals, cupped, slight fragrance; fruit not known; prickles narrow, small, brown; foliage medium, medium green, semi-glossy; bushy, medium growth; [Prominent X Helmut Schmidt]; Horner, Heather M.

'HORBATBEAUT', HT, ly, 1997; (**Battersby Beauty**); flowers full, medium, very dbl., 26–40 petals, borne mostly singly, slight fragrance; foliage medium, light green, semi-glossy; bushy, medium (90cms)growth; [Polar Star X (Elina X Remember Me)]; Horner, Colin P.

'HORBLUSH', HT, w, 1997; (**Helen Margaret**); flowers double, medium, dbl., 15–25 petals, borne in small clusters, moderate fragrance; foliage medium, light green, glossy; compact, medium (80cms)growth; [Champagne Cocktail X Alpine Sunset]; Horner, Colin P.

'HORBONDARC', HT, op, 1997; (**Caroline Clarke**, The Backpackers Rose); flowers full, large, very dbl., 26–40 petals, borne in small clusters, moderate; some prickles; foliage medium, medium green, semi-glossy; bushy, medium (90cms) growth; [(Prominent X Southampton) X New Year]; Horner, Heather M.

'HORBONDSMILE', F, mp, 1996; (**Brave Heart**); flowers high-pointed blooms of classic form, borne in small clusters, dbl., 15–25 petals, 4 in., moderate fragrance; some prickles; foliage large, medium green, semi-glossy; medium (100 cms), upright growth; [(Prominent X Southampton) X Bright Smile]; Horner, Heather M.; Battersby Roses, 1998

'HORBRITESEA', S, w, 1997; (**Forever Free**); flowers full, small, very dbl., 26–40 petals, borne in large clusters, slight fragrance; some prickles; foliage small, medium green, glossy; spreading, low (45cms) growth; [(Anna Ford X Little Darling) X Sea Form]; Horner, Colin P.

'HORCABELLERO', HT, 2000; (**Barry Stephens**); flowers apricot edged red, reverse lighter, dbl., 26–40 petals, 6 in., borne mostly singly, moderate fragrance; prickles moderate; foliage large, dark green, semi-glossy; upright, medium (3.5 ft) growth; [Typhoon X Benson & Hedges Gold]; Horner, Calvin L.; Battersby Roses, 2000

'HORCOCKALORUM', HT, yb, 1997; (**Rebecca Paul**); flowers double, medium, dbl., 15–25 petals, borne in small clusters, slight fragrance; foliage medium, medium green, semi-glossy; bushy, medium (80cms)growth; [Silver Jubilee X Isobel Derby]; Horner, Colin P.

'HORCOED', S, mp, 1997; (**Kathleen Jane**); flowers very full, medium, very dbl., 41 petals, borne in large clusters, slight fragrance; foliage medium, light green, glossy; spreading, medium (4 ft), growth; [Bonica X Leverkusen]; Horner, Colin P.

'HORCOEXIST', S, yb, 1997; (**Starry Eyed**); flowers single, small, single, 4–7 petals, borne in large clusters, slight fragrance; few prickles; foliage small, light green, dull; spreading, bushy, low (90cms) growth; [(Anna Ford X Little Darling) X ((Sea Foam X Little Darling) X (Hamburger Phoenix X Prelude))]; Horner, Colin P.

'HORCOFF', F, rb, 1997; (**Hospitality**); flowers double, medium, dbl., 15–25 petals, borne in small clusters, moderate fragrance; some prickles; foliage medium, medium green, glossy; bushy, medium (100cms) growth; [(Champagne Cocktail X Alpine Sunset) X (Prominent X Southampton)]; Horner, Colin P.

'HORCOFFITUP', S, dy, 1998; (**Celebration 2000**); flowers deep yellow, non-fading, paler reverse, semi-dbl., 8–14 petals, 1.5–2 in., blooms in large clusters, slight fragrance; prickles moderate; foliage small, medium green, very glossy; bushy, tall growth; [(Anytime X Liverpool Echo) X ((Flamenco X Rosa Bella) X Baby Love)]; Horner, Colin P.; Paul Chessum Rose Specialist, 1999

'HORDEN HALL', LCl, w, 1928; flowers pure white, with long yellow stamens, large, single, slight fragrance; [R. wichurana X Frau Karl Druschki]; Conyers

'HOREAGLE', F, pb, 1997; (**Hole in One**); flowers double, small, dbl., 15–25 petals, borne in small clusters, slight fragrance; some prickles; foliage small, medium green, semi-glossy; spreading, medium (90cms) growth; [((Southampton X New Penny) X (White Pet X Stars 'n' Stripes)) X ((Vester X seedling) X Edith Holden))]; Horner, Colin P.

'HORETHEL', HT, pb, 1992; (**Isobel Derby**); flowers peach pink, reverse lighter pink, aging deeper pink, urn-shaped, dbl., 5 in., moderate, fruity fragrance; foliage medium, medium green, glossy; bushy, medium growth; [Champagne Cocktail X [(Honey Favorite X Dr. A.J. Verhage) X Pot 'o Gold]]; Horner, Colin P.; Golden Fields Nursery, 1992

'HORFAIWIL', HT, my, 1998; (**Faith**); flowers medium yellow, lighter reverse, very dbl., 26–40 petals, 4 in., borne mostly singly, moderate fragrance; prickles moderate, medium, curved; foliage medium, medium green, glossy; upright, tall growth; [Golden Future X Solitaire]; Horner, Colin P.; Warley Rose Gardens, Ltd., 2000

'HORFLAME', F, mr, 1997; (**Middlesborough Football Club**); flowers double, large, dbl., 15–25 petals, borne in small clusters, slight fragrance; some prickles; foliage medium, reddish bronze, glossy; upright, medium (60cms) growth; [Alpha X (Old Master X Southampton)]; Horner, Heather M.

'HORFLAN', F, dp, 1990; (**Flanders Field**); bud ovoid; flowers light red aging slightly lighter, urn-shaped, loose, medium, semi-dbl., slight fragrance; foliage medium, medium green, semi-glossy; medium to tall, upright growth; [Prominent X Southampton]; Horner, Heather M., 1992

'HORFLASH', F, yb, 1983; (**Champagne Cocktail**); flowers pale yellow, flecked and splashed pink, yellow reverse, medium, dbl., 20 petals, blooms in clusters of 3 to 5, moderate fragrance; foliage medium, medium green, glossy; bushy growth; GM, Glasgow, 1990; [Old Master X Southampton]; Horner, Colin P., 1985

'HORFLASHROB', F, rb, 1999; (**Valeria Sykes**); flowers red splashed white, reverse silver, single, 5–11 petals, 2 in., borne in small clusters, slight fragrance; few prickles; foliage small, medium green, semi-glossy; bushy, low (2 ft) growth; [Champagne Cocktail X Robin Redbreast]; Horner, Colin P.; Battersby Roses, 2000

'HORFLOSS', F, mp, 1997; (**Flossie**); flowers double, small, dbl., 15–25 petals, borne in large clusters, moderate fragrance; prickles soome; foliage small, medium green, semi-glossy; upright, medium (3ft.) growth; [Coral Reef X (Seedling X Lichtkonigin Lucia)]; Horner, Colin P.

'HORFLOWER', Min, op, 1995; (**Flower Child**); flowers salmon pink, dbl. (15–25 petals), medium blooms; slight fragrance; foliage small, medium green, semi-glossy; compact, tall (40 cms) growth; [Penelope Keith X Gold Bunny]; Horner, Heather M.; Warley Rose Gardens, 1997

'HORFROST', F, w, 1997; (**Frosty Morning**); flowers double, medium, dbl., 15–25 petals, borne in small clusters, slight fragrance; some prickles; foliage medium, medium green, semi-glossy; bushy, medium (100cms) growth; [Sexy Rexy X Princess Michael of Kent]; Horner, Colin P.

'HORGREG', S, w, 1996; (**Greg Moore**); flowers cream tinged pink, small, camellia-shaped blooms) borne in l, dbl., 15–25 petals, 2 in., slight fragrance; some prickles; foliage small, grey green, dull; medium (100 cms), spreading growth; [Bonica X (Lichtkonigin Lucia X Seedling)]; Horner, Colin P.

'HORHARPDOS', Min, ob, 1992; (**Emma Mitchell**); flowers orange/vermillion striped white, light orange reverse, aging light orange, semi-dbl., 12 petals, 1.25 in., slight, fruity fragrance; foliage small, medium green, semi-glossy; bushy, low growth; Horner, Colin P.; Battersby Roses, 1993

'HORHARRYPLUS', F, w, 1988; (**Whitley Bay**); bud ovoid; flowers cream with pink splashes, reverse cream, loose, medium, born, cupped, slight fragrance; ovoid, medium, yellow fruit; prickles medium, brown; foliage medium, medium green, semi-glossy; spreading, medium growth; [Champagne Cocktail X [Colour Wonder X (Vera Dalton X Piccadilly)]]; Horner, Colin P.; Battersby Roses, 1990

'HORIZON', F, or, 1956; bud pointed; flowers geranium-red, open, large, borne in large clusters, dbl., moderate fragrance; foliage leathery; vigorous, upright, bushy growth; [Crimson Glory X Cinnabar]; Tantau, Math.

Horizon 2000, HT, 1990; Delbard, Georges; (Cavriglia)

'HORJACK', HT, ob, 1995; (**Jack Collier**); flowers orange blend, dbl., 15–25 petals, 4 in., borne mostly singly or in small clusters, slight fragrance; foliage medium, medium green, semi-glossy; tall (120 cms), upright growth; [Marjorie May X Remember Me]; Horner, Colin P.; Battersby Roses, 1997

'HORJEMMA', F, my, 1995; (**Jemma Giblin**); dbl., 15–25 petals, 2.5 in., borne in small clusters, moderate fragrance; foliage medium, medium green, semi-glossy; medium (50 cms), bushy growth; [Gingernut X Amanda]; Horner, Colin P.; Battersby Roses, 1997

'HORJOY', F, pb, 1997; (**Our Joy**); flowers candy pink, silver reverse, dbl., 15–25 petals, 3 in., borne in small clusters, moderate fragrance; prickles moderate; foliage medium, semi-glossy; upright, 80cms growth; [Seedling X Seedling]; Horner, Heather M.; Battersby Roses, 1997

'HORKEEPOG', F, w, 1994; (**Susan Jellicoe**); flowers cream/pink blend, blooms borne mostly single and in small clusters, dbl., 26–40 petals, 3–3.5 in., intense fragrance; some prickles; foliage medium, medium green, semi-glossy; medium (100 cms), upright, bushy growth; [Keepsake X Pot O'Gold]; Horner, Colin P.; Horner, 1993

'HORKORBLUSH', HT, r, 1988; (**Georgie Lad**); bud ovoid, red; flowers mahogany-red, yellow at base, reverse lighter red, medium, dbl., cupped, moderate, fruity fragrance; ovoid, medium, medium fruit; prickles small, light brown; foliage medium, medium green, matt; upright, medium growth; [Prominent X (Champagne Cocktail X Alpine Sunset)]; Horner, Colin P.; Battersby Roses, 1990

'HORLEXSTRIP', HT, op, 1991; (**Grace Donnelly**); bud ovoid; flowers pink, orange, yellow striped, urn-shaped, loose, dbl., borne singly or in sprays of 5-9, slight, fruity fragrance; foliage medium, medium green, semi-glossy; bushy, tall growth; [Alexander X [Southampton X [(New Penny X Little White Pet) X Stars 'n' Stripes]]]; Horner, Colin P., 1984; Battersby Roses, 1992

'HORLIGHTS', F, yb, 1987; (**Friction Lights**); flowers canary-yellow edged cherry-red, large, dbl., 15–25 petals, slight fragrance; foliage medium, medium green, semi-glossy; upright growth; [Alexander X Champagne Cocktail]; Horner, Colin P.; Battersby Roses, 1987

'HORLOVEQUEEN', F, ob, 1992; (**Mary Campbell**); flowers orange, blooms borne in small clusters, semi-dbl., 6–14 petals, 1.5–2.75 in., slight fragrance; some prickles; foliage medium, medium green, semi-glossy; medium (80 cms), bushy growth; [Lovers Meeting X Amber Queen]; Horner, Heather M., 1993

'HORMARTIM', F, dr, 1988; (**Margaret Daintry**); flowers crimson-scarlet, large, borne singly and in clusters, dbl., 15–25 petals, intense fragrance; foliage medium, dark green, glossy; bushy growth; [Red Planet X (Blessings X (Parkdirektor Riggers X Honey Favorite))]; Horner, Colin P.; Rosemary Roses, 1988

'HORMASBRICK', F, r, 1992; (**Whickham Highway**); flowers brick red, blooms borne in large clusters, semi-dbl., 6–14 petals, 1.5–2.75 in., slight fragrance; some prickles; foliage medium, medium green, glossy; tall (120 cms), upright growth; [Mary Sumner X (KORp X Southampton)]; Horner, Heather M., 1993

'HORMISLAC', F, dp, 1992; (**Ted Gore**); flowers deep pink, blooms borne in large clusters, dbl., 26–40 petals, 1.5–2.75 in., slight fragrance; some prickles; foliage medium, medium green, semi-glossy; medium (80 cms), bushy growth; [Miss Ireland X Lilac Charm]; Horner, Colin P., 34394

'HORNDON PINK', HT, dp, 1963; flowers old-rose-pink, dbl., 60 petals, 4.5 in., moderate fragrance; foliage dark; [Lady Elgin X Independence]; Barter

'HORNEWGRAM', HT, my, 1988; (**Dame Joyce Frankland**); bud globular, greenish-yellow; flowers urn-shaped, large, borne singly, dbl., 32 petals, slight fragrance; ovoid, large, yellow fruit; prickles straight, medium, light brown; foliage large, medium green, glossy; bushy growth; [(Honey Favorite X Dr. A.J. Verhage) X Pot O'Gold]; Horner, Colin P.

Hornish Plot, HP, rb

'HORODD', F, rb, 1998; (**Oddball**); flowers crimson stripes on buff brown, facing to red stripes on pink, semi-dbl., 8–14 petals, 2.5 in., borne in small clusters, slight fragrance; prickles moderate; foliage medium, medium green, dull; spreading growth; [(Licterloh X (Tall Story X Southampton) X ((New Penny X White Pet) X Stars 'n' Stripes)]; Horner, Colin P.; Paul Chessum Rose Specialist, 1998

'HORPEKBOND', Min, or, 1998; (**Hot Gossip**); flowers luminous orange-red, reverse matt orange, semi-dbl., 8–14 petals, 1.5 in., borne in small clusters, slight fragrance; prickles moderate; foliage medium, medium green, dull; compact, medium (50 cms), dwarf-clustered growth; [Penelope Keith X (Prominent X Southampton)]; Horner, Heather M.; Paul Chessum Rose Specialist, 1998

'HORRIDO'®, F, mr, 1963; (Jockey); bud ovoid; flowers pure blood-red, blooms in large clusters, dbl., cupped; foliage dark, glossy; bushy, low growth; Tantau, Math.

'HORSADDLER', F, ab, 1995; (**Saddler's Gold**); flowers amber gold, blooms borne in small clusters, dbl., 15–25 petals, 2.5 in., slight fragrance; foliage medium, medium green, semi-glossy; medium (50 cms), bushy growth; [Gingernut X Gold Bunny]; Horner, Heather M.; Warley Rose Gardens, 1997

'HORSILBEE', HT, ab, 1988; (**Heather Honey**); bud ovoid, bronze; flowers apricot yellow, reverse apricot urn-shaped, medium, borne usually singly, dbl., 25 petals, moderate, fruity fragrance; globular, medium, yellow fruit; prickles small, greenish-brown; foliage medium, medium green, glossy; bushy, medium growth; [Silver Jubilee X (Honey Favorite X Southampton)]; Horner, Colin P.; LeGrice Roses, 1991

'HORSILKARL', HT, mp, 1997; (**Doris Dowman**); flowers full, large, very dbl., 26–40 petals, borne mostly singly, slight fragrance; some prickles; foliage large, medium green, glossy; compact, medium (3ft.) growth; [Silver Jubilee X (Seedling X Karlsruhe)]; Horner, Colin P.

'HORSTACEY', F, pb, 1996; (**Stacey's Star**); flowers blends of pink, short center petals much darker pink, blooms, dbl., 26–40 petals, 2.5 in., flat, moderate fragrance; some prickles; foliage small, medium green, semi-glossy; low (50 cms), bushy growth; Horner, Colin P.; Battersby Roses, 1998

'HORSTMANN'S BERGFEUER', F, dr, 1954; flowers dark blood-red, well formed, very large, borne in large clusters, dbl.; moderate growth; [World's Fair X Independence]; Horstmann

'HORSTMANN'S JUBILÄUMSROSE', F, op, 1954; flowers pink tinted peach, well formed, large, borne in clusters of, dbl.; foliage glossy, leathery; dwarf growth; [Golden Rain X Alpine Glow]; Tantau, Math.; Horstmann

'HORSTMANN'S LEUCHTFEUER', F, dr, 1954; flowers blood-red, large, borne in large clusters, dbl.; moderate growth; [Red Favorite X Karl Weinhausen]; Tantau, Math.; Horstmann

'HORSTMANN'S ROSENRESLI', F, w, 1955; flowers pure white, large blooms in clusters, dbl., moderate fragrance; bushy growth; [Rudolph Timm X Lavender Pinocchio]; Kordes; Horstmann

'HORSTMANN'S SCHÖNE BRÜNETTE', HT, or, 1955; flowers coppery brick-red, large, dbl.; strong stems; [Independence X Hens Verschuren]; Horstmann

'HORSUN', F, yb, 1986; (**Playgroup Rose**®); flowers yellow, petals edged red, reverse pale yellow, blooms in sprays, dbl., 25 petals, cupped, moderate fragrance; small, globular, orange fruit; large, light brown prickles; foliage medium, light green, glossy; medium, bushy growth; [Prominent X Southampton]; Horner, Heather M.

'HORSUNPEGY', F, op, 1993; (**Marjorie May**); flowers orange/pink blend, blooms borne in small clusters, dbl., 26–40 petals, 3–3.5 in., slight fragrance; some prickles; foliage medium, medium green, semi-glossy; medium (100 cms), upright growth; [Playgroup X Peer Gynt]; Horner, Heather M.; Horner, 1995

'HORSUNSMILE', F, yb, 1993; (**Voice of Thousands**); flowers yellow edged cherry red, dbl., 15–25 petals, 1.5–2.75 in., borne in small clusters, slight fragrance; some prickles; foliage medium, medium green, semi-glossy; medium (80cms), bushy growth; [Playgroup X Bright Smile]; Horner, Heather M.; Horner, 1994

'HORTENSE DE BEAUHARNAIS', HGal, pb; flowers rose edged rosy lilac, dbl.

'HORTENSE VERNET', M, w, 1861; flowers white shaded rose, very dbl.; moderate growth; Moreau et Robert

Horticolor® *see* LAPbal

'HORTICULTOR VIDAL', HT, op, 1952; bud pointed; flowers salmon-pink, large, dbl., 35 petals, intense fragrance; bushy growth; [Mme Butterfly X Federico Casas]; Dot, Pedro

'HORTIFLORA', Pol, mp, 1974; bud oval; flowers full, dbl., 45 petals, 2 in., slight fragrance; Delforge, S.

'HORTROPIC', HT, yb, 1988; (Sir William Leech, **William Leech**); bud ovoid; flowers yellow edged red, reverse light yellow, large, borne usually, dbl., exhibition form, slight fragrance; ovoid, large, orange fruit; prickles medium, red; foliage medium, medium green, glossy; bushy, medium growth; [Royal Dane X Piccadilly]; Horner, Colin P.; Battersby Roses, 1989

'HORTULANUS ALBERT FIET', HT, ab, 1919; flowers apricot and lilac-rose, center coppery orange, dbl., moderate fragrance; [Mme Mélanie Soupert X Mons. Paul Lédé]; Leenders, M.

'HORTULANUS BUDDE', HT, mr, 1919; bud pointed; flowers large, dbl., moderate fragrance; foliage dark; vigorous growth; [Gen. MacArthur X Mme Edouard Herriot]; Verschuren

'HORTULANUS FIET', HT, w, 1919; bud pointed; flowers deep and light cream, open, very large, dbl., slight fragrance; foliage sparse, glossy, dark; vigorous growth; [Cissie Easlea X Golden Star]; Verschuren

'HORTYARD', F, m, 1992; (**Olga Rippon**); flowers purple/mauve, loose, blooms borne in sprays of 5-9, dbl., 2.5 in., slight, spicy fragrance; foliage medium, medium green, semi-glossy; bushy, low growth; [Sexy Rexy X [(INTermezzo X Baby Faurax) X (Tassin X Seedling)]]; Horner, Colin P.

'HORVALKATH', F, yb, 1997; (**Valerie Kathleen**); flowers double, medium, dbl., 15–25 petals, borne in small clusters, moderate fragrance; some prickles; foliage large, dark green, glossy; bushy, medium (2ft. 6in.)growth; [(Silver Jubilee X Pink Favorite) X Amber Queen]; Horner, Colin P.

'HORWANDER', S, mp, 1997; (**Wanderer**); flowers semi double, small, semi-dbl., 8–14 petals, borne in large clusters, slight fragrance; foliage small, dark green, semi-glossy; spreading, low (50-60cms) growth; [Sea Foam X Eye-opener]; Horner, Colin P.

'HORWINGFEL', F, or, 1988; (**Winged Fellowship**); bud ovoid; flowers vermillion orange, reverse lighter, urn-shaped, medium, born, dbl., 28 petals, slight fragrance; globular, medium, red-orange fruit; prickles small, brown; foliage medium, medium green, matt; bushy, medium growth; [Guitare X Prominent]; Horner, Heather M.; Rosemary Roses, 1989

'HOSHIZUKUYO', HT, w, 1990; bud pointed; flowers ivory, large, borne usually singly, dbl., 35–40 petals, exhibition form, intense, fruity fragrance; foliage medium, light green, semi-glossy; medium, upright growth; [(Izayoi X Sodori-Hime) X White Success]; Ohtsuki, Hironaka

Hospitality *see* 'HORCOFF'

'HOSTESS', F, op, 1960; flowers pink tinted salmon, open, large, borne in clusters of 5-7, semi-dbl.; vigorous, upright growth; [Papillon Rose X (Cinnabar X Alain)]; Lens

'HOSTESS GISELA', F, dp, 1973; bud ovoid; flowers medium, dbl., intense fragrance; foliage very glossy; vigorous, upright, bushy growth; [Sympathie X Dr. A.J. Verhage]; Hetzel; GAWA

Hot Chile *see* 'MACBUNBER'

Hot Chocolate *see* SIMcho

Hot Fire *see* INTerbricor

Hot Gossip *see* 'HORPEKBOND'

Hot Gossip *see* JACati

Hot Lips *see* 'TINLIPS'

Hot Lips, F, dp

Hot 'n' Spicy *see* 'MACSODA'

Hot Pants *see* 'SIMHOPAN'

'HOT PEWTER', HT, or, 1978; (Crucencia, Crucenia, Cruzenia); flowers brilliant orange-red, borne usually singly, dbl., 41 petals, exhibition form, slight fragrance; broad, dark prickles; foliage large, mid-green, semi-glossy; bushy growth; [Alec's Red X Red Dandy]; Harkness, R., & Co., Ltd.

Hot Pink Pastel *see* 'WILPINK'

Hot Point *see* BARhop

Hot Princess, HT, lp;, exhibition form; foliage large, dark

'HOT SHOT', Min, or, 1982; (Hotshot); flowers vibrant vermilion, medium, dbl., 28 petals, exhibition form, slight fragrance; foliage small, medium green; upright, bushy growth; AOE, 1984; [Futura X Orange Honey]; Bennett, Cecilia 'Dee'; Tiny Petals Nursery

Hot Spot™ *see* 'CARBANE'

'HOT STUFF', Min, mr, 1978; bud pointed; flowers turkey-red, semi-dbl., 10 petals, 1 in., slight fragrance; foliage tiny; very compact, bushy growth; Lyon

Hot Tamale™ *see* 'JACPOY'

'HOTEL HERSHEY', Gr, or, 1977; bud long, pointed to urn-shaped; flowers salmon orange-red, dbl., 4–4.5 in., exhibition form, slight fragrance; foliage dark, leathery; upright growth; [Queen Elizabeth X Comanche]; Williams, J. Benjamin; Hershey Estates

Hotel Royal, S, mr, 1980; McGredy, Sam IV

'HÔTESSE DE FRANCE', HT, dr, 1962; flowers velvety deep red, medium, cupped; foliage bronze; [Soraya X Unnamed seedling]; Hémeray-Aubert

Hotline™ *see* 'AROMIKEH'

Hotshot *see* **'HOT SHOT'**

'HOUEMILY', HT, lp, 1991; (**Emily Hough**); flowers white, blushing pink toward center, more pink in cooler weather, dbl., 15–25 petals, 3–3.5 in., exhibition form, slight fragrance; some prickles; foliage large, medium green, semi-glossy; medium, upright growth; [Touch of Class sport]; Hough, Robin, 1993

House Beautiful *see* HARbingo

House of York, HT, w, 1998; flowers creamy white, large, exhibition form; Williams, J. Benjamin

Houston *see* **'CATHEDRAL'**

'HOUSTON', HT, dy, 1980; bud ovoid; flowers deep bright yellow, borne singly or 3-4 per cluster, dbl., 38 petals, exhibition form, moderate, tea fragrance; long prickles, hooked downward; foliage leathery, wrinkled, dark; vigorous, upright growth; PP4687; [Summer Sunshine X Unnamed seedling]; Weeks, O.L.

'HOUSTONIAN', Gr, dr, 1962; bud ovoid; dbl., 35 petals, 4 in., moderate fragrance; foliage leathery, glossy, bronze; vigorous, upright, tall growth; [Carrousel X Unnamed seedling]; Patterson; Patterson Roses

'HOVYN DE TRONCHÈRE', T, rb; flowers red with orange depths, bordered silver; Puyravaud, 1899

Howard Florey *see* 'TOMFLO'

'HOWARD JERABEK', S, mp, 1978; flowers pearl-pink, very dbl., 100 petals, moderate, apple-blossom fragrance; foliage large, glossy; vigorous growth; Jerabek, Paul E.

Howard Morrison *see* 'MACCRACKLE'

'HUBAR', HT, dr, 1975; (**Black Night**®); bud long; dbl., 26 petals, 4 in.; foliage dark, leathery; vigorous, upright growth; [Fragrant Cloud X Pharaoh]; Huber

'HUBICKA', HT, pb, 1935; flowers alabaster-white, sometimes rosy, base yellow, very large, dbl., moderate fragrance; foliage dark; bushy growth; [Grete Bermbach sport]; Böhm, J.

'HUDSON', HT, mp; flowers pale rose, well shaped; foliage glossy; Mallerin, C.

Hudson Crimson, HP, dr

Hudson's Bay Rose *see* **R. BLANDA**

'HUETTE'S DAINTY FLORRIE', Min, lp, 1979; dbl., 24 petals, moderate fragrance; dark red prickles; foliage dark green edged dark red; upright, bushy growth; [Sweet and Low X Mary Marshall]; Schwartz, Ernest W.; Men's Garden Club of Virginia

'HUGH DICKSON', HP, mr, 1905; flowers very large, dbl., 38 petals, exhibition form, recurrent bloom, intense fragrance; vigorous growth; [Lord Bacon X Gruss an Teplitz]; Dickson, H.

Hugh Ringold, HT; Gregory, C.; (Cavriglia)

'HUGH WATSON', HP, op, 1905; flowers deep pink tinged salmon and silver-pink, very large, dbl., 24 petals, flat; vigorous growth; Dickson, A.

'HUGO MAWEROFF', HMult, dp, 1910; flowers carmine-pink, small, dbl.; Soupert & Notting; (Sangerhausen)

'HUGO PILLER', HT, ab, 1927; flowers flesh-white, center pale ecru; [Ophelia sport]; Leenders, M.

'HUGO ROLLER', T, yb, 1907; flowers lemon-yellow, edged and suffused crimson, well-formed blooms, dbl., moderate fragrance; foliage small, rich green; weak stems; compact, bushy growth; Paul, W.

'HUGO ROLLER, CLIMBING', Cl T, lp; [Hugo Roller sport]; Rogers, 1932; (Weatherly, L.)

'HUGO SCHLÖSSER', HT, op, 1955; bud long, pointed; flowers salmon-pink, very large, dbl., exhibition form, moderate fragrance; foliage leathery; strong stems; very vigorous, upright, bushy growth; [World's Fair X Peace]; Kordes

Hugs 'n' Kisses™ *see* 'JACIRAPS'

Hugues Aufray *see* DORastri

'HUGUETTE', Min, op; flowers shrimp-pink shaded salmon

'HUGUETTE DESPINEY', HWich, yb, 1911; flowers light buff-yellow, edged red, small, borne in large clusters, very dbl.; vigorous growth; [Marco sport]; Girin

'HUGUETTE DUFLOS', HT, op, 1937; bud pointed, dark pink; flowers satiny pink, touched salmon, large, dbl., 30–40 petals, intense fragrance; foliage dark; vigorous, bushy growth; [Betty Uprichard X Unnamed seedling]; Lille

'HUGUETTE VINCENT', HT, op, 1921; flowers brilliant velvety geranium-red, semi-dbl.; [Mrs Edward Powell seedling X Willowmere]; Chambard, C.

Huilito, B, mp

'HULA GIRL', Min, ob, 1975; bud long, pointed; flowers bright orange, dbl., 45 petals, 1 in., moderate, fruity fragrance; foliage small, glossy, embossed; bushy growth; AOE, 1976; [Miss Hillcrest X Mabel Dot]; Williams, Ernest D.; Mini-Roses;, Sequoia Nursery

'HULA HOOP', HT, pb, 1960; flowers striped bright pink and red, intense fragrance; RULED EXTINCT 11/90; Freud; Horstmann

Hula Hoop *see* 'MORHOOP'

'HULLABALOU', Min, yb, 1990; bud pointed; flowers light orange on yellow with yellow reverse, aging to dull yellow, dbl., 20 petals, moderate fragrance; prickles straight, sparse, small, red; foliage small, medium green when new, aging to maroon, semi-; bushy, low growth; [Rise 'n' Shine X Unknown]; Stoddard, Louis

Hulthemia berberifolia *see* **'HULTHEMIA PERSICA'**

'HULTHEMIA HARDII', ((Cels) Rowley), Sp; (Rosa X hardii, X Hardiilist as X H. hardii); flowers yellow with the crimson eye of hulthemia, 2 in.; foliage 1-7 narrow leaflets; Simplicifoliae; [Hulthemia persica X Rosa clinophylla]; Hardy before 1836

'HULTHEMIA PERSICA', ((Michaux) Bornmueller), Sp, yb; (Hulthemia berberifolia, Lowea berberifolia, Rosa berberifolia, Rosa persica, Rosa simplicifolia); flowers buttercup-yellow with a scarlet eye like a cistus, 1 in. fruit prickly curious xerophyt; foliage leaves simple, bluish-green; Simplicifoliae, (14); 1790

'HULTHEMOSA'; (X Hulthemosa)

Humanity *see* HARcross

'HUMBOLDT', HT, mp, 1922; flowers bright rose-pink, dbl., exhibition form, moderate fragrance; foliage glossy, bronze; vigorous growth; [Ophelia X Unnamed seedling]; Hill, E.G., Co.; Cottage Gardens Co.

'HUMDINGER', Min, op, 1976; bud pointed; flowers orange-red, dbl., 53 petals, 1 in., exhibition form, slight fragrance; foliage glossy, dark; upright, bushy growth; AOE, 1978; [Gold Coin seedling X Unnamed Miniature seedling]; Schwartz, Ernest W.; Nor'East Min. Roses

'HUME'S BLUSH TEA-SCENTED CHINA', T, lp, 1809; dbl., intense fragrance; Hume; A. Hume

'HUMMINGBIRD', HT, ob, 1934; bud pointed; flowers bronze-orange, very dbl., 3 in.; vigorous growth; [Talisman sport]; Scittine; Lainson

'HUMORESKE', Min, lp, 1957; flowers white, tinged pink, small, dbl.; bushy, compact growth; [Midget X Pixie]; Spek; McGredy

'HUMPTY-DUMPTY', Min, lp, 1952; flowers soft carmine-pink, center deeper, blooms in clusters, very dbl.; height 6-8 in; [(R. multiflora nana X Mrs Pierre S. duPont F2) X Tom Thumb]; deVink; T. Robinson, Ltd.

Hungaria, F, mr

Hunslet Moss, M, dp; re-introduced, 1984

'HUNTER', HRg, mr, 1961; (The Hunter); flowers bright crimson, medium, dbl., 43 petals, 2.5 in., borne in clusters, recurrent bloom, moderate, sweet fragrance; foliage rugose, dark, glossy; vigorous (4-5 ft.) growth; [R. rugosa rubra X Independence]; Mattock, R.H.; Mattock

'HUNTER'S MOON', HT, my, 1951; dbl., 30 petals, slight fragrance; foliage glossy, dark; very vigorous growth; [Condesa de Sástago X Gorgeous]; McGredy, Sam IV

Huntingburg, F, lp, 1997; Peters

Huntington Brocade, B, rb, 1973; Robinson

Huntington Pink Tea, T, dp

Huntington Seven Sisters, HMult, mp

Huntington Slater's Crimson China, Ch, mr

Huntington's Hero, S, lp, 1995; Martin

'HUNTSMAN', HT, rb, 1951; flowers spectrum-red, reverse yellow, dbl., 35–40 petals, 5.5 in., moderate fragrance; foliage dark; [The Queen Alexandra Rose seedling X Crimson Glory]; Robinson, H.; J&P

'HURDALSROSEN', S, mp, 1860

Hurdy Gurdy *see* 'MACPLUTO'

'HURON', S, lp, 1932; flowers white flushed pink, semi-dbl., non-recurrent, moderate fragrance; foliage leathery; vigorous (2 1/2 ft.), compact, bushy growth; a good hedge rose; [Pythagoras X R. cinnamomea]; Central Exp. Farm

Huron Sunset *see* 'RENHURON'

'HURRA'®, F, or, 1962; bud pointed; dbl., blooms in clusters; foliage glossy; vigorous, bushy growth; Tantau, Math.

Hurrikan, Pol, mr; flowers large, dbl., slight fragrance; VEG; (Sangerhausen)

'HURST CHARM', LCl, pb, 1936; flowers pink, slightly tinted mauve, large; Hicks

'HURST CRIMSON', Pol, dr, 1933; flowers deep crimson, borne in large trusses; [Ideal sport]; Hicks

'HURST DELIGHT', LCl, w, 1936; flowers clear pale cream, moderate fragrance; vigorous growth; Hicks

'HURST FAVOURITE', LCl, w, 1936; flowers pure ivory-white, very large, borne in clusters, semi-dbl., moderate fragrance; very vigorous, climbing growth; Hicks

'HURST GEM', Pol, ob, 1931; flowers brilliant orange-scarlet, small, semi-dbl.; [Orléans Rose sport]; Hicks

'HURST GLORY', HT, ob, 1936; flowers pale salmon-cerise, flushed yellow, well shaped, large, dbl., moderate fragrance; vigorous growth; Hicks

'HURST SCARLET', HT, dr, 1933; flowers deep scarlet, large, dbl., moderate fragrance; Hicks

Husmoderrose, HSpn, mp

'HUTTON VILLAGE', HT, my, 1973; flowers bright yellow; [Whisky Mac sport]; Deamer; Warley Rose Gardens

'HUX 1', S, mp, 1998; (**Lily Freeman**); flowers medium pink, single, single, 5–11 petals, 5.25 in., flat, borne in small clusters, moderate fragrance; prickles numerous; foliage small, medium green, dull; bushy, medium tall growth; [Schneezwerg X Schneezwerg]; Huxley, Ian

Hviezdoslav *see* **'HWIEZDOSLAV'**

'HVISSINGE-ROSE', F, pb, 1943; flowers pinkish with yellow, borne in clusters, single; vigorous growth; [Orléans Rose X Seedling]; Poulsen, S.

'HWIEZDOSLAV', HT, rb, 1936; (Hviezdoslav); flowers copper-red to orange-copper-red, large, dbl., cupped, intense fragrance; foliage glossy; vigorous, bushy growth; Böhm, J.

'HYBRIDE DI CASTELLO', S, ly, 1920

'HYLO', F, mp, 1961; flowers salmon-pink, medium, borne in clusters, dbl., 28 petals, 2–2.5 in., slight fragrance; bushy growth; [Highlight sport]; Cant, B. R.

Hymenee, S, 1969; Delbard-Chabert; (Cavriglia)

'HYMNE', F, mp, 1964; bud short, pointed; flowers neyron rose, large, borne in large clusters, dbl., 38–52 petals; foliage dark, glossy; vigorous growth; [Miracle X Unnamed seedling]; Verbeek

'HYPACIA', C, pb; (Hypathia); flowers bright rose-pink spotted white, center whitish, large, dbl., cupped, moderate, damask fragrance

Hypathia *see* **'HYPACIA'**

'HYTHE CLUSTER', Pol, dp, 1935; flowers glowing deep pink, borne in large clusters, semi-dbl., cupped; foliage small, glossy, light; vigorous growth; Archer

I

I Love You *see* 'GEELOVE'

'I LOVE YOU', F, dp, 1981; flowers deep pink; [Margaret Thatcher sport]; Takatori, Yoshiho; Japan Rose Nursery

I Promise *see* 'LYOPRI'

'I ZINGARI', HT, ob, 1925; flowers orange-scarlet, corymbs, semi-dbl.; foliage dark; stems claret; Pemberton

'I. X. L.', HMult, dp, 1925; flowers magenta; prickles straight, thornless canes (1 indiam); very vigorous growth; [Tausendschön X Veilchenblau]; Coolidge

'IAN BRINSON', HT, dr, 1945; flowers crimson, compact, dbl., 30 petals, 4–5 in., intense fragrance; foliage dark; [Mrs J.J. Hedley-Willis X J.C. Thornton]; Bees

Iberflora 95® *see* FEbama

Ibica *see* **'IBIZA'**

Ibis *see* 'GODEURPAN'

'IBIZA', HT, w, 1938; (Ibica); flowers well formed, moderate fragrance; erect growth; [Mme Butterfly X Frau Karl Druschki]; Dot, Pedro

Ice Angel *see* JACBER

Ice Breaker *see* 'BRIICE'

Ice Cascade, S, lp, 1999; very dbl., borne in sprays; Williams, J. Benjamin

Ice Cream *see* KORzuri

Ice Crystal *see* 'CLEICE'

Ice Fairy *see* 'SANMED'

Ice Follies, S, w, 1999; flowers white with delicate apricot tint, borne in large sprays, moderate fragrance; foliage dark green; Williams, J. Benjamin

'ICE MAIDEN', F, w, 1977; bud pointed; flowers pure white, semi-formal, semi-dbl., 15 petals, 3.5 in., moderate fragrance; foliage narrow, glossy, light green; vigorous, upright growth; [Iceberg X Iceberg]; Garelya

Ice Meidiland® *see* MEIVAHYN

Ice Princess *see* 'LAVICE'

Ice Queen® *see* 'SAVANICE'

'ICE WHITE', F, w, 1966; (Vision Blanc®); dbl., 25 petals, 3 in., blooms in clusters; foliage glossy; GM, Portland, 1970; [Mme Leon Cuny X (Orange Sweetheart X Cinnabar)]; McGredy, Sam IV; McGredy

Ice White, Climbing *see* MACvisar

Iceberg *see* 'KORBIN'

'ICEBERG, CLIMBING', Cl F, w, 1968; Cant, B. R.

'ICED GINGER', F, ob, 1971; flowers petals palest pink, reverse coppery, dbl., 45 petals, 4.5 in., moderate fragrance; foliage red, veined; [Anne Watkins X ?]; Dickson, A.

'ICED PARFAIT', F, lp, 1972; flowers pale pink, urn-shaped, medium, borne 6 per cluster, dbl., 40 petals, moderate fragrance; straight, red prickles; foliage light green; bushy, compact growth; [Pink Parfait X Iceberg]; Xavier, Sister M.

'ICELAND QUEEN', LCl, w, 1935; flowers creamy white, large, dbl.; vigorous growth; Horvath; Wyant

Ichalkaranji 100, HT, m, 1993; Chiplunkar

ICI Golden Celebration *see* 'PEAQUANT'

'ICO', F, m, 1986; flowers light purple, very dbl., 60–70 petals, globular, slight fragrance; brownish-green prickles curving downward; foliage dark green, glossy; vigorous, upright, bushy growth; [Deep Purple sport]; Patil, B.K.; K.S.G. Son's Roses

Ico Ambassador, HT, my, 1988; Patil, B.K.

'ICO BEAUTY', HT, pb, 1986; flowers rose pink, reverse flushed white, medium, dbl., 25–30 petals, exhibition form, slight fragrance; pale green prickles curving downward; foliage glossy; upright, bushy growth; [Red Planet sport]; Patil, B.K.; K.S.G. Son's Roses

Ico Delight, HT, w, 1989; Patil, B.K.

Ico Deluxe, HT, pb, 1990; Patil, B.K.

Ico Pearl, F, pb, 1990; Patil, B.K.

Ico Talk, F, pb, 1993; Patil, B.K.

Ico Trimurthi, HT, pb, 1989; Patil, B.K.

Icy Tiding, HT, w

Ida Belle® *see* 'MINAYCO'

Ida Elizabeth *see* 'WELIZ'

Ida Hoff, HT, mp; flowers large, dbl.; (Sangerhausen)

'IDA KLEMM', HMult, w, 1907; flowers snow-white, immense clusters; long, strong stems; vigorous growth; [Crimson Rambler sport]; Walter, L.

'IDA MCCRACKEN', HT, op, 1952; flowers salmon and coral, well formed, dbl., 25–30 petals, 4 in., moderate fragrance; foliage leathery, dark; free growth; [Ethel Somerset X Mrs Sam McGredy]; Norman; Harkness

'IDA SCHOLTEN', HT, pb, 1933; flowers pink shaded carmine-red, very large, dbl.; foliage dark, leathery, glossy; bushy, dwarf, compact growth; [Capt. F.S. Harvey-Cant X Gen. MacArthur]; Buisman, G. A. H.

'IDEAL', Pol, mr, 1921; [Miss Edith Cavell sport]; Spek

Ideal Home *see* **'IDYLLE'**

Idée Fixe® *see* 'LENDEC'

Idole, HT, or

'IDUN', F, mp, 1969; flowers neyron rose, large, borne in clusters, dbl., 41 petals, cupped; foliage dark, glossy; bushy growth; [Schneewittchen X Fanal]; Lundstad

'IDYLLE', HT, pb, 1959; (Ideal Home); flowers carmine-pink, base white, well formed, dbl., 25–30 petals, 5 in., moderate fragrance; vigorous, upright growth; [Monte Carlo X Tonnerre]; Laperrière; EFR

'IGA 63', HT, pb, 1963; flowers pink and red; [Confidence X Seedling]; da Silva, Moreira

Iga 83 Munchen *see* 'MEIBALBIKA'®

Iga Erfurt, F, op, 1966; flowers medium, semi-dbl.; GPG Bad Langensalza; (Sangerhausen)

'IGLOO', HT, w, 1969; bud ovoid; flowers medium, dbl.; foliage dark; [Seedling X White Knight]; Verbeek

'IGNASI IGLESIAS', HT, mr, 1934; flowers rose in early season, oriental red in summer, dbl., exhibition form, intense fragrance; foliage wrinkled; vigorous, bushy growth; [Angel Guimera X (Souv.

de Claudius Pernet X Mme Butterfly)]; Dot, Pedro

'IGNIS', S, mr, 1934; flowers fiery red, very large, non-recurrent; foliage leathery, dark; vigorous (3 1/4-6 1/2 ft.) growth; Chotkové Rosarium; Böhm

'IGOR', pb, 1913; flowers carmine-white, medium, dbl., 36 petals, intense fragrance; foliage dark; spreading growth; Geltsendeger

'IKARUGA', HT, yb, 1975; bud ovoid; dbl., 35 petals, 6 in., exhibition form, slight fragrance; vigorous, upright growth; [McGredy's Ivory X Garden Party]; Ito

Ilam *see* 'SIMILAM'

'ILARIA', F, dp, 1962; flowers coral-red, well formed, semi-dbl.; foliage dark; [Cinnabar X Fashion]; Borgatti, G.; Sgaravatti

'ILDIKO', F, mr, 1971; bud ovoid; flowers cherry-red, medium, semi-dbl., cupped, profuse, intermittent bloom; moderate, upright growth; [Mardi Gras X Paprika]; Institute of Ornamental Plant Growing

Ile de France *see* **'ADORATION'**

'ILE DE FRANCE', HWich, rb, 1922; flowers bright scarlet, center white, open, borne in clusters, semi-dbl., slight fragrance; foliage large, leathery, dark; short, strong stems; very vigorous (15-20 ft) growth; [American Pillar Unnamed seedling]; Nonin

Ile de France *see* PEKsolred

'ILLINOIS', Gr, or, 1969; bud long, pointed; flowers large, dbl., 25 petals, moderate fragrance; foliage leathery; vigorous, bushy growth; [Soprano X Tropicana]; Morey, Dr. Dennison; Country Garden Nursery

Illisca® *see* LAPem

Illos®, HT, m

'ILLUMINATION', F, dy, 1970; flowers deep sulfur-yellow, semi-dbl., 12 petals, 3 in., moderate fragrance; foliage glossy, light; free growth; [Clare Grammerstorf X Happy Event]; Dickson, Patrick; A. Dickson

Illumination, F, ab, 1986; Pallek

'ILLUSION', HKor, mr, 1961; flowers blood-red to cinnabar, large blooms in large clusters, dbl., moderate fragrance; foliage leathery, glossy, light green; vigorous growth; Kordes, R.

'ILLUSION', HT, my, 1961; dbl., 50 petals, moderate fragrance; foliage glossy; vigorous growth; [Peace X Unnamed seedling]; Verbeek

Ilmenau, S, mr, 1992; Hetzel

Ilona *see* 'VARLON'

Ilona™ *see* 'MARILO'

'ILSE HABERLAND'®, S, mp, 1956; flowers crimson-pink, very large, dbl., exhibition form, moderate fragrance; foliage glossy; vigorous, upright, bushy growth; Kordes

'ILSE KROHN', LCl, w, 1957; flowers pure white, overlarge, very dbl., exhibition form, non-recurrent, slight fragrance; foliage glossy, leathery; very vigorous growth; [Golden Glow X R. kordesii]; Kordes

'ILSE KROHN SUPERIOR'®, HKor, w, 1964; flowers pure white, dbl., intense fragrance; foliage dark; vigorous (9 ft) growth; Kordes

Ilseta® *see* 'TANATESIL'

Ilsetta *see* 'TANATESIL'

Iluse, HT, ob, 1968; flowers salmon-orange and yellow, large, dbl., slight fragrance; Urban, J.; (Sangerhausen)

Image d'Epinal, LCl, mr, 1997; Croix

Imagination *see* 'WEKMAR'

Imagine *see* 'RENIMAG'

Imagine *see* DORmagi

'IMATRA', F, pb, 1930; flowers pink to white; [Orléans Rose seedling]; Poulsen, S.; Olsson

Imbroglio *see* BENimbro

Imma, HT, lp;, exhibition form

'IMMACULADA GALAN', F, w, 1968; flowers large, borne in trusses, dbl., 60–70 petals, globular; foliage small, blue-gray; very free growth; LeGrice

Immensee® *see* 'KORIMRO'

Immortal Juno *see* 'AUSJUNO'

'IMOGEN', F, lp; Bidwell, 1845; (Weatherly, L.)

Imogene™ *see* 'MINYCO'

'IMP', F, rb, 1971; bud globular; flowers red, reverse silver-pink, small, dbl., slight fragrance; foliage large, dark, leathery; bushy growth; [Daily Sketch X Impeccable]; Dawson, George; Brundrett

Impala *see* 'HERPIM'

Impala *see* 'RULIMPA'

Impatient *see* 'JACDEW'

'IMPECCABLE', HT, dr, 1955; flowers deep velvety red, well-shaped, dbl., intense fragrance; foliage dark; Delbard-Chabert

Imperator *see* 'MEIPERATOR'

'IMPÉRATRICE EUGÉNIE', M, m, 1856; flowers lilac-pink, medium, dbl., moderate fragrance; vigorous growth; Guillot Père

Imperatrice Josephine *see* **'EMPRESS JOSEPHINE'**

Impératrice Rouge *see* **'RED EMPRESS'**

'IMPERIAL', HT, rb, 1957; flowers cardinal-red, reverse golden yellow, well formed; [Geranium X Opera]; da Silva, Moreira

'IMPERIAL GOLD', HT, my, 1962; (Canadiana); bud ovoid; flowers lemon-yellow to indian yellow, dbl., 30–35 petals, 3.5–4.5 in., moderate fragrance; foliage leathery, dark, glossy; vigorous, upright growth; [Charlotte Armstrong X Girona]; Swim, H.C.; C.R. Burr

Imperial Palace *see* POUlchris

'IMPERIAL PINK', HT, mp, 1942; [Royal Beauty sport]; Coddington

'IMPERIAL POTENTATE', HT, mp, 1921; flowers large, dbl., 45 petals, exhibition form; foliage dark, leathery; vigorous growth; GM, Portland, 1921; [Ophelia X Hoosier Beauty]; Clarke Bros.

'IMPERIAL QUEEN', HT, mr, 1962; bud long, pointed; flowers cherry-red, dbl., 21 petals, 4.5–5 in., cupped, moderate fragrance; foliage leathery, glossy; vigorous, compact growth; [Queen Elizabeth X Chrysler Imperial]; Lammerts, Dr. Walter; C.R. Burr

Imperial Rose, B, dp

Imposant®, F, mp

'IMPRESS', HT, rb, 1929; bud ovoid, cardinal-red, shaded orange; flowers salmon-cerise, tinted golden, very large, dbl., 40–45 petals; foliage dark, glossy; vigorous growth; Dickson, A.; Dreer;, Liggit

'IMPROVED CÉCILE BRÜNNER', F, op, 1948; (Rosy Morn); bud long, pointed; flowers salmon-pink, medium, borne in clusters, dbl., 30 petals, exhibition form, slight fragrance; foliage leathery, dull green; very vigorous, upright growth; [Dainty Bess X R. gigantea]; Duehrsen; H&S

'IMPROVED LAFAYETTE', F, mr, 1935; semi-dbl., slight fragrance; foliage soft; vigorous, bushy growth; [E.G. Hill X Seedling]; H&S; Dreer

'IMPROVED PEACE', HT, yb, 1959; bud ovoid; flowers yellow edged and flushed pink, large, dbl., exhibition form, slight fragrance; foliage leathery, wrinkled; very vigorous, bushy growth; [Peace sport]; Dean

'IMPROVED PREMIER BAL', HT, pb; flowers pale cream edged deep pink, large, moderate fragrance; foliage leathery; vigorous growth; Wheatcroft Bros.

'IMPROVED PRINCE PHILIP', Gr, or, 1964; flowers well formed; [Queen Elizabeth X Prince Philip]; Leenders, J.

'IMPROVED UNIVERSAL FAVORITE', HWich, mp, 1901; flowers brilliant pink; Manda, W.A.

'IMPROVED VERDUN', Pol, dp, 1946; flowers vivid carmine-red, borne in clusters, dbl.; foliage leathery; bushy growth; Kluis; Klyn

'IMPULSE', Min, op, 1986; flowers salmon-pink, light yellow reverse, small, dbl., 38 petals, cupped, no fragrance; fruit not observed; long, brownish prickles; foliage small, medium green, semi-glossy; medium, upright growth; [Red Ace X Chris Jolly]; Jolly, Marie; Rosehill Farm

Impulse® *see* 'BARIMP'

In the Mood *see* 'SEAMOOD'

'IN THE PINK', F, mp, 1988; flowers medium, borne in sprays of 4-5, dbl., 60 petals, moderate, musk fragrance; prickles hooked, red; foliage greenish-red, glossy; medium, bushy growth; [Baby Faurax X Unnamed seedling]; Ryan, C.; Melville Nursery, 1988

Ina *see* 'TALINA'

Ina an' Mona®, S, ab, 1992; Jensen

'INANO', HT, dy, 1978; flowers urn-shaped, medium, borne singly, dbl., 40 petals, moderate, fruity fragrance; small, brown prickles; foliage medium, light green; medium, bushy growth; [Doreen X Goldilocks]; Teranishi, K.; Itami Rose Nursery

'INATA', F, op, 1967; flowers pink shaded salmon, open, borne in trusses, semi-dbl.; vigorous growth; [Valeta sport]; deRuiter

'INCA DE MALLORCA', F, mr, 1958; flowers strawberry-red, dbl., 20 petals; strong stems; compact growth; [Soller X Floradora]; Dot, Pedro

Incandescent, HT, lp; globular and short; flowers velvety bright red vermillion-red, dbl., slight; large, reddish dark green, glossy; [Bond Street X Dame de Coeur]; St. Wagner, 1989; Res. Stn. f. Horticulture, Cluj, 1991, Romania

'INCENSE', HT, dr, 1968; flowers deep red, pointed, dbl., intense fragrance; vigorous growth; [(Karl Herbst X New Yorker) X Konrad Adenauer]; LeGrice

Incense Rose *see* **R. PRIMULA**

'INCH'ALLAH', HT, pb, 1944; bud long; flowers bright pink, reverse flesh, stamens yellow, very large, semi-dbl.; vigorous growth; [Pres. Macia X Editor McFarland]; Meilland, F.

Incognito™ *see* 'BRIINCOG'

'INCOMPARABLE', HP, w, 1923; flowers rosy white, large; Giraud, A.

'INCREDIBLE', S, yb, 1984; bud ovoid, pointed; flowers yellow freckled and streaked with orange-red, urn-shaped, large, dbl., 28 petals, repeat bloom, moderate fragrance; awl-like, brown prickles; foliage medium large, leathery, dark olive green, copper tin; vigorous, erect growth; hardy; [Gingersnap X Sevilliana]; Buck, Dr. Griffith J.; Iowa State University

'INDÉFECTIBLE', Pol, mr, 1919; flowers bright clear red, semi-dbl., moderate fragrance; [Annchen Muller seedling]; Turbat

'INDEPENDENCE', F, or, 1951; (Geranium, Kordes' Sondermeldung, Reina Elisenda, Sondermeldung); bud urn-shaped; flowers pure scarlet, blooms in clusters(up to 10), dbl., 35 petals, 4.5 in., cupped, moderate fragrance; foliage glossy, dark; growth moderate; (28); GM, Bagatelle, 1943 GM, NRS, 1950 GM, Portland, 1953; [F2 seedling (Baby Chateau X Crimson Glory)]; Kordes; J&P

'INDEPENDENCE '76', F, mr, 1974; bud short, pointed; flowers high-pointed, dbl., 22 petals, 3–4 in., moderate, tea fragrance; vigorous, upright growth; [Cotillion X Suspense]; Byrum; J.H. Hill Co.

'INDEPENDENCE DAY', HT, ab, 1919; bud pointed; flowers sunflower-gold, stained flame-color and orange-apricot, dbl., exhibition form, intense fragrance; foliage leathery, glossy, dark; vigorous growth; GM, NRS, 1919; [Mme Edouard Herriot X Souv. de Gustave Prat]; Bees

'INDEPENDENCE DAY, CLIMBING', Cl HT, ab, 1930; Brown, W.&J.; E. Murrell

Independence du Luxembourg, HT, w, 1960; flowers creamy white with orange and pink tones, large, dbl., intense fragrance; Lens, Louis

'INDEPENDENCE, CLIMBING', Cl F, or, 1960; Balducci & Figli

'INDIAN CHIEF', HT, rb, 1967; flowers currant-red shaded orange, pointed, dbl., slight fragrance; foliage dark; very free growth; [Tropicana X ?]; Gregory

'INDIAN GODDESS', Cl Min, pb, 1999; semi-dbl., 12–16 petals, 2.5 in., borne singly and in small clusters, no fragrance; few prickles; foliage dark green, semi-glossy, disease-resistant; bushy, medium (4-5 ft) growth; [Crazy Dottie X seedling]; Sridharan, Dr. Lakshmi M.

'INDIAN GOLD', F, yb, 1961; bud ovoid, flushed red; flowers yellow flushed soft pink, borne in clusters, dbl., 30–45 petals, 3.5 in., exhibition form, moderate fragrance; foliage glossy, light green; short stems; upright, compact growth; [Goldilocks X Seedling]; Von Abrams; Peterson & Dering

'INDIAN MAID', HT, ob; bud long, pointed; flowers salmon, reverse bronze-yellow, dbl., exhibition form, intense fragrance; foliage glossy; vigorous growth; [Talisman seedling X Souv. de Claudius Pernet]; Padilla

Indian Meillandina *see* **'CAROL-JEAN'**

'INDIAN PINK', HT, pb, 1975; dbl., 30–36 petals, 5 in., globular; foliage leathery; very vigorous, upright, bushy growth; [Unknown seedling X Orange Tango]; McDaniel, Earl; Carlton Rose Nurseries

Indian Princess *see* 'PIXIPRIN'

Indian Princess *see* **'PRINCESS OF INDIA'**

'INDIAN RED', HT, rb, 1948; bud long, pointed; flowers red shaded deeper, large, dbl., exhibition form, moderate fragrance; foliage small, glossy; vigorous, bushy, upright growth; hardy.; [Pink Princess X Crimson Glory]; Brownell, H.C.

Indian Silk®, MinFl, w, 1991; Lens

Indian Song *see* 'MEIHIMPER'

'INDIAN SUMMER', Cl HT, ob, 1938; bud pointed; flowers orange, streaked red, large, dbl., 25 petals, intense fragrance; foliage dark bronze, leathery, glossy; very vigorous (to 12-18 ft) growth; [Ednah Thomas X Autumn]; Duehrsen; H&S

Indian Summer *see* HARWIGWAM

Indian Summer *see* PEAPERFUME

Indian Sunblaze *see* **'CAROL-JEAN'**

Indian Warrior, HT, dr

'INDIANA', HT, pb, 1907; flowers bright pink, faintly suffused orange, dbl.; bushy growth; [Rosalind X Frau Karl Druschki]; Hill, E.G., Co.

Indiana, F, mp

'INDIANAPOLIS', HT, mp, 1971; flowers deep yellow-pink, medium, very dbl., cupped, slight fragrance; foliage glossy, leathery; moderate, upright growth; [Coloranja X ?]; Schloen, J.; Ellesmere Nursery

'INDICA ALBA', HCh, lp, 1802; (White Daily Rose); flowers very light blush; [Old Blush sport]

Indica Major *see* **'FUN JWAN LO'**

Indica Purpurea, HCh, m; flowers purple/pink, medium, single; Chenault; (Sangerhausen)

'INDIGO', P, m, 1830

Indigoletta, LCl, m

'INDIRA', HT, pb, 1973; bud ovoid; flowers pink, reverse lighter, medium, slight fragrance; foliage soft; vigorous, upright, bushy growth; [Baccará X Prima Ballerina]; Hetzel; GAWA

'INDISPENSABLE', LCl, mp, 1947; bud globular; flowers pink, medium, borne in clusters, dbl., exhibition form, slight fragrance; foliage glossy; moderate, upright, pillar growth; [Roserie sport]; Klyn

'INDRA', LCl, dp, 1937; bud pointed; flowers rose-pink, open, borne in clusters, semi-dbl., slight fragrance; foliage glossy; long stems; very vigorous, climbing growth; [(Ophelia X R. multiflora) X Florex]; Tantau

Indraman, F, or, 1990; Chiplunkar

Indu Singhal, Gr, pb, 1995; Singhal

'INDY 500', Gr, or, 1976; bud tapered; flowers brilliant orange-red, medium-high, dbl., 32 petals, 4.5–5 in., flat, intense fragrance; foliage large, glossy, dark, reddish; vigorous, upright growth; [(Aztec X Queen Elizabeth seedling) X (Independence X Scarlet Knight seedling)]; Williams, J. Benjamin; Krider Nursery

Inermis Morletii *see* **'MORLETII'**

'INFANTANIA', F, w, 1953; flowers snow-white, sometimes tinged green, small, borne in large clusters, dbl., intense fragrance; [Baby Alberic X ?]; Heers; Langbecker

'INFANTE BEATRICE', HT, ob, 1930; flowers orange-yellow, tinted reddish gold, base golden, dbl., intense fragrance; vigorous growth; [Marie Adélaide X Unnamed seedling]; Guillot, M.

'INFANTE MARIA CRISTINA', HT, ob, 1930; bud pointed; flowers coppery, tinged carmine, moderate fragrance; foliage reddish bronze; vigorous growth; Gaujard

Inferno *see* 'AROKUNCE'

Infinity *see* 'RESFINI'

'INGAR OLSSON', F, mr, 1931; flowers brilliant cerise-red, blooms in clusters, semi-dbl., cupped, slight fragrance; foliage leathery; vigorous, rather compact growth; [Else Poulsen X Ophelia]; Poulsen, S.

'INGE HORSTMANN', HT, rb, 1964; bud long, red, reverse white tinged pink; flowers cherry-red, exhibition form, intense fragrance; long stems; vigorous, bushy growth; Tantau, Math.

'INGE PEIN', Pol, mr, 1939; flowers carmine-red, medium, semi-dbl.; Pein; (Sangerhausen)

Inge Schubert® *see* MAUlave

Ingrid Bergman® *see* 'POULMAN'

'INGRID STENZIG', Pol, dp, 1951; (Pink Triumph); flowers rose-pink, buttercup form, small, borne inlarge clusters; [Orange Triumph sport]; Hassefras Bros.; B&A

Ingrid Weibull *see* 'TANWEIEKE'

'INK SPOTS', HT, dr, 1985; flowers medium, dbl., 35 petals, slight fragrance; foliage large, dark, semi-glossy; upright, bushy, spreading growth; PP005855; [Unnamed seedling X Unnamed seedling]; Weeks, O.L.

Inka® *see* TANakni

Inner Glow *see* 'PIXINNER'

Inner Wheel *see* 'FRYJASSO'

'INNISFREE', F, yb, 1964; flowers yellow, orange and pink, blooms in clusters, dbl., 22 petals; vigorous, tall growth; [(Karl Herbst X Masquerade) X Circus]; Dickson, Patrick; A. Dickson

'INNOCENCE', HT, w, 1921; flowers stamens reddish, slightly waved, borne in clusters, semi-dbl., 12 petals, 5 in., moderate fragrance; foliage dark; vigorous growth; Chaplin Bros.

Innocence *see* 'SAVINN'

Innocence, HT, w; Croix

Innocence 96 *see* 'COCORAY'

'INNOCENCE, CLIMBING', Cl HT, w, 1938; Armstrong, J.A.; Armstrong Nursery

Innocencia® *see* 'KORENBON'

Innocent Blush *see* 'RENBLUSH'

Innovation, F, 1973; Combe, M.; (Cavriglia)

Innovation Minijet *see* 'MEIJETTE'

Innoxa Femille *see* 'HARPRINCELY'

'INOA', F, mr, 1958; flowers bright velvety red; vigorous growth; [Gruss an Teplitz X Pioupiou]; Arles; Roses-France

'INSEL MAINAU'®, F, dr, 1959; bud ovoid; flowers deep crimson, large blooms in clusters (up to 5), dbl., slight fragrance; foliage leathery, dark; low, compact growth; ADR, 1960; Kordes, R.

Insolite *see* 'MORCHARI'

'INSPECTEUR JAGOURT', F, m, 1932; bud glowing red; flowers purplish pink to china-rose, large white stamens, large, dbl., 25–30 petals; foliage glossy; vigorous growth; [Mrs Henry Winnett X Eblouissant]; Soupert & Notting

'INSPECTOR ROSE', HT, rb, 1969; flowers maroon-red, reverse yellow, long, pointed, dbl., 35 petals; foliage coppery bronze-red; very free growth; [Piccadilly sport]; Fryers Nursery, Ltd.

'INSPEKTOR BLOHM', HMsk, w, 1942; very dbl., large corymbs, recurrent bloom, intense fragrance; foliage abundant, gray-green; vigorous, well-branched growth; [Joanna Hill X Eva]; Kordes

'INSPIRATION', LCl, mp, 1946; flowers large, semi-dbl., moderate fragrance; foliage large, glossy; moderate growth; [New Dawn X Crimson Glory]; Jacobus; B&A

Inspiration, HT, rb, 1991; Perry

Inspiration *see* 'JACCOH'

'INSTITUTEUR SIRDEY', HT, dy, 1905; Pernet-Ducher

'INSULINDE', HT, yb, 1923; flowers clear yellow shaded golden yellow, dbl., slight fragrance; [Mr Joh. M. Jolles X Melody]; Van Rossem

'INSULINDE', HT, op, 1923; flowers pink and salmon, dbl., moderate fragrance; [Ophelia X Jonkheer J.L. Mock]; Leenders, M.

INTasprint, HT, dy; (**Sprinter**®); Ilsink, 1989

Integrity *see* 'HARVINTAGE'

Intel, Min, m, 1996; Spooner, Raymond A.

INTepin, F, mp, 1995; (**Pink Fire**®); Ilsink

'INTERALL', S, lp, 1979; (**Rosy Cushion**®); flowers small blooms in large clusters, single, 7–8 petals, repeat bloom, slight fragrance; many medium prickles; foliage dark, glossy; vigorous (3-4ft.) groundcover growth; [Yesterday X Unnamed seedling]; Ilsink

Interama *see* 'INTRUMA'

INTeramon, S, w; (**White Diamond**®); Interplant, 1994

'INTERANDER', S, op, 1983; (**Oleander Rose**); flowers salmon-pink, small blooms in large clusters, single, 7 petals, repeat bloom, slight fragrance; very few, small prickles; foliage medium, medium green, semi-glossy; upright growth; [Liverpool Echo X Unnamed seedling]; Interplant

INTerbab, F, lp; (**Yellow Dot**); Ilsink; Interplant, 1995

Interbec *see* **'ESPOIR'**

INTerbricor, S, mr, 1995; (**Hot Fire**); Interplant

'INTERBRONZI', Min, r, 1992; (**Suntan**™); flowers bronze, little fading, urn-shaped, blooms borne in sprays of, dbl., 30 petals, 1.5 in., slight fragrance; foliage small, dark green, glossy; upright, bushy, low (30 cms) growth; [Seedling X The Fairy]; Ilsink, G.P., 1983; Interplant B.V., 1989

'INTERCARP', S, dp, 1983; (**Rosy Carpet**®); flowers deep pink, blooms in clusters, single, 5 petals, repeat blooming, moderate fragrance; many, medium prickles; foliage medium, dark, glossy; spreading (to 4 ft.) growth; [Yesterday X Unnamed seedling]; Interplant, 1984

'INTERCEL', S, dp, 1979; ('INTERCELL', **Red Blanket**®); flowers dull, deep pink, small blooms in small clusters, semi-dbl., repeat bloom, slight fragrance; many medium prickles; foliage dark, glossy; vigorous (to 3-4 ft) groundcover growth; [Yesterday X Unnamed seedling]; Ilsink; Dickson Nurseries, Ltd.

'INTERCHER', S, mr, 1992; (**Chimo**®); flowers red aging dark red, blooms borne singly, single, 5 petals, 1.25 in., cupped, repeat bloom, slight fragrance; foliage medium, medium green, glossy; low (90 cms), spreading growth; [Seedling X Immensee]; Ilsink, Peter; & W. Kordes, 1989;, Interplant B.V., 1989

'INTERCHIMP', S, mp, 1992; (**Pink Chimo**®); single, 5 petals, 1–1.5 in., cupped, borne singly, repeat bloom, no fragrance; foliage medium, medium green, semi-glossy; spreading, low (30 cms) growth; [Unknown seedling X Immensee]; Ilsink, Peter; Interplant B.V., 1990

'INTERCLEM', S, yb, 1982; (**New Face**®); flowers yellow edged pink, small blooms in large clusters, single, 5 petals, repeat bloom, slight fragrance;

many prickles; foliage medium, medium green, semi-glossy; upright growth; Interplant, 1978

'INTERCREAM', F, w, 1986; (**Ivora**); flowers near white, blooms in clusters, dbl., 35 petals, slight fragrance; foliage medium, light green, matt; upright growth; [AmRUda X Unnamed seedling]; Interplant

INTerdain, F, lp, 1996; (**Fellini**); Interplant

INTerdays, HT, lp; (**Tropical Skies**); Ilsink; Interplant, 1997

'INTERDUST', S, lp, 1998; (Stardust, **Sun Runner**); flowers bright yellow, small, single, 4–11 petals, 1.5 in., borne in large clusters, self cleaning, slight fragrance; prickles moderate; foliage small, dark green, glossy; spreading (3 ft.), mounding (18 in), low, growth; PP10240; [seedling X seedling]; Ilsink/ Interplant; Bear Creek Gardens, 1992

'INTERETTE', S, lp, 1977; (Flavia, **Fleurette®**); single, 5 petals, 2 in., blooms in clusters, repeat bloom; few, medium prickles; foliage medium green, glossy; vigorous (to 4 ft) growth; groundcover; [Yesterday X Unnamed seedling]; Interplant

INTereup, S, ob, 1997; (**Euphoria**); Interplant

'INTERFAIR', S, m, 1982; (**Fair Play®**); flowers light violet, blooms in large clusters, semi-dbl., 18 petals, repeat bloom, slight fragrance; dark green prickles; foliage medium, dark, matt; groundcover; vigorous growth; [Yesterday X Seedling]; Interplant, 1977

'INTERFIRE', F, ob, 1992; (Darthuizer Orange Fire, **Orange Fire®**); flowers very bright orange, urn-shaped, blooms borne in sprays, dbl., 15–20 petals, 3.25 in., slight fragrance; foliage large, medium green, matt; upright growth; [Orange Wave X Unnamed seedling]; Ilsink, Peter; Interplant B.V., 1988;, Kordes, 1988;, Pekmez, 1987

Interflora *see* **'INTERVIEW'**

'INTERFLU', Min, w, 1984; (**Fluffy**, Patio Cloud); flowers white, shaded creamy pink, small blooms in clusters, semi-dbl., 15 petals, slight fragrance; few, medium prickles; foliage dark, glossy; groundcover; spreading growth; [Seedling X Nozomi]; Interplant

'INTERFOUR', Min, mp, 1982; (**Petit Four®**); flowers patio, blooms in clusters, semi-dbl., moderate fragrance; many, small prickles; foliage small, medium green, glossy; bushy growth; [Marlena seedling X Unnamed seedling]; Interplant

'INTERGANT', Min, w, 1983; (**Elegant Pearl®**); flowers creamy white, patio, blooms in large clusters, dbl., slight fragrance; few, medium prickles; foliage small, medium green, glossy; bushy growth; [Seedling X Nozomi]; Interplant

'INTERGEORGE', Min, mp, 1992; (Carmela®, **Georgeous®**); semi-dbl., 6–8 petals, 1.25 in., cupped, borne in sprays of 3-8, slight fragrance; foliage small, dark green, glossy; bushy, low (30-40 cms) growth; [Candy Rose X Eyeopener]; Ilsink, Peter; Interplant B.V., 1990

INTerglo, S, pb, 1987; (**Gloire des Anciens**); Interplant

INTergol, Min, dy, 1990; (**Golden Rosamini**); Ilsink

'INTERGRI', F, lp, 1993; (**Silver Dream®**); flowers silver pink, blooms borne in sprays, dbl., 15–25 petals, 1.5–2.75 in., slight fragrance; few prickles; foliage medium, medium green, semi-glossy; medium (60 cms), upright growth; Ilsink, G.P.; Interplant B.V., 1992

'INTERHEL', F, mr, 1979; (**Stadt den Helder**); flowers medium, dbl., 20 petals, slight fragrance; foliage large, dark, matt; upright growth; [Amsterdam X (Olala X Diablotin)]; Interplant

INTerhyro, S, mr; (**Borsalino®**); Interplant

INTerice, HT, op, 1996; (**Coral Ice**); Interplant

INTerim, S, mr, 1991; (**Red Trail**); Ilsink

'INTERJADA', F, dp, 1986; flowers deep pink, blooms in sprays, dbl., 35 petals, slight fragrance; foliage medium, dark, semi-glossy; upright growth; [Unnamed seedling X Unnamed seedling]; Interplant

'INTERLADA', F, lp, 1984; (**Lady of the Dawn®**); flowers large, ruffled, soft cream edged with pink, semi-dbl., borne in large clusters, moderate, fruity fragrance; foliage large, medium green, matte, leathery; upright growth, arching, 4 ft.; PP006068; [INTerdress X Stadt den Helder]; Interplant

'INTERLAV', S, m, 1984; (**Lavender Dream®**); flowers deep lilac pink, semi-dbl., 16 petals, blooms in clusters, repeat bloom, no fragrance; few medium prickles; foliage medium, light green, matte, narrow; bushy, arching canes, (5 ft) growth; PP5916; ADR, 1987; [Yesterday X Nastarana]; Interplant, 1985

'INTERLEER', F, ob, 1982; (**Leersum 700**); flowers light orange-yellow, blooms in clusters, semi-dbl., slight fragrance; many, small prickles; foliage medium, light green, matt; upright growth; [Lichtkonigin Lucia X Marlena]; Interplant, 1979

'INTERLIEN', F, lp, 1985; (**Evelien**); flowers , dbl., 35 petals, blooms in clusters, slight fragrance; foliage medium, medium green, semi-glossy; PP007368; [Seedling X Fresh Pink]; Interplant

'INTERLIGHT', F, ab, 1992; (**Day Light®**, Daylight); flowers apricot-yellow, apricot and light pink reverse, aging to champagne, dbl., 15–25 petals, 3.25 in., no fragrance; foliage medium, dark green, matt; upright, medium (+/- 48 cms) growth; [Seedling X New Year]; Ilsink, G.P., 1985; Interplant B.V., 1991

INTerlis, F, lp, 1995; (**Lydia**); Ilsink

'INTERMEZZO', HT, m, 1963; bud ovoid; flowers deep lavender, medium, dbl., 25 petals, 3.5 in., moderate fragrance; foliage dark, glossy; moderately tall, compact growth; [Grey Pearl X Lila Vidri]; Dot, Simon; McGredy;, Minier

'INTERMINER', Min, lp, 1982; (**Minerette**); semi-dbl., blooms in clusters, slight fragrance; many small prickles; foliage small, medium green, glossy; bushy growth; GM, Baden-Baden, 1984; [Marlena seedling X Unnamed seedling]; Interplant

'INTERMOTO', F, dp, 1985; (**Joy®**); flowers deep pink, small blooms in clusters, dbl., 35 petals, no fragrance; foliage medium, medium green, semi-glossy; upright growth; [AmRUda X Unnamed seedling]; Interplant

INTermug, S, my, 1996; (**Butterflies**); Ilsink

INTermunder, S, mr, 1989; (**Dart's Red Dot®**); Interplant

International Herald Tribune® *see* 'HARQUANTUM'

'INTERNIKI', F, mr, 1993; (**Nikita®**); dbl., 15–25 petals, 1.5 in., borne in sprays, slight fragrance; some prickles; foliage medium, medium green, glossy; medium (45 cms), bushy growth; Ilsink, G.P.; Interplant B.V., 1988

'INTERONLY', HT, dr, 1986; (**Only Love**); flowers large, dbl., 35 petals, no fragrance; foliage medium, dark, semi-glossy; upright growth; PP006698; [Unnamed seedling X Caramba]; Interplant

'INTEROP', S, mr, 1987; (Erica, Eye Opener®, **Eyeopener**, Tapis Rouge); flowers small, semi-dbl., no fragrance; foliage medium, medium green, glossy; spreading growth; [(Unnamed seedling X Eyepaint) X (Unnamed seedling X Dortmund)]; Interplant

'INTERORGE', F, w, 1983; (**Georgette**); flowers large, dbl., 35 petals, slight fragrance; foliage large, medium green, semi-glossy; upright growth; [Unnamed seedling X Bordure Rose]; Interplant

'INTERPEEL', S, dr, 1992; (Ardennes, **Eye Appeal®**); semi-dbl., 8 petals, 2 in., cupped, borne singly, repeat bloom, no fragrance; foliage medium, medium green, glossy; spreading, medium (40-

60 cms) growth; [Eyeopener X Unknown]; Ilsink, Peter; Interplant B.V., 1991

'INTERPINK', S, mp, 1978; (**Pink Star**®); flowers medium, semi-dbl., slight fragrance; foliage medium, light, semi-glossy; spreading growth; [Yesterday X Seedling]; Interplant, 1977

'INTERPOOL', HT, rb, 1960; (**Verdi**); bud ovoid; flowers brick-red, reverse veined darker, medium, dbl., moderate fragrance; foliage dark, glossy; very vigorous, bushy growth; [Cafougnette X Independence]; Dorieux; Pin

'INTERPRINCE', F, w, 1993; (**Princess**®); dbl., 26–40 petals, 1.5–2.75 in., borne in large clusters, slight fragrance; some prickles; foliage medium, dark green, glossy; tall to medium (60 cms), upright growth; [Pink Delight X Seedling]; Ilsink, G.P.; Interplant B.V., 1989

'INTERPUR', F, m, 1993; (**Purple Prince**®); flowers dark mauve to light purple, blooms borne in large clusters, dbl., 26–40 petals, 1.5–2.75 in. slight fragrance; some prickles; foliage medium, dark green, semi-glossy; medium (55 cms), upright growth; Ilsink, G.P.; Interplant B.V., 1991

INTerreflet, S, mr; (**Glowing Cushion**); Ilsink, 1995; Interplant, 1996

'INTERRO', Min, or, 1984; (**Red Jewel**); flowers small, dbl., 20 petals, no fragrance; foliage small, medium green, semi-glossy; bushy growth; PP005769; [Amanda X Unnamed seedling]; Interplant

'INTERROB', MinFl, rb, 1983; (**Robin Red Breast**, Robin Redbreast); flowers dark red, white eye, reverse silver, small blooms in clusters, single, no fragrance; many, medium prickles; foliage small, medium green, glossy; bushy growth; [Unnamed seedling X Eyepaint]; Interplant, 1984

'INTERSINA', HT, lp, 1984; flowers large, dbl., 35 petals, no fragrance; foliage large, medium green, semi-glossy; upright growth; PP005836; [Seedling X Red Success]; Interplant

'INTERSIREE', F, lp, 1993; (**Swing**®) flowers cream pink, blooms borne in sprays, dbl., 15–25 petals, 1.5–2.75 in., no fragrance; few prickles; foliage medium, dark green, glossy; medium (60 cms), upright growth; Ilsink, G.P.; Interplant B.V., 1990

'INTERSMART', S, lp, 1979; (**Smarty**®); single, 7 petals, 2 in., blooms in clusters, repeat bloom; many, small prickles; foliage bright green, matt; groundcover; vigorous growth; [Yesterday X Unnamed seedling]; Ilsink; Dickson Nurseries, Ltd.

INTerstreep, S, pb, 1997; (**Arabesque**); Interplant

'INTERSUM', S, lp, 1981; (**Summerrose**); single, 6–8 petals, 2 in., blooms in large clusters, moderate fragrance; many, long prickles; foliage medium, medium green, matt; groundcover; spreading growth; [Yesterday X Unnamed seedling]; Interplant

'INTERTOR', S, mp, 1987; (**Pink Torch**®, Torche Rose®); flowers small blooms in large pyramidal sprays, single, slight fragrance; foliage medium, medium green, glossy; upright (to 4 ft) growth; [(Mozart X Seedling) X (Seedling X Eyepaint)]; Interplant

'INTERTRA', F, m, 1993; (**Chess**®); flowers deep purple, blooms borne in sprays, dbl., 26–40 petals, 1.5–2.75 in., no fragrance; few prickles; foliage medium, dark green, glossy; medium (55 cms), upright growth; Ilsink, G.P.; Interplant B.V., 1991

INTertropa, HT, lp; (**Temptation**); Interplant, 1995

'INTERTWIK', Min, ly, 1992; (**Twinkle**®); flowers light yellow, does not fade, urn-shaped, blooms borne singly, dbl., 10 petals, 2 in., slight fragrance; foliage medium, medium green, matt; upright, low (40 cms) growth; [McShane X Unknown seedling]; Ilsink, Peter; Interplant B.V., 1989

'INTERTYN', F, op, 1993; (**Sentyna**®); flowers light salmon pink, blooms borne in sprays, dbl., 15–25 petals, 1.5–2.75 in., slight fragrance; few prickles; foliage medium, medium green, glossy; tall (80 cms), upright growth; Ilsink, G.P.; Interplant B.V., 1990

'INTERVAL', HT, w, 1983; (**Porcelina**); flowers cream with pink tint, dbl., 30 petals, borne singly or in small clusters, moderate fragrance; few prickles; foliage large, dark, glossy; upright growth; PP005648; [Seedling X Golden Times]; Interplant

INTervema, HT, mp, 1995; (**Orlando**); Ilsink

'INTERVIEW', HT, dp, 1968; (Interflora); flowers deep pink, large, dbl., 40 petals, exhibition form, slight fragrance; foliage leathery; vigorous, upright growth; [((Baccará X White Knight) X (Baccará X Jolie Madame)) X (Baccará X Paris-Match)]; Meilland

'INTERVILLES'®, LCl, mr, 1968; bud ovoid; flowers large, semi-dbl., cupped, moderate fragrance; foliage dark, glossy; vigorous, climbing growth; [Etendard X Unnamed seedling]; Robichon; Ilgenfritz Nursery

'INTERWAY', HT, w, 1992; (**Milky Way**); flowers white with hint of pink on reverse/edge, urn-shaped, dbl., 20 petals, 4 in., slight fragrance; foliage large, dark green, glossy; upright, medium (100-120 cms) growth; [Esmeralda X True Love]; Ilsink, Peter; Interplant B.V., 1991

INTerwell, S, ly; (**Yellow Fleurette**®); flowers yellow with yellow stamens, semi-dbl., borne in sprays, no fragrance; foliage medium, dark green, glossy; strong, bushy (2.5 ft) growth; Interplant, 1992

'INTIMITÉ', HT, ob, 1956; bud long; flowers golden orange shaded yellow and chamois, open, large, dbl., slight fragrance; foliage glossy, dark; vigorous, bushy growth; [Beauté X Unnamed seedling]; Delforge

Intrepid™ *see* 'PERPID'

Intrepide, LCl, or, 1972; Combe

Intrigue *see* 'JACUM'

Intrigue *see* 'KORLECH'

'INTRUMA', F, dr, 1976; (**Interama**); semi-dbl., 18 petals, 3 in.; foliage large, glossy, dark; bushy growth; [Kohima X (Europeana X Kimona)]; deRuiter

Invention, HT, dp, 1988; Patil, B.K.

Invertilis, HSpn, mr; flowers carmine-red, medium-large, single, moderate fragrance; (Sangerhausen)

Invincible *see* 'RUNTRU'

'INVITATION', HT, ab, 1961; bud long, pointed; flowers rich salmon-pink, base yellow, dbl., 30 petals, 4.5 in., exhibition form, intense, spicy fragrance; foliage leathery, glossy; vigorous, compact, bushy growth; [Charlotte Armstrong X Signora]; Swim & Weeks; C-P

'IOBELLE', HT, pb, 1962; bud ovoid; flowers ivory-white edged and overspread deep pink, large, exhibition form, moderate, fruity fragrance; foliage dark, glossy; vigorous, upright, compact growth; [Dean Collins X Peace]; Buck, Dr. Griffith J.; Iowa State University

'IODE', F, dr, 1974; bud ovoid; flowers large, semi-dbl., cupped, slight fragrance; foliage glossy; vigorous, upright growth; [Lichterloh X Red Pinocchio]; Schloen, J.

'IOLANTHE', HT, rb, 1940; flowers bright red, reverse yellow, large, semi-dbl.; vigorous growth; Gaujard

'ION PHILLIPS', HT, dy, 1934; flowers rich yellow, large, dbl.; vigorous growth; Dickson, A.

'IONA HERDMAN', HT, dy, 1914; flowers brilliant yellow, dbl., slight fragrance; GM, NRS, 1913; McGredy

'IONE', Min, ly, 1989; bud ovoid; flowers white with pale yellow center, medium, borne singly and in small clusters, very dbl., 50 petals, exhibition form, slight fragrance; fruit not observed; prickles very few, very small, light green; foliage medium, medium green, semi-glossy;

bushy, medium growth; [Unnamed seedling X Unnamed seedling]; Jerabek, Paul E., 1990

Ipitombi *see* JACorb

'IPSILANTÉ', HGal, m, 1821; flowers light lilac pink, very large, dbl., quartered; vigorous growth

Iranja® *see* 'LENIRA'

'IRELAND HAMPTON', HT, pb, 1934; flowers flame-pink suffused gold, base gold, large, dbl., cupped, moderate, spicy fragrance; foliage glossy; vigorous, compact growth; [Étoile de Feu X Unnamed seedling]; Hillock

'IRELAND HAMPTON, CLIMBING', Cl HT, pb, 1936; Hillock

'IRENE', F, pb, 1941; flowers rose-white, reverse pure white, semi-dbl., moderate fragrance; [Seedling X Permanent Wave]; Leenders, M.

Irene *see* MEIridol

Irene au Danmark *see* **'IRENE OF DENMARK'**

'IRÈNE BONNET', Cl HT, mp, 1920; flowers hermosa pink, dbl., moderate fragrance; Nabonnand, C.

'IRENE CHURRUCA', HT, ly, 1934; ('GOLDEN MELODY'); bud pointed, yellow; flowers light buff, fading cream, well-formed, large, intense fragrance; [Mme Butterfly X (Lady Hillingdon X Souv. de Claudius Pernet)]; La Florida

'IRENE CURIE', HT, mr, 1952; bud very long, pointed; flowers scarlet, large, dbl., 20 petals; foliage glossy; very vigorous, bushy growth; [Unnamed seedling X Lawrence Johnston (HT)]; San Remo Exp. Sta.

Irene Jane, HWich, m, 1995; Nobbs

'IRENE OF DENMARK', F, w, 1948; (Irene au Danmark, Irene von Danemark); bud pointed; dbl., 40 petals, 3 in., cupped, moderate fragrance; foliage dark; vigorous, upright, bushy growth; (21); [Orléans Rose X (Mme Plantier X Edina)]; Poulsen, S.; C-P, 1950

Irene of Denmark, Climbing, Cl F, lp; [Irene of Denmark sport]; Ruston, D., 1970; (Weatherly, L.)

Irene Smith *see* 'BOSREXEYE'

'IRENE THOMPSON', HT, yb, 1921; flowers deep ruddy gold shaded bronze or coppery, dbl., intense fragrance; GM, NRS, 1919; McGredy

Irene Virag *see* 'ZIPVIR'

Irene von Danemark *see* **'IRENE OF DENMARK'**

'IRENE WATTS', Ch, w, 1896; bud soft apricot-orange; flowers with a button eye, dbl.; Guillot, P.

'IRENE'S BEAUTY', Min, yb, 1999; flowers very colorful, yellow blend, reverse med. yellow, aging to orange-red, dbl., 17–25 petals, 1.5 in., borne in small clusters, no fragrance; few prickles; foliage medium, dark green, semi-glossy; compact, low (1 ft) growth; [Little Darling X Kristin]; Jolly, Betty J.; Langenbach, 1998

'IRENE'S CHOICE', HT, mp, 1978; flowers azalea-pink, full, dbl., 35–40 petals, 4–5 in., slight fragrance; very free growth; [Karl Herbst X Blue Moon]; Ellick; Excelsior Roses

'IRENE'S DELIGHT', HT, lp, 1982; flowers large, dbl., exhibition form, intense fragrance; foliage medium, dark, semi-glossy; upright growth; [Admiral Rodney X Red Lion]; Varney, E.

Irene's Surprise *see* 'KIRBEE'

'IRINA', F, dr, 1969; flowers dark crimson-red, large, semi-dbl.; foliage soft, glossy; moderate, bushy growth; Grabczewski

'IRIS', HSpn, w; dbl.; shining, black fruit; foliage finely divided; dense, shrubby (3-4 ft) growth

Iris *see* FErecha®

'IRIS FOSTER', HT, mp, 1998; flowers full, medium pink, very large, very dbl., 26–40 petals, 6 in., exhibition form, borne mostly singly, slight fragrance; some prickles; foliage large, medium green, dull; vigorous, upright, bushy, floriferous growth; [Gavotte X (Solitaire X Mischief)]; Poole, Lionel

Iris Gee, F, lp, 1987; Gee

'IRIS PATRICIA GREEN', HT, mr, 1928; bud pointed; flowers cherry-red, moderate fragrance; foliage dark; Pemberton

'IRIS SQUIRE', F, mp, 1966; flowers soft rose, 4–5 in.; foliage dull; tall, vigorous growth; [Unnamed seedling X Queen Elizabeth]; Bees

Iris Webb *see* 'CHEWELL'

Irischer Regen *see* **'IRISH MIST'**

'IRISH AFTERGLOW', HT, ob, 1918; flowers very deep tangerine, passing to crushed strawberry; [Irish Fireflame sport]; Dickson, A.

Irish Beauty *see* 'MACEL'

'IRISH BRIGHTNESS', HT, mr, 1904; flowers medium, single; Dickson, A.; (Sangerhausen)

'IRISH CHARITY', HT, rb, 1927; bud intense fiery scarlet with golden sheen; flowers rosy scarlet, dbl.; McGredy; Dreer;, H&S

'IRISH CHARM', HT, ab, 1927; bud pointed; flowers base golden apricot passing to blush-pink, dbl., exhibition form, moderate fragrance; foliage dark, leathery; vigorous growth; McGredy; H&S

'IRISH COURAGE', HT, op, 1927; bud pointed; flowers soft shrimp-pink to salmon, dbl., exhibition form, moderate fragrance; foliage rich green, leathery, glossy; vigorous growth; McGredy

Irish Creme™ *see* 'PERCREME'

'IRISH ELEGANCE', HT, ob, 1905; flowers bronze orange-scarlet, large, single, 5 petals; vigorous growth; (21); Dickson, A.

'IRISH ENGINEER', HT, mr, 1904; flowers dazzling scarlet, large, single; Dickson, A.

'IRISH EYES', F, yb, 1974; bud ovoid; flowers yellow edged red, dbl., 45 petals, 2.5–3 in., exhibition form, slight fragrance; vigorous growth; [Unnamed seedling X Gemini]; Byrum; J.H. Hill Co.

Irish Eyes™ *see* 'DICWITNESS'

'IRISH FIREFLAME', HT, ob, 1914; flowers orange to old-gold, veined crimson, anthers light fawn, single, 5 petals, 5 in., intense fragrance; foliage dark, glossy; compact, bushy growth; GM, NRS, 1912; Dickson, A.

'IRISH FIREFLAME, CLIMBING', Cl HT, ob, 1916; Dickson, A.

'IRISH GLORY', HT, pb, 1900; flowers silvery pink, reverse crimson, large, semi-dbl., 10 petals, moderate fragrance; very vigorous growth; Dickson, A.

'IRISH GOLD', HT, my, 1966; (Grandpa Dickson); bud ovoid; dbl., 33 petals, 7 in., exhibition form, moderate fragrance; foliage dark, glossy, leathery; vigorous, upright, bushy growth; GM, Belfast, 1968 GM, Portland, 1970 GM, RNRS, 1965 GM, The Hague, 1966 PIT, RNRS, 1965; [(Kordes' Perfecta X Governador Braga da Cruz) X Piccadilly]; Dickson, A.; J&P

Irish Gold, Climbing, Cl HT, 1973; Humphreys; (Cavriglia)

Irish Heartbreaker *see* 'SEAHEART'

'IRISH HOPE', HT, dr, 1927; bud pointed; flowers rosy crimson shaded maroon, large, dbl., exhibition form, intense fragrance; foliage dark, leathery; vigorous growth; McGredy; Dreer;, H&S

Irish Hope *see* HARexclaim

'IRISH LADY', Min, pb, 1991; flowers small, dbl., 26–40 petals, slight fragrance; foliage small, dark green, glossy; bushy growth; [Kathy Robinson sport]; Schmidt, Richard; Michigan Mini Roses, 1991

'IRISH MIST', F, op, 1966; (Irischer Regen, Irish Summer®); flowers orange-salmon, well-formed, blooms in clusters, 4.5 in., slight fragrance; foliage dark; dense growth; [Orangeade X Mischief]; McGredy, Sam IV; McGredy

'IRISH MODESTY', HT, op, 1900; flowers light orange-pink, large, single; Dickson, A.; (Sangerhausen)

'IRISH MORN', HT, pb, 1927; flowers pink, center coral, dbl., moderate fragrance; McGredy; Dreer;, H&S

Irish Rich Marbled, HSpn, rb

'IRISH ROVER', HT, op, 1970; flowers salmon-pink, dbl., 36 petals, 4 in., slight fragrance; foliage coppery, dark; vigorous growth; [Violet Carson X Tropicana]; McGredy, Sam IV

Irish Squire, F, dp, 1966; flowers carmine-pink, large, dbl.; Bees of Chester; (Sangerhausen)

Irish Summer® *see* **'IRISH MIST'**

'IRISH SWEETNESS', HT, rb, 1927; bud pointed; flowers crimson suffused scarlet, large, dbl., exhibition form, intense fragrance; foliage dark, leathery; vigorous growth; McGredy; Dreer;, H&S

Irish Wonder *see* 'MACEV'

Irmela, F, W, 1983; Hetzel, K.; (Cavriglia)

'IROQUOIS', S, dp, 1932; flowers deep mauve-pink, semi-dbl., non-recurrent, moderate fragrance; foliage leathery; vigorous, bushy, compact growth; [Pythagoras X R. cinnamomea]; Central Exp. Farm

Irresistible™ *see* 'TINRESIST'

Irwin Lane, B, dr

'ISA', HT, lp, 1931; flowers light pinkish cream, well shaped, dbl., moderate fragrance; vigorous growth; [Abol seedling]; Evans, F.&L.

'ISA MURDOCK', HSpn, w, 1953; flowers white, sometimes tinged with pink, dbl., non-recurrent; many prickles; foliage spinosissima type; height 3 ft; [R. spinosissima altaica X Dbl. white spinosissima]; Skinner

'ISABEL DE ORTIZ', HT, pb, 1962; (Isabel Ortiz); flowers deep pink, reverse silvery, well-formed, dbl., 38 petals, 5 in., moderate fragrance; foliage dark, glossy; vigorous, upright growth; GM, Madrid, 1961 GM, NRS, 1962; [Peace X Kordes' Perfecta]; Kordes, R.; J&P, 1965

'ISABEL LLORACH', HP, yb, 1929; flowers nankeen yellow, tinted red, semi-dbl., moderate fragrance; [Frau Karl Druschki X Benedicte Seguin]; Dot, Pedro

Isabel Ortiz *see* **'ISABEL DE ORTIZ'**

Isabel Renaissance *see* POUlisab

'ISABELLA', HT, ob, 1964; flowers orange, star shaped; [Queen Elizabeth X Pink Lustre]; Leenders, J.

Isabella *see* POUlisab

'ISABELLA', HGal, dp, 1834

'ISABELLA GRAY', N, dy; flowers golden yellow, more fragrant, but otherwise similar to parent; [Chromatella seedling]; Gray, Andrew, 1855

'ISABELLA SKINNER', S, mp; flowers pink, well-formed, dbl., blooms on new wood all summer; bushy growth; [(R. laxa X Tea) X Floribunda]

'ISABELLA SPRUNT', T, my, 1855; flowers sulfur-yellow; [Safrano sport]; Sprunt; Buchanan, 1865

Isabella Sprunt Single Seedling, HMsk, 1980; Worl

'ISABELLE DE FRANCE', HT, or, 1956; bud pointed; flowers vermilion, large, dbl., exhibition form, slight fragrance; vigorous, upright growth; [Peace X (Mme Joseph Perraud X Opera)]; Mallerin, C.; Hémeray-Aubert

'ISABELLE MILNER', HT, ly, 1908; flowers large, dbl., moderate fragrance; Paul, W.; (Sangerhausen)

'ISABELLE NABONNAND', T, 1875; Nabonnand, G.; (Cavriglia)

Isabelle Renaissance *see* POUlisab

Isabel's Jewel *see* 'RENIJEWEL'

'ISELLA', Min, lp, 1973; flowers small, borne singly and several together, dbl., 22 petals, cupped, no fragrance; pale pink prickles; foliage small, green; vigorous, compact, upright growth; [(Baccará X Generosa) X Miss Italia]; Bartolomeo, Embriaco

Isfahan *see* **'ISPAHAN'**

'ISIS', F, w, 1973; flowers ivory-white, dbl., 40–45 petals, 4–5 in., moderate fragrance; compact growth; [Vera Dalton X Shepherdess]; Mattock

Iskra *see* 'MEIHATI'

Iskra 82, LCl, mr

Island of Dreams, F, dy, 1995; Spek

Isle of Man *see* **'MANX QUEEN'**

Isle of Roses *see* 'MORISLE'

'ISOBEL', HT, pb, 1916; bud pointed; flowers light rose-pink, shaded apricot, large, single, 5 petals, cupped, slight fragrance; foliage rich green, soft; GM, NRS, 1915; McGredy

Isobel Champion *see* 'DELGEOT'

Isobel Derby *see* 'HORETHEL'

'ISOBEL HARKNESS', HT, dy, 1957; flowers bright yellow, dbl., 32 petals, 6 in., moderate fragrance; foliage dark, leathery, semi-glossy; vigorous, upright, bushy growth; [McGredy's Yellow X Phyllis Gold]; Norman; Armstrong Nursery;, Harkness

Isolde *see* 'JUDSOLDE'

'ISPAHAN', D, mp; (Isfahan, Pompon des Princes); flowers bright pink, loosely, dbl., blooms over long season, intense fragrance; foliage small

Istropoliteana, S, 1990; Chorvath, F.; (Cavriglia)

Ita Buttrose, HT, op, 1984; Armstrong

'ITALIA', HT, dr, 1959; bud ovoid; flowers cardinal-red and cherry-red, to open, large, dbl., cupped, moderate fragrance; foliage leathery, dark; long, strong stems; bushy, upright growth; [Baccará X Poinsettia]; Valentino

'ITALIAN PINK', S, mp, 1959; bud short, pointed; flowers begonia-pink, open, large, borne in clusters, dbl., moderate fragrance; foliage dark; vigorous, upright, well branched growth; [Cocorico X Yellow Holstein]; Leenders, J.

'ITALIE IMPÉRIALE', HT, m, 1936; flowers purplish garnet-red, large; vigorous growth; Capiago

'ITALIENISCHES DOERFCHEN', F, or, 1967; bud ovoid; flowers open, small, borne in clusters, single, slight fragrance; foliage small, leathery; vigorous, bushy, low growth; [Highlight X ?]; Haenchen, E.; Teschendorff

It's ShowTime™ *see* 'MORJOYBON'

'IVAN MISSON', Pol, lp, 1922; flowers small, dbl.; Soupert & Notting; (Sangerhausen)

'IVANHOE', HT, rb, 1928; bud pointed; flowers brilliant scarlet to rich crimson, large, dbl., exhibition form, intense fragrance; foliage glossy; vigorous growth; Easlea

'IVANY', F, rb, 1975; flowers orange-yellow, shaded pink on petal edges, aging red, dbl., 50 petals, cupped, borne in clusters of 5-30; foliage dark, glossy; vigorous growth; [Masquerade X Rumba]; Staikov, Prof. Dr. V.; Kalaydjiev and Chorbadjiiski

'IVEMAX', HT, my, 1999; (**Delta Dawn**); dbl., 26–40 petals, 5 in., borne mostly singly, intense fragrance; few prickles; foliage medium, medium green, semi-glossy; upright, medium (5 ft) growth; [Out of Africa sport]; Iverson, Halvor

Ivora *see* 'INTERCREAM'

'IVORY', T, w, 1901; flowers ivory-white, large, dbl., moderate fragrance; vigorous growth; [Golden Gate sport]; Dingee & Conard

'IVORY ANN', S, w, 1993; flowers creamy white, blooms borne in small clusters, single, 5 petals, 3–3.5 in., moderate fragrance; foliage large, medium green, glossy; medium, bushy, spreading growth; [Ivory Fashion X ((Valerie Jeanne X Eyepaint) X Twilight Trail)]; Jobson, Daniel J.; Jobson, 1993

Ivory Beauty *see* KORivo

Ivory Buccaneer, HT, w

Ivory Carpet™ *see* 'WILIVORY'

'IVORY CHARM', Cl F, w, 1968; bud ovoid; flowers open, large, semi-dbl., intense fragrance; foliage glossy, leathery; vigorous, climbing growth; [Ivory Fashion sport]; Earing, F.E.

'IVORY FASHION', F, w, 1958; bud ovoid; flowers ivory-white, well-formed, blooms in clusters, semi-dbl., 17 petals, 4–4.5 in., moderate fragrance; foliage leathery; vigorous, upright growth;

AARS, 1959; [Sonata X Fashion]; Boerner; J&P

'IVORY FASHION, CLIMBING', Cl F, w, 1964; Williams, J. Benjamin

Ivory Festival™ *see* 'LAVYIP'

'IVORY FLUSH', F, pb, 1996; flowers orange centered with pink blended edges, fading to pale pink, very dbl., 3.5 in., slight fragrance; some prickles; foliage medium, light green, dull; bushy, medium (26 in patio) growth; Bees of Chester; L W Van Geest Farms, Ltd., 1995

Ivory Grand *see* POUlmount

Ivory Palace *see* 'MORIVORY'

'IVORY QUEEN', HT, w, 1954; bud ovoid; flowers ivory-cream, 5–6 in., moderate fragrance; foliage dark, glossy; vigorous, bushy growth; [Edina X McGredy's Ivory]; Fletcher; Tucker

'IVORY QUEEN', Gr, w, 1965; flowers ivory; [Queen Elizabeth sport]; Delforge

'IVORY SPLENDOR', Min, w, 1991; bud rounded; flowers white, pale yellow center, white reverse, aging white, large, dbl., 35 petals, exhibition form, slight, fruity fragrance; foliage medium, medium green, matt; upright, tall growth; [Rise 'n' Shine X Unnamed seedling]; Gruenbauer, Richard, 1984; Flowers 'n' Friends Miniature Roses, 1993

'IVORY TIP TOP', F, lp, 1976; flowers ivory-pink, full, semi-dbl., 12–16 petals, 2–2.5 in., slight fragrance; low, compact, bushy growth; [Tip Top sport]; Fryers Nursery, Ltd.

'IVORY TOWER', HT, w, 1979; bud very long, pointed; flowers ivory-white, shaded light pink and light yellow, dbl., 35 petals, 5.5 in., exhibition form, moderate fragrance; upright, bushy growth; [Colour Wonder X King's Ransom]; Kordes, R.; Armstrong Nursery

'IVORY TRIUMPH', F, w, 1961; bud pointed; flowers ivory, open, borne in clusters, semi-dbl., 12 petals, 3–4 in., slight fragrance; foliage leathery, light green; upright, compact growth; [Goldilocks X Unnamed seedling]; Von Abrams; Peterson & Dering

'IVRESSE', HT, rb, 1958; flowers clear red, reverse silvery, large; vigorous growth; [Peace X Spectacular]; Combe; Japan Rose Society

'IVY ALICE', HWich, op, 1927; flowers soft pink to blush-salmon, splashed carmine when fading, dbl., cupped, borne in very large clusters, slight fragrance; foliage glossy, light; very vigorous, climbing (6 ft) growth; [Excelsa sport]; Letts

'IVY EVANS', HT, mr, 1926; flowers light cerise, moderate fragrance; [George C. Waud seedling X Gen. MacArthur]; Evans

'IVY MAY', HT, pb, 1925; bud pointed; flowers rose-pink, base and edges amber, dbl., moderate fragrance; foliage dark; vigorous growth; [Mme Butterfly sport]; Beckwith

'IWARA', Misc OGR, w; (R. X iwara, R. yesoensis); flowers small, single; (14); [R. multiflora X R. rugosa]

'IZUMI', HT, m, 1999; flowers lavender blended pale pink, dbl., 40 petals, 5–5.5 in., exhibition form, slight fragrance; foliage bronze dark green; 5 ft growth; [(Intermezzo X Soir d'Automne) X (Intermezzo X Soir d'Automne)]; Ohkawara, Kiyoshi, 1994; Komaba Rose Nursery, 1997

'IZY', HT, pb, 1997; flowers full, large, white stripes frequently appear, very dbl., 26–50 petals, 5–6 in., borne mostly singly, intense fragrance; some prickles; foliage long, narrow, medium, medium green semi-glossy; upright, medium to tall growth; [Sheer Bliss sport]; Ballin, Don & Paula

J

'J. A. Gomis', HT, rb, 1933; flowers crimson-red and yellow, medium, dbl., cupped, slight fragrance; foliage dark, glossy; upright growth; [Sensation X Souv. de Claudius Pernet]; Camprubi, C.

'J. B. Clark', HP, dr, 1905; flowers deep scarlet, shaded blackish crimson, large, dbl., 25 petals, exhibition form, moderate fragrance; very prickly; vigorous (8-10 ft), bushy, almost climbing growth; [Lord Bacon X Gruss an Teplitz]; Dickson, H.

'J. B. Clark, Climbing', Cl HT, mr, 1939; flowers carmine-red, very large, dbl., moderate fragrance; Vogel, M.; (Sangerhausen)

'J. B. M. Camm', B, lp, 1900; flowers large, very dbl., moderate fragrance; Paul, G.; (Sangerhausen)

'J. B. Meilland', HT, ob, 1941; flowers orange, reverse golden yellow, large, very dbl., moderate fragrance; very vigorous growth; [Mme Joseph Perraud X (Charles P. Kilham X Margaret McGredy)]; Meilland, F.

J. Bienfait *see* **'Mr J. Bienfait'**

'J. C. Thornton', HT, mr, 1926; bud pointed; flowers glowing crimson-scarlet, dbl., slight fragrance; foliage light olive-green, glossy, leathery; vigorous, branching growth; GM, NRS, 1928; [K. of K. X Red-Letter Day]; Bees

'J. F. Barry', HT, my, 1912; flowers light daffodil-yellow; [Arthur R. Goodwin sport]; Piper

'J. F. Müller', F, dr, 1929; flowers large; foliage dark; bushy, dwarf growth; [Rodhatte sport]; Muller, J.F.

'J. G. Glassford', HT, dr, 1921; bud pointed; flowers deep crimson, very large, exhibition form, moderate fragrance; very vigorous, branching growth; Dickson, H.

'J. H. Bruce', HT, mr, 1937; flowers crimson-scarlet, over large, dbl., exhibition form; foliage glossy; vigorous, bushy growth; GM, NRS, 1936; [H.V. Machin X Marion Horton]; Bees

'J. H. Pemberton', HT, mr, 1931; flowers scarlet, moderate, damask fragrance; vigorous growth; Bentall

'J. H. van Heyst', HT, yb, 1936; bud pointed; flowers yellowish flesh, reverse pink; vigorous growth; [Comtesse Vandal X Edith Nellie Perkins]; Leenders, M.

'J. K. B. Roos', HT, m, 1933; bud pointed; flowers pale reddish lilac, shaded salmon-flesh, large, dbl., intense fragrance; very vigorous growth; Leenders, M.

'J. K. Tyl', LCl, mp, 1936; flowers bright pink, cactus form, very free bloom; vigorous growth; Brada, Dr.; Böhm

'J. M. López Picó', HT, dr, 1947; bud long, pointed; flowers crimson, large, dbl., exhibition form, slight fragrance; upright growth; [Editor McFarland X Comtesse Vandal]; Camprubi, C.

'J. M. López Picó, Climbing', Cl HT, dr, 1954; Camprubi, C.

J. Michael® *see* 'KINmike'

'J. Michel', HT, dy, 1930; flowers dark golden yellow, large, dbl.; foliage leathery; vigorous growth; RULED EXTINCT 1/87; [Unnamed seedling X The Queen Alexandra Rose]; Felberg-Leclerc

'J. N. Hart', HT, dp, 1924; flowers rose-pink, dbl., moderate fragrance; [George Dickson X Edith Cavell]; Chaplin Bros.

J. Otto Ripoff, HT, lp

'J. Otto Thilow', HT, mp, 1927; bud pointed, well shaped; flowers rich glowing rose-pink, large, dbl., exhibition form; very vigorous growth; [Hadley X Souv. de H.A. Verschuren]; Verschuren; Dreer;, H&S

'J. Otto Thilow, Climbing', Cl HT, mp, 1933; Howard Rose Co.

'J. P. Connell', S, my, 1987; flowers pale, medium yellow at inner petals, yellow-white on upper petals, dbl., exhibition form, borne 1-8 per cluster, repeat bloom, intense, tea fragrance; no prickles; foliage abundant, dark yellow-green, wide ovate, double serr; bushy, winter hardy growth; [Arthur Bell X Von Scharnhorst]; Svedja, Felicitas; Agriculture Canada, 1986

J. R. Byfield, Cl HT, lp; deep red flowers with petals edged purple, slight; strong and long; [Sensation seedling sport]; Clark, A., 1941

'J. S. Baar', HT, dr, 1934; flowers pure dark carmine-red, dbl., intense fragrance; foliage leathery; long, strong stems; vigorous, bushy growth; Mikes Böhm, J.; Böhm

'J. S. Fay', HP, rb, 1905; flowers dark crimson tipped scarlet, dbl.; vigorous growth; [Prince Camille de Rohan X Souv. de Pierre Notting]; Walsh

'J. W. Fargo', sp, (variety of R. arkansana), mp; flowers wild-rose-pink, borne in clusters, non-recurrent; well branched (20 in.) growth

J.C. Flat Pink, HT, lp

J.E. Murphy's Pink Tea, T, lp

'J.F. Bailey', HT, lp; [Frau Karl Druschki X Unknown]; Williams, A., 1911; (Weatherly, L.)

J.F. Quadra, S, mp, 1995; Ogilvie, Ian S.

J.G. Mendel, S, lp; Urban, J., 1989; (Czech Rosa Club)

J.G. Sandberg *see* **'Jonkheer G. Sandberg'**

J.H. Pierneef, F, rb, 1994

'JACable', HT, ob, 1998; (**Spice Twice**™); flowers orange, full, very dbl., 26–40 petals, 5–6 in., borne in large clusters, moderate fragrance; prickles moderate; foliage medium, dark green, semi-glossy; upright, tall (5 ft) growth; [Spirit of Glasnost X Kardinal]; Zary, Dr. Keith W.; Bear Creek Gardens, Inc., 1997

'JACaby', HT, or, 1979; flowers large, slight fragrance; foliage large, medium green; upright, bushy growth; [South Seas X Tonight]; Warriner, William A.; J&P

'JACade', HT, pb, 1984; (Home's Choice, **Olympic Dream**); flowers large, dbl., 35 petals, slight fragrance; foliage large, dark, semi-glossy; upright growth; PP005842; [Unnamed seedling X Unnamed seedling]; J&P; McConnell Nurs., Inc.

JACaebi, HT, dp; (**Rendez-vous**)

Jacaford, S; (Cavriglia)

'JACAGE', HT, yb, 1981; flowers medium, dbl., 35 petals, slight fragrance; foliage medium, dark, leathery; upright, bushy growth; [Unnamed seedling X Spellbinder]; Warriner, William A.; J&P

'JACAIM', HT, ob, 1982; (**Acclaim**); flowers orange, large, dbl., 34 petals, slight fragrance; foliage medium green, semi-glossy; upright, bushy growth; PP005638; [Sunfire X Spellbinder]; Warriner, William A.; J&P, 1986

'JACAIR', S, dp, 1998; (**Watermelon Ice**™); flowers dark lavender pink, lighter reverse, small, single, 5–11 petals, 1.5 in., borne in large clusters, repeats quickly, slight fragrance; prickles moderate, straight; foliage small, dark green, semi-glossy; spreading, bushy, low (1 1/2 ft) growth; PP10229; [The Fairy X Seedling]; Zary, Dr. Keith W.; Bear Creek Gardens, Inc., 1997

'JACAKOR'®, HT, mp, 1985; (**Jacaranda**®, Jackaranda); flowers mauve-pink, large, dbl., 35 petals, intense fragrance; foliage large, medium green; upright growth; PP006316; [(Mercedes X Emily Post) X Unnamed seedling]; Kordes, W.

'JACAL', HT, or, 1984; flowers large, dbl., 35 petals, no fragrance; foliage medium, dark, semi-glossy; upright growth; PP005341; [Spellbinder X Futura]; Warriner, William A.; J&P

'JACALE', F, yb, 1986; (Arc de Triomphe, **Summer Fashion**); flowers petals light yellow edged pink, pink spreading with a, dbl., 20 petals, moderate fragrance; foliage large, medium green, semi-glossy; PP005860; [Precilla X Bridal Pink]; Warriner, William A.; J&P, 1985

'JACALP', HT, or, 1980; flowers brick red, large, dbl., 50 petals, slight fragrance; foliage large, dark, leathery; upright growth; [Unnamed seedling X Medallion]; Warriner, William A.; J&P

'JACAMQUE', F, ab, 1999; (**Amber Waves**™); bud copper/gold, pointed, ovoid; flowers amber yellow, reverse yellow/apricot, opens flat, dbl., 26–40 petals, 3.5–4 in., cupped, borne singly and in small clusters, moderate, fruity fragrance; prickles moderate; foliage medium, dark green, glossy; upright, bushy, medium (3.5 ft) growth; Bronze Medal, Australian Nationall Rose Trials, 1999; [seedling X Amber Queen]; Zary, Dr. Keith W.; Bear Creek Gardens, Inc., 2001

JACams, MinFl, lp; (**Sweet Dreams**); J&P, 1995

'JACANGE', HT, ob, 1984; (**John-Paul II**); flowers very large, dbl., no fragrance; foliage medium, dark, semi-glossy; upright growth; PP005639; [Apricot Parfait X Futura]; J&P; McConnell Nurs., Inc.

'JACANGEL', Gr, m, 1991; (**Spellcaster**); flowers lavender and deep mauve, full with heavy substance, blooms b, very dbl., 26–40 petals, 3–3.5 in., moderate fragrance; some prickles; foliage large, dark green, glossy; tall (145-160 cms), upright, spreading, uniform growth; [Unnamed seedling X Angel Face]; Warriner, William A.; Bear Creek Gardens, 1992

'JACANT', F, r, 1985; (**Topaz**); flowers tan, medium, dbl., 35 petals, slight fragrance; foliage medium, medium green, semi-glossy; upright growth; [Unnamed seedling X Intrigue]; Warriner, William A.; J&P

'JACANTH', HT, mr, 1974; ('JACMANTHA', **Samantha**®); bud ovoid; flowers floratea, medium, dbl., exhibition form, slight fragrance; foliage leathery; vigorous, upright growth; [Bridal Pink X Seedling]; Warriner, William A.; J&P

JACAP, F, ob, 1992; (**Amazon**); Christensen, Jack E.

'JACAPRI', HT, ab, 1991; (**Lucille Ball**); flowers apricot, amber blend, good petal substance, blooms borne mostly singly, dbl., 26–40 petals, 2.75 in., moderate fragrance; some prickles; foliage medium, medium green, semi-glossy; tall (145-160 cms), upright, bushy growth; [Hello Dolly X Unnamed seedling]; Christensen, Jack E.; Bear Creek Gardens, 1993

'JACARA', HT, pb, 1991; (**Spirit of Glasnost**); flowers coral pink and ivory blend, large blooms borne mostly singly, dbl., 26–40 petals, moderate fragrance; foliage medium, dark green, semi-glossy; upright, spreading growth; [Unnamed seedling X Unnamed seedling]; Warriner, William A.; Bear Creek Gardens

Jacaranda® *see* 'JACAKOR'®

'JACARCH', LCl, w, 1992; (**Lace Cascade**); bud pointed, ovoid; flowers icy white, floriferous, dbl., 26–40 petals, 3–3.5 in., borne in small and large clusters, moderate, sweet fragrance; some prickles; foliage large, medium green to dark green, semi-glossy, milde; tall (150-160 cms), upright, spreading growth; PP8689; [Iceberg X Prairie Fire]; Warriner, William A.; Bear Creek Gardens

'JACARE', F, w, 1988; (**Class Act**, First Class, White Magic); flowers loose, borne in sprays of 3-6, semi-dbl., flat, slight, fruity fragrance; prickles long, narrow; foliage medium, dark green, semi-glossy; upright, bushy, medium growth; AARS, 1989 GM, Portland, 1989 Gold Star of the South Pacific, Palmerston North, NZ, 1990; [Sun Flare X Seedling]; Warriner, William A.; J&P

'JACARED', HT, dr, 1978; (**American Pride**); bud pointed, ovoid; dbl., 33 petals, 4–5 in., exhibition form, slight fragrance; foliage large, dark; tall, upright growth; PP006515; Warriner, William A.; J&P, 1974

'JACARET', Min, dr, 1983; (**Cabaret**); bud small; flowers small, dbl., 35 petals, moderate fragrance; foliage small, medium green, semi-glossy; upright, bushy growth; PP005576; [(Fire Princess X Mary DeVor) X (Seedling X Caliente)]; Warriner, William A.; J&P

'JACARINA', HT, lp, 1986; (**Amorous**); flowers shell pink, dbl., 30 petals, exhibition form, borne singly, slight, fruity fragrance; foliage large, medium green, glossy; upright, tall growth; [White Masterpiece X Marina]; Warriner, William A.; J&P

'JACARY', HT, dy, 1983; (**Firstar**, Golden Anniversary, Home's Pride, Olympic Spirit); flowers large, dbl., 35 petals, moderate fragrance; foliage medium, medium green, glossy; bushy, spreading growth; [New Day X Oregold]; Warriner, William A.; J&P

JACATI, Min, lp; (**Hot Gossip**); Zary, Dr. Keith W., 1997

JACAUTE, HT, mp, 1997; (**Sara**); J&P

'JACAVE', F, m, 1983; (**Blue Chip**); flowers medium, dbl., 20 petals, intense fragrance; foliage medium, medium green, semi-glossy; bushy, upright growth; PP005684; [Heirloom X Angel Face]; Warriner, William A.; J&P

JACAY, F, w, 1991; (**Avalanche**); Warriner, William A.

'JACBED', S, mr, 1986; (**Red Rascal**); flowers patio, small, bright red, dbl., 35 petals, 1.5–2 in., cupped, blooms in sprays of 2-5, self cleaning, repeat bloom, slight fragrance; medium, red to brown prickles, hooked downward; foliage small, medium green, semi-glossy; medium, bushy growth; PP006693; [Unnamed seedling X Unnamed seedling]; Warriner, William A.; J&P

JACBER, min, w, 1997; (**Ice Angel**); Zary

'JACBIP', HT, lp, 1990; (**Wedding Pink**); bud ovoid; flowers light pink, reverse slightly paler pink, urn-shaped, large, very dbl., exhibition form, slight fragrance; prickles long, very narrow, reddish-brown; foliage medium, dark green, matt; upright, tall growth; [Bridal White sport]; Nakashima, Tosh; Bear Creek Gardens, 1990

'JACBITOU', HT, yb, 1999; (**Weight Watchers Success**™); flowers light creamy yellow edged pink, dbl., 17–25 petals, 4.5–5 in., exhibition form, borne mostly singly, intense, sweet and spicy fragrance; few prickles; foliage medium, dark green, semi-glossy; upright spreading, tall (4-4.5 ft) growth; [Henry Fonda X French Perfume]; Zary, Dr. Keith W.; Bear Creek Gardens, Inc., 1999

JACblar, HT, m, 1995; (**Evita Bezuidenhout**); J&P

'JACBLEM', HT, my, 1981; (**Emblem**®); dbl., 25 petals, exhibition form, borne singly, no fragrance; straight, long, light green prickles; foliage glossy, dark; upright growth; [Seedling X Sunshine]; Warriner, William A.; J&P

'JACBOW', S, m, 1998; (**Kaleidoscope**®); flowers tan mauve blend, reverse yellow mauve blend, very dbl., 26–40 petals, 3 in.., borne in large clusters, slight, fruity fragrance; many prickles; foliage dark green, glossy; spreading, bushy, hedge, 3 1/2 ft. growth; AARS, 1999; [Pink Pollyanna X Rainbow's End]; Walden, John K.; Bear Creek Gardens, Inc., 1999

JACboy, HT, ob; (**Tango**); J&P, 1992

'JACBRI', F, mp, 1967; (**Bridal Pink**®); bud ovoid, pointed, long; flowers large, pink blended with cream, dbl., 30–35 petals, exhibition form, borne in small clusters or singly, moderate, spicy fragrance; foliage leathery; vigorous, upright, bushy growth; [Summertime seedling X Spartan seedling]; Boerner; J&P

'JACBUB', Min, w, 1989; (**Tiny Bubbles**®); bud ovoid; flowers ivory, near white, fading to white, medium, borne usually singly, dbl., cupped, no fragrance; prickles straight, slightly angled downward, green-yellow; foliage medium, medium green, semi-glossy; upright, spreading, medium growth; [Zorina X Funny Girl]; Warriner, William A.; Bear Creek Gardens, 1990

'JACBUNCH', Min, dp, 1998; (**Raspberry Punch**®); flowers large, deep pink, full, very dbl., 26–40 petals, 2 in., borne in large clusters, slight fragrance; prickles moderate, straight; foliage medium, dark green, semi-glossy; compact, moderate, border plant growth; [Tournament of Roses X Seedling]; Zary, Dr. Keith W.; Bear Creek Gardens, Inc., 1999

'JACBUSH', HT, pb, 1990; (**Barbara Bush**®); flowers salmon pink where exposed to sun, lighter pink to ivory white where unexposed, dbl., 25–30 petals, 5 in., exhibition form, borne mostly singly, moderate, damask fragrance; foliage medium, medium green, glossy; tall, upright, spreading growth; PP007542; [Pristine X Antigua]; Warriner, William A.; Bear Creek Gardens, 1991

'JACBUTE', HT, pb, 1991; (**Abracadabra**); bud pointed, ovoid, deep pink/yellow; flowers purple pink, tan and yellow, large blooms, dbl., 26–40 petals, 4 5–5 in., exhibition form, borne mostly singly, intense fragrance; some prickles; foliage large, dark green, semi-glossy; medium (90-120 cms), upright, bushy growth; PP8590; [White Masterpiece X Tribute]; Warriner, William A.; Bear Creek Gardens, 1993

'JACCART', Min, rb, 1989; (**Cartwheel**); bud ovoid; flowers red and white picotee, small, dbl., 20 petals, cupped, borne singly, slight fragrance; prickles hooked, small, tan; foliage small, medium green, matt; low, bushy growth; PP007508; [Libby X Seedling]; Warriner, William A.; Bear Creek Gardens, 1990

'JACCASP', Min, pb, 1992; (**Happy Trails**); flowers pink with cream center, blooms borne in small clusters, very dbl., 1.5 in., no fragrance; some prickles; foliage small, medium to dark green, glossy; low (20-30 cms), groundcover, spreading 60-90 cms across. growth; [Immensee X Roller Coaster]; Warriner, William A. & Zary, Keith W.; Bear Creek Gardens, 1993

JACchar, Min, lp, 1994; (**Fairy Tale**); J&P

'JACCHARI', S, dp, 1998; (**Candy Mountain**®); flowers deep pink, single, single, 5–11 petals, 1.5–2 in., borne in large clusters, slight fragrance; prickles moderate; foliage small, medium green, dull; bushy, arching, low, 2ft growth; PP10616; [Sweet Chariot X Seedling]; Walden, John K.; Bear Creek Gardens, Inc., 1997

'JACCHRY', HT, dp, 1993; (**Breathless**); flowers deep pink, blooms borne mostly single, dbl., 15–25 petals, 3–3.5 in., moderate fragrance; some prickles; foliage large, medium green, purple-red when young, semi-glo; tall (150 cms), upright growth; [Seedling X Chrysler Imperial]; Warriner, William A.; Bear Creek Gardens, 1993;, Bear Creek Gardens, Inc., 1993, USA

'JACCINQO', Gr, pb, 1998; (**Candelabra**®); bud long, pointed; flowers coral orange, dbl., 20–25 petals, 4 in., borne in small clusters, slight, tea fragrance; prickles moderate; foliage medium, dark green, glossy; bushy, medium tall growth; AARS, J&P, 1999; [Tournament of Roses X Seedling]; Zary, Dr. Keith W.; Bear Creek Gardens, 1999

'JACCLAM', LCl, op, 1976; (**America**®); bud ovoid, pointed; flowers salmon, reverse lighter, imbricated, very dbl., 43 petals, 3.5–4.5 in., borne singly and in small clusters, intense fragrance; foliage large, medium green; climbing (10 - 12 ft.) growth; AARS, 1976; [Fragrant Cloud X Tradition]; Warriner, William A.; J&P

'JACCLIP', Cl Min, mp, 1993; (**Bunny Hop**); flowers moderately full, dbl., 15–25 petals, 1.5 in., borne in large clusters, slight fragrance; some prickles; foliage small, dark green, semi-glossy; upright (180 cms), spreading, climbing growth; [Pink Pollyanna X Seedling]; Zary, Dr. Keith W.; Bear Creek Gardens, 1995;, Bear Creek Gardens, Inc., 1995, USA

'JACCLIST', Cl HT, pb, 1976; (**First Prize, Climbing**); very vigorous, climbing growth; Reasoner; J&P

'JACCLOP', LCl, dr, 1975; (**Tempo**); bud ovoid; flowers deep red, large, very dbl., slight fragrance; foliage large, glossy, dark; climbing growth; [Ena Harkness, Climbing X ?]; Warriner, William A.; J&P

'JACCO', HT, w, 1976; (**Coquette**); bud long; dbl., 30–35 petals, 4 in., exhibition form, slight fragrance; foliage light green, reddish underneath; upright growth; [Seedling X Seedling]; Warriner, William A.; J&P

'JACCOEUR', HT, pb, 1993; (**Gardens of the World**); bud small, ovoid; flowers magenta pink and cream blend, blooms borne mostly single, dbl., 26–40 petals, 3–3.5 in., slight fragrance; few prickles; foliage medium, medium green, matt; medium (120 cms), upright growth; [Dame de Coeur X Sunbright]; Christensen, Jack E.; Bear Creek Gardens, 1991

'JACCOFL', F, ab, 1993; (**Brass Band**®); flowers melon orange and yellow bicolor, dbl., 30–35 petals, 3–3.5 in., blooms in small clusters, slight, fruity fragrance; some prickles; foliage large, dark green, semi-glossy; medium (100 cms), upright, bushy growth; PP9171; AARS, 1995; [Gold Badge X Seedling]; Christensen, Jack E.; Bear Creek Gardens, 1994;, Bear Creek Gardens, Inc., 1994, USA

'JACCOGEL', HT, ob, 1999; (**Orange Passion**®); flowers coral orange, reverse lighter, dbl., 26–40 petals, 4.5–5 in., borne mostly singly, slight fragrance; prickles moderate; foliage large, dark green, upright; upright growth; [Anne Morrow Lindbergh X Angelique]; Zary, Dr. Keith W.; Bear Creek Gardens, Inc., 1999

'JACCOH', HT, op; (**Inspiration**)

'JACCOL', HT, mp, 1990; (**Colette**); bud ovoid, pointed; flowers medium soft pink, same reverse, aging slightly paler, urn-shaped, dbl., 25–30 petals, exhibition form, slight, damask fragrance; foliage large, dark green, semi-glossy; upright, bushy, tall growth; [Lorena X Seedling]; Warriner, William A. & Zary, Keith W., 1986; Bear Creek Gardens, 1990

'JACCOR', HT, mp, 1991; (**Glory Days**); flowers coral pink, large, dbl., 26–40 petals, moderate fragrance; foliage medium, medium green, semi-glossy; tall, upright, bushy growth; [Unnamed seedling X Showstopper]; Warriner, William A.; Bear Creek Gardens

JACcream, HT, yb; (**Angela**); J&P

'JACDAL', F, mp, 1993; (**Great Expectations**); J&P

'JACDASH', HT, dy, 1993; (**Sun Goddess**); dbl., 15–25 petals, 3–3.5 in., borne mostly singly, moderate fragrance; some prickles; foliage medium, dark green, matt; tall, upright, spreading growth; PP007659; [Sunbright X Seedling]; Warriner, William A.; Bear Creek Gardens, 1993

'JACDAZ', S, dr, 1992; (**Chorus Line**); flowers red, cream base, blooms borne in small clusters, dbl., 16–20 petals, 3–3.5 in., moderate fragrance; some prickles; foliage large, dark green, semi-glossy; medium (120-135 cms), upright, bushy, very vigorous growth; [Razzle Dazzle X Seedling]; Zary, Keith & Warriner, William; Bear Creek Gardens

JACdebu, HT, m, 1992; (**Paul Gauguin**); J&P

JACdeep, HT, lp, 1994; (**Kiko**); J&P

JACdeli, HT, yb, 1992; (**Paul Cezanne**); J&P

'JACDER', HT, dr, 1974; (**Red Masterpiece**); flowers deep red, large, dbl., exhibition form, intense fragrance; foliage large, dark, leathery; vigorous, upright growth; [(Siren X Chrysler Imperial) X (Carrousel X Chrysler Imperial)]; Warriner, William A.; J&P

JACdesa, HT, rb, 1992; (**Claude Monet®**)

'JACDEW', F, or, 1982; (**Impatient**); flowers medium, orange-red, semi-dbl., 20–25 petals, 4 in., borne in large clusters, slight fragrance; foliage medium, light green, glossy; upright, bushy growth; PP005122; AARS, 1984; [America X Unnamed seedling]; Warriner, William A.; J&P, 1984

'JACDIS', HT, mp, 1974; (Fragrant Memory, **Jadis**); flowers large, dbl., exhibition form, intense fragrance; foliage large, light, leathery; vigorous, upright, bushy growth; [Chrysler Imperial X Virgo]; Warriner, William A.; J&P

'JACDOR', F, rb, 1991; (**Señorita**); flowers orange-red to red on top of petal with a yellow petal base a, dbl., 26–40 petals, 1.5–2.75 in., slight fragrance; many prickles; foliage medium, dark green, semi-glossy; upright (75 cms), bushy growth; [Unnamed seedling X Matador]; Warriner, William A.; Bear Creek Gardens, 1992

JACdreco, F, m, 1999; (**Greetings**); Zary, Dr. Keith W.

'JACDRIFT', Min, w, 1991; (**Blizzard**); bud pointed; flowers medium, dbl., 26–40 petals, no fragrance; foliage small, dark green, glossy; low, spreading, compact growth; [Petticoat X Orange Honey]; Warriner, William A.; Bear Creek Gardens

JACdrive, S, mr, 1998; (**Red Delicious**); J&P

'JACECOND', HT, dp, 1998; (**Timeless™**); flowers red to deep pink, deep pink reverse, dbl., 17–25 petals, 4.5 in., borne in small clusters, slight fragrance; prickles moderate; foliage medium, dark green, semi glossy; upright, bushy growth; AARS, J&P, 1997; [Spirit of Glasnost X Kardinal]; Zary, Dr. Keith W.; Bear Creek Gardens, Inc., 1997

'JACED', HT, dr, 1984; (Home's Beauty, **Olympic Glory**); flowers medium, dbl., 35 petals, slight fragrance; foliage large, medium green, glossy; upright growth; [Unnamed seedling X Unnamed seedling]; J&P; McConnell Nurs., Inc.

'JACEL', HT, my, 1988; (**Graceland**); flowers medium yellow, aging lighter at margins, loose, medium, dbl., 30–35 petals, cupped, borne singly, no fragrance; prickles medium, reddish-green; foliage medium, dark green, matt, smooth; upright growth; PP006069; GM, The Hague, 1988; [New Day X Unnamed seedling]; Warriner, William A.; Bear Creek Gardens, 1989

'JACERN', HT, my, 1985; (**Radiant Gold**); flowers large, dbl., 35 petals; foliage large, dark, glossy; upright growth; [Precilla X Sunshine]; J&P; McConnell Nurs., Inc.

JACess, F, w, 1997; (**Moonstruck**); J&P

'JACETTE', HT, w, 1986; (**Regatta**); flowers large, dbl., 48 petals, exhibition form, borne singly, intense fragrance; no fruit; medium red to brown prickles, hooked downward; foliage medium, medium green, matt; medium, upright growth; PP005896; [Bernadette X Coquette]; Warriner, William A.; J&P

JACeve, F, lp; (**Venus**); J&P, 1997

'JACEZZ', F, dr, 1998; (**Crimson Lace™**); flowers dark red, old-fashioned, dbl., 17–25 petals, 3–3.5 in., borne in small clusters, slight fragrance; prickles moderate; foliage medium, dark green, glossy; compact, medium growth; [Esprit X Razzle Dazzle]; Zary, Dr. Keith W.; Bear Creek Gardens, Inc., 1997

'JACFED', HT, mp, 1993; very dbl., 3–3.5 in., borne mostly singly, moderate fragrance; some prickles; foliage large, medium green, semi-glossy; tall (150-180 cms), upright, spreading growth; [Grand Masterpiece X First Federal Renaissance]; Warriner, William A.; Bear Creek Gardens, 1992

'JACFISEG', Min, ab, 1992; ('JACFISEQ', **Paper Doll**); flowers light apricot with a hint of pale pink fading to light amber, dbl., 15–25 petals, 1.5–2.75 in., no fragrance; some prickles; foliage small, dark greeen, glossy; low (45-60 cms), upright growth; [Fiddler's Gold X Sequoia Gold]; Zary, Dr. Keith W.; Bear Creek Gardens

'JACFLARE', Min, dy, 1991; (**Miss Daisy**); flowers small, very dbl., slight fragrance; foliage small, dark green, glossy; low, bushy growth; [Unnamed seedling X Sun Flare]; Warriner, William A.; Bear Creek Gardens

'JACFOG', Min, dy, 1992; (**Charm Bracelet**); flowers dark yellow, aging to pink, red blush on petal tips of outer petals, very dbl., 1.5 in., slight fragrance; some prickles; foliage small, dark green, semi-glossy; low (45 cms), bushy, compact growth; [Fool's Gold X Seedling]; Christensen, Jack E.; Bear Creek Gardens, 1992

'JACFON', HT, w, 1993; (**Heaven**); flowers ivory or cream with a light pink blend, blooms borne mostly, dbl., 26–40 petals, 3–3.5 in., intense fragrance; some prickles; foliage large, dark green, matt; tall (150-160 cms), upright, bushy growth; [Honor X First Prize]; Warriner, William A.; Bear Creek Gardens, 1994

'JACFRUIT', HT, lp, 1993; (**Camille**); flowers light pink, near white, blooms borne mostly single, very dbl., 3–3.5 in., intense fragrance; some prickles; foliage large, dark green, semi-glossy; medium (120 cms), upright, bushy growth; [Honor X Fragrant Memory]; Warriner, William A.; Bear Creek Gardens, 1995

'JACFUN', Min, lp, 1982; (**Funny Girl**); flowers small, dbl., 20 petals, slight fragrance; foliage small, medium green, matt; upright, bushy growth; PP005249; [Bridal Pink X Fire Princess]; Warriner, William A.; J&P, 1983

'JACGATE', HT, mr, 1991; (**Brandenburg Gate**); flowers red with light reverse, large, dbl., 26–40 petals, moderate fragrance; foliage large, medium green, matt; tall, upright growth; [Seedling X Madras]; Warriner, William A.; Bear Creek Gardens, 1990

'JACGLOW', HT, w, 1989; (**White Delight**); bud ovoid, pointed; flowers white with pink blush, large, dbl., 38 petals, exhibition form, borne usually singly, slight fragrance; prickles hooked, medium, brown; foliage medium, dark green, matt; upright, medium growth; PP006664; [White Masterpiece X Futura]; Warriner, William A.; J&P, 1990

'JACGOLD', HT, dy, 1982; (**Golden Emblem**); flowers large, dbl., 20 petals, slight fragrance; foliage large, medium green, glossy; upright growth; PP005121; [(Bridal Pink X Dr. A.J. Verhage) X (Golden Sun X South Seas)]; Warriner, William A.; J&P

'JACGRANT', F, ab, 1998; (**Fragrant Apricot™**); bud pointed, oval, apricot blend; flowers light apricot, copper tinted, light coral reverse, very dbl., 26–40 petals, 4 in., borne in large clusters, slight, musk fragrance; prickles moderate; foliage large, dark green, glossy; compact, medium (3 ft.) growth; [Impatient X Amber Queen]; Zary, Dr. Keith W.; Bear Creek Gardens, Inc., 1999

'JACGRAY', HT, lp, 1998; (**Billy Graham**®); bud long, pointed; flowers light pink, full, large, dbl., 26–40 petals, 5 in., exhibition form, borne singly, slight, sweet fragrance; prickles moderate; foliage large, dark green, glossy; upright, tall, 5 ft. growth; [Honor X Color Magic]; Zary, Dr. Keith W.; Bear Creek Gardens, Inc., 1998

'JACGRON', S, dp, 1998; (**Garden Blanket**®); flowers dark pink, white at base, lighter reverse, single, 4–11 petals, 1–1.5 in., borne in large clusters, slight fragrance; prickles moderate; foliage small. dark green, glossy; spreading, bushy, tall (3.5 ft.) growth; [Magic Carpet X Happy Trails]; Walden, John K.; Bear Creek Gardens, Inc., 1998

'JACGUM', Min, dr, 1981; (**Gumdrop**) bud fat, pointed; flowers , dbl., 25 petals, borne 3-12 per cluster; foliage small, semi-glossy; spreading growth; PP005152; [(San Fernando X Bridal Pink) X (Fire Princess X Mary DeVor)]; Warriner, William A.; J&P

'JACHAL', Min, dy, 1999; (**Sun Sprinkles**®); flowers deep yellow, dbl., 26–40 petals, 2–2.5 in., borne mostly singly and in small clusters, slight fragrance; prickles moderate; foliage medium, dark green, glossy; compact, low (18-20 in.) growth; AOE, U.S., 2001; [Yellow Jacket X seedling]; Walden, John K.; Bear Creek Gardens, 2001

JAChart, F, lp; (**Starburst**); J&P, 1997

'JACHEIR', HT, m, 1993; (**Fortune Teller**); flowers deep mauve/purple, blooms borne mostly single, dbl., 26–40 petals, 3–3.5 in., intense fragrance; some prickles; foliage large, dark green, semi-glossy; upright, spreading (150-160 cms) growth; [Seedling X Heirloom]; Warriner, William A. & Zary, Keith W.; Bear Creek Gardens

'JACHILL', Min, mp, 1991; (**Suzy Q**®); bud ovoid; flowers medium shell pink, deeper color at petal margins, medium, very dbl., 40–45 petals, cupped, no fragrance; foliage medium, dark green, glossy; upright, bushy, medium growth; [Rose Hills Red X Baby Ophelia]; Warriner, William A. & Zary, Keith W., 1985; Bear Creek Gardens, 1992

JAChoc, HT, w; (**Suzette van der Merwe**); J&P, 1994

JAChon, HT, mp, 1995; (**Enchanted**); J&P

'JACHOP', HT, dp, 1987; (**Haute Pink**); flowers rose pink fading little, loose, medium, dbl., 25–30 petals, cupped, borne singly, slight fragrance; no fruit; long, narrow, red prickles; foliage medium, medium green, matt; upright, tall growth; PP006653; [Bridal Pink X Grand Masterpiece]; Warriner, William A.; J&P

'JACHOTTA', Min, dp, 1999; (**Bedazzled**®); dbl., 26–40 petals, 2–2.5 in., borne in small clusters, slight fragrance; prickles moderate; foliage dark green, semi-glossy; compact, medium (20-24 in.) growth; [Hot Tamale sport]; Walden, John K.; Bear Creek Gardens, Inc., 2000

'JACHY', HT, dy, 1991; (**Old Faithful**); flowers large, dbl., 26–40 petals, moderate fragrance; foliage large, dark green, semi-glossy; upright, bushy growth; PP006445; [Sunbright X Medallion]; Warriner, William A.; Bear Creek Gardens

'JACHYP', HT, mp, 1991; (**Unforgettable**); flowers , dbl., 26–40 petals, 3–3.5 in., borne mostly singly, intense fragrance; foliage large, dark green, semi-glossy; tall (140 cms), upright, spreading growth; PP006318; [Honor X American Dawn]; Warriner, William A.; Bear Creek Gardens, 1992

'JACIAT', Min, mr, 1988; (**Little Sizzler**®, Patio Jewel); bud ovoid, pointed; flowers large, dbl., 38 petals, cupped, borne usually singly and in sprays of 16-20; prickles hooked downward, reddish-brown; foliage large, dark green, semi-glossy; bushy, medium growth; PP006091; [Unnamed seedling X Funny Girl]; Warriner, William A.; Bear Creek Gardens, 1989

'JACIBE', Min, ab, 1982; (**Caribe**); flowers apricot, semi-dbl., slight fragrance; foliage small, light green, semi-glossy; upright, bushy growth; PP005246; [Bridal Pink X Fire Princess]; Warriner, William A.; J&P, 1983

JACibras, F, dy, 1993; (**Catherine McAuley**); Bear Creek Gardens

'JACICE', HT, w, 1984; (**Moonlight**); flowers medium, dbl., 35 petals, no fragrance; foliage large, medium green, matt; upright growth; PP005743; [Coquette X Unnamed seedling]; Warriner, William A.; J&P

'JACIENT', Gr, mp, 1988; (Berkeley, Poesie, **Tournament of Roses**); flowers light coral pink, reverse deep pink, aging coral pink, large, dbl., exhibition form, no fragrance; large prickles; foliage large, dark green, semi-glossy, disease resistant; upright, bushy, medium growth; PP006725; AARS, 1989; [Impatient X Unnamed seedling]; Warriner, William A.; J&P, 1989

'JACIF', F, ab, 1983; (**Pacifica**®); flowers medium, dbl., 35 petals, slight fragrance; foliage medium, light green, matt; upright, bushy growth; PP005261; [Mercedes X Marina]; Warriner, William A.; J&P

'JACIMIN', Min, w, 1993; (**Small Miracle**); bud plump; flowers white, moderately classic hybrid tea, small, dbl., 15–25 petals, 2 in., exhibition form, borne in small clusters, slight fragrance; few prickles; foliage small, dark green, glossy, resistant to powdery mild; low (50-60 cms), bushy growth; PP8850; [Libby X Sun Flare]; Warriner, William A.; Bear Creek Gardens, 1993

'JACINA', S, dp, 1998; (**Wild Dancer**®); flowers bright magenta-pink, single, single, 4–11 petals, 1.5 in., borne in large clusters, slight fragrance; prickles moderate; foliage small, medium green, semi double; compact, arching, low, 2.5 x 2.5 ft. growth; [China Doll X Sweet Chariot]; Walden, John K.; Bear Creek Gardens, Inc, 1998

'JACINAL', F, mp, 1985; (**Debutante**); flowers small, dbl., 20 petals, slight fragrance; foliage medium, medium green, matt; bushy growth; PP005729; [Bridal Pink X Zorina]; Warriner, William A.; J&P

'JACINK', F, mp, 1978; (**Simplicity**®); bud long, pointed; semi-dbl., 18 petals, 3–4 in., flat, slight fragrance; bushy, upright growth; GM, NZ, 1976; [Iceberg X Unnamed seedling]; Warriner, William A.; J&P, 1979

'JACIRAPS', Min, pb, 1999; (**Hugs 'n' Kisses**®); flowers pink, white, reverse near white, medium, dbl., 17–25 petals, 2.5 in., borne in small to large clusters, moderate fragrance; prickles moderate; foliage large, dark, grey-green, glossy; compact, spreading, medium (16-18 in.) growth; [Small Miracle X seedling]; Walden, John K.; Bear Creek Gardens, Inc., 1999

'JACIRST', HT, op, 1998; (**Artistry**®); flowers coral orange and creamy coral reverse, very dbl., 30–35 petals, 4.5 in., exhibition form, borne singly, full, best form and color in heat, slight fragrance; prickles moderate; foliage large, dark green, semi glossy; upright, bushy, tall growth; PP10230; AARS, J&P, 1997; [Seedling X Seedling]; Zary, Dr. Keith W.; Bear Creek Gardens, Inc., 1997

'JACIS', HT, lp, 1976; (Poesie, **Promise**); bud ovoid; flowers large, dbl., exhibition form, slight fragrance; foliage large, glossy; vigorous growth; GM, Bagatelle, 1976; [South Seas X Peace]; Warriner, William A.; J&P

'JACIV', F, dr, 1974; (**Viva**); bud ovoid; flowers medium, dbl., exhibition form, slight fragrance; foliage glossy, dark; vigorous, upright growth; GM, Portland, 1984; [Unnamed seedling X Unnamed seedling]; Warriner, William A.; J&P

'JACJEL', HT, my, 1984; (**Sunbright**®); bud long, pointed; dbl., 28 petals, 4 in., flat, slight fragrance; upright growth; [Unnamed seedling X New Day]; Warriner, William A.; J&P

'JACJEM', F, my, 1981; (**Sun Flare**, Sunflare); bud pointed; flowers , dbl., 30 petals, flat, borne 3-12 per cluster, no fragrance; reddish prickles; foliage small, glossy; low, compact growth; PP005247; AARS, 1983 GM, Japan, 1981 GM, Portland, 1985; [Sunsprite X Unnamed seedling]; Warriner, William A.; J&P, 1983

'JACJUBEL', F, my, 1998; (**Mirabella**®); flowers medium yellow, very dbl., 41 petals, 4–4.5 in., borne singly and in small clusters, intense fragrance; some prickles; foliage medium, dark green, glossy; medium (80-100 cms), upright, bushy growth; [Sunsprite X Silver Jubilee]; Zary, Dr. Keith W.; Bear Creek Gardens, 1994

Jack Collier *see* 'HORJACK'

Jack Dayson *see* 'KORWILMA'

'JACK FOLLY', HT, w, 1976; bud ovoid; flowers cream color, opening, dbl., 40 petals, 5.5 in., cupped; foliage dull, leathery; spreading growth; Buss, H., Nursery

'JACK FROST', F, w, 1962; bud pointed; flowers white to creamy, sweetheart, medium, dbl., 42 petals, exhibition form, moderate fragrance; foliage dark; vigorous, upright growth; [Garnette X Unnamed seedling]; Jelly; E.G. Hill Co.

'JACK HORNER', Min, mp, 1955; flowers bright pink, dbl., 50 petals, slight fragrance; no prickles; height 4-8 in; [Margo Koster X Tom Thumb]; Robinson, T.

'JACK MCCANDLESS', HT, rb, 1935; bud long, pointed, carmine and yellow; flowers amber-yellow, veined red, dbl., exhibition form, slight fragrance; foliage small, glossy, dark; McGredy

'JACK OF HEARTS', F, 1968; flowers cardinal red, trusses, semi-dbl., cupped, free bloom; foliage dark, glossy; low, bushy growth; Waterhouse Nursery

'JACK O'LANTERN', Gr, yb, 1960; bud ovoid; flowers gold and yellow blend, dbl., 25 petals, 4.5 in., exhibition form, slight fragrance; foliage leathery; vigorous, tall, bushy growth; [Circus X Golden Scepter]; Swim & Weeks

Jack Rose *see* **'GÉNÉRAL JACQUEMINOT'**

Jack Rose *see* **R. MAXIMOWICZIANA JACKII**

JACka, HT, pb, 1994; (**Loving**); J&P

Jackaranda *see* 'JACAKOR'®

JACkari, HT, yb; (**Askari**); J&P

'JACKEE', Min, yb, 1998; (**Bees Knees**®); flowers yellow/pink blend, large, very dbl., 26–40 petals, 2.5 in., borne in large clusters, slight fragrance; prickles moderate; foliage medium, dark green, dull; bushy, tall, 3 ft. growth; [Seedling X Haute Pink]; Zary, Dr. Keith W.; Bear Creek Gardens, Inc., 1998

'JACKIE'®, Min, ly, 1955; flowers straw-yellow changing to white, small, dbl., 60 petals, 1.5 in., exhibition form, moderate fragrance; foliage glossy; vigorous, dwarf (12 in), bushy, spreading growth; [Golden Glow X Zee]; Moore, Ralph S.; Sequoia Nursery

'JACKIE CLARK', HT, dr, 1990; bud pointed; very dbl., 35 petals, 3.5 in., exhibition form, borne singly, intense fragrance; prickles curved down, pink; foliage medium, dark green, matt, disease-resistant; vigorous, upright growth; [White Masterpiece X Red Planet]; Wambach, Alex A., 1985; Alex A. Wambach, 1989

'JACKIE, CLIMBING', Cl Min, ly, 1957; flowers soft yellow to creamy white, dbl., 60 petals, 1–1.5 in., moderate fragrance; foliage semi-glossy, leathery; height to10 ft; [Golden Glow X Zee]; Moore, Ralph S.; Sequoia Nursery

'JACKMAN'S WHITE', HT, w; (Brookdale Giant White); flowers creamy white, very large, exhibition form; vigorous growth; very hardy for this type; Brookdale-Kingsway; Bosley Nursery, 1940

'JACKO', F, or, 1974; (**Sunfire**); bud ovoid; flowers mandarin-red, medium, dbl., exhibition form, slight fragrance; foliage large, leathery; vigorous, upright, bushy growth; [Tropicana X Zorina]; Warriner, William A.; J&P

Jackpot *see* 'MORJACK'

Jack's Fantasy *see* 'BILFAN'

Jackson Square *see* 'TALJAC'

'JACLACE', F, w, 1980; (**French Lace**); bud pointed; flowers ivory, pastel apricot to white, dbl., 30–35 petals, 4.5 in., exhibition form, borne in small clusters, slight, fruity fragrance; small prickles; foliage small, dark; bushy growth; PP4848; AARS, 1982 GM, Portland, 1984; [Dr. A.J. Verhage X Bridal Pink]; Warriner, William A.; J&P

'JACLAF', HT, m, 1998; (**Moon Shadow**®); bud long, pointed, oval; flowers medium lavendar, very dbl., 26–40 petals, 4.5–5 in., borne in small clusters, intense, sweet pea fragrance; few prickles; foliage medium, dark green, semi glossy; upright, tall, bushy, 4.5 ft. growth; PP9538; Warriner, William A.; Bear Creek Gardens, Inc., 1998

'JACLAM', F, dr, 1986; (**Carmen**); flowers deep red, very little fading, small, dbl., 30 petals, exhibition form, borne usually singly, no fragrance; foliage medium, dark green, semi-glossy; upright, bushy growth; PP006070; [Unnamed seedling X Samantha]; Warriner, William A.; J&P

'JACLAV', HT, m, 1990; (Heather, **Silver Anniversary**®); flowers medium lavender, reverse medium lavender, spot of yellow at, dbl., 25–30 petals, exhibition form, intense, damask fragrance; foliage large, very dark green, semi-glossy-glossy; long, straight, sturdy stems; average 24-32 in; long, tall, upright growth; PP007658; [Crystalline X Shocking Blue]; Christensen, Jack E., 1985; Bear Creek Gardens, 1990

JAClay, HT, my; (**Treasure Trove**); Warriner, William A., 1991

'JACLEM', Cl F, my, 1987; (**Sun Flare, Climbing**); [Sun Flare sport]; Warriner, William A.; J&P

'JACLES', HT, rb, 1983; (**Milestone**); flowers medium red, silvery red reverse, opening to coral-pink, dbl., cupped, slight fragrance; foliage large, medium green, semi-glossy; upright growth; [Sunfire X Spellbinder]; Warriner, William A.; J&P, 1985

'JACLEWT', Gr, m, 1999; (**English Perfume**®); flowers lavender blend, very dbl., 41 petals, 5 in., intense fragrance; prickles moderate; foliage medium, dark green, semi-glossy; upright growth; [seedling X Carefree Wonder]; Zary, Dr. Keith W.; Bear Creek Gardens, Inc., 1999

'JACLIANG', Min, m, 1991; (**Plum Dandy**®); bud ovoid, pointed; flowers medium lavender, lighter near petal base, same reverse, fade, very dbl., 35 petals, cupped, moderate, fruity fragrance; foliage medium, medium green, semi-glossy; bushy, spreading, medium growth; [Unnamed seedling X Angel Face]; Warriner, William A.; Bear Creek Gardens

JAClib, Min, pb, 1993; (**Petite Michelle**); J&P

'JACLIN', HT, dr, 1991; (**Patriot**); dbl., 26–40 petals, 3–3.5 in., borne mostly singly, slight fragrance; foliage large, dark green, semi-glossy to glossy; upright (150 cms), spreading growth; [Showstopper X Mister Lincoln]; Warriner, William A.; Bear Creek Gardens

'JACLING', F, pb, 1991; (**First Kiss**); flowers light pink, light yellow blend at base, moderately large, dbl., 15–25 petals, slight fragrance; foliage medium, medium green, matt; bushy, compact growth; [Sun Flare X Simplicity]; Warriner, William A.; Bear Creek Gardens

'JACLION', HT, ab, 1998; (**Grand Impression**®, Monet); flowers yellow/peach/pink blend, apricot yellow reverse, very dbl., 26–40 petals, 6 in., borne in small clusters, slight fragrance; few prickles; foliage large, dark green, semi glossy; upright, tall, 5 ft. growth; [Spirit of Glasnost X Medallion]; Warriner, William A.; Bear Creek Gardens, Inc., 1999

'JACLIP', Min, lp, 1986; (**Le Pink**); flowers crowded, small, dbl., 40 petals, cupped, blooms in sprays of 3-20, slight, damask fragrance; long, thin prickles, numerous on pecuncle; foliage small, light green, matt; very dense, bushy, upright, medium growth; PP006011;

[Unnamed seedling X Watercolor]; Warriner, William A.; J&P

'JACLIP', Min, mp; (**Pink Cloud®**)

'JACLITE', Min, w, 1982; (**Cloud Nine**); flowers medium, semi-dbl.; foliage light green, matt; spreading growth; PP005557; [Bon Bon X Calgold]; Warriner, William A.; J&P

'JACLIV', F, my, 1987; (**Blondie**, 'JACLIY'); flowers medium yellow, fading slightly, medium, borne usually singly, dbl., 25 petals, exhibition form, slight, fruity fragrance; no fruit; straight, medium prickles; foliage medium, light green, semi-glossy; upright, bushy, medium growth; [(Bridal Pink X Golden Wave) X Gold Rush]; Warriner, William A.; J&P

JACLOGO, HT, ob, 1997; (**Saturn**); J&P

'JACLOOM', HT, m, 1972; (**Heirloom**) bud long, pointed; flowers deep lilac, medium, semi-dbl., intense fragrance; foliage leathery; vigorous, upright growth; [Seedling X Seedling]; Warriner, William A.; J&P

'JACLOPI', HT, mp, 1985; (**Leading Lady**); flowers large, dbl., 35 petals, slight fragrance; foliage large, light green, matt; upright growth; PP005859; [Unnamed seedling X Unnamed seedling]; Warriner, William A.; J&P

JACLORNA, HT, dp; (**Sugar Daddy**); J&P, 1992

'JACLOVER', S, m, 1992; (**Magic Carpet**); flowers lavender blooms, small, profuse, semi-dbl., 12 petals, 2 in., borne in clusters, intense, spicy fragrance; foliage medium, dark green, glossy; spreading, low, 18 in., ground cover growth; PP9324; Zary, Keith & Warriner, William

JACMAD, HT, pb; (**Winner's Choice**); J&P, 1992

'JACMAG', HT, pb, 1978; (**Color Magic**); bud long; flowers pale salmon blushing pink, dbl., 25–30 petals, 5 in., flat, borne singly, intense, fruity fragrance; foliage large, dark green; upright growth; AARS, 1978; [Seedling X Spellbinder]; Warriner, William A.; J&P

'JACMANTHA', HT, mr, 1974; ('JACANTH', **Samantha®**); bud ovoid; flowers floratea, medium, dbl., exhibition form, slight fragrance; foliage leathery; vigorous, upright growth; [Bridal Pink X Seedling]; Warriner, William A.; J&P

'JACMAR', F, ob, 1991; (**Kaileen**); flowers bright orange, yellow reverse, blooms borne in small clusters, dbl., 26–40 petals, 1.5–2.75 in., slight fragrance; some prickles, light green to yellow; foliage medium, dark green, glossy; upright (180 cms) growth; [Marina sport]; Nakashima, Tosh; Bear Creek Gardens, 1992

'JACMAS', HT, w, 1969; (**White Masterpiece**); bud long, pointed; flowers very large, large peduncle, dbl., exhibition form, slight fragrance; upright growth; Boerner; J&P

'JACMAT', Min, or, 1981; (**Razzmatazz**); flowers , dbl., exhibition form, in clusters, no fragrance; foliage semi-glossy; upright growth; PP005118; [Zorina X Fire Princess]; Warriner, William A.; J&P

JACmeen, F, op, 1998; (**Good Show**); PPAF; J&P

JACMEM, HT, yb, 1994; (**Mango**); J&P

'JACMENT', F, op, 1998; (**Pure Poetry**); flowers yellow/orange/pink blend, very full, very dbl., 41 petals, 4 in., borne in small clusters, slight fragrance; prickles many, straight; foliage medium, dark green, glossy; compact, 2.5 ft. growth; [Seedling X Tournament of Roses]; Zary, Dr. Keith W.; Bear Creek Gardens, Inc., 1997

'JACMEY', HT, dy, 1987; (**Sun Glory**); flowers medium, borne singly, dbl., 30 petals, exhibition form, slight, fruity fragrance; no fruit; long, red prickles, pointed slightly downwards; foliage medium, medium green, matt; upright growth; PP006612; [Golden Emblem X Unnamed seedling]; Warriner, William A.; J&P

'JACMINNO', Min, ob, 1992; (**Jitterbug**); flowers orange, slightly lighter on reverse, nicely formed, open, dbl., 15–25 petals, 1.5–2.75 in., slight fragrance; some prickles; foliage medium, dark green, glossy; tall (60-75 cms), upright, bushy growth; [Caribe X Impatient]; Warriner, William A.; Bear Creek Gardens, 1993

JACMINT, Min, pb; (**Peppermint Stick**); J&P

'JACMIRYL', Min, my, 1999; (**Lemon Gems™**); flowers rich, deep yellow, dbl., 26–40 petals, 2.5–2.75 in., cupped, borne mostly singly, slight fragrance; prickles moderate; foliage medium, dark green, glossy; upright, spreading, low (18-20 in.) growth; [seedling X Small Miracle]; Walden, John K.; Bear Creek Gardens, 1999

'JACMISS', Min, lp, 1989; (**Miss Perfect™**); bud ovoid, pointed; flowers light pink to near white at edge, reverse lighter, medium, very dbl., cupped, slight fragrance; prickles straight to slightly hooked downward, light yellow; foliage medium, medium green, semi-glossy; spreading, low growth; PP007650; [Over the Rainbow X Lavender Lace]; Warriner, William A.; Bear Creek Gardens

'JACMIY', Min, my, 1988; (**Sunsmile™**); bud ovoid; flowers medium, borne singly and in sprays of 2-3, dbl., 48 petals, exhibition form, slight fragrance; prickles straight to slightly hooked downward, brown; foliage medium, medium green, semi-glossy; bushy, spreading, compact, vigorous growth; [Spanish Sun X Calgold]; Warriner, William A.; Bear Creek Gardens, 1989

'JACMO', Min, ob, 1991; (**Pride 'n' Joy™**); bud ovoid; flowers bright, medium orange, reverse orange and cream, fades to sa, very dbl., 30–35 petals, 2 in., exhibition form, moderate, fruity fragrance; foliage medium, dark green, semi-glossy; bushy, spreading, medium (36 in) growth; PP8578; AARS, 1992; [Chattem Centennial X Prominent]; Warriner, William A., 1983; Bear Creek Gardens, 1992

'JACMOBLI', Min, mp, 1998; (**Barbie®**); flowers medium pink, long lasting, very dbl., 41 petals, 2 in., borne in small clusters, slight fragrance; prickles moderate; foliage medium, dark green, semi glossy; upright, low, suitable for pots growth; [(Seedling X Watercolor) X Red Minimo]; Walden, John K.; Bear Creek Gardens, Inc., 1999

'JACMOCL', Cl Min, ob, 1989; (**Orange Blossom Special™**); bud ovoid, sepals serrated; flowers coral orange, edges darker, reverse lighter coral pink, agin, very dbl., exhibition form, slight fragrance; prickles straight to hooked down, reddish-green; foliage small, dark green, semi-glossy; upright, bushy, tall growth; [Zorina X Andrea]; Warriner, William A.; Bear Creek Gardens, 1990

'JACMUR', HT, w, 1984; (**Fountain Square™**); flowers clear white, dbl., 25–30 petals, 5 in.–5.5 in., exhibition form, borne singly, slight fragrance; foliage large, dark green, semi-glossy; tall growth; PP006805; [Pristine sport]; Humenick, Muriel F.; Fountain Square, Inc., 1986

'JACNEL', HT, dy, 1992; (**Sundance**); very dbl., 3–3.5 in., borne mostly singly, moderate fragrance; some prickles; foliage large, dark green, semi-glossy; tall (180-120 cms), upright growth; [Unnamed seedling X Emblem]; Warriner, William A.; Bear Creek Gardens, 1991

'JACNEPAL', HT, pb, 1999; (**Gemini™**); bud pointed, ovoid; flowers cream, blushing coral pink, dbl., 25–30 petals, 4.5–5 in., exhibition form, usually borne singly, moderate, sweet fragrance; prickles moderate; foliage large, deep green, glossy; long stems; upright, spreading, tall (5.5 ft) growth; AARS, U.S., 2000 Silver Medal - Monza, 1999; [Anne Morrow Lindbergh X New Year]; Zary, Dr. Keith W.; Bear Creek Gardens, 2000

'JACNON', HT, ob, 1985; flowers medium, dbl., slight fragrance; foliage large, medium green, matt; upright, bushy growth; PP005318; [Baccará X Unnamed seedling]; Warriner, William A.; J&P

'JACNOR', HT, 1998; (**Signature®**); flowers deep pink and cream blend, light pink and cream reverse, very dbl., 26–40 petals, 5 in., exhibition form, borne singly, moderate fragrance; many prickles; foliage large, dark green, semi glossy; upright, bushy, tall, 4-5 ft. growth; [Honor X First Federal Renaissance]; PP9539 Warriner, William A.; Bear Creek Gardens, Inc., 1996

JACnu, HT, op, 1993; (**Nuance**); J&P

JACnuel, HT, lp; (**Sterling '95**); J&P, 1995

'JACNUYE', Min, ob, 1998; (**Little Flame™**); flowers brilliant orange, dbl., 17–25 petals, 2 in., borne in large clusters, no fragrance; prickles moderate, straight; foliage medium, dark green, semi-glossy; compact, tall, border (3 ft.) growth; AOE, J&P, 1999; [New Year X Seedling]; Walden, John K.; Bear Creek Gardens, Inc., 1998

Jacob van Ruysdael, F, op, 1997; flowers peach orange, semi-dbl.; upright, compact growth; Williams, J. Benjamin

Jacobite Rose *see* White Rose of York

'JACOLBER', HT, dr, 1998; (**Opening Night™**); flowers bright, deep red, very dbl., 26–40 petals, 4.5 in., exhibition form, borne singly, slight fragrance; prickles moderate, straight; foliage medium, dark green, semi glossy; upright, tall, slightly spreading growth; AARS, AARS, 1998; [Olympiad X Ingrid Bergman]; Zary, Dr. Keith W.; Bear Creek Gardens, Inc., 1998

'JACOLFA', HT, lp, 1999; (**English Sachet™**); flowers cupped, ruffled, almost quartered, delicate petals, very dbl., 40–50 petals, 4–4.5 in., cupped, borne mostly singly, intense, sweet fragrance; prickles moderate; foliage medium, medium green, matte; upright, spreading, tall (5 ft) growth; [Summer Fashion X Silver Jubilee]; Zary, Dr. Keith W.; Bear Creek Gardens, Inc., 1999

'JACOLITE', HT, w, 1980; (**Honor®**, Honour, Michèle Torr®); bud ovoid, pointed; flowers loose, dbl., 23 petals, 5 in., slight fragrance; foliage large, dark; upright growth; AARS, 1980 GM, Portland, 1978; Warriner, William A.; J&P

'JACOLPUR', HT, m, 1999; (**Purple Passion™**); bud pointed, ovoid; flowers rich purple/lavender, reverse lavender, dbl., 26–40 petals, 4.5–5 in., exhibition form, borne mostly singly, intense, lemony fragrance; prickles moderate; foliage large, dark green, glossy; upright, spreading, tall (5 ft) growth; [seedling X seedling]; Zary, Dr. Keith W.; Bear Creek Gardens, Inc., 1999

'JACOPPER', HT, dr, 1999; (City of Newcastle Bicentennary, Five-Roses Rose, Lady in Red, **Veteran's Honor** ™); bud 2 i. furled, pointed; flowers bright red, high centered, large, dbl., 26–40 petals, 5–5.5 in., exhibition form, borne mostly singly, slight, raspberry fragrance; prickles moderate; foliage medium, dark green, semi-glossy; upright, spreading, tall (5 ft) growth; Rose of the Year - 2000, J&P, 2000; [Showstopper X (seedling X Royalty)]; Zary, Dr. Keith W.; Jackson & Perkins, 2000, U.S.

'JACORAL', Min, mp, 1988; (**Coral Sprite™**); bud ovoid, pointed; flowers urn-shaped, medium, borne in sprays of 3-18, very dbl., no fragrance; prickles straight to slightly angled downward, yellow-green; foliage medium, dark green, matt; bushy, spreading, low growth; PP007717; [Merci X Party Girl]; Warriner, William A.; Bear Creek Gardens, 1989

JACorb, HT, ob; (**Ipitombi**); J&P

'JACORCA', Min, op, 1998; (**Tropical Twist™**); flowers coral orange and apricot with yellow reverse, dbl., 17–25 petals, borne mostly singly, floriferous, slight fragrance; prickles moderate; foliage medium, dark green, glossy; compact, tall growth; hedging or border; AOE, ARS, 1997; [Seedling X Pink Pollyanna]; Walden, John K.; Bear Creek Gardens, Inc., 1997

'JACORE', HT, dr, 1984; (**Encore**); flowers large, dbl., 20 petals, no fragrance; foliage medium, dark, semi-glossy; upright growth; PP005658; [Seedling X Samantha]; Warriner, William A.; J&P

'JACORG', Min, ob, 1999; (**Crackling Fire™**); flowers copper orange, reverse deep red orange, dbl., 20–40 petals, 2–2.25 in., borne in small clusters, no fragrance; prickles moderate; foliage medium, dark green, semi-glossy; upright, compact, low (14-18 in.) growth; [seedling X Rainbow's End]; Walden, John K.; Bear Creek Gardens, Inc., 1999

'JACORMIN', Min, or, 1988; (**Vixen**); bud ovoid, pointed, green with reddish-brown; flowers red-orange with yellow base, aging red-orange to pink orange, dbl., 28 petals, cupped, no fragrance; prickles small, straight, reddish-brown; foliage medium, medium green, semi-glossy, small; bushy, spreading, low growth; [Petticoat X Red Minimo]; Warriner, William A.; Bear Creek Gardens, 1990

JACosch, HT, w, 1997; (**Opulence**); J&P

JACosol, HT, my, 1995; (**Moonlight Lady**); J&P

'JACOTTE', LCl, ab, 1920; bud ovoid, orange and yellow; flowers deep coppery yellow, tinted coppery red, open, semi-dbl., 3 in., cupped, moderate fragrance; foliage leathery, glossy, dark; long, strong stems; very vigorous, climbing growth; [R. wichurana X Arthur R. Goodwin]; Barbier

'JACOUCH', Min, ab, 1984; (**Soft Touch**); flowers medium, semi-dbl., slight fragrance; foliage medium, medium green, semi-glossy; PP005578; [Bridal Pink X Fire Princess]; Warriner, William A.; J&P

'JACOUT', F, dp, 1991; (**Neon Lights**); flowers hot magenta pink, moderately large blooms borne in small clusters, dbl., 15–25 petals, moderate fragrance; foliage medium, medium green, semi-glossy; medium, bushy growth; [Intrigue X Impatient]; Warriner, William A.; Bear Creek Gardens, 1992

'JACOYEL', HT, op, 1982; (**Fascination**); bud nearly globular; flowers orange and rose blend, blooms borne usually singly, dbl., 55 petals, exhibition form, no fragrance; long-based prickles, hooked down; foliage very large, semi-glossy; upright, heavy branching growth; GM, NZ, 1976; [Unnamed seedling X Spellbinder]; Warriner, William A.; J&P

'JACPAL', HT, pb, 1991; (**Brigadoon™**); bud ovoid, pointed; flowers cream pink near center, coral pink at petal margin, rose pink edges, very dbl., 25–30 petals, 5 in., exhibition form, borne mostly singly, moderate, tea fragrance; foliage medium, dark green, semi-glossy; upright, spreading, tall growth; PP8591; AARS, 1992; [Seedling X Pristine]; Warriner, William A., 1983; Bear Creek Gardens, 1992

JACpan, F, mp; (**Meteor Shower**); J&P

JACpat, F, op; (**Gnome World**); J&P

'JACPET', Min, w, 1981; (**Petticoat**); bud short; flowers white tinted pink, dbl., 55 petals, slight fragrance; foliage dark, small, pointed; very compact, spreading growth; PP005039; [Bon Bon X Lemon Delight]; Warriner, William A.; J&P

'JACPICL', Cl F, op, 1998; (**Dream Weaver™**); bud short, pointed; flowers coral pink, very dbl., 26–40 petals, 3–3.5 in., rosette, borne in large clusters, slight, rose fragrance; prickles moderate; foliage large, dark green, glossy; tall, spreading, arching, a climber growth; PP9492; [Seedling X Lady of the Dawn]; Zary, Dr. Keith W.; Bear Creek Gardens, Inc., 1997

'JACPICO', HT, w, 1978; (**Pristine®**); bud long; flowers near white, shaded light pink, imbricated, dbl., 28 petals, 5–6 in., slight fragrance; foliage very large, dark; upright growth; Edland Fragrance Medal, ARS, 1979 GM, Portland, 1979; [White Masterpiece X First Prize]; Warriner, William A.; J&P

'JACPIE', HT, mr, 1978; (**Grand Masterpiece**); bud ovoid, pointed; 5 in., exhibition form, slight fragrance; tall, upright growth; [Seedling X Tonight]; Warriner, William A.; J&P

'JACPIF', F, mp, 1988; (**Pleasure**); bud ovoid, pointed; flowers coral pink, reverse lighter, large, dbl., 33 petals, cupped, borne in sprays of 3-7, slight fragrance; globular fruit; prickles slightly hooked downward, reddish-brown; foliage medium, dark green, semi-glossy; low, compact growth; PP007480; AOE, 1990; [(Merci X Faberge) X Intrigue]; Warriner, William A.; Bear Creek Gardens, 1990

'JACPIHI', HT, w, 1998; (**Grand Finale**®); bud long, ovoid; flowers ivory white, full, very dbl., 26–40 petals, 4–4.5 in., exhibition form, borne singly, slight, honeysuckle fragrance; prickles moderate; foliage large, medium green, semi-glossy; upright, medium (4 ft.) growth; Rose of the Year, J&P, 1998; [Honor X Pristine]; Zary, Dr. Keith W.; Bear Creek Gardens, Inc., 1998

JACpik, HT, or, 1994; (**African Sunset**); J&P

JACpin, F, op, 1993; (**Dream Dolly**); J&P

'JACPLY', HT, yb, 1993; (**Somersault**); flowers yellow and rose red/orange blend, blooms borne in small or l, dbl., 26–40 petals, 1.5–2.75 in., slight fragrance; some prickles; foliage small, dark green, glossy; low (45-50cms), upright, bushy growth; Zary, Dr. Keith W.; Bear Creek Gardens

'JACPO', Min, my, 1984; (**Goldpoint**); flowers small, dbl., 20 petals, slight fragrance; foliage small, light green, matt; upright, bushy growth; PP005645; [Rise 'n' Shine X (Faberge X Precilla)]; Warriner, William A.; J&P

'JACPOP', HT, yb, 1985; (**Prince of Peace**); flowers yellow edged with pink, medium, dbl., 35 petals, slight fragrance; foliage medium, medium green, glossy; upright growth; PP005901; [Bridal Pink X ?]; J&P; McConnell Nurs., Inc.

JACpow, HT, w, 1992; (**Fantasy**); J&P

'JACPOY', Min, yb, 1993; (**Hot Tamale**®); flowers yellow-orange blend changing to yellow pink, finishing pink, dbl., 26–40 petals, 1.5–2.75 in., exhibition form, borne singly and in small clusters, slight fragrance; some prickles; foliage small, dark green, semi-glossy; low (36 cms), bushy, compact growth; PP9015; AOE, ARS, 1994; Zary, Dr. Keith W.; Bear Creek Gardens, Inc. (J&P);, Nor'East Miniature Roses

JACpri, Min, ab; (**Joyous Moment**); J&P

'JACPURR', F, m, 1991; (**Purple Tiger**); bud pointed, ovoid; flowers very deep purple with stripes and flecks of white and mauve, dbl., 26–40 petals, 3.5–4 in., borne in small clusters, moderate, damask fragrance; nearly thornless; foliage medium, medium green, glossy; stems very glabrous (shiny); medium (70-90cms), bushy growth; [Intrigue X Pinstripe]; Christensen, Jack E.; Bear Creek Gardens, 1992

'JACPURSH', S, m, 1998; (**Purple Simplicity**); flowers raspberry purple, semi-double, semi-dbl., 12–16 petals, 3 in.., borne in large clusters, slight fragrance; prickles few, straight, thin; foliage medium, medium green, semi-glossy; tall, bushy, for hedges growth; [Seedling X Love Potion]; Zary, Dr. Keith W.; Bear Creek Gardens, Inc., 1999

Jacqueline *see* 'MEIGER'

'JACQUELINE', HT, mr, 1961; bud short, pointed; flowers turkey-red, dbl., 30 petals, 4–4.5 in., exhibition form, moderate fragrance; foliage glossy; vigorous, well-branched growth; [Topper X Unnamed seedling]; J.H. Hill Co.

Jacqueline *see* KORoranki

Jacqueline du Pré® *see* 'HARWANNA'

'JACQUELINE DUFIER', HT, dr, 1957; flowers crimson shaded black, high pointed, dbl., 25 petals, 4 in., slight fragrance; foliage bronze; vigorous growth; [Dicksons Red X Ena Harkness]; Kemp, M.L.

Jacqueline Humery, HMsk, lp, 1995; Lens

Jacqueline Nebout® *see* MEIchoiju

'JACQUELINE STERNOTTE', HT, rb, 1974; bud ovoid; flowers red and white, full, very dbl., 51 petals, 4.5 in., moderate fragrance; Delforge, S.

'JACQUES AMYOT', N, lp; flowers lilac pink, medium, good repeat bloom; tall growth; Varangot, 1844

'JACQUES CARROY', Pol, m, 1929; flowers carmine, slightly tinted purple, center velvety, borne in clusters; Turbat

'JACQUES CARTEAU', HT, ly, 1957; bud long; flowers creamy yellow, medium; foliage glossy; moderate growth; Privat

Jacques Cartier, P, lp, 1868; flowers clear rose, center darker, very dbl., 50 petals, 3.5–4 in., rosette, borne singly and in small clusters, intense, sweet fragrance; foliage medium green, matt; compact, low, branching growth; Moreau et Robert

Jacques Esterel, HT, m; Croix

'JACQUES HACKENBURG', HT, pb, 1919; flowers deep rose-pink and carmine, opening flesh-white, dbl., intense fragrance; [Jonkheer J.L. Mock X Marquise de Sinéty]; Leenders, M.

'JACQUES LATOUCHE', HT, pb, 1935; flowers orange-pink in spring, red in summer, reverse yellow, very large, dbl., cupped, slight fragrance; foliage glossy; vigorous growth; [Souv. de Claudius Pernet X Director Rubió]; Mallerin, C.; H. Guillot

'JACQUES PORCHER', HT, rb, 1914; flowers blended tints of carmine, saffron and deep yellow, dbl., moderate fragrance; Guillot, P.

Jacques Prevert® *see* 'MEIMOUSLIN'

'JACQUES PROUST', Pol, m, 1904; flowers violet-red, small, semi-dbl., moderate fragrance; Robichon; (Sangerhausen)

'JACQUES VINCENT', HT, rb, 1908; flowers coral-red, center golden; vigorous growth; [Mme J.W. Budde X Souv. de Catherine Guillot]; Soupert & Notting

Jacquie Williams *see* 'MORWHEELS'

'JACQUINT', F, ab, 1998; (**Royal Wedding**®); flowers pink amber, very full, very dbl., 41 petals, 3.5–4 in., borne mostly singly, moderate fragrance; moderate, straight prickles; foliage medium, dark green, semi-glossy; compact, old-fashioned growth; [Impatient X Amber Queen]; Zary, Dr. Keith W.; Bear Creek Gardens, Inc., 1998

Jacquot, F, 1973; Dorieux, Francois; (Cavriglia)

JACrain, Min, pb, 1993; (**Kleine Renate**); J&P

JACrare, F, pb; (**Sixth Sense**); J&P

'JACRAW', F, rb, 1992; (**Peppermint Twist**); flowers red/white/pink striped, open slightly, blooms borne in small, very dbl., 3–3.5 in., cupped, slight fragrance; some prickles on peduncle; foliage large, medium green, semi-glossy; medium (90-110 cms), upright, bushy growth; [Pinstripe X Maestro]; Christensen, Jack E.; Bear Creek Gardens, 1992

'JACRAZ', F, rb, 1977; (**Razzle Dazzle**); flowers red, reverse white, dbl., 25 petals, 2.5 in., slight fragrance; foliage dark, leathery; bushy growth; GM, Portland, 1978; Warriner, William A.; J&P

'JACREDEM', Min, dr, 1998; (**Jingle Bells**®); flowers dark red, light red reverse, dbl., 17–25 petals, 2 in., borne in large clusters, slight fragrance; moderate, straight prickles; foliage medium, dark green, semi-glossy; bushy, medium growth, 2 ft., border; PP9325; AOE, J&P, 1995; [Seedling X Seedling]; Zary, Dr. Keith W.; Bear Creek Gardens, Inc., 1997

JACredi, HT, mr, 1997; (**Colinda**)

'JACREF', HT, mr, 1984; (**Limited Edition**); flowers large, dbl., no fragrance; foliage medium, medium green, semi-glossy; upright growth; [Unnamed seedling X Unnamed seedling]; J&P; McConnell Nurs., Inc.

JACREG, F, lp, 1995; (**Jubilation**)

'JACRENEW', Min, mr, 1999; (**Very Cherry**®); flowers fresh, clean lavender, dbl., 17–25 petals, 2–2.5 in., borne in small clusters, slight fragrance; prickles moderate; foliage large, medium green, glossy; upright, compact, globular, medium (20-26 in.) growth; [seedling X New Year]; Walden, John K.; Bear Creek Gardens, Inc., 1999

JACrette, HT, mr, 1997; (**Emperor**); J&P

'JACRIM', F, rb, 1986; (**Target**); flowers white with red petal edges, dbl., 25 petals, exhibition form, borne usually singly, no fragrance; straight, large and small prickles; foliage medium, medium green, semi-glossy; upright, bushy, medium growth; PP006255; [Prominent X Unnamed seedling]; Warriner, William A.; J&P

'JACRINK', Min, dp, 1988; (**Day Glow**); bud ovoid, pointed; flowers deep pink, fading lighter, medium, borne usually singly, very dbl., 60 petals, exhibition form, slight fragrance; prickles straight, short; foliage medium, dark green, matt; low growth; [Petticoat X Red Jewel]; Warriner, William A.; Bear Creek Gardens, 1989

'JACRIS', S, dp, 1987; (**Contessa**); flowers deep pink with yellow center, reverse deep pink, fading slightly, semi-dbl., 6–10 petals, flat, fast cycle, slight, spicy fragrance; straight, long, green-brown prickles; foliage medium, medium green, semi-glossy; upright, tall growth; PP006577; [Sunsprite X Seedling]; Warriner, William A.; J&P

'JACRITE', HT, yb, 1991; (**Rio Samba**®); flowers medium yellow fading to peach-pink, dbl., 15–25 petals, 2.75–3 in., exhibition form, borne singly and small clusters, slight fragrance; some prickles; foliage medium, dark green, matt; medium (110-120 cms), upright, bushy growth; PP8361; AARS, 1993; [Unnamed seedling X Sunbright]; Warriner, William A.; Bear Creek Gardens, 1993

'JACRO', F, dr, 1987; (**Bravado**); flowers , dbl., 26 petals, exhibition form, borne usually singly, slight fragrance; medium, red to brown prickles, hooked downward; foliage medium, dark, semi-glossy; medium, upright growth; PP005978; [Seedling X Gabriella]; Warriner, William A.; J&P

'JACROM', HT, mp, 1983; (**Romance**); flowers medium salmon-pink, large, dbl., 35 petals, slight fragrance; foliage large, medium green, semi-glossy; upright growth; PP005250; [Unnamed variety X Prominent]; Warriner, William A.; J&P

'JACROSE', HT, dp, 1983; (**Tribute**); bud long, pointed; flowers deep pink, loose form, large blooms borne singly, dbl., 30 petals, slight fragrance; reddish prickles, hooked down; foliage large; upright growth; [Unnamed seedling X Unnamed seedling]; Warriner, William A.; J&P

'JACROVE', HT, dr, 1999; (**Perfectly Red**®); dbl., 26–40 petals, 4–4.5 in., borne mostly singly, slight fragrance; prickles moderate; foliage medium, dark green, semi-glossy; upright (5 ft) growth; [seedling X Love]; Zary, Dr. Keith W.; Bear Creek Gardens, Inc., 1999

'JACRULAV', HRg, m, 1999; (**Wild Berry Breeze**®, Wildberry Breeze); flowers lavender-pink, veined, large petals, single, 5–11 petals, 3.5 in., borne in clusters, intense, spicy clove fragrance; round, large, orange fruit; many prickles; foliage medium, dark green, semi-glossy, deeply rugose; upright, bushy, medium (3.5 ft) growth; [Buffalo Gal X seedling]; Zary, Dr. Keith W.; Bear Creek Gardens, Inc., 1999

'JACRUWHI', HRg, w, 1999; (**Wild Spice**); flowers snowy white, delicate, ruffled, single, single, 5–11 petals, 3.5 in., borne in small clusters, intense, clove fragrance; foliage dark green, semi-glossy, deeply rugose; upright, bushy, medium (3.5 ft) growth; [Buffalo Gal X seedling]; Zary, Dr. Keith W.; Bear Creek Gardens, Inc., 1999

'JACRYBI', Min, rb, 1999; (**Petite Perfection**®); bud long, ovoid; flowers bright red/yellow, reverse deep yellow, dbl., 17–25 petals, 2.25–2.5 in., exhibition form, borne singly and in small clusters, slight fragrance; prickles moderate; foliage medium, dark green, glossy; patio mini; upright, compact, medium (16-22 in.) growth; [seedling X seedling]; Walden, John K.; Bear Creek Gardens, Inc., 1999

JACsak, HT, mr, 1994; (**Riviera**); J&P

'JACSAL', F, op, 1980; (**Cherish**); bud short, flat; flowers coral-pink, dbl., 28 petals, 3 in., exhibition form, slight fragrance; foliage large, dark; compact, spreading growth; AARS, 1980; [Bridal Pink X Matador]; Warriner, William A.; J&P

JACsan, Gr, ab, 1994; (**Pearl of Joy**); J&P

'JACSASH', HT, lp, 1999; (**Rose Rhapsody**®); flowers deep, dusty pink, very dbl., 41–50+ petals, 6–6.5 in., borne mostly singly, intense, citrus fragrance; prickles moderate; foliage large, dark green, glossy; upright, spreading, tall (4.5-5 ft) growth; [Fragrant Cloud X Ingrid Bergman]; Zary, Dr. Keith W.; Bear Creek Gardens, Inc., 1999

'JACSAT', LCl, dr, 1992; (**Dynamite**®); flowers bright red, dbl., 30 petals, 4–4.5 in., borne in small clusters, slight, citrus fragrance; some prickles; foliage large, dark green, glossy, resistant to powdery mild; tall (150-185 cms), upright, spreading, arching growth; PP8741; [Seedling X Simpathie]; Warriner, William A.; Bear Creek Gardens

'JACSAY', HT, dr, 1990; (**Avanti**); bud ovoid, pointed; flowers dark red, hint of blue reverse, aging dark, dark red, urn-shaped, large, very dbl., 25–30 petals, exhibition form, slight, fruity fragrance; foliage medium to large, dark green, matt to semi-glossy; upright, bushy, tall growth; [Royalty X Samantha]; Warriner, William A. & Zary, Keith W., 1986; Bear Creek Gardens, 1992

'JACSEDI', F, m, 1993; (**Love Potion**®, Purple Puff); flowers deep clear lavender, dbl., 26–40 petals, 3–3.5 in., borne in small clusters, intense, raspberry fragrance; some prickles; foliage medium, dark green, glossy; medium (90-100 cms), upright, spreading growth; PP9172; [Seedling X Dilly Dilly]; Christensen, Jack E.; Bear Creek Gardens, 1994

'JACSEE', HT, pb, 1998; (**Fragrant Lace**®); flowers cream and lavendar pink with cream and light yellow reverse, full, large, very dbl., 26–40 petals, 5 in.–5.5 in., borne mostly singly, intense fragrance; prickles moderate; foliage large, dark green, glossy; upright growth; [Seedling X Cherry Jubilee]; Zary, Dr. Keith W.; Bear Creek Gardens, 1998

'JACSERAW', F, ob, 1991; (**Orange Splash**); flowers bright orange to orange-red with white and lighter orange st, dbl., 15–25 petals, 3–3.5 in., moderate fragrance; some prickles; foliage large, dark green, glossy; medium (75-90 cms), upright, bushy growth; [Unnamed seedling X Unnamed seedling]; Christensen, Jack E.; Bear Creek Gardens, 1992

'JACSHAQ', HT, pb, 1998; (**Diana, Princess of Wales**®, The Work Continues); bud ovoid, pointed; flowers luminous pink and creamy ivory blend, very dbl., 26–40 petals, 4 in., exhibition form, borne in small clusters, moderate fragrance; prickles moderate; foliage medium, dark green, glossy; upright, 4.5 - 5 ft. growth; [Anne Morrow Lindbergh X Sheer elegance]; Zary, Dr. Keith W.; Bear Creek Gardens, Inc., 1999

'JACSHE', HT, ab, 1986; (**Summer Dream**); flowers apricot pink, blooms borne singly, dbl., 30 petals, exhibition form, slight, fruity fragrance; foliage medium, medium green, matt; upright, tall growth; [Sunshine X Unnamed seedling]; Warriner, William A.; J&P, 1987

'JACSHIP', S, mp, 1989; (**Pink Polyanna**®); bud ovoid; flowers medium pink, reverse slightly darker, aging lighter, small, dbl., exhibition form, repeat bloom, moderate, fruity fragrance; prickles small, straight, red to brown; foliage small, medium green, glossy; upright, bushy, medium growth; [Zorina X Heidi]; Warriner, William A.; Bear Creek Gardens, 1990

'JACSHO', HT, dr, 1981; (**Showstopper**); bud pointed; flowers deep red, spiral blooms borne singly or 3-4 per cluster, dbl., 33 petals, intense fragrance; long prickles; foliage large; strong, upright, bushy growth; [Unnamed seedling X Samantha]; Warriner, William A.; J&P

'JACSHOK', F, m, 1999; (**Cotillion**™); bud pointed, ovoid; flowers lavender, reverse pale lavender, very dbl., 41 petals, 4 in., rosette, borne in large clusters, intense, sweet fragrance; prickles moderate; foliage medium, dark green, glossy; upright, spreading (3.5 ft) growth; GM, Rome, 1998 [seedling X Shocking Blue]; Zary, Dr. Keith W.; Bear Creek Gardens, Inc., 1999

'JACSHUR', S, dr, 1991; (**King Crimson**); flowers large, semi-dbl., 6–14 petals, no fragrance; foliage large, dark green, semi-glossy; tall, upright, bushy growth; [Razzle Dazzle X Unnamed seedling]; Warriner, William A.; Bear Creek Gardens

'JACSIL', F, m, 1986; (**Eloquence**, 'JACAIL', Little Silver); flowers lavender, mini-flora, small, dbl., 20 petals, flat, borne singly and in clusters, slight, spicy fragrance; foliage medium, medium green, matt; upright, bushy growth; PP006010; [(Merci X Faberge) X Angel Face]; Warriner, William A.; J&P

'JACSIM', F, mp, 1991; (**Sweet Inspiration**); flowers medium pink with some cream coloration at petal base on both, dbl., 26–40 petals, 3–3.5 in.; few prickles; foliage medium, medium green, matt; upright (85-100 cms), bushy growth; [Sun Flare X Simplicity]; Warriner, William A.; Bear Creek Gardens, 1993

'JACSIMPL', S, mr, 1991; (**Red Simplicity**); flowers bright red, a bit of blackening near petal edges, moderately, dbl., 15–25 petals, slight fragrance; foliage medium, medium green, semi-glossy; medium, upright, spreading growth; [Unnamed seedling X Sun Flare]; Warriner, William A. & Zary, Keith W.; Bear Creek Gardens, 1992

'JACSLING', HT, mr, 1993; (**Legacy**™); J&P

'JACSMI', Min, my, 1984; (**Smiles**); flowers medium, dbl., 35 petals, slight fragrance; foliage small, light green, semi-glossy; upright, bushy growth; PP005577; [Spanish Sun X Calgold]; Warriner, William A.; J&P

JACsnow, F, w; (**White Simplicity**®); J&P, 1991

'JACSOD', HT, mp, 1985; (**Rhapsody**); flowers medium, dbl., 20 petals, slight fragrance; foliage medium, medium green, matt; [Unnamed variety X Unnamed variety]; Warriner, William A.; J&P

'JACSOS', S, mr, 1991; (**Cherry Jubilee**); flowers light to medium red, moderately medium, dbl., 15–25 petals, no fragrance; foliage medium, medium green, semi-glossy; upright, bushy growth; [Seedling X Simplicity]; Warriner, William A.; Bear Creek Gardens

JACsowe, HT, dp; (**Sowetan Peace**); J&P, 1994

'JACSPIF', F, ly, 1986; (**Satin Doll**); flowers cream to pale apricot; [Pacifica sport]; Warriner, William A.; J&P

JACspri, F, dy, 1992; (**Eldorado**); Rose of The Year, Auckland, NZ, 1994; Christensen, Jack E.; J&P, 1992

'JACSTAND', S, rb, 1999; (**Raspberry Swirl**™); flowers red, white, pink stripe, reverse same, dbl., 17–25 petals, 2 in., borne in large clusters, slight, apple scent fragrance; prickles moderate; foliage small, dark green, glossy; upright, arching, medium (3-4 ft) growth; [seedling X Plum Dandy]; Walden, John K.; Bear Creek Gardens, Inc., 1999

'JACSTINE', HT, yb, 1992; (**Spirit of Peace**); flowers apricot yellow opening with pink tinge where sun strikes the, dbl., 26–40 petals, 3–3.5 in., slight fragrance; some prickles; foliage medium, dark green, semi-glossy; tall (170-180 cms), upright, spreading growth; [Pristine X Unnamed seedling]; Warriner, William A.; Bear Creek Gardens

JACsto, HT, mr, 1993; (**Future Award**); J&P

'JACSTOP', HT, dr, 1990; (**Salsa**); bud ovoid, pointed; flowers urn-shaped, large, very dbl., 30–35 petals, exhibition form, borne singly, moderate, damask fragrance; few prickles; foliage medium, dark green, semi-glossy; upright, tall growth; PP006971; [Showstopper X Unnamed seedling]; Warriner, William A.; Bear Creek Gardens

'JACSUN', Min, dy, 1981; (**Bojangles**); flowers deep yellow, small, dbl., 20 petals, no fragrance; foliage small, light green, semi-glossy; upright, bushy growth; PP004990; [Spanish Sun X Calgold]; Warriner, William A.; J&P, 1983

JACsunov, F, lp; (**Super Nova**); J&P, 1997

'JACSUR', F, op, 1998; (**Bill Warriner**™); bud pointed, ovoid; flowers coral pink, ruffled edges, medium blooms, very dbl., 26–40 petals, 3.5–4 in., borne in large clusters, slight, sweet fragrance; prickles moderate, straight; foliage medium, dark green, semi-glossy; compact growth (3 ft.); PP9494; [Sun Flare X Impatient]; Warriner, William A. & Zary, Keith W.; Bear Creek Gardens, Inc., 1997

'JACSUYEL', S, my, 1999; (**Golden Bounty**™); dbl., 17–25 petals, 2.5 in., borne in large clusters, slight fragrance; few prickles; foliage medium, dark green, semi-glossy to glossy; upright, arching, medium (3-4 ft) growth; [Sun Flare X seedling]; Zary, Dr. Keith W.; Bear Creek Gardens, Inc., 2001

'JACTAFL', F, lp, 1998; (**Bridal Shower**™); flowers light pink, medium, very dbl., 26–40 petals, 4 in., borne in small clusters, slight fragrance; prickles moderate; foliage medium, dark green, glossy; upright, bushy, tall (4 ft.) growth; [Seedling X Sunflare]; Zary, Dr. Keith W.; Bear Creek Gardens, Inc., 1999

'JACTAN', LCl, r, 1986; (**Butterscotch**); flowers tannish-orange, double to loose, dbl., 25 petals, cupped, blooms in clusters of 3-5, slight fragrance; no fruit; foliage medium, medium green, semi-glossy; PP005895; [(Buccaneer X Zorina) X Royal Sunset]; Warriner, William A.; J&P

'JACTIG', HT, ab, 1974; (**Antigua**); bud ovoid; flowers large, dbl., exhibition form, slight fragrance; foliage leathery; vigorous, upright, bushy growth; GM, Geneva, 1972; [South Seas X Golden Masterpiece]; Warriner, William A.; J&P

'JACTIGER', F, ob, 1991; ('JACTANG', **Tiger Tail**); flowers deep orange, white or cream colored stripes, cream colored r, dbl., 26–40 petals, no fragrance; foliage medium, medium green, glossy; medium, bushy growth; [Matangi X Pinstripe]; Christensen, Jack E.; Bear Creek Gardens, 1992

'JACTOP', HT, mr, 1992; (**Legend**); flowers, very dbl., 3–3.5 in., exhibition form, borne mostly singly, intense fragrance; very small prickles on peduncle; foliage medium, dark green, semi-glossy; tall (120-140 cms), upright, bushy growth; PP006092; [Grand Masterpiece X Unnamed seedling]; Warriner, William A.; Bear Creek Gardens

'JACTOU', HT, dy, 1992; (**Midas Touch**™); bud urn-shaped; flowers bright, non-fading yellow, dbl., 15–25 petals, 3.5–4 in., exhibition form, borne mostly singly, moderate, musk fragrance; small prickles on peduncle; foliage large, medium green, matt; tall (150-160 cms), upright, bushy growth; PP8706; [Brandy X Friesensohne]; Christensen, Jack E.; Bear Creek Gardens, 1994

'JACTRED', HT, mr, 1988; (**American Spirit**, Medal of Honor); flowers large blooms, dbl., 35 petals, exhibition form, borne usually singly, no fragrance; foliage medium, medium green, semi-glossy; upright, tall growth; PP006083; [Seedling X American Pride]; Warriner, William A.; J&P

'JACTRIG', HT, m, 1999; (**Fragrant Rhapsody**™); flowers light lavender blend, small, dbl., 20–40 petals, borne mostly singly, intense fragrance; prickles moderate; foliage medium, medium green, glossy; upright, spreading, medium (4 ft) growth; [Intrigue X seedling]; Zary, Dr. Keith W.; Bear Creek Gardens, Inc., 1999

'JACTRO', HT, w, 1985; (**Sheer Bliss**); flowers white with pink center, large, dbl., 35 petals, exhibition form, borne singly, moderate, spicy fragrance; medium brown prickles; foliage medium, medium green, matt; medium, upright, bushy growth; PP006282; AARS, 1987 GM, Japan, 1984; [White Masterpiece X Grand Masterpiece]; Warriner, William A.; J&P, 1987

'JACTU', HT, pb, 1985; (**Headliner**); flowers petals white, blending to deep pink at edges, large, dbl., 40 petals, exhibition form, slight fragrance; foliage large, medium green, glossy; upright growth; PP005340; [Love X Color Magic]; Warriner, William A.; J&P

'JACTUTTI', Min, yb, 1991; (**Tutti-Frutti**); bud ovoid, pointed; flowers yellow with red-orange stripes, medium, dbl., 25–30 petals, cupped, slight fragrance; foliage medium, dark green, semi-glossy; bushy, spreading, medium growth; [Fool's Gold X Pinstripe]; Christensen, Jack E., 1984; Bear Creek Gardens, 1991

'JACTWIN', Gr, rb, 1980; (**Love**); bud short, pointed; flowers bright scarlet red, reverse silvery white, dbl., 35 petals, 3.5 in., exhibition form, borne mostly singly, slight, rose fragrance; foliage dark green, glossy, new growth is red; long stems; upright growth; AARS, 1980 GM, Portland, 1980; [? X Redgold]; Warriner, William A.; J&P

'JACUM', F, m, 1982; (**Intrigue**); flowers reddish purple, large, dbl., 20 petals, intense fragrance; foliage medium, dark, semi-glossy; PP005002; AARS, 1984; [White Masterpiece X Heirloom]; Warriner, William A.; J&P, 1984

'JACUP', Min, mr, 1984; (**Upstart**); flowers small, semi-dbl., slight fragrance; foliage small, medium green, semi-glossy; upright growth; [Merci X Fire Princess]; Warriner, William A.; J&P

JACurp, F, m, 1997; (**Night Star**); J&P

'JACUTE', Min, mp, 1988; (**Confection**); bud ovoid, pointed; flowers pink with yellow to cream base, medium, very dbl., exhibition form, borne usually singly, slight fragrance; prickles straight to slightly hooked, light green; foliage medium, dark green, matt; micro-mini low, bushy growth; PP006518; [Seedling X Seedling]; Warriner, William A.; Bear Creek Gardens

'JACVAL', HT, lp, 1998; (**Ultimate Pink**™); bud pointed, ovoid; flowers large, full, light pink, very dbl., 26–40 petals, 4.5–5 in., borne mostly singly, slight, sweet fragrance; prickles moderate, straight; foliage large, medium green, dull; tall, upright, 4.5 ft, vigorous growth; Rose of the Year, J&P, 1999; [Seedling X Fragrant Memory]; Zary, Dr. Keith W.; Bear Creek Gardens, Inc., 1999

'JACVEN', F, w, 1974; (**Evening Star**®); flowers white, base shading pale yellow, large, dbl., exhibition form, slight fragrance; foliage large, dark, leathery; vigorous, upright, bushy growth; GM, Belfast, 1977 GM, Portland, 1977; [White Masterpiece X Saratoga]; Warriner, William A.; J&P

'JACVEP', HT, pb, 1992; (**Corina**); flowers salmon pink with lighter almost cream reverse, yellow crescent at petal base, dbl., 26–40 petals, 3–3.5 in., exhibition form, moderate fragrance; few prickles; foliage large, dark green, matt; medium (150 cms), upright growth; [Bridal Pink X Kardinal]; Warriner, William A. & Zary, Keith W.; Bear Creek Gardens, 1993

JACvet, HT, dr; (**Velvet Treasure**); J&P, 1992

JACvoo, HT, w, 1996; (**First Affair**); J&P

'JACWADE', Min, w, 1992; (**Snow Shower**); flowers snow white, small, opens flat, very dbl., 1.5 in., flat, borne in small clusters, no fragrance; some prickles; foliage small, medium to dark green, glossy; low (15-20 cms), spreading 60-75 cms across; groundcover; PP9374; [Immensee X Roller Coaster]; Warriner, William A. & Zary, Keith W.; Bear Creek Gardens, 1993

'JACWHIM', Min, w, 1988; (**Shy Girl**); bud ovoid, pointed; flowers medium, very dbl., 80 petals, exhibition form, borne usually singly and in sprays of 2-4, slight fragrance; no fruit; prickles straight, short; foliage medium, dark green, semi-glossy; upright, spreading, low growth; PP006514; [Petticoat X Red Minimo]; Warriner, William A.; Bear Creek Gardens

'JACWHIP', S, w, 1993; (**Snow Angel**); flowers white with a light pink tone and a hint of yellow near petal, very dbl., 1.5–2.75 in., slight fragrance; some prickles; foliage small, dark green, semi-glossy; medium (5 ft), upright, bushy growth; [Pink Pollyanna sport]; Bear Creek Gardens; Bear Creek Gardens, 1993

'JACWHY', F, w, 1970; (**Bridal White**™); flowers ivory-white; [Bridal Pink sport]; Warriner, William A.; J&P

'JACWIG', Min, m, 1998; (**Wild Plum**™); flowers lavender blend, darker reverse, dbl., 17–25 petals, 1.5–2 in., borne in large clusters, slight fragrance; prickles few, straight; foliage medium, dark green, glossy; bushy, tall, hedge type, 3 ft. growth; [Winsome X Elegance]; Zary, Dr. Keith W.; Bear Creek Gardens, Inc., 1995

JACwiht, HT, w, 1995; (**Cathedral Peak**); J&P

'JACWOTTE', HT, ab, 1999; (**Vanilla Perfume**™); bud plump, ovoid; flowers light cream apricot/pink, reverse light apricot, dbl., 26–40 petals, 4–4.5 in., borne mostly singly and in small clusters, intense, sweet, spicy vanilla fragrance; prickles moderate; foliage medium, dark green, semi-glossy to glossy; upright, tall (5 ft) growth; [Anne Morrow Lindbergh X seedling]; Zary, Dr. Keith W.; Bear Creek Gardens, Inc., 1999

'JACYAP', HT, pb, 1993; (**Anne Morrow Lindbergh**); flowers pink, white, yellow blend, dbl., 26–40 petals, 3–3.5 in., borne mostly singly, moderate, some fragrance; some prickles; foliage medium to large, medium green, semi-glossy; tall (150-160 cms), upright growth; [Seedling X Seedling]; Warriner, William A.; Bear Creek Gardens, 1994

'JACYEF', Gr, dy, 1989; (**Shining Hour**™); bud ovoid; flowers deep, bright yellow, medium, borne usually singly and in sprays, dbl., 33 petals, exhibition form, moderate, fruity fragrance; prickles fairly long, hooked downward, red to yellow; foliage large, dark green, semi-glossy; upright, bushy, medium growth; AARS, 1991; [Sunbright X Sun Flare]; Warriner, William A.; Bear Creek Gardens, 1991

'JACYELSH', S, dy, 1998; (**Yellow Simplicity**); flowers deep yellow, medium-sized blooms, 12–16 petals, 3 in.., borne in small clusters, slight fragrance; prickles moderate; foliage medium, dark green, glossy; upright, tall hedge growth; [Seedling X Seedling]; Zary, Dr. Keith W.; Bear Creek Gardens, Inc., 1997

'JACYEM', HT, my, 1993; (**Santa Fe**); flowers rich, medium yellow, blooms, 3–3.5 in., moderate fragrance; few prickles; foliage large, dark green, matt; tall (180-210 cms), upright growth; [Emblem X Seedling]; Zary, Dr. Keith W.; Bear Creek Gardens, 1993

'JACYEPAT', S, dy, 1992; (**Yellow Jacket**); flowers bright yellow, loose, dbl., 26–40 petals, 1.5–2.75 in., borne mostly singly or small clusters, slight fragrance; some prickles; foliage medium, dark green, glossy, very disease resistant; patio; very vigorous, medium (75-90 cms), upright, bushy growt growth; [Unnamed seedling X Unnamed seedling]; Christensen, Jack E.; Bear Creek Gardens, 1992

'JACYES', HT, dy, 1995; (**Henry Fonda**™); bud pointed, ovoid; flowers deep yellow, pointed, dbl., 20–25 petals, 4.5–5 in., exhibition form, borne most singly, slight fragrance; foliage medium, dark green, glossy; tall growth; PP9390; Zary, Dr. Keith W.

JACyesp, F, my, 1997; (**Golden Galaxy**); J&P

'JACYIM', Min, dy, 1989; (**Sunsplash**™); bud ovoid, pointed; flowers deep yellow, aging pale yellow, dbl., 43 petals, exhibition form, borne in sprays of 18-21, slight fragrance; prickles long, slightly hooked downward, red to green-tan; foliage medium, medium green, very glossy, attractive; tall, upright, spreading growth; PP007127; [Rise 'n' Shine X Sun

Flare]; Warriner, William A.; Bear Creek Gardens

'JACYIMP', F, yb, 1999; (**Honey Bouquet**®); bud pointed, ovoid; flowers light yellow, hint of pink, reverse light yellow, dbl., 20–40 petals, 4.5–5 in., borne in small clusters, intense, sweet fragrance; foliage medium, dark green, semi-glossy; upright, spreading, medium, 3.5 ft. growth; [seedling X Amber Queen]; Zary, Dr. Keith W.; Bear Creek Gardens, Inc., 1999

'JACYO', HT, ob, 1989; (**Dynasty**®); bud pointed; flowers bright orange with yellow blending at petal base, aging to coral, dbl., 30 petals, cupped, no fragrance; prickles hooked down slightly, red to yellow-green; foliage medium, medium green, semi-glossy; upright, spreading, tall growth; PP006443; [Seedling X Seedling]; Warriner, William A.; Bear Creek Gardens, 1991

'JACZAP', F, ob, 1999; (**Outrageous**®); dbl., 26–40 petals, 3.5–4 in., borne in small clusters, moderate fragrance; prickles moderate; foliage medium, dark green, dull; bushy, medium (3.5 ft) growth; [Summer Fashion X seedling]; Zary, Dr. Keith W.; Bear Creek Gardens, Inc., 1999

'JACZOR', Gr, dp, 1999; (**Fame!**®); flowers deep shocking pink - almost light red, very dbl., 30–35 petals, 5 in., exhibition form, borne in small clusters, slight fragrance; foliage large, dark green; shrubby, full bush, tall growth; AARS, 1998; [Tournament of Roses X Zorina]; Zary, Dr. Keith W.; Bear Creek Gardens, Inc., 1998

'JACZOTTA', F, pb, 1999; (**Sorbet Bouquet**®); flowers rose pink, reverse light yellow, dbl., 26–40 petals, 4 in., borne in large clusters, slight fragrance; prickles moderate; foliage medium, dark green, glossy; bushy, medium (3.5 ft) growth; [Tournament of Roses X seedling]; Zary, Dr. Keith W.; Bear Creek Gardens, Inc., 2000

JADecosill, HT, dp, 1997; (**Ensa de Rennes**); Adam, M.

Jadis *see* 'JACDIS'

Jaen, HT, pb

'JÄGERBATAILLON', sp, (strain of R. canina), lp; almost thornless; once popular as an understock; Klinken

Jake McIlroy, HT, mp, 1997; Brundrett

'JALACT', Min, dr, 1993; (**Opening Act**); single, 1.5 in., borne in small clusters, no fragrance; some prickles; foliage medium, dark green, glossy; medium (30-35 cms), upright growth; [Anytime X Black Jade]; Jalbert, Brad; Select Roses, 1994

'JALCORAL', Min, op, 1996; (**Super Cascade Coral**); flowers coral, blooms borne in large clusters, very dbl., 1.5 in., slight fragrance; few prickles; foliage medium, dark green, glossy; spreading, medium (12 in), cascading growth; [Orange Honey X Sexy Rexy]; Jalbert, Brad; Select Roses, 1997

'JALPINK', Min, lp, 1996; (**Heaven Scent Pink**); flowers wide blooms, dbl., 26–40 petals, 2 in., flat, intense fragrance; few prickles; foliage medium, medium green, dull, reddish; upright, tall (24 in) growth; [Rise 'n' Shine X Rosemary Harkness]; Jalbert, Brad; Select Roses, 1997

'JALRAY', Min, pb, 1995; (**Reiko**); flowers medium pink, lighter pink reverse, dbl. (15–25 petals), medium (1 in.) blooms borne in small clusters; fragrant; some prickles; foliage medium, medium green, glossy; medium (12 in.), compact, bushy growyh; [Winsome X Springwood® Gold]; Jalbert, Brad; Select Roses, 1995

'JALSTAR', Min, pb, 1995; (**Rising Star**); flowers cream with strong pink blend edges, full (26–40 petals), large (1–2 in.) blooms borne mostly single; slight fragrance; some prickles; foliage medium, medium green, semi-glossy; tall (18 in.), upright, bushy growth; [Sans Souci X Pink Petticoat]; Jalbert, Brad; Select Roses, 1995

'JALSUN', Min, my, 1995; (**Sun chariot**); flowers medium yellow, bright gold stamens, dbl. (15–25 petals), medium blooms borne in small clusters; fragrant; numerous prickles; foliage medium, dark green, matt; medium, upright, tall growth; [Pink Petticoat X Bright Smile]; Jalbert, Brad; Select Roses, 1996

'JALWED', Min, mp, 1995; (**Summer Wedding**); flowers medium pink, full (26–40 petals), medium blooms borne in small clusters; slight fragrance; few prickles; foliage medium, medium green, semi-glossy; compact, bushy, medium growth; [Maurine Neuberger X Sexy Rexy]; Jalbert Brad; Select Roses, 1995

Jaltinskii Suvenir, F, ly, 1956; flowers large, dbl., moderate fragrance; Klimenko, V. N.; (Sangerhausen)

'JALWINE', Min, mr, 1993; (**Cherry Wine**); dbl., 26–40 petals, 1.5 in., borne mostly singly, slight fragrance; some prickles; foliage medium, medium green, matt; medium (30-35 cms), spreading growth; [Dee Bennett X Winsome]; Jalbert, Brad; Select Roses, 1994

'JALWOOD', Min, r, 1995; (**Sandalwood**); flowers russet with lighter edges, blooms borne in small clusters, dbl., 15–25 petals, 1 in., no fragrance; some prickles; foliage medium, medium green, glossy; medium (12 in), compact, bushy growth; [Blushing Blue X Springwood Gold]; Jalbert, Brad; Select Roses

Jam Session *see* 'ZIPJAM'

'JAMAICA', HT, mr, 1965; bud ovoid; flowers cherry-red, large, semi-dbl., cupped, moderate fragrance; foliage dark, glossy, leathery; vigorous, upright growth; [(Charlotte Armstrong X Floradora) X Nocturne]; Lindquist; Howard Rose Co.

'JAMBOREE', F, rb, 1964; flowers cherry-red, reverse lighter, borne in clusters, dbl., 26 petals, 1.5–2 in., flat; foliage glossy, light green; free growth; [Masquerade X ?]; Gregory

'JAMES APPLEBY', F, ob, 1963; flowers orange-scarlet flecked deep crimson, medium, borne in clusters, semi-dbl.; vigorous growth; [Orangeade sport]; Wood

James Biddle *see* 'WILJAME'

James Bond *see* **'JAMES BOND 007'**

'JAMES BOND 007', HT, op, 1966; (James Bond); flowers coral-pink, pointed; foliage glossy; bushy growth; [? X Fragrant Cloud]; Tantau, Math.

James Bougault *see* **'JAMES BOURGAULT'**

'JAMES BOURGAULT', HP, w, 1887; (James Bougault); flowers white, shaded rose; Renault

'JAMES FERRIS', HT, w, 1927; flowers creamy white; vigorous growth; Hall

'JAMES GIBSON', HT, mr, 1928; bud pointed; flowers crimson-scarlet, large, dbl., exhibition form, slight fragrance; foliage dark, leathery; vigorous, bushy growth; GM, NRS, 1929; McGredy

James Mason, HGal, mr, 1982; Beales, Peter

'JAMES MITCHELL', M, dp; flowers pink, small, dbl., early bloom; very heavily mossed stems; Verdier, E., 1861; E. Verdier

James Pereire *see* MEInorep

'JAMES REA', HT, dp, 1930; bud pointed; flowers rich carmine or rose-pink, very large, dbl., exhibition form, intense fragrance; foliage light, leathery; vigorous, bushy growth; GM, NRS, 1929; McGredy

'JAMES SMILE'®, HT, pb, 1987; flowers china-pink, reverse pink-yellow blend, loose, medium, borne, semi-dbl., 40 petals, cupped, moderate, spicy fragrance; rounded, medium, pink-red fruit; rare, small, brown prickles; foliage medium, medium green, semi-glossy; upright, medium growth; [Sea Pearl sport]; Rogin, Josip

'JAMES VEITCH', M, m, 1865; flowers violet-slate shaded fiery red, blooms in corymbs, dbl.; moderately vigorous growth; Verdier, E.

'JAMES WALLEY', HT, ab, 1923; flowers apricot and fiery salmon, large, dbl., moderate fragrance; foliage olive-green, leathery; vigorous growth; [Ophelia X Unnamed seedling]; Easlea

Jamestown, Gr, op, 1999; flowers soft peach-pink, semi-dbl., moderate fragrance; Williams, J. Benjamin

Jamie™ *see* 'MOGAJAM'

Jamie Alexander, S, lp; [Ko's Yellow X Eyepaint]; Hannemann, F., 1995; The Rose Paradise; (Weatherly, L.)

Jamie's Love *see* 'GELJAM'

'JAN ABBING', HT, rb, 1933; bud pointed; flowers salmon-red shaded yellow, large, cupped, slight fragrance; foliage leathery, dark; vigorous growth; [Columbia X Étoile de Hollande]; Tantau

'JAN BÖHM', HP, mr, 1928; flowers velvety fiery red, dbl., 40 petals, recurrent bloom, moderate fragrance; vigorous growth; [Hugh Dickson X King George V]; Böhm, J.

'JAN GUEST', HT, pb, 1975; flowers carmine-pink with yellow reverse, dbl., 43 petals, 4 in.; foliage glossy; vigorous, upright growth; [Fragrant Cloud X Irish Gold]; Guest

'JAN H. MEYER', HT, mr, 1954; flowers cinnabar-red, well formed, large; vigorous growth; [Tawny Gold X Gaudia]; Leenders, M.

'JAN HUS', HT, lp, 1933; flowers large, dbl., moderate fragrance; Böhm, J.; (Sangerhausen)

'JAN SPEK'®, F, dy, 1966; dbl., 44 petals, 3 in., flat, blooms in clusters, slight fragrance; foliage dark, glossy; GM, Belfast, 1968 GM, The Hague, 1970; [Clare Grammerstorf X Faust]; McGredy, Sam IV; McGredy

'JAN STEEN', HT, mr, 1923; flowers brilliant scarlet-red, semi-dbl., slight fragrance; [Mev. Dora van Tets X Gruss an Dresden]; Spek

Jan Steen, S, rb, 1997; bud high; flowers ivory and red striped, single; upright growth; Williams, J. Benjamin

'JAN VAN RIEBEECK', HT, m, 1952; flowers carmine, large, dbl., moderate fragrance; vigorous growth; Leenders, M.

Jan Vermeer, F, m, 1997; flowers purple/burgundy blend, borne in sprays; Williams, J. Benjamin

'JAN WELLUM', Min, dr, 1986; (Vancouver Centennial); flowers large blooms borne singly and in sprays of 3-5, dbl., 20 petals, slight, fruity fragrance; small, globular fruit; small, straight, near-white prickles; foliage medium, medium green, matt; spreading growth; [Unnamed Shrub X Dwarfking '78]; Fischer, C.&H.; Alpenflora Gardens

Jana, HT, dy, 1976; flowers large, dbl.; Cocker; (Sangerhausen)

'JANAL', HT, dp, 1978; bud pointed; flowers deep pink, dbl., 55 petals, exhibition form, intense fragrance; hooked red prickles; foliage bronze-red to dark green; vigorous growth; [Charles Mallerin X (Duet X Kordes' Perfecta)]; Dawson, George; Rainbow Roses

'JANE', HT, ob, 1956; flowers coppery orange suffused pink, well formed, dbl., 24 petals, 3.5 in., moderate fragrance; foliage dark; vigorous, upright growth; [Signora X Mrs Edward Laxton]; Mee; Sanday

Jane Asher *see* 'PEAPET'

'JANE CARREL', HT, yb, 1940; bud pointed; flowers yellow shaded orange, medium, dbl., moderate fragrance; very vigorous, bushy growth; Gaujard

Jane Eyre *see* 'MEHPARK'

Jane Isobella Linton, T, pb; L'Hay

'JANE JACKSON', F, ob, 1976; flowers strawberry-orange, dbl.; low, vigorous, bushy growth; [Tip Top sport]; Jackson, J.R.

Jane Lathrop Stanford, HT, lp

'JANE LAZENBY', F, mp, 1959; flowers rose-pink, dbl., 25 petals, 3.5 in., flat, moderate fragrance; foliage dark; very vigorous, bushy growth; [Alain X Mme Henri Guillot]; McGredy, Sam IV; McGredy

'JANE PAULEY', HT, ob, 1992; flowers orange, reverse orange, aging orange pink, over large blooms borne, dbl., 35 petals, 6.5 in., exhibition form, moderate fragrance; foliage large, medium green, semi-glossy; upright, medium growth; [Elizabeth Taylor X Fortuna]; Weddle, Von C.; Hortico Roses, 1993

'JANE PIEKARSKI', HT, w, 1994; very dbl., 3–3.5 in., borne mostly singly, intense fragrance; many prickles; foliage medium, medium green, matt; medium (24 in), upright growth; [Honor X John F. Kennedy]; Hoshall, Howard

'JANE PROBYN', HT, dp, 1978; flowers deep pink; [Red Devil sport]; Anderson, K.

'JANE ROGERS', HT, ab, 1991; flowers large blooms borne singly, sometimes in small clusters, dbl., 26–40 petals, slight fragrance; foliage medium, medium green, semi-glossy; medium, spreading growth; [Fragrant Cloud X Diamond Jubilee]; Mander, George

'JANE THORNTON', HT, rb; flowers velvety crimson shaded maroon, intense fragrance; Bees, about 1940

'JANET', HT, ab, 1915; flowers golden fawn, shaded copper and rose, fading, dbl., moderate fragrance; GM, NRS, 1916; Dickson, A.

Janet B. Wood, Ayr, w, 1990; Beales, Peter

Janet Bebb *see* 'MEHGOLDIE'

Janet Carnochan, HT, w, 1996; Pallek

'JANET FRAZER', F, op, 1937; flowers shrimp-pink and yellow, blooms in clusters, semi-dbl., 3.5 in., slight fragrance; foliage light green; [Mme Leon Cuny X (Orange Sweetheart X Cinnabar)]; McGredy

'JANET MORRISON', Cl HT, dp, 1936; flowers deep pink, large blooms, semi-dbl., moderate fragrance; long stems; vigorous growth; [Black Boy X Seedling]; Clark, A.; NRS Victoria

'JANET'S PRIDE', HEg, pb, 1892; (Clementine); flowers white, edged carmine-rose, semi-dbl., non-recurrent; vigorous growth; Paul; Paul

'JANETTE MURRAY', HT, op, 1985; flowers orange pink, large blooms borne singly, dbl., 35 petals, slight fragrance; foliage large, medium green, semi-glossy; vigorous growth; [Daily Sketch seedling X Montezuma]; Bell, Ronald J.; Brundrett & Sons

'JANICE', Min, mp, 1971; bud ovoid; flowers small, dbl.; foliage small, glossy, leathery; vigorous, dwarf, upright, bushy growth; [(R. wichurana X Floradora) X Eleanor]; Moore, Ralph S.; Sequoia Nursery

Janice Heyes, HT, lp, 1986; Dawson

'JANICE TELLIAN', Min, op, 1979; bud pointed; flowers light coral-pink, dbl., 40 petals, 1 in., exhibition form, slight fragrance; foliage small; dwarf, bushy, compact growth; [Fairy Moss X Fire Princess]; Moore, Ralph S.; Sequoia Nursery

'JANIDA', HT, rb, 1955; bud long, pointed, blood-red veined maroon; flowers orange, dbl.; [Crimson Glory X Baby Chateau]; Robichon

Janina® *see* TANija

'JANINE ASTLE', F, mp, 1971; flowers clear pink, dbl., 22 petals, 3 in., exhibition form, slight fragrance; foliage bluish green; vigorous, upright growth; [Charlotte Elizabeth X Grand Slam]; Hunt

Janine Herholdt, HT, ab, 1989; Herholdt

'JANINE VIAUD-BRUANT', HT, mr, 1910; flowers large, dbl., moderate fragrance; Viaud-Bruant; (Sangerhausen)

'JANNA', Min, pb, 1970; bud pointed; flowers pink, reverse white, small, dbl.; foliage leathery; dwarf, bushy growth; [Little Darling X (Little Darling X (R. wichurana X Miniatureseedling))]; Moore, Ralph S.; Sequoia Nursery

Janos, Pol, or; flowers medium, dbl.; VEG; (Sangerhausen)

Jan's Wedding *see* 'ADAJAN'

'JANTAR', HT, dr, 1966; bud elongated; flowers dark crimson, shaded darker, large; vigorous, upright growth; Grabczewski

'JANTZEN GIRL', Gr, mr, 1961; bud ovoid; dbl., 4–5 in., exhibition form, blooms in clusters, slight fragrance; foliage glossy; upright growth; [Carrousel X (Chrysler Imperial X Seedling)]; Von Abrams; Peterson & Dering

Japanese Rose *see* **R. RUGOSA**

'JAPONICA' A strain of R. multiflora, probably indistinguishable from it; popular as an understock

Japonica *see* **'MOUSSEUX DU JAPON'**

'JAPONICA THORNLESS'; prickle-free strain of R. multiflora

Jaquenetta, S, ab, 1983; Austin, David

Jardinero Ortiz® *see* DOTraner

Jardins de Bagatelle® *see* 'MEIMAFRIS'

Jardins de France *see* MEIzebul

Jardins de Valloires, Pol, dp, 1994; Eve, A.

Jardins de Viels Maisons *see* MASframb

Jardins de Villandry *see* DELrovrai

Jaris Hudson, Min, mp

Jarlina *see* 'POULENA'

'JARVIS BROOK', HT, rb, 1928; flowers carmine, reverse orange, dbl., moderate fragrance; Low

Jasena, HT, dy, 1983; flowers large, dbl., slight fragrance; Urban, J.; (Sangerhausen)

'JASLOK', HT, m, 1999; flowers light mauve with dark mauve stripes, reverse lighter, very dbl., 41 petals, 4–5 in., borne mostly singly, slight fragrance; prickles moderate; foliage large, medium green, semi-glossy; compact, medium (2.5-3.5 ft.) growth; [Blue Ocean sport]; Chiplunkar, C. R., 1992; KSG's Roses, 1992

Jasnaja Poljana, HT, pb, 1958; flowers salmon-pink with light yellow, large, very dbl., moderate fragrance; Shtanko, E.E.; (Sangerhausen)

Jason *see* 'SHERICE'

Jaunatella, HT, yb

'JAUNÂTRE', HSem, ly; flowers yellowish white, moderate fragrance

'JAUNE BICOLOR', S, yb, 1633; flowers yellow streaked red, medium, single; [R. foetida bicolor sport]

'JAUNE DESPREZ', N, yb, 1830; (Desprez à Fleur Jaunes, Noisette Desprez); flowers warm yellow shaded peach and apricot, dbl., flat, moderate fragrance; vigorous (to 20 ft) growth; [Blush Noisette X Parks' Yellow Tea-scented China]; Desprez

'JAUNE D'OR', T, ob; flowers coppery yellow, large, dbl., globular; Oger, 1863

'JAVA', F, ob, 1955; bud ovoid; flowers orange-red, medium, borne in clusters of 4-5, dbl., 40–45 petals, slight fragrance; foliage bronze; vigorous, upright, bushy growth; GM, Geneva, 1954; [Francais X Unnamed seedling]; Mallerin, C.; EFR

'JAWAHAR', HT, w, 1980; flowers creamy white, blooms borne 2-6 per cluster, dbl., 47 petals, exhibition form, intense fragrance; straight, brown prickles; foliage light green, glossy; vigorous, bushy growth; [Sweet Afton X Delhi Princess]; Division of Vegetable Crops and Floriculture

Jawarosa Mawa, HT, or; flowers large, dbl.; (Sangerhausen)

'JAY JAY', HT, w, 1971; flowers white, edged pink, very large, dbl., exhibition form, slight fragrance; foliage large, dark, leathery; very vigorous, bushy growth; [Peace sport]; Kern Rose Nursery

'JAYANC'®, Min, mp, 1997; (**Ancestry**); flowers semi double, medium, semi-dbl., 8–14 petals, borne in small clusters, moderate fragrance; foliage medium, medium green, glossy; spreading, bushy, medium (30cms) growth; [Seedling X Party Girl]; Jellyman, J.S.

'JAYART'®, Min, pb, 1997; (**Arthur Cox**); flowers double, medium, dbl., 15–25 petals, borne in small clusters, slight fragrance; foliage medium, medium green, semi-glossy; compact, medium growth; [Cotswold Gold X Peaches'n'Cream]; Jellyman, J.S.

Jayatsen, HT, pb, 1998; K&S

'JAYBO', Min, rb, 1988; flowers very bright red, yellow at base, fading topink, urn-shaped, semi-dbl., 22 petals, slight, fruity fragrance; prickles long, pointed, pink; foliage medium, dark green, glossy; bushy, medium growth; [Rise 'n' Shine X Unnamed seedling]; Bridges, Dennis A.; Bridges Roses

'JAYCOT', Min, ob, 1992; (**Cotswold Gold**); flowers orange, blooms borne in small clusters, semi-dbl., 6–14 petals, 1.5 in., slight fragrance; few prickles; foliage small, medium green, semi-glossy; low (20 cms), bushy growth; [Tony Jacklin X Judy Fischer]; Jellyman, J.S., 1993

'JAYDON', F, mr, 1998; (**Beverley Stoop**); flowers velvety red, dbl., 15–25 petals, borne in clusters, slight fragrance; foliage medium, medium green, semi-glossy; upright, compact, 2 1/2 ft. growth; [Seedling X (Tony Jacklin X Andrea)]; Jellyman, J.S.

'JAYEMM', F, mr, 1992; (**Emma Kate**); flowers light red with lighter reverse, full, blooms borne 4-10+ per cluster, dbl., 35 petals, moderate, fruity fragrance; foliage medium, dark green, glossy; upright, bushy, medium growth; [Tony Jacklin X Cairngorm]; Jellyman, J.S., 34001

'JAYGUP', HT, op, 1999; (**Diana H. Gupta**); flowers orange-pink, reverse pale lemon yellow with faint white striping, dbl., 26–40 petals, 3 in., borne in small clusters, slight fragrance; few prickles; foliage medium, medium green, semi-glossy; upright, bushy, medium (3 ft) growth; [Gavotte X Bill Temple]; Jellyman, J.S.

'JAYKOO'®, F, or, 1997; (**kooshti-Bok**); flowers single, medium, single, 4–7 petals, borne in small clusters, slight fragrance; foliage medium, dark green, semi-glossy; upright, spreading, medium to tall (3.5-4ft.)growth; [(Tony Jacklin X seedling) X (Tony Jacklin X Andrea)]; Jellyman, J.S.

'JAYMARJ'®, HT, pb, 1997; (**Marjorie Reid**); flowers double, small, dbl., 15–25 petals, slight fragrance; foliage small, dark green, glossy; low (25 cms) growth; [Seedling X Wee Barbie]; Jellyman, J.S.

'JAYMATT', HT, pb, 1997; (**The Matthew**®); flowers full, medium, very dbl., 26–40 petals, slight fragrance; few prickles; [Silver Jubilee X Pres. Petts]; Jellyman, J.S.

'JAYMER', f, ab, 1994; (**Merrie**); flowers apricot, pale pink, double, blooms borne 6-20 per cluster, 30 petals, moderate fragrance; foliage medium, dark green, semi-glossy; medium, upright, bushy growth; [Minnie Pearl X Seedling]; Jellyman, J.S.

'JAYMOLL', F, pb, 1997; (**Molly Beckley**®); flowers semi double, small, semi-dbl., 8–14 petals, slight fragrance; foliage small, dark green, glossy; spreading, low (1ft.6in.)growth; [(Tony Jacklin X Andrea) X Wee Barbie]; Jellyman, J.S.

Jayne Austin *see* 'AUSBREAK'

'JAYSIL', F, dp, 1997; (**Precious Gift**®); flowers double, medium, dbl., 15–25 petals, borne in small clusters, slight fragrance; few prickles; foliage medium, medium green, semi-glossy; upright, bushy, medium (20cms)growth; [Silver Jubilee X Seedling]; Jellyman, J.S.

'JAYSUN', F, pb, 1992; (**Cotswold Sunset**); flowers gold, pink edge, pale pink, moderately blooms borne in small clusters, dbl., 15–25 petals, 3–3.5 in., exhibition form, slight fragrance; some prickles; foliage medium, dark green, glossy; medium (80-90 cms), upright, compact growth; [Cairngorm X Seedling (Alexander X Wembley)]; Jellyman, J.S., 1993

'JAZZ', F, ob, 1960; flowers orange-yellow flushed crimson, borne in clusters, dbl., 26 petals, 2 in., slight fragrance; foliage dark, glossy; vigorous growth; [Masquerade X Unknown seedling]; deRuiter; Gregory

Jazz *see* POUlnorm

Jazz Club, S, ab, 1995; Lowery/Robinson

Jazz Dancer *see* 'SEATWINKLE'

'JAZZ FEST', F, mr, 1971; bud long, pointed; flowers medium, semi-dbl., slight fragrance; foliage large, leathery; vigorous, upright, bushy growth; [Pink Parfait X Garnette]; Armstrong, D.L.; Armstrong Nursery

Jazz Time™ *see* 'MINACCO'

Jean Adrien Mercer, F, 1991; Dorieux, Francois; (Cavriglia)

'JEAN BACH SISLEY', HCh, pb, 1889; flowers silvery rose, outer petals salmon-rose veined carmine, moderate fragrance; growth moderate; Dubreuil

'JEAN BAKER', HT, w, 1998; flowers white, exhibition blooms, very dbl., 26–40 petals, borne singly, intense fragrance; few prickles; foliage medium, dark green, glossy; upright, tall (4 - 5 ft) growth; [Crystalline X Classic Touch]; Edwards, Eddie

'JEAN BAPTISTE CASATI', HP, dp, 1886; flowers carmine-pink, large, very dbl., moderate fragrance; Schwartz; (Sangerhausen)

'JEAN BODIN', M, lp; flowers light rose-pink, quartered; not very mossy; vigorous growth; Vibert, 1846

'JEAN BOSTICK', HT, my, 1936; (Yellow Condesa de Sastago); flowers deep yellow, sometimes splotched red, large, dbl., 50 petals, globular, intense fragrance; foliage leathery, glossy; very vigorous growth; [Condesa de Sástago sport]; Bostick

'JEAN BROWN', HT, ly, 1930; flowers large, very dbl., moderate fragrance; Evans, F. David; (Sangerhausen)

'JEAN C. N. FORESTIER', HT, rb, 1919; flowers carmine, slightly tinted orange and yellow, very large, dbl.; foliage glossy, bronze; very vigorous growth; GM, Bagatelle, 1919; [Unnamed variety X Mme Edouard Herriot]; Pernet-Ducher

'JEAN CAMPBELL', HT, pb, 1964; flowers blush-pink suffused apricot, well formed, dbl., 28 petals, 4.5 in., moderate fragrance; foliage dark; upright growth; [Ethel Sanday seedling X Lady Sylvia]; Sanday, John

'JEAN CHERPIN', HP, m, 1865; flowers purple/pink, very large, very dbl.; Liabaud; (Sangerhausen)

'JEAN COTE', HT, yb, 1936; flowers old-gold, center deeper, very large, dbl.; foliage brilliant green; vigorous growth; Gaujard; J&P

Jean de la Lune *see* 'DELCRO'

'JEAN DU TILLEUX', HT, mp, 1980; bud long; flowers medium lavender pink, dbl., 30 petals, exhibition form, slight fragrance; slightly hooked red prickles; foliage deep green, waxy; medium, vigorous growth; [King of Hearts X Golden Masterpiece]; Winchel, Joseph F.; Kimbrew-Walter Roses

'JEAN DUCHER', T, op; flowers salmon to peachy pink, globular, moderate fragrance; Ducher, 1873

Jean Galbraith, HT, lp; [Abraham Darby sport]; Nieuwesteeg, J., 1999; (Weatherly, L.)

Jean Gaujard® *see* 'GAUVITOR'

Jean Giono® *see* 'MEIROKOI'

'JEAN GIRIN', HWich, pb, 1910; flowers bright rose-pink, base rosy white, stamens forming a yellow, dbl., profuse bloom, sometimes repeated; vigorous, climbing growth; Girin

'JEAN GUICHARD', HWich, pb, 1905; bud bronzy crimson; flowers copper-pink; vigorous, climbing growth; [R. wichurana X Souv. de Catherine Guillot]; Barbier

'JEAN KATHRYN', F, pb, 1973; flowers neyron rose, reverse white, very full, dbl., 45 petals, 4 in., intense fragrance; foliage light; very vigorous growth; [Memoriam X Gavotte]; Ellick

Jean Kenneally™ *see* 'TINEALLY'

'JEAN LAFITTE', LCl, mp, 1934; bud pointed; flowers willowmere pink, large, dbl., cupped, moderate fragrance; foliage leathery; ; very vigorous, climbing (8-10 ft. very hardy.; [R. setigera seedling X Willowmere]; Horvath; J&P

'JEAN LAPEYRE', S, my, 1960; flowers well formed, recurrent bloom; foliage bronze; vigorous, bushy growth; Gaujard

'JEAN LELIÈVRE', HP, dr, 1879; flowers large, dbl., moderate fragrance; Oger; (Sangerhausen)

'JEAN LHOSTE', HWich, rb, 1926; flowers rosy carmine, base flesh-white, large, borne in clusters, dbl.; [Alexandre Girault X Gerbe Rose]; Congy; Cochet-Cochet

'JEAN LIABAUD', HP, dr, 1875; flowers crimson-maroon, shaded scarlet, large, dbl., 60 petals, some recurrent bloom, moderate fragrance; vigorous growth; [Baron de Bonstetten X ?]; Liabaud

'JEAN LORTHOIS', HT, pb, 1879; flowers rose-pink, center darker, reverse silvery, well formed, large; Ducher, Vve.

'JEAN MACARTHUR', HT, mr, 1942; bud long, pointed, begonia-red; dbl., 30–40 petals, 3.5–4 in., exhibition form, slight fragrance; foliage leathery, wrinkled, dark; vigorous, upright growth; [Joanna Hill X California]; Hill, Joseph H., Co.

Jean Marc, F, or, 1975; Croix; (Cavriglia)

Jean Marc Rosé, F, 1995; Fineschi, G.; (Cavriglia)

Jean Marmoz *see* **'JEAN MERMOZ'**

'JEAN MAYCOCK', HT, mr, 1999; flowers crimson pink, reverse lighter, dbl., 26–40 petals, 3 in., borne in small clusters, slight fragrance; prickles moderate; foliage medium, medium green, semi-glossy; compact, medium (36 in.) growth; [Sexy Rexy X Sharifa Asma]; Rawlins, R.

'JEAN MCGREGOR REID', HT, w, 1962; flowers cream, large, moderate fragrance; foliage dark, glossy; vigorous growth; [Peace sport]; Sunter; Ross & Son

'JEAN MERMOZ', Pol, mp, 1937; (Jean Marmoz); flowers ruddy pink, imbricated, small blooms in long clusters, very dbl., slight fragrance; foliage glossy, dark; vigorous growth; [R. wichurana X Hybrid Tea]; Chenault; Hémeray-Aubert

Jean Monnet® *see* PEKcram

Jean Morrison, HT, or, 1996; Sutherland

'JEAN MURAOUR', F, w, 1935; flowers pure white, center light yellow; [Gruss an Aachen sport]; Vogel, M.

'JEAN NOTÉ', HT, op, 1909; flowers light salmon-pink, large, dbl.; Pernet-Ducher; (Sangerhausen)

'JEAN RAMEAU', B, dp, 1918; flowers iridescent rose; [Mme Isaac Pereire sport]; Darclanne; Turbat

'JEAN RENTON', HT, my, 1940; vigorous growth; Clark, A.

Jean Rex *see* 'REXJEAN'

Jean Rose *see* 'TINJEAN'

'JEAN ROSENKRANTZ', HP, or, 1864; flowers very bright coral red, large, dbl., intense fragrance; Portemer fils

'JEAN SISLEY', HT, m, 1879; flowers lilac-rose, large; moderate growth; [Adam X Emilie Hausburg]; Bennett

'JEAN SOUPERT', HP, dr, 1875; flowers crimson-maroon, large, dbl., moderate fragrance; [Charles Lefebvre X Souv. du Baron de Semur]; Lacharme, F.

'JEAN THOMSON HARRIS', F, op, 1976; flowers salmon, shaded orange, dbl., 30 petals, 4 in.; [(Fragrant Cloud X Heidelberg) X (Heidelberg X Kingcup)]; Cocker

'JEAN WEBB', HT, rb, 1966; flowers cochineal, reverse light bronze, 4 in., moderate fragrance; foliage dark; vigorous growth; [Bettina sport]; Marks

Jeanette Talbot *see* 'WEBEYAM'

'JEANIE', HT, w, 1959; flowers cream to pink, dbl., 66 petals, 4–4.5 in., exhibition form, moderate fragrance; foliage dark; vigorous, spreading growth; [Condesa de Sástago X Mme Edmond Labbe]; Eddie; Wyant

'JEANIE WILLIAMS', Min, rb, 1965; flowers orange-red, reverse yellow, small, dbl., slight fragrance; foliage leathery; vigorous, bushy growth; [Little Darling X Magic Wand]; Moore, Ralph S.; Sequoia Nursery

Jeanine, Gr, op

'JEANINE DEFAUCAMBERGE', Pol, pb, 1931; flowers bright salmon-pink, passing to light pink, peony form, large, very dbl.; foliage slender; [Merveille sport]; Turbat

'JEANINE WEBER', HT, ob, 1954; flowers orange, large, moderate fragrance; vigorous growth; [Soestdijk X Mary Hart]; Leenders, M.

'JEANNE CABANIS', HT, rb, 1922; bud coral-red; flowers bright rose-carmine, reverse silvery, center coppery rose, dbl.; Guillot, P.

'JEANNE CORBOEUF', HT, lp, 1902; flowers large, dbl.; Corboeuf; (Sangerhausen)

'JEANNE D'ARC', A, w, 1818; flowers creamy flesh fading to ivory-white, large, dbl., intense fragrance; dense bush (to 5 ft) growth; Vibert

'JEANNE D'ARC', N, w, 1848; flowers creamy white, dbl.; vigorous growth; Verdier, V.

'JEANNE D'ARC', Pol, w, 1909; flowers pure milky white, moderate fragrance; [Mme Norbert Levavasseur sport]; Levavasseur

'JEANNE DE MONTFORT', M, mp, 1851; bud heavily mossed, dark carmine; flowers clear pink, edged silver, blooms in large clusters, semi-dbl., flat, moderate fragrance; foliage emerald-green; tall, vigorous growth; Robert

'JEANNE DRIVON', Pol, w, 1883; flowers white, faintly shaded pink, very dbl., moderate fragrance; Schwartz, J.

'JEANNE EXCOFFIER', HT, pb, 1921; flowers daybreak-pink, inside buff, dbl.; [Mme Philippe Rivoire X Mme Edouard Herriot]; Buatois

'JEANNE HACHETTE', M, m, 1851; flowers slaty violet; Robert

Jeanne Hachette, HP, lp; flowers carmine-rose; Oger

'JEANNE LAJOIE', Cl Min, mp, 1975; bud long, pointed; dbl., 40 petals, 1 in., exhibition form, slight fragrance; foliage small, glossy, dark, embossed; upright, bushy growth; AOE, 1977; [(Casa Blanca X Independence) X Midget]; Sima; Mini-Roses

'JEANNE LALLEMAND', LCl, pb, 1954; flowers pink, reverse salmon-pink, large, dbl., 40–50 petals, very free, recurrent bloom; very vigorous growth; [Mrs Pierre S. duPont X George Dickson]; Buatois

'JEANNE LASSALLE', LCl, mp, 1936;, borne in clusters of 50-60, free, recurrent boom; foliage broad, light; long, stiff stems; vigorous (5 ft.) growth; Lassalle; Vially

'JEANNE MASSON', HP, lp, 1891; flowers medium, dbl., moderate fragrance; Liabaud; (Sangerhausen)

'JEANNE MERMET', Pol, w, 1909; vigorous growth; Mermet

'JEANNE NICOD', HT, w, 1929; flowers white, center tinted cream, dbl.; Schwartz, A.

'JEANNE RICHERT', HWich, w, 1929; flowers cream, center red-brown, borne in large clusters; foliage glossy; very vigorous, climbing growth; [Leontine Gervais X Unnamed seedling]; Walter, L.

'JEANNE SAULTIER', HT, pb, 1927; flowers salmon-rose, reverse reddish pink, base yellow, dbl., moderate fragrance; [Louise Catherine Breslau X Mme Edouard Herriot]; Laperrière

'JEANNE SURY', HP, dp, 1868; flowers large, dbl., moderate fragrance; Faudon; (Sangerhausen)

'JEANNETTE', HGal, dp; flowers bright light red, fading light rose-pink, dbl.; Descemet, M.

Jeannette, Min, pb, 1997; Spooner, Raymond A.

Jeannette Heller *see* **'WILLIAM R. SMITH'**

'JEANNIE DEANS', HEg, dr, 1895; flowers scarlet-crimson, semi-dbl., very free seasonal bloom; foliage fragrant; vigorous growth; Penzance

'JEANNIE DICKSON', HP, pb, 1890; flowers rose-pink, edged silvery pink, large, dbl., 45 petals, exhibition form, moderate bloom; vigorous growth; Dickson, A.

'JEANNIE SOUPERT', Pol, w, 1913; flowers soft flesh-white, borne in large clusters; moderately vigorous growth; [Mme Norbert Levavasseur X Petite Léonie]; Soupert & Notting

Jeannine *see* 'FRONEEN'

Jeannine Michelle *see* 'FROJEAN'

'JEAN'S DREAM', HT, pb, 1971; flowers orient pink to azalea-pink, dbl., 30 petals, 6 in., exhibition form; foliage large; vigorous, upright, bushy growth; [My Choice X (Seedling X Memoriam)]; Ellick

Jeeper's Creeper *see* 'KORISSEL'

Jeffrey *see* 'RESJEFF'

Jehoca *see* 'TANJEKA'

Jehoka *see* 'TANJEKA'

Jekyll's Own Rambler, Ayr, w

'JELBAR', Min, w, 1980; (**Wee Barbie**); bud globular; flowers cream-white, blooms borne 10-20 per cluster, dbl., 43 petals, moderate fragrance; foliage dark; bushy growth; [Unnamed seedling X Unnamed seedling]; Jellyman, J.S.

'JELCANODIR', F, dy, 1976; flowers aureolin-yellow, full, dbl., 30 petals, 2.5–3 in., almost continuous bloom in glasshouse, slight fragrance; vigorous, upright growth; [Undisseminated seedling 1-61-ys X Golden Garnette]; Jelly; Universal Plants

'JELDANIRAN', F, lp, 1985; (**Mimi Pink**(TM), Mimi Rose); flowers light pink, reverse darker, sweetheart, large, dbl., 35 petals, no fragrance; foliage medium, dark, semi-glossy; upright growth; PP005636; [Unnamed seedling X Misty Pink]; Jelly, Robert G.; E.G. Hill Co., 1985;, Universal Plants, 1980

'JELEIT', F, w, 1971; dbl., 30–35 petals, 3 in., globular, slight fragrance; foliage large, dull, leathery; vigorous, upright growth; [Seventeen X Jack Frost]; Hill, E.G., Co.; Meilland

Jelena de Belder, S, w, 1996; Lens, Louis; (Cavriglia)

'JELICO', HT, or, 1973; flowers vermilion, reverse crimson, dbl., 35 petals, 5 in., cupped, slight fragrance; foliage large, dark; [Baccará X (Forever Yours X Unnamed seedling)]; Jelly; Universal Plants

'JELLO', F, my, 1969; (**Coed**(TM)); bud short, pointed; flowers mini-flora, medium, dbl., exhibition form, slight fragrance; foliage dark, leathery; vigorous, upright growth; [Golden Garnette X Seedling]; Jelly; E.G. Hill Co.

Jelly Bean(TM) *see* 'SAVABEAN'

'JELPIROFOR', HT, w, 1985; (**White Success**); flowers large, dbl., 55 petals, exhibition form, borne singly, no fragrance; medium, straight prickles; foliage large, dark, semi-glossy; upright growth; PP005632; [Bridal Pink X Unnamed seedling]; Jelly; E.G. Hill Co.

'JELRANDOLI', F, rb, 1976; flowers vermilion-red, base cardinal-red, full, dbl., 35 petals, 3–3.5 in., slight fragrance; foliage large; vigorous, upright growth; [San Francisco X Little Leaguer]; Jelly; Universal Plants

'JELROGANOR', F, rb, 1977; (**Charisma**, Surprise Party); bud ovoid; flowers scarlet and yellow, medium, very full, very dbl., 40–50 petals, 2–2.5 in., exhibition form, borne in small clusters, slight, fruity fragrance; foliage medium, dark green, glossy, leathery; vigorous, bushy, upright growth; PP004173; AARS, 1978 GM, Portland, 1976; [Gemini X Zorina]; Hill, E.G., Co.; C-P

'JELWHITE', HT, w, 1969; (**Promise Me**); dbl., 45 petals, 4 in., exhibition form, slight, sweetbriar fragrance; foliage matt, dark, leathery; vigorous, upright growth; [Snowsong Supreme X Unnamed seedling]; Jelly; E.G. Hill Co.

'JEMA', HT, ab, 1981; bud ovoid; flowers large blooms borne singly, dbl., 45 petals, moderate fragrance; small, triangular, straw prickles; foliage medium, light green; tall, vigorous growth; [Helen Traubel X Lolita]; Perry, Astor; Perry Roses, 1982

Jemma Giblin *see* 'HORJEMMA'

Jemne Losos Interflora, HT, lp; Strnad; (Czech Rosa Club)

Jennie Anne™ *see* 'KINJEN'

Jennie June™ *see* 'TANJENJU'

Jennie Robinson *see* 'TROBETTE'

Jennifer™ *see* 'BENJEN'

'JENNIFER', HT, 1954; flowers pale flesh pink, 4–5 in., intense fragrance; foliage dull green; vigorous growth; RULED EXTINCT 1/85 ARM; [(Mrs Henry Bowles X Phyllis Gold) X Edina.]; Fletcher; Tucker

'JENNIFER', F, 1959; RULED EXTINCT 1/85 ARM; [Independence X Fashion]; Fryers Nursery, Ltd.

Jennifer Hart *see* 'AROART'

'JENNIFER JAY', HT, pb, 1977; flowers light to medium pink; [Christian Dior sport]; Thomas, Dr. A.S.

Jennifer Joy *see* 'POULJENJOY'

Jennifer-Betty Kenward *see* 'LANBET'

Jennifer's Rose, HT, mr, 1997; Dawson

'JENNY BROWN', HT, or, 1974; bud long, pointed; flowers salmon-pink, center paler, open, single, 5 petals, 4 in., intense fragrance; foliage glossy; very vigorous growth; [(Pink Favorite X Dorothy Peach) X Dainty Bess]; Parkes, Mrs M.H.; Rumsey

Jenny Butchart, HT, lp; flowers deep orange pink fading to lighter pink, exhibition form, 30–35 petals, slight; medium sized matt green; [Miss Canada X Fragrant Cloud]; Hepworth, George, 1970

Jenny Charlton *see* SIMway

'Jenny Duval' HCh; flowers rosy bush; no longer in cultivation; thought to be Président de Seze

'JENNY FAIR', HT, mp, 1967; flowers pink, globular, moderate fragrance; foliage dark; slender, upright growth; [Tropicana X ?]; Gregory

'JENNY JONES', T, lp; Williams, A., 1890; (Weatherly, L.)

'JENNY LIND', M, mp, 1845; Laffay, M.

Jenny Lynn, HT, pb

'JENNY WREN', F, ab, 1957; bud salmon-red; flowers creamy apricot, reverse pale salmon, small blooms in large, dbl., intense fragrance; foliage dark; [Cécile Brunner X Fashion]; Ratcliffe

'JENNY'S DREAM', HT, dp, 1980; flowers deep pink; [Red Devil sport]; Beckett, Ian

Jenny's Rose *see* CANsit

'JENS MUNK', HRg, mp, 1974; bud ovoid; flowers yellow stamens, dbl., 25 petals, 3 in., intense fragrance; upright, bushy growth; [Schneezwerg X Frau Dagmar Hartopp]; Svedja, Felicitas; Canada Dept. of Agric.

Jericho, HT, 1971; Dorieux, Francois; (Cavriglia)

'JERRY', Gr, pb, 1987; flowers white, flushing carmine red, reverse carmine, grading to white, dbl., cupped, intense, sweet fragrance; small, rarely sets fruit; medium, red-brown prickles, hooked downward; foliage medium, medium green, semi-glossy; bushy, tall growth; Silver, ARC TG, 1986; [Unnamed seedling X Unnamed seedling]; Jerabek, Paul E.

'JERRY DESMONDE', HT, mp, 1959; flowers rose-pink, reverse silvery, well formed, dbl., 50 petals, 5 in.; foliage dark, glossy; vigorous, upright growth; [Lord Rossmore X Karl Herbst]; Norman; Harkness

Jerry-O™ *see* 'SAVAJERRY'

'JERSEY BEAUTY', HWich, ly, 1899; flowers pale yellow, fading white, large blooms in clusters, single, non-recurrent, intense fragrance; foliage very glossy; vigorous, climbing growth; [R. wichurana X Perle des Jardins]; Manda, W.A.

Jersey Cream, S, ly, 1996; Sutherland

'JERSEY QUEEN', HT, ob, 1920; flowers flame-orange, edged rose, reverse lemon, dbl.; [Mme Mélanie Soupert X Queen Mary]; Le Cornu; Jersey Nursery

Jerusalem *see* 'HOLJE'

'JESMOND DENE', F, op, 1975; flowers pastel salmon-pink, dbl., 3.5 in., intense fragrance; vigorous, upright growth; [Arthur Bell X Betty May Wood]; Wood

Jesse's Jewels *see* 'BURJES'

'JESSICA' *see* **Angle**

'JESSICA', HWich, pb, 1910; flowers cream-white, center light rose, large; Walsh

Jessica *see* 'TANJEKA'

'JESSIE', Pol, rb, 1909; flowers bright crimson, fading rose-pink, center white, small, semi-dbl., 1.5 in., slight fragrance; foliage small, soft, glossy; bushy growth; [Phyllis sport]; Merryweather

'JESSIE ANDERSON', S, dp; flowers deep rose, well formed, large, dbl., blooms continuously on new wood; [(Old Crimson China X R. canina) X Souv. d'Alphonse Lavallee]

'JESSIE BROWN', Min, mp, 1978; bud mossy; flowers loosely formed, small, semi-dbl., 15 petals, 1.5 in.; foliage small; bushy growth; [Fairy Moss X Fairy Moss]; Dobbs; Small World Min. Roses

'JESSIE CLARK', LCl, mp, 1915; flowers rosy pink, becoming lighter, overlarge, single, early spring bloom; foliage dark, leathery; very vigorous growth; [R. gigantea X Mme Martignier]; Clark, A.; NRS Victoria

Jessie Mathews *see* 'BEEJES'

'JESSIE SEGRAVE', HT, rb, 1937; flowers scarlet on deep chrome base, with pencil markings on inside, dbl.; vigorous growth; Mee; Beckwith

Jessika® *see* 'TANJEKA'

'JET', HT, dr, 1948; bud long, pointed; flowers red to very dark red, large, dbl., moderate fragrance; foliage glossy, dark; vigorous, bushy, compact growth; [Pink Princess X Crimson Glory]; Brownell, H.C.

'JET FIRE', F, or, 1964; bud ovoid; flowers large, dbl., cupped, moderate, spicy fragrance; foliage dark, glossy; vigorous, upright, bushy growth; [Sumatra X Fashion]; Schloen, J.; Ellesmere Nursery

Jet Flame® *see* 'LENPEN'

Jet Flame Nirpaysage® *see* 'LENPEN'

Jet Spray® *see* 'LENCARA'

'JET TRAIL', Min, w, 1964; bud pointed; flowers white, sometimes tinted pale green, small, dbl., 40 petals; bushy (12-14 in) growth; [Little Darling X Magic Wand]; Moore, Ralph S.; Sequoia Nursery

Jeune Fille *see* **'BROCADE'**

'JEUNE FILLE', F, op, 1964; flowers bright salmon-pink, medium, borne in clusters, dbl., moderate fragrance; foliage leathery; very vigorous, bushy growth; [Rose Gaujard seedling X Vendome]; Gaujard

Jeune France *see* **'YOUNG FRANCE'**

'JEUNE HENRY', P, mp; flowers vivid rose; Vibert, prior to 1815

'JEUNESSE', HT, mp, 1959; flowers bright pink, dbl.; moderately bushy growth; [(Independence X Tonnerre) X Michele Meilland]; Laperrière; EFR

Jeunesse Éternelle *see* **'ETERNAL YOUTH'**

'JEWEL', HT, dr, 1938; flowers velvety red, dbl., 50 petals, 5 in., moderate fragrance; [Better Times sport]; Grillo

Jewel Box *see* 'MORBOX'

Jewel's Delight™ *see* 'MINBACO'

'JEZEBEL', HT, mp, 1964; dbl.; strong stems; [Queen Elizabeth X Pink Lustre]; Leenders, J.

JFK *see* **'JOHN F. KENNEDY'**

Jhb Garden Club *see* 'POULFAN'

'JIAN', Min, mr, 1965; bud ovoid; flowers medium red, reverse lighter, very small, dbl.; foliage narrow, leathery; very vigorous, bushy, dwarf growth; [Juliette X Oakington Ruby]; Williams, Ernest D.; Mini-Roses

'JIHOCESKE SLUNCE', HT, w, 1937; flowers medium, dbl.; Böhm, J.; (Sangerhausen)

'JILL', F, mr, 1939; flowers cerise-scarlet, open, borne in clusters, semi-dbl., slight fragrance; long stems; vigorous, bushy growth; [(Else Poulsen X Seedling) X ((Seedling (single red) X Étoile de Hollande) X Daily Mail Scented Rose)]; LeGrice

Jill Carter *see* 'SHERIJILL'

'JILL DARLING', HT, rb, 1937; flowers rich cerise, reverse cinnamon-yellow, moderate fragrance; foliage glossy; vigorous growth; Austin & McAslan

'JILLIAN LOUISE', Min, w, 1994; flowers white with light pink edge and center, blooms borne in large clusters, dbl., 26–40 petals, 1.5–2.75 in., slight fragrance; some prickles; foliage medium, medium green, semi-glossy; medium to tall, upright growth; [Magic Carrousel sport]; Fairweather, Mrs. P.H.

Jillian McGredy *see* 'MACARNHE'

Jill's Rose *see* GANjil

Jilly Cooper *see* LEGcream

Jilly Jewel *see* BENmfig

Jim Bowie, S, dp, 1996; flowers dark pink to light pink center, semi-dbl., 4 in., borne in sprays, intense fragrance; arching growth; hardy; Williams, J. Benjamin

Jim Dandy™ *see* 'BENJIM'

'JIM INGALL', HT, mr, 1998; flowers medium red, medium sized, full, very dbl., 26–40 petals, 4 in., borne singly, slight fragrance; prickles moderate; foliage medium, dark green, semi-glossy; bushy, upright, 120-150 cms growth; [Debbie Thomas X Maria Theresa]; Thomas, D.

'JIM LOUNSBERY', S, or, 1995; single, 5–7 petals, 1.5–2.75 in., borne in sprays of 5-25, slight fragrance; foliage medium, medium green, matt; upright (120-150 cms), bushy growth; [Liverpool Echo X R. virginiana]; Fleming, Joyce L.; Hortico Roses, 1994

'JIM TODD', HT, rb, 1940; flowers nasturtium-red, reverse touched yellow, large, semi-dbl., cupped; very vigorous, bushy growth; Mallerin, C.; A. Meilland

'JIMINY CRICKET', F, op, 1954; bud ovoid; flowers coral-orange to pink-coral, blooms in clusters, dbl., 28 petals, 3–4 in., cupped, moderate, rose geranium fragrance; foliage glossy; vigorous, upright, bushy growth; AARS, 1955; [Goldilocks X Geranium Red]; Boerner; J&P

Jimmy, Min, ob, 1993; Chandrakant

'JIMMY GREAVES', HT, m, 1971; flowers red-purple, reverse silver, dbl., 55 petals, 5 in., exhibition form, slight fragrance; foliage large; erect, bushy growth; [Dorothy Peach X Prima Ballerina]; Gandy, Douglas L.

Jimmy Savile *see* 'PEAPOLLY'

Jimmy Saville *see* 'PEAPOLLY'

Jim's Fence Corner, Pol, w; (found rose)

Jingle Bells® *see* 'JACREDEM'

'JINGLES', F, mp, 1956; bud ovoid; flowers pink overcast deep rose-pink, to open, dbl., 35–40 petals, 2.5–3 in., cupped, moderate fragrance; foliage leathery, glossy; vigorous, upright, bushy growth; [Goldilocks X Garnette]; Boerner; J&P

Jisraela Amira, HT, 1980; Nevo, Motke

Jitka, F, lp; Strnad; (Czech Rosa Club)

Jitka, F, lp; Urban, J., 1987; (Czech Rosa Club)

'JITRENKA', HWich, lp, 1933; flowers medium, semi-dbl.; Mikes Böhm, J.; (Sangerhausen)

Jitrenka, Pol, dy, 1978; flowers medium, semi-dbl., slight fragrance; Urban, J.; (Sangerhausen)

Jitterbug *see* 'JACMINNO'

'JOAN', HMsk, ob, 1919; bud peach; flowers copper, borne in clusters, semi-dbl.; [Trier X Perle des Jeannes]; Pemberton

'JOAN ALDER', HT, pb, 1950; bud long; flowers salmon-pink tinted mauve; foliage dark; very vigorous growth; Moss

Joan Anderson *see* **'ELSE POULSEN'**

Joan Austin *see* 'MORDEB'

Joan Ball *see* 'TROBALL'

'JOAN BELL', HT, dp, 1985; flowers deep pink; [Portland Trailblazer sport]; Bell, John C.

Joan Brickhill® *see* 'HILGOFAN'

'JOAN CANT', HT, pb, 1929; bud pointed; flowers salmon-pink, reverse brighter, very large, dbl., intense fragrance; foliage light, leathery; vigorous, bushy growth; Cant, B. R.

'JOAN DAVIS', HT, ab, 1927; flowers salmon-apricot shaded cerise-pink, base yellow, dbl., intense fragrance; [Ophelia seedling]; Allen

'JOAN ELIZABETH', HT, yb, 1949; bud long, pointed; flowers golden yellow, reverse flushed pink, well formed, dbl., 30 petals, 5 in., moderate fragrance; foliage dark, glossy; vigorous growth; Fletcher; Tucker

'JOAN FITTALL', HT, yb, 1946; bud long, pointed; flowers bronze and gold fading to pink, open, medium, semi-dbl., slight fragrance; foliage leathery; vigorous, bushy growth; [Luis Brinas X ?]; Moss; F. Mason

Joan Fontaine™ *see* CLEjoan

'JOAN FRUEH', HT, lp, 1924; flowers shell-pink, dbl., moderate fragrance; [Ophelia X General-Superior Arnold Janssen]; Frueh

'JOAN HOWARTH', HT, pb, 1924; flowers shell-pink shaded carmine, very large, dbl., moderate fragrance; [Lyon Rose X Mme Abel Chatenay]; Bees

'JOAN KNIGHT', Cl HT, dr, 1928; vigorous growth; Knight, J.

Joan Kruger, S, w, 1997; Kordes

Joan Longer™ *see* 'WILKTWO'

'JOAN MARGARET DERRICK', Pol, dr, 1953; flowers carmine-red, small, semi-dbl., 15 petals; vigorous growth; [Golden Salmon sport]; Derrick

'JOAN ROSS', HP, pb, 1933; bud pointed; flowers blush, reverse light pink, very large, dbl., profuse, non-recurrent bloom, slight fragrance; vigorous growth; [Frau Karl Druschki X Paul Neyron]; Nicolas

'JOANNA BRIDGE', HT, yb, 1916; flowers canary-yellow shaded strawberry, borne on large trusses, semi-dbl.; vigorous growth; Hicks

'JOANNA HILL', HT, ly, 1928; bud long, pointed; flowers creamy yellow, base flushed orange, large, dbl., 48 petals, moderate fragrance; foliage leathery; vigorous growth; [Mme Butterfly X Miss Amelia Gude]; J.H. Hill Co.

'JOANNA HILL, CLIMBING', Cl HT, yb, 1935; Howard Rose Co.

'JOANNA LUMLEY', HT, ab, 1994; dbl., 26–40 petals, 3–3.5 in., borne singly, slight fragrance; some prickles; foliage medium, dark green, glossy; upright (100 cms) growth; [Chicago Peace X Joanne]; Poole, Lionel; F. Haynes & Partners, 1995

'JOANNA TROUTMAN', HT, ob, 1929; bud pointed; flowers orange-yellow, open, semi-dbl., moderate fragrance; foliage bronze, glossy; [Mme Alexandre Dreux seedling]; Vestal

'JOANNE', HT, op, 1985; flowers medium shrimp pink, large blooms borne singly, dbl., 43 petals, exhibition form, slight fragrance; large, globular, orange fruit; large, dark brown prickles; foliage large, dark, semi-glossy; medium, upright growth; [Courvoisier X Princesse]; Poole, Lionel

'JOANNES GINET', HT, pb, 1929; flowers white, tinted cream, edged oriental red; [The Queen Alexandra Rose sport]; Gaujard

'JOAO MOREIRA DA SILVA', HT, my, 1959;, moderate, damask fragrance; [Mme Marie Curie X Dr. Manuel Alves de Castro]; da Silva, Moreira

'JOAO PEREIRA DA ROSA', HT, rb, 1936; flowers brilliant red shading orange and yellow, large, dbl., cupped; foliage light, soft; vigorous growth; [Angèle Pernet X Mme Méha Sabatier]; da Silva, Moreira

'JOAQUIN ALDRUFEU', HT, m, 1897; flowers garnet-purple, reverse violet to magenta, dbl., slight fragrance; foliage light green; moderate growth; Aldrufeu

'JOAQUIN MIR', HT, dy, 1940; flowers golden yellow, large, dbl., cupped; foliage glossy, dark; upright growth; [Mrs Pierre S. duPont X Senora Gari]; Dot, Pedro

Joaquina Munoz, HT, 1980; Dot, Simon; (Cavriglia)

'JOASINE HANET', P, m; (Glendora, Portland from Glendora); flowers deep rose tinged with violet, medium, dbl., quartered, heavy bloomer, moderate fragrance; very hardy.; Vibert, prior to 1882

'JOCELYN', F, r, 1970; flowers mahogany aging purplish-brown, dbl., 3 in.; LeGrice

Jockey *see* **'HORRIDO'**®

Joconde *see* 'DELJACQ'

'JODRELL BANK', HT, pb, 1967; flowers light pink, reverse rose-pink; [Charles F. Warren sport]; Dale, F.

Joe Gray, S, mp, 1999; flowers pink and lavender; dwarf, creeping, ground cover growth; Williams, J. Benjamin

'JOE LONGTHORNE', HT, op, 1995; flowers orange pink, blooms borne singly, dbl., 26–40 petals, 3–3.5 in., intense fragrance; some prickles; foliage medium, dark green, matt; medium (100 cms), upright growth; [Gavotte X Pot of Gold]; Poole, Lionel; Battersby Roses, 1995

'JOE ROSCOE', HT, mp, 1971; flowers rose-red, dbl., 62 petals, 6 in., moderate fragrance; [Karl Herbst X Tzigane]; Wright & Son

Joe-Joe™ *see* 'KINJOE'

Joella® *see* 'KRITIBAN'

Joëlle® *see* 'KRITIBAN'

'JOFITALI', F, rb, 1976; flowers rose-bengal, center cardinal-red; PP004083; [Sonia sport]; DeWitte; Meilland

Jogan, HT, ab, 1988; Bansal, O.P.

Johann Strauss® *see* 'MEIOFFIC'

'JOHANNA OFMAN', HT, dp, 1962; flowers carmine-pink, large, dbl., intense fragrance; [Pink Sensation sport]; Ofman

'JOHANNA RÖPCKE', HWich, op, 1931; (Johanna Ropke); bud pointed; flowers salmon-pink, resembling ophelia but smaller, small, dbl., cupped, borne in clusters; foliage dark, bronze; very vigorous, climbing growth; [Dorothy Perkins X Ophelia]; Tantau

Johanna Ropke *see* **'JOHANNA RÖPCKE'**

'JOHANNA TANTAU', Pol, w, 1928; flowers white, center pinkish yellow, large, borne in clusters, dbl.; foliage dark, leathery; bushy, dwarf growth; [Dorothy Perkins X Ophelia]; Tantau

'JOHANNES BOETTNER', F, mr, 1943; flowers light crimson, very large, borne in clusters, dbl., exhibition form, slight fragrance; vigorous, bushy, compact growth; [Baby Chateau seedling X Else Poulsen]; Kordes

Johannes XXIII, HT, w

Johannesburg Centennial, HT, dy

Johannesburg Garden Club *see* 'POULFAN'

Johannesburg Sun *see* 'KORDOUBT'

Johannisfeuer® *see* TANeufis

'JOHANNISZAUBER', HT, dr, 1926; bud pointed; flowers dark velvety blood-red, very dbl.; vigorous growth; [Château de Clos Vougeot X Unnamed seedling]; Tantau

'JOHASINE HANET', HP, pb, 1847; flowers pink with red, medium, dbl.; Vibert; (Sangerhausen)

'JOHILLGOLD', Gr, dy, 1992; (**Tropical Sunrise**); flowers deep yellow aging to medium yellow, blooms borne singly and, semi-dbl., 21 petals, 4 in., exhibition form, slight fragrance; foliage medium, medium green, semi-glossy; upright, medium growth; [Golden Fantasie X Unnamed seedling]; Hoy, Lowel L.; DeVor Nurseries, Inc.

'JOHILLGOLF', S, mr, 1992; (**Crimson Delight**); dbl., 2.5 in., borne in sprays of 3-4, no fragrance; foliage medium, medium green, semi-glossy; upright, bushy, low growth; [Volare X Seedling]; Hoy, Lowel L.; DeVor Nurseries, Inc.

'JOHILLSTAR', HT, mr, 1992; (**Classical Velvet**); semi-dbl., 25 petals, 5.25 in., borne in sprays of 4-5, no fragrance; foliage medium, dark green, semi-glossy; upright, bushy, tall growth; [Anniversary X Seedling]; Hoy, Lowel L.; DeVor Nurseries, Inc.

JOHilmar, Gr, op; (**After Glow**)

'JOHN A. ALLISON', Gr, lp, 1973; bud ovoid; flowers large, very dbl., slight fragrance; very vigorous, upright growth; [Queen Elizabeth X Montezuma]; Golik; J. Schloen

John A. Macdonald, Gr, dr

John A. Weall, LCl, dr, 1994

'JOHN ABRAMS', F, op, 1976; flowers vermilion and salmon, semi-dbl., 15 petals, 3 in.; [Vera Dalton X Sarabande]; Sanday, John

'JOHN ALLEN', sp, (form of R. suffulta), lp, 1944;, recurrent bloom; height 18 in.; (28); P.H. Wright; ((collected in southern Saskatchewan))

John Andrews, HT, dr, 1997; flowers large; foliage disease-resistant; Williams, J. Benjamin

John Bradshaw *see* 'HARQUISP'

'JOHN BRIGHT', HP, mr, 1878; flowers bright crimson, medium; Paul & Son

'JOHN C. M. MENSING', HT, mp, 1924; (Pink Ophelia); flowers deep bright rose-pink, open, large, dbl., intense fragrance; vigorous growth; [Ophelia sport]; Eveleens

'JOHN CABOT', HKor, mr, 1978; bud ovoid; flowers fuschia tinted red, opens flat, dbl., 40 petals, 2.5 in., blooms in clusters, moderate fragrance; foliage yellow-green; vigorous, upright, medium growth; [R. kordesii X Unnamed seedling]; Svedja, Felicitas; Canada Dept. of Agric.

'JOHN CANT', S, dp, 1895; flowers carmine-pink, small, semi-dbl., slight fragrance; Cant, B. R.; (Sangerhausen)

'JOHN CHURCH', F, or, 1964; flowers orange-scarlet, well-formed, blooms in clusters, dbl., 30 petals, 3.5 in., moderate fragrance; vigorous growth; [Ma Perkins X Red Favorite]; McGredy, Sam IV; McGredy

John Clare *see* 'AUSCENT'

'JOHN COOK', HT, pb, 1917; bud dark pink; flowers la france pink, reverse very dark, dbl., moderate fragrance; [La France X Unknown seedling]; Krüger; Ketten Bros

John Cramphorn, F, or, 1980; flowers large, dbl.; Kriloff; (Sangerhausen)

'JOHN CRANSTON', M, m, 1861; flowers crimson, shaded purple, expanded, medium, dbl.; vigorous growth; Verdier, E.

'JOHN CRONIN', HT, dp, 1935; flowers deep pink, large, dbl., globular, moderate fragrance; vigorous growth; Clark, A.; NRS Victoria

'JOHN DAVIS', HKor, mp, 1986; flowers medium pink, yellow at base blooms in clusters of up to 17, dbl., 40 petals, 3.5 in., recurrent bloom, intense, spicy fragrance; straight prickles; foliage glossy, leathery; trailing growth; [(R. kordesii X Unnamed seedling) X ?]; Svedja, Felicitas; Agriculture Canada

'JOHN DAVISON', HT, dr, 1919; flowers rich velvety crimson, dbl., moderate fragrance; McGredy

'JOHN DIJKSTRA', F, dr, 1965; (Letkis); bud ovoid; flowers medium, borne in clusters, semi-dbl.; foliage dark; [Olala X Paprika]; Buisman, G. A. H.

John Donne, S, mp

'JOHN DOWNIE', HT, ob, 1921; flowers salmon; [Lyon Rose sport]; Dobbie

'JOHN E. SLEATH', HT, rb, 1937; flowers carmine-red, suffused vermilion-orange; vigorous growth; Mee; Beckwith

'JOHN EDWARD REED', HT, dy, 1950; bud long, pointed; flowers buttercup-yellow, dbl., 32 petals, 5.5–6. in., moderate fragrance; vigorous, upright growth; [Talisman sport]; Reed

'JOHN F. KENNEDY', HT, w, 1965; (JFK, President John F. Kennedy); bud ovoid, tinted greenish; dbl., 48 petals, 5–5.5 in., exhibition form, moderate fragrance; foliage leathery; vigorous growth; [Unnamed seedling X White Queen]; Boerner; J&P

'JOHN FRANKLIN', S, mr, 1980; bud ovoid; flowers vibrant red, large, dbl., 25 petals, borne in clusters, floriferous, moderate fragrance; yellow-green prickles with purple hues; foliage rounded; upright, bushy growth, 3 - 4 ft.; [Lilli Marleen X Unnamed seedling]; Svedja, Felicitas; Agriculture Canada

'JOHN FRASER', M, rb; flowers bright red, shaded crimson and purple, large, dbl.; shy growth; Lévêque, 1861

John Gold Veitch, HP, mr

'JOHN GREENWOOD', F, mr, 1976; flowers bright red, full, dbl., 25 petals, 2.5 in., slight fragrance; foliage large, dark; vigorous, free growth; [Marlena X Fragrant Cloud]; Lea

John Grooms, LCl, pb, 1993; Beales, Peter

'JOHN GROW', M, mp, 1859; flowers clear pink; Laffay, M.

'JOHN H. ELLIS', HT, dp, 1948; flowers deep rose-pink, well formed, large, dbl., 48 petals; free, bushy growth; McGredy

John Harris, HT, pb, 1995; Dawson

'JOHN HART', HT, mp, 1922; flowers cherry-pink, dbl., slight fragrance; Hicks

'JOHN HENRY', HT, mr, 1925; bud rosy scarlet; flowers rich pink, slight fragrance; Beckwith

'JOHN HOPPER', HP, pb, 1862; flowers bright rose edged lilac, center carmine, semi-globular, large, dbl., 70 petals, occasionally recurrent bloom, intense fragrance; vigorous, upright, bushy growth; [Jules Margottin X Mme Vidot]; Ward

John Hughes *see* 'SANPHYLLIS'

John Keats *see* 'MEIROUPIS'

John Kemp, HWich, w

'JOHN KEYNES', HP, dr; flowers red shaded maroon, dbl., 48 petals, intense fragrance; vigorous growth; Verdier E., 1864

'JOHN KIDMAN', HT, my, 1969; bud ovoid; flowers lemon-yellow, camellia form, very large, dbl., intense fragrance; foliage leathery; vigorous, tall, compact growth; [Radar X Allgold]; Fankhauser

'JOHN LAWRENCE', F, my, 1990; bud pointed; flowers canary yellow, fading as it ages, shaped, medium, borne in small cluster, dbl., 18 petals, flat, moderate, spicy fragrance; foliage medium, glossy; medium, upright growth; [Unnamed seedling X Sunsprite]; Bracegirdle, Derek T., 1984

'JOHN MCNABB', HRg, mp, 1932; dbl., profuse midseason bloom, sometimes continuing later; [R. rugosa kamtchatica X R. beggeriana]; Skinner

'JOHN MOORE', HT, yb, 1939; flowers buff shaded gold, well shaped, very large, dbl., 47 petals, exhibition form, moderate fragrance; foliage dark; Gaujard

'JOHN MORLEY', HT, dp, 1945; bud long, pointed; flowers glowing pink, very large, dbl., exhibition form, moderate fragrance; foliage dark, leathery; vigorous, bushy growth; [Joanna Hill X J.C. Thornton]; Duehrsen; California Roses

John Phillip Sousa, S, rb

John Ruskin *see* **'RUSKIN'**

'JOHN RUSSELL', HT, dr, 1924; flowers glowing crimson flushed deeper, well-shaped, large, very dbl.; vigorous growth; GM, Bagatelle, 1924; Dobbie

'JOHN RUSSELL, CLIMBING', Cl HT, dr, 1930; Ketten Bros.

'JOHN S. ARMSTRONG', Gr, dr, 1961; bud ovoid to urn-shaped; dbl., 40 petals, 3.5–4 in., exhibition form, slight fragrance; foliage leathery, semi-glossy, dark; tall, bushy growth; AARS, 1962; [Charlotte Armstrong X Seedling]; Swim, H.C.; Armstrong Nursery

'JOHN S. BLOOMFIELD', HT, ab, 1964; bud ovoid; flowers deep apricot flushed pink, open, large, dbl., slight fragrance; compact growth; [Ma Perkins X Burnaby]; Fankhauser

John Snowball, HT, w

'JOHN SQUARE', Cl HT, dy, 1937; flowers sunflower-yellow, center deeper, very large, dbl., cupped; foliage glossy, dark; vigorous, climbing (6-8 ft in season) growth; [Souv. de Claudius Pernet sport]; Square

'JOHN STUART MILL', HP, mr, 1875; flowers large, dbl.; Turner; (Sangerhausen)

'JOHN WALLACE', Pol, dr, 1941; flowers deep red, open, large, dbl.; foliage large, leathery, glossy; bushy growth; [Marianne Kluis Superior sport]; Kluis; Klyn

'JOHN WATERER'®, HT, dr, 1970; dbl., 44 petals, 4 in., exhibition form, moderate fragrance; [King of Hearts X Hanne]; McGredy, Sam IV; McGredy

'JOHNAGO', HT, pb, 1962; (**Chicago Peace**®); flowers phlox-pink, base canary-yellow, very dbl., 50–60 petals, 5–5.5 in., exhibition form, borne mostly singly, slight fragrance; foliage large, dark green, leathery, glossy; tall growth; GM, Portland, 1961; [Peace sport]; Johnston; C-P;, URS

John-John™ *see* 'WEKGIBOTEX'

Johnnie Walker *see* 'FRYGRAN'

John-Paul II *see* 'JACANGE'

John's Rose *see* 'MEJORIA'

Johnson Street Giant, LCl, mp; (found rose)

'JOIA', F, yb, 1962; flowers yellow shaded carmine; [Seedling X Virgo]; da Silva, Moreira

'JOIE DE VIVRE', HT, pb, 1949; flowers pink, base gold, well shaped, very large, dbl., moderate fragrance; foliage bronze green; moderately vigorous growth; Gaujard; Wheatcroft Bros.

Jo-Jo, Min, dy, 1988; Interplant

'JOKER', HT, ob, 1958; flowers orange-red, reverse lighter, slight fragrance; foliage glossy; [Peace X Karl Herbst]; Lens

'JOLAFT', Min, mp, 1991; (**Afternoon Delight**); dbl., 21 petals, 1.5–2.75 in., borne mostly singly and in small clusters, slight fragrance; few prickles; foliage medium, medium green, matt; tall (45 cms), upright, spreading, vigorous growth; [Party Girl X Fashion Flame]; Jolly, Marie; Rosehill Farm, 1992

'JOLANDA', HT, dp, 1959; bud long, pointed; flowers rose, cupped; vigorous, upright, bushy growth; Malandrone

'JOLCHER', Min, ob, 1991; (**Cherry Bomb**); flowers orange, small, blooms borne mostly singly, very dbl., 1.5 in., no fragrance; few prickles; foliage small, medium green, semi-glossy; medium, upright growth; [Fashion Flame X Sheri Anne]; Jolly, Marie; Rosehill Farm, 1991

'JOLCOL', Min, mp, 1991; (**Color Guard**); dbl., 2–5 cms, borne mostly singly, no fragrance; few prickles; foliage medium, medium green, matt; upright (20 cms), bushy, very hardy growth; [Anita Charles X Poker Chip]; Jolly, Marie; Rosehill Farm, 1993

'JOLI COEUR', F, dr, 1963; bud globular; flowers dark crimson, medium, dbl., moderate fragrance; foliage dark; symmetrical growth; [Rose Gaujard X (Seedling X Josephine Bruce)]; Gaujard

Joli Mome, HT, 1969; Ducher, Ch.; (Cavriglia)

Jolie Comtoise®, F, op, 1991

Jolie Demoiselle, Pol, lp

'JOLIE MADAME', HT, or, 1958; bud ovoid; flowers vermilion-red, dbl., 65 petals, 4–4.5 in., cupped, slight fragrance; foliage leathery, glossy; vigorous, upright, bushy growth; [(Independence X Happiness) X Better Times]; Meilland, F.; C-P, 1960;, URS, 1958

'JOLIE PRINCESSE', F, pb, 1955; flowers pink, shaded ochre, borne on large trusses, dbl., moderate fragrance; foliage leathery, bronze; very vigorous, bushy growth; [Peace X Independence]; Gaujard

Jolie Rose, HT, 1967; Mondial Roses; (Cavriglia)

Jolisquare®, S, mp

'JOLLITY JANE', Pol, m, 1993; flowers lilac-cream blend, blooms borne in large clusters, very dbl., 1.5 in., slight fragrance; few prickles; foliage small, medium green, glossy; tall, upright, bushy growth; [Valerie Jeanne X Yesterday]; Jobson, Daniel J.; Jobson, 1993

'JOLLMEA', F, rb, 1995; (**Atlantic City**); flowers ivory white with dark red blend, full (26–40 petals), large (7+ cms) blooms borne mostly single; slight fragrance; few prickles; foliage medium, dark green, semi-glossy; medium (105 cms), compact growth; [Crimson Glory X Orange Honey]; Jolly, Marie; Paramount Nursery, 1996

'JOLLY', Pol, m, 1934; flowers carmine-purple, center white, borne in clusters, single, slight fragrance; foliage sparse, dark; dwarf growth; [Miss Edith Cavell X Tip-Top]; Leenders, M.

Jolly *see* 'LUKOR'

Jolly Cupido *see* RUIxandra

Jolly Dance, HKor, ob

'JOLLY GOOD', F, pb, 1973; bud ovoid; flowers salmon-pink, dbl., 55 petals, 3.5 in., moderate fragrance; foliage glossy, dark leathery; bushy, compact growth; [Cupid's Charm X Lucky Piece]; Fuller; Wyant

Jolly Good, MinFl, op, 1991; Lens

Jolly Joker, F, w

'JOLLY ROGER', F, or, 1973; bud ovoid, pointed; flowers bright reddish orange, medium, semi-dbl., cupped, slight fragrance; foliage wrinkled; growth moderate, bushy; [Spartan X Angelique]; Armstrong, D.L.; Armstrong Nursery

'JOLSIR', Min, ab, 1992; (**Sir**); semi-dbl., 6–14 petals, 1.5 in., moderate fragrance; few prickles; foliage medium, dark green, semi-glossy; medium growth; [Olympic Gold X Rise 'n' Shine]; Jolly, Marie; Rosehill Farm, 1993

'JOLSON', Min, pb, 1991; (**Sonata in Pink**); flowers medium pink fading to light pink, blooms borne mostly singly, dbl., 24 petals, 1.25 in., slight fragrance; few prickles; foliage medium, medium green, semi-glossy; medium (44 cms), upright growth; [Chris Jolly X Chattem Centennial]; Jolly, Marie; Rosehill Farm, 1992

'JOLTIP', Min, mp, 1987; (**Tipper**™); flowers medium pink, aging lighter, medium, long lasting, dbl., 32 petals, exhibition form, borne singly, slight fragrance; round, orange fruit; prickles bayonet, light brown; foliage medium, medium green, semi-glossy; upright, medium growth; PP007340; AOE, 1989; [Chris Jolly X Chattem Centennial]; Jolly, Marie; Rosehill Farm, 1988

Joly Rose Primaplant, HT, mp, 1963; flowers medium to large, dbl., slight fragrance; Vlaeminck; (Sangerhausen)

'JONDAD', Min, op, 1995; (**For You Dad**); flowers coral, single (5 petals), reflexes to form star, medium blooms borne mostly single; slight fragrance; some prickles; foliage medium, dark green, semi-glossy; upright, tall growth; [Heartbreaker X Seedling]; Jones, Steve

'JONDOROSE', Min, ob, 1998; (**Dorothy Rose**); flowers orange and white, turning orange red and red, striped, single, 4–7 petals, 2 in., borne in small clusters, slight fragrance; foliage medium, medium green, semi glossy; climbing, 6 ft. growth; [Sarabande X Peggy T]; Jones, Steve

Jone Asher, MinFl, 1987; Pearce, C.A.; (Cavriglia)

'JONETSU', HT, mr, 1978; bud pointed; flowers dark scarlet, dbl., 30–35 petals, 4.5–6. in., exhibition form, moderate fragrance; foliage dark, leathery; vigorous growth; [(Kagayaki X Prima Ballerina) X Kagayaki]; Suzuki, Seizo; Keisei Rose Nursery

'JONKHEER G. SANDBERG', HT, my, 1936; (J.G. Sandberg); flowers clear yellow, dbl.; foliage dark, leathery; vigorous growth; [Christine X Mrs Wemyss Quin]; Buisman, G. A. H.; Armstrong Nursery, 1941

'JONKHEER J. L. MOCK', HT, pb, 1910; bud pointed; flowers silvery rose-white, reverse carmine-pink, bluing slightly, dbl., exhibition form, moderate fragrance; foliage dark, leathery; vigorous growth; GM, Bagatelle, 1911; [(Mme Caroline Testout X Mme Abel Chatenay) X Farbenkonigin]; Leenders, M.

'JONKHEER J. L. MOCK, CLIMBING', Cl HT, pb, 1923; Timmermans

'JONKHEER MR G. RUYS DE BEERENBROUCK', HT, ob, 1919; flowers pure orange-yellow fading clear yellow, dbl.; vigorous growth; [Mme Mélanie Soupert X Joseph Hill]; Timmermans

Jonquille *see* 'DELJONQ'

'JONWHITE', Min, w, 1998; (**White Sunshine**); flowers white, single, like my sunshine, single, 4–7 petals; foliage medium, medium green, semi-glossy; compact, bushy, medium, 2 ft. growth; [Sarabande X Peggy T]; Jones, Steve

Jorianda, HT, m

Joro, HT, ob, 1979; RvS-Melle

José Carréras *see* POUlnew

'JOSEF ANGENDOHR', F, dp, 1982; flowers deep pink; [Dame de Coeur sport]; Angendohr, Hans-Werner; Baumschulen Angendohr

'JOSEF PETER', HT, m, 1929; flowers pale blush, reverse mauve-rose, dbl., moderate fragrance; [Ruth X Frank W. Dunlop]; Ketten Bros.

'JOSEF ROTHMUND', HEg, ob, 1940; (Joseph Rothmund); bud small, ovoid, orange-red; flowers light red with pinkish yellow, borne in clusters, very dbl., profuse, non-recurrent bloom, intense fragrance; foliage bronze, leathery; very vigorous growth; [Joanna Hill X Magnifica]; Kordes

'JOSEF STRNAD', HT, rb, 1932; flowers dark red, with traces of yellow, orange and rose, very large, dbl., cupped, moderate fragrance; foliage leathery, glossy, dark, bronze; very vigorous, bushy, branching growth; [Aspirant Marcel Rouyer X Toison d'Or]; Böhm, J.; J&P, 1934

Josefa, HT, 1982; Laperriere, J.; (Cavriglia)

'JOSEFINA DE SALGADO', HT, dp, 1963; (Joséphine de Salgado); flowers bright pink, large, dbl., 30 petals, moderate fragrance; somewhat weak stems; very vigorous growth; [Queen Elizabeth X Peace]; Dot, Simon

'JOSEPH ARLES', Cl HT, rb, 1964; flowers vermilion-red, reverse silvery white, moderate fragrance; foliage leathery; vigorous, climbing growth; [Aloha X Gabychette]; Arles; Roses-France

'JOSEPH BAUD', HT, yb, 1919; flowers golden yellow and orange-yellow, dbl., intense fragrance; [Rayon d'Or X Unnamed seedling]; Gillot, F.

'JOSEPH BILLARD', HWich, mr, 1906; flowers carmine-red, medium, single; Barbier; (Sangerhausen)

'JOSEPH CHAPPAZ', HP, m, 1883; flowers violet-pink, large, dbl.; Schmitt; (Sangerhausen)

'JOSEPH COURBIS', HT, pb, 1958; flowers carthamus-pink to orange-red, dbl., 48 petals; foliage dark, glossy; vigorous, upright growth; [Margaret McGredy X Emma Wright]; Arles; Roses-France

'JOSEPH F. LAMB', S, dr, 1988; bud ovoid, pointed; flowers dark red, reverse lighter, aging darker, loose, medium, dbl., 23 petals, cupped, moderate, fruity fragrance; yellow-orange fruit; prickles awl-like, small, tan toreddish brown; foliage medium, medium green semi-glossy; upright, bushy, low, winter hardy growth; [[Prairie Star X [(Dornroschen X Peace) X Music Maker]] X (Music Maker X Topsi)]; Buck, Dr. Griffith J., 1989

Joseph Goudreau, B, pb

Joseph Guy *see* **'LAFAYETTE'**

'JOSEPH GUY', Pol, dp, 1930; flowers carmine-pink, medium, semi-dbl., slight fragrance; Feldmann; (Sangerhausen)

'JOSEPH GUY', Pol, mp, 1950; flowers large, semi-dbl., slight fragrance; Westhus; (Sangerhausen)

'JOSEPH HILL', HT, pb, 1903; bud pointed; flowers pink shaded salmon, reverse coppery pink, dbl., slight fragrance; Pernet-Ducher

Joseph Klimes, HT, dr, 1985; flowers large, very dbl., slight fragrance; Urban, J.; (Sangerhausen)

'JOSEPH LIGER', HWich, yb, 1909; flowers canary-yellow, edged and washed light pink, reverse cream-white; vigorous, climbing growth; [R. wichurana X Irene Watts]; Barbier

'JOSEPH LOWE', HT, op; flowers salmon-pink; [Mrs W.J. Grant sport]

'JOSEPH PERNET D'ANNEMASSE', HT, op, 1934; bud pointed; flowers salmon, dbl.; foliage glossy, dark, bronze; very vigorous growth; Pernet-Ducher; Gaujard

Joseph Rothmund *see* **'JOSEF ROTHMUND'**

Joseph Sauvageot®, HT, ob, 1989; Sauvageot

'JOSEPHINE', Min, w, 1969; flowers white or soft pink, micro-mini, small, dbl.; foliage small, glossy; dwarf, bushy growth; [(R. wichurana X Carolyn Dean) X Jet Trail]; Moore, Ralph S.; Sequoia Nursery

Joséphine Baker *see* 'MEIMAUR'

Joséphine Beauharnais *see* **'BELLE DE SÉGUR'**

'JOSEPHINE BRUCE', HT, dr, 1949; flowers crimson, dbl., 24 petals, 5–6 in., slight fragrance; foliage dark; vigorous, branching growth; [Crimson Glory X Madge Whipp]; Bees, 1949; Totty, 1953

'JOSEPHINE CARMODY', S, lp, 1995; flowers light pink, double, medium sized, dbl., 15–25 petals, borne in small clusters, slight fragrance; prickles numerous; foliage small, dark green, glossy; bushy, medium growth (36 in.); [Sea Foam sport]; Collins, Frank E. Jr

Joséphine de Beauharnais *see* **'BELLE DE SÉGUR'**

Joséphine de Salgado *see* **'JOSEFINA DE SALGADO'**

Josephine Elizabeth, S, lp; Peden, R., 1999; (Weatherly, L.)

'JOSÉPHINE GUYET', B, dr, 1873; (Mme Joséphine Guyet); flowers deep red, recurrent bloom; Touvais

Joséphine Maltot *see* **'MME BRAVY'**

'JOSÉPHINE MAROT', HT, lp, 1894; flowers white washed pink; Bonnaire

'JOSEPHINE MOREL', Pol, dp, 1892; flowers carmine-pink, small, dbl.; Alégatière; (Sangerhausen)

'JOSEPHINE RITTER', HMult, mp, 1900; flowers medium, dbl.; Geschwind, R.; (Sangerhausen)

'JOSEPHINE SPIECKER', HT, ob, 1939; flowers deep orange to yellow, very dbl., globular; foliage glossy, dark bronze; vigorous, bushy growth; Verschuren-Pechtold; Bentley;, Harkness

'JOSEPHINE THOMAS', HT, ob, 1924; flowers orange-salmon to cream-flesh, very dbl., exhibition form, slight fragrance; foliage leathery; vigorous, bushy growth; H&S; Dreer

'JOSEPHINE VESTAL', HT, lp, 1923; flowers soft pink, dbl., exhibition form, moderate fragrance; very vigorous, bushy growth; [Ophelia X Unnamed seedling]; Hill, E.G., Co.; Vestal

Josephine Wheatcroft *see* **'ROSINA'**

'JOSEPH'S COAT'®, LCl, rb, 1969; flowers yellow and red, repeats, dbl., 23–28 petals, 3 in., blooms in clusters, recurrent bloom, slight fragrance; foliage dark, glossy; vigorous, pillar growth; GM, Bagatelle, 1964; [Buccaneer X Circus]; Armstrong, D.L. & Swim, 1964; Armstrong Nursery, 1964

Josh *see* 'TINJOSH'

Joshua™ *see* 'MOGAJOSH'

Joshua Bradley *see* 'CHRISGOBRO'

'JOSPINK', F, lp, 1998; (**Anne Marie Laing**); flowers light pink, darker reverse, full, camillia form, very dbl., 26–40 petals, 4 in., borne in large clusters, moderate fragrance; few prickles; foliage medium, dark green, dull; upright, tall, 110cms growth; [(Iceberg X Anytime) X Sexy Rexy]; Shipway, John; F. Haynes Partners, 1998

Josyane, S, m, 1980; flowers violet-red, small, dbl.; Lens, Louis; (Sangerhausen)

'JOSYSIGAL', HT, mr, 1975; flowers large, dbl., 60 petals, cupped, intense fragrance; Delforge, S.

Jour de Fête, HT, w, 1968; Lens

Jour des Pères *see* **'VATERTAG'**®

'JOUR D'ETÉ', HT, mr, 1964; bud very long; flowers bright red, open; vigorous growth; [Coup de Foudre X Berthe Mallerin]; Combe; Vilmorin-Andrieux

'JOURNEY'S END', HT, ob, 1978; flowers indian orange, pointed, dbl., 37 petals, 6 in., slight fragrance; foliage large, glossy; vigorous, upright growth; [Doreen X Vienna Charm]; Gandy, Douglas L.

'JOUVENCELLE', HT, pb, 1969; flowers salmon-pink suffused red, dbl., 50 petals; foliage reddish; [Prima Ballerina X Helen Traubel]; Gaujard

'JOVE', F, or, 1968; flowers scarlet, semi-dbl., blooms in clusters, slight fragrance; foliage glossy; low growth; [Vera Dalton X Paprika]; Harkness

'JOVITA'®, F, or, 1975;, blooms in large clusters, slight fragrance; foliage bright green; medium, bushy growth; [Jove X Tip Top]; Harkness; Hauser

'JOVITA PÉREZ', HT, ob, 1929; flowers coppery salmon, shaded coral, dbl., cupped, moderate fragrance; foliage soft, dark; long, strong stems; vigorous, compact growth; [Mme Butterfly X Souv. de Claudius Pernet]; Munné, B.

Joy® *see* 'INTERMOTO'

'JOY', HT, pb, 1929; bud tangerine-red; flowers rose-pink suffused tangerine, base yellow, dbl., exhibition form, moderate fragrance; foliage leathery; vigorous, branching growth; Beckwith

Joy Bells *see* **'JOYBELLS'**

Joy Button *see* LUDpinbu

'JOY O'BRIEN', Gr, op, 1969; flowers pink shaded orange-salmon, large, dbl., moderate fragrance; foliage dark; vigorous growth; [Queen Elizabeth X Seedling]; Verschuren; Stassen

Joy of Health *see* HARxever

Joy Owens *see* 'MACRED'

Joy Pagram, HT, lp, 1995; Dawson

'JOY PARFAIT', F, lp, 1965; [Pink Parfait sport]; McIlroy

'JOYANCE', HT, mr, 1939; flowers velvety red, camellia shape, dbl., 50 petals, 4 in., moderate fragrance; foliage leathery, dark; very vigorous, upright growth; [Regina Elena sport]; Grillo

'JOYBELLS', F, mp, 1961; (Joy Bells); flowers rich pink, camellia-shaped, blooms in clusters, dbl., 30 petals, 3.5 in., moderate fragrance; [Unnamed seedling X Fashion]; Robinson, H.

Joybells *see* 'KORSORB'

'JOYCE', HT, dr, 1953; flowers dark velvety crimson, pointed, medium, dbl., 24 petals, moderate fragrance; foliage leathery; vigorous growth; [George Dickson X Étoile de Hollande]; Cant, F.

Joyce Barden *see* 'ARDOUR'

'JOYCE CLAIRE', F, dp, 1965; bud globular; flowers deep pink, open, small, dbl., slight fragrance; foliage glossy; very vigorous, upright growth; [Queen Elizabeth X ?]; Tonkin

Joyce Edmonds, F, op, 1992; Nieuwesteeg, J.

'JOYCE FAIREY', Cl HT, dp, 1929; flowers soft red; pillar growth; Clark, A.; NRS Victoria

Joyce Hunt, lp; [Scented Bouquet X Stella]; Dawson, George

Joyce Lomax *see* **'SATAN'**

'JOYCE LONGLEY', HT, ob, 1958; flowers in sunset shades; [Opera sport]; Court

'JOYCE NORTHFIELD', HT, ob, 1977; flowers deep orange, high-pointed, dbl., 3–4 in., slight fragrance; foliage dark; vigorous, upright growth; [Fred Gibson X Vienna Charm]; Northfield

'JOYCE RILEY', F, ob, 1978; bud well formed; flowers vermilion, yellow-salmon, full, dbl., 25 petals, 2.5 in., slight fragrance; foliage dark, leathery; vigorous, upright growth; [Paddy McGredy X Arthur Bell]; Wood

'JOYCE ROBINSON', HT, op, 1945; flowers peach-pink, exhibition form, moderate fragrance; foliage dark, leathery; vigorous, bushy growth; [Rose Berkley sport]; Selwood; Rosecraft Nursery

Joycie™ *see* 'MORJOYC'

'JOYENA', LCl, m, 1964; flowers spirea-red, reverse tyrian purple, borne in small clusters, exhibition form, moderate fragrance; foliage dark, bronze, leathery; vigorous growth; [Blossomtime X ?]; Mason, P.G.

'JOYEUX NOËL', HT, or, 1960; bud long, pointed; flowers well formed, medium, dbl., 30–35 petals, moderate fragrance; foliage bronze, leathery; vigorous, bushy growth; [(Floradora X Independence) X (La Vaudoise X Léonce Colombier)]; Delbard-Chabert

'JOYFUL', HT, pb, 1931; bud pointed; flowers pink, reverse streaked red, base red and orange; vigorous growth; Vestal

Joyfulness *see* **'FROHSINN'**®

Joyfulness *see* 'TANSINNROH'

Joyfulness, Climbing, Cl HT, lp; [Joyfulness sport]; Ruston, D., 1988; (Weatherly, L.)

'JOYOUS', F, pb, 1939; flowers rose-pink, reverse slightly darker; vigorous, bushy growth; [Else Poulsen sport]; deRuiter; J&P

'JOYOUS CAVALIER', HT, mr, 1926; bud pointed; flowers brilliant red, open, large, dbl., 25–30 petals, slight fragrance; foliage dark, glossy; very vigorous growth; [Red-Letter Day X Clarice Goodacre]; Archer

Joyous Moment *see* JACpri

Joyride *see* 'RENRIDE'

'JUAN MARAGALL', HT, mr, 1960; flowers bright strawberry-red, large, dbl., 35 petals, intense fragrance; long, strong stems; vigorous, compact growth; [Chrysler Imperial X Buccaneer]; Dot, Simon

'JUAN PICH', HT, m, 1921; flowers purplish wine-red, dbl., moderate fragrance; Leenders, M.

'JUAN QUEVEDO', HT, ly, 1921; flowers cream-yellow, dbl., moderate fragrance; [Entente Cordiale X My Maryland]; Leenders, M.

'JUANA DE DARDER', HT, yb, 1947; flowers deep yellow shaded salmon, cupped, slight fragrance; foliage bright green; strong stems; vigorous growth; [Souv. de Claudius Pernet X (Sensation X Souv. de Claudius Pernet)]; Munné, M.

Juane Adam, F, my, 1997; Adam, M.

'JUANITA', HGal, pb; flowers pink edged paler, medium, dbl.

'JUANITA', Ch, pb, 1885; flowers pink spotted white; Robert

Jubilaire de Masaryk *see* **'MASARYKOVA JUBILEJNI'**

'JUBILANT', F, lp, 1967; flowers flesh pink, blooms in clusters, 2.5 in., moderate fragrance; foliage glossy; [Dearest X Circus]; Dickson, A.

Jubilation *see* 'JACREG'

'JUBILÄUMSROSE', HT, w, 1910; flowers creamy white, large, semi-dbl.; Schmidt, I. C.; (Sangerhausen)

'JUBILÄUMSROSE', HFt, mr, 1929; flowers large, dbl.; Schmidt, I. C.; (Sangerhausen)

'JUBILEE', HP, m, 1897; flowers purple, shaded maroon, large, dbl., some recurrent bloom, moderate fragrance; moderate growth; [Victor Hugo X Prince Camille de Rohan]; Walsh

'JUBILEE', HT, w, 1930; (Allen's Jubilee); bud pointed; flowers cream, tinged salmon-pink and indian yellow, center coral-pink, dbl., exhibition form, intense fragrance; vigorous growth; [Paul's Lemon Pillar X Aspirant Marcel Rouyer]; Allen

Jubilee *see* **'MASARYKOVA JUBILEJNI'**

Jubilee 150 *see* 'MEICLOUX'

'JUBILEE CELEBRATION', F, pb, 1977; flowers pink shaded salmon, dbl., 20 petals, 4 in., intense fragrance; foliage matt, green; growth moderate; [Elizabeth of Glamis X Prima Ballerina]; Smith, E.; Wheatcroft, 1976

Jubilee Sunset *see* 'TALJUB'

'JUBILEJNAJA', HT, pb, 1940; flowers rose pink with dark yellow, medium, semi-dbl., slight fragrance; Kosteckij; (Sangerhausen)

Jubilejni, HT, lp; Urban, J., 1979; (Czech Rosa Club)

Jubileum 110, HT, lp; Urban, J., 1989; (Czech Rosa Club)

'JUDANN', Cl Min, op, 1996; (**Ann's Rose**); flowers light orange pink, blooms borne in small clusters, dbl., 15–25 petals, 1.5 in.; few prickles; foliage medium, light green, glossy; tall (5 ft), upright, pillar (2 ft) growth; Bell, Judy G.; Michigan Mini Roses, 1996

'JUDBABY', MinFl, dp, 1994; (**Betty's Baby**); flowers deep pink with white center, blooms borne in small clusters, semi-dbl., 6–14 petals, 1.5 in., no fragrance; few prickles; foliage small, medium green, semi-glossy; medium (12-14 in), upright, bushy growth; [Twilight Trail X Charmglo]; Bell, Judy G.; Michigan Mini Roses, 1995

'JUDBRIGHT', Min, ob, 1993; (**Bright Sight**); flowers light orange, dbl., 26–40 petals, 1.5–2.5 in., borne in large clusters, no fragrance; few prickles; foliage small, medium green, matt; upright (18 in), bushy growth; [Charmglo X Poker Chip]; Bell, Judy G.; Michigan Mini Roses, 1994;, Michigan Miniature Roses, 1994, USA

'JUDDANCE', Min, m, 1993; (**Folk Dance**); dbl., 26–40 petals, 1.5 in., borne mostly singly, slight fragrance; few prickles; foliage small, dark green, glossy; bushy (41 cms) growth; [Dale's Sunrise X Angel Face]; Bell, Judy G.; Michigan Mini Roses, 1994

'JUDDOLL', Min, ly, 1996; (**Diamond Doll**); flowers light yellow, darker yellow center, quilled reflex, blooms b, dbl., 26–40 petals, 1.5 in., intense fragrance; few prickles; foliage medium, medium green, matt; upright growth; medium (18 in. winter hardy; [Rise 'n' Shine X Angel Face]; Bell, Judy G.; Michigan Mini Roses, 1997

Jude the Obscure *see* 'AUSjo'

'JUDFEST', Min, ab, 1999; (**Bloomfest**™); flowers apricot blend, dbl. (17–25 petals), medium (1 in.) blooms borne in small clusters; no fragrance; few prickles; foliage medium, medium green, dull; compact, low (14 in.) growth; [Rise 'n' Shine X Selected Pollen]; Bell, Judy G.; Michigan Miniature Roses, 1996

Judi Dench *see* 'PEATHUNDER'

'JUDIAN', Min, mr, 1996; (**Here's Ian**); semi-dbl., 6–14 petals, 1.5 in., borne mostly singly, no fragrance; few prickles; foliage small, medium green, semi-glossy, disease resistant; small, compact, rounded growth; low (12 in. winter hardy; [Dale's Sunrise X Seedling]; Bell, Judy G.; Michigan Mini Roses, 1997

'JUDIE DARLING', Cl Min, pb, 1979; flowers marbled pink, reflexed, dbl., 45–50

petals, 1–1.5 in., moderate fragrance; vigorous, climbing growth; Sudol, Julia

Judit, F, lp; cupped, 35 petals, slight; large, dark green, leathery, glossy; [(Frankfurt am Main X Maria Callas) X Dr Faust]; St. Wagner, 1996; Res. Stn. f. Fruit growing, Cluj, 1997, Romania

'Judith', HT, rb, 1938; flowers glowing cerise, reverse golden yellow, dbl., globular, intense, fruity fragrance; foliage glossy, bronze; vigorous, bushy growth; LeGrice

'Judith Ann', HT, dp, 1998; flowers deep pink with medium reverse and silvery sheen, very dbl., 41 petals, 6–7 in., borne singly, candelabras in summer, slight fragrance; prickles moderate; foliage medium, dark green, glossy; upright, medium, 4 - 5 ft. growth; [Red Devil sport]; Schamel. Al; Edmunds Roses, 1998

'Judith Black', HT, dr, 1930; bud pointed; flowers rich dark red flushed fiery red, dbl., globular, moderate fragrance; foliage soft; dwarf growth; Clark, A.; Hazlewood Bros.

'Judith I. B. Hall', HT, pb, 1953; flowers pink, base orange, dbl., 32 petals, 5 in., intense fragrance; very vigorous growth; [Crimson Glory X Sterling]; Balcombe Nursery

'JUDlee', MinFl, ob, 1992; (**Essie Lee**); flowers white with orange picotee down 1/2 of petals, white to light orange reverse, dbl., 26–40 petals, 1.5 in., globular, no fragrance; few prickles; foliage small to medium, dark green, semi-glossy; upright (46 cms), bushy growth; [Tennessee X Tennessee]; Bell, Judy G., 1991; Michigan Mini Roses

'JUDlov', Min, m, 1994; (**Love in Bloom**); flowers mauve to tan with bright yellow stamens, blooms borne in small clusters, single, 5 petals, 1.5 in., slight fragrance; few prickles; foliage medium, dark green, semi-glossy; bushy, spreading (10-12 in) growth; [Angel Face X Unknown]; Bell, Judy G.; Michigan Mini Roses, 1995

'JUDmarie', Min, m, 1995; (**Ashley Marie**); flowers light lavender, moderately full (15–25 petals), small blooms borne mostly single; no fragrance; few prickles; foliage small, dark green, semi-glossy; low (14in.), bushy growth; [Dale's Sunrise X Angel Face]; Bell, Judy G.; Michigan Miniature Roses, 1996

'JUDmon', Min, m, 1995; (**Monday's Child**); flowers lavender and white, moderately full (15–25 petals), small blooms borne in small clusters; no fragrance; few prickles; foliage medium, medium green, glossy; medium (16 in.), upright, bushy growth; [Dale's Sunrise X Angel Face]; Bell, Judy G.; Michigan Miniature Roses, 1996

'JUDmorn', Min, m, 1995; (**Misty Morning**); flowers lavender shaded yellow/ tan, full (26–40 petals), small blooms borne mostly single; slight fragrance; few prickles; foliage medium, medium green, matt; tall (18 in.), upright, bushy growth; [Dale's Sunrise X Angel Face]; Bell, Judy G.; Michigan Miniature Roses, 1996

'JUDnic', Min, op, 1992; (**Baby Dominic**); flowers light orange-coral pink, micro-mini, dime-sized blooms, dbl., 15–25 petals, exhibition form, borne in small clusters, slight fragrance; few prickles; foliage small, light green, matt; low, compact, bushy growth; [Centergold X Cuddles]; Bell, Judy G.; Michigan Mini Roses

'JUDpride', Min, rb, 1996; (**P.J.'s Pride**); bud medium; flowers white with bright pink to red edging shadiing down petals to, exhibition form, no fragrance; some prickles; foliage medium, dark green, dull; medium, upright growth; [American Rose Centennial X Selected Pollen]; Bell, Judy G.; Michigan Mini Roses, 1997

'JUDriv', Min, ob, 1996; (**River City Jubilee**); flowers orange, tight, dbl., 26–40 petals, exhibition form, no fragrance; some prickles; foliage medium, dark green, semi-glossy, very disease-resist; medium (16 in), upright growth; [Jean Kenneally X Rainbow's End]; Bell, Judy G.; Michigan Mini Roses, 1996

'JUDsar', Min, dr, 1996; (**Caesar's Rose**); semi-dbl., 1–1.5 in., borne mostly singly, no fragrance; many dark red, short prickles; foliage small, dark green, matt; upright growth; medium (15-18 in. winter hardy; [Jean Kenneally X Angel Face]; Bell, Judy G.; Michigan Mini Roses, 1997

'JUDsilk', Min, ab, 1996; (**Silk 'n' Satin**); flowers apricot, bright yellow stamens, blooms borne in small clusters, semi-dbl., 8–9 wide petals, 1.5 in., no fragrance; no prickles; foliage small, light green, matt; bushy, spreading growth; low (10 in. winter hardy; [Seedling X Unknown]; Bell, Judy G.; Michigan Mini Roses, 1997

'JUDsolde', MinFl, pb, 1994; (**Isolde**); flowers pink blend (light petals in center shading darker toward edges), dbl., 26–40 petals, 1.5 in., slight fragrance; few prickles; foliage medium, medium green, matt; upright, bushy (14-16 in) growth; [Dale's Sunrise X Charmglo]; Bell, Judy G.; Michigan Mini Roses, 1995

'JUDspunk', Min, mr, 1993; (**Spunky**); dbl., 15–25 petals, 1.5 in., borne in small clusters, no fragrance; many prickles; foliage small, dark green, glossy; medium (16 in), bushy, upright growth; [Dale's Sunrise X J. Michael]; Bell, Judy G.; Michigan Mini Roses, 1994

'JUDvee', Min, dp, 1993; (**Vee Marie**); flowers bright deep pink, slight white on reverse, yellow stamens, dbl., 26–40 petals, 1.5 in., borne in small clusters, no fragrance; few prickles; foliage small, dark green, glossy; low (15 in), bushy, spreading, compact growth; [Dale's Sunrise X Angel Face]; Bell, Judy G.; Michigan Mini Roses, 1993

'JUDvirg', Min, ly, 1999; (**Dorothy Virginia**); flowers medium, dbl., 17–25 petals, borne mostly singly, no fragrance; few prickles; foliage large, dark green, semi-glossy; upright, medium growth; [Loving Touch X Unknown]; Bell, Judy G.; Michigan Mini Roses, 1999

'Judy', HT, mr, 1940; flowers cerise-red, dbl., 55 petals, 4 in., moderate fragrance; [Jewel sport]; Grillo

Judy *see* TANydu

Judy Dench *see* 'PEAthunder'

'Judy Finnigan', HT, op, 1999; flowers orange/peach, classic-shaped, dbl., 26–40 petals, 4–4.5 in., borne in small clusters, moderate fragrance; almost thornless; foliage medium, medium green, semi-glossy, very rain-resista; upright, tall (4 ft) growth; [Hazel Rose X Joe Longthorne]; Poole, Lionel; David Lister, Ltd., 2000

'Judy Fischer', Min, mp, 1968; bud pointed; flowers rose-pink, small, dbl.; foliage dark, bronze, leathery; vigorous, bushy, low growth; AOE, 1975; [Little Darling X Magic Wand]; Moore, Ralph S.; Sequoia Nursery

Judy Garland *see* 'HARking'

'Judy Hart', HT, mp, 1958; bud ovoid; dbl., 30–40 petals, 4–5 in., intense fragrance; foliage leathery; vigorous, bushy growth; [Pink Delight (HT) sport X (Senator X Florex)]; Motose; G.B. Hart

'Judy Robertson', Min, my, 1998; flowers medium yellow, very full, 80 petals, 2 in., exhibition form, borne singly and in small clusters, slight fragrance; no prickles; foliage medium, medium green, semi-glossy; long, thin stems; upright, bushy, 3 ft. growth; [Little Darling X Rise 'n' Shine]; Warner, A.J.

'Jugoslavie', HT, w, 1936; flowers large, dbl., slight fragrance; Böhm, J.; (Sangerhausen)

Juillet® *see* BRIletjui

'Jujnoberejnaia', F, dr, 1955; (River's South Bank); flowers velvety red, well shaped, medium; [Independence X Vaterland (?)]; Klimenko, V. N.

JuJu™ *see* 'BISred'

'JULES', LCl, dr, 1997; flowers dark to medium red, velvety, very full, very dbl., 41 petals, 2.5 in, borne in small clusters, moderate fragrance; foliage medium, light green turning medium green, sem-glossy; spreading, medium growth; [Seedling X Seedling]; Jerabek, Paul E.

'JULES BARIGNY', HP, mr, 1886; flowers large, dbl., intense fragrance; Verdier, E.; (Sangerhausen)

'JULES CLOSEN', HWich, dr, 1935; flowers darker and more, dbl.; [Excelsa sport]; Opdebeeck

'JULES FINGER', T, rb, 1879; flowers vivid red fading light red, shaded silvery, very large, dbl.; vigorous growth; [Catherine Mermet X Mme de Tartas]; Ducher, Vve.

'JULES GAUJARD', HT, ob, 1928; flowers bright orange-red flushed carmine, very large, cupped, intense fragrance; foliage bright green; very vigorous growth; [Jean C.N. Forestier X Unnamed seedling]; Pernet-Ducher; Gaujard

'JULES GIRODIT', HT, op, 1900; flowers light orange-pink, large, dbl.; Buatois; (Sangerhausen)

Jules Jurgensen, B, m, 1879; flowers purple-crimson, large; Schwartz

'JULES LEVACHER', HWich, lp, 1908; flowers creamy light rose pink, medium, semi-dbl.; Barbier; (Sangerhausen)

'JULES MARGOTTIN', HP, mp, 1853; flowers carmine-rose, rather, large, dbl., 90 petals, flat, recurrent bloom, slight fragrance; vigorous growth; [Probably La Reine seedling sport]; Margottin

'JULES MARGOTTIN, CLIMBING', Cl HP, mr, 1874; flowers red with purple tints, large, dbl., intense fragrance; Cranston; (Sangerhausen)

'JULES SEURRE', HP, 1869; Liabaud, I.; (Cavriglia)

'JULES TABART', HT, pb, 1920; flowers silvery salmon-pink, center coppery coral-pink, dbl., moderate fragrance; [Unnamed seedling X Mme Edouard Herriot]; Barbier

'JULES TOUSSAINT', HT, dr, 1900; flowers large, dbl., moderate fragrance; Bonnaire; (Sangerhausen)

'JULESCHILDREN', F, lp, 1998; (**Me Too**); flowers pink-purple, white eye, white reverse, small, semi-dbl., 8–14 petals, .5–1.5 in., no fragrance; prickles moderate; foliage small, light green, dull; upright, medium growth; [MACkaukaup X Nickelodeon]; Muha, Julius; Garden Jules, 1999

'JULESDEAR', HT, dp, 1998; (**Thanx Mom**, Thanx Mum); flowers hot fuschia pink, opens decorative, dbl., 26–40 petals, 4–4.5 in., borne in small clusters, intense, raspberry fragrance; prickles moderate; foliage medium, medium green, semi-glossy; bushy, medium, 3 ft. growth; [Lancome X Stephen*s Big Purple]; Muha, Julius; Hortico, Inc., 1998

'JULESLOVE', HT, lp, 1998; (**Luv Ya**); flowers soft apricot pink, full, very dbl., 26–40 petals, 4–4.5 in., borne mostly singly, moderate fragrance; prickles moderate; foliage medium, medium green, semi-glossy; upright, 3-5 ft. growth; Muha, Julius

Juli de Sala® *see* FEmental

Julia® *see* PEKcouliane

'JULIA ANN BOSTICK', Pol, pb, 1935; flowers apple-blossom-pink, base white, small, single, cupped, moderate fragrance; dwarf growth; [Ideal sport]; Bostick

'JULIA BARTET', HT, yb, 1920; flowers dark canary-yellow, fading pale straw-yellow, dbl., slight fragrance; [Lyon Rose X Georges Schwartz]; Schwartz, A.

'JULIA CLEMENTS', F, mr, 1957; flowers bright red, borne in clusters, single, 3 in.; foliage dark, glossy; very vigorous growth; Wheatcroft Bros.

'JULIA COUNTESS OF DARTREY', HT, pb, 1927; flowers rose pink with golden yellow, very large, dbl., intense fragrance; Hall; (Sangerhausen)

Julia Ferran *see* **'LADY TRENT'**

'JULIA MANNERING', HEg, lp, 1895; flowers pearly pink, yellow stamens blooms borne along the cane, semi-dbl., summer bloom, moderate fragrance; vigorous growth; Penzance

Julia Renaissance *see* POUlheart

'JULIA, COUNTESS OF DARTREY', HT, rb, 1927; bud pointed; flowers tyrian rose, base yellow, very large, dbl., exhibition form, intense fragrance; foliage dark, leathery, glossy; very vigorous growth; GM, NRS, 1925; Hall; McGredy

'JULIANA ROSE', Pol, op, 1920; flowers pale salmon; [Orléans Rose sport]; Den Ouden

'JULIA'S ROSE'®, HT, r, 1976; bud long, pointed; flowers parchment and copper shades, pointed, small, dbl., 22 petals, 2.5 in., slight fragrance; foliage reddish; upright growth; GM, Baden-Baden, 1983; [Blue Moon X Dr. A.J. Verhage]; Wisbech Plant Co.

Julia's Rose, Climbing, Cl HT, lp; [Julia's Rose sport]; St. Kilda's, 1994; (Weatherly, L.)

'JULIE', HT, dr, 1970; bud ovoid; flowers large, dbl., cupped, intense fragrance; foliage dark, soft; upright growth; [Seedling X Red American Beauty]; Kordes, R.; Kordes

Julie Andrews *see* FRYvivacious

Julie Ann™ *see* 'SAVAWEEK'

'JULIE ANNE ASHMORE', HT, yb, 1985; flowers deep yellow, suffused with pink throughout; [Peace sport]; Owen, Fred

Julie Cussons *see* FRYprincess

'JULIE DE MERSAN', M, mp, 1854; (Julie de Mersent); flowers rose shaded blush; Thomas

Julie de Mersent *see* **'JULIE DE MERSAN'**

Julie Delbard® *see* 'DELJULI'

'JULIE D'ÉTANGES', HGal, m; flowers rosy lilac, edged blush, large, dbl., cupped; erect, vigorous growth; Vibert, ca 1834

'JULIE SHARP', F, w, 1976; flowers white, pink edge maturing to scarlet, full, dbl., 25–30 petals, 3.5 in., cupped, slight fragrance; foliage matt green; vigorous, upright growth; [Evelyn Fison sport]; Sharp

'JULIE STRAHL', HT, rb, 1928; flowers nasturtium-red, passing to golden yellow, dbl., intense fragrance; [Lady Greenall X Gorgeous]; Leenders Bros.

Julie Y *see* HARbinger

'JULIEN POTIN', HT, ly, 1927; (Golden Pernet); bud golden, pointed; flowers primrose-yellow, large, dbl., exhibition form, moderate fragrance; foliage bright green; vigorous growth; GM, Portland, 1929; [Souv. de Claudius Pernet X Unnamed seedling]; Pernet-Ducher; Dreer

'JULIEN POTIN, CLIMBING', Cl HT, dy, 1935; Bostick

Julien Renoard, HT, 1987; Dorieux, Francois; (Cavriglia)

'JULIENNE', HT, mp, 1940; flowers silvery pink; [Jewel sport]; Grillo

Julies Choice *see* 'RENJULIE'

Julie's Choice *see* 'RENJULIE'

'JULIET', HP, pb, 1910; bud globular, golden yellow; flowers rich rosy red to deep rose, reverse old-gold, large, dbl., occasionally recurrent bloom, moderate fragrance; foliage curiously curled; vigorous growth; [Capt. Hayward X Soleil d'Or]; Paul, W.

Juliet Ann *see* HARvissa

'JULIET STAUNTON CLARK', HT, w, 1933; flowers white, center blush-white, turning white, large, very dbl., moderate fragrance; [Juliet sport]; Robichon

'JULIETTE', Min, mr; flowers brilliant crimson-scarlet, dbl., 30 petals; foliage bright red in fall; vigorous (10-12 in) growth; Lamb Nursery

'JULIETTE', HGal, m; Miellez, before 1828

'JULIETTE E. VAN BEUNINGEN', HT, dp, 1937; (Mme Juliette); bud pointed; flowers bright pink, open, very large, semi-dbl.; foliage leathery, dark; [Dame Edith Helen X Mrs Sam McGredy]; Buisman, G. A. H.

Julischka® *see* 'TANJUKA'

'JULIUS FABIANICS DE MISEFA', T, mr, 1902; flowers crimson; Geschwind, R.

'JULIUS GOFFERJE', HT, pb, 1930; flowers peach-pink on yellow ground, dbl.; foliage bright green; vigorous growth; Schmidt, J.C.

'JULKLAP', S, op, 1940; flowers light salmon-pink, large, semi-dbl., slight fragrance; Krause; (Sangerhausen)

'JULY GLORY', HWich, dp, 1932; flowers rich rose-pink, small, borne in large clusters, dbl.; foliage glossy; vigorous growth; Chaplin Bros.

Jumpin' Jack, F, op, 1997

Jumping Jack Flash *see* 'TALJUM'

'JUNE', HT, pb, 1937; bud pointed; flowers shell-pink, center darker, well shaped, moderate fragrance; vigorous growth; Archer

'JUNE ABERDEEN', F, ob, 1977; flowers salmon, dbl., 20 petals, 2.5 in., slight fragrance; foliage dark; [Anne Cocker X (Sabine X Circus)]; Cocker

'JUNE BOYD', HT, rb, 1924; flowers salmon-carmine, base yellow, opening to bright peach-blossom, dbl., moderate fragrance; McGredy

'JUNE BRIDE', Gr, w, 1957; bud pointed, greenish white tipped pink; flowers creamy white, blooms in clusters of 3-7, dbl., 30 petals, 4 in., exhibition form, moderate fragrance; foliage leathery, crinkled; vigorous, upright growth; [(Mme Butterfly X New Dawn) X Crimson Glory]; Shepherd; Bosley Nursery

June Bug *see* 'PIXJUN'

'JUNE FLAME', HT, ob, 1949; bud small, tight; flowers bright orange-flame shaded copper, very early bloom; Fletcher; Tucker

June Laver™ *see* 'LAVJUNE'

'JUNE MORN', LCl, rb, 1939; bud ovoid; flowers carmine-red, reverse touched gold, dbl., 5 in., exhibition form, some recurrent bloom; vigorous, climbing (8 ft.) growth; [Mme Gregoire Staechelin X Souv. de Claudius Pernet, Climbing]; Nicolas; J&P

'JUNE OPIE', F, ab, 1958; flowers apricot shaded salmon-pink, borne on trusses, semi-dbl., 3 in., slight fragrance; foliage leathery; very free, upright growth; [Masquerade X Seedling]; Kordes; Morse

'JUNE PARK', HT, dp, 1958; flowers rose-pink, dbl., 40 petals, 4.5–5 in., intense fragrance; foliage dark; vigorous, spreading growth; GM, NRS, 1959; [Peace X Crimson Glory]; Park; Sanday and Nonin

'JUNE PATRICIA', HT, lp, 1966; flowers silvery pink, dbl., 3 in., moderate fragrance; foliage dark, glossy; moderate growth; [Peace X Ena Harkness]; Lens

'JUNE TIME', Min, lp, 1963; flowers light pink, reverse darker, small, in clusters, dbl., 75 petals; foliage glossy; bushy, compact (10-12 in. growth; [(R. wichurana X Floradora) X ((Étoile Luisante seedling X Red Ripples) X Zee)]; Moore, Ralph S.; Sequoia Nursery

'JUNE WAY', HT, mp, 1977; dbl., 33 petals, 4–5 in., cupped, slight fragrance; foliage glossy; spreading growth; [Pink Favorite X Chrysler Imperial]; Atkiss; Wyant

'JUNE WEDDING', HT, w, 1977; bud pointed; flowers white, tinted yellow, dbl., 27 petals, 4 in., exhibition form, slight fragrance; foliage glossy, dark; upright growth; [Bewitched sport]; Graham; South Forrest Rose Nursery

June Whitfield *see* HARchutzpah

'JUNEEN', HT, my, 1967;, exhibition form, slight fragrance; compact growth; [Burnaby X Burnaby seedling]; Mason, P.G.

Junior, HT, 1973; St Zila; (Cavriglia)

'JUNIOR BRIDESMAID', F, mp, 1962; bud short, pointed; flowers sweetheart, dbl., 35 petals, 2 in., slight fragrance; foliage leathery; vigorous, upright growth; [Stoplite X Lovelight]; Jelly; E.G. Hill Co.

'JUNIOR GILBERT', HT, or, 1954;, cupped; vigorous growth; Mallerin, C.; EFR

Junior Miss *see* **'AMERICA'S JUNIOR MISS'**

'JUNIOR MISS', F, pb, 1943; bud well formed; flowers pink and yellow, medium, borne in clusters, semi-dbl., exhibition form, slight fragrance; foliage glossy; vigorous, bushy growth; [Joanna Hill X Heidekind]; Duehrsen; California Roses

'JUNIOR PROM', F, dr, 1962; bud ovoid; flowers crimson, open, dbl., 30–45 petals, 1.5–2 in., moderate fragrance; vigorous, upright growth; [Orange Sweetheart X Lovelight]; Jelly; E.G. Hill Co.

'JUNIOR VAN FLEET', S, lp, 1923; flowers flesh-pink, dbl., non-recurrent, moderate fragrance; [Dr. W. Van Fleet X Frau Karl Druschki]; Kemp, J.A.

'JUNO', HCh, lp, 1847; flowers pale rose, very large, dbl., globular; Laffay, M.

'JUNO', HT, mp, 1950; bud ovoid; flowers soft medium pink, large, dbl., 30 petals, exhibition form, slight fragrance; foliage bright, leathery, wrinkled, glossy; moderate upright, bushy growth; [Duquesa de Peñaranda X Charlotte Armstrong]; Swim, H.C.; Armstrong Nursery

'JUNO', C, lp; flowers blush pink, very dbl., globular; arching shrub growth; L1832

Junon® *see* 'GAUCOVA'

'JUPITER', sp; flowers bright pink, dbl.; shining, black fruit; foliage finely divided; dense, shrubby (3-4 ft) growth

Jupiter, HT, pb

Jupiter, S, mr, 1987; GPG Bad Langensalza

'JUPITER', HP, lp; [Prince Camille de Rohan X Unknown]; Williams, A., 1900; (Weatherly, L.)

Jupon Rose® *see* 'LENNIRO'

Jura® *see* **Berliner Luft**®

Jurassic Pink, HKor, mp

'JUSAMANDA', Min, w, 1991; (**Amanda Kay**); bud ovoid; flowers white to near white, some pink shading, small, dbl., exhibition form, borne singly, intense, spicy fragrance; foliage small, medium green, glossy; bushy, low growth; [Seedling X Seedling]; Justice, Jerry G., 1990; Justice Miniature Roses

'JUSCLARE', Min, w, 1995; (**Nancy Clare**); flowers nearly pure white with pink accent at outer edge, blooms bor, dbl., 18–20 petals, 1.75–2 in., moderate, spicy fragrance; foliage medium, light green, semi-glossy; medium (32 in), upright growth; RR; [Jennifer X Seedling]; Justice, Jerry G.; Justice Miniature Roses, 1995

'JUSELLE', Min, dp, 1991; (**Giselle**); bud small, pointed with medium green sepals; flowers pink outer edges with very light pink (almost white) at midline, semi-dbl., 18 petals, 1.25 in., no fragrance; foliage small, dark green, glossy, disease-resistant; bushy, low growth; [Crazy Dottie X Unnamed seedling]; Justice, Jerry G.; Justice Miniature Roses, 1992

'JUSHAZE', Min, m, 1996; (**Purple Haze**); flowers soft mauve with light tan shading at the throat, blooms born, dbl., 26–40 petals, 2.75 in., intense fragrance; some prickles; foliage medium, medium green, semi-glossy; bushy, medium growth; PPAF; [Twilight Trail X Seedling]; Justice, Jerry G.; Justice Miniature Roses, 1996

'JUSHONEY', Min, ob, 1991; (**My Honey**); bud pointed; flowers orange with light yellow accent at base, acquires pinkish cast, dbl., no fragrance; foliage medium, dark green, semi-glossy; spreading, medium growth; [Orange Honey sport]; Justice, Jerry G., 1990; Justice Miniature Roses

'JUSLATTE', Min, m, 1995; (**Latte**); flowers russet with lavender edge, blooms borne singly, very dbl., 56 petals, 1.5 in., moderate, spicy fragrance; foliage medium, medium green, semi-glossy; medium, bushy growth; RR; [Twilight Trail X Seedling]; Justice, Jerry G.; Justice Miniature Roses, 1995

'JUSLEMON', Min, my, 1996; (**Lemon Meringue**); flowers medium yellow, with white to ivory tips, reverse becomes lig, dbl., 26–40 highly reflexed petals, 1 in.; nearly thornless; foliage medium, dark green, dull; compact (16-20 in) growth; RR; [Rise 'n' Shine X Seedling]; Justice, Jerry G.; Justice Miniature Roses, 1997

'JUSMAIDEN', Min, w, 1995; (**Snow Maiden**); flowers pure white with ivory to bone white throat, reverse pure white, pink tones in cool weather, full (26–40 petals), petals form starlike cluster, medium blooms borne mostly single; slight fragrance; few prickles; foliage medium, dark green, glossy; upright, medium growth; [Snow Twinkle X Seedling]; Justice, Jerry G.; Justice Miniature Roses, 1996

'JUSMICHAEL', Min, dp, 1991; (**Baby Michael**); bud pointed; flowers deep rose pink, reverse medium pink with a touch of silver at base, very dbl., 38–40 petals, exhibition form, slight, spicy fragrance; foliage medium, dark green, semi-glossy; upright, medium growth; [Seedling X Seedling]; Justice, Jerry G., 1990; Justice Miniature Roses

'JUSPROP', Min, pb, 1999; (**Whirlygig**); flowers soft pink and white, reverse darker, single, 5–11 petals, 2 in., slight fragrance; foliage medium, dark green, semi-glossy; upright, medium (20-22 in.) growth; [(seedling X Magic Carrousel) X Unknown]; Justice, Jerry G.; Justice Miniature Roses, 1999

'JUSREGAL', Cl Min, dp, 1999; (**Regal**); flowers old rose pink, reverse same, very dbl., 41 petals, 1.75 in., borne mostly singly, no fragrance; prickles moderate; foliage medium, medium green, glossy; upright, climbing, medium to tall (6 ft.) growth; [Rosanna X Unknown]; Justice, Jerry G.; Justice Miniature Roses, 1999

'JUSSABRA', Min, mp, 1999; (**Sabra**); flowers medium pink, reverse light pink, single, 5–11 petals, 1.75 in., borne mostly singly, no fragrance; foliage large, light green, dull; upright, medium (18-21 in.) growth; [(Fairy Moss X Unknown) X Unknown]; Justice, Jerry G.; Justice Miniature Roses, 1999

Just Buddy™ *see* 'KINBUD'

Just Dreamy *see* 'PERJUSDREAM'

Just For Fun *see* 'STRILDEW'

Just For You *see* 'MORYOU'

Just Happy *see* 'DICTALENT'

Just Jenny, Min, ab, 1993

'JUST JOEY', HT, ob, 1972; flowers buff-orange, classic form, dbl., 30 petals, 5 in., intense fragrance; foliage glossy, leathery; growth moderate; GM, RNRS, 1986 Hall of Fame, WFRS, 1994 James Mason, RNRS, 1986; [Fragrant Cloud X Dr. A.J. Verhage]; Cants of Colchester, Ltd.

'JUST JUDY', HT, lp, 1992; dbl., 26–40 petals, 1.5–2.75 in., borne mostly singly, slight fragrance; some prickles; foliage medium, dark green, semi-glossy; upright (0-8 m) growth; [Mischief X Simba]; Poole, Lionel, 1994

'JUST LUCKY', HT, w, 1984; flowers well-formed, large, dbl., 35 petals, intense fragrance; foliage medium, dark, glossy; bushy growth; [Typhoo Tea X Pascali]; Bridges, Dennis A.; Bridges Roses, 1985

Just Magic *see* 'TROBIC'

Just Peachy *see* 'GELPEACH'

Just Reward *see* 'GUESTALL'

Just Simon *see* 'BURBRAHMVIDYA'

Just William *see* 'DRISCORED'

Justa Little Goofy *see* 'RENGOOFY'

Justin, HT, mr, 1997

'JUSTINA', HT, mp;, moderate fragrance

'JUSTINE', C, m, 1822; flowers pale lilac-pink, medium, very dbl.; Vibert

'JUSTINE', HT, ly, 1935; bud orange; flowers creamy yellow, base dark orange, reverse almost white, large, dbl., 30–35 petals; foliage dark, leathery; vigorous growth; [Joanna Hill X Sweet Adeline]; Hill, Joseph H., Co.

'JUSTINE RAMET', C, m, 1845; flowers purplish-rose, medium, dbl.; Vibert

Justine Silva *see* **'RITA SAMMONS'**

'JUSTINO HENRIQUES', HT, yb, 1926; flowers yellow tinted orange, stamens carmine, dbl., moderate fragrance; [Louise Catherine Breslau sport]; deFreitas; P. Guillot

'JUSTIZRAT DR HESSERT', HT, pb, 1919; bud carmine-red; flowers salmon-pink shaded red and yellow, dbl., moderate fragrance; [Gen. MacArthur X Tip-Top]; Lambert, P.

Jutland, F, dp, 1995

'JUTLANDIA', HWich, mp, 1913; dbl.; vigorous growth; [Mme Norbert Levavasseur X Dorothy Perkins]; Poulsen, D.T.

Jutta, Pol, lp, 1964; flowers large, dbl.; Rupprecht-Radke; (Sangerhausen)

Jutta, Pol, mr, 1970; flowers medium, semi-dbl.; Scholle, E.; (Sangerhausen)

'JUTUL', F, dp, 1982; flowers deep pink, small blooms in clusters of 7-9, semi-dbl., 14 petals, cupped, slight fragrance; curved, red-brown prickles; foliage dark, glossy; vigorous growth; [New Dawn X Moulin Rouge]; Lundstad, Arne; Agricultural University of Norway

'JUWEL', HT, ly, 1911; flowers large, dbl., moderate fragrance; Hinner, W.; (Sangerhausen)

K

K S G Centenary, HT, yb, 1995; K&S

K S R, HT, 1987; Keisei Rose Nurseries, Inc.; (Cavriglia)

K. of K. *see* **'KITCHENER OF KHARTOUM'**

K.A. Viktoria *see* **'KAISERIN AUGUSTE VIKTORIA'**

K.T. *see* 'TALKEV'

K.T. Marshall *see* **'KATHERINE T. MARSHALL'**

Kabuki® *see* 'MEIGOLD'

'KAGAYAKI', F, rb, 1970; (Brilliant Light); flowers brilliant scarlet and yellow, large, dbl., exhibition form, slight fragrance; foliage glossy, dark; vigorous, upright growth; [(Aztec seedling X (Spectacular X Aztec)) X Cover Girl seedling]; Suzuki, Seizo; Keisei Rose Nursery

Kaikoura® *see* 'MACWALLA'

Kaileen *see* 'JACMAR'

Kaina, HT, pb, 1985; G&L

'KAISER WILHELM', T, pb, 1889; flowers golden pink, large, very dbl., moderate fragrance; Drögemüller; (Sangerhausen)

'KAISER WILHELM I', HP, m, 1878; flowers purple/pink, large, dbl., moderate fragrance; Ruschpler; (Sangerhausen)

'KAISER WILHELM II', HT, mr, 1909; flowers medium, semi-dbl., moderate fragrance; Welter; (Sangerhausen)

'KAISERIN AUGUSTE VIKTORIA', HT, w, 1891; (K.A. Viktoria); bud long, pointed; flowers snowy white, center tinted lemon, well-formed, very dbl., 100 petals, intense fragrance; foliage rich green, soft; [Coquette de Lyon X Lady Mary Fitzwilliam]; Lambert, P.

'KAISERIN AUGUSTE VIKTORIA, CLIMBING', Cl HT, w, 1897; Dickson, A.

Kaiserin Farah, HT, dr, 1965; flowers very large, very dbl., intense fragrance; Kordes, W. Söhne; (Sangerhausen)

'KAISERIN FRIEDRICH', T, pb, 1890; flowers golden yellow on pink, large, very dbl., intense fragrance; Drögemüller; (Sangerhausen)

Kaiserin Zita, HT, lp

Kaitlyn Ainsley, HRg, m, 1998; Baskerville

Kakadu, S, lp; Peden, R., 2000; (Weatherly, L.)

Kakwa, S, w, 1973; Wallace

'KALAHARI', HT, op, 1971; flowers salmon-pink, high-pointed, dbl., 25 petals, 4 in., slight fragrance; foliage glossy, dark; [Uncle Walter X (Hamburger Phoenix X Danse de Feu)]; McGredy, Sam IV

'KALAVALLA', F, mp, 1935; dbl., borne in large clusters; vigorous growth; [Else Poulsen X Seedling]; Poulsen, S.

'KALEIDOSCOPE', F, ob, 1972; flowers orange and yellow, dbl., 28 petals, 3 in., slight fragrance; foliage glossy; [Circus X Redgold]; Fryer, Gareth; Fryer's Nursery, Ltd.

Kaleidoscope™ *see* 'JACBOW'

Kalinka *see* 'MEIHARTFOR'

Kalinka, F, 1970; Meilland, L.; (Cavriglia)

Kalinka, Climbing *see* 'MEIHARTFORSAR'

'KALMIA', HWich, w, 1911; flowers white, upper half of petals tinged pink, borne in clusters, single; foliage dark, glossy; vigorous, climbing growth; Walsh

Kalyana, HT, ob, 1995; Kasturi

Kamaladevi Chattopadhayay, HT, op, 1989; Pal, Dr. B.P.

Kambala, HT, my, 1988; Swane

'KAMCHIN', HT, mp, 1972; flowers neyron rose, outside rose, dbl., 45 petals, 4–5 in., cupped, slight fragrance; foliage dull, dark; vigorous growth; [Carina sport]; Kammeraad; Meilland

Kamelia, HT, mr, 1969; flowers large, dbl., moderate fragrance; Urban, J.; (Sangerhausen)

'KAMION', Min, or, 1988; flowers bright orange-red, reverse matte finish, clear, golden stamens, semi-dbl., 12 petals, exhibition form, no fragrance; oval, large, light, light orange fruit; prickles curved, small, reddish; foliage large, dark green, glossy; upright, tall, sturdy growth; [Starina X Unnamed seedling]; Schoen-Jones, Helen; Justice Miniature Roses, 1988

'KAMMERSANGER TERKAL', F, ob, 1971; (Hartina); bud small, globular; flowers pure orange, medium, dbl., slight fragrance; foliage glossy; dwarf, bushy growth; Tantau, Math.; Ahrens & Sieberz

Kammersingerin Perra, S; (Cavriglia)

Kampai *see* **'KAN-PAI'**

'KANA', HT, dr, 1982; flowers large blooms borne 1-3 per stem, dbl., 38 petals, exhibition form, no fragrance; small prickles, slanted downward; foliage medium green; tall, bushy growth; [Ginger Rogers X Chiyo]; Ota, Kaichiro

Kanaal, F, op, 1998; RvS-Melle

Kanak, F, yb, 1988; Sunil Jolly

'KANAKANGI', HT, ab, 1968; bud globular; flowers gold and apricot, open, medium, semi-dbl., intense fragrance; foliage leathery; moderate, bushy, open growth; [Mme Charles Sauvage X ?]; Pal, Dr. B.P.; Indian Agric. Research Inst.

'KANARIE', HT, dy, 1919; flowers clear dark yellow, dbl.; [Golden Star X Melody]; Verschuren

Kanchi, HT, dp, 1976; Viraraghavan, M.S.

'KANEGEM', F, or, 1985; flowers large blooms, dbl., 42 petals, exhibition form, borne 1-7 per cluster, no fragrance; foliage dark, glossy; upright growth; [Ludwigshafen am Rhein X Satchmo]; Rijksstation Voor Sierplantenteelt, 1982

'KAN-PAI', HT, dr, 1983; (Kampai); flowers deep red, blooms borne 1-3 per cluster, dbl., 48 petals, exhibition form, moderate fragrance; foliage medium, dark; upright growth; GM, Rome, 1983; [(Yu-ai X (Happiness X American Beauty)) X Pharoah]; Suzuki, Seizo; Keisei Rose Nursery, 1980

Kantha Selvon *see* 'EVEGAYTIME'

Kanva, HT, op, 1974; Kasturi

Kanyakumari, LCl, op, 1979; Viraraghavan, M.S.

Kapai® *see* 'MACGAM'

Kapiti *see* 'MACGLEMIL'

'KARA', Min, mp, 1972; (Dear One); bud long, mossy; flowers light to medium pink, micro-mini, mini-moss, small, single; foliage small, soft; vigorous, dwarf, bushy growth; [Fairy Moss X Fairy Moss]; Moore, Ralph S.; Sequoia Nursery

'KARDINAL'®, HT, dr, 1934; flowers scarlet-red, sometimes tipped blackish, large, cupped; vigorous, compact, bushy growth; [Château de Clos Vougeot seedling X Unnamed seedling]; Krause

Kardinal™ *see* 'KORLINGO'

Kardinal 85 *see* 'KORLINGO'

'KARDINAL PIFFL', HT, ob, 1925; flowers red-orange, reverse golden yellow, very large, dbl., moderate fragrance; foliage good; GM, Bagatelle, 1926; [(Mme Edouard Herriot X Rayon d'Or) X Mme Charles Lutaud]; Leenders Bros.

'KARDINAL SCHULTE', HT, mr, 1926; flowers brilliant scarlet-red, dbl., moderate fragrance; [(Jonkheer J.L. Mock X Radiance) X Commandeur Jules Gravereaux]; Leenders Bros.

Kardinal Schulte, Climbing, Cl HT, mr; flowers large, dbl., moderate fragrance; (Sangerhausen)

Kardinal, Climbing, Cl HT, lp; [Kardinal sport]; Knight, J., 1999; (Weatherly, L.)

Kardinal's Flame, HT, or

'KAREL HYNEK MÁCHA', HT, dr, 1936; flowers velvety red, well shaped; Brada, Dr.; Böhm

'KAREL IV', HT, mr, 1935; flowers large, dbl., moderate fragrance; Brada, Dr.; (Sangerhausen)

'KAREN', S, w; flowers creamy white flecked red, well-shaped, very dbl., non-recurrent, moderate fragrance; bushy, erect (5 ft.) growth; [R. primula X R. spinosissima cultivar]

'KAREN', F, my, 1961; flowers straw-yellow shaded ochre-yellow, dbl., 30–35 petals, 3 in.; bushy growth; [Goldilocks X Fashion]; Borgatti, G.; Sgaravatti

Karen *see* DEVoran

Karen Blixen *see* 'POULARI'

Karen Julie, HT, or; [Alexander X Weinerwold]; Allender, Robert William, 1979

'KAREN MARIE', HT, yb, 1973; flowers cream and yellow, flushed pink, full, dbl., 40 petals, 4 in., intense fragrance; foliage glossy, dark; free growth; [Kordes' Perfecta sport]; Gates

'KAREN POULSEN', F, mr, 1932; flowers scarlet, single blooms in huge trusses; vigorous growth; (28); GM, NRS, 1933 GM, Portland, 1935; [Kirsten Poulsen X Vesuvius]; Poulsen, S., 1932; J&P, 1933

'KAREN POULSEN, CLIMBING', Cl F, mr; Roger

Karenina, Min, op, 1999

Karen's Cream, HT, lp; [Shocking Blue sport]; Neil, J.; (Weatherly, L.)

'KAREN'S PINK LACE', F, mp, 1998; flowers medium pink, creamy reverse at base, very full, 41 petals, borne mostly singly, moderate fragrance; prickles moderate, medium size; foliage medium, medium green, dull; bushy, medium, 3 - 4 ft. growth; [French Lace sport]; Prevatt, Clarence; Giles Rose Nursery, 1998

Karine, Pol, mp, 1992; Knopf, Ruth

Karine Sauvageot® *see* SAUkar

Karkulka, F, 1986; Urban, J.; (Cavriglia)

Karkulka, HT, lp; Strnad; (Czech Rosa Club)

Karl Fischer, S, or, 1976; Hetzel

Karl Foerster *see* **'KARL FÖRSTER'**

'KARL FÖRSTER', HSpn, w; (Karl Foerster); bud pointed; flowers snow-white, large, dbl., exhibition form, repeat bloom, slight fragrance; foliage wrinkled, light; vigorous (7 ft.) growth; [Frau Karl Druschki X R. spinosissima altaica]; Kordes, 1930

'KARL HERBST', HT, mr, 1950; (Red Peace); flowers dull dark scarlet, well-shaped, large, dbl., 60 petals, intense fragrance; vigorous growth; GM, NRS, 1950; [Independence X Peace]; Kordes

Karl Höchst, F, dp, 1983; flowers carmine-pink, large, dbl., slight fragrance; Hetzel; (Sangerhausen)

Karl Mayer, HT, 1970; Hetzel, K.; (Cavriglia)

'KARL SCHNEIDER', HMult, lp, 1934; flowers medium, semi-dbl.; Vogel, M.; (Sangerhausen)

'KARL WEINHAUSEN', F, dr, 1942; flowers dark red tinted salmon, rosette form, large, borne in clusters, semi-dbl., 20 petals, slight fragrance; vigorous, upright growth; [Baby Chateau X (Heidekind X Ingar Olsson)]; Tantau

Karla *see* FRYtrooper

'KARLEA', HT, pb, 1965; flowers pink and salmon, base yellow, full, dbl., slight fragrance; foliage glossy; vigorous, upright growth; O'Brien

Karl-Heinz Hanisch, HT, w, 1986; flowers cream white to light pink, medium, dbl., intense fragrance; Meilland; (Sangerhausen)

'KARLIAN', HT, mr, 1991; dbl., borne mostly singly, intense fragrance; foliage medium, medium green, semi-glossy; medium, upright, compact growth; [Pristine X Burgund]; Alde, Robert O.

'KARLSRUHE', HKor, dp, 1957; bud ovoid; flowers deep rose-pink, large blooms in large clusters, very dbl., cupped, repeat bloom, slight fragrance; foliage glossy; vigorous, climbing growth; Kordes

Karma *see* 'MACNIC'

Karneol Rose, HT, op, 1964; flowers dark salmon-pink, large, dbl.; Rupprecht-Radke; (Sangerhausen)

Karolina, F, mr, 1994; flowers carmine-red, medium, very dbl.; Hempelmann; (Sangerhausen)

Karoline Reiber, Pol, mr, 1991; flowers carmine-red, small, dbl., intense fragrance; Hetzel; (Sangerhausen)

'KAROLINE SVETLA', Pol, lp, 1937; flowers medium, semi-dbl.; Brada, Dr.; (Sangerhausen)

Karol's Rose, Min, mp

Karolyn *see* **'CORALIN'**

'KAROO', HT, lp; flowers bright pink, large, exhibition form; vigorous growth; RULED EXTINCT 2/88; Elton Farm Nursery

Karoo *see* 'POULKARO'

Karoo Rose *see* 'POULKARO'

Kasachstanskaia Jubilejnaja, HT, dr, 1958; flowers large, very dbl., slight fragrance; Sushkov & Besschetnova; (Sangerhausen)

'KASBAH', HT, dp, 1977; bud pointed; flowers dark pink, reverse lighter, very dbl., 80–110 petals, 3.5 in., cupped, moderate fragrance; foliage leathery; upright growth; [Unnamed seedling X Tropicana]; Takatori, Yoshiho; Japan Rose Nursery

Kasbek, HT, w; flowers medium to large, semi-dbl., slight fragrance; VEG; (Sangerhausen)

'KASHMIR', Cl F, m, 1973; bud ovoid; flowers clear mauve, full, medium, dbl., slight fragrance; foliage glossy, leathery; vigorous, climbing growth; [Magenta X Royal Tan]; Thomson

'KASSEL', LCl, or, 1957; bud ovoid; flowers orange-scarlet, large blooms in clusters to 10, semi-dbl., recurrent bloom, moderate fragrance; foliage dark, glossy; vigorous growth; not dependably hardy.; [Hamburg X Scarlet Else]; Kordes, R.; Morse

Kasteel Van Ooidonk, F, lp; single, 10 petals, flat; foliage matt; good grower growth; RvS-Melle, 1995

Kasteel van Ooidonk, F, 1995; Rijksstation Voor Sierplantenteelt; (Cavriglia)

Kasturi Rangan, HT, m, 1983; Agarwal

'KÄTCHEN VON HEILBRONN', Pol, dr, 1922; flowers very dark red, small, dbl.; [Freudenfeuer X Unnamed variety]; Kiese

'KÄTE BEYER', Pol, op, 1947; flowers salmon-pink, medium, dbl.; Vogel, M.; (Sangerhausen)

'KATE EDWARDS', HT, m, 1963; flowers magenta; [Condessa de Mayalde sport]; Edwards

Kate Emily McCormack *see* 'KENMACHO'

'KÄTE FELBERG', HT, pb, 1930; bud pointed; flowers creamy white, reverse violet-rose, large, dbl., moderate fragrance; vigorous growth; [Unnamed seedling X Mrs Wemyss Quin]; Felberg-Leclerc

'KATE MOULTON, CLIMBING', Cl HT, pb, 1928; [Miss Kate Moulton sport]; Opdebeeck

'KATE MULL', HT, op, 1934; flowers crushed strawberry and coppery rose, very dbl.; Easlea

'KATE RAINBOW', HT, pb, 1935; flowers blend of glowing pinks and gold, well formed, very large, dbl., intense fragrance; foliage glossy, leathery; very vigorous growth; Beckwith

'KÄTE SCHMID', HMult, dp, 1931; flowers deep rose-pink, very large, borne in clusters, dbl., slight fragrance; foliage light; long stems; very vigorous, climbing growth; [Fragezeichen X Tausendschon]; Vogel, R., Jr.; Kordes

Kate Sheppard, F, mp

'KATE SMITH', HT, pb, 1954; bud ovoid; flowers apricot overcast grenadine-pink, dbl., 35–40 petals, 4.5–5 in., exhibition form, intense fragrance; foliage glossy; vigorous growth; [(Break o' Day X Golden Rapture) X Ballet]; Boerner; J&P

Katelyn Ann *see* 'MICKATE'

Katerina, HP, mr, 1955; flowers large, very dbl., moderate fragrance; Kosteckij; (Sangerhausen)

Kateryna *see* CLEkate

Kate's Rose, Pol, m, 1988; [Baby Faurax X Unknown]; Cox, Mrs K., 1988

Katharina Kündgen, S, mp, 1996; dbl., moderate fragrance; Michler, K. H.; (Sangerhausen)

Katharina Sophia, F, lp, 1991; Karwecki

'KATHARINA ZEIMET', Pol, w, 1901; (White Baby Rambler); flowers pure white, small blooms in clusters of 25-50, dbl., moderate fragrance; foliage small, rich green; short stems; dwarf, bushy growth; [Étoile de Mai X Marie Pavie]; Lambert, P.

'KATHARINE PECHTOLD', HT, ob, 1934; bud old-gold and bronzy orange; flowers coppery orange, flushed rose and gold, semi-dbl., moderate, clove pink fragrance; foliage leathery; vigorous, bushy growth; [Roselandia X Charles P. Kilham]; Verschuren-Pechtold; Dreer;, H&S

'KATHARINE WORSLEY', F, mr, 1962; flowers bright oriental red, borne in large, well-spaced clusters, dbl., 28 petals, 3.5 in.; moderate, bushy growth; Waterhouse Nursery

Käthchen, F, mr, 1985; flowers large, dbl.; Hetzel; (Sangerhausen)

'KÄTHE DUVIGNEAU', S, mr, 1942; flowers glistening red tinted salmon, large blooms in clusters of 12-15, semi-dbl., 15 petals, slight fragrance; foliage leathery, glossy, bright green; vigorous, upright growth; [Baby Chateau X R. roxburghii]; Tantau

'KATHE VON SAALFELD', T, 1914; Eibel

'KATHERINE COOK', HT, mr, 1927; flowers cherry-red, dbl., moderate fragrance; [Crusader X Unnamed seedling]; Cook, J.W.

'KATHERINE HARBOUR', HT, mp, 1973; bud ovoid; flowers pink, tinted apricot, open, large, dbl., moderate fragrance; foliage glossy; vigorous, upright growth; [Queen Elizabeth X Comtesse Vandal]; Aloe Vera Nursery

Katherine Loker *see* 'AROKR'

Katherine Mansfield *see* 'MEILANEIN'

'KATHERINE MCCARTY', Min, yb, 1991; bud ovoid; flowers yellow center, then white, with coral edges, reverse same, very dbl., 60 petals, exhibition form, moderate, damask fragrance; foliage medium, dark green, matt; bushy, medium growth; [Poker Chip X Rise 'n' Shine]; Gruenbauer, Richard, 1984; Flowers 'n' Friends Miniature Roses, 1993

Katherine McGredy, HT, 1997; McGredy, Sam IV; (Cavriglia)

'KATHERINE MOCK', HT, my, 1943; [Pres. Herbert Hoover sport]; Mock

'KATHERINE T. MARSHALL', HT, mp, 1943; (K.T. Marshall); flowers deep rose-pink, flushed yellow, dbl., 22 petals, 5 in., cupped, slight, spicy fragrance; foliage leathery; vigorous, upright growth; AARS, 1944; [Unnamed seedling X Chieftain]; Boerner; J&P

'KATHLEEN', HMsk, lp, 1922; flowers blush-pink, small blooms in large clusters, single, recurrent bloom, slight fragrance; vigorous (6 ft.) growth; [Daphne X Perle des Jeannes]; Pemberton

'KATHLEEN', HMult, pb; flowers soft rose with white eye, single; Paul, W., 1908

'KATHLEEN', HT, lp, 1934; bud very long; flowers light yellowish salmon, large, slight fragrance; vigorous, free branching growth; Dickson, A.

'KATHLEEN FERRIER', F, op, 1952; flowers deep salmon-pink, blooms in small clusters, semi-dbl., 18 petals, 2.5 in., moderate fragrance; foliage dark, glossy; vigorous, upright growth; [Gartenstolz X Shot Silk]; Buisman, G. A. H.

'KATHLEEN HARROP', B, lp, 1919; flowers soft shell-pink; (21); [Zephirine Drouhin sport]; Dickson, A.

Kathleen Jane *see* 'HORCOED'

Kathleen Jermyn *see* 'LEGNEED'

'KATHLEEN JOYCE', F, lp, 1970; flowers soft pink, dbl., 30 petals, 4 in., exhibition form, intense fragrance; [Paddy McGredy X Ice White]; McGredy, Sam IV

'KATHLEEN KAYE', HT, mp, 1959; flowers rich rose-pink, dbl., 60 petals, 5 in., intense fragrance; foliage dark; vigorous growth; [Directeur Guerin X Mirandy]; Kemp, M.L.

'KATHLEEN KELLEHAN', S, rb, 1999; (The Nanango Rose); flowers red blend, full (26–40 petals), large (10 cms) bloom borne in large clusters; very fragrant; moderate prickles; foliage medium, medium green, semi-glossy; upright, medium (4 ft) growth; [Lilian Austin X Oklahoma]; Laving, Peter

'KATHLEEN KENNEDY', HT, op, 1939; flowers light salmon-carmine, shaded orange, well-formed, large, moderate fragrance; strong stems; vigorous growth; Dickson, A.

'KATHLEEN KING', HT, dp, 1930; flowers carmine-pink, well formed, large, exhibition form, intense fragrance; vigorous growth; Marriott

Kathleen Kirkham *see* 'KIRKIT'

'KATHLEEN MILLS', HT, pb, 1934; bud long, pointed; flowers pale pink tinted silvery, reverse deep pink, large, semi-dbl., intense fragrance; foliage leathery; vigorous growth; LeGrice

'KATHLEEN NASH', HT, mr, 1944; bud urn shaped; flowers bright cerise, dbl., 28 petals, 4 in., globular, intense fragrance; foliage leathery, dark; vigorous growth; [Pink Delight sport]; Spera; Rose Farms Corp

'KATHLEEN O'ROURKE', HT, op, 1976; flowers soft orange-pink, dbl., 38 petals, 4 in., exhibition form, moderate fragrance; foliage large, matt; [Fragrant Cloud X Red Planet]; Dickson, Patrick; A. Dickson

'KATHLEEN PEDEN', HT, dr, 1959; bud long pointed; flowers crimson, large, dbl., exhibition form, intense fragrance; foliage leathery, dark; very vigorous, upright growth; [Crimson Glory X Charles Mallerin]; Peden, G.H.

Kathleen Rumble *see* 'WELRUM'

'KATHLEEN WIGGIN', Cl HP, w, 1932; bud long pointed, opening one at a time on each cluster; flowers white, sometimes tinged pink, large, borne in clusters, very dbl., slight fragrance; foliage glossy, heavy; long stems; vigorous, climbing (12-15 ft) growth; [Frau Karl Druschki seedling]; Wiggin

Kathleen's Rose *see* KIRkitt

'KATHRYN', S, ab, 1986; bud copper; flowers apricot, aging to buff, copper stamens, open form, dbl., 25 petals, 4–5 in., moderate, fruity fragrance; foliage semi-glossy; vigorous growth; [Wind Chimes X Yellow HT]; Eggeman, H.W.

'KATHRYN BAILEY', F, w, 1992; dbl., 26–40 petals, 1.5–2.75 in., slight fragrance; some prickles; foliage light green, glossy; medium tall, upright growth; [Gene Boerner sport]; Bailey, Dr. Edwin; Bailey's Plant Farm, 1993

'KATHRYN GRAM', F, ab, 1945; bud urn shaped; flowers large, moderate fragrance; bushy, low growth; [Talisman X ?]; Moore, Ralph S.; Sequoia Nursery

Kathryn McGredy *see* MACauclad

Kathryn Morley *see* 'AUSCLUB'

'KATHY', Min, mr, 1970; bud pointed; flowers small, dbl., moderate fragrance; foliage small, leathery; moderate, dwarf, bushy growth; [Little Darling X Magic Wand]; Moore, Ralph S.; Sequoia Nursery

'KATHY FISCUS', F, mp, 1950; bud ovoid; flowers deep flesh-pink, open, medium, borne in clusters, dbl., 45 petals, moderate fragrance; foliage leathery; very vigorous, upright, bushy growth; [Unnamed seedling X Baby Chateau]; Duehrsen; Elmer Roses Co.

Kathy Reid, HRg, m, 1987; Sandbrook

'KATHY ROBINSON', Min, pb, 1968; flowers pink, creamy reverse, dbl., 26 petals, 1 in., exhibition form, slight fragrance; foliage small, glossy, dark, embossed; upright, bushy growth; [Little Darling X Over the Rainbow]; Williams, Ernest D.; Mini-Roses

Kathy's Hunter's Pink, HWich, mp

'KATIE', LCl, mp, 1959; bud long, pointed; flowers reverse darker, large, semi-dbl., 17 petals, cupped, recurrent bloom, intense fragrance; foliage glossy; vigorous growth; [New Dawn X Crimson Glory]; O'Neal; Wyant

Katie Crocker *see* BURbrindley

Katie Pianto's Rose, LCl, pb

Katie's McNear HP, HP, dp; (found rose)

Katrin, HT, op, 1972; flowers salmon-pink, large, dbl.; GPG Bad Langensalza; (Sangerhausen)

'KATRINA', HT, op, 1964; bud pointed; flowers orange-scarlet, medium, dbl., cupped, slight fragrance; foliage glossy, leathery; bushy, vigorous growth; [Baccará sport]; Samtmann Bros.; C-P

Katy Lampkin *see* 'TALKAT'

Katy Road Pink, S, mp

Kauff *see* **'KAUTH'**

'KAUTH', (strain of R. canina), lp; (Kauff); sometimes used as understock

'KAWAMOBLUE', HT, m, 1995; (**Purple Rain**); flowers mauve, full (26–40 petals), large (7 + cms) blooms borne mostle single; very fragrant; few prickles; foliage medium, dark green, semi-glossy; upright, medium growth; [Seedling X Blue Moon]; Kawamoto, Hiromoto; Carlton Rose Nurseries, 1994

'KAY', F, or, 1971; bud ovoid; flowers vermilion, large, dbl., cupped, slight fragrance; foliage large, bronze; vigorous, upright growth; [Queen Elizabeth X Numero Un]; Delforge

'KAY BARNARD', HT, lp, 1974; bud long, pointed; flowers pale pink, exhibition form, moderate fragrance; upright growth; [South Seas X Queen Elizabeth]; Dingle

Kay Denise *see* 'TINKAY'

Kayla *see* 'MORKAY'

'KAZANLIK', D, dp; (R. damascena trigintipetala, Trigintipetala); flowers deep pink, dbl., 30 petals; (28)

'KDE DOMOV MUJ', HMult, dp, 1935; flowers carmine-pink, medium, semi-dbl.; Böhm, J.; (Sangerhausen)

Kean, HGal, m

KEBesa, HT, lp; (**Pretty Woman**); (KEBesa), 1999

'KEELY', Min, or, 1988; bud pointed; flowers bright, orange-red, reverse slightly darker, aging color fading, dbl., 20 petals, exhibition form; prickles straight, pointed, medium, red; foliage medium, medium green, semi-glossy; bushy, medium growth; [Party Girl X Unnamed seedling]; Bridges, Dennis A.; Bridges Roses, 1989

Keepit, HT, lp, 1988; Dawson

Keepsake *see* 'KORMALDA'

'KEEPSAKE', HT, dr, 1941; flowers deep red, moderate fragrance; RULED EXTINCT 7/81; [Anne Leygues seedling]; Clark, A.

'KEES KNOPPERS', Pol, w, 1930; flowers flesh-white, open, large, borne in clusters, semi-dbl.; foliage rich green; vigorous, bushy growth; [Mev. Nathalie Nypels sport]; Leenders, M.

'KEGON', HT, mr, 1976; bud pointed; flowers deep red, dbl., 30 petals, 5 in., exhibition form; foliage leathery; tall growth; [Gruss an Berlin X Christian Dior]; Onodera, Toru F.; S. Onodera

'KEI', Min, lp, 1980; bud ovoid, pointed; dbl., 33 petals, borne 1-3 per cluster, slight fragrance; straight prickles; foliage tiny, dark; very compact, bushy growth; [Unnamed seedling X Unnamed seedling]; Lyon, Lyndon

'KEIBELMI', Min, rb, 1991; (**Shortcake**®); bud ovoid, pointed; flowers red, white reverse, red color pales with age, blues slightly, dbl., 30–35 petals, cupped, no fragrance; foliage large, dark green, glossy; upright, bushy, tall growth; [Unnamed seedling X Unnamed seedling]; Keisei Rose Nurseries, Inc., 1981; Bear Creek Gardens, 1991

'KEIBIAN', HT, yb, 1993; (**French Perfume**®); flowers light yellow with rose pink picotee, very dbl., 40–45 petals, 5–6.5 in., borne mostly singly, intense, fruity fragrance; foliage large, medium green, purple-red when new, semi-gloss; tall (110-150 cms), upright, spreading growth; PP8476; [(Todoroki X Montana) X Seedling]; Suzuki, Seizo; Bear Creek Gardens, 1993

'KEIFUPIE', Min, dp, 1999; (**Hime**); flowers bright purplish pink, dbl., 20–25 petals, 1.5–2 in., flat, moderate fragrance; foliage dark green, semi-glossy; 10 in. growth; GM, Japan Rose Concours, 1997; [Ko's Yellow X seedling]; Hirabayashi, Hiroshi, 1995; Keisei Rose Nurseries, 1997

'KEIFUPIRA', F, dp, 1999; (**Urara**); flowers pale purplish pink, dbl., 25–30 petals, 3–4 in., cupped, no fragrance; foliage dark green, leathery; 2-3 ft growth; GM, Japan Rose Concours, 1995; [Mimi X Minuette]; Hirabayashi, Hiroshi, 1993; Keisei Rose Nurseries, 1995

'KEIMIPIA', Min, lp, 1999; (**Sayokyoku**); flowers soft pink, very dbl., 45–50 petals, 2–2.5 in., exhibition form, slight fragrance; foliage dark green, leathery; 1.5-2 ft growth; GM, Japan Rose Concours, 1994; [Sonia X Petite Folie]; Hirabayashi, Hiroshi, 1992; Keisei Rose Nurseries, 1996

KEImove, Gr, m; (**Prelude**); (KEImove)

KEInoumi, F, dr; (**Scarlet Mimi**®); (KEInoumi), 1992

KEIreb, HT, mr; (**Mahalia**®); (KEIreb), 1992

KEIren, Pol, dp, 1995; (**Fairy-Tale**); Keiren

KEIromo, HT, op; (**Silva**); (KEIromo)

KEItaibu, Gr, dp; (**Laser**®); (KEItaibu)

'KEITH HARDER', HT, lp; Harder, K., 1954; (Weatherly, L.)

Keith Kirsten *see* TANsirk

Keithie, F, rb

'KEITOLI', S, dp, 1984; (Ferdi, **Ferdy**®); flowers deep pink, small, dbl., 20 petals, no fragrance; foliage medium, medium green, matt; spreading (to 5 ft) growth; groundcover; [Unnamed climbing seedling X Petite Folie seedling]; Suzuki, Seizo; Keisei Rose Nursery

KEIzoubo, HT, ob; (**Pareo**®); (KEIzoubo)

'KELDAZ', Min, yb, 1997; (**Dazzler**®); flowers double, medium, white with red edge and yellow base, dbl., 15–25 petals, borne mostly singly, slight fragrance; foliage medium, dark green, semi-glossy; upright, medium growth; [Rainbow's End X Kristin]; Kelly, Martin

Kelleriis-Rose *see* **'POULSEN'S SUPREME'**

'KELLI ANN', Min, w, 1981; flowers small, dbl., 35 petals, slight fragrance; foliage small, medium green, matt; bushy growth; [Patricia Scranton X (Patricia Scranton X Fairy Moss)]; Dobbs, Annette E.

Kelly Country, LCl, pb, 1997; Sutherland

Kelly Reynolds *see* 'REYKELLY'

Kelly-Leigh *see* 'STEKELLY-LEIGH'

Ken Davis *see* 'GELDAVIS'

'KENDAD', F, ab, 1996; (**May Graham**); flowers coppery apricot, fading to pink, blooms borne in small clust, dbl., 26 petals, 1.5–2.75 in., slight fragrance; many prickles; foliage medium, medium green, glossy; medium (3 1/2 ft), bushy growth; [Prominent X Kiskadee]; Kenny, David

'KENDANQU', F, or, 1993; (**Wiggy**); flowers orange red with cerise, dbl., 26–40 petals, 1.5–2.75 in., borne in small clusters, slight fragrance; some prickles; foliage medium, dark green, semi-glossy; tall (90 cms), upright, bushy growth; [(Prominent X Kiskadee) X (Mary Sumner X Kiskadee)]; Kenny, David; Kenny, 1993

'KENFINE', Min, dp, 1997; (**Riverdance**); flowers double, medium, dbl., 15–25 petals, borne in small clusters, moderate fragrance; foliage small, medium green, glossy; bushy, medium (28in.) growth; [MEIdomonac X Freegold]; Kenny, David

'KENMACHO', F, rb, 1999; (**Kate Emily McCormack**); flowers red, pink to white edges, handpainted with white eye, reverse silver, semi-dbl., 8–14 petals, 3 in., borne in large clusters, moderate fragrance; prickles numerous; foliage large, dark green, glossy; spreading, bushy, tall (4 ft) growth; [((Mary Sumner X Kiskadee) X Bassino) X Little Artist]; Kenny, David

'KENMADO', F, op, 1997; (**Sister Joan**); flowers double, large, dbl., 15–25 petals, borne in small clusters, moderate fragrance; some prickles; foliage large, dark green, semi-glossy; upright, bushy, tall (3.5x2.5ft.)growth; [(Mary Sumner X Kiskadee) X Maestro]; Kenny, David

'KENMASNIA', F, yb, 1995; (**Aunty Lil**); Flowers yellow edged pink, fades to pink, moderately full (15–25 petals), medium (4–7 cms.) Blooms borne in large clusters; slight fragrance; some prickles; foliage medium, medium green, glossy; medium, upright growth; [(Friesia X Kiskadee) X (Mary Sumner X Regensberg)]; Kenny, David; Kenny, 1994

'KENMOLL', F, dy, 1999; (**Tessa O'Keeffe**); dbl., 15–25 petals, 3 in., borne in small clusters, moderate fragrance; few prickles; foliage medium, dark green, glossy; bushy, medium (30 in.) growth; [Friesia X Kiskadee]; Kenny, David

'KENMORE', HT, dy, 1971; flowers deep yellow, dbl., 36 petals, 4.5 in., slight fragrance; foliage glossy; [Jane Lazenby X Golden Delight]; Macara; Sanday

Kenny's Rose *see* 'GELKEN'

'KENOBSESS', F, ab, 1999; (**Audrey McCormack**); flowers soft apricot yellow, dbl., 26–40 petals, 3 in., borne in large clusters, moderate fragrance; many prickles; foliage large, medium green, glossy; spreading, bushy, medium (3 ft) growth; [Golden Wedding X Bright Smile]; Kenny, David

'KENORA DUET', F, ab, 1976; bud long, pointed; flowers salmon-pink, open, medium, semi-dbl., slight fragrance; foliage glossy; vigorous, bushy growth; [Duet sport]; Tresise

'KENSINGO', F, op, 1995; (**Mary Swindells**); flowers salmon pink, full (26–40 petals), medium (4–7 cms) blooms borne in small clusters; slight fragrance; many prickles; foliage medium, dark green, glossy; tall, upright, bushy growth; [(Mary Sumner X Kiskadee) X (Mary Sumner X Regenberg)]; Kenny, David; Kenny, 1990

Kent® *see* 'POULCOV'

'KENTFIELD', Pol, mp, 1922; flowers soft cameo-pink to deeper pink, small, dbl., slight fragrance; [Cécile Brunner X Unnamed seedling]; Diener

Kentucky, HT, w

Kentucky Derby *see* 'ARODER'

Kern White, HP, w; (found rose)

'KERRY GOLD', F, yb, 1967; flowers canary-yellow, outer petals veined red, blooms in clusters, 3 in., globular; foliage dark; [Circus X Allgold]; Dickson, A.

'KERRY MACNEIL', F, mr, 1967; flowers bright vermilion; free growth; [Orangeade X Anna Wheatcroft]; Vincent

'KERRYMAN'®, F, pb, 1971; flowers salmon and pink, dbl., 24 petals, 4.5 in., exhibition form, slight fragrance; [Paddy McGredy X (Mme Leon Cuny X Columbine)]; McGredy, Sam IV; McGredy

'KERSBERGEN', Pol, mr, 1927; flowers bright currant-red; [Miss Edith Cavell sport]; Kersbergen

'KESRI', F, or, 1969; bud ovoid; flowers orient orange, open, small, dbl., slight fragrance; foliage leathery; vigorous growth; [Orangeade, Climbing X Open pollination]; Singh; Gopalsinamiengar

Kessy, F, my

'KETJE', HT, mr, 1938; bud pointed; flowers brilliant red mixed dark pink, well formed, very dbl.; foliage bronze; very vigorous growth; [Mrs Sam McGredy X E.G. Hill]; Lens

Kev *see* 'TALKEV'

Kevin, HT, dp, 1997; Twomey, Jerry

'KEW BEAUTY', HT, dr, 1918; flowers crimson, dbl., moderate fragrance; Therkildsen

'KEW RAMBLER', LCl, mp, 1913; flowers pink, center paler, single, borne in clusters; foliage gray-green; height to 15 ft; [R. soulieana X Hiawatha]

'KEWAI', HT, yb, 1997; flowers full, large, very dbl., 26–40 petals, slight fragrance; few prickles; foliage medium, medium green, semi-glossy; upright, bushy, medium (180cms) growth; Ohtsuki, Hironaka

Key Rock Rose, Misc OGR, yb

'KEYSTONE', LCl, dy, 1904; flowers deep lemon-yellow; Dingee & Conard

'KI KI PAQUEL', F, or, 1960; flowers bright brick color; [Super-Congo X Independence]; da Silva, Moreira

'KIA ORA', F, or, 1962; flowers orange-scarlet, large, dbl., 35 petals; foliage glossy, dark; vigorous, bushy, compact growth; [Independence X Independence]; Mason, P.G.

Kia Ora *see* 'SUNCRAL'

Kibo *see* **'KIBOH'**

'KIBOH', F, rb, 1986; (Gypsy, Gypsy Carnival, Kibo, Lovita); flowers orange-red, reverse yellowish, large, dbl., 50 petals, exhibition form, blooms in clusters of 3, slight fragrance; prickles slanted downward; foliage dark, semi-glossy; vigorous, upright growth; PP007139; GM, The Hague, 1985; [Liberty Bell X Kagayaki]; Suzuki, Seizo; Keisei Rose Nursery

'KICKAPOO', HT, rb, 1990; bud ovoid; flowers yellow, outer petals red, white reverse aging to light red blend, dbl., 20 petals, slight fragrance; round, green light yellow fruit; prickles straight, green aging to tan; foliage medium, medium green, dull; bushy, medium, slightly spreading growth; [(Daisy Mae X First Prize) X Unnamed seedling]; Stoddard, Louis, 1991

'KIDDY', Pol, mr, 1967; bud pointed; flowers dark red, becoming brighter, medium, single, cupped, abundant, recurrent bloom; foliage glossy; vigorous, upright growth; [Mme Dieudonne X Seedling]; Delforge

'KIDWAI', HT, mp, 1933; (Kidway); bud long, pointed; flowers salmon-rose, lower half of petals golden yellow, large, semi-dbl., slight fragrance; foliage leathery, dark, bronze, glossy; very vigorous, bushy growth; Pernet-Ducher; Dreer;, Gaujard;, H&S

Kidway *see* **'KIDWAI'**

Kieran Ross Clark *see* 'HORANYME'

'KIESE', HCan, mr, 1910; flowers bright red, semi-dbl.; [Général Jacqueminot X R. canina]; Kiese

'KIFTSGATE', (R. filipes form), w, 1954; (Filipes Kiftsgate, R. filipes 'Kiftsgate'); flowers creamy white, borne in large clusters, single, 5 petals, moderate fragrance; very vigorous, sprawling growth; E. Murrell

Kika *see* 'ORTKEE'

Kiki, HT, 1972; Dorieux, Francois; (Cavriglia)

Kiki Paquel, Gr, 1960; Moreira da Silva, A.; (Cavriglia)

Kiki Rose® *see* 'POULCAT'

Kiko *see* JACdeep

Kilimanjaro *see* 'HERARO'

'KILLARNEY', HT, mp, 1898; bud long, pointed; flowers bright medium pink, loose, large, dbl., intense fragrance; foliage bronze; [Mrs W.J. Grant X Charles J. Grahame]; Dickson, A.

'KILLARNEY BRILLIANT', HT, dp, 1914; bud long, pointed; flowers brilliant pink to rosy carmine, open, dbl., moderate fragrance; foliage rich green, soft; [Killarney sport]; Dickson, A.

'KILLARNEY DOUBLE PINK', HT, mp, 1935; flowers sparkling shell-pink; [Killarney Double White sport]; Vestal

'KILLARNEY DOUBLE WHITE', HT, w, 1912; flowers snowy white; [Killarney sport]; Budlong; A.N. Pierson

'KILLARNEY DOUBLE WHITE, CLIMBING', Cl HT, w, 1935; Howard Rose Co.

'KILLARNEY QUEEN', HT, dp, 1912; flowers tyrian rose, brighter than killarney; [Killarney sport]; Budlong

'KILLARNEY, CLIMBING', Cl HT, mp, 1908; Reinberg

Kilwinning, HSpn, lp;, intense; fern like; [R. spinosissima altaica X Hansen's Yellow]; Wright, Percy H.

'KILWORTH GOLD', HT, dy, 1977; flowers golden yellow; [Whisky Mac sport]; Gandy, Douglas L.

'KILWORTH PRIDE', F, dr, 1955; semi-dbl., blooms in clusters; foliage bronze; dwarf, bushy growth; [Better Times X Floribunda seedling]; deRuiter; Gandy Roses, Ltd.

'KIM', F, my, 1971; flowers mini-flora, dbl., 28 petals, 3 in., slight fragrance; foliage small, light green, matt; dwarf growth; [(Orange Sensation X Allgold) X Elizabeth of Glamis]; Harkness, 1973

'KIM', Pol, mp, 1956; flowers geranium-rose, small, borne in clusters; vigorous growth; [Independence X Salmon Perfection]; Frères, Buyl

'KIMBERLEY ANNE', HT, w, 1983; flowers large, dbl., 35 petals, intense fragrance; foliage medium, medium green, matt; upright growth; Bronze, ARC TG, 1985; [Virgo X Secret Love]; Evans, F. David

Kimberly *see* 'MICKIM'

Kimbo, F, 1977; Pineau; (Cavriglia)

Kimmy, F, 1975; Laperriere, L.; (Cavriglia)

'KIMONO'®, F, pb, 1961; flowers salmon-pink, blooms in broad clusters, dbl., 30 petals, 3 in., moderate fragrance; vigorous, bushy growth; [Cocorico X Frau Anny Beaufays]; deRuiter; Horstmann

Kim's Cream™ *see* 'RUPKIMCRM'

'KINASH', Min, yb, 1985; (**Baby Ashley**™); flowers light yellow, petals edged light pink, dbl., 24 petals, borne singly and in cluster of 3-5, no fragrance; straight, small, light brown prickles; foliage small, dark, matt; bushy growth; King, Gene; AGM Miniature Roses

'KINBARB', Min, ab, 1990; (**Barbara Mandrell**™); bud pointed; flowers dark apricot changing to apricot-pink at 3/4 open, yellow base, aging to white, dbl., 28 petals, exhibition form, slight fragrance; foliage medium, medium green, matt; upright, medium growth; [(Seedling X Party Girl) X Party Girl]; King, Gene, 1980; AGM Miniature Roses, 1991

'KINBAY', Min, ob, 1989; (**Tampa Bay**™); bud pointed; flowers orange, edges aging darker, medium, borne singly, dbl., 18 petals, exhibition form, slight, fruity fragrance; ovoid, orange fruit; prickles straight, red; foliage medium, medium green, matt; upright, medium growth; [(Arthur Bell X Orange Honey) X Baby Diana]; King, Gene; BDK Nursery, 1988

'KINBEE', Min, dr, 1986; (**B. C.**™); flowers dark red, reverse darker, dbl., 24 petals, exhibition form, slight fragrance; globular fruit; long, hooked, brown prickles; foliage small, medium green, matt; medium, upright, bushy growth; [Evelyn Fison X Magic Mist]; King, Gene; AGM Miniature Roses

'KINBELLE', Min, dp, 1988; (**Rosa Belle**™); flowers deep pink to yellow-cream at base, medium, borne usually singly, dbl., 21 petals, exhibition form; no fruit; prickles straight with hook, few, dark brown; foliage medium, medium green, matt; bushy, medium growth; [Vera Dalton X Party Girl]; King, Gene; AGM Miniature Roses

'KINBO', Min, dy, 1989; (**Tobo**™); bud pointed; flowers medium, borne usually singly and in sprays of 2-3, dbl., 32 petals, exhibition form, slight, fruity fragrance; ovoid, green fruit; prickles straight, slight crook, red to brown; foliage medium, light green, matt; upright, bushy, medium growth; [Arthur Bell X Rise 'n' Shine]; King, Gene; AGM Miniature Roses

'KINBOSCO', Min, ob, 1990; (**Boscobel**™); flowers orange to burnt orange, moderately small, borne mostly singly, dbl., 15–25 petals, no fragrance; foliage medium, light green, matt; upright, medium growth; [(Cheers X Rainbow's End) X Breezy]; King, Gene; AGM Miniature Roses

'KINBUD', Min, ly, 1985; (**Just Buddy**™); flowers mini-flora, large blooms borne singly, dbl., 45 petals, exhibition form, slight fragrance; no hips; straight, light brown prickles; foliage medium, light green, matt; medium upright, bushy growth; [New Day X Rise 'n' Shine]; King, Gene; AGM Miniature Roses, 1986

'KINBUFF', Min, ab, 1986; (**Buffy**™); flowers light apricot, reverse deeper, darker towards center, medium, dbl., 20 petals, exhibition form, slight fragrance; oval, green fruit; slightly hooked, light brown prickles; foliage medium green, matt; mini-flora; bushy spreading growth; [Vera Dalton X Party Girl]; King, Gene; AGM Miniature Roses, 1987

'KINCAP', Min, or, 1991; (**Captivation**™); dbl., borne mostly singly, slight fragrance; some prickles; foliage medium, light green, matt; upright (60-70 cms), bushy growth; [(Arthur Bell X Little Jackie) X Little Jackie]; King, Gene; AGM Miniature Roses, 1992

'KINCHA', Min, mr, 1984; (**Charlie**™); flowers small, dbl., 35 petals, exhibition form, no fragrance; foliage small, dark, matt; upright, bushy growth; [Seedling X Big John]; King, Gene

'KINCLAIRE', Min, w, 1989; (**Mollie Claire**™); bud pointed; flowers white, pink edge blushing toward center, reverse white tipped, dbl., 28 petals, exhibition form, no fragrance; no fruit; prickles crooked, very few, white to brown; foliage small, medium green, matt; bushy, low growth; [(Evelyn Fison X Magic Mist) X Baby Diana]; King, Gene; AGM Miniature Roses

'KINCOACH', Min, op, 1986; (**Stagecoach**™); flowers medium salmon pink, medium, borne usually singly or in spray, dbl., 18 petals, exhibition form, slight fragrance; oval, green fruit; straight, white prickles; foliage medium, light green, matt; upright, spreading growth; [Vera Dalton X Orange Honey]; King, Gene; AGM Miniature Roses, 1987

Kind Regards *see* 'PEATIGAR'

Kind Regards *see* PEAtiger

'KINDIDDY', Min, mp, 1991; (**Pink Diddy**®); bud pointed; flowers pink, medium, dbl., 28 petals, cupped, slight, fruity fragrance; foliage small, medium green, semi-glossy; bushy, spreading, low growth; [[(B.C. X Scamp) X Miss Dovey] X Tudelum]; King, Gene, 1986; AGM Miniature Roses, 1990

'KINDIXIE', Min, ob, 1991; (**Dixie Dazzle**®); bud ovoid; flowers orange-red to yellow, small, dbl., 16 petals, exhibition form, no fragrance; foliage small, medium green, semi-glossy; upright, low growth; [(Rainbow's End X Miss Dovey) X Jennie Anne]; King, Gene, 1986; AGM Miniature Roses, 1990

Kindness, HT, dp, 1994

'KINDOV', Min, ab, 1985; (**Miss Dovey**®, Miss Dovie); flowers deep apricot, blooms borne usually singly, dbl., 21 petals, exhibition form, slight fragrance; sets hips; straight, reddish-brown prickles; foliage medium, medium green, semi-glossy; upright, bushy growth; [Anne Harkness X Rise 'n' Shine]; King, Gene; AGM Miniature Roses

'KINFANCY', Min, rb, 1986; (**Fancy Pants**®); flowers deep pink to golden yellow base, edged red, fading deeper pink, dbl., 40 petals, exhibition form, slight, spicy fragrance; fruit not observed; medium, light red prickles, slightly crooked on end; foliage medium, medium green, matt; upright, bushy, medium growth; [Baby Katie X Rose Window]; King, Gene; AGM Miniature Roses, 1987

'KINFANTA', Min, w, 1985; (**Fantasi**®); flowers cream, reverse cream shaded deep pink, mini-flora, large, dbl., 50 petals, exhibition form, slight fragrance; no fruit; medium, hooked, brown prickles; foliage medium, dark, matt; tall, upright growth; King, Gene; AGM Miniature Roses, 1986

King Alexander I *see* **'ROI ALEXANDRE'**

'KING ARTHUR', F, op, 1967; flowers salmon-pink, blooms in clusters, dbl., 3.5 in., slight fragrance; foliage glossy; [Pink Parfait X Highlight]; Harkness

'KING BOREAS', F, my, 1941; flowers pure yellow shading to nearly white, recurved, medium, very dbl., 100 petals, moderate fragrance; vigorous growth; [Golden Glow (LCl) X ?]; Brownell, H.C.

King Crimson *see* 'JACSHUR'

King George IV *see* **'RIVERS' GEORGE IV'**

'KING GEORGE V', HP, dr, 1912; flowers crimson, large, dbl., 40 petals, exhibition form, sparse, intermittent bloom, moderate fragrance; strong stems; very vigorous, open growth; Dickson, H.

King George's Memorial *see* **'PAMÁTNIK KRÁLE JIRÍHO'**

King Henry *see* GENwine

King Hey, S, dr

'KING J.', HRg, w, 1996; flowers yellow stamens, blooms borne in small clusters, dbl., 15–25 petals, good repeat, intense, spicy fragrance; numerous prickles; foliage dark green, glossy, very healthy, disease/insect res; bushy, medium growth; hardy zone 6; [R. rugosa alba X Assiniboine]; Fleming, Joyce L.; Hortico Roses, 1996

'KING MIDAS', LCl, my, 1942; bud long, pointed; flowers clear yellow, blooms in clusters of 4-6, dbl., 20 petals, 4–5 in., cupped, repeat bloom, slight fragrance; foliage large, leathery, dark; vigorous, climbing or pillar (8-10 ft.) growth; Nicolas; J&P

'KING O' KINGS', HT, mr, 1973; dbl., 35–40 petals, 6 in., exhibition form, intense fragrance; foliage dark; [My Love X Duftwolke]; Anderson's Rose Nurseries

'KING OF HEARTS', HT, mr, 1968; bud long, pointed; flowers medium, dbl., exhibition form; foliage dark, leathery; vigorous, bushy growth; [Karl Herbst X Ethel Sanday]; McGredy, Sam IV; Edmunds Roses

'KING OF SCOTS', HSpn, dp; flowers deep pink, dbl., profuse, early bloom; glossy, black fruit; foliage finely divided; dense, shrubby (3-4 ft) growth; Lee, prior to 1848

'KING OF THE PRAIRIES', HSet, mr, 1843; flowers bright red; Feast

King Richard *see* GENruby

King Tut *see* 'CHEWARVEL'

King Tut *see* 'LAVTREK'

King William *see* GENpink

'KINGAROY', HT, dr, 1979; bud long, pointed; flowers urn-shaped blooms borne singly, dbl., 30 petals, slight, fruity fragrance; small prickles; foliage medium, matt; medium growth; [Unnamed seedling X Red Lion]; Perry, Astor, 1981

'KINGCUP', HT, my, 1953; flowers buttercup-yellow, well formed, dbl., 40 petals, 4 in., intense, fruity fragrance; foliage dark, glossy; very free growth; [Mrs Sam McGredy X Ellinor LeGrice]; LeGrice

Kingi *see* 'MURKI'

'KINGIG', Min, mp, 1987; (**Giggles**); flowers light pink, reverse light to dark pink, fading to creamy pink, dbl., 18 petals, exhibition form, slight fragrance; oval, green fruit; slightly crooked, white prickles; foliage medium, medium green, matt; upright, tall growth; [Vera Dalton X Rose Window]; King, Gene; AGM Miniature Roses

'KINGLOW', Min, ob, 1989; (**Glowry**®); bud pointed; flowers bright orange-yellow bicolor, medium, borne singly, dbl., 24 petals, exhibition form, no fragrance; no fruit; prickles straight, red; foliage small, medium green, matt; bushy, low growth; [(Arthur Bell X Orange Honey) X Baby Diana]; King, Gene; AGM Miniature Roses

'KING'S RANSOM'®, HT, dy, 1961; bud ovoid; flowers clear golden yellow, dbl., 38 petals, 5–6. in., exhibition form, moderate fragrance; foliage leathery, glossy; vigorous, upright growth; AARS, 1962; [Golden Masterpiece X Lydia]; Morey, Dr. Dennison; J&P

'KING'S ROW', S, yb, 1965; flowers yellow becoming rose-red, medium, borne in clusters, dbl., moderate fragrance; foliage bronze, leathery; vigorous (4 ft), compact growth; [Easter Parade X Herrenhausen]; Whisler; Germain's

King's Treasure, F, my, 1995; Williams, J. Benjamin

Kingsmead Heritage *see* KORdalsilk

Kingswood College *see* KORloher

'KINGUN', Min, pb, 1990; (**Top Gun**®); bud ovoid; flowers peony pink, yellow base, apricot yellow reverse, medium, bor, dbl., 28 petals, exhibition form, no fragrance; foliage medium, medium green, semi-glossy; tall, upright growth; [(Rainbow's End X Vera Dalton) X Vera Dalton]; King, Gene, 1985; AGM Miniature Roses, 1991

'KINHEART', Min, ob, 1985; (**Heartlight**®); flowers orange-yellow, reverse yellow, mini-flora, large, 16 petals, exhibition form, slight fragrance; straight, light brown prickles; foliage medium, medium green, matt; upright, bushy, medium growth; [Golden Slippers X Rise 'n' Shine]; King, Gene; AGM Miniature Roses

'KINJEN', Min, rb, 1986; (**Jennie Anne**®); flowers red, reverse yellow, fading light yellow with red edge, medium, dbl., 16 petals, exhibition form, no fragrance; oval fruit; straight, small, white prickles; foliage medium, medium green, matt; bushy, medium growth; [Gingersnap X Charmglo]; King, Gene; AGM Miniature Roses, 1987

'KINJOE', Min, ob, 1985; (**Joe-Joe**®); flowers orange-yellow, reverse yellow, blooms borne singly, dbl., 25 petals, cupped, slight fragrance; straight, light yellow to brown prickles; foliage medium, dark, matt; bushy, spreading growth; [Seedling X Rise 'n' Shine]; King, Gene; AGM Miniature Roses

'KINKAKU', HT, dy, 1975; flowers golden yellow, large blooms borne usually singly, dbl., 38 petals, exhibition form, slight fragrance; brown prickles; foliage medium, light green; medium, bushy growth; [Unnamed seedling X Peace]; Okamoto, K.; K. Hirakata Nursery

'KINLIV', Min, ob, 1991; (**Midwest Living**®); bud globular; flowers orange, slight ruffle to petal edges, good substance, holds, dbl., 26–40 petals, 1.5–2.75 in., slight fragrance; few prickles; foliage medium, medium green, semi-glossy; bushy, spreading, medium (46 cms), healthy, vigorous growth; [Evelyn Fison X Party Girl]; King, Gene; Michigan Mini Roses, 1992

'KINLU', Min, rb, 1990; (**Li'l Alleluia**™); flowers wine to silver with yellow base, moderately small, borne mos, dbl., 15–25 petals, moderate fragrance; foliage medium, medium green, matt; upright, tall growth; [[(B.C. X Scamp) X Tamango] X Magic Carrousel]; King, Gene; AGM Miniature Roses

'KINLUM', Min, dp, 1986; (**Tudelum**™); flowers deep pink to lighter shades, petal edges darken on opening, dbl., 30 petals, exhibition form, no fragrance; no fruit; straight, medium, light pink prickles; foliage medium, medium green, matt; upright, bushy, medium, vigorous growth; [Baby Katie X Watercolor]; King, Gene; AGM Miniature Roses, 1987

'KINMAC', Min, dr, 1989; (**Merrimac**™); bud pointed; flowers deeper at tips, medium, borne usually singly and in sprays, 18 petals, exhibition form, slight, fruity fragrance; ovoid, orange fruit; prickles straight, small, red; foliage medium, medium green, matt; upright, medium growth; [(Alain X Scamp) X Lilli Marleen]; King, Gene; AGM Miniature Roses

'KINMIKE', Min, or, 1987; (**J. Michael**™); flowers orange-red, aging lighter, large, borne usually singly, dbl., 18 petals, exhibition form, moderate, fruity fragrance; straight, medium, white-green prickles; foliage large, light green, matt; mini-flora; upright, tall growth; [Poker Chip X Watercolor]; King, Gene; AGM Miniature Roses

'KINNIGHT', Min, dr, 1990; (**After Midnight**™); bud ovoid; flowers dark red, tips show darker edge, outer petals darker, medium, dbl., 28 petals, exhibition form, borne singly, no fragrance; foliage medium, medium green, semi-glossy; medium, upright growth; [(B.C. X Scamp) X Black Jade]; King, Gene, 1986; AGM Miniature Roses, 1991

'KINNOR', Min, mr, 1987; (**New Orleans**™); flowers medium to dark red toward base, non-fading, reverse medium red, dbl., 32 petals, exhibition form, slight, spicy fragrance; oval, small, green fruit; straight, small, red prickles; foliage medium, medium green, matt; bushy, spreading, medium growth; [Evelyn Fison X Magic Mist]; King, Gene; AGM Miniature Roses

'KINPIC', Min, pb, 1990; (**Pink Picotee**®); flowers white to cream with dark pink picotee edge, small, borne mos, dbl., 26–40 petals, no fragrance; foliage medium, light green, matt; upright, spreading, medium growth; [(Vera Dalton X Fancy Pants) X Magic Carrousel]; King, Gene; AGM Miniature Roses

'KINPLEAS', Min, pb, 1986; ('KINPLEASE', **My Pleasure**™); flowers lavender pink, reverse light pink, fading lighter pink, medium, dbl., 45 petals, exhibition form, moderate, fruity fragrance; oval, medium green fruit; straight, white prickles with brown tips; foliage medium, dark green, matt; upright, medium growth; [Lavender Pinocchio X Unnamed seedling]; King, Gene, 1981; AGM Miniature Roses, 1987

'KINPLEASE', Min, pb, 1986; ('KINPLEAS', **My Pleasure**™); flowers lavender pink, reverse light pink, fading lighter pink, medium, dbl., 45 petals, exhibition form, moderate, fruity fragrance; oval, medium green fruit; straight, white prickles with brown tips; foliage medium, dark green, matt; upright, medium growth; [Lavender Pinocchio X Unnamed seedling]; King, Gene, 1981; AGM Miniature Roses, 1987

'KINSANDY', Min, pb, 1990; (**Pink Sandy**™); flowers light pink to light apricot center, aging deep pink edges, s, dbl., 26–40 petals, slight fragrance; foliage small, medium green, matt; bushy, low growth; [(Vera Dalton X Rainbow's End) X Fancy Pants]; King, Gene; AGM Miniature Roses

'KINSCHON', Min, rb, 1986; ('KINSCHOON', **Prairie Schooner**™); flowers red, reverse yellow edged red, dbl., 22 petals, exhibition form, borne singly and in sprays, slight fragrance; no fruit; straight, brown prickles; foliage small, dark, matt; upright growth; PP006748; [Vera Dalton X Sheri Anne]; King, Gene; AGM Miniature Roses

'KINSLAM', Min, dr, 1990; (**Little Slam**®); flowers deep red, moderately small, borne mostly singly, dbl., 15–25 petals, no fragrance; foliage small, medium green, semi-glossy; bushy, low growth; [[(B.C. X Scamp) X Red Ace] X Scamp]; King, Gene; AGM Miniature Roses, 1989

Kinsmen, HT, rb; Delbard

'KINSPEECH', Min, ob, 1986; (**Speechless**™); flowers dark to light orange, fading darker, medium, borne usually singly, dbl., 30 petals, exhibition form, no fragrance; oval, medium, green fruit; medium, hooked, red prickles, with brown tips; foliage dark green, semi-glossy; upright, bushy growth; [Unnamed seedling X Watercolor]; King, Gene; AGM Miniature Roses, 1987

'KINSUN', Min, yb, 1990; (**Dale's Sunrise**®); flowers yellow tipped medium pink, fading pink, small, borne mostly, dbl., 15–25 petals, slight fragrance; foliage small, medium green, semi-glossy; upright, bushy, medium growth; [[(B.C. X Scamp) X Rainbow's End] X Tobo]; King, Gene; AGM Miniature Roses, 1991

'KINTEE', Min, mr, 1988; (**Peggy "T"**™); flowers medium red to white, circular base, reverse white, aging med, single, 5 petals; no fruit; prickles straight with hook, white; foliage medium, medium green, matt; bushy, medium growth; [Poker Chip X Rise 'n' Shine]; King, Gene; AGM Miniature Roses

'KINTENN', Min, op, 1988; (**Tennessee**™); flowers coral to white, reverse light coral to white, aging darker coral, dbl., 18 petals, exhibition form; oval, large, orange fruit; prickles straight, light green; foliage medium, light green, matt; upright, tall growth; [Kiskadee X Orange Honey]; King, Gene; AGM Miniature Roses

'KINTODDY', Min, dr, 1989; (**Hoddy Toddy**™); bud pointed; flowers dark red, petals tipped darker, small, borne usually singly, dbl., 28 petals, cupped, no fragrance; no fruit; prickles straight, very small, red; foliage small, medium green, matt; bushy, low growth; [(Alain X Scamp) X Scamp]; King, Gene; AGM Miniature Roses

'KINUGASA', HT, lp, 1984; flowers large blooms borne usually singly, dbl., 45 petals, exhibition form, moderate, fruity fragrance; brown, hooked prickles; foliage medium, light green; medium, bushy growth; [Michelle Meilland X (Michelle Meilland X Anne Letts)]; Shibata, T.; K. Hirakata Nursery

'KIORA', F, mr, 1989; bud pointed; flowers light vermillion shading to darker petal edges, small, dbl., flat, no fragrance; foliage dark green, shiny, veined; upright, bushy growth; [Liverpool Echo X John Church]; Cattermole, R.F.; South Pacific Rose Nursery

Kir Royal *see* MEInibur

Kirang, HT, pb, 1995; Tejganga

'KIRBEE', F, ab, 1993; (**Irene's Surprise**); dbl., 15–25 petals, 1.5–2.75 in., exhibition form, borne in small clusters, moderate fragrance; foliage medium, medium green, semi-glossy; medium (2 1/2 ft), upright growth; Kirkham, Gordon Wilson; Kirkham, 1993

'KIRBELL', HT, dy, 1995; (**Berenice Neville**); flowers moderately blooms borne in small clusters, dbl., 15–25 petals, moderate fragrance; many prickles; foliage medium, dark green, semi-glossy; medium, upright growth; [Tynwald X Bright Smile]; Kirkham, Gordon Wilson; Kirkham, 1996

KIRbill, F, 1995; (**Margaret's World**); Kirkham, Gordon Wilson; (Cavriglia)

'KIRBRONZE', F, ob, 1993; (**Elizabeth Ann**); flowers bronze and gold, blooms borne in small clusters, dbl., 15–25 petals, 1.5–2.75 in., slight fragrance; some prickles; foliage medium, medium green, glossy; medium, upright growth; [Kathleens Rose X Eurorose]; Kirkham, Gordon Wilson; Kirkham, 1995

'KIRCLOUD', HT, dp, 1995; (**Mary Hilda Law**); flowers deep pink, blooms, dbl., 15–25 petals, 3–3.5 in., intense fragrance; foliage large, dark green, semi-glossy; medium, upright growth; [Lady Helen X Fragrant Cloud]; Kirkham, Gordon Wilson; Kirkham, 1997

'KIRDEX', F, dp, 1985; (**Roccana Diane**); flowers deep pink, medium, dbl., 20 petals, slight fragrance; foliage large, dark, glossy; upright growth; [Pink Favorite X Attraction]; Kirkham, Gordon Wilson

'KIRHOL', HT, w, 1993; (**May Lawlor**); flowers cream and pink, borne mostly single, dbl., 15–25 petals, 2.75–3 in., intense fragrance; many prickles; foliage large, dark green, semi-glossy; tall, upright growth; [Morgengruss X Mary Sumner]; Kirkham, Gordon Wilson

Kiri *see* MATtkiri

'KIRJACK', LCl, or, 1994; (**Daphne Claire Jones**); dbl., 15–25 petals, 3–3.5 in. moderate fragrance; many prickles; foliage large, dark green, semi-glossy; tall, upright growth; [Parkdirecktor Riggers X Seedling]; Kirkham, Gordon Wilson; Kirkham, 1996

'KIRKIT', F, pb, 1985; (**Kathleen Kirkham**); flowers medium, semi-dbl., moderate fragrance; foliage medium, medium green, semi-glossy; upright growth; [Manx Queen X Unnamed seedling]; Kirkham, Gordon Wilson

KIRkitt, F, ab, 1986; (**Kathleen's Rose**); Kirkham, Gordon Wilson

'KIRLON', F, lp, 1993; (**Woman O'th North**); flowers shell pink, blooms borne in small clusters, dbl., 15–25 petals, 1.5–2.75 in., moderate fragrance; few prickles; foliage medium, medium green, semi-glossy; low (1 1/2 x 1 1/2 ft), bushy growth; Kirkham, Gordon Wilson; Kirkham, 1994

KIRlyl, F, mp, 1997; (**Harlow Carr**); Kirkham, Gordon Wilson

KIRmac, F, w, 1989; (**Greenalls Glory**); Kirkham, Gordon Wilson

KIRmelody, HT, 1996; (**Bangor Cathedral**); Kirkham, Gordon Wilson; (Cavriglia)

'KIRPARK', HT, mr, 1983; (**Ted Goves**); flowers medium, dbl., 20 petals, intense fragrance; foliage medium, dark, matt; [Prima Ballerina X Teenager]; Kirkham, Gordon Wilson

'KIRPICE', HT, mr, 1994; flowers medium red, moderately full (15–25 petals), large (7+ cms) blooms borne in small clusters; slight fragrance; few prickles; foliage large, dark green, semi-glossy; low, bushy growth; [Invincible X Red Dandy]; Kirham, Gordon; 1995

KIRpink, F, mp, 1968; (**Owen's Pride**); Kirkham, Gordon Wilson

'KIRRAD', Min, ab, 1993; (**Arejay**); flowers moderately full, small blooms, dbl., 15–25 petals, 1.5 in., borne in large clusters, slight fragrance; few prickles; foliage small, medium green, glossy; low, bushy growth; [Seedling X Seedling]; Kirkham, Gordon Wilson; Kirkham, 1995

KIRsan, LCl, my, 1999; (**Chris**); Kirkham, Gordon Wilson

Kirsch, Climbing *see* 'POULLACK'

KIRscot, MinFl, op, 1989; (**Graceland**); Kirkham, Gordon Wilson

KIRshru, S, ly; (**st Dunstan's Rose**); Kirkham, Gordon Wilson, 1991

'KIRSI', HT, rb, 1967; flowers cream-white edged pink, becoming red over white, large, dbl.; very vigorous growth; [Rose Gaujard sport]; Palmer, H.E.; Edmunds Roses

'KIRSMILE', F, yb, 1994; (**Donald Thomas Heald**); semi-dbl., 6–14 petals, 1.5 in., borne in small clusters, slight fragrance; many prickles; foliage medium, dark green, semi-glossy; patio; low, compact growth; [Seedling X Bright Smile]; Kirkham, Gordon Wilson; Kirkham, 1996

'KIRSTEIN', HT, ob, 1999; (**Sophie Deborah**); dbl., 15–25 petals, 3 in., borne in small clusters, slight fragrance; prickles moderate; foliage medium, dark green, semi-glossy; upright, medium (2.5 ft) growth; [Prunella X Mary Sumner]; Kirkham, Gordon Wilson

Kirsten *see* POUlkir

Kirsten Klein, HMsk, mp, 1995; Scarman

'KIRSTEN POULSEN', F, mr, 1924; flowers bright scarlet, borne in clusters, single, slight fragrance; foliage leathery; vigorous growth; (21); [Orléans Rose X Red Star]; Poulsen, S.

'KIRSTEN POULSEN IMPROVED', F, mr, 1938; flowers scarlet, borne on trusses, single, free bloom; foliage dark; free growth; (21); Radmore

Kirsty Jane *see* 'SIMKAYJAY'

'KIRSUN', S, dy; (**Gold Rush**)

'KIRTYN', S, ab, 1995; (**Heywood**); flowers peach, blooms borne in large clusters, semi-dbl., 6–14 petals, 1.5–2.75 in., slight fragrance; many prickles; foliage medium, medium green, matt; tall, bushy growth; [Tynwald X Bright Smile]; Kirkham, Gordon Wilson

'KIRWIM', HT, mp, 1998; (**Clara's Surprise**); flowers lilac pink, high centered, medium size, dbl., 26–40 petals, exhibition form, borne in small clusters, moderate fragrance; few prickles; foliage medium, dark green, semi-glossy; compact, low growth; [Seaspray X Admiral Rodney]; Kirkham, Gordon Wilson

'KIRWORJACKIE', HT, pb, 1993; (**Worjackie**); flowers full, pink blend, dbl., 26–40 petals, 2.75–3 in., borne mostly singly, moderate fragrance; many prickles; foliage medium, dark green, semi-glossy; medium, upright growth; [Solitaire X Brandy Butter]; Kirkham, Gordon Wilson; Battersby Roses, 1994

Kiska Rose *see* **R. RUGOSA**

'KISKADEE', F, my, 1973; flowers bright yellow, dbl., 25 petals, 3.5 in., exhibition form, slight fragrance; foliage dark; [Arthur Bell X Cynthia Brooke]; McGredy, Sam IV

Kisme, F, pb, 1960; Raffel

'KISMET', HT, my, 1930; flowers clear yellow, center deeper, large, dbl., cupped, intense fragrance; foliage light, glossy; [Talisman sport]; Nicolas; J&P

'KISS', Min, ob, 1980; bud ovoid, pointed; flowers indian orange, reverse lighter, blooms borne1-3 per cluster, dbl., 28 petals, exhibition form, moderate, fruity fragrance; straight prickles; foliage small, medium green; strong, upright growth; [Unnamed seedling X Unnamed seedling]; Lyon, Lyndon; Dreer

Kiss *see* KORikis

Kiss and Tell *see* SEAris

Kiss 'n' Tell *see* 'SEAKISS'

Kiss of Fire, HT, 1969; Laveena Roses; (Cavriglia)

Kiss of Fire, F, rb, 1960; Gaujard

Kiss the Bride *see* 'SEAWHI'

'KISSIN' COUSIN', S, op, 1978; bud ovoid, pointed; flowers pink to coral-pink, dbl., 28 petals, 4–5 in., exhibition form, slight fragrance; foliage large, dark, leathery; vigorous, upright, spreading growth; [((Ophelia X Prairie Princess) X Tiki) X ((Corbeille Royale xAmerican Heritage) X Hawkeye Belle)]; Buck, Dr. Griffith J.; Iowa State University

'KITANA', HRg, m, 1927; flowers deep lavender-pink, semi-dbl., non-recurrent, intense fragrance; profuse, red fruit; [Tetonkaha X Rose Apples]; Hansen, N.E.

'KITCHENER OF KHARTOUM', HT, mr, 1917; (K. of K.); flowers dazzling velvety scarlet, medium, 10 petals, 3 in., intense fragrance; vigorous, branching growth; GM, NRS, 1916; Dickson, A.

'KITTY', Pol, mr, 1925; flowers carmine-red, small, dbl.; Koster; (Sangerhausen)

'KITTY BICE', LCl, lp; [Ophelia, Climbing X Lady Waterlow]; Fitzhardinge, before 1932; (Weatherly, L.)

'KITTY HAWK', Min, pb, 1986; flowers deep pink, reverse lighter pink, mini-flora blooms borne singly, dbl., 29 petals, exhibition form, slight fragrance; medium, long, red prickles; foliage large, medium green, semi-glossy; medium, upright growth; [Watercolor X ?]; Bridges, Dennis A.

'KITTY KINGSBURY', HT, mp, 1930; flowers shell-pink, moderate fragrance; vigorous growth; [Abol seedling]; Evans

'KITTY KININMONTH', LCl, dp, 1922; flowers deep pink, golden stamens, very large, semi-dbl., cupped, some recurrent bloom, slight fragrance; few prickles; foliage dark, wrinkled; vigorous, climbing growth; [Unnamed seedling X R. gigantea]; Clark, A.; Hackett

'KIWI', HT, w, 1989; bud pointed; flowers creamy pink opening to creamy white, reflexed, pointed, medium, dbl., slight fragrance; prickles needle-like, red to light brown; foliage dark green, shiny; tall, upright growth; [Judith Morton X (Pascali X Blue Moon)]; Cattermole, R.F.; South Pacific Rose Nursery

'KIWI BELLE', F, pb, 1983; flowers tan apricot to pink, large blooms in clusters of 3-5, dbl., 40 petals, globular, intense fragrance; brown prickles; foliage bronze green when young, turning light green; upright, spreading growth; [Silent Night X Irish Mist]; Cattermole, R.F.

'KIWI CHARM', HT, pb, 1971; bud ovoid; flowers creamy yellow, edged pink, large, semi-dbl., exhibition form, moderate fragrance; foliage glossy, leathery; vigorous, bushy growth; [Kordes' Perfecta X Champagne]; Lindquist; Bell Roses, Ltd.

'KIWI DELIGHT', HT, yb, 1983; flowers golden yellow, petals edged deep pink, aging pink overall, large, dbl., 40 petals, globular, intense fragrance; gray-brown, very small and large prickles; foliage bronze green turning dark green, glossy; upright, bushy growth; [Peace X (Peer Gynt X Irish Mist)]; Cattermole, R.F.

'KIWI GOLD', HT, yb, 1984; bud pointed; flowers light yellow, reverse edges of petals flushed pink, dbl., 22 petals, exhibition form, intense fragrance; very few, light brown prickles; foliage light green, glossy; upright growth; [(Pink Parfait X Pink Parfait) X Waipounamu]; Cattermole, R.F.

'KIWI QUEEN', HT, yb, 1983; flowers yellow, shaded orange and pink, large, dbl., 44 petals, exhibition form, slight fragrance; light brown prickles; foliage medium green; upright, spreading growth; [Peer Gynt X Command Performance]; Cattermole, R.F.

Kiwi Sunrise *see* 'MACPAURMAR'

'KIYOSUMI', HT, m, 1979; bud slender; flowers clear light purplish blue, dbl., 25 petals, 3 in., exhibition form, slight fragrance; foliage light green, leathery; upright growth; [Unnamed seedling X Sterling Silver]; Onodera, Toru F.; S. Onodera

'KLAUS GROTH', HSpn, ob, 1951; (Claus Groth); flowers salmon-orange shaded apricot-yellow, large, dbl., intense, spinosissima fragrance; large fruit; foliage dark; vigorous, bushy (5 ft) growth; [R.M.S. Queen Mary X R. spinosissima]; Tantau

'KLAUS STÖRTEBEKER', HT, mr, 1962; flowers well-formed, dbl., 40 petals, 5 in.; foliage dark; low, bushy growth; Kordes, R.

Kleine Ballerina, Pol, 1998; Noack, Werner; (Cavriglia)

Kleine Dortmund®, F, rb, 1992; Noack, Werner; (Cavriglia)

'KLEINE ECHO', Pol, mp, 1925; flowers small, semi-dbl.; Kiese; (Sangerhausen)

Kleine Eva, Pol, mp, 1995; Hetzel

'KLEINE LEO', HT, mr, 1921; flowers brilliant red shaded dark red, dbl.; [Farbenkonigin X Gen. MacArthur]; Timmermans

Kleine Regina® *see* HELreg

Kleine Renate *see* JACrain

'KLEINE ROSEL', HWich, m, 1929; flowers dark violet-pink, medium, single; Vogel, M.; (Sangerhausen)

'KLEINER ALFRED', Pol, or, 1904; bud garnet-red; flowers well formed, medium; foliage glossy; dwarf growth; [Anna-Maria de Montravel X Shirley Hibberd]; Lambert, P.

Kleopatra *see* KORverpea

Klerksdorp Horizon *see* 'KORMOWE'

'KLETTERNDE RUBY', Cl Pol, mr, 1946; bud globular; flowers scarlet, small, bornein clusters, dbl., slight fragrance; foliage dark, wrinkled; very vigorous, trailing growth; [Ruby sport]; Kordes

Klima™ *see* 'SAVAKLIM'

Klimentina, HT, mp, 1955; flowers pink with silvery reverse, large, dbl., moderate fragrance; Klimenko, V. N.; (Sangerhausen)

'KLONDYKE', HWich, my, 1911; flowers soft yellow, center deeper, passing to ivory-white, large, dbl.; very vigorous, climbing growth; Paul

'KLONDYKE', HT, my, 1934; flowers clear golden yellow; [Lady Forteviot sport]; LeGrice

Kluis Orange *see* **'KLYN'S ORANGE'**

'KLUIS SCARLET', F, mr, 1931; flowers brilliant red; very free growth; [Lafayette sport]; Kluis, R.

'KLYN'S ORANGE', Pol, or; (Kluis Orange); flowers orange-scarlet, small, borne in dense clusters, dbl.; foliage light green; dwarf growth; Kluis

'KLYN'S YELLOW', HT, my, 1948; bud pointed; flowers clear yellow, open, large, dbl., moderate fragrance; foliage glossy; very vigorous, compact growth; [McGredy's Yellow X ?]; Klyn

Knezna Libuse, HT, lp; Cerveny; (Czech Rosa Club)

Knirps *see* KORverlandus

Knock Out® *see* 'RADRAZZ'

'KOA', Min, op, 1978; bud ovoid; flowers coral-pink, dbl., 44 petals, 1 in., exhibition form, slight fragrance; tall, open, leggy growth; [Persian Princess X Gene Boerner]; Rovinski & Meredith; Kingsdown Nursery & Miniature Plant Kingdom

Koala® *see* MEItapov

Koba *see* 'MEIROVERNA'

Kobold, F, op, 1981; flowers large, semi-dbl.; McGredy, Sam IV; (Sangerhausen)

Kocher Red, LCl, mr

'KO-CHOH', F, ob, 1983; flowers orange, reverse yellow shaded orange, small, borne 3-5 per cluster, dbl., 45 petals, exhibition form, moderate fragrance; small prickles; foliage dark, semi-glossy; low, compact growth; [(Rumba X Olympic Torch) X Allgold]; Suzuki, Seizo; Keisei Rose Nursery

Kogane, F, my

'KOHA', HT, rb, 1983; flowers medium red, reverse lighter, large, dbl., 32 petals, exhibition form, slight fragrance; brown prickles; foliage light, glossy, veined; upright, branching growth; [Silent Night X (Josephine Bruce X Irish Mist)]; Cattermole, R.F.

'KOHARU-BIYORI', F, lp, 1999; flowers pale pink, single, 5 petals, 2 in., moderate fragrance; foliage medium green; 2.5 ft growth; GM, Japan Rose Concours, 1993; [Hanagasumi X seedling]; Yasuda, Yuji, 1994

'KOHIMA', F, rb, 1972; flowers fire-red to orange, reverse cream, very full, dbl., 45–50 petals, 4 in., intense fragrance; vigorous growth; [Orange Sensation X Mischief]; Ellick; Excelsior Roses

'KOH-I-NOOR', HT, pb, 1973; flowers white, center deep pink, very full, dbl., 60 petals, 5 in., intense fragrance; foliage glossy, dark; vigorous growth; [(Memoriam X Peace) X My Choice]; Ellick

Kohsai *see* **'MIKADO'**

Koh-Sai *see* **'MIKADO'**

Koigokoro, HT, dp

Kojack *see* **'MISS BLANCHE'**

Kojo *see* **'SWEET DREAMS'**

'KOJO NO TSUKI', LCl, dy, 1975; bud circular; flowers deep yellow, dbl., 35 petals, 4.5–5 in., exhibition form, slight fragrance; foliage light; vigorous, climbing growth; [(Souv. de Jacques Verschuren X Thais) X Amarillo]; Teranishi, K.; Itami Bara-en

'KOKO', Min, yb, 1982; bud globular; flowers pale yellow, aging orange-red on petal edges, blooms borne singly, dbl., 55 petals, moderate fragrance; thin, triangular, light brown prickles; foliage medium, medium green, matt; low, bushy growth; Meredith, E.A. & Rovinski, M.E.; Casa de Rosa Domingo

Koksijde, F, pb; 19 petals, flat; foliage matt; strong growth; RvS-Melle, 1998

'KOKULENSKY', (strain of R. canina), lp; susceptible to rust; once used as understock; Kokulensky

Kokyu, HT, lp; good size and form growth

'KOLDINGHUS', F, dp, 1968; flowers deep rose, open, dbl., 23 petals, 2–2.5 in., slight fragrance; foliage dark, glossy; upright growth; [Pernille Poulsen X Seedling]; Poulsen, Niels D.; Poulsen

'KOLIBRE', F, mr, 1946; bud long, pointed; semi-dbl., 20 petals, 3 in., small trusses, intense fragrance; foliage bronze; moderately vigorous growth; [Mev. Nathalie Nypels X Unnamed seedling]; Leenders, M.; Longley

'KOLKHOZNITSA', HT, m, 1957; flowers lilac-pink, large, dbl., 45 petals, moderate fragrance; foliage dark; very vigorous, compact growth; [Peace X Mirandy]; Sushkov & Besschetnova

'KÖLN AM RHEIN', HKor, op, 1956; (Cologne); flowers deep salmon-pink, large blooms in clusters, dbl., recurrent bloom, moderate fragrance; foliage dark, glossy; vigorous, climbing growth; Kordes

Kölner Karneval *see* 'KORGI'

Kölner Karneval, Climbing, Cl HT, m; (Cavriglia)

Komala, HT, pb, 1989; K&S

Kombination, F, 1982; VEG; (Sangerhausen)

Komet, HT, mr; flowers vermilion red, medium to large, dbl., slight fragrance; VEG; (Sangerhausen)

'KOMFORT', F, mp, 1967; flowers salmon-pink, large, borne in clusters (to 20), semi-dbl.; foliage glossy; very vigorous, bushy growth; Tantau, Math.

'KOMMERZIENRAT W. RAUTENSTRAUCH', HFt, mp, 1909; flowers pure salmon-pink, center yellow, reverse lighter, borne in clusters, semi-dbl., slight fragrance; [Léonie Lamesch X R. foetida bicolor]; Lambert, P.

'KOMMODORE', F, mr, 1959; (The Commodore); flowers blood-red, well-formed, large blooms in clusters, dbl.; vigorous, low growth; Tantau, Math.

'KONFETTI', HT, pb, 1965; bud globular; flowers claret-rose, reverse light cream, well formed, dbl., 40–45 petals, 5 in.; foliage dark, leathery; strong stems; bushy, upright growth; Tantau, Math.

König Friedrich II von Dänemark, HP, dr; flowers medium, dbl., slight fragrance; (Sangerhausen)

'KÖNIG LAURIN', HT, w, 1910; flowers large, dbl., moderate fragrance; Türke; (Sangerhausen)

König Ludwig-Rose, S, 1994; flowers large, dbl.; Tantau; (Sangerhausen)

Königen, HWich, mp; flowers small, dbl.; (Sangerhausen)

Königin Beatrix® *see* 'HETKORA'

'KÖNIGIN CAROLA', HT, pb, 1904; (Königin Carola von Sachsen); bud pointed; flowers satiny rose-pink, reverse silvery white, very large, dbl., slight fragrance; foliage dark, leathery; vigorous growth; [Mme Caroline Testout X Viscountess Folkestone]; Türke

Königin Carola von Sachsen *see* **'KÖNIGIN CAROLA'**

Königin der Rosen *see* 'KORBICO'

'KÖNIGIN EMMA', HT, lp, 1903; flowers large, dbl.; Verschuren; (Sangerhausen)

'KÖNIGIN JULIANA', F, 1948; Buisman, G. A. H.; (Cavriglia)

'KÖNIGIN LUISE', HT, w, 1927; flowers very large, dbl., exhibition form, slight fragrance; foliage dark, leathery, glossy; vigorous, bushy growth; [Frau Karl Druschki X Sunburst]; Weigand, C.

'KÖNIGIN MARIA THERESE', HT, dp, 1916; flowers carmine-pink, very large, very dbl., intense fragrance; Lambert, P.; (Sangerhausen)

'KÖNIGIN VIKTORIA VON SCHWEDEN', HT, yb, 1919; flowers light saffron-yellow, aging to pale salmon-pink, dbl.; [Mme Segond Weber X Mrs Joseph Hill]; Ries; Teschendorff

'KÖNIGIN VON DÄNEMARK', A, mp; (Belle Courtisanne, Queen of Denmark, Reine du Dänemark); flowers flesh-pink, center darker, medium, very dbl., non-recurrent, intense fragrance; vigorous growth; [Probably R. alba X Damask hybrid]; Booth, 1816

Konigin Wilhelmine, F, mr, 1997; deVor

'KÖNIGIN WILHELMINE', Pol, dp, 1925; flowers small, dbl.; Koster; (Sangerhausen)

Konigliche Hoheit *see* **'ROYAL HIGHNESS'**

Königlicht Hoheit *see* **'ROYAL HIGHNESS'**

'KÖNIGSBERG', HT, mr, 1940; flowers scarlet-red; [Mrs Henry Winnett X Marechal Petain]; Weigand, L.; Teschendorff

'KONINGIN ASTRID', HT, mr, 1935; (Queen Astrid); bud long, pointed, nasturtium-red; flowers reddish apricot and bronze, very large, dbl.; foliage dark, bronze; vigorous growth; Leenders, M.

'KONINGIN JULIANA', F, mp, 1948; flowers salmon-pink becoming light yellow with salmon, large, borne, dbl., blooms late, moderate fragrance; foliage glossy; [Poulsen's Pink X Mrs Pierre S. duPont]; Buisman, G. A. H.

'KONRAD ADENAUER'®, HT, dr, 1955; (Konrad Adenauer Rose®); bud globular; flowers blood-red, dbl., 35 petals, 4 in., cupped, intense fragrance; foliage light green, glossy; vigorous, upright growth; [Crimson Glory X Hens Verschuren]; Tantau, Math.; J&P, 1954

Konrad Adenauer Rose® *see* **'KONRAD ADENAUER'**®

'KONRAD GLOCKER', F, dr, 1962; dbl., 3 in.; foliage dark; vigorous, bushy, low growth; Kordes, R.

Konrad Henkel® *see* 'KORJET'

'KONRAD THÖNGES', HFt, op, 1929; flowers salmon-orange/pink, large, semi-dbl.; Thönges; (Sangerhausen)

'KON-TIKI', F, mr, 1971; bud ovoid; flowers bright red, large, semi-dbl., cupped; foliage glossy, dark; vigorous, upright growth; [Aztec X Paprika]; Institute of Ornamental Plant Growing

Kontrast, HT, rb, 1981; Hchen

Kooiana Butterscotch, HT, lp; [Golden Times sport]; Gibson, P., 1995; Sunrise Flowers, Int.; (Weatherly, L.)

Kooiana Daybreak, HT, ab

Kooiana Moonlight, HT, lp; [Gerdo sport]; Gibson, P., 1995; Sunrise Flowers, Int.; (Weatherly, L.)

Kooiana Watermelon, HT, lp; [Kooiana Daybreak sport]; Gibson, P., 1995; Sunrise Flowers, Int.; (Weatherly, L.)

Kookaburra, S, lp

'KOOPMANNS SPORT VON ELSE POULSEN', Pol, mr, 1940; flowers medium, semi-dbl.; Koopmann; (Sangerhausen)

Kooshti-Bok *see* JAYKOO

'KOOTENAY', HT, lp, 1917; flowers primrose-color; Dickson, A.

Köpenicker Sommer, F, dy, 1968; flowers medium-large, dbl.; Rupprecht-Radke; (Sangerhausen)

'KOPPIES', HT, lp, 1979; bud ovoid; dbl., 35 petals, exhibition form, borne usually singly, moderate, fruity fragrance; short, recurved prickles; foliage medium, matt; vigorous, tall growth; [Tropicana X Wendy Cussons]; Perry, Astor

KORabmask, HT, dy, 1996; (**Harry Oppenheimer**); Kordes

KORaby, HT, mp, 1995; (**Barby**); Kordes

KORacona, F, mp, 1995; (**Georgette**); Kordes

'KORAD', HT, yb, 1970; (Susan, **Susan Massu**); bud ovoid; flowers light yellow and salmon blend, large, dbl., cupped, moderate fragrance; foliage glossy, dark, leathery; vigorous, upright growth; GM, Baden-Baden, 1968; [Colour Wonder X Liberty Bell]; Kordes, R.; Kordes

'KORADES', S, dp, 1993; (**Wild at Heart**); flowers deep pink, large, full, dbl., 15–25 petals, 3–3.5 in., borne in large clusters, slight fragrance; foliage large, medium green, glossy; medium (120-130 cms), bushy, spreading (180 cms) growth; [Bonanza X Seedling]; Kordes, W. Söhne; Bear Creek Gardens, 1994

'KORAL', F, or, 1969; (**Bengali**®); bud ovoid; flowers red-orange, medium, dbl.; foliage dark; [Dacapo X Seedling]; Kordes; Buisman, 1969

Koralie, S, lp

'KORALLE', F, mr, 1942; flowers light red, slight fragrance; [Else Poulsen sport]; Koopman; C-P;, Krause;, Tantau

Koralle, F, op, 1991; Harkness

Korallovyj Sjurpriz, F, op, 1966; flowers luminous coral-red, large, dbl., moderate fragrance; Klimenko, V. N.; (Sangerhausen)

KORalogen, F, ab, 1996; (**Apps Rose**); Kordes

'KORALU', HT, pb, 1957; (**Kordes' Perfecta**, Perfecta); bud urn-shaped; flowers cream tipped and then flushed crimson, suffused yellow, dbl., 68 petals, 4.5–5 in., exhibition form, intense fragrance; foliage dark, leathery, glossy; vigorous, upright growth; GM, NRS, 1957 GM, Portland, 1958 PIT, NRS, 1957; [Golden Scepter X Karl Herbst]; Kordes; A. Dickson, 1957;, J&P, 1958;, McGredy, 1957

KORamator, HT, ab, 1996; (**Marjorie Chase**); Kordes

'KORAMBO', HT, my, 1986; (**Professor Fred Ziady**); flowers clear medium yellow, star-shaped, borne singly, very dbl., 43 petals, moderate fragrance; straight, dark brown prickles; foliage medium green; medium, sturdy growth; [Lusambo X Deep Secret]; Kordes, R.; Ludwigs Roses Pty. Ltd., 1985

KORameget, HT, dy, 1995; (**Egoli**); Kordes

KORamgis, HT, dy, 1996; (**Golden Spire**); Kordes

'KORAMI', F, op, 1964; (Ami des Jardins, **Finale**); flowers salmon-rose, well formed, borne in clusters, dbl., 21 petals, 3.5 in.; foliage light green; low, compact growth; [Nordlicht X Meteor (F)]; Kordes, R.; A. Dickson;, McGredy

'KORAMPA', F, w, 1985; (**Antique Silk**, Champagner®, Kordes' Rose Champagner); flowers near white, flora-tea, large, dbl., 20 petals, slight fragrance; foliage medium, medium green, semi-glossy; upright, bushy growth; PP005411; [Anabell seedling X Seedling]; Kordes, 1982

KORampli, HT, dr, 1996; (**Bles Bridges**); Kordes

KORamvis, HT, ly, 1997; (**Lemon Dream**); Kordes

KORanderer

KORandpunk, HT, m, 1996; (**Madiba**); Kordes

'KORANGELI', HT, ob, 1985; (Angelique®, **Ankori**, Kordes' Rose Angelique); flowers orange, well-formed, large, dbl., 40 petals, slight fragrance; foliage medium, medium green, matt; bushy growth; [Mercedes X Seedling]; Kordes, 1980

'KORANTEL', HT, yb, 1985; (**Tarantella**); flowers creamy yellow, petals edged and marked with pink, large blooms, dbl., 54 petals, exhibition form; light brown prickles; foliage medium, medium green, glossy; upright growth; [Colour Wonder X Wiener Charme]; Kordes, W.; Horstmann, 1979

KORaucher, S, ab, 1996; (**Deloitte & Touche**); Kordes

'KORAV', HT, op, 1985; (**Ave Maria**®, Sunburnt Country); flowers orange-salmon, large, dbl., 35 petals, exhibition form, borne singly, moderate fragrance; small prickles; foliage large, medium green, semi-glossy; upright growth; [Uwe Seeler X Sonia]; Kordes, W., 1981

KORazerka, HT, dr, 1994; (**Ecstasy**); Kordes

KORbacol, HT, lp; (**Texas**); Kordes, 1993

'KORBAD', F, mr, 1980; (**Bad Füssing**®); bud large; flowers , dbl., 23 petals, cupped, blooms in clusters, moderate fragrance; few prickles; foliage glossy, dark; vigorous, upright, bushy growth; [Gruss an Bayern X Seedling]; Kordes, W.

KORbalem, Min, my, 1988; (**Little Lemmy**); Kordes

KORbarkeit, S, lp; (**Yorkshire**); Kordes, 1998

KORbasren, S, pb, 1993; (**Pink Bassino**); Kordes

KORbasren, S, lp; (**St Tiggywinkles**); Kordes

KORbasta, HT, my; Kordes

KORbaxand, HT, yb, 1973; (**Alexandra**); Kordes

'KORBE', HKor, mr, 1959; (Grüss an Heidelberg, **Heidelberg**); flowers bright crimson, reverse lighter, blooms in clusters, dbl., 32 petals, 4 in., exhibition form; foliage glossy, leathery; very vigorous, bushy growth; Kordes, R.; McGredy

'KORBEEN', HT, dy, 1981; (**Goldy**); flowers large, dbl., 35 petals, moderate fragrance; foliage medium, medium green, semi-glossy; upright, bushy growth; [Berolina X Unnamed seedling]; Kordes, W.; Kordes Sons

Korbel Canyon Red China, Ch, mr; (found rose)

Korbel Vinca Purple, LCl, m; (found rose)

'KORBELL', F, ob, 1972; (**Anabell**®, Annabelle, Kordes' Rose Anabel); flowers orange and silvery blend, well-formed, dbl., 30 petals, 4 in., moderate fragrance; foliage small; [Zorina X Colour Wonder]; Kordes; Dicksons of Hawlmark

'KORBELMA', HT, my, 1979; (Goldsmith, **Helmut Schmidt**®, Simba); bud large, long, pointed; flowers clear, even yellow, dbl., 30–40 petals, 4.5–5 in., exhibition form, borne 1-3 per cluster, moderate, sweet tea fragrance; foliage dark green, matt; vigorous, upright, bushy growth; GM, Belgium, 1979 GM, Geneva, 1979; [New Day X Unnamed seedling]; Kordes, W.

KORberis, S, mp, 1998; (**Felicitas**); Kordes

KORberuhig, LCl, dp, 1996; (**Manita**); Kordes

'KORBETEILICH', Gr, dr, 1999; (**Crimson Bouquet**™); flowers dark garnet red, reverse shiny dark red, good substance, dbl., 20–25 petals, 4–4.5 in., exhibition form, borne in large clusters, slight, sweet fragrance; prickles large; foliage large, dark green, glossy; upright, medium (4.5 ft) growth; AARS, U.S., 2000; [Bad Füssing X Ingrid Bergman]; Kordes; Bear Creek Gardens, Inc., 2000

'KORBI', HT, dy, 1961; (Fièvre d'Or, **Golden Giant**®, Goldrausch); flowers rich golden yellow, well-formed, dbl., 45 petals, 5 in., moderate fragrance; foliage dark; vigorous, tall growth; GM, NRS, 1960; Kordes, R.; A. Dickson;, McGredy

'KORBICO', HT, ob, 1964; (**Colour Wonder**, Königin der Rosen, Queen of Roses, Reine des Roses); bud ovoid; flowers orange-coral, reverse cream, large, dbl., 50 petals, slight fragrance; many, large prickles; foliage glossy, bronze; vigorous, bushy growth; ADR, 1964 GM, Belfast, 1966; [Kordes' Perfecta X Tropicana]; Kordes, R.; A. Dickson, 1964;, McGredy, 1964;, Wyant, 1966

'KORBIDO', HT, m, 1966; (**Silver Star**®); flowers lavender, well-formed, dbl., 5 in., intense fragrance; foliage dark; [Sterling Silver X Magenta (F) seedling]; Kordes, R.; McGredy

'KORBIN', F, w, 1958; (Fée des Neiges, **Iceberg**, Schneewittchen); bud long, pointed; flowers pure white, large blooms in clusters, dbl., intense fragrance; foliage light green, glossy; vigorous, upright, bushy growth; GM, Baden-Baden, 1958 GM, NRS, 1958 Hall of Fame, WFRS, 1983; [Robin Hood X Virgo]; Kordes, R.

'KORBIRAC', Gr, ab, 1992; (**Caribbean**™); flowers apricot orange/yellow blend, dbl., 30–40 petals, 3–3.5 in., exhibition form, borne in small clusters, moderate fragrance; many prickles; foliage large, dark green, semi-glossy, has pronounced red; medium (110-125 cms), upright, bushy growth; PP8592; AARS, 1994; [Mercedes X (New Day X Seedling)]; Kordes, W.; Bear Creek Gardens, 1994

'KORBISCH', S, ob, 1983; (**Kordes' Brillant**®, Kordes' Brilliant); flowers orange, large, dbl., 35 petals, slight fragrance; foliage medium, medium green, glossy; upright, bushy growth; [Sympathie seedling X Unnamed seedling]; Kordes, W., 1982

'KORBLUE', F, m, 1974; (**Shocking Blue**®); bud very large; flowers lilac-mauve, dbl., exhibition form, intense fragrance; foliage dark, glossy, leathery; vigorous growth; [Seedling X Silver Star]; Kordes, 1985

KORboden, HT, op, 1997; (**Dawn Haggle**); Kordes

KORbolak, HT, lp, 1991; (**Melody**); Kordes

KORbonnet, HT, mr; (**Wessie Roos**); Kordes, 1993

'KORBOTAF', S, mp, 1985; (**Elveshörn**®); flowers medium, dbl., 35 petals, slight fragrance; foliage medium, dark, semi-glossy; bushy, spreading growth; [The Fairy X Seedling]; Kordes, W.

KORbraufa, S, mr, 1992; (**Godewind**®); Kordes

'KORCAPAS', F, yb, 1964; (**Samba**®); bud globular; flowers golden yellow touched red, becoming fiery red, blooms in clusters; foliage glossy; low, bushy growth; Kordes, R.

KORcelin, Min, ob, 1987; (**Mandarin**); Kordes

KORciate, HT, ob, 1991; (**Felicia Teichmann**); Kordes

KORcilmo, F, w, 1991; (**Eskimo**®); Kordes

'KORCONTA', S, w, 1998; (**Moon River**™); flowers white, single, small, single, 4–11 petals, 2 in., borne in large clusters, slight fragrance; prickles moderate; foliage medium, dark green, glossy; medium tall groundcover, 2 1/2 ft., spreading growth; [Yesterday X (Sea Foam X Red Max Graf)]; Kordes, W.; Bear Creek Gardens, 1996

'KORCOON', F, yb, 1981; (**Coon Carnival**®); bud ovoid; flowers yellow, changing to pink and red, large blooms borne 1-7 per cluster, dbl., 56 petals, slight fragrance; straight, brown prickles; foliage matt, green; medium high, bushy growth; [Seedling X Seedling]; Kordes, R.; Ludwigs Roses Pty. Ltd.

KORcountry, HT, ab, 1994; (**Bernalene**); Kordes

KORcremkis, F, w, 1991; (**Madeo**); Kordes

KORcrisett, HT, or, 1994; (**Calibra**); Kordes

KORdaba, HT, op, 1992; (**Lambada**); Kordes

KORdadel, Min, yb; (**The Valois Rose**); Kordes, 1990

'KORDAISY', F, pb, 1982; (**Daisy Rose**); flowers pink with white eye, small blooms in large clusters, single, 5 petals; foliage medium, medium green, semi-glossy; bushy growth; [Robin Hood X Topsi]; Kordes, W.

KORdalen, LCl, pb, 1988; (**Antike 89**®); flowers creamy white with pink, large, very dbl.; Kordes, W. Söhne; (Sangerhausen)

KORdalsilk, F, ab, 1996; (**Kingsmead Heritage**); Kordes

'KORDALU', F, mr, 1976; (**Andalusien**®); bud long, pointed; flowers red, dbl., 34 petals, 3 in., cupped; vigorous, bushy growth; ADR, 1976; [Seedling X Zorina]; Kordes, 1977

'KORDAPT', Min, lp, 1985; (**Heidekönigin**®, Palissade Rose®, Pheasant); flowers large, dbl., 35 petals, slight fragrance; foliage small, medium green, glossy; groundcover; spreading growth; [Zwerkonig '78 X R. wichurana seedling]; Kordes, W.

'KORDAY', F, dp, 1984; (Angela®, **Angelica**); flowers deep pink, dbl., 35 petals, cupped, borne singly and in clusters, no fragrance; foliage medium, medium green, glossy; bushy growth; ADR, 1982; [Yesterday X Peter Frankenfeld]; Kordes

KORdehei, F, m, 1991; (**Heidelinde**®); Kordes

KORdehei, S, lp; (**Malverns**); Kordes

'KORDEHN', F, or, 1973; (**Mireille Mathieu**); bud ovoid; dbl., 27 petals, 3.5 in., exhibition form, slight fragrance; foliage soft; vigorous, upright, bushy growth; [Duftwolke X Peer Gynt]; Kordes; Dehner & Co.

KORdemas, Min, mp, 1990; (**Pink Button**); Kordes

Kordes' Brillant® *see* 'KORBISCH'

Kordes' Harmonie *see* **Harmonie**®

Kordes' Magenta *see* **'MAGENTA'**

Kordes' Perfecta *see* 'KORALU'

'KORDES' PERFECTA SUPERIOR', HT, mp, 1963; (Perfecta Superior); flowers bright pink; [Kordes' Perfecta sport]; Kordes, 1963; Wyant, 1966

'KORDES' PERFECTA, CLIMBING', Cl HT, pb, 1962; Japan Rose Soc.

Kordes' Rose Anabel *see* 'KORBELL'

Kordes' Rose Angelique *see* 'KORANGELI'

Kordes' Rose Bella Rosa *see* 'KORWONDIS'

Kordes' Rose Champagner *see* 'KORAMPA'

Kordes' Rose Delicia *see* 'KORGATUM'

Kordes' Rose Esmeralda *see* 'KORMALDA'

Kordes' Rose Florentine® *see* **'FLORENTINA'**

Kordes' Rose Holstein *see* 'KORHOLST'

Kordes' Rose Immensee® *see* 'KORIMRO'

Kordes Rose Kardinal *see* 'KORLINGO'

Kordes' Rose Lady Rose *see* 'KORLADY'

Kordes' Rose Pasadena *see* 'KORLAND'

Kordes' Rose Patricia *see* 'KORPATRI'

Kordes' Rose Repandia® *see* 'KORSAMI'

Kordes' Rose Robusta® *see* 'KORGOSA'

Kordes' Rose Sylt® *see* 'KORYLT'

Kordes' Rose Sylvia *see* 'KORLIFT'

Kordes' Rose Weiss Immensee *see* 'KORWEIRIM'

Kordes' Rose Westfalenpark® *see* 'KORPLAVI'

Kordes' Sondermeldung *see* **'INDEPENDENCE'**

KORdialo, MinFl, rb, 1994; (**Festival**); Kordes

'KORDIAM', HT, op, 1987; (Heidi Kabel, **Holsteinperle**®, Testa Rossa); flowers orange pink, large, dbl., no fragrance; foliage medium, medium green, semi-glossy; bushy growth; [Unnamed seedling X Flamingo]; Kordes, W., 1985

KORdibor, S, lp, 1988; (**Rosendorf Sparrieshoop®**); Kordes

'KORDIENA', HT, w, 1996; (**Helen Naude**); flowers white flushed with pink, blooms, very dbl., 4.75 in.; moderate prickles; foliage large, medium green, dull; bushy, medium growth; Kordes, W. Söhne; Ludwigs Roses Pty. Ltd.

KORdisard, S, mr, 1999; (**Girtnerfreude**); Kordes

'KORDODO', F, mr, 1987; (**Larissa®**); flowers medium, dbl., 26–40 petals, no fragrance; foliage small, medium green, semi-glossy; spreading growth; [(Unnamed seedling X Marina) X Rumba]; Kordes, W., 1989

KORdomal, S, w, 1991; (**Gletscherfee®**); Kordes

'KORDOSELBLA', HT, ab, 1987; (**Sunbeam®**); flowers large, dbl., 26–40 petals, slight fragrance; foliage large, dark green, semi-glossy; bushy growth; Kordes, W.

'KORDOUBT', HT, dy, 1987; (**Johannesburg Sun**); flowers deep golden yellow, large, borne singly, dbl., 22 petals, moderate fragrance; prickles concave, brown; foliage glossy, deep green; tall, upright growth; [Unnamed seedling X Unnamed seedling]; Kordes, W. Söhne; Ludwigs Roses Pty. Ltd., 1988

KORdreischi, HT, w, 1997; (**Beverley Watson**); Kordes

KORdrekes, 1998; (**Myrna's Dream**); Kordes

KORdreweer, 1998; (**Arena's Dream**); Kordes

'KORDU', HT, mr, 1969; (**Duftzauber**, Fragrant Charm); bud ovoid; flowers rose-red, large, dbl., exhibition form, intense fragrance; foliage light, soft; moderate, upright growth; [Prima Ballerina X Kaiserin Farah]; Kordes, R.; Kordes

'KORDURBAN', F, rb, 1982; (**Durban July**, Durbankor); flowers yellow, orange to red, medium, semi-dbl.; foliage medium, medium green, semi-glossy; Kordes, W.

Kore, S, yb, 1980; flowers yellowish-pink, large, dbl., intense fragrance; Urban, J.; (Sangerhausen)

'KOREB', F, or, 1962; (**Diamant®**); bud ovoid; flowers bright orange-scarlet, well-formed, large blooms in clusters, dbl., 40 petals, slight fragrance; foliage dark, glossy; vigorous, upright growth; Kordes, R.; A. Dickson;, McGredy

KORechtem, HT, w; (**White Spire**); Kordes, 1994

KORedan, F, mp, 1995; (Bluewunder, Bluhwunder®, **Flower Power**); Kordes; (Cavriglia)

KOReibei, F, dp, 1993; (**Bridget**); Kordes

'KOREIPARK', S, mr, 1983; (**Rheinaupark®**); flowers large, dbl., 20 petals, slight fragrance; foliage large, dark, glossy; upright, bushy growth; [(Gruss an Bayern X Seedling) X R. rugosa seedling]; Kordes; Kordes Roses

KOReklia, HT, ab, 1989; (New Valencia, Valeccia, **Valencia®**, Valencia 89); Kordes, 1989

KORelasting, S, mp, 1999; (**Linderof**); Kordes

KOReledas, S, lp, 1996; (**NDR I Radio Niedersachsen**); flowers medium, semi-dbl., slight fragrance; Kordes, W. Söhne; (Sangerhausen)

KOReledas, F, lp; (**Centenary**, Heartache); Rose of The Year, Auckland, NZ, 1999; Kordes

KORelgas, HT, w, 1992; (**Esme Euvrard**); flowers tends to light pink, petals creped, dbl., 4–4.5 in., borne mostly singly, slight fragrance; upright, tall (4 ft) growth; Kordes

'KORENBON', F, w, 1987; (**Innocencia®**); flowers medium, dbl., 26–40 petals, no fragrance; foliage medium, medium green, matt; bushy growth; [Lorena sport]; Kordes, W., 1986

'KORENLO', F, mp, 1984; (**Lorena®**, Lorina); flowers medium salmon-pink, flora-tea, large, dbl., 35 petals, exhibition form, no fragrance; foliage medium, medium green, semi-glossy; upright growth; PP005679; [Angelique X Seedling]; Kordes, W., 1983

KORenon, HT, dp, 1997; (**Neon**); Kordes

'KORENPI', F, mp, 1987; (**Loretta®**); flowers medium, dbl., 26–40 petals, no fragrance; foliage medium, medium green, matt; flora-tea bushy growth; PP007479; [Lorena sport]; Kordes, W., 1986

KORfachrit, HT, lp, 1992; (**Andrea Stelzer**); Kordes

'KORFALT', F, dy, 1984; (**Goldmarie**, Goldmarie 82®, Goldmarie Nirp®); flowers deep yellow, red on reverse of outer petal, large, dbl., 35 petals, slight fragrance; foliage medium, medium green, glossy; bushy growth; [((Arthur Bell X Zorina) X (Honeymoon X Dr. A.J. Verhage)) x(Seedling X Sunsprite)]; Kordes, W.; Kordes Sons

'KORFAN', HT, pb, 1974; (**Fantasia**); bud ovoid; flowers lilac-red, flora-tea, globular, moderate fragrance; foliage leathery; vigorous growth; [Silver Star X Tradition]; Kordes; Horstmann, 1977

KORfanto, Min, dp, 1989; (**Rosmarin 89®**); Kordes

'KORFARIM', F, ob, 1972; (Esther Ofarim, Esther O'Farim, **Matador**); bud ovoid; flowers light scarlet and orange, reverse gold, medium, dbl., exhibition form, slight fragrance; foliage large, dark, leathery; vigorous growth; [Colour Wonder X Zorina]; Kordes, R.; J&P, 1970

'KORFEE', Min, or, 1979; (Dwarf Fairy, **Zwergenfee®**); bud globular; dbl., 29 petals, 2 in., cupped, moderate fragrance; foliage small, glossy; dwarf, vigorous, upright growth; [Miniature seedling X Traumerei]; Kordes, W.

KORfeimot, HT, pb, 1994; (**Grafin Sonja**); Kordes

KORfeining, S, pb, 1997; (**Dawn Sunsation**); Kordes

KORfeldwo, S, lp; (**Tudor Sunsation**); Kordes, 1997

KORferse, F, or, 1989; (**Coco**); Kordes

'KORFEU', HT, or, 1973; (**Feuerzauber®**, Fire Magic, Magic de Feu); flowers orange-red, reverse lighter, medium to large, dbl., exhibition form; foliage dark, glossy; vigorous, upright growth; [Fragrant Cloud X Unnamed seedling]; Kordes, R., 1974

'KORFI', F, mr, 1976; (**Fiona**); bud long, pointed; dbl., 24 petals, 3 in., exhibition form, slight fragrance; foliage wrinkled; vigorous, upright growth; RULED EXTINCT 9/82 ARM; [Seedling X Prominent]; Kordes

KORfibi, S, dp, 1997; (**Camilla Sunsation**); Kordes

KORfischer, S, m, 1994; (**Hansa-Park®**); Kordes

KORfisro, S, lp; (**Sweet Sunsation**); Kordes

'KORFLAPEI', F, my, 1987; (**Frisco®**, Pamela); flowers medium, dbl., 26–40 petals, slight fragrance; foliage medium, dark green, semi-glossy; mini-flora bushy growth; PP006695; [((New Day X Minigold) X Banzai) X Antique Silk]; Kordes, W., 1986

'KORFLATA', LCl, mr, 1955; (Flame Dance, **Flammentanz®**); flowers crimson, very large blooms in clusters, dbl., exhibition form, non-recurrent, moderate fragrance; foliage dark, leathery; very vigorous (10 ft.) growth; ADR, 1952; [R. eglanteria hybrid X R. kordesii]; Kordes

'KORFLOT', HT, dp, 1958; (**Ballet®**); flowers deep pink, dbl., 52 petals, 5 in., slight fragrance; foliage gray-green; vigorous, bushy growth; [Florex X Karl Herbst]; Kordes, R.; A. Dickson;, McGredy

'KORFLÜG', HT, lp, 1979; (**Flamingo®**, Margaret Thatcher, Porcelain, Veronica, Veronika); bud large, long, pointed; flowers large blooms borne singly, dbl., 24 petals, exhibition form, moderate fragrance; many prickles; foliage matt, green; vigorous, upright, bushy growth; [Unnamed seedling X Lady Like]; Kordes, W.; Kordes, 1978;, Ludwigs Roses, 1983

KORfolga, S, lp; (**Splish Splash**); Kordes, 1996

KORfolklori, HT, rb, 1994; (**Herman Steyn**); Kordes

'KORFRILLA', F, rb, 1987; (**Carnaval**®, Carnival); flowers white with red edges, large, dbl., 26–40 petals, no fragrance; foliage medium, dark green, matt; bushy growth; [Seedling X (Die Krone X Simona)]; Kordes, W., 1986

'KORFULLWIND', S, lp, 1993; (**Baby Blanket**, Oxfordshire, Sommermorgen); flowers moderately full, medium blooms, dbl., 15–25 petals, 1.5–2.75 in., borne in large clusters, slight fragrance; few prickles; foliage small, dark green, glossy; medium (60-75 cms), bushy, spreading (120-150 cms) growth; [Weisse Immensee X Goldmarie]; Kordes, W. Söhne; Bear Creek Gardens;, Bear Creek Gardens, Inc., 1993, USA

KORfungo, S, mr, 1990; (**Royal Bassino**); Kordes

'KORGANE', HT, ob, 1981; (**Las Vegas**®); bud large, pointed; flowers deep orange, reverse lighter, blooms borne 1-3 per cluster, dbl., 26 petals, moderate fragrance; brown prickles; foliage green, slightly glossy; vigorous, upright, bushy growth; GM, Genoa, 1985 GM, Portland, 1988; [Ludwigshafen am Rhein X Feuerzauber]; Kordes, W.

'KORGATER', HT, yb, 1985; ('KORSUN', **Sun City**); flowers deep yellow with red petal edges, red spreading, star-shaped, dbl., 32 petals, slight fragrance; dark brown prickles; foliage deep green, red when young, leathery; upright, tall growth; [((New Day X Minigold) X Unnamed seedling) X MEItakilor]; Kordes, R.; Ludwigs Roses Pty. Ltd.

KORgatine, S, w, 1996; (**Cream Sunsation**); Kordes

'KORGATUM', HT, ly, 1982; (Delicia, **Elegant Beauty**, Kordes' Rose Delicia); flowers light yellow flushed pink, large, dbl., 20 petals; foliage large, dark, matt; upright, bushy growth; [New Day X Seedling]; Kordes, W.

KORgenda, F, m, 1994; (**Shocking Sky**); Kordes

'KORGERA', F, or, 1976; (**Bagheera**); bud ovoid; dbl., 35 petals, 4 in., exhibition form, slight fragrance; foliage glossy, dark; vigorous, upright growth; [Nordia X Seedling]; Kordes, W.

'KORGI', HT, m, 1964; (Blue Girl, Cologne Carnival, **Kölner Karneval**); flowers silvery lilac-lavender, dbl., 35–40 petals, 5.5 in., exhibition form, borne singly, moderate, lightly fruity fragrance; foliage large, deep green; vigorous, bushy growth; GM, Rome, 1964; Kordes, R.

KORgitte, HT, ob, 1986; (**Golden Queen**); Gold Star of the South Pacific, Palmerston North, NZ, 1984; Kordes

'KORGO', F, my, 1963; (Gold Topaz, **Goldtopas**®, Goldtopaz, 'KORTOSSGO'); bud ovoid; flowers amber-yellow, blooms in clusters (up to 10), dbl., 3.5 in., cupped; foliage glossy; vigorous, bushy growth; ADR, 1963; Kordes, R.

'KORGOLD', HT, my, 1977; (Mabella, **New Day**); bud ovoid, pointed; flowers mimosa-yellow, dbl., 30 petals, 4–5 in., exhibition form, intense fragrance; foliage large, light; upright growth; [Arlene Francis X Roselandia]; Kordes, R.; J&P

'KORGOSA', S, mr, 1979; (Kordes' Rose Robusta®, **Robusta**®); bud long, pointed; flowers medium, single, 5 petals, 2.5 in., moderate fragrance; foliage dark, glossy, leathery; very vigorous, upright, bushy growth; ADR, 1980; [Unnamed seedling X R. rugosa]; Kordes, W.

'KORGRAM', S, w, 1983; (**Weisse Max Graf**®); flowers medium, semi-dbl., intense fragrance; foliage small, dark, glossy; spreading (10 ft) growth; groundcover; [Seedling X R. wichurana seedling]; Kordes, W.; Kordes Roses

KORgrapet, F, my; (**Suncluster**); Kordes

KORgrayel, F, m, 1993; (**Saffex Rose**); Kordes

KORgreyel, F, dy; (**The J.S.E. Rose**); Kordes, 1993

'KORGUND', HT, dr, 1977; (**Burgund**®); bud long, pointed; dbl., 30 petals, 4 in., exhibition form, intense fragrance; vigorous, bushy growth; [Henkell Royal X Seedling]; Kordes, W.

'KORGUND', HT, mr, 1983; (Burgund '81, 'KORGUND '81', **Loving Memory**, Red Cedar); flowers large, dbl., exhibition form, slight fragrance; foliage medium green, semi-glossy; upright, bushy growth; [Seedling X Red Planet seedling]; Kordes, W., 1981

'KORGUST', HT, ab, 1987; (Sunsation, **Veldfire**, Wurzburg); flowers orange, reverse chrome-yellow, large, borne singly, dbl., 38 petals, moderate fragrance; prickles concave, yellow-brown; foliage glossy, medium green; upright, well-branched, free-flowering growth; [Unnamed seedling X Unnamed seedling]; Kordes, W. Söhne; Ludwigs Roses Pty. Ltd., 1988

KORhagon, HT, dp, 1992; (**Pretoria**); Kordes

KORhamp, S, mr, 1989; (**Hampshire**); Kordes

'KORHANBU', HT, op, 1985; (**Belami**, 'KORPRILL'); flowers orange pink, large, dbl., 35 petals, moderate fragrance; foliage medium, dark, glossy; upright, bushy growth; [(Prominent X Carina) X Emily Post]; Kordes, W.

KORharment, S, dr, 1993; (**Francois Krige**); Kordes

KORhassi, HRg, dr, 1993; (**Hansaland**®); Kordes

KORhassi, S, lp; (**Charles Notcutt**); Kordes

KORhaugen, S, rb, 1994; (**Carpet of Color**); Kordes

'KORHEIM', S, rb, 1985; (**Morgenrot**®); flowers medium, single, 5 petals, slight fragrance; foliage small, dark, matt; bushy growth; [(Marlena X Europeana) X ((Tropicana X Carina) X (Clare Grammerstorf X Fruhlingsmorgen))]; Kordes, W.

'KORHITOM', Min, dy, 1987; (Perestroika, **Sonnenkind**®); flowers medium, dbl., slight fragrance; foliage small, medium green, semi-glossy; bushy growth; [Unnamed seedling X Goldmarie]; Kordes, W., 1986

'KORHOLST', S, dr, 1987; (City of Birmingham, **Esprit**®, Holstein 87, Kordes' Rose Holstein, Petit Marquis); flowers deep red, aging darker, medium, semi-dbl., 12 petals, flat, borne in sprays of 5-7, repeat bloom, no fragrance; no fruit; medium, tan prickles, slightly down pointed; foliage small, medium green, semi-glossy; upright, bushy, tall growth; PP006117; [Seedling X Chorus]; Kordes, W.; J&P, 1989

KORhood, HT, mr, 1993; (**Rooi Rose**); Kordes

KORhoro, LCl, my, 1988; (**Morgensonne**); Kordes

KORhota, HT, pb; (**Sarie Marais**); Kordes

KORhuba, HT, ab, 1995; (**Glendora**); Kordes

'KORHUG', HT, my, 1987; (**St Hughs**); flowers creamy yellow, large, dbl., 26–40 petals, moderate fragrance; foliage medium, medium green, semi-glossy; upright, bushy growth; [Unnamed seedling X Unnamed seedling]; Kordes, W.; The Rose Nursery, 1986

KORhungen, S, rb, 1998; (**Cambridgeshire**); flowers lavender pink, borne in large clusters; Kordes

KORhurtlen, LCl, lb, 1991; (**Copper Arch**); Kordes

'KORIANT', S, w, 1982; (**Weisse Repandia**®); flowers small, semi-dbl., slight fragrance; foliage small, dark, glossy; spreading (7 ft) growth; groundcover; [The Fairy X R. wichurana seedling]; Kordes, W.; Kordes Roses

KORichard, F, ab, 1990; (**Plisiedame**, Polisiedame); Kordes

'KORicole', F, w, 1985; (**Nicole**); flowers white with pink petal edges, large, dbl., 35 petals, slight fragrance; foliage large, dark, semi-glossy; upright growth; [Seedling X Bordure Rose]; Kordes, W., 1984

KORiganta, HT, ab, 1993; (**Antoinette**); Kordes

KORignale, HT, w; (**Amorosa**); Kordes

KORikis, F, op, 1988; (**Kiss**); Kordes

'KORikon', HT, ly, 1984; (Golden Medaillon, **Limelight**); flowers large, dbl., 35 petals, exhibition form, intense fragrance; foliage medium, dark, semi-glossy; upright, bushy, spreading growth; [Peach Melba X Seedling]; Kordes, W., 1984

'KORilona', F, my, 1976; (**Bellona**); bud ovoid; flowers golden yellow, pointed, dbl., 27 petals, 3 in., slight fragrance; foliage light; very vigorous, upright growth; [New Day X Minigold]; Kordes, R.; J&P

'KORimro', S, lp, 1983; (Grouse, **Immensee**®, Kordes' Rose Immensee®, Lac Rose®); flowers light pink to near white, small, single, moderate fragrance; foliage small, dark, glossy; spreading (to 13 ft) growth; groundcover; GM, RNRS, 1984; [The Fairy X R. wichurana seedling]; Kordes, W., 1982

KORinor, HT, dr, 1987; (**Mandy**); Kordes

'KORinter', LCl, op, 1982; (**Rosanna**); flowers orange pink, large, dbl., 35 petals, exhibition form, moderate fragrance; foliage medium, medium green, glossy; upright growth; [Coral Dawn X Unnamed seedling]; Kordes, W.

'KORissel', S, w, 1993; (**Jeeper's Creeper**); flowers semi-double, blooms borne in large clusters, semi-dbl., 6–14 petals, 1.5–2.75 in., borne in large clusters, slight fragrance; many prickles; foliage medium, dark green, semi-glossy; low (40-50 cms), spreading growth; [Yesterday X Edelweiss]; Kordes, W. Söhne; Bear Creek Gardens, 1994

'KORita', HT, ab, 1978; (**Gitte**, Peach Melba); bud long, pointed; flowers apricot-pink blend, dbl., 33 petals, 4 in., exhibition form, intense fragrance; foliage dark; vigorous, upright, bushy growth; [(Fragrant Cloud X Peer Gynt) X ((Dr. A.J. Verhage X Colour Wonder) X Zorina)]; Kordes, W. Söhne; Horstmann

KORituscha, F, lp; (**Triodene**); Kordes, 1995

'KORiver', S, dr, 1985; (Esterel, **Heidikind**); flowers medium, dbl., 20 petals, slight fragrance; foliage small, medium green, glossy; upright, bushy growth; [The Fairy X Seedling]; Kordes, W.

KORivo, F, w, 1985; (**Ivory Beauty**); Kordes

'KORizont', LCl, dp, 1985; (**Summer Wine**); flowers deep pink, red stamens, large, single, 5 petals, moderate fragrance; foliage large, medium green, semi-glossy; upright growth; Kordes, W.; John Mattock, Ltd

'KORjet', HT, mr, 1983; (Avenue's Red, **Konrad Henkel**®); flowers large, dbl., 35 petals, exhibition form, moderate fragrance; foliage large, medium green, semi-glossy; upright, bushy growth; [Unnamed seedling X Red Planet]; Kordes, W.

'KORjoni', HT, my, 1987; (**Zitronenjette**®); flowers large, dbl., 15–25 petals, intense fragrance; foliage large, medium green, glossy; spreading growth; [Sutter's Gold X Sunblest]; Kordes, W., 1986

KORkalba, Min, w; (**Zepheline**); Kordes, 1994

KORkandel, S, pb; (**Warwickshire**); Kordes, 1991

'KORkeilich', Min, dp, 1987; (**Pepita**); flowers deep pink, small, dbl., 26–40 petals, no fragrance; foliage small, medium green, semi-glossy; spreading growth; [LENpi X (Mercedes X Garnette)]; Kordes, W., 1985

KORkeindor, F, op, 1994; (**Nancy Gardiner**); Kordes

KORkeltin, S, mr; (**Tradition 95**®); Kordes, 1995

KORkilt, HT, lp; (**Susan**); Kordes, 1996

KORkister, Min, ob, 1987; (**Orange Juwel**®); Kordes

KORkojotie, F, lp; (**Toorenburg**); Kordes, 1995

KORkompo, F, op, 1993; (**Anja**); Kordes

KORkonig, Min, dr; (**Zwergkonig 78**); Kordes, 1978

KORkragor, HT, mp, 1994; (**Coral Spire**); Kordes

KORkultop, F, m, 1993; (**Blue Bajou**®); Kordes

KORkuma, Cl F, yb, 1998; (**Goldmarie, Climbing**); Martens

KORkunde, F, lp; (**Toscana**); Kordes, 1991

'KORlady', HT, op, 1979; (Kordes' Rose Lady Rose, **Lady Rose**®); bud long, pointed; dbl., 34 petals, 5 in., exhibition form, moderate fragrance; vigorous, upright, bushy growth; GM, Belfast, 1981; [Seedling X Traumerei]; Kordes, W. Söhne

KORlallal, LCl, dp, 1994; (**Pink Curtain**); Kordes

KORlamber, F, mr, 1993; (**Pixie Hat**); Kordes

'KORland', HT, or, 1982; (Kordes' Rose Pasadena, **Pasadena**®); flowers large, dbl., 35 petals, exhibition form, no fragrance; foliage large, medium green, matt; upright growth; [Mercedes X (Sweet Promise X (Miss Ireland X Zorina))]; Kordes, W., 1980

'KORlanum', S, lp, 1985; (Sommerwind, **Surrey**); flowers medium, semi-dbl., slight fragrance; foliage small, medium green, semi-glossy; bushy growth; [The Fairy X Seedling]; Kordes, W.

KORlaper, F, lp; (**La Perla**); Kordes

KORlarkon, HT, w; (**Delicate Beauty**); Kordes

KORlasche, HT, m, 1993; (**Mary Pope**); Kordes

KORlastine, LCl, lp; (**Summer Breeze**); Kordes

'KORlawe', S, ab, 1969; ('KORwest', **Westerland**®); bud ovoid; flowers apricot-orange, cupped, 20 petals, 3 in., cupped, repeat bloom, intense; foliage large, dark, soft; vigorous, upright, climbing growth; ADR, 1974; [Friedrich Worlein X Circus]; Kordes

'KORlech', F, dr, 1978; (Intrigue, Lavaglow, **Lavaglut**®); dbl., 24 petals, 2.5 in., globular, blooms in clusters, slight fragrance; foliage glossy; vigorous, upright, bushy growth; [Gruss an Bayern X Unnamed seedling]; Kordes, W. Söhne, 1979

'KORleen', F, mr, 1984; (Bavarian Girl, **Schöne Münchnerin**®); dbl., 22 petals, cupped, borne 2-3 per cluster, slight fragrance; medium, green prickles; foliage medium, medium green, semi-glossy; bushy growth; [Sympathie X Tornado]; Kordes, W.; Kordes;, Ludwig Roses

'KORlette', HT, m, 1998; (**Harlequin**™); flowers lavendar pink, white reverse, dbl., 26–40 petals, 4 in., borne mostly singly, slight fragrance; few prickles; foliage medium, dark green, glossy; bushy, low, 3 ft. growth; [Prima Ballerina X Peace]; Kordes, W.; Bear Creek Gardens, 1998

KORlichtung, F, my, 1993; (**Ruth Pennington**); Kordes

'KORlift', HT, op, 1979; (**Congratulations**, Kordes' Rose Sylvia, Sylvia); bud long, pointed; flowers medium pink, dbl., 42 petals, 4.5 in., exhibition form, moderate fragrance; vigorous, upright, bushy growth; ADR, 1977; [Carina X Seedling]; Kordes, W. Söhne

'KORlilub', S, my, 1985; (**Lichtkönigin Lucia**®); semi-dbl., 18 petals, cupped, borne 3-5 per cluster, moderate fragrance; foliage medium, dark, glossy; bushy, tall growth; ADR, 1968; [Zitronenfalter X Clare Grammerstorf]; Kordes, W., 1966

'KORLIMA', F, mr, 1959; (Lili Marlene, **Lili Marleen**®, Lilli Marlene); bud ovoid; dbl., 25 petals, 3 in., cupped, moderate fragrance; foliage leathery; vigorous growth; ADR, 1960 Golden Rose, The Hague, 1966; [(Our Princess X Rudolph Timm) X Ama]; Kordes, R.; J&P, 1961;, McGredy, 1959

'KORLIMIT', HT, mr, 1987; (**Cora Marie** Dallas); flowers large, dbl., 15–25 petals, no fragrance; foliage large dark green, semi-glossy; upright growth; [Ankori X Seedling]; Kordes, W., 1986

KORlinde, HT, ob; (**Taihape Sunset**); Kordes, 1987

'KORLINGO', HT, mr, 1986; (**Kardinal**™, Kardinal 85, Kordes Rose Kardinal); flowers large, bright red, dbl., 30–35 petals, 4.5–5 in., exhibition form, borne singly, long vase life, slight fragrance; foliage medium, dark, semi-glossy; upright growth, medium; PP005846; [Unnamed seedling X Flamingo]; Kordes, W., 1985

'KORLIRUS', F, w, 1985; (Cevennes, **Heidesommer**®); flowers medium, dbl., 20 petals, intense fragrance; foliage small, dark, glossy; upright, bushy growth; [The Fairy X Unnamed seedling]; Kordes, W.

KORlis, HT, mp, 1996; (**Eliza**); Kordes

'KORLITA', HT, ab, 1973; ('LITAKOR', **Lolita**®); flowers golden bronze, dbl., 28 petals, 5 in., moderate fragrance; ADR, 1973; [Colour Wonder X Seedling]; Kordes, R.; Dicksons of Hawlmark, 1972

KORlitze, HT, yb, 1992; (**Lady Mavis Pilkington**); Kordes

KORlodera, S, mp, 1995; (**Pink Spectacle**); Kordes

KORloher, F, mr, 1990; (**Kingswood College**); Kordes

KORlomet, S, lp; (**Vogelpark Walsrode**); Kordes, 1988

'KORLORE', HT, ob, 1977; (**Folklore**®); bud long, pointed; flowers orange, reverse lighter, dbl., 44 petals, 4.5 in., exhibition form, intense fragrance; foliage glossy; very tall and vigorous, upright, bushy growth; [Fragrant Cloud X Seedling]; Kordes; Barni-Kordes

KORlowi, Gr, ob; (**Cathie Irwin**); Kordes

'KORLUDWIG', F, dp, 1975; (Ludwigshafen, **Ludwigshafen am Rhein**®); bud ovoid; flowers deep pink, flora-tea, dbl., 45 petals, 4 in., exhibition form, moderate fragrance; foliage soft; vigorous, upright, bushy growth; ADR, 1973; [Seedling X Pink Puff]; Kordes

KORlunta, Min, lp, 1987; (**Daniela**®); Kordes

KORlupo, LCl, pb, 1986; (**Harlekin**®); Kordes

'KORLYN', HT, w, 1985; (**Lynette**); bud long, pointed; flowers cream blended with coral pink, well-formed, large blooms in, dbl., no fragrance; straight, brown prickles; foliage dark; tall, upright growth; [Clivia X MEItakilor]; Kordes, R.; Ludwigs Roses Pty. Ltd., 1983

KORmador, F, lp; (**Tamara**); Kordes, 1988

'KORMAI', HT, dr, 1969; (**Mainauperle**®); bud ovoid; flowers large, dbl., exhibition form, intense fragrance; foliage large, dark, leathery; vigorous, upright, bushy growth; ADR, 1966; [Seedling X Americana]; Kordes, R.

'KORMALDA', HT, pb, 1981; (Esmeralda®, **Keepsake**, Kordes' Rose Esmeralda); bud ovoid; flowers deep pink blended with lighter pink shades, reflexed, large, dbl., 40 petals, moderate fragrance; large, stout prickles; foliage dark; vigorous, bushy growth; GM, Portland, 1987; Kordes, W. Söhne; John Mattock, Ltd.

KORmarec, S, mr, 1995; (**Sommerabend**®); Kordes

'KORMARIE', S, yb, 1982; (**Bonanza**®, Miss Pam Ayres); flowers yellow tipped red, large, dbl., 20 petals, slight fragrance; foliage medium, dark, glossy; upright growth; ADR, 1984; [Seedling X Arthur Bell]; Kordes, W., 1983

KORmarter, S, ob, 1988; (**Freisinger Morgenrote**®); flowers large, semi-dbl.; Kordes, W. Söhne

KORmarter, LCl, lp; (**Sunrise**); Kordes

KORmasyl, HT, lp, 1987; (**Fee**); Martens

'KORMAT', F, ab, 1980; (**Australian Gold**®, Mona Lisa); bud ovoid; flowers apricot-peach, dbl., 20 petals, borne 5 per cluster, moderate fragrance; red prickles; foliage dark, leathery; bushy growth; Kordes; John Mattock, Ltd, 1985

'KORMATE', HT, ab, 1987; (**Graaff-Reinet**); flowers apricot with orange on petal margin, large, borne in sprays, dbl., 36 petals, moderate fragrance; prickles concave, reddish-brown; foliage dull, medium green; compact, medium, well branched, free flowering growth; [Unnamed seedling X Unnamed seedling]; Kordes, W. Söhne; Ludwigs Roses Pty. Ltd., 1988

'KORMATT', F, or, 1982; (**St Boniface**); flowers patio, medium, dbl., 35 petals, slight fragrance; foliage medium, dark, semi-glossy; [Diablotin X Traumerei]; Kordes, W., 1980

KORmauret, LCl, dr, 1994; (**Red Curtain**); Kordes

'KORMAX', HKor, mr, 1980; (Red Max Graf, **Rote Max Graf**®); bud ovoid; single, 6 petals, blooms in clusters, moderate fragrance; dark brown prickles; foliage small, leathery, matt; vigorous, trailing groundcover growth; GM, Baden-Baden, 1981; [R. kordesii X Seedling]; Kordes, W.

KORmazin, F, my, 1991; (**Sommermond**®); Kordes

KORmeeram, HT, 1997; (**Royal Dream**); Kordes

KORmeita, LCl, mp, 1988; (**Ramira**®); Kordes

KORmeneint, S, mp, 1997; (**Mein Schoner Garten**); Kordes

'KORMETTER', F, op, 1985; (**Anna Livia**, Sandton Smile, Trier 2000); flowers orange pink, large, dbl., 20 petals, slight fragrance; foliage medium, medium green, semi-glossy; bushy growth; GM, Orleans, 1987 Golden Prize, Glasgow, 1991; [(Seedling X Tornado) X Seedling]; Kordes, W.

KORmiach, HT, mr, 1990; (**Liebeszauber**); Kordes

KORmiller, HT, lp, 1979; (**Dream**); Kordes

'KORMIORA', F, or, 1975; (**Luminion**®, Rosi Mittermeier); bud globular; dbl., 34 petals, 3 in., cupped, moderate fragrance; foliage glossy, dark; vigorous, upright, bushy growth; [Hurra X Peer Gynt]; Kordes; Vilmorin

'KORMIXEL', S, mr, 1988; (**Bassino**®); Kordes

KORmollis, F, mr, 1999; (**Garden Party**); Kordes

KORmorlet, LCl, ob, 1987; (**Salita**®); Kordes

'KORMOWE', HT, rb, 1987; (**Klerksdorp Horizon**); flowers tomato-red, reverse golden yellow, large, borne 1-3 per cluster, dbl., 32 petals, no fragrance; prickles concave, reddish-brown; foliage medium green, large; upright, medium, free-flowering growth; [Unnamed seedling X Unnamed seedling]; Kordes, W. Söhne; Ludwigs Roses Pty. Ltd., 1988

'KORMUN', F, mr, 1971; (Baveria, **Grüss an Bayern**®); flowers blood-red, medium, semi-dbl., globular, slight fragrance; foliage dark, leathery; vigorous, upright growth; ADR, 1973; [Messestadt Hannover X Hamburg]; Kordes, R.; Kordes

KORmunde, F, mp, 1975; (**Rosamunde**); Kordes

KORmuse, S, mp; (**Wiltshire**); Kordes, 1993

'KORNACHO', F, mr, 1984; (**Gartenzauber**®, Gartenzauber '84); flowers large, dbl., 35 petals, exhibition form, slight fragrance; foliage medium, dark, semi-glossy; upright growth; [(Unnamed seedling X Tornado) X Chorus]; Kordes, W.; Kordes Sons

KORnagent, F, lp, 1987; (**Florence**); Kordes

KORnanze, HT, my, 1990; (**Golden Medallion**); Kordes

KORnauer, HT, lp, 1989; (**Agnes Bernauer**®); Kordes

'KORNEMARK', HT, op, 1992; (**Louisiana**); Kordes

KORnends, HT, mr; (**Charlotte Searle**); Kordes

'KÖRNER', HMsk, yb, 1914; bud reddish; flowers orange-yellow tinted salmon, borne in clusters, dbl.; [Trier X Eugenie Lamesch]; Lambert, P.

KORnicken, S, op; (**Anjou**); Kordes

KORniebon, LCl, lp, 1998; (**Bonny**); Kordes

KORnieoch, HT, w; (**Table Mountain**); Kordes, 1990

'KORNITA', F, or, 1955; (**Korona**®); flowers orange-scarlet, blooms in large trusses, dbl., 20 petals, 2.5 in., slight fragrance; vigorous, upright growth; GM, NRS, 1954; Kordes; Morse

'KORNITZEL', HT, mr, 1987; (**Boksburg Fantasia**); flowers large, bornein sprays of 1-3, dbl., 23 petals, moderate fragrance; prickles needle point, brown; foliage medium green; tall, well branched growth; [Seedling X Seedling]; Kordes, W. Söhne; Ludwigs Roses Pty. Ltd., 1988

'KORNUMA', F, mr, 1983; (**Heinzelmännchen**®, Red Pixie); flowers large, dbl., 35 petals, slight fragrance; foliage medium, medium green, glossy; bushy growth; [(Satchmo X Unnamed seedling) X (Messestadt Hannover X Hamburg)]; Kordes, W., 1984

'KORO', HT, mr, 1983; flowers scarlet, large, dbl., 32 petals, exhibition form, intense, fruity fragrance; reddish prickles; foliage medium green; upright growth; [Pink Parfait seedling X Red Planet]; Cattermole, R.F.

KORoberfinz, HT, w, 1994; (**Margaret Wasserfall**); Kordes

KORocken, HT, ob; (**Out of Africa**); Kordes

KORofaser, HT, ob, 1994; (**Peach Spire**); Kordes

KORogesa, HT, ab, 1993; (**Helpmekaar Roos**); Kordes

KORoketto, F, m, 1994; (**Futura**); Kordes

'KOROL', HT, ob, 1969; (Coppertone, Old Time, **Oldtimer**); flowers bronze, long, pointed, slight fragrance; Kordes, R.; McGredy

'KOROL', HT, yb, 1968; (**Peer Gynt**®); flowers yellow, outer petals edged red, large, dbl., slight fragrance; vigorous, bushy growth; GM, Belfast, 1970; [Colour Wonder X Golden Giant]; Kordes, R.; McGredy

KORomega, HT, or, 1996; (**Pat's Choice**); Kordes

KORommerla, S, ab, 1997; (**Peach Sunsatien**); Kordes

KORomtar, HT, w, 1997; (**Cream Dream**); Kordes

Korona® *see* 'KORNITA'

'KORONA, CLIMBING', Cl F, or, 1957; Kordes

KORonam, HT, w, 1994; Kordes

'KORONET', HT, ob, 1941; bud globular, lemon shaded orange; flowers orange, reverse deep primrose, dbl., 40–50 petals, 4.5 in., exhibition form; foliage dark, glossy, leathery; vigorous, upright, bushy, open growth; [Julien Potin X Bright Wings]; Mallerin, C.; A. Meilland;, C-P

KORonto, S, m, 1991; (**Burghausen**®); Kordes

KORopas, HRg, m, 1988; (**Sir Henry**®); Kordes

KORoranki, F, ob, 1994; (**Jacqueline**); Kordes

'KOROVO', HT, mp, 1931; bud pointed; flowers peach-blossom-pink and coppery old-rose, large, dbl., 30 petals, moderate fragrance; foliage thick; vigorous growth; [Mrs T. Hillas X Étoile de Hollande]; Leenders, M.

'KOROYNESS', HT, lp, 1985; (**Bride's Dream**, Fairy Tale Queen, Marchenkonigin); bud long, pointed, ovoid; flowers very pale pink, large, full, dbl., 25–30 petals, 5 in., exhibition form, borne singly, slight fragrance; dark brown prickles; foliage large, medium green, matt; tall, upright growth; [Royal Highness X Seedling]; Kordes, R., 1984; Ludwigs Roses Pty. Ltd., 1986

KORozon, HT, pb, 1993; (**Mondiale**®); Kordes

'KORP', Gr, or, 1971; (**Prominent**®); bud long, pointed; dbl., 33 petals, 3.5 in., cupped, slight fragrance; foliage matt; upright growth; AARS, 1977 GM, Portland, 1975; [Colour Wonder X Zorina]; Kordes, R.; Kordes

KORpalmor, F, op, 1997; (**Schloss Balthasar**); Kordes

KORpalud, HT, op, 1997; (**Hamburger Deern**); flowers large, dbl., moderate fragrance; Kordes, W. Söhne; (Sangerhausen)

KORpapie, S, op; (**Salmon Sunsation**); Kordes

KORpapiro, F, ab, 1995; (**Apricot Summer**); Kordes

KORparall, S, mp, 1989; (**Mary Hayley Bell**); Kordes

KORparesni, F, lp, 1997; (**Rosenprofessor Sieber**); flowers medium-large, dbl., slight fragrance; Kordes, W. Söhne; (Sangerhausen)

KORpastato, HT, ly, 1997; (**Sebastian Kneipp**); flowers medium-large, very dbl., intense fragrance; Kordes, W. Söhne; (Sangerhausen)

'KORPATRI', F, ab, 1972; (Kordes' Rose Patricia, **Patricia**); flowers apricot, base shaded gold; GM, Orleans, 1979; [Elizabeth of Glamis sport]; Kordes; Fermor

'KORPEAHN', F, mr, 1986; (Carl PhilipKristian IV, The Times Rose, **Mariandel**®); flowers scarlet crimson-red, medium, semi-dbl., 6–14 petals, slight fragrance; foliage medium, dark green, semi-glossy, disease-resistant; bushy growth; Golden Rose, The Hague, 1990 PIT, RNRS, 1982; [Tornado X Redgold]; Kordes, W.; John Mattock, Ltd, 1985

'KORPEK', HT, mr, 1985; dbl., 20 petals, no fragrance; foliage dark, glossy; upright growth; [Unnamed seedling X Unnamed seedling]; Kordes, W.; Paul Pekmez

'KORPERKI', HT, dr, 1986; (**Trojan Victory**); bud ovoid; flowers deep red, dbl., 50 petals, exhibition form, borne singly, moderate, damask fragrance; medium prickles, hooked downward; foliage medium, medium green, semi-glossy; medium, spreading growth; PP005678; [Unnamed seedling X Uwe Seeler]; Kordes, R.; J&P

'KORPESH', Gr, ob, 1981; (**Shreveport**™); bud ovoid, pointed; flowers orange, dbl., 50 petals, exhibition form, borne 1-3 per cluster, slight, tea fragrance; small prickles, hooked downward; foliage large; tall, upright, vigorous growth; PP005157; AARS, 1982; [Zorina X Uwe Seeler]; Kordes, R.; Armstrong Nursery

KORpinka, S, dp, 1992; (**Sommermarchen**); Kordes

'KORPINROB', S, mp, 1987; (**Pink Robusta**®, The Seckford Rose); flowers large, semi-dbl., 16–14 petals, slight fragrance; foliage large, dark green, glossy; bushy, spreading growth; [(Zitronenfalter X Grammerstorf, Climbing) X Robusta]; Kordes, W., 1986

KORplasina, F, w; (**Vanilla**®); Kordes, 1994

'KORPLAVI', S, ab, 1987; (Chevreuse, Kordes' Rose Westfalenpark®, **Westfalenpark**®); flowers large, dbl., 26–40 petals, moderate fragrance; foliage large, dark green, glossy; bushy, spreading growth; [Unnamed seedling X Las Vegas]; Kordes, W., 1986

'KORPON', F, or, 1970; (**Ponderosa**®); bud globular; flowers red-orange, medium, dbl., cupped, slight fragrance; foliage leathery; vigorous, dwarf, bushy growth; ADR, 1971; [Seedling X Marlena]; Kordes, R.; Kordes

KORpora, HT, dy, 1990; (**Christel von der Post**®); Kordes

'KORPORT', HT, yb, 1984; (**Funkuhr**®, Golden Summers, Laser Beam); flowers yellow, petals edged medium red, aging red, large, dbl., 35 petals, no fragrance; foliage medium, medium green, glossy; upright growth; [Unnamed seedling X Unnamed seedling]; Kordes, W.; Kordes Sons

'KORPRILL', HT, op, 1985; (**Belami**, 'KORHANBU'); flowers orange pink, large, dbl., 35 petals, moderate fragrance; foliage medium, dark, glossy; upright, bushy growth; [(Prominent X Carina) X Emily Post]; Kordes, W.

'KORPRIWA', HT, dy, 1984; (Berolina, **Selfridges**); flowers amber yellow, large, dbl., 35 petals, moderate fragrance; foliage medium, medium green, semi-glossy; ADR, 1986; Kordes, W.; John Mattock, Ltd

KORproa, HT, yb; (**Technikon Pretoria**); Kordes, 1993

KORquermu, F, lp, 1991; (**Queen Mother**); Kordes

KORquick, HT, ob, 1988; (**Oudtshorn Joy**); Kordes

'KORRANTU', F, mr, 1968; (**Travemünde**®); bud ovoid; flowers medium, dbl.; foliage dark; ADR, 1966; [Lilli Marleen X Ama]; Kordes; Buisman

'KORRED', HT, dr, 1983; (Ace of Hearts, **Asso di Cuori**®, Toque Rouge); bud large, ovoid; dbl., 30 petals, cupped, moderate fragrance; dark green prickles; foliage large, dark; bushy growth; Kordes, W.; John Mattock, Ltd, 1981

'KORREI', F, ob, 1974; (Dreaming, 'REIKOR', Reverie, **Träumerei**®); bud long, pointed; flowers orange, medium, dbl., cupped, intense fragrance; foliage leathery; vigorous, upright, bushy growth; [Colour Wonder X Seedling]; Kordes

'KORRESIA', F, dy, 1977; (Friesia, **Sunsprite**); bud ovoid; dbl., 28 petals, 3 in., flat, intense fragrance; foliage light; upright growth; GM, Baden-Baden, 1972 Gold Star of the South Pacific, Palmerston North, NZ, 1975; [Unnamed seedling X Spanish Sun]; Kordes, R.; J&P

'KORRESLI', S, dp, 1987; (**Rosenresli**®); flowers deep pink, large, dbl., 26–40 petals, intense fragrance; foliage medium, dark green, glossy; upright, bushy, possibly climbing growth; ADR, 1984; [(New Dawn X Prima Ballerina) X Unnamed seedling]; Kordes, W., 1986

KORretra, HT, mr, 1989; (**Mother's Value**); Kordes

'KORREV', HT, rb, 1962; (**Neue Revue**®, News Review); flowers yellow-white, touched dark red, well-formed, dbl., 30 petals, 4.5 in., intense fragrance; many large prickles; foliage leathery; upright growth; ADR, 1969; [Colour Wonder X ?]; Kordes

'KORRIGAN', F, lp, 1972; bud globular; flowers medium, dbl., 25 petals, 2–2.5 in., slight fragrance; foliage glossy, dark, leathery; vigorous, bushy growth; [Seedling X (Orléans Rose X Eden Rose)]; Poulsen, Niels D.; Poulsen

KORrohe, F, mr, 1984; (**Heckenfeuer**®); Kordes

KORruge, HRg, yb, 1989; (**Rugelda**®); Kordes

KORrundum, HT, my, 1993; (**Heike**); Kordes

'KORSAKU', HRg, dp, 1998; (Playtime, **Rosalina**™); bud slender, pointed; flowers dark pink with traces of lavendar, dark pink reverse, single, 4–11 petals, 3 in., borne in small clusters, slight fragrance; many prickles; foliage medium, medium green, dull; bushy, medium, for hedging, 3 - 4 ft. growth; PP9011; [The Fairy X Seedling]; Kordes, W.; Bear Creek Gardens, 1992

'KORSAMI', S, lp, 1983; (Kordes' Rose Repandia®, **Repandia**®); flowers small, semi-dbl., moderate fragrance; foliage small, dark, glossy; low, spreading (to 5 ft groundcover) growth; ADR, 1986; [The Fairy X R. wichurana seedling]; Kordes, W.; Kordes Roses, 1982

KORsanter, HT, 1988; Kordes, R.; (Cavriglia)

'KORSCHAPRAT', HT, ob, 1963; (Charme de Vienne, Charming Vienne, Vienna Charm, **Wiener Charme**®); bud pointed; flowers coppery, orange, dbl., 27 petals, 6 in., exhibition form, moderate fragrance; foliage dark; vigorous, tall growth; [Chantré X Golden Sun]; Kordes, R.; J&P, 1965;, McGredy, 1963

'KORSCHLOSS', F, or, 1975; (**Schloss Mannheim**®); flowers red-orange, dbl., globular, slight fragrance; foliage dark, leathery; vigorous, upright, bushy growth; ADR, 1972; [Marlena X Europeana]; Kordes

'KORSCHNUPPE', LCl, dy, 1984; (**Goldener Olymp**®, Olympic Gold); flowers large, dbl., 20 petals, moderate fragrance; foliage large, medium green, matt; upright, bushy (to 7 ft) growth; [Seedling X Goldstern]; Kordes, W.; Kordes Sons

KORschwama, HT, dr, 1992; (**Schwarze Madonna**®); Kordes

KORschwama, HT, lp; (**Barry Fearn**); Kordes

KORsebue, HT, ob, 1995; (**Hanneli Rupert**); Kordes

'KORSEE', F, ob, 1970; (Gitta Grummer, Orange Vilmoria, Orange Vilmorin, Rainer Maria Rilke, Reiner Maria Rilke, **Uwe Seeler**®); bud ovoid; flowers salmon-orange, large, semi-dbl., exhibition form, moderate fragrance; foliage large, glossy, bronze, leathery; vigorous, upright, bushy growth; [Queen Elizabeth X Colour Wonder]; Kordes, R.; Kordes

KORsenter, HT, pb; (**Garden Perfume**); Kordes

KORseubel, F, ab; (**City of Pretoria**); Kordes

'KORSHEL', HT, op, 1976; (**Seashell**); bud short, pointed; flowers burnt-orange, imbricated, dbl., 48 petals, 3–4 in., slight fragrance; upright growth; AARS, 1976; [Unnamed seedling X Colour Wonder]; Kordes, R.; J&P

'KORSICHT', HT, m, 1984; (**Blue River**®); bud large; flowers lilac, shaded deeper at petal edges, large, dbl., 35 petals, exhibition form, intense fragrance; foliage medium, medium green, semi-glossy; upright growth; GM, Baden-Baden; [Blue Moon X Zorina]; Kordes, W.

KORsilan, S, mp, 1988; (**Palmengarten Frankfurt**®); Kordes

KORsion, Gr, my; (**Randfontein Gold**); Kordes

KORsisten, F, op, 1995; (**Bienkie**); Kordes

'KORSITA', HT, dr, 1982; (**Pasita**); flowers bright dark red, flora-tea, blooms borne singly, dbl., 25 petals, slight fragrance; straight, light brown prickles; foliage glossy; medium-high, densely branched growth; [Mercedes X Seedling]; Kordes, R.; Ludwigs Roses Pty. Ltd.

'KORSKIPEI', HT, op, 1987; (**Esther Geldenhuys**); flowers light coral pink, clam-shaped, large, borne singly, dbl., 32 petals, moderate fragrance; prickles concave, yellow-brown; foliage glossy, purple to medium green; vigorous, very tall, well-branched growth; [Seedling X Seedling]; Kordes, W. Söhne; Ludwigs Roses Pty. Ltd., 1988

'KORSOMMER', F, w, 1984; (**Roseromantic**®); flowers light pink to white, large, single, 5 petals, slight fragrance; foliage small, dark, glossy; bushy, spreading growth; GM, Baden-Baden, 1982; [Seedling X Tornado]; Kordes, W.

'KORSONN', F, dy, 1977; (**Sonnenröschen**®); bud ovoid; flowers deep yellow, dbl., 40 petals, 4 in., cupped, slight fragrance; foliage glossy; vigorous, bushy growth; [Arthur Bell X Yellow seedling]; Kordes, W. Söhne, 1978

'KORSORB', HT, ab, 1988; (**Joybells**); Kordes

'KORSPARKO', F, ob, 1984; (**Orange Sparkle**); flowers bright orange, yellow stamens, blooms in clusters of 1-5, semi-dbl., moderate fragrance; straight, brown prickles; foliage very glossy; tall, bushy growth; [(Colour Wonder X Zorina) X Uwe Seeler]; Kordes, R.; Ludwigs Roses Pty. Ltd.

KORspatax, HT, mp, 1994; (**Darling**); Kordes

KORstacha, S, w; (**Weisse Wolke**®); Kordes, 1993

KORstacha, LCl, lp; (**White Cloud**); Kordes

KORstatis, S, pb, 1989; (**Rosenstadt Zweibrucken**®); Kordes

KORstesgli, S, mr, 1998; (**Lancashire**); Kordes

'KORSUN', HT, yb, 1985; ('KORGATER', **Sun City**); flowers deep yellow with red petal edges, red spreading, star-shaped, dbl., 32 petals, slight fragrance; dark brown prickles; foliage deep green, red when young, leathery; upright, tall growth; [((New Day X Minigold) X Unnamed seedling) X MEItakilor]; Kordes, R.; Ludwigs Roses Pty. Ltd.

KORtabo, F, mr; (**Little Red Hedge**); GM, Durbanville, 1983; Kordes

KORtabris, F, pb, 1989; (**Raspberry Ice**, Tabris); Kordes

'KORTAG', HT, ob, 1985; (**Clivia**®); bud ovoid; flowers salmon orange-red blend, blooms borne singly, dbl., 30 petals, exhibition form, moderate fragrance; brown prickles; foliage medium, medium green, matt; upright, bushy growth; [Mercedes X (Sonia X Uwe Seeler)]; Kordes, W., 1979

'KORTAKE', HT, dy, 1987; (**Germiston Gold**); flowers deep golden-yellow, large, borne in sprays of 1-3, dbl., 36 petals, intense fragrance; prickles concave, brown; foliage medium green; medium, well branched, free-flowering growth; [Unnamed seedling X Unnamed seedling]; Kordes, W. Söhne; Ludwigs Roses Pty. Ltd., 1988

KORtaly, S, dy, 1988; (**Lucinde**®); Kordes

KORtanken, F, dr, 1994; (**Domstadt Fulda**®); flowers large, semi-dbl., slight fragrance; Kordes, W. Söhne; (Sangerhausen)

KORtara, F, dp, 1987; (**Schleswig 87**®); Kordes

'KORTAT', HT, dr, 1970; (Rosenthal, **Tatjana**®); bud long, pointed; flowers large, dbl., cupped, intense fragrance; foliage leathery, soft; vigorous, upright growth; [Liebeszauber X Prasident Dr. H.C. Schroder]; Kordes, R.; Kordes

KORtechna, HT, ob, 1992; (**Germania-Africana**); Kordes

KORtelin, F, yb; (**Spotted Gold**); Kordes, 1994

'KORTELLO', S, dr, 1981; (**Ulmer Münster**®); flowers large, dbl., 35 petals, slight fragrance; foliage large, dark, glossy; [Sympathie X Unnamed seedling]; Kordes, W., 1982

'KORTEMBER', HT, op, 1981; (**Harmonie**®); bud long, pointed; flowers deep salmon, dbl., 20 petals, exhibition form, intense fragrance; foliage slightly glossy; vigorous, upright, bushy growth; GM, Baden-Baden, 1981; [Fragrant Cloud X Uwe Seeler]; Kordes, W.

'KORTEMMA', S, dr, 1998; (**Red Ribbons**, Roselina™, Fiery Sunsation, Chilterns); flowers bright red, double, 17–25 petals, 3 in.., borne in large clusters, slight fragrance; prickles moderate; foliage medium, dark green, glossy; low, spreading, 2 ft x 5 ft growth; [Weisse Max Graf X Waltzertraum]; Kordes, W.; Bear Creek Gardens, 1990

KORtenay, S, dp, 1991; (**Hertfordshire**); Kordes

KORtensei, S, mp, 1993; (**Dornroschenschloss Sababurg**®); Kordes, R.

'KORTENSES', Min, or, 1992; (**Sergeant Pepper**); semi-dbl., 6–14 petals, 1.5–2.75 in., borne in small clusters, no fragrance; some prickles; foliage small, dark green, glossy; low (45-60 cms), upright, bushy, spreading growth; [Unnamed seedling X LAVglut]; Kordes, W. Söhne; Bear Creek Gardens, 1992

'KORTERSEN', LCl, dp, 1977; (Netersen, Rosarium Netersen®, **Rosarium Ueteresen**®); bud ovoid; flowers deep pink, very dbl., 142 petals, 3 in., moderate fragrance; foliage large, glossy; vigorous, climbing growth; [Karlsruhe X Seedling]; Kordes, W.

'KORTEXUNG'®, F, mp, 1987; (**Europa**, Fleurop®); flowers medium, dbl., slight fragrance; foliage medium, medium green, matt; bushy growth; PP006513; [(Unnamed seedling X Banzai) X (Mercedes X Carol)]; Kordes, W., 1985

KORtikel, F, dy, 1989; (**Golden Holstein**®); Kordes

'KORTIME', F, my, 1985; (**Golden Times**); flowers medium, dbl., intense fragrance; foliage medium, dark, semi-glossy; bushy growth; [New Day X Minigold]; Kordes, W., 1976

KORtingle, MinFl, mr, 1993; (**Scarlet Patio**); Kordes

KORtionza, S, dy, 1998; (**Postillion**); Kordes

KORtisching, S, lp; (**Yellow Sunsation**); Kordes, 1997

KORtitu, S, mp, 1991; (**Marondo**®); Kordes

'KORTOR', F, or, 1973; (**Tornado**®); flowers medium, semi-dbl., globular, slight fragrance; foliage leathery; vigorous, bushy growth; ADR, 1972; [Europeana X Marlena]; Kordes, R.; Kordes

'KORTOSSGO', F, my, 1963; (Gold Topaz, **Goldtopas**®, Goldtopaz, 'KORGO'); bud ovoid; flowers amber-yellow, blooms in clusters (up to 10), dbl., 3.5 in., cupped; foliage glossy; vigorous, bushy growth; ADR, 1963; Kordes, R.

KORtradkos, F, mr, 1995; (**Parkwood Scarlet**); Kordes

KORtragfei, S, or, 1989; (**Rote Mozart**®); Kordes

'KORTREU', F, or, 1985; (**Fuggerstadt Augsburg**®); flowers medium, semi-dbl., slight fragrance; foliage medium, dark, glossy; upright, bushy growth; [Cordula X Topsi]; Kordes, W.; Kordes Sons

'KORTRI', F, or, 1972; (**Cordula**®); bud globular; flowers red-orange, medium, dbl., slight fragrance; foliage dark, bronze, leathery; vigorous, dwarf, bushy growth; [Europeana X Marlena]; Kordes, R.; Kordes

KORtrolle, F, ob, 1988; (**Bloemfontein**); Kordes

'KORTUEL', F, w, 1989; (**Bella Weiss**®); Kordes

'KORTUNA', HT, op, 1985; (**Fortuna**®); flowers medium salmon-pink, well-formed, large blooms borne singly, dbl., 30 petals, exhibition form, moderate fragrance; foliage medium, medium green, semi-glossy; medium, upright growth; [Sonia X Unnamed seedling]; Kordes, W., 1977

KORturnus, HT, op, 1993; (**Salmon Spire**); Kordes

'KORUB', S, yb, 1971; (**Charivari**); bud ovoid; flowers golden yellow to salmon, large, dbl., cupped, moderate fragrance; foliage glossy; vigorous, upright, bushy growth; [Königin der Rosen X Goldrausch]; Kordes, R.

KORulas, HT, ly, 1997; (**Limona**); Kordes

'KORUMELST', F, mp, 1961; (**Elysium**®); bud pointed; flowers salmon-pink, well-formed, large, dbl., 35 petals, cupped, moderate fragrance; foliage glossy; vigorous, tall, growth; Kordes, R.

'KORVALUE', HT, pb, 1984; (**Woman's Value**); flowers cream with shades of rose and coral, star-shaped blooms born, dbl., 42 petals, slight fragrance; straight, brown prickles; foliage glossy, deep green; medium, bushy, well-branched growth; [((Sonia X ((Dr. A.J. Verhage X Colour Wonder) X Zorina)) X Asso di Cuori]; Kordes, R.; Ludwigs Roses Pty. Ltd.

KORveco, HT, or, 1997; (**Corvette**); Kordes

KORvegata, HT, 1996; (**Rebell 96**); Kordes, R.; (Cavriglia)

'KORVEJOH', F, ob; (**Vera Johns**); flowers purple-violet, dbl., cupped

'KORVERA', Gr, or, 1977; (**Vera Johns**®); bud ovoid, pointed; flowers large, dbl., 40 petals, exhibition form, slight fragrance; foliage glossy, dark, leathery; vigorous, upright growth; [? X Prominent]; Kordes, R.; Ludwigs Roses Pty. Ltd.

KORveril, HT, op, 1990; (**Cadillac**); Kordes

KORverlandus, S, dp, 1997; (**Knirps**); Kordes

KORverpea, HT, rb, 1955; (**'Cleopatra'**, Kleopatra); flowers scarlet, reverse old-gold, well-formed, medium, dbl., 45 petals, moderate fragrance; foliage dark, glossy; vigorous growth; GM, NRS, 1955; [(Walter Bentley X Condesa de Sástago) X Golden Scepter]; Kordes

KORviga, Min, lp; (**Young Mistress**); Kordes, 1988

'KORVILA', F, mp, 1983; (**Vierländerin**); flowers medium salmon-pink, large, dbl., 35 petals, moderate fragrance; foliage medium, medium green, matt; upright growth; [(Zorina X Zorina) X Rosenelfe]; Kordes, W.; Kordes Roses, 1982

KORvolomin, HT, ab, 1994; (**Saturday Star**); Kordes

KORvondra, F, mp, 1990; (**Flirt**); Kordes

'KORWALBE', Min, rb, 1984; (**Maidy**®); flowers medium, dbl., 20 petals, no fragrance; foliage small, medium green, semi-glossy; bushy growth; [Regensberg X Seedling]; Kordes, W.

'KORWEIRIM', S, w, 1983; (Kordes' Rose Weiss Immensee, Lac Blanc, Partridge, **Weisse Immensee**®); bud light pink; flowers small, single, 5 petals, intense fragrance; foliage small, dark, glossy; spreading (10 ft) growth; groundcover; [The Fairy X R. wichurana seedling]; Kordes, W.; Kordes Roses, 1982

'KORWEISO', F, pb, 1983; (**Hannah Gordon**); flowers white with deep pink petal edges, large, dbl., 35 petals, slight fragrance; foliage large, medium green, semi-glossy; upright, bushy growth; [Unnamed seedling X Bordure]; Kordes, W.; John Mattock, Ltd.

'KORWERK', Min, mp, 1982; (Dwarf Queen '82, Zwergkönigin '82); flowers medium, dbl., 35 petals, slight fragrance; foliage small, medium green, glossy; bushy growth; [KORkonig X Sunday Times]; Kordes, W.

'KORWEST', S, ab, 1969; ('KORLAWE', **Westerland**®); bud ovoid; flowers apricot-orange, cupped, 20 petals, 3 in., cupped, repeat bloom, intense; foliage large, dark, soft; vigorous, upright, climbing growth; ADR, 1974; [Friedrich Worlein X Circus]; Kordes

'KORWILMA', HT, rb, 1989; (Jack Dayson, **Perfect Moment**™); bud pointed; flowers red on outer half of petals, yellow on inner, reverse yellow, dbl., 30–35 petals, 4–4.5 in., exhibition form, borne singly and in small clusters, slight fragrance; prickles broad at base, narrowing, hooked down, red to brown; foliage medium, medium green, semi-glossy; upright, bushy, medium growth; PP8007; AARS, 1991; [New Day X Unnamed seedling]; Kordes, W. Söhne; Bear Creek Gardens, 1991

KORwilpa, HGal, mp, 1987; (**Park Wilhelmshohe**); Kordes

'KORWINGS', HT, ob, 1978; (**Soaring Wings**); bud ovoid; flowers deep dusky orange, large, dbl., 64 petals, exhibition form, moderate fragrance; foliage medium green, matt; vigorous, upright, bushy growth; [Colour Wonder X ?]; Kordes, W. Söhne; Ludwigs Roses Pty. Ltd.

KORwisco, S, lp; (**The Compass Rose**); Kordes, 1997

'KORWONDER', F, mp, 1981; (**Bella Rosa**®, Kordes' Rose Bella Rosa, 'KORWONDIS', Toynbee Hall); flowers mini-flora, large, dbl., 34 petals, slight fragrance; foliage small, medium green, glossy; bushy growth; GM, Baden-Baden, 1983 GM, Copenhagen, 1981 GM, Durbanville, 1981; [Seedling X Traümerei]; Kordes, W., 1982

KORworm, F, mp, 1994; (**Romantic Hedgerose**); Kordes

KORxenna, HT, mr, 1988; (**Robina**); Kordes

KORyard, HT, mp; (**Yardley English Rose**); Kordes, 1990

'KORYLT', HKor, dr, 1981; (Kordes' Rose Sylt®, **Sylt**); flowers medium, semi-dbl., slight fragrance; foliage medium, dark, glossy; spreading growth; [R. kordesii X Unnamed seedling]; Kordes, W.

'KORZAUN', HT, dr, 1984; (Duftzauber '84, **Royal William**); flowers large, dbl., 35 petals, exhibition form, moderate fragrance; foliage large, dark, semi-glossy; upright, bushy growth; [Feuerzauber X Unnamed seedling]; Kordes, W.

KORzeito, Min, w, 1991; (**Silk Button**); Kordes

'KORZIMKO', Min, mr, 1987; (**Charmant**®); flowers small, dbl., 40 petals, no fragrance; foliage small, medium green, semi-glossy; bushy growth; [((Seedling X Tornado) X KORkonig) X Trumpeter]; Kordes, W.

KORzimko, HT, op, 1975; (**Charmante**); flowers large, dbl., slight fragrance; Huber; (Sangerhausen)

KORzinta, HT, pb; (**Taubie Kushlik**); Kordes, 1991

'KORZOLA', F, rb, 1986; (**Zola Budd**); flowers white with ruby red, borne 1-3 per cluster, single, 5–7 petals, slight fragrance; straight, light brown prickles; foliage deep green, leathery; vigorous, bushy, densely-branched growth; [Mabella X [(Dr. A.J. Verhage X Colour Wonder) X Zorina]]; Kordes, R.; Ludwigs Roses Pty. Ltd., 1985

KORzuri, HT, w, 1992; (**Memoire**®); GM, Belfast, 1994; Kordes

KORzuri, HT, lp; (**Ice Cream**); Kordes

Ko's Yellow *see* 'MACKOSYEL'

Kosai, F, op

Kosmos, F, dr; flowers medium, dbl.; VEG; (Sangerhausen)

'KOSTER'S ORLÉANS', Pol, mr, 1920; flowers brilliant scarlet-red; [Orléans Rose sport]; Koster, M.

'KOSTER'S TRIUMPH', Pol, or, 1920; flowers small, dbl.; Koster; (Sangerhausen)

'KOSTIOR ARTEKA', F, or, 1955; (Campfire Arteka); flowers coral-red tinted orange, large, slight fragrance; [Independence X ?]; Klimenko, V. N.

'KOTO', HT, dy, 1972; bud ovoid; flowers pure deep yellow, large, dbl., exhibition form, moderate fragrance; foliage glossy, dark, leathery; vigorous, upright growth; [Lydia seedling X Peace seedling]; Suzuki, Seizo; Keisei Rose Nursery

'KOTOBUKI', HT, yb, 1990; bud pointed; flowers creamy to creamy yellow, creamy yellow reverse with light pink, dbl., 30–35 petals, exhibition form, slight fragrance; foliage dark green, glossy; medium growth; [Souma sport]; Ogura Rose Nurseries, 1985

'KOVALAM', HT, w, 1976; bud globular; flowers cream-white, dbl., 20–25 petals, 4 in., cupped, intense fragrance; foliage wrinkled; moderate, bushy growth; [(Amberlight X Traumland) X Western Sun]; Viraraghavan, M.S.; Gopalsinamiengar

'KOYO', F, rb, 1989; bud ovoid; flowers vermillion, reverse orange-yellow, small, borne in sprays of many blooms, dbl., 17–20 petals, cupped, slight fragrance; prickles ordinary; foliage dark green, glossy; spreading growth; [Masquerade X Matador]; Kikuchi, Rikichi, 1990

Koyuki *see* **'SNOW INFANT'**

'KOZA', HRg, dp, 1927; flowers deep pink, semi-dbl.; vigorous (over 7 ft) growth; extremely hardy.; [(R. rugosa X La France) seedling X La Mélusine]; Hansen, N.E.

'KRAKOW', HT, dr; bud ovate; very dbl.; Grabczewski

'KRALJ ALEXANDER I', HT, dr, 1935; flowers velvety blood-red, reflexes fiery red, very large, dbl., intense fragrance; vigorous growth; [Capt. Kilbee Stuart X Jan Bohm]; Böhm, J.

'KRALJ PETAR II', HT, pb, 1936; flowers salmon-pink, reverse carmine, with coppery sheen, moderate fragrance; Brada, Dr.

'KRALJ TOMISLAV', HT, mr, 1931; bud long, pointed; flowers solferino-red, open, very large, dbl., exhibition form; vigorous growth; [Dora Stober X Étoile de Hollande]; Leenders, M.

'KRALJICA MARIJA', HT, ly, 1935; flowers creamy yellow to creamy white, large, dbl., intense fragrance; [Frau Karl Druschki X Golden Ophelia]; Brada, Dr.; Böhm

'KRANENBURG', F, dp, 1963; flowers very dark pink, borne in clusters, dbl., 32 petals; foliage glossy, dark; bushy growth; [Pinocchio X Ma Perkins]; Verschuren, A.; van Engelen

'KRASAVITZA FESTIVALIA', HT, yb, 1955; (Festival Beauty); flowers yellow edged raspberry-red, medium, dbl., 28 petals; foliage glossy, light green; spreading growth; [(Peace X Crimson Glory) X Poinsettia]; Klimenko, V. N.

Krásná Azurea *see* **'GENERÁL STEFÁNIK'**

'KRÁSNÁ USLAVANKA', HT, ob, 1930; flowers orange-rose, reverse dark orange-yellow, moderate fragrance; very vigorous, bushy growth; [Mrs Beckwith X Arthur Cook]; Böhm, J.

'KRASNAIA MOSKVA', HT, dr, 1955; (Red Moscow); flowers dark velvety red, medium, slight fragrance; [Peace X Crimson Glory]; Klimenko, V. N.

Krasni Mak-Sin, F, dr, 1975; flowers dark, velvety red, medium, dbl.; Klimenko, V. N.; (Sangerhausen)

'KRASNOKAMENKA', F, mr, 1955; (Red Stone); flowers crimson-red, medium, semi-dbl., 12 petals; foliage dark, glossy; upright growth; [Independence X Kirsten Poulsen]; Klimenko, V. N.

'KRASNYI MAK', F, mr, 1955; (Red Poppy); flowers scarlet, medium, dbl., 26 petals, slight fragrance; foliage glossy, light green; short stems; upright growth; [Independence X Kirsten Poulsen]; Klimenko, V. N.

Krause's Rote Joseph Guy *see* **'FEUERSCHEIN'**

'KRIBATIS', HT, yb, 1980; (Comtesse d'Alcantara, **Home & Country**); flowers large, moderate fragrance; foliage medium green glossy; upright growth; [Unnamed seedling X Peace]; Kriloff, Michel; Primavera

'KRICARLO', HT, op, 1984; (Marachal Le Clerc, Maréchal le Clerc, **Touch of Class**™); flowers medium pink, shaded coral and cream, large, dbl., 3–35 petals, 4.5–5.5 in., exhibition form, borne mostly singly, slight fragrance; foliage large, dark, semi-glossy; upright, bushy growth; PP005165; AARS, 1986 GM, Portland, 1988; [Micaela X (Queen Elizabeth X Romantica)]; Kriloff, Michel; Armstrong Nursery, 1984

KRIlamy, HT, op; (**Harmonie**); Kriloff

'KRILEVILLE', HT, rb, 1986; (**St Quentin**); flowers red, silver petal edges; foliage dark, glossy; [Unnamed seedling X Tropicana]; Kriloff, Michel

'KRILEXIS', HT, dp, 1986; (**Resurrection**®); bud red; flowers bright pink, large, moderate, sweet fragrance; foliage dark; [Unnamed seedling X Unnamed seedling]; Kriloff, Michel

KRImasar, Cl HT, or, 1984; (**Maryse Kriloff, Climbing**); flowers large, dbl., moderate fragrance; Kriloff; (Sangerhausen)

KRImony, HT, ob, 1994; (**Hello**); Kriloff

KRInico, HT, lp; (**Tresor**); Kriloff, 1996

KRInirosy, HT, op, 1995; (**Helene de Savoie**); Kriloff

'KRIOGA', F, lp, 1986; dbl., 35 petals, blooms in clusters; foliage dense; bushy growth; [Unnamed seedling X Orange Garnet]; Kriloff, Michel

'KRIPRILE', F, yb, 1986; (**Souviens-Toi**); flowers medium yellow, petals edged and washed carmine, large, dbl., 38 petals, moderate fragrance; foliage matt; Kriloff, Michel

Kristall®, F, w, 1979; GPG Bad Langensalza

'KRISTI', HT, mp, 1976; bud ovoid, pointed; flowers clear medium pink, dbl., 45 petals, 5 in., moderate, spicy fragrance; foliage large; vigorous, upright growth; [White Satin X Bewitched]; Swim, H.C. & Ellis, A.E.; Armstrong Nursery

Kristin™ *see* 'BENMAGIC'

'KRISTINA AV TUNSBERG', F, or, 1972; bud long, pointed; flowers large, very dbl., exhibition form, slight fragrance; foliage glossy, dark; vigorous, bushy growth; [Charleston X Toni Lander]; Lundstad; Norges Landbruks-hogskole

Kristo Pienaar *see* HERpot

'KRITIBAN', HT, w, 1986; (Joella®, **Joëlle**®); flowers white, aging light pink, well-formed, dbl.; foliage glossy; [Unnamed seedling X Unnamed seedling]; Kriloff, Michel

'KRIWI', HT, ob, 1968; (**Wizo**®); flowers salmon-red, reverse darker, dbl., 40 petals, exhibition form; foliage reddish-green, matt; [Tropicana X (Gamine X Romantica)]; Kriloff, Michel; Domaine Agricole de Cronenbourg

Kronborg *see* POUltry

Kronenbourg *see* 'MACBO'

Kronenbourg, Climbing, Cl HT, rb, 1973; Kasturi

Kronjuwel, F, mr, 1997; Noack, Werner

'KRONPRINCESSIN VIKTORIA', B, w, 1887; flowers milk-white, center tinted yellow; [Souv. de la Malmaison sport]; Vollert

'KRONPRINSESSE INGRID', F, mp, 1936; (Kronprinzessin Ingrid); bud long, pointed; flowers deep rose-pink, open, semi-dbl., slight fragrance; foliage dark; vigorous growth; [Else Poulsen X Dainty Bess]; Poulsen, S.; C-P, 1942;, Poulsen, 1942

'KRONPRINZESSIN CECILIE', HT, lp, 1907; flowers large, very dbl.; Kiese; (Sangerhausen)

Kronprinzessin Ingrid *see* **'KRONPRINSESSE INGRID'**

Kronprinzessin Victoria *see* HELtoria

Krymchanka, F, m, 1955; flowers dark purple-red, medium, dbl.; Klimenko, V. N.; (Sangerhausen)

Krymskaja Notsch, HT, dr, 1955; flowers large, dbl., intense fragrance; Klimenko, V. N.; (Sangerhausen)

'KSEELITE', Min, pb, 1996; (**Bright Lights**); flowers dark pink with bright yellow stripe on some petals, reverse, dbl., 15–25 petals, slight fragrance; numerous prickles; foliage medium, medium green, semi-glossy; upright, medium (24 in) growth; [Red Delight X Hurdy Gurdy]; Apple, Ken; K&C Roses, 1997

Kühnhilda, S, mp, 1983; flowers medium, single, slight fragrance; Weihrauch; (Sangerhausen)

Kumaradhara, F, pb, 1988; K&S

Kumari, F, mp, 1982; Kasturi

'KUMBAYA', F, my, 1981; bud pointed; flowers bright medium yellow, borne up to 15 per cluster, dbl., 22 petals, slight fragrance; slightly hooked, red prickles; foliage deep green; low, bushy growth; [Chatterbox X Allgold]; Sanday, John; Sanday Roses, Ltd.

Kumiko, HT, dp, 1997; Reuter

'KUNIGUNDE', F, 1960; flowers pink, large, dbl., 35 petals, exhibition form, moderate fragrance; vigorous growth; Horstmann

Kupferkönig, F, op, 1993; flowers coppery pink, medium, dbl.; Kordes, W. Söhne; (Sangerhausen)

Kuroshinju, HT, dr, 1988; Suzuki, Seizo

Kurstadt Baden, HT, pb, 1966; flowers pink and red, large, dbl., slight fragrance; Tantau, Math.; (Sangerhausen)

'KURT SCHOLZ', HT, mr, 1934; bud pointed; flowers blood-red with some crimson, very large, dbl., exhibition form, slight fragrance; foliage leathery; vigorous, bushy growth; [Cathrine Kordes X W.E. Chaplin]; Kordes

Kusum, F, lp, 1978; Padhye

'KUTNO', Pol, m, 1965; flowers lavender-pink, dbl.; growth low; [Margo Koster seedling]; Wituszynski, B.

'KWINANA', F, rb, 1962; bud ovoid; flowers crimson overlaid carmine, open, borne in clusters, single, moderate fragrance; foliage leathery; strong stems; vigorous, tall growth; [Orange Triumph X ?]; Riethmuller; Hazlewood Bros.

'KYNAST', HT, mr, 1917; flowers amaranth-red; Krüger

'KYO-MAIKO', F, op, 1974; bud ovoid; flowers deep bright salmon-orange, small, dbl., cupped, slight fragrance; foliage small, glossy, light; moderate, dwarf, bushy growth; [Sarabande seedling X Ruby Lips seedling]; Suzuki, Seizo; Keisei Rose Nursery

Kyoto 1200, *see* **'ARASHIYAMA'**

'KYRIA', F, lp, 1976; (Pitica); PP004017; [Sonia sport]; Royon

'KYSON', Cl HT, mr, 1940; flowers bright red, open, very large, borne in clusters, single, recurrent bloom, slight fragrance; foliage leathery, glossy, dark; short, strong stems; very vigorous, climbing (7 ft. or more) growth; [New Dawn X Allen Chandler]; Eacott; R. Murrell

L

L. D. Braithwaite® *see* 'AUSCRIM'

'L. E. LONGLEY', HT, mr, 1949; flowers open, large to medium, semi-dbl., moderate fragrance; foliage glossy, dark, bronze; very vigorous, bushy growth; [Pink Princess X Crimson Glory]; Longley, L.E.; Univ. of Minn.

L. G. Harris, HT, 1972; Gregory, C.; (Cavriglia)

'L. J. DE HOOG', HT, mr, 1934; flowers scarlet-red, well formed, large, dbl., slight fragrance; vigorous, bushy growth; [Hadley X Hawlmark Scarlet]; Leenders Bros.

'L. R. MAY', HT, dr, 1935; flowers scarlet-crimson, paling to silvery pink, base orange, well formed; vigorous growth; Chaplin Bros.

La Baraka®, HT, or

'LA BELLA', HT, dr, 1976; bud large, ovoid; dbl., globular, slight fragrance; foliage glossy, dark; very vigorous, upright growth; [Liebeszauber X Herz As]; Kordes; Horstmann

'LA BELLE DISTINGUÉE', HEg, mr; (La Petite Duchesse, Lee's Duchess, Scarlet Sweet Brier); flowers bright crimson, small, dbl., non-recurrent; foliage dainty, fragrant; compact growth

'LA BELLE IRISÉE', HT, ob, 1943; bud ovoid, coppery; flowers clear orange-yellow, medium, dbl., globular, slight fragrance; foliage leathery; dwarf growth; [Mme Joseph Perraud X Seedling]; Gaujard

'LA BELLE MARIE', T, dp; flowers pink, veined darker, reverse deep rose, center incurved, medium, moderate, fruity fragrance; foliage smooth, pointed; vigorous, tall growth; [Old Blush X Mme Laurette Messimy]; Tillotson

'LA BELLE SULTANE', HGal, dr; (Gallica Maheca, Violacea); flowers deep crimson, becoming violet, base white, almost single; Hardy, before 1801; ca. 1795

'LA BELLE SUZANNE', T, lp; flowers light pink suffused white, moderate fragrance; foliage smooth, pointed; vigorous, tall growth; [Old Blush X Mme Laurette Messimy]; Tillotson

La Belle Villageoise *see* **'VILLAGE MAID'**

'LA BICHE', N, w, 1832; flowers white, center flesh, large, very dbl., cupped; Trouillet

La Blancheur *see* **'SODORI-HIMÉ'**

'LA BRILLANTE', HP, mr; flowers bright crimson, well formed, large; Verdier, V., 1861

'LA CAILLE', M, mp, 1857; flowers bright rose-pink; Robert et Moreau

La Canada, Min, dp

'LA CANADIENNE', HT, ob, 1967; bud long, pointed; flowers orange to shrimp and cream, large, semi-dbl., intense fragrance; foliage dark, bronze, glossy; vigorous, bushy growth; [Royal Sunset X Sierra Sunset]; Morey, Dr. Dennison; Country Garden Nursery

'LA CHAMPAGNE', HT, ob, 1919; bud long, pointed; flowers light coppery red, base yellow, edged light pink, large, dbl., 25 petals, globular, moderate fragrance; foliage rich green, leathery; vigorous growth; Barbier

La Coquette de Lyon *see* **'COQUETTE DE LYON'**

'LA COURONNE TENDRE', HGal, lp; flowers flesh-pink, small, dbl.

'LA DETROITE', HT, lp, 1904; flowers large, dbl., intense fragrance; Hopp & Lemke; (Sangerhausen)

'LA DIAPHANE', M, lp, 1900

'LA DUCHESSE', HGal, mp, 1838

La Favorita, HT, or

'LA FAVORITE', HT, mp, 1900; flowers large, dbl.; Schwartz; (Sangerhausen)

'LA FAVORITE', HGal, mr, 1844; Vibert

'LA FIAMMA', HWich, ob, 1909; flowers flame, borne in very large clusters, single; vigorous, climbing (15 ft) growth; Walsh

'LA FLORIDA', HT, mp, 1932; flowers salmon, well formed; foliage glossy; strong stems; very vigorous growth; La Florida

'LA FOLLETTE', LCl, mp, 1910; flowers pink and carmine, large, dbl., moderate fragrance; ; vigorous (to 20 ft. half-hardy; (14); [R. gigantea X ?]; Busby, about 1910

'LA FONTAINE', F, my, 1961; bud pointed, ovoid; flowers barium-yellow, to, borne in clusters, dbl., 20–25 petals, 3.5–4 in., exhibition form, slight fragrance; foliage leathery, dark; vigorous, bushy growth; [Mme Charles Sauvage X Fashion]; Meilland, Mrs. Marie-Louise; C-P

La Fraicheur, HWich, mp

'LA FRANCE', HT, lp, 1867; bud long, pointed; flowers silvery pink, reverse bright pink, large, dbl., 60 petals, intense fragrance; vigorous growth; Guillot et Fils

'LA FRANCE DE '89', HT, rb, 1889; bud long, pointed; flowers bright red, sometimes striped white, very large, dbl.; very vigorous growth; [Reine Marie Henriette X La France]; Moreau et Robert

'LA FRANCE STRIPED', HT, pb, 1956; flowers deep pink to red, and white to blush, moderate fragrance; [La France sport]; Hennessey

'LA FRANCE VICTORIEUSE', HT, dp, 1919; flowers silvery carmine-pink, inside tinted yellow, dbl., moderate fragrance; Gravereaux; M. Guillot

'LA FRANCE, CLIMBING', Cl HT, lp, 1893; Henderson, P.

'LA GIRALDA', HT, mp, 1926; [Frau Karl Druschki X Mme Edouard Herriot]; Dot, Pedro

La Grande Parade, S, mr

'LA JOCONDE', HT, my, 1920; flowers pure golden yellow; RULED EXTINCT 2/88; [Arthur R. Goodwin sport]; Croibier

La Joconde *see* 'DELJACQ'

'LA JOLIE', HT, mr, 1956; bud long, pointed; flowers geranium-red, dbl., 30–35 petals; very vigorous growth; [Independence X Hens Verschuren]; Frères, Buyl

'LA JOLLA', HT, pb, 1954; bud long, pointed; flowers soft pink veined deeper, center cream and gold, dbl., 65 petals, 5 in., exhibition form, moderate fragrance; foliage dark, glossy; upright growth; [Charlotte Armstrong X Contrast]; Swim, H.C.; Armstrong Nursery

'LA LOUISE', HGal, m, 1840; flowers purple/pink, medium, semi-dbl.; Parmentier; (Sangerhausen)

'LA MACULÈE', HGal, pb, 1810; flowers rose pink with carmine striping, medium, semi-dbl.; Dupony, Andre; (Sangerhausen)

'LA MADELON DE PARIS', F, lp, 1962; flowers bright pink, well formed, medium, borne in clusters of 6-8, dbl.; vigorous growth; [Cécile Brunner X Unnamed seedling]; Robichon

'LA MARÉCHALE PÉTAIN', HT, mr, 1927; flowers carmine, dbl., moderate fragrance; [Col. Leclerc X Château de Clos Vougeot]; Sauvageot, H.; F. Gillot

'LA MARNE', Pol, pb, 1915; flowers blush white, edged vivid pink, single blooms in large, loose clusters; vigorous growth; [Mme Norbert Levavasseur X Comtesse du Cayla]; Barbier

La Marseillaise® *see* 'DELGEOT'

La Marylene, F, dr

'LA MASCOTTE', Cl HT, dy, 1933; flowers deep saffron-yellow, passing to straw-yellow, slightly tinted salmon, dbl.; foliage dark, glossy; [Reine Marie Henriette X Laurent Carle]; Schwartz, A.

'LA MÉLUSINE', HRg, dp, 1906; flowers pinkish red, large, borne in large clusters, dbl., intense fragrance; very vigorous growth; Spath

'LA MIE AU ROY', HT, yb, 1927; flowers yellow, salmon and copper, dbl.; [Duchess of Wellington X Pax Labor]; Bernaix, P.

La Mienne *see* **'FLON'**

La Minuette *see* **'LAMINUETTE'**

'LA MORTOLA', (form of R. moschata nepalensis), w, 1954; single, 5 petals, 3 in., blooms in trusses, intense fragrance; foliage downy gray-green; very vigorous, climbing growth; Hanbury; Sunningdale Nursery

La Nationale, HGal, pb; flowers pink and red, medium, dbl.; Roseraie de l'Hay; (Sangerhausen)

'LA NEIGE', M, w, 1905; flowers pure white, medium, dbl.; foliage turns purple; vigorous growth; [Blanche Moreau sport]; Moranville

'LA NEIGE', HGal, w, 1853; Robert

'LA NOBLESSE', C, lp, 1856; flowers soft pink, dbl., later bloom than most centifolias, intense fragrance; bushy growth; Soupert & Notting

La Nymphe, N, mp

'LA PALOMA'®, F, w, 1959; flowers creamy white, well-formed blooms in clusters (to 30), dbl.; foliage dark, glossy; [Yellow Rambler X Goldene Johanna Tantau]; Tantau, Math.

La Paloma '85 *see* 'TANAMOLA'

La Pâquerette *see* **'PÂQUERETTE'**

'LA PARFAITE', HT, op, 1956; bud pointed; flowers bright salmon-pink, large, dbl., 40–50 petals; vigorous, bushy growth; [R.M.S. Queen Mary X Lady Sylvia]; Frères, Buyl

La Parfumee, HT, or

'LA PARISIENNE', HT, or, 1937; bud long, pointed, deep coral-red; flowers orange-coral, open, very large, semi-dbl., slight fragrance; foliage glossy, dark; very vigorous growth; [Lucy Nicolas X Charles P. Kilham]; Mallerin, C.; A. Meilland;, C-P

La Parisienne® *see* 'SUNPARI'

La Passionata *see* 'DELUP'

La Passionata, HT, 1987; Barni, V.; (Cavriglia)

La Passionate *see* DELapo

La Perla *see* KORlaper

'LA PERLE', HWich, ly, 1905; flowers pale yellow to white; Fauque

La Petite Duchesse *see* **'LA BELLE DISTINGUÉE'**

'LA PLUS BELLE DES PONCTUÉES', HGal, pb; flowers deep rose, spotted pale rose, flat; Hébert, Mme., before 1829

'LA POLOGNE', HT, or, 1938; flowers orange-carmine, very large, cupped; foliage bronze; vigorous growth; Chambard, C.

La Presumida *see* **'PRESUMIDA'**

'LA PROMISE', HT, mr, 1956; bud long, pointed; flowers brick-red, semi-dbl.; tall growth; [Seedling X Betty Uprichard]; Frères, Buyl

'LA PUCELLE', HGal, m, 1820; Vibert

La Quatre Saisons Continue, P, mp

'LA RAMÉE', F, dr, 1949; flowers crimson, borne in clusters, semi-dbl., 13 petals; foliage dark, bronze; vigorous growth; [Holstein X Alain]; Meilland, F.

'LA REINE', HP, mp, 1842; (Reine des Francais, Rose de la Reine); flowers glossy rose-pink, large, dbl., 78 petals, cupped, moderate fragrance; vigorous growth; Laffay, M.

La Reine de Hamburg, P, mp

La Reine Victoria *see* **'REINE VICTORIA'**

'LA REVENANTE', HGal, mp, 1825; Miellez

La Roche aux Fees, HT, yb, 1997; Adam, M.

La Rose de Mme Raymond Poincaré *see* **'MME RAYMOND POINCARÉ'**

La Rose de York *see* White Rose of York

'LA ROSE TATOUÉE', HT, m, 1956; (The Rose Tattoo); flowers salmon-pink tipped and spotted lavender, large, semi-dbl., slight fragrance; foliage leathery; very vigorous, upright growth; [Opera seedling X (Opera X Unnamed seedling)]; Gaujard

La Rosée *see* **'DR DEBAT'**

'LA ROSÉE', Pol, w, 1920; flowers sulfur-white, passing to pure white, then to soft pink; Turbat

'LA ROSIERE', HP, 1861; Verdier, E.; (Cavriglia)

La Rosière, HP, lp; Damaizin, 1874

La Rosière, Climbing *see* **'PRINCE CAMILLE DE ROHAN'**

La Rossa *see* BARlar

'LA ROXELANE', HGal, m, 1828; Vibert

La Royale *see* **'GREAT MAIDEN'S BLUSH'**

La Rubanée *see* **'VILLAGE MAID'**

La Sanguine *see* **'SANGUINEA'**

'LA SCALA', HT, rb, 1964; flowers reddish orange, large, dbl., cupped, intense fragrance; foliage leathery; upright, open growth; GM, Rome, 1961; [(Mme Henri Guillot X Mirandy) X Peace]; Lindquist

La Séduisante *see* **'GREAT MAIDEN'S BLUSH'**

La Serenissima, HT; (Italia); (Cavriglia)

La Sevillana® *see* 'MEIGEKANU'

La Sevillana, Climbing *see* MEIgekanusar

'LA SIRÈNE', HP, m, 1867; (La Syrène); flowers dark reddish-purple; Soupert & Notting

'LA SOMME', HT, or, 1919; flowers deep coral-red tinted copper, turning salmon, semi-dbl., moderate fragrance; [Mme Caroline Testout X Rayon d'Or]; Barbier

'LA STUPENDA', HT, pb, 1966; bud long, pointed; flowers pink, reverse darker, large, very dbl., slight fragrance; foliage glossy; upright, bushy growth; [Aztec X First Love]; Taylor, L.R.

La Superba, Gr, 1970; Mansuino; (Cavriglia)

'LA SYLPHIDE', T, m; flowers pink tinted lavender, deeper in fall, semi-dbl.; tall growth; Boyau, 1842

La Syrène *see* **'LA SIRÈNE'**

'LA TENDRESSE', HP, mp, 1864; Oger

'LA TOSCA', HT, pb, 1901; flowers shell-pink, center and reverse darker, dbl., 4 in., moderate fragrance; foliage rich green, leathery; vigorous, bushy growth; [Josephine Marot X Luciole]; Schwartz, Vve.

La Tour d'Argent™ *see* 'DELRICOS'

'LA TOUR D'AUVERGNE', HGal, or, 1842; flowers deep rosy crimson, flecked carmine, incurved, large, dbl.; Vibert

'LA VAUDOISE', HT, mr, 1946; bud oval; flowers blood-red touched brilliant scarlet, medium, dbl., moderate fragrance; foliage leathery; vigorous, bushy growth; Heizmann, E.; A. Meilland

'LA VENDOMOISE', HT, pb, 1907; flowers pink and red, large, dbl., moderate fragrance; Moullière; (Sangerhausen)

'LA VIE', HT, dr, 1931; flowers crimson to scarlet, base orange to carmine, fading to red; [Talisman sport]; Groshens

La Vie en Rose *see* **'VIE EN ROSE'**

'LA VILLE DE BRUXELLES', D, dp, 1849; (Ville de Bruxelles); flowers pink, center incurved, large, dbl., quartered; foliage glossy, light green; height 5 ft; Vibert

La Villiageoise *see* **'VILLAGE MAID'**

La Virginale *see* **'GREAT MAIDEN'S BLUSH'**

'LA VIRGINALE', HP, w, 1858; (Mme Liabaud); flowers white with flesh center, medium, dbl.; delicate habit growth; Lacharme, F.

'LA VIRGINALE', A, w, 1840; Moreau et Robert

'LA VOULZIE', F, dr, 1953; flowers garnet-red, very large, borne in clusters of 8-15, semi-dbl.; very vigorous growth; [Brise Parfumee X Alain]; Robichon

'LABAREDA', F, mr; flowers bright red; da Silva, Moreira

'L'ABONDANCE', N, w, 1877; flowers pure white; Moreau et Robert

Labrador Rose *see* **R. BLANDA**

Lac Blanc *see* 'KORWEIRIM'

'LAC LA NONNE', HRg, dp, 1950; bud pointed, deep red; flowers very deep pink, semi-dbl., 2–3 in., moderate fragrance; foliage light; ;vigorous (7-8 ft. hardy.; [R. rugosa plena X R. acicularis]; Bugnet; P.H. Wright

Lac Majeau, HRg, w; Bugnet

Lac Rose® *see* 'KORIMRO'

Lace Cascade *see* 'JACARCH'

'LACHS', Pol, or, 1943; flowers glowing orange-red; [Dick Koster sport]; Kordes

Lacination, S, mp, 1997; Twomey, Jerry

'LACRE', F, mr, 1963; flowers bright red, loose, semi-dbl.; [Concerto X Seedling]; da Silva, Moreira

Lada, HT, lp; Urban, J., 1978; (Czech Rosa Club)

'LADDIE', HT, or, 1926; flowers deep carmine, flushed orange and scarlet, base orange; bushy growth; McGredy; Beckwith

'LADIES CHOICE', HT, pb, 1969; flowers cerise, reverse silvery, high pointed, large, dbl., intense fragrance; foliage light green; free growth; [Liberty Bell X Prima Ballerina]; Anderson's Rose Nurseries

'LADIES' CHOICE, CLIMBING', Cl HT, pb, 1975; flowers cerise, reverse silvery, high-pointed, large, dbl., intense fragrance; foliage light green; [Ladies' Choice sport]; Anderson's Rose Nurseries

Ladies Home Journal® *see* 'BURWINLADHOM'

Ladies' View *see* 'SEAVIEW'

'L'ADMIRATION', HCh, mp, 1856; flowers medium, single; Robert; (Sangerhausen)

'LADY', HT, mp, 1983; flowers medium, dbl., 35 petals; foliage large, medium green, matt to semi-glossy; upright, compact growth; [Song of Paris X Royal Highness]; Weeks, O.L., 1984

'LADY ALICE STANLEY', HT, pb, 1909; bud pointed; flowers pale flesh-pink, reverse coral-rose, large, dbl., 75 petals, moderate fragrance; foliage rich green, leathery; branching growth; McGredy

'LADY ANDERSON', HT, pb, 1920; flowers coral pink to flesh pink and yellow, dbl.; Hall

'LADY ANN', Min, mp, 1961; bud pointed; flowers rose-pink, dbl., 42 petals, 1.75 in., cupped, moderate fragrance; foliage leathery, glossy, dark; vigorous, bushy, low growth; [(R. wichurana X Floradora) X Little Buckaroo]; Moore, Ralph S.; Sequoia Nursery

'LADY ANN KIDWELL', Pol, dp, 1948; bud pointed; flowers deep pink, star-shaped, medium, dbl., slight fragrance; foliage glossy; vigorous, upright growth; [Cécile Brunner X ?]; Krebs; Marsh's Nursery

'LADY ASHTOWN', HT, pb, 1904; bud pointed; flowers carmine-pink, base yellow, large, dbl., 43 petals, exhibition form, slight fragrance; foliage rich green, soft; vigorous, bushy growth; [Mrs W.J. Grant seedling]; Dickson, A.

'LADY ASHTOWN, CLIMBING', Cl HT, pb, 1909; Bradley

'LADY BAILLIE', HSpn, ly; flowers pale sulfur, semi-dbl., profuse, early bloom; glossy, black fruit; foliage finely divided; dense, shrubby (3-4 ft) growth; Lee, prior to 1848

Lady Banks'Rose *see* **R. BANKSIAE**

'LADY BARBARA', LCl, ob, 1987; flowers tangerine, medium, dbl., 20 petals, moderate fragrance; foliage medium, medium green, semi-glossy; upright growth; [Red Planet X (Elizabeth of Glamis X (Galway Bay X Sutter's Gold))]; Warner, Chris

Lady Barbara Blossom *see* 'BOSHOTEARL'

'LADY BARNBY', HT, pb, 1930; bud pointed; flowers glowing pink, shaded red, large, dbl., exhibition form, moderate fragrance; foliage rich green, leathery; bushy, low growth; Dickson, A.

'LADY BARNETT', HT, dr, 1957; flowers crimson, reverse darker, exhibition form, moderate fragrance; vigorous, upright growth; Verschuren; Blaby Rose Gardens

'LADY BATTERSEA', HT, pb, 1901; (Red Niphetos); flowers cherry-blossom, base orange, dbl.; [Mme Abel Chatenay X Liberty]; Paul

Lady Be Good *see* 'SEAGOOD'

'LADY BEATTY', HT, lp, 1918; flowers blush-pink, well formed, moderate fragrance; vigorous growth; Chaplin Bros.

'LADY BEAUTY', HT, pb, 1984; bud ovoid; flowers light pink flushed yellow, reverse deeper, large, dbl., 33 petals, exhibition form, slight fragrance; few sickle-shaped prickles; foliage medium, medium green, glossy; vigorous, upright growth; [Lady X Princess Takamatsu]; Kono, Yoshito, 1986

'LADY BELPER', HT, ob, 1948; flowers bronze-orange shaded light orange, semi-globular, dbl., 38 petals, 4 in., exhibition form, moderate fragrance; foliage glossy, dark; vigorous growth; [Mev. G.A. van Rossem X Seedling]; Verschuren; Gregory

'LADY BETTY', HT, ab, 1930; bud pointed; flowers apricot-pink, veined red, semi-dbl., exhibition form, slight fragrance; [Sunburst X Mrs Aaron Ward]; Bees

'LADY BEVERLEY', F, lp; flowers clear pink; [Independence sport]; Owens

'LADY BIRD', F, yb, 1966; flowers yellow shaded red at edge, small, borne in clusters, dbl., slight fragrance; foliage leathery; vigorous, bushy growth; [Unnamed seedling X Rumba]; Hill, Joseph H., Co.

'LADY BIRD JOHNSON', HT, or, 1971; bud long, pointed; flowers medium, dbl., moderate fragrance; vigorous, upright growth; [Montezuma X Hawaii]; Curtis, E.C.; Texas Rose Research Foundation

'LADY BISSETT', HT, ob, 1928; flowers bright orange, reverse apricot; Lilley

'LADY BLANCHE', R., w, 1913; flowers snow-white, dbl., borne in large clusters, free bloom, sometimes repeated in fall, intense fragrance; very vigorous, climbing growth; Walsh

'LADY BOUNTIFUL', LCl, mr, 1938; flowers scarlet-rose, center white, open, large, borne in clusters, single; foliage leathery, dark; very vigorous, climbing or trailing growth; [American Pillar seedling]; Tait; B&A

'LADY BRAYE', HT, dp, 1960; flowers deep rose-pink, long, pointed; foliage dark; Verschuren; Gandy Roses, Ltd.

Lady Brisbane *see* **'CRAMOISI SUPÉRIEUR'**

Lady Brisbane, Climbing *see* **'CRAMOISI SUPÉRIEUR, CLIMBING'**

'LADY CAHN', HT, ab, 1937; bud long, pointed; flowers rich apricot-yellow, veined darker, large, dbl., 40–50 petals; long, strong stems; vigorous growth; Gaujard

'LADY CANADA', HT, mp, 1927; flowers bright rose, dbl., moderate fragrance; [Mme Butterfly X Premier]; Dale

'LADY CAROLINA', S, lp, 1990; bud ovoid; flowers blush pink, same reverse, aging to white, decorative, small, 35–40 petals, repeat bloom, slight fragrance; sound, very small, very red fruit; prickles small, hooked, reddish-brown; foliage average, dark, glossy; bushy, spreading, hedge-type, medium growth; [Lady Gay sport]; Jeremias, Lephon L.

'LADY CASTLEREAGH', T, yb; flowers rosy yellow, well formed; vigorous growth

'LADY CATHERINE', HT, or, 1973; bud urn shaped; dbl., 32–45 petals, 3.5–4 in., exhibition form, slight fragrance; foliage leathery; vigorous, upright, compact growth; [Montezuma X Rubaiyat]; Von Koss; Kern Rose Nursery

'LADY CHARLES TOWNSHEND', HT, ob, 193?; flowers orange, overlaid salmon, large, dbl., globular; vigorous, bushy growth; [The Queen Alexandra Rose X Shot Silk]; Daniels Bros.

'LADY CHARMION', HT, mr, 1923; flowers bright cherry-carmine, dbl., moderate fragrance; [Lyon Rose X Gen. MacArthur]; Bees

Lady Clonbrock, N, lp; flowers medium, dbl.; (Sangerhausen)

'LADY CRAIG', HT, ly, 1922; flowers cream-yellow, center apricot-yellow, well formed, large, dbl., moderate fragrance; vigorous, free branching growth; Dickson, H.

'LADY CROMWELL', HT, dr, 1956; flowers crimson, base gold, large, dbl.; foliage bronze; vigorous growth; Verschuren; Gandy Roses, Ltd.

'LADY CUNLIFFE OWEN', HT, pb, 1932; flowers salmon and cream, base yellow, flushed carmine-rose, outer petals reflexed, dbl., exhibition form, moderate fragrance; foliage leathery; vigorous growth; [Mrs A.R. Barraclough sport]; Ley

'LADY CURZON', HRg, mp, 1901; flowers large, single, moderate fragrance prickly; arching stems; vigorous growth; (21); [R. macrantha X R. rugosa rubra]; Turner

Lady D, HT, 1982; Huber, R.; (Cavriglia)

'LADY DALLAS BROOKS', HT, mp, 1955; [Peace sport]; Downes

'LADY DAWSON BATES', HT, yb, 1939; flowers golden yellow, flushed pink, opening, exhibition form, moderate fragrance; vigorous growth; McGredy

Lady Di®, HT, lp, 1982; Huber

'LADY DIANA', HT, lp, 1986; flowers , dbl., 37 petals, exhibition form, blooms in sprays of 3-4, slight fragrance; ovoid, orange fruit; short, hooked prickles; foliage medium, medium green, matt; tall, upright growth; PP005360; [Sonia X Caress]; Hoy, Lowel L.; Joseph H. Hill, Co., 1983

'LADY DIXON', HT, ab, 1919; flowers rich apricot, flushed salmon-pink, dbl., moderate fragrance; Dickson, A.

'LADY DIXON-HARTLAND', HT, pb, 1923; flowers centers deep salmon, outer petals pale pink, exhibition form; vigorous growth; Cant, B. R.

Lady Dumas, HT, lp; (Weatherly, L.)

'LADY DUNCAN', HRg, pb, 1900; flowers rich glowing pink, center and stamens yellow, single, 3 in., non-recurrent; foliage glossy; trailing (6 ft.) growth; [R. wichurana X R. rugosa]; Dawson; Eastern Nursery

'LADY DUNLEATH', HT, w, 1913; flowers ivory-white edged yellow, small, dbl., moderate fragrance; Dickson, A.

'LADY EDGEWORTH DAVID', HT, mp, 1939; bud long, pointed; flowers malmaison rose shaded soft pink, open, large, dbl., moderate fragrance; foliage glossy; vigorous growth; [Unnamed seedling X Betty Uprichard]; Fitzhardinge; Hazlewood Bros.

'LADY ELEANORE', Cl HT, yb, 1923; flowers light yellow-cream splashed rose, center golden to copper; [Gruss an Teplitz X Barbara]; Dreer

Lady Elgin *see* 'MEIMAJ'

'LADY ELPHINSTONE', HT, yb, 1921; flowers indian yellow to clear rose, semi-dbl.; [Mme Edouard Herriot sport]; Dobbie

'LADY EMILY PEEL', N, w, 1862; flowers white, tinged with blush; [Blanche Lafitte X Sappho]; Lacharme, F.

'LADY ENGLISH', HT, or, 1934; flowers bright cerise, center orange, open, very large, dbl., slight fragrance; foliage glossy, bronze; long stems; vigorous growth; Cant, B. R.

Lady Ethel, S, lp; Sutherland, P, 1998; Golden Vale Nursery; (Weatherly, L.)

'LADY EVE', Min, op, 1978; bud globular; flowers creamy white, edged coral-pink, dbl., 40 petals, 1.5–2 in., exhibition form, slight fragrance; vigorous, upright, tall, spreading growth; [Neue Revue X Sheri Anne]; Rovinski & Meredith; Casa de Rosa Domingo

Lady Eve Price *see* **'CAPRICE'**

'LADY EVELYN GUINNESS', HT, mp, 1932; [Ophelia sport]; Evans

'LADY FAIRBAIRN', HT, mp, 1929; flowers bright pink, dbl., 40 petals; vigorous, upright growth; [Mme Abel Chatenay X Seedling]; Clark, A.; NRS New South Wales

'LADY FAIRE', HT, mp, 1907; flowers salmon-pink; [Mrs W.J. Grant sport]; Bentley

'LADY FAIRFAX', HT, ob, 1930; flowers rose and orange-cerise, flushed orange to yellow, well formed, moderate fragrance; foliage light green; long stems; vigorous growth; Cant, F.

'LADY FLORENCE STRONGE', HT, lp, 1925; bud pointed; flowers pale flesh, base pink and gold, very large, dbl., exhibition form, slight fragrance; foliage leathery, glossy; vigorous, bushy growth; McGredy

'LADY FORTEVIOT', HT, yb, 1926; flowers golden yellow to deep apricot, large, dbl., exhibition form, intense fragrance; foliage bronze, glossy; vigorous, bushy growth; (28); GM, NRS, 1927; Cant, B. R., 1928

'LADY FORTEVIOT, CLIMBING', Cl HT, yb, 1935; Howard Rose Co.

'LADY FRASER', HT, mr, 1941; flowers rich red; [War Paint seedling]; Clark, A.

'LADY FROST', HT, dp, 1935; flowers deep rose, very large, dbl., intense fragrance; foliage leathery; vigorous, bushy growth; [Lady Alice Stanley X Dr. Herbert Hawkesworth]; Bees

'LADY GAY', HWich, op, 1905; dbl., blooms in clusters; foliage small, dark, glossy; vigorous, climbing (12-20 ft.) growth; [R. wichurana X Bardou Job]; Walsh

Lady Genevieve, HP, mr

'LADY GEORGIA', HT, pb, 1973; flowers pink, base blending to ivory, full, medium, dbl., moderate fragrance; foliage dark, leathery; very vigorous, bushy growth; [Miss Hillcrest X Peace]; Curtis, E.C.; Kimbrew

Lady Glencora *see* 'ARORAJU'

'LADY GODIVA', HWich, lp, 1908; flowers cameo-pink; [Dorothy Perkins sport]; Paul

'LADY GOWRIE', Cl HT, my, 1938; bud long, pointed; flowers maize and champagne-yellow, large, very dbl., intermittent bloom, moderate fragrance; foliage leathery, glossy, dark; long stems; very vigorous, climbing growth; [Sunburst, Climbing X Rev. F. Page-Roberts]; Fitzhardinge; Hazlewood Bros.

'LADY GRADE', HT, or, 1982; (Lady Kathleen Grade); flowers vermilion, large, dbl., 35 petals, moderate fragrance; foliage large, medium green, semi-glossy; bushy growth; [Tropicana X Unnamed seedling]; Gregory, C.

'LADY GREENALL', HT, yb, 1911; flowers saffron-yellow, edges tinted shell-pink, dbl., moderate fragrance; Dickson, A.

'LADY GREENALL, CLIMBING', Cl HT, yb, 1923; Lippiatt

'LADY GWENDOLINE COLVIN', Cl HT, pb, 1918; flowers apricot-salmon, shaded chrome-yellow, outer petals stained carmine, dbl., moderate fragrance; height 6-10 ft; Chaplin Bros.

'LADY HAILSHAM', HT, or, 1951; bud pointed; flowers orange flushed red, dbl., 30 petals, 3 in., intense fragrance; foliage glossy; vigorous growth; [McGredy's Sunset sport]; Knight's Nursery

'LADY HAMILTON', HSpn, lp; flowers rosy blush, semi-dbl., profuse, non-recurrent bloom; dwarf growth

Lady Harriet *see* 'WILHERB'

'LADY HELEN', HT, mp, 1970; flowers soft clear pink, pointed, dbl., 30 petals, 5 in., intense fragrance; foliage glossy, dark; bushy growth; [Margaret X (McGredy's Ivory X Peace)]; McTeer, Gilbert; Waterhouse Nursery

'LADY HELEN MAGLONA', HT, dr, 1926; bud pointed; flowers bright crimson-red to scarlet-red, center deeper, very large, dbl., exhibition form, intense fragrance; foliage leathery; vigorous, bushy growth; GM, NRS, 1926; Dickson, A.

'LADY HELEN STEWART', HP, dr, 1887; flowers bright crimson shaded scarlet, dbl., intense fragrance; vigorous growth; Dickson, A.

'LADY HILLINGDON', T, yb, 1910; bud long, pointed; flowers deep apricot-yellow, semi-dbl., moderate fragrance; foliage bronze; bushy growth; sometimes as hardy as a Hybrid Tea; (21); [Papa Gontier X Mme Hoste]; Lowe & Shawyer

'LADY HILLINGDON, CLIMBING', Cl T, yb, 1917; Hicks

'LADY HUDSON', HT, ab, 1930; flowers deep apricot, large, dbl.; vigorous growth; Chaplin Bros.

'LADY HUNTINGFIELD', HT, my, 1937; flowers rich golden yellow, reverse lighter, large, dbl., globular, moderate fragrance; long stems; vigorous, bushy growth; [Busybody X ?]; Clark, A.; NRS Victoria

'LADY ILIFFE', HT, mr, 1976; flowers tyrian rose, dbl., 38 petals, 5 in., intense fragrance; foliage large; [Saul X Wendy Cussons]; Gandy, Douglas L.

Lady in Red *see* 'SEALADY'

Lady in Red *see* 'JACOPPER'

'LADY INCHIQUIN', HT, or, 1922; flowers orange-vermilion, large, very dbl., exhibition form, slight fragrance; foliage leathery, rich glossy green; very vigorous, bushy growth; GM, NRS, 1920; Dickson, A.

'LADY JANE', HT, dy, 1992; dbl., 26–40 petals, 3–3.5 in., borne mostly singly, slight fragrance; some prickles; foliage medium, medium green, semi-glossy; medium (75 cms), upright growth; [Dorothe X Helmut Schimdt]; Poole, Lionel, 1993

'LADY JOHNSTONE', LCl, dp, 1922; bud yellow; flowers reddish pink, turning lilac-rose, stamens yellow, large, single, moderate fragrance; vigorous, climbing growth; [R. gigantea X Beauté Lyonnaise]; Nabonnand, P.

Lady Kathleen Grade *see* **'LADY GRADE'**

'LADY KATHRYN', Min, m, 1989; bud pointed; flowers lavender, aging brown, urn-shaped, medium, borne usually singly, dbl., 22 petals, exhibition form, moderate, damask fragrance; round, green-brown fruit; no prickles; foliage medium, medium green, semi-glossy, disease-resistant; upright, medium, vigorous growth; [Lavender Jade X Angel Face]; Jolly, Marie; Rosehill Farm, 1991

'LADY LAUDER', HT, yb, 1931; flowers deep canary-yellow, reverse flushed crimson, dbl., cupped; foliage thick, light; long stems; very vigorous growth; Morse

'LADY LAVENDER', HT, m, 1993; dbl., 26–40 petals, 3–3.5 in., borne mostly singly, moderate fragrance; some prickles; foliage medium, dark green, glossy; bushy (40-48 in) growth; [(Paradise X Seedling) X Swarthmore]; Weeks, O.L.; Estrella Rose Company, 1993

'LADY LAYTON', HT, my, 1932; bud long, pointed; flowers light sunflower-yellow, deepening as it opens, large, moderate fragrance; vigorous growth; [Joanna Hill sport]; Layton

'LADY LECONFIELD', HT, w, 1939; bud long, pointed, cream, flushed pink; flowers cream-white, dbl., 25–30 petals, cupped, intense fragrance; foliage leathery; vigorous, bushy growth; Burbage Nursery; C-P

'LADY LE-RU', HRg, dp, 1963; bud round; flowers deep pink, dbl., 50 petals, 3.5–4 in., cupped, moderate, open (2 1/2 ft. free, non-recurrent bloom, moderate fragrance; thornless; long stems; hardy.; [R. rugosa hybrid X Unknown Hybrid Tea]; Lothrop

'LADY LESLIE', HT, rb, 1929; flowers rosy scarlet to scarlet-carmine suffused saffron-yellow, large, dbl., exhibition form; foliage dark, leathery, glossy; vigorous growth; McGredy

'LADY LIBERTY', HT, w, 1986; flowers clear white, yellow at base, large, dbl., 35 petals, exhibition form, borne usually singly or in small clusters, slight fragrance; globular, medium, medium green fruit; straight, medium, light yellow prickles; foliage medium, medium green, semi-glossy; extremely long stems; upright, profuse growth; PP006142; [Lady Diana sport]; deVor, Tom, 1987

'LADY LIKE', HT, ob, 1971; bud globular; flowers dark orange, large, dbl., slight fragrance; vigorous, upright growth; [Seedling X Tropicana]; Tantau, Math.; Ahrens & Sieberz

Lady Like® *see* TANekily

'LADY LILFORD', HT, my, 1930; flowers clear yellow, center deep golden yellow, slight fragrance; foliage rich green, glossy; vigorous, bushy, branching, compact growth; [Independence Day sport]; Gregory

'LADY LOCH', HT, lp; [Aspasia sport]; Johnson, 1885; (Weatherly, L.)

'LADY LOU', HT, ob, 1948; bud long, pointed; flowers coral-peach, large to medium, dbl., 50 petals, exhibition form, moderate fragrance; foliage glossy, light; vigorous, dwarf growth; [Pink Princess X Shades of Autumn]; Brownell, H.C.

'LADY LUCK', HT, pb, 1956; bud long, pointed; flowers blends of pale to rich pink, dbl., 38 petals, 4–4.5 in., exhibition form, intense, damask fragrance; foliage dark, leathery; vigorous, upright, bushy growth; [Tom Breneman X Show Girl]; Miller, A.J.; Elmer Roses Co.

'LADY MACGREGOR', HT, lp; [Boule de Neige X a Tea]; Williams, A., 1911; (Weatherly, L.)

Lady MacRobert *see* 'COCLENT'

Lady Madeleine *see* 'EVELEINE'

'LADY MANDEVILLE', HT, my, 1941; flowers yellow, flushed amber, well-formed, dbl., 35 petals, 5 in., slight, fruity fragrance; foliage dark, bronze; branching, moderate growth; [Unnamed seedling X Mrs Sam McGredy]; McGredy; J&P

'LADY MANN', HT, pb, 1940; flowers rosy salmon; [Lorraine Lee X ?]; Clark, A.

'LADY MARGARET BOSCAWEN', HT, mp, 1911; flowers large, dbl., intense fragrance; Dickson, A.; (Sangerhausen)

'LADY MARGARET STEWART', HT, ob, 1926; bud long, pointed; flowers golden yellow shaded and streaked orange and red, very large, dbl., exhibition form, moderate fragrance; foliage sage-green, leathery; vigorous, bushy growth; GM, Bagatelle, 1928 GM, NRS, 1926; Dickson, A.

'LADY MARINE', HT, or, 1981; flowers dark orange-red, urn-shaped, borne 1-3 per cluster, dbl., 53 petals, exhibition form, moderate fragrance; medium, reddish prickles; foliage dark, leathery; medium growth; [Unnamed seedling X Tropicana]; DeLashmutt; Roseway Nursery

'LADY MARTHA BRUCE', HT, mp, 1925; flowers pink, outer petals tinged peach-blossom-pink; Ferguson, W.

'LADY MARY', HT, mr, 1999; flowers crimson red, dbl., 40 petals, exhibition form, slight fragrance; foliage medium green; 3 ft growth; [Princess Mikasa X seedling]; Teranishi, K.; Itami Rose Nursery, 1990

'LADY MARY CORRY', T, 1900; Dickson, A.; (Cavriglia)

'LADY MARY ELIZABETH', HT, dp, 1927; bud pointed; flowers brilliant carmine-pink, large, dbl., exhibition form, intense fragrance; vigorous, bushy growth; Dickson, A.

'LADY MARY FITZWILLIAM', HT, lp, 1882; flowers flesh-color, large, globular, intense fragrance; vigorous growth; [Devoniensis X Victor Verdier]; Bennett

'LADY MARY WARD', HT, ob, 1913; flowers orange, shaded deeper, dbl., moderate fragrance; GM, NRS, 1912; McGredy

'LADY MAUREEN STEWART', HT, dr, 1920; flowers velvety blackish scarlet-cerise, reflex orange-maroon, dbl., moderate fragrance; Dickson, A.

Lady Mavis Pilkington *see* KORlitze

'LADY MAYSIE ROBINSON', HT, pb, 1956; flowers deep pink, center white, large, dbl., 22 petals, cupped, moderate fragrance; foliage dark, glossy; vigorous, upright, bushy growth; [Seedling X Peace]; Kordes

'LADY MEDALLIST', Cl HT, lp, 1912; large centered flowers, pink with lighter reverse; dark green foliage; very vigorous early summer climber; Clark, A.

Lady Meilland *see* MEIalzonite

Lady Meillandina *see* 'MEILARCO'

'LADY MILLER', HT, dr, 1940; flowers dark-red, well-formed, moderate fragrance; Clark, A.

Lady Mitchell *see* 'HARYEARN'

'LADY MOND', HT, w, 1920; flowers deep cream, outer petals shaded rose; Paul, W.

'LADY MOYRA BEAUCLERC', HT, mr, 1901; flowers carmine-red, large, dbl.; Dickson, A.; (Sangerhausen)

'LADY MOYRA CAVENDISH', HT, mr, 1939; bud long, pointed; flowers bright strawberry-red, flushed crimson, dbl., exhibition form, slight fragrance; foliage glossy, dark; bushy growth; McGredy

'LADY NUTTING', HT, mp, 1938; flowers soft salmon-pink, large, exhibition form; foliage leathery, dark; vigorous growth; Wheatcroft Bros.

Lady of Hertford *see* 'BOSCONPEA'

'LADY OF SKY', HT, or, 1974; dbl., 35 petals, exhibition form, slight fragrance; foliage dark; vigorous, upright growth; [Queen Elizabeth X ?]; Gregory

'LADY OF STIFFORD', F, or, 1982; [Matador sport]; Warley Rose Gardens, 1981

Lady of the Dawn® *see* 'INTERLADA'

Lady Penelope *see* 'CHEWDOR'

'LADY PENZANCE', HEg, op, 1894; (R. X penzanceana); flowers coppery salmon-pink with yellow stamens, single, summer bloom; foliage dark, fragrant (apple); very vigorous growth; (42); [R. eglanteria X R. foetida bicolor]; Penzance

'LADY PIRRIE', HT, ab, 1910; (Mme Pirrie); bud pointed; flowers apricot-yellow, reverse coppery, large, dbl., moderate fragrance; vigorous, bushy growth; GM, NRS, 1909; Dickson, H.

'LADY PIRRIE, CLIMBING', Cl HT, ab, 1938; Unknown

'LADY PLYMOUTH', T, w, 1914; bud long, pointed; flowers deep ivory-cream, very faintly flushed, large, dbl., moderate fragrance; foliage rich green, leathery; many canes; vigorous, bushy growth; GM, NRS, 1913; Dickson, A.

'LADY R. VERNEY', HT, mp, 1935; flowers medium, dbl., moderate fragrance; Bees of Chester; (Sangerhausen)

Lady Rachel *see* CANdoodle

'LADY RACHEL VERNEY', HT, mp, 1935; flowers rose, base lemon, large, dbl., cupped; foliage glossy, bronze; vigorous, bushy growth; [Annie Laurie X Lord Charlemont]; Bees

'LADY READING', Pol, mr, 1921; flowers clear red; [Ellen Poulsen sport]; Van Kleef

'LADY RHODES', HT, lp; Clark, A., 1934; (Weatherly, L.)

'LADY ROBERTS', T, ab, 1902; flowers rich reddish apricot, base coppery red, edges shaded orange, dbl., moderate fragrance; vigorous growth; [Anna Olivier bud sport]; Cant, F.

Lady Romsey, F, w, 1985; Beales, Peter

Lady Rose® *see* 'KORLADY'

Lady Rose, Climbing *see* ORAkosar

'LADY ROUNDWAY', HT, ab, 1923; flowers bright apricot-orange, fading to creamy buff, open, semi-dbl., slight fragrance; vigorous, but stubby growth; GM, NRS, 1923; Cant, B. R.

Lady Russon *see* **'ORANGE DELBARD'**

'LADY SACKVILLE', HT, w, 1933; bud pointed; flowers pure white, very large, dbl., exhibition form, moderate fragrance; foliage leathery, bronze; very vigorous growth; Cant, B. R.

Lady Sackville *see* **'NIGHT'**

'LADY SETON', HT, lp, 1966; dbl., 35 petals, 4.5 in., intense fragrance; vigorous, tall growth; [Ma Perkins X Mischief]; McGredy, Sam IV; McGredy

'LADY SOMERS', HT, lp, 1930; flowers fresh pink, tinted flesh, dbl., slight fragrance; foliage wrinkled, light; bushy growth; [Comte G. de Rochemur X Scorcher]; Clark, A.; NRS Victoria

'LADY SONIA', S, my, 1961; flowers golden yellow, dbl., 20 petals, 4–4.5 in.; foliage dark; vigorous, upright, branching growth; [Grandmaster X Doreen]; Mattock

'LADY STANLEY', T, dr; bud long; flowers crimson, very large, dbl., moderate fragrance

'LADY STUART', HCh, lp, 1851; flowers flesh pink to blush; foliage leaflets 5-7; Portemer fils

Lady Sunblaze™ *see* 'MEILARCO'

'LADY SUNSHINE', HT, my, 1965; bud ovoid; flowers large, very dbl., intense fragrance; foliage dark; [Belle Étoile X (Michele Meilland X Tawny Gold)]; Lens; Spek

'LADY SUSAN BIRCH', HT, ab, 1934; bud pointed; flowers large, dbl., exhibition form, moderate fragrance; foliage glossy, dark; vigorous, bushy growth; Cant, B. R.

'LADY SUZANNE', HT, w, 1985; flowers creamy white, large, dbl., 32 petals, exhibition form, slight fragrance; foliage large, dark, glossy; bushy growth; [Lady X X Flaming Beauty]; Bridges, Dennis A.; Bridges Roses

'LADY SYDNEY EARDLEY-WILMOT', HT, mp, 1925; flowers coppery reddish salmon, tinted fawn and apricot, semi-dbl., moderate fragrance; Chaplin Bros.

'LADY SYLVIA', HT, lp, 1926; [Mme Butterfly sport]; Stevens, W.

'LADY SYLVIA, CLIMBING', Cl HT, pb, 1933; Stevens, W.; Low

Lady Taylor *see* **'SMITLING'**

'LADY TERVUEREN', Pol, mr, 1969; flowers medium to small, dbl.; foliage dark; [Allotria X Seedling]; Buisman, G. A. H.

'LADY TRENT', HT, ob, 1940; (Julia Ferran); flowers coppery orange, large, dbl., 46 petals, exhibition form, moderate fragrance; foliage dark, glossy; vigorous growth; [Rosieriste Gaston Leveque X Federico Casas]; Dot, Pedro; Wheatcroft Bros.

'LADY URSULA', HT, pb, 1908; flowers pink fading lighter, reverse cameo-pink, base lemon, large, very dbl., exhibition form, slight fragrance; foliage dark, leathery, glossy; vigorous, bushy growth; Dickson, A.

'LADY VENABLES VERNON', HT, lp, 1922; flowers soft flesh-color, overlaid blush, moderate fragrance; [Mrs Amy Hammond X Sir Alexander N. Rochfort]; Jersey Nursery

'LADY VERA', HT, pb, 1974; flowers silvery pink, reverse rose-pink, dbl., slight fragrance; vigorous growth; [Royal Highness X Christian Dior]; Smith, R.W.; Brundrett

'LADY VEREY', HT, mp, 1922; flowers rose-pink, dbl., moderate fragrance; Hicks

'LADY VIOLET ASTOR', HT, dp, 1933; flowers deep rose-pink, over large, dbl., exhibition form, slight fragrance; foliage leathery; vigorous growth; Cant, B. R.

'LADY WAKEFIELD', HT, ab, 1926; flowers bright apricot, moderate fragrance; Cant, B. R.

'LADY WATERLOW', Cl HT, pb, 1903; flowers salmon-pink edged carmine, large, dbl.; vigorous (8-10 ft) growth; [La France de '89 X Mme Marie Lavalley]; Nabonnand, G.

'LADY WENLOCK', HT, lp, 1904; flowers china-pink tinted apricot; Bernaix, P.

'LADY WILLINGDON', HT, lp, 1928; flowers very light pink, large, dbl., moderate fragrance; foliage rich green, glossy; vigorous growth; [Ophelia X Premier]; Dale

'LADY WOODWARD', HT, lp, 1959; bud long, pointed; flowers pink veined, large, dbl., exhibition form, moderate fragrance; foliage dark, glossy; vigorous, upright, bushy growth; [Heinrich Wendland X Elli Knab]; Riethmuller

'LADY WORTHINGTON EVANS', HT, dr, 1926; bud pointed; flowers deep crimson shaded blackish, semi-dbl., exhibition form, moderate fragrance; foliage bronze, leathery; vigorous, bushy growth; GM, NRS, 1926; Dickson, A.

Lady X *see* 'MEIFIGU'

'LADY X, CLIMBING', Cl HT, m, 1976; Takatori, Yoshiho; Japan Rose Nursery

Lady Yvonne *see* 'BOSBIGSOUTH'

'LADY ZIA', HT, or, 1959; flowers light orange-scarlet, well-formed, dbl., 50 petals, 5–6. in., moderate fragrance; foliage dark, glossy; vigorous growth; GM, NRS, 1959; [Peace X Independence]; Park; Harkness

Ladybug *see* 'MORBUG'

'LADYLOVE', HT, mp, 1926; flowers light rose-pink fading hydrangea-pink, flushed apricot, well-formed, dbl., intense fragrance; foliage dark; very vigorous growth; [Opehlia X Unnamed seedling]; McGredy; Beckwith

'LAFAYETTE', F, dp, 1924; (August Kordes, Joseph Guy); flowers bright cherry-crimson, large blooms in clusters (up to 40), semi-dbl., cupped, slight fragrance; foliage rich green, glossy; vigorous, bushy growth; [Rodhatte X Richmond]; Nonin; Dreer;, H&S

Lafayette, Climbing *see* **'AUGUSTE KORDES'**

'L'AFRICAINE', LCl, dr, 1953; flowers garnet shaded coppery, well formed, large, borne in clusters, abundant early bloom, not recurrent; strong stems; very vigorous growth; [Guinee X Crimson Glory]; Mallerin, C.; EFR

'LAFTER', HT, yb, 1948; bud pointed; flowers salmon-yellow, dbl., 23 petals, 4 in., moderate fragrance; vigorous, upright, branching growth; [(V for Victory X (Général Jacqueminot X Dr. W. Van Fleet)) X Pink Princess]; Brownell, H.C.

Lagerfeld™ *see* 'AROLAQUELI'

'LAGERFEUER'®, F, mr, 1958; (Feu de Camp); bud pointed; flowers velvety scarlet, large blooms in clusters, dbl., slight fragrance; foliage leathery, dark; vigorous, upright growth; [Red Favorite X Kathe Duvigneau]; Tantau, Math.

'LAGOON', F, m, 1970; flowers lilac, reverse darker, gold stamens, single, 7 petals, 2.5 in., moderate fragrance; foliage glossy; [Lilac Charm X Sterling Silver]; Harkness, 1973

'LAGUNA', HT, or, 1974; bud large, long, pointed; dbl., cupped, moderate fragrance; foliage glossy; vigorous, upright growth; [Hawaii X Orange Delbard]; Kordes; Horstmann

Laguna, HT, lp; flowers yellow with pink edges

Laguna Palace *see* POUlagun

Lahar, F, my, 1991; IARI

'LAILA', F, ob, 1968; flowers bright orange, well formed, borne on trusses, dbl., slight fragrance; foliage glossy; vigorous growth; [Orangeade seedling]; Abdullah

L'Aimant *see* HARzola

'LAKE COMO', F, m, 1968; flowers lilac, blooms in trusses, semi-dbl., intense fragrance; [Lilac Charm X Sterling Silver]; Harkness

'LAKELAND', HT, lp, 1976; flowers soft shell-pink, dbl., 36 petals, 5–6. in., slight fragrance; [Fragrant Cloud X Queen Elizabeth]; Fryer, Gareth; Fryer's Nursery, Ltd.

Lakeland Princess *see* 'POULLACK'

Lakeland's Pride *see* **'DOUBLE FEATURE'**

'LAL', HT, pb, 1933; bud long, pointed; flowers deep salmon pink, suffused yellow, intense fragrance; foliage dark; vigorous growth; [Commonwealth (HT) X Florence L. Izzard]; Easlea

'LAL, CLIMBING', Cl HT, mp, 1937; flowers very large, dbl., intense fragrance; Vogel, M.; (Sangerhausen)

'LALIMA', HT, mr, 1978; flowers large blooms borne singly, dbl., 50 petals, exhibition form, intense fragrance; foliage large; vigorous, upright growth; [Picture X Jour d'Ete]; Pal, Dr. B.P.; K.S.G. Son's Roses

'LALLITA', HT, mp, 1929; flowers rose, very large, dbl., slight fragrance; foliage rich green, leathery; long, strong stems; [Pres. Briand X ?]; Mallerin, C.

'LAMARQUE', N, w, 1830; flowers pure white, center lemon-yellow, blooms in clusters, dbl., intense fragrance; vigorous, climbing growth (long, trailing shoots).; [Blush Noisette X Parks' Yellow Tea-scented China]; Marechal

'LAMARTINE', HT, op, 1943; flowers pearly pink shaded orange, dbl.; Meilland, F.

Lamb Chop, S, lp; Peden, R., 1997; (Weatherly, L.)

Lambada *see* KORdaba

'LAMBERT CLOSSE', S, mp, 1994; very dbl., 53 petals, 3–3.5 in., borne mostly singly, slight; some prickles; foliage medium, light green, glossy; medium (85 cms), upright growth; [Arthur Bell X John Davis]; Ogilvie, Ian S.; Agriculture Canada, 1994

'L'AMI E. DAUMONT', HP, mr, 1903; flowers carmine-red, large, very dbl.; Vilin; (Sangerhausen)

'LAMIA', HT, or, 1918; bud rich apricot; flowers intense reddish orange, semi-dbl.; vigorous growth; GM, NRS, 1918; Easlea

'LAMINUETTE', F, rb, 1969; (La Minuette, Minuette); bud ovoid, pointed; flowers petals ivory-white, tipped red, heart-shaped, medium, dbl., slight fragrance; foliage glossy, dark green; bushy, branching, strong (2 ft) growth; [Peace X Rumba (F)]; Lammerts, Dr. Walter; DeVor Nurseries, Inc.

'LAMLAM', HT, yb, 1965; (**American Heritage**®); bud long, pointed; flowers ivory and salmon blend, becoming salmon, large, dbl., exhibition form; foliage dark, leathery; tall growth; AARS, 1966; [Queen Elizabeth X Yellow Perfection]; Lammerts, Dr. Walter; Germain's

'LAMOTTE SANGUIN', HP, mr, 1869; flowers very large, dbl.; Vigneron; (Sangerhausen)

'LAMPION', F, or, 1957; bud pointed; flowers blood-red shaded orange, open, large, borne in large clusters, single, 5 petals, slight fragrance; foliage dark, leathery, glossy; dwarf, bushy growth; [Fanal X Kathe Duvigneau]; Tantau, Math.

'LAMPLIGHTER', Cl HT, dy, 1948; flowers blend similar to talisman and autumn, large, dbl., globular, intense fragrance; foliage leathery; very vigorous, climbing (10-15 ft) growth; [Talisman X Gold Rush]; Duehrsen; California Roses

'LAMPLIGHTER', HT, pb, 1950; flowers salmon-rose, reverse gold, very large, dbl., 37 petals, exhibition form, slight fragrance; foliage bronze; vigorous growth; [Sam McGredy X Unnamed seedling]; McGredy, Sam IV

'LAMPLIGHTER', HT, yb, 1959; bud long, pointed; flowers mimosa-yellow, dbl., 55–60 petals, 4–4.5 in., exhibition form, moderate fragrance; foliage dark, semi-glossy, leathery; strong stems; vigorous, upright growth; [Peace X Yellow Perfection]; Hill, Joseph H., Co.

Lampo, F; Cazzaniga, F. G.; (Cavriglia)

'LANBET', HT, or, 1985; (**Jennifer-Betty Kenward**); flowers medium, dbl., 35 petals, slight fragrance; foliage medium, medium green, glossy; spreading growth; [Mildred Reynolds X Whisky Mac]; Sealand Nurseries, Ltd.

'LANCASHIRE', HT, mr, 1950; flowers fiery red, intense fragrance; [Christopher Stone X Seedling]; Wright, R. & Sons

Lancashire *see* KORstesgli

'LANCASHIRE LASS', HT, mr, 1939; bud well shaped, crimson; flowers scarlet-cerise, semi-dbl., moderate fragrance; foliage glossy, dark; vigorous growth; Archer

Lancashire Life *see* 'RUILANCA'

'LANCASTER', Cl HT, mr, 1962; bud ovoid; flowers medium rose-red, dbl., 35–40 petals, 5 in., exhibition form, intense fragrance; foliage leathery, dark; vigorous growth; [Pres. Eisenhower sport]; Hicks, S.J.; C-P

'LANCASTRIAN', HT, mr, 1965; flowers crimson-scarlet, dbl., 40 petals, 3.5–4 in., intense fragrance; foliage light green, glossy; vigorous, upright growth; [Ena Harkness X ?]; Gregory

'LANCE HIRD', HT, mp, 1973; flowers deep pink, reverse lighter, full, dbl., 20–30 petals, 4 in., slight fragrance; free growth; [Seedling X Greetings]; Wood; Wood Roses

'LANCIER', HT, dr, 1959; flowers crimson-red; [Karl Herbst X Unnamed seedling]; Mallerin, C.; Vilmorin-Andrieux

Lancéme® *see* 'DELBOIP'

Land of the Long White Cloud *see* 'MACLANOFLON'

'LANDEN', HT, ly, 1988; (**Denman**); flowers creamy yellow, urn-shaped, large, borne in sprays of 2-3, dbl., 35 petals, intense fragrance; prickles medium, red; foliage large, dark green, glossy; upright growth; [Mildred Reynolds X Arthur Bell]; Sealand Nurseries, Ltd.

Lander Gold *see* 'MICGOLD'

'LANDIA', HT, ab, 1988; (Anne Diamond); bud pointed; flowers apricot, reverse pink, aging apricot, urn-shaped, medium, bo, dbl., 38 petals; prickles slightly hooked, brown; foliage medium, dark green, semi-glossy; upright, bushy growth; [Mildred Reynolds X Arthur Bell]; Sealand Nurseries, Ltd.

LANdisney, HT, w, 1987; (**Snow White**); Sealand Nurseries, Ltd.

Landmark Rose, S, mp, 1996

Landora *see* **'SUNBLEST'**

Landora, Climbing *see* ClinORA

'LANDOUR', Cl HT, w, 1978; bud oval; dbl., 40 petals, 6–6.5 in., exhibition form, intermittent bloom, moderate, fruity fragrance; foliage large, glossy, dark, leathery; upright growth; [Peace, Climbing self seedling]; Thakur; Doon Valley Roses

Landrover, HRg, mr, 1995

Landscape Splendor *see* CLEland

Landwirtschaftsrat Hubert Schilling, Pol, or, 1964; flowers medium, semi-dbl.; Pieper; (Sangerhausen)

Lane *see* **'LANEII'**

'LANEII', M, mr, 1845; (Lane, Lane's Moss); bud large, globular, well-mossed; flowers rosy crimson, occasionally tinted purple, large, dbl., globular; foliage large, bright green, 5 leaflets; robust growth; Laffay, M., 1845; Int. into England by Lane & Son, ca. 1846

Lane's Moss *see* **'LANEII'**

Langdale Chase *see* FRYrhapsody

'LANGFORD', LCl, dp; flowers nearly red, dbl., free bloom; needs protection in cold areas; [R. setigera seedling]; Preston; Central Exp. Farm

Langford Light *see* 'LANNIE'

'LANGLEY', HT, dy, 1942; bud streaked red; flowers clear deep yellow, edges flushed old-gold, dbl., 50 petals, 4–5 in., moderate fragrance; foliage bronze, dark; [Mrs Sam McGredy X Phyllis Gold]; Eacott; Ley

'LANGLEY GEM', F, dp, 1939; flowers scarlet-cerise, open, borne in clusters, single; foliage leathery, bronze; strong stems; vigorous growth; [Karen Poulsen X R. moyesii]; Eacott; R. Murrell

'LANICAN', HT, yb, 1988; (**Great Expectations**); flowers light pink, reverse light yellow, aging fading slightly, well-formed, dbl., 55 petals, exhibition form, moderate fragrance; prickles long, pointed, medium, red; foliage medium, medium green, semi-glossy, clean; upright growth; [Rosenella X Cassandra]; Sealand Nurseries, Ltd.

'LANKEN', HT, op, 1985; (**Felicity Kendal**); flowers salmon-orange, large, dbl., 35 petals; foliage large; bushy growth; [Fragrant Cloud X Mildred Reynolds]; Sealand Nurseries, Ltd.

'LANNIE', Min, w, 1985; (**Langford Light**); flowers bright yellow stamens, small, semi-dbl., moderate fragrance; foliage medium, dark, matt; bushy growth; [Ballerina X Little Flirt]; Sealand Nurseries, Ltd., 1984

'LANPIPE', F, dy, 1985; (**St Bruno**); flowers large, dbl., 35 petals, intense fragrance; foliage medium, medium green, semi-glossy; bushy growth; Edland Fragrance Medal, ARS, 1986; [Arthur Bell X Zambra]; Sealand Nurseries, Ltd.

Lantern *see* 'MACLANTER'

'LANTOR', HT, pb, 1985; (**Torvill & Dean**); flowers pink, yellow reverse, medium, dbl., 35 petals, slight fragrance; foliage medium, dark, semi-glossy; upright growth; [Irish Gold X Alexander]; Sealand Nurseries, Ltd., 1984

Lanvin™ *see* 'AROLEMO'

'LANWOOL', F, ab, 1985; (**Chelsea Gold**); flowers medium, dbl., 20 petals, slight fragrance; many prickles; foliage medium, medium green, semi-glossy; bushy growth; [Arthur Bell X Elizabeth of Glamis]; Sealand Nurseries, Ltd.

LAPad, F, or, 1971; (**Porthos**®); Laperrière

LAPam, F, my; (**Golden Flame**®); Laperrière

LAParan, F, 1958; (**Chantefleur**); Laperriere, L.; (Cavriglia)

'LAPAV', HT, ob, 1967; (**Mme L. Pradel**®); flowers clear orange, medium, dbl., exhibition form, slight fragrance; foliage bronze, glossy, leathery; vigorous, upright growth; [Magicienne X Seedling]; Laperrière; EFR

LAPbal, HT, yb, 1989; (**Horticolor**®); Laperrière

'LAPBAU', HT, my, 1991; (**Bellissima**®); Laperrière

LAPbel, HT, mr, 1988; (**Duchesse de Savoie**®); Laperriere, L.

LAPbu, HT, op, 1988; (**Ecole d'Ecully**); Laperrière

'LAPCAL', HT, rb, 1966; (**Fanette**®); flowers red, reverse white, medium, dbl., 36 petals, exhibition form, slight fragrance; foliage dark, glossy; vigorous, bushy growth; [Jeunesse X Souv. du President Plumecocq]; Laperrière; EFR

'LAPCI', F, mr, 1958; (**Bel Ami**®); flowers large blooms in clusters of 7-8, dbl.; bushy growth; [Michèle Meilland X Tonnerre]; Laperrière; EFR

LAPcli, F, mr, 1997; (**Hillary First Lady**); Laperrière

LAPdade, HT, mr, 1984; (**Ferline**®); Laperrière

LAPdal, HT, yb, 1990; (**Roxane**®); Golden Rose, Geneva, 1990; Laperrière

'LAPDEF', HT, dy, 1986; (**Banco 86**®); Laperrière

LAPdev, F, or, 1991; (**Plein Soleil**®); Laperrière

'LAPDI', HT, or, 1966; (**Renouveau de Provins**®); bud ovoid; flowers geranium color, open, very large, dbl., 36 petals; foliage glossy; vigorous, bushy growth; [Magicienne X Numero Un]; Laperrière; EFR

LAPdil, HT, op, 1988; (**Blondie**); Laperrière

LAPduf, HT, 1989; (**Don Bosco**); Laperriere, L.; (Cavriglia)

'LAPDUL', HT, pb, 1967; (**Régence**®); bud globular; flowers flesh color edged bright pink, open, dbl.; foliage glossy; vigorous, bushy growth; Laperrière; EFR

LAPed, HT, rb; (**Cabaret**®); Laperrière, 1992; (CRL)

LAPej, HT, rb, 1992; (**Republique de Genève**); Laperière

LAPem, HT, dp, 1974; (**Illisca**®); Laperrière

LAPjaminal, F, ob, 1983; (**Magic Fire**®); Laperrière

'LAPLAM', HT, ob, 1969; (**Mme Franck Augis**®); bud ovoid; flowers clear orange, medium, dbl., exhibition form, slight fragrance; foliage glossy; vigorous, bushy growth; [Magicienne X Seedling]; Laperrière; EFR

LAPlical, HT, rb, 1981; (**Mamy Laperrière**); Laperrière; (Cavriglia)

'LAPMAU', HT, m, 1968; (**Ludmilla**®); bud ovoid; flowers medium, semi-dbl., slight fragrance; foliage glossy; vigorous, upright growth; [(Peace X Independence) X Heure Mauve]; Laperrière; EFR

LAPmiravi, HT, mr, 1982; (**Carmosine**); Laperriere, L.

LAPmisal, HT, op; (**Mistraline**®); Laperrière

LAPnat, HT, op, 1971; (**Domila**®); Laperrière

'LAPNEUF', F, or, 1967; (**Magic Fire**®); bud pointed; flowers bright orange-red, small, semi-dbl., exhibition form; foliage glossy; vigorous, bushy growth; Laperrière; EFR

LAPniro, F, ob; (**D'Artagnan**); Laperrière

'LAPON', HT, mr, 1968; (**Ursula**®); bud ovoid; flowers clear red, open, medium, dbl., slight fragrance; foliage glossy; vigorous, upright growth; [Jeunesse X (Peace X Independence)]; Laperrière; EFR

'LAPOUDA', HT, ob, 1978; (**Magicienne 78**®); Laperrière

LAPovi, F, dp; (**Capitole**®); Laperrière

Lapponia® *see* 'TANNIPOLA'

LAPruni, HT, w, 1974; (**Banquise**); Laperrière

'LAPWON', F, ob, 1965; (**Athos**®); flowers bright orange, large, dbl., 28 petals; foliage dark; bushy growth; [Coup de Foudre X Soleil]; Laperrière; EFR

Laque de Chine, HMsk, 1991; Lens

'LARA', HT, op, 1967; flowers salmon-carmine, well-formed, large, dbl., 38 petals, slight fragrance; foliage dark, glossy; vigorous, tall growth; [Tropicana X Romantica]; Kriloff, Michel; Cramphorn's Nursery

Larado *see* 'SPOLAR'

L'Argentee, HT, rb

'LARGO D'HAENDEL', HT, dp, 1947; (Handel's Largo); flowers reddish apricot to salmon-carmine; Mallerin, C.; URS

Larissa® *see* 'KORDODO'

L'Arlésienne *see* **'FRED EDMUNDS'**

'LARRY BURNETT', HSpn, w, 1925; flowers blush-white, center deeper, semi-dbl., 3.5 in., cupped, profuse bloom, intense fragrance; foliage small, rich green, soft; short stems; vigorous, bushy, spreading growth; very hardy.; [R. acicularis X R. spinosissima]; Skinner

Larry's Surprize *see* 'RENLARRY'

Las Vegas® *see* 'KORGANE'

'LAS VEGAS', HT, pb, 1957; bud long, pointed; flowers salmon-pink, reverse darker, dbl., 25–30 petals, 5 in., exhibition form, slight fragrance; foliage leathery; vigorous, bushy growth; RULED EXTINCE 6/80; [Charlotte Armstrong X Mission Bells]; Whisler; Germain's

Las Vegas, HT, 1974; Kordes, R.; (Cavriglia)

'LAS-CASES', B, dp, 1828; Vibert

Laser™ *see* KEItaibu

Laser Beam *see* 'KORPORT'

'LASKER', HT, pb, 1979; bud very long; dbl., 40 petals, exhibition form, borne singly, moderate fragrance; curved prickles; foliage glossy; tall growth; [South Seas X Oregold]; Perry, Astor

'LASSIE', HT, pb, 1946; bud long, pointed; flowers shell-pink, base canary-yellow, dbl., exhibition form, moderate fragrance; foliage leathery; vigorous, compact growth; [Picture sport]; Tuttle Bros. Nursery

Lasting Impression *see* 'MICIMP'

Lasting Peace *see* MEImurge

'LATAN', HT, 1979; bud long; flowers light lavender-tan, dbl., 43 petals, 4.5–5.5 in., exhibition form, intense fragrance; foliage leathery; upright, bushy growth; [Lady X X Bronze Masterpiece]; Williams, J. Benjamin; J.B. Williams & Associates

'LATHOM CHAPEL', F, w, 1973; flowers buff-cream, blended to white, full, dbl., 20–25 petals, 4 in., intense fragrance; vigorous growth; [Orange Sensation X Sutter's Gold]; Ellick

'LATHOM PARK', F, dp, 1972; flowers carmine, full, dbl., 40 petals, 4 in., slight fragrance; foliage glossy, light; vigorous growth; [Orange Sensation X Ballet]; Ellick

'LATHOM SUNRISE', F, ob, 1973; flowers orange-flame, full, dbl., 25–30 petals, 4 in., intense fragrance; vigorous growth; [(Orange Sensation X Ballet) X Sutter's Gold]; Ellick

'LATHOM SUNSET', HT, ob, 1973; flowers orange-yellow, pink and red, full, dbl., 20–25 petals, 4.5 in., moderate fragrance; vigorous growth; [Karl Herbst X Mischief]; Ellick

'LATHOM TWILIGHT', HT, m, 1973; flowers deep purple to indian lake, full, dbl., 28–30 petals, 5 in., intense fragrance; foliage dark; vigorous growth; [Karl Herbst X Mischief]; Ellick

Latte *see* 'JUSLATTE'

Laughter Lines *see* 'DICKERRY'

'LAURA', HT, op, 1969; flowers coral-pink, large, dbl., exhibition form, moderate fragrance; foliage leathery; vigorous, upright, bushy growth; [((Happiness X Independence) X Better Times) X (Baccará X WhiteKnight)]; Meilland, Mrs. Marie-Louise; C-P

Laura™ *see* 'MEIVOUPLIX'

Laura® *see* 'MEIDRAGELAC'

'LAURA', C, mp, 1848

Laura '81 *see* 'MEIDRAGELAC'

Laura Anne *see* 'COCCLARION'

Laura Ashley *see* 'CHEWHARLA'

Laura Chantal *see* 'TOMTAL'

Laura Clements *see* CLEspirit

Laura Ford® *see* 'CHEWARVEL'

Laura Jane, HT, lp, 1990; Gandy

Laura Louisa, LCl, lp

Laura Louise, HKor, op, 1995; Riches

'LAURA ROSE ELLIOTT', HT, dp, 1995; flowers deep pink with light reverse, blooms borne mostly single, very dbl., 4–4.75 in., moderate fragrance; numerous prickles; foliage medium, medium green, dull; medium (60-70 cms), upright growth; [Gavotte X City of Gloucester]; Thomas, D.

'LAURA TOWILL', HT, yb, 1929; flowers copper-yellow, becoming copper-pink, semi-dbl., slight fragrance; [Phantom X Buttercup]; Towill

Laura's Red Koster, Pol, mr, 1992

Laure Charton, F, or

'LAURÉ DAVOUST', HMult, lp, 1834; (Marjorie W. Lester); flowers clear pink, fading to flesh, then white, small blooms in clusters, dbl., cupped, non-recurrent; Laffay, M.

'LAURÉ SOUPERT', HMult, ly, 1927; flowers yellowish white to pure white, small blooms in clusters of 80, dbl., recurrent bloom, intense fragrance; foliage small, glossy; strong stems; vigorous, climbing or trailing growth; [Tausendschön X George Elger]; Soupert & Notting; Notting

'LAURE WATTINNE', HT, mp, 1902; flowers large, dbl., moderate fragrance; Soupert & Notting; (Sangerhausen)

'LAUREATE', F, m, 1989; bud pointed; flowers lavender with pink highlights, aging lighter, urn-shaped, medium, dbl., exhibition form, slight, fruity fragrance; fruit not observed; prickles straight, medium, red; foliage medim, dark green, glossy, disease-resistant; upright, bushy, medium, winter-hardy growth; [Baby Talk X Angel Face]; Jobson, Daniel J.

Laurel Louise, MinFl, ab, 1999

'LAURELLE', F, m, 1966; flowers lavender-pink, base veined lemon, borne in clusters, dbl., moderate fragrance; foliage glossy; vigorous, bushy growth; [Rumba X Lavender Princess]; Harris, J.R.; Paulen Park Nursery

Lauren Amy, HT, or, 1995; Dawson

Lauren Elizabeth *see* 'ORTLAE'

Laurence Olivier, F, 1990; Meilland, Alain A.; (Cavriglia)

'LAURENT CARLE', HT, mr, 1907; flowers brilliant velvety carmine, open, large, dbl., intense fragrance; foliage rich green, soft; bushy growth; Pernet-Ducher

'LAURENT CARLE, CLIMBING', Cl HT, mr, 1923; Rosen, L.P.

'LAURENT CARLE, CLIMBING', Cl HT, mr, 1924; Mermet

'LAURENT DE RILLÈ', HP, mr, 1885; flowers large, dbl., moderate fragrance Lévêque; (Sangerhausen)

Laurette, HRg, dp, 1999; Hortico

'LAURIE', HT, lp, 1970; flowers creamy pale pink, tipped salmon, full, dbl., 28 petals, 5 in., moderate fragrance; foliage matt green; vigorous growth; [Princesse sport]; Scott, D.H.

Lausitz, S, mp, 1959; flowers medium, semi-dbl., slight fragrance; Berger, W.; (Sangerhausen)

Lava Gold, F, ob, 1999; Williams, J. Benjamin

'LAVABY', Min, lp, 1982; (**Baby Face**); bud tapered; flowers light pink, very small, dbl., 35 petals, blooms in sprays of many, slight; foliage small, light green, matt; compact growth; [Popcorn X Popcorn]; Laver, Keith G.

'LAVACEK', Min, mr, 1984; ('LAVCALE', 'LAVCALI', **Mountie**™); dbl., 35 petals, exhibition form, no fragrance; foliage small, dark, semi-glossy; bushy growth; PP006054; [Party Girl X Dwarfking '78]; Laver, Keith G.; Springwood Roses, 1985

'LAVACRE', Min, ly, 1987; ('LAVCREAM', **Potluck Cream**™); flowers light yellow center, cream edges, reverse cream with yellow, dbl., 30 petals, exhibition form, no fragrance; no fruit; prickles slender, brown; foliage medium, medium green, matt; bushy, low, very free flowering growth; [Cornsilk X Unnamed seedling]; Laver, Keith G.; Springwood Roses, 1988

Lavaglow *see* 'KORLECH'

Lavaglut® *see* 'KORLECH'

'LAVALIER', Min, dp, 1989; bud pointed; flowers deep mauve-pink, reverse medium pink, small, borne singly, dbl., 50 petals, cupped, no fragrance; round, red fruit; prickles straight, small, light brown; foliage small, medium green, matt, disease-resistant; upright, bushy, low, prolific growth; [Loving Touch X (Honest Abe X Unnamed seedling)]; Laver, Keith G.; Springwood Roses

'LAVALOT', Min, mp, 1992; (**Potluck Pink**™); flowers rose pink, blooms borne mostly single, dbl., 26–40 petals, 1.5 in., slight fragrance; few prickles; foliage small, medium green, matt; low (25-30 cms), compact growth; [June Laver X Julie Ann]; Laver, Keith G.; Springwood Roses

'LAVALUCK', Min, dp, 1989; bud pointed; flowers vivid, deep cherry pink, aging fuchsia, urn-shaped, small, b, dbl., 28 petals, no fragrance; ovoid, orange-red fruit; no prickles; foliage small, medium green, matt, long sepals, disease resi; bushy, low growth; [Blueblood X Julie Ann]; Laver, Keith G.; Springwood Roses

'LAVAMAZE', Min, lp, 1989; bud ovoid; flowers light pink with deeper center, reverse light pink, full, small, very dbl., 80 petals, exhibition form, moderate fragrance; round, orange-red fruit; prickles straight, narrow, red; foliage small, medium green to red, disease resistant; bushy, low growth; [Loving Touch X PotluckTM]; Laver, Keith G.; Springwood Roses

Lavande, F, m; Rennie, Bruce F.

'LAVARU', Min, dr, 1993; (**Springwood Ruby**); dbl., 26–40 petals, 1.5 in., borne in small clusters, no fragrance; some prickles; foliage medium, medium green, matt; low (20-25 cms), compact growth; [(Breezy X June Laver) X (June Laver X Ontario Celebration)]; Laver, Keith G.; Springwood Roses

'LAVAWAY', Min, w, 1989; (**Snowbound**); bud ovoid; flowers ivory, reverse white, aging white, urn-shaped, small, borne, dbl., 55 petals, slight fragrance; round, yellow-orange fruit; prickles straight, narrow, greenish-brown; foliage small, dark green, matt, disease-resistant; bushy, low, compact growth; [Tabris X June Laver]; Laver, Keith G.; Springwood Roses, 1990

'LAVBERT', Min, dr, 1995; (**Springwood**® **Red Victor**); flowers dark red, very full (41+ petals), small (0–4 cms) blooms borne 3–10 per cluster; no fragrance; some prickles; foliage medium, medium green, semi-glossy; medium (30–33cms), upright, compact growth; [Seedling X Seedling]; Laver, Keith G.; Springwood Roses, 1995

'LAVBLU', Min, mr, 1982; (**Blueblood**™); flowers velvety texture, small, dbl., 55 petals; foliage small, deep green, semi-glossy; spreading growth; [Dwarfking '78 X Hokey Pokey]; Laver, Keith G.

'LAVBOUND', Min, op, 1989; bud pointed; flowers coral to orange-pink, outer petals pink-apricot, reverse pink, dbl., 53 petals, exhibition form, no fragrance; foliage medium, medium green, semi-glossy, disease-resistant; upright, bushy, low, prolific growth; [June Laver X Black Jade]; Laver, Keith G.; Springwood Roses

'LAVBRIC', Min, lp, 1995; (**Show Carpet**); Flowers light pink, dbl. (15–25 petals), small, micro-mini blooms borne in large clusters; slight fragrance; foliage small, dark green, glossy; upright growth; [Seedling X Seedling]; Laver, Keith G.; Springwood Roses, 1996

'LAVBRUN', Min, ob, 1995; flowers orange, blooms borne mostly single, dbl., 26–40 petals, no fragrance; some prickles; foliage medium, medium green, semi-glossy; bushy, medium growth; [Seedling X Apricot Doll]; Laver, Keith G.; Springwood Roses, 1995

'LAVCALE', Min, mr, 1984; ('LAVACEK', 'LAVCALI', **Mountie**™); dbl., 35 petals, exhibition form, no fragrance; foliage small, dark, semi-glossy; bushy growth; PP006054; [Party Girl X Dwarfking '78]; Laver, Keith G.; Springwood Roses, 1985

'LAVCALI', Min, mr, 1984; ('LAVACEK', 'LAVCALE', **Mountie**™); dbl., 35 petals, exhibition form, no fragrance; foliage small, dark, semi-glossy; bushy growth; PP006054; [Party Girl X Dwarfking '78]; Laver, Keith G.; Springwood Roses, 1985

'LAVCAP', Min, op, 1997; (**Peach Festival**™); flowers very full, medium, very dbl., 41 petals, borne mostly singly, slight fragrance; few prickles; foliage medium, medium green, glossy; upright, bushy, medium growth; [Seedling X Painted Doll]; Laver, Keith G.

'LAVCHIP', Min, m, 1993; (**Blue Ice**); flowers lavender, reflex upon opening, blooms borne mostly singly, dbl., 26–40 petals, 1.5–2.75 in., moderate fragrance; some prickles; foliage small, medium green, semi-glossy; low (25-30 cms), upright, bushy growth; [June Laver X Apricot Doll]; Laver, Keith G.; Springwood Roses, 1993

'LAVCHRO', Min, dy, 1993; (**Gala Gold**); dbl., 26–40 petals, 1.5 in., borne mostly singly, no fragrance; some prickles;

foliage small, medium green, matt; low (24-30 cms), upright, bushy, compact growth; [Golden Promise X (June Laver X Tut's Treasure)]; Laver, Keith G.; Springwood Roses, 1994

'LAVCLASS', Min, mp, 1999; (**Springwood Classic**®); dbl., 26–40 petals, 2 in., borne in small clusters, no fragrance; few prickles; foliage medium, medium green, semi-glossy; bushy, medium (10-12 in.) growth; [seedling X seedling]; Laver, Keith G.; Springwood Roses, 1999

'LAVCLO', Min, rb, 1994; (**Fiesta Clown**); flowers scarlet with orange-yellow reverse, bicolor, pointed, recurved, dbl., 26–40 petals, 1.5–2.75 in., intense fragrance; some prickles; foliage medium, dark green, semi-glossy; medium (24-30 cms), upright growth; [(June Laver X Seedling) X (Painted Doll X June Laver)]; Laver, Keith G.; Springwood Roses, 1994

'LAVCOAT', F, yb, 1991; (**Colour Parade**); flowers yellow turning to red, small, blooms borne in small clusters, dbl., 26–40 petals, 1.5 in., slight fragrance; few prickles; foliage medium, medium green, glossy; low (45 cms), bushy growth; [Breezy X Julie Ann]; Laver, Keith G.; Springwood Roses, 1992

'LAVCOM', Min, ob, 1993; (**Silken Laumann**); flowers luminous orange, yellow reverse, blooms borne in large cluster, dbl., 26–40 nicely recurved petals, 1.5–2.75 in., exhibition form, slight fragrance; few prickles; foliage medium, medium green, matt; medium (45 cms), bushy growth; [June Laver X Potluck Red]; Laver, Keith G.; Springwood Roses, 1994

'LAVCREAM', Min, ly, 1987; ('LAVACRE', **Potluck Cream**®); flowers light yellow center, cream edges, reverse cream with yellow, dbl., 30 petals, exhibition form, no fragrance; no fruit; prickles slender, brown; foliage medium, medium green, matt; bushy, low, very free flowering growth; [Cornsilk X Unnamed seedling]; Laver, Keith G.; Springwood Roses, 1988

'LAVCREME', Min, ly, 1988; (**Crème Glacée**); bud pointed; flowers urn-shaped, small, borne singly, dbl., 23 petals, slight fragrance; globular, orange fruit; prickles pointed and straight out, white-beige; foliage small, medium green, matt; bushy growth; [June Laver X Summer Butter]; Laver, Keith G.; Springwood Roses, 1990

'LAVDANCE', Min, pb, 1993; (**Striped Pet**); flowers variable pink stripes on white, blooms borne mostly single, dbl., 15–25 petals, 1.5–2.75 in., moderate fragrance; few prickles; foliage medium, dark green, matt; low (25-30 cms), upright, compact growth; [June Laver X (June Laver X Ontario Celebration)]; Laver, Keith G.; Springwood Roses, 1993

'LAVDOLL', Min, ab, 1990; (**Apricot Doll**); bud ovoid; flowers apricot, yellow center, reverse lighter, aging light apricot, dbl., 30–35 petals, borne usually singly and in sprays of 1-4, moderate fragrance; foliage small, medium green, matt; spreading, low growth; [Painted Doll X Painted Doll]; Laver, Keith G., 1986; Springwood Roses, 1991

'LAVDUSK', Min, dp, 1992; (**Springwood Pink**®); flowers deep pink, dbl., 15–25 petals, 1.5 in., borne in small clusters, slight fragrance; some prickles; foliage medium, medium green, matt; suitable for pots; low (30-40 cms), spreading growth; [Maurine Neuberger X (June Laver X Ontario Celebration)]; Laver, Keith G.; Springwood Roses

'LAVEENA', HT, yb, 1969; bud long, pointed; flowers yellow tinged pinkish, large, dbl., exhibition form; foliage glossy; vigorous, upright, compact growth; [Kiss of Fire X ?]; Laveena Roses

'LAVENDALE', Min, m, 1989; bud pointed; flowers deep lavender, medium, borne singly, dbl., 40 petals, exhibition form, intense fragrance; round, greenish-brown fruit; no prickles; foliage medium, medium green, semi-glossy; upright, bushy, medium, vigorous growth; [Lavender Jade X Angel Face]; Jolly, Marie; Rosehill Farm, 1990

'LAVENDER BIRD', HT, m, 1964; bud long, spiral, pointed; flowers lavender-pink, well formed, large, dbl., moderate, lavender fragrance; vigorous growth; Herholdt, J.A.; Herholdt's Nursery

Lavender Blue *see* 'LEOLAVBLU'

'LAVENDER CHARM', HT, m, 1964; flowers persian lilac, dbl., 45–50 petals, 4.5–5 in., cupped, intense fragrance; foliage dark, leathery; vigorous, bushy growth; [Brownie X Sterling Silver]; Boerner; J&P

Lavender Cover *see* POUlrust

Lavender Crystal *see* ASAlav

Lavender Delight *see* 'MORORCHERI'

Lavender Dream® *see* 'INTERLAV'

Lavender Friendship, S, m, 1984; Verschuren

'LAVENDER GARNETTE', F, m, 1958; bud globular; flowers lavender, open, medium, dbl., 35–45 petals, moderate fragrance; foliage leathery; vigorous, bushy growth; [Grey Pearl seedling X Garnette]; Boerner; J&P

'LAVENDER GIRL', F, m, 1958; flowers rosy purple, reverse magenta, changing to lavender, dbl., 35–42 petals, 3.5 in., cupped, moderate, spicy fragrance; dwarf, bushy growth; [Fantastique X (Ampere X (Charles P. Kilham X Capucine Chambard))]; Meilland, F.; C-P

Lavender Jade® *see* 'BENALAV'

'LAVENDER JEWEL', Min, m, 1978; bud pointed; flowers clear lavender-mauve, dbl., 38 petals, 1 in., exhibition form, slight fragrance; foliage dark; compact, bushy growth; [Little Chief X Angel Face]; Moore, Ralph S.; Sequoia Nursery

'LAVENDER LACE', Min, m, 1968; flowers lavender, small, dbl., exhibition form, moderate fragrance; foliage small, glossy; vigorous, bushy dwarf growth; AOE, 1975; [Ellen Poulsen X Debbie]; Moore, Ralph S.; Sequoia Nursery

'LAVENDER LACE, CLIMBING', Cl Min, m, 1971; Rumsey, R.H.

'LAVENDER LADY', F, m, 1956; flowers pastel mauve, large, borne in clusters, semi-dbl., moderate fragrance; vigorous, upright growth; [Unnamed seedling X Lavender Pinocchio]; LeGrice

Lavender Lady, HT, m, 1996; Williams, J. Benjamin

'LAVENDER LASSIE', HMsk, m, 1960; flowers lilac-pink, blooms in large clusters, dbl., 3 in., intense fragrance; very vigorous, tall growth; Kordes; Morse

'LAVENDER LOVE', F, m, 1964; flowers lavender, medium; low growth; [Fashion X Floradora]; Daugherty; Wyant

Lavender Meillandina, Min, m

Lavender Midinette, Min, m; Moore

Lavender Mist *see* 'TINVIOLET'

'LAVENDER MIST', LCl, m, 1981; (Mystic Mauve); bud ovoid; dbl., 35 petals, borne mostly 3 per cluster, slight, tea fragrance; medium prickles, hooked downward; foliage large; vigorous growth; long arching canes; [Angel Face X Allspice]; Christensen, Jack E.; Armstrong Nursery

'LAVENDER PEARL', HT, m, 1976; flowers lavender center shading to pearl, full, dbl., 40 petals, 3.5 in., moderate, lemon fragrance; foliage small; moderate, bushy growth; [Blue Moon X Grandpa Dickson]; Shaw

Lavender Pink Parfait, Pol, m

'LAVENDER PINOCCHIO', F, m, 1948; bud ovoid, light chocolate-olive-brown; flowers pink-lavender, blooms in clusters, dbl., 28 petals, 3–3.5 in., moderate fragrance; vigorous, bushy, compact growth; [Pinocchio X Grey Pearl]; Boerner; J&P

'LAVENDER PRINCESS', F, m, 1959; bud ovoid; flowers lavender, lightly overcast purplish lilac, open, borne in large clusters, dbl., 25 petals, 3.5–4 in., moderate, fruity fragrance; foliage leathery; vigorous, upright growth; [World's Fair seedling X Lavender Pinocchio seedling]; Boerner; J&P

'LAVENDER QUEEN', HT, m, 1951; bud pointed, touched red; flowers pinkish, lavender, dbl., 20–35 petals, 5–6 in., cupped, moderate fragrance; vigorous growth; Raffel; Port Stockton Nursery

Lavender Simplex® *see* 'MINTCC'

Lavender Spoon *see* 'SPOLAV'

Lavender Star® *see* 'MINAUCO'

Lavender Sunblaze®, Min, m, 1993; flowers small, lavender-mauve, exhibition, dbl., 25–30 petals, 1–1.5 in., exhibition form, borne singly and in small clusters, slight fragrance; some prickles; foliage small, deep green, glossy; bushy, compact, medium, 18 on. growth; Meilland

Lavender Sweetheart® *see* 'WILLASH'

'LAVENDULA', F, m, 1965; flowers lavender, blooms in clusters, 4 in., intense fragrance; foliage dark; [Magenta (F) X Sterling Silver]; Kordes, R.; McGredy

Laverne's Greenhouse Climber, LCl, lb

Laverne's Little Red Creeper, Min, rb

'LAVFAL', Min, rb, 1997; (**Fall Festival**®); flowers full, medium, very dbl., 26–40 petals, borne mostly singly, no fragrance; some prickles; foliage medium, dark green, semi-glossy; upright, bushy, medium growth; [Seedling X Springwood Red Victor]; Laver, Keith G.

'LAVFAN', Min, dr, 1998; (**Fancy Potluck**); flowers dark red, full, good substance, 26–40 petals, .5–1 in., borne mostly singly, no fragrance; few prickles; foliage small, dark green, matte; low (15-20 cms), compact growth; [Seedling X Antique Gold]; Laver, Keith G.; Springwood Roses, 1998

'LAVFIRE', Min, or, 1999; (**Flame Bouquet**®); flowers brilliant orange red very long lasting, dbl., 26–40 petals 2.5–3 in., borne mostly singly, no fragrance; few prickles; foliage medium medium green, semi-glossy; upright medium (12 in.) growth; Laver, Keith G.; Springwood Roses, 1999

'LAVFLAME', Min, or, 1995; (**Flaming Potluck**); flowers flaming orange-scarlet, deeper center, blooms borne mostly singly, dbl., 26–40 petals, 1.5 in., slight fragrance; some prickles; foliage medium, dark green, semi-glossy; medium (30-40 cms), upright, bushy growth; Laver, Keith G.; Springwood Roses, 1994

'LAVFLUSH', Min, op, 1993; (**Double Date**); flowers orange pink, blooms borne mostly single, very dbl., 1.5 in., slight fragrance; some prickles; foliage small, medium green, semi-glossy; low, upright, bushy growth; [(Breezy X June Laver selfed) X Seedling]; Laver, Keith G.; Springwood Roses

'LAVFUHR', Min, mp, 1994; (**Dainty Bouquet**); flowers clear pink, blooms borne mostly single, very dbl., 1.5 in., slight fragrance; some prickles; foliage medium, medium green, semi-glossy; medium (20 cms), upright, spreading growth; [Seedling X Pink Bouquet]; Laver, Keith G.; Springwood Roses, 1994

'LAVFUN', Min, pb, 1993; (**Blessed Event**); flowers white suffused pink, blooms borne singly and in small clusters, dbl., 26–40 petals, 1.5–2.75 in., slight fragrance; some prickles; foliage medium, dark green, matt; low (25-30 cms), upright growth; [Enjoy X (June Laver X Party Girl)]; Laver, Keith G.; Springwood Roses, 1993

'LAVGLO', Min, my, 1985; (**Potluck Yellow**®, Spicy Minijet, Yellow Mini-Wonder); dbl., 20 petals, no fragrance; foliage small, medium green, semi-glossy; bushy growth; [Rise 'n' Shine X Lemon Delight]; Laver, Keith G., 1986

'LAVGOLD', Min, my, 1991; (**Potluck Gold**®); flowers small, dbl., 20 petals, flat, borne in sprays of 3-5, no fragrance; foliage small, medium green, semi-glossy; bushy, low, very dwarf growth; [Dorola X Julie Ann]; Laver, Keith G., 1987; Springwood Roses

'LAVHOMME', Min, w, 1986; (**Bonhomme**); [Blueblood sport]; Laver, Keith G.

'LAVICE', Min, lp, 1983; (**Ice Princess**); flowers light pink, light yellow stamens, finishing white, medium, dbl., no fragrance; foliage small, medium green, matt; [Unnamed pink seedling X Lemon Delight]; Laver, Keith G.

'LAVICTOR', Min, op, 1991; (**Royal Victoria**); flowers coral pink, white reverse, small blooms borne mostly singly, very dbl., moderate fragrance; foliage small, medium green, semi-glossy; tall, upright, bushy growth; [(Painted Doll X June Laver) X Mountie]; Laver, Keith G.; Springwood Roses

'LAVINA', HT, mp, 1962; flowers buff-pink, large, dbl., 60 petals, exhibition form, moderate fragrance; foliage dark, leathery; vigorous growth; [Queen Elizabeth X Anne Letts]; Reynolds, W.H.

Lavina, HT, lp; Strnad; (Czech Rosa Club)

Lavinia *see* 'TANKLEWI'

'LAVINIA HARRISON', HT, mp, 1988; flowers medium, shell pink with light undertones, large, dbl., 30–35 petals, borne in sprays of 1-3, slight fragrance; prickles pyramid, red-brown; foliage red to light green, aging dark green; vigorous growth; [Duet X Unnamed seedling]; Harrison, G.

'LAVISLE', Min, m, 1995; (**Springwood**® **Mauvette**); flowers deep mauve, full (26–40 petals), medium blooms borne mostly single; slight fragrance; numerous prickles; foliage medium, dark green, glossy, spreading, bushy growth; [Blue Ice X Seedling]; Laver, Keith G.; Springwood Roses, 1996

'LAVJACK', Min, or, 1989; (Orange Mini-Wonder, **Potluck Orange**®); bud ovoid; flowers intense, dark orange, compact, full, small, dbl., 24 petals, borne in sprays, no fragrance; fruit not observed, fruit; prickles straight, pointed, light brown; foliage small, medium green, matt; bushy, low, symmetrical growth; PP007326; [Julie Ann X PotluckTM]; Laver, Keith G., 1990

'LAVJOY', Min, pb, 1989; (**Enjoy**); bud pointed; flowers blush pink edged deeper pink, reverse white, small, borne si, dbl., 22 petals, exhibition form, no fragrance; ovoid, orange fruit; prickles straight, very small, sparse, light brown; foliage small, medium green, disease resistant; upright, low growth; [(Moulin Rouge X Unnamed seedling) X Party Girl]; Laver, Keith G.; Springwood Roses, 1990

'LAVJUNE', Min, dy, 1987; (**June Laver**®); flowers dark yellow, aging cream, large, dbl., 20–25 petals, exhibition form, borne usually singly or in small sprays, no fragrance; rounded, light, light orange-red fruit; prickles small, short, green; foliage large, dark green, matt; bushy, medium, compact growth; PP006859; [Helmut Schmidt X Gold Mine]; Laver, Keith G., 1985; Springwood Roses, 1988

'LAVKIN', Min, dy, 1993; (**Yellow Bouquet**); flowers deep yellow, dbl., 15–25 petals, 1.5 in., borne in small clusters, slight fragrance; few prickles; foliage small, medium green, matt; low (15-20 cms), compact growth; [June Laver X Potluck Gold]; Laver, Keith G.; Springwood Roses, 1994

'LAVLAD', Min, dp, 1993; (**Rosy Forecast**); flowers deep pink, blooms borne in small clusters, dbl., 15–25 petals, 1.5 in., moderate fragrance; few prickles; foliage small, dark green, matt; low (15-18 cms), compact growth; [(June Laver X Ontario Celebration) X Potluck Purple]; Laver, Keith G.; Springwood Roses, 1994

'LAVLEMO', Min, my, 1989; bud ovoid; flowers lemon yellow, deeper in center, medium, borne usually singly, dbl., 28 petals, exhibition form; ovoid, green fruit; prickles beige; foliage small, medium green, matt; upright, bushy, medium growth; [Dorola X Genevieve]; Laver, Keith G.

'LAVLINC', Min, dr, 1984; (**Small Slam**); flowers small, dbl., slight fragrance; foliage small, dark, semi-glossy; bushy growth; [Nic Noc X Party Girl]; Laver, Keith G.

'LAVLINGER', Min, ly, 1991; (**Living Bouquet**, Yellow Festival); bud ovoid; flowers full, blooms borne in sprays of 3-5, very dbl., 40–50 petals, flat, no fragrance; foliage medium, light green, semi-glossy; bushy, compact, semi-upright growth; [Loving Touch X (Dorola X Genevieve)]; Laver, Keith G., 1986; Springwood Roses

'LAVLOW', Min, ob, 1991; flowers orange yellow, small blooms borne in small clusters, dbl., no fragrance; few prickles; foliage small, medium green, semi-glossy; low, compact growth; [(Painted Doll X June Laver) X Potluck Yellow (TM)]; Laver, Keith G.; Springwood Roses

'LAVLUV', Min, dy, 1992; (**Golden Promise**); dbl., 26–40 petals, 1.5 in., borne mostly singly, no fragrance; few prickles; foliage small, dark green, glossy; low (30 cms), bushy growth; [June Laver X Potluck Gold]; Laver, Keith G.; Springwood Roses, 1993

'LAVMILNE', Min, lp, 1987; (**Alexander Milne**); flowers light pink, reverse cream, small, borne singly, dbl., 35 petals, exhibition form, no fragrance; rounded, orange fruit; straight, light brown prickles; foliage small, medium green, semi-glossy; bushy, low growth; [Corn Silk X Ice Princess]; Laver, Keith G.; Springwood Roses

'LAVMIN', Min, dr, 1988; (**Potluck Red**™); bud pointed; flowers deep red, compact, full, small, borne singly, dbl., 35 petals, no fragrance; globular, orange-red fruit; no prickles; foliage small, dark green, young edged in red, matt; upright, low growth; [Breezy X June Laver]; Laver, Keith G.; Springwood Roses, 1989

'LAVMING', Min, rb, 1999; (**Striped Festival**™); flowers very rich stripe, deeper than most, dbl., 17–25 petals, 2.5 in., exhibition form, borne mostly singly, slight fragrance; prickles moderate; foliage medium, medium green, semi-glossy; easy-growing, upright, medium (12-15 in.) growth; Laver, Keith G.; Springwood Roses, 1999

'LAVMOTH', Min, pb, 1988; (**Fair Genie**); bud pointed; flowers light orange in center with pink outer petals, reverse pink, dbl., 33 petals, exhibition form, moderate fragrance; ovoid, orange fruit; prickles slender, straight, almost white, translucent; foliage small, medium green, glossy; upright, bushy, low, strong growth; [Breezy X June Laver]; Laver, Keith G.; Springwood Roses, 1990

'LAVMOUNT', Min, or, 1983; (**Ontario Celebration**); flowers small, dbl., 35 petals, moderate fragrance; foliage small, deep reddish green, semi-glossy; spreading, compact growth; [Nic-Noc X (Party Girl X Queen of the Dwarfs)]; Laver, Keith G., 1984

'LAVNEW', Min, op, 1995; (**Springwood Beauty**™); flowers orange pink, blooms borne mostly single, dbl., 26–40 petals, no fragrance; few prickles; foliage small, medium green, semi-glossy; bushy, low growth; [Seedling X Apricot Doll]; Laver, Keith G.; Springwood Roses, 1995

'LAVNOVA', Min, m, 1995; (**Potluck Blue**™); flowers lavender, reverse white, double, blooms borne mostly, single, 15–25 petals, no fragrance; some prickles; foliage medium, medium green, semi-glossy; upright, bushy, medium growth; [Blue Ice X Seedling]; Laver, Keith G.; Springwood Roses, 1995

'LAVONT', Min, op, 1995; (**Touche**); flowers orange-pink, very full (41 + petals), medium blooms borne in small clusters; very fragrant; numerous prickles; foliage medium, medium green, semi-glossy; miniature moss; bushy, medium growth; [Seedling X Seedling]; Laver, Keith G.; Springwood Roses, 1996

'LAVPAINT', Min, ob, 1985; (**Painted Doll**™); flowers orange, reverse yellow, small, dbl., 35 petals, slight fragrance; foliage small, light green, matt; bushy growth; [Party Girl X Dwarfking '78]; Laver, Keith G.

'LAVPERT', F, w, 1991; (**Snow Princess**); dbl., 26–40 petals, 1.5–2.75 in., borne in large clusters, no fragrance; some prickles; foliage large, dark green, glossy; bushy, low (45 cms), compact growth; [Regensberg X June Laver]; Laver, Keith G.; Springwood Roses, 1992

'LAVPET', Min, dy, 1992; ('LAVROY', **Perfect Potluck**™); flowers golden yellow, blooms borne in small clusters, dbl., 15–25 petals, 1.5 in., slight fragrance; few prickles; foliage medium, dark green, glossy; low (30 cms), compact growth; [Showbound X (Party Girl X June Laver)]; Laver, Keith G.; Springwood Roses

'LAVPLAT', Min, m, 1988; (**Platinum Lady**); bud pointed; flowers light lavender, reverse white, borne singly, very slow to open, dbl., 21 petals, exhibition form, no fragrance; ovoid, orange fruit; prickles slender "v" shape, red; foliage small, dark green, young edged in red, semi-glossy; upright, bushy, medium growth; [Unnamed seedling (lavender) X Lavender Jade]; Laver, Keith G.; Springwood Roses, 1989

'LAVPLEASE', Min, mp, 1992; (**Pleasantly Pink**); flowers clear pink, blooms borne in small clusters, very dbl., 1.5–2.75 in., moderate fragrance; some prickles; foliage medium, medium green, matt; medium (30-35 cms), bushy growth; [(June Laver X Rosamini Red) X (Dwarfking X Julie Ann)]; Laver, Keith G.; Springwood Roses

'LAVPOP', Min, w, 1993; (**Party Popcorn**); dbl., 26–40 petals, 1.5 in., borne mostly singly, slight fragrance; few prickles; foliage small, medium green, matt; medium (20-25 cms), bushy growth; [(June Laver X Painted Doll) X Popcorn]; Laver, Keith G.; Springwood Roses, 1993

'LAVPOT', Min, dr, 1984; (**Potluck**); flowers small, dbl., 20 petals, no fragrance; foliage small, medium green, glossy; compact, bushy, tiny growth; [Dwarfking '78 X Little Liza]; Laver, Keith G.

'LAVPUP', Min, dr, 1992; (**Potluck Purple**™); flowers fuchsia, blooms borne in small clusters, dbl., 15–25 petals, 1.5 in., no fragrance; few prickles; foliage small, medium green, matt; micro-mini; low (25-30 cms), compact growth; [(June Laver X Painted Doll) X Springwood Purple]; Laver, Keith G.; Springwood Roses

'LAVPURE', Min, w, 1995; (**Potluck® Frosty**); flowers white, full (26–40 petals), medium (4–7 cms) blooms borne mostly single; no fragrance, some prickles; foliage small, medium green, semi-glossy; low (20–30 cms), bushy, spreading growth; [Seedling X Potluck Yellow]; Laver, Keith G.; Springwood Roses, 1996

'LAVPURR', Min, dp, 1990; (**Springwood Purple**™); flowers fuchsia, loose, sweetheart, medium to large, dbl., 25–30 petals, borne in sprays of 3, no fragrance; foliage large, medium green, semi-glossy; bushy, spreading, medium, very compact growth; [June Laver X (Small Slam X Mountie)]; Laver, Keith G., 1988; Springwood Roses, 1991

'LAVQUEST', Min, lp, 1990; (**Pink Bouquet**, Pink Festival); bud pointed; flowers white to luminous blush pink, white reverse, aging white, very dbl., 50–60 petals, flat, no fragrance; foliage small, dark green, matt; bushy, spreading, low growth; [Loving Touch X Ontario Celebration]; Laver, Keith G., 1987; Springwood Roses, 1991

'LAVRANGE', Min, ob, 1992; (**By Design**, 'LAVSIGN'); flowers orange, blooms borne in small clusters, dbl., 26–40 petals, 1.5–2.75 in., exhibition form, slight fragrance; some prickles; foliage medium, medium green, semi-glossy; medium (40-50 cms), bushygrowth; [(Breezy X Julie Ann) X (June Laver X Painted Doll)]; Laver, Keith G.; Springwood Roses

'LAVRED', Min, dr, 1988; (**Springwood Red**™); flowers deep red, aging slightly lighter, urn-shaped, small, borne u, dbl., 40 petals, no fragrance; ovoid, orange fruit; prickles slightly recurved, light yellow-green; foliage small, medium green, matt; bushy, low growth; [Small Slam X Mountie]; Laver, Keith G.; Springwood Roses, 1990

'LAVRIDE', Min, dr, 1991; (Red Rider, **Red Rover**); bud pointed; flowers loose blooms borne usually singly, dbl., exhibition form, no fragrance; foliage small to medium, dark green, semi-glossy to glossy; bushy, low growth; [Mountie sport]; Laver, Keith G., 1988; Springwood Roses, 1990

'LAVROSY', Min, mp, 1995; (**Rosy Potluck**); dbl., 26–40 petals, 1.5 in., borne in small clusters, moderate fragrance; few prickles; foliage small, dark green, semi-glossy; low (20-30 cms), compact growth; Laver, Keith G.; Springwood Roses, 1995

'LAVROY', Min, dy, 1992; ('LAVPET', **Perfect Potluck**™); flowers golden yellow, blooms borne in small clusters, dbl., 15–25 petals, 1.5 in., slight fragrance; few prickles; foliage medium, dark green, glossy; low (30 cms), compact growth; [Showbound X (Party Girl X June Laver)]; Laver, Keith G.; Springwood Roses

'LAVSANS', Min, dp, 1986; (**Sans Souci**); flowers deep fuchsia pink, fading slightly, medium, borne usually singly, dbl., exhibition form, slight fragrance; oblong, narrow, red fruit; many, very fine, light brown prickles; foliage medium, light green, matt; bushy, medium growth; [Rise 'n' Shine X Ontario Celebration]; Laver, Keith G.; Springwood Roses

'LAVSAPH', Min, dr, 1995; (**Black Sapphire**); dbl., 26–40 petals, borne mostly singly; few prickles; foliage medium, medium green, glossy; patio, upright, medium growth; [Royal Victoria X Seedling]; Laver, Keith G.; Springwood Roses, 1995

'LAVSASK', Min, rb, 1995; (**Maverick**); flowers striped red and white, double, blooms borne mostly, single, 15–25 petals, slight fragrance; some prickles; foliage medium, medium green, dull; upright, bushy, medium growth; [Striped Pet X Apricot Doll]; Laver, Keith G.; Springwood Roses, 1995

'LAVSAT', Min, mp, 1997; (**Springwood Pink Satin**®); flowers full, small, very dbl., 26–40 petals, borne in small clusters, intense fragrance; few prickles foliage medium, medium green, semi-glossy; compact, bushy, low growth [Pleasantly Pink X Springwood Roses]; Laver, Keith G.

'LAVSCENT', Min, or, 1987; (**Springwood Coral**™); flowers coral-pink, turning slightly white at stem, loose, large bo, dbl., 23 petals, intense fragrance round, light orange fruit; prickles curved, red; foliage large, light green matt; bushy, medium growth; [Helmut Schmidt X PotluckTM]; Laver, Keith G. 1986; Springwood Roses, 1990

'LAVSER', Min, pb, 1998; (**Serene Bouquet**); flowers pink with white reverse, 26–40 petals, 1.5–2 in., borne in sprays of 5, slight fragrance; some prickles; foliage medium, medium green, matte; upright, vigorous growth; [Seedling X Antique Gold]; Laver, Keith G.; Springwood Roses, 1998

'**LAVSHO**', Min, lp, 1984; flowers small blooms in large clusters, dbl., 20 petals, slight fragrance; foliage small, light green, glossy; spreading growth; [Mighty Mouse X Fairy Rose]; Laver, Keith G.

'LAVSHRIMP', Min, mp, 1987; (**Wilfrid H. Perron**); flowers clear shrimp-pink, aging paler, rosea, large, borne singly, very dbl., 70–80 petals, no fragrance; round, very rare, green-red fruit; prickles straight out, very slender, greenish-red; foliage very small, medium green, matt; bushy, low growth; [(Dwarfking '78 X Baby Katie) X Painted Doll]; Laver, Keith G., 1988

'LAVSIGN', Min, ob, 1992; (**By Design**, 'LAVRANGE'); flowers orange, blooms borne in small clusters, dbl., 26–40 petals, 1.5–2.75 in., exhibition form, slight fragrance; some prickles; foliage medium, medium green, semi-glossy; medium (40-50 cms), bushygrowth; [(Breezy X Julie Ann) X (June Laver X Painted Doll)]; Laver, Keith G.; Springwood Roses

'**LAVSNO**', Min, w, 1984; dbl., no fragrance; foliage small, medium green, matt; upright, bushy growth; [Ice Princess X Sue Lawley]; Laver, Keith G.

'LAVSNOW', Min, w, 1991; (**Springwood White**™, White Festival); very dbl., 1.5–2.75 in., borne mostly singly, no fragrance; few prickles; foliage small, medium green, matt; low (22 cms), bushy growth; [Loving Touch X June Laver]; Laver, Keith G.; Springwood Roses

'LAVSON', Min, dr, 1997; (**Potluck Crimson**®); flowers very full, medium, very dbl., 41 petals, borne mostly single and in small clusters, no fragrance; few prickles; foliage medium, medium green, semi-glossy; compact, low growth; [June Laver Sport X Springwood Red Victor]; Laver, Keith G.

'LAVSTAR', Min, m, 1989; (**Sweet Butterfly**); bud rounded; flowers mauve pink, pointed, open, loose, medium, star-like form, bo, semi-dbl., 12–15 petals, flat, intense fragrance; globular, yellow fruit; prickles short, pointed, beige; foliage medium, medium green, matt; bushy, low growth; [(Dwarfking X Baby Katie) X (Small Slam X Mountie)]; Laver, Keith G.; Springwood Roses

'LAVSUM', Min, rb, 1997; (**Summer Festival**™); flowers very full, medium, very dbl., 41 petals, borne singly and in small clusters, moderate fragrance; some prickles; foliage medium, dark green, semi-glossy; upright, compact, bushy, medium growth; [Seedling X Painted Doll]; Laver, Keith G.

'LAVTIKI', Min, my, 1999; (**Lemon Bouquet**™); flowers medium yellow, reverse light yellow, dbl., 26–40 petals, 2–2.5 in., exhibition form, borne mostly singly, slight fragrance; no prickles; foliage medium, dark green, glossy; upright, medium (12-15 in.) growth; Laver, Keith G.; Springwood Roses, 1999

'LAVTIPAWS', Min, ob, 1994; (**Tiger Paws**™); flowers orange with yellow reverse, blooms borne mostly singly, dbl., 26–40 petals, 1.5–2.75 in., no fragrance; foliage medium, medium green, matt; medium (20 cms), compact growth; [Painted Doll X Seedling]; Laver, Keith G.; Springwood Roses, 1994

'LAVTIQUE', Min, yb, 1995; (**Antique Gold**); flowers deep chrome yellow tipped with red, dbl. (15–25 petals), small blooms borne mostly single; slight fragrance; some prickles; foliage medium, light green, semi-glossy; spreading, medium growth; [Seedling X Seedling]; Laver, Keith G.; Springwood Roses, 1996

'LAVTREK', Min, dy, 1989; (**King Tut**); bud pointed; flowers rich, deep yellow, reverse medium yellow, small, borne singly, dbl., 45 petals, exhibition form, moderate fragrance; ovoid, orange-red fruit; prickles curved down, light brown; foliage small, dark green, disease resistant; upright, low, compact growth; [June Laver X Genevieve]; Laver, Keith G.; Springwood Roses

'LAVTRICE', Min, or, 1994; (**Orange Zest**); very dbl., 1.5 in., borne mostly singly, slight fragrance; few prickles; foliage small, dark green, semi-glossy; upright, bushy (25 cms) growth; [Seedling X Painted Doll]; Laver, Keith G.; Springwood Roses, 1994

'LAVTRU', Min, dy, 1997; (**True Gold** ™); flowers very full, medium, very dbl., 41 petals, borne mostly singly, no fragrance; few prickles; foliage medium, medium green, dull; upright, bushy, medium growth; [Antique Gold X Yellow Bouquet]; Laver, Keith G.

'LAVTUCH', Min, ob, 1998; (**Painter's Touch**); flowers luminous orange blend, very dbl., 26–40 petals, 1.5–3 in., borne mostly singly, moderate fragrance; prickles moderate; foliage large, medium green, matte; bushy, medium tall growth; [Seedling X Antique Gold]; Laver, Keith G.; Springwood Roses, 1998

'LAVTYNINE', Min, my, 1989; (**Springwood Gold**™); bud pointed; flowers deep, buttery yellow, reverse lighter, medium, borne usually, dbl., 20 petals, exhibition form, slight fragrance; globular, light orange fruit; prickles very narrow, straight, green; foliage small, medium green, semi-glossy; upright, bushy, medium growth; [Rise 'n' Shine X June Laver]; Laver, Keith G.; Springwood Roses

'LAVWYTE', Min, w, 1985; (**Potluck White**™); dbl., 35 petals, no fragrance; foliage small, light green, glossy;

spreading growth; [Baby Katie X Mountie]; Laver, Keith G., 1987

'LAVYIP', Min, w, 1997; (**Ivory Festival**®); flowers full, large, very dbl., 26–40 petals, borne singly and in small clusters, intense fragrance; foliage medium, dark green, glossy; compact, bushy, medium growth; [Seedling X Living Bouquet]; Laver, Keith G.

Lawinia® *see* 'TANKLEWI'

'LAWRENCE JOHNSTON', LCl, my, 1923; (Hidcote Yellow); flowers large, semi-dbl., blooms in clusters, repeat bloom, moderate fragrance; very vigorous, climbing (to 30 ft.) growth; [Mme Eugene Verdier X R. foetida persiana]; Ducher

'LAWRENCE JOHNSTON', HT, yb, 1946; bud pointed; flowers yellow edged red and salmon, large, semi-dbl., moderate fragrance; foliage light green; vigorous, upright, bushy growth; GM, Rome, 1954; [Souv. de Denier van der Gon X Brazier]; San Remo Exp. Sta.

'LAWRENCE OF ARABIA', HT, yb, 1938; flowers indian yellow, flushed coppery rose, large, dbl.; vigorous, bushy growth; Dickson, A.

Laxton's Pink Delight *see* **'PINK DELIGHT'**

'LAXTON'S STANDARD', HT, mp, 1926; flowers clear cerise-pink; Laxton Bros.

Lazy Daze *see* 'TALDAZ'

'LE CHAMOIS', F, 1954; Combe, M.; (Cavriglia)

'LE CID', HRg, mr, 1908; flowers dazzling crimson, large; vigorous growth

Le Cid *see* DORpark

'LE DROIT HUMAIN', HMult, pb, 1907; flowers pink with darker carmine-pink, medium, very dbl.; Vilin; (Sangerhausen)

Le Grande Capitain, B, dr; very dbl.

'LE HAVRE', HP, mr, 1870; flowers vermilion, imbricated, large; vigorous growth; Eudes

Le Jacobin *see* **'MARCEL BOURGOUIN'**

'LE LOBÈRDE', mp, 1800; flowers medium, very dbl.; (Sangerhausen)

'LE LOIRET', Pol, mp, 1920; flowers very brilliant pink to salmon-rose, borne in clusters of 10-15; Turbat

'LE MEXIQUE', HWich, lp, 1912; flowers pale silvery rose, medium, borne in clusters, semi-dbl., slight fragrance; vigorous, climbing growth; [Dorothy Perkins X Marie Pavie]; Schwartz, A.

'LE NANKIN', T, yb, 1871; flowers yellow shaded coppery, well formed, moderate fragrance; Ducher

'LE PACTOLE', T, ly; flowers pale yellow, large, dbl.; [Lamarque X Yellow Tea]; Miellez, before 1841

Le Petit Orange, Pol, ob; Adam

'LE PHOENIX', HGal, mp, 1843; flowers medium carmine-pink, large, very dbl., moderate fragrance; Vibert; (Sangerhausen)

Le Pink *see* 'JACLIP'

Le Poilu, Cl M, lp, 1900; (**'CUMBERLAND BELLE'**); flowers silvery pink, well mossed, small, very dbl., intense fragrance; [Princesse Adélaide sport]; Dreer

'LE PONCEAU', Pol, dr, 1912; flowers deep garnet-red, small, semi-dbl.; [Gruss an Teplitz X Mme Norbert Levavasseur]; Hémeray-Aubert

'LE PROGRÈS', HT, my, 1903; very vigorous growth; Pernet-Ducher

'LE RÊVE', LCl, ly, 1923; bud pointed; flowers pale yellow, large, semi-dbl., moderate fragrance; foliage rich green, glossy; vigorous, climbing growth; [Mme Eugene Verdier X R. foetida persiana]; Pernet-Ducher

'LE RIGIDE', HMult, mp, 1920; flowers neyron pink, borne in clustersof 25-30, semi-dbl.; vigorous growth; Turbat

'LE RIRE NIAIS', C, mp, 1810; Dupony, Andre

'LE ROITELET', B, mp, 1869; flowers small, dbl.; Soupert; (Sangerhausen)

Le Rosier du Jardin ou Cecile, S, op

Le Rosier Eveque, HGal, m; flowers dark violet-red, medium, dbl.; (Sangerhausen)

Le Rosier Pompon Blanc *see* **'ROSE DE MEAUX WHITE'**

Le Rouge et Le Noir *see* DELcart

'LE VÉSUVE', Ch, pb, 1825; flowers carmine shading to pink, large, very dbl.; vigorous growth; (14); Laffay, M.

Lea, F, lp; Urban, J., 1988; (Czech Rosa Club)

'LÉA', HGal, lp; (Lee); flowers blush, shaded flesh, expanded, large, very dbl., abundant early bloom; Vibert, before 1828

'LEA ANN', Gr, w, 1976; bud long, pointed; flowers clear white, dbl., 25–30 petals, 4 in., exhibition form, moderate fragrance; foliage soft; vigorous growth; [Queen Elizabeth X Ivory Fashion]; Patterson; Patterson Roses

'LEA RUBRA', C, mr, 1856

'LEADER', HT, mr, 1924; [Premier sport]; Hill, E.G., Co.

Leading Lady *see* 'JACLOPI'

'LEADING LADY', HT, lp, 1935; flowers flesh-pink, flushed peach-blossom, large, dbl., exhibition form, intense fragrance; foliage deep green, leathery; vigorous, bushy growth; RULED EXTINCT 1/85; GM, NRS, 1934; Dickson, A.

'LEAGUE OF NATIONS', HT, m, 1929; flowers reddish lilac, shaded salmon-pink, semi-dbl., moderate fragrance; [Frau Felix Tonnar X Solliden]; Leenders, M.

Leah, Pol, mr, 1970; Holzman

'LEALAND JEWEL', S, dp, 1963; flowers deep pink, fading quickly, semi-dbl.; (14); [Athabasca seedling]; Erskine

Leana *see* ROElanda

Leander® *see* 'AUSLEA'

LeAnn Rimes *see* 'HARZIPPEE'

Leaping Salmon *see* 'PEAMIGHT'

LEBbet, F, dp; (**Favorite**); (LEBbet)

Lebensfreude, HT, or, 1979; flowers large, dbl.; GPG Bad Langensalza; (Sangerhausen)

'LECBELPA', HT, dr, 1995; (**Belvedere Park**); flowers dark red, edging to black, blooms borne mostly single, dbl., 26–40 petals, 3.5 in., slight fragrance; few prickles; foliage medium, dark green, glossy; medium (4 ft), spreading, bushy growth; [Swarthmore X First Prize]; LeCroy, Jack; K&M Nursery, 1995

'LECBLU', F, or, 1996; (**Blushing June**); flowers orange, vivid stamens, large blooms borne mostly single, semi-dbl., 6–14 petals, slight; some prickles; foliage medium, medium green, semi-glossy; upright, medium growth; [Altissimo X Frances Ashton]; LeCroy, Jack

'LECOCQ-DUMESNIL', HP, mr, 1882; flowers carmine-red, very large, very dbl., moderate fragrance; Verdier, E.; (Sangerhausen)

'LEDA', S, lp, 1960; flowers pale pink, shallow, borne in clusters of 4-5, cupped, intense fragrance; large, apple-like, red fruit; foliage dark; bushy (5 ft) growth; Skinner

'LÉDA', D, w; (Painted Damask); flowers white to blush, edged crimson, medium, dbl., sometimes recurrent bloom, moderate fragrance; foliage dark; compact (3 ft.) growth; hardy.

Lee *see* **'LÉA'**

Leea Rubra, C, lp; flowers medium, dbl., moderate fragrance; (Sangerhausen)

'LEECAL', Min, my, 1992; (**Colleen Little**); dbl., 26–40 petals, 1.5 in., exhibition form, borne usually singly or in sprays of 5-7, slight fragrance; some prickles; foliage small, dark green, semi-glossy, disease resistant; medium (26-30 cms), upright, bushygrowth; [Luis Desamero X Cheyenne]; Little, Lee W.; Oregon Miniature Roses, 1992

Leeds Castle *see* 'TANRUPEZA'

'LEEFIR', LCl, rb, 1992; (**Flashfire**); flowers brilliant coppery red-orange with yellow eye, blooms borne in clusters, single, 5 petals, 3–3.5 in., slight fragrance; some prickles; foliage large, dark green, glossy, disease resistant; tall (210 cms), upright, spreading growth; [Altissimo X Playboy]; Little, Lee W.; Heirloom Old Garden Roses, 1992

'LEENDERS' BERGFEUER', F, mr, 1959; bud long; flowers scarlet, large, semi-dbl., moderate fragrance; foliage dark, glossy; [Independence X Fashion]; Leenders, J.

'LEENDERS' FLAMINGO', F, ob, 1960; bud ovoid; flowers bright coral-peach, open, large, borne in clusters, semi-dbl., moderate fragrance; vigorous, bushy growth; [Cocorico X Ma Perkins]; Leenders, J.

'LEENDERS' PINK', F, mp, 1959; flowers bright pink, large, borne in heavy clusters, dbl., slight fragrance; foliage dark, leathery, glossy; [Goldilocks X Mrs Inge Poulsen]; Leenders, J.

'LEEPEL', Cl HT, mp, 1992; (**Royal Porcelain**); dbl., 26–40 petals, 3–3.5 in., exhibition form, borne in small clusters on strong upright laterals, slight fragrance; some prickles; foliage large, disease resistant; tall (260 cms), upright, spreading growth; [Pele X Altissimo]; Little, Lee W.; Heirloom Old Garden Roses, 1992

Leerder's Harlequin *see* **'THE JESTER'**

Leersum 700 *see* 'INTERLEER'

Lee's Crimson Perpetual *see* **'ROSE DU ROI'**

Lee's Duchess *see* **'LA BELLE DISTINGUEE'**

'LEESHO', LCl, pb, 1992; (**Rainbow Shower**); flowers shrimp pink suffused with yellow and darker pink edge, aging, single, 5 petals, 3–3.5 in., slight fragrance; some prickles; foliage medium, medium green, glossy, disease resistant; tall (210 cms), upright, spreading growth; [Altissimo X Playboy]; Little, Lee W.; Heirloom Old Garden Roses, 1992

'LEGACY', HT, pb, 1963; flowers rose-pink streaked white, large, dbl., intense fragrance; foliage leathery; moderate growth; [Mrs Bryce Allan sport]; Hamilton; Junior Legacy Club

Legacy™ *see* 'JACSLING'

'LEGACY JUBILEE', HT, yb, 1974; bud long, pointed; flowers yellow, edged red, large, dbl., intense fragrance; foliage large, glossy; vigorous, upright growth; [Great Venture X Fred Streeter]; Dawson, George; Neil

LEGcream, F, w, 1998; (**Jilly Cooper**); LeGrice

Legend *see* 'JACTOP'

'LEGENDARY', HT, lp, 1962; bud long, pointed; flowers soft pink, dbl., 55 petals, 5.5 in., exhibition form, moderate fragrance; foliage leathery; vigorous, upright growth; Von Abrams; Peterson & Dering

'LEGGA', HT, my, 1979; (**Dorothe**®, Midas); bud pointed; dbl., 48 petals, borne singly, slight fragrance; large, curved, light brown prickles; foliage glossy, medium green; vigorous, upright growth; [Irish Gold X Dr. A.J. Verhage]; LeGrice, E.B., 1980

'LEGGAB', S, w, 1981; (**Pearl Drift**®); flowers white flushed pink, blooms borne several together, semi-dbl., 18 petals, slight fragrance; small, light brown prickles; foliage glossy, reddish to dark green; vigorous, compact, spreading growth; [Mermaid X New Dawn]; LeGrice, 1980

'LEGGLOW', HT, ob, 1981; (**Can Can**, Can-Can); flowers orange, large, dbl., 24 petals, intense fragrance; foliage large, dark, semi-glossy; bushy growth; [Just Joey X (Superior X Mischief)]; LeGrice, E.B., 1982

'LEGGREY', F, m, 1989; (**Nimbus**); bud pointed; flowers lilac-gray, medium, borne in sprays, very dbl., cupped, slight fragrance; foliage medium, medium green, semi-glossy; bushy, medium growth; [Grey Dawn X Unnamed seedling]; LeGrice, E.B. Roses

'LEGION', HT, dp, 1920; (American Legion); flowers deep cerise-red, dbl., moderate fragrance; [Milady seedling X Hadley]; Towill

Légion d'Honneur® *see* 'DELSAMAR'

'LEGNEED', HT, or, 1988; (**Kathleen Jermyn**); flowers urn-shaped, loose, medium, borne usually singly, dbl., slight fragrance; foliage large, dark green, semi-glossy; bushy, medium growth; [Royal Dane X Alexander]; LeGrice, E.B.; E.B. LeGrice Roses, Ltd.

'LEGNEWS', F, m, 1968; (**News**®); flowers red-purple, blooms in trusses, semi-dbl., moderate fragrance; foliage olive-green, glossy; GM, RNRS, 1970; [Lilac Charm X Tuscany Superb]; LeGrice

LEGram, LCl, dp, 1990; (**Pink Fountain**); LeGrice

'LEGSWEET', F, pb, 1983; (**Deb's Delight**); flowers silvery salmon-pink blend, patio, medium, dbl., 35 petals, moderate fragrance; foliage medium, medium green, semi-glossy; low, bushy growth; [Tip Top X Seedling]; LeGrice, E.B., 1982

Leicester Abbey *see* 'NOSAB'

'LEIGH ANN'™, Min, pb, 1986; flowers pink, reverse cream, yellow center, fading to light pink, medium, dbl., 34 petals, exhibition form, slight fragrance; round, red-orange fruit; bayonet-shaped, pinkish brown prickles; foliage medium, medium green, semi-glossy; upright, medium, vigorous growth; [Poker Chip X Rise 'n' Shine]; Jolly, Marie; Rosehill Farm, 1987

Leigh-Lo *see* 'HARPURL'

Leila *see* 'BERLEI'

'LEILA FRANCIS', HT, pb, 1948; bud ovoid; flowers two-toned pink, large, dbl., exhibition form, intense fragrance; foliage leathery; vigorous, bushy growth; [Earl Haig X Crimson Glory]; Francis; F. Mason

Leila Verde, F, m

'LEIPZIG', HMsk, or, 1939; flowers orange-scarlet, open, borne in clusters, semi-dbl., recurrent bloom; foliage leathery, glossy, wrinkled; long, strong stems; vigorous, bushy growth; [Eva X Mermaid]; Kordes

'LEITRIM GLORY', HT, dy, 1976; flowers deep yellow, full, dbl., 28–30 petals, 5 in., moderate fragrance; foliage bronze, matt green; [Whisky Mac sport]; Hughes

'LELE', HT, dr, 1939; bud pointed; flowers dark red, veined, very large, dbl., 30 petals, moderate fragrance; foliage bright green; strong stems; very vigorous growth; [Marquise d'Andigne X Pres. Herbert Hoover]; San Remo Exp. Sta.

'LELIA LAIRD', Min, ob, 1979; bud long, pointed; flowers orange-red with yellow eye and reverse, blooms borne 1-4 per cluster, dbl., 38 petals, exhibition form, moderate, tea fragrance; long, thin, red prickles; foliage medium green with red edging; upright growth; [Contempo X Sheri Anne]; Bennett, Cecilia 'Dee'; Tiny Petals Nursery, 1980

'LEMANIA', HT, dr, 1937; bud almost black; flowers velvety blackish red, well formed, very large, dbl., intense fragrance; vigorous growth; Heizmann, E.; A. Meilland

'LEMBUT', Min, w, 1980; (**Buttons**); bud globular; dbl., 35 petals, borne singly, no fragrance; no prickles; foliage very tiny, smooth; compact growth; [Seedling X Seedling]; Lemrow, Dr. Maynard W.

'LEMCAR', Min, m, 1985; (**Carl's Rose**); flowers small blooms in sprays, dbl., 35 petals, exhibition form, no fragrance; foliage small, medium green, semi-glossy; low, upright growth; Lemrow, Dr. Maynard W.

'LEMFALL', Cl Min, w, 1988; (**Snowfall**™); flowers decorative, medium, dbl., cupped, borne usually singly or in sprays of 2-5, no fragrance; no fruit; prickles short, pointed, small; foliage medium, medium green, semi-glossy; spreading, tall, profuse, hardy growth; PP006706; [Jeanne Lajoie X Unnamed seedling]; Lemrow, Dr. Maynard W.; Nor'East Min. Roses, 1988

'LEMLIL', Min, pb, 1984; (**Sealily**); flowers white, petals edged pink, small, dbl., 35 petals, no fragrance; foliage medium, dark, semi-glossy; [Max Colwell X ?]; Lemrow, Dr. Maynard W.

'LEMMAR', Min, pb, 1981; (**Maryellen**); flowers deep pink, yellow center, blooms borne 3 or more per cluster, single, 5 petals, no fragrance; triangle-shaped prickles; foliage small, deep green; upright growth; [Darling Flame X Unnamed seedling]; Lemrow, Dr. Maynard W.

'LEMON BEAUTY', HT, w, 1932; flowers creamy white, quickly fading to paper-white, base lemon-yellow, moderate, fruity fragrance; foliage light; Cant, B. R.

Lemon Blush *see* SIElemon

Lemon Bouquet™ *see* 'LAVTIKI'

'LEMON CHIFFON', HT, my, 1954; bud long, pointed; flowers lemon-yellow, to open, dbl., 40 petals, 3–4 in., exhibition form, intense, spicy fragrance; compact, bushy growth; RULED EXTINCT 9/86; [Soeur Thérèse X Golden Dawn]; Swim, H.C.; Arp Nursery Co.

'LEMON DELIGHT', Min, my, 1978; bud mossy, long, pointed; flowers mini-moss, small, semi-dbl., 10 petals, 1.5 in., slight fragrance; bushy, upright growth; [Fairy Moss X Goldmoss]; Moore, Ralph S.; Sequoia Nursery

Lemon Dream *see* KORamvis

'LEMON DROP', Min, ly, 1954; dbl., 0.75 in.; very prickly; foliage very small; dwarf 6 in) growth; [(R. wichurana X Floradora) X Zee]; Moore, Ralph S.; Sequoia Nursery

Lemon Drop *see* 'WEKYEGI'

'LEMON ELEGANCE', HT, my, 1960; bud long, pointed; flowers lemon-yellow, well-shaped, dbl., 38 petals, 4.5–5 in., moderate fragrance; foliage leathery; vigorous, tall growth; Jones; Hennessey

Lemon Fluff *see* 'CURLEM'

Lemon Gems™ *see* 'JACMIRYL'

'LEMON GLOW', HT, my, 1964; bud long, pointed; flowers lemon-yellow, dbl., 55 petals, 6–7 in., exhibition form, moderate fragrance; foliage soft; vigorous, upright growth; [Sunlight X Golden Masterpiece]; Schwartz, Ernest W.; Wyant

Lemon Honey *see* DICkindlel

'LEMON ICE', HT, my, 1960; bud pointed; flowers lemon-yellow, dbl., 65 petals, 4 in., slight fragrance; foliage leathery, dark; vigorous, bushy, compact growth; [Leonard Barron sport]; Kern Rose Nursery

Lemon Ice, Min, lp; [Oz Gold X Lemon Delight]; Hannemann, F., 1990; The Rose Paradise; (Weatherly, L.)

Lemon Light, LCl, ly

Lemon Meringue *see* 'JUSLEMON'

'LEMON MIST', Min, w, 1991; bud ovoid; flowers white with yellow center, aging white, medium, very dbl., 50 petals, exhibition form, slight, fruity fragrance; foliage medium, medium green, matt; upright growth; [Rise 'n' Shine X Unnamed seedling]; Gruenbauer, Richard, 1984; Flowers 'n' Friends Miniature Roses, 1993

'LEMON OPHELIA', HT, my, 1922; flowers lemon-yellow, dbl., moderate fragrance; [Ophelia sport]; Leenders, M.

Lemon Pearl *see* 'MICPEARL'

Lemon Pillar *see* **'PAUL'S LEMON PILLAR'**

'LEMON QUEEN', Cl HT, w, 1912; flowers creamy white, large, dbl., moderate fragrance; Hobbies; (Sangerhausen)

Lemon Rosamini, Min, ly

'LEMON SHERBET', HT, ly, 1973; bud ovoid; flowers white, center light yellow, dbl., 35 petals, 4 in., exhibition form, slight fragrance; foliage large, leathery; upright growth; [Florence sport]; Kern Rose Nursery

'LEMON SPICE', HT, ly, 1966; bud long, pointed; flowers large, dbl., exhibition form, intense fragrance; foliage dark, leathery; vigorous, spreading growth; [Helen Traubel X Seedling]; Armstrong, D.L.; Armstrong Nursery

'LEMON SURPRISE', F, my, 1978; flowers lemon, semi-dbl., 12 petals, moderate, spicy fragrance; foliage glossy; low, upright growth; [Allgold X Elizabeth of Glamis]; Slack

Lemon Swirl *see* 'RENSWIRL'

Lemon Time, HT, my, 1985; Kasturi

Lemon Twist *see* 'FOUTWIST'

'LEMON YELLOW', F, my, 1977; flowers lemon-yellow blooms, dbl., 21 petals, 2 in., slight fragrance; foliage dark; bushy growth; [Orange Sensation X King's Ransom]; Gandy, Douglas L.

'LEMONADE', F, w, 1974; flowers cream, dbl., 25 petals, 4.5 in., slight fragrance; foliage matt; [Nancy West sport]; Haynes Roses

'LEMSEA', Min, mp, 1976; (**Seabreeze**); bud short, pointed; dbl., 35 petals, 1 in., slight fragrance; foliage light to medium green; very vigorous, bushy growth; [White Fairy X ?]; Lemrow, Dr. Maynard W.; Nor'East Min. Roses

'LEMSUR', Min, ob, 1984; (**Surf's Up**); flowers orange, yellow reverse, small blooms borne singly, dbl., 35 petals, exhibition form, no fragrance; foliage small, medium green, semi-glossy; upright growth; [Avandel X Unnamed seedling]; Lemrow, Dr. Maynard W.

'LEMSWI', Min, rb, 1987; (**Swiss Lass**); flowers red, reverse almost all white, medium, very dbl., slight fragrance; foliage small, medium green, semi-glossy; upright growth; [Unnamed seedling X Unnamed seedling]; Lemrow, Dr. Maynard W.

'LEN 1', S, pb, 1985; (**Bouquet Fait**); flowers medium pink, white eye blooms in clusters of 1-5, semi-dbl., 2 in., non-recurrent, intense fragrance; foliage grayish green, hairy; bushy (to 5 ft.) growth; [R. mollis X Complicata]; Lens, Louis

'LEN 2', Gr, dr, 1966; (**Epoca**); flowers very dark red, large blooms in sprays of 3-18, dbl., 45 petals, exhibition form, no fragrance; brown-green prickles; foliage very dark; upright, bushy growth; [Seedling X Seedling]; Lens, Louis

'LEN 3', F, op, 1980; (**Favori**, Favorite); flowers light salmon-pink, urn-shaped blooms in clusters of 8-24, dbl., 23 petals, 3 in., intense fragrance; hooked, brownish-green prickles; foliage dark; large, bushy growth; [Unnamed seedling X Unnamed seedling]; Lens, Louis

'LEN 4', Min, m, 1984; (**Jet Spray**®, 'LENCARA'); flowers purple-pink, rosette, small, dbl., 21 petals, slight fragrance; no prickles; foliage very small, dark; bushy, spreading growth; [New Penny X Unnamed seedling]; Lens, Louis

Len Turner *see* 'DICJEEP'

'LENA', T, ab, 1906; flowers apricot, edged yellow; Dickson, A.

Lena, HT, or, 1981; deRuiter

'LENANA', S, lp, 1984; (**Flash**); dbl., 22 petals, 1.5 in., blooms in clusters of 3-50, slight fragrance; hooked, green prickles; foliage small, dark; bushy, spreading growth; [R. multiflora nana X Unnamed seedling]; Lens, Louis

'LENBLANK', Min, w, 1985; ('LENVIR', **White Dream**®); flowers small blooms in clusters of 3-28, dbl., 36 petals, exhibition form, slight fragrance; hooked, greenish-brown prickles; low, bushy growth; [Unnamed seedling X Unnamed seedling]; Lens, 1982

'LENBRAC', HRg, mp, 1987; (**Pink Surprise**); Lens

'LENBRU', HT, dy, 1980; (**Bruocsella**®); flowers golden yellow, urn-shaped, large, dbl., 35 petals, exhibition form, borne singly or in small clusters, intense fragrance; dark green prickles; upright, bushy growth; [Peace X Golden Garnette]; Lens, Louis

'LENCAN', Min, ly, 1984; (**Petit Canard**®); flowers small blooms in clusters of 12-22, dbl., 20 petals, exhibition form, slight fragrance; few, green prickles; foliage small; bushy growth; GM, Paris, 1984; [Rosina X (Le Vesuve X Belle Étoile)]; Lens

'LENCARA', Min, m, 1984; (**Jet Spray**®, 'LEN 4'); flowers purple-pink, rosette, small, dbl., 21 petals, slight fragrance; no prickles; foliage very small, dark; bushy, spreading growth; [New Penny X Unnamed seedling]; Lens, Louis

'L'ENCHANTRESSE', HGal, mp, 1829; (Grande Henriette); flowers clear pink, fading with age, large, dbl.; François, 1824

'LENDA', Pol, rb, 1980; (**Anda**); flowers dark red, white eye, single, 5 petals, 2 in., blooms in clusters of 18-24, moderate, fruity fragrance; purple prickles; bushy growth; [(Britannia X R. moschata) X (Little Angel X Europeana)]; Lens

'LENDADI', S, m, 1985; (**Violine**); flowers lilac, pink, white blend blooms in clusters of 5-28, dbl., 20 petals, 2 in., moderate, fruity fragrance; few, small, brown prickles; foliage light green; upright, bushy growth; [(Little Angel X Picasso) X Skyrocket]; Lens

'LENDEC', F, or, 1980; (**Idée Fixe**®); flowers light orange-red, blooms in clusters of 3-18, 15 petals, 2 in., slight fragrance; hooked, light red prickles; foliage small, dark; low, bushy growth; [Unnamed seedling X (Unnamed seedling X Floradora)]; Lens, Louis

'LENDORI', F, op, 1985; (**Erna Doris**); flowers medium salmon pink, blooms in clusters of 3-12, dbl., 24 petals, exhibition form, slight fragrance; small, hooked, brown-green prickles; foliage small, medium green; upright, bushy growth; [Little Angel X Elizabeth of Glamis]; Lens, Louis

LENfil, S, lp, 1988; (**Fil d'Ariadne**®); Lens

'LENGRA', S, w, 1985; (**Pleine de Grâce**); flowers creamy white, small blooms in clusters of 12-24, single, 5 petals, non-recurrent, intense fragrance; foliage yellowish-green; bushy, spreading growth; [Ballerina X R. filipes]; Lens, 1983

'LENI NEUSS', HT, pb, 1933; bud pointed; flowers hydrangea-pink, reverse reddish old-rose, very large, dbl., moderate fragrance; very vigorous, bushy growth; [Lilly Jung X Baronesse M. van Tuyll van Serooskerken]; Leenders, M.; C-P

Lenidora, S, 1990; Lens, Louis; (Cavriglia)

'LENIP', HT, w, 1963; (Blanche Pasca®, **Pascali**®); flowers creamy white, well-formed, medium, dbl., 30 petals; foliage dark; vigorous, bushy growth; All-American Rose Selection, AARS, 1969 GM, Portland, 1967 GM, The Hague, 1963 Hall of Fame, WFRS, 1991; [Queen Elizabeth X White Butterfly]; Lens; A. Dickson, 1963;, Armstrong Nursery, 1968

'LENIRA', F, or, 1984; (**Iranja**®); flowers orange-red blooms in clusters of 3-24, dbl., 35 petals, 2 in., no fragrance; hooked, red-green prickles; foliage small, brilliant green; low, bushy growth; [Little Angel X (Floradora X Angelina Louro)]; Lens, Louis

'LENISUR', S, lp, 1985; (**Merle Blanc**); flowers blush pink blooms in clusters of 3-22, dbl., 22 petals, 2 in., recurrent bloom, intense fragrance; hooked, brownish-green prickles; foliage dark; tall, spreading growth; [Ballerina X Surf Rider]; Lens

'LENIVILL', S, dr, 1983; (**Ville d'Ettelbruck**); flowers deep red blooms in clusters of 5-18, dbl., 20 petals, 2 in., moderate, fruity fragrance; hooked, green prickles; foliage dark, leathery; upright, bushy growth; [Satchmo X Skyrocket]; Lens

Lenka, F, lp; Vecera, L., 1968; (Czech Rosa Club)

'LENLIT', F, lp, 1984; (**Little Shrimp**®); flowers light shrimp-pink, rosette blooms in clusters of 3-24, dbl., 40 petals, 2 in., slight fragrance; hooked, brownish-red prickles; foliage small; bushy growth; [Little Angel X (Little Angel X Spartan)]; Lens, Louis

'LENMACRA', S, lp, 1984; (**Maria Teresa**®); dbl., 28 petals, cupped, blooms in clusters of 32, recurrent bloom, intense fragrance; hooked, brownish-green prickles; foliage light green; bushy growth; [Unnamed seedling X R. macrantha]; Lens, Louis

'LENMAR', Min, w, 1980; (**Maria Mathilda**®); flowers white, shaded pink, small blooms in clusters of 3-24, dbl., 24 petals, intense fragrance; hooked, brownish-red prickles; foliage very dark, glossy; upright, bushy growth; Golden Rose, The Hague, 1981 Rose d'or a La Haye, 1981; [Unnamed Min seedling X (New Penny X Jour de Fete)]; Lens, Louis

'LENMAU', F, m, 1984; (**Bluesette**®); flowers lilac, blooms in clusters of 3-18, dbl., 50 petals, 2 in., flat, slight fragrance; dark green prickles; bushy growth; [Little Angel X (Westmauve X Blue Diamond)]; Lens, Louis

'LENMOBAR', S, mr, 1984; (**Vivaldi**); single, 5 petals, 1 in., blooms in clusters of 7-50; very hooked, greenish-brown prickles; foliage small; bushy, spreading growth; [R. multiflora seedling X (Unnamed seedling X Robin Hood)]; Lens

'LENMOBRI', S, pb, 1983; (**Rush**®); flowers pink, white eye blooms in clusters of 3-32, single, 5 petals, 2 in., recurrent bloom, moderate, fruity fragrance; hooked, brownish-green prickles; foliage light green; upright, bushy growth; GM, Bagatelle, 1986 GM, Monza, 1982 GM, Rome, 1982; [(Ballerina X Britannia) X R. multiflora]; Lens

'LENMOL', Min, m, 1980; (**Si Bemol**®); flowers lilac blue and white, small blooms in clusters of 3-24, dbl., 30 petals, slight fragrance; few, green prickles; foliage dark; bushy growth; [(Little Angel X Le Vesuve) X Mr Bluebird]; Lens, 1983

'LENMOR', S, pb, 1984; (**Schubert**®); flowers pink, white-eyeblooms in clusters of 7-60, single, 5 petals, 1 in., recurrent bloom, slight fragrance; very hooked, greenish-brown prickles; foliage small; bushy, spreading growth; [Ballerina X R. multiflora]; Lens

LENneel, HMsk, mp, 1997; (**Heavenly Pink**); Lens

'LENNIRO', F, op, 1982; (**Jupon Rose**®); flowers light salmon pink, blooms in clusters of 3-24, dbl., 28 petals, cupped, slight fragrance; hooked, reddish-green prickles; foliage dark; upright, bushy growth; [Little Angel X Pernille Poulsen]; Lens, Louis

'LENNOP', Min, mr, 1984; (**Little Pim**®); flowers rosette, small blooms in clusters of 3-32, dbl., 24 petals, no fragrance; hooked, greenish-brown prickles; foliage dark; bushy growth; [Unnamed Miniature seedling X Ruth Leuwerick]; Lens, Louis

'LENOBIT', HMsk, pb, 1982; (**Poesie**); flowers white shaded pink blooms in clusters of 5-32, dbl., 20 petals, 2 in., intense fragrance; hooked, light brown prickles; foliage large, leathery, dark; bushy, spreading growth; [Ballerina X Moonlight]; Lens, Louis, 1985

'LENOR', F, ob, 1980; (**Festivity**); flowers orange blooms in clusters of 3-24, dbl., 24 petals, 3 in., slight fragrance; hooked, green prickles; foliage dark; bushy growth; [Unnamed seedling X Unnamed seedling]; Lens, Louis

'LENORAN', HT, lp, 1984; (**Eurosong**®); dbl., 38 petals, exhibition form, borne singly or in three's, slight fragrance; large, brownish-red prickles; foliage dark reddish-green; upright growth; [(Queen Elizabeth X Seedling) X Queen Elizabeth]; Lens, Louis

'LENPAC', HMsk, lp, 1984; (**Focus**); flowers small blooms in clusters of 3-50, dbl., 20 petals, recurrent bloom, intense fragrance; hooked, brownish-green prickles; foliage small, dark; spreading growth; [Marie Pavie X Unnamed seedling]; Lens, Louis

LENpaga, S, mr; (**Harley**®); Lens

'LENPAS', Min, w, 1984; (**Pascaline**®); flowers urn-shaped, blooms in clusters of 3-24, dbl., 35 petals, exhibition form, no fragrance; slightly hooked, brownish-red prickles; foliage dark gray-green; upright, bushy growth;

[Unnamed Miniature seedling X (New Penny X Jour de Fete)]; Lens

LENpaya, S, w; (**White Spray**®); Lens, 1980

'LENPEN', Min, m, 1984; (**Jet Flame**®, Jet Flame Nirpaysage®); flowers lavender-purple, rosette, micro-mini, small blooms in clusters, dbl., 30 petals, no fragrance; no prickles; foliage very small, brilliant dark green; bushy, spreading growth; GM, Paris, 1984; [New Penny X Violet Hood]; Lens, Louis

'LENPIC', F, op, 1982; (**Tache de Beauté**®); dbl., 22 petals, 2 in., exhibition form, blooms in clusters of 3-24, slight fragrance; hooked, brown prickles; foliage dark; upright, bushy growth; [Little Angel X Picasso]; Lens

LENpil, Min, mp, 1982; (**Pink Delight**®); Lens

'LENPLAT', S, pb, 1982; (**Tapis Volant**®); flowers white-pink blend blooms in clusters of 7-35, semi-dbl., 2 in., recurrent bloom, moderate, fruity fragrance; hooked, reddish-brown prickles; foliage reddish green; groundcover; spreading growth; GM, Kortrijk, 1987; [(R. luciae X ?) X (R. multiflora X Ballerina)]; Lens

'LENPON', Min, or, 1980; (**Asterix**®); flowers small, dbl., 65 petals, flat, blooms in clusters of 12-22, no fragrance; hooked, brown prickles; bushy growth; [(Little Red X (Little Angel X Robin Hood)) X Idée Fixe]; Lens

'LENRAG', F, r, 1980; (**Ragtime**); flowers brownish red, blooms in clusters of 7-18, dbl., 25 petals, exhibition form, slight fragrance; foliage brownish-green; low, bushy growth; [Little Angel X Goldtopas]; Lens, Louis

'LENRAMP', S, m, 1985; (**Running Maid**®); flowers lilac-red, white eye blooms in clusters of 3-32, single, 5 petals, 2 in., recurrent bloom, intense fragrance; hooked, brown prickles; foliage reddish-green; spreading growth; groundcover; [R. multiflora X (R. wichurana X Violet Hood)]; Lens, 1982

'LENRAV', HMsk, pb, 1988; (**Ravel**®); Lens

LENray, S, pb, 1980; (**Pink Spray**®); Lens

'LENSIM', F, m, 1982; (**Eagle Wings**®); flowers white, shaded lilac blooms in clusters of 3-24, single, 5 petals, 2 in., moderate, spicy fragrance; hooked, brown prickles; foliage very dark; bushy growth; [Seedling X Picasso]; Lens, Louis

'LENSUN', Min, yb, 1983; (**Sunnyside '83**); flowers yellow, spotted red, small blooms in clusters of 3-22, dbl., 20 petals, moderate, fruity fragrance; small, hooked, green prickles; foliage small; bushy growth; [Little Angel X (Rosina X Unnamed seedling)]; Lens, Louis

'LENTINI', F, lp, 1985; (**Tineke van Heule**®); 7–12 petals, blooms in clusters of 3-12, slight fragrance; few, reddish-brown prickles; foliage large, leathery, greenish-brown; upright, bushy growth; [Unnamed seedling X City of Belfast]; Lens

'LENVIE', Pol, dp, 1980; (**View**®); flowers deep pink, blooms in clusters of 3-32, 7–15 petals, 2 in., no fragrance; hooked, brown prickles; foliage greenish-brown; low, bushy growth; GM, The Hague, 1978; [Britannia X R. multiflora]; Lens

'LENVIR', Min, w, 1985; ('LENBLANK', **White Dream**®); flowers small blooms in clusters of 3-28, dbl., 36 petals, exhibition form, slight fragrance; hooked, greenish-brown prickles; low, bushy growth; [Unnamed seedling X Unnamed seedling]; Lens, 1982

'LENWAY', HT, yb, 1982; (**Waverland**®); flowers petals deep yellow, edged orange-red, large, borne singly, dbl., 30 petals, exhibition form, intense fragrance; brown prickles; foliage dark, leathery; upright, bushy growth; GM, Monza, 1980; [Peer Gynt X Peace]; Lens

'LENWICH', S, w, 1985; (**Green Snake**®, 'LENWIGA', Serpent Vert®); flowers pure white, small blooms in clusters of 3-24, single, 5 petals, non-recurrent; hooked, light brownish-green prickles; foliage small, spoon-shaped; groundcover growth; [R. arvensis X R. wichurana]; Lens, Louis, 1987

'LENWIGA', S, w, 1985; (**Green Snake**®, 'LENWICH', Serpent Vert®); flowers pure white, small blooms in clusters of 3-24, single, 5 petals, non-recurrent; hooked, light brownish-green prickles; foliage small, spoon-shaped; groundcover growth; [R. arvensis X R. wichurana]; Lens, Louis, 1987

Lenzburger Duft, HT, mr, 1985; Schaiber

'LEOBRETAK', HT, dp, 1982; (**Breathtaking**); bud ovoid, long, pointed; flowers deep pink, dbl., 40 petals, exhibition form, intense, damask fragrance; foliage medium to large, medium green, semi-glossy; upright, bushy growth; [Pink Silk seedling X (Pink Parfait X Wendy Cussons)]; Leon, Charles F., Sr.

'LEOCOK', Min, pb, 1985; (**Cockle Shells**); flowers pale yellow tinged deep pink, medium, dbl., 35 petals, slight fragrance; foliage medium; bushy, spreading growth; [(Kathy Robinson X Seedling) X (Janna X Seedling)]; Leon, Charles F., Sr.

'LEOCREL', HT, dr, 1990; (**Crimson Elegance**); bud rounded; flowers crimson/scarlet, crimson reverse, aging darker red, dbl., 40 petals, exhibition form, borne usually singly, moderate fragrance; foliage medium, medium green, semi-glossy; upright, bushy, tall growth; [Big Red X Swarthmore]; Leon, Charles F., Sr.; Oregon Grown Roses, 1990

'LEOEXBEAU', HT, op, 1990; (**Exotic Beauty**); bud rounded; flowers orange blend, reverse pink blend, large blooms borne usually singly, dbl., exhibition form, slight, fruity fragrance; foliage large, dark green, glossy; upright, bushy, medium growth; [Silver Jubilee (surmised) X Undetermined seedling]; Leon, Charles F., Sr.; Oregon Grown Roses, 1990

'LEOFIRE', Min, or, 1985; (**Forest Fire**); flowers medium, dbl., slight fragrance; foliage medium, medium to dark, semi-glossy; bushy growth; [((Sheri Anne X Starina) X (Persian Princess X Starina)) X ((Sheri Anne X Persian Princess) X Starina)]; Leon, Charles F., Sr.

'LEOGLO', HT, ab, 1985; (**Glorious**); flowers medium pink tinted apricot, well-formed, large, dbl., 37 petals, slight fragrance; foliage medium to large, medium green, semi-glossy; upright, bushy growth; [Unnamed seedling X Mirato]; Leon, Charles F., Sr.

'LEOLAVBLU', HT, m, 1982; (**Lavender Blue**); flowers lavender, darker edges, large, dbl., exhibition form, intense fragrance; foliage medium, dark, semi-glossy; upright, bushy growth; [Silver Star X Kolner Karneval]; Leon, Charles F., Sr.

'LEOLOYROS', HT, mr, 1990; (**Loyal Rosarian**); bud ovoid; flowers medium red becoming darker red as it ages, large blooms borne usually singly, dbl., 36–42 petals, exhibition form, intense fragrance; large fruit, germinates rapidly; foliage medium, leathery, green; upright, bushy, tall growth; [Red Planet X Red Devil (surmised)]; Leon, Charles F., Sr.; Oregon Grown Roses, 1990

'LÉON CHENAULT', HT, pb, 1931; flowers carmine-rose shaded salmon, base deeper, very large, dbl., intense fragrance; foliage dark; very vigorous growth; Pernet-Ducher; Gaujard

'LÉON DELAVILLE', HP, dr, 1885; flowers large, dbl.; Verdier, E.; (Sangerhausen)

Leon Lecomte, D, dp

'LÉON RENAULT', HP, mr, 1878; flowers large, very dbl.; Lédéchaux; (Sangerhausen)

'LÉON ROBICHON', HP, w, 1901; flowers large, dbl., moderate fragrance; Robichon; (Sangerhausen)

'LEONARD BARRON', HT, op, 1931; bud pointed; flowers salmon-copper and shell-pink, very large, dbl., moderate fragrance; foliage leathery; bushy growth; David Fuerstenberg Prize, ARS, 1933; [Schoener's Nutkana X Souv. de Mme Boullet]; Nicolas; C-P

Leonard Dudley Braithwaite *see* 'AUSCRIM'

Leonardo de Vinci® *see* 'MEIDEAURI'

'LÉONCE COLOMBIER', HT, mr, 1943; bud oval; flowers brilliant geranium-red, stamens yellow, medium, dbl., cupped, slight fragrance; foliage leathery; very vigorous, bushy growth; [Charles P. Kilham X (Charles P. Kilham X Capucine Chambard)]; Meilland, F.; A. Meilland

'LEONHART', HT, mr, 1982; (**Heart Throb**); dbl., 37 petals, exhibition form, borne singly and several together, moderate, damask fragrance; foliage medium to large, medium green, leathery; vigorous, bushy, tall growth; [Norita X ((Norita X Unnamed seedling) X Papa Meilland)]; Leon, Charles F., Sr.

Leonidas *see* MEIcofum

'LEONIE', F, mp, 1964; (Dazzler); dbl.; vigorous growth; [Queen Elizabeth X Circus]; Leenders, J.

'LÉONIE LAMBERT', HT, lp, 1913; flowers silver-pink, shaded yellow and flesh, well shaped, large, moderate fragrance; very vigorous growth; [Frau Karl Druschki X Prince de Bulgarie]; Lambert, P.

'LÉONIE LAMESCH', Pol, ob, 1899; flowers light coppery red, center yellow, edges flecked darker, semi-dbl.; foliage rich green, soft; vigorous, bushy growth; [Aglaia X Kleiner Alfred]; Lambert, P.

Leonie's Appoline, HT, dp

'LÉONOR DE MARCH', HT, dr, 1957; bud long, pointed; flowers deep blood-red, large, dbl., moderate fragrance; foliage glossy; upright growth; GM, Rome 1958; [J.M. Lopez Pico X Poinsettia] Camprubi, C.

'LÉONORA', HT, mr, 1921; flowers brilliant velvety red, center brighter, dbl.; Paul W.

'LEONORO', HT, my, 1982; (**Solid Gold**); flowers golden yellow, large, borne 3-6 per cluster, dbl., 36 petals, intense fragrance; foliage dark, glossy; vigorous growth; [(Royal Gold seedling X Golden Giant) X (Bright Gold X PhyllisGold seedling)]; Leon, Charles F., Sr.

'LÉONTINE CONTENOT', HT, yb, 1935; bud pointed; flowers sunflower-yellow bordered pink and yellow, large, dbl., 45–50 petals, exhibition form; foliage quaker green; vigorous growth; [Joanna Hill X Souv. de Claudius Pernet]; Ketten Bros.

'LÉONTINE GERVAIS', LCl, ab, 1903; bud coppery red; flowers salmon-orange and yellow, large blooms in clusters of 3-10, non-recurrent, moderate fragrance; foliage dark, glossy; very vigorous, climbing growth; [R. wichurana X Souv. de Catherine Guillot]; Barbier

'LÉOPOLD I', HP, dr, 1863; flowers deep red, well formed, large, dbl.; [Général Jacqueminot seedling]; Van Assche

'LÉOPOLD LAMBOTTE', HT, mr, 1944; bud long, pointed; flowers scarlet-red, moderate fragrance; very tall growth; [Grenoble X National Flower Guild]; Meilland, F.

'LEOPOLD RITTER', HMult, mr, 1900; Geschwind, R.

'LÉOPOLDINE D'ORLÉANS', HSem, w, 1828; flowers white shaded rose, medium, dbl.; Jacques

'LEORA STEWART', HT, op;, long continued bloom, intense fragrance; Wilber

'LEOSILMU', HT, m, 1982; (**Silvery Moon**); flowers lilac, large, dbl., 35 petals, intense fragrance; foliage medium, medium green, leathery, matt; upright growth; [Silver Star X Blue Moon]; Leon, Charles F., Sr.

'LEOSTRA', Min, w, 1985; (**Strawberry Sundae**); flowers ivory flushed pink, medium, dbl., 35 petals, slight fragrance; foliage medium, light to medium green, semi-glossy; bushy, spreading growth; [(Kathy Robinson X Seedling) X (Kathy Robinson X Seedling)]; Leon, Charles F., Sr.

'LEOTILDE MINGUEZ', HT, rb, 1961; flowers carmine, shaded vermilion and yellow, large, dbl., 30 petals; upright growth; [Pres. Herbert Hoover X Vicky Marfa]; Dot, Pedro

'LEOTOELG', Min, w, 1988; (**Touch of Elegance**); bud pointed; flowers white with creamy yellow center, large, bornein sprays of 1-, dbl., exhibition form, slight fragrance; foliage medium, medium green, semi-glossy; upright, tall growth; [(Gavotte X Buccaneer) X Unnamed Miniature seedling]; Leon, Charles F., Sr.; John Carrigg, 1987

'LEOVELV', HT, rb, 1988; (**Touch of Velvet**); bud pointed, long; flowers magenta red with lighter tones, borne usually singly and in, dbl. 31 petals, exhibition form, moderate fragrance; foliage medium, dark green, semi-glossy; upright, tall growth; [(First Prize X Gypsy) X Unnamed seedling]; Leon, Charles F., Sr.; John Carrigg, 1987

'LEPRECHAUN', F, yb, 1971; bud ovoid; flowers red, reverse yellow, small, dbl., globular, moderate fragrance; foliage glossy; growth moderate, upright; [(Easter Parade X Masquerade) X Little Darling]; Adams, M.R.

'LES AMIS DE LILLE', HT, pb, 1928; bud lincoln red on pale buff ground; flowers pale buff, edges and reverse suffused salmon-pink; [Golden Emblem X Pres. Bouche]; Ketten Bros.

'LES AMIS DE TROYES', HT, mp, 1935; flowers china-pink, passing to carmine-pink, base ochre-yellow, well-formed, dbl.; vigorous growth; [Feu Joseph Looymans X Unnamed seedling]; Vially

Les Amoureux de Peynet® *see* MEItobla

'LES ROSATI', HFt, mr, 1906; flowers bright carmine; Gravereaux

Les Saisons d'Italie, HGal, mr; flowers carmine-red, medium, very dbl., intense fragrance; (Sangerhausen)

'LES SJULIN', Gr, pb, 1981; flowers coral pink, reverse light red, urn-shaped to imbricated blooms, dbl., 28 petals, moderate, old rose fragrance; awl-like prickles; foliage medium, dark olive green, leathery; erect, bushy growth; [Country Dancer X ((Dornroschen X Peace) X Pink Peace)]; Buck, Dr. Griffith J.; Iowa State University

'LES SYLPHIDES', HT, op, 1960; flowers pink shading to yellow and orange, well shaped, intense fragrance; foliage reddish; long stems; very vigorous growth; [Margaret sport]; Watkins Roses

Lesbos, LCl, lp; Peden, R., 1998; (Weatherly, L.)

Lesdain, HT, ob, 1989; RvS-Melle

'LESJA JEAN', HT, lp, 1990; bud pointed; flowers large blooms borne usually singly, very dbl., 37 petals, exhibition form, moderate fragrance; prickles medium, light; foliage medium, medium green, semi-glossy; upright, medium growth; [Unnamed seedling X Unnamed seedling]; Weddle, Von C., 1989

Lesja Ukrajinka, HT, lp, 1964; flowers medium, dbl., moderate fragrance; Lempickij; (Sangerhausen)

Lesley Anne *see* 'FRYMINILES'

'LESLEY DUDLEY', HT, op, 1932; (Leslie Dudley); flowers warm carmine-pink shaded orange, well shaped, large, dbl., intense fragrance; foliage dark; vigorous, bushy growth; GM, NRS, 1931; McGredy

'LESLEY JOHNS', HT, dp, 1972; flowers deep pink, high-pointed, dbl., 32 petals, 5–6 in., moderate fragrance; foliage dark, glossy; [Soraya X ?]; Gregory

Leslie *see* 'TINLESLIE'

Leslie Dudley *see* **'LESLEY DUDLEY'**

'LESLIE EVANS', HT, dr, 1927; flowers rich dark velvety crimson, semi-dbl.; vigorous growth; [J.B. Clark X Red-Letter Day]; Evans; Beckwith

'LESLIE G. HARRIS', HT, mr, 1970; flowers crimson-scarlet, pointed, dbl., 27 petals, 4 in., intense fragrance; moderate growth; Gregory

'LESLIE HOLLAND', HT, dr, 1911; flowers deep velvety crimson, large, dbl., moderate fragrance; GM, NRS, 1909; Dickson, H.

Leslie Johns, HT, 1972; Gregory, C.; (Cavriglia)

'LESLIE PIDGEON', HT, ob, 1922; flowers orange-buff, suffused terra-cotta, semi-dbl., moderate fragrance; Dickson, H.

'LESLIE WHEAL', F, yb, 1977; bud ovoid; flowers cream-yellow, small, dbl., 25–35 petals, 1.5 in., cupped; foliage light green; compact growth; [Zorina sport]; Buss; H. Buss Nursery

Leslie's Dream *see* 'DICJOON'

'LESSING', HMsk, rb, 1914; flowers reddish rose streaked white, center citron-yellow, small, dbl., moderate fragrance; foliage large, light; [Trier X Entente Cordiale]; Lambert, P.

'LESTRA HIBBERD', HT, dy, 1935; bud dark yellow; flowers amber-yellow to orange-yellow, large, dbl., moderate fragrance; foliage leathery; strong stems; vigorous, compact growth; [Joanna Hill X Sweet Adeline]; Hill, Joseph H., Co.

Letchworth Garden City® *see* 'HARKOVER'

'LETITIA', HT, dr, 1949; bud long, pointed; flowers crimson, dbl., 35 petals, 5–6 in.; foliage dark; very vigorous growth; [Crimson Glory X Southport]; Bees

Letizia, Gr, 1962; Cazzaniga, F. G.; (Cavriglia)

Letizia *see* BARitzia

Letizia Bianca, 1963; Mansuino; (Cavriglia)

Letkis *see* **'JOHN DIJKSTRA'**

Letkiss, F, mr

LETlyda, S, w, 1994; (**Lyda Rose**); Lettunich

'L'ETNA', Ch, mp, 1825; flowers pink, becoming brighter, medium, very dbl.; Laffay, M.

LETrob, HMsk, ab, 1996; (**Doctor Robert Korns**); Lettunich

LETsilk, HCh, mp, 1992; (**Mateo's Silk Butterflies**); Lettunich

Lettle Len, Min, ab

Lettle Mermaid, Cl Min, ly

'LETTY COLES', T, w, 1876; flowers white, center pink; [Mme Mélanie Willermoz sport]; Keynes

'LEUCHTFEUER', HCh, mr, 1909; (President Taft); flowers bright red, moderate fragrance; [Gruss an Teplitz X Cramoisi Superieur]; Türke

'LEUCHTSTERN', HMult, pb, 1899; flowers deep rose pink, center white, blooms in clusters, single, slight fragrance; height 8-10 ft; [Daniel Lacombe X Crimson Rambler]; Schmidt, J.C.

Leventina®, HT, op, 1984; Huber

'LEVERKUSEN', HKor, ly, 1954; bud long, pointed; flowers large blooms on long sprays, dbl., exhibition form, recurrent bloom, slight fragrance; foliage glossy, light green; vigorous, creeper or pillar (to 8 ft) growth; [R. kordesii X Golden Glow]; Kordes

Leverson Gower *see* **'LEWESON GOWER'**

Leveson Gower *see* **'LEWESON GOWER'**

Levianthan, lp; [Great Venture X Fritz Thiedemann]; Dawson, George

'LEWESON GOWER', B, op, 1845; (Leverson Gower, Leveson Gower, Red Malmaison, Souvenir de la Malmaison Rouge); flowers rose shaded salmon, very large, dbl., cupped; Béluze

'LEWISTON', HT, my, 1981; flowers pointed, large blooms borne singly, dbl., 40 petals, exhibition form, intense, fruity fragrance; foliage large, glossy; vigorous, tall growth; [Red Lion X King's Ransom]; Perry, Astor, 1982

'LEXIE', Pol, lp; Miers, A., 1935; (Weatherly, L.)

Lexington *see* POUlgode

'LEY'S PERPETUAL', Class unknown; flowers pale yellow, dbl., cupped, moderate fragrance; vigorous, climbing (to 15 ft.) growth

'LI BURÉS', HT, rb, 1929; flowers rose-red and yellow mixed (very variable), dbl., cupped, slight fragrance; very vigorous, bushy growth; [Château de Clos Vougeot X Souv. de Claudius Pernet]; Dot, Pedro; C-P

Li Schluter, HT; Moreira da Silva, A.; (Cavriglia)

Lia, Min, or

Liaison, Gr, or, 1996; Williams, J. Benjamin

Liane®, LCl, ob, 1989; Cocker

Lianne Reynolds *see* 'REYLIANNE'

'LIBBY', Min, rb, 1978; bud ovoid; flowers white, edged red, high-pointed, dbl., 20 petals, 1–1.5 in., flat; foliage glossy; upright, compact growth; [Overture X Perla de Alcanada]; Rovinski; Casa de Rosa Domingo

'LIBBY'S GOLD', HT, yb, 1977; bud long, slender; dbl., 40 petals, 5 in., slight fragrance; foliage dark; upright, compact growth; [Apollo sport]; Carpenter

'LIBERTÉ', F, dr, 1946; (Crimson Masse); flowers brilliant crimson-red, large, dbl.; [Florentina X Seedling]; Leenders, M.

'LIBERTY', HT, dr, 1900; flowers brilliant velvety crimson, large, dbl., intense fragrance; foliage dark; vigorous growth; [Mrs W.J. Grant X Charles J. Grahame]; Dickson, A.

Liberty Bell *see* **'FREIHEITSGLOCKE'**

'LIBERTY BELL', HT, pb, 1963; (Freiheitsglocke); flowers deep pink, silver reverse, large, dbl.; Kordes, R.; A. Dickson;, McGredy

Liberty Miss *see* 'ZIPLIB'

'LIBERTY, CLIMBING', Cl HT, dr, 1908; May

'LIBIA', HT, w, 1934; flowers milk-white, center creamy yellow, large, very dbl.; vigorous growth; Borgatti, G.

Libre Ingenue *see* DELciste

'LIBRETTO', HT, dp, 1966; bud ovoid; flowers pink-red, medium, borne in clusters, dbl.; foliage dark; [Elli Knab X Seedling]; Verschuren, A.; van Engelen

Libretto, LCl, w

Lichtblick, S, yb, 1972; flowers yellow and pink, large, dbl.; VEG; (Sangerhausen)

'LICHTERLOH'®, F, mr, 1955; bud ovoid; flowers velvety blood-red, blooms in clusters, semi-dbl., slight fragrance; foliage leathery, dark, glossy; vigorous, upright (3 ft) growth; [Red Favorite X New Dawn]; Tantau, Math.

Lichtkönigin Lucia® *see* 'KORLILUB'

Licorice Twist *see* DEVlicor

'LÍDA BAAROVÁ', HT, mp, 1934; ('LIDA PAAR'); bud pointed; flowers rosy salmon, large, dbl., cupped; foliage glossy; vigorous, bushy growth; [Ophelia seedling]; Böhm, J.

'LIDA PAAR' *see* **'LÍDA BAAROVÁ'**

Lida O *see* 'BISYEL'

'L'IDÉAL', N, mr, 1887; flowers geranium to turkey-red, base indian yellow, semi-dbl.; vigorous, climbing growth; Nabonnand, G.

Lidice, F, or, 1961; flowers large, dbl., moderate fragrance; Wheatcroft; (Sangerhausen)

Lidka, HT, lp; Urban, J., 1985; (Czech Rosa Club)

Lidka Böhm *see* **'LIDKA BÖHMOVA'**

'LIDKA BÖHMOVA', HT, mp, 1929; (Lidka Böhm); flowers salmon-pink, tips veined reddish, base golden yellow, large, dbl., intense fragrance; foliage soft, bronze; very vigorous, bushy growth; [Una Wallace sport]; Böhm, J.

Lido di Roma *see* 'DELGAP'

'LIEBESBOTE', HT, dr, 1935; bud pointed; flowers velvety dark red, very large, dbl., exhibition form, intense fragrance; foliage soft; very vigorous, bushy growth; [Hadley X Miss C.E. van Rossem]; Weigand, C.; Pfitzer

Liebeserklärung, HT, mr, 1968; flowers large, dbl.; Horstmann; (Sangerhausen)

Liebesglut *see* **'CRIMSON KING'**

Liebeslied *see* **'LOVE SONG'**

Liebeslied, F, pb, 1990; Noack, Werner

Liebestraum *see* **'RED QUEEN'**

'LIEBESZAUBER', HT, mr, 1959; (Charme d'Amour); flowers velvety red, large, cupped, intense fragrance; foliage dark; vigorous, upright growth; [Detroiter X Crimson King]; Kordes

Liebeszauber *see* KORmiach

Lied *see* **'BOUQUET'**

Lien Budde *see* **'CAROLINE BUDDE'**

Liesbeth Canneman, F, dy, 1997; Interplant

'LIESBETH VAN ENGELEN', HT, w, 1960; flowers creamy white edged lilac-pink, dbl., moderate fragrance; foliage dark, glossy; upright growth; [Briarcliff X Unnamed seedling]; Verschuren, A.; van Engelen

'LIEUTENANT CHAURÉ', HT, dr, 1909; flowers velvety crimson-red, shaded garnet, very large, dbl., cupped, moderate fragrance; foliage rich green, leathery; vigorous, bushy growth; [Liberty X Étoile de France]; Pernet-Ducher

'LIEUTENANT COLONEL A. FAIRRIE', HT, my, 1930; bud pointed; flowers primrose, base deep yellow large, dbl., exhibition form, moderate, apple fragrance; [Rev. F. Page-Roberts X Mme Ravary]; Bees

'LIEUTENANT COLONEL DESMAIRES', F, ob, 1967; bud ovoid; flowers orange and yellow, borne in clusters, dbl., moderate fragrance; foliage leathery; vigorous, low growth; [Rondo X Fashion seedling]; Boerner; J&P

'LIEVEN GEVAERT', HT, mr, 1974; bud ovoid; flowers full, dbl., 30 petals, 4 in., moderate fragrance; foliage dark; Delforge, S.

'LIFEBOAT JUBILEE', HT, mr, 1974; flowers scarlet, shaded crimson, classic form, dbl., 20 petals, 4 in.; [Karl Herbst X (Karl Herbst X Crimson Glory)]; Sanday, John

Lifestyle *see* 'MORDARAIN'

Lifestyle™, Min, pb

'LIFIRANE', HT, mp, 1976; flowers neyron rose, full, dbl., 20–25 petals, 3 in., exhibition form, slight fragrance; foliage dull; vigorous, upright growth; PP004034; [Sweet Promise sport]; Zwemstra; Meilland

'LIGHT EDITOR MCFARLAND', HT, lp, 1950; [Editor McFarland sport]; Thomasville Nursery

'LIGHT TOUCH', S, pb, 1997; flowers double, medium, dbl., 15–25 petals, borne in large clusters, slight fragrance; foliage medium, light green, very glossy; bushy tall (2m) growth; [Cousin Essie X Open Pollination]; Weatherly, Lila

Lights of Broadway™ *see* 'SAVALIGHTS'

Ligia, HT; Moreira da Silva, A.; (Cavriglia)

Lijang Rose, Misc OGR, yb

'LIJNBAANROOS', F, ly, 1961; bud yellow; flowers creamy yellow, borne in clusters, semi-dbl.; foliage dark; vigorous, upright growth; [Schneewittchen X Koningin Juliana]; Buisman, G. A. H.

Li'l Alleluia™ *see* 'KINLU'

Li'l Rip™ *see* 'MOGARIP'

Li'l Touch™ *see* 'MOGATOUCH'

'LILA', Pol, lp; Miers, A., 1935; (Weatherly, L.)

Lila Queen of Bermuda, HT, m, 1970; flowers lilac-pink, large, dbl.; Strnad, J.; (Sangerhausen)

'LILA TAN', HT, m, 1961; flowers violet-cobalt, medium, dbl., 30 petals; vigorous, spreading growth; [Grey Pearl X Simone]; Dot, Pedro

'LILA VIDRI', HT, m, 1958; bud pointed; flowers lilac, dbl., 30 petals, exhibition form, moderate fragrance; strong stems; upright, compact growth; [(Unnamed seedling X Prelude) X Rosa de Friera]; Dot, Simon

Lila Zwerg *see* HELzwerg

Lilac Airs *see* 'SANLILAC'

'LILAC CHARM', F, m, 1962; bud pointed; flowers pastel mauve, anthers golden, filaments red, blooms in clusters, single, 5–8 petals, 4 in., flat, moderate fragrance; foliage dark; upright, compact growth; GM, NRS, 1931; LeGrice; Wayside Gardens Co., 1952

Lilac Chrysler Imperial, HT, lp; Strnad; (Czech Rosa Club)

'LILAC DAWN', F, m, 1964; bud pointed; flowers lavender-pink to lilac, blooms in clusters, dbl., 43 petals, 2.5 in., moderate, lilac fragrance; foliage leathery, light green; vigorous, bushy growth; [Lavender Pinocchio X Frolic]; Swim & Weeks; Armstrong Nursery

Lilac Joseph Guy, lp; Strnad; (Czech Rosa Club)

Lilac Minimo, Min, m, 1997

Lilac Mystery, S, m, 1994

Lilac Pink Moss, M, dp

Lilac Queen Elizabeth, Gr, m

Lilac Rose *see* 'AUSLILAC'

'LILAC ROSE', HT, lp, 1962; flowers pink tinted lilac, dbl., 28 petals, 5 in., moderate fragrance; foliage dark; vigorous, upright growth; [Karl Herbst X Chrysler Imperial]; Sanday, John

Lilac Rose™, S, pb

'LILAC SNOW', HT, w, 1985; flowers white, strong lilac tinge to inside petals; [Admiral Rodney sport]; Williams, H.; P. Loaney

'LILAC TIME', Min, m, 1955; flowers lilac-pink to light red, becoming lilac tinted, small, dbl.; dwarf (10 in) growth; [Violette X Zee]; Moore, Ralph S.; Sequoia Nursery

'LILAC TIME', HT, m, 1956; flowers lilac, dbl., 33 petals, 3.5 in., exhibition form, moderate fragrance; foliage light green; growth moderate; [Golden Dawn X Luis Brinas]; McGredy, Sam IV

Lilet Dot, F, 1987; Dot, Simon; (Cavriglia)

'LILETTE MALLERIN', HT, m, 1937; bud pointed, yellow; flowers mauve-red to mauve pink, deepening in cooler weather, reverse yellow, dbl., cupped, slight fragrance; foliage glossy; vigorous growth; [Charles P. Kilham X Unnamed R. foetida bicolor seedling]; Mallerin, C.; A. Meilland

'LILETTE MALLERIN IMPROVED', HT, m, 1942; bud pointed; flowers mauve-red, reverse yellow, large, dbl., 25 petals, exhibition form; foliage bronze, glossy; vigorous, upright, bushy growth; [Lilette Mallerin sport]; J&P

Lili Marlene *see* 'KORLIMA'

Lili Marlene, Climbing *see* 'PEKLIMASAR'

'LILIAN', HT, dy, 1931; flowers golden yellow, very large, dbl., cupped, intense fragrance; foliage bronze, glossy; vigorous, bushy growth; RULED EXTINCT 12/85; Cant, B. R.

Lilian Austin® *see* 'AUSMOUND'

Lilian Baylis *see* HARdeluxe

Lilian Baylis *see* HARdelux

'LILIAN BOOTLE', HT, dp, 1973; flowers cerise, dbl., 35 petals, 4.5–5 in., exhibition form, moderate fragrance; foliage matt green; compact, bushy growth; [Margaret X Seedling (red)]; Bootle

'LILIAN NORDICA', HT, w, 1898; flowers large, dbl., moderate fragrance; vigorous growth; [Margaret Dickson X Mme Hoste]; Walsh

'LILIANA', Gr, w, 1977; flowers large, dbl., 50 petals; foliage light green, leathery; vigorous, tall, upright growth; [Queen Elizabeth X Unnamed seedling]; Staikov, Prof. Dr. V.; Kalaydjiev and Chorbadjiiski

'LILIANA', HT, pb, 1956; flowers soft pink, reverse carmine; RULED EXTINCT 12/85; GM, Geneva, 1956; [Edith Krause X Fashion]; Camprubi, C.

Liliana *see* POUlsyng

'LILIBET', F, lp, 1953; (Fairy Princess); bud ovoid; dbl., 30 petals, 2.5–3.5 in., blooms in clusters, moderate, spicy fragrance; foliage glossy; low, bushy growth; AARS, 1954; [Floradora X Pinocchio]; Lindquist; Howard Rose Co.

'LILLAN', Min, pb, 1958; flowers rose pink, white center, small, semi-dbl.; very dwarf growth; [Ellen Poulsen X Tom Thumb]; deVink

'LILLEBROR', Min, mr, 1978; dbl.; dwarf, bushy growth; Hubner; O. Baum

Lilli Marleen® *see* 'KORLIMA'

Lilli Marleen, Climbing *see* 'PEKLIMASAR'

Lilli Marlene *see* 'KORLIMA'

Lilli Marlene, Climbing *see* 'PEKLIMASAR'

'LILLI VON POSERN', HT, mp, 1910; flowers large, dbl.; Kiese; (Sangerhausen)

Lillian, F, ob, 1997; Williams, J. Benjamin

'LILLIAN FRENCH', HT, pb, 1985; flowers light to medium pink, reverse deep pink, large blooms borne, very dbl., 50–120 petals, quartered, moderate fragrance; large prickles pointed downward; foliage medium, medium green, semi-glossy; upright, bushy growth; [Miss Hillcrest X Bishop Darlington]; French, Richard

'LILLIAN GIBSON', S, mp, 1938; flowers rose-pink, dbl., 40 petals, 3 in., non-recurrent, intense fragrance; vigorous growth; very hardy.; [R. blanda X Red Star]; Hansen, N.E.

'LILLIAN GISH', F, ly, 1985; flowers very pale yellow, large blooms borne singly and in small clusters, dbl., 35 petals, slight fragrance; few, straight prickles; foliage medium, medium green, semi-glossy; upright growth; [Queen Elizabeth X Allgold]; French, Richard

'LILLIAN GOMEZ-MENA', Cl HT, pb, 1927; flowers salmon-cream, reverse carmine, dbl., intense fragrance; Chambard, C.

'LILLIAN MOORE', HT, dy, 1917;, moderate fragrance; Dickson, H.

'LILLIE BELL', HT, dp, 1979; bud ovoid to pointed; flowers deep pink to cerise-red, ruffled, dbl., 34 petals, 4.5–5 in., moderate fragrance; foliage large; upright growth; [Pink Peace X Miss All-American Beauty]; Williams, J. Benjamin; Central Mississippi Rose Soc.

'LILLIE DAWBER', HT, dr, 1952; flowers scarlet overlaid crimson, dbl., 25 petals, 5 in., cupped, moderate fragrance; foliage dark; free growth; Kordes

'LILLIPUT', Cl Pol, mr, 1897; flowers carmine-red, small, dbl.; Paul, G.; (Sangerhausen)

'LILLY JUNG', HT, my, 1925; flowers golden yellow, dbl., moderate fragrance; Leenders, M.

Lily Bertschinger®, F, w, 1980; Hauser

'LILY DE GERLACHE', HT, dp, 1971; bud long, pointed; flowers rose-red, large, dbl., cupped, intense fragrance; foliage glossy, bronze, leathery; growth moderate, upright, bushy; [Kordes' Perfecta X Prima Ballerina]; Institute of Ornamental Plant Growing

Lily Freeman *see* 'HUX 1'

'LILY KEMP', HT, dp, 1928; flowers deep cherry-cerise, dbl., moderate fragrance; [Mme Butterfly X Capt. Ronald Clerk]; Morse

'LILY MERTSCHERSKY', N, m, 1878; flowers violet-red; Nabonnand

'LILY PONS', HT, ly, 1939; flowers yellow center, shading to white outer petals, large, dbl., 50 petals, exhibition form, moderate fragrance; foliage glossy; [Glenn Dale X Stargold]; Brownell, H.C.

'LILY THE PINK', HT, lp, 1992; very dbl., 3–3.5 in., borne mostly singly, slight fragrance; some prickles; foliage medium, medium green, semi-glossy; tall (1000 cms), upright growth; [Seedling X Seedling]; Scrivens, Len, 1991

'LILY VAN OOST', HT, mp, 1938; flowers very large, dbl.; Verschuren-Pechtold; (Sangerhausen)

'LILY WHITE', HT, w, 1950; (Starbright); bud ovoid; dbl., 40 petals, 4.5 in., exhibition form, slight fragrance; foliage glossy; vigorous, upright growth; [Starlite sport]; Hartgerink; J&P

'LIMBURGIA', HT, rb, 1921; flowers glowing deep carmine, reverse lilac-white, dbl., moderate fragrance; Leenders, M.

Lime Kiln, HWich, w, 1970

Limelight *see* 'KORIKON'

'LIMELIGHT', F, r, 1959; flowers vermilion splashed golden, borne in clusters, semi-dbl., slight fragrance; vigorous growth; RULED EXTINCT 5/85; [Enterprise sport]; Morse

Limelight *see* 'WEKAMANDA'

Limerick *see* 'ZIPLIME'

Limited Edition *see* 'JACREF'

Limona *see* KORulas

'LINA', Pol, lp, 1930; flowers creamy light pink, medium, dbl.; Vogel, M.; (Sangerhausen)

'LINA SCHMIDT-MICHEL', HT, yb, 1906; flowers yellowish-pink, large, semi-dbl.; Lambert, P.; (Sangerhausen)

'LINA VOGEL', HFt, ob, 1936; flowers orange with golden and light yellow shades, medium, single; Vogel, M.; (Sangerhausen)

Linburn Light, F, mp; Andersons

Lincoln Cathedral *see* 'GLANLIN'

Lincolnshire Poacher *see* 'GLAREABIT'

'LINDA', HT, yb, 1997; flowers full, medium, very dbl., 26–40 petals, borne in small clusters, slight fragrance; foliage medium, medium green, semi-glossy; upright, medium (3.5ft.) growth; [Champion X (Pink Favourite X Golden Autumn)]; Bracegirdle, A.J.

'LINDA ANN', Gr, mp, 1996; flowers medium pink, deeper pink than parents, more petals than parent, dbl., 26–40 petals, slight fragrance; moderate prickles; foliage medium, medium green, semi-glossy; upright, tall growth; [Queen Elizabeth X Katherine T. Marshall]; Dobbs, Annette E.

Linda Buford *see* 'TINBUFORD'

Linda Campbell *see* 'MORTEN'

'LINDA CHRISTINE', F, pb, 1980; flowers cerise, silver reverse; [Molly McGredy sport]; Taylor, W.J.

'LINDA GUEST', HT, ab, 1984; flowers large, dbl., 20 petals, exhibition form, slight fragrance; foliage medium, medium green, matt; bushy growth; [((Golden Jewel X Mischief) X Red Planet) X Valencia]; Guest, M.M.

'LINDA LOU'™, F, ob, 1987; flowers orange-yellow-red, medium, borne in sprays of 1-8, dbl., 30 petals, exhibition form, moderate fragrance; foliage medium, dark green; medium growth; Silver, ARC TG, 1987; [Golden Slippers X Roman Holiday]; Harvey, R.E.; Kimbrew Walter Roses, 1988

Linda Mary *see* 'BOSELADNEE'

'LINDA PORTER', HT, op, 1957; (Miguel Aldrufeu); bud ovoid; flowers salmon-pink, dbl., 55 petals, 5–6 in., globular, intense fragrance; foliage leathery; vigorous, upright growth; [Senateur Potie X Poinsettia]; Dot, Pedro; B&A

Linda Thomson *see* 'TOMTWO'

Linda Vista, F, 1974; de Ruiter, G.; (Cavriglia)

Linda's Lipstick *see* 'SHERILIP'

'LINDBERGH', Pol, mr, 1927; flowers bright geranium-red; [Orléans Rose sport]; Croibier

Lindee, Pol, w

L'Indéfrisible *see* **'PERMANENT WAVE'**

Linden Heath, S, mp

Linderof *see* KORelasting

'LINDLIFFE', F, op, 1984; **(Guyscliffe)**; flowers orange-salmon, blooms borne 3-5 per cluster, semi-dbl., 12 petals, cupped, intense fragrance; light brown prickles; foliage light green; tall, densely branched growth; Lindner, Richard; Ludwigs Roses Pty. Ltd.

'LINDSEY', HT, op, 1972; flowers salmon-pink, shaded copper, pointed, 3–4 in., moderate fragrance; foliage large; free growth; [Whisky Mac sport]; Watkins Roses

Line & Twine, HP, lp

'LINETTE', Pol, dp, 1922; bud reddish apricot; flowers shrimp-carmine-pink, passing to soft rose-pink, large; Turbat

'L'INGENUE', HGal, w, 1848

Lingo Musk, N, pb

'LINRICK', HT, dp, 1978; (**Friendship**®); bud ovoid; flowers deep pink, dbl., 23 petals, 5.5–6 in., cupped, intense fragrance; foliage large, dark; vigorous, upright growth; AARS, 1979; [Fragrant Cloud X Miss All-American Beauty]; Lindquist; C-P

'LINRO', F, rb, 1966; (**Roman Holiday**); bud ovoid; flowers orange turning blood-red, baseyellow, blooms in clusters, dbl., 28 petals, exhibition form, moderate fragrance; foliage dark, leathery; vigorous, bushy, low growth; AARS, 1967; [(Pinkie X Independence) X Circus]; Lindquist; Howard Rose Co.

'LINVILLE', Min, w, 1989; bud pointed; flowers light pink, aging white, large, borne usually singly, dbl., 28 petals, exhibition form, slight, fruity fragrance; prickles straight, medium, deep pink; foliage medium, medium green, semi-glossy; upright, medium growth; PP007737; [Unnamed seedling X Unknown seedling]; Bridges, Dennis A.; Bridges Roses, 1990

'LIOLÀ', HT, w, 1960; flowers ivory edged crimson, dbl.; very vigorous growth; [Peace X (Baiser X Marguerite Chambard)]; Giacomasso

'LION DES COMBATS', HP, mr, 1850; flowers carmine-red, large, dbl.; Lartay; (Sangerhausen)

'LIONEL BARRYMORE', F, my, 1956; bud urn shaped; flowers yellow, center deeper very dbl., 3–3.5 in., cupped, intense fragrance; foliage dark; strong stems bushy growth; [Duchess of Atholl X Orange Everglow]; Silva; Booy Rose Nursery

Lionheart *see* 'MINHEART'

Lions International *see* FRYcharm

Lippe-Detmold®, HT, mp, 1985; Noack Werner

'LIPSIANA', Pol, 1952; Cazzaniga, F. G. (Cavriglia)

'LIPSTICK', F, rb, 1940; flowers deep cerise shaded salmon, blooms in clusters, semi-dbl., 2 in., cupped, slight fragrance; foliage glossy, dark; vigorous (3 ft), bushy growth; Verschuren; Dreer

Lipstick 'n' Lace *see* 'CLELIPS'

Lisa, F, 1988; Zandri, R.; (Cavriglia)

Lisa, HSpn, lp; [Single Cherry X Primula]; Sutherland, P, 1996; Golden Vale Nursery; (Weatherly, L.)

'LISA COLFAX', F, ob, 1975; bud ovoid; flowers vermilion, overlaid brown, medium, dbl., exhibition form, intense fragrance; foliage leathery; very vigorous growth; [Mignonne X Sherry]; Parkes, Mrs M.H.; Rumsey

Lisa Kent, HT, dp, 1995; Dawson

'LISA MAREE', HT, dp, 1989; flowers deep pink, reverse lighter; [Esther Geldenhuys sport]; Cowper, Mrs. Maree; Cherry Wood's Nursery, 1989

'LISBETH PRIM', HT, or, 1934; flowers coppery red, fading lighter, large, dbl., intense fragrance; vigorous growth; [Hadley X Lady Inchiquin]; Felberg-Leclerc

'LISBETH STELLMACHER', Pol, or, 1919; bud coppery orange-red; flowers coppery red with golden yellow, striped pink, small, dbl., intense fragrance; [Aglaia X Marie van Houtte]; Lambert, P.

'LISBETH VON KAMECKE', HMult, lp, 1910; flowers small, semi-dbl.; Kiese; (Sangerhausen)

'LISBOA', HT, dr, 1953; flowers red shaded darker, large; very vigorous growth; [Barcelona X Crimson Glory]; da Silva, Moreira

'LISBOA DE 1947', HT, 1947; Moreira da Silva, A.; (Cavriglia)

'LISE CHIAVASSA', HT, mr, 1931; flowers carmine, very large, dbl., cupped, intense fragrance; foliage glossy; vigorous growth; [Mme Philippe Rivoire X Yves Druhen]; Buatois

Lise Palais *see* **'OPAL'**

Liselle *see* 'RULIS'

Liselotte Hollweg, F, mr, 1962

Lisette, HT, ly; Select Roses, B.V.

'LISETTE DE BERANGER', HP, dp; Guillot et Fils, 1867

Lisima *see* GAUzimi

'LISSY HORSTMANN', HT, mr, 1943; flowers brilliant scarlet-crimson, large, dbl., 28 petals, cupped; foliage leathery; vigorous growth; [Hadley X Heros]; Tantau; C-P

'LITAKOR', HT, ab, 1973; ('KORLITA', **Lolita**®); flowers golden bronze, dbl., 28 petals, 5 in., moderate fragrance; ADR, 1973; [Colour Wonder X Seedling]; Kordes, R.; Dicksons of Hawlmark, 1972

Literary Giant, HT, pb, 1991; Clements, John K.

'LITLIN', Min, w, 1983; (**Autumn Frost**); flowers medium, dbl., 35 petals, exhibition form, slight fragrance; foliage medium, medium green, semi-glossy; bushy growth; [Seedling X Seedling]; Rose, Euie; Little Gems Mini Rose Nursery

Little Amigo *see* 'TINAMIGO'

'LITTLE ANGEL', F, mr, 1961; flowers salmon becoming scarlet, borne in large clusters, dbl., 30 petals; foliage dark, glossy; moderate growth; Verschuren; Blaby Rose Gardens

Little Artist® *see* 'MACMANLY'

Little Ballerina *see* 'CURBAL'

'LITTLE BEAUTY', F, mp, 1935; (Crown of Jewels); flowers bright cerise-pink, fading to pink, small, very dbl., slight fragrance; foliage leathery; vigorous, bushy growth; [E.G. Hill seedling X Polyantha]; Howard, F.H.; Dreer

'LITTLE BETTY', S, lp, 1940; flowers soft pink, small, non-recurrent; height 3 ft; [Betty Bland X R. nitida]; Wright, Percy H.

Little Bit o' Sunshine *see* **'BIT O' SUNSHINE'**

Little Bo-Peep *see* 'POULLEN'

'LITTLE BREEZE', Min, ob, 1981; bud long, slender; flowers orange-red, fading to pink, loose blooms borne singly, semi-dbl., 17 petals, moderate fragrance; straight, gray prickles; foliage large, dark, glossy; vigorous growth; [Anytime X Elizabeth of Glamis]; McCann, Sean

'LITTLE BRIDESMAID', HT, lp; flowers shrimp-pink, well shaped; Archer

Little Brother *see* 'PIXBRO'

'LITTLE BUCKAROO', Min, mr, 1956; flowers bright red, small, dbl., 23 petals, moderate, fresh apple fragrance; foliage bronze, glossy, leathery; height 14-16 in; [(R. wichurana X Floradora) X (Oakington Ruby X Floradora)]; Moore, Ralph S.; Sequoia Nursery

'LITTLE CAMEO', F, pb, 1981; bud medium, pointed; flowers pink blend with cream outer petals, dbl., 26 petals, borne mostly singly, slight fragrance; straight, short, red prickles; foliage medium green, leathery; vigorous, upright growth; PP005381; [Lara X Unnamed seedling]; Strahle, Robert; Carlton Rose Nurseries

Little Carol *see* 'TINCAROL'

'LITTLE CHAMELEON', Min, pb, 1977; bud pointed; flowers rose-pink to red, small, dbl., 30 petals, 1.5 in., slight fragrance; upright growth; [Little Amy X ?]; Lyon

Little Charm *see* 'LYOLIT'

'LITTLE CHIEF', Min, dp, 1971; bud long, pointed; flowers deep pink, small, semi-dbl.; foliage small, glossy, leathery; growth moderate, dwarf, bushy; [Cotton Candy X Magic Wand]; Moore, Ralph S.; Sequoia Nursery

'LITTLE COMPTON CREEPER', LCl, dp, 1938; flowers deep rose-pink, borne in open clusters, single; yellow-orange red fruit; foliage glossy, dark; Brownell, H.C.

Little Crimson *see* 'SAVACRIM'

'LITTLE CURT', Min, dr, 1971; bud long, pointed; flowers deep velvety red, medium, semi-dbl.; foliage dark, leathery; vigorous, upright, bushy growth; [Unnamed red Cl F X Westmont]; Moore, Ralph S.; Sequoia Nursery

'LITTLE DARLING', F, yb, 1956; bud ovoid; flowers blend of yellow and soft salmon-pink, well-formed, dbl., 27 petals, 2.5 in., moderate, spicy fragrance; foliage dark, glossy, leathery; very vigorous, spreading growth; David Fuerstenberg Prize, ARS, 1964 GM, Portland, 1958; [Capt. Thomas X (Baby Chateau X Fashion)]; Duehrsen; Elmer Roses Co.

Little Debbie, MinFl, op

Little Devil *see* 'DELPO'

Little Devil, Climbing *see* 'DELPOSAR'

Little Diamond *see* 'SUNDIA'

Little Diane, Min, op

'LITTLE DICKENS', Min, ob, 1979; bud ovoid; flowers orange-red and yellow, small, dbl., 25 petals, cupped, slight fragrance; foliage small; compact growth; [(Ma Perkins X Sheri Anne) X Over the Rainbow]; Schwartz, Ernest W.; Nor'East Min. Roses

'LITTLE DORRIT', Pol, mp, 1930; flowers glowing pink; [Coral Cluster sport]; Reeves

'LITTLE DORRIT, CLIMBING', Cl Pol, mp, 1935; Letts

'LITTLE DOT', Pol, lp, 1889; flowers soft pink, flaked deeper on the outer petals clusters; very dwarf growth; Bennett

'LITTLE EMBERS', F, rb, 1988; flowers cerise to yellow, reverse yellow with cerise on tips, aging, dbl., 25 petals, exhibition form; no fruit; prickles long, narrow, small, light green; foliage medium, medium green, matt; upright, low growth; [Rise 'n' Shine X Prominent]; McFarland, John

Little Eskimo *see* 'MORWIT'

Little Fire *see* **'OGONIOK'**

'LITTLE FIREBALL', Min, or, 1968; bud ovoid; flowers bright coral-red, small, dbl., moderate fragrance; foliage small, glossy; bushy, compact, low growth; [(R. wichurana X Floradora) X New Penny]; Moore, Ralph S.; Mini-Roses;, Sequoia Nursery

Little Flame(TM) *see* 'JACNUYE'

'LITTLE FLIRT', Min, rb, 1961; bud pointed; flowers orange-red, reverse yellow, small, dbl., 42 petals, 1.5 in., moderate fragrance; foliage light green; vigorous, bushy (12-14 in.) growth; [(R. wichurana X Floradora) X (Golden Glow (LCl) X Zee)]; Moore, Ralph S.; Sequoia Nursery

'LITTLE GEM', M, dp, 1880; bud heavily mossed; flowers bright deep pink, small blooms in clusters, dbl., intense fragrance; dwarf, compact growth; Paul, W.

Little Gem *see* 'SUNSALM'

'LITTLE GEM', Pol, lp; Alderton, 1930; (Weatherly, L.)

'LITTLE GIRL', Cl Min, op, 1973; bud long, pointed; flowers coral-salmon-pink, medium, dbl.; foliage glossy; bushy, climbing growth; [Little Darling X Westmont]; Moore, Ralph S.; Sequoia Nursery

Little Green Snake, S, w, 1996; Lens

'LITTLE GUY', Min, mr, 1980; flowers medium red, reverse lighter, small blooms in clusters of 3-7, semi-dbl., no fragrance; no fruit; foliage small, medium green, semi-glossy; low, bushy growth; [Magic Wand X Violette]; Moore, Ralph S.; Moore Min. Roses

Little Huzzy *see* 'TINHUZZY'

Little Jackie(TM) *see* 'SAVOR'

Little Jewel *see* 'COCABEL'

Little Jewel *see* 'SUNJEW'

'LITTLE JOE', HT, dr, 1921; flowers crimson-red, single; [Red-Letter Day X H.V. Machin]; Looymans; Prior

'LITTLE JOHN', Min, rb, 1959; bud pointed; flowers scarlet, center white, medium, borne in clusters, semi-dbl., exhibition form, slight fragrance; foliage soft; vigorous, bushy (12-15 in.) growth; [Baby Masquerade sport]; Mason, F.

'LITTLE JOKER', Min, pb, 1958; flowers rose-pink, center cream, well-shaped; bushy growth; Spek

'LITTLE JUAN', Min, mr, 1966; flowers medium red, reverse lighter, small, dbl., slight fragrance; foliage small, leathery; vigorous, dwarf growth; [Juliette X Seedling]; Williams, Ernest D.; Mini-Roses

'LITTLE JULIET', HT, yb, 1924; flowers apricot and peach on yellow ground, dbl., slight fragrance; [F.J. Looymans seedling X ?]; Looymans; Prior

'LITTLE LADY', Pol, w, 1967; flowers blush to ice-white, small, dbl., 70 petals, slight fragrance; dwarf growth; [Schneewittchen (F) X Baby Faurax]; Harkness

'LITTLE LEAGUER', F, dr, 1962; bud ovoid; dbl., 28–40 petals, 2–2.5 in., cupped, slight fragrance; vigorous, upright growth; [Garnette seedling X Yuletide]; Jelly; E.G. Hill Co.

Little Lemmy *see* KORbalem

Little Len, MinFl, ab, 1987; Hatfield & Buckley

Little Lighthouse *see* 'CLELIGHT'

'LITTLE LINDA', Min, ly, 1976; bud high-pointed; flowers micro-mini, semi-dbl., 17 petals, 1 in., exhibition form, slight fragrance; compact growth; [Gold Coin seedling X Seedling]; Schwartz, Ernest W.; Nor'East Min. Roses

'LITTLE LIZA', Min, mp, 1975; bud mossy; flowers rose-pink, semi-dbl., 15 petals, .5–1 in.; low, very compact growth; [Fairy Moss X Fairy Moss]; Saville, F. Harmon; Nor'East Min. Roses

'LITTLE LOVE', Min, mp, 1977; bud pointed; flowers rose-pink, open, small, dbl., 25–30 petals, 1.5 in., slight fragrance; foliage tiny; very low, compact growth; [Little Amy X Unnamed seedling]; Lyon; L. Lyon Greenhouses

Little Mabel, Min, mr

Little Magician(TM) *see* 'WILMAG'

Little Marvel *see* 'RUIGERDAN'

Little Meghan *see* 'SPOMEG'

'LITTLE MELODY', Min, pb, 1979; bud globular; flowers soft peach-yellow blend, dbl., 38 petals, exhibition form, moderate fragrance; light brown prickles, curved downward; foliage medium green; compact, bushy growth; [Neue Revue X Sheri Anne]; Strawn, Leslie E.; Tiny Petals Nursery

Little Mermaid *see* 'MORWIMER'

'LITTLE MIKE', Min, dr, 1967; bud ovoid; flowers deep red, small, dbl., exhibition form; foliage dark, glossy, leathery; vigorous, dwarf growth; [((R. wichurana X Floradora) X Seedling) X Little Buckaroo]; Moore, Ralph S.; Sequoia Nursery

Little Miss Muffet *see* POUlense

'LITTLE MISS MUFFETT', F, pb, 1940; bud pointed, cerise; flowers bright rose-pink, reverse deeper, open, large, borne in clusters, semi-dbl., slight fragrance; strong stems; vigorous, bushy growth; [Else Poulsen X Étoile de Hollande]; LeGrice; C-P

'LITTLE NELL', HT, yb, 1933; flowers deep cream, center apricot, reverse primrose, well shaped, intense fragrance; foliage glossy, dark; vigorous growth; Archer

Little Nugget *see* 'SUNGOLD'

Little One *see* 'SANONE'

Little Opal *see* 'SUNPAT'

Little Paradise(TM) *see* 'WEKLIPS'

Little Peaces *see* 'TINPEACES'

Little Pearl *see* 'SUNPEARL'

Little Pim® *see* 'LENNOP'

'LITTLE PINK', F, pb, 1965; flowers pink, center creamy pink, small, very dbl., exhibition form, slight fragrance; foliage glossy; vigorous, compact growth; [Little Darling X ?]; Castleberry

Little Pink Hedge, F, pb, 1992

Little Pioneer *see* 'CURNEER'

'LITTLE POOH', Min, or, 1979; bud slim, pointed; single, 5–6 petals, flat, borne 5-6 per cluster, moderate fragrance; triangular prickles; foliage heavy, thick; upright growth; [Anytime X Unnamed seedling]; Fong, William P.

'LITTLE PRANCER', Cl HT, mr, 1996; flowers red, lighter reversee, blooms borne mostly singly, dbl., 15–25 petals, 3 in., moderate fragrance; moderate prickles; foliage medium, medium green striped with red, dull; upright, tall (7-9 ft.) growth; [Don Juan X Joseph's Coat]; Thurman, Robert R.; Thurman's Nursery

Little Prince *see* 'COCCORD'

'LITTLE PRINCESS', Pol, ob, 1937; flowers pale salmon-coral, well-formed, dbl.; vigorous, bushy growth; Knight, G.; Beckwith

'LITTLE PRINCESS', F, mp, 1955; flowers rose-pink, reverse lighter, medium, dbl., 50–75 petals, 2 in., cupped, moderate fragrance; long stems; vigorous growth; [Garnette sport]; Sodano, J.

Little Princess *see* **'PIXIE'**

Little Purple, F, m

Little Rambler *see* 'CHEWRAMB'

'LITTLE RASCAL', Min, mr, 1981; flowers medium red, shading to yellow at base, reverse lighter, dbl., 34 petals, exhibition form, slight fragrance; slightly hooked prickles; foliage tiny, light green; compact, bushy growth; [Sher Anne X Rise 'n' Shine]; Jolly, Betty J. Rosehill Farm

Little Rascal *see* PEAalamo

'LITTLE RED', Min, or, 1975; bud ovoid flowers red-orange, pompon shape small, semi-dbl., 18–22 petals, 1.5 in., slight fragrance; foliage glossy compact growth; [New Penny X Coloranja]; Lens; Spek

Little Red Devil *see* 'AROVIDIL'

Little Red Hedge *see* KORtabo

'LITTLE RED MONKEY', F, mr, 1954; flowers bright red, short stiff petals, borne singly and in sprays, very dbl.; bushy growth; [Donald Prior X Seedling]; Ratcliffe

Little Red Run-Around, HCh, dr

Little Russel *see* 'TROBRIC'

Little Sapphire *see* 'SUNSAP'

'LITTLE SCOTCH', Min, ly, 1958; bud long; flowers straw-yellow to white, dbl., 55 petals, 1.5 in., moderate fragrance; foliage leathery; vigorous, bushy (12 in) growth; [Golden Glow (LCl) X Zee]; Moore, Ralph S.; Sequoia Nursery

'LITTLE SHOWOFF', Cl Min, yb, 1960; bud pointed; flowers bright yellow, sometimes tinted red, dbl., 30 petals, 1–1.5 in., exhibition form, moderate fragrance; upright (to 4 ft.) growth; [Golden Glow (LCl) X Zee]; Moore, Ralph S.; Sequoia Nursery

Little Shrimp® *see* 'LENLIT'

Little Silver *see* 'JACSIL'

'LITTLE SIR ECHO', Min, mp, 1977; bud long, pointed; dbl., 48 petals, 1–1.5 in., exhibition form, moderate fragrance; foliage matt, green; compact, upright growth; [Ma Perkins X Baby Betsy McCall]; Schwartz, Ernest W.; Nor'East Min. Roses

'LITTLE SIR ECHO, CLIMBING', Cl Min, mp, 1985; [Little Sir Echo sport]; Watterberg, Leah

Little Sister *see* 'PIXSIS'

Little Sizzler™ *see* 'JACIAT'

Little Slam® *see* 'KINSLAM'

'LITTLE SMILES', Min, yb, 1978; bud ovoid; flowers chinese yellow vermilion, semi-dbl., 12 petals, 1 in., moderate, spicy fragrance; compact growth; [Q17a X Redgold]; Lyon

Little Squirt *see* 'TINSQUIRT'

Little Star Rose *see* 'BENSTAR'

'LITTLE STARBURST'™, Min, yb, 1991; flowers orange-red washing on golden yellow, very small blooms, dbl., slight fragrance; foliage small, medium green, semi-glossy; low, bushy growth; PP006821; [Prominent X Rise 'n' Shine]; Williams, J. Benjamin, 1988

'LITTLE STEPHEN', HT, mp, 1971; flowers glowing pink, dbl., 40 petals, 3 in., exhibition form, moderate fragrance; foliage matt; upright growth; [Gavotte X Seedling]; Sanday, John

Little Stripes *see* 'SHERILOWSTRI'

'LITTLE SUNSET', Min, pb, 1967; flowers salmon-pink on yellow, star-shaped blooms in clusters; foliage small, light green; [Seedling X Tom Thumb]; Kordes

'LITTLE SUNSHINE', Pol, my, 1915; flowers creamy yellow, varying to deep golden yellow, occasionally flecked crimson, dbl., 1.5–2 in.; dwarf growth; [R. multiflora nana X Soleil d'Or]; Cumming; A.N. Pierson

Little 't' *see* 'TRAT'

Little Tease *see* 'ZIPTEASE'

Little Tiger *see* 'MORSHAKI'

Little Tommy Tucker™ *see* 'TUCTOMMY'

Little Tyke *see* 'RENYKE'

Little Vegas *see* 'SPOVEGAS'

'LITTLE WALLACE', F, pb, 1952; bud very long; flowers light pink, center yellow, blooms in clusters, dbl., 25 petals, exhibition form, slight fragrance; foliage leathery, dark; vigorous growth; [Elfe sport]; Beall; Beall Greenhouse Co.

Little White Pet *see* **'WHITE PET'**

Little White Spray®, Min, w, 1991; Lens

Little Wings™, Min, ly

Little Woman *see* 'DICLITTLE'

'LITTLE WONDER', F, or, 1975; bud pointed; dbl., 22–28 petals, 3.5–4 in., moderate, spicy fragrance; foliage small; upright to spreading growth; [Duftwolke X Ena Harkness]; Huber

Little Woods Rose *see* **R. GYMNOCARPA**

'LITTLEST ANGEL', Min, my, 1976; bud short, pointed; flowers medium to deep yellow, micro-mini, dbl., 28 petals, .5 in., exhibition form, slight fragrance; foliage small; low, compact, bushy growth; [Gold Coin seedling X Unnamed Miniature seedling]; Schwartz, Ernest W.; Nor'East Min. Roses

Littlest Spartan *see* 'GRULIT'

Live Wire™ *see* 'SAVAWIRE'

'LIVELY', HT, dp, 1959; flowers rose pink, dbl., 32 petals, 4–6. in., intense fragrance; foliage dark, glossy; vigorous, compact, low growth; [Wellworth X Ena Harkness]; LeGrice

'LIVELY LADY', F, or, 1969; flowers vermilion, large, dbl., slight fragrance; foliage dark, glossy; [Elizabeth of Glamis X Tropicana]; Cocker

Liverpool® *see* **'LIVERPOOL ECHO'**

Liverpool Daily Post *see* FRYtrooper

'LIVERPOOL ECHO', F, op, 1971; (Liverpool®); flowers salmon, dbl., 23 petals, 4 in., exhibition form, slight fragrance; foliage light; tall growth; GM, Portland, 1979; [(Little Darling X Goldilocks) X Munchen]; McGredy, Sam IV; McGredy

Liverpool Remembers *see* 'FRYSTAR'

'LIVERTON LADY', Cl HT, mp, 1978; semi-dbl., 16–20 petals, 3–4 in., slight fragrance; foliage small, glossy; tall, climbing growth; [Bantry Bay X Sympathie]; Warner; Bradley Nursery

Livia *see* MEIkola

Livin' Easy™ *see* 'HARWELCOME'

'LIVING', HT, rb, 1957; flowers reddish orange, reverse copper streaked red, dbl., 24 petals, 5–6. in., exhibition form, moderate, spicy fragrance; foliage leathery, semi-glossy; vigorous growth; [Charlotte Armstong X Grande Duchesse Charlotte]; Lammerts, Dr. Walter; Consolidated Nursery

Living Bouquet *see* 'LAVLINGER'

'LIVING CORAL', HT, op, 1974; bud ovoid; flowers soft coral-pink, dbl., 28 petals, 5 in., exhibition form, moderate, fruity fragrance; foliage glossy; moderate growth; [Queen of Bermuda X Tropicana]; Golik; Dynarose

'LIVING FIRE', F, ob, 1972; flowers orange, suffused orange-red, rosette form, dbl., 33 petals, 2.5 in., moderate fragrance; foliage dark; [Tropicana X Unknown seedling]; Gregory, 1973

Living Well, S, mp, 1999

Liz *see* MEIrebuc

Liza® *see* BARliz

Lizabeth's Lullabye *see* 'WIRCUROB'

'LIZZIE MOLK', S, pb, 1983; flowers light pink, white center, blooms in clusters, dbl., 20 petals, does not repeat, intense fragrance; foliage medium, blue-green, glossy; spreading growth; Rusnock, Ann M.

'LJUBA RIZZOLI'®, HT, dr, 1980; (Via Romana); bud pointed; dbl., 25 petals, cupped, borne singly, intense fragrance; purple-green prickles; foliage large, dark, matt; upright, bushy growth; Dot, Simon; Rose Barni-Pistoia

'LLEIDA', HT, rb, 1936; flowers bright red, reverse yellow, intense fragrance; very vigorous growth; [Edouard Renard X Condesa de Sástago]; Dot, Pedro; A. Meilland

Lloyd Center Supreme *see* 'TWOLOY'

Lloyds of London *see* CANlloyd

Lloyds Supreme, HT, mp; [Brion X Silver Jubilee]; Twomey, Jerry, 1995

Loads of Pink *see* 'FIALOPI'

'LOBO', HT, rb, 1987; flowers red-purple, reverse white, large, dbl., 26–40 petals, exhibition form, intense fragrance; foliage large, dark green, matt; upright growth; [Kordes' Perfecta X Gavotte]; Perry, Astor, 1989

'L'OBSCURITÉ', M, dr, 1848; flowers dark garnet-crimson, large, semi-dbl.; Lacharme, F.

'LOCARNO', Pol, or, 1926; flowers large, borne in huge clusters; vigorous, bushy growth; [Orléans Rose sport]; deRuiter

'LOCOMOTION', HT, yb, 1965; flowers salmon tinted peach, becoming citron-yellow; foliage dark, leathery; vigorous, bushy growth; [R.M.S. Queen Mary X Lady Sylvia]; Verschuren, H.A.M.; Ravensberg

'LODESTAR', HT, pb, 1953; bud ovoid to globular; flowers buff-pink, center peach, dbl., 35–40 petals, 4.5–5 in., intense, spicy fragrance; foliage leathery; bushy growth; [Diamond Jubilee seedling X Serenade]; Boerner; J&P

'LODEWIJK OPDEBEEK', HT, dr, 1921; flowers oxblood-red, reverse rose, dbl., moderate fragrance; [Jonkheer J.L. Mock X Mev. Dora van Tets]; Leenders, M.

Lodovico, F, 1960; Cazzaniga, F. G.; (Cavriglia)

'LOETA LIGGETT', HT, pb, 1984; flowers light pink, darker pink reverse, often with salmon center; [Duet sport]; Liggett, Myron T.

'LOHENGRIN', HT, lp, 1903; flowers large, dbl.; Kiese; (Sangerhausen)

Lois *see* 'TINLOIS'

'LOIS CROUSE', HT, lp, 1937; bud pointed; flowers light pink, suffused salmon, peony form, large, dbl.; foliage dark; vigorous, bushy growth; [Mme Butterfly seedling]; Moore, Ralph S.; Brooks & Son

'LOIS MANEY', LCl, op, 1953; flowers salmon-pink, dbl., 4–5 in., abundant, nonrecurrent bloom, slight fragrance; foliage leathery; very vigorous (25 ft.) growth; [R. maximowicziana pilosa X Templar]; Maney; Iowa State College

Lois Wilson, HT, pb, 1998

Lola de Porcioles, HT; Camprubi, C.; (Cavriglia)

Lola Montes, HT, mr, 1969; Dot

'LOLA VENDRELL', HT, pb, 1957; flowers soft pink streaked deeper, well formed, large, dbl., 40 petals, moderate fragrance; foliage glossy; long stems; vigorous growth; [Serafina Longa X Mme Kriloff]; Bofill; Toree Blanca

'LOLETTE DUPAIN', Cl Pol, pb, 1918; flowers yellowish rose, reverse silvery rose, borne in clusters, dbl., sometimes recurrent bloom; vigorous growth; [Casimir Moulle X Mme Norbert Levavasseur]; Lottin

'LOLI CREUS', HT, dp, 1953; flowers carmine, well formed, large; foliage glossy; strong stems; very vigorous growth; [Cynthia X Manuelita]; Dot, Pedro

Lolita® *see* 'LITAKOR'

'LOLITA', HT, w, 1937; flowers white, center cream, overlarge, dbl., globular; foliage leathery; vigorous growth; [Frau Karl Druschki X Unnamed Hybrid Tea]; Croibier

'LOLLIPOP', Min, mr, 1959; flowers bright red, dbl., 35 petals, 1–1.5 in., slight fragrance; foliage glossy; vigorous (14 in), bushy growth; [(R. wichurana X Floradora) X Little Buckaroo]; Moore, Ralph S.; Sequoia Nursery

'LOLLO', HT, rb, 1949; bud pointed; flowers purplish red, reverse crimson-carmine, well formed, dbl., 32–36 petals, moderate fragrance; foliage dark; strong stems; vigorous, bushy growth; [Lele X Crimson Glory]; San Remo Exp. Sta.

'LONDON PRIDE', HT, op, 1954; flowers deep salmon-pink shaded coral, exhibition form, moderate fragrance; upright growth; Ratcliffe

'LONDON STARLETS', HT, ob, 1959; flowers orange to orange-red, well shaped, large, dbl., 50 petals, moderate fragrance; long, strong stems; vigorous growth; [Mission Bells X Jiminy Cricket seedling]; Maarse, G.

'LONDON TOWN', HT, op, 1955; flowers salmon-pink, base buff, well formed, medium, moderate, fruity fragrance; vigorous growth; [Peace X Charles Gregory]; Letts

'LONE STAR', HT, mr, 1925; flowers velvety cardinal-red, large, semi-dbl., moderate fragrance; vigorous, upright growth; [Étoile de France seedling]; Buller; Hillje

'LONE STAR STATE', HT, pb, 1943; flowers darker than parent, irregularly striped white; [Texas Centennial sport]; Collins

'LONETTE CHENAULT', Pol, lp, 1925; flowers creamy rose pink, medium, dbl.; Chenault; (Sangerhausen)

Long Island® *see* 'CASLON'

'LONG JOHN SILVER', LCl, w, 1934; bud pointed; flowers silvery white, very large blooms in clusters, dbl., cupped, moderate fragrance; foliage large, leathery; vigorous, climbing growth; [R. setigera seedling X Sunburst]; Horvath; J&P

Long Tall Sally ™ *see* 'WEKAJAZOUL'

Long White Cloud *see* 'MACLANOFLON'

'LONGCHAMP', F, mr, 1960; flowers bright red cerise, medium, borne in clusters, semi-dbl.; foliage glossy; vigorous, symmetrical growth; Laperrière; EFR

Longchamp 80, S, 1980; Laperrière; (Cavriglia)

'LONGFORD', S, dp, 1938; flowers carmine-pink, small to medium, dbl.; Central Exp. Farm; (Sangerhausen)

Longleat *see* 'MACINCA'

'LONGWOOD', LCl, mp, 1914; semi-dbl.; [American Pillar X Unnamed seedling]; Wintzer

'LONGWORTH RAMBLER', LCl, mr, 1880; flowers light crimson, semi-dbl.; vigorous growth; Liabaud

Looks Like Fun *see* 'SEAFUN'

Looping® *see* 'MEIROVONEX'

'LORD ALLENBY', HT, dr, 1923; flowers bright crimson, very large, dbl., exhibition form, slight fragrance; foliage rich green, leathery; dwarf, sturdy growth; Dickson, A.

'LORD BACON', HP, dr, 1883; flowers deep crimson shaded scarlet, large, dbl., globular; vigorous growth; Paul & Son

'LORD BADEN-POWELL', HT, ob, 1937; bud orange striped red; flowers saffron-yellow, large, very dbl., moderate fragrance; vigorous, bushy, compact growth; Leenders, M.

Lord Byron *see* 'MEITOSIER'

'LORD CALVERT', HT, dr, 1919; flowers dark velvety red, dbl., intense fragrance; [Radiance X Hoosier Beauty]; Cook, J.W.

'LORD CASTLEREAGH', HT, dr, 1927; bud pointed; flowers dark blackish crimson, open, semi-dbl., intense fragrance; foliage dark, leathery; vigorous, bushy growth; Dickson, A.

'LORD CHARLEMONT', HT, dr, 1922; bud long, pointed; flowers clear deep crimson, well shaped, large, dbl., exhibition form, intense fragrance; foliage dark, leathery; bushy growth; McGredy

John-John™
(WEKgibotex)
Floribunda, medium yellow, 2000
Hybridized by Tom Carruth
[(Seedling x Sunprite) x (Old Master x Texas)]
Photo by Gene Sasse; Courtesy Weeks Roses

Betty Boop™
(WEKPLAPIC)
Floribunda, red blend, 1999
Hybridized by Tom Carruth
[Playboy x Picasso]
AARS 1999
Photo courtesy Gene Sasse/Weeks Roses
™ of the Hearst Corporation

Fourth of July™
(WEKROALT)
Large Flowered Climber, red blend, 1999
Hybridized by Tom Carruth
[Roller Coaster x Altissimo]
AARS 1999
Photo courtesy Gene Sasse/Weeks Roses

Scentimental™
(WEKplapep)
Floribunda, red blend, 1997
Hybridized by Tom Carruth
[Playboy x Peppermint Twist]
AARS 1997
Photo courtesy Gene Sasse/Weeks Roses

Tootsie
(GREtoots)
Floribunda, pink blend, 1991
Hybridized by Chris Greenwood
[Angel Face x Old Master]
Photo by Gene Sasse

Smooth Prince
(HADprince)
Hybrid Tea, medium red, 1990
Hybridized by Harvey Davidson
[Smooth Sailing x Old Smoothie]
Western Sun Roses
Photo courtesy Harvey Davidson

White Treasure
(HADtreasure)
Hybrid Tea, white, 2002
Hybridized by Harvey Davidson
[Smooth Sailing x Sunset Jubilee]
Western Sun Roses
Photo courtesy Harvey Davidson

Olde Romeo
(HADromeo)
Hybrid Tea, dark red, 1993
Hybridized by Harvey Davidson
[Smooth Sailing x (Old Smoothie x ((Polly x Peace) x Simon Bolivar))]
Western Sun Roses
Photo courtesy Harvey Davidson

Pink Empress
(HADemp)
Hybrid Tea, medium pink, 2002
Hybridized by Harvey Davidson
[Smooth Sailing x Medallion]
Western Sun Roses
Photo courtesy Harvey Davidson

Overnight Scentsation™
(SAVanight)
Miniature, medium pink, 1998
Hybridized by F. Harmon Saville
[Taxi x Lavender Jade]
Photo courtesy Nor'East Miniature Roses, Inc.

Scentsational™
(SAVAMOR)
Minature, mauve, 1995
Hybridized by F. Harmon Saville
[Lavender Jade x Silverado]
Photo courtesy Nor'East Miniature Roses, Inc.

'LORD CHARLEMONT, CLIMBING', Cl HT, cr, 1932; Hurcombe

'LORD CHELMSFORD', HT, lp; Williams, F., 1906; (Weatherly, L.)

'LORD CLYDE', HP, dr, 1863; flowers bright crimson, well formed, large; Paul & Son

Lord Don, S, rb, 1993

'LORD FAIRFAX', HT, dp, 1925; flowers cherry-rose-pink, moderate fragrance; foliage leathery; long stems; good habit growth; Gray, W.R.

'LORD FREDERICK CAVENDISH', HP, mr; flowers bright scarlet; Frettingham, 1884

Lord Gold® *see* DELgold

Lord Houghton of Sowerby *see* 'HARTUBOND'

'LORD KITCHENER', HT, dp, 1918; flowers bright carmine-rose, dbl., intense fragrance; Chaplin Bros.

'LORD LAMBOURNE', HT, yb, 1925; bud pointed; flowers buttercup-yellow, edged carmine-scarlet, very large, dbl., exhibition form, moderate fragrance; foliage light, leathery, glossy; very vigorous, bushy growth; McGredy

'LORD LONSDALE', HT, ab, 1933; bud pointed; flowers deepest orange-yellow, large, dbl., exhibition form, moderate fragrance; foliage glossy, light; vigorous, bushy growth; GM, NRS, 1931; Dickson, A.; Dreer;, H&S

'LORD LOUIS', HT, dp, 1981; bud pointed; flowers light crimson, blooms borne several together, dbl., 30 petals, moderate fragrance; foliage mid-green, glossy; vigorous growth; [Pink Favorite X Seedling]; Gregory, C.; C. Gregory & Sons

'LORD MACAULAY', HP, dr, 1874; flowers crimson, globular; Paul, W.

'LORD PENZANCE', HEg, yb, 1894; flowers soft rosy yellow, paler at base, yellow stamens, blooms in clusters, single, summer bloom, moderate fragrance; foliage small, dark, fragrant; very vigorous growth; [R. eglanteria X Harison's Yellow]; Penzance

'LORD RAGLAN', HP, dr, 1854; flowers bright velvety crimson, very large, very dbl.; [Geant des Batailles seedling]; Guillot Père

'LORD ROSSMORE', HT, w, 1930; flowers creamy white shaded rose toward edge, large, dbl., exhibition form, moderate fragrance; foliage dark olive-green, leathery; vigorous growth; GM, NRS, 1928; Hall; McGredy

Lord Scarman, HGal, pb, 1996; Scarman

'LORD STAIR', HT, dr, 1930; bud pointed; flowers velvety crimson-scarlet, large, dbl., exhibition form, moderate fragrance; foliage dark, leathery; very vigorous, bushy growth; Smith, T.

'LORD WORTHINGTON', HT, mp, 1928; flowers medium, dbl.; Dickson, A.; (Sangerhausen)

Lordly Oberon *see* 'AUSRON'

L'Oréal Trophy *see* 'HARLEXIS'

'LOREE', F, w, 1969; bud long, pointed; flowers white tinted pink, small, dbl., cupped, slight fragrance; foliage leathery; vigorous, bushy, compact growth; [Frolic X ?]; Pal, Dr. B.P.; Son

'LORELEI', S, ab, 1947; bud ovoid; flowers peach-pink, center yellow, semi-dbl., 3–4 in., non-recurrent, intense fragrance; foliage leathery, dark; vigorous, upright growth; [Joanna Hill X Harison's Yellow]; Fisher, R.C.

'LORELEY', Pol, lp, 1913; flowers creamy rose pink, medium, dbl.; Kiese; (Sangerhausen)

Loreley 82®, HT, dr, 1982; Noack, Werner

Lorena® *see* 'KORENLO'

Lorenz Schwamborn, Pol, mr, 1960; flowers medium, semi-dbl.; Schmid, P.; (Sangerhausen)

'LORENZO PAHISSA', HT, op, 1941; flowers coral, very large, dbl., moderate fragrance; foliage abundant; very vigorous, upright growth; GM, Bagatelle, 1941; [Seedling X Mari Dot]; Pahissa

Loretta® *see* 'KORENPI'

'LORETTO', LCl, pb, 1922; flowers reddish, center white, blooms in clusters, semi-dbl., moderate fragrance; few prickles; vigorous, climbing growth; [Jersey Beauty X Seedling]; Clark, A.; Brundrett

'LORI ANN', F, yb, 1990; flowers bright amber yellow, flushed orange and pink, reverse yellow, dbl., 20 petals, cupped, moderate, fruity fragrance; foliage medium, medium green, semi-glossy, slightly elongate; slightly spreading, medium growth; [Judy Garland X Sutter's Gold]; Schneider, Peter, 1987

'LORI NAN', Min, dp, 1965; bud globular; flowers rose-red, small, dbl.; foliage glossy, leathery; moderate growth; [(R. wichurana X Floradora) X (Seedling X Zee)]; Moore, Ralph S.; Sequoia Nursery

Lorina *see* 'KORENLO'

'LORNA', HT, op, 1936; bud pointed; flowers salmon, dbl., exhibition form; foliage leathery, glossy, light; vigorous, bushy growth; Cant, B. R.

'LORNA ANDERSON', HT, mr, 1940; flowers well-formed, moderate fragrance; Clark, A.

'LORNA DOONE', F, mr, 1972; dbl., 24 petals, 4 in., slight fragrance; foliage dark, glossy; [Red Dandy X Lilli Marleen]; Harkness

'LORNA MAY', F, rb, 1958; flowers crimson-red, center white, large trusses, single, 2 in.; foliage light green; vigorous growth; [Poulsen's Pink X Kathe Duvigneau]; deRuiter; Blaby Rose Gardens

'LORRAINE', HT, op, 1945; flowers salmon-carmine touched red, large, dbl.; vigorous growth; [Peace X Mme Mallerin]; Meilland, F.

'LORRAINE LEE', T, pb, 1924; bud pointed; flowers rosy apricot-pink, dbl., cupped, moderate fragrance; foliage rich green, leathery, glossy; vigorous growth; (14); [Jessie Clark X Capitaine Millet]; Clark, A.; Hackett

'LORRAINE LEE, CLIMBING', Cl T, pb, 1932; McKay

'LORRAINE STEBBINGS', Min, w, 1985; flowers small blooms borne singly, dbl., 23 petals, slight, honey fragrance; foliage small, glossy; upright, bushy growth; [Cinderella X Popcorn]; Morey, Dr. Dennison

'LOS ANGELES', HT, op, 1916; bud pointed; flowers coral-pink, base gold, large, dbl., intense fragrance; foliage leathery; vigorous, spreading growth; GM, Bagatelle, 1918; [Mme Segond Weber X Lyon Rose]; Howard, F.H.; H&S

'LOS ANGELES BEAUTIFUL', Gr, yb, 1967; flowers yellow blended with coral and scarlet, medium, dbl., exhibition form, slight fragrance; foliage dark, leathery; vigorous, upright, compact growth; [Queen Elizabeth X Rumba]; Lammerts, Dr. Walter; Germain's

Los Angeles Frost, HT, lp, 1989; Robinson

'LOS ANGELES, CLIMBING', Cl HT, pb, 1925; H&S

'LOS TEJAS', HT, mr, 1968; bud globular; flowers large, dbl., exhibition form, moderate fragrance; foliage leathery; vigorous, upright growth; [Chrysler Imperial X Happiness]; Patterson; Patterson Roses

Lost in Paradise, Min, m, 1997; Spooner, Raymond A.

'LOTTE GÜNTHART', HT, mr, 1964; bud ovoid; flowers bright red, peony form, large, very dbl., 90 petals; foliage leathery; tall, upright, bushy growth; [Queen Elizabeth X Bravo]; Armstrong, D.L.; Armstrong Nursery

'LOTTIE FORSTER', Pol, lp; Matthews, W.J., 1915; (Weatherly, L.)

'LOU-CELINA', S, dr, 1979; flowers large, dbl., 50 petals, repeat bloom, moderate fragrance; foliage medium, dark red-green, semi-glossy; upright growth; [(Venture X ((Cecilia X China Belle) X Suzanne)) X (Paula X Soeur Kristin)]; James, John; Historical Roses, 1986

'L'OUCHE', HCh, pb, 1901; flowers rose shaded yellow; Buatois, 1891

'LOUIS BARBIER', HFt, ob, 1909; flowers coppery orange with dark yellow striping, medium, semi-dbl.; Barbier; (Sangerhausen)

Louis Bernard, HT, or, 1969; Croix

'LOUIS BOURGOIN', HT, lp, 1921; flowers flesh-pink passing to silvery pink, dbl., slight fragrance; [Jonkheer J.L. Mock X Frau Karl Druschki]; Gillot, F.

Louis Bugnet, HRg, w, 1960; Bugnet

'LOUIS CALLA', HP, m, 1885; flowers purple/pink, large, dbl.; Verdier, E.; (Sangerhausen)

Louis de Funès® *see* 'MEIRESTIF'

Louis d'Or, HT, 1965; Dorieux; (Cavriglia)

'LOUIS FAURAX', HT, ob, 1941; flowers coppery-salmon with dark yellow, large, dbl.; Gaujard; (Sangerhausen)

'LOUIS GIMARD', M, mp, 1877; flowers bright pink, very large blooms, dbl.; prickles well-mossed; long stems; vigorous growth; Pernet Père

'LOUIS JOLLIET', HKor, mp, 1991; flowers medium, dbl., 26–40 petals, slight fragrance; foliage medium, medium green, semi-glossy; spreading, low growth; Ogilvie, Ian S.; Agriculture Canada, 1992

'LOUIS KAHLE', HT, mr, 1922; flowers bright cherry-red, intense fragrance; [Lieutenant Chaure X Étoile de France]; Kiese

'LOUIS PAJOTIN', HT, op, 1940; flowers coral, stamens yellow, large, very dbl.; long, stiff stems; very vigorous growth; [Souv. de Claudius Pernet X Margaret McGredy]; Mallerin, C.; Meilland, A.

'LOUIS PAJOTIN, CLIMBING', Cl HT, op, 1959; Pajotin-Chédane

Louis Philippe *see* **'GRANDISSIMA'**

'LOUIS PHILIPPE', Ch, rb, 1834; flowers dark crimson with edges of center petals blush, aging crimson, dbl., globular; bushy growth; Guérin

Louis Philippe d'Angers, HGal, lp; Hardy, 1824

'LOUIS PHILIPPE D'ORLEANS', HP, 1884; Verdier, E.; (Cavriglia)

'LOUIS PHILIPPE, CLIMBING', Cl Ch, rb

Louis Rambler, HMult, 1997; Lens, Louis; (Cavriglia)

Louis Riel *see* 'ZUBLOU'

'LOUIS RÖDIGER', LCl, ob, 1935; bud pointed; flowers orange shaded yellow and red, open, very large, semi-dbl., moderate, fruity fragrance; foliage leathery, wrinkled, dark; very vigorous, climbing trailing growth; [Daisy Hill X (Charles P. Kilham X Mev. G.A. van Rossem)]; Kordes

'LOUIS SAUVAGE', HWich, m, 1914; flowers purple/red, small, dbl.; Turbat; (Sangerhausen)

'LOUIS VAN HOUTTE', HP, dr; flowers crimson-maroon, well-formed, large, dbl., 40 petals, some recurrent bloom, intense fragrance; [Général Jacqueminot X ?]; Lacharme, F., 1869

Louis van Till, HGal, mp; flowers luminous pink, very dbl., intense fragrance; (Sangerhausen)

'LOUIS VAN TYLE', m; flowers light crimson, shaded black or purple, small, semi-dbl.

'LOUIS WALTER', HT, ab, 1938; flowers golden orange-yellow, large, dbl.; foliage clear green, glossy; vigorous growth; [Mrs Pierre S. duPont seedling X Charles P. Kilham]; Mallerin, C.

'LOUIS XII', T, m; flowers violet, cupped

'LOUIS XIV', HP, dr, 1859; flowers dark crimson, medium, dbl., 25 petals, intense fragrance; foliage sparse; moderate growth; Guillot et Fils

'LOUISA JANE', HT, m, 1976; flowers center soft white, shading to mauve-pink; [Baronne Edmond de Rothschild sport]; Ross, A., & Son; A. Ross & Son

Louisa Jane Morris *see* 'BOSADSON'

'LOUISA SCHULTHEIS', HT, op, 1925; flowers pink and salmon, center darker, dbl., 32–36 petals, slight fragrance; [Golden Ophelia X Ruhm von Steinfurth]; Schultheis, A.

Louisa Stone *see* HARbadge

'LOUISE', HT, mp, 1924; bud pointed; flowers rose-pink shaded cerise, large, dbl., intense fragrance; foliage dark, leathery; vigorous growth; [Isobel X Unnamed variety]; Prince

'LOUISE ABDY', F, mp, 1964; flowers pink, base yellow, single, 6 petals, 4 in., moderate fragrance; foliage dark; vigorous growth; [Donald Prior X McGredy's Yellow]; Abdy

'LOUISE BALDWIN', HT, ob, 1919; bud pointed; flowers rich orange, tinted soft apricot, dbl., exhibition form, intense fragrance; vigorous growth; McGredy

'LOUISE CATHERINE BRESLAU', HT, op, 1912; flowers shrimp-pink shaded reddish coppery orange, reverse chrome-yellow, dbl., moderate fragrance; foliage dark, bronze, leathery; bushy growth; [Unnamed variety X Soleil d'Or seedling]; Pernet-Ducher

'LOUISE CATHERINE BRESLAU, CLIMBING', Cl HT, op, 1917; Kordes

Louise Clements *see* CLElou

'LOUISE CRETTÉ', HP, w, 1915; flowers snow-white, center creamy white, well-formed, dbl., 55 petals, 6–7 in., exhibition form, moderate fragrance; foliage dark; vigorous, bushy growth; [Frau Karl Druschki X Kaiserin Auguste Viktoria, Climbing]; Chambard, C.

'LOUISE CRINER', HT, w, 1919; flowers snow-white, center creamy, dbl.; few thorns; [Unknown seedling X Louise Crette]; Chambard, C.

'LOUISE D'ARZENS', N, w; Lacharme, F., 1861

'LOUISE DE VILMORIN', HT, ab, 1944; bud ovoid; flowers orange-yellow, overlarge, very dbl., slight fragrance; foliage glossy, dark; dwarf growth; Gaujard

'LOUISE ESTES', HT, pb, 1991; bud pointed; flowers pink blend, reverse white, aging medium pink, medium, dbl., 35 petals, 4 in., exhibition form, quick repeat, moderate, fruity fragrance; foliage medium, medium green, matt, disease-resistant; upright, medium growth; Silver, ARC TG, 1991; [Unnamed seedling X Miss Canada]; Winchel, Joseph F.; Coiner Nursery, 1992

Louise Gardner *see* 'MACERUPT'

'LOUISE GAUJARD', HT, op, 1941; flowers coppery pink shaded coral, open, very large, very dbl., cupped, moderate fragrance; foliage light green; vigorous, upright growth; [Mme Joseph Perraud X Seedling]; Gaujard

'LOUISE HOPKINS', HMult, lp, 1923; flowers white, center shell-pink, very dbl., 200–225 petals, non-recurrent; vigorous, climbing (20 ft. or more) growth; [Trier sport]; Hopkins

'LOUISE JOLY', HT, or, 1922; flowers coral-red, shaded shrimp-pink, dbl., moderate fragrance; [Unnamed seedling X Mme Edouard Herriot]; Buatois

'LOUISE KRAUSE', HT, ob, 1930; flowers reddish orange, passing to golden yellow, large, dbl., slight fragrance; foliage dark, glossy; vigorous, bushy growth; [Mrs Beckwith X Souv. de H.A. Verschuren]; Krause

'LOUISE LE CARDONNEL', HT, yb, 1939; bud globular; flowers yellow tinted coral, very large, dbl.; stiff stems; very vigorous growth; [Unnamed seedling X Mev. G.A. van Rossem]; Mallerin, C.; Meilland, A.

'LOUISE MACK', Min, dr, 1992; dbl., 30 petals, 1.25 in., exhibition form, borne usually singly, slight fragrance; foliage medium, dark green, semi-glossy; medium, spreading growth; [Unknown X Unknown]; Jerabek, Paul E., 1984

'LOUISE MÉHUL', HGal, pb; flowers light red, spotted white, large, flat; Parmentier

'LOUISE ODIER', B, dp, 1851; (Mme de Stella); flowers bright rose-pink, well-formed, full, dbl., cupped, intense fragrance; prickles very few; foliage medium green, matt; tall growth; (28); Margottin

'LOUISE PERNOT', HT, lp, 1903; flowers large, dbl.; Robichon; (Sangerhausen)

'LOUISE PEYRONY', HP, dp, 1844; flowers deep pink shaded carmine, very large, dbl., moderate fragrance; moderate growth; [La Reine seedling]; Lacharme, F.

'LOUISE PIGNÉ', T, mp, 1905; flowers china-pink, base buff-yellow, petals crinkled, very large, very dbl., moderate fragrance; very vigorous growth; [Mme Eugene Resal X Mme Lombard]; Pigné

'LOUISE VERGER', M, lp, 1860; flowers bright pink, medium, dbl.; Robert et Moreau

'LOUISE WALTER', Pol, pb, 1909; flowers white and flesh-pink, open, dbl.; foliage small, rich green; very dwarf growth; (14); [Tausendschön X Rosel Dach]; Walter, L.; P. Lambert

'LOUISIANA', HT, w, 1969; flowers cream/ white, dbl., 38 petals, 3–4 in., exhibition form, slight fragrance; foliage dark, leathery; upright growth; [Unnamed seedling X Unnamed seedling]; Weeks

Louisiana *see* 'KORNEMARK'

Louisiana Lady® *see* 'TALLOU'

'LOUISIANA PURCHASE', HT, dp, 1954; bud long, pointed; flowers rich cerise, dbl. 20–25 petals, 4–5 in., cupped, intense damask fragrance; foliage dark, leathery; very vigorous, upright growth [Charlotte Armstrong X Piccaninny] Swim, H.C.; Stark Bros.

Louisiana, Climbing, Cl HT, w, 1985; Weeks

'LOUISVILLE LADY', HT, pb, 1986; flowers bright pink, silver reverse, medium, dbl., 35 petals, exhibition form, intense fragrance; foliage medium, dark, semi-glossy; bushy growth; [Osiria X Unnamed seedling]; Weddle, Von C.

Louita, F, rb

Louksor *see* 'DELCRAFT'

'LOUMUSKOKA', HT, w, 1999; (**Muskoka Moonlight**); flowers near white, medium, dbl., 26–40 petals; some prickles; foliage medium, medium green, glossy; vigorous, spreading, tall growth; very hardy; [Pristine X Lichtkönigin Lucia]; Lougheed, Larry; Lougheed Hybrid Roses, 1998

Louqsor® *see* 'DELCRAFT'

'LOURDES', HT, dy, 1959; bud ovoid; flowers golden yellow, dbl., 35–50 petals, 4–5 in., moderate fragrance; foliage leathery, glossy, dark; vigorous growth; hardy for the class.; [V for Victory X (New Dawn sport X Copper Glow sport)]; Brownell, H.C.; Stern's Nursery

Lourdes Arroyo® *see* FEprogo

Lou's Sonoma Rose, HCh, dp; (found rose)

'LOUVIOLA', Gr, pb, 1997; (**Viola Lougheed**); flowers full, large, very dbl., 26–40 petals, borne in large clusters, moderate fragrance; many prickles; foliage medium, medium green, semi-glossy; tall, spreading growth; [Pristine X Peter Frankenfeld]; Lougheed, Larry

Louvre® *see* 'DELFAT'

'LOVABLE', HT, lp, 1979; bud long, pointed; dbl., 30 petals, 4.5–6 in., exhibition form, moderate fragrance; vigorous, upright, bushy growth; [(Helen Traubel X Michele Meilland) X ((Blanche Mallerin X Peace) X (Peace X Virgo))]; Leon, Charles F., Sr.

'LOVANIA', S, or, 1978; bud ovoid; flowers bright red-orange, pompon shape, small, semi-dbl., 18–22 petals, 1.5 in., recurrent bloom, moderate, fruity fragrance; foliage dark; vigorous, upright, climbing growth; [Robin Hood X (New Penny X Coloranja)]; Lens, 1977

Love *see* 'JACTWIN'

'LOVE', Cl HT, or, 1935; flowers scarlet, open, semi-dbl., moderate fragrance; foliage dark; long stems; vigorous, climbing (6 1/2-10 ft), bushy growth; RULED EXTINCT 1/79; [Hadley X Ami Quinard]; Caron, B.; H. Guillot

'LOVE AFFAIR', HT, dr, 1970; bud short pointed; flowers brilliant red, large, very dbl., exhibition form, moderate fragrance; foliage large, dark, leathery; vigorous, upright growth; [Undisseminated red seedling X Forever Yours]; Jelly; E.G. Hill Co.

'LOVE BUG', Min, ob, 1986; flowers orange, white base, reverse orange veining to white base, fades slightly, semi-dbl., 14 petals, exhibition form, slight fragrance; short, small, pink prickles, hooked slightly downward; foliage medium, medium green, semi-glossy; bushy, medium growth; [Heartland X Unnamed seedling]; Bridges, Dennis A., 1987

'LOVE CALL', HT, pb, 1989; bud ovoid; flowers deep pink, lighter at base, reverse light pink, large, dbl., 50 petals, exhibition form, moderate, damask fragrance; prickles pointed slightly downward, medium, light green; foliage medium, dark green, glossy; upright, tall growth; [Lady X X Wini Edmunds]; Bridges, Dennis A.; Bridges Roses, 1990

Love Dove *see* 'LYOVE'

Love in Bloom *see* 'JUDLOV'

Love In Bloom, Min, m

Love Knot *see* 'CHEWGLORIOUS'

'LOVE LETTER'®, F, w, 1977; flowers creamy white, dbl., 30–35 petals, 3–3.5 in., cupped, intense fragrance; foliage glossy, dark; vigorous, bushy, upright growth; [Pink Parfait X Rosenelfe]; Lens, 1980

Love Magic, HT, dr

Love Me *see* 'WALLOVE'

Love Note *see* 'ZIPNOTE'

Love Potion™ *see* 'JACSEDI'

'LOVE SONG', HT, pb, 1955; (Liebeslied); bud ovoid; flowers neyron rose, reverse yellow, dbl., 45 petals, 4.5–5 in., cupped, intense fragrance; foliage dark, glossy; vigorous, upright growth; [Peace X Orange Nassau]; Fisher, G.; C-P

Love Story® *see* 'TANVERY'

'LOVE TOKEN', F, mp, 1964; flowers peach-pink, well formed, dbl., 28 petals, 2 in.; foliage dark, glossy; vigorous growth; Gregory

Love Torch™ *see* 'MINABECO'

Love 'Ya Dad *see* 'TINDAD'

Love You *see* 'NEMLOVE'

Lovebird *see* FRAnmilro

Loveglo™ *see* 'MINJCO'

'LOVELIEST', HT, lp, 1956; bud ovoid; flowers clear rose-pink, large, dbl., exhibition form, intense fragrance; foliage leathery, light green; vigorous, upright growth; [Charlotte Armstrong X Juno]; Leon, Charles F., Sr.

'LOVELIGHT', F, mp, 1959; bud short, pointed; flowers large, dbl., 33 petals, exhibition form, slight fragrance; vigorous, upright growth; [Garnette X Unnamed seedling]; Jelly; E.G. Hill Co.

'LOVELINESS', HWich, lp, 1935; flowers pale pink, white spot at base, round, large, borne in large clusters, dbl.; foliage light; vigorous, climbing growth; Chaplin Bros.

'LOVELY', HT, dp, 1936; bud pointed; flowers carmine-pink, large, semi-dbl., exhibition form, moderate, violet fragrance; foliage soft; vigorous, compact growth; H&S; Dreer

Lovely Dream *see* HILraz

Lovely Fairy *see* SPEvu

Lovely Girl *see* MEljasonl

Lovely Jubilee, HT, m, 1992; Interplant

Lovely Lady *see* 'DICJUBELL'

'LOVELY LADY', HT, mr, 1934; bud long, pointed; flowers pure rose-red, large, very dbl., exhibition form, intense fragrance; foliage leathery; long, strong stems; very vigorous, open hait growth; RULED EXTINCT 10/86; [Better Times sport]; Asmus

Lovely Lorrie, Min, mp, 1998; Moore

Lovely Louise, F, ab

Lovely Pink Hybrid Musk, HMsk, mp; (found rose)

Lovely Rosamini, Min, lp

'LOVERLY', F, ab, 1986; flowers medium, semi-dbl., moderate fragrance; foliage medium, medium green, semi-glossy; bushy growth; Silver, ARC TG, 1985; [Restless Native X Apricot Nectar]; Stoddard, Louis

'LOVERS' MEETING', HT, ob, 1980; bud pointed; flowers bright orange, blooms borne singly and in clusters, dbl., 25 petals, exhibition form, moderate fragrance; short prickles; foliage bronze; strong, upright growth; [Unnamed seedling X Egyptian Treasure]; Gandy, O.L.

Lovers Only *see* 'SEALOVE'

Love's Spring *see* 'MACIVY'

Loving *see* JACka

Loving Memory *see* 'KORGUND '81'

'LOVING SON', HT, w, 1994; very dbl., 1.5–2.75 in., borne mostly singly, moderate fragrance; some prickles; foliage medium, medium green, semi-glossy; medium, bushy growth; [Honor X Captain Harry Stebbings]; Mengel, Russell R.; Mengel, 1994

'LOVING TOUCH'™, Min, ab, 1983; flowers mini-flora, blooms borne usually singly, exhibition form, 25 petals, exhibition form, slight; globular fruit; foliage medium, medium green, semi-glossy; bushy, spreading growth; AOE, 1985; [Rise 'n' Shine X First Prize]; Jolly, Nelson F.; Rosehill Farm, 1982

Lovita *see* **'KIBOH'**

'LOVITA', HT, mr, 1965; bud ovoid; flowers bright red, large, dbl.; foliage dark; vigorous, upright growth; [Baccará X (Independence X Peace)]; Meilland; C-P, 1967;, Moerheim, 1965

Lovita, F, ab, 1989

Lowburn Rose, Misc OGR, dp

Lowea berberifolia *see* **'HULTHEMI PERSICA'**

'LOWELL THOMAS', HT, dy, 1943; (Botaniste Abrial); bud long, pointed; flowers rich yellow, dbl., 38 petals, 4–4.5 in., exhibition form; foliage leathery; vigorous, upright, bushy, compact growth; AARS, 1944 GM, Portland, 1944; [Mme Mélanie Soupert X Nonin]; Mallerin, C.; A. Meilland, 1943;, C-P, 1943

'LOWELL THOMAS, CLIMBING', Cl HT, dy, 1954; Armstrong, J.A.; Armstrong Nursery

Loyal Rosarian *see* 'LEOLOYROS'

'LOYALIST', Misc OGR, lp, 1773; flowers pale rose fading almost white, very dbl., cupped, blooms well in june; repeats little, moderate fragrance; bushy, densely branched growth; [Possibly R. damascena X R. virginiana]

Luarca, HT; Dot, Simon; (Cavriglia)

Lübeck *see* **'HANSESTADT LÜBECK'**

Lübecker Rotspon® *see* 'TANTIDE'

'LUBOV CHEVTSOVA', HT, lp, 1956; flowers light pink, base tinted orange, large (5 in); foliage glossy; strong stems; upright growth; [Cathrine Kordes X Peace]; Sushkov, K. L.

'LUBRA', HT, dr, 1938; bud long, pointed; flowers dark crimson, dbl., exhibition form, intense fragrance; foliage leathery, dark; vigorous growth; [Ophelia, Climbing seedling X Black Boy]; Fitzhardinge; Hazlewood Bros.

Luc Steeno, HT, mp, 1997; RvS-Melle

'LUC VARENNE', F, mr, 1959; bud oval; flowers scarlet, open, medium, borne in clusters, semi-dbl.; foliage dark, glossy; vigorous, bushy growth; [Alain X Montrouge]; Delforge

'LUCDOD', S, mp, 1981; (**Rendez-vous**); flowers medium, semi-dbl., intense fragrance; foliage medium, medium green, matt; bushy growth; [R. wichurana X Alain Blanchard]; Lucas, C.C.

Luce di Todi™ *see* 'WILLUCE'

'LUCENS ERECTA', S, op, 1921; Paul & Son

Lucetta® *see* 'AUSEMI'

Luchian, F, lp; globular; flowers large, velvety vermillion red, dbl., 35 petals, slight; [Paprika X Coup de Foudre]; Palocsay and St. Wagner, 1971; Res. Stn. f. Horticulture, Cluj, 1972, Romania

Lucia®, S, my

Lucia Cotarelo® *see* FEtural

'LUCIA ZULOAGA', HT, rb, 1932; flowers velvety brownish scarlet with a golden undertone, open, very large, semi-dbl., moderate, fruity fragrance; foliage glossy, dark; [Duquesa de Peñaranda X F. Cambo]; Dot, Pedro; C-P, 1934

'LUCIE DUPLESSIS', M, mp; flowers rosy white; Robert, 1853

'LUCIE FERNAND-DAVID', HT, w, 1924; flowers white, center slightly tinted cream, dbl., intense fragrance; Chambard, C.

'LUCIE MARIE', HT, yb, 1930; bud pointed; flowers buttercup-yellow veined apricot-orange and shaded salmon-cerise, dbl., exhibition form, moderate fragrance; foliage dark, leathery, glossy; vigorous, compact, bushy growth; Dickson, A.

'LUCIE NICOLAS MEYER', HT, dp, 1922; flowers dark pink, edged lighter, dbl., moderate fragrance; [Jonkheer J.L. Mock X Unnamed seedling]; Gillot, F.

'LUCIE PETZKA', Pol, dr, 1953; flowers medium, dbl.; Petzka; (Sangerhausen)

'LUCIEN CHAURÈ', HT, lp, 1913; flowers large, dbl.; Soupert & Notting; (Sangerhausen)

'LUCIEN DE LEMOS', HT, lp, 1905; flowers large, dbl., moderate fragrance; Lambert, P.; (Sangerhausen)

'LUCIEN DURANTHON', HP, mr, 1894; flowers large, dbl.; Bonnaire; (Sangerhausen)

Lucie's Dream *see* 'MEICITREM'

'LUCIFER', Pol, or, 1931; flowers small, dbl.; Easlea; (Sangerhausen)

'LUCILE', HWich, lp, 1911; flowers flesh-pink, base tinged rosy salmon, borne in clusters, dbl.; foliage large, rich green, glosy; vigorous, climbing growth; Walsh

'LUCILE BARKER', HT, ab, 1922; bud pointed; flowers apricot-yellow, semi-dbl., exhibition form, moderate fragrance; foliage bronze; vigorous growth; Hicks

'LUCILE DUPLESSIS', HGal, dp, 1836; Vibert

'LUCILE HILL', HT, dp, 1939; bud long, pointed, rose-red; flowers spinel-pink, very large, dbl., slight fragrance; foliage leathery; long stems; very vigorous, compact growth; [(Senior X De Luxe) X Sweet Adeline]; Hill, Joseph H., Co.

'LUCILE RAND', HT, pb, 1930; bud pointed; flowers brilliant carmine, shaded yellow and orange, very large, dbl., intense fragrance; very vigorous, semi-climbing, bushy growth; Pernet-Ducher; Gaujard

'LUCILE SUPREME', HT, dp, 1941; flowers rose-red, dbl., 25–35 petals, 6 in.; [Lucile Hill sport]; Hill, Joseph H., Co.

'LUCILLA', F, ab, 1992; semi-dbl., 6–14 petals, 1.5 in., borne in small clusters, slight fragrance; some prickles; foliage small, medium green, semi-glossy; low (35 cms), compact growth; [Conservation sport]; Stainthorpe, Eric; Battersby Roses, 1993

Lucille Ball *see* 'JACAPRI'

Lucille McWherter *see* 'WELLSNED'

'LUCILLE ROSS', LCl, w, 1940; flowers white, center golden, open, semi-dbl., slight fragrance; vigorous, climbing (20-40 ft) growth; [Dr. W. Van Fleet sport]; Ross; Roselawn Gardens

'LUCINDA', HT, dp, 1927; [Columbia sport]; Heacock

Lucinde® *see* KORtaly

'LUCIOLE', T, dp; bud long; flowers carmine-rose, base coppery yellow, large, intense fragrance; Guillot et Fils, 1887

'LUCIOLE', HWich, rb, 1923; flowers bright scarlet, center white, small, single, slight fragrance; vigorous, climbing growth; [Hiawatha X Unknown seedling]; Nonin

Lucious *see* 'RESJUC'

'LUCKY', F, dr, 1962; flowers deep velvety red, semi-dbl., 13 petals, 3 in.; [Goldilocks X Independence]; Leenders, J.

'LUCKY BEAUTY', HT, pb, 1970; bud ovoid; flowers pink, yellow reverse, large, very dbl., 90 petals, exhibition form, moderate fragrance; vigorous, upright, bushy growth; [Kordes' Perfecta X Lucky Piece]; Fuller; Wyant

'LUCKY CHARM', F, yb, 1961; flowers bright yellow tipped red, borne in clusters of 6-10, semi-dbl., 18 petals, 3.5–4 in., intense fragrance; foliage glossy; vigorous growth; Robinson, H.; Lowe

Lucky Charm *see* 'MORAIN'

'LUCKY CHOICE', HT, w, 1985; flowers creamy white, flushed light pink in center, large blooms borne 1-3 per stem, dbl., 35 petals, exhibition form, slight fragrance; few, small, slender prickles; foliage large, medium green; upright growth; [American Heritage X Sodori-Hime]; Ota, Kaichiro

Lucky Lady *see* 'ARMLU'

Lucky Me *see* SOCluck

'LUCKY PIECE', HT, pb, 1962; flowers copper, pink and gold blend; [Peace sport]; Gordon; Wyant

'LUCKY STAR', HT, yb, 1936; flowers golden yellow, suffused flame-scarlet; [Souvenir sport]; Armacost; Armacost & Royston

Lucretia, F, mr

'LUCY', Cl HT, dp, 1936; flowers brilliant carmine, well formed, large; foliage glossy; very vigorous, climbing growth; Williams, A.

Lucy *see* BARluc

Lucy *see* 'MORLUCY'

'LUCY ASHTON', HEg, w, 1894; flowers pure white, edged pink, single, seasonal bloom; foliage dark, fragrant; vigorous growth; (42); Penzance

'LUCY BERTRAM', HEg, rb, 1895; flowers dark shining crimson, center white, single, seasonal bloom; foliage dark, fragrant; very vigorous growth; Penzance

'LUCY CONSTABLE', HT, pb, 1924; flowers silver pink, reverse deep salmon pink, moderate fragrance; Lilley

'LUCY CRAMPHORN', HT, or, 1960; (Maryse Kriloff); flowers signal-red, well-formed dbl., 40 petals, 5 in., moderate fragrance; foliage glossy; vigorous upright growth; [Peace X Baccará]; Kriloff, Michel; Cramphorn's Nursery

'LUCY MARGUERITE', HT, yb, 1978; flowers buttercup-yellow, diffused red, full, dbl., 35–40 petals, 4 in., moderate fragrance; foliage small, glossy, dark; [Val De Mosa X Denise-Anne]; Ellick; Excelsior Roses

'LUCY NICOLAS', HT, r, 1935; bud pointed; flowers coppery salmon, large, dbl., exhibition form; foliage glossy, bronze; very vigorous growth; [Odette Foussier X Cécile Walter]; Mallerin, C.; C-P;, H. Guillot

'LUCY THOMAS', Cl HP, mp, 1924; flowers pink, center brighter, semi-dbl., rarely recurrent bloom; [Ulrich Brunner Fils X Georg Arends]; Nabonnand, P.

'LUCYLE', HT, mp, 1933; flowers glowing pink, very large, dbl., intense fragrance; foliage leathery, dark; vigorous, bushy growth; Vestal

'LUDEK PIK', HP, mr, 1933; flowers large, very dbl.; Böhm, J.; (Sangerhausen)

Ludmilla® *see* 'LAPMAU'

LUDpinbu, Min, ab, 1993; (**Joy Button**); Taschner

LUDsporcoma, HT, rb, 1995; (**Flames 'n' Sparks**); Taschner

LUDsportiana, HT, dp, 1995; (**Alberton Amor**); Taschner

LUDswenic, HT, ob, 1995; (**Nicolette®**); Taschner

Ludvik Vecera, LCl, mr, 1981; flowers medium, dbl., moderate fragrance; Vecera, L.; (Sangerhausen)

'LUDWIG MÖLLER', HP, ab; flowers bright amber-yellow, fading white; [Frau Karl Druschki X Marechal Niel seedling]; Kiese, 1915

'LUDWIG OPPENHEIMER', HT, dr, 1932; bud pointed; flowers crimson-scarlet, large, dbl., moderate fragrance; foliage dark; vigorous growth; [Villa Pia X Capitaine Georges Dessirier]; Leenders Bros.

Ludwigshafen *see* 'KORLUDWIG'

Ludwigshafen am Rhein® *see* 'KORLUDWIG'

'LUIS BRINAS', HT, ob, 1934; bud long, pointed; flowers rose-orange, large, dbl., cupped, moderate fragrance; foliage soft; vigorous growth; GM, Bagatelle, 1932 GM, Portland, 1934; [Mme Butterfly X Federico Casas]; Dot, Pedro; C-P

Luis Desamero *see* 'TINLUIS'

'LUISA FERNANDA DE SILVA', HT, rb, 1946; flowers reddish passing to purplish and then yellowish red, base yellow, moderate fragrance; foliage dark, glossy; vigorous growth; Dot, Pedro

Luisa Stone, S, 1997; Harkness, R.; (Cavriglia)

Luise Kiese *see* **'FRAU LUISE KIESE'**

'LUISE KÜHNEL', HT, dy, 1937; flowers medium, dbl.; Kühnel; (Sangerhausen)

'LUISE LILIA', HT, dr, 1912;, intense fragrance; moderate growth; Lambert, P.

Luisella Angelini, HT, 1980; Dot, Simon; (Cavriglia)

'LUKOR', HT, rb, 1973; (Jolly, **Lustige**); bud ovoid; flowers copper-red, reverse yellow, large, dbl., cupped, moderate fragrance; foliage large, glossy, leathery; vigorous, upright growth; [Peace X Brandenburg]; Kordes

'LULLABY', Pol, w, 1953; bud ovoid; flowers white, center flushed pink, blooms in loose clusters, dbl., 75 petals, 1.5–2 in., cupped, slight fragrance; foliage dark, leathery; vigorous, bushy, compact growth; [(R. soulieana X Mrs Joseph Hiess) X Cécile Brunner]; Shepherd; Bosley Nursery

'LULLABY', F, dp, 1957; flowers rich rose-pink, camellia shape, borne in clusters (to 28), semi-dbl., 3 in., flat; vigorous growth; Bishop; Baker's Nursery

'LULU', HT, op, 1919; bud very long, pointed, deep orange-red; flowers salmon-pink, large, single, 8 petals, slight fragrance; foliage glossy; bushy, compact growth; Easlea

'LULU', F, ob, 1973; flowers orange-pink, medium, dbl., exhibition form, slight fragrance; foliage glossy, dark, bronze; vigorous, upright, bushy growth; [Zorina X Seedling]; Kordes

'LULUETTE', Min, mp, 1986; flowers small blooms borne singly and in small sprays, single, no fragrance; small, straight, red prickles; foliage small, medium green, semi-glossy; low, compact, tiny growth; [Fairy Moss X Fairy Moss]; Fischer, C.&H.; Alpenflora Gardens

Lumen®, HT, ab, 1990; Sauvageot

'LUMIÈRE', HT, yb, 1944; flowers golden yellow suffused capucine-red, large; foliage dark, leathery; vigorous growth; Mallerin, C.; A. Meilland

'LUMINA', F, or, 1955; flowers orange-scarlet, rosette shape, blooms in large trusses, dbl., 25 petals, 2.5 in.; foliage dark; vigorous growth; [Fanal X Alpine Glow]; Tantau, Math.

'LUMINATOR', Pol, dr, 1938; flowers scarlet-crimson; vigorous growth; [Lady Reading sport]; Smith, J.; Eddie

Luminion® *see* 'KORMIORA'

'LUMINOSA', HT, or, 1964; bud long, pointed; flowers vermilion, large, dbl., 52 petals, exhibition form; foliage dark, glossy; strong stems; vigorous, upright, bushy growth; Mondial Roses

'LUMINOUS', Pol, or, 1932; flowers brilliant scarlet tinted orange, small, dbl., cupped, slight fragrance; foliage small, light, wrinkled; dwarf growth; [Gloria Mundi sport]; deRuiter; J&P

'LUM'S DOUBLE WHITE', HT, w, 1930; flowers pure white, overlarge, semi-dbl., exhibition form, moderate fragrance; foliage dark, leathery; vigorous growth; [Killarney Double White sport]; Lum; Totty

'LUNA', HT, ly, 1918; bud pointed; flowers pale yellow, large, dbl., exhibition form, intense fragrance; foliage dark; strong stems; very vigorous, bushy growth; [Harry Kirk X Sunburst]; Poulsen, S.

'LUNA PARK', LCl, or, 1964; flowers red shaded orange, large, recurrent bloom; vigorous, climbing growth; [Gladiator X Unnamed seedling]; Croix, P.

Luna Rossa, HT; Delbard-Chabert; (Cavriglia)

Lund's Jubiläum *see* **'BABY BLAZE'**

Lundy's Lane Yellow, HT, ob; (found rose)

'LUNELLE', HT, lp, 1955; (George Sand); flowers pale pink, very large, dbl., 50 petals, exhibition form, intense fragrance; vigorous growth; [Young France X Signora]; Meilland, F.

'LURAY', F, mr, 1958; flowers rose-red, medium, dbl., 40–45 petals, 2–3 in., moderate fragrance; foliage leathery, glossy; vigorous, bushy growth; [Patty's Pink sport]; Masek; Carlton Rose Nurseries

Lusambo *see* 'MEILUSAM'

Lustige *see* 'LUKOR'

'LUSTRE', HT, mp, 1926; flowers rose-pink, large, semi-dbl., moderate fragrance; [Ophelia X Hoosier Beauty]; Hill, E.G., Co.; Amling Bros.

'LUSTRE D'ENGLISE', HGal, mp, 1813; flowers medium pink with lighter reverse, small, dbl., intense fragrance; Roseraie de l'Hay; (Sangerhausen)

Lustrous *see* **'CELEBRITY'**

'LUTETIA', HT, yb, 1961; flowers coppery yellow, medium, dbl., 28 petals; vigorous, upright growth; [The Optimist X Tudor]; deRuiter

'LUTHER RUSSELL', F, op, 1956; flowers salmon, borne several together, semi-dbl.; foliage olive-green; [Korona sport]; Morse

Lutin *see* 'MEIRADIA'

Lutin *see* MEIsecaso

Luv Ya *see* 'JULESLOVE'

'LUVVIE', Min, pb, 1979; bud ovoid; flowers soft to deep coral pink, micro-mini, very small, dbl., 38 petals, exhibition form, slight fragrance; straight, thin, red prickles; foliage small, deep green with red peduncle and petiole; low, bushy growth; [Little Darling X Over the Rainbow]; Bennett, Cecilia 'Dee'; Tiny Petals Nursery, 1980

Luxembourg *see* **'FANNY BLANKERS-KOEN'**

'LUXEMBOURG', M, dr; flowers crimson, not very mossy; vigorous growth

Luxembourg *see* **'MARIE ADÉLAÏDE'**

'LUXEMBOURG, CLIMBING', Cl HT, op, 1932; flowers pinkish orange, base deep yellow; very vigorous, climbing growth; [Marie Adélaide sport]; Wight

Luxor, HT, mp

'LUXURY', HT, rb, 1968; flowers light red, reverse white, large, dbl., exhibition form, moderate fragrance; foliage soft; moderate growth; [Unnamed seedling X Suspense]; Patterson; Patterson Roses

Lybelle, LCl, mp, 1984; flowers large, dbl.; Noack, Werner; (Sangerhausen)

Lybie *see* 'MARLYB'

'LYCORIS', M, pb; flowers light rosy red spotted white, large, dbl., flat

Lyda Rose *see* LETlyda

'LYDIA', HT, dy, 1949; bud long, pointed; flowers intense saffron-yellow, medium, very dbl., exhibition form, moderate fragrance; foliage dark, leathery, glossy; vigorous, bushy growth; [Phyllis Gold X Seedling]; Robinson, H.; Baker's Nursery

'LYDIA'®, S, ob, 1973; (Clubrose Lydia); bud ovoid; flowers deep orange, reverse yellow, medium, semi-dbl., cupped, moderate fragrance; foliage glossy, dark, leathery; very vigorous, upright, climbing growth; [Seedling X Circus]; Kordes

'LYDIA', HT, op, 1933; bud pointed; flowers bright orange-rose, large, very dbl., moderate fragrance; foliage glossy; vigorous, bushy growth; [Briarcliff X Florex]; Verschuren; Dreer;, H&S

Lydia *see* INTerlis

Lydia, HT, op; Croix

'LYDIA GRIMM', HT, ly, 1907; flowers large, dbl., moderate fragrance; Geduldig; (Sangerhausen)

Lydia Morris, S, m

Lydie, F, ob, 1993; Harkness

Lykke Dazia, HT, ob

Lykke White Pet, Pol, w

'LYKKEFUND', LCl, w, 1930; flowers cream color, blooms in clusters, semi-dbl., intense fragrance; no prickles; foliage dark, glossy; vigorous growth; [Seedling from R. helenae, Barbier's form, possibly crossedwith Zephirine Drouhin sport]; Olsen

Lyla Barbour, HT, lp; Watt, Mrs B., 1970; (Weatherly, L.)

Lynda, HT, 1970; Mondial Roses; (Cavriglia)

'LYNDA HURST', HT, mp, 1938; flowers large, dbl.; vigorous growth; [Mme Abel Chatenay X Seedling]; Clark, A.; NRS Victoria

Lyndarajha, HT, 1982; Dorieux; (Cavriglia)

Lynette *see* 'KORLYN'

Lynn Anderson *see* 'WEKJOE'

'LYNN ANNE', Min, ob, 1981; bud short, pointed; flowers orange-yellow blend, blooms borne singly and in large sprays, dbl., 38 petals, exhibition form, moderate fragrance; long, thin prickles; foliage medium; vigorous, upright, compact growth; [Rise 'n' Shine X Sheri Anne]; Saville, F. Harmon; Nor'East Min. Roses

Lynne Elizabeth™ *see* 'BYRLYNNE'

Lynne Gold™ *see* 'MORLYN'

'LYOAD', Min, or, 1984; (**Added Touch**); bud small; flowers orange-red, touch of yellow in the center and back of petals, dbl., 20 petals, moderate fragrance; foliage small, medium green, semi-glossy; upright, bushy growth; [Dandy Lyon X Seedling]; Lyon

'LYOCA', Min, ab, 1982; (**Care Deeply**); flowers medium, dbl., 35 petals, moderate fragrance; foliage medium, medium green, semi-glossy; upright, bushy growth; [Honey Hill X Seedling]; Lyon, Lyndon; L. Lyon Greenhouses

'LYOCH', Min, dr, 1982; (**Cherokee Fire**); flowers deep red, medium, semi-dbl., no fragrance; foliage medium green, semi-glossy; upright, bushy growth; [Merry Christmas X Seedling]; Lyon, Lyndon; L. Lyon Greenhouses

'LYOCL', Min, mp, 1983; (**Classic Love**); flowers small blooms in clusters, dbl., 20 petals, no fragrance; foliage small, medium green, semi-glossy; upright, bushy growth; [Baby Betsy McCall X Seedling]; Lyon, Lyndon; L. Lyon Greenhouses

'LYOCO', Min, ab, 1982; (**Coral Fantasy**); flowers medium, dbl., 34 petals, moderate fragrance; foliage medium green, semi-glossy; vigorous, upright, bushy growth; [Dandy Lyon X Seedling]; Lyon, Lyndon; L. Lyon Greenhouses

'LYODIT', Min, dr, 1986; (**Ditto**); flowers very small blooms borne usually singly, semi-dbl., 14 petals, no fragrance; no prickles; foliage small, medium green, matt; low, bushy growth; [Baby Betsy McCall X Seedling]; Lyon, Lyndon; M.B. Farm Min. Roses, Inc.

'LYODRO', Min, my, 1986; (**Sun Drops**); dbl., 35 petals, cupped, borne usually singly, slight fragrance; very small, reddish prickles; foliage medium, medium green, matt; low, bushy growth; [Unnamed seedling X Redgold]; Lyon; M.B. Farm Min. Roses, Inc.

'LYOET', Min, dr, 1982; (**Velvet Dreams**); flowers small, semi-dbl., slight fragrance; foliage small, medium green, semi-glossy; very miniature, upright, bushy growth; [Unnamed seedling X Unnamed seedling]; Lyon

'LYOFI', Min, dr, 1982; (**Night Fire**); flowers deep red, petals often edged black, small, dbl., 20 petals, moderate fragrance; foliage medium, dark, semi-glossy; upright, bushy growth; [Unnamed seedling X Unnamed seedling]; Lyon, Lyndon

'LYOFIN', Min, ab, 1986; (**Fine Touch**); flowers pale apricot, deeper in center, large blooms borne singly, dbl., 37 petals, exhibition form, slight fragrance; globular, medium, orange fruit; few, reddish prickles; foliage medium, medium green, matt; medium, upright growth; [Honey Hill X ?]; Lyon; M.B. Farm Min. Roses, Inc.

'LYOFOR', Min, mr, 1986; (**Forevermore**); flowers large blooms borne singly, dbl., 38 petals, exhibition form, intense, spicy fragrance; globular, medium, orange-red fruit; few, reddish, small prickles; foliage medium, medium green, semi-glossy; medium, upright growth; [Unnamed seedling X ?]; Lyon; M.B. Farm Min. Roses, Inc.

'LYOGI', Min, pb, 1982; (**Giggles**); bud pointed; flowers medium pink, white center, small, semi-dbl., slight fragrance; foliage small, medium green, semi-glossy; very small, upright, bushy growth; [Unnamed seedling X Unnamed seedling]; Lyon, Lyndon; L. Lyon Greenhouses

'LYOHI', Min, mr, 1983; (**High Style**); flowers cardinal red, blooms in clusters, dbl., 35 petals, slight fragrance; foliage medium, medium green, semi-glossy; upright, bushy growth; [Unnamed seedling X Unnamed seedling]; Lyon

'LYOLIT', Min, mp, 1984; (**Little Charm**); flowers small, semi-dbl., no fragrance; foliage small, medium green, semi-glossy; upright, bushy growth; [Unnamed seedling X Unnamed seedling]; Lyon, Lyndon

'LYOLITE', Min, ab, 1986; (**Peach Elite**); flowers peach, to loose form, large blooms borne singly, dbl., 26 petals, exhibition form, no fragrance; globular, medium, orange-red fruit; small prickles; foliage large, medium green, matt; upright growth; [Dandy Lyon X ?]; Lyon; M.B. Farm Min. Roses, Inc.

'LYOMA', Min, dr, 1983; (**Magic Splendor**); flowers small blooms in clusters, dbl., 35 petals, no fragrance; foliage medium, medium green, semi-glossy; upright, bushy growth; [Baby Betsy McCall X Seedling]; Lyon, Lyndon

'LYON RAMBLER', HMult, dp, 1909; flowers bright rose-pink, flushed carmine; [Crimson Rambler seedling]; Dubreuil

'LYON ROSE', HT, op, 1907; flowers shrimp-pink, center coral-red shaded yellow, large, dbl., 44 petals, moderate fragrance; (28); GM, Bagatelle, 1909; [Mme Mélanie Soupert X Soliel c'Or seedling]; Pernet-Ducher

'LYON ROSE, CLIMBING', Cl HT, op, 1924; Ketten Bros.

'LYONFARBIGE DRUSCHKI', HP, yb, 1928; flowers yellowish-pink, large, dbl., slight fragrance; Rosarium Sangerhausen; (Sangerhausen)

'LYONNAIS', HP, mp, 1872; flowers large, dbl.; Lacharme, F.; (Sangerhausen)

'LYOPIN', Min, mp, 1984; (**Pink Charmer**); flowers medium, dbl., 20 petals, moderate fragrance; foliage small, medium green, semi-glossy; upright, bushy growth; [Baby Betsy McCall X Unnamed seedling]; Lyon, Lyndon

'LYOPR', Min, my, 1982; (**Precious Moments**); flowers medium, dbl., 35 petals, moderate fragrance; foliage medium, medium green, semi-glossy; upright, bushy growth; [Dandy Lyon X Unnamed seedling]; Lyon

'LYOPRI', Min, pb, 1986; (**I Promise**); flowers pink, reverse pale yellow, medium, borne usually singly, dbl., 28 petals, exhibition form, moderate, spicy fragrance; globular, very small, red-gold fruit; few, small prickles; foliage medium, medium green, semi-glossy; low, bushy growth; [Unnamed seedling X Unnamed seedling]; Lyon; M.B. Farm Min. Roses, Inc.

'LYOPRO', Min, dr, 1985; (**Promenade**); flowers very dark red, informal, ruffled, small blooms borne singly, dbl., 35 petals, no fragrance; foliage medium, dark, matt; bushy growth; [Red Can Can X Unnamed seedling]; Lyon; M.B. Farm Min. Roses, Inc.

'LYORA', Min, ob, 1985; (**Orange Sherbert**); flowers orange, small blooms borne singly, dbl., 35 petals, cupped, moderate fragrance; foliage medium, medium green, matt; upright growth; [Dandy Lyon X ?]; Lyon; M.B. Farm Min. Roses, Inc.

'LYOREN', Min, mr, 1986; (**Renegade**); dbl., 52 petals, cupped, borne usually singly, no fragrance; small, green prickles; foliage medium, medium green, matt; medium, upright growth; [Unnamed seedling X Unnamed seedling]; Lyon; M.B. Farm Min. Roses, Inc.

'LYOSHY', Min, mp, 1985; (**Shy Beauty**); flowers small blooms borne singly, dbl., 55 petals, slight fragrance; foliage medium, medium green, semi-glossy; bushy growth; [Unnamed seedling X ?]; Lyon; M.B. Farm Min. Roses, Inc.

'LYOSS', Min, mr, 1982; (**Endless Love**); flowers medium, dbl., 35 petals, no fragrance; foliage medium, dark, semi-glossy; upright, bushy growth; [Red Can Can X Seedling]; Lyon, Lyndon

'LYOSUN', Min, yb, 1984; (**Sun Sparkle**); flowers yellow blended red, small, dbl., 20 petals, moderate fragrance; foliage small, medium green, semi-glossy; upright, bushy growth; [Dandy Lyon X Unnamed seedling]; Lyon

'LYOTER', Min, mr, 1984; (**Cheery Chatter**); flowers medium, dbl., 20 petals, moderate fragrance; foliage small, medium green, semi-glossy; upright, bushy growth; [Dandy Lyon X Seedling]; Lyon, Lyndon; L. Lyon Greenhouses

'LYOTO', Min, mr, 1982; (**Twinkle Toes**); flowers small, semi-dbl., no fragrance; foliage small, medium green, semi-glossy; very compact growth; [Merry Christmas X Unnamed seedling]; Lyon

'LYOVE', Min, w, 1982; (**Love Dove**); flowers near white, medium, dbl., 35 petals, exhibition form, intense fragrance; foliage medium, medium green, semi-glossy; low growth; [Unnamed seedling X Unnamed seedling]; Lyon, Lyndon

'LYOWE', Min, mr, 1982; (**Wild One**); flowers medium red, lighter center, medium, semi-dbl., no fragrance; foliage medium, dark, semi-glossy; bushy, upright growth; [Merry Christmas X Unnamed seedling]; Lyon

'LYRIC', S, mp, 1951; bud ovoid; flowers rose-pink, blooms in large clusters, dbl., 28 petals, cupped, moderate fragrance; foliage leathery; vigorous (4 ft), upright growth; good for hedge and border; [Sangerhausen X Seedling]; deRuiter; J&P

'LYRICAL', F, pb, 1989; bud pointed; flowers salmon pink shading to cream at base, medium, semi-dbl., 11 petals, flat, borne in sprays of 5-7, moderate, spicy fragrance; fruit not observed; prickles hooked downward, red; foliage medium, dark green, glossy, disease-resistant; upright, bushy, low, winter-hardy growth; [Snow White X Yesterday]; Jobson, Daniel J.

'LYS ASSIA', F, mr, 1958; flowers deep orange-scarlet, borne in small clusters, semi-dbl., 20 petals, 4 in., exhibition form, slight fragrance; foliage glossy; vigorous, upright, bushy growth; [Spartan X Hens Verschuren]; Kordes, R.

Lysa, HT, w, 1989; RvS-Melle

'LYSBETH-VICTORIA', F, lp, 1978; flowers light shell-pink, semi-dbl., 11 petals, 4.5 in., slight fragrance; foliage matt; growth moderate; [Pink Parfait X Nevada]; Harkness

M

M Boncenne *see* **'MONS. BONCENNE'**

'M. A. KEESSEN', Pol, dp, 1923; flowers darker; [Ellen Poulsen sport]; Keessen

'M. A. WILLETT', HT, op, 1959; bud globular; flowers pink to salmon-pink, base coppery, large, dbl., moderate fragrance; foliage leathery; moderate, upright growth; [Picture X Tahiti]; Kernovski, V.R.; Langbecker

'M. B.', HT, rb, 1941; flowers red shading to pink, overlaid orange and yellow, open, large, semi-dbl., 25 petals; long stems; vigorous, compact, upright growth; [Dr. W. Van Fleet hybrid X Frau Karl Druschki hybrid]; Brownell, H.C.

'M. GEIER', HT, dp, 1929; flowers fiery dark carmine, shaded darker, dbl.; [Augustus Hartmann X Admiral Ward]; Felberg-Leclerc

'M. H. WALSH', HP, mr, 1905; flowers carmine-red, large, dbl., moderate fragrance; Dickson, A.; (Sangerhausen)

'M. KOCKEL', Pol, lp, 1926; flowers small, dbl.; Kockel; (Sangerhausen)

'M. S. HERSHEY', HT, dr, 1941; bud long, pointed; flowers velvety crimson, dbl., 33 petals, 4–4.5 in., cupped, slight fragrance; bushy growth; [Seedling X E.G. Hill]; Coddington; C-P

'MA FIANCÉE', HT, dr, 1922; flowers dark crimson, often nearly black, dbl., intense fragrance; [Gen. MacArthur seedling X Red-Letter Day]; Van Rossem

'MA FILLE', HT, ob, 1960; flowers orange, large, dbl., 45 petals, slight fragrance; foliage glossy; vigorous, bushy, symmetrical growth; [Berthe Mallerin X Unnamed seedling]; Mallerin, C.; EFR

'MA MIE', HT, yb, 1955; flowers yellow, edge tinted pink, well formed, very large, dbl., 50 petals; foliage bright green; very vigorous, upright growth; [Peace X Unnamed seedling]; Laperrière; EFR

Ma Pâqueretta *see* **'PÂQUERETTE'**

'MA PERKINS', F, pb, 1952; bud ovoid; flowers sparkling salmon shell-pink, dbl., 25 petals, 3.5 in., cupped, moderate fragrance; foliage rich green, glossy; vigorous, bushy growth; AARS, 1953; [Red Radiance X Fashion]; Boerner; J&P

'MA PONCTUÉE', M, pb, 1847; (Ponctuée); flowers rose, spotted white, medium, semi-dbl., recurrent bloom; moderate growth; Moreau et Robert

'MA SURPRISE', S, w, 1872; flowers white tinged salmon, large, dbl., moderate fragrance; [Probably R. roxburghii X R. odorata]; Guillot et Fils

'MA TULIPE', HT, mr, 1899; flowers carmine-red, large, semi-dbl., moderate fragrance; Bonnaire; (Sangerhausen)

Maaike®, HT, pb

Maaseik 750, HMsk, pb, 1994; Lens

'MAB GRIMWADE', HT, yb, 1937; flowers rich chrome, shaded yellow, center apricot, dbl.; [Souv. de Gustave Prat X ?]; Clark, A.; NRS Victoria

'MABEL DOT', Min, or, 1966; flowers rose-coral, small blooms in clusters, dbl.; foliage small, bronze; [Orient X Perla de Alcanada]; Dot; Minier

'MABEL DREW', HT, w, 1911; flowers deep cream, center canary-yellow, borne in clusters, dbl., moderate fragrance; GM, Bagatelle, 1913 GM, NRS, 1910; Dickson, A.

'MABEL FRANCIS', HT, my, 1943; flowers rose-pink, dbl., 35–40 petals, 5 in., moderate fragrance; foliage light green; free growth; [Leading Lady X Southport]; Bees

'MABEL JACKSON', HT, ab, 1924; flowers apricot and pink, dbl., intense fragrance; [Edith Part X Queen Mary]; Easlea

'MABEL LYNAS', HT, dr, 1926; bud pointed; flowers crimson-scarlet, base yellow, large, dbl., moderate fragrance; foliage glossy; strong stems; very vigorous, bushy growth; GM, NRS, 1924; McGredy

'MABEL MORRISON', HP, w, 1878; flowers flesh-white, becoming pure white, sometimes tinged pink in autumn, dbl., 30 petals, cupped, seasonal bloom; stout, erect growth; [Baroness Rothschild sport]; Broughton

'MABEL MORSE', HT, my, 1922; bud pointed; flowers bright golden yellow, well shaped, large, dbl.; foliage dark, bronze; bushy growth; GM, NRS, 1921; McGredy

'MABEL MORSE, CLIMBING', Cl HT, my, 1931; Moulden

'MABEL MORSE, CLIMBING', Cl HT, my, 1932; Ley

'MABEL PRENTICE', HT, mp, 1923; flowers clear rose-pink; Lippiatt

'MABEL STEWART', LCl, dr, 1942; bud almost black; flowers velvety crimson, semi-single; Clark

'MABEL TURNER', HT, pb, 1923; bud pointed; flowers blush, center and reverse rosy carmine, very large, exhibition form, slight fragrance; foliage olive-green; vigorous growth; Dickson, H.

Mabella *see* 'KORGOLD'

'MABELLE STEARNS', S, mp, 1938; (Maybelle Stearns); flowers peach-pink with silvery reflex, blooms in clusters, dbl., 55 petals, recurrent bloom, intense fragrance; foliage small, glossy, dark; height 2 ft., spreading (6-8 ft.) growth; hardy.; [Mrs F.F. Prentiss X Souv. de Georges Pernet]; Horvath; Wayside Gardens Co.

Mableton Crimson China, Ch, rb

'MACAFT', HT, op, 1979; (Margaret Trudeau, **Sweepstakes**); bud long, pointed; flowers coral-orange to salmon-pink, imbricated, dbl., 35 petals, 3.5–5 in., exhibition form, moderate fragrance; upright, spreading, bushy growth; [Prima Ballerina X Ginger Rogers]; McGredy, Sam IV; Armstrong Nursery

'MACAI', F, pb, 1961; (**Daily Sketch**); flowers petals silver, edged deep pink, well-formed, dbl., 46 petals, 3.5 in., blooms in clusters, moderate fragrance; foliage dark; vigorous, bushy growth; GM, NRS, 1960; [Ma Perkins X Grand Gala]; McGredy, Sam IV; McGredy

'MACAL', F, op, 1977; (**Captain Cook**); flowers orange-salmon, 10 petals, 4 in., moderate fragrance; foliage glossy; [Irish Mist X Seedling]; McGredy, Sam IV; Mattock

'MACAMSTER', HT, ab, 1998; (**Matawhero Magic**); flowers apricot blend, double, semi-dbl., 17–25 petals, 4 in., borne mostly singly, intense fragrance; prickles moderate; foliage large, medium green, glossy; bushy, 120 cms growth; [Spek's Centennial X New Year]; McGredy, Sam IV

'MACANAT', Min, mp, 1981; (**Hi-de-hi**); flowers small, dbl., 20 petals, moderate fragrance; foliage small, dark, glossy; bushy growth; [Anytime X Gartendirektor Otto Linne]; McGredy, Sam IV

'MACANGEL', Min, w, 1981; (**Angelita**®, 'MACANGELI', Snowball); flowers small, very dbl., slight fragrance; foliage small, dark, glossy; spreading growth; PP005849; Gold Star of the South Pacific, Palmerston North, NZ, 1983; [Moana X Snow Carpet]; McGredy, Sam IV, 1982

'MACAR', HT, rb, 1960; (**Piccadilly**®); flowers scarlet, base and reverse gold, high-pointed, dbl., 28 petals, 4.5–5 in.; foliage dark, glossy; vigorous, upright, branching growth; GM, Madrid, 1960 GM, Rome, 1960; [McGredy's Yellow X Karl Herbst]; McGredy, Sam IV; McGredy

'MACARNHE', F, lp, 1998; (**Jillian McGredy**); flowers light pink, double, semi-dbl., 17–25 petals, 3.5 in., borne in small clusters, moderate fragrance; prickles moderate; foliage medium, medium green, semi-glossy; bushy, 120 cms growth; [Sexy Rexy X Lagerfeld]; McGredy, Sam IV

'MACAROM', HT, ly, 1988; (**Goldie**); flowers pale yellow, large, dbl., 26–40 petals, intense fragrance; foliage large, medium green, glossy; upright growth; [Unnamed seedling X Golden Gate]; McGredy, Sam IV

Macartney Rose *see* **R. BRACTEATA**

'MACATSAN', S, mr, 1992; ('MACCATSAN', **Phantom**); flowers scarlet red, blooms, semi-dbl., 6–14 petals, 3–3.5 in., slight fragrance; foliage large, medium green, semi-glossy; spreading (80 cms) growth; [Pandemonium X Eyeopener]; McGredy, Sam IV; McGredy Roses International

'MACAUCK', HT, mr, 1982; (**Olympiad**™, Olympiode®); flowers brilliant medium red, large, non-fading, dbl., 30–35 petals, 4.5–5 in., exhibition form, borne mostly singly, slight fragrance; foliage large, medium green, matte; upright, bushy growth, vigorous; PP005519; AARS, 1984 GM, Portland, 1985; [Red Planet X Pharaoh]; McGredy, Sam IV, 1983; Armstrong Nursery, 1984

MACauclad, HT, mp, 1998; (**Kathryn McGredy**); flowers medium pink, double, semi-dbl., 17–25 petals, 4.5 in., borne mostly singly, slight fragrance; slight prickles; foliage medium, medium green, glossy; moderate (110 cms) growth; [City of Auckland X Lady Rose]; McGredy, Sam IV, 1996; McGredy, Sam, 1996, NZ

MACausal, S, pb, 1994; (**Pink Magic Carpet**™); McGredy, Sam IV

'MACBA', LCl, op, 1966; (**Galway Bay**®); flowers salmon-pink, well-formed, blooms in clusters, 3.5 in., slight fragrance; [Heidelberg X Queen Elizabeth]; McGredy, Sam IV

'MACBERN', HT, my, 1975; (Yellow Wonder, **Young Quinn**®); bud small, short; dbl., 28 petals, borne singly, slight fragrance; foliage heavily veined, dark; upright, tall growth; GM, Belfast, 1978; [Peer Gynt X Kiskadee]; McGredy, Sam IV

'MACBETH', HT, dr, 1921; bud pointed; flowers deep crimson, shaded darker, large, dbl., exhibition form, intense fragrance; foliage dark, bronze; vigorous, bushy, compact growth; [Richmond X Admiral Ward]; Bees

'MACBIGMA', F, pb, 1988; (Red Hot, **Redhot**); flowers hand-painted, medium, dbl., 15–25 petals, slight fragrance; foliage small, medium green, semi-glossy; bushy growth; [Eyepaint X Ko's Yellow]; McGredy, Sam IV

'MACBIPI', Min, mp, 1978; (**Moana**®); flowers rose-pink, small, dbl., 35 petals, slight fragrance; foliage glossy, dark; [Seedling X New Penny]; McGredy, Sam IV

'MACBO', HT, rb, 1966; (**Flaming Peace**, Kronenbourg); flowers bright medium red, reverse straw-yellow veined red; [Peace sport]; McGredy, Sam IV; Country Garden Nursery, 1966

'MACBUCPAL', HT, w, 1987; (**Auckland Metro**, Métro); flowers large, dbl., 26–40 petals, intense fragrance; foliage large, dark green, semi-glossy; bushy growth; [Sexy Rexy X (Seedling X Ferry Porsche)]; McGredy, Sam IV; McGredy Roses International, 1988

'MACBUNBER', S, or, 1998; (**Hot Chile**); flowers orange-red, semi-double, semi-dbl., 12–16 petals, 2.75 in., borne in small clusters, slight fragrance; prickles moderate; foliage small, dark green, glossy; spreading, low, bushy, 30 cms growth; [Trumpeter X Eyeopener]; McGredy, Sam IV; McGredy, Sam, 1995, NZ

'MACCA', LCl, ly, 1963; (**Casino**®, Gerbe d'Or); flowers soft yellow, well-formed, large, dbl., recurrent bloom, moderate fragrance; foliage dark, glossy; vigorous (10 ft.) growth; GM, NRS, 1963; [Coral Dawn X Buccaneer]; McGredy, Sam IV; McGredy

'MACCARLTO', F, mp, 1991; (Northland, **Sweet Gesture**); dbl., 26–40 petals, 3 in., slight fragrance; foliage large, medium green, semi-glossy; ultra healthy, bushy(100 cms) growth; [(Sexy Rexy X New Year) X West Coast]; McGredy, Sam IV; McGredy Roses International, 1992

'MACCARPE', Min, w, 1980; (Blanche Neige, **Snow Carpet**®); flowers small, dbl., 55 petals, slight fragrance; spreading growth; GM, Baden-Baden, 1982; [New Penny X Temple Bells]; McGredy, Sam IV

'MACCARRIB', F, lp, 1992; (**Rexy's Baby**); dbl., 26–40 petals, 1.5–2.75 in., slight fragrance; foliage small, medium green, glossy; bushy (90 cms) growth; [Sexy Rexy X [Freude X ((Anytime X Eyepaint) X Stars 'n' Stripes)]]; McGredy, Sam IV; McGredy Roses International

'MACCATSAN', S, mr, 1992; ('MACATSAN', **Phantom**); flowers scarlet red, blooms, semi-dbl., 6–14 petals, 3–3.5 in., slight fragrance; foliage large, medium green, semi-glossy; spreading (80 cms) growth; [Pandemonium X Eyeopener]; McGredy, Sam IV; McGredy Roses International

'MACCHE', Gr, ab, 1978; ('MACCHEUP', **Sundowner**); flowers golden orange, dbl., 35 petals, 4 in., exhibition form, intense fragrance; foliage leathery; tall, upright growth; AARS, 1979; [Bond Street X Peer Gynt]; McGredy, Sam IV; Edmunds Roses

'MACCHEUP', Gr, ab, 1978; ('MACCHE', **Sundowner**); flowers golden orange, dbl., 35 petals, 4 in., exhibition form, intense fragrance; foliage leathery; tall, upright growth; AARS, 1979; [Bond Street X Peer Gynt]; McGredy, Sam IV; Edmunds Roses

'MACCHOME', HT, or, 1994; (**Big Daddy**®, Rita MacNeil); flowers scarlet-orange, blooms, dbl., 26–40 petals, 3 in., slight fragrance; foliage large, medium green, semi-glossy; bushy (100 cms) growth; [Howard Morrison X Mme Delbard]; McGredy, Sam IV; McGredy Roses International, 1994

'MACCHOPSU', Gr, or, 1998; (**Painted Desert**); flowers orange red, double, dbl., 17–25 petals, 4 in., borne in small clusters, slight fragrance; prickles moderate; foliage large, medium green, bushy growth; upright, bushy, 110 cms. growth; [Louise Gardner X Mme. Delbard]; McGredy, Sam IV, 1997; McGredy, Sam, 1997, NZ

'MACCI', F, or, 1968; (**City of Belfast**®); flowers bright red, blooms in trusses, dbl., cupped; foliage glossy; GM, Belfast, 1970 GM, RNRS, 1967 GM, The Hague, 1976 Gold Star of the South Pacific, Palmerston North, NZ, 1969 PIT, RNRS, 1967; [Evelyn Fison X (Circus X Korona)]; McGredy, Sam IV; McGredy

'MACCLACK', HT, ob, 1991; (**Paddy Stephens**); flowers moderately large blooms, dbl., 15–25 petals, slight fragrance; foliage large, dark green, red when young; bushy (100 cms) growth; [Solitaire X [[(Tombola X (Elizabeth of Glamis X (Circus X Golden Fleece))) X Mary Sumner] X Unknown]]; McGredy, Sam IV; McGredy Roses International

'MACcloseup', HT, ab, 1998; (**Carmel Sunset**); flowers apricot blend, double, dbl., 17–25 petals, 4.5 in., borne mostly singly, moderate fragrance; prickles moderate; foliage large, medium green, dull; bushy, tall (120 cms) growth; [Freude X Silver Jubilee]; McGredy, Sam IV, 1997; McGredy, Sam, 1997, NZ

'MACcolumb', HT, mp, 1998; (**Carolyn**); flowers medium pink, full, very dbl., 26–40 petals, 4 in., borne mostly singly, intense fragrance; prickles moderate; foliage large, light green, dull; bushy, medium, 110 cms growth; [((Courvoisier X Arthur Bell) X Traumerei) X (Tojo X Vienna Woods)]; McGredy, Sam IV; McGredy, Sam, 1995, NZ

'MACcompu', Gr, ob, 1989; (**Today**™); bud ovoid; flowers light orange blending to yellow, urn-shaped, medium, dbl., 33 petals, cupped, borne usually singly, slight fragrance; globular, medium, orange-red fruit; prickles recurved, medium, brown; foliage medium, dark green, glossy; bushy, medium growth; PP007202; [[Typhoo Tea X (Yellow Pages X Kabuki)] X [(Yellow Pages X Kabuki) X (MACjose X Typhoon)]]; McGredy, Sam IV; Co-Operative Rose Growers, 1990

'MACcourlod', Min, pb, 1984; (**Ragtime**); flowers small, dbl., slight fragrance; foliage small, dark, matt; bushy growth; [Mary Sumner X Unnamed seedling]; McGredy, Sam IV; McGredy Roses International, 1982

'MACcrackle', HT, dr, 1983; (**Howard Morrison**); flowers large, dbl., 35 petals, slight fragrance; foliage medium, medium green, semi-glossy; upright, bushy growth; [Unnamed seedling X Unnamed seedling]; McGredy, Sam IV, 1982

'MACcricke', Min, dp, 1989; (**Marriotta**); flowers deep pink, small blooms with side buds and sprays, dbl., 20 petals, slight fragrance; foliage small, medium green, semi-glossy, resistant to black; bushy growth; [Seaspray X Little Artist]; McGredy, Sam IV; McGredy Roses International, 1989;, Oregon Mini. Roses, 199

'MACdeepo', HT, my, 1985; (**Guadalajara**); flowers large, dbl., 24 petals, exhibition form, borne singly, moderate fragrance; fruit never observed; large, deltoid, red to brown prickles; foliage large, medium green, semi-glossy; medium, upright growth; (28); PP006263; [New Day X Yellow Bird]; McGredy, Sam IV; Roses by Fred Edmunds, 1984

'MACdub', LCl, mr, 1975; (**Dublin Bay**®); bud ovoid; dbl., 25 petals, 4.5 in., moderate fragrance; climbing growth; Rose of The Year, Auckland, NZ, 1993; [Bantry Bay X Altissimo]; McGredy, Sam IV; McGredy Roses International, 1974

'MACel', F, op, 1964; (**Elizabeth of Glamis**®, Irish Beauty); flowers light orange-salmon, blooms in clusters, dbl., 35 petals, 4 in., flat, intense fragrance; vigorous, compact bushy growth; GM, NRS, 1963 PIT, NRS, 1963; [Spartan X Highlight]; McGredy, Sam IV; Edmunds, 1965;, McGredy, 1964

'MACerupt', HT, yb, 1987; (**Louise Gardner**); flowers large, dbl., 15–25 petals, slight fragrance; foliage medium, medium green, matt; upright, bushy growth; [Freude X Sunblest seedling]; McGredy, Sam IV; McGredy Roses International, 1988

'MACesp', F, rb, 1974; (**Old Master**); flowers carmine, white eye and reverse, semi-dbl., 15 petals, 4.5 in., slight fragrance; foliage semi-glossy, medium green; vigorous, bushy growth; [(Maxi X Evelyn Fison) X (Orange Sweetheart X Fruhlingsmorgen)]; McGredy, Sam IV

'MACev', F, mr, 1962; (**Evelyn Fison**, Irish Wonder); flowers scarlet, blooms in broad clusters, dbl., 3 in., slight fragrance; foliage dark, glossy; compact, bushy growth; GM, NRS, 1963; [Moulin Rouge X Korona]; McGredy, Sam IV; J&P, 1964;, McGredy, 1962

'MACeye', F, rb, 1975; (Eye Paint, **Eyepaint**®, Tapis Persan); bud ovoid; flowers bright red, whitish eye, gold stamens, single, 5–6 petals, 2.5 in., slight fragrance; foliage small, dark; tall, bushy growth; GM, Baden-Baden, 1974 GM, Belfast, 1978; [Seedling X Picasso]; McGredy, Sam IV

MacFarlane's Own, F, pb, 1999; Williams, J. Benjamin

'MACfirinlin', F, mr, 1998; (Fiesta, **War Dance**); flowers medium red, hand-painted in cool weather, semi-dbl., 12–16 petals, 2.5 in, borne in small clusters, slight fragrance; prickles moderate; foliage small, medium green, semi-glossy; compact, low (30 cms) growth; [Howard Morrison X Sue Lawley]; McGredy, Sam IV; McGredy, 1997

'MACfirwal', S, rb, 1988; (Rock 'n' Roll, Tango, **Stretch Johnson**); flowers orange blend, medium, semi-dbl., 6–14 petals, slight fragrance; foliage large, medium green, semi-glossy; bushy growth; PP007472; GM, RNRS, 1988; [Sexy Rexy X Maestro]; McGredy, Sam IV; McGredy Roses International

'MACfrabro', Min, ob, 1985; (**Firefly**); flowers small, dbl., 20 petals, slight fragrance; foliage small, dark, glossy; bushy growth; [Mary Sumner X Ko's Yellow]; McGredy, Sam IV

'MACfreego', Min, dy, 1983; (Free Gold®, **Freegold**®, Penelope Keith); flowers deep yellow, gold reverse, small, dbl., 20 petals, exhibition form, moderate fragrance; foliage small, light green, semi-glossy; upright growth; [Seaspray X Dorola]; McGredy, Sam IV

MACfrothy, Min, w, 1992; (**Frothy**); McGredy, Sam IV

'MACgam', F, or, 1977; (**Kapai**®); flowers medium, dbl., 30 petals, 3.5 in., intense fragrance; foliage small; low, bushy growth; [Madame Bollinger X Tombola]; McGredy, Sam IV

'MACgem', HT, yb, 1979; (**Benson & Hedges Gold**®); bud ovoid; flowers deep golden yellow flushed coppery red, medium, dbl., 33 petals, moderate fragrance; bushy growth; Gold Star of the South Pacific, Palmerston North, NZ, 1978; [Yellow Pages X (Arthur Bell X Cynthia Brooke)]; McGredy, Sam IV; Mattock

'MACgenev', HT, lp, 1989; (Aotearoa-New Zealand, **New Zealand**); flowers large, soft creamy pink, dbl., 34 petals, 4.5–5 in., exhibition form, borne singly, intense, honeysuckle fragrance; foliage large, medium green, semi-glossy; upright, medium growth; PP8279; GM, Portland, USA, 1995 Rose of The Year, Auckland, NZ, 1990; [Harmonie X Auckland Metro]; McGredy, Sam IV, 1991

MACghovie, S, dr; (**Chateau Canon**); McGredy, Sam IV

'MACglemil', F, mp, 1991; (**Kapiti**); semi-dbl., 6–14 petals, 1.5–2.75 in., slight fragrance; foliage medium, medium green, semi-glossy; spreading growth (to 60 cms.); [Sexy Rexy X Eyeopener]; McGredy, Sam IV; McGredy Roses International, 1992

'MACgolold', F, ab, 1997; (**Golden Oldie**); flowers apricot blend, dbl., 17–25 petals, 2.75 in., borne in large clusters, slight fragrance; prickles moderate; foliage medium, dark green, semi-glossy; bushy, low (50 cms) growth; [Maiden Voyage X Orange Honey]; McGredy, Sam IV; McGredy, 1997

'MACgoofy', HT, rb, 1990; (**Papageno**®); flowers red blend (striped), large blooms, dbl., 15–25 petals, slight fragrance; foliage large, light green, matt; upright growth; [Freude X [(Anytime X Eyepaint) X Stars 'n' Stripes]]; McGredy, Sam IV; McGredy Roses International, 1989

MACgremli, HT, rb; (**Tiger**); McGredy, Sam IV, 1990

'MACGUTSY', Min, pb, 1983; (**Academy®**); flowers patio, small, dbl., 20 petals, slight fragrance; foliage small, medium green, semi-glossy; bushy growth; [Anytime X Matangi]; McGredy, Sam IV, 1982

'MACHA', LCl, rb, 1965; (Haendel®, **Handel®**); flowers cream edged red, dbl., 22–30 petals, 3.5 in., recurrent bloom, slight, honey fragrance; foliage glossy, olive-green; blooms on old and new wood, climber growth; GM, Portland, 1975; [Columbine X Heidelberg]; McGredy, Sam IV; McGredy

'MACHADEN', HT, dr, 1986; (**Red Perfection™**); flowers dark red, reverse slightly lighter, fading purple-red, medium, semi-dbl., 50–60 petals, exhibition form, moderate, old rose fragrance; globular, large, dark red fruit; slightly recurved, average, green prickles; foliage large, dark green, semi-glossy; tall growth; PP005426; [Karma X Arturo Toscanini]; McGredy, Sam IV; Co-Operative Rose Growers, 1987

'MACHAHEI', Gr, or, 1998; (**Reba McEntire**, Spirit of Hope); flowers double, orange red blooms, dbl., 17–25 petals, 4–4.5 in., exhibition form, borne in small clusters, slight fragrance; prickles moderate; foliage large, dark green, glossy; bushy, medium (110cms) growth; Gold Star of the South Pacific, Palmerston North, NZ, 1994; [(Howard Morrison X Red Perfection) X Maiden Voyage]; McGredy, Sam IV; Edmunds Roses, 1997;, McGredy, 1997

MAChana, Min, my, 1983; McGredy, Sam IV

'MACHILVER', F, lp, 1998; (**Dame Cath**); flowers small, light pink, semi-dbl., 12–16 petals, 2.75 in., borne in small clusters, slight fragrance; prickles moderate; foliage medium, dark green, dull, bushy, low (50 cm) growth; [Chaumant X (Seaspray X Freegold)]; McGredy, Sam IV; McGredy, 1997

Macho Man *see* 'PERMACH'

'MACHOMAI', Min, dr, 1984; (**Grandee**); flowers gold stamens, medium, semi-dbl., moderate fragrance; foliage medium, medium green, semi-glossy; upright growth; [Regensberg X Ko's Yellow]; McGredy, Sam IV; Oregon Miniature Roses

'MACHORO', HT, mr, 1990; (Placido Domingo, **Velvet Ruby**, Waikato); flowers moderate, large blooms, 15–25 petals, moderate fragrance; foliage large, medium green, matt; bushy growth; [Candella X Auckland Metro]; McGredy, Sam IV; McGredy Roses International

'MACHOROWHEN', HT, or, 1993; (**Sophileo**); dbl., 15–25 petals, 1.5–2.75 in., slight fragrance; foliage medium, medium green, semi-glossy; medium, bushy growth; McGredy, Sam IV; C & K Jones;, Golden Fields Nursery

'MACINCA', Min, or, 1978; (Longleat, **Wanaka®**, Young Cale); flowers small, dbl., 40 petals, slight fragrance; foliage light green; low, bushy growth; [Anytime X Trumpeter]; McGredy, Sam IV; McGredy Roses International

'MACINJU', HT, rb, 1981; ('MACKINJU', **Maestro®**); bud ovoid; flowers medium red, painted white, reverse lighter red and white, dbl., 28 petals, slight fragrance; narrow, red prickles; foliage matt, olive green; upright, bushy growth; [Picasso seedling X Unnamed seedling]; McGredy, Sam IV, 1980

'MACIO', F, op, 1964; (**Violet Carson**); flowers peach-pink, reverse silvery, well-formed, blooms in large clusters, dbl., 35 petals, moderate fragrance; foliage dark, glossy; compact, bushy growth; [Mme Leon Cuny X Spartan]; McGredy, Sam IV; McGredy

'MACIR', HT, or, 1961; (**Miss Ireland**); flowers orange-red, yellow reverse, dbl., 37 petals, 5 in., moderate fragrance; foliage dark; vigorous, bushy growth; [Tzigane X Independence]; McGredy, Sam IV; McGredy

'MACIVY', F, ab, 1994; (Love's Spring, **Singin' in the Rain**, Spek's Centennial); flowers apricot/copper, very dbl., 25–30 petals, 1.5–2.75 in., borne in large clusters, moderate, sweet musk fragrance; some prickles; foliage medium, dark green, glossy; medium, (3-4 ft) upright growth, free branching; PP8362; AARS, 1995; [Sexy Rexy X Pot O'Gold]; McGredy, Sam IV; Edmunds Roses, 1994

'MACJILLI', Gr, op, 1987; (**Profile**); flowers orange blend, large, dbl., 26–40 petals, slight fragrance; foliage large, medium green, semi-glossy; upright growth; [Freude X (Unnamed Arthur Bell seedling X Sunsong)]; McGredy, Sam IV; McGredy Roses International, 1988

'MACJOCEL', F, w, 1987; (Dresselhuys, **Petite Penny**); flowers small, semi-dbl., 6–14 petals, moderate fragrance; foliage small, medium green, semi-glossy; bushy growth; [(R. crepuscule X Unnamed seedling) X Royal Occasion]; McGredy, Sam IV; McGredy Roses International, 1988

'MACJULIAT', HT, r, 1990; (Old Spice, Siegfried Sassoon, **Spiced Coffee**, Vidal Sassoon); flowers pale lavender with brown overtones, large blooms, dbl., 15–25 petals, intense fragrance; foliage medium, medium green, matt; upright growth; [Harmonie X Big Purple]; McGredy, Sam IV; McGredy Roses International, 1991

'MACKATI', F, m, 1990; (**Old Port**); dbl., 26–40 petals, moderate fragrance; foliage medium, medium green, matt; bushy growth; [[(Anytime X Eyepaint) X Purple Splendour] X Big Purple]; McGredy, Sam IV; McGredy Roses International, 1991

'MACKEPA', S, ob, 1979; (Blushing Maid, **Rocky™**); bud ovoid; flowers coral orange, reverse whitish, medium, dbl., 25 petals; vigorous, tall, bushy growth; [Liverpool Echo X (Evelyn Fison X (Orange Sweetheart X Fruhlingsmorgen))]; McGredy, Sam IV; McGredy Roses International

'MACKINJU', HT, rb, 1981; ('MACINJU', **Maestro®**); bud ovoid; flowers medium red, painted white, reverse lighter red and white, dbl., 28 petals, slight fragrance; narrow, red prickles; foliage matt, olive green; upright, bushy growth; [Picasso seedling X Unnamed seedling]; McGredy, Sam IV, 1980

'MACKOSRED', Min, mr, 1982; (**Paragon**); dbl., 35 petals, moderate fragrance; foliage small, dark, glossy; bushy growth; [Ko's Yellow X Little Artist]; McGredy, Sam IV

'MACKOSYEL', Min, yb, 1978; (**Ko's Yellow**); flowers yellow, edges marked red, fading to cream, classic form, medium, dbl., 39 petals; foliage dark, glossy; bushy growth; [(New Penny X Banbridge) X (Border Flame X Manx Queen)]; McGredy, Sam IV; McGredy Roses International

'MACKUNG', HT, my, 1980; (**Austragold®**); McGredy, Sam IV

'MACLANOFLON', HT, w, 1998; (**Full Sail**, Land of the Long White Cloud, Long White Cloud); flowers white, full, large, very dbl., 26–40 petals, 4.5 in., exhibition form, borne mostly singly, intense fragrance; few prickles; foliage large, dark green, glossy; bushy, medium, 110 cm growth; [New Zealand sport]; McGredy, Sam IV; Edmunds' Roses, 1998;, McGredy, 1998

'MACLANTER', Gr, ob, 1998; (Lantern, **Octoberfest®**, Oktoberfest); bud long, pointed; flowers blended autmnal colors, cream to red-orange, semi-dbl., 127–25 petals, 5–5.5 in., exhibition form, borne in small clusters, slight, fruity fragrance; prickles moderate; foliage large, dark green, glossy; reddish new growth, tall, upright, 6 ft; [Louis Gardner X New Zealand]; McGredy, Sam IV

'MACLAPAZ', F, r, 1986; (**Beachcomber**); flowers medium, semi-dbl., 6–14 petals, slight fragrance; foliage medium, medium green, semi-glossy; bushy growth; Gold Star of the South Pacific, Palmerston North, NZ, 1985; [Seedling X Colorbreak]; McGredy, Sam IV

'MACLEIDE', F, rb, 1998; (**Wizard**); flowers orange-red with white eye, semi-dbl., 12–16 petals, 2.5 in., borne in large clusters, slight fragrance; prickles moderate; foliage small, medium green, semi-glossy; bushy, low, 30 cm growth; [Seedling X Genesis]; McGredy, Sam IV; McGredy, 1996

'MACLOCKER', F, pb, 1992; (Chaumant, Chaumont, **Maiden Voyage**, Oriola); dbl., 15–25 petals, 1.5–2.75 in., borne in small clusters, slight fragrance; foliage medium, medium green, glossy; medium (3 1/2'), bushy growth; [Sexy Rexy X New Year]; McGredy, Sam IV; Cooperative Rose Growers, 1992

'MACLOUPRI', LCl, or, 1998; (**Too Hot To Handle**); flowers orange-red, double, dbl., 17–25 petals, 2.75 in., borne in small clusters, slight fragrance; prickles moderate; foliage large, dark green, glossy; upright, tall, 150 cm growth; [Waiheke X Eyeopener]; McGredy, Sam IV; McGredy, 1995

'MACMAN', F, rb, 1974; (**Matangi**®); bud ovoid; flowers orange-red, silver eye and reverse, dbl., 30 petals, 3.5 in., slight fragrance; foliage small; bushy growth; GM, Belfast, 1976 GM, Portland, 1982 GM, Rome, 1974 PIT, RNRS, 1974; [Unnamed seedling X Picasso]; McGredy, Sam IV

'MACMANLEY', Min, rb, 1982; (**Little Artist**®, 'MACMANLY', Top Gear); flowers open with hand-painted marks becoming solid medium red, off-white in base half, semi-dbl., slight fragrance; foliage small, medium green, semi-glossy; upright growth; [Eyepaint X Ko's Yellow]; McGredy, Sam IV

MACmatan, HT, mr; (**Ten Ten CFRB**); McGredy, Sam IV, 1987

'MACMELAN', Min, op, 1988; (**Melanie**); bud ovoid; flowers light salmon-pink, reverse lighter, small, borne in sprays o, dbl., 23 petals, cupped, slight fragrance; prickles straight, red-brown; foliage small, medium green, semi-glossy; bushy, low growth; [Seaspray X Wanaka]; McGredy, Sam IV; Sealand Nursery, 1989

'MACMI', HT, op, 1961; (**Mischief**); flowers salmon-pink, dbl., 28 petals, 4 in., moderate fragrance; foliage light green; vigorous, upright growth; GM, NRS, 1961 GM, Portland, 1965 PIT, NRS, 1961; [Peace X Spartan]; McGredy, Sam IV; McGredy

'MACMIGMOU', F, rb, 1980; (**Mighty Mouse**, Painted Star); flowers scarlet, white eye, blooms borne 20 per cluster, 7–12 petals, slight fragrance; sets hips, orange-red fruit; slightly hooked, red prickles; foliage dark, very long, pointed; tall, spreading growth; [Anytime X Eyepaint]; McGredy, Sam IV

Macmillan Nurse *see* 'BEAMAC'

'MACMINMO', Min, rb, 1987; (Minnie Mouse, **Roller Coaster**); flowers red blend (striped), semi-dbl., 6–14 petals, slight fragrance; foliage small, medium green, glossy; upright growth; PP007319; [(Anytime X Eyepaint) X Stars 'n' Stripes]; McGredy, Sam IV; McGredy Roses International, 1988

'MACMISECH', F, op, 1978; (**Taupo**®); flowers light rose-salmon, flora-tea, dbl., 40 petals, 5 in.; tall growth; [Liverpool Echo X Irish Mist]; McGredy, Sam IV; McGredy

'MACMO', F, rb, 1969; (**Molly McGredy**); flowers medium red, reverse silver, well-formed, large blooms in trusses, dbl., 35 petals, slight fragrance; foliage dark, glossy; GM, Belfast, 1971 GM, Portland, 1971 GM, RNRS, 1968 PIT, RNRS, 1968; [Paddy McGredy X (Mme Leon Cuny X Columbine)]; McGredy, Sam IV; McGredy

'MACMOOBLU', F, m, 1998; (**Moody Blues**); flowers mauve, full, large, very dbl., 26–40 petals, 2.75 in., moderate fragrance; foliage large, medium green, semi-glossy; bushy, medium, 100 cm growth; [Sexy Rexy X Blue Nile]; McGredy, Sam IV; McGredy, 1991

MACmoodre, F, m, 1996; (**Moody Dream**); McGredy, Sam IV

'MACMOSCO', Min, rb, 1988; (**Whistle Stop**); flowers red blend (striped), small, semi-dbl., 6–14 petals, slight fragrance; foliage small, medium green, semi-glossy; bushy growth; [Mighty Mouse X Hurdy Gurdy]; McGredy, Sam IV, 1989

'MACMOTA', HT, mr, 1983; (**Crimson Tide**); flowers crimson, large blooms borne singly, dbl., 35 petals, exhibition form, slight fragrance; foliage large, dark, leathery; upright growth; [Seedling X Seedling]; McGredy, Sam IV; Roses by Fred Edmunds

'MACMOUHOO', HT, w, 1988; (Christine Horbiger, Foster's Melbourne Cup, Foster's Wellington Cup, **Mount Hood**); flowers medium, dbl., 40–45 petals, moderate; foliage large, medium green, semi-glossy; bushy growth; Gold Star of the South Pacific, Palmerston North, NZ, 1991; [Sexy Rexy X Pot O'Gold]; PP9095; McGredy, Sam IV; McGredy Roses International, 1991

'MACMU', HT, op, 1968; (**Muria**); bud long, pointed; flowers salmon-orange, classic form, dbl., 40 petals, 4 in., slight fragrance; foliage light green; bushy growth; [Miss Ireland X Tropicana]; McGredy, Sam IV; Spek

'MACNARU', HT, mp, 1988; ('MACNAURU', 'MACSATUR', **Penthouse**); flowers large, dbl., 15–25 petals, moderate fragrance; foliage large, medium green, matt; bushy growth; [Unnamed seedling X Ferry Porsche]; McGredy, Sam IV

'MACNAURU', HT, mp, 1988; ('MACNARU', 'MACSATUR', **Penthouse**); flowers large, dbl., 15–25 petals, moderate fragrance; foliage large, medium green, matt; bushy growth; [Unnamed seedling X Ferry Porsche]; McGredy, Sam IV

'MACNAURU', HT, mp, 1987; ('MACNGAURU', Metropolitain, Penthouse, **West Coast**®); flowers large, dbl., 15–25 petals, slight fragrance; foliage large, light green, matt; bushy growth; [((Yellow Pages X Kabuki) X Golden Gate) X (Unnamed Poulsen seedling X Picasso)]; McGredy, Sam IV; McGredy Roses International, 1986

'MACNECTA', Min, or, 1978; (**Otago**®); flowers medium, dbl., 35 petals, exhibition form, moderate fragrance; bushy growth; [Anytime X Minuette]; McGredy, Sam IV; McGredy Roses International

'MACNEW', Min, pb, 1982; ('MACNEWING', **Seaspray**); flowers pale pink flushed red, medium, semi-dbl., moderate fragrance; foliage medium, medium green, matt; bushy growth; [Anytime X Moana]; McGredy, Sam IV; John Mattock, Ltd

'MACNEWEYE', Gr, ob, 1983; (Arcadian, 'MACNEWYE', **New Year**®); flowers orange and gold blend, medium, dbl., 20 petals, slight fragrance; foliage large, dark, glossy; upright growth; PP005428; AARS, 1987; [Mary Sumner X Unnamed seedling]; McGredy, Sam IV, 1982

'MACNEWING', Min, pb, 1982; ('MACNEW', **Seaspray**); flowers pale pink flushed red, medium, semi-dbl., moderate fragrance; foliage medium, medium green, matt; bushy growth; [Anytime X Moana]; McGredy, Sam IV; John Mattock, Ltd

'MACNEWYE', Gr, ob, 1983; (Arcadian, 'MACNEWEYE', **New Year**®); flowers orange and gold blend, medium, dbl., 20 petals, slight fragrance; foliage large, dark, glossy; upright growth; PP005428; AARS, 1987; [Mary Sumner X Unnamed seedling]; McGredy, Sam IV, 1982

'MACNGAURU', HT, mp, 1987; ('MACNAURU', Metropolitain, Penthouse, **West Coast**®); flowers large, dbl., 15–25 petals, slight fragrance; foliage large, light green, matt; bushy growth; [((Yellow Pages X Kabuki) X Golden Gate) X (Unnamed Poulsen seedling X Picasso)]; McGredy, Sam IV; McGredy Roses International, 1986

'MACNIC', HT, mr, 1981; (Karma, **Pounder Star**); bud long, pointed; dbl., 20 petals, exhibition form, moderate, old rose to spicy fragrance; small, curved prickles; foliage dark, glossy; upright, medium growth; [John Waterer X Kalahari]; McGredy, Sam IV; Roses by Fred Edmunds, 1982

'MACNICKEL', Min, rb, 1989; (**Nickelodeon**); flowers small, semi-dbl., 6–14 petals, slight fragrance; foliage small, dark green, semi-glossy; patio; bushy growth; [Roller Coaster X [Freude X [(Anytime X Eyepaint) X Stars 'n'Stripes]]]; McGredy, Sam IV, 1991

'MACNON', HT, mp, 1965; (**Shannon**®); bud ovoid; dbl., 58 petals, 5 in.; foliage dark, rounded; vigorous growth; [Queen Elizabeth X McGredy's Yellow]; McGredy, Sam IV; McGredy

'MACOBORN', HT, pb, 1995; (**Maggie Barry**); flowers salmon edged orange, blooms, dbl., 26–40 petals, 3 in., slight fragrance; foliage large, light green, matt; bushy (100 cms) growth; [Louise Gardner X West Coast]; McGredy, Sam IV; McGredy Roses International, 1993

MACoffer, HT, op, 1988; (**City of Christchurch**); McGredy, Sam IV

'MACON', HT, mr, 1963; (**Uncle Walter**); flowers crimson-scarlet, dbl., 30 petals, 5 in., exhibition form; foliage leathery, coppery; vigorous, tall growth; [Detroiter X Heidelberg]; McGredy, Sam IV; McGredy

'MACORANLEM', S, ob, 1994; (**Oranges 'n' Lemons**); flowers striking orange and yellow striped, dbl., 26–40 petals, 3–3.5 in., borne in small clusters, slight, fruity fragrance; some prickles; foliage medium to large, dark red when new, turning dark gre; tall, spreading, very vigorous growth; PP9191; [New Year X (Freude X Seedling)]; McGredy, Sam IV; Weeks Roses, 1995

'MACPA', F, mp, 1962; (**Paddy McGredy**); bud ovoid; flowers deep rose-pink, blooms in clusters, dbl., 33 petals, 4 in., cupped, moderate fragrance; foliage leathery; vigorous, bushy growth; GM, NRS, 1961; [Spartan X Tzigane]; McGredy, Sam IV; J&P

'MACPANDEM', F, yb, 1988; (Claire Rayner, **Pandemonium**); flowers yellow and red stripes, dbl., 26–40 petals, slight fragrance; foliage small, medium green, glossy; patio; bushy growth; [New Year X ((Anytime X Eyepaint) X Stars 'n' Stripes)]; McGredy, Sam IV

'MACPARLEZ', Min, ab, 1984; (**Baby Sunrise**, Gold Fever); bud small; flowers copper apricot, small, semi-dbl., slight fragrance; foliage small, medium green, semi-glossy; bushy growth; PP007031; [Dorola X Moana]; McGredy, Sam IV; John Mattock, Ltd

'MACPAURMAR', Min, ob, 1993; (**Kiwi Sunrise**, MACtaumaruniu); flowers florescent orange with yellow eye, light yellow to white accents, dbl., 15–25 round petals, 2 in., borne in small clusters, moderate, spicy fragrance; few prickles; foliage medium, dark green, glossy; medium (24-30 in), bushy growth; [Orange Honey X Pandemonium]; McGredy, Sam IV; Justice Miniature Roses, 1994

'MACPIC', F, pb, 1971; (**Picasso**); flowers deep pink, petal edges lighter, white eye and reverse, semi-dbl., 18 petals, 3 in.; foliage small; GM, Belfast, 1973 Gold Star of the South Pacific, Palmerston North, NZ, 1971; [Marlena X (Evelyn Fison X (Frulingsmorgen X Orange Sweetheart))]; McGredy, Sam IV; McGredy

MACpinderal, Min, mp, 1987; (**Fresh Pink**); McGredy, Sam IV

'MACPIOPI', HT, mp, 1995; (**Graeme Douglas**); dbl., 26–40 petals, 3 in., slight fragrance; foliage large, medium green, semi-glossy; bushy (100 cms) growth; [Silver Jubilee X (Harmonie X Auckland Metro)]; McGredy, Sam IV; McGredy Roses International, 1994

'MACPLUTO', Min, rb, 1986; (**Hurdy Gurdy**); flowers dark red with white stripes, small, dbl., 26–40 petals, slight fragrance; foliage small, medium green, semi-glossy; mini-flora; upright growth; [Matangi X Stars 'n' Stripes]; McGredy, Sam IV

'MACPONUI', HT, pb, 1988; (**Pinky**); flowers large, dbl., 26–40 petals, slight fragrance; foliage large, dark green, glossy; upright growth; [Freude X Typhoo Tea]; McGredy, Sam IV

'MACPOW', HT, rb, 1977; (**Chivalry**®); flowers red, yellowish reverse, large, dbl., 35 petals; foliage glossy, dark; [Peer Gynt X Brasilia]; McGredy, Sam IV; Mattock

'MACRAIDA', HT, mp, 1982; (**Rainy Day**®); flowers large, dbl., 20 petals, slight fragrance; foliage large, dark, semi-glossy; bushy growth; [Trumpeter X Typhoon]; McGredy, Sam IV; McGredy Roses International

'MACRAMAR', HT, lp, 1998; (**Tess**); flowers large, light pink, very full, very dbl., 41 petals, 4.5 in., borne mostly singly, slight fragrance; prickles moderate; foliage large, medium green, semi-glossy; bushy, tall (120 cms) growth; [((Sexy Rexy X New Year) X (Freude X Courvoisier X Arthur Bell)) X Dreaming]; McGredy, Sam IV; McGredy, 1997

'MACRAT', F, rb, 1978; (**Priscilla Burton**®); flowers deep carmine pink and white blend (with variable combinations), semi-dbl., 10 petals, 2.5 in., moderate fragrance; foliage glossy, dark; PIT, RNRS, 1976; [Old Master X Unnamed seedling]; McGredy, Sam IV; Mattock

'MACRED', HT, mr, 1977; (**Joy Owens**); bud ovoid; 4 in., exhibition form; foliage very dark; moderate, bushy growth; [Electron X Pharaoh]; McGredy, Sam IV; McGredy Roses International

'MACREDPARAP', S, dr, 1994; (**Trinity**); dbl., 15–25 petals, 1.5 in., slight fragrance; foliage small, dark green, semi-glossy; spreading (90 cms) growth; [Sexy Rexy X Eyeopener]; McGredy, Sam IV; McGredy Roses International, 1994

'MACREDSAUS', S, mr, 1999; (**Red Magic Carpet**™); semi-dbl., 12–16 petals, 3 in., borne in small clusters, moderate fragrance; foliage medium, semi-glossy; spreading, low (12 in.); groundcover growth; [Sexy Rexy X Eyeopener]; McGredy, Sam IV, 1994

'MACRELEA', HT, my, 1984; (Chardonnay, Chardony, 'MACRELEASE', **Nobilo's Chardonnay**®); flowers orange yellow, large, dbl., 35 petals, slight fragrance; foliage small, light green, glossy; bushy growth; [Freude X (Wienerwald X Benson & Hedges Gold)]; McGredy, Sam IV

'MACRELEASE', HT, my, 1984; (Chardonnay, Chardony, 'MACRELEA', **Nobilo's Chardonnay**®); flowers orange yellow, large, dbl., 35 petals, slight fragrance; foliage small, light green, glossy; bushy growth; [Freude X (Wienerwald X Benson & Hedges Gold)]; McGredy, Sam IV

'MACRENO', HT, ob, 1994; (**Cardinal's Rose**, Susan Devoy); flowers ornage blend, blooms, dbl., 26–40 petals, 3–3.5 in., slight fragrance; foliage large, medium green, matt; bushy (100 cms) growth; [Waiheke X Las Vegas]; McGredy, Sam IV; McGredy Roses International, 1994

'MACRERO', HT, mr, 1977; (**Bulls Red**); bud ovoid; dbl., 28 petals, 4 in.; foliage dark; tall, upright growth; [Sympathie X Irish Rover]; McGredy, Sam IV; McGredy Roses International

'MACREXY', F, mp, 1984; (Heckenzauber, **Sexy Rexy**®); flowers medium pink to light pink, dbl., blooms in large clusters, slight fragrance; foliage small, light green, glossy; bushy growth; PP006713; Gold Star of the South Pacific, Palmerston North, NZ, 1984 Golden Prize, Glasgow, 1989 Rose of The Year, Auckland, NZ, 1991; [Seaspray X Dreaming]; McGredy, Sam IV

'MACROS', Min, mr, 1976; (Rose Baby, **Royal Salute**); flowers rose-red, small, dbl., 30 petals, 1.5 in.; foliage small, dark; [New Penny X Marlena]; McGredy, Sam IV; Mattock

'MACSATUR', HT, mp, 1988; ('MACNARU', 'MACNAURU', **Penthouse**); flowers large, dbl., 15–25 petals, moderate fragrance; foliage large, medium green, matt; bushy growth; [Unnamed seedling X Ferry Porsche]; McGredy, Sam IV

'MACSEATRI', HT, pb, 1987; (**Happy Days**); flowers medium, dbl., 15–25 petals, slight fragrance; foliage medium, medium green, matt; bushy growth; [(Unnamed Poulsen seedling X Picasso) X Paradise]; McGredy, Sam IV; McGredy Roses International, 1988

'MACSEE', F, dy, 1970; (**Courvoisier**®); dbl., 49 petals, 3.5 in., exhibition form, intense fragrance; foliage glossy, dark; [Elizabeth of Glamis X Casanova]; McGredy, Sam IV; McGredy

'MACSEV', HT, op, 1978; (**Christina Atherton**); flowers salmon-pink, classic form, dbl., 33 petals, 4 in., slight fragrance; free growth; [Tiki X Seedling]; McGredy, Sam IV

'MACSHANA', Min, dy, 1982; (Benson & Hedges Special, **Dorola**®, Parkay); flowers mini-flora, medium, dbl., 26 petals, moderate fragrance; foliage small, medium green, semi-glossy; bushy growth; [Darling Flame X New Day]; McGredy, Sam IV; McGredy Roses International, 1983

'MACSINGAL', Min, op, 1988; (**Singalong**); flowers salmon-orange and yellow, medium, dbl., 26–40 petals, slight fragrance; foliage small, medium green, glossy; patio; bushy growth; [(Anytime X Eyepaint) X New Year]; McGredy, Sam IV

'MACSINGAP', HT, or, 1987; (**Cup Final**); flowers large, dbl., 15–25 petals, slight fragrance; foliage medium, medium green, semi-glossy; upright growth; [Benson & Hedges Gold X (Kalahari X Papa Meilland)]; McGredy, Sam IV; McGredy Roses International, 1988

'MACSODA', F, or, 1991; (**Hot 'n' Spicy**); semi-dbl., 6–14 petals, borne in small clusters, slight fragrance; foliage medium, dark green, glossy; bushy growth; [Mary Sumner X Precious Platinum]; McGredy, Sam IV; Co-Operative Rose Growers, 1990

'MACSPASH', F, rb, 1980; (**Sue Lawley**); flowers medium red, petals edged light pink all around, blooms borne, 19 petals, slight fragrance; small, straight prickles; foliage matt, red when young; bushy growth; Gold Star of the South Pacific, Palmerston North, NZ, 1981; [((Little Darling X Goldilocks) X (Evelyn Fison X (Coryana X Tantau's Triumph)) X (((John Church X Elizabeth of Glamis)) x]; McGredy, Sam IV

'MACSPEEGO', HT, rb, 1990; (**Candella**®, Eternally Yours); flowers moderately large blooms, dbl., 15–25 petals, slight fragrance; foliage large, dark green, glossy; bushy growth; [Howard Morrison X Esmeralda]; McGredy, Sam IV; McGredy Roses International

'MACSPICE', Min, m, 1983; ('MACSPIKE'); flowers small, semi-dbl., slight fragrance; foliage small, medium green, semi-glossy; spreading growth; [Anytime X Gartendirektor Otto Linne]; McGredy, Sam IV

'MACSPIKE', Min, m, 1983; (**'MACSPICE'**); flowers small, semi-dbl., slight fragrance; foliage small, medium green, semi-glossy; spreading growth; [Anytime X Gartendirektor Otto Linne]; McGredy, Sam IV

'MACSTEWAR', F, or, 1986; (**Bradley Craig**®, Bradley Graig); flowers scarlet, medium, dbl., 15–25 petals, slight fragrance; foliage medium, medium green, glossy; bushy growth; [Tojo X Montana]; McGredy, Sam IV

'MACSTRA', F, or, 1976; (**Mary Sumner**); flowers medium, semi-dbl., 15 petals, 3 in., slight fragrance; tall, upright growth; [Seedling X Seedling]; McGredy, Sam IV; McGredy Roses International

'MACSUPBOW', Gr, m, 1998; (**Cologne**, Super Bowl); flowers mauve, double, large, dbl., 17–25 petals, 4.5 in., borne in small clusters, intense fragrance; prickles moderate; foliage large, medium green, matte; bushy, medium, 110 cm growth; [Harmonie X Lagerfeld]; McGredy, Sam IV; McGredy, 1996

'MACSUPCAT', HT, yb, 1989; (**Miriam**); flowers medium, dbl., 20 petals, slight fragrance; foliage medium, dark green, semi-glossy; bushy growth; [Sexy Rexy X Yabadabadoo]; McGredy, Sam IV, 1990

'MACSWEETWA', HT, op, 1987; (Freshie, **Sweetwaters**); flowers medium salmon-pink, large, dbl., 26–40 petals, slight fragrance; foliage large, dark green, glossy; upright, tall, very big growth; McGredy, Sam IV; McGredy Roses International, 1986

'MACTANE', HT, ob, 1981; (**City of Auckland**); flowers large, dbl., intense fragrance; foliage medium green, semi-glossy; bushy growth; [Benson & Hedges Gold X Whisky Mac]; McGredy, Sam IV; McGredy Roses International, 1982

'MACTAURANG', HT, ob, 1998; (Marvelle, **Tropical Sunset** ™); bud big, pointed, yellow with orange-red; flowers yellow and pink blend with stripes, dbl., 17–25 petals, 4 in., borne mostly singly, slight; prickles moderate; foliage large, light green, matte; bushy, tall (4 ft) growth; [Louise Gardner X (Auckland Metro X Stars 'n' Stripes seedling)]; McGredy, Sam IV; McGredy, 1995

'MACTEL', LCl, mr, 1972; (**Grand Hotel**®, Grandhotel®); flowers scarlet, 4 in., exhibition form, repeat bloom; foliage dark; ADR, 1977; McGredy, Sam IV

'MACTEMAIK', F, rb, 1998; (**Michelangelo**, The Painter); flowers red blend with stripes, double, dbl., 17–25 petals, 4 in., borne in small clusters, slight fragrance; prickles moderate; foliage medium, medium green, matte; tall, bushy, 120 cm growth; [Louise Gardner X (Auckland Metro X Stars 'n' Stripes seedling)]; McGredy, Sam IV; McGredy, 1995

'MACTENNI', Min, op, 1978; (**Takapuna**®); flowers light peach-pink, medium, dbl., 37 petals, intense fragrance; tall, spreading growth; [New Penny X ((Clare Grammerstorf X Cavalcade) X Elizabeth of Glamis)]; McGredy, Sam IV; McGredy Roses International

'MACTITIR', HT, mr, 1990; (**Hoagy Carmichael**); flowers large blooms, dbl., 26–40 petals, moderate fragrance; foliage medium, dark green, matt; upright, bushy growth; [(Sir Harry Pilkington X Elegy) X Pounder Star]; McGredy, Sam IV; McGredy Roses International, 1990;, Roses by Fred Edmunds, 1990

'MACTRAMPOL', F, w, 1998; (**Remuera**); flowers white, very full, small blooms, very dbl., 26–41 petals, 2.75 in., borne in large clusters, slight fragrance; prickles moderate; foliage small, medium green, semi-glossy; bushy, low, 40 cm growth; Gold Star of the South Pacific, Palmerston North, NZ, 1989; [Seaspray X Sexy Rexy]; McGredy, Sam IV; McGredy, 1991

'MACTRUM', F, or, 1977; (**Trumpeter**®); bud ovoid; dbl., 39 petals, 3.5 in., cupped, slight fragrance; foliage medium green, glossy; bushy, compact growth; GM, NZ, 1977 GM, Portland, 1977 James Mason, RNRS, 1991; [Satchmo X Seedling]; McGredy, Sam IV; McGredy Roses International

'MACTUAL', HT, dr, 1998; (**Romeo**); flowers dark red, full, large, very dbl., 26–40 petals, 4.5 in., borne mostly singly, moderate fragrance; prickles moderate; foliage large, dark green, semi-glossy; bushy, tall (110 cms) growth; [Howard Morrison X Harmonie]; McGredy, Sam IV; McGredy, 1995

'MACTURANG', HT, pb, 1998; (**Maria McGredy**); flowers pink blend, large, full, very dbl., 26–40 petals, 4.5 in., borne mostly singly, slight fragrance; prickles moderate; foliage large, medium green, semi-glossy; bushy, tall (120 cms) growth; [Freude X Remember Me]; McGredy, Sam IV; McGredy, 1997

'MACULTRA', F, r, 1983; (**Brown Velvet**, Colorbreak); flowers orange, tinged brownish, medium, dbl., 35 petals, slight fragrance; foliage medium, dark, glossy; upright growth; Gold Star of the South Pacific, Palmerston North, NZ, 1979; [Mary Sumner X Kapai]; McGredy, Sam IV, 1982

MACvisar, Cl F, w; (**Ice White, Climbing**); McGredy, Sam IV

'MACVOLAR', F, or, 1987; (**Volare**); flowers medium, dbl., 15–25 petals, slight fragrance; foliage large, medium green, glossy; upright growth; [Julischka X Matangi]; McGredy, Sam IV; McGredy Roses International, 1988

'MACWAIKE', Gr, op, 1986; (**Waiheke**®, Waikiki); flowers coral-pink fading lighter, medium, semi-dbl., 30 petals, exhibition form, borne in sprays of 5-9, slight, spicy fragrance; ovoid, small, tan-orange fruit; small, green prickles; foliage medium, dark green, glossy; upright, bushy growth; PP005429; [Tony Jacklin X Young Quinn]; McGredy, Sam IV, 1985; Co-Operative Rose Growers, 1987

'MACWAIRAR', HT, my, 1998; (**Apertif**); flowers medium yellow, large, double, dbl., 17–25 petals, 4 in., borne mostly singly, slight; prickles moderate; foliage large, light green, matte; bushy, medium, 110cm growth; [Solitaire X Sunbright]; McGredy, Sam IV

'MACWAIWER', Min, or, 1987; (Tara, **Tara Allison**); flowers small, semi-dbl., 6–14 petals, slight fragrance; foliage small, medium green, semi-glossy; bushy growth; [Wanaka X Eyepaint]; McGredy, Sam IV; Justice Miniature Roses, 1986

'MACWALLA', Min, ob, 1978; (**Kaikoura**®); flowers orange, patio, medium, dbl., 27 petals, slight fragrance; foliage glossy, dark; vigorous, bushy growth; [Anytime X Matangi]; McGredy, Sam IV

'MACWEEMAT', Min, rb, 1978; (Waitemata, **Waitmata**®, Wee Matt); flowers medium, dbl., 42 petals, slight fragrance; foliage light green, glossy; bushy growth; [Wee Man X Matangi]; McGredy, Sam IV; McGredy Roses International, 1980

'MACWHAKA', S, mp, 1987; (**Gwen Swane**); flowers medium, dbl., 26–40 petals, slight fragrance; foliage small, medium green, matt; spreading growth; [MACbroey X Snow Carpet]; McGredy, Sam IV; McGredy Roses International, 1988

'MACWHENU', HT, ob, 1981; (**Derek Nimmo**); bud ovoid; flowers orange-red, silvery reverse, borne 1-3 per cluster, dbl., 30 petals, moderate fragrance; red prickles; foliage medium green; vigorous growth; [Seedling X Seedling]; McGredy, Sam IV; John Mattock, Ltd.

'MACWHITOUT', Min, w, 1988; (**Whiteout**); flowers small, dbl., 20 petals, slight fragrance; foliage small, medium green, semi-glossy; bushy growth; [Sexy Rexy X Popcorn]; McGredy, Sam IV; Justice Miniature Roses, 1990

'MACWOODMA', F, rb, 1988; (**Redwood**); flowers hand-painted, medium, semi-dbl., 14 petals, slight fragrance; foliage large, dark green, glossy; bushy growth; [Old Master X Wienerwald]; McGredy, Sam IV

'MACWOOHERM', F, mr, 1998; (**Red Point**) flowers medium red, double, dbl., 17–25 petals, 3.5 in., borne in small clusters, slight fragrance; prickles moderate; foliage large, medium, glossy; bushy, medium, 110 cm growth; [Maiden Voyage X Eyeopener]; McGredy, Sam IV; McGredy, 1993

'MACWYNPIN', F, mp, 1994; (**Manapouri**); dbl., 15–25 petals, 1.5–2.75 in., slight fragrance; foliage medium, medium green, semi-glossy; patio; bushy (35 cms) growth; [MACredparap X Moody Blues]; McGredy, Sam IV; McGredy Roses International, 1994

'MACWYNSCAR', F, or, 1994; (**Hauraki**); dbl., 15–25 petals, 1.5–2.75 in., slight fragrance; foliage small, medium green, semi-glossy; patio; bushy (30 cms) growth; Rose of The Year, Auckland, NZ, 1998; [Trumpeter X Kapiti]; McGredy, Sam IV; McGredy Roses International, 1994

'MACYABA', HT, dy, 1981; (**Yabadabadoo**®); flowers large, dbl., 20 petals, slight fragrance; foliage medium, medium green, semi-glossy; bushy growth; [Yellow Pages X Bonfire Night]; McGredy, Sam IV

'MACYEFRE', HT, yb, 1987; (Chartreuse, **Solitaire**); flowers yellow tinted pink, reverse yellow, fading without blanching, dbl., 25 petals, cupped, slight fragrance; no fruit; slightly hooked, large, reddish-brown prickles; foliage medium, dark green, semi-glossy; bushy, strong growth; PIT, RNRS, 1985; [Freude X Benson & Hedges Gold]; McGredy, Sam IV, 1978; Sealand Nursery, 1987

'MACYELKIL', HT, dy, 1998; (**Gold Heart**, Heart of Gold); flowers deep yellow, dbl., 17–25 petals, 4 in., borne in small clusters, moderate fragrance; prickles moderate; foliage medium, medium green, glossy; upright, very tall, 160 cm growth; [Solitaire X Remember Me]; McGredy, Sam IV; McGredy, 1994

'MACYOUMIS', F, pb, 1979; (Buffalo Bill, **Regensberg**®); flowers petals pink, edged white, white eye, yellow stamens, dbl., 21 petals, 4.5 in., cupped, moderate fragrance; low, bushy growth; GM, Baden-Baden, 1980; [Geoff Boycott X Old Master]; McGredy, Sam IV; McGredy Roses International

Mada *see* **'SPECTACULAR'**

'MADAM PRESIDENT', F, pb, 1975; (Madame President); bud long, pointed; flowers blend of pink shades, dbl., 70 petals, 4 in., exhibition form, slight fragrance; bushy growth; [Seedling X Handel]; McGredy, Sam IV; Avenue Nursery

Madam Speaker *see* MEIzuzes

'MADAME BOLLINGER', F, op, 1972; flowers deep orange-salmon, dbl., 25 petals, 3 in., exhibition form, slight fragrance; free growth; [(Little Darling X Goldilocks) X Bobbie Lucas]; McGredy, Sam IV; McGredy

Madame Delbard® *see* 'DELADEL'

Madame Gilberte Dubois, HT, op; 30 petals, flat; foliage dense, luxurious; RvS-Melle, 1990

Madame President *see* **'MADAM PRESIDENT'**

Madame Seneclause, HT, lp

'MADAME STOLZ', D, ly, 1848

'MADAME VERBELEN', HT, dr, 1973; flowers full, dbl., 65 petals, 4.5 in., cupped, intense fragrance; Delforge, S.

'MADAME VIOLET', HT, m, 1981; flowers lavender, dbl., 45 petals, exhibition form, borne singly, no fragrance; medium, reddish light-green prickles; foliage medium, medium green, semi-glossy; tall, upright growth; PP006315; Teranishi, K., 1985; Itama Rose Nursery

'MADAME VIOLET, CLIMBING', Cl HT, m, 1999; flowers pale violet, very dbl., 45 petals, 5.5 in., exhibition form, no fragrance; growth to 9-12 ft.; Itami Rose Nursery, 1992

Madames de Villeparisis, M, lp; [Henri Martin sport]; Morley, Dr B., 1988; (Weatherly, L.)

'MADCAP', F, mr, 1954; bud urn-shaped; flowers scarlet-red, dbl., 20–25 petals, 4.5 in., cupped, moderate fragrance; foliage leathery; bushy growth; [Independence X Crimson King]; Kordes; Stark Bros.

'MADCAP', HT, rb, 1955; bud dark red; flowers flame-scarlet striped yellow, or orange without stripe; [Grand Duchesse Charlotte sport]; Ratcliffe

'MADDALENA', HT, op, 1934; bud pointed to ovoid; flowers salmon-pink, reverse begonia-rose, very large, dbl., 30–32 petals, cupped, moderate fragrance; foliage light green, glossy, leathery; long stems; very vigorous, upright, bushy growth; [Julien Potin X J.C. Thornton]; San Remo Exp. Sta.

'MADELEINE FAIVRE', HT, lp, 1902; flowers large, very dbl.; Buatois; (Sangerhausen)

'MADELEINE LEMAIRE', HWich, op, 1923; flowers bright salmon-pink, medium, semi-dbl.; [Mrs F.W. Flight X Unnamed seedling]; Nonin

'MADELEINE MONOD', HT, op, 1939; bud long; flowers salmon-carmine, very large, dbl., moderate fragrance; foliage bronze; vigorous growth; Chambard, C.

'MADELEINE PACAUD', HT, op, 1922; flowers silvery rose, tinted salmon; Chambard, C.

Madeleine Rivoire®, HT, lp, 1988; Orard

'MADELEINE SELTZER', HMult, ly, 1926; (Madeleine Selzer, Yellow Tausendschön); flowers pale lemon, fading white; few prickles; vigorous growth; [Tausendschön X Mrs Aaron Ward]; Walter, L.

Madeleine Selzer *see* **'MADELEINE SELTZER'**

'MADELEINE WEIDERT', HMult, mp, 1928; flowers rose-pink; [Tausendschön X Rosel Dach]; Walter, L.

'MADELINE', HEg, w; (Emmeline); flowers creamy white edged pink, semi-dbl.

'MADELINE CORREY', HT, dp, 1971; bud globular; flowers cerise, full, medium, dbl., moderate, lemon fragrance; foliage glossy, dark; moderate growth; [Minnie Watson X Sterling Silver]; Watson

Madeline Spezzano *see* 'TINMAD'

Madelon *see* 'RUIMEVA'

Madelon de Paris, F, 1961; Robichon, M.; (Cavriglia)

'MADELYN LANG', Cl Min, dp, 1970; bud slightly ovoid; flowers deep pink, dbl., 40 petals, 1 in.; foliage small, glossy, dark, embossed; upright growth; [Little Darling X Little Chief]; Williams, Ernest D.; Mini-Roses

'MADEMOISELLE', F, dp, 1950; bud ovoid; flowers rose-red, dbl., 50–60 petals, 3.5 in.; very vigorous, branching growth; [Goldilocks X Marionette]; Boerner; J&P

Mademoiselle *see* MEIplovon

'MADEMOISELLE DE DINANT', F, w, 1966; bud globular; flowers creamy white, open, dbl., 22 petals, 3–3.5 in., intense, fruity fragrance; foliage light green; vigorous, spreading growth; [Purpurine X Lavender Pinocchio]; Lens

Mademoiselle Jacqueline, Pol, dp, 1985; Weihrauch

Madeo *see* KORcremkis

'MADETTE', HT, op, 1922; flowers coppery orange-pink; Guillot, P.

'MADGE ELLIOTT', HT, lp, 1964; flowers light pink, center shaded apricot, medium, dbl., slight fragrance; foliage dark, leathery; very vigorous, upright growth; [Queen Elizabeth X ?]; Darvall; G. Knight

'MADGE PRIOR', F, pb, 1934; flowers brilliant claret, white eye, borne in large clusters, single; foliage dark; vigorous growth; Prior

'MADGE TAYLOR', HT, dp, 1930; flowers deep pink, large, dbl., globular, slight fragrance; foliage light; vigorous growth; [Rhea Reid seedling X ?]; Clark, A.; Hazlewood Bros.

'MADGE WHIPP', HT, or, 1936; flowers bright scarlet, dbl., moderate fragrance; foliage leathery; vigorous, bushy growth; [Lady Charmion X J.C. Thornton]; Bees

'MADGE WILDFIRE', HT, or, 1932; bud pointed; flowers indian red, very large, very dbl., exhibition form, moderate fragrance; foliage leathery; vigorous growth; GM, NRS, 1933; Dobbie

Madhatter *see* 'TINHAT'

'MADHOSH', HT, rb, 1975; bud globular; flowers deep magenta-red, streaked mauve, dbl., 45 petals, 4.5 in., slight fragrance; foliage leathery; vigorous, bushy, compact growth; IARI

'MADHUMATI', HT, mp, 1973; flowers large blooms borne singly, dbl., 55 petals, exhibition form, intense fragrance; brown prickles; foliage medium, medium green, smooth; vigorous, upright, bushy growth; [General MacArthur X Unnamed seedling]; Pal, Dr. B.P.; IARI

'MADHURA', F, yb, 1979; bud pointed; dbl., 70 petals, exhibition form, borne 3-6 per cluster, moderate fragrance; foliage glossy; vigorous, upright, bushy growth; [Kiss of Fire X Goudvlinder]; Pal, Dr. B.P.; K.S.G. Son's Roses

Mädi, Pol, or, 1969; flowers medium-large, semi-dbl.; GPG Bad Langensalza; (Sangerhausen)

Madiba *see* KORandpunk

'MADISON', T, w, 1912; flowers pure white; [(Perle X The Bride) X Meteor (N)]; Brant-Hentz

Madison, S, lp, 1998; flowers very small, borne in large clusters; groundcover (2 ft) growth; Poulsen

Madlenka, F, lp; Vecera, L., 1975; (Czech Rosa Club)

'MADOKA', F, rb, 1977; dbl., 25 petals, 2–2.5 in., exhibition form, slight fragrance; foliage glossy, dark; upright growth; [(Zambra X Peace) X Cherry Brandy]; Teranishi, K.; Itami Bara-en

Madona *see* MEInopia

'MADONNA', HT, w, 1908; [Mlle Alice Furon X Marie van Houtte]; Cook, J.W.

'MADRAS', HT, pb, 1981; bud ovoid; flowers rose with yellow and light pink reverse, borne singly, dbl., 48 petals, moderate fragrance; prickles hooked downward; foliage large, leathery; medium, spreading growth; [Unnamed seedlng X Unnamed seedling]; Warriner, William A.; J&P

'MADRIGAL', HT, dp, 1950; bud long, pointed; flowers brilliant salmon-pink flushed coppery, very large, dbl., moderate, spicy fragrance; vigorous growth; [(Mme Joseph Perraud seedling X Unnamed seedling) X (Mme Joseph Perraud seedling X R. foetida bicolor)]; Gaujard

Madrigal *see* HARextra

'MADY', Cl HT, ly, 1925; flowers large, dbl.; Gemen & Bourg; (Sangerhausen)

Maestro® *see* 'MACKINJU'

'MAESTRO', HT, dr, 1957; bud oval; flowers deep velvety red, open, large, dbl., intense fragrance; foliage dark, glossy; vigorous, bushy growth; RULED EXTINCT 2/81; [Crimson Glory X Charles Mallerin]; Delforge

Mafalda di Savoia *see* BARsav

'MAGALI', HT, dp, 1952; flowers carmine, open, medium, dbl., 35–40 petals, slight fragrance; foliage abundant, leathery; very vigorous, upright, bushy growth; [Charles P. Kilham X Brazier]; Mallerin, C.; Meilland-Richardier

Magali *see* MEIgoufin

'MAGALI BONNEFON', HT, pb, 1916; flowers pink, reverse bright salmon-pink, semi-dbl., moderate fragrance; [Mme Abel Chatenay sport]; Nabonnand

Maganwilare, HT, pb, 1995; Huber

'MAGDALENA DE NUBIOLA', HT, op, 1932; flowers salmon-rose, semi-dbl., slight fragrance; [Li Bures X Mari Dot]; Dot, Pedro; C-P

Magdeburg, HT, 1988; VEG; (Sangerhausen)

'MAGENTA', Pol, m, 1916; flowers violet-red, semi-dbl.; dwarf growth; Barbier

'MAGENTA', F, m, 1954; (Kordes' Magenta); bud ovoid; flowers rosy magenta to soft deep mauve, large blooms in large clusters, dbl., intense fragrance; foliage dark, leathery; vigorous, upright, bushy growth; [Yellow Floribunda X Lavender Pinocchio]; Kordes

'MAGENTA', HT, mr, 1934; bud pointed; flowers crimson-carmine, large, dbl., intense fragrance; foliage glossy, dark; vigorous, bushy growth; Leenders, M.

Magenta Buttons, LCl, dp

Magenta Diadem *see* TANdelbel

Maggie, B, mr

Maggie Barry *see* 'MACOBORN'

Magia Nera®, HT, dr

'MAGIC', HT, pb, 1954; bud globular; flowers silver-pink, open, dbl., 75 petals, 5.5 in., moderate fragrance; foliage leathery; very vigorous, upright growth; RULED EXTINCT 11/88; [Thornless Beauty sport]; Grillo

'MAGIC', HT, mr, 1988; bud pointed; flowers medium, borne usually singly, dbl., 24 petals, exhibition form; pear shaped, medium, orange fruit; prickles slender, straight, medium, light green; foliage medium, medium green, matt; upright, tall growth; [MACvolar X Tonight]; Strahle, Robert; Carlton Rose Nurseries, 1987

Magic Baby *see* 'GELMAGIC'

Magic Blanket™ *see* 'TANIGINO'

'MAGIC CARPET', LCl, yb, 1941; flowers yellow, splashed orange, scarlet and rose, large, borne in clusters, semi-dbl., exhibition form; foliage small, bronze, leathery, glossy; strong stems; vigorous, climbing or trailing growth; [Coral Creeper X Stargold]; Brownell, H.C.

Magic Carpet *see* 'JACLOVER'

Magic Carrousel® *see* 'MORROUSEL'

'MAGIC CHARM', HT, mp, 1966; flowers medium, dbl., exhibition form; vigorous growth; [Mount Shasta X Granada]; Aufill

Magic de Feu *see* 'KORFEU'

'**MAGIC DRAGON**', Cl Min, dr, 1969; bud short, pointed; flowers small, dbl.; foliage leathery; very vigorous, upright growth; [((R. wichurana X Floradora) X Seedling) X Little Buckaroo]; Moore, Ralph S.; Sequoia Nursery

Magic East *see* 'VIRSPLASH'

Magic Fire® *see* 'LAPNEUF'

Magic Fire® *see* LAPjaminal

Magic Forez, HT, 1970; Croix; (Cavriglia)

Magic Lantern *see* 'STANQUELI'

Magic Medley *see* 'ABHAYA'

Magic Meidiland™ *see* 'MEIBONRIB'

Magic Meillandecor® *see* 'MEIBONRIB'

'**MAGIC MIST**', Min, mr, 1980; bud pointed; flowers medium red, veined darker, blooms borne usually singly, dbl., 47 petals, exhibition form, moderate fragrance; thin, tan prickles; foliage small, medium to dark green, glossy; bushy, spreading growth; [Tom Brown X Little Chief]; Williams, Ernest D.; Mini-Roses

'**MAGIC MOMENT**', HT, mr, 1964; bud pointed; flowers scarlet to geranium-red, large, dbl., exhibition form, moderate fragrance; Frères, Buyl; Cuthbert

Magic Moments, HT, op, 1983; Martin

'**MAGIC MOON**', Gr, ob, 1970; bud ovoid; flowers deep salmon, reverse silver, medium, very dbl., slight fragrance; foliage large, leathery; vigorous, upright growth; [Little Darling X Golden Scepter]; Schwartz, Ernest W.; Wyant

'**MAGIC MOUNTAIN**', F, yb, 1973; bud ovoid; flowers large, dbl., exhibition form, slight fragrance; foliage glossy, dark, leathery; vigorous, bushy growth; [Circus X Texan]; Armstrong, D.L.; Armstrong Nursery

'**MAGIC RED**', F, mr, 1942; flowers large blooms in clusters, dbl., 45 petals, globular, slight fragrance; foliage leathery, glossy; vigorous, bushy growth; [Henri Panthier X Dance of Joy]; Kordes; J&P

Magic Silver, HT, m, 1998; Kordes

Magic Splendor *see* 'LYOMA'

Magic Sunblaze *see* SCHanbiran

Magic Sunset, F, ob, 1995; Tantau

'**MAGIC TOUCH**', HT, lp, 1974; bud ovoid; flowers soft pink, dbl., 32 petals, 5 in., exhibition form, moderate fragrance; foliage glossy; moderate growth; [Tropicana X Queen of Bermuda]; Golik; Dynarose

'**MAGIC WAND**', Cl Min, dp, 1957; flowers light red, blooms in clusters, semi-dbl., 20 petals, 1 in.; orange fruit; foliage small, dark; arching (to 4 ft) growth; [Eblouissant X Zee]; Moore, Ralph S.; Sequoia Nursery

Magician™ *see* 'MINIMAG'

'**MAGICIENNE**', HT, or, 1957; flowers geranium-red, well formed, dbl., 30 petals; foliage bronze; dwarf, bushy growth; [Comtesse Vandal X (Peace X Independence]; Laperrière; EFR

Magicienne 78® *see* 'LAPOUDA'

Magie des Jardins *see* **Gartenzauber®**

Magie d'Orient, HMsk, 1991; Lens

Magitta® *see* 'TANATTIGAM'

Magma, HT, or, 1998; Kordes

'**MAGNA CHARTA**', HP, mp, 1876; (Casper); flowers bright pink, suffused carmine, very large, dbl., globular, some recurrent bloom, moderate fragrance; foliage thick, rich green; vigorous, compact growth; Paul, W.

'**MAGNAFRANO**', HT, mr, 1900; flowers rich crimson-rose, large, dbl., intense fragrance; [Magna Charta X Safrano]; Van Fleet; Conard & Jones

Magneet *see* **'FEUERWERK'®**

Magnet, F, mr; flowers luminous red, large, dbl., slight fragrance; VEG; (Sangerhausen)

'**MAGNIFICA**', HEg, m, 1916; (R. eglanteria duplex, R. rubiginosa magnifica); flowers purplish red, semi-dbl.; [Lucy Ashton self seedling]; Hesse

'**MAGNIFICENCE**', HT, op, 1954; flowers salmon-pink tinted yellow, large, dbl., exhibition form, moderate fragrance; foliage dark; vigorous, bushy growth; [Peace seedling]; Gaujard; McGredy

'**MAGNIFIQUE**', Pol, lp, 1928; flowers clear pink, open, large, borne in clusters, semi-dbl., cupped; foliage rich green, glossy; vigorous growth; [Orléans Rose sport]; deRuiter; Sliedrecht & Co.

Magnolia Rose *see* **'DEVONIENSIS'**

Magnolia Springs™ *see* 'TALSPRINGS'

'**MAGNOLIJA**', HP, w, 1940; flowers creamy white, large, dbl.; Kosteckij; (Sangerhausen)

'**MAGRANA**', F, or, 1954; flowers well formed, dbl.; vigorous growth; [Méphisto X Alain]; Dot, Pedro

'**MAHADEV**', F, or, 1975; bud long, pointed; flowers full, open, small, dbl., 20 petals, 1.5 in., slight fragrance; foliage glossy; vigorous, tall, bushy growth; [Seedling X (Unknown X Unknown)]; Viraraghavan, M.S.

'**MAHAGONA**', HT, pb, 1956; (Mahogany); bud pointed; flowers dull orange-scarlet, open, very large, moderate fragrance; foliage leathery, wrinkled; very vigorous, upright, bushy growth; [Golden Rapture X Hens Verschuren]; Kordes; McGredy

'**MAHAJA**', HT, dp, 1936; bud long, pointed; flowers deep rose-pink, center rose-red, base yellow, dbl., 4 in., exhibition form, moderate fragrance; foliage glossy; long stems; vigorous growth; [Rose Hill sport]; Carbaugh; Johnstown Greenhouses

Mahalaxmi, HT, yb, 1988; Patil, B.K.

Mahalia™ *see* KEIreb

'**MAHARAJAH**', HP, mr, 1931; flowers velvety crimson, golden anthers, large blooms in clusters, semi-dbl., moderate fragrance; foliage dark, bronze, leathery; Cant, B. R.

Maharani, HT, pb, 1996

Maharishi, HT, or, 1995; Tejganga

Maheka *see* **'PURPUREA'**

'**MAHINA**', HT, ab, 1952; flowers reddish apricot, reverse golden yellow, large, dbl., 35 petals, moderate fragrance; foliage leathery; very vigorous, bushy growth; GM, Bagatelle, 1952; [Peace X Fred Edmunds]; Meilland, F.

Mahogany *see* **'MAHAGONA'**

Maia, HT, 1986; Barni, V.; (Cavriglia)

'**MAID MARIAN**', HT, pb, 1920; flowers carmine-rose, reverse silvery pink, dbl.; Therkildsen

'**MAID MARION**', HWich, w, 1909; flowers white, tipped pink, center filled with yellow stamens, slightly incurved, single; foliage large, glossy; vigorous growth; Walsh

'**MAID MARION**', HMsk, w, 1930; flowers white opening blush, borne in very large clusters, semi-dbl.; vigorous (3-4 ft) growth; Pemberton; Bentall

Maid Marion *see* **'RED IMP'**

'**MAID OF GOLD**', Cl HT, my, 1936; bud globular, reddish; flowers golden yellow, large, very dbl., profuse, intermittent bloom; foliage glossy; vigorous, climbing (12 ft.), compact growth; [Golden Emblem, Climbing X ?]; Raffel; Port Stockton Nursery

Maid of Honor *see* **'MAID OF HONOUR'**

'**MAID OF HONOUR**', HT, yb, 1986; (Maid of Honor); flowers yellow, light pink center, large blooms borne singly and in, dbl., moderate fragrance; foliage large, dark, semi-glossy; tall, upright growth; [Folklore X Unnamed seedling]; Weddle, Von C.; Hortico Roses, 1984

'**MAID OF HONOUR**', F, op, 1951; (Schleswig); bud long, pointed; flowers salmon-pink, open, large, borne in trusses, single, moderate fragrance; foliage leathery, glossy, light green; vigorous, upright growth; (28); RULED EXTINCT 1/86; [Crimson Glory X Holstein]; Kordes; Morse

'**MAID OF KENT**', HT, op, 1929; flowers soft salmon-pink, moderate fragrance; [Ophelia X Mrs W.J. Grant]; Archer

'**MAID OF ORLEANS**', F, ob, 1977; flowers orange-flame blend, full, dbl., 40 petals, 4 in., slight fragrance; foliage glossy, dark; very free growth; [Val De Mosa X Tropicana]; Ellick

Maiden Voyage *see* 'MACLOCKER'

Maiden's Blush *see* **'GREAT MAIDEN'S BLUSH'**

'MAIDEN'S BLUSH', HAlba, w, 1797; (Small Maiden's Blush); flowers soft blush, medium, dbl., globular, non-recurrent, moderate fragrance; vigorous (to 8 ft.) growth; [Presumed to be a natural hybrid of R. alba X R. centifolia]; Kew

Maiden's Blush *see* **'WILLIAM R. SMITH'**

Maidi, HT, pb

Maids of Jubilee *see* 'TALMAID'

Maidy® *see* 'KORWALBE'

'MAIGOLD', S, dy, 1953; flowers bronze-yellow, semi-dbl., 14 petals, 4 in., cupped, non-recurrent, intense fragrance; foliage glossy; bushy (5 ft.), pillar or shrub growth; [Poulsen's Pink X Fruhlingstag]; Kordes

Maile's Double Pink OGR, Misc OGR, mp; (found rose)

Maile's Pink Quill, Misc OGR, lp; (found rose)

Mainauperle® *see* 'KORMAI'

'MAINZ', HT, my, 1930; flowers citron-yellow, very large, dbl., moderate fragrance; [Kardinal Piffl sport]; Leenders Bros.

Mainzer Fastnacht *see* 'TANSI'

Mainzer Fastnacht, Climbing *see* **'BLUE MOON, CLIMBING'**

Mainzer Rad *see* **'MAINZER WAPPEN'**

'MAINZER WAPPEN', S, or, 1963; (Mainzer Rad); bud pointed; flowers red tinted orange, blooms in clusters (up to 20), dbl., 25 petals, 3.5 in., moderate fragrance; foliage dark; bushy, upright (5 ft) growth; Kordes, R.

Mainzer Wappen, S, or

'MAISIE', HT, op, 1926; flowers salmon-orange and pink, medium, semi-dbl.; Dickson, A.; (Sangerhausen)

'MAISIE GOWIE', HT, ?, 1968;, intense fragrance; foliage dark, leathery; bushy growth; Gowie

Maisieres, F, mr, 1993; 29 petals, flat; foliage dense dark green; RvS-Melle, 1992

'MAISON PERNET-DUCHER', HT, my, 1934; bud pointed; flowers golden yellow, veined copper, large, dbl.; foliage glossy, dark; very vigorous, bushy growth; Pernet-Ducher; Gaujard

Maitland White, T, w

Maiwunder, S, ly, 1966; Kordes

'MAJA', F, mr, 1959; flowers cinnabar-red, borne in broad clusters, dbl., 30 petals, cupped; vigorous, bushy growth; [Independence X Signal Red]; deRuiter; Horstmann

Maja Mauser *see* **'CARA MIA'**

'MAJA MAUSER', F, or, 1970; flowers dark orange-red, dbl., 20 petals, 4 in., slight fragrance; foliage glossy, dark; bushy, upright growth; [Evelyn Fison X Seedling]; Poulsen, Niels D.; Poulsen

Maja Oetker *see* **'CORONATION GOLD'**

'MAJESTADE', HT, w, 1957; flowers cream-white, base deep yellow, well formed; very vigorous growth; [Mme Marie Curie X Peace]; da Silva, Moreira

Majesté® *see* 'DORMA'

'MAJESTIC', HT, or, 1955; bud long, purplish; flowers cinnabar-red, open, large, dbl., moderate fragrance; vigorous growth; [Peace X Unnamed seedling]; Gaujard

'MAJESTUEUSE', HGal, mp, 1811; Guerrapain

Majeure *see* **'TANYA'**

Majolika® *see* TANkeijoli

'MAJOR FRANK HAYES', HT, dr, 1934; flowers crimson, center darker, very large, dbl., exhibition form, intense fragrance; foliage leathery, dark; vigorous growth; [Joan Howarth X J.C. Thornton]; Bees

'MAJOR SHELLEY', HT, mr, 1939; bud pointed; flowers rich crimson-scarlet, dbl., 35 petals, 5–6. in., exhibition form; foliage leathery, dark; very vigorous, bushy growth; [Mrs J.D. Eisele X Crimson Glory]; Howard, F.H.; H&S

'MAJORCA', HT, mr, 1938; (Majorica); flowers scarlet, large, dbl., cupped; foliage glossy, bronze; vigorous, bushy growth; [Aribau X Angels Mateu]; Dot, Pedro; C-P, 1941

'MAJORCA', HT, op, 1958; bud short, thick; flowers rose-salmon, dbl., 52 petals; moderate growth; Frères, Buyl

Majorette® *see* 'MEIDAD'

Majorette® *see* 'MEIPIESS'

Majorette 86 *see* 'MEIPIESS'

Majorica *see* **'MAJORCA'**

Make A Wish *see* 'MEHPAT'

Make Believe™ *see* 'MORMAKE'

Make Mine Sherry™, Min, rb

'MALA RUBINSTEIN', HT, mp, 1971; flowers high-pointed, dbl., 45 petals, 5.5 in., moderate fragrance; foliage large, matt; Edland Fragrance Medal, ARS, 1972 GM, Belfast, 1973; [Sea Pearl X Colour Wonder]; Dickson, A.

'MALAGA', Cl HT, dp, 1971; flowers reddish pink, classic form, dbl., 36 petals, 4.5 in., intense fragrance; foliage glossy, dark; [(Hamburger Phoenix X Danse de Feu) X Copenhagen]; McGredy, Sam IV

'MALAGUENA', S, mp, 1976; bud ovoid, long, pointed; dbl., 28 petals, 4.5 in., cupped, slight fragrance; foliage large, dark, leathery; erect, bushy growth; [Tickled Pink X Country Music]; Buck, Dr. Griffith J.; Iowa State University

Malahat *see* 'BETAHAT'

Malak, HT, pb, 1988; Kumar

Malakarsiddha, HT, lp, 1987; Patil, B.K.

'MÄLAR-ROS', HT, dr, 1932; bud pointed; flowers glowing ruby-red indoors, dark blood-red with crimson outdoors, dbl., exhibition form, moderate fragrance; very vigorous, bushy growth; [Hadley X Fragrance]; Kordes

'MALCAIR', LCl, or, 1959; (**Danse des Sylphes®**, Grimpant Danse des Sylphes); flowers rich red suffused geranium-red, blooms in large clusters, globular; foliage glossy; very vigorous growth; [Spectacular X (Peace X Independence)]; Mallerin, C.; URS

'MALCOLM', HT, dp, 1979; flowers carmine, full, dbl., 35–40 petals, 4–5 in., slight fragrance; foliage large, glossy, dark; vigorous, free growth; [Hector Deane X Chopin]; Ellick; Excelsior Roses

Malcolm Sargent *see* 'HARWHARRY'

Maleica *see* TANcelia

'MALEK-ADEL', HGal, lp; (Melik El Adel); flowers soft pink, dotted white, large

Malene, HT, 1980; Laperrière; (Cavriglia)

'MALESHERBES', HGal, m, 1834; flowers purple, spotted; Vibert

'MALIA', Min, mp, 1992; very dbl., 1.5 in., borne singly, highly recurrent bloom, no fragrance; some prickles; foliage small, medium green, semi-glossy; bushy (100-150 cms) growth; [Seedling (Pink 1172) X Rosa Maria]; Mansuino, Dr. Domenico, 1988

'MALIBU', F, or, 1959; bud pointed, ovoid; flowers coral-orange-red, becoming lighter, dbl., 35 petals, 4–4.5 in., exhibition form, intense fragrance; foliage leathery; vigorous, bushy, upright growth; [Charlotte Armstrong X Independence]; Morey, Dr. Dennison; J&P

Malibu Lake Shady Pink, HT, dp, 1994

Malicorne *see* 'DELMATOR'

Malindall, F, mr, 1996; Noack, Werner

Malindi, F, mr, 1974; flowers medium, semi-dbl.; Noack, Werner; (Sangerhausen)

'MALINOVKA', F, dp, 1956; (Raspberry Wine); flowers raspberry-red, medium, slight fragrance; [Staatsprasident Pats X Independence]; Sushkov, K. L.

'MALJOANNA', HT, mp, 1994; (**Querida**); dbl., 26–40 petals, 3–3.5 in., borne mostly singly, slight fragrance; few prickles; foliage large, dark green, semi-glossy; upright (5 ft.) growth; PPAF; [Silver Anniversary X Unknown]; Maltagliati, Mark G.; Meilland-Star Roses, Inc., 1995

'MALKARSIDDHA', HT, lp, 1989; [Century Two sport]; Patil, B.K.; K.S.G. Son's Roses, 1987

MALloeur, LCl, mr; (**Coup de Coeur**)

'MALMESBURY', HT, my, 1980; bud pointed; dbl., 23 petals, borne singly and several together, slight fragrance; slightly hooked prickles; foliage medium green; compact, bushy growth; [Vera Dalton X Parasol]; Sanday, John

'MALMIYA', Gr, ly, 1991; (**Ayako**™); flowers pale yellow cream, large, dbl., 26–40 petals, slight fragrance; foliage medium, medium green, semi-glossy; medium, upright growth; [Sonia sport]; Maltagliati, Mark G.; Nino-miya Nursery Co., 1990

'MALNINO', Gr, w, 1989; (**Saint Louis**); flowers white to ivory; PP006908; [Lifirane sport]; Maltagliati, Mark G.; Conard-Pyle Co., 1989

MALren, HT, yb, 1993; (**Nerene**); Malanseuns

'MALSO', F, or, 1958; (**Soleil**®); flowers bright vermilion, blooms in clusters, dbl., 29 petals, 3 in.; foliage clear green; dwarf, bushy growth; Mallerin, C.; EFR

'MALTON', HCh, mr, 1830; (Fulgens); flowers bright crimson; (28); Guérin

'MALVA', F, m, 1935; flowers mauve, center white, open, large, semi-dbl.; foliage large, glossy, light; very vigorous, bushy growth; Leenders, M.

Malvena, HT, lp; Dawson, George, 1971; (Weatherly, L.)

Malverns *see* KORdehei

'MALVINA', C, lp, 1841; flowers pale pink, edged lighter, well-formed, large blooms in clusters, very dbl.; Verdier, V.

Malwa 94, HT, pb, 1994; Gokhale

Mama, HT, dr, 1991; flowers large, very dbl., moderate fragrance; Wänninger, Franz; (Sangerhausen)

'MAMA DE MEYER', HT, w, 1931; flowers cream, center salmon, well formed, large, dbl.; vigorous growth; [Duchess of Wellington X Aspirant Marcel Rouyer]; Lens

'MAMA LAMESCH', HT, op, 1922; flowers orange-rose, center deeper, reverse reddish rose, dbl., moderate fragrance; [Frau Oberprasident von Grothe X Mme Edouard Herriot]; Lambert, P.

'MAMA LOOYMANS', HT, dp, 1910; flowers carmine-pink, medium, semi-dbl.; Leenders, M.; (Sangerhausen)

Mama Mia *see* 'ZIPMIA'

'MAMA PECHTOLD', HT, 1938; bud long, pointed; flowers rosy salmon, dbl.; foliage bronze; long, strong stems; vigorous growth; [Katherine Pechtold X Briarcliff]; Pechtold; op

Mamaia, Pol, or, 1972; semi-dbl.; GPG Bad Langensalza; (Sangerhausen)

'MAMAN', HT, or, 1963; flowers orange-vermilion, well formed, dbl., 30–40 petals, 4–5 in., slight fragrance; foliage bright green; bushy growth; GM, Geneva, 1962; [(Rome Glory X La Vaudoise) X Impeccable]; Delbard-Chabert

Maman Chérie *see* 'DELMANCHE'

'MAMAN COCHET', T, pb, 1893; bud pointed; flowers pale pink, center deeper, base lemon-yellow, dbl., 4 in., exhibition form, moderate fragrance; foliage dark, leathery; vigorous, bushy growth; [Marie van Houtte X Mme Lombard]; Cochet, S.

'MAMAN COCHET, CLIMBING', Cl T, pb, 1909; Upton, 1909 & H&S, 1915

'MAMAN DENTAL', HT, mp, 1921; flowers pure rose-pink; [Mme Caroline Testout sport]; Dental

'MAMAN GENEVIÈVE', F, or, 1960; flowers red tinted orange; bushy growth; Hémeray-Aubert

'MAMAN LEVAVASSEUR', Pol, dp, 1907; (Baby Dorothy); flowers bright crimson-pink, dbl.; [Mme Norbert Levavasseur sport]; Levavasseur

'MAMAN LYLY', HT, lp, 1911; flowers large, semi-dbl.; Soupert & Notting; (Sangerhausen)

Maman Pineau *see* 'GODILOFTER'

'MAMAN TURBAT', Pol, pb, 1911; flowers china-rose shaded lilac, reverse almost white, large, borne, semi-dbl.; foliage dark, soft; bushy growth; [Mme Norbert Levavasseur X Katharina Zeimet]; Turbat

'MAMBO', F, op, 1971; bud ovoid; flowers salmon-pink, medium, dbl., slight fragrance; dwarf, bushy growth; [Tropicana X Zorina]; Tantau, Math.

'MAMBO', Gr, mr, 1960; bud ovoid; flowers currant-red to cardinal-red, medium, dbl., exhibition form, slight fragrance; foliage dark, leathery; vigorous, tall, bushy growth; [Charlotte Armstrong X Unnamed seedling]; Swim & Weeks

'MAMIE', HT, pb, 1901; flowers pink and red, large, dbl., moderate fragrance; Dickson, A.; (Sangerhausen)

'MAMIE SERPA', F, w, 1955; bud ovoid, creamy; flowers open, dbl., 50 petals, 2.5 in., slight fragrance; foliage dark, soft; vigorous, bushy, compact growth; [Goldilocks X Snowbird]; Serpa

'MAMILLE', HT, my, 1976; bud long, pointed; dbl., 35 petals, 4.5 in., exhibition form, moderate fragrance; foliage soft; vigorous, upright growth; [Peer Gynt X Valencia]; Kordes; Vilmorin

'MAMITA', HT, mr, 1958; flowers garnet-red, large, intense fragrance; vigorous growth; [Dicksons Red X Unamed seedling]; Robichon

Mamy Blue® *see* 'DELBLUE'

Mamy Laperrière *see* LAPlical

'MANA', F, yb, 1989; flowers creamy yellow, edges flushed pink, hybrid-tea shaped, large, dbl., moderate fragrance; prickles light brown; foliage medium green, glossy, veined; upright growth; [Liverpool Echo X Arthur Bell]; Cattermole, R.F.; South Pacific Rose Nursery

Manaia *see* 'SUNMANI'

'MANAMSUN', Min, ob, 1996; (**Amber Sunset**); flowers orange with orange and dark yellow reverse, star-shaped, dbl., 26–40 petals, 1.75–2 in., exhibition form, no fragrance; some prickles; foliage medium, dark green, glossy, disease resistant; very vigorous, medium (40-50 cms), bushy growth; [June Laver X Rubies 'n' Pearls]; Mander, George; Select Roses, 1996

Manapouri *see* 'MACWYNPIN'

Manasi, F, lp, 1991; IARI

'MANBERYL', Min, yb, 1995; (**Golden Beryl**); flowers deep yellow brushed orange inside of petals, blooms borne in, dbl., 15–25 petals, 1.75–2 in., slight fragrance; few prickles; foliage medium, medium green, glossy; medium (35-40 cms), bushy growth; [June Laver X Rubies 'n' Pearls]; Mander, George; Oregon Miniature Roses, 1995;, Select Roses, 1995

Manchester Guardian Angel, N, w

Manchu Rose *see* **R. XANTHINA**

'MANCLASSIC', F, op, 1996 (**Buffy Sainte-Marie**, Valia Balkanska); flowers blend of salmon, orange and pink with yellow center and yell, dbl., 15–25 petals, 3.25–4.25 in., slight fragrance; some prickles; foliage medium, dark green, glossy, disease-resistant; very vigorous, medium (70-90 cms), bushy growth; [June Laver X Rubies 'n' Pearls]; Mander, George; Select Roses, 1996

'MANCORAL', Min, op, 1995; (**Coral 'n' Gold**); flowers coral/salmon/pink blend inside, yellow centre, yellow cream reverse, some yellow stripes inside petals, full (26–40 petals) medium (4–6 cms) blooms borne in small clusters; no fragrance; few prickles; foliage medium, medium green, semi-glossy; bushy, medium (40–45 cms) growth; [June Laver X Rubies 'n' Pearls]; Mander, George; Select Roses, 1995–96

'MANDALAY', HT, my, 1942; (Helvétia); flowers clear yellow, open, dbl., 40

petals, 6 in., moderate fragrance; foliage leathery; long stems; very vigorous, upright growth; [Soeur Thérèse X Feu Joseph Locymans]; Mallerin, C.; A. Meilland;, J&P

'MANDARIN', F, mr, 1951; bud ovoid; flowers mandarin-red, blooms in large clusters, semi-dbl., 18 petals, 3–4 in., exhibition form, moderate fragrance; foliage leathery, glossy; vigorous (3 1/2-4 ft), upright growth; [Lilette Mallerin sport X Red F seedling]; Boerner; J&P

'MANDARIN', HT, my, 1946; flowers golden yellow, intense fragrance; foliage glossy; vigorous growth; [Betty Uprichard X Ville de Paris]; Robichon

Mandarin *see* KORcelin

Mandarin Delight *see* 'FOUMANDE'

Mandarine *see* **'MANDRINA'**

'MANDA'S TRIUMPH', HWich, w, 1899; flowers pure white, borne in clusters of 10-12, very dbl.; very vigorous growth; [R. wichurana X Paquerette]; Horvath; W.A. Manda

'MANDRINA', F, or, 1964; (Mandarine); dbl., 30 petals, 2.5–3 in., blooms in clusters; foliage dark; vigorous, bushy growth; [Moulin Rouge X Frau Anny Beaufays]; deRuiter; Carlton Rose Nurseries

Mandryka, S, or, 1992; flowers medium red with orange; Mekdeci-Olsen

'MANDY', F, ly, 1963; flowers creamy yellow to soft peach, large, dbl.; foliage coppery; very free growth; [Pinocchio X Sweet Repose]; Robinson, H.

Mandy *see* KORinor

'MANDY JO', HT, pb, 1970; flowers pale biscuit, flushed light pink, full, dbl., 22 petals, 4.5 in., slight fragrance; foliage very glossy, bright green; vigorous growth; [Pink Favorite sport]; Abrahams

Mandy Singleton, HT, pb, 1995; Dawson

'MANECA', HT, lp, 1929; dbl., moderate fragrance; Pernet-Ducher; Gaujard

'MANETTE', HGal, rb; flowers rosy crimson marbled with purple, green eye, small, blooms l, very dbl., cupped; vigorous, upright growth; Lecoffé, before 1824; Prior to 1848

'MANETTI', N, lp, 1835; (R. chinensis manettii, R. manettii, R. X noisettiana manettii); 2 in.; red shoots; vigorous shrub growth; Rivers; Rivers, 1835, from S. Manetti, Monza Bot. Gdn., It

'MANGLOW', Min, rb, 1996; (**Glowing Amber**); flowers scarlet red with deep yellow reverse and yellow center, dbl., 26–40 petals, 1.5–2 in., borne mostly singly, slight fragrance; some prickles; foliage medium, dark green, glossy; medium (40-50 cms), bushy growth; [June Laver X Rubies 'n' Pearls]; Mander, George; Select Roses, 1996

Mango® *see* 'DORAPRI'

Mango *see* 'JACMEM'

'MANHALE', Min, m, 1996; (**Haleakala**); flowers velvety dark ruby-purple inside with yellow center, dbl., 15–25 petals, 1.75–2 in., slight fragrance; some prickles; foliage medium, dark green, glossy; medium (40-50 cms), bushy growth; [Rubies 'n' Pearls X June Laver]; Mander, George; Select Roses, 1996

'MANHATTAN', HT, or, 1936; (Passion Rose); bud urn shaped; flowers jasper-red to coral-red, base yellow, very large, dbl., cupped, intense fragrance; very vigorous growth; [Souvenir sport]; Asmus

Manhatten Blue, HT, m

'MANIFESTO', HT, op, 1920; flowers flesh-pink, tinged salmon, dbl., moderate fragrance; McGredy

Manilla, HT; Kordes, R.; (Cavriglia)

'MANILLE', HT, yb, 1951; flowers golden yellow edged red, very large, dbl., 40–50 petals, globular, intense fragrance; foliage leathery, bronze; vigorous, upright growth; [Pierre X Lumiere]; Mallerin, C.; Vilmorin-Andrieux

Manit *see* 'UMSDAD'

Manita *see* KORberuhig

'MANITOU', HT, rb, 1957; flowers coppery red, reverse golden yellow, well formed, very large, dbl.; upright, bushy growth; Swim, H.C.; Kordes

Manja *see* **'SIX FLAGS'**

Manja Böhm *see* **'MÁNJA BÖHMOVÁ'**

'MÁNJA BÖHMOVÁ', HMult, w, 1925; (Manja Böhm); flowers greenish white; [Tausendschön sport]; Böhm, J.

Manja Mourier *see* **'SIX FLAGS'**

'MANJANA', F, op, 1969; flowers salmon-pink-apricot, exhibition form; free growth; [Orange Sensation X (Pink Parfait X Lavender Pinocchio)]; deRuiter

Manmatha, F, m, 1989; Pal, Dr. B.P.

'MANNEQUIN', F, mp, 1961;, slight fragrance; vigorous, upright growth; [(Peace X Cinnabar) X Fashion]; Lens

'MANNHEIM'®, S, dr, 1959; flowers crimson, large blooms in clusters, dbl., recurrent bloom; upright (3 ft.), bushy growth; [Rudolph Timm X Fanal]; Kordes, R., 1958

'MANNING'S BLUSH', HEg, w; flowers white faintly flushed pink, very small, dbl.; foliage fragrant (apple); height 4-5 ft

Mannington Cascade, HWich, w, 1998; Walpol & Lowne

Manola *see* **'FRED CRAMPHORN'**

'MANON', HT, ab, 1924; flowers yellow mixed with apricot, semi-dbl.; [Christine X Mrs Farmer]; Bernaix, P.

Manora® *see* TANaronam

'MANORSUN', Min, ob, 1996; (**Orange Sunset**); flowers salmon-orange inside, some yellow stripes, orange-yellow reverse, dbl., 26–40 petals, 1.5–2 in., slight fragrance; some prickles; foliage medium, dark green, glossy; medium (40-50 cms), bushy growth; [June Laver X Rubies 'n' Pearls]; Mander, George; Select Roses, 1996

'MANOSQUE', HT, mr, 1958; flowers well formed, intense fragrance; moderately vigorous growth; Frères, Buyl

Manou, Gr, 1979; Meilland, L.; (Cavriglia)

Manou Meilland® *see* 'MEITULIMON'

Manou Meilland, Climbing *see* MEItulimonsar

'MANPETALS', Min, op, 1996; (**Glowing Petals**); flowers blend of salmon orange and pink, dark yellow reverse, dbl., 26–40 petals, 2–2.5 in., no fragrance; some prickles; foliage large, dark green, glossy; medium (45-60 cms), bushy growth; [June Laver X Rubies 'n' Pearls]; Mander, George; Select Roses, 1996

'MANPRINCESS', HT, lp, 1991; flowers large blooms borne mostly singly, very dbl., slight fragrance; foliage medium, light green, semi-glossy; tall, upright growth; [(Tiffany X Pascali) X Super Sun]; Mander, George

'MANPURPEARL', F, m, 1992; flowers purple/cream bicolor, mauve color intensifies with sun and a, dbl., 26–40 petals, 3–3.5 in., slight fragrance; very few prickles; foliage medium, dark green, semi-glossy; medium (80-100 cms), bushy growth; [Rise 'n' Shine X MANpurple]; Mander, George

'MANPURPLE', HT, m, 1991; flowers purple/cream bicolor, large blooms borne mostly singly, dbl., 15–25 petals, slight fragrance; foliage medium, dark green, glossy; medium, bushy growth; [Mount Shasta X Super Sun]; Mander, George

'MANPURSUN', Min, m, 1992; (**Purple Sunset**); flowers purple/cream bicolor, very attractive bicolor combination, b, dbl., 15–25 petals, 1.5 in., no fragrance; few prickles; foliage small, dark green, glossy; low (35-40 cms), upright growth; [Rise 'n' Shine X MANpurple]; Mander, George

'MANRUPEARL', Min, m, 1992; (**Rubies 'n' Pearl**); flowers purple/cream bicolor, very attractive, very dark purple on i, dbl., 15–25 petals, 1.5 in., slight fragrance; few prickles; foliage small, medium green, semi-glossy; low (40-50 cms), upright, bushy growth; [Rise 'n' Shine X MANpurple]; Mander, George

'MANSCARLET', F, rb, 1993; (**Scarlet Pearl**); flowers scarlet red with white eye and white reverse, blooms borne i, single, 5 petals, 1.5–2.75 in., slight fragrance; many prickles on basals, few on laterals; foliage medium, dark green, glossy; medium (90-100 cms), bushy, spreading, very vigorous growth; [Pink Meidiland X Seedling]; Mander, George; Christie Nursery, Ltd., 1993

'MANSOPAS', Cl F, op, 1991; flowers medium pink shaded coral and cream, blooms borne in small clusters, dbl., 26–40 petals, slight fragrance; foliage medium, dark green, glossy; tall, upright, bushy growth; [Shades of Pink X Pascali]; Mander, George

'MANSTAR', Min, ob, 1999; (**Amber Star**, Brittany's Glowing Star); flowers amber/orange, reverse golden yellow, dbl., 26–40 petals, 1.75–2.25 in., borne in small clusters, slight fragrance; prickles moderate; foliage medium, dark green, glossy; upright, tall (15-20 in.) growth; [Glowing Amber sport]; Mander, George & Pazdzierski, Jim; Select Roses, 2000;, The Mini Rose Garden, 1999

'MANSUINO ROSE', HT, mr, 1964; (Generosa); flowers crimson to spirea-red, small to blooms, dbl., 35 petals, cupped, slight fragrance; foliage small, dark; thin stems; Mansuino, Q.; Carlton Rose Nurseries

'MANTEAU POURPRE', HGal, 1823; flowers medium, dbl.; Vibert; (Sangerhausen)

Manteo *see* 'BRIMAN'

'MANTOPAS', Min, dy, 1995; (**Golden Topas**); dbl., 15–25 petals, 1.75–2 in., borne in small clusters, slight fragrance; some prickles; foliage medium, medium green, semi-glossy; medium (35 cms), bushy growth; [June Laver X Rubies 'n' Pearls]; Mander, George; Select Roses, 1995

'MANU MUKERJI', HT, my, 1972; [Fragrant Cloud sport]; Friends Rosery

Manuel Canovas *see* MASpagui

'MANUEL P. AZEVEDO', HT, mr, 1911; flowers carmine-red, large, dbl., intense fragrance; Soupert & Notting; (Sangerhausen)

'MANUEL PINTO D'AZEVEDO', HT, pb, 1954; flowers deep rose-pink, reverse lighter, dbl., 48 petals, 4 in., moderate fragrance; foliage dark, glossy, leathery; vigorous, upright growth; [Unnamed seedling X Peace]; da Silva, Moreira

'MANUELA'®, HT, mp, 1968; flowers large, dbl., 30 petals, exhibition form, moderate fragrance; foliage glossy; vigorous, upright, bushy growth; Tantau, Math.

'MANUELITA', HT, or, 1947; flowers large; very vigorous growth; [Cynthia X Vive la France]; Dot, Pedro

Manureva *see* SAUdril

'MANX QUEEN', F, ob, 1963; (Isle of Man); flowers rich gold flushed bronze red, blooms in large clusters, semi-dbl., 18 petals, moderate fragrance; foliage dark; bushy, compact growth; [Shepherd's Delight X Circus]; Dickson, Patrick; A. Dickson

Many Happy Returns *see* 'HARWANTED'

'MANY MOONS', F, my, 1985; flowers deep yellow to light yellow, large, dbl., 35 petals, moderate fragrance; foliage medium, medium green, matt; upright, bushy, tall, arching growth; [Chinatown X Maigold]; Stoddard, Louis

'MANY SUMMERS', HT, ob, 1975; flowers orange-copper, full, dbl., 30 petals, 6 in., intense fragrance; vigorous growth; [Arthur Bell X Belle Blonde]; Fryer, Gareth; Fryer's Nursery, Ltd.

Many Thanks, Min, ab; [Mary Marshall sport]; Geytenbeek, 1976

Manyo, F, ab

'MAORI DOLL', Min, yb, 1977; flowers buff, yellow center; [Yellow Doll sport]; Bell Roses; Sequoia Nursery

'MAORI LULLABY', F, pb, 1963; flowers carmine-rose, base light yellow to white, borne in clusters, semi-dbl., 10–12 petals, 3–3.5 in.; foliage bronze, leathery; bushy, low, compact growth; [Traumland X Unnamed seedling]; Mason, P.G.

'MAORI MOON', F, pb, 1974; bud pointed; flowers pink, center cream, semi-dbl., 10 petals, 3.5 in., cupped, slight fragrance; foliage light, aging darker; upright, compact, bushy growth; [Bengali X (Pink Parfait X King Boreas)]; Clayworth

'MAORILANDER', HT, mr, 1956; bud ovoid, pointed; flowers crimson, lighter reverse, large, dbl., exhibition form, moderate fragrance; foliage bronze, leathery; bushy growth; [Crimson Glory X Peace]; Mason, P.

Marachal Le Clerc *see* 'KRICARLO'

Marama *see* 'SIMARAMAM'

'MARANTA', F, yb, 1974; bud globular; flowers yellow, red, medium, semi-dbl., slight fragrance; foliage soft; upright, bushy growth; Tantau, Math.; Ahrens & Sieberz

'MARATHON', F, op, 1956; flowers salmon-rose, borne in clusters, semi-dbl., 4 in.; moderate, compact growth; [Fashion X Hybrid Tea seedling]; Mondial Roses

'MARBEL', HT, lp, 1999; (**Belle**); flowers light pink, reverse medium pink, dbl., 17–25 petals, 4.5 in., borne mostly singly, slight fragrance; prickles moderate; foliage small, dark green, semi-glossy; upright, medium (2.5 ft) growth; [First Prize X Handel]; Hiltner, Martin

'MARBRÉE', P, rb, 1858; flowers red marbled white, large, dbl.; Robert et Moreau

'MARC GUILLOT', F, dr, 1955; bud ovoid; flowers dark scarlet-red, medium, dbl., 25–35 petals, cupped, slight fragrance; foliage dull green, leathery; vigorous, upright growth; [Happiness X Demain]; Mallerin, C.; EFR

'MARCEL BOIVIN', HT, op, 1954; flowers coral-pink, base yellow, very large, dbl., 50–60 petals, moderate fragrance; vigorous growth; [Souv. de Claudius Pernet X Château de Clos Vougeot]; Buatois

'MARCEL BOURGOUIN', HGal, m; (Le Jacobin); flowers velvety rich scarlet-purple mottled violet; Corboeuf-Marsault, 1899; Corboeuf, 1899

Marcel Pagnol *see* 'MEISOYRIO'

Marcel Pajotin, F, 1976; Kriloff, Michel; (Cavriglia)

Marcel Turbat, LCl; Moreira da Silva, A.; (Cavriglia)

'MARCELINE', HT, rb, 1928; flowers crimson, edge and reverse violet-rose, dbl.; [Frau Karl Druschki X Yves Druhen]; Buatois

Marcel Pagnoe *see* 'MEISOYRIO'

Marcella Baldge *see* **'SOUN DE MARCELLE BALAGE'**

'MARCELLE AUCLAIR', F, or, 1964; flowers medium, borne in clusters of 5-8, semi-dbl., exhibition form, intense fragrance; foliage glossy, leathery; vigorous, upright growth; [Soleil de Lyon X Seedling]; Robichon; Ilgenfritz Nursery

'MARCELLE GRET', HT, my, 1947; bud long, pointed; flowers saffron-yellow, dbl., 28 petals, 6 in., moderate fragrance; foliage dark; vigorous growth; GM, Geneva, 1948; [Peace X Prinses Beatrix]; Meilland, F.

'MARCELLE GRET, CLIMBING', Cl HT, my, 1957; Frères, Brenier

'MARCELLE PETIT', F, op, 1958; flowers glowing salmon-pink; foliage glossy; vigorous growth; [Pinocchio X Independence]; Arles; Roses-France

'MÄRCHEN', Pol, lp, 1927; flowers light rose pink with white, small, single; Kiese; (Sangerhausen)

Marchenkonigin *see* 'KOROYNESS'

'MÄRCHENLAND', F, ob, 1951; (Exception); flowers bright rose tinted salmon, large blooms in clusters of 40, semi-dbl., 18 petals, moderate fragrance; foliage dark; vigorous, upright growth; [Swantje X Hamburg]; Tantau, 1946

'MÄRCHENTAG', HT, lp, 1929; flowers creamy pink, large, dbl.; von Württemberg, Herzogin Elsa; (Sangerhausen)

'MARCHESA BOCCELLA', HP, lp, 1842; (Marquise Boccella, Marquise Bocella);

flowers delicate pink, edges almost blush, compact, large; petals smaller than other HP's, dbl.; stiff, erect stems; dwarf, robust habit growth; Desprez

'Marchioness of Dufferin', HP, mp, 1891; flowers very large, dbl.; Dickson, A.; (Sangerhausen)

'Marchioness of Linlithgow', HT, dr, 1929; flowers deep blackish crimson, open, large, dbl.; foliage soft, bronze; vigorous growth; GM, NRS, 1930; Dobbie

'Marchioness of Londonderry', HP, lp, 1893; flowers pale pink, very large, dbl., 50 petals, exhibition form, moderate fragrance; very vigorous growth; Dickson, A.

'Marchioness of Lorne', HP, pb, 1889; flowers rich rosy pink shaded darker, large, dbl., cupped, intense fragrance; vigorous growth; Paul, W.

'Marchioness of Ormonde', HT, ly, 1918; flowers clear wheat-straw color, center deep honey-yellow, dbl.; Dickson, H.

'Marchioness of Salisbury', HT, dr, 1890; Pernet

'Marcia', HT, op, 1952; flowers pink to coral, base yellow, well shaped, large; foliage glossy; vigorous, upright growth; [Étoile de Hollande X Raffel's Yellow]; Raffel; Port Stockton Nursery

'Marcia Coolidge', HT, pb, 1927; flowers very light pink, reverse darker, stamens dark crimson, semi-dbl., intense fragrance; [Gen. MacArthur seedling]; Coolidge

'Marcia Gandy', HT, rb, 1957; flowers crimson to rose-red, reverse rose-opal, intense fragrance; vigorous growth; Verschuren; Gandy Roses, Ltd.

'Marcia Stanhope', HT, w, 1922; flowers pure white, large, dbl., 25 petals, globular, intense fragrance; foliage leathery; GM, NRS, 1924; [Frau Karl Druschki seedling]; Lilley

'Marco', HWich, w, 1905; flowers white, center coppery; [R. wichurana X Souv. de Catherine Guillot]; Guillot, P.

'Marco Polo', HT, lp, 1971; flowers soft dawn-pink, large, dbl., exhibition form, moderate fragrance; foliage glossy; vigorous, upright growth; [Memoriam X Elizabeth Fankhauser]; Fankhauser

Marco Polo *see* 'MEIpaleo'

'Mardi Gras', HT, dr, 1953; bud ovoid; flowers deep velvety red, dbl., 33 petals, 5 in., exhibition form, moderate fragrance; foliage leathery; vigorous, upright, bushy growth; GM, Baden-Baden, 1953; [Crimson Glory X Poinsettia]; Jordan, G.L.; J&P

'Mardi Gras, Climbing', Cl HT, dr, 1956; Kordes

'Marechal Carmona', HT, op; flowers salmon-pink with reddish tones; da Silva, Moreira

'Maréchal Davoust', M, mp, 1853; flowers bright rose, large, cupped; Robert

'Maréchal de Villars', B, dp; flowers deep rose shaded violet, , cupped

Marechal du Palais, B, lp; Béluze, 1845

'Maréchal Foch', Pol, dp, 1918; (Red Orléans Rose); flowers cherry-red to pink, open, borne in compact clusters, semi-dbl., moderate fragrance; vigorous, bushy growth; [Orléans Rose sport]; Levavasseur

Maréchal le Clerc *see* 'KRIcarlo'

'Maréchal Lyautey', HT, dr, 1931; flowers deep red, large, very dbl., exhibition form, intense fragrance; foliage thick, dark, bronze; very vigorous, bushy growth; [Hadley X Laurent Carle]; Croibier

'Maréchal Niel', N, my, 1864; bud long, pointed; flowers golden yellow, large blooms, dbl., intense fragrance; foliage rich green; weak stems; very vigorous, climbing growth; (14); Pradel

'Maréchal Pétain', HT, yb, 1926; flowers soft pink on yellow ground; Reymond

Maréchal Stalin, HT, mr, 1965; flowers carmine-red, large, dbl., moderate fragrance; Mélichar; (Sangerhausen)

'Marella', HT, pb, 1961; bud globular; flowers pink-red, large, dbl., 35 petals, slight fragrance; foliage glossy, leathery; vigorous growth; [(Happiness X Independence) X Better Times]; Meilland, Mrs. Marie-Louise; URS

Maren *see* 'WILgmar'

Marfa, HT, w, 1975; dbl., slight fragrance; Shtanko, E.E.; (Sangerhausen)

'Marfil', HT, w, 1962; flowers ivory-white, large; strong stems; tall growth; [White Knight X Angelis]; Dot, Pedro

Marga, HRg, lp, 1979; flowers medium, semi-dbl., slight fragrance; Rieksta; (Sangerhausen)

'Marga Weil', HT, or, 1938; flowers salmon-red, large, dbl., slight fragrance; Weil; (Sangerhausen)

'Margaret', HT, pb, 1954; flowers bright pink, reverse silvery pink, well-shaped, dbl., 70 petals, moderate fragrance; vigorous growth; GM, NRS, 1954; [May Wettern seedling X Souv. de Denier van der Gon]; Dickson, A.

'Margaret Amos', HT, dp, 1952; flowers strawberry-red, large, dbl., 25 petals, exhibition form, slight fragrance; foliage dark reddish green; very free growth; [McGredy's Scarlet seedling]; McGredy, Sam IV

'Margaret Anderson', LCl, w, 1931; (Broomfield Novelty); flowers centers deep cream, outer petals cream-flesh, very large, dbl., recurrent bloom, moderate fragrance; foliage thick, leathery; very vigorous, climbing growth; Thomas; H&S

Margaret Anne, HT, lp; Matthews, W.J., 1977; (Weatherly, L.)

'Margaret Anne Baxter', HT, w, 1927; bud pointed; flowers white, sometimes tinted flesh, large, very dbl., 88 petals, moderate fragrance; foliage thick, leathery, glossy, bronze; vigorous, bushy growth; GM, NRS, 1927; [Harry Kirk seedling]; Smith, T.

'Margaret Belle Houston', HT, dr, 1929; flowers velvety crimson, very large, dbl., moderate fragrance; foliage light, leathery; Vestal

Margaret Bushby, S, lp; [R. macrantha X Unknown]; Weatherly, Lila, 1999; Prophyl Pty, Ltd.; (Weatherly, L.)

'Margaret Chase Smith', HT, dr, 1966; bud long, pointed; flowers large, dbl., moderate fragrance; vigorous, upright growth; [Red Duchess X Queen Elizabeth]; Brownell, H.C.

'Margaret Clara', HT, ab, 1999; dbl., 26–40 petals, 2 in., borne mostly singly, moderate fragrance; prickles moderate; foliage medium, dark green, glossy, impervious to rain; upright, medium (3.5 ft) growth; [Solitaire X (Alexander X Remember Me)]; Jones, L.J.

Margaret Daintry *see* 'HORmartim'

'Margaret Dickson', HP, w; flowers white, center pale flesh, well-formed, large, dbl., 65 petals, cupped, occasionally recurrent bloom, slight fragrance; foliage dark; vigorous growth; [Lady Mary Fitzwilliam X Merveille de Lyon]; Dickson, A., 1892

'Margaret Dickson Hamill', HT, yb, 1915; flowers straw-yellow, flushed salmon, petals shell-shaped, large, dbl., slight fragrance; foliage dark, leathery; bushy growth; GM, Bagatelle, 1917 GM, NRS, 1914; Dickson, A.

'Margaret Egerton', HT, dp, 1931; flowers rosy cerise passing to carmine, base yellow, well formed, moderate fragrance; vigorous growth; Chaplin Bros.

'Margaret Elbogen', Pol, w, 1936; flowers pinkish white, intense fragrance; very vigorous growth; Brada, Dr.; Böhm

'Margaret Fleming', HRg, mp, 1995; flowers very clear pink, no lavender tones, blooms borne up to 5 per cluster, semi-dbl., 6–14 petals, 1.5 in., no fragrance; foliage medium tolarge, medium green, semi-glossy, somewhat; bushy (up to 120 cms) growth; [R. rugosa alba X Masquerade]; Fleming, Joyce L.; Hortico Roses, 1994

'Margaret Herbert', HT, lp, 1956; Harrison

'MARGARET HORTON', HT, ab, 1921; bud pointed; flowers apricot-yellow, open, large, dbl., exhibition form; foliage leathery, glossy, light; vigorous growth; Hicks

'MARGARET ISABEL', HT, w, 1983; flowers oyster white, cream center, delicately flushed pink petal edges, dbl., 45 petals, exhibition form, intense fragrance; red-brown prickles; upright growth; [Strawberry Ice X Redgold]; Summerell, B.L.

'MARGARET JEAN', Min, dp, 1978; bud ovoid; flowers deep pink, blooms borne 4-8 per cluster, dbl., 40 petals, slight fragrance; very few prickles; foliage small, dark; bushy, vigorous growth; [Fairy Moss X Fairy Moss]; Dobbs, Annette E.; Small World Min. Roses

'MARGARET LAW', F, ob, 1981; flowers luminous orange, pink on outer petals, borne 3-23 per cluster, dbl., 25 petals, exhibition form, intense fragrance; hooked, red prickles; foliage dark, glossy; vigorous growth; [Princess Michiko X Gold Gleam]; McTeer, Gilbert; Bridgemere Nursery

'MARGARET M. WYLIE', HT, pb, 1921; (Mrs A.J. Wylie); flowers flesh, edges heavily flushed deep rosy pink, dbl., intense fragrance; GM, NRS, 1920; Dickson, H.

Margaret McDowell *see* 'SEADOW'

'MARGARET MCGREDY', HT, or, 1927; flowers orange-scarlet, large, dbl., 35 petals, exhibition form, moderate fragrance; foliage light, leathery, glossy; vigorous growth; GM, NRS, 1925; McGredy

'MARGARET MCGREDY, CLIMBING', Cl HT, or, 1936; Dixie Rose Nursery

'MARGARET MERCER', HT, yb, 1977; flowers pale yellow, edged light pink, full, very dbl., 53–65 petals, 4.5 in., moderate fragrance; foliage glossy, dark; vigorous growth; [Pink Favorite X Peace]; Mercer

Margaret Merril® *see* 'HARKULY'

'MARGARET MOLYNEUX', HT, my, 1909; flowers canary-yellow, moderate fragrance; Dickson, A.

'MARGARET MOORE JACOBS', HT, mp, 1968; bud ovoid; flowers large, dbl., cupped, intense fragrance; foliage glossy; vigorous, bushy growth; [Tiffany X Pink Masterpiece]; Fuller; Wyant

Margaret O *see* 'ORTMAR'

'MARGARET ROBERTS', F, dp, 1976; flowers cerise, full, dbl., 20–25 petals, 3.5 in., intense fragrance; foliage small, dark; moderate, free growth; [Elizabeth of Glamis X Wendy Cussons]; Wood

'MARGARET RUTH', HT, w, 1969; flowers creamy white, center pale pink, large, dbl., exhibition form, intermittent bloom, slight fragrance; foliage glossy, leathery; vigorous, upright, bushy growth; [Anne Letts X Christian Dior]; Taylor, L.R.

Margaret Sharpe, T, mp

'MARGARET SPAULL', HT, ob, 1928; flowers variable orange and lilac, dbl., moderate fragrance; [Ophelia X Unnamed seedling]; Cant, B. R.

Margaret Telfer *see* 'SEATEL'

Margaret Thatcher *see* 'KORFLÜG'

Margaret Thatcher *see* 'TAKSUN'

Margaret Trudeau *see* 'MACAFT'

'MARGARET TURNBULL', Cl HT, yb, 1931; flowers soft pink on amber ground, large, dbl., cupped, slight fragrance; foliage wrinkled, light; vigorous, pillar growth; Clark, A.; NRS New South Wales

'MARGARET VAN ROSSEM', HT, op, 1946; flowers coppery salmon, center old-gold; moderate growth; Van Rossem

Margaret von Hessen, HT, 1973; Hetzel, K.; (Cavriglia)

Margaret Wasserfall *see* KORoberfinz

Margaret Watson, HT, pb

'MARGARETE GNAU', HT, ob, 1930; flowers creamy white on orange ground, very large, dbl., exhibition form, slight fragrance; foliage leathery; very vigorous growth; [Mrs Charles Lamplough X Souv. de H.A. Verschuren]; Krause

'MARGARETE HERBST', Pol, dr, 1934; flowers dark blood-red, dbl., late bloom; recurrent; foliage ruby-red when young; vigorous, bushy growth; Herbst

Margarete Krüger, HT, or; flowers salmon-red, large, semi-dbl., slight fragrance; (Sangerhausen)

'MARGARETHA MÜHLE', HT, pb, 1925; flowers clear satiny pink with silvery reflex, dbl., moderate fragrance; [Mme Caroline Testout X Mrs W.J. Grant]; Mühle; F.J. Grootendorst

'MARGARETHE VAN DE MANDERE', F, dp, 1952; flowers raspberry-red, large, single; very vigorous growth; Leenders, M.

'MARGARET'S CHOICE', Gr, or, 1999; flowers orange-vermillion, reverse whitish, stiff petals, dbl., 26–40 petals, 4.5 in., borne mostly singly, moderate fragrance; prickles moderate; foliage medium, dark green, semi-glossy; upright, tall (3-5 ft) growth; Tough, Ian Murray

Margaret's World *see* KIRbill

'MARGARITA RIERA', HT, op, 1924; flowers brilliant rose-salmon, base yellow, dbl., slight fragrance; [Mme Ravary X Mme Edouard Herriot]; Dot, Pedro

'MARGE', S, pb, 1999; flowers med. pink, creamy white at base, deep pink splotches on outer petals, dbl., 26–40 petals, 3 in., borne mostly singly, moderate fragrance; prickles moderate; foliage medium, medium green, dull; spreading, medium to tall (5 ft) growth; Jerabek, Paul E.

'MARGHERITA CROZE', HT, m, 1914; flowers carmine-purple changing to purple-rose, base shaded deep rose, dbl., moderate fragrance; [Étoile de France X Earl of Warwick]; Ketten Bros.

Margie™ *see* 'MARMIE'

'MARGIE BURNS', HT, mp, 1968; flowers rose-pink, reverse tyrian rose, large, dbl., exhibition form, moderate fragrance; foliage leathery; very vigorous, upright, bushy growth; Carrigg

'MARGO KOSTER', Pol, ob, 1931; (Sunbeam); flowers salmon; [Dick Koster sport]; Koster, D.A.

'MARGO KOSTER SUPERIOR', Pol, ob, 1956; flowers deep salmon-pink; [Dick Koster sport]; Koster, D.A.

'MARGO KOSTER, CLIMBING', Cl Pol, ob, 1962; (Sunbeam, Climbing); Golie; Crombie Nursery

'MARGO'S BABY', Pol, yb, 1987; flowers creamy ivory, salmon edges, aging dark edge, small, dbl., 30 petals, cupped, no fragrance; prickles few, medium, light green; foliage small, medium green, glossy; upright, bushy, low, vigorous growth; winter hardy.; [Margo Koster sport]; Partain, Joe L., 1988

'MARGO'S SISTER', Pol, lp, 1954; flowers shell pink; [Margo Koster sport]; Ratcliffe

'MARGOT AMOS', HT, op; bud high, pointed; flowers coral-pink flushed strawberry, large; foliage glossy, bronze

'MARGOT ANSTISS', HT, lp, 1947; flowers glossy satin-pink, dbl., 40–45 petals, 6 in., moderate fragrance; vigorous, branching growth; Norman; Harkness

'MARGOT ASQUITH', HT, mr, 1934; flowers shining cerise-red, well formed, moderate fragrance; very vigorous growth; [Betty Uprichard X K. of K.]; Prince

'MARGOT FONTEYN', HT, op, 1964; flowers salmon-orange, dbl., 40 petals, 4 in., intense fragrance; very free growth; [Independence X Ma Perkins]; McGredy, Sam IV; Fisons Horticulture

'MARGRATEN', F, mr, 1949; flowers currant-red, borne in trusses, semi-dbl., 16 petals, 3–4 in., flat, slight fragrance; foliage light green; vigorous, bushy growth; [Donald Prior X World's Fair]; Leenders, M.

'MARGRETHE MÖLLER', HT, dp, 1914; flowers deep cerise-rose, well formed, dbl., moderate fragrance; weak necks; [Lady Mary Fitzwilliam X Seedling]; Poulsen, D.T.

Margriet Hermans, HT, yb, 1994; RvS-Melle

'MARGUERITE AMIDIEU DE CLOS', HT, my, 1926; flowers buttercup-yellow, dbl., moderate fragrance; [Souv. de Claudius Pernet X Golden Emblem]; Ketten Bros.

Marguerite Anne *see* 'COCREDWOOD'

'MARGUERITE BRASSAC', HP, dr, 1874; flowers large, dbl.; Brassac; (Sangerhausen)

'MARGUERITE CARELS', Cl HP, mp, 1922; flowers neyron pink, center darker, large, dbl., recurrent bloom; vigorous, climbing growth; [Frau Karl Druschki X Gen. MacArthur]; Nabonnand, P.

'MARGUERITE CHAMBARD', HT, or, 1928; bud pointed; flowers geranium-red to vermilion, very large, exhibition form, slight fragrance; foliage dark; very vigorous, bushy growth; Chambard, C.

Marguerite d'Autricho *see* DORtiche

'MARGUERITE DE ROMAN', HP, w, 1882; Schwartz

Marguerite Defforey, HT, 1973; Orard, Joseph; (Cavriglia)

'MARGUERITE DESRAYAUX', N, mp, 1906; flowers large, semi-dbl.; Nabonnand; (Sangerhausen)

'MARGUERITE GUILLARD', HP, w, 1915; flowers stamens yellow, semi-dbl., 20 petals, flat; [Frau Karl Druschki sport]; Chambard, C.

'MARGUERITE GUILLOT', HT, ly, 1902; flowers very large, dbl., moderate fragrance; Guillot; (Sangerhausen)

'MARGUERITE HEITZMANN', HT, or, 1930; flowers salmon-pink, stamens golden yellow, very large, dbl., cupped, moderate fragrance; foliage leathery; very vigorous growth; [Frau Karl Druschki X Mme Edouard Herriot]; Buatois

'MARGUERITE HILLING', HMoy, mp, 1959; (Pink Nevada); [Nevada sport]; Hilling

'MARGUERITE JAMAIN', HP, lp, 1873; flowers large, very dbl.; Jamain, H.; (Sangerhausen)

'MARGUERITE MOULIN', HT, pb, 1938; flowers lilac-pink, center salmon, large, very dbl., cupped, moderate fragrance; vigorous, bushy growth; [(Mme Edouard Herriot X Mrs Aaron Ward) X Mme Caroline Testout]; Moulin; Hamonière

'MARGUERITE ROSE', Pol, mp, 1905; flowers medium, dbl.; Robichon; (Sangerhausen)

'MARGY', Pol, mr, 1936; flowers brilliant red, open, semi-dbl., moderate, spicy fragrance; foliage small, soft; bushy growth; Sauvageot, H.; C-P

'MARI CAROLYN', HT, m, 1999; dbl., 17–25 petals, 4–5 in., borne mostly singly, intense fragrance; few prickles; foliage medium, medium green, semi-glossy; bushy, medium (5 ft) growth; [Heirloom X Chrysler Imperial]; Sproul, James A.; Roses By Design, 1999

'MARI DOT', HT, ab, 1927; flowers bright salmon to salmon-pink, base yellow, borne in clusters, moderate fragrance; foliage glossy; long, strong stems; very vigorous growth; [O. Junyent X Jean C.N. Forestier]; Dot, Pedro; C-P

'MARIA', HT, rb, 1974; flowers red, yellow reverse, large blooms borne singly, dbl., 60 petals, moderate, tea fragrance; foliage dark, glossy; vigorous growth; [Rina Herholdt X Unnamed seedling]; Staikov, Prof. Dr. V.; Kalaydjiev and Chorbadjiiski

'MARIA', F, or, 1965; flowers orange-scarlet, borne in clusters, single, 10 petals, 3 in., slight fragrance; foliage very large, dark, leathery; vigorous, upright growth; [Unknown seedling X Border Beauty]; Gregory

'MARIA ANTONIA CAMPRUBI', HT, mp, 1956; flowers soft carmine, well formed, large, intense fragrance; vigorous growth; [Peace X Rosa Munne]; Munné, M.

'MARIA BURNETT', Min, yb, 1980; bud globular; flowers light yellow, edged pink, small, dbl., 70 petals, globular, slight fragrance; foliage light green; low growth; [Little Darling X Patricia Scranton]; Miniature Plant Kingdom

Maria Callas *see* 'MEIDAUD'

Maria Callas, Climbing *see* 'MEIDAUDSAR'

Maria Carta, HT, 1976; Zandri, R.; (Cavriglia)

'MARIA CHAVARRI DE SALAZAR', HT, lp, 1935; flowers large, dbl.; Dot, Pedro; (Sangerhausen)

'MARIA CINTA', HT, op, 1967; bud long, pointed; flowers coral, large, dbl., 28 petals, moderate fragrance; foliage glossy, bronze; low, compact growth; [Duet X (Soraya X Chrysler Imperial)]; Dot, Simon; Minier

'MARIA DE MELLO', HT, dr, 1935; bud pointed; flowers bright velvety carmine-purple, large, dbl., 40–45 petals, cupped; foliage bronze quaker green; vigorous, branching growth; [Mme Gabriel Hanra X Mrs John Bell]; Ketten Bros.

'MARIA DELFORGE', Pol, mp, 1959; bud oval; flowers pink, rosette form, medium, borne in clusters, dbl., slight fragrance; foliage glossy; vigorous, bushy, low growth; [Orange Triumph X Ma Perkins]; Delforge

'MARIA GRAEBNER', S, mp; (R. X mariaegraebneriae);, a few borne all summer; subglobose, red fruit; foliage orange and red in fall; 5 ft. growth; (28); [R. palustris (?) X R. virginiana]

'MARIA GUARRO', HT, dp, 1935; bud pointed; flowers pink in spring, blood-red in summer, large, very dbl., exhibition form, slight fragrance; foliage glossy, dark; very vigorous growth; [Château de Clos Vougeot X Li Bures]; Dot, Pedro

Maria Hofker, HT, my, 1993; Interplant

'MARIA ISABEL', HT, mr; flowers deep strawberry-red, base yellow, exhibition form; foliage dark; vigorous growth; Camprubi, C.

'MARIA LEONIDA', HBc, w, 1829; (R. X leonida); flowers flesh, anthers purplish, very dbl., intense, tea fragrance; (14); Lemoyne

'MARIA LIESA', HMult, dp, 1925; flowers carmine-pink, small, single; Bruder Alfons; (Sangerhausen)

'MARIA LISA', HWich, pb, 1936; flowers clear rose, center white, stamens yellow, open, small, single, profuse, non-recurrent bloom; foliage dark, leathery; very vigorous, climbing or trailing growth; Liebau

'MARIA MAASS', HT, w, 1927; flowers very large, very dbl., intense fragrance; Maass; (Sangerhausen)

Maria Mathilda® *see* 'LENMAR'

Maria McGredy *see* 'MACTURANG'

'MARIA PERAL', HT, yb, 1941; flowers yellow suffused red, large, dbl.; very vigorous growth; Dot, Pedro

'MARIA REID', HT, dp, 1924; flowers dark rose-pink tinted peach, base yellow, dbl.; [Mme Caroline Testout X George C. Waud]; Ferguson, W.

'MARIA SERRAT', HT, op, 1946; flowers salmon-pink, base yellow, reverse deep yellow; [Mrs Pierre S. duPont X Baronesa de Ovilar]; Munné, M.

'MARIA STERN', HT, ob, 1969; bud pointed; flowers orange, large, dbl., 43 petals, globular, moderate fragrance; vigorous, upright growth; [Tip Toes X Queen Elizabeth]; Brownell, H.C.; Stern's Nursery

Maria Teresa® *see* 'LENMACRA'

'MARIA TERESA BORDAS', HT, mp, 1953; bud ovoid; flowers rose-pink, very large, very dbl., exhibition form, moderate fragrance; foliage dark, glossy; very vigorous, upright, bushy growth; [Sensation X Peace]; Bordas

Maria Teresa de Esteban *see* 'DOTRAMES'

'MARIALE', HT, ab, 1956; flowers orange-yellow, semi-dbl., slight fragrance; vigorous growth; [Souv. de Jacques Verschuren sport]; de Boer

'MARIAN ANDERSON', HT, dp, 1964; flowers deep pink, large, semi-dbl., exhibition form, moderate, spicy fragrance; foliage glossy; vigorous, tall, compact growth; [Queen Elizabeth X Merry Widow]; Lammerts, Dr. Walter; Germain's

'MARIAN COLTHORPE', HT, op, 1946; flowers coral shaded lemon and pink; Wheatcroft Bros.

Mariandel® *see* 'KORPEAHN'

Mariandel 92, F, lp; [Ko's Yellow X Eyepaint]; Hannemann, F., 1992; (Weatherly, L.)

'MARIANNA ROLFS', Cl HT, lp, 1926; flowers silvery pink, semi-dbl.; Walter, L.

'MARIANNE', HT, yb, 1933; flowers mixture of copper-yellow, pink and red, large, exhibition form, moderate fragrance; vigorous, bushy growth; [Sybil X Sunstar]; Krause

Marianne *see* MEIklusy

'MARIANNE KLUIS', Pol, mr, 1942; flowers carmine-red, small, dbl.; Kordes, W. Söhne; (Sangerhausen)

'MARIANNE KLUIS SUPERIOR', Pol, mr, 1930; flowers deep violet-red; [Greta Kluis Superior sport]; Kluis & Koning

'MARIANNE PFITZER', HT, lp, 1902; flowers very large, dbl.; Jacobs; (Sangerhausen)

'MARIANNE POWELL', HT, dr, 1986; flowers large, very dbl., 50 petals, moderate fragrance; foliage large, dark green, glossy; upright growth; [Kerryman X Red Dandy]; Powell, G.

Marianne Tudor *see* FRYmartor

'MARIANO VERGARA', T, mr, 1896; flowers magenta-red with vermilion reflections, large, dbl.; vigorous growth; Aldrufeu

Maria-Theresa, HT, ab, 1994; Fryer, Gareth

Maribel *see* 'COCDANA'

'MARIBELL', HT, rb, 1987; flowers carmine red, reverse silverwhite, large, dbl., 40–45 petals, slight fragrance; upright, vigorous growth; [(Sea Pearl X Zorina) X Lovita]; Gressard, J.; Delbard Roses, 1988

'MARICA', HT, mp, 1964; flowers bright pink, large, dbl., slight fragrance; strong stems; vigorous, symmetrical growth; Mondial Roses

'MARICA', Cl Min, pb, 1989; flowers pink, reverse white; [Ginza Komachi sport]; Kono, Yoshito

Marie *see* 'RESmar'

'MARIE ACCARIE', N; Guillot et Fils, 1872

'MARIE ADÉLAÏDE', HT, yb, 1912; (Grande Duchesse de Luxembourg, Luxembourg); bud pointed; flowers coppery yellow, center deeper, large, dbl., exhibition form, slight fragrance; foliage bronze, soft; vigorous, spreading growth; [Mme J.W. Budde X Lyon Rose]; Soupert & Notting

'MARIE ANTOINETTE', HGal, m; flowers lilac-rose, large, dbl.; Vibert, 1829

'MARIE ANTOINETTE', HT, mp, 1968; bud long, pointed; flowers pink, reverse darker, large, dbl., cupped, slight fragrance; foliage dark, glossy; very vigorous, upright growth; [Queen Elizabeth X Chrysler Imperial]; Armstrong, D.L.; Armstrong Nursery

Marie Antoinette Rety, HT, 1983; Orard; (Cavriglia)

'MARIE BAUMANN', HP, mr, 1863; flowers carmine-red, large, dbl., 55 petals, globular, moderate fragrance; foliage dark; vigorous growth; [Alfred Colomb seedling]; Baumann

'MARIE BRISSONET', Pol, lp, 1913; flowers flesh-rose borne in pyramidal clusters of 75-100; Turbat

'MARIE BUGNET', S, w, 1963; bud long, pointed; flowers snow white, dbl., 3 in., recurrent bloom, intense fragrance; foliage light green, rugose; vigorous (3 ft.), bushy, compact growth; [(Therese Bugnet X Seedling) X F.J. Grootendorst]; Bugnet; Skinner

Marie Casant *see* **'MARYTJE CAZANT'**

Marie Christina, MinFl, lp, 1988

Marie Curie *see* MEIlomit

'MARIE DE BLOIS', M, mp; flowers pink tinted lighter, large; Moreau et Robert, 1852

'MARIE DE BOURGOGNE', M, mp, 1853; flowers large, dbl.; Robert

'MARIE DE SAINT JEAN', P, w, 1869; (Marie de st Jean); flowers large, dbl.; Damaizin

Marie de st Jean *see* **'MARIE DE SAINT JEAN'**

'MARIE DERMAR', N, ly, 1889; flowers medium, dbl., moderate fragrance; Geschwind, R.; (Sangerhausen)

'MARIE DIETRICH', LCl, yb, 1928; flowers yellowish red, passing to white; [Leontine Gervais X Eugenie Lamesch]; Walter, L.

'MARIE D'ORLÉANS', T, mp; flowers bright pink shaded darker, large, dbl., flat; vigorous growth; Nabonnand, G., 1884

'MARIE DOUGHERTY', F, dr, 1976; bud long, slender; single, 5 petals, 3.5 in., slight, tea fragrance; foliage glossy; moderate, bushy growth; [Sarabande X Sarabande]; Linscott

'MARIE DUTOUR', F, op, 1962; flowers reddish salmon, well formed, large; [Aloha X (Gloire du Midi X Edith de Martinelli)]; Arles; Roses France

'MARIE EADS', Pol, mr, 1991; bud rounded; flowers medium red with blue overtones, reverse light red, small blooms, dbl., 25 petals, repeat bloom, no fragrance; foliage medium, medium green, semi-glossy with fringed stipu; bushy, low growth; [Baby Faurax X Verdun]; Eads, C.E.

'MARIE ELIZABETH', F, yb, 1965; flowers yellow shaded rose-pink, dbl., 28 petals, 3 in., flat, intense fragrance; foliage dark, heavily veined; vigorous growth; [Clare Grammerstorf X Cavalcade]; McGredy, Sam IV; McGredy

'MARIE FAIST', HT, op, 1925; flowers shell-pink tinted salmon, center darker, with orange, dbl., intense fragrance; [Mme Edmond Rostand X Mrs T. Hillas]; Berger, V.; Faist

'MARIE GIRARD', HT, lp, 1898; flowers large, dbl., moderate fragrance; Buatois; (Sangerhausen)

'MARIE GOUCHAULT', HWich, mr, 1927; flowers clear red passing to brilliant salmon-rose, small, borne in, dbl., sometimes recurrent bloom; very vigorous growth; Turbat

'MARIE GREENE', HT, mr, 1941; flowers rich red, moderate fragrance; Clark, A.

'MARIE GUILLOT', T OGR), w, 1874; flowers white, tinged yellow, large, dbl., intense fragrance; strong stems; vigorous growth; Guillot et Fils

'MARIE GUILLOT', T, lp; Guillot et Fils, 1873

'MARIE GUILLOT, CLIMBING', Cl T, w, 1898; Dingee & Conard

Marie Henriette Grafin Chotek, HMult, dr

'MARIE HENRIETTE GRÄFIN CHOTEK', HMult, dr, 1911; flowers dark crimson, large, borne in clusters, dbl.; very vigorous, climbing growth; Lambert, P.

'MARIE HENRY', HT, ly, 1900; flowers large, dbl., moderate fragrance; Buatois; (Sangerhausen)

Marie Laforet *see* DORlain

'MARIE LAMBERT', T, w, 1886; (Snowflake, White Hermosa); flowers pure white; [Mme Bravy sport]; Lambert, E.

'MARIE LAVIER', HT, yb, 1935; bud brownish yellow; flowers reddish nankeen yellow to salmon-yellow, rather large, dbl., moderate fragrance; vigorous, bushy growth; [Souv. de Claudius Pernet X Mme Edouard Herriot]; Buatois

Marie Legonde, T, w

'MARIE LEONIDA', HBr, 1832; Moldenke; (Cavriglia)

'MARIE LOUISE', D, mp; flowers mauve-pink, large, very dbl., intense fragrance; bushy, shrubby (about 4 ft) growth

'MARIE LOUISE MATHIAN', HT, ly, 1912; flowers large, dbl., intense fragrance; Fugier; (Sangerhausen)

'MARIE LÜNNEMANN', HT, my, 1920; flowers clear pink, dbl., moderate fragrance; [Pharisaer X Laurent Carle]; Timmermans

'MARIE MAASS', HT, w, 1928; flowers pure white to ivory-white, very large, dbl., intense fragrance; vigorous, bushy growth; [Kaiserin Auguste Viktoria X Marechal Niel]; Maass

'MARIE MENUDEL', HP, op, 1927; flowers rose-pink, tinted salmon, large, dbl., moderate fragrance; Barbier

Marie Nordlinger, M, lp; [Henri Martin X Unknown]; Morley, Dr B., 1988; (Weatherly, L.)

Marie Palit, HT, pb, 1988; Datt, Braham

Marie Pavic *see* **'MARIE PAVIÉ'**

'MARIE PAVIÉ', Pol, w, 1888; (Marie Pavic); flowers white, center flesh, blooms in clusters, dbl., 2 in.; no prickles; foliage large, rich green; vigorous, bushy growth; Allégatière

'MARIE ROBERT', P, m, 1850; flowers lilac-pink, medium, dbl.; Robert et Moreau; (Sangerhausen)

'MARIE SCHMITT', HT, lp, 1910; flowers large, dbl.; Schmitt; (Sangerhausen)

Marie Shields™ *see* 'MORMARI'

'MARIE TUDOR', HGal, dp

'MARIE VAN HOUTTE', T, pb, 1871; (Mlle Marie van Houtte); flowers deep cream, tinged pink, base buff-yellow, large, very dbl., exhibition form, moderate fragrance; foliage rich green, leathery; vigorous, bushy, sprawling growth; (14); [Mme de Tartas X Mme Falcot]; Ducher

'MARIE VAN HOUTTE, CLIMBING', Cl T, pb, 1936; Thomasville Nursery

'MARIE VERBRUGH', F, op, 1954; flowers yellow-salmon, reverse coral, well formed, large, dbl., intense fragrance; bushy growth; [Ambassadeur Nemry X Souv. de Claudius Pernet]; Leenders, M.

'MARIE YOUNG', HT, or; flowers brilliant orange-red, well formed

'MARIE ZAHN', HT, lp, 1887; flowers large, dbl.; Müller, Dr. F.; (Sangerhausen)

Marie-Antoinette Rety, HT, op

'MARIE-CHANTAL', F, mp, 1959; flowers bright pink, open, large, semi-dbl., slight fragrance; foliage glossy; vigorous, bushy growth; [Peace X Fernand Arles]; Gaujard

'MARIE-CLAIRE', HT, or, 1938; bud deep orange-red; flowers golden coral-red, passing to orange-yellow, large, dbl., moderate fragrance; foliage bronze, glossy; strong stems; very vigorous, cmpact growth; [(Charles P. Kilham X Duquesa de Peñaranda) X (Charles P. Kilham X Margaret McGredy]; Meilland, F.

'MARIE-CLAIRE, CLIMBING', Cl HT, or, 1944; (Grimpant Marie-Claire); Meilland, F.

'MARIE-FRANCE', HT, my, 1957; bud apricot shaded orange; flowers pure yellow, large, semi-dbl.; very vigorous growth; [Feu Pernet-Ducher X Léonce Colombier]; Dorieux; Pin

Marie-Francoise Saignes, HT, mr

'MARIE-JEANNE', Pol, w, 1913; flowers pale blush-cream borne in clusters of 40-60; no prickles; height 2-3 ft; Turbat

Marieken, F, ab, 1987; RvS-Melle

'MARIELLE', F, mp, 1963; flowers deep rosy pink, blooms in large clusters, dbl., 25 petals, 4 in., moderate fragrance; foliage dark; vigorous growth; [Independence X F seedling]; deRuiter

Marie-Louise Marjan, HT, ab, 1999; Kordes

'MARIE-LOUISE PONCET', HT, op, 1929; flowers coppery rose to pale coppery pink, reverse carmine-salmon; Gaujard

'MARIE-LOUISE SONDAZ', HT, w, 1967; flowers cream shaded red, very large, dbl., moderate fragrance; foliage dark; very vigorous, upright growth; [Rose Gaujard X Peace]; Gaujard

Marie-Louise Velge, F, lp, 1997; 20 petals, 4 in., flat; foliage dense dark green; GM, The Hague, 1996 Golden Rose, Geneva, 1997; RvS-Melle, 1995

Marie-Louise Velge, F, 1997; Institute of Ornamental Plant Growing; (Cavriglia)

'MARIE-ROSE', Pol, mp, 1930; flowers ruddy pink, carnation shaped, large, borne in clusters of 30, dbl.; foliage glossy; very vigorous growth; [Marie-Jeanne sport]; Truffaut, T.A.; Turbat

'MARIE-ROSE BESSON', HT, lp, 1939; bud long, pointed, yellow, tinted coral; flowers light pink, tinted coral-orange, large, dbl., slight fragrance; foliage glossy; long stems; vigorous growth; [Souv. de Claudius Pernet X Unnamed seedling]; Mallerin, C.; A. Meilland

'MARIE-ROSE TOUSSAINT', HT, lp, 1946; flowers satiny pink, very large, moderate fragrance; foliage dark, leathery; strong stems; vigorous, bushy growth; Gaujard

Marietta® *see* 'TANATIRAM'

'MARIETTA SILVA TAROUCOVÁ', HMult, mp, 1925; flowers bright rose, large, borne in clusters; foliage rich green; very vigorous, climbing growth; [Colibri X Crimson Rambler]; Zeman

Marie-Victorin, S, lp, 1999; L'Assomption

Marie-Victorin *see* 'AC MARIE-VICTORIN'

'MARIGOLD', HT, op, 1955; bud long, pointed; flowers salmon-yellow lightly washed pink, dbl., 52 petals, 6 in., exhibition form, intense fragrance; foliage leathery, glossy, bright green; vigorous, upright growth; [Peace X Mme Joseph Perraud]; Lens

'MARIJKE KOOPMAN', HT, mp, 1979; bud long, pointed; dbl., 25 petals, borne 3-5 per cluster, moderate fragrance; red prickles; foliage dark, leathery; vigorous, medium-tall, upright growth; GM, The Hague, 1978; Fryer, Gareth; Fryer's Nursery, Ltd.

Mariko *see* 'DEVDORADO'

'MARILO', Min, ab, 1999; (**Ilona**™); flowers orange/apricot striped, reverse lighter, single, 5–11 petals, 1 in., borne in small clusters, no fragrance; prickles moderate; foliage small, dark green, near rugose, disease-resistant; compact, low (18-24 in.) growth; [Altissimo X Roller Coaster]; Martin, Robert B., Jr.

'MARILYN', Min, lp, 1955; flowers light pink, base purplish, small blooms in clusters, dbl., 60 petals; very compact growth; [Perla de Montserrat X Bambino]; Dot, M.

'MARILYN', HT, ab, 1954; bud long, pointed; flowers apricot-pink veined red, intense fragrance; foliage dull, green; vigorous growth; [May Wettern X Phyllis Gold]; Fletcher; Tucker

'MARILYN GOWIE', F, ?, 1968;, borne in trusses; foliage bronze; bushy growth; Gowie

'MARIMBA'®, F, mp, 1965; [Garnette sport]; Dekkers; Verbeek

Marime *see* 'MORMARME'

Marina® *see* 'RINAKOR'

'MARINA FONTCUBERTA', HT, dp, 1924; flowers brilliant carmine, center rose-carmine, dbl., intense fragrance; [Entente Cordiale X Laurent Carle]; Dot, Pedro

Marina Marini® *see* BARmar

Marinette *see* 'AUSCAM'

Mario Clemente, HT; Moreira da Silva, A.; (Cavriglia)

Mario Lanza *see* 'HORAARDVARK'

Mariolina, S, 1989; Embriaco, B.; (Cavriglia)

'MARION', F, op, 1956; (Salmon Sensation); flowers pink tinted salmon, medium, semi-dbl.; bushy growth; [Duchess of Rutland X Fashion]; deRuiter

'MARION CRAN', HT, or, 1927; bud buttercup-yellow, flushed cerise; flowers scarlet veined orange and yellow, dbl., exhibition form; foliage bronze, leathery, glossy; very vigorous, bushy growth; McGredy

'MARION DINGEE', HT, dr, 1889; flowers crimson; [((Comtesse de Caserta X Général Jacqueminot) X Marechal Niel)x (Pierre Notting X Safrano)]; Cook, J.W.

Marion Foster *see* 'MEICAPULA'

Marion Harkness® *see* 'HARKANTABIL'

Marion Hess, HT, dr, 1981; Hetzel

'MARION HORTON', HT, my, 1929; flowers primrose-yellow; [Gorgeous X Sunstar]; Bees

'MARION LAWRIE', HT, mp, 1976; flowers pink, base gold, very full, very dbl., 75 petals, 4 in., slight fragrance; foliage dark, leathery; [Kordes' Perfecta sport]; Lawrie

Marion Manifold *see* **'MISS MARION MANIFOLD'**

'MARION R. HALL', HT, dp; flowers bright cerise-red, well formed, semi-dbl., 20 petals, 5 in.; vigorous growth; [Crimson Glory X Sterling]; Balcombe Nursery

'MARION RICH', HT, yb, 1997; flowers double, medium, dbl., 15–25 petals, borne mostly singly, slight fragrance; foliage medium, dark green, glossy; upright, medium (3-3.5ft.) growth; [Solitaire X Remember Me]; Skinner, A.W.

'MARIONETTE', F, w, 1944; bud cream-yellow; flowers small, borne in clusters, dbl., 25–30 petals, 1.5 in., moderate fragrance; vigorous, bushy growth; [Pinocchio sport]; deVor, Paul F.; J&P

'MARIPOSA', Pol, or, 1927; flowers deeper orange-red; [Orange King sport]; Allen

Mariposa Gem *see* 'MORMAGEM'

Marisa(TM) *see* 'ORTISA'

Mariska *see* VALkita

'MARISTA', HT, rb, 1975; flowers medium red, whitish reverse, large blooms borne singly, dbl., 75 petals, moderate, tea fragrance; foliage dark, glossy; upright growth; [Sarah Arnot X Rina Herholdt]; Staikov, Prof. Dr. V.; Kalaydjiev and Chorbadjiiski

'MARISTELLA', HT, 1952; Giacomasso; (Cavriglia)

'MARITA', F, ob, 1961; flowers copper-orange, heavily veined yellow, medium, borne in trusses, dbl., 30–40 petals; foliage coppery; growth very free, straggly; [Masquerade X Serenade]; Mattock

Maritime Bristol *see* 'SANTANG'

'MARITIME HEIR', S, dp, 1986; flowers lavender pink, carnation-like, large, dbl., occasionally recurrent bloom, intense fragrance; fine prickles; foliage small, light green, disease resistant; upright, bushy, very hardy growth; [Therese Bugnet X R. nitida]; James, John

Marjan, HT, ob; Select Roses, B.V.

'MARJOLIN', Ch, dr; flowers deep crimson-purple, cupped

'MARJOLINE', HT, rb, 1949; flowers cardinal-red, reverse indian yellow, dbl., 40 petals, 6 in., cupped, slight fragrance; vigorous, upright growth; [Boudoir X Léonce Colombier]; Meilland, F.

'MARJORIE', HT, w, 1895; flowers white with pink, large, dbl., moderate fragrance; Dickson, A.; (Sangerhausen)

'MARJORIE ANDERSON', F, mp, 1973; dbl., 26 petals, 5.5 in., moderate fragrance; foliage very large, matt; [Fragrant Cloud X Sea Pearl]; Dickson, Patrick; Dicksons of Hawlmark

'MARJORIE ATHERTON', HT, my, 1977; bud ovoid; dbl.; foliage light, leathery; vigorous, upright, bushy growth; [Mt. Shasta X Peace]; Bell, Ronald J.; Brundrett

'MARJORIE BULKELEY', HT, yb, 1921; flowers buff, flushed rose-pink, passing to silvery pink, dbl., moderate fragrance; GM, NRS, 1920; Dickson, H.

Marjorie Chase *see* KORamator

'MARJORIE CONN', F, lp, 1981; bud ovoid; semi-dbl., 14 petals, blooms in clusters of 5-7, slight fragrance; brown prickles, hooked down; foliage dark; bushy growth; [Bon Bon X Unnamed seedling]; Berry, Howard

'MARJORIE ELLICK', F, w, 1978; flowers rowanberry, very full, very dbl., 95 petals, 4–6. in.; foliage glossy, light; very vigorous, free growth; [(Spion-Kop X Ena Harkness) X (Sam Ferris X Karl Herbst)]; Ellick; Excelsior Roses

Marjorie Fair(R) *see* 'HARHERO'

'MARJORIE FOSTER', HWich, dr, 1934; flowers deep blood-red, small, dbl.; vigorous growth; Burbage Nursery

'MARJORIE LEGRICE', HT, ob, 1949; flowers orange and yellow, pointed, dbl., 30 petals, 5 in., moderate fragrance; foliage glossy; vigorous growth; [Mrs Sam McGredy X President Plumecocq]; LeGrice

'MARJORIE LEGRICE, CLIMBING', Cl HT, ob, 1956; Tantau, Math.

Marjorie Marshall *see* HARdenier

Marjorie May *see* 'HORSUNPEGY'

'MARJORIE PROOPS', HT, dr, 1969; flowers crimson, dbl., exhibition form, moderate fragrance; [Red Dandy X Ena Harkness]; Harkness

Marjorie W. Lester *see* **'LAURÉ DAVOUST'**

'MARJORY PALMER', Pol, mp, 1936; flowers rich pink, blooms in clusters, dbl., intense fragrance; bushy, compact growth; [Jersey Beauty X ?]; Clark, A.; NRS Victoria

Marjorie Reid *see* 'JACMARJ'

Mark 1 *see* 'SAVAMARK'

Mark One(TM) *see* 'SAVAMARK'

'MARK SULLIVAN', HT, op, 1942; (Président Chaussé); flowers gold flushed and veined rose, dbl., 33 petals, 4–4.5 in., exhibition form, moderate fragrance; foliage dark, leathery, glossy; vigorous, upright. growth; [Luis Brinas X Brazier]; Mallerin, C.; A. Meilland;, C-P

'MARK SULLIVAN, CLIMBING', Cl HT, ob

Markgräfin Wilhelmine, F, lp, 1995; flowers large, dbl.; Tantau; (Sangerhausen)

'MARLENA', F, mr, 1964; flowers crimson scarlet, patio, blooms in clusters, semi-dbl., 18 petals, flat; low, compact bushy growth; ADR, 1964 GM, Baden-Baden, 1962 GM, Belfast, 1966; [Gertrud Westphal X Lilli Marleen]; Kordes, R.; A. Dickson, 1964;, J&P, 1967;, McGredy, 1964

Marlowe Soft Orange, LCl, ab

'MARLYB', HT, yb, 1999; (**Lybie**); flowers yellow blend, reverse medium yellow, dbl., 17–25 petals, 4 in., borne in small clusters, moderate fragrance; few prickles; foliage medium, dark green, semi-glossy; upright, medium (3 ft) growth; [Lynn Anderson X ((Carefree Beauty X Picasso) X seedling)]; Hiltner, Martin

Marlyn *see* TANliram

Marlyse(R) *see* PEKomecli

'MARMALADE', HT, ob, 1977; bud long, pointed; flowers bright orange, reverse deep yellow, dbl., 30 petals, 5 in., intense, tea fragrance; foliage large, glossy, dark; upright growth; [Arlene Francis X Bewitched]; Swim, H.C. & Ellis, A.E.; Armstrong Nursery

'MARMALADE MIST'(TM), HT, op, 1990; bud pointed; flowers medium salmon pink, with lighter salmon pink reverse, aging, dbl., 25 petals, cupped, slight, fruity fragrance; deep plum and brown prickles; foliage large, dark green, semi-glossy; upright, medium growth; PP007549; Lammerts, Dr. Walter, 1981; DeVor Nurseries, Inc., 1991

Marmalade Skies(TM) *see* MEIMONBLAN

'MARMIE', Cl Min, rb, 1999; (**Margie**(TM)); flowers dard red with white eye, reverse lighter, dbl., 17–25 petals, 2 in., borne in small clusters, slight fragrance; prickles moderate; foliage medium, dark green, semi-glossy; upright, climbing (5-7 ft); shade tolerant growth; [Roller Coaster X Roller Coaster]; Martin, Robert B., Jr.

'MARMION', HT, pb, 1934; flowers pale rose flushed orange, reverse salmon-pink, dbl., moderate fragrance; vigorous, branching growth; Dobbie

Marondo(R) *see* KORtitu

Marovilla, F, or, 1992

'MARQUES DE NARROS', HT, ob, 1951; flowers salmon-pink, pointed; thornless; La Florida

'MARQUESA DE AGUILAR', HT, mr, 1955; bud ovoid; flowers cardinal-red to begonia pink, large, dbl., 60 petals, exhibition form, moderate fragrance; foliage glossy; vigorous growth; [Comtesse Vandal X Caprice]; Bofill; Torre Blanca

'**MARQUESA DE BOLARQUE**', HT, my, 1945; bud long, pointed; flowers lemon-yellow, large, dbl., exhibition form, slight fragrance; foliage dark, glossy; vigorous growth; [Shot Silk X Julien Potin]; Camprubi, C.

'**MARQUESA DE CASA VALDÉS**', HT, or, 1955; bud pointed; flowers scarlet-red slightly shaded orange, large, dbl., 35–40 petals, exhibition form, moderate fragrance; foliage dark; very vigorous, compact growth; [Peace X Poinsettia]; Dot, Pedro

'**MARQUESA DE GOICOERROTEA**', HT, ab, 1947; flowers amber-yellow, well formed, slight fragrance; upright growth; [Eclipse X Joanna Hill]; Dot, Pedro

Marquesa de Urquijo *see* '**PILAR LANDECHO**'

Marquesa de Urquijo, Climbing *see* '**PILAR LANDECHO, CLIMBING**'

'**MARQUESA DEL VADILLO**', HT, pb, 1945; (Spanish Main); flowers neyron pink, reverse silvery pink, open, 5 in., moderate fragrance; foliage glossy, dark; upright growth; [Girona X Condesa de Sástago]; Dot, Pedro

Marquise Boccella *see* '**MARCHESA BOCCELLA**'

Marquise Bocella *see* '**MARCHESA BOCCELLA**'

'**MARQUISE D'ANDIGNÉ**', HT, dr, 1927; flowers velvety scarlet-crimson, dbl., intense fragrance; [(Lieutenant Chaure X George C. Waud) X Laurent Carle]; Leenders Bros.

'**MARQUISE DE BALBIANO**', B, m, 1855; Lacharme, F.

'**MARQUISE DE BARBENTANE**', HT, ab, 1928; flowers apricot-yellow, shaded orange and sunflower-yellow, dbl.; [Mrs Farmer X Severine]; Fugier

'**MARQUISE DE CASTELLANE**', HP, dp, 1869; flowers dark rose-pink, well formed, large, dbl., moderate fragrance; moderate growth; Pere Pernet

'**MARQUISE DE GANAY**', HT, mp, 1910; flowers silvery rose, dbl., moderate fragrance; [Liberty X La France]; Guillot, P.

'**MARQUISE DE SINÉTY**', HT, yb, 1906; flowers golden yellow, shaded bronzy red, dbl., moderate fragrance; GM, Bagatelle, 1907; Pernet-Ducher

'**MARQUISE DE SINÉTY, CLIMBING**', Cl HT, yb, 1912; Griffon

'**MARQUISE DE VIVENS**', T, dp, 1886; flowers carmine, base yellowish, large, dbl.; Dubreuil

'**MARQUISE LITTA DE BRÉTEUIL**', HT, mr, 1893; Pernet-Ducher

Marquisette® *see* DUCmar

'**MARR**', Pol, op; flowers coral overcast orange, borne in clusters, 1 in., cupped; thornless; rangy habit (10-12 in.) growth

'**MARRAKECH**', HT, dr, 1945; flowers oxblood-red shaded bright scarlet, well formed, very large, dbl.; upright, vigorous growth; [Rome Glory X Tassin]; Meilland, F.

Marriotta *see* 'MACCRICKE'

Marry Me™ *see* 'DICWONDER'

'**MARS**', HT, op, 1927; flowers deep coral, moderate fragrance; Chaplin Bros.

'**MARSHALL P. WILDER**', HP, dr, 1885; flowers bright deep red, large, dbl., globular, intense fragrance; vigorous, tall growth; [Général Jacqueminot X ?]; Ellwanger & Barry

Marsyanka, HT, dp, 1958; flowers carmine-pink, medium, semi-dbl.; (Sangerhausen)

'**MARTA**', Min, or, 1983; flowers small, dbl., 20 petals, slight fragrance; foliage small, light green, matt; upright growth; [Persian Princess X Anytime]; Dobbs; Small World Min. Roses

Marta Salvador® *see* FEdugia

'**MARTHA**', Pol, pb, 1906; flowers coppery rose, blooms in clusters of 7-20, dbl.; dwarf growth; [Thalia X Mme Laurette Messimy]; Lambert, P.

'**MARTHA**', B, pb, 1912; flowers pale coral, dbl., borne in clusters of 1-5, moderate fragrance; few thorns; tall, climbing growth; Knudson

'**MARTHA BUGNET**', HRg, mr, 1959; bud long, pointed; flowers purplish red, open, large, semi-dbl., abundant, recurrent bloom, intense fragrance; very large fruit; foliage dark rugosa type; weak stems; vigorous, bushy (5-6 ft.) tall and broad growth; [(R. rugosa kamtchatica X R. amblyotis X R. rugosa plena) X F.J. Grootendorst]; Bugnet

'**MARTHA DREW**', HT, w, 1919; flowers creamy white, center rose, dbl., moderate fragrance; GM, NRS, 1919; McGredy

Martha Ford, HMult, ab, 1987; Nobbs

Martha Gonzales, Ch, mr

'**MARTHA KELLER**', Pol, lp, 1912; flowers medium, very dbl.; Walter; (Sangerhausen)

'**MARTHA KORDES**', F, dp, 1941; flowers light capucine-red, becoming pink, open, medium, borne in clusters, semi-dbl., slight fragrance; vigorous, upright growth; [Hedwig Fulda X Holstein]; Kordes

'**MARTHA LAMBERT**', Pol, rb, 1939; flowers brilliant scarlet with small yellow eye, small, borne in clusters, single, recurrent bloom, slight fragrance; foliage glossy; strong stems; vigorous, bushy growth; [Frans Leddy X Paul's Scarlet Climber]; Lambert, P.; C-P

'**MARTHA'S CHOICE**', HT, dp, 1977; bud high-centered; flowers deep pink, full, dbl., 37 petals, 5 in., slight fragrance; [Gavotte X Prima Ballerina]; Bailey

Martha's Vineyard *see* 'POULANS'

'**MARTHE ANCEY**', HT, w, 1932; (Mme Martha Ancey); flowers cream, tinted salmon-pink, center with straw-yellow reflections, dbl.; vigorous growth; [Souv. de Claudius Pernet X Mme Mélanie Soupert]; Schwartz, A.

'**MARTHE CAHUZAC**', Pol, ly, 1902; flowers medium, dbl.; Ketten, Gebrüder; (Sangerhausen)

'**MARTIAN GLOW**', F, mr, 1972; flowers red, reverse lighter, blooms in clusters, semi-dbl., 10 petals, 1.5 in., slight fragrance; foliage semi-glossy, medium green; vigorous, spreading growth; [Joseph's Coat X Dorothy Wheatcroft]; Gandy, Douglas L.

'**MARTIAN SUNRISE**', S, or, 1979; bud ovoid; flowers urn-shaped, blooms borne singly or 3-5 per cluster, dbl., 43 petals, exhibition form, repeats well, slight fragrance; semi-hooked prickles; foliage light to medium green, semi-glossy; upright, bushy growth; [Paddy McGredy X Heidelberg]; Taylor, Thomas E.

'**MARTIN FAASSEN**', HT, dp, 1965; bud ovoid; flowers pink-red, medium, borne in clusters, dbl.; foliage dark; [Baccará X Seedling]; Verbeek

'**MARTIN FROBISHER**', HRg, lp, 1968; bud ovoid; flowers light pink, center darker, medium, dbl., intense fragrance; foliage light green; vigorous, tall growth; (14); [Schneezwerg X ?]; Svedja, Felicitas; Canada Dept. of Agric.

'**MARTIN LIEBAU**', HP, mp, 1930; flowers large, dbl., moderate fragrance; Kiese; (Sangerhausen)

Martin Martin *see* 'SEAMAR'

Martina®, HT, w, 1989; Noack, Werner

Martina, F, lp; Urban, J., 1987; (Czech Rosa Club)

Martine Guillot *see* MASmabay

'**MARTINE HÉMERAY**', F, or, 1958; flowers china-red, well formed, dbl., slight fragrance; foliage leathery; [(Orange Triumph X Mme Edouard Herriot) X Unnamed seedling]; Gaujard, R.; Hémeray-Aubert

Martini, HT, 1967; Delforge; (Cavriglia)

'**MARTONE**', Min, pb, 1989; bud pointed; flowers medium pink, lighter at base, reverse light pink, medium, dbl., 30 petals, exhibition form, slight, damask fragrance; prickles straight, medium, deep pink; foliage medium, medium green, semi-glossy; upright, medium growth; [Queen City X Unknown seedling]; Bridges, Dennis A.; Bridges Roses, 1990

Marty *see* 'PEASWEET'

Marty's Triumph® *see* 'BISMAR'

'MARUSHKA', HT, dy, 1984; dbl., 35 petals, exhibition form, borne 1-3 per stem; foliage dark, leathery, glossy; vigorous, bushy growth; [Kabuki X Unnamed seedling]; Staikov, Prof. Dr. V.; Kalaydjiev and Chorbadjiiski

Marvelle *see* 'MACTAURANG'

Marvie *see* 'GELMAT'

'MARVLOUS', HT, mr, 1937; flowers crimson, open, large, dbl., intense fragrance; foliage leathery; very vigorous growth; Cant, B. R.

'MARY', HT, ob, 1931; flowers buff and orange, moderate fragrance; vigorous growth; Bentall

'MARY', Pol, op, 1947; flowers orange-cerise, borne in small clusters; vigorous growth; [Orange Triumph sport]; Qualm; Spek

'MARY ADAIR', Min, ab, 1966; flowers buffy apricot, small, dbl., moderate fragrance; foliage light green, soft; vigorous, bushy, dwarf growth; [Golden Glow (LCl) X Zee]; Moore, Ralph S.; Sequoia Nursery

'MARY ANN', F, dp, 1959; bud ovoid; flowers rose-red, dbl., 60–78 petals, 2 in., exhibition form, slight fragrance; foliage dark; upright growth; [Garnette sport]; Restani

'MARY BARNARD', F, op, 1978; bud ovoid; flowers deep salmon-pink, semi-dbl., 18 petals, 3.5 in., intense fragrance; foliage dark; low, vigorous growth; [(Karl Herbst X Sarabande) X Ernest H. Morse]; Sanday, John

'MARY BEAUFORT', HT, lp, 1969; flowers light peach-pink, well shaped, small, intense fragrance; compact, low growth; [Gavotte X (Ethel Sanday X Crimson Glory)]; Sanday, John

'MARY BELL', Min, w, 1987; flowers small, dbl., 26–40 petals, moderate fragrance; foliage medium, medium green, semi-glossy, disease resistant; bushy, hardy growth; [Cherish X Rise 'n' Shine]; Bell, Charles E., Jr.; Kimbrew Walter Roses, 1987

'MARY BOSTOCK', HT, lp, 1952; flowers shell-pink tinted white, dbl., 60 petals; strong stems; vigorous growth; Clark, A.

Mary Bradby® *see* 'BYRBRADBY'

'MARY BRUNI', Pol, lp, 1914; flowers light creamy rose pink, small, dbl.; Gratama; (Sangerhausen)

Mary Burke, Min, lp; Hannemann, F., 1990; The Rose Paradise; (Weatherly, L.)

Mary Campbell *see* 'HORLOVEQUEEN'

'MARY CARVER', HT, lp, 1950; bud globular; flowers shell-pink, dbl., 80 petals, 5 in., cupped, intense fragrance; foliage leathery, light green; vigorous, upright growth; [Red Radiance sport]; Chick

Mary Casant *see* **'MARYTJE CAZANT'**

Mary Cave, F, 1993; Harkness; (Cavriglia)

'MARY CLARK', Min, lp, 1981; flowers medium, dbl., 55 petals, slight fragrance; straight, light yellow prickles; foliage mid-green; [Janna X Gene Boerner]; Hooper, John C.

'MARY CLAY', HT, dr, 1951; flowers blood-red, dbl., 40–45 petals, 6 in., intense fragrance; foliage very heavy, dark; very free growth; [Kardinal X Crimson Glory]; Kordes; Morse

'MARY CORELLY', HP, op, 1901; flowers dark salmon-pink, medium, dbl.; Prince; (Sangerhausen)

'MARY DELAHUNTY', HT, dr, 1990; bud pointed; flowers large, borne usually singly, dbl., 35–40 petals, moderate, damask fragrance; foliage dark green, glossy; bushy, tall growth; [(Daily Sketch X Impeccable) X Red Planet]; Bell, Ronald J., 1985; Treloar Roses Pty. Ltd., 1990

'MARY DEVOR', F, mr, 1967; (Douchka); flowers cardinal-red, sweetheart, medium, dbl., 35 petals, moderate fragrance; foliage leathery; vigorous, upright growth; [Christian Dior X Rumba]; Lammerts, Dr. Walter; Amling-DeVor Nursery

Mary Donaldson *see* 'CANANA'

'MARY DUTTON', HT, op, 1949; bud long, pointed; flowers salmon-pink, dbl., 40 petals, 6 in., slight fragrance; foliage glossy; very vigorous growth; [Crimson Glory X Mrs Sam McGredy]; Bees

Mary Edith *see* 'TALMAR'

'MARY EGERTON', F, rb, 1982; flowers white, petals edged orange-red, spreading with age, reverse, dbl., 35 petals; foliage large, dark, glossy; upright, bushy growth; [Fragrant Cloud X Prominent]; Lea, R.F.G.

'MARY ELIZABETH', Min, lp, 1982; flowers small, dbl., 29 petals, slight fragrance; foliage small, light green, matt; bushy growth; [Fairy Moss X Fairy Moss]; Dobbs; Small World Min. Roses

'MARY FLEMING', F, mp, 1995; bud opening to very pale pink; semi-dbl., 6–14 petals, 1.5–2.75 in., borne 2-15 per cluster, moderate fragrance; foliage medium, medium green, matt; bushy (90-100 cms), spreading growth; [Marchenland X Golden Salmon Superieur]; Fleming, Joyce L.; Hortico Roses, 1994

Mary Gammon *see* FRYsweetie

Mary Gamon, Min, 1993; Fryer, Gareth; (Cavriglia)

'MARY GUTHRIE', Pol, mp, 1929; flowers rich pink, small blooms in large clusters, single, moderate fragrance; foliage light; bushy (2 1/2 ft) growth; [Jersey Beauty X Scorcher]; Clark, A.; NRS Victoria

'MARY HART', HT, mr, 1931; [Talisman sport]; Hart, G.B.

'MARY HART, CLIMBING', Cl HT, mr, 1937; Western Rose Co.

'MARY HART, CLIMBING', Cl HT, mr, 1942; Meilland, F.

Mary Hayley Bell *see* KORparall

'MARY HAYWOOD', Min, mp, 1957; flowers bright pink, base white, dbl., 50 petals, 1 in., moderate fragrance; foliage glossy; very compact (10 in), bushy growth; [(R. wichurana X Floradora) X Oakington Ruby]; Moore, Ralph S.; Sequoia Nursery

'MARY HELEN TANNER', HT, dp, 1932; flowers carmine, stems white or pinkish; [Templar sport]; Tanner

'MARY HICKS', HMult, dr, 1927; flowers deep scarlet, borne in clusters, semi-dbl., moderate fragrance; foliage light; vigorous, climbing growth; Hicks

Mary Hilda Law *see* 'KIRCLOUD'

'MARY HILL', HT, w, 1916; flowers cream, center deep orange, dbl.; [Ophelia X Sunburst]; Hill, E.G., Co.

Mary Hill *see* 'MORHILL'

Mary Jean *see* 'HARYEN'

'MARY JO', HT, my, 1958; bud short, pointed; flowers maize-yellow, to open, medium, dbl., 30–35 petals, 3.5 in., exhibition form, slight fragrance; foliage leathery; strong stems; vigorous, upright, bushy growth; [Unnamed seedling X Orange Delight]; Hill, Joseph H., Co.

Mary Johnston *see* 'WAMMARY'

'MARY KATE', Min, dp, 1977; bud mossy; flowers light red, dbl., 29 petals, 1 in., flat, slight fragrance; foliage dark, soft; vigorous, upright growth; [Fairy Moss X Fairy Moss]; Dobbs; Small World Min. Roses

Mary Kay® *see* 'MINOCO'

'MARY KITTEL', HT, dr, 1975; bud pointed; dbl., 35 petals, 5 in., exhibition form, intense fragrance; foliage large, glossy; vigorous growth; [(Chrysler Imperial X Night 'n' Day) X Night 'n' Day]; Harvey, R.E.; Kimbrew-Walter Roses

'MARY L. EVANS', HRg, dp, 1936; flowers deep wild-rose-pink, non-recurrent; very similar to tekonkaha but growth more spreading; [Hansa X R. macounii]; Wright, Percy H.

'MARY LAWRANCE'S SHELL ROSE', Ch, mr, 1799

Mary Lee Johnson Richards, HT, lp

Mary Louise *see* 'BUSMALOU'

Mary Louise *see* 'TAYMAR'

'MARY LOVETT', LCl, w, 1915; flowers snow-white, large, dbl., some repeat bloom, moderate fragrance; vigorous, climbing (to 10-12 ft.) growth; [R. wichurana X Kaiserin Auguste Viktoria]; Van Fleet; J.T. Lovett

Mary Lynn *see* 'RENLYNN'

Mary Lyon® *see* 'WILMTHO'

Mary MacKillop, HT, pb, 1989; Swane

Mary Magdalene *see* 'AUSJOLLY'

'MARY MALVA', HT, lp, 1971; bud long, pointed; dbl., 18–24 petals, 3.5–4.5 in., exhibition form, moderate, spicy fragrance; foliage dark, leathery; upright growth; [(Pascali X Charlotte Armstrong) X Lilac Charm]; Lens, 1972

'MARY MANGANO', Min, op, 1989; flowers deep coral pink; [Petite Folie sport]; Papandrea, John T.

Mary Manners, HRg, w, 1970; Leicester

Mary Margaret *see* 'TALMAG'

'MARY MARGARET MCBRIDE', HT, mp, 1942; bud long, pointed; flowers salmon-pink, dbl., 42 petals, 4–5 in., exhibition form, moderate fragrance; foliage dark, leathery, glossy; vigorous, upright, bushy growth; AARS, 1943 David Fuerstenberg Prize, ARS, 1945; [Sunkist X Olympiad]; Nicolas; J&P

'MARY MARQUES', HT, or, 1955; bud long, pointed; flowers orange-red tinted yellow, medium, dbl., 40 petals, exhibition form, slight fragrance; upright growth; [Mediterranea X Suzanne Balitrand]; Bofill; Torre Blanca

'MARY MARSHALL', Min, ob, 1970; bud long, pointed; flowers orange, base yellow, small, dbl., cupped, moderate fragrance; foliage small, leathery; vigorous, dwarf, bushy growth; AOE, 1975; [Little Darling X Fairy Princess]; Moore, Ralph S.; Sequoia Nursery

Mary Marshall, Climbing *see* 'MINICO'

Mary Maud *see* 'GELMAUD'

Mary May *see* 'SHERISKEP'

'MARY MCHUTCHIN', Cl Pol, mr, 1935; flowers crimson, borne in clusters, semi-dbl., cupped, slight fragrance; foliage large, leathery; vigorous, climbing (6-8 ft) growth; Cant, B. R.

'MARY MERRYWEATHER', HT, dy, 1925; bud pointed; flowers deep golden yellow, semi-dbl., cupped, moderate fragrance; foliage glossy; vigorous, bushy growth; [Marquise de Sinéty X Lady Hillingdon]; Merryweather

'MARY MINE', Gr, op, 1973; flowers salmon-pink to light rose, large, dbl., 27 petals, slight fragrance; [Queen Elizabeth X Buccaneer]; Harkness

'MARY MONRO', HT, pb, 1921; flowers carmine-pink, flushed saffron-yellow, dbl., moderate fragrance; Pemberton

'MARY MULLIGAN', HT, or, 1944; bud long, pointed; flowers flame, dbl., cupped, moderate fragrance; vigorous growth; Mallerin, C.; A. Meilland

'MARY MURRAY', HT, ab, 1930; flowers deep apricot-yellow becoming lighter, dbl., 30 petals; foliage glossy, bronze; vigorous growth; Prior

'MARY NISH', HT, w, 1928; (White Radiance); flowers white, center tinted shell-pink, very large, dbl., moderate fragrance; foliage rich green, soft, glossy; vigorous, bushy growth; [Red Radiance sport]; Pacific Rose Co.

Mary of Penola *see* 'TOMMOP'

'MARY PAULINE', HT, w, 1998; flowers pale pink, double (15–25 petals), exhibition, large blooms borne mostly single; some prickles; foliage large, medium green, semi-glossy; upright, tall growth; cut flowers; [Queen Elizabeth X Chablis]; McCruiry, Curtis

'MARY PICKFORD', HT, ab, 1923; bud pointed, orange-yellow; flowers pale yellow, center deeper, large, dbl., intense fragrance; foliage bronze; vigorous, bushy growth; [Grange Colombe seedling X Souv. de Claudius Pernet]; H&S

Mary Pilkington *see* 'EVEPILK'

'MARY POPE', F, yb, 1965; flowers golden yellow suffused pink, edged darker, borne in clusters, dbl., 25 petals, 3 in.; foliage glossy, dark; vigorous growth; [Seedling X Independence]; Sanday, John

Mary Pope *see* KORlasche

'MARY POPPINS', HT, mp, 1967; [Hallmark sport]; Morey, Dr. Dennison; Country Garden Nursery

'MARY QUEEN OF SCOTS', HSpn, w

'MARY RAND', HT, w, 1965; flowers cream, edged rose-pink, well formed, 3.5 in.; foliage glossy, light green; vigorous growth; [Caprice X Scandale]; Latham

'MARY RATCLIFFE', HT, pb, 1958; flowers soft pink, reverse darker, moderate fragrance; Ratcliffe

'MARY ROBERTSON', Pol, mr, 1969; bud ovoid; flowers medium, dbl.; foliage dark; [Paprika X Seedling]; Buisman, G. A. H.

Mary Rose® *see* 'AUSMARY'

'MARY RUSSELL', HT, mr, 1940; flowers well formed, large; Clark, A.

Mary Sampere de Guanabara, F; Moreira da Silva, A.; (Cavriglia)

Mary Sheffield *see* 'DEBRAH'

Mary Sumner *see* 'MACSTRA'

Mary Swindells *see* 'KENSINGO'

Mary Taylor, HT, lp; Taylor, L.R., 1966; (Weatherly, L.)

Mary Thomson, S, lp; Thomson, G.A.; (Weatherly, L.)

'MARY WALLACE', LCl, mp, 1924; bud long, pointed; flowers warm rose pink, very large, semi-dbl., cupped, seasonal bloom, moderate fragrance; foliage glossy, rich green; vigorous, climbing (8-12 ft) growth; [R. wichurana X Pink Hybrid Tea]; Van Fleet; American Rose Society

Mary Warm, Cl Min, or, 1997

'MARY WARREN', Cl HT, mp, 1931; flowers pink, open, large, semi-dbl., recurrent bloom, slight fragrance; foliage soft, large, dark; vigorous, pillar growth; [Mrs Frank Guthrie X Scorcher]; Clark, A.; NRS New South Wales

'MARY WASHINGTON', N, w; flowers white tinted pink, fading to white, dbl., borne in clusters of 7-11, moderate fragrance; very vigorous growth; Registered by Frank L. Ross, Nashville, Tenn

Mary Webb® *see* 'AUSWEBB'

'MARY WHEATCROFT', HT, ob, 1945; bud high-pointed; flowers deep copper; foliage bronze; [Mrs Sam McGredy X Princess Marina]; Robinson, H.; Wheatcroft Bros.

Mary Wise *see* 'HERMAWI'

Mary Woodcock, HMsk, mp, 1989; Stydd

'MARY, COUNTESS OF ILCHESTER', HT, mp, 1909; (Countess Mary of Ilchester, Countess of Ilchester); flowers deep rose-pink, open, large, dbl., moderate fragrance; foliage rich green, leathery, glossy; very vigorous, bushy growth; Dickson, A.

Marybeth® *see* 'MINIBETH'

'MARYCKA MAGDANOVÁ', Pol, dr, 1938; flowers small, semi-dbl.; Böhm, J.; (Sangerhausen)

Maryellen *see* 'LEMMAR'

Mary-Jean *see* 'HARYEN'

'MARYKE-MARIKA', F, ob, 1973; bud medium, ovoid; flowers orange, base yellow, dbl., globular, slight fragrance; foliage glossy; vigorous, upright growth; [Colour Wonder X Zorina]; Kordes

Marylea Johnson Richards *see* 'WILLEA'

'MARYLÈNE', HT, mp, 1965; bud long, pointed; flowers pearl-pink, medium, dbl., moderate fragrance; foliage dark, glossy; upright growth; [Mignonne X Queen Elizabeth]; Gaujard

Marylka, Min, 1967; Wituszynski, B.; (Cavriglia)

'MARY'S FAVORITE', F, dp, 1994; flowers deep pink, blooms borne in small clusters, dbl., 15–25 petals, 3–3.5 in., slight fragrance; few prickles; foliage large, medium green, semi-glossy, disease resistant; medium (48-54 in), upright, compact growth; [Pink Favorite X Unknown]; Hemphill, William J.

'MARY'S PINK', HT, mp, 1952; bud long, pointed; flowers neyron rose, dbl., 32 petals, 4–4.5 in., slight fragrance; foliage leathery; vigorous, compact, bushy growth; [Better Times sport]; Spanbauer

'MARYSA', Pol, w, 1936; flowers pure white, moderate, lily-of-the-valley fragrance; vigorous growth; Brada, Dr.; Böhm

Maryse Kriloff *see* **'LUCY CRAMPHORN'**

Maryse Kriloff, Climbing *see* KRImasar

'MARYTJE CAZANT', Pol, ab, 1927; (Marie Casant, Mary Casant); bud globular; flowers coral-pink, blooms in large clusters; dwarf growth; [Jessie sport]; van Nes

'MASARYKOVA JUBILEJNI', HT, dr, 1931; (Jubilaire de Masaryk, Jubilee, Masaryk's Jubilee, Masaryk's Jubileums-Rose); flowers velvety red shaded black, very large, single, slight fragrance; foliage dark, glossy; vigorous growth; [Blanta sport]; Böhm, J.

Masaryk's Jubilee *see* **'MASARYKOVA JUBILEJNI'**

Masaryk's Jubileums-Rose *see* **'MASARYKOVA JUBILEJNI'**

MAScanta, S, my, 1997; (**Claudia Cardinale**, MAScatna); Guillot-Massad, 1997

Mascara® *see* MEIdalnu

MAScatna, S, my, 1997; (**Claudia Cardinale**, MAScanta); Guillot-Massad, 1997

'MASCOTTE', HT, mp, 1951; bud pointed; flowers hermosa pink, dbl., 4 in., moderate fragrance; foliage dark; vigorous growth; [Michele Meilland X Pres. Herbert Hoover]; Meilland, F.

Mascotte '77® *see* 'MEITILOLY'

MASdogui, S, op, 1995; (**Sonia Rykiel**); Guillot-Massad, 1995

MASdomo, S, pb, 1996; (**Belle de Dom**); Guillot-Massad

MASelgi, S, w, 1998; (**Eliane Gillet**); Guillot-Massad

MASflodel, S, m, 1997; (**Florence Delattre**); Guillot-Massad

MASflodel, S, ab, 1997; (**Morabito**); Guillot-Massad

MASframb, S, mp, 1998; (**Jardins de Viels Maisons**); Guillot-Massad

'MASKED BALL', HT, pb, 1966; bud globular; flowers scarlet and gold, large, dbl., moderate fragrance; foliage dark, glossy; vigorous, bushy growth; [Masquerade X Peace]; Schwartz, Ernest W.; Wyant

MASlako, S, dr, 1996; (**Anne Laure**); Guillot-Massad

MASlupau, S, ab, 1997; (**Mme Paule Massad**); Guillot-Massad

MASmabay, S, w, 1996; (**Martine Guillot**); Guillot-Massad

MASmarti, S, mp, 1996; (**Orsola Spinola**); Guillot-Massad

MASpagui, S, w, 1995; (**Manuel Canovas**); Guillot-Massad

MASpaujeu, S, ab, 1997; (**Paul Bocuse**); Guillot-Massad, 1999

'MASQUERADE', F, rb, 1949; bud small, ovoid, yellow; flowers bright yellow turning salmon-pink and then dark red, blooms, semi-dbl., 17 petals, 2.5 in., slight fragrance; foliage leathery, dark; vigorous, bushy, compact growth; (28); GM, NRS, 1952; [Goldilocks X Holiday]; Boerner; J&P

'MASQUERADE, CLIMBING', Cl F, rb, 1958; Dillian; Gregory

'MASSABIELLE', HT, w, 1958; flowers well formed, large; upright growth; Guillot, M.

Massara, F, 1995; Sauvageot; (Cavriglia)

'MASTER DAVID', HT, pb, 1949; flowers pink, reverse carmine, pointed, dbl., 25 petals, 4 in., moderate fragrance; almost thornless; vigorous growth; Cox

'MASTER HUGH', (R. macrophylla seedling), dp, 1970; flowers rich rose-pink, blooms in clusters, single, 5 petals, 2.5–3 in., slight fragrance; [R. macrophylla seedling]; Mason, L.M.; Sunningdale Nursery

'MASTER JOHN', Cl HT, or, 1944; flowers fiery orange-red, base gold, large, dbl., globular, profuse spring bloom, then scattered until fall, intense fragrance; foliage glossy, dark; very vigorous, climbing growth; [Ednah Thomas X Golden Rapture]; Duehrsen; California Roses

MASversi, S, lp; (**Versigny**); Guillot-Massad, 1998

Matador *see* 'KORFARIM'

'MATADOR', HT, dr, 1935; bud long, pointed; flowers scarlet-crimson, shaded darker, large, very dbl., cupped, moderate fragrance; foliage leathery, dark; vigorous, bushy growth; GM, Portland, 1940; [Charles P. Kilham seedling X Étoile de Hollande]; Van Rossem; J&P

'MATADOR, CLIMBING', Cl HT, dr, 1938; Western Rose Co.

Matangi® *see* 'MACMAN'

Matangi, Climbing, Cl F, ob, 1985; Chiplunkar

Matawhero Magic *see* 'MACAMSTER'

Matchball®, HMsk, w, 1990; Lens

'MATCHLESS', HT, dp, 1926; flowers cerise-pink, dbl.; upright, bushy growth; [Premier sport]; Duckham-Pierson Co.

'MATDOVE', S, mp, 1998; (**Dove Dale**); flowers medium pink, very full, very dbl., 41 petals, 3.0in., borne in small clusters, slight fragrance; prickles moderate, medium, straight; foliage medium, dark green, glossy; spreading, low, 3 ft. x 6 ft. growth; [Seedling X Seedling]; Mather, Wendy

Mateo's Silk Butterflies *see* LETsilk

Mater Casta, S; (Cavriglia)

'MATEUS ROSE', HT, pb, 1973; flowers peach, reverse pale pink, dbl., 20 petals, 4 in., slight fragrance; [Pink Parfait X Mme Butterfly]; Winship

Mathé Altéry *see* **'DRESDEN'**

'MATHILDE KÄRGER', Pol, dp, 1929; flowers carmine-pink, medium, dbl.; Kärger; (Sangerhausen)

'MATI BRADOVÁ', HT, dp, 1934; flowers dark rose-pink, sometimes almost carmine, very large, dbl., moderate fragrance; foliage glossy, dark; very vigorous, bushy growth; [Gorgeous X Gen. MacArthur]; Brada, Dr.; Böhm

Matilda® *see* 'MEIBEAUSAI'

Matilda™ *see* 'MEISOYRIS'

'MATILDA CAMPBELL', HT, dp, 1952; bud ovoid; flowers bengal rose, very dbl., 5–6.5 in., exhibition form, intense fragrance; foliage glossy; vigorous, bushy growth; Campbell

'MATSON MODESTY', HT, lp, 1946; flowers pale pink, dbl., 60 petals, 4 in., moderate fragrance; vigorous growth; [Mrs Sam McGredy X Heinrich Wendland]; Prosser

'MATSUO-HIME', HT, lp, 1999; flowers pale pink, dbl., 31–33 petals, 5.5 in., exhibition form, slight fragrance; foliage medium green; 4.5 ft growth; SM, Japan Rose Concours, 1995; [Hatakeyama X Yonina]; Shimizu, Junji, 1991

Matsuri, LCl, rb

'MATTCHE', Min, mr, 1982; (**Chelsea Pensioner**); flowers scarlet, patio, small, dbl., 20 petals, slight fragrance; foliage small, dark, semi-glossy; bushy growth; [Gold Pin seedling X Seedling]; Mattock, John, Ltd.; John Mattock, Ltd.

MATtdor, S, w, 1990; (**Northamptonshire**); Mattock

Matterhorn® *see* 'ARMMA'

'MATTGRO', F, mp, 1983; (**Pink Wave**); flowers soft medium pink, medium, semi-dbl., moderate fragrance; foliage medium, medium green, semi-glossy;

groundcover; spreading growth; [Moon Maiden X Eyepaint]; Mattock, John, Ltd.

'MATTHEW DUCKETT', HT, or, 1999; dbl., 17–25 petals, 2 in., borne mostly singly, moderate fragrance; few prickles; foliage medium, medium green, bronze when new, semi-glossy; upright, tall growth; [Solitaire X (Alexander X Remember Me)]; Jones, L.J.

'MATTHEWS SURPRISE', T, lp; Matthews, H., 1889; (Weatherly, L.)

Matthias Meilland® *see* 'MEIFOLIO'

Mattinata, HT; Voightlander; (Cavriglia)

'MATTJO', HT, w, 1983; (**Snowman**); flowers large, dbl., moderate fragrance; foliage large, dark, semi-glossy; upright, bushy growth; [(Peer Gynt X Isis) X Lady Seton]; Mattock, John, Ltd.

MATtkiri, HT, dp, 1994; (**Kiri**); Matthews

MATtlace, F, lp, 1998; (**Anniversary**); Gold Star of the South Pacific, Palmerston North, NZ, 1997; Matthews

'MATTNOT', LCl, lp, 1987; (**Elizabeth Heather Grierson**); flowers soft pink, medium, dbl., 15–25 petals, moderate fragrance; foliage medium, dark green, semi-glossy; upright growth; [Bonfire Night X Dreaming Spires]; Mattock, John, Ltd.; The Rose Nursery, 1986

'MATTSUN', F, ab, 1979; (**Young Venturer**); bud ovoid; flowers rich apricot, dbl., 30 petals, 4.5 in., intense fragrance; foliage dark, glossy, leathery; vigorous, upright growth; [Arthur Bell X Cynthia Brooke]; Mattock

'MATTWYT', HT, ly, 1979; (**Tynwald**); bud ovoid; flowers cream, center yellow, dbl., 60 petals, 5 in., moderate fragrance; bushy, upright growth; [Peer Gynt X Isis]; Mattock

Matty *see* 'DEVHAUTE'

'MATURITY', HT, mp, 1973; flowers rose-pink, full, pointed, dbl., 50 petals, 7 in.; foliage large, dark; free growth; [Duftwolke X Lively]; LeGrice

'MAUD', HT, pb, 1921; flowers salmon-pink; Paul, W.

'MAUD BETTERTON', HT, mp, 1960; flowers rose-pink, medium, dbl.; vigorous growth; Gregory

'MAUD COLE', F, m, 1968; flowers mauve-purple, dbl., moderate fragrance; foliage dark, glossy; [Lilac Charm X Africa Star]; Harkness

'MAUD CUMING', HT, op, 1923; bud pointed; flowers coral-pink, shaded peach and orange, very large, dbl., exhibition form, slight fragrance; foliage dark, glossy; vigorous, bushy growth; Dickson, A.

'MAUD DAWSON', HT, or, 1915; flowers large, dbl., intense fragrance; Dickson, A.; (Sangerhausen)

'MAUD E. GLADSTONE', Pol, pb, 1926; bud pointed; flowers malmaison pink, shaded coral and chrome-yellow, small, dbl., globular, moderate fragrance; foliage rich green, leathery, glossy; vigorous, bushy growth; [Orléans Rose X Edward VII]; Bees

'MAUD NASH', HT, mr, 1942; flowers rich red with fire; upright growth; Clark, A.

Maud Nunn *see* 'DRISCONUN'

Maude Sumner, F, ob, 1994

MAUkeole, F, op; (**Vent des Indes**®)

MAUlave, F, rb; (**Inge Schubert**®); (MAUlave), 1994

'MAUNA LOA', HT, mr, 1937; flowers bright red, large, dbl.; foliage heavy; H&S

'MAUPERTUIS', M, mp, 1868; flowers rosy-pink; Moreau et Robert

Maureen Lipman *see* FRAntier

'MAUREEN MACNEIL', F, op, 1967; flowers salmon, borne in clusters; moderate growth; [Anna Wheatcroft X Orangeade]; Vincent

'MAUREEN THOMPSON', HT, dr, 1949; bud pointed; dbl., 35 petals, 3 in., moderate fragrance; foliage leathery; very vigorous growth; Cant, B. R.; Bosley Nursery

Maurice *see* **'MCGREDY'S TRIUMPH'**

'MAURICE BERNARDIN', HP, mr, 1861; bud vermilion; flowers bright crimson, moderately large blooms in clusters, dbl., intense fragrance; vigorous growth; [Général Jacqueminot X ?]; Granger

Maurice Chevalier® *see* 'DELTRE'

'MAURICE LEPELLETIER', HP, dr; flowers vermilion red; Moreau et Robert, 1868

'MAURICE NOYELLE', HT, 1951; Buatois; (Cavriglia)

'MAURICE VILMORIN', HP, rb, 1868; flowers garnet-crimson to purple-maroon, dbl., profuse bloom, sometimes repeated, moderate fragrance; Lédéchaux

'MAURICETTE SISTAU', Pol, w, 1925; flowers pure white to rosy white, large, borne in clusters of 25-50, dbl.; thornless; Turbat

Maurine, F, mp, 1991; Poulsen

Maurine Neuberger™ *see* 'SPOMAUR'

Mauve Mallerin *see* **'SIMONE'**

'MAUVE MELODEE'®, HT, m, 1962; bud long, pointed, purple; flowers rose-mauve, semi-dbl., 17–25 petals, 4.5–5 in., moderate fragrance; foliage dark, leathery; vigorous, upright growth; [Sterling Silver X Unnamed seedling]; Raffell; Port Stockton Nursery

Maverick *see* 'LAVSASK'

Maverick, HT, lp; Edwards, Eddie

'MAVIS CAMPBELL', HT, dp, 1942; bud long, deep pink; vigorous, tall growth; Clark, A.

'MAVOURNEEN', S, rb, 1984; bud medium-large, ovoid, pointed; flowers medium red, white reverse, large blooms borne 1-10 per cluster, dbl., 23 petals, repeat bloom, slight fragrance; awl-like, tan prickles; foliage large, leathery, semi-glossy, dark; erect, bushy growth; [(Tickled Pink X Prairie Princess) X El Catala]; Buck, Dr. Griffith J.; Iowa State University

'MAX COLWELL', Min, or, 1969; bud long, pointed; flowers orange-red to red, small, dbl., 25 petals, 1.5 in., slight fragrance; foliage leathery; bushy, spreading growth; [Unnamed red F seedling X (Little Darling X Unnamed Min seedling)]; Moore, Ralph S.; Sequoia Nursery

'MAX GRAF', HRg, pb, 1919; flowers bright pink, center golden, medium, single, non-recurrent; foliage glossy, rugose; vigorous, bushy, trailing growth; valuable hardy groundcover; (14); [Probable hybrid of R. rugosa X R. wichurana]; Bowditch

'MAX HAUFE', HEg, lp, 1939; bud long, pointed, dark pink; flowers large, semi-dbl., seasonal bloom; foliage large, leathery, light; very vigorous (5-7 ft), trailing growth; [Joanna Hill X R. eglanteria hybrid]; Kordes

'MAX HESDÖRFFER', HT, lp, 1903; flowers large, dbl., moderate fragrance; Jacobs; (Sangerhausen)

'MAX KRAUSE', HT, yb, 1930; flowers reddish orange, opening golden yellow, very large, dbl., moderate fragrance; foliage dark, glossy; vigorous growth; [Mrs Beckwith X Souv. de H.A. Verschuren]; Krause; J&P

'MAX KRAUSE SUPERIOR', HT, op, 1940; flowers coppery pink, very large, very dbl., moderate fragrance; Heizmann, E.; (Sangerhausen)

'MAX KRAUSE, CLIMBING', Cl HT, yb, 1940; da Silva, Moreira

'MAX SCHMELING', F, or, 1973; bud ovoid; flowers large, dbl., slight fragrance; foliage large, glossy; upright, bushy growth; [Unknown X Unknown]; Tantau, Math.; Ahrens & Sieberz

'MAX SINGER', HMult, dr, 1885; flowers medium, dbl.; Lacharme, F.; (Sangerhausen)

'MAX VOGEL', HT, op, 1929; flowers coppery orange, large, very dbl., intense fragrance; foliage bronze; [Fritz Maydt X Lilly Jung]; Leenders, M.

'MAXI', F, rb, 1971; flowers red, white eye, semi-dbl., 12 petals, 3 in., slight fragrance; free growth; [(Evelyn Fison X (Tantau's Triumph X R. macrophylla coryana)) x(Hamburger Phoenix X Danse de Feu)]; McGredy, Sam IV; McGredy

'MAXIM', F, op, 1961; bud ovoid; flowers salmon-pink, large, bornein broad clusters, dbl.; foliage leathery; vigorous, bushy growth; Tantau, Math.

Maxim *see* TANmixa

Maxima *see* **'ALBA MAXIMA'**

Maxima Regina *see* 'SULTANA'

'MAXIME CORBON', HWich, rb, 1918; flowers dark coppery red turning apricot-yellow, borne in clusters, dbl., abundant seasonal bloom, moderate fragrance; foliage rich green, glossy, leathery; vigorous, climbing and trailing growth (8-18 ft.); [R. wichurana X Léonie Lamesch]; Barbier

'MAXIMIN CHABUEL', HT, op, 1943; Mallerin, C.; A. Meilland

'MAXINE', F, rb, 1958; bud globular, creamy pink; flowers camellia-red becoming maroon flecked pink and white, rosette, dbl., 37–40 petals, flat, moderate fragrance; foliage leathery, glossy; very vigorous, low, bushy growth; [Pinocchio X Crimson Glory]; Silva; Booy Rose Nursery

Maxine, S, dp, 1999; Williams, J. Benjamin

'MAXISTAR', HT, or, 1975; bud round; very dbl., 70–80 petals, 5.5–6 in.; foliage dark, leathery; vigorous, upright growth; [Duftwolke X Pharaon]; Huber

'MAY BANKS', T, ly, 1938; flowers lemon-yellow; [Lady Hillingdon sport]; Banks

May Choice, HT, 1958; LeGrice, E.B.; (Cavriglia)

May Graham *see* 'KENDAD'

May Lawlor *see* 'KIRHOL'

May Lawlor

May Lyon *see* 'COCBAY'

'MAY MARTIN', HT, ly, 1918; flowers pure canary-yellow, center darker, semi-dbl.; [Ophelia sport]; Martin & Forbes Co.

'MAY MILLER', HT, op, 1911; flowers coppery pink, large, dbl.; Hill, E.G., Co.; (Sangerhausen)

'MAY QUEEN', HWich, mp, 1898; flowers pink, very dbl., quartered, moderate, fruity fragrance; height to 25 ft; [R. wichurana X Champion of the World]; W.A. Manda

'MAY QUEEN', HWich, lp, 1898; flowers lilac-pink, large, semi-dbl., flat, profuse bloom, occasionally repeated, intense fragrance; foliage glossy; vigorous, climbing or groundcover growth

'MAY ROBINSON', F, op; flowers bright salmon-pink

'MAY TAYLOR', HT, dp, 1966; flowers deep rose-pink, large, dbl., exhibition form, slight fragrance; foliage soft; vigorous, bushy growth; [Tassin X Ballet]; Taylor, L.R.

'MAY WETTERN', HT, mp, 1928; bud pointed; flowers rosy pink, large, dbl., exhibition form, moderate fragrance; foliage rich green, leathery; vigorous, bushy growth; GM, NRS 1928; Dickson, A.

'MAY WOOLLEY', F, ab, 1976; flowers bronze-apricot to peach, full, dbl., 25–25 petals, 2.5–3 in., cupped, intense fragrance; foliage small, glossy; moderate, free growth; [Fairlight X Arthur Bell]; Wood

'MAYA', F, or, 1986;, blooms in clusters; foliage dense, bright; low growth; [Zorina X Lara]; Kriloff, Miche

'MAYA LEE', HT, or, 1992; very dbl., 50 petals, 3–3.5 in., borne mostly singly, intense fragrance; many prickles; foliage medium, medium green, semi-glossy; long stems; medium (100+ cms), upright growth; [Unknown X Unknown]; Jerabek, Paul E.

Maybelle Stearns *see* **'MABELLE STEARNS'**

'MAYDAY', F, lp, 1957; bud ovoid; flowers white overcast pink, borne in clusters, dbl., 25–30 petals, 3 in., cupped, moderate fragrance; foliage rich green, leathery; vigorous, bushy growth; [(Pinocchio seedling X Unnamed Hybrid Tea) X Fashion]; Boerner; J&P

'MAYET', HT, mr, 1951; flowers crimson-red passing to neyron pink, base yellow, well formed, dbl., 30 petals, intense fragrance; foliage olive-green; vigorous growth; [Condesa de Sástago X Mme Henri Guillot]; Dot, Pedro

'MAYFAIR', HT, dp, 1935; flowers deep pink, well formed, large, moderate fragrance; foliage bronze; very vigorous growth; Bentall

'MAYFLOWER', HT, pb, 1958; flowers light cerise, reverse silvery pink, dbl., 30 petals, 4 in., intense fragrance; foliage glossy; vigorous growth; [Eden Rose X ?]; Gregory

'MAYLINA', Cl HT, w, 1916; (Mrs Charles E.F. Gersdorff); flowers silvery white, reverse killarney pink to shell-pink, very large, dbl., cupped, abundant, intermittent bloom, moderate, spicy fragrance; foliage large, soft; vigorous, climbing (15 ft.), fairly compact growth; [White climbing rose X Killarney]; Gersdorff

'MAYOR BAKER', HT, or, 1928; flowers terra-cotta to scarlet, base light orange, semi-dbl., moderate fragrance; vigorous growth; [Mons. Paul Lédé X Hadley]; Thomas

'MAYOR CERMÁK', HT, dr, 1932; flowers very dark red, shaded purple, large; vigorous, branching growth; [Mrs Henry Winnett X Vaterland]; Böhm, J.; J&P, 1934

Mayor of Casterbridge *see* 'AUSBRID'

Maysa, HT; (Cavriglia)

'MAYTIME', S, pb, 1975; bud ovoid, pointed; flowers carmine-rose, base yellow, shallowly-cupped, single, 6–10 petals, 3.5–4 in., cupped, repeat bloom, moderate fragrance; foliage dark, leathery; upright, bushy growth; [Elegance (HT) X Prairie Princess]; Buck, Dr. Griffith J.; Iowa State University

'MAYTIME', LCl, lp, 1953; flowers flesh-pink, reverse rose-pink, borne in clusters of 5, single, 5–6 in., profuse, non-recurrent bloom, slight fragrance; foliage leathery; vigorous growth; (21); [R. maximowicziana pilosa X Betty Uprichard]; Maney; Iowa State College

Maywonder, Pol, or, 1968; Grootendorst

'MAYWOOD', HT, mr, 1924; flowers bright red, dbl., moderate fragrance; [Charles K. Douglas X (Killarney X Ophelia)]; Hill, Joseph H., Co.; Amling Bros.

'MAYWOOD RED', HT, mr, 1923;, moderate fragrance; [Premier X Seedling]; Hill, E.G., Co.; J.H. Hill Co.

'MAZEPPA', HGal, rb; flowers red, edged and marbled with white, medium, dbl.

'MAZOWSZE', HT, dp, 1966; bud oblong; flowers deep pink edged lighter, dbl.; foliage leathery; very vigorous growth; [Marella seedling]; Grabczewski

'MAZURKA', F, mp, 1965; bud ovoid; flowers pink, medium, borne in clusters, dbl., slight fragrance; foliage dark; Verbeek

Mazurka *see* MEItune

'MAZZINI', HT, w, 1925; flowers blush-white suffused pink, dbl., moderate fragrance; [Mme Butterfly X Gladys Holland]; Easlea

McCallun House, T, ly

McCartney Rose *see* 'MEIZELI'

McClinton Tea, T, mp

McGinnis Tea, T, ab

'McGREDY'S CORAL', HT, op, 1936; flowers coral-pink, overlaid salmon, shaded copper, large, exhibition form; foliage dark cedar-green; very vigorous, branching growth; McGredy

'McGREDY'S GEM', HT, lp, 1933; bud pointed; flowers creamy pink, base yellow, deepening to rose-pink edges, dbl., 25–40 petals, cupped, slight fragrance; very vigorous growth; McGredy

'McGREDY'S IVORY', HT, w, 1930; (Portadown Ivory); bud long, pointed; flowers creamy white, base yellow, large, dbl., 28 petals, exhibition form, moderate, damask fragrance; foliage dark, leathery, glossy; vigorous growth; GM, NRS, 1928; [Mrs Charles Lamplough X Mabel Morse]; McGredy; Dreer

'McGREDY'S IVORY, CLIMBING', Cl HT, w, 1939; Raffel; Port Stockton Nursery

'McGREDY'S ORANGE', HT, ob, 1936; (Golden Thoughts, Morning Glory,

Sunglow); flowers deep indian yellow, reverse orange, flushed salmon, dbl., exhibition form, moderate fragrance; foliage dark, bronze; vigorous growth; [Mrs Sam McGredy sport]; McGredy

'McGREDY'S PEACH', HT, op, 1933; bud pointed; flowers creamy yellow, washed salmon, very large, cupped, slight fragrance; foliage glossy, dark; vigorous growth; GM, NRS, 1932; McGredy

'McGREDY'S PERFECTION', HT, lp, 1936; flowers large, dbl.; McGredy; (Sangerhausen)

'McGREDY'S PILLAR', HT, or, 1935; flowers terra-cotta, moderate fragrance; McGredy

'McGREDY'S PINK', HT, lp, 1936; flowers bright rose, outer petals pearly cream and pink, base saffron, intense fragrance; foliage dark; vigorous, branching growth; McGredy; J&P

'McGREDY'S PRIDE', HT, op, 1936; bud long, pointed; flowers orange and salmon-pink, flushed saffron-yellow, reverse yellow, dbl., moderate fragrance; GM, NRS, 1936; [Angèle Pernet X Mrs Charles Lamplough]; McGredy; J&P

'McGREDY'S SALMON', HT, ab, 1940; bud pointed; flowers apricot-salmon, dbl., slight fragrance; foliage dark, wrinkled; strong stems; vigorous, compact growth; [Mrs Henry Morse X Unnamed seedling]; McGredy; J&P

'McGREDY'S SCARLET', HT, mr, 1930; flowers medium, dbl., 35 petals, exhibition form, slight fragrance; foliage leathery, glossy; vigorous growth; McGredy

'McGREDY'S SUNSET', HT, ob, 1936; bud long, pointed; flowers chrome-yellow shading to scarlet, reverse clear buttercup-yellow, dbl., 40 petals, globular, moderate fragrance; foliage glossy, bronze; vigorous growth; [Margaret McGredy X Mabel Morse]; McGredy; J&P

'McGREDY'S SUNSET, CLIMBING', Cl HT, ob, 1957; Shamburger, P.

'McGREDY'S TRIUMPH', HT, dp, 1934; (Maurice); flowers soft rose flushed orange, very large, dbl., exhibition form, moderate fragrance; foliage dark reddish bronze, glossy; strong stems; vigorous, branching growth; (28); GM, NRS, 1932; [Admiration X Seedling]; McGredy; J&P

'McGREDY'S TRIUMPH, CLIMBING', Cl HT, dp, 1948; Simmonds Nursery

'McGREDY'S WONDER', HT, ob, 1934; flowers coppery orange, flushed orange-red, reverse orange-red, large, semi-dbl., cupped, moderate, fruity fragrance; foliage glossy, olive-green; vigorous growth; McGredy

'McGREDY'S YELLOW', HT, my, 1933; bud long, pointed; flowers bright buttercup-yellow, large, dbl., 30 petals, cupped, slight fragrance; foliage glossy, bronze; vigorous growth; GM, NRS, 1930 GM, Portland, 1956; [Mrs Charles Lamplough X (The Queen Alexandra Rose X J.B. Clark)]; McGredy, 1934

'McGREDY'S YELLOW, CLIMBING', Cl HT, my, 1937; Western Rose Co.

McMillan's Pink *see* **'AFFIRM'**

McMillan's Yellow, HT, my

'ME DARLING', F, pb, 1971; flowers cream suffused pink, dbl., 37 petals, 3.5–4 in., exhibition form, slight fragrance; foliage glossy, light; [Evelyn Fison X Dearest]; Anderson's Rose Nurseries

Me Too *see* 'JULESCHILDREN'

'MEADOW DANCER', Min, lp, 1991; bud ovoid; flowers medium, dbl., 45 petals, cupped, no fragrance; foliage medium, medium green, semi-glossy; spreading, tall growth; [Judy Fischer sport]; Gruenbauer, Richard, 1984; Flowers 'n' Friends Miniature Roses, 1990

'MEADOW RUBY', S, mr, 1978; flowers large blooms borne singly, dbl., 40 petals, exhibition form, repeat bloom, moderate fragrance; long, red prickles; foliage leathery; vigorous, upright growth; [Prairie Princess X (Queen Elizabeth X Borealis)]; James, John

'MEBAN', HT, w, 1955; (Message, **White Knight**); bud long, pointed; flowers clear white, dbl., 33 petals, 4 in., exhibition form; foliage leathery, light green; vigorous, upright growth; AARS, 1958; [(Virgo X Peace) X Virgo]; Meilland, F.; C-P, 1957;, URS, 1955

'MÉCÈNE', HGal, pb, 1845; flowers white, striped with rose, compact, medium, dbl.; shoots very smooth; erect, moderate growth; Vibert

'MECHAK', HT, dr, 1979; flowers very large, dbl., 20 petals, exhibition form, no fragrance; brown prickles; foliage medium, dark, smooth; upright growth; [Samourai X Unnamed seedling]; Pal, Dr. B.P.; Arand Roses

'MECHTILDE VON NEUERBURG', HEg, mp, 1920; semi-dbl., 1 in., blooms in corymbs; large, crimson fruit; foliage typical sweetbriar; vigorous (10' x 10') growth; very hardy.; Boden

'MEDA', LCl, op, 1942; flowers shrimp-pink, open, imbricated, dbl., 40–60 petals, 3.5–4 in., intense fragrance; foliage large, leathery; long stems; very vigorous, climbing (10-12 ft) growth; [(R. setigera X Mme Butterfly) X Golden Dawn]; Horvath; Wayside Gardens Co.

Medal of Honor *see* 'JACTRED'

'MEDALLION'®, HT, ab, 1973; bud long, pointed; flowers light apricot, very large, dbl., moderate fragrance; foliage large, leathery; vigorous, upright growth; AARS, 1973 GM, Portland, 1972; [South Seas X King's Ransom]; Warriner, William A.; J&P

Medeo, F, w

Medialis *see* HELmedia

'MEDIATOR', HT, op, 1949; flowers coral-pink shading to salmon base, medium, dbl., 45 petals, exhibition form; foliage soft, dark green; vigorous, bushy growth; Totty

Medima®, HT, ob, 1993; Poulsen

'MEDINA', HT, w, 1918; [Sunburst sport]; White Bros.; American Bulb Co., 1923;, White Bros., 1923

'MEDITERRANEA', HT, pb, 1943; flowers carmine with yellow, passing to pink with white markings, dbl., 40 petals, 5 in., exhibition form, moderate fragrance; upright growth; [Signora sport]; Dot, Pedro

'MEDLEY', HT, lp, 1962; bud pointed; flowers bright salmon-pink, reverse flushed yellow, semi-dbl., 35–40 petals, 5–5.5 in., cupped, moderate fragrance; foliage leathery; vigorous, upright growth; [Unnamed Hybrid Tea seedling X Pageant]; Boerner; J&P

Medusa, S, m, 1996; Noack, Werner

Méduse® *see* 'GAUTARA'

Meerzicht Glory *see* **'ORANGE DELIGHT'**

'MEG', Cl HT, ab, 1954; flowers salmon-apricot, stamens red, blooms in large clusters, single, 10 petals, 5.5 in., recurrent bloom, moderate fragrance; foliage dark, glossy; vigorous growth; GM, NRS, 1954; [Probably Paul's Lemon Pillar X Mme Butterfly.]; Gosset; Harkness

Meg '81 *see* **'STRING OF PEARLS'**

'MEG MERRILIES', HEg, dp, 1894; flowers rosy crimson, single, summer bloom, moderate fragrance; foliage very fragrant; very vigorous (10 ft.) growth; [R. eglanteria X HP or B]; Penzance; Keynes, Williams & Co.

'ME-GAMI', HT, op, 1980; flowers medium salmon-pink, large, dbl., 33 petals, exhibition form, moderate fragrance; foliage dark, leathery; upright growth; [Unnamed seedling X Fragrant Cloud]; Suzuki, Seizo; Keisei Rose Nursery

'MEGAN', HT, pb, 1981; bud pointed; flowers white with pink petal edges, blooms borne 3-5 per cluster, exhibition form, moderate fragrance; broad, slightly hooked prickles; foliage medium green, glossy; upright, strong growth; Bronze, ARC TG, 1981; [Daily Sketch X Unnamed seedling]; Adams, Dr. Neil D.

Megan *see* 'DEVNOVIA'

Megan Dolan *see* 'DEVNOVIA'

'MEGAN LOUISE', HT, pb, 1981; bud ovoid; flowers silvery pink, deep pink petal edges, large blooms borne 1-5, dbl., 48 petals, exhibition form, intense fragrance; red-brown prickles; foliage matt, green, tough; short, bushy growth; [Red Lion X Silver Lining]; Erich Welsh Roses; Australian Rose Society & Roy Rumsey Ltd.

'MEGASTAR', HT, or, 1980; bud globular; dbl., 40 petals, cupped, borne singly, intense, fruity fragrance; curved yellow-brown prickles; foliage light green; Sohne, W. Kordes; Rose Barni-Pistoia

'MEGERSAR', Cl HT, or, 1965; (**Baccará, Climbing**®); URS

Meggie, Min, lp

Meghan's Arrival *see* 'BULLMEG'

'MEGIDDO', F, or, 1970; dbl., 25 petals, 4.5 in., slight fragrance; foliage large, olive-green; upright growth; [Coup de Foudre X S'Agaro]; Gandy, Douglas L.

'MEHAMBER', F, ab, 1996; (**Amber Nectar**); flowers amber/apricot with amber yellow reverse, blooms borne in large clusters, dbl., 26–40 petals, 2 in., slight fragrance; moderate prickles; foliage medium, medium green, glossy; upright, medium (60 cms) growth; [Alexander X Sweet Magic]; Mehring, Bernhard F.; Henry Street Nurseries, 1997

'MEHBRONZE', Min, r, 1996; (**Free As Air**); flowers orange red/russet with light orange pink reverse, blooms bor, semi-dbl., 8–14 petals, 1.5 in., no fragrance; few prickles; foliage medium, dark green, glossy; compact, medium (35 cms) growth; PVRO 5448; [Anna Ford X Brown Velvet]; Mehring, Bernhard F.; Mehring, 1996

'MEHBROWN', F, mr, 1997; (**Garry Brown**); flowers double, medium, dbl., 15–25 petals, borne in small clusters, intense fragrance; some prickles; foliage large, dark green, glossy; upright, medium (2-6in.)growth; [Roger Lamberlin X Arthur Bell]; Mehring, Bernhard F.

'MEHER', HT, mp, 1998; flowers medium pink, lighter reverse, very dbl., 26–40 petals, 3–3.5 in., borne mostly singly, moderate fragrance; numerous small and large prickles; foliage medium, medium green, semi-glossy; upright, tall, 5 ft. growth; [American Heritage X Paradise]; Pavri, Nadir J.

'MEHGOLDIE', F, ly, 1997; (**Janet Bebb**); flowers full, medium, dbl., 15–25 petals, borne in small clusters, slight fragrance; foliage large, medium green, glossy; compact, medium (100cms) growth; [Golden Holstein X Seafarer]; Meahring, Bernhard F.

MEHnan, Min, yb, 1998; (**Frechdachs**); Mehring, Bernhard F.

'MEHPARK', LCl, mp, 1998; (**Jane Eyre**); flowers medium pink, double, dbl., 15–25 petals, 4.5 in., borne in small clusters, slight fragrance; prickles moderate; foliage large, medium green, semi-glossy; upright, climbing, medium, 300cms growth; [(Westfalenpark Dortmund X (Anna Ford X Frank Naylor)) X (Westerland X ((Casino X Mermaid) X (Vesper X Picasso)))]; Mehring, Bernhard F.; Apuldram Roses, 1998

'MEHPAT', Min, pb, 1996; (**Make A Wish**); flowers salmon pink, light pink reverse, blooms borne in large clusters, very dbl., 2 in., moderate fragrance; moderate prickles; foliage meidum, dark green, glossy; bushy, medium (35 cms) growth; [Robin Redbreast X Amber Queen]; Mehring, Bernhard F.; Henry Street Nurseries, 1996

'MEHPIC', F, yb, 1998; (**Golden Hope**); flowers yellow, pink edge, reverse amber with pink edge, dbl., 15–25 petals, 2.75 in., borne in small clusters, slight fragrance; no prickles; foliage large, medium green, semi-glossy; upright, medium, 80 cm growth; [Kronprinzessin Victoria von Schweden X Goldene Holstein]; Mehring, Bernhard F.; Henry Street Nurseries, 1998

'MEHREX', F, rb, 1993; (**Sarah Jo**); flowers red with white eye, salmon reverse, blooms borne in small clusters, dbl., 15–25 petals, 1.5–2.75 in., slight fragrance; some prickles; foliage large, dark green, glossy; medium (60 cms), upright, bushy growth; [Sheri Ann X Dortmund]; Mehring, Bernhard F.; Mehring, 1994

'MEHSHERRY', S, dp, 1995; (**The Holt**); flowers fuchsia pink, dbl. (15–25 petals), small blooms borne in large clusters; slight fragrance; some prickles; foliage medium, dark green, glossy; bushy (100 cms) growth; [Sexy Rexy X Robin Redbreast]; Mehring, B. F.; Mehring, 1994-95

'MEIALATE', S, yb, 1998; (**Mystic Meidiland**®); flowers coppery peach fading to butter cream yellow, semi-dbl., 8–14 petals, 1–1.5 in., borne in small clusters, no fragrance; prickles moderate; foliage medium, dark green, semi-glossy; compact, medium, 3 ft. growth; Selection Meilland; Conard-Pyle Co., 1997

'MEIALFI', F, ob, 1961; (**Zambra**®); bud ovoid; flowers orange, reverse yellow, blooms in clusters, semi-dbl., 13 petals, 2.5–3 in., flat, slight, sweetbriar fragrance; foliage leathery, glossy, light green; vigorous, well-branched growth; GM, Bagatelle, 1961 GM, Rome, 1961; [(Goldilocks X Fashion) X (Goldilocks X Fashion)]; Meilland, Mrs. Marie-Louise; C-P, 1964;, URS, 1961

'MEIALFISAR', Cl F, ob, 1969; (**Zambra, Climbing**®); Meilland; URS

MEIalzonite, HT, op, 1986; (**Lady Meilland**); GM, Durbanville, 1982 Gold Star of the South Pacific, Palmerston North, NZ, 1982; Meilland

'MEIARLO', HT, or, 1962; (**Allegro**®); dbl., 30 petals, 3.5–4.5 in., exhibition form, slight fragrance; foliage leathery, glossy; vigorous, bushy growth; GM, Rome, 1962 GM, The Hague, 1962; [(Happiness X Independence) X Soraya]; Meilland, Alain A.; C-P, 1964;, URS, 1962

MEIbalani, S, mp, 1980; (**Rétro**); flowers large, dbl., slight fragrance; Meilland; (Sangerhausen)

'MEIBALBIKA'®, F, mr, 1984; (Iga 83 Munchen, Meilland Rosiga '83, Munchen 83, **Rose Iga**, The Wyevale rose); flowers large, dbl., 20 petals; foliage medium, dark, semi-glossy; PP006281; [MEIgurami X (Cruosa X City of Leeds)]; Meilland, Mrs. Marie-Louise; Meilland & Son, 1981

'MEIBARKE', Min, rb, 1988; (**Debut**™, Douce Symphonie, Sweet Symphony); flowers luminous scarlet blending to cream to yellow at base, aging, dbl., 15–18 petals, exhibition form, no fragrance; ovoid, few, dull orange-red fruit; prickles slender, few, straw; foliage medium, dark green, semi-glossy; bushy growth; PP006791; AARS, 1989; [Coppelia X Magic Carrousel]; Selection Meilland; C-P, 1989

'MEIBEAUSAI', F, w, 1988; (Charles Aznavour, **Matilda**®, Pearl of Bedfordview, Seduction); flowers edged pink, large, dbl., 15–20 petals, no fragrance; foliage meidum, dark green, semi-glossy; upright, low, compact, proliferous growth; PP007667; GM, Bagatelle, 1987 GM, Courtrai, 1987; [MEIgurami X Nirvana]; Meilland, Alain A.; SNC Meilland & Cie

'MEIBEKARB', HP, mr; (**Cherry Sunblaze**™); flowers purple-violet, dbl., cupped; [Debut sport]

'MEIBELUXEN', S, dr, 1979; (**Fiona**®); flowers small, dbl., 20 petals, slight fragrance; foliage small, dark, semi-glossy; spreading growth; [Sea Foam X Picasso]; Meilland, Mrs. Marie-Louise; Meilland Et Cie, 1982

MEIbicmarj, HT, pb, 1996; (**Elegance**); Meilland

MEIbigoud, HT, mr, 1995; (**Ambassador**); flowers velvety red, frilly; Meilland

'MEIBIL', HT, mp, 1959; (**Pink Peace**®); flowers dusty pink, dbl., 58 petals, 4.5–6 in., intense fragrance; foliage leathery; vigorous, tall, bushy growth; GM, Geneva, 1959 GM, Rome, 1959; [(Peace X Monique) X (Peace X Mrs John Laing)]; Meilland, F.; C-P;, URS

'MEIBILSAR', Cl HT, mp, 1968; (**Pink Peace, Climbing**®); Meilland; URS

'MEIBINOSOR', F, or, 1976; (**Concertino**®); flowers cherry-red, medium, dbl., 20 petals, cupped, slight fragrance; foliage matt, dark; vigorous, bushy growth; [((Fidélio X Fidélio) X

(Zambra X Zambra)) X Marlena]; Meilland, Mrs. Marie-Louise; Meilland

'MEIBIRANDA', HT, rb, 1978; (**Candia**™); flowers red and yellow blend, large, dbl., 38 petals, exhibition form, slight fragrance; foliage large, light green; PP004705; [Matador X (Tropicana X Flirt)]; Meilland, Mrs. Marie-Louise; Meilland Et Cie

'MEIBLAM', Min, w, 1984; (**White Meillandina**, Yorkshire Sunblaze); flowers medium, semi-dbl., no fragrance; foliage small, light green, semi-glossy; bushy growth; [Katharina Zeimet X White Gem]; Meilland, Mrs. Marie-Louise; Meilland Et Cie, 1983

'MEIBLERI', HT, w, 1985; (**Alliance**); flowers large, semi-dbl., no fragrance; foliage medium, dark, matt; upright growth; [Rustica X Youki San]; Meilland, Mrs. Marie-Louise, 1981; Meilland Et Cie, 1984

MEIbojat, HT, dy, 1987; (**Aurea**, MEIbosat); Meilland

'MEIBOKARB', Min, mr, 1993; (**Cherry Sunblaze**®); flowers small, bright red, nonfading, very dbl., 40–25 petals, 1–1.5 in., borne in small clusters, no fragrance; some prickles; foliage medium, dark green, semi-glossy; medium (35 cms), bushy growth; PP8448; [Coppelia X Magic Carrousel]; Hutton, R.J.; Conard-Pyle, 1993, USA

MEIbolnay, HT, lp; (**Top Secret**); Meilland, 1996

'MEIBONRIB', S, mp, 1994; (**Magic Meidiland**™, Magic Meillandecor®); flowers red purple, lighter centers, dbl., 17–25 petals, 1.5–2.75 in., flat, borne in small clusters, no fragrance; some prickles; foliage small, dark green, glossy; medium to low, spreading, growth; groundcover; PP9469; [R. sempervirens X (Milrose X MEIdomonac)]; Meilland, Alain A.; SNC Meilland & Cie, 1992

MEIbonver, HT, w; (**White Majesty**); Meilland, 1992

MEIborfil, HT, m; (**Silver Anniversary**); Meilland

MEIbosat, HT, dy, 1987; (**Aurea**, MEIbojat); Meilland

'MEIBRICO', HT, op, 1986; (**Bettina '78**); flowers coral, dbl., 30 petals, 4 in., slight fragrance; foliage dark, leathery; vigorous growth; [(Jolie Madame X Sunlight) X (Lady Elgin X Dr. A.J. Verhage)]; Paolino; URS, 1974

'MEIBRINPAY', HT, ab, 1996; (**Abbaye de Cluny**™, Romantic Seranade); dbl., 26–40 petals, 4.25 in., borne mostly singly, slight fragrance; moderate prickles; foliage medium, dark green, semi-glossy; bushy, medium (90 cms) growth; Eurosa, Chinensis; US PPAF; GM, Belfast, 1995; [Just Joey X (MEIrestif X MEInarual)]; Meilland International SA, 1992; The Conard-Pyle Co., 1993; (Hay)

'MEIBURENAC', Min, w, 1978; (**Swany**®); bud ovoid; flowers pure white, large, very dbl., 95 petals, cupped; foliage glossy, bronze; very vigorous, spreading growth; [R. sempervirens X Mlle Marthe Carron]; Meilland, Mrs. Marie-Louise; Meilland, 1977

'MEIBURGANA', HT, mr, 1981; (**Miriana**); flowers large, dbl., no fragrance; foliage medium, dark, semi-glossy; upright growth; [((Seedling X Independence) X Suspense) X (((Alain X R. mutabilis) X Caprice) X Pharaoh)]; Meilland, Mrs. Marie-Louise; Meilland Et Cie, 1982

'MEIBYBA', Min, or, 1965; (**Baby Baccará**®); flowers orange-scarlet, small, dbl., 1.5 in., slight fragrance; foliage dark; [Callisto X Perla de Alcañada]; Meilland, Alain A.; URS;, Wheatcroft Bros.

'MEIBYSTER', HT, or, 1971; (Atoll, **Clarita**); flowers vermilion, dbl., 30–35 petals, 5 in., exhibition form, slight fragrance; foliage dark; very vigorous, upright growth; GM, Geneva, 1971 GM, Lyon, 1971; [Tropicana X (Zambra X Romantica)]; Meilland

'MEICAPINAL', HT, pb, 1982; (Aachener Dom, Panthere Rose, **Pink Panther**®); flowers petals silvery pink, edged deep pink, large, dbl., no fragrance; foliage medium, semi-glossy; upright growth; [MEIgurami X MEInaregi]; Meilland, Mrs. Marie-Louise; Meilland Et Cie, 1981

'MEICAPULA', Gr, mp, 1981; (**Fiorella**, Fiorella '82, Marion Foster); flowers large, dbl., 35 petals, slight fragrance; foliage medium, dark, matt; upright growth; [(Queen Elizabeth X Nirvana) X (Tropicana X MEInaregi)]; Meilland, Mrs. Marie-Louise; Meilland Et Cie

'MEICARI', HT, dr, 1957; (**Champs-Elysées**®, 'MEICARL'); flowers rich crimson-red, large, dbl., 35 petals, cupped, slight fragrance; vigorous, bushy growth; GM, Madrid, 1957; [Monique X Happiness]; Meilland, F.; URS

'MEICARL', HT, dr, 1957; (**Champs-Elysées**®, 'MEICARI'); flowers rich crimson-red, large, dbl., 35 petals, cupped, slight fragrance; vigorous, bushy growth; GM, Madrid, 1957; [Monique X Happiness]; Meilland, F.; URS

'MEICARLSAR', Cl HT, dr, 1969; (**Champs-Elysées, Climbing**®); Meilland; URS

'MEICAUF', HT, dr, 1994; (**Amalia**™); dbl., 26–40 petals, 3–3.5 in., borne mostly singly, no fragrance; many prickles; foliage medium, dark green, semi-glossy; medium (100-120 cms), upright growth; PP007718; [(Queen Elizabeth X Karl Herbst) X Papa Meilland]; Meilland, Alain A.; Selection Meilland, 1989;, The Conard-Pyle Co., 1990

MEIcelna, HT, mp; (**Tino Rossi**®); Meilland, 1990

'MEICESAR', HT, dr, 1963; ('MEISAR', **Papa Meilland**®); bud pointed; flowers dark velvety crimson, large, dbl., 35 petals, exhibition form, intense fragrance; foliage leathery, glossy, olive-green; vigorous, upright growth; GM, Baden-Baden, 1962 Hall of Fame, WFRS, 1988 James Alexander Gamble Fragrance Medal, ARS, 1974; [Chrysler Imperial X Charles Mallerin]; Meilland, Alain A.; URS

'MEICHAM', HT, pb, 1964; (**Silva**®); bud long, pointed; flowers yellowish salmon shaded bright rose, dbl., 38 petals, 5.5 in., exhibition form, slight fragrance; foliage dark, glossy, leathery; vigorous, upright growth; GM, The Hague, 1964; [Peace X Confidence]; Meilland, Alain A.; URS

'MEICHANSO', HT, or, 1978; (**Parador**®, Tchin-Tchin®); dbl., 20 petals, 3.5 in., cupped; very vigorous growth; GM, Tokyo, 1978; [((Sarabande X Meikim) X (Alain X Orange Triumph)) X Diablotin]; Paolino; URS

MEIchansosar, Cl F, lp; (**Tchin-Tchin, Climbing**); Meilland, 1995

MEIchavrin, LCl, w, 1998; (**Mon Jardin et Ma Maison**); Meilland

'MEICHEST', F, or, 1964; (**Fidélio**®); bud long, pointed; flowers medium, dbl., 35 petals, exhibition form, slight fragrance; foliage leathery; vigorous, upright growth; [(Radar X Caprice) X Fire King]; Meilland, Alain A.; URS

'MEICHIM', HT, mp, 1963; (**Carina**®); dbl., 40 petals, 5 in., exhibition form, moderate fragrance; foliage leathery; upright, bushy growth; ADR, 1966; [White Knight X (Happiness X Independence)]; Meilland, Alain A.; C-P, 1963;, J.H. Hill Co., 1964;, URS, 1963

'MEICHIMSAR', Cl HT, mp, 1968; (**Carina, Climbing**®, Grimpant Carina); Meilland; URS

MEIchoiju, F, mp, 1989; (**Jacqueline Nebout**®); Meilland

'MEICHONAR', S, w, 1996; (**Pearl Sevillana**™); flowers pearly white, blooms borne in small clusters, semi-dbl., 8–14 petals, slight fragrance; moderate prickles; foliage medium, dark green, dull; bushy, medium (4-5 ft) growth; US PPAF; [(MEIdomonac X Paseali) X Edelweiss]; Meilland International SA; The Conard-Pyle Co., 1996

'MEICIJAS', HT, yb, 1991; (**Texas Girl**™); flowers creamy yellow, large, dbl., 26–40 petals, slight fragrance; foliage large, medium green; tall, upright growth; PP007784; [Lovely Girl sport]; Meilland, Alain A.; The Conard-Pyle Co., 1991

'MEICITREM', Min, my, 1995; (**Lucie's Dream**); very dbl., 1.5–2.75 in., borne in small clusters, moderate, lemony fragrance; some prickles; foliage medium, medium green, semi-glossy; medium (25-30 cms), bushy growth; [(Yellow Pages X Gold Badge) X Lemon Delight]; Meilland, Alain A.; SNC Meilland & Cie, 1994

'MEIcloux', F, ob, 1985; (Chacok, Fakir, Jubilee 150, **Pigalle**®, Pigalle 84); flowers yellow blended with orange and orange-red, large, dbl., 40 petals, no fragrance; foliage medium, medium green, semi-glossy; bushy growth; [Frenzy X ((Zambra X Suspense) X King's Ransom)]; Meilland, Mrs. Marie-Louise; Meilland Et Cie, 1984

MEIcobius, HT, lp; (**Terracotta**); Meilland, 1999

MEIcobuis, HT, lp; (**Terracotta**); Meilland, 1999

MEIcofum, HT, rb, 1995; (**Leonidas**); Meilland

'MEIcoublan', S, w, 1986; (Alba Meidiland®, Blanc Meillandécor, **White Meidiland**®); flowers large, very dbl., no fragrance; foliage medium, dark green glossy; spreading growth; [Temple Bells X MEIgurami]; Meilland, Mrs Marie-Louise, 1987

'MEIcri', Min, ob, 1958; (**Cricri**®, Cri-Cri, Gavolda); flowers salmon shaded coral, small, dbl.; foliage leathery; dwarf, very bushy growth; [(Alain X Independence) X Perla de Alcañada]; Meilland, F.; URS

'MEIcubasi', Min, dr, 1982; (Scarlet Meillandina, **Scarlet Sunblaze**®); flowers medium, dbl., 20 petals, no fragrance; foliage medium, dark, matt; bushy growth; PP004681; [Tamango X (Baby Bettina X Duchess of Windsor)]; Meilland, Mrs. Marie-Louise; C-P, 1980

'MEIcupag', Min, my, 1994; (**Golden Sunblaze'**®); bud short, plump, ovoid; flowers bright golden yellow, like a zinnia, very dbl., 40–45 petals, 1.5 in., cupped, blooms in small clusters, slight fragrance; few prickles; foliage medium, dark green, glossy; medium (15 in), upright, bushy growth; PP8493; [(Rise 'n' Shine X Mark One) X Yellow Meillandina]; Selection Meilland; The Conard-Pyle Co., 1993

MEIcurbos, F, op; (**Zambra**®, Zambra 92); GM, Bagatelle, 1992; Meilland, 1992

'MEIdacinu', Min, or, 1977; (**Baby Bettina**); bud ovoid; flowers vermilion, reverse carmine, deep, mini-flora, 15–20 petals, cupped, borne 3-11 per cluster, slight fragrance; foliage matt, dense; vigorous growth; [(Callisto X Perla de Alcañada) X Starina]; Meilland, Mrs. Marie-Louise; Meilland Et Cie

'MEIdad', HT, op, 1967; (**Majorette**®, Minna Lerche Lerchenborg®); bud pointed; flowers coppery salmon, large, dbl., exhibition form, slight fragrance; foliage dark, leathery; vigorous, upright growth; GM, Bagatelle, 1966; [Zambra X Fred Edmunds]; Meilland; URS

'MEIdali', HT, pb, 1968; (**Arianna**®); flowers carmine-rose suffused coral, large, dbl., 35 petals, exhibition form, slight fragrance; foliage dark, leathery; vigorous, upright, open growth; GM, Bagatelle, 1965 GM, Rome, 1965 GM, The Hague, 1965; [Charlotte Armstrong X (Peace X Michèle Meilland)]; Meilland, Mrs. Marie-Louise; URS

MEIdalnu, HT, m, 1992; (**Mascara**®); Meilland

'MEIdanclar', Min, dp, 1991; (**Candy Sunblaze**®, Romantique Meillandina); flowers deep pink, blooms, very dbl., slight fragrance; foliage medium, dark green, glossy; tall, upright growth; PP007621; [Lady Sunblaze sport]; Selection Meilland, 1988; The Conard-Pyle Co., 1992

'MEIdanego', Min, dr, 1984; (**Belle Meillandina**®, Belle Sunblaze); [Meillandina sport]; Meilland, Mrs. Marie-Louise; Meilland Et Cie, 1980

'MEIdanover', Min, ob, 1979; (**Colibre 79**®); Meilland

'MEIdanu', F, dr, 1967; (**Tamango**®); flowers large blooms in large clusters, dbl., 35 petals, slight fragrance; [(Alain X Mutabilis) X (Radar X Caprice)]; Meilland, Mrs. Marie-Louise; Wheatcroft Bros.

MEIdarwet, S, pb, 1993; (**Relax Meillandecor**); Meilland

MEIdarwet, S, lp; (**Relax Meidiland**); Meilland

'MEIdaud', HT, dp, 1965; (Maria Callas, **Miss All-American Beauty**); bud ovoid; flowers large, dbl., 55 petals, cupped, intense fragrance; foliage leathery; vigorous, bushy growth; AARS, 1968 GM, Portland, 1966; [Chrysler Imperial X Karl Herbst]; Meilland, Mrs. Marie-Louise; C-P, 1967;, Wheatcroft Bros, 1965

'MEIdaudsar', Cl HT, dp, 1969; (Maria Callas, Climbing, **Miss All-American Beauty, Climbing**); Meilland; URS

'MEIdeauri', F, lp, 1994; (**Leonardo de Vinci**®); flowers bengal pink, blooms borne in small clusters, very dbl., 3–3.5 in., slight fragrance; some prickles; foliage medium, dark green, glossy; bushy (70-110 cms) growth; [Sommerwind X (Milrose X Rosamunde)]; Meilland, Alain A.; SNC Meilland & Cie, 1993

'MEIdeauri', mp, 1998; (**Leonardo da Vinci** TM); flowers medium pink, rosette blooms, very dbl., 25–30 petals, rosette, borne singly and in small clusters, slight fragrance; foliage medium, mediu green; climbing (3-4 ft) growth; Meilland

MEIdenji, LCl, mr, 1989; (**Cassandre**®); Meilland

'MEIdesi', Min, mp, 1965; (**Mimi**); bud ovoid; dbl., 33 petals, 1–1.5 in., cupped, blooms in clusters, slight fragrance; foliage leathery; vigorous, bushy (14 in.) growth; [Moulin Rouge X (Fashion X Perla de Montserrat)]; Meilland, Mrs. Marie-Louise; URS

Meidiland Alba *see* 'MEIflopan'

'MEIdinro', HT, rb, 1990; (Altesse, **Anticipation**, Rodin); flowers red with silver/ white reverse, medium, semi-dbl., 35 petals, exhibition form, borne singly, slight fragrance; foliage medium, medium green, semi-glossy; bushy growth; [Seedling X Seedling]; Meilland, 1980; Co-Operative Rose Growers, 1991

'MEIdipser', Min, dp, 1994; (**Spot Meillandina**®); flowers rose bengal, blooms borne in small clusters, very dbl., 74 petals, 1.5 in., no fragrance; few prickles; foliage medium, dark green, semi-glossy; low (20-25 cms), bushy growth; [(Orange Sunblaze X Pink Symphony) X Red Minimo]; Meilland, Alain A.; SNC Meilland & Cie, 1992

'MEIdirapo', HT, dr, 1975; flowers red to purple, full, dbl., 25 petals, 3.5–4 in., cupped, slight fragrance; vigorous growth; [(Queen Elizabeth X (Peace X Michele Meilland)) X (Baccará X Seedling)]; Paolino; Meilland

'MEIdo', Min, or, 1961; (**Scarlet Gem**®, Scarlet Pimpernel); bud ovoid; flowers orange-scarlet, dbl., 58 petals, 1 in., cupped, slight fragrance; foliage dark, glossy, leathery; bushy, dwarf (12-15 in) growth; [(Moulin Rouge X Fashion) X (Perla de Montserrat X Perla de Alcanada)]; Meilland, Alain A.; C-P

'MEIdomonac', S, mp, 1985; (**Bonica**®, Bonica '82, Demon); flowers medium pink center, lighter at edges, medium, dbl., no fragrance; foliage small, dark, semi-glossy; bushy growth; PP005105; AARS, 1987 ADR, 1983; [(R. sempervirens X Mlle Marthe Carron) X Picasso]; Meilland, Mrs. Marie-Louise; C-P, 1987;, Meilland Et Cie, 1981

'MEIdona', HT, w, 1965; (Mme Neige, **Youki San**®); flowers large, dbl., 40 petals, intense fragrance; foliage light green; tall growth; GM, Baden-Baden, 1964; [Lady Sylvia X White Knight]; Meilland; URS;, Wheatcroft Bros.

'MEIdonfe', Min, dp, 1998; (**Raspberry Sunblaze**); flowers near red purple (deep pink), double (15–25 petals), medium (2 in) blooms borne in small clusters; no fragrance; moderate prickles; foliage medium, medium green, semi-glossy; bushy, medium (24 in.) growth, pot forcing; [(Orange Sunblaze X Prince Meillandina) X Red

Minimo]; Selection Meilland; The Conard Pyle Co. 1998

MEIdorsun, HT, dy, 1997; (**Concorde**); Meilland

'MEIDRAGELAC', HT, or, 1985; (Laura '81, **Natilda**); flowers orange-red, lighter reverse, large, dbl., 30 petals, slight fragrance; foliage small, dark, semi-glossy; medium growth; GM, Japan, 1981; [(Pharaoh X Colour Wonder) X ((Suspense X Suspense) X King's Ransom)]; Meilland, Mrs. Marie-Louise, 1974; Meilland Et Cie, 1981

'MEIDRAGELAC', HT, ob; (**Laura**®)

MEIdrimy, LCl, lp, 1995; (**Domaine de Courso**); Meilland

'MEIDUJARAN', F, mr, 1980; (**Pimlico**, Pimlico '81); flowers large, dbl., 35 petals, no fragrance; foliage large, dark, glossy; bushy growth; GM, Belfast, 1983; [(Tamango X Fidélio) X (Charleston X Lilli Marleen)]; Meilland, Mrs. Marie-Louise; Meilland Et Cie

'MEIELEC', Gr, mr, 1966; (Samourai, **Scarlet Knight**); bud ovoid; flowers crimson-scarlet, dbl., 4–5 in., cupped, slight fragrance; foliage leathery; vigorous, upright, bushy growth; AARS, 1968 GM, Madrid, 1966; [(Happiness X Independence) X Sutter's Gold]; Meilland, Mrs. Marie-Louise; C-P, 1967;, URS, 1966

'MEIELPA', LCl, mr, 1967; (**Benvenuto**®); flowers rose-red, medium, semi-dbl., recurrent bloom, slight fragrance; vigorous, climbing growth; [(Alain X Guinée) X Cocktail]; Meilland; URS

'MEIESTHO', F, or, 1966; (**Sangria**®); semi-dbl., in large trusses; vigorous growth; GM, Geneva, 1966 GM, The Hague, 1966; [Fire King X (Happiness X Independence)]; Meilland, Mrs. Marie-Louise; URS

'MEIFAN', HT, rb, 1960; (**Suspense**); bud ovoid; flowers turkey-red, reverse yellow-ochre, dbl., 58 petals, 4.5–5 in., exhibition form, slight fragrance; foliage leathery, dark, glossy; vigorous, upright, bushy growth; [Henri Mallerin X (Happiness X Floradora)]; Meilland, F.; C-P, 1961;, URS, 1960

MEIfarent, HT, my, 1986; (**Solidor**®); Meilland

'MEIFERJAC', Min, or, 1996; (**Autumn Sunblaze**™); flowers bright, non-fading orange, very dbl., 40–45 petals, 1.5 in., borne in small clusters, no fragrance; some prickles; foliage medium, medium green, semi-glossy; medium (24 in), bushy growth; neat tolerant; PP9562; [(Bonfire Night X MEIninrut) X Orange Jewel]; Selection Meilland; Conard-Pyle Co., 1995

MEIfersi, HT, mr, 1986; (**Michel Lis le Jardinier**®); Meilland

'MEIFIGA', HT, or, 1967; (**Pharaoh**, Pharaon®); bud ovoid; flowers bright orange-red, dbl., 5 in., exhibition form, moderate fragrance; foliage dark, glossy, leathery; vigorous, upright growth; GM, Belfast, 1969 GM, Geneva, 1967 GM, The Hague, 1967; [(Happiness X Independence) X Suspense]; Meilland, Mrs. Marie-Louise; C-P, 1969;, URS, 1967

'MEIFIGU', HT, m, 1965; (**Lady X**); bud long, pointed; flowers large, dbl., exhibition form, slight fragrance; foliage leathery; vigorous, upright growth; GM, Portland, 1968; [Seedling X Simone]; Meilland, Mrs. Marie-Louise; C-P, 1966

'MEIFIKALIF', HT, or, 1980; (**Allegeo '80**); flowers large, dbl., 35 petals, no fragrance; foliage medium, medium green, semi-glossy; [(Diorette X Tropicana) X (Unnamed seedling X (Diorette X Tropicana))]; Meilland, Mrs. Marie-Louise; Meilland Et Cie

'MEIFINARO', Min, yb, 1982; (**Air France**, Air France Meillandina, American Independence, Rosy Meillandina); flowers yellow, pink petal edges, small, dbl., no fragrance; foliage small, dark, matt; bushy growth; PP005180; [Minijet X (Darling Flame X Perle de Montserrat)]; Meilland, Mrs. Marie-Louise, 1983

'MEIFLAROL', F, ob, 1986; (**Pareo**®); Meilland

'MEIFLOPAN', S, w, 1987; (**Alba Meidiland**™, Alba Meillandécor, Meidiland Alba); flowers medium, very white, dbl., no fragrance; foliage medium, medium green, glossy; spreading growth, vigorous; PP006891; [R. sempervirens X Marthe Carron]; Meilland, Mrs. Marie-Louise, 1985

'MEIFLOREM', F, or, 1970; (**My Fancy**); flowers dutch vermilion, imbricated, dbl., 25–30 petals, 3 in., slight fragrance; foliage dull, dark; vigorous, upright, bushy growth; [(Dany Robin X Fire King) X Rumba]; Meilland; URS

'MEIFLUNEY', Min, dp, 1984; flowers deep pink, medium, dbl., no fragrance; foliage medium, medium green, matt; bushy growth; [(Alain X Fashion) X (Rumba X (Zambra X Cinderella))]; Meilland, Mrs. Marie-Louise; Meilland Et Cie

'MEIFOLIO', F, mr, 1988; (**Matthias Meilland**®); flowers large, dbl., 15–25 petals, no fragrance; foliage medium, dark green, glossy, disease resistant; upright, floriferous growth; [(Mme Charles Sauvage X Fashion) X (Poppy Flash X Parador)]; Meilland, Mrs. Marie-Louise; SNC Meilland & Cie, 1985

'MEIFOTA', HT, mp, 1981; (**Celica**); flowers large, dbl., 20 petals, slight fragrance; foliage medium, dark, semi-glossy; upright growth; PP005683; [Unnamed seedling X (((Zambra X (Baccará X White Knight)) X Golden Garnette) X Unnamed seedling)]; Meilland, Mrs. Marie-Louise; Meilland Et Cie

'MEIFOUR', HT, dp, 1968; (Anne Marie, **Anne Marie Trechslin**®); bud long, pointed; flowers deep pink, large, dbl., exhibition form, intense fragrance; foliage dark, leathery; vigorous, branching growth; [Sutter's Gold X (Demain X Peace)]; Meilland; URS

'MEIFRISON', HT, op, 1968; (**Feria**); flowers coral suffused pink, large, dbl., globular, moderate fragrance; foliage leathery; upright growth; [(Grand Gala X Premier Bal) X Love Song]; Meilland; URS

MEIfructoz, F, ob, 1984; (**Fruité**); Meilland, L.; (Cavriglia)

'MEIFRUIJE', Min, yb, 1994; (**Apricot Sunblaze**®); flowers yellow edged in bright orange, very dbl., 25–40 petals, 1.5 in., exhibition form, borne mostly singly, no fragrance; few prickles; foliage small, dark green, glossy; medium, bushy growth, compact; PP9033; [(Mark One X Yellow Meillandina) X Gold Badge]; Meilland, Alain A.; The Conard-Pyle Co., 1994

'MEIGABI', Min, or, 1965; ('MEIGALI', **Starina**®); flowers orange-scarlet, small, dbl.; foliage glossy; vigorous, dwarf growth; ADR, Germany, 1971 GM, Japan, 1968 Miniature Rose Hall of Fame, ARS, 1999; [(Dany Robin X Fire King) X Perla de Montserrat]; Meilland, Mrs. Marie-Louise; C-P

'MEIGAFOR', HT, w, 1995; (**Confidence**™); Meilland

'MEIGALI', Min, or, 1965; ('MEIGABI', **Starina**®); flowers orange-scarlet, small, dbl.; foliage glossy; vigorous, dwarf growth; ADR, Germany, 1971 GM, Japan, 1968 Miniature Rose Hall of Fame, ARS, 1999; [(Dany Robin X Fire King) X Perla de Montserrat]; Meilland, Mrs. Marie-Louise; C-P

'MEIGALIL', HT, mr, 1971; (**Galileo**); flowers currant-red to cherry-red, dbl., 30 petals, 5 in., globular; foliage large, glossy; vigorous, upright growth; [Ma Fille X Love Song]; Meilland

'MEIGANDOR', HT, ab, 1977; (**Sabrina**); flowers apricot-orange, large, dbl., 35 petals; upright growth; PP004520; [(Sweet Promise X Golden Garnette) X ((Zambra X Suspense) X (King's Ransom X Whisky Mac))]; Meilland, Mrs. Marie-Louise; Meilland Et Cie

'MEIGAVESOL', F, ob, 1982; (**Tequila**®); flowers light yellow, overlaid with orange, stained carmine on outer, dbl., 20 petals, slight fragrance; foliage medium, dark, matt; bushy growth; [Poppy Flash X (Rumba X (Meikim X Fire King))]; Meilland, Mrs. Marie-Louise; Meilland Et Cie

'MEIGEKANU', F, or, 1978; (**La Sevillana**®); bud conical; flowers vermilion, medium, semi-dbl., 13 petals; foliage bronze; vigorous, bushy growth; PP006384; ADR, 1979; [((MEIbrim X Jolie Madame) X (Zambra X Zambra)) X ((Tropicana xTropicana) X (Poppy Flash X Rusticana))]; Meilland, Mrs. Marie-Louise; Meilland, 1982

MEIgekanusar, Cl F, or, 1997; (**La Sevillana**, **Climbing**); Meilland

'MEIGENE', Gr, mp, 1964; (**Garden State**); bud ovoid, pointed; flowers rose-pink, dbl., 42 petals, 3.5–4 in., moderate fragrance; foliage leathery; vigorous, tall, bushy growth; [(Happiness X Independence) X White Knight]; Meilland, Alain A.; C-P

'MEIGER', HT, or, 1954; (**Baccará**®, Jacqueline); flowers bright orange-red, dbl., 75 petals, 3 in., cupped; foliage dark, leathery; bushy, upright growth; [Happiness X Independence]; Meilland, F.; C-P, 1957;, URS, 1954

'MEIGERIUM', HT, or, 1975; flowers light vermilion-red, full, dbl., 35 petals, 4–4.5 in.; foliage glossy; very vigorous growth; [(Romantica X Tropicana) X ((Show Girl X Baccará) X Romantica)]; Paolino; URS

'MEIGEROKA', F, mp, 1984; (**Pink La Sevillana**®, Pink La Sevilliana, Pink Sevilliana, Rosy La Sevilliana); PP007117; GM, Baden-Baden, 1985 GM, Durbanville, 1986; [La Sevillana sport]; Meilland, Mrs. Marie-Louise; Meilland Et Cie, 1985

MEIgerokosar, Cl F, mp, 1993; (**Rosy La Sevillana**, **Climbing**); Meilland

MEIglassol, Min, lp; (**Tropico Sunblaze**); Meilland, 1995

'MEIGLUSOR', Min, dp, 1988; (**Concertino**™); flowers cardinal-pink, medium, dbl., 26–40 petals, no fragrance; foliage small, medium green, semi-glossy; bushy, compact growth; [(Anytime X Julita) X Lavender Jewel]; Meilland, Alain A.; SNC Meilland & Cie

'MEIGOLD', HT, dy, 1968; (Golden Prince, **Kabuki**®); flowers medium, dbl., 45 petals, exhibition form, moderate fragrance; foliage bronze, glossy, leathery; vigorous, upright growth; [(Monte Carlo X Bettina) X (Peace X Soraya)]; Meilland, Mrs. Marie-Louise; C-P;, URS

MEIgosar, Cl F, ob; (**Prince Igor**, **Climbing**); Meilland

MEIgoufin, F, 1986; (**Magali**); Meilland, Alain A.; (Cavriglia)

MEIgovin, Min, w, 1991; (**Snow Meillandina**®); Meilland

MEIgrelou, F, lp; (**Grisbi**); Meilland

'MEIGRISO', HT, rb, 1968; (Baronne de Rothschild, **Baronne Edmond de Rothschild**®); flowers ruby-red, whitish reverse, large, dbl., 40 petals, exhibition form, intense fragrance; foliage glossy, leathery; vigorous growth; GM, Lyon, 1968 GM, Rome, 1968; [(Baccará X Crimson King) X Peace]; Meilland; URS, 1969

'MEIGRISOSAR', Cl HT, rb, 1974; (Baronne de Rothschild, Climbing, **Baronne Edmond de Rothschild**, **Climbing**®, Grimpant Baronne de Rothschild); Meilland, Mrs. Marie-Louise

MEIgro, Cl F, my, 1991; (**Gold Badge**, **Climbing**); Meilland

'MEIGRONURI', F, my, 1978; (**Gold Badge**™, Gold Bunny®, Rimosa 79); bud conical; flowers lemon-yellow, dbl., 38 petals, 3 in., cupped; vigorous growth; PP004625; [Poppy Flash X (Charleston X Allgold)]; Paolino; Meilland

MEIgrouge, HT, mr; (**Preference**); Meilland

MEIgurami, F, op, 1976; (**Coppélia 76**); flowers medium, dbl., slight fragrance; Meilland; (Sangerhausen)

MEIhailoil, S, ab, 1995; (**Concerto**); Meilland

'MEIHAITI', Cl F, mr, 1970; (Iskra, 'MEIHATI', **Sparkling Scarlet**); flowers scarlet-red, small, semi-dbl., 13 petals, 2.5 in., moderate, fruity fragrance; foliage large; GM, Paris, 1969; [Danse des Sylphes X Zambra]; Meilland; URS

MEIhaitoil, S, lp; (**Romantic Sunrise**); Meilland

'MEIHAND', F, or, 1957; ('MEIRABANDE', **Sarabande**®); flowers light orange-red, stamens yellow, blooms in large trusses, semi-dbl., 13 petals, 2.5 in., cupped, slight fragrance; foliage semi-glossy; low, bushy growth; AARS, 1960 GM, Bagatelle, 1957 GM, Geneva, 1957 GM, Portland, 1958 GM, Rome, 1957; [Cocorico X Moulin Rouge]; Meilland, F.; C-P, 1959;, URS, 1957

'MEIHANDSAR', Cl F, or, 1968; (**Sarabande**, **Climbing**®); GM, Japan, 1968; Meilland; URS

'MEIHARTFOR', F, lp, 1970; (Kalinka, **Pink Wonder**); flowers imbricated, dbl., 28 petals, 3 in., intense fragrance; foliage large, glossy, leathery; vigorous, upright growth; GM, Belfast, 1972 GM, Madrid, 1969; [Zambra X (Sarabande X (Goldilocks X Fashion))]; Meilland; URS

'MEIHARTFORSAR', Cl F, lp, 1976; (Kalinka, Climbing, **Pink Wonder**, **Climbing**); Meilland, Mrs. Marie-Louise; Meilland

'MEIHATI', Cl F, mr, 1970; (Iskra, 'MEIHAITI', **Sparkling Scarlet**); flowers scarlet-red, small, semi-dbl., 13 petals, 2.5 in., moderate, fruity fragrance; foliage large; GM, Paris, 1969; [Danse des Sylphes X Zambra]; Meilland; URS

'MEIHELVET', Gr, pb, 1974; (**Sonia**, Sonia Meilland, Sweet Promise); bud long; flowers pink suffused coral to yellow, dbl., 30 petals, 4–4.5 in., exhibition form, intense, fruity fragrance; foliage glossy, dark, leathery; [Zambra X (Baccará X White Knight)]; Meilland; C-P

'MEIHELVETSAR', Cl Gr, pb, 1976; (Grimpant Sonia Meilland®, Sonia Meilland, Climbing, **Sonia**, **Climbing**, Sweet Promise, Climbing); Meilland, Mrs. Marie-Louise; Meilland

'MEIHERODE', Min, ob, 1968; (**Petite Folie**®); flowers vermilion, reverse carmine, small blooms in trusses, dbl., globular, slight, fruity fragrance; foliage leathery; vigorous growth; GM, Japan, 1969; [(Dany Robin X Fire King) X (Cricri X Perla de Montserrat)]; Meilland; URS

'MEIHIGOR', F, rb, 1970; (**Frenzy**, Prince Igor®); flowers nasturtium-red, reverse yellow, rounded, dbl., 25 petals, 2 in., moderate, fruity fragrance; foliage matt; vigorous, bushy growth; [(Sarabande X Dany Robin) X Zambra]; Meilland

'MEIHIMPER', HT, pb, 1971; (**Indian Song**, Preziosa); flowers rose, reverse gold, dbl., 40 petals, 5 in., exhibition form, slight fragrance; foliage glossy, dark; vigorous, upright growth; [(Radar X Karl Herbst) X Sabrina]; Meilland

MEIhirvin, HT, mr, 1988; (**Charlotte Rampling**®); Meilland

'MEIHIVANO', F, mp, 1971; (**Sweet Song**); dbl., 35 petals, 4.5 in., moderate, fruity fragrance; bushy growth; [Fidélio X Bettina]; Meilland

'MEIHOURAG', LCl, ob, 1991; (**Arielle Dombasie**®); Meilland

MEIhuterb, F, yb, 1995; (**Galaxy**); Meilland

MEIjadel, LCl, pb, 1989; (**Fandango**); Meilland

'MEIJALITA', F, or, 1977; (**Chorus**®, 'MEIJULITO', 'MEIMORE'); flowers vermilion-red, dbl., 35 petals, 4 in., slight, fruity fragrance; foliage glossy; vigorous growth; ADR, 1977; [Tamango X (Sarabande X Zambra)]; Paolino; URS, 1975

MEIjasonl, HT, ly; (**Lovely Girl**); Meilland

'MEIJENOR', HT, or, 1955; ('MEJENOR', **Soraya**®); bud pointed; flowers orange-red reverse crimson-red, large, dbl., 30 petals, cupped, slight fragrance; foliage glossy; vigorous, bushy growth; [(Peace X Floradora) X Grand'mere Jenny]; Meilland, F.; URS

'MEIJENORSAR', Cl HT, or, 1960; (Grimpant Soraya®, 'MEJENORSAR', **Soraya**, **Climbing**®); Barni, V.; URS

'MEIJETTE', Min, my, 1986; (French Liberty, **Innovation Minijet**); flowers medium, dbl., slight fragrance; foliage small, light green, glossy; bushy growth;

PP006177; [(Rumba X Carol Jean) X (Zambra X Darling Flame)]; Meilland, Mrs. Marie-Louise; SNC Meilland & Cie, 1987

'MEIJIDIRO', Min, mp, 1982; (**Pink Meillandina**®, Pink Sunblaze™); PP004961; [Orange Sunblaze sport]; Meilland, Mrs. Marie-Louise; Meilland Et Cie, 1980

'MEIJIKATAR', Min, or, 1982; (Orange Meillandina, **Orange Sunblaze**™, Sunblaze); flowers non-fading deep orange, very dbl., 40 petals, 1.5–2 in., cupped, borne 1-3 per cluster, slight fragrance; straw-brown prickles; foliage small, light green, matt; upright, bushy (15 in.) growth; drops spent blooms; PP004682; [Parador X (Baby Bettina X Duchess of Windsor)]; Meilland, Mrs. Marie-Louise; C-P, 1981

'MEIJIKATARSAR', Cl Min, or, 1986; (Grimpant Orange Meillandina®, **Orange Meillandina, Climbing**, Orange Sunblaze, Climbing); flowers medium, dbl., 26–40 petals, no fragrance; foliage small, medium green, matt; upright growth; PP006817; [Orange Meillandina sport]; Meilland, Mrs. Marie-Louise

MEIjulitasar, Cl F, or, 1986; (**Chorus, Climbing**); Meilland

'MEIJULITO', F, or, 1977; (**Chorus**®, 'MEIJALITA', 'MEIMORE'); flowers vermilion-red, dbl., 35 petals, 4 in., slight, fruity fragrance; foliage glossy; vigorous growth; ADR, 1977; [Tamango X (Sarabande X Zambra)]; Paolino; URS, 1975

'MEIKANS', F, or, 1959; (**Fire King**); bud ovoid; flowers fiery scarlet, blooms in clusters, dbl., 48 petals, 2.5 in., exhibition form, moderate, musk fragrance; foliage dark, leathery; vigorous, upright, bushy growth; AARS, 1960; [Moulin Rouge X Fashion]; Meilland; C-P, 1959;, URS, 1958

MEIkiji, F, w, 1989; (**Destiny**); Meilland

'MEIKINOSI', HT, yb, 1978; (Parador, **Tchin-Tchin**®); flowers chrome-yellow, dbl., 30 petals, 4.5 in., cupped; vigorous growth; [((Zambra X Suspense) X King's Ransom) X (Kabuki X Dr. A.J. Verhage)]; Paolino; URS

MEIklusy, F, rb, 1993; (**Marianne**); Meilland

MEIkola, HT, mp; (**Livia**); Meilland

'MEIKROTAL', S, mr, 1987; (**Scarlet Meidiland**™, Scarlet Meillandécor); flowers light cherry red, reverse dark carmine pink, small, dbl., 15–25 petals, blooms in clusters of 10-15, no fragrance; foliage medium, dark green, glossy; spreading growth; PP006087; [MEItiraca X Clair Matin]; Meilland, Mrs. Marie-Louise; SNC Meilland & Cie, 1985

MEIkruza, HT, w; (**Arianna**); Meilland

MEIkyster, F, dp; (**Trudy Mimi**™); Meilland, 1992

'MEILANEIN', HT, m, 1974; (**Charles de Gaulle**®, Katherine Mansfield); flowers lilac, dbl., 38 petals, 3.5–4 in., cupped, intense fragrance; vigorous growth; [(Sissi X Prelude) X (Kordes' Sondermeldung X Caprice)]; Meilland, Mrs. Marie-Louise; URS

'MEILANODIN', F, or, 1976; (**Paso Doble**); flowers geranium-red, 9 petals, 3 in., slight fragrance; vigorous growth; Paolino; URS

'MEILARCO', Min, lp, 1986; (Lady Meillandina, **Lady Sunblaze**™, Peace Meillandina®, Peace Sunblaze); bud plump, pointed; flowers pale orient pink to light coral pink, medium, very dbl., 40 petals, 1.5 in., exhibition form, borne singly, no fragrance; foliage small, dark green, glossy; bushy growth; PP006170; [(Fashion X Zambra) X Belle Meillandina]; Meilland, Mrs. Marie-Louise, 1987

MEIlasso, F, pb, 1989; (**Cocorico**®); Meilland

'MEILAUR', F, mr, 1971; (Florian, **Tender Night**); dbl., 25 petals, 4 in., slight, fruity fragrance; foliage matt, dark; upright growth; GM, Rome, 1971; [Tamango X (Fire King X Banzai)]; Meilland

'MEILAURSAR', Cl F, mr, 1976; (Florian, Climbing, **Tender Night, Climbing**); Meilland, Mrs. Marie-Louise

'MEILAVIO', HT, dr, 1998; (**Traviata**); flowers dark red, very full (41+ petals), large (10–11 cms) blooms borne mostly single; slight fragrance; moderate prickles; large foliage, dark green, glossy; upright, medium (60–80 cms) growth; [(Porta Nigra X Paola) X William Shakespeare]; Selection Meilland; The Conard Pyle Co., 1997

'MEILÉNA', F, or, 1971; (**Poppy Flash**, Rusticana); flowers vermilion, dbl., 20 petals, 3 in., slight, fruity fragrance; vigorous, bushy growth; GM, Geneva, 1970 GM, Rome, 1972; [(Dany Robin X Fire King) X (Alain X Mutabilis)]; Meilland, 1972

MEIlenangal, HT, 1981; (**Mitzi**); Meilland; (Cavriglia)

'MEILÉNASAR', Cl F, or, 1975; (**Poppy Flash, Climbing**, Rusticana, Climbing); Paolino; URS

MEIliaxi, S, lp, 1989; (**Schuss**®); Meilland

MEIlicafal, HT, 1983; (**Cannes Festival 83**); Meilland, Alain A.; (Cavriglia)

'MEILIDER', LCl, dp, 1985; (All In One, **Exploit**®, Grimpant Exploit®); flowers deep pink, medium, dbl., 20 petals, no fragrance; foliage small, medium green, matt; very vigorous, spreading, climbing growth; [Fugue X Sparkling Scarlet]; Meilland, Mrs. Marie-Louise, 1983

'MEILIE', HT, mr, 1958; (**Christian Dior**); bud ovoid, pointed; flowers clear true red, high centered, very dbl., 45–50 petals, 4–4.5 in., exhibition form, borne mostly singly, slight, soft spice fragrance; foliage leathery, glossy; long stems; vigorous, upright, bushy growth; AARS, 1962 GM, Geneva, 1958; [(Independence X Happiness) X (Peace X Happiness)]; Meilland, F.; C-P, 1961;, URS, 1958

'MEILIMONA', HT, rb, 1961; (**Banzai**); bud ovoid; flowers red, pink and cream blend, dbl., 30 petals, 3.5 in., exhibition form, slight fragrance; foliage leathery, dark; vigorous, bushy growth; [Radar X Caprice (Meilland)]; Meilland, Mrs. Marie-Louise; C-P

'MEILIMONA', HT, my, 1976; (**Banzai**, Banzai '76); flowers canary-yellow, full, dbl., 35 petals, 4 in., cupped, slight fragrance; vigorous growth; PP004013; [Coed X ((Seedling X Seedling) X Verla)]; Paolino; Meilland

'MEILISIA', HT, mp, 1968; ('MEILISTA', **Princess Margaret of England**, Princesse Margaret d'Angleterre®); flowers phlox-pink, large, dbl., exhibition form, slight fragrance; foliage leathery; vigorous, upright growth; GM, Portland, 1977; [Queen Elizabeth X (Peace X Michele Meilland)]; Meilland, Mrs. Marie-Louise; URS, 1969

'MEILISTA', HT, mp, 1968; ('MEILISIA', **Princess Margaret of England**, Princesse Margaret d'Angleterre®); flowers phlox-pink, large, dbl., exhibition form, slight fragrance; foliage leathery; vigorous, upright growth; GM, Portland, 1977; [Queen Elizabeth X (Peace X Michele Meilland)]; Meilland, Mrs. Marie-Louise; URS, 1969

'MEILIVAR', HT, dy, 1997; (**Gina Lollobrigida**); flowers full, large, very dbl., 70–90 petals, borne mostly singly, moderate fragrance; some prickles; foliage medium, medium green, semi-glossy; upright, medium (5ft.) growth; [MEIdragelac X MEIkinosi]; Meilland International SA

Meilland Decor Arlequin® *see* 'MEIZOURAYOR'

Meilland Decor Rose *see* 'MEITURAPHAR'

Meilland Rosiga '83 *see* 'MEIBALBIKA'®

Meillandina *see* 'MEIROV'

'MEILMERA', Min, w, 1996; (**Bridal Sunblaze**®); very dbl., 30–40 petals, 2 in., borne in small clusters, slight fragrance; few prickles; foliage large, dark green, dull; compact, medium (18 in) growth; PP10002; [(MEIringa X Schneewitchen) X (MEIzogrel X MEIlarco)]; Meilland International SA; The Conard-Pyle Co., 1996

MEIloise, HT, op, 1985; (**Reve de Paris**®); Meilland

MEIlomit, F, ob, 1996; (**Marie Curie**); Meilland

MEIlomit, S, lp; (**Romantic Dreams**); Meilland

'MEILONTIG', S, w, 1987; (**Repens Meidiland**); flowers medium, single, 5 petals, no fragrance; foliage medium, light green, glossy; spreading, strong growth; PP006598; [Swany X New Dawn]; Meilland, Mrs. Marie-Louise, 1985

'MEILOTUP', HT, dr, 1986; (**Royal Velvet**™); flowers large blooms borne usually singly, dbl., 28 petals, exhibition form, slight fragrance; globular, small, medium green fruit; large, straw-colored prickles; medium, bushy growth; [(Exciting X Suspense) X Duke of Windsor]; Meilland, Mrs. Marie-Louise; Wayside Gardens Co.

MEIlsinplox, HT, my, 1994; (**Golden Nugget**™); Meilland

'MEILUCCA', Min, or, 1971; (**Darling Flame**, Minuette, Minuetto); flowers mandarin-red to vermillion-red, yellow anthers, small, dbl., 25 petals, 1.5 in., globular, slight, fruity fragrance; foliage glossy, dark; vigorous growth; [(Rimosa X Josephine Wheatcroft) X Zambra]; Meilland

'MEILUCRE', HT, or, 1971; (Arturo Toscanini, **Elegy**); flowers vermilion, dbl., 30 petals, 5.5 in., globular, slight fragrance; foliage semi-matt, dark; vigorous growth; [((Happiness X Independence) X Sutter's Gold) X ((Happiness xIndependence) X Suspense))]; Meilland

'MEILUMINAC', HT, dr, 1979; (President L. Senghor®, **Président Leopold Senghor**®); bud conical; flowers large, dbl., 25 petals, cupped; foliage glossy, dark; vigorous, bushy growth; Meilland, Mrs. Marie-Louise; Meilland

MEIluminacsar, Cl HT, dr; (**President Leopold Senghor, Climbing**); Meilland

MEIlupin, HT, lp; (**Starlite**); Meilland, 1989

'MEILUSAM', HT, mr, 1973; (Lusambo, **Red Rock**, Red Rocky); flowers cherry-red, open, imbricated, dbl., 35 petals, 5 in., slight fragrance; very vigorous growth; [(Royal Velvet X Chrysler Imperial) X Pharaoh]; Meilland; URS

'MEILUTIDA', F, or, 1977; (**Muchacha**®); bud conical; flowers brilliant vermilion, shallowly-cupped, medium, semi-dbl., 12 petals, cupped; foliage dark; very vigorous, semi-shrub growth; Golden Rose, The Hague, 1976; [(Frenzy X Frenzy) X (Sangria X Sangria)]; Meilland, Mrs. Marie-Louise; Meilland

'MEIMAFRIS', HT, w, 1986; (**Jardins de Bagatelle**®, Sarah); flowers large, very dbl., intense fragrance; foliage large, medium green, semi-glossy; upright growth; GM, Genoa, 1987 Golden Rose, Geneva, 1984; [(Queen Elizabeth X Eleg) X MEIdragelac]; Meilland, Mrs. Marie-Louise; SNC Meilland & Cie, 1987

'MEIMAGARMIC', HT, w, 1982; (Grace Kelly, Preference, Princess Grace, Princess of Monaco, **Princesse de Monaco**®); flowers petals cream, edged pink, large, dbl., 35 petals, exhibition form, moderate fragrance; foliage large, dark, glossy; upright, bushy growth; PP005067; [Ambassador X Peace]; Meilland, Mrs. Marie-Louise; Meilland Et Cie, 1981

'MEIMAGUL', Min, rb, 1994; (**Gypsy Sunblaze**™); flowers red/yellow bicolor, blooms borne mostly single, dbl., 15–25 petals, 1.5 in., no fragrance; some prickles; foliage small, light green, matt; medium, bushy growth; [(Bonfire Night X MEIalfi) X Tapis Jaune]; Meilland, Alain A.; The Conard-Pyle Co., 1993

'MEIMAINGER', HT, ob, 1987; (Bushveld Dawn, **Cary Grant**™); flowers vivid orange blend, lighter at petal base, very dbl., 35–40 petals, 5 in., exhibition form, borne singly, open slowly, intense, spicy fragrance; ovoid, green red-orange fruit; slightly recurved, light green-straw prickles; foliage medium, dark green, glossy; upright, medium growth; PP006792; [(Pharaoh X Königin der Rosen) X ((Zambra X Suspense) X King'sRansom)]; Meilland, Mrs. Marie-Louise; Conard-Pyle, 1989;, Wayside Gardens Co., 1989

'MEIMAJ', HT, yb, 1954; (**Lady Elgin**, Thaïs); bud ovoid; flowers buff-yellow washed pink, large, dbl., 40 petals, cupped, moderate fragrance; foliage dark, leathery; vigorous, upright, bushy growth; [Mme Kriloff X (Peace X Geneve)]; Meilland, F.; C-P, 1957;, URS, 1954

'MEIMAL', Min, ob, 1958; (Colibre, **Colibri**); bud ovoid; flowers bright orange-yellow, small blooms in clusters, dbl., slight fragrance; foliage glossy; bushy growth; Golden Rose of The Hague, The Hague, 1962; [Goldilocks X Perla de Montserrat]; Meilland, F.; URS

'MEIMALYNA', HT, mr, 1984; (New Rouge Meilland, **Rouge Meilland**); flowers large, dbl., no fragrance; foliage large, dark, semi-glossy; upright growth; [((Queen Elizabeth X Karl Herbst) X Pharoah) X Antonia Ridge]; Meilland, Mrs. Marie-Louise; Meilland Et Cie, 1982

MEImanoir, LCl, mp, 1994; (**Blossom Magic**); Meilland

'MEIMAUR', HT, dr, 1973; (Joséphine Baker, **Velvet Flame**); dbl., 30 petals, 5–5.5 in., slight fragrance; foliage dark; vigorous growth; [Tropicana X Papa Meilland]; Meilland; URS

'MEIMEX', HT, r, 1959; (**Fantan**); bud urn-shaped; flowers burnt-orange to yellow-ochre, dbl., 48 petals, 3.5 in., cupped, slight fragrance; foliage leathery; moderate growth; [(Pigalle X Prelude) X Self]; Meilland, F.; C-P;, URS

'MEIMICK', S, rb, 1961; (**Cocktail**®); bud pointed; flowers geranium red, base primrose yellow, medium, single, 5 petals, 2.5 in., borne in clusters, slight, spicy fragrance; foliage leathery, glossy; vigorous, semi-climbing growth; shrub or hedge; [(Independence X Orange Triumph) X Phyllis Bide]; Meilland, F.; C-P, 1961;, URS, 1957

'MEIMIT', HT, lp, 1956; (**Grace de Monaco**®); flowers light rose-pink, well-formed, large, dbl., intense fragrance; foliage leathery; vigorous, bushy growth; [Peace X Michele Meilland]; Meilland, F.; URS

'MEIMODAC', S, mp, 1994; (**Royal Bonica**™); dbl., 15–25 petals, 1.5–2.75 in., borne in small clusters, slight fragrance; some prickles; foliage medium, medium green, semi-glossy; medium, upright growth; [MEIdomonac sport]; Meilland, Alain A.; The Conard-Pyle Co., 1993

'MEIMONBLAN', F, ob, 1999; (**Marmalade Skies**™); flowers medium, dbl., 17–25 petals, borne in large clusters, slight fragrance; prickles moderate; foliage medium, dark green, glossy; bushy, medium (3-4 ft) growth; AARS, U.S., 2001; [(Tamango X Parador) X Patricia]; Selection Meilland; Conard-Pyle, 2001

'MEIMONT', LCl, mp, 1960; (**Clair Matin**®, Grimpant Clair Matin); bud pointed; cupped, 15 petals, 2–3 in., cupped, blooms in rounded clusters, moderate, sweetbriar fragrance; foliage dark, leathery; vigorous (10-12 ft), well-branched growth; GM, Bagatelle, 1960; [Fashion X ((Independence X Orange Triumph) X Phyllis Bide)]; Meilland, Mrs. Marie-Louise; C-P, 1963;, URS, 1960

'MEIMORE', F, or, 1977; (**Chorus**®, 'MEIJALITA', 'MEIJULITO'); flowers vermilion-red, dbl., 35 petals, 4 in., slight, fruity fragrance; foliage glossy; vigorous growth; ADR, 1977; [Tamango X (Sarabande X Zambra)]; Paolino; URS, 1975

'MEIMOUSLIN', Gr, mr, 1993; (**Cardinal Song**™); very dbl., 3–3.5 in., borne mostly singly, slight fragrance; some prickles; foliage large, dark green, glossy; medium, semi-erect growth; [Olympiad X (Michel Lis Le Jardinier X Red Lady)]; Selection Meilland; The Conard-Pyle Co., 1992

MEImucas, HT, op, 1995; (**Arioso**); Meilland

MEImuide, F, op, 1990; (**Holstentor**); flowers salmon-pink, large, dbl.; Meilland; (Sangerhausen)

MEImurge, Gr, ob, 1997; (**Lasting Peace**); Meilland

Mein München, F, yb, 1987; flowers yellow and red, large, dbl., slight fragrance; Cocker; (Sangerhausen)

'MEIN RUBIN', HT, dr, 1984; flowers urn-shaped, well-formed, blooms borne usually singly, dbl., 35 petals; medium, lavender prickles; foliage medium, medium green, semi-glossy; medium, bushy growth; [(Helene Schoen X Charles Mallerin) X (Helene Schoen X CharlesMallerin)]; Teranishi, K.; Itami Rose Nursery

Mein Schoner Garten *see* KORmeneint

'MEINABRON', Min, w, 1988; (**Alba Meillandina**®); flowers medium, very dbl., no fragrance; foliage small, medium green, semi-glossy; bushy growth; PP007347; [MEIdonq X (Darling X Jack Frost)]; Meilland, Mrs. Marie-Louise; SNC Meilland & Cie, 1987

MEInagre, F, rb; (**Wapiti**®); Meilland, 1988

MEInalpir, HT, pb, 1997; (**Blue Bell**); Meilland

MEInartemi, F; Meilland; (Cavriglia)

'MEINARVAL', HT, my, 1974; (**Sun King**, Sun King '74); flowers large, dbl., 30 petals, exhibition form; foliage small; vigorous, upright growth; [(Soroya X Signora) X King's Ransom]; Paolino; URS, 1972

'MEINASTUR', HT, or, 1975; (**Alpha**); flowers bright vermilion, dbl., 20 petals, 4 in.; foliage leathery; [((Show Girl X Baccará) X Romantica) X (Romantica X Tropicana)]; Paolino; URS

'MEINATAC', HT, lp, 1972; (**Susan Hampshire**); flowers light fuchsia-pink, dbl., 40 petals, 5.5 in., globular, intense fragrance; foliage matt; vigorous, upright growth; [(Monique X Symphonie) X Miss All-American Beauty]; Paolino; URS

Meindeert Hobbema, S, rb, 1997; Williams, J. Benjamin

'MEINEBLE', S, rb, 1989; (**Red Meidiland**™, Rouge Meillandécor®); bud conical; flowers red with white eye, medium, single, 5 petals, cupped, borne in sprays of 7-15, repeat bloom, no fragrance; globular, small, red fruit; prickles gray-brown; foliage medium, dark green, glossy, disease resistant; spreading, medium, very winter hardy growth; PP007116; [Sea Foam X (Picasso X Eyepaint)]; Meilland, Alain A.; The Conard-Pyle Co., 1989

MEInececa, F, mr; (**Tempo**); Meilland, 1978

'MEINEREAU', HT, yb, 1996; (**Centennial Star**™); very dbl., 4.75 in., moderate fragrance; foliage large, dark green, semi-glossy; medium growth; [(Peace X Landora) X Kings Ransom]; Meilland International SA; Conard-Pyle Co., 1996

'MEINEYTA', Min, ob, 1994; (**Cumba Meillandina**®, Fiery Sunblaze); very dbl., 1.5–2.75 in., borne in small clusters, no fragrance; some prickles; foliage large, medium green, semi-glossy; medium, bushy growth; [Orange Honey X (Darling Flame X Tapis Jaune)]; Meilland, Alain A.; SNC Meilland & Cie, 1992

'MEINIACIN', HT, mp, 1991; (Fragrant Lady, **Perfume Beauty**™); bud conical; flowers rose bengal, blooms borne singly, very dbl., cupped, moderate fragrance; foliage large, dark green; upright, tall growth; PP007819; [MEIlista X (Carina X Silvia)]; Meilland, Alain A., 1990; The Conard-Pyle Co., 1990

MEInibur, LCl, ab, 1995; (**Kir Royal**); Meilland

MEInical, HT, dr, 1999; (**Edith Piaf**); Meilland

'MEINIMO', HT, lp, 1994; (21 Again!, Penny Coelen, Prestige de Lyon, **Regatta**™); dbl., 26–40 petals, 3–3.5 in., borne mostly singly, moderate fragrance; few prickles; foliage large, dark green, matt; medium, bushy growth; PP8390; Geneva, 1989; [MEIgurami X (MEInaregi X MEIdragelac)]; Meilland, Alain A.; SNC Meilland & Cie, 1992;, The Conard-Pyle Co., 1994

'MEININRUT', Min, or, 1994; (**Orange Symphonie**®); flowers orange vermillion, blooms borne mostly single or in small clusters, dbl., 26–40 petals, 1.5–2.75 in., no fragrance; some prickles; foliage medium, dark green, semi-glossy; low (35-45 cms), bushy growth; [(Baby Bettina X Anytime) X Meteor]; Meilland, Alain A.; SNC Meilland & Cie, 1993

MEInirlo, S, mr, 1992; (**Rote Woge**); Meilland

MEInitper, HT, ab, 1986; (**Caramella**®); Meilland

'MEINIVOZ', HT, yb, 1994; (**Paul Ricard**, Paul Richard, Spirit of Peace); flowers amber yellow, blooms, very dbl., 3–3.5 in., intense fragrance; foliage large, medium green, matt; tall (110-120 cms), upright growth; PPAF; GM, Rome, 1991; [(Hidalgo X Mischief) X Ambassador]; Meilland, Alain A.; SNC Meilland & Cie, 1990;, The Conard-Pyle Co., 1994

'MEINOFRAI', Min, dr, 1990; (**Red Mini-Wonder**™); bud rounded; flowers currant red, cardinal red reverse, aging to dark red, small, very dbl., 40–43 petals, cupped; prickles small, green to tan; foliage small, medium green, very dense, semi-glossy; bushy, low growth; PP007384; [(Anytime X Parador) X Mogral]; Selection Meilland, 1987; The Conard-Pyle Co., 1989

MEInoiral, S, mp, 1990; (**Deborah**®); Meilland

'MEINOMAD', HT, yb, 1992; (**Desert Peace**); flowers yellow tinged with red, blooms borne mostly singly, dbl., 15–25 petals, 1.5–2.75 in., slight fragrance; foliage large, dark green, glossy; tall (120-130 cms), upright growth; [(Sonia X Rumba) X (Piccadilly X Chicago Peace)]; Selection Meilland; C-P

MEInopia, HT, pb, 1992; (**Madona**); Meilland

MEInorep, HT, pb, 1995; (**James Pereire**); Meilland

'MEINRONSSE', HT, ob, 1992; (Christoph Colombus, Christophe Colomb, **Christopher Columbus**, Cristobal Colon, Cristoforo Colombo, 'MEIRONSSE'); flowers orange blend/copper, blooms borne mostly singly, dbl., 26–40 petals, 3–3.5 in., slight fragrance; many prickles; foliage large, dark green, semi-glossy; medium (120-140 cms), upright growth; GM, Durbanville, 1990; [MEIgurani X (Ambassador X MEInaregi)]; Selection Meilland; France, 1990;, Germany;, Italy, 1991;, Spain, 1991;, The Conard-Pyle

'MEINUSIAN', F, mr, 1998; (**Francois Rabelais**™); flowers bright red, very full, very dbl., 41 petals, 2–2.5 in., borne in small clusters, slight fragrance; prickles moderate; foliage medium, dark green, glossy; bushy, medium, 3 ft. growth; [(MEIchanso X MEIfolio) X KORlima]; Selection Meilland; Conard-Pyle Co., 1996

MEInustrel, F; Meilland; (Cavriglia)

'MEINUZETEN', HT, ob, 1979; (**Ambassador**®); bud conical; flowers orange-red, reverse blended with golden yellow (orange-apricot), large, dbl., 33 petals, 4 in., cupped; foliage dark, glossy; PP004224; [Seedling X Whisky Mac]; Meilland; C-P

'MEIOFFIC', F, pb, 1994; (Johann Straus, **Johann Strauss**®, Sweet Sonata); flowers orient pink slightly suffused with aureolin yellow, blooms b, very dbl., 3–3.5 in., slight fragrance; few prickles; foliage medium, dark green, semi-glossy; low (50-60 cms), bushy, compact growth; [Flamingo X (Pink Wonder X Tip Top)]; Meilland, Alain A.; SNC Meilland & Cie, 1993

MEIonagre, HT, mr, 1995; (**Michaelhouse Centenary**); Meilland

MEIouscki, S, dr, 1996; (**Crimson Meillandecor**); Meilland

'MEIPAL', HT, op, 1953; (**Bettina**®, MEpal); flowers salmon-orange, veined, well-formed, dbl., 37 petals, 4 in., moderate fragrance; foliage dark, glossy, bronze; vigorous growth; [Peace X (Mme Joseph Perraud X Demain)]; Meilland, F.; URS

'MEIPALEO', HT, my, 1994; (**Marco Polo**); dbl., 26–40 petals, 3–3.5 in., borne mostly singly, moderate fragrance; some prickles; foliage medium green, semi-glossy; tall, upright growth; [Ambassador X (Sunlight X Oregold)]; Meilland, Alain A.; SNC Meilland & Cie, 1992;, The Conard-Pyle Co., 1994

'MEIPARADON', HT, mr, 1976; (**Antonia Ridge**®); flowers cardinal-red, dbl., 30 petals, 4–4.5 in., exhibition form, slight fragrance; vigorous growth; [(Chrysler Imperial X Karl Herbst) X Seedling]; Paolino; URS

'MEIPARNIN', HT, pb, 1998; (**Honore de Balzac**®, Romantic Days); flowers light creamy yellow suffused with carmine, very full, 41 petals, 4.5 in., borne in large clusters, moderate fragrance; prickles moderate; foliage medium, medium green, semi-glossy; upright, bushy, medium, 3 ft. growth; [(Marion Foster X KORax) X Lancome]; Selection Meilland; Conard-Pyle Co., 1996

MEIparos, HT, mp, 1999; (**Perfumella**); Meilland

MEIpazdia, HT, w, 1998; (**Poker**); Meilland

'MEIPELTA', S, dp, 1994; (Cyclamen Meillandecor, **Fuchsia Meidiland**™, Fuchsia Meillandecor®); flowers deep pink, open flat, semi-dbl., 6–14 petals, 1.5–2.75 in., borne in large clusters, slight fragrance; few prickles; foliage medium, light green, glossy; low, spreading growth; PP8839; [MEItiraca X Clair Matin]; Meilland, Alain A.; The Conard-Pyle Co., 1993

MEIpeluj, LCl, lp; (**Sorbet**); Meilland, 1993

'MEIPERATOR', F, mr, 1971; (**Crimson Wave**, Imperator); flowers cardinal-red, shaded cherry, dbl., 25–30 petals, 4–5 in., slight, apple fragrance; foliage large, semi-matt, dark; vigorous, upright growth; [Zambra X (Sarabande X (Goldilocks X Fashion))]; Meilland

'MEIPIERAR', HT, w, 1993; (Cameo Cream, **Caroline de Monaco**); flowers cream white, blooms borne mostly single, very dbl., 3–3.5 in., no fragrance; foliage large, dark green, semi-glossy; medium, bushy growth; [Chicago Peace X Tchin-Tchin]; Meilland, Alain A., 1989; Conard-Pyle, 1993, USA;, The Conard-Pyle Co., 1993

'MEIPIESS', Min, mr, 1986; (**Majorette**®, Majorette 86); flowers medium cardinal-red, moderately medium, dbl., 15–25 petals, no fragrance; fruit produced in autumn; foliage small, medium green, semi-glossy; mini-flora; bushy growth; [Magic Carrousel X (Grumpy X Scarletta)]; Meilland, Mrs. Marie-Louise, 1985

'MEIPINJID', Min, mp, 1985; (**Classic Sunblaze**®, Duc Meillandina, Duke Meillandina); flowers rich pink, very double, big, very dbl., 40 petals, 1.5–2 in., borne mostly singly, slight fragrance; foliage medium, dark green; short, bushy, 15 in. growth; PP005958; [Pink Sunblaze sport]; Meilland, Mrs. Marie-Louise; SNC Meilland & Cie

'MEIPITAC', S, pb, 1990; (**Carefree Wonder**™); bud pointed; flowers medium pink with light pink reverse, aging to medium pink, large, dbl., 26 petals, cupped, borne in sprays of 1-4, slight fragrance; oval, reddish-brown fruit; prickles narrow, reddish; foliage medium, medium green, semi-glossy; bushy, medium growth, suitable for hedges.; hardy; PP007783; AARS, 1991; [(Prairie Princess X Nirvana) X (Eyepaint X Rustica)]; Selection Meilland, 1978; The Conard-Pyle Co., 1991

'MEIPLATIN', S, lp, 1989; (**Pearl Meidiland**™, Perle Meillandécor®); bud ovoid; flowers light ochre pink, aging white, medium, dbl., 30 petals, flat, borne in sprays of 3-15, repeat bloom, no fragrance; globular, small dish, reddish fruit; prickles small, reddish-brown; foliage medium, dark green, glossy; spreading, low growth; PP006807; [(Sea Foam X MEIsecaso) X Sea Foam]; Meilland, Alain A.; The Conard-Pyle Co., 1989

MEIplovon, HT, mp; (**Mademoiselle**); Meilland

'MEIPOBIL', HT, dp, 1987; (**Rendez-vous**); Meilland

'MEIPONAL', Min, ob, 1986; (Sunblaze, **Sunny Meillandina**®, Sunny Sunblaze); flowers yellow-orange, medium, dbl., 35 petals, no fragrance; foliage small, dark, matt; bushy growth; PP006810; [(Sarabande X Moulin Rouge) X (Zambra X Meikim)]; Meilland, Mrs. Marie-Louise; Meilland, 1985

'MEIPOPUL', S, mp, 1994; (**Douceur Normande**®, Sandton City, Goose Fair); flowers scarlet pink, blooms borne in small clusters, single, 5 petals, 1.5–2.75 in., no fragrance; many prickles; foliage small, medium green, semi-glossy; medium (70-90 cms), bushy growth; [(Immensee X Green Snake) X (Temple Bells X Red Cascade)]; Meilland, Alain A.; SNC Meilland & Cie

'MEIPOQUE', S, pb, 1984; (**Pink Meidiland**®, Schloss Heidegg); flowers deep pink, white eye, medium, single, 5 petals, no fragrance; foliage small, medium green, semi-glossy; bushy growth; PP005956; ADR, 1987; [Anne de Bretagne X Nirvana]; Meilland, Mrs. Marie-Louise; Meilland & Son, 1985

'MEIPOTAL', S, pb, 1994; (Bingo Meidiland®, Bingo Meillandecor®, **Carefree Delight**™); flowers carmine pink with white eye, blooms borne in large clusters, single, 5 petals, 1.5–2.75 in., no fragrance; many prickles; foliage small, dark green, glossy; medium (70-80 cms), bushy growth; PPAF; AARS, 1996; [(Eyepaint X Nirvana) X Smarty]; Meilland, Alain A.; SNC Meilland & Cie, 1991;, The Conard-Pyle Co., 1994

MEIpraserpi, HT, op, 1981; (**Catherine Deneuve**®); Meilland

'MEIPSIDUE', S, mr, 1999; (**Fire Meidiland**™); flowers fire engine red, opens flat, dbl., 17–25 petals, 2.5–3 in., borne in small clusters; foliage dark green, glossy; mounding, ground cover growth; Meilland

MEIpsilon, HT, rb; (**Yakimour**®); Meilland, 1980

'MEIPUMA', F, rb, 1975; (**Scherzo**®); flowers bright scarlet, reverse white and crimson, spiraled, dbl., 40 petals, 3.5 in., exhibition form; foliage dark; vigorous, bushy growth; GM, Belfast, 1975; [Tamango X Frenzy]; Paolino; URS

MEIqualis, HT, dr, 1995; (**Grand Gala**); Meilland

'MEIRABANDE', F, or, 1957; ('MEIHAND', **Sarabande**®); flowers light orange-red, stamens yellow, blooms in large trusses, semi-dbl., 13 petals, 2.5 in., cupped, slight fragrance; foliage semi-glossy; low, bushy growth; AARS, 1960 GM, Bagatelle, 1957 GM, Geneva, 1957 GM, Portland, 1958 GM, Rome, 1957; [Cocorico X Moulin Rouge]; Meilland, F.; C-P, 1959;, URS, 1957

'MEIRADIA', Min, mp, 1971; (Lutin, **Rosy Gem**); [Scarlet Gem sport]; Meilland, 1973

'MEIRANDIVAL', HT, yb, 1980; [Meibiranda sport]; Meilland, Mrs. Marie-Louise; Meilland Et Cie

'MEIRANOGA', Min, or, 1975; (**Brilliant Meillandina**); semi-dbl., 15 petals, cupped, blooms in clusters of 3-20, no fragrance; foliage dark, semi-glossy; [Parador X (Baby Bettina X Duchess of Windsor)]; Meilland, Mrs. Marie-Louise; Meilland Et Cie

'MEIRANOVI', S, rb, 1982; (**Candy Rose**®); flowers deep pink, reverse medium red, medium, dbl., 20 petals, no fragrance; foliage small, medium green, semi-glossy; spreading growth; PP006385; [(R. sempervirens X Mlle Marthe Carron) X ((Lilli Marleen X Evelyn Fison) X (Orange Sweetheart X Fruhlingsmorgen))]; Meilland, Mrs. Marie-Louise, 1980; Meilland Et Cie, 1983

MEIrasimac, S, mr, 1983; (**Royal Show**); GM, Durbanville, 1984; Meilland

MEIrebuc, F, w, 1995; (**Liz**); Meilland

MEIrecrom, HT, ob, 1999; (**Exotica**); Meilland

'MEIRED', HT, mr, 1972; (**Visa**); flowers turkey-red, dbl., 38 petals, 5 in., exhibition form, slight fragrance; foliage large, leathery; vigorous, upright growth; [(Baccará X Queen Elizabeth) X Lovita]; Meilland

MEIreibat, F, 1993; (**Concorde**); Meilland, Alain A.; (Cavriglia)

'MEIRESTIF', HT, ob, 1987; (Charleston 88, **Louis de Funès**®); flowers orange capucine reverse cadmium yellow, large, dbl., 15–25 petals, slight fragrance; foliage medium, dark green, glossy; upright, strong growth; [(Ambassador X Whisky Mac) X (Arthur Bell X Kabuki)]; Meilland, Mrs. Marie-Louise, 1983

'MEIREVOLT', HT, my, 1994; (**Toulouse Lautrec**®); very dbl., 3–3.5 in., borne mostly singly, no fragrance; some prickles; foliage medium, medium green, glossy; medium (70-80 cms), bushy growth; [Ambassador X (King's Ransom X Sunblest)]; Meilland, Alain A.; SNC Meilland & Cie, 1992

MEIrgano, HT, mr, 1992; (**Porto Nigra**); Meilland

'MEIRIANOPUR', F, or, 1979; (**Orange Bunny**®); bud pointed; flowers orange-red, reverse darker, cup-shaped, borne 1-25 per cluster, semi-dbl., 13 petals, slight fragrance; foliage bronze, matt, very dense; bushy growth; [Scherzo X (Sarabande X Frenzy)]; Meilland, Mrs. Marie-Louise; Meilland Et Cie

'MEIRIDGE', F, yb, 1963; (**Charleston**); bud pointed; flowers yellow flushed crimson, becoming crimson, blooms in clusters, dbl., 20 petals, 3 in., slight fragrance; foliage dark, leathery, glossy; upright, compact growth; [Masquerade X (Radar X Caprice)]; Meilland, Alain A.; URS;, Wheatcroft & Sons;, Wheatcroft Bros.

MEIridol, F, dp, 1983; (**Irene**); Meilland

'MEIRIDORIO', HT, yb, 1984; (**Gilbert Bécaud**); flowers orange and yellow blend, large, dbl., 45 petals, slight fragrance; foliage bronze, matt; upight growth; [(Peace X Mrs John Laing) X Bettina]; Meilland, Mrs. Marie-Louise; Meilland Et Cie, 1979

'MEIRIENTAL', mr, 1997; (**Paprika**™); flowers bright red, large, dbl., 20–25 petals, borne in clusters, slight fragrance; foliage large, deep green, leathery, glossy; climbing (8–10 ft.) growth; PP9537; [Marchenland X Red Favorite]; Meilland

'MEIRIGALU', HT, rb, 1968; (**Colorama**, Colourama, Dr R. Maag); bud ovoid; flowers red and yellow, large, dbl., cupped, moderate fragrance; foliage very glossy; vigorous, upright, bushy growth; [Suspense X Confidence]; Meilland, Mrs. Marie-Louise; C-P

'MEIRILOCRA', HT, ab, 1984; (**Capella**, Zambra '80); flowers large, dbl., 35 petals, slight fragrance; foliage medium, dark, matt; PP004868; [Unnamed seedling X Banzai]; Meilland, Mrs. Marie-Louise; Meilland Et Cie, 1979

'MEIRINGA', HT, w, 1975; (**Carte Blanche**™); flowers flora-tea, dbl., 38 petals, 4 in.; foliage matt; vigorous growth; [(Carina X White Knight) X Jack Frost]; Paolino; URS

'MEIRINLOR', HT, my, 1975; semi-dbl., 15–18 petals, 4 in.; foliage dark; [Golden Garnette X ((Golden Garnette X Bettina) X Dr. A.J. Verhage)]; Paolino; URS

'MEIRISOURU', F, lp, 1977; (**Nirvana**®); bud ovoid; flowers large, dbl., 20 petals, cupped; foliage glossy; bushy growth; GM, Geneva, 1975; [(Pink Wonder X Kalinka) X Centenaire de Lourdes]; Meilland, Mrs. Marie-Louise; Meilland

MEIrivoui, Min, mp, 1997; (**Belle Symphonie**); Meilland

'MEIROBIDOR', HT, dy, 1980; (Carte d'Or, **Supra**™); flowers flora-tea, large, dbl., 20 petals, no fragrance; foliage medium, dark, semi-glossy; upright growth; PP005042; [((Zambra X (Baccará X White Knight) X Golden Garnette)) X Unnamed seedling]; Meilland, Mrs. Marie-Louise; Meilland Et Cie

'MEIRODIUM', HT, rb, 1976; (**Red Success**); flowers blood-red, base cardinal-red, very full, dbl., 40–45 petals, 4–4.5 in., slight fragrance; foliage large; vigorous, upright growth; PP004037; [(Tropicana X MEIalto) X ((MEIbrem X Zambra) X Tropicana)]; Paolino; URS

'MEIROKOI', HT, yb, 1998; (**Jean Giono**™, Romantic Moments); flowers sunny yellow, veined and edged in orange/apricot, very dbl., 26–40 petals, 3–4 in., borne in small clusters, moderate fragrance; prickles numerous; foliage medium, medium green, glossy; upright, bushy, medium, 90 cms growth; [(Yakimour X Landora) X Graham Thomas]; Selection Meilland; Conard-Pyle Co., 1996

MEIrolange, HT, mp, 1999; (**Paganini**); Meilland

'MEIROLOUR', F, mr; (**Concerto**™)

MEIrolyz, F, ob, 1999; (**Fire King**); Meilland

MEIromar, HT, w, 1995; (**Gin Fizz**); Meilland

'MEIRONSSE', HT, ob, 1992; (Christoph Colombus, Christophe Colomb, **Christopher Columbus**, Cristobal Colon, Cristoforo Colombo, 'MEINRONSSE'); flowers orange blend/copper, blooms borne mostly singly, dbl., 26–40 petals, 3–3.5 in., slight fragrance; many prickles; foliage large, dark green, semi-glossy; medium (120-140 cms), upright growth; GM, Durbanville, 1990; [MEIgurani X (Ambassador X MEInaregi)]; Selection Meilland; France, 1990;, Germany;, Italy, 1991;, Spain, 1991;, The Conard-Pyle

MEIrotego, Min, mp, 1977; (**Minijet**®); Meilland

'MEIROUPIS', LCl, mp, 1996; (Colette™, John Keats); very dbl., 3–3.5 in., borne in small clusters, intense fragrance; numerous prickles; foliage small, dark green, glossy; spreading, medium (2 m) growth; US PPAF; [(MEIbeluxen X Friesia) X Prairie Princess]; Meilland International SA; The Conard-Pyle Co., 1995

MEIroupis, HT, lp, 1996; (**Colette**™); flowers pink, medium sized, old-fashioned blooms, dbl., 25 petals, 3 in., cupped, intense fragrance; foliage large, medium green; climbing growth; PP9994; [(Centenaire de Lourdes X Pocasso) X MEIhatiti]; Meilland; Meilland, 1998

MEIrouve, HT, ab, 1997; (**Alphonse Daudet**); Meilland

'MEIROV', Min, mr, 1975; (**Meillandina**); flowers currant-red, then imbricated, small, dbl., 20 petals, 1.5 in., cupped; foliage matt; vigorous growth; [Rumba X (Dany Robin X Fire King)]; Paolino; URS

'MEIROVERNA', HT, or, 1979; (**Koba**); flowers scarlet, medium, dbl., 40 petals, cupped; foliage dark; very vigorous, upright growth; [((Unnamed seedling X Rouge Meilland) X Independence) X QueenElizabeth]; Meilland, Mrs. Marie-Louise; Meilland

'MEIROVONEX', LCl, ob, 1977; (**Looping**®); bud conical; flowers orange-coral, medium, dbl., 40 petals, cupped, spring bloom, slight fragrance; foliage dark; vigorous, climbing growth; [((Zambra X Zambra) X (Malcair X Danse des Sylphes) X (Cocktailx Cocktail)) X Royal Gold]; Meilland, Mrs. Marie-Louise; Meilland

'MEIROZRUG', HRg, mp, 1994; (**Turbo**™, Turbo Meidiland®, Turbo Rugostar); flowers fuchsia pink, blooms, dbl., 15–25 petals, 3–3.5 in., no fragrance; many prickles; foliage medium, light green, semi-glossy; tall (130-140 cms), bushy growth; [(Frau Dagmar Hastrup X Manou Meilland) X Pink Grootendorst]; Meilland, Alain A.; SNC Meilland & Cie, 1993

MEIruchka, F, rb, 1999; (**Bolchoi**); Meilland

MEIrulex, MinFl, lp; (**Fuchsia Sunblaze**); Meilland

'MEIRUMOUR', S, rd, 1995; (Cherry Meidiland, **Cherry Meillandecor**®); flowers medium red with white eye, single (5 petals), medium (4–7 cms) blooms borne in small clusters; no fragrance;

many prickles; foliage medium, dark green, semi-glossy; GM Geneva, 1994; [MEIpoque X (Regensberg X Fair Play)]; Meilland, 1994; The Conard- Pyle Co.

'MEIRUTRAL', Min, dr, 1988; (**Prince Meillandina**®, Prince Sunblaze, Red Sunblaze®); flowers dark currant-red, medium, dbl., 15–25 petals, 1.5–2 in., cupped, borne in large clusters, no fragrance; foliage medium, dark green, semi-glossy; bushy growth, branching, 15 in.; PP007021; [Parador X Mogral]; Meilland, Alain A.; SNC Meilland & Cie

'MEIRYPOUX', HT, dr, 1990; (**Rotary Rose**); bud pointed; flowers , dbl., 27 petals, exhibition form, borne usually singly, no fragrance; prickles large, red; foliage medium green; vigorous, upright growth; PP007408; [(Mister Lincoln X Pres. Leopold Senghor) X Karl Herbst]; Meilland, Alain A.; Rotary Rose Co., 1989

'MEISADINA', HT, or, 1975; (**Golestan**); bud tapering; flowers vermilion, small, dbl., 25 petals, exhibition form; foliage glossy, dark; vigorous, upright growth; [(Tropicana X Tropicana) X ((Seedling X Rouge Meilland) xIndependence)]; Meilland, Mrs. Marie-Louise; Meilland

'MEISAR', HT, dr, 1963; ('MEICESAR', **Papa Meilland**®); bud pointed; flowers dark velvety crimson, large, dbl., 35 petals, exhibition form, intense fragrance; foliage leathery, glossy, olive-green; vigorous, upright growth; GM, Baden-Baden, 1962 Hall of Fame, WFRS, 1988 James Alexander Gamble Fragrance Medal, ARS, 1974; [Chrysler Imperial X Charles Mallerin]; Meilland, Alain A.; URS

MEIsardan, S, yb, 1993; (**Cesar** ™, Romantic Occasion); flowers creamy yellow with pink, very large, dbl., slight fragrance; Meilland; (Sangerhausen)

'MEISARSAR', Cl HT, dr, 1970; (Grimpant Papa Meilland®, **Papa Meilland, Climbing**®); Stratford; Rumsey

MEIsazy, F, lp, 1997; (**Gala Charles Aznavour**); Meilland

MEIsecaso, S, pb; (**Lutin**); Meilland

'MEISELGRA', Min, mp, 1990; (**Pink Mini-Wonder**™); bud rounded; flowers light rose bengal, reverse pale rose bengal, aging pale rose, dbl., 39–42 petals, cupped, no fragrance; prickles small, very few, green, aging tan; foliage medium green, semi-glossy; bushy, medium growth; PP007361; [(Anytime X Parador) X Mogral]; Selection Meilland, 1987; The Conard-Pyle Co., 1989

'MEISHUTO', Min, op, 1996; (**Salmon Sunblaze**™); flowers salmon pink, full, dbl., 35–40 petals, 1.5–2 in., borne in small clusters, slight fragrance; few prickles; foliage small, dark green, semi-glossy; compact, low (12-16 in) growth; US PPAF; [MEIchanso X (MEIlarco X KORwerk)]; Meilland International SA; The Conard-Pyle Co., 1997

MEIsionver, HT, rb, 1998; (**Caprice de Mielland**); Meilland

'MEISOCRAT', F, mp, 1996; (**Guy de Maupassant**™, Romantic Fragrance); very dbl., 3–3.5 in., borne in small clusters, intense fragrance; moderate prickles; foliage large, dark green, glossy; bushy, tall (70-90 cms) growth; US PPAF; [(MEIturaphar X Mrs John Laing) X Egekov]; Meilland International SA; The Conard-Pyle Co., 1995

'MEISOGREL', Min, w, 1988; ('MEIZOGREL', Spot Minijet, **White Mini-Wonder**™); flowers small, clear white, thin petals, very dbl., 40 petals, cupped, borne in small clusters, no fragrance; foliage small, medium green, matt; short, bushy growth, 12 in.; PP007276; [White Gem X Cinderella]; Meilland, Alain A.; SNC Meilland & Cie

MEIsolroz, S, mp, 1990; (**Flamingo Meidiland**); Meilland

'MEISOYRIS', Gr, dr, 1994; (**Matilda**™, Marcel Pagnol, Velveteen); dbl., 26–40 petals, 1.5–2.75 in., moderate fragrance; some prickles; foliage large, light green, glossy; medium, semi-erect growth; PPAF; [Fragrant Cloud X (Oklahoma X Royal William)]; Meilland, Alain A.; SNC Meilland & Cie, 1993;, The Conard-Pyle Co., 1994

MEIspola, HT, w; (**Preview**); Meilland

MEIspreyp, F, my, 1994; (**Mimi Golden**™); Meilland

'MEISTERSTUCK', F, mr, 1940; bud ovoid; flowers velvety crimson, large, borne in clusters, dbl., cupped, moderate fragrance; foliage dark, glossy; upright, bushy growth; [Holstein X Kardinal]; Kordes

MEIstocko, S, dp, 1991; (**Flash Meidiland**); flowers medium-large, single; Meilland; (Sangerhausen)

MEIsunaj, F, dp, 1989; (**Mme Fernandel**®); Meilland

'MEITABIFOB', HT, my, 1977; (**Cocktail '80**); bud elongated; flowers, dbl., 20 petals, 5 in., exhibition form, slight fragrance; foliage dark; very vigorous growth; Meilland, Mrs. Marie-Louise; Meilland

'MEITALBAZ', F, my, 1994; (**Anthony Meilland**); dbl., 26–40 petals, 3–3.5 in., borne in small clusters; some prickles; foliage large, medium green, semi-glossy; medium, bushy growth; PP8449; [Sunblest X MEIlenangal]; Meilland, Alain A.; SNC Meilland & Cie, 1990;, The Conard-Pyle Co., 1994

'MEITAM', LCl, dr, 1958; (**Fugue**®); bud globular; dbl., 30 petals, blooms in clusters, slight fragrance; foliage leathery, glossy; vigorous growth; GM, Madrid, 1958; [Alain X Guinee]; Meilland, Mrs. Marie-Louise; URS

MEItanet, F, lp; (**Starlite**); Meilland, 1995

MEItapov, LCl, w, 1993; (**Koala**®); Meilland

'MEITARAS', HT, pb, 1963; (**Swarthmore**); flowers shades of pink blended, dbl., 50 petals, 4 in., exhibition form, slight fragrance; foliage dark, leathery; vigorous, tall, bushy growth; [(Independence X Happiness) X Peace]; Meilland, Alain A.; C-P

'MEITEBROS', HT, lp, 1998; (**Frederic Mistral**®); flowers venetian pink, reverse suffused rose, full, very dbl., 26–40 petals, 4.5 in., borne mostly singly, intense fragrance; prickles moderate; foliage large, dark green, semi-glossy; upright, tall, 6 ft. growth; [(Perfume Delight X Prima Ballerina) X The McCartney Rose]; Selection Meilland; Conard-Pyle Co., 1995

MEItelov, HT, lp; (**Michelangelo**); Meilland, 1997

'MEITIFRAN', Min, rb, 1987; (**Baron Meillandina**®, Baron Sunblaze); flowers white with red edges, umbrella-shaped, medium, dbl., no fragrance; foliage small, medium green, glossy; bushy growth; PP006818; [Magic Carrousel X ((Alain X R. mutabilis) X (Medar X Caprice))]; Meilland, Mrs. Marie-Louise; Meilland Et Cie, 1989

'MEITILOLY', HT, yb, 1976; (**Mascotte '77**®); flowers yellow, edged cardinal-red, dbl., 40 petals, 4.5 in., slight fragrance; foliage glossy; vigorous growth; GM, Belfast, 1979; [(MEIrendal X (Rim X Peace)) X Peace]; Paolino; URS

'MEITINIROL', HT, or, 1977; (**Galia**®); dbl., 38 petals, cupped, borne singly, no fragrance; foliage matt, dense; vigorous growth; [MEIretni X Elegy]; Meilland, Mrs. Marie-Louise; Meilland Et Cie

MEItinor, F, mp; (**Cindy**); Meilland

MEItixia, HT, mr, 1989; (**Revolution Francaise**®); Meilland

MEItobla, F, dp, 1992; (**Les Amoureux de Peynet**®); Meilland

'MEITOFLAPO', F, mp, 1980; flowers large, dbl., 29 petals, no fragrance; foliage small, dark, semi-glossy; [(Jack Frost X (Zambra X (Baccará X White Knight))) X ((Zambrax (Baccará X White Knight)) X Unnamed seedling)]; Meilland, Mrs. Marie-Louise; Meilland Et Cie

'MEITOIFAR', HT, mp, 1994; (**Auguste Renoir**®); flowers old form, blooms borne mostly single, very dbl., 3–3.5 in., moderate fragrance; many prickles; foliage medium, medium green, semi-dull; medium, bushy growth; [(Versailles X Pierre de Ronsard) X Kimono];

Meilland, Alain A.; SNC Meilland & Cie, 1992

'MEITOLEIL', Min, my, 1994; (**Gold Symphonie**); very dbl., 1.5–2.75 in., borne mostly singly or in small clusters, no fragrance; some prickles; foliage medium, dark green, semi-glossy; medium (40-50 cms), bushy growth; [(Rise 'n' Shine X Yellow Meillandina) X Gold Badge]; Meilland, Alain A.; SNC Meilland & Cie, 1993

MEItomkin, Min, lp; (**Symphonie Lumiere**); Meilland, 1995

'MEITONJE', Min, lp, 1987; (Pink Symphonie®, **Pink Symphony**, Pretty Polly, Sweet Sunblaze®); flowers light cardinal pink, medium, unfading, dbl., 26–40 petals, 1.5–2 in., flat, borne singly and in small clusters, slight fragrance; foliage medium, dark green, glossy, disease resistant; vigorous, bushy growth, 18 in.; PP007277; GM, Glasgow, 1992; [Darling Flame X Air France]; Meilland, Mrs. Marie-Louise

MEItorpo, F, mr, 1992; (**Sangria**); Meilland

'MEITOSIER', LCl, ab, 1996; (Lord Byron, **Polka**™, Polka 91, Scented Dawn); flowers ivory yellow, large blooms, very dbl., 30–35 petals, borne in small clusters, intense fragrance; moderate prickles; foliage large, medium green, semi-glossy; upright, climbing, tall (12 ft) growth; PP009233; [Golden Showers X Lichtkonigin Lucia]; Meilland International SA; The Conard-Pyle Co., 1991

MEItrino, Min, mr; (**Tonic Meillandina**®); Meilland, 1994

'MEITRISICAL', Min, yb, 1982; (**Yellow Meillandina**, Yellow Sunblaze); flowers yellow, petals edged pink, dbl., 20 petals, moderate fragrance; foliage medium, dark, semi-glossy; bushy growth; [(Poppy Flash X (Charleston X Allgold)) X Gold Coin]; Meilland, Mrs. Marie-Louise; Meilland Et Cie, 1980

'MEITROGANA', HT, my, 1977; (**Vabene**); bud conical; flowers medium, dbl., 30 petals, exhibition form, slight fragrance; very vigorous, upright growth; [Arthur Bell X (MEIgold X Kabuki)]; Meilland, Mrs. Marie-Louise; Meilland

'MEITULANDI', HT, mr, 1979; (**Hidalgo**®, Michel Hidalgo); bud conical; flowers currant-red, very large, dbl., 30 petals, cupped, intense fragrance; foliage matt, bronze; vigorous, upright growth; [((Queen Elizabeth X Karl Herbst) X (Lady X X Pharaon)) X (MEIcesar X Papa Meilland)]; Meilland, Mrs. Marie-Louise; Meilland

'MEITULIMON', HT, m, 1979; (**Manou Meilland**®); bud conical; flowers mauve-pink, medium, dbl., 50 petals, cupped, slight fragrance; foliage glossy, dark; vigorous, bushy growth; Gold Star of the South Pacific, Palmerston North, NZ, 1980; [(Baronne Edmond de Rothschild X Baronne Edmond de Rothschild)x (Ma Fille X Love Song)]; Meilland, Mrs. Marie-Louise, 1980

MEItulimonsar, Cl F, dp, 1997; (**Manou Meilland, Climbing**); Meilland

MEItune, F, mp, 1995; (**Mazurka**); Meilland

'MEITURAPHAR', S, dp, 1979; (**Anne de Bretagne**®, Decor Rose, Meilland Decor Rose); bud conical; flowers deep pink, shallow-cupped, dbl., 20 petals, cupped; foliage semi-glossy; vigorous, upright growth; [(Malcair X Danse des Sylphes) X ((Zambra X Zambra) X Centenaire de Lourdes)]; Meilland, Mrs. Marie-Louise, 1976

'MEITURUSA', Min, w, 1976; (**White Gem**®); bud long; flowers soft ivory, shaded pale tan, very dbl., 90 petals, 1.5 in., slight fragrance; foliage large, glossy, dark; upright, bushy growth; [Darling Flame X Jack Frost]; Meilland; C-P

MEItylpic, HT, dy, 1999; (**Solidor**); PPAF; Meilland

MEIvahyn, S, w, 1996; (**Ice Meidiland**™); flowers white, opens flat, dbl., 25–30 petals, 2.5–3 in., pompon, borne in clusters, no fragrance; foliage medium, medium green, glossy; compact, mounding, ground cover growth; Meilland

MEIvaleir, LCl, yb, 1998; (**Michka**); Meilland

'MEIVANAMA', HT, or, 1979; (**Ondella**); bud conical; flowers vermilion, large, dbl., 33 petals; foliage dark; vigorous, upright growth; [(Elegy X Arturo Toscanini) X (Peace X Demain)]; Meilland, Mrs. Marie-Louise; Meilland

'MEIVESTAL', HT, dr, 1988; (Dreams Come True, **Senator Burda**®, Spirit of Youth, Victor Hugo®); flowers brilliant currant-red, large, dbl., 26–40 petals, intense fragrance; foliage large, medium green, semi-glossy; upright, strong, floriferous growth; [(Karl Herbst X (Royal Velvet X Suspense)) X Erotika]; Meilland, Mrs. Marie-Louise; SNC Meilland & Cie

MEIvibroum, HRg, lp, 1996; (**Pinky Rugostar**); Meilland

'MEIVILANIC', F, yb, 1981; (**Rustica**®, Stadt Basel, Ville de Bâle); bud very long; flowers yellow-peach blend, reverse buff yellow-orange, blooms borne, dbl., 35 petals, cupped, slight fragrance; foliage dark, semi-matt; half upright growth; [(Queen Elizabeth X Seedling) X Sweet Promise]; Meilland, Mrs. Marie-Louise; Meilland Et Cie

'MEIVILDO', HT, dp, 1985; (Queen Adelaide, The Royal Brompton Rose, **Yves Piaget**®); flowers deep pink, large, very dbl., globular, intense fragrance; foliage medium, dark, semi-glossy; upright growth; PP006895; Golden Rose, Geneve, 1998; [((Pharaoh X Peace) X (Chrysler Imperial X Charles Mallerin)) xTamango]; Meilland, Mrs. Marie-Louise; Meilland Et Cie, 1983

MEIviola, HT, dr, 1998; (**Traviata**™, Xaviere); flowers dark red, very full, very dbl., 41 petals, 4–4.5 in., borne mostly singly, slight fragrance; prickles moderate; foliage large, dark green, glossy; upright, medium, 80 cm growth; [(Porta Nigra X Paolo) X William Shakespeare]; Selection Meilland; Conard-Pyle Co., 1997

'MEIVIOLIN', LCl, pb, 1987; (Eden, Eden Climber™, Eden Rose 88, Grimpant Pierre de Ronsard®, **Pierre de Ronsard**®); flowers cream white suffused with carmine pink, old rose shape, large, very dbl., 40–55 petals, 4.5–5 in., borne in clusters, repeat bloom, slight fragrance; foliage medium, deep green, semi-glossy; spreading growth, climbing habit; frost resistant; PP006892; [(Danse des Sylphes X Haender) X Pink Wonder, Climbing]; Meilland, Mrs. Marie-Louise, 1985

'MEIVIRGI', Gr, my, 1959; (**Golden Girl**); bud pointed; flowers golden yellow, dbl., 45 petals, 4–4.5 in., exhibition form, moderate fragrance; foliage leathery, light green; upright, vigorous, bushy growth; [(Joanna Hill X Eclipse) X Michele Meilland]; Meilland; C-P

MEIvostro, Min, pb, 1995; (**Carinita**); Meilland

'MEIVOUPLIX', HT, dy; (**Laura**™); Meilland

MEIvraivou, Min, dr; (**Tilt Symphonie**®); Meilland, 1994

MEIvrita, HT, mp, 1999; (**Seduction**); Meilland

MEIvrofix, F, mp; (**Zurella**™); Meilland

MEIwaton, LCl, dp, 1994; (**Gites de France**®); Meilland

'MEIWONDER', Pol, mr, 1965; bud globular; flowers small, dbl., borne in clusters; [Marianne Kluis Superior sport]; Grootendorst, F.J.

'MEIVAMO',HT, dp, 1995 ; (**Paris de Yves St. Laurent**™);flowers deep pink, full (26–40 petals), large (7 + cms) blooms; slight fragrance; some prickles; foliage large, dark green, glossy; upright growth;[Silva sport]; Meiland, 1992; The Conard Pyle Co.

MEIxerul, Min, ab, 1991; (**Peche Meillandina**®); Meilland

MEIxetal, S, lp, 1988; (**Denise Grey**®); Meilland

MEIyacom, HT, mp, 1997; (**Claudia**); Meilland

MEIzalitaf, LCl, yb; (**Spectra**); Meilland, 1983

MEIzebul, F, mp, 1998; (**Jardins de France**); GM, Bagatelle, 1998 GM,

Dublin, 1998 GM, Geneve, 1998; Meilland

'MEIZELI',HT, mp, 1995; (**The McCartney Rose**™, McCartney Rose, Paul McCartney,Sweet Lady, The MacCartney Rose);flowers medium pink, full (26-40 petals), large (7 + cms) blooms; very fragrant; foliage large, medium green, semi-glossy; bushy, medium growth; GM, Geneva, 1988; GM LeRoeix, 1988, GM, Monza,1988; GM Paris, 1988;[(Nirvana X Papa Meilland) X First Prize]; Meilland, Alain; The Conard Pyle Co.

'MEIZOELO', Gr, yb, 1999; (**Glowing Peace** ™); flowers yellow and orange blend, medium, dbl., 26–40 petals, borne in small clusters, slight fragrance; prickles moderate; foliage medium, dark green, glossy; upright, bushy, medium growth; AARS, U.S., 2001; [Sun King X Roxane]; Selection Meilland; Conard-Pyle, 2001

'MEIZOGREL', Min, w, 1988; ('MEISOGREL', Spot Minijet, **White Mini-Wonder**®); flowers small, clear white, thin petals, very dbl., 40 petals, cupped, borne in small clusters, no fragrance; foliage small, medium green, matt; short, bushy growth, 12 in.; PP007276; [White Gem X Cinderella]; Meilland, Alain A.; SNC Meilland & Cie

'MEIZOURAYOR', S, rb, 1986; (**Decor Arlequin**, **Meilland Decor Arlequin**®); flowers strawberry-red and yellow, medium, semi-dbl., 18 petals, cupped; foliage dark; very vigorous, upright growth; GM, Rome, 1975; [((Zambra X Zambra) X (Suspense X Suspense)) X Arthur Bell]; Meilland, Mrs. Marie-Louise; Meilland Et Cie, 1977

MEIzuzes, HT, rb, 1996; (**Bolcholi**, Madam Speaker); Meilland

'MEJAKKA', S, rb, 1997; (**Rosy Wings**); flowers semi double, large, semi-dbl., 8–15 petals, borne in large clusters, slight fragrance; foliage large, medium green, glossy; spreading, tall (6ft.) growth; [Dornroschen X Golden Wings]; Mekdeci, John

'MEJENOR', HT, or, 1955; ('MEIJENOR', **Soraya**®); bud pointed; flowers orange-red reverse crimson-red, large, dbl., 30 petals, cupped, slight fragrance; foliage glossy; vigorous, bushy growth; [(Peace X Floradora) X Grand'mere Jenny]; Meilland, F.; URS

'MEJENORSAR', Cl HT, or, 1960; (Grimpant Soraya®, 'MEIJENORSAR', **Soraya, Climbing**®); Barni, V.; URS

'MEJORIA', S, ob, 1995; (**John's Rose**); flowers brilliant orange to fluorescent pink, blooms borne in small, dbl., 15–25 petals, 3–3.5 in., no fragrance; some prickles; foliage large, dark green, glossy; tall, upright growth; [[(Tropicana X Queen of the Lakes) X Seedling] X Golden Wings]; Mekdeci, John; Hortico Roses, 1995

'MEKKADA', S, ob, 1997; (**Butterflies**); flowers single, medium, single, 4–7 petals, borne in large clusters, moderate fragrance; foliage medium, medium green, semi-glossy; bushy (5ft.) growth; [Dornroschen X Golden Wings]; Mekdeci, Dr. Anthony Casimir

Melanie *see* 'MACMELAN'

'MELANIE', S, dr, 1946; flowers deep red, semi-dbl., non-recurrent; foliage reddish; hardy (to about -15).; [R. rubrifolia X Gruss an Teplitz]; Wright, Percy H.

'MELANIE', HT, dp, 1958; bud long; flowers carmine-pink, dbl.; vigorous growth; [Sterling X Mme Auguste Chatain]; Combe; Japan Rose Society

'MELANIE DE MONTJOIE', HSem, w; flowers pure white, large; vigorous growth; Jacques

Melanie Lemaire *see* **'HERMOSA'**

'MÉLANIE SOUPERT', T, w; [Gloire de Dijon seedling]; Nabonnand, G., 1882

'MELANIE WALDOR', M, ab, 1865; Moreau et Robert

'MELBA', F, yb, 1963; flowers soft cream and peach, well formed, small, borne in clusters, dbl.; foliage light green; spreading growth; [Masquerade X Independence seedling]; Sanday, John

Meletta, Min, lp; flowers light orange, exhibition form; Giles, Diann

Melflor, F, lp, 1988; RvS-Melle

'MELGLORY', F, mr, 1982; flowers large blooms in clusters of 4-16, dbl., 23 petals, no fragrance; foliage dark, matt; upright growth; [Lilli Marleen X Patricia]; Rijksstation Voor Sierplantenteelt

'MELGOLD', HT, dy, 1980; flowers large blooms in cluster of 1-6, dbl., 62 petals, cupped, no fragrance; prickles reddish-green; foliage dark, matt; upright growth; [Sunblest X Souv. de Jacques Verschuren]; Rijksstation Voor Sierplantenteelt

Melica, HT, lp; Hicl, 1991; (Czech Rosa Club)

Melik El Adel *see* **'MALEK-ADEL'**

Meli-Melo *see* ORAstrip

Melina *see* 'TANEMA'

Melinda *see* 'RULIMPA'

Melinda Claire *see* 'TAYMEL'

Melinda Gainsford *see* 'JACYAP'

'MELINDA MARIE', F, lp, 1972; bud ovoid; flowers clear light pink, dbl., 20–25 petals, 3 in., flat, slight fragrance; foliage dark, leathery; bushy growth; [Sarabande X Sarabande]; Linscott

'MÉLISANDE', HT, lp, 1964; bud long; flowers soft pink, reverse darker, well formed, large, dbl., moderate fragrance; foliage dark; long, strong stems; vigorous, symmetrical growth; Mondial Roses

'MELISSA', HT, mp, 1975; dbl., 27–32 petals, 4 in., globular, slight fragrance; foliage leathery; very vigorous, upright, bushy growth; PP004082; [Seedling 1968-1 X Sedling 167-d]; McDaniel, Earl; Carlton Rose Nurseries

Melissa, F, lp, 1996; Noack, Werner

Melissa Joyce *see* 'RENHOM'

'MELISSA MCCARTNEY', F, dp, 1991; bud ovoid; flowers holds color well, does not fade, blooms borne usually singly, very dbl., 36 petals, 5 in., exhibition form, slight fragrance; foliage medium, dark green, glossy, slight red tinge to new; bushy, low (70 cms) growth; [Cherish sport]; Troyer, Ray & Pat

'MELITA', HWich, mp, 1934; flowers carnation-pink, large, dbl.; foliage glossy, light; vigorous, climbing growth; GM, NRS, 1933; [Thelma sport]; Easlea

'MELLE FISCHER', Pol, mp, 1914; flowers small, dbl.; Pfitzer; (Sangerhausen)

Mellow, HT, ab, 1995; Dawson

'MELLOW GLOW', HT, pb, 1989; bud ovoid; flowers medium pink, reverse creamy yellow, medium, borne usually singly, dbl., 52 petals, exhibition form, intense, damask fragrance; prickles pointed slightly downward, large, yellow; foliage medium, medium green, semi-glossy; bushy growth; [Thriller X Wild Cherry]; Bridges, Dennis A.; Bridges Roses, 1990

'MELLOW YELLOW', HT, my, 1968; flowers sunflower-yellow, edged pink, urn shaped; free growth; [Piccadilly sport]; Waterhouse Nursery

Melmore Terrace, HSem, w

Melodie®, F, mr, 1980; Noack, Werner

Melodie Parfumee *see* 'DORIENT'

'MELODY', HT, pb, 1946; bud urn-shaped; flowers deep pink, edged lighter, large, dbl., 35 petals, moderate fragrance; low, bushy growth; [Joanna Hill X Miss C.E. van Rossem]; Lammerts, Dr. Walter; Armstrong Nursery

Melody *see* KORbolak

Melody Lane *see* 'MICLANE'

Melody Maker *see* 'DICQUEEN'

Melody Marshall *see* 'MORMELMA'

Melody Parfumee™ *see* 'DORIENT'

Melody Queen, HT, pb, 1998; Ghosh

'MELO-MELO-DAY', Cl F, mp, 1955; bud ovoid; flowers cameo-pink, dbl., 40 petals, 2–2.5 in., cupped, moderate fragrance; climbing (10-15 ft) growth; [Demure sport]; Motose

Melonda, HT, ob, 1974; flowers yellowish salmon-orange, very large, dbl., slight fragrance; J&P; (Sangerhausen)

Melpink, F, mp

'MELROSE', HT, rb, 1963; flowers creamy white flushed cherry-red, dbl., 35 petals, 3.5 in., moderate fragrance; foliage dark, leathery; vigorous, bushy growth; [Silver Lining X E.G. Hill]; Dickson, A.

Melrose, F, dp, 1990; 24 petals, 3 in.; GM, Durbanville, 1994 GM, Geneve, 1992; RvS-Melle, 1992

'MELVENA', HT, op, 1971; bud ovoid; flowers salmon-pink, full, medium, dbl., moderate fragrance; foliage leathery; vigorous, upright growth; [Daily Sketch X Impeccable]; Dawson, George

'MELVIN', F, rb, 1980; bud pointed; flowers ivory, flushed red, aging darker red, blooms borne 1-11 per, dbl., 30 petals, exhibition form, slight fragrance; triangular, hooked prickles; foliage medium green; medium, dense growth; Bronze, ARC TG, 1980; [Unnamed seedling X Unnamed seedling]; Jerabek, Paul E.

'MÉMÉ BUY', HT, or, 1935; flowers coppery coral-red, lightly streaked golden yellow, large, very dbl., cupped; foliage bronze; very vigorous growth; Chambard, C.

'MÉMÉE ARLES', HT, dr, 1955; flowers deep red tinted vermilion; foliage glaucous green; [Peace X Emma Wright]; Arles; Roses-France

'MÉMÉE AZY', HT, pb, 1921; flowers pink shaded carmine, bordered whitish, stamens orange-yellow, dbl., slight fragrance; [Étoile de France X Le Progres]; Gillot, F.

'MÉMÉE CHANTEUR', F, dr, 1959; bud ovoid; flowers deep crimson, large, borne in clusters, dbl., cupped, slight fragrance; foliage dark, leathery; vigorous, upright growth; [Karl Herbst X Pioupiou]; Arles; Roses-France

Memento® *see* 'DICBAR'

Memoire® *see* KORzuri

Memorial Day Rose, Misc OGR, mp

Memorial Rose *see* **R. WICHURANA**

'MEMORIAM', HT, lp, 1961; bud long, pointed; flowers pastel pink to nearly white, dbl., 55 petals, 6 in., exhibition form, moderate fragrance; foliage dark, leathery; moderately tall growth; GM, Portland, 1960; [(Blanche Mallerin X Peace) X (Peace X Frau Karl Druschki)]; Von Abrams; Peterson & Dering

'MEMORIES', HT, dy, 1977; bud short, pointed, ovoid; flowers empire-yellow, full, small, dbl., 25–30 petals, 2.5–3 in., exhibition form, continuous bloom in greenhouse, intense, tea fragrance; foliage large, glossy; vigorous, upright, bushy growth; [Spanish Sun X Hoosier Gold]; Byrum; J.H. Hill Co.

Memories *see* CLEjoy

'MEMORY', HT, pb, 1932; bud pointed; flowers light pink to deeper pink, base yellow, very large, dbl., exhibition form, intense fragrance; foliage rich green, leathery; vigorous, compact growth; GM, NRS, 1932; Cant, B. R.; J&P

Memory Bells *see* 'POULMA'

'MEMORY LANE', Min, lp, 1973; bud ovoid; flowers rose-pink, small, very dbl., slight fragrance; foliage leathery; vigorous, dwarf, bushy growth; [(Pinocchio X William Lobb) X Little Chief]; Moore, Ralph S.; Sequoia Nursery

Memory of D. M. Roy, HT, w, 1998; Ghosh

Memory Lane *see* 'PEAVOODOO'

Memphis, HT, dp; Select Roses, B.V.

'MEMPHIS HERITAGE', Gr, pb, 1969; bud long, pointed; flowers pink blended with gold, large, dbl., exhibition form, moderate fragrance; very vigorous, compact growth; [Queen Elizabeth X Happiness]; Patterson; Patterson Roses

Memphis Queen *see* 'WELQUEEN'

'MÉNAGE', A, w, 1847; flowers flesh, medium, dbl., cupped; Vibert

Menaka Durga Roy *see* 'DRISCOROY'

'MENAP', HT, my, 1955; (**Belle Blonde**); flowers yellow, center darker, well-formed, moderate fragrance; foliage glossy; bushy growth; [Peace X Lorraine]; Meilland, F.; URS

'MENDEL', F, mr, 1946; bud pointed; flowers cherry-red, borne in trusses, semi-dbl., 15 petals, 4 in., moderate fragrance; foliage reddish green; vigorous, bushy growth; [Florentina X Unnamed seedling]; Leenders, M.; Longley

Mendocino Delight, HWich, w; Demits

Mendocino Gothic, HMsk, w; Demits, 1993

Mendocino Headlands, HWich, lp; (found rose)

Mendocino Pale Pink China, HCh, lp

Mendocino Smiles, HWich, pb; (found rose)

Menja, HMsk, mp, 1960; Petersen

'MENNIE D'AGNIN', F, mr, 1962; flowers vermilion-red, large; [Independence X Fashion]; Orard, Joseph

'MENTOR', S, mp, 1959; bud globular; flowers light pink, reverse darker, medium, dbl., recurrent bloom, intense fragrance; foliage dark, glossy; very vigorous (6 ft.), bushy, upright growth; [Tallyho X New Dawn]; Wyant

Menuett, HT, yb, 1988; flowers dark yellow with carmine pink, large, dbl.; GPG Bad Langensalza; (Sangerhausen)

Menuett, HT, ob, 1961; flowers orange and yellow, large, dbl., slight fragrance; Hill, S. H.; (Sangerhausen)

'MENUT', Min, dp, 1956; flowers carmine, small, semi-dbl., 20 petals, abundant, intermittent bloom; dwarf, bushy, compact growth; [Rouletii X Perla de Alcanada]; Dot, Simon

MEpal, HT, op, 1953; (**Bettina**®, 'MEIPAL'); flowers salmon-orange, veined, well-formed, dbl., 37 petals, 4 in., moderate fragrance; foliage dark, glossy, bronze; vigorous growth; [Peace X (Mme Joseph Perraud X Demain)]; Meilland, F.; URS

'MEPALSAR', Cl HT, op, 1958; (**Bettina, Climbing**®, Grimpant Bettina); GM, Geneva, 1959; Meilland, F.; URS

Mephisto, Pol, or, 1985; flowers large, semi-dbl.; GPG Bad Langensalza; (Sangerhausen)

'MÉPHISTO', F, or, 1951; flowers geranium-red, medium, borne in clusters, semi-dbl.; foliage leathery; vigorous, upright, bushy growth; [Francais X Unnamed seedling]; Mallerin, C.; EFR

'MERCATOR', HT, dr, 1962; flowers deep red, large, dbl., intense fragrance; foliage dark; [Chrysler Imperial X Tango]; Delforge

Mercedes® *see* 'MERKOR'

'MERCEDES', HGal, lp, 1847; flowers white and lilac, changing to pale pink, large, dbl.; Vibert

'MERCEDES GALLART', Cl HT, dp, 1932; flowers deep pink, base yellow, very large, dbl., recurrent bloom, intense fragrance; foliage glossy; very vigorous, climbing growth; [Souv. de Claudius Denoyel X Souv. de Claudius Pernet]; Munné, B.; J&P

'MERCEDES JUNCADELLA', HT, op, 1933; flowers salmon-orange; [Frau Karl Druschki X Angèle Pernet]; Munné, B.

'MERCEDES MENDOZA', HT, or, 1962; flowers large, dbl., 30 petals; foliage glossy; vigorous growth; [Asturias X Grand'mere Jenny]; Dot, Simon

Mercedes, Climbing, Cl HT, lp; [Mercedes sport]; Ruston, D., 1982; (Weatherly, L.)

'MERCI', F, mr, 1974; bud ovoid, long, pointed; flowers medium, dbl., slight fragrance; foliage dark, leathery; vigorous growth; Warriner, William A.; J&P

'MERCURIUS', LCl, op, 1940; flowers light coral-pink, large, borne in clusters, semi-dbl., cupped, profuse seasonal bloom; foliage glossy; vigorous, climbing growth; [(Doubloons X Damask) X Clio]; Horvath; Wayside Gardens Co.

'MERCURY', F, or, 1967; flowers orange-scarlet edged mahogany, borne in clusters; foliage dark; low, bushy growth; [Independence X Paprika]; Sanday, John

Mercy *see* SIMmer

Mère de la Patrie *see* **'MOEDER DES VADERLANDS'**

'MEREDITH', LCl, lp, 1984; flowers light pink to almost white, large blooms, dbl., 20 petals, exhibition form, repeat blooming, moderate fragrance; foliage large, dark, semi-glossy; long stems; upright (to 12ft.) growth; [(Charlotte Armstong X New Dawn) X Araby]; Thomson, Richard

Meredith Anne *see* 'TINMERE'

'MEREDITH HUGHES', Min, op, 1984; flowers medium coral pink, medium, dbl., exhibition form, slight fragrance; foliage medium, medium green, glossy; bushy growth; [Anne Scranton X Patricia Scranton]; Dobbs, Annette E.

Meridian, HT, mr; flowers luminous red, very large, dbl.; VEG; (Sangerhausen)

Meridiana® *see* BARmeri

Meriggio *see* BARafne

Merindah, HMult, mp, 1994; Sutherland

Merindah Red, HMult, mr, 1995; Sutherland

'MERIT', F, mp, 1952; bud short, pointed; flowers brilliant rose, reverse lighter, base white, borne in clusters, dbl., 50–90 petals, 1.5–2 in., flat, intense fragrance; foliage leathery, dark; very vigorous, bushy growth; [Garnette sport]; Domilla; Twin Nursery

Merit *see* 'SPOMERIT'

Merko *see* 'MERKOR'

'MERKOR', F, or, 1974; (**Mercedes®**, Merko); bud ovoid; flowers bright scarlet, dbl., 33 petals, exhibition form, slight fragrance; foliage large, leathery; [Anabell X Seedling]; Kordes, R.; J&P, 1975

Merle Blanc *see* 'LENISUR'

'MERLIN', F, pb, 1967; flowers yellow, pink and red, blooms in clusters, dbl., 2.5 in., slight fragrance; foliage glossy; [Pink Parfait X Circus]; Harkness

Merlin, S, lp, 1991; Poulsen

'MERMAID', HBc, ly, 1918; flowers creamy yellow, amber stamens, single, 5 petals, 5–6 in., dependably recurrent, moderate fragrance; foliage dark, glossy; vigorous, climbing, pillar or trailer (6-9 ft.) growth; tender incold regions; (14); GM, NRS, 1917; [R. bracteata X Double yellow Tea rose]; Paul, W.

Merrie *see* 'JAYMER'

'MERRIE MISS', HT, mp, 1966; bud long, ovoid; flowers soft rose-pink, dbl., 60 petals, 4 in., exhibition form, intense fragrance; foliage dark, matt; upright, bushy growth; [Pink Favorite X Margaret]; Fuller; Wyant

Merrimac™ *see* 'KINMAC'

'MERRIMENT', Min, yb, 1986; flowers bright yellow tipped red, reverse yellow edged red, 8 petals, slight fragrance; medium, long, pink prickles; foliage large, medium green, semi-glossy; tall upright growth; [Rise 'n' Shine X ?]; Bridges, Dennis A.

Merrouw Daendels *see* **'MRS HENRI DAENDELS'**

'MERRY CHRISTMAS', Min, mr, 1977; bud pointed; flowers currant-red, open, single, 5 petals, 2 in., slight fragrance; foliage dark; upright, compact, branching growth; [Red Can Can X ?]; Lyon; L. Lyon Greenhouses

'MERRY ENGLAND', HP, mr, 1897; flowers satiny light red, shaded carmine, dbl., 50 petals, intense fragrance; vigorous growth; Harkness

Merry Go Round, F, w

'MERRY HEART', Gr, rb, 1960; bud ovoid; flowers orient red, borne in small clusters, dbl., 30 petals, 3.5–4 in., exhibition form, slight fragrance; foliage glossy, dark; vigorous, upright growth; [El Capitan X Seedling]; Swim, H.C.; C.R. Burr

'MERRY WIDOW', Gr, dr, 1958; bud long, pointed; flowers velvety crimson, dbl., 23 petals, 6 in., cupped, intense, spicy fragrance; foliage dark, glossy; vigorous growth; [Mirandy X Grande Duchesse Charlotte]; Lammerts, Dr. Walter; Germain's

Merryglo™ *see* 'MINIMERR'

'MERRY-GO-ROUND', HT, op, 1950; bud pointed, becoming urn shaped, orange; flowers orange and pink blend, dbl., 25 petals, 5 in.; foliage dark, leathery; tall, compact growth; [Talisman X R.M.S. Queen Mary]; Fisher, G.; Arnold-Fisher Co.

'MERRYWEATHER'S CRIMSON', HT, mr, 1958; flowers bright crimson, well formed, moderate fragrance; strong stems; vigorous, bushy growth; Merryweather

Mers du Sud *see* **'SOUTH SEAS'**

Merveille de la Brie, HWich, mr

'MERVEILLE DE LYON', HP, w, 1882; flowers pure white, tinted satiny rose, dbl., 4 in., cupped, some recurrent bloom; vigorous growth; [Baroness Rothschild sport]; Pernet Père

'MERVEILLE DES BLANCHES', HP, w, 1894; flowers medium-large, dbl., moderate fragrance; Pernet; (Sangerhausen)

'MERVEILLE DES JAUNES', Pol, yb, 1920; flowers bright coppery golden yellow, borne in clusters, dbl.; dwarf growth; Turbat

'MERVEILLE DES ROUGES', Pol, rb, 1911; flowers deep velvety crimson, center whitish, borne in large clusters, semi-dbl., cupped; dwarf growth; Dubreuil

Merveilleuse, HT; Dot, Simon; (Cavriglia)

Mervelle, HT, 1995; McGredy, Sam IV; (Cavriglia)

'MERYL JANE GASKIN', HT, pb, 1948; flowers shell-pink edged deeper pink, well formed, dbl., 48 petals, 5–6 in., intense fragrance; free, branching growth; [Rose Berkeley sport]; Mee; Fryer's Nursery, Ltd.

Message *see* 'MEBAN'

Message, Climbing *see* **'WHITE KNIGHT, CLIMBING'**

Messagere *see* 'HERTIE'

Messara *see* SAUdive

'MESSESTADT HANNOVER', F, mr, 1962; (Hannover); flowers large blooms in clusters, dbl.; foliage light green, glossy; moderate growth; Kordes, R.

Messidor, HT, dy

'MESSIRE', LCl, mr, 1963; flowers bright red, borne in clusters of 6-7, semi-dbl., 10–15 petals, 3 in., profuse, repeated bloom; foliage bronze; moderately vigorous growth; [Unnamed seedling X Spectacular]; Laperrière; EFR

Messire Delbard® *see* 'DELSIRE'

Metanoia®, LCl, ob

'METEOR'®, F, or, 1959; flowers orange-scarlet, patio, blooms in clusters (up to 10), dbl., 40 petals, 3 in., cupped; foliage light green; vigorous, bushy, low growth; ADR, 1960; [Feurio X Gertrud Westphal]; Kordes, R.; A. Dickson;, McGredy

'METEOR', HT, mr, 1887; flowers crimson-carmine, open, dbl., recurrent bloom, slight fragrance; few thorns; foliage small, soft; dwarf growth; Bennett

'METEOR', N, dp, 1887; flowers deep rose tinted carmine-purple, large, dbl., moderate fragrance; vigorous growth; Geschwind, R.

Meteor, B, dp

Meteor Shower *see* JACpan

'METEOR, CLIMBING', Cl HT, mr, 1901; Dingee & Conard

'METIS', S, mp, 1967; (Simonet); bud ovoid; flowers soft rose, dbl., 35 petals, 2.5–3 in., flat, spring bloom, slight fragrance; foliage small, glossy; [R. nitida X Therese Bugnet]; Harp; Morden Exp. Farm

Métro *see* 'MACBUCPAL'

'METROPOLE', HT, mp, 1961; dbl., 45 petals, 4–6 in., globular, moderate fragrance; foliage matt, green; vigorous growth; [Sidney Peabody X Peace]; deRuiter; Blaby Rose Gardens

Metropolitain *see* 'MACNGAURU'

METset, F, m; (**Cristian**); (METset)

'MEVROUW A. DEL COURT VAN KRIMPEN', HT, lp, 1917; flowers flesh-white and pale pink, tinted copper, dbl., moderate fragrance; [Unnamed seedling X Prince de Bulgarie]; Leenders, M.

'MEVROUW A. H. DE BEAUFORT', HT, pb, 1934; flowers clear salmon-pink, large, dbl., moderate fragrance; foliage bronze, glossy; vigorous growth; [Morgenglans X Gooiland Beauty]; Van Rossem

'MEVROUW AMÉLIE MÜLLER', HT, yb, 1927; flowers old-gold, shaded orange, dbl., moderate fragrance; [Golden Ophelia X Golden Emblem]; Verschuren

'MEVROUW BOREEL VAN HOGELANDEN', HT, w, 1918; flowers flesh-white shaded carmine and pink, dbl., intense fragrance; [Mme Leon Pain X Mme Antoine Mari]; Leenders, M.

'MEVROUW C. VAN MARWIJK KOOY', HT, w, 1921; flowers white, center indian yellow, sometimes coppery orange, dbl., moderate fragrance; [Mme Caroline Testout X Mrs Aaron Ward]; Leenders, M.

'MEVROUW D. A. KOSTER', Pol, mr, 1934; flowers bright red; [Dick Koster sport]; Koster, D.A.

'MEVROUW DORA VAN TETS', HT, mr, 1913; flowers velvety deep crimson, dbl., moderate fragrance; GM, Bagatelle, 1914; [Farbenkonigin X Gen. MacArthur]; Leenders, M.

'MEVROUW DR L. CROBACH', HT, pb, 1928; flowers carmine, base salmon, dbl., moderate fragrance; [Pink Pearl X Red Star]; Leenders, M.

'MEVROUW G. A. VAN ROSSEM', HT, ob, 1929; (Mrs G.A. van Rossem); flowers orange and apricot on golden yellow, veined red, reverse often dark bronze, dbl., intense fragrance; foliage very large, dark, bronze, leathery; vigorous growth; (28); [Souv. de Claudius Pernet X Gorgeous]; Van Rossem; C-P

'MEVROUW G. A. VAN ROSSEM, CLIMBING', Cl HT, ob, 1937; (Mrs G.A. van Rossem, Climbing); Gaujard

'MEVROUW G. DE JONGE VAN ZWYNSBERGEN', HT, pb, 1923; flowers pale flesh, center flesh-pink and salmon, dbl.; [Mme Mélanie Soupert X George C. Waud]; Leenders, M.

'MEVROUW H. CREMER', HT, mr, 1932; flowers large, dbl., intense fragrance; Buisman, G. A. H.; (Sangerhausen)

'MEVROUW HENRI DAENDELS', HT, op, 1931; flowers light salmon, large, very dbl., intense fragrance; Buisman, G. A. H.; (Sangerhausen)

'MEVROUW L. C. VAN GENDT', HT, ab, 1925; flowers salmon-apricot on yellow ground, dbl., moderate fragrance; [Unnamed seedling X Golden Emblem]; Van Rossem

'MEVROUW LALA PHILIPS', HT, or, 1931; flowers brilliant orange-toned shrimp-red, well formed, large, dbl.; foliage dark; vigorous growth; [Elvira Aramayo sport]; Leenders Bros.

'MEVROUW NATHALIE NYPELS', Pol, mp, 1919; (Nathalie Nypels); flowers rose-pink, medium, semi-dbl., intense fragrance; dwarf, spreading growth; (14); [Orléans Rose X (Comtesse du Cayla X R. foetida bicolor)]; Leenders, M.

Mevrouw S. van den Bergh, Jr. *see* **'MRS VANDENBERGH'**

'MEVROUW SMITS GOMPERTZ', HT, op, 1917; flowers yellowish salmon and coppery orange shaded lilac, dbl., intense fragrance; [Lady Wenlock X (Mme J.W. Budde X Souv. de Catherine Guillot)]; Leenders, M.

Mevrouw van Straaten van Nes *see* **'PERMANENT WAVE'**

'MEVROUW WELMOET VAN HEEK', HT, dp, 1933; flowers carmine-red, well formed; foliage dark; vigorous, bushy growth; Buisman, G. A. H.

'MEXICALI ROSE', F, yb, 1957; bud short, pointed, yellow suffused red; flowers deep yellow turning deep rose-pink, then cerise-red, open, dbl., 65–70 petals, 3 in., slight, spicy fragrance; foliage dark; vigorous, upright growth; [(Herrenhausen seedling X Golden Rapture) X Easter Parade]; Whisler; Germain's

Mexican Festival, S, dp, 1996; Dickson, Patrick

'MEXICANA', HT, rb, 1966; bud ovoid; flowers red, reverse silvery, large, dbl., 33 petals, exhibition form, moderate fragrance; foliage dark, glossy, leathery; vigorous, upright growth; [Kordes' Perfecta X Unnamed seedling]; Boerner; J&P

'MEXICO', F, or, 1944; bud globular, deep carmine; flowers deep scarlet, suffused orange, borne in clusters, semi-dbl., 18 petals, 4 in., cupped, slight fragrance; foliage leathery; strong stems; vigorous, bushy growth; [Baby Chateau X Helgoland]; Krause; J&P

Mexico, Min, rb

'MEYPINK', HT, pb, 1971; (**Portrait**, Stéphanie de Monaco); bud ovoid; flowers medium pink and light pink blend, medium, dbl., moderate fragrance; foliage glossy, dark; upright, bushy growth; AARS, (First amateur winner), 1972; [Pink Parfait X Pink Peace]; Meyer, C.; C-P

'MHAIRI'S WEDDING', HT, w, 1983; dbl., 35 petals, flat, moderate fragrance; foliage large, dark, matt; upright growth; [Dalvey X Virgo]; MacLeod, Major C.A.; Christie Nursery, Ltd.

Mia, HT, lp; Urban, J., 1975; (Czech Rosa Club)

'MIA MAID', HT, mp, 1953; bud ovoid; flowers phlox-pink, open, dbl., 40–50 petals, 3.5–4 in., moderate fragrance; foliage leathery, glossy; upright, compact growth; [Charlotte Armstrong X Signora]; Swim, H.C.; Mt. Arbor Nursery

'MIA SNOCK', HT, ly, 1925; flowers lemon-yellow, dbl.; [Mrs T. Hillas X Mrs Wemyss Quin]; Leenders, M.

'MIAMI', HT, yb, 1953; flowers orange, veined, reverse yellow, long, pointed, dbl., 25 petals, 5 in., slight fragrance; foliage dark; very vigorous, branching growth; [Mme Joseph Perraud X Fred Edmunds]; Meilland, F.; C-P

'MIAMI HOLIDAY', Min, rb, 1976; bud pointed; flowers red, reverse yellow, dbl., 60 petals, 1–1.5 in., moderate fragrance; foliage small, glossy; upright, bushy growth; [Seedling X Over the Rainbow]; Williams, Ernest D.; Mini-Roses

Miami Playa, HT; Dot; (Cavriglia)

'MICADA', Min, pb, 1992; (**Orlando Sunshine**); flowers pink and yellow blend, blooms borne mostly singly, dbl., 15–25 petals, 1.5–2.75 in., slight fragrance; some prickles; foliage large, medium green, semi-glossy; medium (18-24"), bushy growth; [Unnamed Miniature seedling X Unnamed Miniature seedling]; Williams, Michael C.; The Mini Rose Garden

'MICAELA', M, lp, 1864; flowers blush, compact, medium, very dbl.; erect growth; Moreau et Robert

'MICAËLA', HT, or, 1986; (Trudor); flowers orange-red, fading to dark carmine red, blooms in clusters; vigorous growth; [Manola X Unnamed seedling]; Kriloff, Michel

'MICAM', Min, rb, 1989; (**Carolina Morning**, 'MICAR'); bud pointed; flowers red with yellow center, aging red, small, borne singly and in small clusters, dbl., 20 petals, exhibition form, no fragrance; round, orange fruit; prickles straight, small, red; foliage small, medium green, semi-glossy; bushy, medium growth; [Rise 'n' Shine X Rainbow's End]; Williams, Michael C.; The Rose Garden & Mini Rose Nursery, 1990

'MICANGEL', Min, ab, 1996; (**Angel's Blush**); flowers apricot, darkest color is on the very edge of the petals, bl, dbl., 15–25 petals, 1.5–2.75 in., slight fragrance; few prickles; foliage small, medium green, semi-glossy; medium (50 cms), upright growth; [Seedling X Selected Pollen]; William, Michael C.; The Mini Rose Garden, 1997

'MICAR', Min, rb, 1989; (**Carolina Morning**, 'MICAM'); bud pointed; flowers red with yellow center, aging red, small, borne singly and in small clusters, dbl., 20 petals, exhibition form, no fragrance; round, orange fruit; prickles straight, small, red; foliage small, medium green, semi-glossy; bushy, medium growth; [Rise 'n' Shine X Rainbow's End]; Williams, Michael C.; The Rose Garden & Mini Rose Nursery, 1990

'MICAROLINE', Min, rb, 1998; (**Sweet Caroline**); flowers white with light to medium red on outer petals, dbl., 17–25 petals, 1.5 in., exhibition form, borne mostly singly, no fragrance; prickles moderate, small; foliage medium, dark green, semi-glossy; tall, 36 in., slightly spreading growth; [Seedling X Select Pollen]; Williams, Michael C.; The Mini Rose Garden, 1999

'MICAUTUMN', MinFl, yb, 1999; (**Autumn Splendor**); exhibition form, 26–40 petals, 2 in., exhibition form, borne mostly singly, slight; foliage large, medium green, semi-glossy; upright, tall (30-40 in.) growth; Williams, Michael C.; The Mini Rose Garden, 1999

'MICCOUNTRY', Min, pb, 1996; (**Old Country Charm**); flowers medium to dark pink with salmon, blooms borne mostly singly, dbl., 26–40 petals, no fragrance; foliage medium, dark green, dull; upright, medium growth; [Seedling X Selected pollen]; Williams, Michael C.; The Mini Rose Garden, 1996

'MICDEB', Min, dp, 1991; (**Debidue®**); flowers magenta, blooms borne mostly singly, dbl., 26–40 petals, 1.5 in., slight fragrance; some prickles; foliage medium, dark green, semi-glossy; medium (12 cms), upright growth; AOE, 1992; [Jazz Fest X Party Girl]; Williams, Michael C.; The Rose Garden & Mini Rose Nursery, 1992

'MICEVEN', Min, pb, 1989; (**Evening Shadows**); bud pointed; flowers medium, borne usuallysingly, dbl., 20 petals, exhibition form, slight, fruity fragrance; globular, light orange fruit; prickles slight downward curve, small, red; foliage large, medium green, semi-glossy; bushy, medium growth; [Tiki X Party Girl]; Williams, Michael C.; The Rose Garden & Mini Rose Nursery, 1990

'MICEXPLORE', Min, op, 1992; (**Explorer's Dream**); flowers deep orange pink, just a touch of yellow at base of each petal, dbl., 15–25 petals, 1.5 in., no fragrance; some prickles; foliage medium, dark green, semi-glossy; medium (50 cms), upright growth; [Miniature seedling X Homecoming]; Williams, Michael C.; The Rose Garden & Mini Rose Nursery

'MICFINEST', Min, ob, 1996; (**Finest Hour**); flowers very bright orange with yellow at base, creamy orange yellow, very dbl., no fragrance; some prickles; foliage medium, medium green, semi-glossy; upright, medium growth; [Glowry X Selected pollen]; Williams, Michael C.; The Mini Rose Garden, 1996

'MICGOLD', Min, my, 1986; (**Lander Gold**); dbl., 36 petals, exhibition form, borne usually singly, moderate, fruity fragrance; no fruit; very few, small, straight prickles; foliage medium, medium green, semi-glossy; medium, bushy growth; [Rise 'n' Shine X Little Jackie]; Williams, Michael C.; The Rose Garden & Mini Rose Nursery

Michael Bates' Moss, M, mp

Michael Crawford *see* POUlvue

'MICHAEL JON HALVORSON', HT, lp, 1999; dbl., 17–25 petals, 3 in., borne in small clusters, slight fragrance; prickles moderate; foliage medium, medium green, semi-glossy; spreading, tall (6 ft) growth; Adams, Dr. Neil D.

'MICHAEL LEEK', F, or, 1995; flowers bright orange-red, blooms borne 2-8 per cluster, semi-dbl., 5–8 petals, 1.5–2.75 in., slight fragrance; foliage medium, dark green, glossy; bushy (75-90 cms) growth; [Liverpool Echo X Traumerei]; Fleming, Joyce L.; Hortico Roses, 1994

'MICHAEL SAUNDERS', HT, pb, 1879; flowers deep pink and coppery red, well formed, medium, moderate fragrance; moderate growth; [Adam X Mme Victor Verdier]; Bennett

Michaelhouse Centenary *see* MEIonagre

'MICHEL BONNET', B, pb, 1864; flowers bright rose; vigorous growth; Guillot Père

Michel Hidalgo *see* 'MEITULANDI'

Michel Lis le Jardinier® *see* MEIfersi

Michelangelo *see* 'MACTEMAIK'

Michelangelo *see* MEItelov

'MICHÈLE', F, op, 1970; flowers deep salmon-pink, large, dbl., 25 petals, moderate fragrance; foliage light green; bushy growth; [Seedling X Orange Sensation]; deRuiter

Michele Bross, F, 1975; Dorieux; (Cavriglia)

'MICHÈLE MEILLAND', HT, lp, 1945; flowers light pink shaded lilac, center salmon, large, dbl.; vigorous growth; [Joanna Hill X Peace]; Meilland, F.

'MICHÈLE MEILLAND, CLIMBING', Cl HT, lp, 1951; (Grimpant Michèle Meilland); Meilland, F.

Michèle Torr® *see* 'JACOLITE'

'MICHELINE', HT, dp, 1953; flowers deep salmon-pink, base darker, dbl., 30–35 petals; foliage dark; [Edouard Renard X Luis Brinas]; de Basso, Mata

Michelle, F, rb, 1998

Michelle Joy *see* AROshrel

Michelle Wright, F, ab, 1998

'MICHIGAN', HT, pb, 1948; flowers salmon-carmine, reverse indian yellow, very large, dbl.; bushy growth; [Mme Joseph Perraud X Vive la France]; Mallerin, C.; URS

Michka *see* MEIvaleir

'MICHOME', Min, mp, 1989; (Homecoming, **Home-Coming**); bud pointed; flowers medium pink with slightly darker petal edges, medium, dbl., 35 petals, exhibition form, no fragrance; no fruit; prickles straight, green; foliage medium, medium green, semi-glossy; upright, tall growth; PP007454; [Tiki X Party Girl]; Williams, Michael C.; The Rose Garden & Mini Rose Nursery

'MICIMP', Min, pb, 1998; (**Lasting Impression**); flowers medium pink with darker pink edges, dbl., 17–25 petals, 1.5 in., borne mostly singly, no fragrance; no prickles; foliage medium, dark green, semi-glossy; compact, medium, 20 in. growth; [Seedling X Select Pollen]; Williams, Michael C.; The Mini Rose Garden, 1998

Mick Micheyl, Gr; (Cavriglia)

'MICKATE', F, mp, 1992; (**Katelyn Ann**); flowers deep pink, reverse pale pink with hint of yellow, aging pale, very dbl., 60 petals, 2.5 in., cupped, intense fragrance; foliage medium, dark green, glossy; medium, upright growth; [Unknown X Unknown]; Williams, Michael C., 1987; Roses Unlimited, 1992

'MICKI', HT, or, 1982; flowers light tangerine, dbl., 36 petals, exhibition form, moderate fragrance; small, brown prickles; foliage dark, glossy, leathery; vigorous, upright growth; [Fragrant Cloud X Tropicana]; James, John

'MICKIM', Min, ab, 1986; (**Kimberly**); flowers urn-shaped, blooms borne usually singly, dbl., 24 petals, exhibition form, moderate, fruity fragrance; globular, small, orange fruit; no prickles; foliage medium, medium green, semi-glossy; upright, medium growth; [Party Girl X Sheri Anne]; Williams, Michael C.; The Rose Garden & Mini Rose Nursery

'MICKY', HT, op, 1951; bud long, pointed; flowers coral-pink, single, 9 petals, slight fragrance; few thorns; foliage leathery; very vigorous, upright, bushy growth; [Lulu X Vesuvius]; Houghton, D.; Elmer Roses Co.

Micky, Cl Min, mr, 1995; Warner, Chris

'MICLANE', Min, m, 1991; (**Melody Lane**); dbl., 26–40 petals, 1.5 in., borne in small clusters, no fragrance; foliage small, medium green, semi-glossy; low (6 cms) growth; [Lavender Jewel X Party Girl]; Williams, Michael C.; The Rose Garden & Mini Rose Nursery, 1992

'MICLEE', Min, yb, 1991; (**Brenda Lee**); bud pointed; flowers red edge to yellow base, urn-shaped, small, dbl., 20 petals, exhibition form; foliage small, medium green, semi-glossy; bushy, low, compact growth; [Rise 'n' Shine X Rainbow's End]; Williams, Michael C., 1988; The Rose Garden & Mini Rose Nursery

'MICLEE', Min, yb, 1989; (**Virginia Lee**); bud ovoid; flowers creamy yellow with pink border, reverse creamy yellow, aging, dbl., 30 petals, cupped, no fragrance; globular, green orange-yellow fruit; prickles straight, green; foliage medium, dark green, glossy; slightly spreading, medium growth; [Rise 'n' Shine X Baby Katie]; Williams, Michael C.; The Rose Garden & Mini Rose Nursery

'MICMAC', S, w;, borne in clusters, non-recurrent; foliage deep purplish red; open habit (4 ft.) growth; hardy.; [R. rubrifolia X R. rugosa]; Central Exp. Farm

'MICMAG', Min, rb, 1988; ('MICMAGIC', **Mini Magic**); flowers white with red edges, reverse more red color, small, dbl., 35 petals, cupped, no fragrance; prickles straight, light green; foliage small, medium green, semi-glossy; slightly spreading, low growth; [Baby Katie sport X Watercolor.]; Williams, Michael C.; The Rose Garden & Mini Rose Nursery

'MICMAGIC', Min, rb, 1988; ('MICMAG', **Mini Magic**); flowers white with red edges, reverse more red color, small, dbl., 35 petals, cupped, no fragrance; prickles straight, light green; foliage small, medium green, semi-glossy; slightly spreading, low growth; [Baby Katie sport X Watercolor.]; Williams, Michael C.; The Rose Garden & Mini Rose Nursery

'MICODESSA', Min, m, 1998; (**Odessa**); flowers full, mauve, exhibition blooms, very dbl., 26–40 petals, 1.5 in., borne mostly singly, no fragrance; prickles few, small; foliage medium, dark green, matte; upright, tall (30 in) growth; [Jean Kenneally X Select Pollen]; Williams, Michael C.; The Mini Rose Garden, 1998

'MICONE', Min, mr, 1989; (**Bill Cone**); bud ovoid; flowers medium, borne usually singly, dbl., 35 petals, cupped, no fragrance; round, green orange-yellow fruit; prickles light green, globular; foliage medium, medium green, semi-glossy; upright, tall growth; [Heartland X Anita Charles]; Williams, Michael C.; The Rose Garden & Mini Rose Nursery

'MICPAL', Min, ob, 1992; (**Palmetto Sunrise**); flowers orange with a yellow base, yellow reverse, opening orange, m, dbl., 15–25 petals, 1.5–2.75 in.; some prickles; foliage medium, medium green, semi-glossy; upright (50 cms) growth; AOE, 1993; [Orange Honey X Unnamed Miniature seedling]; Williams, Michael C.; The Rose Garden & Mini Rose Nursery, 1993

'MICPEARL', Min, ly, 1998; (**Lemon Pearl**); flowers light yellow, very full, very dbl., 41 petals, 1–1.5 in., borne mostly singly, no fragrance; prickles few, small, straight; foliage medium, light green, matte; compact, medium (22 in.) growth; [Seedling X Select Pollen]; Williams, Michael C.; The Mini Rose Garden, 1998

'MICPIE', Min, op, 1988; (**Pierrine**); flowers medium pink, reverse slightly lighter, medium, borne singly, dbl., 40 petals, exhibition form, slight, damask fragrance; round, green orange-yellow fruit; prickles curved down slightly, light green; foliage medium, medium green, semi-glossy serrated edges; upright, medium growth; [Tiki X Party Girl]; William, Michael C.; The Rose Garden & Mini Rose Nursery

'MICPRIDE', Min, pb, 1991; (**Founder's Pride**); bud pointed; flowers deep pink, white center, mostly white reverse, aging deep pink, dbl., 24 petals, exhibition form, slight, spicy fragrance; foliage large, dark green, semi-glossy; upright growth; [Seedling X Party Girl]; Williams, Michael C., 1987; The Rose Garden & Mini Rose Nursery

'MICRUGOSA', S, lp; (R. vilmorinii, R. wilsonii, R. X micrugosa); single, 3 in.; orange-red fruit; depressed-globose prickles, about 1 3/4 in. diameter; (14); [R. roxburghii X R. rugosa]

'MICRUGOSA ALBA', S, w; flowers otherwise similar to micrugosa, single; [R. roxburghii X R. rugosa]

'MICSHADE', Min, yb, 1989; ('MICSHADY', **Shady Charmer**); bud ovoid; flowers light yellow base with light pink edges, aging cream, medium, dbl., 43 petals, exhibition form, slight, spicy fragrance; no fruit; prickles straight, very few, small, lightgreen; foliage medium, dark green, semi-glossy; bushy, medium growth; [Party Girl X Anita Charles]; Williams, Michael C.; The Rose Garden & Mini Rose Nursery, 1990

'MICSHADY', Min, yb, 1989; ('MICSHADE', **Shady Charmer**); bud ovoid; flowers light yellow base with light pink edges, aging cream, medium, dbl., 43 petals, exhibition form, slight, spicy fragrance; no fruit; prickles straight, very few, small, lightgreen; foliage medium, dark green, semi-glossy; bushy, medium growth; [Party Girl X Anita Charles]; Williams, Michael C.; The Rose Garden & Mini Rose Nursery, 1990

'MICSOX', Min, rb, 1986; (**Sox**); flowers medium red, small white eye blooms borne usually singly, single, 7 petals, 1 in., no fragrance; small, globular, orange fruit; small, straight prickles; foliage medium, dark; medium, upright growth; [Baby Katie X Angel Darling]; Williams, Michael C.; The Rose Garden & Mini Rose Nursery

'MICSTEAL', Min, mp, 1995; (**Stolen Dream**); flowers medium pink, touch of yellow at base of petals, dbl., 15–25 petals, 1.5–2.75 in., no fragrance; foliage small, dark green, semi-glossy; upright (55 cms) growth; Williams, Michael C.; The Mini Rose Garden, 1995

'MICSURF', Min, pb, 1988; (**Surfside**); flowers medium pink to creamy yellow to creamy white, medium, borne, dbl., 24 petals, exhibition form, no fragrance; round, medium, green orange-yellow fruit; prickles curved down, very few, red-green; foliage large, light green, semi-glossy; upright, medium growth; [Tiki X Party Girl]; Williams, Michael C.; The Rose Garden & Mini Rose Nursery

'MICTWIST', Min, ob, 1996; (**Tangerine Twist**); flowers medium orange with yellow, light yellow reverse edged orange, dbl., 15–25 petals, no fragrance; few prickles; foliage medium, medium green, semi-glossy; upright, medium growth; [Pierrine X Selected pollen]; Williams, Michael C.; The Mini Rose Garden, 1996

'MICURIN', LCl, mr, 1936; flowers bright red, very large, borne in large clusters, semi-dbl., globular, abundant seasonal bloom; foliage dark, soft; vigorous, climbing growth; Böhm, J.

'MICVOYAGE', Min, yb, 1996; (**Voyager**); flowers yellow blend, at times touch of pink, blooms borne mostly singly, dbl., 26–40 petals, 1.5–2.75 in., no fragrance; foliage medium, dark green, semi-glossy; medium (50 cms), upright growth; [Seedling X Selected Pollen]; Williams, Michael C.; The Mini Rose Garden, 1997

Midas *see* 'LEGGA'

Midas Touch™ *see* 'JACTOU'

Middlesbrough Football Club *see* 'HORFLAME'

Middlesbrough Pride *see* 'NOSMID'

Middlesex County *see* 'BOSANNE'

Middo, F, 1971; Laperrière; (Cavriglia)

'MIDGET', Min, mr, 1941; bud pink; flowers carmine-red, micro-mini, dbl., 20 petals, .5 in., slight fragrance; foliage fern-like; dwarf growth; [Ellen Poulsen X Tom Thumb]; deVink; C-P

Midget Gem®, S, ly, 1994; Dickson, Patrick

Midi®, HT, op

'MIDINETTE', HT, mp, 1962; flowers cyclamen-pink, well formed, dbl., moderate fragrance; vigorous growth; [Pink Spiral X Seedling]; Delforge

'MIDJET', Min, 1941; deVink, J.; (Cavriglia)

'MIDNIGHT', HT, dr, 1956; bud urn shaped; flowers currant-red to cardinal-red, dbl., 23–30 petals, 3.5–4.5 in., exhibition form, intense fragrance; foliage dark, glossy; long strong stems; vigorous, compact growth; [Gay Lady X Texas Centennial]; Swim, H.C.; Armstrong Nursery

Midnight Magic™ *see* 'WILMNMG'

Midnight Rambler *see* 'TALMID'

'MIDNIGHT RENDEZVOUS', Min, dr, 1985; flowers small, dbl., 27 petals, slight fragrance; foliage small, dark, semi-glossy; bushy growth; [Scarlet King X Big John]; Hardgrove, Donald L.; Rose World Originals

'MIDNIGHT SUN', HT, dr, 1921; flowers deep crimson flushed velvety black, semi-dbl.; [Star of Queensland X Red-Letter Day]; Grant; Kershaw

'MIDNITE SUN', HT, yb, 1955; flowers buttercup-yellow edged red, becoming lighter, dbl., 40 petals, 5 in., moderate fragrance; bushy growth; [Sutter's Gold X R. wichurana seedling]; Brownell, H.C.

'MIDSUMMER', HT, mp, 1947; semi-dbl., 20 petals, 4–5 in., exhibition form, moderate, fruity fragrance; foliage leathery, dark; [Heinrich Wendland X Lady Sylvia]; Prosser

Midtown Mini-Climber, Cl Min, ob

Midwest Living® *see* 'KINLIV'

'MIE', Min, mr, 1980; bud long, pointed; semi-dbl., 18 petals, cupped, borne several together, slight fragrance; tiny, curved prickles; foliage very tiny, medium green; very compact growth; [Red Can Can X Unnamed seedling]; Lyon

'MIEN DE JONGE', F, mr, 1969; bud ovoid; flowers scarlet, large, semi-dbl., cupped, moderate fragrance; vigorous, bushy growth; [Sumatra X Seedling]; Verschuren, A.; Stassen

'MIES BOUWMAN', F, op, 1973; bud round; flowers salmon-reverse yellow, slight fragrance; foliage glossy, dark; vigorous, bushy growth; [Lijnbaanroos X Unknown seedling]; Buisman, G. A. H.

'MIESZKO', Pol, lp, 1966; flowers pale pink, small; foliage small, light green; low growth; Grabczewski

'MIEZE', Pol, ob, 1909; flowers orange-yellow, borne in small clusters; vigorous growth; very hardy.; [Petite Léonie X R. foetida bicolor]; Lambert, P.

'MIEZE SCHWALBE', Pol, mr, 1927; flowers rose-red; [Frau Rudolf Schmidt sport]; Lohse & Schubert

Mighty Moe *see* 'MOEMIGHTY'

Mighty Mouse *see* 'MACMIGMOU'

'MIGNARDISE', Pol, mr, 1971; bud ovoid; flowers open, large, single, slight fragrance; foliage large, leathery; vigorous, upright growth; [Luc Varenne X Seedling]; Delforge

Mignon *see* **'MLLE CÉCILE BRÜNNER'**

Mignon, Climbing *see* **'MLLE CÉCILE BRÜNNER, CLIMBING'**

'MIGNONETTE', Pol, lp, 1880; flowers rose, sometimes blush white bordered with wine-red spots, dbl., 1 in., blooms in short, full panicles of 50, resembling the Chinas; hooked, red prickles; foliage dark green above, reddish beneath, glossy, 5-7 leafl; branches dark red; very dwarf, bushy growth; (14); [Dbl. Fld. Multiflora X probably China or Tea]; Guillot et Fils

'MIGNONNE', HT, 1962; bud long, pointed; flowers bright salmon-pink, large, dbl., 80 petals, moderate fragrance; foliage leathery; very vigorous growth; [Mme Butterfly X Fernand Arles]; Gaujard; Ilgenfritz Nursery, 1966

Miguel Aldrufeu *see* **'LINDA PORTER'**

Mijayima, F, dy, 1992; flowers large, dbl.; Tagashira, Kazuso; (Sangerhausen)

'MIKADO', Ch, dy, 1929; flowers deep golden yellow; vigorous growth; RULED EXTINCT 2/87; Dobbie

'MIKADO'™, HT, rb, 1987; (Kohsai, Koh-Sai); flowers brilliant luminous light scarlet, suffused with yellow at base, dbl., 30–35 petals, exhibition form, slight fragrance; rounded, dull orange-red fruit; slightly recurved, medium green prickles, tinged purple; foliage medium, medium green, glossy; upright, tall growth; PP006470; AARS, 1988; [Duftwolke X Kagayaki]; Suzuki, Seizo, 1976; C-P, 1987;, Keisei Rose Nursery, 1987

'MIKAGURA', HT, dp, 1999; flowers bright velvet red, dbl., 30 petals, exhibition form, slight fragrance; foliage medium green; 4.5 ft growth; BM, Japan Rose Concours, 1995; [Christian Dior X seedling]; Hayashi, Shunzo, 1992

Mike Thompson *see* 'SHERIRED'

'MIKE'S OLD-FASHIONED PINK', S, lp, 1999; dbl., 26–40 petals, 2.5 in., borne in small clusters, slight fragrance; prickles moderate; foliage large, dark green, semi-glossy; bushy, medium (4 ft) growth; [Heritage X William Baffin]; Lowe, Malcolm

Mikheil, F, m, 1998

'MIKULÁS ALES', HT, lp, 1936; flowers rosy salmon-white, intense fragrance; very vigorous growth; Bojan

Miky Tea *see* 'SHERIMIKY'

'MIL LAMP', HT, w, 1991; flowers white with pink border, moderately medium, dbl., 15–25 petals, intense fragrance; foliage medium, medium green, semi-glossy; bushy growth; [Admiral Rodney sport]; Colclasure, C.E., 1989

'MILADY', HT, dp, 1914; flowers deep pink, dbl., exhibition form, moderate fragrance; foliage small, soft; dwarf, sparse bloom. growth; [Richmond X J.B. Clark]; Towill; A.N. Pierson

'MILAGROS DE FONTCUBERTA', HT, m, 1968; bud pointed; flowers violet-mauve, medium, dbl., 50 petals, moderate fragrance; upright, compact growth; [(Sterling Silver X Intermezzo) X (Sterling Silver X Simone)]; Dot, Simon; Rosas Dot

'MILAN', F, rb, 1986; flowers red, silver or yellow reverse, medium, dbl., 28 petals, slight fragrance; foliage medium, medium green, semi-glossy; upright, bushy growth; [Unnamed seedling X Unnamed seedling]; Jerabek, Paul E.

'MILANO', HWich, op, 1923; bud very long, pointed; flowers nasturtium-pink on indian yellow ground, borne in large clusters, dbl., moderate fragrance; foliage glossy; vigorous, climbing (10 ft) growth; Ingegnoli

'MILBEN', Min, pb, 1988; (**Dr B. Benacerraf**); flowers white to light pink, dark pink borders, opening to light pink, dbl., 15–20 petals, slight fragrance; foliage ovoid, medium green, matt, disease resistant; vigorous, compact growth; [(Double Delight X Simplex) X Magic Carrousel]; Miller, F.

'MILDEWFREE ELSE POULSEN', F, pb, 1937; flowers pink slightly tinged yellow, medium, single; [Else Poulsen X Dainty Bess]; Poulsen, S.

'MILDRED CANT', HT, mr, 1935; bud pointed; flowers bright crimson, very large, dbl., exhibition form, intense fragrance; foliage leathery; vigorous, open growth; Cant, B. R.

'MILDRED GRANT', HT, w, 1901; flowers silvery white, tinted pink at edge, dbl., exhibition form; very large, vigorous growth; GM, NRS, 1898; [Niphetos X Mme Mélanie Willermoz]; Dickson, A.

'MILDRED REYNOLDS', F, mr, 1966; flowers cardinal-red, large, borne in large, compact clusters, 3 in., cupped, slight fragrance; foliage glossy; [Peace X Seedling]; Dorieux; Bees

Mildred Scheel *see* **'DEEP SECRET'**

Milena, HT, 1971; Vecera, L.; (Cavriglia)

Milestone *see* 'JACLES'

Milevsko, S, mp, 1980; flowers medium, dbl.; Vecera, L.; (Sangerhausen)

Mili de Vega, HT; Moreira da Silva, A.; (Cavriglia)

'MILKANA', HT, rb, 1974; flowers brick-red, outer petals shaded pink, large, dbl., 50 petals; foliage dark, glossy; vigorous, upright, tall growth; [Tallyho X Spartan]; Staikov, Prof. Dr. V.; Kalaydjiev and Chorbadjiiski

'MILKMAID', N, w, 1925; flowers white tinted fawn, small blooms in clusters, semi-dbl., moderate fragrance; foliage rich green; very vigorous, climbing growth; [Crepuscule X Unnamed variety]; Clark, A.; Brundrett

Milky Way *see* 'INTERWAY'

'MILKY WAY', HWich, w, 1900; flowers pure white, tips lightly tinged pink, stamens yellow, very large, semi-dbl., seasonal bloom, moderate fragrance; foliage glossy; vigorous, climbing growth; RULED EXTINCT 9/86; Walsh

'MILKYWAY', Gr, w, 1986; flowers yellow stamens, loose, medium, borne singly, semi-dbl., 21 petals, cupped, slight, fruity fragrance; straight, small, greenish-yellow prickles; foliage small, medium green, semi-glossy; spreading, medium growth; [Unnamed seedling X Unnamed seedling]; Ohlson, John

'MILLEDGEVILLE', (natural variation of R. setigera), mp, 1842; flowers carmine

Millefleurs, Min, ob

Millennium® *see* 'GRALOVE'

Millennium® *see* 'PERMILL'

'MILLER HAYES', HP, mr, 1873; flowers large, dbl., moderate fragrance; Verdier, E.; (Sangerhausen)

Millgrove, HMult, lp, 1998

'MILLICENT', S, mp; flowers light coral-red fading to flesh-pink, reverse yellowish, non-recurrent; flattened, globe shape, light red fruit; foliage dark green veined red-brown; medium-tall growth; hardy.; Central Exp. Farm

'MILLIE', Pol, rb, 1937; flowers light cherry-red, base yellow, borne in clusters; foliage dark; strong stems; Russ

'MILLIE PERKINS', HT, mr, 1960; flowers velvety red, well formed, large, dbl.; long, strong stems; vigorous growth; [Ena Harkness X Chrysler Imperial]; Maarse, G.

Millie Walters® *see* 'MORMILLI'

Millionaire *see* PEAzara

'MILORD', F, dr, 1961; flowers large, borne in clusters, dbl.; foliage dark; vigorous growth; [Opera X Ville de Gand]; Gaujard

'MILORD', HT, mr, 1962; flowers crimson-scarlet, well formed, dbl., 35 petals, 5–6. in., intense fragrance; foliage dark; upright growth; [Rubaiyat X Karl Herbst]; McGredy, Sam IV; McGredy

'MILOU', F, dr, 1964; flowers dark vermilion-red, medium, semi-dbl.; vigorous, low growth; Mondial Roses

Milrose® *see* 'DELBIR'

'MILSWEET', F, ab, 1989; (**Elsie Warren**); flowers apricot with lemon yellow eye, medium, dbl., 15–25 petals, moderate fragrance; foliage large, medium green, semi-glossy; upright growth; [Arthur Bell X Arthur Bell]; Milner, William; Battersby Roses, 1990

Milva® *see* 'TANAVLIM'

Mimi *see* 'MEIDESI'

'MIMI COERTSE', Gr, mp, 1963; bud pointed; flowers bright rose-pink, dbl., 4 in., exhibition form; foliage glossy; upright growth; [Queen Elizabeth X Constantia]; Herholdt, J.A.; Herholdt's Nursery

Mimi Golden® *see* MEIspreyp

Mimi Pearl®, F, w

Mimi Pink® *see* 'JELDANIRAN'

'MIMI PINSON', Pol, dp, 1919; flowers clear crimson, passing to purplish rose and then to neyron purple; Barbier

Mimi Rose *see* 'JELDANIRAN'

'MIMOLLET', F, lp, 1975; bud pointed; flowers bright light pink, dbl., 40 petals, 3–4 in., exhibition form; foliage glossy; upright growth; [(Queen Elizabeth X Ethel Sanday) X Zambra]; Ota, Kaichiro

'MIN JURGENSEN', HT, w, 1973; flowers center buff, aging to pink edges, full, medium, dbl., slight fragrance; foliage dark, soft; very vigorous, upright growth; [Pascali sport]; Bauer; Trewallyn Nursery

'MINAACO', Min, dp, 1985; (**Cindy®**); flowers deep pink, well-formed, small, dbl., slight fragrance; foliage small, dark, glossy; [Tom Brown X Over the Rainbow]; Williams, Ernest D.; Mini-Roses

'MINABBCO', Min, mr, 1991; (**Redglo®**); flowers very color fast, velvety knapp, outstanding substance, dbl., 26–40 petals, 1.5 in., exhibition form, slight fragrance; some prickles; foliage small, dark green, semi-glossy; low (36 cms), bushy growth; [Starburst X Over the Rainbow]; Williams, Ernest D.; Mini-Roses, 1992

'MINABCCO', Min, m, 1991; (**Space Walk®**); flowers mauve-tan/yellow, russet reverse, blooms borne mostly singly, dbl., 26–40 petals, 1.5 in., intense fragrance; few prickles; foliage small, dark green, semi-glossy; low (30 cms), bushy growth; [Unnamed seedling X Twilight Trail]; Williams, Ernest D.; Mini-Roses, 1992

'MINABCO', Cl Min, yb, 1985; (**Touch o' Midas®**); flowers deep yellow, petals edged deep pink, small blooms borne singly, dbl., 35 petals, slight fragrance; foliage small, dark, semi-glossy; upright (to 5 ft) growth; [Little Darling X Over the Rainbow]; Williams, Ernest D.; Mini-Roses

'MINABDCO', Min, pb, 1991; (**Pretty Woman**); flowers pink with yellow and orange, deeper at edges, aging darker, dbl., 26–40 petals, 1.5 in., exhibition form, moderate fragrance; few prickles; foliage small, medium green, glossy; low (30 cms), bushy growth; [Tom Brown X Over the Rainbow]; Williams, Ernest D.; Mini-Roses, 1992

'MINABECO', Min, rb, 1992; (**Love Torch®**); bud long, pointed; flowers red, orange and yellowblend, holds color well, heavy substance, dbl., 26–40 petals, 1.5 in., slight fragrance; some prickles; foliage small, dark green, glossy; low (30 cms), upright, bushy growth; hardy.; [Starburst X Over the Rainbow]; Williams, Ernest D.

'MINACCO', Cl Min, dp, 1985; (**Jazz Time®**); flowers deep pink, small blooms in clusters, dbl., 35 petals, slight fragrance; foliage small, dark, semi-glossy; upright, bushy (4-5 ft) growth; [Little Darling X Little Chief]; Williams, Ernest D.; Mini-Roses, 1986

'MINACO', Min, m, 1982; (**Angelglo®**); flowers lavender, micro-mini, small, dbl., slight fragrance; foliage small, dark, semi-glossy; bushy growth; PP005356; [Angel Face X (Angel Face X Over the Rainbow)]; Williams, Ernest D.; Mini-Roses

'MINACON', Min, rb, 1995; (**Space Probe**); flowers red edges on white petals, blooms, dbl., 15–25 petals, 1.5 in., moderate fragrance; some prickles; foliage small, medium green, semi-glossy; medium, upright, bushy growth; [Little Darling X Over the Rainbow]; Williams, Ernest D.; K&C Roses, 1995

'MINADCO', Min, m, 1986; (**Royalglo®**); flowers non-fading, small, dbl., 40 petals, exhibition form, borne usually singly, intense, damask, sweet fragrance; no fruit; few, long, light tan prickles; foliage small, dark green, semi-glossy; bushy, medium growth; PP006673; [Angel Face X Anita Charles]; Williams, Ernest D.; Mini-Roses

'MINAECO', Min, ab, 1986; (**Apricot Charm®**); flowers apricot, reverse slightly deeper, blending to yellow at base, dbl., 42 petals, exhibition form, borne usually singly, slight fragrance; no fruit; few, long, thin, short, light tan prickles; foliage small, dark green, glossy; bushy, spreading, medium growth; [Gingersnap X Anita Charles]; Williams, Ernest D.; Mini-Roses

'MINAFCO', Min, pb, 1986; (**Orange Love®**); flowers orange, reverse deeper orange to dark red, fading lighter, dbl., 45 petals, exhibition form, slight fragrance; no fruit; very few, short, light tan prickles; foliage small, dark green, semi-glossy; upright, bushy, medium growth; PP006546; [Tom Brown X Over the Rainbow]; Williams, Ernest D.; Mini-Roses

'MINAGCO', Min, r, 1986; (**Honey 'n' Spice®**); flowers tan with red highlights, reverse deeper tan with more red, small, dbl., 45–49 petals, borne usually singly, moderate, fresh honey fragrance; no fruit; few, short, thin, light tan prickles; foliage small, medium green, semi-glossy; bushy, medium growth; [Tom Brown X Over the Rainbow]; Williams, Ernest D.; Mini-Roses

'MINAHCO', Min, dy, 1986; (**Summer Beauty**™); flowers small, borne usually singly or in sprays of 3-5, dbl., exhibition form, moderate fragrance; no fruit; very few, thin, short, light prickles; foliage small, dark green, semi-glossy; upright, bushy, medium growth; [Unnamed seedling X Unnamed seedling]; Williams, Ernest D.; Mini-Roses

'MINAICO', Cl Min, mr, 1986; (**Red Delight**™); flowers medium red, yellow at base, small, dbl., 33 petals, exhibition form, borne singly and in sprays, intense, spicy fragrance; round, red fruit; few, small, tan prickles, slanted downwards; foliage small, dark green, semi-glossy; tall growth; PP006893; [Golden Song X Magic Mist]; Williams, Ernest D.; Mini-Roses, 1987

'MINAJCO', Min, m, 1987; (**Purple Majesty**™); flowers mauve, edged red, reverse mauve, small, dbl., 40 petals, exhibition form, borne usually singly; fruit not observed; tan, needle-like prickles, dilated at base; foliage small, medium green, semi-glossy; bushy, medium growth; PP006894; [Tom Brown X Black Jack]; Williams, Ernest D.; Mini-Roses

'MINAKCO', Min, lp, 1987; (**Pink Kiss**™); flowers light pink, non-fading, small, borne usually singly or in spr, dbl., 40–45 petals, exhibition form, no fragrance; fruit not observed; tan-red prickles, dilated at base; foliage small, medium green, semi-glossy; bushy, medium growth; [Tom Brown X Anita Charles]; Williams, Ernest D.; Mini-Roses

'MINALCO', Min, ob, 1987; (**Orange Star**™); flowers orange, reverse orange with yellow at base, non-fading, high, dbl., 35 petals, exhibition form, intense, damask fragrance; fruit not observed; tan prickles, dilated at base; foliage small, medium green, semi-glossy; bushy, spreading, medium growth; [Unnamed Miniature seedling X Unnamed Miniature seedling]; Williams, Ernest D.; Mini-Roses

'MINAMCO', Min, mr, 1987; (**Ethel Orr**™); flowers small, borne usually singly or in sprays of 3-5, dbl., 35 petals, exhibition form, slight fragrance; no fruit; few, small, tan prickles; foliage small, medium green, glossy; upright, bushy, medium, profuse growth; [Miniature seedling X Big John]; Williams, Ernest D.; Mini-Roses

'MINANCO', Min, or, 1987; (**Oriental Simplex**™); flowers bright orange-red, reverse creamy yellow, aging deeper, non-, single, 5 petals, flat, no fragrance; fruit not observed; tan prickles, declining, dilated at base; foliage small, medium green, glossy; upright, bushy, medium growth; [(Starburst X Over the Rainbow) X Little Chief]; Williams, Ernest D.; Mini-Roses

'MINAPCO', Min, r, 1988; (**Amberglo**™); flowers small, dbl., 34 petals, intense fragrance; foliage small, dark green, semi-glossy; bushy, sturdy growth; [Tom Brown X Twilight Trail]; Williams, Ernest D.

'MINAQCO', Min, w, 1988; (**White Charm**™); bud long; flowers small, dbl., 33 petals, exhibition form, intense fragrance; foliage small, medium green, glossy; bushy growth; [Tom Brown X Over the Rainbow]; Williams, Ernest D.

'MINARCO', Min, mp, 1988; (**Evelyn Rogers**™); flowers small, dbl., 34 petals, slight fragrance; foliage small, medium green, glossy; upright, bushy growth; [Tom Brown X Over the Rainbow]; Williams, Ernest D.

'MINASCO', Min, dy, 1988; (**The Garden Editor**®); flowers small, dbl., 33 petals, moderate fragrance; foliage small, medium green, glossy; upright, bushy growth; [Gold Badge X Yellow Jewel]; Williams, Ernest D.

'MINATCO', Min, mp, 1988; (**Ta Ta**); flowers small, dbl., 33 petals, no fragrance; foliage small, medium green, semi-glossy, dense; bushy growth; [Tom Brown X Over the Rainbow]; Williams, Ernest D.

'MINAUCO', Min, m, 1988; (**Lavender Star**™); flowers mauve blended lavender-tan, small, single, 5 petals, intense fragrance; foliage small, dark green, semi-glossy; upright, bushy, dense growth; [Unnamed seedling X Lavender Simplex]; Williams, Ernest D., 1989

'MINAVCO', Min, r, 1989; (**Sue Jo**™); flowers russet, outside petals tan-lavender, inside petals golden-am, dbl., 33 petals, intense fragrance; foliage small, medium green, glossy; upright, bushy growth; PP007606; [Tom Brown X Twilight Trail]; Williams, Ernest D.; Mini-Roses

'MINAWCO', Min, r, 1989; (**Suntan Beauty**™); flowers tan with rosy highlights, small, dbl., 33–45 reflexed petals, exhibition form, intense fragrance; foliage small, dark green, glossy, disease-resistant; bushy growth; PP007596; [(Angel Face X Golden Angel) X Yellow Jewel]; Williams, Ernest D.; Mini-Roses

'MINAXCO', Min, m, 1989; (**Smoke Signals**™); flowers smoky, lavender-gray, small, dbl., 33 petals, exhibition form, intense fragrance; foliage small, medium green, glossy; upright, bushy growth; PP007604; [[Tom Brown X (Rise 'n' Shine X Watercolor)] X Twilight Trail]; Williams, Ernest D.; Mini-Roses

'MINAYCO', Min, m, 1989; (**Ida Belle**™); flowers lavender with blends of amber, very long-lasting, small, dbl., 33 petals, exhibition form, intense fragrance; foliage small, medium green, glossy; bushy growth; PP007605; [[Tom Brown X ((Rise 'n' Shine) X Watercolor)] X Twilight Trail]; Williams, Ernest D.

'MINAZCO', Min, r, 1989; (**Sands of Time**™); flowers small, dbl., 32 petals, exhibition form, intense fragrance; foliage small, medium green, glossy; upright, bushy growth; PP007607; [Tom Brown X Twilight Trail]; Williams, Ernest D.; Mini-Roses

'MINBACO', Min, mp, 1989; (**Jewel's Delight**™); flowers small, dbl., 33 petals, moderate fragrance; foliage small, medium green, glossy; bushy growth; [Tom Brown X Twilight Trail]; Williams, Ernest D.

'MINBCO', Min, dy, 1982; (**Great Day**); bud long, pointed; flowers well-formed, small, dbl., 35 petals, moderate fragrance; foliage small, light green, glossy; bushy growth; PP005358; [(Little Darling X Gold Coin) X (Little Darling X Gold Coin)]; Williams, Ernest D.; Mini-Roses

'MINCCO', Min, pb, 1982; (**Peggy's Delight**); flowers deep pink, white reverse, small, dbl., slight fragrance; foliage small, dark, glossy; bushy growth; PP005357; [Little Darling X Over the Rainbow]; Williams, Ernest D.; Mini-Roses

'MINDCO', Min, or, 1982; (**Orange Charm**); flowers well-formed, small, dbl., slight fragrance; foliage small, medium green, glossy; bushy growth; [Starburst X Over the Rainbow]; Williams, Ernest D.; Mini-Roses

'MINDOR', F, yb, 1976; flowers deep yellow, slightly flushed orange, urn shaped, dbl., 30–60 petals, 3 in.; foliage dark; vigorous, low growth; [Dr. Faust X Mme Lucky]; Station Exp. de Roses

'MINDREAM', Min, m, 1992; (**Twilight Dream**); flowers good color stability, heavy substance, blooms borne mostly singly, very dbl., 1.5 in., exhibition form, good repeat, intense fragrance; few prickles; foliage small, dark green, glossy; low (30 cms), bushy growth; hardy; [Tom Brown X Twilight Beauty]; Williams, Ernest D.

'MINECO', Min, pb, 1982; (**Spring Beauty**); flowers pastel pink and yellow blend, small, dbl., 35 petals, slight fragrance; foliage small, medium green, glossy; upright growth; [Little Darling X Over the Rainbow]; Williams, Ernest D.; Mini-Roses

Minerette *see* 'INTERMINER'

Miners Cottage, Misc OGR, dp

'MINERVE', HT, yb, 1947; flowers chrome-yellow shaded red, high pointed, dbl., 36 petals, moderate fragrance; foliage thick, dark; vigorous, branching growth; [Peace X Prinses Beatrix]; Meilland, F.

Minette, A, lp

'MINFCO', Cl Min, pb, 1982; (**Break o' Dawn**); flowers white with deep pink edges, white reverse, small, dbl., 35 petals, slight fragrance; foliage small, dark, glossy; upright, bushy, climbing growth; [Little Darling X Over the Rainbow]; Williams, Ernest D.; Mini-Roses

'MINFIRE', Min, or, 1998; (**Watchfire**); flowers bright coral red with deep yellow base, full, very dbl., 35 petals, 1.25–1.5 in., exhibition form, borne mostly singly, slight fragrance; prickles moderate; foliage medium, medium green, semi-glossy; upright, compact, bushy, medium (18 in.) growth; [Sue Jo X Twilight Trail]; Williams, Ernest D.; Texas Mini Roses, 1997

'MINFRIV', Min, mr, 1998; (**Frivolous**); flowers cherry red, broad petals, good substance, single, 8–10 petals, 1.5 in., borne mostly singly, no fragrance; almost thornless; foliage medium, semi-glossy; upright, medium (18 in.) growth; [Twilight Trail X Seedling]; Williams, Ernest D.; Texas Mini Roses, 1997

'MING TOY', F, dp, 1947; bud globular; flowers deep rose-pink, large, loose trusses, dbl., 50 petals, 2 in., moderate fragrance; foliage leathery, dark; vigorous, upright, bushy growth; Krebs; H&S

'MINGCO', Min, rb, 1983; (**Fiesta Time**®); flowers yellow to orange to red, reverse yellow, small, dbl., slight fragrance; foliage small, dark, glossy; bushy growth; [Starburst X Over the Rainbow]; Williams, Ernest D.; Mini-Roses

Minha Mulher, HT, 1969; Moreira da Silva, A.; (Cavriglia)

'MINHCO', Min, mp, 1983; (**DeGrazia's Pink**); flowers heavy substance, small, dbl., slight fragrance; foliage small, dark, glossy; bushy growth; [(Seedling X Over the Rainbow) X (Seedling X Over the Rainbow)]; Williams, Ernest D.; Mini-Roses

'MINHEART', Min, r, 1998; (**Lionheart**); flowers russet with coral edge, deep yellow reverse, dbl., 25 petals, 1.2in., borne mostly singly, moderate fragrance; prickles moderate; foliage medium, dark green, semi-glossy; upright, bushy, medium growth; [Seedling X Twilight Trail]; Williams, Ernest D.; Texas Mini Roses, 1997

Mini Lights *see* 'DICMOPPET'

Mini Magic *see* 'MICMAGIC'

Mini Magic, Min, lp; Walsh, 1998; (Weatherly, L.)

Mini Mercedes, Min, mr

Mini Metro *see* 'RUFIN'

Mini Nicole, Min, pb

'MINI PINK MELODIES', Min, mp, 1991; flowers moderately small, dbl., 15–25 petals, slight fragrance; foliage small, medium green, semi-glossy; bushy growth; [Unnamed seedling X Unnamed seedling]; Spooner, Raymond A.

'MINIATURE', Pol, lp, 1884; flowers pink, becoming yellowish white, very small, very dbl., intense fragrance; moderate growth; Alégatière

'MINIBETH', Min, m, 1994; (**Marybeth**®); flowers pastel mauve, very double, blooms borne mostly, single, exhibition form, moderate fragrance; few prickles; foliage small, medium green, semi-glossy; medium (18-20 in), upright growth; [Seedling X Twilight Trail]; Williams, Ernest D.; Texas Mini Roses, 1994

'MINIBID', Min, m, 1994; (**Forbidden**®); flowers medium lavender with dark lavender on edges, dbl., 26–40 petals, 1.5 in., moderate fragrance; few prickles; foliage small, dark green, semi-glossy; low (14-16 in), bushy growth; [Seedling X Twilight Trail]; Williams, Ernest D.; Texas Mini Roses, 1994

'MINICIN', Min, r, 1993; (**Cinnamon Delight**®); flowers russet with deep yellow base, dbl., 26–40 petals, 1.5 in., borne mostly single, intense fragrance; few prickles; foliage small, dark green, semi-glossy; medium (40 cms), upright growth; [Seedling X Twilight Trail]; Williams, Ernest D.; Mini Roses of Texas, 1993

'MINICO', Cl Min, ob, 1983; (**Mary Marshall, Climbing**); Williams, Ernest D.; Mini-Roses, 1982

'MINIDOR', Min, yb, 1996; (**Fall Splendor**); flowers rich yellow with orange blush on edges, blooms borne mostly, dbl., 26–40 petals, 1.25–1.5 in., moderate fragrance; some prickles; foliage medium, dark green, semi-glossy; medium (16 in), compact, bushy growth; [Rise 'n' Shine X Twilight Trail]; Williams, Ernest D.; Texas Mini Roses, 1996

'MINIFIRE', Min, rb, 1993; (**Dragon's Fire**®); flowers bright red with deep yellow at base of petals, yellow reverse, dbl., 26–40 petals, 1.5 in., moderate fragrance; few prickles; foliage small, medium green, semi-glossy; medium (35 cms), upright, compact growth; [Starburst X Twilight Trail]; Williams, Ernest D.; Mini Roses of Texas, 1993

'MINIGOLD', F, my, 1970; flowers pointed, dbl., 35 petals, 2.5 in., moderate fragrance; foliage glossy, dark; [Whisky Mac X Zorina]; Tantau, Math.

Minijet® *see* MEIrotego

Minilights *see* 'DICMOPPET'

'MINIMAG', Min, m, 1993; (**Magician**®); flowers mauve with deeper petal edges, blooms borne mostly single, very dbl., 1.5–2.75 in., intense fragrance; few prickles; foliage small, medium green, semi-glossy; medium (45 cms), upright, bushy growth; [Seedling X Twilight Trail]; Williams, Ernest D.; Mini Roses of Texas, 1993

'MINIMERR', Min, m, 1995; (**Merryglo**®); flowers medium lavender, blooms borne mostly single, dbl., 26–40 petals, 1.25 in., intense fragrance; some prickles; foliage medium, medium green, semi-glossy; medium (18 in), bushy growth; [Seedling X Twilight Trail]; Williams, Ernest D.; Texas Mini Roses, 1995

'MINIMOON', Min, mr, 1995; (**Red Moon**™ (USA)); dbl., 26–40 petals, 1–1.5 in., borne in small clusters, no fragrance; some prickles; foliage medium, medium green, semi-glossy; upright, bushy, climbing (5 ft) growth; [Red Delight X Twilight Trail]; Williams, Ernest D.; Texas Mini Roses, 1995

'MINIMYS',Min, m, 1995; (**Mysterious**); flowers mauve, exhibition, full (26–40 petals), medium blooms borne mostly single; fragrant; some prickles; foliage medium, medium green, semi-glossy; upright (16–18 in.) bushy growth; [Seedling X Twlight Trail]; Williams, Ernest; Texas Mini Roses,1995

'MINPUM',Min, r, 1995; (**Pumpkin Frost**®); flowers russet, dbl. (15–25 petals), medium (1-1 1/4 in.) blooms borne small clusters; fragrant; some prickles; foliage medium, dark green, semi-glossy; low (12in.), compact growth; [Seedling X Twlight Trail]; Williams, Ernest; Texas Mini Roses 1995

'MINI-POUL'®, Min, pb, 1978; flowers yellow and deep pink blend, small, dbl., 25 petals, no fragrance; foliage small, dark, glossy; compact growth; [Darling Flame X Unnamed seedling]; Poulsen, Niels D.; D.T. Poulsen

'MINIRASP', Min, mr, 1994; (**Raspberry Beauty**®); flowers non-fading raspberry red, blooms borne mostly single, dbl., 26–40 petals, 1.5 in., exhibition form, slight fragrance; few prickles; foliage small, medium green, semi-glossy; medium (16-20 in), bushy growth; [Angel Face X Anita Charles]; Williams, Ernest D.; Texas Mini Roses, 1994

Minirosa, Min, lp; flowers small, semi-dbl., slight fragrance; VEG; (Sangerhausen)

Minirot, Min, m; flowers purple/pink with cream reverse, small, semi-dbl., slight fragrance; VEG; (Sangerhausen)

'MINIRYL', Min, m, 1993; (**Royal Lady**®); flowers lavender pink, blooms borne mostly single, dbl., 26–40 ruffled petals,

1.5 in., intense fragrance; few prickles; foliage small, medium green, semi-glossy; medium (40 cms), bushy growth; [Angel Face X Twilight Beauty]; Williams, Ernest D.; Mini Roses of Texas, 1993

'MINISA', HRg, mr, 1927; flowers deep crimson, semi-dbl., 17 petals, bloom repeats, intense fragrance; very hardy.; [R. rugosa X Prince Camille de Rohan]; Hansen, N.E.

'MINISHOW', Min, dr, 1996; (**Showdown**); dbl., 26–40 petals, 1.25 in., borne mostly singly, no fragrance; some prickles; foliage dark green, semi-glossy; medium (16-18 in), upright, bushy growth; [Rise 'n' Shine X Twilight Trail]; Williams, Ernest D.; Texas Mini Roses, 1996

'MINISONG', Min, ob, 1996; (**Morning Song**); flowers orange with yellow base, blooms borne mostly single, dbl., 26–40 petals, 1.25 in., slight fragrance; some prickles; foliage medium, medium green, semi-glossy; medium (16-18 in), upright, bushy growth; [Sue Jo X Twilight Trail]; Williams, Ernest D.; Texas Mini Roses, 1996

'MINISTER AFRITSCH', F, or, 1964; flowers large, dbl.; vigorous, bushy growth; [Unknown seedling X Signalfeuer]; Tantau, Math.; Starkl

'MINISTER LUNS', F, dp, 1968; bud ovoid; flowers pink-red, medium, semi-dbl.; foliage dark; [Marchenland X Florence Mary Morse]; Wijnhoven

Minister Rasín *see* **'MINISTRE DES FINANCES RASÍN'**

'MINISTORM', Min, m, 1994; (**Stormy Weather**®); flowers medium lavender, blooms borne mostly single, dbl., 26–40 petals, 1.5 in., exhibition form, moderate fragrance; few prickles; foliage small, medium green, semi-glossy; tall (24-30 in), upright growth; [Seedling X Twilight Trail]; Williams, Ernest D.; Texas Mini Roses, 1994

'MINISTRE DES FINANCES RASÍN', HT, dp, 1930; (Minister Rasín); bud long; flowers carmine-rose, large, dbl.; vigorous growth; [Mme Maurice de Luze X Hadley]; Böhm, J.

'MINITOC',Min, r, 1995; (**Touch O' Cloves**);flowers russet, full (26–40 petals), medium (1-1/4 in.) blooms borne mostly single; fragrant; few prickles; foliage medium, dark green, semi-glossy; upright (18 in.)bushy growth; [Seedling X Twlight Trail]; Williams, Ernest; Texas Mini Roses 1995

'MINITOK', Min, dy, 1993; (**Golden Token**®); foliage small, medium green, glossy; meidum (40 cms), upright, bushy growth; [Flowers deep golden yellow, very full (41+ petals), small (0-4cms) blooms borne in small clusters; slight fragrance; few pr]; Williams, Ernest D.; Mini Roses of Texas, 1993

'MINIUMBER', Min, r, 1996; (**Umberglo**®); flowers russet with red blush on edges, blooms borne mostly singly, dbl., 26–40 petals, 1–1.25 in., moderate fragrance; few prickles; foliage medium, dark green, semi-glossy; low (12-14 in), compact, bushy growth; [Seedling X Twilight Trail]; Williams, Ernest D.; Texas Mini Roses, 1996

'MINIVIG', Min, w, 1995; (**Vigilance**®); flowers white, full (26–40 petals), medium ($1^1/_2$ - $1^3/_4$ in.) blooms borne in small clusters; no fragrance; numerous prickles; foliage medium, medium green, glossy; spreading (6ft.), bushy, climbing growth; [Jeanne Lajoie sport]; Williams, Ernest; Texas Mini Roses, 1995

'MINIVISION', Min, mp, 1996; (**Heavenly Vision**); flowers clear, medium pink, blooms borne mostly single, dbl., 15–25 petals, 1.5 in., slight fragrance; few prickles; foliage medium green, semi-glossy; medium (16-18 in), upright, bushy growth; [Angel Face X Tom Brown]; Williams, Ernest D.; Texas Mini Roses, 1996

'MINIWACK', Min, pb, 1993; (**Paddywack**®); dbl., 26–40 petals, 1.5 in., borne mostly singly, slight fragrance; few prickles; foliage small, medium green, semi-glossy; medium (50 cms), upright, compact growth; [Tom Brown X Over the Rainbow]; Williams, Ernest D.; Mini Roses of Texas, 1993

'MINIWIN',Min, ly, 1995; (**Winter Wheat**®); flowers light yellow, full (26–40 petals), medium (1 1/4 in.) blooms borne mostly single; slight fragrance; some prickles; foliage medium; medium green, semi-glossy; upright (16–18 in.), bushy growth; [Seedling X Twlight Trail]; Williams, Ernest; Texas Mini Roses, 1995

'MINIWON', Min, r, 1994; (**Wonderglo**®); flowers unusual russet with red highlights, very dbl., 26–40 petals, 1.5 in., borne mostly singly, moderate fragrance; few prickles; foliage small, medium green, semi-glossy; medium (18 in), bushy growth; [Seedling X Twilight Trail]; Williams, Ernest D.; Texas Mini Roses, 1994

'MINJCO', Min, op, 1983; (**Loveglo**®); flowers light coral pink and cream, small, dbl., exhibition form, moderate fragrance; foliage small, dark, glossy; bushy growth; PP005677; [Little Darling X Over the Rainbow]; Williams, Ernest D.; Mini-Roses

'MINKCO', Min, dr, 1983; (**Black Jack**®); flowers deep red, golden stamens, small, dbl., 40 petals, exhibition form, slight fragrance; foliage small, dark, semi-glossy; bushy growth; PP005671; [Tom Brown X Over the Rainbow]; Williams, Ernest D.; Mini-Roses

'MINLCO', Min, w, 1983; (**Arctic Snow**®); flowers small, dbl., exhibition form, slight fragrance; foliage small, dark, glossy; upright, bushy growth; [Miniature seedling X Over the Rainbow]; Williams, Ernest D.; Mini-Roses

'MINMCO', Min, ob, 1983; (**Spring Melody**®); flowers orange, small, dbl., 35 petals, slight fragrance; foliage small, bronze green, semi-glossy; bushy growth; [Little Darling X Over the Rainbow]; Williams, Ernest D.; Mini-Roses

'MINNA', HEg, w, 1895; flowers pure white, semi-dbl.; very vigorous growth; Penzance

'MINNA', F, mp, 1930; flowers rosy pink, large, borne in clusters, dbl., slight fragrance; foliage rich green, leathery; bushy, dwarf growth; [Gruss an Aachen sport]; Kordes

Minna Kordes *see* **'WORLD'S FAIR'**

Minna Lerche Lerchenborg® *see* 'MEIDAD'

'MINNCO', Min, mp, 1983; (**Pink Carpet**®); flowers small, dbl., blooms in large clusters, no fragrance; foliage small, light to medium green, glossy; low, spreading (to 6 ft hanging basket) growth; PP005606; [Red Cascade X Red Cascade]; Williams, Ernest D.; Mini-Roses

'MINNEHAHA', HWich, lp, 1905; flowers pink fading white, small blooms in large clusters, dbl., non-recurrent, slight fragrance; foliage small, glossy, dark; climbing (15-20 ft.) growth; [R. wichurana X Paul Neyron]; Walsh

Minnesota Mo, M, dr

'MINNIE', Min, rb; bud long, pointed; flowers red and yellow blend, dbl., 40 petals, 1 in., exhibition form, moderate fragrance; foliage small, glossy; upright growth; [Starburst X Over the Rainbow]; Williams, Ernest D.; Mini-Roses, 1977

'MINNIE DAWSON', HMult, w, 1896; flowers pure white, borne in very large clusters; [Dawson X R. multiflora]; Dawson

'MINNIE FRANCIS', T, dp, 1905; flowers deep pink, open; vigorous growth; Griffing Nursery

'MINNIE MARCUS', HT, pb, 1984; flowers large, dbl., 35 petals, slight fragrance; foliage large, dark, glossy; upright growth; [Queen Elizabeth X Windsounds]; Rodgers, Shafner R.; Neiman-Marcus

Minnie Mouse *see* 'MACMINMO'

Minnie Pearl® *see* 'SAVAHOWDY'

Minnie Pearl White, Min, lp; [Minnie Pearl sport]; Lee, 1993; (Weatherly, L.)

'MINNIE SAUNDERS', HT, mr, 1921; flowers bright red, single; vigorous growth; Hicks

'MINNIE WATSON', HT, lp, 1965; bud globular; semi-dbl., slight fragrance; foliage glossy; compact, bushy growth; [Dickson's Flame X Dickson's Flame]; Watson

'MINOCO', Min, lp, 1984; (**Mary Kay**®); flowers small, dbl., 35 petals, slight fragrance; foliage small, dark, glossy; upright, bushy growth; PP005631; [Tom Brown X Over the Rainbow]; Williams, Ernest D.; Mini-Roses

Minou, F; Moreira da Silva, A.; (Cavriglia)

Minouchette, LCl, 1971; Croix; (Cavriglia)

'MINPCO', Min, yb, 1984; (**Partyglo**®); flowers yellow with pink petal edges, yellow reverse, small, dbl., 35 petals, exhibition form, slight fragrance; foliage small, dark, semi-glossy; PP005880; [Little Darling X Over the Rainbow]; Williams, Ernest D.; Mini-Roses

'MINPEEP', Cl Min, mr, 1992; (**Peep-Eye**®); flowers very distinct shade of medium red, with large clusters of go, dbl., 25 petals, 1.5 in., slight fragrance; some prickles; foliage small, dark green, glossy; tall (4'+), upright, bushygrowth; hardy (both heat & cold).; [Red Delight X Unnamed seedling]; Williams, Ernest D.

'MINPROP', Min, mr, 1998; (**Prophecy**); flowers rose red, heat stable, medium, very dbl., 26–40 petals, 1–1.25 in., exhibition form, borne mostly singly, slight fragrance; prickles moderate; foliage medium, medium green, semi-glossy; upright, bushy, medium (18 in.) growth; [Red Delight X Twilight Trail]; Williams, Ernest D.; Texas Mini Roses, 1997

'MINQCO', Min, mr, 1984; (**Red Love**®); flowers spiral form, small, dbl., 35 petals, slight fragrance; foliage small, dark, semi-glossy to glossy; upright, bushy growth; [Tom Brown X Over the Rainbow]; Williams, Ernest D.; Mini Roses

'MINRCO', Min, pb, 1984; (**Cheryl's Delight**®); flowers medium pink, reverse white, small, dbl., 35 petals, slight fragrance; foliage small, dark, semi-glossy; bushy growth; PP005910; [Little Darling X Over the Rainbow]; Williams, Ernest D.; Mini-Roses

'MINRID', Min, yb, 1998; (**Dreamrider**); flowers med. yellow, pink edge, reverse yellow, very dbl., 26–40 petals, 1.75 in., borne mostly singly, slight fragrance; few prickles; foliage medium, medium green, semi-glossy; upright, bushy, medium (18 in) growth; [Sue Jo X Twilight Trail]; Williams, Ernest D.; Texas Mini Roses, 1997

'MINSCO', Min, mp, 1984; (**Pink Bounty**®); flowers small, dbl., 35 petals, exhibition form, slight fragrance; foliage small, medium green, semi-glossy; very dense, bushy growth; [Tom Brown X Over the Rainbow]; Williams, Ernest D.; Mini-Roses

'MINSCO', Min, m, 1985; ('MINXCO', **Twilight Trail**®); flowers lavender-tan, small, dbl., 35 petals, intense fragrance; foliage small, dark, semi-glossy; upright, bushy growth; PP006198; [Angel Face X Anita Charles]; Williams, Ernest D.; Mini-Roses

'MINSOUNDS', Min, mp, 1998; (**Heartsounds**); flowers clear medium pink, cream reverse, very dbl., 30–35 petals, 1–1.25 in., borne mostly singly, slight fragrance; few prickles; foliage medium, medium green, semi-glossy; long stems; upright, bushy, medium (18 in) growth; [Rise 'n' Shine X Twilight Trail]; Williams, Ernest D.; Texas Mini Roses, 1997

'MINSTREL', F, mr, 1967; flowers scarlet, semi- borne in clusters, single, 2.5 in.; foliage dark, glossy; very free growth; [Independence X Paprika]; Sanday, John

Mint Julep *see* 'AROGRESH'

'MINTCO', Min, m, 1984; (**Lavender Simplex**®); flowers lavender, purple stamens, small blooms borne singly, single, 5 petals, moderate fragrance; foliage small, dark, semi-glossy; upright, bushy growth; [Angel Face X Yellow Jewel]; Williams, Ernest D.; Mini-Roses

'MINUCO', Min, mr, 1984; (**Royal Carpet**®); flowers small, dbl., blooms in clusters, no fragrance; foliage small, dark, semi-glossy; groundcover spreading growth; PP005909; [Red Cascade X Red Cascade]; Williams, Ernest D.; Mini-Roses

'MINUET', HT, my, 1930; flowers open, large, semi-dbl., slight fragrance; foliage leathery; vigorous growth; [Joanna Hill sport]; Thompson's, J.H., Sons

Minuette *see* 'MEILUCCA'

Minuette *see* **'LAMINUETTE'**

Minuette, Climbing, Cl F; (Cavriglia)

Minuetto *see* 'MEILUCCA'

'MINVCO', Min, dr, 1984; (**Red Pendant**®); flowers small, dbl., 35 petals, no fragrance; foliage small, medium green, very glossy; low, spreading growth; [Red Cascade X Red Cascade]; Williams, Ernest D.; Mini Roses

'MINWCO', Min, lp, 1985; (**Beautyglo**®); flowers small, dbl., 35 petals, exhibition form, slight fragrance; foliage small, dark, semi-glossy; bushy growth; PP006199; [Tom Brown X Black Jack]; Williams, Ernest D.; Mini-Roses

'MINX', F, ob, 1955; bud globular; flowers orange-pink, borne in pyramidal clusters, dbl., 75–80 petals, 2.5 in., flat, moderate fragrance; strong stems; vigorous, upright growth; [Pinocchio seedling X Garnette]; Boerner; J&P

'MINXCO', Min, m, 1985; ('MINSCO', **Twilight Trail**®); flowers lavender-tan, small, dbl., 35 petals, intense fragrance; foliage small, dark, semi-glossy; upright, bushy growth; PP006198; [Angel Face X Anita Charles]; Williams, Ernest D.; Mini-Roses

'MINYCO', Min, yb, 1985; (**Imogene**®); flowers yellow, marked red, well-formed, small, dbl., 45 petals, slight fragrance; foliage small, dark, glossy; upright, bushy growth; [Little Darling X Over the Rainbow]; Williams, Ernest D.; Mini-Roses

'MINZCO', Min, dy, 1985; (**Fiddler's Gold**®); flowers small, dbl., moderate fragrance; foliage small, dark, glossy; bushy growth; [Tom Brown X Golden Angel]; Williams, Ernest D.; Mini-Roses

'MIO MAC', F, dy, 1973; bud ovoid; flowers coppery yellow, medium, dbl., slight fragrance; foliage glossy; growth moderate, upright, bushy; Tantau, Math.; Horstmann

Mirabella® *see* 'JACJUBEL'

Mirabilia, F, 1960; Cazzaniga, F. G.; (Cavriglia)

'MIRACLE', F, op, 1962; flowers soft coral blooms in large trusses, semi-dbl., 16–35 petals, 3 in., exhibition form; foliage glossy; vigorous growth; GM, Bagatelle, 1958; [Unnamed seedling X Fashion]; Verbeek; Ilgenfritz Nursery

'MIRAGAIA', HT, pb, 1958; flowers lilac-pink; [Peace X Coimbra]; da Silva, Moreira

'MIRAGE', Gr, op, 1966; bud ovoid; flowers bright salmon shaded red, medium, dbl., intense fragrance; foliage dark; very vigorous, bushy growth; [Peace X Circus]; Gaujard

Mirakel, F, yb; flowers dark yellow with orange and red, medium, dbl., moderate fragrance; VEG; (Sangerhausen)

'MIRALDA', HCh, m; flowers dark crimson-purple, nearly black, small, very dbl.; vigorous, branching growth; Vibert, ca 1830

Miramar *see* **'ALMIRANTE AMÉRICO TOMÁS'**

'MIRAMAR', F, or, 1956; bud globular; flowers cinnabar shaded coppery, large; foliage dark; very vigorous growth; [Opera seedling X Seedling]; Gaujard

'MIRANDA', P, mp, 1869; flowers satiny pink, very dbl.; de Sansal

'MIRANDA JANE', F, op, 1972; flowers salmon, dbl., 20 petals, 4 in., exhibition form, moderate fragrance; [Orange Sensation X Red Dandy]; Cocker

Mirandolina®, Min, w

'MIRANDY', HT, dr, 1945; flowers garnet-red, aging darker, dbl., 45 petals, 5–6. in., globular, intense, damask fragrance; foliage leathery; vigorous, upright, bushy growth; AARS, 1945; [Night X Charlotte Armstrong]; Lammerts, Dr. Walter; Armstrong Nursery

'MIRANDY, CLIMBING', Cl HT, dr, 1961; Moore

'MIRATO', HT, mp, 1974; bud ovoid; dbl., intense fragrance; foliage large, glossy; upright, bushy growth; [Unknown X Unknown]; Tantau, Math.

Mirato *see* TANotari

'MIREILLE', HT, mr, 1952; flowers coppery crimson-red, dbl., 26 petals, 4 in., moderate fragrance; foliage rich green; vigorous growth; [Opera X Unnamed seedling]; Gaujard

Mireille Mathieu *see* 'KORDEHN'

Mirella® *see* BARnec

Miriam *see* 'MACSUPCAT'

'MIRIAM', HT, my, 1919; flowers nasturtium-yellow, dbl., slight fragrance; GM, NRS, 1919; Pemberton

Miriam Wilkins, HP, lp

'MIRIAM'S CLIMBER', LCl, op, 1950; bud ovoid, apricot-pink; flowers peach-pink, lighter at tip, large, bornein clusters, dbl., intense fragrance; foliage dark, soft; very vigorous, climbing (25 ft) growth; [Seedling X Edith Nellie Perkins]; Rosen, H.R.

Miriam's Pink Powderpuff, HP, pb

Miriam's Single White Wichurana, HWich, w; (found rose)

Miriana *see* 'MEIBURGANA'

'MIRZA', F, ob, 1974; bud ovoid; dbl., 35 petals, 4 in., cupped, moderate fragrance; foliage glossy; vigorous, upright, bushy growth; [Zorina X Samba]; Kordes; Willemse

Mischief *see* 'MACMI'

'MISILJURIKA', F, mr, 1995; (Preservation); flowers bright red, blooms borne in small clusters, very dbl., 2.75in., slight fragrance; some prickles; foliage medium, medium green, glossy; medium, upright growth; [Silver Jubilee X Paprika]; Bossom, W.E.

'MISS ADA', F, lp, 1998; flowers light pink, darker edges, reverse darker pink, single, 4–7 petals, 2 in.–4 in., borne singly and in small clusters, slight fragrance; prickles few, brown; foliage medium, medium green, semi-glossy; upright, bushy, medium growth; [Playgirl sport]; Pawlikowski, Martin & Elaine; Giles Rambli' Roses, 1998

'MISS AGNES C. SHERMAN', T, rb, 1901; flowers rose, salmon and red; Nabonnand

'MISS ALICE DE ROTHSCHILD', T, ly, 1910; flowers light canary-yellow, center deeper, dbl., moderate fragrance; Dickson, A.

Miss All American Dream™ *see* 'WILDREM'

'MISS ALL AUSTRALIAN BEAUTY', HT, dp, 1969; bud ovoid; flowers light red, reverse darker, medium, dbl., slight fragrance; upright growth; [Aztec X Impeccable]; Armbrust; Langbecker

Miss All-American Beauty *see* 'MEIDAUD'

Miss All-American Beauty, Climbing *see* 'MEIDAUDSAR'

Miss All-American Dream™ *see* 'WILDREM'

'MISS AMELIA GUDE', HT, dy, 1921; flowers deep yellow center shading to cream, dbl., 35–40 petals, moderate fragrance; foliage dark; vigorous growth; [Columbia X Sunburst]; Lemon

'MISS AMERICA', HT, pb, 1938; flowers light pink, flushed salmon and gold, dbl., 65 petals, 6 in., moderate fragrance; foliage dark, leathery; vigorous growth; [Joanna Hill X S.M. Gustave V]; Nicolas; J&P

'MISS ANNAMARIE BALLY', HT, or, 1926; flowers reddish copper, reverse suffused whitish fawn; [Aspirant Marcel Rouyer X Lamia]; Easlea

Miss Annie Crawford *see* **'ANNIE CRAWFORD'**

Miss Atwood, T, ab

'MISS AUSTRALIA', HT, mp, 1933; flowers pink, center salmon, very large, dbl., 50 petals, globular, intense fragrance; foliage thick; very vigorous, bushy growth; [Dame Edith Helen X Mme Segond Weber]; Knight, G.

'MISS BLANCHE', HT, w, 1980; (Kojack); bud long; flowers urn-shaped to blooms borne usually singly, dbl., 38 petals, exhibition form, no fragrance; straight, reddish prickles; foliage large, dark, leathery; upright growth; [Evening Star X Coquette]; Warriner, William A.; J&P

'MISS BRISBANE', HT, lp, 1953; flowers shell-pink, medium, dbl., cupped, moderate fragrance; foliage light green; vigorous growth; [? X The Doctor]; Ulrick, L.W.

'MISS C. E. VAN ROSSEM', HT, mr, 1919; bud long, pointed; flowers crimson-scarlet shaded carmine and black, open, medium to small, semi-dbl., cupped, slight fragrance; foliage leathery, bronze, dark; vigorous, bushy growth; [Leuchtfeuer (HCh) X Red-Letter Day]; Verschuren

'MISS CALIFORNIA', HT, dp, 1933; flowers deep glowing pink; [Dame Edith Helen sport]; Smith, J.

'MISS CANADA', HT, pb, 1963; bud ovoid; flowers rose-madder, reverse silver, large, dbl., exhibition form, slight fragrance; foliage glossy, leathery; vigorous, upright, spreading growth; [Peace X Karl Herbst]; Blakeney; Eddie

'MISS CAROLINE', T, pb, 1997; flowers very full, medium, very dbl., 50–60 petals, borne in small clusters, moderate fragrance; foliage medium, medium green, semi-glossy; bushy, medium (4-5ft.) growth; [Duchesse de Brabant sport]; Knopf, Ruth

'MISS CLIPPER', HT, pb, 1942; bud long, pointed to ovoid; flowers pale salmon-pink shaded yellow, dbl., 25–30 petals, 3.5–4 in., exhibition form, moderate, spicy fragrance; foliage glossy, light; strong stems; vigorous, upright, bushy growth; [Angèle Pernet X Pres. Herbert Hoover]; Lammerts, Dr. Walter; Armstrong Nursery

'MISS CONNER', HT, ly, 1920; flowers canary-yellow on lemon-yellow, dbl., moderate fragrance; Dickson, A.

'MISS CYNTHIA FORDE', HT, dp, 1909; flowers deep brilliant rose-pink, reverse lighter, large, dbl., moderate fragrance; long stems; vigorous, bushy growth; GM, NRS, 1909; Dickson, H.

Miss Daisy *see* 'JACFLARE'

'MISS DELIGHTFUL', F, my, 1966; flowers bright yellow, rosette form, borne in heavy clusters, dbl., 30 petals, 3 in., moderate fragrance; foliage glossy; vigorous, upright growth; [(Masquerade X Seedling) X Golden Scepter]; Sanday, John

Miss Dior *see* HARencens

Miss Dovey™ *see* 'KINDOV'

Miss Dovie *see* 'KINDOV'

'MISS EDITH CAVELL', Pol, dr, 1917; (Edith Cavell, Nurse Cavell); flowers scarlet-crimson overlaid velvety crimson; [Orléans Rose sport]; deRuiter; Spek

'MISS ENGLAND', HT, w, 1936; flowers creamy, very large, dbl., moderate fragrance; foliage leathery, dark; vigorous, compact growth; Cant, B. R.

'MISS ETHEL RICHARDSON', HP, w, 1897; flowers white tinged pink, very large, dbl.; Dickson, A.; (Sangerhausen)

'MISS EVELYN DAVEY', T, lp; [Souv de S.A. Prince X Niphetos]; Williams, A., 1905; (Weatherly, L.)

Miss Finland, F, ob

Miss Flippins™ *see* 'TUCKFLIP'

'MISS FLORA MITTEN', LCl, lp, 1913; flowers soft pink, stamens yellow, single, 3 in.; vigorous growth; [R. wichurana X R. canina]; Lawrenson

'MISS FRANCE', Gr, or, 1955; (Pretty Girl); flowers bright scarlet, large, dbl., globular, moderate fragrance; foliage bronze; vigorous growth; [Peace X Independence]; Gaujard

'MISS G. MESMAN', Cl Pol, mr; (Baby Rambler, Climbing); [Mme Norbert Levavasseur climbing sport]

'MISS G. RADCLIFFE', HT, lp; [Pharisaer X Unknown]; Williams, A., 1900; (Weatherly, L.)

'MISS GEORGIE', HT, ab, 1980; dbl., 35 petals, borne usually singly, moderate fragrance; long prickles; foliage large, semi-glossy; upright, compact growth; [South Seas X Unnamed seedling]; Warriner, William A.; J&P

Miss Harp *see* 'TANOLG'

'MISS HAWAII', HT, w, 1976; flowers creamy white, dbl., 30–35 petals, 5.5–6 in., exhibition form, intense fragrance; foliage leathery; very vigorous growth; [Hawaii sport]; Payne; Lone Star Nursery

'MISS HELYETT', LCl, pb, 1909; (Miss Heylett); flowers bright carmine-pink, center yellowish salmon-pink, open, large, dbl., slight fragrance; vigorous, climbing (10-12 ft) growth; [R. wichurana X Ernest Metz]; Fauque

'MISS HENRIETTE TERSTEEG', HT, pb, 1922; flowers flesh and salmon-pink, dbl., slight fragrance; [Mme Abel Chatenay seedling X Mrs Joseph Hill]; Van Rossem

Miss Heylett *see* **'MISS HELYETT'**

'MISS HILLCREST', HT, or, 1969; flowers large, dbl., exhibition form, intense, fruity fragrance; foliage glossy; vigorous, tall growth; [Peace X Hawaii]; Curtis, E.C.; Kimbrew

Miss Hillcrest *see* 'POULSALFAI'

'MISS HOUSE', HP, w, 1838; flowers satin white; House

'MISS HUNTINGTON', HT, ob, 1971; bud ovoid; flowers bright orange, large, dbl., exhibition form, moderate fragrance; foliage leathery; vigorous, upright growth; [Ma Perkins X San Francisco]; Patterson; Patterson Roses

Miss Ireland *see* 'MACIR'

Miss Italia, S, 1963; Mansuino; (Cavriglia)

'MISS JEKYLL', Ayr, w; flowers light rose pink, dbl.; (14)

'MISS JOAN', Cl HT, op, 1943; flowers copper-bronze and salmon-pink, large, dbl., globular, intense fragrance; foliage dark, glossy; very vigorous, climbing growth; [Ednah Thomas X Golden Dawn]; Duehrsen; California Roses

'MISS KATE MOULTON', HT, pb, 1906; flowers rosy pink shaded rosy salmon, dbl.; [Mme Caroline Testout X (La France X Mrs W.J. Grant)]; Monson; Minneapolis Floral Co.

'MISS KATE SESSIONS', LCl, pb, 1953; flowers deep rose-pink on white base, reverse shell-pink, open, large, dbl., profuse, intermittent bloom, moderate fragrance; foliage light green, leathery; moderate, climbing growth; [Heart of Gold X Ednah Thomas]; Hieatt

'MISS KOGANEI', Cl Min, rb, 1985; flowers medium red, white eye, yellow stamens, single, 7–10 petals, cupped; few prickles; foliage small; vigorous (to 3 ft) growth; [Nozomi X ?]; Asano, S.

'MISS LESLIE', HT, w, 1998; flowers white with light pink edges, full, very dbl., 26–40 petals, exhibition form, borne mostly singly, slight fragrance; prickles moderate; foliage medium, medium green, semi-glossy; upright, tall growth; [Rina Hugo sport]; Garrett, Troy O.

Miss Liberté *see* 'AROVULE'

Miss Liberty *see* 'AROVULE'

'MISS LIBERTY', LCl, mp, 1956; bud ovoid; flowers tyrian rose, borne in large clusters, semi-dbl., 15–20 petals, 3.5 in., cupped, repeat bloom, moderate fragrance; foliage dark, leathery; strong stems; vigorous, climbing (10-12 ft.) growth; [New Dawn X World's Fair, Climbing]; Boerner; Stuart

'MISS LOLITA ARMOUR', HT, or, 1919; flowers deep coral-red suffused coppery red, base yellow, very large, dbl., cupped, intense fragrance; GM, Bagatelle, 1921; Howard, F.H.; H&S

'MISS LOLITA ARMOUR, CLIMBING', Cl HT, or; H&S, about 1925

Miss Lowe *see* **'MISS LOWE'S VARIETY'**

'MISS LOWE'S VARIETY', Ch, mr; (Miss Lowe); flowers bright red, single, recurrent bloom; dwarf growth; (14); [Possibly Slater's Crimson China sport]

'MISS M. J. SPENCER', HT, my, 1920; flowers clear bright golden yellow, dbl.; Dickson, H.

'MISS M. ROBERTSON', Pol, lp; [Mlle Cécile Brünner X Unknown]; Williams, A., 1902; (Weatherly, L.)

'MISS M.J.F. GOSTLING', Pol, lp; [Miss Pollock X Unknown]; Williams, A., 1911; (Weatherly, L.)

'MISS MAINE 1999', S, dp, 1999; flowers deep pink, reverse medium pink, dbl., 26–40 petals, 2 in., borne in small clusters, slight fragrance; many prickles; foliage small, medium green, semi-glossy; bushy, medium (4 ft) growth; hardy to zone 4; [Henry Kelsey sport]; Law, Stephen

'MISS MARION MANIFOLD', Cl HT, mr; (Marion Manifold); flowers velvety scarlet, shaded crimson, large, dbl., globular, moderate fragrance; foliage large, leathery; vigorous, climbing (12 ft) growth; Adamson, 1913; Brundrett

Miss Mary *see* 'GELMISS'

'MISS MAY MARRIOTT', HT, ab, 1917; flowers glowing apricot; [Mme Edouard Herriot sport]; Robinson, T.

'MISS MAY THURLOWE', T, lp; Williams, A., 1904; (Weatherly, L.)

'MISS MAYNE', HT, lp; [Marie van Houtte X General Jacqueminot]; Williams, A., 1900; (Weatherly, L.)

'MISS MIDDLETON', F, op, 1963; flowers coral-pink, borne in small clusters, dbl., 20 petals, 2.5–3 in., cupped; foliage light green; low, bushy growth; [Independence X Masquerade]; Hill, A.

'MISS M'LISS', F, mp, 1958; bud short pointed; flowers phlox-pink, dbl., 25–30 petals, 2.5–3 in., exhibition form, slight fragrance; very vigorous, upright growth; [Garnette X Garnette seedling]; Jelly; E.G. Hill Co.

'MISS MODESTO', HT, my, 1934; flowers pure yellow, very large, dbl., exhibition form, moderate fragrance; [Rev. F. Page-Roberts sport]; Brooks, L.L.; Brooks & Son

'MISS MUFFETT', Min, mp, 1955; bud rather mossy; flowers apple-blossom-pink, compact growth, dbl., 80 petals, 4–6 in.; [Baby Bunting X Tom Thumb]; Robinson, T.

'MISS MURINE', F, dp, 1937; flowers deep pink, base yellow, small, dbl.; upright, bushy growth; [Unnamed seedling X Cécile Brunner]; Fitzgerald

'MISS N. ROBERTSON', Pol, lp; [Mlle Cécile Brünner X Unknown]; Williams, A., 1902; (Weatherly, L.)

'MISS P. WILLIAMS', Pol, lp; [Perle d'Or X Unknown]; Williams, A., 1902; (Weatherly, L.)

Miss Pam Ayres *see* 'KORMARIE'

Miss Pearl® *see* 'HOOPEARL'

Miss Perfect® *see* 'JACMISS'

'MISS PERSONALITY', F, pb, 1972; bud ovoid; flowers cerise-pink, white eye, open, medium, semi-dbl., slight fragrance; foliage small, dark, leathery; vigorous, bushy growth; [Pink Parfait X (Ophelia X Parkdirektor Riggers)]; Sherwood; F. Mason

'MISS POLLOCK', P, lp; Williams, A., 1906; (Weatherly, L.)

Miss Prissy, HT, w, 1995; Wells

Miss Reus, HT, op; Dot

Miss Rita, F, lp, 1986; Robertson

Miss Rose, Gr; Dorieux; (Cavriglia)

'MISS ROSE HILL', HT, lp; Williams, A., 1913; (Weatherly, L.)

'MISS ROWENA THOM', HT, pb, 1927; flowers fiery rose and rosy mauve, center washed gold, dbl., 50 petals, 6 in., intense fragrance; vigorous growth; [Radiance X Los Angeles]; H&S

'MISS ROWENA THOM, CLIMBING', Cl HT, pb, 1937; van Barneveld; California Roses

Miss Schweiz *see* TANziewsim

'MISS TALMADGE', HT, dy, 1927; dbl., moderate fragrance; [Constance sport]; Pacific Rose Co.

'MISS UNIVERSE', HT, rb, 1956; bud long, pointed; flowers orange-red, reverse tinted copper, large, moderate fragrance; foliage dark; vigorous growth; [(Peace X Seedling) X Seedling]; Gaujard

'MISS WILLMOTT', HT, ly, 1917; flowers soft sulfur-cream, edges flushed pale pink, well formed, dbl., moderate fragrance; vigorous growth; GM, NRS, 1916; McGredy

'MISS WINDSOR', HT, mr, 1967; flowers medium, dbl., exhibition form, intense fragrance; foliage light green, leathery; very vigorous, upright growth; [Tropicana sport]; Heron

'MISS WINIFRED DENHAM', T, lp; [Marie van Houtte X Unknown]; Williams, A., 1911; (Weatherly, L.)

'MISSION BELLS', HT, pb, 1949; bud long, pointed; flowers vermilion-pink, dbl., 43 petals, 5 in., exhibition form, moderate fragrance; foliage dark, soft; vigorous, bushy growth; AARS, 1950; [Mrs Sam McGredy X Malar-Ros]; Morris; Germain's

'MISSION SUPREME', HT, ab, 1981; bud pointed; flowers pale peach pink to apricot, blooms borne singly, dbl., 30 petals, moderate fragrance; straight, red-brown prickles; foliage deep green; vigorous, bushy, medium growth; [City of Glouchester X Unnamed seedling]; Sanday, John

'MISSISSIPPI', HT, dr, 1976; bud pointed; flowers deep red, dbl., 38 petals, 4.5–5 in., moderate, damask fragrance; foliage dark; upright growth; [Charlotte Armstrong X Mister Lincoln]; Williams, J. Benjamin

Mississippi *see* **'MME CHARLES SAUVAGE'**

'MISSISSIPPI RAINBOW', HT, yb, 1977; flowers full, dbl., 41 petals, 4 in., slight fragrance; foliage thick, glossy; upright growth; Graham; South Forrest Rose Nursery

'MISSY', Min, mr, 1978; bud pointed; flowers cardinal-red, small, dbl., 22 petals, 1.5 in., slight fragrance; foliage small; compact, bushy growth; [Seedling X Seedling]; Lyon

'MISSY ELEANOR', S, pb, 1999; flowers light pink, yellow eye, reverse medium pink, single, 5–11 petals, 2.5 in., borne in large clusters, moderate fragrance; prickles moderate; foliage medium, light green, glossy; bushy, medium (4-6 ft) growth; [Anna Ford X Ballerina]; Watson, Thomas L, & Glenda, 1996

'MISTEE', Min, ly, 1979; bud long, pointed; flowers white, tinted yellow, small, dbl., 28 petals, 1.5 in., flat, moderate fragrance; foliage small; bushy, upright growth; [Little Darling X Peachy White]; Moore, Ralph S.; Sequoia Nursery

'MISTER AMERICA', LCl, dr, 1974; flowers blood-red, very large, dbl., exhibition form, abundant bloom, moderate fragrance; foliage leathery; very vigorous, climbing growth; [(Paul's Scarlet Climber X Golden Climber) X Pinocchio]; Zombory; General Bionomics

Mister Chips, HT, 1970; Dickson, A.; (Cavriglia)

'MISTER LINCOLN'®, HT, dr, 1964; bud urn-shaped; dbl., 35 petals, 4.5–6 in., exhibition form, intense fragrance; foliage leathery, matt, dark; vigorous growth; AARS, 1965; [Chrysler Imperial X Charles Mallerin]; Swim & Weeks; C-P

'MISTER LINCOLN, CLIMBING', Cl HT, dr, 1974; Ram

Mister Otis *see* 'CURMIST'

'MISTER SOFTEE', HT, w, 1964; flowers creamy white edged pink, open, dbl., 40 petals, 4.5 in., moderate fragrance; foliage glossy, dark; very free growth; Morton's Rose Nursery

'MISTICA', HT, m, 1966; bud ovoid; flowers lilac, very large, dbl., cupped, intense fragrance; foliage dark; vigorous growth; [Sterling Silver X Intermezzo]; Dot; Minier

Mistigri *see* 'TANNIMOLL'

Mistraline® *see* LAPmisal

'MISTRESS BOSANQUET', B, lp, 1832; (Mrs Bosanquet); flowers rosy flesh, very dbl.; vigorous growth; Laffay, M.

'MISTRESS PAT', Pol, mp, 1928; flowers chatenay pink, moderate fragrance; Lilley

Mistress Quickly *see* 'AUSKY'

'MISTY', HT, w, 1965; bud ovoid, pointed; flowers creamy white, to formal, dbl., 35 petals, 4 in., cupped, moderate, tea fragrance; foliage large, leathery; vigorous, upright growth; [Mount Shasta X Matterhorn]; Armstrong, D.L.; Armstrong Nursery

'MISTY DAWN', Min, w, 1979; bud ovoid, pointed; flowers pure white, dbl., 33 petals, 1 in., cupped; foliage small, dark; vigorous, compact, spreading growth; [Charlie McCarthy X Unnamed seedling]; Schwartz, Ernest W.; Nor'East Min. Roses

'MISTY DELIGHT', HT, mp, 1990; dbl., 26–40 petals, borne mostly singly; foliage large, dark green, glossy; upright growth; [Blue Wonder X Unnamed seedling]; Christensen, Jack E.; Vaughan's Seed Co., 1991

'MISTY GOLD', F, my, 1954; bud ovoid, pointed; flowers empire-yellow, dbl., 45–50 petals, 3.5–4 in., cupped, moderate fragrance; foliage glossy; vigorous growth; [Unnamed Floribunda (Yellow, semi-dbl., self seedling)]; Boerner; Stark Bros.

'MISTY MORN', HT, ly, 1949; flowers pale lemon-yellow, pointed, large, dbl., 45 petals, slight fragrance; foliage dark; [Unnamed seedling X Mrs Charles Lamplough]; McGredy

Misty Morning *see* 'JUDMORN'

Misty Pink, F, mp

Misty Veil, S, mp, 1999; Williams, J. Benjamin

'MITCHELTONII', S, mp, 1966; (Rosa Xmitcheltonii); flowers pink, small, single, profuse spring bloom; thornless; foliage small, leathery; very vigorous, climbing growth; [R. multiflora X I.X.L.]; Armbrust

Mitsouko® *see* 'DELNAT'

'MITZI', HT, yb, 1956; flowers pearly tints flushed mauve-rose, well formed, dbl., 35 petals, moderate fragrance; foliage dark; strong stems; upright, bushy growth; GM, Rome, 1956; [(Peace X Mme Joseph Perraud) X (Mrs Pierre S. duPont X Mrs John Laing)]; Meilland, F.; URS

Mitzi *see* MEIlenangal

Miwaku, HT, w, 1988; Suzuki, Seizo

Mix 'n' Match™ *see* 'CHEWILY'

Mixed Marriage, F, lp; [Bridal Pink sport]; Ruston, D., 1987; (Weatherly, L.)

'MIYABI', HT, w, 1977; bud ovoid; flowers near white, dbl., 30 petals, 4–4.5 in., exhibition form, slight fragrance; upright growth; [(Amatsu-Otome X Samba) X (Kordes' Perfecta X American Heritage)]; Teranishi, K.; Itami Bara-en, 1976

'MIYAGINO', Cl Min, pb, 1978; bud rounded; flowers light pink, small, single, 5 petals, flat; foliage tiny; [Nozomi X Unnamed seedling]; Onodera, Toru F.; S. Onodera

Mizar *see* BARmiz

Mlada, F, yb, 1983; flowers yellow and red, medium, dbl., slight fragrance; Urban, J.; (Sangerhausen)

'MLLE ALICE FURON', HT, ly, 1896; flowers yellowish white; [Lady Mary Fitzwilliam X Mme Chedane-Guinoisseau]; Pernet-Ducher

'MLLE ALICE LEROY', m, 1842; flowers light violet-pink, medium, dbl.; Vibert; (Sangerhausen)

'Mlle Alice Rousseau', Pol, lp, 1903; flowers small, dbl.; Vilin; (Sangerhausen)

'Mlle Andree Worth', B, lp, 1890; Lévêque

'Mlle Annie Wood', HP, mr, 1866; (Annie Wood); flowers clear red, large, dbl., recurrent bloom, moderate fragrance; Verdier, E.

'Mlle Augustine Guinoisseau', HT, lp, 1889; flowers large, dbl.; Guinoisseau; (Sangerhausen)

'Mlle Bep van Rossem', HT, my, 1926; flowers deep canary-yellow, dbl.; [Unnamed seedling X Souv. de Claudius Pernet]; Van Rossem

Mlle Berthe Leveque *see* **'Mlle Berthe Levet'**

'Mlle Berthe Levet', HP, mp; (Mlle Berthe Leveque)

'Mlle Blanche Lafitte', B, w, 1851; flowers whitish-rose, dbl., blooms in clusters, repeats in autumn; vigorous growth; Pradel

'Mlle Bonnaire', HP, w, 1859; flowers white, center sometimes pink, dbl.; growth moderate; Pernet Père

'Mlle Cécile Brünner', Pol, lp, 1881; (Cécile Brünner, Mignon, Mme Cécile Brünner, Sweetheart Rose); bud long, pointed; flowers bright pink on yellow ground, small blooms in clusters, dbl., moderate; very few prickles; foliage sparse, soft, dark, 3-5 leaflets; growth like a tea, dwarf; (14); [Said to be Dbl. fld. Multiflora X Souv. d'un Ami]; Ducher, Vve.; Pernet-Ducher

'Mlle Cécile Brünner, Climbing', Cl Pol, lp, 1894; (Cécile Brünner, Climbing, Mignon, Climbing, Mme Cécile Brünner, Climbing, Sweetheart Rose, Climbing); Hosp

Mlle Cécile Brünner, Climbing, Cl Pol, lp, 1904; Ardagh

'Mlle Claire Andruejol', HT, lp, 1920; flowers pale pink tinted carmine, dbl., slight fragrance; [Comte G. de Rochemur X Mme Maurice de Luze]; Schwartz, A.

'Mlle Claire Jacquier', N, ly, 1888; flowers small to medium, dbl., moderate fragrance; Bernaix, A.; (Sangerhausen)

'Mlle Claire Truffaut', B, lp, 1887; flowers medium, dbl.; Verdier, E.; (Sangerhausen)

'Mlle Claudine Perreault', T, mp, 1885; flowers rose flesh, center darker, very large, dbl., free bloom; Lambert, E.

'Mlle Danielle Dumur', HT, lp, 1910; flowers large, dbl., moderate fragrance; Laroulandie; (Sangerhausen)

Mlle de Dinant, F, 1966; Lens, Louis; (Cavriglia)

'Mlle de Meux', HT, dp, 1901; flowers carmine-pink, medium, dbl.; Berland; (Sangerhausen)

'Mlle de Morlaincourt', HT, pb, 1934; flowers pink and yellow; [Cécile Walter X Korovo]; Walter, L.; Amis des Roses

'Mlle de Sombreuil', T, w, 1851; Robert

'Mlle Emilienne Moreau', HT, yb, 1920; flowers dark yellow and creamy pink, medium, semi-dbl.; Verschuren; (Sangerhausen)

'Mlle Eugénie Verdier', HP, mp, 1869; flowers clear silvery pink, reverse silvery white, large, dbl., 40 petals, moderate fragrance; vigorous, upright growth; Guillot et Fils

'Mlle Eugénie Verdier', HP, lp; Schwartz, 1872

'Mlle Fernande Dupuy', Pol, dp, 1899; flowers carmine-pink, small, dbl.; Vigneron; (Sangerhausen)

'Mlle Franziska Kruger', T, op; flowers coppery yellow and pink, center often green, large blooms, very dbl., moderate fragrance; weak stems; ; hardy for this class; (14); [Catherine Mermet X Gén. Schablikine]; Nabonnand, G., 1879

'Mlle Geneviève Godard', T, yb, 1889; flowers yellow with pink and orange, medium, dbl.; Godard; (Sangerhausen)

'Mlle Hélène Gambier', HT, op, 1895; (Helen Gambier); flowers salmon-pink to coppery rose, large, dbl., intense fragrance; Pernet-Ducher

'Mlle Henriette Martin', HT, w, 1936; flowers white, shaded ivory, edges lightly tinted pale pink, moderate fragrance; Reymond; Vially

'Mlle Honorine Duboc', HP, dp, 1894; flowers very large, dbl.; Duboc; (Sangerhausen)

'Mlle Irene Hennessy', HT, or, 1923; flowers bright vermilion-orange, dbl., moderate fragrance; [George C. Waud X Unnamed seedling]; Guillot, P.

'Mlle Jeanne Lenail', Pol, mr, 1924; flowers bright ruby-red shaded carmine, large, dbl.; [Mrs W.H. Cutbush X Mme Taft]; Schwartz, A.

'Mlle Jeanne Philippe', T, yb, 1898; Godard

'Mlle la Comtesse de Leusee', T, pb, 1878; Nabonnand

Mlle Lenari la Granada, HT, 1982; Dot, Simon; (Cavriglia)

'Mlle Louise Chabier', HP, lp, 1867; flowers large, dbl.; Cochet, S.; (Sangerhausen)

'Mlle Madelaine de la Roche', T, 1890; Corboeuf; (Cavriglia)

'Mlle Madeleine de Vauzolles', B, lp, 1881; Vigneron

'Mlle Marcelle Gaugin', Pol, lp, 1910; flowers creamy pink, small, dbl.; Corboeuf; (Sangerhausen)

'Mlle Marie Dauvesse', HP, mp; Vigneron, 1861

'Mlle Marie Drivon', B, pb, 1887; flowers medium, very dbl.; Schwartz; (Sangerhausen)

'Mlle Marie Gaze', N, pb, 1892; flowers yellowish-pink with dark yellow, medium, dbl.; Godard; (Sangerhausen)

'Mlle Marie Louise Bourgeois', M, lp, 1891; flowers light pink, edged white; Corboeuf

'Mlle Marie Magat', HP, mr, 1889; flowers carmine-red, large, dbl.; Liabaud; (Sangerhausen)

'Mlle Marie Mascuraud', HT, w, 1909; flowers white tinted flesh; Bernaix, P.

'Mlle Marie Moreau', T, w; flowers silver-white flushed crimson, well formed; Nabonnand, G., 1879

Mlle Marie van Houtte *see* **'Marie van Houtte'**

'Mlle Marthe Carron', HWich, w, 1931; flowers white, slightly tinted pink on opening, borne in clusters of 40-50; vigorous growth; [R. wichurana sport X R. wichurana]; Mermet

'Mlle Marthe Moisset', HT, my, 1935; flowers chrome-yellow on ochre ground, very large, cupped, moderate fragrance; foliage glossy, dark; vigorous, erect, bushy growth; [Mme Henri Queuille X Unnamed seedling]; Ducroz

'Mlle Mathilde Lenaerts', T, lp, 1880; flowers medium, dbl., moderate fragrance; Levet; (Sangerhausen)

'Mlle Maurand', HP, lp; flowers pale flesh

'Mlle Renée Denis', HP, w, 1906; flowers white with light pink, dbl.; Chédane-Guinoisseau; (Sangerhausen)

Mlle Rosa Bonheur *see* **'Rosa Bonheur'**

'Mlle Simone Beaumez', HT, w, 1907; flowers flesh-white, center sometimes tinted saffron-yellow, dbl., moderate fragrance; Pernet-Ducher

'Mlle Sontag', HGal, dp; flowers deep pink, reverse pale blush

'Mlle Stella Mallerin', HT, w, 1926; bud pointed; flowers white, center slightly shaded cream, very large, dbl., cupped, moderate fragrance; Chambard, C.

'Mlle Thérese Levet', HP, mp, 1864; flowers large, very dbl., moderate fragrance; Levet; (Sangerhausen)

'Mlle Yvette Bouquil', F, yb, 1955; flowers yellow edged orange; bushy growth; Privat

'MME A. BOUCHAYER', HT, op, 1927; flowers shrimp-pink, base indian yellow; Siret-Pernet

'MME A. CHATAIN', HT, 1940; Mallerin, C.; (Cavriglia)

'MME A. GALLAND', HT, pb, 1928; flowers rose-pink, shaded shrimp-pink, dbl., intense fragrance; [Pharisaer X (Constance X Unnamed Hybrid Tea]; Mallerin, C.; Laperrière

'MME A. LABBEY', HP, pb, 1843; flowers pink and lilac, medium or small

'MME A. LERCHE', HT, mp, 1928; flowers china-rose shaded carmine, reverse silvery rose, dbl.; [Mme L. Hot X Mrs Henry Winnett]; Bernaix, P.

Mme A. Meilland *see* **'PEACE'**

Mme A. Meilland, Climbing *see* **'PEACE, CLIMBING'**

'MME A. ROURE', HT, yb, 1932; flowers brilliant chrome-yellow shaded salmon, to amaranth-red, dbl., moderate fragrance; [Wilhelm Kordes X Mev. G.A. van Rossem]; Lens

'MME ABEL CHATENAY', HT, pb, 1895; bud pointed; flowers pale pink, center deeper, reverse carmine-pink, dbl., 3 in., moderate fragrance; foliage bronze when young; [Dr. Grill X Victor Verdier]; Pernet-Ducher

'MME ABEL CHATENAY, CLIMBING', Cl HT, pb, 1917; Page; Easlea

'MME ACHILLE FOULD', T, yb, 1903; flowers yellow, shaded carmine rose and salmon, very large, dbl., globular; Lévêque

'MME ACHILLE VILLEY', HT, or, 1939; flowers coral-red, tinted yellow, dbl.; foliage dark; very vigorous growth; [Charles P. Kilham X Mrs Aaron Ward]; Colombier

Mme Adelaide Ristori, B, dp; flowers cerise and fawn; Praden, 1861

'MME ADOLPHE DAHAIR', T, ly, 1901; flowers large, dbl., moderate fragrance; Puyravaud; (Sangerhausen)

'MME ADOLPHE LAFONT', HT, ab, 1921; flowers deep apricot-red tinted buff, semi-dbl., intense fragrance; [Joseph Hill sport]; Croibier

'MME AGATHE NABONNAND', T, lp, 1886; (Agathe Nabonnand); flowers rosy flesh tinted amber, petals shell-like, very large; Nabonnand, G.

'MME ALBERT BARBIER', HP, ob, 1925; flowers salmon, tinted nankeen yellow, center darker, large, dbl., 50 petals, cupped, recurrent bloom, slight fragrance; vigorous, bushy growth; [Frau Karl Druschki X ?]; Barbier; Dreer

'MME ALBERT BERNARDIN', T, w; flowers white, shaded with carmine, center yellow; Mari, 1905

'MME ALBERT GILLES', HT, op, 1934; flowers light coral-pink, very large, very dbl.; [Unnamed variety X Jean C.N. Forestier]; Guillot, H.; Mallerin

'MME ALÉGATIÈRE', Pol, dp, 1888; flowers carmine-pink, medium, dbl.; Alégatière; (Sangerhausen)

'MME ALEXANDRE', HT, rb, 1926; flowers crimson-vermilion-red, shaded velvety purple, dbl.; Walter, L.

'MME ALEXANDRE CHARVET', HT, or, 1943; bud long, pointed; flowers orange-red, edged lilac, medium, dbl., cupped, slight fragrance; foliage leathery; moderate growth; [Charles P. Kilham X (Charles P. Kilham X Mme Joseph Perraud)]; Meilland, F.; A. Meilland

'MME ALEXANDRE DREUX', HT, my, 1921; bud pointed; flowers golden yellow, large, dbl.; [Rayon d'Or X Primerose]; Soupert & Notting

'MME ALFRED CARRIÈRE', N, w, 1879; flowers pale pinkish white, large, dbl., globular, recurrent bloom, intense fragrance; vigorous, climbing growth; (21); Schwartz, J.

'MME ALFRED DE ROUGEMONT', HP, lp, 1862; flowers white tinted pink, medium; vigorous growth; [Blanche Lafitte X Sappho (Vibert's)]; Lacharme, F.

'MME ALFRED PONNIER', HT, w, 1920; Bernaix, P.

'MME ALFRED SCHISSELÉ', HT, m, 1930; flowers lilac-white, center coppery orange, large, dbl., moderate fragrance; foliage bronze; [Frau Felix Tonnar X Angèle Pernet]; Leenders, M.

'MME ALICE GARNIER', HWich, pb, 1906; flowers bright rose, center yellow to light pink, small, intense, sweet fragrance; foliage dark, glossy; [R. wichurana X Mme Charles]; Fauque

'MME ANCELOT', HRg, mp, 1901; Müller, Dr. F.

'MME ANDRÉ CHARMET', HT, mp, 1921; flowers carnation-pink, dbl., moderate fragrance; [Mme Mélanie Soupert X Mme Maurice de Luze]; Croibier

'MME ANDRÉ DE HALLOY', HT, ob, 1929; flowers orange, reverse salmon-pink, very dbl., slight fragrance; [Gloire de Hollande X Benedicte Seguin]; Ketten Bros.

'MME ANDRÉ DULIN', HT, op, 1959; flowers bright coppery pink, large, dbl., moderate fragrance; vigorous, bushy growth; [Opera X Ville de Gand]; Gaujard

'MME ANDRÉ GILLIER', HT, rb, 1934; bud pointed; flowers coppery red, shaded yellow, semi-dbl., cupped, slight fragrance; [Padre X Unnamed seedling]; Reymond; Vially

'MME ANDRÉ LEROY', HP, mp, 1864; flowers salmon-rose, large, dbl.; vigorous growth; Trouillard

'MME ANDRÉ SAINT', HP, w, 1926; flowers milk-white to pure white, center clear chamois, dbl., moderate fragrance; few thorns; stocky growth; [Frau Karl Druschki X Benedicte Seguin]; Barbier

'MME ANGÉLIQUE VEYSSET', HT, lp, 1890; flowers pink tinted white, large, dbl.; Veysset

'MME ANNETTE AYNARD', HT, w, 1919; flowers milk-white edged pink, passing to amber-yellow, dbl., moderate fragrance; [Mme Caroline Testout X Prince de Bulgarie]; Leenders, M.

'MME ANTH. KLUIS', Pol, op, 1924; flowers salmon-pink tinted orange, large, borne in clusters, semi-dbl.; Kluis; Kluis & Koning

'MME ANTOINE MARI', T, pb, 1901; flowers rosy flesh, shaded lilac and rose, large, dbl., moderate fragrance; (14); Mari, A.

'MME ANTOINE MONTAGNE', HT, lp, 1930; flowers flesh-pink, with reflexes of old ivory; Richardier

'MME ANTOINE REBE', T, rb, 1900; Laperrière

'MME ARMAND SOUZY', HT, rb, 1945; flowers geranium-red and saffron-yellow, very large; [Charles P. Kilham X (Charles P. Kilham X Margaret McGredy)]; Meilland, F.

'MME ARTHAUD', HT, ob, 1938; flowers deep orange, large, very dbl., slight fragrance; vigorous growth; [Charles P. Kilham X (K. of K. X Mari Dot)]; Mallerin, C.

'MME ARTHUR OGER', Cl B, mp, 1899; flowers brilliant pink, very large, dbl.; very vigorous, climbing growth; Oger

'MME ARTHUR ROBICHON', Pol, dp, 1912; flowers carmine-pink, small, dbl.; Robichon; (Sangerhausen)

'MME AUGUSTE CHATAIN', HT, op, 1940; bud very long; flowers coral-salmon, well formed, large, dbl.; very vigorous growth; GM, Bagatelle, 1939; Mallerin, C.; A. Meilland

'MME AUGUSTE CHOUTET', LCl, ob, 1901; flowers orange-yellow; [William Allen Richardson X Kaiserin Auguste Viktoria]; Godard

'MME AUGUSTE NONIN', HWich, dp, 1914; flowers deep shell-pink, center white, small, borne in clusters, semi-dbl., cupped, slight fragrance; foliage large, glossy; very vigorous, climbing(15-20 ft), open growth; [Dorothy Perkins X Blush Rambler (?)]; Nonin

'MME AUGUSTINE HAMMOND', HT, mp, 1897; flowers very large, dbl.; Vigneron; (Sangerhausen)

'MME AUTRAND', HT, ob, 1922; flowers coppery orange, dbl., intense fragrance; [Mme Caroline Testout X Prince de Bulgarie]; Leenders, M.

'MME AZELIE IMBERT', T, 1870; Levet, F.; (Cavriglia)

'MME BALLU', HRg, m, 1901; flowers violet-pink, medium, semi-dbl.; Roseraie de l'Hay; (Sangerhausen)

'MME BARDOU JOB', HT, ly, 1913; flowers canary-yellow, center chrome-yello, semi-dbl., slight fragrance; [Prince de Bulgarie sport]; Dubreuil

'MME BARTHÉLEMY LEVET', T, ly; flowers canary-yellow; Levet, A., 1879

'MME BAULOT', HP, mp, 1885; flowers large, dbl.; Lévêque; (Sangerhausen)

'MME BÉRARD', Cl T, ob; flowers salmon-yellow shaded salmon-rose, large, dbl., cupped, moderate fragrance; [Mme Falcot X Gloire de Dijon]; Levet, F., 1870

'MME BERKELEY', T, pb, 1898; Bernaix, A.

'MME BERTHE DE FORGE', HT, op, 1935; flowers orange-coral, tinted coppery salmon, very large, cupped; foliage bronze; Chambard, C.

'MME BLONDEL', HT, mp, 1899; flowers very large, dbl., moderate fragrance; Veysset; (Sangerhausen)

'MME BLYTHA PEARKES', HT, ab, 1968; flowers light apricot flushed yellow and pink, medium, dbl.; bushy, compact growth; [Karl Herbst X Lady Hillingdon]; Blakeney; Eddie

'MME BOLL', HP, dp, 1859; flowers carmine-rose, large, recurrent bloom; foliage 5 leaflets per leaf; vigorous growth; Boll, Daniel; Boyeau

'MME BOLLAERT', HT, rb, 1938; bud long, coppery red; flowers carmine, shaded nasturtium-red, very large, dbl., cupped, intense fragrance; strong stems; very vigorous, bushy growth; [Ami F. Mayery X Unnamed seedling]; Chambard, C.

Mme Bollinger, F, rb, 1972; flowers coppery red with dark yellow, medium, semi-dbl.; McGredy, Sam IV; (Sangerhausen)

'MME BRAVY', T, w, 1846; (Adèle Pradel, Alba Rosea, Danzille, Joséphine Maltot, Mme de Sertot, Mme Denis); flowers creamy white shaded blush, dbl., moderate fragrance; Guillot Père

Mme Bruno Coquatrix *see* PINturnia

'MME BUTTERFLY', HT, lp, 1918; flowers light creamy pink; [Ophelia sport]; Hill, E.G., Co.

'MME BUTTERFLY, CLIMBING', Cl HT, lp, 1926; Smith, E.P.

'MME BYRNE', N, w, 1840; flowers cream color, center rose, large, dbl.; [Lamarque seedling]; Buist

'MME C. CHAMBARD', HT, lp, 1911; flowers rosy flesh-pink, shaded salmon, base yellow, dbl., 72 petals, moderate fragrance; [Frau Karl Druschki X Prince de Bulgarie]; Chambard, C.

'MME C. LIGIER', T, pb, 1900; flowers pink with darker red, large, very dbl., intense fragrance; Berland; (Sangerhausen)

'MME C. RICHARDIER', HT, my, 1924; flowers yellow, passing to clear yellow, dbl.; Richardier

'MME CAMILLE', T, lp, 1871; flowers aurora-pink, veined, with white reflections, well formed, large, dbl., moderate fragrance; Guillot et Fils

'MME CAMILLE LAURENS', HT, or, 1956; flowers crimson-red tinted orange, well formed, moderate fragrance; foliage bright green; strong stems; [Peace X Happiness]; Dorieux; Pin

'MME CARISTIE MARTEL', HT, ly, 1916; flowers pure sulfur-yellow, center deeper, 5–6 in.; GM, Bagatelle, 1917; Pernet-Ducher

'MME CARNOT', N, yb, 1889; flowers golden yellow tinged orange, center darker, large, very dbl., globular, moderate fragrance; vigorous, climbing growth; [William Allen Richardson seedling]; Moreau et Robert

'MME CAROLINE KÜSTER', N, yb, 1872; flowers yellow and orange blend, medium, dbl., moderate fragrance; Pernet; (Sangerhausen)

'MME CAROLINE SCHMITT', N, yb, 1878; (Caroline Schmitt); flowers salmon-yellow, dbl., recurrent bloom; [Solfaterre seedling]; Schmitt

'MME CAROLINE TESTOUT', HT, mp, 1890; bud pointed; flowers bright satiny rose, center darker, edged soft carmine-pink, dbl., moderate fragrance; foliage rich green, soft; vigorous, bushy growth; (28); [Mme de Tartas X Lady Mary Fitzwilliam]; Pernet-Ducher

'MME CAROLINE TESTOUT, CLIMBING', Cl HT, mp, 1901; Chauvry

'MME CECCALDI', HT, op, 1938; bud long; flowers salmon-carmine-pink, shaded vermilion, very large, dbl.; strong stems; very vigorous growth; [Soeur Thérèse X Unnamed seedling]; Chambard, C.

Mme Cécile Brünner *see* **'MLLE CÉCILE BRÜNNER'**

Mme Cécile Brünner, Climbing *see* **'MLLE CÉCILE BRÜNNER, CLIMBING'**

'MME CECILE MORAND', HP, 1890; Corboeuf; (Cavriglia)

'MME CÉLINE NOIREY', T, lp, 1868; flowers soft pink, reverse purple, large, very dbl., moderate fragrance; Guillot et Fils

'MME CÉLINE TOUVAIS', HP, mp, 1859; flowers large, dbl.; Touvais; (Sangerhausen)

'MME CÉSAR BRUNIER', HP, mp, 1888; flowers large, very dbl., moderate fragrance; Bernaix, A.; (Sangerhausen)

'MME CHABAN DELMAS', HT, mr, 1957; flowers bright red, well formed, large; foliage bright green; Privat

'MME CHAMOUTON-MURGUE', HT, op, 1925; flowers orange-carmine, shaded vermilion, very large, dbl., cupped, moderate fragrance; foliage dark; vigorous, erect, branching growth; [Unnamed seedling X Mrs Edward Powell]; Chambard, C.

'MME CHARLES', T, yb, 1864; flowers yellow, center salmon; [Mme Damaizin X ?]; Damaizin

'MME CHARLES ALLIZON', HT, w, 1928; flowers rosy white, edges tinted yellow, dbl.; [Mme Vittoria Gagniere X Lady Pirrie]; Schwartz, A.

'MME CHARLES BALTET', B, lp, 1865; Verdier, E.

'MME CHARLES DÉTREAUX', B, mr; flowers bright carmine-red, large

'MME CHARLES DUBREUIL', HT, pb, 1911; flowers salmon-rose, reverse shaded carmine, dbl.; [Pharisaer sport]; Guillot, P.

'MME CHARLES FREDERIC WORTH', HRg, dp, 1889; (C.F. Worth); flowers rosy carmine, fading, large, borne in large clusters, semi-dbl., profuse early bloom, but sparse in summer and fall, moderate fragrance; vigorous growth; Schwartz, Vve.

'MME CHARLES GUILLAUD', HT, op, 1943; flowers orange-pink with fiery tints; Mallerin, C.

'MME CHARLES GUILLOT', HT, pb, 1943; flowers carmine-pink and orange, large, dbl.; Mallerin, C.; (Sangerhausen)

'MME CHARLES HAAS', HT, w, 1930; flowers amber-white, tinted flesh-white, large, dbl., 60–70 petals, exhibition form, moderate fragrance; foliage leathery; vigorous, free branching growth; [Mme Abel Chatenay X Golden Emblem]; Ketten Bros.

'MME CHARLES JOLY', HT, op, 1942; flowers bright salmon shaded coppery yellow; Chambard, C.

'MME CHARLES LEJEUNE', LCl, lp, 1924; flowers soft pink; [Mme Caroline Testout X Lady Ashtown]; Vandevelde

'MME CHARLES LUTAUD', HT, yb, 1912; flowers chrome-yellow, blending to rosy scarlet at edge, dbl.; GM, Bagatelle, 1913; [Unnamed seedling X Marquise de Sinéty]; Pernet-Ducher

'MME CHARLES LUTAUD, CLIMBING', Cl HT, yb, 1922; Guillot, P.

'MME CHARLES MAGNY', HT, rb, 1941; flowers coppery red and golden yellow, very large, very dbl., cupped, moderate fragrance; vigorous, bushy growth; [Mme Joseph Perraud X Seedling]; Gaujard

'MME CHARLES MALLERIN', HT, ob, 1939; flowers orange-salmon, large, dbl., cupped, slight fragrance; foliage leathery, dark; vigorous growth; [Lucy Nicolas X Brazier]; Mallerin, C.; A. Meilland;, C-P

'MME CHARLES MEURICE', HP, dr, 1878; flowers velvety dark red, well formed, large, dbl.; bushy growth; Meurice

'MME CHARLES ROUVEURE', HT, dy, 1946; dbl., 36 petals, 4–5 in., moderate, fruity fragrance; foliage dark; free growth; Mallerin, C.; A. Meilland

'MME CHARLES SAUVAGE', HT, yb, 1949; (Mississippi); flowers yellow tinted saffron, center orange-yellow, well-shaped, dbl., 30 petals, 5 in., slight fragrance; bushy growth; [Julien Potin X Orange Nassau]; Mallerin, C.; URS

'MME CHARLES SINGER', T, dp, 1916; flowers garnet, becoming dark velvety purple-garnet, dbl.; vigorous growth; Nabonnand, C.

'MME CHARLES TRUFFANT', HP, lp, 1878; flowers satiny rose, large; vigorous growth; Verdier, E.

'MME CHARLES WOOD', HP, mr, 1861; (Dinsmore); flowers fiery scarlet, large, dbl., 45 petals, moderate fragrance; moderate growth; Verdier, E.

Mme Chatelaine de Lullier, HT, 1987; Meilland, L.; (Cavriglia)

'MME CHÉDANE-GUINOISEAU', T, ly, 1880; flowers sulphur yellow; [Safrano X ?]; Lévêque

'MME CHEINE DUGUY', HT, pb, 1929; flowers cerise-red, shaded scarlet, dbl.; Schwartz, A.

'MME CHEVALIER', B, mp, 1886; Pere Pernet

'MME CHIANG KAI-SHEK', HT, ly, 1942; bud long, pointed; flowers lemon yellow, turning lighter, dbl., 27 petals, 5–5.5 in., exhibition form, moderate fragrance; foliage leathery, glossy, dark; vigorous, upright, compact growth; AARS, 1944; [Joanna Hill X Sir Henry Segrave]; Duehrsen; H&S

'MME CLARA D'ARCIS', HT, pb, 1931; (Clara d'Arcis); flowers brilliant rose-pink, base yellow, large, dbl., moderate, spicy fragrance; foliage dark, leathery; [Julien Potin X Unnamed seedling]; Gaujard; C-P

'MME CLAUDE OLIVIER', HT, op, 1939; bud long, pointed; flowers coral, tinted nasturtium-yellow, very large, dbl., exhibition form, slight fragrance; foliage leathery; strong stems; very vigorous, bushy growth; [Soeur Thérèse X Unnamed R. foetida bicolor seedling] Mallerin, C.; A. Meilland

'MME CLÉMENCE JOIGNEAUX', HP, pb, 1861; flowers pink and red, very large, dbl.; Liabaud; (Sangerhausen)

'MME CLEMENCE MARCHIX', HT, 1899; Bernaix, P.; (Cavriglia)

'MME COCHET-COCHET', HT, op, 1934; bud very long; flowers coppery rose-pink, tinted coral, large, dbl., 30 petals, cupped, moderate fragrance; foliage glossy; vigorous growth; GM, Bagatelle, 1932; [Mrs Pierre S. duPont X Cécile Walter]; Mallerin, C.; C-P

'MME COLETTE MARTINET', HT, yb, 1915; flowers old-gold shaded orange-yellow, dbl., moderate fragrance; Pernet-Ducher

'MME CONSTANS', HWich, lp, 1900

'MME CONSTANT SOUPERT', T, yb, 1905; flowers yellow shaded peach; [Souv. de Pierre Notting X Duchesse Marie Salviati]; Soupert & Notting

'MME CORDIER', HP, m, 1903; flowers bright lilac-pink, large, dbl.; Leroy

'MME CORNÉLISSEN', B, lp, 1865; flowers large, very dbl., moderate fragrance; Cornélissen; (Sangerhausen)

'MME COUIBES', HT, ab, 1938; flowers salmon, center apricot, edged lighter, passing to golden coral, dbl., moderate fragrance; foliage fresh green; long stems; vigorous, bushy growth; [Charles P. Kilham X Rochefort]; Meilland, F.

Mme Crego, HT, w

'MME CRESPIN', HP, m, 1862; flowers rose shaded violet; Damaizin

'MME CREUX', T, op, 1890; flowers light salmon-pink, large, dbl., moderate fragrance; Godard; (Sangerhausen)

'MME CROIBIER', HT, op, 1935; (Mme J. B. Croibier, Mme Jean Croibier); flowers bright salmon, open, very large, dbl.; foliage leathery; very vigorous growth; GM, Rome, 1936; [Unnamed seedling X Mme Nicolas Aussel]; Gaujard

'MME CROMBEZ', T, op, 1888; flowers rosy buff tinted bronze, well formed, dbl., moderate fragrance; Nabonnand, G.

'MME CUNISSET CARNOT', HT, 1900; Buatois; (Cavriglia)

'MME CUSIN', T, rb; flowers crimson, center yellowish white, well formed, dbl.; Guillot et Fils, 1882

'MME DAILLEUX', HT, pb, 1901; flowers salmon pink with dark yellow, large, dbl., intense fragrance; Buatois; (Sangerhausen)

'MME DAMAIZIN', T, w, 1858; flowers creamy white shaded salmon, poorly formed, very large, dbl.; Damaizin

'MME D'ARBLAY', HMsk, w, 1835; flowers soft flesh changing to white, borne in large clusters, semi-dbl., moderate fragrance; [R. multiflora X R. moschata]; Wells

'MME DAVID', T, lp, 1895; flowers pale flesh, center darker, very dbl., flat, moderate fragrance; Pernet Père

'MME DE BEAUVOIRE', HT, lp, 1922; flowers pinkish white, center pale pink, dbl.; [Mme Vittoria Gagniere X Lady Ashtown]; Schwartz, A.

'MME DE CARBUCCIA', HT, mr, 1941; bud long, pointed; flowers bright crimson-red, medium, dbl., intense fragrance; foliage leathery; very vigorous, upright growth; [Admiral Ward X Mme Méha Sabatier]; Kriloff; A. Meilland

'MME DE LA RÔCHE-LAMBERT', M, m, 1851; flowers dark reddish purple, large, dbl., globular; moderate growth; Robert

'MME DE POMPADOUR', HT, rb, 1945; flowers coppery red and bright yellow, medium, borne, dbl., globular, slight fragrance; foliage dark, glossy; short stems; vigorous, bushy growth; Gaujard

Mme de Sancy de Parabère *see* **'MME SANCY DE PARABÈRE'**

Mme de Sertot *see* **'MME BRAVY'**

Mme de Sevigne, S, lp, 1998; Delbard

'MME DE SÉVIGNÉ', B, pb, 1874; flowers bright rose in center, edges lighter, large blooms in clusters, dbl.; vigorous growth; Moreau et Robert

Mme de Soubeyran *see* **'MME SOUBEYRAN'**

'MME DE ST JOSEPH', T, lp; flowers fawn shaded salmon, large, dbl., intense fragrance; moderate growth

Mme de Stella *see* **'LOUISE ODIER'**

'MME DE TARTAS', T, lp, 1859; (Mme de Thartas); flowers flush-pink, large, dbl., cupped; vigorous, sprawling growth; Bernède

Mme de Thartas *see* **'MME DE TARTAS'**

'MME DE TROTTER', HP, pb, 1854; flowers pink and red, medium, dbl.; Granger; (Sangerhausen)

'MME DE VATRY', T, dp, 1855; flowers deep pink, center lighter, large, dbl., recurrent bloom, moderate fragrance; vigorous growth; Guérin

Mme de Villars, M, mp; Béluze, 1855

'MME DE WATTEVILLE', T, yb, 1883; flowers lemon edged pink, large, dbl.; Guillot et Fils

Mme Delaunay, HT, 1963; Privat, J.; (Cavriglia)

'MME D'ENFERT', B, w, 1904; Vilin

Mme Denis *see* **'MME BRAVY'**

Mme Denise Cassegrain, Misc OGR, w

'MME DENISE GALLOIS', HT, pb, 1941; flowers salmon shaded yellow; Sauvageot, H.

'MME DÉSIRÉ GIRAUD', HP, pb; flowers blush-white striped rose; [Baronne Prevost sport]; Van Houtte, 1854

'MME DESMARS', HT, yb, 1929; bud long; flowers golden yellow, tinted nasturtium-red, very large, dbl., exhibition form, slight fragrance; foliage glossy; vigorous, bushy growth; [Ophelia X Constance]; Mallerin, C.; H. Guillot, 1932

'MME DESMARY', HT, yb, 1950; flowers ochre-yellow tinted orange, dbl., 30–35 petals; foliage bright green; vigorous growth; [Aspirant Marcel Rouyer X Emma Wright]; Moulin-Epinay; Vilmorin-Andrieux

'MME DESPREZ', B, mp, 1831; flowers rosy lilac, dbl., cupped; Desprez

'MME DESPREZ', Ch, w

'MME DEVOUCOUX', T, my, 1874; flowers bright yellow, well formed, dbl.; Ducher, Vve.

'MME D'HÉBRAY', HGal, w, 1820; flowers white with pink stripes, large, dbl.; Pradel; (Sangerhausen)

'MME DIDKOWSKY', HT, rb, 1943; flowers fiery red, reverse golden yellow; Mallerin, C.; A. Meilland

'MME DIEUDONNÉ', HT, rb, 1949; (Mme L. Dieudonné); flowers rose-red, reverse gold, dbl., 30 petals, 4–5 in., exhibition form, moderate fragrance; foliage dark, glossy; vigorous growth; [(Mme Joseph Perraud X Brazier) X (Charles P. Kilham X Capucine Chambard)]; Meilland, F.

'MME DIEUDONNÉ, CLIMBING', Cl HT, rb, 1959; Anderson

Mme Dimitriu® *see* 'DELCRIP'

'MME DORÉ', B, lp, 1863; dbl., moderate fragrance; medium growth; Fontaine

'MME DRIOUT', Cl T, pb, 1902; flowers bright rose, striped carmine, large, dbl.; very vigorous, climbing growth; Thirat, J.

'MME DUBOST', B, lp, 1890; flowers flesh, center rose; Pernet Père

'MME DUBROCA', T, pb, 1882; flowers salmon shaded carmine, large, dbl.; Nabonnand, G.

'MME E. ROCQUE', HMult, m, 1918; flowers violet, sometimes striped white, passing to amethyst, borne, dbl.; [Veilchenblau X Reine des Violettes]; Lottin

'MME E. TERRACOL', HT, yb, 1940; bud globular; flowers pure yellow, base orange, very large, dbl., slight fragrance; foliage leathery; very vigorous, upright growth; [Julien Potin X Soeur Therese]; Meilland, F.; A. Meilland

'MME EDMÉE METZ', HT, pb, 1901; flowers rosy carmine shaded salmon, dbl.; [Mme Caroline Testout X Ferdinand Jamin]; Soupert & Notting

'MME EDMOND FABRE', HP, mp, 1884; flowers large, dbl., moderate fragrance; Verdier, E.; (Sangerhausen)

'MME EDMOND GILLET', HT, yb, 1921; flowers reddish nankeen yellow, slightly shaded carmine at tips, dbl.; [Mme Edmond Rostand X Marquise de Sinéty]; Pernet-Ducher

'MME EDMOND LABBÉ', HT, rb, 1938; bud pointed; flowers orange-red, reverse golden yellow, large, dbl., slight fragrance; long stems; very vigorous growth; GM, Bagatelle, 1938; [Souv. de Claudius Pernet X Unnamed Hybrid Tea seedling]; Mallerin, C.; A. Meilland

'MME EDMOND RAYNAL', HT, ly, 1927; flowers yellowish cream, center salmon, dbl., intense fragrance; Sauvageot, H.; F. Guillot

'MME EDMOND ROSTAND', HT, pb, 1912; flowers pale flesh, center shaded salmon and reddish orange-yellow, dbl., moderate fragrance; [Unnamed variety X Prince de Bulgarie]; Pernet-Ducher

'MME EDOUARD ESTAUNIÉ', HT, yb, 1936; bud long; flowers nankeen yellow, center reddish, edges and reverse flesh-pink, dbl., moderate fragrance; long stems; very vigorous growth; [Seabird X Souv. de Claudius Pernet]; Buatois

'MME EDOUARD HERRIOT', HT, ob, 1913; (Daily Mail Rose); bud pointed; flowers coral-red shaded yellow and bright rosy scarlet, passing to, semi-dbl., moderate fragrance; foliage bronze, glossy; vigorous, spreading, branching growth; (28); GM, NRS, 1913; [Mme Caroline Testout X HT]; Pernet-Ducher

'MME EDOUARD HERRIOT, CLIMBING', Cl HT, ob, 1921; (Daily Mail Rose, Climbing); Ketten Bros.

'MME EDOUARD ORY', M, dp, 1854; flowers bright carmine-pink, dbl., globular; Robert

'MME ELIE DUPRAZ', HT, mr, 1948; bud large; flowers brilliant red, medium, dbl., exhibition form, moderate fragrance; foliage dark, glossy; very vigorous, bushy growth; Gaujard

'MME ELISA DE VILMORIN', HT, mr; (Mme Eliza de Vilmorin); flowers dark carmine, large, dbl., 30 petals, sparse bloom; upright, bushy growth; Lévêque, 1867

Mme Eliza de Vilmorin *see* **'MME ELISA DE VILMORIN'**

'MME EMILE DALOZ', HT, pb, 1934; flowers satiny purplish pink, reverse bright rose-pink, very large, dbl., globular, intense fragrance; foliage leathery, glossy; very vigorous, bushy growth; [Frau Karl Druschki X Souv. de Georges Pernet]; Sauvageot, H.; C-P;, Sauvageot

'MME EMILE MAYEN', HT, ly, 1924; flowers sulfur-yellow passing to cream, dbl.; Chambard, C.

Mme Emile Seneclauze, HT, 1966; Croix, P.; (Cavriglia)

'MME EMILE THIERRARD', HT, yb, 1919; flowers chamois-yellow and pink, stamens pure yellow, dbl., moderate fragrance; [Mrs Aaron Ward X Joseph Hill]; Turbat

'MME EMILIE CHARRON', T, mp, 1895; flowers china-pink, large, cupped; very vigorous growth; Perrier

'MME EMILIE LAFON', HT, mr, 1905; flowers large, dbl.; Moranville; (Sangerhausen)

'MME EMILIE VAN DER GOES', HT, op, 1925; bud pointed, orange-yellow and rosy shadings; flowers more pink than bud, large, semi-dbl., moderate fragrance; foliage bronze, leathery; bushy growth; [Columbia X Irish Fireflame]; Verschuren

'MME ERNEST CALVAT', B, mp, 1888; (Mme Ernst Calvat, Pink Bourbon); flowers pink shaded darker; [Mme Isaac Pereire sport]; Schwartz, Vve.

'MME ERNEST CHARLES', HT, or, 1933; bud large, long; flowers coral-red, reverse shrimp-red, dbl., cupped, moderate fragrance; foliage leathery, glossy, bronze; vigorous, bushy growth; [Mme Edmond Rostand X Severine]; Buatois

'MME ERNEST LEVAVASSEUR', HP, mr, 1900; flowers carmine-red, very large, very dbl., moderate fragrance; Vigneron; (Sangerhausen)

'MME ERNEST PERRIN', T, op, 1900; flowers light orange-pink, very large, dbl.; Schwartz; (Sangerhausen)

'MME ERNEST PIARD', HT, dp, 1888; flowers carmine-pink, very large, dbl.; Bonnaire; (Sangerhausen)

Mme Ernst Calvat *see* **'MME ERNEST CALVAT'**

'MME ETIENNE', T, mp, 1887; flowers rose, well formed; vigorous growth; Bernaix, A.

'MME ETIENNE LEVET', HP, mr, 1871; flowers carmine-red, large, dbl., moderate fragrance; Levet; (Sangerhausen)

'MME EUGÈNE MALLET', N, yb, 1875; flowers pink and yellow, dbl., moderate fragrance; moderate, climbing growth; Nabonnand, G.

Mme Eugène Marlitt *see* **'EUGÈNE E. MARLITT'**

'MME EUGÈNE MOREAU', HT, my, 1925; dbl.; vigorous growth; Richardier

'MME EUGÈNE PICARD', HT, my, 1932; [Ariel sport]; Gillot, F.; C-P

'MME EUGÈNE RÉSAL', Ch, pb, 1894; flowers bright pink shaded reddish orange, base yellow; [Mme Laurette Messimy sport]; Guillot, P.

'MME EUGÈNE VERDIER', HP, lp; flowers silvery pink, large, dbl., globular, moderate fragrance; vigorous growth; Verdier, E., before 1866

'MME EUGÉNIE BOULLET', HT, pb, 1897; flowers pink tinted yellow, moderate fragrance; Pernet-Ducher

'MME EUGÈNIE FRÈMY', HP, dp, 1885; flowers carmine-pink, large, very dbl., moderate fragrance; Verdier, E.

'MME FALCOT', T, my, 1858; flowers nankeen yellow passing to clear yellow, large, dbl., moderate fragrance; Guillot et Fils

'MME FANNY DE FOREST', N, lp, 1882; flowers very large, dbl.; Schwartz; (Sangerhausen)

'MME FANNY PAUWELS', T, 1885; Soupert, C.; (Cavriglia)

'MME FAURAX-LILLE', HT, mr, 1933; bud pointed; flowers bright vermilion-red passing to geranium-red, large, dbl.; [Cuba X Sir David Davis]; Reymond; Vially

'MME FEARNLEY SANDER', HT, m, 1921; flowers carmine, deepening to purple, base yellow, dbl.; [Gen. MacArthur X Rayon d'Or]; Ketten Bros.

'MME FERDINAND JAMAIN', HP, lp; bud globular; flowers deep pink, large, dbl., 50 petals, cupped, intense fragrance; Lédéchaux, 1875

'MME FERNAND GENTIN', HT, op, 1939; bud long, pointed; flowers copper shaded coral, large, semi-dbl., cupped, slight fragrance; vigorous growth; [Unnamed seedling X Brazier]; Mallerin, C.; A. Meilland

'MME FERNAND GREGH', HT, yb, 1955; flowers canary-yellow shaded coppery, large, slight fragrance; foliage glossy; vigorous growth; [Padre X Madeleine Pacaud]; Robichon

Mme Fernandel® *see* MEIsunaj

'MME FILLION', HP, op, 1865; flowers salmon-pink, large, dbl.; Gonod; (Sangerhausen)

'MME FOJO', HT, ob, 1937; flowers orange, well formed, large; strong stems; vigorous growth; Dot, Pedro; H. Guillot

'MME FOREST', HT, mp, 1928; [Lieutenant Chaure X Mrs George Shawyer]; Walter, L.

'MME FOUREAU', N, dy, 1913; flowers medium, dbl.; Viaud-Bruant; (Sangerhausen)

'MME FRACULY', HT, dy, 1929; flowers deep golden yellow, dbl.; foliage dark, glossy; long, strong stems; vigorous growth; [Ophelia X Constance]; Siret-Pernet

'MME FRANCISQUE FAVRE', Pol, 1915; Dubreuil; (Cavriglia)

Mme Franck Augis® *see* 'LAPLAM'

'MME FRANÇOIS BOLLEZ', HT, op, 1934; flowers coral-pink, center brighter, tinted orange, very large, dbl., moderate fragrance; foliage dark, bronze; very vigorous growth; Gillot, F.

'MME FRANÇOIS GRAINDORGE', Pol, pb, 1922; flowers dark reddish pink shaded magenta, base tinged lilac, large, ; vigorous growth; Grandes Roseraies

'MME FRANÇOIS HOT', HT, pb, 1928; flowers salmon shaded coppery rose, edged lighter, base salmon-yellow; [Lady Pirrie X Mme de Bauvoire]; Schwartz, A.

'MME FRANÇOIS PITTET', N, w, 1878; Lacharme, F.

'MME FRANÇOIS ROYET', HMult, mr, 1926; flowers bright red; [Crimson Rambler X Général Jacqueminot]; Royet

Mme Frank Augis, HT, 1969; Laperrière; (Cavriglia)

'MME G. FOREST-COLCOMBET', HT, dp, 1928; flowers deep carmine, strongly tinted scarlet; [Hadley sport]; Mallerin, C.; Grandes Roseraies

'MME G. HEKKENS', HT, dr, 1929; flowers velvety dark red (nearly as dark as Chateau de Clos Vougeot), single, intense fragrance; foliage glossy; [Gloire de Holland X Hawlmark Crimson]; Faassen-Hekkens

'MME GABRIEL HANRA', HT, rb, 1929; flowers strawberry-red, shaded carmine-purple, dbl., slight fragrance; [The Adjutant X K. of K.]; Ketten Bros.

'MME GABRIEL LUIZET', HP, lp, 1877; flowers light silvery pink, edged lighter, large, dbl., 34 petals, cupped, non-recurrent, moderate fragrance; vigorous growth; Liabaud

'MME GASTON ANNOUILH', N, w, 1899; flowers medium, dbl., intense fragrance; Chauvry; (Sangerhausen)

'MME GASTON DOUMERGUE', Pol, lp, 1934; flowers soft salmon; [Probably Gloria Mundi sport]; Levavasseur

'MME GASTON MESTREIT', Pol, lp, 1922; flowers very soft flesh-white, borne in clusters; [Jeanny Soupert X Katharina Zeimet]; Soupert & Notting

'MME GASTON NOCTON', Pol, w, 1928; flowers white, center flesh-pink, opening pure white, borne in clusters; dwarf growth; [Amaury Fonseca X Jeanny Soupert]; Soupert & Notting

'MME GEORGES BÉNARD', HT, dp, 1900; flowers carmine-pink with silvery reverse, large, dbl.; Corboeuf; (Sangerhausen)

'MME GEORGES BRÉDIF', HT, rb, 1955; flowers red mottled garnet, well formed, moderate fragrance; strong stems; Privat

'MME GEORGES BRUANT', HRg, w, 1887; bud pointed; flowers loose, large blooms in clusters, recurrent bloom, moderate fragrance; (14); [R. rugosa X Sombreuil]; Bruant

Mme Georges Clenet, HT, 1980; Godin, M.; (Cavriglia)

'MME GEORGES COZON', HT, pb, 1929; flowers shrimp-pink, reverse yellow; [Mme Charles Lutaud X Unnamed Hybrid Tea seedling]; Laperrière

Mme Georges Delbard® *see* 'DELADEL'

'MME GEORGES DELBARD', HT, or, 1959; flowers bright red suffused orange, dbl., 35–45 petals, 5 in., moderate fragrance; foliage dull green; long, strong stems; very vigorous growth; RULED EXTINCT 4/85; [Impeccable X Mme Robert Joffet]; Delbard-Chabert

Mme Georges Delbard, HT, 1979; Delbard-Chabert; (Cavriglia)

'MME GEORGES DROIN', HT, op, 1930; flowers orange-shrimp-pink; foliage bronze; very vigorous growth; Gaujard, Jules

'MME GEORGES LANDARD', HT, lp, 1925; foliage glossy; [Mme Abel Chatenay X Lyon Rose]; Walter, L.; Lamesch

'MME GEORGES PETIT', HT, mr, 1928; bud pointed; flowers bright purple-red to velvety crimson-red, large, dbl., exhibition form, intense fragrance; very vigorous growth; [Gen. MacArthur X Mme Edouard Herriot]; Ketten Bros.

Mme Georges Renoard *see* 'DELRENO'

'MME GHYS', HMult, m, 1912; flowers lilac-rose, borne in clusters; height 6-10ft; [Crimson Rambler X ?]; Ghys; Decault

'MME GILBERTE JANAUD', F, mp, 1957; flowers bright pink tinted salmon, well-formed, very dbl.; dwarf growth; Privat

'MME GILLET LAFOND', HT, pb, 1930; flowers salmon-white, center old-rose, large, semi-dbl., moderate fragrance; foliage light, leathery; vigorous growth; [Frau Felix Tonnar X Angèle Pernet]; Leenders, M.

'MME GINA DEMOUSTIER', HT, mr, 1920; flowers pure garnet-red, dbl., intense fragrance; [Étoile de France X Unnamed variety]; Laperrière

'MME GRÉGOIRE STAECHELIN', LCl, pb, 1927; (Spanish Beauty); flowers delicate pink, reverse stained crimson, ruffled, very large, non-recurrent, moderate fragrance; pear-shaped fruit; foliage heavy, large, dark; vigorous (13-14 ft.) growth; GM, Bagatelle, 1927 John Cook Medal, ARS, 1929; [Frau Karl Druschki X Château de Clos Vougeot]; Dot, Pedro; C-P, 1929

'MME GUIMET', HT, dp, 1942; flowers large, dbl., slight fragrance; Laperrière; (Sangerhausen)

'MME GUSTAVE METZ', HT, lp, 1906; flowers large, dbl.; Lamesch; (Sangerhausen)

'MME GUSTAVE SOUPERT', HT, pb, 1928; flowers purplish pink, center brighter, reverse silvery carmine, dbl., moderate fragrance; [Augustus Hartmann X Souv. de Georges Pernet]; Soupert & Notting

'MME HARDY', D, w, 1832; flowers pure white, occasionally tinged flesh-pink, green pip, large, dbl., cupped, borne in clusters, intense fragrance; vigorous growth; Hardy

Mme Hébert *see* **'PRÉSIDENT DE SÈZE'**

'MME HECTOR LEUILLOT', HT, yb, 1903; bud pointed; flowers golden yellow on carmine ground, dbl., moderate fragrance; Pernet-Ducher

'MME HÉLÈNE DAPPLES', HT, mr, 1932; flowers large, semi-dbl., intense fragrance; Heizmann, E.; (Sangerhausen)

'MME HÉLÈNE DUCHÉ', HT, pb, 1921; flowers soft rose with silvery reflexes, edged carmine, dbl.; [Mme Caroline Testout X Reine Emma des Pays-Bas]; Buatois

'MME HÉLÈNE PARMENTIER', HT, or, 1935; bud long; flowers clear nasturtium-red, shaded orange, passing to pink, semi-dbl., cupped, moderate fragrance; foliage glossy, wrinkled, bronze; long stems; vigorous, bushy growth; [Unnamed seedling X Angèle Pernet]; Sauvageot, H.

'MME HENRI BONNET', HT, pb, 1948; bud ovoid; flowers deep salmon-pink suffused golden orange, large, dbl., exhibition form, moderate fragrance; foliage leathery; vigorous, upright, compact growth; [Elite X Unnamed seedling]; Boerner; J&P

'MME HENRI FONTAINE', HT, lp, 1914; flowers creamy pink, large, dbl., moderate fragrance; Guillot; (Sangerhausen)

'MME HENRI GRAVEREAUX', HT, yb, 1926; flowers coppery yellow tinted bronze-yellow, veined orange, dbl., moderate fragrance; [Mrs Aaron Ward X Unnamed seedling]; Barbier

'MME HENRI GRIMM', HT, lp, 1934; bud long; flowers pinkish white tinged carmine, edged crimson, base yellow, dbl., exhibition form, intense fragrance; foliage leathery, dark; vigorous, bushy growth; [Mme Charles Detreaux X Mme Edouard Herriot]; Buatois

'MME HENRI GUILLOT', HT, rb, 1938; flowers orange-coral-red, large, dbl., 25 petals, slight fragrance; foliage glossy; vigorous, bushy growth; GM, Bagatelle, 1936 GM, Portland, 1939; [Rochefort X Unnamed R. foetida bicolor seedling]; Mallerin, C.; A. Meilland;, C-P

'MME HENRI GUILLOT, CLIMBING', Cl HT, rb, 1942; Meilland, F.

'MME HENRI GUILLOT, CLIMBING', Cl HT, rb, 1947; van Barneveld; C-P

'MME HENRI LAFOREST', HT, my, 1942; bud pointed; flowers golden yellow, very large, semi-dbl., moderate fragrance; foliage glossy; very vigorous, upright growth; Gaujard

'MME HENRI LUSTRE', HT, rb, 1924; flowers purplish garnet tinted currant-red, very large, dbl., exhibition form, slight fragrance; foliage leathery; very vigorous, bushy growth; [Mme Edouard Herriot X Yves Druhen]; Buatois

'MME HENRI PATÉ', HT, ly, 1929; bud pointed; flowers sulfur-yellow, very large, semi-dbl., intense fragrance; foliage bronze; very vigorous, bushy growth; [Souv. de Claudius Pernet X Unnamed seedling]; Pernet-Ducher; Gaujard

'MME HENRI PELLEY', HT, ly, 1928; flowers transparent cream-yellow; Richardier

'MME HENRI PEREIRE', HP, mr, 1887; flowers crimson-red, moderate fragrance; Vilin

'MME HENRI QUEUILLE', HT, op, 1928; bud long; flowers bright shrimp-pink, center deeper, reverse coppery gold, semi-dbl., moderate fragrance; foliage bronze; very vigorous growth; Pernet-Ducher; Gaujard

'MME HENRI QUEUILLE', HFt, or, 1928; flowers light orange/red, large, dbl., intense fragrance; Gaujard; (Sangerhausen)

'MME HENRI THIEBAUT', HT, op, 1931; bud pointed; flowers salmon-coral base coral-orange, very large, dbl., intense fragrance; foliage bright green; strong stems; vigorous growth; Chambard, C.

'MME HERMANN HAEFLIGER', S, dr, 1951; flowers large; foliage dark; vigorous growth; [R. foetida bicolor seedling X Charles P. Kilham]; Hauser

'MME HERRIOT PANACHÉE', HT, yb, 1921; flowers coral and golden yellow; [Mme Edouard Herriot sport]; Cassegrain

'MME HIDE', HT pb, 1997; flowers blush edged pink, moderately full (15–25 petals), large (7 cms) blooms; slight fragrance; some prickles; foliage large, medium green, matt; compact (1.2 m) growth; [Pink Picnica X Garden Party]; Ohta, Kaichiro

'MME HIPPOLYTE DUMAS', HT, pb, 1924; flowers flesh tinted salmon-pink, base yellow, dbl., slight fragrance; Guillot, P.

Mme Hippolyte Jamain, HP, lp; Garçon, 1871

'MME HONORÉ DEFRESNE', Cl T, my, 1886; flowers golden yellow, large, dbl., intense fragrance; vigorous growth; [Mme Falcot seedling sport]; Levet, F.

'MME HOSTE', T, w, 1887; flowers yellowish-white, imbricated, large, globular; vigorous growth; Guillot et Fils

'MME ISAAC PEREIRE', B, dp, 1881; flowers deep rose-pink shaded purple, large, dbl., moderate fragrance; vigorous growth; Garçon

Mme J. B. Croibier *see* **'MME CROIBIER'**

'MME J. BONNAIRE PIERRE', HP, 1892; Bonnaire; (Cavriglia)

'MME J. P. SOUPERT', HT, w, 1900; flowers very large, very dbl., moderate fragrance; Soupert & Notting; (Sangerhausen)

'MME J.M. FRUCTUS', HT, pb, 1935; flowers satiny carmine, shaded salmon, base yellow, well formed, dbl., cupped; foliage dark; very vigorous growth; Chambard, C.

'MME J.W. BÜDDE', HT, dp, 1906; flowers bright carmine; Soupert & Notting

'MME JACQUES PRIVAT', HT, mr, 1959; bud long; flowers red, slight fragrance; foliage glossy; vigorous growth; Privat, J.

Mme Jean Croibier *see* **'MME CROIBIER'**

'MME JEAN DEMESHAYNE', HT, op, 1934; flowers yellowish salmon-pink, large, very dbl., slight fragrance; Nicolas; (Sangerhausen)

'MME JEAN DUPUY', T, yb, 1902; bud long, pointed; flowers golden yellow washed pink, large, dbl., moderate fragrance; vigorous growth; Lambert, P.

'MME JEAN EVERAERTS', HP, dr, 1907; flowers large, dbl., moderate fragrance; Geduldig; (Sangerhausen)

'MME JEAN GAUJARD', HT, yb, 1938; bud long, pointed; flowers creamy yellow, reverse shaded orange and carmine-pink, very, dbl.; foliage leathery, light; very vigorous, bushy growth; [Julien Potin X Unnamed seedling]; Gaujard; J&P

'MME JEAN PAQUEL', HT, ly, 1934; bud long, pointed; flowers yellow, passing to cream; foliage glossy; [Alice Stern X Lilly Jung]; Walter, L.; Amis des Roses

Mme Jean Philippe, T, yb

'MME JEAN RATY', HT, w, 1932; bud pointed; flowers amber-white, edges tinted peach-blossom, well formed, large, dbl., moderate fragrance; [Mme Abel Chatenay X Unnamed seedling]; Ketten Bros.

'MME JEANNINE JOUBERT', B, dp, 1877; Margottin fils

'MME JENNY', HMult, mp; flowers satiny rose, small, borne in clusters, cupped, moderate fragrance; foliage dark; vigorous, climbing growth; Nonin, 1925

'MME JENNY GILLEMOT', HT, ly, 1905; bud pointed; flowers light saffron-yellow, large, dbl., exhibition form, moderate fragrance; Pernet-Ducher

'MME JOANNES BEURRIER', HT, op, 1942; bud pointed; flowers bright orange-pink, reddish reflections, very large, very dbl., moderate fragrance; foliage glossy, bronze; very vigorous, bushy growth; Gaujard

'MME JOSEPH BONNAIRE', HT, pb, 1891; Bonnaire

'MME JOSEPH COMBET', HT, pb, 1894; flowers pink with yellow and white, large, very dbl., intense fragrance; Bonnaire; (Sangerhausen)

'MME JOSEPH JULLIEN', HT, pb, 1938; flowers coppery carmine, very large, dbl., cupped, moderate fragrance; foliage slightly bronze; [Ami F. Mayery X Unnamed seedling]; Chambard, C.

'MME JOSEPH PERRAUD', HT, yb, 1934; (Sunburst); flowers yellow, center deeply tinted coppery, large, dbl., 33 petals, moderate fragrance; foliage glossy; GM, Bagatelle, 1934; [Julien Potin X Unnamed seedling]; Gaujard

'MME JOSEPH PERRAUD, CLIMBING', Cl HT, yb, 1945; Marsh's Nursery

'MME JOSEPH SCHWARTZ', T, w, 1880; (White Duchesse de Brabant); flowers white washed flesh-pink, medium, dbl.; vigorous growth; [Probably a Duchesse de Brabant sport]; Schwartz, J.

Mme Joséphine Guyet *see* **'JOSÉPHINE GUYET'**

'MME JULES BOUCHÉ', HT, w, 1911; bud pointed; flowers white, center shaded primrose or pale blush, large, dbl., 34 petals, moderate fragrance; tall growth; [Pharisaer X Unknown seedling]; Croibier

'MME JULES BOUCHÉ, CLIMBING', Cl HT, w, 1938; California Roses

'MME JULES FINGER', HT, lp, 1893; Guillot

'MME JULES FONTAINE-LAMARCHE', HT, mr, 1936; flowers velvety scarlet-red, large, dbl.; [Sensation X E.G. Hill]; Soupert & Notting

'MME JULES GOUCHAULT', Pol, op, 1913; flowers bright pink, tinted coral and orange, borne in clusters, dbl., cupped, moderate fragrance; bushy growth; [Maman Turbat X George Elger]; Turbat; Teschendorff

'MME JULES GRAVEREAUX', Cl T, ab, 1901; flowers flesh, shaded peach or yellow, very dbl., recurrent bloom, slight fragrance; foliage large, dark, glossy; height 3-4 ft., bushy growth; [Reve d'Or X Viscountess Folkestone]; Soupert & Notting

'MME JULES GROLEZ', HT, mp, 1896; bud long, pointed; flowers bright china-rose, dbl., exhibition form, moderate fragrance; bushy growth; [Triomphe de l'Exposition X Mme Falcot]; Guillot, P

'MME JULES GUÉRIN', HT, ly, 1931; flowers deep cream, very large, dbl., intense fragrance; foliage bronze; Gaujard

'MME JULES THIBAUD', Pol, op; flowers coral-pink; [Cécile Brunner sport]

'MME JULES WALTHERY', HT, my, 1924; flowers outer petals becoming white; Allen

'MME JULIE DARAN', HP, mr, 1861; flowers crimson; Touvais

'MME JULIEN POTIN', HRg, lp, 1913; flowers pure flesh-pink, large, dbl., flat, recurrent bloom; foliage very leathery; vigorous growth; [R. rugosa X Gloire de Dijon]; Gravereaux; Cochet-Cochet

Mme Juliette *see* **'JULIETTE E. VAN BEUNINGEN'**

Mme Juliette Guillot, F, dp

'MME KAHN', HT, ly, 1939; flowers canary-yellow, reverse slightly reddish; vigorous growth; [Charles P. Kilham X Ville de Paris]; Colombier

'MME KAREN SHADISH', F, mp, 1999; dbl., 17–25 petals, 2–3 in., borne in small clusters, slight fragrance; prickles moderate; foliage small, medium green, semi-glossy; upright, medium (2-3 ft) growth; [Pink Petticoat X Mary Rose]; Chaney, William

'MME KASTLER', HT, mr, 1934; bud long, pointed; flowers large, dbl.; very vigorous growth; [Mme Adele Gance X Mme Caroline Testout]; Walter, L.; Amis des Roses

'MME KLATZ', M, dp; flowers deep pink, quite large

'MME KNORR', P, mp, 1855; dbl., moderate fragrance; Verdier, V.

'MME KRILOFF', HT, yb, 1944; flowers clear saffron-yellow, veined reddish orange, large, dbl., globular, moderate fragrance; foliage leathery; vigorous, bushy growth; GM, Bagatelle, 1944 GM, NRS, 1948; [Peace X Signora]; Meilland, F.; A. Meilland

Mme L. Dieudonné *see* **'MME DIEUDONNÉ'**

'MME L. HOT', HT, pb, 1926; flowers reddish salmon, shaded salmon-rose and chrome-yellow, dbl.; [Gorgeous X Rosomane Narcisse Thomas]; Bernaix, P.

Mme L. Ladoire, F, op

'MME L. LAPERRIERE', HT, 1951; Laperrière; (Cavriglia)

Mme L. Pradel® *see* 'LAPAV'

'MME LA COLONELLE DESMAIRES', F, dr, 1969; flowers large, dbl., slight fragrance; foliage soft, bronze; moderate growth; [Mercator X Alain]; Delforge

'MME LA GÉNÉRALE ARDOUIN', HT, rb, 1927; flowers coppery carmine shaded chrome-yellow, dbl., moderate fragrance; Chambard, C.

'MME LAFFAY', HP, mr, 1839; flowers bright crimson, dbl., intense fragrance; very vigorous growth; Laffay, M.

'MME LAJOTTE', HT, pb; flowers bright salmon, reverse yellow, large, dbl.; very vigorous, bushy growth; [Rome Glory X Marie-Rose Toussaint]; Gaujard

Mme Lambard *see* **'MME LOMBARD'**

'MME LANDEAU', rb, 1875; flowers light red with white striping, medium, dbl.; Moreau et Robert; (Sangerhausen)

'MME LANQUETIN', HT, ab; flowers apricot; Ofman; Pin

'MME LAURAS', HT, mr, 1954; flowers bright red, medium, dbl., slight fragrance; foliage leathery; very vigorous growth; [Rome Glory X Mme Elie Dupraz seedling]; Gaujard

'MME LAURENT SIMONS', T, op, 1894; flowers coppery-pink, large, very dbl., moderate fragrance; Lévêque; (Sangerhausen)

'MME LAURETTE MESSIMY', Ch, dp, 1887; bud long; flowers rose-pink, base shaded yellow, large, dbl.; vigorous growth; (14); [Rival de Paestum X Mme Falcot]; Guillot et Fils

'MME LAURIOL DE BARNY', B, lp, 1868; flowers silvery pink, large, dbl., quartered, rarely repeats, moderate, fruity fragrance; height 5-6 ft; Trouillard

'MME LE GUELINEL', HT, mr, 1959; flowers large, intense fragrance; [Mme Kriloff X Marrakech]; Gaujard, R.; G. Truffaut

'MME LEFÈBVRE', HP, mp, 1886; flowers large, dbl.; Moreau et Robert; (Sangerhausen)

'MME LEGRAND', pb, 1863; flowers pink and red, large, dbl.; Fontaine; (Sangerhausen)

'MME LEGRAS DE ST GERMAIN', A, w, 1846; flowers white, center rich cream, large, dbl.; few prickles; foliage gray-green; vigorous (6-7 ft) growth

'MME LEMESLE', HP, m, 1892; flowers purple/violet, medium, dbl.; Moreau et Robert; (Sangerhausen)

'MME LÉON CUNY', HT, mr, 1955; bud long, pointed; flowers bright red, veined purple, very large, dbl., exhibition form, moderate fragrance; foliage dark, bronze; upright growth; [Peace X Seedling]; Gaujard

'MME LÉON FÉVRIER', T, pb, 1884; flowers silvery rose shaded crimson, well formed, dbl., intense fragrance; Nabonnand, G.

'MME LÉON GUINOTTE', HT, pb, 1924; flowers glistening pink shaded yellow; [Mme Edouard Herriot X Old Gold]; Verschuren

'MME LÉON PAIN', HT, pb, 1904; bud pointed; flowers silvery flesh-pink, center orange-yellow, reverse salmon-pink, dbl., 45 petals, moderate fragrance; vigorous, bushy growth; [Mme Caroline Testout X Souv. de Catherine Guillot]; Guillot, P.

'MME LÉON PIN', HT, ob, 1954; flowers soft orange, medium, dbl., slight fragrance; foliage dark, leathery; upright growth; [Louise de Vilmorin X Capucine seedling]; Gaujard

'MME LÉON SIMON', HT, pb, 1909; flowers dark pink with cream, large, dbl.; Lambert, P.; (Sangerhausen)

'MME LÉON TROUSSIER', HT, op, 1941; flowers coral, base golden yellow; Mallerin, C.; A. Meilland

'MME LÉON VOLTERRA', HT, mp, 1958; flowers salmon-pink, well shaped, dbl., intense fragrance; vigorous growth; GM, Bagatelle, 1957; [Ambassadeur Nemry X Tawny Gold]; Leenders, M.

'MME LÉONCE COLOMBIER', HT, ly, 1926; flowers center straw-yellow passing to white, reverse light rose, dbl.; Richardier

'MME LEOPOLD DUPUY', HT, mp, 1912; flowers large, dbl., intense fragrance; Robichon; (Sangerhausen)

Mme Liabaud *see* La Virginale

'MME LIERVAL', HP, pb, 1868; Fontaine

'MME LINE RENAUD', HT, mr, 1956; flowers velvety red, intense fragrance; foliage bright green; [Crimson Glory X Unnamed seedling]; Mondial Roses

'MME LOEBEN SELS', HT, w, 1879; flowers silvery white shaded rose, large, dbl., flat; moderate growth; Soupert & Notting

'MME LOMBARD', T, op, 1878; (Mme Lambard); flowers rosy salmon, center darker, sometimes rosy flesh, large, very dbl., moderate fragrance; vigorous growth; [Mme de Tartas seedling]; Lacharme, F.

Mme Louis Armand, HT, 1974; Croix; (Cavriglia)

Mme Louis Ladoire, F; Arles, F.; (Cavriglia)

'MME LOUIS LAPERRIÈRE', HT, mr, 1951; flowers rich scarlet, well-formed, medium, dbl., 48 petals, intense fragrance; foliage dark; upright, bushy growth; GM, Bagatelle, 1950; [Crimson Glory X Seedling]; Laperrière; EFR

Mme Louis Lens *see* **'WHITE BRIARCLIFF'**

'MME LOUIS LENS, CLIMBING', Cl HT, w, 1935; [White Briarcliff sport]; Lens

'MME LOUIS LÉVÊQUE', HP, dp, 1864; flowers carmine-rose, very large, dbl.; moderate growth; Lévêque

'MME LOUIS LÉVÊQUE', M, mp; flowers brilliant salmon-pink, large, dbl., globular, sometimes blooms in fall, moderate fragrance; prickles well mossed; Lévêque, 1898

'MME LOUIS LÉVÊQUE', T, yb, 1892; flowers yellow washed pink, dbl.; moderate growth; Lévêque

'MME LOUIS RICARD', HP, dp, 1904; flowers carmine-pink, medium, dbl.; Boutigny; (Sangerhausen)

'MME LOUIS RICART', B, lp, 1892; Duboc

'MME LOUISA COINTREAU', HT, mr, 1957; flowers garnet-red, dbl., slight fragrance; vigorous growth; [Crimson Glory X Symbole]; Robichon

Mme Louise Guillot, F, 1955; Guillot, M.; (Cavriglia)

'MME LOUISE PIRON', HP, lp, 1903; Piron-Medard

'MME LOUISE TRÉMEAU', HT, op, 1931; flowers pink, shaded nasturtium-red, center brighter, open, large, dbl., cupped, slight fragrance; vigorous growth; [(Frau Karl Druschki X Mme Edouard Herriot) X (Mrs Edward Powell X R. foetida bicolor)]; Mallerin, C.; C-P

'MME LUCIEN CHAURE', HP, 1884; Vigneron; (Cavriglia)

'MME LUCIEN PERRIER', HT, rb, 1938; bud long; flowers coppery red, lighter on opening, open, very large, dbl., moderate fragrance; foliage glossy, dark; vigorous growth; Gaujard

'MME LUCIEN VILLEMINOT', HRg, lp, 1901; flowers pale pink, large, dbl., globular, moderate fragrance; vigorous growth; [Conrad Ferdinand Meyer X Belle Poitevine]; L'Hay

'MME LUREAU-ESCALAIS', HP, mp, 1886; flowers large, dbl.; Verdier, E.; (Sangerhausen)

'MME MALLERIN', HT, rb, 1924; flowers crimson-scarlet shaded vermilion, intense fragrance; [Mrs Edward Powell X Unnamed seedling]; Chambard, C.

Mme Marcel Astic, F, 1978; Orard, Joseph; (Cavriglia)

'MME MARCEL DELAUNEY', HT, pb, 1916; flowers pale pink or soft rose, shaded hydrangea-pink, dbl., moderate fragrance; GM, Bagatelle, 1915; Leenders, M.

'MME MARCEL FAUNEAU', HP, mr, 1886; flowers carmine-red, large, dbl.; Vigneron; (Sangerhausen)

'MME MARGOTTIN', T, my, 1866; flowers lemon-yellow, large, dbl.; Guillot et Fils

'MME MARGUERITE LAGIÈRES', HT, w, 1955; flowers cream, base golden yellow, very large, dbl.; foliage glossy; strong stems; vigorous growth; Privat

'MME MARGUERITE MARSAULT', HP, m, 1894; flowers reddish-violet, large, dbl.; Corboeuf; (Sangerhausen)

'MME MARIE CROIBIER', HT, dp, 1901; flowers large, dbl.; Croibier; (Sangerhausen)

'MME MARIE CURIE', HT, dy, 1943; (Québec); flowers clear yellow, dbl., 25 petals, 5 in., exhibition form, slight fragrance; foliage leathery, dark; vigorous, bushy, compact growth; AARS, 1944; Gaujard; J&P

'MME MARIE EBERLIN', HT, w, 1923; flowers cream, passing to white, base light yellow, dbl.; [Comtesse Melanie de Pourtales X Capt. Christy]; Walter, A.; Bacher

'MME MARIE LAVALLEY', HT, pb, 1881; flowers bright rose tinted white; Nabonnand, G.

Mme Marie Van Houtte, HP, lp

'MME MARIUS DÉVIGNE', HT, pb, 1930; bud long; flowers salmon-pink, reverse vivid carmine, large, globular; Reymond

Mme Martha Ancey *see* **'MARTHE ANCEY'**

'MME MARTIGNIER', Cl T, m; flowers red tinted purplish, on yellow ground; Dubreuil, 1903

'MME MASSON', HP, dr, 1856; flowers pure crimson-rose, free bloom during long season, moderate fragrance; vigorous, compact growth; Masson

'MME MASSOT', B, w, 1856; Lacharme, F.

'MME MAURICE BAUDOT', HT, ob, 1941; flowers bright orange, medium, semi-dbl., exhibition form, moderate fragrance; foliage dark; very vigorous, bushy growth; [Mme Joseph Perraud X Seedling]; Gaujard

'MME MAURICE CAPRON', HT, ab, 1914; flowers deep apricot-yellow tinted salmon; Guillot, P.

'MME MAURICE CAZIN', HT, rb, 1931; bud pointed; flowers dark scarlet, reverse clear reddish crimson, large, dbl.; [Gen. MacArthur X Hadley]; Schwartz, A.

'MME MAURICE DE LUZE', HT, pb, 1907; flowers rose-pink, center carmine; [Mme Abel Chatenay X Eugene Furst]; Pernet-Ducher

'MME MAURICE FENAILLE', HT, 1904; Boutigny; (Cavriglia)

Mme Maurice Genevoix, F, op, 1966; Hémeray-Aubert

'MME MAURICE KUPPENHEIM', T, ab, 1877; flowers salmon-yellow, large, dbl.; Ducher, Vve.

'MME MAURIN', T, ob, 1853; flowers creamy white, shaded with salmon, large, dbl.; Guillot Père

'MME MÉHA SABATIER', HT, rb, 1916; flowers deep red, with white stripes in some petals, dbl.; [Seedling X Château de Clos Vougeot]; Pernet-Ducher

'MME MÉLANIE SOUPERT', HT, yb, 1905; bud pointed; flowers salmon yellow, suffused pinkand carmine, large semi-dbl., moderate fragrance; vigorous growth; Pernet-Ducher

'MME MÉLANIE WILLERMOZ', T, lp, 1849; flowers white tinted pink, moderate fragrance; Lacharme, F.

'MME MERCIER DE MOLIN', HT, rb, 1921; flowers fiery red, tinted crimson, edges slightly tinged rose-pink, dbl., moderate fragrance; [Comte G. de Rochemur X Liberty]; Schwartz, A.

'MME MICHEL DUFAY', HMoy, rb, 1932; flowers maroon, reverse purplish garnet, large, dbl., cupped, non-recurrent, slight fragrance; foliage dark; very vigorous, bushy growth; [George Dickson X R. moyesii]; Sauvageot, H.

'MME MILLERAND', HT, w, 1926; flowers rosy white, shaded salmon, dbl.; [Pharisaer X Mme Henriette Schissele]; Walter, L.

'MME MINIVER', HT, ob, 1947; bud pointed; flowers orange-red tinted apricot, base yellow, large, dbl.; vigorous growth; [Joanna Hill X Charles P. Kilham]; Vilmorin-Andrieux

'MME MOISANS', LCl, mp, 1955; flowers hortensia-pink, dbl., recurrent bloom, moderate fragrance; foliage glossy; [Lady Sylvia X Unnamed seedling]; Robichon; Pin

'MME MORAND ANDRÉE', HT, dr, 1957; flowers blackish red, very large; foliage dark; vigorous growth; Privat

'MME MOREAU', M, pb, 1872; flowers rose, edged white, large, dbl.; Moreau et Robert

'MME MOSER', HT, pb, 1889; flowers pink to rosy white, large, dbl., intense fragrance; Vigneron

'MME MOUSEUR-FONTAINE', HT, yb, 1931; bud pointed; flowers sulfur-yellow, center saffron-yellow, large, dbl., slight fragrance; [Sunburst X Primerose]; Soupert & Notting

Mme Neige *see* 'MEIDONA'

'MME NERARD', B, pb, 1938; Nerard

Mme Neumann *see* **'HERMOSA'**

'MME NICOLAS AUSSEL', HT, ob, 1930; bud pointed; flowers salmon shaded carmine and ochre, large, dbl.; foliage dark; vigorous growth; GM, Portland; Pernet-Ducher; Gaujard

'MME NICOLAS BOUDLER', HT, pb, 1934; bud round; flowers rose, reverse shaded yellow, large, dbl.; [Souv. de Georges Pernet X Gloire de Dijon]; Boudler

'MME NOBÈCOURT', B, lp, 1893; flowers very large, dbl., intense fragrance; Moreau et Robert; (Sangerhausen)

'MME NOËL', HT, rb, 1939; flowers vermilion-red, reverse carmine-yellow, very large, dbl., cupped; foliage dark; vigorous growth; Chambard, C.

'MME NOËL LE MIRE', HT, rb, 1934; flowers brilliant crimson-red with yellow reflections, borne in clusters, semi-dbl., cupped, moderate fragrance; strong stems; dwarf growth; [George Dawson X Dance of Joy]; Sauvageot, H.

'MME NORBERT LEVAVASSEUR', Pol, mr, 1903; (Red Baby Rambler); flowers crimson-red, center lighter, bluing badly, small blooms in l, semi-dbl., cupped, slight fragrance; foliage glossy, dark; bushy, dwarf growth; (14); [Crimson Rambler X Gloire des Polyantha]; Levavasseur

'MME NORMAN', HP, w, 1867; flowers medium, dbl.; Guillot; (Sangerhausen)

'MME OCKER FERENCZ', T, yb, 1892; flowers canary and carmine; Bernaix, A.

'MME OFMAN', HT, or, 1954; bud long, pointed; flowers large, very dbl., exhibition form, moderate fragrance; foliage dark, glossy; strong stems; vigorous, upright, bushy growth; Ofman

'MME OLYMPE TERETSCHENKO', B, 1882; Lévêque; (Cavriglia)

'MME ORÈVE', HT, ob, 1926; flowers rose-salmon, center coppery salmon, moderate fragrance; Chambard, M.

'MME P. DOITHIER', HT, pb, 1920; flowers glossy pink, shaded shrimp-pink; Chambard, M.

Mme P. Euler *see* **'MME PIERRE EULER'**

Mme P. Olivier *see* **'SHINING STAR'**

'MME PAQUEL', HT, my, 1945; flowers chrome-yellow, well formed, dbl.; foliage wood and reddish; Mallerin, C.; A. Meilland

'MME PAUL BOUJU', HT, ob, 1930; bud long, pointed; flowers carmine-orange, very large, cupped; foliage bronze; Chambard, C.

'MME PAUL DURINGE', HT, mp, 1934; flowers deep coral, large; very vigorous growth; Chambard, C.

Mme Paul Euler *see* **'MME PIERRE EULER'**

'MME PAUL LACOUTIÈRE', HT, op, 1897; flowers coppery-pink, large, semi-dbl., intense fragrance; Buatois; (Sangerhausen)

'MME PAUL MARCHANDEAU', HT, w, 1928; flowers white, base deep yellow, dbl., moderate fragrance; Barbaras

'MME PAUL OLLIVARY', HT, dp, 1924; bud pointed; flowers coppery salmon, reverse shaded yellow, nearly large, single; [Mme Mélanie Soupert X Emma Wright]; Schwartz, A.

'MME PAUL PARMENTIER', HT, ob, 1919; flowers salmon-yellow shaded flesh, copper and daybreak-pink, dbl.; [Le Progres X Lyon Rose]; Gillot, F.

'MME PAUL ROTTIER', F, rb, 1957; flowers red shaded orange, borne in large clusters; moderate growth; Buisman, G. A. H.

Mme Paula Guisez *see* **'HEAT WAVE'**

Mme Paule Massad *see* MASlupau

'MME PAULINE LABONTÉ', T, op, 1852; flowers salmon-pink, large, dbl., moderate fragrance; vigorous growth; Pradel

'MME PERNET-DUCHER', HT, dy, 1891; flowers medium, very dbl., moderate fragrance; Pernet-Ducher; (Sangerhausen)

'MME PH. PLANTAMOUR', HRg, mr, 1900; flowers large, semi-dbl.; (Sangerhausen)

'MME PHILIPPE RIVOIRE', HT, ab, 1908; flowers apricot-yellow, center nankeen yellow, reverse red, large, dbl., globular; vigorous growth; Pernet-Ducher

'MME PIERRE COCHET', N, yb, 1891; flowers saffron-yellow, shaded scarlet, center apricot, dbl., intense fragrance; vigorous, climbing growth; [Reve d'Or seedling]; Cochet, S.

'MME PIERRE EULER', HT, rb, 1907; (Mme P. Euler, Mme Paul Euler); flowers silvery vermilion-pink, dbl., moderate fragrance; [Antoine Rivoire X Killarney]; Guillot, P.

'MME PIERRE FORESTIER', HT, ob, 1933; flowers orange, shaded shrimp-carmine to satiny china-pink, very large, cupped, moderate fragrance; Chambard, C.

'MME PIERRE GUILLOT', HT, 1928; Guillot; (Cavriglia)

'MME PIERRE KOECHLIN', HT, op, 1936; flowers salmon-pink, very large, dbl., exhibition form, slight fragrance; [Unnamed seedling X The Queen Alexandra Rose]; Sauvageot, H.; C-P

'MME PIERRE OGER', B, pb, 1878; flowers blush, reverse tinged rosy lilac; (28); [Reine Victoria sport]; Oger; C. Verdier

'MME PIERRE PERNY', T, ly, 1880; flowers large, semi-dbl.; Nabonnand; (Sangerhausen)

Mme Pirrie *see* **'LADY PIRRIE'**

'MME PIZAY', HT, op, 1920; flowers light salmon; [Unnamed variety X Mme Mélanie Soupert]; Chambard, C.

'MME PIZAY, CLIMBING', Cl HT, op, 1920; flowers light salmon-pink, very large, dbl.; Chambard, C.; (Sangerhausen)

'MME PLANTIER', HAlba, w, 1835; flowers creamy white changing to pure white, green pip, blooms in cl, very dbl., flat, non-recurrent, moderate fragrance; foliage small, 7 leaflets; vigorous, spreading, bushy (about 5 ft.) growth; (21); [Thought to be R. alba X R. moschata]; Plantier

'MME PLATZ', dp, 1864; flowers carmine-pink, medium, dbl., intense fragrance; Moreau et Robert; (Sangerhausen)

'MME PLUMECOCQ', HT, dy, 1931; flowers golden yellow, center brighter; [Roselandia X Ville de Paris]; Lens

'MME PLUMECOCQ', HT, pb, 1954; flowers bright pink, reverse silvery, very large, dbl., exhibition form; foliage bronze; vigorous, upright growth; [Peace X Seedling]; Gaujard

Mme Poincaré *see* **'MME RAYMOND POINCARÉ'**

'MME PORTIER DUREL', LCl, 1910; Durel, P.; (Cavriglia)

'MME PROSPER LAUGIER', HP, mr, 1875; flowers carmine-red, large, dbl., moderate fragrance; Verdier, E.; (Sangerhausen)

'MME RAMBAUX', HP, dp, 1881; flowers very large, very dbl., moderate fragrance; Rambaux; (Sangerhausen)

'MME RAOUL FAURAN', HT, m, 1934; flowers carmine shaded velvety purple, reverse light purple, very large, semi-dbl., cupped, moderate fragrance; foliage leathery, dark; vigorous, bushy growth; Sauvageot, H.

'MME RAVARY', HT, ob, 1899; flowers orange-yellow, dbl., moderate fragrance; vigorous, bushy growth; Pernet-Ducher

'MME RAYMOND CHEVALIER-APPERT', HT, rb, 1917; flowers cerise-red, edged lighter, dbl., intense fragrance; [Gen. MacArthur X Richmond]; Guillot, P.

Mme Raymond Gaujard *see* **Olympiad**®

'MME RAYMOND POINCARÉ', HT, pb, 1919; (La Rose de Mme Raymond Poincaré, Mme Poincaré); flowers pale pink and salmon, center yellow, dbl., slight fragrance; GM, Bagatelle, 1915; Gravereaux; Kieffer & Sons

'MME RÉMOND', T, yb; flowers sulfur-yellow, edged red, very dbl., moderate fragrance; Lambert, E., 1883

'MME RENAHY', HP, dp, 1889; (Mme Renshy); flowers carmine; Guillot et Fils

'MME RENARD', HP, op, 1872; flowers salmon-pink, very large, dbl.; Moreau et Robert; (Sangerhausen)

'MME RENÉ ANDRÉ', HT, lp, 1906; flowers large, dbl., slight fragrance; (Sangerhausen)

Mme René Cassin *see* 'DELTAF'

'MME RENÉ COLLETTE', HT, pb, 1909; flowers yellowish-pink with red, large, dbl., intense fragrance; Gamon; (Sangerhausen)

'MME RENÉ COTY', HT, rb, 1955; bud globular; flowers persian red, reverse yellow, very large, dbl., moderate fragrance; foliage glossy, leathery; long, strong stems; vigorous, upright growth; GM, Bagatelle, 1954; [Peace X Brazil]; Meilland, F.; URS

'MME RENÉ LEFÈVRE', HT, yb, 1938; bud long, pointed, sulphur-yellow, edged cerise; flowers golden yellow, flushed carmine, open, large, semi-dbl.; foliage leathery, bronze; [Elizabeth of York X Mme Henri Pate]; Robichon

'MME RENÉ TRUCHOT', HT, 1952; Orard, Joseph; (Cavriglia)

'MME RENEE GRAVEREAUX', HRg, lp, 1906; Gravereaux

'MME RENÉE OBERTHÜR', HT, w, 1908; flowers large, dbl., moderate fragrance; Vigneron; (Sangerhausen)

Mme Renshy *see* **'MME RENAHY'**

'MME ROBERT', HT, yb, 1917; flowers nankeen yellow with chamois reflexes, dbl.; Chambard, C.

'MME ROBERT FORTIN', HT, m, 1935; flowers carmine-purple, large, dbl., cupped, intense fragrance; very vigorous growth; [Mme Caroline Testout X Yves Druhen]; Buatois

'MME ROBERT JOFFET', F, rb, 1956; flowers salmon to geranium, reverse carmine-pink, well formed, large, semi-dbl.; Delbard-Chabert

'MME ROBERT MARTIN', HT, pb, 1943; bud long, pointed; flowers pink, center coral, very large, very dbl., slight fragrance; foliage leathery; vigorous, bushy growth; [Charles P. Kilham X Mme Joseph Perraud]; Meilland, F.; A. Meilland

'MME ROBERT PERRIER', HT, ob; flowers coppery, stamens saffron, large, semi-dbl.; Orard, Joseph

'MME ROBERTE HUET', HT, rb, 1960; flowers velvety scarlet, reverse raspberry-red, base gold, dbl., slight fragrance; foliage glossy, slightly bronze; strong stems; vigorous growth; Hémeray-Aubert

'MME ROCHEFONTAINE', HP, pb; flowers rosy flesh to clear pink, large, dbl., moderate fragrance

'MME ROGER DOUINE', HT, rb, 1926; flowers crimson, shaded scarlet, dbl.; [Souv. de Claudius Denoyel X Mme Edouard Herriot]; Reymond

'MME ROGER VERLOMME', HT, pb, 1951; flowers ochre edged flesh-pink, very large, very dbl.; very vigorous growth; GM, Geneva, 1951; Mallerin, C.

'MME ROSA MONNET', HP, dp, 1885; Monnet

'MME ROSE CHÉRIE', mp, 1850; flowers medium, dbl.; Laffay, M.; (Sangerhausen)

'MME ROSE ROMARIN', T, rb, 1888; flowers light coppery red with salmon, large, semi-dbl.; Nabonnand; (Sangerhausen)

'MME ROUDILLON', HP, mr, 1903; flowers very large, very dbl., intense fragrance; Vigneron; (Sangerhausen)

Mme Royet, LCl, m

'MME RUAU', HFt, pb, 1909; flowers orange/pink with dark yellow, large, dbl.; Gravereaux; (Sangerhausen)

'MME S. CROZA', HT, ab, 1935; bud long; flowers flesh-pink, very large, dbl., exhibition form; foliage leathery; long stems; vigorous growth; [Sunburst seedling]; Laperrière

Mme Sachi, HT, w

'MME SANCY DE PARABÈRE', Bslt, lp, 1874; (Mme de Sancy de Parabère); flowers clear soft pink, outer petals r than inner ones, which tend, 5 in., very early, nonrecurrent bloom, slight fragrance; no prickles; very hardy.; (14); Bonnet

'MME SAPORTAS', HGal, dp; flowers bright rosy red, large, dbl., intense fragrance

'MME SCHMITT', HT, ab, 1922; flowers salmon-pink, shaded peach-blossom-pink, dbl.; Schwartz, A.

'MME SCHULTZ', N, yb; flowers pale yellow, center darker, medium, dbl., intense fragrance; vigorous growth; Béluze, 1856

'MME SCHWALLER', HT, mp, 1886; flowers large; vigorous growth; Bernaix, A.

'MME SCIPION COCHET'®, T, pb, 1872; flowers pale pink to white, center yellow, large, dbl., cupped, moderate fragrance; vigorous growth; [Anna Olivier X Duchesse de Brabant]; Bernaix, A.

'MME SCIPION COCHET', HP, m; flowers purplish pink edged soft pink, center petals wrinkled, dbl., cupped; vigorous growth; Cochet, S., 1872

'MME SEGOND WEBER', HT, op, 1907; flowers clear salmon-pink, open, dbl., moderate fragrance; long, strong stems; bushy growth; GM, Bagatelle, 1909; [Antoine Rivoire X Souv. de Victor Hugo]; Soupert & Notting

'MME SEGOND WEBER, CLIMBING', Cl HT, op, 1911; Ardagh

'MME SEGOND WEBER, CLIMBING', Cl HT, op, 1929; Reymond

'MME SEGOND WEBER, CLIMBING', Cl HT, lp; Ardagh, 1911; (Weatherly, L.)

'MME SOLEDAD DE AMPUERA DE LEGUIZAMON', HT, pb, 1928; flowers hydrangea-pink, reverse carmine-rose, dbl., intense fragrance; [General-Superior Arnold Janssen X Mrs E.G. Hill]; Soupert & Notting

Mme Solvay, LCl, dp, 1992; Eve, A.

Mme Sophie Charlotte, A, lp, 1986; Weihrauch

'MME SOPHIE FROPPOT', HP, mp, 1876; flowers large, dbl.; Levet; (Sangerhausen)

'MME SOPHIE STERN', HP, dp, 1887; flowers large, dbl.; Lévêque; (Sangerhausen)

'MME SOUBEYRAN', HP, dp; (Mme de Soubeyran); flowers bright rose; Gonod, 1872

'MME SOUCHET', B, pb, 1843; Souchet

'MME SOUPERT', M, mr, 1851; flowers well mossed, red; Soupert & Notting, 1871

Mme Souveton, P, pb

'MME SPOTTI', HT, m, 1955; flowers pink striped mauve, very dbl.; foliage glossy; very vigorous growth; Privat

'MME STEINBACH', HT, pb, 1934; bud very long; flowers coppery pink tinted coral, very large, dbl.; [Mrs Pierre S. duPont X Cécile Walter]; Caron, B.

'MME SUZANNE HERVÉ', HP, dr, 1936; bud very long, pointed; flowers velvety red, heavily streaked maroon, intense fragrance; vigorous growth; [Baron Girod de l'Ain seedling]; Hervé; Vially

'MME TAFT', Pol, dp, 1909; flowers ruby-pink medium, dbl.; vigorous, bushy growth; [Crimson Rambler X Mme Norbert Levavasseur]; Levavasseur

'MME TAHA HUSSEIN', HT, rb, 1939; flowers indian red, reverse darker, semi-dbl.; vigorous growth; [Charles P. Kilham X Betty Uprichard]; Colombier

Mme Teresa Estaban *see* **'DOTRAMES'**

'MME THÉODORE DELACOURT', HT, mp, 1913; flowers large, very dbl.; Pernet-Ducher; (Sangerhausen)

'MME THÉVENOT', HP, dr, 1878; flowers large, very dbl., slight fragrance; Jamain, H.; (Sangerhausen)

'MME THIBAUT', HP, lp, 1889; flowers large, dbl.; Lévêque; (Sangerhausen)

'MME THIERS', B, pb, 1873; Pradel

'MME TIRET', HRg, mr, 1901; flowers large, semi-dbl.; Roseraie de l'Hay; (Sangerhausen)

'MME TONY BABOUD', HT, my, 1895; flowers medium golden yellow, large, semi-dbl.; Godard; (Sangerhausen)

'MME TRIFLE', N, yb, 1869; Levet

'MME TRIFLE', Cl T, lp; flowers flesh pink, very dbl.; tall, vigorous, climbing growth; [Gloire de Dijon X Unknown]

'MME TROTTIER', HT, yb, 1937; flowers yellowish flesh, large, very dbl.; vigorous, bushy growth; Leenders, M.

Mme Tsiranana, HT, mr, 1971; Croix

'MME VAN DE VOORDE', HT, mr, 1928; bud pointed; flowers brilliant scarlet, large, semi-dbl., cupped, intense fragrance; very vigorous growth; [Mme Méha Sabatier X K. of K.]; Mallerin, C.; C-P

'MME VANNIER', HT, or; flowers coppery red, large, dbl., cupped; Orard, Joseph

Mme Verbelen, HT, 1973; Delforge; (Cavriglia)

'MME VERDIER', HP, ab, 1840; flowers pale flesh; foliage gray-green; Verdier, V.

'MME VERRIER-CACHET', HP, mr, 1895; flowers carmine-red, very large, dbl., intense fragrance; Chédane-Guinoisseau; (Sangerhausen)

'MME VICTOR BOZZOLA', HT, mp, 1935; bud pointed; flowers bright coral-pink, well formed, large; [Kardinal Piffl X Mme Edouard Herriot]; Soupert, C.

'MME VICTOR LOTTIN', HWich, rb, 1921; flowers dark red, shaded crimson; vigorous, climbing growth; Lottin

'MME VICTOR RAULT', HT, w, 1920; flowers white tinted salmon, center yellow, dbl.; [Mme Mélanie Soupert X Lyon Rose]; Croibier

'MME VICTOR VERDIER', HP, mr, 1863; flowers clear light crimson, large, dbl., 75 petals, flat, seasonal bloom, intense fragrance; vigorous growth; [Senateur Vaisse X ?]; Verdier, E.

'MME VIDOT', HP, ab, 1854; flowers flesh-white; Couturier; E. Verdier

'MME VIGER', HT, lp, 1901; flowers creamy pink, very large, dbl., moderate fragrance; Jupeau; (Sangerhausen)

'MME VILLATE', HT, ob, 1936; flowers orange with yellow, dbl.; foliage glossy; [Korovo X Unnamed seedling]; Walter, L.

'MME VINCENT AURIOL', HT, yb, 1951; bud long; flowers golden coral, peony form, large, dbl., slight fragrance; foliage glossy; vigorous growth; GM, Bagatelle, 1948; [Trylon X Unnamed seedling]; Caron, B.; URS

'MME VIRGILIO PIROLA', HT, mr, 1939; bud long, pointed; flowers very bright red, dbl., moderate fragrance; foliage dark; vigorous, bushy growth; GM, Bagatelle, 1939; [Charles P. Kilham X Étoile de Hollande]; Lens

'MME VISSEAUX', HT, op, 1936; bud long; flowers orange-pink, base yellow, dbl., 4–5 in., slight fragrance; [Odette Foussier X Elvira Aramayo]; Mallerin, C.; C-P

'MME VITTORIA GAGNIÈRE', HT, w, 1909; flowers white tinted pink; [Anna Chartron X Mrs W.J. Grant]; Schwartz, Vve.

Mme W. Baumann, HT, 1975; Sauvageot; (Cavriglia)

'MME W.C. WHITNEY', HT, lp, 1894; flowers flesh-pink, intense fragrance

'MME WAGRAM, COMTESSE DE TURENNE', T, pb, 1894; flowers bright satiny rose suffused with darker rose, very large, moderate fragrance; vigorous growth; Bernaix, A.

'MME WALTER BAUMANN', HT, rb, 1934; bud long; flowers carmine, base yellow, open, dbl.; [Mlle Franziska Kruger X Gwynne Carr]; Reymond; Vially

'MME WELCH', T, ob, 1878; flowers pale yellow, well formed, large, very dbl.; [Devoniensis X Souv. d'un Ami]; Ducher, Vve.

'MME WILLIAM PAUL', M, dp; flowers bright rose, large, dbl., cupped, recurrent bloom; Moreau et Robert, 1869

'MME YORK', HP, dr, 1881; flowers large, dbl., intense fragrance; Moreau et Robert; (Sangerhausen)

'MME YVES LATIEULE', HT, my, 1949; (Nankin); flowers primrose-yellow, dbl., 70 petals, 5 in., slight fragrance; foliage glossy, dark; vigorous growth; GM, NRS, 1950; [Mme Joseph Perraud X Léonce Colombier]; Meilland, F.

Mme Yvette Gayraud, HT, ob; Wisbech Plant Co.

Mme Yvonne Chaverot, HT, 1976; Orard, Joseph; (Cavriglia)

'MME ZÖETMANS', D, w, 1830; flowers pale flesh, dbl.; Marest

Mme. Hide, HT, pb, 1997; flowers double, large, dbl., 15–25 petals, slight fragrance; some prickles; foliage large, medium green, matt; compact (1.2m) growth; [Pink Picnica X Garden Party]; Ohta, Kaichiro

'MME. LA GÈNÈRALE PAUL DE BENOIST', Cl T, op, 1901; flowers salmon-pink, very large, very dbl.; Berland; (Sangerhausen)

Moana® *see* 'MACBIPI'

Moane st Yellow Tea, T, my; (found rose)

Mobile Jubilee *see* 'TALMOBILE'

'MOCK'S ROSA DRUSCHKI', Cl HP, mp, 1935; flowers large, dbl.; Mock; (Sangerhausen)

'MODEL OF PERFECTION', F, ob, 1977; bud globular; flowers yellow, pink and orange, large, dbl., 28 petals, slight fragrance; [Zorina X Arthur Bell]; Dickson, A.

Modern Art® *see* 'POULART'

'MODERN TIMES', HT, rb, 1956; flowers red striped pink; [Better Times sport]; Verbeek; Minier

'MODESTY', HT, pb, 1916; bud pointed; flowers white, center rose-pink, large, dbl., exhibition form, intense fragrance; GM, NRS, 1915; McGredy

'MOEALPHA', Min, ob, 1998; (**Alpha Moe**); flowers cream white with orange edges, single, 6–11 petals, 1.5 in., borne mostly singly, slight fragrance; few prickles; foliage medium, medium green, semi-glossy; upright, medium (18 in) growth; [Pink Petticoat X Seedling]; Moe, Mitchie; Mitchie's Roses & More, 1999

'MOEANNE', Min, r, 1999; flowers pale russet, reverse lighter, yellow base, dbl., 17–25 petals, 1–2 in., exhibition form, slight fragrance; few prickles; foliage medium, medium green, semi-glossy; upright, medium (15 in.) growth; [Fairhope X Wistful]; Moe, Mitchie; Mitchie's Roses & More, 1999

'MOEDER DES VADERLANDS', F, mr, 1956; (Mère de la Patrie, Mother's Country); flowers bright vermilion-red; strong stems; vigorous growth; [Ambassadeur Nemry X Cinnabar]; Leenders, M.

'MOEJO-NETSU', Min, dp, 1998; flowers magenta pink, dbl., 15–25 petals, 1–1.5 in., exhibition form, borne mostly singly, slight fragrance; no prickles; foliage medium, dark green, semi-glossy; upright, tall, 18 in. growth; [Klima X Seedling (mauve)]; Moe, Mitchie; Mitchie's Roses & More, 1999

'MOEMARILYN', Min, ab, 1999; flowers apricot, reverse lighter, dbl., 17–25 petals, 1–2 in., exhibition form, slight fragrance; few prickles; foliage medium, dark green, semi-glossy; upright, tall (18 in.) growth; [Sheri Anne X Wistful]; Moe, Mitchie; Mitchie's Roses & More, 1999

'MOEMIGHTY', Min, ob, 1999; (**Mighty Moe**); flowers orange, reverse lighter, semi-dbl., 12–16 petals, .5 in., exhibition form, slight fragrance; few prickles; foliage small, medium green, dull; compact, medium (12 in.) growth; [Luis Desamero X seedling]; Moe, Mitchie; Mitchie's Roses & More, 1999

'MOGADOR', P, dr; (Roides Pourpres); flowers deeper in color, repeats sparingly; [Rose du Roi sport]; Descemet, M. (?), ca 1819

'MOGAJAM', Min, lp, 1992; (**Jamie**™); flowers full, outer petals horizontal at exhibition stage, 26–40 petals, 1.5–2.75 in., exhibition form, borne in small clusters, slight fragrance; some prickles; foliage medium, medium green, semi-glossy; medium (45-50 cms), spreading growth; [Party Girl X Fairlane]; Moglia, Thomas; Gloria Dei Nursery, 1993

'MOGAJOSH', Min, pb, 1989; (**Joshua**™); bud pointed; flowers clear, deep pink with yellow at base, reverse lighter to white, dbl., 20 petals, exhibition form, slight fragrance; fruit not observed; prickles hooked, small, red; foliage medium, medium green, semi-glossy; long stems; upright, medium growth; [Loving Touch X Rainbow's End]; Moglia, Thomas; Gloria Dei Nursery

'MOGALINE', Min, op, 1992; (**Party Line**™); flowers coral, blooms borne in clusters of 6 or more on basal canes, dbl., 15–25 petals, 1.5 in., slight fragrance; few prickles; foliage small, medium green, matt; medium (45-50cms), compact growth; [Party Girl X Fairlane]; Moglia, Thomas; Gloria Dei Nursery, 1993

'MOGARIP', Min, w, 1992; (**Li'l Rip**™); flowers white with yellow center, blooms borne in small clusters, dbl., 26–40 petals, 1.5 in., slight fragrance; few prickles; foliage small, medium green, matt; medium (35-45 cms), narrow, upright growth; [Loving Touch X Cupcake]; Moglia, Thomas; Gloria Dei Nursery, 1993

'MOGATOUCH', Min, ab, 1992; (**Li'l Touch**™); flowers apricot, yellow stamens showing, moderately open, circular o, dbl., 15–25 petals, 1.5 in., slight fragrance; some prickles; foliage medium, medium green, matt; medium (40-50 cms), upright growth; [Loving Touch X Prima Donna]; Moglia, Thomas; Gloria Dei Nursery, 1993

Mohak, HT, ob, 1993; Chiplunkar

'MOHAWK', S, m; flowers brighter than aster-purple, center white, single, profuse, non-recurrent bloom; foliage dull green; rounded, dwarf growth; hardy.; [R. rubrifolia X R. rugosa]; Central Exp. Farm

'MOHICAN', F, mr, 1961; flowers dark rose-red; [Garnette sport]; Thompson's, J.H., Sons

'MOHICAN', HT, mr, 1937; bud long, pointed; flowers cherry-red, large, dbl., slight fragrance; foliage leathery, glossy, dark; long stems; very vigorous growth; [Unnamed variety X Briarcliff]; J.H. Thompson's Sons

'MOHINI', F, ob, 1970; bud long, pointed; flowers chocolate-brown, base tinged yellow, full, medium, dbl.; foliage glossy, dark; moderate, bushy growth; [Sea Pearl X Shola]; Division of Vegetable Crops and Floriculture

Mohykan, HT, lp; Strnad, 1972; (Czech Rosa Club)

'MOÏSE', HGal, m, 1828; flowers rosy-carmine, shaded purplish slate, expanded, large, dbl.; moderate upright growth; Parmentier

'MOJAVE', HT, ob, 1954; bud long; flowers apricot-orange tinted red, prominently veined, dbl., 25 petals, 4–4.5 in., exhibition form, moderate fragrance; foliage glossy; vigorous, upright growth; AARS, 1954 GM, Bagatelle, 1953 GM, Geneva, 1953; [Charlotte Armstrong X Signora]; Swim, H.C.; Armstrong Nursery

'MOJAVE, CLIMBING', Cl HT, ob, 1964; Trimper, K.; Ruston

'MOJE HAMMARBERG'®, HRg, m, 1931; flowers reddish violet, large blooms, dbl., recurrent bloom, intense fragrance; large, red fruit; short, weak stems; vigorous growth; hardy.; Hammarberg; Stockholm Stads Gatukontor

Moksha, HT, op; flowers large, intense fragrance; (Sangerhausen)

Molde® *see* 'TANNIMOLL'

'MOLHIP', Min, w, 1986; (**High Point**); [Helen Boehm sport]; Molder, W.A.; Rose Acres

Molineux™ *see* 'AUSMOL'

Molitor® *see* 'GAUTIRA'

Mollie Claire™ *see* 'KINCLAIRE'

'MOLLY ABDY', HT, rb, 1957; bud pointed; flowers deep scarlet to lighter red, medium, dbl., moderate fragrance; vigorous growth; [Southport X ?]; Abdy; Shepperson

Molly Beckley™ *see* 'JAYMOLL'

'MOLLY BISHOP', HT, op, 1951; bud long, pointed; flowers large, dbl., exhibition form, intense fragrance; foliage leathery, rich green; vigorous growth; Robinson, H.

'MOLLY BLIGH', HT, pb, 1917; flowers deep pink, base deep orange, moderate fragrance; Dickson, A.

'MOLLY DARRAGH', HT, pb, 1930; bud pointed; flowers bright old-rose, base orange-yellow, very large, dbl., exhibition form, moderate fragrance; McGredy

'MOLLY DOYLE', F, pb, 1963; flowers old-rose-pink, base silvery white, single, 6 petals, 5 in., intense fragrance; foliage light green; low, bushy growth; [Étoile de Hollande X Queen Elizabeth]; Barter

'MOLLY KIRBY', F, or, 1980; flowers vermilion; [Matangi sport]; Wilson, George D.

Molly McGredy *see* 'MACMO'

'MOLLY SHARMAN-CRAWFORD', T, w; bud long, pointed; flowers greenish white, becoming whiter, very dbl., exhibition form, moderate fragrance; foliage sparse, rich green; bushy growth; Dickson, A., 1909

'MOLODOST', HT, lp, 1956; (Youth); dbl., 37 petals, 5 in.; foliage dark; strong stems; vigorous, upright growth; [Staatsprasident Pats X Cathrine Kordes]; Sushkov, K. L.

'MOLODOST MIRA', HT, mr, 1955; (Early Peace); flowers coral-red, well formed, large, dbl., 50 petals, moderate fragrance; foliage dark; vigorous, well branched growth; [Peace X (Crimson Glory X Poinsettia)]; Klimenko, V. N.

'MOLSUNLEM', HT, lp, 1986; (**Sunday Lemonade**); dbl., 30 petals, exhibition form, borne usually singly, intense, fruity fragrance; medium, amber prickles, hooked downward; foliage medium, dark, matt; upright, medium-tall growth; [Lemon Spice sport]; Molder, W.A.; Rose Acres

'MOMO', HT, dp, 1987; flowers deep pink, large, borne usually singly, dbl., 35–40 petals, slight fragrance; medium, copper fruit; prickles pointed, straight; foliage medium, dark green; [(Picnic X Kordes' Perfecta) X Christian Dior]; Ota, Kaichiro, 1988

Momo, LCl, dr, 1995; Noack, Werner

Mom's Fancy *see* 'RENMOM'

Momy, F, dp

Mon Amour *see* 'COGAMO'

'MON AMOUR, CLIMBING', Cl HT, pb, 1966; bud ovoid; flowers phlox-pink, reverse silver-rose, large, dbl., cupped, profuse, intermittent bloom, moderate fragrance; foliage dark, leathery; very vigorous growth; [Peace, Climbing X Caprice]; Coggiatti, Stelvio

Mon Cheri™ *see* 'AROCHER'

Mon Jardin et Ma Maison *see* MEIchavrin

'MON PAYS', Gr, w, 1983; (Rigobec); flowers large, dbl., slight fragrance; foliage medium to large, medium green, glossy; upright growth; [Iceberg X Peace]; Gailloux, Gilles

'MON PETIT', Min, dp, 1947; flowers light red, dbl., 80 petals; foliage pointed; dwarf, compact growth; [Merveille des Rouges X Pompon de Paris]; Dot, Pedro; URS

Mon Tresor *see* **'RED IMP'**

'MONA', F, mr, 1957; flowers light crimson, center lighter, borne in large clusters, dbl., 2 in., exhibition form, slight fragrance; foliage light green; vigorous, upright, bushy growth; [Rudolph Timm X Fanal]; Kordes, R.; Dehner & Co.

Mona Lisa *see* 'KORMAT'

'MONA LISA', Cl HT, pb, 1956; bud ovoid; flowers warm pink overcast cameo-pink, borne singly and in clusters, dbl., 35–40 petals, 4–4.5 in., cupped, free, recurrent boom, intense fragrance; foliage dark, leathery; vigorous growth; bushy pillar or climbing (8-10 ft.); [Mrs Sam McGredy X ((Mrs Sam McGredy X (Seedling X Capt. Thomas))]; Malandrone; Armstrong Nursery

Mona Lisa, F, my, 1979; Kordes

'MONA ROSETTE', F, pb, 1963; bud ovoid; flowers pink and white striped, medium, in clusters, dbl., cupped; foliage dark, leathery; bushy, low growth; [Pink Rosette sport]; Manski; Paulen Park Nursery

'MONA RUTH', Min, mp, 1959; dbl., 30 petals, 1–1.5 in., slight fragrance; foliage leathery; vigorous (12-14 in), bushy growth; [((Soeur Thérèse X Skyrocket) X (Seedling X Red Ripples)) X Zee]; Moore, Ralph S.; Sequoia Nursery

'MONARCH', HT, mp, 1926; bud pointed; flowers silvery pink, very large, dbl., exhibition form, moderate fragrance; Dobbie

'MÖNCH', F, mp, 1952; flowers pink, large large corymbs, dbl.; foliage dark, leathery; vigorous, upright growth; [Karl Weinhausen seedling]; Tantau

'MONCTON', HRg, lp, 1977; bud ovoid; flowers medium, dbl., 20 petals, 2 in., intense fragrance; foliage gray-green; upright, bushy growth; [Schneezwerg X R. chinensis]; Svedja, Felicitas; Canada Dept. of Agric.

Monday's Child, *see* **'JUDMON'**

'MONDIAL PINK', HT, pb, 1965; flowers pink, base yellow, medium, exhibition form, slight fragrance; foliage glossy; Mondial Roses

Mondiale® *see* KORozon

Mondovision *see* 'DELBRAT'

Monet *see* 'JACLION'

Monet Rose, HT, m

'MONETA', F, pb, 1969; bud ovoid; flowers pinkish red, small, very dbl.; foliage dark; [Garnette sport]; Freytag; Santhof

'MONETTE', HWich, w, 1921; flowers small, borne in clusters, very dbl., intense fragrance; vigorous, climbing growth; [R. wichurana X Yvonne Rabier]; Hémeray-Aubert

'MONETTE', Pol, rb, 1922; flowers fiery red, passing to rose, with many white streaks, borne i; [Phyllis X Unnamed seedling]; Turbat

Mongioia, Gr, 1966; Mansuino; (Cavriglia)

Monica® *see* 'TANAKNOM'

Monica Astmann, F, or; flowers light orange, large, dbl., moderate fragrance; (Sangerhausen)

Monika® *see* 'TANAKNOM'

'MONIQUE', HT, op, 1949; flowers medium salmon-pink, well-shaped, large, dbl., 25 petals, intense fragrance; vigorous, upright growth; GM, NRS, 1950; [Lady Sylvia X Unnamed seedling]; Paolino; URS

Monique van Honsebrouck, Gr; Delforge, H.; (Cavriglia)

Monna Lisa®, HT, w

Monna Lisa, Climbing®, Cl HT, w

'MONS A. MEILLE', B, dr, 1889; flowers very large, dbl., intense fragrance; Moreau et Robert; (Sangerhausen)

'MONS CHÉDANE-GUINOISSEAU', HRg, lp, 1895; flowers very large, dbl.; Chédane-Guinoisseau; (Sangerhausen)

'MONS CORDEAU', B, dp, 1892; flowers deep red with violet, very dbl., intense fragrance; very thorny; Moreau et Robert

'MONS DE MONTIGNY', HP, dp, 1857; flowers carmine-pink, large, dbl.; Paillet; (Sangerhausen)

'MONS DE MORAND', HP, 1891; Schwartz, J.; (Cavriglia)

'MONS EDOUARD DETAILLE', HP, m, 1893; flowers purple/pink, large, dbl.; Gouchault; (Sangerhausen)

'MONS ERNEST DUPRÉ', HP, mr, 1904; flowers large, dbl.; Boutigny; (Sangerhausen)

'MONS FRAISSENON', HT, dp, 1911; flowers large, dbl., moderate fragrance; Gamon; (Sangerhausen)

'MONS FRANCISQUE RIVE', HP, 1893; Schwartz, J.; (Cavriglia)

'MONS HÉLYE', HRg, lp, 1900; flowers creamy pink, medium, semi-dbl.; Morlet; (Sangerhausen)

'MONS HOSTE', HP, mr, 1884; flowers large, dbl.; Liabaud; (Sangerhausen)

'MONS JULES DEROUDILHE', HP, m, 1886; flowers purple/pink, medium, dbl.; Liabaud; (Sangerhausen)

'MONS JULES LEMAITRE', HP, dp, 1890; flowers carmine-pink with lighter reverse, very large, dbl., moderate fragrance; Vigneron; (Sangerhausen)

'MONS JULES MAQUINANT', HP, mr, 1882; flowers large, dbl.; Vigneron; (Sangerhausen)

'MONS LAURIOL DE BARNY', HP, 1894; Trouillard; (Cavriglia)

'MONS LE CAPITAINE LOUIS FRERE', HP, dr, 1883; Vigneron

'MONS MATHIEU BARON', HP, dr, 1886; flowers dark violet-red, large, dbl., moderate fragrance; Schwartz; (Sangerhausen)

'MONS. BONCENNE', HP, dr, 1864; (M Boncenne); flowers deep crimson, medium, dbl.; moderate growth; Liabaud

Mons. E.Y. Teas *see* **'E. Y. TEAS'**

'MONS. LOUIS RICARD', HP, m, 1894; flowers blackish purple, shaded vermilion, large, dbl., cupped, moderate fragrance; vigorous growth; [Simon de St. Jean X Abel Carrière]; Boutigny

'MONS. PAUL LÉDÉ', HT, pb, 1902; flowers carmine-pink, shaded yellow, large, dbl., cupped, moderate fragrance; Pernet-Ducher

'MONS. PAUL LÉDÉ, CLIMBING', Cl HT, pb, 1913; Low

Mons. Pélisson *see* **'PÉLISSON'**

'MONS. ROSIER', T, pb, 1887; flowers rose and yellow; Nabonnand

'MONS. TILLIER', T, op, 1891; flowers rosy flesh, shaded salmon-rose and purple-rose, fairly large, dbl.; vigorous growth; Bernaix, A.

Monsieur Cordeau, B, dp; Moreau et Robert, 1892

Monsieur Lede *see* **'PAUL LEDE, CLIMBING'**

Monsieur Paul Lede *see* **'PAUL LEDE, CLIMBING'**

Mont Blanc, HRg, lp, 1986; semi-dbl.; numerous hips; 3 ft. growth; very hardy; Baum

'Mont Hamel', LCl, mp, 1937; semi-dbl., globular, borne in clusters; foliage glossy; very vigorous, climbing growth; Constantin

'Mont-à-Camp', HT, yb, 1928; flowers pure yellow, reverse shaded orange, very large, dbl.; vigorous, upright growth; Delobel

Montagny *see* **'Flaminaire'**

Montagut, HT, 1991; Dot, Simon; (Cavriglia)

Montana® *see* **'Royal Occasion'**

Montauban de Bretagne® *see* ADAmonbu

Montblanc, HRg, w, 1984; Baum

'Monte Carlo', HT, ob, 1949; flowers indian yellow suffused russet-orange, well shaped, dbl., 45 petals, 5 in., moderate fragrance; foliage glossy; vigorous growth; GM, NRS, 1950; [Peace X Unnamed seedling]; Meilland, F.; URS

Monte Cassino, HRg, dr, 1987; semi-dbl., moderate fragrance; 3 ft growth; Baum

'Monte Cristo', HP, dr, 1861; flowers large, dbl., 50 petals, globular; low growth; Fontaine

'Monte Igueldo', HT, dr, 1944; flowers velvety dark red, well formed, open, intense fragrance; upright, open habit growth; [Étoile de Hollande X Majorca]; La Florida

'Monte Nevoso', HT, 1929; Cazzaniga, F. G.; (Cavriglia)

Monte Rosa, HRg, mp, 1984; flowers large, semi-dbl., moderate fragrance; numerous hips; 3 ft. growth; Baum

'Monte Toro', HT, pb, 1962; flowers strawberry-pink shaded red, large, dbl., moderate fragrance; [Berthe Mallerin X Grand'mere Jenny]; Dot, Simon

'Montecito', S, w, 1930; single, 5 petals, 6 in.; foliage medium green, disease-resistant; vigorous, climbing (to 50 ft.) growth; Franceshi-Fenzi

'Monterey', HT, rb, 1933; flowers light gold flushed rose, veined and edged crimson, reverse lighter, dbl., moderate, fruity fragrance; [The Queen Alexandra Rose sport]; Lester Rose Gardens

'Monterosa', HT, pb, 1952; flowers deep rose, edged lighter, peony form, dbl., 40 petals; very vigorous growth; GM, Rome, 1952; [Elettra X Superba (HT)]; Giacomasso

'Montesquieu', HT, pb, 1959; flowers deep rose shaded crimson, dbl., 45 petals, exhibition form, moderate fragrance; strong stems; very vigorous growth; [Loli Creus X Tahiti]; Dot, Simon

Montesuma Rosice, HT, lp; Strnad; (Czech Rosa Club)

'Montezuma', Gr, op, 1955; bud urn-shaped; dbl., 36 petals, 3.5–4 in., exhibition form, slight fragrance; foliage leathery, semi-glossy; very vigorous, compact growth; GM, Geneva, 1955 GM, NRS, 1956 GM, Portland, 1957; [Fandango X Floradora]; Swim, H.C.; Armstrong Nursery

Montezuma, Climbing, Cl HT, lp; [Montezuma sport]; Langbecker, 1970; (Weatherly, L.)

'Monthly Rambler', LCl, dr, 1926; flowers brilliant crimson-red, large, borne in clusters, semi-dbl., blooms throughout summr and on young wood in autumn, moderate fragrance; [R. wichurana X Old Crimson China]; Laxton Bros.

'Monthyon', HGal, m; (Montigny); flowers slaty pink blotched purple, large, dbl., globular

Montigny *see* **'Monthyon'**

Montigny-le-Tilleul, S, dr; very dbl., 60 petals, flat; foliage dense dark green; RvS-Melle, 1995

'Montijo', F, dr, 1954; bud ovoid; flowers crimson-red, medium, dbl., 40 petals; very vigorous growth; [Méphisto X Magrana]; Dot, Pedro

'Montmartre', F, m, 1955; bud ovoid; flowers bright purplish red, large, dbl., moderate fragrance; foliage dark, bronze; very vigorous growth; [Peace X Seedling]; Gaujard

'Montparnasse, Climbing', Cl HT, ab, 1980; flowers large, dbl., 34 petals, exhibition form, slight fragrance; foliage coppery green, leathery, semi-glossy; bushy, vigorous growth; [Montparnasse sport]; Kodoya, Y.; Ogura Rose Nursery

Montreal® *see* 'GAUzeca'

Montresor *see* **'Red Imp'**

'Montrose', HT, dp, 1916; flowers deep pink; [Unnamed red seedling X Laurent Carle]; Cook, J.W.

Montrose *see* 'TALmontrose'

Montrose Tea, T, lp

'Montrouge', F, rb, 1956; flowers clear red, center tinted copper, medium, semi-dbl., slight fragrance; foliage glossy; vigorous, bushy growth; [Peace X Alain seedling]; Gaujard

'Montseny', HT, op, 1944; flowers salmon-pink, well formed, dbl., 40 petals, exhibition form, intense fragrance; foliage glossy; upright growth; [Mme Butterfly X Jean C.N. Forestier]; Dot, Pedro

'Montserrat', HT, op, 1954; flowers orange-salmon-pink, large, dbl., cupped, moderate fragrance; foliage dark, glossy; very vigorous, upright growth; [Comtesse Vandal X Angels Mateu]; Camprubi, C.

'Montuna', HP, lp; [Prince Camille de Rohan X Unknown]; Luke, E.R., 1928; (Weatherly, L.)

Monty's White, T, w

Monument, HT, mr, 1969; flowers large, dbl., moderate fragrance; Urban, J.; (Sangerhausen)

'Monviso', HT, pb, 1955; bud pointed; flowers white streaked pink, large, dbl., slight fragrance; strong stems; [Julien Potin X Monterosa]; Giacomasso

'Monymusk', HMsk, my, 1954; flowers clear yellow, large, semi-dbl., recurrent bloom, moderate fragrance; vigorous (3-4 ft.) growth; [Pax X Phyllis Gold]; Ratcliffe

Monza, HT, 1965; Leenders; (Cavriglia)

'Mood Music', Min, op, 1977; bud mossy; flowers orange to orange-pink, flat to rounded, mini-moss, dbl., 45 petals, 1 in., slight fragrance; foliage small; upright, bushy growth; [Fairy Moss X Goldmoss]; Moore, Ralph S.; Sequoia Nursery

Moody Blues *see* 'MACmooblu'

Moody Dream *see* MACmoodre

'Moon Glow', LCl, pb, 1937; flowers creamy primrose, center soft yellow, dbl., 60–75 petals, 3 in., moderate fragrance; strong stems; vigorous growth; [Glenn Dale X Mrs Arthur Curtiss James]; Brownell, H.C.

Moon Magic *see* 'DELcro'

'Moon Maiden', F, ly, 1970; flowers creamy yellow, dbl., 55 petals, 3.5–4 in., moderate fragrance; foliage dark; [Fred Streeter X Allgold]; Mattock

Moon Mist™ *see* 'PIXimis'

Moon River™ *see* 'KORconta'

Moon River *see* BENlavscent

Moon Shadow™ *see* 'JAClaf'

'Moonbeam', HT, dy, 1950; flowers deep golden yellow, large, exhibition form, moderate fragrance; foliage dark, glossy; RULED EXTINCT 4/92; [Unnamed seedling X McGredy's Yellow]; Robinson, H.; Baker's Nursery

Moonbeam *see* 'AUSbeam'

'Moonbeam, Climbing', Cl HT, dy, 1955; Kordes

Moondance, *see* **'MEInivoz'**

'Moondrops', HT, op, 1965; bud ovoid; flowers rose-amaranth shaded salmon, open, single, intense, spicy fragrance; vigorous growth; [Sunny Boy X Seedling]; Delforge

Moondrops 85, HT, 1985; Delforge; (Cavriglia)

Moonlight *see* 'JACICE'

'MOONLIGHT', HMsk, ly, 1913; flowers lemon-white, prominent yellow stamens, blooms in small clusters, single, repeat bloom, moderate fragrance; foliage dark, glossy; vigorous (4-5 ft.) bushy growth; GM, NRS, 1913; [Trier X Sulphurea]; Pemberton

Moonlight and Roses *see* 'BRIROSE'

Moonlight Bay *see* 'RENMOBAY'

Moonlight Fragrance, LCl, ly

Moonlight Lady *see* 'SOCALP'

Moonlight Lady *see* JACosol

'MOONLIGHT MAGIC', HT, m, 1991; flowers lavender, moderately blooms borne mostly singly, dbl., slight fragrance; foliage medium, medium green, matt; medium, bushy growth; [Unknown seedling X Unknown seedling]; Burks, Larry; Co-Operative Rose Growers

'MOONLIGHT MIST', HT, w, 1965; bud pointed; flowers ivory-white, very large, dbl., slight fragrance; very vigorous growth; [Buccaneer X Golden Harvest]; Armbrust; Langbecker

'MOONLIGHT NIAGARA'®, HT, or, 1987; flowers orange, reverse pale orange, loose, star-shaped, medium, bor, very dbl., 60 petals, exhibition form, moderate, spicy fragrance; rounded, medium, orange-red fruit; semi-thick, medium, brow-red prickles; foliage large, dark green, matt; upright, tall growth; [Red Queen sport]; Rogin, Josip

'MOONLIGHT SERENADE', HT, yb, 1994; flowers yellow with red fringe, blooms, dbl., 26–40 petals, 3–3.5 in., moderate fragrance; many prickles; foliage large, medium green, semi-glossy; medium (120cms), spreading growth; GM, JRC, 1991; [Garden Party X Jana]; Ohata, Hatsuo

'MOONLIGHT SONATA', HT, ab, 1966; bud ovoid; flowers large, dbl., cupped, moderate fragrance; foliage dark, glossy; vigorous, upright, bushy growth; [Diamond Jubilee seedling X (Goldilocks X Orange Nassau)]; Boerner; J&P

'MOONRAKER'®, F, ly, 1968; flowers pale yellow to white, large blooms in clusters, dbl., slight fragrance; foliage light green; [Pink Parfait X Highlight]; Harkness

'MOONRISE', HT, ly, 1997; flowers full, medium, very dbl., 26–40 petals, borne mostly singly, no fragrance; foliage medium, medium green, semi-glossy; upright, low (2ft.) growth; [(Lady X X Antigua) X Midas Touch]; McMillan, Thomas G.

'MOONSPRITE', F, ly, 1956; bud ovoid; flowers creamy white, center pale gold, blooms in clusters, dbl., 80 petals, 2–2.5 in., cupped, intense fragrance; foliage leathery, semi-glossy; dwarf, bushy growth; GM, Baden-Baden, 1955 GM, Rome, 1956; [Sutter's Gold X Ondine]; Swim, H.C.; Armstrong Nursery

Moonstone™ *see* 'WEKCRYLAND'

Moonstruck *see* JACess

Moorcap *see* **'RED CASCADE'**

'MOORCAR', Min, rb, 1972; (**Magic Carrousel**®, 'MORROUSEL'); flowers petals white, edged red, small, dbl., exhibition form, slight fragrance; foliage small, glossy, leathery; vigorous, bushy growth; AOE, ARS, 1975 Miniature Rose Hall of Fame, ARS, 1999; [Little Darling X Westmont]; Moore, Ralph S.; Sequoia Nursery

Moore's Classic Perpetual *see* 'MORCLASSIC'

Moore's Pink Perpetual *see* 'MORPINKPER'

Moore's Yellow, Min, my

'MOOSE RANGE', HRg, mr, 1944; [Hansa X Mary L. Evans]; Wright, Percy H.

Morabito *see* MASflodel

'MORAIN', Min, yb, 1989; (**Lucky Charm**); bud ovoid; flowers yellow, reverse tinting red, aging to pink to red, lightly s, dbl., 50–60 petals, exhibition form, no fragrance; no fruit; prickles straight, small, brown; foliage small, medium green, matt; bushy, spreading, low growth; [Rumba X Pinstripe]; Moore, Ralph S.; Sequoia Nursery

'MORALBUQUE', Min, rb, 1997; (**Albuquerque Enchantment**); flowers double, medium, dbl., 15–25 petals, borne in small clusters, slight fragrance; foliage medium, medium green, semi-glossy; medium (30-45cms) bushy, spreading growth; [Poker Chip X Cherry Magic]; Moore, Ralph S.

'MORALERT', Min, mr, 1990; (**Red Alert**); bud pointed; flowers medium red, slightly lighter reverse, aging similar, medium, dbl., 35 petals, exhibition form, slight fragrance; foliage medium, medium green, semi-glossy; upright, bushy, medium growth; [Orangeade X Rainbow's End]; Moore, Ralph S., 1986; Sequoia Nursery, 1991

'MORANIUM', Min, or, 1983; (**Autumn Fire**); flowers small, semi-dbl., no fragrance; foliage small, medium green, semi-glossy; bushy, very spreading growth; [Little Chief X Anytime]; Moore, Ralph S., 1982

'MORANYBLAC', Min, dr, 1993; (**Red Cameo**); flowers dark red, does not burn, blooms borne mostly single, semi-dbl., 6–14 petals, 1.5–2.75 in., slight fragrance; few prickles; foliage medium, medium green, semi-glossy; medium (34-38 cms), upright, bushy growth; [Anytime X Black Jade]; Moore, Ralph S.; Sequoia Nursery, 1994

'MORANYFACE', Min, ob, 1992; (**Halo Star**); flowers reddish on outside, orange to pink on inside, reddish lavender, single, 5 petals, 1.5 in., no fragrance; foliage small, medium green, matt; medium (35-40 cms), upright, bushy growth; [Seedling (Anytime X Angel Face) X Seedling (Anytime X Angel Face)]; Moore, Ralph S.; Sequoia Nursery, 1993

'MORART', Cl Min, ob, 1989; (**Work of Art**); bud short; flowers orange blend, reverse slightly more yellow, urn-shaped, medium, dbl., 35 petals, exhibition form, slight fragrance; globular, medium, orange fruit; prickles slender, nearly straight, average, brown, few; foliage medium, bronze aging to medium green, semi-glossy; upright, spreading, tall growth; PP007617; [Unnamed yellow Climbing Miniature seedling X Gold Badge]; Moore, Ralph S.; Sequoia Nursery

Morava, HT, 1970; Strnad; (Cavriglia)

'MORBERG', Min, or, 1981; (**Ann Moore**); bud long, pointed; flowers , dbl., 30 petals, exhibition form, borne usually singly, moderate fragrance; long prickles; foliage leathery, semi-glossy; vigorous, bushy, upright growth; [Little Darling X Fire Princess]; Moore, Ralph S.

'MORBERK', Min, m, 1988; (**Berkeley Beauty**™); flowers picotee white, edged lavender to pink (striped), small, dbl., exhibition form, moderate fragrance; prickles varying green to brown; foliage medium, medium green, semi-glossy, abundant; bushy, medium, neat growth; [Pink Petticoat X Make Believe]; Moore, Ralph S.; Sequoia Nursery, 1987

'MORBLACK', Min, dr, 1982; (**Don Marshall**); flowers medium red, reverse blackish-red, small, dbl., 35 petals, exhibition form, slight fragrance; foliage small, dark, matt; bushy, spreading growth; [Baccará X Little Chief]; Moore, Ralph S.; Moore Min. Roses

'MORBOUQUET', S, mp, 1991; (**Grandma's Pink**); very dbl., borne in small clusters, no fragrance; foliage medium, medium green, matt; medium, upright, bushy growth; [Shakespeare Festival X Marchioness of Londonderry]; Moore, Ralph S.; Sequoia Nursery

'MORBOX', Min, pb, 1984; (**Jewel Box**); flowers light to deep pink blend, reverse lighter, small, dbl., 20 petals, slight fragrance; foliage small, medium green, semi-glossy; bushy growth; [Avandel X Old Master]; Moore, Ralph S.; Moore Min. Roses, 1983

'MORBOY', Min, mr, 1990; (**Billy Boy**); bud pointed, short; flowers small blooms borne usually singly, dbl., 15–18 petals, cupped, slight fragrance; prickles slender, straight, small, brownish; foliage small, medium green, matt, dense;

bushy, low to medium, compact growth; [Anytime X Happy Hour]; Moore, Ralph S.; Sequoia Nursery

'MORBRIGHTS', Min, ob, 1999; (**Spotlight**); flowers bright orange blended rose with white star-shaped eye, single, 5 petals, 2–2.5 in., borne mostly singly, no fragrance; foliage medium, medium green, semi-glossy; bushy, compact (12-18 in) growth; [Orangeade X Little Artist]; Moore, Ralph S.; Sequoia Nursery, 1999

'MORBROWN', Min, ab, 1993; (**Apricot Twist**); flowers moderately full, medium sized blooms, dbl., 15–25 petals, 1.5–2.75 in., borne in small clusters, slight fragrance; few prickles; foliage small, medium green, semi-glossy; low (30-32 cms), bushy, compact growth; [Golden Angel X Sequoia Gold]; Moore, Ralph S.; Sequoia Nursery, 1994

'MORBUG', Min, mr, 1992; (**Ladybug**); semi-dbl., 6–14 petals, 1.5 in., borne in small clusters, fast repeat, slight fragrance; few prickles; foliage small, medium green, semi-glossy; low (30-40 cms), bushy, compact growth; [Sheri Anne X Cherry Magic]; Moore, Ralph S.; Sequoia Nursery, 1993

'MORCADE', Cl Min, mp, 1981; (**Pink Cascade**); flowers small, dbl., 35 petals, slight fragrance; foliage small, medium green, matt to semi-glossy; spreading (5-7 ft; similar to red cascade) growth; [(R. wichurana X Floradora) X Magic Dragon]; Moore, Ralph S.; Moore Min. Roses

'MORCALYN', Min, dp, 1986; (**Coral Cameo**); flowers deep pink, borne singly and in small sprays, dbl., exhibition form, slight fragrance; small, globular, orange fruit; very few prickles; foliage small, medium green, semi-glossy; medium, upright, bushy growth; [Little Darling X Anytime]; Moore, Ralph S.; Moore Min. Roses, 1982

'MORCANA', Min, mp, 1980; (**Antique Rose**®, 'MORCARA'); bud pointed; flowers rose pink, mini-flora, medium, dbl., 38 petals, exhibition form, borne usually singly, slight fragrance; straight, brown prickles; foliage dark, semi-glossy; vigorous, upright growth; [Baccará X Little Chief]; Moore, Ralph S.

'MORCARA', Min, mp, 1980; (**Antique Rose**®, 'MORCANA'); bud pointed; flowers rose pink, mini-flora, medium, dbl., 38 petals, exhibition form, borne usually singly, slight fragrance; straight, brown prickles; foliage dark, semi-glossy; vigorous, upright growth; [Baccará X Little Chief]; Moore, Ralph S.

'MORCARLET', Min, mr, 1988; (**Scarlet Moss**®); flowers intense, scarlet-red, open, small, semi-dbl., borne in sprays of 3-10, no fragrance; round elonngated, orange fruit; prickles slender, various, green to brown; foliage medium, medium green, glossy, leathery; upright, bushy, tall growth; PP007128; [(Dortmund X Unnamed mossed Miniature seedling) X (Dortmund X Unnamed mossed Miniature striped seedling)]; Moore, Ralph S.; Sequoia Nursery

'MORCAS', Min, w, 1991; (**White Rain**); dbl., 15–25 petals, 1.5 in., borne in large clusters, moderate fragrance; some prickles; foliage small, light green, semi-glossy; low (22 cms), spreading, groundcover growth; [Papoose X Renae]; Moore, Ralph S.; Sequoia Nursery, 1992

'MORCAT', Min, mp, 1986; (**Copy Cat**); [Beauty Secret sport]; Moore, Ralph S.; Moore Min. Roses, 1985

'MORCEBRU', Min, lp, 1981; (**Baby Cécile Brunner**); flowers soft pink, small, slight fragrance; foliage small, medium green, matt; upright, bushy growth; [Cécile Brunner, Climbing X Fairy Princess]; Moore, Ralph S.; Moore Min. Roses

'MORCEEDEE', S, pb, 1990; (**Cee Dee Moss**); bud pointed; flowers pink with occasional white striped, lighter reverse, ages lighter, semi-dbl., 15–25 petals, 2.5–2.75 in., cupped, slight fragrance; foliage medium, light green, glossy; bushy, spreading, medium growth; [Carolyn Dean X Seedling]; Moore, Ralph S., 1981; Sequoia Nursery

'MORCHARI', Min, m, 1984; (Insolite, **Sweet Chariot**™); flowers lavender to purple blend, small, dbl., blooms in clusters of 5-20, intense fragrance; foliage small, medium green, matt; upright growth; PP005975; [Little Chief X Violette]; Moore, Ralph S.; Moore Min. Roses

'MORCHARLIE', Min, rb, 1996; (**Charlie Brown**); flowers red and white stripe, blooms borne mostly single or clusters, semi-dbl., 8–14 petals, 1–1.25 in., slight fragrance; few prickles; foliage small, medium green, semi-glossy; compact, bushy, low (12-15 in) growth; [Seedling X Pinstripe]; Moore, Ralph S.; Sequoia Nursery, 1997

'MORCHEER', Min, dr, 1986; (**Cheerleader**™); flowers small, very dbl., blooms in sprays of 5-10, no fragrance; no fruit; foliage small, medium green, semi-glossy; bushy, spreading growth; PP005977; [Fairy Moss X Orange Honey]; Moore, Ralph S.; Moore Min. Roses, 1985

'MORCHERMAG', Min, dr, 1988; (**Cherry Magic**™); flowers deep red, reverse lighter red with silver sheen, aging lighter, dbl., 25 petals, exhibition form, no fragrance; round, small, orange red fruit; prickles short, small, brown; foliage small, medium green, matt; bushy, spreading, low growth; PP007061; [Anytime X Lavender Jewel]; Moore, Ralph S.; Sequoia Nursery

'MORCHICK', Min, mp, 1990; (**Chick-a-dee**); bud pointed; flowers medium pink with occasional white stripes, reverse similar, dbl., 40–50 petals, exhibition form, slight fragrance; prickles small, hooked downward, brownish; foliage small, medium green, matt to semi-glossy; bushy, low, compact, rounded growth; [Cécile Brunner X (Dortmund X (Fairy Moss X (Little Darling X Ferdinand Pichard)))]; Moore, Ralph S.; Sequoia Nursery

'MORCINGOLD', Min, my, 1995; (**Cinderella Gold**); flowers medium yellow, moderately full (15–25 petals), small (0–4 cms) blooms borne mostly in small clusters; slight fragrance; no prickles; small foliage, light green, semi-glossy; low (15 cms), bushy growth; [Cal Poly X Cal Poly]; Moore, Ralph S.; Sequoia Nursery, 1996

'MORCLASSIC', S, dp, 1999; (**Moore's Classic Perpetual**); very dbl., 41 petals, 3 in., borne in small clusters, moderate fragrance; few prickles; foliage medium, medium green, semi-glossy; dwarf hybrid perpetual, bushy, compact (16-24 in.) growth; [Anytime X Paul Neyron]; Moore, Ralph S.; Sequoia Nursery, 1999

'MORCLILAV', Cl Min, m, 1992; (**Orchid Jubilee**); flowers mauve blend, holds color well in heat, blooms borne in small, dbl., 15–25 petals, 1.5–2.75 in., profuse repeat bloom, no fragrance; few prickles; foliage medium, medium green, matt; tall (2 meters), upright, climbing growth; [Seedling (Little Darling X Yellow Magic) X Make Believe]; Moore, Ralph S.; Sequoia Nursery, 1993

'MORCLIM', Cl Min, yb, 1983; (**High Stepper**); flowers yellow overlaid pink, reverse yellow, small, dbl., slight fragrance; foliage medium, dark, semi-glossy; bushy, spreading (to 5 ft; needs support) growth; [(Little Darling X Yellow Magic) X Magic Wand]; Moore, Ralph S.; Moore Min. Roses

'MORCLIP', Cl F, mp, 1993; (**Playgirl, Climbing**); single, 5 petals, 3–3.5 in., borne in small clusters, slight fragrance; foliage medium, medium green, semi-glossy; tall (2-3 m), upright, spreading, climbing growth; [Playgirl sport]; Moore, Ralph S.; Sequoia Nursery, 1995

'MORCOAT', Min, rb, 1992; (**Tag-a-long**); flowers contrast of reddish lavender against white, blooms, somewhat, semi-dbl., 6–14 petals, 1.5–2.75 in., no fragrance; few prickles; foliage medium, medium green, new bronze, matt; medium (30-40 cms), upright, bushy growth; [Seedling (Little Darling X Yellow Magic) X Make Believe]; Moore, Ralph S.; Sequoia Nursery, 1993

'MORcofair', Cl S, pb, 1995; (**Coral Fairy**); single, 5 petals, 1.5 in., borne in large clusters, slight fragrance; few prickles; foliage small, dark green, semi-glossy; tall (3-4 m), upright, bushy, spreading growth; [(R. wichurana X Floradora) X Hallelujah]; Moore, Ralph S.; Sequoia Nursery, 1995

'MORcojo', Min, pb, 1984; (**Country Joy**); flowers light pink, yellow reverse, small, dbl., slight fragrance; foliage small, medium green, matt; compact growth; [Pinocchio X Yellow Jewel]; Moore, Ralph S.; Moore Min. Roses

'MORdan', Min, mr, 1980; (**Red Wagon**); dbl., 23 petals, borne1-3 or more per cluster, slight fragrance; brown prickles; foliage glossy; vigorous, bushy, rounded growth; [Little Darling X Little Chief]; Moore, Ralph S.; Moore Min. Roses

'MORdarain', Min, pb, 1992; (**Lifestyle**); flowers color holds well, blooms mostly singly, dbl., 15–25 petals, 1.5–2.75 in., slight fragrance; few prickles; foliage medium, medium green, semi-glossy; medium (35-45 cms), upright, bushy, rounded growth; [Little Darling X Rainbow's End]; Moore, Ralph S.; Sequoia Nursery, 1993

'MORdarcrest', F, pb, 1994; (**Elegant Design**); flowers pink, crested sepals, blooms borne in smallclusters, dbl., 15–25 petals, 1.5–2.75 in., slight fragrance; few prickles; foliage medium, medium green, matt; medium to tall (45-60cms), upright, bushy growth; [Little Darling X Crested Jewel]; Moore, Ralph S.; Sequoia Nursery, 1995

'MORdashin', Min, yb, 1991; (**Southern Delight**); flowers yellow edged with red, aging to pink and yellow, blooms born, dbl., 15–25 petals, 1.5–2.75 in., slight fragrance; few prickles; foliage medium, medium green, semi-glossy; medium to tall (22-24 cms), upright, bushy growth; [Little Darling X Rise 'n' Shine]; Moore, Ralph S.; Sequoia Nursery, 1992

'MORday', Min, yb, 1984; (**Sunday Brunch**); flowers soft creamy yellow, petal tips becoming pink, which spreads, dbl., 20 petals, moderate fragrance; foliage small to medium, medium green, semi-glossy; vigorous, upright, bushy growth; [Rumba X Peachy White]; Moore, Ralph S.; Moore Min. Roses, 1983

'MORdeb', Min, pb, 1981; (**Joan Austin**); bud pointed; flowers light to medium pink, white stripes, dbl., 38 petals, exhibition form, borne singly, sometimes 3 or more per cluster, intense fragrance; foliage small, medium green, semi-glossy to matt; very bushy, compact growth; [Avandel X Unnamed seedling]; Moore, Ralph S.; Moore Min. Roses

'Morden Amorette', S, dp, 1977; bud pointed; flowers deep pink, dbl., 28 petals, 3 in.; foliage dark; [(Independence X (Donald Prior X R. arkansana)) X (Fire King x(J.W. Fargo X Assiniboine))]; Marshall, H.H.; Agriculture Canada

'Morden Blush', S, lp, 1988; flowers light pink, fading to ivory, small, flat, dbl., 51 petals, flat, borne in sprays of 1-5, repeat bloom; prickles straight; foliage medium, medium green, matt; bushy, low growth, 3 - 4 ft.; hardy; PP8054; [(Prairie Princess X Morden Amorette) X (Prairie Princess X (White Bouquet X (R. arkansana X Assiniboine)))]; Collicutt, L.M. & Marshall, H.H.; Agriculture Canada

'Morden Cardinette', S, mr, 1980; bud ovoid; flowers cardinal red, blooms borne 1-5 per cluster, occasionally 15, dbl., 25 petals, slight fragrance; foliage 7 leaflets, globuous, dark; Marshall, H.H.; Agriculture Canada

'Morden Centennial', S, mp, 1980; bud ovoid; flowers medium-pink, shapely, very dbl., 40 petals, borne 1-15 per cluster, repeat bloom, slight fragrance; slightly recurved prickles; foliage 7 leaflets, dark, slightly glossy; typical shrub growth, 3 - 4 ft.; [Prairie Princess X (White Bouquet X (J.W. Fargo X Assiniboine))]; Marshall, H.H.; Agriculture Canada

'Morden Fireglow', S, or, 1989; bud pointed; flowers brilliant red-orange, reverse red, loose, medium, dbl., 28 petals, 3 in.., cupped, borne in sprays, repeat bloom, slight fragrance; globular, reflexed calyx fruit; prickles slight downward curve, tan; foliage medium, medium green, matte; bushy, low, medium growth, 4 ft.; PP8060; [Unnamed seedling X Morden Cardinette]; Collicutt, L.M. & Marshall, H.H.; Agriculture Canada, 1991

'Morden Ruby', S, pb, 1977; bud ovoid; very dbl., 2.5–3.5 in., heavy bloom; vigorous, irregular growth; [Fire King X (J.W. Fargo X Assiniboine)]; Marshall, H.H.; Agriculture Canada

Morden Snow Beauty *see* **'Morden Snowbeauty'**

'Morden Snowbeauty', S, w, 1998; (Morden Snow Beauty); flowers white, medium, semi-dbl., 8–14 petals, 3 in., borne in small clusters, slight fragrance; prickles small; foliage medium, medium green, semi-glossy; compact, low growth; [(Prairie Princess X Morden Amorette) X (Mount Shasta X (Adelaide Hoodless X R. arkansana))]; Davidson, C.G. & Colicutt, L.M.; Agriculture & Agri-Food Canada, 1997

'MORdora',Cl Min, mr, 1995; (**Sequoia Ruby**);flowers medium red, moderately full (15–25 petals), medium (2 in.) Blooms borne in small clusters; some prickles; foliage medium, dark green, matt; tall (5 ft.), arching shrub, upright, bushy, spreading growth; [(Little Darling X Yellow Magic) X Floradora]; Moore, Ralph, S.; Sequoia Nursery, 1996

'MORdort', Min, mr, 1991; (**Sincerely Yours**); semi-dbl., 6–14 petals, 1.5–2.75 in., borne in small clusters, no fragrance; some prickles; foliage medium, medium green, semi-glossy; medium (20-24 cms), bushy growth; AOE, 1992; [Sheri Anne X Dortmund]; Moore, Ralph S.; Sequoia Nursery, 1992

'More Vale Pride', Gr, mp, 1957; [Ma Perkins X Overloon]; Ulrick, L.W.

'MORecli', Min, ly, 1984; (**Baby Eclipse**™); bud small; flowers miniflora, small, semi-dbl., slight fragrance; foliage small, medium green, matt; bushy, spreading growth; AOE, 1984; [(R. wichurana X Floradora) X Yellow Jewel]; Moore, Ralph S.; Moore Min. Roses

'MORed', Min, lp, 1988; (**Edna Marie**); flowers very soft pink, soft yellow base, aging becomes near white, dbl., 20 petals, exhibition form, slight, fruity fragrance; prickles small, brown; foliage small, light green, semi-glossy; upright, bushy, medium growth; [Pinocchio X Peachy White]; Moore, Ralph S.; Sequoia Nursery

'MORedfar',Pol, mr, 1995; (**Red Fairy**); flowers deep pink to red, moderately full (15–25 petals), small (0–4 cms.) blooms borne in large clusters of 10–25; no fragrance; few prickles; foliage medium, medium green, semi-glossy; upright, bushy, spreading (45–60 cms) growth; [Simon Robinson X Unknown]; Moore, Ralph S.; Sequoia Nursery, 1996

'MORelfire', Min, pb, 1982; (**Pink Elf**); flowers medium pink, blended with yellow, small, semi-dbl., no fragrance; foliage small, medium green, matt to semi-glossy; upright, bushy growth; [Ellen Poulsen X Fire Princess]; Moore, Ralph S.; Moore Min. Roses

Morena®, Min, op, 1983; Cocker

'MOReny', Min, mp, 1989; (**Renny**); bud pointed; flowers medium rose pink, reverse lighter, old fashioned, medium, bo, dbl., 25 petals, moderate fragrance; no fruit; no prickles; foliage medium, medium green, matt; upright, bushy, low growth; [Anytime X Renae]; Moore, Ralph S.; Sequoia Nursery

'MOReyes', Min, pb, 1998; (**Halo Karol**); flowers dark pink-lavendar base creates halo effect, small, single, 6–14 petals, 1–1.5 in., borne singly, no fragrance; few prickles; foliage medium, medium green, semi-glossy; bushy, compact growth; [(Anytime X Angel Face) X (Anytime X Angel Face)]; Moore, Ralph S.; Sequoia Nursery, 1998

Morey's Pink, S, mp, 1994; Morey

Morey's Salmon, HT, pb, 1953; Morey, Dr. Dennison

'MORFAIR', Min, lp, 1980; (**Pink Cherub**); bud ovoid; flowers medium to light pink, often lighter at tips, dbl., 43 petals, slight fragrance; straight, small prickles; foliage small, medium green, matt; compact, very bushy growth; [Fairy Moss X Fairy Moss]; Moore, Ralph S.; Moore Min. Roses

'MORFAIRPOL', Pol, lp, 1999; (**Fair Molly**); flowers white blend, semi-dbl., 6–14 petals, 1–1.5 in., borne in small clusters, slight fragrance; few prickles; foliage small, medium green, semi-glossy; bushy, spreading, medium (16-24 in.); landscape shrub growth; [(R. polyantha nana X Unknown) X Fairy Moss]; Moore, Ralph S.; Sequoia Nursery, 1999

'MORFEAT', Min, w, 1980; (**White Feather**); flowers loose form, small blooms in clusters of 3-5, semi-dbl., 15 petals, slight fragrance; no fruit; very few prickles; foliage small, light green, semi-glossy; medium, bushy growth; [(R. wichurana X Floradora) X Peachy White]; Moore, Ralph S.; Moore Min. Roses

'MORFEVER', Min, my, 1990; (**Gold Fever**); bud pointed; flowers medium yellow, aging lighter, dbl., 40–50 petals, exhibition form, borne usually singly or in sprays of 3-5, moderate, spicy fragrance; round, small, orange fruit; prickles slender, straight, medium to long, brownish; foliage medium, medium green, semi-glossy; upright, bushy, medium growth; [Sheri Anne X Gold Badge]; Moore, Ralph S.; Sequoia Nursery

'MORFING', Min, ob, 1990; (**Fingerpaint**); bud short, pointed; flowers orange blend, yellow base, with light yellow reverse, aging, semi-dbl., 12–14 petals, flat, no fragrance; prickles small, brownish, straight; foliage medium, medium green, semi-glossy; bushy, spreading, low-medium growth; [Orangeade X Little Artist]; Moore, Ralph S.; Sequoia Nursery

'MORFIRE', Min, yb, 1986; (**Ring of Fire**™); flowers yellow blended orange, reverse yellow, fading lighter, imbricated, very dbl., 60 petals, slight fragrance; no fruit; slender, sharp pointed, medium, green to brown prickles; foliage medium, medium green, semi-glossy; upright, bushy, vigorous growth; PP006618; AOE, 1987; [Pink Petticoat X Gold Badge]; Moore, Ralph S.

'MORFLASH', F, ob, 1993; (**Blastoff**™); flowers scarlet orange, white reverse, petals ruffled, very dbl., 35–40 petals, 1.5–2 in., quartered, borne in large clusters, slight, slight spice fragrance; many prickles; foliage medium, dark green, semi-glossy; medium (100-120 cms), upright growth; PP9405; [Orangeade X Little Artist]; Moore, Ralph S.; Weeks Roses, 1995

'MORFLASH', HP, ob; flowers purple-violet, dbl., cupped

'MORFREE', Min, w, 1983; (**Softee**); flowers creamy white, small, dbl., 35 petals, slight fragrance; no prickles; foliage small to medium, medium green, matt; bushy, spreading growth; [Unnamed seedling X Unnamed seedling]; Moore, Ralph S.; Moore Min. Roses, 1982

'MORGA', HRg, mp; dbl., blooms in clusters

'MORGAL', Min, dr, 1980; (**Galaxy**™); bud long, pointed; flowers deep velvety red, borne 3, sometimes 5-10 per cluster, dbl., 23 petals, exhibition form, slight fragrance; prickles slightly curved; foliage small to medium; vigorous, bushy, upright growth; [Fairy Moss X Fairy Princess]; Moore, Ralph S.; Sequoia Nursery

Morgan Hill Red, HT, dr

Morgans Spring, Ch, lp

'MORGEL', Cl Min, op, 1987; (**Angel Pink**, Pink Angel); flowers pink to soft coral pink, holds color, dbl., exhibition form, borne in sprays of 3-7, slight fragrance; no fruit; medium, brown prickles, slightly hooked downwards; foliage medium, light green, semi-glossy; upright, tall, climbing (5-7 ft) growth; [Little Darling X Eleanor]; Moore, Ralph S.

'MORGENGLANS', HT, op, 1916; flowers salmon-flesh, semi-dbl., slight fragrance; Van Rossem

'MORGENGRÜSS', HKor, op, 1962; (Morning Greeting); bud ovoid; flowers light pink tinted orange-yellow, large blooms in clusters, dbl., intense fragrance; foliage glossy, light green; very vigorous (13-14 ft), bushy growth; Kordes, R.

'MORGENLUFT', F, pb, 1962; flowers dark pink, reverse lighter, borne in clusters, dbl., 55–60 petals; foliage dark buff-green; upright, bushy growth; [La France X Unnamed seedling]; Verschuren, A.; van Engelen

Morgenrot® *see* 'KORHEIM'

'MORGENRÖTE', HT, mp, 1951; RULED EXTINCT 12/85; Burkhard

Morgensen *see* **'MORGENSONNE'**

'MORGENSONNE', S, my, 1954; (Morgensen, Morning Sun); bud ovoid; flowers golden yellow, very large, moderate fragrance; foliage glossy, light green; very vigorous, upright, bushy growth; Kordes

Morgensonne *see* KORhoro

Morgenstern®, S, ly, 1989; Liebig

'MORGLO', Min, or, 1981; (**Gypsy Fire**); flowers small blooms in clusters, semi-dbl., no fragrance; foliage small, medium green, semi-glossy to glossy; upright, bushy growth; [(R. wichurana X Carolyn Dean) X Fire Princess]; Moore, Ralph S.; Moore Min. Roses

'MORGOGARD', Min, my, 1988; (**Golden Gardens**); bud ovoid; flowers bright, clear medium yellow, reverse slightly lighter, informal, dbl., 28 petals, cupped, no fragrance; no fruit; prickles slender, inclined downward, small, brownish; foliage medium, medium green, semi-glossy; upright, medium growth; [(Little Darling X Yellow Magic) X Gold Badge]; Moore, Ralph S.; Sequoia Nursery, 1989

'MORGOLDART', Min, pb, 1993; (**Splish Splash**); dbl., 26–40 petals, 1.5–2.75 in., borne in small clusters, slight fragrance; few prickles; foliage medium, medium green, semi-glossy; medium (30 cms), bushy, spreading growth; [Sequoia Gold X Little Artist]; Moore, Ralph S.; Sequoia Nursery, 1994

'MORGOLDPOLY', Pol, my, 1999; (**Polly Sunshine**); dbl., 26–40 petals, 2–3 in., borne in small clusters; few prickles; foliage medium, medium green, glossy; bushy (16-24 in.) growth; [Golden Angel X seedling]; Moore, Ralph S.; Sequoia Nursery, 1999

'MORGRAPES', S, m, 1999; (**Vineyard Song**); dbl., 15–25 petals, 1–1.5 in., borne in large clusters, moderate fragrance; some prickles; foliage dark green, glossy; medium (12-16 in.) growth; Moore, Ralph S.; Sequoia Nursery, 1999

'MORHALFIRE', Min, rb, 1995; (**Halo**® **Fire**); flowers bright orange red with darker red halo, semi-dbl. (6–14 petals), medium (4–7 cms) blooms borne in small clusters; some prickles; foliage small, medium green, matt; medium (35–45 cms), upright, bushy growth; [Orangeade X Seedling]; Moore, Ralph S.; Sequoia Nursery, 1996

'MORHILL', Min, pb, 1990; (**Mary Hill**); bud pointed; flowers medium pink, yellow reverse, aging lighter, decorative, medium, dbl., 30–35 petals, exhibition form, moderate, fruity fragrance; prickles slender, straight, small, brownish; foliage medium, medium green, semi-glossy; upright, bushy, medium growth; [Little Darling X Golden Angel]; Moore, Ralph S.; Sequoia Nursery

'MORHONEY', Min, my, 1983; (**Sun Honey**); [Orange Honey sport]; Moore, Ralph S.; Moore Min. Roses

'MORHOOP', F, pb, 1990; (**Hula Hoop**); bud pointed; flowers white with pink to red edge, similar reverse, aging less intense, semi-dbl., 15 petals, 3–3.5 in., flat, borne in sprays of 5-15, no fragrance; foliage medium, medium green, matt; upright, medium growth;

[(Dortmund X Unnamed seedling) X Self]; Moore, Ralph S.; Sequoia Nursery, 1991

'MORHOOT', Min, rb, 1990; (**Hoot Owl**); bud pointed; flowers red with white eye, small, borne usually singly or in sprays, single, 5 petals, no fragrance; foliage small, medium green, semi-glossy; bushy, low growth; [Orangeade X Little Artist]; Moore, Ralph S., 1987; Sequoia Nursery, 1991

Moriah *see* 'GANHOL'

Moriah, HT, ob

Morinaye, F, rb

'MORISLE', Min, yb, 1993; (**Isle of Roses**); dbl., 26–40 petals, 1.5–2.75 in., borne in small clusters, no fragrance; few prickles; foliage medium, medium green, semi-glossy; medium to tall (18-24 in), upright, bushy growth; [Pink Petticoat X Gold Badge]; Moore, Ralph S.; Sequoia Nursery, 1994

'MORIVORY', Min, w, 1990; (**Ivory Palace**); bud ovoid; flowers ivory white, white reverse, aging similar color, medium, very dbl., exhibition form, borne in sprays of 3-8, slight fragrance; foliage medium, medium green, semi-glossy; bushy, medium growth; [Sheri Anne X Pinocchio]; Moore, Ralph S., 1982; Sequoia Nursery, 1991

'MORJACK', Min, dy, 1984; (**Jackpot**); flowers small, dbl., moderate fragrance; foliage small, medium green, matt; vigorous, bushy, spreading growth; [Little Darling X Sunspray]; Moore, Ralph S.; Moore Min. Roses, 1985

'MORJOYBON', HT, ab, 1996; (**It's Show-Time**™); flowers reverse or slightly lighter, blooms borne mostly singly, dbl., 15–25 petals, 4–4.5 in.; foliage medium, dark green, semi-glossy; upright, medium (3-5 ft) growth; [Joycie X Bon Silene]; Moore, Ralph S.; Spring Hill Nurseries Co., 1997

'MORJOYC', Min, ob, 1988; (**Joycie**™); flowers orange-apricot, reverse lighter, small, dbl., exhibition form, borne singly or in small clusters, moderate, fruity fragrance; globular, orange fruit; prickles slender, small, brown; foliage small, medium green, semi-glossy; bushy, medium growth; PP007055; [(Little Darling X Yellow Magic) X Gold Badge]; Moore, Ralph S.; Sequoia Nursery

'MORKAY', Min, pb, 1994; (**Kayla**); dbl., 15–25 petals, 3–3.5 in., borne mostly single but some in small clusters, slight fragrance; few prickles; foliage large, medium green, semi-glossy; tall (45-55 cms), upright, bushy growth; [Sheri Ann X Violette]; Moore, Ralph S.; Sequoia Nursery, 1995

'MORKINSHINE', Cl Min, my, 1990; (**Rise 'n' Shine, Climbing**); bud pointed; flowers medium yellow, aging lighter, small, borne usually singly or, dbl., 35 petals, exhibition form, slight fragrance; round, orange fruit; prickles straight, inclined slightly downward, brown; foliage medium, medium green, semi-glossy; upright, spreading, tall, climbing. growth; [Rise 'n' Shine sport]; King, Gene; Sequoia Nursery, 1990

'MORKITA', Min, dr, 1988; (**Chiquita**™); flowers rich, dark red with fluorescent glow, borne singly, semi-dbl., 20 petals, flat, slight fragrance; few to none, fruit; prickles hooked, short, brown; foliage small, medium green, matt; upright, bushy, medium growth; [Anytime X Happy Hour]; Moore, Ralph S.; Sequoia Nursery

'MORLAVMAG', Min, m, 1994; (**Surprise Surprise**); flowers mauve-lavender with occasional petals being striped red, or, semi-dbl., 6–14 petals, 1.5 in., no fragrance; few prickles; foliage small, dark green, semi-glossy; low (12-16 in), bushy, compact growth; [Cherry Magic sport]; Moore, Ralph S.; Sequoia Nursery, 1994

'MORLEM', Min, ob, 1983; (Sun Gold, **Sungold**); flowers yellow overlaid with orange on outer half of petals, reverse, dbl., 20 petals, no fragrance; foliage small, medium green, matt to semi-glossy; upright, bushy growth; [Rumba X Lemon Delight]; Moore, Ralph S.; Moore Min. Roses, 1982

'MORLETII', Bslt, m, 1883; (Inermis Morletii); flowers magenta, small to blooms in clusters, slight fragrance; branches plum-colored, without prickles; 5 ft. growth; very hardy; (14); Morlet

'MORLIDAY', Min, dr, 1982; (**Holiday Cheer**); flowers small blooms in clusters, dbl., 35 petals, no fragrance; foliage small, dark, matt; upright, bushy growth; [Red Pinocchio X Little Chief]; Moore, Ralph S.; Moore Min. Roses

'MORLIYEL', Min, yb, 1980; (**Sierra Sunrise**); bud medium, pointed; flowers soft yellow, petals tipped pink, borne singly or 3-5 per cluster, dbl., 43 petals, exhibition form, slight fragrance; long, straight prickles; foliage medium, medium green; vigorous, bushy, upright growth; Gold Star of the South Pacific, Palmerston North, NZ, 1984; [Little Darling X Yellow Magic]; Moore, Ralph S.; Sequoia Nursery

'MORLIZ', Min, or, 1986; (**Spanish Dancer**); flowers small blooms in sprays of 5-7, semi-dbl., 18 petals, slight fragrance; no fruit; small, brown prickles; foliage medium, medium green, semi-glossy; medium, bushy, spreading growth; [Sarabande X Little Chief]; Moore, Ralph S.; Moore Min. Roses, 1980

'MORLOGEN', Min, yb, 1984; (**Namib Sunrise**); flowers yellow blended with coral pink, blooms in clusters of 15, dbl., 64 petals, slight fragrance; straight, light brown prickles; foliage light; dense growth; [Rumba X Yellow Jewel]; Moore, Ralph S.; Ludwigs Roses Pty. Ltd.

'MORLU', F, mp, 1992; (**Vintage Visalia**); flowers reverse of outer petals deeper pink than inside surface bea, very dbl., 2.75–3 in., slight fragrance; few prickles; foliage large, medium green, semi-glossy; medium (50-60 cms), upright, bushy growth; [Pink Petticoat X Lulu]; Moore, Ralph S.; Sequoia Nursery, 1993

MORlucy, Min, mp, 1997; (**Lucy**); Moore, Ralph S.

'MORLYN', Min, my, 1983; (**Lynne Gold**™); flowers micro-mini, small, dbl., 20 petals, slight fragrance; foliage small, medium green, semi-glossy; bushy, spreading growth; PP005742; [Ellen Poulsen X Yellow Jewel]; Moore, Ralph S.; Moore Min. Roses

'MORMAGEM', S, rb, 1995; (**Mariposa Gem**); flowers red/yellow blend, blooms borne in small clusters, semi-dbl., 6–14 petals, 1.5–2.75 in., slight fragrance; few prickles; foliage medium, medium green, semi-glossy; upright, bushy, spreading(1 1/2 m) growth with arching canes; [Little Darling X Magic Wand]; Moore, Ralph S.; Sequoia Nursery, 1995

'MORMAKE', Min, m, 1986; (**Make Believe**™); flowers mauve and white blend, reverse red-purple, blooms in sprays, 10 petals, no fragrance; small, globular, orange fruit; very few, hooked, dark brown prickles; foliage medium, dark, semi-glossy; upright, bushy growth; [Anytime X Angel Face]; Moore, Ralph S., 1980; Moore Min. Roses, 1985

'MORMARI', Min, mp, 1988; (**Marie Shields**™); flowers medium pink, reverse pink veined white, small, very dbl., exhibition form, borne usually in sprays or clusters, slight fragrance; prickles slightly hooked, small, brown; foliage small, medium green, semi-glossy; bushy, medium growth; [Avandel X (Rumba X Unnamed Floribunda mossed seedling)]; Moore, Ralph S.; Sequoia Nursery

'MORMARME', Min, mr, 1992; (**Marime**); flowers excellent medium red, good form, blooms borne mostly singly, semi-dbl., 6–14 petals, 1.5–2.75 in., no fragrance; few prickles; foliage medium, medium green, semi-glossy; medium (35-45 cms), upright, bushy growth; [Anytime X Happy Hour]; Moore, Ralph S.; Sequoia Nursery

'MORMELMA', Min, ly, 1993; (**Melody Marshall**); dbl., 15–25 petals, 1.5–2.75 in., borne mostly singly, slight fragrance; no prickles; foliage medium, medium green, semi-glossy; medium (25-35 cms), upright, bushy growth; [Mary Marshall sport]; Moore, Ralph S.; Sequoia Nursery, 1989

'MORMILLI', Min, op, 1983; (**Millie Walters**®); flowers deep coral pink, dbl., 45 petals, slight fragrance; foliage small, medium green, matt; upright, bushy growth; PP005741; [Little Darling X Galaxy]; Moore, Ralph S.; Moore Min. Roses, 1984

'MORMINT', Min, my, 1991; (**Hall of Flowers**); bud pointed; flowers lemon yellow, similar reverse, aging slightly lighter, dbl., exhibition form, slight fragrance; foliage medium, medium green, semi-glossy; upright, bushy, medium growth; [Avandel X Gold Badge]; Moore, Ralph S., 1983; Sequoia Nursery

'MORMITCHELL', Min, ly, 1996; (**Annie R. Mitchell**); flowers light yellow to white, reverse similar, blooms borne in small clusters, dbl., 26–40 petals, 1.5–2 in., moderate fragrance; few prickles; foliage medium, medium green, semi-glossy; spreading, bushy, medium (15-18 in) growth; [Mary Hill sport]; Moore, Ralph S.; Sequoia Nursery, 1996

'MORMUM', Min, rb, 1986; (**Strange Music**, Striped Meillandina); bud mossy; flowers red, striped white, reverse near white, mini-moss, blooms in, dbl., 50 petals, exhibition form, no fragrance; small to medium, straight, brown prickles; foliage medium, medium green, matt; bushy, spreading growth; [Little Darling X (Fairy Moss X (Little Darling X Ferdinand Pichard))]; Moore, Ralph S.; Moore Min. Roses, 1982

'MORMURI', HBc, lp, 1989; (**Muriel**); bud ovoid; flowers light to medium pink, aging slightly lighter, large, borne s, semi-dbl., 15 petals, flat; short, oval, large, prickly, orange fruit; prickles sharp, pointed, average, brown; foliage large, medium green, semi-glossy; spreading, tall growth; [R. bracteata X Guinee]; Moore, Ralph S.; Sequoia Nursery

'MORMYVAL', Min, dr, 1975; (**My Valentine**®); flowers deep red, dbl., 65 petals, 1 in., exhibition form; foliage small, glossy, tinted bronze; vigorous, bushy growth; [Little Chief X Little Curt]; Moore, Ralph S.; Sequoia Nursery

'MORNET', Min, op, 1991; (**Annette Dobbs**); bud pointed; flowers coral red, lighter reverse, ages slightly lighter, with peta, semi-dbl., 15 petals, flat, borne in sprays of 3-5, no fragrance; foliage medium, medium green, semi-glossy; bushy, medium growth; [Anytime X Playgirl]; Moore, Ralph S., 1986; Sequoia Nursery, 1990

Morning Blush *see* **'DUQUESA DE PEÑARANDA'**

Morning Blush, A, ly, 1988; Sievers

'MORNING DAWN', LCl, lp, 1955; flowers silvery rose flushed salmon, dbl., 63 petals, 5 in., exhibition form, moderate, spicy fragrance; foliage dark, glossy, leathery; vigorous, pillar (6-8 ft) growth; [New Dawn seedling X R.M.S. Queen Mary]; Boerner; J&P

Morning Glory *see* **'MCGREDY'S ORANGE'**

'MORNING GLORY', Cl HT, rb, 1937; flowers carmine, reverse sulfur-yellow stained carmine, well formed, dbl., moderate fragrance; very vigorous, compact growth; [Portadown Sally sport]; Beckwith

'MORNING GLOW', Min, ly, 1999; flowers light yellow, slight blushing on top surface in cool weather, dbl., 17–25 petals, 1.2in., exhibition form, borne mostly singly, no fragrance; few prickles; foliage large, medium green, semi-glossy; upright, tall (24 in.) growth; [Tidewater X Kristin]; Jolly, Betty J.; Langenbach, 1998

Morning Greeting *see* **'MORGENGRÜSS'**

Morning Has Broken *see* CLEwedding

Morning in Moscow *see* **'UTRO MOSKVY'**

'MORNING JEWEL'®, LCl, mp, 1968; flowers large, semi-dbl., recurrent bloom, moderate fragrance; foliage glossy; ADR, 1975; [New Dawn X Red Dandy]; Cocker

'MORNING JOY', HT, ab, 1968; flowers creamy amber, reverse flushed coppery pink, well-formed, dbl., slight fragrance; foliage light green; vigorous, upright growth; [Mischief sport]; Williamson

Morning Light *see* HARlecho

'MORNING MIST', HT, m, 1950; flowers lavender tinted gray, semi-dbl., 15–20 petals, 3.5 in., exhibition form, moderate fragrance; moderately vigorous, upright growth; Fisher, G.; Arnold-Fisher Co.

Morning Mist *see* 'AUSFIRE'

Morning Skies *see* 'GELSKIES'

Morning Song *see* 'MINISONG'

Morning Star, S, yb, 1998

'MORNING STARS', S, w, 1949; bud ovoid; flowers clusters, dbl., 3 in., cupped, free, recurrent boom, moderate fragrance; foliage glossy; upright, bushy, compact growth; [(New Dawn X Autumn Bouquet) X (New Dawn X Inspiration)]; Jacobus; B&A

Morning Sun *see* **'MORGENSONNE'**

Morning Sun *see* AROsumo

'MORNITA', Min, op, 1981; (**Anita Charles**®); bud pointed; flowers bright pink, reverse lighter pink-yellow blend, dbl., 43 petals, exhibition form, borne singly, sometimes 2-3 per cluster, moderate, tea fragrance; straight prickles; foliage small, leathery, matt; vigorous, upright, spreading growth; [Golden Glow (Brownell) X Over the Rainbow]; Moore, Ralph S.

Moroccan Rose, P, dp

'MOROCCO', HT, mr, 1961; bud pointed; flowers velvety red, dbl., 30 petals, 5 in., exhibition form, moderate fragrance; foliage glossy; vigorous, upright growth; [Carrousel X Charles Mallerin]; Von Abrams; Peterson & Dering

'MOROLÉ', Min, r, 1990; (**Café Olé**); bud pointed; flowers medium to large blooms, borne singly or in sprays of 3-5, very dbl., 40–50 + petals, cupped, moderate, spicy fragrance; round, orangish fruit; prickles slender, hooked downward, brown; foliage medium to large, medium green, dull to semi-glossy; upright, bushy, tall, vigorous growth; [Winter Magic sport]; Moore, Ralph S.; Sequoia Nursery

Morongo, S, lp; Peden, G.H., 1996; (Weatherly, L.)

'MORORCHERI', Min, m, 1993; (**Lavender Delight**); flowers semi double, medium, semi-dbl., 6–14 petals, 1.5–2.75 in., borne in small clusters, slight fragrance; few prickles; foliage medium, medium green, semi-glossy; medium, upright, bushy growth; [Orangeade X Cherry Magic]; Moore, Ralph S.; Sequoia Nursery, 1994

'MOROSE', Min, rb, 1987; (**Rose Gilardi**®); flowers red and pink striped, reverse similar, aging well, informal, semi-dbl., 12–15 petals, slight fragrance; slender, straight, small to medium, brownish prickles; foliage small, medium green, semi-glossy; bushy, spreading, medium growth; [Dortmund X ((Fairy Moss X (Little Darling X Ferdinand Pichard)) X Unnamed seedling)]; Moore, Ralph S., 1986

'MORPALE', Min, rb, 1984; (**Painter's Palette**); bud mossy; flowers creamy white, striped deep pink to deep red, blooms borne 3-5 per cluste, dbl., 72 petals, slight fragrance; needle straight, brown prickles; foliage deep green; shrubby growth; Moore, Ralph S.; Ludwigs Roses Pty. Ltd.

'MORPAPPLAY', S, rb, 1988; (Creepy, Glowing Garpet, Highveld Sun, **Ralph's Creeper**); flowers dark orange-red, bright yellow eye, reverse bright yellow to, semi-dbl., 15–18 petals, repeat bloom, moderate, apple-blossom fragrance; round, orange-red fruit; prickles normal, intermediate, tall; foliage small, dark green, matt; spreading, low growth; PP006548; [Papoose X Playboy]; Moore, Ralph S.; Armstrong Nursery, 1987

'MORPETTI', Min, rb, 1993; (**Hoot 'n' Holler**); flowers pleasing blend of red changing to lavender, blooms borne in, semi-dbl., 6–14 petals, 1.5–2.75 in., no fragrance; few prickles; foliage medium, medium green, matt; upright, bushy growth; [Pink Petticoat X Make Believe]; Moore, Ralph S.; Sequoia Nursery, 1993

Morpho, F; Leenders; (Cavriglia)

'MORPICO', Min, rb, 1991; (**Circus Clown**); semi-dbl., 6–14 petals, 1.5 in., borne mostly singly or in small clusters, slight fragrance; few prickles; foliage small, medium green, semi-glossy; low (18-22 cms), bushy, compact growth; [Pink Petticoat X Make Believe]; Moore, Ralph S.; Sequoia Nursery, 1992

'MORPINKPER', S, pb, 1999; (**Moore's Pink Perpetual**); flowers pink blend, very dbl., 41 petals, 3–3.5 in., borne in small clusters, moderate fragrance; many prickles; foliage medium, medium green, semi-glossy; dwarf hybrid perpetual, bushy, compact (16-24 in.) growth; [seedling X Paul Neyron]; Moore, Ralph S.; Sequoia Nursery, 1999

'MORPINTS', Min, rb, 1985; (**Pinstripe**®); flowers red with white stripes, well-formed, small, dbl., 35 petals, slight fragrance; foliage small, medium green, semi-glossy; low, mounded habit growth; [Pinocchio X Seedling]; Moore, Ralph S.; Armstrong Nursery, 1986

'MORPLAG', F, mp, 1986; (**Playgirl**®); flowers yellow stamens, blooms borne usually singly and in sprays, single, slight fragrance; no fruit; few, brown prickles; foliage medium, medium green, semi-glossy; upright, bushy growth; [Playboy X Angel Face]; Moore, Ralph S.; Moore Min. Roses

'MORPLATI', F, or, 1989; (**Playtime**®); bud pointed; flowers vibrant orange-red, aging slightly darker, medium, borne usu, single, 5 petals, flat, slight fragrance; round, medium, orange-red fruit; prickles straight, slightly hooked, medium, light brown; foliage medium, dark green, semi-glossy; upright, bushy, medium growth; [Playboy X Old Master]; Moore, Ralph S.; Sequoia Nursery, 1990

'MORPLAYGOLD', Min, ob, 1997; (**Playgold**); flowers semi double, small, semi-dbl., 6–14 petals, borne in small clusters, slight fragrance; foliage small, medium, green, glossy; low (30-45 min.) growth; [Playgold X Sequoia Gold]; Moore, Ralph S.

'MORPLUM', Min, lp, 1994; (**Sugar Plum**); flowers lavender purple opening to expose a silver shading on inside, semi-dbl., 6–14 petals, 1.5–2.75 in., no fragrance; no prickles; foliage medium, medium green, semi-glossy; medium (30-40 cms), bushy, spreading growth; [Anytime X Angel Face]; Moore, Ralph S.; Sequoia Nursery, 1995

'MORPOLY', Min, my, 1991; (**Cal Poly**); flowers long lasting color, non-fading, blooms borne in small clusters, dbl., 15–25 petals, 1.5–2.75 in., slight fragrance; few prickles; foliage medium, medium green, semi-glossy; medium (24 cms), upright, bushy growth; AOE, 1992; [(Little Darling X Yellow Magic) X Gold Badge]; Moore, Ralph S.; Sequoia Nursery, 1992

'MORPOODLE', Min, pb, 1991; (**Pink Poodle**); flowers pink changing to lighter pink to white, typical china characteristics, very dbl., 1.5–2.75 in., moderate fragrance; few prickles; foliage small, medium green, matt; low (26-30 cms), upright growth; [(Little Darling X Yellow Magic) X Old Blush]; Moore, Ralph S.; Sequoia Nursery, 1992

'MORPORC', Min, lp, 1994; (**English Porcelain**); flowers light pink, moderately full (15–25 petals), medium (4–7 cms) blooms borne mostly single; slight fragrance; no prickles; foliage small, medium green, matt; low to medium (28–34 cms), bushy, spreading, compact growth; [Pink Petticoat X Happy time]; Moore, Ralph S.; Sequoia Nursery, 1995

MORposa, S, rb, 1997; (**Royal Midinette**); Moore

'MORPREPEN', Min, ob, 1994; (**Pretty Penny**); semi-dbl., 6–14 petals, 1.5–2.75 in., borne mostly singly or in small clusters, no fragrance; no prickles; foliage medium, medium green, matt; medium (28-34 cms), bushy, spreading, compact growth; [Seedling X Self]; Moore, Ralph S.; Sequoia Nursery, 1995

'MORPROM', Min, dp, 1989; (**Prom Date**); bud short, pointed; flowers deep pink, aging lighter, medium, borne singly, profuse, dbl., 38 petals, globular, no fragrance; globular, orange fruit; prickles straight, pointed, medium, gray to brown; foliage medium, medium green, matt; upright, bushy, medium growth; [Sheri Anne X (Unnamed seedling X Fairy Moss)]; Moore, Ralph S.; Sequoia Nursery

'MORPUFF', Cl HBc, lp, 1990; (**Pink Powderpuff**); bud pointed; flowers light pink, aging slightly lighter, old garden rose type, large, very dbl., exhibition form, repeat bloom, intense, damask fragrance; foliage large, medium green, semi-glossy; tall, spreading, climbing growth; [Lulu X Muriel]; Moore, Ralph S., 1964

'MORQUAKE', Min, rb, 1983; (**Earthquake**®); flowers striped red and yellow, reverse yellow, small, dbl., no fragrance; foliage small, medium green, semi-glossy; upright, bushy growth; PP005791; [Golden Angel X Seedling]; Moore, Ralph S.; Moore Min. Roses, 1984

'MORRAINBOW', Min, pb, 1994; (**Halo Rainbow**); flowers pink edging with center of each petal creamy white and base, single, 5 petals, 1.5–2.75 in., slight fragrance; no prickles; foliage medium, medium green, semi-glossy; medium (28-35 cms), bushy, spreading growth; [Seedling X Make Believe]; Moore, Ralph S.; Sequoia Nursery, 1995

'MORREDFAR', Pol, mr, 1996; (**Red Fairy**); flowers cherry red, dbl., 24–30 petals, 1.5 in., borne in conical sprays, somewhat pendulous, slight fragrance; some prickles; foliage long, bright medium green, semi-glossy; wiry stems; medium, semi-cascading growth; PP10150; [Simon Robinson X Simon Robinson]; Moore, Ralph S.

'MORROUSEL', Min, rb, 1972; (**Magic Carrousel**®, 'MOORCAR'); flowers petals white, edged red, small, dbl., exhibition form, slight fragrance; foliage small, glossy, leathery; vigorous, bushy growth; AOE, ARS, 1975 Miniature Rose Hall of Fame, ARS, 1999; [Little Darling X Westmont]; Moore, Ralph S.; Sequoia Nursery

'MORRUBI', Min, mr, 1986; ('MORUBY', **Ruby Magic**); flowers small to blooms borne usually singly and in sprays of 3-5, dbl., 20 petals, slight fragrance; usually no fruit; brown prickles; foliage small to medium, medium green, semi-glossy; medium, upright, bushy growth; [Orangeade X Pinstripe]; Moore, Ralph S.; Moore Min. Roses

Morsdag *see* **'MOTHERSDAY'**

Morsdag Alba, Pol, w

Morsdag Red, Pol, mr

'MORSEA', Min, mp, 1986; (**Chelsea**); flowers small blooms borne in clusters of 5 or more, dbl., 25 petals, cupped, slight fragrance; few, globular, medium, orange fruit; foliage small to medium, medium green, semi-glossy; upright, bushy growth; [(Little Darling X Yellow Magic) X Crested Jewel]; Moore, Ralph S.; Sequoia Nursery

'MORSEGOLD', Min, my, 1986; (**Sequoia Gold**®); flowers medium yellow, fading lighter, medium, dbl., 30 petals, exhibition form, borne usually singly, moderate, fruity fragrance; round, orange fruit; slender, medium, pale green-brown prickles; foliage medium, medium green, glossy; bushy, spreading growth; PP006617; AOE, 1987 Rose of The Year, Auckland, NZ, 1995; [(Little Darling X Lemon Delight) X Gold Badge]; Moore, Ralph S.

'MORSEMURI',S, pb, 1995; (**Star Magic**); flowers pink blend, single (5 petals, small (0–4 cms) blooms borne in small clusters; no fragrance; some prickles; foliage small, dark green, glossy, groundcover, medium (3ft. X 9 ft.), spreading compact growth; [Sequoia Gold X MORmuri]; Moore, Ralph, S.; Sequoia Nursery, 1996

'MORSEWEL', Min, mr, 1989; (**Sequoia Jewel**); bud rounded; flowers medium, borne in sprays of 3-7, dbl., 33 petals, cupped, slight fragrance; round, medium, orange-red fruit; prickles straight, short, brown; foliage medium, medium green, matt; upright, bushy, medium growth; [Sheri Anne X Paul Neyron]; Moore, Ralph S.; Sequoia Nursery, 1990

'MORSHACK', Min, ob, 1987; (**Phyllis Shackelford**®); flowers orange, fading pink, small borne usually singly, dbl., 20 petals, exhibition form, moderate, fruity fragrance; rounded, medium, orange fruit; few, medium, brownish prickles, slightly hooked downwards; foliage small, medium green, semi-glossy; upright, bushy, medium growth; [Anytime X Gold Badge]; Moore, Ralph S.

'MORSHAKI', Min, rb, 1989; (**Little Tiger**); bud short, pointed; flowers red, yellow and white stripes of varying patterns, reverse with more yellow, very dbl., exhibition form, no fragrance; prickles average, slender, inclined downward, brown; foliage small, medium green, matt; bushy, low, rounded growth; [Golden Angel X Pinstripe]; Moore, Ralph S.; Sequoia Nursery

'MORSHAKRUG', S, pb, 1992; (**Yesterday's Garden**); very dbl., 1.5–2.75 in., borne in small clusters, no fragrance; foliage medium, medium green, semi-glossy; tall to medium (over 50 cms), upright, bushy, spreading gro growth; [Shakespeare Festival X Belle Poitevine]; Moore, Ralph S.; Sequoia Nursery, 1993

'MORSHARON', S, w, 1996; (**Sharon's Delight**); single, 4–7 petals, 3.5 in., borne in small clusters, slight fragrance; few prickles; foliage medium, medium green, semi-glossy; bushy, medium (2-3 ft) growth; [Golden Angel X Safrano]; Moore, Ralph S.; Sequoia Nursery, 1996

'MORSHEFRAN', Min, w, 1989; (**New Adventure**); bud pointed; flowers creamy white, medium, borne singly and in sprays of 3-5, dbl., 25 petals, flat, no fragrance; globular, orange fruit; prickles short, pointed, brownish-gray; foliage small, medium green, matt; upright, bushy, medium growth; [Sheri Anne X Safrano]; Moore, Ralph S.; Sequoia Nursery

'MORSHERI', Min, or, 1973; (**Sheri Anne**); bud long, pointed; flowers orange-red, base yellow, semi-dbl., 17 petals, 1–1.5 in., moderate fragrance; foliage glossy, leathery; upright, bushy growth; AOE, 1975; [Little Darling X New Penny]; Moore, Ralph S.; Sequoia Nursery

'MORSHODOT', Min, dr, 1996; (**Cherry Hi**); flowers dark red, reverse slightly lighter, blooms borne in small clusters, very dbl., 1.5 in., no fragrance; few prickles; [Show 'n' Tell X Sincerely Yours]; Moore, Ralph S.; Sequoia Nursery, 1997

'MORSHOOK', Cl Min, rb, 1990; (**Earthquake, Climbing**); bud rounded; flowers red/yellow stripes, yellow reverse, aging similar, urn-shape, dbl., no fragrance; foliage small, medium green, semi-glossy; upright, tall growth; [Earthquake sport]; Moore, Ralph S.; Sequoia Nursery, 1991

'MORSIXTY', Min, m, 1996; (**Diamond Anniversary**); flowers mauve, reverse mauve or slightly lighter, blooms borne in small clusters, dbl., 26–40 petals, 1.5 in., slight fragrance; few prickles; [Joycie X Cherry Magic]; Moore, Ralph S.; Sequoia Nursery, 1997

'MORSNO', Min, w, 1987; (**Snow Twinkle**®); flowers small, borne usually singly, dbl., exhibition form, slight fragrance; round, orange fruit; small, brown prickles; foliage small, medium green, matt; bushy, medium growth; [(Little Darling X Yellow Magic) X Magic Carrousel]; Moore, Ralph S.

'MORSNOP', Min, my, 1993; (**Buttermint**); flowers moderately full, dbl., 15–25 petals, 1.5–2.75 in., borne mostly singly, slight fragrance; few prickles; foliage medium, medium green, semi-glossy; medium to tall (36-40 cms), upright, bushy growth; [Pink Petticoat X Gold Badge]; Moore, Ralph S.; Sequoia Nursery, 1994

'MORSTAR', HRg, mp, 1990; (**Star Delight**, Starry Eyed); bud pointed; flowers rose pink, white base, silvery pink reverse, medium, borne u, single, 5 petals, flat, repeat bloom, slight fragrance; foliage medium, olive to bluish-green, semi-glossy; upright, bushy, tall growth; [Yellow Jewel X Rugosa Magnifica]; Moore, Ralph S., 1989; Sequoia Nursery

'MORSTOCK', Min, yb, 1999; (**Woodstock**); dbl., 26–40 petals, 1–1.5 in., borne mostly singly, slight fragrance; few prickles; foliage small, medium green, semi-glossy; low (8-12 in.), bushy, compact growth; [seedling X Clytemnestra]; Moore, Ralph S.; Sequoia Nursery, 1999

'MORSTRORT', LCl, pb, 1999; (Cl. Shadow Dancer, **Shadow Dancer**®); flowers striped, swirled two-tone pink, reverse same, dbl., 17–25 petals, 3.5–4 in., borne in large clusters, slight fragrance; prickles moderate; foliage medium, dark green, holly-like; climbing, tall (8-10 ft) growth; [Dortmund X (Dortmund X Unknown)]; Moore, Ralph S.; Weeks Roses, 1998

MORsuise, Min, yb, 1997; (**Halo Suise**); Moore

'MORSUN', Min, my, 1981; (**Sweet Sunshine**); flowers small, dbl., 20 petals, intense fragrance; foliage small, medium green, semi-glossy; upright, bushy growth; [Rumba X Yellow Jewel]; Moore, Ralph S.; Moore Min. Roses

'MORSUNRISE', Min, gls, 1997; (**Halo Sunrise**); Moore

'MORSWEET', LCl, mp, 1988; (**Crested Sweetheart**); flowers medium rose pink, large, borne in sprays of 3-5, very dbl., cupped, intense, damask fragrance; prickles small, gray to brown; foliage large, medium green, matt, rugose; upright, tall growth; [Little Darling X Crested Moss]; Moore, Ralph S.; Wayside Gardens Co., 1988

'MORSWISS', Min, ob, 1990; (**Two-Timer**); bud ovoid; flowers orange-red with lighter shades of orange and white striped, very dbl., exhibition form, no fragrance; foliage medium, medium green, semi-glossy; bushy, spreading, low growth; [Orangeade X Pinstripe]; Moore, Ralph S., 1982; Sequoia Nursery, 1991

MORsycheek, S, op; (**Rosy Cheeks**); Moore

'MORTEN', HRg, mr, 1990; (**Linda Campbell**, Tall Poppy); bud pointed; flowers medium red, blooms slightly lighter reverse, aging medium red, semi-dbl., 25 petals, cupped, fast repeat, no fragrance; foliage large, dark green, semi-glossy; upright, bushy, medium growth; [Anytime X Rugosa Magnifica]; Moore, Ralph S., 1985; Wayside Gardens Co., 1991

'MORTODAY', Min, op, 1994; (**Halo Today**); flowers have distinct pink/lavender area at base of each petal, semi-dbl., 6–14 petals, 1.5–2.75 in., no fragrance; no prickles; foliage medium, medium green, semi-glossy; low (16-18 in), upright, bushy growth; [(Anytime X Gold Badge) X (Anytime X Lavender Jewel)]; Moore, Ralph S.; Sequoia Nursery, 1994

'MORTOPAZ',Min, ab, 1995; (**Peach Candy**);flowers soft peach, moderately full (15–25 petals), small (0–4 cms) blooms borne in small clusters; slight fragrance; few prickles; foliage small, light green, matt; medium (24–30 cms), upright, bushy growth; [Sheri Anne X Topaz Jewel]; Moore, Ralph, S.; Sequoia Nursery, 1996

'MORTORCH', Min, or, 1986; (**Torch of Liberty**®); flowers orange-red, silver reverse, small, dbl., 20 petals, slight fragrance; foliage small, medium green, semi-glossy; upright, bushy growth; PP006254; [Orangeade X Golden Angel]; Moore, Ralph S.; Moore Min. Roses, 1985

'MORTREAT', Min, yb, 1986; (**Double Treat**); bud mossy; flowers bright red and orange-yellow, striped, mini-moss, dbl., cupped, borne usually singly, slight fragrance; small, globular with numerous spines, orange fruit; slender, brown prickles; foliage small to medium, medium green, semi-glossy; medium, upright, bushy growth; [Arizona X ((Fairy Moss X Fairy Moss) X (Little Darling X Ferdinand Pichard))]; Moore, Ralph S.; Sequoia Nursery, 1985

'MORTRIP', Min, rb, 1980; (**Crazy Quilt**); bud pointed; flowers red and white striped, borne 1-3 or more per cluster, dbl., flat; small, straight prickles; foliage medium green; coopact, bushy growth; [Little Darling X Seedling]; Moore, Ralph S.; Sequoia Nursery

'MORTWIST', Min, ob, 1986; (**Orange Twist**); flowers tannish-orange; [Sungold sport]; Moore, Ralph S.; Moore Min. Roses, 1985

'MORTWISTER', Cl Min, rb, 1997; (**Twister**); flowers full, medium, very dbl., 26–40 petals, borne in small clusters, slight fragrance; few prickles; foliage medium, medium green, dull; upright, climbing tall (3-ft.) growtb growth; [Little Darling X Little Magic]; Rekdeci, John Jason

'MORUBY', Min, mr, 1986; ('MORRUEI', **Ruby Magic**); flowers small to blooms borne usually singly and in sprays of 3-5, dbl., 20 petals, slight fragrance; usually no fruit; brown prickles; foliage small to medium, medium green, semi-glossy; medium, upright, bushy growth; [Orangeade X Pinstripe]; Moore, Ralph S.; Moore Min. Roses

'MORUNUSUAL', F, pb, 1999; (**Most Unusual Day**); dbl., 15–25 petals, 1.5–2.5 in., borne in small clusters, no fragrance; few prickles; foliage medium, medium green, semi-glossy; medium (16-24 in.), upright, bushy growth; [Show 'n' Tell X Blastoff]; Moore, Ralph S.; Sequoia Nursery, 1999

'MORVANDEL', Min, yb, 1977; (**Avandel**); bud long, pointed; flowers pink-yellow blend, dbl., 23 petals, 1–1.5 in., cupped, moderate, fruity fragrance; foliage medium to dark green, leathery; upright, bushy growth; AOE, 1978; [Little Darling X New Penny]; Moore, Ralph S.; Sequoia Nursery

'MORVI', Min, m, 1991; (**Vi's Violet**); flowers soft lavender, small, dbl., slight fragrance; foliage small, medium green, matt; upright, bushy, compact growth; [Unnamed seedling X Angel Face]; Moore, Ralph S.

'MORWATEYE', Min, pb, 1992; (**Halo Dolly**); flowers bicolor, reddish outside, pink inside with reddish lavender, semi-dbl., 6–14 petals, 1.5 in., no fragrance; few prickles; foliage medium, medium green, semi-glossy; medium (30-45 cms), upright, bushy, rounded growth; [Anytime X Seedling (Anytime X Angel Face)]; Moore, Ralph S.; Sequoia Nursery, 1993

'MORWHEELS', Min, rb, 1997; (**Jacquie Williams**); flowers full, medium, very dbl., 26–40 petals, borne mostly single but some clusters, slight fragrance; foliage medium, medium green, semi-glossy, ; low to medium (30-45 cms) growth; [(Yellow Jewel X Tamango) X Strawberry Ice]; Moore, Ralph S.

'MORWHIT', Min, w, 1981; (**Little Eskimo**, 'MORWIT'); bud long, pointed; flowers near white, blooms borne 3-7 per cluster, sometimes singly, dbl., 55 petals, slight fragrance; long, slender prickles; foliage small, semi-glossy, leathery; vigorous, bushy, upright growth; [(R. wichurana X Floradora) X Jet Trail]; Moore, Ralph S.; Moore Min. Roses

'MORWHY', Min, rb, 1983; (**Why Not**, Why Not?); flowers medium red, yellow eye, small, single, 7 petals, slight fragrance; foliage small, medium green, matt to semi-glossy; upright, bushy growth; PP005676; [Golden Angel X Unnamed seedling]; Moore, Ralph S.; Moore Min. Roses

'MORWIMER', Cl Min, ly, 1995; (**Little Mermaid**); flowers light yellow, single (5 petals), medium (4–7 cms) blooms borne in small clusters, no fragrance; some prickles; foliage small, medium green, semi-glossy; bushy, spreading growth; [Mermaid sport] Moore, Ralph S.; Sequoia nursery, 1995

'MORWINGS', S, ly, 1989; (**Yellow Butterfly**); bud pointed; flowers light yellow, aging to near white, informal, medium, borne i, single, 5 petals, no fragrance; ovoid, small, yellowish-orange fruit; prickles slender, inclined downward, small, brownish; foliage medium, light green, semi-glossy; bushy, spreading, medium, clean growth; [Ellen Poulsen X Yellow Jewel]; Moore, Ralph S.; Sequoia Nursery

'MORWINST', Min, rb, 1994; (**Secret Recipe**); flowers combination of burgundy, red and white stripes, blooms borne, semi-dbl., 6–14 petals, 1.5–2.75 in., slight fragrance; some prickles; foliage medium, medium green, semi-glossy; upright, bushy (30-40 cms) growth; [Little Darling X Seedling]; Moore, Ralph S.; Sequoia Nursery, 1995

'MORWIT', Min, w, 1981; (**Little Eskimo**, 'MORWHIT'); bud long, pointed; flowers near white, blooms borne 3-7 per cluster, sometimes singly, dbl., 55 petals, slight fragrance; long, slender prickles; foliage small, semi-glossy, leathery; vigorous, bushy, upright growth; [(R. wichurana X Floradora) X Jet Trail]; Moore, Ralph S.; Moore Min. Roses

'MORWOOD', Min, or, 1984; (**Redwood Empire**); flowers small, dbl., 20 petals, slight fragrance; foliage small, medium green, semi-glossy; upright, bushy growth; [Rumba X Sheri Anne]; Moore, Ralph S.; Moore Min. Roses, 1983

'MORYEARS', S, w, 1999; (**Out of Yesteryear**); flowers white blend, very dbl., 41 petals, 3 in., borne in large clusters, moderate fragrance; many prickles; foliage medium, dark green, glossy; upright, spreading (4-6 ft) growth; [Golden Angel X Muriel]; Moore, Ralph S.; Sequoia Nursery, 1999

'MORYELRUG', HRg, my, 1987; (Gelbe Dagmar Hastrup, Rustica 91, **Topaz Jewel**®, Yellow Fru Dagmar Hartopp); flowers medium yellow fading to cream, large, dbl., cupped, borne in sprays of 5-8, repeat bloom, moderate, fruity fragrance; no fruit; slender, fairly long, brown prickles in variable sizes; foliage large, medium green, rugose, matt; upright, bushy, spreading, vigorous growth; PP006793; [Golden Angel X Belle Poitevine]; Moore, Ralph S.; Moore Miniature Roses, 1987;, Wayside Gardens, 1987

'MORYETTEM', Min, m, 1986; (**Saint Mary**®); flowers deep red purple, small, dbl., cupped, slight fragrance; no fruit; small, brown prickles; foliage small, medium green, semi-glossy; medium, upright, bushy growth; [Little Chief X Angel Face]; Moore, Ralph S.; Moore Min. Roses

'MORYOU', Min, dp, 1990; (**Just For You**); bud pointed; flowers dark pink to light red, lighter reverse, aging lighter, medium, dbl., 35 petals, exhibition form, slight fragrance; foliage medium, medium green, semi-glossy; bushy, medium growth; AOE, 1991; [Orangeade X Rainbow's End]; Moore, Ralph S., 1986; Sequoia Nursery, 1991

Mosaik, F, rb, 1978; flowers red with dark yellow, medium, dbl., slight fragrance; VEG; (Sangerhausen)

'MOSAÏQUE', F, yb, 1960; flowers light yellowish pink, becoming red, open, semi-dbl.; foliage light green; low, bushy growth; [(Alain X Cinnabar) X Circus]; Lens

'MOSCHATA GRANDIFLORA', HMsk, w; flowers prominent golden stamens, large, single, intense fragrance; very vigorous growth; [R. moschata X R. multiflora]; Bernaix, A., ca. 1867

Moscow Morn *see* **'UTRO MOSKVY'**

'MOSEL', HMult, m, 1920; flowers bluish violet, center reddish violet, dbl., sometimes recurrent bloom, slight fragrance; [Mme Norbert Levavasseur X Trier]; Lambert, P.

'MOSELLIED', HMsk, m, 1932; flowers purplish red, center white, stamens golden yellow, petals he, single, non-recurrent, intense fragrance; foliage dark; vigorous (about 6 1/2 ft.), broad growth;

[(Geheimrat Dr. Mittweg X Tip-Top) X (Chamisso X Parkzierde)]; Lambert, P.

'MOSS MAGIC', Min, mp, 1977; bud cupped, mossy; flowers medium to dark pink, circular, dbl., 48 petals, 1 in., flat, profuse bloom, repeating well; foliage dark; spreading growth; [Fairy Moss X ?]; Sudol, Julia

Moss Rose *see* **'CENTIFOLIA MUSCOSA'**

'MOSSMAN', S, mp, 1954; flowers heavily mossed, pale dusty pink, very dbl., 2.5 in.; vigorous (4 ft.) growth; [(R. acicularis X R. rugosa) X Moss rose]; Skinner

Mossy de Meaux *see* **'MOSSY ROSE DE MEAUX'**

'MOSSY GEM', Min, mp, 1984; bud small, mossy; flowers medium pink, outer petals fading, small, dbl., slight fragrance; foliage medium, dark, semi-glossy; bushy growth; [Heidi X Violette]; Kelly, Martin

'MOSSY ROSE DE MEAUX', M, mp; (Mossy de Meaux); [Rose de Meaux sport]

Most Unusual Day *see* 'MORUNUSUAL'

Moth, S, m, 1983; Austin, David

'MOTHER', HT, pb, 1939; flowers white edged bright pink, intense fragrance; [Memory sport]; Cant, B. R.

Mother and Baby® *see* 'SANBABY'

'MOTHER MARIE', Pol, w, 1954; flowers white, center green, borne in clusters, dbl., 40 petals, 1.5–2 in., flat; foliage light green; [Garnette sport]; Podesta

Mother Theresa, HT, w, 1994; IARI

Mother's Country *see* **'MOEDER DES VADERLANDS'**

Mothers Day *see* **'MOTHERSDAY'**

'MOTHER'S DAY', HT, w, 1937; flowers white, at times shaded pink, very large; very vigorous growth; Knight, G.

'MOTHER'S DAY ORANGE SPORT', Pol, ob, 1958; flowers orange-yellow; [Mothersday sport]; Klyn

Mother's Love *see* 'TINLOVE'

Mother's Value *see* KORretra

'MOTHERSDAY', Pol, dr, 1949; (Fête des Mères, Morsdag, Mothers Day, Muttertag); flowers deep red, blooms in clusters (up to 20), dbl., globular; foliage glossy, small; dwarf growth; [Dick Koster sport]; Grootendorst, F.J.

'MOTHERSDAY, CLIMBING', Cl Pol, dr, 1956; (Muttertag, Climbing); Kordes

Motiv, HT, mp, 1983; flowers large, dbl.; VEG; (Sangerhausen)

'MOTREA', F, mr, 1968; bud ovoid; flowers light red, small, very dbl.; foliage dark; [Coronet sport]; Mosselman & Terreehorst

'MOTYLEK', HT, w, 1954; (Butterfly Papilio); flowers cream, base yellow, medium, dbl., 40–45 petals, exhibition form, moderate fragrance; moderate growth with thin, strong shoots; [Freiburg II X Golden Dawn]; Shtanko, I.

'MOULIN ROUGE'®, F, mr, 1952; (Sans souci); dbl., 20–25 petals, 2 in., cupped, blooms in clusters, slight fragrance; foliage glossy; upright, very bushy growth; GM, Geneva, 1952 GM, NRS, 1952 PIT, NRS, 1952; [Alain X Orange Triumph]; Meilland, F.; URS

'MOULIN ROUGE, CLIMBING', Cl F, mr, 1957; Mondial Roses; URS

Mouna®, HT, mr

'MOUNT EVEREST', S, mp

'MOUNT HOBBEN', HT, mr, 1981; flowers bright scarlet, large blooms borne 1-3 per stem, dbl., 30 petals, exhibition form, no fragrance; many, small prickles; foliage medium green; tall, upright growth; [Duftzauber X Christian Dior]; Ota, Kaichiro

'MOUNT HOMAN', HT, ab, 1983; bud ovoid; flowers large blooms borne 1-3 per stem, dbl., exhibition form, slight fragrance; large prickles; foliage dark, semi-glossy; tall, upright, bushy growth; [Ginger Rogers X Sunblest]; Ota, Kaichiro

Mount Hood *see* 'MACMOUHOO'

Mount Nelson, F, w, 1998

'MOUNT SHASTA', Gr, w, 1963; bud long, pointed; dbl., 4.5–5 in., cupped, moderate fragrance; foliage leathery, gray-green; vigorous, upright growth; [Queen Elizabeth X Blanche Mallerin]; Swim & Weeks; C-P

Mount Shasta, Climbing, Cl HT, lp; [Mount Shasta sport]; Knight, G., 1968; (Weatherly, L.)

'MOUNTAIN HAZE', F, m, 1967; flowers lavender, reverse silver, medium, dbl., 55–60 petals, exhibition form, moderate fragrance; foliage dark, leathery; vigorous, low, spreading growth; [Amy Vanderbilt X Lilac Time]; Morey, Dr. Dennison; Country Garden Nursery

Mountain Mignonette, Pol, lp

'MOUNTAIN MIST', S, m, 1989; bud pointed; flowers purple-lavender shading to white at base, small, borne in sprays, single, 5 petals, flat, intense, multiflora, vanilla fragrance; globular, red fruit; prickles very few, red; foliage small, light green, semi-glossy; spreading, medium (4x6'), prolific growth; [Yesterday X Yesterday]; Jobson, Daniel J.

'MOUNTAIN MUSIC', S, pb, 1984; flowers pink and yellow blend, blooms borne 5-15 per cluster, dbl., 23 petals, cupped, repeat bloom, moderate fragrance; awl-like, tan prickles; foliage leathery, semi-glossy; vigorous, erect, bushy growth; hardy.; [Sevilla X Tom Brown]; Buck, Dr. Griffith J.; Iowa State University

Mountain Rose *see* **'R. WOODSII'**

Mountain Side, S

Mountain Snow *see* 'AUSSNOW'

'MOUNTAINEER', HT, op, 1963; bud long, pointed, orange-red; flowers orange-pink, reverse creamy yellow, dbl., 35 petals, 5 in., exhibition form, moderate fragrance; foliage glossy; vigorous growth; [Mrs Sam McGredy X Crimson Glory]; Wyant

Mountbatten® *see* 'HARMANTELLE'

Mountie™ *see* 'LAVCALI'

'MOUNTJOY', (natural variation of R. setigera), lp; flowers blush, center darker; (found in Ohio)

'MOURNE GOLD', HT, yb, 1980; flowers deep yellow, pink flush; [Whisky Mac sport]; Kane Brothers

Mouse *see* 'SEAMOU'

'MOUSHA', HT, lp, 1974; bud long, pointed; dbl., 24 petals, 4 in., exhibition form; foliage glossy, dark; vigorous, upright growth; [Königin der Rosen X King's Ransom]; Kordes; Willemse

Mousseau Ancien *see* **'COMMUNIS'**

Mousseline *see* **'ALFRED DE DALMAS'**

Mousseux Ancien *see* **'CENTIFOLIA MUSCOSA'**

'MOUSSEUX DU JAPON', M, m; (Japonica, Moussu du Japon, Muscosa Japonica); bud and stems heavily mossed; flowers purplish rose, quickly fading to lavender, many stamens, semi-dbl.; growth moderate; (Japan), before 1900

Moussu du Japon *see* **'MOUSSEUX DU JAPON'**

'MOUSSUE DES QUATRE SAISON BLANC', w, 1835; flowers medium, dbl.; (Sangerhausen)

Moussue Partout *see* **'ZOÉ'**

Movie Star *see* TANeivom

'MOZART', HMsk, pb, 1937; flowers deep pink with large white eye, small blooms in clusters, single, repeat bloom, moderate fragrance; vigorous, trailing or bushy growth; [Robin Hood X Rote Pharisaer]; Lambert, P.

'MR BLUEBIRD', Min, m, 1960; bud ovoid; flowers lavender-blue, semi-dbl., 15 petals, 1.25 in.; foliage dark; compact, bushy (10-14 in) growth; [Old Blush X Old Blush]; Moore, Ralph S.; Sequoia Nursery

'MR CHIPS', HT, yb, 1970; flowers yellow and orange, high-pointed, large, dbl.; foliage glossy; [Irish Gold X Miss Ireland]; Dickson, A.

Mr E.E. Greenwell *see* 'HARJOOBILY'

'MR FAITHFUL', F, mp, 1968; dbl., blooms in trusses, moderate fragrance; [Pink Parfait X Self]; Harkness

Mr J C B *see* DICsun

'MR J. BIENFAIT', HT, mr, 1923; (J. Bienfait); flowers brick-red; [Mme Leon Pain X Red-Letter Day]; Van Rossem

Mr J.C.B., F, dy, 1993; Dickson, Patrick

'MR JOH. M. JOLLES', HT, ly, 1920; flowers clear creamy yellow, shaded apricot and golden yellow, dbl.; [Frau Karl Druschki X Mrs Joseph Hill]; Van Rossem

Mr Leigh, F, mp

Mr McCawber *see* 'WHICAW'

Mr Nash, LCl, my

'MR PAT', HT, dr, 1966; bud globular; flowers large, dbl., exhibition form, intense, spicy fragrance; foliage leathery; compact growth; [Red Jacket X Mirandy]; Patterson; Patterson Roses

Mr Pomeroy's Rose, S, w

'MR STANDFAST', HT, w, 1968; flowers cream, large, dbl., moderate fragrance; foliage glossy; [Dr. A.J. Verhage X Kordes' Perfecta]; Harkness

'MR TALL', HT, op, 1958; bud long, pointed; flowers salmon-pink, open, single, 7 petals, 4 in., intense, cinnamon fragrance; foliage dark, leathery; vigorous, upright growth; [Vogue X Grande Duchesse Charlotte]; Wyant

'MRIDULA', HT, w, 1975; bud long, pointed; flowers white, center soft pink, dbl., 35 petals, 3.5 in., exhibition form, moderate fragrance; foliage large, light, soft; very vigorous, upright, compact growth; [Queen Elizabeth X Sir Henry Segrave seedling]; IARI

Mrilliant Hit, Min, ob

'MRINALINI', HT, mp, 1973; bud long, pointed; flowers very large blooms borne singly, dbl., exhibition form, moderate fragrance; foliage medium, medium green, soft; vigorous, bushy growth; [Pink Parfait X Christian Dior]; IARI

'MRS A. GORDON', Cl T, lp; Fell, J.B., 1916; (Weatherly, L.)

Mrs A. Hudig *see* **'FLAMING JUNE'**

'MRS A.J. ALLEN', HT, pb, 1930; flowers rich pink, base lemon-yellow, tipped almost white, large, dbl., slight fragrance; [Richmond X Souv. de Mme Boullet]; Allen

Mrs A.J. Wylie *see* **'MARGARET M. WYLIE'**

'MRS A.R. BARRACLOUGH', HT, mp, 1926; bud long, pointed; flowers bright carmine pink, large, dbl., exhibition form, slight fragrance; vigorous growth; McGredy

'MRS A.R. BARRACLOUGH, CLIMBING', Cl HT, mp, 1935; Fryers Nursery, Ltd.

'MRS A.W. ATKINSON', HT, w, 1918; flowers ivory-white, dbl.; Chaplin Bros.

'MRS AARON WARD', HT, yb, 1907; bud long, pointed; flowers yellow, occasionally washed salmon (quite variable), dbl., exhibition form, moderate fragrance; dwarf, compact growth; Pernet-Ducher

'MRS AARON WARD, CLIMBING', Cl HT, yb, 1922; Dickson, A.

'MRS ALBERT NASH', HT, dr, 1929;, moderate fragrance; Clark, A.; NRS Victoria

'MRS ALFRED TATE', HT, rb, 1909; flowers coppery red shaded fawn, base shaded ochre, moderate fragrance; McGredy

'MRS ALFRED WEST', HT, op, 1922; flowers salmon-pink, dbl., intense fragrance; Cant, F.

Mrs Alston's Rose, Pol, rb, 1940; large trusses of reddish pink, semi-double blooms with lighter eye; Clark

'MRS AMBROSE RICARDO', HT, yb, 1914; flowers deep honey-yellow, overlaid brighter yellow, dbl., moderate fragrance; McGredy

'MRS AMY HAMMOND', HT, pb, 1911; flowers cream and amber, sometimes flushed pink, base apricot, dbl., moderate fragrance; GM, NRS, 1910; [Mme Abel Chatenay seedling]; McGredy

'MRS ANDREW CARNEGIE', HT, w, 1913; flowers white, center lightly tinted lemon-yellow, dbl., slight fragrance; GM, NRS, 1912; [Niphetos X Frau Karl Druschki]; Cocker

'MRS ANNE DAKIN', LCl, pb, 1972; flowers salmon-pink, reverse cream, dbl., 3.5 in., slight fragrance; foliage glossy; moderate, climbing growth; Holmes, R.; Albrighton Roses

Mrs Annie Beaufays *see* **'FRAU ANNY BEAUFAYS'**

'MRS ANTHONY SPALDING', HT, or, 1934; flowers strawberry-red flushed orange, reverse shaded orange, well s, dbl.; vigorous growth; McGredy

'MRS ANTHONY WATERER', HRg, dr, 1898; flowers deep crimson, semi-dbl., intense fragrance; vigorous growth; hardy; (21); [R. rugosa X Général Jacqueminot]; Waterer

'MRS ARCHIE GRAY', HT, ly, 1914; flowers deep creamy yellow, opening light canary-yellow; GM, NRS, 1913; Dickson, H.

'MRS ARNOLD BURR', HT, w, 1945; flowers pure white, large, very dbl., exhibition form, moderate fragrance; foliage glossy; vigorous, upright growth; [Peace X ?]; Burr, A.

'MRS ARTHUR CURTISS JAMES', LCl, my, 1933; (Golden Climber); bud long, pointed; flowers large, semi-dbl., 18 petals, non-recurrent, moderate fragrance; foliage glossy; vigorous, climbing growth; [Mary Wallace X Seedling]; Brownell, H.C.; J&P

'MRS ARTHUR E. COXHEAD', HT, dr, 1911; flowers claret-red, shaded brighter red, dbl., moderate fragrance; McGredy

'MRS ARTHUR JOHNSON', HT, yb, 1920; flowers rich orange-yellow to chrome-yellow, slight fragrance; McGredy

'MRS ARTHUR MOORE', HT, lp; [Victor Hugo X C.J. Graham]; Moore, A.K., 1911; (Weatherly, L.)

'MRS ARTHUR ROBERT WADDELL', HT, pb, 1909; bud long, pointed; flowers reddish salmon, reverse rosy scarlet, fading, open, large, semi-dbl., slight, apricot fragrance; vigorous growth; Pernet-Ducher

'MRS ATLEE', HT, op, 1926; flowers silvery pink, shaded soft salmon, moderate fragrance; Chaplin Bros.

'MRS B. STORY', T, lp; [Mme Lombard X Unknown]; Williams, A., 1905; (Weatherly, L.)

'MRS B.R. CANT', T, mp, 1901; (Mrs Benjamin R. Cant); flowers silvery rose, base suffused buff, reverse deep rose, dbl., cupped, moderate fragrance; vigorous growth; Cant, B. R.

'MRS B.R. CANT, CLIMBING', Cl T, mp, 1960; Hjort

Mrs Baker, HP, mr; Turner, 1876

'MRS BAYARD THAYER', HT, pb, 1915; flowers clear silvery pink, reverse deep rose; [Mrs Charles E. Russell sport]; Waban Conservatories

'MRS BEATTY', HT, my, 1926; flowers marechal niel yellow, well formed, moderate fragrance; foliage bronze; vigorous growth; GM, NRS, 1925; Cant, B. R.

'MRS BECKWITH', HT, my, 1922; bud long, pointed; flowers buttercup-yellow, edged lighter, open, semi-dbl., slight fragrance; vigorous growth; GM, NRS, 1923; Pernet-Ducher

'MRS BELMONT TIFFANY', HT, yb, 1918; flowers golden yellow, base apricot-orange; [Sunburst sport]; Budlong; A.N. Pierson

Mrs Benjamin R. Cant *see* **'MRS B.R. CANT'**

'MRS BERTRAM J. WALKER', HT, dp, 1915; flowers bright cerise-pink, dbl.; GM, NRS, 1914; Dickson, H.

Mrs Billy Crick, LCl, mp, 1995; Scarman

'MRS BLAMIRE YOUNG', HT, op, 1932; flowers salmon pink, slightly more petals than parent; [Una Wallace sport]; Young

Mrs Bosanquet *see* **'MISTRESS BOSANQUET'**

'MRS BREEDLOVE', HT, pb, 1947; bud ovoid; flowers pink, base yellow, very large, very dbl., moderate fragrance;

foliage leathery, glossy, dark; vigorous, bushy growth; [Golden Dawn sport]; Breedlove; Breedlove Nursery

'MRS BROWNELL', HT, rb, 1942; flowers red to pink and coppery orange, medium, semi-dbl., moderate, spicy fragrance; Brownell, H.C.

'MRS BRYCE ALLAN', HT, dp, 1916; flowers rose-pink, very dbl., intense fragrance; GM, NRS, 1916; Dickson, A.

'MRS BULLEN', HT, rb, 1917; flowers crimson, shaded yellow, passing to carmine, dbl.; Pernet-Ducher

'MRS C. V. HAWORTH, CLIMBING', Cl HT, ab, 1932; flowers dark apricot with pink, large, semi-dbl., intense fragrance; Cant, F.; (Sangerhausen)

'MRS C.E. BERNAYS', HT, lp; [Mme Lombard X Unknown]; Williams, A., 1905; (Weatherly, L.)

'MRS C.E. PRELL', HT, rb, 1938; bud long, pointed; flowers dark cerise, reverse pink, stamens golden, very large, dbl.; long stems; very vigorous growth; [Gustav Grunerwald X Betty Uprichard]; Fitzhardinge; Hazlewood Bros.

Mrs C.J. Bell *see* **'MRS CHARLES BELL'**

'MRS C.L. FITZGERALD', F, pb, 1937; bud pointed; flowers deep pink, base apricot-yellow, small, dbl., slight fragrance; foliage glossy; upright growth; [Unnamed seedling X Cécile Brunner]; Fitzgerald

'MRS C.V. HAWORTH', HT, ab, 1919; flowers cinnamon-apricot, passing to buff, semi-dbl., intense fragrance; [Mrs Wemyss Quin X Hugh Dickson]; Dickson, A.

'MRS C.W. DUNBAR-BULLER', HT, dp, 1919; flowers deep rosy carmine, large, dbl., intense fragrance; Dickson, A.

'MRS C.W. EDWARDS', HT, rb, 1924; bud pointed; flowers crimson-carmine, base yellow, reverse veined yellow, over large, dbl., exhibition form, moderate fragrance; foliage dark, glossy; long stems; very vigorous growth; McGredy

'MRS C.W. THOMPSON', HMult, dp, 1920; flowers deep pink, quilled, small, borne in clusters, dbl., good seasonal bloom; vigorous, climbing growth; U.S. Dept. of Agric.; Storrs & Harrison Co.

'MRS CALVIN COOLIDGE', HT, yb, 1924; bud long, pointed; flowers golden yellow, deepening to rich orange, semi-dbl., moderate fragrance; [Ophelia sport]; U.S. Cut Flower Co.; F.R. Pierson

'MRS CAMPBELL HALL', T, ab, 1914; flowers soft creamy buff, edged or suffused rose, center warm salmon, exhibition form, intense fragrance; foliage leathery, dark; vigorous growth; Hall; A. Dickson

Mrs Cecily McMullen *see* 'BOSEMWINE'

'MRS CHAPLIN', HT, pb, 1918; flowers creamy pink, base shaded yellow; Chaplin Bros.

'MRS CHARLES BELL', HT, op, 1917; (Mrs C.J. Bell, Salmon Radiance, Shell-Pink Radiance); flowers shell-pink, shaded soft salmon; [Red Radiance sport]; Bell, Mrs. C.J.; A.N. Pierson

'MRS CHARLES BELL, CLIMBING', Cl HT, op, 1929; Thomasville Nursery

'MRS CHARLES CURTIS-HARRISON', HT, m, 1910; flowers purple/pink, large, dbl., moderate fragrance; Dickson, A.; (Sangerhausen)

'MRS CHARLES E. PEARSON', HT, ob, 1913; flowers orange-apricot, flushed fawn and yellow, well formed, moderate fragrance; GM, NRS, 1912; McGredy

'MRS CHARLES E. RUSSELL', HT, dp, 1914; bud long, pointed; flowers rosy carmine, large, dbl., globular, moderate fragrance; foliage leathery; long stems; vigorous growth; [Mme Caroline Testout X (Mme Abel Chatenay X Marquise Litta de Breteuil)]; Montgomery, A.; Waban Conservatories

Mrs Charles E.F. Gersdorff *see* **'MAYLINA'**

'MRS CHARLES H. RIGG', HT, my, 1946; bud long, pointed; flowers lemon-yellow, large, dbl., moderate fragrance; vigorous growth; McGredy

'MRS CHARLES HUNTER', HT, dp, 1912; flowers carmine-pink, large, dbl., intense fragrance; Paul, W.; (Sangerhausen)

'MRS CHARLES LAMPLOUGH', HT, ly, 1920; flowers pale lemon-yellow, very large, dbl., moderate fragrance; vigorous growth; GM, NRS, 1919; [Frau Karl Druschki X ?]; McGredy

'MRS CHARLES REED', HT, w, 1914; flowers pale cream, tinted deep peach, base soft golden yellow, dbl., intense fragrance; Hicks

'MRS CHARLES STEWARD', HT, mp, 1959; flowers bright pink, large, dbl., intense fragrance; foliage dark; vigorous growth; Verschuren; Gandy Roses, Ltd.

'MRS CHARLES TENNANT', HT, yb, 1936; flowers clear primrose, shaded rich canary-yellow, moderate fragrance; foliage bronze; vigorous growth; Cant, F.

'MRS CHARLOTTE GUILFOYLE', HT, lp; [M. Berard X Unknown]; Johnson, 1885; (Weatherly, L.)

'MRS CLAUDE AVELING', HT, rb, 1929; flowers scarlet-cerise, tinted orange, base buttercup-yellow, anther, semi-dbl., moderate, fruity fragrance; [The Queen Alexandra Rose X Gorgeous]; Bees

'MRS CLEMENT YATMAN', HT, dr, 1927; flowers deep crimson, dbl., intense fragrance; Hicks

'MRS COCKER', HP, mp, 1899; flowers large, dbl., intense fragrance; Cocker; (Sangerhausen)

'MRS COLVILLE', HSpn, m; flowers bright crimson-purple, white eye, single; vigorous (4 ft) growth; [Probably R. spinosissima X R. pendulina]

'MRS CORNWALLIS WEST', HT, w, 1911; flowers white, center blush, very large, imbricated, globular; vigorous growth; GM, NRS, 1910; Dickson, A.

'MRS COURTNEY PAGE', HT, ob, 1922; bud long, pointed; flowers orange-cerise, shaded carmine, very large, dbl., exhibition form, moderate fragrance; strong stems; GM, NRS, 1922; McGredy

'MRS CURNOCK SAWDAY', HT, mp, 1920; flowers satiny pink, dbl., slight fragrance; Hicks

'MRS D. A. KOSTER', Pol, 1934; Koster, D.A.; (Cavriglia)

'MRS DAN PROSSER', HT, rb, 1946; flowers red shaded gold, dbl., 60 petals, 5.5 in., exhibition form; foliage glossy, dark red when young; [Mrs Sam McGredy X Heinrich Wendland]; Prosser

'MRS DAVID BAILLIE', HT, dp, 1912; flowers carmine, penciled deeper, dbl., moderate fragrance; Dickson, H.

'MRS DAVID MCKEE', HT, ly, 1904; flowers creamy yellow, well formed, large; dwarf, compact growth; [Frau Karl Druschki X Kaiserin Auguste Viktoria]; Dickson, A.

Mrs DeGraw *see* **'CHAMPION OF THE WORLD'**

Mrs Doreen Pike *see* 'AUSDOR'

'MRS DOUGLAS COPLAND', HT, mp, 1945; dbl.; vigorous growth; Clark, A.

'MRS DUDLEY CROSS', T, yb; (Dudley Cross); flowers pale yellow, tinted in autumn with crimson, dbl., slight fragrance; no prickles; vigorous growth; Paul, W., 1908

'MRS DUDLEY FULTON', Pol, w, 1931; flowers silver-white, large, borne in large clusters, single; dwarf growth; (14); [Dorothy Howarth X Perle d'Or]; Thomas; Armstrong Nursery

'MRS DUNLOP BEST', HT, ab, 1916; (Cleveland II); bud long, pointed; flowers reddish apricot, base coppery yellow, large, dbl., 28 petals, moderate fragrance; foliage leathery, bronze, glossy; vigorous growth; Hicks

'MRS DUNLOP BEST, CLIMBING', Cl HT, ab, 1933; Rosen, L.P.

'MRS E. ALFORD', HT, lp, 1913; flowers large, dbl.; Lowe; (Sangerhausen)

'MRS E. CLAXTON', HT, pb, 1928; flowers light pink, shaded salmon and carmine; Cant, F.

'MRS E. GALLAGHER', HT, dr, 1924; flowers dark crimson, moderate fragrance; McGredy

'MRS E. TOWNSHEND', HT, 1911; Guillot; (Cavriglia)

'MRS E. WILLIS', HT, lp; semi-dbl., moderate fragrance; growth moderate; [Mme Segond Weber seedling]; Weightman

'MRS E. WILLIS', HT, op, 1923; flowers light salmon-pink, large, dbl., moderate fragrance; Poulsen; (Sangerhausen)

'MRS E. WILLIS, CLIMBING', Cl HT, lp, 1948; Wilson

'MRS E. WOOD', HT, yb, 1934; bud very long; flowers light buff-yellow, tinted yellowish salmon, becoming cream-yellow, intense fragrance; vigorous growth; Dickson, A.

'MRS E.G. HILL', HT, w, 1906; flowers alabaster-white, reverse rose-coral, well shaped, large, dbl., moderate fragrance; vigorous growth; [Mme Caroline Testout X Liberty]; Soupert & Notting

'MRS E.J. HUDSON', HT, mp, 1923; flowers bright pink, dbl., slight fragrance; [Mrs W.J. Grant seedling]; Lilley

'MRS E.J. MANNERS', HT, dr, 1938; flowers deep velvety crimson, becoming darker, well shaped; long, strong stems; vigorous, branching growth; Burbage Nursery

'MRS E.M. GIBSON', HT, dr, 1940; [Countess of Stradbrooke X ?]; Clark, A.

'MRS E.M. GILMER', HT, mr, 1927; (Dorothy Dix); [Seedling X Crusader]; Cook, J.W.

'MRS E.T. STOTESBURY', HT, pb, 1918; flowers light cream-pink, reverse dark pink, very dbl.; [(Joseph Hill X My Maryland) X Milady]; Towill

'MRS E.W. STERLING', HT, mp, 1916; flowers rose-pink; [Antoine Rivoire X Unnamed pink seedling]; Cook, J.W.

'MRS EDITH STANLEY', HT, w, 1919; flowers creamy white, shaded indian yellow; Easlea

'MRS EDWARD J. HOLLAND', HT, op, 1909; flowers salmon-rose, large, moderate fragrance; moderately vigorous growth; GM, NRS, 1909; McGredy

'MRS EDWARD LAXTON', HT, ob, 1935; flowers flaming orange and old-rose, large, very dbl., exhibition form; foliage leathery, dark; very vigorous growth; [Mrs Henry Bowles X Shot Silk]; Laxton Bros.

'MRS EDWARD POWELL', HT, dr, 1911; flowers velvety crimson, dbl., moderate fragrance; Bernaix, P.

'MRS ELISHA HICKS', HT, w, 1919; flowers flesh, nearly white, dbl., intense fragrance; [Frau Karl Druschki X Mme Gabriel Luizet]; Hicks

'MRS ERSKINE PEMBROKE THOM', HT, my, 1926; bud long, pointed; flowers clear yellow, well-formed, large, dbl., 40 petals, moderate fragrance; vigorous growth; [Grange Colombe X Souv. de Claudius Pernet]; Howard, F.H.; H&S

'MRS ERSKINE PEMBROKE THOM, CLIMBING', Cl HT, my, 1933; Dixie Rose Nursery

'MRS EVELINE GANDY', HT, dr, 1959; flowers dark velvety crimson-scarlet, well-formed, large, dbl., 50 petals, intense fragrance; Verschuren; Gandy Roses, Ltd.

'MRS F. MILLAR', HT, lp; Williams, A., 1904; (Weatherly, L.)

'MRS F.F. PRENTISS', LCl, lp, 1925; flowers pale pink, large, dbl.; vigorous, climbing growth; extremely hardy.; [(R. setigera X R. wichurana) X Lady Alice Stanley]; Horvath

'MRS F.J. JACKSON', HT, mr, 1933; bud pointed; flowers cerise, dbl., slight fragrance; foliage leathery, bronze; vigorous growth; LeGrice

'MRS F.J. KNIGHT', HT, mr, 1928; flowers velvety scarlet, moderate fragrance; [Lord Charlemont sport]; Knight, J.

'MRS F.R. PIERSON', HT, dr, 1926; bud pointed; flowers crimson, shaded scarlet, very large, dbl., intense fragrance; [Premier sport]; Pierson, F.R.

'MRS F.W. FLIGHT', LCl, dp, 1905; flowers rose-pink, open, borne in clusters, semi-dbl., non-recurrent; foliage large, rich green, soft; pillar (6-8ft.) growth; [Crimson Rambler X ?]; Cutbush

'MRS F.W. SANFORD', HP, lp, 1898; flowers pink tinged white; [Mrs John Laing sport]; Curtis

'MRS F.W. WOODROFFE', T, lp; Williams, A., 1904; (Weatherly, L.)

'MRS FARMER', HT, yb, 1918; bud pointed; flowers yellow, reverse reddish apricot, dbl., moderate fragrance; Pernet-Ducher

'MRS FOLEY HOBBS', T, pb, 1910; flowers soft ivory-white, edges tinged clear pink, dbl., moderate fragrance; vigorous, upright growth; Dickson, A.

'MRS FORDE', HT, pb, 1913; flowers carmine-rose on soft rose-pink, base chrome-yellow, dbl., moderate fragrance; GM, NRS, 1913; Dickson, A.

'MRS FRANCIS KING', HT, ly, 1934; flowers cream-white shaded straw-yellow, large, dbl., 80 petals, exhibition form, slight fragrance; foliage leathery; vigorous, bushy growth; [Lady Lilford X Leonard Barron]; Nicolas; J&P

'MRS FRANK GUTHRIE', HT, lp, 1923; flowers deep flesh in autumn, pale in summer, semi-dbl., moderate fragrance; foliage dark, leathery; [R. gigantea seedling X ?]; Clark, A.; Hazelwood Bros.

'MRS FRANK J. USHER', HT, yb, 1920; flowers rich yellow, edged rosy carmine, very dbl.; weak stems growth; [Queen Mary X Unnamed variety]; Dobbie

'MRS FRANK SCHRAMM', HT, pb, 1934; flowers bright glowing rose-pink, reverse slightly lighter, dbl., moderate fragrance; foliage leathery, dark; very large, long stems; very vigorous growth; [Briarcliff sport]; Schramm

'MRS FRANK SERPA', Pol, dp, 1954; flowers deep pink, borne in clusters; dwarf (18 in) growth; [Rouletii X China Doll]; Serpa; Port Stockton Nursery

'MRS FRANK VERDON', HT, my, 1935; flowers creamy yellow, very large, dbl., slight fragrance; foliage leathery, dark; vigorous growth; [Joan Horton X Marion Horton]; Bees

'MRS FRANK WORKMAN', HT, dp, 1911; flowers bright rose-pink; Dickson, H.

'MRS FRANKLIN D. ROOSEVELT', HT, dy, 1933; bud long, pointed; flowers golden yellow, large, dbl., globular, moderate fragrance; foliage glossy; very vigorous growth; [Talisman sport]; Traendly & Schenck

'MRS FRANKLIN DENNISON', HT, w, 1915; bud long, pointed; flowers porcelain-white, veiled primrose-yellow, base ochre, very large, dbl., exhibition form, moderate fragrance; McGredy

'MRS FRED COOK', HT, ob, 1920; flowers light terra-cotta, edged silvery white, dbl.; Easlea

'MRS FRED DANKS', HT, m, 1951; bud long, pointed; flowers pink tinted lilac, large, semi-dbl., 15 petals, moderate fragrance; foliage leathery; very vigorous, upright, pillar growth; Clark, A.; NRS Victoria

'MRS FRED H. HOWARD', HT, ob, 1926; flowers orange-apricot edged straw-yellow, dbl., intense fragrance; Dobbie

'MRS FRED L. LAINSON', HT, pb, 1934; bud long, pointed; flowers deep pink, almost red, base yellow and bronze, reverse orange, dbl., moderate fragrance; foliage leathery, glossy, bronze; vigorous growth; [Talisman sport]; Scittine; Lainson

'MRS FRED POULSOM', HT, mp, 1920; flowers vivid pink; thorny; vigorous growth; [Edith Part X Unnamed variety]; Therkildsen

'MRS FRED SEARL', HT, lp, 1917; flowers very large, dbl., moderate fragrance; Dickson, A.; (Sangerhausen)

'MRS FRED STRAKER', HT, yb, 1910; flowers light orange-yellow, medium, dbl., moderate fragrance; Dickson, A.; (Sangerhausen)

'MRS FREDERICK W. VANDERBILT', HT, or, 1912; flowers deep orange-red, shaded apricot, dbl.; GM, NRS, 1913; McGredy

'MRS G. A. WHEATCROFT', HT, lp, 1926; flowers light creamy pink, medium, dbl., moderate fragrance; Wheatcroft; (Sangerhausen)

'MRS G. PAYNE', T, lp; Williams, A., 1911; (Weatherly, L.)

Mrs G.A. van Rossem *see* **'MEVROUW G. A. VAN ROSSEM'**

Mrs G.A. van Rossem, Climbing *see* **'MEVROUW G. A. VAN ROSSEM, CLIMBING'**

'MRS G.A. WHEATCROFT', HT, pb, 1926; flowers coppery pink to silver-rose at tips, reverse soft salmon-pink, dbl., intense fragrance; [Lady Pirrie sport]; Wheatcroft Bros.

'MRS G.M. SMITH', HT, dr, 1935; bud long, pointed; flowers deep crimson, dbl., cupped, slight fragrance; vigorous growth; [Red-Letter Day X Mrs J.J. Hedley-Willis]; Bees

'MRS GEORGE B. EASLEA', HT, dp, 1939; flowers sparkling carmine-pink, well formed, very large, exhibition form, intense fragrance; strong stems; Easlea

'MRS GEORGE C. THOMAS', HMsk, op, 1925; flowers salmon-pink, center orange, semi-dbl., 2–2.5 in., repeat bloom, moderate fragrance; climbing growth to 10ft.; [Mme Caroline Testout, Climbing X Moonlight]; Thomas; B&A

'MRS GEORGE GEARY', HT, ob, 1929; bud pointed; flowers orange-cerise, shaded cardinal, very large, dbl., 35–40 petals, exhibition form, intense fragrance; vigorous growth; [Red-Letter Day X Mrs Wemyss Quin]; Burbage Nursery

'MRS GEORGE MARRIOTT', HT, w, 1918; bud long, pointed; flowers deep cream and pearl, suffused rose, very large, dbl., exhibition form, moderate fragrance; GM, NRS, 1917; McGredy

'MRS GEORGE SHAWYER', HT, pb, 1911; flowers rosy pink, reverse pale pink, very large, dbl., slight fragrance; [Mme Hoste X Joseph Lowe]; Lowe & Shawyer

'MRS GEORGE SHAWYER, CLIMBING', Cl HT, pb, 1918; Lindquist, E.J.

'MRS GEORGIA CHOBE', HT, lp, 1937; bud long; flowers large, dbl., exhibition form; foliage leathery, light; vigorous growth; [Miss Rowena Thom X Renault]; H&S

'MRS GRAHAM HART', Cl T, lp; Williams, A., 1900; (Weatherly, L.)

'MRS H. COBDEN TURNER', HT, rb, 1948; flowers cherry-cerise flushed orange, base yellow, dbl., 26 petals, 3–4 in., intense fragrance; foliage glossy; moderately vigorous growth; [Ophelia X Unnamed seedling]; Mee; Fryer's Nursery, Ltd.

'MRS H. J. HEDLEY-WILLIS', HT, dr, 1930; flowers medium, semi-dbl., intense fragrance; Bees of Chester; (Sangerhausen)

'MRS H.D. GREENE', HT, rb, 1918; flowers reddish bronze, becoming flame and coppery pink, moderate fragrance; [Joseph Hill sport]; Easlea

'MRS H.G. JOHNSTONE', HT, pb, 1930; bud long; flowers rose-pink, base and edges rose, very large, dbl., exhibition form, slight fragrance; vigorous growth; [Mme Caroline Testout X Mrs George Shawyer]; Bees

'MRS H.L. WETTERN', HT, mp, 1922; flowers vivid pink, dbl., moderate fragrance; McGredy

'MRS H.M. EDDIE', HT, w, 1932; flowers creamy white, passing to purest white, dbl., 40–45 petals, 5–6 in., exhibition form, slight fragrance; foliage dark, leathery, glossy; vigorous, bushy growth; [Mrs Charles Lamplough X Mev. G.A. van Rossem]; Eddie

'MRS H.M. EDDIE, CLIMBING', Cl HT, w, 1944; Eddie

'MRS H.P. ABBOT', T, lp; [Mme Lombard X Unknown]; Williams, A., 1901; (Weatherly, L.)

'MRS H.R. DARLINGTON', HT, ly, 1919; bud long, pointed; flowers clear creamy yellow, well formed, balling in wet weather, moderate fragrance; long stems; vigorous growth; GM, NRS, 1919; McGredy

Mrs Harkness *see* **'PAUL'S EARLY BLUSH'**

'MRS HAROLD ALSTON', Cl HT, mp, 1940; Clark, A.

'MRS HAROLD BIBBY', HT, pb, 1936; flowers soft pink veined red, outer petals silvery pink, moderate fragrance; foliage dark; Bees

'MRS HAROLD BROCKLEBANK', HT, w, 1907; flowers creamy white, center buff, base soft yellow, dbl., moderate fragrance; Dickson, A.

'MRS HAROLD BROOKES', HT, mr, 1931; flowers very bright red, large, dbl., cupped, moderate fragrance; foliage light; vigorous, bushy growth; [Frau Oberhofgartner Singer X Firebrand]; Clark, A.; NRS Victoria

'MRS HENRI DAENDELS', HT, ab, 1931; (Merrouw Daendels); flowers apricot shaded orange, reverse violet-pink, large, very dbl., intense fragrance; foliage bronze; strong stems; very vigorous growth; [Mrs Henry Bowles X Rev. F. Page-Roberts]; Buisman, G. A. H.

'MRS HENRI DAENDELS, CLIMBING', Cl HT, ab, 1950; Buisman, G. A. H.

'MRS HENRY BALFOUR', HT, pb, 1918; flowers ivory-white, base primrose, edge penciled rose, like a picotee, moderate fragrance; McGredy

'MRS HENRY BOWLES', HT, mp, 1921; flowers rosy-pink, flushed salmon, well-formed, large, dbl., 50 petals, exhibition form; foliage dark, glossy; vigorous growth; [Lady Pirrie X Gorgeous]; Chaplin Bros.

'MRS HENRY BOWLES, CLIMBING', Cl HT, mp, 1929; Dobbie

'MRS HENRY MORSE', HT, pb, 1919; bud long, pointed; flowers cream, tinted rose, marked and veined red, large, dbl., exhibition form, moderate fragrance; dwarf growth; GM, NRS, 1919; McGredy

'MRS HENRY MORSE, CLIMBING', Cl HT, pb, 1929; Chaplin Bros.

'MRS HENRY WINNETT', HT, dr, 1917; bud long, pointed; flowers deep rich red, large, dbl., exhibition form, moderate fragrance; foliage leathery; vigorous growth; [Mrs Charles E. Russell X Mrs George Shawyer]; Dunlop

'MRS HENRY WINNETT, CLIMBING', Cl HT, dr, 1930; Bernaix, P.

'MRS HERBERT CARTER', HT, ab, 1934; flowers apricot-yellow, center deeper, veined bronze, well formed, l, dbl., moderate fragrance; vigorous growth; Cant, F.

'MRS HERBERT DOWSETT', HT, pb, 1928; flowers several shades deeper, otherwise similar to parent; [Los Angeles sport]; Easlea

'MRS HERBERT HAWKSWORTH', T, r, 1912; flowers ecru on milk-white, dbl., moderate fragrance; Dickson, A.

'MRS HERBERT HOOVER', HT, mr, 1928; bud long, pointed; flowers rich velvety red, dbl., intense fragrance; foliage dark, leathery; vigorous growth; [Ophelia X Hoosier Beauty]; Coddington

'MRS HERBERT NASH', HT, dr, 1925; flowers scarlet-crimson shaded deep crimson, overlarge, moderate fragrance; long, strong stems; vigorous growth; Chaplin Bros.

'MRS HERBERT STEVENS', HT, w, 1910; bud long, pointed; dbl., exhibition form, moderate fragrance; foliage light; vigorous, bushy growth; [Frau Karl Druschki X Niphetos]; McGredy

'MRS HERBERT STEVENS, CLIMBING', Cl HT, w, 1922; (Grimpant Mrs Herbert Stevens, Stevens, Climbing); Pernet-Ducher

'MRS HILTON BROOKS', HT, yb, 1929; flowers saffron-yellow, base deeper, suffused pinkand carmine, dbl., moderate fragrance; Cant, F.

Mrs Holz Climber #1, LCl, m

Mrs Holz climber #2, LCl, lp

'MRS HORNBY LEWIS', HT, ob, 1921; flowers orange-yellow, very dbl., moderate fragrance; [Gorgeous X Mme Mélanie Soupert]; Hicks

'MRS HOVEY', HSet, w, 1850; flowers blush to almost white; [R. setigera X ?]; Pierce

'MRS HUGH DETTMANN', Cl HT, ab, 1930; flowers bright apricot-yellow; good pillar growth; Clark, A.; NRS Victoria

'MRS HUGH DICKSON', HT, yb, 1915; flowers deep cream, heavily suffused orange and apricot, dbl., 44 petals; GM, NRS, 1916; Dickson, H.

'MRS INGE POULSEN', F, pb, 1949; (Fru Inge Poulsen); flowers pink with a yellow center, open, medium, semi-dbl., slight fragrance; foliage light, matt; vigorous, compact (80 cms) growth; [Poulsen's Pink X Seedling]; Poulsen, S.; McGredy, 1952;, Poulsen, 1949

Mrs Iris Clow *see* 'HARBRITE'

'MRS J. HEATH', HT, yb, 1924; flowers maize-yellow, tinted peach-red, center yellow, dbl., moderate fragrance; McGredy

'MRS J. WYLIE', HT, lp, 1923; flowers silvery blush-pink; Dickson, H.

'MRS J.C. AINSWORTH', HT, mp, 1918; flowers rose-pink, very dbl., moderate fragrance; [Mrs Charles E. Russell sport]; Clarke Bros.

'MRS J.C. MANIFOLD', HP, lp; [Prince Camille de Rohan X Unknown]; Adamson, 1915; (Weatherly, L.)

'MRS J.D. EISELE', HT, pb, 1933; bud long, pointed; flowers brilliant cherry-pink, center shaded scarlet, very large, dbl., intense fragrance; vigorous growth; [Premier Supreme X McGredy's Scarlet]; Howard, F.H.; H&S

'MRS J.D. RUSSELL', HT, dr, 1930; flowers deep crimson, center maroon, almost black in certain lights, dbl., cupped, intense fragrance; [Prince Camille de Rohan X Mrs Aaron Ward]; Bees

'MRS J.F. REDLY', HT, lp; flowers pale flesh-pink, center tinted salmon, large, dbl., 40 petals, lightly repeats in late summer.; vigorous growth

'MRS J.J. HEDLEY-WILLIS', HT, dr, 1929; bud pointed; flowers dark crimson, center almost plum-black, dbl., exhibition form, intense fragrance; [Admiral Ward X Richmond]; Bees

'MRS J.T. MCINTOSH', HT, ab, 1935; flowers creamy apricot, center deeper apricot, base golden yellow, dbl.; vigorous growth; McIntosh; Brundrett

'MRS JAMES GARNER', HT, ab, 1931; flowers buff, base orange, moderate fragrance; Cant, F.

'MRS JAMES LYNAS', HT, pb, 1914; flowers pearly pink, reverse and edges flushed rosy peach, dbl., moderate fragrance; GM, NRS, 1913; Dickson, H.

'MRS JAMES SHEARER', HT, w, 1923; flowers pure white, base yellow, large, dbl., exhibition form, moderate fragrance; [Unnamed seedling X Mme Colette Martinet]; Ferguson, W.

'MRS JAMES WILLIAMSON', HT, mp, 1922; flowers clear pink, dbl., moderate fragrance; Dickson, H.

'MRS JEANNETTE G. LEEDS', HT, rb, 1942; bud globular, jasper-red and apricot-buff; flowers venetian pink, dbl., 50–60 petals, 5–6 in., slight fragrance; foliage dark, leathery; very vigorous, upright growth; [Joanna Hill X R.M.S. Queen Mary]; Hill, Joseph H., Co.

'MRS JENNIE DEVERMAN', HT, pb, 1933; flowers cerise edged silvery, base tinted gold; [Pres. Herbert Hoover sport]; Deverman

'MRS JOHN BELL', HT, dr, 1928; bud long; flowers carmine, large, dbl., cupped, moderate fragrance; Bell, Judy G.; Dobbie

'MRS JOHN COOK', HT, w, 1920; bud long, pointed; flowers white, suffused soft pink, deepening in cool weather, large, dbl., cupped, moderate fragrance; [Ophelia X Unnamed seedling]; Cook, J.W.; A.N. Pierson

'MRS JOHN FOSTER', HT, dr, 1915; flowers large, dbl., moderate fragrance; Hicks; (Sangerhausen)

'MRS JOHN INGLIS', HT, dr, 1920; flowers rich crimson, dbl., intense fragrance; McGredy

'MRS JOHN LAING', HP, mp, 1887; bud pointed; flowers soft pink, large, dbl., 45 petals, recurrent bloom, moderate fragrance; foliage light; vigorous, rather dwarf growth; (28); [Francois Michelon X ?]; Bennett

'MRS JOHN MCNABB', S, w, 1941; bud with very long sepals; very dbl., non-recurrent, moderate fragrance; few prickles; foliage large, dark, slightly rugose, prickly underneath; height 5 ft; [R. beggeriana X R. rugosa]; Skinner

'MRS JOHN R. ALLAN', HT, lp, 1920; flowers soft rosy pink, reverse darker, dbl.; GM, NRS, 1920; Dickson, H.

Mrs Jones *see* 'DELGE'

'MRS JOSEPH H. WELCH', HT, mp, 1911; flowers brilliant rose-pink, large, dbl., slight fragrance; vigorous growth; GM, NRS, 1910; McGredy

'MRS JOSEPH H. WELCH, CLIMBING', Cl HT, mp, 1922; Perkins, H.S.

'MRS JOSEPH HIESS', Pol, lp, 1943; flowers mary wallace pink, base white, borne in clusters of 3-16, dbl., 40 petals, cupped, slight fragrance; foliage leathery; strong stems; vigorous, upright, bushy, compact growth; [Roserie X ?]; Shepherd; Klyn

Mrs K.B. Sharma, HT, w, 1989; Pal, Dr. B.P.

Mrs Keays' Red Damask, HCh, mr

Mrs Keays' The Bishop, HCh, m

'MRS L.B. CODDINGTON', HT, mp, 1931; bud long, pointed; flowers large, dbl., moderate fragrance; foliage leathery; vigorous growth; [Templar X Souv. de Claudius Pernet]; Coddington

'MRS L.B. COPELAND', F, pb, 1934; flowers salmon-pink, veined, base yellow, small, dbl., globular, intense fragrance; foliage dark, glossy; vigorous growth; [Unnamed seedling X Cécile Brunner]; Fitzgerald

'MRS LESLIE MOSS', Cl HT, dp, 1944; flowers carmine-pink, medium, semi-dbl.; foliage leathery; vigorous, climbing growth; Moss; Mason, F.

Mrs Lincoln, HT, dr, 1997; Williams, J. Benjamin

'MRS LOVELL SWISHER', HT, pb, 1926; bud pointed; flowers salmon-pink, edged flesh, well-formed, dbl., moderate fragrance; foliage bronze; very vigorous growth; [Unnamed seedling X Souv. de Claudius Pernet]; Howard, F.H.; H&S

'MRS LOVELL SWISHER, CLIMBING', Cl HT, pb, 1930; H&S

'MRS LUTHER BURBANK', HT, mp, 1954; bud long, pointed; flowers rose-pink, dbl., 34 petals, 4–4.5 in., cupped, intense, spicy fragrance; foliage leathery; vigorous growth; [Christopher Stone X Charlotte Armstrong]; Swim, H.C.; Stark Bros.

'MRS M.H. HORVATH', HT, ly, 1940; bud long, pointed; flowers pale yellow, open, large, dbl., 40 petals; foliage glossy; long stems; very vigorous, upright growth; [(Mme Butterfly X ?) X Souv. de Claudius Pernet]; Horvath; Wayside Gardens Co.

'MRS M.H. WALSH', HWich, w, 1913; flowers pure snow-white, small, borne in clusters, dbl.; foliage large, glossy; very vigorous, trailing growth; Walsh

'MRS M.J. GILLON', HT, dr, 1974; bud ovoid; flowers red-purple, dbl., 37 petals, 4 in., globular, moderate fragrance; foliage glossy, dark; upright growth; [Tropicana X Prima Ballerina]; Rijksstation Voor Sierplantenteelt

'MRS MABEL V. SOCHA', HT, my, 1935; bud pointed; flowers pure lemon-yellow, dbl., intense fragrance; long, strong stems; [Seedling X Souv. de Claudius Pernet]; H&S

'MRS MACDONALD'S ROSE', HRg, dp;, non-recurrent; [Apparently R. rugosa plena X R. acicularis]; Reid; P.H. Wright

'MRS MACKELLAR', HT, my, 1915; flowers deep citron or pure canary, passing to primrose, dbl., moderate fragrance; Dickson, A.

'MRS MARY D. WARD', HT, pb, 1927; flowers shell-pink and gold, reverse ivory-white, moderate fragrance; [Double Ophelia X Souv. de Claudius Pernet]; Ward, F.B.

Mrs Mary Thomson *see* 'TOMONE'

Mrs Matso's Moss, M, dr

'MRS MATTHEWS', HT, lp; Matthews, W.J., 1923; (Weatherly, L.)

'MRS MAUD ALSTON', F, lp; Clark, A., 1940; (Weatherly, L.)

'MRS MAYNARD SINTON', HT, pb, 1909; flowers silvery white suffused pink, very large, moderate fragrance; moderately vigorous growth; GM, NRS, 1909; McGredy

'MRS MINA LINDELL', (form of R. macounii), lp, 1927; semi-dbl., 10–12 petals, non-recurrent; height to 4 ft; hardy.; N.E. Hansen; (found in S. Dakota)

'MRS MINIVER', HT, mr, 1944; (Souv de Louis Simon); flowers scarlet-crimson, reverse slightly darker, dbl., 20 petals, 5.5–6 in., cupped, moderate fragrance; foliage soft; vigorous, upright, bushy, compact growth; Chambard, C.; J&P

'MRS MONA HUNTING', HT, yb, 1916; flowers chamois-yellow, opening fawn, dbl.; Dickson, H.

'MRS MOORFIELD STOREY', HT, lp, 1915; flowers shell-pink, center deeper, dbl.; [Gen. MacArthur seedling X Joseph Hill]; Waban Conservatories

'MRS MUIR MACKEAN', HT, dr, 1912; flowers carmine-crimson, large, exhibition form, intense fragrance; vigorous growth; McGredy

'MRS MURRAY ALLISON', HT, pb, 1925; flowers rose-pink, base carmine, dbl., slight fragrance; Prior

'MRS MYLES KENNEDY', T, pb, 1906; flowers silvery white tinted buff, center and reverse pink, large, dbl.; vigorous growth; Dickson, A.

'MRS NIEMINEN', HT, rb, 1954; flowers blood-red shaded scarlet, well-formed, large, very dbl., slight fragrance; bushy, compact growth; [Hens Verschuren X Poinsettia]; Frères, Buyl

'MRS NORMAN WATSON', Cl HT, dp, 1930; flowers deep cherry-pink, large; very vigorous, pillar growth; [Radiance X Gwen Hash]; Clark, A.; Geelong Hort. Soc.

'MRS NORRIS M. AGNEW', HT, ob, 1934; bud pointed; flowers orange-cerise, large, dbl., slight fragrance; foliage leathery, light; very vigorous growth; [J.C. Thornton X Florence L. Izzard]; Bees

Mrs O.G. Orpen, HGal, dp

'MRS OAKLEY FISHER', HT, dy, 1921; flowers deep orange-yellow, blooms in clusters, single, moderate fragrance; foliage dark, bronze, glossy; vigorous growth; Cant, B. R.

'MRS OLIVE SACKETT', F, mr, 1931; flowers bright red, well formed, semi-dbl., 2–2.5 in.; foliage bronze in autumn; vigorous growth; [Else Poulsen sport]; Wirtz & Eicke; Spath

'MRS OLIVER AMES', HT, my, 1941; bud long, pointed; flowers lemon-yellow, dbl., 63 petals, 3.5–4 in., globular, intense fragrance; foliage leathery; long stems; vigorous growth; [Max Krause X Julien Potin]; Verschuren; Dreer

'MRS OLIVER MEE', HT, rb, 1948; flowers scarlet shaded gold, dbl., 35 petals, 5–6 in., exhibition form, moderate fragrance; foliage glossy, bronze; very vigorous growth; [Mrs Charles Lamplough X Edith Mary Mee]; Mee; Fryer's Nursery, Ltd.

'MRS OSWALD LEWIS', HT, yb, 1936; flowers soft canary-yellow, outer petals edged flame, well formed, dbl., intense fragrance; long stems; vigorous growth; GM, Bagatelle, 1935; Cant, F.

'MRS OSWALD SMEATON', HT, pb, 1932; flowers ivory-cream, center and petal tips pink, very large, dbl., 50–60 petals, moderate fragrance; long stems; vigorous growth; Easlea

'MRS PAT', Pol, lp, 1928; flowers small, dbl., moderate fragrance; Lilley; (Sangerhausen)

'MRS PAUL', B, lp, 1891; flowers blush-white shaded rosy peach, large; Paul & Son

'MRS PAUL GOUDIE', HT, yb, 1932; (Shining Sun); flowers deep buttercup-yellow, edged carmine-scarlet, very large, dbl., moderate, fruity fragrance; McGredy

Mrs Paul Hvid, B, w

'MRS PAUL J. HOWARD', Cl HT, rb, 1938; bud long, pointed; flowers brilliant crimson, reverse flame-red, dbl., 30 petals, 5 in., moderate, spicy fragrance; foliage large, bronze, leathery; long stems; very vigorous, climbing (12-15 ft) growth; [Miss Rowena Thom X Paul's Lemon Pillar]; Howard, F.H.; H&S

'MRS PAUL M. PIERSON', HT, pb, 1930; bud long, pointed; flowers soft pink, reverse brighter, very large, dbl., exhibition form, moderate fragrance; [Premier sport]; Pierson, P.M.

'MRS PAUL R. BOSLEY', HT, my, 1941; flowers apricot-yellow; [Mme Joseph Perraud sport]; Bosley Nursery

'MRS PERCY V. PENNYBACKER', HT, pb, 1929; flowers peach-pink shaded silver, dbl., cupped, moderate fragrance; [Mme Butterfly seedling]; Vestal

'MRS PETER BLAIR', HT, yb, 1906; flowers lemon-chrome, center golden yellow, well formed, large; GM, NRS, 1906; Dickson, A.

'MRS PHILIP RUSSELL', HT, dr, 1927; bud long, pointed; flowers dark red, shaded black, semi-dbl., slight fragrance; foliage glaucous green; vigorous, semi-climbing growth; pillar or large bush; [Hadley X Red Letter Day]; Clark, A.; Hackett

'MRS PIERCE', HSet, mp, 1850; flowers pink, well formed; Pierce

'MRS PIERRE S. DUPONT', HT, my, 1929; bud long, pointed, reddish gold; flowers golden yellow, becoming lighter, dbl., 40 petals, moderate, fruity fragrance; foliage rich green; moderate growth; GM, Bagatelle, 1929; [(Ophelia X Rayon d'Or) X (Ophelia X (Constance X Souv. de Claudius Pernet))]; Mallerin, C.; C-P

'MRS PIERRE S. DUPONT, CLIMBING', Cl HT, my, 1933; Hillock

'MRS PRENTISS NICHOLS', HT, dp, 1923; flowers brilliant deep pink, dbl., moderate fragrance; [Ophelia X Seedling]; Scott, R.

Mrs Price, Pol, pb

'MRS R. D. MCCLURE', HT, op, 1913; flowers salmon-pink, very large, dbl.; Dickson, H.; (Sangerhausen)

'MRS R.B. MCLENNAN', HT, pb, 1924; flowers satiny rose suffused yellow, very dbl., intense fragrance; [George C. Waud X Mme Caristie Martel]; Easlea

'MRS R.B. MOLONEY', HT, mr, 1925; flowers brilliant carmine-red, dbl., intense fragrance; McGredy

'MRS R.C. BELL', HT, mr, 1920; flowers bright red; [Gen. MacArthur X Château de Clos Vougeot]; Clark, A.

'MRS R.G. SHARMAN-CRAWFORD', HP, pb, 1894; flowers rosy pink, outer petals tinted flesh, large, dbl., 75 petals, cupped, recurrent bloom, moderate fragrance; vigorous growth; Dickson, A.

'MRS R.M. FINCH', Pol, mp, 1923; flowers rosy pink, becoming lighter, medium blooms in large clusters, dbl.; bushy growth; [Orléans Rose X ?]; Finch

'MRS R.M. FINCH, CLIMBING', Cl Pol, mp

'MRS R.M. KING', HT, lp; [Mme Abel Chatenay X W.R. Smith]; Harrison, A., 1927; (Weatherly, L.)

'MRS RAMON DE ESCOFET', HT, dr, 1919; flowers flame-crimson, large, dbl.; Easlea

'MRS REDFORD', HT, ab, 1919; flowers bright apricot-orange, semi-dbl., intense fragrance; foliage holly-like; GM, NRS, 1917; McGredy

'MRS REYNOLDS HOLE', T, pb, 1900; flowers carmine shaded purple rose, reverse carmine, large, dbl.; Nabonnand, P.&C.

'MRS RICHARD TURNBULL', LCl, w, 1945; flowers handsome stamens, very large, single; Clark, A.

'MRS RICHARDS', F, pb, 1968; flowers pink tinged apricot, blooms in trusses, single, moderate fragrance; foliage dark, glossy; [Ann Elizabeth X Circus]; Harkness

'MRS ROBERT BACON', HT, ab, 1934; flowers golden apricot shading to coral, base yellow, large, very dbl., cupped, intense fragrance; [Talisman sport]; Bertanzel

'MRS ROBERT GARRETT', HT, mp, 1900; [Comtesse de Caserta X Mme Eugene Verdier]; Cook, J.W.

'MRS ROBERT MITCHELL', HT, pb, 1926; flowers salmon-rose, overlaid coppery pink; [St. Helena X Mrs Redford]; Jersey Nursery

'MRS ROBERT PEARY', Cl HT, w, 1898; [Kaiserin Auguste Viktoria sport]; De Voecht & De Wilde; Dingee & Conard

'MRS ROSALIE WRINCH', LCl, lp, 1915; flowers shell-pink, large, semi-dbl.; vigorous, pillar (5-8 ft) growth; [Frau Karl Druschki X Hugh Dickson]; Brown, W.&J.

'MRS ROY GREEN', HT, dr, 1940; flowers very large; long stems; Clark, A.

'MRS RUSSELL GRIMWADE', T, mp, 1938; flowers fuchsia-pink; [Lorraine Lee sport]; Grimwade

'MRS S. PATON', HT, ob, 1928; bud long, pointed; flowers orange-carmine, base orange, large, dbl., slight fragrance; McGredy

Mrs S. Peters, HT, dp

'MRS S.K. RINDGE', HT, yb, 1919; bud long, pointed; flowers deep golden yellow, suffused soft pink with age, over large, semi-dbl., cupped, moderate fragrance; long stems; [Rayon d'Or X Frau Karl Druschki]; H&S

'MRS S.T. WRIGHT', T, ob, 1914; flowers old gold, center suffused rose-pink on orange, dbl., moderate fragrance; [Harry Kirk sport]; Dickson, A.

'MRS S.W. BURGESS', HT, ab, 1925; flowers apricot-yellow, base deeper, dbl., moderate fragrance; [Mme Mélanie Soupert X Joseph Hill]; Burgess, S.W

'MRS SAM MCGREDY', HT, op, 1929; bud pointed; flowers scarlet-copper-orange, reverse heavily flushed red, large, dbl., 40 petals, exhibition form, moderate fragrance; foliage glossy, reddish bronze; vigorous growth; GM, NRS, 1929 GM, Portland, 1956; [(Donald Macdonald X Golden Emblem) X (Seedling X The Queen Alexandra Rose)]; McGredy

'MRS SAM MCGREDY, CLIMBING', Cl HT, op, 1937; (Geneviève Genest); Buisman, G. A. H.

'MRS SAM MCGREDY, CLIMBING', Cl HT, ob, 1938; (Genevieve Genest); Guillaud

'MRS SAM MCGREDY, CLIMBING', Cl HT, ob, 1940; Western Rose Co.

'MRS SPENCER BROWNE', HT, lp; Harrison, A., 1928; (Weatherly, L.)

'MRS STEWART CLARK', HT, mp, 1906; flowers large, dbl., moderate fragrance; Dickson, A.; (Sangerhausen)

'MRS T. HILLAS', HT, my, 1913; flowers pure chrome-yellow, dbl.; Pernet-Ducher

'MRS T.B. DOXFORD', HT, rb, 1932; bud pointed; flowers salmon-carmine to peach-blossom-pink, reverse old-rose, large, dbl., exhibition form, slight fragrance; Dickson, A.

'MRS T.J. ENGLISH', HT, ab, 1922; flowers apricot and amber, tinted salmon-flesh, heavily veined, moderate fragrance; English

'MRS TALBOT O'FARRELL', HT, rb, 1926; flowers cerise flushed bronze, reverse old-gold, dbl., moderate fragrance; McGredy

'MRS THEODORE ROOSEVELT', HT, lp, 1903; flowers well shaped, very large, dbl., moderate fragrance; vigorous, bushy growth; [La France seedling]; Hill, E.G., Co.

'MRS THEODORE SALVESEN', HT, op, 1922; flowers salmon-pink; Dobbie

'MRS THEONVILLE VAN BERKEL', HT, pb, 1935; bud long, pointed; flowers pink, reverse flushed yellow, large, dbl., exhibition form, intense fragrance; foliage leathery, dark; vigorous growth; [Briarcliff X Mrs Sam McGredy]; Buisman, G. A. H.

'MRS TOM PAUL', HT, yb, 1920; flowers saffron-yellow, suffused pink, dbl.; Dickson, H.

'MRS TOM SMITH', HT, mr, 1924; flowers glowing cerise, dbl., moderate fragrance; Smith, T.

'MRS TOM WHITEHEAD', HT, ab, 1938; flowers cream, center apricot-orange, outer petals veined, very large, dbl., exhibition form; very vigorous growth; [Mrs Charles Lamplough X Unnamed seedling]; Whitehead; Beckwith

'MRS TRESHAM GILBEY', HT, op, 1923; bud pointed; flowers coral-rose shaded salmon, very large, dbl., moderate fragrance; GM, NRS, 1923; [Waltham Flame X Edith Cavell (HT)]; Chaplin Bros.

'MRS TRESHAM GILBEY, CLIMBING', Cl HT, op, 1938; flowers light salmon-pink, very large, dbl., moderate fragrance; Vogel, M.; (Sangerhausen)

'MRS U.M. ROSE', HT, dp, 1931; bud long, pointed; flowers cerise-pink, semi-dbl., cupped, intense fragrance; Vestal

'MRS VAN BERESTEYN-FROWEIN', HT, op, 1935; flowers salmon, large, very dbl., moderate fragrance; foliage leathery, light; vigorous growth; [Souv. de Claudius Pernet X Mrs Henry Bowles]; Buisman, G. A. H.

Mrs Van Nes *see* **'PERMANENT WAVE'**

'MRS VANDENBERGH', HT, mr, 1938; (Mevrouw S. van den Bergh, Jr.); flowers bright red, semi-dbl., exhibition form; foliage dark; strong stems; vigorous growth; [E.G. Hill X Étoile de Hollande]; Buisman, G. A. H.

Mrs Verschuren *see* **'R.M.S. QUEEN MARY'**

Mrs Villotti's Pink, Pol, lp

'MRS W. EWART', T, lp; [Souv de Catherine Guillot X Unknown]; Williams, A., 1909; (Weatherly, L.)

'MRS W.A. LINDSAY', HT, pb, 1920; flowers peach-pink, center golden yellow, dbl., moderate fragrance; Dickson, H.

'MRS W.E. NICKERSON', HT, pb, 1927; bud pointed; flowers silvery pink deeply shaded old-gold and salmon, large, dbl., moderate fragrance; McGredy; Dreer;, H&S

'MRS W.H. CUTBUSH', Pol, dp, 1907; flowers deep pink; [Mme Norbert Levavasseur sport]; Levavasseur

'MRS W.J. GRANT', HT, lp, 1895; (Belle Siebrecht); bud long, pointed; flowers light pink fading purplish, dbl., moderate fragrance; [La France X Lady Mary Fitzwilliam]; Dickson, A.

'MRS W.J. GRANT, CLIMBING', Cl HT, lp, 1899; (Belle Siebrecht, Climbing); Hill, E.G., Co.

'MRS W.R. GROVES', HT, dr, 1941; flowers deep red; foliage good; Clark, A.

'MRS WAKEFIELD CHRISTIE-MILLER', HT, pb, 1909; flowers blush, shaded salmon, reverse vermilion-rose, dbl.; foliage light, leathery; dwarf growth; McGredy

'MRS WALLACE H. ROWE', HT, m, 1912;, exhibition form; vigorous growth; McGredy

'MRS WALTER BRACE', HT, dp, 1939; flowers vivid cerise-rose-pink, slightly larger, of better shape; [Picture sport]; Beckwith

'MRS WALTER BURNS', F, mp, 1978; bud ovoid; flowers patio, very dbl., 2.5 in., flat; foliage fragrant (musky), medium, matt, dark; compact, bushy growth; Harkness

'MRS WALTER EASLEA', HT, dr, 1910; flowers crimson, large, dbl., moderate fragrance; Dickson, A.

'MRS WALTER JONES', HT, rb, 1930; flowers brilliant coral-red, shaded orange, dbl.; Cant, B. R.

'MRS WALTER T. SUMNER', HT, dp, 1920; flowers carmine to deep rose-pink, large, semi-dbl., 12–18 petals, intense fragrance; [Ophelia X Hadley]; Clarke Bros.

'MRS WALTER T. SUMNER, CLIMBING', Cl HT, dp, 1932; Hazlewood Bros.

'MRS WARREN E. LENON', HT, dr, 1924; bud long, pointed; flowers crimson, large, dbl., globular, intense fragrance; vigorous growth; [Hoosier Beauty X Premier]; Hill, E.G., Co.; Vestal

'MRS WARREN G. HARDING', HT, dp, 1923; [Columbia sport]; Pierson, A.N.

'MRS WEMYSS QUIN', HT, dy, 1914; dbl., moderate fragrance; bushy, branching growth; GM, Bagatelle, 1916; Dickson, A.

'MRS WHITMAN CROSS', Cl HT, ob, 1943; bud long, pointed; flowers orange-apricot, overlaid pinkish, reverse sometimes striped, semi-dbl., moderate fragrance; foliage glossy, soft; height 8-9 ft as climbing or pillar, upright.; [Nanjemoy X Marion Cran]; Cross, C.W.; B&A

'MRS WILLIAM C. EGAN', HT, lp, 1922; bud long, pointed; flowers soft pink, dbl., 35 petals, moderate fragrance; H&S

'MRS WILLIAM C. EGAN, CLIMBING', Cl HT, lp, 1933; Howard Rose Co.

'MRS WILLIAM FIFE', HT, pb, 1926; flowers soft rose-pink, flushed blush-pink, dbl., moderate fragrance; Dobbie

'MRS WILLIAM R. HEARST', HT, dp, 1915; flowers clear dark pink; [My Maryland sport]; Pierson, A.N.

'MRS WILLIAM SERGENT', HT, ab, 1923; flowers apricot and peach, edges flushed rose-pink, dbl.; Dickson, H.

'MRS WILLIAM SPROTT', HT, my, 1938; bud long, pointed; flowers large, dbl., slight fragrance; foliage glossy, bronze; vigorous, compact growth; [Portadown Glory X Mrs Sam McGredy]; McGredy

'MRS WM. G. KONING', Pol, w, 1917; flowers pure white, open, borne in clusters, dbl., slight fragrance; vigorous, bushy growth; [Louise Walter sport]; Kluis & Koning

Mrs Woods' Lavender-Pink Noisette, N, m

Ms Mary *see* 'TINMARY'

Mt Vernon Pompon, LCl, w

Mt. Hood *see* 'MACMOUHOO'

'MT. RAI', HT, w, 1987; flowers white, pale cream at center, aging pale pink at edges, urn-shaped, dbl., 42 petals, exhibition form, slight fragrance; medium, yellow-green fruit; prickles medium, downward-shaped, slightly dented, green; foliage dark green, semi-glossy; bushy, medium growth; [(Golden Scepter X Narcissus) X Ginger Rogers]; Ota, Kaichiro, 1988

'MT. ST HELENS', Gr, dp, 1981; bud long, pointed; flowers deep pink, urn-shaped blooms borne 1-3 per cluster, dbl., 35 petals, slight, spicy fragrance; bronze prickles; foliage dark, leathery; growth to 4-5 ft.; [Unnamed seedling X Queen Elizabeth]; Northwest Rose Growers

'MT. TARA', HT, or, 1987; flowers large, borne usually singly, dbl., 35 petals, exhibition form, slight fragrance; medium, yellow-green fruit; prickles medium, downward-shaped, slightly dented, green; foliage dark green, semi-glossy; bushy, medium growth; [Ginger Rogers X Duftzauber]; Ota, Kaichiro, 1988

'MUCABA', HT, pb, 1962; flowers pink stained carmine; [Marechal Carmona X Suzon Lotthe]; da Silva, Moreira

Muchacha® *see* 'MEILUTIDA'

Mudgee Red, HT, mr; Dot

'MÜHLE HERMSDORF', HWich, w, 1928; flowers pure white, all open at once, dbl.; [R. wichurana X Gruss an Zabern]; Dechant

'MUHME LÖFFLER', HT, dr, 1940; flowers large, dbl., intense fragrance; Krause; (Sangerhausen)

'MULBARTON', LCl, mr, 1965; dbl., 4.5 in., moderate fragrance; foliage coppery; vigorous growth; [Paul's Lemon Pillar X (Ena Harkness X Richmond)]; Hooney

Mulbry Rose, F, ob, 1998

Mullard Jubilee *see* **'ELECTRON'**®

Mullem, HT, op, 1997; 29 petals, 4 in., cupped, intense fragrance; foliage dense dark green; RvS-Melle, 1996

Multiflora Japonica *see* **R. MULTIFLORA**

'MULTIFLORA NANA PERPÉTUELLE', Pol, lp, 1893; flowers small, single; Lille; (Sangerhausen)

Multiflore de Vaumarcus, N, lp, 1875; Menet

'MULTNOMAH', HT, rb, 1948; bud long, pointed; flowers carmine, base gold, open, semi-dbl., 18–22 petals, 4–5 in., slight fragrance; foliage glossy, dark; vigorous, upright, compact growth; hardy in Pacific Northwest.; [Contrast X Charlotte Armstrong]; Swim, H.C.; Peterson & Dering

'MUM MUM', Min, w, 1986; flowers pure white, small blooms in clusters of 3-25, dbl., 55 petals, cupped, no fragrance; short, curved, light green prickles; foliage light green, semi-glossy; low, upright, bushy, spreading growth; [Unnamed seedling X Unnamed seedling]; McDaniel, Earl; McDaniel's Min. Roses

'MÜNCHEN', HMsk, dr, 1940; bud long, pointed; flowers scarlet-crimson, large blooms in clusters, semi-dbl., repeat bloom, slight fragrance; foliage dark, glossy; very vigorous, trailing growth; [Eva X Reveil Dijonnais]; Kordes

Munchen 83 *see* 'MEIBALBIKA'®

München Kindl *see* 'DICKIMONO'

'MÜNCHENER FASCHING', S, mr, 1963; bud ovoid; flowers bright red, blooms in large clusters (up to 40), dbl., 30 petals; foliage dark, glossy; vigorous, bushy (6-6 1/2 ft.) growth; Kordes, R.

'MUNCHKIN', Min, pb, 1986; flowers pink edging, white center, reverse slight pink edged on white, dbl., 22 petals, exhibition form, slight fragrance; short, pointed, small, tan prickles; foliage medium, medium green, semi-glossy; bushy, low growth; [Watercolor X Unnamed seedling]; Bridges, Dennis A., 1987

Munchner Herz®, F, ab, 1990; Cocker

Münchner Kindl, F, op, 1984; flowers large, dbl.; Dickson, Patrick; (Sangerhausen)

Munot®, Gr, or, 1983; Huber

'MUNRO'S IMPROVED PREMIER', HT, dp, 1927; Munro

'MUNSTER', S, pb, 1958; flowers soft pink shaded deeper, blooms in trusses, dbl., 28 petals, 3.5 in., exhibition form, slight fragrance; foliage light green; Kordes; McGredy

Munsterland®, S, pb, 1986; Noack, Werner

'MURASAKI NO SONO', F, m, 1984; flowers silver lilac, blooms borne 1-5 per cluster, semi-dbl., 13 petals, flat, slight fragrance; prickles slanted downward; foliage light green; vigorous, tall growth; [Tasogare X Unnamed seedling]; Kobayashi, Moriji

'MURCA', HT, dr, 1986; (**Plum Cake**); flowers deep plum red, pointed, borne in sprays of 3-7, dbl., 27 petals, almost everblooming, slight fragrance; pointed, brown prickles; foliage medium, dark green, glossy; bushy growth; [(Chanelle X Sabine) X Pompadour]; Murray, Nola

'MURFRI', F, dp, 1986; (**Frilly Dilly**); flowers light magenta red, pointed, small, borne in sprays of 5-7, dbl., 25 petals, slight fragrance; pointed, brown prickles; foliage large, medium green, flat; upright growth; [Red Lion X Magenta]; Murray, Nola

'MURGO', F, yb, 1983; (**Hipango**); flowers deep yellow, petals edged orange, reverse yellow, shapely, large, dbl., 21 petals, slight fragrance; foliage medium, medium to light green; vigorous growth; [Smiley X Una Hawken]; Murray, Nola

'MURHA', HT, mp, 1986; (**Aroha**); flowers medium, soft pink, elongated blooms, dbl., 43 petals, borne in sprays of 3-5, slight fragrance; pointed, brown prickles; foliage medium, light green; tall growth; [Rifleman X Pascali]; Murray, Nola

Muria *see* 'MACMU'

Muriel *see* 'HARVOOL'

Muriel *see* 'MORMURI'

'MURIEL', HT, dr, 1929; bud long, pointed; flowers brilliant velvety scarlet, semi-dbl.; strong stems; Archer

'MURIEL ARMITAGE', HT, dp, 1972; flowers deep rose-pink, very full, very dbl., 60–70 petals, 5 in., moderate fragrance; foliage large, glossy, dark; vigorous, upright growth; [(Orange Sensation X Ballet) X Mischief]; Ellick; Excelsior Roses

'MURIEL DICKSON', HT, ob, 1915; flowers reddish copper; GM, NRS, 1913; Dickson, H.

Muriel Linton, T, dr; Hay

'MURIEL MOORE', HT, w, 1916; [My Maryland sport]; Moore, F.M.

'MURIEL O'LEARY', HT, pb, 1979; bud long; flowers light and deep pink, well-formed, dbl., 27 petals, 3 in., slight fragrance; foliage large, glossy; vigorous, upright growth; [Honey Favorite X Rose Gaujard]; Murray, Nola

'MURIEL PASQUILL', HT, pb, 1927; flowers strawberry-pink, reverse golden yellow, semi-dbl., intense fragrance; [Padre sport]; Pasquill

'MURIEL WILSON', T, ly, 1923; flowers rich lemon-cream, dbl., slight fragrance; GM, NRS, 1921; Hall; Prince

'MURKI', F, pb, 1983; (**Kingi**); flowers yellow flushed rose pink, large, dbl., 35 petals, slight fragrance; large, red prickles; foliage dense, dark; bushy growth; [Liverpool Echo X Una Hawken]; Murray, Nola

Murmure, LCl, 1971; Croix; (Cavriglia)

Murphy's Law *see* 'SEAMURP'

'MUR-RAY', HT, pb, 1936; flowers darker pink, tinged salmon; [Briarcliff sport]; Murray

'MURRAY HILL', HT, my, 1939; [Joanna Hill sport]; Coddington

'MURSE', HT, lp, 1986; (**Sedgebrook**); flowers very pale pink, almost white, classic form, dbl., 47 petals, slight fragrance; pointed, brown prickles; foliage large, medium green, flat; low growth; [(Chanelle X Prima Ballerina) X Deep Secret]; Murray, Nola

'MURTA', Min, dp, 1983; (**Tasman**); flowers deep pink, small, dbl., 17 petals, flat, no fragrance; foliage medium, dark, leathery; compact growth; [Smiley X ((Pink Parfait X Una Hawken) X Anytime)]; Murray, Nola

'MUSASHINO', HT, op, 1989; bud ovoid; flowers orange-pink, changing to pink, urn-shaped, large, borne usually singly, dbl., 35 petals, exhibition form; few prickles; foliage medium green, oblong, semi-glossy; upright, medium growth; [Garden Party X Daimonji]; Takahashi, Takeshi

Muscosa Japonica *see* **'MOUSSEUX DU JAPON'**

Muscosa Rubra *see* **'OLD RED MOSS'**

'MUSCOSA SIMPLEX', M, mp; (R. centifolia andrewsii, R. muscosa simplex); single; (28)

'MUSETTE', F, dr, 1936; flowers glowing light crimson, open, large, borne in clusters, single, slight fragrance; foliage leathery, wrinkled, dark; strong stems; bushy growth; [Ingar Olsson X Johanniszauber]; Tantau

'MUSIC MAKER', S, lp, 1973; flowers medium, dbl., exhibition form, moderate fragrance; foliage glossy, light, leathery; vigorous, dwarf, upright, bushy growth; Buck, Dr. Griffith J.; Iowa State University

Music Man *see* 'BROMAN'

'MUSICALE', Gr, rb, 1964; bud long, pointed; flowers straw-yellow changing to cherry red, small, dbl., moderate fragrance; foliage glossy, bronze; moderate, bushy growth; [(Bravo X Nellie E. Hillock) X lobelle]; Buck, Dr. Griffith J.; Iowa State University

'MUSICIAN', S, rb, 1953; flowers bicolor, nearer red than yellow, with gray tones, borne in s, semi-dbl., 20 petals, non-recurrent; foliage modified rugose; The first bicolor hardy to -50.; [Hansa X Hazeldean]; Wright, Percy H.

Musikantenland, F, mp, 1993; Hetzel

Musimara®, LCl, mr

Musk Rose *see* **R. MOSCHATA**

'MUSKETEER', HMsk, w; flowers stamens bright yellow, semi-dbl., spring bloom; vigorous (20-25 ft.) growth; Lester Rose Gardens

Musketeer *see* 'WILFOLK'

Musketeer, Gr, op, 1997; Williams, J. Benjamin

Muskoka Moonlight *see* 'LOUMUSKOKA'

Mustang *see* **'COLORANJA'**

'MUTABILIS', Ch, yb; (R. chinensis mutabilis, R. mutabilis, Tipo Ideale); flowers sulfur-yellow, changing to orange, red and finally crimson, 2 in.; (14)

Mutace Caramba, HT; (Cavriglia)

'MUTTER BRADA', HT, ab, 1934; flowers apricot-yellow and red, varying, intense fragrance; vigorous growth; [Lady Craig X Freifrau Ida von Schubert]; Brada, Dr.

'MUTTER BRADA II', HT, dp, 1934; flowers large, dbl., intense fragrance; Brada, Dr.; (Sangerhausen)

'MUTTER BRADA III', HT, lp, 1934; flowers large, dbl., slight fragrance; Brada, Dr.; (Sangerhausen)

Muttergruss®, F, dr

Muttertag *see* **'MOTHERSDAY'**

Muttertag, Climbing *see* **'MOTHERSDAY, CLIMBING'**

'MY ANGEL', F, dp, 1987; flowers deep pink, medium, borne singly or in sprays of 6-10, dbl., 15–25 petals, slight fragrance; foliage medium, medium green, matt; upright, hardy growth; [Pink Parfait X Roman Holiday]; Pencil, Paul S., 1988

'MY BABY', Min, dp, 1967; bud ovoid; flowers deep pink, small, very dbl., cupped, slight fragrance; foliage small, glossy, dark; moderate, dwarf growth; [Cinderella X R. rouletti]; Quackenbush

'MY CHOICE', HT, pb, 1958; flowers pink, reverse pale yellow, dbl., 33 petals, 4.5–5 in., intense, damask fragrance; foliage leathery; vigorous, upright growth; GM, NRS, 1958 GM, Portland, 1961; [Wellworth X Ena Harkness]; LeGrice

My Delight *see* 'TINMYDE'

'MY DREAM', HT, dp, 1970; bud long, pointed; flowers deep pink, medium, dbl., exhibition form, slight fragrance; foliage glossy; upright growth; [Pink Favorite X Karl Herbst]; Winchel, Joseph F.; Country Garden Nursery

'MY FAIR LADY', F, pb, 1959; flowers rose-pink, reverse darker, large, borne in large clusters, semi-dbl.; foliage glossy; tall growth; Wheatcroft Bros.

My Fancy *see* 'MEIFLOREM'

'MY FRIEND', F, lp, 1956; bud ovoid; flowers apple-blossom-pink, dbl., 30 petals, 2.5 in., flat, slight fragrance; dwarf, bushy growth; [Garnette seedling X Summer Snow sport]; Motose

'MY GAL GALE', F, lp, 1958; flowers soft pink, borne in clusters, very dbl.; dwarf growth; [Pinocchio X Morning Star]; Marsh; Marsh's Nursery

'MY GINA', S, mp, 1973; [Dorothy Wheatcroft sport]; Shortland

'MY GIRL', F, ob, 1964; (Cunosa); flowers deep salmon, to open, large, borne in clusters, dbl., 30 petals, cupped; foliage dark; vigorous growth; GM, The Hague, 1963; [Dacapo X Floribunda seedling]; deRuiter; Gregory

My Granny, S, mp, 1991; Poulsen

'MY GUY', F, or, 1986; flowers bright orange-red, fading darker, medium, borne in sprays of, semi-dbl., 13 petals, cupped, moderate fragrance; round, medium, light, light green fruit; medium, light brown, slightly hooked prickles; foliage medium, medium green, semi-glossy; bushy, medium growth; [Rosalynn Carter X Dorothy Wheatcroft]; Milner, William, 1979

My Honey *see* 'JUSHONEY'

'My Joy', HT, mp, 1976; [Red Devil sport]; Wood

'My Lady', HT, ab, 1956; bud pointed; flowers apricot flushed gold, dbl., 46 petals, 5 in., exhibition form, moderate fragrance; foliage dark, leathery; very vigorous, bushy growth; [Seedling X Peace]; Robinson, H.

My Little Boy, Min, my, 1982; Burrows, Steven

My Love® *see* 'COGAMO'

'My Love', HT, dr, 1960; flowers deep red, large, dbl., 45 petals, exhibition form, intense fragrance; foliage dark; vigorous growth; [Bayadere X Ena Harkness]; Anderson's Rose Nurseries

My Love, Climbing, Cl HT, dr

'My Lucky Starr', HT, mr, 1987; flowers large, dbl., 26–40 petals, no fragrance; foliage large, dark green, semi-glossy, leathery; spreading growth; [Secret Love X Karl Herbst]; Evans, F. David

'My Maryland', HT, pb, 1908; flowers bright salmon-pink, edged paler, large, dbl., intense fragrance; vigorous growth; [Madonna X Enchanter]; Cook, J.W.

My Own *see* 'COOSYL'

My Pleasure® *see* 'KINPLEASE'

'My Prayer', HT, op, 1953; bud long, pointed; flowers peach-pink, dbl., 25–30 petals, 5 in., exhibition form, intense fragrance; foliage leathery; vigorous, upright growth; [Seedling X The Doctor]; H&S

My Sunshine *see* 'TINSHINE'

My Valentine® *see* 'MORMYVAL'

My Way, HT, w, 1988; Simpson

'Myfanwy Welbourne', HT, pb, 1999; dbl., 26–40 petals, 4.5 in., borne mostly singly, intense fragrance; prickles moderate; foliage medium, dark green, semi-glossy; upright, bushy, medium (3 ft.) growth; [(Hazel Rose X Cardiff Bay) X Darling Jenny]; Poole, Lionel; David Lister Roses, 2000

'Myra', LCl, w, 1926; flowers creamy white, very large, borne in clusters of 2-3, dbl., slight fragrance; foliage dark, bronze; vigorous, climbing growth; RULED EXTINCT 1/92; [Dr. W. Van Fleet X Lady Roberts]; Wilber

Myra *see* 'BATTOO'

Myra, HT, pb

Myra Stegmann, Gr, op, 1991

Myriam® *see* 'COCGRAND'

'Myrianthes Renoncule', HSem, pb; flowers pale peach, large, dbl., free bloom; not dependably hardy.

Myrna's Dream *see* KORdrekes

Myrniong, HT, lp; Bell, Ronald J., 1994; (Weatherly, L.)

Myrrh-scented Rose *see* **'Splendous'**

'Mystère', HT, dp, 1969; bud long; flowers scarlet pink, dbl., 45 petals; RULED EXTINCT 10/86; [Premiere Ballerine X Femina seedling]; Gaujard

Mysterious *see* 'MINIMYS'

'Mysterium', F, yb, 1963; bud long, pointed; flowers golden yellow striped scarlet, borne in clusters (up to 20), dbl., 25 petals, 3 in., slight fragrance; foliage glossy; bushy, spreading, low growth; [Masquerade X Kordes' Perfecta]; Kordes, R.

Mystery *see* 'HERMESY'

Mystic *see* POUlor

Mystic, HT, m; Dot

'Mystic Gem', HT, dp, 1959; bud long, pointed; flowers reddish pink, large to medium, dbl., exhibition form, slight fragrance; foliage leathery; vigorous, upright growth; [Rod Stillman X Bravo]; Armbrust; Langbecker

Mystic Mauve *see* **'Lavender Mist'**

Mystic Meidiland® *see* 'MEIALATE'

Mystique® *see* 'DEVSTICA'

Mystique, HT, lp; [Mysty sport]; Neil, J., 1989; (Weatherly, L.)

N

N S J Pink Moss, M, dp

N. L. Chrestensen, Pol, or, 1968; flowers medium, semi-dbl.; GPG Bad Langensalza; (Sangerhausen)

'NAARDEN RED', HT, rb, 1932; bud pointed; flowers crimson-red shaded orange, large, dbl., exhibition form, intense fragrance; [Étoile de Hollande X Charles P. Kilham]; Van Rossem

Naas Botha *see* TANrosilb

Nabab, F, 1959; Delbard; (Cavriglia)

'NACHA POBEDA', HT, dr; (Our Victory); flowers dark velvety red; Costetske

Nachez *see* 'POULLEN'

Nachsommer, LCl, mp, 1990; flowers large, dbl.; Wänninger, Franz; (Sangerhausen)

Nachtfalter, HT, dr, 1971; flowers large, dbl., intense fragrance; Baum, Oswald; (Sangerhausen)

'NADIA', HT, dr, 1956; foliage light green; [Mme G. Forest-Colcombet X Seedling]; Delforge

'NADINE', F, dr, 1962; bud long, pointed; flowers maroon-red, dbl., 38 petals, 3 in., cupped, intense fragrance; foliage bronze, soft; vigorous, bushy growth; [Red Pinocchio X Unnamed seedling]; Schwartz, Ernest W.; Wyant

Nadine, HT, ob; Croix

Nadja, HT, mr, 1964; flowers large, dbl.; Rupprecht-Radke; (Sangerhausen)

Naema, LCl, lp, 1998; Delbard

Naia, F; Gaujard; (Cavriglia)

Naina, HT, lp, 1998; Kordes

'NAÏR', HFt, yb, 1936; bud long, yellow and carmine; flowers vermilion-red, reverse yellow, stamens yellow, large, semi-dbl., recurrent bloom; vigorous, bushy growth; [R. foetida bicolor X R. wichurana seedling]; Chambard, C.

'NAKBET', HT, my, 1985; (**Golden Phoenix**); [Bettina sport]; Nakashima, Tosh

'NAMBOUR', F, w, 1953; flowers white flushed pink, medium, borne in clusters, very dbl.; foliage bronze; very vigorous growth; [Yvonne Rabier X Mrs Tom Henderson]; Ulrick, L.W.

'NAMENLOSE SCHÖNE', T, w, 1886; flowers white tinted blush, well formed, large, moderate fragrance; Deegen

Namib Sunrise *see* 'MORLOGEN'

'NAN ANDERSON', F, op, 1970; flowers deep pink, coral sheen, dbl., 30 petals, 2.5–3 in., slight fragrance; foliage glossy, dark; low, bushy growth; Anderson's Rose Nurseries

'NAN POOLE', Min, pb, 1983; bud globular; flowers yellow and pink blend, giving an overall coral-pink effect, dbl., 25 petals, exhibition form, moderate fragrance; thin prickles; foliage small, dark, semi-glossy; low, spreading growth; [Unnamed seedling X Libby]; Meredith, E.A. & Rovinski, M.E.; Casa de Rosa Domingo

'NANA MOUSKOURI', F, w, 1975; flowers well-formed, dbl., 30 petals, 2.5 in., moderate fragrance; [Redgold X Iced Ginger]; Dickson, A.

'NANCY', F, mr; flowers bright red, very small, single; foliage very glossy

'NANCY', HT, dr, 1930; flowers bright scarlet-crimson, semi-dbl., slight fragrance; Ferguson, W.

'NANCY', HT, my, 1934; flowers chamois-yellow, semi-dbl., cupped, slight fragrance; [Mrs T. Hillas X Souv. de Claudius Pernet]; Mallerin, C.; H. Guillot

Nancy, F, or; Croix

Nancy Bennett *see* 'BOSPARDON'

'NANCY BERGH', F, pb, 1968; bud ovoid; flowers pink, reverse silver-pink, open, large, semi-dbl., intense fragrance; foliage dark, glossy, leathery; vigorous, open growth; [Independence X Impeccable]; Fankhauser; A. Ross & Son

Nancy Clare *see* 'JUSCLARE'

'NANCY ELIZABETH', HT, w, 1947; bud long, pointed; flowers cream, large, dbl., exhibition form, moderate fragrance; foliage glossy; vigorous, upright growth; [Korovo X Florinda Norman Thompson]; Mason, F.

Nancy Gardiner *see* KORkeindor

'NANCY HALL', Min, pb, 1972; flowers peachy; [Mary Adair sport]; Moore, Ralph S.; Mini-Roses

'NANCY HAYWARD', LCl, mr, 1937; flowers rich bright cerise, large, single; very vigorous, climbing growth; [Jessie Clark X ?]; Clark, A.; NRS Victoria

'NANCY LEE', HT, pb, 1879; flowers deep pink, intense fragrance; weak growth; [Mme Bravy X Edward Morren]; Bennett

'NANCY PRETTY', HWich, pb, 1917; flowers pink, reverse lighter, small, bornein clusters, dbl.; [Dorothy Perkins X Ellen Poulsen]; MacLellan

'NANCY REAGAN', HT, or, 1967; flowers orange-scarlet, dbl., 28 petals, 5–5.5 in., exhibition form, intense fragrance; foliage dark, bronze, glossy, leathery; vigorous, tall growth; [Orange Delight X Hawaii]; Morey, Dr. Dennison; Country Garden Nursery

Nancy Shaw *see* 'WIWANCY'

'NANCY STEEN', F, pb, 1976; flowers blush pink, center pale cream, dbl., 30 petals, 3.5 in., flat, moderate fragrance; foliage glossy, dark, bronze, leathery; [Pink Parfait X (Ophelia X Parkdirektor Riggers)]; Sherwood; F. Mason

'NANCY WEST', F, yb, 1970; flowers medium yellow, suffused peach; [Elizabeth of Glamis sport]; Haynes

'NANCY WILSON', HT, mp, 1940; Clark, A.

'NANCY WILSON, CLIMBING', Cl HT, mp, 1959; Campton; Hazlewood Bros.

'NANCY'S KEEPSAKE', HT, lp, 1995; flowers pale pink, blooms borne mostly single, dbl., 26–40 petals, 1.5–2.75 in., slight fragrance; some prickles; foliage medium, medium green, glossy; medium (2 1/2 ft), upright growth; [Keepsake sport]; Grierson, Mrs. Nancy; Battersby Roses, 1995

'NANDA', HT, mp, 1959; bud ovoid to urn shaped, claret-red; flowers light rose, large; vigorous growth; [Dame Edith Helen X Eternal Youth]; Sartore

'NANDINI', HT, pb, 1983; bud long, pointed; flowers pink with creamy white reverse, blooms borne singly, dbl., 46 petals, exhibition form, intense fragrance; brown-gray, hooked prickles; foliage large, medium green, leathery; compact growth; [Kiss of Fire X ?]; Pal, Dr. B.P.; Laveena Roses

'NANETTE', LCl, w, 1926; flowers creamy white, large, borne in clusters, dbl.; Hicks

'NANJEMOY', Cl HT, mp, 1937; bud long, pointed; flowers pink, open, large, semi-dbl., free, intermittent bloom; foliage dark; strong stems; vigorous, climbing growth; [Mme Gregoire Staechelin X Bloomfield Comet]; Cross, C.W.

Nankin *see* **'MME YVES LATIEULE'**

'NANO NAGLE', HT, dr, 1997; flowers full, medium, very dbl., 26–40 petals, borne mostly singly, slight fragrance; foliage medium, medium green, semi-glossy; medium (114 cms) upright growth; [Papa Meilland X Northern Lights]; Xavier, Sister M.

'NANTUCKET', HT, ab, 1972; flowers peach apricot, large, dbl., exhibition form, slight fragrance; foliage leathery; vigorous, upright growth; [Chantré sport]; Kern Rose Nursery

'NAOMI', HT, ob, 1926; flowers coppery buff, dbl., moderate fragrance; RULED EXTINCT 1/88; Pemberton

'NAOMI', HT, dr, 1988; flowers dark red, shaded very dark in center, fading to red-purple, dbl., 32 petals, exhibition form, slight fragrance; prickles flat, medium, medium green; foliage large, dark green, semi-glossy; tall, spreading, upright growth; [Red Lion X Unnamed seedling]; Poole, Lionel, 1989

Naomi, HT, pb, 1998

'NAOMI REBECCA', HT, pb, 1996; flowers deep pink, reverse pale pink, large blooms borne mostly singly, dbl., 26–40 petals, slight fragrance; some prickles; foliage large, dark green, dull; upright, bushy, tall growth; [Solitaire X Gavotte]; Poole, Lionel

Napa Valley *see* 'POULINO'

Napoleon, HCh, pb, 1835; Laffay, M.

'NAPOLÉON', HGal, m; flowers bright rose, shaded purple, very large, dbl.; erect, vigorous growth; Hardy, before 1814

'NARCISSE', T, my, 1859; (Enfant de lyon); flowers large, dbl.; Avoux

'NARCISSE DE SALVANDY', HGal, mp; flowers deep rose-pink, prominent yellow stamens, large blooms in clusters, flat; large, spreading growth; Van Houtte, before 1847

Nardy, N, dy, 1888; Nabonnand

'NARITA KORINKAKU', HT, dy, 1978; bud pointed; flowers light orange, dbl., 35–40 petals, 5–6 in., exhibition form, intense fragrance; foliage glossy, dark; upright growth; [Burnaby X Montparnasse]; Kikuchi, Rikichi

'NARMADA LAHARI', HT, pb, 1979; bud ovoid; very dbl., 90 petals, borne singly, moderate fragrance; beak-shaped prickles; foliage small, green; bushy, dwarf growth; [Shree Dayananda sport]; Hardikar, Dr. M.N.

'NARRE FRAGRANCE', HT, ob, 1942; flowers golden orange; [Portadown Fragrance sport]; Brundrett

'NARRE PEACE', HT, yb, 1960; [Peace sport]; Brundrett

'NARROW WATER', N, lp; flowers darker pink; Daisy Hill Nursery, 1883

Nartaki, Pol, m, 1977; Kasturi

'NARVIK', HT, op, 1960; bud long, pointed; flowers salmon-pink to coppery, large, dbl., slight fragrance; foliage leathery; long, strong stems; vigorous growth; [Unnamed seedling X Praline]; Robichon

'NARZISSE', HT, my, 1942; bud long, pointed; flowers apricot to maize-yellow, large, dbl., 23 petals, exhibition form, moderate fragrance; foliage dark, leathery; vigorous, upright growth; [Golden Rapture X Golden Glory]; Krause; C-P

'NASCAPEE', S, w, 1946; flowers open, single, 5 petals, free, recurrent boom, slight fragrance; tall, vigorous growth; hardy.; [(Ross Rambler X (R. rugosa X R. eglanteria)) X ?]; Preston; Central Exp. Farm

'NASE NÁRODNI', HP, m, 1935; flowers purple/pink with white, medium, dbl., intense fragrance; Böhm, J.; (Sangerhausen)

Naseby Rose, HGal, m

'NASTARANA', N, w, 1879; (Persian Musk Rose, R. moschata nastarana, R. pissartii); flowers white tinged pink, blooms in clusters on new wood, semi-dbl., 2 in., recurrent bloom; very vigorous growth; (14); [Probably an early R. chinensis X R. moschata hybrid]

Natacha *see* **'CARA MIA'**

Natali® *see* 'TANROTREILI'

Natali *see* 'TANNALI'

Natalie, Pol, ob, 1972; flowers orange to orange-red, with dark yellow, medium, semi-dbl.; GPG Bad Langensalza; (Sangerhausen)

'NATALIE BOETTNER', HT, my, 1909; flowers sulfur-yellow, passing to cream-yellow, tinted flesh, dbl.; [Frau Karl Druschki X Goldelse]; Boettner

'NATALIE WARD', F, mp, 1974; bud ht type; flowers pink, base yellow, dbl., 20 petals, 2–2.5 in., bloom repeats quickly, slight fragrance; foliage light; bushy growth; [Pink Parfait X ?]; Thomas, R.; Pattollo's Nursery

'NATALKA', HT, lp; Costetske

Natascha *see* 'HARWHARRY'

'NATASHA', F, my, 1974; flowers lemon yellow, fading white, large blooms in clusters of 7-13, very dbl., 135 petals; foliage dark, glossy; bushy growth; [Highlight X Masquerade]; Staikov, Prof. Dr. V.; Kalaydjiev and Chorbadjiiski

'NATASHA MARIA', F, or, 1995; flowers bright, intense, vibrant orange-red, yellow base, blooms bor, semi-dbl., 6–14 petals, 1.5–2.75 in., moderate fragrance; foliage medium, dark green, glossy; 60-80 cms growth; [Wilfrid H. Perron X Golden Olymp]; Fleming, Joyce L.; Hortico Roses, 1994

'NATASHA MONET', HT, m, 1993; flowers very light lavender, dbl., 26–40 petals, 3–3.5 in., borne mostly singly, slight fragrance; some prickles; foliage medium, medium green, semi-glossy; tall, upright, bushy growth; [Crystalline sport]; Pawlikowski, Martin & Elaine

Natchez *see* 'POULLEN'

Natchitoches Noisette, N, lp

Nathalie, S, mp

Nathalie Nypels *see* **'MEVROUW NATHALIE NYPELS'**

Nathalie's High Hills, HCh, dp

Natilda *see* 'MEIDRAGELAC'

'NATIONAL BEAUTY'™, HT, mr, 1983; [The Alamo sport]; Burks, Joe J.; J.B. Williams & Associates

'NATIONAL EMBLEM', HT, dr, 1915; flowers velvety dark crimson, edged vermilion, dbl., exhibition form, moderate fragrance; McGredy

'NATIONAL FLOWER GUILD', HT, mr, 1927; flowers pure scarlet-red, large, dbl.; long stems; very vigorous growth; [(Capt. F. Bald X K. of K.) X Mme Van de Voorde]; Mallerin, C.; C-P, 1930

'NATIONAL TRUST', HT, dr, 1970; (Bad Nauheim, Bad Naukeim); flowers bright red, classic form, dbl., 53 petals, 4 in.; [Evelyn Fison X King of Hearts]; McGredy, Sam IV; McGredy

National Velvet™ *see* 'BURALP'

Nationalstolz, HT, mr, 1970; flowers large, dbl.; McGredy, Sam IV; (Sangerhausen)

'NATIVE WEDDING', S, lp, 1979; bud short, pointed; semi-dbl., 14 petals, cupped, borne singly, repeat bloom, slight fragrance; straight prickles; foliage broad, waved, medium green, semi-glossy, 5-7 leaflet; upright, medium growth; [Restless Native X (Mount Shasta X R. suffulta)]; Stoddard, Louis

Natural Beauty *see* 'ROGSCRIV'

Naughty But Nice *see* 'TINNAUGHTY'

'NAUGHTY NANCY', F, pb, 1970; flowers cream to red, semi-dbl., 12 petals, 2 in., slight fragrance; foliage dull, matt green; moderate growth; Cants of Colchester, Ltd.

Naughty Patricia *see* 'BILPAT'

'NAUTILUS', F, op, 1960; flowers coral-salmon, open, borne in clusters, dbl., 30 petals, slight fragrance; vigorous growth; [Signal Red X Fashion]; deRuiter

'NAVAJO', HT, dp, 1958; bud long, pointed; flowers dark red to rose-red, dbl., 35–40 petals, 4.5–5 in., exhibition form, intense fragrance; foliage leathery; vigorous, upright growth; [Hortulanus Budde X E.G. Hill]; Malandrone; J&P

Navarro Rambler, LCl, pb

Navarro Ridge, N, lp

'NAVID', F, or, 1985; flowers patio, medium, semi-dbl., slight fragrance; foliage medium, medium green, semi-glossy; bushy growth; [Roydon Hall X Trumpeter]; Payne, A.J.

'NAVIGATOR', LCl, mp, 1925; bud globular; flowers radiance pink, large, dbl., 25 petals, cupped; foliage soft; long stems; very vigorous, climbing growth; [R. soulieana X Radiance]; Verhalen; Verhalen Nursery Co., 1943

'NAV-SADABAHAR', F, pb, 1980; bud pointed; flowers deep pink striped white, blooms borne 15 per cluster, dbl., 20 petals, no fragrance; straight, pink to brown prickles; foliage medium, green; medium, spreading, bushy growth; [Sadabahar sport]; Division of Vegetable Crops and Floriculture

'NAYIKA', HT, pb, 1975; bud pointed; flowers delft-rose, reverse and base darker, very dbl., 45 petals, 4 in., exhibition form, slight, tea fragrance; foliage leathery; moderate, upright, bushy growth; Pal, Dr. B.P.; Anand Roses

'NAZNEEN', HT, lp, 1969; bud ovoid; flowers very soft pink, large, very dbl., exhibition form, moderate fragrance; foliage light green, glossy; vigorous, upright growth; [Queen Elizabeth X ?]; Pal, Dr. B.P.; Son

'NAZR-E-NAZAR', HT, mp, 1968; bud pointed; flowers light pink edges flushed darker, large, dbl., exhibition form, slight fragrance; foliage leathery; vigorous, upright, compact growth; [Clovelly X ?]; Singh

NDR I Radio Niedersachsen *see* KOReledas

'NE PLUS ULTRA', N, w; flowers creamy white, moderate fragrance

'NÉALA', HGal, dp, 1822; flowers deep rose edged lighter, medium, dbl.; Vibert, 1845 (?)

Near You *see* 'SEANEAR'

Nearly Black, HT, dr

'NEARLY WILD', F, mp, 1941; bud small, long, pointed; flowers rose-pink, single, white eye, single, 5 petals, moderate, sweet fragrance; bushy growth, 2 - 3 ft.; [Dr. W. Van Fleet X Leuchtstern]; Brownell, H.C.

Nearly Wild Gallica, HGal, dp

'NEARLY WILD, CLIMBING', Cl F, mp, 1962; Burks; Co-Operative Rose Growers

'NÉBULEUSE'®, F, dr, 1971; flowers deep crimson, medium, dbl.; vigorous growth; [Ritz (F) X Lilli Marleen]; Gaujard

Nedbank Rose *see* DELverjaune

'NEDERLAND', HT, dr, 1919; flowers deep glowing red, dbl., 60 petals; [General-Superior Arnold Janssen X George C. Waud]; Verschuren

Neela, HT, m, 1989; K&S

Neelakanti, F, m, 1993; Chiplunkar

'NEELAMABARI', F, dr, 1975; flowers deep red, large, dbl., 35 petals; foliage dark, glossy; vigorous, compact, bushy growth; [Blue Moon X Africa Star]; IARI

Neena, HT, lp, 1990; Bulsara

'NEERVELT', Cl HT, mr, 1910; flowers carmine-red, large, dbl.; Verschuren; (Sangerhausen)

Nefertiti *see* 'VIRAMBER'

Negridte, F, dr, 1972; flowers medium, dbl.; Urban, J.; (Sangerhausen)

'NEHRU CENTENARY', HT, dr, 1989; bud pointed; flowers dark red, reverse deeper, large, borne singly, 60 petals, exhibition form, slight fragrance; prickles hooked, brown; foliage very large, dark green, dense; tall, upright growth; [Christian Dior X Avon]; IARI

'NEIGE D'AVRIL', HMult, w, 1908; flowers pure white, stamens yellow, nearly large, borne in pyramidal, dbl., early, seasonal bloom; very vigorous, climbing (to 8 ft.) growth; Robichon

Neige de Printemps, HMsk, w, 1991; Lens

Neige d'Ete, HMsk, w, 1991; flowers creamy white, medium-large, semi-dbl., moderate fragrance; Lens, Louis; (Sangerhausen)

Neige d'Ete *see* 'TANOTAX'

'NEIGE PARFUM', HT, w, 1942; flowers white, sometimes tinted cream, large, dbl., intense fragrance; foliage leathery; vigorous growth; [Joanna Hill X (White Ophelia X Unnamed seedling)]; Mallerin, C.; A. Meilland;, J&P

'NEIGE ROSE', LCl, pb, 1955; flowers center deep pink, becoming lighter at petal edges, large, slight fragrance; very vigorous growth; Delbard-Chabert

'NEIGES D'ÉTÉ', Min, w, 1984; (Rigobec 2, Summer's Snow); flowers small, dbl., 37 petals, no fragrance; foliage medium, medium green, glossy; upright growth; [Unnamed Baby Masquerade sport X Baby Masquerade seedling]; Gailloux, Gilles

Neisse, S, pb, 1959; flowers yellowish-pink, medium, dbl.; Berger, W.; (Sangerhausen)

'NEJENKA', F, m, 1955; (Tender One); flowers purplish pink, large, dbl., 57 petals, moderate fragrance; foliage dark, glossy; upright growth; Klimenko, V. N.

Nelie Niel, HT, w

Nelkenrose *see* **'F. J. GROOTENDORST'**

'NELL GWYN', HT, or, 1968; flowers orange-copper, globular, moderate fragrance; upright growth; [Tzigane sport]; Cobley; Blaby Rose Gardens

Nella Martinetti®, HT, mp, 1990; Meilland

'NELLIE CHARLTON', HT, mp, 1923; flowers silvery pink, reverse salmon-pink, dbl., intense fragrance; [Mme Abel Chatenay seedling]; Lilley

'NELLIE E. HILLOCK', HT, mp, 1934; flowers silvery pink, base deep gold, reverse old-rose, peony form, dbl., 60 petals, cupped, moderate fragrance; foliage leathery, dark; low, spreading growth; [Golden Dawn X Unnamed seedling]; Hillock

'NELLIE E. HILLOCK, CLIMBING', Cl HT, mp, 1948; Buck, Dr. Griffith J.; Lester Rose Gardens

'NELLIE MAUD POWELL', HT, yb, 1977; flowers yellow edged red, very dbl., 75–80 petals, 4–5 in., intense fragrance; foliage glossy; vigorous growth; [Columbine sport X Kordes' Perfecta]; Powell

Nellie Niel, HT, w, 1975; Dawson

'NELLIE PARKER', HT, w, 1916; flowers creamy white, center darker, well shaped, large, dbl.; vigorous, upright growth; GM, NRS, 1916; Dickson, H.

'NELLY CUSTIS', HMsk, w; dbl., borne in clusters of 7-9, moderate fragrance; vigorous, upright growth

'NELLY VERSCHUREN', HT, my, 1918; flowers clear yellow, intense fragrance; [Unnamed seedling X Duchess of Wellington]; Verschuren

'NÉMÉSIS', Ch, dr, 1836; flowers purplish crimson, pompon form, dbl.; dwarf growth; (14); Bizard

'NEMLOVE', S, mp, 1999; (**Love You**); .75–1 in., borne in large clusters, slight fragrance; prickles moderate; foliage medium, medium green, glossy; bushy, low (2 ft) growth; Nemko, Martin; Bailey Nurseries, 1999

'NÉNETTE LEYDIER', Pol, mr, 1924; flowers crimson-scarlet, base lighter; Richardier

'NENIKUJAKU', F, pb, 1984; flowers pink, blended with yellow and red, blooms in clusters, semi-dbl., 15 petals, cupped, moderate fragrance; vigorous growth; [Masquerade X Matador]; Kikuchi, Rikichi

'NENITA', F, w, 1962; flowers large, dbl., 22 petals, moderate fragrance; foliage light green; vigorous, low growth; [Seedling X Virgo]; da Silva, Moreira

'NEON', F, ob, 1971; flowers intense orange, dbl., 26 petals, 2 in.; compact growth; Waterhouse Nursery

'NEON', HT, mr, 1936; flowers crimson-scarlet, large, very dbl., moderate fragrance; vigorous, branching growth; Nicolas; Beckwith

Neon *see* KORenon

Neon Lights *see* 'JACOUT'

'NEPTUNIA', HT, pb, 1998; flowers pink blending to white, classic, very large, very dbl., 26–40 petals, 6 in., exhibition form, borne mostly singly, slight fragrance; some prickles; foliage large, dark green, semi-glossy; upright, medium growth; [(Precious Platinum X Silver Jubilee) X Seedling]; Poole, Lionel

Nerene *see* MALren

'NERISSA', HT, ly, 1912; flowers cream-yellow, shaded white, dbl., slight fragrance; Paul, W.

'NÉRON', HGal, rb, 1841; flowers crimson, blotched and marbled violet; Laffay, M.

Nerone, HT; Cazzaniga, F. G.; (Cavriglia)

Nervi, HT; Mansuino; (Cavriglia)

'NESTOR', HGal, mr; flowers crimson, large, dbl.; very vigorous growth; Vibert, 1834

'NESTOR BOLDERDIJK', HT, pb, 1938; bud long, pointed; flowers pale ecru, reverse yellowish salmon, base golden, very large, dbl.; foliage glossy; vigorous growth; [Comtesse Vandal X Pres. Macia]; Leenders, M.

Netersen *see* 'KORTERSEN'

Netravathy, F, mp, 1975; Kasturi

'NETUJOH', HT, dr, 1999; flowers deep red, dbl., 30–35 petals, 4–5 in., exhibition form, slight fragrance; foliage dark green, semi-glossy; upright (3-4 ft) growth; GM, Japan Rose Concours, 1993; Hirabayashi, Hiroshi, 1991; Keisei Rose Nurseries, Inc., 1993

Netujou, HT, dr

Neue Revue® *see* 'KORREV'

Neues Europa *see* **'NOUVELLE EUROPE'**

Neus® *see* FEsaru

'NEUTRON', (form of R. rugosa), m, 1984; flowers purple, shallow- blooms borne singly or few per cluster, semi-dbl., 10 petals, 3 in., cupped, repeat bloom, intense fragrance; straight, gray prickles; foliage thick, rugose, shining, dark; upright, dense growth; [Grown from neutron-irradiated seed of R. rugosa]; Lundstad, Arne; Agricultural University of Norway; (grown from irradiated seed)

'NEVADA', HMoy, w, 1927; bud ovoid, pink or apricot; flowers white, reverse sometimes splashed carmine, large blooms on s, single, heavy bloom usually repeated; vigorous (7 ft.), shrubby growth; (28); Dot, Pedro

Nevertheless *see* 'HENNEV'

Never Forgotten *see* 'GREGART'

'NEVILLE CHAMBERLAIN', HT, ob, 1940; bud ovoid; flowers salmon, center orange, dbl., 26 petals, 4 in., exhibition form; foliage bronze; vigorous, tall growth; GM, Portland, 1941; [Charles P. Kilham X Mrs Sam McGredy]; Lens; J&P

Neville Gibson *see* 'HARPORTLY'

Nevis Moss, M, mp

New Adventure *see* 'MORSHEFRAN'

New Antique *see* 'BRONEW'

New Arrival, MinFl, or, 1998

New Ave Maria, HT, ob

New Beginning™ *see* 'SAVABEG'

'NEW CASTLE', Min, rb, 1984; flowers red, white reverse, small, dbl., 20 petals, no fragrance; foliage medium, dark, semi-glossy; upright growth; [Watercolor X Unnamed seedling]; Bridges, Dennis A.

'NEW CENTURY', HRg, pb, 1900; flowers flesh-pink, center light red, edges creamy, dbl., intermittent bloom, moderate fragrance; foliage wrinkled, light, tough; vigorous (4-5 ft.), bushy growth; [R. rugosa alba X Clotilde Soupert]; Van Fleet; Conard & Jones

'NEW COLUMBIA', HT, mp, 1924; flowers true pink, deepening to glowing pink; [Columbia sport]; Hill, E.G., Co.

'NEW DAILY MAIL', F, dr, 1972; (Pussta); bud globular; flowers large, semi-dbl.; vigorous, upright, bushy growth; [Letkis X Walzertraum]; Tantau, Math.

'NEW DAILY MAIL, CLIMBING', Cl F, dr, 1989; (Pussta, Climbing); [New Daily Mail sport]; Patil, B.K.; K.S.G. Son's Roses, 1987

'NEW DAWN', LCl, lp, 1930; (Everblooming Dr W. Van Fleet, The New Dawn); flowers same as parent except repeat blooming, very dbl., 35–40 petals, borne singly and in clusters, moderate, sweet fragrance; foliage medium, dark green; climbing to 20 ft. growth; Hall of Fame, WFRS, 1997; [Dr. W. Van Fleet sport]; Somerset Rose Nursery; Somerset Rose Nursery, 1930

New Dawn Rouge *see* **'ÉTENDARD'**

New Day *see* 'KORGOLD'

New Europe *see* **'NOUVELLE EUROPE'**

New Face® *see* 'INTERCLEM'

'NEW GOLD', Min, dy, 1977; bud long, pointed; flowers buttercup-yellow, open, dbl., 20 petals, 2–2.5 in., intense fragrance; foliage small, very dark; vigorous, upright growth; [Yellow Jewel X Allgold]; Lyon; L. Lyon Greenhouses

'NEW HAVEN QUEEN', HT, ob, 1939; bud pointed; flowers tangerine, dbl., 40 petals, 3.5 in., moderate fragrance; foliage leathery; vigorous growth; [Token sport]; Grillo

'NEW HOPE', Min, w, 1988; bud pointed; flowers creamy white, slight pink edge, medium, borne usually singly, exhibition form; prickles straight, pointed, medium, medium red; foliage medium, dark green, semi-glossy; bushy, medium growth; [Party Girl X Unnamed seedling]; Bridges, Dennis A.; Bridges Roses, 1989

New Horizon *see* DICplay

'NEW LOOK', F, rb, 1961; flowers maroon, reverse silver, blooms in clusters, dbl., 50 petals, 3 in., moderate fragrance; foliage glossy, coppery; vigorous, bushy growth; [Charles Gregory seedling X Orange Triumph seedling]; Gaujard; Gandy Roses, Ltd.

'NEW LOVE', HT, rb, 1968; flowers cardinal-red, reverse golden yellow, large, dbl., exhibition form; foliage glossy, bronze; vigorous, bushy growth; [South Seas X Coronado]; Morey, Dr. Dennison; Country Garden Nursery

'NEW MEXICO', HT, mr, 1967; flowers medium, dbl., exhibition form, moderate fragrance; foliage bronze, leathery; very vigorous growth; [Mount Shasta X Granada]; Aufill

'NEW MOONLIGHT', HT, my, 1935; flowers very large, dbl.; foliage glossy; long stems; very vigorous growth; [Sun Gold X Joseph Hill]; Elmer's Nursery

New Orange, Min, ob

'NEW ORLEANS', HT, yb, 1966; bud ovoid; flowers yellow, veined light red, large, dbl., globular, moderate fragrance; foliage glossy; vigorous, upright growth; RULED EXTINCT 2/87; [Peace sport]; Tate

New Orleans™ *see* 'KINNOR'

New Peace™ *see* 'ARONEWP'

'NEW PENNY', Min, or, 1962; bud short, pointed; flowers orange-red to coral-pink, small, dbl., 20 petals, 1.5 in., moderate fragrance; foliage leathery, glossy; bushy, dwarf (10 in) growth; [(R. wichurana X Floradora) X Unnamed seedling]; Moore, Ralph S.; Sequoia Nursery

'NEW PLANET', HT, MP, 1930; flowers bright rose-pink, center light pink, dbl., intense fragrance; [Premier sport]; Cleveland Cut-Flower Co.

New Pristine *see* 'KORDIENA'

New Rouge Meilland *see* 'MEIMALYNA'

New Star, HT, my, 1987; Keisei Rose Nurseries, Inc.

'NEW STYLE', HT, mr, 1962; bud oval; flowers crimson flushed brighter, large, dbl., 25 petals, slight fragrance; foliage leathery, glossy; long stems; very vigorous, bushy growth; [(Happiness X Independence) X Peace]; Meilland, Alain A.; URS

New Valencia *see* KOReklia

'NEW WORLD', F, rb, 1945; flowers velvety red and crimson, reverse lighter, dbl., intense fragrance; foliage soft, glossy; bushy growth; (28); [Crimson Glory X Château de Clos Vougeot]; Jacobus; B&A

New Year® *see* 'MACNEWYE'

New Year, Climbing *see* 'BURYEAR'

'NEW YORKER', HT, mr, 1947; flowers velvety bright scarlet, dbl., 35 petals, 4–4.5 in., exhibition form, moderate, fruity fragrance; vigorous, bushy growth; [Flambeau X Seedling]; Boerner; J&P

'NEW YORKER, CLIMBING', Cl HT, mr, 1951; Boerner

New Zealand *see* 'MACGENEV'

Newbury Angel *see* 'BARDORG'

Newcomer *see* **'NOVITCHKOVA'**

Newport *see* 'POULMA'

'NEWPORT FAIRY', HWich, pb, 1908; (Newport Rambler); flowers very deep rosy pink, white eye, golden stamens, small blooms, single; [R. wichurana X Crimson Rambler]; Gardner; Roehrs

Newport Rambler *see* **'NEWPORT FAIRY'**

News® *see* 'LEGNEWS'

News Review *see* 'KORREV'

'NEWSACE', HT, ab, 1961; bud ovoid; flowers light apricot, center darker, dbl., 70 petals, 4 in., moderate fragrance; foliage glossy; strong stems; vigorous, upright growth; [Horace McFarland X Good News]; Wyant

'NEWSGATE', HT, op, 1964; flowers rose and persimmon-orange, base yellow, reverse darker, dbl., 50–58 petals, 4 in.; foliage dark, glossy; vigorous growth; [Paulien Verbeek X (Jolie Madame X Baccará)]; Verbeek

Ngarla, P, lp; Riethmuller, 1955; (Weatherly, L.)

'NIAGARA', HT, mr, 1952; bud long, pointed; flowers bright red, dbl., exhibition form, intense, fruity fragrance; [Crimson Glory X Seedling]; Davis; Garden Town Nursery

'NIAGARA MIST', HT, pb, 1968; bud ovoid; flowers light pink, base yellow, large, dbl., exhibition form, moderate fragrance; foliage glossy; vigorous, upright growth; [Tiffany X Mrs A.R. Barraclough]; Davis; Wyant

Niagara Pride *see* 'DELDAL'

'NIAGARA SUNSHINE', HT, dy, 1969; flowers chrome-yellow, large, dbl., globular; foliage glossy, light green; vigorous, upright growth; Davis; Wyant

Niccolo Paganini®, F, mr, 1991; Meilland

Nice Day *see* 'CHEWSEA'

'NICHOLAS SWEETBRIAR', (strain of R. eglanteria), lp; flowers small, 1.5 in.; abundant, red fruit; prickles very thorny; foliage fragrant; height 6-8 ft.; Nicholas; Univ. of Neb., 1959; (probably brought from Virginia to Kansas before 1875)

Nickelodeon *see* 'MACNICKEL'

'NICKY', F, or, 1970; bud medium, pointed; flowers large blooms in clusters, semi-dbl., cupped, no fragrance; foliage medium, light green, sparse, leathery; vigorous, upright, bushy growth; [Cyclamen X Fire King]; Institute of Ornamental Plant Growing

'NIC-NOC', Min, mr, 1978; flowers medium red, lighter reverse, blooms in clusters, dbl., 20 petals, cupped, slight fragrance; foliage small, dark, semi-glossy; spreading growth; [Anytime X Gruss an Bayern]; Poulsen, Niels D.; D.T. Poulsen

'NICOLA', F, dp, 1980; flowers deep rose pink, blooms borne 6-10 per cluster, single, 8 petals, slight fragrance; green prickles; bushy growth; [Unnamed seedling X Unnamed seedling]; Gandy, Douglas L.; Gandy Roses, Ltd.

Nicolas, S, lp; flat, 18 petals, none; L'Assomption, 1996

Nicole *see* 'KORICOLE'

'NICOLE', HT, yb, 1931; bud long, pointed; flowers yellow, center coppery, shaded carmine, intense fragrance; very vigorous growth; RULED EXTINCT 11/80; Gaujard; C-P

Nicole, HT, rb, 1998; Barni, V.

'NICOLE DEBROSSE', HT, dr, 1962; flowers dark red, shaded scarlet; vigorous growth; [Unnamed seedling X Baccará]; Croix, P.; Minier

'NICOLE, CLIMBING', Cl HT, yb, 1933; Kordes

'NICOLETTA', HT, mp, 1969; bud ovoid; flowers pink, medium, dbl.; foliage dark; [Carla sport]; deRuiter

Nicolette® *see* LUDswenic

Nicolina *see* 'ZIPNIC'

'NIDA SENFF', Pol, mp, 1946; flowers soft rosy pink, borne in large clusters; Kersbergen

Nigel Hawthorne *see* 'HARQUIBBLER'

'NIGGER BOY', HT, dr, 1933; bud long, pointed; flowers very dark velvety blackish maroon, dbl., 56 petals, exhibition form, intense fragrance; foliage thick, glossy, bronze; low, compact growth; [Hadley X Yves Druhen]; Knight, G.

'NIGHT', HT, dr, 1930; (Lady Sackville); bud long, pointed; flowers deepest blackish crimson, shaded maroon, dbl., exhibition form, intense fragrance; foliage dark, glossy; bushy growth; McGredy

Night Fire *see* 'LYOFI'

Night Flight *see* 'DELRIO'

Night Hawk *see* 'DOHHAWK'

'NIGHT LADY', Min, rb, 1983; flowers red, white reverse, medium, borne singly, dbl., 36 petals, exhibition form, intense fragrance; foliage medium, dark, semi-glossy; upright growth; [Unnamed seedling X Libby]; Meredith, E.A. & Rovinski, M.E.; Casa de Rosa Domingo

Night Light® *see* 'POULLIGHT'

Night Music *see* 'ZIPMUSIC'

'NIGHT 'N' DAY', HT, dr, 1968; bud pointed; flowers large, dbl., moderate fragrance; foliage dark, leathery; vigorous, tall, bushy growth; [(World's Fair X Chrysler Imperial) X Happiness]; Swim & Weeks

'NIGHT SONG', S, dr, 1984; flowers large blooms borne 1-10 per cluster, dbl., 33 petals, repeat bloom, moderate fragrance; awl-like, tan prickles; foliage medium-large, dark bronze green, semi-glossy; compact, erect, bushy growth; hardy.; [(Rosali X Music Maker) X Meisterstuck]; Buck, Dr. Griffith J.; Iowa State University

Night Star *see* JACurp

'NIGHT TIME', HT, dr, 1975; bud long, pointed; flowers dark black-red, dbl., 39 petals, exhibition form, intense fragrance; foliage dark, leathery; vigorous growth; [Forty-niner X Oklahoma]; Weeks

'NIGHT, CLIMBING', Cl HT, dr, 1936; Armstrong, J.A.; Armstrong Nursery

Nighthawk™ *see* 'DOHHAWK'

Nightingale *see* 'HERGALE'

Night's Musk, LCl, w

'NIGRETTE', HT, dr, 1934; flowers blackish maroon or plum color, varying with season and weath, dbl.; bushy growth; [Château de Clos Vougeot X Lord Castlereagh]; Krause; C-P

'NIGRITELLA', F, dr, 1953; flowers red shaded darker, dbl.; dwarf, bushy growth; Cazzaniga, F. G.

Nikita® *see* 'INTERNIKI'

'NIKITSKAJA ROSOWAJA', HP, mp, 1937; flowers medium, very dbl., intense fragrance; Kosteckij; (Sangerhausen)

'NIKKI', F, ob, 1981; flowers vermilion, white eye and reverse, medium, semi-dbl., no fragrance; foliage medium, medium green, semi-glossy; bushy growth; [Dusky Maiden X Eyepaint]; Bracegirdle, A.J.

Nil Bleu *see* 'DELNIBLE'

'NIL DESPERANDUM', HT, ob, 1979; flowers indian orange, full, dbl., 25–30 petals, 4 in.; foliage large, light matt green; very vigorous growth; [Gavotte X Montezuma]; Ellick; Excelsior Roses

'NILES COCHET', T, rb, 1906; (Red Maman Cochet); flowers cherry-red on outer petals, lighter within; [Maman Cochet sport]; California Nursery Co.

'NILSSON GUY', F, dp, 1930; flowers deep rose-pink, open, large, borne in clusters, semi-dbl.; [Lafayette sport]; Leenders, M.

'NIMBLE', S, pb, 1995; Flowers hand-painted pink stippling to white centre , white reverse, semi-dbl.(6–14 petals), large (4 in.) blooms borne in small and large clusters; slight fragrance; some prickles; foliage medium, medium green, dull, medium (5ft, X 5ft.), upright, bushy growth; [Ivory Fashion X ((Valerie Jeanne X Eyepaint) X Twlight Trail)]; Jobson, Daniel J.; Jobson, 1995

Nimbus *see* 'LEGGREY'

'NIMES', F, rb, 1970; flowers vermilion, reverse gold, dbl.; [Pampa X Piccadilly]; Gaujard

'NINA MARSHALL', HT, dp, 1966; flowers cerise, base gold, semi-dbl., cupped, slight fragrance; foliage glossy; moderate growth; [Serenade X Queen o' the Lakes]; Golik; Ellesmere Nursery

'NINA POULSEN', F, mr, 1940; flowers clear red, semi-dbl.; [Grethe Poulsen X Hybrid Tea (red)]; Poulsen, S.; Poulsen

'NINA ROSA', HT, op, 1946; flowers coppery pink shaded yellow, very large, dbl., moderate fragrance; [Frank Reader X Condesa de Sástago]; Robichon

'NINA WEIBULL'®, F, dr, 1962; flowers medium, dbl.; foliage dark; compact, bushy growth; [Fanal X Masquerade]; Poulsen, S.

Nina Weibull, Climbing, Cl F, dr

Ninetta® *see* 'TANATTENIN'

Ninfea, HT, 1957; Borgatti, G.; (Cavriglia)

'NINIE VANDEVELDE', Pol, mp, 1924; flowers salmon; Vandevelde

'NINON VALLIN', HT, ab, 1936; flowers apricot, reverse fresh yellow, large, dbl.; foliage bright green; very vigorous growth; Gaujard

'NIOBE', F, w, 1942; flowers white, center sometimes flushed light pink; [Rosenelfe sport]; J&P

'NIOUMIYA', Min, w, 1999; flowers ivory white, dbl., 35 petals, 1.5 in., exhibition form, borne 2-5 per cluster, moderate fragrance; foliage medium, dark green; vigorous, bushy, compact (10 in.) growth; [seedling X Hatuzakura]; Yamazaki, Kazuko, 1989; Takii & Co., 1995

'NIPHETOS', T, w, 1843; bud pointed; flowers large, globular, intense fragrance; Bougere

'NIPHETOS, CLIMBING', Cl T, w, 1889; Keynes, Williams & Co.

Nipper *see* HAReco

'NIPPY', HT, yb, 1932; flowers canary-yellow, reverse splashed red, moderate, fruity fragrance; foliage dark; Cant, B. R.

NIPRventyel, HT, lp; (**Versilia**); NIRP, 1996

Niramol *see* 'UMSNIRA'

Nirpette Pink & White, Min, pb, 1997

NIRPnufdeu, HT, yb; (**Ambiance**); Carlton, 1999;, NIRP, 1995

Nirvana® *see* 'MEIRISOURU'

'NISETTE', F, dp, 1967; flowers pink-red, borne in trusses, globular; foliage small, dark; moderate growth; [Garnette sport]; van't Kruis; DeRuiters Nieuwe Rozen B.V.

Nishikie *see* **'NISHIKI-E'**

'NISHIKI-E', F, ob, 1981; (Nishikie); flowers orange-yellow, blooms borne 2-5 per stem, dbl., 38 petals, exhibition form; small prickles slanted downward; foliage dark, semi-glossy; upright growth; [(Sarabande X Amanogawa) X Kagayaki]; Suzuki, Seizo; Keisei Rose Nursery

'NITA', Min, ab, 1987; flowers apricot, lighter apricot reverse, blooms borne singly, dbl., 55 petals, exhibition form, slight fragrance; few, light green prickles; foliage medium, dark, semi-glossy; medium, upright, bushy growth; [Unnamed seedling X Unknown Miniature]; McDaniel, Earl; McDaniel's Min. Roses

'NITOUCHE'®, F, pb, 1974; flowers silvery, deep salmon pink reverse, dbl., 25 petals, 4 in., slight fragrance; foliage glossy, dark; bushy, upright growth; [Seedling X Whisky Mac]; Poulsen, Niels D.; Poulsen

Nivaida, HT, ab, 1997

Nivea *see* **'AIMÉE VIBERT'**

'NIVEA', HT, w, 1949; bud long, pointed; flowers medium, dbl., exhibition form, intense fragrance; foliage sparse; dwarf growth; [Nuria de Recolons X Blanche Mallerin]; Dot, Pedro

Nivescens, HT, lp; flowers medium, semi-dbl., slight fragrance; (Sangerhausen)

NOAbell, F, mp, 1997; (**Bluhendes Barock**); Noack, Werner

Noack's Uberraschung®, HT, yb, 1985; Noack, Werner

NOAdort, F, mp, 1991; (**Schone Dortmunderin**®); Noack, Werner

NOAfeuer, F, dr, 1995; (**Heidefeuer**); Noack, Werner

NOAgut, Gr, op, 1997; (**Focus**); Noack, Werner

'NOAH', HT, ob, 1985; flowers orange; [Dr. A.J. Verhage sport]; Nevo, Motke; Maoz Haim Rose Nursery, 1976

NOAhan, S, w, 1997; (**Hannover's Weisse**); Noack, Werner

NOAmet, F, mp, 1997; (**Rose Minarett**); Noack, Werner

NOArda, S, w, 1998; (**Danica**); Noack, Werner

NOAre, S, mr, 2001; (**Red Flower Carpet**); Noack, Werner

NOAschnee, F, lp; (**White Flower Carpet**); Noack, Werner, 1991

NOAsun, S, dy, 1997; (**Celina**); Noack, Werner

'NOATRAUM', S, dp, 1989; (**Flower Carpet**™, Heidetraum™, Pink Flower Carpet); bud globular; flowers deep pink, reverse lighter, small, profuse, semi-dbl., 15 petals, cupped, borne in sprays, slight fragrance; globular, small, light red fruit; prickles crooked, dark; foliage small, dark green, glossy, disease resistant; vigorous, hardy, low, spreading growth; PP007282; GM, Glasgow, 1993 GM, The Hague, 1990; [Immensee X Amanda]; Noack, Werner; Pan-Am Northwest, Inc., 1991

NOAtwi, S, lp; (**Twilight**); Noack, Werner, 1997

'NOBAL', HMult, w, 1985; (**Thornfree Wonder**); flowers peach pink, fading to white, blooms in clusters (up to14), semi-dbl., 14 petals, cupped, summer bloom, slight fragrance; no prickles; long canes; spreading growth; Nobbs, Kenneth J.

'NOBAM', S, w, 1985; (**Daughter Margaret**); flowers peach, fading to white, borne in clusters of 2-6, dbl., 48 petals, repeat bloom, slight fragrance; broad, pink prickles; foliage typical china; compact growth; [Mutabilis X Cornelia]; Nobbs, Kenneth J.

'NOBILITY', HT, lp, 1961; (Elle); bud ovoid; flowers ivory lightly overcast pink, center deeper, dbl., 35–40 petals, 5–5.5 in., exhibition form, moderate fragrance; foliage leathery; vigorous, upright growth; [Peace seedling X Peace]; Boerner; J&P

Nobilo's Chardonnay® *see* 'MACRELEASE'

Noble Antony *see* 'AUSWAY'

Noble Hit *see* POUlnola

'NOBLESSE', HT, or, 1969; flowers very large, dbl., 28 petals, slight fragrance; foliage glossy; moderate growth; [Coloranja X Coloranja]; Spek; A. Dickson

Noblesse *see* TANselbon

'NOCTURNE', HT, dr, 1947; bud long, pointed; dbl., 24 petals, 4.5 in., cupped, moderate, spicy fragrance; foliage leathery, dark; vigorous, upright, bushy growth; AARS, 1948; [Charlotte Armstrong X Night]; Swim, H.C.; Armstrong Nursery

'NOCTURNE', F, dr; dbl.; moderate growth; Archer

'NOCTURNE, CLIMBING', Cl HT, dr, 1955; Armstrong, J.A.; Armstrong Nursery

'NOËLLA NABONNAND', Cl T, dr, 1901; flowers velvety crimson-red, large, semi-dbl., moderate fragrance; (21); [Reine Marie Henriette X Bardou Job]; Nabonnand, G.

'NOELLA VIREBENT', Cl T, lp, 1922; flowers flesh-pink, center brighter, semi-dbl., moderate fragrance; few thorns; foliage dark, glossy; very vigorous growth; [R. gigantea X Archiduc Joseph]; Nabonnand, P.

Noelle Marie, Min, w

'NOGAWA', HT, w, 1989; bud ovoid; flowers cream, fringed with pink, large, borne usually singly, dbl., 30 petals, exhibition form, moderate fragrance; prickles almost right-angled to stem; foliage medium, dark green, semi-glossy; bushy, tall growth; [Garden Party X Kordes' Perfecta]; Takahashi, Takeshi

Noisette Desprez *see* **'JAUNE DESPREZ'**

Noisette Garden Pink Tea, T, mp

'NOISETTE ROSE'; (R. indica noisettiana, R. moschata autumnalis, R. X noisettiana)

'NOKOMIS', HWich, dp, 1918; flowers dark rose-pink, larger than lady gay or dorothy perkins, bor, dbl., intense fragrance; foliage light, glossy; vigorous, climbing growth; [R. wichurana X Comte Raimbaud]; Walsh

Nola Emily, HT, mp; [Peter Benjamin sport]; Allender, Robert William, 1990

'NOLDANYU', F, my, 1997; (**Sunny Sam**); flowers very full, medium, very dbl., 41 petals, borne in small clusters, slight fragrance; foliage medium, medium green, semi-glossy; upright, compact, low (2.5-3ft.)growth; [(Gold Bunny X Gold Bunny) X Seedling]; Nolan, Gordon D.

'NON PLUS ULTRA', HMult, dr, 1904; (Weigand's Crimson Rambler); flowers small blooms in clusters, dbl.; vigorous growth; [Crimson Rambler X Blanche Rebatel]; Weigand, C.

'NONA', HT, op, 1924; bud long, pointed; flowers flame and pink, open, semi-dbl., moderate fragrance; [Mme Edouard Herriot X Constance]; Easlea

'NONIN', HT, yb, 1938; flowers golden yellow, tinted coral-orange, very large, dbl., slight fragrance; foliage glossy; vigorous growth; [Souv. de Claudius Pernet X Unnamed seedling]; Mallerin, C.; A. Meilland

Nonino, F, mr, 1997

Nora *see* 'BERNORA'

'NORA CUNINGHAM', Cl HT, lp, 1920; flowers flesh-pink, center paler, large, semi-dbl., cupped, free bloom, sometimes recurrent, moderate fragrance; foliage wrinkled, light; vigorous growth, long stems; climbing; [Gustav Grunerwald seedling]; Clark, A.; Hackett

'NORA HENSLOW', HT, dp, 1925; flowers crimson-cerise, single, moderate fragrance; [Mme Mélanie Soupert X Gen. MacArthur]; Evans; Beckwith

'NORA HOOKER', F, mr, 1970; dbl., 22 petals, 4 in., slight fragrance; free growth; [Queen Elizabeth sport]; Hooker; Harkness

'NORA JOHNSON', HMoy, dp, 1957; flowers cerise, small, slight fragrance; small, bright fruit; arching wands growth; [Believed to be R. willmottiae X R. moyesii]; Sunningdale Nursery

'NORA POWER', Cl HT, lp; Dearing, 1919; (Weatherly, L.)

Norah Cruickshank, HT, op

'NORAH LONGLEY', Pol, ob, 1948; flowers flame-orange, borne in trusses; foliage bright green; vigorous, branching growth; [Cameo sport]; Longley

Nordfeuer®, F, mr, 1986; Noack, Werner

'NORDHAUSEN', S, dp, 1940; flowers carmine-pink, large, semi-dbl., slight fragrance; Krause; (Sangerhausen)

Nordia, F, or, 1967; flowers medium, dbl.; Poulsen, D.T.; (Sangerhausen)

'NORDIC CHANT', HT, mp, 1974; bud long, pointed; flowers salmon-pink, dbl., 40 petals, 4.5 in., exhibition form, moderate, spicy fragrance; foliage glossy, light; vigorous growth; [Tropicana X Queen of Bermuda]; Golik; Dynarose

'NORDLICHT', HT, or, 1910; flowers coppery red; [Mme Caroline Testout X Luciole]; Kiese

'NORDLICHT', F, or, 1957; (Northlight); flowers deep cinnabar-red, large, borne in small clusters, dbl., exhibition form, slight fragrance; foliage leathery; very vigorous, low, bushy growth; [Bergfeuer X Gertrud Westphal]; Kordes, R.; R. Schmidt

Nordstern, F, 1964; Kordes, R.; (Cavriglia)

Norfolk *see* POUlfolk

'NORFOLK HARMONY', HT, dp, 1940; bud long, pointed; flowers rosy cerise, large, dbl., exhibition form, slight fragrance; foliage glossy, dark; very vigorous, tall growth; [Comtesse Vandal X Mrs Sam McGredy]; LeGrice

'NORIDA', F, dp, 1967; flowers light crimson-scarlet, medium, dbl.; [(Pinocchio X Pinocchio) X Elsinore]; Poulsen; DeVor Nursery;, McGredy

Norita® *see* 'COMSOR'

Norita-Schwarze Rose® *see* 'COMSOR'

'NORMA', HT, mr, 1976; flowers brilliant red, large, dbl., 50 petals; [Clio X Credo]; Gaujard

'NORMA BENNETT', F, dr, 1958; flowers crimson; [Florence Mary Morse X Border Queen]; Bennett, H.; Waikato Rose Soc.

Norma Major, HT, mp

Norma Margaret *see* 'FRONORM'

'NORMAN', HT, mr, 1934; flowers well formed, bright scarlet-red, moderate fragrance; vigorous growth; Dickson, A.

'NORMAN HARTNELL', HT, mr, 1964; flowers crimson-red, well-formed, large, dbl., 21 petals; foliage dark; very vigorous growth; [Ballet X Detroiter]; Kordes, R.; Wheatcroft Bros.

'NORMAN LAMBERT', HT, ob, 1926; bud long, pointed; flowers deep salmon-orange, suffused bronze and yellow, base lighter, dbl., exhibition form, slight fragrance; GM, NRS, 1924; McGredy

'NORMAN ROGERS', HT, dp, 1933; flowers deep rose-pink, base yellow, large; Chaplin Bros.

Normandie *see* 'CHEWARVEL'

'NORMANDIE', HWich, mp, 1929; flowers salmon-pink; vigorous growth; Nonin

'NORRIS PRATT'®, HT, my, 1964; flowers bright yellow, large; foliage leathery; growth moderate; [Mrs Pierre S. duPont X Marcelle Gret]; Buisman, G. A. H.

'NORRKÖPING', HT, mr, 1961; bud pointed; flowers scarlet, reverse darker; long stems; very vigorous growth; [Karl Herbst X (Baccará X Golden Sun)]; Poulsen, S.

'NORSEMAN', F, mp, 1963; flowers pink, medium, dbl., intense fragrance; foliage soft; vigorous, upright growth; [Unnamed seelding X Pinocchio]; Von Abrams

'NORTH STAR', HT, lp, 1964; bud ovoid; flowers light silvery pink, dbl., 60 petals, 6 in., intense fragrance; foliage glossy; vigorous, medium growth; [Marcia Stanhope X Peace]; Golik; Ellesmere Nursery

Northamptonshire *see* MATtdor

'NORTHERN DANCER', HT, ob, 1965; bud ovoid; flowers orange-yellow, edges flushed pink, large, dbl., slight fragrance; foliage dark, glossy, leathery; vigorous, tall, compact growth; [Tzigane sport]; Schloen; Ellesmere Nursery

Northern Gold, HT, my

'NORTHERN LIGHTS'®, HT, yb, 1969; flowers lemon-cream, tinted pink, dbl., 50 petals, 5 in., intense fragrance; [Fragrant Cloud X Kingcup]; Cocker, 1971

Northern Lights, S, mp, 1997; Noack, Werner

'NORTHERN STATES', HSpn, w, 1952; bud long, pointed; flowers white tinged pink and yellow, open, large, single, 5 petals, profuse, non-recurrent bloom, moderate fragrance; foliage leathery; bushy, compact (2 1/2 ft.) growth; makes a good hedge; hardy; [R. spinosissima X Irish Charm]; Shepherd; Kern Rose Nursery

Northern Yellow, HGal, yb

Northland *see* 'MACCARLTO'

'NORTHLANDER', S, mp, 1985; single, 5 petals, 4 in., borne singly and in clusters of 3, repeat bloom, moderate fragrance; foliage medium, dark, matt; vigorous, upright (to 8 ft.) growth; [Baronne Prevost X ((Magnifica X Joanna Hill) X (Blanche Mallerin X R. laxa))]; James, John

Northlight *see* **'NORDLICHT'**

'NORTHUMBERLAND W.I.', HT, ab, 1988; flowers deep apricot, reverse lighter, medium, dbl., 26–40 petals, slight fragrance; foliage large, dark green, glossy; bushy growth; [Silver Jubilee X Doris Tysterman]; Thompson, Robert; Battersby Roses, 1988

Northwest Sunset, MinFl, op

'NORWICH CASTLE', F, ob, 1980; flowers orange, borne 3-5 per cluster, dbl., 30 petals, slight, fruity fragrance; wedged prickles; foliage medium green, shiny, smooth; vigorous, upright growth; [(Whisky Mac X Arthur Bell) X Unnamed seedling]; Beales, Peter, 1979

Norwich Cathedral *see* 'BEACATH'

'NORWICH CERISE', HT, mr, 1962; flowers cerise, dbl., 25–30 petals, 4–5 in., moderate fragrance; vigorous growth; [Bettina sport]; Morse

'NORWICH GOLD', S, ob, 1962; bud well-formed; flowers yellow shaded orange, large, dbl., 55 petals, moderate fragrance; vigorous, upright growth; Kordes; Morse

'NORWICH PINK', HKor, dp, 1962; flowers bright cerise, semi-dbl., 16 petals, 4 in., moderate fragrance; vigorous, pillar, well-branched growth; Kordes; Morse

'NORWICH SALMON', HKor, op, 1962; flowers salmon-pink, blooms in clusters, dbl., 30 petals, moderate fragrance; foliage glossy; vigorous, pillar, well-branched growth; Kordes; Morse

'NORWICH UNION', F, my, 1975; flowers, 3 in., cupped, intense fragrance; foliage glossy, leathery; [Arthur Bell X (Unnamed seedling X Allgold)]; Beales, Peter, 1976

'NOSAB', HT, mr, 1986; (**Leicester Abbey**); flowers large, very dbl., moderate fragrance; foliage large, dark green, glossy; bushy growth; [(Gavotte X E.H. Morse seedling) X Erotika]; Greensitt, J.A.; Nostell Priory Rose Gardens

'NOSCHAL', HT, dy, 1986; (**Challis Gold**); flowers medium, dbl., 15–20 petals, slight fragrance; foliage medium, medium green, glossy; spreading growth; [Gold Dot sport]; Greensitt, J.A.; Nostell Priory Rose Gardens

'NOSCOOK', HT, lp, 1986; (**Catherine Cookson**); flowers large, dbl., 26–40 petals, moderate fragrance; foliage large, dark green, glossy; bushy growth; [Gavotte X King's Ransom]; Greensitt, J.A.; Nostell Priory Rose Gardens

'NOSMAN', HT, pb, 1986; (**City of Manchester**); flowers large, dbl., moderate fragrance; foliage large, dark green, semi-glossy; bushy growth; [Gavotte X Red Lion]; Greensitt, J.A.; Nostell Priory Rose Gardens

'NOSMID', HT, pb, 1986; (**Middlesbrough Pride**); flowers medium, dbl., 26–40 petals, moderate fragrance; foliage medium, medium green, semi-glossy; bushy growth; [Prima Ballerina X E.H. Morse]; Greensitt, J.A.; Nostell Priory Rose Gardens, 1984

'NOSSA SENHORA DE FÁTIMA', HT, rb; flowers deep red, reverse golden yellow; da Silva, Moreira

'NOSSHEF', HT, yb, 1986; (**City of Sheffield**); flowers medium, dbl., 15–25 petals, intense fragrance; foliage medium, medium green, matt; spreading growth; [Diorama sport]; Greensitt, J.A.; Nostell Priory Rose Gardens

'NOSSUN', HT, lp, 1986; (**Sunderland Supreme**); flowers medium, dbl., 26–40 petals, slight fragrance; foliage meidum, medium green, semi-glossy; upright growth; [Paul Neyron X Royal Highness]; Greensitt, J.A.; Nostell Priory Rose Gardens, 1980

Nostalgia™ *see* 'SAVARITA'

Nostalgie *see* TANeiglat

'NOSTARN', HT, yb, 1986; (**Arnold Greensitt**); flowers medium, dbl., 26–40 petals, intense fragrance; foliage large, light green, matt; bushy growth; [E.H. Morse X Summer Sunshine]; Greensitt, J.A.; Nostell Priory Rose Gardens

'NOSTPRI', HT, mp, 1986; (**Priory Pride**); flowers medium, dbl., 26–40 petals, moderate fragrance; foliage medium, medium green, semi-glossy; bushy growth; [Pink Peace X Chicago Peace]; Greensitt, J.A.; Nostell Priory Rose Gardens, 1981

'NOSTRAD', HT, w, 1986; (Radiant Super Glaze, **Radiant Superglaze**); flowers near white, large, very dbl., moderate fragrance; foliage large, dark green, glossy; [Gavotte X Erotika]; Greensitt, J.A.; Nostell Priory Rose Gardens, 1982

'NOSTROS', HT, mp, 1986; (**Priory Rose**); flowers medium, dbl., 26–40 petals, intense fragrance; foliage medium, medium green, semi-glossy; upright growth; [Unnamed seedling X Unnamed seedling]; Greensitt, J.A.; Nostell Priory Rose Gardens, 1976

'NOSWAN', HT, dp, 1986; (**Pride of Wansbeck**); flowers light red, medium, dbl., 26–49 petals, moderate fragrance; foliage medium, medium green, semi-glossy; bushy growth; [Christian Dior X Unnamed seedling]; Greensitt, J.A.; Nostell Priory Rose Gardens, 1979

Not *see* **'TONNERRE'**®

Notre Dame, LCl, lp; Peden, R., 1997; (Weatherly, L.)

Notre Pere, F, 1982; Croix; (Cavriglia)

'NOTTINGHAM', HT, yb, 1938; flowers clear yellow, center tinted orange, moderate fragrance; vigorous growth; Robinson, H.; Wheatcroft Bros.

'NOTTINGHAM FOREST', F, mr, 1971; dbl., 28 petals, 3.5 in.; foliage dark; moderate, bushy growth; [Metropole X Diamant]; deRuiter; Geo. deRuiter

'NOTTURNO'®, HT, dr, 1981; flowers dark purplish red, dbl., 35 petals, intense fragrance; reddish, hooked prickles; foliage large, dark; upright, bushy growth; [Papa Meilland X Unnamed seedling]; Rose Barni-Pistoia

Nouveau Monde, HCh, mp

Nouveau Rouge, HGal, dr, 1813

'NOUVEAU VULCAIN', HGal, m; flowers dark purple, medium, very dbl.

'NOUVELLE ETOILE', HT, yb, 1966; flowers creamy yellow, edged carmine-red, well shaped, dbl., 40–48 petals; free growth; [Chic Parisien X Provence]; Delbard-Chabert; Cuthbert

'NOUVELLE EUROPE', F, or, 1964; (Neues Europa, New Europe); flowers bright orange, medium, dbl., moderate fragrance; foliage dark; vigorous, bushy growth; ADR, 1964; [Miss France X Vendome]; Gaujard

'NOUVELLE PIVOINE', HGal, m; flowers violet tinted, center vivid red, large

'NOUVELLE TRANSPARENTE', HGal, dp, 1835; flowers rosy crimson, large, dbl., intense fragrance; Miellez, ca 1835

'NOVA', F, or, 1968; semi-dbl., blooms in clusters, slight fragrance; foliage dark, glossy; [Anne Elizabeth X Paprika]; Harkness

Nova, S, mp

'NOVA LUX', HT, yb, 1955; flowers chrome-yellow with red reflections; foliage glossy; very vigorous growth; [Julien Potin X Sensation]; Aicardi, D.; Giacomasso

'NOVA RED', Min, mr, 1964; bud pointed; flowers crimson, small blooms in clusters, semi-dbl., 10 petals; low (12 in) growth; [Unnamed seedling X Little Buckaroo]; Moore, Ralph S.; Sequoia Nursery

'NOVA ZEMBLA', HRg, w, 1907; flowers light pink to white; [Conrad Ferdinand Meyer sport]; Mees

Novaia *see* ORAgofe

November Rain, HT, lp; Urban, J.; (Czech Rosa Club)

'NOVITCHKOVA', HT, ob; (Newcomer); flowers orange-yellow, medium, slight fragrance; foliage dark, leathery; low growth; Novitchkov

'NOWETA', F, mr, 1960; bud ovoid; flowers rose-red, medium, borne in clusters, dbl., moderate fragrance; foliage leathery; vigorous, upright, bushy growth; [Spice X Garnette seedling]; Boerner; J&P

'NOZOMI', Cl Min, lp, 1968; (Heideroslein Nozomi); flowers pearl-pink, blooms in trusses, single, flat, slight fragrance; foliage small, glossy; trailing growth; [Fairy Princess X Sweet Fairy]; Onodera, Toru F.

Nu Gold, Min, my

Nuage Parfume *see* 'TANELLIS'

Nuage Parfume, Climbing *see* 'COLFRAGRASAR'

'NUANCE', Min, or, 1992; dbl., 26–40 petals, 1.5 in., borne mostly singly, slight fragrance; few prickles; foliage small, medium green, matt; medium, upright growth; [Pierrine sport]; White, Al; Giles Rose Nursery, 1991

Nuance *see* JACnu

'NUBIAN', LCl, dr, 1937; flowers dark velvety red, borne in huge clusters, dbl., exhibition form, sometimes recurrent bloom, slight fragrance; foliage large, leathery; vigorous (6-8 ft.) growth; B&A

'NUBIENNE', HCh, dr, 1825; flowers medium, dbl.; Laffay, M.; (Sangerhausen)

'NUGGET', F, my, 1973; bud ovoid; flowers small, very dbl., exhibition form, slight fragrance; foliage large, glossy, dark; vigorous growth; [Yellow Pinocchio X Unnamed seedling]; Warriner, William A.; J&P

'NUGGETS', HT, dy, 1941; bud short, pointed, buff-yellow; flowers pale orange-yellow, open, small, semi-dbl., 15–20 petals, 2–3 in., slight fragrance; foliage small, dark, leathery; [Joanna Hill X Unnamed seedling]; Hill, Joseph H., Co.

Nuit d'Orient *see* 'STEBIGPU'

'NUITS DE YOUNG', M, dr, 1845; (Old Black); flowers reddish purple shading to dusky violet-maroon, well-mossed, ; Laffay, M.

'NUMA FAY', HT, pb, 1938; flowers salmon-pink, edged pale pink, well-formed, large, dbl.; vigorous growth; Richard; A. Meilland

'NUMÉRO UN', HT, or, 1961; flowers scarlet-red passing to vermilion-red, dbl., 35–40 petals, 5 in., globular; foliage bronze, glossy; vigorous, bushy, symmetrical growth; Mallerin, C.; EFR

'NUNTIUS PACELLI', HT, w, 1929; flowers white, center cream, large, dbl., intense fragrance; [Mrs David McKee X British Queen]; Leenders Bros.; C-P

'NUNTIUS SCHIOPPA', HT, my, 1931; flowers golden yellow, sometimes washed peach-blossom-pink; [Los Angeles sport]; Leenders Bros.

'NUR MAHAL', HMsk, mr, 1923; flowers bright crimson, blooms in clusters, semi-dbl., recurrent bloom, moderate, musk fragrance; foliage small; strong stems; vigorous bush or pillar growth; (21); [Château de Clos Vougeot X Hybrid Musk seedling]; Pemberton

Nuria de Recolona *see* **'NURIA DE RECOLONS'**

'NURIA DE RECOLONS', HP, w, 1933; (Nuria de Recolona); flowers well-formed very short peduncle, very dbl., slight fragrance; foliage dense; [Canigo X Frau Karl Druschki]; Dot, Pedro

'NURJEHAN', HT, dp, 1980; bud long, pointed; flowers deep pink, blooms borne singly, dbl., 50 petals, exhibition form, intense fragrance; straight prickles; foliage medium, dark green, coppery when young; medium, spreading growth; [Sweet Afton X Crimson Glory]; Division of Vegetable Crops and Floriculture

Nurse Cavell *see* **'MISS EDITH CAVELL'**

Nursing Centenary®, HT, w

Nutka Rose *see* **R. NUTKANA**

'NUTKHUT', F, or, 1969; bud long, pointed; flowers coral-red, small, very dbl., globular, slight fragrance; foliage leathery; very vigorous, bushy, open growth; [Rumba X Cocorico]; Pal, Dr. B.P.; Son

'NUTNEYRON', S, mp; semi-dbl., occasionally repeats sparingly; height 4 ft; [Paul Neyron X R. nutkana]; Schoener

'NUTZWEDEL', F, mr, 1937; flowers light crimson; [Else Poulsen sport]; Schmidt, K.; Kordes

'NYMPH', F, op, 1953; flowers coral-salmon, borne in trusses, dbl., 30 petals, 3 in., moderate fragrance; foliage dark, glossy; very free growth; [Fashion X Unnamed seedling]; Dickson, A.

'NYMPHE EGERIA', HMult, 1839; Geschwind, R.; (Cavriglia)

'NYMPHE TEPLA', HSet, 1886; Geschwind, R.; (Cavriglia)

'NYMPHENBURG', HMsk, op, 1954; flowers salmon-pink shaded orange, very large blooms in clusters (up, semi-dbl., flat, recurrent bloom, moderate fragrance; foliage large, glossy; upright growth; [Sangerhausen X Sunmist]; Kordes; Morse

'NYPELS PERFECTION', Pol, pb, 1930; flowers hydrangea-pink, shaded deep pink, large blooms in clusters, semi-dbl.; vigorous, bushy growth; Leenders, M.

'NYVELDT'S WHITE', HRg, w, 1955; flowers snow-white, large, single; orange-red fruit; [(R. rugosa rubra X R. cinnamomea) X R. nitida]; Nyveldt

O

O Sole Mio® *see* 'DELOSOL'

'O. JUNYENT', HT, mr, 1924; (Olegario Junyent); flowers coral-red, base yellow, semi-dbl.; [Frau Karl Druschki X Mme Edouard Herriot]; Dot, Pedro

'OAKINGTON RUBY', Min, mr, 1933; bud deep crimson; flowers ruby-crimson, white-eye, dbl., 1–1.5 in.; dwarf (1 ft or less) growth; (14); Bloom

'OAKLEY', HT, pb, 1937; flowers bright rose, base deep red, tipped flesh-pink, large, intense fragrance; vigorous growth; Fairhead

'OAKMONT', HP, pb, 1893; flowers deep pink, reverse lighter, blooms in clusters, recurrent bloom; May

Oasis Sunset *see* 'SUNMANI'

'OBBO', HT, lp; [Alfred Colomb sport]; (Weatherly, L.)

Obélisque® *see* 'DELMOT'

Oberbürgermeister Boock, Pol, dr, 1964; flowers medium, dbl.; GPG Bad Langensalza; (Sangerhausen)

'OBERBÜRGERMEISTER DR KÜLB', HT, op, 1931; flowers flame-colored, passing to salmon; [Roselandia sport]; Nauheimer

'OBERBÜRGERMEISTER DR. TROENDLIN', HT, lp, 1904; flowers large, very dbl.; Kaiser; (Sangerhausen)

'OBERBÜRGERMEISTER HEIMERICH', HP, mp, 1929; flowers fresh rose, some petals with reverse lighter, dbl., slight fragrance; [Frau Karl Druschki X Souv. de Claudius Pernet]; Weigand, C.

'OBERGÄRTNER BURGNER', HT, mr, 1934; flowers medium, semi-dbl., slight fragrance; Burgner; (Sangerhausen)

'OBERGÄRTNER WIEBICKE', F, mr, 1950; bud long, pointed; flowers light red, open, very large, borne in clusters, semi-dbl., moderate fragrance; foliage glossy, light green; vigorous, bushy growth; [Johannes Boetnner X Magnifica]; Kordes

'OBERHOFGÄRTNER A. SINGER', HP, mr, 1904; flowers carmine, center darker, dbl., 40 petals; dwarf, compact growth; [Mme Caroline Testout X Marie Baumann]; Lambert, H.

'OBERLEHRER L. BURKHARDT', HT, dp, 1939; flowers carmine-pink, large, dbl., moderate fragrance; Burkhardt; (Sangerhausen)

'OBERLEUTNANT IMMELMANN', HT, lp, 1936; flowers soft yellowish pink, center deeper, petals incurved, somewha, globular; Henniger

'OBERON', F, ab, 1955; flowers salmon-apricot blooms in trusses, dbl., 38 petals, 2–2.5 in., slight fragrance; bushy growth; [Nymph X Seedling]; Dickson, A.

Obsession *see* 'DEVTINTA'

Ocarina *see* 'OCARU'

'OCARU', Min, mp, 1978; (**Angela Rippon**®, Ocarina); flowers salmon-pink, patio, small, moderate fragrance; dwarf, compact growth; [Rosy Jewel X Zorina]; deRuiter; Fryer's Nursery, Ltd.

Ocho, HT, ab, 1983; Suzuki, Seizo

'OCOOCH MOUNTAIN ROSE', S, mp, 1981; bud small, pointed; single, 5 petals, borne 3-4 per cluster, intense, spicy fragrance; straight, fine prickles; foliage 9 leaflet, small, slightly rugose, medium green; arching growth; Hall, William W.

Octandre, T, pb

Octavia Hill *see* 'HARZEAL'

'OCTAVIE', HGal, lp; flowers light pink, edged blush, open, medium, dbl.; vigorous, branching growth; Coquerel, before 1829

Octavius Weld, T, pb

'OCTET', S, m, 1977; bud narrow, pointed; flowers pale purple, single, 5 petals, 3 in., prolific bloom in summer only, slight fragrance; foliage gray-green; extrememly vigorous growth; [R. rudiuscula X R. subglauca]; Rowley; Royal National Rose Soc.

'OCTOBER', HT, ob, 1980; bud long, pointed; flowers rich salmon orange, blooms borne singly and 2-3 per cluster, dbl., 30 petals, exhibition form, moderate fragrance; long, hooked prickles; foliage leathery, dark; tall, upright growth; [Unnamed seedling X Unnamed seedling]; Weeks, O.L.

Octoberfest™ *see* 'MACLANTER'

Oddball *see* 'HORODD'

Odee Pink, T, lp

Odense City, S, yb, 1996

Odeon *see* 'GAUFRARNER'

'ODÉRIC VITAL', HP, lp, 1858; [Baronne Prevost sport]; Oger

Odessa *see* 'MICODESSA'

Odette, F, mp, 1972; flowers medium, dbl., intense fragrance; GPG Bad Langensalza; (Sangerhausen)

'ODETTE CHÈNE', HT, mp, 1940; flowers pink, base coral; vigorous growth; [Richmond, Climbing X Charles P. Kilham]; Colombier

'ODETTE FOUSSIER', HT, mp, 1924; flowers salmon-pink, inside chrome on yellow ground, dbl., intense fragrance; Chambard, C.

'ODETTE FOUSSIER, CLIMBING', Cl HT, mp, 1929; Chambard, C.

'ODETTE JOYEUX', LCl, op, 1959; bud globular, coral-orange; flowers pink to lilac-pink, large, dbl., cupped, abundant, recurrent bloom, intense fragrance; foliage leathery, glossy; very vigorous growth; [Lady Sylvia X Unnamed seedling]; Robichon

Odine *see* **'ONDINE'**

'ODORATA', HT, pb, 1928; flowers carmine-pink, reverse white edged pink, base golden yellow, dbl., intense fragrance; [Sunburst X Ma Fiancee]; Van Rossem

Odorata 22449 *see* **'FUN JWAN LO'**

Odyssée *see* 'GAUZOMI'

'ILLET', C, lp, 1800; flowers bright pink, medium, dbl., moderate fragrance; vigorous growth; Dupony, Andre

Illet Double, HGal, m, 1835; Prévost

'ILLET FLAMAND', HGal, pb, 1845; flowers pale pink striped white and brighter pink, medium, very dbl., flat, intense fragrance; very vigorous growth; Vibert

'ILLET PANACHÉE', M, pb, 1888; (Striped Moss); flowers pale pink striped deep pink, petals quilled, small, flat; Verdier, C.

'ILLET PARFAIT', HGal, pb; flowers blush, striped light and dark red, medium, dbl., flat; dwarf growth; Foulard, ca 1841

'OEKONOMIERAT ECHTERMEYER', HT, dp, 1913; flowers very large, dbl., moderate fragrance; Lambert, P.; (Sangerhausen)

Officinalis *see* **R. GALLICA OFFICINALIS**

Offrandé, HT, 1982; Gaujard; (Cavriglia)

Offranville, F, dp, 1993; Eve, A.

'OGNEWAJA', HT, mr, 1937; flowers medium, dbl.; Kosteckij; (Sangerhausen)

Ogni Jalty, F, mr, 1955; flowers medium, dbl., slight fragrance; Klimenko, V. N.; (Sangerhausen)

'OGONIOK', F, ob, 1955; (Little Fire); flowers fiery orange edged darker, medium, slight fragrance; Sushkov, K. L.

Oh Darlin *see* 'RENDAR'

Oh La La *see* **'OLALA'**®

Oh My God *see* 'WEKJOE'

'OH MY STARS', Min, ly, 1996; dbl., 26–40 petals, exhibition form, borne mostly singly, slight fragrance; moderate prickles; foliage medium, medium green, dull; upright, tall (2-2 1/2 ft) growth; [Giggles sport]; Bough, Melvin; Giles Rose Nursery

'OH-CHOH', HT, yb, 1983; flowers yellow tinted rose, aging red, large, dbl., 38 petals, exhibition form, moderate fragrance; small prickles, slanted downward; foliage dark, semi-glossy; upright growth; [(Rumba X Olympic Torch) X Wisbech Gold]; Suzuki, Seizo; Keisei Rose Nursery

'OHIO', S, mr, 1949; flowers bright red, semi-dbl., recurrent bloom; height 4 ft; hardy.; [R. soulieana X Gruss an Teplitz seedling]; Shepherd

'OHIO BELLE', LCl, lp, 1974; bud globular; flowers medium, dbl., 53 petals, 3 in., repeat bloom, slight fragrance; foliage glossy, dark; [New Dawn X ?]; Jerabek, Paul E.; Wyant

'OHL', HGal, m; flowers violet-purple, center bright red, large, dbl.; vigorous growth; Vibert, before 1843

Ohlala *see* **'OLALA'**®

Ohshima Rose *see* 'COCHUSTER'

'OIRASE', HT, mr, 1973; bud ovoid; dbl., 24 petals, 5.5 in., exhibition form, below-average bloom continuity, moderate fragrance; foliage glossy, dark; vigorous, upright growth; [Red Lion X Christian Dior]; Ito

'OISEAU BLEU', HT, m, 1970; flowers mauve-rose, dbl., 30 petals, 4–4.5 in., intense fragrance; foliage large, glossy, dark; vigorous growth; Poulsen, Niels D.; Vilmorin-Andrieux

'OISEAU DE FEU', F, mr, 1956; flowers scarlet-red, to open, medium borne in clusters, dbl., 35 petals, 2 in., cupped, slight fragrance; bushy growth; [Chant Indou X Peace]; Mallerin, C.

'OJIBWAY', LCl, w, 1946; bud pointed; semi-dbl., 12–15 petals, 3 in., borne in clusters, free, non-recurrent bloom, slight fragrance; foliage dark; vigorous, spreading growth; hardy.; [Ross Rambler X ((R. rugosa X R. eglanteria) X ?)]; Preston; Central Exp. Farm

'OKAGA', HRg, dp, 1927; flowers deep pink, semi-dbl., non-recurrent; low, bushy growth; very hardy.; [Alika X Tetonkaha]; Hansen, N.E.

'OKLAHOMA', HT, dr, 1964; bud ovoid, long, pointed; flowers very dark red, dbl., 48 petals, 4–5.5 in., exhibition form, intense fragrance; foliage leathery, dark, matt; vigorous, bushy growth; GM, Japan, 1963; [Chrysler Imperial X Charles Mallerin]; Swim & Weeks; Weeks Wholesale Rose Growers

'OKLAHOMA, CLIMBING', Cl HT, dr, 1972; Ross, A., & Son

'OKLAHOMA, CLIMBING', Cl HT, dr, 1968; Swim & Weeks; Weeks Wholesale Rose Growers

'OKRESNI HEITMAN CUBR', HT, mp, 1933; flowers large, dbl.; Böhm, J.; (Sangerhausen)

Oksana, HT, pb, 1995; Singer, Steven

Oktoberfest *see* 'MACLANTER'

'OLALA'®, F, mr, 1956; (Oh La La, Ohlala); bud pointed; flowers blood-red, center lighter, large blooms in clusters to 25, semi-dbl., slight fragrance; foliage leathery, dark, glossy; vigorous, bushy, upright growth; GM, Baden-Baden, 1955; [Fanal X Crimson Glory]; Tantau, Math.

'OLAVE BADEN-POWELL', HT, mr, 1972; flowers scarlet, 5 in., slight fragrance; foliage dark, leathery; Tantau, Math.; Harry Wheatcroft Gardening

'OLAVUS', M, op, 1932; flowers salmon-pink, medium, semi-dbl.; Nielsen; (Sangerhausen)

Old Black *see* **'NUITS DE YOUNG'**

'OLD BLUSH', Ch, mp; (Common Blush China, Common Monthly, Old Pink Daily, Old Pink Monthly, Parsons' Pink China); flowers two-tone pink, blooms in loose sprays, semi-dbl., dependably recurrent, slight fragrance; vigorous, upright growth; (14); Parsons, 1752; Int. into Sweden in 1752 and into England before 1

Old Blush Single, Ch, pb

'OLD BLUSH, CLIMBING', Cl Ch, mp

Old Castilian *see* **'AUTUMN DAMASK'**

Old Country Charm *see* 'MICCOUNTRY'

Old Crimson China *see* **'SLATER'S CRIMSON CHINA'**

Old Danish, LCl, mp

Old Faithful *see* 'JACHY'

'OLD FASHION RED', HT, mr, 1947; flowers spectrum-red fading blush, large, very dbl., exhibition form, moderate fragrance; foliage glossy; vigorous, bushy growth; [Pink Princess X Crimson Glory]; Brownell, H.C.

Old Fashioned Girl *see* 'TINOLD'

Old Fashioned Lady, S, w

Old Flame *see* DICkitty

Old Gay Hill Red China, Ch, mr

'OLD GLORY', HT, mp, 1940; flowers clear brilliant pink, large, dbl., moderate fragrance; RULED EXTINCT 2/88; [Briarcliff sport]; Hausermann

Old Glory™ *see* 'BENDAY'

'OLD GOLD', HT, or, 1913; flowers vivid reddish orange, shaded coppery red and apricot, semi-dbl., 10 petals, moderate fragrance; foliage dark; short stems; GM, NRS, 1912; McGredy

Old Homestead, HCh, m

Old John™ *see* DICwillynilly

Old Lavender, Ayr, m

Old Lavender Rambler, HWich, m

Old Lilac *see* 'AUSLILAC'

Old Master *see* 'MACESP'

Old Nanaimo Rose, Misc OGR, mp

Old Pink Daily *see* **'OLD BLUSH'**

Old Pink Monthly *see* **'OLD BLUSH'**

Old Pink Moss *see* **Communis**

Old Port *see* 'MACKATI'

Old Red, HCh, mr

'OLD RED BOURSAULT', Bslt, mr; flowers pale red, boursault type, poorly formed, borne in large clusters, semi-dbl.; nearly round fruit; very vigorous growth

'OLD RED MOSS', M, mr; (Muscosa Rubra); flowers carmine-red, medium, very dbl., heavy bloom; non-recurrent, moderate fragrance; vigorous growth

Old Smokey, F, lp; Murley, J.J., 1976; (Weatherly, L.)

'OLD SMOOTHIE', HT, mr, 1970; flowers large, very dbl., exhibition form, slight fragrance; foliage large, glossy, leathery; vigorous, upright growth; [Night 'n' Day X (First Love seedling X (Queen Elizabeth X Chrysler Imperial))]; Weeks

Old Spanish Rose *see* **'RUSSELLIANA'**

Old Spice *see* 'MACJULIAT'

Old Stone School, Misc OGR, mp

Old Time *see* 'KOROL'

Old Velvet Moss *see* **'WILLIAM LOBB'**

Old White Moss, M, w

Old Yellow Scotch, HSpn, dy

'OLDCASTLE', HT, mp, 1985; flowers large blooms borne singly, dbl., 55 petals, exhibition form, moderate fragrance; foliage large, dark, glossy; medium, upright growth; [Queen Elizabeth X Charlotte Armstrong]; LeMire, Walter; Roses by Walter LeMire

'OLDE ENGLISH', F, dp, 1974; flowers light red, dbl., 30–35 petals, 3–4 in., slight fragrance; foliage glossy, bronze; free growth; [(Floradora X Independence) X Siren]; Orard, Joseph; Harry Wheatcroft Gardening

Olde Fragrance *see* 'TANSCHAUBUD'

Olde Lace *see* 'HADLACE'

Olde Romeo *see* 'HADROMEO'

Olde Sweetheart™ *see* 'HADSWEE'

Olde Tango™ *see* 'HADANGO'

'OLDEN DAYS', Pol, mr, 1989; (Olden Times); flowers small, borne in sprays of 36, semi-dbl., 10 petals, slight fragrance; prickles brown; foliage small, light green, ellipical, slightly serrated, ma; bushy, compact growth; [Unnamed pink Polyantha Paul Carpel sport]; Vash, Ernest J.; Historical Roses, 1987

Olden Times *see* **'OLDEN DAYS'**

Oldtimer *see* 'KOROL'

'OLÉ', Gr, or, 1964; bud well-shaped; flowers medium, dbl., 50 petals, exhibition form, moderate fragrance; foliage glossy; vigorous growth; [Roundelay X El Capitan]; Armstrong, D.L.; Armstrong Nursery

Olé, Climbing™ *see* 'AROHAICLO'

Oleander Rose *see* 'INTERANDER'

Olegario Junyent *see* **'O. JUNYENT'**

Olga, HT, 1988; Delforge; (Cavriglia)

Olga, HT; Mansuino; (Cavriglia)

Olga Rippon *see* 'HORTYARD'

Olga Tschechowa, HT, ly, 1978; flowers creamy yellow, very large, very dbl., intense fragrance; Cocker; (Sangerhausen)

OLIjbrau, F, r; (**Estelle**); (OLIjbrau), 1999

OLIjcrem, F, w; (**Sahara**); (OLIjcrem), 1997

OLIjfaon, F, pb; (**Evolution**); (OLIjfaon)

OLIjglu, HT, lp; (**Toscanini**)

OLIjkroet, HT, rb; (**Faria**); (OLIjkroet)

OLIjplam, HT, mr, 1999; (**Red Berlin**)

OLIjsab, HT, ob, 1999; (**Orange Flame**)

OLIjzouc, HT, dp, 1997; (**Daytona**); (OLIjzouc)

OLItel, F, pb; (**Super Disco**)

Olive *see* 'HARPILLAR'

'OLIVE COOK', HT, w, 1934; flowers white, base faintly tinged lemon-yellow, large, exhibition form, moderate fragrance; foliage glossy; vigorous growth; Cant, F.

Olive Elsie *see* 'BOSSEXEYE'

'OLIVE MCKENZIE', HT, yb, 1970; bud long, pointed; flowers orange-yellow, marked red, large, dbl., moderate fragrance; foliage large, dark; vigorous, upright, bushy growth; [Daily Sketch X Manitou]; Dawson, George; Brundrett

'OLIVE MOORE', HT, lp, 1927; flowers pale rose-pink, reflexed silver-pink, intense fragrance; Allen

'OLIVE PERCIVAL', HT, rb, 1948; bud long, pointed; flowers intense cherry-red, base gold, semi-dbl., 14–20 petals, 3.5–4 in., cupped, intense fragrance; foliage leathery, bronze; very vigorous, upright, free branching growth; [California X Eternal Youth]; Howard, P.J.

Olive Taylor *see* 'PEAP'

'OLIVE WHITTAKER', HT, op, 1920; flowers rich coppery rose to cerise and salmon; Easlea

'OLIVER DELHOMME', HP, mr, 1861; flowers carmine-red, medium, dbl., moderate fragrance; Verdier, V.; (Sangerhausen)

'OLIVER MEE', HT, mp, 1927; flowers deep salmon tinted fawn, becoming deep salmon-pink, large, dbl., exhibition form, moderate fragrance; Dickson, S.

Oliver Twist *see* 'SABBYRON'

'OLIVERS', HT, m, 1963; flowers purple to pink, medium, semi-dbl., moderate fragrance; [Pres. Herbert Hoover X ?]; Oliver, H.

Olivet, HMult, dp, 1892; Vigneron

Olivia® *see* TANavilo

'OLIVIER METRA', HP, 1885; Verdier, E.; (Cavriglia)

'OLLIE', HT, mr, 1982; bud globular, pointed; dbl., 45 petals, borne singly, intense fragrance; red brown prickles; foliage large, leathery, rounded; vigorous, tall growth; [(Pink Garnette X Pink Hat) X ((Frau Karl Druschki X McGredy'sYellow) X (Baronne Prevost X Gruss an Teplitz))]; James, John

'OLWYN', HT, mp, 1946; bud long, pointed; flowers pink, medium, exhibition form, slight fragrance; foliage leathery; vigorous, bushy growth; Bird; F. Mason

Olympe, HT, 1974; Gaujard; (Cavriglia)

'OLYMPE FRESCENCY', T, w; flowers white, becoming yellow; Damaizin, 1859

'OLYMPIA', HT, mr, 1935; flowers bright red, large, dbl., moderate fragrance; long, strong stems; vigorous growth; [Johanniszauber X Hadley]; Tantau

'OLYMPIA', HT, my, 1955; bud bright yellow, well formed; bushy, semi-upright growth; [Eclipse X Seedling]; Delforge

Olympiad® *see* 'MACAUCK'

'OLYMPIAD', HT, rb, 1931; (Mme Raymond Gaujard); bud long, pointed; flowers blood-red, shaded copper and yellow, large, dbl., intense fragrance; RULED EXTINCT 11/82; GM, Bagatelle, 1930; Pernet-Ducher; Dreer;, Gaujard

'OLYMPIAD, CLIMBING', Cl HT, rb, 1938; Raffel; Port Stockton Nursery

'OLYMPIC CHAMPION', HT, yb, 1995; flowers yellow/pink blend, very dbl., 1.5–2.75 in., slight fragrance; some prickles; foliage medium, dark green, semi-glossy; medium (70 cms), bushy growth; [Champion sport]; Pugh, D. J.; Pugh, 1995

'OLYMPIC CHARM', HT, lp, 1963; flowers silvery, reverse bright pink, well formed, dbl., 65 petals, 3.5 in., intense fragrance; foliage dark, glossy; [Grand'mere Jenny X Claude]; Barter

Olympic Dream *see* 'JACADE'

'OLYMPIC FLAME', F, or, 1962; bud pointed; flowers orange-vermilion, blooms in clusters, single, 8 petals, 3 in.; foliage glossy, bright green; vigorous, upright, bushy growth; Brett

Olympic Glory *see* 'JACED'

Olympic Gold *see* 'KORSCHNUPPE'

'OLYMPIC GOLD', Min, ly, 1983; flowers mini-flora, medium, dbl., 31 petals, exhibition form, slight fragrance; foliage medium, medium green, semi-glossy; upright, bushy growth; [Rise 'n' Shine X Bonny]; Jolly, Nelson F.; Rosehill Farm

Olympic Spirit *see* 'JACARY'

Olympic Spirit *see* 'PEAPRINCE'

'OLYMPIC STAR', HT, mp, 1959; flowers cerise-pink; vigorous growth; [Picture sport]; Trebbin

'OLYMPIC TORCH', HT, rb, 1966; (Seika, Sei-Ka); bud long, pointed; flowers white and red, becoming all red, medium, dbl., exhibition form; foliage glossy, bronze, leathery; vigorous growth; Gold Star of the South Pacific, Palmerston North, NZ, 1970; [Rose Gaujard X Crimson Glory]; Suzuki, Seizo; Keisei Rose Nursery

'OLYMPIC TRIUMPH', F, rb, 1973; flowers red and yellow, dbl., 4.5 in., globular; foliage sage green; upright growth; [Shiralee X Apricot Nectar]; Dickson, Patrick; Dicksons of Hawlmark

Olympiode® *see* 'MACAUCK'

Olympisches Feuer® *see* 'TANOLFEU'

Olympisches Feuer 92 *see* TANolfeu92

Olympus *see* BARoly

'OMAR KHAYYÁM', D, lp, 1893; flowers center incurved, small, very dbl., quartered, moderate fragrance; foliage small, downy; dense, prickly growth (to 3 ft)

'OMAR PACHA', B, mr, 1863; flowers bright cherry, recurrent bloom, slight fragrance; vigorous growth; Pradel

'OMBRÉE PARFAITE', HGal, m, 1823; flowers variable, light pink to deep purple, often in the same flower, dbl.; Vibert

Ombretta, S, 1991; Embriaco, B.; (Cavriglia)

'OMEGA', HT, mp, 1974; (**Emily Post**); flowers soft medium pink, dbl., 48 petals, 3.5–4 in., exhibition form, moderate fragrance; upright, bushy growth; [Eternal Sun X Carina]; Byrum; J.H. Hill Co.

Omega, HT, 1975; Hill, J. H.; (Cavriglia)

Omi Oswald®, HMsk, ly, 1988; Lens

Omni, S, yb, 1988; Lens

'OMPHALE', HGal, 1845; Vibert, J. P.; (Cavriglia)

Omul, S, dy, 1974; flowers dark golden yellow, large, dbl.; GPG Bad Langensalza; (Sangerhausen)

On Fire, S, lp; Peden, R., 1999; (Weatherly, L.)

Onda Rose®, S, mp

Ondella *see* 'MEIVANAMA'

'ONDINE', HT, w, 1936; (Odine); bud pointed; flowers creamy white, slightly tinted pink, high pointed, large, dbl., 20–25 petals; foliage dark gray-green; long, strong stems; very vigorous growth; [Louise Criner X Souv. de Claudius Pernet]; Ketten Bros.

'O'NEAL'S BEQUEST', S, yb, 1986; flowers yellow with pink petal edges, medium, semi-dbl., slight fragrance; foliage large, medium green, glossy; upright growth; O'Neal, Conrad; Paul Jerabek

O'Neal's Bequest, S, yb

'O'NEAL'S WHITE', LCl, w, 1961; flowers large; vigorous growth; [Blossomtime X New Dawn]; O'Neal; Wyant

Onex, HGal, dr

Onkaparinga *see* 'TOMRING'

'ONKEL SVEND', F, rb, 1978; flowers medium red and silvery red blend, large blooms in sprays of, dbl., 23 petals, moderate, spicy fragrance; foliage medium, dark, matt; low, bushy, spreading growth; [Sonia X Ernest H. Morse]; Poulsen, Niels D.; D.T. Poulsen

Only Love *see* 'INTERONLY'

'ONLY YOU', HT, mr, 1970; flowers rose-carmine; [Carina sport]; Vahldiek-Bissingen; URS

Ontario Celebration *see* 'LAVMOUNT'

'ONYX FLAMBOYANT', HT, lp, 1965; flowers peach to shell-pink, 4 in., exhibition form, slight fragrance; foliage serrated; vigorous, bushy growth; [Sultane X Queen Elizabeth]; Delbard-Chabert; Cuthbert

'OOMASTON PRIDE', HT, yb, 1954; flowers buttercup-yellow suffused deep carnation-pink, large, dbl.; foliage dark, glossy; very vigorous growth; [Haisha sport]; C.W.S. Ltd. Hort. Dept.; Co-op. Wholesale Soc.

Oonagh *see* 'SEAOONA'

'OOR WULLIE', Min, mp, 1978; flowers salmon-pink, full, medium, dbl., 24 petals, slight fragrance; foliage glossy, leathery; free growth; [New Penny sport]; Anderson's Rose Nurseries

Opa Pötschke *see* **'PRECIOUS PLATINUM'**

'OPAL', HT, mp, 1934; (Lise Palais); flowers salmon to opal, dbl., 45 petals, 5.5 in., exhibition form, intense fragrance; vigorous growth; [Julien Potin X R. foetida bicolor seedling]; Gaujard; J&P, 1941

'OPAL BRÜNNER', Cl F, lp; bud very small; flowers blush-pink to pale rose, borne in large clusters, dbl., slight, musk fragrance; tall, pillar (10 ft) growth; Marshall, O.C., about 1948

'OPAL GOLD', HT, ab, 1936; flowers apricot, changing to pink; [Joanna Hill sport]; Bate

'OPAL JEWEL', Min, mp, 1962; bud ovoid; flowers pink, center darker, dbl., 45 petals, 1 in., slight fragrance; foliage leathery; vigorous, compact (8-10 in) growth; [Mothersday X Rosy Jewel]; Morey, Dr. Dennison; J&P

'OPAL OF ARZ', HT, ob, 1938; flowers variable orange shades, dbl., moderate fragrance; [Mary Wallace seedling]; Brownell, H.C.

'OPALINE', HT, pb, 1922; flowers pale pink, shaded carmine and old-gold; [Louise Catherine Breslau X Frau Karl Druschki]; Lippiatt

Open Arms *see* 'CHEWPIXEL'

Open d'Australie® *see* DORsafr

Open Secret *see* 'WEEPAIN'

Opening Act *see* 'JALACT'

Opening Night™ *see* 'JACOLBER'

'OPÉRA', HT, rb, 1950; bud long, pointed; flowers light scarlet-red, base yellow, dbl., 6 in., moderate fragrance; foliage leathery, light green; vigorous, erect growth; GM, NRS, 1949; [La Belle Irisee X Unnamed seedling]; Gaujard; J&P

'OPERA, CLIMBING', Cl HT, rb, 1956; (Grimpant Opera); Armbrust

Operettenrose®, HT, lp

'OPHELIA', HT, lp, 1912; bud long, pointed; flowers salmon-flesh, center tinted light yellow, dbl., 28 petals, moderate fragrance; foliage leathery; vigorous growth; (28); [Parentage unknown. Perhaps a chance seedling of Antoine Rivoire. sport]; Paul, W.

'OPHELIA QUEEN, CLIMBING', Cl HT, lp, 1923; [Ophelia sport]; Westbury Rose Co.

'OPHELIA SUPREME', HT, lp, 1917; flowers light rose-pink, center darker; [Ophelia sport]; Dailledouze Bros.

'OPHELIA, CLIMBING', Cl HT, lp, 1920; Dickson, A.

'OPHIRIE', N, op, 1841; flowers reddish copper, reverse rosy and fawn, very dbl., cupped, moderate fragrance; vigorous growth; Goubault

'OPORTO', HT, dr, 1930; flowers port-wine-red, petals very leathery, dbl., slight fragrance; [Château de Clos Vougeot X Betty Uprichard]; Allen

Opulence *see* JACosch

Opus, Pol, or; flowers medium to large, semi-dbl., slight fragrance; VEG; (Sangerhausen)

Or Hatabor, HT, dy, 1989; Fischel

Ora Kingsley *see* CLEhonor

ORAblan, HT, w; (**Therese de Lisieux**®); Orard, 1992

ORAcreps, HT, pb, 2000; (**Hope**); Orard, P.

'ORADIT', Cl F, op, 1983; (**Edith de Martinelli, Climbing**®); Orard, Joseph; Pekmez

ORAdon, HT; (**Triolet**); Orard, 1995; (Cavriglia)

'ORADOUR', F, dr, 1955; dbl.; [Happiness X Demain]; Mallerin, C.; EFR

Orafe, HT, pb, 1997

ORAgofe, F, rb, 1997; (**Novaia**); Orard

ORAju, HT, 1993; (**Tassili**); Orard; (Cavriglia)

ORAkah, HT, m; (**A Caen la Paix**®); Eurosa, Chinensis; Orard, Joseph, 1994; (Hay)

ORAkosar, Cl HT, or, 1993; (**Lady Rose, Climbing**); Orard

ORAlamar, HT, lp; (**Ville de Moulins**); Orard, 1997

'ORAMONT', F, or, 1983; (**Orangina**®); [Royal Occasion sport]; Orard, Joseph; Paul Pekmez

ORAnat, HT, dr, 1997; (**Abbe Lemire**); Orard

Orange Adam®, F, ob, 1993; Adam, M.

Orange Appeal *see* 'RENAPP'

'ORANGE BEAUTY', Cl F, op, 1961; flowers orange to salmon, blooms in clusters, dbl., exhibition form, slight fragrance; foliage glossy; vigorous (6-8 ft) growth; [Little Darling X Gertrude Raffel]; Raffel; Port Stockton Nursery

Orange Blossom, S, lp; [R. brunonii X Unknown]; McLeod, J., 1990; (Weatherly, L.)

Orange Blossom Special™ *see* 'JACMOCL'

'ORANGE BOUQUET', F, ob, 1972; flowers coppery orange, turning pink, rosette form, dbl., 25 petals, 2.5 in., moderate fragrance; foliage dark; upright, free growth; [Masquerade X Paris-Match]; Northfield

Orange Bunny® *see* 'MEIRIANOPUR'

Orange Butterfly, S, or, 1997; Williams, J. Benjamin

'ORANGE CASCADE', Cl Min, ob, 1979; bud pointed; flowers yellow-orange, dbl., 20 petals, 1 in., moderate fragrance; foliage small, fern-like; slender, willowy growth; [Unnamed yellow seedling X Magic Wand]; Moore, Ralph S.; Sequoia Nursery

Orange Charm *see* 'MINDCO'

Orange Chateau *see* **'HATUZAKURA'**

'ORANGE CHATEAU', HT, ab, 1999; flowers apricot orange, dbl., 30 petals, 5.5 in., moderate fragrance; 3.5 ft growth; winter hardy; heat tolerant; Teranishi, K.; Itami Rose Nursery, 1997

'ORANGE CHEER', Pol, ob, 1937; flowers clear orange, almost, dbl.; vigorous growth; Letts

'ORANGE CHIFFON', F, ob, 1966; flowers orange-salmon, reverse silvery orange, medium, dbl., cupped; foliage dark, leathery; vigorous, bushy growth; [Little Darling X Goldmarie]; Mease; Wyant

Orange Class, Min, ob

'ORANGE COMBE', HT, ob, 1956; flowers orange, very dbl., globular; vigorous, upright growth; [Charlotte Armstrong X Unnamed seedling]; Combe; Delbard

'ORANGE CUP', F, or, 1965; bud pointed; flowers orange-scarlet, open, medium, borne in clusters, single, intense fragrance; foliage dark, glossy; vigorous, compact growth; [Cocorico X ?]; Pal, Dr. B.P.; Indian Agric. Research Inst.

'ORANGE DARLING', Cl Min, ob, 1979; flowers orange, small, dbl., 32 petals, 1.5 in., slight fragrance; foliage dark, leathery; climbing or pillar growth; Sudol, Julia

'ORANGE DAWN', F, ob, 1973; flowers orange, tinted vermilion, full, dbl., 30–35 petals, 4 in., intense fragrance; foliage glossy, dark; vigorous growth; [Orange Sensation X Sutter's Gold]; Ellick

'ORANGE DELBARD', HT, or, 1959; (Lady Russon); flowers bright orange, well formed, large, dbl., moderate fragrance; strong stems; vigorous growth; [Impeccable X Mme Robert Joffet]; Delbard-Chabert

'ORANGE DELIGHT', HT, ob, 1950; (Meerzicht Glory, Queen Juliana); bud urn shaped; flowers orange, reverse veined red, dbl., 30 petals, 5.5–6 in., cupped, intense fragrance; foliage glossy, dark; vigorous, tall growth; Verschuren-Pechtold; J&P

'ORANGE DELIGHT, CLIMBING', Cl HT, ob, 1957; Verschuren-Pechtold

Orange d'Éte *see* DORnice

'ORANGE DOT', HT, ob, 1963; bud long; flowers carthamus-red, large, dbl., 50 petals, moderate fragrance; bushy growth; [Chrysler Imperial X Soraya]; Dot, Pedro

'ORANGE DROP', Min, ob, 1988; flowers soft orange, white at base, reverse soft orange edging veining, semi-dbl., 20 petals, exhibition form; prickles long, very pointed, medium, light green; foliage medium, medium green, semi-glossy; bushy, medium, vigorous growth; [Heartland X Unnamed seedling]; Bridges, Dennis A.; Bridges Roses

'ORANGE ELF', Cl Min, ob, 1959; bud pointed; flowers orange, fading lighter, small, dbl., 25 petals, slight fragrance; trailer or groundcover; vigorous, climbing growth; [Golden Glow (LCl) X Zee]; Moore, Ralph S.; Sequoia Nursery

'ORANGE ELIZABETH OF GLAMIS', F, ob, 1974; flowers orange-flame, dbl., 30 petals, 3 in., intense fragrance; foliage glossy; compact, bushy growth; [Elizabeth of Glamis sport]; Thames Valley Rose Growers

'ORANGE EVERGLOW', LCl, ob, 1942; flowers orange slightly shaded red and yellow; [Copper Glow sport]; Brownell, H.C.

'ORANGE FESTIVAL', HT, op, 1961; flowers pink to coral-orange, small; [Souv. de Jacques Verschuren X Serenade]; Leenders, J.

Orange Fire® *see* 'INTERFIRE'

'ORANGE FIRE', Min, op, 1974; bud short, pointed; flowers orange, pink, carmine, rose, dbl., 40 petals, 1 in.; foliage very glossy, leathery; upright growth; [(R. wichurana X Floradora) X Fire Princess]; Moore, Ralph S.; Sequoia Nursery

'ORANGE FLAME', HT, or, 1963; bud ovoid, pointed; dbl., 33 petals, 4.5–5 in., exhibition form, moderate fragrance; foliage leathery, glossy; vigorous growth; [Monte Carlo X Radar]; Meilland, Mrs. Marie-Louise; C-P

Orange Flame, F, ob, 1983; Williams, J. Benjamin

Orange Flame *see* OLIjsab

Orange Flare, F, ob

'ORANGE GARNET', F, or, 1965; bud ovoid; flowers small, dbl.; foliage dark, leathery; vigorous, upright, bushy growth; [(Garnette X Circus) X Spartan]; Swim & Weeks

'ORANGE GLORY', HT, ob, 1936; flowers orange, open, large, semi-dbl.; foliage leathery, light; vigorous growth; [Unnamed seedling X Charles P. Kilham]; Leenders, M.

'ORANGE GLOW', Pol, ob, 1936; flowers bright golden orange, borne in large trusses; Verschuren; Dreer

'ORANGE GOLIATH', HT, op, 1975; bud long, pointed; flowers copper-orange, dbl., 26 petals, 8 in., moderate fragrance; foliage matt, green; vigorous growth; [Beauté X Serenade]; Gandy, Douglas L.

'ORANGE HONEY', Min, ob, 1979; bud pointed; flowers orange-yellow, dbl., 23 petals, 1.5 in., exhibition form, moderate, fruity fragrance; foliage matt, green; bushy, spreading growth; [Rumba X Over the Rainbow]; Moore, Ralph S.; Sequoia Nursery

'ORANGE ICE', F, or, 1963; bud ovoid; flowers light orange-red, blooms in clusters, dbl., 25 petals, 3 in., cupped, slight fragrance; foliage leathery, dark; vigorous, upright, bushy growth; [Fashion X Sumatra]; Thomson, R.; Tillotson

Orange Ice, F, ob, 1999; Williams, J. Benjamin

Orange Ilseta® *see* 'TANILVOBA'

'ORANGE JADE', Min, or, 1991; dbl., 26–40 petals, borne mostly singly, slight fragrance; foliage medium, dark green, semi-glossy; upright, medium growth; [Black Jade sport]; Cole, Catherine W.; East Tennessee Miniature Roses, 1990

Orange Juice *see* 'ARORAJU'

Orange Juwel® *see* KORkister

'ORANGE KING', Pol, ob, 1922; flowers light coral-red, fading greenish, open, borne in clusters, dbl., 0.75 in.; foliage glossy, dark; bushy, dwarf growth; [Orléans Rose sport]; Cutbush

Orange Kordana, Min, or; Kordes

'ORANGE KORONA', F, or, 1959; flowers orange-scarlet, well-formed, blooms in clusters, dbl., 20 petals, 4 in., slight fragrance; foliage olive-green; vigorous, upright growth; [Bergfeuer X Independence]; Morse

Orange Love™ *see* 'MINAFCO'

'ORANGE MARMALADE', Min, ob, 1991; flowers bronze with yellow blend, small, dbl., borne mostly singly, moderate fragrance; foliage medium, light green, matt; low, upright, bushy growth; PP006820; [Gingersnap X Orange Honey]; Williams, J. Benjamin, 1988

'ORANGE MARVEL', Pol, op, 1928; flowers salmon-orange; [Miss Edith Cavell sport]; Van der Vis

'ORANGE MASTERPIECE', F, or, 1970; flowers small, borne in trusses, dbl.; foliage dark; bushy growth; [Seedling X Orange Sensation]; deRuiter

Orange Meillandina *see* 'MEIJIKATAR'

Orange Meillandina, Climbing *see* 'MEIJIKATARSAR'

Orange Mini-Wonder *see* 'LAVJACK'

'ORANGE MIST', F, ob, 1957; flowers orange-salmon to yellow-orange, dbl., 4 in., moderate fragrance; dwarf, bushy growth; [Ma Perkins X Unnamed seedling]; Boerner; J&P

'ORANGE MORSDAG', Pol, ob, 1956; (Orange Mothersday); flowers deep orange; [Mothersday sport]; Grootendorst, F.J.

Orange Mothersday *see* **'ORANGE MORSDAG'**

Orange Muttertag *see* **'VATERTAG'**®

'ORANGE NASSAU', HT, ob, 1941; bud long, pointed; flowers two-toned coppery orange, reverse yellow, dbl., cupped, moderate fragrance; foliage leathery; [Mev. G.A. van Rossem X Unnamed seedling]; Verschuren; Dreer

'ORANGE NYMPH', F, or, 1960; flowers medium, borne in small clusters, dbl.; [Nymph sport]; Tulp; G. Maarse

'ORANGE PARFAIT', HT, ob, 1982; bud ovoid, pointed; flowers orange, yellow reverse, large blooms borne singly, dbl., 45 petals, exhibition form, slight, tea fragrance; long, narrow prickles, hooked downward, reddish at base; foliage medium to large, bronze green, leathery; vigorous growth; Weeks, O.L.; Weeks Wholesale Rose Growers

Orange Parfait, Min, ob, 1998; Moore

Orange Passion™ *see* 'JACCOGEL'

Orange Pastel™ *see* 'WILORPAT'

'ORANGE PERFECTION', Pol, or, 1927; [Ideal sport]; Spek

'ORANGE PIXIE', Min, or, 1978; bud ovoid, pointed; flowers bright orange-red, dbl., 48 petals, 1 in., exhibition form; foliage small, glossy, leathery; bushy, compact, upright growth; [Little Chief X Fire Princess]; Moore, Ralph S.; Sequoia Nursery

'ORANGE QUEEN', Pol, op, 1923; flowers salmon-orange; [Orléans Rose sport]; van Nes

'ORANGE RAPTURE', HT, op, 1935; bud long, pointed; flowers orange with pink, large, dbl., exhibition form, moderate fragrance; [Rapture sport]; Schmidt, K.; Kordes

Orange Rosamini® *see* 'RUISETO'

'ORANGE ROSETTE', Pol, or, 1941; flowers scarlet-orange, borne in clusters, dbl., 30 petals, 1–1.25 in.; vigorous, compact growth; [Gloire du Midi sport]; deRuiter; J&P

'ORANGE RUFFELS', HT, ob, 1952; bud long, pointed; flowers orange to saffron-yellow, dbl., 4–5 in., exhibition form, moderate fragrance; foliage dark, glossy; vigorous, compact growth; [(Dr. W. Van Fleet X Général Jacqueminot) X Lafter]; Brownell, H.C.

'ORANGE RUMBA', F, ob, 1962; flowers orange; [Rumba sport]; Zieger; C-P

Orange Sauvageot, F, ob, 1987; Sauvageot

'ORANGE SCEPTER', HT, yb, 1953; flowers orange-yellow with dark golden yellow, large, dbl.; Verschuren-Pechtold; (Sangerhausen)

Orange Scepter, HT, ob

'ORANGE SCHOON', HT, ob, 1938; flowers pure orange; [Katharine Pechtold sport]; Lens

Orange Seabreeze, MinFl, op, 1996; Williams, J. Benjamin

'ORANGE SENSATION'®, F, or, 1961; dbl., 24 petals, 3 in., blooms in clusters, moderate fragrance; foliage dark; vigorous, bushy growth; GM, NRS, 1961 Golden Rose, The Hague, 1968; deRuiter; Gregory

Orange Sherbert *see* 'LYORA'

'ORANGE SILK', F, or, 1968; flowers orange-vermilion, shallow-, large blooms in large clusters, cupped, slight fragrance; foliage dark, glossy; [Orangeade X (Ma Perkins X Independence)]; McGredy, Sam IV; Gregory

'ORANGE SMOKE', F, ob, 1964; flowers orange, with blue haze in cool weather, 3–4 in.; very vigorous growth; [Orange Ruffels X (Eva X Guinee)]; Hennessey

Orange Sovereign, Pol, ob

Orange Sparkle *see* 'KORSPARKO'

'ORANGE SPARKS', HT, ob, 1969; flowers orange-vermilion blend, striped gold, medium, dbl.; foliage glossy, bronze; vigorous, bushy growth; [Cherry Brandy sport]; Vasishth

'ORANGE SPECIAL', F, or, 1957; flowers orange-scarlet, small, borne in large trusses, dbl., 30–35 petals, 1.5 in., cupped; dwarf, bushy growth; [Salmon Perfection sport]; deRuiter; Blaby Rose Gardens

'ORANGE SPICE', Min, ob, 1980; bud ovoid, pointed; flowers orange, deeper on petal edges, blooms borne singly or severa, dbl., 23 petals, cupped, moderate fragrance; recurved prickles; foliage small, medium green; bushy, upright growth; [Unnamed seedling X Unnamed seedling]; Lyon

Orange Spice *see* 'SOCOTRA'

Orange Splash *see* 'JACSERAW'

Orange Splendor, F, ob, 1998; Select Roses, B.V.

Orange Star™ *see* 'MINALCO'

'ORANGE STARINA', Min, ob, 1981; flowers light orange; [Starina sport]; Graff, Roy; Mini Roses

Orange Sunblaze™ *see* 'MEIJIKATAR'

Orange Sunblaze, Climbing *see* 'MEIJIKATARSAR'

Orange Sunset *see* 'MANORSUN'

'ORANGE SUNSHINE', Min, ob, 1968; flowers orange; [Bit o' Sunshine sport]; Moore, Ralph S.; Sequoia Nursery

Orange Supreme, HT, ob

Orange Surprise, F, ob, 1998

'ORANGE SWEETHEART', F, op, 1952; bud ovoid; dbl., 20–25 petals, 3–3.5 in., cupped, slight, fruity fragrance; foliage dark; upright, bushy growth; [Pinocchio seedling X Fashion]; Boerner; J&P

Orange Symphonie® *see* 'MEININRUT'

Orange Tango *see* 'DANORANG'

'ORANGE THÉRÈSE', HT, ob, 1943; flowers orange-yellow; [Soeur Thérèse sport]; Howard Rose Co.

'ORANGE TRIUMPH'®, Pol, mr, 1937; flowers small blooms in clusters, semi-dbl., cupped, slight fragrance; foliage glossy; compact, bushy growth; GM, NRS, 1937; [Eva X Solarium]; Kordes; Dreer

'ORANGE TRIUMPH IMPROVED', Pol, or, 1960; flowers orange, borne in large clusters, dbl.; [Orange Triumph sport]; Cant, F.

'ORANGE TRIUMPH SUPERBA', Pol, ob, 1953; flowers clear orange; [Orange Triumph sport]; Maarse, J.D.

'ORANGE TRIUMPH, CLIMBING', Cl Pol, mr, 1945; Leenders, M., 1945; Koopmann, 1948

Orange Twist *see* 'MORTWIST'

Orange Unique, HT, lp; flowers medium orange, reverse rose pink, exhibition form; long stems

'ORANGE VELVET', LCl, or, 1986; flowers bright orange-red, reverse orange-red to orange pink, fading, dbl., 37 petals, exhibition form, moderate, damask fragrance; medium, rounded, medium bright orange fruit; foliage large, dark green, glossy, dark waxed; upright, tall, vigorous growth; [Tropicana, Climbing X Swarthmore]; Williams, J. Benjamin

Orange Vilmoria *see* 'KORSEE'

Orange Vilmorin *see* 'KORSEE'

Orange Zest *see* 'LAVTRICE'

'ORANGEADE'®, F, or, 1959; flowers bright orange-red, blooms in clusters, semi-dbl., slight fragrance; foliage dark; very vigorous, bushy growth; GM, NRS, 1959 GM, Portland, 1965; [Orange Sweetheart X Independence]; McGredy, Sam IV; McGredy

'ORANGEADE, CLIMBING', Cl F, or, 1964; Waterhouse Nursery

'ORANGE-RED SUPREME', F, or, 1958; bud ovoid; flowers open, borne in clusters, dbl., 18–24 petals, 2.5–3 in., moderate fragrance; foliage glossy, wrinkled; bushy growth; [Spice X Garnette seedling]; Boerner; J&P

Oranges 'n' Lemons *see* 'MACORANLEM'

Orangina® *see* 'ORAMONT'

'ORANIËN', F, or, 1962; flowers orange-red, sometimes lined yellow, base light yellow, large, dbl., 42 petals; foliage dark; upright, compact, symmetrical growth; [Highlight X Unnamed seedling]; Verschuren, A.; van Engelen

Oranzova Garnette, F, lp; Strnad; (Czech Rosa Club)

ORAred, F, mr, 1991; (**Romy Schneider**®); Orard

ORAreg, F, lp, 1997; (**Chateau de Bagnols**); Orard

ORAroro, HT, ly; (**Sourire d'Enfant**®); Orard, 1989

ORAstrip, HT, rb, 1998; (**Meli-Melo**); Orard

'ORATAM', S, yb, 1939; bud globular; flowers pink edged copper-pink, base and reverse yellow, large, dbl., non-recurrent, intense, damask fragrance; foliage leathery, dark, matt, yellow-green; bushy growth; [R. damascena X Souv. de Claudius Pernet]; Jacobus; B&A

Oratia Maid, F, lp, 1986; Garelja, Anita

Oratorio *see* 'WARDIDO'

ORAuna, F, pb, 1992; (**Cyclamen La Sevillana**); Orard

ORAval, HT, lp; (**Ville de Bar sur Seine**); Orard, 1996

ORAwest, S, yb, 1999; (**President Wilson**); Orard

Orchid Jubilee *see* 'MORCLILAV'

Orchid Lace *see* BENorchid

'ORCHID MASTERPIECE', HT, m, 1960; bud ovoid, deep orchid; flowers lavender-orchid, large, dbl., 68 petals, moderate, fruity fragrance; foliage leathery, dark; vigorous growth; [Golden Masterpiece X Grey Pearl seedling]; Boerner; J&P

Oregold *see* 'TANOLG'

'OREGON CENTENNIAL', HT, mr, 1959; bud pointed; flowers rose-red, dbl., 30–35 petals, 4–5 in., exhibition form, moderate, fruity fragrance; foliage dark; long, strong stems; vigorous, upright, bushy growth; [Charles Mallerin X (Charles Mallerin X Chrysler Imperial)]; Von Abrams; Peterson & Dering

'OREGON OPHELIA', HT, mp, 1921; flowers salmon, edged pink, base yellow, dbl.; [Ophelia sport]; Clarke Bros.

Oregon Rainbow *see* 'CLERAIN'

Oreste Sgaravatti, HT; Sgaravatti, A.; (Cavriglia)

Orfeo, LCl, 1963; Leenders, J.; (Cavriglia)

'ORFEO, CLIMBING', LCl, dr, 1963; flowers deep red, large, dbl., moderate fragrance; [Curly Pink X Guinee]; Leenders, J.

Organdi, HT, dr

'ORGUEIL DE LYON', HP, or, 1886; flowers vermilion; Besson

'ORIANA', HT, rb, 1970; flowers cherry-red, reverse white, dbl., 38 petals, 5 in., slight fragrance; foliage glossy, dark; Tantau, Math.; Wheatcroft & Sons

'ORIENT', HT, dr, 1959; bud pointed; flowers crimson becoming vermilion, reverse vermilion with yellow base, dbl., 22 petals; [Queen of Bermuda X Henri Mallerin]; Dot, Simon

'ORIENT EXPRESS', HT, or, 1978; flowers deep orange-red, reverse lighter, dbl., 40 petals, 4–5 in., intense fragrance; foliage bronze; vigorous growth; [Sunblest X Seedling]; Wheatcroft

Orient Silk *see* **'AHIMSA'**

'ORIENT SPICE', HT, m, 1996; (Rajni); flowers pinkish lilac, reverse same, blooms borne mostly single, dbl., 26–40 petals, 5 in., intense fragrance; moderate prickles; foliage large, medium green, semi-glossy; bushy, medium growth; [Violaine X Margaret Merrill]; Viraraghavan, M.S.; Anand Roses, 1984

'ORIENTAL CHARM', HT, mr, 1960; bud globular; semi-dbl., 11 petals, 3–4 in., slight fragrance; foliage leathery, glossy, dark; vigorous, upright, bushy growth; [(Charlotte Armstrong X Gruss an Teplitz) X (Mme Butterfly xFloradora)]; Duehrsen; Elmer Roses Co.

Oriental Dawn *see* **'ASAGUMO'**

Oriental Emporer, LCl, or, 1999; Clements, John K.

'ORIENTAL GLAMOUR', F, mr, 1972; flowers orient red, semi-dbl., 15 petals, 4–5 in., exhibition form, slight fragrance; foliage bronze; free growth; [Coup de Foudre X Tropicana]; Gandy, Douglas L.

Oriental Palace *see* POUlorin

'ORIENTAL QUEEN', Ch, or, 1926; flowers brilliant orange-scarlet, base yellow, fading to vivid carmine, dbl., 26 petals; few thorns; foliage dark; McGredy; Beckwith

Oriental Simplex™ *see* 'MINANCO'

'ORIENTALE', Cl HT, ob, 1946; flowers coral, becoming old-rose, free, recurrent boom, moderate fragrance; foliage glossy; very vigorous growth; not dependably hardy.; [George Dickson X Mrs Pierre S. duPont]; Robichon

'ORIFLAMME', HMult, pb, 1914; flowers deep rose-pink, suffused coppery gold, dbl., 55 petals, 2.5 in., moderate fragrance; vigorous, climbing growth; Paul

Origami™ *see* 'AROCHARM'

'ORIHIME', HT, pb, 1983; flowers light pink flushed rose pink, reverse pearly light pink; [Confidence sport]; Itami Rose Nursery

O'Rilla *see* 'RESILLA'

'ORIMANDA', HT, dr, 1975; bud long, pointed; dbl., 27 petals, 3.5 in., exhibition form, slight fragrance; foliage glossy, dark; vigorous, upright, bushy growth; [Duftwolke X Seedling]; Kordes; Willemse

'ORINDA', S, ly, 1922; flowers deep cream, medium, dbl., profuse, non-recurrent bloom; foliage soft, dark; vigorous (5 ft.), bushy growth; [Harison's Yellow X ?]; Central Exp. Farm

Oriola *see* 'MACLOCKER'

'ORIOLE', HMult, my, 1912; flowers marechal niel yellow, dbl.; vigorous, climbing growth; [Aglaia seedling]; Lambert, P.

'ORION', F, mr, 1968; flowers scarlet, medium, dbl., slight fragrance; foliage glossy; [Pink Parfait X Red Dandy]; Harkness

Orione®, HT, ob

Orlando *see* INTervema

Orlando Sunshine *see* 'MICADA'

'ORLÉANS IMPROVED', Pol, mp, 1931; flowers vivid rose-pink; [Orléans Rose sport]; Norfolk Nursery

'ORLÉANS ROSE', Pol, rb, 1909; flowers vivid rosy crimson, center white, semi-dbl., repeat bloom, slight fragrance; foliage glossy; vigorous, bushy growth; (14); [Thought to be Mme Norbert Levavasseur seedling sport]; Levavasseur

'ORLÉANS ROSE, CLIMBING', Cl Pol, rb, 1913; Levavasseur

'ORMISTON ROY', S, dy, 1953; flowers large, single, non-recurrent; bushy (3 ft.) growth; [R. spinosissima X R. xanthina]; Doorenbos

Ornella Muti®, HT, pb

'ORNEMENT DE LA NATURE', HGal, m, 1826; Hardy, before 1814

Ornement des Bosquets, HSem, dp; Jamain, H., 1860

'ORNEMENT DU LUXEMBOURG', HP, m, 1840; flowers violet/red, small, dbl., intense fragrance; Hardy; (Sangerhausen)

'ORPHELINE DE JUILLET', HGal, m; flowers crimson-purple, base of petals fiery red with occasional str, very dbl.; erect, moderate growth; Paul, before 1836; Prior to 1837

'ORPHEUS', HT, op, 1963; flowers salmon-orange, medium, dbl., 35–40 petals;

foliage glossy, dark; long stems; vigorous growth; [Montezuma X Unnamed seedling]; Verschuren, A.; Stassen

'ORPINGTON GEM', HT, or, 1953; bud long, pointed; flowers coppery orange shaded gold, dbl., 32 petals, 3 in., intense fragrance; foliage dark, leathery; vigorous growth; [Princess Marina X Alamein]; Buckwell

'ORPINGTON JEWEL', HT, op, 1953; bud very long; flowers coral-pink shaded coppery, base orange, medium size, dbl., 30 petals, 3 in., intense fragrance; foliage dark, leathery; vigorous growth; [Princess Marina X Picture]; Buckwell

Orsen's Big Red, HT, mr, 1990

Orsola Spinola *see* MASmarti

'ORTANA', HT, w, 1994; (**Anna Katherine**); flowers white blend, moderately full, dbl., 15–25 petals, 1.5–2.75 in., borne mostly singly, intense fragrance; some prickles; foliage large, medium green, matt; tall, upright growth; [Osiana X Seedling]; Ortega, Carlos, 1994; Aebi Nursery, 1994

'ORTCAL', HT, rb, 1995; (**California Glory**); very dbl., 3–3.5 in., borne mostly singly, intense fragrance; many prickles; foliage medium, dark green, matt; tall (150 cms), upright growth; [Osiana X Kardinal]; Ortega, Carlos; Aebi Nursery

'ORTISA', HT, mp, 1996; (**Marisa**™); flowers medium pink, lighter pink reverse, classic, urn-shaped bloom, dbl., 15–25 petals, 5 in., no fragrance; moderate prickles; foliage medium, dark green, semi-glossy; upright, tall (5 ft) growth; [Paris d'Yves St. Laurent sport]; Ortega, Carlos; Aebi Nursery, 1996

'ORTKEE', HT, ly, 1993; (**Kika**); dbl., 15–25 petals, 3–3.5 in., borne mostly singly, intense fragrance; no prickles; foliage medium, dark green, semi-glossy; medium (150 cms), upright growth; [JACice X Seedling]; Ortega, Carlos; Aebi Nursery, 1993

'ORTLAE', HT, m, 1991; (**Lauren Elizabeth**); flowers mauve blend, large blooms borne mostly singly, dbl., intense fragrance; foliage large, dark green, matt; tall, upright growth; [Moonlight X Unnamed seedling]; Ortega, Carlos, 1992

'ORTMAR', HT, w, 1995; (**Margaret O**); very dbl., 1.5–2.75 in., borne mostly singly, intense fragrance; some prickles; foliage medium, dark green, matt; tall (180 cms), upright growth; [Pristine X Lauren Elizabeth]; Ortega, Carlos; Aebi Nursery, 1995

'ORTSAR', HT, m, 1997; (**Sterling Star**™); flowers very full, large, very dbl., 41 petals, borne mostly singly, intense fragrance; foliage medium, medium green, dull; upright, tall (8-10ft.) growth; [Lauren Elizabeth X Sterling Silver]; Ortega, Carlos

'ORTSEA', HT, pb, 1997; (**Chelsea**™); flowers double, medium, dbl., 15–25 petals, borne mostly singly, slight fragrance; foliage small, dark green, semi-glossy; upright, medium (4-5ft.)growth; [Lauren Elizabeth X Paul's Pink]; Ortega, Carlos

'ORTVIC', HT, w, 1997; (**Victoria's Song**™); flowers double, large, dbl., 15–25 petals, borne mostly singly, intense fragrance; foliage medium, medium green, semi-glossy; upright, tall (6.5-7ft.) growth; [Moonlight X Seedling]; Ortega, Carlos

Oscar, S, pb, 1994; Hauser

'OSCAR CHAUVRY', N, mp, 1900; flowers large, dbl., moderate fragrance; Chauvry; (Sangerhausen)

'OSCAR II ROI DE SUÈDE', HP, mr, 1890; flowers large, very dbl., moderate fragrance; Soupert & Notting; (Sangerhausen)

'OSCAR LECLERC', M, 1853; Moreau et Robert; (Cavriglia)

Osiana, HT, ab, 1988; PP007660; Tantau

'OSIRIA', HT, rb, 1978; bud long, pointed; flowers dark red, white reverse, blooms, dbl., 50 petals, 4.5 in., exhibition form, intense fragrance; short stems; vigorous, upright, bushy growth; [Snowfire X Unnamed seedling]; Kordes, W.; Willemse

Osjen, HT, dp; flowers carmine-pink, medium, dbl., moderate fragrance; (Sangerhausen)

Oskar Cordel *see* **'OSKAR KORDEL'**

'OSKAR KORDEL', HP, mp, 1897; (Oskar Cordel); flowers carmine, large, dbl., 40 petals, cupped, moderate fragrance; vigorous, compact growth; [Merveille de Lyon X Andre Schwartz]; Lambert, P.

'OSKAR SCHEERER', S, dr, 1961; flowers velvety dark red, large blooms in large clusters, dbl.; vigorous (6 ft), well-branched growth; Kordes, R.

'OSMUNDA', Pol, dr, 1923; flowers dark carmine; [Jessie sport]; Holland (?)

'OSTARA', F, mr, 1964; flowers bright red, medium, borne in clusters, dbl.; foliage dark; [Highlight X Valeta]; deRuiter

Ostrava, S, lp; Vecera, L., 1971; (Czech Rosa Club)

'OSWALD SIEPER', HT, w, 1933; bud long, pointed; flowers creamy white, very large, dbl., exhibition form, intense fragrance; foliage glossy; vigorous growth; [Mrs Charles Lamplough X Ville de Paris]; Krause; C-P

Otago® *see* 'MACNECTA'

Othello *see* 'AUSLO'

'OTHELLO', F, or, 1961; flowers light orange-red, medium, dbl., 22 petals, 2.5 in., cupped; [Ma Perkins X Cocorico]; Leenders, J.

'OTHELLO', HT, mr, 1963; flowers velvety bright red, large, dbl., 48–60 petals, moderate fragrance; foliage dark; vigorous, upright, bushy growth; [New Yorker X Unnamed seedling]; Verschuren, A.; Stassen

'OTOHIME', HT, or, 1977; (Fancy Princess); bud pointed; dbl., 58 petals, 6–6.5 in., exhibition form, moderate fragrance; foliage large, glossy, dark; vigorous growth; [(Hawaii X Tropicana) X (Tropicana X Peace)]; Keisei Rose Nurseries, Inc.

'OTOME PINK', HT, pb, 1967; flowers salmon-pink, reverse buff-yellow; [Amatsu-Otome sport]; Rumsey, R.H.

'OTTO KRAUSE', HT, dy, 1931; flowers coppery yellow, large, dbl., exhibition form, slight fragrance; foliage soft, glossy, bronze, dark; vigorous growth; [Mme Caroline Testout X Souv. de Claudius Pernet]; Weigand, C.

'OTTO MILLER', HT, mr, 1968; bud long, pointed; flowers deep red, dbl., 6 in., exhibition form, slight fragrance; foliage glossy, leathery; vigorous, upright growth; [(Chrysler Imperial X Independence) X (Mrs Charles Russell X Happiness)]; Morey, Dr. Dennison; Country Garden Nursery

'OTTO VON BISMARCK', HT, mp, 1908; flowers large, dbl., moderate fragrance; Kiese; (Sangerhausen)

Oudtshorn Joy *see* KORquick

'OUMA SMUTS', F, ly, 1950; flowers straw-yellow shaded rosy flesh, intense fragrance; [Egalite X Vanessa]; Leenders, M.

'OUR ALLIES', Pol, lp; Matthews, W.J., 1915; (Weatherly, L.)

'OUR ANNIE', HT, op, 1937; flowers vivid orange-cerise, large, dbl., globular, moderate fragrance; foliage glossy; very vigorous growth; Letts

'OUR BOB', HT, mr, 1928; flowers velvety red, center golden, intense fragrance; Dawes

Our Coral Pearl *see* 'DOSPEARL'

Our Diana *see* 'TWOEX'

Our Indira, HT, w, 1998; Viraraghavan, M.S.

Our Joy *see* 'HORJOY'

Our Jubilee *see* 'COCCAGES'

'OUR LADY', F, mr, 1956; flowers carmine, moderate fragrance; vigorous growth; [Seedling X Soestdijk]; Leenders, M.

Our Love *see* 'ANDOUR'

Our Molly *see* 'DICREASON'

'OUR PEARL', LCL, pb, 1993; flowers white with pink edges, blooms borne in small clusters, dbl., 13–18 petals, 1.5–2.75 in., slight fragrance; some prickles; foliage medium, medium green, matt; medium (7 ft), bushy, spreading growth; Bronze, ARC TG, 1993; Jerabek, Paul E.

'OUR PRINCESS', F, dr, 1949; (Crimson Glow); flowers deep velvety crimson, borne in large trusses, semi-dbl., slight fragrance; foliage glossy, dark; vigorous growth; (28); [Donald Prior X Orange Triumph]; Robinson, H.; Baker's Nursery

Our Rodeo, F, rb, 1997; Kordes

'OUR ROSAMOND', HT, pb, 1983; flowers silver and pink blend, large blooms borne singly, dbl., 35 petals, exhibition form, slight fragrance; foliage medium, medium green, glossy; upright growth; [Daily Sketch seedling X Red Planet]; Bell, Ronald J.; Treloar Roses Pty. Ltd.

'OUR SHIRLEY', HT, w, 1989; bud pointed; flowers medium, borne singly, dbl., 28 petals, no fragrance; prickles long, slender, red; foliage small, medium, dark green; medium, upright growth; [Judith Morton X Sylvia]; Wilson, George D.

'OUR SWEET ANN', F, op, 1975; flowers warm pink, with orange glow, full, dbl., 28 petals, 3.5–4 in., intense fragrance; foliage dark, leathery; vigorous, upright growth; [Queen Elizabeth X Elizabeth of Glamis]; Horsfield

Our Town *see* 'ZIPTOWN'

Our Victory *see* **'NACHA POBEDA'**

Out of Africa *see* KORocken

Out of Canada, HT, pb, 1998; Williams, J. Benjamin

Out of Yesteryear *see* 'MORYEARS'

Outrageous® *see* 'JACZAP'

'OUTSIDER', F, mr, 1956; bud ovoid; flowers bright blood-red, open, medium, borne in clusters, semi-dbl., slight fragrance; foliage glossy; vigorous, upright growth; [Fanal X Red Favorite]; Tantau, Math.

'OVATION', HT, or, 1977; bud ovoid; dbl., 25 petals, 4–4.5 in., exhibition form; foliage dark, leathery; vigorous, upright growth; [First Prize X Unnamed seedling]; Weeks

'OVER THE RAINBOW', Min, rb, 1972; flowers red, yellow reverse, small, dbl., exhibition form, slight fragrance; foliage leathery; vigorous, bushy growth; AOE, 1975; [Little Darling X Westmont]; Moore, Ralph S.; Sequoia Nursery

'OVER THE RAINBOW, CLIMBING', Cl Min, rb, 1974; Rumsey, R.H.

'OVERLOON', F, dp, 1949; flowers rosy pink, medium, dbl.; vigorous growth; [Irene X Hebe (F)]; Leenders, M.

Overnight Scentsation *see* 'SAVANIGHT'

'OVERTON ON DEE', HT, pb, 1976; flowers pink, reverse cream, very full, dbl., 40 petals, 4.5 in., slight fragrance; [(Ballet X Gavotte) X Orange Sensation]; Ellick

'OVERTURE', F, m, 1960; flowers lilac-lavender, well formed, semi-dbl., 18 petals, 3–3.5 in., moderate fragrance; foliage dark; vigorous, low growth; [(Seedling X Lavender Pinocchio) X Prelude]; LeGrice

'OVID', HSet, 1890; Geschwind, R.; (Cavriglia)

Owen's Pride *see* KIRpink

Oxbow *see* 'BERBOW'

'OXFAM', HT, pb, 1973; flowers deep pink, shaded lilac, dbl., 40 petals, 5 in., intense fragrance; foliage dark; upright growth; [Fragrant Cloud X Blue Moon]; Cobley; Harry Wheatcroft Gardening

'OXFORD', HT, yb, 1930; flowers deep warm peach, tipped orange-gold, well-shaped, large, moderate fragrance; Prince

Oxfordshire *see* 'KORFULLWIND'

Oyster Pearl, HT, lp; [Bewitched sport]

Oz Baby, Min, lp; [Poker Chip X Oz Gold]; Hannemann, F., 1990; The Rose Paradise; (Weatherly, L.)

Oz Gold, Min, ob, 1981; McGredy, Sam IV

P

P.J.'s Pride *see* 'JUDPRIDE'

P.M. Leenders *see* **'PRESIDENT MACIA'**

Pablito, Min, mp

Pablo Diez, HT; Moreira da Silva, A.; (Cavriglia)

Pace Setter *see* 'SAVAPACE'

Pacemaker *see* 'HARNOBLE'

Pacesetter *see* 'SAVAPACE'

Pachinko *see* 'SUNPACH'

'PACIFIC', HT, lp, 1927; flowers soft pink, semi-dbl.; [Los Angeles sport]; Pacific Rose Co.

'PACIFIC', F, mr, 1958; flowers bright red, semi-dbl.; foliage bright green; [Alain X Chanteclerc]; Gaujard

'PACIFIC', HT, lp; [Catherine Kordes X Rose Marie]; Heers, C W, 1948; (Weatherly, L.)

'PACIFIC BELLE', HT, mp, 1989; bud pointed; flowers light pink, reverse deeper pink, reflexed, pointed, medium, dbl., 31 petals, slight fragrance; prickles pointed, light brown; foliage light green, veined, glossy; bushy, branching, upright growth; [(Unnamed Peer Gynt seedling) X Josephine Bruce]; Cattermole, R.F.

'PACIFIC DARLING', Min, pb, 1991; flowers large blooms borne in large clusters, semi-dbl., 6–14 petals, slight fragrance; foliage large, dark green, glossy; low, upright, bushy, compact growth; [Winifred Coulter X Unnamed seedling]; Sudol, Julia

'PACIFIC PRINCESS', HT, mp, 1989; bud tapering; flowers medium, borne usually singly and in sprays of 3-4, very dbl., 56 petals, globular, intense fragrance; foliage dark green, semi-glossy, veined; upright, branching growth; [Pink Parfait X Red Planet]; Cattermole, R.F.; South Pacific Rose Nursery, 1988

Pacific Serenade® *see* 'SAVANADE'

'PACIFIC SUNSET', HT, pb, 1977; bud deep pointed; flowers pink to orange-scarlet, semi-dbl., 12–16 petals, 5 in., exhibition form, slight fragrance; foliage dark; upright growth; [Mme Henri Guillot X California]; Fong; United Rose Growers

'PACIFIC TRIUMPH', Pol, mp, 1949; flowers salmon-pink, intense fragrance; growth like parent; [Orange Triumph sport]; Heers; Pacific Nursery

Pacifica® *see* 'JACIF'

'PACOIMA', HT, lp, 1927; flowers flesh tinted yellow, dbl.; [William F. Dreer sport]; Pacific Rose Co.

Paddy McGredy *see* 'MACPA'

Paddy Stephens *see* 'MACCLACK'

Paddywack® *see* 'MINIWACK'

Padmavathi, HT, pb, 1995; K&S

'PADRE', HT, rb, 1921; flowers bright coppery scarlet, flushed yellow, well-formed, semi-dbl., moderate, fruity fragrance; vigorous growth; Cant, B. R.

'PADRE AMÉRICO', HT, mr, 1956; flowers carmine-red, intense fragrance; [Crimson Glory X Peace]; da Silva, Moreira

'PADRE CRUZ', HT, w, 1956; flowers rosy white edged ruby; [Branca X Peace]; da Silva, Moreira

'PADRE MAÑANET', HT, dr, 1957; flowers bright purple-garnet, reverse crimson-red, large, dbl., 30 petals, exhibition form, moderate fragrance; strong stems; upright growth; [Charles Mallerin X (Satan X Mirandy)]; Dot, Pedro

'PAEONIA', HP, dr, 1855; flowers crimson, large, dbl., intense fragrance; Lacharme, F.

Paeonia, HT, lp; Strnad; (Czech Rosa Club)

Paeonienrose, HGal, mp

'PAGAN BEAUTY', F, pb, 1965; flowers bright vermilion, reverse salmon-pink, dbl., 42 petals, 4–5 in., globular; foliage dark, leathery; vigorous growth; [Montezuma X Seedling]; Verschuren; Blaby Rose Gardens

Paganini, HMsk, rb, 1989; Lens

Paganini *see* MEIrolange

'PAGEANT', HT, rb, 1953; bud ovoid; flowers red, reverse yellow, dbl., 35–40 petals, 4.5–5 in., exhibition form, moderate fragrance; foliage leathery; vigorous, upright growth; [Unnamed Hybrid Tea seedling X Orange Nassau seedling]; Boerner; J&P

'PAGLIACCI', Gr, yb, 1969; bud pointed; flowers yellow, becoming cerise, medium, semi-dbl., moderate fragrance; foliage glossy, bronze; vigorous, upright growth; Von Abrams; Edmunds Roses

'PAGODA', S, dr; flowers dark velvet red with showy gold stamens, single (5 petals), small blooms borne in large clusters; slight fragrance; no prickles; foliage medium, medium green, semi-glossy; low, spreaing bushy growth; [Anytime X Mountain Mist]; Jobson, Daniel J.; Jobson, 1995

Pahadi Dhun, HT, m, 1981; Pal, Dr. B.P.

Paheka, HT, ly, 1980; Dawson

Pailine®, HT, ob; flowers yellow with pink edges

'PAINT BOX', F, yb, 1963; bud ovoid; flowers red and golden yellow, becoming deep red, blooms in clusters, semi-dbl., 3 in., flat, slight fragrance; foliage dark; vigorous, upright growth; [Seedling X St. Pauli]; Dickson, Patrick; A. Dickson

'PAINTBRUSH', Min, ly, 1975; bud mossy; flowers soft yellow to white, mini-moss, small, 8–10 petals, 1.5 in.; foliage small, glossy, leathery; vigorous, upright, bushy growth; [Fairy Moss X Goldmoss]; Moore, Ralph S.; Sequoia Nursery

Painted Damask *see* **'LÉDA'**

'PAINTED DESERT', HT, op, 1965; bud long, pointed; flowers pink and copper, open, dbl., moderate fragrance; foliage leathery; vigorous, upright growth; [Talisman sport]; Lone Star Rose Nursery

Painted Desert *see* 'MACCHOPSU'

Painted Doll® *see* 'LAVPAINT'

'PAINTED LADY', HT, mp, 1931; bud long, pointed; flowers bright cerise, base old-gold, large, dbl., 42 petals; [(Crusader X Premier) X Julien Potin]; Ward, F.B.

'PAINTED LADY', HT, yb, 1979; bud pointed; flowers cream and gold, red tipped, dbl., 4 in., slight fragrance; foliage glossy, bronze; bushy growth; Herholdt, J.A.

Painted Melody, HT, yb, 1993; K&S

Painted Moon *see* 'DICPAINT'

'PAINTED SPAIN', Min, ob, 1995; flowers hand-painted orange stippled with white, white reverse, dbl. (15–25 petals), medium blooms borne in small clusters; slight fragrance; numerous prickles; foliage medium, medium green , semi-glossy; tall(3 ft), almost climbing, upright growth; [Rise 'n' Shine X (Queen Elizabeth X Eyepaint)];Jobson, Daniel J.; Jobson, 1995

Painted Star *see* 'MACMIGMOU'

Painter's Palette *see* 'MORPALE'

Painter's Touch *see* 'LAVTUCH'

Paint-Pot *see* 'TROBGLOW'

Paisley Anniversary *see* 'ANDPAI'

Pal, HT, ob

'PALACKY', HT, ob, 1936; flowers orange-yellow, open, large, single, intense fragrance; foliage glossy; vigorous growth; [Mme Mélanie Soupert X Sunburst]; Böhm, J.

'PALADIN', F, or, 1960; flowers open, borne in clusters, semi-dbl., 2.5 in.; moderate growth; [Signal Red X Fashion]; deRuiter

Palais de Laeken, HGal, lp; flowers medium, dbl.; (Sangerhausen)

'PALATINO', F, mr, 1956; dbl.; very vigorous growth; Frères, Buyl

'PALE HANDS', HT, ab, 1965; bud pointed; flowers ivory-white to buff and peach, very large, dbl., moderate fragrance; foliage leathery; vigorous, upright growth; [McGredy's Ivory X ?]; Pal, Dr. B.P.; Indian Agric. Research Inst.

'PALE MOON', HT, my, 1967; flowers large, dbl., exhibition form, moderate fragrance; foliage glossy; vigorous, compact growth; [Ma Perkins X Peace]; Patterson; Patterson Roses

'PALEFACE', Gr, w, 1959; bud ovoid; flowers nearly white, base naples yellow, semi-dbl., 15–25 petals, 3–5 in., exhibition form, slight fragrance; foliage leathery, semi-glossy; vigorous, upright growth; [Joanna Hill X Unnamed seedling]; Lindquist; Howard Rose Co.

'PALETTE', F, yb, 1960; bud ovoid; flowers yellow to salmon-pink and then red, blooms in clusters, dbl., slight fragrance; foliage glossy; compact growth; [Masquerade X High Noon]; Leenders, J.

Palissade Rose® *see* 'KORDAPT'

Paljas®, HT, pb

Pallas *see* 'HARVESTAL'

'PALLAS', HT, dr, 1975; flowers garnet-red, reverse paler, dbl., 40 petals, exhibition form, intense, spicy fragrance; foliage large, dull, dark; tall growth; [Chrysler Imperial X Shannon]; Murray & Hawken; Kennedy

'PALLIDA', Ch, mp, 1789; (Bengal Ordinaire); flowers clear rose; Kerr

'PALM SPRINGS', F, rb, 1965; bud ovoid; flowers oriental red to light bronze, center yellowish pink, dbl., 50 petals, intense fragrance; foliage bronze; vigorous, upright growth; [Oriental Charm X Circus]; Duehrsen; Elmer Roses Co.

Palmengarten Frankfurt® *see* KORsilan

Palmetto Sunrise *see* 'MICPAL'

Palmira Bastos, HT; Moreira da Silva, A.; (Cavriglia)

'PALMYRE', P, lp, 1844; flowers pale pink, portland type, often repeats; Laffay, M.

Palocsay Rudolf, HT, lp; long; flowers well shaped suitable for glasshouse as cut flower, exhibition form, 30 petals, moderate; leaves medium, medium green, glossy; semi-vigorous upright; [(Charles P Kilham X R. harisonii) X Ville de Paris]; St. Wagner, 1950; Res. Stn. f. Horticulture, Cluj, 1955, Romania

'PALOK', HR, rb, 1993,(**Brisbane Blush**); flowers cerise with silvery cream base, full (26–40 petals), large (7+ cms) blooms borne in small clusters; fragrant; few prickles; foliage medium, medium green, semi-glossy; medium upright growth; [(Golden Slippers X Lavendule) X Prima Ballerin]; Long, Peter; Queensland Rose Society, 1993

'PALOMA', HT, w, 1968; bud urn-shaped; flowers large, dbl., exhibition form; foliage leathery, olive-green; vigorous, bushy growth; GM, Portland, 1971; [Mount Shasta X White Knight]; Swim & Weeks; Weeks Wholesale Rose Growers

'PALOMA BLANCA', S, w, 1984; flowers ivory-white, blooms borne 3-10 per cluster, dbl., 35 petals, cupped, repeat bloom, moderate fragrance; awl-like, tan prickles; foliage leathery, dark olive green; erect, spreading, bushy, compact growth; hardy.; [Vera Dalton X ((Pink Princess X Lillian Gibson) X (Florence Mary Morse X (Josef Rothmund X R. laxa))]; Buck, Dr. Griffith J.; Iowa State University

'PALOMA FALCÓ', HT, op, 1930; flowers coral-salmon, dbl., moderate fragrance; [Li Bures X Château de Clos Vougeot]; Dot, Pedro; C-P

'PAM', F, yb, 1962; bud ovoid; flowers yellow edged red, becoming deep red, small, semi-dbl., cupped, slight fragrance; moderate, bushy growth; [Masquerade X ?]; Annabel

'PAMÁTNIK KOMENSKÉHO', LCl, mp, 1936; flowers salmon-pink, large, free early bloom, moderate fragrance; vigorous, climbing growth; Bojan; Böhm

'PAMÁTNIK KRÁLE JIRÍHO', HT, dr, 1936; (King George's Memorial); bud pointed; flowers deep crimson, marked velvety purple, large, dbl., intense fragrance; foliage glossy, dark; vigorous growth; [Gorgeous X Gen. MacArthur]; Böhm, J.

Pamela *see* 'KORFLAPEI'

'PAMELA', HT, yb, 1924; flowers canary-yellow and bright blush-pink shaded deep carmine, dbl., moderate fragrance; Therkildsen

Pamela Ann, HT, mr, 1998; Tudor

Pamela Joy, S, lp; [Oz Gold X Eye Paint]; Hannemann, F., 1999; The Rose Paradise; (Weatherly, L.)

'PAMELA TRAVERS', HT, mp, 1966; flowers large, dbl., intense fragrance; foliage leathery; vigorous, upright growth; [Pink Favorite X Queen Elizabeth]; Morey, Dr. Dennison; Country Garden Nursery

'PAMELA'S CHOICE', HT, my, 1966; flowers golden yellow, 4.5 in., slight fragrance; foliage bronze; [Piccadilly sport]; Bardill Nursery

Pamina®, HT, dp, 1991; Liebig

'PAMPA', F, mr, 1971; bud pointed; flowers brilliant vermilion-red, dbl.; foliage dark; [Colisee X Atlantic]; Gaujard

Pampa, HT, dy, 1988; K&S

Pam's Passion *see* 'GELPAM'

Pam's Pink, Ch, pb, 1880

'PAN AMERICA', HT, ob, 1941; flowers deep orange suffused tawny yellow, becoming light gold, open, dbl., 30–40 petals, 5.5 in., moderate fragrance; foliage glossy, bronze; long stems; vigorous, upright, bushy, open, growth; [Heinrich Wendland sport X Max Krause]; Boerner; J&P

Panache *see* 'POULTOP'

'PANACHÉ', Gr, mp, 1959; flowers light salmon-pink, well formed, large, intense fragrance; vigorous growth; [Rubin (HT) X Cinnabar]; Lens

'PANACHEÉ À FLEURS DOUBLES', HGal, pb, 1839; flowers rose pink with violet striping, medium, dbl., moderate fragrance; Viert; (Sangerhausen)

Panachée a Fleurs Plaines, HGal, m, 1839; Vibert

'PANACHÉE D'ANGERS', P, lp, 1879; flowers anthers golden, semi-dbl.; prickles spiny; vigorous growth; Moreau et Robert

'PANACHEÉ DE BORDEAUX', HP, rb, 1898; flowers red with white striping, large, dbl.; Duprat; (Sangerhausen)

'PANACHÉE DE LYON', P, pb, 1895; flowers pink, variegated crimson; [Rose du Roi sport]; Dubreuil

'PANACHÉE D'ORLÉANS', HP, pb, 1854; flowers blush-white striped deep rose; [Baronne Prevost sport]; Dauvesse

Panachee Double *see* **'VILLAGE MAID'**

Panachée Double *see* **'PERLE DES PANACHÉES'**

'PANAMA', HT, lp, 1913; flowers flesh edged lighter, dbl., slight fragrance; [Frau Karl Druschki X Unnamed seedling (pink)]; Cook, J.W.

'PANAME', LCl, mp, 1959; flowers bright pink, reverse apricot tinted salmon, large, dbl., free, recurrent boom, slight fragrance; long, strong stems; well branched growth; [Spectacular X Unnamed seedling]; Delbard-Chabert

Panchganga, HT, pb, 1993; Patil, B.K.

'PANCHU', F, mr, 1966; bud globular; flowers ruby-red, open, medium, semi-dbl.; foliage leathery; very vigorous, upright, compact growth; Pal, Dr. B.P.; Indian Agric. Research Inst.

Pandemonium *see* 'MACPANDEM'

Pandora *see* 'HARWINNER'

'PANDORA', HT, my, 1947; bud ovoid; flowers cream to deep yellow, dbl., 35–55 petals, 5 in., exhibition form, moderate fragrance; foliage dark, semi-glossy; very vigorous, tall growth; [Golden Rapture X R.M.S. Queen Mary]; Barké; Arnold-Fisher Co.

'PANEERA', HT, w, 1983; flowers large, dbl., 35 petals, moderate fragrance; foliage medium, dark, glossy; upright growth; [Unnamed X Unnamed]; Fumagalli, Niso

'PANIA', HT, lp, 1968; flowers well-formed, slight fragrance; foliage leathery; Gold Star of the South Pacific, Palmerston North, NZ, 1969; [Paddy McGredy X (Kordes' Perfecta X Montezuma)]; McGredy, Sam IV

'PANORAMA', HT, mp, 1943; flowers rose-pink, reverse silvery pink, dbl., 30 petals, 5 in., cupped, moderate fragrance; foliage glossy; vigorous, upright, compact growth; [Mrs A.R. Barraclough X Unnamed seedling]; McGredy; J&P

Panorama *see* **'PANORAMA HOLIDAY'**

'PANORAMA HOLIDAY', F, mp, 1973; (Panorama); flowers rose, pointed, dbl., 34 petals, 3.5 in., moderate fragrance; foliage glossy, dark; [Queen Elizabeth X Unknown seedling]; Gregory

Panthere Rose *see* 'MEICAPINAL'

'PANTOMIME', F, dp, 1965; flowers deep pink, in clusters, dbl., 30 petals, 4 in., exhibition form, moderate, fruity fragrance; [Ma Perkins X Karl Herbst]; McGredy, Sam IV; McGredy

Paola *see* 'TANALOAP'

'PAPA GONTIER', T, pb, 1883; bud long, pointed; flowers bright pink, reverse carmine-red, large, semi-dbl., intermittent bloom, moderate fragrance; foliage rich green; vigorous, bushy growth; (21); Nabonnand, G.

'PAPA GONTIER, CLIMBING', Cl T, pb, 1904; Chevrier

'PAPA GOUCHAULT', HMult, dr, 1922; bud long, pointed; flowers pure crimson-red, open, borne in clusters of 10-20, dbl., slight fragrance; foliage large, glossy; long stems; very vigorous, climbing growth; [Rubin X Unnamed seedling]; Turbat

'PAPA HÉMERAY', Ch, rb, 1912; flowers red, center white, blooms in clusters, single; very vigorous growth; Hémeray-Aubert

'PAPA HENDRICKX', HT, or, 1964; bud long, pointed; flowers vermilion-orange, large, dbl., 45–50 petals, exhibition form; foliage bronze; very vigorous, upright growth; [(Jolie Madame X R. rugosa rubra) X Unnamed seedling]; Mondial Roses

'PAPA JOAO XXIII', HT, lp, 1963; flowers pearl-pink; [Plaisir de France X La Jolla]; da Silva, Moreira

'PAPA KLEIN', HT, or, 1934; flowers reddish coppery orange, passing to salmon-pink, imbricated, dbl., moderate fragrance; foliage cedar-green; very vigorous growth; [Margaret Spaull X Norman Lambert]; Ketten Bros.

'PAPA LAMBERT', HT, mp, 1899; flowers large, dbl., intense fragrance; Lambert, P.; (Sangerhausen)

Papa Meilland® *see* 'MEISAR'

Papa Meilland, Climbing® *see* 'MEISARSAR'

Papa Pirosha, HT, my, 1990; Bulsara

'PAPA ROUILLARD', HWich, mr, 1923; flowers bright carmine, borne in long clusters of 15-25, dbl., abundant seasonal bloom; thornless; foliage rich green, glossy; long stems; very vigorous, climbing growth; [Leontine Grevais X Unnamed seedling]; Turbat

'PAPA SCHNEIDER', HT, dr, 1961; flowers dark red-purple, large, dbl., moderate fragrance; foliage glossy; vigorous, upright growth; [Crimson Glory X Unnamed seedling]; Kriloff, Michel

Papagena, F, yb; flowers light yellow with carmine-red, medium, semi-dbl., slight fragrance; VEG; (Sangerhausen)

Papageno® *see* 'MACGOOFY'

Paper Doll *see* 'JACFISEQ'

Papi Delbard *see* DELaby

Papilio, HT, 1955; Aicardi, D.; (Cavriglia)

'PAPILLON', T, pb, 1881; flowers coppery salmon rose, medium, semi-dbl.; Nabonnand

Papillon, Ch, mr, 1900; Dubourg

Papillon *see* TANollipa

'PAPILLON ROSE', F, mp, 1956; flowers pink tinted salmon, blooms in clusters, dbl., exhibition form, intense fragrance; vigorous, bushy growth; [White Briarcliff X (Lady Sylvia X Fashion)]; Lens

'PAPOOSE', Cl Min, w, 1955; bud pointed; single, 1 in., blooms in clusters; foliage small, fern-like, semi-glossy; vigorous, spreading (to 3-4 ft) growth; trailer or groundcover; [R. wichurana X Zee]; Moore, Ralph S.; Sequoia Nursery

Paprika® *see* 'TANPRIK'

Paprika™ *see* 'MEIriental'™

Papst Johannes XXIII, HT, w, 1963; flowers large, dbl., slight fragrance; Brauner; (Sangerhausen)

'PÂQUERETTE', Pol, w, 1875; (La Pâquerette, Ma Pâqueretta); flowers pure white, with imbricated petals, blooms in broad clusters, very dbl., 1 in., cupped, slight fragrance; very few prickles; foliage glossy, 3-5 leaflets; stems bright green; dwarf (12-15 in), bushy growth; (14); Guillot et Fils

'PARA TI', Min, w, 1949; (For You, Pour Toi, Wendy); flowers white, base tinted yellow, semi-dbl.; foliage glossy; very bushy (6-8 in) growth; [Eduardo Toda X Pompon de Paris]; Dot, Pedro; A. Meilland, 1946

'PARADE', LCl, dp, 1953; bud ovoid; flowers deep rose-pink, dbl., 33 petals, 3.5 in., cupped, moderate fragrance; foliage glossy; vigorous growth; [New Dawn seedlng X World's Fair, Climbing]; Boerner; J&P

'PARADE MARSHAL', F, mr, 1974; flowers full, dbl., 25–30 petals, 2–2.5 in., exhibition form, slight fragrance; vigorous growth; [Little Leaguer X Gemini]; Byrum; J.H. Hill Co.

'PARADIS', HT, mr, 1944; bud pointed; flowers clear red, medium, dbl., cupped, slight fragrance; foliage glossy; vigorous growth; Gaujard

'PARADISE', HWich, pb, 1907; flowers rose-pink, center white, tips of petals notched, large bloom, single, non-recurrent, slight fragrance; foliage glossy; vigorous, climbing (10-15 ft.) growth; RULED EXTINCT 1/79; Walsh

Paradise™ *see* 'WEZIP'

'PARADISE RAMBLER', HWich, pb, 1907; flowers pink with white, large, single, slight fragrance; Walsh; (Sangerhausen)

Paradise, Climbing, Cl HT, m, 1985; Weeks

Parador® *see* 'MEICHANSO'

Parador *see* 'MEIKINOSI'

'PARAGLIDER', S, ob, 1984; bud ovoid, pointed; flowers light pink, reverse orange-red, medium-large blooms borne 1-, dbl., 28 petals, cupped, repeat bloom, slight fragrance; awl-like, tan prickles; foliage medium, dark green tinted copper, leathery; vigorous, bushy, spreading growth; hardy.; [(Country Dancer X Carefree Beauty) X Alexander]; Buck, Dr. Griffith J.; Iowa State University

Paragon *see* 'MACKOSRED'

'PARAMOUNT', HT, ab, 1950; bud long, pointed; flowers orange-salmon-buff, becoming, dbl., 30 petals, 4–5 in., exhibition form, slight fragrance; foliage glossy; very vigorous, upright, bushy growth; [Charlotte Armstrong X Glowing Sunset]; Swim, H.C.; Paramount Nursery

'PARASOL', HT, dy, 1964; flowers rich yellow, dbl., 28 petals, 5 in., moderate fragrance; foliage dark; compact growth; [Peace X Ethel Sanday seedling]; Sanday, John

Paray, F, mr

Parc des Princes® *see* DORpurp

'PARDINAS BONET', HT, yb, 1931; flowers deep yellow, reverse red, dbl., intense fragrance; [La Giralda X Souv. de Claudius Pernet]; Dot, Pedro; C-P

Paree Pink, Pol, mp

Paree Red, Pol, mr

Paree Salmon, Pol, op

Paree White, Pol, w

'PAREL VAN AALSMEER', HT, dr, 1941; (Perle von Aalsmeer); flowers deep red; [Better Times sport]; Verschuren

Pareo® *see* 'MEIFLAROL'

Pareo® *see* KEIzoubo

'PARFAIT', F, mr, 1975; flowers sweetheart; PP004046; [Minuette sport]; Knight, C.; DeVor Nurseries, Inc.

Parfum d'Armor® *see* ADAnuamn

Parfum de Franche-Comte, HT, op, 1989; Sauvageot

'PARFUM DE LA NEIGE', HT, w, 1939; flowers large, dbl., intense fragrance; Mallerin, C.; (Sangerhausen)

Parfum de l'Hay *see* **'ROSE À PARFUM DE L'HAY'**

Parfum LIffreen, F, m, 1995

'PARIS', Pol, mr, 1929; flowers bright red; vigorous growth; deRuiter

Paris 2000 *see* 'DELSAMO'

Paris de Yves st Laurent™ *see* 'MEIVAMO'

Paris Pink *see* 'ARODOUSNA'

Paris Red *see* **'ROUGE DE PARIS'**

'PARIS SUPERIOR', Pol, mr; flowers have more lasting color than parent; [Paris sport]; deRuiter

'PARISER CHARME'®, HT, mp, 1965; bud ovoid; flowers pink, well-formed, blooms in clusters (to 10), dbl., 28 petals, 5 in., intense fragrance; foliage dark, glossy; vigorous, upright growth; ADR, 1966; Tantau, Math.

'PARISH LIFE', Gr, or, 1999; flowers orange, red edged, reverse orange, fades to light pink, dbl., 26–40 petals, 4 in., borne mostly singly, slight fragrance; few prickles; foliage medium, medium green, semi-glossy; upright, tall (5 ft) growth; [Piccadilly X Marijke Koopman]; Wilson, George D.; Holy Family Catholic Church, 2000

'PARIS-MATCH', HT, dp, 1957; flowers carmine to rose, center darker, dbl.; foliage leathery; vigorous growth; GM, Bagatelle, 1956; [Independence X Grand'mere Jenny]; Meilland, F.; URS

'PARK AVENUE', HT, dr, 1962; bud ovoid; flowers cardinal-red, open, dbl., 38–48 petals, 3–4 in., moderate fragrance; foliage leathery, dark; vigorous, upright growth; [Yuletide X San Fernando]; Jelly; E.G. Hill Co.

Park Jewell *see* **'PARKJUWEL'**

Park Place *see* AROcruby

'PARK ROYAL', F, ob, 1967; bud ovoid; flowers coral-pink, center white, reverse silver, open, medium, dbl., slight fragrance; foliage light green; vigorous, upright growth; [Tropicana X Shepherd's Delight]; Eddie, J.H.; Sheridan Nursery

Park Wilhelmshohe *see* KORwilpa

Parkay *see* 'MACSHANA'

'PARKDIREKTOR RIGGERS'®, HKor, dr, 1957; bud long, pointed; flowers velvety crimson, blooms in clusters (up to 50), semi-dbl., recurrent bloom, slight fragrance; foliage dark, glossy, leathery; very vigorous, climbing growth; ADR, 1960; [R. kordesii X Our Princess]; Kordes, R.

'PARKFEUER', HFt, or; flowers bright scarlet, single, non-recurrent; vigorous (6-8 ft.) growth; Lambert, P., 1906

Parkjewel *see* **'PARKJUWEL'**

'PARKJUWEL', S, lp, 1956; (Park Jewell, Parkjewel); bud ovoid; flowers very large, very dbl., cupped, non-recurrent, intense fragrance; foliage leathery, wrinkled, light green; vigorous (4 ft.), bushy growth; [Independence X Red Moss]; Kordes; Morse Roses, 1950

'PARKLANE', HT, my, 1961; bud ovoid; flowers canary-yellow, dbl., 25–35 petals, 4.5–5 in., exhibition form, slight fragrance; foliage glossy; vigorous, upright growth; [Peace X Dawn]; Jelly; E.G. Hill Co.

Parkprinzessin, HT, mp

Parkrose, S, dr; flowers large, dbl., moderate fragrance; Elvinge, H.; (Sangerhausen)

'PARKS' YELLOW TEA-SCENTED CHINA', T, my, 1824; (R. indica ochroleuca, R. X odorata ochroleuca); flowers pale yellow, dbl.; Int. into England in 1824

'PARKSIDE ROSE', S, dp; flowers deep pink, single; vigorous (6 ft) growth

Parkstone Pride, S; Delforge; (Cavriglia)

Parkstown's Pride *see* 'TANDIRPKRAP'

Parkstrauchrose, S, rb, 1975; flowers dark red with golden yellow, small to medium, semi-dbl., slight fragrance; Scholle, E.; (Sangerhausen)

Parkwood Scarlet *see* KORtradkos

'PARKZAUBER', S, dr, 1956; bud long, pointed; flowers dark crimson, large, dbl., non-recurrent, moderate fragrance; foliage dark, leathery; very vigorous (4 ft.), upright, bushy growth; [Independence X Nuits de Young]; Kordes

'PARKZIERDE', B, dr, 1909; flowers scarlet crimson, petals shell-shaped, dbl., non-recurrent, moderate fragrance; long stems; very vigorous growth; Lambert, P.

Parly 2, HT, lp; flowers large, dbl., slight fragrance; (Sangerhausen)

Parme *see* **'SIMONE'**

'PARMELIA', HT, mr, 1957; bud long, pointed; flowers flamingo-red, dbl., 24–30 petals, 5–5.5 in., exhibition form; foliage dark, glossy, leathery; very vigorous, upright, compact growth; [Mme Chiang Kai-shek sport]; Lennard

Parmentier, HP, mp; Guillot et Fils, 1860

'PARSIFAL', F, op, 1968; flowers coral-salmon, well shaped, dbl., 34 petals; vigorous, bushy growth; [Dacapo X Ballade]; deRuiter

Parsons' Pink China *see* **'OLD BLUSH'**

Parthenon® *see* 'DELBRO'

Partridge *see* 'KORWEIRIM'

'PARTY DOLL', F, ab, 1958; bud ovoid; flowers pink, dbl., 40 petals, 2.5 in., cupped, moderate fragrance; foliage leathery, glossy; vigorous, bushy growth; [Goldilocks seedling X Fashion]; Boerner; J&P

'PARTY DRESS', HT, ab, 1961; flowers deep apricot shaded buff-peach, dbl., 25 petals, 5 in., exhibition form, moderate fragrance; foliage glossy; vigorous, bushy, compact growth; [Gay Crusader X Unnamed seedling]; Robinson, H.; Lowe

'PARTY GIRL'™, Min, yb, 1979; bud long, pointed; flowers soft apricot-yellow, high centered, dbl., 23 petals, 1–1.5 in., exhibition form, borne mostly singly, moderate, spicy fragrance; compact, bushy growth; PP4598; AOE, ARS, 1981 Miniature Rose Hall of Fame, ARS, 1999; [Rise 'n' Shine X Sheri Anne]; Saville, F. Harmon

Party Line™ *see* 'MOGALINE'

'PARTY PINK', F, mp, 1956; flowers bright pink, borne in clusters, dbl., 30–50 petals, 1–1.5 in., moderate fragrance; moderate growth; [Unnamed seedling X Pinocchio seedling]; Raffel; Port Stockton Nursery

Party Popcorn *see* 'LAVPOP'

'PARTY TIME', HT, yb, 1986; flowers lemon yellow with pink overlay, reverse lemon yellow, fading, dbl., 45 petals, cupped, moderate, fruity fragrance; medium, reddish prickles, hooked downwards; foliage medium, medium green, semi-glossy; upright, medium growth; PP006457; [Perfume Delight X Half Time]; Weeks, O.L.

Party Trick® *see* DICparty

Partyglo® *see* 'MINPCO'

'PARURE', HT, dp, 1965; flowers carmine-rose, 5 in., cupped, slight fragrance; foliage serrated; tall to moderate growth; [Michele Meilland X Chic Parisien]; Delbard-Chabert; Cuthbert

Parure d'Or® *see* 'DELMIR'

Parvifolia *see* **'BURGUNDIAN ROSE'**

'PARWANA', F, yb, 1974; bud pointed; flowers golden yellow, edged plum-red, open, dbl., 37 petals, 3 in., cupped, slight fragrance; foliage glossy, dark; vigorous growth; [Unknown X Unknown]; Pal, Dr. B.P.

Pasadena® *see* 'KORLAND'

'PASADENA', HT, yb, 1927; flowers golden yellow, edged flame, dbl., moderate fragrance; RULED EXTINCT 1/82; [The Queen Alexandra Rose sport]; Coolidge

'PASADENA TOURNAMENT', F, mr, 1942; (Red Cécile Brünner); bud long, pointed; flowers velvety red, small, dbl., 36 petals, cupped, moderate fragrance; foliage bronze; long stems; very vigorous, bushy growth; [Cécile Brunner X Unnamed seedling]; Krebs; Marsh's Nursery

'PASADENA TOURNAMENT, CLIMBING', Cl F, mr, 1945; (Red Cécile Brünner, Climbing); Marsh's Nursery

Pascali® *see* 'LENIP'

'PASCALI, CLIMBING', Cl HT, w, 1978; Anderson's Rose Nurseries

Pascaline® *see* 'LENPAS'

Pasita *see* 'KORSITA'

Paso Doble *see* 'MEILANODIN'

Passaya, HT, w

'PASSION', Gr, mr, 1954; bud long, pointed; flowers scarlet-cerise, dbl., 36 petals, 4 in., moderate fragrance; foliage dark; very vigorous, bushy growth; [Peace X Alain]; Gaujard

Passion de J. Renoard®, HT, m

Passion Rose *see* **'MANHATTAN'**

Passion's Flame *see* 'RENPAFLAME'

'PASSPORT', Cl HT, dr, 1940; flowers well-formed, moderate fragrance; vigorous growth; Clark, A.

'PASTEL', HT, ab, 1961; bud long, pointed; flowers creamy yellow and soft pink, dbl., 25 petals, 5 in., exhibition form, moderate fragrance; foliage glossy, wrinkled; strong stems; vigorous, upright growth; [(Sutter's Gold X Unnamed seedling) X Fred Edmunds]; Von Abrams; Peterson & Dering

Pastel Delight, HT, lp, 1984; Kasturi

'PASTEL PRINCESS', HT, pb, 1995; flowers two-tone pink, dbl. (15–25 petals), medium blooms borne mostly single; fragrant; some prickles; foliage medium, medium green, semi-glossy; upright, tall growth; [Seedling X Seedling]; Perry, Astor; Certified Roses Inc., 1995

Pastel Tower, HT, lp; [Eiffel Tower sport]; Ruston, D., 1968; (Weatherly, L.)

Pastelina *see* 'SUNPASTE'

'PASTEUR', HT, pb, 1973; bud long; flowers brilliant pink, flushed red, dbl.; upright growth; [Firmament X Femina]; Gaujard

Pastorale® *see* 'POURALE'

'PASTORELLA', HT, 1953; Meilland, F.; (Cavriglia)

Pastourelle *see* 'POURALE'

'PASTOURELLE', HT, op, 1952; flowers salmon-pink, very large, very free bloom; RULED EXTINCT 12/85; GM, Rome, 1953; [Comtesse Vandal X Étoile d'Or]; Robichon

Pasture Rose *see* **R. CAROLINA**

Pat Austin® *see* 'AUSMUM'

Pat James *see* HARyoricks

'PAT NIXON', F, dr, 1972; bud ovoid; flowers large, dbl., moderate fragrance; foliage large, glossy, dark; vigorous, upright, bushy growth; [Tamango X (Fire King X Banzai)]; Meilland; Stuart

'PAT PHOENIX', HT, yb, 1964; flowers cream to yellow, flushed pink, base yellow, dbl., 40 petals, 4.5–5 in., moderate fragrance; foliage dark, leathery; very vigorous growth; [(Wellworth X Clarice Goodacre) X Peace]; Latham

'PAT STEWART', HT, dp, 1976; flowers cerise to deep rose-pink, full, dbl., 26 petals, 5–5.5 in., slight fragrance; foliage dark; vigorous, upright growth; [Red Devil X Honey Favorite]; Stewart, G.

Patchwork, HT, rb

'PATCHWORK QUILT', Min, ob, 1990; bud ovoid; flowers orange-yellow-pink blend, aging light orange, urn-shaped, medium, dbl., 60 petals, slight fragrance; foliage small, medium green, matt; upright, bushy, medium growth; [Rise 'n' Shine X Dandy Lyon]; Jolly, Marie, 1985; Rosehill Farm, 1991

Pathfinder *see* 'CHEWPOBEY'

'PATIENCE', HT, ob, 1927; bud pointed; flowers scarlet-carmine shaded orange and orange-scarlet, large, dbl., exhibition form, moderate fragrance; GM, NRS, 1926; McGredy

'PATIENCE STRONG', HT, mr, 1969; flowers crimson-scarlet, pointed, large, dbl.; foliage dull, gray-green; free growth; [Basildon Belle X Red Dandy]; Trew, C.; Basildon Rose Gardens

'PATIENCE, CLIMBING', Cl HT, ob, 1935; Shamburger, C.S.

Patio Charm *see* CHEwapri

Patio Cloud *see* 'INTERFLU'

Patio Dance® *see* 'WILPADA'

Patio Delight, F, op

Patio Flame *see* 'POULCOT'

Patio Gem *see* 'SUNSALM'

Patio Gold® *see* 'WILPAGO'

Patio Honey *see* CHEwpiwitel

Patio Jewel *see* 'JACIAT'

'PATIO JEWEL', MinFl, m, 1975; bud pointed; flowers purple to clear amethyst, open, single, 5–7 petals, 2.5 in., slight fragrance; foliage leathery; very vigorous growth; [Europeana X Angel Face]; Williams, J. Benjamin

Patio Orange, MinFl, or; Harkness

'PATIO PATTY', MinFl, yb, 1975; bud pointed; flowers yellow, washed peach and orange, semi-dbl., 16 petals, 2 in., globular, intense fragrance; foliage small, reddish-green; [(Circus X The Optimist) X (Little Darling X Starina)]; Williams, J. Benjamin

'PATIO PEARL', MinFl, pb, 1975; flowers light pearl-pink, base deeper, semi-dbl., 18–20 petals, 1–1.5 in., exhibition form, slight fragrance; foliage small, glossy, dark; vigorous growth; [Fairy Queen X The Optimist]; Williams, J. Benjamin

Patio Pearl *see* POUlpearl

Patio Prince *see* 'POULCRIM'

Patio Princess *see* POUlholm

Patio Queen *see* 'CHEWSEA'

'PATIO RIBBON', MinFl, dr, 1975; bud ovoid; flowers velvety bright dark scarlet, semi-dbl., 16 petals, 2.5–3 in., cupped, moderate fragrance; foliage dark; strong growth; [Europeana X Red Favorite]; Williams, J. Benjamin

Patio Snow® *see* 'WILPASN'

'PATRICE', HT, w, 1985; flowers large, dbl., 25 petals, exhibition form, slight fragrance; foliage medium, dark, semi-glossy; upright, bushy growth; [(Baccará X Generosa) X Zecchino d'Oro]; Bartolomeo, Embriaco

Patricia *see* 'KORPATRI'

'PATRICIA', HT, mr, 1932; flowers carmine flecked pink, base orange-yellow, moderate fragrance; foliage glossy, dark; vigorous growth; Chaplin Bros.

'PATRICIA ANNE', HT, dp, 1962; flowers deep pink, reverse silvery pink, 4–5 in., moderate fragrance; foliage dark, glossy; vigorous growth; [Kordes' Perfecta sport]; Buzza

'PATRICIA C. OPPMANN', F, w, 1981; bud pointed; flowers very light yellow, blooms borne 1-7 per cluster, dbl., 43 petals, cupped, slight fragrance; slightly hooked prickles; foliage glossy, medium green; vigorous, upright, dense growth; Bronze, ARC TG, 1984; [Unnamed seedling X Unnamed seedling]; Jerabek, Paul E.

'PATRICIA DAWN', HT, or, 1998; flowers orange-red, lighter reverse, ehibition form, very dbl., 41 petals, 4 in., exhibition form, borne singly, moderate fragrance; prickles moderate; foliage medium, medium green, dull; upright, medium growth; [unknown sport]; Austin, M.L.

'PATRICIA HARKNETT', HT, dp, 1961; flowers deep pink, dbl., 23 petals, 4 in., moderate fragrance; foliage bronze-red; vigorous growth; [Lady Sylvia sport]; Harknett

'PATRICIA HYDE', F, mp, 1969; flowers medium, semi-dbl.; [Ann Elizabeth X Red Dandy]; Harkness

'PATRICIA MACOUN', LCl, w, 1945; dbl., slight fragrance; foliage dark, glossy; ; hardy.; [R. helenae X ?]; Central Exp. Farm

'PATRICIA MILLER', HT, w, 1978; bud ovoid; flowers pure white, open, large, semi-dbl., slight fragrance; foliage leathery; vigorous, bushy growth; [Queen Elizabeth sport X Queen Elizabeth sport]; Miller, J.; Treloar Roses Pty. Ltd.

'PATRICIA NEAL', HT, mr, 1963; flowers bright cerise, large, dbl., exhibition form; foliage dark, leathery; vigorous, tall growth; [Better Times sport]; Macres; Paragon Greenhouses

'PATRICIA PIESSE', HT, lp, 1971; flowers luminous light pink, medium, dbl., 50–60 petals, exhibition form, moderate fragrance; foliage glossy; vigorous, upright growth; [Elizabeth Fankhauser X Memoriam]; Fankhauser

'PATRICIA SCRANTON', Cl Min, yb, 1977; bud pointed; flowers light yellow, streaked red, small, semi-dbl., 17 petals, 1.5 in., slight fragrance; foliage glossy, dark, soft; [Fairy Moss X Fairy Moss]; Dobbs; Small World Min. Roses

'PATRICIA WATKINS', F, lp, 1947; flowers bright pink, medium, borne in trusses, single, 5 petals, slight fragrance; foliage dark; vigorous growth; [Karen Poulsen sport]; Watkins Roses

Patricia Weston *see* 'REYPAT'

'PATRICIAN', HT, mr, 1977; bud ovoid, pointed; flowers cardinal red, dbl., 28 petals, 4–5 in., exhibition form, intense fragrance; foliage large, dark; very upright growth; [Fragrant Cloud X Proud Land]; Warriner, William A.; J&P

'PATRICK ANDERSON', HT, dp, 1938; bud long, pointed; flowers deep rose-pink, large, dbl., exhibition form, intense fragrance; foliage leathery; vigorous growth; [John Henry X Portadown Fragrance]; McGredy; J&P

'PATRICK VINCENT', HT, dr, 1967; flowers crimson, intense fragrance; foliage dark; free growth; [Mirandy X F.W. Alesworth]; Vincent

Patriot *see* 'JACLIN'

Patriot Flame *see* 'CLESCRUB'

Pat's Choice *see* KORomega

Pat's Delight *see* 'WILRSS'

'PATSY', HT, w, 1930; bud pointed; flowers pure white, open, very large, dbl., cupped, slight fragrance; Dickson, H.; Morse

Patsy Cline™ *see* 'AROCUMU'

'PATTY LOU', Min, pb, 1953; (Petite); bud ovoid; flowers rose pink, reverse silvery pink, dbl., 55 petals, 1 in., moderate fragrance; foliage small; dwarf (10-12 in), bushy growth; [Oakington Ruby X Oakington Ruby]; Moore, Ralph S.; Sequoia Nursery

Patty Sue *see* 'TINPAT'

'PATTY'S PINK', F, lp, 1960; bud long, pointed; flowers rose-opal, reverse camellia-rose, dbl., 40–45 petals, 2.5 in., exhibition form, slight fragrance; foliage leathery, glossy; very vigorous, upright, compact growth; [(Cécile Brunner X Mrs R.M. Finch) X Self]; Spanbauer

'PATTY'S RED', F, mr, 1968; flowers cherry-red; [Patty's Pink sport]; Paulen Park Nursery

'PAUL BIGOT', HWich, dp, 1924; flowers bright rose, shaded vermilion, borne in clusters of 5-10; vigorous, climbing growth; Turbat

Paul Bocuse *see* MASpaujeu

'PAUL BOUCLAINVILLE', HT, yb, 1930; flowers carmine on yellow ground, reverse pinkish white tinted yellow, semi-dbl., cupped, moderate fragrance; very vigorous growth; [Mme Charles Detreaux X Mme Edouard Herriot]; Buatois

'PAUL BUATOIS', Cl HT, dr, 1931; flowers velvety red, base yellow passing to purplish carmine, very large, dbl., cupped, moderate fragrance; foliage leathery; very vigorous, climbing growth; [Marie Baumann X Mme Edouard Herriot]; Buatois

'PAUL BUNYAN', Gr, dr, 1961; bud long, pointed; flowers deep red, dbl., 55 petals, 5 in., exhibition form, slight fragrance; foliage leathery; very vigorous, upright growth; [Charles Mallerin X Carrousel]; Von Abrams; Peterson & Dering

Paul Cezanne *see* JACdeli

'PAUL CRAMPEL', Pol, or, 1930; flowers deep orange-scarlet, brighter and larger than gloria mundi, dbl.; (14); Kersbergen

'PAUL CRAMPEL, CLIMBING', Cl Pol, or, 1934; Appleton

'PAUL CRAMPEL, CLIMBING', Cl Pol, or, 1934; Vially

'PAUL CRAMPEL, CLIMBING', Cl Pol, or, 1937; Tantau

'PAUL DAUVESSE', HMult, my, 1933; bud long, golden yellow; flowers bright canary-yellow, large, borne in clusters of 4-8, dbl.; vigorous, climbing growth; Barbier

Paul de Fontainne *see* **'DEUIL DE PAUL FONTAINE'**

'PAUL DE LA MEILLERAYE', HP, mr, 1863; flowers carmine-red with lighter reverse, very large, dbl.; Guillot; (Sangerhausen)

'PAUL DÉLÉPINE', Pol, mp, 1933; flowers brilliant rose-pink, borne in clusters, dbl., globular; foliage leathery, glossy; vigorous growth; [Yvonne Rabier X Dorothy Perkins]; Délépine; Pajotin-Chédane

'PAUL DUVIVIER', HT, mr, 1932; flowers carmine, base yellow, dbl., intense fragrance; [Constance X Pax Labor]; Laperrière

Paul Fromont *see* **'BOUDOIR'**

Paul Gauguin *see* JACdebu

Paul Gold, HT, my, 1998; Poulsen

'PAUL HARRIS', HT, dy, 1993; dbl., 26–40 petals, 3–3.5 in., borne mostly singly, slight fragrance; some prickles; foliage medium, medium green, matt; medium (40-48 in), bushy growth; [(Summer Sunshine X Georgia) X Seedling]; Weeks, O.L.; Estrella Rose Company, 1993

Paul Holtge *see* **'GRACE MOORE'**

'PAUL KADOLOZIGUE', HMult, lp, 1912; flowers medium, dbl.; Lambert, P.; (Sangerhausen)

'PAUL KRAMPEL', Pol, 1930; Kersbergen; (Cavriglia)

'PAUL LAFONT', HT, ly, 1920; flowers golden yellow to white tinted yellow, dbl., slight fragrance; [Mme Maurice Capron X Unnamed seedling]; Guillot, P.

'PAUL LEDE, CLIMBING', Cl T, ab, 1913; (Monsieur Lede, Monsieur Paul Lede); pink, large and pointed; flowers yellowish buff, center flushed carmine, outer petals salmon buff, exhibition form, intense, lush, fruity fragrance; small dark green semi-glossy; vigorous to 15 feet; Lowe

'PAUL LUCCHINI', HT, dr, 1931; flowers purplish garnet, shaded velvety red, dbl., cupped, intense fragrance; foliage bronze, leathery; vigorous growth; [Rhea Reid X Yves Druhen]; Buatois

Paul McCartney *see* 'MEIZELI'

'PAUL MONNIER', HT, yb, 1902; flowers light yellow shaded salmon; Buatois

'PAUL NABONNAND', T, mp; flowers satiny rose, large, dbl., cupped; vigorous growth; Nabonnand, G., 1876

'PAUL NEYRON', HP, mp, 1869; flowers clear pink to rose-pink, very large, dbl., 50 petals, cupped, occasionally recurrent bloom, moderate fragrance; foliage large, rich green; vigorous growth; [Victor Verdier X Anna de Diesbach]; Levet, A.

'PAUL NOËL', LCl, pb, 1913; flowers old rose and pale yellow, blooms in clusters of 4-6, 2–3 in.; vigorous, climbing growth; [R. wichurana X Mons. Tillier]; Tanne

'PAUL PERRAS', HP, lp, 1870; (The Crepe Rose); flowers pale rose, compact, large, very dbl.; sets many hips; vigorous, pillar growth; Levet

'PAUL PLOTON', HWich, mr, 1910; flowers bright red, blooms late; [R. wichurana X Mme Norbert Levavasseur]; Barbier

Paul Potter, S, rb, 1997; Williams, J. Benjamin

'PAUL REVERE', HT, mr, 1940; flowers carmine, opening scarlet-crimson, dbl., 24–30 petals, 4–5 in., cupped, moderate fragrance; vigorous, upright growth; [Talisman sport]; Roland

Paul Ricard® *see* 'MEINIVOZ'

'PAUL RICAULT', C, mp, 1845; flowers rose-pink, dbl., quartered, intense fragrance; height to 5 ft; Portemer fils

Paul Richard *see* 'MEINIVOZ'

Paul Shirville *see* 'HARQUETERWIFE'

'PAUL TRANSON', LCl, op, 1900; flowers bright pink, large blooms in clusters of 3-5, dbl., slight fragrance; foliage glossy, dark; short, strong stems; height 10 ft; [R. wichurana X L'Ideal]; Barbier

'PAUL VERDIER', HP, dp, 1866; flowers bright rose; Verdier, C.

'PAUL VOGEL', HT, mp, 1932; flowers medium, dbl.; Vogel, M.; (Sangerhausen)

'PAULA', Gr, or, 1980; flowers dusty salmon; [Radiation induced Queen Elizabeth sport]; James, John

'PAULA ANNE CREASEY', F, rb, 1999; flowers upper 2/3 medium red, lower 1/3 white, reverse light pink, dbl., 26 petals, 2.5 in., borne in small clusters, slight fragrance; prickles moderate; foliage medium, medium green, dull; compact, low (2.5 ft) growth; [Tamango X Anne Scranton]; Dobbs, Annette E., 1998

'PAULA CLEGG', HT, mr, 1919; flowers bright scarlet; [Kaiserin Auguste Viktoria X ?]; Kiese

'PAULA MAYER', HT, pb, 1929; flowers silvery carmine-pink, reverse yellowish pink, semi-dbl., intense fragrance; [Mme Edmee Metz X Betty Uprichard]; Leenders Bros.

'PAULA MEIDINGER', HT, lp, 1923; flowers large, dbl., slight fragrance; Meidinger; (Sangerhausen)

Paula Scholle, S, mp, 1975; flowers medium, semi-dbl., slight fragrance; Scholle, E.; (Sangerhausen)

Paula Scholle II, F, op, 1980; flowers large, dbl.; Scholle, E.; (Sangerhausen)

'PAULE DELAVEY', Pol, ly, 1957; flowers creamy yellow; vigorous, bushy growth; Privat

'PAULETTE', F, ly, 1934; flowers saffron-yellow to yellowish white, borne in clusters, very dbl., cupped, intense fragrance; foliage leathery, glossy; vigorous growth; [Leontine Gervais X Paul Monnier]; Buatois

'PAULETTE', HT, dp, 1946; flowers bright rosy scarlet, center tinted salmon, well formed, large, very dbl., slight fragrance; foliage rich green; tall growth; [Peace X Signora]; Meilland, F.

'PAULETTE BENTALL', HMult, m, 1916; flowers purple/pink, small, semi-dbl.; Bentall; (Sangerhausen)

'PAULETTE BUFFET', HT, lp, 1921; flowers pale flesh-pink, reverse silvery pink, dbl., slight fragrance; [Jonkheer J.L. Mock X Unnamed seedling]; Gillot, F.

'PAULETTE COQUELET', HT, mr, 1947; bud long; flowers salmon-red tinted bright coral, large, dbl.; very vigorous growth; [Unnamed seedling X Daniel]; Mallerin, C.; URS

'PAULIEN VERBEEK', HT, ob, 1958; flowers orange-yellow, large, dbl., 55 petals; very vigorous growth; Verbeek

'PAULII', S, w; (R. rugosa repens alba, R. X paulii); flowers yellow stamens, borne in corymbose, single, 2.5 in.; prostrate shrub growth; (14); [R. arvensis X R. rugosa]

'PAULII ROSEA', S, pb; (R. rugosa repens rosea, R. X paulii rosea); flowers medium pink, white eye, yellow stamens, single; [Possibly a Paulii sport]

'PAULINE DAWSON', HWich, dp, 1916; flowers deep pink, large, single; vigorous, climbing growth; Dawson; Eastern Nursery

'PAULINE LANCEZEUR', HP, dr, 1854; (Pauline Lansezeur); flowers crimson shaded violet, large, dbl., recurrent bloom; Lancezeur

Pauline Lansezeur *see* **'PAULINE LANCEZEUR'**

'PAUL'S CARMINE PILLAR', Cl HT, mr, 1895; (Carmine Pillar); bud long, pointed; flowers carmine-red, open, large, single, very early bloom, slight fragrance; foliage rich green; vigorous growth; Paul

Paul's Double Musk, S, w

'PAUL'S EARLY BLUSH', HP, lp, 1893; (Mrs Harkness); flowers blush; [Heinrich Schultheis sport]; Paul

'PAUL'S HIMALAYAN MUSK RAMBLER', HMsk, lp; flowers blush-lilac-pink, rosette, blooms in clusters; thread-like stems; growth to 30 ft.

'PAUL'S LEMON PILLAR', Cl HT, ly, 1915; (Lemon Pillar); bud pale lemon-yellow; flowers pale sulfur-yellow to almost white, very large blooms, dbl., non-recurrent, intense fragrance; long, strong stems; vigorous growth; not dependably hardy.; GM, NRS, 1915; [Frau Karl Druschki X Marechal Niel]; Paul

Paul's Perpetual White *see* **'PAUL'S SINGLE WHITE PERPETUAL'**

'PAUL'S PINK', HT, lp, 1978; flowers soft pink, small, dbl., 28 petals, intense fragrance; vigorous growth; [Snowsong Supreme X Pink Puff]; deVor, Paul F.; DeVor Nurseries, Inc.

'PAUL'S SCARLET CLIMBER', LCl, mr, 1916; flowers vivid scarlet, shaded bright crimson, blooms in large clusters, semi-dbl., sometimes slightly recurrent, slight fragrance; vigorous, climbing or pillar growth; very hardy; (21); GM, Bagatelle, 1918 GM, NRS, 1915; [Paul's Carmine Pillar X Reve d'Or]; Paul, W.

'PAUL'S SINGLE WHITE PERPETUAL', HP, w, 1883; (Paul's Perpetual White); flowers pure white, solitary or in small clusters, single; foliage light green; vigorous growth; Paul

'PAULSPRIDE', LCl, yb, 1996; flowers yellow with pink edge changing to mostly pink, reverse medium, dbl., 26–40 petals, 3.5 in., moderate fragrance; few prickles; foliage medium, medium green, semi-glossy; spreading, medium (8 ft) growth; Jerabek, Paul E.

Pavane, HT, 1969; Herholdt, J.A.; (Cavriglia)

Pavarotti *see* RUItulvi

'PAVILLON DE PREGNY', N, pb; Guillot, 1863

'PAVILLON DE PREGNY', N, m, 1863; flowers light violet-pink, medium, dbl., moderate fragrance; Guillot; (Sangerhausen)

'PAW MAW', Gr, m, 1994; flowers mauve turning red at edges of petals, blooms borne 1-5 per cluster, dbl., 25–30 petals, 3–3.5 in., moderate fragrance; some prickles; foliage large, medium green, semi-glossy; tall, upright growth; Jerabek, Paul E.

Paws *see* 'BEAPAW'

'PAX', HMsk, w, 1918; flowers pure white, prominent golden anthers, blooms in clusters, semi-dbl., 3–4 in., recurrent bloom, intense fragrance; foliage large, leathery; long, strong stems; vigorous (4 ft.), bushy growth; (21); GM, NRS, 1918; [Trier X Sunburst]; Pemberton

'PAX', F, w, 1946; flowers white tinted greenish, large, semi-dbl., moderate fragrance; [Irene X Mme Alexandre Dreux]; Leenders, M.

'PAX AMANDA', S, lp, 1937; flowers light pink, turning to white, borne in clusters, semi-dbl., 17 petals, free, non-recurrent bloom; thornless; vigorous (7 ft), growth; very hardy.; [Frau Georg von Simson X R. blanda]; Hansen, N.E.

'PAX APOLLO', S, dp, 1938; flowers deep pink, borne in large clusters, semi-dbl., 14 petals, non-recurrent; thornless; vigorous (7 ft), growth; very hardy.; [R. sempervirens X R. blanda]; Hansen, N.E.

'PAX IOLA', S, lp, 1938; flowers clear shell-pink, passing to nearly white, large, borne in l, semi-dbl., 25 petals, non-recurent bloom; thorness; vigorous, pillar, growth; very hardy.; [Anci Bohmova X R. blanda]; Hansen, N.E.

'PAX LABOR', HT, yb, 1918; flowers pale golden yellow, slightly shaded coppery carmine, passing, dbl., globular; vigorous growth; [Beauté de Lyon X Unnamed seedling]; Chambard, C.

'PAX LABOR, CLIMBING', Cl HT, yb, 1929; Gaujard

'PAYABLE', F, rb, 1992; (**Anusheh**); flowers red with yellow reverse, dbl., 26–40 petals, 1.5–2.75 in., borne in large clusters, slight fragrance; many prickles; foliage medium, dark green, glossy; medium, upright growth; [Len Turner X Seedling]; Payne, A.J.

'PAYSAGISTE FAURE-LAURENT', HT, ob, 1947; flowers orange, reverse orange-yellow, large, dbl., slight fragrance; foliage glossy, dark; vigorous growth; Gaujard

'PAZ VILA', HT, mr, 1931; flowers large, dbl., moderate fragrance; foliage glossy; vigorous growth; [Jovita Perez X Jean C.N. Forestier]; Munné, B.

PEAalamo, F, dp, 1998; (**Little Rascal**); Pearce

'PEACAP', F, ab, 1981; (**Ragtime**); flowers apricot pink, blooms borne 10-15 per cluster, dbl., 65 petals, slight fragrance; large, straight, red prickles; foliage small, mid-green, glossy; bushy growth; [Vesper X Aloha]; Pearce, C.A.; Limes Rose Nursery

'PEACE', HT, yb, 1945; (Gioia, Gloria Dei, Mme A. Meilland); bud ; flowers golden yellow edged rose-pink, dbl., 43 petals, 6 in., exhibition form, slight fragrance; foliage large, very dark, leathery, glossy; very vigorous, tall, bushy growth; All-American Rose Selection, AARS, 1946 GM, NRS, 1947 GM, Portland, 1944 GM Certificate, ARS, 1947 Golden Rose, The Hague, 1965 Hall of Fame, WFRS, 1976; [((George Dickson X Souv. de Claudius Pernet) X (Joanna Hill xCharles P. Kilham)) X Margaret McGredy]; Meilland, F.; C-P, 1945, U.S.

Peace, T, ly; Piper, 1902

Peace Maker, F, 1990; Tagashira, Kazuso; (Sangerhausen)

Peace Meillandina® *see* 'MEILARCO'

Peace Sunblaze *see* 'MEILARCO'

'PEACE, CLIMBING', Cl HT, yb, 1950; (Gioia, Climbing, Gloria Dei, Climbing, Mme A. Meilland, Climbing); Brandy; C-P

'PEACE, CLIMBING', Cl HT, yb, 1951; Kordes

'PEACEFUL', HT, pb, 1956; bud globular; flowers deep coral rose pink, reverse lighter, dbl., 50 petals, 5.5–6 in., cupped, moderate fragrance; foliage leathery; vigorous, upright growth; [Unnamed seedling X Peace]; Boerner; J&P

Peacekeeper *see* 'HARBELLA'

'PEACEPORT', HT, op, 1960; flowers deep orange-pink; [Peace sport]; Rokos; Wyant

'PEACH BEAUTY', HT, pb, 1970; bud ovoid; flowers peach-pink, large, dbl., moderate fragrance; foliage large, leathery; vigorous, bushy growth; [Ma Perkins X Polynesian Sunset]; Boerner; Thomasville Nursery

Peach Blossom *see* 'AUSBLOSSOM'

'PEACH BLOSSOM', F, dp, 1932; bud orange-red; flowers soft carmine-rose, medium, borne in clusters of 6-12, semi-dbl.; vigorous growth; Chaplin Bros.

'PEACH BRANDY', Min, ab, 1978; bud pointed; flowers small, dbl., 23 petals, exhibition form, moderate fragrance; compact, bushy growth; Schwartz, Ernest W.; Bountiful Ridge Nursery;, Gulf Stream Nursery

Peach Candy *see* 'MORTOPAZ'

'PEACH DREAM', F, op, 1996; bud opening to peach orange center with pink edging; dbl., 15–25 petals, 2.5 in., exhibition form, borne in small clusters, slight fragrance; few prickles; foliage medium, dark green, glossy; upright, medium (26 in.) growth; patio; Bees of Chester; L W Van Geest Farms, Ltd., 1995

Peach Elite *see* 'LYOLITE'

Peach Festival™ *see* 'LAVCAP'

Peach Fuzz™ *see* 'WEKHELEN'

'PEACH GLOW', F, pb, 1960; bud ovoid; flowers golden coral, base pink, dbl., 30 petals, 3 in., cupped, moderate, spicy fragrance; foliage leathery; vigorous, upright, compact growth; [Goldilocks X Fashion]; Boerner; J&P

Peach Melba *see* 'KORITA'

'PEACH MELBA', HT, yb, 1960; flowers yellow marked flame and pink, moderate fragrance; vigorous growth; [Golden Scepter X Hazel Alexander]; Dicksons of Hawlmark

Peach Parfait™ *see* 'CARFAIT'

'PEACH SHERBET', F, pb, 1999; flowers salmon pink, dbl., 30 petals, 4 in., exhibition form, slight fragrance; [Sonia X seedling]; Teranishi, K.; Itami Rose Nursery, 1996

Peach Silks *see* 'CLEPEACH'

Peach Spire *see* KORofaser

Peach Sunsation *see* 'KOROMMERLA'

Peach Surprise *see* 'POULRISE'

'PEACH TREAT', HT, pb, 1968; bud ovoid; flowers peach-pink, large, very dbl., intense fragrance; foliage leathery; vigorous, bushy growth; [Beauté X Kordes' Perfecta]; Fuller; Wyant

'PEACHBLOW', HT, lp, 1942; bud long, pointed; flowers large, dbl., moderate fragrance; foliage glossy; vigorous growth; [Mme Butterfly X Yellow seedling]; Coddington; C-P

'PEACHES AND CREAM', HT, ob, 1936; flowers salmon shaded gold and rose-pink, very dbl., slight fragrance; bushy growth; RULED EXTINCT 4/77; [Seedling X Miss Rowena Thom]; H&S; Dreer

'PEACHES 'N' CREAM', Min, pb, 1976; bud tapering; flowers light peach-pink blend, dbl., 52 petals, 1 in., exhibition form, slight fragrance; foliage dark; upright, spreading growth; AOE, 1977; [Little Darling X Magic Wand]; Woolcock; Pixie Treasures Min. Roses

'PEACHY', Min, pb, 1964; flowers pink tinted yellow, small, dbl., 50 petals, moderate fragrance; foliage light green, soft; vigorous, bushy (12 in) growth; [Golden Glow (LCl) X Zee]; Moore, Ralph S.; Sequoia Nursery

'PEACHY KEEN', Min, ab, 1979; bud long, pointed; flowers soft apricot pink, 18–20 petals, 1 in., slight fragrance; bushy, spreading growth; [Little Darling X Sheri

Anne]; Bennett, Cecilia 'Dee'; Tiny Petals Nursery

Peachy Pink Magic Carpet *see* POUlor

'PEACHY WHITE', Min, w, 1976; bud long, pointed; flowers near white, often tinted pink, small, semi-dbl., 18 petals, 1.5 in., moderate fragrance; foliage leathery; upright, bushy growth; AOE, 1976; [Little Darling X Red Germain]; Moore, Ralph S.; Nor'East Min. Roses

'PEACOCK', HT, op, 1985; bud ovoid, pointed; flowers orange-coral, reverse cream, blooms borne usually singly, dbl., 5.5 in., exhibition form, moderate, spicy fragrance; large, globular, orange-red fruit; foliage large, dark, semi-glossy; vigorous, upright, bushy growth; [Red Queen sport X Red Queen]; LeMire, Walter

'PEAFEVER', F, dp, 1982; (**Fever**); flowers deep pink, large, dbl., intense fragrance; foliage large, dark; upright growth; Pearce, C.A.; Limes Rose Nursery

'PEAFIRST', F, dp, 1981; (**Dr McAlpine**, Seafirst); flowers deep rose-pink, patio, large blooms borne 1-10 per cluster, dbl., 30 petals, exhibition form, intense fragrance; straight, red prickles; foliage dark; low, bushy growth; Pearce, C.A.; Limes Rose Nursery, 1983

'PEAHAZE', F, ob, 1982; (**Geraldine**); flowers orange, medium, dbl., 20 petals, slight fragrance; foliage medium, light green, semi-glossy; upright growth; [Unnamed seedling X Unnamed seedling]; Pearce, C.A.; Limes Rose Nursery, 1984

'PEAHIGH', F, lp, 1987; (Daily Telegraph, **Gilda**); flowers pale shell pink, moderately medium, dbl., 15–25 petals, intense fragrance; foliage medium, medium green, matt; upright, spreading growth; [Unnamed seedling X Unnamed seedling]; Pearce, C.A.; The Limes New Roses

'PEAK PERFORMANCE', HT, dr, 1991; flowers large blooms borne mostly singly, dbl., slight fragrance; foliage medium, dark green, matt; tall, bushy, spreading growth; [Unknown seedling X Unknown seedling]; Burks, Larry; Co-Operative Rose Growers

'PEAMAX', F, or, 1986; (Dame Vera, **Dame Vera Lynn**); flowers brick red, large, dbl., 20 petals, slight fragrance; foliage medium, dark, semi-glossy; upright growth; [Seedling X Seedling]; Pearce, C.A.; Limes Rose Nursery

'PEAMIGHT', LCl, op, 1983; (**Leaping Salmon**); flowers salmon pink, large, dbl., 20 petals, moderate fragrance; foliage large, medium green, semi-glossy; upright (to 8-10 ft.) growth; PP007196; [((Vesper X Aloha) X (Paddy McGredy X Maigold)) X Prima Ballerina]; Pearce, C.A.; Limes Rose Nursery, 1986

'PEANOB', Min, pb, 1983; (**Bush Baby**); flowers small, dbl., 35 petals, slight fragrance; foliage small, medium green, matt; bushy growth; Pearce, C.A.; Limes Rose Nursery

'PEANUT', Min, dr, 1985; (**Pot Black**); flowers small, dbl., no fragrance; foliage small, medium green, matt; bushy growth; [Unnamed seedling X Unnamed seedling]; Pearce, C.A.; Limes Rose Nursery

Peanut Butter & Jelly *see* 'PIXICHIP'

'PEAP', Min, rb, 1988; (**Olive Taylor**); flowers vermillion red with yellow eye, small, dbl., 20 petals, slight fragrance; foliage medium, dark green, glossy; bushy growth; [Unnamed seedling X Unnamed seedling]; Pearce, C.A.; The Limes New Roses

PEApatio, MinFl, pb, 1992; (**Royal Flush**); Pearce

PEAPERFUME, HT, ob, 1991; (**Indian Summer**); Pearce

'PEAPET', Min, mr, 1988; (**Jane Asher**); flowers scarlet, aging slightly paler, rounded, small, borne in sprays, very dbl., no fragrance; prickles straight, average, red; foliage small, medium green, semi-glossy; patio; bushy, low growth; [Unnamed seedling X Unnamed seedling]; Pearce, C.A.; The Limes New Roses, 1987

'PEAPOLLY', F, ob, 1988; (**Jimmy Savile**, Jimmy Saville); flowers coppery-orange, aging lighter, loose, medium, borne usually, dbl., 25–30 petals, cupped, moderate, fruity fragrance; foliage medium, medium green, matt; bushy, low growth; [Unnamed seedling X Unnamed seedling]; Pearce, C.A.; Rearsby Roses, Ltd., 1988

'PEAPOST', F, mr, 1989; (**Carol Ann**); bud rounded; flowers scarlet, loose, medium, borne in sprays of 3-15, dbl., 42 petals, slight fragrance; prickles straight, large, red; foliage large, medium green, semi-glossy; spreading, medium growth; [Geraldine X Unnamed seedling]; Pearce, C.A.; Rearsby Roses, Ltd., 1990

'PEAPRINCE', F, ob, 1988; (**Olympic Spirit**); flowers orange, reverse yellow, aging orange-brown, medium, borne in, semi-dbl., cupped, slight, fruity fragrance; prickles pointed, medium, red; foliage small, dark green, glossy; bushy, low growth; [Unnamed seedling X Unnamed seedling]; Pearce, C.A.; Rearsby Roses, Ltd., 1988

'PEAPROOF', F, op, 1989; (**Freddy**); bud ovoid; flowers deep coral pink, aging pales slightly, urn-shaped, medium, dbl., 25 petals, borne in sprays of 3-21, slight fragrance; rare fruit; prickles hooked, medium, red; foliage medium, medium green, matt; bushy, even growth; [Unnamed seedling X Unnamed seedling]; Pearce, C.A.; Rearsby Roses, Ltd., 1989

'PEAQUANT', Gr, dy, 1988; (**ICI Golden Celebration**); flowers golden yellow, large, dbl., 26–40 petals, moderate fragrance; foliage medium, medium green, glossy, mildew resistant; bushy, healthy growth; [Unnamed seedling X Unnamed seedling]; Pearce, C.A.; The Limes New Roses

'PEARICH', HT, ob, 1992; (**Alistair Sheridan**); flowers orange, reverse orange, aging pinky orange, urn-shaped, dbl., 4.75 in., borne usually singly, slight fragrance; foliage large, medium green, glossy; upright growth; [F seedling (HO79) X HT seedling (315)]; Pearce, C.A., 1986; The Limes New Roses, 1991

'PEARL', HT, lp, 1879; flowers small, very dbl., moderate fragrance; weak growth; [Adam X Comtesse de Serenye]; Bennett

'PEARL', HT, w, 1933; flowers white, shaded pink, moderate fragrance; Bentall

Pearl, HMult, w; Turner, 1915

Pearl *see* 'WEKPEARL'

'PEARL', P, lp; Matthews, D.W., 1915; (Weatherly, L.)

Pearl Anniversary *see* WHItsun

'PEARL COSTIN', HT, yb, 1959; bud long, pointed; flowers light yellow, center pink, large, semi-dbl., cupped, slight fragrance; foliage leathery, wrinkled; vigorous, upright, bushy growth; [Elli Knab X Amy Johnson]; Reithmuller

'PEARL DAWN', Min, mp, 1975; bud short, pointed; flowers micro-mini, dbl., 38 petals, 1 in.; very compact, bushy growth; [(Cécile Brunner X Perla de Montserrat) X Perla de Montserrat]; Saville, F. Harmon; Nor'East Min. Roses

Pearl Drift® *see* 'LEGGAB'

'PEARL HARBOR', HT, pb, 1943; bud long, pointed; flowers light pink reverse china-rose, dbl., 45 petals, 3–3.5 in., exhibition form, intense fragrance; foliage leathery, dark; very vigorous, upright growth; [Unnamed variety X Miss Rowena Thom]; Howard, F.H.; H&S

Pearl Meidiland™ *see* 'MEIPLATIN'

'PEARL OF BALTIMORE', HT, lp, 1925; flowers shell-pink, center deeper, very dbl., moderate fragrance; [Ophelia X Glorified La France]; Cook, J.W.

Pearl of Bedfordview *see* 'MEIBEAUSAI'

Pearl of Canada *see* **'PERLA DE ALCAÑADA'**

Pearl of Joy *see* JACsan

'PEARL S. BUCK', HT, yb, 1940; bud long, pointed, deep orange; flowers golden yellow suffused apricot, dbl., 45 petals,

4.5 in., moderate fragrance; foliage leathery, dark; long stems; vigorous, bushy growth; [Joanna Hill X Étoile d'Or]; Kordes; J&P

Pearl Sevillana® *see* 'MEIchonar'

'Pearl Wilson Kissel', HT, dp, 1954; flowers bright red, dbl., 30–40 petals, 4.5 in., moderate fragrance; foliage leathery; vigorous, bushy growth; [Red Columbia X Chrysler Imperial]; Kissel and Motose

'Pearlie Mae', Gr, ab, 1981; bud ovoid, pointed; flowers yellow blended with pink, reverse pink, blooms borne 1-8 per, dbl., 35 petals, cupped, moderate fragrance; awl-shaped prickles; foliage leathery, semi-glossy, dark olive green, tinted copp; erect, bushy growth; [Music Maker X (Queen Elizabeth X Country Music)]; Buck, Dr. Griffith J.; Iowa State University

Pearly Drift, S, 1981; LeGrice, E.B.; (Cavriglia)

Pearly Gates® *see* 'WEKmeyer'

Pearly Gates, Climbing *see* 'WEKmeyer'

Pearly King *see* GENdee

'Pearly Peace', HT, lp, 1959; flowers soft pearl-pink; [Peace sport]; Fryers Nursery, Ltd.

'Pearly Queen', F, lp, 1963; flowers well formed, dbl., 22 petals, 4 in., moderate fragrance; vigorous growth; [Queen Elizabeth sport]; North Hill Nursery

'Pearly Shell', HT, pb, 1972; flowers shell pink, center cream, large, dbl., 30 petals, exhibition form, moderate fragrance; foliage glossy; vigorous, upright growth; [Pink Parfait X Michele Meilland]; Sherwood

Pearly Shores *see* EUGhien

'Pearly White', LCl, w, 1942; bud long, pointed; flowers white tinted pearl, open, semi-dbl., 21 petals; long, strong stems; vigorous, climbing (to 20 ft.), upright growth; [Glenn Dale X Mrs Arthur Curtiss James self seedling]; Brownell, H.C.

'PEArobin', F, ob, 1991; (**Gary Lineker**); flowers luminous orange, yellow reverse, single to blooms borne in small clusters, semi-dbl., 1.5–2.75 in., slight fragrance; many prickles; foliage medium, medium green, glossy; medium (50-90 cms), upright growth; [Seedling X Seedling]; Pearce, C.A.; Rearsby Roses, Ltd., 1991

'PEAroyal', F, dy, 1992; (**Goldfinger**, William David); dbl., 15–25 petals, 1.5–2.75 in., borne in small clusters, slight fragrance; some prickles; foliage medium, dark green, glossy; low (40 cms), compact growth; Pearce, C.A.; Hewlett-Packard, 1992

'PEAshine', F, pb, 1992; (**Anthea Fortescue**); flowers pink, yellow center, buff reverse, urn-shaped, blooms borne, dbl., 2.5 in., slight fragrance; foliage small, medium green, glossy, immune to powdery milde; low, spreading growth; [F seedling (P135) X F seedling (N147)]; Pearce, C.A., 1987; The Limes New Roses, 1991

'PEAspecial', Min, lp, 1989; (**Audrey Gardner**); bud rounded; flowers shell pink, reverse slightly darker, aging same, large, born, dbl., 45 petals, cupped, no fragrance; no fruit; prickles pointed, red; foliage small, medium green, semi-glossy; bushy, medium growth; [Unnamed seedling X Unnamed seedling]; Pearce, C.A.; Rearsby Roses, Ltd., 1990

'PEAsweet', F, mr, 1991; (**Marty**); flowers dusty red, blooms borne in small clusters, very dbl., 1.5–2.75 in., exhibition form, slight fragrance; few prickles; foliage small, medium green, semi-glossy; low (40-60 cms), bushy growth; [Seedling X Sweetheart]; Pearce, C.A.; Rearsby Roses, Ltd., 1991

'Peat Fire Flame', HT, op, 1985; flowers pale orange, reverse salmon-pink, medium, dbl., 20 petals, moderate fragrance; foliage medium, medium green, glossy; upright growth; [Red Planet X Bonnie Anne]; MacLeod, Major C.A.

'PEAthunder', F, dr, 1995; (Judy Dench, **Red Velvet**); flowers deep red with a hint of brown, blooms borne in large cluster, dbl., 26–40 petals, 1.5–2.75 in., slight fragrance; foliage medium, dark green, semi-glossy; low (50 cms), bushy, spreading growth; Pearce, C.A.; Limes New Roses, 1996;, Rearsby Roses, Ltd., 1996

'PEAtigar', F, mr, 1956; (**Kind Regards**); flowers crimson-red, very large, borne in clusters, dbl.; vigorous, bushy growth; Kordes

PEAtiger, F, mr, 1995; (**Kind Regards**); Pearce

PEAtrophy, F, lp; (**Charisma**); Pearce

Peaudouce *see* 'DICjana'

PEAvandyke, F, lp; (**Stardust**); Pearce, 1995

PEAvenus, S, lp; (**St Katherine's**); Pearce, 1995

PEAvesta, F, ob, 1996; (**Ginger Toddler**); Pearce

'PEAvoodoo', F, ab, 1994; (**Memory Lane**); flowers apricot pink, blooms borne in small clusters, dbl., 26–40 petals, 3–3.5 in., slight fragrance; some prickles; foliage large, dark green, semi-glossy; upright (60 cms) growth; [PEApost X Seedling]; Pearce, C.A.; The Limes New Roses, 1995

PEAwinner, F, dr, 1995; (**Ruby Celebration**); Pearce

PEAxanadu, S, mp, 1996; (**Pink Midinette**); Pearce

PEAxanthous, MinFl, yb, 1997; (**Bronze Baby**); Pearce

PEAxi, Cl HT, dr, 1977; (**'Bright Fire'**); flowers crimson, shaded darker, semi-dbl., 15 petals, 4.5 in., slight fragrance; foliage glossy, dark; free growth; [Parkdirektor Riggers X Guinee]; Pearce

PEAyetti, MinFl, mp, 1998; (**Dream Lover**); Pearce

PEAzara, F, dp, 1998; (**Millionaire**); Pearce

'Pebble Mill', F, rb, 1973; flowers magenta, reverse spirea-red, dbl., 28 petals, 3.5 in., flat, moderate fragrance; foliage dark; [Paddy McGredy X Unknown seedling]; Gregory

Peche Meillandina® *see* MEIxerul

Pechtold's Flame *see* **'Tudor'**

'Pechtold's Triumph', F, dr, 1961; flowers oxblood-red, medium, borne in large clusters, semi-dbl.; [Red Favorite X Frensham]; Verschuren-Pechtold

'Pedrálbes', HT, w, 1935; flowers cream to pure white, dbl., 30 petals, 5 in., exhibition form, slight fragrance; foliage dark, glossy; very vigorous growth; [Frau Karl Druschki X Souv. de Mme Boullet]; Camprubi, C.; J&P

'Pedro Veyrat', HT, ab, 1933; bud long, pointed; flowers large, dbl., cupped, moderate fragrance; [Li Bures X Benedicte Seguin]; Dot, Pedro; C-P

'Pedrus Aquarius', HT, w, 1999; flowers white/ivory, dbl., 26–40 petals, 6 in., borne mostly singly, slight fragrance; prickles moderate; foliage medium, dark green, glossy; upright, bushy, medium (3 ft.) growth; [Solitaire X Joe Longthorne]; Poole, Lionel; David Lister Roses, 1999

Peek a Boo *see* 'DICgrow'

'Peep o' Day', F, op, 1972; flowers salmon, shaded orange, large, dbl., 28 petals, slight fragrance; foliage dark; [(Pink Parfait X Highlight) X Orion]; Harkness; Mason, 1973

Peep-Eye® *see* 'MINpeep'

'Peeping Tom', S, dp, 1968; flowers deep pink, pointed, large, dbl., recurrent bloom, slight fragrance; foliage medium, medium green, matt; vigorous, tall growth; [Kordes' Perfecta X Parade]; MacLeod

Peer Gynt® *see* 'KORol'

'Peerless', HT, mr, 1935; bud long, pointed; flowers bright velvety scarlet-carmine, large, dbl., 38–40 petals; [Better Times sport]; Hill, Joseph H., Co.

Pegasus *see* 'AUSmoon'

'Peggy', HT, dp, 1934; bud long, pointed; flowers deep rose, dbl., exhibition form, intense fragrance; foliage glossy, light; vigorous growth; [Ophelia X Red-Letter Day]; Bees

Peggy "T"™ *see* 'KINTEE'

'PEGGY A. SMITH', HT, pb, 1936; flowers pink with yellow tints, medium, dbl.; Smith, A.; (Sangerhausen)

'PEGGY ANN LANDON', LCl, ob, 1938; flowers yellow-orange, becoming lighter, dbl., 3.5–5 in., exhibition form, moderate fragrance; foliage large, leathery, glossy, dark; long, strong stems; very vigorous, climbing (15-25 ft) growth; [Glenn Dale X ((Unnamed seedling X Unnamed seedling) X Mary Wallace)]; Brownell, H.C.

'PEGGY ASTBURY', HT, ab, 1920; flowers soft amber to light yellow; Easlea

'PEGGY BELL', HT, ab, 1929; Clark, A.; NRS New South Wales

'PEGGY ENGLAND', HT, ab, 1923; flowers cream-apricot, sometimes tinted carmine; Lilley

'PEGGY GRANT', Min, lp, 1954; flowers shell-pink, small, dbl., 25 petals; foliage light green; dwarf (5-6 in), bushy growth; [(Robinette X Mons. Tillier) X Zee]; Moore, Ralph S.; Sequoia Nursery

Peggy Jane™ *see* 'SAVAPEG'

Peggy Joan Reynolds *see* 'REYPEG'

Peggy Lee *see* 'AROFEIGEL'

'PEGGY M', S, pb, 1996; flowers pink gradually turning white on outer petals, reverse yellow, very dbl., 75 petals, 2.5–3 in., moderate fragrance; numerous prickles; foliage medium, light green, glossy; upright, medium (5 ft) growth; Jerabek, Paul E.

'PEGGY NETHERTHORPE', HT, mp, 1974; dbl., 35 petals, 5 in., slight fragrance; foliage light; [(Voeux de Bonheur X Chic Parisien) X (Michele Meilland X Mme-Joseph Perraud)]; Delbard; Harry Wheatcroft Gardening

'PEGGY NEWTON', F, my, 1957; flowers primrose-yellow, borne in clusters, dbl., 40–50 petals, 2.5–3 in., globular, slight fragrance; foliage small, leathery, glossy; dwarf, spreading growth; [Golden Glow X Goldilocks]; Boerner; Stuart

Peggy Rockefeller™ *see* 'WILACE'

Peggy's Delight *see* 'MINCCO'

'PEINTRE RENOIR', HT, mp, 1925; flowers large, dbl.; Roseraie de St. Jean; (Sangerhausen)

PEKaledon, HT, dp, 1993; (**Femnet**); Pekmez, Paul

'PEKALI', HT, ob; flowers orange, large, dbl., 35 petals, no fragrance; foliage large, glossy; upright growth; [Unnamed seedling X Marina]; Pekmez, Paul

'PEKAMECEL', F, or, 1984; dbl., 20 petals, slight fragrance; foliage dark; upright growth; [Unnamed seedling X Unnamed seedling]; Pekmez, Paul

'PEKATAN', HT, pb, 1984; (**Clo-Clo**); dbl., 20 petals, slight fragrance; foliage dark; upright growth; [Emily Post X Bellona]; Pekmez, Paul

PEKcoucan, HT, or; (**Star 2000**®); Pekmez, Paul

PEKcougel, HT, lp, 1990; (**Anna**); Pekmez

PEKcoujenny, HT, dr, 1988; (**First Red**®); Pekmez, Paul

PEKcouliane, HT, ob; (**Julia**®); (PEIcouliane)

PEKcourofondu, HT, lp; (**Vogue**); Pekmez, Paul, 1997

PEKcram, F, mp, 1988; (**Jean Monnet**®); Pekmez, Paul

'PEKGOLD', HT, dy, 1984; flowers medium, moderate fragrance; foliage dark, glossy; upright growth; [Unnamed seedling X Unnamed seedling]; Pekmez, Paul

'PEKINOIS', Pol, dr, 1975; (Tapis Afghan); bud round; semi-dbl., 15 petals, 1–2 in., cupped; dwarf growth; [Marlena X Lampion]; Pekmez, Paul

'PEKLICAN', HT, dy, 1984; flowers large, dbl., 35 petals, moderate fragrance; foliage medium, light green, matt; bushy growth; [Unnamed seedling X Bellona]; Pekmez, Paul

'PEKLIMASAR', Cl F, mr, 1983; (Grimpant Lilli Marleen®, Lili Marlene, Climbing, **Lilli Marleen**, **Climbing**, Lilli Marlene, Climbing); Pekmez, Paul

'PEKLIPINK', F, mp, 1984; (**Claridge**); flowers large, dbl., 35 petals, moderate fragrance; foliage medium, light green, semi-glossy; upright growth; [Unnamed seedling X Unnamed seedling]; Pekmez, Paul

PEKmiwhite, S, lp, 1991; (**Perce Neige**®); Pekmez, Paul

'PEKOMEGIR', F, mp, 1984; (**President Souzy**); dbl., 20 petals, moderate fragrance; upright growth; [Emily Post X Unnamed seedling]; Pekmez, Paul

PEKomecli, HT, pb, 1985; (**Marlyse**®); (PEIomecli)

PEKoubo, HT, op, 1997; (**Bo**); Pekmez, Paul

PEKsolred, S, mr, 1992; (**Ile de France**); Pekmez, Paul

PEKtarampe, S, dr, 1992; (**Alsace**); Pekmez, Paul

PEKwhina, HT, lp; (**Virginia**); Pekmez, Paul, 1994

'PELÉ', Cl HT, w, 1979; bud ovoid; flowers large blooms borne 1-3 per cluster, dbl., 35 petals, repeat bloom, slight, soft, fruity fragrance; triangular, hooked prickles; foliage medium green; upright growth with long canes; Bronze, ARC TG, 1980; [Unnamed seedling X Unnamed seedling]; Benardella, Frank A.

'PÉLISSON', M, dr, 1848; (Mons. Pélisson); flowers velvety red, turning purple; Vibert

'PELTON LONNEN', HT, my, 1975; flowers pure yellow, full, dbl., 28–30 petals, cupped, moderate fragrance; foliage light; bushy, compact growth; [Whisky Mac sport]; Wood

'PEMBERTON'S WHITE RAMBLER', HMult, w, 1914; flowers rosette form, borne in clusters, dbl.; Pemberton

'PEMBRIDGE', HT, ob, 1934; flowers richer, deeper orange-yellow; [Roselandia sport]; Stevens, E.

'PENELOPE', HMsk, lp, 1924; flowers shell-pink fading to white, center lemon, blooms in clusters, semi-dbl., recurrent bloom, moderate fragrance; foliage dark; shrubby growth; (21); GM, NRS, 1925; [Ophelia X Unnamed seedling or possibly William Allen Richardson or Trier]; Pemberton

'PENELOPE', T, rb, 1906; flowers dark red, center creamy white, high pointed, dbl.; Williams, A.

Pénélope® *see* 'GAUBIROC'

Penelope Keith *see* 'MACFREEGO'

'PENELOPE PLUMMER', F, dp, 1970; bud orange; flowers vivid flamingo-pink, semi-dbl., 16 petals, 4 in., flat, slight fragrance; foliage dark; moderate growth; [Anna Wheatcroft X Dearest]; Beales, Peter; Intwood Lane Nursery, 1971

'PENELOPE, CLIMBING', Cl T, rb, 1932; Rosen, L.P.

'Penkala Slavoljub' *see* 'ROGSLAV'

'PENNANT', LCl, mp, 1941; flowers begonia-pink, dbl., profuse, non-recurrent bloom; [Flying Colours X Lorraine Lee]; Clark, A.

'PENNSYLVANIA', HT, pb, 1934; bud long, pointed; flowers salmon-pink, center apricot, outer petals striped dark pink, semi-dbl., exhibition form, intense fragrance; [Joanna Hill sport]; Neuner

'PENNSYLVANIAN', HT, ob, 1953; bud pointed; flowers apricot-orange, dbl., 25–30 petals, 4–5 in., moderate fragrance; upright, bushy growth; [Luna X (Mrs Pierre S. duPont X Mrs Sam McGredy)]; Ohlhus; C-P

'PENNY', F, rb, 1973; flowers strawberry-red, base orange, rosette form, semi-dbl., 17 petals, 2 in., moderate fragrance; dwarf growth; [Sarabande X Circus]; Sanday, John

Penny Annie *see* 'BISFRA'

Penny Candy™ *see* 'SAVPLENTI'

Penny Coelen *see* 'MEINIMO'

Penny Heyns *see* TANpen

Penny Lane *see* 'TALPEN'

Penny Lane *see* HARdwell

Pensioners Voice *see* FRYrelax

Penthouse *see* 'MACSATUR'

Penthouse *see* 'MACNGAURU'

Peon *see* **'TOM THUMB'**

'PEONY OF FRAGRANCE', HP, mp, 1933; flowers pink, peony-like, large, recurrent bloom, moderate fragrance; Pahissa

'PEOPLE', F, dr, 1956; (The People); flowers crimson shaded pink, large, borne in large trusses, dbl., 26 petals, flat, slight fragrance; foliage light green; vigorous, bushy growth; GM, NRS, 1955; [Cinnabar X (Kathe Duvigneau X Cinnabar)]; Tantau, Math.

People's Princess *see* 'SUNTICK'

'PEPE', HT, rb, 1961; flowers flame, base and reverse gold, dbl., 4 in.; foliage dark, glossy; vigorous growth; [Amor X Sutter's Gold]; deRuiter; Blaby Rose Gardens

Pepino® *see* TANipep

Pepita *see* 'KORKEILICH'

'PÉPITA', HGal, pb; flowers soft rosy pink striped white; Moreau

Pépite *see* 'SOUR DE J. CHABERT'

'PEPPER POT', F, pb, 1973; flowers rose-pink to red, splashed yellow, dbl., 24 petals, 3 in., exhibition form, slight fragrance; foliage light; free growth; [Circus X Seedling]; Fryers Nursery, Ltd.

'PEPPERMINT', F, rb, 1964; bud ovoid; flowers red, reverse cream, medium, dbl., cupped; foliage leathery; vigorous, bushy growth; [Jingles X Jingles]; Boerner; J&P

'PEPPERMINT CANDY', Min, w, 1991; flowers ivory with red blend, dbl., 26–40 petals, 1.5 in., borne mostly singly, intense fragrance; foliage small, dark green, glossy; medium (12-28 inches), upright, bushy growth; PP006819; [Rose Parade X Easter Morning]; Williams, J. Benjamin, 1988

Peppermint Delight, LCl, rb

Peppermint Ice *see* 'BOSGREEN'

'PEPPERMINT PATTY', Min, rb, 1990; bud pointed; flowers white with red edging, same reverse, aging darker red, urn-shaped, dbl., 33 petals, slight, fruity fragrance; oblong, orange when ripe fruit; prickles straight, very few, tan; foliage medium, medium green, semi-glossy; very hardy, upright growth; [Libby X Libby]; Gruenbauer, Richard, 1984; Richard Gruenbauer

Peppermint Stick *see* JACmint

'PEPPERMINT STRIPE', Min, rb, 1991; flowers moderately small, dbl., 15–25 petals, intense fragrance; foliage small, medium green, semi-glossy; bushy growth; [Roller Coaster X Unnamed seedling]; Spooner, Raymond A.

Peppermint Swirl™ *see* 'DEVMENTA'

Peppermint Twist *see* 'JACRAW'

Pepperoni®, S, or, 1994; Poulsen

Per Chance *see* 'FLOPER'

Perce Neige® *see* PEKmiwhite

Perception *see* 'HARZIPPEE'

'PERCHÈ SI?', HT, rb, 1956; flowers carmine-red, reverse silvery white, well formed; foliage dark, glossy; very vigorous growth; [Peace X Crimson Glory]; Giacomasso

'PERCREME', HT, w, 1999; (**Irish Creme**™); flowers light beige, reverse same, dbl., 26–40 petals, 4 in., borne mostly singly, moderate fragrance; prickles moderate; foliage medium, medium green, dull; upright, medium (5 ft) growth; [Butterscotch X Mandelon]; Perry, Astor; Certified Roses, 2000

Percussion *see* 'ZIPCUSS'

'PERCY IZZARD', HT, yb, 1936; flowers maize-yellow, reverse buff flushed warm rose, large, dbl., exhibition form, slight fragrance; foliage leathery; vigorous growth; [May Wettern X Barbara Richards]; Robinson, H.; Wheatcroft Bros.

'PERCY PILCHER', F, op, 1961; flowers salmon-orange, large, borne in large clusters, dbl.; Verschuren

'PERCY THROWER', HT, mp, 1964; flowers rose-pink, well-formed, dbl., 28 petals, 4–5 in., moderate fragrance; foliage glossy; vigorous, tall growth; [La Jolla X Karl Herbst]; Lens; Gregory

Percy Thrower, Climbing, Cl HT, mp, 1978; Stoneham

Perdita® *see* 'AUSPERD'

'PERDITA', F, lp; Bidwell, 1845; (Weatherly, L.)

Perestroika *see* 'KORHITOM'

Perfect Hit *see* POUlfect

Perfect Moment™ *see* 'KORWILMA'

Perfect Peace *see* **'DOROTHY GOODWIN'**

Perfect Potluck™ *see* 'LAVROY'

Perfecta *see* 'KORALU'

'PERFECTA', Pol, dr, 1920; flowers scarlet-crimson under glass, dark crimson in the open, large, dbl.; vigorous growth; [Ellen Poulsen X Merveille des Rouges]; Spek

Perfecta Superior *see* **'KORDES' PERFECTA SUPERIOR'**

'PERFECTION', HT, mp, 1925; [Columbia sport]; Stielow Bros.

'PERFECTION', Pol, op, 1932; flowers softer coral-pink; [Marytje Cazant sport]; Prior

'PERFECTION', F, pb, 1957; flowers pink turning red, reverse white; [Masquerade X Self]; Ulrick, L.W.

'PERFECTION DE MONTPLAISIR', T, 1871; Levet, A.; (Cavriglia)

'PERFECTION DES BLANCHES', N, w, 1873; flowers medium, borne in corymbs, dbl.; vigorous growth; Schwartz, J.

Perfectly Red™ *see* 'JACROVE'

'PERFEITA', F, or; flowers orange and carmine; [Cocorico X Vogue]; da Silva, Moreira

'PERFUME', HT, dr, 1929; bud long, pointed; flowers velvety, fiery deep crimson, semi-dbl., 15 petals, intense fragrance; vigorous growth; Marriott; Beckwith

Perfume, HT, lp, 1995; Spek

Perfume Beauty™ *see* 'MEINIACIN'

'PERFUME DE MULTIFLORE', HT, m, 1935; flowers purple/pink, large, dbl., moderate fragrance; Pahissa; (Sangerhausen)

'PERFUME DELIGHT', HT, mp, 1973; bud long, pointed; flowers deep pink, large, dbl., cupped, intense fragrance; foliage large, leathery; vigorous, upright, bushy growth; AARS, 1974; [Peace X ((Happiness X Chrysler Imperial) X El Capitan)]; Weeks; C-P

Perfume Simplex, Min, yb

Perfume Tiger™ *see* 'WINPTIG'

Perfumed Bride, HT, pb, 1994; Bell

Perfumella *see* MEIparos

'PERGOLÈSE', P, m, 1860; flowers bright purplish crimson, shading to lilac, medium, very dbl., occasionally recurrent bloom, moderate fragrance; Moreau et Robert

'PERJUSDREAM', HT, yb, 1999; (**Just Dreamy**); flowers yellow-gold, reverse apricot gold, dbl., 17–25 petals, 3.5–4 in., borne mostly singly, slight fragrance; few prickles; foliage medium, medium green, semi-glossy; upright, bushy, medium growth; [Folklore X Golden Gate]; Perry, Astor; Certified Roses, 2000

'PERKY', Min, dp, 1958; bud pointed; dbl., 1 in., intense fragrance; foliage glossy; very bushy (12 in), compact growth; [(R. wichurana X Floradora) X Oakington Ruby]; Moore, Ralph S.; Sequoia Nursery

'PERLA DE ALCAÑADA', Min, dp, 1944; (Baby Crimson, Pearl of Canada, Perle de Alcañada, Wheatcroft's Baby Crimson); bud small, ovoid; flowers carmine, semi-dbl., 18 petals; foliage dark, glossy; dwarf, very compact (6-10 in) growth; [Perle des Rouges X Rouletti]; Dot, Pedro; A. Meilland

'PERLA DE ALCAÑADA, CLIMBING', Cl Min, dp; Dot, Pedro

'PERLA DE MONTSERRAT', Min, pb, 1945; bud small; flowers hermosa pink edged pearl, blooms in clusters, semi-dbl., 18 petals; dwarf, very compact growth; [Cécile Brunner X Rouletti]; Dot, Pedro; A. Meilland

Perla d'Oriolo, S, 1992; Embriaco, B.; (Cavriglia)

'PERLA ROSA', Min, mp, 1946; bud well-formed; flowers bright pink, very dbl.; very compact (6-8 in) growth; [Perle des Rouges X Rouletti]; Dot, Pedro

'PERLA ROSA, CLIMBING', Cl Min, mp, 1947; Dot, Pedro

'PERLE', Pol, w, 1920; Easlea

'PERLE ANGEVINE', Pol, lp, 1920; flowers pale rose, small, borne in clusters, dbl.; few thorns; [Jeanne d'Arc X Mrs W.H. Cutbush]; Délépine

Perle Blanche *see* 'DELANCHE'

Perle d'Amour, B, lp; Weihrauch

Perle d'Angers, B, lp, 1879; Moreau et Robert

Perle d'Anjou, Min, mp

Perle de Alcañada *see* **'PERLA DE ALCAÑADA'**

'PERLE DE LYON', T, dy, 1872; flowers apricot yellow, large, dbl.; Ducher

Perle de Veissenstein *see* **'PERLE VON WEISSENSTEIN'**

Perle des Blanches, N, w, 1872; Lacharme, F.

'PERLE DES JARDINS', T, ly, 1874; flowers straw-yellow, large, dbl., globular, intense fragrance; foliage dark; slender growth; [Mme Falcot seedling sport]; Levet, F.

'PERLE DES JARDINS, CLIMBING', Cl T, ly, 1890; Henderson, J.

'PERLE DES JAUNES', T, dy, 1913; flowers golden yellow, large, dbl., moderate fragrance; Reymond; (Sangerhausen)

Perle des Neiges, HMult, w; Dubreuil, 1902

'PERLE DES PANACHÉES', HGal, m, 1845; (Panachée Double); flowers white striped with lilac and violet or rose; Vibert

'PERLE DES ROUGES', Pol, dr, 1896; flowers velvety crimson; Dubreuil

'PERLE D'OR', Pol, yb, 1884; (Yellow Cécile Brünner); flowers golden pink, blooms in clusters, very dbl., intense fragrance; foliage rich green, soft; height 3 ft; (14); [Polyantha X Mme Falcot]; Rambaux; Dubreuil

'PERLE D'OR, CLIMBING', Cl Pol, yb; (Yellow Cécile Brünner, Climbing)

Perle du Lac *see* DORouvi

Perle du Lac Annecy *see* DORouvi

Perle Meillandécor® *see* 'MEIPLATIN'

Perle Noire® *see* 'DELURT'

'PERLE ORLÉANAISE', Pol, dp, 1913; flowers carmine-pink, medium, dbl.; Duveau; (Sangerhausen)

'PERLE VOM WIENERWALD', HMult, pb, 1913; flowers carmine-rose, reverse soft rose-pink, borne in clusters of 5, semi-dbl., 2 in.; vigorous, climbing growth; [Helene X Crimson Rambler]; Praskac; Teschendorff

Perle von Aalsmeer *see* **'PAREL VAN AALSMEER'**

'PERLE VON BRITZ', HMult, lp, 1910; flowers small, semi-dbl.; Kiese; (Sangerhausen)

'PERLE VON GODESBERG', HT, ly, 1902; flowers cream, shaded lemon; [Kaiserin Auguste Viktoria sport]; Schneider

'PERLE VON HEIDELBERG', HT, mr, 1905; flowers silvery pink, intense fragrance; vigorous growth; [La France sport]; Scheurer

'PERLE VON HOHENSTEIN', Pol, lp, 1923; flowers carmine-red, small, borne in clusters, semi-dbl.; [Freudenfeuer X Unnamed seedling]; Kiese

'PERLE VON REMAGEN', HT, lp, 1957; bud pointed; flowers soft pink, reverse tinted creamy white, large, dbl., exhibition form, moderate fragrance; foliage glossy, leathery; long stems; vigorous, upright growth; [R.M.S. Queen Mary X Peace]; Burkhard; Kordes;, Kordes & Tantau

'PERLE VON WEISSENSTEIN', HGal, m; (Perle de Veissenstein); flowers brownish, center purple; Schwartzkopf, 1773

Perl-Ilseta® *see* 'TANILSEPO'

Perlmutt-Prinzessin, S, lp, 1990; flowers small, very dbl.; Wänninger, Franz; (Sangerhausen)

'PERMACH', HT, m, 1999; (**Macho Man**); flowers lavender, medium, dbl., 17–25 petals, borne mostly singly, moderate fragrance; few prickles; foliage medium, medium green, dull; bushy growth; Perry, Astor; Certified Roses, 2000

'PERMANENT WAVE', F, mr, 1932; (Duchess of Windsor, L'Indéfrisible, Mevrouw van Straaten van Nes, Mrs Van Nes, Van Nes); flowers bright carmine, petals wavy, large blooms in clusters, semi-dbl., slight fragrance; foliage glossy, dark; vigorous, bushy growth; GM, Bagatelle, 1933 GM, Rome, 1934; [Else Poulsen sport]; Leenders, M.; J&P, 1935

Permeate Peace, HT, pb, 1990; Dawson

'PERMILL', HT, mr, 1997; (**Millennium**™); flowers full, large, very dbl., 26–40 petals, borne mostly singly, slight fragrance; foliage medium, medium green, semi-glossy; upright, medium (5ft. 6 in.) growth; [Fire Magic X Precious Platinum]; Perry, Astor

Permoser, HT, yb; flowers light yellow with carmine edges, large, dbl., moderate fragrance; Institut für Obstbau Dresden Pillnitz; (Sangerhausen)

'PERNETIANA'; Pernet-Ducher

'PERNILLE POULSEN'®, F, mp, 1965; flowers light pink, blooms in clusters, semi-dbl., 18 petals, 3.5 in., moderate fragrance; foliage pointed, light green; [Ma Perkins X Columbine]; Poulsen, Niels D.; McGredy

'PERNILLE POULSEN, CLIMBING', Cl F, mp, 1980; Poulsen, Niels D.; Vilmorin-Andrieux

'PERO D'ALENQUER', F, dr; flowers dark velvety red, center lighter; [Unnamed seedling X Alain]; da Silva, Moreira

'PERPETUAL RED', S, mr, 1955; flowers bright red, open, medium, semi-dbl., recurrent bloom; foliage abundant; very vigorous growth; [Gruss an Teplitz X Unnamed seedling]; Gaujard

Perpetual White Moss *see* **'QUATRE SAISONS BLANC MOUSSEUX'**

'PERPID', HT, dr, 1998; (**Intrepid**™); flowers dark velvety red, exhibition, medium, very dbl., 26–40 petals, 4–4.5 in., exhibition form, borne singly, moderate fragrance; prickles moderate; foliage medium, dark green, dull; medium, upright, 5 ft. growth; [Karl Herbst X Burgundy]; Perry, Astor; Certified Roses Inc., 1999

'PERROQUET', LCl, dy, 1957; flowers large, slight fragrance; foliage glossy; vigorous growth; Robichon

'PERROQUET', F, rb, 1960; flowers red, reverse yellow, becoming dark red, well formed; foliage bronze; vigorous growth; [Peace X (Cinnabar X Circus)]; Lens

Persane, HT, 1984; Dorieux; (Cavriglia)

'PERSEPOLIS', HT, dy, 1986; (Truper); dbl., 25 petals, intense, tea fragrance; foliage clear green, glossy; [Unnamed seedling X Unnamed seedling]; Kriloff, Michel

'PERSIAN CARPET', F, rb, 1995; flowers rich crimson with yellow base of petal and reverse, blooms, semi-dbl., 10 petals, 1.5–2.75 in., slight fragrance; foliage medium, dark green, glossy; upright (up to 70 cms) growth; [Masquerade X Traumerei]; Fleming, Joyce L.; Hortico Roses, 1994

'PERSIAN DELIGHT', F, rb, 1995; flowers rich crimson with yellow base of petal and reverse, semi-dbl. (10 petals), medium (4–7 cms) blooms; slight fragrance;

foliage medium, dark green, glossy; upright growth;[Masquerade X Traumerei]; Fleming, Joyce; Hortico Roses, 1994

Persian Musk Rose *see* **'NASTARANA'**

'PERSIAN PRINCESS', Min, or, 1970; flowers coral-red, small, dbl., moderate fragrance; foliage leathery; vigorous, bushy, dwarf (12-14 in.) growth; [Baccará X Eleanor]; Moore, Ralph S.; Sequoia Nursery

Persian Yellow Rose *see* **R. FOETIDA PERSIANA**

'PERSONALITY', HT, yb, 1960; bud ovoid; flowers golden yellow splashed red, open, dbl., 35–40 petals, 4.5 in., intense fragrance; foliage leathery, glossy; vigorous, upright growth; [Peace X Sutter's Gold]; Morey, Dr. Dennison; J&P

'PERSUADER', HT, mp, 1960; flowers bright pink, large, dbl., 30 petals, exhibition form, intense fragrance; vigorous growth; [Golden Scepter X Southport]; Eacott

'PERSUASION', F, lp, 1972; flowers pale blush-pink, base orange, dbl., 30 petals, 3.5 in., exhibition form, moderate fragrance; foliage glossy; [Vera Dalton X Tropicana]; Sanday, John

'PERSUE DE GOSSART', C, dp; flowers velvety cerise, with garnet reflexes, dbl.

'PERSUS', HT, dy, 1998; (**Colossus**™); flowers golden yellow, full, medium sized, very dbl., 26–40 petals, borne in small clusters, moderate fragrance; foliage medium, medium green, glossy; upright, medium, 5 ft. growth; [Folklore X Golden Gate]; Perry, Astor; Certified Roses, Inc., 1999

'PERTWENTYFIRST', HT, dy, 1999; (**Twenty First Century**); flowers large, dbl., 26–40 petals, borne mostly singly, slight fragrance; few prickles; foliage medium, dark green, semi-glossy; upright, tall (5 ft) growth; [Butterscotch X Mandelon]; Perry, Astor

Pet, lp; [(Betty Neuss X) X Silver Lining]; Dawson, George

'PETER BENJAMIN', HT, ab, 1978; flowers light apricot pink, borne 2-3 per cluster, dbl., 40 petals, exhibition form, moderate fragrance; pear-shaped, red prickles; foliage light; medium-tall growth; [Benjamin Franklin X Peter Frankenfeld]; Allender, Robert William

'PETER FRANKENFELD'®, HT, dp, 1966; flowers rose-pink, well-shaped, large, slight fragrance; Kordes, R.; A. Dickson

'PETER FRANKENFELD, CLIMBING', Cl HT, dp, 1975; Allen, L.C.; Welsh

Peter Goldman *see* 'DICNAME'

'PETER LAMBERT', HMult, dp, 1936; flowers carmine-pink, medium, dbl.; Vogel, M.; (Sangerhausen)

'PETER LAWSON', HP, dr, 1862; flowers deep scarlet shaded purple; Thomas

Peter Long, HT, or

'PETER MAY', HT, mr, 1958; flowers scarlet, large, dbl., intense fragrance; foliage leathery; long stems; very vigorous growth; [The Doctor X New Yorker]; Verschuren; Blaby Rose Gardens

'PETER PAN', HT, dr, 1935; flowers dark crimson, large; vigorous growth; Knight, G.

Peter Pan *see* **'PRESUMIDA'**

Peter Pan *see* 'CHEWPAN'

Peter Pan *see* 'SUNPETE'

'PETER PIPER', HT, or, 1969; flowers chinese orange, urn shaped, large, semi-dbl.; bushy growth; [Piccadilly sport]; Waterhouse Nursery

'PETER ROSEGGER', LCl, op, 1914; flowers coral-pink, rosette form, borne in clusters of 5-15, dbl., repeat bloom; foliage dark; vigorous, climbing growth; [Geheimrat Dr. Mittweg X Tip-Top]; Lambert, P.

Peter Wessel *see* 'TANTIDE'

'PETER'S BRIARCLIFF', HT, mp, 1940; flowers true unshaded rose-pink, long pointed, well formed, intense fragrance; foliage dark; vigorous, free growth; J&P

Pétillante *see* **'ROTER CHAMPAGNER'**

Petit Canard® *see* 'LENCAN'

'PETIT CONSTANT', Pol, mr, 1899; flowers nasturtium-red, small, dbl., intense fragrance; vigorous growth; [Mignonette X Luciole]; Soupert & Notting

Petit Four® *see* 'INTERFOUR'

'PETIT FRANÇOIS', F, or, 1957; semi-dbl.; foliage glossy; very dwarf growth; [Alain X (Brazier X Léonce Colombier)]; Dorieux; Pin

'PETIT JEAN', HT, ob, 1926; flowers deep orange-buff, shaded to yellow, edged peach-pink, well s, dbl., moderate fragrance; [White Killarney X Sunburst]; Vestal

'PETIT LOUIS', HWich, op, 1912; flowers shrimp-pink, small, very dbl., slight fragrance; vigorous, climbing growth; [Dorothy Perkins X Seedling]; Nonin

Petit Marquis *see* 'KORHOLST'

'PETIT POUCET', F, mr, 1955; flowers bright red, center tinted yellow, petals wavy, single; dwarf growth; [Cocorico X Unnamed seedling]; Combe

'PETIT PRINCE', Pol, mr, 1956; flowers geranium-red, borne in clusters of 8-10; very dwarf growth; Laperrière; EFR

Petit Rat de l'Opera®, S, pb, 1990; Lens

'PETIT RENÉ', HWich, mr, 1925; flowers brilliant red, small, borne in clusters, very dbl., intense fragrance; vigorous, climbing growth; Nonin

Petit Sam®, F, dp; Adam

Petit Serpent, S, w, 1994; Lens

Petite *see* **'PATTY LOU'**

'PETITE CARROUSEL', Min, w, 1985; flowers ivory, coral pink petal edges, small blooms borne singly and, slight fragrance; foliage small, dark, semi-glossy; upright, tiny growth; [Thought to be Magic Carrousel X ?]; Michelis, Dorothy, 1991

'PETITE DE HOLLANDE', C, mp; (Petite Junon de Holland, R. centifolia minor); flowers rose pink, small blooms in clusters, dbl., moderate fragrance; moderate growth

Petite de Orlanaise *see* **'PETITE ORLÉANAISE'**

Petite de Terre Franche, HMsk, pb, 1996; Louette

'PETITE ECOSSAISE', HSpn, lp; flowers flesh, small, non-recurrent; moderate growth; Vibert, before 1835

Petite Folie® *see* 'MEIHERODE'

Petite Francoise, Pol, lp, 1915; Gravereaux

Petite Fredaine®, Min, op, 1991; Lens

'PETITE JEANNE', HWich, mr, 1912; flowers currant-red; [Dorothy Perkins seedling]; Nonin

Petite Junon de Holland *see* **'PETITE DE HOLLANDE'**

'PETITE LÉONIE', Pol, w, 1893; flowers pinkish white, small; [Mignonette X Duke of Connaught]; Soupert & Notting

'PETITE LISETTE', C, dp, 1817; flowers rich rose, pompon-shaped, 1 in.; foliage matt, green; height 3-4 ft; Vibert

'PETITE LOUISE', LCl, op; flowers salmon-pink

Petite Michelle *see* JAClib

'PETITE ODETTE', HWich, lp, 1923; very dbl., borne in clusters, slight fragrance; vigorous, climbing growth; [Lady Godiva X Unnamed seedling]; Nonin

'PETITE ORLÉANAISE', C, mp; (Petite de Orlanaise); flowers pink, pompon-shaped, small, dbl.; vigorous, almost climbing. growth

Petite Penny *see* 'MACJOCEL'

Petite Perfection™ *see* 'JACRYBI'

Petite Pink Scotch, S, mp

Petite Red Scotch *see* **'DOUBLE DARK MARBLED'**

Petite Reine *see* **'QUEENIE'**

Petito, F, yb; Agel

'PETRA', F, mr, 1974; bud medium, globular; flowers blood-red, dbl., slight fragrance; foliage dark, leathery;

moderate, bushy growth; [Seedling X Taora]; Kordes; Dehner & Co.

'PETRINE', HT, ob, 1921; flowers coral-red, shaded chrome-yellow, dbl., slight fragrance; [Old Gold X Mme Edouard Herriot]; Therkildsen

Petro, F, w

'PETRONELLA', HT, rb, 1980; flowers pointed blooms borne singly, dbl., 49 petals, moderate fragrance; straight, red prickles; foliage dark, glossy; compact growth; [Gail Borden X (Dalvey X Fragrant Cloud)]; MacLeod, Major C.A.; Christie Nursery, Ltd.

Petticoat *see* 'JACPET'

Petticoat Lane *see* 'SOCAMP'

'PETULA CLARK', HT, mr, 1963; flowers clear red, well-formed, dbl., exhibition form; foliage bronze; vigorous, bushy growth; [Purpurine X Lavender Pinocchio]; Lens

'PETULA CLARK, CLIMBING', Cl HT, mr, 1967; Lens

Pfaffstädt, HP, ly; flowers large, dbl., moderate fragrance; von Württemberg, Herzogin Elsa; (Sangerhausen)

Pfälzer Gold® *see* 'TANÄLZERGO'

'PFANDER'S CANINA', (strain of R. canina), lp; Pfander, prior to 1954

'PFUSS PFREE', F, mp, 1988; bud ovoid, pointed; flowers light pink, reverse medium pink, full, small, borne in spray, dbl., 25 petals, cupped, slight fragrance; round, deep orange fruit; prickles straight, tan-brown; foliage medium, medium green, very glossy, blackspot resista; spreading growth; [Sea Foam X (Unnamed Restless Native seedling X Europeana)]; Stoddard, Louis, 1989

Phab Gold *see* FRYbountiful

'PHAENOMEN', HT, op, 1934; (Phenomenon); flowers slightly rosy salmon, reverse light rosy red, very large, dbl.; foliage dark; very vigorous growth; Chotkové Rosarium; Böhm

Phalaenopsis, F, dp

'PHANTASY', HT, lp, 1927; flowers medium light pink, base yellow, dbl., moderate fragrance; [Lady Alice Stanley X Royal (or Priscilla)]; Dunlop

'PHANTOM', HT, ob, 1920; flowers coppery yellow, center lighter, semi-dbl., moderate fragrance; RULED EXTINCT 4/92; [(Joseph Hill X My Maryland seedling) X Lady Hillingdon]; Towill

Phantom *see* 'MACCATSAN'

Pharaoh *see* 'MEIFIGA'

Pharaon® *see* 'MEIFIGA'

Phare® *see* 'DELGO'

'PHARISÄER', HT, op, 1903; bud long, pointed; flowers rosy white, shaded salmon, blooms, dbl., exhibition form, moderate fragrance; foliage bronze; long, weak stems; vigorous growth; [Mrs W.J. Grant seedling]; Hinner, W.

Pheasant *see* 'KORDAPT'

'PHEIDIPPIDES', F, mr, 1979; flowers currant-red, full, dbl., 35 petals, 4 in.; foliage light green; moderately vigorous, low growth; [Sam Ferris X Chopin]; Ellick; Excelsior Roses

'PHELAN'S FLAG', HCh, rb, 1952; bud ovoid, greenish white flushed purplish red; flowers unique virgin blooms (1) red, (2) pink, (3) white, (4) red e, dbl., 75–100 petals, 2.5–3 in., quartered, semi-recurrent bloom, intense fragrance; foliage dark; vigorous (6 ft or more) growth, with tendency to climb; Phelan

'PHÉNICE', HGal, dp, 1843; flowers reddish rose, spotted, medium, dbl.; Vibert

Phenomenon *see* **'PHAENOMEN'**

Philadelphia *see* **'PHILADELPHIA RAMBLER'**

'PHILADELPHIA RAMBLER', HMult, dr, 1904; (Philadelphia); flowers scarlet-crimson, center lighter, borne in gigantic clusters, dbl., midseason bloom; vigorous, climbing growth; [Crimson Rambler X Victor Hugo]; Van Fleet; Conard & Jones

'PHILÉMON COCHET', B, mp, 1895; flowers rose; Cochet, Sc.

'PHILIBERT BOUTIGNY', HP, mp; flowers silvery rose, very large; vigorous growth

'PHILIP HARVEY', F, ob, 1975; flowers salmon-red, shaded orange, dbl., 25 petals, 5 in., moderate fragrance; foliage glossy, dark; [Fragrant Cloud X Circus]; Harkness

Philippa Pirette, S, pb

'PHILIPPE', F, op, 1959; (Bel Ami); bud oval; flowers peach-salmon, open, medium, borne in clusters, semi-dbl., 15 petals; foliage glossy; moderate growth; [Cognac X Fashion]; Delforge

Philippe Noiret *see* 'MEIZCELE'

'PHILIPPE PÉTAIN', HT, dp, 1940; flowers velvety carmine with coppery reflections, not turning blue, , cupped, slight fragrance; foliage bright chive-green; Nabonnand, C.

'PHILIPPE RIVOIRE', HT, op, 1941; flowers large, dbl.; Gaujard; (Sangerhausen)

Phillipa *see* POUlheart

Phillipp Melanchthon, F, dr, 1994; Hetzel

'PHILLIPP PAULIG', HP, dr, 1908; flowers large, dbl., moderate fragrance; Lambert, P.; (Sangerhausen)

Philomene, C, m; Hardy

'PHILOMÈNE POLLAERT', HT, or, 1925; flowers crimson tinted orange, dbl., intense fragrance; [Gen. MacArthur seedling X Old Gold]; Pollaert

Phil's Chromatella, N, dy

Phil's Hot Pink Perpetual Damask, D, mp

Phloxy®, F, rb

Phoebe *see* 'HARVANDER'

'PHOEBE', HT, w, 1922; flowers cream-white, sometime pure white, dbl., slight fragrance; GM, NRS, 1921; [Ophelia X Verna Mackay]; Cant, B. R.

Phoebe's Choice™ *see* 'BILICE'

Phoebe's Frilled Pink *see* **'FIMBRIATA'**

'PHOEBUS', HP, mp, 1837; flowers bright pink, shading lighter, large, dbl., moderate fragrance; bushy growth

Phoenix *see* 'HARVEE'

'PHOENIX', HT, dp, 1973; flowers light cerise, large, dbl., exhibition form, moderate fragrance; foliage large, glossy, leathery; vigorous, upright growth; [Manitou X Grand Slam]; Armstrong, D.L.; Armstrong Nursery

'PHOENIX FIRST', F, dr, 1959; bud ovoid; flowers dark red shaded black, small, borne in clusters, very dbl., slight fragrance; foliage leathery; bushy growth; [Our Princess X Pompon Beauty]; Kernovski, V.R.; Langbecker

'PHOTOGENIC', S, w, 1999; flowers white with pink edge, reverse white with narrow pink edge, single, 10 petals, 2.5–3 in., borne in small clusters, moderate fragrance; foliage large, medium green, semi-glossy; upright, tall (10 ft) growth; Jerabek, Paul E.; Freedom Gardens, 1998

'PHYLLIS', Pol, mr, 1908; flowers bright red, small, borne in large clusters, dbl., recurrent bloom; [Mme Norbert Levavasseur X ?]; Merryweather

'PHYLLIS BIDE', Cl Pol, yb, 1923; flowers pale gold, shaded pink, almost small blooms in long, loose clusters, dbl., dependably recurrent; height 6 ft; (14); GM, NRS, 1924; [Perle d'Or X Gloire de Dijon]; Bide

'PHYLLIS BURDEN', HT, op, 1935; bud long, pointed; flowers shrimp-pink and orange, large, dbl.; foliage glossy, light; very vigorous growth; Cant, B. R.

'PHYLLIS GOLD', HT, my, 1935; flowers butter-yellow, edged lighter, dbl., exhibition form, slight fragrance; foliage rich olive-green; very vigorous, branching growth; GM, NRS, 1933; [Lady Florence Stronge X Julien Potin]; Robinson, H.; Wheatcroft Bros.

'PHYLLIS GOLD, CLIMBING', Cl HT, my, 1949; Fryers Nursery, Ltd.

'PHYLLIS LUCAS', HT, ob, 1961; flowers orange shaded bronze, well shaped, moderate fragrance; [Bettina sport]; Wheatcroft Bros.

'PHYLLIS POYSER', HT, op, 1962; bud long, pointed; flowers ornage-pink, large, dbl., 50 petals, exhibition form, intense, spicy fragrance; foliage light green, soft, elongated; vigorous, compact, bushy growth; [Golden Sun X Spartan]; Fankhauser

Phyllis Shackelford™ *see* 'MORSHACK'

'PHYNELIA', HT, mr, 1928; flowers crimson-cerise, borne in clusters; Reeves

'PIA BERGHOUT', Pol, mp, 1967; flowers pink, medium, borne in clusters, dbl.; foliage dark; [Saskia X Seedling]; Buisman, G. A. H.

'PICADOR', F, mr, 1964; flowers cherry-red, borne in clusters, dbl.; foliage glossy, dark; upright, compact growth; [Oranien sport]; Verschuren, A.; van Engelen

Picaninni *see* WRIpic

'PICARDY', HT, dp, 1967; flowers rose-bengal, globular, moderate fragrance; foliage dark, glossy; free growth; [Rose Gaujard X Bayadere]; Trew, C.; Willik Bros.

Picasso *see* 'MACPIC'

'PICAYUNE', Ch, lp; flowers light pink to white, small blooms in clusters, dbl.

Piccadilly® *see* 'MACAR'

'PICCADILLY SUNSET', HT, ob, 1970; flowers orange flushed apricot, reverse gold, high pointed, dbl., 34 petals, 4–5 in., slight fragrance; foliage glossy, dark; moderate, upright growth; [Piccadilly sport]; Goodwin; The Valley Nursery

'PICCADILLY, CLIMBING', Cl HT, rb, 1963; Mized, 1963 & Sutton, 1972

'PICCANINNY', HT, dr, 1941; bud long, pointed; flowers stamens yellow, large, single, 5–6 petals, intense fragrance; foliage dark, glossy; vigorous growth; [Night X Sanguinaire]; Lammerts, Dr. Walter; Armstrong Nursery

'PICCIOLA INA', Pol, op, 1937; flowers pure salmon, borne in clusters of 10-50; vigorous, bushy growth; Giacomasso

Piccola *see* 'TANOLOKIP'

'PICCOLA ES', S, lp, 1983; flowers medium, dbl., 20 petals, slight fragrance; foliage medium, medium green, semi-glossy; bushy growth; [Unnamed X Unnamed]; Fumagalli, Niso

Piccolo® *see* 'TANOLOKIP'

'PICCOLO', F, dr, 1957; bud ovoid; flowers velvety dark red, open, borne in clusters, dbl., slight fragrance; foliage dark, leathery, glossy; moderate, bushy growth; RULED EXTINCT 4/85; [Red Favorite X Kathe Duvigneau]; Tantau, Math.

'PICCOLO PETE', S, mr, 1984; 7–10 petals, 4 in., flat, borne 1-10 per cluster, repeat bloom, moderate fragrance; slightly hooked, tan prickles; foliage large, leathery, dark olive green; upright, bushy growth; hardy.; [Carefree Beauty X ((Peace X Dornroschen) X Country Music)]; Buck, Dr. Griffith J.; Iowa State University

'PICK ME UP', HT, mr, 1941; flowers well-shaped; Clark, A.

Pickering Baby Faurax, Pol, m

Pickering Four Seasons Rose, P, dp

Pickering Old Red China, Ch, dp

Pickering Red, HP, dr

Pickwick *see* 'SABCLIVE'

'PICNIC', F, or, 1966; (Pic-Nic); bud short, pointed; 3–3.5 in., exhibition form, slight fragrance; foliage leathery; upright growth; [South Seas X Unnamed seedling]; Warriner, William A.; J&P

Pic-Nic *see* **'PICNIC'**

'PICO', HT, rb, 1962; bud pointed; flowers salmon-red and canary-yellow, dbl.; foliage glossy; bushy growth; Frères, Buyl

Picobello, F, or, 1998

'PICOTEE', F, rb, 1960; flowers white, edged red, blooms in clusters, dbl., 24 petals, 2.5–3.5 in., cupped, slight fragrance; foliage dark, glossy; vigorous, upright, bushy growth; [Little Darling X Gertrude Raffel]; Raffel; Port Stockton Nursery

Picpa, HT, 1974; Dorieux; (Cavriglia)

'PICTURE', HT, lp, 1932; dbl., exhibition form, slight fragrance; foliage glossy, dark; vigorous growth; McGredy

'PICTURE PAGE', HT, lp, 1953; flowers peach-pink shading to flesh-pink, base yellow, well formed, dbl., 23–25 reflexed petals, 4 in., slight fragrance; foliage dark; very free growth; [Picture X Mme Butterfly]; Jordan, B.L.

Picture Perfect *see* **'ELSIE MELTON'**

'PICTURE, CLIMBING', Cl HT, lp, 1942; Swim, H.C.; Armstrong Nursery

'PICTURESQUE', HT, lp, 1950; bud long; flowers pale pink with pronounced red veining, large, dbl., 30–35 petals, exhibition form, slight fragrance; foliage leathery; vigorous, upright, bushy growth; [Mrs H.M. Eddie X Mrs Sam McGredy]; Eddie

Pie IX *see* **'PIUS IX'**

Piece d'Or, S, dy

'PIED PIPER', F, mr, 1969; flowers small, dbl., globular; foliage leathery; moderate, low growth; [Garnette X Moulin Rouge]; Lindquist; Howard Rose Co.

Pierette, HRg, dp, 1987; flowers carmine-pink, large, semi-dbl., intense fragrance; Uhl, J.; (Sangerhausen)

Pierette Pavement, HRg, lp, 1987; semi-dbl., moderate fragrance; spreading; ground cover (2.5 ft) growth; Uhl, J.

'PIERRE', HT, yb, 1945; bud large, ovoid; flowers deep golden yellow edged red, very dbl.; foliage glossy; vigorous, upright, bushy growth; [Soeur Thérèse X Lumiere]; Mallerin, C.; A. Meilland

Pierre, HT, ab

'PIERRE AGUETANT', HT, my, 1938; bud long, pointed; flowers chamois-yellow, open, very large, dbl., moderate fragrance; foliage leathery; vigorous growth; Gaujard

Pierre B., HT, lp; [Dr A.J. Verhage sport]; Bell, Ronald J., 1982; (Weatherly, L.)

'PIERRE BREDY', HT, rb, 1958; flowers currant-red, reverse silvery; foliage dark; low growth; [Peace X Scheherazade]; Arles; Roses-France

'PIERRE CARO', HP, dr, 1879; flowers medium, dbl., intense fragrance; Levet; (Sangerhausen)

'PIERRE CORMIER', Pol, or, 1926; flowers brilliant scarlet-red, center lighter, borne in clusters of; dwarf growth; Turbat

Pierre de Ronsard® *see* 'MEIVIOLIN'

'PIERRE DE ST CYR', B, lp, 1838; flowers glossy pale pink, large, very dbl., cupped; vigorous growth; Plantier

'PIERRE GAUJARD', HT, rb, 1944; bud pointed; flowers fiery shades, very large, dbl., slight fragrance; foliage glossy; vigorous growth; Gaujard

'PIERRE GUILLOT', HT, rb, 1879; flowers red with white, large, dbl., moderate fragrance; Guillot; (Sangerhausen)

'PIERRE NOTTING', HP, dr; flowers large, dbl., 42 petals, globular, not often recurrent, moderate fragrance; upright growth; [Alfred Colomb seedling]; Portemer fils, 1862

Pierre Troisgros® *see* DORtive

'PIERRE WATTINE', HT, dp, 1901; flowers large, dbl., moderate fragrance; Soupert & Notting; (Sangerhausen)

'PIERRETTE', HT, rb, 1931; flowers blackish red streaked white, passing to steel-blue, small, very dbl., cupped, slight fragrance; foliage small; short stems; dwarf growth; [Felix Laporte sport]; Tantau

'PIERRETTE', HT, or, 1945; flowers bright copper red, flecked brighter, large, dbl., 25 petals, intense fragrance; upright, bushy growth; [Texas Centennial sport]; Tantau

Pierrine *see* 'MICPIE'

'PIERROT', F, rb, 1971; bud ovoid; flowers white marked red, becoming red, dbl., 30 petals, 2.5 in., cupped, moderate, spicy fragrance; foliage leathery; vigorous, compact growth; [Poupee X Fillette]; Lens

'PIERSON'S PINK', HT, dp, 1950; bud long, pointed; flowers rose-pink, dbl., 32

petals, 5 in., exhibition form, slight fragrance; very vigorous, upright growth; [Better Times sport]; Pierson, A.N.

'**PIET RETIEF**', F, dr, 1950; flowers dark morocco-red; [Irene X Donald Prior]; Leenders, M.

'**PIET SAVERYS**', F, ob, 1955; flowers orange, semi-dbl.; bushy growth; [Independence X Border King]; Frères, Buyl

Pietermaritzburg, F, mp

'**PIGALLE**', HT, m, 1951; flowers reddish violet, dbl., 4 in., slight fragrance; foliage bronze; bushy growth; [Fantastique X Boudoir]; Meilland, F.

Pigalle® *see* 'MEICLOUX'

Pigalle 84 *see* 'MEICLOUX'

'**PIGALLE, CLIMBING**', Cl HT, m; Roses-France

'**PIGMY GOLD**', F, dy, 1953; bud ovoid; flowers golden yellow, small, dbl., moderate fragrance; foliage dark, glossy; dwarf growth; Boerner; J&P

'**PIGMY LAVENDER**', F, m, 1961; bud ovoid; flowers lavender tinted pink, dbl., 30–35 petals, 2 in., cupped, moderate fragrance; foliage leathery, dark; short, strong stems; vigorous, dwarf growth; [Lavender Pinocchio seedling X Unnamed Hybrid Tea seedling]; Boerner; J&P

'**PIGMY RED**', F, dr, 1953; bud ovoid; flowers deep red, white eye, small blooms in clusters, dbl., moderate fragrance; foliage glossy; dwarf growth; [Chatter X Red Pinocchio]; Boerner; J&P

'**PIKE'S PEAK**', S, pb, 1940; bud long, pointed; flowers light bright red, center yellow, fading white, blooms in clusters, semi-dbl., 13 petals, 3.5 in., non-recurrent; foliage light, wrinkled; very vigorous (6 ft.), bushy growth; [R. acicularis X Hollywood]; Gunter; B&A

'**PILAR DE ARBURUA**', HT, op; flowers salmon, large, dbl., exhibition form, slight fragrance; foliage dark reddish green; free growth; [Comtesse Vandal X Fashion]; Camprubi, C.

'**PILAR DOT**', Min, op, 1964; flowers coral, well-formed, small; vigorous, well-branched growth; [Orient X Perla de Alcanada]; Dot, Pedro

'**PILAR LANDECHO**', HT, ob, 1940; (Marquesa de Urquijo); bud long, pointed; flowers yellow, reverse dark coral-orange, large, dbl., exhibition form, slight fragrance; foliage dark, leathery; vigorous growth; GM, Bagatelle, 1938; [(Sensation X Julien Potin) X Feu Joseph Looymans]; Camprubi, C.; A. Meilland

'**PILAR LANDECHO, CLIMBING**', Cl HT, ob, 1954; (Marquesa de Urquijo, Climbing); Folgado, Comes

Pilarcitos, N, w

'**PILARÍN VILELLA**', HT, or, 1936; flowers lacquer-red, large, dbl., cupped, intense fragrance; foliage dark; vigorous growth; [Mrs Pierre S. duPont X Lucia Zuloaga]; Dot, Pedro

'**PILGRIM**', HT, dr, 1970; bud ovoid; flowers large, dbl., cupped, moderate fragrance; foliage dark, leathery; vigorous, upright, bushy growth; [Unnamed seedling X Chrysler Imperial]; Armstrong, D.L.; Armstrong Nursery

'**PILGRIM**', HT, pb, 1920; bud long, pointed; flowers silvery pink reverse clear rose-pink, large, dbl., exhibition form, moderate fragrance; foliage leathery, rich green; vigorous, bushy growth; Montgomery Co.; A.N. Pierson

'**PÍLINA MATA**', HT, ob, 1934; flowers orange-yellow; [Souv. de Claudius Pernet X Los Angeles]; Munné, B.

Pillar Box *see* 'CHEWAZE'

Pillar BoxWardlip *see* 'CHEWAZE'

'**PILLAR OF FIRE**', Cl F, or, 1963; bud short, ovoid; flowers coral-red, blooms in cluster, dbl., 33 petals, 2–2.5 in., cupped, slight fragrance; foliage leathery, matt; vigorous growth; [Floradora, Climbing sport]; Shamburger, P.; C-P

Pillar of Gold *see* '**E. VEYRAT HERMANOS**'

'**PILLAR STRATFORD**', Cl HT, lp, 1946; flowers silvery pink, base deeper, large, very dbl., intense fragrance; vigorous, upright (8 ft) growth; [Stratford sport]; Watkins, A.F.; Krider Nursery

Pillnitzer Marcellina, Pol, dr; semi-dbl.; Institut für Obstbau Dresden Pillnitz; (Sangerhausen)

Pillow Fight™ *see* 'WEKPIPOGOP'

'**PILLOW TALK**', F, m, 1980; bud short, ovoid; flowers reddish lavender, blooms borne singly or several together, dbl., exhibition form, moderate, tea fragrance; long prickles, hooked downward; foliage medium, thin, leathery, dark; low to medium, rounded growth; [Plain Talk X Angel Face]; Weeks, O.L.

Pilurett, F, or, 1966; flowers medium, semi-dbl.; Schmadlak, Dr.; (Sangerhausen)

Pimlico *see* 'MEIDUJARAN'

Pimlico '81 *see* 'MEIDUJARAN'

'**PIMPANT**', F, or, 1963; flowers bright orange-red, borne in clusters of 7-8, dbl., 25–30 petals; moderate, bushy growth; [Unnamed seedling X Soleil]; Laperrière; EFR

'**PIMPERNELL**', F, or, 1954; flowers turkey-red, open, borne in clusters, semi-dbl.; foliage dark; [Unnamed seedling X (Poulsen's Pink X Golden Dawn)]; LeGrice

Pimpernelle *see* DELdog

'**PINAFORE**', Pol, ly, 1959; flowers pale yellow to white, tinged pink, borne in large, rounded clusters, single, 1.5–2 in., flat; foliage glossy; low, bushy, compact growth; [China Doll X Mrs Dudley Fulton]; Swim, H.C.; Roseway Nursery

Pinal, Pol, mr; flowers medium, single; Institut für Obstbau Dresden Pillnitz; (Sangerhausen)

'**PIÑATA**', LCl, yb, 1978; bud ovoid; flowers yellow overlaid vermilion, dbl., 28 petals, 3 in., exhibition form, slight fragrance; semi-climbing growth; Suzuki, Seizo; J&P

PINault, F, 1982; (**Rêve de Deauville**, Revue de Dauville); Pineau; (Cavriglia)

Pincushion, S, mp, 1996; Kordes

'**PINEAPPLE POLL**', F, ob, 1970; flowers orange-yellow, flushed red, dbl., 30 petals, 2.5 in., moderate fragrance; foliage glossy; [Orange Sensation X Circus]; Cocker

'**PINEHURST**', Min, pb, 1988; flowers light pink, fading lighter at base, reverse light pink to cream, very dbl., 60 petals, moderate, fruity fragrance; prickles straight, medium, pink; foliage medium, dark green, semi-glossy; bushy, medium, vigorous, neat growth; [Rise 'n' Shine X Unnamed seedling]; Bridges, Dennis A.; Bridges Roses

Pine-Scented Rose *see* **R. GLUTINOSA**

'**PINGAN**', F, w, 1980; bud ovoid, pointed; flowers white with pink petal edges, blooms borne 4-6 per cluster, semi-dbl., 13 petals, moderate fragrance; long prickles; foliage large, leathery; upright growth; [Unnamed seedling X (Ivory Fashion X Little Darling)]; Fong, William P.; Del Rose Nursery

'**PINK ALICIA**', HT, mp, 1968; flowers pink, pointed, moderate fragrance; spreading growth; [Duftwolke X Gavotte]; Dale, F.

Pink Angel *see* 'MORGEL'

'**PINK ANGEL**', Min, mp, 1982; [Starina sport]; Hunton, Claude B.

Pink Angel, HT, lp; Parkes, Mrs M.H., 1977; (Weatherly, L.)

'**PINK ANNE**', F, mp, 1951;, borne in trusses, slight fragrance; (28); [Anne Poulsen sport]; Cant, B. R.

Pink Arctic *see* '**SHOW GARDEN**'

Pink Avalanche *see* 'WILPAVL'

Pink Babyflor *see* TANybab

Pink Bassino *see* KORbasren

'**PINK BEAUTY**', HT, mp, 1919; flowers clear pink, large, semi-dbl., cupped, intense fragrance; [Ophelia X My Maryland]; Cook, J.W.

'PINK BEDDER', HT, mp, 1920; flowers rose-pink, center yellow, borne in clusters; Paul, W.

'PINK BELLE', HT, dp, 1974; flowers deep rose pink, large, dbl., 28 petals, moderate fragrance; foliage dark; RULED EXTINCT 11/83; [Fragrant Cloud X Elizabeth of Glamis]; Harkness; Morse Roses

Pink Bells® *see* 'POULBELLS'

'PINK BLUSH', HT, lp, 1974; bud long, pointed; dbl., 30 petals, 4–5 in., cupped, slight fragrance; foliage large, leathery; very free growth; [Bridal Pink X Seedling]; Warriner, William A.; J&P

'PINK BOUNTIFUL', F, mp, 1945; bud short, pointed; dbl., 55 petals, 3 in., blooms in clusters, moderate fragrance; foliage dark, leathery; vigorous, upright, much-branched growth; [Juanita X Mrs R.M. Finch]; Hill, Joseph H., Co.; J&P;, J.H. Hill Co.

Pink Bounty™ *see* 'MINSCO'

'PINK BOUQUET', F, mp, 1954; flowers china-rose pink, slightly tinted yellow, to ovoid, dbl., 60–75 petals, 3–4 in., exhibition form, moderate fragrance; foliage leathery; upright, open, compact growth; RULED EXTINCT 11/90; [Curly Pink X Free Gold]; Brownell, H.C.

Pink Bouquet *see* 'LAVQUEST'

Pink Bouquet, HWich, pb; Treasure

Pink Bourbon *see* **'MME ERNEST CALVAT'**

'PINK BROCADE', F, lp, 1977; flowers very full, very dbl., 70 petals, 3 in., slight fragrance; foliage glossy; small, compact growth; [Spartan X Lilli Marlene]; Bees

Pink Bunting *see* 'TALPIN'

'PINK BUTTERFLY', HT, dp, 1926; flowers bright cerise, base light buff; [Mme Butterfly sport]; Brown, A.C.

Pink Button *see* KORdemas

'PINK CAMEO', Cl Min, mp, 1954; (Cameo, Climbing); flowers rose-pink, center darker, blooms in clusters (to 20), dbl., 23 petals, 1.25 in., slight fragrance; foliage small, glossy, rich green; height 3-5 ft; [(Soeur Thérèse X Skyrocket) X Zee]; Moore, Ralph S.; Sequoia Nursery

Pink Candy, HT, lp; [Candy Stripe sport]; Thomas, before 1992; (Weatherly, L.)

Pink Cardinal Hume, S, mp, 1998; Rupert, Kim L.

Pink Carpet™ *see* 'MINNCO'

Pink Cascade *see* 'MORCADE'

'PINK CASCADE', Pol, lp, 1945; flowers la france pink, small, borne in clusters, very dbl., slight fragrance; foliage glossy; vigorous, bushy growth; RULED EXTINCT 12/81; [Mrs Dudley Fulton X Tom Thumb]; Lammerts, Dr. Walter; Univ. of Calif.

'PINK CAVALCADE', F, pb, 1955; flowers deep pink, reverse light yellow to white, borne in pyramidal, dbl., 28–32 petals, 2.5–3 in., cupped, moderate fragrance; foliage leathery, glossy, bronze; vigorous, bushy growth; [Cavalcade sport]; Shamburger, C.S.; Stuart

'PINK CHAMELEON', Pol, pb, 1945; bud rose-red; flowers venetian pink, darkening to purple, open, small, borne in clusters, single, abundant, recurrent bloom, slight fragrance; foliage dark, glossy; very vigorous, bushy growth; [Mrs Dudley Fulton X R. chinensis mutabilis]; Lammerts, Dr. Walter; Univ. of Calif.

'PINK CHAMPAGNE', HT, mp, 1956; bud long, pointed; dbl., 32–48 petals, 5–6 in., exhibition form, moderate fragrance; foliage leathery; vigorous, bushy growth; [Unnamed seedling X Pink Bountiful]; Jelly; E.G. Hill Co.

Pink Champagne *see* CLEcham

Pink Charles Austin, S, pb, 1992; Austin, David

'PINK CHARM', F, dp, 1938; flowers deep clear pink, very dbl., slight fragrance; foliage leathery; short stems; bushy growth; Kordes; Dreer

Pink Charmer *see* 'LYOPIN'

'PINK CHARMING', HT, lp, 1953; flowers pale pink, loosely formed, dbl., 36 petals, 5.5 in., moderate fragrance; vigorous growth; Leenders, M.

'PINK CHATEAU', HT, w, 1999; flowers white, center soft pink, dbl., 35 petals, 5 in., exhibition form, no fragrance; 4.5 ft growth; [(Sheer Bliss X seedling) X seedling]; Teranishi, K., 1997; Itami Rose Nursery, 1997

Pink Cherokee *see* **'ANEMONE'**

Pink Cherokee *see* R. X anemonoides

Pink Cherub *see* 'MORFAIR'

'PINK CHIFFON', F, lp, 1956; bud ovoid; dbl., 53 petals, 3.5–4 in., cupped, intense fragrance; foliage glossy; vigorous, bushy growth; [Fashion X Fantasia]; Boerner; J&P

Pink Chimo® *see* 'INTERCHIMP'

'PINK CHRISTIAN DIOR', HT, dp, 1966; bud ovoid; flowers light red and deep pink, large, dbl., exhibition form, slight fragrance; foliage glossy; very vigorous, upright growth; [Christian Dior sport]; Chang, Chi-Shiang

'PINK CLOUD', LCl, mp, 1952; bud ovoid; flowers rich pink, large blooms in clusters of 5-20, dbl., 28 petals, cupped, recurrent bloom, moderate fragrance; foliage glossy; vigorous, climbing (6-8 ft.) growth; [New Dawn X New Dawn seedling]; Boerner; J&P

Pink Cloud® *see* 'JACLIP'

'PINK CLOUDS', Cl Min, dp, 1956; flowers deep rose pink, small to blooms in clusters, single, intense fragrance; foliage dark, glossy; vigorous (5-8 ft) growth; [Oakington Ruby X R. multiflora]; Moore, Ralph S.; Sequoia Nursery

'PINK CLUSTER', F, op, 1938; flowers salmon-pink, shaded gold, borne in clusters; Morse

'PINK COCKTAIL', HT, pb, 1963; flowers light pink, reverse darker, dbl., 38 petals, 5 in., slight fragrance; foliage dark, glossy; very free growth; [Queen Elizabeth X Claude]; Barter

Pink Cottage *see* 'CHEWILY'

Pink Cover®, S, mp; Poulsen

Pink Crumble Bar, HT, lp; [Crumble Bar sport]; Thomson, G.A.; (Weatherly, L.)

Pink Crystal *see* WEKslats

Pink Curtain *see* KORlallal

Pink Cushion, Pol, mp

Pink Danyland®, S, dp

'PINK DAWN', HT, mp, 1935; bud long, pointed; flowers large, dbl., 60 petals, exhibition form, intense fragrance; foliage soft; vigorous growth; [Joanna Hill X Seedling]; H&S; Dreer

'PINK DAWN, CLIMBING', Cl HT, mp, 1941; H&S

'PINK DELIGHT', Pol, dp, 1922; (Laxton's Pink Delight); flowers rose-pink, single, moderate fragrance; Laxton Bros.

'PINK DELIGHT', HT, op, 1936; (Vierlanden); bud long, pointed; flowers deep salmon-pink, large, dbl., exhibition form, intense fragrance; foliage leathery, light green; vigorous growth; [Senator X Florex]; Kordes; J&P

Pink Delight® *see* LENpil

Pink Devil, HT, mp

Pink Diadem® *see* TANspolett

'PINK DIAMOND', HT, lp, 1942; bud long pointed; flowers shell-pink, base yellow, dbl., 38 petals, 3.5–4 in., cupped; foliage leathery; vigorous, upright growth; Howard, F.H.; Diamond State Nursery

'PINK DIANE', F, dp, 1959; bud ovoid; flowers deep rose-pink, dbl., 50–55 petals, 3.5–4 in., cupped, intense, fruity fragrance; foliage leathery, glossy, dark; vigorous, upright growth; [Rosenelfe sport]; Sodano, A.

Pink Diddy™ *see* 'KINDIDDY'

'PINK DIËLMA', HT, mp, 1969; bud ovoid; flowers pink, medium, very dbl.; foliage dark; [Furore sport]; Tas

'PINK DON JUAN', LCl, mp, 1996; semi-dbl., 6–14 petals, 1.5–2.75 in., borne in small clusters, slight fragrance; some prickles; foliage medium, medium green, semi-glossy; tall, upright, spreading, climbing growth; [Don Juan sport]; Nelson, Brian; Certified Roses, Inc., 1997

'PINK DREAM', HT, pb, 1951; flowers pink, inside rosy white, dbl., 45 petals, 6–7 in., exhibition form, slight fragrance; foliage dark; vigorous growth; [Mrs Sam McGredy X R.M.S. Queen Mary]; McGredy, Sam IV

Pink Drift *see* 'POULCAT'

'PINK DRUSCHKI', HT, mp, 1949; flowers bright pink, well formed, dbl., 27 petals, 5 in., slight fragrance; very vigorous growth; Longley

'PINK DUCHESS', HT, dp, 1959; bud ovoid; flowers rose-red, dbl., 38 petals, 5–6 in., cupped, moderate fragrance; foliage glossy; vigorous, upright, bushy growth; [Peace seedling X Unnamed seedling]; Boerner; J&P

'PINK ELEGANCE', HT, lp, 1960; bud long, pointed; flowers bright pink, medium, dbl., exhibition form, slight fragrance; foliage glossy; very vigorous growth; [White Butterfly X Baccará]; Hoefer; Carlton Rose Nurseries

Pink Elf *see* 'MORELFIRE'

Pink Elizabeth Arden *see* **'GEISHA'**®

'PINK EMPEROR', HT, mp, 1958; flowers clear pink, dbl., 50–60 petals, 4–5 in.; foliage glossy; vigorous, bushy growth; [Peace X Crimson Glory]; Jones; Hennessey

Pink Empress *see* 'HADEMP'

'PINK EUTIN', F, mp, 1962; [Eutin sport]; Lindquist; Howard Rose Co.

Pink Fairy, Pol, mp

Pink Fantasy, HT, mp, 1995; K&S

'PINK FAVORITE', HT, mp, 1956; (Pink Favourite); bud pointed; flowers neyron rose, dbl., 25 petals, 3–4 in., cupped, slight fragrance; foliage very glossy, bright green; vigorous, upright, bushy growth; GM, Portland, 1957; [Juno X (Georg Arends X New Dawn)]; Von Abrams; Peterson & Dering

Pink Favourite *see* **'PINK FAVORITE'**

Pink Festival *see* 'LAVQUEST'

Pink Fire® *see* INTepin

'PINK FLAIR', F, mp, 1966; bud urn shaped; flowers small, dbl.; foliage leathery; vigorous, bushy growth; [Verona X Escort]; Swim & Weeks; Carlton Rose Nurseries

'PINK FLAMINGO', HT, pb, 1957; flowers rose-pink tinted lighter, dbl., 50 petals, 2.5–3.5 in., intense fragrance; foliage dark, leathery; dwarf, bushy growth; [Golden Dawn sport]; Kern Rose Nursery

'PINK FLORADORA', F, op, 1951; bud ovoid; flowers shrimp-pink, dbl., 35–40 petals, 3 in., cupped, slight fragrance; foliage glossy, light green; bushy growth; [Floradora sport]; Shamburger, P.

Pink Flower Carpet *see* 'NOATRAUM'

Pink Flurries, F, lp, 1999

'PINK FORMAL', S, pb, 1978; bud pointed; flowers bright coral-pink, loosely-ruffled, dbl., 23 petals, 3.5–4 in., moderate fragrance; foliage leathery; vigorous, upright growth; [(Queen Elizabeth X Gladiator) X (Aztec X Little Darling)]; Williams, J. Benjamin; Krider Nursery

Pink Fountain *see* LEGram

'PINK FRAGRANCE', HT, mp, 1956; bud long, pointed; flowers rose-pink, open, borne in pyramidal clusters, dbl., 78–85 petals, 4.5–5 in., moderate, spicy fragrance; foliage glossy, leathery; very vigorous, bushy, upright growth; [(Orange Triumph X Golden Rapture) X Peace]; deRuiter; Ilgenfritz Nursery

Pink Friendship, S, mp, 1986; Verschuren

'PINK FRILLS', F, lp, 1954; [Garnette sport]; Carlton Rose Nursery

Pink Fringe®, F, mp, 1990; Interplant

'PINK FROST', HT, mp, 1954; bud ovoid; flowers rose, dbl., 38–45 petals, 4–5 in., exhibition form, intense fragrance; foliage glossy, leathery; vigorous, bushy growth; [Charlotte Armstrong X Texas Centennial]; Swim, H.C.; Arp Nursery Co.

'PINK FROSTFIRE', Min, lp, 1968; [Frostfire sport]; Moore, Ralph S.; Sequoia Nursery

Pink Frosting *see* 'RENFROST'

'PINK GARLAND', S, mp, 1935; flowers clear pink, open, semi-dbl., 3–3.5 in., non-recurrent, intense fragrance; upright (3 ft.) growth; [R. blanda X R. spinosissima cultivar]; Skinner

'PINK GARNETTE', F, dp, 1950; flowers tyrian rose; [Garnette sport]; Schneeberg

'PINK GARNETTE', F, dp, 1951; flowers deep pink; [Garnette sport]; Boerner; J&P

'PINK GARNETTE SUPREME', F, mp, 1946; flowers rose-pink, borne in clusters, dbl., 35–40 petals, 2.5 in., cupped, moderate fragrance; foliage glossy, leathery; vigorous, upright growth; [Garnette Supreme sport]; Perkins, C.H.; J&P

'PINK GEM', HT, lp, 1949; bud long, pointed; flowers pearl-pink, dbl., 30 petals, 4–5 in., flat, intense fragrance; foliage bluish green; vigorous, tall growth; Fletcher; Tucker

'PINK GEM', Min, lp, 1974; bud ovoid; flowers rose-pink, full, medium, dbl., slight fragrance; foliage soft; moderate, upright, bushy growth; [Scarlet Gem sport]; Meilland; C-P

Pink Gift, Pol, lp

Pink Gin, HT, mp; flowers medium, dbl.; (Sangerhausen)

'PINK GLORY', HT, mp, 1960; bud long, pointed; flowers clear pink, dbl., 25 petals, 5 in., exhibition form, intense fragrance; foliage leathery; vigorous, upright growth; [Ernie Pyle seedling X Peace]; Boerner; J&P

'PINK GLOW', HT, lp, 1951; bud globular; flowers rose-pink, dbl., 40 petals, 5–5.5 in., cupped, moderate fragrance; foliage leathery, glossy; vigorous, upright growth; [((Schoener's Nutkana X Unknown seedling) X Mrs Pierre S. duPont) X Home Sweet Home]; Boerner; J&P

'PINK GOLDEN DAWN', HT, mp, 1938; flowers large, dbl., moderate fragrance; [Golden Dawn sport]; Bostick

'PINK GOWN', HT, lp, 1971; flowers light clear pink, dbl., 20 petals, 4 in.; foliage matt green; tall, very free growth; [Vera Dalton X Tropicana]; Sanday, John

'PINK GROOTENDORST', HRg, mp, 1923; flowers clear pink, carnation style petal edges, very dbl., borne in small clusters; foliage medium, dark green, rugose; tall growth; [F.J. Grootendorst sport]; Grootendorst, F.J.

'PINK GRÜSS AN AACHEN', F, op, 1929; flowers light salmon-pink; [Gruss an Aachen sport]; Kluis & Koning

'PINK HADLEY', HT, mp, 1928; flowers clear rose pink; [Hadley sport]; Knight, G.

'PINK HAT', F, pb, 1980; bud pointed; flowers light pink, center deep pink, blooms borne singly or 3-7 per cluster, dbl., 24 petals, exhibition form, slight fragrance; red prickles; foliage red turning dark green, glossy; vigorous, upright, bushy growth; James, John

'PINK HEATHER', Min, lp, 1959; flowers lavender-pink to white, very small blooms in clusters, dbl., 45 petals; foliage very small, glossy; vigorous (10-12 in), bushy growth; [(R. wichurana X Floradora) X (Violette X Zee)]; Moore, Ralph S.; Sequoia Nursery

'PINK HEDGE', HRg, mp, 1956; red fruit; foliage small, bronze; [(R. rugosa rubra X R. cinnamomea) X R. nitida]; Nyveldt

Pink Heidelberg, HKor, mp

Pink Hit *see* 'POULINK'

Pink Hit *see* POUltipe

'PINK HONEY', Min, mp, 1988; flowers medium honey-pink, yellow at base, reverse light yellow at base, semi-dbl., 20 petals, exhibition form, slight, fruity fragrance; prickles long, pointed, medium, pink; foliage large, medium green, semi-glossy; bushy, tall growth;

[Summer Spice X Unnamed seedling]; Bridges, Dennis A.; Bridges Roses

Pink Ice *see* 'ANDPIN'

Pink Iceberg® *see* 'PROBERG'

Pink Ilseta, Gr, 1986; Tantau; (Cavriglia)

Pink Jacqueline, HT; (Cavriglia)

'PINK JENNY', HT, mp, 1961; flowers rose-pink; [Grand'mere Jenny sport]; Ruston, D.

'PINK JEWEL', F, mp, 1940; flowers arbutus-pink, center camellia-pink, semi-dbl., cupped, slight fragrance; foliage leathery; vigorous growth; [Crimson Glory X Holstein]; Kordes; Dreer

'PINK JOY', Min, dp, 1953; flowers deep pink, well-shaped, dbl., 30 petals, 1 in., moderate, sweet violet fragrance; dwarf (12in), bushy growth; [Oakington Ruby X Oakington Ruby]; Moore, Ralph S.; Sequoia Nursery

Pink Kardinal, HT, dp; Stratford, 1995

'PINK KAREN POULSEN', F, mp, 1936; [Karen Poulsen sport]; Poulsen, S.

'PINK KEY', HT, mp, 1920; [Francis Scott Key sport]; Pierson, F.R.

Pink Kiss® *see* 'MINAKCO'

'PINK KOSTER', Pol, mp; [Margo Koster sport]; Koster, D.A.

Pink La Sevillana® *see* 'MEIGEROKA'

Pink La Sevilliana *see* 'MEIGEROKA'

'PINK LACE', F, pb, 1961; flowers light pink, becoming darker and then red, medium (2-2 1/2"), dbl., 20 petals, flat, slight fragrance; foliage dark; vigorous growth; [The Optimist X Korona]; Watkins Roses

'PINK LADY', HT, mp, 1947; bud urn-shaped; flowers soft pink, reverse darker, semi-dbl., 3.5 in., slight fragrance; medium, vigorous growth; [Pink Delight sport]; Wiltgen; Premier Rose Gardens

'PINK LAFAYETTE', F, mp, 1925; flowers clear rose-pink; [Lafayette sport]; Griffin

'PINK LÉDA', D, mp; flowers light to medium pink, crimson edging on petals; [Leda sport or vice-versa sport]

Pink Lemonade *see* 'AROFRICHEE'

Pink Licorice *see* 'RESINK'

'PINK LORRAINE', HT, w, 1972; flowers pink, paling to white, base cream, very full, dbl., 40 petals, 4.5 in., slight fragrance; foliage matt green; vigorous, upright growth; [Milord X Farah]; Williams, J. Benjamin

Pink Love, HT, mp, 1998

'PINK LUSTRE', HT, lp, 1957; bud ovoid; dbl., 48 petals, 5 in., exhibition form, intense fragrance; foliage dark, glossy, leathery; vigorous, upright growth; [Peace X Dame Edith Helen]; Verschuren; J&P

Pink Macartney Rose, HBc, lp

'PINK MAGIC', HT, dp, 1952; bud long, pointed, spinel-red; flowers phlox-pink, dbl., 40–45 petals, 4–6 in., intense fragrance; foliage leathery; vigorous, upright growth; [Better Times sport (white) X Snow White]; Hill, Joseph H., Co.

Pink Magic, HMsk, pb, 1990; Lens

Pink Magic Carpet® *see* MACausal

'PINK MAIDEN', F, mp, 1965; bud ovoid; flowers large blooms in clusters, dbl., slight fragrance; foliage dark; [Spartan seedling X Queen Elizabeth]; Boerner; Spek

Pink Maman Cochet, T, mp

'PINK MANDY', Min, mp, 1974; bud globular; dbl., 40 petals, 1 in.; foliage very glossy, leathery; low, bushy, spreading growth; [Ellen Poulsen X Little Chief]; Moore, Ralph S.; Sequoia Nursery

'PINK MARVEL', F, dp, 1958; flowers spirea-red, borne in clusters, dbl., 45–55 petals, 2.5 in., flat; foliage leathery; bushy, compact growth; [Rosemary Rose X (Cécile Brunner X Floribunda seedling)]; deRuiter; C-P

Pink Masquerade, S, op; single, 5 petals, moderate; [Suzanne X Red Dawn]; Simonet

'PINK MASTERPIECE', HT, pb, 1962; bud ovoid, pointed; flowers la france pink, tinted shrimp-pink, dbl., 38 petals, 6 in., exhibition form, moderate fragrance; foliage leathery; vigorous, upright growth; [Serenade seedling X Kate Smith]; Boerner; J&P

Pink Meidiland® *see* 'MEIPOQUE'

Pink Meillandina® *see* 'MEIJIDIRO'

Pink Melody, HT, op, 1993; K&S

'PINK MERINGUE', S, lp, 1999; very dbl., 41 petals, 3.5 in., borne in small clusters, slight fragrance; prickles numerous; foliage medium, medium green, semi-glossy; upright, tall (6 ft) growth; [Mary Rose sport]; Bennett, Frank David

Pink Mermaid, LCl, mp, 1940

'PINK METEOR', F, mp, 1964; [Meteor sport]; Timmerman's Roses

Pink Midinette *see* PEAxanadu

Pink Minimo, Min, mp

Pink Mini-Wonder® *see* 'MEISELGRA'

'PINK MIST', HT, mp, 1959; bud pointed, ovoid; flowers phlox-pink, becoming darker, dbl., 25–35 petals, 4–5 in., exhibition form, intense fragrance; foliage dark, leathery; vigorous, bushy growth; [Red Better Times sport]; Hill, Joseph H., Co.

Pink Mist, S, op

'PINK MONTEZUMA', Gr, lp, 1964; flowers light pink, reverse blush-pink, large, dbl., exhibition form, slight fragrance; foliage dark, leathery; vigorous, bushy growth; [Montezuma sport]; Williams, J. Benjamin

Pink Moss *see* **COMMUNIS**

Pink Mothersday, Pol, mp

Pink Mystery, Misc OGR, mp, 1997; Lens

Pink 'n' Pretty, HT, dp

Pink 'n' White®, Min, pb

Pink Nevada *see* **'MARGUERITE HILLING'**

Pink Nevada *see* 'SANCHARM'

'PINK NYMPH', F, mp, 1959; flowers clear pink; [Nymph sport]; Koster, D.A.

'PINK NYMPH', Min, mp; dbl.; (14); [R. chinensis minima sport]

Pink Ocean® *see* 'HAVINK'

Pink Olympiad, HT, mp, 1995; Humenick, Muriel F.

Pink Ophelia *see* **'JOHN C. M. MENSING'**

'PINK OPHELIA', HT, mp, 1916; flowers rose-pink; [Ophelia sport]; Breitmeyer

Pink Osiana *see* TANpinaiso

Pink Panther® *see* 'MEICAPINAL'

Pink Parade®, Min, mp

'PINK PARASOL', HT, mp, 1950; bud long, pointed; flowers clear pink, dbl., 25–30 petals, 6 in., slight fragrance; foliage leathery, dark; very vigorous growth; [Rapture X Rome Glory]; Fisher, G.; Arnold-Fisher Co.

'PINK PARFAIT', Gr, pb, 1960; bud ovoid to urn-shaped; flowers outer petals medium pink, center blended pale orange, dbl., 23 petals, 3.5–4 in., exhibition form, slight fragrance; foliage leathery, semi-glossy; vigorous, upright, bushy growth; AARS, 1961 GM, Baden-Baden, 1959 GM, NRS, 1962 GM, Portland, 1959; [First Love X Pinocchio]; Swim, H.C.; Armstrong Nursery

'PINK PASSION', Min, pb, 1978; bud pointed; flowers shell-pink and ivory, small, dbl., 26 petals, 1.5 in., exhibition form, moderate fragrance; upright, bushy growth; [Sweet and Low X ?]; Schwartz, Ernest W.; Bountiful Ridge Nursery

Pink Patio, MinFl, lp, 1997

Pink Pavement, S, op, 1991; flowers salmon pink, semi-dbl., moderate fragrance; compact, spreading (2.5 ft) growth; Baum

Pink Peace® *see* 'MEIBIL'

Pink Peace, Climbing® *see* 'MEIBILSAR'

'PINK PEARL', Cl HT, op, 1913; flowers pink shaded salmon, single; [Irish Elegance X Una]; Hobbies

'PINK PEARL', HT, mp, 1924; flowers neyron pink, base salmon, dbl., intense fragrance; long stems; vigorous growth; [Ophelia X (Gen. MacArthur X Marie van Houtte)]; Leenders, M.

'PINK PEARL, CLIMBING', Cl HT, mp, 1933; Dixie Rose Nursery

'PINK PERFECTION', HT, mp, 1927; flowers clear rose-pink, dbl., 45 petals, moderate fragrance; [Premier X Baroness Rothschild]; Ward, F.B.

Pink Perfection, HT, mp, 1998

'PINK PERFEKTA', HT, pb, 1962; flowers deep pink, often edged red; [Kordes' Perfecta sport]; Ross, A., & Son; A. Ross & Son

Pink Perfume, HT, 1955; Robinson, H.; (Cavriglia)

'PINK PERPÉTUÉ', LCl, mp, 1965; flowers bright rose-pink, semi-globular, blooms in clusters, dbl., 32 petals, recurrent bloom, moderate fragrance; foliage glossy, light green; vigorous growth; [Spectacular X New Dawn]; Gregory

'PINK PET', Ch, mp, 1928; flowers bright pink, dbl.; Lilley

'PINK PETTICOAT', Min, pb, 1979; bud pointed; flowers creamy white, edged coral-pink, mini-flora, dbl., 33 petals, 1.5–2 in., exhibition form, slight fragrance; foliage glossy, dark; tall, upright growth; AOE, 1980; [Neue Revue X Sheri Anne]; Strawn; Pixie Treasures Min. Roses

Pink Picotee® *see* 'KINPIC'

Pink Pillar, LCl, pb, 1940; Brownell, H.C., 1940

Pink Pirouette *see* HARboul

Pink Pixie *see* **'PIXIE ROSE'**

Pink Polyanna™ *see* 'JACSHIP'

Pink Poodle *see* 'MORPOODLE'

Pink Popcorn™ *see* 'WILPOP'

Pink Porcelain *see* 'TINPORCE'

Pink Posy *see* 'COCANELIA'

'PINK POWDER PUFF', F, mp, 1965; bud ovoid; flowers soft pink, open, medium, dbl.; foliage leathery; vigorous, upright, compact growth; Pal, Dr. B.P.; Indian Agric. Research Inst.

Pink Powderpuff *see* 'MORPUFF'

'PINK PRIDE', HT, lp, 1959; flowers silvery pink, dbl., 30–35 petals, 5–6 in., exhibition form, intense fragrance; foliage light green; free growth; [May Wettern X Peace]; Fletcher

'PINK PRINCESS', HT, pb, 1939; flowers deep rose pink to yellowish, large, dbl., exhibition form, intense fragrance; foliage dark, leathery, glossy; vigorous, bushy growth; [(Dr. W. Van Fleet X Général Jacqueminot) X Break o' Day]; Brownell, H.C.

Pink Princess, Min, mp

'PINK PROFUSION', HSet, pb, 1938; flowers pale flesh-pink, reverse deep coral-rose, small blooms in clusters, dbl., 80 petals, globular, non-recurrent, slight fragrance; foliage large, glossy; long, strong stems; very vigorous growth; [Mrs F.F. Prentiss X Lady Alice Stanley]; Horvath; Wayside Gardens Co.

'PINK PROSPERITY', HMsk, lp, 1931; flowers light pink, blooms in large trusses, moderate, musk fragrance; very vigorous growth; Bentall

'PINK PUFF', F, lp, 1965; flowers soft pink, dbl., moderate fragrance; foliage leathery; vigorous, upright growth; [Pinocchio seedling X (Red Pinocchio seedling X Garnette)]; Boerner; J&P

Pink Quill *see* 'WEEQUILL'

Pink Radiance *see* **'RADIANCE'**

Pink Radiance, Climbing *see* **'RADIANCE, CLIMBING'**

Pink Radiance, Climbing *see* **'RADIANCE, CLIMBING'**

Pink Reflection®, Min, pb

'PINK REVELATION', F, lp, 1979; [Summer Snow sport]; Schramm, D.

'PINK RHAPSODY', LCl, op, 1973; flowers deep vermilion-pink, full, dbl., 35–40 petals, 4 in., slight fragrance; very vigorous growth; [(Heidelberg X Bonn) X Pink Parfait]; Ellick

'PINK RHYTHM', Min, op, 1990; bud pointed; flowers coral pink, white center, medium pink reverse, aging light p, semi-dbl., 18 petals, exhibition form, moderate, fruity fragrance; foliage medium, medium green, matt; upright, spreading, medium growth; [Party Girl X Fashion Flame]; Jolly, Marie, 1984; Rosehill Farm, 1991

'PINK RIBBON', Min, lp, 1966; flowers soft pink, small, dbl., slight fragrance; foliage glossy, light green; vigorous, bushy, dwarf growth; [(R. wichurana X Floradora) X Magic Wand]; Moore, Ralph S.; Sequoia Nursery

'PINK RIPPLES', F, lp, 1956; flowers pink, open, borne in clusters, semi-dbl., 15–20 wavy petals, 2.5–3 in., slight fragrance; very vigorous, upright, compact growth; [Red Ripples sport]; Sanders, H.T.

'PINK ROAMER', HWich, pb, 1897; flowers pink, center white, single; vigorous growth; [R. wichurana X Cramoisi Superieur]; Horvath; W.A. Manda

Pink Robin, S, pb, 1992; Lens

Pink Robusta® *see* 'KORPINROB'

'PINK ROCKET', S, pb, 1949; bud ovoid; flowers deep pink suffused copper, single and open, large, semi-dbl., profuse, non-recurrent bloom, slight fragrance; foliage glossy, bronze, dark; very vigorous, upright growth; hardy.; [Skyrocket seedling]; Longley, L.E.; Univ. of Minn.

'PINK ROSETTE', F, lp, 1948; bud small, ovoid; flowers soft pink, rosette form, blooms in clusters, dbl., 50 petals, 2 in., cupped, slight fragrance; foliage leathery, dark; vigorous, dwarf, bushy growth; Krebs; H&S

Pink Roundelay, HT, lp; [Roundelay sport]; Taylor, L.R., 1968; (Weatherly, L.)

'PINK ROVER', Cl HT, 1891; Paul, W.; (Cavriglia)

'PINK ROYAL', HT, mp, 1928; flowers glowing pink, dbl., moderate fragrance; [Columbia X Mme Butterfly]; Vestal

'PINK RUBY', HT, mp, 1966; flowers high pointed, 4–4.5 in., moderate fragrance; foliage light green; vigorous growth; [Rubaiyat sport]; Anstiss

'PINK RUFFLES', F, mp, 1977; bud pointed; flowers rose-pink, ruffled, single, 5–7 petals, 3 in., slight, tea fragrance; foliage very glossy, dark; very bushy growth; [Frolic self seedling X Pinafore]; Ellis; Mansion Nursery

Pink Ruffles, HT, dp, 1996; Williams, J. Benjamin

Pink Sachet, HT, dp

Pink Sandy™ *see* 'KINSANDY'

'PINK SATIN', F, mp, 1945; bud large; flowers clean rose-pink, showy stamens, open, semi-dbl., cupped, moderate fragrance; foliage soft, light green; very vigorous, bushy, compact growth; [Indiana X William F. Dreer]; Cross, Mrs. C.W.; B&A

'PINK SATIN', F, lp, 1974; bud ovoid; dbl., 30–35 petals, 4–5 in., exhibition form; foliage large; vigorous, very free growth; [Seedling X Bridal Pink]; Warriner, William A.; J&P

Pink Scotch, HSpn, mp

Pink Secret *see* 'TROBINA'

'PINK SEMI', (form of R. laxa), lp; resembles R. blanda

Pink Sensation, HT, mp

Pink Sevilliana *see* 'MEIGEROKA'

'PINK SHADOW', Min, dp, 1977; bud plump, pointed; flowers dusty pink, reverse darker, dbl., 52 petals, 1 in., moderate fragrance; foliage glossy; bushy, spreading growth; [Over the Rainbow X Over the Rainbow]; Williams, Ernest D.; Mini-Roses

Pink Sheri *see* 'RENERI'

'PINK SHOWERS', Cl HT, lp, 1974; flowers large, dbl.; vigorous, tall, climbing growth; [Carla X Golden Showers]; Verschuren, Ted; Verschuren

Pink Showers, Pol, mp, 1968; Ramson

'PINK SILK', HT, mp, 1972; flowers carmine-rose, dbl., 42 petals, 4 in., exhibition form, moderate fragrance; Gold Star of the South Pacific, Palmerston North, NZ, 1974; [Pink Parfait X Unknown seedling]; Gregory

'PINK SNOW', Cl F, pb, 1981; flowers light pink shading to white in center, deeper on edges of petals; [Summer Snow, Climbing sport]; Reed, Harry, Jr.

'PINK SOUPERT', Pol, mp, 1896;, in clusters; vigorous growth; [Clotilde Soupert seedling]; Dingee & Conard

Pink Spectacle *see* KORlodera

'PINK SPICE', HT, lp, 1962; bud long, pointed; flowers light pink flushed yellow, dbl., 30 petals, 5 in., exhibition form, intense fragrance; foliage leathery, light green; vigorous, upright growth; Von Abrams; Peterson & Dering

'PINK SPIRAL', HT, dp, 1953; flowers deep china-rose-pink, high pointed, dbl., 4 in.; foliage cedar-green; very vigorous growth; McGredy, Sam IV

Pink Spire, HT, mp

'PINK SPLENDOUR', HT, mp, 1951; flowers rose-pink, dbl., 48 petals, 6–7 in., intense fragrance; foliage dark; very free growth; [Sam McGredy X Crimson Glory]; McGredy, Sam IV

Pink Spray® *see* LENray

Pink Star® *see* 'INTERPINK'

Pink Strike, Min, lp, 1998; Laver, Keith G.

Pink Sublime, HMsk, pb

Pink Summer Snow, Cl F, lp; [Summer Snow sport]; Hannemann, F., 1987; (Weatherly, L.)

Pink Sunblaze™ *see* 'MEIJIDIRO'

'PINK SUPREME', HT, lp, 1964; dbl., 23 petals, 4–5 in., intense fragrance; foliage light green; vigorous, upright, well-branched growth; GM, Belfast, 1967 GM, Geneva, 1965; [Amor X Peace]; deRuiter

'PINK SURPRISE', Min, mp, 1980; bud ovoid, pointed; dbl., 48 petals, borne singly or several together, moderate fragrance; curved prickles; foliage tiny, medium green; compact, bushy growth; [Unnamed seedling X Unnamed seedling]; Lyon

Pink Surprise, S, dp, 1975; Ottawa

Pink Surprise *see* 'LENBRAC'

Pink Sweetheart™ *see* 'WILPINK'

Pink Symphonie® *see* 'MEITONJE'

Pink Symphony *see* 'MEITONJE'

'PINK TALISMAN', HT, 1943; flowers orange to pink with pink predominating, large; upright growth; [Talisman sport]; Howard Rose Co.

Pink Tapestry, F, dp, 1997; Williams, J. Benjamin

'PINK TINGLE', Min, mp, 1978; bud ovoid; flowers dawn-pink, to recurved, dbl., 20 petals, 1 in., cupped, slight fragrance; foliage tiny; compact growth; Lyon

Pink Torch® *see* 'INTERTOR'

Pink Traumland, F, mp, 1996; flowers luminous pink, medium-large, dbl.; Tantau; (Sangerhausen)

Pink Treasure *see* 'HADPLEASURE'

Pink Triumph *see* **'INGRID STENZIG'**

'PINK TRIUMPH', Min, mp, 1983; flowers small blooms borne usually singly, dbl., 48 petals, moderate fragrance; foliage small, medium green, semi-glossy; upright, bushy growth; [Operetta X Bonny]; Jolly, Nelson F.; Rosehill Farm

Pink Twister, S, mp

Pink Van Fleet, LCl, mp

'PINK VOGUE', F, mp, 1960; bud pointed; flowers pink, medium, borne in clusters, semi-dbl., moderate fragrance; moderate, upright growth; [Vogue sport]; Kelleher; Hazlewood Bros.

Pink Wave *see* 'MATTGRO'

Pink Wings, HT, lp

Pink Winks *see* 'FLOWIN'

Pink Wonder *see* 'MEIHARTFOR'

Pink Wonder, Climbing *see* 'MEIHARTFORSAR'

'PINK-A-BOO', F, mp, 1961; bud ovoid; dbl., 33 petals, 3–3.5 in., blooms in large clusters, moderate fragrance; foliage leathery; vigorous, upright growth; [Spartan X Pink Garnette]; Boerner; J&P

Pink-A-ling *see* 'ZIPPINK'

'PINKCES', LCl, yb, 1962; bud pointed; flowers light yellow to pink, open, dbl., 30 petals, 3.5–4 in., abundant, non-recurrent bloom, moderate fragrance; foliage leathery; vigorous, tall, compact growth; [Doubloons sport]; Schmalz & Limpert

'PINKERTON', HT, mp, 1949; bud long, pointed; flowers pink shaded deeper, dbl., 30 petals, 4 in., moderate fragrance; foliage light green; vigorous growth; Eacott

'PINKIE', Pol, mp, 1947; flowers neyron rose, blooms in large trusses, semi-dbl., 16 petals, 1.75–2.5 in., cupped, intense fragrance; foliage soft, glossy, mostly 7 leaflet leaves; dwarf, bushy growth; (21); AARS, 1948; [China Doll X ?]; Swim, H.C.; Armstrong Nursery

'PINKIE, CLIMBING', Cl Pol, mp, 1952; Dering; Armstrong Nursery

Pinky *see* 'MACPONUI'

Pinky Rugostar *see* MEIvibroum

'PINOCCHIO', F, op, 1940; (Rosenmärchen); flowers pink suffused salmon, edged deeper, blooms in long sprays, dbl., 30 petals, 2 in., cupped, moderate, fruity fragrance; foliage leathery; vigorous, bushy growth; (28); GM, Portland, 1942; [Eva X Golden Rapture]; Kordes; J&P, 1942

Pinocchio *see* DELki

Pinocchio *see* 'SUNPINO'

'PINOCCHIO, CLIMBING', Cl F, pb, 1951; (Rosenmärchen, Climbing); Parmentier, J.; J&P

'PINSON', LCl, my, 1909; flowers chamois-yellow, tinted rosy white, very large, borne in clusters, semi-dbl., early; vigorous, climbing growth; [R. wichurana X Souv. de Catherine Guillot]; Barbier

Pinstripe™ *see* 'MORPINTS'

'PINTA', HT, w, 1973; flowers creamy white, dbl., 23 petals, 3.5 in., moderate, sweetbriar fragrance; foliage dark, matt; [Ena Harkness X Pascali]; Beales, Peter

PINturnia, Gr, 1978; (**Mme Bruno Coquatrix**); Pineau; (Cavriglia)

'PIN-UP', F, dp, 1959; flowers deep china-rose, medium, borne in clusters, semi-dbl., 16–18 wavy petals; foliage light green; low, bushy growth; [Else Poulsen X ?]; Fletcher

'PINWHEEL', Min, pb, 1977; flowers pink and yellow blend; [Jeanie Williams sport]; Moore, Ralph S.; Sequoia Nursery

'PIONEER', HT, mr, 1970; flowers very full, dbl., 50 petals, 4 in., intense fragrance; foliage small, dark; very free growth; LeGrice

'PIONERKA', F, ob, 1955; (Woman Pioneer); flowers reddish orange, medium, slight fragrance; vigorous growth; Sushkov, K. L.

'PIOUPIOU', F, mr, 1958; flowers cerise-red, large; very vigorous, bushy, low growth; [Alain X (Alain X Unnamed seedling)]; Arles; Roses-France

'PIP', Min, mr, 1977; bud pointed; flowers small, semi-dbl., 12–15 petals, cupped, slight fragrance; foliage tiny; very compact, bushy growth; RULED EXTINCT 7/90; [N74 X Unnamed seedling]; Lyon

'PIP', HT, w, 1990; bud ovoid; flowers white rimmed pink, white reverse, aging same, medium, borne, dbl., 30 petals, exhibition form, slight fragrance; foliage medium, light to medium green, matt; bushy, medium growth; [(Poker Chip X Helmut Schmidt) X [(Rise 'n' Shine X Granada) x Handel]]; Stoddard, Louis, 1991

'PIPE DREAMS', S, mp, 1984; flowers medium pink, reverse darker, blooms borne 5-10 per cluster, dbl., 28 petals, 5 in., cupped, repeat bloom, intense, clove fragrance; awl-shaped, red-brown prickles; foliage leathery, medium olive green; bushy, compact, upright growth; hardy.; [Countryman X ((Meisterstuck X Prairie Princess) X (Tickled Pink X Prairie Princess))]; Buck, Dr. Griffith J.; Iowa State University

'PIPPA'S SONG', S, mp, 1984; flowers imbricated, blooms borne 3-10 per cluster, 3 in., repeat bloom, slight fragrance; awl-shaped, tan prickles; foliage medium, leathery, dark olive green; shrubby, erect, bushy growth; hardy.; [Prairie Princess X (Queen Elizabeth X (Morning Stars X Suzanne))]; Buck, Dr. Griffith J.; Iowa State University

Pippy *see* BARpipp

'PIPSQUEAK', Min, m, 1986; flowers pinkish mauve, quilled, small blooms in sprays of 5-20, dbl., 40 petals, flat, no fragrance; no fruit; very small, brown prickles, hooked downward; foliage small, light green, matt; bushy, spreading growth; [Blue Mist X Snow Magic]; Dobbs; Port Stockton Nursery

'PIRATE GOLD', F, dy, 1972; dbl., globular, slight fragrance; foliage glossy, leathery; vigorous, upright growth; [Golden Wave X Seedling]; deRuiter; Carlton Rose Nurseries

'PIRBRIGHT', F, mp, 1948; semi-dbl., borne in trusses, slight fragrance; foliage dark; vigorous growth; Norman

Piro 3, S, lp, 1995; Stritzke

'PIROJA', HT, m, 1999; flowers deep mauve, reverse same, dbl., 26–40 petals, 2.5–3.5 in., borne in small clusters, moderate fragrance; few prickles; foliage medium, medium green, semi-glossy; upright, medium (3-3.5 ft) growth; [Blue Moon X Heirloom]; Chiplunkar, C. R.; KSG's Roses, 1992

Pirol *see* TANlorip

'PIRON MÉDARD', HP, mp, 1906; flowers large, dbl.; Piron-Médard; (Sangerhausen)

'PIRONIA', Gr, op, 1977; (**Centenaire du Vesinet**); bud pointed; flowers salmon-orange, dbl., 32 petals, 3 in., slight fragrance; vigorous, upright growth; [Sonia X Prominent]; Pineau; Searn

Pironti Arabian, HT; Pironti, N.; (Cavriglia)

Pironti Blithe, F, 1970; Pironti, N.; (Cavriglia)

Pironti Critow, Cl HT; Pironti, N.; (Cavriglia)

Pironti Konrad, Gr, 1971; Pironti; (Cavriglia)

Pironti Tornedos, F, 1971; Pironti; (Cavriglia)

Pirontina, LCl, 1975; Pironti; (Cavriglia)

Piroschka® *see* 'TANPIKA'

'PIROUETTE', F, pb, 1968; flowers pink tinged orange-red, open, medium, dbl., moderate fragrance; [Ma Perkins X Radar]; Fankhauser; A. Ross & Son

Pirouette, S, rb, 1984; Lens

Pisen, HT, lp; Brabec, 1980; (Czech Rosa Club)

'PISMI', HT, rb, 1972; bud ovoid; flowers solferino-purple to spirea-red, large, dbl., cupped, moderate fragrance; foliage large, dark, leathery; vigorous, upright bushy growth; [Dame Edith Helen X Rote Rapture]; Pecollo

Pitica *see* **'KYRIA'**

'PITICA', F, lp, 1976; flowers full, dbl., 25–30 petals, 4 in., exhibition form, slight fragrance; [Sweet Promise sport]; Royon; Universal Plants

'PITTSBURGH', HP, lp, 1929; flowers flesh-pink, base yellow, very large, dbl., 25 petals, globular, slight fragrance; foliage leathery; long stems; very vigorous growth; [(R. gigantea seedling X Frau Karl Druschki) X Mrs John Laing]; Schoener; B&A

'PIUS IX', HP, mp, 1849; (Pie IX, Pius the Ninth, Pope Pius IX); flowers violet-rose, very dbl., flat, intense fragrance; vigorous growth; Vibert

Pius the Ninth *see* **'PIUS IX'**

'PIUS XI', HT, w, 1925; (Pope Pius XI); bud long, pointed; flowers cream-white, center cream-yellow, very large, dbl., moderate fragrance; [Ophelia X Unnamed seedling]; Leenders, M.

'PIXANAH', Min, ab, 1990; (**Savannah Miss**); bud pointed; flowers dark apricot, edges lighter, medium apricot reverse, fading, dbl., 40–50 petals, exhibition form, moderate fragrance; globular, medium green fruit; prickles needle-shaped, light tan; foliage medium, medium green, semi-glossy; bushy, tall growth; [Pounder Star X Ann Moore]; Chaffin, Lauren M.; Pixie Treasures Min. Roses

'PIXANNE', Min, mp, 1993; (**Rebecca Anne**); very dbl., 1.5–2.75 in., borne mostly singly, slight fragrance; some prickles; foliage medium, medium green, semi-glossy; medium (40-45 cms), upright, bushy, vigorous growth; [Gene Boerner X Pink Petticoat]; Chaffin, Lauren M.; Pixie Treasures Min. Roses, 1993

'PIXARLA', Min, w, 1990; (**Starla**); bud pointed; flowers medium, borne singly, semi-dbl., 20–25 petals, exhibition form, slight, fruity fragrance; prickles needle-shaped, tan; foliage medium, medium green, semi-glossy; bushy, medium growth; [Honor X Rainbow's End]; Chaffin, Lauren M.; Pixie Treasures Min. Roses, 1991

'PIXBRO', Min, mr, 1998; (**Little Brother**); flowers medium red, full, very dbl., 26–40 petals, 2 in., borne mostly singly, slight fragrance; no prickles; foliage medium, medium green, semi-glossy; compact, medium growth; [Pink Petticoat X Happy Hour]; Chaffin, Lauren M.; Pixie Treasures Roses, 1998

'PIXCHEK', Min, mr, 1990; (**Checkers**); bud pointed; flowers medium red, aging darker, blooms borne singly, semi-dbl., 20 petals, exhibition form, slight, spicy fragrance; rarely forms fruit; prickles needle-shaped, tan; foliage small, medium green, semi-glossy; bushy, low growth; [Deep Purple X Happy Hour]; Chaffin, Lauren M.; Pixie Treasures Min. Roses

'PIXHAR', Min, ob, 1998; (**Summer Harvest**); flowers very bright orange blend, gold stamens, single, 5 petals, 2 in., borne in small clusters, slight fragrance; prickles moderate, medium, straight; foliage large, dark green, glossy; bushy, tall (24 in.) growth; [Ann Moore X Sequoia Gold]; Chaffin, Lauren M.; Pixie Treasures Roses, 1998

'PIXHONEY', Min, ly, 1998; (**Honey Butter**); flowers creamy yellow center, near white outer edges, very dbl., 26–40 petals, 2–2.5 in., borne mostly singly, slight fragrance; prickles few, small, straight; foliage medium, medium, semi-glossy; bushy, medium growth; [Deep Purple X Rainbow's End]; Chaffin, Lauren M.; Pixie Treasures Roses, 1998

'PIXICHAR', Min, lp, 1992; (**Real Charmer**); dbl., 26–40 petals, 3–3.5 in., borne in small clusters, moderate fragrance; foliage large, medium green, semi-glossy; tall (45 cms), upright growth; [Gene Boerner X Crissy]; Chaffin, Lauren M.; Pixie Treasures Min. Roses

'PIXICHIP', Min, r, 1991; (Chipmonk, **Chipmunk**, Peanut Butter & Jelly); flowers tannish brown blending with mauve when full blown, dbl., 26–40 petals, 1.5–2.75 in., borne mostly singly, moderate fragrance; few prickles; foliage medium, medium green, semi-glossy; medium (30 cms), bushy, neat, compact growth; [Deep Purple X Rainbow's End]; Chaffin, Lauren M.; Pixie Treasures Min. Roses, 1992

'PIXIE', Min, w, 1940; (Little Princess, Princesita); flowers white, center light pink, small blooms, dbl., 55 petals, slight fragrance; foliage very small, soft; short stems; dwarf, compact growth; [Ellen Poulsen X Tom Thumb]; deVink; C-P

Pixie Delight℠ *see* 'WILBENTUR'

'PIXIE GOLD', Min, my, 1961; flowers mimosa-yellow, semi-dbl., 11 petals, 1–1.5 in.; foliage very small, thin, dark; bushy growth; [Perla de Montserrat X (Rosina X Eduardo Toda)]; Dot, Pedro; C-P

Pixie Hat *see* KORlamber

Pixie Hedge *see* **'BABY JAYNE'**

Pixie Pearl *see* **'ESTRELLITA'**

'PIXIE PEARL', Min, w; flowers pearly white, small, borne in clusters, very dbl.; vigorous, compact growth; Lamb Nursery

'PIXIE ROSE', Min, dp, 1961; (Pink Pixie); flowers deep pink, blooms in irregular clusters, dbl., 43 petals, 1 in., exhibition form; foliage very small, dark; dwarf, much-branched growth; [Perla de Montserrat X Coralín]; Dot, Pedro; C-P

'PIXIE, CLIMBING', Cl Min, w, 1964; Ruston, D.

'PIXIFULL', Min, or, 1986; (**Fullerton Centennial**); flowers orange-red with white base, bright, fading darker, medium, semi-dbl., 20–25 petals, exhibition form, borne singly or in sprays of 3-5, slight fragrance; occaisionally forms globular, medium green fruit; needle-declining, sparse, light tan prickles; foliage small, medium green, semi-glossy, disease-resistant; upright, bushy, medium, neat, symmetrical growth; [Orange Honey X Rise 'n' Shine]; Chaffin, Lauren M., 1983; Friends of the Fullerton Arboretum, 1987

'PIXIHON', Min, ab, 1987; (**Honey Bear**); flowers deep apricot, yellow base, reverse light apricot, aging cream, very dbl., 50–55 petals, cupped; needle-like, straight, light tan prickles; foliage small, medium green, semi-glossy, disease resistant; bushy, low compact growth; PP006844; [Rise 'n' Shine X Holy Toledo]; Chaffin, Lauren M.; Pixie Treasures Min. Roses

'PIXIMIS', Min, w, 1985; (**Moon Mist**™); flowers ivory tinged pale pink, well-formed, small, dbl., 20 petals, moderate fragrance; foliage small, medium green, semi-glossy; bushy growth; [The Optimist X Darling Flame]; Strawn, Leslie E.; Pixie Treasures Min. Roses

'PIXINK', Min, mp, 1989; (**Tickle Me Pink**); bud pointed; flowers medium pink to soft cream atbase, reverse cream blush, aging, dbl., moderate, spicy fragrance; globular, rare, medium green fruit; prickles needle-like, light tan; foliage small, medium green, semi-glossy; bushy, low growth; [Osiria X Magic Carrousel]; Chaffin, Lauren M.; Pixie Treasures Min. Roses

'PIXINNER', Min, rb, 1990; (**Inner Glow**); bud ovoid; flowers red with golden yellow base, yellow reverse, red veining towards outer edges, dbl., 40 petals, exhibition form, slight fragrance; round, medium green fruit; prickles hooked, tan; foliage medium, medium green, semi-glossy; bushy, medium growth; [Ann Moore X Rainbow's End]; Chaffin, Lauren M.; Pixie Treasures Min. Roses, 1991

'PIXIPRIN', Min, r, 1982; (**Indian Princess**); flowers tan flushed orange, reverse burnt umber, shaded garnet-brown, dbl., 20 petals, intense fragrance; foliage small, dark, semi-glossy; upright growth; [Yellow Jewel X Golden Cougar]; Strawn, Leslie E.; Pixie Treasures Min. Roses

'PIXIREE', Min, ab, 1985; (**Free Spirit**); flowers small, dbl., 20 petals, slight fragrance; foliage small, medium green, semi-glossy; upright growth; PP006188; [Prominent X Gold Pin]; Strawn, Leslie E.; Pixie Treasures Min. Roses, 1984

'PIXISLIP', Min, m, 1991; (**Silver Slippers**); bud slender; flowers unique silvery lavender, blooms borne mostly singly, dbl., 26–40 petals, 1.5–2.75 in., moderate fragrance; foliage medium, medium green, semi-glossy; medium (30 cms), bushy growth; [Deep Purple X Jennifer]; Chaffin, Lauren M.; Pixie Treasures Min. Roses, 1992

'PIXISUN', Min, my, 1985; (**Sunshine Girl**); flowers multi-star shaped, small, dbl., 35 petals, exhibition form, slight fragrance; foliage small, medium green, semi-glossy; upright growth; [Sunsilk X Rise 'n' Shine]; Woolcock, Edward P.; Pixie Treasures Min. Roses

'PIXITER', Min, dy, 1982; (**Tiger Butter**); flowers small, dbl., 20 petals, slight fragrance; foliage small, dark, semi-glossy; upright, bushy growth; [Sunblest X Over the Rainbow]; Strawn, Leslie E.; Pixie Treasures Min. Roses

'PIXIWOW', Min, ob, 1985; (**Wow!**); flowers bright orange, small, dbl., 35 petals, slight fragrance; foliage small, dark, glossy; bushy growth; PP006469; [Ann Cocker X Unnamed seedling]; Strawn, Leslie E.; Pixie Treasures Min. Roses

'PIXJUN', Min, ob, 1994; (**June Bug**); flowers bright medium orange, slightly lighter reverse, yellow base, dbl., 26–40 petals, 1.5 in., slight fragrance; some prickles; foliage medium, dark green, semi-glossy; medium (12 in), bushy, compact growth; [(Prominent X Orange Honey) X Ann Moore]; Chaffin, Lauren M.; Pixie Treasures Min. Roses, 1994

'PIXKICK', F, ob, 1995; (**Side Kick**); flowers orange red, very full (41+ petals), medium (4–7 cms) blooms borne mostly single; slight fragrance; few prickles; foliage medium, medium green, semi-glossy; medium (30 cms), bushy growth; [Dandenong X Ann Moore]; Chaffin, Lauren M.; Pixie Treasures Miniature Rose Nursery, 1996

'PIXRIC', F, rb, 1996; (**Ric Rac**); flowers white edged red, non-fading color, small, long lasting bloom, dbl., 26–40 petals, 2.5–3 in., fast repeat, slight fragrance; moderate prickles; foliage small, medium green, semi-glossy; upright, low (18-20 in.) growth; [Hannah Gordon X Pink Petticoat]; Chaffin, Lauren M.; Pixie Treasures Min. Roses, 1996

'PIXSAR', F, ab, 1998; (**Sarah Elizabeth**); flowers apricot blend, ogr form, very dbl., 41 petals, 2.5–3 in., rosette, borne in small clusters, slight fragrance; prickles few, medium, straight; foliage medium, dark green, glossy; upright, medium growth; [City of Auckland X Seedling]; Chaffin, Lauren M.; Pixie Treasure Roses, 1998

'PIXSIS', Min, mp, 1995; (**Little Sister**); flowers medium pink, very full (41+ petals), medium (4–7 cms) blooms borne mostly single, occasional side buds; slight fragrance; few prickles; foliage medium, medium green, semi-glossy; medium (30cms), bushy growth; [Gene Boerner X Crissy]; Chaffin, Lauren M.; Pixie Treasures Miniature Rose Nursery, 1996

'PIXSNO', F, rb, 1996; (**Sno Cone**); flowers white edged red, blooms borne in small clusters upright, wa, dbl., 15–25 petals, 2.5–3 in., slight fragrance; moderate prickles; foliage small, medium green, semi-glossy, disease-resistant; bushy, low (18 in) growth; [Hannah Gordon X Pink Petticoat]; Chaffin, Lauren M.; Pixie Treasures Min. Roses, 1996

'PIXSTA', Min, w, 1993; (**Christa**); dbl., 26–40 petals, 1.5–2.75 in., exhibition form, borne mostly singly, slight fragrance; some prickles; foliage medium, medium green, semi-glossy, disease-resistant; long stems; tall (45 cms), upright, vigorous growth; [Honor X Rise 'n' Shine]; Chaffin, Lauren M.; Pixie Treasures Min. Roses, 1993

'PIXSUN', F, ob, 1994; (**Sunny Side Up**); flowers blended soft yellow to tangerine, blooms borne in small clusters, semi-dbl., 6–14 petals, 3–3.5 in., moderate fragrance; some prickles; foliage medium, dark green, glossy; medium (2-3 ft), bushy growth; [City of Auckland X Rainbow's End]; Chaffin, Lauren M.; Pixie Treasures Min. Roses, 1994

'PIXZIG', Min, rb, 1993; (**Zig-Zag**); flowers white with dark red picotee edge, blooms borne mostly single, very dbl., 26–40 petals, 1.5–2.75 in., exhibition form, slight fragrance; some prickles; foliage similar to peaches 'n' cream, medium, medium green; low (30 cms), upright, bushy, compact growth; [Osiria X Pink Petticoat]; Chaffin, Lauren M.; Pixie Treasures Min. Roses, 1993

'PIZZAZZ', Min, rb, 1991; bud ovoid; flowers various yellows, pink hues, aging to orange-red, blooms born, dbl., 50 petals, 1.5 in., slight fragrance; foliage medium, medium green, semi-glossy; upright, spreading, medium growth; [Orange Honey X Loving Touch]; Jolly, Marie; Rosehill Farm, 1992

'PIZZICATO', S, lp, 1962; bud ovoid, long, pointed; flowers light salmon-rose, dbl.,

40 petals, 3–4 in., cupped, moderate fragrance; foliage leathery, bronze; vigorous, upright (3-4 ft), bushy. growth; [Florence Mary Morse X (Josef Rothmund X R. laxa)]; Buck, Dr. Griffith J.; Iowa State University

Placerville White Noisette, N, w

'Placet', S, w, 1981; bud small, ovoid; single, 5 petals, borne 40 or more per cluster, all summer bloom, moderate fragrance; foliage soft green, 7 leaflet; bushy growth; Bevan, Mrs. Ruth M.

'Placida', HT, mr, 1959; bud ovoid; flowers crimson, large, dbl., slight fragrance; foliage glossy, dark; vigorous, bushy, compact growth; [Crimson Glory X Una Wallace]; Cayzer

Placido Domingo *see* 'MACHORO'

'Plain Talk', F, mr, 1964; semi-dbl., blooms in clusters, slight fragrance; foliage dark, leathery; vigorous, bushy, low growth; [Spartan X Garnette]; Swim & Weeks; Weeks Wholesale Rose Growers

'Plaisante', F, mp, 1957; flowers bright pink, base and reverse lighter, semi-dbl., moderate fragrance; low growth; [Borderer X Unnamed seedling]; Riethmuller

Plaisanterie, HMsk, 1996; Lens; (Cavriglia)

'Plaisir de France', HT, dp, 1952; flowers bright deep coppery pink, dbl., 35 petals, 6 in., moderate fragrance; foliage glossy, dark; vigorous growth; [Peace X Unnamed seedling]; Gaujard

Plameny, HT, mr, 1970; flowers large, dbl.; Urban, J.; (Sangerhausen)

'Plamya Vostoka', F, or, 1955; (Flame of the East); flowers fiery red tinted darker, medium, dbl., 30 petals, moderate fragrance; medium growth; [Independence X Kirsten Poulsen]; Klimenko, V. N.

Planten un Blomen *see* **'CHERRIO'**

Platinum Lady *see* 'LAVPLAT'

'Plato', HSpn, m; flowers red-violet, dbl., non-recurrent; glossy, black fruit; foliage finely divided; dense, shrubby (3-4 ft.) growth; Vibert

Play Rose, F, dp, 1989; flowers carmine-pink, large, dbl.; Meilland; (Sangerhausen)

'Playboy'®, F, rb, 1976; (Cheerio); flowers scarlet, gold eye, single, 3.5 in., slight fragrance; foliage glossy, dark; GM, Portland, 1989; [City of Leeds X (Chanelle X Piccadilly)]; Cocker

Player *see* 'SUNPLAY'

Playfair *see* 'WIWAIT'

'Playful', HT, ab, 1970; flowers light apricot, edged pink; [Mischief sport]; Watson

Playgirl™ *see* 'MORPLAG'

Playgirl, Climbing *see* 'MORCLIP'

Playgold *see* 'MORPLAYGOLD'

Playgroup Rose® *see* 'HORSUN'

Playmate *see* 'WEOKAY'

Playtime™ *see* 'MORPLATI'

Playtime *see* 'KORSAKU'

Pleasant Hill Cemetery, N, w

Pleasantly Pink *see* 'LAVPLEASE'

Pleasure *see* 'JACPIF'

'Plein Ciel', HT, mp, 1964; bud pointed; flowers large, dbl., exhibition form, moderate fragrance; foliage dark, glossy; vigorous, well branched growth; [Seedling X Golden Rapture]; Mondial Roses

Plein Soleil® *see* LAPdev

Pleine de Grâce *see* 'LENGRA'

Pleins Feux 92 *see* DORtomyl

'Plena', HSem, lp; (Double White Noisette); flowers flesh, dbl.; Laffay, M.

'Plentiful', F, dp, 1961; bud globular; flowers deep pink, blooms in large clusters, dbl., 75 petals, 3–4 in., flat; foliage light green; vigorous, bushy growth; LeGrice

Plisiedame *see* KORichard

Plnokvety Super Star, HT, lp; Strnad; (Czech Rosa Club)

'Plomin', F, dp, 1951; flowers dark rose, well formed, large, borne in large clusters, semi-dbl., intense fragrance; foliage leathery; bushy, dwarf growth; [(Johanna Tantau X Karen Poulsen) X Stammler]; Tantau

'Pluie de Feu', LCl, or, 1964; flowers bright scarlet, borne in clusters, semi-dbl.; vigorous growth; Mondial Roses

'Plukovnik Svec', HT, dr, 1935; (Colonel Svec); bud long (over 3 in.), pointed; flowers blood-red, large, semi-dbl., moderate fragrance; foliage small; vigorous, bushy growth; [Pres. Jac. Smits X K. of K.]; Böhm, J.

Plum Brandy, Min, m, 1974

Plum Cake *see* 'MURCA'

Plum Crazy *see* 'AROGRAJU'

Plum Dandy™ *see* 'JACLIANG'

'Plum Duffy', Min, m, 1978; bud ovoid; flowers deep plum, small, dbl., 25 petals, 1.5 in., exhibition form, slight fragrance; foliage dark; compact growth; [Magic Carrousel X Magic Carrousel]; Bennett, Cecilia 'Dee'; Tiny Petals Nursery

Plum Pudding™ *see* 'WILPLPD'

'Pluton', HGal, m, 1843; (Rose Pluton); Vibert, 1843

'Plzen', HT, dr, 1930; (City of Pilsen, Stadt Pilsen); bud very long; flowers dark blood-red, very large, semi-dbl.; very vigorous growth; [Étoile de Hollande X Macbeth]; Böhm, J.

'Pobjeditel', HT, dp, 1940; flowers large, dbl., slight fragrance; Kosteckij; (Sangerhausen)

Poblet, HT, ob; Dot

Pocahontas, HT, ob, 1998; Williams, J. Benjamin

'Poco', Min, pb, 1986; flowers pink, white center, fades slightly, urn-shaped, medium, born, dbl., 29 petals, slight fragrance; medium, light green prickles, pointed downwards; foliage medium, medium green, semi-glossy; bushy, medium growth; [Heartland X Unnamed seedling]; Bridges, Dennis A., 1987

'Podruga', HT, mp, 1939; (Girl Friend); flowers bright pink, base dark red, oval, irregular shape; low, rather weak growth; [Vaterland X Mme Edouard Herriot]; Costetske

'Poëma', Cl Pol, mp, 1933; flowers bright pink, passing to la france pink, borne in clusters of, recurrent bloom; foliage bright, dark; vigorous (5-10 ft.) growth; [Tausendschön X Farbenkonigin]; Brada, Dr.; Böhm

'Poente', F, ob; flowers orange and carmine; [Pinocchio X Goldilocks]; da Silva, Moreira

Poesie *see* 'LENOBIT'

Poesie *see* 'JACIS'

Poesie *see* 'JACIENT'

Poesie, S, w, 1997

'Poète Jean du Clos', HT, op, 1919; flowers pink, shaded salmon, dbl., intense fragrance; [Le Progres X Lyon Rose]; Gillot, F.

Poetry *see* 'ZIPOET'

Poetry in Motion *see* 'HARELAN'

Pohadka Maje, HT, lp; Certek, 1978; (Czech Rosa Club)

Poiana, HT, dy, 1972; flowers large, dbl.; GPG Bad Langensalza; (Sangerhausen)

'Poinsettia', HT, mr, 1938; bud long, pointed; flowers bright scarlet, large, dbl., 28 petals, moderate fragrance; foliage glossy; vigorous, compact growth; GM, Portland, 1940; [(Mrs J.D. Eisele X Vaterland) X J.C. Thorton]; H&S; Dreer

'Poinsettia, Climbing', Cl HT, mr, 1950; Thompson, D.L.; Rosemont Nursery Co.

Point Clear *see* 'TALPOI'

Point de Avignon, F, or

Point du Jour® *see* 'CANDIDE'

Poker® *see* 'HAVAPS'

Poker *see* MEIpazdia

'POKER CHIP', Min, rb, 1979; bud pointed; flowers red with yellow reverse, medium, dbl., 28 petals, exhibition form, intense fragrance; large fruit; foliage glossy, dark; vigorous, compact growth; [Sheri Anne X (Yellow Jewel X Tamango seedling)]; Saville, F. Harmon

'POKORNYANA', S, lp; (R. scopulosa, R. X pokornyana); 1.75 in.; foliage leaflets doubly serrate; (40); [R. canina X R. glauca]

Pol Robson, F, dr, 1955; flowers medium, very dbl., slight fragrance; Klimenko, V. N.; (Sangerhausen)

'POLAR BEAR', HP, w, 1934; flowers white tinted blush, becoming pure white, large, very dbl., globular, recurrent bloom, intense fragrance; foliage large, leathery, wrinkled; vigorous, bushy growth; [Schoener's Nutkana X New Century]; Nicolas; J&P

Polar Star *see* 'TANLARPOST'

Polareis *see* STRonin

'POLARIS', LCl, w, 1939; flowers pure snow-white, open, borne in clusters, dbl., profuse, non-recurrent bloom, intense fragrance; foliage glossy, light; very vigorous, climbing (12-15 ft.) growth; [(R. wichurana X R. setigera) X R. foetida bicolor]; Horvath; Wayside Gardens Co.

Polarsonne *see* STRolon

Polarstern® *see* 'TANLARPOST'

'PÔLE NORD', HT, w, 1944; bud long, pointed; flowers pure white, base tinted greenish; vigorous growth; Mallerin, C.; A. Meilland

Polestar *see* **'POLSTJÄRNAN'**

'POLIARCHUS', S, w; flowers cream flushed salmon, fading rapidly, non-recurrent; spreading shrub (4 ft.) growth; hardy.; [Seedling X R. Xharisonii]; Central Exp. Farm

'POLINA', HT, dp, 1984; flowers deep pink, large, dbl., 38 petals; foliage dark, leathery; bushy growth; [Mistica X Chenon]; Staikov, Prof. Dr. V.; Kalaydjiev and Chorbadjiiski

Polisiedame *see* KORichard

Polka™ *see* 'MEITOSIER'

'POLKA', F, mp, 1959; bud ovoid; dbl., 42 petals, 3–3.5 in., exhibition form, blooms in clusters, moderate fragrance; foliage leathery; vigorous, bushy growth; [Moulin Rouge X Fashion]; Meilland, Mrs. Marie-Louise; URS

Polka *see* 'MEITOSIER'

Polka 91 *see* 'MEITOSIER'

'POLKA DOT', Min, w, 1956; flowers ivory white, small, dbl.; foliage dark, leathery; vigorous, dwarf (10 in), bushy growth; [Golden Glow X Zee]; Moore, Ralph S.; Sequoia Nursery

'POLKA TIME', S, op, 1984; bud ovoid, pointed; flowers salmon tinted yellow, veined pink, imbricated, blooms borne, dbl., 38 petals, .5 in., cupped, repeat bloom; moderate fragrance; awl-like, tan prickles; foliage dark, leathery; low, bushy, compact growth; hardy.; [Bonfire Night X Countryman]; Buck, Dr. Griffith J.; Iowa State University

'POLLENTIA', HT, dp, 1942; bud large, long pointed; flowers satiny strawberry-red, dbl., 40 petals, exhibition form, intense fragrance; foliage dark, glossy; upright, compact growth; Dot, Pedro

'POLLINIANA', Misc OGR, w; (R. X polliniana); flowers white to pink, large; long stems; (21); [R. arvensis X R. gallica]

'POLLMERIANA', (strain of R. canina), lp; almost thornless; very vigorous growth; Pollmer

Polluce® *see* BARpoll

'POLLY', HT, w, 1927; bud long, pointed; flowers cream, center tinted pink or light orange, fading white, large, dbl., 38 petals, exhibition form, intense fragrance; [Ophelia seedling X Mme Colette Martinet]; Beckwith

'POLLY FLINDERS', Min, w, 1954; flowers cream tinted copper-orange, dbl., 30 petals, 1 in.; foliage veined red; [Little Princess (Pol) X Fashion]; Robinson, T.

'POLLY PERKINS', HT, op, 1967; flowers orange-cerise, pointed, moderate fragrance; foliage dark; very free growth; Gregory

Polly Sunshine *see* 'MORGOLDPOLY'

Polo *see* TANolop

Polo Club™ *see* 'AROTIGY'

'POLONAISE', S, dp, 1984; bud ovoid, pointed; flowers deep pink, imbricated blooms borne 5-10 per cluster, dbl., 43 petals, 3 in., repeat bloom; awl-like, red-brown prickles; foliage leathery, dark green with reddish veination; dwarf, upright, bushy growth; hardy.; [San Francisco X Prairie Princess]; Buck, Dr. Griffith J.; Iowa State University

'POLONAISE', HT, mr, 1961; flowers bright red, dbl., 40–50 ruffled petals, 6 in., exhibition form, moderate fragrance; foliage glossy; vigorous, upright, compact growth; RULED EXTINCT 1/85; [Carrousel X (Chrysler Imperial X Seedling)]; Von Abrams; Peterson & Dering

'POLSTJÄRNAN', LCl, w, 1937; (Polestar, The Polar Star, The Wasa Star, Wasastiernan, White Rose of Finland, White Star of Finland); flowers pure white, very small, blooms in clusters, non-recurrent; very vigorous (to 18 ft.) growth; very hardy.; Wasast-jarna

'POLY PEACE', F, yb, 1959; flowers golden yellow edged rose-pink, large, semi-dbl., exhibition form, slight fragrance; foliage dark, glossy; strong stems; very vigorous, tall, bushy growth; [Masquerade X Peace]; Leenders, J.

'POLY PRIM', F, dy, 1953; flowers well shaped, borne in clusters, dbl., 70 petals, 3 in., moderate fragrance; foliage dark, glossy; very vigorous, bushy growth; GM, NRS, 1954; [Goldilocks X Golden Rapture]; Eddie

'POLYANA', S, dp, 1925; [R. rugosa X Polyantha (Mme Norbert Levavasseur?)]; Skinner

'POLYANTHA GRANDIFLORA', HMult, w; (Gentiliana); single; 1888

'POLYBAG JOSHI', HT, lp, 1999; flowers light pale pink, reverse deeper, very dbl., 41 petals, 4–5 in., borne mostly singly, moderate fragrance; many prickles; foliage large, medium green, dull, leathery; spreading, medium (3-4 ft) growth; [First Prize X (First Prize X Perfume Delight)]; Chiplunkar, C. R.; KSG's Roses, 1995

Polygold® *see* 'TANLYPOLO'

'POLYNESIAN PEARL', Gr, lp, 1977; bud ovoid; flowers shell-pink, small, dbl., 35–59 petals, 2 in., exhibition form, moderate, fruity fragrance; foliage leathery; upright growth; [Floribunda X Tropicana]; Takatori, Yoshiho; Japan Rose Nursery

'POLYNESIAN SUNSET', HT, op, 1965; bud long; flowers coral-orange, dbl., 6 in., exhibition form, moderate, fruity fragrance; foliage leathery; vigorous, bushy growth; [Diamond Jubilee seedling X Hawaii]; Boerner; J&P

Pomona *see* FRYyeh

'POMPADOUR', HT, dp, 1979; bud pointed; flowers deep pink, dbl., 43 petals, 5.5 in., exhibition form, intense, fruity fragrance; foliage dark, glossy; bushy, compact growth; [Molly McGredy X Prima Ballerina]; Murray, Nola; Rasmussen's

'POMPADOUR RED', F, dp, 1951; bud globular; flowers rose-red, dbl., 30–35 petals, 3.5 in., cupped, moderate, fruity fragrance; vigorous growth; [Orange Triumph X Anne Poulsen]; deRuiter; J&P

Pompeii, HT, rb

Pompon *see* **'POMPON PANACHÉE'**

'POMPON', C, w; flowers white, streaked lilac-pink, small, dbl.

'POMPON', HGal, dr, 1835; flowers brilliant crimson; Joly

'POMPON BEAUTY', F, or, 1949; flowers scarlet, borne in large trusses, very dbl.; vigorous growth; (28); GM, NRS, 1950; [Polyantha seedling X Hybrid Tea seedling]; deRuiter; Spek

'POMPON BLANC PARFAIT', A, w; (Blanc Parfait); flowers blush-white, pompon form; almost without prickles; compact growth; Verdier, E., 1876; 1876

Pompon de Bourgogne *see* **'BURGUNDIAN ROSE'**

Pompon de Bretagne, S, dp, 1997

Pompon de Burgogne *see* **'BURGUNDIAN ROSE'**

Pompon de Panachée *see* **'POMPON PANACHÉE'**

'POMPON DE PARIS', Ch, mp, 1839; bud very pointed; flowers bright pink, very small, dbl.

'POMPON DE PARIS, CLIMBING', Cl Ch, mp

'POMPON DE SAINT-FRANÇOIS', C, dp; flowers deep violet-pink, small, dbl.; low, bushy growth

Pompon des Princes *see* **'ISPAHAN'**

Pompon Diadem *see* TANmop

'POMPON PANACHÉE', HGal, w; (Pompon, Pompon de Panachée); flowers cream to white, striped dark pink, well-formed, dbl., flat; foliage very small; wiry, erect growth; Robert et Moreau, 1856 (?)

Pompon Perpetual *see* **'BERNARD'**

Pompon Rose *see* **'ROSE DE MEAUX'**

'POMPON ROUGE', F, mr, 1971; bud ovoid; flowers brilliant red, full, medium, very dbl.; foliage soft; dwarf, bushy growth; [Reverence X Miracle]; Delforge

Pompon Spong, C, m, 1820; flowers small, very dbl., moderate fragrance; Spong, about 1820; (Sangerhausen)

'PONCHEAU-CAPIAUMONT', HCh, dr; (Rosa Poncheaux, Rose Ponceau, Rose Poncheaux);, borne in clusters of 3; foliage curiously stiff, thick, deeply veined

Ponctuée *see* **'MA PONCTUÉE'**

Ponderosa® *see* 'KORPON'

Pont d'Avignon, F, or; Dot

'PONTBRIANT', HT, mp, 1948; flowers bright pink, very large, dbl., slight fragrance; foliage leathery; vigorous growth; [Mme Joseph Perraud X Kidwai]; Gaujard

'PONTCARRAL', HT, dp, 1943; bud long, pointed; flowers strawberry-red, reverse dull yellow, medium, semi-dbl., cupped, slight fragrance; foliage leathery; moderate, bushy growth; [Charles P. Kilham X (Charles P. Kilham X Margaret McGredy)]; Meilland, F.; A. Meilland

'PONTE D'ARRABIDA', HT, op, 1963; flowers salmon-pink veined deep pink; [Grand'mere Jenny X Seedling]; da Silva, Moreira

Ponte Rosa, LCl, mr

Ponte Salazar, F; Moreira da Silva, A.; (Cavriglia)

Pontevedra, HT, 1969; Moreira da Silva, A.; (Cavriglia)

'POORNIMA', HT, ly, 1971; bud long, pointed; flowers large blooms borne singly, dbl., 46 petals, exhibition form, moderate fragrance; foliage medium, medium green, smooth; bushy, tall growth; [Fernand Arles X Unnamed seedling]; Pal, Dr. B.P.; K.S.G. Son's Roses

'POPCORN', Min, w, 1973; bud ovoid; flowers pure white, semi-dbl., 13 petals, 1 in., moderate, honey fragrance; foliage glossy; upright growth; [Katharina Zeimet X Diamond Jewel]; Morey, Dr. Dennison; Pixie Treasures Min. Roses

Pope Pius IX *see* **'PIUS IX'**

Pope Pius XI *see* **'PIUS XI'**

'POPPET', F, mp, 1979; dbl., 60 petals, 3.5 in., cupped, slight fragrance; foliage light green, matt; upright growth; [Spartan X Arthur Bell]; Bees

'POPPING WHITE', Min, w, 1977; bud ovoid; semi-dbl., 12 petals, 1 in.; vigorous, compact growth; Lyon

Poppius, HSpn, mp; Steinberg, 1838

'POPPY', HT, op, 1939; flowers coral-pink, passing to shell-pink, petals serrated, dbl., moderate fragrance; vigorous growth; Archer

'POPPY', F, or, 1960; flowers scarlet, open, borne in clusters, semi-dbl., 2–2.5 in.; very vigorous growth; [Cocorico X Geranium Red]; Soenderhousen; Hoersholm Nursery

Poppy Flash *see* 'MEILÉNA'

Poppy Flash, Climbing *see* 'MEILÉNASAR'

Pops *see* 'TINPOPS'

Popsy, S, lp; Peden, G.H.; (Weatherly, L.)

Porcelain *see* 'KORFLÜG'

'PORCELAIN', F, lp, 1985; flowers shell pink over cream, medium, semi-dbl., slight fragrance; foliage small, dark, semi-glossy; upright, bushy growth; [Dainty Maid X ((Tropicana X Anna Wheatcroft) X (Tropicana X Anna Wheatcroft))]; Gobbee, W.D.

'PORCELAIN PRINCESS', Min, pb, 1990; bud rounded; flowers cream with pale pink edging, aging light pink, medium, borne, dbl., 35 petals, exhibition form, no fragrance; oblong, orange-red fruit; prickles straight, tan; foliage medium, dark green, semi-glossy; bushy, medium growth; [Libby X Unnamed seedling]; Gruenbauer, Richard, 1984; Richard Gruenbauer

Porcelaine de Chine, S, 1996; Lens; (Cavriglia)

Porcelina *see* 'INTERVAL'

Port St John, HT, 1974; Herholdt; (Cavriglia)

Porta Nigra, HT, dr, 1992; flowers large, dbl., moderate fragrance; Meilland; (Sangerhausen)

Porta Westfalica®, F, dr

'PORTADOWN', HT, dr, 1928; bud long, pointed; flowers velvety deep crimson, large, dbl., exhibition form, slight fragrance; vigorous, bushy growth; GM, NRS, 1929; McGredy

'PORTADOWN BEDDER', HT, op, 1929; bud long, pointed; flowers scarlet-cerise on orange ground, reverse orange-yellow flush, dbl., exhibition form, moderate fragrance; foliage glossy, dark; vigorous, bushy growth; McGredy

'PORTADOWN FRAGRANCE', HT, op, 1931; bud long, pointed; flowers brilliant orange-salmon-pink, flushed orange-scarlet, dbl., exhibition form, intense fragrance; foliage thick, bronze; low, sprawling growth; GM, NRS, 1928; McGredy

'PORTADOWN GLORY', HT, my, 1932; flowers clear canary-yellow, well formed, large, dbl., moderate fragrance; foliage glossy, bright; vigorous growth; GM, NRS, 1933; McGredy

Portadown Ivory *see* **'MCGREDY'S IVORY'**

'PORTADOWN SALLY', HT, mr, 1931; bud long, pointed; flowers crimson-carmine, base yellow, reverse sulfur-yellow, semi-dbl., exhibition form, moderate fragrance; foliage thick, glossy; vigorous growth; McGredy

Portail Rouge, LCl, 1973; Combe, C.; (Cavriglia)

'PORTCHESTER PINK', HT, dp, 1978; flowers deep rose-pink, intense fragrance; [Red Devil sport]; Poole

Portho, Climbing, Cl F, or, 1976; Laperrière

Porthos® *see* LAPad

'PORTIA', HT, lp, 1910; flowers pale rose, center yellow; Paul, W.

'PORTIA', HT, mr, 1921; flowers nasturtium-red, paling to pink, dbl., moderate fragrance; [Bridesmaid X Sunburst]; Bees

'PORTLAND', HT, dp, 1958; flowers rose-madder, large, dbl., moderate fragrance; foliage dark, glossy; vigorous, symmetrical growth; Lowe

Portland Dawn *see* 'SEATIP'

Portland from Glendora *see* **'JOASINE HANET'**

'PORTLAND PINK', F, dp, 1957; flowers glowing deep pink; [Pinocchio sport]; Murrel, E.

Portland Rose *see* **'DUCHESS OF PORTLAND'**

Portland Rose Festival *see* 'DORJURE'

'PORTLAND TRAILBLAZER', HT, dr, 1975; (Big Chief); flowers crimson, dbl., 28 petals, 5.5 in., exhibition form, moderate fragrance; foliage large, matt, brittle; [Ernest H. Morse X Red Planet]; Dickson, A.; Edmunds Roses, 1978

Portmeirion, S, mp, 1988; Austin, David

'PORTO', HT, dr, 1934; flowers deep garnet tinted bright scarlet, large, dbl., slight fragrance; foliage leathery; vigorous growth; [Capt. F. Bald X Mrs Edward Powell]; Mallerin, C.

Porto Nigra *see* MEIrgano

'PORTOFINO', F, rb, 1964; flowers red, reverse deep yellow; low growth; [Miramar X Seedling]; Delforge

Portrait *see* 'MEYPINK'

'PORTRAIT OF JENNY', F, dr, 1951; flowers crimson, borne in trusses, dbl., 30 petals, 2 in.; foliage dark; vigorous, bushy growth; [Donald Prior seedling]; Hope

'PORTUGAL PINK', HT, dp, 1959; bud globular; flowers deep pink, large, dbl., cupped, intense fragrance; foliage leathery; vigorous, upright growth; [Independence X Seedling]; Mondial Roses

Post Office, F, rb, 1997; J&P

Postillion *see* KORtionza

'POSTILLON', HT, ab, 1962; flowers coppery yellow, large, dbl.; foliage glossy; moderate growth; [Peace X Seedling]; Verbeek

'POSY', F, pb, 1951; flowers clear pink, reverse deeper, blooms in large clusters, dbl., 30 petals, 3 in., slight fragrance; foliage dark; dwarf growth; [Rosenelfe X Dusky Maiden]; LeGrice

Pot Black *see* 'PEANUT'

Pot o' Gold *see* 'DICDIVINE'

Potch Pearl *see* HERtroci

'POTIFAR', HT, ab, 1979; (Wieteke van Dordt); flowers pale peachy orange, urn-shaped, large blooms borne usually singly, dbl., 25 petals, slight fragrance; foliage large, leathery, dark, matt; bushy growth; [Royal Dane X Pjerrot]; Poulsen, Niels D.; A. Grumet, 1979;, Jan Spek, 1982

Potluck *see* 'LAVPOT'

Potluck Blue® *see* 'LAVNOVA'

Potluck Cream® *see* 'LAVCREAM'

Potluck Crimson® *see* 'LAVSON'

Potluck Frosty® *see* 'LAVPURE'

Potluck Gold® *see* 'LAVGOLD'

Potluck Orange® *see* 'LAVJACK'

Potluck Pink® *see* 'LAVALOT'

Potluck Purple® *see* 'LAVPUP'

Potluck Red® *see* 'LAVMIN'

Potluck White® *see* 'LAVWYTE'

Potluck Yellow® *see* 'LAVGLO'

Potter & Moore *see* AUSpot

Potton Heritage *see* 'HARSPRICE'

'POUBICARBE', F, ab, 1974; (**Chipie**); flowers rose-begonia to apricot, open, dbl., 24 petals, 4 in.; foliage light green; bushy growth; [((Elizabeth of Glamis X (Heidelberg X 8366-2)) X ((Pernille Poulsen X (Danish Gold X Mischief))]; Poulsen, Niels D.; Vilmorin-Andrieux

'POUDEX', S, lp, 1994; (**Newport**); flowers pink, old fashioned, dbl., 26–49 petals, 2–2.5 in., pompon, borne singly and in small clusters, slight fragrance; foliage medium green; compact, bushy growth; PP9261; Olesen, 1994

POUlagun, MinFl, op, 1999; (**Laguna Palace**); Poulsen

POUlalo, S, op, 1998; (**Coral Border**); Poulsen

'POULANDER', F, dr, 1986; (America's Choice, **H. C. Andersen**®, Hans Christian Andersen, Touraine); flowers large, semi-dbl., cupped, borne in sprays of 1-25, slight fragrance; foliage medium, dark green, glossy; bushy, tall growth; PP006265; [Royal Occasion X Unnamed seedling]; Olesen, Pernille & Mogens N., 1979; Poulsen Roser ApS, 1986

'POULANS', S, dp, 1995; (**Martha's Vineyard**); flowers hot pink, blending to white at center, semi-dbl., 6–14 petals, 2 in., borne in small clusters, slight fragrance; few prickles; foliage small, light green, semi-glossy; spreading (36 in) growth, tall, vigorous; PP9664; Olesen, Pernille & Mogens N.; DeVor Nurseries, Inc., 1995;, Young's American Rose Nursery, 1995

'POULAPS', S, ly, 1985; (**Armorique Nirpaysage**); flowers creamy yellow, urn-shaped, large blooms in clusters of 1-5, dbl., 20 petals, moderate fragrance; foliage large, leathery, dark, glossy; tall, upright, bushy growth; [Unnamed seedling X Kalahari]; Olesen, Pernille & Mogens N.; Poulsen's Roses

'POULAPS', S, w; (**White Nights**)

'POULARI', HT, w, 1994; (**Karen Blixen**, Silver Anniversary); very dbl., 3–3.5 in., borne in small clusters, slight fragrance; some prickles; foliage large, medium green, glossy; medium (60-80 cms), upright growth; Poulsen Roser APS; Cants of Colchester Ltd., 1995

'POULART', HT, rb, 1985; (**Modern Art**®, Prince de Monaco); flowers medium red, dark red petal edges, spreading with age, revers, dbl., 25 petals, slight fragrance; foliage medium, dark, matt; upright, bushy growth; GM, Rome, 1984; [Unnamed seedling X Unnamed seedling]; Olesen, Pernille & Mogens N.; D.T. Poulsen, 1983

POUlave, S, ab, 1991; (**Sussex**); groundcover growth; Poulsen

POUlbella, F, 1994; (**Berleburg**, Bewitched); Poulsen; (Cavriglia)

'POULBELLS', Min, dp, 1983; (**Pink Bells**®); flowers deep pink, small, dbl., 35 petals, slight fragrance; foliage small, medium green, semi-glossy; groundcover; spreading growth; [Mini-Poul X Temple Bells]; Poulsen Roser APS; John Mattock, Ltd

POUlberin, S, m, 1997; (**Bering Renaissance**); Poulsen

'POULBERO', Gr, ob, 1991; (**Solitude**®); flowers orange-yellow bicolor, blooms borne in large clusters, dbl., 26–40 petals, 1.5–2.75 in., slight fragrance; foliage medium, medium green, semi-glossy; medium (110 cms), bushy growth; [Selfridges X Unnamed seedling]; Olesen, Pernille & Mogens N.; Conard-Pyle Co., 1992

POUlbico, F, pb, 1992; (**Christopher Columbus**, POUlstripe); Poulsen

POUlbo, LCl, w, 1999; (**Bolero**); Olesen, 1998

POUlbon, Min, dy, 1992; (**Bonbon Hit**®); Poulsen

POUlbright, F, 1985; Poulsen; (Cavriglia)

POUlbufi, S, ob; (**Bushfire**); Poulsen

POUlbut, S, mp, 1998; (**Butterflies**); Poulsen

POUlcape, MinFl; (**Summer Palace**); Poulsen, 1995; (Cavriglia)

'POULCAT', Min, lp, 1985; (**Caterpillar**®, Kiki Rose®, Pink Drift); flowers small blooms in large trusses, semi-dbl., no fragrance; foliage small, dark, glossy; groundcover; low, spreading growth; [Temple Bells X Seedling]; Olesen, Pernille & Mogens N.; D.T. Poulsen, 1984

POUlchris, MinFl, mr, 1996; (**Imperial Palace**); Poulsen

'POULCLAUS', Min, dr, 1991; (**Santa Claus**); bud long, well-formed; flowers velvety dark red, large, dbl., 15–25 petals, 1.5–2.75 in., exhibition form, borne mostly singly, slight fragrance; few prickles; foliage medium, dark green, glossy; medium (50-60 cms), upright, bushy growth; PP9063; [Unnamed Floribunda seedling X Unnamed Miniature seedling]; Olesen, Pernille & Mogens N.; Weeks Roses, 1995

'POULCLIMB', LCl, pb, 1999; (**Berries 'n' Cream**®, Calypso®, Cl. Berries 'n' Cream); flowers pink striped white, reverse same, dbl., 26–40 petals, 4–5 in., borne in large clusters, moderate, apple fragrance; few prickles; foliage large, medium green, glossy; climbing, tall (10-12 ft) growth; PP10639; [Evita X seedling]; Olesen; Weeks Roses, 1998

'POULCOT', Min, ab, 1985; (**Apricot Medinette**, Apricot Midinette, Patio Flame); flowers apricot-orange, well-formed, small, dbl., 20 petals, blooms in clusters, slight fragrance; brown prickles; foliage leathery, glossy; spreading growth; [Mini-Poul X Mary Sumner]; Olesen, Pernille & Mogens N.; Ludwigs Roses Pty. Ltd., 1983

'POULCOV', S, w, 1988; (**Kent**®, Pyrenees, Sparkley, White Cover); flowers clear white, bright yellow stamens, semi-dbl., 15–25 petals, 2 in., borne singly and in clusters, slight fragrance; foliage medium green; arching, compact, spreading, low growth; Certificate of Merit, RNRS, 1992 GM, Baden-Baden, 1990 PIT, RNRS, 1990; Olesen, 1988

POUlcrack, F, ob, 1989; (**Crackerjack**); Olesen

'POULCRIM', Min, dr, 1984; (**Crimson Medinette**, Crimson Midinette, Patio Prince); flowers small blooms in clusters of 3-5, dbl., 33 petals, no fragrance; straight brown prickles; foliage small, leathery; dense growth; [Seedling X Pygmae]; Olesen, Pernille & Mogens N.; Ludwigs Roses Pty. Ltd.

POUlcub, MinFl, lp; (**Tiger Cub**); Poulsen, 1996

POUldace, MinFl, lp; (**Sundance Palace**); Poulsen, 1999

'POULDAVA', S, ab, 1995; (**Augusta**); flowers peach, blooms, dbl., 15–25 petals, 2–3 in., slight fragrance; some prickles; foliage small, medium green, dull; low (24 in), spreading growth; Olesen, Pernille & Mogens N.; DeVor Nurseries, Inc., 1995;, Young's American Rose Nursery, 1995

POUldel, Min, w, 1993; (**Pure Hit**); Poulsen

POUldiam, S, w, 1998; (**Diamond Border**); Poulsen

POUldom, F, dy, 1994; (**Gold Reef**); Poulsen

POUldotage, HT, pb; (**Addo Heritage**); Poulsen

POUldra, S, dp, 1997; (**Princesse Alexandra**); Poulsen

POUldrik, F, op, 1998; (**Frederiksborg**); Poulsen

'POULDRON', S, op, 1999; (**Countess Celeste**™, Coral Palace); flowers coral pink, reverse same, old-fashioned form, very dbl., 41 petals, 2.5–3 in., borne in small clusters, moderate fragrance; prickles moderate; foliage medium, dark green, glossy; compact, rounded low (20-24 in.) growth; [Queen Margrethe X seedling]; Poulsen Roser APS; Weeks Roses, 1998

'POULDUCE', HT, my, 1994; (**Tivoli Gardens**); flowers soft yellow, blooms borne in small clusters, dbl., 26–40 petals, 3–3.5 in., slight fragrance; some prickles; foliage large, dark green, semi-glossy; tall (80-100 cms), spreading growth; Olesen, Pernille & Mogens N.; Cants of Colchester Ltd., 1995

POUlduff, HT, mr, 1997; (**Courage**, Roy Castle Rose); Poulsen

POUleas, Min, 1995; (**Easy Cover**); Poulsen; (Cavriglia)

POUledge, F, op, 1999; (**Everglades**); Poulsen

POUlelap, MinFl, mp, 1997; (**Queen's Palace**); Poulsen

'POULEMB', S, w, 1995; (**Cliffs of Dover**); flowers small, white, single, bright yellow stamens, single, 5–10 petals, 1–1.5 in., borne in clusters, no fragrance; few prickles; foliage small, medium green, glossy; low (36 in), arching, spreading growth; PP9650; Olesen, Pernille & Mogens N.; DeVor Nurseries, Inc., 1995;, Young's American Rose Nursery, 1995

'POULENA', HT, yb, 1976; (**Bellevue**®, Jarlina); flowers dark yellow and apricot, edged red, dbl., 23 petals, 6 in.; foliage glossy, leathery; vigorous, upright growth; [(Tropicana X Piccadilly) X Fru Jarl]; Poulsen, Niels D.; Poulsen

POUlense, Cl Min, ly, 1995; (**Little Miss Muffet**); Olesen

'POULETTA', HT, ab, 1984; (**Henrietta**); flowers urn-shaped, large, borne singly, dbl., 25 petals, slight, fruity fragrance; foliage large, dark, glossy; vigorous, upright, bushy growth; [Unnamed seedling X (Pink Nordia X Sonny Boy)]; Olesen, Pernille & Mogens N.; Poulsen's Roses

'POULFAIR', S, my, 1988; (**Yellow Fairy**®); flowers small, semi-dbl., 6–14 petals, slight fragrance; foliage small, light green, semi-glossy; spreading, compact, hardy, abundant growth; GM, Madrid, 1988; [Texas X The Fairy]; Olesen, Pernille & Mogens N.; Poulsen Roser ApS

'POULFAN', S, lp, 1995; (**Cape Cod**, Jhb Garden Club, Johannesburg Garden Club); flowers soft pink, blooms borne in large clusters, single, 5 petals, 2–3 in., no fragrance; few prickles; foliage small, medium green, semi-glossy; medium (36-42 in), spreading growth; PP9641; Silver, Bundesgartenshau GER, 1991; Olesen, Pernille & Mogens N.; DeVor Nurseries, Inc., 1995;, Young's American Rose Nursery, 1995

POUlfancy, Min, mr; (**Fancy Hit**); Poulsen

POUlfect, Min, ab; (**Perfect Hit**); Poulsen

POUlfeld, F, w; (**Grasten**); Poulsen

'POULFI', Min, dp, 1985; (**Alfi**™, Alfie™, Elfin); flowers soft lavendar to pink, small, very dbl., 40–45 petals, blooms in large clusters and cadelabras, no fragrance; foliage small, light green, glossy; compact growth, free flowering; PP005058; [Mini-Poul X Harriet Poulsen]; Olesen, Pernille & Mogens N.; Poulsen's Roses, 1981

POUlfiry, Min, ob; (**Fiery Hit**); Poulsen

'POULFLASH', HT, my, 1984; (**Flashdance**); flowers urn-shaped, large blooms borne singly, dbl., slight fragrance; foliage large, medium green, semi-glossy; upright growth; [Berolina X Unnamed seedling]; Olesen, Pernille & Mogens N.; Poulsen Roser ApS

POUlfolk, S, my, 1990; (**Norfolk**); Poulsen

POUlfre, Min, dp, 1990; (**Fresh Hit**); Poulsen

POUlgan, MinFl, dy, 1992; (**City Lights**); Poulsen

'POULGAV', F, ob, 1988; (Bucks Fizz, **Gavno**); flowers orange, medium, dbl., 20 petals, no fragrance; foliage medium, dark green, glossy; bushy growth; [Unnamed seedling X Mary Sumner]; Olesen, Pernille & Mogens N.; Poulsen Roser ApS

POUlgewfa, S, lp, 1992; (**Gwen Fagan**); Poulsen

POUlgode, F, my; (**Yellow Cover**); Poulsen, 1993

POUlgode, S, ly; (**Lexington**); Olesen, 1993

POUlgrad, MinFl, dr, 1999; (**Grand Palace**); Poulsen

POUlgren, F, w, 1992; (**Snow Gosling**); Poulsen

POUlgret, MinFl, w, 1996; (**Hampton Palace**); Poulsen

POUlharm, S, or, 1999; (**Charming Cover**); Poulsen

POUlheart, S, lp, 1996; (**Julia Renaissance**, Phillipa); Olesen, 1996

'POULHIT', Min, dr, 1984; (**Red Hit**); flowers small, dbl., 20 petals, no fragrance; foliage small, dark, matt; low, bushy, compact growth; [Mini-Poul X Unnamed seedling]; Olesen, Pernille & Mogens N.; Poulsen's Roses

POUlhit, Min, my; (**Sun Hit**); Poulsen, 1994

POUlholm, S, op, 1989; (**Patio Princess**); Olesen

POUlijill, S, lp; (**Bella**); Poulsen

'POULINK', Min, mp, 1988; (**Pink Hit**); flowers small, semi-dbl., 6–14 petals, no fragrance; foliage small, medium green, semi-glossy; bushy, compact, even, abundant growth; [Unnamed seedling X Unnamed seedling]; Olesen, Pernille & Mogens N.; Poulsen Roser ApS, 1986

'POULINO', S, mr, 1995; (**Napa Valley**); flowers bright red, semi-double, 2", semi-dbl., 6–14 petals, 2 in., borne in clusters, slight fragrance; foliage small, dark green, glossy; low (36 in), spreading growth; PP9554; Olesen, Pernille & Mogens N.; DeVor Nurseries, Inc., 1995;, Young's American Rose Nursery, 1995

POUlisab, S, dr; (**Isabel Renaissance**, Isabella); Olesen, 1995

POUlisab, S, dr, 1995; (**Isabelle Renaissance**); Poulsen

'POULIT', Min, mp, 1994; (**Baby Grand**®); flowers clear pink, dbl., 25–40 petals, 1.5 in., quartered, borne in small clusters, slight fragrance; some prickles; foliage small, medium green, matt, very disease resistant; low (25-30 cms), bushy, compact, rounded growth; [Egeskov X Seedling]; Poulsen Roser APS; Weeks Roses, 1995

'POULJENJOY', Min, lp, 1985; (**Jennifer Joy**); flowers mini-flora, blooms in clusters, dbl., 75 petals, no fragrance; foliage light green; compact, low, bushy growth; [Mini-Poul X Unnamed seedling]; Olesen, Pernille & Mogens N.; Ludwigs Roses Pty. Ltd., 1983

POUljill, S, my, 1995; (**Bella Renaissance**); Poulsen

'POULJUB', Min, yb, 1978; (**Poulsen's Jubilaeumsrose**); flowers deep pink and yellow blend, blooms in clusters, dbl., 28 petals, slight fragrance; foliage dark, glossy; upright growth; [Darling Flame X Unnamed seedling]; Poulsen, Niels D.

'POULKARO', F, or, 1988; (**Karoo**, Karoo Rose); flowers orange-red, aging watermelon-pink, medium, borne in sprays, dbl., 32 petals, no fragrance; prickles concave, yellow-brown; foliage medium, medium green, very hardy; neat, medium, densely branched growth; [Unnamed seedling X Unnamed seedling]; Poulsen Roser APS; Ludwigs Roses Pty. Ltd., 1988

POUlkef, HT, 1997

POUlkir, HT, lp, 1985; (**Kirsten**); Poulsen

'POULLACK', Cl HT, rb, 1982; (Kirsch, Climbing, Lakeland Princess, **Vintage Wine**); flowers burgundy red, straw-yellow reverse, large, dbl., 40 petals, exhibition form, slight, old rose fragrance; prickles red when new, then tan; foliage large, moss green; pillar (8-10 ft) growth; [Royal Dane X Arthur Bell]; Poulsen, Niels D.; Roses by Fred Edmunds, 1983

POUlland, S, w; (**White Magic Carpet**®); Poulsen, 1994

'POULLEN', Min, mp, 1994; (**Little Bo-Peep**, Nachez); dbl., 26–40 petals, 1.5 in., borne in small clusters, slight fragrance; few prickles; foliage small, dark green; patio; low (65 cms), compact growth; PP9273; PIT, RNRS, 1991; Olesen, Pernille & Mogens N.; DeVor Nurseries, Inc., 1994;, Young's American Rose Nursery, 1994

POUllen, S, lp; (Gentle Cover, Little Bo-Peep, **Natchez**); Olesen, 1994

'POULLIGHT', LCl, dy, 1985; (**Night Light**®); flowers large blooms in sprays of 1-5, dbl., 27 petals, moderate fragrance; large, dark red prickles; foliage large, dark, glossy; bushy growth; [Westerland X Pastorale]; Poulsen, Niels D.; D.T. Poulsen, 1982

'POULLOW', Min, dy, 1988; (**Yellow Hit**); flowers small, dbl., 15–25 petals, slight fragrance; foliage small, dark green, semi-glossy; bushy, compact, even growth; PP006860; GM, International Exhibition, 1986; [Unnamed seedling X Unnamed seedling]; Olesen, Pernille & Mogens N.; Poulsen Roser ApS

POUlly, F, 1991; Poulsen; (Cavriglia)

'POULMA', S, lp, 1994; **'Newport'** very dbl., 3–3.5 in., borne in large clusters, no fragrance; some prickles; foliage medium, medium green, glossy; low (2 ft), spreading growth; Olesen, Pernille & Mogens N.; DeVor Nurseries, Inc., 1994;, Young's American Rose Nursery, 1994

'POULMAN', HT, dr, 1984; (**Ingrid Bergman**®); flowers large, clear bright red, very dbl., 35–40 petals, 5 in., exhibition form, borne singly, slight, spice fragrance; foliage medium, dark, semi-glossy; upright growth, vigorous, compact; PP006264; GM, Belfast, 1985 GM, Madrid, 1986 Golden Rose, The Hague, 1987; [Unnamed seedling X Unnamed seedling]; Olesen; John Mattock, Ltd, 1985

POUlmanti, MinFl, dp, 1999; (**Romantic Palace**); Poulsen

POUlmar, Min, ob, 1994; (**Rosy Hit**); Poulsen

POUlmax, F, lp, 1995; (**Fredensborg**); Poulsen

POUlming, Min, lp, 1994; (**Harmony Parade**); Olesen

POUlmo, Min, mr, 1991; (**Scarlet Hit**); Poulsen

POUlmount, Gr, ly, 1996; (**Ivory Grand**); flowers ivory white, large, dbl., exhibition form; Poulsen

POUlmulti, S, w, 1992; (**Avon**, Sunnyside); bud blush pink; flowers pearl white, dbl.; compact, spreading (1 ft x 2.5 ft) growth; Rose of The Year, Auckland, NZ, 1996; Poulsen

POUlna, S, lp, 1997; (Helena, **Helena Renaissance**); Olesen, 1997

POUlnew, HT, lp; (**Jose Carreras**); Golden Rose, Hradec Kralove, Czech Republic, 1999; Olesen, Pernille & Mogens N.

'POULNINO', S, mr, 1995; (**Napa Valley**); flowers medium red, semi-dbl. (6–14 petals), medium (2 in.) blooms; slight fragrance; foliage small, dark green, glossy; low (36 in.), spreading growth; [Seedling X Seedling]; Olesen, Pernille & Mogens N.; DeVor Nursery; Youngs' American Rose Nursery, 1995

'POULNISH', Min, my, 1988; (**Sunbird**®); flowers medium yellow opening to light yellow, fading to light pastel, dbl., 25–30 petals, exhibition form, no fragrance; ovoid, few, pale orange-red fruit; foliage medium, dark green, matt, convex, straw to light mar; bushy, medium growth; [Mini-Poul X Unnamed seedling]; Poulsen Roser APS; Conard-Pyle Co., 1987

POUlnoeu, MinFl, lp; (**Salmo**); Poulsen

POUlnoev, S, op, 1995; (**Salmo**); Olesen

POUlnola, Min, mr; (**Noble Hit**); Poulsen

POUlnorm, LCl, lp; (**Jazz**); Poulsen

POUlnows, Min, w; (**Snow Hit**); Poulsen

POUlnoz, S, mp, 1988; (**Essex**); single; low, spreading growth; Poulsen

POUloma, F, mp, 1992; (**Bossa Nova**); Poulsen

'POULONI', HT, op, 1977; (Benoni '75, **Vision**); flowers medium salmon red, well-formed, large, dbl., 22 petals, borne 1-3 per stem, slight fragrance; foliage dark, glossy; vigorous, spreading growth; PP006939; Poulsen, Niels D.; Poulsen's Roses, 1990

POUlor, S, m; (**Mystic**, Supreme Cover); Olesen, 1994

POUlor, S, lp, 1994; (**Peachy Pink Magic Carpet**); Poulsen

POUlorin, MinFl, op, 1999; (**Oriental Palace**); Poulsen

POUlosy, S, mp, 1998; (**Rosy Border**); Poulsen

POUlover, LCl, lp, 1994; (**Rosenholm**); Poulsen

POUlpearl, F, lp, 1990; (**Patio Pearl**); Olesen

POUlpol, HT, ob, 1997; Poulsen

POUlprima, Min, m; (**Purple Parade**); Poulsen

'POULPYG', S, ab, 1995; (**Central Park**); flowers peach, bright yellow stamens, fades white, dbl., 15–25 petals, 2 in., borne in small clusters, slight fragrance; few prickles; foliage small, medium green, glossy; low (2 ft), spreading growth; PP9665; Olesen, Pernille & Mogens N.; DeVor Nurseries, Inc., 1995;, Young's American Rose Nursery, 1995

POUlrang, Min, ob, 1995; (**First Hit**); Poulsen

POUlreb, F, my, 1995; (**Eurostar**); Olesen

'POULRED', Min, mr, 1983; (**Red Bells**®); flowers small, dbl., 35 petals, slight fragrance; foliage small, medium green, semi-glossy; groundcover; spreading growth; [Mini-Poul X Temple Bells]; Poulsen Roser APS; John Mattock, Ltd.

POUlrek, F, lp, 1995; (**Crystal Palace**); flowers creamy pink, medium-large, dbl.; Poulsen Roser APS; (Sangerhausen)

POUlria, S, lp; (**Velvet Cover**); Poulsen, 1995

POUlriber, F, op, 1998; (**Riberhus**); Poulsen

POUlrijk, S, mp, 1993; (**Devon**®); Poulsen

POUlrim, HT, 1996; (**Flora Danica**); Poulsen Roser APS; (Sangerhausen)

POUlrim, HT, lp; (**Garden News**, Spellbound); Poulsen

POUlrine, F, mp; (**Bangsbo**); Poulsen

'POULRISE', HT, yb, 1994; (**Peach Surprise**); flowers peach and cream, blooms borne mostly single, dbl., 26–40 petals, 3–3.5 in., moderate fragrance; many prickles; foliage large, dark green, glossy; spreading (80-100 cms) growth; [Freude sport]; Poulsen Roser APS; Cants of Colchester Ltd., 1995

POUlrohill, F, mp; (**Rose Hill**); Poulsen

POUlrougel, Min, or; (**Absolute Hit**); Poulsen

'POULRUSSET', Min, or, 1983; (**Russet Beauty**); flowers small blooms in clusters of 3-15, dbl., 46 petals; straight, brown prickles; foliage matt; compact, bushy growth; [Mini-Poul X Unnamed seedling]; Olesen, Pernille & Mogens N.; Ludwigs Roses Pty. Ltd.

POUlrust, S, m, 1994; (**Lavender Cover**); Poulsen

POUlsail, MinFl, lp; (**Versailles Palace**); Poulsen, 1999

'POULSALFAI', S, op, 1993; (**Miss Hillcrest**); Poulsen

'POULSEN'S BEDDER', F, lp, 1948; (Poulsen's Grupperose); flowers clear pink, blooms in trusses, semi-dbl., 3 in., slight fragrance; foliage bronze; vigorous, upright growth; GM, Portland; [Orléans Rose X Talisman]; Poulsen, S.; C-P, 1952;, McGredy, 1948;, Poulsen, 1948

'POULSEN'S COPPER', F, op, 1940; flowers rose-pink to orange, base yellow, borne in clusters, dbl., 25–35 petals, 3–3.5 in., cupped, slight, spicy fragrance; foliage small, light green; vigorous growth; [Grethe Poulsen X Souv. de Claudius Pernet]; Poulsen, S.; C-P, 1940;, McGredy, 1946;, Poulsen, 1940

'POULSEN'S CRIMSON', F, dr, 1950; single, 6 petals, blooms in clusters, slight fragrance; foliage medium, dark, matt; vigorous, bushy growth; [Orange Triumph X (Betty Uprichard X Johanniszauber)]; Poulsen, S.; McGredy;, Poulsen

'POULSEN'S DELIGHT', F, lp, 1948; (Fru Julie Poulsen); flowers apple-blossom-pink, borne in trusses, single, 5–7 petals; foliage dark, glossy; vigorous, upright growth; [Else Poulsen X Seedling]; Poulsen, S.; McGredy

'POULSEN'S FAIRY', F, my, 1940; (Rodovrerosen); single, borne in large clusters; very vigorous, tall growth; [Orléans Rose X Dainty Bess]; Poulsen, S.; Poulsen

Poulsen's Grupperose *see* **'POULSEN'S BEDDER'**

Poulsen's Jubilaeumsrose *see* 'POULJUB'

'POULSEN'S PARK ROSE', S, lp, 1953; flowers silvery pink, well-shaped, blooms in trusses, dbl., 4–5 in.; vigorous (6 x 6 ft) growth; [Great Western X Karen Poulsen]; Poulsen

'POULSEN'S PEACH', F, ab, 1948; flowers peach, large blooms in clusters, semi-dbl., cupped, slight fragrance; foliage medium, medium green, semi-glossy; medium, bushy growth; Poulsen, S.

'POULSEN'S PEARL', F, lp, 1949; flowers pearly pink, blooms in trusses, single, 5 petals; foliage light; vigorous growth; GM, RNRS; [Else Poulsen X Seedling]; Poulsen, S.; Poulsen's Roses, 1948

'POULSEN'S PINK', F, lp, 1939; flowers soft pink with yellow base, blooms in clusters, semi-dbl., cupped, slight fragrance; foliage glossy, light green; vigorous growth; (28); [Golden Salmon X Yellow HT]; Poulsen; C-P, 1942;, Poulsen, 1942

'POULSEN'S SCARLET', F, dp, 1941; flowers bright rose, blooms in clusters, dbl., 30 petals, 2.5 in., slight fragrance; bushy growth; [D.T. Poulsen X Red HT]; Poulsen, S.

'POULSEN'S SUPREME', F, mp, 1945; (Kelleriis-Rose); semi-dbl., 3 in., borne in trusses; foliage light green; very free growth; (28); [Poulsen's Pink X Seedling]; Poulsen, S.; McGredy, 1953

'POULSEN'S YELLOW', F, my, 1938; bud ovoid; semi-dbl., blooms in clusters, intense fragrance; foliage glossy; (21); GM, NRS, 1937; [Mrs W.H. Cutbush X Gottfried Keller]; Poulsen, S.; C-P, 1939

POUlshine, Min, lp, 1988; (**Rutland**); PP009457; Poulsen

POUlsiana, F, dy, 1999; (**Atlantis Palace**); Poulsen

POUlsint, Min, mr, 1998; (**Red Paillette**); Poulsen

POUlske, F, rb, 1996; (**Arlette**); Poulsen

'POULSKOV', S, lp, 1994; (Enchantment, **Queen Margrethe**™, Dronning Margrethe); flowers pastel pink, old-fashioned form, dbl., 50–40 petals, 1.5–2.75 in., quartered, borne in small clusters, moderate, apple fragrance; few prickles; foliage small, medium green, glossy, clean; low to medium (45-50 cms), bushy, compact, rounded growth; PP9062; Gold Star of the South Pacific, Palmerston North, NZ, 1992; [Seedling X Egeskov]; Olesen; NZ, 1994;, Weeks Roses, 1995

POUlsol, HT, dy, 1999; (**Solo Mio**); Poulsen

POUlspan, S, my, 1992; (**Cottage Maid**, Enchantment); Rose of The Year, Auckland, NZ, 1997; Poulsen

POUlstar, MinFl, lp; (**Starlight Parade**); Poulsen, 1991

POUlstripe, F, pb, 1992; (**Christopher Columbus**, POUlbico); Poulsen

POUlsue, S, lp; (**Susan**); Poulsen, 1997

POUlsum, F, op, 1995; (**Dawn Weller**); Poulsen

POUlsyng, S, lp, 1995; (**Clair Renaissance**, Liliana); Olesen, 1995

'POULTAL', HT, dp, 1983; (**Eva Gabor**, Sentimental); flowers deep pink, large, dbl., 40 petals, moderate fragrance; foliage large, medium green, glossy; vigorous, upright, bushy growth; [Seedling X Seedling]; Olesen, Pernille & Mogens N.; Roses by Fred Edmunds

POUlterp, S, w, 1999; (**Brilliant Cover**); Poulsen

'POULTEX', Min, my, 1984; (Golden Piccolo, **Texas**); flowers mini-flora, small, semi-dbl., slight fragrance; foliage small, medium green, matt; PP006261; [Seedling X Seedling]; Poulsen Roser APS; John Mattock, Ltd

POUlthe, MinFl, lp; (**Topkapi Palace**); Poulsen, 1996

POUltika, LCl, mp, 1997; (**Flamenco**); Poulsen

POUltime, LCl, pb, 1997; (**Ragtime**); Poulsen

POUltimes, F, my, 1993; Poulsen

POUltin, MinFl, lp; (**Violet Hit**); Poulsen, 1995

POUltipe, MinFl, lp; (**Pink Hit**); Poulsen

'POULTOP', Min, ob, 1991; (**Carrot Top**, Panache); flowers orange, clear lasting color, moderately well-formed, blooms, dbl., 15–25 petals, 1.5 in., prolific bloom, slight fragrance; few prickles; foliage medium, medium green, matt; low (40-50 cms), bushy, compact growth; [Seedling (F) X Seedling (Min)]; Olesen, Pernille & Mogens N.; Weeks Roses, 1994

'POULTRESS', HT, dp, 1988; (**Cherry Velvet**®); flowers deep pink, large, dbl., 20 petals, moderate fragrance; foliage large, medium green, glossy; upright, vigorous growth; [Vision X Seedling]; Olesen, Pernille & Mogens N.; Poulsen Roser ApS, 1986

POUltry, F, dr, 1995; (**Kronborg**); Poulsen

POUltumb, S, lp; (**Tumbling Waters**); Poulsen, 1998

'POULURT', S, my, 1995; (**Aspen**, Gwent, Sun Cover; Gold Magic Carpet); dbl., 15–25 petals, 1–2 in., borne in small clusters; few prickles; foliage medium, dark green, semi-glossy; low (14-16 in), spreading growth; PP9637; 2nd prize, Copenhagen, 1992 Certificate of Merit, Belgium, 1992; Olesen, Pernille & Mogens N.; DeVor Nurseries, Inc., 1995;, Young's American Rose Nursery, 1995

'POULUSA', S, w, 1999; (**Princess Marianna**™); flowers ivory white, old-fashioned form, very dbl., 41 petals, 2.5–3 in., borne in small clusters, moderate fragrance; prickles moderate; foliage medium, dark green, glossy; compact, bushy, rounded, low (20-24 in.) growth; PP10638; [seedling X Queen Margrethe]; Poulsen Roser APS; Weeks Roses, 1998

'POULVANG', HT, dr, 1988; (**Solvang**); flowers large, semi-dbl., 6–14 petals, slight fragrance; prickles average; foliage large, dark green, glossy; bushy, vigorous growth; [Vision X Unnamed seedling]; Olesen, Pernille & Mogens N.; Poulsen Roser ApS, 1987

POUlvic, Min, mr; (**Victory Parade**®); Poulsen, 1999

'POULVISION', F, dp, 1975; (**Dolly**®, Springs 75); dbl., 20 petals, 2.5–3 in.; foliage glossy, dark; bushy growth; ADR, 1987 GM, Baden-Baden, 1973; [(Nordia X Queen Elizabeth) X (Seedling X Mischief)]; Poulsen, Niels D.; Poulsen

'POULVITA', Min, w, 1984; (**Evita**); flowers white with touch of pale pink, small blooms in clusters, dbl., 20 petals, slight fragrance; foliage small, dark, glossy; bushy growth; [Mini-Poul X Seedling]; Olesen, Pernille & Mogens N.; Poulsen

POUlvue, HT, ob; (Michael Crawford, **Victor Borge**); flowers pink and yellow blend, dbl., 50 petals, exhibition form; medium growth; Cert. of Merit, Belfast, 1992; Olesen, 1991

POUlwarb, S, ly, 1993; (**Fairy Moon**); Olesen

'POULWEE', Min, dy, 1990; (**Rainbow Yellow Parade**™, Yellow Parade™); bud cupped, globular; flowers bright yellow, aging slightly, small, borne singly, dbl., 28–30 petals, cupped, slight, damask fragrance; no prickles; foliage small, dark green, semi-glossy; upright, low, compact growth; GM, The Hague, 1988; [Unnamed seedling X Texas]; Olesen, Pernille & Mogens N., 1985; DeVor Nurseries, Inc., 1990

POUlweet, S, lp; (**Sweet Cover**); Poulsen, 1999

'POULWHITE', Min, w, 1983; (**White Bells**®); dbl., 35 petals, slight fragrance; foliage small, medium green, semi-glossy; groundcover; spreading growth; [Mini-Poul X Temple Bells]; Poulsen Roser APS; John Mattock, Ltd, 1980

'POULXAS', Min, ab, 1994; (**Gingerbread Man**™); flowers deep, long-lasting apricot amber, blooms borne in small clusters, dbl., 25–40 petals, 1.5 in., quartered, slight fragrance; some prickles; foliage small, dark green, semi-glossy; medium (40-45 cms), upright, bushy, vigorous growth; [Seedling X Texas]; Poulsen Roser APS; Weeks Roses, 1995

POUlzazz, MinFl, rb, 1996; (**Pzazz**); Poulsen

POUlzin, MinFl, rb, 1999; (**Prince Palace**); Poulsen

'POUMIDOR', HT, ob, 1971; (**Royal Dane**®, Troika); flowers orange, outer petals red, classic form, 6 in., intense fragrance; foliage large, dark, glossy, leathery; vigorous, upright, bushy growth; [(Tropicana X (Baccará X Princesse Astrid)) X Hanne]; Deutsche Rosen-registerstelle; N. Poulsen

Pounder Star *see* 'MACNIC'

'POUPÉE', F, lp, 1965; flowers flesh-pink, dbl., 25–30 petals, 3–3.5 in., cupped, slight fragrance; foliage glossy; vigorous, upright growth; Lens

Pour Toi *see* **'PARA TI'**

'POUR VOUS MADAME', F, dp, 1961; flowers rose-red, reverse lighter, open, dbl., 25 petals, 4 in., slight fragrance; foliage dark; vigorous growth; Gaujard; Gandy Roses, Ltd.

'POURALE', S, pb, 1970; (**Pastorale**®, Pastourelle); flowers deep pink, yellow reverse, urn-shaped, large blooms in spray, dbl., 25 petals, slight fragrance; dark green prickles; foliage large, leathery, dark, glossy; very vigorous, upright, bushy growth; [Unnamed seedling X Royal Dane]; Poulsen, Niels D.

'POURPRE', Ch, m, 1827; flowers deep purple; Vibert

'POURPRE DU LUXEMBOURG', M, m, 1848; flowers pink, opening to lilac-pink; Hardy

Poustinia, F, ab, 1994; Orye

'POWDER PUFF', F, lp, 1959; flowers creamy light pink, becoming darker, open, medium, dbl., slight fragrance; foliage glossy; bushy growth; [Masquerade X Masquerade seedling]; Mason, A.L.; F. Mason

'POYNTZFIELD', HT, ab, 1977; dbl., 35 petals, 5 in., slight fragrance; foliage glossy, reddish-green; vigorous growth; [Bonsoir X Percy Thrower]; Simpson, J.W.

'POZNAN', HT, dr, 1966; bud ovoid; flowers vivid dark crimson, well-formed, large, dbl.; vigorous growth; Grabczewski

'PRAECOX', (form of R. sericea), w

PRAeddy, HT, lp; (**Teddy**); Pratt, Florence, 1994

Prairie Belle *see* **'QUEEN OF THE PRAIRIES'**

'PRAIRIE BREEZE', S, m, 1978; bud ovoid, pointed; flowers tyrian purple, dbl., 25 petals, 4.5 in., cupped, repeat bloom, moderate, spicy fragrance; foliage olive-green, leathery; upright, bushy, spreading growth; [Dornroschen X (Josef Rothmund X R. laxa)]; Buck, Dr. Griffith J.; Iowa State University

'PRAIRIE CHARM', S, op, 1959; flowers bright salmon-coral, semi-dbl., free, non-recurrent bloom; foliage light green; stems arching; height 4 ft.; hardy on prairies.; [Prairie Youth X Prairie Wren]; Morden Exp. Farm

'PRAIRIE CLOGGER', S, mr, 1984; bud medium-small, ovoid, pointed; 8–10 petals, 2.5 in., cupped, borne 1-10 per cluster, repeat bloom, moderate fragrance; awl-like, tan prickles; foliage leathery, dark olive green; vigorous, erect, bushy growth; [Carefree Beauty X (Marlena X Pippa's Song)]; Buck, Dr. Griffith J.; Iowa State University

'PRAIRIE DAWN', S, mp, 1959; flowers glowing pink, dbl., 2–2.5 in., repeat bloom on current season's wood; foliage dark, glossy; upright (5 ft.) growth; hardy on prairies.; [Prairie Youth X (Ross Rambler X (Dr. W. Van Fleet X R. spinosissima altaica))]; Modern Exp. Farm

'PRAIRIE FIRE', S, mr, 1960; bud pointed; flowers bright red, base white, blooms in clusters of 35-50, semi-dbl., 9 petals, 2.5–3 in., recurrent bloom, moderate fragrance; foliage glossy, dark; long stems; very vigorous, tall growth; [Red Rocket X R. arkansana]; Phillips; Univ. of Minn.

'PRAIRIE FLOWER', S, rb, 1975; bud ovoid, pointed; flowers cardinal-red, center white, single, 7 petals, 2–3 in., flat, repeat bloom, slight, old rose fragrance; foliage dark, leathery; erect, bushy growth; [(Rose of Tralee X Queen Elizabeth) X (Morning Stars X Suzanne)]; Buck, Dr. Griffith J.; Iowa State University

'PRAIRIE HARVEST', S, ly, 1985; flowers medium yellow, imbricated, blooms borne 1-15 per cluster, dbl., 43 petals,

4–5 in., repeat bloom, moderate fragrance; awl-like, tan prickles; foliage leathery, glossy, medium, dark; upright, bushy growth; hardy.; [Carefree Beauty X Sunsprite]; Buck, Dr. Griffith J.; Iowa State University, 1984

'PRAIRIE HERITAGE', S, op, 1978; bud ovoid, pointed; flowers peach to coral-pink, dbl., 50 petals, 4–5 in., cupped, repeat bloom, moderate fragrance; foliage dark, leathery; vigorous, upright, spreading, bushy growth; [(Vera Dalton X Prairie Princess) X (Apricot Nectar X Prairie Princess)]; Buck, Dr. Griffith J.; Iowa State University

'PRAIRIE JOY', S, mp, 1990; flowers medium pink, aging light to medium pink, borne singly or in, dbl., 30–40 petals, sparse, repeat bloom, slight fragrance; foliage medium, medium green, matt, high resistance to powde; tall, bushy growth; [Prairie Princess X Morden Cardinette]; Collicutt, Lynn M.; Agriculture Canada, 1990

'PRAIRIE LASS', S, pb, 1978; bud ovoid, pointed; flowers claret-rose to rose-red, dbl., 28 petals, 3.5 in., repeat bloom, moderate, spicy fragrance; foliage dark, leathery; vigorous, upright, spreading, bushy growth; [(Hawkeye Belle X Vera Dalton) X (Dornroschen X (World's Fair xApplejack))]; Buck, Dr. Griffith J.; Iowa State University

'PRAIRIE MAID', S, w, 1959; flowers cream, dbl., 25 petals, intermittent bloom, moderate fragrance; compact (4 ft.) growth; hardy on prairies.; [(Ophelia X Turkes Rugosa Samling) X R. spinosissima altaica]; Morden Exp. Farm

'PRAIRIE MOON', LCl, ly, 1953; bud ovoid, deep yellow; flowers creamy yellow, large, borne in clusters of 4-5, dbl., 30–35 petals, abundant, non-recurrent bloom, moderate fragrance; foliage glossy, dark, leathery; vigorous, climbing (15 ft.) growth; [R. maximowicziana pilosa X Autumn]; Maney; Iowa State College

'PRAIRIE PINKIE', S, op, 1958; flowers deep coral-pink, dbl., midsummer bloom, intense fragrance; foliage like r spinossissima; upright (2 1/2 ft.), bushy growth; Skinner

'PRAIRIE PRINCESS', S, op, 1972; bud ovoid, long, pointed; flowers light coral-pink, large, semi-dbl., repeat bloom, slight fragrance; foliage large, dark, leathery; vigorous, upright growth; [Carrousel X (Morning Stars X Suzanne)]; Buck, Dr. Griffith J.; Iowa State University

'PRAIRIE RED', Min, rb, 1980; flowers red-yellow blend, blooms borne singly or several together, semi-dbl., 18 petals, moderate fragrance; brownish prickles curved downward; foliage tiny, medium green; bushy, upright growth; [Unnamed seedling X Unnamed seedling]; Lyon

Prairie Rose *see* **R. SETIGERA**

'PRAIRIE SAILOR', HSpn, yb, 1946; flowers golden yellow deeply edged bright red, single, profuse, non-recurrent bloom; vigorous (6 ft.) growth; hardy on prairies.; Morden Exp. Farm

Prairie Schooner™ *see* 'KINSCHOON'

'PRAIRIE SQUIRE', S, mp, 1984; dbl., 20 petals, 4–5 in., cupped, borne 1-10 per cluster, repeat bloom, slight fragrance; awl-like, tan prickles; foliage leathery, dark green, tinted copper; vigorous, erect, bushy, spreading growth; hardy.; [Countryman X Carefree Beauty]; Buck, Dr. Griffith J.; Iowa State University

'PRAIRIE STAR', S, ly, 1975; bud ovoid, pointed; flowers pale chrome-yellow, tinted pink, dbl., 54 petals, 3.5–4 in., cupped, repeat bloom, moderate, green apple fragrance; foliage dark, leathery; vigorous, erect, bushy growth; [Tickled Pink X Prairie Princess]; Buck, Dr. Griffith J.; Iowa State University

'PRAIRIE SUNSET', S, yb, 1984; flowers deep pink, reverse yellow, urn-shaped, blooms borne 5-10 per, dbl., 38 petals, 4 in., repeat bloom, moderate fragrance; tan prickles; foliage moderately large, leathery, dark; erect, bushy, slightly spreading growth; hardy.; [Bonfire Night X (Music Maker X Athlone)]; Buck, Dr. Griffith J.; Iowa State University

'PRAIRIE VALOR', S, mr, 1984; dbl., 38 petals, 4–5 in., cupped, borne 3-5 per cluster, repeat bloom, moderate, damask fragrance; awl-like, brown prickles; foliage leathery, dark; upright, bushy growth; hardy.; [((Dornroschen X (Josef Rothmund X R. laxa)) X Rose du Roi a Fleurs Pourpres) X (Music Maker X Topsi)]; Buck, Dr. Griffith J.; Iowa State University

'PRAIRIE WREN', S, mp, 1946; flowers rich pink, large, semi-dbl., heavy, non-recurrent bloom; very hardy.; Morden Exp. Farm

'PRAIRIE YOUTH', S, op, 1948; flowers pure salmon-pink, blooms in clusters, semi-dbl., repeat bloom, slight fragrance; vigorous (6 ft.) growth; completely hardy on prairies.; [((Ross Rambler X Dr. W. Van Fleet) X R. suffulta) X ((Dr. W. Van Fleet X Turkes Rugosa Samling) X R. spinosissima altaica)]; Morden Exp. Farm

'PRAISE OF JIRO', F, or, 1959; bud ovoid; flowers large, borne in clusters, dbl., 30 petals, slight fragrance; foliage leathery; vigorous, upright, bushy growth; [Korona X Spartan]; Kordes, R.

'PRALINE', HT, dr, 1955; flowers carmine-purple to crimson, slight fragrance; foliage glossy; [Camelia X Unnamed seedling]; Robichon

'PRARIE SUNRISE', S, ab, 1997; flowers very full, very dbl., 41 petals, borne in small clusters, intense fragrance; foliage large, medium green, glossy; glossy, compact, bushy, medium growth; [Friesia X Freckle Face]; Buck, Dr. Griffith J.

'PRÄSENT', F, ab, 1969; (Present); flowers yellow-orange, reverse yellow, large, dbl., 27 petals, exhibition form, moderate fragrance; foliage dark, glossy, leathery; vigorous, upright, bushy growth; [Highlight X Allgold]; Haenchen, E.; Teschendorff

'PRÄSIDENT DR H.C. SCHRÖDER', HT, mr, 1959; flowers velvety deep red, large, dbl.; strong stems; [New Yorker X Seedling]; Kordes, R.

'PRÄSIDENT HINDENBURG', Pol, mr, 1927; flowers carmine to deep red, base white, dbl.; [Greta Kluis X ?]; Bom; P. Lambert

Praterstern®, HT, yb

Prato Rosso, Min, 1984; Motta; (Cavriglia)

'PRATTIGOSA', HRg, mp, 1953; bud long, pointed, red; flowers pink, overlarge, single, slight fragrance; foliage leathery, light green; vigorous (3 1/2 ft), upright, bushy growth; [R. prattii X R. rugosa alba]; Kordes

Pray, Gr, 1987; Delforge; (Cavriglia)

'PRECILLA', HT, dy, 1973; bud ovoid; flowers deep golden yellow, medium, dbl., cupped, moderate fragrance; foliage dark, leathery; vigorous, upright growth; [Peer Gynt X Seedling]; Kordes

Preciosa, HT, pb, 1972; flowers pink and yellow, large, dbl.; Meilland; (Sangerhausen)

'PRECIOUS', HT, pb, 1985; flowers blend of light, medium and deep pink, reverse medium pink, l, dbl., 35 petals, exhibition form, moderate fragrance; very few prickles; foliage medium, medium green, semi-glossy; upright, bushy growth; [(Briarcliff X Carla seedling) X Pink Parfait]; Dawson, Charles P.

Precious Child *see* 'GEAAURA'

Precious Gift™ *see* 'JAYSIL'

Precious Michelle, HT, lp

Precious Moments *see* 'LYOPR'

'PRECIOUS PLATINUM', HT, mr, 1974; (Opa Pötschke, Red Star); flowers cardinal-red, medium to large, dbl., exhibition form, slight fragrance; foliage glossy, leathery; [Red Planet X Franklin Engelmann]; Dicksons of Hawlmark

'PRÉCOCE', M, dp, 1843; flowers rosy red, sometimes spotted, medium, dbl., very early bloom; Vibert

'PREDICTION', HT, mp, 1974; flowers luminous pink, dbl., 35 petals, 4 in., exhibition form, slight fragrance; foliage leathery; moderate growth; [Queen of

Bermuda X Golden Giant]; Golik; Dynarose

Preference *see* 'MEIMAGARMIC'

Preference *see* MEIgrouge

'PREFET LIMBOURG', HP, dr, 1878; flowers large, dbl., slight fragrance; Margottin; (Sangerhausen)

Prekrasnaja Rossijanka, HT, mr, 1966; flowers large, dbl., intense fragrance; Klimenko, V. N.; (Sangerhausen)

Prelud, F, lp; Urban, J., 1989; (Czech Rosa Club)

Prelude *see* **'SUGAR PLUM'**

Prelude *see* KEImove

'PRÉLUDE', HT, m, 1954; flowers lilac-mauve, well formed, medium, dbl., 25 petals, moderate fragrance; foliage dense; strong stems; vigorous, bushy growth; [Fantastique X (Ampere X (Charles P. Kilham X Capucine Chambard))]; Meilland, F.; URS

Preludium, F, lp; [Poker Chip X Unknown]; Hannemann, F., 1989; The Rose Paradise; (Weatherly, L.)

'PREMA', F, pb, 1970; flowers soft pink, petals edged deep pink, medium, dbl., exhibition form; foliage leathery; vigorous, upright growth; [Sea Pearl X Shola]; Division of Vegetable Crops and Floriculture

'PREMIER', HT, mr, 1918; bud long, pointed; flowers dark velvety rose-rev, veined darker, reverse lighter, open, dbl., intense fragrance; foliage rich green, leathery; vigorous growth; [Ophelia seedling X Mrs Charles E. Russell]; Hill, E.G., Co.

Premier Amour *see* **'FIRST LOVE'**

'PREMIER BAL', HT, w, 1955; flowers ivory edged cyclamen-rose, picotee effect, dbl., 45 petals, intense fragrance; vigorous, bushy growth; [(Fantastique X Caprice) X Peace]; Meilland, F.; URS

'PREMIER BAL', HT, w, 1950; flowers ivory edged carmine, dbl., 30–40 petals, cupped, intense fragrance; NOW DISCARDED; [Fantastique X Caprice]; Meilland, F.

'PREMIER SUPREME', HT, dp, 1927; bud long, pointed; flowers deep rose-pink, almost scarlet, very large, dbl., exhibition form, moderate fragrance; foliage dark, leathery; long stems; very vigorous growth; [Premier sport]; Zieger

'PREMIER, CLIMBING', Cl HT, mr, 1927; Vestal

Première Ballerine *see* **'PRIMA BALLERINA'**®

Présence *see* 'DELPRAT'

Present *see* **'PRÄSENT'**

'PRÉSENT FILIAL', HT, pb, 1956; bud long; flowers old rose tinted coppery yellow, center salmon-pink, reverse, very dbl., 85 petals; foliage dense; [Verschuren's Pink X Unnamed seedling]; Delbard-Chabert

Preservation *see* 'BOSILJURIKA'

Preservation *see* **'MISILJURIKA'**

President *see* **'ADAM'**

'PRESIDENT', T, 1860; Paul; (Cavriglia)

President Armand Zinsch *see* DELzinsch

'PRESIDENT BOONE', HT, dr, 1936; bud long, pointed; flowers scarlet-crimson, large, dbl., globular, intense fragrance; foliage leathery; stems sometimes weak; vigorous growth; [Unnamed seedling X Miss Rowena Thom]; Howard, F.H.; H&S

'PRÉSIDENT BOUCHÉ', HT, or, 1917; flowers coral-red, shaded carmine, large to medium; vigorous growth; [Unnamed seedling X Lyon Rose]; Pernet-Ducher

'PRÉSIDENT BRAY', HT, op, 1954; flowers salmon-pink veined pink; strong stems; vigorous growth; Privat

'PRÉSIDENT BRIAND', HP, op, 1929; flowers pink suffused salmon, overlarge, dbl., globular, moderate fragrance; foliage wrinkled; strong stems; vigorous, compact growth; [(Unnamed Frau Karl Druschki seedling X Lyon Rose) X (Frau KarlDruschki X Willowmere)]; Mallerin, C.; C-P

Président Charles Hain *see* **'AMELIA EARHART'**

Président Chaussé *see* **'MARK SULLIVAN'**

'PRÉSIDENT CHÉRIOUX', HT, rb, 1923; flowers red and salmon-pink, shaded yellow, dbl., slight fragrance; GM, Bagatelle, 1923; Pernet-Ducher

'PRÉSIDENT COCHET-COCHET', HT, dr, 1937; flowers deep garnet-red, tinted scarlet, very large, dbl., slight fragrance; foliage leathery; very vigorous growth; [Grenoble X Unnamed seedling]; Mallerin, C.

'PRESIDENT COOLIDGE', LCl, dr, 1925; flowers glowing crimson, dbl.; very hardy.; [(R. setigera X R. wichurana) X Château de Clos Vougeot]; Horvath

President Cosnier, HT, mp, 1994; Orard

'PRÉSIDENT DE LA ROCHTERIE', B, m, 1891; flowers dark purple/red, very large, dbl., intense fragrance; Vigneron; (Sangerhausen)

'PRÉSIDENT DE SÈZE', HGal, m; (Mme Hébert); flowers magenta center, paler at edges, cup-shaped, becoming convex, very dbl.; Hébert, Mme., 1828

'PRÉSIDENT DEVILLE', HT, mr, 1929; dbl., moderate fragrance; GM, Bagatelle, 1929; [Fritz Maydt X Mme J.W. Budde]; Leenders, M.

'PRÉSIDENT DUHEM', Pol, mr, 1930; flowers bright red, dbl.; foliage dark; vigorous growth; Reymond

'PRÉSIDENT DUTAILLY', HGal, m, 1888; flowers reddish purple, dbl., intense fragrance; Dubreuil

'PRESIDENT EISENHOWER', HT, mr, 1953; bud ovoid; flowers rose-red, dbl., 35–40 petals, 4–5 in., exhibition form, intense fragrance; foliage dark, leathery; vigorous, bushy growth; [Unnamed seedling (Captivator X Red Delicious) sport]; Hill, Joseph H., Co.; C-P

'PRESIDENT F.A. DES TOMBE', HT, ab, 1925; flowers apricot on golden yellow ground, reverse peach, dbl., moderate fragrance; [Mr Joh. M. Jolles X Golden Emblem]; Van Rossem

'PRÉSIDENT FÉRIER', HT, op, 1938; flowers reddish coppery pink, base tinted yellow, very dbl., intense fragrance; foliage bright green; very vigorous growth; Gaujard

'PRESIDENT FRANKLIN D. ROOSEVELT', HT, or, 1933; flowers velvety scarlet, large, dbl., 35–40 petals, moderate fragrance; [Templar sport]; Traendly & Schenck; S. Reynolds

President Gaupin, HT, dr

President Gausen, B, mr, 1862; Pradel

'PRÉSIDENT GEORGES FEUILLET', LCl, yb, 1954; flowers sulfur-yellow edged vermilion, reverse saffron, semi-dbl.; Vially

President Heldar Aliyev *see* 'COCOSIMBER'

'PRÉSIDENT HENRI QUEUILLE', HT, mr, 1952; flowers carmine-red, very large, dbl., intense fragrance; very vigorous, upright growth; [Rome Glory X Unnamed seedling]; Gaujard

'PRESIDENT HERBERT HOOVER', HT, pb, 1930; bud long, pointed; flowers orange, rose and gold, reverse lighter, large, dbl., 25 petals, moderate, spicy fragrance; foliage leathery; vigorous, tall growth; Gertrude M. Hubbard, ARS, 1934 John Cook Medal, ARS, 1935; [Sensation X Souv. de Claudius Pernet]; Coddington; Totty

'PRESIDENT HERBERT HOOVER, CLIMBING', Cl HT, pb, 1937; Cant, B. R.

'PRESIDENT HOOVER, CLIMBING', Cl HT, pb, 1931; Dixie Rose Nursery

'PRÉSIDENT J.B. CROIBIER', HT, ob, 1940; bud well shaped, pure orange; flowers dark orange, reverse yellow; foliage light green; very vigorous growth; [Mrs Pierre S. duPont X Talisman]; Colombier

'PRESIDENT JAC. SMITS', HT, dr, 1928; (President Jacob Smith); bud long, pointed; flowers brilliant dark red, large, semi-dbl., moderate fragrance; foliage bronze, leathery; very vigorous growth; [Étoile de Hollande X K. of K.]; Verschuren; Dreer

President Jacob Smith *see* **'PRESIDENT JAC. SMITS'**

President John F. Kennedy *see* **'JOHN F. KENNEDY'**

President Kekkenen, F, 1983; de Ruiter, G.; (Cavriglia)

President L. Senghor® *see* 'MEILUMINAC'

Président Leopold Senghor® *see* 'MEILUMINAC'

President Leopold Senghor, Climbing *see* MEIluminacsar

'PRESIDENT LINCOLN', HP, dr; flowers dark red shaded crimson, large, dbl.; Granger, 1863

'PRESIDENT MACIA', HT, lp, 1933; (P.M. Leenders); flowers light pink, darker veining, dbl., 25 petals, 6 in., intense fragrance; foliage dark; vigorous, bushy growth; [(Ophelia X Gloire de Hollande) X (Ophelia X Sensation)]; Leenders, M.

'PRÉSIDENT MAGNAUD', Ch, dr, 1916; flowers small, semi-dbl.; Nabonnand, C.

'PRÉSIDENT MOREL JOURNEL', HT, rb, 1934; flowers scarlet, reverse yellow, large, cupped, moderate fragrance; foliage bronze; vigorous growth; [Mrs Edward Powell X R. foetida bicolor hybrid]; Chambard, C.

Président Nomblot *see* **'HORACE MCFARLAND'**

'PRÉSIDENT PACAUD', HT, yb, 1946; flowers ochre-yellow shaded dark carmine-red and copper, well formed, dbl.; foliage glossy; [Mme Joseph Perraud X Unnamed seedling]; Sauvageot, H.; Sauvageot

'PRÉSIDENT PARMENTIER', HT, ab, 1926; flowers apricot-pink, dbl., slight fragrance; [Col. Leclerc X Le Progres]; Sauvageot, H.; Cochet-Cochet

President Pats *see* **'STAATSPRÄSIDENT PÄTS'**

'PRÉSIDENT PAUL MARTIN', HT, my, 1942; flowers purplish carmine-red, large, semi-dbl.; vigorous growth; [Charles K. Douglas X Souv. de Georges Pernet]; Moulin-Epinay; Hamonière

'PRÉSIDENT PAULMIER', Pol, dr, 1932; flowers pure blood-red, passing to garnet, well formed, borne in clusters, very dbl.; dwarf growth; Turbat

'PRÉSIDENT PLUMECOCQ', HT, yb, 1931; (Director Plumecock); flowers coppery buff and deep salmon, large, dbl., 34 petals, cupped, moderate, fruity fragrance; Gaujard; C-P

'PRÉSIDENT POINCARÉ', HT, 1920; flowers reddish magenta, center crimson, shaded yellow, reverse brighter, dbl., moderate fragrance; Grandes Roseraies

'PRÉSIDENT SCHLACHTER', HP, m, 1877; flowers purple/violet, large, dbl.; Verdier, E.; (Sangerhausen)

'PRÉSIDENT SEIZE', F, dp, 1958; bud long; flowers light red becoming darker, semi-dbl., 10–15 petals, 4 in.; foliage dark; strong stems; upright growth; Delbard-Chabert

President Smith *see* **'WILLIAM R. SMITH'**

'PRESIDENT SONO', HT, ly, 1972; bud pointed; flowers cream-yellow, dbl., 45–50 petals, exhibition form, intense fragrance; foliage dark; very large (5-6 in) upright growth; [Burnaby X Montparnasse]; Kikuchi, Rikichi

President Souzy *see* 'PEKOMEGIR'

President Taft *see* **'LEUCHTFEUER'**

'PRÉSIDENT VAN OOST', HT, my, 1934; flowers golden yellow, edged deeper, large, dbl.; foliage glossy; long, strong stems; vigorous growth; [Souv. de Claudius Pernet X Ville de Paris]; Lens

President Viard, HT, mr

'PRÉSIDENT VIGNET', HT, dr, 1911; flowers deep carmine-red, large, dbl.; Pernet-Ducher

'PRÉSIDENT VIGNET, CLIMBING', Cl HT, dr, 1942; flowers large, dbl., slight fragrance; Vogel, M.; (Sangerhausen)

'PRESIDENT W.H. TAFT', HT, op, 1908; bud long, pointed; flowers salmon-pink, large, dbl., exhibition form, moderate fragrance; foliage glossy; long, strong stems; vigorous growth; McCullough

President Willermoz, HP, mp; flowers birght rose; Ducher, 1867

'PRESIDENT WILSON', HT, op, 1918; flowers shrimp-pink, dbl.; Easlea

President Wilson *see* ORAwest

President Wm.R. Smith *see* **'WILLIAM R. SMITH'**

Presidente, HT, or; Croix

'PRESIDENTE CARMONA', HT, dr, 1937; flowers blackish crimson, shaded salmon, large, dbl., cupped, intense fragrance; foliage soft; vigorous, bushy growth; [Hortulanus Budde X Château de Clos Vougeot]; da Silva, Moreira

'PRESIDENTE CRAVEIRO LOPES', HT, rb, 1954; flowers cherry, reverse yellow, medium, dbl., 33 petals, moderate fragrance; foliage dark; strong stems; very vigorous, upright growth; [Sirena X Peace]; da Silva, Moreira

'PRESIDENTIAL', Gr, lp, 1960; bud long, pointed; flowers light crimson, reverse china-rose, dbl., 5.5 in., cupped, slight fragrance; foliage leathery, glossy; vigorous, tall growth; [Charlotte Armstrong X (Charlotte Armstrong X Floradora)]; Lammerts, Dr. Walter; Germain's

'PRESQUE PARTOUT', M, dp; flowers rose, medium, dbl., cupped; vigorous growth

Press and Journal *see* COClion

Pressin, HT, 1975; Dorieux; (Cavriglia)

Prestance, HT, 1985; Dorieux; (Cavriglia)

'PRESTATYN ROVER', HT, op, 1929; flowers salmon-pink, reverse darker, dbl., slight fragrance; [Alfred Colomb X Mrs Wemyss Quin]; Lavender

'PRESTIGE', S, mr, 1957; flowers light crimson, large, semi-dbl., recurrent bloom; foliage dense, dark; vigorous, bushy growth; [Rudolph Timm X Fanal]; Kordes, R.; Morse

Prestige de Bellegarde, F, mr, 1992; Eve, A.

Prestige de Lyon *see* 'MEINIMO'

Prestige de Seine-et-Marne, F, yb, 1992; Eve, A.

'PRESUMIDA', Min, yb, 1948; (Baby Talisman, La Presumida, Peter Pan); flowers pumpkin-yellow to white, center yellowish, small, dbl.; dwarf growth; [Eduardo Toda X Pompon de Paris]; Dot, Pedro

'PRETORIA', HT, or, 1953; bud long; flowers copper-red to orange, large; very vigorous growth; GM, Rome, 1953; Moro

Pretoria *see* KORhagon

'PRETTY BABY', Min, pb, 1982; flowers light pink, reverse creamy yellow, medium, dbl., 35 petals, exhibition form, slight fragrance; foliage small, medium green, semi-glossy; bushy, spreading growth; [Baby Katie X Unnamed Miniature seedling]; Jolly, Betty J.; Rosehill Farm

Pretty Celine *see* 'RENCEL'

Pretty Girl *see* **'MISS FRANCE'**

Pretty in Pink® *see* 'DICUMPTEEN'

Pretty Jessica *see* 'AUSJESS'

Pretty Lady *see* 'SCRIVO'

Pretty 'n' Single *see* 'UMSPRETTY'

Pretty Penny *see* 'MORPREPEN'

'PRETTY PINK', LCl, dp, 1968; flowers deep pink, open, medium, dbl., recurrent bloom, intense fragrance; foliage glossy; very vigorous, climbing growth; [New Dawn X Spartan]; Patterson; Patterson Roses

Pretty Pink, HMsk, pb, 1992; Lens

Pretty Pink *see* BARprett

'PRETTY PLEASE', Min, lp, 1985; flowers urn-shaped, small blooms borne usually singly, dbl., 25 petals, no fragrance; very thin, long, straight, light red prickles; foliage medium, light green, semi-glossy; medium, bushy growth; [Bride's White X Unnamed Miniature seedling]; Epperson, Richard G.

Pretty Polly *see* 'MEITONJE'

'PRETTY POLY', F, mr, 1954; flowers cherry-red, camellia shaped, borne in clusters, flat; vigorous, upright growth;

[Mme Butterfly X Our Princess]; Bishop; Baker's Nursery

Pretty Tiny *see* 'RENPINY'

Pretty Woman *see* 'MINABDCO'

Pretty Woman *see* KEBesa

'PREUSSEN', HT, dr, 1920; flowers glowing dark blood-red, dbl., intense fragrance; [Farbenkonigin X Richmond]; Löbner; Kordes

Prevalent, F; Verbeek; (Cavriglia)

Preview *see* MEIspola

'PREVUE', S, w, 1978; flowers medium, semi-dbl., no fragrance; foliage medium, medium green, semi-glossy; bushy growth; [(Tausendschon X (Perle d'Or X Old China)) X Safrano]; James, John

Preziosa *see* 'MEIHIMPER'

Pride Meidiland *see* MEIrumour

Pride 'n' Joy™ *see* 'JACMO'

'PRIDE OF CANADA', HT, dp, 1981; bud long, ovoid; flowers deep pink, blooms borne in pairs, dbl., 34 petals, exhibition form, moderate fragrance; small, brown prickles; foliage dark; tall growth; [Ena Harkness X Charlotte Armstrong]; Collins; Pan American Bulb Co.

Pride of Daylesford, Hmult, lp; Huxley, Ian, 1994; (Weatherly, L.)

Pride of England *see* HARencore

'PRIDE OF HURST', Pol, op, 1926; flowers coral-pink, small, very dbl.; [Coral Cluster sport]; Hicks

Pride of Ichalkaranji, HT, yb, 1989; Patil, B.K.

'PRIDE OF LEICESTER', HT, pb, 1960; flowers rose-pink, base yellow, large, dbl., exhibition form, intense fragrance; foliage light green; vigorous growth; [R.M.S. Queen Mary X Seedling]; Verschuren; Blaby Rose Gardens

Pride of Maldon *see* 'HARWONDER'

Pride of Mountbarker *see* 'TOMMOUNT'

Pride of Nagpur, HT, dr, 1983; Datt, Braham

'PRIDE OF NEW CASTLE', HT, dr, 1930; flowers deep velvety crimson, large, dbl., intense fragrance; vigorous, bushy growth; [Hoosier X Beauty unnamed seedling]; Hill, E.G., Co.; Heller Bros.

'PRIDE OF NEWARK', F, lp, 1966; flowers shell-pink, dbl., 4 in., cupped, intense fragrance; foliage glossy, bronze, leathery; very vigorous, upright, compact growth; [Joanna Hill X The Fairy]; Morey, Dr. Dennison; Country Garden Nursery

'PRIDE OF OAKLAND', F, mp, 1977; bud pointed; flowers rosette form, dbl., 26 petals, 2.5 in., moderate, spicy fragrance; upright, spreading growth; [Pinocchio X China Doll]; Lindquist; Howard of Hemet

'PRIDE OF PACIFIC', F, op, 1957; flowers salmon-pink, borne in clusters of 4-7, dbl.; symmetrical growth; [Pinocchio X Maxine]; Silva

Pride of Park, F, ob

'PRIDE OF REIGATE', HP, rb, 1884; flowers carmine, striped and mottled white; [Comtesse d'Oxford sport]; Brown, J.

'PRIDE OF REIGATE, CLIMBING', Cl HP, rb, 1941; flowers red with white striping, large, dbl., moderate fragrance; Vogel, M.; (Sangerhausen)

'PRIDE OF RUNCORN', HT, lp; Williams, A., 1901; (Weatherly, L.)

'PRIDE OF SUNNYBANK', Gr, mp, 1957; [Ma Perkins X Charlotte Armstrong]; Ulrick, L.W.

'PRIDE OF WALTHAM', HP, mp, 1881; flowers silvery rose, large, dbl., moderate fragrance; vigorous growth; [Comtesse d'Oxford sport]; Paul, W.

Pride of Wansbeck *see* 'NOSWAN'

'PRIDE OF WASHINGTON', HSet, m, 1849; flowers rosy violet, very dbl.; Pierce

Pride of Nagpur, HT, dr

Prima *see* 'HARWANTED'

'PRIMA BALLERINA'®, HT, dp, 1957; (Première Ballerine, Primaballerina®); bud long, pointed; flowers cherry-pink, medium to large, dbl., 20 petals, intense fragrance; foliage leathery, light green; [? X Peace]; Tantau, Math.

Prima Ballerina *see* **'SOLISTKA BALETA'**

'PRIMA DONNA'™, Gr, dp, 1984; (Toboné); flowers deep fuchsia pink, large, dbl., 27 petals, exhibition form, slight fragrance; foliage large, medium green, semi-glossy; bushy, spreading growth; AARS, 1988; [(Unnamed seedling X Happiness) X Prominent]; Shirakawa, Takeshi; Tosh Nakashima, 1983

'PRIMA DONNA', HT, mp, 1944; bud pointed; flowers rich deep salmon-pink tinted buff, dbl., 45 petals, 5 in., exhibition form, slight fragrance; foliage glossy; vigorous, upright, open growth; RULED EXTINCT 1/85; [Heinrich Wendland X Unnamed seedling]; Dickson, A.; J&P

Primaballerina® *see* **'PRIMA BALLERINA'**®

'PRIMAROSA', HT, dp, 1950; bud long; flowers carmine, base yellow, streaked orange, very large; strong stems; Giacomasso

'PRIMAVERA', HT, op, 1936; flowers salmon-pink, well formed, dbl., moderate, musk fragrance; vigorous growth; [Julien Potin X Sensation]; Aicardi, D.; Robichon

'PRIME TIME', Min, my, 1984; flowers medium, dbl., 35 petals, exhibition form, slight fragrance; foliage small, medium green, semi-glossy; upright, bushy growth; [Picnic X Rise 'n' Shine]; Hardgrove, Donald & Mary; Rose World Originals

'PRIMEROSE', HT, my, 1913; flowers melon-yellow, deeper in autumn, shaded apricot, dbl., moderate fragrance; [Mme Mélanie Soupert X Mrs Peter Blair]; Soupert & Notting

'PRIMEROSE SISTAU', Pol, pb, 1925; flowers carmine, shaded yellow, medium, borne in clusters of 5-6, cupped; few thorns; half dwarf growth; Turbat

Primevere, LCl, yb

'PRIMEVÈRE', LCl, yb, 1929; (Primrose); flowers primrose-yellow to canary-yellow, large blooms in clusters, dbl., non-recurrent, slight fragrance; foliage rich green, glossy; long stems; very vigorous, climbing and trailer growth; [R. wichurana X Constance]; Barbier; Dreer, 1930

Primo Passo *see* BARprimo

'PRIMO PREMIO', HT, pb, 1964; flowers rose suffused yellow, well formed, medium; Giacomasso

Primo Sole® *see* BARpris

Primrose *see* **'PRIMEVÈRE'**

'PRIMROSE BEDDER', F, my, 1956; flowers primrose-yellow, borne in large trusses, dbl., slight fragrance; foliage light green; Kordes; Morse

'PRIMROSE PET', Pol, lp; Williams, A., 1913; (Weatherly, L.)

Primrose Queen, HT, my

Primrose Sistau, Pol, mp

Primula, HT, 1956; Aicardi; (Cavriglia)

Prince Abricot *see* 'AROSNAP'

'PRINCE ALBERT', HP, rb, 1841; flowers carmine-rose changing to dark crimson, large, globular; Laffay, M.

Prince Albert, B, lp; Laffay, M., 1852

'PRINCE ARTHUR', HP, mr, 1875; (Triomphe de Caen); flowers deep crimson, medium, dbl., 55 petals, intense fragrance; vigorous growth; [Général Jacqueminot X ?]; Cant, B. R.

'PRINCE BERNHARD', HT, mr, 1937; bud pointed; flowers shining red, shaded strawberry-red, well formed, large, dbl., intense fragrance; foliage dark; vigorous growth; GM, Bagatelle, 1937; [Matador seedling]; Van Rossem; J&P, 1941

'PRINCE CAMILLE DE ROHAN', HP, dr, 1861; (La Rosière, Climbing); flowers very deep velvety crimson-maroon, well-formed, large blooms, very dbl., cupped, sometimes recurrent bloom, intense fragrance; rather weak stems; vigorous, upright growth; [Possibly

Général Jacqueminot X Geant des Batailles hybrid]; Verdier, E.

'PRINCE CHARLES', B, m, 1842; flowers deep red-purple, fading to lavender, veined magenta, base of, semi-dbl., non-recurrent, intense fragrance; very few prickles; foliage large, dark; vigorous (to 5 ft.) growth

Prince Charles, Gr, dr, 1992

Prince Charles d'Aremberg, HP, dp, 1877; Soupert & Notting

'PRINCE CHARLES DU LUXEMBOURG', Ch, dp; flowers bright carmine, cupped

'PRINCE CHARLIE', HT, op, 1932; bud long, pointed; flowers coral-pink, overlaid salmon, base orange, large, dbl., cupped, moderate fragrance; foliage leathery, bronze tinted; vigorous, bushy growth; Dobbie

'PRINCE CHARMING', Min, mr, 1953; flowers bright crimson, dbl., 1 in.; foliage tinted red; dwarf (8-12 in) growth; [Ellen Poulsen X Tom Thumb]; deVink; T. Robinson, Ltd.

'PRINCE CHARMING', S, lp, 1958; flowers pale blush-pink, dbl., prolonged bloom, intense fragrance; upright (2 1/2 ft), bushy growth; Skinner

'PRINCE DAMASK', HT, dr, 1954; flowers maroon, medium, dbl., 30 petals, intense fragrance; foliage small, dark; [Guinee X ?]; LeGrice

'PRINCE DE BULGARIE', HT, lp, 1900; bud long, pointed; flowers silvery flesh, center deeper, shaded salmon and saffron-yellow, dbl., cupped, moderate fragrance; foliage bright green; vigorous growth; Pernet-Ducher

Prince de Monaco *see* 'POULART'

'PRINCE DE PORCIA', HP, dr, 1865; flowers large, dbl.; Verdier, E.; (Sangerhausen)

'PRINCE DE VAUDEMENT', M, 1854; Moreau et Robert; (Cavriglia)

'PRINCE ENGELBERT CHARLES D'ARENBERG', HT, or, 1909; flowers scarlet, dbl., moderate fragrance; [Richmond X Étoile de France]; Soupert & Notting

Prince Eugène *see* **'EUGÈNE DE BEAUHARNAIS'**

Prince Eugene de Beauharnais, HP, dr, 1864; Moreau et Robert

'PRINCE FÉLIX DE LUXEMBOURG', HT, mr, 1930; flowers carmine-red, shaded purplish, large, dbl., 30–35 petals, moderate fragrance; foliage dark; very vigorous, bushy growth; [Gen. MacArthur X George C. Waud]; Ketten Bros.

Prince Frédéric *see* **'PRINCE FREDERICK'**

'PRINCE FREDERICK', HGal, m, 1840; (Prince Frédéric); flowers large, very dbl.; Parmentier; (Sangerhausen)

'PRINCE HENRI', F, mr, 1954; flowers large, dbl.; Reuter, A.; (Sangerhausen)

'PRINCE HENRI D'ORLÉANS', HP, dp, 1886; flowers large, dbl.; Verdier, E.; (Sangerhausen)

'PRINCE HENRY', F, mp, 1926; bud long, pointed; flowers bright pink, passing to blush-pink, clusters of 18-20; [St. Helena X Unnamed seedling]; Easlea

Prince Igor® *see* 'MEIHIGOR'

Prince Igor, Climbing *see* MEIgosar

'PRINCE JEAN DE LUXEMBOURG', Pol, w, 1926; flowers pure white, very small, borne in immense clusters of 100-150, dbl.; moderately dwarf growth; [Jeanny Soupert X Miniature]; Soupert & Notting

Prince Meillandina® *see* 'MEIRUTRAL'

Prince Napoleon, B, pb, 1864; Pernet

'PRINCE NOIR', HP, m, 1854; flowers dark velvety crimson-purple, not, very dbl.; moderate growth; Boyau

'PRINCE OF DENMARK', HT, mp, 1964; flowers rose-pink, dbl., 4 in.; free growth; [Queen Elizabeth X Independence]; McGredy, Sam IV; Fisons Horticulture

Prince of Orange, HT, ob; Williams, A., 1900

Prince of Peace *see* 'JACPOP'

'PRINCE OF WALES', HT, mr, 1921; bud long, pointed; flowers brilliant cherry-scarlet, semi-dbl., intense fragrance; Easlea

Prince Orange, F, op

Prince Palace *see* POUlzin

Prince Philip *see* **'TUDOR PRINCE'**

Prince Regent *see* GENpen

Prince Sunblaze *see* 'MEIRUTRAL'

Prince Tango® *see* 'DELGUS'

Prince Wasiltchikoff *see* **'DUCHESS OF EDINBURGH'**

'PRINCE YUGALA', HT, dr, 1923; bud long, pointed; flowers deep velvety maroon, dbl., slight fragrance; Cant, F.

'PRINCEPS', LCl, mr, 1942; flowers very large, non-recurrent, moderate fragrance; pillar growth; Clark, A.

Princesa Carmen de Borbon, F, 1979; Dot, Simon; (Cavriglia)

Princesa Sofia, HT, 1972; Dot, Simon; (Cavriglia)

Princesita *see* **'PIXIE'**

Princess® *see* 'INTERPRINCE'

'PRINCESS ALEXANDRA', F, yb, 1962; flowers creamy yellow suffused carmine-pink, medium, borne in trusse, dbl., 30–35 petals, moderate fragrance; foliage dark, glossy; very vigorous growth; [Masquerade X Seedling]; Cobley

Princess Alexandra *see* POUldra

Princess Alice *see* 'HARTANNA'

'PRINCESS ANGELA', HT, lp, 1991; flowers large blooms bornemostly singly, very dbl., slight fragrance; foliage medium, dark green, semi-glossy; tall, upright, bushy growth; [Pristine X Granada]; Alde, Robert O.

'PRINCESS ANGELINE', HT, mp, 1945; flowers pink, becoming open, dbl., 35–50 petals, 4.5–5.5 in., cupped, intense, damask fragrance; foliage leathery; vigorous, tall growth; [Charlotte Armstrong X Times Square]; Swim, H.C.; Peterson & Dering

'PRINCESS BONNIE', HT, dr; flowers vivid crimson; [Bon Silène X William Francis Bennett]; Dingee & Conard, 1895

'PRINCESS CHICHIBU', F, pb, 1971; dbl., 30 petals, 2.5–3 in., slight fragrance; foliage glossy, dark; [(Vera Dalton X Highlight) X Merlin]; Harkness

'PRINCESS ELIZABETH OF GREECE', HT, my, 1926; flowers golden yellow shaded terra-cotta, well shaped, exhibition form, moderate fragrance; Chaplin Bros.

'PRINCESS FAIR', HT, ab, 1962; flowers light apricot, large, dbl., cupped, moderate fragrance; foliage glossy; vigorous, upright growth; [Queen Elizabeth X (Crimson Glory, Climbing X Happiness)]; Moren; J&P

Princess Grace *see* 'MEIMAGARMIC'

'PRINCESS HOHENZOLLERN', T, rb; flowers peach-red to crimson, well formed, very large, dbl.

Princess Margaret, HT, or; Poulsen

Princess Margaret of England *see* 'MEILISTA'

'PRINCESS MARGARET OF ENGLAND, CLIMBING', Cl HT, mp, 1969; Meilland; URS

'PRINCESS MARGARET ROSE', HT, pb, 1933; bud long, pointed; flowers glowing pink suffused orange, large, dbl., cupped, moderate fragrance; foliage leathery; vigorous growth; [Unnamed seedling of Los Angeles parentage X Unnamed seedling]; Cant, B. R.

Princess Marianna™ *see* 'POULUSA'

'PRINCESS MARIE', HSem, 1829; Jacques; (Cavriglia)

'PRINCESS MARINA', HT, ab, 1938; flowers apricot, shaded salmon and copper, well shaped, dbl., slight fragrance; stiff stems; vigorous growth; Robinson, H.; Port Stockton Nursery

'PRINCESS MAY', HT, mp, 1893; flowers large, dbl., moderate fragrance; Paul, W.; (Sangerhausen)

Princess Michael of Kent® *see* 'HARLIGHTLY'

'PRINCESS MICHIKO', F, ob, 1966; flowers coppery orange, yellow eye, in clusters, semi-dbl., 15 petals, 3 in., cupped;

foliage glossy; bushy growth; [Circus X Spartan]; Dickson, A.

'PRINCESS MIKASA', HT, dp, 1983; flowers deep pink, urn-shaped, blooms borne singly, dbl., 45 petals, no fragrance; small, light green prickles; foliage large, medium green, semi-glossy; tall, bushy growth; [(Red Lion X Samantha) X (Red Lion X Samantha)]; Teranishi, K.; Itami Rose Nursery

'PRINCESS NAGAKA', HT, rb, 1922; flowers fiery red, shaded yellow, small, semi-dbl., slight fragrance; Pemberton

'PRINCESS OF INDIA', HT, rb, 1980; (Indian Princess); flowers outer petals deep red, inner petals carmine, large, dbl., 35 petals, exhibition form, intense fragrance; brown to gray prickles; foliage medium to large, dark, smooth; upright, bushy growth; [Tropicana X Granada]; Pal, Dr. B.P.; K.S.G. Son's Roses

Princess of Monaco *see* 'MEIMAGARMIC'

Princess of Nassau *see* **'PRINCESSE DE NASSAU'**

Princess of Wales, HP, lp; Laxton, 1871

Princess of Wales *see* HARdinkum

'PRINCESS OF WALES', HP, lp; flowers vivid crimson; Paul, W., 1864

'PRINCESS PEARL', F, w, 1960; flowers white center blush, borne in clusters, dbl., 40 petals, 3 in., slight fragrance; foliage leathery, dark; moderate growth; [Carol Amling sport]; Beldam Bridge Nursery

'PRINCESS ROYAL', HT, mp, 1935; flowers rose-pink to hydrangea-pink, petals shell-shaped, very large, intense fragrance; foliage dark, glossy; vigorous growth; RULED EXTINCT 4/92; Dickson, A.

Princess Royal *see* 'DICROYAL'

'PRINCESS TAKAMATSU', HT, pb, 1974; bud globular; dbl., 50 petals, 6 in., exhibition form, slight fragrance; vigorous, upright growth; [Bonsoir X Christian Dior]; Kono, Yoshito

'PRINCESS VAN ORANGE', Cl Pol, or, 1935; (Princesse d'Orange, Prinses van Orange); [Gloria Mundi, Climbing sport]; deRuiter; J&P

Princess Verona *see* **'PRINCESSE VERONA'**

'PRINCESS VICTORIA', HT, or, 1920; flowers glowing scarlet-crimson shaded orange, passing to carmine, b, dbl., intense fragrance; GM, NRS, 1920; McGredy

'PRINCESS WHITE', F, w, 1956; bud pointed; flowers medium, dbl., flat, moderate fragrance; foliage leathery; strong stems; vigorous, upright, bushy growth; [Unnamed orange seedling X Demure]; Boerner; J&P

'PRINCESSE', HT, or, 1964; flowers geranium-red, dbl., 45 petals, 4–5 in., globular; vigorous, upright growth; [(Peace X Magicienne) X (Independence X Radar)]; Laperrière; EFR

'PRINCESSE ADÉLAIDE', M, lp, 1845; flowers pale pink, dbl., moderate fragrance; foliage dark, often variegated; vigorous growth; Laffay, M.

Princesse Alexandra *see* POUldra

'PRINCESSE ALICE', M, m, 1853; flowers violet-rose, not very mossy; [Luxembourg seedling]; Paul, W.

'PRINCESSE ALICE DE MONACO', T, yb, 1893; flowers cream-yellow edged pink; Weber

'PRINCESSE AMÉDÉE DE BROGLIE', HT, or, 1936; bud very long, fiery red; flowers nasturtium-red, deeper in autumn, well formed, large; foliage dark, glossy; vigorous growth; GM, Bagatelle, 1935; [Charles P. Kilham seedling X Colette Clément]; Mallerin, C.; Meilland

'PRINCESSE BACCHIOCHI', mp, 1866; flowers medium, dbl.; Moreau et Robert; (Sangerhausen)

Princesse Béatrice *see* **'PRINSES BÉATRIX'**

'PRINCESSE DE BASSARABA DE BRANCOVAN', T, pb, 1900; flowers flesh-pink, shaded copper; Bernaix, A.

'PRINCESSE DE BÉARN', HP, mr, 1885; flowers velvety poppy-red, large, dbl., globular; vigorous growth; Lévêque

Princesse de Joinville, B, mp

Princesse de Lamballe *see* **'PRINCESSE LAMBALLE'**

Princesse de Monaco® *see* 'MEIMAGARMIC'

'PRINCESSE DE NASSAU', N, ly, 1835; (Princess of Nassau); bud yellowish, tinted pink; flowers creamy yellow, fading to cream, blooms in large clusters, very dbl., cupped; branches grow in zig-zag fashion growth; Laffay, M.

Princesse de Parme, M, lp; [William Lobb X Unknown]; Morley, Dr B., 1988; (Weatherly, L.)

'PRINCESSE DE SAGAN', Ch, dr, 1887; flowers crimson shaded with purple, medium, dbl., cupped; vigorous growth; Dubreuil

Princesse de Vaudemont, lp; flowers medium, dbl., moderate fragrance; Vibert, about 1825; (Sangerhausen)

Princesse des Roses *see* **'PRINCIPESSA DELLE ROSE'**

Princesse d'Orange *see* **'PRINCESS VAN ORANGE'**

Princesse d'Orient, HT; Poulsen; (Cavriglia)

'PRINCESSE ÉTIENNE DE CROY', T, yb, 1898; flowers violet rose, dbl.; Ketten Bros.

'PRINCESSE GHIKA', T, mr, 1922; flowers brilliant red with dark reflexes, large, dbl.; [Gén. Schablikine X Papa Gontier]; Nabonnand, P.

'PRINCESSE ITA', HT, yb, 1943; bud oval; flowers indian yellow edged red, open, medium, semi-dbl., slight fragrance; foliage leathery; vigorous, bushy growth; [Julien Potin X Charles P. Kilham]; Meilland, F.; A. Meilland

'PRINCESSE JAUNE', HT, my, 1945; flowers citron-yellow, large, dbl.; Fessel; A. Meilland

'PRINCESSE JOSÉPHINE-CHARLOTTE', Pol, op, 1945; flowers bright pink suffused apricot-salmon, small, semi-dbl., cupped, slight fragrance; very vigorous, bushy growth; [Orange Triumph sport]; Lens

'PRINCESSE L. TROUBEZKOI', HP, 1878; Lévêque; (Cavriglia)

'PRINCESSE LAMBALLE', A, w; (Princesse de Lamballe); flowers white, sometimes tinted flesh, medium, dbl.; vigorous, compact, branching growth

'PRINCESSE LILIANE', HT, mr, 1954; flowers blood-red, dbl., 25–30 petals, 5 in.; foliage dark; vigorous, upright growth; [Happiness X Hens Verschuren]; Frères, Buyl; Morse

'PRINCESSE LOUISE', HSem, w, 1829; flowers creamy white, back petals shaded with rose, large, dbl., cupped; Jacques

'PRINCESSE LOUISE', HP, w, 1869; flowers blush, medium, globular; vigorous growth; [Mme Vidot X Virginal]; Laxton; G. Paul

'PRINCESSE LOUISE', Cl HT, m, 1924; bud long, pointed; flowers rich purple, center tinted brilliant garnet, large, semi-dbl., moderate fragrance; vigorous, climbing growth; [La France de '89 X Victor Hugo]; Nabonnand, P.

Princesse Margaret d'Angleterre® *see* 'MEILISTA'

'PRINCESSE MARGARETHA', Pol, lp, 1932; flowers pale pink, blooms in clusters, dbl., no fragrance; foliage medium, medium green, semi-glossy; medium, bushy growth; Poulsen, S.

'PRINCESSE MARIE', HSem, mp, 1829; flowers bright pink, blooms in very large clusters; Jacques

Princesse Marie Adelaide de Luxembourg, Pol, w, 1895; Soupert & Notting

'PRINCESSE MARIE CLOTILDE NAPOLÉON', Pol, w, 1924; flowers white shaded pink, base pale yellow, large, dbl.; vigorous growth; Opdebeeck

'PRINCESSE MARIE DAGMAR', T, 1919; Lévêque; (Cavriglia)

'PRINCESSE MARIE DOLGOROUKY', HP, lp, 1878; flowers light pink, often striped with white, very large, dbl.; Gonod; (Sangerhausen)

'PRINCESSE MARIE JOSÉ', Pol, w, 1924; flowers white shaded rose, base yellow, large, dbl.; vigorous growth; Opdebeeck

'PRINCESSE MARIE JOSÉ', HT, or, 1925; bud long, pointed; flowers orange-scarlet, dbl., moderate, fruity fragrance; Klettenberg-Londes; F.J. Grootendorst

'PRINCESSE MARIE-ASTRID', HT, mp, 1964; bud globular; flowers deep rose to camellia-pink, large, dbl., slight fragrance; moderate, bushy growth; [(Mme Edouard Herriot X R. rugosa rubra) X La Jolie]; Mondial Roses

'PRINCESSE MARIE-CHRISTINE', HT, dr, 1955; bud ovoid to long; flowers blood-red, dbl., 25–30 petals; vigorous, bushy growth; [Poinsettia X Ena Harkness]; Frères, Buyl

'PRINCESSE PAOLI', HT, mp, 1966; flowers bright pink, dbl., 5 in., exhibition form, intense fragrance; foliage glossy; vigorous growth; Mondial Roses

Princesse Queen, Pol, pb

'PRINCESSE STÉPHANIE', T, op, 1880; flowers salmon-yellow, large, dbl., moderate fragrance; very vigorous growth; [Gloire de Dijon seedling]; Levet, A.

'PRINCESSE STÉPHANIE DE BELGIQUE', HT, dp, 1929; bud long, pointed; flowers carmine, center deeper, dbl., intense fragrance; [Gen. MacArthur X Unnamed seedling]; Soupert & Notting; C. Soupert

'PRINCESSE VERA ORBELIONI', HT, op, 1909; flowers light salmon-pink, large, dbl.; Schwartz; (Sangerhausen)

'PRINCESSE VERONA', S, mp, 1984; (Princess Verona); dbl., 30 petals, 4 in., cupped, borne 1-15 per cluster, repeat bloom, slight fragrance; awl-like, tan prickles; foliage leathery, dark olive green; upright, bushy, compact growth; hardy.; [Verona X Prairie Princess]; Buck, Dr. Griffith J.; Iowa State University

'PRINCESSE YVONNE GHIKA', HT, w, 1927; flowers white, center salmon, dbl., moderate fragrance; [Stadtrat Glaser seedling]; Mühle

'PRINCIPE DE NAPOLI', HT, my, 1937; bud long; flowers large, dbl., intense fragrance; foliage clear green; vigorous growth; Aicardi Bros.

'PRINCIPE DE PIEMONTE', HT, dr, 1929; flowers crimson-red, large, intense fragrance; [Mrs Edward Powell X Gen. MacArthur]; Giacomasso

'PRINCIPESSA DELLE ROSE', HT, mr, 1953; (Princesse des Roses); bud long, pointed; flowers lilac-rose, large, dbl., 30–40 petals, cupped, moderate fragrance; very vigorous, upright growth; [Julien Potin X Sensation]; Aicardi, D.; V. Asseretto

Prins Claus *see* 'RUPRINS'

'PRINS HAMLET', HT, lp, 1927; bud long, pointed; flowers light pink, base yellow, open, large, semi-dbl., exhibition form, intense fragrance; foliage light, leathery; vigorous, bushy growth; [Ophelia sport]; Mohr

'PRINS WILLEM-ALEXANDER', F, ob, 1970; flowers coral-vermilion, 2–3 in.; foliage dark, red when young; bushy growth; [Tropicana X Europeana]; Verschuren

'PRINSES BÉATRIX', HT, ob, 1940; (Princesse Béatrice); bud long, pointed; flowers terra-cotta, shaded light apricot, large, dbl., cupped, intense, fruity fragrance; foliage leathery, bronze; vigorous, bushy growth; [Heinrich Wendland X Max Krause]; Busiman; A. Meilland;, Morse

'PRINSES CHRISTINA', F, mr, 1945; bud large; flowers clear carmine-red, borne in clusters, dbl.; very vigorous growth; [Lafayette X Donald Prior]; Buisman, G. A. H.

'PRINSES JULIANA', HT, dr, 1918; bud long, pointed; flowers deep crimson-red, shaded darker, dbl., intense fragrance; foliage dark; vigorous growth; [Gen. MacArthur X Marie van Houtte]; Leenders, M.

Prinses van Orange *see* **'PRINCESS VAN ORANGE'**

'PRINSESSE ASTRID AF NORGE', F, ob, 1958; flowers bright orange, semi-dbl.; vigorous, upright growth; [Pinocchio X Pinocchio]; Poulsen, S.

'PRINSESSE MARGRETHE', HT, op, 1963; flowers dark salmon-orange, dbl., slight fragrance; strong stems; vigorous, upright growth; [Queen Elizabeth X (Independence X Golden Scepter)]; Poulsen, S.

'PRINTEMPS', HT, pb, 1948; bud long; flowers old-rose tinted light red, reverse yellow, dbl., 4 in., moderate fragrance; foliage glossy, dark; very vigorous growth; [Trylon X Brazier]; Mallerin, C.; URS

'PRINTEMPS FLEURI', HMult, m, 1922; flowers bright purple passing to carmine-pink, stamens yellow, mediu, semi-dbl., slight fragrance; [Étoile Luisante X Unnamed seedling]; Turbat

'PRINZ HAMLET', HT, lp, 1927; flowers large, dbl., moderate fragrance; Mohr; (Sangerhausen)

'PRINZ HIRZEPRINZCHEN', HMult, m, 1912; flowers medium, dbl., moderate fragrance; Geschwind, R.; (Sangerhausen)

'PRINZ MAX ZU SCHAUMBURG', HT, op, 1934; flowers salmon-pink, large, dbl., intense fragrance; vigorous growth; [Frau Karl Druschki X Lyon Rose]; Schaumburg-Lippe

'PRINZESSIN BATHILDIS ZU SCHAUMBURG-LIPPE', HT, lp, 1929; flowers large, dbl.; von Württemberg, Herzogin Elsa; (Sangerhausen)

'PRINZESSIN HILDEGARD', HT, my, 1917; flowers bright yellow, fading to cream-yellow, dbl., moderate fragrance; [Frau Karl Druschki X Friedrich Harms]; Lambert, P.

'PRINZESSIN IRRLIEB', F, mr, 1963; flowers velvety bright red, borne in large clusters, dbl.; moderate, symmetrical growth; Kordes, R.

'PRINZESSIN LUDWIG VON BAYERN', HMult, mp, 1911; flowers medium, dbl.; Brög; (Sangerhausen)

'PRINZESSIN M. VON ARENBERG', HT, w, 1928; flowers rosy white, shaded pale rose-pink, dbl., intense fragrance; [Ophelia X Los Angeles]; Leenders, M.

Prinzessin Maria Teresa, F, lp, 1982; flowers medium, dbl.; Lens, Louis; (Sangerhausen)

'PRINZESSIN TATIANA WASILTCHIKOFF', HT, lp, 1941; flowers flesh-pink, dbl.; long stems; upright growth; Spath

'PRIOR M. OBERTHAU', HMult, dp, 1923; flowers small, dbl.; Bruder Alfons; (Sangerhausen)

Priory Pride *see* 'NOSTPRI'

Priory Rose *see* 'NOSTROS'

'PRISCILLA', HT, mp, 1922; bud long, pointed; flowers pink, outer petals rose-pink, very large, dbl., cupped, moderate fragrance; foliage leathery, glossy; vigorous growth; [Unnamed seedling X Ophelia]; Montgomery Co.; A.N. Pierson

Priscilla Burton® *see* 'MACRAT'

'PRISSY MISSY', Min, mp, 1965; flowers medium pink, reverse lighter, small, very dbl., moderate, spicy fragrance; vigorous, bushy, dwarf growth; [Spring Song X Seedling]; Williams, Ernest D.; Mini-Roses

Pristine® *see* 'JACPICO'

Pristine Pavement, S, w, 1990; semi-dbl., moderate fragrance; foliage glossy; compact, mounding (3 ft) growth; Baum

Priti, Pol, lp, 1971; Kasturi

Priub *see* 'UMSPRIUB'

Prive™, Gr, mp

'PRIVET IZ ALMA-ATY', HT, op, 1958; (Greetings from Alma-Aty); flowers pink tinted orange, large, dbl., 60 petals, moderate fragrance; foliage dark, glossy; very vigorous growth; [Independence X Peace]; Sushkov & Besschetnova

Priyadarshini, HT, pb, 1988; IARI

'PRIYATAMA', HT, pb, 1981; (The Beloved); bud ovoid; flowers pink, paler reverse, blooms borne singly, dbl., 35 petals, exhibition form; red prickles; foliage glossy, slightly wrinkled; bushy growth; [Inge Horstmann X Picasso]; Viraraghavan, M.S.; K.S.G. Son's Roses

'PROBERG', F, pb, 1995; (**Pink Iceberg**™); flowers medium pink blend, darkens in cool weather, dbl., 15–25 petals, slight; large, light green, glossy; upright, rounded, medium; [Iceberg sport]; Weatherly, Lila, 1998; Swane's Nursery, 1999, AUS;, Week's Roses, 1999, USA

'PROBRIL', F, pb, 1999; (**Brilliant Pink Iceberg**™); flowers deep pink, reverse white, dbl., 17–25 petals, 3.5–4 in., borne in large clusters, moderate fragrance; few prickles; foliage large, light green, glossy; upright, rounded, medium (30-36 in.) growth; [Pink Iceberg sport]; Weatherly, Lila; Swane Bros. Pty. Ltd., 1999

'PROBUZENI', S, lp, 1935; flowers medium, very dbl., moderate fragrance; Böhm, J.; (Sangerhausen)

'PRODANÁ NEVESTA', S, w, 1934; flowers snow-white, center tinted yellow, borne inlarge clusters, non-recurrent, intense fragrance; foliage light, glossy; vigorous growth; Brada, Dr.; Böhm

Prodigieuse, HT, 1969; Croix; (Cavriglia)

Prof O. L. Kunz, S, mp, 1980; Hetzel

Prof Reynaldo Santos, HT; Moreira da Silva, A.; (Cavriglia)

'PROF. ALFRED DUFOUR', F, mr, 1969; flowers bright red, medium, semi-dbl., slight fragrance; foliage light green; compact growth; [Paprika X Coup de Foudre]; Cazzaniga, F. G.

'PROF. BENTO CARQUEJA', HT, op, 1936; bud long, pointed; flowers pink and salmon, shaded coral-red, large, dbl., exhibition form; foliage soft; vigorous growth; [Ophelia X Mme Edouard Herriot]; da Silva, Moreira

'PROF. C.S. SARGENT', LCl, my, 1903; flowers yellow, center deeper, fading to cream, dbl., early bloom; foliage small, ornamental; vigorous growth; [R. wichurana X Souv. d'Auguste Metral]; Hoopes, Bro. & Thomas

'PROF. CHRIS BARNARD', HT, mr, 1970; flowers blood-red, pointed, dbl., 40 petals, 4 in., slight fragrance; foliage glossy; upright growth; [Ena Harkness X Karl Herbst]; Fishner, P.; Eden Rose Nursery

'PROF. COSTA LEITE', HT, my, 1955; very vigorous growth; [Peace X Julien Potin]; da Silva, Moreira

'PROF. DR VON BECK', HT, pb, 1927; flowers bright rose-pink, reverse carmine-red, dbl., moderate fragrance; [Mme Abel Chatenay X Farbenkonigin]; Ries

'PROF. LEITE PINTO', HT, mr, 1960; flowers cherry-red, center ochre; [Buccaneer X La Jolla]; da Silva, Moreira

'PROF. N.E. HANSEN', HRg, dr, 1892; flowers rich velvety red; Budd

'PROFESSEUR BÉRARD', HT, dp, 1930; flowers bright purple-carmine, base yellow, petals laciniated, dbl., intense fragrance; foliage leathery, dark; very vigorous growth; [Hadley X The Queen Alexandra Rose]; Laperrière

Professeur Christian Cabrol, HMsk, pb, 1993; Lefebvre

'PROFESSEUR DÉAUX', HT, ly, 1935; flowers light yellow streaked chamois, dbl.; foliage glossy, dark, bronze; long stems; very vigorous growth; Pernet-Ducher; Gaujard

'PROFESSEUR ÉMILE PERROT', D, mp; Brought from Persia by Prof. Perrot, 1930; Turbat

'PROFESSEUR GANIVIAT', T, pb, 1890; flowers salmon carmine, shaded carmine purple, large, very dbl.; Perrier

Professeur Jean Bernard® *see* 'DELJABER'

'PROFESSOR BARANOV', HT, yb, 1947; flowers large, dbl.; Vogel, M.; (Sangerhausen)

Professor Boesman, F, mp, 1994; dbl.; GM, Courtrai, 1992; RvS-Melle, 1991

'PROFESSOR BORZA', HT, or, 1940; flowers medium, dbl.; Ambrosi; (Sangerhausen)

'PROFESSOR DR. HANS MOLISCH', Cl HT, mp, 1923; flowers large, dbl.; Mühle; (Sangerhausen)

Professor Dr. Kurth, HT, mr, 1960; flowers large, dbl.; Schmid, P.; (Sangerhausen)

'PROFESSOR ERICH MAURER', Cl HT, mp, 1939; flowers large, dbl.; Tepelmann; (Sangerhausen)

Professor Fred Ziady *see* 'KORAMBO'

'PROFESSOR GNAU', HT, w, 1928; bud long, pointed; flowers creamy white, intense fragrance; [Oskar Cordel X Unnamed seedling]; Tantau

'PROFESSOR IBRAHIM', LCl, mp, 1937; flowers rose-pink, base yellow, large blooms in clusters of 4-5, intense fragrance; very vigorous, climbng (over 13 ft) growth; [Daisy Hill X Talisman]; Krause

Professor Knöll, HT, mr, 1964; flowers very large, very dbl., intense fragrance; Berger, W.; (Sangerhausen)

Professor Panalla, HT, 1988; Dot, Simon; (Cavriglia)

'PROFESSOR SCHMEIL', HFt, yb, 1925; flowers light orange-yellow, large, semi-dbl.; Kröger; (Sangerhausen)

Profil, HT, op, 1985; flowers coral pink, large, dbl., slight fragrance; VEG; (Sangerhausen)

Profile *see* 'MACJILLI'

Profondo Rosso®, F, dr

Profumo Romano, HT; Zandri, R.; (Cavriglia)

'PROFUSION', HT, dp, 1939; bud long, pointed; flowers carmine, base orange-yellow, dbl., moderate fragrance; vigorous growth; Dickson, A.

'PROFUSION', HT, op, 1944; bud very long; flowers orange-salmon and carmine; foliage glossy; [Mme Henri Guillot X Signora]; Meilland, F.

Progress *see* **'FORTSCHRITT'**

Prohadka Maje, HT; (Cavriglia)

'PROLET', S, pb, 1975; flowers brick, shaded deep pink, medium, very dbl., 85 petals, non-recurrent; foliage light green; vigorous, upright, branching growth; [General Stefanik X Bonn]; Staikov, Prof. Dr. V.; Kalaydjiev and Chorbadjiiski

'PROLIFERA DE REDOUTÉ', C, mp

'PROLIFÈRE', M, dp; flowers deep-rose, too full to open well at times, large, cupped; vigorous growth; (England), Prior to 1800

Prom Date *see* 'MORPROM'

Prom Night *see* 'ZIPPRO'

Promenade *see* 'LYOPRO'

'PROMETHEAN', S, rb, 1985; flowers medium red, reverse darker, white eye, medium, semi-dbl., 12 petals, repeat bloom, slight fragrance; foliage medium, light green, smooth, matt; bushy growth; [(Blanche Mallerin X Pink Hat) X R. multibracteata]; James, John

Prominent® *see* 'KORP'

Promise *see* 'JACIS'

'PROMISE', Cl Ch, op, 1929; (Chin Chin); flowers salmon-pink, stamens bright yellow, borne in large clusters, single, non-recurrent, moderate fragrance; foliage almost evergreen; vigorous, climbing growth; [Mme Eugene Resal sport]; Cant, F.

Promise Me *see* 'JELWHITE'

'PROMOTION', F, mp, 1966; bud ovoid; flowers pink, medium, borne in clusters, dbl., moderate fragrance; foliage dark; [Seedling X Miracle]; Verbeek

Prophecy *see* 'MINPROP'

Prophyta™, HT, ab; de Ruiter, G.

Proserpine, B, m, 1841; Lebourg

'PROSPER LAUGIER', HP, mr, 1883; flowers scarlet-red, large, dbl., 30 petals, cupped, recurrent bloom; tall growth; Verdier, E.

'PROSPERITY', HMsk, w, 1919; bud creamy white flushed pale pink; flowers ivory white, small to blooms in large clusters, semi-dbl., recurrent bloom, moderate fragrance; foliage glossy; vigorous, pillar (6-8 ft) (21) growth; [Marie-Jeanne X Perle des Jardins]; Pemberton

Prospero® *see* 'AUSPERO'

'PROTEIFORMIS', S, w; (R. X heterophylla, R. X protieformis); foliage a curiosity of great botanical interest, the very di; (14); [R. rugosa alba X ?]; 1894

Proud Bride®, S, ab, 1992; Austin, David

'PROUD HERITAGE', Min, dr, 1986; flowers patio, large blooms, dbl., 28 petals, exhibition form, blooms in sprays of 2-3, slight fragrance; globular, orange fruit; light green to brown prickles, hooked downward; foliage medium, medium green, matt; upright, bushy growth; PP006561; [Red Beauty X Big John]; Jolly, Nelson F.; Rosehill Farm

'PROUD LAND', HT, dr, 1969; bud urn-shaped; flowers deep red, large, dbl., 60 petals, moderate fragrance; foliage dark, leathery; vigorous, upright growth; [Chrysler Imperial X Unnamed seedling]; Morey, Dr. Dennison; J&P

Proud Mary *see* 'TWOHAVE'

Proud Titania *see* 'AUSTANIA'

'PROVENCE', HT, ob, 1945; flowers orange, tinted copper and pink, well-formed, moderate fragrance; Paolino; A. Meilland

Provence *see* DORlina

Provence Bergsoe, C, mp

Provence Pink, C, mp, 1759

Provence Rose *see* **'CABBAGE ROSE'**

Province d'Anjou, F, 1967; Croix; (Cavriglia)

Province Panachee, B, pb

'PROVINS RENONCULÉ', HGal, mp; (Renoncule); flowers rose; Dupony, Andre, before 1810

Provins Rose *see* **R. GALLICA**

'PRUDENCE', Cl HT, op, 1938; flowers salmon-pink, large, semi-dbl., cupped, recurrent bloom, intense fragrance; very vigorous, climbing (10 ft.) growth; [Warrawee X Souv. de Claudius Pernet]; Fitzhardinge; Hazlewood Bros.

Prudence Rosser, HP, pb

'PRUDHOE PEACH', HT, ab, 1970; flowers apricot-peach, full, dbl., 25 petals; foliage glossy, dark, leathery; free growth; [Piccadilly sport]; Wood; Homedale Nursery

Pruhonice, S, mp, 1973; flowers large, dbl.; Vecera, L.; (Sangerhausen)

Prunella, F, op

Psyche, HMult, mp, 1899; Paul

'PTERAGONIS', ly, 1937; flowers creamy yellow, medium, single, moderate fragrance; Krause; (Sangerhausen)

'P'TIT PACHA', F, rb, 1965; bud pointed; flowers geranium-red edged dark red, large, semi-dbl.; moderate growth; GM, Madrid, 1963; [Unnamed seedling X Coup de Foudre]; Combe

PTMR, HT, 1975; Kurowski, L.; (Cavriglia)

'PUANANI', F, lp, 1997; flowers single, medium, single, 4–7 petals, borne in small clusters, moderate fragrance; foliage medium, medium green, semi-glossy; bushy, medium (3ft.) growth; [Playgirl sport]; Belendez, Kitty

Pubta, Climbing, Cl F, dr, 1987; Patil, B.K.

'PUCCINI', HT, rb, 1968; flowers red, veined yellow, full, dbl., 26–33 petals, 4–6. in., intense fragrance; foliage large, light; very vigorous growth; [Opera X Teenager]; Ellick

Puccini, HMsk, lp, 1984; Lens

'PUCCINI'S DAUGHTER', F, ob, 1972; flowers orange-flame, reverse streaked yellow, full, dbl., 20–25 petals, 4 in., intense fragrance; foliage glossy, dark; very vigorous, upright growth; [(Puccini X Peace) X Orange Sensation]; Ellick; Excelsior Roses

Pucelle de Lille, M, dp; Miellez, before 1860

'PUCK', HT, mr, 1921; bud long, pointed; flowers cherry-crimson, dbl., moderate fragrance; [Lyon Rose X Gen. MacArthur]; Bees

'PUCK', F, or, 1960; bud globular; flowers cinnabar-red, large, dbl., cupped, slight fragrance; foliage leathery; moderate growth; [Pour Toi X Margaret McGredy]; Leenders, J.

'PUCK PULLING', Cl Min, lp, 1997; flowers single, small, single, 4–7 petals, borne in large clusters, slight fragrance; foliage medium, medium green, semi-glossy; spreading, tall (8ft.) growth; [Seedling X Seedling]; Pulling-Smith, Mrs. Pam

Pucker Up *see* 'TINPUCK'

Pudsey Bear *see* BEDchilc

Puerta del Sol® *see* 'DELGLAP'

Puerto Rico *see* 'DELSOB'

Puerto Rico, T, w

Puerto Rico Shell Pink Noisette, N, lp

Pulnor, LCl, mr

Pulsar®, HT, mr

'PUMILA', Min, dp; bud long, pointed; flowers bright pink to red, dbl.; almost thornless; dwarf (8-10 in) growth

Pumila, HGal, dp, 1879

Pumphouse Climber, LCl, mr

Pumpkin Frost™ *see* 'MINIPUM'

'PUNCH', F, dp, 1960; flowers raspberry-red, open, medium, borne in clusters, semi-dbl.; foliage glossy; moderate, bushy growth; [Luc Varenne X Lafayette]; Delforge

Punkin *see* 'TINKIN'

Puppy Love *see* 'SAVAPUP'

Pur Caprice *see* DELjavert

Pur Sang *see* **'WIENER WALZER'**

Pure Bliss™ *see* 'DICTATOR'

Pure Hit *see* POUldel

Pure Love™ *see* 'BURLOV'

Pure Magic, MinFl, ob, 1999

Pure Poetry *see* 'JACMENT'

Puregold *see* **'FANION'**

'PUREZZA', LCl, w, 1961; (R. banksiae 'Purezza'); dbl., 1–2 in., blooms in clusters of 39-50, recurrent bloom, moderate fragrance; foliage leathery; very vigorous growth; GM, Rome, 1960; [Tom Thumb X R. banksiae lutescens]; Mansuino, Q.

'PURITAN', HT, w; flowers large, moderate fragrance

Puritas, HT, 1969; Borgatti, G.; (Cavriglia)

'PURITY', LCl, w, 1917; flowers pure unshaded white, well-formed, large, semi-dbl., slight fragrance; heavy prickles; foliage light; vigorous, climbing growth; [Unnamed seedling X Mme Caroline Testout]; Hoopes, Bro. & Thomas

'PURPLE BEAUTY', HT, m, 1979; flowers red-purple, dbl., 30 petals, 5 in., exhibition form, moderate fragrance; foliage leathery; vigorous, upright growth; [Eminence X Tyrius]; Gandy, Douglas L.

'PURPLE BENGAL', Ch, dr, 1827; (Violet Bengal); flowers maroon shaded darker; (14); Vibert

Purple Boursault from Rockville, Bslt, m

Purple Buttons™, S, m, 1993; Rupert, Kim L.

Purple Carpet, S, m, 1999; Clements, John K.

Purple Cloud, HT, m, 1993; Keisei Rose Nurseries, Inc.

Purple Dawn *see* 'BRIDAWN'

'PURPLE EAST', HMult, m, 1900; flowers crimson-purple, blooms in clusters, semi-dbl., early; [Crimson Rambler X Beauté Inconstante]; Paul

'PURPLE ELF', Min, m, 1967; flowers fuchsia-purple, small, dbl., 43 petals; foliage glossy; dwarf, bushy (10 in) growth; [Violette X Zee]; Moore, Ralph S.; Sequoia Nursery

'PURPLE FANTASY', Min, m, 1982; flowers deep purple, small, dbl., no fragrance; foliage small, medium green, matt;

bushy growth; [Blue Mist X Snow Magic]; Dobbs, Annette E.

Purple Haze *see* 'JUSHAZE'

'PURPLE HEART', HT, m, 1946; bud long, pointed; flowers dahlia-purple to blackish red-purple, large, semi-dbl., cupped, intense fragrance; upright, bushy growth; [Crimson Queen X Self]; Moore, Ralph S.; Sequoia Nursery

Purple Heart™ *see* 'WEKBIPUHIT'

'PURPLE IMP', Min, m, 1967; bud ovoid; flowers magenta to purple, small, dbl.; foliage small, narrow, glossy; vigorous, very compact growth; [Baby Faurax X Red Imp]; Wiliams, E.D.; Mini-Roses

Purple Majesty™ *see* 'MINAJCO'

Purple Noisette *see* **'PURPUREA'**

Purple Parade *see* POUlprima

Purple Passion™ *see* 'JACOLPUR'

Purple Pavement *see* **Rotesmear**

Purple Popcorn™ *see* 'WILPURP'

Purple Prince® *see* 'INTERPUR'

Purple Puff *see* 'JACSEDI'

Purple Rain *see* 'KAWAMOBLUE'

Purple Reign *see* 'WEKPLUROCO'

Purple Simplicity *see* 'JACPURSH'

Purple Spires, F, m, 1997; Hortico

'PURPLE SPLENDOUR', F, m, 1976; flowers glowing purple, dbl., 26 petals, 4 in., slight fragrance; foliage dark; upright growth; [News X Overture]; LeGrice

Purple Sunset *see* 'MANPURSUN'

Purple Tiger *see* 'JACPURR'

Purple Times, F, m, 1999

'PURPUREA', (R. roxburghii hybrid), m; (Maheka, Purpuria); flowers purple-crimson or purplish-rose; [R. roxburghii seedling]; Buist, prior to 1844

'PURPUREA', N, dr, 1822; (Purple Noisette); flowers purple-crimson, poorly formed, semi-dbl.; Laffay, M.

Purpurea di Bologna, B, lp; [Variegata di Bologna sport]; Ruston, D., 1970; (Weatherly, L.)

'PURPUREA RUBRA', M, m; flowers violet-purple, well mossed, large, dbl., intense fragrance

Purpuria *see* **'PURPUREA'**

'PURPURINE', F, m; bud long; flowers fuchsia-purple to rhodamine-purple, dbl., slight fragrance; foliage dark; vigorous, bushy growth; [(Peace X Unnamed seedling) X Fashion]; Lens; Galan

Purpurtraum, HWich, dr

'PUSA CHRISTINA', HT, mp, 1976; bud globular; very dbl., 50 petals, 3 in., exhibition form, slight fragrance; foliage soft; vigorous, upright growth; IARI

'PUSA SONIA', HT, dy, 1968; bud long, pointed; flowers golden yellow, large, dbl., 24 petals, moderate fragrance; foliage leathery; vigorous, upright growth; [McGredy's Yellow X ?]; IARI

'PUSA SONORA', HT, pb, 1984; flowers rose pink, deeper reverse, medium, single, 5 petals, no fragrance; foliage medium green; bushy, upright growth; [Queen Elizabeth X First Prize]; Division of Floriculture and Landscaping

Pushkala, Min, w, 1973; Kasturi

Pushkarini, F, w, 1990; Chiplunkar

Pussta *see* **'NEW DAILY MAIL'**

Pussta, Climbing *see* **'NEW DAILY MAIL, CLIMBING'**

'PYE COLOUR', F, mr, 1972; flowers turkey-red, ovate, dbl., 30 petals, 2 in., slight fragrance; foliage leathery; free growth; [Marlena X Elizabeth of Glamis]; Dickson, A.

'PYGMAE'®, Min, or, 1977; (Pygmy); bud globular; flowers bright orange-red, yellow center, blooms in clusters, semi-dbl., 13 petals, 1 in.; foliage small, glossy, dark; low, compact, spreading, bushy growth; [Anytime X Minuette]; Poulsen, Niels D.; D.T. Poulsen

Pygmy *see* **'PYGMAE'**®

Pyrenees *see* 'POULCOV'

Pyrenees Rose *see* **R. PENDULINA PYRENAICA**

'PYTHAGORAS', HSpn, pb; flowers light pink, flecked deep pink, semi-dbl., early bloom; glossy, black fruit; foliage finely divided; dense, shrubby (3-4 ft) growth

Pzazz *see* POUlzazz

Q

QE2, HT, lp; [Queen Elizabeth sport]; Ruston, D.; Weatherly, L.)

Quadra, HKor, lp; quartered, 60 petals, slight; L'Assomption, 1981

'Quadroon', S, dr; flowers rich dark red, small, single, non-recurrent; growth rather poor; [Supposedly Hansa X (Hansa X R. nitida)]; Wright, Percy H.

'Quaker Beauty', HT, ab, 1936; bud long, pointed; flowers glowing apricot, large, exhibition form; [Joanna Hill sport]; Brookins

'Quaker Maid', F, or, 1959; bud ovoid; flowers nasturtium-red, borne in clusters, semi-dbl., 18–20 petals, 1.5–2 in., flat, moderate, spicy fragrance; foliage leathery; vigorous, upright, bushy growth; [Orange Sweetheart X Pinocchio]; Hill, Joseph H., Co.

Quaker Star *see* 'DICperhaps'

'Quantock Star', HT, or, 1968; flowers vermilion shaded pink, small, dbl., moderate fragrance; foliage variegated; free growth; [Tropicana sport]; Heard

'Quatre Saisons Blanc Mousseux', M, w; (Perpetual White Moss, Quatre Saisons Blanche, Rosier de Thionville); bud very mossy; dbl., blooms in large clusters, repeats sparingly in fall; [Autumn Damask sport]; Laffay, M., prior to 1837

Quatre Saisons Blanche *see* **'Quatre Saisons Blanc Mousseux'**

Quatre Saisons d'Italie, P, dp; Dupony, Andre, before 1815

Quatre Saisons Rose, LCl, mp

Québec *see* **'Mme Marie Curie'**

Queen Adelaide *see* 'MEIvildo'

'Queen Alexandra', HMult, mp; flowers rosy white, semi-dbl., large corymbs; Veitch, 1915

Queen Alexandra *see* **'The Queen Alexandra Rose'**

'Queen Ann', HT, dr, 1949; bud long, pointed; flowers crimson, medium, dbl., exhibition form; strong stems; vigorous growth; [Better Times sport]; Spandikow

Queen Astrid *see* **'Koningin Astrid'**

'Queen Beatrice', HT, lp, 1909; flowers bright silvery pink, large to medium, very dbl., moderate fragrance; vigorous growth; [Mme Abel Chatenay X Liberty]; Kramer

Queen Beatrix *see* 'HETKORA'

'Queen Bee', S, dr, 1984; dbl., 35 petals, 4–5 in., exhibition form, borne 5-8 per cluster, repeat bloom, moderate, old rose fragrance; awl-like, brown prickles; foliage large, leathery, dark olive green; erect, bushy, branching growth; hardy.; [(Rosali X Music Maker) X (Square Dancer X Tatjana)]; Buck, Dr. Griffith J.; Iowa State University

Queen Charlotte *see* 'HARubondee'

'Queen City', Min, ob, 1986; flowers orange, yellow base, fading lighter, medium, borne usually s, dbl., 20 petals, exhibition form, slight fragrance; long, straight, medium, light colored prickles; foliage medium, medium green, semi-glossy; bushy, medium growth; [Rise 'n' Shine X Unnamed seedling]; Bridges, Dennis A., 1987

'Queen Dina', Gr, dr, 1964; flowers deep scarlet, large, dbl.; foliage leathery; very vigorous, upright, bushy growth; [Cocorico X Geranium Red]; Soenderhousen; Hoersholm Nursery

'Queen Dorothy Bell', HT, dr, 1940; flowers velvety scarlet, very dbl., globular, intense fragrance; foliage light, leathery; vigorous growth; [Oswald Sieper sport]; Stell; Stell Rose Nursery

'Queen Elizabeth'®, Gr, mp, 1954; (Queen of England, The Queen Elizabeth Rose); bud pointed; dbl., 38 petals, 3.5–4 in., exhibition form, borne singly and in clusters, moderate fragrance; foliage dark, glossy, leathery; very vigorous, upright, bushy growth; AARS, US, 1955 Gertrude M. Hubbard, ARS, 1957 GM, NRS, 1955 GM, Portland, 1954 Hall of Fame, WFRS, 1978 PIT, NRS, 1955; [Charlotte Armstrong X Floradora]; Lammerts, Dr. Walter; Germain's

'Queen Elizabeth, Climbing', Cl Gr, mp, 1957; (Grimpant Queen Elizabeth, The Queen Elizabeth Rose, Climbing); Whisler; Germain's, 1957;, Wheatcroft Bros., 1960

'Queen Esther', HT, w, 1984; flowers cream, pale pink petal edges, spiraled, large, dbl., 35 petals, slight fragrance; many prickles; foliage medium, medium green, matt; [Golden Masterpiece X Peer Gynt]; Poole, Lionel

'Queen Fabiola', Gr, or, 1961; (Fabiola); [Montezuma sport]; Hazenberg; Delbard

'Queen Frances Connally', HT, mr, 1939; flowers spectrum-red, base lemon-chrome, reverse yellow edged red; [Katharine Pechtold sport]; Stell; Stell Rose Nursery

'Queen Gertrude Anne Windsor', HT, dp, 1936; flowers darker; [Francis Scott Key sport]; Dixie Rose Nursery

Queen Juliana *see* **'Orange Delight'**

'Queen Louise Boren', HT, op, 1935; flowers pink suffused salmon, large, dbl., intense fragrance; very vigorous growth; [(Emile Charles X La France) X Marechal Niel]; Nicolas; Dixie Rose Nursery

'Queen Lucia', F, op, 1954; flowers salmon-pink, pompon form, borne in large trusses, dbl.; vigorous growth; [Pinocchio X Tapis Rose]; Maarse, G.

'Queen Mab', Ch, ab, 1896; flowers soft rosy apricot, center shaded orange, reverse tinted rose, dbl.; Paul, W.

'Queen Margaret Hunt', HT, dr, 1936; bud long, pointed, spiral; flowers dark velvety crimson-maroon, stamens golden yellow, large, cupped, intense fragrance; foliage leathery; very vigorous growth; [Templar X Ami Quinard]; Nicolas; Dixie Rose Nursery

Queen Margrethe™ *see* 'POUlskov'

'Queen Marie', HT, pb, 1925; flowers rose-pink, reverse deeper pink, base bronze yellow, dbl., moderate fragrance; [Mme Butterfly X Lamia]; Chervenka

'Queen Marie of Jugoslavia', HT, my, 1935; flowers bright yellow, slightly flushed pink; [Mme Butterfly sport]; Hicks

'Queen Mary', HT, yb, 1913; flowers bright canary yellow, shaded red, moderate fragrance; GM, NRS, 1913; Dickson, A.

'QUEEN MARY', HT, lp; [Frau Karl Druschki X Unknown]; Williams, A., 1911; (Weatherly, L.)

Queen Mother *see* KORquermu

Queen Nefertiti® *see* AUSap

'QUEEN O' THE LAKES', HT, dr, 1949; bud ovoid, long, pointed; flowers large, dbl., exhibition form, moderate fragrance; foliage glossy; vigorous, bushy growth; [Pink Princess X Crimson Glory]; Brownell, H.C.

'QUEEN O' THE LAKES, CLIMBING', Cl HT, dr, 1965; Brownell, H.C.

'QUEEN OF BATH', HT, dy, 1931; bud long, pointed; flowers deep buttercup-yellow, outer edged chrome-yellow, dbl., exhibition form, intense fragrance; foliage thick, glossy, bronze; vigorous growth; [Souv. de Claudius Pernet X Cleveland]; Bees

Queen of Beauty and Fragrance *see* **'SOUR DE LA MALMAISON'**

'QUEEN OF BEDDERS', B, dp, 1876; flowers deep carmine, well-shaped; dwarf, compact growth; [Sir Joseph Paxton seedling]; Noble

'QUEEN OF BERMUDA', Gr, or, 1956; bud ovoid; flowers orange-vermilion, blooms in small clusters, dbl., 35 petals, 4 in., exhibition form, moderate, fruity fragrance; foliage glossy, bronze; vigorous, bushy growth; [(Independence X Orange Triumph) X Bettina]; Bowie; Bermuda Rose Nursery

'QUEEN OF BOURBONS', B, pb, 1834; (Bourbon Queen, Reine des Iles Bourbon, Souv de la Princesse de Lamballe); flowers fawn and rose, cupped, intense fragrance; Mauget

Queen of Colors *see* **'FARBENKÖNIGEN'**

Queen of Denmark *see* **'KÖNIGIN VON DÄNEMARK'**

Queen of England *see* **'QUEEN ELIZABETH'**®

'QUEEN OF FRAGRANCE', HT, lp, 1915; flowers shell-pink, tipped silver, well shaped, large, dbl., intense fragrance; foliage soft; dwarf growth; Paul, W.

Queen of Hearts *see* **'DAME DE COEUR'**

'QUEEN OF HEARTS', Cl HT, mp, 1920; bud globular; flowers rich pink, large, dbl., cupped, moderate fragrance; foliage dark; very vigorous, climbing growth; [Gustave Grunerwald X Rosy Morn]; Clark, A.; NRS Victoria

Queen of Pearl, HWich, lp, 1898; Van Fleet

Queen of Roses *see* 'KORBICO'

'QUEEN OF THE BELGIANS', Ayr, w; flowers small, very dbl.

'QUEEN OF THE BELGIANS', HT, op, 1916; bud long, pointed; flowers salmon-pink, semi-dbl.; GM, NRS, 1915; Hicks

'QUEEN OF THE DWARFS', Min, dp, 1955; (Dwarf Queen, Zwergkönigin); flowers deep pink, dbl.; foliage rather coarse; height 10-12 in; Kordes

'QUEEN OF THE MUSKS', HMsk, pb, 1913; bud coppery red; flowers deep blush and white, blooms in panicles, recurrent bloom, intense fragrance; foliage dark ivy-green; Paul

Queen of the Night, HT; Sgaravatti, A.; (Cavriglia)

'QUEEN OF THE PRAIRIES', HSet, pb, 1843; (Beauty of the Prairies, Prairie Belle); flowers bright pink, frequently striped white, large blooms in clust, dbl., globular, moderate fragrance; foliage large; vigorous, climbing growth; hardy.; [Probably R. setigera X China]; Feast

Queen of the Violets *see* **'REINE DES VIOLETTES'**

'QUEEN THORNLESS, CLIMBING', Cl HT, dp; bud pointed; flowers deep pink, dbl., 20–40 petals, 3–4 in., exhibition form, intermittent bloom, moderate fragrance; thornless; foliage leathery, dull green; strong stems; vigorous, arching growth; Kittle; Lincoln Nursery Co.

'QUEEN VICTORIA', HP, lp, 1850; flowers blush-pink, large, dbl.; [La Reine seedling]; Fontaine; A. Paul

'QUEEN WILHELMINA', HT, ob, 1942; flowers brilliant orange, base light orange-yellow; [Hinrich Gaede sport]; Deverman

'QUEENIE', F, lp, 1962; (Petite Reine); bud ovoid; dbl., 33 petals, 4 in., cupped, blooms in clusters, moderate fragrance; foliage leathery; vigorous, upright, bushy growth; [Pinocchio seedling X Spartan]; Boerner; J&P

'QUEENIE ROBINSON', HT, op, 1924; bud long, pointed; flowers orange-cerise to flame-pink, semi-dbl., moderate fragrance; Easlea

'QUEENIE'S LOVE', HT, dp, 1968; bud globular; flowers begonia-rose, medium, dbl., moderate fragrance; foliage leathery; vigorous growth; [Libretto X Mme Butterfly]; Verschuren, A.; Stassen

'QUEEN'S KNIGHT', S, mp, 1979; bud short, conical; 6–10 petals, cupped, borne 3 per cluster, profuse for 6 weeks; nonrecurrent, slight fragrance; hooked prickles; foliage semi-glossy, deep green, 7 leaflets; erect arches, self-supporting growth; [Don Juan X R. laxa (Retzius)]; Stoddard, Louis

Queen's Palace *see* POUlelap

'QUEEN'S SCARLET', Ch, mr, 1880; (Red Hermosa, Rote Hermosa); flowers rich velvety scarlet, small, dbl., moderate fragrance; foliage small; bushy, compact growth; Hallock & Thorpe

'QUEEN'S VISIT', HT, dr, 1955; bud long, pointed; flowers dark velvety red veined darker, medium, dbl., exhibition form, moderate fragrance; foliage leathery; vigorous growth; [Crimson Glory sport X Crimson Glory]; Viney; Wynne

'QUEENSLAND BEAUTY', HT, pb, 1934; flowers coppery pink; [Golden Dawn sport]; Alderton & Williams

'QUEENSTOWN', HT, dr, 1991; bud pointed; flowers deep crimson to ruby red, blooms borne up to 6 per cluster, dbl., 32 petals, 5 in., cupped, slight fragrance; foliage leathery, mid-green, semi-glossy; upright, spreading growth; [Silent Night X Josephine Bruce]; Cattermole, R.F., 1975; South Pacific Rose Nursery, 1991

Querida *see* 'MALJOANNA'

Quicksilver *see* 'AROSTAL'

Quiet Reflections, T, dy; Hay

Quiet Time *see* 'TINQUIET'

Quinella, HT, ob

Quintet, HT, op

Quito, HT, yb, 1999

'QUO VADIS?', HT, pb, 1961; bud oval; flowers silvery pink becoming red, large, dbl., 50–60 petals, slight fragrance; very vigorous growth; [Peace X (Baiser X ?)]; Giacomasso

R

R. abietorum *see* **R. GYMNOCARPA**

R. abyssinica *see* **R. MOSCHATA ABYSSINICA**

R. ACICULARIS, (Lindley), Sp, dp, 1805; (Arctic Rose, R. acicularis carelica, R. acicularis taquetii, R. carelica, R. fauriei, R. korsakoviensis, R. sayi, R. stricta); flowers deep rose solitary, 1–1.5 in., spring bloom, moderate fragrance; usually pyriform, 1/2-1 in. long fruit; 3 ft. growth; Eurosa, Cinnamomeae (28, 42, 56)

R. acicularis bourgeauiana *see* **R. ACICULARIS SAYI**

R. acicularis carelica *see* **R. ACICULARIS**

R. acicularis engelmannii *see* **R. X ENGELMANNII**

R. ACICULARIS FENNICA, (Lallemant), Sp, dp; (R. acicularis gmelinii, R. gmelinii); (28)

R. acicularis gmelinii *see* **R. ACICULARIS FENNICA**

R. ACICULARIS NIPPONENSIS, ((Crépin) Koehne), Sp, dp, 1894; (R. nipponensis); 1.5 in.; (14, 28)

R. ACICULARIS SAYI, (Rehder), Sp, dp; (R. acicularis bourgeauiana, R. bourgeauiana, R. sayi); 2.5 in.; usually globular fruit; (42, 56)

R. ACICULARIS SAYI PLENA, (Lewis), Sp, dp

R. acicularis taquetii *see* **R. ACICULARIS**

R. adenosepala *see* **R. WOODSII HISPIDA**

R. afzeliana *see* **R. DUMALIS**

R. AGRESTIS, (Savi), Sp, lp; (R. sepium); flowers pale pink or whitish, small; Eurosa, Caninae (35, 42)

R. agrestis inodora *see* **R. INODORA**

R. alba Allioni *see* **R. SEMPERVIRENS**

R. alba incarnata *see* **'GREAT MAIDEN'S BLUSH'**

R. alba maxima *see* **'ALBA MAXIMA'**

R. alba rubicanda plena *see* **'GREAT MAIDEN'S BLUSH'**

R. alba rubicunda *see* **'GREAT MAIDEN'S BLUSH'**

R. alba semi-plena *see* Alba semi-plena

R. alba suaveolens *see* Alba suaveolens

R. ALBERTII, (Regel), Sp, w; 1.5 in.; 3 ft. growth; Eurosa, Cinnamomeae; 1877

R. albicans *see* **R. ELYMAITICA**

R. alpina *see* **R. PENDULINA**

R. altaica *see* **R. SPINOSISSIMA ALTAICA**

R. AMBLYOTIS, (C.A. Meyer), Sp, mr; 2 in., solitary or few; subglobose or pyriform, 1/2-1 in., red fruit; Eurosa, Cinnamomeae (14)

R. amoyensis *see* **R. CYMOSA**

R. andegavensis *see* **R. CANINA ANDEGAVENSIS**

R. ANDREAE, (Lange), Sp; (14)

R. ANEMONEFLORA, (Fortune), Sp, w; (R. sempervirens anemoniflora, R. triphylla); flowers corymbose, dbl., 1 in.; foliage leaflets 3, lanceolate; climbing growth; Eurosa, Synstelae (14); 1844

R. angustiarum *see* **R. ARKANSANA**

R. apennina *see* **R. SERAFINII**

R. ARKANSANA, (Porter), Sp, mp; (Arkansas Rose, R. angustiarum, R. arkansanoides, R. heliophila, R. pratincola, R. rydbergii, R. suffulta); 1.5 in., in corymbs; 1.5 ft. growth; Eurosa, Cinnamomeae (28, 14)

R. ARKANSANA ALBA, ((Rehder) Lewis), Sp, w; (R. arkansanoides alba, R. heliophila alba, R. pratincola alba, R. suffulta alba); 1901

R. ARKANSANA PLENA, (Lewis), Sp, lp

R. arkansanoides *see* **R. ARKANSANA**

R. arkansanoides alba *see* **R. ARKANSANA ALBA**

R. ARVENSIS, (Hudson), Sp, w; (Field Rose, R. repens, R. serpens, R. silvestris); 1.5–2 in., in few-flowered corymbs, summer bloom, no fragrance; ovoid fruit; foliage deciduous; creeping growth; Eurosa, Synstylae (14)

R. ARVENSIS AYRESHIREA, (Seringe); (Ayrshire Rose, R. arvensis capreolata, R. capreolata)

R. arvensis capreolata *see* Ayrshire Rose

R. atrovirens *see* **R. SEMPERVIRENS**

R. aurantiaca *see* **R. FOETIDA BICOLOR**

R. austriaca *see* **R. GALLICA**

R. austriaca pygmaea *see* **R. GALLICA PUMILA**

R. bakerii *see* **R. X ENGELMANNII**

R. balearica *see* **R. SEMPERVIRENS**

R. balsamea *see* **'TACKHOLMII'**

R. BANKSIAE, (Aiton fil), Sp; (Banks' Rose, Banksian Rose, Lady Banks'Rose, R. banksiana); flowers white or yellow on slender pedicels in many-flowered umbels, 1 in., early spring bloom, slight fragrance; foliage evergreen; climbing (20 ft. or more) growth; Eurosa, Banksianae (14)

R. banksiae 'Purezza' *see* **'PUREZZA'**

R. banksiae alba *see* **R. BANKSIAE BANKSIAE**

R. banksiae alba-plena *see* **R. BANKSIAE BANKSIAE**

R. BANKSIAE BANKSIAE, Sp, w; (Banksiae Alba, R. banksiae alba, R. banksiae alba-plena, White Banksia, White Lady Banks' Rose); dbl.; (14); William Kerr, 1807

R. BANKSIAE LUTEA, (Lindley), Sp, ly; (R. banksiae luteaplena, Yellow Lady Banks' Rose); dbl.; (14); 1824

R. banksiae luteaplena *see* **R. BANKSIAE LUTEA**

R. BANKSIAE LUTESCENS, (Voss), Sp, my; single; (14); ca. 1870

R. BANKSIAE NORMALIS, (Regel), Sp, w; single; (14); 1796, but not distributed until 1877

R. banksiana *see* **R. BANKSIAE**

R. BANKSIOPSIS, (Baker), Sp, mr; 1 in., borne in corymbs; Eurosa, Cinnamomeae (14); 1907

R. BEGGERIANA, (Schrenk ex Fischer & Meyer), Sp, w; (R. cinnamomea sewerzowii, R. lehmanniana, R. regelii, R. silverhjelmii); 1.5 in., borne in corymbs, late summer bloom.; fruit small, globular, without sepals; dense (5 ft.) growth; Eurosa, Cinnamomeae (14); 1868

R. belgica *see* R. X damascena

R. BELLA, (Rehder & Wilson), Sp, mp; flowers pink, solitary, 1.75–2 in.; ovoid, 3/4 in long, scarlet fruit; growth to 8 ft.; Eurosa, Cinnamomeae (28); 1910

R. BELLA PALLENS, (Rehder & Wilson), Sp, lp; flowers pale pink; 1910

R. bengalensis *see* **'SLATER'S CRIMSON CHINA'**

R. bicolor *see* **R. FOETIDA BICOLOR**

R. bifera *see* **'AUTUMN DAMASK'**

R. bifera semperflorens *see* **'AUTUMN DAMASK'**

R. BILLOTIANA, (Crépin), Sp; fruit like R. sherardii; similar to R. canina; Eurosa, Caninae (28)

R. BLANDA, (Aiton), Sp, mp; (Hudson's Bay Rose, Labrador Rose, R. fraxinifolia, R. gratiosa, R. solandri, R. subblanda, R. virginiana blanda); flowers pink borne usually several on smooth peduncles, single, 2–2.5 in., spring bloom; globular, sometimes elongated fruit; usually no prickles; 5 ft. growth; Eurosa, Cinnamomeae (14, 21, 28); 1773

R. blanda *see* **R. NITIDA**

R. BLANDA CARPOHISPIDA, ((Schuette) Lewis), Sp, mp; (R. blanda hispida); prickles densely covered with fine bristles; (14)

R. blanda hispida *see* **R. BLANDA CARPOHISPIDA**

R. BLANDA WILLMOTTIANA, (Baker), Sp, mp; flowers bright coral-pink; stems red

R. blinii *see* **R. MULTIFLORA CARNEA**

R. blondaeana *see* **R. CANINA BLONDAEANA**

R. bourgeauiana *see* **R. ACICULARIS SAYI**

R. boursaultii *see* **'BOURSAULT ROSE'**

R. BRACTEATA, (Wendland), Sp, w; (Chickasaw Rose, Macartney Rose, R. lucida, R. macartnea); 2–2.75 in., borne 1 or few on short stalks, spring-fall bloom; large, globular, tomentose, orange-red fruit; foliage half-evergreen, bright green, somewhat glossy above, ; climbing growth; Eurosa, Bracteata (14); 1793

R. bracteata *see* **R. WICHURANA**

R. bracteata alba odorata, R. microphylla alba odorata *see* **'ALBA ODORATA'**

R. braunii *see* **R. X INVOLUTA**

R. BRITZENSIS, (Koehne), Sp, lp; flowers pale pink changing to white borne 1-2 together, 3–4 in.; ovoid, brown fruit; 6 ft. growth; Eurosa, Caninae (35); 1901

R. brownii *see* **R. MOSCHATA NEPALENSIS**

R. brunonii *see* **R. MOSCHATA NEPALENSIS**

R. burgundensis *see* **'BURGUNDIAN ROSE'**

R. burgundica *see* **'BURGUNDIAN ROSE'**

R. calabrica *see* **R. GLUTINOSA**

R. calendarum *see* **'SUMMER DAMASK'**

R. CALIFORNICA, (Chamisso & Schlechtendahl), Sp, mp; (California Wild Rose, R. gratissima); 1.5 in., borne few or several on slender pedicels in dense corymbs, summer bloom; globose-ovoid, usually with a prominent neck fruit; usually has paired infrastipular prickles, which are often c; up to 8 ft. growth

R. CALIFORNICA NANA, (Bean), Sp, mp; habit very dwarf growth

R. CALIFORNICA PLENA, (Rehder), Sp, 1894; dbl.; (14); Geschwind

R. californica ultramontana *see* **R. WOODSII ULTRAMONTANA**

R. calva *see* **R. MULTIFLORA CALVA**

R. camellia *see* **R. LAEVIGATA**

R. campanulata *see* **R. X FRANCOFURTANA**

R. CANINA, (Linnaeus), Sp, lp; (Brier Bush, Dog Rose, R. leucantha, R. pseudoscabrata, R. sphaerica, R. surculosa); flowers white or pinkish blooms solitary or in few-flowered corymbs, 2 in., summer bloom; ovoid, glabrous (see endpapers), orange-red scarlet fruit; 10 ft. growth; Eurosa, Caninae (35, 42, 34)

R. CANINA ANDEGAVENSIS, ((Bastard) Desportes), Sp; (R. andegavensis); foliage variation with glabrous, eglandular, but glandular; (35)

R. CANINA ANDERSONII, S, mp, 1912; flowers medium, single; Hillier; (Sangerhausen)

R. CANINA BLONDAEANA, ((Ripart) Rouy), Sp; (R. blondaeana); foliage leaflet teeth glandular, bi-serrate; (42)

R. canina borboniana *see* **'BOURBON ROSE'**

R. canina coriifolia *see* **R. CORIIFOLIA**

R. canina dumetorum *see* **R. CORYMBIFERA**

R. CANINA EXILIS, ((Crépin) Keller), Sp, mp; (R. exilis); 1 in.; low growth

R. CANINA FROEBELII, (Christ), Sp, w; (R. coriifolia froebelii, R. froebelii, R. laxa); flowers small; (35)

R. CANINA INERMIS, (hort.), Sp; almost unarmed; vigorous; dog rose popular as an understock

R. canina insignis *see* **R. CANINA SPURIA**

R. CANINA LUTETIANA, ((Léman) Baker), Sp; (R. lutetiana); (35)

R. CANINA SPURIA, ((Puget) Wolley-Dod), Sp; (R. canina insignis); large fruit; foliage serrations almost entirely simple; (35)

R. canina tomentella *see* **R. OBTUSIFOLIA**

R. capreolata *see* Ayrshire Rose

R. carelica *see* **R. ACICULARIS**

R. carnea *see* **'GREAT MAIDEN'S BLUSH'**

R. CAROLINA, (Linnaeus), Sp, mp; (Carolina Rose, Pasture Rose, R. humilis, R. parviflora, R. pensylvanica, R. pratensis, R. virginiana humilis); flowers bright pink, 2 in., summer bloom; 3-6 ft.growth; Eurosa, Carolinae

R. carolina *see* **R. PALUSTRIS**

R. CAROLINA ALBA, (Rehder), Sp, w; (R. lyonii alba, R. virginiana alba); (28)

R. CAROLINA GLANDULOSA, ((Crépin) Farwell), Sp, mp; (R. mexicana, R. parviflora glandulosa, R. sarrulata); foliage leaflets glandular-serrate leaf-stalk glandular; (28)

R. carolina grandiflora *see* **R. LAXA**

R. carolina inermis *see* **R. PALUSTRIS INERMIS**

R. carolina lyonii *see* **R. CAROLINA VILLOSA**

R. carolina nuttalliana *see* **R. PALUSTRIS NUTTALLIANA**

R. CAROLINA PLENA, ((Marshall) Doris Lynes), Sp, mp; (R. pennsylvanica plena); dbl.

R. carolina setigera *see* **R. NITIDA SPINOSA**

R. CAROLINA TRILOBA, ((Watson) Rehder), Sp, mp; (R. humilis triloba); flowers bright pink, petals 3-lobed

R. CAROLINA VILLOSA, ((Best) Rehder), Sp, mp; (R. carolina lyonii, R. humilis villosa, R. lyonii, R. pusilla); foliage pubescent beneath; (28)

R. caroliniana *see* **R. PALUSTRIS**

R. caryophyllacea *see* **R. INODORA**

R. cathayensis *see* **R. MULTIFLORA CATHAYENSIS**

R. cathayensis platyphylla *see* **'SEVEN SISTERS'**

R. CAUDATA, (Baker), Sp, mr; 2 in., borne in few-flowered corymbs; ovoid fruit; 12 ft. growth; Eurosa, Cinnamomeae (14, 28); ca. 1896

R. centifolia *see* **'CABBAGE ROSE'**

R. centifolia albo-muscosa *see* **'SHAILER'S WHITE MOSS'**

R. centifolia andrewsil *see* **'MUSCOSA SIMPLEX'**

R. centifolia cristata *see* **'CRESTED MOSS'**

R. centifolia minor *see* **'PETITE DE HOLLANDE'**

R. centifolia muscosa *see* **'CENTIFOLIA MUSCOSA'**

R. centifolia muscosa *see* Communis

R. centifolia muscosa cristata *see* **'CRESTED MOSS'**

R. centifolia parvifolia *see* **'BURGUNDIAN ROSE'**

R. centifolia pomponia *see* **'ROSE DE MEAUX'**

R. centifolia sancta *see* **St. John's Rose**

R. centifolia simplex *see* **'CIUDAD DE OVIEDO'**

R. centifolia variegata *see* **'VILLAGE MAID'**

City of Dunedin
(REYcidin)
Mini-Flora, yellow blend, 1995
Hybridized by Ted Reynolds
[(Rise 'n' Shine x seedling) x Rainbow's End]
Photo courtesy Rakaia Roses, Canterbury, New Zealand

Shelley Higgins
(REYshelley)
Hybrid Tea, apricot blend, 1998
Hybridized by Ted Reynolds
[Westerland x seedling]
Photo courtesy Rakaia Roses, Canterbury, New Zealand

Kelly Reynolds
(REYkelly)
Hybrid Tea, pink blend, 1999
Hybridized by Ted Reynolds
[Perfume Delight x seedling]
Photo courtesy Rakaia Roses, Canterbury, New Zealand

City of Invercargill
(REYinver)
Hybrid Tea, deep pink, 1996
Hybridized by Ted Reynolds
[Westerland x Perfume Delight]
Photo courtesy Rakaia Roses, Canterbury, New Zealand

Reynolds Rugosa
(REYrug)
Hybrid Rugosa, medium red, 1999
Hybridized by Ted Reynolds
[double rugosa x double rugosa]
Photo courtesy Rakaia Roses, Canterbury, New Zealand

All That Jazz™
(TWOadvance)
Shrub, orange-pink, 1991
Hybridized by Jerry Twomey
[Gitte x Seedling]
AARS 1992
Photo courtesy Devor Nurseries, Inc.

Sheer Elegance™
(TWObe)
Hybrid Tea, orange-pink, 1989
Hybridized by Jerry Twomey
[Pristine x Fortuna]
AARS 1991
Photo courtesy Devor Nurseries, Inc.

All American Bride
(WILbrid)
Hybrid Tea, white, 2000
Hybridized by J. Benjamin Williams
[White Masterpiece x Miss All-American Beauty]
Photo courtesy W. Atlee Burpee Co.

Scarlet Star
(WILSTAR)
Large Flowered Climber, medium red, 2000
Hybridized by J. Benjamin Williams
[Red Fountain x Mister Lincoln]
Photo courtesy Conard-Pyle Co.

Delaney Sisters™
(WILdels)
Grandiflora, pink blend, 1997
Hybridized by J. Benjamin Williams
[Handel x Love]
Photo courtesy Conard-Pyle Co.

Rose Parade
Floribunda, pink blend, 1974
Hybridized by J. Benjamin Williams
[Sumatra x Queen Elizabeth]
AARS 1975
Photo courtesy J. B. Williams & Associates

Amber Flash™
(WILDAK)
Mini-Flora, orange blend, 1982
Hybridized by J. Benjamin Williams
[Zorina x Starina]
Photo courtesy Conard-Pyle Co.

Patio Patty
Mini-Flora, yellow blend, 1975
Hybridized by J. Benjamin Williams
[(Circus x The Optimist) x (Little Darling x Starina)]
Photo courtesy J. B. Williams & Associates

Stardance™
(WILblank)
Mini-Flora, white, 1982
Hybridized by J. Benjamin Williams
[Ma Perkins x (Charlie McCarthy x Easter Morning)]
Photo courtesy Conard-Pyle Co.

Bellissima®
(SUNLAMPO)
Floribunda, orange blend, 1999
Hybridized by Frank B. Schuurman
[Lambada X Pot O'Gold]

Gold Strike®
(SUNLUCK)
Hybrid Tea, dark yellow, 1999
Hybridized by Frank B. Schuurman
[Seedling X Seedling]

Citronella®
(SUNSIEN)
Hybrid Tea, light yellow, 1999
Hybridized by Frank B. Schuurman
[La Parisienne sport]

La Parisienne®
(SUNPARI)
Hybrid Tea, apricot blend, 1999
Hybridized by Frank B. Schuurman
[Kia Ora X Texas]

Delilah®
(SUNDEL)
Hybrid Tea, mauve, 1999
Hybridized by Frank B. Schuurman
[Seedling X Osiana]

Photo courtesy Franks Roses

Peace
(Gioia; Gloria Dei; Mme A. Meilland)
Hybrid Tea, yellow blend, 1945
[((George Dickson x Souvenir de Claudius Pernet) x (Joanna Hill x Charles P. Kilharm)) x Margaret McGredy)]
Hybridized by Francis Meilland; Introduced 1945
GM, Portland, 1944; GM, NRS, 1947; AARS, 1946;
ARS National Gold Medal Certificate, 1947; Golden Rose of the Hague, 1965
Elected to WFRS Hall of Fame, 1976
Photo courtesy of Rich Baer

Queen Elizabeth
(Queen of England; The Queen Elizabeth Rose)
Grandiflora, medium pink, 1954
Hybridized by Dr. W. E. Lammerts, U.S.A; Introduced 1954
[Charlotte Armstrong x Floradora]
GM, Portland, 1954; GM, NRS, 1955; AARS, 1955;
NRS PIT, 1955; ARS (Gertrude M. Hubbard), 1957;
ARS National Certificate, 1960;
Elected to WFRS Hall of Fame, 1978
Photo courtesy of Rich Baer

Fragrant Cloud
(TANellis)
(Duftwolke; Nuage Parfumé)
Hybrid Tea, orange red, 1967
Hybridized by Mathias Tantau; Introduced 1963
[Seedling x Prima Ballerina]
GM, NRS, 1963; GM, Portland, 1967; NRS PIT, 1964;
ARS James Alexander Gamble Rose Fragrance Medal, 1969;
Elected to WFRS Hall of Fame, 1981
Photo courtesy of Tom Hickey

Iceberg
(KORbin)
(Fée des Neiges; Schneewittchen)
Floribunda, white, 1958
Hybridized by Reimer Kordes; Introduced 1958
[Robin Hood x Virgo]
GM, NRS, 1958; GM, Baden-Baden, 1958;
Elected to WFRS Hall of Fame, 1983
Photo courtesy of Rich Baer

Double Delight™
(ANdeli)
Hybrid Tea, red blend, 1977
Hybridized by Herbert Swim & Arnold Ellis, 1977; Introduced 1977
[Granada x Garden Party]
GM, Baden Baden, 1976; GM, Rome, 1976; AARS, 1977
ARS James Alexander Gamble Rose Fragrance Medal, 1986;
Elected to WFRS Hall of Fame, 1985
Photo courtesy of Paul Jerabek

Papa Meilland®
(MEIsar; MEIcesar)
Hybrid Tea, dark red, 1963
Hybridized by Alain Meilland; Introduced 1963.
[Chrysler Imperial x Charles Mallerin]
GM, Baden Baden, 1962;
ARS James Alexander Gamble Rose Fragrance Medal, 1974;
Elected to WFRS Hall of Fame, 1988
Photo courtesy of Paul Jerabek

Pascali®
(LENip)
(Blanche Pasca)
Hybrid Tea, white, 1963
Hybridized by Louis Lens; Introduced 1963
[Queen Elizabeth x White Butterfly]
GM, The Hague, 1963; GM, Portland, 1967; AARS, 1969
Elected to WFRS Hall of Fame, 1991
Photograph courtesy of Ron Shaw

Just Joey
Hybrid Tea, orange blend, 1972
Hybridized by Cants of Colchester; Introduced 1972
[Fragrant Cloud x Dr. A. J. Verhage]
GM, James Mason, 1986
Elected to WFRS Hall of Fame, 1994
Photo courtesy of Rich Baer

New Dawn
(Everblooming Dr. W. Van Fleet; The New Dawn)
Large flowered Climber, light pink, 1930
Somerset Rose Nursery; Introduced 1930
[Dr. W. Van Fleet sport]
Elected to WFRS Hall of Fame, 1997
Photo courtesy of P. A. Haring

AMERICAN ROSE SOCIETY
Miniature Rose Hall of Fame

Starina
(MELgabi)
Miniature, orange-red, 1965
Hybridized by Meilland
[(Dany Robin x Fire King) x Perla de Montserrat]
Elected to Hall of Fame, 1999.
Photo courtesy of Rich Baer

Magic Carrousel
(MORrousel)
Miniature, red blend, 1972
Hybridized by Ralph Moore
[Little Darling x Westmont]
ARS Award of Excellence, 1975
Elected to Hall of Fame, 1999
Photo courtesy of A. William Nelson

Beauty Secret
Miniature, medium red, 1965
Hybridized by Ralph Moore.
[Little Darling x Magic Wand]
ARS Award of Excellence, 1975
Elected to Hall of Fame, 1999
Photo courtesy of P. A. Haring

Rise 'n' Shine
Miniature, medium yellow, 1977
Hybridized by Ralph Moore
[Little Darling x Yellow Magic]
ARS Award of Excellence, 1978
Elected to Hall of Fame, 1999
Photo courtesy of Al Schamel

Party Girl
Miniature, yellow blend, 1979
Hybridized by F. Harmon Saville
[Rise 'n' Shine x Sheri Anne]
ARS Award of Excellence, 1981
Elected to Hall of Fame, 1999
Photo courtesy A. William Nelson

R. CERASOCARPA, (Rolfe), Sp, w; 1–1.25 in.; globose, over 1/3 in., deep red fruit; Eurosa, Synstylae (14)

R. charbonneaui *see* **R. LONGICUSPIS**

R. CHAVINII, (Rapin ex Reuter), Sp, mp; (R. montana chavinii); 2 in.; ovoid, setose fruit; 6-10 ft. growth; Eurosa, Caninae (42); [Perhaps a derivative of R. canina X R. montana]

R. cherokeensis *see* **R. LAEVIGATA**

R. CHINENSIS, (Jacquin), HCh; (Bengal Rose, China Rose, R. chinensis indica, R. indica, R. indica vulgaris, R. nankinensis, R. sinica); flowers crimson or pink, rarely whitish, borne usually several, less, 2 in., recurrent bloom, no fragrance; obovoid or turbinate, about 3/4 in. long fruit; foliage evergreen or partially so; long stems, rarely short-stemmed; Eurosa, Chinensis (14, 21, 28)

R. chinensis fragrans *see* **R. X ODORATA**

R. chinensis indica *see* **R. CHINENSIS**

R. CHINENSIS LONGIFOLIA, ((Willdenow) Rehder), Sp; (R. indica longifolia, R. longifolia); foliage leaflets very long and narrow, like a willow; Int. 1820, but not now in cult.

R. chinensis manettii *see* **'MANETTI'**

R. CHINENSIS MINIMA, HCh; (Angel Rose, R. indica pumila, R. laurentiae, R. lawranceana, R. semperflorens minima); flowers white, pink or red, petals often pointed, semi-dbl., 1.5 in.; height variable; Eurosa, Chinensis

R. chinensis mutabilis *see* **'MUTABILIS'**

R. chinensis pseudindica *see* **'FORTUNE'S DOUBLE YELLOW'**

R. chinensis semperflorens *see* **'SLATER'S CRIMSON CHINA'**

R. chinensis serratipetala *see* **'SERRATIPETALA'**

R. CHINENSIS SPONTANEA, (Rehder & Wilson), Sp, dr; flowers deep red or pink, borne usually solitary, single

R. chinensis viridiflora *see* **'GREEN ROSE'**

R. chlorophylla *see* **R. FOETIDA**

R. CINNAMOMEA, (Linnaeus), Sp, m; (Cinnamon Rose, R. collincola, R. spinosissima); flowers purple borne solitary or few on short, naked pedicels, 2 in., moderate fragrance; depressed-globular, scarlet fruit; 6 ft. growth; Eurosa, Cinnamomeae (14, 28)

R. cinnamomea *see* **R. PENDULINA**

R. CINNAMOMEA PLENA, (Weston), Sp, m; (Double Cinnamon, R. foecundissima, R. majalis, Rose du Saint Sacrament, Stevens Rose, Whitsuntide Rose); dbl.

R. cinnamomea sewerzowii *see* **R. BEGGERIANA**

R. CLINOPHYLLA, (Thory), Sp, w; (R. involucrata, R. lindleyana, R. lyellii); tender.; Eurosa, Bracteata (14); Int. prior to 1817

R. collina, (Jacquin), Misc OGR, mp; flowers rose-colored, sepals shorter

R. collina *see* **R. CORYMBIFERA**

R. collincola *see* **R. CINNAMOMEA**

R. coreana *see* **R. MAXIMOWICZIANA JACKII**

R. CORIIFOLIA, (Fries), Sp, lp; (R. canina coriifolia, R. frutetorum, R. watsonii); flowers pink, short-pediceled, with large bracts; foliage leaflets gray-green, downy; 5 ft. growth; Eurosa, Caninae (35)

R. coriifolia froebelii *see* **R. CANINA FROEBELII**

R. coronata *see* **R. X INVOLUTA**

R. coruscans *see* **R. RUGOSA CHAMISSONIANA**

R. CORYMBIFERA, (Borkhausen), Sp, w; (R. canina dumetorum, R. collina, R. dumetorum, R. saxatilis, R. taurica); flowers light pink to white, borne 1-many, 1.75–2 in.; ovoid to subglobose, 3/4 in. fruit; foliage leaflets downy; Eurosa, Caninae (35, 42)

R. corymbosa *see* **R. PALUSTRIS**

R. CORYMBULOSA, (Rolfe), Sp, rb; flowers red with a white eye, borne in dense, umbel-like corymbs, .75–1 in.; globose fruit; 6 ft. growth; Eurosa, Cinnamomeae (14); 1908

R. CROCACANTHA, (Boulenger), Sp, w; .5 in., borne in many-flowered panicles; globose, 1/3in., red fruit; prickles curved, yellow; Eurosa, Synstylae (14); 1917

R. cursor *see* **R. SETIGERA TOMENTOSA**

R. cuspidata *see* **R. TOMENTOSA**

R. CYMOSA, (Trattinnick), Sp, w; (R. amoyensis, R. esquirolii, R. fragariaeflora, R. indica, R. microcarpa, R. sorbiflora); flowers small, borne in many-flowered corymbs; small globose., red fruit; Eurosa, Banksianae (14); Int. 1904?

R. dalmatica *see* **R. GLUTINOSA DALMATICA**

R. damascena bifera *see* **'AUTUMN DAMASK'**

R. damascena rubrotincta *see* **'HEBE'S LIP'**

R. damascena semperflorens *see* **'AUTUMN DAMASK'**

R. damascena trigintipetala *see* **'KAZANLIK'**

R. damascena variegata *see* **'YORK AND LANCASTER'**

R. damascena versicolor *see* **'YORK AND LANCASTER'**

R. DAVIDII, (Crépin), Sp, lp; (Father David's Rose); flowers pink borne in corymbs, 1.5–2 in.; oblong-ovoid or ovoid, long-necked, 1/2-3/4 in. long, scarlet fruit; 10 ft. growth; Eurosa, Cinnamomeae (28); 1908

R. DAVIDII ELONGATA, (Rehder & Wilson), Sp, lp; (R. parmentieri); large, to 1 in. long fruit; 1908

R. davidii persotosa *see* **R. PERSETOSA**

R. DAVURICA, (Pallas), Sp, m; (R. willdenowii); ovate fruit; straight instead of curved prickles and smaller leaflets; Eurosa, Cinnamomeae (14); 1910

R. dawsoniana *see* **R. MULTIFLORA**

R. deserta *see* **R. WOODSII**

R. dijoniensis *see* **'ROSE DE MEAUX'**

R. dimorpha *see* **R. TOMENTOSA**

R. diversifolia *see* **'SLATER'S CRIMSON CHINA'**

R. doniana *see* **R. X INVOLUTA**

R. DUMALIS, (Bechstein), Sp; (R. afzeliana, R. glauca, R. gypsicola, R. reuteri); heavily prickled; foliage bluish green glaucous leaflets; Eurosa, Caninae (35)

R. dumetorum *see* **R. CORYMBIFERA**

R. ECAE, (Aitchison), Sp, dy; (R. xanthina, R. xanthina ecae); flowers short-stalked, 1–1.25 in., borne solitary; obovoid, 1/3-1/2 in, long fruit; upright (4 ft.) growth; Eurosa, Pimpinellifoliae (14); 1880

R. ecae *see* **R. PRIMULA**

R. EGLANTERIA, (Linnaeus), Sp, lp; (Eglantine, R. rubiginosa, R. suavifolia, R. walpoleana, Sweet Brier Rose, Sweetbriar); flowers pink borne solitary or in few-flowered corymbs, single, 2 in.; many prickles; foliage glandular and fragrant (apple); vigorous, to 8 ft. growth; Eurosa, Caninae (35, 42)

R. eglanteria *see* **R. FOETIDA**

R. eglanteria duplex *see* **'MAGNIFICA'**

R. eglanteria punicea *see* **R. FOETIDA BICOLOR**

R. ehrrhartiana *see* **'BURGUNDIAN ROSE'**

R. ELASMACANTHA, (Trautvetter), Sp, my; (R. pimpinellifolia elasmacantha); flowers light yellow, borne solitary; low glabrous shrub growth; Eurosa, Pimpinellifoliae (28)

R. ELEGANTULA, (Rolfe), Sp, mp; flowers pink, with golden anthers, small, free bloomer; summer; 5-10 ft. growth; Eurosa, Cinnamomeae (14)

R. elliptica *see* **R. INODORA**

R. elongata *see* **R. PALUSTRIS**

R. ELYMAITICA, (Boissier & Haussknecht), Sp, mp; (R. albicans); flowers pink, small, borne 1-3 together; small, globular fruit; dwarf (3 ft) growth; Eurosa, Cinnamomeae (14, 28); 1900

R. engelmannii *see* **R. X ENGELMANNII**

R. ernestii *see* **R. RUBUS**

R. ernestii nudescens *see* **R. RUBUS NUDESCENS**

R. esquirolii *see* **R. CYMOSA**

R. exilis *see* **R. CANINA EXILIS**

R. FARGESII, (Boulenger), Sp, w; flowers paniculate, 1.75 in.; Eurosa, Synstylae

R. fargesii *see* **R. MOYESII FARGESII**

R. fargesii *see* **R. MOYESII**

R. FARRERI, (Stapf), Sp, w; flowers pale pink to white, 1.25–1.5 in.; ovoid to ellipsoid, red fruit; Eurosa, Pimpinellifoliae

R. FARRERI PERSETOSA, (Stapf), Sp, mp; (Threepenny Bit Rose); flowers pink, smaller; prickles whole plant finely bristly; 1914

R. faureri *see* **R. MAXIMOWICZIANA**

R. fauriei *see* **R. ACICULARIS**

R. FEDTSCHENKOANA, (Regel), Sp, w; 1.5–2 in., borne 1-4 together; Eurosa, Cinnamomeae (28); 1876

R. fendleri *see* **R. WOODSII FENDLERI**

R. fenestrata *see* **R. SETIGERA**

R. ferox *see* **R. GLUTINOSA**

R. ferox *see* **R. HORRIDA**

R. ferox *see* **R. RUGOSA**

R. ferruginea *see* R. glauca

R. FILIPES, (Rehder & Wilson), Sp, w; 1 in., borne in large, loose corymbs, moderate fragrance; globose, 1/3-1/2 in., scarlet fruit; 15 ft. growth; Eurosa, Synstylae (14); 1908

R. filipes 'Brenda Colvin' *see* **'BRENDA COLVIN'**

R. filipes 'Kiftsgate' *see* **'KIFTSGATE'**

R. fimbriatula *see* **R. WOODSII**

R. floribunda *see* **R. HELENAE**

R. floribunda *see* **R. MICRANTHA**

R. florida *see* **R. MULTIFLORA CARNEA**

R. foecundissima *see* **R. CINNAMOMEA PLENA**

R. FOETIDA, (Herrmann), Sp, my; (Austrian Briar, Austrian Brier Rose, Austrian Yellow Rose, R. chlorophylla, R. eglanteria, R. lutea); flowers bright yellow borne sometimes several, but without bracts o, single, 2–2.5 in., summer bloom, moderate, sickly sweet fragrance; globular, rarely sets fruit; 10 ft. growth; Eurosa, Pimpinellifoliae (28); Possibly int. prior to 1542

R. FOETIDA BICOLOR, ((Jacquin) Willmott), Sp, rb; (Austrian Copper Rose, R. aurantiaca, R. bicolor, R. eglanteria punicea, R. lutea punicea, R. luteabicolor, R. punicea); flowers orange-scarlet within, yellow reverse; (28); [R. foetida sport]; Int. prior to 1590

R. foetida harisonii *see* **'HARISON'S YELLOW'**

R. FOETIDA PERSIANA, ((Lemaire) Rehder), Sp, my; (Persian Yellow Rose, R. hemisphaerica plena, R. lutea persiana, R. lutea plena); dbl.; (28); 1837

R. FOLIOLOSA, (Nuttall ex Torrey & Gray), Sp, mp; flowers rose borne solitary or in few-flowered clusters, 1.5 in., spring bloom; globose fruit; 1 1/2 ft. growth; Eurosa, Carolinae (14)

R. FOLIOLOSA ALBA, ((Bridwell) Rehder), Sp, w; (14)

R. FORRESTIANA, (Boulenger), Sp, dp; flowers rose borne solitary or few, 1 in.; Eurosa, Cinnamomeae

R. fortuneana *see* **'FORTUNIANA'**

R. fortuniana *see* **'FORTUNE'S DOUBLE YELLOW'**

R. fragariaeflora *see* **R. CYMOSA**

R. fragrans *see* **R. PALUSTRIS**

R. franchetii *see* **R. LUCIAE**

R. franchetii paniculigera *see* **R. MULTIFLORA**

R. francofurtana *see* **'EMPRESS JOSEPHINE'**

R. francofurtensis *see* **R. X FRANCOFURTANA**

R. fraxinifolia *see* **R. BLANDA**

R. fraxinifolia *see* **R. PENDULINA**

R. freudiana *see* **'DUPONTII'**

R. froebelii *see* **R. CANINA FROEBELII**

R. frutetorum *see* **R. CORIIFOLIA**

R. fujisanensis *see* **R. LUCIAE**

R. GALLICA, (Linnaeus), HGal, dp; (French Rose, Provins Rose, R. austriaca, R. grandiflora, R. olympica, R. rubra, R. sylvatica); flowers deep pink to crimson, on stout pedicels, 2–3 in., summer bloom; subglobose or turbinate, brick-red fruit; 5 ft. growth; Eurosa, Gallicanae (28)

R. GALLICA AGATHA, ((Thory) Loiseleur), HGal, lp; (Agatha); flowers inner petals concave, dbl.

R. gallica centifolia *see* **'CABBAGE ROSE'**

R. GALLICA CONDITORUM, (Dieck), HGal, dr; (Conditorum, Tidbit Rose); flowers dark red-purple, paler reverse, perhaps identical with Parkinson's Hungarian Rose, semi-dbl.; (28)

R. gallica damascena *see* **'SUMMER DAMASK'**

R. gallica grandiflora *see* **'ALIKA'**

R. GALLICA HAPLODONTA, ((Borbas) Braun), Sp, dp; foliage wild variety with simply serrate and woolly styles

R. gallica macrantha *see* **'GALLICA MACRANTHA'**

R. gallica maxima *see* **R. GALLICA OFFICINALIS**

R. GALLICA OFFICINALIS, Sp, dp; ((The) Apothecary's Rose of Provins, (The) Apothecary's Rose of Provins, Apothecary's Rose, Double French Rose, Officinalis, R. gallica maxima, R. gallica plena, R. officinalis, R. provincialis, R. X centifolia provincialis, Red Rose of Lancaster, Rose of Provins); flowers deep pink, yellow stamens, semi-dbl.; few prickles; foliage dark; branching growth

R. gallica plena *see* **R. GALLICA OFFICINALIS**

R. GALLICA PUMILA, ((Jacquin) Seringe), Sp, dp; (R. austriaca pygmaea, R. humilis, R. pumila); flowers red, single; dwarf growth; (28)

R. gallica rosa mundi *see* **R. GALLICA VERSICOLOR**

R. gallica variegata *see* **R. GALLICA VERSICOLOR**

R. GALLICA VELUTINAEFLORA, Sp; (R. velutinaeflora); flowers described by déséglise as deep velvety red, late spring; pyriform, reddish-orange fruit

R. GALLICA VERSICOLOR, Sp, pb; (R. gallica rosa mundi, R. gallica variegata, R. mundi, Rosa Mundi); flowers striped white, pink and red, yellow stamens, semi-dbl.; (28); [**R. GALLICA OFFICINALIS** sport]

R. gallicaremensis *see* **'BURGUNDIAN ROSE'**

R. gebleriana *see* **R. LAXA**

R. gechouitangensis *see* **R. X ODORATA**

R. gentili, Sp, w

R. GENTILIANA, (Léveillé & Vaniot), Sp, w; (R. polyantha grandiflora, R. wilsonii); flowers creamy white, borne in dense clusters, semi-dbl.; red fruit; red glandular branches; vigorous shrub growth; Eurosa, Synstylae (14); 1907

R. gentiliana *see* **R. HENRYI**

R. gentilis *see* **R. PENDULINA GENTILIS**

R. germanica *see* **R. X FRANCOFURTANA**

R. GIGANTEA, (Collett), Sp, w; (R. macrocarpa, R. X odorata gigantea, R. xanthocarpa); flowers creamy white, 4–5 in.; vigorous growth; climbing to 50 ft.; Eurosa, Chinensis (14); 1889

R. GIGANTEA ERUBESCENS, (Focke), Sp, mp; (R. X odorata erubescens)

R. GIRALDII, (Crépin), Sp, pb; flowers pink with white center borne solitary or few, 1 in.; ovoid, 1/3-1/2 in., scarlet fruit; 6 ft. growth; Eurosa, Cinnamomeae (14)

R. giraldii *see* **R. GLAUCA**

R. GIRALDII VENULOSA, (Rehder & Wilson), Sp, pb; 1907

R. glabrata *see* **R. X SPINULIFOLIA**

R. glandulosa *see* **R. PENDULINA**

R. glauca *see* **R. DUMALIS**

R. GLAUCA, (Pourret), Sp, mp; (R. ferruginea, R. giraldii, R. ilseana, R. lurida, R. rubrifolia); flowers pink borne 1-3 together, 1.5 in., spring bloom; subglobose, (see endpapers-rubrifolia), scarlet fruit; foliage reddish; 6 ft. growth; Eurosa, Caninae (28)

R. glaucodermis *see* **R. GYMNOCARPA**

R. glaucophylla *see* **R. HEMISPHAERICA**

R. GLOMERATA, (Rehder & Wilson), Sp, w; 1 in., borne in dense corymbs, moderate fragrance; subglobose, about 1/3 in., orange-red fruit; 20 ft. growth; Eurosa, Synstylae; 1908

R. GLUTINOSA, (Sibthorp & Smith), Sp, mp; (Pine-Scented Rose, R. calabrica, R. ferox, R. libanotica, R. pulverulenta, R. pustulosa); flowers pink, small, moderate, pine-scented fragrance; globose, small (see endpapers) fruit; dwarf, glandular growth; Eurosa, Caninae (35, 42); 1821

R. GLUTINOSA DALMATICA, ((Kerner) Borbas), Sp, mp; (R. dalmatica); ellipsoid fruit; (35, 42)

R. gmelinii *see* **R. ACICULARIS FENNICA**

R. godetii *see* **R. MARGINATA GODETII**

R. GRACILIFLORA, (Rehder & Wilson), Sp, mp; flowers pale rose borne solitary, but numerous along the stem, 1.5 in.; Eurosa, Pimpinellifoliae; 1908

R. grandiflora *see* **R. GALLICA**

R. grandiflora *see* **R. SPINOSISSIMA ALTAICA**

R. granulosa *see* **R. MAXIMOWICZIANA**

R. gratiosa *see* **R. BLANDA**

R. gratissima *see* **R. CALIFORNICA**

R. graveolens *see* **R. INODORA**

R. grevillii *see* **R. MULTIFLORA CARNEA**

R. grosseserrata *see* **R. MACOUNII**

R. GYMNOCARPA, (Nuttall ex Torrey & Gray), Sp, lp; (Little Woods Rose, R. abietorum, R. glaucodermis); flowers pale pink borne solitary, on short lateral branchlets, 1 in., early summer; small, globose, orange-red fruit; 10 ft. growth; Eurosa, Cinnamomeae (14); 1893

R. gypsicola *see* **R. DUMALIS**

R. hackeliana *see* **R. HECKELIANA**

R. hakonensis *see* **R. LUCIAE**

R. HAWRANA, (Kmet), Sp, lp; 2 in.; globose, densely bristly fruit; Eurosa, Caninae (28)

R. headleyensis *see* **'HEADLEYENSIS'**

R. HECKELIANA, (Trattinnick), Sp, lp; (R. hackeliana); flowers pink, small, borne usually solitary; dwarf growth; Eurosa, Caninae

R. HELENAE, (Rehder & Wilson), Sp, w; (R. floribunda, R. moschata helenae, R. moschata micrantha); 1.5 in., borne in many-flowered umbel-like corymbs 3-6 in diam, late spring, moderate fragrance; ovoid or oblong-obovoid, about 1/2 in., scarlet fruit; 15 ft. growth; Eurosa, Synstylae (14); 1907

R. heliophila *see* **R. ARKANSANA**

R. heliophila alba *see* **R. ARKANSANA ALBA**

R. HEMISPHAERICA, (Hermann), Sp, my; (R. glaucophylla, R. rapinii, R. sulphurea, Sulphur Rose); flowers sulfur-yellow, nodding, borne usually solitary, dbl., cupped, late spring; foliage bluish-green; Eurosa, Pimpinellifoliae (28); Int. before 1625

R. hemisphaerica plena *see* **R. FOETIDA PERSIANA**

R. HEMISPHAERICA RAPINII, ((Boissier) Rowley), Sp, my; (R. rapinii); (28); 1933

R. HEMSLEYANA, (Täckholm), Sp, lp; (R. macrophylla, R. setipoda); flowers corymbose, 2 in., late spring; ovoid, about 1 in., fruit; stems long, with a distinct neck; Eurosa, Cinnamomeae (42); 1904

R. HENRYI, (Boulenger), Sp, w; (R. gentiliana); flowers corymbose, 1.5 in., summer bloom, moderate fragrance; globose, about 1/3 in. fruit; Eurosa, Synstylae (14); 1907

R. heterophylla *see* **R. MOLLIS**

R. hibernica *see* **'TACKHOLMII'**

R. hillieri *see* **'HILLIER ROSE'**

R. hirtula *see* **R. ROXBURGHII HIRTULA**

R. hispanica *see* **R. POUZINII**

R. hispida *see* **R. POMIFERA**

R. HISPIDA, (Sims, not others), Sp, my; (R. lutescens, R. spinosissima hispida); flowers sulfur-yellow, rather, 2.5–3 in.; (28, 29); Prior to 1781

R. hoffmeisteri *see* R. macrophylla

R. HOLODONTA, (Stapf), Sp, lp; (R. moyesii rosea); foliage leaves coarsely serrate; Eurosa, Cinnamomeae (28); 1908

R. hookeriana *see* R. macrophylla

R. HORRIDA, (Fischer), Sp, w; (R. ferox); 1.5 in., borne solitary or few; subglobose, 1/3-1/2 in. across, dark red fruit; prickly; foliage very small, roundish leaflets; low shrub growth; Eurosa, Caninae (35); 1796

R. hudsoniana *see* **R. PALUSTRIS**

R. hudsoniana scandens *see* **R. PALUSTRIS SCANDENS**

R. HUGONIS, (Hemsley), Sp, my; (Father Hugo Rose, Father Hugo's Rose, Goolden Rose of China, R. xanthina); single, 2.5 in., borne solitary, on slender glabrous pedicels, early spring; depressed-globose, deep scarlet fruit; 6 ft. branches drooping growth; Eurosa, Pimpinellifoliae (14); 1899

R. humilis *see* **R. CAROLINA**

R. humilis *see* **R. GALLICA PUMILA**

R. humilis *see* **R. MARGINATA**

R. humilis grandiflora *see* **R. LAXA**

R. humilis lucida *see* **R. VIRGINIANA**

R. humilis triloba *see* **R. CAROLINA TRILOBA**

R. humilis villosa *see* **R. CAROLINA VILLOSA**

R. hystrix *see* **R. LAEVIGATA**

R. illinoensis *see* **R. SPINOSISSIMA**

R. ilseana *see* **R. GLAUCA**

R. incarnata *see* **'GREAT MAIDEN'S BLUSH'**

R. inconsiderata *see* **R. POUZINII**

R. indica *see* **R. CHINENSIS**

R. indica *see* **R. CYMOSA**

R. indica cruenta *see* **'SANGUINEA'**

R. indica fragrans *see* **R. X ODORATA**

R. indica longifolia *see* **R. CHINENSIS LONGIFOLIA**

R. indica noisettiana *see* **'NOISETTE ROSE'**

R. indica ochroleuca *see* **'PARKS' YELLOW TEA-SCENTED CHINA'**

R. indica odorata *see* **R. X ODORATA**

R. indica odoratissima *see* **R. X ODORATA**

R. indica pumila *see* **R. CHINENSIS MINIMA**

R. indica semperflorens *see* **'SLATER'S CRIMSON CHINA'**

R. indica vulgaris *see* **R. CHINENSIS**

R. inermis *see* **R. X FRANCOFURTANA**

R. INODORA, (Fries), Sp, w; (R. agrestis inodora, R. caryophyllacea, R. elliptica, R. graveolens, R. klukii); flowers white or pink, 1–1.5 in.; ovoid, bright red fruit; Eurosa, Caninae (35, 42)

R. inodora *see* **R. OBTUSIFOLIA**

R. intermedia *see* **R. MULTIFLORA**

R. involucrata *see* **R. CLINOPHYLLA**

R. irridens *see* **R. LONGICUSPIS**

R. jackii *see* **R. MAXIMOWICZIANA JACKII**

R. jackii pilosa *see* **R. MAXIMOWICZIANA PILOSA**

R. jasminoides *see* **R. LUCIAE**

R. jundzilli godetii *see* **R. MARGINATA GODETII**

R. jundzillii *see* **R. MARGINATA**

R. kamtchatica *see* **R. RUGOSA KAMTCHATICA**

R. kelleri *see* **R. MAXIMOWICZIANA JACKII**

R. kentuckensis *see* **R. SETIGERA TOMENTOSA**

R. klukii *see* **R. INODORA**

R. KORDESII, (Wulff), (classification as "species" disputed), dp; semi-dbl., cupped; elliptic-ovoid fruit; foliage shining green; (28); 1952; (a variety that arose in cultivation as a result of spontaneous chromosome doubling (see Max Graf))

R. KOREANA, (Komarov), Sp, lp; flowers white, flushed pink, borne solitary, but numerous along the, 1 in.; ovoid, 1/2 in. long, orange-red fruit; Eurosa, Pimpinellifoliae (14); 1917

R. korsakoviensis *see* **R. ACICULARIS**

R. LAEVIGATA, (Michaux), Sp, w; (Cherokee Rose, R. camellia, R. cherokeensis, R. hystrix, R. nivea, R. sinica, R. ternata, R. triphylla); flowers white, rarely rose, borne solitary, 2.5–3.5 in., early spring, moderate fragrance; large, obovoid, bristly, yellow, later turning brown fruit; foliage glossy, leaflets 3; tall, climbing growth; Eurosa, Laevigatae (14); 1759

R. LATIBRACTEATA, (Boulenger), Sp, mp;, in two's or corymbose; foliage bracts broad, leaflets 7, up to 1 in long; 1936

R. laurentiae *see* **R. CHINENSIS MINIMA**

R. lawranceana *see* **R. CHINENSIS MINIMA**

R. LAXA, (Lindley), Sp, mp; (R. carolina grandiflora, R. humilis grandiflora, R. lindleyi, R. obovata); flowers bright pink, 2 in.; foliage larger than r; Eurosa, Carolinae (28)

R. laxa *see* **R. CANINA FROEBELII**

R. LAXA, (Retzius), Sp, w; (R. gebleriana, R. soongarica); flowers small; oblong-ovoid, small. fruit; foliage small, light green; Eurosa, Cinnamomeae (28)

R. lebrunei *see* **R. MULTIFLORA CARNEA**

R. lehmanniana *see* **R. BEGGERIANA**

R. LESCHENAULTII, (Wight & Arnott), Sp, w; (R. moschata leschenaultii); flowers larger, in few-flowered corymbs; Eurosa, Synstylae (14)

R. leucantha *see* **R. CANINA**

R. l'heritierana *see* **'BOURSAULT ROSE'**

R. libanotica *see* **R. GLUTINOSA**

R. lindleyana *see* **R. CLINOPHYLLA**

R. lindleyi *see* **R. LAXA**

R. linkii *see* **R. MULTIFLORA**

R. LONGICUSPIS, (Bertoloni), Sp, w; (R. charbonneaui, R. irridens, R. lucens, R. willmottiana); flowers petals silky outside, borne corymbose, 1.5 in.; ovoid, scarlet or orange-red fruit; foliage half-evergreen; 20 ft. growth; tender.; Eurosa, Synstylae (14)

R. longifolia *see* **R. CHINENSIS LONGIFOLIA**

R. lucens *see* **R. LONGICUSPIS**

R. LUCIAE, (Franchet & Rochebrune), Sp, w; (R. franchetii, R. fujisanensis, R. hakonensis, R. jasminoides, R. luciae fujisanensis); flowers smaller; habit more upright growth; Somewhat less hardy; Eurosa, Synstylae (14); 1880

R. luciae *see* **R. WICHURANA**

R. luciae fujisanensis *see* **R. LUCIAE**

R. luciae wichurana *see* **R. WICHURANA**

R. luciaetaquetiana *see* **R. WICHURANA**

R. lucida *see* **R. BRACTEATA**

R. lucida *see* **R. VIRGINIANA**

R. lucida alba *see* **R. VIRGINIANA**

R. lucida plena *see* **R. X RAPA**

R. lurida *see* **R. GLAUCA**

R. lutea *see* **R. FOETIDA**

R. lutea hoggii *see* **'HARISON'S YELLOW'**

R. lutea persiana *see* **R. FOETIDA PERSIANA**

R. lutea plena *see* **R. FOETIDA PERSIANA**

R. lutea punicea *see* **R. FOETIDA BICOLOR**

R. luteabicolor *see* **R. FOETIDA BICOLOR**

R. lutescens *see* R. hispida

R. lutetiana *see* **R. CANINA LUTETIANA**

R. lyellii *see* **R. CLINOPHYLLA**

R. lyonii *see* **R. CAROLINA VILLOSA**

R. lyonii alba *see* **R. CAROLINA ALBA**

R. macartnea *see* **R. BRACTEATA**

R. macdougalii *see* **R. NUTKANA HISPIDA**

R. MACOUNII, (Greene), Sp, lp; (R. grosseserrata, R. naiadum, R. subnuda, R. woodsii); flowers pale pink, small; depressed-globose fruit; low growth; Eurosa, Cinnamomeae (14, 21); Int. prior to 1826

R. macounii *see* **R. WOODSII**

R. macrantha *see* **'GALLICA MACRANTHA'**

R. MACRANTHA, (hort., not Desportes), Sp, lp; flowers blush-pink, large, single; 3/4 in., subglobose, dull red fruit; lax, trailing shrub growth

R. macrocarpa *see* **R. GIGANTEA**

R. macrophylla *see* **R. HEMSLEYANA**

R. MACROPHYLLA, (Lindley), Sp, dp; (R. hoffmeisteri, R. hookeriana); flowers red borne 1-3 together, 2 in.; oblong-ovoid, 1-1 1/2 in. long, red fruit; Eurosa, Cinnamomeae (14, 28); 1818

R. macrophylla acicularis *see* **R. PERSETOSA**

R. macrophylla crasseaculeata *see* R. setipoda

R. MACROPHYLLA DONCASTERII (DONCASTERII), Sp, dp, 1930; flowers deep pink; hips flagon-shaped, large, red; foliage purplish-green; stems plum colored; arching growth (6 ft.); Hurst, C.C.; Int. E. Doncaster of J. Burrell & Co., ca. 1930

R. macrophylla gracilis *see* **R. PERSETOSA**

R. macrophylla rubrostaminea *see* **R. MOYESII**

R. majalis *see* **R. CINNAMOMEA PLENA**

R. manca *see* **R. NUTKANA**

R. manettii *see* **'MANETTI'**

R. MARGINATA, (Wallroth), Sp, lp; (R. humilis, R. jundzillii, R. trachyphylla, R. zagrabiensis); flowers pink borne solitary or few, 2–2.5 in.; globose to ellipsoid, scarlet fruit; Eurosa, Caninae (42)

R. MARGINATA GODETII, ((Grenier) Rehder), Sp, lp; (R. godetii, R. jundzilli godetii); flowers small; low shrub growth

R. MARRETII, (Léveillé), Sp, mp; (R. rubrostipullata); flowers pink borne usually 3-6, 1.5–2 in.; subglobose, 1/3-1/2 in. fruit; stems dark purple; Eurosa, Cinnamomeae (14)

R. maximilianii *see* **R. WOODSII**

R. MAXIMOWICZIANA, (Regel), Sp, w; (R. faureri, R. granulosa); 1.5 in., borne in many-flowered corymbs; partially climbing. growth; Eurosa, Synstylae (14)

R. MAXIMOWICZIANA JACKII, ((Rehder) Rehder), Sp, w; (Jack Rose, R. coreana, R. jackii, R. kelleri); prickles without bristles; (14); 1905

R. MAXIMOWICZIANA PILOSA, ((Nakai) Nakai), Sp, w; (R. jackii pilosa); flowers leaf and stalks pubescent; (14)

R. melina *see* **R. X ENGELMANNII**

R. mexicana *see* **R. CAROLINA GLANDULOSA**

R. MICRANTHA, (Smith), Sp, lp; (R. floribunda, R. nemorosa, R. rubiginosa, R. rubiginosa nemoralis); flowers pale pink, small; 6 ft. growth; Eurosa, Caninae (35, 42)

R. microcarpa *see* **R. CYMOSA**

R. microcarpa *see* **R. MULTIFLORA**

R. microphylla *see* **R. ROXBURGHII**

R. microphyllahirtula *see* **R. ROXBURGHII HIRTULA**

R. MINUTIFOLIA, (Englemann), Sp; flowers pink or white short-pediceled, 1 in., early spring; subglobose, hispid fruit; 4 ft. growth; Minutifoliae (Hesperhodos), (14)

R. mirifica *see* **R. STELLATA MIRIFICA**

R. mitissima *see* **R. SPINOSISSIMA INERMIS**

R. mohavensis *see* **R. WOODSII**

R. mokanensis *see* **R. WICHURANA**

R. MOLLIS, (Smith), Sp, dp; (R. heterophylla, R. mollissima, R. villosa, R. villosa mollissima); flowers deep pink, rarely white borne 1-3 together, 1.5–2 in., summer bloom; small fruit; foliage branchlets purple, more or less bloomy; 4 ft. growth; Eurosa, Caninae (28); 1818

R. mollissima *see* **R. MOLLIS**

R. MONTANA, (Chaix), Sp, lp; flowers pink borne solitary or few, 1.5 in.; oblong-ovoid, 3/4 in. fruit; vigorous growth; Eurosa, Caninae (42)

R. montana chavinii *see* **R. CHAVINII**

R. MONTEZUMAE, (Humboldt & Bonpland ex Thory), Sp, dp; flowers pale red, 1.5 in., june; 3 ft. growth; Eurosa, Caninae (35)

R. MOSCHATA, (Herrmann), Sp, w; (Musk Rose, R. ruscinonensis); 1.5–2 in., borne usually in 7-flowered corymbs, summer-fall bloom, moderate, musk fragrance; ovoid, small fruit; Eurosa, Synstylae (14); 1540

R. MOSCHATA ABYSSINICA, ((Lindley) Rehder), Sp, w; (R. abyssinica); prickles more prickly; tender growth; (14)

R. moschata autumnalis *see* **'NOISETTE ROSE'**

R. moschata flore semipleno *see* **R. MOSCHATA PLENA**

R. moschata helenae *see* **R. HELENAE**

R. moschata leschenaultii *see* **R. LESCHENAULTII**

R. moschata micrantha *see* **R. HELENAE**

R. moschata nastarana *see* **'NASTARANA'**

R. MOSCHATA NEPALENSIS, (Lindley), Sp, w; (Himalayan Musk Rose, R. brownii, R. brunonii); foliage like r moschata but dull green, downy; (14); 1822

R. MOSCHATA PLENA, (Weston), Sp, w; (R. moschata flore semipleno); dbl.; (14)

R. MOYESII, (Hemsley & Wilson), Sp, mr; (R. fargesii, R. macrophylla rubrostaminea); flowers deep blood-red through deep rose to light pink borne solitary, 1.75–2.5 in., summer bloom; oblong-ovoid, narrowed into a neck, 2-2 1/2 in. long, deep orange-red fruit; 10 ft. growth; Eurosa, Cinnamomeae (42); 1894, Re-int. 1903

R. MOYESII FARGESII, (Rolfe), Sp, mr; (R. fargesii); foliage leaflets smaller, broad-oval to suborbicular; (28)

R. moyesii rosea *see* **R. HOLODONTA**

R. MULLIGANII, (Boulenger), Sp, w; flowers corymbose, 2 in.; ovoid, 1/2 in. long fruit; Eurosa, Synstylae (14); 1917

R. MULTIBRACTEATA, (Hemsley & Wilson), Sp, lp; (R. reducta); flowers pink corymbose, with numerous crowded bracts, sometimes few, 1.25 in.; ovoid, 1/3-1/2 in. long, orange-red fruit; 6 ft. growth; Eurosa, Cinnamomeae (28); 1910

R. MULTIFLORA, (Thunberg), Sp, w; (Multiflora Japonica, R. dawsoniana, R. franchetii paniculigera, R. intermedia, R. linkii, R. microcarpa, R. multiflora thunbergiana, R. polyantha, R. thunbergii, R. thyrsiflora, R. wichurae); flowers usually white, or more, borne in many-flowered pyramidal cor, 0.75 in., summer bloom; small, globular fruit; recurving or climbing branches. growth; Eurosa, Synstylae (14, 28); ca. 1810

R. MULTIFLORA CALVA, (Franchet & Savatier), Sp, w; (R. calva)

R. MULTIFLORA CARNEA, (Thory), Sp, lp; (R. blinii, R. florida, R. grevillii, R. lebrunei, R. multiflora plena, R. rubeoides); dbl.; (14); 1804

R. MULTIFLORA CATHAYENSIS, (Rehder & Wilson), Sp, lp; (R. cathayensis, R. uchiyamana); flowers pink borne in rather corymbs, 1.5 in., flat; (14); 1907

R. multiflora dawsoniana *see* **R. MULTIFLORA ROSEIFLORA**

R. MULTIFLORA NANA HORT., (hort.), Pol; flowers white to pink, single to very small, in clusters, dbl., recurrent bloom; bushy, dwarf (1-2 ft.) growth

R. multiflora platyphylla *see* **'SEVEN SISTERS'**

R. multiflora plena *see* **R. MULTIFLORA CARNEA**

R. MULTIFLORA ROSEIFLORA, ((Focke) Rehder), Sp, lp; (R. multiflora dawsoniana); semi-dbl., 1 in.

R. multiflora thunbergiana *see* **R. MULTIFLORA**

R. MULTIFLORA WATSONIANA, ((Crépin) Matsumura), Sp, lp, 1870; (Bamboo Rose, R. watsoniana); (14); Int from a Japanese garden in 1870

R. mundi *see* **R. GALLICA VERSICOLOR**

R. muriculata *see* **R. NUTKANA**

R. MURIELAE, (Rehder & Wilson), Sp, w; 1 in., borne in 3-7 flowered corymbs; ellipsoid, 1/2-3/4 in., orange-red fruit; growth to 8 ft.; Eurosa, Cinnamomeae (28); 1904

R. muscosa *see* **'CENTIFOLIA MUSCOSA'**

R. muscosa alba *see* **'SHAILER'S WHITE MOSS'**

R. muscosa simplex *see* **'MUSCOSA SIMPLEX'**

R. mutabilis *see* **'MUTABILIS'**

R. myriacantha *see* **R. SPINOSISSIMA MYRIACANTHA**

R. myriadenia *see* **R. YAINACENSIS**

R. naiadum *see* **R. MACOUNII**

R. nankinensis *see* **R. CHINENSIS**

R. NANOTHAMNUS, (Boulenger), Sp, mp; Eurosa, Cinnamomeae (14); 1935

R. nemorosa *see* **R. MICRANTHA**

R. nipponensis *see* **R. ACICULARIS NIPPONENSIS**

R. NITIDA, (Willdenow), Sp, mp; (R. blanda, R. redutea rubescens, R. rubrispina, Shining Rose); flowers bright pink solitary, or in few-flowered corymbs on slender, 1–2 in., summer bloom; 1 1/2 ft. growth; Eurosa, Carolinae (14)

R. NITIDA SPINOSA, (Lewis), Sp, mp; (R. carolina setigera); stems with enlarged prickles new england & e canada

R. nivea *see* **R. LAEVIGATA**

R. NUTKANA, (Presl), Sp, mp; (Nutka Rose, R. manca, R. muriculata, R. spaldingii); flowers pink usually solitary, 2–2.5 in., summer bloom; globose, without a neck (see endpapers) fruit; 5 ft. growth; Eurosa, Cinnamomeae (42); ca. 1876

R. NUTKANA HISPIDA, (Fernald), Sp, mp; (R. macdougalii)

R. nuttalliana *see* **R. PALUSTRIS NUTTALLIANA**

R. obovata *see* **R. LAXA**

R. OBTUSIFOLIA, (Desvaux), Sp; (R. canina tomentella, R. inodora, R. tomentella); flowers white or rose, 1-few, 1.25 in.; ovoid, 1/2-3/4 in. fruit; foliage leaflets 1 1/2 in long, tip rounded, 5-7 long; 12 ft. growth; Eurosa, Caninae (35)

R. ochroleuca *see* **R. SPINOSISSIMA LUTEOLA**

R. officinalis *see* **R. GALLICA OFFICINALIS**

R. olympica *see* **R. GALLICA**

R. omeiensis *see* **R. SERICEA OMEIENSIS**

R. omeiensis chrysocarpa *see* **R. SERICEA CHRYSOCARPA**

R. omeiensis pteracantha *see* **R. SERICEA PTERACANTHA**

R. omissa *see* **R. SHERARDII**

R. orbicularis *see* **R. SERTATA**

R. oreophila *see* **R. X ENGELMANNII**

R. ORIENTALIS, (Dupont ex Seringe), Sp, lp; flowers pink, solitary, short-pediceled; dwarf growth; Eurosa, Caninae (35)

R. oulengensis *see* **R. X ODORATA**

R. oxyacanthos *see* **R. X KOCHIANA**

R. oxyodon *see* **R. PENDULINA OXYODON**

R. PALUSTRIS, (Marshall), Sp, mp; (R. carolina, R. caroliniana, R. corymbosa, R. elongata, R. fragrans, R. hudsoniana, R. pensylvanica, R. salicifolia, R. virginiana, Swamp Rose); flowers pink usually corymbose, 2 in., spring-summer bloom; growth to 8 ft.; Eurosa, Carolinae (14, 28); 1726

R. PALUSTRIS INERMIS, ((Regel) Lewis), Sp, mp; (R. carolina inermis); no prickles; (14)

R. PALUSTRIS NUTTALLIANA, (Rehder), Sp, mp; (R. carolina nuttalliana, R. nuttalliana); (42)

R. PALUSTRIS SCANDENS, Sp, mp; (R. hudsoniana scandens, Swamp Rose); dbl., spring bloom

R. parmentieri *see* **R. DAVIDII ELONGATA**

R. parviflora *see* **R. CAROLINA**

R. parviflora glandulosa *see* **R. CAROLINA GLANDULOSA**

R. parvifolia *see* **'BURGUNDIAN ROSE'**

R. PENDULINA, (Linnaeus), Sp, dp; (Alpine Rose, R. alpina, R. cinnamomea, R. fraxinifolia, R. glandulosa); flowers pink usually solitary or 2-5 together, single, 2 in., spring bloom; fruit usually nodding, oblong or ovoid, with an elongated neck, scarlet; 3 ft growth; Eurosa, Cinnamomeae (28)

R. PENDULINA GENTILIS, ((Sternberg) R. Keller), Sp, dp; (R. gentilis); flowers deep pink; low shrub with bristly branches growth

R. PENDULINA OXYODON, ((Boissier) Rehder), Sp, mp; (R. oxyodon); flowers rose, solitary; (28)

R. PENDULINA PYRENAICA, ((Gouan) R. Keller), Sp, mp; (Pyrenees Rose, R. pyrenaica); foliage leaves glandular; dwarf growth; (28)

R. pennsylvanica *see* R. virginiana

R. pennsylvanica plena *see* **R. CAROLINA PLENA**

R. pensylvanica *see* **R. CAROLINA**

R. pensylvanica *see* **R. PALUSTRIS**

R. PERSETOSA, (Rolfe), Sp, dp; (R. davidii persotosa, R. macrophylla acicularis, R. macrophylla gracilis); flowers deep pink in large panicles, 1 in.; Eurosa, Cinnamomeae (14); 1895

R. phoenica *see* **R. PHOENICIA**

R. PHOENICIA, (Boissier), Sp, w; (R. phoenica);, in many-flowered pyramidal corymbs; not hardy.; Eurosa, Synstylae (14); ca. 1885

R. pimpinellifolia *see* R. spinosissima

R. pimpinellifolia *see* **R. SPINOSISSIMA PIMPINELLIFOLIA**

R. pimpinellifolia elasmacantha *see* **R. ELASMACANTHA**

R. pimpinellifolia tuschetica *see* **R. TUSCHETICA**

R. pimpinellifoliainermis *see* **R. SPINOSISSIMA INERMIS**

R. PINETORUM, (A. Heller), Sp, dp; flowers deep rose, usually solitary, floral tube usually not glandular, 1.5 in.; foliage leaflets doubly serrate, with gland-tipped teeth; 3 ft growth; Eurosa, Cinnamomeae

R. PISOCARPA, (Gray), Sp, mp; flowers pink on short pedicels in several to many-flowered corymbs, 1 in., summer bloom; globose, with a very short neck (see endpapers). fruit; Eurosa, Cinnamomeae (14, 21); Int. ca. 1882

R. pisocarpa ultramontana *see* **R. WOODSII ULTRAMONTANA**

R. pissartii *see* **'NASTARANA'**

R. platyphylla *see* **'SEVEN SISTERS'**

R. poetica *see* **R. WOODSII FENDLERI**

R. polyantha *see* **R. MULTIFLORA**

R. polyantha grandiflora *see* **R. GENTILIANA**

R. polyantho *see* **'SUMMER DAMASK'**

R. POMIFERA, (Herrmann), Sp, mp; (Apple rose, R. hispida, R. villosa); flowers pink 1-3 together, 1.5–2 in., summer bloom; ovoid or subglobose, to 1 in. fruit; 6 ft. growth; Eurosa, Caninae (28); 1771

R. pomifera duplex *see* **'DUPLEX'**

R. pomponia *see* **'BURGUNDIAN ROSE'**

R. pomponia *see* **'ROSE DE MEAUX'**

R. POUZINII, (Trattinnick), Sp, mp; (R. hispanica, R. inconsiderata, R. rubiginosa parvifolia); flowers pink, small, 1-few; ellipsoid, small fruit; 3-6 ft. growth; Eurosa, Caninae (42)

R. PRAELUCENS, (Byhouwer), Sp, lp; near to r. roxburghii, but less bristly, leaflets larger, fe; Microphyllae (Platyrhodon)

R. pratensis *see* **R. CAROLINA**

R. pratincola *see* **R. ARKANSANA**

R. pratincola alba *see* **R. ARKANSANA ALBA**

R. PRATTII, (Hemsley), Sp, mp; flowers pink 1-3 together, 0.75 in.; subglobose to ovoid, 1/4-1/3 in., scarlet fruit; growth to 8 ft.; Eurosa, Cinnamomeae (14)

R. PRIMULA, (Boulenger), Sp, ly; (Incense Rose, R. ecae, R. sweginzowii, R. xanthina, R. xanthina normalis); flowers yellowish white, very early spring; red prickles; foliage small, stiff, remarkably incense-scented; thin, flexible stems; 2-4 ft. growth; Eurosa, Pimpinellifoliae (14); 1910

R. procera *see* White Rose of York

R. prostrata *see* **R. SEMPERVIRENS PROSTRATA**

R. provincialis *see* **R. GALLICA OFFICINALIS**

R. pseudindica *see* **'FORTUNE'S DOUBLE YELLOW'**

R. pseudoscabrata *see* **R. CANINA**

R. pubescens *see* **R. RUGOSA PLENA**

R. pulchella *see* **R. SPINOSISSIMA**

R. pulchella *see* **'ROSE DE MEAUX'**

R. pulverulenta *see* **R. GLUTINOSA**

R. pumila *see* **R. GALLICA PUMILA**

R. punicea *see* **R. FOETIDA BICOLOR**

R. pusilla *see* **R. CAROLINA VILLOSA**

R. pustulosa *see* **R. GLUTINOSA**

R. pyrenaica *see* **R. PENDULINA PYRENAICA**

R. pyrifera *see* R. woodsii

R. rapinii *see* **R. HEMISPHAERICA**

R. rapinii *see* **R. HEMISPHAERICA RAPINII**

R. reclinata *see* **'BOURSAULT ROSE'**

R. reducta *see* **R. MULTIBRACTEATA**

R. redutea rubescens *see* **R. NITIDA**

R. regeliana *see* **R. RUGOSA**

R. regelii *see* **R. BEGGERIANA**

R. remensis *see* **'BURGUNDIAN ROSE'**

R. repens *see* **R. ARVENSIS**

R. reuteri *see* **R. DUMALIS**

R. rouletii *see* **'ROULETII'**

R. ROXBURGHII, (Trattinnick), Sp, mp; (Burr Rose, Chestnut rose, Chinquapin Rose, R. microphylla, R. roxburghii plena); bud prickly, like a chestnut burr; flowers lilac pink borne often solitary, on short pedicels, dbl., 2–2.5 in., spring-summer bloom; depressed-globose, 1-1 1/2 in. fruit; 6 ft. growth; Microphyllae (Platyrhodon), (14)

R. ROXBURGHII HIRTULA, ((Regel) Rehder & Wilson), Sp, mp; (R. hirtula, R. microphyllahirtula); flowers lilac pink; foliage pubescent beneath; Microphyllae (Platyrhodon), (14); Prior to 1880

R. ROXBURGHII NORMALIS, (Rehder & Wilson), Sp, lp; (Single Chestnut Rose); single; Microphyllae (Platyrhodon), (14); 1908

R. roxburghii plena *see* **R. ROXBURGHII**

R. rubella *see* **R. X REVERSA**

R. rubeoides *see* **R. MULTIFLORA CARNEA**

R. rubicans *see* **'GREAT MAIDEN'S BLUSH'**

R. rubifolia *see* **R. SETIGERA TOMENTOSA**

R. rubiginosa *see* **R. EGLANTERIA**

R. rubiginosa *see* **R. MICRANTHA**

R. rubiginosa magnifica *see* **'MAGNIFICA'**

R. rubiginosa nemoralis *see* **R. MICRANTHA**

R. rubiginosa parvifolia *see* **R. POUZINII**

R. rubra *see* **R. GALLICA**

R. rubrifolia *see* **R. GLAUCA**

R. rubrispina *see* **R. NITIDA**

R. rubrostipullata *see* **R. MARRETII**

R. RUBUS, (Léveillé & Vaniot), Sp, w; (Blackberry Rose, R. ernestii); 1–1.33 in., borne in dense corymbs, moderate fragrance; subglobose, about 1/3 in., dark scarlet fruit; climbing, 20 ft. growth; Eurosa, Synstylae (14); 1907

R. RUBUS NUDESCENS, ((Stapf) Rowley), Sp, w; (R. ernestii nudescens)

R. RUDIUSCULA, (Greene), Sp, mp; flowers pink corymbose, few, 2 in.; globose, 1/2 in., red fruit; prickles very bristly; Eurosa, Carolinae (28); 1917

R. RUGOSA, (Thunberg), Sp, m; (Hedgehog Rose, Japanese Rose, Kiska Rose, R. ferox, R. regeliana, Ramanas Rose, Rugosa Rose, Tomato Rose); flowers purple or white borne solitary or few, 2.5–3.5 in., spring-fall bloom; depressed-globose, to 1 in., brick-red fruit; foliage rugose, shining, darkgreen; 6 ft. growth; Eurosa, Cinnamomeae (14)

R. RUGOSA ALBA, ((Ware) Rehder), Sp, w; (R. rugosa albiflora); (14)

R. rugosa albiflora *see* R. RUGOSA ALBA

R. RUGOSA ALBO-PLENA, (Rehder), Sp, w; dbl.

R. rugosa calocarpa *see* 'CALOCARPA'

R. RUGOSA CHAMISSONIANA, (C.A. Meyer), Sp; (R. coruscans); almost free of bristles

R. RUGOSA KAMTCHATICA, ((Ventenat) Regel), Sp; (R. kamtchatica); (14); ca. 1770

R. RUGOSA PLENA, (Regel), Sp, m; (R. pubescens, R. rugosa rebro-plena); flowers purple, dbl.; very hardy shrub on the prairies, where it is called Empress of the North; (14)

R. rugosa rebro-plena *see* R. RUGOSA PLENA

R. rugosa repens alba *see* 'PAULII'

R. rugosa repens rosea *see* 'PAULII ROSEA'

R. RUGOSA ROSEA, (Rehder), Sp, mp; (14)

R. RUGOSA RUBRA, (hort., not Rehder), Sp, m; (R. rugosa typica); flowers brilliant magenta-purple the largest single-flowered form; very vigorous growth; (14)

R. RUGOSA RUGOSA, Sp, m; (R. rugosa thunbergiana); flowers purple; (14)

R. rugosa scabrosa *see* 'SCABROSA'

R. rugosa thunbergiana *see* R. RUGOSA RUGOSA

R. rugosa typica *see* R. RUGOSA RUBRA

R. ruscinonensis *see* R. MOSCHATA

R. rydbergii *see* R. ARKANSANA

R. sabinii *see* R. X INVOLUTA

R. salicifolia *see* R. PALUSTRIS

R. SALICTORUM, (Rydberg), Sp, mp; flowers pink, borne in corymbs, 1.5 in.; globose, about 1/2 in., fruit; prickles nearly spineless; foliage leaflets ovate oblong, mostly 2 in long, 5-7; growth to about 12 ft.; Eurosa, Cinnamomeae (14)

R. sancta *see* R. WOODSII

R. sandbergii *see* R. WOODSII

R. sarrulata *see* R. CAROLINA GLANDULOSA

R. SATURATA, (Baker), Sp, dr; flowers dark red, anthers purple borne solitary, 2 in.; globose-ovoid, 3/4 in. long, coral-red fruit; growth to 8 ft.; Eurosa, Cinnamomeae (28); 1907

R. saxatilis *see* R. CORYMBIFERA

R. sayi *see* R. ACICULARIS

R. sayi *see* R. ACICULARIS SAYI

R. scandens *see* R. SEMPERVIRENS SCANDENS

R. scopulosa *see* 'POKORNYANA'

R. semperflorens *see* 'AUTUMN DAMASK'

R. semperflorens *see* 'SLATER'S CRIMSON CHINA'

R. semperflorens minima *see* R. CHINENSIS MINIMA

R. SEMPERVIRENS, (Linnaeus), Sp, w; (R. alba Allioni, R. atrovirens, R. balearica); 2 in., borne in few-flowered, rarely many-flowered corymbs, summer bloom, slight fragrance; subglobose or ovoid, ., orange-red fruit; foliage evergreen; Eurosa, Synstylae (14, 21, 28)

R. sempervirens anemoniflora *see* R. ANEMONEFLORA

R. SEMPERVIRENS PROSTRATA, (Desvaux), Sp, w; (R. prostrata); ovoid fruit

R. SEMPERVIRENS SCANDENS, ((Miller) De Candolle), Sp, w; (R. scandens); subglobose fruit

R. sepium *see* R. AGRESTIS

R. SERAFINII, (Viviani), Sp, lp; (R. apennina, R. seraphinii); flowers pink, solitary, very short-stalked, 1 in.; globose-ovoid fruit; 1-3 ft. growth; Eurosa, Caninae (35)

R. seraphinii *see* R. SERAFINII

R. seraphinii *see* R. SICULA

R. SERICEA, (Lindley), Sp, w; (R. tetrapetala, R. wallichii); 4–5 petals, 1–1.5 in., early spring; globose or turbinate fruit; 12 ft. growth; Eurosa, Pimpinellifoliae (14); 1822

R. SERICEA CHRYSOCARPA, ((Rehder) Rowley), Sp, w; (R. omeiensis chrysocarpa); bright yellow fruit

R. SERICEA DENUDATA, (Franchet), Sp, w; branches unarmed

R. SERICEA HOOKERI, (Regel), Sp, w; branches glandular

R. SERICEA OMEIENSIS, ((Rolfe) Rowley), Sp, w; (R. omeiensis); single; foliage leaflets more numerous; (14); 1901

R. SERICEA POLYPHYLLA, (Geier), Sp, w; foliage leaflets numerous; (14)

R. SERICEA PTERACANTHA, (Franchet), Sp, w; (R. omeiensis pteracantha); single; stems withlarge, wing-like prickles, deep red and semi-trans; (14); 1890

R. serpens *see* R. ARVENSIS

R. SERTATA, (Rolfe), Sp, mp; (R. orbicularis, R. webbiana); flowers rose or rose-purple solitary on short branchlets, 2–2.5 in.; 5 ft. growth; Eurosa, Cinnamomeae (14); 1904

R. SETIGERA, (Michaux), Sp, dp; (Prairie Rose, R. fenestrata, R. trifoliata); flowers deep rose, fading to whitish in few-flowered corymbs, 2 in., summer bloom, slight fragrance; globular, 1/3 in. fruit; 6 ft.branches recurving or climbing growth; Eurosa, Synstylae (14); 1810

R. SETIGERA INERMIS, (Palmer & Steyermark), Sp, dp; no prickles; foliage leaves glabrous; 1923

R. SETIGERA SERENA, (Palmer & Steyermark), Sp, dp; no prickles; foliage leaves pubescent; (14); 1924

R. SETIGERA TOMENTOSA, (Torrey & Gray), Sp, dp; (R. cursor, R. kentuckensis, R. rubifolia); foliage leaves tomentose beneath; (14); 1800

R. setigera variegata, Sp, pb

R. setipoda *see* R. HEMSLEYANA

R. SETIPODA, (Hemsley & Wilson), Sp, lp; (R. macrophylla crasseaculeata); flowers pale pink in loose corymbs, 2 in., summer bloom; oblong-ovoid with a narrow neck, about 1 in. long, deep red fruit; foliage sweetbriar-scented; 10 ft. growth; Eurosa, Cinnamomeae (28, 42); 1895

R. SHERARDII, (Davies), Sp, dp; (R. omissa, R. subglobsa); flowers deep pink, 1.5–2 in.; ovoid or pear-shaped, 1/2-3/4 in. across fruit; Eurosa, Caninae (28)

R. SHERARDII PERTHENSIS, (Harrison), Sp, dp; remarkable variety with densely glandular hispid fruit; (35)

R. sibirica *see* R. SPINOSISSIMA ALTAICA

R. SICULA, (Trattinnick), Sp; (R. seraphinii); flowers deep red to whitish solitary, rarely to 4, short-stalked, 1–1.25 in.; small, globose fruit; low, 1 1/2-2 ft. growth; Eurosa, Caninae (35)

R. silverhjelmii *see* R. BEGGERIANA

R. silvestris *see* R. ARVENSIS

R. sinica *see* R. CHINENSIS

R. sinica *see* R. LAEVIGATA

R. SINOWILSONII, (Hemsley), Sp, w; 1.5 in., in many-flowered corymbs; subglobose, small, red fruit; vigorous growth; climbing to 50 ft.; half hardy.; Eurosa, Synstylae (14); 1904

R. solandri *see* R. BLANDA

R. SONOMENSIS, (Greene), Sp, mp; flowers bright pink in dense few-flowered corymbs, 1–1.5 in.; 1 ft. growth; Eurosa, Cinnamomeae

R. soongarica *see* R. LAXA

R. sorbiflora *see* R. CYMOSA

R. SOULIEANA, (Crépin), Sp, w; flowers corymbose, 1.5 in.; ovoid or subglobose, 1/3-1/2 in. long, orange-red fruit; 12 ft. growth; Eurosa, Synstylae (14); 1896

R. spaldingii *see* R. NUTKANA

R. sphaerica *see* R. CANINA

R. spinosissima *see* R. CINNAMOMEA

R. SPINOSISSIMA, (Linnaeus), Sp, w; (Burnet Rose, R. illinoensis, R. pimpinellifolia, R. pulchella, Scotch Rose); flowers cream, but white, yellow, pink or purple in garden forms so, 1.25–2 in., spring bloom; globular, black fruit; 3-4 ft. growth; Eurosa, Pimpinellifoliae (28)

R. SPINOSISSIMA ALTAICA, ((Willdenow) Bean), Sp, w; (Altaica, R. altaica, R. grandiflora, R. sibirica, R. spinossissima baltica); flowers large; (28); ca. 1820

R. SPINOSISSIMA ANDREWSII, (Willmott), Sp, mr; dbl.; (28)

R. SPINOSISSIMA BICOLOR, (Andrews), Sp, pb; flowers pale pink becoming cream flushed with pink, later rotate, semi-dbl., cupped; (28)

R. SPINOSISSIMA FULGENS, (Bean), Sp, mp; flowers bright rose; (28)

R. spinosissima hispida *see* R. hispida

R. SPINOSISSIMA INERMIS, ((De Candolle) Rehder), Sp, lp; (R. mitissima, R. pimpinellifoliainermis, R. spinosissima mitissima); branches almost unarmed

R. SPINOSISSIMA LUTEA, (Bean), Sp, my; flowers bright yellow; (28)

R. SPINOSISSIMA LUTEOLA, (Andrews), Sp, ly; (R. ochroleuca); flowers pale yellow, 2 in.; (28)

R. spinosissima mitissima *see* **R. SPINOSISSIMA INERMIS**

R. SPINOSISSIMA MYRIACANTHA, ((De Candolle) Koehne), Sp, w; (R. myriacantha); flowers white, blushed, small; (28); Prior to 1820

R. SPINOSISSIMA NANA, (Andrews), Sp, w; semi-dbl., 2 in.; (28)

R. SPINOSISSIMA PIMPINELLIFOLIA, ((Linnaeus) Hooker), Sp, w; (R. pimpinellifolia); flowers smooth pedicels; (28)

R. spinossissima baltica *see* **R. SPINOSISSIMA ALTAICA**

R. SPITHAMEA, (S. Watson), Sp; (California Ground Rose); flowers usually solitary floral tube often covered with glandtippe; 1/3 - 1 ft. growth; Eurosa, Cinnamomeae (28); Rarely cult.

R. SPITHAMEA SONOMIENSIS, Sp; foliage leaflets often double serrate, with glandtipped tee

R. STELLATA, (Wooten), Sp, m; (R. vernonii); flowers deep rose-purple, solitary, 1.75–2.25 in.; turbinate, galbrous or puberulent with scattered short bristles; prickles young floral branches with white to yellow, occasio; foliage leaflets 3-5; 2 ft. growth; Minutifoliae (Hesperhodos), (14); 1902

R. STELLATA ERLANSONIAE, (Lewis), Sp, m; Minutifoliae (Hesperhodos)

R. STELLATA MIRIFICA, ((Greene) Cockerell), Sp, m; (R. mirifica, The Gooseberry Rose); prickles floral branches gladrous with many internodal gland; 4 ft. growth; Minutifoliae (Hesperhodos), (14); 1916

R. STELLATA MIRIFICA MIRIFICA, (Lewis), Sp, m; Minutifoliae (Hesperhodos)

R. STELLATA STELLATA, (Lewis), Sp, m; few internodal prickles; Minutifoliae (Hesperhodos)

R. stricta *see* **R. ACICULARIS**

R. STYLOSA, (Desvaux), Sp, w; (R. systyla); flowers white or light pink, 1.5–2 in.; (35, 42, 28, 34)

R. suavifolia *see* **R. EGLANTERIA**

R. subblanda *see* **R. BLANDA**

R. subglobsa *see* **R. SHERARDII**

R. subglosa *see* **R. TOMENTOSA SUBGLOBOSA**

R. subnuda *see* **R. MACOUNII**

R. SUBSERRULATA, (Rydberg), Sp, mp; flowers rose solitary, 2 in.; 2 1/2 ft. growth; Eurosa, Carolinae (14, 28)

R. suffulta *see* **R. ARKANSANA**

R. suffulta alba *see* **R. ARKANSANA ALBA**

R. sulphurea *see* **R. HEMISPHAERICA**

R. surculosa *see* **R. CANINA**

R. sweginzowii *see* **R. PRIMULA**

R. SWEGINZOWII, (Koehne), Sp, mp; flowers pink 1-3 together, 1.75 in.; oblong, 1 in. long fruit; Eurosa, Cinnamomeae (42); 1909

R. sweginzowii inermis *see* **R. WARDII**

R. SWEGINZOWII MACROCARPA, (hort.), Sp, mp

R. sylvatica *see* **R. GALLICA**

R. systyla *see* **R. STYLOSA**

R. taquetii *see* **R. WICHURANA**

R. taurica *see* **R. CORYMBIFERA**

R. terebinthinacea *see* **R. X MARCYANA**

R. ternata *see* **R. LAEVIGATA**

R. tetrapetala *see* **R. SERICEA**

R. thea *see* **R. X ODORATA**

R. thoryi *see* **'SEVEN SISTERS'**

R. thunbergii *see* **R. MULTIFLORA**

R. thyrsiflora *see* **R. MULTIFLORA**

R. tomentella *see* **R. OBTUSIFOLIA**

R. TOMENTOSA, (Smith), Sp, lp; (R. cuspidata, R. dimorpha); flowers pale pink, on longer pedicels; smaller fruit; foliage grayish green, downy; Eurosa, Caninae (35)

R. TOMENTOSA SUBGLOBOSA, ((Smith) Carion), Sp, lp; (R. subglosa); subglobose fruit; prickles hooked

R. tongtchouanensis *see* **R. X ODORATA**

R. trachyphylla *see* **R. MARGINATA**

R. trifoliata *see* **R. SETIGERA**

R. triphylla *see* **R. ANEMONEFLORA**

R. triphylla *see* **R. LAEVIGATA**

R. turbinata *see* **R. X FRANCOFURTANA**

R. TURKESTANICA, (Regel), Sp, my; subglobose fruit; Eurosa, Pimpinellifoliae

R. TUSCHETICA, (Boissier), Sp, mp; (R. pimpinellifolia tuschetica); ovate (not obovate) fruit; differs from R. glutinosa in the leaflets; Eurosa, Caninae; 1945

R. uchiyamana *see* **R. MULTIFLORA CATHAYENSIS**

R. ultramontana *see* **R. WOODSII ULTRAMONTANA**

R. unguicularis *see* R. webbiana

R. usitatissima *see* White Rose of York

R. velutinaeflora *see* **R. GALLICA VELUTINAEFLORA**

R. vernonii *see* **R. STELLATA**

R. villosa *see* **R. MOLLIS**

R. villosa *see* **R. POMIFERA**

R. villosa *see* **R. X INVOLUTA**

R. villosa mollissima *see* **R. MOLLIS**

R. vilmorinii *see* **'MICRUGOSA'**

R. virginiana *see* **R. PALUSTRIS**

R. VIRGINIANA, (Miller), Sp, mp; (R. humilis lucida, R. lucida, R. lucida alba, R. pennsylvanica, Virginia Rose); flowers bright pink usually few or solitary, 2 in., summer bloom; fruit remaining plump until next spring (see endpapers); 6 ft. growth; Eurosa, Carolinae (28); Prior to 1807

R. virginiana alba *see* **R. CAROLINA ALBA**

R. virginiana blanda *see* **R. BLANDA**

R. virginiana humilis *see* **R. CAROLINA**

R. VIRGINIANA LAMPROPHYLLA, (Rehder), Sp, mp; foliage leaves lustrous; 1881

R. virginiana plena *see* **R. X RAPA**

R. virginiana plena *see* **'ROSE D'AMOUR'**

R. viridiflora *see* **'GREEN ROSE'**

R. vorbergii *see* **'VORBERGII'**

R. wallichii *see* **R. SERICEA**

R. walpoleana *see* **R. EGLANTERIA**

R. WARDII, (Mulligan), Sp, w; (R. sweginzowii inermis);, up to 3 together; almost unarmed

R. WARDII CULTA, (Mulligan), Sp, w; flowers smaller pedicels often glandular, 1.25 in.; (42)

R. watsoniana *see* **R. MULTIFLORA WATSONIANA**

R. watsonii *see* **R. CORIIFOLIA**

R. webbiana *see* **R. SERTATA**

R. WEBBIANA, (Wallich ex Royle), Sp, mp; (R. unguicularis); flowers large, mostly solitary; ovoid (see endpapers). fruit; Eurosa, Cinnamomeae (14); 1879

R. wichurae *see* **R. MULTIFLORA**

R. WICHURANA, (Crépin), Sp, w; (Memorial Rose, R. bracteata, R. luciae, R. luciae wichurana, R. luciaetaquetiana, R. mokanensis, R. taquetii); 1.5–2 in., in few-flowered pyramidal corymbs, late summer, moderate fragrance; ovoid, to 1/2 in. fruit; foliage half evergreen; prostrate with creeping branches growth; Eurosa, Synstylae (14); 1891

R. willdenowii *see* **R. DAVURICA**

R. WILLMOTTIAE, (Hemsley), Sp, m; flowers rose-purple, solitary, short-stalked, on short lateral branc, 1–1.25 in., spring bloom; subglobose, 1/3-1/2 in. long, bright orange-red fruit; foliage fern-like; 5-10 ft. growth; Eurosa, Cinnamomeae (14); 1904

R. willmottiana *see* **R. LONGICUSPIS**

R. wilsonii *see* **'MICRUGOSA'**

R. wilsonii *see* **R. GENTILIANA**

R. wilsonii *see* **R. X INVOLUTA WILSONII**

R. woodsii *see* **R. MACOUNII**

R. WOODSII, (Lindley 1820), Sp, mp; (Mountain Rose, R. deserta, R. fimbriatula, R. macounii, R. maximilianii, R. mohavensis, R. pyrifera, R. sandbergii, R. woodsii mohavensis); flowers pink, rarely white corymbose or solitary, on very short, sm, 1.5–2 in., summer bloom; globose, with a short neck (see endpapers) fruit; 3 ft. growth; Eurosa, Cinnamomeae (14)

R. WOODSII FENDLERI, ((Crépin) Rehder), Sp, mp; (R. fendleri, R. poetica); (14)

R. WOODSII HISPIDA, (Lewis), Sp, dp; (R. adenosepala); flowers deep pink; hips glandular; (14)

R. woodsii mohavensis *see* R. woodsii

R. WOODSII ULTRAMONTANA, (Lindley (Watson) Taylor & Ma), Sp, mp; (R. californica ultramontana, R. pisocarpa ultramontana, R. ultramontana); flowers clusters of 3-10, 2 in.; subglobose, small fruit; 10 ft. growth; Eurosa, Cinnamomeae (14)

R. X ALBA, (Linnaeus), A, w; (Bonnie Prince Charlie's Rose, Jacobite Rose, La Rose de York, R. procera, R. usitatissima, White Rose of York); dbl., borne usually several together, summer bloom, moderate fragrance; ovoid (see endpapers), scarlet fruit; 6 ft. growth; (42); **[R. GALLICA X R. CANINA]**

R. X anemonoides *see* **'ANEMONE'**

R. X aschersoniana *see* **'ASCHERSONIANA'**

R. X borboniana *see* **'BOURBON ROSE'**

R. X calocarpa *see* **'CALOCARPA'**

R. X cantabrigiensis *see* **'CANTABRIGIENSIS'**

R. X centifolia bullata *see* **'BULLATA'**

R. X centifolia minima *see* **'ROULETII'**

R. X centifolia provincialis *see* **R. GALLICA OFFICINALIS**

R. X cooperi *see* **'COOPER'S BURMESE'**

R. X coryana *see* **'CORYANA'**

R. X damascena *see* **'SUMMER DAMASK'**

R. X DULCISSIMA, (Lunell), S; [R. blanda X R. woodsii]

R. X dupontii *see* **'DUPONTII'**

R. X ENGELMANNII, (S. Watson), S, dp; (R. acicularis engelmannii, R. bakerii, R. engelmannii, R. melina, R. oreophila); (42); [Naturally-occurring hybrid between R. nutkana and R. acicularis]; 1891

R. X fortuniana *see* **'FORTUNIANA'**

R. X FRANCOFURTANA, (Muenchhausen), Misc OGR, m; (R. campanulata, R. francofurtensis, R. germanica, R. inermis, R. turbinata); flowers purple borne 1-3 together, semi-dbl., 2–3 in., summer bloom, slight fragrance; turbinate fruit; 6 ft. growth; (21, 28); [Probably R. cinnamomea X R. gallica]; Possibly int. prior to 1629

R. X harisonii *see* **'HARISON'S YELLOW'**

R. X harisonii vorbergii *see* **'VORBERGII'**

R. X heterophylla *see* **'PROTEIFORMIS'**

R. X hibernica *see* **'HIBERNICA'**

R. X highdownensis *see* **'HIGHDOWNENSIS'**

R. X INVOLUTA, (Smith), Misc OGR, w; (R. braunii, R. coronata, R. doniana, R. sabinii, R. villosa); flowers whitish; (42); [R. spinosissima X R. tomentosa]

R. X INVOLUTA WILSONII, ((Borrer) Baker), Misc OGR; (R. wilsonii); vigorous, shrubby growth; (42)

R. X iwara *see* **'IWARA'**

R. X JACKSONII, (Willmott), S, mr; flowers bright crimson, very free bloomer; (14); [R. rugosa X R. wichurana]

R. X KOCHIANA, (Koehne), S, dp; (R. oxyacanthos); flowers deep rose borne solitary or 2-3, 1.5 in.; (14); [Possibly an F2 from the cross R. palustris X R. spinosissima]

R. X KOEHNEANA, S, m; flowers purplish red, larger; (21); [R. carolina X R. rugosa]

R. X leonida *see* **'MARIA LEONIDA'**

R. X MALYI, (Kerner), S, mr; flowers bright red; foliage similar to spinosissima; (28); [Possibly R. pendulina X R. spinosissima]

R. X MARCYANA, (Bouller), S, m; (R. terebinthinacea); flowers pink to light purple, 2.5–3 in.; long-stalked; low (2 ft., rarely 4 ft.) growth; (21); [R. gallica X R. tomentosa]

R. X mariae-graebneriae *see* **'MARIA GRAEBNER'**

R. X micrugosa *see* **'MICRUGOSA'**

R. X noisettiana *see* **'NOISETTE ROSE'**

R. X noisettiana manettii *see* **'MANETTI'**

R. X ODORATA, (Sweet); (R. chinensis fragrans, R. gechouitangensis, R. indica fragrans, R. indica odorata, R. indica odoratissima, R. oulengensis, R. thea, R. tongtchouanensis); flowers white, light pink or salmon-pink or yellowish singly or 2-3, dbl., 2–3.5 in.; short stems; Eurosa, Chinensis (14, 21, 28)

R. X odorata erubescens *see* **R. GIGANTEA ERUBESCENS**

R. X odorata gigantea *see* **R. GIGANTEA**

R. X odorata ochroleuca *see* **'PARKS' YELLOW TEA-SCENTED CHINA'**

R. X odorata pseudindica *see* **'FORTUNE'S DOUBLE YELLOW'**

R. X paulii *see* **'PAULII'**

R. X paulii rosea *see* **'PAULII ROSEA'**

R. X penzanceana *see* **'LADY PENZANCE'**

R. X pokornyana *see* **'POKORNYANA'**

R. X polliniana *see* **'POLLINIANA'**

R. X protieformis *see* **'PROTEIFORMIS'**

R. X pruhoniciana *see* **'HILLIER ROSE'**

R. X PTERAGONIS, (Krause), S, my; flowers rich yellow, numerous, 1.5 in.; (14); [R. hugonis X R. sericea]

R. X pteragonis cantabrigiensis *see* **'CANTABRIGIENSIS'**

R. X RAPA, (Bosc), Misc OGR, mp; (R. lucida plena, R. virginiana plena); flowers bright pink, dbl.; (21); [R. virginiana X Unknown diploid]; Before 1820

R. X REVERSA, (Waldstein & Kitaibel), Misc OGR, mr; (R. rubella); obovoid, pendulous, scarlet fruit; foliage dark green; (28); [R. pendulina X R. spinosissima]

R. X richardii *see* **St. John's Rose**

R. X RUBROSA, (Preston), S, mp; flowers large; (28); [R. glauca X R. rugosa]

R. X ruga *see* **'RUGA'**

R. X SALAEVENSIS PERRIERI, ((Verlot) Christ), S; flowers rose-purple, borne 1-3 together; [R. dumalis X R. pendulina]

R. X scharnkeana *see* **'SCHARNKEANA'**

R. X spaethiana *see* **'SPAETHIANA'**

R. X SPINULIFOLIA, (Dematra), S, mp; (R. glabrata); 2–2.5 in., summer bloom; (28); [R. pendulina X R. tomentosa(?)]

R. X tackholmii *see* **'TACKHOLMII'**

R. X waitziana *see* **'WAITZIANA'**

R. X waitziana macrantha *see* **'GALLICA MACRANTHA'**

R. X warleyensis *see* **'WARLEYENSIS'**

R. X wintoniensis *see* **'WINTONIENSIS'**

R. xanthina *see* **R. ECAE**

R. xanthina *see* **R. HUGONIS**

R. xanthina *see* **R. PRIMULA**

R. XANTHINA, (Lindley), Sp, my; (Manchu Rose, R. xanthinoides); flowers short-stalked, dbl., 1.75 in., solitary; 10 ft. growth; Eurosa, Pimpinellifoliae (14); 1906

R. xanthina Allard *see* **'ALLARD'**

R. xanthina 'Canary Bird' *see* **'CANARY BIRD'**

R. xanthina ecae *see* **R. ECAE**

R. xanthina normalis *see* **R. PRIMULA**

R. XANTHINA SPONTANEA, (Rehder), Sp, my; single; (14); 1907

R. xanthinoides *see* **R. XANTHINA**

R. xanthocarpa *see* **R. GIGANTEA**

R. yainacencis *see* **R. YAINACENSIS**

R. YAINACENSIS, (Greene), Sp; (R. myriadenia, R. yainacencis); flowers more numerous; smaller fruit; (28)

R. yesoensis *see* **'IWARA'**

R. zagrabiensis *see* **R. MARGINATA**

'R.G. CASSON', HT, pb, 1923; flowers rose and copper, moderate fragrance; Cant, B. R.

'R.M.S. QUEEN MARY', HT, mp, 1937; (Mrs Verschuren); flowers salmon-pink suffused orange, very large, dbl., cupped; foliage leathery; vigorous growth; [Briarcliff X Mrs Sam McGredy]; Verschuren; Dreer;, J&P

'R.S. HUDSON', HT, yb, 1939; flowers yellow tinged red, large, well-formed; vigorous, upright growth; Wheatcroft Bros.

'R.W. PROCTOR', F, dp, 1947; flowers deep rose-pink, borne in trusses, semi-dbl., 10–12 petals, 4 in., slight fragrance; foliage glossy; vigorous growth; [Anne Poulsen sport]; Proctor

Raalte's Golden Harvest, S, dy, 1992; Interplant

'RAAT-KI-RANI', HT, mr, 1975; bud pointed; flowers velvety crimson-red, full, dbl., 30 petals, 4.5 in., slight fragrance; foliage glossy; vigorous, upright growth; [Seedling X Samourai seedling (open-pollinated)]; IARI

'RABBIE BURNS', HT, mr, 1959; bud long, pointed; flowers bright light red, semi-dbl., intense fragrance; foliage dark; vigorous, upright growth; [Ena Harkness X Sutter's Gold]; Arnot; Croll

Rachel *see* 'BOOYOL'

'RACHEL', HT, ob, 1929; flowers orange-buff, flushed carmine, large, dbl., exhibition form; foliage dark; very vigorous growth; RULED EXTINCT 5/84; Pemberton

Rachel Bowes Lyon *see* 'HARLACAL'

'RACHEL CRAWSHAY', HT, op, 1977; flowers pink to orange-salmon, dbl., 30 petals, 5 in., slight fragrance; foliage olive-green; [Fragrant Cloud X Mary Mine]; Harkness

'RACHEL TOWNSEND', HT, yb, 1963; flowers golden yellow tipped carmine, dbl., 30 petals, 4–4.5 in.; foliage dark, glossy; free growth; [Sultane sport]; Townsend; Townsend & Son

'RACHELLE', F, mp, 1976; bud long; flowers french rose, nearly, dbl., 30 petals, 2–3 in., flat, slight fragrance; upright growth; [Antigua X Unnamed cultivar]; Warriner, William A.; J&P

Racy Lady® *see* 'DICWAFFLE'

'RADA', HT, dr, 1975; flowers large, dbl., 75 petals; foliage medium green, glossy; [Baccará X Unnamed seedling]; Staikov, Prof. Dr. V.; Kalaydjiev and Chorbadjiiski

'RADAMES', HT, dr, 1983; flowers large, dbl., 35 petals, intense fragrance; foliage large, medium green, glossy; upright growth; [Unnamed X Unnamed]; Fumagalli, Niso

'RADAR', HT, or, 1953; bud long; flowers light geranium-red, well formed, large, dbl., 45 petals, moderate fragrance; vigorous growth; [Charles Mallerin X Independence]; Meilland, F.; URS

'RADAR ITALIANA', HT, or, 1944; bud pointed; flowers nasturtium-red edged rose, large, dbl., 32–34 petals, moderate fragrance; foliage bright green; vigorous, bushy growth; [Souv. de Denier van der Gon X Brazier]; San Remo Exp. Sta.

'RADAR, CLIMBING', Cl HT, or, 1959; Meilland, Mrs. Marie-Louise; URS

'RADCLIFFE FLAME', HT, mr, 1986; flowers medium, dbl., 26–40 petals, no fragrance; foliage medium, dark green, matt; upright growth; [Alec's Red X Grandpa Dickson]; Thomson, Colin, 1987

'RADIANCE', HT, lp, 1908; (Pink Radiance); bud globular; flowers rose-pink, reverse lighter, large, dbl., 23 petals, cupped, intense, damask fragrance; foliage leathery; vigorous growth; [Enchanter X Cardinal]; Cook, J.W.; P. Henderson

'RADIANCE, CLIMBING', Cl HT, lp, 1926; (Pink Radiance, Climbing); Griffing, W.D.

'RADIANCE, CLIMBING', Cl HT, lp, 1928; (Pink Radiance, Climbing); Catt

'RADIANT', HT, or, 1962; bud spiral; flowers orange-flame, reverse shaded red, dbl., 30 petals, 4–5 in., slight fragrance; foliage glossy; very free growth; RULED EXTINCT 9/87; [Mrs Sam McGredy X Fantasia]; Fletcher; Tucker

Radiant® *see* 'BENRAD'

'RADIANT BEAUTY', HT, mr, 1934; flowers deeper crimson than parent, not turning blue, dbl., moderate fragrance; [Francis Scott Key sport]; Cleveland Cut-Flower Co.

'RADIANT GLOW', F, or, 1953; bud pointed; flowers bright orange-salmon tinted peach, medium, borne singly and, dbl., cupped; foliage leathery, light green; vigorous, upright growth; [Pinocchio X ?]; Quinn; Roseglen Nursery

Radiant Gold *see* 'JACERN'

Radiant Super Glaze *see* 'NOSTRAD'

Radiant Superglaze *see* 'NOSTRAD'

'RADIATION', F, mr, 1960; flowers open, borne in clusters, semi-dbl., 3–3.5 in.; foliage glossy; vigorous growth; [Poulsen's Pink X Pompadour Red]; deRuiter

'RADIEUSE', HT, mr, 1955; bud long; flowers clear red, well formed, large, dbl., 30–35 petals; vigorous, bushy growth; Laperrière; EFR

'RADIMAN', F, ly, 1975; dbl., 50 petals, cupped, blooms in clusters of 5-35, moderate, tea fragrance; foliage dark, glossy; vigorous growth; [Highlight X Masquerade]; Staikov, Prof. Dr. V.; Kalaydjiev and Chorbadjiiski

'RADIO', HT, yb, 1937; flowers yellow slightly tinted pink, striped and marked rose, large, dbl., 50 petals, cupped, moderate, spicy fragrance; foliage wrinkled, light; vigorous growth; [Condesa de Sástago sport]; Dot, Pedro; C-P

'RADIO LANCASHIRE', F, or, 1987; flowers orange-red, reverse slightly lighter, rosette, medium, borne, dbl., 38 petals, slight, fruity fragrance; round, orange fruit; triangle-shaped, brown prickles; foliage medium, dark green, glossy; upright, medium growth; [Dusky Maiden X Matangi]; Bracegirdle, A.J.; Rosemary Roses, 1987

Radio Times *see* **'AUSSAL'**

Radiosa, HT, 1956; Aicardi; (Cavriglia)

'RADIUM', HT, rb, 1922; flowers carmine, shaded coppery red; [Beauté Lyonnaise X Capt. Hayward]; Lippiatt

Radka, HT, dy, 1975; flowers large, dbl., moderate fragrance; Vecera, L.; (Sangerhausen)

'RADOME', F, lp, 1966; flowers pale rose; [Queen Elizabeth sport]; Nicol; Minier

Radox Bouquet *see* 'HARMUSKY'

'RADRAZZ', S, rb, 1999; (**Knock Out**®); flowers medium, light red to deep pink, single, 5–11 petals, 1.5 in., borne in small clusters, slight, tea fragrance; prickles moderate; foliage large, medium green, semi-glossy, very resistant to; bushy, rounded, medium (3 ft) growth; AARS, U.S., 2000; [(Carefree Beauty X Unknown) X (Razzle Dazzle X Unknown)]; Radler, William; Conard-Pyle, 2000

'RADWAY CHARM', HT, dp, 1959; flowers deep pink, base deep yellow, semi-dbl., 20 petals, moderate fragrance; vigorous growth; [Christopher Stone X McGredy's Wonder sport]; Waterhouse Nursery

'RADWAY GLOW', F, op, 1960; flowers coral-pink, borne in large, open clusters, semi-dbl., 14 petals, slight fragrance; foliage dull green; vigorous growth; Waterhouse Nursery

'RADWAY JEWEL', F, dr, 1960; flowers yellow, becoming orange and deep red, medium, borne in large, dbl., 40 petals; foliage light green; Waterhouse Nursery

'RADWAY PINK', F, mp, 1965; flowers rose-pink, borne in clusters, dbl., 42 petals, 3.5 in.; foliage dull green; very free growth; [Margaret X Korona]; Waterhouse Nursery

'RADWAY SCARLET', F, mr, 1963; low growth; [Karl Weinhausen X Seedling]; Waterhouse Nursery

'RADWAY SUNRISE', S, ob, 1962; flowers yellow, shading through orange to red, blooms in clusters, single, 7 petals, 3.5–4 in., recurrent bloom, slight fragrance; foliage dark, glossy; very vigorous growth; [Masquerade X ?]; Waterhouse Nursery

'RAE DUNGAN', HT, yb, 1971; bud long pointed; flowers creamy yellow, edged deep pink, large, dbl., moderate fragrance; foliage dark; vigorous growth [Daily Sketch X Fred Streeter]; Dawson, George

Raf Braeckman, F, ob, 1985; RvS-Melle

'RAFAELA G. DE PEÑA', Pol, ob, 1938; flowers pure orange, dbl.; foliage bright green, leathery; vigorous, bushy growth; Dot, Pedro

'RAFFEL'S PRIDE', HT, mr, 1937; bud small; flowers oriental red, reverse gold-splashed, open, dbl.; vigorous growth; [Talisman X ?]; Raffel; Port Stockton Nursery

'RAFFEL'S YELLOW', HT, my, 1942; flowers pure yellow, well shaped, large, dbl., 25–35 petals; foliage bronze turning very dark; vigorous growth; [Probably Mrs Beatty X Maid of Gold]; Raffel; Port Stockton Nursery

'RAFFLES BRUCE', HT, ab, 1943; flowers apricot and gold, well shaped, 3–4 in.; foliage dark; compact growth; GM, NRS, 1943; [Mrs Sam McGredy X Aureate]; Bees

Ragamuffin, Min, yb

Ragazzina, F; (Cavriglia)

Ragged Robin *see* **'GLOIRE DES ROSOMANES'**

'RAGGEDY ANN', F, mr, 1956; bud short, pointed; semi-dbl., 13 petals, blooms in clusters, slight fragrance; foliage dark, leathery; vigorous, upright growth; [Garnette X Sister Kenny]; Hill, Joseph H., Co.

Ragtime *see* 'LENRAG'

Ragtime *see* 'MACCOURLOD'

Ragtime *see* 'PEACAP'

Ragtime *see* POUltime

'RAINBOW', T, pb, 1889; flowers pink striped carmine; [Papa Gontier sport]; Sievers

Rainbow, S, ob

Rainbow Bliss™ *see* 'DEVDICHA'

Rainbow Cerise™ *see* 'DEVCLAVEL'

Rainbow Crimson™ *see* 'DEVMESI'

Rainbow Delight, Min, mr, 1998

Rainbow Eclipse™ *see* 'DEVECLIPSAR'

Rainbow Gold™ *see* 'DEVORO'

Rainbow Hot Pink™ *see* 'DEVIENTE'

Rainbow Magic™ *see* 'DICXPLOSION'

Rainbow Pink™ *see* 'DEVROSADO'

Rainbow Red™ *see* 'DEVROJO'

Rainbow Robe, HT, m, 1991; Kordes

Rainbow Shower *see* 'LEESHO'

Rainbow Stanford™ *see* 'DEVRICO'

Rainbow Sunrise™ *see* 'DEVAURORA'

Rainbow Surprise™ *see* 'DEVPRESA'

Rainbow Warrior *see* CHRisgood

Rainbow Yellow™ *see* 'DEVAMARILLO'

Rainbow Yellow Parade™ *see* 'POULWEE'

Rainbow's End, Climbing *see* 'SAVACLENA'

Rainbow's End™ *see* 'SAVALIFE'

Raindrops™ *see* 'SAVARAIN'

Raindrops, Min, pb, 1997; Williams, J. Benjamin

Rainer Maria Rilke *see* 'KORSEE'

Rainy Day® *see* 'MACRAIDA'

Raissa®, HT, mp, 1990; Cocker

'RAJA OF NALAGARH', HT, or, 1977; (Raja Surendra Singh of Nalagarh); bud pointed; flowers red, dbl., 32 petals, 4 in., exhibition form, slight fragrance; foliage dark, leathery; vigorous, upright growth; [Samourai X Montezuma]; Pal, Dr. B.P.; Gopalsinamiengar

Raja Surendra Singh of Nalagarh *see* **'RAJA OF NALAGARH'**

'RAJBALA', F, pb, 1975; bud pointed; flowers open, semi-dbl., 10 petals, 4 in.; foliage large, glossy, light; very vigorous, upright, compact growth; [Delhi Princess X Seedling]; Pal, Dr. B.P.; Indian Agric. Research Inst.

'RAJKUMARI', HT, dp, 1975; bud pointed; flowers deep fuchsine-pink, full, very dbl., 70 petals, 4 in.; foliage glossy, light; vigorous, compact growth; [(Charles Mallerin X Delhi Princess) X Seedling]; IARI

Rajni *see* **'ORIENT SPICE'**

'RAKETA', HT, op, 1952; (Rocket); flowers golden orange-pink, dbl., 50 petals, 5 in.; foliage reddish bronze; vigorous, upright, compact growth; [Narzisse X Comtesse Vandal]; Shtanko, E.E.

Rakete, HT, mr, 1972; flowers large, dbl.; GPG Bad Langensalza; (Sangerhausen)

'RAKTAGANDHA', HT, or, 1975; bud long, pointed; flowers vermilion, dbl., 35 petals, 3 in., exhibition form, slight fragrance; foliage glossy; vigorous, upright growth; [Christian Dior X Carrousel seedling]; IARI

Rakthima, HT, mr, 1991; IARI

'RALLYE', Pol, r, 1966; bud ovoid; flowers cognac color with pink, large, borne in clusters, dbl., moderate fragrance; foliage dark, glossy; vigorous, upright growth; [Cognac X Fashion]; Delforge

Ralph®, HT, ab, 1993; deVor

'RALPH LEIGHTY', F, w, 1971; bud ovoid; flowers near white, medium, dbl., exhibition form, slight fragrance; foliage light, soft; moderate, upright growth; [Gene Boerner sport]; Leighty; McFadden

Ralph Moore *see* 'SAVARALPH'

'RALPH TIZARD', F, ob, 1979; (Ralph Tizzard); flowers pure salmon, pointed, dbl., 28 petals, 4 in., moderate fragrance; foliage dark; vigorous growth; [Vera Dalton X Tropicana]; Sanday, John

Ralph Tizzard *see* **'RALPH TIZARD'**

Ralph's Creeper *see* 'MORPAPPLAY'

Ramanas Rose *see* **R. RUGOSA**

Ramapo *see* 'ZIPRAM'

'RAMAT-GAN', F, dy, 1972; bud urn-shaped; flowers deep lemon-yellow, to urn-shaped, dbl., 25 petals, 1 in., cupped, moderate fragrance; foliage light; moderate growth; [Golden Masterpiece X Zorina]; Holtzman; Holtzman Rose Nursery

'RAMBLING RECTOR', HMult, w; semi-dbl., blooms in large clusters, moderate fragrance; vigorous growth

Ramira® *see* KORmeita

'RAMÓN BACH', HT, ob, 1938; flowers bright orange, edged lighter, reverse reddish gold, stamens, dbl., 80 petals, globular, intense, fruity fragrance; foliage glossy, dark; vigorous growth; [Luis Brinas X Condesa de Sástago]; Dot, Pedro; C-P

'RAMONA', S, mr, 1913; (Red Cherokee); flowers carmine-crimson, single; [Anemone sport]; Dietrich & Turner

'RAMPA PAL', HT, pb, 1975; bud ovoid; flowers fuchsine-pink, reverse lighter, very dbl., 60 petals, 4.5 in., exhibition form, slight fragrance; foliage glossy; moderate, upright, bushy growth; Pal, Dr. B.P.

'RAMPANT', HSem, w, 1830;, profuse bloom, sometimes in autumn; Jacques

'RAN', S, mp, 1972; bud globular; flowers pink, open, small, semi-dbl., slight fragrance; foliage small, light, soft; vigorous, upright growth; [R. cinnamomea X R. helenae hybrida]; Lundstad; Norges Landbruks-hogskole

'RANDALL', HGal, mp; flowers an old variety taken by early pioneers to western canada, dbl.; not as hardy as Alika.

Randfontein Gold *see* KORsion

Randilla Jaune®, Min, my

Randilla Rose®, Min, mp

Randilla Rouge®, Min, mr

Rangatarang, HT, pb, 1993; Chiplunkar

Ranger *see* BARger

Rangila, HT, rb, 1995; Tejganga

Rangitoto *see* 'SUNTOTO'

'RANGOLI', F, op, 1978; bud tapered; flowers coral-pink, dbl., 25 petals, 3–3.5 in., exhibition form, slight, fruity fragrance; foliage glossy; dwarf, bushy growth; [Golden Slippers sport]; Thakur; Doon Valley Roses

'RANGSHALA', HT, ab, 1969; bud pointed; flowers apricot, shaded peach and amber yellow, full, open, medium, dbl.; foliage glossy; moderate growth; [Margaret Spaull X Open pollination]; IARI

'RANJANA', HT, or, 1975; bud pointed; flowers rose-opal, full, dbl., 38 petals, 4.5 in., intense fragrance; foliage dark, leathery; very vigorous, upright, bushy growth; [Samourai X ?]; Pal, Dr. B.P.; Anand Roses

'RANKENDE JOHANNA TANTAU', Cl Pol, 1942; Tantau, Math.; (Sangerhausen)

'RANKENDE MINIATURE', Cl Pol, lp, 1908; flowers very small, very dbl., moderate fragrance; Lambert, P.; (Sangerhausen)

'RANKENDE TESCHENDORFF'S JUBILÄUMSROSE', Cl Pol, dp, 1930; flowers medium, dbl., slight fragrance; Teschendorff; (Sangerhausen)

'RANUNCULUS MUSK CLUSTER', HMsk, w; flowers pure white, very dbl.; very vigorous growth

'RAPHAEL', M, w, 1856; flowers well mossed, pinkish white, slightly recurrent bloom; Robert

Raphaela *see* TANelephar

'RAPPERSWIL'®, HT, or, 1975; bud globular; dbl., 30 petals, 4 in., cupped, intense fragrance; foliage dark, leathery; [Fragrant Cloud X Ena Harkness]; Huber

Rapsodia®, F, ob

'RAPTURE', HT, pb, 1926; flowers deeper pink, flushed gold at base; (28); [Mme Butterfly sport]; Traendly & Schenck

'RAPTURE, CLIMBING', Cl HT, pb, 1933; Dixie Rose Nursery

'RAQUEL MELLER', HT, lp, 1957; flowers soft pink, large, dbl., cupped, intense fragrance; foliage glossy; GM, Geneva, 1956; [Edith Krause X Fashion]; Camprubi, C.

Rare Edition, F, rb, 1982; Kasturi

Rashmi, Pol, dr, 1977; Kasturi

Raspberry Beauty™ *see* 'MINIRASP'

'RASPBERRY DELIGHT', HT, rb, 1979; flowers medium red on outer petals shading to creamy coral on inner, dbl., 30 petals, exhibition form, moderate, raspberry fragrance; hooked prickles; foliage medium green, semi-glossy; upright, bushy growth; [Carrousel X First Prize]; Taylor, Thomas E.

Raspberry Ice *see* 'ZIPBERRY'

Raspberry Ice *see* 'KORTABRIS'

Raspberry Punch™ *see* 'JACBUNCH'

Raspberry Red, F, mr, 1995

Raspberry Ripple, HSet, pb

Raspberry Ripple, F, rb, 1998

Raspberry Rose, F, mr

Raspberry Ruffles *see* 'TALRAS'

Raspberry Sunblaze® *see* 'MEIDONFE'

Raspberry Swirl™ *see* 'JACSTAND'

Raspberry Swirl, HT, lp; flowers cream goin to yellow throat, with outer half of petal clear raspberry, exhibition form, moderate fragrance; Edwards, Eddie

Raspberry Wine *see* **'MALINOVKA'**

'RASSVET', HT, pb, 1955; (Daybreak); flowers soft pink, base creamy yellow, medium, very dbl., 48 petals; foliage dark; spreading growth; [Peace seedling]; Klimenko, V. N.

Ratgeber Rose, HMult, mr; flowers medium, dbl., slight fragrance; (Sangerhausen)

'RATHERNICE', HT, ob, 1957; flowers coppery orange; foliage coppery; moderate growth; Bishop; Baker's Nursery

Ratio Times, HP; flowers purple-violet, dbl., cupped

'RAUBRITTER', S, lp, 1936; dbl., globular, in clusters, moderate fragrance; foliage leathery, wrinkled; vigorous, climbing growth; (21); [Daisy Hill X Solarium]; Kordes

Ravel® *see* 'LENRAV'

Ravel *see* RUIsteenka

Raven™ *see* FRYtrooper

Ravensberg®, S, mr, 1986; Noack, Werner

'RAVENSWOOD', Min, yb, 1992; flowers gold, outer petals flushed pink, blooms borne 3-4 per cluster, semi-dbl., 18–20 petals, .5 in., slight fragrance; foliage healthy; upright, bushy growth; [Rise 'n' Shine X Unknown]; Catt, Graeme Charles, 1989; F.D. Catt Wholesale Nursery

'RAVENSWOOD VILLAGE', HT, ly, 1996; flowers light yellow, deeper in autumn, very large, blooms borne most, dbl., 26–40 petals, exhibition form, slight fragrance; some prickles; foliage medium, medium green, glossy; upright, medium growth; [Golden Splendour X Queen Esther]; Poole, Lionel; F. Haynes & Partners, 1997

'RAVING BEAUTY', HT, mp, 1948; bud ovoid, rose-red; flowers tyrian rose, dbl., 35–40 petals, 3–4 in., globular, intense fragrance; foliage leathery, dark; vigorous, upright growth; Hill, Joseph H., Co.

'RAWANA', HT, dy, 1999; (**Golden Wonders 99**); dbl., 15–25 petals, 3.5 in., borne in small clusters, slight fragrance; prickles moderate; foliage medium, medium green, semi-glossy; bushy (2.5 ft) growth; [(Gold Bunny X (Baby Love X seedling)) X Golden Celebration]; Rawlins, R.

'RAY ADELINE', HT, my, 1989; bud pointed; flowers medium, borne usually singly and in sprays of 2-4, dbl., 40 petals, exhibition form, moderate fragrance; small, light green fruit; prickles straight, hooked down, light green; foliage medium, dark green, semi-glossy; upright growth; [First Prize X King's Ransom]; Bevard, Harry D., 1990

'RAY BUNGE', Cl HP, dp, 1927; flowers dark rose, reverse lighter, borne singly or clusters of 7-8, dbl., 30–40 petals, 4–5 in., prolific early spring bloom, repeating later in season, slight fragrance; foliage dark; very long stems; very vigorous (15 1/2 ft) growth; quite hardy.; [Paul Neyron sport]; Bunge; Andrews Nursery Co., 1959

Ray of Hope *see* COCnilly

Ray of Sunshine *see* 'COCCLARE'

'RAYMOND', HT, pb, 1917; bud long, pointed; flowers peach-blossom-pink, center salmon-carmine, dbl.; [Rayon d'Or seedling]; Pernet-Ducher

Raymond Carver, S, my, 1999; Horner

'RAYMOND CHENAULT', HKor, mr, 1960; flowers bright red, blooms in clusters, semi-dbl., 16 petals, 4 in., moderate fragrance; foliage dark, glossy; vigorous (9-12 ft) growth; [R. kordesii X Montezuma]; Kordes, R.

Raymond Kopa *see* 'WEKJUVOO'

'RAYMOND PRIVAT', Pol, m, 1935; flowers violet, borne in clusters, dbl.; vigorous growth; Privat

Raymond's Pearl, F, yb, 1997; flowers very full, medium, very dbl., 41 petals, borne in small clusters, no fragrance; some prickles; foliage medium, medium green, semi-glossy; upright, medium (75cms) growth; [Amber Queen X Brown Velvet]; Rawlins, R.

'RAYON D'OR', HT, my, 1910; flowers golden yellow, dbl.; [Mme Mélanie Soupert X Soleil d'Or]; Pernet-Ducher

'RAYON D'OR', HT, my, 1962; flowers well formed; Combe

Razzle Dazzle *see* 'JACRAZ'

Razzle Dazzle *see* FRYbright

Razzmatazz *see* 'JACMAT'

'REA SILVIA', HT, or, 1958; flowers fiery red; [Baiser X (Peace X seedling)]; Giacomasso

'REACLIBAL', LCl, pb, 1997; flowers semi double, medium, semi-dbl., 8–15 petals, borne in large clusters, moderate fragrance; some prickles; foliage medium, light green, semi-glossy; upright (15ft.) growth; [Ballerina sport]

Real Charmer *see* 'PIXICHAR'

Rea's Rose, HT, ab, 1997

Reba McEntire *see* 'MACHAHEI'

'REBECCA', HT, pb, 1930; bud large, long, pointed; flowers silvery pink, reverse salmon-pink, dbl.; foliage dark, leathery; vigorous growth; Pemberton; Bentall

Rebecca® *see* 'TANREKTA'

Rebecca Anne *see* 'PIXANNE'

'REBECCA CLAIRE', HT, op, 1981; flowers coppery orange edged light coral, blooms borne singly and in, dbl., 28 petals, intense fragrance; medium brown prickles; foliage medium green, semi-glossy; vigorous, bushy growth; Edland Fragrance Medal, ARS, 1980 GM, RNRS, 1980 PIT, RNRS, 1980; [Blessings X Redgold]; Law, M.J., 1986

'REBECCA GUE', B, mp, 1982; flowers distinction from parent not described; [Mme Ernst Calvat sport]; Gue, Derek J.

Rebecca Paul *see* 'HORCOCKALORUM'

'REBECCA'S DELIGHT', F, mp, 1974; flowers soft salmon pink, shaded darker, large, dbl., 24 petals, slight fragrance; foliage light green; vigorous, upright growth; [(Pink Parfait X Highlight) X Circus]; Harkness; Morse Roses

Rebekah, HT, pb

Rebell *see* 'FAYBELL'

Rebell 96 *see* KORvegata

Rebellastar, HT, 1970; Hetzel; (Cavriglia)

'RECOMPENSE', HT, dp, 1957; flowers deep pink, well shaped, medium, moderate fragrance; moderate growth; [Charles P. Kilham X Polly]; Ratcliffe

Reconciliation *see* HARtillery

Record *see* **'DÉCOR'**

'RECUERDO DE ANGEL PELUFFO', HT, rb, 1928; flowers cardinal-red, center garnet-red, dbl., moderate fragrance; [Mme Edouard Herriot X Elvira Aramayo]; Soupert & Notting

'RECUERDO DE ANTONIO PELUFFO', T, yb, 1910; flowers light yellow, edged pink, dbl.; [Mme Mélanie Soupert X Mme Constant Soupert]; Soupert & Notting

'RECUERDO DE BLAS MUNNÉ', Cl HT, rb, 1948; flowers carmine-red shaded cerise red, slight fragrance; foliage dark; long, strong stems; very vigorous growth; [Maria Serrat X Recuerdo del Doctor Ferran]; Munné, M.

'RECUERDO DE FELIO CAMPRUBI', HT, rb, 1931; flowers crimson suffused pink, reverse yellow suffused red, large, dbl., intense fragrance; vigorous growth; [Hugh Dickson X Souv. de Claudius Pernet]; Camprubi, C.

'RECUERDO DEL DOCTOR FERRARI', HT, rb, 1935; bud long, pointed; flowers scarlet-crimson, shaded fiery red, open, large, very dbl., intense fragrance; foliage dark; vigorous growth; [Sensation X K. of K.]; Munné, B.

Red Ace *see* **'AMRUDA'**

'RED ACE', Min, mr, 1980; dbl., 23 petals, exhibition form, borne usually singly, slight fragrance; long, thin, straight prickles; low, compact, bushy growth; [Rise 'n' Shine X Sheri Anne]; Saville, F. Harmon; Nor'East Min. Roses

'RED ADMIRAL', F, mr, 1940; flowers scarlet, borne in clusters; Archer

Red Admiral, HT, mr

Red Alert *see* 'MORALERT'

'RED AMERICAN BEAUTY', HT, mr, 1959; bud ovoid; flowers scarlet overcast rose-red, dbl., 30–35 petals, 4.5–5 in., exhibition form, intense fragrance; foliage leathery, dark; long stems; vigorous, upright, bushy growth; [Happiness X San Fernando]; Morey, Dr. Dennison; J&P

'RED ARROW', Min, mr, 1962; dbl., 40 petals, 1.25 in., exhibition form, blooms in clusters, slight fragrance; foliage leathery; vigorous (12-18 in) growth; [(R. wichurana X Floradora) X Seedling]; Moore, Ralph S.; Sequoia Nursery

Red Azteca *see* 'SELAZTECA'

Red Baby, S, 1992; Embriaco, B.; (Cavriglia)

Red Baby Rambler *see* **'MME NORBERT LEVAVASSEUR'**

Red Bailer, B, mr

Red Ballerina *see* 'HARHERO'

'RED BALLERINA', S, mr, 1976; flowers bright crimson, full, small, semi-dbl., 10 petals, slight fragrance; foliage glossy; [Ballerina X Evelyn Fison]; Fryer, Gareth; Fryer's Nursery, Ltd.

'RED BEAUTY', Min, dr, 1981; flowers dark red, yellow hinge, small to medium, dbl., 35 petals, exhibition form, slight fragrance; foliage small, dark, glossy; bushy growth; [Starburst X Over the Rainbow]; Williams, Ernest D.; Mini-Roses

'RED BEAUTY', HT, mr, 1929; [Matchless sport]; Dunlop; Liggit

Red Belinda, HMsk, mr

Red Bells® *see* 'POULRED'

Red Berlin *see* OLIjplam

'RED BETTER TIMES', HT, mr, 1937; flowers bright clear red; [Better Times sport]; Asmus

'RED BIRD', HT, mr, 1957; flowers bright red; [Better Times sport]; Manda, Jr., E.A.; J&P

Red Bird *see* 'GELBIRD'

Red Blanket® *see* 'INTERCELL'

Red Blush, A, m, 1988; Sievers

'RED BOY', HT, or, 1939; bud long, pointed; flowers fiery orange-red, fading to dominant pink, single and open, semi-dbl., slight fragrance; foliage dark, glossy; vigorous, bushy growth; [Charles K. Douglas X Pres. Herbert Hoover]; Hansen, N.J.; B&A

Red Brigand *see* 'SANTOR'

'RED BUTTON', Min, dr, 1978; bud short, pointed; flowers deep red, small, dbl.; foliage very small, glossy; bushy, compact, spreading growth; [(R. wichurana X Floradora) X Magic Dragon]; Moore, Ralph S.; Sequoia Nursery

'RED CAMELLIA', F, or, 1943; bud small; flowers orange-scarlet, dbl., 20–30 petals, slight fragrance; foliage dark, leathery; vigorous, upright, bushy growth; [Baby Chateau X Folkestone]; Krause

Red Camellia, HT, mr

Red Cameo *see* 'MORANYBLAC'

'RED CAN CAN', Min, mr, 1977; bud pointed; flowers cardinal-red, dbl., 28 petals, 2 in., moderate, fruity fragrance; foliage dark; vigorous, upright growth; [Unnamed seedling X ?]; Lyon

Red Candy, F, lp; short; flowers simple light red with white eye, flat, 8 petals, borne in clusters, slight; medium large, dark green, glossy; [Candy Rose chance seedling sport]; St. Wagner,

1993; Res. Stn. f. Fruit Growing, Cluj, 1995, Romania

Red Cap, HT, dr

'RED CARPET', Cl F, dr, 1971; bud ovoid; flowers dark scarlet, overlaid darker, medium, dbl., globular; foliage large, dark, leathery; very vigorous, climbing growth; [Don Juan X Red Favorite]; Williams, J. Benjamin

'RED CARROUSEL', Min, dp, 1984; flowers deep pink; [Magic Carrousel sport]; Rumsey, R.H.; Roy H. Rumsey, Pty. Ltd.

Red Carrousel, Min, dp

'RED CASCADE', Cl Min, dr, 1976; (Moorcap); bud pointed; flowers deep red, dbl., 40 petals, 1 in., cupped, slight fragrance; foliage small, leathery; prostrate, bushy growth; AOE, 1976; [(R. wichurana X Floradora) X Magic Dragon]; Moore, Ralph S.; Sequoia Nursery

Red Cécile Brünner *see* **'PASADENA TOURNAMENT'**

Red Cécile Brünner, Climbing *see* **'PASADENA TOURNAMENT, CLIMBING'**

Red Cedar *see* 'KORGUND '81'

'RED CHATEAU', HT, mr, 1999; flowers crimson red, dbl., 35 petals, exhibition form, slight fragrance; 3.5 ft growth; [(Samantha X seedling) X seedling]; Teranishi, K.; Itami Rose Nursery, 1997

'RED CHEER', Min, mr, 1975; flowers cherry-red, full, dbl., 45 petals, 1 in., slight fragrance; foliage very small, dark; compact, bushy growth; [Unnamed seedling X Unnamed seedling]; Lyon

'RED CHEERFUL', HT, dr, 1951; bud pointed; flowers deep red, dbl., 30–40 petals, 4.5 in., exhibition form, moderate fragrance; foliage glossy, leathery, veined red; vigorous, upright growth; [Better Times sport]; Blixen; Woodlawn Gardens

Red Cherokee *see* **'RAMONA'**

'RED CHIEF', HT, mr, 1967; flowers large, dbl., 35 petals, exhibition form, intense fragrance; foliage leathery; vigorous, upright, bushy growth; [Seedling X Chrysler Imperial]; Armstrong, D.L.; Armstrong Nursery

'RED COAT', F, mr, 1981; (Redcoat); bud pointed; semi-dbl., 10 petals, borne 1-5 per cluster, slight fragrance; hooked, brown prickles; foliage dark; bushy growth; [Unnamed seedling X Golden Showers]; Austin, David, 1973

'RED COLUMBIA', HT, mr, 1920; flowers rich velvety scarlet, large, dbl., intense fragrance; foliage leathery; very vigorous growth; [Columbia sport]; Hill, Joseph H., Co.

Red Cottage *see* DICxtol

'RED CROSS', HT, ob, 1916; flowers orange-crimson-scarlet, moderate fragrance; Dickson, A.

Red Cross, HT, dr, 1998; Meilland

Red Curtain *see* KORmauret

'RED CUSHION', F, dr, 1966; bud pointed; flowers small blooms in clusters, semi-dbl., slight fragrance; foliage dark, glossy, leathery; vigorous, bushy growth; [Circus X Ruby Lips]; Armstrong, D.L.; Armstrong Nursery

Red Dagmar *see* SPEruge

Red Damask, HGal, mr; flowers medium, dbl., intense fragrance; (Sangerhausen)

'RED DANDY', F, mr, 1959; flowers cherry-red, dbl., 40 petals, 3 in., moderate fragrance; vigorous, upright growth; [Ena Harkness X Karl Herbst]; Norman; Harkness

'RED DAWN', S, mr, 1957; flowers deep rose-red, well-formed, recurrent bloom; hardy.; [New Dawn X ?]; Simonet; Skinner

Red Ded, Min, dr, 1980; flowers dark scarlet-red, small, very dbl.; Cocker; (Sangerhausen)

'RED DELICIOUS', HT, mr, 1942; bud carmine, shaded oxblood-red; flowers brilliant rose-red, dbl., 30–35 petals, 4–6 in., intense fragrance; foliage leathery, dark; weak necks; vigorous, upright growth; [Rome Glory X Chieftain]; Hill, Joseph H., Co.

Red Delicious *see* JACdrive

'RED DELIGHT', HT, mr; flowers bright red; RULED EXTINCT 11/86; [Pink Delight sport]; Avansino; Mortensen, 1935

Red Delight™ *see* 'MINAICO'

'RED DET', Min, mr, 1978; dbl., flat, blooms in clusters, slight fragrance; foliage small, deep green, glossy; vigorous, bushy growth; [(Marlena X Kim) X Little Buckaroo]; Harkness; Rosen-Union

Red Det 80®, Min, or, 1980; Cocker

Red Devil *see* 'DICAM'

Red Diadem® *see* TANetorde

Red Dorothy Perkins *see* **'EXCELSA'**

'RED DRAGON', F, mr, 1970; single, 5 petals, 3.5–4 in., slight fragrance; vigorous growth; [Anna Wheatcroft X Unnamed seedling]; Cants of Colchester, Ltd.

Red Druschki *see* **'RUHM VON STEINFURTH'**

'RED DUCHESS', HT, mr, 1942; bud long, pointed; dbl., 35–45 petals, 3.5–5 in., exhibition form, moderate fragrance; foliage glossy, bronze; long, strong stems; vigorous, upright, bushy growth; [Pink Princess X Crimson Glory]; Brownell, H.C.

'RED DUCHESS, CLIMBING', Cl HT, mr, 1955; Brownell, H.C.

'RED ECHO', Pol, or, 1932; flowers vermilion tinted crimson, dbl., cupped, slight fragrance; foliage dark, wrinkled; dwarf growth; [Echo X Unnamed Hybrid Tea (red)]; Kluis & Koning; J&P

Red Elegance ™ *see* BALIgance

'RED ELF', Min, dr, 1949; bud ovoid; flowers dark crimson, dbl., 23 petals, .75–1 in., slight fragrance; foliage soft, tiny; vigorous, bushy, dwarf growth; [Eblouissant X Tom Thumb]; deVink; C-P

'RED ELLEN POULSEN', Pol, mr, 1918; (Dunkelrote Ellen Poulsen); [Ellen Poulsen sport]; Poulsen, S.

'RED EMBER', HT, or, 1953; flowers flame, well formed, dbl., 30 petals, 4–5 in., moderate fragrance; foliage glossy, dark; free growth; Cant, F.

'RED EMBLEM', F, dr, 1958; bud ovoid; flowers deep red, open, dbl., 30–35 petals, 2.5 in., moderate fragrance; foliage leathery; upright, bushy growth; [Garnette seedling X Pageant]; Boerner; J&P

Red Emperor *see* 'BROEMP'

'RED EMPRESS', LCl, mr, 1956; (Impératrice Rouge, Robur); bud ovoid; flowers cardinal-red, to loosely, blooms borne singly or 2 per stem, dbl., 33 petals, 3.5 in., exhibition form, recurrent bloom, intense fragrance; foliage leathery; vigorous, climbing growth; [(Holstein X Decor) X Self]; Mallerin, C.; C-P;, EFR

'RED ENSIGN', HT, mr, 1947; flowers crimson, dbl., 40–45 petals, 4–5 in., exhibition form, intense, damask fragrance; foliage dark; very vigorous growth; GM, NRS, 1943; [Crimson Glory X Southport]; Norman; Harkness

'RED EXPLORER', Cl Pol, dr, 1938; flowers deep brilliant crimson, borne in large clusters, recurrent bloom; vigorous, climbing growth; [Miss Edith Cavell sport]; Penny

Red Fairy *see* 'MOREDFAR'

Red Fan, HT, or

Red Favorite *see* 'TANSCHWEIGRU'

Red Favorite, Climbing

Red Festival *see* 'DELFESROU'

'RED FINCH', Pol, mp, 1937; flowers cerise, more, more open, more than parent, borne in clusters, dbl., cupped; [Mrs R.M. Finch sport]; Stielow, F.C.

'RED FLARE', LCl, dr, 1954; bud globular; flowers carmine overcast spectrum-red, borne singly and clusters, dbl., 40 petals, 4–4.5 in., cupped, profuse bloom repeated sparingly, moderate fragrance; foliage dark, glossy; height 7-8 ft; [Reine Marie Henriette X Paul's

Scarlet Climber seedling]; Mansuino, Dr. Andrea; J&P

Red Fleurette, S, mr

Red Flower Carpet *see* NOAre

'Red Flush', Min, mr, 1978; bud ovoid; flowers small, dbl., 53 petals, 1.5 in., cupped; foliage green, matt; very compact growth; AOE, 1979; Schwartz, Ernest W.; Nor'East Min. Roses

'Red Fountain', LCl, dr, 1975; bud ovoid, pointed; flowers scarlet, medium, dbl., 20–25 petals, cupped, blooms in clusters, intense fragrance; foliage large, dark, leathery; very vigorous, climbing growth; [Don Juan X Blaze]; Williams, J. Benjamin; C-P

Red France *see* FAZcanne

Red Friendship, S, mr, 1986; Verschuren

'Red Frost', HT, rb, 1993; flowers medium red, suffused white, shaded light yellow at base, blo, dbl., 26–40 petals, 3–3.5 in., exhibition form, slight fragrance; some prickles; foliage medium, dark green, glossy; medium (90-120 cms), bushy growth; [First Prize X Seedling]; Lienau, David W.; Trophy Roses, Ltd., 1993

Red Garnette *see* **'Garnette'**

Red Garter *see* 'BROgar'

'Red Germain', Min, rb, 1975; bud long, pointed; flowers red, reverse lighter, dbl., 25 petals, 1 in., flat, slight fragrance; foliage small, leathery; vigorous, bushy growth; [(R. wichurana X Floradora) X (Oakington Ruby X Floradora)]; Moore, Ralph S.; Sequoia Nursery

'Red Globe', F, mr, 1971; bud ovoid; flowers full, large, dbl.; foliage soft; moderate, bushy growth; Delforge

'Red Glory', F, mr, 1958; bud ovoid, pointed; flowers cherry to rose-red, blooms in rounded clusters, semi-dbl., 11 petals, 2.5–3.5 in., cupped, slight fragrance; foliage leathery, semi-glossy; very vigorous, tall, bushy growth; good as fence or hedge; [Gay Lady X (Pinocchio X Floradora)]; Swim, H.C.; Armstrong Nursery

Red Gruss an Coburg *see* **'Clotaria'**

'Red Guard', HT, rb, 1935; flowers dark blood-red, well formed, very dbl.; very vigorous growth; Verschuren

'Red Halo', F, mr, 1968; flowers crimson, borne in trusses, flat, slight fragrance; foliage glossy; vigorous, bushy growth; [Tabarin X Karl Herbst]; Oliver, W.G.

'Red Hedge', HRg, mr, 1958; small, red fruit; upright, compact growth; [(R. rugosa rubra X R. cinnamomea) X R. nitida]; Nyveldt

Red Hermosa *see* **'Queen's Scarlet'**

'Red Hill', HT, mr, 1941; flowers well-formed, large; [E.G. Hill seedling]; Clark, A.

Red Hit *see* 'POUlhit'

'Red Hoover', HT, rb, 1937; flowers brilliant red, center salmon-red; [President Herbert Hoover sport]; Lens

Red Hot *see* 'MACbigma'

'Red Imp', Min, dr, 1951; (Maid Marion, Mon Tresor, Montresor); bud ovoid; flowers deep crimson, micro-mini, very dbl., 54 petals, .75–1 in., flat, slight fragrance; upright, bushy, dwarf (9 in) growth; [Ellen Poulsen X Tom Thumb]; deVink; C-P

'Red Jacket', HT, mr, 1950; bud ovoid, pointed; flowers urn-shaped, dbl., 21 petals, 3.75 in., flat, slight fragrance; foliage leathery; upright, bushy growth; [World's Fair X Mirandy]; Swim, H.C.; Stuart

Red Jewel *see* 'INTerro'

Red Joy, HT, mr, 1997

Red Kelly, LCl, lp; Sutherland, P, 1999; (Weatherly, L.)

'Red Knight', LCl, dr, 1964; flowers maroon, medium, very vigorous growth, semi-dbl., globular, recurrent bloom; [Dr. Huey sport]; Booy, P.J.; Booy Rose Nursery

Red La France *see* **'Duchess of Albany'**

Red Lady®, HT, mr

'Red Lion', HT, mr, 1964; flowers red becoming rose-red, dbl., 38 petals, 5 in., exhibition form; [Kordes' Perfecta X Detroiter]; McGredy, Sam IV; Spek

Red Love™ *see* 'MINqco'

'Red Magic', Min, mr, 1977; bud pointed; flowers deep cherry-red, open, medium, semi-dbl., 10–15 petals, slight fragrance; vigorous, upright growth; [Red Can Can X ?]; Lyon; L. Lyon Greenhouses

Red Magic Carpet™ *see* 'MACredsaus'

'Red Maid', F, or, 1975; flowers coral-red, semi-dbl., 18 petals, 3 in.; [Vera Dalton X Stephen Langdon]; Sanday, John

Red Malmaison *see* **'Leweson Gower'**

Red Maman Cochet *see* **'Balduin'**

Red Maman Cochet *see* **'Niles Cochet'**

Red Marechal Niel *see* **'Grossherzog Ernst Ludwig'**

'Red Margo Koster', Pol, mr

'Red Martini', HT, dr, 1967; bud ovoid; flowers large, dbl., slight fragrance; foliage bronze; vigorous, bushy growth; [Chrysler Imperial X Seedling]; Delforge

'Red Masquerade', F, mr, 1962; bud pointed; flowers red becoming darker, borne in clusters, single, 10 petals, 2.5–3 in.; vigorous growth; [Masquerade X Independence]; Hill, A.

Red Masterpiece *see* 'JACder'

Red Matheson, F, mr

Red Max Graf *see* 'KORmax'

Red Meidiland™ *see* 'MEIneble'

Red Meillandina®, Min, mr, 1982; Meilland

Red Mercedes, F, mr

Red Minimo™ *see* 'RUImired'

Red Mini-Wonder™ *see* 'MEInofrai'

Red Moon™ *see* 'MINimoon'

Red Moscow *see* **'Krasnaia Moskva'**

Red Moss *see* **'Henri Martin'**

'Red Moss', M, mr; flowers heavily mossed, reddish rose, large

Red Moss Rambler, HWich, mr, 1988; Moore

'Red Moss, Climbing', Cl M, mr; Foote; B&A

Red 'n' White Glory *see* **'Candystick'**

'Red Nearly Wild', F, mr, 1960; bud globular; flowers rose-red, small, borne in clusters, single, cupped; foliage soft; vigorous, bushy growth; [Nearly Wild X Unnamed seedling]; Brownell, H.C.; Brownell

Red Nella, HT, dp, 1991; Allender, Robert William

Red Nelly, S, m

Red New Dawn *see* **'Étendard'**

Red Niphetos *see* **'Lady Battersea'**

'Red Opal', F, mr, 1968; bud pointed; flowers red, reverse cerise, borne in small clusters, flat; foliage small, dark; free growth; [Karl Herbst X Korona]; Northfield

Red Orléans Rose *see* **'Maréchal Foch'**

Red Paillette *see* POUlsint

'Red Parade', F, mr, 1972; bud ovoid; flowers carmine-red, medium, dbl., exhibition form, moderate fragrance; foliage glossy, bronze; vigorous, bushy growth; [(Frolic X Peace) X Texan]; Patterson; Patterson Roses

Red Parfum®, Cl F, mr, 1972; Eve, A.

Red Pastel *see* 'WILrib'

Red Peace *see* **'Karl Herbst'**

'Red Pearl', HT, mr, 1970; flowers bright red, pointed, dbl., 40 petals, 5 in., moderate fragrance; foliage large, dark; very vigorous growth; [Josephine Bruce X Kordes' Perfecta]; Watkins Roses

Red Pendant™ *see* 'MINvco'

Red Perfection™ *see* 'MACHaden'

'RED PETTICOAT', F, mr, 1967; flowers blood-red, short petaled, borne in trusses, dbl., flat; foliage dark, glossy; very vigorous growth; [Buisman's Triumph X Lilli Marlene]; Watkins Roses

'RED PINOCCHIO', F, dr, 1947; flowers velvety carmine-red, blooms in clusters, dbl., 28 petals, 3 in., cupped, moderate fragrance; vigorous, bushy growth; (28); [Unnamed yellow Pinocchio seedling X Donald Prior]; Boerner; J&P

'RED PISTOLS', Min, dy, 1982; flowers deep golden yellow, small, 15–20 petals, no fragrance; brown, needle-like prickles; foliage medium green, matt; upright, bushy growth; [Rise 'n' Shine X Unnamed seedling]; Ballmer, Gordon W.; Biotika International

Red Pixie *see* 'KORNUMA'

'RED PLANET', HT, dr, 1970; flowers crimson, dbl., 30 petals, 5.5–6 in., intense fragrance; foliage glossy; GM, RNRS, 1969 PIT, RNRS, 1969; [Red Devil X Seedling]; Dickson, Patrick; A. Dickson

'RED PLUME', F, or, 1966; flowers bright scarlet, rosette form, borne in clusters, dbl., 35 petals, 3 in.; foliage glossy; vigorous, bushy growth; [Masquerade X Independence seedling]; Sanday, John

Red Point *see* 'MACWOOHERM'

Red Poppy *see* **'KRASNYI MAK'**

'RED PREMIER', HT, dp, 1924; bud long, pointed; flowers bright carmine, large, dbl., intense fragrance; foliage leathery; vigorous growth; [Premier sport]; Scott, R.

'RED PRIDE', F, mr, 1968; bud ovoid; flowers small, dbl.; Verbeek

Red Prince® *see* **'FOUNTAIN'**

Red Prolific *see* 'DELIFIC'

'RED PROVENCE', C, mr; flowers clear crimson-red, large, cupped, intense fragrance; low, spreading growth

'RED QUEEN', HT, mr, 1968; (Liebestraum); bud ovoid; flowers large, dbl.; foliage dark; vigorous, upright growth; [Colour Wonder X Liberty Bell]; Kordes; Buisman;, McGredy

'RED RADIANCE', HT, dp, 1916; flowers light crimson; [Radiance sport]; Gude Bros.

'RED RADIANCE, CLIMBING', Cl HT, dp, 1927; Pacific Rose Co.

'RED RADIANCE, CLIMBING', Cl HT, dp, 1929; Catt

Red Rascal *see* 'JACBED'

Red Recker, HT, dr, 1986; Kasturi

'RED REFLECTION', HT, mr, 1964; bud ovoid, pointed; 4–5 in., exhibition form, slight fragrance; foliage large, leathery; upright growth; [Tropicana X Living]; Warriner, William A.; J&P

Red Ribbon™ *see* 'WILRIB'

Red Ribbons *see* 'KORTEMMA'

'RED RIDER', Cl F, rb, 1970; flowers red, base yellow, full, small, dbl., 50 petals, 2–3 in.; foliage large, glossy; vigorous growth; [Circus X Danse de Feu]; Guest

Red Rider *see* 'LAVRIDE'

'RED RIDING HOOD', Min, dr, 1955; flowers brilliant dark red; [Red Imp sport]; Robinson, T.

Red Riding Hood *see* **'RÖDHÄTTE'**

'RED RIPPLES', F, dr, 1942; (Willi Maass); bud globular; flowers deep red, petals wavy, small blooms in large clusters, semi-dbl., slight fragrance; foliage leathery, glossy, wrinkled; upright, bushy growth; [Hamburg X Anne Poulsen]; Krause; C-P

Red River, F; Kordes; (Cavriglia)

Red Robin *see* **'GLOIRE DES ROSOMANES'**

'RED ROBIN', HT, rb, 1940; flowers red, tending toward scarlet; vigorous growth; Brownell, H.C.

Red Robin, S, rb, 1992; Lens

Red Rock *see* 'MEILUSAM'

'RED ROCKET', S, mr, 1949; bud ovoid; flowers crimson, large blooms in clusters, semi-dbl., non-recurrent, slight fragrance; foliage large, dark, glossy, bronze; vigorous, upright growth; [Skyrocket seedling]; Longley, L.E.; Univ. of Minn.

Red Rocky *see* 'MEILUSAM'

Red Roma, HT, mr; flowers large, dbl.; (Sangerhausen)

Red Rosamini® *see* 'RUIREDRO'

'RED ROSE MARIE', HT, rb, 1938; bud ovoid; flowers cerise-red, large, dbl., exhibition form, slight fragrance; foliage dark, leathery; vigorous, bushy growth; [Rose Marie sport]; Mordigan Evergreen Nursery

Red Rose of Lancaster *see* **R. GALLICA OFFICINALIS**

Red Roulette, Min, mr

Red Rover *see* 'LAVRIDE'

'RED RUFFLES', F, dr, 1960; bud pointed; flowers dull dark red, borne in large clusters, dbl., 2.5–3 in., cupped, slight fragrance; foliage dark, glossy; vigorous, bushy, compact growth; [Improved Lafayette X Carrousel]; Von Abrams; Peterson & Dering

Red Rugostar, HRg, mr, 1995; flowers carmine-red, medium-large, semi-dbl.; Meilland; (Sangerhausen)

'RED RUM', F, mr, 1976; flowers red, shaded scarlet, dbl., 24 petals, 2.5 in., slight fragrance; foliage dark; vigorous growth; [Handel X Arthur Bell]; Bees

'RED SARONG', HT, dr, 1974; bud ovoid; flowers deep red, dbl., 36 petals, 5 in., exhibition form, moderate fragrance; foliage dark, leathery; tall growth; [Baccará X Golden Showers]; Golik; Dynarose

Red Satin *see* 'DEVPAJARO'

Red Sea® *see* BARred

Red Shadows™ *see* 'SAVMORE'

Red Simplicity *see* 'JACSIMPL'

'RED SKELTON', HT, or, 1968; bud long, pointed; flowers vermilion, large, dbl., exhibition form, moderate fragrance; foliage bronze, leathery; vigorous, upright growth; [Rose Queen X Charlotte Armstrong]; Whisler, D.; Germain's

Red Smith's Parish, T, rb, 1997; flowers double, small, dbl., 15–25 petals, borne in small clusters, slight fragrance; foliage medium, light green, semi-glossy; bushy, medium (4-6ft.) growth; [Smith's Parish sport]; Manners, Dr. Malcolm

Red Soldier *see* **'FUSILIER'**

'RED SPARKLER', HT, rb, 1967; flowers red, striped pink and white, dbl., 55 petals, 4–4.5 in., cupped, intense, damask fragrance; foliage dark, leathery; upright growth; [Scarlet Royal X Rouge Mallerin]; Buck, Dr. Griffith J.; Iowa State University

Red Spectacle, S, dr, 1993; Poulsen

'RED SPICE', F, dr, 1958; bud short, flat topped; dbl., 80–85 petals, 2–2.5 in., cupped, moderate fragrance; foliage dark, leathery; vigorous, bushy growth; [Spice X Garnette]; Boerner; J&P

'RED SPLENDOR', HT, mr, 1954; flowers bright red, dbl., 55 petals, 6 in., moderate fragrance; foliage leathery; vigorous, upright growth; [Joyance sport]; Grillo

Red Splendour *see* 'DAVONA'

'RED SPRITE', F, mr, 1974; flowers glowing red, dbl., 35 petals, 2–2.5 in., slight fragrance; foliage small, glossy, dark; low growth; LeGrice

Red Star *see* **'PRECIOUS PLATINUM'**

Red Stone *see* **'KRASNOKAMENKA'**

Red Storm, Min, dr

Red Success *see* 'MEIRODIUM'

Red Summer, F, or, 1996; Noack, Werner

'RED SUMMIT', Min, dr, 1981; bud ovoid, pointed; dbl., 33 petals, borne usually singly, slight fragrance; long, thin, straight prickles; low, compact, bushy growth; [Unnamed seedling X Unnamed seedling]; Lyon

Red Sunblaze® *see* 'MEIRUTRAL'

'RED SWEETHEART', F, rb, 1944; bud pointed; flowers crimson-carmine, dbl., 25 petals, 1–1.5 in., exhibition form, moderate, spicy fragrance; foliage small, glossy; strong stems; vigorous, bushy growth; [Intercrossing of Cécile Brunner seedlings sport]; Krebs; Marsh's Nursery

'RED TAG', Min, rb, 1978; bud ovoid, pointed; flowers medium red, white reverse, dbl., 48 petals, 1 in., exhibition form, moderate fragrance; foliage small, glossy, dark; upright, spreading growth; [Seedling X Over the Rainbow]; Williams, Ernest D.; Mini-Roses

'RED TALISMAN', HT, rb, 1931; flowers deep cerise, base yellow; [Talisman sport]; Amling Bros.

'RED TALISMAN, CLIMBING', Cl HT, rb

'RED TAUSENDSCHÖN', HMult, rb, 1931; flowers vivid red with white eye, but varies considerably; [Tausendschön sport]; Walter, L.

'RED TORNADO', HT, mr, 1999; flowers medium red, reverse light red, dbl., 26–40 petals, 5–5.5 in., borne mostly singly, slight fragrance; prickles moderate; foliage medium, dark green, glossy; upright, medium (4 ft) growth; [Dorothy Anne X Classic Touch]; McCall, Sharan

Red Trail *see* INTerim

'RED TRIUMPH', Pol, or, 1956; [Orange Triumph sport]; Morse

Red Trooper, F, dr

Red Velvet *see* 'PEATHUNDER'

'RED VELVET', F, mr, 1940; bud ovoid; flowers vivid crimson, semi-dbl., cupped, slight fragrance; vigorous, compact growth; Kordes; Dreer

Red Velvet *see* 'PEATHUNDER'

Red Velvet *see* TANorelav

Red Wagon *see* 'MORDAN'

'RED WAND', Cl Min, mr, 1964; flowers light crimson-red, small, dbl.; vigorous, climbing (3 1/2-4 ft) growth; [((R. wichurana X Floradora) X Orange Triumph) X Unknown Miniature]; Moore, Ralph S.; Sequoia Nursery

Red Wand, S, mr, 1994; Kordes

'RED WAVE', F, mr, 1964; flowers tulip shaped, small; will bloom in pot; low growth; [Carolyn Dean X Unnamed Floribunda seedling]; Moore, Ralph S.; Sequoia Nursery

'RED WINE', HT, mr, 1970; dbl., 22 petals, 3.5 in., exhibition form, blooms in clusters, slight fragrance; foliage medium green, matt; upright growth; [Lilac Rose X Seedling]; Sanday, John

'RED WING', S, my; flowers bright golden yellow; large, winged red prickles; [Probably R. sericea pteracantha X R. hugonis]

'RED WINGS', F, dr, 1958; bud ovoid; flowers rich dark red, semi-dbl., 20 petals, 3.5 in., cupped, moderate fragrance; foliage dark, leathery; vigorous, upright, bushy growth; [(Improved Lafayette X Herrenhausen) X Lavender Pinocchio]; Boerner; J&P

'RED WONDER', F, dr, 1954; bud globular; dbl., 28 petals, 3–3.5 in., cupped, blooms in large clusters, moderate fragrance; foliage leathery, semi-glossy; vigorous, bushy growth; [Better Times X Floribunda seedling]; deRuiter; C-P

Red Yesterday *see* 'HARHERO'

Red Zinnia, Min, dr

'REDCAP', F, mr, 1954; bud ovoid; semi-dbl., 18 petals, 2.5 in., exhibition form, blooms in clusters; foliage leathery, semi-glossy; vigorous, upright, bushy growth; [World's Fair X Pinocchio]; Swim, H.C.; Armstrong Nursery

'REDCLIFFE', F, mr, 1975; flowers bright crimson, semi-dbl., 15 petals, 3.5 in., slight fragrance; [Seedling X Sarabande]; Sanday, John

Redcoat *see* **'RED COAT'**

'REDCRAZE', HT, ob, 1959; bud long, pointed; flowers orange-scarlet, medium, dbl., exhibition form, slight fragrance; foliage glossy, dark; vigorous, bushy growth; [Independence X Independence]; Doley

Reddy Teddy *see* 'SUNHINO'

Redglo® *see* 'MINABBCO'

Redgold *see* 'DICOR'

'REDGOLD, CLIMBING', Cl F, yb, 1974; Lynch; J&P

Redgold, Climbing *see* 'DICORSAR'

'REDHEAD', Cl Min, dr, 1956; flowers blood-red; height 2 1/2 ft; [((Soeur Therese X Skyrocket) X (Seedling X Red Ripples)) X Zee]; Moore, Ralph S.; Sequoia Nursery

Redhot *see* 'MACBIGMA'

'REDHOTS', Min, dr, 1982; dbl., 22 petals, exhibition form, borne singly, slight fragrance; foliage small, long, medium green, finely serrated; very small, tiny growth; [Unnamed seedling X Darling Flame]; Meredith, E.A. & Rovinski, M.E.; Rosa de Casa Domingo

Rediffusion Gold *see* 'HARQUORGOLD'

'REDIPUGLIA', HT, w, 1933; bud long, pointed; flowers pinkish white, reverse golden rose, base chrome-yellow, large, dbl.; vigorous growth; Ingegnoli

'REDLAND COURT', F, ab, 1982; flowers soft apricot, small, dbl., 20 petals, slight fragrance; foliage small, medium green, glossy; bushy growth; [Red Maid X Sarabande]; Sanday, John; Sanday Roses, Ltd.

Redlands Century® *see* 'BURCEN'

'RED-LETTER DAY', HT, dp, 1914; flowers velvety rose-red, white streak in center of inner petals, st, semi-dbl., 3.5 in.; foliage glaucous sage-green; short stems; vigorous growth; (28); GM, NRS, 1913; Dickson, A.

Redneck Girl *see* 'GELRED'

'REDONDA', HT, mr, 1968; bud globular; flowers large, dbl., moderate fragrance; foliage leathery, wrinkled; vigorous, upright growth; [Queen Elizabeth X Happiness]; Patterson; Patterson Roses

Redoute® *see* 'AUSPALE'

'REDWAY', HT, mr, 1951; flowers crimson-carmine, exhibition form, moderate fragrance; foliage leathery; moderate, upright growth; [Ena Harkness X Unnamed seedling]; Waterhouse Nursery

Redwood *see* 'MACWOODMA'

Redwood Empire *see* 'MORWOOD'

Referenz, F, mp, 1983; dbl., slight fragrance; VEG; (Sangerhausen)

'REFLECTION', HT, yb, 1952; bud long, pointed; flowers amber, reverse streaked scarlet, moderate fragrance; vigorous growth; [Mme Henri Guillot sport]; Ratcliffe

Reflection, HT, lp; Brady, M., 1995; (Weatherly, L.)

'REFLETS', F, or, 1964; flowers large, single; upright growth; [Gertrud Westphal X Sarabande]; Croix, P.

'REFRESHER', LCl, w, 1929; flowers pure white, blooms in clusters, single, early; vigorous, climbing growth; Clark, A.; Hazlewood Bros.

'REFULGENCE', HEg, mr, 1909; flowers scarlet, aging to crimson, large, semi-dbl.; foliage very fragrant; vigorous growth; (41); Paul, W.

'REG WILLIS', HT, dp, 1966; flowers deep rose-pink, base yellow, well-formed, 4.5 in., slight fragrance; [Golden Masterpiece X Karl Herbst]; McGredy, Sam IV

Regal *see* 'JUSREGAL'

'REGAL GOLD', HT, dy, 1958; bud pointed; flowers clear golden yellow, open, dbl., 40–45 petals, 5 in., moderate fragrance; foliage glossy, leathery; vigorous, upright, bushy growth; [Golden Rapture sport]; Dale; Amling-DeVor Nursery

'REGAL LADY', Min, pb, 1989; bud pointed; flowers white, edged pink, borne usually singly and in sprays of 3-5, dbl., 27 petals, exhibition form, slight fragrance; round fruit; no prickles; foliage medium, medium green, semi-glossy; upright, spreading, tall growth; [Sassy Lassie X First Prize]; Jolly, Nelson F.; Rosehill Farm, 1991

'REGAL PINK', HT, mp, 1980; [Red Queen sport]; Thomas, Dr. A.S.

Regal Red *see* COCfoster

'REGALIA', HT, rb, 1964; flowers cherry-red, reverse silver, well-formed, dbl., 60 petals, 5.5 in.; foliage bronze; [Rose Gaujard X Seedling]; Robinson, H.

Regatta® *see* 'MEInimo'

Regatta *see* 'JACette'

Régence® *see* 'LAPdul'

Regensberg® *see* 'MACyoumis'

Regente Agricola, HT; Moreira da Silva, A.; (Cavriglia)

'REGIERUNGSRAT ROTTENBERGER', HMult, mp, 1926; flowers medium, dbl.; Praskac; (Sangerhausen)

'REGINA', HWich, w, 1916; flowers creamy white tipped pink large panicles, single; foliage large, glossy; vigorous, climbing growth; Walsh

Regina, HT, rb

'REGINA BADET', HRg, dp; flowers deep pink, large, very dbl., repeat bloom, intense fragrance; foliage dark; bushy growth; [(Général Jacqueminot X Empereur du Maroc) X R. rugosa]

'RÉGINA DE ALVÉAR', HT, w, 1922; flowers white, center slightly shaded pink, dbl., slight fragrance; [Mme Mélanie Soupert X Mme Segond Weber]; Sauvageot, H.

'REGINA ELENA', HT, mp, 1938; bud long, pointed; flowers darker rose-pink than briarcliff, dbl., 50 petals, 5 in., moderate fragrance; foliage dark; [Briarcliff sport]; Grillo

Regina Louise *see* CLEconcert

'REGINA PACIS', HT, w, 1945; 5 in., exhibition form, intense fragrance; foliage soft; branching growth; [Nuria de Recolons X Ibiza]; Dot, Pedro

'REGINALD FERNYHOUGH', HT, op, 1949; flowers pink lightly suffused orange, long, pointed, dbl., 35 petals, 5 in., intense fragrance; foliage olive-green; very vigorous growth; [Southport X Unnamed seedling]; Bees

Regine® *see* 'HEFqueen'

Regine Crespin *see* DELcreme

'REGULUS', T, pb, 1860; flowers bright coppery pink, medium, dbl., flat, intense fragrance; Moreau et Robert

'REICHSPRÄSIDENT VON HINDENBERG', HP, pb, 1933; flowers dark pink to carmine, veined lighter, reverse darker, very dbl., 6 in., moderate; foliage broad; very vigorous, bushy growth; [Frau Karl Druschki X Graf Silva Tarouca]; Lambert, P.

Reigen, F, ob, 1984; flowers medium, dbl., slight fragrance; GPG Bad Langensalza; (Sangerhausen)

'REIKO', HT, op, 1963; bud long, pointed; flowers bright coral-peach, dbl., 25 petals, 3.5–4 in., exhibition form, slight fragrance; foliage dark, leathery; vigorous, upright growth; [Spartan X Fred Streeter]; Teranishi, K.; Itami Rose Nursery

Reiko *see* 'JALray'

'REIKOR', F, ob, 1974; (Dreaming, 'KORrei', Reverie, **Träumerei**®); bud long, pointed; flowers orange, medium, dbl., cupped, intense fragrance; foliage leathery; vigorous, upright, bushy growth; [Colour Wonder X Seedling]; Kordes

'REIMS', HT, yb, 1924; bud long, pointed; flowers bright nankeen yellow shaded fiery red, orange-apricot and c; Barbier

Reina Elisenda *see* **'INDEPENDENCE'**

'REINA MARIA CRISTINA', T, ob, 1894; flowers orange-yellow, very dbl., globular; foliage bronze; very vigorous, compact growth; Aldrufeu

Reinaerdiana, F, 1974; Delforge; (Cavriglia)

Reine Amelia, HGal, mp

'REINE ASTRID', HT, rb, 1937; bud long, pointed; flowers bright deep coppery red, reverse golden yellow, dbl., cupped, moderate fragrance; foliage glossy; vigorous growth; Gaujard

Reine Blanche *see* **'HEBE'S LIP'**

'REINE BLANCHE', M, w; flowers pure white, large, dbl., flat; Moreau et Robert, 1858

Reine Chabeau®, HMsk, w, 1994; Lens

'REINE DE CASTILLE', B, mp, 1863; bud rose-white; flowers bright rose, well-formed; vigorous growth; Pernet

Reine de Perse, HGal, lp; flowers light cream-pink, small to medium, very dbl., intense fragrance; Roseraie de l'Hay; (Sangerhausen)

Reine de Perse, HGal, w

'REINE DE PORTUGAL', T, yb; flowers coppery yellow shaded pink, large, very dbl.; Guillot et Fils, 1867

Reine de Saxe, C, mp; flowers medium, dbl., moderate fragrance; Roseraie de l'Hay; (Sangerhausen)

'REINE DES AMATEURS', HGal, m; flowers clear lilac, edged pale pink, well shaped, very large; Hébert, Mme., before 1832

'REINE DES BELGES', HCh, w, 1867;, sparse bloom; vigorous growth; [Globe Hip X R. chinensis]; Cochet

Reine des Bordures *see* **'BORDER QUEEN'**

'REINE DES CENTFEUILLES', C, mp, 1824; flowers clear pink, reflexed, very large, dbl.; very vigorous growth

Reine des Couleurs *see* **'FARBENKÖNIGEN'**

Reine des Francais *see* **'LA REINE'**

Reine des Iles Bourbon *see* **'QUEEN OF BOURBONS'**

Reine des Mousseuses, M, lp; Moreau et Robert, 1860

Reine des Neiges *see* **'FRAU KARL DRUSCHKI'**

Reine des Roses *see* KORbico

Reine des Vierges, B, lp; Béluze, 1884

'REINE DES VIOLETTES', HP, m, 1860; (Queen of the Violets); flowers violet-red, large, dbl., 75 petals, recurrent bloom, intense fragrance; foliage sparse; [Pius IX seedling]; Millet-Malet

'REINE D'ESPAGNE', HP, mr, 1861; flowers brilliant red; foliage light; bushy growth; Fontaine

Reine du Danemark, A, pb, 1816; Booth

Reine du Dänemark *see* **'KÖNIGIN VON DÄNEMARK'**

Reine du Forez, HT, 1969; Croix; (Cavriglia)

'REINE ELISABETH', HT, mr, 1955; flowers velvety blood-red; [Princesse Liliane X Ena Harkness]; Frères, Buyl

'REINE ELIZABETH', Pol, mr; flowers crimson, pompon shape, borne in clusters; compact, dwarf growth

'REINE ELIZABETH', HT, rb, 1925; bud long, pointed; flowers coral-red, changing to prawn-red, tinted yellow, dbl., intense fragrance; Opdebeeck

'REINE EMMA DES PAYS-BAS', T, my, 1879; flowers coppery yellow; Nabonnand, G.

Reine France, HT, op; flowers large, dbl., moderate fragrance; Delbard-Chabert; (Sangerhausen)

Reine Marguerite *see* **'TRICOLORE'**

Reine Marguerite d'Italie, HT, mr, 1904; Soupert & Notting

'REINE MARIA DE ROUMANIE', HT, w, 1927; bud long, pointed; flowers marble-white, center yellow, dbl.; [Stadtrat Glaser seedling]; Mühle

'REINE MARIA PIA', T, dp, 1880; flowers deep pink, center crimson, large, moderate fragrance; very vigorous growth; [Gloire de Dijon seedling]; Schwartz, J.

'REINE MARIE HENRIETTE', Cl HT, mr, 1878; flowers pure cherry-red, large, dbl., moderate fragrance; vigorous, climbing growth; [Mme Berard X Général Jacqueminot]; Levet, F.

'REINE OLGA DE WURTEMBERG', N, mr, 1881; flowers bright red, large, dbl., moderate fragrance; very vigorous, pillar growth; Nabonnand, G.

'REINE VICTORIA', B, mp, 1872; (La Reine Victoria); flowers rich pink, well-formed, dbl., cupped, repeat bloom; foliage soft green; slender, upright (to 6 ft) growth; Schwartz, J.

Reiner Maria Rilke *see* 'KORsee'

'REINHARD BÄDECKER', HP, my, 1918; flowers clear golden yellow, dbl., slight fragrance; [Frau Karl Druschki X Rayon d'Or]; Kordes

Reinhard Pusch, HT, 1992; flowers large, dbl.; Tagashira, Kazuso; (Sangerhausen)

'REJOICE', Gr, pb, 1985; flowers salmon-pink blended with yellow, large blooms in clusters of, dbl., 40 petals, exhibition form, moderate fragrance; foliage large, medium green, glossy; tall, upright, bushy growth; Gold (First Gold Winner), ARC TG, 1985; [Little Darling X Color Magic]; McMillan, Thomas G.

Rejouissence, HT, 1977; Croix; (Cavriglia)

Rekordblueher, F, 1965; Tantau; (Cavriglia)

'REKORDBLÜHER', F, mp, 1965; flowers rose-pink, borne in large trusses, semi-dbl., 2–3 in.; foliage dark, glossy; bushy growth; Tantau, Math.

'REKTOR FOERSTER', HT, pb, 1936; flowers solferino-pink, flushed yellow, large, dbl., exhibition form; foliage leathery; vigorous growth; [Golden Ophelia X Mme Caroline Testout]; Weigand, C.; Pfitzer

Relax, F, or, 1979

Relax Meidiland *see* MEIdarwet

Relax Meillandecor *see* MEIdarwet

Release®, HT, yb

'RELIEF', HT, w, 1919; flowers ivory-white center yellowish pink; [Kaiserin Auguste Viktoria X Sunburst]; Verschuren

'REMBRANDT', P, or, 1883; flowers vermilion; Moreau et Robert

'REMBRANDT', HT, ob, 1914; flowers salmon, tinted orange-red, well-formed, large, dbl., 40 petals; foliage leathery; vigorous growth; [Frau Karl Druschki X Lyon Rose]; Van Rossem

Rembrandt van Ryn, S, dr, 1997; Williams, J. Benjamin

Remember Me® *see* 'COCDESTIN'

'REMEMBRANCE', HT, yb, 1953; bud ovoid; flowers soft yellow edged dawn-pink, dbl., 50 petals, 4–5 in., exhibition form, profuse bloom, moderate fragrance; foliage leathery, glossy, coppery green; vigorous, upright growth; RULED EXTINCT 4/92; [Fred Howard X Unnamed seedling]; Howard, A.P.; H&S

Remembrance *see* 'HARXAMPTON'

Remuera *see* 'MACTRAMPOL'

Remy Martin *see* 'STARQUELI'

'RENACK', Min, lp, 1995; (**Wee Ack**); dbl., 15–25 petals, 1 in., borne mostly singly, slight fragrance; few prickles; foliage small, light green, semi-glossy; medium (15-18 in), upright growth; [Seedling X Pink Sheri]; Rennie, Bruce F.; Rennie Roses International, 1994

'RENAE', Cl F, mp, 1954; bud pointed; dbl., 43 petals, 2.5 in., blooms in clusters, recurrent bloom, intense fragrance; foliage small, glossy; vigorous, climbing growth; [Étoile Luisante X Sierra Snowstorm]; Moore, Ralph S.; Armstrong Nursery

Renaissance® *see* 'GAUDINO'

'RENAISSANCE'®, HT, ob, 1986; flowers brilliant orange, dbl., exhibition form, moderate fragrance; rounded fruit; small prickles; foliage large; [Unnamed seedling X Pampa]; Gaujard, Jean

Renaissance *see* HARzart

'RENAL', Min, w, 1995; (**Grandpa Alex**); dbl., 26–40 petals, 1.5 in., borne in small clusters, moderate fragrance; few prickles; foliage medium, dark green, semi-glossy; medium (15-18 in), bushy growth; [Hap Renshaw X Summer Scent]; Rennie, Bruce F.; Rennie Roses International, 1993

'RENANGEL', F, m, 1992; (**Blue Angel**); dbl., 15–25 petals, 1.5–2.75 in., borne mostly singly, moderate fragrance; few prickles; foliage medium, dark green, semi-glossy; medium, bushy growth; [Lavonde X Shocking Blue]; Rennie, Bruce F.; Rennie Roses International

'RENAPP', Min, or, 1995; (**Orange Appeal**); very dbl., 2.75 in., borne mostly singly, slight fragrance; some prickles; foliage large, medium green, semi-glossy; tall (24 in), spreading, bushy growth; [Justa Little Goofy X Seedling]; Rennie, Bruce F.; Rennie Roses International, 1992

Renata Tebaldi *see* DELredi

'RENATE', HT, w, 1925; flowers cream-white, dbl., intense fragrance; [Kaiserin Auguste Viktoria seedling]; Berger, V.

'RENBABY', Min, lp, 1993; (**Dream Baby**); dbl., 15–25 petals, 1.5 in., borne mostly singly, slight fragrance; few prickles; foliage small, medium green, semi-glossy; medium, bushy growth; [Party Girl X Silver Phantom]; Rennie, Bruce F.; Rennie Roses International, 1993

'RENBLONDE', Min, yb, 1993; (**California Blonde**); flowers moderately full, dbl., 15–25 petals, 1.5 in., slight fragrance; few prickles; foliage medium, medium green, semi-glossy; medium, upright growth; [Party Girl X Tooth of Time]; Rennie, Bruce F.; Rennie Roses International, 1995

'RENBLUE', Min, m, 1988; (**Blushing Blue**); flowers lavender with magenta edge, yellow-tan at base, medium, dbl., 30–35 petals, exhibition form, borne usually singly, repeat bloom, no fragrance; rounded, medium, orange fruit; prickles straight, medium, transparent yellow-green; foliage medium, medium green, semi-glossy; upright, bushy growth; [Shocking Blue X Twilight Trail]; Rennie, Bruce F.; Rennie Roses International, 1989

'RENBLUSH', Min, lp, 1989; (**Innocent Blush**); bud ovoid; flowers pale blush pink, reverse white, aging white, medium, dbl., 40 petals, exhibition form, borne usually singly and in sprays of 5-7, no fragrance; fruit not observed; prickles straight, medium, yellow; foliage medium, medium green, matt; upright, medium growth; [Paul Shirville X Party Girl]; Rennie, Bruce F.; Rennie Roses International, 1990

'RENBOY', Min, rb, 1995; (**Augie Boy**); dbl., 15–25 petals, 1–1.5 in., borne mostly singly, moderate fragrance; few prickles; foliage medium, medium green, glossy; low (12-15 in), compact growth; [Party Girl X California Dreaming]; Rennie, Bruce F.; Rennie Roses International, 1994

'RENBUN', Min, m, 1995; (**Easter Bunny**); single, 5 petals, 2 in., borne mostly singly, intense fragrance; few prickles; foliage large, medium green, semi-glossy; tall (2 ft), upright growth; [Lavonde X Seedling]; Rennie, Bruce F.; Rennie Roses International, 1992

'RENCEL', Min, pb, 1995; (**Pretty Celine**); dbl., 15–25 petals, 1.25 in., borne mostly singly, no fragrance; some prickles; foliage medium, medium green, semi-glossy; medium (15 in), upright, bushy growth; [Strawberry Delight X Strawberry Delight]; Rennie, Bruce F.; Rennie Roses International, 1994

Rendan, F, rb

'RENDAR', Min, ab, 1995; (**Oh Darlin**); semi-dbl., 6–14 petals, 1.25 in., borne mostly singly, moderate fragrance; few prickles; foliage small, dark green, glossy; low (10-12 in), compact growth; [Pink Sheri X Hap Renshaw]; Rennie, Bruce F.; Rennie Roses International, 1992

Rendez Vous 81, HT, 1981; Meilland, L.; (Cavriglia)

Rendez-vous *see* **'DAY OF TRIUMPH'**

Rendez-vous *see* 'LUCDOD'

Rendez-vous *see* 'MEIPOBIL'

Rendez-vous *see* 'JACAEBI'

'RENDUDE', Min, m, 1995; (**Cool Dude**); dbl., 15–25 petals, 1 in., borne mostly singly, moderate fragrance; no prickles; foliage small, medium green, semi-glossy; low (10-12 in), compact growth; [Seedling X Blushing Blue]; Rennie, Bruce F.; Rennie Roses International, 1995

'RENÉ ANDRÉ', HWich, ab, 1901; bud coppery; flowers saffron-yellow, becoming pale pink and carmine, blooms in clusters, semi-dbl., 2–2.5 in., moderate fragrance; very vigorous growth; [R. wichurana X L'Ideal]; Barbier

'RENÉ BUATOIS', LCl, mp, 1936; flowers ruddy pink; [Leontine Gervais sport]; Buatois

'RENÉ D'ANJOU', M, dp, 1853; flowers deep pink, veined, small, dbl., globular; Robert

'RENÉ JAVEY', HT, ab, 1934; flowers clear apricot, reverse salmon-pink shaded yellow, passing to, dbl.; foliage clear bronze green; Gillot, F.

'RENÉE BRIGHTMAN', HT, rb, 1936; flowers brilliant scarlet, shaded orange, reverse striped yellow, dbl., moderate fragrance; vigorous growth; [Emma Wright seedling]; Hurran

Renee Columb *see* GAUvila

'RENÉE DANIELLE', LCl, my, 1913; flowers jonquil-yellow to golden yellow, passing to white, medium, b, sometimes blooms again in autumn; vigorous, climbing growth; Guillot, P.

'RENÉE WILMART-URBAN', HT, pb, 1908; bud long, pointed; flowers salmon-flesh, edged carmine, dbl., slight fragrance; Pernet-Ducher

'RENEEN', Min, r, 1987; (**Caribbean Queen**); flowers apricot-copper, reverse yellow, blooms borne singly, dbl., 33 petals, exhibition form, slight fragrance; ovoid, yellow fruit; medium, orange-red prickles; foliage small to medium, medium green, semi-glossy; vigorous, bushy growth; [Sunday Brunch X Gold Mine]; Rennie, Bruce F.; Rennie Roses International

Renegade *see* 'LYOREN'

Renegade, MinFl, lp, 1995; Spooner, Raymond A.

'RENERI', Min, lp, 1985; (**Pink Sheri**); [Sheri Anne sport]; Rennie, Bruce F.

'RENERRY', Min, pb, 1989; (**Strawberry Delight**); bud pointed; flowers white with medium pink freckles, reverse white with pink edg, dbl., 28 petals, exhibition form, no fragrance; fruit not observed; prickles straight, medium, brown; foliage medium, dark green, semi-glossy; spreading, medium growth; [Little Darling X California Dreaming]; Rennie, Bruce F.; Rennie Roses International, 1990

Rene's Rose, HT, pb, 1995; Dawson

'RENFER', Min, or, 1988; (**Surfer Girl**); flowers light orange-red, salmon, reverse lighter, medium, borne usua, dbl., 25–30 petals, exhibition form, moderate, spicy fragrance; round, small, yellow-orange fruit; prickles hooked, medium, transparent yellow-brown; foliage medium, medium green, matt; bushy, medium growth; [Pink Sheri X Paul Shirville]; Rennie, Bruce F.; Rennie Roses International

'RENFIRE', Min, ob, 1988; (**Touch of Fire**); bud ovoid; flowers orange-yellow, reverse light orange, aging apricot-orange, u, dbl., 25 petals, moderate, spicy fragrance; no fruit; prickles straight, small, orange-red; foliage small, medium green, semi-glossy; bushy growth; [Tangerine Mist X California Girl]; Rennie, Bruce F.; Rennie Roses International, 1989

'RENFRAGOBSES', Min, m, 1995; (**Fragrant Obsession**); dbl., 15–25 petals, 1.5 in., borne mostly singly, intense fragrance; few prickles; foliage medium, medium green, semi-glossy; medium (18 in), upright growth; [Lavonde X Blushing Blue]; Rennie, Bruce F.; Rennie Roses International, 1995

'RENFROST', HT, mp, 1992; (**Pink Frosting**); dbl., 26–40 petals, 3–3.5 in., borne mostly singly, intense fragrance; some prickles; foliage large, medium green, matt; tall, upright growth; [Unnamed seedling X Prima Ballerina]; Rennie, Bruce F.; Rennie Roses International

'RENGOOFY', F, or, 1992; (**Justa Little Goofy**); dbl., 26–40 petals, 1.5–2.75 in., borne mostly singly, moderate fragrance; some prickles; foliage medium, medium green, semi-glossy; medium, bushy growth; [Unnamed seedling X Unnamed seedling]; Rennie, Bruce F.; Rennie Roses International

'RENHAP', Min, ab, 1991; (**Hap Renshaw**); flowers moderately small blooms borne mostly singly, dbl., slight fragrance; foliage small, dark green, semi-glossy; upright growth; [Party Girl X Lavonde]; Rennie, Bruce F.; Rennie Roses International

'RENHEART', Min, dr 1995; (**Young Hearts**); flowers dark red, dbl. (15–25 petals), medium (1.5 in.) blooms borne mostly single; fragrant; some prickles; foliage medium, dark green, glossy; medium (18 in.), upright growth; [California Dreaming X Royal City]; Rennie, B.F.; Rennie Roses International, 1993

'RENHOM', Min, m, 1988; (Melissa Joyce, **Silver Phantom**); bud pointed; flowers silver-lavender, large, borne usually singly, dbl., 33 petals, exhibition form, slight, licorice fragrance; globular, medium, yellow-orange fruit; prickles hooked, medium, light tan-red to dark; foliage medium, dark green, semi-glossy; upright, bushy, tall growth; [Shocking Blue X Angelglo]; Rennie, Bruce F.; Rennie Roses International, 1989

'RENHURON', Min, ly, 1990; (**Huron Sunset**); flowers small, borne in small clusters, dbl., 26–40 petals, no fragrance; foliage small, medium green, semi-glossy; bushy growth; [Party Girl X Golden Rule]; Rennie, Bruce F.; Rennie Roses International, 1991

Renica *see* 'TANREKTA'

'RENIJEWEL', Min, ob, 1995; (**Isabel's Jewel**); semi-dbl., 6–14 petals, 1.25 in., borne mostly singly, moderate fragrance; some prickles; foliage medium, medium green, glossy; medium (15-18 in), spreading, bushy growth; [Forever Mine X Fragrant Morning]; Rennie, Bruce F.; Rennie Roses International, 1993

'RENIMAG', Min, pb, 1995; (**Imagine**); dbl., 26–40 petals, 1 in., borne in clusters, slight fragrance; few prickles; foliage small, medium green, semi-glossy; low (12-15 in), compact growth; [Forever Mine X Pink Sheri]; Rennie, Bruce F.; Rennie Roses International, 1995

'RENIRL', Min, ab, 1986; (**California Girl**); flowers apricot, reverse yellow blend, blooms borne singly, dbl., 28 petals, exhibition form, moderate, fruity fragrance; orange-red fruit; small prickles; foliage medium to large, medium green, semi-glossy; bushy, spreading growth; [Julie Ann X Red Love]; Rennie, Bruce F.; Rennie Roses International, 1987

'RENIST', Min, ob, 1987; (**Tangerine Mist**); flowers orange, medium, borne singly, semi-dbl., 15 petals, exhibition form, moderate, fruity fragrance; round, small, orange fruit; prickles straight, medium, red-brown; foliage medium, light green, glossy; upright growth; [(Avandel X Unnamed seedling) X Gold Mine]; Rennie, Bruce F.; Rennie Roses International, 1988

'RENITY', Min, dr, 1989; (**Royal City**); bud ovoid; flowers urn-shaped, medium, borne singly, dbl., 23 petals, exhibition form, no fragrance; globular, medium, orange-red fruit; prickles hooked, medium, yellow; foliage medium, medium green, semi-glossy; upright, tall growth; [Goldmarie X Pink Sheri]; Rennie, Bruce F.; Rennie Roses International, 1990

'RENJULIE', Min, ly, 1990; (**Julie's Choice**); flowers small, borne in small clusters, dbl., 26–40 petals, slight fragrance; foliage small, medium green, glossy; bushy growth; [Unnamed seedling X Unnamed seedling]; Rennie, Bruce F.; Rennie Roses International, 1991

'RENJULIE', Min, ly, 1990; (**Julies Choice**); Rennie, Bruce F.

'RENLARRY', Min, dy, 1990; (**Larry's Surprize**); flowers small, borne mostly singly, semi-dbl., 6–14 petals, slight fragrance; foliage small, light green, matt; bushy (15 in) growth; [Golden Rule X Rise 'n' Shine]; Rennie, Bruce F.; Rennie Roses International, 1991

'RENLOVER', Min, lp, 1995; (**Dream Lover**); dbl., 26–40 petals, 1–1.5 in., borne mostly singly, moderate, raspberry fragrance; some prickles; foliage medium, medium green, glossy; medium (18 in), upright growth; [Pink Sheri X Innocent Blush]; Rennie, Bruce F.; Rennie Roses International, 1995

'RENLYNN', HT, dr, 1987; (**Mary Lynn**); flowers maroon red, reverse lighter, aging darker, medium, borne usually singly, semi-dbl., 15 petals, exhibition form, intense, damask fragrance; oblong, medium, yellow fruit; prickles hooked, medium, red-brown; foliage medium, medium green, semi-glossy; upright, tall growth; [(Electron X Watercolor) X Lavonde]; Rennie, Bruce F.; Rennie Roses International, 1988

'RENMER', Min, op, 1988; (**Endless Summer**); bud pointed; flowers shrimp-pink, reverse light pink, small, borne singly, dbl., 33 petals, exhibition form, slight, spicy fragrance; no fruit; prickles straight, small, transparent to brown; foliage small, dark green, semi-glossy; low, bushy growth; [Paul Shirville X California Dreaming]; Rennie, Bruce F.; Rennie Roses International, 1990

'RENMERCY', Min, mr, 1995; (**Tender Mercy**); dbl., 15–25 petals, 1–1.5 in., borne mostly singly, slight fragrance; few prickles; foliage small, medium green, glossy; low (12-15 in), compact growth; [California Dreaming X Seedling]; Rennie, Bruce F.; Rennie Roses International, 1992

'RENMINE', Min, ob, 1993; (**Forever Mine**); dbl., 26–40 petals, 1.5 in., slight fragrance; some prickles; foliage small, medium green, matt; medium, bushy growth; [Hap Renshaw X Party Girl]; Rennie, Bruce F.; Rennie Roses International, 1994

'RENMOBAY', F, w, 1995; (**Moonlight Bay**); very dbl., 3.25 in., borne mostly singly, no fragrance; some prickles; foliage small, light green, semi-glossy; medium (3 ft), upright growth; [(Paul Shirville X Lavonde) X Party Girl]; Rennie, Bruce F.; Rennie Roses International, 1991

'RENMOM', Min, lp, 1995; (**Mom's Fancy**); flowers slight moss on buds, blooms borne in large clusters, dbl., 15–25 petals, 1.25 in., slight fragrance; some prickles; foliage small, medium green, semi-glossy; low (12-15 in), compact growth; [Pink Sheri X Hap Renshaw]; Rennie, Bruce F.; Rennie Roses International, 1994

'RENMORNING', Min, my, 1990; (**Fragrant Morning**); flowers small, borne mostly singly, dbl., 26–40 petals, intense fragrance; foliage small, light green, semi-glossy; upright (18 in) growth; [Sunsprite X Unnamed Miniature seedling]; Rennie, Bruce F.; Rennie Roses International, 1991

Renny *see* 'MORENY'

'RENO', Gr, mp, 1957; bud urn-shaped; flowers coral-salmon, wavy petals, blooms in clusters, 16–20 petals, 3–3.5 in., globular, slight fragrance; foliage bronze; vigorous growth; [Mrs Sam McGredy X Mme Henri Guillot]; Silva; Booy Rose Nursery

'RENOIR', Gr, pb, 1982; bud plump, pointed; flowers flesh pink with peach tones toward base, borne 1-4 per cluster, semi-dbl., 18 petals, repeat bloom, moderate, tea fragrance; curved prickles; foliage 7-9 leaflet, dark, smooth; vigorous, upright to arching growth; Hall, William W.

Renoncule *see* **'PROVINS RENONCULÉ'**

'RENONCULE', Pol, dp, 1913; flowers deep pink tinted lighter, buttercup form; Barbier

'RENONCULE PONCTUÉE', HGal, mr, 1835; flowers small, dbl.; Vibert; (Sangerhausen)

Renouveau de Provins® *see* 'LAPDI'

'RENOVE', Min, mr, 1987; (**California Dreaming**); flowers medium red, reverse lighter, blooms borne usually singly, dbl., 28 petals, exhibition form, moderate, fruity fragrance; globular, yellow-orange fruit; tiny, maroon prickles; foliage medium, dark, semi-glossy; medium, bushy growth; [Julie Ann X Black Jack]; Rennie, Bruce F.; Rennie Roses International

'RENOWN', HT, or, 1927; flowers glowing orange-cerise, shaded cardinal, dbl., intense fragrance; [Red-Letter Day X Mrs Wemyss Quin]; Burbage Nursery

'RENPAFLAME', Min, dp, 1995; (**Passion's Flame**); flowers deep pink, blooms borne in small clusters, semi-dbl., 6–14 petals, 1.5 in., slight fragrance; some prickles; foliage medium, medium green, semi-glossy; tall (18-24 in), spreading growth; Rennie, Bruce F.; Rennie Roses International, 1993

'RENPIE', Min, ly, 1988; (**Cutie Pie**); bud ovoid; flowers light to medium yellow, small, dbl., 23 petals, exhibition form, borne usually singly and in small clusters, moderate, fruity fragrance; globular, very small, yellow orange fruit; prickles straight, small, yellow-red; foliage small, dark green, glossy; bushy, low growth; [Tangerine Mist X California Girl]; Rennie, Bruce F.; Rennie Roses International, 1989

'RENPINY', Min, w, 1995; (**Pretty Tiny**); dbl., 15–25 petals, 1 in., borne in small clusters, slight fragrance; few prickles; foliage small, medium green, semi-glossy; low (8-10 in), compact growth; Rennie, Bruce F.; Rennie Roses International, 1995

'RENRIDE', Min, mp, 1995; (**Joyride**); very dbl., 2.75 in., borne mostly singly, slight fragrance; few prickles; foliage large, medium green, semi-glossy; tall (24 in), upright growth; [Seedling X Innocent Blush]; Rennie, Bruce F.; Rennie Roses International, 1994

'RENRUBY', Min, dr, 1988; (**Ruby Tuesday**); flowers dark, velvety red, reverse lighter, aging darker, pom-pom, s, dbl., 25–30 petals, cupped, no fragrance; round, small, red-orange fruit; prickles slightly hooked, small, brownish; foliage small, dark green, semi-glossy; bushy, low growth; [Pink Sheri X Black Jack]; Rennie, Bruce F.; Rennie Roses International, 1989

'RENSAND', F, op, 1990; (**Coral Sand**); flowers coral, large, borne singly, dbl., 26–40 petals, intense fragrance; foliage medium, medium green, semi-glossy; bushy growth; [Paul Shirville X Shocking Blue]; Rennie, Bruce F.; Rennie Roses International, 1991

'RENSCENT', F, pb, 1988; (**Soft Scent**); bud ovoid; flowers pink blend, reverse light pink, sweetheart, urn-shaped, medium, dbl., 48 petals, exhibition form, borne usually singly, intense, spicy fragrance; elongated, medium, orange fruit; prickles straight, medium, red; foliage medium, dark green, semi-glossy; bushy, medium growth; [Paul Shirville X California Girl]; Rennie, Bruce F.; Rennie Roses International, 1990

'RENSECOBSES', Min, lp, 1995; (**Secret Obsession**); dbl., 26–40 petals, 3 in., borne mostly singly, slight fragrance; few prickles; foliage large, medium green, semi-glossy; tall (24-30 in), upright growth; [Seedling X Innocent Blush]; Rennie, Bruce F.; Rennie Roses International, 1994

'RENSILEN', Min, yb, 1995; (**Golden Silence**); dbl., 15–25 petals, 1.5 in., borne mostly singly, slight fragrance; some prickles; foliage medium, dark green, semi-glossy; medium (15-18 in), upright growth; Rennie, Bruce F.; Rennie Roses International, 1995

'RENSUM', Min, lp, 1989; (**Summer Scent**); bud ovoid; flowers medium pink with white center, reverse lighter, medium, born, dbl., 30 petals, exhibition form, moderate, spicy fragrance; fruit not observed; prickles straight, small, few, brown; foliage medium, dark green, glossy; bushy, low growth; [Paul Shirville X Unnamed seedling]; Rennie, Bruce F.; Rennie Roses International, 1990

'RENSUN', Min, ab, 1988; (**California Sun**); flowers golden apricot, opening to golden center, reverse lighter, dbl., 30–35 petals, exhibition form, no fragrance; rounded, medium, yellow-orange fruit; prickles straight, medium, pinking; foliage large, medium green, matt; patio; upright, tall growth; [Shocking Blue X Rise 'n' Shine]; Rennie, Bruce F.; Rennie Roses International, 1989

'RENSWEET', Min, mp, 1993; (**Sweet Scent**); dbl., 26–40 petals, 1.5 in., moderate fragrance; few prickles; foliage small, medium green, semi-glossy; medium, upright growth; [Party Girl X Silver Phantom]; Rennie, Bruce F.; Rennie Roses International, 1994

'RENSWIRL', Min, ly, 1992; (**Lemon Swirl**); dbl., 15–25 petals, 1.5–2.75 in., borne in small clusters, slight fragrance; few prickles; foliage medium, light green, glossy; tall, upright growth; [Unnamed seedling X Sunsprite]; Rennie, Bruce F.; Rennie Roses International

'RENTDEL', Min, m, 1995; (**Timeless Delight**); dbl., 15–25 petals, 2.25 in., borne mostly singly, moderate fragrance; some prickles; foliage large, medium green, dull; tall (24-36 in), upright growth; [Silver Phantom X Innocent Blush]; Rennie, Bruce F.; Rennie Roses International, 1996

'RENTIME', Min, w, 1989; (**Tooth of Time**); bud pointed; flowers white with cream, urn-shaped, small, borne usually singly an, dbl., 25 petals, slight, spicy fragrance; fruit not observed; prickles hooked downward, medium, light yellow; foliage small, dark green, semi-glossy; bushy, medium growth; [Party Girl X Paul Shirville]; Rennie, Bruce F.; Rennie Roses International, 1990

'RENUGGLE', Min, mp, 1995; (**Snuggles**); dbl., 15–25 petals, 1.5 in., borne in small clusters, no fragrance; some prickles; foliage medium, medium green, semi-glossy; medium (15-18 in), bushy growth; Rennie, Bruce F.; Rennie Roses International, 1993

'RENURF', Min, w, 1987; (**California Surf**); flowers creamy white with peachy-pink edging, urn-shaped, small, dbl., 20 petals, borne usually singly, slight fragrance; round, small, yellow-orange fruit; prickles straight, small, reddish-brown; foliage small, medium green, semi-glossy; upright growth; [Seedling X Seedling]; Rennie, Bruce F.; Rennie Roses International, 1988

'RENWINNIE', Min, lp, 1991; (**Winnie Renshaw**); flowers moderately small blooms borne in small clusters, dbl., moderate fragrance; foliage small, medium green, semi-glossy; medium growth; [Lavonde X Party Girl]; Rennie, Bruce F.; Rennie Roses International

'RENYKE', Min, pb, 1986; (**Little Tyke**); flowers medium pink, white reverse, loose form small blooms borne singly, dbl., 25 petals, no fragrance; small fruit; small, reddish prickles; foliage small, medium green, matt; small, bushy, very compact growth; [Julie Ann X Red Love]; Rennie, Bruce F.; Rennie Roses International

'RENYOUNG', Min, w, 1993; (**Young 'n' Innocent**); flowers full, small, white, very dbl., 26–40 petals, 1.5 in., borne in small clusters, moderate fragrance; some prickles; foliage small, medium green, semi-glossy; low, spreading growth; [Hap Renshaw X Tooth of Time]; Rennie, Bruce F.; Rennie Roses International, 1994

'RENYOURS', Min, w, 1993; (**Tenderly Yours**); flowers small, white, full, dbl., 26–40 petals, 1.5 in., slight fragrance; some prickles; foliage small, medium green, semi-glossy; low, spreading growth; [Party Girl X Tooth of Time]; Rennie, Bruce F.; Rennie Roses International, 1993

'RENZIG', Min, r, 1995; (**Ziggy Stardust**); dbl., 15–25 petals, 1.25 in., borne mostly singly, slight fragrance; few prickles; foliage medium, medium green, semi-glossy; medium (15 in), upright growth; [Silver Phantom X Party Girl]; Rennie, Bruce F.; Rennie Roses International

Repandia® *see* 'KORSAMI'

Repartee, HSpn, w

Repelsteeltje *see* **'RUMPELSTILZCHEN'**

Repens Meidiland *see* 'MEILONTIG'

Rephidy, HT, pb

Republique de Geneve *see* LAPEJ

'RESAND', HT, lp, 1996; (**Gene Sandberg**); flowers light pink with a dark pink line around outer edge of each p, dbl., 26–40 petals, 4–6. in., slight fragrance; moderate prickles; foliage medium, dark green, dull; upright, medium (2-3 ft) growth; [Touch of Elegance X First Prize]; Sheldon, John, Jennifer & Robyn

'RESBOBIE', HT, pb, 1996; (**Bobbie Vesely**); flowers medium pink blend, reverse medium pink with yellowish tinge, dbl., 26–40 petals, slight fragrance; moderate prickles; foliage medium, medium green, dull; upright, medium growth; [Pristine X Gold Medal]; Sheldon, John, Jennifer & Robyn; Sheldon, 1995

'RESCARD', HT, ly, 1995; (**Christmas Card**); flowers light yellow, fading to white on outer petals, full (26–40 petals), large blooms borne mostly single; fragrant; foliage medium, medium green, matt; upright; medium growth [Spirit of Glasnost X Lanvin]; Sheldon, John & Robin; Sheldon Manor, 1995

'RESCLARA', HT, ab, 1995; (**Clara**); flowers unusual apricot, full (26–40 petals), medium blooms borne mostly single; slight fragrance; foliage medium, medium green, matt; upright, medium growth; [Nantucket X Lovely Lady]; Sheldon, John & Robin; Sheldon Manor, 1995

'RESFINI', HT, mp, 1995; (**Infinity**); flowers medium pink full (26–40 petals), medium blooms borne mostly single; slight fragrance; foliage medium; medium green, matt; upright, tall growth; [Kordes Perfecta X Prima Donna]; Sheldon, John & Robin; Sheldon Manor, 1995

'RESGOLD', HT, my, 1996; (**Doris Reese**); bud urn-shaped with tips of petals folding downward; flowers medium yellow, center of rose darker than outer petals, dbl., 26–40 petals, 4 in., exhibition form, moderate fragrance; moderate prickles; foliage medium, dark green, semi-glossy; upright, medium (3 ft) growth; [Seedling X Lanvin]; Sheldon, John, Jennifer & Robyn

'RESHONOR', HT, w, 1995; (**Sheldon's Honor**); flowers white, full (26–40 petals), large blooms borne in small clusters; slight fragrance; foliage medium, medium green, matt; upright, medium growth; [Sheer Bliss X Anastasia]; Sheldon, John & Robin; Sheldon Manor, 1995

'RESHOPE', HT lp, 1995; (**Hopeful**); flowers light pink, full (26–40 petals), medium blooms borne mostly single; slight fragrance; foliage medium, medium green, matt; upright, medium growth; [Sheer Bliss X Elizabeth Taylor]; Sheldon, John & Robin; Sheldon Manor, 1995

'RESILLA', HT, r, 1996; (**O'Rilla**); dbl., 26–40 petals, 5–7 in., exhibition form, slight fragrance; moderate prickles; foliage medium, dark green, semi-glossy; upright, medium (3-4 ft) growth; [Lanvin X First Prize]; Sheldon, John, Jennifer & Robyn

'RESJEFF', HT, my, 1995; (**Jeffery**); flowers yellow, full (26–40 petals), medium blooms borne in small clusters; slight fragrance; foliage medium, light green, matt; upright, medium growth; [Polarstern X (Spirit of Glasnost X Lanvin)]; Sheldon, John & Robin; Sheldon Manor, 1995

'RESLAND', HT, ob, 1995; (**Dreamland**); flowers orange blend tipped with darker orange, full (26–40 petals), medium blooms borne mostly single; slight fragrance; numerous prickles; foliage medium, medium green, matt; upright, medium growth; [Summer Dream X Lanvin]; Sheldon, John & Robin; Sheldon Manor, 1995

'RESLINE', HT w, 1995; (**White House**); flowers white with pink tinge, full (26–40 petals), medium blooms borne mostly single; slight fragrance; foliage medium, medium green, matt; upright, medium growth; [Pristine X (Spirit of Glasnost X Lanvin)]; Sheldon, John & Robin; Sheldon Manor, 1995

'RESLINK', HT, pb, 1995; (**Pink Licorice**); flowers pink blend, full (26–40 petals), medium blooms borne mostly single; licorice fragrance; foliage medium, medium green, matt; upright, medium growth; [Osiria X Elegant Beauty]; Sheldon, John & Robin; Sheldon Manor, 1995

'RESLUC', HT, mr, 1995; (**Lucious**); flowers intense med. red, double, dbl., 15–25 petals, borne mostly singly, slight fragrance; foliage medium, dark green, semi-glossy; upright growth; Sheldon, John & Robin

'RESMAR', HT, w, 1995; (**Marie**); flowers white, full (26–40 petals), medium blooms borne mostly single; slight fragrance; foliage medium, medium green, mett; upright, medium growth; [Sheer Bliss X Headliner]; Sheldon, John & Robin; Sheldon Manor, 1995

'RESOLUT', F, or, 1962; bud oval; dbl., 30 petals, 4–5 in., borne in large sprays, slight fragrance; foliage glossy; upright, bushy growth; Tantau, Math.

'RESONE', HT, m, 1991; (**Violet Dawson**); flowers mauve blend, blooms borne in small clusters, dbl., moderate fragrance; foliage medium, dark green, disease-resistant; medium, bushy growth; winter hardy.; [Paradise X Unnamed seedling]; Sheldon, John & Jennifer

'RESPAD', HT, pb, 1995; (**Granpa Dan**); flowers pink blend, full (26–40 petals), medium blooms borne mostly single; slight fragrance; foliage medium, medium green, matt; upright, extremely vigorous, medium growth; [Pristine X Touch of Class]; Sheldon, John & Robin; Sheldon Manor, 1995

'RESPAR', HT, pb, 1995; (**Granpa Ray**); flowers light pink blend, full (26–40 petals), medium blooms borne mostly single; slight fragrance; foliage medium, medium green, matt; upright, medium growth; [Pristine X Touch of Class]; Sheldon, John & Robin; Sheldon Manor, 1995

'RESPAT', HT, pb, 1995; (**Granpa Toni**); flowers light pink blend, full (26–40 petals), medium blooms borne mostly single; slight fragrance; foliage medium, medium green, matt; upright, medium growth; [Pristine X Touch of Class]; Sheldon, John & Robin; Sheldon Manor, 1995

'RESPLENDA', HT, w, 1974; bud ovoid; dbl., 30 petals, 4 in., moderate, fruity fragrance; foliage glossy; moderate growth; [Queen of Bermuda X Golden Giant]; Golik; Dynarose

'RESPLENDENT', HP, lp; Williams, A., 1896; (Weatherly, L.)

'RESPOND', HT, pb, 1995; (**Respond**); flowers dark pink blend, full (26–40 petals), large blooms borne mostly single; slight fragrance; foliage medium, medium green, matt; upright, medium growth; [Sheer Elegance X Seedling]; Sheldon, John & Robin; Sheldon Manor, 1995

Respond *see* 'RESPOND'

Responso, F, yb, 1981; flowers medium, dbl.; GPG Bad Langensalza; (Sangerhausen)

'RESROB', HT, m; (**Robin**)

'RESRUTH', HT, yb; (**Ruthe**)

'RESSOX', HT, mr, 1995; (**Roxie**); flowers medium red, full (26–40 petals), medium blooms borne mostly single; slight fragrance; numerous prickles; foliage medium, medium green, matt; upright, medium growth; [Sheer Bliss X Headliner]; Sheldon, John & Robin; Sheldon Manor, 1995

Ressins Etienne Gautier *see* DORster

Rest in Peace *see* BEDswap

'RESTFUL', HT, my, 1995; (**Restful**); flowers yellow, full (26–40 petals), large blooms borne mostly single; slight fragrance; foliage medium, medium green, matt; upright, tall growth; [Miyabi X Lanvin]; Sheldon, John & Robin; Sheldon Manor, 1995

Restful *see* 'RESTFUL'

Restless, HT, mr, 1938; flowers dark red, semi-dbl.; almost thornless; Clark

'RESTLESS NATIVE', S, or, 1973; bud ovoid; flowers orange-scarlet, base white, 8–12 petals, 2.5 in., cupped; foliage large, dark, leathery; vigorous, compact, bushy growth; [Orangeade X R. carolina]; Stoddard, Louis

'RESTO', HT, pb, 1995; (**Thoughtful**); flowers dark pink blend, full (26–40 petals), large blooms borne mostly single; slight fragrance; foliage medium, dark green, matt; upright, medium growth; [Pristine X Seedling]; Sheldon, John & Robin; Sheldon Manor, 1995

Results *see* 'RESULTS'

'RESULTS', HT, pb, 1995; (**Results**); flowers red, white, pink, purple hues depending on temperature and sunlight, dbl. (15–25 petals), large blooms borne mostly single; slight fragrance; foliage medium, dark green, matt; upright, tall growth; [Pristine X Seedling]; Sheldon, John & Robin; Sheldon Manor, 1995

'RESUNC', HT, mr, 1995; (**Uncle Steve**); flowers medium red, full (26–40 petals), large blooms borne mostly single; slight fragrance; foliage medium, dark green, matt; upright, medium growth; [Pristine X Touch of Class]; Sheldon, John & Robin; Sheldon Manor, 1995

Resurrection® *see* 'KRILEXIS'

'RETINA', Min, mr, 1980; flowers medium red, reverse lighter, base gold, blooms borne singly, dbl., 43 petals, slight, fruity fragrance; straight, long, tan prickles; foliage green, matt; bushy, spreading growth; [Unnamed seedling X Over the Rainbow]; Williams, Ernest D.; Kimbrew-Walter Roses

Rétro *see* MEIbalani

Rétro Blanc, S, w

'REUS', HT, m, 1949; bud long, pointed; flowers magenta, large, dbl., exhibition form, moderate fragrance; [Cynthia X Manuelita]; Dot, Pedro

'REV T.C. COLE', Cl T, lp; [Chromatella X Marechal Neil]; Cole, Rev T.C., 1880; (Weatherly, L.)

'REV. DAVID R. WILLIAMSON', HT, mr, 1904; flowers dark crimson, globular, moderate fragrance; Dickson, A.

'REV. F. PAGE-ROBERTS', HT, yb, 1921; bud long, pointed; flowers yellow shaded red, large, dbl., intense fragrance; GM, NRS, 1920; [Queen Mary X Unnamed seedling]; Cant, B. R.

'REV. F. PAGE-ROBERTS, CLIMBING', Cl HT, yb, 1931; Beverley; W.B. Clarke

Rev. Floris Ferwerda *see* **'BEN STAD'**

'REV. H. D'OMBRAIN', B, mr, 1863; flowers brilliant red, recurrent bloom; vigorous growth; Margottin

'REV. JAMES SPRUNT', Cl HCh, mr, 1856; flowers crimson-red, larger; [Cramoisi Superieur, Climbing sport]; Sprunt; P. Henderson, 1858

'REV. WILLIAMSON', HT, or, 1921; bud long, pointed; flowers coral-red, shaded carmine, dbl.; Pernet-Ducher

'RÊVE DE CAPRI', HT, op, 1953; flowers salmon-orange and yellow; foliage bronze; compact, low growth; [Pres. Herbert Hoover X Seedling]; Frères, Buyl

Rêve de Deauville *see* PINault

Reve de Paris® *see* MEIloise

Rêve de Valse®, F, mr

'RÊVE D'HÉLÈNE', HT, mp, 1959; flowers bright pink edged silvery, large; foliage clear green; free growth; [Michele Meilland X Unnamed seedling]; Orard, Joseph

'RÊVE D'OR', N, my, 1869; flowers buff-yellow tinted lighter, dbl., moderate fragrance; foliage rich green; vigorous, climbing growth; [Mme Schultz seedling]; Ducher, Vve.

Rêve d'un Soir, HT, m, 1997; Croix

'RÊVE ROSE', F, mp, 1950; dbl., 60 petals, borne in large clusters; Mallerin, C.; Vilmorin-Andrieux

'RÉVEIL', HT, yb, 1924; flowers golden yellow, reverse striped red, dbl., moderate fragrance; [Mr Joh. M. Jolles X Mrs Wemyss Quinn]; Van Rossem

'RÉVEIL DIJONNAIS', Cl HT, rb, 1931; flowers cerise, large golden yellow center, reverse yellow streaked, semi-dbl., 13 petals, 5 in., cupped, moderate fragrance; foliage thick, glossy, bronze; short stems; vigorous, climbing growth; GM, Portland, 1929; [Eugene Furst X Constance]; Buatois

'REVEILLE', HT, pb, 1941; flowers light salmon-buff-pink, center deeper pink, open, large, dbl., 40–50 petals, cupped, slight fragrance; foliage dark, leathery; vigorous, upright, bushy, compact growth; [Kidwai X Golden Main]; Nicolas; J&P

'REVELATION', HT, pb, 1938; bud long; flowers thulite-pink to rose-red, dbl., 35 petals, 4.5–5 in., exhibition form, moderate fragrance; foliage dark; very vigorous growth; [Briarcliff sport]; Witter; Evans City Cut Flower Co.

Revelry® *see* 'SWISHREV'

'REVENANTE', HGal, pb; flowers light rose-pink, edged lilac; Miellez

'RÉVÉRENCE', F, mr, 1962; flowers geranium-red, carnation form, borne in large clusters, dbl., 26 petals, 2 in.; foliage dark, glossy; vigorous, upright growth; [Orange Triumph X Unnamed seedling]; Delforge

'REVEREND ALAN CHEALES', HP, mr, 1898; flowers very large, dbl.; Paul, G.; (Sangerhausen)

'REVERIE', HT, ab, 1925; flowers apricot, reverse shrimp-pink veined rose, dbl.; [Mme Mélanie Soupert X Jean C.N. Forestier]; Ketten Bros.

Reverie *see* 'REIKOR'

'REVIEW', HT, mp, 1951; flowers bright rose, reverse flesh, dbl., 50 petals, 6–7 in., moderate fragrance; foliage dull, green; medium, dwarf growth; [Mrs Henry Bowles X Trigo]; Fletcher; Tucker

'REVIVAL', HT, ab, 1980; flowers light apricot; [Folklore sport]; Rose Barni-Pistoia, 1979

Revolution, Pol, or, 1972; flowers medium, semi-dbl.; GPG Bad Langensalza; (Sangerhausen)

Revolution Francaise® *see* MEItixia

Revue de Dauville *see* PINault

'REWARD', HT, yb, 1934; bud long, pointed; flowers clear yellow shaded peach, medium, moderate fragrance; foliage dark, glossy; vigorous, bushy growth; Dickson, A.

'REX', HT, mr, 1959; bud long, pointed; flowers velvety cardinal-red, large, semi-dbl., cupped, slight fragrance; foliage leathery, glossy; vigorous, upright growth; [Senior X Better Times]; Spanbauer

'REX ANDERSON', HT, w, 1938; flowers ivory-white, very well-formed, large, dbl., shy bloom, moderate, fruity fragrance; foliage gray-green; vigorous growth; [Florence L. Izzard X Mrs Charles Lamplough]; McGredy; J&P

'REXJEAN', S, dr, 1999; (**Jean Rex**); flowers dark red (3 shades), center vein darkest, reverse dark red, very dbl., 41 petals, 4–4.5 in., borne in small clusters, intense fragrance; prickles numerous; foliage medium, dark green, matte; bushy, very vigorous, strong, medium (4 ft) growth; [Prospero sport]; Rex, Dr. Robert W.; Hortico, Inc., 1995

Rexy's Baby *see* 'MACCARRIB'

'REYCANTPRI', Min, r, 1992; (**Canterbury Pride**); very dbl., 1.5 in., slight fragrance; foliage small, medium green, semi-glossy; low (28 cms), upright growth; [Hot Chocolate X Miniature seedling]; Reynolds, Ted; Reynolds Roses

'REYCIDIN', Min, yb, 1995; (**City of Dunedin**); flowers old gold changing to yellow then white with red edge, full (26–40 petals), small blooms borne in small clusters; slight fragrance; few prickles; foliage medium, medium green, glossy; upright, medium growth; [(Rise 'n' Shine X Seedling) X Rainbow's End]; Reynolds, Ted; Ted Reynolds Roses International, 1995

'REYFAYE', Cl F, dp, 1992; (**Faye Reynolds**); flowers deep pink, blooms borne in small clusters, dbl., 26–40 petals, 3–3.5 in., moderate fragrance; some prickles; foliage sage green, semi-glossy; medium (6' 6"), upright growth; [Westerland X Gingersnap]; Reynolds, Ted; Reynolds Roses

'REYINVER', HT, dp, 1996; (**City of Invercargill**); bud very dark red; flowers deep pink, large blooms borne mostly single, very dbl., moderate fragrance; few prickles; foliage medium, medium green, glossy, disease-free; upright, medium growth; [Westerland X Perfume Delight]; Reynolds, Ted; Ted Reynolds Roses International, 1996

'REYKELLY', HT, pb, 1999; (**Kelly Reynolds**); flowers apricot peach pink blend, large, dbl., 26–40 petals, exhibition form, borne mostly singly, slight fragrance; few prickles; foliage large, medium green, glossy; upright, medium growth; [Perfume Delight X seedling]; Reynolds, Ted; Ted Reynolds Roses International, 1999

'REYLIANNE', F, lp, 1996; (**Lianne Reynolds**); flowers light pink, china porcelain appearance, blooms borne in small clusters, dbl., 26–40 petals, no fragrance; some prickles; foliage small, medium green, glossy; compact, medium growth; [Gloire de Dijon X Hot Pewter]; Reynolds, Ted; Ted Reynolds Roses International, 1997

Reynolda House *see* 'WILANGIL'

'REYNOLDS HOLE', HT, mp, 1862; flowers medium, very dbl.; Standisch; (Sangerhausen)

Reynolds Rugosa *see* 'REYRUG'

'REYPAT', LCl, lp, 1992; (**Patricia Weston**); dbl., 26–40 petals, 3–3.5 in., borne in small clusters, moderate fragrance; some prickles; foliage large, dark green, glossy; medium (2 m), upright growth; [Westerland X Perfume Delight]; Reynolds, Ted; Reynolds Roses

'REYPEG', S, mp, 1992; (**Peggy Joan Reynolds**); single, 5 petals, 1.5 in., borne in small clusters, very remontant, almost perpetual, slight fragrance; few prickles; foliage medium, light green, semi-glossy, disease-resistant; upright (150 cms) growth; very hardy; (14); [Parentage unknown. (Possibly HCh X R. gigantea)]; Reynolds, Ted, 1993

'REYRED', HT, mr, 1992; (**Ted's Red**); dbl., 26–40 petals, 3–3.5 in., borne mostly singly, slight fragrance; few prickles; foliage medium, medium, semi-glossy; medium, upright growth; [Alec's Red X Seedling]; Reynolds, Ted; Reynolds Roses

'REYRUG', HRg, mr, 1999; (**Reynolds Rugosa**); [double rugosa X double rugosa]; Reynolds, Ted; Ted Reynolds Roses International, 1999

'REYSEL', HMsk, dp, 1992; (**Selina**); flowers deep pink, blooms borne in small clusters, single, 5 petals, 1.5 in., slight fragrance; few prickles; foliage small, medium green, semi-glossy; medium, spreading growth; [Cornelia X Trier]; Reynolds, Ted; Reynolds Roses, 1992

'REYSHELLEY', HT, ab, 1998; (**Shelley Higgins**); flowers apricot-peach blend, double, dbl., 15–25 petals, exhibition form, borne mostly singly, intense, spicy fragrance; some prickles; foliage medium, medium green, semi-glossy; upright, medium growth; [Westerland X Seedling]; Reynolds, Ted; Ted Reynolds Roses International, 1998

'RHAPSODY', HT, ob, 1951; bud long, pointed; flowers orange, reverse terra-cotta, medium, 20 petals, exhibition form; foliage leathery; vigorous, bushy growth; RULED EXTINCT 1/85; [Lulu X Cecil]; Houghton, D.; Elmer Roses Co.

Rhapsody *see* 'JACSOD'

Rhapsody in Blue *see* 'FRANTASIA'

'RHEA REID', HT, mr, 1908; bud long, pointed; flowers crimson-red, dbl., exhibition form, moderate fragrance; foliage soft; GM, Bagatelle, 1908; [American Beauty seedling X Red seedling]; Hill, E.G., Co.

'RHEA REID, CLIMBING', Cl HT, mr, 1914; California Nursery Co.

Rheda-Wiedenbruck, F, mp

Rheinaupark® *see* 'KOREIPARK'

'RHEINGOLD', HT, my, 1934; flowers golden yellow, large, very dbl., intense fragrance; foliage leathery, light; dwarf, bushy, compact growth; [Mrs T. Hillas X Mabel Morse]; Leenders, M.; J&P

'RHODE ISLAND RED', LCl, dr, 1957; dbl., 33 petals, 4–5 in., cupped, moderate fragrance; foliage glossy; [Everblooming

Pillar No. 73 X Unnamed seedling]; Brownell, H.C.

Rhodes Rose, HMult, w

'RHODOLOQUE JULES GRAVEREAUX', T, 1908; Fontes

'RHODOPHILE GRAVEREAUX', HFt, yb, 1900; flowers dark yellow with pink, large, semi-dbl.; Pernet-Ducher; (Sangerhausen)

'RHONA', HT, mp, 1984; flowers large, dbl., 20 petals, intense fragrance; foliage medium, medium green, semi-glossy; bushy growth; [Anne Letts X (Dainty Maid X Pink Favorite)]; Gobbee, W.D.

'RHONDA', LCl, mp, 1968; bud globular; flowers carmine-rose, large, dbl., slight fragrance; foliage dark, glossy; vigorous, climbing growth; [New Dawn X Spartan]; Lissemore; C-P

Rhotare, Pol, mr; flowers medium, semi-dbl.; (Sangerhausen)

'RIA WENNING', HT, dp, 1932; bud long, pointed; flowers carmine, semi-dbl., moderate fragrance; vigorous growth; [Mme Maurice de Luze X Red Star]; Leenders, M.

Ribambelle, F, 1972; Croix; (Cavriglia)

'RIBATEJO', F, w, 1962; flowers white, center deep yellow; [Virgo X Seedling]; da Silva, Moreira

Riberhus *see* POUlriber

Ric Rac *see* 'PIXRIC'

Ricarda®, F, op, 1989; Noack, Werner

Riccordo di Fernando Scarlatti, HP, dr; flowers large, dbl., intense fragrance; (Sangerhausen)

'RICCORDO DI GEO CHAVEZ', HT, dp, 1911; flowers large, dbl.; Bonfiglio, A.; (Sangerhausen)

'RICCORDO DI GIOSUE CARDUCCI', HT, lp, 1909; flowers very large, very dbl.; Bonfiglio, A.; (Sangerhausen)

Riccordo di Giovanni Spotti, HT, mr; flowers large, dbl., slight fragrance; (Sangerhausen)

Rich and Rare *see* 'SEARICH'

'RICHARD', HT, dr, 1979; bud long, pointed; flowers spinel-red, dbl., 25–30 petals, 4–5 in., intense fragrance; foliage light green; bushy, upright growth; [Gavotte X Memoriam]; Ellick; Excelsior Roses

Richard Buckley *see* 'SMITSHORT'

'RICHARD E. WEST', HT, my, 1924; bud long, pointed; flowers large, dbl., intense fragrance; foliage leathery; vigorous growth; Dickson, A.

'RICHARD HAYES', S, mp, 1973; 4 in., moderate fragrance; foliage light; vigorous, tall, upright growth; [Fred Loads sport]; Holmes, R.A.; Wonnacott

Richard Strauss®, S, rb, 1989; Noack, Werner

Richard Tauber®, HT, rb, 1986; Pouw

'RICHARDSON WRIGHT', HT, pb, 1931; flowers pearl with carmine dashes and lemon reflexes, large, globular, moderate fragrance; foliage dark, leathery; vigorous growth; [Radiance X Ville de Paris]; C-P

'RICHELIEU', HCh, m, 1845; flowers violet-pink, large, very dbl.; Verdier, V.; (Sangerhausen)

'RICHMOND', HT, mr, 1905; (Everblooming Jack Rose); bud long, pointed; flowers bright scarlet, varying greatly at times, dbl., moderate, damask fragrance; vigorous growth; [Lady Battersea X Liberty]; Hill, E.G., Co.

'RICHMOND, CLIMBING', Cl HT, mr, 1912; Dickson, A.

Ricky® *see* HAUric

Ricordo di Leone Sgaravatti, HT; Sgaravatti, A.; (Cavriglia)

'RIDGEWAY', HT, pb, 1953; flowers salmon-pink shaded apricot, well shaped, dbl., moderate fragrance; foliage rather sparse; strong stems; very vigorous growth; [Princess Marina X Vanessa]; Ratcliffe

'RIEDER'S SOLIN', Cl HT, mr, 1930; flowers large, dbl., moderate fragrance; Rieder; (Sangerhausen)

'RIFLEMAN', HT, or, 1979; bud ovoid; flowers light vermilion, dbl., 35–43 petals, 3.5 in., exhibition form, slight fragrance; tall, very vigorous, bushy growth; [Tropicana X Orange Sensation]; Murray, Nola

'RIGAUDON', F, dr, 1957; flowers dark red tinted geranium, dbl.; very dwarf, dense growth; [(Independence X Unnamed seedling) X Unnamed seedling]; Combe; Japan Rose Society

'RIGHT ROYAL', HT, mp, 1979; flowers silvery pink, dbl., 30 petals, 4 in., slight fragrance; foliage dark; moderate growth; [Scented Air X Anne Letts]; Hawken, Una

Rigobec *see* **'MON PAYS'**

Rigobec 2 *see* **'NEIGES D'ÉTÉ'**

Rigobec 3 *see* **'DANSE AZTEQUE'**

'RIGOLETTO', F, ab, 1954; flowers apricot-yellow tinted copper, well formed, large, semi-dbl., intense fragrance; vigorous growth; [Floribunda seedling X Souv. de Claudius Pernet]; Leenders, M.

'RIJSWIJK', F, ob, 1964; bud deep yellow; flowers orange-yellow, borne in clusters, semi-dbl.; foliage dark, glossy; vigorous growth; [Goldmarie X Fata Morgana]; Buisman, G. A. H.

'RIKSBYGGEROSEN', F, or, 1969; flowers open, semi-dbl., 23 petals, 3 in.; foliage glossy, light green; low, compact growth; [Irish Wonder X Seedling]; Poulsen, Niels D.; Poulsen

Rilla, HT, r

Rim *see* **'HAPPINESS'**

'RIMA', HT, lp, 1964; flowers light silvery pink; [Prima Ballerina sport]; Samuels

'RIMOSA', F, my, 1958; flowers indian yellow to citron-yellow, well-formed, medium, borne i, semi-dbl., 18–25 petals, moderate fragrance; foliage leathery; upright, symmetrical, compact growth; [Goldilocks X Perla de Montserrat]; Meilland, F.; URS

Rimosa 79 *see* 'MEIGRONURI'

'RINA', HT, w, 1999; flowers cream white blended pink, dbl., 32 petals, 5 in., exhibition form, intense fragrance; foliage medium, medium green, semi-glossy; tall (6 ft) growth; [Royal Highness X Shizunomai]; Yonzda, Kazuo, 1997

Rina Herholdt *see* 'HERANI'

'RINA HERHOLDT, CLIMBING', Cl HT, pb, 1973; Arora, Bal Raj; The Rosery, India

Rina Hugo *see* DORvizo

'RINAKOR', F, ob, 1974; (**Marina®**); bud long, pointed; flowers orange, base yellow, dbl., moderate fragrance; foliage glossy, dark, leathery; vigorous, upright growth; AARS, 1981; [Colour Wonder X Seedling]; Kordes, 1975

Rinascimento® *see* BARinas

Ring of Fire™ *see* 'MORFIRE'

'RINGFIELD', HT, or, 1977; flowers deep vermilion, full, dbl., 30 petals, 5 in., slight fragrance; vigorous growth; [Ernest H. Morse X Fragrant Cloud]; Plumpton, E.

'RINGLET', Cl HT, pb, 1922; flowers white, tipped pink and lilac, small blooms in clusters, single; vigorous, climbing growth; [Ernest Morel X Betty Berkeley]; Clark, A.; Brundrett

'RIO GRANDE', HT, dr, 1973; bud ovoid; flowers velvety dark red, medium, dbl., moderate fragrance; foliage soft; moderate, upright, bushy growth; [Unknown X Unknown]; Tantau, Math.; Ahrens & Sieberz

'RIO RITA', HT, w, 1931; bud long, pointed; flowers white tinged pink, large, dbl., intense fragrance; foliage glossy; vigorous growth; [Mme Butterfly X Premier]; Hill, Joseph H., Co.

'RIO RITA', Cl HT, w, 1935; flowers velvety scarlet-crimson, very large, dbl., cupped, intense fragrance; foliage glossy; very vigorous growth; [E.G. Hill sport]; Elmer's Nursery

Rio Samba™ *see* 'JACRITE'

'RIPPLES', F, m, 1971; flowers lilac-lavender, semi-dbl., 3.5 in., slight fragrance; foliage small, green, matt; [(Tantau's Surprise X Marjorie LeGrice) X (Unnamed seedling X Africa Star)]; LeGrice

'RISE 'N' SHINE', Min, my, 1977; (Golden Meillandina, Golden Sunblaze); bud long, pointed; flowers rich medium yellow, small, dbl., 35 petals, 1.5 in., exhibition form, moderate fragrance; bushy, upright growth; AOE, ARS, 1978 Miniature Rose Hall of Fame, ARS, 1999; [Little Darling X Yellow Magic]; Moore, Ralph S.; Sequoia Nursery

Rise 'n' Shine, Climbing *see* 'MORKINSHINE'

Rising Star *see* 'JALSTAR'

Rising Star *see* HAReast

'RISING SUN', HT, rb, 1924; bud long, pointed; flowers rich copper, base old-gold, cactus dahlia form; Hicks

'RISQUÉ', Gr, rb, 1985; flowers medium red, light yellow reverse, medium, dbl., 20 petals, no fragrance; foliage medium, dark, semi-glossy; upright, slightly spreading growth; [Bob Hope X Unnamed seedling]; Weeks, O.L.; Weeks Wholesale Rose Growers

'RITA', F, mp, 1960; flowers rich pink, blooms in large clusters, dbl., moderate fragrance; foliage glossy; vigorous growth; [Karl Herbst X Pinocchio]; Fryers Nursery, Ltd.

Rita Applegate *see* 'TINRITA'

Rita Bugnet, HRg, w, 1958; Bugnet

'RITA JACKSON', HT, rb, 1964; flowers large, dbl., exhibition form, slight fragrance; upright, bushy growth; [Tzigane sport]; Jackson, F.

Rita Levi Montalcini® *see* BARlev

Rita MacNeil *see* 'MACCHOME'

'RITA SAMMONS', Pol, pb, 1925; (Justine Silva); flowers deep rose-pink, opening pink, edged lighter; [Cécile Brunner sport]; Clarke, B.; Clarke Bros.

Ritausma, HRg, lp

Ritchie's Red Climber, Cl HT, dr

Rite Brite *see* 'GELRITE'

Rittenhouse™ *see* 'WILHKPK'

Ritter Taler, HT, rb

'RITTER VON BARMSTEDE', HKor, mp, 1959; 20 petals, 2 in., blooms in clusters of 30-40; foliage glossy; vigorous (10-15 ft) growth; Kordes

'RITZ', LCl, mr, 1955; flowers velvety red blooms in clusters, semi-dbl., 11 petals, 3 in., recurrent bloom; vigorous (8-9 ft) growth; [R. setigera X ?]; Horvath; Wyant

'RITZ', F, mr, 1961; flowers bright scarlet, semi-dbl., 16 petals, 3.5 in., slight fragrance; foliage dark, glossy; vigorous, well branched growth; Gaujard

'RIVA LIGURE', HT, dp, 1947; bud pointed; flowers velvety carmine, dbl., 24–35 petals, intense fragrance; foliage dark; vigorous, bushy growth; [Unnamed seedling X Crimson Glory]; San Remo Exp. Sta.

'RIVAL', HT, mr, 1954; flowers cherry-scarlet, loosely formed, 4–5 in., flat; foliage bronze; free growth; [Southport X The Rev. W.S. Crawford]; Fletcher; Tucker

'RIVAL DE PAESTUM', T, w; bud tinged pink; flowers white, base blush and ivory, dbl.; foliage dark; moderate growth; Béluze, 1841

Rival de Paestum, Ch, w, 1863; Paul

River City Jubilee *see* 'JUDRIV'

Riverdance *see* 'KENFINE'

Riverdance, Min, pb, 1998; Laver, Keith G.

Riverina Sunset, T, mp; Hay

'RIVERS', HP, mr, 1832; flowers bright crimson, large, borne in corymbs, dbl., recurrent bloom; vigorous growth; Laffay, M.

'RIVERS' GEORGE IV', HCh, dr; (George IV, King George IV); flowers vivid crimson, shaded with dark purple, loosely, dbl., cupped, non-recurrent; branching shoots tinged with purple; vigorous, growth; [Thought to be Damask X China]; Rivers, ca. 1817

'RIVERS' MUSK', HMsk, mp; flowers rosy buff, small, intense fragrance; Rivers, prior to 1925

River's South Bank *see* **'JUJNOBEREJNAIA'**

Riverview Centennial, HT, dr, 1980

'RIVIERA', HT, or, 1939; bud globular, yellow; flowers orange-scarlet, reverse lighter, base yellow, open, large, dbl., cupped; foliage glossy, wrinkled, dark; very vigorous, bushy growth; [Luis Brinas X Catalonia]; Dot, Pedro; J&P, 1940

Riviera *see* JACsak

Riviere de Diamant *see* BRIsiness

Road to Freedom *see* 'FRANLAC'

'ROADMAN', HT, w, 1977; bud pointed; flowers near white, dbl., 40–50 petals, 6 in., exhibition form; foliage leathery; upright growth; [Hawaii X Kordes' Perfecta]; Ota, Kaichiro

'ROAMING', HT, dp, 1970; flowers reddish pink shades, pointed, dbl., 24 petals, 3.5 in.; foliage green, matt; [Vera Dalton X Tropicana]; Sanday, John

Rob in des Bois *see* **'ROBIN HOOD'**

Rob Roy® *see* 'COROB'

Robbie Burns *see* 'AUSBURN'

Robe de Neige, S, 1995; Lens; (Cavriglia)

Robe de Soie, S, lp, 1996; Lens

'ROBE D'ETÉ', F, or, 1966; flowers salmon-orange, dbl., 22–24 petals, 3–3.5 in., exhibition form, slight fragrance; foliage bronze; bushy, upright growth; [(Chatelaine X Mannequin) X (Montezuma X Floradora)]; Lens

Robe Fleuri, S, 1995; Lens; (Cavriglia)

Robe Rose, S, 1995; Lens; (Cavriglia)

'ROBERT', P, pb, 1856; flowers carmine, marbled white; Robert

Robert Aliano *see* 'WILALNO'

'ROBERT BETTEN', HT, mr, 1920; flowers clear dark carmine-red, not turning blue, dbl.; [Frau Karl Druschki X Corallina]; Schmidt, J.C.

'ROBERT BLAND', S, dp, 1960; flowers open, small, dbl., non-recurrent; thornless; foliage richgreen; vigorous, bushy growth; quite hardy.; [((Hansa X R. macounii) X Betty Bland) X (R. blanda X Betty Bland)]; Wright, Percy H.

'ROBERT COTTON', HT, w, 1968; bud ovoid; flowers white, edges flushed pink, large, dbl., slight fragrance; foliage glossy, serrated, leathery; moderate growth; [Marcia Stanhope X Karl Herbst]; Golik; Ellesmere Nursery

'ROBERT DE BRIE', HP, m, 1860; flowers violet-pink with white stripes, large, dbl.; Granger; (Sangerhausen)

'ROBERT DUBOL', HT, mr, 1946; flowers warm orient red, stamens golden, very dbl., exhibition form; Sauvageot, H.; Sauvageot

'ROBERT DUNCAN', HP, pb, 1897; flowers purplish pink, sometimes flamed brilliant red, well-formed, dbl., 70 petals, repeat bloom, moderate fragrance; vigorous growth; Dickson, A.

'ROBERT F. KENNEDY', HT, mr, 1968; flowers scarlet, exhibition form, intense fragrance; foliage dark; very vigorous, upright growth; [Chrysler Imperial X Ena Harkness]; Takatori, Yoshiho; Parnass Rose Nursery

'ROBERT HUEY', HT, dp, 1911; bud long, pointed; flowers carmine edged lighter, bluing slightly, dbl., moderate fragrance; moderate growth; Dickson, A.

'ROBERT LE DIABLE', HGal, m; flowers scarlet-pink aging to deep purple, center often green, dbl.; low, lax growth

'ROBERT LÉOPOLD', M, pb, 1941; flowers salmon-flesh-pink edged light carmine, large, dbl.; Buatois

'ROBERT PERPÉTUEL', P, m, 1856; flowers violet-pink, medium, dbl.; Robert; (Sangerhausen)

Robert Stolz, F, dr, 1974; flowers medium, semi-dbl.; deRuiter; (Sangerhausen)

'ROBERTA BONDAR', LCl, my, 1993; dbl., 26–40 petals, 3–3.5 in., borne 1-8 per truss, moderate fragrance; foliage large, dark green, semi-glossy; climbing (70cms) growth; [King's Ransom X Buff Beauty]; Fleming, Joyce L.; Hortico Roses, 1993

'ROBESPIERRE', Pol, pb, 1975; bud full; flowers small, very dbl., 88 petals, 2.5 in., cupped, moderate fragrance; foliage bronze; Delforge, S.

'ROBIN', Min, mr, 1956; bud urn-shaped; flowers rich red, blooms in clusters of 15, dbl., 65 petals, 1.25 in., flat; foliage leathery, green, matt; vigorous, dwarf (12 in), bushy growth; [Perla de Montserrat X Perla de Alcanada]; Dot, Pedro; C-P

Robin *see* 'RESROB'

'ROBIN HOOD', HMsk, mr, 1927; (Rob in des Bois); flowers cherry-red, blooms in large clusters, recurrent bloom; vigorous (4-5 ft.), dense, compact growth; (14); [Seedling X Miss Edith Cavell]; Pemberton

'ROBIN HOOD', HT, mr, 1912; flowers soft bright rosy scarlet, changing to bright scarlet-crimson, dbl., intense fragrance; Hill, E.G., Co.

Robin Red Breast *see* 'INTERROB'

Robin Redbreast *see* 'INTERROB'

Robina *see* KORxenna

'ROBINETTE', HMult, rb, 1943; flowers amaranth-red, white eye, open, small, borne in clusters, single, intense fragrance; foliage glossy; height 10-12 ft; [Hiawatha X Hiawatha]; Moore, Ralph S.; Hennessey

Robur *see* **'RED EMPRESS'**

Robusta® *see* 'KORGOSA'

Robusta, B, mr, 1877; Soupert & Notting

Robusta, Cl T, mp

Roby, HMult, rb; Guillot, 1912

Roccana Diane *see* 'KIRDEX'

Roche Centenary *see* DICvintage

'ROCHEFORT', HT, ab, 1936; flowers large, dbl., intense, fruity fragrance; foliage leathery; vigorous growth; GM, Portland, 1935; [Mrs Pierre S. duPont X Charles P. Kilham]; Mallerin, C.; C-P

'ROCHELLE HUDSON', HT, rb, 1937; bud long, pointed; flowers carmine, base yellow, orange undertone deepening with age, semi-dbl., slight, fruity fragrance; foliage dark; vigorous growth; [Isobel X Mme Edouard Herriot]; Moore, Ralph S.; Brooks & Son

'ROCHESTER', F, ab, 1934; flowers buff, reverse orange-carmine, medium, dbl., moderate fragrance; foliage leathery; vigorous, bushy growth; [Echo X Rev. F. Page-Roberts]; Nicolas; J&P

Rochester Cathedral *see* 'HARROFFEN'

Rocio Elias® *see* FEtrone

Rock Hill Peach, T, pb

Rock 'n' Roll *see* 'MACFIRWAL'

Rocket *see* **'RAKETA'**

'ROCKET', HT, mr, 1935; flowers brilliant scarlet, reverse crimson, large, dbl., exhibition form, intense fragrance; foliage leathery, dark, bronze; very vigorous growth; [Dame Edith Helen X Scorcher]; Nicolas; J&P

Rocketeer *see* 'CHEWALLOP'

Rockin' Robin™ *see* 'WEKBOROCO'

Rocky™ *see* 'MACKEPA'

'ROCOCO', F, mr, 1964; flowers scarlet, borne in clusters, semi-dbl., 15 petals, 3 in.; free growth; [Moulin Rouge X Fire Opal]; McGredy, Sam IV; Spek

'ROD STILLMAN', HT, lp, 1948; flowers light pink, base flushed orange, large, dbl., 35 petals, intense fragrance; foliage dark; vigorous growth; [Ophelia X Editor McFarland]; Hamilton

Roddy MacMillan *see* 'COCARED'

'RODEO', F, or, 1960; flowers bright scarlet, borne in clusters (up to 10), dbl., 3 in., slight fragrance; foliage light green; bushy, low growth; [Obergärtner Wiebicke X Spartan]; Kordes, R.; McGredy

Rodeo Drive™ *see* 'AROCORE'

'RÖDHÄTTE', F, mr, 1912; (Red Riding Hood); flowers clear cherry-red, large blooms in large clusters, semi-dbl.; foliage rich green; bushy, compact growth; [Mme Norbert Levavasseur X Richmond]; Poulsen, D.T.; Poulsen

'RÖDHÄTTE, CLIMBING', Cl F, mr, 1925; Grootendorst, F.J.

Rodin *see* 'MEIDINRO'

Rodinghausen®, S, mr, 1987; Noack, Werner

Rodovrerosen *see* **'POULSEN'S FAIRY'**

Rody *see* TANydor

Roedean *see* BENcamelia

ROElanda, HT, w, 1993; (**Leana**)

'ROELOF BUISMAN', HT, mr, 1964; bud ovoid; flowers bright pure red, wellformed, large; vigorous, upright, bushy growth; Kordes, R.

'ROGER BOUDOU', Pol, mr, 1957; flowers very bright red; [Lafayette seedling]; Privat

'ROGER LAMBELIN', HP, rb, 1890; flowers bright crimson fading maroon, petals margined white, very di, dbl., recurrent bloom; vigorous growth; (28); [Fisher Holmes sport]; Schwartz, Vve.

'ROGER LAMBELIN STRIPED', HP, rb, 1953; flowers deep maroon to pink stripes on white ground, recurrent bloom; [Roger Lambelin sport]; Hennessey

Roger Secretrain, LCl; Moreira da Silva, A.; (Cavriglia)

'ROGSCRIV', HT, pb, 1999; (**Natural Beauty**); flowers pink/apricot blend, reverse pink blend, dbl., 26–40 petals, 3 in., rosette, borne in small clusters, intense fragrance; few prickles; foliage medium, medium green, semi-glossy; compact, medium (3 ft) growth; [Pretty Lady X Silver Jubilee]; Scrivens, Len; John Tooby & Co., Ltd., 1998

'ROGslav', HT, ab 1995; (**Penkala Slavoljub**); flowers apricot-orange striped, full (26–40 petals), medium (4–7 cms) blooms borne mostly single; few prickles; foliage medium, dark green, semi-glossy; medium (100cms), bushy growth; [Ambassador sport]; Rogin, Josip; Rogin, 1995

'ROGZAG', HT, pb 1995; (**Zagreb**, Zagreb 900); flowers pink striped, full (26–40 petals), medium (4–7 cms) blooms borne mostly single; very fragrant; few prickles; foliage large, medium green, matt; tall (150cms), upright growth; [Eiffel Tower sport]; Rogin, Josip; Rogin, 1995

'ROI ALBERT', HT, pb, 1925; bud long, pointed; flowers bright carmine-rose, center tinted scarlet, intense fragrance; [Laurent Carle seedling X Mme Abel Chatenay]; Klettenberg-Londes

'ROI ALEXANDRE', HT, ob, 1937; (King Alexander I, S.M. Alexander I); flowers coppery orange, tinted salmon, over large, dbl., moderate fragrance; foliage leathery, glossy, bronze; very vigorous growth; Gaujard

Roi de Nains, Min, mr

'ROI DE PAYS-BAS', C, dp, 1848; (Roi des Pays-Bas); flowers deep pink, large, dbl., cupped

Roi de Siam, Cl T, mr, 1825; Laffay, M.

'ROI DES AUNES', S, dp, 1885; (Erlkönig); flowers carmine tinted red, large, dbl., globular, non-recurrent; very vigorous growth; Geschwind, R.

'ROI DES BENGALES', F, rb, 1958; flowers grenadine-red; vigorous, low growth; [(Hermosa X Gruss an Teplitz) X Independence]; Arles; Roses-France

Roi des Bordures *see* **'BORDER KING'**

'ROI DES CRAMOISIS', HCh, mr; flowers bright red, cupped

Roi des Pays-Bas *see* **'ROI DE PAYS-BAS'**

Roi des Rois, F, 1955; Delbard-Chabert; (Cavriglia)

'ROI SOLEIL', HT, my, 1962; bud long, pointed; flowers citron-yellow; very vigorous growth; GM, Madrid, 1963; [Peace X Independence seedling]; Dorieux; Le Blévenec

Roides Pourpres *see* **'MOGADOR'**

'ROKLEA'®, HT, ob, 1975; flowers bright orange, large, dbl., 20 petals, exhibition form, moderate fragrance; foliage large, dark, semi-glossy; upright growth; Tantau, Math.

Rokoko® *see* TANokor

Rokoko, HRg, ob, 1985; Baum

'**ROLAND**', F, ob, 1961; flowers salmon-orange-red; low growth; [Karl Weinhausen X Independence]; Leenders, J.

Roland Garros® see BRIgarold

Roletta, F, op, 1983; flowers salmon-orange/pink, large, dbl.; GPG Bad Langensalza; (Sangerhausen)

Roller Coaster see 'MACMINMO'

Roma, HT, lp; [Duet sport]; Spronk, 1970; (Weatherly, L.)

Roma di Notte, HT, 1980; Zandri, R.; (Cavriglia)

Romaggi Plot Bourbon, B, pb

'**ROMÁN**', HT, mp, 1961; flowers nilsson pink, becoming hermosa pink, large, dbl., 35 petals, intense fragrance; [Asturias X Rosa de Friera]; Dot, Simon

'**ROMAN FESTIVAL**', F, pb, 1968; bud ovoid; flowers coral, base yellow, medium, borne in clusters, exhibition form, slight fragrance; foliage dark, glossy; vigorous, low, compact growth; [Queen Elizabeth X Sumatra]; Williams, J. Benjamin

Roman Holiday see 'LINRO'

'**ROMAN TRIUMPH**', F, mr, 1977; flowers large, semi-dbl., 13 petals, slight fragrance; foliage glossy; upright, bushy growth; [Jove X City of Leeds]; Harkness

'**ROMANA**', HT, rb, 1938; bud long, pointed; flowers rose-red to light purple, open, large, dbl.; foliage leathery, glossy, dark; very vigorous growth; [Better Times sport]; Ringdahl

Romana, HT, lp; Vecera, L., 1975; (Czech Rosa Club)

Romance see 'JACROM'

'**ROMANCE**', HT, my, 1931; bud long, pointed; flowers golden yellow, shading toward lemon, open, large, dbl., moderate fragrance; foliage thick; very vigorous growth; RULED EXTINCT 11/82; [Souv. de Claudius Pernet X Buttercup seedling]; Towill

'**ROMANCE**', Cl HT, lp, 1933; flowers shell-pink, fading to pale blush, large, not too full, intense fragrance; vigorous, climbing growth; RULED EXTINCT 11/82; [Isa sport]; Beckwith

Romance see 'TANEZAMOR'

Romanina, F, 1974; Zandri, R.; (Cavriglia)

Romantic Days see 'MEIPARNIN'

Romantic Dreams see MEIlomit

Romantic Fragrance see 'MEISOCRAT'

Romantic Hedgerose see KORworm

Romantic Moments see 'MEIROKOI'

Romantic Occasion see MEIsardan

Romantic Palace see POUlmanti

Romantic Seranade see 'MEIBRINPAY'

Romantic Sunrise see MEIhaitoil

'**ROMANTICA**', HT, mp, 1962; bud oval; flowers phlox-pink, large, dbl., 40 petals, exhibition form, slight fragrance; foliage leathery, glossy; very vigorous, upright growth; [Baccará X White Knight]; Meilland, Mrs. Marie-Louise; URS

Romantica 76®, HT, op

Romantique Meillandina see 'MEIDANCLAR'

'**ROMANY**', F, op, 1965; flowers salmon, well formed, borne in clusters, 3.5 in., slight fragrance; free growth; [Orangeade X Mischief]; McGredy, Sam IV; Geest Industries

Romanze see 'TANEZAMOR'

'**ROME GLORY**', HT, mr, 1937; (Gloire de Rome, Gloria di Roma, Glory of Rome); bud ovoid; flowers scarlet, reverse cerise, dbl., 55 petals, 4–5 in., globular, moderate fragrance; vigorous, bushy growth; [Dame Edith Helen X Sensation]; Aicardi, D.; J&P

Rome Glory, Climbing, Cl HT, mr; (Cavriglia)

'**ROMEO**', HT, yb, 1918; flowers indian yellow, suffused coppery pink, slight fragrance; [Edith Part seedling X Ophelia]; Therkildsen

'**ROMEO**', LCl, dr, 1919; flowers deep red, well formed, borne in small clusters, dbl.; vigorous, climbing growth; Easlea

Romeo see 'MACTUAL'

Romina see TANanim

Romstar, HT, lp; long; flowers medium red, velvetyt, high centered, dbl., 34 petals, moderate; medium, medium-green, semi-glossy; [Baccara X Coronado]; St. Wagner, 1989; Res. Stn. f. Horticulture, Cluj, 1991, Romania

Romy, HT, w; Select Roses, B.V.

Romy Schneider® see ORAred

'**RON WEST**', HT, pb, 1985; flowers white with deep pink petal edges; [Admiral Rodney sport]; West, Ronald

Ronald George Kent see 'BRANDYPINK'

'**RONALD HEALY**', HT, pb, 1932; bud long, pointed; flowers old-rose, shaded salmon and yellow, dbl., exhibition form, moderate fragrance; foliage glossy; bushy growth; Dobbie

'**RONALD TOOKE**', HT, dr, 1927; flowers deep blackish crimson; [Col. Oswald Fitzgerald sport]; Morse

Roncalli, S, rb, 1997; Noack, Werner

'**RONDE ENDIABLÉE**', F, or, 1963; flowers geranium-red, edged darker, large, semi-dbl.; foliage dark, glossy; moderate growth; GM, The Hague, 1964; Combe

'**RONDO**', HT, or, 1955; bud ovoid; semi-dbl., 15–20 petals, 3.5–4 in., exhibition form, moderate, fruity fragrance; foliage dark, leathery; vigorous, upright, compact growth; [Danzig X (Crimson Glory X Floradora sister seedling)]; Tantau, Math.; J&P

'**RONNY TEMMER**', F, mp, 1974; bud ovoid; flowers full, dbl., 44 petals, 3.5 in., moderate fragrance; Delforge, S.

'**RONSARD**', HT, rb, 1937; bud long, pointed, yellowish edged red; flowers brilliant red, reverse yellow and cream, semi-dbl., cupped, slight fragrance; foliage leathery, dark; bushy growth; [Conrad Ferdinand Meyer X R. foetida bicolor seedling]; Gaujard; J&P

Rooi Rose see KORhood

'**ROOSENDAAL**', F, mr, 1965; flowers medium, dbl.; foliage dark; [Gartendirektor Glocker X Alpine Glow]; Buisman, G. A. H.

'**ROQUEBRUNE**', HT, ob, 1959; bud oval; flowers ochre-yellow edged orange, medium, dbl., slight fragrance; foliage dark, glossy; strong stems; moderate, bushy growth; Delforge

'**RORO**', HT, 1954; San Remo Exp. Sta.; (Cavriglia)

'**RORY CARLTON**', HT, dp, 1996; flowers deep pink blooms borne mostly single, dbl., 26–40 petals, slight fragrance; few prickles; foliage medium, dark green, semi-glossy; medium (100-120 cms), upright growth; [Sylvia X Royal Highness]; Macredie, W.R.

'**ROSA**'

Rosa Belle™ see 'KINBELLE'

Rosa berberifolia see '**HULTHEMIA PERSICA**'

'**ROSA BONHEUR**', M, mp; (Mlle Rosa Bonheur); flowers pink or bright rose, large, dbl.; moderate growth; Laffay, M., 1871

'**ROSA D'ABRIL**', HT, dp, 1948; flowers carmine, large, very dbl., globular; very vigorous growth; Dot, Pedro

'**ROSA DE FRIERA**', HT, pb, 1956; flowers violet-pink with carmine reflections, large, dbl., 35 petals, intense fragrance; strong stems; vigorous growth; [Rosa Gallart X Paulette]; Dot, Pedro

'**ROSA GALLART**', HT, mp, 1935; bud long, pointed; flowers rose-pink, large, dbl., cupped, moderate fragrance; foliage glossy; vigorous growth; [Unnamed variety X (Li Bures X Rose Marie)]; Dot, Pedro

'**ROSA GRÜSS AN AACHEN**', F, yb, 1930; flowers satiny yellowish pink; [Gruss an Aachen sport]; Spek

Rosa Kaiserin, HT, mp; flowers large, dbl., slight fragrance; (Sangerhausen)

'**ROSA MAMIE**', HT, mr, 1956; bud ovoid; flowers bright rose, large, dbl., 35 petals, moderate fragrance; long stems; vigorous growth; [Rome Glory sport]; Asseretto, V.

'ROSA MONNET', HP, 1866; Monnet; (Cavriglia)

Rosa Mundi *see* **R. GALLICA VERSICOLOR**

'ROSA MUNNÉ', HT, rb, 1952; flowers red to saffron-pink, large, dbl., intense fragrance; foliage clear green; very vigorous growth; [Maria Serrat X Paz Vila]; Munné, M.

'ROSA ORANGE TRIUMPH', Pol, mp, 1943; flowers small, semi-dbl., slight fragrance; Verschuren; (Sangerhausen)

Rosa persica *see* **'HULTHEMIA PERSICA'**

Rosa Poncheaux *see* **'PONCHEAU-CAPIAUMONT'**

Rosa Prominent, Gr, 1982; Sieber; (Cavriglia)

Rosa simplicifolia *see* **'HULTHEMIA PERSICA'**

Rosa Sinfonie, F, mp; flowers medium, semi-dbl., slight fragrance; VEG; (Sangerhausen)

Rosa Stern, HT, mp; flowers large, dbl.; (Sangerhausen)

'ROSA TRAUM', F, mp, 1974; bud pointed; dbl., 32 petals, 2.5 in., exhibition form, slight fragrance; foliage glossy; vigorous, upright, bushy growth; [Duftwolke X Seedling]; Kordes; Dehner & Co.

Rosa Union, HT; Dot; (Cavriglia)

'ROSA VERSCHUREN', HT, 1904; Verschuren; (Cavriglia)

'ROSA VOLLENDUNG', F, pb, 1943; bud long, pointed; flowers salmon-pink, reverse capucine-red, large, borne in clusters, dbl., slight fragrance; foliage leathery, wrinkled; vigorous, bushy growth; [Crimson Glory X Else Poulsen]; Kordes

Rosa X hardii *see* **'HULTHEMIA HARDII'**

Rosa Xmitcheltonii *see* **'MITCHELTONII'**

Rosa Zwerg, HRg, mp, 1984; Baum

'ROSABEL WALKER', HT, mr, 1922; bud long, pointed; flowers brilliant velvety crimson, dbl., 3.5 in., moderate fragrance; very vigorous, bushy, spreading growth; Cant, F.

Rosabell *see* 'COCCELESTE'

'ROSABELLA', HT, mr, 1941; flowers red with cerise reflections, very large; foliage abundant; long stems; [Mrs J.D. Russell X Julien Potin]; Giacomasso

'ROSABELLA', F, op, 1955; flowers salmon-pink shaded orange, well shaped, borne in very large, very dbl.; vigorous growth; [Pinocchio seedling]; Maarse, G.

'ROSABELLE BARNETT', F, op, 1970; flowers coral, dbl., 32 petals, 2.5 in., globular, slight fragrance; foliage dark; very free growth; [Tropicana X ?]; Gregory

Rosabunda, F, lp; cylindrical and short; cupped, 30 petals, flowers in clusters of 3-9 florets, moderate; dark green, medium large, glossy; [Frankfurt am Main X Maria Callas]; St. Wagner, 1978; Res. Stn. f. Hortiuclture, Cluj, 1978, Romania

'ROSADA', Min, op, 1950; (Rosata); flowers peach edged pink, dbl., 25 petals, cupped; foliage small, glossy; compact (7-8 in) growth; [Perla de Alcanada X Rouletii]; Dot, Pedro; URS

Rosadora, F, lp; rosette, 25 petals, slight; leaves small, medium green, glossy; spreading habit; [Candy Rose X Yesterday]; St. Wagner, 1995; Res. Stn. f. Fruit Growing, Cluj, 1996, Romania

Rosagold, F, lp; medium long; flowers yellow-apricot, cupped, 35 petals, borne in clusters, intense, very; medium large, light green, glossy; vigorous upright tall growth; [Vigorosa X Allgold]; St. Wagner, 1995; Res. Stn. f. Fruit Growing, Cluj, 1996, Romania

'ROSALBA', HT, pb, 1934; bud long, pointed, streaked carmine; flowers lilac-pink, center shaded salmon-pink, large, dbl.; vigorous growth; [Souv. de Claudius Pernet X Willowmere]; Borgatti, G.

'ROSALEDA', HT, yb, 1958; flowers yellow and white, well formed, large, dbl., moderate fragrance; foliage dark; strong stems; upright, bushy growth; [Monte Carlo X Michele Meilland]; da Silva, Moreira

'ROSALEEN', HMsk, dr, 1933; dbl., large clusters, recurrent bloom; Bentall

'ROSALEEN DUNN', HT, mr, 1942; flowers crimson-red, dbl., 22 petals, 5.5 in., cupped, intense fragrance; foliage dark; vigorous growth; McGredy; J&P

Rosaletta *see* FRYstassi

Rosali® *see* 'TANILASOR'

Rosali *see* 'TANLI'

Rosali '83® *see* 'TANILASOR'

'ROSALIA', HT, dp, 1954; bud long, pointed, light cerise; long stems; Cant, F.

Rosalie *see* **'ROSELLA SWEET'**

Rosalie Coral *see* 'CHEWALLOP'

'ROSALIE RICHARDSON', HT, lp, 1932; flowers soft pink, intense fragrance; Evans

Rosalina™ *see* 'KORSAKU'

'ROSALIND', HT, op, 1918; bud bright coral; flowers apricot-pink becoming shell-pink, dbl., moderate fragrance; [Ophelia sport]; Pierson, F.R.

Rosalind, HT, pb, 1993; Bell

'ROSALIND ORR ENGLISH', HT, op, 1905; flowers salmon-pink, large, dbl.; Hill, E.G., Co.; (Sangerhausen)

'ROSALIND RUSSELL', HT, mp, 1950; bud long, well formed; flowers bright pink, dbl., 45–50 petals, 5 in., exhibition form, moderate fragrance; foliage dark, leathery; very vigorous, upright growth; [Briarcliff X Regina Elena]; Grillo

'ROSALINDA', HT, dp, 1945; flowers carmine, well formed, large, very dbl., moderate fragrance; very vigorous growth; [Editor McFarland X Comtesse Vandal]; Camprubi, C.

Rosalinda, F, lp; flowers medium pink borne in clusters, cupped, 39 petals, slight; medium green, semi-glossy; vigorous; [(Frankfurt am Main X Maria Callas) X Foc de Tabara]; St. Wagner, 1992; Res. Stn. f. Horticulture, Cluj, 1994, Romania

'ROSALINDE', F, mp, 1944; flowers clear pink, blooms in large clusters, dbl.; Krause

Rosalita, HMsk, 1997; Lens; (Cavriglia)

'ROSALPINA', HT, pb, 1953; flowers coppery pink, very large, dbl., 50 petals; foliage glossy; strong stems; very vigorous growth; [Signora X Unnamed variety]; Giacomasso

Rosalynn Carter *see* 'RUPRINS'

Rosamaria, S, 1974; Mansuino; (Cavriglia)

'ROSAMOND', HT, op, 1927; flowers orange-salmon, stamens golden, single; [Red-Letter Day X R. foetida bicolor]; Burbage Nursery

'ROSAMUNDE', F, pb, 1941; flowers salmon-carmine, reverse hydrangea-pink, large, dbl., moderate fragrance; [Seedling X Permanent Wave]; Leenders, M.

Rosamunde *see* KORmunde

Rosanna *see* 'KORINTER'

'ROSANNA', HT, mp, 1959; bud urn shaped; flowers spinel-rose, open, medium to large, slight fragrance; foliage leathery; vigorous, upright, compact growth; RULED EXTINCT 4/77; [Baccará X Gruss an Coburg]; Valentino

Rosarama, HT, op, 1988; Martin

Rosario® *see* TANoras

Rosarito, HT; Moreira da Silva, A.; (Cavriglia)

Rosarium den Blakken, F, lp, 1998; RvS-Melle

Rosarium Dortmund, LCl, pb, 1995; Noack, Werner

Rosarium Netersen® *see* 'KORTERSEN'

Rosarium Ueteresen® *see* 'KORTERSEN'

Rosary *see* **'ROSERIE'**

Rosata *see* **'ROSADA'**

Rosazwerg *see* **Dwarf Pavement**

Röschen Albrecht, F, or, 1981; flowers medium, semi-dbl., slight fragrance; Tantau; (Sangerhausen)

Rose 2000 *see* 'COCQUETRUM'

Rose à Feuilles de Laitue *see* **'BULLATA'**

'ROSE À FEUILLES LUISANTES', M, lp, 1843; flowers soft pink edged blush, medium, borne in clusters, dbl., globular, moderate fragrance; foliage glossy; branching growth; Vibert

Rose à Parfum de Bulgarie, D, mp; flowers medium, dbl., intense fragrance; (Sangerhausen)

Rose a Parfum de Grasse, D, mp; flowers medium, dbl., intense fragrance

'ROSE À PARFUM DE L'HAY', HRg, mr, 1901; (Parfum de l'Hay); flowers cherry-carmine-red, turning blue in heat, large, dbl., globular, recurrent bloom, intense fragrance; foliage nottypically rugose; vigorous (5 ft.) growth; (21); [(Summer Damask X Général Jacqueminot) X R. rugosa]; Gravereaux

'ROSE AIMÉE', HT, m, 1955; flowers gold, flushed and splashed crimson, well formed, dbl., moderate fragrance; very vigorous, bushy growth; [Peace X Seedling]; Gaujard

Rose Angle *see* **'ANGLE'**

'ROSE ANGLE', HEg, mp; (Rose Angle Blush); flowers bright lilac rose, cupped; foliage very fragrant; very vigorous growth; Martin, prior to 1848

Rose Angle Blush *see* Rose Angle

Rose Anil, Gr, m, 1998; Viraraghavan, M.S.

'ROSE ANNE', Cl HT, ob, 1938; flowers orange-apricot, base deeper yellow, semi-dbl., cupped, recurrent bloom; foliage large, glossy; very vigorous, climbing (15 ft.) growth; RULED EXTINCT 4/77; [Francesca X Margaret McGredy]; Thomas; Armstrong Nursery

'ROSE APPLES', HRg, dp, 1906; flowers carmine-rose, large, semi-dbl., moderate fragrance; vigorous growth; Paul

Rose Baby *see* 'MACROS'

'ROSE BAMPTON', HT, dp, 1940; bud carmine; flowers bright china-red, camellia form, dbl., 50 petals, 5.5 in., moderate fragrance; foliage dark; vigorous growth; [Charles P. Kilham X Margaret McGredy]; Van Rossem; J&P

'ROSE BANSAL', HT, yb, 1972; flowers yellow, petal edges sometimes crimson; [Ambossfunken sport]; Friends Rosery

'ROSE BENARY', HT, mp, 1908; flowers large, dbl., moderate fragrance; Lambert, P.; (Sangerhausen)

'ROSE BERKLEY', HT, op, 1928; (Souv de Rose Berkley); bud long, pointed; flowers deep rosy salmon-pink suffused orange, base orange, large, exhibition form, slight fragrance; foliage rich green, glossy; very vigorous growth; McGredy

Rose Bicolore, HT, ob; Croix

Rose Blanche, HT, w; Croix

Rose Bleue, HT, m

'ROSE BOWL', HT, mr, 1961; bud urn shaped; flowers bright red, large, dbl., 35 petals, exhibition form, intense fragrance; foliage leathery; long, strong stems; vigorous, upright, bushy growth; [Mardi Gras X Chrysler Imperial]; Morey, Dr. Dennison; J&P

'ROSE BRADWARDINE', HEg, mp, 1894; flowers clear rose-pink, blooms in graceful clusters, single, seasonal bloom; foliage very fragrant; vigorous growth; Penzance

'ROSE BRUFORD', HT, ab, 1961; flowers creamy peach, shaded rosy bronze; [Soraya sport]; Wheatcroft Bros.

Rose Cascade *see* DELcouro

Rose Céleste® *see* 'DELROCELES'

'ROSE CHARM', HT, mp, 1934; flowers sanguineous pink; [Talisman sport]; Scittine; Lainson

'ROSE CHEAL', HT, mp, 1970; flowers rose-pink, spiral form, dbl., 42 petals, 4–5 in., moderate fragrance; foliage glossy, dark; free growth; [New Style X Scarlet Queen Elizabeth]; Herincx

Rose Cherry Pastel, Min, mr

Rose Chou, C, mp

Rose Csárdás, F, mr, 1965; flowers medium, dbl.; Berger, W.; (Sangerhausen)

'ROSE D'AMOUR', Misc OGR, dp; (R. virginiana plena, st Mark's Rose, The st Mark's Rose); bud deep pink with long sepals; flowers deep pink, outer petals fading to pale pink receptacle wide, dbl., summer bloom, moderate fragrance; prickles of mixed sizes, scattered; foliage leaflets 5-7, rich green, marked with red on leaf st; lax growth to 10 ft.

'ROSE D'AMOUR', HT, rb, 1936; flowers brown-red, reverse yellow, open, large, dbl.; foliage glossy, dark; vigorous growth; Gaujard

Rose d'Annecy®, S, mp; (ADAfetap); Golden Rose, Geneva, 1995; Adam

'ROSE DAWN', HT, pb, 1924; bud long, pointed; flowers soft shell-pink, base yellow, large, dbl., exhibition form, intense fragrance; foliage light; vigorous growth; [(Joseph Hill X Mrs George Shawyer seedling) X Ophelia]; Towill

'ROSE DE FRANCE', HT, op, 1942; bud ovoid; flowers brilliant salmon tinted orange, very large, dbl., slight fragrance; foliage glossy; dwarf growth; Gaujard

Rose de la Reine *see* **'LA REINE'**

'ROSE DE LOS ANDES', Gr, op, 1974; bud ovoid; flowers pink-salmon-peach, dbl., 44–46 petals, 4–4.5 in.; vigorous growth; [Duet sport]; Gutierrez

'ROSE DE LYON', HT, ob, 1945; bud pointed; flowers orange-yellow, medium, dbl., moderate fragrance; foliage glossy; Gaujard

'ROSE DE MEAUX', C, mp; (De Meaux, Pompon Rose, R. centifolia pomponia, R. dijoniensis, R. pomponia, R. pulchella); dbl., 1.5 in.; dwarf growth; (28)

'ROSE DE MEAUX WHITE', C, w; (De Meaux White, Le Rosier Pompon Blanc, White de Meaux); flowers white with pink centers

Rose de Montfort, F, dp, 1993

Rose de Puteaux, P, mp

Rose de Rennes® *see* ADAharos

'ROSE DE RESCHT', P, dp; flowers bright fuchsia-red, fading with lilac tints, rosette form, very dbl., long-season bloom, intense, damask fragrance; foliage dense; vigorous, compact growth; re-introduced, 1950

Rose de Schelfhout, HGal, lp; Parmentier, 1840

Rose des Cisterciens *see* DELarle

'ROSE DES MAURES', HGal, dr; (Sissinghurst Castle); flowers deep plum-crimson, yellow stamens, semi-dbl., 17 petals, 2.5 in., moderate fragrance; foliage thin; vigorous growth; Sackville-West, 1947; Re-int. 1947

'ROSE DES PEINTRES', C, mp; (Centifolia Major); flowers bright pink, dbl., moderate fragrance; vigorous growth

Rose des Quatre Saisons *see* **'AUTUMN DAMASK'**

'ROSE D'ESPÉRANCE', HT, dr, 1918; flowers medium-large, dbl.; Verschuren; (Sangerhausen)

'ROSE D'EVIAN', T, pb, 1895; flowers pink, center carmine; Bernaix, A.

'ROSE D'HIVERS', D, w; flowers whitish, center shell-pink, well-shaped

'ROSE D'OR', HT, dy, 1941; bud long, pointed; flowers intense yellow, dbl., 35 petals, 4.5–5 in., slight fragrance; foliage dark, bronze, glossy; very vigorous growth; [Julien Potin X Unnamed seedling]; Gaujard; J&P

Rose d'Or de Montreux *see* ADAtonysil

Rose d'Orsay, S, mp; flowers medium, dbl., moderate fragrance; (Sangerhausen)

'ROSE DOT', HT, rb, 1962; flowers red, reverse white, large, dbl., 35 petals, intense fragrance; foliage dark; vigorous, upright growth; [Baccará X Peace]; Dot, Simon

'ROSE DU BARRI', HT, pb, 1940; flowers salmon-pink, reverse carmine, large, single, moderate fragrance; vigorous growth; Archer

'ROSE DU BENGALE', HCh, 1759; Jacquin; (Cavriglia)

Rose du Ciel® *see* 'DELFROR'

'ROSE DU MAÎTRE D'ÉCOLE', HGal, m; flowers soft pink to lilac, very large, dbl., flat, intense fragrance; Miellez, 1840

'ROSE DU PRINCE', HT, mr, 1959; bud long, pointed; flowers strawberry-rose, large, dbl., slight fragrance; foliage leathery, light green; vigorous, upright growth; [Blanche Mallerin X Profusion]; Dorieux; Pin

'ROSE DU ROI', P, mr, 1815; (Lee's Crimson Perpetual); flowers bright red shaded violet, large, semi-dbl., remontant bloom, intense fragrance; foliage clear green, slightly fluted; vigorous growth; Lelieur; Souchet

'ROSE DU ROI À FLEURS POURPRES', P, m; flowers purplish, medium, very dbl., intense fragrance; (28); [Rose du Roi sport]; (France), before 1819; 1819

'ROSE DU ROI PANACHÉ', P, rb; flowers pale flesh striped crimson; [Rose du Roi sport]

Rose du Saint Sacrament *see* **R. CINNAMOMEA PLENA**

Rose Edouard, B, mp, 1818; Perichon/ Neumann

Rose Elf *see* **'ROSENELFE'**

Rose Ellen, S, lp, 1995; Lykke

'ROSE EUTIN', F, mp, 1958; flowers rose-pink; [Eutin sport]; Hennessey

Rose Festival, S, mr; [Black Beauty X Candy Cane]; Sutherland, P, 1996; Golden Vale Nursery

'ROSE FUKUOKA', HT, pb, 1983; flowers blend of light, medium and deep pink, (deeper at petal edges, dbl., 27 petals, exhibition form, slight fragrance; foliage small, dark, glossy; compact growth; [(Utage X Kordes' Perfecta) X Miss Ireland]; Ota, Kaichiro

Rose Garnette, Pol, mp

Rose Gaujard® *see* 'GAUMO'

'ROSE GAUJARD, CLIMBING', Cl HT, rb, 1964; (Grimpant Rose Gaujard®); Nakashima, Tosh

Rose Gilardi™ *see* 'MOROSE'

'ROSE HANNES', HT, w, 1982; (Hannes); flowers large, dbl., moderate fragrance; foliage medium, medium green, glossy; upright growth; [Pascali X Unnamed seedling]; Wheatcroft, Christopher; Timmerman's Roses

'ROSE HILL', HT, dp, 1928; bud long, pointed; flowers darker than briarcliff, very large, dbl., exhibition form, moderate fragrance; [Columbia sport]; Hill, Joseph H., Co.

Rose Hill *see* POUlrohill

'ROSE HILLS RED', Min, dr, 1978; bud pointed; flowers deep red, small, dbl., 30 petals, 1.5 in.; foliage glossy, leathery; vigorous, upright growth; [(R. wichurana X Floradora) X Westmont]; Moore, Ralph S.; Sequoia Nursery

Rose Iga *see* 'MEIBALBIKA'®

'ROSE IMPÉRIALE', HT, ob, 1942; flowers flame and gold, over large, very dbl., globular, moderate fragrance; foliage dark, glossy; vigorous growth; GM, Bagatelle, 1941; Gaujard

Rose Magic *see* 'GELLANDO'

'ROSE MARIE', HT, mp, 1918; flowers clear rose-pink, very large, dbl., cupped, moderate fragrance; foliage glossy, dark; vigorous growth; [Hoosier Beauty X Sunburst]; Dorner

Rose Marie, F, m

'ROSE MARIE REID', HT, mp, 1956; bud globular; flowers neyron rose, dbl., 48 petals, 5–6 in., cupped, moderate fragrance; foliage dark, leathery; vigorous growth; [Charlotte Armstrong X Katherine T. Marshall]; Whisler; Germain's

'ROSE MARIE, CLIMBING', Cl HT, mp, 1927; Pacific Rose Co.

Rose Mary, HT, mp

'ROSE MERK', HT, or, 1931; flowers bright geranium-red, moderate fragrance; vigorous growth; Cant, F.

Rose Minarett *see* NOAmet

'ROSE MOET', HT, my, 1962; flowers golden yellow, large, dbl.; foliage leathery; long, strong stems; Dorieux; Pin

'ROSE NABONNAND', T, pb, 1883; flowers salmon-pink tinted yellow, large, intense fragrance; Nabonnand, G.

'ROSE NOBLE', HT, lp, 1927; bud long, pointed; flowers silvery pink, semi-dbl., moderate fragrance; [Mme Caroline Testout X ?]; Mühle

'ROSE NUGGETS', Min, lp, 1991; bud pointed; flowers light pink, aging lighter, self-cleaning, many petaloids, bl, dbl., 30 petals, 1.5 in., exhibition form, no fragrance; foliage small, dark green, glossy, frequently 7 leaflets; very full, bushy, low growth; [Red Ace sport]; Michelis, Dorothy; Justice Miniature Roses, 1992

Rose O'Bree, F, lp; O'Bree, Nancy, 1997; (Weatherly, L.)

Rose Odyssey 2000 *see* 'BALPARTY'

Rose Oeilletde Saint Arquey *see* **'SERRATIPETALA'**

'ROSE OF CLIFTON', F, yb, 1978; bud pointed; flowers gold edged peach-pink, dbl., 27 petals, 3.5 in., slight fragrance; foliage dark, matt; vigorous, upright growth; [Vera Dalton X Parasol]; Sanday, John

'ROSE OF FREEDOM', HT, mr, 1948; bud ovoid; dbl., 50 petals, 3.5–4.5 in., intense fragrance; foliage leathery, dark; vigorous, upright, bushy growth; [Charlotte Armstrong X Night]; Swim, H.C.; Mt. Arbor Nursery

'ROSE OF LIDICE', HT, yb, 1961; flowers lemon-yellow shaded poppy red; [Tzigane sport]; Wheatcroft Bros.

Rose of Narromine *see* 'WEKDYKSTRA'

Rose of Provins *see* **R. GALLICA OFFICINALIS**

'ROSE OF TORRIDGE', HT, dp, 1961; flowers deep pink, dbl., 36 petals, 6 in., moderate fragrance; foliage glossy; vigorous growth; [Karl Herbst X Pink Charming]; Allen, E.M.

'ROSE OF TRALEE', S, op, 1964; flowers deep pink shaded salmon, blooms in small clusters, dbl., 35 petals, 4 in., slight fragrance; foliage dark; very vigorous, bushy growth; [Leverkusen X Korona]; McGredy, Sam IV

Rose A Wagga Wagga *see* 'JACDASH'

'ROSE OPAL', HT, dp, 1962; flowers pink-opal, well formed, dbl., 25–30 petals, 4–5 in., moderate fragrance; foliage dark (reddish when young); vigorous, upright, bushy growth; [Wellworth X Independence]; LeGrice; Wayside Gardens Co.

'ROSE PARADE', F, pb, 1974; bud ovoid; flowers coral-peach to pink, large, dbl., cupped, moderate fragrance; foliage large, glossy; vigorous, bushy, compact growth; AARS, 1975; [Sumatra X Queen Elizabeth]; Williams, J. Benjamin; Howard Rose Co.

Rose Pluton *see* **'PLUTON'**

Rose Ponceau *see* **'PONCHEAU-CAPIAUMONT'**

Rose Poncheaux *see* **'PONCHEAU-CAPIAUMONT'**

'ROSE QUEEN', HT, or, 1962; bud globular; dbl., 38 petals, 4–5 in., exhibition form, moderate fragrance; foliage leathery, dark; vigorous, upright, bushy growth; [Chrysler Imperial X Unnamed seedling]; Whisler; Germain's

Rose Rhapsody™ *see* 'JACSASH'

'ROSE SHERBET', F, dp, 1962; bud globular; flowers deep rose-pink, medium, dbl., intense fragrance; foliage glossy; vigorous, upright, open growth; [Gruss an Teplitz X ?]; Pal, Dr. B.P.; Indian Agric. Research Inst.

Rose Two Thousand *see* 'COCQUETRUM'

'ROSE VALMAE', S, mr, 1981; flowers decorative blooms borne 3 per cluster, dbl., 40 petals, slight fragrance; foliage light green, 7 leaflet; upright, medium growth; Watts, Mrs. M.A.

Rose Verte, HT, 1989; Croix; (Cavriglia)

'ROSE WINDOW', Min, ob, 1978; bud ovoid, pointed; flowers orange, yellow, and red blend, 15–20 petals, 1 in., slight fragrance; foliage small, dark; spreading, bushy growth; [Seedling X Over the Rainbow]; Williams, Ernest D.; Mini-Roses

Rosea, F, 1974; Delforge; (Cavriglia)

Rosea Centfeuilles, HP, dp; Touvais

'ROSEANNA', Cl Min, dp, 1976; bud long, pointed; flowers deep pink, dbl., 1–1.5 in., exhibition form; foliage glossy; tall, semi-climbing growth; [Little Darling X Seedling]; Williams, Ernest D.; Mini-Roses

'ROSEATE', Cl HT, mr, 1931; flowers large, semi-dbl., cupped, slight fragrance; foliage wrinkled; vigorous, climbing growth; good as a pillar; Clark, A.; Ivanhoe Hort. Soc.

Rosebud, S, pb, 1997

'ROSECARPE', HT, yb, 1993; flowers cream shaded pink, blooms borne mostly single, dbl., 26–40 petals, 3–3.5 in., slight fragrance; some prickles; foliage large, dark green, semi-glossy; medium (100 cms), upright growth; [Selfridges X Mischief]; Poole, Lionel; Battersby Roses, 1994

Rosée du Matin *see* **'CHLORIS'**

'ROSEE O'BREE', F, mp, 1998; flowers medium pink, reverse light pink, multi-colored, double (15–25 petals), small (2.5 cms) blooms borne in large clusters; slight fragrance; moderate prickles; foliage medium green, semi-glossy; spreading, medium (7 ft) rambling growth; recurrent; hedge; [Parentage unknown]; O'Bree, Nancy

'ROSEFORD', LCl, lp, 1981; bud ovoid; flowers light pink, slightly darker reverse, blooms borne 6 per cluster, dbl., 28 petals, cupped, repeats well, slight fragrance; straight, red prickles; foliage dark, semi-glossy; bushy (can be grown as shrub) growth; Bronze, ARC TG, 1986; [Unnamed seedling X Unnamed seedling]; Jerabek, Paul E.

'ROSEGLEN BOUQUET', F, pb, 1953; flowers ivory-white edged deep pink, small, very dbl., exhibition form, slight fragrance; foliage dark, glossy, leathery; vigorous, upright growth; [Pinocchio X ?]; Quinn; Roseglen Nursery

'RÖSEL DACH', Pol, mr, 1907; flowers cherry-rose, borne in upright clusters, dbl.; dwarf, bushy growth; Walter, L.

'ROSEL VOGEL', S, m, 1938; flowers small, dbl., slight fragrance; Vogel, M.; (Sangerhausen)

'ROSELAND ROSETTE', F, dp, 1952; bud globular; flowers deep rose-pink, small, dbl.; vigorous, bushy growth; [Crimson Rosette sport]; Houghton, T.B.

'ROSELANDIA', HT, my, 1924; flowers darker and larger; [Golden Ophelia sport]; Stevens, W.; Low

'ROSELANDIA, CLIMBING', Cl HT, my, 1933; Lens

'ROSELETTE', F, mp, 1957; flowers soft salmon-pink; foliage bright green; very vigorous growth; [Orange Triumph X Alain]; Lens

Roselina™ *see* 'KORTEMMA'

Roseline de Kersa, HT, dp, 1997

'ROSELLA', F, ab, 1972; (**Bobby Dazzler**); flowers blush, shaded rich apricot, dbl., 50 petals, 3 in., slight fragrance; foliage small, matt; [(Vera Dalton X Highlight) X (Ann Elizabeth X Circus)]; Harkness, 1973

'ROSELLA', F, pb, 1930; flowers salmon-rose-pink, ruffled, borne in large trusses, semi-dbl.; vigorous growth; [Else Poulsen sport]; Prior

'ROSELLA', Cl HT, pb, 1931; bud long, pointed; flowers velvety carmine, base yellow, orange undertone, borne in clusters, single, 2.75 in., slight fragrance; foliage large, glossy; vigorous (8 ft or more) growth; [Mme Edouard Herriot X Roger Lambelin]; Dot, Pedro; C-P

'ROSELLA GRACE CREMER', F, dp, 1957; bud conical to ovoid; flowers dark tyrian rose, reverse lighter, rosette center, medium, dbl., 35–40 petals, moderate fragrance; very vigorous, bushy growth; [Garnette sport]; Cremer

'ROSELLA SWEET', HT, op, 1930; (Rosalie); flowers nasturtium-yellow suffused salmon-pink, large, intense fragrance; Pernet-Ducher; Dreer

Rosellana® *see* BARros

'ROSEMARIE HINNER', HT, mp, 1949; bud long, pointed; flowers large, exhibition form, moderate fragrance; foliage leathery, light green; very vigorous, upright, bushy growth; [Frau Karl Druschki X (Ellen X Una Wallace)]; Hinner, P.; Bauské Bros. & Hinner

'ROSE-MARIE VIAUD', HMult, m, 1924; (Rosemary Viaud); flowers same as parent except bluer and more, dbl.; [Veilchenblau seedling]; Igoult; Viaud-Bruant

'ROSEMARY', Cl HT, lp, 1925; bud long, pointed; flowers pink shaded old-gold, very dbl., exhibition form, free, recurrent boom, moderate fragrance; foliage leathery; pillar growth; Cant, F.

'ROSEMARY', Min, mp; flowers apple-blossom-pink

'ROSEMARY', F, dp, 1955; bud ovoid; flowers rose-pink, borne in clusters, dbl., 55–60 petals, 2.5 in., flat, moderate fragrance; foliage leathery; vigorous, bushy growth; [Garnette sport]; Schenkel; Amling-DeVor Nursery

'ROSEMARY CLOONEY', F, ab, 1985; flowers creamy apricot, blooms in clusters of 3-6, dbl., 20 petals, slight fragrance; small, red prickles; foliage small, medium green, matt; upright, bushy growth; [Vera Dalton X Elizabeth of Glamis]; French, Richard

'ROSEMARY DUNCAN', F, my, 1964; flowers yellow, edged lighter, medium, dbl., 40 petals; foliage glossy; vigorous, bushy growth; [Pinocchio sport]; Duncan

'ROSEMARY EDDIE', F, mp, 1956; flowers bright pink, well formed; [New Dawn X Fashion]; Eddie

'ROSEMARY GANDY', F, yb, 1957; (Flash); flowers yellow and coppery, medium, semi-dbl., cupped, moderate fragrance; foliage glossy, bronze; bushy growth; [Tabarin seedling X Jolie Princesse seedling]; Gaujard

Rosemary Harkness *see* 'HARROWBOND'

'ROSEMARY ROSE', F, dp, 1954; flowers deep pink, camellia-shaped, blooms in large trusses, dbl., moderate fragrance; foliage coppery; vigorous, bushy growth; GM, NRS, 1954 GM, Rome, 1954; [Gruss an Teplitz X Floribunda seedling]; deRuiter; Gregory

'ROSEMARY STONE', HT, mp, 1970; flowers pink, very dbl., exhibition form, slight fragrance; foliage leathery; moderate, upright growth; [Waltzing Matilda sport]; Stone

Rosemary Viaud *see* **'ROSE-MARIE VIAUD'**

'ROSEMARY, CLIMBING', Cl HT, lp, 1920; flowers very large, dbl.; Dingee & Conard; (Sangerhausen)

Rosenau, F, dr, 1961; Eggert

Rosendel, F, m

Rosendorf Schmitshausen, LCl, dr, 1977; Cocker

Rosendorf Sparrieshoop® *see* KORdibor

Rosendorf Steinfurth, S, mp, 1993; Schultheis

'ROSENDORF UFHOVEN', HEg, dr, 1949; flowers crimson, very large blooms, very dbl., globular, non-recurrent, slight fragrance; foliage leathery, glossy, dark; weak stems; vigorous, upright, bushy growth; [Gen. MacArthur X Magnifica]; Kordes, W. Söhne

'ROSENELFE', F, mp, 1939; (Rose Elf); bud long, pointed; dbl., 2.5 in., exhibition form, blooms in clusters, moderate fragrance; foliage leathery, glossy, light; vigorous, bushy growth; [Else Poulsen X Sir Basil McFarland]; Kordes; Dreer

'ROSENELLA', HT, mp, 1964; flowers clear pink, reverse brighter, intense fragrance; Mondial Roses

Rosenfee®, F, lp, 1967; Boerner

Rosenfest, S, op, 1981; GPG Bad Langensalza

Rosenholm *see* POUlover

'ROSENKAVALIER', F, dp, 1963; flowers deep rose, borne in clusters, dbl., 45 petals, intense fragrance; foliage dark; vigorous, upright, bushy growth; Verschuren, A.; Stassen

Rosenkavalier, F, or, 1995; Huber

Rosenkreis Neunkirchen, F, mr, 1995; flowers luminous red, semi-dbl., slight fragrance; Michler, K. H.; (Sangerhausen)

'ROSEN-LAMBERT', LCl, dr, 1937; flowers oxblood-red, very large, dbl.; foliage large, bronze, leathery; vigorous, climbing (9 ft) growth; [Fragezeichen X American Pillar]; Vogel, M.; P. Lambert

Rosenmärchen *see* **'PINOCCHIO'**

Rosenmärchen, Climbing *see* **'PINOCCHIO, CLIMBING'**

'ROSEN-MÜLLER', S, mr, 1940; flowers medium, single, slight fragrance; Vogel, M.; (Sangerhausen)

'ROSENPFARRER MEYER', HT, or, 1930; flowers coral-red passing to prawn-red, semi-dbl., slight fragrance; very vigorous growth; [Mme Edouard Herriot X Louise Catherine Breslau]; Soupert & Notting

Rosenprinz, S, dp, 1990; flowers small, dbl., slight fragrance; very low, almost dwarf growth; Wänninger, Franz; (Sangerhausen)

'ROSENPRINZESSIN', HT, mp, 1975; bud ovoid; dbl., 3.5–5 in., moderate fragrance; foliage glossy; upright growth; [Nordia X Sans Souci]; Hetzel

Rosenprinzessin *see* HELprinzess

Rosenprinzessin Andrea, HT, mr, 1993; Hetzel

Rosenprofessor Sieber *see* KORparesni

Rosenreigen®, S, lp

Rosenresli® *see* 'KORRESLI'

Rosenrot®, HT, mr, 1978; Tantau

Rosenstadt Zweibrucken® *see* KORstatis

Rosentanz, S, rb, 1989; flowers dark red with white, large, semi-dbl.; Fryer, Gareth; (Sangerhausen)

Rosenthal *see* 'KORTAT'

Rosenwalzer®, S, pb, 1993; Dickson, Patrick

'ROSENWUNDER', HEg, dp, 1934; bud long, pointed; flowers rose-red, very large, semi-dbl., cupped, non-recurrent, slight fragrance; foliage large, leathery, glossy, wrinkled; very vigorous, bushy growth; (29); [W.E. Chaplin X Magnifica]; Kordes

Rosenzauber, F, lp, 1991; flowers medium, semi-dbl., slight fragrance; Wänninger, Franz; (Sangerhausen)

Roseraie de Blois *see* DORfuri

'ROSERAIE DE L'HAY', HRg, dr, 1901; flowers crimson-red changing to rosy magenta, large, dbl., recurrent bloom, intense fragrance; foliage rugose; vigorous (4-5 ft.) growth; (14); Cochet-Cochet

'ROSERIE', HMult, mp, 1917; (Rosary); flowers tyrian pink, base white, darkening with age, open, semi-dbl., 3.75 in., slight fragrance; thornless; foliage large, rich green, leathery, glossy; very vigorous, climbing growth; [Tausendschön sport]; Witterstaetter

Roseromantic® *see* 'KORSOMMER'

Rosetime *see* 'BENTEM'

'ROSETONE', Min, mp, 1977; bud ovoid, pointed; dbl., 60 petals, 1 in., exhibition form, slight fragrance; foliage small, dark, leathery; very bushy growth; [Dream Dust X Little Chief]; Moore, Ralph S.; Sequoia Nursery

Rosetta *see* 'SPOROSE'

'ROSETTE', Pol, mr, 1926; flowers fuchsia-red, medium, dbl.; semi-dwarf growth; Grandes Roseraies

'ROSETTE', HT, or, 1934; flowers rose red, shaded orange, base yellow, well formed, large, dbl., moderate fragrance; vigorous, bushy growth; Dickson, A.

'ROSETTE', F, op; flowers peach-pink, borne in clusters, dbl.; vigorous growth; Archer

'ROSETTE DE LA LÉGION D'HONNEUR', HT, op, 1896; flowers salmon-pink, medium, semi-dbl., moderate fragrance; Bonnaire; (Sangerhausen)

'ROSETTE DELIZY', T, yb, 1922; flowers yellow, apricot reflexes, outer petals dark carmine, well-formed, dbl.; vigorous growth; [Gen. Gallieni X Comtesse Bardi]; Nabonnand, P.

Roseville College *see* **The Roseville College Rose**

'ROSEWOOD', HT, dp, 1998; flowers neon pink, double, medium-sized, dbl., 15–25 petals, 4–4.5 in., blooms in small clusters, moderate fragrance; prickles moderate; foliage medium, medium green, semi-glossy; compact, medium growth; [Kardinal sport]; Prescott, Cheryl

Roseworld *see* 'SIMILAM'

Rosi Mittermeier *see* 'KORMIORA'

Rosie™ *see* 'BENROS'

Rosie Larkin *see* FRYyippee

'ROSIE O'BREE', F, mp, 1998; flowers medium pink, lighter reverse, multi-clored, dbl., 15–25 petals, .75–2.5 in., borne in large clusters, slight fragrance; prickles moderate; foliage medium green, semi-glossy; spreading, rambling, medium growth; [Seedling sport]; O'Bree, Nancy

Rosie O'Donnell *see* 'WEKWINWIN'

Rosier de Damas, D, mp; flowers medium, dbl., moderate fragrance; (Sangerhausen)

Rosier de Thionville *see* **'QUATRE SAISONS BLANC MOUSSEUX'**

Rosier des Parfumeurs, HGal, lp, 1800; Cochet

Rosier d'Or *see* 'DELPLI'

'ROSIER GLORIETTE', Min, op; flowers orange-salmon, rosette form

'ROSIÉRISTE GASTON LÉVÊQUE', HT, dp, 1932; bud long, pointed; flowers brilliant carmine, large, dbl., cupped, intense fragrance; foliage dark; very vigorous growth; [Mme Butterfly X Jean C.N. Forestier]; Dot, Pedro

'ROSIERISTE MAX SINGER', HMult, 1885; Lacharme, F.; (Cavriglia)

'ROSIÉRISTE PAJOTIN-CHÉDANE', Pol, dr, 1934; flowers deep red, with white, thread-like markings, borne in cluster, semi-dbl., cupped, slight fragrance; foliage leathery; strong stems; vigorous, bushy growth; Délépine; Pajotin-Chédane

Rosika *see* 'HARMUSKY'

Rosilia, Cl Min, or, 1991; flowers small to medium, semi-dbl.; Warner; (Sangerhausen)

'ROSINA', Min, my, 1951; (Josephine Wheatcroft, Yellow Sweetheart); flowers sunflower-yellow, small blooms in clusters, semi-dbl., 16 petals, slight fragrance; foliage glossy, light; dwarf, compact growth; [Eduardo Toda X Rouletii]; Dot, Pedro; URS, 1935

'ROSINE', HT, pb, 1935; bud citron-yellow with red; flowers flesh-pink, base clear salmon, very dbl.; foliage bronze; Lens

'ROSITA', HT, lp, 1956; bud long; flowers soft pink, large, dbl., moderate fragrance; foliage glossy; vigorous growth; [The Doctor X Unnamed seedling]; Delforge

'ROSITA MAURI', HT, 1914; Ketten Bros.; (Cavriglia)

Rosita Missoni *see* BARmiss

'ROSLAW', HT, yb, 1985; (**D. H. Lawrence**); [Gay Gordons sport]; McCarthy, Mrs. Rosemary; Rosemary Roses

'ROSLYN', HT, dy, 1929; bud long, pointed, deep orange; flowers golden yellow, reverse darker orange, large, semi-dbl., slight fragrance; vigorous, compact growth; [Souv. de Claudius Pernet X Buttercup]; Towill

'ROSLYN, CLIMBING', Cl HT, dy, 1937; Vestal

Roslyne *see* 'DELBARA'

'ROSMARI', HT, pb, 1962; bud pointed; flowers pink, reverse purplish, dbl., 30 petals; strong stems; vigorous, upright growth; [Vigoro X First Love]; Dot, Simon

'ROSMARIN'®, Min, pb, 1965; flowers light pink, reverse light red, small, globular, slight fragrance; foliage glossy, light green; dwarf, bushy growth; [Tom Thumb X Dacapo]; Kordes, R.

Rosmarin 89® *see* KORfanto

Rosnella, Gr, 1958; Moro, L.; (Cavriglia)

'ROSOMANE NARCISSE THOMAS', T, rb, 1908; flowers crimson suffused apricot-yellow; Bernaix, P.

'ROSORUM', F, or, 1959; flowers orange-scarlet, borne in large clusters, semi-dbl.; foliage dark, glossy; vigorous, upright growth; [Buisman's Triumph X Alpine Glow]; Buisman, G. A. H.

'ROSS GOWIE', HT, dr, 1968; flowers conical, large, slight fragrance; vigorous, tall growth; Gowie; Gandy Roses, Ltd.

'ROSS RAMBLER', LCl, w, 1938;, blooms all summer; height 9 ft; P.H. Wright

'ROSSANA', HT, dr, 1958; flowers blood-red, well shaped, large, dbl.; vigorous, spreading growth; [Orange Delight X Tudor]; Frères, Buyl

Rosso Giacomasso, HT, 1967; Giacomasso; (Cavriglia)

'ROSTOCK', HMsk, lp, 1937; bud long, pointed; flowers very large blooms in clusters, dbl., exhibition form, recurrent bloom, slight fragrance; foliage leathery, glossy, dark; very vigorous, bushy growth; [Eva X Louise Catherine Breslau]; Kordes

'ROSWYTHA', F, mp, 1968; bud ovoid; flowers pink, small borne in clusters, very dbl.; foliage dark; [Carol sport]; van der Meyden

Rosy, Min, mp, 1993; Chandrakant

Rosy Ann, Min, op, 1994; Lens

Rosy Border *see* POUlosy

Rosy Carpet® *see* 'INTERCARP'

'ROSY CHEEKS', HT, rb, 1975; flowers red, reverse yellow, dbl., 35 petals, 7 in., intense fragrance; foliage dark, glossy; [? X Irish Gold]; Anderson's Rose Nurseries, 1976

Rosy Cheeks *see* MORsycheek

Rosy Cheeks, **Climbing** *see* 'ANDROS'

Rosy Cushion® *see* 'INTERALL'

'ROSY DAWN', Min, yb, 1982; bud ovoid; flowers yellow, petals edged deep pink, medium, dbl., 28 petals, exhibition form, slight, tea fragrance; foliage small, medium green, glossy; PP005274; [Magic Carrousel X Magic Carrousel]; Bennett, Cecilia 'Dee'; Tiny Petals Nursery

Rosy Forecast *see* 'LAVLAD'

Rosy Future *see* 'HARWADEROX'

Rosy Gem *see* 'MEIRADIA'

'ROSY GLOW', HT, mp, 1947; bud long, pointed; flowers hermosa pink, dbl., 25–35 petals, 4–5 in., exhibition form, intense fragrance; foliage leathery; vigorous, bushy growth; [Better Times sport]; Hill, Joseph H., Co.

Rosy Hit *see* POUlmar

'ROSY JEWEL', Min, pb, 1958; bud ovoid; flowers rose-red, reverse lighter, center white, dbl., 25 petals, 1 in., moderate fragrance; low (6-8 in), compact growth; [Dick Koster sport X Tom Thumb]; Morey, Dr. Dennison; J&P

Rosy Koster, Pol, mp

Rosy La Sevillana, **Climbing** *see* MEIgerokosar

Rosy La Sevilliana *see* 'MEIGEROKA'

'ROSY LIFE', HT, mp, 1970; flowers sparkling pink, full, dbl., 25 petals, 3–4 in., slight fragrance; foliage glossy; vigorous growth; [Walko X Souv. de J. Chabert]; Delbard; Laxton & Bunyard Nursery

'ROSY MANTLE', LCl, mp, 1968; flowers large, dbl., moderate fragrance; foliage dark, glossy; [New Dawn X Prima Ballerina]; Cocker

Rosy Meillandina *see* 'MEIFINARO'

Rosy Minimo, Min, mp

Rosy Morn *see* **'IMPROVED CÉCILE BRÜNNER'**

'ROSY MORN', LCl, mp, 1914; flowers large; [Frau Karl Druschki seedling]; Clark, A.

'ROSY MORN', HP, op, 1878; flowers soft peach shaded salmon-pink, well formed, very large, slight fragrance; very vigorous growth; Paul, W.

'ROSY MORN', Pol, mp, 1930; flowers rose-pink, large, borne in large clusters, moderate fragrance; vigorous growth; Burbage Nursery

Rosy Outlook *see* 'WEKTORROC'

Rosy Potluck *see* 'LAVROSY'

Rosy Purple, HMsk, m, 1995; Lens

Rosy Star, HT, mp

'ROSY WINGS', HT, dp, 1962; flowers pink shaded light red, dbl., 35 petals, 2–2.5 in., globular, moderate fragrance; foliage dark, glossy; vigorous growth; [Pink Spiral X Unnamed seedling]; Delforge

Rosy Wings *see* 'MEJAKKA'

'ROTARIAN', HT, mr, 1921; bud long, pointed; flowers bright cherry-crimson, dbl., 35–40 petals, moderate fragrance; [Ophelia X ?]; Lemon

'ROTARY JUBILEE', HT, yb, 1971; bud ovoid; flowers yellow blend, creamy yellow, large, dbl., exhibition form, moderate fragrance; foliage large, glossy, dark, leathery; vigorous, bushy growth; [Queen Elizabeth X Peace]; Lindquist; Bell Roses, Ltd.

'ROTARY PRESIDENT', F, or, 1975; flowers deep vermilion, dbl., 4 in., intense fragrance; foliage small, plum color, turning dark green; dwarf, bushy growth; [Fairlight X Summer Holiday]; Wood

Rotary Rose *see* 'MEIRYPOUX'

'ROTARY-LYON', HT, yb, 1936; bud long, pointed, well formed, golden pink; flowers old-gold shaded carmine, reverse yellow, stamens deep yellow, dbl., cupped, moderate fragrance; foliage dark; very vigorous growth; Chambard, C.

Rote Apart, S, mr, 1991; Uhl, J.

'ROTE BETTER TIMES', HT, mr, 1940; flowers large, dbl., intense fragrance; Noack, Werner; (Sangerhausen)

'ROTE CENTIFOLIE', C, dr, 1938; flowers medium, very dbl., intense fragrance; Krause; (Sangerhausen)

'ROTE ELSE POULSEN', F, dp, 1934; (21); [Else Poulsen sport]; Koopmann; Tantau Roses

Rote Flamme®, LCl, dr, 1967; Kordes

'ROTE GABRIELLE PRIVAT', Pol, mr, 1940; [Gabrielle Privat sport]; Koopmann

Rote Hannover, S, mr, 1998

Rote Hermosa *see* **'QUEEN'S SCARLET'**

Rote Hiroshima, HT, mr, 1990; flowers carmine-red, large, dbl., moderate fragrance; Tagashira, Kazuso; (Sangerhausen)

Rote Max Graf® *see* 'KORMAX'

'ROTE MEVROUW G.A. VAN ROSSEM', HT, or, 1934; flowers nasturtium-red, reverse lightly tinted yellow, open, large, dbl., intense fragrance; foliage glossy, dark; vigorous growth; [Mev. G.A. van Rossem sport]; Kordes

Rote Mozart® *see* KORtragfei

'ROTE PERLE', F, dr, 1962; flowers deep red, medium, dbl., 25 petals, 1.5 in.; foliage dark; vigorous, bushy growth; Tantau, Math.

'ROTE PHARISÄER', HT, mr, 1927; bud very long, pointed; flowers well shaped, large, dbl., intense fragrance; foliage reddish green, leathery; very vigorous growth; [Pharisaer X George C. Waud]; Hinner, W.

'ROTE RAPTURE', HT, mr, 1934; flowers bright cherry-red; [Rapture sport]; Weber, J.

'ROTE TESCHENDORFFS JUBILÄUMSROSE', Pol, dp, 1930; flowers dark crimson-pink to light red, large, borne in large clusters, dbl.; vigorous growth; Grunewald; Teschendorff

Rote Woge *see* MEInirlo

Rotelfe® *see* 'TANRECKTOR'

'ROTELFE', HT, dr, 1922; flowers very dark red; RULED EXTINCT 4/85; [Château de Clos Vougeot X Ulrich Brunner Fils]; Tantau

'ROTER CHAMPAGNER', HT, dp, 1963; (Pétillante); bud pointed; flowers red champagne color, large, dbl., slight fragrance; foliage bright green; long, strong stems; vigorous, upright growth; Tantau, Math.

Roter Kobold®, F, dr, 1994; Interplant

'ROTER SCHMETTERLING', HFt, mr, 1935; flowers medium, semi-dbl., slight fragrance; Vogel, M.; (Sangerhausen)

Roter Stern, HT, mr, 1958; Meilland

'ROTES MEER', Pol, or, 1960; dbl., borne in large clusters; foliage glossy, dark; upright, bushy, symmetrical growth; [Orange Triumph X Unnamed seedling]; Verschuren, A.; van Engelen

Rotesmeer, HRg, m, 1986; (**Purple Pavement**); Baum

Rotfassade, LCl, mr, 1997; Noack, Werner

'ROTORUA', F, or, 1963; bud ovoid; flowers orange-scarlet, borne in clusters, dbl., 27 petals, 3.5 in., exhibition form, slight fragrance; foliage leathery, dark; short, strong stems; very vigorous, bushy growth; [Independence X Spartan]; McGredy, Sam IV; Avenue Nursery

'ROTRAUT', Pol, dr, 1931; dbl., borne in clusters; vigorous growth; [Miss Edith Cavell sport]; Grunewald

'ROTROU', M, m, 1848; flowers reddish-violet, medium, dbl.; Vibert; (Sangerhausen)

'ROTTKÄPPCHEN', Pol, mr, 1887; flowers medium-large, dbl.; Geschwind, R.; (Sangerhausen)

'ROUGE', Pol, mr, 1934; flowers brilliant scarlet-crimson, open, semi-dbl.; foliage leathery; strong stems; dwarf growth; Verschuren; J&P

Rouge Adam, HT, 1996; Adam, M.; (Cavriglia)

Rouge Admirable, HGal, dp, 1825; Vibert

'ROUGE ANGEVINE', HP, mr, 1907; flowers large, dbl.; Chedane-Pajotin; (Sangerhausen)

Rouge Baiser ™, HT, lp; flowers red to reddish orange, exhibition form, exhibition form; greenhouse variety; Delbard

'ROUGE CHAMPION', HT, mr, 1956; flowers bright red, dbl., 40–50 petals; very vigorous growth; [Happiness X Princesse Liliane]; Frères, Buyl

'ROUGE DE PARIS', F, mr, 1958; (Paris Red); flowers light red, dbl., 35–45 petals, 3 in.; foliage purplish; vigorous, bushy growth; [Floradora seedlings X Orange Triumph seedlings]; Delbard-Chabert; Stark Bros., 1964

'ROUGE DE PARME', HT, pb, 1963; flowers mauve-pink, large, dbl., exhibition form; foliage dark, dull; [(Peace X Fred Edmunds) X Buccaneer]; Dorieux; Le Blévenec

Rouge Dorieux® *see* 'DORMAL'

'ROUGE DOT', HT, mr, 1962; flowers currant-red, dbl., 40 petals, slight fragrance; vigorous growth; [Baccará X Lydia]; Dot, Simon

Rouge et Or *see* 'DICOR'

Rouge et Or, Climbing *see* 'DICORSAR'

Rouge Koster, Pol, mr

'ROUGE MALLERIN', HT, mr, 1934; (Henri Mallerin); bud long, pointed; flowers brilliant red, large, dbl., exhibition form, intense, damask fragrance; compact growth; [Mme Van de Voorde X Lady Maureen Stewart]; Mallerin, C.; C-P

Rouge Marbree, B, rb

Rouge Meilland *see* **'HAPPINESS'**

Rouge Meilland *see* 'MEIMALYNA'

Rouge Meilland, Climbing *see* **'HAPPINESS, CLIMBING'**

Rouge Meillandécor® *see* 'MEINEBLE'

Rouge Pineau, F, 1980; Pineau; (Cavriglia)

Rouge Prolific *see* 'DELIFIC'

'ROUGEMOSS', F, or, 1972; bud ovoid; flowers medium, dbl., moderate fragrance; foliage leathery; vigorous, dwarf growth; [Rumba X Moss hybrid]; Moore, Ralph S.; Sequoia Nursery

'ROULETII', HCh, mp; (Centifolia Minima, R. rouletii, R. X centifolia minima); flowers rose-pink, dbl., .5 in., recurrent bloom; dwarf growth; Correvon, 1922

Rouletii, Climbing, Cl HCh, mp

Round Robin, Min, lp; [Avandel X Oz Gold]; Hannemann, F., 1990; The Rose Paradise; (Weatherly, L.)

Roundabout *see* 'ZIPROUND'

'ROUNDELAY', Gr, dr, 1954; bud ovoid; flowers medium-large, dbl., 38 petals, exhibition form, moderate fragrance; foliage dark; very vigorous growth; GM, Geneva, 1954; [Charlotte Armstrong X Floradora]; Swim, H.C.; Armstrong Nursery

Roundelay, Climbing, Cl Gr, dr, 1970

Roussillon®, F, mr

'ROWENA', HWich, dp, 1912; flowers carmine, changing to mauve-pink, small, borne in clusters, dbl.; vigorous, climbing growth; Paul, W.

'ROXANA', HT, yb, 1933; bud large, long, pointed; flowers orange-yellow and copper, open, semi-dbl., moderate fragrance; foliage glossy, light; vigorous, bushy growth; Dickson, A.

Roxana, HT, lp; Urban, J., 1986; (Czech Rosa Club)

Roxane® *see* LAPdal

Roxburghe Rose *see* COCember

Roxie *see* 'RESSOX'

Roxie Baby™ *see* 'TUCKROX'

Roxie's Romance, Min, or

Roy Black *see* 'POULARI'

Roy Castle Rose *see* POUlduff

Roy Rumsey *see* **'FIREFALL'**

'ROYAL AIR FORCE', HT, m, 1969; flowers light lavender-blue, dbl., 35 petals, 6 in., exhibition form, slight fragrance; foliage dark, matt; Laperrière

'ROYAL ALBERT HALL', HT, rb, 1972; flowers wine-red, reverse gold, dbl., 32 petals, 5 in., intense fragrance; foliage dark; [Fragrant Cloud X Postillon]; Cocker

'ROYAL AMERICA', LCl, w, 1994; flowers ivory, blooms borne in small clusters, semi-dbl., 6–14 petals, 1.5–2.75 in., slight fragrance; foliage medium, light green, matt; tall, bushy, spreading growth; [America sport]; Cooper, Curt; Certified Roses, Inc., 1997

Royal Amethyst™ *see* 'DEVMORADA'

Royal Anniversary *see* 'HARBONNY'

Royal Ascot® *see* 'DELSAB'

Royal Baby *see* 'DEBRAD'

Royal Bassino *see* KORfungo

'ROYAL BATH & WEST', Gr, lp, 1977; flowers pastel pink, dbl., 20 petals, 4 in., intense fragrance; foliage green, matt; [Unnamed seedling X Prima Ballerina]; Sanday, John

'ROYAL BEAUTY', HT, dr, 1940; flowers dark velvety red, dbl., 25 petals, 5 in., exhibition form, moderate, spicy fragrance; foliage bronze; [Better Times sport]; Coddington

Royal Blush *see* SIEroyal

Royal Bonica™ *see* 'MEIMODAC'

'ROYAL BRIGHT', HT, my, 1976; bud ovoid; dbl., 40 petals, 6 in., exhibition form, slight fragrance; vigorous, upright growth; [Garden Party X (Bronze Masterpiece X Memoriam)]; Kono, Yoshito

'ROYAL CANADIAN', HT, mr, 1968; bud ovoid; flowers scarlet, large, dbl., 35 petals, cupped, moderate fragrance; foliage glossy, leathery; vigorous, upright growth; [Seedling X Talisman]; Morey, Dr. Dennison; J&P

Royal Carpet® *see* 'MINuco'

'**Royal Chinook**', HT, mr, 1939; flowers brilliant rose-red; [Rapture sport]; Chase

Royal City *see* 'RENity'

'**Royal Cluster**', HMult, lp, 1899; flowers small, single; Conard & Jones; (Sangerhausen)

Royal Dane® *see* 'POUmidor'

'**Royal Dawn**', LCl, mp, 1962; flowers coral-pink, large, dbl., 35 petals, moderate fragrance; foliage dark, glossy; vigorous, climbing growth; [Royal Sunset X Aloha]; Morey, Dr. Dennison; J&P

Royal Delight *see* 'AROrerov'

Royal Dot, HT, pb

Royal Dream *see* KORmeeram

'**Royal Edward**', S, mp, 1994; dbl., 15–25 petals, 1.5 in., borne in large clusters, slight fragrance; few prickles; foliage small, medium green, glossy; low (45 cms), bushy growth; [R. kordesii X [(R. kordesii X (Red Dawn X Suzanne)o.p. X Zeus]]; Ogilvie, Ian S.; Agriculture Canada, 1994

'**Royal Flare**', HT, rb, 1988; flowers red and white blend, large, borne singly, exhibition form, slight fragrance; prickles down-turned, dark green; foliage medium, dark green, disease resistant; upright, medium growth; [Pristine X Standout]; Wambach, Alex A.

'**Royal Flush**', LCl, pb, 1970; bud ovoid; flowers cream, edges blending pink, medium, semi-dbl., cupped, repeat bloom, moderate fragrance; foliage dark, leathery; vigorous, upright, climbing growth; [Little Darling X Suspense]; Fuller; Wyant

Royal Flush *see* PEApatio

Royal Gem, F, lp; Brooks, M.L., 1959; (Weatherly, L.)

'**Royal Gold**', LCl, my, 1957; flowers golden yellow, blooms borne singly and in clusters of 3-7, dbl., 35 petals, 4 in., cupped, moderate, fruity fragrance; foliage glossy; vigorous, pillar (5-7 ft) growth; [Goldilocks, Climbing X Lydia]; Morey, Dr. Dennison; J&P

'**Royal Highness**', HT, lp, 1962; (Konigliche Hoheit, Königlicht Hoheit); bud long, pointed; flowers soft light pink, dbl., 43 petals, 5–5.5 in., exhibition form, intense fragrance; foliage dark, glossy, leathery; tender, upright, bushy growth; AARS, 1963 David Fuerstenberg Prize, ARS, 1964 GM, Madrid, 1962 GM, Portland, 1960; [Virgo X Peace]; Swim & Weeks; C-P

Royal Lady® *see* 'MINiryl'

'**Royal Lavender**', LCl, m, 1961; bud ovoid; flowers lavender tinted gray and pink, blooms in clusters, dbl., 35 petals, 3 in., cupped, intense fragrance; foliage leathery; vigorous (6-9 ft) growth; [Lavender Queen X Amy Vanderbilt]; Morey, Dr. Dennison; J&P

'**Royal Lustre**', HT, op; flowers orange-buff overlaid salmon, dbl., 45 petals, 6–7 in., slight fragrance; foliage coppery; very free growth; [Mrs Sam McGredy X Crimson Glory]; McGredy

Royal Mail *see* 'ANDroi'

'**Royal Marbrée**', HGal, m; flowers lilac and purple, marbled (striped) pink, medium, very dbl.; Moreau-Robert (?), before 1837

Royal Meillandina *see* 'SCHobitet'

Royal Midinette *see* MORposa

'**Royal Occasion**', F, or, 1974; (Montana®); bud long, pointed; flowers luminous orange-scarlet, dbl., 20 petals, 3 in., cupped, slight fragrance; foliage glossy, large, leathery; upright, compact, strong (3 ft) growth; ADR, 1974; [Walzertraum X Europeana]; Tantau, Math.

'**Royal Perfection**', HT, or, 1964; flowers orange-coral, full, dbl., 20 petals, 4.5 in., slight fragrance; vigorous growth; [Rome Glory X Bayadere]; Delbard; Laxton & Bunyard Nursery

Royal Philarmonic, HT, 1997; Harkness, R.; (Cavriglia)

Royal Philharmonic *see* HARdeed

Royal Porcelain *see* 'LEEpel'

'**Royal Princess**', Min, pb, 1981; bud ovoid; flowers pink and yellow blend, blooms borne 1-3 per cluster, dbl., 65 petals, moderate, fruity fragrance; hooked, bronze prickles; foliage dark, leathery; upright, bushy growth; [Unnamed seedling X Unnamed seedling]; Lyon

'**Royal Queen**', Gr, w, 1965; flowers greenish white; [Queen Elizabeth sport]; Verschuren

'**Royal Red**', HT, mr, 1924; bud pointed; flowers intense crimson-scarlet, very large, dbl., moderate fragrance; foliage leathery; vigorous growth; Hill, E.G., Co.

'**Royal Red**', HT, dr, 1956; flowers deep scarlet, well formed; moderate growth; [Happiness sport]; Lowe

'**Royal Robe**', S, mr, 1946; flowers crimson, almost purple, rather large, semi-dbl., non-recurrent; vigorous growth; [(R. rugosa X Hybrid Perpetual) X (R. multiflora X R. blanda)]; Wright, Percy H.

Royal Romance *see* 'RUlis'

'**Royal Romance**', HT, my, 1974; flowers bright lemon-yellow, full, dbl., 30 petals, 5 in., moderate fragrance; foliage leathery; very free growth; [Pink Parfait X Unknown seedling]; Fryer, Gareth; Fryer's Nursery, Ltd.

'**Royal Rose**', HT, mp, 1999; flowers rose pink, very dbl., 80 petals, exhibition form, slight fragrance; 4.5 ft growth; [Happiness X Peter Frankenfeld]; Teranishi, K.; Itami Rose Nursery, 1987

'**Royal Ruby**', Min, mr, 1972; flowers red, base white, small, dbl., globular, slight fragrance; foliage small, dark, leathery; vigorous, upright growth; [Garnette X (Tom Thumb X Ruby Jewel)]; Morey, Dr. Dennison; Pixie Treasures Min. Roses

Royal Salute *see* 'MACros'

'**Royal Scarlet**', HT, mr, 1966; bud ovoid; flowers scarlet-red, large, dbl., exhibition form, moderate fragrance; foliage glossy; vigorous, bushy growth; [McGredy's Scarlet X Christian Dior]; Kraus; Wyant

'**Royal Scarlet Hybrid**', LCl, mr, 1926; Chaplin Bros.

'**Royal Scot**', HT, yb, 1928; flowers golden yellow, edged crimson, open, semi-dbl., moderate fragrance; foliage dark; vigorous growth; Dobbie

'**Royal Show**', HT, mr, 1973; flowers pointed, dbl., 27 petals, 4 in., slight fragrance; foliage glossy; vigorous growth; [Queen Elizabeth X ?]; Gregory

Royal Show *see* MEIrasimac

Royal Smile, HT, w, 1980; Beales, Peter

Royal Success® *see* 'AROrerov'

Royal Sunblaze® *see* 'SCHobitet'

'**Royal Sunset**', LCl, ab, 1960; bud ovoid; dbl., 20 petals, 4.5–5 in., cupped, moderate, fruity fragrance; foliage leathery; vigorous (6 ft) growth; GM, Portland, 1960; [Sungold X Sutter's Gold]; Morey, Dr. Dennison; J&P

'**Royal Tan**', HT, m, 1955; flowers pale purple feathered violet and chocolate, high-pointed, 5 in., slight fragrance; foliage dark; vigorous growth; [Charles P. Kilham X Mrs Sam McGredy]; McGredy, Sam IV

Royal Touch *see* 'ANDroy'

Royal Velvet® *see* 'MEIlotup'

'**Royal Velvet**', HT, dr, 1959; flowers rich, velvety cardinal-red, blooms, dbl., 55–65 petals, 4–5 in.; vigorous, tall growth; RULED EXTINCT 3/86 ARM.; [(Happiness X Independence) X (Happiness X Floradora)]; Meilland, F.; C-P

Royal Victoria *see* 'LAVictor'

'**Royal Visit**', HT, op, 1939; flowers deep tangerine-orange, reverse coral passing to apricot, dbl., slight fragrance; foliage leathery, glossy, dark; bushy growth; [Picture X Mrs Pierre S. duPont]; Eddie

Royal Volunteer *see* 'COCdandy'

Royal Wedding® *see* 'JACquint'

'ROYAL WELCOME', HT, rb, 1955; flowers velvety dark red, reverse deep rosy pink, well formed, intense fragrance; foliage glossy; vigorous growth; [Crimson Glory X Peace]; Homan; G.A. Williams

Royal William *see* 'KORZAUN'

'ROYAL WILLIAM, CLIMBING', Cl HT, dr, 1995; flowers other growth similar to parent; [Royal William sport]; Newman, F. S.; Abbey Rose Gardens

Royal Worcester® *see* 'TROBROY'

Royale *see* **'BELLE ÉPOQUE'**

Royale Perfection, HT, or, 1964; flowers large, dbl., moderate fragrance; Delbard-Chabert; (Sangerhausen)

'ROYALET', HT, mr, 1968; flowers crimson-red, dbl., 35–40 petals, 4.5–5.5 in., slight fragrance; foliage dark; upright, bushy growth; [Unnamed seedling X Chrysler Imperial]; Herholdt, J.A.

Royalglo™ *see* 'MINADCO'

'ROYALIST', HT, mp, 1954; flowers deep rose-pink, large, dbl., 28 petals, exhibition form, intense fragrance; foliage dull green; very free growth; [(Billy Boy X Blossom) X Mrs Redford]; McGredy, Sam IV

'ROYALTY', HT, dr, 1976; bud globular, pointed; flowers deep red, dbl., 20 petals, 4–4.5 in., slight fragrance; foliage leathery; vigorous growth; PP004057; [Forever Yours X Love Affair]; Jelly; E.G. Hill Co.

'ROYDEN', F, my, 1989; flowers bright gold, fading quickly to white, large, borne in sprays, dbl., 40 petals, cupped, intense fragrance; foliage light green, glossy; tall, upright growth; [Liverpool Echo X Arthur Bell]; Cattermole, R.F.; South Pacific Rose Nursery

'ROYDON HALL', F, mr, 1983; flowers large, dbl., 35 petals, no fragrance; foliage medium, medium green, semi-glossy; bushy growth; [City of Leeds X (Paprika X Rose Gaujard)]; Scrivens, Len

'ROZE KONINGIN', HT, mr, 1938; bud long, pointed; flowers medium red, passing to pink, well formed large, dbl., moderate fragrance; vigorous growth; [Lady Sylvia seedling X Étoile de Hollande]; Lens

'ROZENAMATEUR A. BOK', F, mr, 1961; flowers bright red, borne in large clusters, single; foliage dark; vigorous, upright growth; [Prinses Christina X Kathe Duvigneau]; Buisman, G. A. H.

Rozorina *see* **'ZORINA'**

Roztomila, F, lp; Urban, J., 1988; (Czech Rosa Club)

Ruatara, LCl, dp, 1995; Nobbs

'RUBAIYAT', HT, dp, 1946; bud long, pointed; flowers rose-red, reverse lighter, dbl., 25 petals, 4.5–5 in., exhibition form, intense fragrance; foliage dark, leathery; vigorous, upright growth; AARS, 1947 GM, Portland, 1945; [(McGredy's Scarlet X Mrs Sam McGredy) X (Seedling X Sir BasilMcFarland)]; McGredy; J&P

'RUBAN ROUGE', F, dr, 1957; bud globular; flowers deep velvety red, borne in clusters; foliage dark, leathery; vigorous growth; [Alain X Cinnabar]; Lens

'RUBELLA', F, dr, 1972; flowers deep red, large, dbl., 24 petals; bushy growth; [Kimono X Lilli Marleen]; deRuiter

'RUBENS', T, w, 1859; flowers white, shaded with rose, center bronzy yellow, large, dbl., cupped; Robert

'RUBENS', HT, mr, 1972; bud pointed; flowers vermilion red, large, dbl.; vigorous growth; [Rose Gaujard X Miss France]; Gaujard

'RÜBEZAHL', HT, m, 1917; flowers large, dbl., slight fragrance; Krüger; (Sangerhausen)

Rubies 'n' Pearl *see* 'MANRUPEARL'

'RUBIN', HMult, mr, 1899; flowers crimson to spinel-red, medium, borne in clusters, semi-dbl., moderate fragrance; height 10-12 ft; [Daniel Lacombe X Fellemberg]; Schmidt, J.C.

'RUBIN', HT, or, 1956; flowers orange-red, base yellow, large, semi-dbl., slight fragrance; vigorous growth; [Mme Henri Guillot X Grande Duchesse Charlotte]; Lens

'RUBIN', F, dr, 1962; bud long, pointed; flowers deep red, borne in broad clusters, dbl.; foliage dark; bushy, compact, low growth; Kordes; Dehner & Co.

Rubina®, F, dr, 1989; Dickson, Patrick

'RUBINETTE', F, dr, 1971; flowers deep red, dbl., 28 petals, 3.5 in., slight fragrance; foliage dark; vigorous, upright growth; [(Mandrina X Baccará) X (Mandrina X Baccara)]; deRuiter; Carlton Rose Nurseries

Rubino, S, 1966; Mansuino; (Cavriglia)

'RUBIS', Pol, rb, 1926; flowers bright ruby-red, center white, medium, borne in clusters; dwarf growth; [Merveille des Rouges X Jessie]; Nonin

'RUBIS', Cl HT, mr, 1948; flowers blood-red, well formed, large, intense fragrance; foliage bronze; vigorous growth; [Unnamed climber X Mme G. Forest-Colcombet]; Mallerin, C.; URS

'RUBOR', HT, mp, 1947; flowers neyron pink, dbl., moderate fragrance; foliage dark; very vigorous growth; [Cynthia X Director Rubió]; Dot, Pedro

RUbrispa, Min, ob; (**Bright Spark**); Ru?

Rubrotincta *see* **'HEBE'S LIP'**

Rubus coronarius, Rubus rosaefolius coronarius *see* **'EASTER ROSE'**

'RUBY', Pol, mr, 1932; flowers glowing scarlet; deRuiter; Sliedrecht & Co.

Ruby Anniversary *see* 'HARBONNY'

Ruby Celebration *see* PEAwinner

'RUBY DEE', Gr, mr, 1968; bud ovoid; flowers large blooms in clusters, dbl., exhibition form, moderate fragrance; foliage dark, glossy; vigorous, upright growth; [Queen Elizabeth X Happiness]; Patterson; Patterson Roses

'RUBY GEM', HT, mr, 1962; flowers bright red, large, dbl., 28 petals, slight fragrance; [Seedling X Red Favorite]; Leenders, J.

'RUBY GLOW', LCl, mr, 1955; bud short, oval; flowers spectrum-red, open, borne in clusters, semi-dbl., 13–15 petals, 3 in., moderate fragrance; foliage glossy; moderate pillar (7 ft) growth; [Dream Girl X New World]; Jacobus; B&A

'RUBY JEWEL', Min, mr, 1958; bud ovoid; flowers ruby-red, reverse lighter, open, dbl., 35–40 petals, .5 in., moderate fragrance; foliage glossy; low (6-8 in), compact growth; [Dick Koster sport X Tom Thumb]; Morey, Dr. Dennison; J&P

'RUBY LIPS', F, mr, 1958; bud ovoid, pointed; flowers bright cardinal-red, loose form, blooms in clusters, semi-dbl., 18 petals, 3 in., slight fragrance; foliage semi-glossy; vigorous, semi-spreading, bushy growth; [World's Fair X Pinocchio]; Swim, H.C.; Armstrong Nursery

Ruby Magic *see* 'MORUBY'

'RUBY MANWARING', HT, dp, 1932; flowers rich rosy cerise; [Betty Uprichard sport]; Longley

'RUBY PENDANT', Min, m, 1979; bud pointed; flowers red-purple, borne singly, dbl., 28 petals, exhibition form, slight fragrance; needle-shaped prickles; foliage reddish-green; vigorous growth; [(Lotte Gunthart X Salvo) X Baby Betsy McCall]; Strawn, Leslie E.; Pixie Treasures Min. Roses

'RUBY PRINCESS', HT, mr, 1949; flowers velvety red, dbl., 50 petals, 5 in., moderate fragrance; [Jewel sport]; Grillo

'RUBY QUEEN', HWich, dp, 1899; flowers deep rose-pink, reverse lighter, small, borne in clusters of, dbl., non-recurrent, slight fragrance; height 10 ft; [R. wichurana X Queen's Scarlet]; Van Fleet; Conard & Jones

Ruby Rain, Cl Min, mr

'RUBY RING', HWich, rb, 1926; flowers white with ruby edge, borne in sprays, dbl.; Clark, A.; (Sangerhausen)

Ruby Star *see* 'UMSTAR'

'RUBY SUPERIOR', Pol, mr; flowers have more lasting color; [Ruby sport]; deRuiter

'RUBY TALISMAN', HT, mr, 1935; flowers rich ruby-red, more shapely than parent, with reflexed petal; [Talisman sport]; Eddie

Ruby Treasure, Min, lp; [Oz Gold X Lemon Delight]; Hannemann, F., 1990; The Rose Paradise; (Weatherly, L.)

Ruby Tuesday *see* 'RENRUBY'

'RUBY ULRICK', F, dp, 1953; flowers deep pink, base white, borne in clusters, dbl.; foliage leathery; very vigorous growth; [Mrs Tom Henderson X Gloria Mundi]; Ulrick, L.W.

Ruby Velvet *see* 'HINRUBY'

'RUBY WEDDING', HT, dr, 1979; flowers deep red, dbl., 44 petals, 3–3.5 in.; vigorous, spreading growth; [Mayflower X ?]; Gregory

'RUDELSBURG', HMult, dp, 1919; flowers shining carmine-rose, blooms in clusters; no prickles; vigorous, climbing growth; Kiese

'RUDI KORTE', Pol, or, 1929; flowers medium, semi-dbl.; Kersbergen; (Sangerhausen)

'RUDOLA', F, mr, 1973; bud ovoid; flowers light geranium-red, open, large, semi-dbl., slight fragrance; foliage leathery; very vigorous, bushy growth; [Dacapo X Kimono]; deRuiter

'RUDOLF ALEXANDER SCHRÖDER', HT, w, 1930; bud long, pointed; flowers white, center tinted lemon, large, dbl., exhibition form, intense fragrance; foliage leathery, light; vigorous growth; [Mrs Herbert Stevens X Pius XI]; Kordes

'RUDOLF SCHMIDTS JUBILÄUMSROSE', F, my, 1955; flowers golden yellow, large, borne in clusters, semi-dbl., exhibition form, moderate fragrance; foliage light green, glossy; very vigorous, upright, bushy growth; Kordes; R. Schmidt

'RUDOLF SCHOCK', HT, mr, 1968; flowers bright currant-red, large, dbl., moderate fragrance; foliage dark, leathery; vigorous growth; [Josephine Bruce X Seedling]; Verschuren, A.; Stassen

'RUDOLF VON BENNIGSEN', LCl, mp, 1932; bud pointed; flowers rosy-pink, edges fading, large, borne in loose trusses of 5-, recurrent bloom; foliage broad, dark, glossy; vigorous, bushy, semi-climbing growth; [(Geheimrat Dr. Mittweg X Souv. de Paul Neyron) X Mrs Joseph Hill]; Lambert, P.

'RUDOLPH KLUIS', Pol, mr, 1922; bud globular; flowers pure vermilion-red, dbl., moderate fragrance; foliage rich green, glossy; bushy growth; [Ellen Poulsen sport]; Kluis & Koning

'RUDOLPH KLUIS SUPERIOR', Pol, mr, 1928; flowers glowing scarlet; more compact growth than Rudolph Kluis.; [Ellen Poulsen sport]; Kluis

'RUDOLPH TIMM', F, rb, 1951; flowers white, reverse red, open, small, borne in trusses to 40, semi-dbl., 15–20 petals, 2 in., moderate fragrance; foliage glossy, light green; very free growth; (28); [(Johannes Boettner X Magnifica) X (Baby Chateau X Else Poulsen)]; Kordes; Wheatcroft Bros.

'RUDOLPH VALENTINO', HT, op, 1929; flowers lively shrimp-pink or coral-red, suffused golden coppery, dbl., moderate fragrance; vigorous growth; Pernet-Ducher; Dreer

'RUFFLES', HT, mp, 1994; dbl., 26–40 petals, 1.5–2.75 in., borne mostly singly, moderate fragrance; some prickles; foliage large, medium green, tall growth; [Seedling X Alec's Red]; Perry, Astor; Certified Roses, Inc., 1997

Ruffles 'n' Flourishes *see* 'CLERUFF'

'RUFIN', Min, op, 1982; (Finnstar®, **Finstar**, Mini Metro); flowers orange-salmon, small, dbl., 20 petals, slight fragrance; foliage small, medium green, semi-glossy; bushy growth; [Minuette X Seedling]; deRuiter, George; Fryer's Nursery, Ltd., 1980

'RUFUS', Pol, mr, 1925; flowers intense scarlet, borne in clusters, dbl.; compact growth; [Orléans Rose sport]; Allen

'RUGA', Ayr, w; (R. X ruga); flowers flesh changing to creamy white, large blooms in several-flow, dbl., cupped; trailing growth; (14); [Thought to be R. arvensis X China or Tea]

Rugelda® *see* KORruge

'RUGGELDA', HT, dy, 1982; (**Golden Days**, 'RUGOLDA'); flowers large, dbl., 35 petals, slight fragrance; foliage large, medium green, semi-glossy; bushy growth; [Peer Gynt X Seedling]; deRuiter, George; Fryer's Nursery, Ltd., 1980

'RUGOLDA', HT, dy, 1982; (**Golden Days**, 'RUGGELDA'); flowers large, dbl., 35 petals, slight fragrance; foliage large, medium green, semi-glossy; bushy growth; [Peer Gynt X Seedling]; deRuiter, George; Fryer's Nursery, Ltd., 1980

'RUGOSA COPPER', S, ob, 1955; flowers coppery orange, large, recurrent bloom; vigorous growth; [Conrad Ferdinand Meyer X Unnamed seedling]; Gaujard

'RUGOSA MAGNIFICA', HRg, m, 1905; flowers reddish lavender, gold stamens, dbl., moderate fragrance; orange-red fruit; vigorous, spreading growth; very hardy.; Van Fleet

Rugosa Ottawa, HRg, lp; flat, 5 petals, intense; L'Assomption, 1969

Rugosa Rose *see* **R. RUGOSA**

'RUGOTIDA', HRg, 1950; Darthuis; (Cavriglia)

Rugspin, HRg, dr, 1966; Petersen

'RUGUL', Min, my, 1973; (Gulletta, **Tapis Jaune®**); flowers small, dbl., 20 petals, 1.5 in.; foliage small, glossy, dark; low, compact growth; [Rosy Jewel X Allgold]; deRuiter

'RÜHKOR', HT, w, 1984; (Athena); flowers large, dbl., 35 petals, moderate fragrance; foliage large, medium green, matt; upright, bushy growth; [Unnamed seedling X Helmut Schmidt]; Kordes, W., 1981

'RUHM DER GARTENWELT', HP, dr, 1904; flowers large, dbl.; [American Beauty X Fracis Dubreuil]; Jacobs

'RUHM VON STEINFURTH', HP, mr, 1920; (Red Druschki); bud long, pointed; flowers large, dbl., 34 petals, cupped, intense fragrance; foliage dark, leathery; vigorous growth; [Frau Karl Druschki X Ulrich Brunner Fils]; Weigand, C.; H. Schultheis

'RUIBLUN', Min, m, 1983; (Azulabria, **Blue Peter**, Bluenette); flowers lilac-purple, patio, small, semi-dbl., slight fragrance; foliage small, light green, semi-glossy; bushy growth; [Little Flirt X Seedling]; deRuiter, George; Fryer's Nursery, Ltd.

RUIcharo, Min, ob; (**Charming Rosamini**); deRuiter

'RUICO', F, ob, 1968; (**Fresco®**); flowers orange, reverse golden yellow, well-formed, 3 in., slight fragrance; foliage dark, glossy; vigorous, bushy growth; [Metropole X Orange Sensation]; deRuiter

RUIdriko, HT, w; (**Vivaldi** TM); deRuiter, 1988

RUIfarol, Min, mp, 1989; (**Favourite Rosamini**); deRuiter

RUIflami, Min, mr, 1987; (**Flaming Rosamini**); deRuiter

'RUIGERDAN', Min, or, 1988; (**Little Marvel**); flowers bright orange-scarlet, small, semi-dbl., 6–14 petals, slight fragrance; foliage small, medium green, semi-glossy; bushy growth; PP007572; [Unnamed seedling X Unnamed seedling]; deRuiter, George; Fryer's Nursery, Ltd., 1987

'RUILANCA', F, or, 1984; (**Lancashire Life**); flowers scarlet, blooms in clusters, semi-dbl., slight fragrance; foliage medium, medium green, semi-glossy; bushy growth; [Robert Stolz X Diablotin]; deRuiter, George; Fryer's Nursery, Ltd.

'RUIMEVA', HT, or, 1983; (**Madelon**); flowers medium, dbl., 20 petals, slight fragrance; foliage medium, medium green, semi-glossy; upright growth; PP005820; [Varlon X MEIgenon]; deRuiter, George; Fryer's Nursery, Ltd., 1981

'RUIMIRED', Min, dr, 1991; (**Red Minimo®**); flowers moderately small, dbl., 15–25 petals, borne in small clusters, no fragrance; foliage small, dark green, semi-glossy; low, bushy growth; PP005770; [Unnamed seedling X Unnamed seedling]; deRuiter, George; Bear Creek Gardens, 1987

'RUIREDRO', Min, dr, 1988; (**Red Rosamini**®); flowers clear crimson, white at base, dbl., 30–40 petals, 1–1.5 in., cupped, floriferous, borne in small clusters, slight fragrance; few prickles; foliage medium, medium green, matt; bushy, medium growth, 15 in; PP005976; [Unnamed seedling X Unnamed seedling]; deRuiter, George; Conard-Pyle Co., 1987

RUIrupo, Min, mr, 1989; (**Scarlet Rosamini**); deRuiter

RUIsalro, Min, op; (**Salmon Rosamini**); deRuiter

'RUISETO', Min, or, 1988; (**Orange Rosamini**®); semi-dbl.; medium growth; PP006236; [Unnamed seedling X Red Rosamini]; deRuiter, George; DeRuiters Nieuwe Rozen B.V.

RUIsteenka, HT, m, 1994; (**Ravel**); deRuiter

RUItulvi, HT, pb, 1993; (**Pavarotti**); deRuiter

RUIxandra, Min, rb, 1992; (**Jolly Cupido**); deRuiter

RUIzesac, HT, pb, 1989; (**Astra**); deRuiter

'RUKHSAAR', HT, w, 1969; flowers cream, center shell-pink, medium, dbl., exhibition form, slight fragrance; foliage glossy; moderate growth; [Virgo X Open pollination]; Singh; Gopalsinamiengar

'RULIMPA', HT, rb, 1980; (Impala, **Melinda**); flowers scarlet, gold reverse, large, dbl., 35 petals, moderate fragrance; foliage medium, dark, glossy; upright growth; [Whisky Mac X Criterion]; deRuiter, George; Fryer's Nursery, Ltd.

'RULIS', HT, ob, 1982; (**Liselle**, Royal Romance); flowers orange-peach, well-formed, large, dbl., 35 petals, moderate fragrance; foliage large, dark, semi-glossy; bushy growth; [Whisky Mac X Matador]; deRuiter, George, 1980; Fryer's Nursery, Ltd., 1981

Rum Butter *see* 'TALRUM'

Rum Candy *see* 'UMSRUM'

'RUMBA'®, F, rb, 1958; bud ovoid; flowers poppy-red, center yellow, blooms in clusters, dbl., 35 petals, 2–2.5 in., cupped, slight, spicy fragrance; foliage dark, glossy, leathery; vigorous, bushy growth; [Masquerade X (Poulsen's Bedder X Floradora)]; Poulsen, S.; C-P, 1962;, McGredy, 1960

'RUMBA', HT, op, 1956; bud long, pointed; flowers salmon-orange, intense fragrance; long stems; [Talisman X ?]; Faassen-Houba

'RUMBA, CLIMBING', Cl F, rb, 1972; Bansal, O.P.; Bansal Roses

'RUMPELSTILZCHEN', S, dr, 1956; (Repelsteeltje); flowers deep red, small, single; dwarf growth; deRuiter; Willicher Baumschulen

Running Maid® *see* 'LENRAMP'

'RUNTRU', F, dr, 1983; (Fennica, **Invincible**); bud large; flowers large, dbl., 20 petals, slight fragrance; foliage large, medium green, glossy; upright growth; [Rubella X National Trust]; deRuiter; Fryer's Nursery, Ltd., 1982

'RUPERT BROOKE', HT, pb, 1928; flowers fawn-pink to cream, dbl., moderate fragrance; [Miss Cynthia Forde X Mrs Wemyss Quin]; Easlea

'RUPFESPIN', S, mp, 1991; (**Festival Pink**™); bud pointed; flowers clear medium pink, white petal base, golden stamens, single, 5–7 petals, 1.5–2.75 in., borne in small and large clusters, slight fragrance; many prickles; foliage medium, bright green, glossy; tall (120-150 cms), bushy, spreading growth; [Festival Fanfare sport]; Rupert, Kim L., 1992

'RUPGOLJUL', HT, my, 1992; (**Golden Julia**™); bud long, pointed; flowers medium golden mustard, opening medium golden yellow, fading, dbl., 18–22 petals, 3–4 in., flat, slight fragrance; few, yellow prickles; foliage medium, medium green, matt; green wood; medium (75-90 cms), upright, bushy growth; [Julia's Rose sport]; Rupert, Kim L.

'RUPKIMCRM', F, w, 1991; (**Kim's Cream**™); bud ovoid, pointed; flowers cream with coffee and gold tints, opening cream, tint remains in cool weather, dbl., 26–40 petals, 3–3.5 in., flat, intense, spicy fragrance; some prickles; foliage leathery, medium, medium green, semi-glossy; low (45-50 cms), bushy, compact, vigorous growth; [Lavender Pinocchio X Lavender Pinocchio]; Rupert, Kim L., 1992

'RUPRINS', Gr, or, 1979; (Prins Claus, **Rosalynn Carter**); flowers coral-red with orange tones, dbl., 30 petals, 3.5–4 in., exhibition form, moderate, spicy fragrance; tall, vigorous, bushy, upright growth; [Unnamed seedling X Scania]; deRuiter; C-P, 1978

'RURAL RHYTHM', S, lp, 1984; dbl., 30 petals, 4 in., cupped, borne 1-5 per cluster, repeat bloom, moderate, myrrh fragrance; awl-like, tan prickles; foliage dark, leathery; erect, bushy, spreading growth; hardy.; [Carefree Beauty X The Yeoman]; Buck, Dr. Griffith J.; Iowa State University

'RURITANIA', HT, or, 1972; bud ovoid; flowers full, medium, very dbl., intense fragrance; foliage dark, leathery; vigorous, upright growth; [Miss Hillcrest X Hawaii]; Curtis, E.C.; Kimbrew

'RUSALKA', LCl, dp, 1934; flowers carmine to sunset-rose, base yellow, very large, semi-dbl., exhibition form, profuse, repeated bloom; foliage glossy, light; climbing growth; [Tausendschön X Farbenkonigin]; Brada, Dr.; Böhm

Rush® *see* 'LENMOBRI'

Rushing Stream *see* 'AUSTREAM'

'RUSKIN', HRg, dr, 1928; (John Ruskin); bud ovoid; flowers deep crimson, large, dbl., 50 petals, cupped, sparingly recurrent, intense fragrance; foliage large, rich green, leathery; vigorous, bushy (4-5 ft.) growth; (21); [Souv. de Pierre Leperdrieux X Victor Hugo]; Van Fleet; American Rose Society

'RUSSELL SUPREME', HT, lp, 1927; [Mrs Charles E. Russell sport]; Pacific Rose Co.

'RUSSELLIANA', HMult, m; (Old Spanish Rose, Russell's Cottage RoseScarlet Grevillea, Souv de la Bataille de Marengo); flowers magenta-crimson fading to mauve, blooms in clusters, dbl., flat, moderate fragrance; foliage coarse; height to 20 ft

Russell's Cottage RoseScarlet Grevillea *see* **'RUSSELLIANA'**

Russet Beauty *see* 'POULRUSSET'

Rustica® *see* 'MEIVILANIC'

'RUSTICA', HFt, yb, 1929; flowers straw-yellow and gold, center apricot, reverse citron-yellow, semi-dbl., non-recurrent bloom, moderate fragrance; (28); RULED EXTINCT 4/81; [Mme Edouard Herriot X Harison's Yellow]; Barbier

Rustica 91 *see* 'MORYELRUG'

Rusticana *see* 'MEILÉNA'

Rusticana, F, lp; short; flowers pale pink, flat, 15 petals, borne in clusters, slight; medium large, dark green, glossy; ground cover type; [Candy Rose X Yesterday]; St. Wagner, 1994; Res. Stn. f. Fruit Growing, Cluj, 1995, Romania

Rusticana, Climbing *see* 'MEILÉNASAR'

Ruston's Blush, LCl, lp; [R. gigantea hybrid sport]; Morley, Dr B., 1985; (Weatherly, L.)

'RUTH', HT, ob, 1921; flowers orange flushed carmine, large, dbl., exhibition form, moderate fragrance; vigorous, bushy growth; Pemberton

'RUTH', HGal, mr, 1947; flowers less bright than alika, more double than alika, dbl., non-recurrent; upright growth (to 7 ft.); (21); [Mary L. Evans X Alika]; Wright, Percy H.

'RUTH ALEXANDER', LCl, ob, 1937; bud long, pointed; flowers orange, base yellow, large, semi-dbl., exhibition form, moderate fragrance; foliage leathery, glossy, bronze; vigorous, climbing growth; [Myra X Constance Casson]; Wilber; Bertsch

Ruth Harker *see* 'HARPOOH'

'RUTH HEWITT', F, w, 1963; dbl., 36 petals, 4 in., cupped, borne in open clusters, slight fragrance; foliage dark; vigorous, compact growth; [Unnamed seedling X Queen Elizabeth]; Norman; Harkness

Ruth Knopf, LCl, mp

Ruth Knopf's Pink Poly, Pol, lp

'RUTH LEUWERIK', F, mr, 1961; flowers bright red, blooms in clusters, dbl., 30 petals, 3 in., moderate fragrance; foliage bronze; vigorous, bushy growth; [Kathe Duvigneau X Rosemary Rose]; deRuiter; Gandy Roses, Ltd.

Ruth Pennington *see* KORlichtung

'RUTH SHAMBURGER', F, lp, 1934; [Kirsten Poulsen sport]; Shamburger, C.S.

Ruth Staley *see* 'TINSTALEY'

'RUTH TURNER', Pol, dp, 1941; flowers rose-pink, small, borne in clusters of 5-10, dbl., exhibition form, moderate fragrance; foliage glossy, pointed; vigorous, bushy growth; [Étoile Luisante X Sierra Snowstorm]; Moore, Ralph S.; California Roses

'RUTH VESTAL', Cl T, w, 1908; (Climbing Bride); flowers snow-white, well formed, very large, recurrent bloom, moderate fragrance; vigorous growth; [The Bride climbing sport]; Vestal

'RUTH WARNER', F, m, 1978; flowers deep lavender, medium, borne singly and in clusters up to 5, dbl., 48 petals, exhibition form, intense fragrance; very few prickles; foliage dark, smooth; upright growth; Bronze, ARC TG, 1983; [Little Darling X Angel Face]; Warner, A.J.

Ruth Westwood's Single Pink, Pol, lp

Ruth Woodward *see* 'DICPLEASANT'

'RUTH, CLIMBING', Cl HT, ob; flowers light orange shaded to red at edges, very large, very dbl., recurrent bloom, intense fragrance; foliage very large, glossy, olive-green; vigorous, climbing growth; not hardy.

'RUTHCHEN', Pol, mr, 1937; flowers carmine-red, medium, dbl.; Vogel, M.; (Sangerhausen)

Ruthe *see* 'RESRUTH'

Ruthie *see* 'TINRUTH'

Ruth's Bright Pink Eglantine, HEg, dp

Ruth's German Rose, HCh, m

Ruth's Pink Musk, HMsk, lp

Ruth's Steeple Rose, HCh, lp

Ruth's Wagon Wheel Bright Pink, HT, mp

'RUTILANT', HT, mr, 1960; bud ovoid; flowers carthamus-red, open, medium, semi-dbl., slight fragrance; foliage glossy; vigorous, bushy growth; [Gloire de Cibeins X Independence]; Arles; Roses-France

Rutland *see* POUlshine

'RUTRULO', HT, w, 1979; (True Love, **Yorkshire Bank**); dbl., 36 petals, borne 3-5per cluster, moderate fragrance; pointed, brown prickles; foliage bright green; vigorous, medium, bushy growth; GM, Geneva, 1979 Gold Star of the South Pacific, Palmerston North, NZ, 1979; [Pascali X Peer Gynt]; deRuiter, George; Fryer's Nursery, Ltd.

Ruyton, HT, mp, 1989; (Ruyton Girl's School); Bell, R.

Ruzova Lavina, F, lp; Strnad, 1976; (Czech Rosa Club)

'RUZYNE', HT, rb, 1936; flowers carmine-red marked rose, large, moderate fragrance; Mikes Böhm, J.

'RYOKKOH', F, ly, 1999; (Chidori); flowers light yellow green turning to green, dbl., 15–20 petals, 1–2 in., no fragrance; foliage dark green, leathery; 3-4 ft growth; GM, Baden-Baden, 1989; [Bridal Pink X (seedling X Tasogare)]; Suzuki, Seizo, 1986; Keisei Rose Nurseries, 1991

Ryokukou, F, ly

S

S. Antonio di Padova, HT, 1960; Sgaravatti, A.; (Cavriglia)

'S.&M. PERRIER', HT, pb, 1936; flowers very pale rose-pink, with orange glow, large, dbl., 60 petals, exhibition form, slight fragrance; foliage glossy; vigorous growth; [Magdalena de Nubiola X Pres. Cherioux]; Mallerin, C.; C-P

'S.A. PRINCE YOUSSOF KAMAL', HP, rb, 1922; bud long, pointed; flowers crimson, streaked brilliant scarlet, large, semi-dbl., moderate fragrance; [Souv. de Mme Chedane-Guinoisseau X Ulrich Brunner Fils]; Nabonnand, P.

S.H. Freedom *see* 'TWOFREE'

S.M. Alexander I *see* **'ROI ALEXANDRE'**

'S.M. GUSTAVE V', HP, dr, 1922; bud long, pointed, brilliant crimson; flowers carmine, reflexes crimson, well formed, very large, dbl., 30 petals, cupped, moderate fragrance; vigorous growth; [Frau Karl Druschki X Avoca]; Nabonnand, P.

'S.S. PENNOCK', HT, pb, 1922; flowers light rose-pink, with sulfur-yellow sheen (under glass, clea, dbl., moderate fragrance; [Lieutenant Chaure X Mrs George Shawyer]; Kordes

'S.S. PENNOCK, CLIMBING', Cl HT, pb, 1932; Lens; J&P

S.W.A.L.K. *see* 'SEAWALK'

'S'AAGARÓ', HT, mr, 1959; bud long, pointed; flowers geranium-red, large, dbl., 30 petals, exhibition form, slight fragrance; foliage dark, glossy; compact growth; [Angels Mateu X (Radar X Grand'mere Jenny)]; Dot, Simon

'SAARBRÜCKEN', S, mr, 1959; flowers scarlet-red, blooms in clusters (up to 20), semi-dbl., 3 in.; foliage dark; vigorous, bushy (5 ft) growth; Kordes

Saarlandwelle, HT, or

'SABAUDIA', HT, pb, 1934; flowers pink with gold reflections, base light chrome-yellow, large, dbl., cupped, moderate fragrance; foliage leathery, glossy, dark; vigorous, bushy growth; [R. foetida hybrid X Harison's Yellow]; Cazzaniga, F. G.

'SABBELIEF', Min, rb, 1996; (**Artful Dodger**); flowers cherry red, cream reverse, dbl., 15–25 petals, 1 in., borne in large clusters, slight fragrance; some prickles; foliage medium, dark green, glossy; tall (45 cms), upright, spreading, bushy growth; [Sheri Anne X Richard Buckley]; Burrows, Steven; Burrows Roses, 1997

'SABBYRON', Min, w, 1996; (**Oliver Twist**); flowers creamy white, blooms borne in large clusters, dbl., 15–25 petals, 1 in., slight fragrance; some prickles; foliage medium, dark green, glossy; tall (45-60 cms), upright, spreading, bushy growth; [Sheri Anne X Richard Buckley]; Burrows, Steven; Burrows Roses, 1996

SABchurchill, HT, lp; (**Sweet Bouquet**); Burrows, 1996

'SABCLIVE', Min, w, 1998; (**Pickwick**); flowers cream and white, double, medium, dbl., 15–25 petals, borne in small clusters, slight fragrance; prickles moderate; foliage medium, medium green, semi-glossy; compact, bushy, medium growth; [Buttons X Richard Buckley]; Burrows, Steven; Burrows Roses, 1998

'SABINE', HT, ob, 1958; flowers salmon, well formed, dbl., 40 petals; vigorous growth; [Dame Edith Helen X Seedling]; Frères, Buyl

'SABINE', HT, dp, 1962; (Sabine Sinjen); bud pointed; flowers cherry-red, medium, dbl., 30 petals, exhibition form, intense fragrance; foliage dark, glossy; upright, bushy growth; Tantau, Math.

Sabine, HT, pb; RvS-Melle

Sabine Rancy, HT, 1972; Croix; (Cavriglia)

Sabine Ruf *see* HELsabin

Sabine Sinjen *see* **'SABINE'**

'SABINIA', HT, pb, 1940; bud ovoid; flowers dark eglantine-pink passing to reddish yellow, large, dbl., exhibition form, moderate fragrance; foliage dark, leathery; very vigorous, upright growth; GM, Rome, 1939; [Julien Potin X Sensation]; Aicardi, D.; Giacomasso

Sable Chaud *see* 'DELSOB'

Sable Cloud *see* 'DELSOB'

Sabra *see* 'JUSSABRA'

'SABRE DANCE', F, rb, 1975; flowers scarlet, reverse gold, dbl., 40 petals, 4 in., exhibition form, slight fragrance; foliage dark; vigorous growth; [Mildred Reynolds X Arthur Bell]; Bees

Sabrina *see* 'MEIGANDOR'

'SABRINA', HT, rb, 1960; flowers crimson, reverse amber-yellow marked crimson, dbl., 35 petals, 5 in., exhibition form, intense fragrance; foliage dark, leathery; vigorous, bushy growth; [Grand Gala X Premier Bal]; Meilland, Mrs. Marie-Louise; URS

'SACAJAWEA', F, mr, 1982; flowers small blooms borne singly, dbl., slight fragrance; foliage medium, medium green, smooth; upright growth; [(Baccará X Garnet) X ?]; Elliott, Charles P.

Sacha *see* SPEKES

Sachalin, HRg, pb, 1988

Sachet® *see* 'SAVASACH'

Sächs. Lichtenstein, S, lp, 1996; flowers medium-large, semi-dbl.; Noack, Werner; (Sangerhausen)

'SACHSENGRUSS', HP, lp, 1912; (Tendresse); flowers soft flesh, well-formed, very large; vigorous growth; [Frau Karl Druschki X Mme Jules Gravereaux]; Neubert; Hoyer & Klemm

Sachsenrose, Pol, dp; flowers carmine-pink, small, semi-dbl.; (Sangerhausen)

'SACLAY', F, pb, 1970; flowers opal, full, dbl., 20–22 petals, 2–3 in., cupped, slight fragrance; foliage dull, sage-green; bushy growth; [Miss France sport]; Inst. National Agronomique; Commissariat a l'Energie Atomique

Sacramento, HT, dr, 1981; flowers dark velvety red, medium, dbl.; GPG Bad Langensalza; (Sangerhausen)

'SADABAHAR', F, mp, 1970; bud pointed; flowers pink, open, medium, semi-dbl.; foliage glossy, dark; very vigorous, bushy growth; [Frolic X Open pollination]; IARI

Sadabahar, Climbing, Cl F, mp, 1991; IARI

'SADARANGA', HT, mp, 1977; bud ovoid; very dbl., 60 petals, 4.5 in., globular, slight fragrance; foliage glossy; vigorous, upright, compact growth; [Kronenbourg X Peace]; Hardikar, Dr. M.N.

Saddler's Gold *see* 'HORSADDLER'

Sadie *see* DESmother

Sadie Hawkins *see* 'SPOHAWK'

Sadler *see* 'BISCOF'

Sadlers Wells, S, pb, 1983; Beales, Peter

'SAFARI', F, my, 1966; flowers large, semi-dbl., 22 petals, cupped, blooms in trusses; vigorous, compact growth; Tantau, Math.

Safari, F, ab, 1998; Kordes

Saffex Rose *see* KORgrayel

'SAFFO', HT, 1934; Aicardi, D.; (Cavriglia)

'SAFRANO', T, ab, 1839; bud pointed; flowers saffron and apricot-yellow, large, semi-dbl., moderate fragrance; vigorous growth; Beauregard

'SAGA', F, w, 1973; flowers cream-white, blooms in clusters, semi-dbl., 12 petals, 2.5 in., slight fragrance; [Rudolph Timm X (Chanelle X Piccadilly)]; Harkness, 1974

Saga Holiday *see* 'HARZEST'

'SAHARA', HT, w, 1956; bud golden yellow; flowers creamy buff, well shaped, rather late bloom; very vigorous, upright growth; [Mrs Wemyss Quinn X Mev. G.A. van Rossem]; Stevenson; Waterer

Sahara *see* OLIjcrem

Sahara, S, yb, 1996; Tantau

'SAÏD', F, rb, 1964; bud ovoid; flowers cinnabar-red, reverse dark red, medium, borne in clusters, dbl., 28 petals, cupped; short stems; [Aloha X Pioupiou]; Arles; Roses-France

'SAIGON', HT, yb, 1943; flowers dark yellow with coppery pink tints, large, dbl.; Gaujard; (Sangerhausen)

'SAILOZ MOOKHERJEA', F, or, 1974; flowers cadmium-orange to orange-vermilion, dbl., 20 petals, 3 in., exhibition form, slight fragrance; very vigorous growth; [Unknown X Unknown]; Pal, Dr. B.P.

'SAINT ALBAN', F, mr, 1977; semi-dbl., 15 petals, 3 in., slight fragrance; foliage glossy; dwarf growth; [Marlena X ?]; Harkness

'SAINT FIACRE', Pol, or, 1965; bud ovoid; flowers geranium-red, borne in clusters, single, cupped; vigorous, upright growth; [Reverence X Sumatra]; Delforge

Saint Louis *see* 'MALNINO'

Saint Mary® *see* 'MORYETTEM'

Saint Patrick *see* 'WEKAMANDA'

Saint Patrick, HT, lp; Strickland, Frank A., 1991

'SAINT PRIST DE BREUZE', HCh, dp, 1838; flowers medium, dbl.; Desprez; (Sangerhausen)

Sainte-Genevieve, S, w, 1987; Guimont, G.

Saint-Exupéry® *see* 'DELVOR'

Saint-Fiacre d'Orleans, Pol, op, 1997; Eve, A.

Saint-Victor, HT, mr, 1979; Croix

Saint-Vincent® *see* DELtrap

Saison Bluete, HT, dy

'SAI-UN', HT, ob, 1980; flowers yellow-orange shaded deep orange-red, large, dbl., 50 petals, moderate fragrance; prickles slanted downward; foliage dark, glossy; upright growth; [(Miss All-American Beauty X Kagayaki) X Unnamed seedling]; Suzuki, Seizo; Keisei Rose Nursery

Sakura Hime, Min, lp, 1990

'SAKURA-GASUMI', F, op, 1999; flowers pale pink, turning deeper, dbl., 20–25 petals, 1.5–2 in., slight fragrance; foliage dark green, semi-glossy; dwarf, compact (1 ft) growth; GM, Japan Rose Concours, 1990; Suzuki, Seizo, 1988; Keisei Rose Nursery, 1990

'SALAM ALEIK', HT, mr; flowers pure red, very large, dbl.

'SALAMANDER', HP, mr, 1891; flowers bright scarlet-crimson, large, dbl., cupped, non-recurrent; upright growth; Paul, W.

'SALAMBO', HT, mr, 1959; flowers carthamus-red; Kordes, R.; Vilmorin-Andrieux

Salammbo® *see* DELperl

'SALAZAR', HT, rb; flowers bright velvety red with salmon reflections, petals waved; da Silva, Moreira

'SALDEN MONARCH', HT, mp, 1968; flowers large, dbl.; foliage glossy; vigorous growth; [Ballet X ?]; Carrigg

'SALET', M, mp, 1854; bud well mossed; flowers rosy pink, very large, dbl., flat, some recurrence; vigorous growth; Lacharme, F.

Salita® *see* KORmorlet

'SALLIE LEWIS', HT, ab, 1924; bud long, pointed; flowers creamy apricot, center deeper, dbl., intense fragrance; [Mme Charles Lutaud X Gladys Holland]; Morse

'SALLY', F, ab, 1981; flowers pale apricot, fading to ivory, with pink veins; [Elizabeth of Glamis sport]; Walker, B.; Homeric Rose Nurseries

'SALLY', HT, mp, 1938; bud long, pointed; flowers pink, dbl., exhibition form; foliage leathery; long stems; vigorous growth; RULED EXTINCT 3/81; [Mrs Charles Russell X Mme Butterfly]; Spandikow; Eddie

'SALLY ALDER', HT, mp, 1944; bud long, pointed; flowers pink, medium, semi-dbl., exhibition form; foliage dark; moderate, bushy growth; [Portadown X Seedling]; Moss; F. Mason

Sally Forth, S, lp; Peden, R., 1999; (Weatherly, L.)

'SALLY HOLMES'®, S, w, 1976; bud pointed, apricot; flowers creamy white, hydrangea-like clusters, single, 5–8 petals, 3.5–4 in., blooms in large clusters, slight fragrance; few prickles; foliage dark, glossy; straight, long stems; vigorous, bushy growth; GM, Baden-Baden, 1980; [Ivory Fashion X Ballerina]; Holmes, R.; Fryer's Nursery, Ltd.

'SALLY MAC', F, ab, 1981; flowers apricot pink blend, yellow at base, blooms borne 4 per cluster, dbl., 40 petals, moderate fragrance; hooked, red-brown prickles; foliage medium, dark, glossy; vigorous, upright growth; [(Joyfulness X Paddy McGredy) X (Circus X Joyfulness)]; McCann, Sean; Hughes Roses

'SALLY PIGTAIL', HT, pb, 1959; flowers cream flushed rose-pink, high pointed, large, dbl., 32 petals, moderate fragrance; free growth; [Wilfred Pickles X Karl Herbst]; Mee

'SALLY TITE', HT, dr, 1930; flowers glowing crimson, over large, dbl., exhibition form; Dickson, S.; Armstrong Nursery;, B&A

Sally's Rose *see* 'CANREM'

Salmo *see* POUlnoev

Salmo *see* POUlnoeu

Salmon, Cl Min, op, 1994

'SALMON ARCTIC', LCl, lp, 1954; (Everblooming Pillar No. 83); flowers dawn-pink, very dbl., 90–100 petals, 3.5–4.5 in.; growth like a hybrid tea, followed by 4-5 ft. canes; [Unnamed seedling X Break o' Day, Climbing]; Brownell, H.C.

'SALMON BEAUTY', Pol, op, 1929; flowers peach-pink, passing to soft salmon, medium, borne in cluster, very dbl., slight fragrance; dwarf growth; [Orange King sport]; Wezelenburg

Salmon Button, Min, ob

Salmon Charm *see* **'GINGER ROGERS'**

Salmon Dream, MinFl, op, 1999

Salmon Drops, HT, op, 1999; Williams, J. Benjamin

'SALMON GLORY', Pol, op, 1937; flowers pinkish salmon; [Gloria Mundi sport]; deRuiter

'SALMON GLOW', F, op, 1970; flowers salmon, full, dbl., 25 petals, 3 in., slight fragrance; foliage matt green; compact, bushy growth; [Vera Dalton X Seedling]; Sanday, John

'SALMON MARVEL', F, op, 1958; flowers orange-salmon, rosette shape, borne in trusses, dbl., 44 petals, 2.5–3 in.; foliage dark, glossy, crinkled; bushy, upright growth; [Red Pinocchio X Signal Red]; deRuiter; Blaby Rose Gardens

Salmon Midinette, Cl Min, op, 1996; Moore

'SALMON PERFECTION', F, rb, 1952; flowers scarlet-red shaded orange, medium, borne in trusses, dbl., 25 petals, cupped; foliage dark, leathery; vigorous growth; deRuiter; A. Dickson;, McGredy

Salmon Pink Garnette, Pol, op

'SALMON QUEEN', Pol, op, 1923; flowers deep salmon; [Juliana Rose sport]; Den Ouden

Salmon Radiance *see* **'MRS CHARLES BELL'**

Salmon Rosamini *see* RUIsalro

Salmon Sensation *see* **'MARION'**

Salmon Sensation, F, 1956; de Ruiter, G.; (Cavriglia)

Salmon Sorbet, F, pb

Salmon Spire *see* KORturnus

Salmon Splash, ob, 1997; K&S

'SALMON SPRAY', F, pb, 1923; flowers light salmon-pink, reverse carmine, borne in clusters, semi-dbl., cupped, moderate fragrance; foliage rich green, leathery; very vigorous, bushy growth; [Orléans Rose, Climbing X Midnight Sun]; Grant; Kershaw

'SALMON SPRITE', F, op, 1964; flowers salmon suffused strawberry, blooms in clusters to 15, dbl., 40 petals, 3 in., intense fragrance; upright growth; [Unnamed seedling X Jiminy Cricket]; LeGrice

Salmon Sunblaze™ *see* 'MEIshuto'

Salmon Sunsation *see* KORpapie

Salmon Symphony, HT, op

'SALMONE', Gr, lp, 1968; bud ovoid; flowers medium, borne in clusters; foliage dark; [Queen Elizabeth sport]; Faassen-Houba

'SALMONEA', Pol, op, 1927; flowers salmon-pink; Cutbush

'SALMON-KING', F, rb, 1972; flowers red, inside orange-red, full, high pointed, dbl., 28–30 petals, 5 in., intense fragrance; foliage glossy; vigorous, bushy growth; [Show Girl X Orange Sensation]; Ellick; Excelsior Roses

'SALOMÉ', HT, op, 1945; bud ovoid; flowers salmon-flesh-pink, medium, dbl., moderate fragrance; foliage glossy; very vigorous, upright growth; Gaujard

'SALOMÉ', F, pb, 1958; flowers rose-orange, dbl.; vigorous growth; Frères, Buyl

'SALOU', F, pb, 1971; bud long, pointed; flowers deep pink to flesh-pink, open, large, dbl., moderate fragrance; foliage dark, leathery; very vigorous, upright growth; [Astoria X Tiffany]; Delforge

Salsa *see* 'JACstop'

'SALTAIRE', HT, dr, 1925; bud long, pointed; flowers deep velvety crimson, dbl., intense fragrance; Dickson, A.

'SALTATINA', F, or, 1974; bud ovoid; flowers orange-copper to red, open, dbl., 18–22 petals, 2.5–3 in., moderate, spicy fragrance; foliage dark; bushy, spreading growth; [Zorina X Fillette]; Lens

'SALTWELL PARK', F, mr, 1976; flowers very deep scarlet, full, dbl., 20 petals, 3 in., slight fragrance; foliage dark; moderate, upright growth; [Paddy McGredy X Arabian Nights]; Wood

Salut, HT, op, 1987; flowers salmon-orange, large, dbl.; VEG; (Sangerhausen)

Salut à la Suisse *see* 'TANschweigru'

'SALUTATION', LCl, op, 1972; flowers salmon, full, small, dbl., 38 petals, 2.5 in., intense fragrance; foliage glossy, dark; very vigorous growth; [Salute X Mme Isaac Pereire]; Oliver, W.G.

'SALUTE', F, rb, 1958; flowers cherry and ochre bicolor, small borne in trusses, semi-dbl., 20 petals, 1.5 in.; foliage dark, leathery; free growth; [Masquerade X Lady Sylvia]; McGredy, Sam IV; McGredy

'SALVO', HT, mr, 1959; flowers velvety crimson, dbl., 45 petals, 5 in., moderate fragrance; vigorous, bushy growth; [Happiness X Grand Gala]; Herholdt, J.A.; Herholdt's Nursery

Salza, S, dp, 1956; flowers carmine-pink, medium, dbl.; Berger, W.; (Sangerhausen)

Salzagold, HT, dy

Salzajubiläum, F, or, 1982; flowers salmon-red, large, semi-dbl.; GPG Bad Langensalza; (Sangerhausen)

Salzaperle, HT, mp, 1977; flowers large, dbl., moderate fragrance; GPG Bad Langensalza; (Sangerhausen)

Salzaquelle, HT, m, 1977; flowers light lilac, large, dbl., intense fragrance; GPG Bad Langensalza; (Sangerhausen)

Salzburg, S, op, 1967; flowers medium, dbl.; Kordes, W. Söhne; (Sangerhausen)

Sam *see* 'STEsam'

Sam Buff *see* **'AFTERGLOW'**

'SAM FERRIS', F, ob, 1967; flowers orange-scarlet, borne in trusses, dbl., slight fragrance; vigorous growth; [Heidleberg sport]; Ellick

Sam Houston Rose, S, pb, 1996; Antique Rose Emporium

'SAM MCGREDY', HT, yb, 1937; flowers dark cream, base sunflower-yellow, large, exhibition form, slight fragrance; foliage dark, leathery; vigorous growth; GM, NRS, 1935; [Delightful X Mrs Charles Lamplough]; McGredy

Sam Tucker, N, my

'SAMANDI', F, w, 1997; flowers double, dbl., 15–25 petals, borne in small clusters, no fragrance; foliage medium; medium (1m) growth; [Sexy Rexy X Brownie]; Rawlins, R.

Samantha® *see* 'JACmantha'

Samantha®, HT, dr, 1984; Noack, Werner

Samantha Ruth *see* 'WELsam'

Samara®, HT, dy, 1989; Noack, Werner

Samaritan® *see* 'HARverag'

Samba® *see* 'KORcapas'

'SAMMETGLUT', F, dr, 1943; bud long, pointed; flowers crimson, open, very large, borne in clusters, semi-dbl., slight fragrance; foliage dark, leathery; very vigorous, upright, bushy growth; [Holstein X Kardinal]; Kordes

'SAMMY', HMsk, dr, 1921; bud small; flowers carmine, borne in erect clusters, single, recurrent bloom, slight fragrance; foliage glossy, bronze; very vigorous, bushy growth; [(Trier X Watta?) X Gruss an Teplitz]; Pemberton

Samoa *see* **'FRED CRAMPHORN'**

'SAMOA SUNSET', HT, ob, 1965; flowers chrome-yellow and copper-orange, open, large, dbl., moderate fragrance; foliage glossy, bronze; vigorous, bushy, compact growth; [McGredy's Sunset sport]; Lone Star Rose Nursery

Samourai *see* 'MEIelec'

Samourai, Climbing *see* **'SCARLET KNIGHT, CLIMBING'**

Sampson *see* 'SAVasam'

Samptosa, F, dr, 1969; flowers large, dbl.; Institut für Obstbau Dresden Pillnitz; (Sangerhausen)

Samson™ *see* 'SAVasam'

'SAMUEL MARSDEN', Ch, mp, 1989; [Slater's Crimson China sport]; Nobbs, Kenneth J., 1987

'SAMUEL PEPYS', HT, w, 1934; flowers white, center slightly shaded cream, very large, dbl., globular, moderate fragrance; foliage leathery; bushy growth; Cant, B. R.

'SAN ANTONIO', Gr, or, 1967; bud ovoid; flowers large, dbl.; foliage dark, glossy, leathery; very vigorous, upright, bushy growth; [Roundelay X El Capitan]; Armstrong, D.L.; Armstrong Nursery

'SAN DIEGO', HT, my, 1968; bud ovoid; flowers light yellow, large, dbl., 50 petals, exhibition form, moderate fragrance; foliage leathery; vigorous, upright, bushy, compact growth; [Helen Traubel X Tiffany]; Armstrong, D.L.; Armstrong Nursery

San Diego, HT, pb, 1993; deVor

'SAN FERNANDO', HT, mr, 1948; flowers scarlet, large, dbl., 30 petals, exhibition form, intense fragrance; vigorous, upright growth; AARS, 1948; [Heart's Desire X (Crimson Glory X Poinsettia)]; Morris; Western Rose Co.

'SAN FERNANDO, CLIMBING', Cl HT, mr, 1951; Whisler; Germain's

'SAN FRANCISCO', HT, mr, 1962; bud ovoid; flowers signal-red, dbl., 40 petals, 4–5 in., exhibition form, moderate fragrance; foliage dark, glossy, leathery; vigorous, compact, branching growth; [Dean Collins X Independence]; Lammerts, Dr. Walter; Amling-DeVor Nursery;, Germain's

'SAN GABRIEL', HT, op, 1947; bud long, pointed; flowers deep salmon-pink, dbl., 45 petals, 5.5 in., exhibition form, moderate fragrance; foliage dark, leathery; vigorous, bushy growth; [Poinsettia X Unnamed variety]; Morris; Germain's

San Gerolamo, HGal, dp

'SAN JOAQUIN', HT, ob, 1939; flowers glowing orange; [Talisman sport]; Moore, Ralph S.

San Jordi, HT; Camprubi, C.; (Cavriglia)

'SAN JOSÉ', HT, ob, 1933; bud ovoid; flowers orange-salmon suffused gold, very large, dbl., moderate fragrance; foliage glossy; very vigorous growth; Denoyel, Vve.; J&P

San Jose Sunshine *see* 'FOUSUN'

San Juan Settler, B, pb

San Leandro Dark Red HT, HT, dr

'SAN LUIS REY', HT, dy, 1947; bud ovoid; flowers deep saffron-yellow, dbl., 40 petals, 4.5 in., cupped, intense fragrance; foliage dark, glossy; bushy growth; [Lady Forteviot X Pedralbes]; Morris; Germain's

San Rafael Rose *see* **'FORTUNE'S DOUBLE YELLOW'**

San Valentin, Pol, 1971; Dot, Simon; (Cavriglia)

SANaran, HT, w, 1995; (**Aorangi**); Sandbrook

'SANBABY', HT, dr, 1988; (**Mother and Baby**®); flowers dark red with a velvet sheen, some black marking, reverse de, dbl., 48 petals, quartered; prickles barbed, light brown; foliage medium, medium green, semi-glossy; low growth; [Bristol X (Lilac Rose X (Magenta X Crimson Glory))]; Sanday, John; John Sanday Roses, Ltd.

Sanbi, HT, ab

'SANCHARM', HMoy, lp, 1985; (**Pink Nevada**); flowers pale lilac pink; [Nevada sport]; Sanday, John, 1970

'SANCOL', F, dr, 1990; (**Edward Colston**); bud rounded; flowers dark red with medium red reverse, aging dark red, very dbl., 40–45 petals, cupped, borne in sprays of 3-5, slight, fruity fragrance; no fruit; prickles barbed, red; foliage medium, medium green, matt; upright, medium growth; [Vera Dalton X Stephen Langdon]; Sanday, John, 1982; John Sanday Roses, Ltd.

'SANCTUS', HKor, w, 1978; flowers pure white, large blooms borne singly, dbl., 36 petals, exhibition form, repeat bloom, moderate fragrance; large, straight, rust-brown prickles; foliage large, dark, leathery; vigorous, upright, tall growth; [Borealis X Borealis]; James, John

Sandalwood *see* 'JALWOOD'

'SANDAR', HT, rb, 1946; flowers carmine, center yellow, large, dbl., 25 petals; foliage dark; very vigorous growth; [Charles P. Kilham X Lleida]; Laperrière

'SANDAYA', S, ly, 1992; (**Enhance**); flowers soft apricot yellow, blooms borne in large clusters, dbl., 15–25 petals, 1.5 in., moderate fragrance; some prickles; foliage medium, medium green, glossy; low (25 cms), spreading growth; [Malmesbury X The Fairy]; Sanday, John; John Sanday Roses Ltd.

Sandberg, HT, pb

'SANDER'S WHITE RAMBLER', HWich, w, 1912; flowers rosette form, small blooms in large clusters, moderate fragrance; foliage bright green, glossy; vigorous growth; Sander & Sons

'SANDKOR', HT, op, 1981; (**Sandra**Rs); flowers salmon, large, dbl., 35 petals, exhibition form, slight fragrance; foliage medium, medium green, matt; upright growth; [Mercedes X Unnamed seedling]; Kordes, W.

'SANDOKAN', F, rb, 1977; (**Sanmez**); dbl., 27–32 petals; Sande; Pekmez

'SANDOLINA DE MAJOR', HT, dp, 1977; bud ovoid; flowers deep pink, very dbl., 57 petals, 4 in., cupped, moderate fragrance; vigorous, upright growth; [Dr. A.J. Verhage X Königin der Rosen]; Kordes, W. Söhne; Willemse

SandraRs *see* 'SANDKOR'

Sandra Kim®, S, mp

Sandra May Williamson, HT, mp, 1988; Williamson

Sandrina, S, 1996; Embriaco, B.; (Cavriglia)

Sandrina, HT, yb, 1997; Kordes

Sandrine, F, lp, 1975; Eve, A.

'SANDRINGHAM', F, my, 1955; dbl., 30 petals, 2.5 in., blooms in large clusters, moderate fragrance; foliage light green, glossy; vigorous, tall growth; Kordes; Morse

'SANDRINGHAM CENTENARY', HT, op, 1980; (Sandringham Century); bud pointed; flowers deep salmon-pink, blooms borne singly, dbl., 22 petals, moderate fragrance; foliage dark, glossy; vigorous, upright growth; [Queen Elizabeth X Baccará]; Wisbech Plant Co., 1981

Sandringham Century *see* **'SANDRINGHAM CENTENARY'**

Sands of Time™ *see* 'MINAZCO'

Sandton City *see* 'MEIPOPUL'

Sandton Smile *see* 'KORMETTER'

Sandy, HT, op, 1981; flowers coppery-pink, large, dbl., slight fragrance; Wisbech Plant Co.; (Sangerhausen)

'SANGERHAUSEN', HMsk, dp, 1938; bud long, pointed; flowers deep pink, large blooms in large clusters, semi-dbl., cupped, repeat bloom, slight fragrance; foliage large, leathery, wrinkled; vigorous (4-5 ft.), bushy growth; (28); [Ingar Olsson X Eva]; Kordes

Sangria® *see* 'MEIESTHO'

Sangria *see* MEItorpo

'SANGUINAIRE', HRg, pb, 1933; bud long, pointed; flowers brilliant oxblood-red with orange, stamens yellow, open, semi-dbl., 18 petals, 4 in., recurrent bloom, slight fragrance; foliage glossy; short stems; shrub or pillar (6 1/2 ft) grwth growth; [Bergers Erfolg X Capt. Ronald Clerk]; Gillot, F.

'SANGUINEA', Ch, dr; (Bengal Cramoisi Double, Blood-red China Rose, La Sanguine, R. indica cruenta); flowers velvety, vivid purple crimson, petals concave with white bas, very dbl., globular; foliage branches, leaves, and flowered stalks very purple; weak, spreading growth

Sanka, HT, or, 1986; (Enchantment); Keisei

'SANKT ANTON', Pol, mr, 1971; bud long, pointed; flowers brilliant red, open, small, single, profuse, continuous bloom, slight fragrance; foliage leathery; vigorous, upright growth; [Alain X Seedling]; Delforge

Sanktflorian, F, mr, 1971; flowers medium to large, dbl.; Meilland; (Sangerhausen)

'SANLILAC', HT, lp, 1988; (**Lilac Airs**); flowers soft, lilac pink, yellow at base, reverse silver at base, large, very dbl., 36 petals, exhibition form, intense, damask fragrance; prickles long, fairly straight, medium brown, matt; foliage large, medium green, semi-glossy;

bushy, tall, strong growth; [Fred Gibson X Whisky Mac]; Sanday, John; John Sanday Roses, Ltd.

'SANMAR', F, or, 1983; (**Bright Eyes**); flowers yellow stamens, medium, dbl., slight fragrance; foliage medium, dark, matt; bushy growth; [Seedling X Circus]; Sanday, John; John Sanday Roses, Ltd.

'SANMED', S, w, 1984; (**Ice Fairy**); [The Fairy sport]; Sanday, John; Sanday Roses, Ltd.

Sanmez *see* 'SANDOKAN'

'SANOLENCE', HT, or, 1985; (**Benevolence**); flowers large, dbl., 35 petals, moderate fragrance; long, narrow prickles; foliage large, dark, semi-glossy; bushy growth; [Vera Dalton X Seedling]; Sanday, John, 1986

'SANONE', S, rb, 1988; (**Little One**); flowers red with white eye, borne in sprays of 12-20, single, 5 petals, repeat bloom, no fragrance; prickles average, light brown; foliage medium, medium green, glossy; bushy, low growth; [(Sarabande X Unnamed seedling) X Circus]; Sanday, John; John Sanday Roses, Ltd.

'SANPHYLLIS', F, ab, 1986; (**John Hughes**); flowers soft apricot, medium, semi-dbl., 6–14 petals, slight fragrance; foliage medium, dark green, glossy; bushy growth; [City of Gloucester X Bristol Post]; Sanday, John

'SANREMO', HT, mr, 1962; bud ovoid, pointed; flowers cardinal-red, dbl., 20–30 petals, 5.5–6 in., exhibition form, moderate fragrance; foliage glossy; strong stems; vigorous, upright, bushy growth; [(Pink Delight X Rome Glory) X Baccará]; Mansuino, Q.; Carlton Rose Nurseries

'SANROC', F, or, 1985; (**Corsair**); flowers medium, dbl., 20 petals, slight fragrance; foliage medium, dark, glossy; bushy growth; [(Vera Dalton X Stephen Langdon) X Fiesta Flame]; Sanday, John; Sanday Roses, Ltd.

'SANROZO', HT, pb, 1986; (**Esperanto Jubileo**); flowers cream, edged deep rose pink, yellow suffused at base, large, dbl., slight fragrance; foliage medium, medium green, matt; bushy growth; [Gavotte X Piccadilly]; Sanday, John, 1987

Sans Sépales, lp; flowers medium, dbl.; (Sangerhausen)

Sans Souci *see* 'LAVSANS'

Sans souci *see* **'MOULIN ROUGE'**®

Sans Souci, Pol, or, 1960; flowers small, dbl.; Schmid, P.; (Sangerhausen)

Sans Souci *see* BARsan

'SANSPIC', HT, ab, 1985; (**Apricot Spice**); flowers medium, dbl., 35 petals, slight fragrance; foliage medium, medium green, matt; bushy growth; [City of Gloucester X Seedling]; Sanday, John

'SANTA ANITA', HT, mp, 1932; bud long, pointed; dbl., 22 petals, 3.5–4 in., exhibition form, moderate fragrance; foliage light; vigorous, bushy growth; [(Mrs J.D. Eisele X Unnamed seedling) X E.G. Hill]; Howard, F.H.; H&S

'SANTA ANITA, CLIMBING', Cl HT, mp, 1946; Howard, F.H.; H&S

'SANTA CATALINA', Cl F, lp, 1970; flowers medium, semi-dbl., 18 petals, 3.5 in., slight fragrance; [Paddy McGredy X Heidelberg]; McGredy, Sam IV; McGredy

Santa Claus *see* 'POULCLAUS'

Santa Fe *see* 'JACYEM'

Santa Fe, HT, ly; flowers daffodil yellow, exhibition form

'SANTA FÉ', HT, op, 1967; flowers deep salmon-pink, reverse lighter, large, dbl.; [Mischief X Tropicana]; McGredy, Sam IV; McGredy

'SANTA MARIA', F, mr, 1969; flowers scarlet, blooms in trusses, slight fragrance; foliage small; [Evelyn Fison X (Ma Perkins X Moulin Rouge)]; McGredy, Sam IV; McGredy

'SANTA RITA', HT, op, 1961; flowers salmon-pink, well formed, large; vigorous growth; [Independence X Papillon Rose]; Lens

Santa Rosa *see* **'BURBANK'**

'SANTA ROSA', F, op, 1954; bud ovoid; flowers salmon-pink tipped brick, small, dbl., globular, intense fragrance; foliage glossy, soft, bronze; vigorous, bushy growth; [Golden Salmon X Pinocchio]; Silva

Santa Rosa, HCh, dp, 1899; Burbank

'SANTA TEREZA D'AVILA', HT, op, 1959; flowers salmon-pink and orange, reverse gold, well-formed, moderate fragrance; foliage glossy; vigorous growth; GM, Madrid, 1959; [Monte Carlo X Michele Meilland]; da Silva, Moreira

Santana® *see* 'TANKLESANT'

'SANTANG', HT, ob, 1983; (**Maritime Bristol**); flowers tangerine, large, dbl., 35 petals, moderate fragrance; foliage medium, dark, semi-glossy; upright, bushy growth; [City of Gloucester X Unnamed seedling]; Sanday, John; Sanday Roses, Ltd.

Santiago® *see* ADAmona

'SANTOR', F, or, 1984; (**Red Brigand**); flowers medium, dbl., 20 petals, slight fragrance; short, slightly hooked prickles; foliage large, dark, semi-glossy; bushy growth; [Vera Dalton X Stephen Langdon]; Sanday, John; Sanday Roses, Ltd.

'SANWIFE', HT, rb, 1988; (**Wiltshire Pride**); flowers scarlet with yellow at base, reverse yellow with red edging, dbl., 28 petals, exhibition form, slight fragrance; prickles slightly hooked, dark brown; foliage large, dark green, semi-glossy; upright, tall growth; [Bristol X Piccadilly]; Sanday, John; John Sanday Roses, Ltd., 1989

'SAOHIME', HT, lp, 1999; flowers clean, light pink, dbl., 38 petals, exhibition form, slight fragrance; foliage medium green, half-leathery; 3.5 ft growth; BM, Japan Rose Concours, 1992; [Gavotte X Bridal Robe]; Hayashi, Shunzo, 1990

'SAONARA', HT, mr, 1962; flowers geranium-red, dbl.; foliage dark; strong stems; vigorous growth; [Baccará X Peace]; Borgatti, G.; Sgaravatti

Saphir *see* **'SONG OF PARIS'**

Saphir *see* TANrikas

'SAPHO', HT, ob, 1933; flowers salmon, tinted coppery, very large, dbl., slight fragrance; foliage leathery, glossy, dark; very vigorous growth; Gaujard

'SAPPHO', T, pb; flowers fawn shaded pink, center yellow; vigorous growth; Paul & Son, 1888

Sappho, A, w, 1817

Sara *see* JACautel

Sarabande® *see* 'MEIRABANDE'

Sarabande, Climbing® *see* 'MEIHANDSAR'

'SARAGAT', HT, rb, 1968; flowers deep strawberry to cream, large, dbl., globular, intense fragrance; foliage leathery; very vigorous growth; [Seedling X Seedling]; Malandrone

Sarah *see* 'MEIMAFRIS'

Sarah, Min, ab; [Poker Chip X Holy Toledo]; Hannemann, F., 1991; The Rose Paradise

'SARAH ARNOT', HT, mp, 1957; flowers warm rose-pink, dbl., 25 petals, 4.5 in., moderate fragrance; foliage leathery; vigorous, upright growth; GM, NRS, 1958; [Ena Harkness X Peace]; Croll

'SARAH BACHERACH', F, mp, 1965; flowers pink, medium, dbl., intense fragrance; foliage dark; [Harmonie X Buisman's Triumph]; Buisman, G. A. H.

'SARAH BERNHARDT', Cl HT, dr, 1906; flowers scarlet-crimson, very large, semi-dbl.; Dubreuil

'SARAH COVENTRY', F, dr, 1956; bud ovoid; flowers cardinal-red, open, medium, borne in irregular clusters, dbl., 40–45 petals, cupped, moderate fragrance; foliage bright green; vigorous, compact, bushy growth; [Red Pinocchio X Garnette]; Boerner; Stuart

'SARAH DARLEY', HT, dy, 1938; flowers clear deep golden yellow, well shaped;

foliage dark; vigorous growth; Wheatcroft Bros.

Sarah Elizabeth *see* 'PIXSAR'

'SARAH HILL', F, mp, 1956; bud short, pointed; flowers phlox-pink, open, dbl., 25–30 petals, 2.5–3 in., slight, spicy fragrance; foliage dark, leathery; vigorous, upright, bushy growth; [Garnette X Pink Bountiful]; Hill, Joseph H., Co.

'SARAH JANE', HT, pb, 1971; flowers rose-pink, edged deeper, reverse white, full, dbl., 35 petals, 4.5–5 in., slight fragrance; foliage glossy; very vigorous, tall growth; [Rose Gaujard sport]; Heath, W.L.

Sarah Jo *see* 'MEHREX'

'SARAH LYNN', HT, mp, 1991; dbl., borne mostly singly, slight fragrance; foliage large, dark green, semi-glossy; strong stems; vigorous, medium, compact growth; Bronze, ARC TG, 1991; [Seedling X Prima Ballerina]; Ohlson, John

'SARAH MAUD', HT, dy, 1934; bud long; flowers golden yellow, large; foliage bright, dark; very vigorous growth; Mallerin, C.

Sarah Philp, Min, ab

Sarah Robinson *see* 'TROBINETTE'

'SARAH VAN FLEET', HRg, mp, 1926; flowers wild-rose-pink, large, semi-dbl., cupped, recurrent bloom, intense fragrance; foliage leathery, rugose; compact (6-8 ft.) growth; (14); Van Fleet; American Rose Society

'SARAH WRIGHT', HT, dp, 1927; bud long, pointed; flowers rose-pink, dbl., slight fragrance; [Ophelia X Emma Wright]; Morse

Sarah, Duchess of York *see* 'DICRACER'

Sarah's Tea, T, pb

'SARAJEAN', Min, pb, 1979; bud pointed; flowers peach-pink, dbl., 50 petals, 1 in., globular, moderate fragrance; foliage small, bronze-green, glossy; upright, bushy growth; [Seedling X Over the Rainbow]; Williams, Ernest D.; Mini-Roses

'SARATOGA', F, w, 1963; bud ovoid; flowers gardenia-shaped, blooms in irregular clusters, dbl., 33 petals, 4 in., intense fragrance; foliage glossy, leathery; vigorous, upright, bushy growth; AARS, 1964; [White Bouquet X Princess White]; Boerner; J&P

'SARDANE', F, rb, 1962; bud long; flowers coral-red tinted silvery, dbl., 35–40 petals, 3 in.; foliage bright green; very bushy, compact growth; Laperrière; EFR

'SARGENT', LCl, pb, 1912; flowers apple-blossom-rose to pale pink, base amber-yellow, borne in clusters, semi-dbl., 3 in.; height 10-12 ft; [(R. wichurana X Crimson Rambler) X Baroness Rothschild]; Dawson; Eastern Nursery

'SARIE', HT, mp, 1977; dbl., 35 petals, 4.5–5 in., slight fragrance; foliage glossy, bright green; vigorous growth; [Pink Favorite X Nightingale]; Herholdt, J.A.

Sarie Marais *see* KORhota

'SARIE MAREIS', F, mr, 1950; flowers glowing scarlet; [Irene X Donald Prior]; Leenders, M.

Saris Sepals, M, mp, 1839

Saroda *see* **'SHARADA'**

Saroja, F, pb, 1984; Pal, Dr. B.P.

Sarong *see* 'GLANLIN'

'SAROOR', F, or, 1969; bud ovoid; flowers orange-scarlet, open, large, dbl., slight fragrance; foliage large, leathery; upright growth; [Gertrud Westphal X Open pollination]; Singh; Gopalsinamiengar

Sasa *see* DELrima

Saskabec *see* **'TANT MIEUX'**

'SASKIA', F, mp, 1961; single, borne in large clusters; foliage dark; bushy growth; [Pinocchio X Gartenstolz]; Buisman, G. A. H.

Sassy™ *see* 'HILTACO'

'SASSY LASSY', Min, yb, 1975; bud pointed; flowers yellow and pink blend, dbl., 28 petals, 1 in., slight fragrance; foliage small, bronze, glossy, leathery; upright, spreading growth; [Seedling X Over the Rainbow]; Williams, Ernest D.; Mini-Roses

'SATAN', HT, dr, 1939; (Joyce Lomax, Satanas); flowers very dark red, reverse lighter, nearly, large, dbl., exhibition form, moderate fragrance; foliage leathery; bushy growth; [(Mme Edouard Herriot X Angèle Pernet) X Mari Dot]; Pahissa; J&P

Satanas *see* **'SATAN'**

'SATCHMO'®, F, or, 1970; flowers bright scarlet, dbl., 25 petals, 3 in., exhibition form, slight fragrance; GM, The Hague, 1970; [Evelyn Fison X Diamant]; McGredy, Sam IV; McGredy

Satellite *see* 'DELSATEL'

'SATELLITE'®, HT, dr, 1958; bud pointed; flowers deep crimson, dbl., 30 petals, 4.5 in., exhibition form, intense fragrance; foliage dark, glossy; vigorous, upright, compact growth; [Editor McFarland X William Harvey]; Priestly

Satin, HT, dp, 1996

Satin Doll *see* 'JACSPIF'

Satina® *see* TANinat

'SATINETTE', Pol, w, 1971; bud ovoid; flowers large, dbl., cupped, slight fragrance; foliage large, light, leathery; vigorous, upright growth; [Maria Delforge X Irene of Denmark]; Delforge

'SATINGLO', F, mp, 1954; bud globular; flowers glowing coral, borne in clusters, dbl., 45–50 petals, 2.5 in., cupped, moderate fragrance; foliage leathery, glossy; vigorous growth; [Pinocchio seedling X Vogue]; Boerner

'SATISFACTION', HT, dp, 1956; flowers carmine-pink, large, dbl., intense fragrance; [Parfum X Unnamed seedling]; Verbeek

'SATMIR', HT, dr, 1956; flowers carmine-red, dbl.; long, strong stems; very vigorous growth; [Satan X Mirandy]; Dot, Pedro

Saturday®, HT, pb, 1990; McGredy, Sam IV

Saturday Star *see* KORvolomin

Saturn *see* JAClogo

'SATURNIA', HT, rb, 1936; bud long, pointed; flowers bright scarlet with gold, large, dbl., 20 petals, cupped, moderate, fruity fragrance; foliage dark, glossy; vigorous growth; GM, Portland, 1938 GM, Rome, 1933; [Julien Potin X Sensation]; Aicardi, D.; J&P;, Robichon

Satvika, HT, yb; flowers occasionally striped

SAUbaflor, F, lp, 1990; (**Heriflor**); Sauvageot

SAUbima, F, op, 1994; (**Chateau de Filain**); Sauvageot

SAUblim, HT, mr, 1991; (**Atida**); Sauvageot

SAUbord, HT, dp, 1992; (**Gustave Courbet**®); Sauvageot

'SAUCY SUE', F, mp, 1974; flowers full, dbl., 20–25 petals, 3.5 in., moderate fragrance; free growth; [Pink Parfait X Europeana]; Lowe

'SAUDADE D'ANIBAL DE MORAIS', HT, dr, 1935; bud velvety crimson; flowers salmony crimson-red, dbl.; vigorous growth; [Sir David Davis X Pres. Jac. Smits]; da Silva, Moreira

SAUdero, F, 1995; (**Dance of Joy 95**); Sauvageot, H.; (Cavriglia)

SAUdime, F, rb, 1999; (**Amours de Savernem**); Sauvageot

SAUdive, F, 1995; (**Messara**); Sauvageot; (Cavriglia)

SAUdora, F, pb, 1997; (**Eclipse**); Golden Rose, Geneva, 1996; Sauvageot

SAUdril, F, lp, 1997; (**Manureva**); Sauvageot

SAUkar, F, op, 1989; (**Karine Sauvageot**®); Sauvageot

'SAUL', HT, dp, 1970; flowers light rose-madder, dbl., 28 petals, 4 in., moderate fragrance; foliage large, dark; upright growth; [Tropicana X Sterling Silver]; Gandy, Douglas L.

Saumonia, HT, op, 1998

'SAVABE', Min, dy, 1994; (**Glory Be**); dbl., 15–25 petals, 1.5 in., borne singly or in small clusters, no fragrance; foliage small, dark green, semi-glossy; medium (16-20 in), upright, bushygrowth; [Party Girl X Sonnenkind]; Saville, F. Harmon; Nor'East Min. Roses, 1995

'SAVABEAN', Min, rb, 1981; (**Jelly Bean**®); bud ovoid, pointed; flowers red-yellow blend, micro-mini blooms borne 1-6 per cluster, dbl., 20 petals, exhibition form, moderate, spicy fragrance; no prickles; foliage small; very compact, tiny growth; [Unnamed seedling X Poker Chip]; Saville, F. Harmon; Nor'East Min. Roses, 1982

'SAVABEAR', Min, r, 1989; (**Teddy Bear**®); bud urn shaped; flowers terra-cotta, reverse lighter, aging to mauve-pink, dbl., 28 petals, exhibition form, borne singly, slight fragrance; ovoid, grayed-orange fruit; prickles slight downward curve, medium, purple; foliage medium, dark green, semi-glossy; upright, bushy, medium, vigorous growth; PP007424; [Sachet X Rainbow's End]; Saville, F. Harmon; Nor'East Min. Roses, 1990

'SAVABEG', Min, ob, 1988; (**New Beginning**®); flowers bright orange-yellow bicolor, decorative, medium, very dbl., 40–50 petals, borne usually singly, no fragrance; no fruit; very few prickles; foliage medium, medium green, semi-glossy; bushy, compact growth; PP006707; AARS, 1989; [Zorina X Unnamed seedling]; Saville, F. Harmon; Nor'East Min. Roses, 1989

'SAVABET', Min, mr, 1985; (**Single's Better**®); bud mossy; flowers medium red, yellow hinge, mini-moss, small, single, 5 petals, slight fragrance; foliage medium, medium green, semi-glossy; bushy growth; [(Yellow Jewel X Tamango) X ((Little Chief X Sarabande) X LemonDelight)]; Saville, F. Harmon; Nor'East Min. Roses

'SAVABINO', Min, or, 1997; (**Bambino**); flowers full, small, micro mini, vibrant orange, very dbl., 26–40 petals, borne in small clusters, no fragrance; foliage small, medium green, semi-glossy changing to matte; compact, bushy, low (to 14in.) growth; [Sequoia Gold X Sparks]; Saville, F. Harmon

'SAVABREZ', Min, ob, 1984; (**Breezy**®); bud small; flowers bright orange-red, yellow reverse, small blooms in sprays, dbl., 20 petals, exhibition form, moderate fragrance; foliage small, medium green, semi-glossy; upright growth; [Sheri Anne X Seedling]; Saville, F. Harmon; Nor'East Min. Roses

'SAVACALL', Min, or, 1984; (**Total Recall**®); flowers medium, dbl., no fragrance; foliage small, medium green, semi-glossy; bushy growth; [Zorina X Baby Katie]; Saville, F. Harmon; Nor'East Min. Roses

'SAVACAMP', Min, mr, 1984; (**Scamp**®); flowers micro-mini, small, dbl., 35 petals, exhibition form, slight fragrance; foliage small, medium green, semi-glossy; compact, bushy growth; [Baby Katie X (Yellow Jewel X Tamango)]; Saville, F. Harmon; Nor'East Min. Roses

'SAVACENT', Min, dy, 1981; (Atkins Beauty, **Center Gold**®); bud pointed; flowers deep yellow, sometimes near white, borne singly or up to 12, dbl., 60 petals, exhibition form, moderate, spicy fragrance; long, thin prickles, slanted downward; foliage glossy, textured; upright, compact growth; AOE, 1982; [Rise 'n' Shine X Kiskadee]; Saville, F. Harmon; American Rose Foundation

'SAVACHASE', Min, rb, 1988; (**Chasin' Rainbows**®, Chasing Rainbows); bud ovoid; flowers very brilliant yellow, edged red with scarlet becoming more red with age, dbl., 21 petals, exhibition form, slight, spicy fragrance; no fruit; prickles long, thin, angled, light brown; foliage small, dark green, semi-glossy; micro-mini bushy, low growth; PP007058; [Zorina X Rainbow's End]; Saville, F. Harmon; Nor'East Min. Roses, 1990

'SAVACHILD', Min, pb, 1991; (**Child's Play**®); flowers porcelain pink/white bicolor, pink edges, semi-dbl., 1.5 in., exhibition form, borne singly or in sprays of 3+, moderate, sweet fragrance; foliage medium, dark green, matt; upright, medium growth; PP8175; AARS, 1993 AOE, 1993; [(Yellow Jewel X Tamango) X Party Girl]; Saville, F. Harmon; Nor'East Min. Roses

'SAVACIN', Min, r, 1986; (**Cinnamon Toast**®); flowers russet brown, small, dbl., 28 petals, exhibition form, slight fragrance; small, red prickles; foliage small, medium green, semi-glossy; low, upright, bushy growth; [Zorina X (Sheri Anne X Glenfiddich)]; Saville, F. Harmon; Nor'East Min. Roses

'SAVACLEND', Min, yb, 1999; (**Climbing Rainbow's End**®); flowers yellow blooms suffused with red at edges and bloom ages, gradually turni, dbl., 25–40 petals, 1.5 in., exhibition form, borne in small clusters, slight fragrance; prickles moderate; foliage small, medium green, semi-glossy; short, thick stems; climbing to 10 ft. growth; [Sport of Rainbow's End sport]; Saville, F. Harmon

'SAVACLOUD', Min, w, 1988; (**White Cloud**®); flowers white with pale pink; PP006876; [Buttons 'n' Bows sport]; Saville, F. Harmon; Nor'East Min. Roses

'SAVACOOK', Min, ab, 1995; (**Fortune Cookie**®); flowers apricot blend, moderately full (15–26 petals), small blooms borne singly and in small clusters; no fragrance; no prickles; foliage small. Medium green, semi-glossy; medium (16–18 in.) upright; [Baby Katie X Mazurka]; Saville, F. Harmon; Nor'East Miniature Roses, 1996

'SAVACOP', Min, ob, 1988; (**Copper Sunset**®); bud pointed; flowers coppery-orange, flushed orange-red, reverse medium red, dbl., 21 petals, exhibition form, slight fragrance; ovoid, orange fruit; prickles long, thin, slanted, gray-red; foliage medium, dark green, semi-glossy; upright, medium, angular growth; PP007032; [Acey Deucy X Rainbow's End]; Saville, F. Harmon

'SAVACRIM', Min, dr, 1994; (**Little Crimson**); dbl., 15–25 petals, 1.5 in., borne mostly singly, no fragrance; few prickles; foliage small, dark green, semi-glossy; medium (16-20 in), upright, bushy growth; [Teddy Bear X (Zorina X Baby Katie)]; Saville, F. Harmon; Nor'East Min. Roses, 1995

'SAVADAY', Min, op, 1988; (**Heavenly Days**®, 'SAVAHE'); flowers glowing indian-orange, reverse lemon yellow, flushed fire-red, dbl., 28–32 petals, cupped, no fragrance; foliage medium, medium green, glossy, underside matt; bushy, medium, compact growth; PP006808; AOE, 1988; [Unnamed Climbing Yellow Miniature seedling X (Sheri Anne X Glenfiddich)]; Saville, F. Harmon; Nor'East Min. Roses

'SAVADEE', Min, ob, 1988; (**Dee Bennett**®); flowers yellow and orange becoming orange-yellow, medium, dbl., 25 petals, exhibition form, borne usually singly, slight, fruity fragrance; no fruit; prickles long, thin, curved, gray-orange; foliage medium, dark green, semi-glossy; bushy, medium growth; PP006951; AOE, 1989; [Zorina X (Sheri Anne X (Yellow Jewel X Tamango))]; Saville, F. Harmon; Nor'East Min. Roses, 1989

'SAVADEN', Min, ob, 1995; (**Denver's Dream**®); flowers copper orange with red reverse, dbl., 15–25 petals, 1.5 in., cupped, borne mostly singly, no fragrance; foliage medium, dark green, semi-glossy; medium (16-20 in), upright growth; PP9435; [Gingersnap X Klima]; Saville, F. Harmon; Nor'East Min. Roses, 1994

'SAVADI', Min, or, 1986; (**Baby Diana**®); flowers orange-red, yellow reverse, small, dbl., 20 petals, exhibition form, borne usually singly, moderate fragrance; long, thin, hooked, brown prickles; foliage small, medium green, semi-glossy; bushy growth; PP005957; [Zorina X (Sheri Anne X Glenfiddich)]; Saville, F. Harmon; Nor'East Min. Roses

'SAVADOLL', Min, ob, 1979; (**Calico Doll**); bud ovoid, pointed; flowers orange, striped yellow, semi-dbl., 18 petals, 1–1.5 in., cupped; foliage dark; compact growth; [Rise 'n' Shine X Glenfiddich]; Saville, F. Harmon; Nor'East Min. Roses

'SAVADREAM', Min, mp, 1990; (**Dreamer**®); bud ovoid, pointed; flowers dusty pink, medium, borne singly or in sprays of 3-5, dbl., 20 petals, cupped, no fragrance; foliage medium, dark green, semi-glossy; upright, bushy, medium growth; PP007757; [Baby Katie X Shocking Blue]; Saville, F. Harmon; Nor'East Min. Roses, 1991

'SAVAFACE', Min, dp, 1991; (**Happy Face**®); bud ovoid; flowers clear rosy pink, blooms borne usually singly or in sprays of 3-5, very dbl., 35–40 petals, cupped, no fragrance; foliage medium, dark green, glossy; bushy, medium, compact growth; [(Sheri Anne X Rise 'n' Shine) X Mountie]; Saville, F. Harmon, 1986; Nor'East Min. Roses

'SAVAFIRE', Min, rb, 1991; (**Fireworks**®); flowers brilliant orange-yellow bicolor, dbl., 35 petals, 1.5 in., exhibition form, profuse, borne singly or in small clusters, slight fragrance; foliage medium, dark green, semi-glossy; bushy, medium growth; PP8182; [(Rise 'n' Shine X Sheri Anne) X Rainbow's End]; Saville, F. Harmon; Nor'East Min. Roses, 1992

'SAVAFISH', Min, ob, 1992; (**Gone Fishin'**®); flowers bright orange, blooms borne singly and in sprays of 4-10, dbl., 28–35 petals, 1.25 in., cupped, slight fragrance; foliage medium, dark green, glossy, very disease resistant; bushy, medium growth; [Fairlane X Zorina]; Saville, F. Harmon; Nor'East Min. Roses, 1993

'SAVAGALA', Min, dp, 1999; (**Gala**); dbl., 17–25 petals, 1.5 in., borne mostly single and in small clusters, loose sprays, no fragrance; prickles moderate; foliage medium, dark green, semi-glossy; upright, spreading, medium (15-16 in) growth; [High Jinks X seedling]; Saville, F. Harmon; Nor' East Miniature Roses, 2000

'SAVAGEN', Min, yb, 1982; (**Genevieve**®); flowers yellow, streaked scarlet, scarlet increasing with age, small, dbl., 35 petals, moderate, spicy fragrance; foliage medium, medium green, semi-glossy; upright, bushy growth; [Unnamed Miniature Cl. X Unnamed Miniature seedling]; Saville, F. Harmon; Nor'East Min. Roses, 1983

'SAVAGOOD', Min, my, 1991; (**Good Morning America**®); bud ovoid, urn-shaped; flowers urn-shaped, large blooms, sunshine yellow, slight red edges, very dbl., 45–50 petals, 1.5 in., exhibition form, borne mostly singly, moderate, fruity fragrance; foliage medium, dark green, semi-glossy; long stems; upright, bushy, tall growth; PP007761; AOE, 1991; [Fantasia X Rainbow's End]; Saville, F. Harmon, 1984; Nor'East Min. Roses

'SAVAHALO', Min, my, 1991; (**Golden Halo**®); bud ovoid, pointed; flowers bright yellow, medium, dbl., 24–26 petals, cupped, profuse, borne mostly dingly, moderate fragrance; foliage medium, medium green, semi-glossy; upright, bushy growth; PP007761; AOE, 1991; [Arthur Bell X Rainbow's End]; Saville, F. Harmon, 1985; Nor'East Min. Roses

'SAVAHE', Min, op, 1988; (**Heavenly Days**®, 'SAVADAY'); flowers glowing indian-orange, reverse lemon yellow, flushed fire-red, dbl., 28–32 petals, cupped, no fragrance; foliage medium, medium green, glossy, underside matt; bushy, medium, compact growth; PP006808; AOE, 1988; [Unnamed Climbing Yellow Miniature seedling X (Sheri Anne X Glenfiddich)]; Saville, F. Harmon; Nor'East Min. Roses

'SAVAHOLD', Min, my, 1996; (**Behold**®); bud medium; flowers clear, bright medium yellow, reverse lighter, holds color, dbl., 15–25 petals, borne mostly singly, no fragrance; few prickles; foliage medium, medium green, semi-glossy; upright, compact, medium (22-26 in) growth; [(Rise 'n' Shine X Sheri Ann) X (Heideroslein X Nozomi)]; Saville, F. Harmon; Nor'East Min. Roses, 1997

'SAVAHOWDY', Min, pb, 1982; (**Minnie Pearl**®); flowers light pink, reverse darker, yellow base, exhibition form, borne mostly singly, slight fragrance; few prickles; foliage small, medium green, semi-glossy; thin, wiry stems; upright growth; PP5097; [(Little Darling X Tiki) X Party Girl]; Saville, F. Harmon; Nor'East Min. Roses

'SAVAJERRY', Min, mr, 1998; (**Jerry-O**®); flowers light ot medium red, large blooms, dbl., 25 petals, 1.5 in., exhibition form, borne singly, intense fragrance; Saville, F. Harmon

'SAVAJINKS', Min, pb, 1992; (**High Jinks**®); dbl., 25–32 petals, 1.5 in., exhibition form, borne singly and in sprays of 4-30, slight fragrance; foliage medium, dark green, semi-glossy; upright, bushy, medium to tall growth; [Rise 'n' Shine X Sheri Anne]; Saville, F. Harmon; Nor'East Min. Roses, 1993

'SAVAKLIM', Cl Min, dy, 1993; (**Klima**®); semi-dbl., 6–14 petals, 1.5 in., no fragrance; few prickles; foliage small, medium green, semi-glossy; tall (4-5ft), upright, climbing growth; [Ferris Wheel X COCagold]; Saville, F. Harmon; Nor'East Min. Roses, 1993

'SAVALAV', Min, m, 1994; (**Vista**®); flowers soft lavender, moderately excellent form, dbl., 15–25 petals, 1.5 in., exhibition form, borne mostly singly, slight fragrance; some prickles; foliage medium, medium green, semi-glossy; medium, compact growth; PP9031; [Sachet X Copper Sunset]; Saville, F. Harmon; Nor'East Min. Roses, 1994

'SAVALIFE', Min, yb, 1984; (**Rainbow's End**®); bud small; flowers deep yellow, red petal edges, aging red all over, dbl., 35 petals, exhibition form, borne in small clusters, no fragrance; foliage small, dark, glossy; upright, bushy growth, 18 in.; PP5482; AOE, 1986; [Rise 'n' Shine X Watercolor]; Saville, F. Harmon; Nor'East Min. Roses

'SAVALIGHTS', Min, rb, 1993; (**Lights of Broadway**®); flowers yellow with broad red edge, yellow reverse, very dbl., 1.5 in., borne mostly singly, no fragrance; some prickles; foliage medium, medium green, semi-glossy; medium (16-20 in), upright, bushy growth; PP8918; [(Tamango X Yellow Jewel) X Party Girl]; Saville, F. Harmon; Nor'East Min. Roses, 1994

'SAVALOT', Min, ob, 1984; (**Cheers**®); flowers orange-red, cream reverse, dbl., 20 petals, slight fragrance; foliage small, medium green, semi-glossy; compact, bushy growth; [Poker Chip X Zinger]; Saville, F. Harmon; Nor'East Min. Roses

'SAVALUCK', Min, ob, 1987; (**Happy Go Lucky**®); flowers brilliant orange-yellow blend, small, dbl., 17–24 petals, exhibition form, borne usually singly, slight, sweet fragrance; no fruit; long, thin pointed prickles; foliage small, dark green, semi-glossy; bushy, medium growth; PP006506; [Cheers X (Sheri Ann X (Yellow Jewel X Tamango))]; Saville, F. Harmon; Nor'East Min. Roses

'SAVALUTE', Min, yb, 1998; (**Absolutely**®); flowers pale to medium yellow, dusted apricot pink, lighter reverse, borne singly and in small clusters, floriferous, slight fragrance; few prickles; foliage medium, medium green, semi-glossy; upright, compact, spreading, bushy, vigorous growth; [Seedling X Rise 'n' Shine]; Saville, F. Harmon; Nor'East Miniature Roses, 1998

'SAVAMAE', Min, ab, 1986; (**Ellamae**®); flowers , dbl., 35 petals, exhibition form, borne singly and in sprays of 3-5, moderate fragrance; long, thin prickles; foliage medium, dark, glossy; medium, upright, bushy growth; PP006053; [Zorina X (Sheri Anne X Glenfiddich)]; Saville, F. Harmon; Nor'East Min. Roses

'SAVAMARK', Min, or, 1982; (Apricot Sunblaze, Mark 1, **Mark One**®); flowers brilliant orange-red, blooms borne singly and up to 10 per cluster, dbl., 43 petals, cupped, moderate, spicy fragrance; long, thin prickles, soft on peduncles; foliage very glossy; compact, bushy growth; [Sheri Anne X Glenfiddich]; Saville, F. Harmon; Nor'East Min. Roses

'SAVAMIST', Min, ab, 1987; (**Apricot Mist**®); flowers apricot with tones of pink and yellow, medium, very dbl., 40–45 petals, exhibition form, borne singly, slight fragrance; no fruit; long, thin, straight, brown prickles; foliage small, dark green, glossy; bushy, low, profuse, compact, symmetrical growth; PP006507; [Fantasia X Baby Katie]; Saville, F. Harmon; Nor'East Min. Roses

'SAVAMOR', Min, m, 1995; (**Scentsational**®); flowers light mauve, large blooms, dbl., exhibition form, borne singly on long stems, intense fragrance; Saville, F. Harmon

'SAVANADE', Min, dy, 1997; (**Pacific Serenade**®); flowers double, medium, bright yellow, dbl., 15–25 petals, cupped, borne mostly singly, moderate fragrance; no prickles; foliage medium, medium green, semi-glossy; upright, compact, spreading, bushy, vigorous growth; [Cal Poly X New Zealand]; Saville, F. Harmon

'SAVANHOUR', Min, mr, 1983; (**Happy Hour**®); flowers bright medium red, yellow eye, small, dbl., 20 petals, moderate fragrance; foliage small, dark, glossy; bushy, spreading growth; PP005449; [(Tamango X Yellow Jewel) X Zinger]; Saville, F. Harmon; Nor'East Min. Roses

'SAVANICE', Min, w, 1991; (**Ice Queen**®); bud ovoid, pointed; flowers medium, white, fully double, very dbl., 60 petals, exhibition form, borne mostly singly, no fragrance; foliage medium, dark green, semi-glossy; bushy, medium growth; PP007771; [Cupcake sport]; Saville, F. Harmon, 1983; Nor'East Min. Roses

'SAVANIGHT', MinFl, mp, 1997; (**Overnight Scentsation**); flowers very full, large, very dbl., 41 petals, borne mostly singly, intense fragrance; foliage medium, medium to dark green, semi-glossy; upright, medium (2ft.) growth; [Taxi X Lavender Jade]; Saville, F. Harmon

'SAVANIN', Min, rb, 1986; flowers medium red with yellow center, moderately small, dbl., 15–25 petals, slight fragrance; foliage small, medium green, semi-glossy; bushy growth; [Rise 'n' Shine X Zinger]; Saville, F. Harmon; SNC Meilland & Cie, 1986

'SAVANNAH', HT, ab, 1980; bud ovoid, pointed; flowers soft apricot, borne singly or 2-3 per cluster, dbl., 32 petals, cupped, slight, spicy fragrance; long prickles, hooked down; foliage moderately leathery; vigorous, upright growth; [Unnamed seedling X Arizona]; Weeks, O.L.

Savannah Miss *see* 'PIXANAH'

Savanne, HT, dy, 1985; Noack, Werner

'SAVAPACE', Min, w, 1979; (Pace Setter, **Pacesetter**); bud long, pointed; flowers pure white, dbl., 46 petals, 1.5 in., exhibition form, borne in small clusters, needs disbudding, moderate fragrance; few prickles; foliage dark, matt; vigorous, compact growth, tall; PP4513; AOE, 1981; [Ma Perkins X Magic Carrousel]; Schwartz, Ernest W.; Nor'East Min. Roses

'SAVAPAINT', Min, rb, 1982; (**Flying Colors**®); flowers red and yellow blend, aging to pink and white, micro-mini, semi-dbl., slight fragrance; foliage medium green, semi-glossy; upright, bushy growth; [(Yellow Jewel X Tamango) X Sheri Anne]; Saville, F. Harmon; Nor'East Min. Roses, 1983

'SAVAPEG', Min, lp, 1986; (**Peggy Jane**®); flowers light pink, with lighter petal edges and base; PP005999; [Starina sport]; Utz, Peggy L.; Nor'East Min. Roses

'SAVAPIE', Min, rb, 1991; (**Something Else**®); flowers white with very contrasting red edge, blooms borne usually singly, dbl., cupped, slight fragrance; foliage small, medium green, semi-glossy; upright, bushy, medium growth; [(Yellow Jewel X Tamango) X Party Girl]; Saville, F. Harmon; Nor'East Min. Roses, 1985

'SAVAPIECE', Min, mr, 1984; (**Centerpiece**®, Centre Piece); flowers deep medium red, small, dbl., 35 petals, exhibition form, slight fragrance; foliage small, dark, semi-glossy; bushy growth; PP007769; AOE, 1985; [(Sheri Anne X Tamango) X (Sheri Anne X (Yellow Jewel X Tamango))]; Saville, F. Harmon; Nor'East Min. Roses, 1985

'SAVAPPLE', Min, ab, 1999; (**Applause**); flowers medium coral, reverse lighter, dbl., 17–25 petals, 1.2in., slight fragrance; prickles moderate; foliage small, medium green, semi-glossy; upright, compact, low growth; AOE, ARS, 2000; [Sequoia Gold X Sparks]; Saville, F. Harmon; Nor'East Miniature Roses, 2000

'SAVAPUP', Min, ob, 1978; (**Puppy Love**); bud pointed; flowers pink, coral, orange blend, small, dbl., 23 petals, 1.5 in., exhibition form, slight fragrance; foliage matt; upright, compact growth; AOE, 1979; [Zorina X Unnamed seedling]; Schwartz, Ernest W.; Nor'East Min. Roses

'SAVARAIN', Min, m, 1989; (**Raindrops**®); bud ovoid; flowers light mauve-purple, light yellow at base, reverse lighter, dbl., 24 petals, borne in small clusters, slight fragrance; fruit not observed; prickles straight, slanted downward, small, gray-red; foliage small, dark green, semi-glossy; upright, medium growth; PP007748; [Sachet X Rainbow's End]; Saville, F. Harmon; Nor'East Min. Roses, 1990

'SAVARALPH', Min, mr, 1999; (**Ralph Moore**); flowers medium, dbl., 17–25 petals, borne mostly singly, slight fragrance; few prickles; foliage medium, dark green, semi-glossy; upright, compact, well-branched, medium (16 in) growth; AOE, ARS, 2000; [Sachet X seedling]; Saville, F. Harmon; Nor'East Miniature Roses, 2000

'SAVAREND', Min, dp, 1987; (**Adam's Smile**®, 'SAVASMILE'); flowers deep pink, medium, dbl., 23–27 petals, exhibition form, borne usually singly and in sprays of 3-5, no fragrance; no fruit; prickles long, thin, pointed slightly downward, gray-red; foliage medium, medium green, semi-glossy; upright, bushy, medium growth; PP006570; [(Rise 'n' Shine X Sheri Anne) X Rainbow's End]; Saville, F. Harmon; Nor'East Min. Roses, 1991

Savaria, Pol, 1972; Mark; (Cavriglia)

'SAVARITA', Min, mp, 1989; (**Nostalgia**®); bud ovoid; flowers medium pink, reverse lighter, aging lighter, medium, borne u, dbl., 38 petals, cupped, no fragrance; fruit not observed; prickles thin, straight, medium, gray-purple to brown; foliage medium, medium green, semi-glossy; spreading, low growth; [Rita X (Rise 'n' Shine X Sheri Anne)]; Saville, F. Harmon; Nor'East Min. Roses, 1990

'SAVARS', Min, pb, 1991; (A.R.S. Centennial, **American Rose Centennial**®); bud ovoid; flowers creamy white, edged soft pink, very dbl., 50–55 petals, 1.5 in., exhibition form, borne singly or in sprays of 3-5, slight fragrance; foliage medium, dark green, semi-glossy; bushy, medium growth; PP8181; [High Spirits X Rainbow's End]; Saville, F. Harmon; Nor'East Min. Roses, 1992

'SAVASACH', Min, m, 1986; (**Sachet**®); flowers lavender, yellow stamens, urn-shaped to loose, mini-flora, dbl., 30 petals, intense, damask fragrance; medium, thin prickles; foliage small, dark, semi-glossy; medium upright, bushy growth; PP005967; [Unnamed yellow Miniature X Shocking Blue]; Saville, F. Harmon; Nor'East Min. Roses

'SAVASAM', Min, mr, 1986; (Sampson, **Samson**®); flowers brilliant scarlet red, urn-shaped, blooms in sprays of 20-60, dbl., 28 petals, slight fragrance; long, thin, prickles; foliage large, dark, semi-glossy; tall, upright growth; [Sheri Anne X (Yellow Jewel X Tamango)]; Saville, F. Harmon; Nor'East Min. Roses

'SAVASCHOOL', Min, dy, 1993; (**School Days**); dbl., 26–40 petals, 1.5 in., borne mostly singly, no fragrance; few prickles; foliage medium, medium green, semi-glossy; medium (14-18 in), bushy, compact growth; [Klima X Sonnenkind]; Saville, F. Harmon; Nor'East Min. Roses, 1994

'SAVASEAT', Min, pb, 1996; (**Seattle Scentsation**™); flowers yellow, apricot base and mauve-pink blooms, dbl., 15–25 petals, 1.5–2.75 in., borne singly and in small clusters, intense fragrance; no prickles; foliage small, dark green, semi-glossy; medium (26-30 in), upright, bushy, vigorous growth; Named in honor of ARS Spring Convention, Seattle, WA, ARS, 1996; [Lavender Jade X New Zealand]; Saville, F. Harmon; Nor'East Min. Roses, 1996

'SAVASHEL', Min, pb, 1988; (**Shelly Renee**™); flowers shrimp pink, reverse peach, aging light pink to white; foliage small, dark green, semi-glossy; PP006952; [Spice Drop sport]; Saville, F. Harmon; Nor'East Min. Roses, 1989

'SAVASILK', Min, ly, 1982; (**Cornsilk**™); flowers light pastel yellow, small, dbl., 40 petals, exhibition form, borne mostly singly, moderate fragrance; foliage medium, medium green, semi-glossy; vigorous, bushy growth; PP5164; AOE, 1983; [Rise 'n' Shine X Sheri Anne]; Saville, F. Harmon; Nor'East Min. Roses

'SAVASMILE', Min, dp, 1987; (**Adam's Smile**™, 'SAVAREND'); flowers deep pink, medium, dbl., 23–27 petals, exhibition form, borne usually singly and in sprays of 3-5, no fragrance; no fruit; prickles long, thin, pointed slightly downward, gray-red; foliage medium, medium green, semi-glossy; upright, bushy, medium growth; PP006570; [(Rise 'n' Shine X Sheri Anne) X Rainbow's End]; Saville, F. Harmon; Nor'East Min. Roses, 1991

'SAVASORC', Min, mr, 1994; (**Sorcerer**™); flowers bright, medium red, dbl., 15–25 petals, 1.5 in., borne mostly singly, floriferous, no fragrance; some prickles; foliage small, medium green, semi-glossy; medium (16-20 in), upright, bushy growth; [Ginger Snap X Rainbow's End]; Saville, F. Harmon; Nor'East Min. Roses, 1995

'SAVASPARK', Min, mr, 1995; (**Sparks**); flowers medium red, semi-dbl. (6–14 petals), small blooms borne mostly in small to large clusters; no fragrance; few prickles; foliage small, medium green, semi-glossy; medium (14–16 in.), upright, compact bush; [(Zorina X Baby Katie) X Red Minimo]; Saville, F. Harmon; Nor'East Miniature Roses

'SAVASPIR', Min, mr, 1983; (**High Spirits**™); flowers small, dbl., 35 petals, exhibition form, blooms in sprays, slight fragrance; foliage small, dark, semi-glossy; upright growth; PP005450; [Sheri Anne X Tamango]; Saville, F. Harmon; Nor'East Min. Roses

'SAVASUN', Min, dy, 1986; (**Sunny Day**™); flowers bright yellow, fading lighter, slight red blush on edges, dbl., 30 petals, exhibition form, borne in small clusters, slight, spicy fragrance; no fruit; long, thin prickles, angled slightly downwards; foliage small, dark green, semi-glossy; upright, bushy, low growth; PP6213; [Golden Slippers X Rise 'n' Shine]; Saville, F. Harmon; Nor'East Min. Roses, 1987

'SAVASWEET', Min, op, 1982; ('SAVSWET', **Spice Drop**™); flowers light salmon-pink, micro-mini, small, dbl., 35 petals, exhibition form, slight fragrance; foliage small, medium green, glossy; tiny growth; [(Sheri Anne X Glenfiddich) X (Unnamed moss seedling X (Sarabande X Little Chief))]; Saville, F. Harmon; Nor'East Min. Roses

'SAVATHREE', Min, mr, 1982; (**Acey Deucy**™); flowers small, dbl., 20 petals, exhibition form, moderate fragrance; foliage small, medium green, semi-glossy; bushy growth; [(Yellow Jewel X Tamango) X Sheri Anne]; Saville, F. Harmon; Nor'East Min. Roses

'SAVAVAL', Min, dp, 1980; (**Valerie Jeanne**); bud globular; flowers deep magenta pink, blooms borne 1-20 per cluster, dbl., 58 petals, exhibition form, slight fragrance; long, straight, thin prickles; foliage very glossy; vigorous, upright growth; AOE, 1983; [Sheri Anne X Tamango]; Saville, F. Harmon; Nor'East Min. Roses

'SAVAVEL', Min, mr, 1993; (**Velvet Touch**); flowers small, medium red, dbl., 15–25 petals, 1.5 in., borne mostly singly, slight fragrance; some prickles; foliage medium, medium green, semi-glossy; medium, bushy, compact growth; [Rainbow's End X Acey Deucy]; Saville, F. Harmon; Nor'East Min. Roses, 1994

'SAVAWED', Min, mp, 1985; (**Wedded Bliss**™); semi-dbl., blooms in clusters, slight fragrance; foliage small, medium green, glossy; groundcover; very spreading growth; [(Yellow Jewel X Tamango) X Nozomi]; Saville, F. Harmon; Nor'East Min. Roses

'SAVAWEEK', Min, or, 1984; (**Julie Ann**™); flowers small, brilliant vermillion orange, dbl., 20 petals, 1–1.5 in., exhibition form, borne mostly singly, moderate fragrance; prickles moderate; foliage small, medium green, semi-glossy; upright, bushy growth; PP5415; AOE, 1984; [Zorina X Poker Chip]; Saville, F. Harmon; Nor'East Min. Roses

'SAVAWHOOP', Min, rb, 1991; (**Whoopi**™); flowers red and white blend, ages as red edges suffused more, dbl., 28–32 petals, 1.5 in., exhibition form, borne mostly singly or in small clusters, slight, spicy fragrance; foliage medium, dark green, semi-glossy; bushy, medium growth; PP8176; [(Yellow Jewel X Tamango) X Party Girl]; Saville, F. Harmon; Nor'East Min. Roses, 1992

'SAVAWIN', Min, m, 1984; (**Winsome**™); flowers lilac-lavender with red tinge, mini-flora, dbl., exhibition form, borne singly and in sprays, no fragrance; foliage medium, dark, semi-glossy; upright, bushy growth; PP5691; AOE, 1985; [Party Girl X Shocking Blue]; Saville, F. Harmon; Nor'East Min. Roses, 1985

'SAVAWIRE', Min, dp, 1988; (**Live Wire**™); flowers deep cardinal red-pink, reverse darker, aging lighter, loose, semi-dbl., 27–30 petals, cupped, no fragrance; no fruit; prickles thin, straight, dark gray-purple; foliage small, medium green, semi-glossy; micro-mini; bushy, low, compact growth; [(Rise 'n' Shine X Sheri Anne) X Rainbow's End]; Saville, F. Harmon; Nor'East Min. Roses

'SAVAWIST', Min, m, 1994; (**Wistful**); flowers medium mauve, good form, dbl., 15–25 petals, 1.5 in., borne singly, repeats well, slight fragrance; some prickles; foliage small, dark green, semi-glossy; medium (16-22 in), upright, bushy growth; [Sachet X Rainbow's End]; Saville, F. Harmon; Nor'East Min. Roses, 1995

Save the Children *see* 'HARTRED'

'SAVEMBER', Min, or, 1994; (**Ember**); dbl., 15–25 petals, 1.5 in., borne mostly singly, no fragrance; few prickles; foliage small, medium green, semi-glossy; medium (16-22 in), upright, bushy, compact growth; [Copper Sunset X (Zorina X Baby Katie)]; Saville, F. Harmon; Nor'East Min. Roses, 1995

'SAVERNE', HT, rb, 1937; flowers nasturtium tinted brownish red, reverse tinted yellow, large, dbl.; foliage bright, bronze; Heizmann, E.; A. Meilland

'SAVINN', Min, w, 1997; (**Innocence**); flowers full to very full, medium, pure ivory white, very dbl., 26–41 petals, borne mostly singly, slight fragrance; few prickles; foliage medium, dark green, glossy; upright, spreading, bushy, medium (30 in.) growth; [SAVajinks X SAValite]; Saville, F. Harmon

Savkar, HT, mr, 1988; Patil, B.K.

'SAVMORE', Min, dr, 1984; (**Red Shadows**™); flowers small, dbl., exhibition form, no fragrance; foliage small, medium green, semi-glossy; bushy, slightly spreading growth; [Tamango X Sheri Anne]; Saville, F. Harmon; Nor'East Min. Roses

'SAVOIA', HT, m, 1937; bud pointed, rosy lilac; foliage glossy; vigorous growth; [Julien Potin X Sensation]; Aicardi, D.; Giacomasso

'SAVOR', Min, ob, 1982; (**Little Jackie**™); flowers light orange-red, yellow reverse, small, dbl., 20 petals, intense fragrance; foliage medium, medium green, semi-glossy; vigorous growth; AOE, 1984; [(Prominent X Sheri Anne)

X Glenfiddich]; Saville, F. Harmon; Nor'East Min. Roses

Savoy Hotel *see* 'HARVINTAGE'

'SAVPLENTI', Min, ob, 1981; (**Penny Candy**®); bud ovoid; flowers orange-yellow blend, micro-mini, blooms borne singly or seve, dbl., 30 petals, cupped, slight fragrance; long, thin prickles; foliage small; tiny, compact growth; [Rise 'n' Shine X Sheri Anne]; Saville, F. Harmon; Nor'East Min. Roses

'SAVROJET', Min, dp, 1986; flowers deep pink, moderately small, dbl., 15–25 petals, no fragrance; foliage small, medium green, semi-glossy; bushy growth; [(Tamango X Yellow Jewel) X Watercolor]; Saville, F. Harmon; SNC Meilland & Cie, 1985

'SAVSAY', Min, op, 1982; (**Heartland**®); bud short, pointed; flowers orange-red, blooms in clusters, dbl., 38 petals, exhibition form, slight fragrance; long, thin prickles; vigorous, upright growth; [Sheri Anne X Watercolor]; Saville, F. Harmon; Nor'East Min. Roses

'SAVSWAT', Min, dp, 1982; (**Windjammer**); flowers deep pink, reddened by sun, small, dbl., 35 petals, slight fragrance; foliage medium, medium green, semi-glossy; [Sheri Anne X Watercolor]; Saville, F. Harmon; Nor'East Min. Roses

'SAVSWET', Min, op, 1982; ('SAVASWEET', **Spice Drop**®); flowers light salmon-pink, micro-mini, small, dbl., 35 petals, exhibition form, slight fragrance; foliage small, medium green, glossy; tiny growth; [(Sheri Anne X Glenfiddich) X (Unnamed moss seedling X (Sarabande X Little Chief))]; Saville, F. Harmon; Nor'East Min. Roses

'SAVYK', Min, dy, 1999; (**Y2K**); flowers deep yellow with coating of coral, reverse medium yellow, dbl., 17–25 petals, 1.5–2 in., borne mostly singly, slight fragrance; few prickles; foliage medium, dark green, glossy; bushy, low (12-15 in.) growth; [Cal Poly X New Zealand]; Saville, F. Harmon; Nor'East Miniature Roses, 2000

Saya, Min, ob

Sayokyoku *see* 'KEIMIPIA'

'SAYONARA', HT, yb, 1959; bud long, pointed; flowers yellow blend tinted pink, dbl., 50 petals, 5 in., moderate fragrance; foliage leathery; vigorous, upright growth; [Sunnymount sport]; Grillo

'SAZANAMI', Min, lp, 1984; flowers soft pink, small blooms borne 2-5 per cluster, dbl., 60 petals, flat, moderate fragrance; prickles slanted downward; foliage dark, semi-glossy; bushy growth; GM, Japan, 1980; [Yorokobi seedling X Yellow Doll]; Suzuki, Seizo; Keisei Rose Nursery, 1982

Sázava, S, mr, 1964; flowers medium, semi-dbl., slight fragrance; Pajer, J.; (Sangerhausen)

'SCABRATA', S, m; flowers rich pinkish purple, large, single; foliage downy; [R. corymbifera (?) X R. gallica]

'SCABROSA', HRg, m, 1950; (R. rugosa scabrosa); flowers mauve-pink, stamens light sulfur, blooms in clusters of 5 or, single, 5 petals, 5 in., recurrent bloom, moderate, carnation fragrance; large, bright red fruit; foliage light, glossy, soft; very bushy (5 ft.) growth; Harkness

Scala *see* **'CLUBROSE SCALA'**

Scala *see* GAUesca

Scaldia, HT, 1987; Delforge; (Cavriglia)

Scamp® *see* 'SAVACAMP'

'SCANDALE', HT, pb, 1958; bud pink; flowers crimson shaded coppery, dbl., 25 petals, 4 in., moderate fragrance; foliage glossy, dark; long stems; vigorous, upright growth; [Peace X Opera]; Gaujard

'SCANDIA', Pol, op, 1951; flowers salmon-orange; [Margo Koster sport]; van de Water; van Nes

Scandia, HT, m, 1996; Cooper

Scandica, F, mr

'SCANIA', F, dr, 1965; flowers deep red, well-formed, 3.5 in.; foliage matt; [Cocorico X Seedling]; deRuiter

Scar P97, Misc OGR, lp, 1997; Scarman

'SCARAMOUCHE', F, mr, 1968; bud long, pointed; flowers coral-red, large, dbl., 30–35 petals, exhibition form, slight fragrance; foliage leathery; vigorous, compact, tall growth; [Ma Perkins X Duet]; Fankhauser

'SCARLANO', HSet, mr, 1938; (Faust); flowers cerise-red, open, borne in clusters, semi-dbl., sometimes recurrent bloom, slight fragrance; foliage leathery, dark; short, strong stems; bushy (2 1/2 ft.) growth; [(R. setigera X Papoose) X Paul's Scarlet Climber]; Horvath; Wayside Gardens Co.

'SCARLET ADVENTURER', HT, dr, 1957; flowers scarlet, medium, dbl.; foliage dark; Lowe

'SCARLET BEAUTY', HT, dr, 1934; flowers crimson-scarlet, very large, dbl., slight fragrance; vigorous growth; [Mme Butterfly X Premier Supreme]; Vestal

'SCARLET BEDDER', HT, ob, 1927; flowers rich orange-scarlet; dwarf growth; [Mme Edouard Herriot X Gen. MacArthur]; Henderson, W.H.

'SCARLET BETTY UPRICHARD', HT, dr, 1930; bud long, pointed, shaded black; flowers intense scarlet, semi-dbl., cupped, intense fragrance; foliage thick, light; vigorous growth; [Betty Uprichard seedling]; Allen

'SCARLET BUTTON', Pol, dr, 1932; flowers brilliant scarlet; [Locarno sport]; Dreer

'SCARLET CRAMPEL', Pol, dr; flowers scarlet; [Paul Crampel sport]

'SCARLET ELSE', F, dr; flowers scarlet, borne in trusses, semi-dbl., 10–12 petals, 3–4 in., slight fragrance; foliage leathery; very free growth; [Else Poulsen X Hybrid Tea seedling (red)]; Kordes, about 1925

'SCARLET EMPEROR', HT, mr, 1961; flowers scarlet-red, large, dbl., exhibition form; vigorous, upright growth; [Karl Herbst X Fandango]; LeGrice; Wayside Gardens Co.

Scarlet Fire *see* **'SCHARLACHGLUT'**

'SCARLET FLAME', HT, mr, 1934; flowers brilliant red, petals recurved and fringed, large, dbl., moderate fragrance; very vigorous growth; Burbank; Stark Bros.

'SCARLET GARNETTE', F, mr, 1971; flowers scarlet, small, dbl., 45 petals, globular, slight fragrance; foliage dark; vigorous growth; [Garnette sport]; Newberry

Scarlet Gem® *see* 'MEIDO'

'SCARLET GLORY', HT, ob, 1925; bud long, pointed; flowers orange-scarlet, dbl., exhibition form, slight fragrance; foliage leathery, rich green; Dickson, A.

'SCARLET GLOW', HT, dr, 1945; flowers brilliant velvety scarlet; [Briarcliff sport]; Sodano, A.; St. Leonards Farms

Scarlet Glow *see* **'SCHARLACHGLUT'**

Scarlet Hit *see* POUlmo

Scarlet Knight *see* 'MEIELEC'

'SCARLET KNIGHT, CLIMBING', Cl Gr, mr, 1972; (Samourai, Climbing); [Scarlet Knight sport]; Jack; A. Ross & Son

'SCARLET LADY', Min, mr, 1990; bud ovoid; flowers medium red, aging light pink, large, borne singly, dbl., 30 petals, exhibition form, no fragrance; foliage medium, dark green, semi-glossy; upright, medium growth; [Anita Charles X Chris Jolly]; Jolly, Nelson F., 1983; Rosehill Farm, 1991

'SCARLET LEADER', Pol, ob, 1927; flowers brilliant orange-scarlet, large, borne in clusters, dbl.; Wezelenburg

'SCARLET MARINER', F, mr, 1972; flowers bright scarlet-red, medium, dbl., exhibition form, moderate fragrance; foliage leathery; vigorous, bushy growth; [Unnamed seedling X Showboat]; Patterson; Patterson Roses

'SCARLET MARVEL', F, mr, 1958; flowers orange-scarlet, borne in clusters, dbl., 45–50 petals, 2.5 in., flat; foliage leathery; vigorous, compact growth; [Alain X Floribunda seedling]; deRuiter; C-P

Scarlet Meidiland® *see* 'MEIKROTAL'

Scarlet Meillandécor *see* 'MEIKROTAL'

Scarlet Meillandina *see* 'MEIcubasi'

Scarlet Mimi® *see* KEInoumi

Scarlet Moss® *see* 'MORcarlet'

Scarlet Patio *see* KORtingle

Scarlet Pavement, S, lp, 1991; bud pink; flowers light red, semi-dbl.; numerous, dark red fruit; spreading (2.5 ft) growth; Uhl, J.

Scarlet Pearl *see* 'MANscarlet'

Scarlet Pimpernel *see* 'MEIdo'

'SCARLET QUEEN', F, dr, 1939; bud long, pointed; flowers pure scarlet, open, large, borne in clusters, dbl., slight fragrance; foliage glossy, leathery, bronze; long stems; vigorous, bushy growth; [Dance of Joy X Crimson Glory]; Kordes; Morse

Scarlet Queen Elizabeth® *see* 'DICel'

'SCARLET RIBBON', Cl Min, dr, 1961; bud ovoid; flowers red, sometimes almost maroon, dbl., 50 petals, 1.25 in., exhibition form, slight fragrance; vigorous (3 ft) growth; [((Soeur Therese X Wilhelm) X (Seedling X Red Ripples)) X Zee]; Moore, Ralph S.; Sequoia Nursery

Scarlet Rosamini *see* RUIrupo

'SCARLET ROYAL', HT, mr, 1963; flowers scarlet, base yellow, well formed; vigorous growth; [Karl Herbst X Independence]; Park; Tantau Roses

'SCARLET RUFFLES', Min, or, 1990; bud pointed; flowers orange-red with yellow eye, aging to light red, loose, large, semi-dbl., 8 petals, intense, spicy fragrance; oblong, gold and orange fruit; prickles straight, slightly hooked downwards, tan; foliage medium, dark green, glossy; spreading, medium growth; [Poker Chip X Zinger]; Gruenbauer, Richard, 1984; Richard Gruenbauer

'SCARLET SENSATION', LCl, dr, 1954; (Everblooming Pillar No.73); bud high pointed, crimson; flowers rose-madder, borne in clusters, dbl., 35 petals, 3.5–4.5 in., free bloom, moderate fragrance; growth like a hybrid tea, followed by 4-5 foot canes; [Unnamed seedling X Queen o' the Lakes]; Brownell, H.C.

Scarlet Showers, LCl, mr

Scarlet Spreader® *see* 'WILscarlet'

Scarlet Star *see* 'WILstar'

Scarlet Sunblaze® *see* 'MEIcubasi'

'SCARLET SUNSET', F, or, 1970; flowers small, borne on trusses, semi-dbl., 12 petals; foliage dark, leathery; moderate bushy growth; [Orange Sensation X Seedling]; deRuiter; Geo. deRuiter

Scarlet Sweet Brier *see* **'LA BELLE DISTINGUÉE'**

'SCARLET TRIUMPH', F, rb, 1951; bud ovoid; flowers deep scarlet, base yellow, small, borne in clusters, semi-dbl., cupped, slight fragrance; foliage glossy, light green; vigorous, bushy growth; [Orange Triumph sport]; Poulter

'SCARLET WAVES', F, mr, 1961; flowers bright scarlet, ruffled petals; tall growth; [(Florence Mary Morse X Border Queen) X Mrs Inge Poulsen]; Bennett, H.; Pedigree Nursery

'SCARLET WONDER', F, or, 1958; flowers bright orange-scarlet, borne in clusters, semi-dbl., 3 in., flat, slight fragrance; foliage dark, glossy; vigorous growth; [Signal Red X Fashion]; deRuiter; Blaby Rose Gardens

'SCARLETINA', Min, or, 1985; flowers spiraled, blooms borne usually singly, dbl., moderate fragrance; foliage small, medium green, semi-glossy; upright, bushy growth; [Futura X Poker Chip]; Hardgrove, Donald L.; Rose World Originals

'SCARLETT O'HARA', Pol, rb, 1947; bud ovoid; flowers brilliant red, overcast orange, open, large, borne in clusters, dbl., slight fragrance; foliage leathery; vigorous, bushy growth; RULED EXTINCT 2/88; [Gloria Mundi sport]; Klyn

Scarlett O'Hara® *see* 'AROresas'

Scarletta®, Min, mr, 1972; deRuiter

Scarman's Crimson China, HCh, mr, 1995; Scarman

Scentasia *see* 'SUNscent'

'SCENTED AIR', F, op, 1965; flowers salmon-pink, well-formed, blooms in clusters, dbl., 5 in., intense fragrance; foliage very large; vigorous growth; GM, Belfast, 1967 GM, The Hague, 1965; [Spartan seedling X Queen Elizabeth]; Dickson, Patrick; A. Dickson

Scented Bouquet, lp; [My Choice X Great Venture]; Dawson, George

'SCENTED BOWL', HT, dr, 1965; bud ovoid; flowers bright red, medium, dbl., intense fragrance; foliage glossy; open, upright growth; [Gen. MacArthur X ?]; Pal, Dr. B.P.; Indian Agric. Research Inst.

Scented Dawn *see* 'MEItosier'

'SCENTED STAR', HT, lp, 1973; flowers coral-pink, classic form, dbl., 30–35 petals, 5–5.5 in., intense fragrance; foliage dark; moderate growth; [Fragrant Cloud X Spek's Yellow]; Lowe

Scentimental® *see* 'WEKplapep'

Scent-Sation *see* FRYromeo

Scentsational® *see* 'SAVamor'

Scepter'd Isle *see* 'AUSland'

'SCEPTRE', HT, ob, 1923; flowers bright flame, base shaded orange, reverse dull yellow; low growth; McGredy

Schackenborg, HT, dr, 1997; Poulsen

'SCHAFFNERS ERFOLG', F, mr; dbl., 30 petals, 2 in., borne in trusses; vigorous, bushy growth; [Red Favorite X Fanal]; Tantau

SCHanbiran, Min, rb, 1984; (**Magic Sunblaze**); Schwartz

'SCHARLACHGLUT', S, dr, 1952; (Scarlet Fire, Scarlet Glow); flowers scarlet-crimson, single, 5 in.; foliage dull green; vigorous (4-5 ft), dense, spreading growth; [Poinsettia X Alika]; Kordes

'SCHARNKEANA', S, m; (R. X scharnkeana); flowers rose-purple, 1-5 together; 3 ft. growth; [R. californica X R. nitida]

'SCHÉHÉRAZADE', HT, mr, 1942; bud large, oval; flowers fiery red, dbl., cupped, slight fragrance; foliage dark, glossy; vigorous, bushy growth; GM, Rome, 1940; Mallerin, C.; A. Meilland

Scherzo® *see* 'MEIpuma'

'SCHIEHALLION', HT, mr, 1982; dbl., 50 petals, 4.5 in., intense fragrance; foliage medium, medium green, glossy; upright growth; [Red Planet X Bonnie Anne]; MacLeod, Major C.A.

'SCHILLER', HMult, mp, 1913; flowers clear pink, medium, borne in clusters; [Trier X Lady Mary Fitzwilliam]; Lambert, P.

Schleswig *see* **'MAID OF HONOUR'**

Schleswig 87® *see* KORtara

'SCHLESWIG-HOLSTEIN', HT, rb, 1921; flowers reddish yellow; [Mme Edouard Herriot sport]; Engelbrecht

Schloss Balthasar *see* KORpalmor

Schloss Dryburg, Cl HT, dy, 1969; flowers luminous golden yellow, large, dbl.; GPG Bad Langensalza; (Sangerhausen)

'SCHLOSS FRIEDENSTEIN', HMult, m, 1915; flowers reddish-violet, small, dbl., moderate fragrance; Schmidt, I. C.; (Sangerhausen)

Schloss Heidegg *see* 'MEIpoque'

Schloss Mannheim® *see* 'KORschloss'

'SCHLOSS MORITZBURG', F, dr, 1967; flowers medium, semi-dbl., cupped, slight fragrance; foliage dark, leathery; vigorous growth; [Donald Prior X ?]; Haenchen, E.; Teschendorff

'SCHLOSS SEUSSLITZ', HSpn, ly; flowers very large, semi-dbl., very early bloom; moderate growth; Dechan, 1933

Schlösser Brilliant, Climbing *see* **'DETROITER, CLIMBING'**

Schlosser's Brilliant *see* **'DETROITER'**

Schlossgarten, S, mp, 1989; flowers large, dbl.; GPG Bad Langensalza; (Sangerhausen)

'SCHMETTERLING', HFt, yb, 1905; flowers yellow and red, medium, semi-dbl., moderate fragrance; Müller, Dr. F.; (Sangerhausen)

'SCHMID'S IDEAL', (strain of R. canina), lp; thorny; used as an understock; R. Schmidt

'SCHMID'S REKORD', S, 1930; Schmid, R.; (Cavriglia)

'SCHNEEBALL', HWich, w, 1905; flowers medium, semi-dbl., moderate fragrance; Weigand, C.; (Sangerhausen)

Schneeberg, S, w

Schnee-Eule, HRg, w, 1989; flowers large, semi-dbl., intense fragrance; Uhl, J.; (Sangerhausen)

Schneekonigin®, *see* 'TANINGO'

Schneekonigin® *see* 'TANOTAX'

'SCHNEEKOPF', HP, w, 1903; flowers snow-white or light pink, regular form, recurrent bloom; vigorous growth; [Mignonette X Souv. de Mme Sablayrolles]; Lambert, P.

Schneekoppe, HRg, w, 1984; (**Snow Pavement**); Baum

Schneekusschen®, Min, w, 1993; Kordes

'SCHNEELICHT', HRg, w, 1894; flowers pure white, large blooms in clusters, single; makes an impenetrable, prickly hedge; very vigorous, climbing growth; very hardy.; Geschwind, R.

'SCHNEEPRINZESSIN', HT, w, 1946; bud long; flowers large, intense fragrance; vigorous growth; Meilland, F.; Pfitzer

'SCHNEESCHIRM', F, w, 1946; flowers white, center tinted rose-yellow, large blooms in clusters, single, slight fragrance; foliage dark; [Johanna Tantau X (Karen Poulsen X Stammler)]; Tantau

Schneesturm® *see* TANmurse

Schneewalzer *see* 'TANSCHNEEWA'

Schneewalzer *see* TANrazlaw

Schneeweisschen® *see* TANensch

Schneewittchen *see* 'KORBIN'

'SCHNEEWITTCHEN', Pol, ly, 1901; flowers yellow fading to white; [Aglaia X (Paquerette X Souv. de Mme Levet)]; Lambert, P.

Schneewolke, S, w, 1996; Noack, Werner

'SCHNEEZWERG', HRg, w, 1912; (Snow Dwarf, Snowdwarf); flowers snow-white, stamens golden yellow, blooms in clusters of 3-1, semi-dbl., flat, recurrent bloom; abundant, small, red fruit; prickles spiny; foliage glossy, rugose; vigorous (3-4 ft.), growth; (14); [Possibly R. rugosa X Polyantha hybrid]; Lambert, P.

'SCHOBITET', Min, my, 1987; (Royal Meillandina, **Royal Sunblaze®**); flowers lemon yellow, medium, full, quilled petals, very dbl., rosette, borne in small clusters, slight fragrance; foliage medium, medium green, semi-glossy; mini-flora, bushy (15 in) growth; PP005690; [Unnamed seedling X Unnamed seedling]; Schwartz, Ernest W.; SNC Meilland & Cie, 1984

'SCHOENER'S MUSK', HMsk, w; flowers milk-white, medium enormous trusses; vigorous, pillar growth; Schoener

'SCHOENER'S NUTKANA', S, mp, 1930; flowers clear rose-pink, blooms in arching canes, single, 4 in., non-recurrent, moderate fragrance; few prickles; vigorous, shrub (4-6 ft.) growth; [R. nutkana X Paul Neyron]; Schoener; C-P

Scholle's Golden Moss, M, dy, 1985; flowers amber-yellow, medium, dbl., moderate fragrance; Scholle, E.; (Sangerhausen)

'SCHÖN INGEBORG', HP, lp, 1921; flowers large, dbl.; Kiese; (Sangerhausen)

Schone Berlinerin® *see* TANrised

Schone Dortmunderin® *see* NOAdort

Schöne Münchnerin® *see* 'KORLEEN'

'SCHÖNE VON HOLSTEIN', Pol, mp, 1919; flowers pure hermosa pink, dbl.; [Orléans Rose X ?]; Tantau

'SCHÖNE VON KAISERSLAUTERN', F, op, 1957; bud long, pointed; flowers orange tinted salmon-red, dbl., 30 petals, 4 in., exhibition form, free, intermittent bloom, intense fragrance; foliage dark, leathery; strong stems; very vigorous, upright, bushy growth; [R.M.S. Queen Mary X Obergärtner Wiebicke]; Kordes

'SCHÖNE VON MARQUARDT', HWich, rb, 1928; flowers bright dark red, variegated with white; vigorous, climbing growth; [Sodenia sport]; Clauberg

'SCHONERTS MEISTERKLASSE', HT, or, 1952; flowers salmon and coral-red, well formed, very large, dbl., moderate fragrance; vigorous growth; Leenders, M.

School Days *see* 'SAVASCHOOL'

'SCHOOLGIRL', LCl, ab, 1964; flowers orange-apricot, well-formed, 4 in., recurrent bloom; [Coral Dawn X Belle Blonde]; McGredy, Sam IV; McGredy

Schubert® *see* 'LENMOR'

Schultheis' American Beauty, HP, dp

'SCHUPRAK', HT, pb, 1989; (**Susan Elizabeth**); bud ovoid; flowers cream tipped with very deep purple-pink, reverse cream to pa, dbl., 40 petals, exhibition form, slight, fruity fragrance; fruit not observed; prickles slightly recurved, sparse, red; foliage medium, medium green, semi-glossy; upright, medium growth; [Pristine X Akebono]; Schlueter, Barry; Schlueter Rose Culture

Schuss® *see* MEIliaxi

'SCHWABENLAND', HRg, mp, 1928; flowers amaranth-pink, open, large, dbl., profuse, repeated bloom, moderate fragrance; foliage large, rich green, leathery; vigorous (3 ft.) growth; [R. rugosa seedling X Elizabeth Cullen]; Berger, V.; Pfitzer

Schwabenmadel, F, dr

'SCHWÄBISCHE HEIMAT', HT, dr, 1934; flowers deep amaranth-red, dbl.; [Jonkheer J.L. Mock sport]; Pfitzer

Schwanensee *see* **'SWAN LAKE'**

Schwarzaldmadel, F, 1977; Hetzel, K.; (Cavriglia)

Schwarze Madonna® *see* KORschwama

Schwarzer Samt, HKor, dr, 1969; flowers dark velvety red, medium, semi-dbl.; VEG; (Sangerhausen)

Schwarzwaldmadel®, F, ob, 1977; Hetzel

'SCHWEIZER GOLD', HT, ly, 1975; (Swiss Gold); bud ovoid; dbl., exhibition form, moderate fragrance; foliage large, light; vigorous growth; GM, Baden-Baden, 1972; [Peer Gynt X King's Ransom]; Kordes; Horstmann

Schweizer Gold, HT, lp; Urban, J.; (Czech Rosa Club)

Schweizer Grüss *see* 'TANSCHWEIGRU'

Schweizer Grüss, Climbing

Schweizer Woche®, HT, w, 1989; Huber

'SCHWERIN', HMsk, dp, 1937; bud long, pointed; flowers light crimson, open, large clusters, cupped, profuse, intermittent bloom, moderate, musk fragrance; foliage large, leathery, glossy, bronze; long, strong stems; very vigorous, bushy growth; [Eva X D.T. Poulsen]; Kordes

'SCINTILLATION', S, lp, 1967; flowers pink, blooms in large clusters, semi-dbl., intense fragrance; foliage dark, matt; vigorous growth; [R. macrantha X Vanity]; Austin, David; Sunningdale Nursery

Scipion Cochet, T, dp

Scipion Cochet, HP, dr

Scipion Cochet, T, pb, 1893

'SCOOP JACKSON', Gr, dr, 1983; (Canterbury, Fidelity); flowers large, dbl., 20 petals, slight fragrance; foliage glossy; upright growth; [Kalahari X John Waterer]; McGredy, Sam IV; Roses by Fred Edmunds

'SCORCHER', Cl HT, dr, 1922; flowers brilliant scarlet-crimson, semi-dbl., 4 in., non-recurrent, slight fragrance; foliage large, wrinkled; vigorous, climbing or pillar (to 10 ft.) growth; [Mme Abel Chatenay X Seedling]; Clark, A.; Hackett

'SCOTCH BLEND', HT, pb, 1975; bud long, pointed; dbl., 35 petals, 5–6 in., exhibition form, slight fragrance; foliage dark, leathery; upright growth; [Queen Elizabeth X Peace]; J&B Roses; Eastern Roses

Scotch Heather, HSpn, yb, 1996; Sutherland

Scotch Rose *see* R. spinosissima

Scotch Yellow, HT, my, 1991; Mercer

Scotland Yard, HT, ab

Scotland's Trust *see* 'COCLANDS'

Scott William, F, lp; [Oz Gold X Gold Bunny]; Hannemann, F., 1992; The Rose Paradise; (Weatherly, L.)

Scottish Soldier, S, yb, 1999; Williams, J. Benjamin

Scottish Special *see* 'COCDAPPLE'

Scottish Tartan, F, yb

'SCOTT'S COLUMBIA', HT, dp, 1928; flowers clear bright pink; [Columbia sport]; Scott, R.

Scout *see* **'BOY SCOUT'**

'SCRABO', F, op, 1968; flowers light salmon-pink, large, dbl., exhibition form, moderate fragrance; free growth; [Celebration X Elizabeth of Glamis]; Dickson, A.

'SCRIVLEO', HT, ly, 1997; (**Albert Weedall**); flowers double, large, dbl., 15–25 petals, borne in large clusters, intense fragrance; some prickles; foliage medium, medium green, semi-glossy; upright, tall (5ft.) growth; [Seedling X Seedling]; Scrivens, Len

'SCRIVLUV', Min, dy, 1992; (**Baby Love**™); bud small, pointed; flowers buttercup yellow, single, 5 petals, 1.5 in., borne mostly singly, slight, licorice fragrance; some prickles; foliage small, medium green, semi-glossy; low (400 cms), compact growth; [Sweet Magic X Seedling]; Scrivens, Len

'SCRIVO', F, lp, 1997; (**Pretty Lady**); flowers semi double, medium, semi-dbl., 8–15 petals; many prickles; foliage medium, dark green, semi-glossy; medium (4ft.) growth; [(seedling X (R. davidii elongata X seedling)) X ((Troika X Alpine Sunset) X Freedom)]; Scrivens, Len

Scudbuster *see* 'CLESCRUB'

'SEA FOAM'®, S, w, 1964; flowers white to cream, blooms in clusters, dbl., recurrent bloom, slight fragrance; foliage small, glossy, leathery; vigorous, climbing trailer, semi-prostrate growth; David Fuerstenberg Prize, ARS, 1968 GM, Rome, 1963; [((White Dawn X Pinocchio) X (White Dawn X Pinocchio)) X (WhiteDawn X Pinocchio)]; Schwartz, Ernest W.; C-P

Sea Foam Rosa, S, 1971; Zandri, R.; (Cavriglia)

'SEA MIST', HT, w, 1960; bud long, pointed; flowers cream, center golden, large, semi-dbl., exhibition form, slight fragrance; foliage leathery; vigorous growth; [Helen Traubel X Golden Harvest]; Armbrust; Langbecker

'SEA NYMPH', Min, pb, 1986; flowers petals blush pink, edged coral, large blooms borne singly, dbl., 48 petals, exhibition form, no fragrance; medium, curved prickles; foliage medium, medium green, semi-glossy; upright, bushy growth; [Unnamed seedling X Unnamed seedling]; McDaniel, Earl; McDaniel's Min. Roses

'SEA OF FIRE', F, or, 1954; (Feuermeer); flowers orange-scarlet, blooms in small clusters, semi-dbl., 3 in., slight fragrance; foliage dark, leathery; vigorous, upright, bushy growth; [(Baby Chateau X Else Poulsen) X Independence]; Kordes

'SEA OF TRANQUILITY', HT, ab, 1991; bud classical; flowers delicate light pink in bud, opening to apricot pink blush, l, very dbl., 40–45 petals, 4.75 in., exhibition form, slight, sweet fragrance; foliage light green, small, disease-resistant; upright (up to 5 ft), non-spreading growth; [Sylvia sport]; Keene's Rose Nursery

'SEA PEARL', F, pb, 1964; (Flower Girl); bud long, pointed; flowers soft pink, reverse flushed peach and yellow, well-formed, bl, dbl., 24 petals, 4.5 in.; foliage dark; upright, bushy growth; [Kordes' Perfecta X Montezuma]; Dickson, Patrick; A. Dickson

'SEA SPRAY', HMsk, w, 1923; flowers stone-white, flushed pink, borne in clusters, non-recurrent; height 3-5 ft; RULED EXTINCT 12/82; Pemberton

'SEABASKET', Min, rb, 1995; (**Flower Basket**); flowers hand-painted carmine red to pink stripes, fades to white with dark red bud, prominent yellow stamens, dbl. (15–25 tightly quilled petals), large blooms borne mostly single; fragrant, numerous prickles; foliage small, medium green, semi-glossy; bushy, very low growth; [Rose Gilardi X Ain't Misbehavin']; McCann, Sean; Justice Miniature Roses, 1996

'SEABET', Min, pb, 1993; (**Elizabeth Munn**); flowers mainly pink with lighter reverse, dbl., 15–25 petals, flat, borne in small clusters; some prickles; foliage medium, medium green, semi-glossy; medium, upright growth; McCann, Sean; McCann, 1993

'SEABIRD', HT, my, 1913; flowers primrose-yellow; vigorous growth; Dickson, H.

'SEABLA', Min, dr, 1990; (**Ain't Misbehavin'**); single, 5 petals, slight fragrance; foliage small, green, semi-glossy; bushy growth; [Oonagh X (Pot Black X Black Jade)]; McCann, Sean, 1991

Seabreeze *see* 'LEMSEA'

'SEACATCH', Min, rb, 1995; (**Dreamcatcher**); flowers medium to dark red striped with white, blooms borne singly, dbl., 35–40 petals, 2–2.5 in., moderate, damask fragrance; foliage medium, medium green, glossy; tall, spreading growth; RR; [Lady in Red X [(Rose Gilardi X Wit's End) X (Remember Me X Stars 'n' Stripes)]]; McCann, Sean; Justice Miniature Roses, 1995

'SEACLOAK', Min, dr, 1999; (**Velvet Cloak**); flowers burgundy red, very prominent yellow stamens, dbl., 17–25 petals, 2 in., borne mostly singly, slight fragrance; prickles moderate; foliage medium, medium green, semi-glossy; bushy, medium (20-24 in) growth; [Lady in Red X Ain't Misbehavin]; McCann, Sean; Justice Miniature Roses, 1998

'SEACOAT', Min, yb, 1996; (**Dreamcoat**); flowers medium to dark yellow with scarlet accent, fades to reddish, dbl., 15–25 petals, 2.5 in., slight fragrance; foliage large, dark green, semi-glossy; upright, tall (4-6 ft) growth; RR; [Lady in Red X Joseph's Coat]; McCann, Sean; Justice Miniature Roses, 1997

'SEADARBY', Min, ob, 1996; (**Darby O'Gill**); flowers orange with creamy white striping, blooms borne mostly singly, dbl., 18 petals, 1.5 in., slight fragrance; some prickles; foliage medium, dark green, glossy; upright, medium (24-26 in) growth; PPAF; [Tattooed Lady X Lady in Red]; McCann, Sean; Justice Miniature Roses, 1996

'SEADARK', Min, dr, 1996; (**Dark Mirage**); flowers velvety dark red with very dark shading at tip, reverse dark, dbl., 26–40 petals, 2 in., exhibition form, intense fragrance; foliage medium, dark green, glossy, highly serrated; long stems; medium (18-20 in) growth; RR; [Lady in Red X Fountain]; McCann, Sean; Justice Miniature Roses, 1997

'SEADEEP', Min, rb, 1993; (**Tattooed Lady**); flowers red, yellow reverse, yellow stripes and veins, blooms borne, dbl., 15–25 petals, 1.5–2.75 in., slight fragrance; many prickles; foliage medium, dark green, bronze, glossy; medium (12-14 in), upright growth; [Siobhan X (Picasso X Near You)]; McCann, Sean; Justice Miniature Roses, 1994

'SEADOT', Min, ob, 1988; (**Crazy Dottie**); flowers orange-red with star shaped copper center, small, single, 5 petals, slight fragrance; foliage small, medium green, semi-glossy; bushy growth; [Rise 'n' Shine X (Sheri Anne X Picasso)]; McCann, Sean

'SEADOW', HT, mr, 1992; (**Margaret McDowell**); dbl., 26–40 petals, 1.5–2.75 in., slight fragrance; some prickles; foliage medium, medium green, semi-glossy; medium (36"), upright growth; [[Ruby Wedding X (Seedling X

Oonagh)] X (Seedling X Oonagh)]; McCann, Sean, 1993

'SEAEUS', F, pb, 1999; (**Eustacia**); flowers handpainted with patches of pink and red, yellow center, dbl., 15–25 petals, 3 in., borne in small clusters, slight fragrance; prickles numerous; foliage medium, medium green, glossy; upright, medium (30 in) growth; [(Copper Pot X Maxi) X Picasso]; McCann, Sean

'SEAFAIR', HT, ab, 1959; bud long, pointed; flowers deep apricot, large, dbl., 45 petals, exhibition form, slight fragrance; foliage glossy; vigorous, upright, symmetrical growth; RULED EXTINCT 4/86; [Charlotte Armstrong X Signora]; Von Abrams; Peterson & Dering

Seafarer *see* 'HARTILION'

Seafirst *see* 'PEAFIRST'

'SEAFUN', Min, pb, 1997; (**Looks Like Fun**); flowers very full, medium, very dbl., 41 petals, borne mostly singly, moderate fragrance; foliage medium, medium green, semi-glossy; spreading, bushy, medium (16in.)growth; [Lady in Red X (Rose Gilardi X Seedling)]; McCann, Sean

'SEAGER WHEELER', HSpn, lp, 1947; dbl., non-recurrent; height 6 ft.; hardy.; [R. spinosissima altaica X ?]; Wheeler; P.H. Wright

'SEAGOLD', Min, my, 1987; (**Gold Country**); flowers small blooms borne singly, dbl., 20 petals, exhibition form, intense fragrance; foliage small, light green, semi-glossy; bushy growth; [Rise 'n' Shine X (Rise 'n' Shine X Casino)]; McCann, Sean

'SEAGOOD', Min, mp, 1990; (**Lady Be Good**); dbl., 15–25 petals, slight fragrance; foliage small, medium green, semi-glossy; bushy growth; [Kiss 'n' Tell X (Irish Mist X Matangi)]; McCann, Sean, 1991

'SEAGRU', Min, w, 1983; (**Blushing Groom**); flowers near white, small, dbl., 35 petals, slight fragrance; foliage small, dark, semi-glossy; upright growth; [Rise 'n' Shine X Karl Herbst]; McCann, Sean

'SEAGULL', HWich, w, 1907; flowers pure white, stamens golden, blooms in large clusters, single; very vigorous growth; Pritchard

'SEAHEART', Cl Min, rb, 1990; (**Irish Heartbreaker**); flowers small blooms, dbl., 26–40 petals, slight fragrance; foliage medium, medium green, semi-glossy; upright growth; [Rise 'n' Shine X (Oonagh X Siobhan)]; McCann, Sean, 1991

'SEAKISS', Min, ab, 1985; (**Kiss 'n' Tell**); flowers small, dbl., 35 petals, flat, slight fragrance; foliage small, medium green, semi-glossy; bushy growth; [Rise 'n' Shine X (Sally Mac X New Penny)]; McCann, Sean, 1989

'SEALADY', Min, rb, 1988; (**Lady in Red**); flowers red with touch of white at base of petals, small, dbl., 15–25 petals, slight fragrance; foliage small, medium green, semi-glossy; bushy growth; [Rise 'n' Shine X Siobhan]; McCann, Sean, 1990

'SEALARK', HT, rb, 1998; (**Big Jim Larkin**); flowers red with silver reverse, silvers flecks on top surface, very dbl., 26–40 petals, 4.5–5 in., borne in large clusters, moderate fragrance; foliage large, dark green, semi-glossy; free blooming, spreading, medium (4 ft.) growth; [Lady in Red X Old Master]; McCann, Sean; Hughes Roses, 1999

'SEALIFE', Min, rb, 1996; (**High Life**); flowers medium red with white at throat, reverse white with red, fad, dbl., 26–40 petals, 2.5 in., exhibition form, slight fragrance; foliage large, medium green, dull; upright, tall (30 in.) growth; RR; [Lady in Red X Oriana]; McCann, Sean; Justice Miniature Roses, 1997

Sealily *see* 'LEMLIL'

'SEALING WAX', HMoy, mp; flowers large; abundant, bright red fruit; height 8 ft; Royal Hort. Soc.

'SEALOVE', Min, rb, 1989; (**Lovers Only**); flowers cherry red, with straw-yellow bicolor, reverse fades to cream, dbl., 15–25 petals, slight fragrance; prickles pubescent on peduncle; foliage medium, medium green, glossy; upright, compact growth; [Rise 'n' Shine X Siobhan]; McCann, Sean

'SEAMAR', HT, w, 1985; (**Martin Martin**); flowers cream, petals tipped red, large, exhibition form; foliage large, dark, semi-glossy; upright growth; [Jimmy Greaves X Irish Gold]; McCann, Sean

'SEAMOM', Min, m, 1990; (**Stolen Moment**); flowers mauve blend, small, semi-dbl., 6–14 petals, slight fragrance; foliage small, medium green, semi-glossy; bushy growth; [Kiss 'n' Tell X (Aunty Dora X Charles de Gaulle)]; McCann, Sean

'SEAMOOD', Min, yb, 1988; (**In the Mood**); flowers yellow streaked pink, small, dbl., 20 petals, slight fragrance; foliage small, medium green, semi-glossy; bushy growth; [Rise 'n' Shine X Unnamed seedling]; McCann, Sean

'SEAMOU', Min, pb, 1999; (**Mouse**); flowers pink yellow at base, changes to red-pink, dbl., 15–25 petals, 1 in., borne in small clusters, slight fragrance; few prickles; foliage small, medium green, semi-glossy; bushy, medium (18 in.) growth; [Portland Dawn X Siobhan]; McCann, Sean

'SEAMURP', HT, mp, 1982; (**Murphy's Law**); flowers medium pink, lighter reverse, large, dbl., 35 petals, slight fragrance; foliage medium, dark, matt; bushy growth; [Fragrant Cloud X (Prima Ballerina X Gavotte)]; McCann, Sean

Sean, S, lp; [Mme Segond Weber sport]; Sutherland, P; Golden Vale Nursery, 2000; (Weatherly, L.)

'SEANEAR', Min, w, 1990; (**Near You**); flowers moderately blooms, dbl., 15–25 petals, slight fragrance; foliage small, medium green, semi-glossy; spreading growth; [Rise 'n' Shine X (Elina X Royal Gold)]; McCann, Sean

'SEAODD', Min, rb, 1992; (**Alice Faye**); flowers full circle of red with distinct yellow eye from the bottom view, dbl., 22–24 petals, 2.5 in., no fragrance; foliage large, medium green, semi-glossy; upright, tall growth; [Seedling X Seedling]; McCann, Sean; Justice Miniature Roses, 1992

'SEAOONA', HT, mr, 1990; (**Oonagh**); flowers large blooms, dbl., 26–40 petals, slight fragrance; foliage medium, medium green, semi-glossy; very spreading growth; [Matangi X Gavotte]; McCann, Sean

Search for Life™ *see* 'WILSFOL'

'SEARICH', Min, rb, 1987; (**Rich and Rare**); flowers scarlet, reverse white veined red, small blooms borne singly, dbl., 35 petals, exhibition form, slight fragrance; foliage small, dark, semi-glossy; bushy growth; [(Rise 'n' Shine X Siobhan) X Beauty Secret]; McCann, Sean

SEAris, Min, lp; (**Kiss and Tell**); McCann, Sean

'SEARODNEY', Min, w, 1991; (**Admirable**); bud ovoid, sharply pointed sepals; flowers soft pink with darker accent, reverse ivory with pink accent, dbl., 2.5 in., exhibition form, intense, fruity fragrance; foliage large, medium green, semi-glossy; long, straight stems; upright, tall growth; [Seedling X Admiral Rodney]; McCann, Sean; Justice Miniature Roses, 1992

'SEARONAN', Min, op, 1996; (**Street Wise**); flowers orange and pink, reverse yellow at throat, blooms borne most, very dbl., 60 petals, intense fragrance; numerous prickles; foliage medium, medium green, semi-glossy; upright, medium growth; PPAF; [Rise 'n' Shine X Siobhan]; McCann, Sean; Justice Miniature Roses, 1996

Seashell *see* 'KORSHEL'

'SEASHELL', HWich, mp, 1916; flowers large, borne in large clusters, semi-dbl.; foliage glossy; Dawson

'SEASIO', F, rb, 1984; (**Siobhan**); flowers red, reverse yellow, medium, dbl., 20 petals, exhibition form, slight fragrance; foliage medium, dark bronze, semi-glossy; upright growth; [Maxi X Copper Pot]; McCann, Sean

SEAsol, Min, pb; (**Solitude**); McCann, Sean

'SEASOON', Min, ly, 1992; (**Someday Soon**); flowers light yellow framed in creamy white outer petals, reverse cream, dbl., 52 petals, 1.5 in., exhibition form, slight, spicy fragrance; foliage medium, light green, matt, highly serrated; upright, medium growth; [Unnamed Miniature seedling X Antique Silk]; McCann, Sean; Justice Miniature Roses, 1993

Seaspray *see* 'MACNEWING'

'SEASWAN', Min, w, 1988; (Swan Song, **Swansong**); flowers small, dbl., 15–25 petals, intense fragrance; foliage small, medium green, semi-glossy; bushy growth; [(Rise 'n' Shine X Party Girl) X Margaret Merril]; McCann, Sean

'SEATEL', Min, w, 1992; (**Margaret Telfer**); dbl., 15–25 petals, 1.5–2.75 in., borne in small clusters, slight fragrance; some prickles; foliage small, medium green, semi-glossy; low (14"), upright growth; [Kiss the Bride X Margaret Merril]; McCann, Sean

'SEATIP', Min, pb, 1988; (**Portland Dawn**); flowers veined pink and bronze, small, dbl., 20 petals, slight fragrance; foliage small, medium green, semi-glossy; bushy growth; [Rise 'n' Shine X (Copper Pot X Maxi)]; McCann, Sean, 1989

'SEATOFF', Min, ob, 1992; (**Toffee**); flowers bright orange, yellow eye, dark orange stamens, reverse oran, semi-dbl., 18 petals, 1 in., slight, fruity fragrance; foliage medium, medium green, semi-glossy, disease-resistant; bushy, medium growth; [Bloomsday X Unnamed Min seedling]; McCann, Sean; Justice Miniature Roses, 1993

Seattle Scentsation™ *see* 'SAVASEAT'

'SEATWINKLE', Min, ob, 1999; (**Jazz Dancer**); flowers orange and yellow, reverse orange, dbl., 26–40 petals, 1.75 in., borne mostly singly, slight fragrance; prickles moderate; foliage medium, dark green, glossy; compact, medium (14-16 in) growth; McCann, Sean; Justice Miniature Roses, 1999

'SEAVIEW', Min, yb, 1990; (**Ladies' View**); flowers moderately small, dbl., 15–25 petals, slight fragrance; foliage small, medium green, semi-glossy; bushy growth; [You 'n' Me X Amber Queen]; McCann, Sean

'SEAVIN', Min, mr, 1999; (**True Vintage**); flowers medium red to fuchsia, reverse silver/light red, medium, semi-dbl., 12–16 petals, borne mostly singly, slight fragrance; many prickles; foliage medium, medium green, semi-glossy; compact, medium (16-18 in) growth; [Rose Gilardi X Stolen Moment]; McCann, Sean; Justice Miniature Roses, 1999

'SEAWALK', Cl Min, mr, 1999; (**S.W.A.L.K.**); flowers barn red, dbl., 26–40 petals, 2.25 in., borne mostly singly, slight fragrance; prickles moderate; foliage medium, dark green, glossy; climbing (6-7 ft) growth; trellis or fence for support; McCann, Sean; Justice Miniature Roses, 1998

'SEAWHI', Min, w, 1987; (**Kiss the Bride**); flowers small blooms borne singly and in clusters of 4-6, dbl., 20 petals, moderate, spicy fragrance; foliage medium, medium green, semi-glossy; bushy growth; [Rise 'n' Shine X White Bouquet]; McCann, Sean, 1990

'SEAWISH', Min, rb, 1996; (**Wishful Thinking**); flowers medium red, large light yellow eye, heavy mass of yellow sta, dbl., 15–25 highly reflexed petals, 1.5 in., intense fragrance; foliage medium, dark green to light orange, glossy; upright, medium (16-20 in) growth; RR; [Seedling X Wit's End]; McCann, Sean; Justice Miniature Roses, 1997

'SEAWIT', Min, rb, 1988; (**Wit's End**); flowers red with yellow reverse, small, borne in sprays of 3-5, dbl., 20 petals, slight fragrance; foliage small, medium green, semi-glossy; bushy growth; [Rise 'n' Shine X Siobhan]; McCann, Sean, 1989

'SEAYOU', Min, w, 1985; (**You 'n' Me**); flowers white, light apricot center, small, dbl., exhibition form, slight fragrance; foliage small, medium green, semi-glossy; bushy growth; [Avandel X Party Girl]; McCann, Sean, 1987

Sebago *see* **'TYLER'**™

Sebastian Kneipp *see* KORpastato

Sebastian Schultheis, Gr, dp

'SECOND CHANCE', HT, pb, 1988; bud ovoid; flowers coral orange, tips orange-red, reverse deep pink, aging medi, dbl., 35 petals, exhibition form, moderate, damask fragrance; foliage medium, medium green to maroon, semi-glossy, a bit r; bushy, slightly spreading growth; [Carefree Beauty X Sonia]; Stoddard, Louis, 1991

Secret™ *see* 'HILAROMA'

Secret Garden Musk Climber, LCl, w

'SECRET LOVE', HT, dr, 1973; bud ovoid; flowers deep red, large, dbl., exhibition form, moderate fragrance; foliage leathery; moderate, upright, bushy growth; [Unnamed seedling X Unnamed seedling]; Armstrong, D.L.; Armstrong Nursery

Secret Obsession *see* 'RENSECOBSES'

Secret Recipe *see* 'MORWINST'

'SECRÉTAIRE BELPAIRE', HT, op, 1934; flowers brilliant salmon-pink, large, dbl.; foliage bright bronze; vigorous growth; [Angèle Pernet X Mme Edouard Herriot]; Lens

'SECRETAIRE J. NICOLAS', HP, m, 1883; flowers purple; Schwartz

'SECRETARIS ZWART', HT, dp, 1918; flowers bright rose, reverse silvery rose; [Gen. MacArthur X Lyon Rose]; Van Rossem

Sedgebrook *see* 'MURSE'

Seduction *see* 'MEIBEAUSAI'

'SEDUCTION', Pol, mp, 1927; flowers peach-blossom-pink, large, borne in clusters of 50-60, dbl.; few thorns; dwarf growth; Turbat

Seduction *see* MEIvrita

'SEEFELD', F, rb, 1958; bud oval; flowers red tinted lighter, open, medium, borne in large clusters, dbl.; foliage glossy; very vigorous, bushy growth; [Fashion X Orange Triumph]; Delforge

'SEFTOPOLIS', Gr, dp, 1977; flowers deep pink, large, dbl., 36 petals; foliage glossy, leathery; vigorous growth; [Queen Elizabeth X Unnamed seedling]; Staikov, Prof. Dr. V.; Kalaydjiev and Chorbadjiiski

Segovia, HT, 1964; Combe, M.; (Cavriglia)

'SÉGUIER', HGal, m, 1853; flowers purple/violet with white stripes, medium, dbl.; Robert; (Sangerhausen)

Sehnsucht®, F, dr

Seika *see* **'OLYMPIC TORCH'**

Sei-Ka *see* **'OLYMPIC TORCH'**

Seiko, HT, dy

'SEION', HT, my, 1987; flowers large, dbl., 35–40 petals, exhibition form, moderate fragrance; medium, pale, pale orange fruit; few prickles; foliage light green; upright, medium growth; [Grandpa Dickson X Sunblest]; Yokota, Kiyoshi, 1988

'SEISHO', HT, pb, 1999; flowers soft pink, deep pink petal edge, dbl., 30–35 petals, 5.5 in., exhibition form, moderate fragrance; foliage dark green; 5 ft growth; [Michele Meilland X Maria Teresa Bordas]; Ohkawara, Kiyoshi, 1990; Komaba Rose Nursery, 1992

'SEKEL', S, yb, 1984; flowers light yellow, bright red blend blooms in clusters of 21, semi-dbl., 11 petals, 3.5 in., slight fragrance; foliage dark, leathery; vigorous, upright growth; [Lichterloh X Zitronenfalter]; Lundstad, Arne; Agricultural University of Norway

'SEKI-YOH', HT, or, 1975; flowers large, dbl., 52 petals, exhibition form, slight fragrance; foliage dark, leathery, semi-glossy; upright, compact growth; [Miss France X Christian Dior]; Suzuki, Seizo; Keisei Rose Nursery

'SELANDIA', LCl, mp, 1913; dbl.; vigorous growth; [Mme Norbert Levavasseur X Dorothy Perkins]; Poulsen, D.T.

'SELAZTECA', HT, mr, 1990; (**Red Azteca**); bud pointed, tapering, slender; flowers bright red, no fading, large, borne singly, dbl., 30–35 petals, cupped; prickles reddish, with yellow tip; foliage large, dark green, glossy; upright, tall growth; [Unnamed seedling X Unnamed seedling]; Select Roses, B.V.; DeVor Nurseries, Inc.

Selena ™, HT, lp; deVor

Selfridges *see* 'KORPRIWA'

Selina *see* 'REYSEL'

'SELSTAR', HT, my, 1989; (**Christina**); bud pointed; flowers bright lemon-yellow, does not fade, to, large, borne singly, dbl., 43 petals, cupped, no fragrance; prickles declining, light red; foliage large, dark green, glossy; upright, tall growth; [Eliora X Unnamed seedling]; Select Roses, B.V.; DeVor Nurseries, Inc., 1991

'SELVETTA'®, HT, pb, 1982; flowers light pink, petals edged salmon, large, dbl., 40 petals, no fragrance; foliage medium, dark, matt; upright growth; Cazzaniga-Como; Rose Barni-Pistoia

'SELWYN BIRD', HT, or, 1969; flowers salmon-cerise, dbl., 35 petals, exhibition form, moderate fragrance; foliage dark, glossy; vigorous growth; [Fragrant Cloud X Stella]; Cocker

'SELWYN TOOGOOD', Min, mp, 1983; flowers small, dbl., 33 petals, exhibition form, slight fragrance; light moss on stems; foliage small, light green; bushy, upright growth; [Heidi X ?]; Eagle, Barry & Dawn; Southern Cross Nursery

'SEMI', (R. laxa form), w, 1913; flowers small, blooms all summer; bright red fruit; tall (8 ft.) growth; very hardy.; (grown from seed collected in Siberia by N.E. Hansen)

Semillante, HT, 1976; Combe, M.; (Cavriglia)

'SEMIRAMIS', HT, yb, 1957; bud pointed; flowers rose, center buff or amber, dbl., 35–45 petals, 5 in., exhibition form, intense, raspberry fragrance; foliage dark, glossy; vigorous growth; [Capistrano X (Peace X Crimson Glory)]; Motose

'SÉNATEUR AMIC', LCl, mr, 1924; bud long, pointed; flowers brilliant carmine, nearly, single; very vigorous growth; [R. gigantea X Gen. MacArthur]; Nabonnand, P.

Senateur Lafollette, S, ob

'SÉNATEUR MASCURAUD', HT, ly, 1909; flowers light yellow, center darker, dbl., moderate fragrance; Pernet-Ducher

'SÉNATEUR POTIÉ', HT, ob, 1937; flowers orange-yellow, large, semi-dbl., cupped; foliage glossy, bronze; vigorous growth; [Mme Butterfly X Carito MacMahon]; Dot, Pedro

'SÉNATEUR VAISSE', HP, rb, 1859; flowers red, shaded darker, large, dbl., 32 petals, occasionally recurrent bloom, intense fragrance; upright growth; [? X Général Jacqueminot]; Guillot Père

'SENATOR', HT, mr, 1926; bud long, pointed; flowers brilliant scarlet, dbl., intense fragrance; [Red Columbia X Premier]; Florex Gardens

Senator Burda® *see* 'MEIVESTAL'

'SENATOR JOE T. ROBINSON', HT, dr, 1938; bud long, pointed; flowers dark crimson, semi-dbl., cupped, intense fragrance; foliage leathery; vigorous growth; [Harvard X David O. Dodd]; Vestal

'SENECA QUEEN', HT, pb, 1965; flowers apricot-pink, reverse darker, dbl., 50 petals, 6 in., exhibition form, moderate fragrance; foliage leathery; vigorous growth; [(Serenade seedling X Fashion) X Golden Masterpiece]; Boerner; J&P

'SÉNÉGAL', Cl HT, dr, 1944; flowers very dark red; very vigorous growth; Mallerin, C.; A. Meilland

'SENFF', (strain of R. canina), lp; almost thornless; sometimes used as understock; Senff

Sengodea, HT, pb

'SENHORA DA GRAÇA', HT, rb; flowers red with carmine reflections; da Silva, Moreira

'SENIOR', HT, rb, 1932; bud pointed; flowers scarlet-crimson, open, dbl., moderate fragrance; foliage thick; long stems; very vigorous growth; [Richmond X Général Jacqueminot]; Spanbauer; Hill Floral Products Co.

'SENIOR PROM', HT, dp, 1964; bud long, pointed; flowers deep pink, dbl., 38 petals, 4.5 in., exhibition form; foliage dark, glossy; vigorous, upright growth; [Pink Princess X Queen Elizabeth]; Brownell, H.C.; Brownell

'SEÑOR PHILIPPE', LCl, m; flowers lilac-pink, center paler, dbl.; vigorous growth

'SEÑORA DE BORNAS', HT, mr, 1955; flowers vermilion-red, medium, very dbl., cupped, intense fragrance; foliage glossy; vigorous, upright growth; [J.M. Lopez Pico X Concerto]; Camprubi, C.

'SEÑORA DE CARULLA', F, mr, 1961; flowers cerise-red; GM, Madrid, 1961; Torre Blanca

'SEÑORA GARI', HT, yb, 1935; bud long, pointed; flowers deep orange-yellow, very large, dbl., exhibition form, moderate fragrance; sprawling growth; [Mari Dot X Constance]; Dot, Pedro; C-P

'SEÑORA LEON DE AUJURIA', HT, ob, 1935; flowers orange; foliage glossy; vigorous growth; La Florida

Señorita *see* 'JACDOR'

'SEÑORITA CARMEN SERT', HT, yb, 1917; flowers indian yellow, shaded pale pink, edged bright carmine, dbl.; [Marquise de Sinéty seedling]; Pernet-Ducher

'SEÑORITA DE ALVAREZ', HT, mp, 1931; bud long, pointed; flowers glowing salmon, very large, single, cupped, intense fragrance; foliage leathery, dark; vigorous growth; GM, NRS, 1930; Cant, B. R.

Sensass *see* 'DELMOUN'

Sensass Delbard® *see* 'DELMOUN'

'SENSATION', HT, mr, 1922; bud long, pointed; flowers scarlet-crimson, open, dbl., 36 petals, 5 in., moderate fragrance; foliage dark; free, branching growth; [Hoosier Beauty X Premier]; Hill, Joseph H., Co.

Sensation, HT, lp

Sensucht, F, 1978; Noack, Werner; (Cavriglia)

'SENTA SCHMIDT', Pol, ob, 1930; flowers coppery orange, semi-dbl.; foliage small, soft, light; dwarf growth; [Suzanne Turbat sport]; Schmidt, R.

Senteur Royale *see* 'TANSCHAUBUD'

Sentimental *see* 'POULTAL'

'SENTINEL', HT, pb, 1934; flowers velvety cerise, reverse silvery cerise, large, dbl., cupped, intense fragrance; foliage glossy; vigorous growth; Clark, A.; Wyant

'SENTRY', HT, dr, 1948; bud long, pointed; flowers clear crimson, dbl., 25 petals, 4–5 in., intense fragrance; dwarf, compact growth; Fletcher; Tucker

Sentyna® *see* 'INTERTYN'

Seppenrade, S, ob, 1970; flowers salmon-orange, large, dbl., moderate fragrance; Scholle, E.; (Sangerhausen)

Seppenrader Elfe, HKor, ly, 1975; flowers medium, dbl., slight fragrance; Scholle, E.; (Sangerhausen)

'SEPTEMBER DAYS', Min, dy, 1976; bud pointed; flowers reflexed, micro-mini, small, dbl., 40 petals, 1.5 in., exhibition form, moderate fragrance; foliage glossy; upright, compact growth; [Rise 'n' Shine X Yellow Jewel]; Saville, F. Harmon; Days Inn

September Eighteenth *see* 'WILSEPT'

'SEPTEMBER MORN', HT, pb, 1913; flowers flesh-pink, center deeper, large, moderate fragrance; vigorous growth; [Mme Pierre Euler sport]; Dietrich & Turner

'SEPTEMBER SONG', Gr, ab, 1981; bud ovoid; dbl., 28 petals, cupped, borne singly and in clusters of 5-8, moderate, fruity fragrance; thin, awl-like prickles; foliage dark, tinted with copper, semi-glossy, leathery; erect, bushy growth; [(Vera Dalton X Prairie Princess) X (Apricot Nectar X Prairie Princess)];

Buck, Dr. Griffith J.; Iowa State University

'SEPTEMBER WEDDING', HT, mp, 1964; bud ovoid; flowers deep pink, reverse darker, large, dbl., exhibition form, moderate fragrance; foliage dark, glossy; vigorous, tall, compact growth; [Montezuma sport]; Schloen, J.; Ellesmere Nursery

'SEQUOIA', HT, ob, 1939; flowers ripe pumpkin-flesh shaded apricot, dbl., globular, slight fragrance; foliage leathery, bronze; vigorous growth; Verschuren-Pechtold; Dreer

Sequoia Gold™ *see* 'MORSEGOLD'

Sequoia Jewel *see* 'MORSEWEL'

Sequoia Ruby *see* 'MORDORA'

'SEQUOIA, CLIMBING', Cl HT, ob, 1940; Swim, H.C.; Armstrong Nursery

'SERAFINA LONGA', HT, dp, 1933; flowers old-rose, heavily veined, well formed, large, moderate fragrance; vigorous growth; [Mme Butterfly X Mme Abel Chatenay]; La Florida

Seraphim *see* **'SERAPHINE'**

'SERAPHINE', HSet, pb, 1840; (Seraphim); flowers soft pink, center darker, very dbl.; Prince Nursery

Serena, HT, 1955; Lens; (Cavriglia)

'SERENADE', HT, ob, 1949; bud ovoid; flowers coral-orange, dbl., 28 petals, 4–4.5 in., cupped, slight fragrance; foliage glossy, leathery; vigorous, upright growth; [Sonata X R.M.S. Queen Mary]; Boerner; J&P

Serenata *see* BARser

'SERENDIPITY', S, ob, 1978; bud ovoid, pointed; flowers orange to buttercup-yellow, dbl., 23 petals, 4–5 in., cupped, moderate fragrance; foliage dark, glossy, leathery; vigorous, upright, spreading, bushy growth; [(Western Sun X Carefree Beauty) X (Apricot Nectar X Prairie Princess)]; Buck, Dr. Griffith J.; Iowa State University

'SERENE', HT, w, 1940; bud long, pointed, light buff; flowers shining silvery white, open, very large, dbl., 30–40 petals; foliage sparse, soft; vigorous, upright growth; Mallerin, C.; C-P

Serene Bouquet *see* 'LAVSER'

'SERENELLA', HT, 1954; Cazzaniga, F. G.; (Cavriglia)

'SERENISSIMA', LCl, m, 1980; bud pointed; flowers light lilac, dbl., 38 petals, recurrent bloom, intense fragrance; straight, reddish-green prickles; foliage large, light green, matt; upright growth; Takatori, Yoshiho; Rose Barni-Pistoia

Serenité *see* GAUvera

'SÉRÉNITÉ', HT, my, 1946; bud pointed; flowers chrome-yellow, large; foliage reddish; stiff stems; very vigorous growth; RULED EXTINCT 7/86; Gaujard

'SÉRÉNITÉ'®, HT, my, 1986; flowers coppery-yellow, borne singly, dbl., exhibition form, intense fragrance; rounded fruit; large prickles; foliage large, medium green; tall growth; [John Armstrong X Tanagra]; Gaujard, Jean

Serenity, F, lp; [Playgirl sport]; Snetsinger

'SERGE BASSET', HT, dr, 1918; flowers brilliant garnet-red, dbl.; Pernet-Ducher

Sergeant Pepper *see* 'KORTENSES'

'SERGENT ULMANN', HT, dr, 1930; bud long, pointed; flowers deep garnet, lightened with scarlet, open, very large, semi-dbl., slight fragrance; foliage leathery, bronze; vigorous growth; [Grenoble X Mme Van de Voorde]; Mallerin, C.

Serpent Rose®, S, pb, 1994; Lens

Serpent Vert® *see* 'LENWIGA'

'SERRATIPETALA', Ch, pb, 1912; (R. chinensis serratipetala, Rose Oeilletde Saint Arquey); Vilfray

Set of Gold, HT, yb, 1994; Delbard

'SETINA', Cl Ch, lp, 1879; (Hermosa, Climbing); [Hermosa, Climbing sport]; P. Henderson

'SETINA', B, lp; Henderson, 1859

'SETSUKO', HT, w, 1999; flowers cream white, blended red in petal edge, dbl., 37 petals, 5 in., slight fragrance; foliage dark green, half leathery; 4.5 ft growth; [Marchenkonigin X Hakuchoh]; Sasaki, Keiji, 1996

'SEVEN SEAS', F, m, 1973; flowers lilac, dbl., 26 petals, 4 in., moderate fragrance; foliage large, glossy; [Lilac Charm X Sterling Silver]; Harkness

'SEVEN SISTERS', HMult, pb, 1817; (R. cathayensis platyphylla, R. multiflora platyphylla, R. platyphylla, R. thoryi, Seven Sisters Rose); flowers pale rose to crimson, much larger than r; (14); 1817

Seven Sisters Rose *see* **'SEVEN SISTERS'**

'SEVENTEEN', F, pb, 1959; bud ovoid; flowers pink-coral, borne in pyramidal clusters, dbl., 20–25 petals, 3.5 in., cupped, intense fragrance; foliage dark, leathery; vigorous, upright growth; [Pinocchio seedling X Fashion seedling]; Boerner; J&P

'SEVENTH HEAVEN', HT, dr, 1966; flowers large, dbl., exhibition form; foliage glossy; upright, bushy growth; [Unnamed seedling X Chrysler Imperial]; Armstrong, D.L. & Swim; Armstrong Nursery

'SÉVERINE', HT, rb, 1918; flowers coral-red, passing to shrimp-red, semi-dbl.; foliage bronze; vigorous growth; Pernet-Ducher

'SEVERN VALE', HT, ob, 1967; flowers salmon, 4.5 in., exhibition form; free growth; [Beauté sport]; Sanday, John

'SEVILLIANA', S, pb, 1976; bud ovoid, pointed; flowers light claret-rose, yellow from base, semi-dbl., 15 petals, 3.5–4 in., cupped, moderate, spicy fragrance; foliage tinted copper, leathery; upright, bushy growth; [(Vera Dalton X Dornroschen) X ((World's Fair X Floradora) X Applejack)]; Buck, Dr. Griffith J.; Iowa State University

Sexy Rexy® *see* 'MACREXY'

'SFINGE', HT, rb, 1954; flowers deep red edged rose; long stems; vigorous, upright growth; [Julien Potin X Sensation]; Aicardi, D.; Giacomasso

'SHABNAM', F, w, 1976; bud ovoid; flowers white, center pinkish, open, small, very dbl., 85 petals, 2 in.; foliage soft; upright, open growth; [Baby Sylvia X ?]; IARI

'SHADES OF AUTUMN', HT, rb, 1943; flowers red to pink with some yellow, center yellow, large, dbl., intense fragrance; foliage glossy, leathery; vigorous, compact, upright, bushy growth; [Golden Glow X Condesa de Sástago]; Brownell, H.C.

'SHADES OF PINK'™, F, pb, 1985; flowers pink with white eye, imbricated, in clusters of 10-50, dbl., 33 petals, 3.5 in., no fragrance; curved, hooked prickles; foliage medium green, glossy; bushy, upright growth; [Robin Hood X Pascali]; Mander, George

Shadow, HT, dr, 1966; Dawson

'SHADOW DANCE', F, rb, 1969; bud ovoid; flowers silver-pink edged red, small, dbl., 30–35 petals, intense fragrance; foliage glossy, leathery; vigorous, low, compact growth; [Pink Parfait X Crimson Glory]; Fankhauser

Shadow Dancer™ *see* 'MORSTRORT'

Shady Charmer *see* 'MICSHADY'

'SHADY FLAME', Min, or, 1981; dbl., 35 petals, exhibition form, borne usually singly, slight fragrance; straight prickles; foliage small, light green; compact, bushy growth; [Prominent X Zinger]; Jolly, Betty J.; Rosehill Farm

'SHADY LADY', Min, dy, 1981; flowers yellow-orange, blooms borne mostly singly, dbl., 35 petals, exhibition form, moderate fragrance; no prickles; foliage tiny, green; upright, bushy growth; [(Prominent X Zinger) X Puppy Love]; Jolly, Betty J.; Rosehill Farm

'SHADY LANE', HT, pb, 1990; bud ovoid; flowers deep pink to white base, aging deeper pink, large, borne usu, dbl., 35 petals, exhibition form, intense, damask fragrance; prickles medium, pointed downwards, light green; foliage medium, dark green, semi-glossy; upright, medium growth; [Thriller X

Just Lucky]; Bridges, Dennis A.; Bridges Roses

Shafter *see* **'DR HUEY'**

'SHAIDA', HT, or, 1976; bud long, pointed; flowers red-orange-salmon, dbl., 28–32 petals, 3.5–4.5 in., exhibition form, slight fragrance; foliage dark, leathery; vigorous, upright growth; [(Fandango X Fillette) X Coloranja]; Lens

'SHAILER'S PROVENCE', C, lp; (Gracilis); flowers lilac-pink, base white, inner petals rolled and wrinkled, bl, dbl., cupped, moderate fragrance; foliage small; vigorous (4-5 ft) growth; Shailer, Prior to 1799

'SHAILER'S WHITE MOSS', M, w; (Blanche Mousseuse, Clifton Moss, R. centifolia albo-muscosa, R. muscosa alba, White Bath); flowers medium, dbl.; (28); Shailer, Cult. 1810

'SHAKESPEARE FESTIVAL', Min, my, 1979; bud pointed; flowers clear yellow, dbl., 45 petals, 1.5 in., exhibition form, moderate, tea fragrance; foliage green, matt; bushy, compact growth; [Golden Angel X Golden Angel]; Moore, Ralph S.; Sequoia Nursery

Shakespeare Garden Eglantine, HEg, w

Shaleen Surtie-Richards, HT, mp, 1996

'SHALIMAR', HWich, yb, 1914; flowers creamy blush, picotee edge of bright rose-pink, borne in imm; vigorous, climbing growth; [Minnehaha sport]; Burrell

'SHALOM'®, F, or, 1973; (Flammenmeer); bud globular; dbl., 23 petals, 3.5–4 in.; foliage dark; vigorous, upright growth; [Korona seedling X Korona seedling]; Poulsen, Niels D.; Poulsen

'SHANDON', HT, dr, 1899; flowers large, dbl., intense fragrance; Dickson, A.; (Sangerhausen)

Shanghai Autumn, HT, yb

'SHANGRI-LA', HT, mp, 1945; bud long, pointed; flowers silvery pink, open, dbl., 35 petals, 3.5–4.5 in., intense fragrance; foliage leathery; long stems; very vigorous, upright, bushy growth; [Mrs J.D. Eisele X Pres. Herbert Hoover]; Howard, F.H.; H&S

Shankar Jaikishan, HT, mr, 1998; Shastri

Shannie *see* 'GELSHAN'

Shannon® *see* 'MACNON'

Shantaraj, HT, dr, 1998; K&S

Shanthi Pal, HT, op, 1989; Pal, Dr. B.P.

Shantung® *see* 'DELCHINE'

'SHARADA', Gr, lp, 1983; (Saroda); [Queen Elizabeth sport]; Gupta, Dr. M.N., Datta, Dr. S.K. & Nath, P.; National Botanical Research Institute

'SHARI', HT, pb, 1992; flowers growth, and foliage same as swarthmore color is lighter, dbl., 26–40 petals, slight fragrance; foliage medium, medium green, matt; upright (185 cms) growth; [Sweetie Pie sport]; Perry, Astor; Hortico Roses, 1993

Sharifa Asma™ *see* 'AUSREEF'

'SHARON', HT, lp, 1962; bud long, pointed; flowers soft pink edged lighter, base light yellow, reverse darker pink, dbl., 35–40 petals, 5.5 in., exhibition form; foliage leathery, dull; vigorous, upright growth; [Golden Rapture X Happiness]; Spandikow

Sharon Anne *see* 'BOSLORVET'

'SHARON FRANCES', HT, pb, 1999; flowers light pink blending to peach center, dbl., 26–40 petals, 5.5–6 in., borne mostly singly, slight fragrance; some prickles; foliage large, dark green, glossy; upright, tall (3 ft) growth; [Tom Foster X Ravenswood Village]; Poole, Lionel

'SHARON LORRAINE', F, rb, 1978; bud pointed; flowers ivory yellow, aging red, reverse red, blooms borne 30-36 per cluster, dbl., 40 petals; long, red prickles; foliage dark, semi-glossy; medium, upright (2-2 1/2 ft) growth; [Unnamed seedling X Unnamed seedling]; Bossom, W.E.

'SHARON LOUISE', HT, w, 1968; bud ovoid; flowers near white, center pale pink, medium, dbl., slight fragrance; foliage dark, leathery; vigorous, tall, bushy growth; [Queen Elizabeth X Virgo]; Parkes, Mrs M.H.

'SHARON MUXLOW', Min, op, 1980; bud globular; flowers bright coppery orange, dbl., 25 petals, 1 in., flat; foliage leathery; upright growth; [Anytime X Persian Princess]; Dobbs; Min. Plant Kingdom

Sharon's Delight *see* 'MORSHARON'

Sharon's Love, S, lp, 1998; Lens

'SHASTA', F, w, 1962; bud pointed; dbl., 23 petals, 4 in., blooms in clusters, moderate fragrance; foliage leathery; vigorous, bushy growth; [Paul's Lemon Pillar X Fashion]; Schwartz, Ernest W.; Wyant

Shatadhara, F, mp, 1991; Chiplunkar

'SHAYELA', F, yb, 1999; (**Swonderful**); flowers yellow-orange, reverse lighter, similar to little darling, dbl., 26–40 petals, 1.5 in., exhibition form, borne in small clusters, moderate fragrance; prickles moderate; foliage small, dark green, glossy; upright, medium (4 ft); patio growth; [Pink Petticoat X Redgold]; Schakelford, Grace, 1998

'SHE', F, or, 1962; flowers salmon-opal, base lemon, borne in clusters, semi-dbl., 19 petals, 2.5 in., slight fragrance; moderate, bushy growth; [(Independence X Fashion) X Brownie]; Dickson, Patrick; A. Dickson

'SHEELAGH BAIRD', Pol, pb, 1934; flowers shell-pink, overlaid rich rose pink, base yellow, large bloom, dbl.; vigorous growth; Cant, F.

Sheer Bliss *see* 'JACTRO'

Sheer Delight *see* 'HARWAZZLE'

Sheer Elegance™ *see* 'TWOBE'

Sheer Elegance, Climbing *see* 'DEVBILL'

Sheer Grace, HT, pb, 1995; Patil, B.K.

'SHEER STRIPES', S, pb, 1999; flowers dark pink and white stripes, single, 5–11 petals, 3 in., borne in small clusters, slight fragrance; few prickles; foliage medium, medium green, semi-glossy; upright, tall (6 ft) growth; [Hurdy Gurdy X Heritage]; Lowe, Malcolm

'SHEERWATER', HT, ly, 1977; flowers beige to cream, veined carmine, full, dbl., 35 petals, 5 in., moderate fragrance; foliage dark, matt green; free growth; [My Choice X Premier Bal]; Plumpton, E.

Sheffield Pride *see* 'WEBCYRENE'

'SHEILA', Pol, op, 1930; orange-salmon flowers; Walsh, J.; Beckwith

'SHEILA BELLAIR', HT, op, 1937; flowers salmon-pink, large, semi-dbl.; bushy growth; [Miss Mocatta X ?]; Clark, A.; NRS Victoria

'SHEILA FLEMING', Gr, mp, 1995; flowers prominent stamens, blooms borne 15-20 per cluster, single, 5 petals, 1.5–2.75 in., moderate fragrance; foliage medium, medium green, matt; upright (120 cms), bushy growth; 1st prize, Unnamed Seedling, Toronto Show, 1992; [Marchenland X Montezuma]; Fleming, Joyce L.; Hortico Roses, 1994

Sheila MacQueen *see* HARwotnext

'SHEILA MITCHELL', HT, yb, 1998; flowers yellow shading to pink, reverse lemon yellow, very full, very dbl., 41 petals, 5 in., exhibition form, borne mostly singly, moderate fragrance; prickles moderate; foliage large, medium green, semi-glossy; compact, medium (3 ft.) growth; [Trumpeter X Grandpa Dickson]; Mitchell, Harold V.; Nicholas Maple, 2000

'SHEILA WILSON', HT, mr, 1910; flowers light scarlet; vigorous growth; Hall; A. Dickson

Sheila's Perfume *see* 'HARSHERRY'

'SHELBY WALLACE', Cl Pol, op, 1929; flowers light salmon-pink, small, semi-dbl., slight fragrance; [Cécile Brunner, Climbing seedling]; Moore, Ralph S.

Sheldon's Honor, *see* 'RESHONOR'

Shell Beach *see* 'TROBEACH'

'SHELL QUEEN', Gr, lp, 1961; flowers shell pink, fading white; [Queen Elizabeth sport]; Allen, L.C.

'SHELLBROOK ROSE', (form of R. acicularis), dr; long, bottle-shaped fruit; (found near Shellbrook, Sask., Canada)

Shelley Higgins *see* 'REYSHELLEY'

Shell-Pink Radiance *see* **'MRS CHARLES BELL'**

'SHELLY', HT, pb, 1987; flowers pale pink with cyclamen-pink shading, reverse fleck & stripe, dbl., 20–25 petals, slight, sweet fragrance; prickles slightly hooked, beige-cream; foliage dark green, glossy, disease resistant; medium growth; [Francine sport]; Melville Nurseries Pty., Ltd., 1988

Shelly Renee® *see* 'SAVASHEL'

'SHENANDOAH', LCl, dr, 1935; bud long, pointed; flowers crimson, large, semi-dbl., exhibition form, intense fragrance; foliage large, glossy; vigorous, climbing (10 ft) growth; [Étoile de Hollande X Schoener's Nutkana]; Nicolas; C-P

'SHEPHERDESS', F, yb, 1967; flowers yellow flushed salmon, blooms in clusters, dbl., 3.5–4 in., slight fragrance; foliage dark, glossy, leathery; vigorous growth; [Allgold X Peace]; Mattock

'SHEPHERD'S DELIGHT', F, rb, 1956; flowers flame and yellow, blooms in trusses, semi-dbl., 15 petals, 3 in., slight fragrance; foliage dark; vigorous growth; GM, NRS, 1958; [Masquerade seedling X Joanna Hill]; Dickson, A.

Sheri Anne *see* 'MORSHERI'

'SHERICE', HT, mp, 1999; (**Jason**); flowers deep pink, reverse paler, medium, dbl., 15–25 petals, borne mostly singly, slight fragrance; few prickles; foliage medium, dark green, dull; upright (4 ft) growth; TGC, RNRS, 1989; [Silver Jubilee X (Redgold X Golden Slippers)]; Sheridan, John

Sheridan Pink, F, dp, 1998

'SHERIGREY', HT, yb, 1998; (**Dorothy Lloyd**); flowers grey with brown center, reverse grey, dbl., 15–25 petals, 4 in., borne in small clusters, slight fragrance; prickles few, medium, straight; foliage medium, medium green, semi-glossy; bushy, medium (2.5 ft) growth; [Cream Peach X Paradise]; Sheridan, John

'SHERIJILL', F, yb, 1993; (**Jill Carter**); flowers yellow edged red, blooms borne in small clusters, dbl., 15–25 petals, 1.5 in., moderate fragrance; few prickles; foliage small, dark green, semi-glossy; medium, upright growth; [Sheila's Perfume X Darling Flame]; Sheridan, John; Sheridan Nursery

'SHERIJOE', HT, dp, 1999; (**Smokey Joe's Cafe**); dbl., 15–25 petals, 3.5–4 in., borne in small clusters, moderate fragrance; few prickles; foliage medium, medium green, semi-glossy; upright, medium (2.5 ft) growth; [Red Planet X Pretty Lady]; Sheridan, John

'SHERILIP', F, mr, 1998; (**Linda's Lipstick**); flowers scarlet red, white reverse, hand painted, semi-dbl., 8–14 petals, 3.0in., borne in large clusters, slight fragrance; prickles moderated, pointed; foliage medium, dark green, semi-glossy; upright, tall growth; [Seedling X Picasso]; Sheridan, John

'SHERILOWSTRI', Min, yb, 1999; (**Little Stripes**); flowers yellow with orange stripes, reverse yellow, dbl., 15–25 petals, 2 in., borne in small clusters, slight fragrance; few prickles; foliage medium, semi-glossy; patio; compact, low (18 in) growth; [Summer Tan X seedling]; Sheridan, John

'SHERIMIKY', HT, pb, 1999; (**Miky Tea**); flowers pink blend, reverse cream, semi-dbl., 8–14 petals, 4 in., borne in small clusters, intense fragrance; few prickles; foliage medium, light green, semi-glossy; upright, low (2 ft) growth; [Cream Peach X Double Delight]; Sheridan, John

'SHERIRED', HT, mr, 1997; (**Mike Thompson**); flowers full, large, very dbl., 26–40 petals, borne mostly singly, slight fragrance; some prickles; foliage large, dark green, semi-glossy; medium, bushy growth; [(Silver Jubilee X seedling) X (Red Planet X Pharoah)]; Thompson, Mike

'SAVRISCENT', Min, rb, 1995; (**Betty Wilson**); flowers red,reverse yellow, double (15–25 petals),medium (2 in.) blooms borne in small clusters; very fragrance; few, small, hooked prickles; foliage medium, light green, glossy; medium (2ft.) growth; patio; [Sheila's Perfume X (Little Darling X Seedling)]; Sheridan, John

'SHERISILVER', HT, mp, 1999; (**Emma May**); flowers light pink, reverse deeper pink, dbl., 15–25 petals, 5–6 in., exhibition form, borne mostly singly, slight fragrance; prickles moderate; foliage large, dark green, semi-glossy; upright, medium (3.5 ft) growth; TGC, RNRS, 1989; [Silver Jubilee X Dr A. J. Verhage]; Sheridan, John

'SHERISKEP', F, rb, 1999; (**Mary May**); flowers red on white, reverse white, double (15–26 petals), medium (3 in.) blooms borne in small clusters; very fragrant; moderate prickles; foliage small, dark green, semi-glossy; bushy (2 ft) growth; garden decorative; [Daily Sketch X (Cream Peach X Seedling)]; Sheridan, John

Sherrill Anne® *see* 'BYRSHERRILL'

'SHERRY', F, r, 1960; flowers dark sherry color, blooms in clusters, semi-dbl., 14 petals, 2.5 in., slight fragrance; foliage dark; vigorous growth; [Independence X Orange Sweetheart]; McGredy, Sam IV; McGredy

Shi Tz-mei *see* **'TURNER'S CRIMSON RAMBLER'**

Shikou, HT, m

'SHIMMERING DAWN', F, mp, 1965; bud rose-pink; flowers blush-pink, borne in clusters, 3 in., moderate fragrance; foliage dark, glossy; very vigorous growth; Verschuren

'SHIMMERING SILK', HT, rb, 1968; flowers cerise tinted silvery pink, large, dbl.; foliage dark; vigorous, upright growth; [Ena Harkness X Molly Doyle]; Barter

Shimsha, HT, pb, 1976; Kasturi

Shine On® *see* DICtalent

Shining Coral *see* 'HADCORAL'

Shining Flare *see* 'HADFLARE'

Shining Hour® *see* 'JACYEF'

Shining Rose *see* **R. NITIDA**

Shining Ruby *see* 'HADRUBY'

'SHINING STAR', HT, my, 1945; (Mme P. Olivier, Trylon); bud long, pointed; flowers vivid chrome-yellow, large, dbl., moderate, fruity fragrance; foliage dark, leathery; [Soeur Thérèse X Feu Pernet-Ducher]; Mallerin, C.; C-P

Shining Sun *see* **'MRS PAUL GOUDIE'**

'SHINING SUN', HT, yb, 1932; bud long, pointed, golden yellow splashed scarlet; flowers yellow deepening to reddish center, dbl., intense fragrance; foliage thick, bronze; vigorous growth; [Charles P. Kilham X Julien Potin]; Van Rossem

'SHINJU', HT, lp, 1976; bud large, ovoid; flowers light pink, paler at petal edges, large blooms borne 1-3 per cluster, dbl., 28 petals, exhibition form, moderate fragrance; many, medium prickles, slanted downward; foliage medium green, leathery; vigorous, upright growth; [Royal Highness X Garden Party]; Harada, Toshiyuki

Shinsei *see* **'SHIN-SEI'**

'SHIN-SEI', HT, dy, 1979; (Shinsei); bud pointed; flowers well-formed, large blooms borne singly, dbl., 38 petals, exhibition form, moderate fragrance; prickles slanted downward; foliage medium large, medium green, glossy; upright, bushy growth; [(Ethel Sanday X Lydia) X Koto]; Suzuki, Seizo; Keisei Rose Nursery, 1978

'SHIN-SETSU', LCl, w, 1972; bud ovoid; flowers white, center soft cream, large, very dbl., exhibition form, moderate fragrance; foliage glossy, dark; very vigorous, climbing growth; [(Blanche Mallerin X Neige Parfum) X New Dawn seedling]; Suzuki, Seizo; Keisei Rose Nursery

'SHIRALEE', HT, yb, 1965; flowers yellow flushed orange, dbl., 36 petals, 5.5 in., exhibition form, moderate fragrance; vigorous, tall growth; GM, Japan, 1964; [Seedling X Kordes' Perfecta]; Dickson, Patrick; A. Dickson

Shire County *see* 'HARSAMY'

'SHIRLEY', HT, rb, 1933; bud shaded russet; flowers light prawn-red, base yellow, intense fragrance; vigorous growth; Dickson, A.

'SHIRLEY HIBBERD', T, my; flowers small; Levet, F., 1873

'SHIRLEY HOLMES', HT, ob, 1958; bud long, pointed; flowers golden orange, dbl., 30 petals, 4 in., intense fragrance; vigorous, bushy growth; [McGredy's Yellow X Ethel Sanday]; Mee; Edenvale Nursery

'SHIRLEY LAUGHARN', HT, yb, 1974; bud ovoid; flowers creamy yellow, edged pink, dbl., 35 petals, 5 in., moderate, fruity fragrance; foliage dark; very vigorous, upright, slightly spreading growth; [Granada X Garden Party]; Swim, H.C.; Laugharn

'SHIRLEY ROSE', HT, pb, 1966; flowers cream and carmine, loose, 5 in., slight fragrance; foliage dark; upright growth; [Eden Rose sport]; Lawrence

Shirley Spain *see* 'COCHAROD'

'SHIRLEY TEMPLE', HT, ly, 1936; flowers light yellow, edged lemon-yellow; [Joanna Hill sport]; Engle; Wyant

Shirpa, LCl, op, 1976; Eve, A.

Shiun *see* **'SHI-UN'**

'SHI-UN', HT, m, 1984; (Shiun); flowers deep lilac purple, reverse deeper, blooms borne singly, dbl., 33 petals, exhibition form, moderate fragrance; prickles slanted downward; foliage dark, leathery, semi-glossy; vigorous, upright, bushy growth; [(Blue Moon X Twilight) X (Red American Beauty X Happiness)]; Suzuki, Seizo; Keisei Rose Nursery

'SHIZU NO MAI', HT, lp, 1990; flowers light pastel pink, large, borne usually singly, dbl., 32–35 petals, exhibition form, slight fragrance; foliage medium, dark green, semi-glossy; sturdy, upright growth; [Jana X Madame Violet]; Ohtsuki, Hironaka, 1986

Shleby Belogorsky, HMsk, w, 1996

Shobha, HT, ob, 1988; Friends Rosery

'SHOCKING', Pol, rb, 1967; flowers dark red, reverse lighter, borne in clusters, 3 in., exhibition form, slight fragrance; foliage dark, glossy; very low, erect growth; [Red Favorite X Alain]; Hémeray-Aubert; McGredy

Shocking Blue® *see* 'KORBLUE'

'SHOCKING PINK', HT, mp, 1968; flowers large, dbl., cupped, intense fragrance; foliage leathery; vigorous, upright growth; [Pink Sensation sport]; McCannon; DeVor Nurseries, Inc.

Shocking Sky *see* KORgenda

'SHOLA', F, or, 1969; bud pointed; flowers sparkling orient red, full, medium, dbl.; foliage leathery; vigorous, dwarf growth; [Anna Wheatcroft X Open pollination]; IARI

Shona *see* 'DICDRUM'

'SHOOTING STAR', Min, yb, 1972; bud ovoid; flowers yellow, tipped red, small, dbl., cupped, slight fragrance; foliage small, light, soft; vigorous, dwarf growth; [Rumba X (Dany Robin X Perla de Montserrat)]; Meilland; C-P

'SHOOTOUT', Min, ab, 1981; bud ovoid; flowers apricot-orange, blooms borne singly, dbl., 28 petals, slight fragrance; long, triangular, light green prickles; foliage medium green; vigorous, bushy growth; [Tiki X Darling Flame]; Borst, Jim; Kimbrew-Walter Roses

Shoreline Tawny, N, ly

Short 'n' Sweet *see* 'TINSHORT'

Shortcake™ *see* 'KEIBELMI'

'SHOT SILK', HT, pb, 1924; flowers cherry-cerise, shading to golden yellow at base, dbl., 27 petals, exhibition form, intense fragrance; foliage glossy, slightly curled; vigorous growth; GM, NRS, 1923; [Hugh Dickson seedling X Sunstar]; Dickson, A.

'SHOT SILK, CLIMBING', Cl HT, pb, 1931; Knight, C.

Show, HT, lp, 1995; RvS-Melle

Show Carpet, *see* 'LAVBRIC'

'SHOW GARDEN', LCl, mp, 1954; (Pink Arctic); flowers crimson to rose-bengal, then magenta, dbl., 43 petals, 4–5 in.; growth like a hybrid tea, followed by 4-5 ft. canes; [Unnamed seedling X Queen o' the Lakes]; Brownell, H.C.

Show Garden, LCl, mp

'SHOW GIRL', HT, mp, 1946; bud long, pointed; flowers rose-pink, 15–20 petals, 3.5–4.5 in., exhibition form, moderate fragrance; foliage leathery; vigorous, upright, bushy growth; GM, NRS, 1950; [Joanna Hill X Crimson Glory]; Lammerts, Dr. Walter; Armstrong Nursery

Show Girl, HT, mp

'SHOW GIRL, CLIMBING', Cl HT, mp, 1949; Chaffin; Armstrong Nursery

Show 'n' Tell *see* 'FOUTELL'

Show Off *see* 'AROWAGO'

Showbiz *see* 'TANWEIEKE'

Showbiz, Climbing, Cl F, mr, 1995; Tejganga

'SHOWBOAT', F, yb, 1965; bud ovoid; flowers deep yellow on pink to cream ground, dbl., 30 petals, 2.5 in., moderate fragrance; foliage leathery; moderate, bushy growth; [Carrousel X Unnamed seedling]; Patterson; Patterson Roses

Showbound™, Min, op

'SHOWCASE', Min, ab, 1986; flowers small, dbl., 35 petals, slight fragrance; foliage small, light green, semi-glossy; bushy growth; [Rise 'n' Shine X Over the Rainbow]; Stoddard, Louis

Showdown *see* 'MINISHOW'

'SHOWER OF GOLD', HWich, my, 1910; flowers golden yellow, rapidly fading to pale yellow, rosette-shaped, dbl.; foliage glossy, fern-like; very vigorous, climbing growth; [Jersey Beauty X Instituteur Sirdey]; Paul

Showoff *see* 'AROWAGO'

'SHOWOFF', LCl, mr, 1952; flowers scarlet, medium to large, borne in large clusters, very dbl., cupped, profuse, repeated bloom; vigorous growth; [Blaze sport]; Moffet; Earl May Seed Co.

'SHOWPIECE', HT, mr, 1958; flowers bright scarlet, moderate fragrance; Ratcliffe

Showqueen, HT, mp, 1997; Williams, J. Benjamin

Showstopper *see* 'JACSHO'

'SHOWTIME', HT, mp, 1969; flowers large, dbl., exhibition form, moderate, fruity fragrance; foliage glossy, leathery; vigorous, bushy growth; [Kordes' Perfecta X Granada]; Lindquist; Howard Rose Co.

Showy Gold, F, dy

Showy Pavement, S, mp, 1990; flowers large, semi-dbl., intense fragrance; low, arching, spreading (2 ft) growth; Baum

'SHREE DAYANANDA', HT, dp, 1979; bud ovoid; flowers deep pink, very dbl., 90 petals, moderate fragrance; beak-shaped prickles; foliage small, green; bushy, dwarf growth; [Scarlet Knight X Festival Beauty]; Hardikar, Dr. M.N.

Shreveport™ *see* 'KORPESH'

Shrewsbury Show *see* FRYshrewby

Shreyasi, HT, m, 1992; IARI

Shri Swamy Samarth, LCl, dp, 1992; Patil, B.K.

'SHROPSHIRE LASS', S, lp, 1968; flowers blush-pink, single, 5 petals, 5 in., summer bloom, moderate fragrance; [Mme Butterfly X Mme Legras de St. Germain]; Austin, David

Shrubby Pink *see* **'SUNDAY TIMES'**

Shugetsu *see* **'SHU-GETSU'**

'SHU-GETSU', HT, dy, 1982; (Shugetsu); flowers large, dbl., 38 petals, exhibition form, moderate fragrance; large, straight prickles; foliage large, dark, glossy; upright, bushy growth; [Seiko

X King's Ransom]; Suzuki, Seizo; Keisei Rose Nursery

Shunpo, HT, lp, 1987; Suzuki, Seizo

'SHUN'YO', HT, ob, 1982; flowers light yellow orange, large, dbl., 33 petals, exhibition form, slight fragrance; sickle-shaped prickles; foliage glossy; vigorous, upright growth; [(Golden Sun X Summer Holiday) X (Garden Party X Narzisse)]; Kono, Yoshito

Shuo *see* **'SHU-OH'**

'SHU-OH', HT, or, 1982; (Shuo); flowers medium, cupped, moderate fragrance; prickles slanted downward; foliage dark, semi-glossy; upright growth; [San Francisco X Pharaoh]; Suzuki, Seizo; Keisei Rose Nursery

Shurpee, HT, mr

Shuzao Red, S, mr; (China), 1991

Shy Beauty *see* 'LYOSHY'

Shy Girl *see* 'JACWHIM'

Shy Maiden *see* 'ANDSHY'

'SI', Min, w; flowers rosy white, micro-mini, small; dwarf growth; [Perla de Montserrat X (Anny X Tom Thumb)]; Dot, Pedro

Si Bemol® *see* 'LENMOL'

'SIBELIUS', HT, mr, 1958; flowers velvety crimson, dbl., exhibition form; vigorous growth; [New Yorker X Étoile de Hollande]; Verschuren; Blaby Rose Gardens

Sibelius, HMsk, m, 1984; Lens

Sibilla® *see* BARsib

Sibylle, F, op

Siddartha, HT, rb, 1973; Kasturi

Side Kick, *see* 'PIXKICK'

'SIDI-BRAHIM', HT, dy, 1953; flowers jonquil-yellow, dbl., slight fragrance; strong stems; [Feu Pernet-Ducher X Mme Rene Lefevre]; Robichon

'SIDNEY PEABODY', Gr, dp, 1955; flowers reddish pink, well formed, borne in clusters, dbl., 36 petals, 3.5 in., slight fragrance; foliage dark, leathery; very vigorous growth; [Rome Glory X Unknown Floribunda seedling]; deRuiter; Gandy Roses, Ltd.

Sidonie *see* **'SYDONIE'**

Siegeslied *see* **'BOUQUET'**

'SIEGESPERLE', Pol, ly, 1915; flowers small, semi-dbl.; Kiese; (Sangerhausen)

Siegfried Sassoon *see* 'MACJULIAT'

Sieguier, HGal, m, 1853; Robert

SIElemon, A, ly, 1988; (**Lemon Blush**); Sievers

SIEroyal, A, lp, 1988; (**Royal Blush**); Sievers

'SIERRA DAWN', HT, pb, 1967; bud long, pointed; flowers bright pink blend, large, dbl., exhibition form, moderate fragrance; foliage dark, bronze, leathery; vigorous, upright, bushy growth; [Helen Traubel X Manitou]; Armstrong, D.L.; Armstrong Nursery

'SIERRA GLOW', HT, pb, 1942; bud ovoid; flowers shrimp-pink, reverse strawberry-pink, dbl., 33 petals, 3.5–4.5 in., exhibition form, moderate fragrance; foliage leathery; vigorous, bushy, spreading growth; [Crimson Glory X Soeur Thérèse]; Lammerts, Dr. Walter; Armstrong Nursery

'SIERRA GOLD', HT, my, 1960; bud urn shaped; flowers indian yellow, dbl., 45–55 petals, 4–4.5 in., exhibition form, slight fragrance; foliage leathery, semi-glossy; upright, compact growth; [Queen Elizabeth X Tawny Gold]; Lammerts, Dr. Walter; Amling-DeVor Nursery

Sierra Lynn *see* 'GELLYNN'

'SIERRA SNOWSTORM', S, ly, 1936; bud long, pointed, cream and yellow; flowers open, small borne in clusters, single, 2 in., abundant, recurrent bloom, intense fragrance; foliage large, leathery, glossy, light; vigorous (5-6 ft.), bushy, arching growth; [Gloire des Rosomanes X Dorothy Perkins]; Moore, Ralph S.; Henderson's Exp. Gardens and Brooks & Son

Sierra Sun, HT, rb

Sierra Sunrise *see* 'MORLIYEL'

'SIERRA SUNSET', LCl, yb, 1961; bud pointed; flowers blend of yellow, peach, orange and red, in clusters, dbl., 40 petals, 6 in., exhibition form, intense, fruity fragrance; foliage glossy; vigorous (6-8 ft) growth; [(Capt. Thomas X Joanna Hill) X Mme Kriloff]; Morey, Dr. Dennison; J&P

Sieska Fervid, F; Klimenko, V. N.; (Cavriglia)

SIEson, A, dr, 1988; (**Crimson Blush**); Sievers

'SIESTA', F, or, 1966; flowers light vermilion on cream base, large, dbl.; foliage glossy, leathery; very free growth; [Sarabande X Dany Robin]; Meilland; URS

'SIF', LCl, ob, 1969; flowers tangerine-orange, large, dbl., 25 petals, cupped; foliage dark, glossy; vigorous, climbing growth; [Traumland X Royal Gold]; Lundstad

Sight Saver *see* FRYaffair

Sightsaver *see* FRYaffair

Siglinde, F, 1972; Noack, Werner; (Cavriglia)

'SIGNAL RED', Pol, mr, 1949; flowers scarlet, rosette-shaped, blooms in large trusses, 2 in.; foliage glossy, bronze; vigorous, bushy growth; (28); [DeRuiter's Herald X Polyantha seedling]; deRuiter; Gandy Roses, Ltd.

Signal Rot, F, 1982; VEG; (Cavriglia)

'SIGNALFEUER'®, F, or, 1959; (Fire Signal); flowers cinnabar to orange, blooms in clusters, dbl.; bushy growth; [Lumina X Cinnabar seedling]; Tantau, Math.

Signature® *see* 'JACNOR'

'SIGNE RELANDER', HRg, dr, 1928; flowers bright dark red, small blooms in clusters, dbl., recurrent bloom; vigorous (6 1/2 ft.) growth; [R. rugosa hybrid X Orléans Rose]; Poulsen, S.; Poulsen

'SIGNET', HT, dp, 1938; bud long, pointed; flowers deep pure pink, large, dbl., exhibition form, intense fragrance; foliage leathery; very vigorous growth; [Premier X Talisman]; Montgomery Co.

'SIGNORA', HT, ob, 1936; (Signora Piero Puricelli); bud long, pointed; flowers orange-apricot, suffused gold, outer petals magenta-pink, la, dbl., 27 petals, cupped, moderate fragrance; foliage glossy; vigorous growth; GM, Portland, 1937; [Julien Potin X Sensation]; Aicardi, D.; J&P

Signora Maria Sgaravatti, HT; Sgaravatti, A.; (Cavriglia)

Signora Piero Puricelli *see* **'SIGNORA'**

Sika®, HT, op, 1983; flowers large, dbl.; Cocker; (Sangerhausen)

Sika *see* DORto

Sila, HT, op, 1983; Cocker

'SILBERLACHS', F, op, 1944; (Silver Salmon); flowers pale salmon, shell shaped, borne in clusters of 3-10, single, 6–8 petals, slight fragrance; foliage light green; upright, bushy growth; [Rosenelfe X Hamburg]; Tantau

Silberzauber, F, mp, 1985; flowers large, dbl.; GPG Bad Langensalza; (Sangerhausen)

'SILENT NIGHT', HT, yb, 1969; flowers creamy yellow suffused pink, well-formed, slight fragrance; GM, Geneva, 1969; [Daily Sketch X Hassan]; McGredy, Sam IV

Silhouette *see* 'TANOLG'

'SILHOUETTE'®, HT, pb, 1984; (Silver Medal); bud long, pointed; flowers white, petals edged pink, dbl., 25 petals, 4–5 in.; foliage light green; upright growth; Warriner, William A.; J&P, 1980

Silk Button *see* KORzeito

Silk Hat™ *see* 'AROSILHA'

Silk 'n' Satin *see* 'JUDSILK'

'SILK SASH', HT, pb, 1971; flowers pastel pink, reverse silvery pink, very large, very dbl., 85 petals, exhibition form, slight fragrance; foliage dark, leathery; vigorous, upright growth; [Memoriam X Elizabeth Fankhauser]; Fankhauser

Silken Laumann *see* 'LAVCOM'

Silky Mist *see* SIMsilko

Silva® *see* 'MEICHAM'

Silva *see* KEIromo

'SILVA GRAÇA', HT, ob, 1956; flowers salmon and yellow shaded pink, well shaped; [Michele Meilland X Comtesse Vandal]; da Silva, Moreira

'SILVABELLA', HT, pb, 1967; flowers deep pink, reverse carmine-pink, large, dbl., cupped, slight fragrance; foliage leathery; upright, bushy growth; [Dr. Debat X Eden Rose]; Guiseppe, M.

Silver Angel, *see* 'GELANGEL'

Silver Anniversary *see* 'POULARI'

Silver Anniversary™ *see* 'JACLAV'

Silver Anniversary *see* MEIborfil

'SILVER BEAUTY', F, m, 1956; flowers silvery rose-madder; vigorous, bushy growth; [La France seedling]; Verschuren; Gandy Roses, Ltd.

'SILVER BELL', F, lp, 1976; (Gin no suzu); bud ovoid; flowers soft light pink, small, dbl., 35 petals, 2 in., exhibition form; foliage glossy; vigorous, upright growth; [Gene Boerner X Unnamed seedling]; Takatori, Yoshiho; Japan Rose Nursery

'SILVER CHARM', F, m, 1968; flowers lavender-blue, large blooms in large trusses, semi-dbl., moderate fragrance; foliage dark; vigorous, low growth; [Lilac Charm X Sterling Silver]; LeGrice

Silver Cloud, F, r, 1990; Moore

'SILVER COLUMBIA', HT, pb, 1924; flowers clear silver-pink; [Columbia sport]; Leonard

'SILVER DAWN', LCl, w, 1942; flowers creamy white, center deeper, stamens golden yellow, open, large, very dbl., 90–100 petals; foliage large, leathery, glossy, dark; vigorous (10-12 ft), compact, climbing growth; [Silver Moon X Silver Moon]; Zombory

Silver Dream® *see* 'INTERGRI'

'SILVER ENCHANTMENT', HT, rb, 1971; flowers rich red, reverse silver, conical, large, dbl., 36 petals, moderate fragrance; foliage dark; very free growth; [Tropicana X ?]; Gregory

Silver Fox *see* 'DEVSPILIO'

'SILVER GEM', F, m, 1969; flowers silver-mauve, large, borne in trusses, semi-dbl.; foliage dark, leathery; bushy growth; deRuiter

'SILVER JUBILEE'®, HT, pb, 1978; flowers silvery pink, reverse darker, dbl., 33 petals, 5 in., exhibition form, slight fragrance; foliage glossy; vigorous growth; GM, Belfast, 1980 GM, RNRS, 1977 PIT, RNRS, 1977; [((Highlight X Colour Wonder) X (Parkdirektor Riggers X Piccadilly)) X Mischief]; Cocker

'SILVER JUBILEE', HT, yb, 1937; flowers light golden yellow, base chrome, edged canary, very large, dbl., moderate fragrance; foliage very large, glossy; vigorous growth; RULED EXTINCT 4/77; Dickson, A.

'SILVER JUBILEE, CLIMBING', Cl HT, pb, 1983; Cocker

'SILVER LADY', F, m, 1991; flowers pale lavender, with a hint of darker lavender on edges of petals, dbl., 15–25 petals, 3–3.5 in., no fragrance; some prickles; foliage medium, medium green, semi-glossy; medium (90 cms), upright, bushy growth; [Azure Sea X Unnamed seedling]; Taylor, Franklin "Pete" & Kay; Taylor's Roses, 1993

'SILVER LINING', HT, pb, 1958; flowers silvery rose, dbl., 30 petals, 5 in., intense fragrance; vigorous growth; GM, NRS, 1958 GM, Portland, 1964; [Karl Herbst X Eden Rose seedling]; Dickson, A.

Silver Lining, Climbing *see* 'DICSILVING'

Silver Medal *see* **'SILHOUETTE'**

'SILVER MOON', LCl, w, 1910; bud long, pointed; flowers creamy white, base amber, stamens darker, blooms in clusters, dbl., 20 petals, 4.5 in., non-recurrent, slight fragrance; foliage large, dark, leathery, glossy; very vigorous (to 20 ft or more) growth; [Reputedly (R. wichurana X Devoniensis) X R. laevigata]; Van Fleet; P. Henderson

Silver Peach, Min, lp; [Silver Jubilee X Oz Gold]; Hannemann, F., 1989; The Rose Paradise; (Weatherly, L.)

Silver Phantom *see* 'RENHOM'

Silver Pink®, F, pb, 1990; Lens

'SILVER PRINCESS', HT, ob, 1934; bud salmon-yellow to delicate pink; flowers almost ivory-white, tinged pink, very large, semi-dbl.; vigorous growth; Burbank; Stark Bros.

Silver Queen, HP, pb; Paul, W., 1886

Silver Queen, HT, w

Silver River®, S, pb, 1989; Lens

Silver Salmon *see* **'SILBERLACHS'**

'SILVER SHADOWS', HT, m, 1984; flowers light blue-lavender, large blooms borne 3-5 per cluster, dbl., 33 petals, cupped, repeat bloom, intense fragrance; awl-like, tan prickles; foliage large, dark, leathery; upright, bushy growth; hardy.; [((Soir d'Automne X Music Maker) X Solitude) X ((Blue Moon X Tom Brown) X Autumn Dusk]; Buck, Dr. Griffith J.; Iowa State University

Silver Slippers *see* 'PIXISLIP'

'SILVER SPOON', HT, m, 1985; flowers large, dbl., 35 petals, no fragrance; foliage large, medium green, semi-glossy; upright, bushy, spreading growth; [Louisiana X Unnamed seedling]; Weeks, O.L.; Weeks Wholesale Rose Growers

Silver Star® *see* 'KORBIDO'

'SILVER TIPS', Min, pb, 1961; bud pointed; flowers pink, reverse and tips silvery, becoming soft lavender, dbl., 50 petals, 1 in., slight fragrance; foliage leathery; vigorous, bushy (10-12 in) growth; [(R. wichurana X Floradora) X Lilac Time (Min)]; Moore, Ralph S.; Sequoia Nursery

'SILVER WEDDING', HT, w, 1921; flowers almost identical to parent; [Ophelia sport]; Amling Co.

Silver Wedding, HT, w, 1976; Gregory

'SILVER WINGS', HT, w, 1943; bud long, pointed; flowers ivory-white, large, dbl., exhibition form, slight fragrance; foliage glossy; vigorous, upright, bushy growth; McGredy

Silverado *see* 'AROGREWOD'

'SILVERELDA', HT, pb; flowers veined silvery buff-pink, edged lighter, base yellow, dbl., 35 petals, slight fragrance; foliage glossy; vigorous, compact growth; [Heinrich Wendland X Nancy Wilson]; Riethmuller

Silverhill *see* 'TALSILVER'

Silvery Moon *see* 'LEOSILMU'

'SILVIA', HT, yb, 1920; (Yellow Ophelia); flowers sulfur-yellow shading to white; [Ophelia sport]; Pierson, F.R.

'SILVIA LEYVA', HT, mr, 1933; flowers cardinal-red, large, dbl., cupped, moderate fragrance; foliage glossy; vigorous growth; [Mrs C.W. Edwards X Mari Dot]; Dot, Pedro; C-P

'SIMARAMAM', HT, w, 1979; (**Marama**); bud pointed; dbl., 45 petals, exhibition form, intense fragrance; dark brown prickles; foliage large, medium green, semi-glossy; bushy, medium growth; [Lady Helen X Unnamed seedling]; Simpson, J.W.

Simba *see* 'KORBELMA'

SIMcho, F, rb, 1986; (**Hot Chocolate**); Gold Star of the South Pacific, Palmerston North, NZ, 1986; Simpson

'SIMEROSE', HT, rb, 1939; bud long; flowers nasturtium-red, reverse golden yellow; vigorous growth; [Charles P. Kilham X (Charles P. Kilham X Margaret McGredy)]; Meilland, F.

'SIMETNA', F, r, 1979; (**Tane**); bud pointed; dbl., 38 petals, exhibition form, borne 4-10 per cluster, no fragrance; short, straight, dark brown prickles; foliage small, dark, very glossy; vigorous, upright growth; [(Orangeade X Megiddo) X (Unnamed seedling X Jocelyn)]; Simpson, J.W.

Simfonia, HT, lp; high-centered, exhibition form, 29 petals, mostly single or in clusters; dark green, semi-glossy; vigorous upright; [Mount Shasta X Pascali]; St. Wagner, 1977; Res. Stn. f. Horticulture, Cluj, 1977, Romania

'SIMHAHA', HT, yb, 1979; (**Hamish**); bud ovoid; flowers foliage medium green, dbl., 50 petals, exhibition form; dark brown prickles; vigorous, medium, upright to bushy growth; [Fairy Dancers X Diamond Jubilee]; Simpson, J.W.

'SIMHOPAN', HT, or, 1979; (**Hot Pants**); bud high-pointed; dbl., 48 petals, exhibition form, borne usually 3 per cluster, no fragrance; few prickles; foliage medium green; spreading growth; [Gypsy Moth X Princesse]; Simpson, J.W.

'SIMILAM', HT, mr, 1994; (Ilam, **Roseworld**); dbl., 26–40 petals, 3–3.5 in., borne mostly singly, slight fragrance; few prickles; foliage medium, medium green, semi-glossy; medium, upright growth; [Melina X Mme G. Delbard]; Simpson, Nola; Simpson, 1994

Simina, F, lp; flowers clear pink, cupped, 40 petals, borne in clusters, moderate; large, dark green, glossy; vigorous tall upright habit; [Vigorosa X Allgold]; St. Wagner, 1996; Res. Stn. f. Fruit Growing, Cluj, 1997, Romania

'SIMJEZBEL', HT, dp, 1980; (**Anne Kercher**); bud long, pointed; flowers deep pink, dbl., 30 petals, exhibition form, borne singly or 3 per cluster; straight prickles; foliage large, dense, medium green; vigorous, very bushy, compact growth; [First Prize X Seedling]; Simpson, J.W.

'SIMKAYJAY', F, or, 1979; (**Kirsty Jane**); bud ovoid; flowers fluted blooms borne 3-10 per cluster, dbl., 28 petals, no fragrance; straight brown prickles; foliage medium, glossy; bushy, vigorous growth; [Orangeade seedling X Megiddo]; Simpson, J.W.

SIMmer, HT, dr, 1997; (**Mercy**); Simpson

'SIMON BOLIVAR', HT, or, 1966; bud ovoid; flowers bright orange-red, large, dbl., 40 petals, exhibition form, slight fragrance; foliage dark, glossy, leathery; vigorous, upright, bushy growth; [Roundelay X El Capitan]; Armstrong, D.L.; Armstrong Nursery

'SIMON DE ST JEAN', HP, m, 1861; flowers velvety purple; Liabaud

Simon Dot *see* 'DOTSURODO'

'SIMON FRASER', S, mp, 1992; semi-dbl., 6–14 petals, 1.5–2.75 in., borne in small clusters, slight fragrance; some prickles; foliage dark green, semi-glossy; low (60 cms), upright growth; extremelywinter hardy.; Ogilvie, Ian S.; Agriculture Canada, 1992

Simon Robinson *see* 'TROBWICH'

Simona *see* 'TANMOSINA'

'SIMONE', HT, w, 1924; flowers flesh-white, center deeper pink, passing to creamy white, dbl., moderate fragrance; [Mme Caroline Testout X Paul Monnier]; Buatois

'SIMONE', HT, m, 1957; (Mauve Mallerin, Parme); bud ovoid; flowers pastel lilac, dbl., 50 petals, 6 in., exhibition form, moderate fragrance; foliage leathery, dark, glossy; vigorous growth; [(Peace X Independence) X Grey Pearl]; Mallerin, C.; Hémeray-Aubert

'SIMONE DAMAURY', HT, dr, 1925; bud long, pointed; flowers brilliant crimson, semi-dbl., intense fragrance; [Liberty X Gen. MacArthur]; Soupert & Notting

'SIMONE DE CHEVIGNÉ', HT, pb, 1924; flowers flesh-pink, shaded yellow, dbl.; Pernet-Ducher

'SIMONE DE NANTEUIL', HT, pb, 1925; bud long, pointed; flowers rosy white, tinted carmine-pink, reverse flesh, dbl., moderate fragrance; [Ophelia X Mme Vittoria Gagniere]; Schwartz, A.

'SIMONE GUÉRIN', HT, my, 1929; flowers coral-yellow, large, semi-dbl., exhibition form, slight fragrance; foliage glossy, dark; vigorous growth; [Constance X Unnamed Hybrid Tea seedling]; Mallerin, C.

'SIMONE LABBÉ', HT, ab, 1922; bud long, pointed; flowers apricot-yellow, passing to clear saffron-yellow, dbl., intense fragrance; [Le Progres X Lady Greenall]; Ketten Bros.

'SIMONE MAYERY', HT, yb, 1937; bud long, pointed, cream-yellow, shaded carmine-pink; flowers cream, center dark yellow, very large, cupped; foliage bronze; vigorous growth; Chambard, C.

Simone Merieux® *see* 'GAUTARA'

Simonet *see* **'METIS'**

'SIMONET'S DOUBLE PINK RUGOSA', HRg, mp; flowers pink, smaller than hansa, very dbl., non-recurrent; hardy.; [R. macounii X Mme Georges Bruant (?)]; Simonet; P.H. Wright

'SIMPALNO', HT, rb, 1979; (**Happy Day**, Velvet Lustre); bud pointed; dbl., 30 petals, exhibition form, borne singly, moderate fragrance; brown prickles; foliage large, dark, semi-glossy; strong, upright, medium growth; [First Prize X Gypsy Moth]; Simpson, J.W.

Simple Pleasures™ *see* 'TALSIMPLE'

'SIMPLE SIMON', Min, mp, 1955; flowers carmine-rose, base yellow, dbl.; [(R. multiflora nana X Mrs Pierre S. duPont)F2 X Tom Thumb]; deVink; T. Robinson, Ltd.

'SIMPLEX', Min, w, 1961; bud long, pointed, apricot; single, 5 petals, 1.5 in., blooms in small clusters, slight fragrance; foliage leathery; vigorous, bushy (12-14 in.) growth; [(R. wichurana X Floradora) X Seedling]; Moore, Ralph S.; Sequoia Nursery

Simplex Multiflora, HMult, pb, 1905

Simplicity™ *see* 'JACINK'

'SIMPLICITY', HT, w, 1909; flowers pure white, open, single, slight fragrance; foliage glossy, rich green; Dickson, H.

Simplicity, F, op, 1990; Lens

'SIMPLY DIVINE', F, rb, 1978; semi-dbl., 18 petals, 3.5 in., slight fragrance; foliage dark; [Elizabeth of Glamis X Evelyn Fison]; Anderson's Rose Nurseries

Simply Elegant *see* 'BUSIMEL'

Simply Irresistible *see* 'BURBRAHMVIDYA'

'SIMPSON'S RED', HT, dr, 1977; flowers medium, dbl., 40 petals, 3.5 in., slight fragrance; [Red Lion X Grande Amore]; Simpson, J.W.; Lower Rangitikei Rose Soc.

SIMsilko, HT, lp, 1989; (**Silky Mist**); Simpson

'SIMTEONO', HT, pb, 1979; (**Te Moana**); bud pointed; dbl., 38 petals, exhibition form, borne mostly singly, intense fragrance; brown prickles; foliage dark, semi-glossy; medium, bushy, upright growth; Gold Star of the South Pacific, Palmerston North, NZ, 1981; [(Gypsy Moth X Percy Thrower) X First Prize]; Simpson, J.W.

SIMway, HT, w; (**Jenny Charlton**); Simpson

'SINBAD', HT, op, 1964; flowers salmon-pink; [Pink Lustre X Circus]; Leenders, J.

Sincera *see* **'AMISTAD SINCERA'**

Sincerely Mine, Min

Sincerely Yours *see* 'MORDORT'

'SINCERITY', HT, ob, 1940; bud long, pointed; flowers flesh, shaded amber and orange, large, dbl., slight fragrance; foliage dark, leathery; vigorous, compact growth; [Comtesse Vandal seedling X Mrs Sam McGredy]; LeGrice

'SINDOOR', F, or, 1980; bud long, pointed; dbl., 23 petals, borne singly or 15 per cluster, no fragrance; straight prickles, bending downward; foliage large, glossy, coppery when young; upright, tall growth; [Sea Pearl X Suryodaya]; Division of Vegetable Crops and Floriculture

Singalong *see* 'MACSINGAL'

Singin' in the Rain *see* 'MACIVY'

'SINGLE BLISS', Min, pb, 1980; bud short, pointed; flowers deep pink and white, blooms borne 5-30 per cluster, single, 5 petals, flat, slight fragrance; straight prickles; foliage very small, very glossy; very compact, bushy growth; [Seabreeze X Baby Betsy McCall]; Saville, F. Harmon; Nor'East Min. Roses

Single Charm *see* 'UMSCHARM'

Single Cherry, S, pb; flowers purple/pink with dary yellow, medium, single; (Sangerhausen)

Single Chestnut Rose *see* **R. ROXBURGHII NORMALIS**

Single Pink, Ch, pb

Single's Better® *see* 'SAVABET'

Sinoia, Pol, ob

Sinsetu, LCl, w

Siobhan *see* 'SEASIO'

'SIOUX BEAUTY', S, rb, 1927; flowers bright rose, center dark crimson, very dbl., 100 petals, non-recurrent bloom, moderate fragrance; hardy.; [Tetonkaha X American Beauty]; Hansen, N.E.

Sir *see* 'JOLSIR'

'SIR ALEXANDER N. ROCHFORT', HT, ob, 1917; flowers flesh, center darker; [Lady Alice Stanley X Marquise de Sinéty]; Le Cornu

'SIR ARTHUR STREETON', HT, mp, 1940; flowers pink, well formed; Clark, A.

'SIR BASIL MCFARLAND', HT, ob, 1931; bud long, pointed; flowers orange-salmon-pink, flushed yellow, dbl., exhibition form, moderate fragrance; foliage thick; vigorous growth; McGredy

'SIR BILLY BUTLIN', HT, or, 1969; (Sir William Butlin); flowers red-orange, large, dbl., 30 petals, cupped, slight fragrance; foliage bronze; vigorous, upright growth; [Bettina X Majorca]; Gandy, Douglas L.

Sir C.V. Raman, HT, or, 1989; Pal, Dr. B.P.

'SIR CEDRIC MORRIS', LCl, w, 1980; bud small, globular; single, 5 petals, borne 20-40 per cluster, summer bloom only in large quantity, slight fragrance; large prickles; foliage elongated, large; very vigorous growth; [R. glauca X Unnamed seedling]; Morris, Sir Cedric; Peter Beales, 1979

Sir Clough *see* AUSclough

'SIR DALLAS BROOKS', HT, mr, 1962; bud long, pointed; flowers large, dbl., exhibition form, moderate fragrance; foliage leathery; vigorous, upright, open growth; [Ena Harkness X Charles Mallerin]; Smith, R.W.; T.G. Stewart

'SIR DAVID DAVIS', HT, rb, 1926; bud long, pointed; flowers deep glowing crimson, base light yellow, dbl., exhibition form, intense fragrance; foliage dark, leathery; vigorous growth; McGredy

'SIR DAVID REID', HT, mr, 1941; flowers crimson-red, large, dbl., 25 petals, flat, moderate fragrance; Dickson, A.

Sir Edward Elgar *see* 'AUSPRIMA'

Sir Edward's Rose, HT, mp, 1998; Dawson

Sir Frederick Ashton, HT, w, 1987; Beales, Peter

'SIR GALAHAD', F, dp, 1967; flowers deep pink, medium, dbl., slight fragrance; foliage glossy; [Pink Parfait X Highlight]; Harkness

'SIR HARRY', HT, my, 1989; bud ovoid; flowers medium, borne usually singly, dbl., 30 petals, exhibition form, slight, fruity fragrance; round, orange fruit; prickles triangle, brown; foliage medium, dark green, glossy; bushy, medium growth; [Irish Gold X (Pink Favorite X Golden Autumn)]; Bracegirdle, A.J.; Gregory's Roses, 1989

Sir Harry Pilkington *see* 'TANEMA'

Sir Henry® *see* KORopas

'SIR HENRY SEAGRAVE', HT, ly, 1932; bud long, pointed; flowers light primrose-yellow, large, dbl., exhibition form, intense fragrance; foliage leathery; GM, NRS, 1932; Dickson, A.

'SIR JOSEPH PAXTON', B, dr, 1852; flowers deep red tinted violet, well-formed, medium; Laffay, M.

'SIR LANCELOT', F, ab, 1967; flowers apricot-yellow, semi-dbl., 3–4 in., slight fragrance; foliage light green, glossy; [Vera Dalton X Woburn Abbey]; Harkness

'SIR MATTHEW NATHAN', HT, lp; [Rhea Reid X Laurent Carle]; Harrison, A., 1927; (Weatherly, L.)

Sir Neville Marriner *see* 'GLANMUSIC'

'SIR ROBERT DUFF', Cl T, lp; [Gloire de Dijon X Unknown]; Johnson, 1893; (Weatherly, L.)

'SIR ROWLAND HILL', HP, m, 1888; flowers dark purple/red, large, very dbl., moderate fragrance; Mack; (Sangerhausen)

'SIR THOMAS LIPTON', HRg, w, 1900; bud ovoid; flowers very white, round and cupped, dbl., cupped, recurrent bloom, intense fragrance; foliage dark, leathery; vigorous (6-8 ft.), bushy growth; [R. rugosa alba X Clotilde Soupert]; Van Fleet; Conard & Jones

Sir Walter Raleigh® *see* 'AUSSPRY'

'SIR WALTER RALEIGH', HT, ob, 1919; bud long, pointed; flowers coppery reddish salmon, reverse deep crimson, dbl., moderate fragrance; [Lady Pirrie X Unnamed seedling]; Le Cornu

Sir Wilfrid Laurier, LCl, lp

Sir William Butlin *see* **'SIR BILLY BUTLIN'**

Sir William Leech *see* 'HORTROPIC'

'SIR WINSTON CHURCHILL', HT, op, 1955; flowers salmon-pink shaded orange, dbl., 48 petals, 5 in., exhibition form, moderate fragrance; foliage dark, glossy; very vigorous growth; GM, NRS, 1955; [Unnamed seedling X Souv. de Denier van der Gon]; Dickson, A.

'SIREN', F, or, 1953; bud ovoid; flowers bright scarlet-red, blooms in clusters, semi-dbl., 18 petals, 3–3.5 in., cupped, moderate fragrance; foliage leathery; vigorous, compact growth; GM, NRS, 1952; [(Baby Chateau X Else Poulsen) X Independence]; Kordes; J&P

'SIRENA', HT, rb, 1941; bud ovoid; flowers cardinal-red, center shaded yellow, very large, dbl., cupped, intense fragrance; foliage leathery; very vigorous growth; [Saturnia X Anemone]; Aicardi, D.; Giacomasso

Sirenella, HT; (Cavriglia)

Sirenetta®, Min, ab

'SIRIUS', Cl HT, mr, 1939; bud long, pointed; flowers cherry-red, center lighter, large, dbl., intermittent bloom, intense fragrance; vigorous, climbing (15 ft), open habit growth; [Unnamed seedling X Lubra]; Fitzhardinge; Hazlewood Bros.

'SIROCCO', F, or, 1971; bud globular; flowers vermilion-red, large, very dbl., cupped, slight fragrance; foliage dark, leathery; vigorous, bushy growth; [Independence X Orangeade]; Delforge

Sis *see* 'BRISIS'

'SISI KETTEN', Pol, dp, 1900; flowers medium, dbl.; Ketten, Gebrüder; (Sangerhausen)

Sissi *see* 'TANSI'

Sissi, Climbing *see* **'BLUE MOON, CLIMBING'**

Sissinghurst Castle *see* **'ROSE DES MAURES'**

Sister Joan *see* 'KENMADO'

'SISTER KENNY', F, mr, 1954; bud ovoid; flowers scarlet, medium, borne in clusters, semi-dbl., 10–12 petals, flat, slight fragrance; foliage leathery; very vigorous, upright growth; [Baby Chateau X Red Delicious]; Hill, Joseph H., Co.; H&S

'SISTER SUSAN', HT, r; bud ovoid, yellow stained crimson; flowers cream tinted flesh, center copper-orange, base yellow, medium, dbl., 48 petals, exhibition form, slight fragrance; foliage leathery, glossy, dark; vigorous, spreading growth; [Mrs Sam McGredy X Soeur Therese]; Oliver, F.

Sister Thérèse *see* **'SOEUR THÉRÈSE'**

'SITPRETTY', F, w, 1999; flowers white with shrimp pink edges, reverse darker, dbl., 17–25 petals, 3.5 in., borne in small clusters, no fragrance; prickles moderate; foliage medium dark green, glossy; bushy, shrub-like, tall (5-6 ft) growth; [Nicole sport]; Sitton, John; Edmunds' Roses, 2000

'SITTING BULL', HSet, dp; flowers deep pink, semi-dbl., non-recurrent; height 6-8 ft; not hardy.; Horvath

Sitting Pretty *see* 'TINSIT'

'SIX FLAGS', Gr., mr, 1962; (Manja, Manja Mourier); flowers cherry-red, dbl., 25–30 petals, 3.5–4 in., exhibition form; foliage slender, leathery, semi-glossy; vigorous, upright, spreading growth; [First Love X Roundelay]; Swim, H.C.; Five M Nursery;, Poulsen

Six Fours Auri, Misc OGR, w

Sixth Sense *see* JACrare

Sizzle Pink™ *see* 'DEVCALI'

'SIZZLER', Min, or, 1975; dbl., 28 petals, 1–1.5 in., exhibition form, moderate, spicy fragrance; vigorous, upright, spreading growth; [Sheri Anne X Prominent]; Saville, F. Harmon; Nor'East Min. Roses

'SJOUKJI DIJKSTRA', Pol, mr, 1966; bud ovoid; flowers scarlet, medium, borne in large clusters, semi-dbl.; compact growth; [Chatter X Paprika]; Buisman, G. A. H.

'SKAGGARAK', F, mr, 1970; 18–20 petals, 3.5–6 in., slight fragrance; [Irish Wonder X Seedling]; Poulsen

'SKARLAGEN GENERAL MACARTHUR', HT, dr, 1930; flowers scarlet, large, dbl., slight fragrance; foliage medium, dark, semi-glossy; upright growth; [General MacArthur X Unnamed seedling]; Poulsen, S.; Poulsen's Roses

'SKINNER'S RAMBLER', LCl, lp, 1955; bud long, pointed; flowers pale pink, open, small, single, 5 petals, borne in clusters of 10-40, slight fragrance; foliage soft green; very vigorous, climbing (20 ft. annually) growth; [R. maximowicziana X ?]; Skinner; Univ. of NH

'SKOGUL', F, dp, 1969; flowers rose-madder, open, borne in clusters, semi-dbl., 18 petals; foliage dark, glossy; vigorous growth; [Lichterloh X Lumina]; Lundstad

Skvost, HT; (Cavriglia)

'SKYLARK', F, mp, 1958; bud urn shaped; flowers pink, reverse darker, borne in clusters of 3-15, dbl., 55–65 petals, 2.5 in., exhibition form; foliage glossy; very vigorous, upright growth; [Carol Amling sport]; deVor, Paul F.; Amling Bros.

Skyline *see* TANilyks

Skyliner, LCl, 1980; Delforge; (Cavriglia)

'SKYLON', HT, rb, 1952; bud very long, pointed; flowers orange shaded peach, veined red, dbl., 26 petals, 5.5–6 in.; foliage glossy; very free growth; [Mme Henri Guillot sport]; Lowe

'SKYROCKET', HMsk, dr, 1934; (Wilhelm); bud long, pointed; semi-dbl., blooms in clusters of 50, recurrent bloom, moderate fragrance; foliage large, glossy, leathery; vigorous, bushy (6-8 ft.) growth; (28); [Robin Hood (HMsk) X J.C. Thornton]; Kordes

'SLATER'S CRIMSON CHINA', Ch, mr; (Chinese Monthly Rose, Crimson China Rose, Old Crimson China, R. bengalensis, R. chinensis semperflorens, R. diversifolia, R. indica semperflorens, R. semperflorens); flowers cherry red, usually solitary; slender stems; (14, 21); ca. 1790

'SLATS', HT, mp, 1999; dbl., 26–40 petals, 6 in., borne mostly singly, moderate fragrance; prickles moderate; foliage large, medium green, semi-glossy; upright, tall (5-6 ft) growth; [Louise Estes X Great Scott]; Edwards, Eddie; Johnny Becnel Show Roses, 1999

Slava™ *see* 'WILSLAV'

'SLÁVA BÖHMOVA', HT, rb, 1930; bud long; flowers salmon-red, base golden yellow, large, dbl.; foliage blood-red, mahonia-like; [Covent Garden X Golden Emblem]; Böhm, J.

'SLAVIA', Cl Pol, w, 1934; flowers rosy white, borne in large clusters, recurrent bloom, intense fragrance; vigorous growth; [Tausendschön X Seedling]; Brada, Dr.; Böhm

'SLÁVUSE', HT, w, 1936; flowers cactus dahlia form; Brada, Dr.

'SLEEPING BEAUTY', HT, w, 1966; bud long, pointed; flowers large, dbl., exhibition form, moderate fragrance; foliage leathery; vigorous, upright growth; [(Frau Karl Druschki X Rex Anderson) X Virgo]; Morey, Dr. Dennison; Country Garden Nursery

'SLEEPY', Pol, mp, 1955; (Baldwin); flowers very small blooms in trusses, dbl.; [(Orange Triumph X Golden Rapture) X Polyantha seedling]; deRuiter; Gregory & Willicher Baumschulen

'SLEEPY TIME', Min, op, 1973; bud long, pointed; flowers soft peach to soft salmon-pink, small, dbl., slight fragrance; foliage small, light, leathery; vigorous, dwarf, upright, bushy growth; [Ellen Poulsen X Fairy Princess]; Moore, Ralph S.; Sequoia Nursery

'SLEIGH BELLS', HT, w, 1950; bud ovoid, cream; flowers white, center creamy, open, large, dbl., 40 petals, intense fragrance; foliage leathery, glossy; very vigorous, upright growth; [Capt. Thomas X Eternal Youth]; Howard, P.J.; H&S

Sleigh Bells, Climbing, Cl HT, w, 1995; Wilkins

Slot van Laarne, S, lp, 1993; RvS-Melle

'SMALL FANTASY', Min, lp, 1982; flowers small, dbl., slight fragrance; foliage small, light green, semi-glossy; bushy growth; [Snow Magic X Blue Mist]; Dobbs, Annette E.

Small Maiden's Blush *see* Maiden's Blush

Small Miracle *see* 'JACIMIN'

Small Slam *see* 'LAVLINC'

'SMALL TALK', F, my, 1963; dbl., 33 petals, 2.5 in., exhibition form, blooms in large clusters, slight fragrance; foliage leathery, glossy, dark; compact, low growth; [Yellow Pinocchio X Circus]; Swim & Weeks; Weeks Roses

'SMALL VIRTUE', Min, w, 1986; flowers small blooms borne singly and in spraysof 2-7, dbl., 48 petals, exhibition form, slight fragrance; fruit not observed; medium pink prickles; foliage small, medium green, semi-glossy; low, bushy growth; [Party Girl X Snow Bride]; Jolly, Marie; Rosehill Farm, 1983

'SMALL WONDER', Min, pb, 1983; flowers light pink, darker reverse, small blooms borne singly, dbl., exhibition form, no fragrance; foliage small, medium green, semi-glossy; upright, bushy growth; [Futura X Orange Honey]; Hardgrove, Donald L.; Rose World Originals

'SMALL WORLD', Min, or, 1975; bud pointed; flowers rich orange-red, dbl., 21 petals, 2.5 in., flat; foliage small, glossy, leathery; dwarf, compact growth; [Little Chief X Fire Princess]; Moore, Ralph S.; Sequoia Nursery

Smarty® *see* 'INTERSMART'

Smerisal, HT, op

'SMILES', F, lp, 1937; flowers light salmon-pink, blooms in clusters, semi-dbl., slight fragrance; foliage leathery; bushy growth; [Echo X Rev. F. Page-Roberts]; Nicolas; J&P

Smiles *see* 'JACSMI'

'SMILEY', F, ob, 1979; bud pointed; flowers orange, apricot, yellow, shapely, dbl., 21 petals, 2.5 in.; foliage matt green; tall, vigorous, upright growth; [Arthur Bell X Little Darling]; Murray, Nola

'SMILING THROUGH', F, or, 1976; flowers orange to red, dbl., 27 petals, 3 in., slight fragrance; foliage glossy, light; [Orange Sensation X Mme Louis Laperriere]; Anderson's Rose Nurseries

Smiling Wings, F, dp

'SMITFIRST', F, ab, 1986; (**Summer Sérénade**®); flowers apricot to gold, fading to cream, urn-shaped, medium, semi-dbl., 25 petals, slight fragrance; round, small, orange fruit; medium, pointed, brown prickles; foliage medium, dark green, glossy; upright medium growth; GM, Bagatelle, 1986 GM, Geneva, 1986; [(Unnamed seedling X Zambra) X Baby Bio]; Smith, Edward; Wheatcroft Roses, 1987

Smith's Parish *see* **'FORTUNE'S FIVE-COLORED ROSE'**

'SMITLING', F, or, 1983; (Lady Taylor); flowers vermilion, patio, medium, dbl., 35 petals, slight fragrance; foliage medium, medium green, matt; bushy growth; [Elizabeth of Glamis X Topsi]; Smith, Edward; C. Gregory & Sons, 1984

'SMITS' BRIAR', HRg, ?; [R. rugosa X R. canina]; Smits

'SMITSBLANC', F, w, 1984; (**Fragrant Mist**); flowers well-formed, medium, dbl., 35 petals, intense fragrance; foliage medium, medium green, matt; upright growth; [Elizabeth of Glamis X Jubilee Celebration]; Smith, Edward

'SMITSHORT', F, mp, 1994; (**Richard Buckley**); flowers pale salmon pink, blooms borne in small clusters, dbl., 15–25 petals, 1.5–2.75 in., slight fragrance; some prickles; foliage medium, medium green, matt; low (45-55 cms), bushy, compact growth; [Lady Taylor X Regensberg]; Smith, Edward; Smith, 1992

Smoke Signals™ *see* 'MINAXCO'

Smokey Joe, ob, 1998; Williams, J. Benjamin

Smokey Joe's Cafe *see* 'SHERIJOE'

'SMOKY', HT, rb, 1968; flowers smoky oxblood-red shaded burgundy, medium, dbl.; vigorous, upright growth; Combe; J&P

'SMOKY MOUNTAIN', Min, m, 1988; bud pointed; flowers deep mauve, medium, borne usually singly, dbl., 24 petals, exhibition form, slight fragrance; prickles straight, pointed, small, tan; foliage medium, dark green, semi-glossy; upright, medium growth; [Black Jade X Unnamed seedling]; Bridges, Dennis A.; Bridges Roses, 1989

Smooth Angel *see* 'HADANGEL'

Smooth Lady *see* 'HADLADY'

Smooth Melody *see* 'HADMELODY'

Smooth Perfume *see* 'HADPERFUME'

Smooth Prince *see* 'HADPRINCE'

Smooth Princess, HT, w, 1999; Davidson

Smooth Romance™ *see* 'HADROMANCE'

'SMOOTH SAILING', Gr, w, 1977; flowers cream color, dbl., 30 petals, 4 in., slight fragrance; foliage dark, glossy; tall growth; [Little Darling X Pink Favorite]; Davidson, Harvey D.; Burgess Seed & Plant Co.

Smooth Satin *see* 'HADSATIN'

'SMOOTH TALK', F, w, 1987; flowers ivory-white, aging bright white, medium, abundant, dbl., 25 petals, exhibition form, borne in clusters, slight fragrance; prickles pointed, small, yellow-brown; foliage medium, medium green, semi-glossy; upright, bushy, low, compact growth; PP007156; [Unnamed seedling X Unnamed seedling]; Weeks, O.L.; Weeks Wholesale Rose Growers, 1988

Smooth Velvet *see* 'HADVELVET'

'SMUTS MEMORY', HT, dr, 1950; flowers brilliant deep crimson-red; [Sensation X World's Fair]; Leenders, M.

'SNAPPIE', HT, rb, 1989; bud pointed; flowers medium red with yellow at base, reverse creamy-yellow, aging, dbl., 52 petals, exhibition form; prickles curved slightly downward, medium, deep pink; foliage medium, dark green, semi-glossy; upright, medium growth; [Lady X X Wini Edmunds]; Bridges, Dennis A.; Bridges Roses, 1990

'SNEEZY', Pol, dp, 1955; (Bertram); flowers neyron rose, small, single; deRuiter; Gregory & Willicher Baumschulen

'SNEHURKA', Pol, w, 1937; flowers small, dbl.; Böhm, J.; (Sangerhausen)

Snehurka, Min, lp; Urban, J.; (Czech Rosa Club)

Snehurka, Min, lp; Vik, 1950; (Czech Rosa Club)

'SNEPRINCESSE', Pol, w, 1953; bud ovoid; flowers pure white, small, borne in clusters, dbl., cupped, slight fragrance; foliage light green, glossy; moderate, upright, bushy growth; [Mothersday sport]; Bang

'SNEPRINSESSE'®, Pol, w, 1946; [Dick Koster sport]; Grootendorst, F.J.

'SNO', Min, w, 1982; dbl., 48 petals, exhibition form, blooms in clusters of 1-3, intense fragrance; foliage narrow, light green, matt; sprawling growth; [Unnamed seedling X Gold Pin]; Meredith, E.A. & Rovinski, M.E.; Casa de Rosa Domingo

Sno Cone *see* 'PIXSNO'

Snodoll *see* 'TRASNO'

Snookie *see* 'TINSNOOK'

'SNOOPY', HT, rb, 1973; flowers carmine, reverse silver, urn shaped, very large, dbl., 28 petals, moderate, spicy fragrance; foliage large, glossy; [Paddy McGredy X Rose Gaujard]; Cadle's Roses

'SNÖVIT', Pol, w, 1946; flowers pure white; [Dick Koster sport]; Grootendorst, F.J.

Snow Angel *see* 'JACWHIP'

'SNOW BALLET'®, S, w, 1977; (Claysnow, Snowballet®); flowers pure white, dbl., 45 petals, 4 in., cupped, slight fragrance; foliage small, dark, glossy; GM, Baden-Baden, 1980; [Sea Foam X Iceberg]; Clayworth; Harkness New Roses, Ltd., 1978

Snow Bear *see* 'ADASNOW'

'SNOW BRIDE', Min, w, 1982; (Snowbride); dbl., 20 petals, exhibition form, slight fragrance; foliage medium, medium green, semi-glossy; bushy growth; PP005579; AOE, 1983; [Avandel X Zinger]; Jolly, Betty J.; Rosehill Farm

Snow Carpet® *see* 'MACCARPE'

Snow Cloud *see* TANinaso

'SNOW CREAM', HT, w, 1986; flowers cream, yellow stamens, blooms in clusters of 4-6, semi-dbl., 11 petals, slight fragrance; medium, red, hooked prickles; foliage medium, dark, glossy; medium, upright growth; [Garden Party X Portrait]; Bridges, Dennis A.; Bridges Roses

'SNOW CRYSTAL', HT, w, 1978; flowers large, dbl., 20 petals, slight fragrance; foliage large, light green, semi-glossy; [Sonia X Pascali]; Verschuren, Ted; H.A. Verschuren

Snow Drop, Min, w, 1996; Welsh

Snow Dwarf *see* **'SCHNEEZWERG'**

'SNOW FAIRY', F, w, 1963; bud pointed; semi-dbl., 18 petals, cupped, blooms in clusters, moderate fragrance; foliage dark, glossy, leathery; vigorous, bushy growth; [Virgo X Katharina Zeimet]; Camprubi, C.; C-P

'SNOW FLURRY', F, w, 1952; bud small, long, pointed, tinted pink in cool weather; flowers open, borne in clusters, semi-dbl., moderate fragrance; foliage leathery, glossy, dark; vigorous, bushy growth; [Unnamed seedling X Red Ripples]; Moore, Ralph S.; Marsh's Nursery

Snow Goose *see* 'AUSPOM'

Snow Gosling *see* POUlgren

'SNOW HEDGE', HRg, w, 1963; flowers pure white, medium, single; red fruit; upright, compact growth; [(R. rugosa rubra X R. cinnamomea) X R. nitida]; Nyveldt

Snow Hit *see* POUlnows

'SNOW INFANT', Min, w, 1989; (Koyuki); bud ovoid; flowers light green, reverse white, aging white, small, borne in sprays, dbl., 40 petals, no fragrance; ovoid, small, red fruit; prickles small, light green; foliage small, light green, semi-glossy; upright, medium growth; [Katharina Zeimet X Unnamed seedling]; Yamasaki, Kazuko

Snow Kiss, Min, w, 1997; Kordes

'SNOW MAGIC', Min, w, 1976; bud short pointed, light pink; flowers medium, dbl., 40 petals, slight fragrance; foliage small, soft; bushy growth; [From unnamed seedling sport]; Moore, Ralph S.; Sequoia Nursery

Snow Maiden, *see* 'JUSMAIDEN'

Snow Meillandina® *see* MEIgovin

Snow Owl *see* UHLensch

Snow Pavement *see* Schneekoppe

Snow Princess *see* 'LAVPERT'

Snow Queen *see* **'FRAU KARL DRUSCHKI'**

Snow Rambler, HMult, w, 1995; Nieuwesteeg, J.

Snow Ruby, MinFl, rb, 1996

Snow Shower *see* 'JACWADE'

'SNOW SPRAY', F, w, 1957; flowers pure white, stamens yellow, dbl., moderate fragrance; dwarf growth; [Gartendirektor Otto Linne X Gartendirektor Otto Linne]; Riethmuller

Snow Twinkle® *see* 'MORSNO'

'SNOW WHITE', HT, w, 1941; bud long, pointed; flowers large, dbl., exhibition form, moderate fragrance; foliage leathery, dark; vigorous, bushy, compact growth; [Joanna Hill X White Briarcliff]; Hill, Joseph H., Co.

'SNOW WHITE', Min, w, 1955; flowers white tinted blush, micro-mini, blooms in clusters, dbl., 1 in.; low (3-6 in), spreading growth; [Little Princess (Pol) X Baby Bunting]; Robinson, T.

'SNOW WHITE', Pol, w; [Dick Koster sport]

'SNOW WHITE', HT, w, 1938; flowers well formed, large, intense fragrance; [White Ophelia X Nuria de Recolons]; Dot, Pedro

Snow White *see* LANdisney

'SNOW WHITE, CLIMBING', Cl HT, w; Cant, F.

'SNOW WONDER', Min, w, 1980; bud ovoid, pointed; dbl., 48 petals, borne 3-5 per cluster, slight fragrance; tiny, curved prickles; foliage tiny, medium green; vigorous, bushy growth; [Red Can Can X Baby Betsy McCall]; Lyon

Snowball *see* 'MACANGELI'

'SNOWBALL', Pol, w, 1901; dbl.; Walsh

Snowballet® *see* **'SNOW BALLET'**®

'SNOWBANK', F, w, 1937; flowers white tinted blush, changing to white, dbl., 30 petals, 3.5 in., cupped, slight fragrance; foliage leathery, dark; dwarf growth; [Mrs E.P. Thom X Gloria Mundi]; Nicolas; J&P

'SNOWBELT', Pol, w, 1997; flowers double, medium, dbl., 15–25 petals, borne in large clusters, slight fragrance; some prickles; foliage medium, light green turning medium green, semi-gloss; medium, bushy growth; [Seedling X Seedling]; Jerabek, Paul E.

'SNOWBIRD', HT, w, 1936; bud long, pointed; flowers white, center creamy, dbl., exhibition form, intense fragrance; foliage leathery; vigorous, compact, bushy growth; [Chastity X Louise Crette]; Hatton; C-P

'SNOWBIRD, CLIMBING', Cl HT, w, 1949; Weeks

Snowblush *see* 'CLESNOW'

Snowbound *see* 'LAVAWAY'

Snowbride *see* **'SNOW BRIDE'**

'SNOWDANCE', F, w, 1971; dbl., 36 petals, 3 in., slight fragrance; moderate, bushy growth; [Orange Sensation X Iceberg]; deRuiter

Snowdon, HRg, w, 1989; Austin, David

'SNOWDRIFT', HWich, w, 1913; flowers pure white, small blooms in clustersof 20-30, dbl.; foliage very large, light; vigorous (8-12 ft) growth; RULED EXTINCT 1/85; Walsh

Snowdrift® *see* 'WILSNOD'

Snowdrop *see* 'AMORU'

Snowdwarf *see* **'SCHNEEZWERG'**

Snowfall® *see* 'LEMFALL'

'SNOWFIRE', HT, rb, 1970; bud ovoid; flowers bright red, reverse white, large, dbl., slight fragrance; foliage large, dark, leathery; vigorous growth; [Detroiter X Liberty Bell]; Kordes, R.; J&P

Snowflake *see* **'MARIE LAMBERT'**

'SNOWFLAKE', T, w, 1890; Strauss & Co.

'SNOWFLAKE', HWich, w, 1922; flowers pure white, blooms in clusters, dbl., intense fragrance; foliage dark, glossy; vigorous, climbing growth; GM, NRS, 1921; Cant, F.

'SNOWFLAKE', Min, w, 1977; [Chipper sport]; Ludwig Roses

Snowflake *see* 'SUNSNOFLE'

'SNOWFLAKES', Min, w, 1954; flowers small, dbl.; dwarf (6 in) growth; [(R. wichurana X Floradora) X Zee]; Moore, Ralph S.; Sequoia Nursery

Snowgoose *see* 'BARSHIFLO'

'SNOWHITE CLIMBER', LCl, w, 1938; flowers large, dbl.; long, strong stems; vigorous climbing growth; Burbank; Stark Bros.

'SNOWLINE'®, F, w, 1970; (Edelweiss); dbl., 31 petals, 3 in., slight fragrance; ADR, 1970; Poulsen, Niels D.; McGredy

Snowman *see* 'MATTJO'

'SNOWSONG', HT, w, 1961; bud long, pointed; flowers white, base yellow, semi-dbl., 20 petals, 4.5 in., slight, sweetbriar fragrance; foliage dark, glossy; moderately vigorous, upright growth; [Snow White (HT, Hill) X Snow White seedling]; Jelly; E.G. Hill Co.

'SNOWSONG SUPREME', HT, w, 1965; bud long, pointed; flowers white, base greenish, medium, exhibition form, moderate fragrance; moderate growth; [Snowsong X White Butterfly]; Jelly; E.G. Hill Co.

'SNOWSTORM', HMsk, w, 1907; flowers pure white, small blooms in clusters of 6-20, semi-dbl., recurrent bloom; vigorous growth; [R. moschata X Climber]; Paul

Snuggles *see* 'RENUGGLE'

So In Love *see* 'UMSLOVE'

Soaring Flight, LCl, pb, 1996; Clements, John K.

Soaring Wings *see* 'KORWINGS'

'SOBHAG', F, ob, 1972; bud urn shaped; flowers orange, open, medium, semi-dbl., moderate fragrance; foliage glossy, dark, leathery; vigorous, bushy growth; [Orangeade X ?]; Bansal, O.P.; Bansal Roses

'SOCALP', Min, w, 1986; (**Moonlight Lady**); flowers creamy white, center buff pink, small blooms borne 1-3 per cluster, dbl., 52 petals, exhibition form, slight fragrance; narrow, red prickles; foliage medium, dark, semi-glossy; upright growth; [Pink Petticoat X Pink Petticoat]; Eagle, Barry & Dawn; Southern Cross Nursery, 1987

'SOCAMP', Min, pb, 1985; (**Petticoat Lane**); flowers medium pink, deeper in center, light pink reverse, small blooms, dbl., 28 petals, exhibition form, slight fragrance; foliage small, dark, semi-glossy; upright, very small growth; [Pink Petticoat X Pink Petticoat]; Eagle, Barry & Dawn; Southern Cross Nursery

'SOCAPAN', Min, yb, 1985; (**Calumet**); flowers creamy yellow, petals edged with pink, blooms borne 1-3 per, dbl., 30 petals, exhibition form, moderate fragrance; very few, small, red prickles; foliage medium green, semi-glossy; upright, bushy growth; [Golden Angel X Golden Angel]; Eagle, Barry & Dawn; Southern Cross Nursery

Society Special *see* 'CHEWSOS'

SOCluck, MinFl, mp, 1997; (**Lucky Me**); Eagle

'SOCOTRA', Min, ob, 1985; (**Orange Spice**); flowers orange, yellow reverse, small blooms in clusters of 1-3, dbl., 25 petals, slight fragrance; very few prickles; foliage small, medium green; short, compact, very bushy growth; [Over the Rainbow X Over the Rainbow]; Eagle, Barry & Dawn; Southern Cross Nursery

'SOCRATES', T, pb, 1858; flowers deep rose tinged fawn, large, dbl.; Moreau et Robert

'SOCSPO', Min, ab; (**Apricot Mist**)

'SODENIA', HWich, rb, 1911; flowers bright carmine changing to deep pink, small, borne in large, dbl.; vigorous, climbing growth; Weigand, C.

'SODORI-HIMÉ', HT, w, 1975; (La Blancheur); dbl., 30 petals, 4.5 in., exhibition form, slight fragrance; foliage dark; bushy growth; [White Knight X White Prince]; Onodera, Toru F.; S. Onodera

'SOESTDIJK', F, ob, 1949; flowers deep orange, base buttercup-yellow, medium, semi-dbl., moderate fragrance;

vigorous, compact growth; [Vanessa X Seedling]; Leenders, M.

'SOEUR KRISTIN', HKor, my, 1978; flowers large blooms borne singly, dbl., 20 petals, repeat bloom, slight fragrance; foliage medium, medium green, glossy; semi-spreading growth; [Blanche Mallerin X (R. kordesii X (Van Bergen X Soeur Thérèse))]; James, John

'SOEUR MARIE-ANGE', F, w, 1957; flowers snow-white, very dbl.; bushy, dwarf growth; Privat

Soeur Marthe, M, dp, 1848; Vibert

'SOEUR THÉRÈSE', HT, yb, 1931; (Sister Thérèse); bud long, pointed; flowers golden yellow flushed and edged carmine, large, dbl., 25 petals, cupped, slight fragrance; foliage leathery, bronze; vigorous, bushy growth; (28); [(Général Jacqueminot X Juliet) X Souv. de Claudius Pernet]; Gillot, F.; C-P

'SOEUR THÉRÈSE, CLIMBING', Cl HT, yb, 1953; Shira

Soft Scent *see* 'RENSCENT'

Soft Steps *see* 'FLOSOF'

Soft Touch *see* 'JACOUCH'

Softee *see* 'MORFREE'

Softee, Climbing, Cl Min, lp; [Softee sport]; Hannemann, F., 1992; The Rose Paradise; (Weatherly, L.)

Softly Softly *see* 'HARKOTUR'

Softy *see* TANpretty

Softy, Pol, lp, 1992; flowers small, dbl.; Tantau; (Sangerhausen)

'SOGNO', HT, mr, 1973; bud ovoid, globular; flowers orient red to geranium-like, open, medium, dbl., cupped, slight fragrance; foliage large, dark, leathery; vigorous, upright, bushy growth; Calvino

'SOIR D'AUTOMNE', HT, m, 1966; bud long, pointed; flowers violet, medium, dbl., cupped, moderate fragrance; foliage light green, leathery; vigorous growth; [(Sterling Silver X Intermezzo) X (Sterling Silver X Simone)]; Dot; Minier

Soir de Fete, HT, my; Croix

Soir d'Eté, F, 1977; Croix; (Cavriglia)

Soiree, S, op, 1993; (Soiree de Bonheur); Croix; (Cavriglia)

'SOLAR FLAIR', Min, rb, 1996; flowers bright crimson with yellow reverse, small blooms borne mostl, semi-dbl., 6–14 petals, no fragrance; some prickles; foliage small, dark green, glossy; compact, low growth; [Rise 'n' Shine X Captivation]; Tucker, Robbie

Solar Flash, HT, lp

Solarflare, *see* 'CAPFLARE'

Solarflash *see* 'CAPFLARE'

'SOLARIA', HT, or, 1976; bud long, pointed; dbl., cupped; foliage dark; upright growth; [(Anabell X Zorina) X Seedling]; Kordes, W. Söhne; Barni

'SOLARIUM', HWich, w, 1925; flowers velvety vermilion-red, stamens yellow, very large, borne in, single; foliage rich glossy green; very vigorous, climbing growth; Turbat

'SOLDIER BOY', LCl, mr, 1953; flowers scarlet, large, single, repeat bloom; vigorous, pillar growth; [Unnamed seedling X Guinee]; LeGrice

'SOLDIER'S PRIDE', HT, dr, 1987; flowers medium, dbl., 26–40 petals, slight fragrance; foliage medium, dark green, semi-glossy; upright growth; [Red Planet X Clare]; MacLeod, Major C.A.

Sole di San Remo, S, 1975; Mansuino; (Cavriglia)

Soleil® *see* 'MALSO'

'SOLEIL D'ANGERS', HFt, yb, 1910; flowers orange-yellow, medium, very dbl.; Détriché; (Sangerhausen)

'SOLEIL DE FRANCE', HT, yb, 1931; flowers sun-yellow, center reddish, large, dbl., moderate fragrance; foliage thick; vigorous growth; [Souv. de Claudius Pernet X Unnamed seedling]; Mermet; J&P

'SOLEIL DE LYON', F, op, 1955; flowers salmon-pink; vigorous growth; Robichon; Pin

Soleil de Minut, HT; Delbard-Chabert; (Cavriglia)

'SOLEIL DE RUSTICA', LCl, dy; flowers golden yellow, large; Cognet

Soleil d'Été *see* **'SUMMER SUNSHINE'**

'SOLEIL D'OR', HFt, yb, 1900; bud long, pointed; flowers orange-yellow to ruddy gold, shaded nasturtium-red, large, dbl., moderate fragrance; foliage rich green; vigorous growth; (28); [Said to be F2 seedling of Antoine Ducher X R. foetida persiana.]; Pernet-Ducher

'SOLEIL D'ORIENT', Cl HT, rb, 1935; bud long, pointed; flowers indian red, shaded yellow, very large, dbl., abundant, non-recurrent bloom; foliage glossy; long stems; vigorous, climbing (6 ft.) growth; [Frau Karl Druschki X Mme Edouard Herriot]; Croibier

'SOLEIL LEVANT', LCl, ob, 1956; flowers orange, borne in clusters, single; [Spectacular X Unnamed seedling]; Mondial Roses

Soleillade *see* DORcast

'SOLFATERRE', N, my, 1843; flowers large, dbl., slight fragrance; vigorous growth; [Lamarque seedling]; Boyau

Solid Gold *see* 'LEONORO'

Solidor® *see* MEIfarent

Solidor *see* MEItylpic

Soliman, F, 1975; VEG; (Cavriglia)

'SOLISTKA BALETA', HT, yb, 1955; (Prima Ballerina); flowers soft lemon-yellow edged pink, large, moderate fragrance; [(Peace X Crimson Glory) X Poinsettia]; Klimenko, V. N.

Solitaire *see* 'MACYEFRE'

'SOLITAIRE', F, mp, 1970; flowers coral-pink, reverse silvery, dbl., 25 petals, 3 in., moderate fragrance; foliage glossy, dark; vigorous growth; [Queen Elizabeth X Elysium]; Cants of Colchester, Ltd.

'SOLITAIRE, CLIMBING', Cl, F, 1997; flowers double, medium, dbl., 15–25 petals, moderate fragrance; many prickles; foliage medium, medium green, semi-glossy; upright (7ft.) growth; [Solitaire sport]; Earnshaw, Ronald

Solitude™ *see* 'POULBERO'

Solitude *see* SEAsol

'SOLLER', HT, my, 1949; bud long, pointed; flowers overlarge, very dbl.; foliage glossy, bronze; bushy growth; [Eduardo Toda X Senateur Potie]; Dot, Pedro

'SOLLIDEN', HT, rb, 1924; bud long, pointed; flowers carmine, reverse shaded ochre, open, large, semi-dbl., slight fragrance; foliage dark; vigorous growth; [(Mme Mélanie Soupert X George C. Waud) X Mme Edouard Herriot]; Leenders, M.

'SOLO'®, Cl HT, mr, 1956; flowers dark red, well-formed, large blooms in large clusters, dbl., slight fragrance; foliage dark, leathery; vigorous growth; [Crimson Glory X ?]; Tantau, Math.

Solo Mio, *see* **Sophia Renaissance**

Solo Mio *see* POUlsol

'SOLUS', S, or, 1967; flowers bright orange-scarlet, blooms in clusters, semi-dbl., 19 petals, 2.5 in., slight fragrance; foliage dark, glossy, leathery; very vigorous growth; [Kathleen Ferrier X Dickson's Flame]; Watkins Roses

Solvang *see* 'POULVANG'

'SOMA', HT, m, 1980; bud long, pointed; flowers mauve blend, blooms borne 8 per cluster, dbl., 40 petals, no fragrance; straight prickles; foliage dark, leathery; bushy growth; [Chandrama X Surekha]; Division of Vegetable Crops and Floriculture

'SOMBRERO', F, pb, 1962; flowers cream flushed pink, well-formed, dbl., 25 petals, 4 in., slight fragrance; foliage light green; vigorous growth; [Masquerade X Rubaiyat]; McGredy, Sam IV; McGredy

'SOMBREUIL', Cl T, w, 1850; flowers creamy white, often tinted pink, well-formed, large, dbl., flat; vigorous growth; [Gigantesque seedling]; Robert

Someday Soon *see* 'SEASOON'

Somersault *see* 'JACPLY'

Something Else® *see* 'SAVAPIE'

Something for Judy *see* 'TINJUDY'

Sommerabend® *see* KORmarec

Sommerduft® *see* 'TANFUDERMOS'

Sommerfreude®, F, lp

Sommerlachen, F, ob, 1972; flowers coppery orange and dark golden yellow, large, dbl., intense fragrance; GPG Bad Langensalza; (Sangerhausen)

Sommerliebe, F, ob, 1986; flowers coppery-pink, large, dbl., intense fragrance; GPG Bad Langensalza; (Sangerhausen)

Sommerliebe, HT, or, 1975; flowers large, dbl., slight fragrance; Pörschmann; (Sangerhausen)

Sommermarchen *see* KORpinka

'SOMMERMÄRCHEN', F, dp, 1945; flowers dark rose, shell shaped, large, borne in clusters of 12-18, single, slight fragrance; foliage light green, leathery; upright, vigorous, bushy growth; [Prof. Gnau X Baby Chateau]; Tantau

Sommermelodie®, S, lp

Sommermond® *see* KORmazin

Sommermorgen *see* 'KORFULLWIND'

'SOMMERNACHTSTRAUM', Pol, rb, 1944; flowers carmine-red with white, medium, single; Krause; (Sangerhausen)

Sommerschirm, F, 1993; Tantau; (Cavriglia)

Sommerschnee, F, w; flowers large, dbl.; VEG; (Sangerhausen)

Sommertag, HT, or, 1969; flowers large, very dbl.; Gregory; (Sangerhausen)

'SOMMERTRAUM', F, pb, 1960; flowers pink, red and yellow, borne in clusters, semi-dbl.; foliage dark; [Masquerade X Unnamed seedling]; Verschuren, A.; van Engelen

Sommerwind *see* 'KORLANUM'

'SOMMET', LCl, or, 1960; flowers well formed, dbl.; foliage bright green; vigorous growth; Mallerin, C.; Hémeray-Aubert

'SOMNIP', HT, m, 1996; (**Blackberry Nip**); flowers deep purple, blooms borne mostly single, very dbl., intense fragrance; some prickles; foliage medium, medium green, semi-glossy; upright, medium growth; [Deep Secret X Old Port]; Somerfield, Rob

'SONATA', HT, pb, 1942; bud long, pointed; flowers red becoming lively pink, reverse darker, dbl., 30 petals, 4 in., exhibition form, moderate, fruity fragrance; foliage glossy; vigorous, bushy growth; Van Rossem; J&P

Sonata *see* 'ZIPSON'

Sonata in Pink *see* 'JOLSON'

'SONATINA', F, lp, 1982; flowers small, dbl., 35 petals, slight fragrance; foliage medium, dark, semi-glossy; [Red Maid X Sarabande]; Sanday, John; Sanday Roses, Ltd.

Soncy, T, ly

Sondermeldung *see* **'INDEPENDENCE'**

'SONG BIRD', Gr, w, 1978; [Tammy sport]; Ryan, C.; J. Hill Co.

'SONG OF PARIS', HT, m, 1964; (Saphir); flowers silvery lavender, dbl., 4 in., exhibition form, moderate fragrance; foliage leathery; upright growth; [(Holstein X Bayadere) X Prelude]; Delbard-Chabert; Armstrong Nursery

'SONG OF SONGS', HT, dr, 1998; flowers dark red, full, classic shape, very dbl., 26–40 petals, 5.5 in., exhibition form, borne mostly singly, intense fragrance; some prickles; foliage medium, dark green, semi-glossy; upright, bushy, medium (3.5 ft) growth; [Adrienne Berman X (Royal William X Gabi)]; Poole, Lionel

Songfest *see* 'ZIPSON'

Sonia *see* 'MEIHELVET'

'SONIA', HT, rb, 1938; flowers cherry-red, center orange, camellia form, semi-dbl., slight fragrance; foliage glossy, dark; short stems; vigorous, bushy growth; hardy.; [(R. multiflora X R. canina) X Hortulanis Budde]; Horvath; Wayside Gardens Co.

Sonia Meilland *see* 'MEIHELVET'

Sonia Meilland, Climbing *see* 'MEIHELVETSAR'

Sonia Rykiel *see* MASdogui

Sonia, Climbing *see* 'MEIHELVETSAR'

Sonja 92, Gr, mp

'SONJA HENIE', HT, pb, 1949; bud long, pointed to ovoid; flowers pink, reverse darker, very large, very dbl., exhibition form; foliage leathery, dark; vigorous, upright growth; [Briarcliff Supreme X Rosemarie Hinner]; Hinner, P.; Bauské Bros. & Hinner

Sonja Horstmann, HT, mr

Sonja Supreme, Gr, ab

Sonne der Freundschaft, F, dy, 1978; flowers golden yellow, medium, dbl., slight fragrance; GPG Bad Langensalza; (Sangerhausen)

'SONNENGOLD', HT, dy, 1936; bud long, pointed; flowers golden yellow, large, dbl., exhibition form, slight fragrance; foliage leathery, glossy, light; bushy growth; [Lilian X Sir Basil McFarland]; Kordes

Sonnenkind® *see* 'KORHITOM'

'SONNENLICHT', HFt, my, 1913; flowers canary-yellow, semi-dbl., non-recurrent, moderate fragrance; vigorous growth; [Lady Mary Fitzwilliam X Harison's Yellow]; Krüger; Kriese

Sonnenröschen® *see* 'KORSONN'

Sonnenschirm® *see* TANmirsch

Sonnenuntergang, F, 1970; deRuiter; (Cavriglia)

'SONNET', HT, lp, 1961; bud ovoid; flowers light salmon-pink, dbl., 58 petals, 5 in., cupped, moderate fragrance; foliage leathery; vigorous, upright growth; [Golden Masterpiece X Spartan]; Boerner; J&P

Sonny, S, lp; Peden, G.H., 1995; (Weatherly, L.)

'SONNYCHILD', HT, yb, 1949; (Zonnekeild); flowers yellow, reverse edged scarlet, semi-dbl., 20 petals, 6 in., moderate fragrance; foliage glossy; vigorous growth; Lowe

'SONOMA', F, mp, 1973; flowers medium salmon-pink, medium, dbl., exhibition form, moderate fragrance; foliage leathery; vigorous, upright, bushy growth; [Sumatra X Circus]; Armstrong, D.L.; Armstrong Nursery

'SONORA', F, yb, 1962; bud ovoid; flowers buff-yellow flushed to pink, dbl., 30 petals, 3.5–4 in., cupped, moderate fragrance; foliage leathery; vigorous, upright growth; [Orange Mist X Mayday]; Boerner; J&P

Sonora Sunset *see* DEVmauve

'SONRISA', HT, dr, 1969; flowers deep crimson-red, medium-large, dbl., 48 petals, exhibition form, intense, damask fragrance; foliage dark, leathery; vigorous, upright growth; [Mister Lincoln X Night 'n' Day]; Swim & Weeks; Weeks Wholesale Rose Growers

'SOPHIA', HT, pb, 1988; flowers medium pink with persimmon-orange, large, borne usually singly, dbl., 27 petals, exhibition form, moderate, fruity fragrance; no fruit; no prickles; foliage large, dark green, glossy; upright, tall growth; [First Prize X Dolce Vita]; Weddle, Von C.; Hortico Roses, 1987

'SOPHIA FLEUR', F, ly, 1978; flowers creamy yellow; [Elizabeth of Glamis sport]; Timmerman's Roses

'SOPHIA LOREN'®, HT, mr, 1967; flowers well-formed, large, dbl., 33 petals; foliage glossy; vigorous, upright growth; Tantau, Math.

Sophia Renaissance, S, my; (Solo Mio)

'SOPHIA SMITH', HT, m, 1996; flowers pinkish mauve with salmon center, reverse pinkish mauve, very dbl., 3.25–4 in., slight fragrance; moderate prickles; foliage medium, dark green; upright, medium (5 ft) growth; [Lady X X (Blue Moon X Great News)]; Linck, Robert G.; Linck, 1996

Sophia's Song™ *see* 'DEVLOREN'

Sophie, F, lp, 1972; Eve, A.

'Sophie de Bavière', A, mp; flowers clear pink, very regular, medium, dbl.; Vibert

'Sophie de Marsilly', M, pb, 1863; flowers pink with light white stripes, dbl., globular; more prickly than most albas; vigorous, upright growth; Moreau et Robert; Prior to 1848

Sophie Deborah *see* 'KIRstein'

'Sophie MacKinnon', HT, dr, 1937; flowers deep red, large, dbl.; vigorous growth; [John Cromin X ?]; Clark, A.; NRS Victoria

'Sophie Ortlieb', HP, mp, 1933; bud long, pointed; flowers silvery pink, very large, dbl.; foliage wrinkled; very vigorous, open habit growth; [Georg Arends X Unnamed seedling]; Walter, L.

'Sophie Thomas', Cl HT, my, 1931; (Bloomfield Loveliness); flowers deep yellow, passing to lighter yellow but not cream, large, dbl., moderate fragrance; foliage good; long, strong stems; vigorous growth; [Seedling climbing X Los Angeles]; Thomas; H&S

'Sophie's Perpetual', Ch, pb; flowers pale pink, overlaid with deep pink and cerise red, globular; few prickles; foliage dark green; growth to 8 ft.; Paul, W., prior to 1905; Re-int. Humphrey Brooke, 1960

Sophileo *see* 'MAChorowhen'

'Sophisticated Lady', Gr, lp, 1985; dbl., 25 petals, exhibition form, borne usually singly, slight, fruity fragrance; medium, hooked, dull red prickles; foliage medium, medium green, semi-glossy; upright, tall growth; [Queen Elizabeth X Arlene Francis]; Epperson, Richard G.

'Sophocle', HT, m, 1974; flowers purple-red, large; [Rose Gaujard X Credo]; Gaujard

Sophy's Rose *see* 'AUSlot'

'Soprano', F, or, 1961; flowers large, dbl., moderate fragrance; vigorous growth; [Mannequin X Aztec]; Lens

Soraya® *see* 'MEjenor'

Soraya, Climbing® *see* 'MEjenorsar'

Sorbet *see* MElpeluj

Sorbet Bouquet™ *see* 'JACzotta'

Sorbet Framboise® *see* DELsorb

Sorcerer™ *see* 'SAVasorc'

'Sorcier', LCl, ob, 1958; flowers bright orange, recurrent bloom; foliage bronze; vigorous growth; [Seedling X Spectacular]; Hémeray-Aubert

'Soroptimist', HT, dy, 1958; flowers orange-yellow, well formed, dbl., 40 petals, 4 in.; foliage glossy; vigorous growth; [Golden Scepter X Seedling]; Verbeek

Soroptimist International *see* 'BENstar'

'Souchet', B, m, 1842; flowers purple/pink, large, dbl., intense fragrance; Souchet; (Sangerhausen)

Soukara-Ibara *see* **'Turner's Crimson Rambler'**

'Souma', HT, mp, 1977; dbl., 45 petals, 6 in., exhibition form, slight fragrance; foliage glossy, dark; vigorous, upright growth; Souma Rose Soc.

'Soupert et Notting', M, dp, 1874; flowers deep pink, very large, globular, recurrent bloom, intense fragrance; foliage 5 leaflets per leaf; dwarf growth; Pernet Père

'Source d'Or', LCl, yb, 1913; flowers amber-yellow, edged creamy yellow, large, borne in clusters, very dbl., abundant, non-recurrent bloom, intense fragrance; foliage glossy, dark; short stems; height 6-8 ft; Turbat

Sourire d'Antan®, S, dp; Lens, 1988

'Sourire de France', HT, ob, 1940; bud oval; flowers orange, base ochre-yellow, open, medium, dbl., slight fragrance; foliage leathery, glossy; vigorous, bushy growth; [Ampere X (Charles P. Kilham X Capucine Chambard)]; Meilland, F.; A. Meilland

Sourire d'Enfant® *see* ORAroro

Sourire d'Orchidee, S, lp; Golden Rose, Geneva, 1985; Croix, 1985

Sourire Rosel, HMsk, lp; Lens, 1996

'Sousse', HT, lp, 1942; flowers large, dbl.; Meilland, F.; (Sangerhausen)

Sousyun, F, w

South Moon *see* 'GELmoon'

'South Orange Perfection', HWich, pb, 1899; flowers blush-pink, turning white, rosette form borne in clusters, moderate fragrance; ; very hardy.; [R. wichurana X Cramoisi Superieur]; Horvath; W.A. Manda

South Pacific™ *see* 'ARONesuf'

'South Seas', HT, op, 1962; (Mers du Sud); bud ovoid; flowers coral-pink, dbl., 48 petals, 6–7 in., cupped, moderate fragrance; foliage leathery; vigorous, upright growth; [Rapture X HT, Climbing seedling]; Morey, Dr. Dennison; J&P

'Southampton', F, ab, 1971; (Susan Ann); dbl., 28 petals, 3 in., moderate fragrance; foliage glossy; GM, Belfast, 1974; [(Ann Elizabeth X Allgold) X Yellow Cushion]; Harkness, 1972

'Southend Jubilee', F, dp, 1965; flowers deep pink, borne in clusters, semi-dbl., 2.5 in., flat; very vigorous, bushy growth; McCreadie

'Southern Aurora', F, rb, 1999; flowers red, reverse orange, dbl., 17–25 petals, 3 in.; few prickles; foliage medium, medium green, semi-glossy; upright, medium (4.5 ft) growth; [Redgold sport]; Le Fevre, Ian

'Southern Beauty', HT, pb, 1926; flowers deep rose-pink, edged light pink; [Columbia sport]; Rowe

'Southern Belle', HT, pb, 1981; bud ovoid, long, pointed; flowers deep pink, white reverse, spiraled blooms, dbl., 28 petals, borne singly, slight fragrance; long, narrow prickles; foliage large, semi-glossy; medium, upright, spreading growth; PP005077; [Pink Parfait X Phoenix]; Swim, H.C. & Ellis, A.E.; Armstrong Nursery

Southern Charm *see* 'BRIcharm'

'Southern Cross', HT, mp, 1931; flowers pink, dbl., globular, moderate fragrance; bushy growth; [Joseph Hill X Gen. MacArthur]; Clark, A.; Ballarat Hort. Soc.

Southern Cross, F, lp; [Redgold sport]; Jack, 1977

Southern Delight *see* 'MORdashin'

'Southern Lady', HT, pb, 1988; bud pointed; flowers light pink, center flesh tones, reverse slightly darker pink, dbl., 40 petals, exhibition form, moderate fragrance; prickles downward pointed, medium, tan; foliage medium, dark green, semi-glossy; upright, medium growth; [Lady X X Flaming Beauty]; Bridges, Dennis A.; Bridges Roses, 1989

'Southern Spring', Min, m, 1988; bud ovoid; flowers light mauve shaded pink, reverse lighter mauve edges, darker, dbl., 65 petals, moderate fragrance; prickles slightly downward pointed, medium, tan; foliage medium, dark green, matt; bushy, low growth; [Twilight Trail X Unnamed seedling]; Bridges, Dennis A.; Bridges Roses, 1989

Southern Sun *see* HERbeau

Southern Sunset, HT, ob, 1994 (Tamrabarani); flowers silky orange with yellow reverse and copper overtones, slight fragrance; foliage mildew resistant; 2.5 ft growth; Viraraghavan, M.S.

'Southport', HT, mr, 1933; bud long, pointed; flowers bright scarlet, semi-dbl., 18 petals, cupped, moderate fragrance; vigorous growth; GM, NRS, 1931; [(George Dickson X Crimson Queen) X Souv. de George Beckwith]; McGredy

'Southport, Climbing', Cl HT, mr, 1946; Howard Rose Co.

Souv d'Adele Launay, B, lp; Moreau et Robert, 1872

Souv d'Adolphe de Charvoik, HWich, lp; flowers small, semi-dbl.; (Sangerhausen)

'SOUV D'ADOLPHE THIERS', HP, mr, 1877; flowers very large, dbl., moderate fragrance; Moreau et Robert; (Sangerhausen)

'SOUV D'ADOLPHE TURC', Pol, op, 1924; (Souvenir de Adolphe Turc); flowers clear salmon-pink; Turc, 1926

'SOUV D'AIMÉE TERREL DES CHENES', HCh, op, 1897; flowers coppery-pink, small, dbl., slight fragrance; Schwartz; (Sangerhausen)

'SOUV D'ALEXANDRE BACOT', HT, mr, 1958; flowers geranium-red, large, dbl., 50 petals, slight fragrance; very vigorous growth; [Crimson Glory X Independence]; Arles; Roses-France

'SOUV D'ALEXANDRE BERNAIX', HT, dr, 1926; flowers crimson-scarlet, shaded darker, very large, dbl., cupped; foliage purplish green; vigorous growth; [Étoile de Hollande X Gen. MacArthur]; Bernaix, P.

'SOUV D'ALEXANDRE HARDY', HP, mr, 1898; flowers large, dbl.; Lévêque; (Sangerhausen)

'SOUV D'ALINE FONTAINE', HP, lp, 1879; flowers large, very dbl.; Fontaine; (Sangerhausen)

'SOUV D'ALPHONSE LAVALLÉE', HP, dr, 1884; flowers dark velvety crimson to maroon, moderate fragrance; height to 6 ft; Verdier, C.; C. Verdier

'SOUV D'ANDRÉ RAFFY', HP, dr, 1899; flowers large, dbl., intense fragrance; Vigneron; (Sangerhausen)

'SOUV D'ANGÈLE OPDEBEECK', HT, my, 1926; bud long, pointed; flowers canary-yellow, dbl., moderate fragrance; [Golden Ophelia X Golden Emblem]; Verschuren

'SOUV D'ANNE FRANK', F, ob, 1960; flowers orange tinted yellow and salmon, open, medium, borne in clusters, semi-dbl., 17 petals; foliage glossy, dark; moderate, bushy growth; [Reve de Capri X Chanteclerc]; Delforge

'SOUV D'ANNE-MARIE', HT, op, 1902; flowers yellowish salmon-pink, large, dbl.; Ketten, Gebrüder; (Sangerhausen)

Souv d'Anselme, B, dp; flowers large, dbl.; (Sangerhausen)

'SOUV D'ANTONIN PONCET', HT, dr, 1921; bud long, pointed; flowers carmine, flecked paler, dbl.; [Mme Maurice de Luze X Lady Ashtown]; Schwartz, A.

'SOUV D'ARTHUR DE SANSAL', HP, mp, 1876; flowers large, dbl., intense fragrance; Guénoux; (Sangerhausen)

'SOUV D'AUGUSTE LEGROS', T, mr, 1890; flowers red-crimson, large; Bonnaire

'SOUV D'AUGUSTE MÉTRAL', HT, dr, 1895; flowers crimson; Guillot, P.

'SOUV DE AMAND OPDEBEECK', HT, yb, 1936; bud long, pointed; flowers yellowish apricot-pink, edged pink, very large, dbl., exhibition form, moderate fragrance; foliage leathery, bronze; very vigorous growth; Belge; Opdebeeck

'SOUV DE BEN-HUR', HT, mr, 1960; bud pointed; flowers crimson-scarlet, dbl., 5 in., intense fragrance; very vigorous growth; [Ena Harkness X Charles Mallerin]; Verschuren; Blaby Rose Gardens

'SOUV DE BÉRANGER', HP, mp, 1857; flowers rose, large, dbl.; moderate growth; Bruant

'SOUV DE BERTRAND GUINOISSEAU', HP, m, 1895; flowers large, very dbl., moderate fragrance; Chédane-Guinoisseau; (Sangerhausen)

Souv de Brod *see* **'ERINNERUNG AN BROD'**

'SOUV DE CATHERINE FONTAINE', HT, yb, 1934; bud long, pointed; flowers brownish yellow, center brick-red, reverse coral-red, very large, moderate fragrance; [Souv. de Jean Soupert X Mme Edouard Herriot]; Soupert, C.

'SOUV DE CATHERINE GUILLOT', T, rb, 1895; flowers coppery carmine, center shaded orange, large, dbl., intense fragrance; weak growth; Guillot, R.

'SOUV DE CHARLES GOUVERNEUR', HT, pb, 1927; bud long, pointed; flowers flesh-pink, center salmon-orange, very large, moderate fragrance; Chambard, C.

'SOUV DE CHARLES LAEMMEL', HT, yb, 1919; flowers golden yellow, streaked orange and shaded pink, intense fragrance; [Frau Karl Druschki X Soleil d'Or]; Gillot, F.

'SOUV DE CHRISTOPHE COCHET', HRg, mp, 1894; flowers large, semi-dbl., moderate fragrance; Cochet-Cochet

'SOUV DE CLAUDE VIALLY', HT, rb, 1931; bud long, pointed; flowers light red, tinted pink and aurora; Reymond

'SOUV DE CLAUDIUS DENOYEL', Cl HT, dr, 1920; (Denoyel); bud long, pointed; flowers rich crimson-red, tinted scarlet, large, dbl., cupped, sparse, recurrent bloom, moderate fragrance; vigorous, climbing growth; [Château de Clos Vougeot X Commandeur Jules Gravereaux]; Chambard, C.

'SOUV DE CLAUDIUS PERNET', HT, my, 1920; bud long, pointed; flowers pure sunflower-yellow, center deeper, large, dbl., 28 petals, moderate fragrance; foliage glossy, rich green; long, strong stems; vigorous growth; GM, Bagatelle, 1920; [Constance X Unnamed variety]; Pernet-Ducher

'SOUV DE CLAUDIUS PERNET, CLIMBING', Cl HT, my, 1925; Western Rose Co.

'SOUV DE CLAUDIUS PERNET, CLIMBING', Cl HT, my, 1932; Schmidt, J.C.

'SOUV DE CLAUDIUS PERNET, CLIMBING', Cl HT, my, 1933; Gaujard

'SOUV DE CLERMONDE', HT, pb, 1925; flowers salmon-rose, center darker, shaded yellow, semi-dbl., moderate fragrance; Pernet-Ducher

'SOUV DE DENIER VAN DER GON', HT, rb, 1935; bud long, pointed; flowers reddish yellow to golden yellow, large, dbl., moderate fragrance; very vigorous growth; [Roselandia X Souv. de Claudius Pernet]; Verschuren-Pechtold

'SOUV DE E. GUILLARD', HT, rb, 1912; flowers reddish yellow shaded coppery carmine, large, dbl., moderate fragrance; very vigorous, branching growth; GM, Bagatelle, 1914; [Beauté Inconstante X Le Progres]; Chambard, C.

'SOUV DE F. BOHÉ', HT, op, 1922; bud long, pointed; flowers orange-salmon, dbl., 35 petals; [Willowmere X Unnamed seedling]; Chambard, C.

'SOUV DE FRANCIS BORGES', HT, ob, 1932; flowers flesh, center orange, very large, moderate fragrance; vigorous growth; [Mme Leon Pain X Unnamed seedling]; Chambard, C.

'SOUV DE FRANÇOIS GAULAIN', T, rb, 1889; flowers crimson and violet; Guillot

'SOUV DE FRANÇOIS GRAINDORGE', HT, yb, 1928; bud long, pointed; flowers ochre-yellow, base indian yellow, dbl., moderate fragrance; [Benedicte Sequin X Lady Hillingdon]; Grandes Roseraies

'SOUV DE FRANÇOIS MERCIER', HT, ob, 1923; bud long, pointed; flowers light coppery rose, edged deeper, dbl., moderate fragrance; [Antoine Rivoire sport]; Laperrière

'SOUV DE FRANÇOIS RICHARDIER', HT, rb, 1923; flowers bright carmine-pink, tinted cherry; Richardier

'SOUV DE GABRIEL LUIZET', HT, yb, 1922; flowers sulfur-yellow tinted salmon, passing to deep rich yellow and, dbl.; [Mme Mélanie Soupert seedling X Lyon Rose]; Croibier

'SOUV DE GABRIELLE DREVET', T, op; flowers salmon-pink, base coppery, large, well formed, intense fragrance; Guillot et Fils, 1885

'SOUV DE GASTON COMMAGÈRES', HT, ob, 1954; flowers orange-ivory veined yellow, edge veined pink, very dbl.; vigorous, bushy growth; Privat

'SOUV DE GEORGE BECKWITH', HT, pb, 1919; bud long, pointed; flowers shrimp-pink, tinted chrome-yellow, base deeper, very large, dbl., 55 petals, globular, moderate fragrance; foliage glossy; very vigorous growth; [Unnamed variety X Lyon Rose]; Pernet-Ducher

'SOUV DE GEORGE KNIGHT', HT, my, 1926; flowers nanking yellow; [Rayon d'Or sport]; Knight, J.

'SOUV DE GEORGES PERNET', HT, op, 1921; flowers medium salmon-pink, large, dbl., 31 petals, intense fragrance; foliage dark, bronze; vigorous growth; GM, Bagatelle, 1921; [Unnamed seedling X Mme Edouard Herriot]; Pernet-Ducher

'SOUV DE GEORGES PERNET, CLIMBING', Cl HT, op, 1927; Pernet-Ducher

'SOUV DE GILBERT NABONNAND', T, pb, 1920; flowers in summer: yellow base edged carmine-pink, dbl., intense fragrance; vigorous growth; (28); Nabonnand, C.

'SOUV DE GREGOIRE BORDILLON', HP, 1860; Moreau et Robert; (Cavriglia)

'SOUV DE GUSTAVE PRAT', HT, ly, 1909; flowers pure light sulfur-yellow, dbl.; Pernet-Ducher

'SOUV DE GUSTAVE SCHICKELÉ', HT, yb, 1927; bud long, pointed; flowers chrome-yellow, reverse bright rosy scarlet, shaded apricot, dbl.; [Mme Edouard Herriot X Duchess of Wellington]; Ketten Bros.

'SOUV DE H.A. VERSCHUREN', HT, yb, 1922; bud long, pointed; flowers yellow to orange-yellow, well-formed, large, dbl., 38 petals, moderate fragrance; vigorous growth; [Unnamed seedling X Golden Ophelia]; Verschuren

'SOUV DE HENRI FAASSEN', HT, pb, 1929; bud long, pointed; flowers deep pink, base orange-yellow, open, large, semi-dbl., intense fragrance; foliage bronze; very vigorous growth; [Betty Uprichard X (Mrs George Shawyer X Los Angeles)]; Faassen-Hekkens

'SOUV DE HENRI VENOT', HT, mr, 1931; flowers brilliant red, very dbl., intense fragrance; foliage dark; vigorous growth; [Lord Charlemont X Red Star]; Lens

'SOUV DE J. CHABERT', F, mr, 1956; (Bobbie Robbie, Pépite); flowers well-formed, large blooms in clusters of 3-6, dbl.; foliage dark, bronze; vigorous, low growth; [Francais X Unnamed seedling]; Delbard-Chabert

'SOUV DE J. MERMET', HWich, dr, 1934; flowers silvery carmine, large, borne in clusters, recurrent bloom; foliage bright, bronze; very vigorous growth; Mermet

Souv de J.B. Guillot, T, rb; Guillot, 1897

'SOUV DE J.B. WEIBEL', HT, dr, 1930; flowers carmine, very large, dbl., cupped, slight fragrance; foliage dark; very vigorous, bushy growth; [Unnamed seedling of Mrs Bullen X Edouard Mignot]; Sauvageot, H.; C-P, 1933;, F. Gillot, 1930

'SOUV DE JACQUES VERSCHUREN', HT, op, 1950; bud long, pointed; flowers apricot-salmon, large, dbl., moderate fragrance; foliage leathery, dark; vigorous, bushy growth; [Katharine Pechtold X Orange Delight]; Verschuren-Pechtold

'SOUV DE JEAN CROIBIER', HT, pb, 1921; flowers bright salmon-pink shaded chamois, center coral-red shaded yellow, dbl., slight fragrance; [Mme Mélanie Soupert X Lyon Rose]; Croibier

'SOUV DE JEAN GINET', HT, mr, 1935; flowers scarlet-red, base coppery, reverse golden yellow; Brenier, E.C.; Buatois

'SOUV DE JEAN SOUPERT', HT, my, 1929; bud long, pointed; flowers golden yellow, very large, semi-dbl., cupped, intense fragrance; foliage bronze; vigorous growth; [Ophelia X Feu Joseph Looymans]; Soupert & Notting

'SOUV DE JEANNE BALANDREAU', HP, mr, 1899; Vilin

'SOUV DE JOHN E. KNIGHT', HT, ob, 1928; flowers terra-cotta, salmon-pink and flesh streaked yellow, moderate fragrance; foliage dark; vigorous growth; Knight, J.

'SOUV DE JOSEFINA PLÀ', HT, mr, 1929; flowers bright red, large, semi-dbl., moderate fragrance; vigorous growth; [Étoile de Hollande X Mme Butterfly]; Munné, B.

'SOUV DE JOSEPH BESSON', HT, rb, 1931; flowers reddish orange, base yellow; Brenier, E.C.

'SOUV DE JULES NICOLAS MATHIEU LAMARCHE', Pol, rb, 1934; flowers cardinal-red, base yellow, very small large panicles, single; [Eblouissant X Petit Constant]; Soupert, C.

Souv de la Bataille de Marengo *see* **'RUSSELLIANA'**

'SOUV DE LA MALMAISON', B, lp, 1843; (Queen of Beauty and Fragrance); flowers creamy flesh, center rosy shaded, very large, dbl., quartered, repeat bloom, intense, spicy fragrance; dwarf, bushy (2 ft.) growth; (21); [Mme Desprez X Tea]; Béluze

'SOUV DE LA MALMAISON, CLIMBING', Cl B, lp, 1893; Bennett

'SOUV DE LA PRINCESSE AMÉLIE DES PAYS-BAS', HP, m, 1873; flowers large, dbl.; Liabaud; (Sangerhausen)

Souv de la Princesse de Lamballe *see* **'QUEEN OF BOURBONS'**

'SOUV DE LA REINE D'ANGLETERRE', HP, mp, 1855; flowers bright pink, large, dbl., recurrent bloom; very vigorous growth; [La Reine X ?]; Cochet Freres

'SOUV DE LAFFAY', HP, mr, 1878; flowers crimson-red; Verdier, E.

'SOUV DE LAURENT GUILLOT', HT, 1894; Bonnaire; (Cavriglia)

'SOUV DE L'AVIATEUR MÉTIVIER', HWich, ly, 1913; flowers clear yellow, passing to white, dbl.; vigorous, climbing growth; [R. wichurana X Mme Ravary]; Tanne

'SOUV DE L'AVIATEUR OLIVIER DE MONTALENT', HWich, rb, 1913; flowers dull rose, base salmon, borne in clusters, dbl., profuse bloom, rarely recurrent; foliage very glossy, dark; vigorous, climbing growth; [R. wichurana X Anna Olivier]; Tanne

'SOUV DE L'EXPOSITION DE BORDEAUX', HMult, mr, 1906; flowers medium, dbl.; Puyravaud; (Sangerhausen)

'SOUV DE LILETTE', HT, w, 1937; bud very long; flowers snow-white, large, intense fragrance; foliage slightly bronze; compact, bushy growth; Chambard, C.

Souv de l'Imperatrice Josephine *see* **'EMPRESS JOSEPHINE'**

Souv de Louis Amade *see* DELilac

Souv de Louis Simon *see* **'MRS MINIVER'**

'SOUV DE LUCIE', N, dp, 1893; flowers medium, dbl.; Schwartz; (Sangerhausen)

'SOUV DE LUCIENNE VALAYER', HT, pb, 1938; bud long; flowers soft pink, shaded light salmon, very large, cupped, moderate fragrance; vigorous, bushy growth; Chambard, C.

'SOUV DE MADELEINE ROUILLON', HT, ob, 1929; bud long, pointed; flowers orange, base yellow, dbl., moderate fragrance; [Manon X Elvira Aramayo]; Bernaix, P.

Souv de Malmedy, HGal, lp; Scarman, 1996

'SOUV DE MAMAN CORBOEUF', HP, mp, 1900; flowers medium, dbl.; Corboeuf; (Sangerhausen)

Souv de Marcel Proust, HT, 1992; Delbard; (Cavriglia)

'SOUV DE MARCELLE BALAGE', HT, pb, 1930; (Marcella Baldge); bud long, pointed; flowers satiny flesh-pink, center slightly tinted salmon, very large, dbl., cupped, intense fragrance; foliage dark; [Willowmere X Mme Pizay]; Bernaix, P.

'SOUV DE MARIA CLOTILDE', HT, ly, 1934; bud long; flowers amber-white, often passing to light yellow, center sometimes, dbl., moderate fragrance; vigorous growth; [Mme Abel Chatenay sport]; Carneiro; Ketten Bros.

'SOUV DE MARIA DE ZAYAS', HT, mr, 1906; flowers large, dbl., moderate fragrance; Soupert & Notting; (Sangerhausen)

'SOUV DE MARIE FINON', HT, ab, 1924; flowers apricot-yellow passing to clear yellow shaded salmon, dbl., intense fragrance; Croibier

'SOUV DE MARIE THÉRÈSE PRIVAT', Pol, mr, 1935; flowers bright vermilion, well formed, borne in clusters of 40-50, very dbl.; foliage bright green; dwarf, good habit growth; Privat

'SOUV DE MARIE-THÉRÈSE', HT, w; flowers ivory-white, slight fragrance; low growth; Chabanat; Roses-France

'SOUV DE MARQUES LOUREIRO', HT, rb, 1912; flowers light red shading to rose, tinted yellow and purple, dbl.; [Mons. Paul Lédé X Mme Hoste]; Ketten Bros.

Souv de Maurice Chevalier *see* 'DELSOUCHE'

Souv de McKinley, HT, lp; Godard, 1902

'SOUV DE MLLE JULIET DE BRICARD', Pol, dp, 1934; flowers pale rosy pink, very dbl., globular, slight fragrance; foliage glossy, dark; vigorous, bushy growth; [Cécile Brunner X Yvonne Rabier]; Délépine; Pajotin-Chédane

'SOUV DE MME A. HESS', HT, pb, 1936; bud long, pointed; flowers shrimp-pink, center deep coral, very large, cupped, moderate fragrance; foliage bronze; very vigorous, bushy growth; [Unnamed seedling X Ami F. Mayery]; Chambard, C.

'SOUV DE MME ACHILLE VAN HERREWEGHE', HT, dr, 1936; flowers carmine-red; vigorous growth; Van Herreweghe-Coppitters

Souv de Mme Auguste Charles, B, mp; Moreau et Robert, 1866

'SOUV DE MME AUGUSTINE GILLOT', HT, op, 1920; bud long, pointed; flowers salmony flesh-pink, base salmon-yellow, reverse silvery flesh, slight fragrance; [Frau Karl Druschki X Lyon Rose]; Gillot, F.

Souv de Mme Berthier, HP, lp; Berthier/ Liabaud, 1881

'SOUV DE MME BOULLET', HT, dy, 1921; bud long, pointed; flowers well-formed, large, dbl.; vigorous, spreading growth; [Sunburst X Unnamed variety]; Pernet-Ducher

'SOUV DE MME BOULLET, CLIMBING', Cl HT, dy, 1930; H&S

Souv de Mme Breuil, B, dp

'SOUV DE MME C. CHAMBARD', HT, op, 1931; flowers coral-rose-pink, center flushed gold, large, semi-dbl., cupped, moderate fragrance; vigorous growth; Chambard, C.; C-P

'SOUV DE MME C. CHAMBARD, CLIMBING', Cl HT, op, 1935; Armstrong, J.A.

'SOUV DE MME CANEL', HT, rb, 1932; flowers carmine-orange, large, very dbl.; foliage pointed, bronze; robust, bushy growth; Gillot, F.

'SOUV DE MME CHÉDANE-GUINOISSEAU', HP, mr, 1900; flowers bright geranium red, very large, dbl.; Chédane-Guinoisseau

Souv de Mme de Corval, HP, dy; Gonod, 1867

'SOUV DE MME DURAND', HT, yb, 1954; flowers naples yellow, base bright yellow, edges veined pink, dbl., globular; foliage glossy; strong stems; Privat

'SOUV DE MME EUGÈNE VERDIER', HT, w, 1894; flowers white, base saffron-yellow; [Lady Mary Fitzwilliam X Mme Chedane-Guinoisseau]; Pernet-Ducher

'SOUV DE MME F. ZURICH', HT, mp, 1910; flowers silvery pink, large, dbl.; Puyravaud; (Sangerhausen)

'SOUV DE MME GAUTHIER-DUMONT', HT, mr, 1921; bud long, pointed; flowers scarlet; Guillot, P.

'SOUV DE MME H. THURET', HP, pb, 1922; flowers salmon-pink, center shrimp-red, edged chrome, well-formed, cupped; foliage rich green; very vigorous growth; [Frau Karl Druschki X Lyon Rose]; Texier; P. Nabonnand

'SOUV DE MME HENNECOURT', HP, mp, 1869; flowers large, dbl.; Cochet, S.; (Sangerhausen)

'SOUV DE MME JULES PAGES', HMult, pb, 1937; flowers deep pink shaded orange and red, open, small, dbl., free, intermittent bloom, slight fragrance; vigorous, climbing or pillar growth; [Phyllis Bide X Eblouissant]; Reiter; Stocking

'SOUV DE MME KREUGER', HT, op, 1919; flowers pure salmon-orange, passing to coppery pink, dbl.; [Mme Mélanie Soupert X Willowmere]; Chambard, C.

'SOUV DE MME L'ADVOCAT', N, op, 1899; flowers coppery pink, medium, dbl., moderate fragrance; Veysset; (Sangerhausen)

'SOUV DE MME LEFÈBVRE', HT, rb, 1929; flowers oriental red, passing to pink, golden yellow and red; vigorous growth; Richardier

'SOUV DE MME LÉONIE VIENNOT', Cl T, yb, 1898; flowers yellow shaded pink, very dbl.; vigorous growth; Bernaix, A.

'SOUV DE MME LEVET', T, ob, 1891; flowers orange-yellow; Levet, F.

'SOUV DE MME LOUISE CRETTÉ', HT, yb, 1924; flowers golden yellow, shaded coral and red, dbl.; [Mme Edouard Herriot seedling]; Cretté; C. Chambard

'SOUV DE MME MORIN-LATUNE', HT, dp, 1920; bud long, pointed; flowers cream-rose, dbl.; Bernaix, P.

'SOUV DE MME PIDOUX', HT, yb, 1926; bud long, pointed; flowers chrome-yellow, reverse pink, cupped; [Unnamed seedling X Mrs Aaron Ward]; Chambard, C.

'SOUV DE MME ROBERT', HP, lp, 1879; flowers large, dbl.; Moreau et Robert; (Sangerhausen)

'SOUV DE MME SABLAYROLLES', T, ab, 1891; flowers apricot-pink; Bonnaire

'SOUV DE MME SALATI-MONGELLAZ', HP, dp, 1937; bud long, pointed; flowers satiny rose-pink, very large, dbl.; foliage dark; very vigorous growth; [Frau Karl Druschki X Unnamed seedling]; Croibier

'SOUV DE MONS ROUSSEAU', HP, mr, 1861; flowers medium, dbl.; Fargeton; (Sangerhausen)

'SOUV DE NORAH LINDSAY', LCl, mp; flowers bright pink, stamens golden, very large, single; vigorous, pillar growth

'SOUV DE NUNGESSER', HT, dr, 1927; flowers brilliant deep carmine-red, dbl., moderate fragrance; [Mme Maurice de Luze X Laurent Carle]; Croibier

'SOUV DE PAPA CALAME', HT, dp, 1922; flowers large, dbl., moderate fragrance; Guillot; (Sangerhausen)

'SOUV DE PAUL GRANDCLAUDE', HT, yb, 1923; bud large, long, pointed; flowers yellow, shaded pink and in autumn light brown, semi-dbl.; [Mme Mélanie Soupert X Beauté de Lyon]; Sauvageot, H.

'SOUV DE PAUL NEYRON', T, yb, 1871; flowers salmon-yellow edged pink, semi-globular, large, intense fragrance; very vigorous growth; [Ophirie seedling]; Levet, A.

'SOUV DE PAUL RAUDNITZ', HWich, lp, 1910; flowers small, very dbl.; Cochet-Cochet; (Sangerhausen)

'SOUV DE PHILÉMON COCHET', HRg, w, 1899; flowers white, center rose; [Blanc Double de Coubert sport]; Cochet-Cochet

Souv de Pierre Dupuy, B, lp; Levet, 1876

'SOUV DE PIERRE GUILLOT', HT, ob, 1928; bud long, pointed; flowers coral-orange blend, dbl.; [Marie Adélaide X Unnamed seedling]; Guillot, M.

'SOUV DE PIERRE KETTEN', HT, rb, 1928; bud long, pointed; flowers bright rose, inside pink, base chrome-yellow, dbl., 30–35 petals, intense fragrance; [Mme Mélanie Soupert X Pilgrim]; Ketten Bros.

'SOUV DE PIERRE LEPERDRIEUX', HRg, mr, 1895; flowers bright red; large fruit; very vigorous growth; Cochet-Cochet

'SOUV DE PIERRE NOTTING', T, yb, 1902; bud long, pointed; flowers sunflower-yellow tinted apricot and coppery yellow, edged rose, very dbl., slight fragrance; foliage rich green, soft; [Marechal Niel X Maman Cochet]; Soupert & Notting

'SOUV DE PIERRE VIBERT', M, rb; flowers dark red, shaded carmine and violet, large, dbl., sometimes recurrent bloom; moderate growth; Moreau et Robert, 1867

'SOUV DE PROSPER FRAISSENON', HT, m, 1927; bud long, pointed; flowers geranium-red tinted violet; Richardier

'SOUV DE R.B. FERGUSON', HT, pb, 1922; flowers shell-pink and apricot, sometimes shaded rose-pink and cerise, slight fragrance; [Unnamed seedling X Constance]; Ferguson, W.

'SOUV DE RAYMOND GAUJARD', HT, yb, 1943; bud ovoid; flowers golden yellow, reverse often veined red, open, medium, dbl., moderate fragrance; foliage dark, glossy; vigorous, upright growth; Gaujard

'SOUV DE RENÉ GROGNET', HT, ob, 1921; bud long, pointed; flowers coppery orange-yellow, shaded carmine; Chambard, C.

Souv de Rose Berkley *see* **'ROSE BERKLEY'**

Souv de Rose-Marie, HMsk, lp; Lens, 1998

'SOUV DE S.A. PRINCE', T, w, 1889; (The Queen); [Souv. d'un Ami sport]; Prince

'SOUV DE ST ANNE'S', B, lp; flowers blush-pink, nearly, single, recurrent bloom; (21); [Souv. de la Malmaison sport]; Hilling, 1950

'SOUV DE THÉRÈSE LEVET', T, dr, 1886; [Adam X ?]; Lovet, A.

'SOUV DE VICTOR HUGO', T, pb, 1886; flowers china-pink, center salmon-pink, large, dbl., intense fragrance; [Duchesse de Brabant X Regulus]; Bonnaire

'SOUV DE VICTOR HUGO', HP, lp, 1885; flowers very large, very dbl.; Pernet; (Sangerhausen)

'SOUV DE VICTOR LANDEAU', B, mr, 1890; bud rose; flowers vivid red; Moreau et Robert

'SOUV DE VICTOR LANDEAU', HP, pb; flowers pale pink to mauve rose, large, dbl., globular; Moreau et Robert, 1890

'SOUV DE WILLIAM WOOD', HP, dr, 1864; flowers dark velvety red, large, dbl., 37 petals, cupped, sometimes recurrent bloom, moderate fragrance; vigorous growth; Verdier, E.

'SOUV DE YEDDO', HRg, lp, 1874; flowers large, dbl.; Morlet; (Sangerhausen)

'SOUV DE. H.A. VERSCHUREN, CLIMBING', Cl HT, yb, 1927; H&S

'SOUV D'ELISE VARDON', T, w, 1855; flowers creamy white, center yellowish, very large, dbl., globular; Marest

'SOUV D'EMILE MAYRISCH', HT, dr, 1932; flowers dark crimson-garnet, well formed, large, dbl., 35–40 petals; stiff stems; vigorous growth; Ketten Bros.

'SOUV D'EMMANUEL BUATOIS', HT, rb, 1932; bud long, pointed; flowers coral-red shaded shrimp-red, reverse clear rose, base golden, dbl., moderate fragrance; foliage leathery, bronze; vigorous growth; [Mme Edouard Herriot X Souv. de Claudius Pernet]; Buatois

'SOUV D'ERNEST THÉBAULT', HWich, dr, 1921; dbl., borne in clusters of 10-20; vigorous, climbing growth; Thebault

Souv di Castagneto, HP, lp; Scarman, 1988; re-introduced by Scarman, 1988

'SOUV DU BARON DE SÉMUR', HP, m, 1874; flowers deep purple-red shaded fiery red and black, large, dbl.; vigorous growth; [Charles Lefebvre X ?]; Lacharme, F.

'SOUV DU CAPITAINE CRÉMONA', HT, rb, 1928; flowers salmon-carmine over yellow ground, dbl., moderate fragrance; [Admiration X Gorgeous]; Bernaix, P.

'SOUV DU CAPITAINE FERNAND JAPY', HT, mr, 1922; dbl., slight fragrance; [Le Progres X Les Rosati]; Sauvageot, H.

'SOUV DU CAPITAINE FERRAND', HT, rb, 1939; bud long, well formed; flowers nasturtium-red, reverse golden yellow; erect, vigorous growth; Gaujard

'SOUV DU DOCTEUR ALBERT REVERDIN', HT, rb, 1930; bud long, pointed; flowers brilliant carmine, shaded vermilion, large, very dbl., moderate fragrance; foliage dark; strong stems; vigorous growth; [George C. Waud X Mrs Edward Powell]; Bernaix, P.

'SOUV DU DOCTEUR JAMAIN', HP, dr, 1865; flowers plum shaded deep crimson, dbl., moderate fragrance; moderate growth; [Charles Lefebvre X ?]; Lacharme, F.

'SOUV DU LIEUTENANT BUJON', B, mr, 1891; flowers very large, dbl., moderate fragrance; Moreau et Robert; (Sangerhausen)

'SOUV DU PAPA CALAME', HT, pb, 1921; bud long, pointed; flowers carmine-pink, reverse silvery pink, stamens salmon, dbl.; [Jonkheer J.L. Mock sport]; Gillot, F.

'SOUV DU PRESIDENT CARNOT', HT, lp, 1894; bud long, pointed; flowers flesh-pink, center shell-pink, moderate fragrance; [Unnamed seedling X Lady Mary Fitzwilliam]; Pernet-Ducher

'SOUV DU PRÉSIDENT CARNOT, CLIMBING', Cl HT, lp, 1926; Grandes Roseraies

'SOUV DU PRÉSIDENT DAUREL', HT, dp, 1906; flowers large, dbl.; Chauvry; (Sangerhausen)

Souv du President Lincoln, B, mp; Moreau et Robert, 1865

'SOUV DU PRÉSIDENT PLUMECOCQ', HT, rb, 1958; flowers bright red, reverse marked silvery, large, dbl., moderate fragrance; foliage bronze; very vigorous, well branched growth; [Unnamed seedling X Peace]; Laperrière; EFR

'SOUV DU REVEREND PÉRE PLANQUE', HT, ob, 1932; flowers orange-chrome-yellow, fading lighter; foliage very glossy; very vigorous growth; [Unnamed seedling X Souv. de Georges Pernet]; Bel

'SOUV DU ROSIÉRISTE GONOD', HP, mr, 1889; flowers large, dbl., intense fragrance; Ducher; (Sangerhausen)

Souv du Rosieriste Rambaux, T, pb; Rambaux, 1883

'SOUV DU SÉNATEUR BAZIRE', HMult, m, 1918; flowers violet, center violet-rose, borne in clusters of 25-50, semi-dbl.; [Veilchenblau X Bordeaux]; Lottin

'SOUV DU SERGENT CRETTÉ', HT, yb, 1921; bud long, pointed; flowers coppery golden yellow, shaded carmine, very large, cupped; foliage bronze; very vigorous growth; [Mme Mélanie Soupert X Unnamed seedling]; Chambard, C.

'SOUV D'UN AMI', T, lp, 1846; flowers pale rose tinged salmon, very large, dbl., cupped, intense fragrance; vigorous growth; Belot-Defougere

'SOUV OF MIAMI', HT, dr, 1925; Cook, J.W.

'SOUV OF STELLA GRAY', T, ob, 1907; flowers deep orange splashed apricot, salmon and crimson, small to medium, semi-dbl., moderate fragrance; Dickson, A.

'SOUV OF THE OLD ROSE GARDEN', HT, pb, 1929; flowers silvery pink, brighter inside, very large, dbl., globular, intense fragrance; foliage glossy, light; strong stems; vigorous growth; GM, NRS, 1928; Cant, B. R.

'SOUV OF WOOTTON', HT, mr, 1888; flowers rich velvety red, very large, dbl., intense fragrance; foliage dark, leathery; vigorous growth; [Bon Silène X Louis van Houtte]; Cook, J.W.

'SOUV OF WOOTTON, CLIMBING', Cl HT, mr, 1899; Butler

'SOUVENANCE', HT, m, 1965; bud very long, pointed; flowers deep lavender-lilac, dbl., 25–30 petals, 3.5–4.5 in., exhibition form, intense fragrance; foliage dark; moderately vigorous growth; [Unnamed seedling X Sterling Silver]; Lens

'SOUVENIR', HT, dy, 1930; bud pointed; flowers golden yellow, dbl., 36–42 petals, moderate fragrance; foliage glossy; vigorous growth; [Talisman sport]; Pierson, A.N.

Souvenir *see* EuroGROOT

Souvenir de Adolphe Turc *see* **'SOUV D'ADOLPHE TURC'**

Souvenir de la Malmaison Rouge *see* **'LEWESON GOWER'**

Souvenirs de Marcel Proust *see* DELpapy

Souviens-Toi *see* 'KRIPRILE'

'SOVEREIGN', HT, dy, 1922; flowers deep yellow and old-gold, open, cupped, intense, fruity fragrance; foliage glossy, dark, bronze; vigorous growth; [Queen Mary X Unnamed seedling]; Cant, B. R.

'SOVRANA', HT, ob, 1930; flowers orange streaked yellow, well formed, large; very vigorous, upright growth; [Julien

Potin X Signora seedling (yellow)]; Aicardi, D.; Giacomasso

Sowetan Peace *see* JACsowe

Sox *see* 'MICSOX'

'SOYÉCOURT', HT, rb, 1921; flowers blood-red, overlaid orange; [Gen. MacArthur X George C. Waud]; Jersey Nursery

'SPACE GIRL', LCl, dr, 1966; flowers dark crimson, very large, dbl., moderate fragrance; foliage dark; vigorous, climbing growth; [Queen Elizabeth X Étoile de Hollande]; Barter

Space Invader *see* 'DICROCKY'

Space Odyssey® *see* 'WEKSNACARE'

Space Probe *see* 'MINACON'

Space Walk® *see* 'MINABCCO'

'SPAETHIANA', S, m; (R. X spaethiana); flowers purple, large, corymbose; (21); [R. palustris X R. rugosa]

Spangles *see* 'GANSPA'

Spanish Beauty *see* **'MME GRÉGOIRE STAECHELIN'**

Spanish Dancer *see* 'MORLIZ'

Spanish Enchantress, S, lp

'SPANISH EYES', F, op, 1981; flowers orange-salmon, reverse yellow; [Prominent sport]; Takatori, Yoshiho; Japan Rose Nursery

'SPANISH GOLD', HT, dy, 1960; flowers straw-yellow; tall, bushy growth; Fletcher; Tucker

Spanish Main *see* **'MARQUESA DEL VADILLO'**

'SPANISH ORANGE', F, ob, 1966; flowers orange, small borne in clusters, very dbl., 1.5 in., moderate fragrance; foliage dark, glossy; very free growth; deRuiter; Gregory

'SPANISH RHAPSODY', S, pb, 1984; flowers deep pink, tinted orange and freckled red, shallow-cupped, large blooms, dbl., 30 petals, cupped, repeat bloom, moderate, raspberry fragrance; awl-like prickles; foliage medium-large, leathery, dark olive green; upright, bushy, branching growth; hardy.; [Gingersnap X Sevilliana]; Buck, Dr. Griffith J.; Iowa State University

'SPANISH SUN', F, dy, 1966; flowers large, dbl., intense fragrance; foliage glossy, leathery; vigorous, bushy growth; [Yellow Pinocchio X Golden Garnette climbing seedling]; Boerner; J&P

'SPANKY', Min, yb, 1986; flowers bright yellow, shaded pink to red, reverse yellow, aging pink, dbl., exhibition form, slight fragrance; few, pink, small prickles; foliage large, medium green, semi-glossy; medium, upright growth; [Rise 'n' Shine X ?]; Bridges, Dennis A.; Bridges Roses

'SPARKELS', HT, rb, 1956; bud urn shaped; flowers red striped white, medium, dbl., 35–50 petals, exhibition form, intense fragrance; foliage leathery; [Briarcliff sport]; Webber

'SPARKIE', Min, mr, 1957; flowers bright red, becoming darker, blooms in clusters, single, 6 petals; foliage glossy; vigorous (12-16 in) growth; [(R. wichurana X Floradora) X Little Buckaroo]; Moore, Ralph S.; Sequoia Nursery

'SPARKLE', HT, pb, 1949; bud pointed, ovoid; flowers white turning cream, rose-pink and yellow toward center, large, dbl., exhibition form, moderate fragrance; foliage glossy, light green; vigorous, upright, bushy growth; [Pink Princess X Shades of Autumn]; Brownell, H.C.

'SPARKLE PLENTY', F, ob, 1976; bud ovoid; flowers bright orange, to open, dbl., 25 petals, 3 in., exhibition form, moderate fragrance; foliage dark, leathery, wrinkled; vigorous growth; [Ma Perkins X Engagement]; Patterson; Patterson Roses

Sparkler *see* 'DELECLA'

'SPARKLER', Pol, mr, 1929; [Golden Salmon sport]; deRuiter; Sliedrecht & Co.

Sparkley *see* 'POULCOV'

'SPARKLING BURGUNDY', F, dr, 1965; flowers burgundy-red, open, medium, borne in clusters, dbl., moderate fragrance; foliage dark, leathery, glossy; vigorous, upright growth; [Queen Elizabeth X Carrousel]; Williams, J. Benjamin

Sparkling Orange® *see* 'DEVILK'

Sparkling Scarlet *see* 'MEIHATI'

Sparks® *see* 'SAVASPARK'

'SPARRIESHOOP', S, lp, 1953; bud long, pointed; flowers gold stamens, single, 4 in., repeat bloom, intense fragrance; foliage leathery; very vigorous (5 ft.), upright, bushy growth; GM, Portland, 1971; [(Baby Chateau X Else Poulsen) X Magnifica]; Kordes

'SPARTAN', F, or, 1955; (Aparte); bud pointed; flowers orange-red to reddish coral, blooms borne singly and in clusters, dbl., 30 petals, 3–3.5 in., exhibition form, intense fragrance; foliage dark, leathery, glossy; vigorous, bushy growth; David Fuerstenberg Prize, ARS, 1957 GM, ARS, 1961 GM, NRS, 1954 GM, Portland, 1955 PIT, NRS, 1954; [Geranium Red X Fashion]; Boerner; J&P

'SPARTAN BLAZE', Min, rb, 1991; bud ovoid; flowers red with yellow reverse, blooms borne usually singly, dbl., exhibition form, slight fragrance; foliage medium, medium green, semi-glossy; upright growth; [Poker Chip X Rise 'n' Shine]; Gruenbauer, Richard, 1984; Flowers 'n' Friends Miniature Roses, 1991

'SPARTAN DAWN', Min, yb, 1990; bud rounded; flowers yellow with orange edges, reverse yellow to cream to orange, dbl., 30 petals, no fragrance; round, flat on top, orange-red fruit; few or no prickles; foliage medium, medium green, matt; bushy, medium growth; [Rise 'n' Shine X Hokey Pokey]; Gruenbauer, Richard, 1984; Richard Gruenbauer

Spartan II, F; Moreira da Silva, A.; (Cavriglia)

'SPARTAN, CLIMBING', Cl F, or, 1960; (Grimpant Spartan); Kordes

'SPARTAN, CLIMBING', Cl F, or, 1964; Martinez; J & P

Späth 250® *see* **'SPÄTH'S JUBILÄUM'**

'SPÄTH'S JUBILÄUM', F, or, 1970; (Späth 250®); bud ovoid; dbl., 22 petals, 2.5 in., exhibition form, slight fragrance; foliage dark, soft; vigorous, upright, bushy growth; [Castanet X Seedling]; Kordes

'SPEAKER SAM', HT, yb, 1962; flowers light yellow edged red; [Peace sport]; Dean; Arp Nursery Co.

Spearmint, Min, w

SPEcawijk, HT, mr, 1996; (**Aruba**); Spek

'SPECIAL ANGEL', Min, m, 1992; flowers mauve, tipped in pink, blooms borne mostly singly, dbl., 15–25 petals, 1.5 in., exhibition form, slight fragrance; few prickles; foliage medium, dark green, semi-glossy; tall (60-70 cms), upright, bushy growth; [Jean Kenneally X (Rise 'n' Shine X Acey Deucy)]; Stoddard, Louis; Bridges Roses, 1992

Special Effects, F, lp; flowers orange and white stripes, intense fragrance; [Scentimental sport]; Desamero, Luis

'SPECIAL GUEST', HT, mr, 1992; flowers crimson, blooms borne mostly singly, very dbl., 3–3.5 in., slight fragrance; many prickles; foliage medium, medium green, semi-glossy; medium (70 cms), bushy growth; [Jan Guest sport]; Guest, M.M., 1990

'SPECIAL MERIT', HT, mr, 1990; bud pointed; flowers large, borne singly, very dbl., 30 petals, exhibition form; foliage medium, dark green, semi-glossy; medium, upright, bushy growth; [Seedling X First Prize]; Wambach, Alex A.

Special Occasion *see* FRYyoung

Speckled Delight, HT, lp; Patel, 1998

'SPECTABILE', HSem, m; (Spectabilis); flowers bright rosy lilac, petals "curiously incised" (notched?), cupped, occasionally repeats in autumn; vigorous, climbing growth; [Possibly R. sempervirens X Noisette hybrid]

Spectabilis *see* **'SPECTABILE'**

'SPECTACULAR', LCl, or, 1953; (Danse du Feu, Mada); bud ovoid; flowers scarlet-red, blooms in clusters, dbl., 33 petals, cupped, recurrent bloom, moderate fragrance; foliage glossy, bronze; vigorous, climbing (8-10 ft.) growth; [Paul's Scarlet Climber X Unnamed R. multiflora seedling]; Mallerin, C.; EFR, 1953;, J&P, 1956

Spectra *see* MEIzalitaf

Speechless® *see* 'KINspeech'

Speelwark, HT, lp; Kordes, 1999

SPEhewe, HT, lp; (**Tiamo**); Spek, 1996

SPEkes, HT, mr; 1996; (**Sacha**, Our Sacha); Spek

Spek's Centennial *see* 'MACivy'

Spek's Improved, HRg, lp

Spek's Yellow *see* **'Golden Scepter'**

Spellbinder *see* 'WARdido'

'Spellbound', HT, dr, 1949; flowers deep velvety red, becoming darker, dbl., 25–30 petals, 5.5–6 in., moderate fragrance; thornless; foliage dark; very vigorous, upright growth; [Better Times sport]; Sodano, J.; Amling-DeVor Nursery

Spellbound *see* POUlrim

Spellcaster *see* 'JACangel'

'Spencer', HP, lp, 1892; flowers soft pink; [Merveille de Lyon sport]; Paul, W.

'Spencer's Delight', HT, ab, 1998; flowers coral pink, reverse soft amber, full, very dbl., 26–40 petals, 3.5–4.5 in., exhibition form, borne mostly singly, moderate fragrance; prickles moderate; foliage medium, dark green, semi-glossy; long, straight stems; vigorous, upright, tall (48 - 62 in.) growth; [Maid of Honor sport]; Spencer, Keith V.

Speragina, T, lp; Hay

SPErien, S, lp; (**Fairy Queen**); Spek

SPEruge, HRg, mr, 1997; (**Red Dagmar**); Spek

Spes, HT, mp, 1970; flowers large, very dbl., intense fragrance; Urban, J.; (Sangerhausen)

SPEvu, Pol, dp, 1990; (**Lovely Fairy**); Vurens-Spek

'Sphinx', HT, lp, 1967; flowers bright pink tinted lighter, very large, dbl., moderate fragrance; foliage leathery; vigorous, bushy growth; [Rose Gaujard X Gail Borden]; Gaujard

'Spice', F, or, 1954; bud globular; flowers scarlet-red, blooms in clusters, dbl., 53 petals, 3 in., cupped, moderate fragrance; foliage glossy; vigorous, compact, upright growth; [(Goldilocks X Floradora) X Unnamed Floribunda seedling]; Boerner; J&P

Spice, HCh, lp

Spice Drop® *see* 'SAVswet'

Spice 'n' Nice, HT, lp; Twomey, Jerry, 1997

Spice Twice® *see* 'JACable'

Spiced Coffee *see* 'MACjuliat'

'SPIcup', Min, mp, 1981; (**Cupcake**®); bud ovoid; flowers clear medium pink, long lasting, dbl., 60 petals, exhibition form, borne 1-5 per cluster, no fragrance; no prickles; foliage glossy; compact, bushy growth; PP4835; AOE, 1983; [Gene Boerner X (Gay Princess X Yellow Jewel)]; Spies, Mark C.; Nor'East Min. Roses

Spicy Minijet *see* 'LAVglo'

Spielplatz DRS®, HT, yb; Huber, 1987

'Spinning Wheel'®, S, rb, 1991; flowers cherry red with ivory striping, large blooms borne in large, dbl., slight fragrance; foliage large, dark green, semi-glossy; tall, upright, spreading growth; [Handel X Love]; Williams, J. Benjamin, 1992

'Spion-Kop', F, mr, 1968; flowers signal-red, full, dbl., 30–35 petals, 3–5 in., slight fragrance; foliage light, bronze, matt green; vigorous growth; [Evelyn Fison X Orange Sensation]; Ellick

Spirit, HT, lp

'Spirit of '76', Gr, or, 1971; bud ovoid; flowers medium, dbl., moderate fragrance; foliage large, glossy, dark; vigorous, upright growth; [Queen Elizabeth X San Francisco]; Whisler, D.; Gro-Plant Industries

'Spirit of Canada', F, dr, 1995; flowers dark red enriched with flashes of scarlet, prominent stamens, semi dbl. (6–14 petals), medium blooms borne in small and large clusters; numerous prickles; foliage medium, medium green, semi-glossy; upright, bushy, tall growth; [Bambula X RedMax]; Fleming, Joyce; Hortico Roses, 1995

Spirit of Glasnost *see* 'JACara'

Spirit of Hope *see* 'MAChahei'

Spirit of Peace *see* 'MEInivoz'

Spirit of Peace *see* 'JACstine'

Spirit of Pentax *see* 'HARwex'

Spirit of Tollcross *see* HAReverso

Spirit of Youth *see* 'MEIvestal'

'Spitfire', Min, mr, 1986; flowers large blooms borne singly, dbl., 38 petals, exhibition form, no fragrance; thin, light green prickles; foliage medium, medium green, matt; upright, bushy growth; [Unnamed seedling X Unnamed seedling]; McDaniel, Earl; McDaniel's Min. Roses

'Spitfire', HT, mr, 1943; bud oxblood-red; flowers carmine, open, small, dbl., 25–30 petals, 2–3 in., moderate fragrance; foliage dark, leathery; short stems, vigorous, upright, much branched growth; RULED EXTINCT 6/86; [Better Times X Colleen Moore]; Hill, Joseph H., Co.

'Spitfire Improved', HT, dr, 1949; bud medium, short-pointed, ovoid; flowers velvety dark carmine, semi-dbl., 4–5 in., globular, intense fragrance; foliage leathery, dark; vigorous, upright, bushy growth; RULED EXTINCT 6/86; [Spitfire sport]; Hill, Joseph H., Co.

Spitzenschleier, S, lp; flowers small, single; VEG; (Sangerhausen)

Splash, HT, lp; flowers white with splashes of pink; [Moonstone sport]; Becnel, Johnny

'Splendens', HGal, mr; (Frankfurt); flowers glistening crimson, semi-dbl.

'SPLendens', Gr, mr; (Frankfurt, R. gallica Splendens)

'Splendid Garland', HSem, lp, 1835; flowers medium, dbl.; Wells; (Sangerhausen)

'Splendor', HT, or, 1933; bud long, pointed; flowers orange-carmine, dbl., cupped, moderate fragrance; foliage leathery, glossy; vigorous growth; [La Marechale Petain X Souv. de Claudius Pernet]; Sauvageot, H.; C-P

'Splendor', HT, mr, 1940; bud long; flowers rose-red, reverse slightly lighter, pointed, dbl., 4–5 in., intense fragrance; foliage dark; [Better Times sport]; Abrams, Von

Splish Splash *see* 'MORgoldart'

Splish Splash *see* KORfolga

'SPOboom', Min, rb, 1992; (**Boomerang**); flowers red and white, blooms borne in small clusters, dbl., 26–40 petals, 1.5–2.75 in., no fragrance; some prickles; foliage medium, medium green, semi-glossy; medium, bushy growth; AOE, 1993; [Seedling X Seedling]; Spooner, Raymond A.; Oregon Miniature Roses, 1993

'SPOchey', Min, ab, 1985; (**Cheyenne**); flowers golden apricot, blooms borne singly, dbl., 20 petals, intense fragrance; foliage medium, medium green, semi-glossy; upright growth; [Rise 'n' Shine X Center Gold]; Spooner, Raymond A.; Oregon Miniature Roses

'SPOdarc', Min, w, 1991; (**Darcelle**); flowers small, blooms borne in sprays of 3-5, single, 5 petals, 1.5 in., slight fragrance; foliage small, medium green, semi-glossy; bushy, low (13 cms), very compact growth; [(Whistle Stop X Popcorn) X Nozomi]; Spooner, Raymond A.; Oregon Miniature Roses

'SPOdash', Min, mr, 1993; (**Dasher**); dbl., 26–40 petals, 1.5–2.75 in., borne mostly singly, slight fragrance; some prickles; foliage medium, medium green, semi-glossy; medium (40 cms), upright, bushy growth; [Maurine Neuberger X Seedling]; Spooner, Raymond A.; Oregon Miniature Roses, 1994

'SPODAY', Min, pb, 1995; (**Alisha**); flowers soft pink with white margin on outer edge, blooms borne singly, very dbl., 1.5 in., no fragrance; few prickles; foliage medium, medium green, matt; medium (20 in), bushy growth; [Marriotta X Seedling]; Spooner, Raymond A.; Oregon Miniature Roses, 1995

'SPODE', HT, ob, 1972; flowers orange-scarlet, flushed cream, long, pointed, dbl., 35 petals, 6 in., intense fragrance; foliage glossy, dark; [Diorama X Fragrant Cloud]; Fryers Nursery, Ltd.

'SPOGLO', Min, or, 1985; (**Superglo**); flowers small, semi-dbl., slight fragrance; foliage small, medium green, semi-glossy; bushy growth; [((Prominent X Rise 'n' Shine) X Trumpeter) X Chattem Centennial]; Spooner, Raymond A.

'SPOGRET', Min, dp, 1993; (**Gretta**); flowers deep pink, blooms borne mostly single, very dbl., 1.5–2.75 in., moderate fragrance; some prickles; foliage medium, dark green, glossy; bushy (44 cms) growth; Spooner, Raymond A.; Oregon Miniature Roses

'SPOHAWK', Min, pb, 1989; (**Sadie Hawkins**); bud globular; flowers medium pink with white stripes, reverse white, aging lighter, dbl., 16 petals, moderate, fruity fragrance; globular, medium green fruit; no prickles; foliage medium, medium green, matt; bushy, medium growth; [Roller Coaster X Unnamed seedling]; Spooner, Raymond A.; Oregon Miniature Roses, 1990

'SPOLAR', Min, mr, 1991; (**Larado**); flowers bright, clear red with small white eye in center, compact, semi-dbl., 6–14 petals, 1.5 in., no fragrance; foliage small, medium green, semi-glossy; micro-mini; low (15 cms), bushy growth; [Tobo Yellow X Seedling]; Spooner, Raymond A.; Oregon Miniature Roses, 1992

'SPOLAV', Min, m, 1995; (**Lavender Spoon**); single, 5 petals, 1.5–2.75 in., borne in sprays of 3-7, slight fragrance; some prickles; foliage small, dark green, semi-glossy; medium (12-14 in), bushy growth; [(Black Jade X Blue Peter) X Seedling]; Spooner, Raymond A.; Oregon Miniature Roses, 1995

'SPOMAUR', Min, mr, 1989; (**Maurine Neuberger**™); bud pointed; flowers medium, borne singly, dbl., 30 petals, exhibition form, moderate fragrance; ovoid, light green fruit; prickles needle-like, light brown; foliage medium, medium green, matt; upright, medium growth; PP007772; [(Prominent X Zinger) X Centerpiece]; Spooner, Raymond A.; Oregon Miniature Roses

'SPOMEG', Min, dy, 1993; (**Little Meghan**); flowers bright yellow, blooms borne mostly single, dbl., 15–25 petals, 1.5 in., slight fragrance; no prickles; foliage small, light green, matt; low, compact growth; [Good Morning America X Seedling]; Spooner, Raymond A.; Oregon Miniature Roses, 1994

'SPOMERIT', Min, ob, 1989; (**Merit**); bud pointed; flowers brilliant orange, yellow base, reverse yellow, small, dbl., 17 petals, exhibition form, no fragrance; no fruit; prickles needle-like, brown; foliage small, dark green, semi-glossy; bushy, low growth; [(Prominent X Zinger) X Unnamed Miniature seedling]; Spooner, Raymond A.; Oregon Miniature Roses

'SPONG', C, mp; flowers rose-pink, richer in center, rosette form, larger than rose, early bloom; dwarf, compact (to 4 ft) growth; 1805

'SPOROSE', Min, rb, 1991; (**Rosetta**); flowers white center, red on outer half of petals, blooms borne singly, dbl., 26–40 petals, 1.5 in., no fragrance; foliage small, medium green, semi-glossy, disease-resistant; bushy (34 cms) growth; [Scamp X Seedling]; Spooner, Raymond A.; Oregon Miniature Roses, 1992

Sporting Duo *see* HARbar

Spot Meillandina® *see* 'MEIDIPSER'

Spot Minijet *see* 'MEIZOGREL'

Spot 'o Gold *see* 'WEESPOT'

'SPOTLIGHT', HT, my, 1969; flowers peach and gold, globular, moderate fragrance; foliage dark; free growth; [Seedling X Piccadilly]; Dickson, A.

Spotlight *see* 'MORBRIGHTS'

Spotted Gold *see* KORtelin

'SPOVEGAS', Min, mr, 1985; (**Little Vegas**); flowers small, dbl., 20 petals, exhibition form, slight fragrance; foliage small, medium green, semi-glossy; bushy growth; [(Rise 'n' Shine X Prominent) X Unnamed red seedling.]; Spooner, Raymond A.; Oregon Miniature Roses, 1984

'SPRAY CÉCILE BRÜNNER', Pol, pb, 1941; bud long, pointed; flowers bright pink on yellow, edged clear pink, center yellowish, dbl., moderate fragrance; foliage sparse, soft, dark; bushy growth; [Cécile Brunner sport]; Howard Rose Co.

Spreeathen, HT, mr, 1968; flowers large, dbl.; Rupprecht-Radke; (Sangerhausen)

Spreeglut, S, dr; flowers medium, semi-dbl.; Kopenick; Institut für Zierpflanzenbau Berlin, 1985; (Sangerhausen)

Spring *see* **'WESNIANKA'**

Spring Beauty *see* 'MINECO'

Spring Bouquet *see* 'TINSPRING'

Spring Break *see* 'STRAHEIDI'

Spring Fever *see* 'AROCANT'

Spring Fragrance *see* **'VESENII AROMAT'**

'SPRING FROLIC', Min, my, 1978; bud ovoid; flowers pompon-shaped, dbl., 70 petals, 1 in., intense fragrance; low, compact, spreading growth; [(Little Darling X Gold Coin) X Golden Angel]; Williams, Ernest D.; Mini-Roses

'SPRING FROST', Min, w, 1978; flowers pure white, ruffled, semi-dbl., 17 petals, 1 in., slight fragrance; foliage small; low, very compact growth; Schwartz, Ernest W.; Bountiful Ridge Nursery

Spring Gold *see* **'FRÜHLINGSGOLD'**®

Spring Hill's Freedom *see* 'TWOFREE'

Spring Hill's Pink Freedom, S, lp; Spring Hill, 1997

Spring Melody™ *see* 'MINMCO'

Spring Morning *see* **'FRÜHLINGSMORGEN'**

'SPRING SONG', Min, pb, 1957; flowers pink tinted salmon, small blooms in clusters, dbl.; foliage glossy; bushy growth; [(R. wichurana X Floradora) X Thumbelina]; Moore, Ralph S.; Sequoia Nursery

'SPRING SONG', S, dp, 1954; flowers rich carmine-pink, semi-dbl., moderate fragrance; vigorous, tall growth; [Gartendirektor Otto Linne seedling]; Riethmuller

Spring Time™ *see* 'WILSPRING'

Springfields *see* 'DICBAND'

Springhill Freedom *see* 'TWOFREE'

'SPRINGMOON', Min, ly, 1994; (**Full Moon**); flowers creamy light yellow, blooms opening to old fashion form, dbl., 26–40 petals, 1–1.5 in., slight fragrance; few prickles; foliage medium, medium green, semi-glossy, disease-resistant; medium (20 in), spreading, rounded growth; [Avandel X Olympiad]; Sproul, James A.

'SPRINGPROMISE', Min, dr, 1994; (**Crimson Promise**); flowers antique, smoky dark red, blooms borne mostly single, dbl., 26–40 petals, 1.25–1.75 in., no fragrance; few prickles; foliage medium, dark green, semi-glossy, disease-resistant; medium (20 in), bushy growth; [Avandel X Chrysler Imperial]; Sproul, James A.

Springs 75 *see* 'POULVISION'

'SPRINGTIME', F, lp, 1935; flowers wild-rose-pink, center white, borne in clusters, semi-dbl., cupped, slight fragrance; foliage leathery; bushy growth; RULED EXTINCT 7/90; [Miss Rowena Thom X Unnamed seedling]; Howard, F.H.; Dreer

'SPRINGVALE', HT, w, 1975; flowers white to pink, very dbl., 50 petals, 4 in., exhibition form, slight fragrance; foliage dark; [Mme A. Meilland sport]; Miller, J.

Springwood Beauty® *see* 'LAVNEW'

Springwood Classic® *see* 'LAVCLASS'

Springwood Coral® *see* 'LAVSCENT'

Springwood Gold® *see* 'LAVTYNINE'

Springwood Mauvette®, *see* 'LAVISLE'

Springwood Pink® *see* 'LAVDUSK'

Springwood Pink Satin® *see* 'LAVSAT'

Springwood Purple® *see* 'LAVPURR'

Springwood Red® *see* 'LAVRED'

Springwood Red Victor®, *see* 'LAVBERT'

Springwood Ruby *see* 'LAVARU'

Springwood White® *see* 'LAVSNOW'

'SPRINT', F, mr, 1961; bud oval; flowers bright red, open, borne in clusters, semi-dbl., 12 petals, 2.5 in., slight fragrance; foliage glossy, dark; vigorous growth; [Jolie Princesse X Chanteclerc]; Gaujard

Sprinter® *see* INTasprint

Spun Glass® *see* 'DEVLASS'

'SPUN GOLD', HT, my, 1941; flowers gold, dbl., 27 petals, 4.5 in., exhibition form; foliage glossy, leathery; vigorous, bushy, fairly compact growth; [Unnamed seedling X Portadown Glory]; McGredy; J&P

Spunglass *see* 'DEVLASS'

Spunky *see* 'JUDSPUNK'

'SPUTNIK', F, or, 1958; semi-dbl., large trusses of 25-30; low, bushy growth; [Jiminy Cricket X Gloria Mundi]; Maarse, G.

'SQUARE DANCER', S, dp, 1973; bud ovoid; flowers deep pink, large, dbl., cupped, repeat bloom, moderate fragrance; foliage large, dark, leathery; vigorous, upright, bushy growth; [Meisterstuck X ((World's Fair X Flordora) X Applejack)]; Buck, Dr. Griffith J.; Iowa State University

'SQUATTER'S DREAM', S, my, 1923; flowers medium yellow, becoming lighter, semi-dbl., cupped, recurrent bloom, moderate fragrance; no prickles; foliage dark bronze-green; dwarf, bushy growth; [Seedling from R. gigantea seedling sport]; Clark, A.; Hackett

'SRDCE EUROPY', HWich, dp, 1937; flowers small, single; Böhm, J.; (Sangerhausen)

'SREBRA', F, pb, 1975; flowers pink, shaded mauve, blooms in clusters of 5-20, dbl., moderate, tea fragrance; foliage dark, glossy; bushy growth; [Highlight X Masquerade]; Staikov, Prof. Dr. V.; Kalaydjiev and Chorbadjiiski

Srinivasa, HT, rb; Kasturi, 1969

St Alban's Gem, S, lp

St Boniface *see* 'KORMATT'

St Bruno *see* 'LANPIPE'

St Catherine, LCl, lp

St Cecilia *see* 'AUSMIT'

St Christopher *see* HARcogent

St Clair's Rose, HT, lp

St David's, Ch, rb

St Dunstan's Rose *see* KIRshru

St George Tucker, Misc OGR, lp

St Helena *see* 'CANLISH'

'ST HELENA', HT, ly, 1912; flowers creamy yellow, center pinkish; RULED EXTINCT 6/83; GM, RNRS, 1912; Cant, B. R.

St Helena Texas Centennial, HT, lp

'ST HILAIRE', (R. blanda form), mp; no prickles

St Hildas, HT, lp

St Hughs *see* 'KORHUG'

'ST INGEBERT', HP, w, 1926; bud long, pointed; flowers white, center yellowish and reddish, large, dbl., 62 petals, slight fragrance; vigorous growth; [Frau Karl Druschki X Mme Mélanie Soupert]; Lambert, P.

St John *see* 'HARBILBO'

St John's College, F, lp; J&P, 1998

'ST JOHN'S ROSE', S, lp; (R. centifolia sancta, R. sancta, R. X richardii); flowers rose, 2–2.5 in.; (21); [R. gallica X R. phoenicia]

St Katherine's *see* PEAvenus

'ST LUCIA', HT, mp, 1973; bud ovoid; flowers pink, medium, dbl., moderate fragrance; foliage glossy; moderate, upright, bushy growth; [Unknown X Unknown]; Tantau, Math.; Ahrens & Sieberz

St Margaret, HT, lp; Bell, Ronald J., 1998; (Weatherly, L.)

St Mark's Rose *see* **'ROSE D'AMOUR'**

'ST NICHOLAS', D, dp, 1950; flowers rich pink, semi-dbl.; vigorous, erect growth; Hilling; Hilling, 1950

St Patrick® *see* 'WEKAMANDA'

'ST PAULI', F, yb, 1958; flowers golden yellow edged red to pink, becoming fused, open, borne, semi-dbl., 15 petals, 3 in., slight fragrance; foliage dark, glossy; very vigorous, upright, bushy growth; [Masquerade X Karl Herbst]; Kordes, R.

St Piers *see* HARentrap

St Piers *see* HARentrap

'ST PRIST DE BREUZE', Ch, rb, 1838; flowers rich deep crimson with rose center, medium, dbl., globular; Desprez

St Quentin *see* 'KRILEVILLE'

St Swithun *see* 'AUSWITH'

St Thomas China, Ch, m

St Tiggywinkles *see* KORbasren

St Victor, HT, 1979; Croix; (Cavriglia)

'ST WILFRID'S HOSPICE', F, w, 1994; flowers medium red, blooms borne in small clusters, semi-dbl., 6–14 petals, 1.5–2.75 in., slight fragrance; some prickles; foliage small, medium green, semi-glossy; medium, bushy growth; [Len Turner sport]; Sawday, Mrs. D.R.; Apuldram Roses, 1995

St Wilfrid's Hospice, F, mr

'STAATSPRÄSIDENT PÄTS', HT, my, 1937; (President Pats); bud long, pointed; flowers amber-yellow, very varying, very large, dbl., intense fragrance; foliage leathery, dark; vigorous growth; [Ophelia X Souv. de Claudius Pernet]; Weigand, C.; Spath

'STACCATO', F, mr, 1959; bud ovoid; flowers vermilion-red, open, medium, semi-dbl., 14–20 petals; moderately vigorous, bushy growth; [Cantate sport]; van de Water; Klyn

'STACEY', F, yb, 1965; flowers maize-yellow tinted blush, borne in clusters, dbl., 25–30 petals, 4–5 in., moderate fragrance; foliage dark, glossy, leathery; vigorous growth; Verschuren, A.; Blaby Rose Gardens

'STACEY SUE', Min, lp, 1976; bud short, pointed; flowers soft pink, dbl., 60 petals, 1 in., slight fragrance; foliage small, glossy; bushy growth; [Ellen Poulsen X Fairy Princess]; Moore, Ralph S.; Sequoia Nursery

Stacey's Star *see* 'HORSTACEY'

'STAD DARMSTADT', Pol, mr, 1966; bud ovoid; flowers bright red, small, borne in clusters, semi-dbl.; foliage dark; [Red Favorite X The Doctor]; deRuiter

'STAD DEN HAAG', F, mr, 1969; flowers scarlet, open, large, semi-dbl., 20 petals, slight fragrance; foliage glossy; free growth; [Evelyn Fison X (Spartan X Red Favorite)]; McGredy, Sam IV

Stadt Basel *see* 'MEIVILANIC'

Stadt den Helder *see* 'INTERHEL'

Stadt Eltville® *see* TANelliv

'STADT ESSEN', Pol, mr, 1937; flowers medium, dbl.; Tantau, Math.; (Sangerhausen)

Stadt Ettelbrück, S, mr, 1981; flowers medium-large, semi-dbl.; Lens, Louis; (Sangerhausen)

Stadt Hockenheim, HMsk, lp; Weihrauch

'STADT KIEL'®, S, mr, 1962; flowers cinnabar-red, blooms in large clusters, dbl., 30 petals, slight fragrance; foliage dark; very vigorous, bushy growth; Kordes, R.

Stadt Luzern, F, op; Huber, 1967

Stadt Pilsen *see* **'PLZEN'**

Stadt Pottrop, F, 1961; deRuiter; (Cavriglia)

'STADT ROSENHEIM'®, S, or, 1961; dbl., blooms in clusters (up to 10), moderate fragrance; foliage glossy, light green; vigorous, upright growth; ADR, 1960; Kordes, R.

Stadt Wurzburg, HT, 1987; Kordes, R.; (Cavriglia)

'STADTRAT F. KÖHLER', HT, 1906; Geduldig; (Sangerhausen)

'STADTRAT GLASER', HT, yb, 1910; flowers sulfur-yellow edged soft red; Kiese

'STADTRAT MEYN', Pol, mr, 1919; flowers luminous brick-red, large, dbl.; [Orléans Rose X ?]; Tantau

Stagecoach™ *see* 'KINCOACH'

Stainless Steel™ *see* 'WEKBLUSI'

Stamela, F; (Holland); (Cavriglia)

'STÄMMLER', HP, mp, 1933; flowers rose-pink, large, very dbl., exhibition form, intense fragrance; foliage glossy; vigorous growth; [Victor Verdier X Arabella]; Tantau; C-P

'STANDING', Pol, dr, 1969; flowers open, medium, dbl., slight fragrance; foliage dark, glossy; vigorous growth; [Atlantic X Seedling]; Delforge

Standing Ovation™ *see* 'TUCSTAND'

'STANDOUT', HT, rb, 1977; bud ovoid, pointed; flowers red, white reverse, dbl., 40 petals, 3–4 in., exhibition form, slight, tea fragrance; vigorous, upright growth; [Tiffany X Suspense]; Weeks; Weeks Wholesale Rose Growers

Stanley Duncan, Min, ab

'STANLEY GIBBONS', HT, op, 1976; flowers salmon-orange, pointed, dbl., 28 petals, 4 in., moderate fragrance; foliage glossy, dark; [Fragrant Cloud X Papa Meilland]; Gregory

'STANLEY MATTHEWS', HT, dr, 1964; flowers crimson-scarlet, dbl., 26 petals, 4.5–5 in., moderate fragrance; foliage glossy, light green; vigorous growth; [(Independence X Crimson Glory) X Happiness]; Latham

'STANWELL PERPETUAL', HSpn, w, 1838; flowers blush, medium, dbl., recurrent bloom, slight fragrance; many prickles; foliage very small; moderate, spreading growth; (28); [Thought to be a repeat-blooming Damask X Spinosissima]; Lee

'STANZA', F, lp, 1969; bud ovoid; flowers light pink, center darker, open, medium, semi-dbl., slight fragrance; foliage glossy, dark; very vigorous, bushy growth; Pal, Dr. B.P.; Gopalsinamiengar

Star 2000® *see* PEKcoucan

Star Burst, S, 1994; Ilsink, Peter; (Cavriglia)

Star Child® *see* 'DICMADDER'

Star Delight *see* 'MORSTAR'

Star Gazer, HT, pb

Star Magic *see* 'MORSEMURI'

'STAR OF BETHLEHEM', HWich, w, 1947; flowers stamens red, single, 5 petals, 3–3.5 in., moderate bloom; foliage glossy; [Innocence X Silver Moon]; Fisher, R.C.

'STAR OF PERSIA', HFt, my, 1919; flowers bright yellow, stamens golden, medium, semi-dbl.; vigorous (8-10 ft) growth; [R. foetida X Trier]; Pemberton

'STAR OF QUEENSLAND', HT, w, 1909; flowers creamy white, medium, dbl., 40 petals, exhibition form; foliage grayish green; vigorous, bushy growth; [Étoile de France X Earl of Dufferin]; Williams, A.

'STAR OF THAILAND', HT, w, 1977; flowers creamy white, medium, dbl., 40 petals, exhibition form; foliage grayish green; vigorous, bushy growth; [Mount Shasta X Pascali]; Chinprayoon; Chavalit Nursery

Star of Tokio, Min, 1984; Dot, Simon; (Cavriglia)

'STAR OF WALTHAM', HP, mr, 1875; flowers carmine-crimson, semi-globular, medium, dbl., moderate fragrance; smooth green wood with occasional red prickles; foliage very large; dwarf to medium growth; Paul, W.

Star Performer *see* 'CHEWPEARL'

'STAR TWINKLE', Min, pb, 1978; bud pointed; flowers pink, coral and orange, single, 5 petals, 1 in.; foliage glossy; bushy, compact growth; [Fairy Moss X Fire Princess]; Moore, Ralph S.; Sequoia Nursery

Starbright *see* **'LILY WHITE'**

'STARBRIGHT', HT, w, 1962; bud ovoid; dbl., 43 petals, 4 in., exhibition form, moderate fragrance; foliage glossy, dark; vigorous, upright growth; [Princess White X Hybrid Tea seedling]; Boerner; J&P

'STARBURST', Gr, rb, 1969; bud ovoid; flowers red and yellow, medium, dbl., exhibition form, slight fragrance; foliage glossy; vigorous, bushy growth; [Zambra X Suspense]; Meilland, Mrs. Marie-Louise; C-P

Starburst *see* JAChart

Stardance™ *see* 'WILBLANK'

'STARDUST', HT, w, 1939; flowers rose-red; RULED EXTINCT 5/92; [Better Times sport]; Florex Gardens

Stardust *see* 'DEVSTAR'

Stardust *see* 'INTERDUST'

Stardust *see* PEAvandyke

'STARFIRE', Gr, mr, 1958; bud urn-shaped; dbl., 28 petals, 5 in., exhibition form, blooms in clusters, moderate fragrance; foliage glossy; vigorous, tall, bushy growth; AARS, 1959; [Charlotte Armstrong X (Charlotte Armstrong X Floradora)]; Lammerts, Dr. Walter; Germain's

Starfire, Climbing, Cl F, lp; [Starfire sport]; Jack, V., 1972; (Weatherly, L.)

'STARGAZER', F, ob, 1977; flowers orange-red, yellow eye, patio, small to blooms in large clusters, single, 9 petals, slight fragrance; foliage medium, medium green, matt; medium, bushy growth; [Marlena X Kim]; Harkness

'STARGLO', Min, w, 1973; bud long, pointed; flowers small, dbl., exhibition form, slight fragrance; foliage small, leathery; vigorous, bushy growth; AOE, 1975; [Little Darling X Jet Trail]; Williams, Ernest D.; Mini-Roses

'STARGOLD', HT, my, 1936; bud long, pointed; flowers yellow, often splashed red, dbl., 32 petals, exhibition form, intense fragrance; foliage dark, glossy; long, strong stems; branching, upright growth; [Mary Wallace self X (Unnamed seedling X Unnamed seedling)]; Brownell, H.C.; Inter-State Nursery

Starina® *see* 'MEIGALI'

'STARINA, CLIMBING', Cl Min, or, 1981; [Starina sport]; Asami, Hitoshi

Starion® *see* DORdeli

'STARK WHITECAP', F, w, 1959; flowers pure white, borne in clusters, dbl., 30–35 petals, 3 in., cupped, moderate fragrance; foliage glossy; vigorous, upright growth; [Glacier X Garnette]; Boerner; Stark Bros.

'STARKRIMSON', HT, dr, 1961; flowers scarlet-red, dbl., 48 petals, 4.5 in., cupped, intense fragrance; foliage leathery; vigorous, upright growth; [Happiness X San Fernando]; Morey, Dr. Dennison; Stark Bros.

Starla *see* 'PIXARLA'

'STARLET', F, my, 1957; bud pointed; dbl., 60 petals, 2.5 in., exhibition form, blooms in clusters, slight fragrance; foliage dark, leathery, glossy; vigorous, compact, upright growth; [Goldilocks X Unnamed seedling]; Swim, H.C.; Armstrong Nursery

Starlight *see* 'AROLAQUELI'

'STARLIGHT', HT, my, 1934; flowers orange and buff, moderate fragrance; vigorous growth; RULED EXTINCT 3/87; Wood & Ingram

'STARLIGHT', HT, my, 1987; flowers medium, dbl., 26–40 petals, slight fragrance; foliage large, light green, matt; bushy growth; [Devotion X Benson & Hedges Gold]; Guest, M.M.

Starlight Express *see* TRObstar

Starlight Parade *see* POUlstar

Starlina, Min, mp

'STARLITE', HT, w, 1942; bud long, pointed; flowers clear white, center tinted cream, large, dbl., 40 petals, exhibition form; foliage dark, soft; vigorous, upright, bushy growth; RULED EXTINCT 3/87; [Unknown seedling X White Briarcliff]; Nicolas; J&P

Starlite *see* MEIlupin

Starlite *see* MEItanet

'STARQUELI', Gr, ab, 1995; (**Magic Lantern,** Remy Martin®); flowers copper-orange-gold, full (26–40 petals), large (7+ cms) blooms borne in small clusters; slight fragrance; some prickles; foliage medium green, semi-glossy; upright, bushy (150 cms) growth; [Gold Medal sport]; Royon, Rene; Bear Creek Gardens, 1994

Starry Bouquet, S, lp; Clements, John K., 1999

Starry Eyed *see* 'HORCOXIST'

Starry Eyed *see* 'MORSTAR'

'STARS 'N' STRIPES', Min, rb, 1975; bud long, pointed; flowers evenly striped red and white, blooms in clusters, dbl., 21 petals, exhibition form, moderate, sweet fragrance; foliage light to medium green; bushy, upright growth; [Little Chief X (Little Darling X Ferdinand Pichard)]; Moore, Ralph S.; Sequoia Nursery, 1976

'STARSCENT', HT, pb, 1976; flowers rich pink, base yellow, dbl., 20 petals, 4.5 in., intense fragrance; foliage dull, blue-green; upright growth; [Silver Star X Peter Frankenfeld]; Parkes, Mrs M.H.; Rumsey

'STARSHINE', HT, yb, 1971; flowers pink and yellow blend, very large, dbl., globular, slight fragrance; foliage glossy; vigorous, bushy growth; [Peace X Condesa de Sástago]; Hamm; Five M Nursery

Starstruck *see* 'TINSTAR'

Start, F, or; flowers medium, semi-dbl., slight fragrance; VEG; (Sangerhausen)

'STARTLER', HT, or, 1955; flowers orange-scarlet, base rich yellow, well formed; vigorous growth; [Hector Deane X Mary Wheatcroft]; Robinson, H.

'STATE OF MAINE', HT, w, 1930; flowers white, center slightly tinged greenish, large, very dbl., exhibition form, slight fragrance; [Lady Ursula sport]; de Bree

'STATELY', HT, lp, 1930; bud long, pointed; flowers pale flesh, center deeper, well-shaped, large, dbl., exhibition form, slight fragrance; vigorous growth; [Souv. de Gustave Prat X Seedling]; Clark, A.; Hackett

'STATESMAN', HT, dr, 1991; flowers large blooms borne mostly singly, dbl., slight fragrance; foliage medium, medium green, matt; medium, bushy growth; [Unknown seedling X Unknown seedling]; Burks, Larry; Co-Operative Rose Growers

'STEADFAST', HT, ob, 1939; flowers amber, flushed yellow and pink, semi-dbl.; vigorous growth; [Mme Auguste Choutet X ?]; Clark, A.; T.G. Stewart

'STEBIGPU', HT, m, 1985; (Big Purple, Nuit d'Orient, **Stephen's Big Purple**, Stephens' Rose Big Purple); flowers purple, large, dbl., 35 petals, intense fragrance; foliage large, medium green, matt; upright growth; PP006262; [Unnamed seedling X Purple Splendour]; Stephens, Pat; McGredy Roses International, 1986

'STEEPLE ROSE', C, dp; bud full, globular, large; flowers rich rosy crimson with large green foliage; [Perhaps Prolifera de Redoute sport]

'STEFANOVITCH', Cl HT, dr, 1943; flowers large, dbl., recurrent bloom, intense fragrance; vigorous growth; [Lemania sport]; Meilland, F.

Steffi Graf *see* HELgraf

'STEINO', HT, my, 1965; bud ovoid; dbl., slight fragrance; foliage dark; [Golden Rapture sport]; Stein

'STEKELLY-LEIGH', Min, yb, 1981; (**Kelly-Leigh**); flowers white and gold, medium, semi-dbl., no fragrance; foliage medium, medium green, glossy; bushy growth; [Ko's Yellow X Unnamed seedling]; Stephens, Paddy

'STELLA'®, Gr, w, 1958; bud ovoid; flowers blush, edged deep pink on outer petals, in clusters, dbl., 36 petals, 4–5 in., exhibition form, slight fragrance; foliage dark, leathery; vigorous, upright growth; GM, NRS, 1960; [Horstmann's Jubilaumsrose X Peace]; Tantau, Math.

'STELLA DI BOLOGNA', HT, mr, 1909; flowers large, dbl., moderate fragrance; Bonfiglio, A.; (Sangerhausen)

'STELLA DUCE', F, dr, 1956; flowers dark crimson-red, stamens deep gold, moderate fragrance; vigorous growth; Leenders, M.

'STELLA ELIZABETH', Min, w, 1983; flowers creamy white, petals edged light pink, small, dbl., 35 petals, slight fragrance; foliage small, medium green, semi-glossy; bushy growth; [Unnamed seedling X Unnamed seedling]; Moore, Ralph S.; Roy H. Rumsey, Pty. Ltd.

'STELLA MATTUTINA', HT, lp, 1951; flowers flesh-pink, well formed, large, dbl., 35 petals; foliage glossy; vigorous growth; [Numa Fay X Asso Francesco Baracca]; Giacomasso

'STELLA PACIS', HT, ob, 1946; flowers combination of orange, salmon, red and yellow, large; vigorous growth; [Julien Potin X Mme G. Forest-Colcombet]; Giacomasso

'STELLA POLARIS', HRg, w, 1890; flowers silvery white, large, single, recurrent bloom; foliage dark; vigorous (4 ft.) growth; Jensen

'STELLMACHER', S, mr, 1937; flowers bright red, well formed umbels, dbl., abundant, recurrent bloom, moderate fragrance; hardy.; [D.T. Poulsen X Stammler]; Tantau

Stelvio Coggiatti *see* BARstel

'STEPHAN', Pol, mp, 1967; bud ovoid; flowers cyclamen-pink, large, borne in clusters, dbl., slight fragrance; foliage glossy; vigorous, upright growth; [Maria Delforge X Arc-en-Ciel]; Delforge

'STEPHANIE', Cl HT, lp, 1973; flowers open, large, semi-dbl., profuse, repeated bloom, slight fragrance; foliage glossy; vigorous, climbing growth; [Coral Dawn X Titian]; Gatty, Joseph

'STEPHANIE ANN', Min, mp, 1978; bud long, pointed; flowers neyron rose, semi-dbl., 16 petals, 1 in., slight fragrance; foliage tiny; upright, bushy growth; Lyon

Stéphanie de Monaco *see* 'MEYPINK'

'STEPHANIE DIANE', HT, mr, 1971; flowers scarlet-red, dbl., 50 petals, 5 in., exhibition form; moderately vigorous, upright growth; [Fragrant Cloud X Cassandra]; Bees

Stephanie Jo, HT, lp; [Sylvia sport]; Allender, Robert William, 1995; (Weatherly, L.)

'STEPHEN FOSTER', LCl, mp, 1950; bud ovoid; flowers pink, recurved, semi-dbl., 20 petals, 3.5–4 in.; vigorous, upright growth, spreading to 15 ft. or more; [Unnamed seedling X Black Knight]; Rosen, H.R.

'STEPHEN LANGDON', F, dr, 1969; flowers deep scarlet, large, semi-dbl., slight fragrance; foliage dark; vigorous, compact growth; [Karl Herbst X Sarabande]; Sanday, John

Stephen's Big Purple *see* 'STEBIGPU'

Stephens' Rose Big Purple *see* 'STEBIGPU'

Steppin' Out *see* 'BRISTEP'

'STEREO', F, or, 1972; bud long, pointed; flowers pure orange, open, dbl., 18–24 petals, 3–3.5 in., slight fragrance; foliage brownish, leathery; bushy growth; [Baccará X (Ole X Independence)]; Lens; Spek

'STERKMANNS', HGal, mr, 1847; flowers large, dbl.; Vibert; (Sangerhausen)

'STERLING', HT, mp, 1933; bud long, pointed; flowers brilliant pink, base yellow, large, dbl., 35 petals, moderate fragrance; foliage glossy; vigorous growth; Gertrude M. Hubbard, ARS, 1939 GM, Portland, 1938; [Mme Butterfly X Unnamed seedling]; Hill, E.G., Co.

Sterling '95 *see* JACnuel

'STERLING SILVER', HT, m, 1957; bud long, pointed; flowers lilac, becoming lighter, dbl., 30 petals, 3.5 in., exhibition form, intense fragrance; foliage dark, glossy; vigorous, upright growth; [Seedling X Peace]; Fisher, G.; J&P

'STERLING SILVER, CLIMBING', Cl HT, m, 1963; Miyawaki

Sterling Star™ *see* 'ORTSAR'

'STERN VON PRAG', HRg, dr, 1924; flowers dark blood-red, large, dbl., moderate fragrance; foliage large, dark; very vigorous, bushy growth; [R. rugosa seedling X Edward Mawley]; Berger, V.; Faist

Sternenflor®, S, w; Schultheis, 1989

'STESAM', F, r, 1981; (**Sam**); flowers russet, reverse cream, dbl., 20 petals, slight fragrance; foliage medium, dark, glossy; bushy growth; [Brown Eye X Unnamed seedling]; Stephens, Paddy

'STETEAW', HT, rb, 1985; (**Te Awamutu Centennial**); flowers white-edged medium red, dbl., 35 petals, slight fragrance; foliage dark, glossy; bushy growth; [Strawberry Ice X Unnamed seedling]; Stephens, Pat

Steve Silverthorne *see* 'FULSTEVE'

Stevens Rose *see* **R. CINNAMOMEA PLENA**

Stevens, Climbing *see* **'MRS HERBERT STEVENS, CLIMBING'**

Stevie *see* 'BUSTEV'

'STEYL RAMBLER', HMult, mr, 1915; flowers medium, dbl.; Leenders, M.; (Sangerhausen)

'STIEDWARD', HT, mp, 1993; (**Hilton Edward**); flowers lavender pink, blooms borne mostly single, dbl., 15–25 petals, 3–3.5 in., intense fragrance; some prickles; foliage large, dark green, semi-glossy; tall, upright growth; [Sweet Afton X News]; Stibbard, Robert H.; Tallisker Nursery Enterprises, 1995

Stifontein Rose *see* TANnietnof

'STIRLING CASTLE', F, mr, 1978; flowers bright scarlet, 18–20 petals, 3 in., slight fragrance; foliage matt; low, compact growth; [(Anne Cocker X Elizabeth of Glamis) X (Orange Sensation X TheOptimist)]; Cocker

'STOCKTON BEAUTY', HT, pb, 1948; bud long, pointed; flowers deep salmon-pink, base yellow, open, large, dbl., moderate fragrance; foliage soft; very vigorous, bushy growth; [Banner sport]; Raffel; Port Stockton Nursery

'STOCKTON RED', Cl HT, mr, 1962; flowers clear red, well formed; very long stems, vigorous growth; Raffel; Port Stockton Nursery

'STOKES', HT, op, 1982; flowers peach, reverse salmon, large, dbl., 35 petals, exhibition form, slight, fruity fragrance; foliage medium green, matt; medium-tall growth; [Susan Massu X Yellow seedling]; Perry, Astor; Perry Roses

Stolen Dream *see* 'MICSTEAL'

Stolen Moment *see* 'SEAMOM'

Stonelea Rambler, HMsk, lp; Nieuwesteeg, J., 1994

'STOPLITE', F, mr, 1955; flowers rose-red, medium, dbl., 45–55 petals, moderate fragrance; vigorous, upright growth; [Garnette X Unnamed seedling]; Jelly; E.G. Hill Co.

'STORMLY', HT, op, 1966; flowers coral, reverse red-pink, oval, cupped; foliage dull; tall growth; [(Eclipse X Michele Meilland) X Baccará]; Caranta, M.&H.

Stormy Weather™ *see* 'MINISTORM'

'STRAHEIDI', HT, dp, 1993; (**Spring Break**); flowers deep florescent pink, dbl., 15–25 petals, 1.5–3 in., exhibition form, borne mostly singly, no fragrance; some prickles; foliage medium, medium green, matt; medium (100-300 cms), upright growth; ARC Test Gardens - Silver Certificate, 1995; [Secret Love X KORlingo]; Strickland, Frank A.; Strickland, 1994

Strange Music *see* 'MORMUM'

'STRATFORD', HT, op, 1936; bud long-pointed, ovoid; flowers luminous pink tinted salmon, very dbl., intense fragrance; foliage leathery; long stems; very vigorous, bushy growth; [(Emile Charles X La France) X Marechal Niel]; Nicolas; Dixie Rose Nursery

'STRATOSFÉRA', C, dp, 1934; flowers large, very dbl., moderate fragrance; Böhm, J.; (Sangerhausen)

Strauchmaskerade, S, lp

Strawberries and Cream *see* 'GEESTRAW'

'STRAWBERRY', HT, mr, 1950; flowers strawberry, reverse deep flesh, dbl., 35–40 petals, 5.5–6 in.; foliage light green; vigorous, compact growth; [McGredy's Pink X Phyllis Gold]; Fletcher; Tucker

'STRAWBERRY BLONDE', Gr, or, 1966; bud urn-shaped; flowers light orange-red, medium, dbl., 25 petals, exhibition form, moderate, spicy fragrance; foliage dark, leathery; very vigorous, upright, bushy growth; [Ma Perkins X Spartan]; Armstrong, D.L.; Armstrong Nursery

'STRAWBERRY CREAM', F, pb, 1972; flowers strawberry-pink, streaked cream, very full, dbl., 45 petals, 3 in., intense fragrance; foliage light; vigorous, low growth; Ellick; Excelsior Roses

'STRAWBERRY CRUSH', F, mr, 1974; dbl., 3 in., slight fragrance; foliage red when young; [Bridal Pink X Franklin Engelmann]; Dickson, A.

Strawberry Delight *see* 'RENERRY'

'STRAWBERRY FAIR', F, mr, 1966; flowers scarlet, dbl., 2 in., blooms in clusters, slight fragrance; foliage dark; vigorous, bushy growth; [Orangeade X ?]; Gregory

Strawberry Fayre *see* AROwillip

Strawberry Ice *see* 'DELBARA'

'STRAWBERRY ICE', F, w, 1975; flower petals cream-white, edged pink, dbl., 22 petals, 3–3.5 in.; foliage dark; compact, bushy growth; GM, Madrid, 1974; [((Goldilocks X Virgo) X (Orange Triumph X Yvonne Rabier)) X Fashion]; Bees

Strawberry Kiss *see* 'CLEKISS'

'STRAWBERRY SOCIAL', F, mr, 1979; bud ovoid; dbl., 23 petals, exhibition form, borne 3-7 per cluster, slight fragrance; hooked prickles; foliage medium green, semi-glossy; upright to spreading, bushy growth; [Pink Parfait X Chrysler Imperial]; Taylor, Thomas E.

Strawberry Sundae *see* 'LEOSTRA'

'STRAWBERRY SWIRL', Min, rb, 1978; bud ovoid, pointed; flowers red mixed with white, mini-moss, dbl., 48 petals, 1 in., exhibition form; bushy growth; [Little Darling X Unnamed Miniature seedling]; Moore, Ralph S.; Sequoia Nursery

Strawrose, Pol, lp; Schneider, 1998

Street Wise *see* 'SEARONAN'

Stretch Johnson *see* 'MACFIRWAL'

'STRILDEW', HT, lp, 1996; (**Just For Fun**); flowers light pink, outer petals sometimes white with inner petals l, very dbl., 6 in., slight fragrance; moderate prickles; foliage medium, medium green, dull; very long stems, upright, medium (4-5 ft) growth; [Secret Love X First Prize]; Strickland, Frank A.

Strilli, F, my

'STRING OF PEARLS', F, w, 1985; (Meg '81); flowers near white, medium, semi-dbl., slight fragrance; foliage medium, dark, semi-glossy; dwarf, bushy growth; [Meg, Climbing X Sunsprite]; Gandy's Roses, Ltd.; Rearsby Roses, Ltd.

String of Pearls, Min, op; deRuiter, 1991

String of Pearls, S, lp; Williams, J. Benjamin

Striped Fairy Rose, Min, rb

Striped Festival™ *see* 'LAVMING'

Striped Meillandina *see* 'MORMUM'

Striped Moss *see* **'ILLET PANACHÉE'**

Striped Pet *see* 'LAVDANCE'

'STRIPED RADIANCE', HT, pb, 1919; flowers distinctly striped white; [Radiance sport]; Vestal

'STRIPED TEXAS CENTENNIAL', HT, ob, 1950; flowers orange-red with dark yellow striping, large, dbl., moderate fragrance; Heckmann; (Sangerhausen)

'STRIPEZ', HT, pb, 1942; flowers striped and variegated; [Better Times sport]; Janssen; Premier Rose Gardens

Stroke-o-Luck, F, lp

'STROLLER', F, rb, 1968; bud loose; flowers cerise, reverse gold, large blooms in trusses, dbl.; foliage matt; [Manx Queen X Happy Event]; Dickson, A.

STRolon, HRg, mp, 1991; (**Polarsonne**); Strobel

Stromboli, S, dr

STRonin, HRg, w, 1991; (**Polareis**); Strobel

'STRYKER', HT, pb, 1994; flowers pink and white, veining in outer petals, blooms borne mostly, dbl., 26–40 petals, 3–3.5 in., slight fragrance; no prickles; foliage medium, dark green, glossy; tall, bushy, upright growth; [Pristine X Suffolk]; Edwards, Eddie; Testing at J&P and Weeks

'STUART'S QUARRY', HT, pb, 1977; bud high centered; flowers pale bluish pink, reverse silver, full, dbl., 30 petals, 4 in., bloom repeats quickly, moderate fragrance; foliage small, glossy, dark; [Silver Lining X Anne Letts]; Bailey

'STUDIENRAT SCHLENZ', Cl HT, mp, 1926; flowers rose-pink, reverse pink, large, dbl., non-recurrent, moderate fragrance; height 6 1/2-8 ft; [Mrs Aaron Ward X Frau Karl Druschki]; Lambert, P.

'STURDY GERTIE', HT, op, 1949; flowers coral-pink, medium, dbl., 30 petals, moderate, raspberry fragrance; foliage small, dark, glossy, somewhat ribbed; [Pink Princess X Peace]; Taylor, C.A.

'STUTTGART', HT, my, 1928; bud long, pointed; flowers pure yellow, medium, moderate fragrance; [Edith Cavell X Mrs Franklin Dennison]; Berger, V.; Pfitzer

'STYLISH', HT, pb, 1953; flowers rose-pink, base yellow, 5–6 in., exhibition form, moderate fragrance; foliage dark, glossy; vigorous growth; Robinson, H.

'SU EXCELENCIA SEÑORA DE FRANCO', F, mr, 1956; bud long, pointed; flowers medium, dbl., cupped, slight fragrance; foliage dark, glossy; vigorous, upright growth; [Cocorico X Independence]; Camprubi, C.

'SUBARU', F, mr, 1977; bud pointed; flowers small, single, 5 petals, 1.5 in., slight fragrance; foliage dark; low growth; [Masquerade X Permanent Wave]; Kikuchi, Rikichi

'SUBLAEVIS', Sp, mp

'SUBLIME', HT, op, 1931; bud long, pointed; flowers orange-salmon, suffused scarlet, semi-dbl., moderate fragrance; long stems; very vigorous growth; [Talisman sport]; Amling Co.

'SUBSTITUT JACQUES CHAPEL', HT, pb, 1922; flowers peach-blossom-pink, edged rose-pink, base shaded citron-yellow, intense fragrance; [Mme Mélanie Soupert X Lyon Rose]; Bernaix, P.

'SUCCES', HCan, lp, 1954; flowers pinkish white, blooms in trusses, single; foliage dark, glossy; [R. canina seedling]; deRuiter

'SUCCÈS FOU', HT, mr, 1963; bud long; flowers deep cherry-red, well formed; vigorous growth; [Walko X Souv. de J. Chabert]; Delbard-Chabert

Sue Belle *see* 'TALSUE'

Sue Betts *see* 'BETSUE'

Sue Hipkin *see* HARzazz

Sue Jo™ *see* 'MINAVCO'

Sue Lawley *see* 'MACSPASH'

'SUE LEAT', HT, pb, 1983; flowers medium pink on outer half of petals, golden yellow on lower, dbl., 20 petals, moderate fragrance; brown prickles; upright growth; [Golden Slippers X Anne Letts]; Summerell, B.L.

Sue Ryder® *see* 'HARLINO'

'SUFFOLK', HT, w, 1983; flowers white with pink petal tips, urn-shaped, large, dbl., exhibition form, slight fragrance; foliage large, medium green, matt; upright growth; [Garden Party X Unnamed yellow seedling]; Perry, Astor; Perry Roses, 1984

Sugandharaj, HT, lp; Datt, Braham, 1992

'SUGANDHINI', HT, lp, 1969; bud pointed; flowers full, medium, dbl., intense fragrance; foliage glossy, light; bushy growth; [Margaret Spaull X ?]; IARI

'SUGAR BABE', Min, lp, 1978; bud long, pointed; semi-dbl., 16 petals, 1 in., slight fragrance; foliage bronze; compact growth; Lyon

Sugar Baby *see* TANragus

'SUGAR BEAR', Min, w, 1986; flowers medium, borne singly, dbl., 18 petals, exhibition form, slight fragrance; straight, pointed, medium, light green prickles; foliage medium, medium green, semi-glossy; upright, medium growth; [Heartland X Unnamed seedling]; Bridges, Dennis A., 1987

'SUGAR CANDY', Pol, mp, 1951; flowers pink, very small, dbl., profuse, repeated bloom; compact growth; Proctor

Sugar Daddy *see* JAClorna

'SUGAR ELF', Cl Min, pb, 1974; bud long, pointed; flowers pink and gold blend, semi-dbl., 15 petals, 1 in., slight fragrance; foliage glossy, leathery; bushy, spreading growth; Moore, Ralph S.; Sequoia Nursery

Sugar Magnolia™ *see* 'TALSUGAR'

Sugar 'n' Spice *see* 'TINSPICE'

'SUGAR PLUM', HT, mp, 1954; (Prelude); bud ovoid; flowers tyrian rose, dbl., 55 petals, 4–5 in., exhibition form, moderate, spicy fragrance; foliage dark, glossy, leathery; vigorous, upright, bushy growth; [Crimson Glory X Girona]; Swim, H.C.; Breedlove Nursery

Sugar Plum *see* 'MORPLUM'

Sugar Plum Fairy, Min, lp; [Sweet Chariot X seedling]; Hannemann, F., 1995; The Rose Paradise; (Weatherly, L.)

'SUGAR SWEET', F, pb, 1973; flowers soft pink, reverse yellow and pink, semi-dbl., 13 petals, 3–4 in., exhibition form, moderate fragrance; foliage green, matt; Edland Fragrance Medal, ARS, 1974; [Wendy Cussons X Prima Ballerina]; Sanday, John

Suitor, Pol, mp; light pink flowers; low bush; Clark, A., 1942

'SUKUMARI', F, w, 1983; flowers rosy white; [America's Junior Miss sport]; Gupta, Dr. M.N. & Datta, Dr. S.K.; National Botanical Research Institute

'SULCOVA KLADENSKA', Pol, lp, 1936; flowers bright pink; vigorous growth; Sulc

Sulphur Rose *see* **R. HEMISPHAERICA**

'SULPHUREA', T, ly, 1900; flowers sulfur-yellow; Paul, W.

Sulphureux, T, lp; flowers sulphur-yellow; Ducher, 1869

'SULTAN OF ZANZIBAR', HP, m; flowers purplish maroon, medium, globular; Paul & Son, 1876

Sultana, Gr, lp; (Maxima Regina)

'SULTANE', HT, rb, 1946; bud long, pointed; flowers vermilion, reverse gold, dbl., 5 in., slight fragrance; foliage leathery, bronze; [J.B. Meilland X Orange Nassau]; Meilland, F.

Suma *see* HARsuma

'SUMA NO URA', HT, mr, 1980; flowers crimson, blooms borne usually singly, dbl., 40 petals, exhibition form, no fragrance; large, reddish-brown prickles; foliage small, medium green, semi-glossy; medium, bushy growth; [(Rob Roy X Himatsuri) X Unnamed seedling.]; Teranishi, K.; Itami Rose Nursery

'SUMATRA', F, or, 1956; bud ovoid; flowers signal-red, blooms in pyramidal clusters, dbl., 26 petals, 3 in., globular, moderate fragrance; foliage leathery; moderate, upright growth; [Olga X Fashion]; Mallerin, C.; C-P

Summer Beauty® *see* 'MINAHCO'

'**SUMMER BLOSSOM**', F, mp, 1971; flowers soft geranium-pink, open, 2–3 in.; upright, bushy growth; [Orange Sensation X Kimono]; deRuiter

Summer Blush, A, dp; Sievers, 1988

Summer Breeze *see* 'TALSUM'

Summer Breeze *see* KORlastine

Summer Breeze, F, lp; Meilland, 1987

'**SUMMER BUTTER**', Min, dy, 1979; bud ovoid, pointed; dbl., 22 petals, 1–1.5 in., cupped, intense, spicy fragrance; foliage glossy; vigorous, compact growth; [Arthur Bell X Yellow Jewel]; Saville, F. Harmon; Nor'East Min. Roses

Summer Cloud *see* 'ZIPCLOUD'

Summer Crest *see* 'HERCRES'

'**SUMMER DAMASK**', D; (Damask Rose, R. belgica, R. calendarum, R. gallica damascena, R. polyantho, R. X damascena); flowers red, pink or white, sometimes striped, borne usually in corymbs, dbl., summer bloom, intense fragrance; obovoid fruit; 5 ft. growth; Eurosa, Gallicanae (28, 35); Int. to Eu. from Asia Minor in 16th Century

'**SUMMER DAWN**', Pol, mp, 1950; flowers soft rose-pink; foliage small, dainty; [Margo Koster sport]; Proctor

'**SUMMER DAYS**', HT, ly, 1976; flowers pale yellow, full, dbl., 36 petals, 3.5 in., exhibition form, moderate fragrance; vigorous growth; [Fragrant Cloud X Dr. A.J. Verhage]; Bees

Summer Dream *see* 'JACSHE'

Summer Dream *see* FRYmaxicot

'**SUMMER EVENING**', Min, ob, 1999; flowers orange and white, reverse white, dbl., 26–40 petals, 1.75 in., exhibition form, borne mostly singly, no fragrance; few prickles; foliage medium, dark green, semi-glossy; upright, medium (14 in) growth; [Tennessee X Kristin]; Jolly, Betty J.; Langenbach, 1998

Summer Fantasy *see* 'ZIPFAN'

Summer Fashion *see* 'JACALE'

Summer Festival® *see* 'LAVSUM'

Summer Fever, HT, my

'**SUMMER FIELDS**', F, dp, 1971; flowers light scarlet, semi-dbl., 17 petals, 5 in., moderate fragrance; foliage red when young, then green, matt; [Tropicana X Goldmarie]; Mattock

Summer Fragrance *see* 'TANFUDERMOS'

'**SUMMER FROST**', F, w, 1962; bud ovoid; dbl., 30–35 petals, 4.5 in., cupped, moderate fragrance; foliage leathery, dark; vigorous, bushy, compact growth; [Princess White X Golden Masterpiece]; Boerner; Home Nursery Products Corp.

'**SUMMER GLORY**', HT, my, 1961; flowers lemon, well formed, large; upright growth; [Dr. van Rijn X Seedling]; Leenders, J.

Summer Harvest *see* 'PIXHAR'

'**SUMMER HOLIDAY**'®, HT, or, 1967; flowers vermilion, dbl., 48 petals, exhibition form, moderate fragrance; foliage semi-glossy; very vigorous growth; [Tropicana X ?]; Gregory

'**SUMMER JOY**', HWich, dp, 1911; bud pure white; flowers dark rose-pink, borne in clusters, dbl.; foliage large, glossy; vigorous, climbing (18-20 ft) growth; Walsh

Summer Lady® *see* TANydal

Summer Love *see* 'FRANLUV'

Summer Madness® *see* 'FOUMAD'

'**SUMMER MAGIC**', Min, op, 1989; bud ovoid; flowers light pink blended apricot, reverse cream center, pink edging, very dbl., 150 petals, exhibition form, slight, fruity fragrance; no prickles; foliage small, medium green, matt; bushy, spreading, medium growth; [Fashion Flame X Anita Charles]; Jolly, Marie; Rosehill Farm, 1991

'**SUMMER MEETING**', F, my, 1968; flowers large, dbl., 45 petals, blooms in trusses, slight fragrance; foliage glossy; compact, bushy growth; [Selgold X Circus]; Harkness

Summer Morning *see* 'KORFULLWIND'

Summer Palace *see* POUlcape

'**SUMMER PERFUME**', LCl, mr, 1962; flowers vermilion-red, semi-dbl., 19 petals, intense fragrance; [Coral Dawn X Cocorico]; Leenders, J.

'**SUMMER PROMISE**', HT, ly, 1969; flowers very large, dbl., exhibition form, slight fragrance; foliage glossy; vigorous growth; Von Abrams; United Rose Growers

'**SUMMER QUEEN**', HT, w, 1964; flowers ivory-white, center tinted pink; [Queen Elizabeth sport]; Delforge

'**SUMMER RAINBOW**', HT, pb, 1966; bud ovoid; flowers pink, reverse yellow, large, dbl., exhibition form, slight fragrance; foliage glossy, leathery; vigorous, bushy growth; [Peace X Dawn]; Jelly; C-P

Summer Scent *see* 'RENSUM'

Summer Sérénade® *see* 'SMITFIRST'

'**SUMMER SNOW**', F, w, 1938; 3.5 in., blooms in large clusters, heavy, recurrent bloom, slight fragrance; foliage light green; bushy growth; [Summer Snow, Climbing sport]; Perkins, C.H.; J&P

Summer Snow, F, lp

'**SUMMER SNOW, CLIMBING**', Cl F, w, 1936; semi-dbl., 2 in., cupped, blooms in large clusters, sparse recurrent bloom, slight fragrance; foliage leathery; vigorous, pillar (8-10 ft) growth; [Tausendschön seedling]; Couteau; J&P

Summer Snowflake, Pol, lp

'**SUMMER SONG**', F, ob, 1962; (Chanson d'Été); flowers orange and yellow, large blooms in clusters, semi-dbl., 12 petals, slight fragrance; foliage glossy; low, bushy growth; [Seedling X Masquerade]; Dickson, Patrick; A. Dickson

'**SUMMER SPICE**', Min, ab, 1983; flowers light apricot, medium, dbl., 35 petals, moderate fragrance; foliage medium, medium green, semi-glossy; spreading growth; [Sheri Anne X Unnamed seedling]; Bridges, Dennis A.

Summer Sun *see* 'ADASUN'

Summer Sunrise, S, mp

Summer Sunset *see* 'BRISUN'

'**SUMMER SUNSHINE**', HT, dy, 1962; (Soleil d'Été); bud ovoid; dbl., 25 petals, 3.5–5 in., exhibition form, slight fragrance; foliage leathery, dark, semi-glossy; vigorous, upright, well-branched growth; [Buccaneer X Lemon Chiffon]; Swim, H.C.; Armstrong Nursery

'**SUMMER SURPRISE**', Min, mp, 1991; bud rounded; flowers pink with yellow stamens, slightly darker reverse, aging lighter, very dbl., 60 petals, no fragrance; foliage small, medium green, matt; bushy, low growth; [Libby X Unnamed seedling]; Gruenbauer, Richard, 1984; Flowers 'n' Friends Miniature Roses, 1990

Summer Sweet, S, pb

'**SUMMER SYMPHONY**', Min, yb, 1984; flowers large, dbl., exhibition form, intense fragrance; foliage medium, medium green, semi-glossy; bushy, spreading growth; [Lady Eve X Little Darling]; Hardgrove, Donald & Mary; Rose World Originals

'**SUMMER TAN**', F, lp, 1979; flowers flesh-pink; vigorous, compact growth; [Golden Slippers X Picasso]; Sheridan, John

Summer Wedding *see* 'JALWED'

'**SUMMER WIND**', S, op, 1975; bud ovoid, pointed; single, 3.5–4 in., flat, moderate, spicy clove fragrance; foliage dark, leathery; moderately vigorous, erect, bushy growth; [(Fandango X Florence Mary Morse) X Applejack]; Buck, Dr. Griffith J.; Iowa State University

Summer Wine *see* 'KORIZONT'

Summerdale *see* 'TALDAL'

Summerrose *see* 'INTERSUM'

Summer's Snow *see* **'NEIGES D'ÉTÉ'**

'SUMMERTIME', HT, mp, 1957; bud ovoid; flowers cameo-pink overcast rose-pink, borne in irregular clusters, dbl., 65–70 petals, 4–4.5 in., exhibition form, intense fragrance; foliage glossy, olive-green; bushy growth; [Diamond Jubilee X Fashion]; Boerner; J&P

Summerwind, S, 1985; Kordes, R.; (Cavriglia)

'SUMMERWINE', HT, mp, 1974; bud ovoid, pointed; dbl., 45 petals, 4–6 in., slight fragrance; foliage leathery; [Tiffany X South Seas]; Warriner, William A.; J&P

'SUN BLUSH', HT, yb, 1981; dbl., 37 petals, exhibition form, borne singly or 4 per cluster, slight fragrance; broad based, dark redprickles; foliage mid-green; upright growth; [Circus X Summer Sunshine]; Anderson's Rose Nurseries

Sun Chariot *see* 'JALSUN'

Sun City *see* 'KORSUN'

Sun Cover *see* 'POULURT'

Sun Drops *see* 'LYODRO'

Sun Flare *see* 'JACJEM'

Sun Flare, Climbing *see* 'JACLEM'

Sun Flare, Climbing™ *see* 'BURYELLOW'

Sun Glory *see* 'JACMEY'

'SUN GLOW', HT, op, 1934; bud long, pointed; flowers dark coral-pink, large, dbl., exhibition form, moderate fragrance; foliage glossy; long, strong stems; very vigorous growth; [Talisman sport]; Florex Gardens

'SUN GOD', HT, pb, 1930; flowers shrimp-pink, yellow, and orange-copper, very dbl., moderate fragrance; foliage leathery, bronze; short stems; dwarf growth; [Unnamed seedling X Mme Edouard Herriot]; Klyn; Wayside Gardens Co.

Sun Goddess *see* 'JACDASH'

'SUN GOLD', HT, my, 1935; flowers pure yellow, large, dbl., slight fragrance; foliage glossy, dark; very vigorous growth; RULED EXTINCT 3/83; Elmer's Nursery

Sun Gold *see* 'MORLEM'

Sun Hit *see* 'POULHIT'

Sun Honey *see* 'MORHONEY'

Sun King *see* 'MEINARVAL'

'SUN KING'®, HT, my, 1954; bud long, pointed; flowers bright lemon-yellow, dbl., 45 petals, 3.5 in., exhibition form, moderate fragrance; foliage dark, glossy, leathery; vigorous, upright growth; [Peace X Duchesse de Talleyrand]; Meilland, F.; C-P

Sun King '74 *see* 'MEINARVAL'

Sun Runner *see* 'INTERDUST'

Sun Sparkle *see* 'LYOSUN'

Sun Sprinkles™ *see* 'JACHAL'

'SUN UP', HT, lp, 1951; flowers china-pink, large, dbl., moderate fragrance; vigorous growth; [Break o' Day sport]; Brownell, H.C.

'SUN VALLEY', HT, dy, 1952; (Yukon); bud ovoid; flowers golden yellow, open, dbl., 30–35 petals, 4–5 in., intense, spicy fragrance; foliage dark, leathery; vigorous, upright growth; [Soeur Thérèse X Mark Sullivan]; Whisler; Germain's

'SUNALAD', Min, 1999; (**Aladdin**); dbl., 26–40 petals, 1.75–2 in., borne in large clusters, moderate fragrance; prickles moderate; foliage large, dark green, glossy; spreading, tall growth; [Tinkerbell X Texas]; Schuurman, Frank B.; Franko Roses New Zealand, 1995

'SUNAMBro', Gr, ab, 1999; (**Ambrosia**); dbl., 26–40 petals, 2.75–3 in., borne in small clusters, moderate fragrance; prickles moderate; foliage large, medium green, semi-glossy; bushy, medium growth; [Sexy Rexy X New Year]; Schuurman, Frank B.; Franko Roses New Zealand, 1996

Sunanda, HT, lp; Ghosh, 1998

'SUNAUCK', HT, ab, 1999; (**Barossa Dream**); flowers apricot, reverse light apricot, dbl., 17–25 petals, 4.5–5 in., borne mostly singly, slight fragrance; prickles moderate; foliage large, medium green, semi-glossy; bushy, tall (3-4 ft) growth; [City of Auckland X New Year]; Schuurman, Frank B.; Franko Roses New Zealand, Ltd., 1995

Sunbeam *see* **'MARGO KOSTER'**

'SUNBEAM', Min, my, 1957;, intense fragrance; dwarf (14-18 in) growth; [(Tom Thumb X Polly Flinders) X Golden Scepter]; Robinson, T.

Sunbeam® *see* 'KORDOSELBLA'

Sunbeam, Climbing *see* **'MARGO KOSTER, CLIMBING'**

Sunbird™ *see* 'POULNISH'

Sunbrid *see* 'JACPOY'

Sunblaze *see* 'MEIJIKATAR'

Sunblaze *see* 'MEIPONAL'

'SUNBLEST', HT, dy, 1970; (Landora); dbl., 38 petals, 5 in., slight fragrance; foliage glossy; GM, Japan, 1971 Gold Star of the South Pacific, Palmerston North, NZ, 1971; [Seedling X King's Ransom]; Tantau, Math.

Sunblest, Climbing, Cl HT, dy, 1979; Orard, Joseph; (Cavriglia)

'SUNBONNET', F, dy, 1967; bud pointed; flowers bright greenish yellow, medium, dbl., exhibition form, moderate fragrance; foliage dark, leathery; vigorous, low, bushy growth; [Arlene Francis X (Circus X Sweet Talk)]; Swim & Weeks; Weeks Wholesale Rose Growers

'SUNBONNET SUE', S, yb, 1984; flowers yellow stippled with scarlet, large blooms borne 1-10 per cluster, dbl., 25 petals, repeat bloom, moderate fragrance; small, awl-like, brown prickles; foliage large, leathery, semi-glossy, medium green; upright, bushy growth; hardy.; [Gold Dot X Malaguena]; Buck, Dr. Griffith J.; Iowa State University

Sunbright® *see* 'JACJEL'

Sunburnt *see* 'GELBURN'

Sunburnt Country *see* 'KORAV'

Sunburst *see* **'MME JOSEPH PERRAUD'**

'SUNBURST', HT, dy, 1912; bud long, pointed; flowers variable yellow, toward orange, dbl., cupped; GM, NRS, 1912; Pernet-Ducher

'SUNBURST JEWEL', HT, rb, 1979; bud pointed; flowers red, reverse pale gold, dbl., 35 petals; bushy growth; [Muchacha X Unnamed seedling]; Herholdt, J.A.

'SUNBURST, CLIMBING', Cl HT, dy, 1914; Howard Rose Co.

'SUNBURST, CLIMBING', Cl HT, dy, 1915; Low

Suncharm *see* HARfab

Suncharm *see* HARfab

Suncluster *see* KORgrapet

'SUNCOR', Min, op, 1999; (**Coralitos**); dbl., 17–25 petals, 1 in., borne in small clusters, no fragrance; prickles moderate; foliage small, medium green, semi-glossy; spreading, low (12-16 in) growth; Schuurman, Frank B.; Franko Roses New Zealand, Ltd., 1996

'SUNCRAL', Gr, op, 1999; (**Kia Ora**); 4 in., borne in large clusters, slight fragrance; few prickles; foliage medium green, glossy; bushy, medium growth; [Innocendi X Firefly]; Schuurman, Frank B.; Franko Roses New Zealand, Ltd., 1994

'SUNCREDEL', Gr, lp, 1983; (**Cream Delight**, Darling); [Sonia sport]; Schuurman, Frank B.

Sundance *see* 'JACNEL'

'SUNDANCE', F, yb, 1954; flowers orange-yellow changing to bright rose-pink, blooms in trusses, dbl., 22 petals, 2.5 in., slight fragrance; foliage light green, matt, thin; vigorous, tall (4 ft), upright, uneven growth; GM, NRS, 1954; [Poulsen's Supreme X Eugene Furst]; Poulsen, S.; McGredy;, Poulsen

Sundance Palace *see* POUldace

Sundancer *see* DEVsunset

'SUNDAY BEST', LCl, rb, 1924; bud long, pointed; flowers brilliant red, center white, single blooms in clusters, long seasonal bloom, slight fragrance; foliage wrinkled; vigorous, climbing growth; [Frau Karl Druschki X

Unnamed seedling]; Clark, A.; NRS Victoria

Sunday Brunch *see* 'MORDAY'

Sunday China *see* 'GELDAY'

Sunday Lemonade *see* 'MOLSUNLEM'

'SUNDAY PRESS', HT, mr, 1970; dbl., 53 petals, 4.5 in., exhibition form; Kordes; McGredy

'SUNDAY TIMES', F, dp, 1972; (Shrubby Pink); flowers deep rosy pink, medium, dbl., globular, slight fragrance; foliage light; moderate, dwarf growth; McGredy, Sam IV; Kordes

'SUNDEL', HT, lp, 1999; (**Delilah**®); dbl., 17–25 petals, 4 in., borne mostly singly, no fragrance; few prickles; foliage large, medium green, glossy; upright, tall (3-4 ft) growth; [seedling X Osiana]; Schuurman, Frank B.; Franko Roses New Zealand, Ltd., 1995

Sunderland Supreme *see* 'NOSSUN'

'SUNDIA', Min, or, 1999; (**Little Diamond**); dbl., 26–40 petals, 1 in., borne in small clusters, no fragrance; few prickles; foliage medium, dark green, glossy; upright, medium (12-18 in) growth; [Sexy Rexy X Firefly]; Schuurman, Frank B.; Franko Roses New Zealand, Ltd., 1993

Sundial™ *see* 'AROLYME'

'SUNDOHA', HT, lp, 1999; (**Double Happy**); flowers bi-color red and yellow, dbl., 17–25 petals, 3.5 in., borne mostly singly, no fragrance; prickles moderate; foliage medium, dark green, semi-glossy; upright, medium (30-36 in.) growth; [Louise Gardner X Goldmarie]; Schuurman, Frank B.; Franko Roses New Zealand, Ltd., 1999

'SUNDOWN', HT, mp, 1934; bud globular; flowers pink, medium, dbl., 35–40 petals, 3.5 in., intense fragrance; foliage glossy; very vigorous growth; [Talisman sport]; Scittine; Lainson

Sundowner *see* 'MACCHEUP'

Sundra® *see* 'GAISU'

Sundream, F, 1971; Leenders; (Cavriglia)

'SUNDUST', Min, yb, 1977; bud pointed; flowers light apricot, small, dbl., 23 petals, 1.5 in., moderate, fruity fragrance; foliage small, light; low, bushy, compact growth; [Golden Glow X Magic Wand]; Moore, Ralph S.; Sequoia Nursery

Sunfire *see* 'JACKO'

Sunflare *see* 'JACJEM'

'SUNGIRL', F, yb, 1981; flowers deep yellow, light red on petal tips, dbl., 25 petals, no fragrance; reddish-green prickles; foliage medium, deep green, glossy; upright growth; [Charleston X Unnamed seedling]; Rose Barni-Pistoia

'SUNGLO', F, or, 1976; bud ovoid; flowers open, dbl., 25 petals, 3 in., flat, slight, fruity fragrance; [Woburn Abbey sport]; Kuramoto, H.; J&P

Sunglow *see* **'MCGREDY'S ORANGE'**

'SUNGOLD', Min, my, 1992; (**Little Nugget**); flowers golden yellow, patio, small, dbl., 15–25 petals, 1.5 in., slight fragrance; foliage small, medium green, semi-glossy; bushy growth; [Lorena X Firefly]; Schuurman, Frank B.; Riverland Nurseries, Ltd., 1991

'SUNGOLD', Cl HT, my, 1939; bud long, pointed; flowers bright golden yellow, large, dbl.; foliage glossy, dark; vigorous (15-18ft) growth; RULED EXTINCT 3/83; [Margaret Anderson X Souv. de Claudius Pernet, Climbing]; Thomas; Armstrong Nursery

Sungold *see* 'MORLEM'

'SUNHAP', Gr, dr, 1999; (**Happy Go Lucky**); flowers dark red, reverse medium red, dbl., 17–25 petals, 3 in., borne in large clusters, slight fragrance; prickles moderate; foliage medium, medium green, semi-glossy; spreading, medium (30-36 in) growth; [Happy Days X Only Love]; Schuurman, Frank B., 1992; Franko Roses New Zealand, Ltd., 1998

'SUNHIFI', F, lp, 1999; (**High Five**); dbl., 26–40 petals, 3.25 in., borne in large clusters, slight fragrance; prickles moderate; foliage medium, dark green, glossy; spreading, medium (30-36 in) growth; [Jacaranda X French Lace]; Schuurman, Frank B.

'SUNHINO', Min, lp, 1999; (**Reddy Teddy**); dbl., 17–25 petals, 1.5 in., borne mostly singly, no fragrance; prickles moderate; foliage small, medium green, semi-glossy; compact, low (8-12 in) growth; [Sexy Rexy X Firefly]; Schuurman, Frank B.

Suni® *see* BARsus

'SUNIL GAVASKAR', F, m, 1985; dbl., 60 petals, 3 in., cupped, blooms in clusters of 7 or more, no fragrance; light green, needle-like prickles; foliage large, dark, leathery; dwarf, bushy growth; [(First Rose Convention X Scarlet Knight) X Ena Harkness]; Hardikar, Dr. M.N.

'SUNJEW', Min, ab, 1999; (**Little Jewel**); dbl., 17–25 petals, 1.5–2 in., borne in small clusters, slight fragrance; few prickles; foliage small, medium green, semi-glossy; upright growth; [Innocenti X Firefly]; Schuurman, Frank B.; Franko Roses New Zealand, Ltd., 1991

Sunking™, F, dy

'SUNKISSED', HT, my, 1980; bud ovoid, pointed; flowers yellow-orange, blooms borne singly, no fragrance; long, light green prickles, hooked down; foliage leathery; upright growth; [Minigold X Precilla]; Tantau, Math.; J&P

'SUNKIST'™, HT, yb, 1932; flowers orange-copper; [Joanna Hill sport]; Hill, E.G., Co.

'SUNLAMPO', F, lp, 1999; (**Bellisima**®); dbl., 26–40 petals, 2.5 in., borne mostly singly, slight fragrance; few prickles; foliage medium, dark green, glossy; upright, medium growth; [Lambada X Pot O' Gold]; Schuurman, Frank B.; Franko Roses New Zealand, Ltd., 1998

'SUNLIGHT', HT, my, 1956; (Grisbi); bud ovoid; dbl., 45 petals, 4 in., exhibition form, moderate fragrance; foliage leathery; vigorous, upright, bushy growth; [(Eclipse X Ophelia) X Monte Carlo]; Meilland, F.; C-P, 1958;, URS, 1956

'SUNLIGHT, CLIMBING', Cl HT, my, 1963; (Grisbi, Climbing); Meilland, Mrs. Marie-Louise; URS

'SUNLIT', HT, ab, 1937; flowers rich apricot, dbl., globular; compact growth; Clark, A.; NRS Victoria

'SUNLOCK', Min, lp, 1999; (**Goldilocks**); flowers medium yellow, reverse light yellow, dbl., 26–40 petals, 1.5 in., borne in large clusters, moderate fragrance; prickles moderate; foliage large, medium green, glossy; spreading, tall growth; [Scentasia X seedling]; Schuurman, Frank B.; Franko Roses New Zealand, Ltd., 1998

'SUNLUCK', HT, lp, 1999; (**Gold Strike**®); dbl., 26–40 petals, 4.5 in., borne in small clusters, slight fragrance; prickles moderate; foliage large, dark green, glossy; upright, tall growth; Schuurman, Frank B.; Franko Roses New Zealand, Ltd., 1997

Sunmaid®, Min, ob; Spek, 1972

'SUNMANI', HT, ob, 1999; (**Manaia**, Oasis Sunset); flowers orange, reverse apricot, dbl., 17–25 petals, 4.5–5 in., borne mostly singly, slight fragrance; prickles moderate; foliage large, medium green, semi-glossy; bushy, tall (3.5-4 ft) growth; Schuurman, Frank B.; Franko Roses New Zealand, Ltd., 1994

'SUNMIST', F, ly, 1940; bud pointed; flowers clear light sulfur-yellow, open, semi-dbl., slight fragrance; foliage leathery; vigorous growth; [(Eva X Golden Rapture) X Hede]; Kordes; Dreer

'SUNNINGDALE', HT, mp, 1929; bud long, pointed; flowers reddish carmine passing to cherry-pink, dbl., exhibition form, moderate fragrance; Hicks

'SUNNY', F, my, 1952; bud pointed, yellow overlaid red; flowers lighter yellow, open, small, semi-dbl., moderate fragrance; foliage leathery, glossy; vigorous, upright, bushy growth; [Unnamed seedling X Goldilocks]; Moore, Ralph S.; Marsh's Nursery

Sunny Afternoon *see* 'TALSUN'

'SUNNY BOY', HT, pb, 1961; flowers buff-pink, tinted gold and lilac, center darker, well formed, moderate fragrance; foliage bronze; [Mme Butterfly X The Queen]; Delforge

'SUNNY CALIFORNIA', HT, my, 1936; bud long, pointed; flowers yellow like Ville de Paris, very large, dbl., cupped; [Feu Joseph Looymans sport]; Hanshaw, E.

Sunny Child® *see* 'HAVYCHI'

Sunny Day™ *see* 'SAVASUN'

'SUNNY DAYS', HT, my, 1939; bud long, pointed; flowers chrome-yellow at edges, large, dbl., exhibition form, slight fragrance; foliage leathery, dark; vigorous growth; RULED EXTINCT 11/86; Verschuren; Dreer

'SUNNY HONEY', F, yb, 1972; flowers yellow-pink blend, reverse red, dbl., 22 petals, 4 in., moderate fragrance; foliage large, dark; [Happy Event X Elizabeth of Glamis]; Dickson, Patrick

'SUNNY JERSEY', HT, ab, 1928; flowers bronze, apricot, salmon and orange; [Mme Edouard Herriot sport]; Le Cornu

'SUNNY JUNE', S, dy, 1952; bud pointed; flowers deep canary-yellow, blooms in large clusters, single, 5 petals, 3–3.5 in., cupped, slight, spicy fragrance; foliage dark, glossy; upright, compact, pillar or shrub (8 ft) growth; [Crimson Glory X Capt. Thomas]; Lammerts, Dr. Walter; Descanso Distributors

'SUNNY MAID', F, my, 1949; flowers bright yellow, borne in clusters, single, 6–8 petals, 4 in.; foliage glossy, light green; very vigorous, upright growth; [(Golden Rapture X Fred Walker) X Unnamed seedling]; Fletcher; Tucker

Sunny Meillandina® *see* 'MEIPONAL'

'SUNNY MORNING', Min, my, 1974; bud long, pointed; flowers medium to creamy yellow, dbl., 35 petals, 1–1.5 in., flat, moderate fragrance; bushy, upright growth; [Golden Glow (LCl) X Peachy White]; Moore, Ralph S.; Sequoia Nursery

Sunny Sam *see* 'NOLDANYU'

'SUNNY SAN JOAQUIN', F, rb, 1962; flowers cerise-red, reverse tinted ivory well shaped, borne in large clusters, slight fragrance; tall growth; [Little Darling X Gertrude Raffel]; Raffel; Port Stockton Nursery

Sunny Side Up *see* 'PIXSUN'

'SUNNY SOUTH', HT, pb, 1918; flowers pink flushed carmine, base yellow, large, semi-dbl., cupped, moderate fragrance; foliage rich green; very vigorous growth; [Gustav Grunerwald X Betty Berkeley]; Clark, A.; NRS Victoria

Sunny Sunblaze *see* 'MEIPONAL'

'SUNNY TODAY', F, dy, 1970; flowers clear deep yellow, medium, dbl., exhibition form, moderate fragrance; foliage glossy, dark; moderate, bushy growth; [(Summer Sunshine X Gold Cup) X Isobel Harkness]; Whisler, D.; Gro-Plant Industries

Sunnybrook, HT, my

'SUNNYDEW', Min, dy, 1978; bud ovoid; semi-dbl., 18 petals, 1 in., exhibition form, slight fragrance; well-branched, spreading growth; [Yellow Doll X Unnamed seedling]; Schwartz, Ernest W.; Bountiful Ridge Nursery

'SUNNYMOUNT', HT, my, 1936; bud sharp pointed but short; flowers buttercup-yellow, large, dbl., 50 petals, intense fragrance; foliage leathery; vigorous growth; [Joanna Hill sport]; Grillo

'SUNNYSIDE'®, Min, yb, 1963; flowers yellow, becoming pink and then red; vigorous growth; [(Purpurine X Miniature seedling) X Rosina]; Lens

'SUNNYSIDE', HT, op, 1949; bud pointed, orange; flowers pink shaded orange, dbl., 24–30 petals, 4.5–5 in., exhibition form, slight fragrance; very vigorous growth; [Yellow Gloria sport]; Andre; Andre Greenhouses

Sunnyside *see* POUlmulti

Sunnyside '83 *see* 'LENSUN'

'SUNPACH', F, ob, 1999; (**Pachinko**); semi-dbl., 12–16 petals, 3 in., borne in large clusters, slight fragrance; prickles moderate; foliage medium, dark green, glossy; spreading, medium (30-36 in) growth; [Alexander X Golden Emblem]; Schuurman, Frank B.; Franko Roses New Zealand, Ltd., 1997

'SUNPARI', HT, ab, 1999; (**La Parisienne**); dbl., 26–40 petals, 4.5 in., borne mostly singly, slight fragrance; prickles moderate; foliage large, dark green, glossy; upright, tall growth; [Kia Ora X Texas]; Schuurman, Frank B.; Franko Roses New Zealand, Ltd., 1996

'SUNPASTE', F, w, 1992; (**Pastelina**); flowers near white, blooms, dbl., 26–40 petals, 1.5–2.75 in., slight fragrance; foliage medium, medium green, glossy; upright growth; [White Dream X Freegold]; Schuurman, Frank B.; Riverland Nurseries, Ltd., 1991

'SUNPAT', Min, lp, 1992; (**Little Opal**); flowers patio, dbl., 15–25 petals, 1.5 in., slight fragrance; foliage small, medium green, glossy; upright growth; Rose of The Year, Auckland, NZ, 1992; [White Dream X Dicky Bird]; Schuurman, Frank B.; Riverland Nurseries, Ltd., 1991

'SUNPEARL', Min, lp, 1992; (**Little Pearl**); dbl., 26–40 petals, 1.5 in., slight fragrance; few prickles; foliage small, medium green, glossy; medium (50 cms), upright growth; [Innocent X MACfrabro (Firefly)]; Schuurman, Frank B.; Riverland Nurseries, Ltd.

'SUNPETE', F, lp, 1999; (**Peter Pan**); dbl., 26–40 petals, 2 in., borne in large clusters and sprays, no fragrance; prickles moderate; foliage medium, dark green, glossy; upright, medium growth; [Pink Delight X seedling]; Schuurman, Frank B.; Franko Roses New Zealand, Ltd., 1993

'SUNPINO', Min, lp, 1999; (**Pinocchio**); dbl., 17–25 petals, 2–2.5 in., borne in large clusters, no fragrance; prickles moderate; foliage large, dark green, semi-glossy; patio; upright, medium growth; [Tinkerbell X Chess]; Schuurman, Frank B.; Franko Roses New Zealand, Ltd., 1998

'SUNPLAY', HT, w, 1999; (**Player**); dbl., 17–25 petals, 4.5–5 in., borne mostly singly, no fragrance; many prickles; foliage large, dark green, glossy; upright, tall (4.5-5 ft) growth; [Raspberry Ice X Karen]; Schuurman, Frank B.; Franko Roses New Zealand, Ltd., 1993

'SUN-RAY', HT, my, 1932; flowers golden yellow, semi-dbl., moderate fragrance; vigorous growth; Bentall

'SUNRISE', HT, mp, 1939; bud long, pointed; flowers salmon, open, large, dbl.; foliage glossy, bronze; vigorous, bushy growth; [Pilarin Vilella X Rosa Gallart]; Dot, Pedro

Sunrise *see* KORmarter

'SUNRISE-SUNSET', HT, pb, 1971; flowers blended pink, cream and lavender, large, dbl., exhibition form, slight fragrance; foliage glossy, light to dark, leathery; vigorous growth; [Tiffany X (Unnamed seedling X Happiness)]; Swim & Weeks; Weeks Wholesale Rose Growers

'SUNSALM', Min, op, 1992; (Little Gem, **Patio Gem**); flowers moderately patio, small, dbl., 15–25 petals, 1.5 in., moderate fragrance; foliage small, medium green, matt; spreading growth; [Sexy Rexy X Firefly]; Schuurman, Frank B.; Riverland Nurseries, Ltd., 1991

'SUNSAP', Min, pb, 1999; (**Little Sapphire**); flowers medium pink, reverse light pink, dbl., 17–25 petals, 2.5–3 in., borne in small clusters, slight fragrance; prickles moderate; foliage medium, medium green, semi-glossy; bushy, medium (16-20 in) growth; [White Dream X seedling]; Schuurman, Frank B.; Franko Roses New Zealand, Ltd., 1993

Sunsation *see* 'WEKCELPEP'

Sunsation *see* 'KORGUST'

'SUNSCENT', Min, ly, 1999; (**Scentasia**); dbl., 26–40 petals, 1.5–2 in., borne in large clusters, moderate fragrance; few prickles; foliage large, medium green, semi-glossy; spreading, tall (20-24 in) growth; [Tinkerbell X Little Nugget];

Schuurman, Frank B.; Franko Roses New Zealand, Ltd., 1997

Sunseeker *see* 'DICRACER'

'SUNSET', T, dy, 1883; flowers orange-yellow; [Perle des Jardins sport]; Henderson, P.

'SUNSET', Cl HT, yb, 1947; flowers peach and apricot, reverse yellow, dbl., 45–60 petals, 5 in., cupped, slight fragrance; vigorous growth; [Faience sport]; Marsh; Marsh's Nursery

Sunset Boulevard *see* HARbabble

Sunset Celebration™ *see* 'FRYXOTIC'

'SUNSET GLORY', HT, ab, 1947; bud ovoid; flowers golden yellow overlaid dusty rose-pink, dbl., 35–40 petals, 4–4.5 in., cupped, moderate, fruity fragrance; vigorous, upright, compact growth; [McGredy's Sunset sport]; Boerner; J&P

Sunset Glow *see* **'CANARIENVOGEL'**

'SUNSET JUBILEE', HT, pb, 1973; bud ovoid; flowers medium pink, tinted lighter, large, dbl., exhibition form, slight fragrance; foliage large, light, leathery; vigorous, upright, bushy growth; [Kordes' Perfecta X Pink Duchess seedling]; Boerner; J&P

Sunset Song *see* 'COCASUN'

Sunset Strip *see* 'AROCORE'

'SUNSHINE', HT, yb, 1918; bud long, pointed; flowers golden yellow, shaded apricot, dbl., moderate fragrance; Chaplin Bros.

'SUNSHINE', Pol, ob, 1927; bud ovoid; flowers golden orange, small blooms in clusters, dbl., moderate fragrance; foliage glossy; dwarf growth; (14); [George Elger X William Allen Richardson]; Robichon; Cutbush

Sunshine, LCl, lp

Sunshine Girl *see* 'PIXISUN'

Sunshine Princess *see* 'ANDSUN'

'SUNSHOWER', F, my, 1974; bud ovoid; flowers bright yellow, ruffled, semi-dbl., 16 petals, 2.5 in.; foliage glossy, light; moderate growth; [Golden Showers X Prairie Fire]; Golik; Dynarose

'SUNSIEN', HT, lp, 1999; (**Citronella**®); dbl., 26–40 petals, 4.5–5 in., borne mostly singly, slight fragrance; prickles moderate; foliage large, medium green, glossy; upright, tall growth; [La Parisienne sport]; Schuurman, Frank B.; Franko Roses New Zealand, Ltd., 1998

'SUNSILK', F, my, 1974; flowers lemon-yellow, dbl., 30 petals, 5 in., slight fragrance; foliage dark; GM, Belfast, 1976; [Pink Parfait X Redgold seedling]; Fryer, Gareth; Fryer's Nursery, Ltd.

Sunsmile™ *see* 'JACMIY'

'SUNSNOFLE', Min, lp, 1999; (**Snowflake**); dbl., 26–40 petals, 1 in., borne in small clusters, no fragrance; prickles moderate; foliage small, light green, semi-glossy; compact, low growth; Schuurman, Frank B.; Franko Roses New Zealand, Ltd., 1996

'SUNSONG', Gr, ob, 1966; (Dorrit); bud globular; flowers orange to coral blend, informal, dbl., 63 petals, 3 in., slight, tea fragrance; foliage glossy, leathery; upright, bushy growth; [Folie d'Espagne X (Zambra X Danish Pink)]; Soenderhousen; Armstrong Nursery

Sunsplash™ *see* 'JACYIM'

'SUNSPOT', F, my, 1965; flowers mimosa-yellow, large, borne in clusters, dbl., cupped; foliage leathery; vigorous, bushy growth; [Golden Anniversary X Masquerade]; Fisher, G.; C-P

Sunspray™ *see* 'ARORASP'

Sunsprite *see* 'KORRESIA'

'SUNSPRITE, CLIMBING', Cl F, dy, 1989; [Sunsprite sport]; Kroeger, Henry, 1990

'SUNSTAR', HT, or, 1921; bud long, pointed; flowers orange and red, semi-dbl., intense fragrance; foliage light green; vigorous, bushy growth; GM, Bagatelle, 1923 GM, NRS, 1917; [Mrs C.V. Haworth X Hugh Dickson]; Dickson, A.

'SUNSTAR, CLIMBING', Cl HT, or, 1925; Dickson, A.

'SUNSTONE', HT, yb, 1957; bud long; flowers bright yellow splashed red, 6 in., slight fragrance; foliage glossy, bronze; strong stems; vigorous, tall growth; [Bridget X Marcelle Gret]; Fletcher; Tucker

'SUNSTRIKE', F, dy, 1974; bud small, pointed, ovoid; flowers medium, dbl., exhibition form, slight fragrance; foliage large, light, leathery; very vigorous, bushy growth; [Spanish Sun X (Buccaneer X Zorina)]; Warriner, William A.; J&P

'SUNSTRUCK', Min, yb, 1978; bud long, pointed; flowers aureolin-yellow and signal-red, dbl., 20 petals, 2 in., intense fragrance; foliage dark; upright growth; [Redgold X ?]; Lyon

'SUNTAN', HT, dy, 1939; flowers deep orange-yellow, fading slowly, large, dbl., 35 petals; foliage leathery, dark, bronze; strong stems; RULED EXTINCT 4/92; [Nanjemoy X Mrs Pierre S. duPont]; Hansen, N.J.; B&A

'SUNTAN', HT, lp, 1940; bud long, pointed; flowers buff-salmon, base yellow, reverse light coral-red, dbl., 25 petals, 3.5 in., cupped, moderate fragrance; vigorous, upright growth; RULED EXTINCT 4/92; [Mrs Franklin D. Roosevelt sport]; Yoder

Suntan™ *see* 'INTERBRONZI'

Suntan Beauty™ *see* 'MINAWCO'

'SUNTHUMB', Min, lp, 1999; (**Thumbelina**); dbl., 17–25 petals, 1.75 in., borne in small clusters, moderate fragrance; prickles moderate; foliage medium, medium green, glossy; patio; compact, medium growth; [Champagner X Tinkerbell]; Schuurman, Frank B.; Franko Roses New Zealand, Ltd., 1998

'SUNTICK', HT, lp, 1999; (People's Princess, **Tickled Pink**); dbl., 26–40 petals, 4.5–5 in., borne mostly singly, intense fragrance; few prickles; foliage large, medium green, glossy; upright, tall growth; [Jacaranda X Gold Medal]; Schuurman, Frank B.; Franko Roses New Zealand, Ltd., 1997

'SUNTINK', F, lp, 1999; (**Tinkerbell**); dbl., 26–40 petals, 1.5 in., borne in large clusters and sprays, moderate fragrance; prickles moderate; foliage medium, dark green, glossy; upright, medium growth; [White Dream X Evelien]; Schuurman, Frank B.; Franko Roses New Zealand, Ltd., 1993

'SUNTOTO', HT, m, 1992; (**Rangitoto**); flowers medium, very dbl., 1.5–2.75 in., slight fragrance; foliage medium, medium green, glossy; upright growth; [Champagne X Chantilly Lace]; Schuurman, Frank B.; Riverland Nurseries, Ltd., 1990

SUNwend, F, lp, 1999; (**Wendy**); dbl., 26–40 petals, 1.5 in., borne in large clusters and sprays, no fragrance; many prickles; foliage medium, medium green, glossy; upright, medium growth; Schuurman, Frank B.

'SUPER ARIBAU', HT, dr, 1960; flowers crimson-carmine, reverse silvery crimson, large, dbl., 24 petals, intense fragrance; strong stems; upright, vigorous growth; [Aribau X Director Rubió]; Dot, Pedro

Super Bowl *see* 'MACSUPBOW'

Super Cascade Coral *see* 'JALCORAL'

'SUPER CHIEF', HT, mr, 1976; bud globular; flowers bright red, aging darker, very dbl., 60–65 petals, 5 in., exhibition form, slight fragrance; foliage dark; vigorous growth; [Queen Elizabeth X Happiness]; Patterson; Patterson Roses

'SUPER CONGO', HT, dr, 1950; bud ovoid; flowers velvety dark blood-red, medium, dbl., 30 petals, slight fragrance; foliage dull green; upright growth; [Congo X Léonce Colombier]; Meilland, F.

Super Derby *see* 'BURSTEIN'

Super Disco *see* OLItel

Super Dorothy® *see* HELdoro

Super Elfin *see* 'HELKLEGER'

Super Excelsa® *see* HELexa

Super Fairy *see* 'HELSUFAIR'

'SUPER GAMUSIN', HT, m, 1962; flowers chamois to mauve; [Grey Pearl X Tristesse]; Dot, Pedro

Super Gold™ *see* 'WINGOLD'

'SUPER HARRINGTON', HT, mr, 1938; flowers medium, dbl., moderate fragrance; Brown, J.; (Sangerhausen)

Super Nova *see* JACsunov

Super Pink®, LCl, pb; Lens, 1992

Super Prince Tango, F, 1971; Delbard; (Cavriglia)

Super Sparkle *see* 'HELHEIN'

Super Star *see* 'TANORSTAR'

'SUPER STAR SUPREME', HT, or, 1982; flowers deeper color; [Tropicana sport]; U.S. Patent Sales, Inc.

Super Star, Climbing *see* 'TANGOSTAR'

'SUPER SUN', HT, my, 1967; flowers maize-yellow; [Piccadilly sport]; Bentley

Super Swany®, S, w

'SUPER TABARIN', F, ob, 1963; flowers coppery, reverse orange, semi-dbl.; vigorous growth; [Faust X Tabarin]; Gaujard

Super Tan, Min, op, 1995; flowers coppery orange, small to medium, semi-dbl.; Interplant; (Sangerhausen)

'SUPERB', HT, lp, 1924; flowers pale pink, tinted blush; [Mme Caroline Testout X Willowmere]; Evans

'SUPERB', Pol, dr, 1927; flowers crimson, dbl.; deRuiter

Superb Tuscan *see* **'TUSCANY SUPERB'**

Superb Tuscany *see* **'TUSCANY SUPERB'**

'SUPERB, CLIMBING', Cl Pol, dr, 1933; Guillot, P.

'SUPERBA', HSet, lp, 1843; flowers blush, borne in large clusters; [R. setigera seedling]; Feast

'SUPERBA', HMoy, dr; flowers dark crimson, large, dbl.; no fruit; [Charles P. Kilham X R. moyesii]; Van Rossem

'SUPERBA', (R. setigera seedling), w

'SUPERBA', HT, dp, 1940; flowers intense rose with yellow reflections, overlarge, intense fragrance; long stems; [Julien Potin X Seedling]; Aicardi, L.; Giacomasso

Supercandy, Gr, 1967; Mansuino; (Cavriglia)

Super-Dupont *see* **'DOLLY MADISON'**

'SUPERGA', HT, lp, 1956; flowers rose-pearl, high pointed, slight fragrance; foliage dark, glossy; long stems; very vigorous growth; [Savoia sport]; Giacomasso

Superglo *see* 'SPOGLO'

'SUPER-HARRINGTON', HT, dr, 1938; bud oval; flowers dark crimson, mottled scarlet, medium, dbl., moderate fragrance; foliage leathery; vigorous, upright growth; [Tassin X Victoria Harrington]; Brown, John; A. Meilland

'SUPERIOR', F, ab, 1968; flowers apricot-yellow, blooms in trusses, semi-dbl., slight fragrance; foliage dark, glossy; vigorous, low growth; [Masquerade X Amberlight]; LeGrice

Supra™ *see* 'MEIROBIDOR'

'SUPREME', HT, my, 1963; flowers lemon-yellow, well formed, dbl., 40–45 petals, 4.5–5 in., moderate fragrance; foliage dark, leathery, glossy; vigorous, tall growth; [Golden Masterpiece X Ethel Sanday]; LeGrice

Supreme Cover *see* POUlor

'SUPREME RENDEZVOUS', HT, pb, 1961; flowers cream to pale pink, edged carmine; [Day of Triumph sport]; Langbecker

'SUPREME'S SISTER', F, mp, 1960; bud red; flowers coppery salmon to salmon-pink, medium, dbl., 42 petals; foliage leathery, bronze; low, bushy growth; Jones; Hennessey

Supriya, HT, pb; San, 1982

'SURABHI', HT, mp, 1975; bud long, pointed; flowers phlox-pink, very dbl., 80 petals, exhibition form, intense fragrance; foliage glossy, dark; upright, compact growth; [Oklahoma X Delhi Princess]; IARI

'SURAIN', HT, mp, 1985; flowers large, dbl., 35 petals, exhibition form, slight fragrance; upright growth; [Unnamed seedling X Forever Yours]; Elliott, Charles P.

Surbrunnsgatan, HP, lp

'SUREKHA', HT, or, 1969; bud pointed; flowers coral-red, full, large, dbl.; foliage dark, leathery; very vigorous, upright growth; [Queen Elizabeth X Open pollination]; IARI

'SURF RIDER', S, w, 1968; flowers creamy white, blooms in large trusses, dbl., 25 petals, moderate, musk fragrance; foliage bright, glossy; very vigorous, erect growth; [Ivory Fashion X Ballerina]; Holmes, R.A.; Fryer's Nursery, Ltd.

Surfer Girl *see* 'RENFER'

Surf's Up *see* 'LEMSUR'

Surfside *see* 'MICSURF'

'SURKHAB', HT, m, 1976; bud globular; flowers tyrian purple, reverse silvery amaranth, very dbl., 76 petals, 4 in., cupped, intense fragrance; foliage leathery; very vigorous, upright, compact growth; Pal, Dr. B.P.; Anand Roses

Surkhab, HT, lp; Pal, Dr. B.P., 1980

'SURPASSE TOUT', HGal, mr; (Cérisette la Jolie); flowers rosy crimson, fading cerise-pink, dbl.; height 3-4 ft; Hardy, before 1823

Surpassing Beauty of Woolverstone, HP, dr; re-introduced, 1980

'SURPRISE', HT, mp, 1925; flowers salmon-pink, dbl., cupped, slight fragrance; [Frau Karl Druschki X Mme Edouard Herriot]; Van Rossem

Surprise Party *see* 'JELROGANOR'

Surprise Surprise *see* 'MORLAVMAG'

Surrey *see* 'KORLANUM'

'SURVILLE', HT, mr, 1924; flowers indian red, shaded cerise, dbl.; [Mme Edouard Herriot seedling]; Croibier

Survivor, S, dr; Ottawa, 1975

'SURYA KIRAN', F, or, 1979; bud pointed; 19 petals, borne 2-14 per cluster, slight fragrance; hooked prickles; foliage large, glossy; tall, upright, vigorous, open growth; [Flamenco X Orangeade]; Pal, Dr. B.P.; Friends Rosery

Surya Shikha, HT, lp; Diby

'SURYODAYA', F, ob, 1968; bud ovoid; flowers bright orange, medium, semi-dbl.; foliage dark, leathery; very vigorous, upright growth; [Orangeade X ?]; IARI

'SUSAN', F, mr, 1955; flowers crimson, well-formed blooms in large clusters, semi-dbl.; foliage dark; bushy, compact growth; [Donald Prior X Our Princess]; Robinson, H.

Susan *see* 'KORAD'

Susan *see* POUlsue

Susan *see* KORkilt

Susan Ann *see* **'SOUTHAMPTON'**

'SUSAN BECKWITH', HT, mp, 1998; bud short, squat; flowers medium pink, apricot base, opens flat, very dbl., 26–40 petals, 4–4.5 in., flat, borne in clusters, intense fragrance; prickles moderate; foliage medium, medium green, semi-glossy; bushy, medium (2.5 ft.) growth; [Dr. Sybil Johnston X Seedling]; Beckwith, R.; Battersby Roses, 1998

Susan Blixen *see* 'POULARI'

Susan Devoy *see* 'MACRENO'

Susan Elizabeth *see* 'SCHUPRAK'

Susan Hampshire *see* 'MEINATAC'

'SUSAN HAYWARD', Cl HT, lp, 1967; flowers light pink, center and reverse darker, large, dbl., exhibition form; foliage dark, glossy; vigorous, climbing, open habit growth; [Fontanelle X Gen. MacArthur]; Hayward

Susan Irvine *see* 'TOMSUE'

Susan Irvine Cream, T, lp

Susan Irvine Pink Tea, T, lp

Susan Jellicoe *see* 'HORKEEPOG'

'SUSAN LOUISE', S, lp; bud very long, pointed, deep pink; flowers flesh-pink, semi-dbl., recurrent bloom, slight fragrance; vigorous (4-5 ft.), bushy growth; [Belle Portugaise seedling]; Adams, Dr. Neil D.; Stocking

Susan Massu *see* 'KORAD'

'SUSAN NOEL', Min, ab, 1986; flowers light apricot, urn-shaped, small blooms borne usually singly, very dbl., 80 petals, slight fragrance; small, globular fruit; small, slightly hooked, gray-orange prickles; foliage small, medium green, semi-glossy; medium, upright growth; [Rise 'n' Shine X Orange Honey]; Jolly, Nelson F.; Rosehill Farm

'SUSAN SCHNEIDER', S, lp, 1999; flowers med. yellow, turning orange pink, then red, reverse med. yellow, semi-dbl., 8–14 petals, 3 in., borne in small clusters, slight fragrance; prickles moderate, tan; foliage medium, medium green, glossy; spreading, medium (5 ft) growth; Jerabek, Paul E.; Freedom Gardens, 1998

'SUSANA MARCHAL', HT, or, 1953; bud pointed; flowers coral-red, large, semi-dbl., 18 petals, exhibition form; foliage clear green; strong stems; upright growth; [Cynthia X Vive la France]; Dot, Pedro

'SUSANE DOT', Cl HT, dr, 1963; flowers crimson, large, dbl., 32 petals, moderate fragrance; vigorous growth; [Queen Elizabeth X Peace, Climbing]; Dot, Simon

'SUSANNA', Pol, lp, 1914; flowers small, dbl.; Weigand, C.; (Sangerhausen)

Sushma, F, m; Kasturi, 1975

'SUSIE', HT, mp, 1955; flowers pink, base yellow, dbl., 45–50 petals, 4–6 in., exhibition form, slight fragrance; foliage leathery; vigorous, bushy growth; RULED EXTINCT 11/90; [Mme Butterfly X Sally]; Spandikow

'SU-SPANTU', F, m, 1962; flowers purplish pink, borne in clusters of 7-8, dbl., 25 petals; upright growth; [Alain X Independence]; Borgatti, G.; Sgaravatti

Suspense *see* 'MEIFAN'

Sussex *see* POUlave

'SUTTER'S GOLD', HT, ob, 1950; bud orange overlaid indian red; flowers golden orange, often with red on outer petals, dbl., 33 petals, 4–5 in., exhibition form, intense fragrance; foliage dark, leathery; very vigorous, upright growth; AARS, 1950 GM, Bagatelle, 1948 GM, Geneva, 1949 GM, Portland, 1946 James Alexander Gamble Fragrance Medal, ARS, 1966; [Charlotte Armstrong X Signora]; Swim, H.C.; Armstrong Nursery

'SUTTER'S GOLD, CLIMBING', Cl HT, ob, 1950; (Grimpant Sutter's Gold); Weeks; Armstrong Nursery

'SUTTON PLACE'®, Gr, op, 1991; flowers light pink with salmon pink edge, large blooms borne mostly, single, intense fragrance; foliage large, dark green, semi-glossy; tall, upright, bushy growth; [Queen Elizabeth X (Carla X Command Performance)]; Williams, J. Benjamin, 1992

'SUZAN BALL', F, ab, 1957; bud conical; flowers salmon-coral, to loosely open, borne in clusters, dbl., 25 petals, 2.5–3.5 in., exhibition form, moderate, spicy fragrance; foliage dark; vigorous, upright growth; [Tom Breneman X Fashion]; Warriner, William A.; H&S

'SUZANNE', HSpn, op, 1950; flowers pale coral-pink, very dbl., recurrent bloom; foliage small, dark; height 4 ft; [Second generation R. laxa X R. spinosissima]; Skinner

'SUZANNE ALBRAND', Pol, mr, 1930; flowers bright red, large, borne in large clusters; foliage glossy; vigorous growth; Turbat

'SUZANNE BALITRAND', HT, or, 1942; flowers coral edged fiery red, stamens yellow; Mallerin, C.; A. Meilland

'SUZANNE BIDARD', Pol, mp, 1914; flowers small, dbl.; Vigneron; (Sangerhausen)

'SUZANNE BLANCHET', T, lp, 1885; flowers flesh-pink, large, intense fragrance; Nabonnand, G.

'SUZANNE CARROL OF CARROLTON', HP, mp, 1924; flowers light satiny rose and salmon, large, semi-dbl.; vigorous growth; [Frau Karl Druschki X Mme Gabriel Luizet]; Nabonnand, P.

Suzanne Dolard, Pol, dr, 1966; flowers medium, semi-dbl.; Gaujard; (Sangerhausen)

'SUZANNE HESTER', Cl HT, w, 1950; bud ovoid; flowers white, base tinted yellow, dbl., 30 petals, exhibition form, intense fragrance; foliage glossy; vigorous growth; [Marie Maass sport]; Hester

'SUZANNE MEYER', Pol, w, 1926; flowers white shaded soft pink, center bright rose-pink, medium; medium growth; [Tausendschön X Rosel Dach]; Walter, L.

'SUZANNE MICHELA', HT, my, 1932; bud long, pointed; flowers pure chrome-yellow, very large, dbl., cupped, moderate fragrance; very vigorous growth; [Mme la Generale Ardouin X Unnamed seedling]; Chambard, C.

'SUZANNE MILLER', Pol, mr, 1927; flowers clear bright cherry-red, medium, borne in clusters, dbl.; low, bushy growth; Wezelenburg

'SUZANNE TURBAT', Pol, ob, 1919; flowers coral-red, shaded shrimp-pink, medium, borne in clusters of, dbl., cupped, slight fragrance; dwarf growth; [Petit Constant seedling X Unnamed seedling]; Turbat

'SUZANNE VILLAIN', HT, pb, 1935; flowers peach-blossom-pink suffused cherry-red, reverse salmon-pink, dbl., 60–65 petals; foliage rich green; vigorous, bushy growth; [(Rev. Williamson X Gorgeous) X Mrs John Bell]; Ketten Bros.

'SUZANNE WOOD', HP, mp, 1869; flowers large, dbl.; Verdier, E.; (Sangerhausen)

'SUZANNE-MARIE RODOCANACHI', HP, mp, 1883; flowers dark rosy cerise, shaded and bordered lighter, well formed, dbl., 45 petals, globular, profuse bloom, somewhat recurrent; [Victor Verdier seedling]; Lévêque

Suzette van der Merwe *see* JAChoc

Suzon, LCl, pb; Eve, 1994

'SUZON LOTTHÉ', HT, pb, 1951; bud peach; flowers pearl-pink, flushed deeper toward edge, dbl., 60 petals, 4 in., exhibition form, intense, damask fragrance; foliage dark; very vigorous growth; [Peace X (Signora X Mrs John Laing)]; Meilland, F.; C-P

'SUZON LOTTHÉ, CLIMBING', Cl HT, pb, 1964; Trimper, K.; Ruston

Suzy *see* 'BRIsuzy'

Suzy, LCl, yb; Eve, 1994

Suzy Q® *see* 'JAChill'

'SVATOPLUK CECH', LCl, dy, 1936; flowers orange-yellow, recurrent bloom, intense fragrance; vigorous growth; Brada, Dr.; Böhm

'SVATY VÁCLAV', HT, w, 1936; flowers white, center yellowish, large, intense fragrance; Berger, A.

Svensk Pimpinelle, HSpn, w

'SVETLA ALBENA', F, yb, 1974; flowers opening yellow orange, aging red, blooms in large clusters, very dbl., 85 petals; foliage dark; vigorous, upright growth; [Highlight X Masquerade]; Staikov, Prof. Dr. V.; Kalaydjiev and Chorbadjiiski

Svetlana, F, lp; Urban, J., 1975; (Czech Rosa Club)

Svitani, HT, lp; Kavka, 1962; (Czech Rosa Club)

'SVORNOST', Pol, dr, 1935; flowers pure dark red, with fiery streaks, borne in large clusters; [Orléans Rose sport]; Böhm, J.

'SWAGATAM', HT, ab, 1989; flowers light pink, reverse apricot; [Surkhab sport]; Patil, B.K.; K.S.G. Son's Roses, 1987

'SWAMI', HT, lp, 1973; bud long, pointed; very dbl., 100 petals, 5 in., exhibition form, moderate fragrance; very vigorous, upright growth; [Scarlet Knight X Festival Beauty]; Hardikar, Dr. M.N.

Swamp Rose *see* **R. PALUSTRIS**

Swamp Rose *see* **R. PALUSTRIS SCANDENS**

Swan *see* 'AUSwhite'

'SWAN LAKE', LCl, w, 1968; (Schwanensee); flowers white, center tinged pinkish, well-formed, large, dbl., 50 petals, slight fragrance; [Memoriam X Heidelberg]; McGredy, Sam IV

Swan Song *see* 'SEASWAN'

'SWANSDOWN', HT, w, 1928; bud large, long, pointed; flowers creamy white, center deeper cream, spiral form, dbl., slight fragrance; foliage olive-green; fairly vigorous growth; Dickson, A.

Swansong *see* 'SEASWAN'

'SWANTJE', F, w, 1936; flowers snow-white, very large, borne in clusters, dbl.; foliage glossy, light; very vigorous, bushy growth; [Johanna Tantau X (Prof. Gnau X Joanna Hill)]; Tantau

Swany® *see* 'MEIBURENAC'

'SWANY RIVER', HT, w, 1981; (Fredagh of Bellinchamp); flowers large, dbl., moderate fragrance; foliage medium, dark, semi-glossy; upright growth; [Pascali X Unnamed seedling]; Herholdt, J.A.

Swarthmore *see* 'MEITARAS'

'SWARTHMORE, CLIMBING', Cl HT, pb, 1973; Thomas, Dr. A.S.; A. Ross & Son

Swashbuckle, HT, or; Dawson, before 1990

'SWATI', Pol, w, 1968; bud pointed; flowers white, edged deep pink, open, small, semi-dbl.; foliage dark, leathery; moderate, bushy growth; [Winifred Coulter X ?]; IARI

'SWEDISH DOLL', Min, op, 1976; bud long, pointed; flowers small, dbl., 28 petals, 1.5 in., slight fragrance; foliage glossy, leathery; vigorous, upright, branched growth; [Fire King X Little Buckaroo]; Moore, Ralph S.; Sequoia Nursery

Sweepstakes *see* 'MACAFT'

'SWEET', Min, my, 1980; bud long, pointed; dbl., 50 petals, borne 1-3 per cluster, moderate fragrance; tiny, curved prickles; foliage tiny, glossy, deep green; compact, bushy growth; [Unnamed seedling X Unnamed seedling]; Lyon

'SWEET ADELINE', HT, mp, 1929; bud long, pointed; flowers rose-pink, large, semi-dbl., moderate fragrance; [Rapture X Souv. de Claudius Pernet]; Hill, Joseph H., Co.

Sweet Adeline, S, lp; Erskine

'SWEET AFTON', HT, w, 1964; flowers near white, reverse pale pink, dbl., 4.5–5 in., exhibition form, intense fragrance; foliage leathery; tall, spreading, bushy growth; [(Charlotte Armstrong X Signora) X (Alice Stern X Ondine)]; Armstrong, D.L. & Swim; Armstrong Nursery

'SWEET ALLISON', MinFl, rb, 1985; flowers orange-red shading to yellow or white, reverse cream, deeper, dbl., 35 petals, slight fragrance; foliage medium, medium green, semi-glossy; upright, bushy growth; [Bonny X (Tiki X Seedling)]; Jolly, Nelson F.; Rosehill Farm

Sweet Amazone, HT, op

'SWEET AMY', HT, lp, 1999; flowers medium pink, reverse lighter, dbl., 26–40 petals, 4–5 in., borne mostly singly, intense fragrance; prickles moderate; foliage medium, dark green, glossy; upright, medium (5 ft) growth; [Crystalline X Stainless Steel]; Edwards, Eddie; Johnny Becnel Show Roses, 1999

'SWEET AND LOW', F, pb, 1962; bud globular; flowers salmon-pink, center lighter, small, borne in large clusters, dbl., 36 petals, slight fragrance; foliage glossy, bronze; compact, low growth; [Pinocchio X Sweet Fairy]; Schwartz, Ernest W.; Wyant

Sweet Bouquet *see* SABchurchill

Sweet Bouquet *see* SABchurchill

'SWEET BRIAR QUEEN', F, mp, 1960; flowers rose-pink, borne in clusters, dbl., 45–50 petals, 2 in., flat, moderate fragrance; foliage maroon when young, leathery; vigorous, branching growth; [Rosemary sport]; Schenkel; H.R. Schenkel, Inc.

Sweet Brier Rose *see* **R. EGLANTERIA**

Sweet Butterfly *see* 'LAVSTAR'

'SWEET CARESS', F, mp, 1958; bud ovoid; flowers rose-pink, large, dbl., 55–60 petals, moderate fragrance; foliage leathery, glossy; vigorous, upright growth; [Pigmy Red X Demure]; Boerner; J&P

Sweet Caroline *see* 'MICAROLINE'

Sweet Chariot™ *see* 'MORCHARI'

'SWEET CHARITY', HT, w, 1967; flowers waxy white, sometimes marked pink, medium, semi-dbl., moderate fragrance; foliage dark, glossy; vigorous, upright growth; Park, F.; Moulton-Jones

'SWEET CHERRY', HT, mr, 1960; (Cherry Glow); flowers cherry-red and yellow, large, exhibition form; [Margaret X Karl Herbst]; Dickson, A.

Sweet Cory, Pol, lp

Sweet Cover *see* POUlweet

Sweet Diana, HT, lp; Hortico, 1999

Sweet Dream *see* 'FRYMINICOT'

'SWEET DREAMS', HT, pb, 1977; (Kojo); bud long, pointed; flowers light pink, reverse deep pink, dbl., 45–50 petals, 6 in., exhibition form, slight, tea fragrance; foliage dark; vigorous, very upright growth; [Jrn. No. 8 seedling X Royal Highness]; Takatori, Yoshiho; Japan Rose Nursery

Sweet Dreams *see* JACams

Sweet Ecstasy *see* 'HINSWEET'

'SWEET FAIRY', Min, lp, 1946; flowers micro-mini, dbl., 57 petals, 1 in., cupped, moderate fragrance; foliage small, dark; vigorous, dwarf (6-8 in) growth; [Tom Thumb X Unnamed seedling]; deVink; C-P

Sweet Frederique, S, lp; Interplant, 1997

Sweet Gesture *see* 'MACCARLTO'

Sweet Hannah, Min, lp; Moore, 1998

'SWEET HARMONY', F, pb, 1962; flowers canary-yellow edged crimson-pink, borne in clusters, dbl., 36 petals, 3.5 in., moderate fragrance; foliage glossy, light green; vigorous growth; [Peace X Masquerade]; Gaujard; Gandy Roses, Ltd.

'SWEET HOME', HT, dp, 1969; flowers deep pink, large, dbl., 35 petals, exhibition form, slight fragrance; foliage leathery; vigorous growth; [(Jolie Madame X Baccará) X (Baccará X Jolie Madame)]; Meilland, Mrs. Marie-Louise; URS

Sweet Home Alabama *see* 'TALSWEET'

Sweet Honesty, Min, w

Sweet India, HT, lp; Tejganga, 1996

Sweet Inspiration *see* 'JACSIM'

'SWEET INTERLUDE', HT, pb, 1989; bud pointed; flowers white center with pink edging, large, borne singly, slow opening, dbl., 32 petals, exhibition form, intense fragrance; prickles moderate, brown; foliage medium green, disease-resistant; upright, bushy, medium growth; [Pristine X Olympiad]; Wambach, Alex A.

Sweet Juliet *see* 'AUSLEAP'

'SWEET KERI', HT, lp, 1988; flowers light pale pink, large, borne singly, dbl., 44 petals, exhibition form, no fragrance; prickles triangular, red; foliage medium green, large; medium, upright growth; [Judith Marten X Folklore]; Wilson, George D.

Sweet Lady *see* 'MEIZELI'

'SWEET LAVENDER', HMult, pb, 1912; flowers blush, edged mauve, borne in large clusters, single; Paul

'SWEET LAVINIA', HT, my, 1967;, exhibition form, intense fragrance; foliage glossy; [Golden Scepter X Peace]; McTeer, F.

'SWEET LELANIE', HT, lp, 1967; flowers light pink, base yellow, large, dbl., globular, moderate fragrance; foliage wrinkled, bronze; vigorous, upright growth; [Charlotte Armstrong X Seedling.]; Duehrsen; Elmer Roses Co.

'SWEET LOVE', HT, mp, 1980; flowers medium-large blooms borne 1-7 per cluster, very dbl., 90 petals, flat, no fragrance; red prickles; foliage matt, dark; upright growth; [Queen Elizabeth X Duke of Windsor]; Rijksstation Voor Sierplantenteelt

Sweet Magic *see* 'DICMAGIC'

'SWEET MAID', HT, lp, 1950; bud pointed; flowers porcelain-pink, semi-dbl., intense fragrance; Moss

'SWEET MANDARIN', Min, ob, 1978; bud pointed; flowers light orange, small, semi-dbl., 16 petals, flat, moderate fragrance; foliage small, soft; compact growth; [Sweet and Low X Gypsy Moth seedling]; Schwartz, Ernest W.; Bountiful Ridge Nursery

Sweet Marie, Min, ob

Sweet Marvel, Min, lp; deRuiter, 1992

Sweet Melody, F; Leenders; (Cavriglia)

Sweet Melody *see* FISmelody

'SWEET MEMORIE', HT, mp, 1937; flowers pink, base yellow, reverse purplish, large, semi-dbl., cupped, intense fragrance; foliage leathery; vigorous growth; [Mrs C.W. Edwards X Rose Marie]; Hieatt

Sweet Memories *see* WHAmemo

Sweet Memories, MinFl, lp; Whartons Roses, 1995

'SWEET MEMORY', F, lp, 1991; bud pointed; flowers urn-shaped, dbl., 30 petals, 4 in., exhibition form, moderate fragrance; foliage medium, medium green, matt; bushy, medium growth; [Hana-Gasumi X (Lady X X Paradise)]; Yasuda, Yuji, 1992

Sweet Mimi® *see* 'HAUMI'

Sweet Mystery *see* 'FLOMYST'

'SWEET 'N' PINK', HT, dp, 1976; bud ovoid, pointed; flowers deep pink, dbl., 48 petals, 4–4.5 in., exhibition form, intense fragrance; foliage dark; upright, branching growth; [(Prima Ballerina X Seedling) X Unnamed seedling]; Weeks; Weeks Wholesale Rose Growers

Sweet Nell *see* 'COCAVOTER'

Sweet Pea, Pol, lp; Antique Rose Emporium, 1998

Sweet Perfume, F, or

Sweet Petite *see* FRYxquisite

Sweet Pickins *see* 'TINSWEET'

'SWEET PRINCE', HT, pb, 1988; flowers deep pink, petals edged medium red blooms borne singly, dbl., 20 petals, 3.5 in., intense fragrance; heavily prickled; foliage large, dark, matt; bushy growth; [Honey Favorite X Granada]; Stoddard, Louis

Sweet Promise *see* 'MEIHELVET'

Sweet Promise, Climbing *see* 'MEIHELVETSAR'

'SWEET RASPBERRY', Min, dp, 1984; flowers deep purplish pink, urn-shaped, small, dbl., flat, slight fragrance; foliage small, medium green, matt; bushy growth; [Little Rascal X Cinderella]; Jolly, Nelson F.; Rosehill Farm

Sweet Repose *see* **'THE OPTIMIST'**

Sweet Revelation *see* HARzazz

Sweet Revenge *see* 'TINREVENGE'

Sweet Rosamini, Min, lp

Sweet Salmon, F; Leenders; (Cavriglia)

Sweet Scent *see* 'RENSWEET'

Sweet Sensation, F, lp

'SWEET SEVENTEEN', HT, lp, 1923; semi-dbl., moderate fragrance; foliage light, wrinkled; bushy, dwarf growth; [Frau Karl Druschki X Bardou Job]; Clark, A.; NRS Victoria

'SWEET SEVENTEEN', Gr, mp, 1960; flowers pink, small; vigorous growth; [Fred Howard X Cocorico]; Leenders, J.

'SWEET SHADDOW', HT, mr, 1963; flowers scarlet, edged darker, dbl., 22 petals, 4.5 in., moderate fragrance; foliage dark; vigorous growth; [Queen Elizabeth X Étoile de Hollande]; Barter

Sweet Shirley *see* 'CLESHIR'

'SWEET SIXTEEN', HT, pb, 1943; bud long, pointed; flowers salmon-pink, base yellow, 16–20 petals, 4–5 in., intense fragrance; vigorous, upright, bushy growth; [Mrs Sam McGredy X Pres. Herbert Hoover]; Lammerts, Dr. Walter; Armstrong Nursery

Sweet Sonata *see* 'MEIOFFIC'

Sweet Song *see* 'MEIHIVANO'

Sweet Success, Gr, lp; Pottschmidt, 1995

'SWEET SUE', Min, lp, 1979; bud ovoid; flowers very small blooms in clusters of 10-25, semi-dbl., 13 petals, exhibition form; foliage small, dark; bushy growth; [Pink Ribbon X Pink Ribbon]; Bennett, Cecilia 'Dee'; Tiny Petals Nursery

'SWEET SUE', HT, op, 1940; bud long, pointed, flame to blood-red; flowers coral-pink, stamens maroon, single, cupped, moderate, spicy fragrance; vigorous growth; [Joanna Hill X Night]; Lammerts, Dr. Walter; Armstrong Nursery

'SWEET SULTAN', Cl HT, dr, 1958; (Swet Sultain); flowers crimson shaded maroon, blooms in trusses, single, 5 petals, 4 in., recurrent bloom, intense fragrance; vigorous, pillar growth; [Independence X Honour Bright]; Eacott; LeGrice Roses

Sweet Sunblaze® *see* 'MEITONJE'

Sweet Sunsation *see* KORfisro

Sweet Sunshine *see* 'MORSUN'

'SWEET SURRENDER', HT, mp, 1983; flowers medium silvery pink, large, dbl., 40 petals, cupped, slight, tea fragrance; foliage dark, leathery; AARS, 1983; [Unnamed seedling X Tiffany]; Weeks, O.L.; Weeks Wholesale Rose Growers

Sweet Symphony *see* 'MEIBARKE'

'SWEET TALK', F, ly, 1964; flowers lemon to white, blooms in clusters, dbl., moderate fragrance; foliage light green, leathery; low, uniform, bushy growth; [Frolic X Lavender Pinocchio]; Swim & Weeks; Weeks Wholesale Rose Growers

'SWEET THOUGHTS', HT, lp, 1965; 5 in., slight fragrance; vigorous growth; [Kordes' Perfecta X Show Girl]; Cant, F.

'SWEET TWILIGHT', Cl HT, ly, 1954; flowers buff-yellow, center darker, dbl., 25 petals, 4–5.5 in., exhibition form, abundant, recurrent bloom, moderate, fruity fragrance; vigorous (15-30 ft.) growth; [Tawny Gold sport]; Motose

'SWEET VELVET', F, mr, 1969; flowers large, dbl., moderate fragrance; foliage small, leathery, glossy; moderate growth; [Leverkusen X S'Agaro]; Martin, J.; Gandy Roses, Ltd.

Sweet Vibes *see* 'GELSWEET'

'SWEET VIVID', Min, pb; flowers pink, yellow center, well-formed, dbl., moderate fragrance; vigorous growth

'SWEET VIVIEN', F, pb, 1961; bud ovoid; flowers pink, center light yellow, blooms in clusters, semi-dbl., 17 petals, 3 in., slight fragrance; large, pear-shaped fruit; foliage small, dark, glossy; short stems; very compact, bushy growth; [Little Darling X Odorata (HT)]; Raffel; Port Stockton Nursery

Sweet Wonder, MinFl, lp

Sweetbriar *see* **R. EGLANTERIA**

Sweetcorn, Min, lp; Hannemann, F., 1991; The Rose Paradise; (Weatherly, L.)

Sweetheart *see* 'COCAPEER'

'SWEETHEART', HWich, w, 1901; bud rose-pink; very dbl., 2.5 in., moderate fragrance; foliage glossy, dark; vigorous, climbing growth; RULED EXTINCT 4/80; [R. wichurana X Bridesmaid]; Walsh

Sweetheart Rose *see* **'MLLE CÉCILE BRÜNNER'**

Sweetheart Rose, Climbing *see* **'MLLE CÉCILE BRÜNNER, CLIMBING'**

'SWEETIE PIE', HT, lp, 1971; flowers light pink, outer edges darker; [Swarthmore sport]; Hyde

'SWEETNESS', HT, mp, 1919; flowers intense rose-pink, shaded scarlet, well formed, moderate fragrance; McGredy

'SWEETNESS', HT, ab, 1937; bud long, pointed; flowers apricot-yellow shaded pink, spiral, reflexed petals, large, dbl., moderate fragrance; foliage glossy; very vigorous growth; GM, NRS, 1935; Dickson, A.

Sweetnesse *see* TANsemil

Sweetwaters *see* 'MACSWEETWA'

Swet Sultain *see* **'SWEET SULTAN'**

Swift, S, w, 1992; dbl.; groundcover; low (1-1.5 ft) growth; Poulsen

Swing® *see* 'INTERSIREE'

'SWINGER', Min, my, 1984; flowers, dbl., 35 petals, exhibition form, borne usually singly, slight fragrance; foliage medium, medium green, semi-glossy; spreading growth; PP006560; [Anita Charles X Orange Honey]; Jolly, Nelson F.; Rosehill Farm

'SWINGTIME', F, rb, 1961; bud pointed; flowers crimson, reverse buff-white, medium, dbl., 40 petals, cupped, moderate, sweetbriar fragrance; vigorous, upright growth; [Little Darling X (Red Pinocchio X Masquerade)]; Morgan; Carlton Rose Nurseries

'SWISHREV', HT, dr, 1998; (**Revelry**™); flowers dark red to purple, full, very dbl., 26–40 petals, 4 in., borne moslty single, moderate fragrance; foliage medium, dark green, semi-glossy; upright, tall growth; [Pristine X National Velvet]; Swisher, Stan; Certified Roses, Inc., 1998

'SWISS BLISS', Min, my, 1988; flowers aging lighter, decorative, small, borne singly or in, dbl., 45–50 petals, slight fragrance; prickles reddish-brown; foliage small, medium green, matt, serrated; upright, bushy, medium growth; [Anytime X Unnamed seedling]; Lemrow, Dr. Maynard W.

Swiss Fire® *see* 'HUBAR'

Swiss Gold *see* **'SCHWEIZER GOLD'**

Swiss Lass *see* 'LEMSWI'

Swonderful *see* 'SHAYELA'

'SWORD OF HOPE', HT, mr, 1964; bud long, pointed; dbl., 45 petals, 5–6 in., exhibition form, moderate fragrance; foliage glossy; vigorous, upright growth; Von Abrams; Peterson & Dering

'SYBIL', HT, mp, 1921; bud long, pointed; flowers silvery salmon-rose, center orange-salmon, very dbl., moderate fragrance; [Sunburst X Mary, Countess of Ilchester]; Bees

Sybil Hipkin, HT, lp; Dawson, before 1979

'SYBIL THORNDIKE', HT, mr, 1975; flowers full, dbl., 25 petals, 5 in., slight fragrance; foliage dark; [Liebeszauber X Seedling]; Kordes; Harry Wheatcroft Gardening

'SYDNEY', HCh, mp, 1977; bud ovoid; flowers phlox-pink, dbl., 20 petals, 3–3.5 in., cupped, intense fragrance; foliage rugose, yellow-green; bushy, spreading growth; [Old Blush X Frau Dagmar Hartopp]; Svedja, Felicitas; Canada Dept. of Agric.

'SYDNEY', HP, lp; [Prince Camille de Rohan sport]; Moore, A.K., 1908; (Weatherly, L.)

'SYDONIE', HP, mp, 1846; (Sidonie); flowers rose, medium, very dbl., quartered; red prickles; vigorous growth; Dorisy

Sylphide®, F, dp

Sylt *see* 'KORYLT'

Sylvaine, HT, lp

'SYLVAN SUNSET', HT, pb, 1979; bud pointed; dbl., 28 petals, exhibition form, slight fragrance; hooked prickles; foliage medium green, semi-glossy; vigorous, upright, tall, bushy growth; [Swarthmore X First Prize]; Taylor, Thomas E.

Sylvana, HT, lp; Huber, 1997

'SYLVANDER', S, my; flowers clear yellow, large, single; dwarf (2-3 ft) growth; [R. X harisonii open-pollinated seedling]; Central Exp. Farm

Sylvia *see* 'KORLIFT'

'SYLVIA', HWich, ly, 1911; flowers pale lemon-yellow, passing to white, dbl., moderate fragrance; moderate growth; Paul, W.

'SYLVIA', HT, my, 1949; flowers buttercup-yellow, dbl., 25 petals, moderate, fruity fragrance; RULED EXTINCT 1/79; [Golden Gleam sport]; Foster

'SYLVIA DOT', F, lp, 1965; bud pointed; flowers salmon, medium, dbl., 30 petals, exhibition form, slight fragrance; foliage glossy, bronze; upright, dense growth; [Queen Elizabeth X Orient]; Dot, Simon; Minier

'SYLVIA GROEN', HT, pb, 1935; flowers coral-rose, shaded crimson; vigorous, upright growth; [Pres. Herbert Hoover sport]; Groen

'SYLVIA LOUISE', HT, op, 1973; flowers pink and orange blend; foliage glossy; [Ena Harkness, Climbing seedling]; Shaw, H.C.W.

Sylvie Briant, HT, op

Sylvie Leblanc, F, w

Sylvie Vartan®, Pol, lp; Eve, 1969

Sylvor, HT, 1971; Croix; (Cavriglia)

Symbol®, F, mr

'SYMBOL MIRU', HWich, w, 1937; flowers medium, dbl.; Böhm, J.; (Sangerhausen)

'SYMBOLE', S, mr, 1945; bud thick, pointed; flowers carmine, well shaped, very large, slight fragrance; long, strong stems; vigorous growth; [Mev. G.A. van Rossem X Roseraie de l'Hay]; Robichon; Vilmorin-Andrieux

'SYMPATHIE'®, HKor, mr, 1964; flowers velvety dark red, well-formed, intense fragrance; foliage glossy, dark; very vigorous (9-12 ft) growth; ADR, 1966; Kordes, R.

'SYMPHONETTE', Min, pb, 1973; bud ovoid; flowers light pink, reverse deep pink, small, dbl., moderate fragrance; foliage small, glossy, leathery; vigorous, dwarf, bushy growth; [Cécile Brunner X Cinderella]; Morey, Dr. Dennison; Pixie Treasures Min. Roses

'SYMPHONIE', HT, pb, 1951; bud pointed, ovoid to globular; flowers shades of pink, broad petals, veined carmine-pink, dbl., 25 petals, 4.5–5 in., exhibition form; foliage glossy, leathery; vigorous, upright, bushy growth; GM, NRS, 1949; [Peace X (Signora X Mrs John Laing)]; Meilland, F.; C-P

Symphonie Lumiere *see* MEItomkin

'SYMPHONIE, CLIMBING', Cl HT, pb; Elmer's Nursery, about 1952

Symphony *see* 'AUSLETT'

'SYMPHONY', HP, lp, 1935; bud long, pointed; flowers large, dbl., exhibition form; foliage leathery, dark; [Frau Karl Druschki X Souv. de Claudius Pernet]; Weigand, C.; P.J. Howard

'SYR', S, dr, 1977; bud pointed, ovoid; dbl., 23 petals, 4 in., exhibition form, slight fragrance; foliage glossy, dark; very vigorous, upright growth; [Stadt Rosenheim X Sangerhausen]; Lundstad; Agricultural University of Norway

'SYRACUSE', HT, mr, 1930; bud crimson; flowers scarlet-crimson, large, dbl., slight fragrance; foliage dark, leathery; vigorous growth; [Mme G. Forest-Colcombet X Aspirant Maumejean]; Mallerin, C.; C-P

Syrikit, HT, 1969; Mondial Roses; (Cavriglia)

'SYRINGA', HWich, w, 1931; flowers small, single; Browning; (Sangerhausen)

T

T. B. McQuesten, F, lp

'T.F. Crozier', HT, dy, 1918; flowers deep canary-yellow, fading, dbl.; Dickson, H.

'T.V. Times', HT, dr, 1970; flowers crimson, ovate, dbl., 35 petals, 5.5 in., moderate fragrance; foliage very large, dull; very free growth; [Gallant X (Brilliant X Seedling)]; Dickson, A.

'Ta Nase Pisnicka Ceska', Pol, mp, 1938; flowers small, dbl.; Böhm, J.; (Sangerhausen)

Ta Ta *see* 'MINatco'

'Tabarin', F, mp, 1956; bud oval; flowers salmon-pink flushed yellow and copper-orange to red, semi-dbl., 3–4 in., slight fragrance; foliage light green; vigorous, bushy growth; [Opera X Masquerade]; Gaujard

Table Mountain *see* KORnieoch

'Tablers' Choice', F, rb, 1974; flowers deep red, yellow reverse, dbl., 38 petals, 4 in., slight fragrance; foliage glossy; [(Ann Elizabeth X Allgold) X (Tropicana X Piccadilly)]; Harkness

Taboo™ *see* 'TANelorak'

Tabris *see* 'KORtabris'

'Tabriz', HT, lp, 1986; flowers soft pink, large, slight fragrance; few prickles; [Belle Epoque X Unnamed seedling]; Kriloff, Michel

Tache de Beauté® *see* 'LENpic'

'Tackholmii', S, mp; (R. balsamea, R. hibernica, R. X tackholmii); flowers pink, solitary or few, early flowering; vigorous, erect, shrub growth; (56)

'Tacoma', HT, ab, 1921; bud long, pointed; flowers flaming apricot, dbl., moderate fragrance; [Mme Butterfly X Honeymoon]; Chervenka

'Taconis'®, F, or, 1968; flowers orange-scarlet, in trusses, slight fragrance; foliage glossy; vigorous growth; GM, Baden-Baden, 1968; [Ruth Leuwerik X City of Nottingham]; deRuiter

'Taffeta', HT, pb, 1947; bud urn-shaped; flowers pink-yellow blend, dbl., 20 petals, 3–3.5 in., moderate fragrance; foliage leathery, glossy, bronze-green; vigorous, upright growth; AARS, 1948; [Mrs Sam McGredy X Pres. Herbert Hoover]; Lammerts, Dr. Walter; Armstrong Nursery

Taffeta *see* CHEwecrest

'Taffeta, Climbing', Cl HT, pb, 1954; Armstrong, J.A.; Weeks Roses

Taffy *see* 'BAKtaf'

Tag-a-long *see* 'MORcoat'

'Tagore', F, my, 1955; flowers yellow-ochre, reverse shaded orange, bornein clusters; compact growth; [Pinocchio seedling]; Maarse, G.

'Tahiti', HT, yb, 1947; flowers amber-yellow suffused carmine, 6 in., intense fragrance; foliage glossy, dark; [Peace X Signora]; Meilland, F.

'Tahore', HT, mp, 1958; flowers rose, large, dbl., 45–50 petals, exhibition form, moderate fragrance; foliage glossy; strong stems; vigorous, upright growth; [Peace X Seedling]; Frères, Buyl

Taifun *see* **'Typhoon'**

Taiga® *see* 'TANtiga'

'Tai-Gong', F, or, 1974; (Apachi); bud long, pointed; flowers orange to red, dbl., 28 petals, 3.5 in., exhibition form, slight fragrance; foliage glossy, dark; vigorous, upright growth; [Klaus Stortebeker X Seedling]; Kordes

Taihape Sunset *see* KORlinde

Taischa, HT, m; Robinson, 1993

'Taj Mahal', HT, dp, 1972; flowers deep pink, large, dbl., exhibition form, moderate fragrance; foliage large, leathery; vigorous, upright, bushy growth; [Manitou X Grand Slam]; Armstrong, D.L.; Armstrong Nursery

'Takao', HT, yb, 1975; flowers deep yellow, aging scarlet, blooms borne usually singly, dbl., 33 petals, exhibition form, moderate, damask fragrance; deep brown prickles; foliage medium, medium green; medium, bushy growth; [(Masquerade X Lydia) X (Montezuma X Miss Ireland)]; Okamoto, K.; K. Hirakata Nursery

Takapuna® *see* 'MACtenni'

'Takatori Prominent', F, op, 1981; flowers orange-salmon; [Prominent sport]; Takatori, Yoshiho; Japan Rose Nursery

'TAKsun', F, rb, 1983; (**Margaret Thatcher**); flowers striped red and white; [Bridal Pink sport]; Takatori, Yoshiho & Sunao; Japan Rose Nursery

'TALalconbury', F, dp, 1996; (**Alconbury**); flowers deep pink with touch of violet, blooms borne single and in s, very dbl., 1.5–2.75 in., exhibition form, slight fragrance; few prickles; foliage medium, dark green, semi-glossy; tall (48 in), upright, bushy growth; Taylor, Franklin "Pete" & Kay; Taylor's Roses, 1997

'TALalf', Min, op, 1991; (**Alfie Luv**); flowers orange-pink with lavender hue, reverse same, yellow base, dbl., 15–25 petals, 1.5–2.75 in., borne mostly singly, slight fragrance; some prickles; foliage medium, medium green, semi-glossy; upright, medium (60 cms) growth; [Azure Sea X Seedling]; Taylor, Franklin "Pete" & Kay; Taylor's Roses, 1992

'TALali', Min, pb, 1991; (**Alice Lee**); bud pointed; flowers pink with color getting lighter towards edge of petals, giving a lavende, dbl., 15–18 petals, exhibition form, slight fragrance; foliage medium, medium green, semi-glossy; upright, bushy, medium growth; [Azure Sea X Seedling]; Taylor, Franklin "Pete" & Kay, 1986; Taylor's Roses, 1990

'TELbaby', Min, rb, 1995; (**Baby Cakes**); flowers white, edged red, opening to red and white apperance, full (26–40 petals), small blooms borne mostly single; no fragrance; few prickles; foliage medium, medium green, semi-glossy; upright, bushy, medium; [Party Girl X Admiral Rodney]; Taylor, Pete & Kay; Taylor's Roses, 1996

'TALber', Min, mr, 1993; (**Berry Berry Red**); flowers medium red, moderately full, medium sized, dbl., 15–25 petals, 1.5–2.75 in., borne mostly singly, slight fragrance; few prickles; foliage medium, medium green, semi-glossy; medium (24 in), upright, bushy growth; [Party Girl X Papa Meilland]; Taylor, Franklin "Pete" & Kay; Taylor's Roses, 1993;, Taylors Roses, 1993, USA

'TALblo', Min, pb, 1993; (**Bloomtown**); flowers pink petals with slightly darker edges, reverse cream, dbl., 15–25 petals, 1.5–2.75 in., small clusters, no fragrance; some prickles; foliage medium, medium green, semi-glossy; medium (24 in), upright, spreading growth; [Party Girl X Andrea]; Taylor, Franklin "Pete" & Kay; Taylor's Roses, 1993;, Taylors Roses, 1993, USA

'TALBRO',Min, ab, 1995; (**Brown Sugar**); flowers apricot, moderately full (15–25 petals), medium (4–7 cms) blooms borne mostly single; slight fragrance; some prickles; foliage medium, dark green, glossy; upright, bushy, medium (30 in.) growth; [Party Girl X Julia's Rose]; Taylor, Pete & Kay; Taylor's Roses, 1996

'TALCAJ', Min, rb, 1993; (**Cajun Dancer**); flowers red with white eye, light reverse, semi-dbl., 6–14 petals, 1.5–2.75 in., borne singly and in large clusters, slight fragrance; some prickles; foliage medium, dark green, semi-glossy; tall (36 in), upright, bushy growth; [Party Girl X Unknown]; Taylor, Franklin "Pete" & Kay; Taylor's Roses, 1993;, Taylors Roses, 1993, USA

'TALCAN', Min, lp, 1999; (**Candy Corn**); flowers orange-yellow, single, 5–11 petals, 1.2in., borne mostly singly, moderate fragrance; few prickles; foliage medium, medium green, semi-glossy; upright, bushy, medium (18 in) growth; Taylor, Franklin "Pete" & Kay; Taylor's Roses, 1998

'TALCHELSEA', Min, mr, 1991; (**Chelsea Belle**); bud ovoid; flowers medium red with white base, whitish reverse, ages lighter, dbl., 28–30 petals, exhibition form, moderate fragrance; foliage medium, medium green, semi-glossy; upright, bushy, medium growth; [Azure Sea X Party Girl]; Taylor, Franklin "Pete" & Kay, 1987; Taylor's Roses, 1990

'TALCHRIS', Min, rb, 1996; (**Father Christmas**); flowers red with white base and white reverse, blooms borne in small clusters, dbl., 26–40 petals, 1.5 in., slight fragrance; some prickles; foliage small, dark green, semi-glossy; low (18 in), upright, bushy growth; [Party Girl X Seedling]; Taylor, Franklin "Pete" & Kay; Taylor's Roses, 1997

'TALCLOUSSEAU',HT, mr, 1995; (**Clousseau**); flowers medium red, cream reverse full (26–40 petals), large (7+ cms) blooms borne mostly single; slight fragrance; some prickles; foliage medium, dark green, semi-glossy; upright, bushy, medium (48 in.) growth; [Azure Sea X Marijke Koopman]; Taylor, Pete & Kay; Taylor's Roses, 1996

'TALDAL', Min, dy, 1991; (**Summerdale**); bud pointed; flowers deep yellow, aging lighter, loose, medium, borne usually singly, very dbl., 30–35 petals, cupped, no fragrance; foliage small, medium green, matt; bushy, low growth; [Rise 'n' Shine X Unnamed seedling]; Taylor, Franklin "Pete" & Kay, 1986; Taylor's Roses, 1990

'TALDAZ', Min, m, 1991; (**Lazy Daze**); flowers light lavender, reverse slightly darker, white base, blooms, very dbl., 1.5–2.75 in., blooms sometimes quarters, slight fragrance; some prickles; foliage small, medium green, semi-glossy; low (30 cms), compact growth; [Azure Sea X Unnamed seedling]; Taylor, Franklin "Pete" & Kay; Taylor's Roses, 1992

'TALDEARPRU', F, pb, 1995; (**Dear Prudence**); flowers light pink blushing with darker pink edges, moderately full (15–25 petals), medium (4–7 cms) blooms borne in large clusters; slight fragrance; some prickles; foliage medium, medium green, semi-glossy; upright, bushy, medium; [Seedling X Seedling]; Taylor, Pete & Kay; Taylor's Roses,1996

'TALDON', F, m, 1991; (**Donna Jean**); flowers mauve with white eye, blooms borne singly and in small clusters, semi-dbl., 6–14 petals, no fragrance; foliage medium, dark green, semi-glossy; medium, upright, bushy growth; [Azure Sea X Party Girl]; Taylor, Franklin "Pete" & Kay; Taylor's Roses

'TALDOR', Min, pb, 1991; (**Dora Delle**); flowers light pink with lavender hue, lighter in center, reverse cream, dbl., exhibition form, moderate fragrance; foliage medium, medium green, semi-glossy; medium, upright growth; [Azure Sea X Jean Kenneally]; Taylor, Franklin "Pete" & Kay; Taylor's Roses, 1991

'TALDRE', Min, ob, 1992; (**Dreamsicle**); flowers creamy white edged with orangish pink edges, reverse same, dbl., 15–25 petals, 1.5–2.75 in., slight fragrance; some prickles; foliage small, medium green, semi-glossy; medium (36 cms), compact growth; [Poker Chip X Party Girl]; Taylor, Franklin "Pete" & Kay; Taylor's Roses, 1993

'TALEND', Min, ab, 1993; (**Endeavour**); flowers apricot with yellow base, reverse slightly darker, color fading, dbl., 26–40 petals, 1.5–2.75 in., slight fragrance; few prickles; foliage medium, medium green, semi-glossy; medium, upright growth; [Party Girl X Azure Sea]; Taylor, Franklin "Pete" & Kay; Taylor's Roses, 1993

'TALFAIRHOPE', Min, ly, 1989; (**Fairhope**); bud pointed; flowers soft, light pastel yellow, reverse same, aging same, color holds well, dbl., 16–28 petals, exhibition form, slight fragrance; round, small, green fruit; prickles straight, medium, red; foliage medium, medium green, semi-glossy; upright, bushy, medium growth; [Azure Sea X Unnamed seedling]; Taylor, Franklin "Pete" & Kay; Taylor's Roses

'TALFAT', Min, m, 1991; (**Fat Tuesday**); flowers lavender with darker edges, blending to lighter center, reverse lighter, dbl., 26–40 petals, 1.5–2.75 in., no fragrance; some prickles; foliage medium, medium green, semi-glossy; tall (76 cms), upright growth; [Azure Sea X Lavender Jewel]; Taylor, Franklin "Pete" & Kay; Taylor's Roses, 1992

'TALFLORA', Min, rb, 1996; (**Flora Bama**); flowers medium red blending to cream center, cream reverse, blooms b, dbl., 26–40 petals, 1.5–2.75 in.; few prickles; foliage medium, dark green, semi-glossy; medium (24 in), upright, spreading growth; [Party Girl X Poker Chip]; Taylor, Franklin "Pete" & Kay; Taylor's Roses, 1997

'TALGOO', Min, dy, 1992; (**Good Day Sunshine**); flowers bright yellow, reverse same, blooms borne mostly singly, very dbl., 1.5–2.75 in., exhibition form, slight fragrance; few prickles; foliage medium, medium green, semi-glossy; medium (60cms), upright, bushy, spreading growth; [Party Girl X Elina]; Taylor, Franklin "Pete" & Kay; Taylor's Roses, 1993

'TALGUL', Min, pb, 1991; (**Gulf Breeze**); flowers creamy getting darker pink toward edges, reverse same, very dbl., 1.5–2.75 in., slight fragrance; some prickles; foliage medium, medium green, semi-glossy; upright, medium (35 cms), bushy growth; [Baby Katie X Poker Chip]; Taylor, Franklin "Pete" & Kay; Taylor's Roses, 1992

'TALHEATHER', Min, mp, 1989; (**Heather Leigh**); bud pointed; flowers medium pink, reverse slightly darker, aging lighter, holds color well, dbl., 35–40 petals, exhibition form, borne singly, no fragrance; no fruit; prickles straight, small, reddish-brown; foliage medium, medium green, semi-glossy; upright, bushy, medium growth; [Azure Sea X Unnamed Miniature seedling]; Taylor, Franklin "Pete" & Kay; Taylor's Roses

'TALHON', Min, rb, 1996; (**Honky Tonk**); flowers red with darker edges, white reverse, blooms borne mostly singly, dbl., 15–25 petals, 1.5–2.75 in., slight fragrance; many prickles; foliage medium, dark green, glossy; medium (30 in), upright, bushy growth; [Party Girl X Seedling]; Taylor, Franklin "Pete" & Kay; Taylor's Roses, 1997

Talia, S, 1990; Barni, V.; (Cavriglia)

'TALINA', Min, w, 1994; (**Ina**); flowers white with pinkish edge, moderately good form, blooms borne, dbl., 15–25 petals, 1.5–2.75 in., slight fragrance; few prickles; foliage medium, medium green, semi-glossy; medium (30 in), upright, bushy growth; [Party Girl X Fairhope]; Taylor, Franklin "Pete" & Kay; Taylor's Roses, 1993

'TALISMAN', HT, yb, 1929; bud pointed; flowers golden yellow and copper, medium, dbl., 25 petals, flat, moderate fragrance; foliage light green, leathery, glossy; vigorous growth; (28); Gertrude M. Hubbard, ARS, 1929 John Cook Medal, ARS, 1932; [Ophelia X Souv. de Claudius Pernet]; Montgomery Co.

'TALISMAN NO. 5', HT, yb; flowers deeper color, free bloom; good growth

'TALISMAN, CLIMBING', Cl HT, yb, 1930; Western Rose Co.

'TALISMAN, CLIMBING', Cl HT, yb, 1932; Dixie Rose Nursery

'TALJAC', Min, dr, 1995; (**Jackson Square**); flowers dark red, full (26–40 petals), medium (4–7 cms) blooms borne mostly single; slight fragrance; some prickles; foliage medium, medium green, semi-glossy; upright, bushy, medium (30 in.) growth; [Jean Kenneally X Black Jade]; Taylor, Pete & Kay; Taylor's Roses,1996

'TALJUB', Min, ob, 1991; (**Jubilee Sunset**); flowers bright vivid orange, yellow eye, reverse creamy yellow, dbl., 15–25 petals, 1.5–2.75 in., slight fragrance; some prickles; foliage medium, medium green, semi-glossy; low (40 cms), upright, bushy growth; [Baby Katie X Poker Chip]; Taylor, Franklin "Pete" & Kay; Taylor's Roses, 1992

'TALJUM', Min, yb, 1992; (**Jumping Jack Flash**); flowers yellow edged with deep pink to red, cream reverse, tipped with deep pink, dbl., 15–25 petals, 1.5–2.75 in., exhibition form, slight fragrance; few prickles; foliage medium, medium green, semi-glossy; medium (42 cms), upright, bushy growth; [Party Girl X Poker Chip]; Taylor, Franklin "Pete" & Kay; Taylor's Roses, 1993

'TALKAT', Min, mp, 1993; (**Katy Lampkin**); flowers medium pink, cream base, reverse same, dbl., 15–25 petals, 1.5 in., borne singly and in small clusters, slight fragrance; few prickles; foliage small, medium green, semi-glossy; low (18 in), compact growth; [Winsome X Admiral Rodney]; Taylor, Franklin "Pete" & Kay; Taylor's Roses, 1993

'TALKEV', Min, dy, 1991; (K.T., **Kev**); dbl., borne mostly singly, no fragrance; foliage medium, medium green, semi-glossy; medium, upright growth; [Azure Sea X Party Girl]; Taylor, Franklin "Pete" & Kay; Taylor's Roses

Tall Poppy *see* 'MORTEN'

Tall Story® *see* 'DICKOOKY'

'TALLOU', Min, yb, 1997; (**Louisiana Lady**™); flowers full, medium, very dbl., 26–40 petals, borne mostly singly, slight fragrance; foliage medium, medium green, semi-glossy; upright, bushy, medium (24 in.)growth; [Seedling X Seedling]; Taylor, Franklin "Pete" & Kay

Tallulah *see* 'BEELAH'

'TALLYHO', HT, dp, 1948; bud urn-shaped; flowers deep pink, dbl., 35 petals, 3.5–4 in., exhibition form, intense, spicy fragrance; foliage leathery; vigorous, upright, bushy growth; (28); AARS, 1949 David Fuerstenberg Prize, ARS, 1951; [Charlotte Armstrong X Unnamed seedling]; Swim, H.C.; Armstrong Nursery

'TALLYHO, CLIMBING', Cl HT, dp, 1952; Armstrong, J.A.; Armstrong Nursery

'TALMAG', Min, op, 1993; (**Mary Margaret**); flowers coral pink, creamy yellow base, reverse same, blooms borne singly, dbl., 26–40 petals, 1.5 in., slight fragrance; few prickles; foliage medium, medium green, semi-glossy; low (18 in), spreading growth; [Party Girl X Watercolor]; Taylor, Franklin "Pete" & Kay; Taylor's Roses, 1993

'TALMAID', Min, pb, 1989; (**Maids of Jubilee**); bud pointed; flowers bright deep pink with cream base in center, reverse cream and pink blend, dbl., 25–30 petals, exhibition form, no fragrance; prickles yellowish-green; foliage medium, dark green, semi-glossy; upright, bushy, medium growth; [Azure Sea X Unnamed Miniature seedling]; Taylor, Franklin "Pete" & Kay; Taylor's Roses

'TALMAR', Min, w, 1991; (**Mary Edith**); flowers white, tipped with pink edges, flower turns darker pink, tints with age, dbl., moderate fragrance; foliage medium, medium green, semi-glossy; low, upright, bushy growth; [Azure Sea X Unnamed seedling]; Taylor, Franklin "Pete" & Kay; Taylor's Roses

'TALMID', Cl Min, rb, 1993; (**Midnight Rambler**); flowers red with white eye, blooms borne in small clusters, semi-dbl., 6–14 petals, 1.5–2.75 in., no fragrance; some prickles; foliage medium, medium green, semi-glossy; medium (30 in), spreading growth; [Party Girl X Andrea]; Taylor, Franklin "Pete" & Kay; Taylor's Roses, 1993

'TALMOBILE', Min, op, 1994; (**Mobile Jubilee**); flowers light salmon blending to darker edges, yellow base, reverse, dbl., 26–40 petals, 1.5–2.75 in., slight fragrance; some prickles; foliage medium, medium green, semi-glossy; tall (90 cms), upright, bushy growth; [Party Girl X Gitte]; Taylor, Franklin "Pete" & Kay; Taylor's Roses, 1994

'TALMONTROSE', Min, mp, 1994; (**Montrose**); flowers bright yellow stamens, blooms borne single and in large clusters, dbl., 15–25 petals, 1.5–2.75 in., moderate fragrance; some prickles; foliage medium, dark green, semi-glossy; medium (60 cms), upright, bushy growth; [Party Girl X Andrea]; Taylor, Franklin "Pete" & Kay; Taylor's Roses, 1994

'TALPEN', Min, lp, 1992; (**Penny Lane**); flowers cream base blending to light pink, reverse same, blooms born, dbl., 26–40 petals, 1.5 in., exhibition form, no fragrance; few prickles; foliage small, medium green, semi-glossy; medium (38 cms), upright, bushy growth; [Party Girl X Maids of Jubilee]; Taylor, Franklin "Pete" & Kay; Taylor's Roses, 1993

'TALPIN', Min, mp, 1993; (**Pink Bunting**); flowers pink, creamy yellow base blending to pink edges around petal, dbl., 15–25 petals, 1.5 in., no fragrance; few prickles; foliage small, dark green, semi-glossy; low (12 in), spreading, compact growth; Taylor, Franklin "Pete" & Kay; Taylor's Roses, 1993

'TALPOI', Min, w, 1991; (**Point Clear**); flowers white, reverse same, light yellow base, blooms borne mostly, dbl., 26–40 petals, 1.5 in., slight fragrance; some prickles; foliage small, medium green, semi-glossy; low (40 cms), upright growth; [Azure Sea X Party Girl]; Taylor, Franklin "Pete" & Kay; Taylor's Roses, 1992

'TALRAS', F, dp, 1991; (**Raspberry Ruffles**); bud pointed; flowers deep pink with white eye, deep pink reverse, aging lighter, dbl., 18–20 petals, exhibition form, intense fragrance; foliage medium, medium green, semi-glossy; upright, bushy, medium growth; [Garnette X Unnamed seedling]; Taylor, Franklin "Pete" & Kay, 1987; Taylor's Roses, 1990

'TALRUM', Min, ab, 1991; (**Rum Butter**); bud pointed; flowers apricot with yellow base-bloom gives a lavender cast blended, dbl., 12–16 petals, exhibition form, no fragrance; foliage medium, medium green, semi-glossy; upright, bushy, medium growth; [Azure Sea X Unnamed seedling]; Taylor, Franklin "Pete" & Kay, 1987; Taylor's Roses, 1990

'TALSILVER', Min, pb, 1996; (**Silverhill**); flowers pink with mauve tint, blending to cream center and cream reverse, dbl., 26–40 petals, 1.5–2.75 in., exhibition form, slight fragrance; some prickles; foliage medium, medium green, semi-glossy; medium (36 in), upright, bushy growth; [Azure Sea X Seedling]; Taylor, Franklin "Pete" & Kay; Taylor's Roses, 1997

'TALSIMPLE', Min, lp, 1999; (**Simple Pleasures**™); flowers red petals, white eye, yellow stamens, single, 5–11 petals, 1.75 in., borne mostly singly, no fragrance; few prickles; foliage medium, medium green, semi-glossy; upright, spreading, medium (18 in.) growth; Taylor, Franklin "Pete" & Kay; Taylor's Roses, 1998

'TALSPI', Min, ob, 1993; (**Cajun Spice**); flowers bright orange, white eye, light reverse, semi-dbl., 6–14 petals, 1.5–2.75 in., borne mostly singly, no fragrance; some prickles; foliage medium, medium green, semi-glossy; medium (24 in), upright, bushy growth; [Party Girl X Baby Diana]; Taylor, Franklin "Pete" & Kay; Taylor's Roses, 1993;, Taylors Roses, 1993, USA

'TALSPRINGS', Min, lp, 1999; (**Magnolia Springs**™); flowers white with pink edge, greenish guard petals, very dbl., 41 petals, 2 in., borne in large clusters, no fragrance; few prickles; foliage medium, medium green, semi-glossy; upright, bushy, medium (18 in) growth; Taylor, Franklin "Pete" & Kay; Taylor's Roses, 1998

'TALSUE', Min, ob, 1992; (**Sue Belle**); flowers orange with yellow eye, yellow stamens, aging to pink, semi-dbl., 6–14 petals, 1.5–2.75 in., slight fragrance; some prickles; foliage small, medium green, semi-glossy; medium (40 cms), upright, bushy growth; [Party Girl X Poker Chip]; Taylor, Franklin "Pete" & Kay; Taylor's Roses, 1993

'TALSUGAR', F, lp, 1999; (**Sugar Magnolia**™); flowers white, burgundy pistil and stamens, set against fleshtone petals, dbl., 17–25 petals, 2.75 in., borne in small clusters, no fragrance; few prickles; foliage medium, dark green, semi-glossy; upright, bushy (3 ft) growth; Taylor, Franklin "Pete" & Kay; Taylor's Roses, 1998

'TALSUM', Min, op, 1991; (**Summer Breeze**); flowers light orange-pink, ages lighter, yellow base, reverse slight, dbl., 26–40 petals, 1.5 in., slight fragrance; some prickles; foliage medium, medium green, semi-glossy; low (40 cms), spreading growth; [Baby Katie X Poker Chip]; Taylor, Franklin "Pete" & Kay; Taylor's Roses, 1992

'TALSUN', Min, yb, 1995; (**Sunny Afternoon**); flowers light yellow, outer edges pinkish/apricot tinge, blooms born, dbl., 15–25 petals, 1.5 in., slight fragrance; some prickles; foliage medium, medium green, semi-glossy; medium, upright, bushy growth; [Party Girl X Elina]; Taylor, Franklin "Pete" & Kay; Taylor's Roses, 1995

'TALSWEET', Min, pb, 1995; (**Sweet Home Alabama**); flowers magenta pink with white reverse, white eye, very dbl., 26–40 petals, 1.5–3 in., borne mostly singly, slight fragrance; some prickles; foliage medium, medium green, semi-glossy; upright, bushy, medium growth; [Party Girl X Azure Sea]; Taylor, Franklin "Pete" & Kay; Taylor's Roses, 1996

'TALTRICK', Min, ob, 1996; (**Trick Or Treat**); flowers orange blending to yellow base, very dbl., 41 petals, 1.5 in., flat, borne mostly singly, slight fragrance; few prickles; foliage small, medium green, semi-glossy; medium (24 in), upright, bushy growth; [Seedling X Seedling]; Taylor, Franklin "Pete" & Kay; Taylor's Roses, 1997

'TALUPTOWN', Min, m, 1991; (**Uptown**); flowers lavender, edges darker violet, reverse lighter, yellow base, dbl., 26–40 petals, 1.5–2.75 in., exhibition form, no fragrance; some prickles; foliage medium, medium green, semi-glossy; medium (46 cms), upright, bushy growth; [Azure Sea X Party Girl]; Taylor, Franklin "Pete" & Kay; Taylor's Roses, 1992

'TALWIL', Min, yb, 1992; (**Wildest Dreams**); flowers yellow blending to deep pink edges, aging lighter, reverse yellow, dbl., 15–25 petals, 1.5–2.75 in., no fragrance; few prickles; foliage small, medium green, semi-glossy; medium (40 cms), upright, bushy growth; [Poker Chip X Unknown]; Taylor, Franklin "Pete" & Kay; Taylor's Roses, 1993

'TAM O'SHANTER', F, yb, 1969; flowers yellow and red, borne in clusters, dbl., slight fragrance; low, compact growth; [Orange Sensation X Circus]; Cocker

'TAMA', HT, op, 1989; bud pointed; flowers salmon-pink, fringed with pink, reverse creamy yellow, urn-shaped, dbl., 35 petals, exhibition form, slight fragrance; prickles small, lower part hollow; foliage dark green; upright, tall growth; [Red Lion X Garden Party]; Sato, Kohi

Tamango® *see* 'MEIDANU'

'TAMARA', HT, mr, 1966; flowers geranium-red, dbl., exhibition form; foliage dark; very vigorous growth; [Parel van Aalsmeer X Independence]; Mondial Roses

Tamara *see* KORmador

'TAMARISK', HT, mp, 1954; flowers clear rose-pink, camellia shaped, intense fragrance; foliage leathery, dull green; very vigorous growth; [Mme Butterfly X Signora]; Ratcliffe

'TAMBOURINE', F, rb, 1958; flowers cherry-red, base yellow, reverse light burnt-orange, borne i, dbl., 25 petals, 3.5 in., slight fragrance; foliage dark, veined; vigorous, tall, bushy growth; [Independence seedling X Karl Herbst]; Dickson, A.

Tambourine *see* HARdolly

Tambrabarani *see* **Southern Sunset**

'TAMMY', Pol, lp, 1972; flowers medium, dbl., exhibition form, moderate fragrance; foliage large, leathery; vigorous, upright, bushy growth; [Seventeen X Jack Frost]; Byrum

Tamora *see* 'AUSTAMORA'

Tamouree, F; Combe; (Cavriglia)

Tampa Bay™ *see* 'KINBAY'

'TAMPICO', HT, or, 1976; bud ovoid, long, pointed; flowers coral-pink, large, dbl., slight fragrance; foliage large, leathery; vigorous, upright growth; [South Seas X Hawaii]; Warriner, William A.; J&P

'TAMRABARANI', HT, ob, 1997; (**Southern Sunset**); flowers full, large, very dbl., 26–40 petals, borne mostly singly, slight fragrance; many prickles; foliage large, dark green, glossy; upright, compact, medium (2.5ft.) growth; Viraraghavan, M.S.

'TAM-TAM', F, my, 1961; bud ovoid; flowers yellow flushed pink, dbl., 35–40 petals, 3 in., cupped, moderate fragrance; foliage glossy; vigorous, upright growth; RULED EXTINCT 1/88; [Goldilocks seedling X Demure seedling]; Boerner; J&P

'TAM-TAM', Min, lp, 1988; flowers inner petals upright, borne in large clusters, dbl., 18–20 petals, flat, slight fragrance; elongated, orange fruit; foliage small, light green; [(Nozomi X Nozomi) X Bo-Peep]; Onodera, Toru F.

Tan Cho, HT, lp; Keisei, 1994

TANadac, F, mr, 1994; (**Escada**); Tantau

'TANAGNARAT', F, mr, 1982; (**Taranga**®); flowers medium, semi-dbl., no fragrance; foliage medium, medium green, semi-glossy; bushy growth; Tantau, Math., 1981

Tanagra® *see* 'GACTA'

'TANAKINOM', HT, pb, 1985; (Monica®, **Monika**®, 'TANAKNOM'); flowers medium, dbl., 35 petals, slight fragrance; foliage medium, dark, glossy; upright growth; Tantau, Math., 1986

'TANAKNI', HT, op, 1978; (**Inka**®); bud pointed; flowers salmon, large, dbl., slight fragrance; foliage large, glossy; upright, bushy growth; Tantau, Math., 1981

TANAKNI, HT, dy, 1993; (**Inka**®); Tantau

'TANAKNOM', HT, pb, 1985; (Monica®, **Monika**®, 'TANAKINOM'); flowers medium, dbl., 35 petals, slight fragrance; foliage medium, dark, glossy; upright growth; Tantau, Math., 1986

'TANAL', F, or, 1958; (**Allotria**®); flowers orange-scarlet, dbl., 3 in., flat, blooms in large clusters; foliage dark, glossy; vigorous growth; [Fanal X Cinnabar seedling]; Tantau, Math.

TANalam, HT, lp; (**Unforgettable**); Tantau, 1996

TANalednev, HT, lp; (**Velvet Star**); Tantau, 1997

TANallepal, HT, mr, 1994; (**Acapella** R); bud long, pointed; flowers cherry red, reverse silver, large, very dbl., intense fragrance; foliage dark green, leathery, glossy; upright, vigorous growth; Tantau, Math.; (Sangerhausen)

'TANALLET', S, mr, 1985; (**Castella**®); flowers medium, semi-dbl., no fragrance; foliage large, dark, semi-glossy; upright growth; Tantau, Math., 1984

'TANALOAP', HT, mr, 1982; (**Paola**); flowers large, dbl., 20 petals, slight fragrance; foliage large, dark, matt; Tantau, Math., 1981

TANaluma, HT, mr, 1932; (**'AMULETT'**); bud ovoid; flowers fiery red, very dbl., exhibition form, moderate fragrance; foliage dark, glossy; vigorous, bushy growth; [Mrs Henry Winnett X Johanniszauber]; Tantau; C-P

'TANÄLZERGO', HT, dy, 1982; (**Pfälzer Gold**®); flowers large, dbl., 20 petals, no fragrance; Tantau, Math., 1981

'TANAMOLA', F, w, 1985; (La Paloma '85, **The Dove**); flowers medium, dbl., 20 petals, no fragrance; foliage medium, light green, glossy; upright growth; Tantau, Math.

TANanairan, F, mr, 1996; (**Austriana**); flowers blood red, medium-large, dbl.; Tantau, Math.; (Sangerhausen)

TANanilov, HT, pb, 1997; (**Violina**®); bud long, pointed, dark pink; dbl., borne mostly singly, intense fragrance; foliage large, medium green, glossy; upright, strong (3 ft) growth; Tantau

TANanim, HT, dp, 1994; (**Romina**); Tantau

TANaronam, HT, mr, 1992; (**Manora**®); Tantau

'TANATESIL', F, mp, 1985; (**Ilseta**®, Ilsetta); flowers medium, dbl., 35 petals, no fragrance; foliage medium, medium green, matt; upright growth; PP005707; Tantau, Math., 1983

'TANATIRAM', F, yb, 1985; (**Marietta**®); flowers medium, dbl., 20 petals, no fragrance; foliage medium green, glossy; upright growth; Tantau, Math.

'TANATTENIN', F, dp, 1985; (**Ninetta**®); flowers deep pink, medium, dbl., 20 petals, slight fragrance; foliage medium green, semi-glossy; upright growth; Tantau, Math.

'TANATTIGAM', HT, mr, 1982; (**Magitta**®); flowers large, dbl., 20 petals, slight fragrance; foliage medium, medium green, semi-glossy; upright growth; Tantau, Math., 1981

TANavilo, HT, mr, 1993; (**Olivia**®); Tantau

'TANAVLIM', F, m, 1983; (**Milva**®); flowers medium, dbl., 35 petals, slight fragrance; foliage medium, medium green, semi-glossy; upright growth; Tantau, Math.

'TANBA', Min, rb, 1956; (Baby Carnaval, Baby Carnival, Baby Mascarade, Baby Maskarade, Baby Maskerade, **Baby Masquerade**®, 'TANBAKEDE'); bud ovoid; flowers yellow aging red, dbl., 23 petals, 1 in., slight, fruity fragrance; foliage leathery; vigorous, compact (8 in. tall) growth; [Tom Thumb X Masquerade]; Tantau, Math.; J&P, 1955

'TANBAKEDE', Min, rb, 1956; (Baby Carnaval, Baby Carnival, Baby Mascarade, Baby Maskarade, Baby Maskerade, **Baby Masquerade**®, 'TANBA'); bud ovoid; flowers yellow aging red, dbl., 23 petals, 1 in., slight, fruity fragrance; foliage leathery; vigorous, compact (8 in. tall) growth; [Tom Thumb X Masquerade]; Tantau, Math.; J&P, 1955

'TANBEEDEE', F, ob, 1971; (**Belinda**); bud ovoid; flowers copper to orange, medium, dbl., moderate fragrance; vigorous, upright, bushy growth; [Seedling X Zorina]; Tantau, Math.

'TANBLADY', HT, dr, 1976; (**Black Lady**®); bud globose; flowers blackish-red, medium, dbl., intense fragrance; foliage matt; bushy growth; Tantau, Math.

'TANCA', HT, rb, 1966; (**Caramba**®); bud pointed; flowers bright red, reverse white, well-formed, large, dbl., 45 petals; foliage glossy; upright growth; Tantau, Math.

'TANCARY', HT, yb, 1976; (**Canary**); flowers yellow, petals marked orange, well-formed, medium, dbl., moderate fragrance; Tantau, Math.; Krussmann, 1972

TANcelia, F, w, 1988; (**Maleica**); Tantau

'TANCOFEUMA', F, or, 1982; (**Cosima**®); flowers medium, dbl., 20 petals, no fragrance; foliage large, dark, glossy; Tantau, Math., 1981

TANcreif, F, w, 1988; (**Elfe**); Tantau

'TANCRESSOR', HT, mr, 1991; (**Celebrate America**™); dbl., 45–55 petals, 4–4.75 in., exhibition form, borne usually singly, slight fragrance; foliage large, medium green, semi-glossy; upright, spreading, medium growth; [Seedling X Seedling]; Rosen Tantau; Bear Creek Gardens, 1990

TANdalaum, F, mp; (**Traumland**®); Tantau, 1988

TANdelbel, F, m, 1994; (**Magenta Diadem**); Tantau

'TANDINADI', F, my, 1977; (**Diana**®); bud globular; flowers medium, dbl., slight fragrance; foliage medium, glossy; vigorous, upright growth; Tantau, Math.

TANdirpkrap, HT, mr; (**Parkstown's Pride**); Tantau

TANdomo, Min, mr, 1994; (**Domino**); Tantau

'TANDUFT', HT, dy, 1981; (Duftgold, **Fragrant Gold**, 'TANDUGOFT'); flowers large, semi-dbl., moderate fragrance; foliage medium, dark, glossy; upright growth; Tantau, Math., 1982

'TANDUGOFT', HT, dy, 1981; (Duftgold, **Fragrant Gold**, 'TANDUFT'); flowers large, semi-dbl., moderate fragrance; foliage medium, dark, glossy; upright growth; Tantau, Math., 1982

Tane *see* 'SIMETNA'

Tanecnice, F, lp; Strnad, 1976; (Czech Rosa Club)

TANedallab, F, lp, 1991; (**Ballade**); Tantau

TANegnaro, Min, or, 1994; (**Ceres**); Tantau

TANeiglat, HT, 1995; (**Nostalgie**); Tantau; (Sangerhausen)

TANeitber, F, dy, 1987; (**Bernstein-Rose**®); Tantau

TANeivom, HT, op, 1995; (**Movie Star**); Tantau

TANekily, HT, pb, 1989; (**Lady Like**®); Tantau

'TANEKILY'

'TANELAIGIB', F, pb, 1988; (**Abigaile**®); Tantau

TANelephar, HT, mr, 1994; (**Raphaela**); Tantau

'TANELFE', HT, lp, 1985; (**Elfe**®); flowers medium, dbl., 20 petals, slight fragrance; foliage medium, medium green, matt; upright growth; Tantau, Math., 1972

TANeliet, HT, dp, 1996; (**African Queen**); Tantau

TANellelog, F, lp; (**Goldquelle**); Tantau, 1988

'TANELLIS', HT, or, 1967; (Duftwolke, **Fragrant Cloud**, Nuage Parfume); bud ovoid; flowers coral-red becoming geranium-red, well-formed, dbl., 28–35 petals, 5 in., exhibition form, borne mostly singly, intense, damask-fruity-spicy-citrus-sharp fragrance; prickles moderate; foliage dark, glossy; vigorous, upright growth; GM, NRS, 1963 GM, Portland, 1963 Hall of Fame, WFRS, 1981 James Alexander Gamble Fragrance Medal, ARS, 1969 PIT, NRS, 1964; [Seedling X Prima Ballerina]; Tantau, Math., 1963; J&P, 1968

TANelliv, F, mr; (Glow Achievement, **Stadt Eltville**®); GM, Durbanville, 1989; Tantau, 1990

'TANELORAK', HT, dr, 1993; (Barkarole, Grand Chateau, **Taboo**™); bud pointed, ovoid, almost black; flowers deepest red, velvety, dbl., 26–40 petals, 4–4.5 in., exhibition form, borne mostly singly, moderate, sweet citrus fragrance; some prickles; foliage large, dark green, semi-glossy; tall (170 cms), upright, bushy growth; PP007665; Evers, Hans & Rosen Tantau; Bear Creek Gardens, 1994

'TANEMA', HT, mr, 1974; (Melina, **Sir Harry Pilkington**); flowers well-formed, dbl., 30 petals, 4–5 in., slight fragrance; foliage dark; [Inge Horstmann X Sophia Loren]; Harry Wheatcroft Gardening

TANemrac, HT, or, 1986; (**Carmen**); Tantau

TANensch, Min, w, 1992; (**Schneeweisschen**®); Tantau

'TANERISED', HT, lp, 1985; (**Desirée**®); flowers large, dbl., 35 petals, slight fragrance; foliage large, dark, semi-glossy; upright growth; Tantau, Math., 1986

'TANESSENOR', HT, or, 1989; (**Baronesse**®); Tantau

TANetee, HT, lp; (**Tea Time**); Tantau, 1994

TANetorde, F, dr, 1991; (**Red Diadem**®); Tantau

TANettola, HT, mr, 1997; (**Charlotte**); Tantau

TANeufis, S, mr, 1988; (**Johannisfeuer**®); flowers blood red, semi-dbl., borne in sprays; foliage large, dark green; bushy, upright, strong (4.5 ft) growth; Tantau

'TANEZAMOR', S, mp, 1985; (Romance, **Romanze**); flowers medium, dbl., 20 petals, slight fragrance; foliage medium, dark, semi-glossy; bushy growth; ADR, 1986 GM, Baden-Baden, 1985; Tantau, Math.; Tantau Roses, 1984

'TANFIFUM', HT, m, 1978; (**Blue Parfum**®, Blue Perfume, 'TANIFUME', 'TANTIFUM'); bud ovoid; flowers mauve-blush, large, dbl., intense fragrance; foliage glossy; bushy, upright growth; Tantau, Math.

'TANFUDERMOS', HT, dr, 1985; (**Sommerduft**®, Summer Fragrance); flowers medium-large, dbl., 20 petals, moderate fragrance; foliage medium, dark, semi-glossy; upright growth; Tantau, Math., 1986

Tangeglow *see* 'DICMADDER'

'TANGER', HT, rb, 1949; bud ovoid; flowers crimson, reverse yellow, large, dbl., 60 petals, moderate fragrance; moderate, compact growth; [Condesa de Sástago X Peace]; Dot, Pedro

'TANGERINE', HT, op, 1942; bud globular, coral-red; flowers light salmon-orange, dbl., 30–35 petals, 3.5–4 in., slight fragrance; foliage dark, leathery; strong stems; vigorous, very upright, well branched growth; [Capt. Glisson X R.M.S. Queen Mary]; Hill, Joseph H., Co.

'TANGERINE CONTEMPO', F, ob, 1983; flowers tangerine orange; [Contempo sport]; Gupta, Dr. M.N. & Datta, Dr. S.K.; National Botanical Research Institute

Tangerine Mist *see* 'RENIST'

Tangerine Twist *see* 'MICTWIST'

'TANGIER', HT, rb, 1955; bud long, pointed; flowers peach-red to scarlet-copper, medium, dbl., 60 petals, exhibition form, slight fragrance; vigorous, bushy growth; [Katharine Pechtold X R.M.S. Queen Mary]; Parmentier, J.

'TANGO', HT, mr, 1937; flowers scarlet, reverse bronze, base shaded old-gold; vigorous growth; [Seedling X Talisman]; H&S

'TANGO', HT, mr, 1955; bud long; flowers tango-red, open, large, dbl., moderate fragrance; foliage dark, glossy; vigorous, bushy growth; [The Doctor X Karl Herbst]; Delforge

Tango *see* JACboy

Tango *see* 'MACFIRWAL'

Tango Rose *see* 'DELTANGA'

TANgolca, HT, dy, 1988; (**Golden Monica**®); Tantau

'TANGOSAR', Cl HT, or, 1971; (Super Star, Climbing, 'TANGOSTAR', **Tropicana, Climbing**); Boerner; J&P

'TANGOSTAR', Cl HT, or, 1971; (Super Star, Climbing, 'TANGOSAR', **Tropicana, Climbing**); Boerner; J&P

'TANIA VERSTAK', HT, mr, 1962; flowers rich red, large, dbl., slight fragrance; foliage light green, soft; vigorous, upright growth; [Charlotte Armstrong X Happiness]; Armbrust; Langbecker

'TANIBARA', F, ob, 1986; (**Arabia**®); Tantau

'TANIFUME', HT, m, 1978; (**Blue Parfum**®, Blue Perfume, 'TANFIFUM', 'TANTIFUM'); bud ovoid; flowers mauve-blush, large, dbl., intense fragrance; foliage glossy; bushy, upright growth; Tantau, Math.

TANija, HT, ob, 1974; (**Janina**®); Tantau

'TANILASOR', F, mp, 1983; (**Rosali**®, Rosali '83®); flowers medium, dbl., 20 petals, no fragrance; foliage medium, medium green, glossy; bushy growth; Tantau, Math.

'TANILSEPO', F, w, 1985; (**Perl-Ilseta**®); flowers pearly white; [Ilseta sport]; Tantau, Math., 1984

'TANILVOBA', F, ob, 1985; (**Orange Ilseta**®); flowers orange; [Ilseta sport]; Tantau, Math.

TANilyks, HT, my, 1991; (**Skyline**); Tantau

TANinaso, S, w, 1993; (**Snow Cloud**); Tantau

TANinaso, MinFl, lp; (**Harewood**); Tantau

TANinat, S, mp, 1992; (**Satina**®); Tantau

TANipep, F, lp, 1987; (**Pepino**®); Tantau

TANireb, HT, mp, 1994; (**Belle of Berlin**); Tantau

TANiripsa, F, w, 1997; (**Aspirin**); flowers medium-large, dbl., moderate fragrance; Tantau, Math.; (Sangerhausen)

'TANITO', F, w, 1974; (**Akito**); bud ovoid; flowers , dbl., blooms in clusters, slight fragrance; foliage medium, medium green; upright, bushy growth; [Zorina X Nordia]; Tantau, Math.

'TANJEKA', HT, op, 1971; (Jehoca, Jehoka, Jessica, **Jessika**®); bud long, pointed; flowers peach-salmon, medium, moderate fragrance; vigorous, upright growth; [Colour Wonder X Piccadilly]; Tantau, Math.

'TANJENJU', HT, mr, 1989; (**Jennie June**™); bud ovoid; flowers medium red, reverse lighter with some bluing, large, dbl., 50 petals, exhibition form, borne usually singly, slight fragrance; prickles short, narrow, hooked down, red-green; foliage large, medium green, semi-glossy; upright, spreading, medium growth; [Unnamed seedling X Unnamed seedling]; Rosen Tantau; Bear Creek Gardens, 1990

'TANJUKA', F, mr, 1974; (**Julischka**®); bud long, pointed; flowers bright red, medium, semi-dbl., slight fragrance; foliage glossy, bronze; Gold Star of the South Pacific, Palmerston North, NZ, 1976; Tantau, Math.; Horstmann

TANkalgic, HT, dr, 1997; (**Black Magic**); Tantau

'TANKANUSCH', F, or, 1978; (**Anuschka**®); bud ovoid; flowers large, dbl., 23 petals, slight fragrance; foliage large; bushy, upright growth; Tantau, Math., 1977

TANkeijoli, F, lp, 1988; (**Majolika**®); Tantau

'TANKENFRAM', F, or, 1982; (**Frankenland**®); bud medium, pointed; flowers brilliant scarlet, patio, borne in clusters of many, dbl., 23 petals, exhibition form, no fragrance; straight, brown-red prickles; foliage large, dark, semi-glossy; bushy, low growth; Tantau, Math.; Tantau Roses, 1978

'TANKLESANT', LCl, mr, 1985; (**Santana**®); flowers medium, dbl., 20 petals, no fragrance; foliage large, medium green, glossy; upright growth; Tantau, Math., 1984

'TANKLEVI', LCl, mp, 1982; (Lavinia, **Lawinia**®, 'TANKLEWI'); flowers large, dbl., 20 petals, cupped, moderate fragrance; foliage large, medium green, semi-glossy; spreading (to 8 ft) growth; Tantau, Math., 1980

TANkobi, MinFl, pb, 1986; (**Bubikopf**®); Tantau, Math.; (Cavriglia)

TANkorab, HT, my, 1990; (**Barock**); Tantau

'TANKY', HT, yb, 1967; (Whisky, **Whisky Mac**); bud ovoid; flowers bronze-yellow, well-formed, large, dbl., 30 petals, intense fragrance; foliage glossy; vigorous, upright, bushy growth; Tantau, Math.

'TANLARPOST', HT, w, 1982; (Polar Star, **Polarstern**®); flowers pale yellow to white, medium, dbl., 35 petals, exhibition form, no fragrance; foliage medium, medium green, matt; bushy growth; ROTY, 1985; Tantau, Math.

TANledolg, MinFl, dy, 1959; (Bijou d'Or, **'GOLDEN JEWEL'**, Goldjuwel); flowers golden yellow, blooms in clusters (up to 10), dbl., 3 in., moderate fragrance; foliage dark, glossy; vigorous, bushy growth; [Goldilocks X Masquerade seedling]; Tantau, Math.

TANledolg, Min, dy, 1995; (**Bijou d'Or**); Tantau

TANlewu, Min, dy, 1993; (**Goldjuwel**); Tantau

'TANLI', F, mp, 1971; (**Rosali**); bud long, pointed; flowers large, dbl., moderate fragrance; vigorous, upright growth; RULED EXTINCT 4/85; [Seedling X Junior Miss]; Tantau, Math.

'TANLILIDA', F, dr, 1977; (**Dalli Dalli**®); bud ovoid; flowers medium, dbl., cupped, slight fragrance; foliage very glossy; upright, bushy growth; ADR, 1975; Tantau, Math.

TANliram, HT, mp, 1997; (**Marlyn**); Tantau

TANlorip, Pol, dy, 1994; (**Pirol**); flowers small, dbl.; Tantau; (Sangerhausen)

'TANLYPOLO', F, dy, 1982; (**Polygold**®); flowers medium, semi-dbl., no fragrance; foliage medium, light green, glossy; bushy growth; Tantau, Math., 1978

TANmeda, F, mp, 1986; (**Diadem**); GM, Durbanville, 1987; Tantau

TANmirsch, S, my, 1993; (**Sonnenschirm**®); Tantau

TANmirson, S, lp; (**Broadlands**); Tantau

TANmixa, HT, rb, 1993; (**Maxim**); Tantau

TANmop, F, mp; (**Pompon Diadem**); Tantau

'TANMOSINA', HT, ab, 1979; (**Simona**); flowers large, dbl., 20 petals, moderate fragrance; foliage large, medium green, matt; upright growth; Tantau, Math.

TANmurse, S, w, 1990; (**Schneesturm**®); Tantau

TANmurse, S, lp; (**Blenheim**); Tantau

'TANNACHT', HT, m, 1965; (Blue Monday, **Blue Moon**®, Mainzer Fastnacht, Sissi, 'TANSI'); bud long, pointed; flowers lilac, dbl., 40 petals, 4 in., intense fragrance; vigorous growth; ADR, 1964 GM, Rome, 1964; [Sterling Silver seedling X Seedling]; Tantau, Math., 1964

'TANNALI', F, dr, 1973; (**Natali**); bud ovoid; flowers small, semi-dbl.; foliage glossy; dwarf, upright growth; RULED EXTINCT 4/85; [Unknown X Unknown]; Tantau, Math.; Horstmann

TANnietnof, HT, ob; (**Stifontein Rose**); Tantau, 1992

'TANNIMOLL', F, or, 1964; (Mistigri, **Molde**®); dbl., blooms in clusters; foliage dark, glossy; bushy, low, compact. growth; GM, Baden-Baden, 1964; Tantau, Math.

'TANNIPOLA', F, op, 1978; (**Lapponia**®); bud broadly ovoid; flowers medium salmon-pink, dbl., 25 petals, slight fragrance; bushy, upright growth; Tantau, Math.

'TANNIREB', HT, mp, 1985; flowers large, dbl., 20 petals, slight fragrance; foliage medium, dark, semi-glossy; upright growth; Tantau, Math.

TANnus, HT, ob, 1991; (**Capri Sun**); Tantau

'TANOB', F, op, 1955; (**Hobby**); flowers coral-pink, large blooms in open clusters, dbl., slight fragrance; foliage dark; upright, bushy growth; [Red Favorite X Kathe Duvigneau]; Tantau, Math.

TANogrew, Min, ob, 1997; (**Clementine**); Tantau

TANokor, S, ly, 1987; (**Rokoko**®); flowers light yellow with rose tints, large, dbl.; foliage dark green; wide, bushy, well-branched (4.5 ft) growth; Tantau

'TANOLFEU', F, ob, 1971; (**Olympisches Feuer**®); bud ovoid; flowers orange, medium, dbl., slight fragrance; upright growth; [Ahoi X Signalfeuer]; Tantau, Math.; Dehner & Co.

TANolfeu92, F, or, 1992; (**Olympisches Feuer 92**); Tantau

'TANOLG', HT, dy, 1975; (Anneliesse Rothenberger, Miss Harp, **Oregold**, Silhouette); flowers large, dbl., exhibition form, slight fragrance; foliage large, glossy, dark; vigorous, upright, bushy growth; AARS, 1975; [Piccadilly X Colour Wonder]; Tantau, Math.; J&P, 1970

TANollipa, HT, my, 1997; (**Papillon**); Tantau

'TANOLOKIP', F, or, 1985; (Piccola, **Piccolo**®); flowers patio, medium, dbl., 20 petals, no fragrance; foliage large, dark, glossy; upright growth; Tantau, Math., 1984

TANolop, HT, w, 1997; (**Polo**); Tantau

TANopel, F, op, 1963; (**Tip Top**); flowers salmon-pink, large, dbl., moderate fragrance; Tantau, Math.; (Sangerhausen)

TANoras, S, mp, 1993; (**Rosario**®); Tantau

TANorelav, HT, dr, 1994; (**Red Velvet**); Tantau

'TANORSTAR', HT, or, 1960; (Super Star, **Tropicana**); bud pointed; flowers coral-orange, well-formed, dbl., 33 petals, 5 in., intense, fruity fragrance; foliage dark, glossy, leathery; vigorous, upright growth; AARS, 1963 GM, ARS, 1967 GM, NRS, 1960 GM, Portland, 1961 PIT, NRS, 1960; [(Unnamed X Peace) X (Unnamed X Alpine Glow]; Tantau, Math.; J&P, 1962

TANotari, S, mp, 1990; (**Mirato**); Tantau

'TANOTAX', S, dp, 1998; (**Footloose**™, Mirato, Neige d'Ete, Schneekonigin®); flowers dark pink, medium sized, double, flat, borne in large clusters, no fragrance; prickles moderate, straight; foliage small, dark green, glossy; spreading, bushy, medium growth; [Seedling X Seedling]; Evers, Hans & Rosen Tantau; Bear Creek Gardens, 1990

TANotax, MinFl, lp; (**Chatsworth**); Tantau

TANpen, F, ly, 1997; (**Penny Heyns**); Tantau

'TANPIKA', HT, mp, 1972; (**Piroschka**®); bud long, pointed; flowers large, dbl., moderate fragrance; moderate, upright growth; [Fragrant Cloud X Dr. A.J. Verhage]; Tantau, Math.

TANpinaiso, HT, lp, 1997; (**Pink Osiana**); Tantau

TANpretty, HT, lp, 1990; (**Softy**); flowers small, dbl.; Tantau; (Sangerhausen)

'TANPRIK', F, or, 1958; (**Paprika**®); bud long, pointed; flowers brick red, blooms in large clusters, semi-dbl., 3.5 in., slight fragrance; foliage leathery, glossy, olive-green; vigorous, upright growth; GM, NRS, 1959 Golden Rose, The Hague, 1961; [Marchenland X Red Favorite]; Tantau, Math.

TANragus, Min, dp; (**Sugar Baby**); flowers small, dbl.; Tantau, 1997; (Sangerhausen)

'TANRAUSCH', F, mp, 1972; (**Duftrausch**); bud globular; semi-dbl., intense fragrance; foliage soft; upright growth; RULED EXTINCT 4/85 ARM; [Unknown X Unknown]; Tantau, Math.; Horstmann

'TANRAVENS', F, or, 1993; (**Wild West**); flowers orange-red, full, very dbl., 26–40 petals, 1.5–2.75 in., borne in large clusters, slight fragrance; small, soft, glandular prickles on peduncles; foliage large, medium green, glossy, resistant to powdery mi; medium (85-95 cms), bushy growth; [Showbiz X Seedling]; Evers, Hans & Rosen Tantau; Bear Creek Gardens, 1993

TANrazlaw, LCl, w, 1987; (**Schneewalzer**); Tantau

'TANRECKERTOR', F, mr, 1985; (**Rotelfe**®, 'TANRECKTOR'); flowers medium, semi-dbl., no fragrance; foliage small, medium green, semi-glossy; groundcover; spreading growth; Tantau, Math.

'TANREKTA', HT, rb, 1970; (**Rebecca**®, Renica); bud ovoid; flowers red, yellow reverse, well-formed, large, dbl., slight fragrance; vigorous, upright growth; [Konfetti X Piccadilly]; Tantau, Math.

TANrif, HT, mp, 1997; (**First Lady**); Tantau

TANrikas, HT, op, 1989; (**Saphir**); Tantau

TANrised, HT, mp, 1986; (**Schone Berlinerin**®); Tantau

TANrolfy, Min, m, 1992; (**Babyflor®**); Tantau

TANrosilb, F, pb, 1990; (**Naas Botha**); Tantau

'TANROTREILI', F, mp, 1981; (**Natali®**); flowers large, dbl., 20 petals, slight fragrance; foliage medium, medium green, matt; Tantau, Math.

'TANROWISA', HT, pb, 1982; (**Wimi®**); flowers shiny pink, silver reverse, well-formed, large, dbl., 20 petals, intense fragrance; foliage large, dark, semi-glossy; bushy growth; Tantau, Math., 1983

'TANRUPEZA', S, m, 1998; (**Leeds Castle**); flowers carmine purple, reverse medium pink, single (4–7 petals), medium (6 cms) blooms borne in large clusters; no fragrance; moderate prickles; foliage medium, dark green, glossy; spreading, low (30–40 cms) growth; groundcover [Seedling X Seedling]; Rosen Tantau; Eurosa, 1998

'TANRYRANDY', HT, ob, 1985; (**Cherry Brandy '85**); flowers orange, large, dbl., 35 petals, no fragrance; foliage large, medium green, glossy; upright growth; Tantau, Math.

'TANSARAS', HT, lp, 1999; (**Century Sunset**, Herz As); dbl., 26–40 petals, 4.5–5 in., borne mostly singly, slight fragrance; prickles moderate; foliage large, dark green, semi-glossy; upright, medium (2-2.5 ft) growth; Tantau

'TANSCHAUBUD', HT, mp, 1985; (**Duftrausch®**, Olde Fragrance, Senteur Royale); flowers large, dbl., intense fragrance; foliage medium, medium green, semi-glossy; upright growth; Tantau, Math., 1986

'TANSCHNEEWA', F, w, 1972; (**Schneewalzer**); bud ovoid; flowers large, dbl.; moderate, upright growth; Tantau, Math.

'TANSCHWEIGRU', F, mr, 1954; (Holländerin, **Red Favorite**, Red Favourite, Salut à la Suisse, Schweizer Grüss); bud ovoid; flowers velvety oxblood-red, blooms in trusses, semi-dbl., 13 petals, 2.5 in., slight fragrance; foliage dark, leathery, glossy; vigorous, bushy growth; ADR, 1950; [Karl Weinhausen X Cinnabar]; Tantau, Math.; C-P

TANselbon, HT, op, 1989; (**Noblesse**); Tantau

TANsemil, HT, lp; (**Sweetnesse**); Tantau, 1997

'TANSENFRIE', S, dy, 1982; (**Friesensöhne®**); flowers medium, semi-dbl., no fragrance; foliage medium, light, glossy; PP005706; Tantau, Math.; Tantau Roses, 1981

TANseus, HT, lp; (**The Robe**); Tantau, 1995

'TANSI', HT, m, 1965; (Blue Monday, **Blue Moon®**, Mainzer Fastnacht, Sissi, 'TANNACHT'); bud long, pointed; flowers lilac, dbl., 40 petals, 4 in., intense fragrance; vigorous growth; ADR, 1964 GM, Rome, 1964; [Sterling Silver seedling X Seedling]; Tantau, Math., 1964

'TANSINNROH', HT, ab, 1984; (Frohsinn '82, **Joyfulness**); flowers apricot and orange blend, large, dbl., 35 petals, slight fragrance; foliage large, dark, glossy; Tantau, Math., 1982

TANsirk, F, or, 1990; (**Keith Kirsten**); Tantau

TANspolett, F, dp, 1994; (**Pink Diadem®**); Tantau

'TANT MIEUX', S, ob, 1984; (Saskabec); flowers tangerine, yellow center, blooms in clusters, single, 5 petals, no fragrance; foliage medium, medium green, glossy; bushy growth; [R. gallica X Maria Stern]; Wright, Percy H.; Gilles Gailloux

'TANTA', F, or, 1968; (**Taora®**); bud long, pointed; flowers large, dbl.; foliage glossy; very vigorous, bushy growth; ADR, 1969; [Fragrant Cloud X Schweizer Gruss]; Tantau, Math.

TANtaatsyr, HT, ob; (**Vrystaat**); Tantau, 1992

Tantaliser, F, lp

'TANTALLON', HT, mp, 1934; flowers salmon-pink, base yellow, reverse cerise-red touched orange, dbl.; foliage glossy, dark; dwarf, stocky growth; Dobbie

Tantarra, HT, lp; Gardner, 1996

'TANTASCH', F, dy, 1998; (Castle Howard Tercentenary, Golden Jet, **Goldschatz**); flowers deep yellow, dbl., 15–25 petals, 4 in., borne in small clusters, slight fragrance; prickles moderate; foliage large, medium green, glossy; bushy, medium (70cm) growth; [Seedling X Seedling]; Rosen Tantau; Eurosa, 1999

'TANTAU'S DELIGHT', F, or, 1951; bud long, pointed; flowers open, large, borne in clusters, semi-dbl., slight fragrance; foliage dark, glossy; vigorous, bushy growth; [Cinnabar X Kathe Duvigneau]; Tantau

'TANTAU'S SURPRISE', F, mr, 1951; (Tantau's Ueberraschung); flowers blood-red, large, borne in clusters of 5-8, dbl., 40 petals, slight fragrance; foliage dark, glossy; upright growth; [Bouquet X Hamburg]; Tantau

Tantau's Triumph *see* **'CINNABAR'**

Tantau's Triumph, Climbing, Cl Pol, mr, 1957; flowers medium, semi-dbl.; Rosarium Sangerhausen; (Sangerhausen)

Tantau's Ueberraschung *see* **'TANTAU'S SURPRISE'**

Tante Frieda, HT, mr, 1995; flowers large, dbl., intense fragrance; Michler, K. H.; (Sangerhausen)

'TANTERN', HKor, my, 1966; (Gold Star, **Goldstern®**); bud long, pointed; flowers golden yellow, large blooms in clusters, recurrent bloom; foliage glossy; vigorous, bushy (7-8 ft.) growth; Tantau, Math.

'TANTIDE', F, dr, 1988; (**Glad Tidings**, Lübecker Rotspon®, Peter Wessel); bud ovoid; flowers bright crimson, medium, dbl., 20 petals, cupped, borne in sprays, no fragrance; foliage medium, medium green, semi-glossy; upright, medium growth (2 ft); GM, Durbanville, 1988 ROTY, St. Albans, UK, 1989; [Unnamed seedling X Unnamed seedling]; Tantau, R.; Wheatcroft Ltd., 1989

'TANTIFUM', HT, m, 1978; (**Blue Parfum®**, Blue Perfume, 'TANFIFUM', 'TANIFUME'); bud ovoid; flowers mauve-blush, large, dbl., intense fragrance; foliage glossy; bushy, upright growth; Tantau, Math.

'TANTIGA', F, mp, 1972; (**Taiga®**); bud globular; flowers pink, medium, dbl.; vigorous, upright growth; [Geisha X Junior Miss]; Tantau, Math.

'TANTIVVY', Pol, mr, 1954; flowers crimson-scarlet, rosette form, semi-dbl.; compact, bushy growth; [Gloire du Midi sport]; Ratcliffe

TANttelos, S, dp, 1994; (**Violett Satina**); flowers medium, dbl.; Tantau; (Sangerhausen)

TANtumleh, HT, dr, 1996; (**Helmut Kohl**); flowers very large, semi-dbl., slight fragrance; Tantau, Math.; (Sangerhausen)

'TANVERY', HT, ob, 1974; (**Love Story®**); bud ovoid; flowers orange, large, dbl., cupped, slight fragrance; Tantau, Math.; Horstmann, 1972

'TANWEIEKE', F, mr, 1983; (Bernhard Daneke Rose, Ingrid Weibull, **Showbiz**); flowers bright medium red, medium, dbl., 20 petals, no fragrance; foliage medium, dark, semi-glossy; bushy growth; AARS, 1985; Tantau, Math., 1981

'TANYA', HT, ob, 1959; (Majeure); bud pointed; flowers deep orange to apricot-orange, dbl., 48 petals, 5 in., exhibition form, moderate fragrance; foliage leathery, glossy; vigorous, upright growth; [Peace X (Peace X Orange Nassau)]; Combe; J&P;, Vilmorin-Andrieux

'TANYA KIM', HT, my, 1975; flowers clear yellow, full, dbl., 24 petals, 4.5 in., intense fragrance; foliage glossy; [Whisky Mac sport]; Marks

'TANYAB', F, my, 1990; (**Bayerngold®**); Tantau

TANybab, Min, dp, 1993; (**Pink Babyflor**); Tantau

TANydal, HT, pb; (**Summer Lady**®); Tantau, 1991

TANydor, S, mr, 1995; (**Rody**); flowers medium, semi-dbl.; Tantau; (Sangerhausen)

TANydu, HT, my, 1997; (**Judy**); Tantau

'TANYELLDA', F, dy, 1972; (**Yellow Belinda**); bud long, pointed; flowers orange-yellow, medium, dbl.; [Tanbeedee sport]; Tantau, Math.

'TANZAHDE', S, mr, 1998; (**Beaulieu**); flowers medium red, single, 4–7 petals, 2.5 in, borne in large clusters, no fragrance; prickles moderate; foliage medium, dark green, glossy; low, spreading (30 cms) growth; [Seedling X Seedling]; Rosen Tantau; Eurosa, 1998

TANzecon, HT, ob, 1992; (**Constanze**); Tantau

TANziewsim, HT, dr, 1996; (**Miss Schweiz**); flowers very dark red, large, dbl.; Tantau; (Sangerhausen)

Taora® *see* 'TANTA'

'TAPESTRY', HT, rb, 1958; bud pointed; flowers red, yellow and pink, dbl., 38 petals, 4.5–6 in., exhibition form, moderate, spicy fragrance; foliage glossy; upright, bushy growth; [Peace X Mission Bells]; Fisher, G.; C-P

Tapis, Gr, lp

Tapis Afghan *see* **'PEKINOIS'**

'TAPIS BLANC', Pol, w, 1927; flowers pure white, center tinted cream, large, borne in clusters, dbl.; bushy, dwarf growth; Turbat

Tapis de Soie *see* **'WEE MAN'**

Tapis Jaune® *see* 'RUGUL'

Tapis Persan *see* 'MACEYE'

'TAPIS ROSE', F, mp, 1950; bud ovoid; flowers rose-pink, borne in clusters, dbl., 40 petals, 2–2.5 in., exhibition form, moderate fragrance; foliage leathery, dark; vigorous, upright, bushy growth; (27); [Pinocchio X Mme Jules Gouchault]; Meilland, F.; C-P

Tapis Rouge *see* 'INTEROP'

'TAPIS ROUGE', F, mr; more compact and dwarf growth

Tapis Rustic, S, lp

Tapis Volant® *see* 'LENPLAT'

Tapti, Gr, lp; Kasturi, 1978

Tara *see* 'MACWAIWER'

Tara Allison *see* 'MACWAIWER'

'TARA RED', HT, dr, 1974; bud ovoid; flowers carmine-red, becoming very dark, dbl., 38 petals, 6 in., exhibition form, moderate fragrance; foliage leathery; tall growth; [Peace X Charles Mallerin]; Golik; Dynarose

Taranga® *see* 'TANAGNARAT'

Tarantella *see* 'KORANTEL'

'TARANTELLA'®, HT, dy, 1936; flowers deep golden yellow, large, dbl., cupped, slight fragrance; foliage light, wrinkled; vigorous, bushy growth; [Charles P. Kilham X Pres. Herbert Hoover]; Tantau

'TARANTELLE', HT, or, 1965; flowers well-formed, dbl., 40 petals; foliage bronze; vigorous, bushy growth; [(Beauté X Ma Fille) X Magicienne]; Laperrière; EFR

'TARDE GRIS', HT, m, 1966; (Evening Light); flowers soft lilac, large, dbl., 25 petals, moderate fragrance; foliage dark; upright growth; [(Sterling Silver X Intermezzo) X (Sterling Silver X Simone)]; Dot, Simon; Roses Dot

Target *see* 'JACRIM'

'TARO', HT, my, 1977; bud oval; flowers high pointed, dbl., 35 petals, 5.5 in.; foliage dark, leathery; vigorous, upright growth; [Edith Krause X Narzisse]; Ota, Kaichiro

Tarragona *see* **'DIPUTACION DE TARRAGONA'**

'TARRAGONA', HT, ab, 1945; flowers apricot-yellow, open, dbl., 25–35 petals, exhibition form, slight fragrance; foliage glossy; upright growth; [Duquesa de Peñaranda X Federico Casas]; Dot, Pedro

Tarrawarra, Pol, pb; Nieuwesteeg, J., 1992

'TARTUFE', F, mr, 1956; flowers dark geranium-red, dbl., 28 petals; moderate growth; [Pompadour Red X Independence]; Frères, Buyl

Taryn, Min, op

'TARZAN', LCl, mr, 1955; flowers coppery carmine, dbl., free, recurrent bloom, moderate fragrance; foliage bronze; very vigorous growth; Delbard-Chabert

Tasja, HT, lp; RvS-Melle, 1995

Tasman *see* 'MURTA'

Tasman Bay, LCl, w; Pratt, Florence, 1994

'TASOGARE', F, m, 1977; bud cupped; semi-dbl., 18 petals, 2.5–4 in., flat, moderate fragrance; foliage glossy; vigorous, spreading growth; [Gletscher X (Sterling Silver X Gletscher)]; Kobayashi, Moriji

'TASSIE PRINCESS', F, lp, 1999; dbl., 15–25 petals, 2–2.5 in., borne in small clusters, no fragrance; few prickles; foliage small, medium green, semi-glossy; bushy, low (36 in) growth; Elszele, L. H.

Tassili *see* ORAju

'TASSIN', HT, mr, 1942; bud ovoid; flowers large, dbl., intense fragrance; foliage leathery; vigorous, upright growth; [National Flower Guild X Lemania]; Meilland, F.; A. Meilland

'TASSIN, CLIMBING', Cl HT, mr; da Silva, Moreira

'TASTE OF HONEY', F, ob, 1977; bud ovoid; flowers honey color, aging pink, open, large, dbl., intense fragrance; foliage leathery; vigorous, upright growth; [Elizabeth of Glamis sport]; Young, S.A.

'TATA CENTENARY', HT, rb, 1974; flowers deep carmine, striped pale yellow, reverse buff; [Pigalle sport]; Pradhan, Rauf, Murmu, Ghosh

Tata Centenary, **Climbing**, Cl HT, lp; K&S, 1995

Tatjana® *see* 'KORTAT'

Tattletale *see* 'ZIPTALE'

Tattooed Lady *see* 'SEADEEP'

Taubie Kushlik *see* KORzinta

'TAUNUSBLÜMCHEN', HMult, m, 1904; flowers violet-pink, small, semi-dbl.; Weigand, C.; (Sangerhausen)

Taupo® *see* 'MACMISECH'

'TAURANGA CENTENNIAL', F, or, 1982; flowers orange-red, center yellow, large blooms in cluster of 3, dbl., 22 petals, slight fragrance; light brown prickles; upright growth; [Arthur Bell X (Alexander X Sympathy)]; Wareham, Phyllis e.

'TAURINIA', HT, mp, 1942; bud ovoid; flowers bright salmon-pink, large, very dbl., cupped, moderate fragrance; foliage leathery; very vigorous growth; [Julien Potin X Sensation]; Aicardi; Giacomasso

'TAURO', F, dr, 1958; bud globular; flowers oxblood-red, large, borne in clusters, dbl.; foliage glossy; upright growth; [Poinsettia X Alain]; Bofill; Torre Blanca

'TAURUS', Min, or, 1978; bud pointed, ovoid; flowers vermilion, full, dbl., 35–40 petals, 2–3 in.; foliage dark; compact, upright growth; [Baby Masquerade X (Spion-Kop X Evelyn Fison)]; Ellick; Excelsior Roses

'TAUSENDSCHÖN', HMult, pb, 1906; (Thousand Beauties); flowers deep rose-pink, center white, large blooms in large clusters, dbl., cupped, slight fragrance; no prickles; foliage soft; climbing (8-10 ft.) growth; [Daniel Lacombe X Weisser Herumstreicher]; Schmidt, J.C.

'TAWNY GOLD', HT, dy, 1951; bud pointed; flowers tawny gold, well-shaped, dbl., 25 petals, 5 in., exhibition form, intense fragrance; foliage leathery; bushy, compact growth; [Vanessa X Burgemeester van Oppen]; Leenders, M.; J&P

'TAWNY SUPERIOR', HT, my, 1958; flowers golden yellow; Leenders, M.

'TAXI', F, rb, 1969; flowers mandarin red, reverse soft rose, pointed, large, dbl.;

foliage dark, matt; RULED EXTINCT 12/85; [Isabelle de France X Basildon Belle]; Trew, C.; Basildon Rose Gardens

'TAXI', HT, dr, 1985; flowers large blooms borne usually singly, dbl., 35 petals, exhibition form, intense fragrance; foliage large, dark, matt; vigorous, upright growth; [Fragrant Cloud X Gisselfeld]; Poulsen, Niels D.; Poulsen's Roses, 1978

'TAY CAITLIN', Min, ab, 1997; flowers double, small, dbl., 15–25 petals, .5–1 in., borne mostly singly, slight fragrance; some prickles; foliage medium, medium green, semi-glossy; upright, medium (16-22in.) growth; [Queen City sport]; Garrett, Troy O.

'TAYBIRD', HT, yb, 1992; (**Bird of Fire**); flowers yellow blending to reddish edges, reverse yellow, aging lighter, exhibition form, 25–30 petals, 4–4.75 in., exhibition form, moderate; foliage medium, dark green, semi-glossy, leathery, reddish w; upright, bushy, medium growth; [Fragrant Delight X Sunsilk]; Taylor, Thomas E., 1993; National Kirtland's Warbler Recovery Team

Taylors Gold *see* 'HORBARJEFF'

'TAYMAR', HT, pb, 1996; (**Mary Louise**); flowers light pink, reverse medium pink, form similar to first prize, dbl., 15–25 petals, 5 in., borne mostly singly, slight fragrance; moderate prickles; foliage medium, medium green, glossy, similar to honey favor; compact, bushy, medium (3-4 ft) growth; [Honey Favorite X First Prize]; Taylor, Thomas E.

'TAYMEL', Min, m, 1990; (**Melinda Claire**); bud pointed; flowers reddish magenta, yellow stamens, small to blooms borne usually singly, semi-dbl., 15 petals, exhibition form, slight fragrance; globular, medium, orange fruit; prickles curved downward, small, widely spaced, brownish; foliage medium, medium green, semi-glossy; [Charmglo X Charmglo]; Taylor, Thomas E., 1988; Michigan Mini Roses, 1990

Tchin-Tchin® *see* 'MEICHANSO'

Tchin-Tchin® *see* 'MEIKINOSI'

Tchin-Tchin, Climbing *see* MEIchansosar

Te Awamutu Centennial *see* 'STETEAW'

Te Kawanata Hou, S, lp; Nobbs, 1995

'TE KOWHAI', F, dr, 1960; flowers dark velvety red, reverse white overlaid red; [(Florence Mary Morse X Border Queen) X Confidence]; Bennett, H.; Pedigree Nursery

Te Moana *see* 'SIMTEONO'

'TEA PARTY', Min, ab, 1972; flowers orange-apricot to pink, small, dbl.; foliage small, light; vigorous, dwarf, bushy growth; [(R. wichurana X Floradora) X Eleanor]; Moore, Ralph S.; Sequoia Nursery

'TEA RAMBLER', HMult, op, 1904; flowers soft salmon-pink, blooms in clusters, dbl., early, moderate fragrance; vigorous, climbing growth; [Crimson Rambler X Tea]; Paul

Tea Rose *see* **'R. X ODORATA'**

'TEA TIME', F, mp, 1960; bud ovoid; flowers clear pink, medium, borne in clusters, dbl., 60–65 petals, exhibition form, moderate fragrance; foliage leathery; vigorous, bushy growth; [Floribunda seedling X Demure]; Boerner; J&P

Tea Time *see* TANetee

'TEACHER'S PET', HT, ab, 1979; bud ovoid, pointed; flowers apricot, dbl., 30 petals, exhibition form, slight fragrance; straight prickles; foliage dark green with red mid-rib and edges; upright, bushy growth; [First Prize X Bonsoir]; Taylor, Thomas E.

Tear Drop *see* 'DICOMO'

'TEASE', F, ob, 1967; bud ovoid, coppery; flowers golden yellow, dbl., cupped, moderate fragrance; foliage leathery; vigorous, upright growth; [Yellow Pinocchio X Fashion seedling]; Boerner; J&P

Teasing Georgia *see* 'AUSBAKER'

Technikon Pretoria *see* KORproa

Ted Allen *see* 'GUESGENUS'

'TED CARNAC', LCl, 1968; vigorous growth; Gowie

Ted Gore *see* 'HORMISLAC'

Ted Goves *see* 'KIRPARK'

'TEDCASTLE', HP, m, 1940; flowers purple, open, very dbl., cupped, slight fragrance; foliage glossy; upright, bushy growth; [Ulrich Brunner Fils X Paul Neyron]; Tedcastle

Teddy *see* PRAeddy

Teddy Bear™ *see* 'SAVABEAR'

Teddy's Coat, F, dy

Ted's Red *see* 'REYRED'

'TEENAGER', HT, yb, 1958; flowers yellow shaded pink and carmine, loosely formed, dbl., 25 petals, 4–5 in., intense fragrance; foliage dark, leathery; vigorous, upright growth; [Ena Harkness X Sutter's Gold]; Arnot; Croll

'TEENAGER', F, pb, 1958; bud ovoid; flowers la france pink, lightly overcast shrimp-pink, dbl., 55–60 petals, 3.5 in., exhibition form, intense fragrance; foliage leathery; strong stems; vigorous, upright growth; [Demure X Demure]; Boerner; J&P

Teeny Bopper *see* 'TINBOPPER'

Teeny-Weeny *see* **'BUTTONS 'N' BOWS'**

'TEGALA', S, dp, 1927; flowers deep pink, semi-dbl., non-recurrent; height 5 ft; very hardy.; [Tetonkaha X Alika]; Hansen, N.E.

'TEIDE', LCl, mr, 1948; flowers carmine, well formed; foliage reddish; [Texas Centennial X Guinee]; Dot, M.

Telford's Promise *see* 'CHEWOZ'

'TELSTAR', F, ob, 1962; flowers orange to orange-buff, blooms in clusters (up to 5), semi-dbl., 14 petals, 3.5 in., moderate fragrance; foliage dark; vigorous, upright growth; [Flash (F) X Masquerade]; Gandy, Douglas L.

'TEMNO', HT, m, 1936; flowers dark maroon, fairly large, dbl.; vigorous growth; Böhm, J.; J&P

'TEMPER TANTRUM', Min, dr, 1986; flowers white center, medium, borne usually singly or in sprays of 2, dbl., 30 petals, exhibition form, slight fragrance; no fruit; no prickles; foliage small, medium green, semi-glossy; upright, medium growth; [Red Beauty X Libby]; Jolly, Marie; Rosehill Farm, 1988

'TEMPERAMENT', F, mr, 1957; (Vivacity); flowers light scarlet, open, borne in clusters, dbl., 24 petals, 2.5 in., slight fragrance; foliage glossy, leathery; vigorous, upright growth; [Fanal X Karl Weinhausen]; Tantau, Math.

'TEMPEST, CLIMBING', LCl, rb, 1962; bud pointed; flowers yellow, pink and orange-red, borne in clusters, dbl., 26 petals, 2.5–3 in., intense fragrance; foliage dark, glossy; vigorous growth; [Masquerade X Unnamed seedling]; Verschuren; Blaby Rose Gardens

Tempi Moderni *see* BARdem

'TEMPIE LEE', HT, pb, 1991; dbl., 26–40 petals, 1.5–2.75 in., exhibition form, borne mostly singly, slight fragrance; some prickles; foliage medium, medium green, semi-glossy, problems with blackspot; long canes; vigorous, medium, spreading growth; [Elizabeth Taylor sport]; Whittington, J.O., Sr.; Edmunds' Roses, 1998

'TEMPLAR', HT, mr, 1924; flowers bright red, dbl., globular; foliage dark, leathery; vigorous growth; GM, NRS, 1925; [Premier X Seedling]; Montgomery Co.; A.N. Pierson

'TEMPLE BELLS', Cl Min, w, 1971; flowers small, single, 7 petals, 2 in., slight fragrance; foliage small, glossy; [R. wichurana X Blushing Jewel]; Morey, Dr. Dennison; McGredy

'TEMPLE FLAME', F, or, 1965; flowers orange flushed red, open, medium, semi-dbl.; foliage leathery; very vigorous, upright, compact growth; [Orangeade X ?]; Pal, Dr. B.P.; Indian Agric. Research Inst.

Tempo *see* 'JACCLOP'

Tempo *see* MEInececa

'TEMPTATION', LCl, mr, 1950; bud ovoid, carmine-red; flowers rose-red, open, large, dbl., 40 petals, free, recurrent boom, moderate fragrance; foliage large, glossy; vigorous, climbing growth; good pillar; very hardy.; [(New Dawn X Crimson Glory) X Dream Girl]; Jacobus; B&A

Temptation *see* 'TROPAT'

Temptation *see* INTertropa

'TEMPTRESS', HT, mr, 1948; bud long, pointed, crimson; flowers carmine, medium, dbl., 35–40 petals, cupped, moderate fragrance; foliage leathery, dark; vigorous, upright, bushy growth; [(Lucile Hill X Chieftain) X Unnamed seedling]; Hill, Joseph H., Co.

Temptress *see* 'WAMTRESS'

'TEMPUS-FUGIT', HT, mr, 1977; flowers cardinal-red, full, dbl., 40–45 petals, 3–4 in., moderate fragrance; foliage glossy, dark; very free growth; [Chopin X Heidelberg]; Ellick

Ten Sisters *see* **'TURNER'S CRIMSON RAMBLER'**

Ten Ten CFRB *see* MACmatan

Tender Blush, A, lp; Sievers, 1988

'TENDER LOVE', Min, lp, 1980; bud slender, tapering; dbl., 23 petals, borne mostly singly, sometimes 3-5 per cluster, slight fragrance; prickles dilated at base; foliage glossy, medium green; tall, upright growth; [Doris Ashwell X Sheri Anne]; Strawn, Leslie E.; Pixie Treasures Min. Roses

Tender Love, F, lp, 1997; flowers very full, exhibition, medium, very dbl., 41 petals, 2.75 in., borne in large clusters, no fragrance; some prickles; foliage medium, light green, glossy; upright, medium (90cms.) growth; [Sexy Rexy X Seedling]; Bossom, W.E.

'TENDER LOVING CARE' *see* BOSpeabay

Tender Mercy *see* 'RENMERCY'

Tender Night *see* 'MEILAUR'

Tender Night, Climbing *see* 'MEILAURSAR'

Tender One *see* **'NEJENKA'**

Tenderly, HT, 1984; Delforge; (Cavriglia)

Tenderly Yours *see* 'RENYOURS'

Tendresse *see* 'DELJOFEM'

Tendresse *see* **'SACHSENGRUSS'**

Tendresse *see* DELtendre

'TENEREZZA', HT, rb, 1973; bud ovoid, long, pointed; flowers spinel-rose to cherry-red, medium, dbl., cupped, moderate fragrance; foliage large, dark, leathery; vigorous, upright, bushy growth; [Unnamed seedling X Ninfa Rossa]; Calvino

'TENERIFE', HT, op, 1972; flowers deep orange, yellowish reverse, pointed, dbl., 45 petals, 5 in., intense fragrance; foliage glossy; [Fragrant Cloud X Piccadilly]; Timmerman's Roses

Tennessee™ *see* 'KINTENN'

Tennessee, F, ob; Select Roses, 1991

'TENNESSEE BELLE', LCl, mp; flowers bright rosy blush, large, borne in clusters, dbl., moderate fragrance; vigorous, climbing growth

Ténor *see* 'DELCAP'

'TENTATION', HT, op, 1968; bud pointed; flowers carmine-pink to orange, large, dbl., slight fragrance; foliage glossy; vigorous, upright growth; [Touggout X Flaminaire]; Dorieux; Vilmorin

Tenth Rose Convention, HT, dy; Patil, B.K., 1990

'TEODORA', HT, mr, 1983; flowers large, dbl., 30 petals, cupped, slight fragrance; reddish-green prickles; foliage large, brownish-green, matt; upright growth; [Honey Favorite X Command Performance]; Stoddard, Louis; Rose Barni-Pistoia

Tequila® *see* 'MEIGAVESOL'

Tequila Sunrise *see* 'DICOBEY'

Tequilla Sunset™ *see* 'WILSUN'

'TERESA FINOTTI-MASIERI', HT, m, 1929; bud long, pointed; flowers brilliant purple-rose, rosy scarlet reflexes, bordered yellow; [George C. Waud X Souv. de Gabriel Luizet]; Ketten Bros.

'TERESA OZORES', Cl HT, lp, 1957; bud ovoid; flowers soft pearly pink, large, dbl., cupped, intense fragrance; vigorous growth; [Frau Karl Druschki X Lady Sylvia]; Camprubi, C.

Terra Jubilee, F, lp; Zary, Dr. Keith W., 1996

Terracotta *see* MEIcobius

Terrell Anne *see* 'GRETAG'

'TERRY O'', F, rb, 1976; flowers dark red, reverse copper yellow; [Redgold sport]; Hughes; Hughes Roses

Tersicore, S, 1985; Barni, V.; (Cavriglia)

'TESCHENDORFFS JUBILÄUMSROSE', Pol, dr, 1928; flowers vivid rosy crimson, darker than Orleans Rose and more brilliant; [Orléans Rose sport]; Teschendorff

'TESCO BERNSTEIN', F, dy, 1972; bud ovoid; flowers brownish yellow, open, medium, semi-dbl., globular, slight fragrance; foliage glossy, dark, leathery; vigorous, bushy growth; [Cognac X Allgold]; Haenchen, E.; Teschendorff

'TESCO BLICKFANG', F, or, 1972; bud ovoid; flowers dark vermilion-red, large, dbl., slight fragrance; foliage dark, soft; vigorous, upright, bushy growth; [Dicksons Flame X Circus]; Haenchen, E.

'TESCO BRENNPUNKT', F, or, 1972; bud ovoid; flowers bright vermilion-red, open, medium, semi-dbl., slight fragrance; foliage dark, soft; vigorous, upright growth; [Highlight X Dicksons Flame]; Haenchen, E.; Teschendorff

'TESCO GOLDTEPPICH', F, my, 1972; bud ovoid; flowers medium, semi-dbl., cupped, slight fragrance; foliage glossy, dark, leathery; vigorous, bushy growth; [Cognac X Allgold]; Haenchen, E.; Teschendorff

'TESCO LICHTBLICK', S, ob, 1972; flowers pink and orange-pink on yellow ground, large, dbl., exhibition form, abundant, continuous bloom, slight fragrance; foliage glossy, leathery; vigorous, upright growth; [Yellow Holstein X Heidelberg]; Haenchen, E.; Teschendorff

'TESCO ROMANZE', F, pb, 1972; bud globular; flowers pink, center white, open, small, semi-dbl., slight fragrance; foliage small, glossy, dark, leathery; vigorous, dwarf, bushy growth; [Marchenland X ?]; Haenchen, E.; Teschendorff

'TESORINO'®, Min, lp, 1982; flowers small, dbl., 20 petals, no fragrance; light green prickles; foliage small, light green, matt; groundcover; spreading growth; [Unnamed X Unnamed]; Rose Barni-Pistoia

'TESS', HMult, mp, 1939; flowers neyron pink, passing to deeper pink, center brighter, large, dbl., slight fragrance; foliage large, rich green; strong stems; very vigorous growth; RULED EXTINCT 11/91; Beckwith

Tess *see* 'CLETEST'

Tess *see* 'MACRAMAR'

Tess of the D'Urbervilles *see* 'AUSMOVE'

Tessa O'Keeffe *see* 'KENMOLL'

Testa Rossa *see* 'KORDIAM'

'TETON BEAUTY', HRg, dp, 1927; flowers rich pink to crimson, dbl., cupped, recurrent bloom, moderate fragrance; foliage rugosa-like; very hardy.; [Tetonkaha X American Beauty]; Hansen, N.E.

'TETONKAHA', HRg, dp, 1912; flowers deep rich pink, medium, semi-dbl., profuse, non-recurrent bloom, intense fragrance; vigorous (6 ft tall and wide) growth; hardy.; [R. macounii X Rugosa hybrid]; Hansen, N.E.

Teutonia®, HT, dp; Noack, Werner, 1985

'TEWANTIN', Cl HT, dp, 1953; bud globular; flowers deep pink, large, dbl., moderate fragrance; long stems; vigorous (12 ft) growth; [Editor McFarland X Self]; Ulrick, L.W.

'TEXAN', F, mr, 1956; (The Texan); bud ovoid; flowers rose-red, borne singly and in clusters, dbl., 23 petals, 3.5–4 in., exhibition form, moderate fragrance; foliage dark, leathery; vigorous, upright growth; [Improved

Lafayette X Peace]; Lindquist; Howard Rose Co.

Texas *see* 'POULTEX'

Texas *see* KORbacol

'TEXAS CENTENNIAL', HT, rb, 1935; flowers vermilion-red with some gold, center lighter; GM, Portland, 1935; [Pres. Herbert Hoover sport]; Watkins, A.F.; Dixie Rose Nursery

'TEXAS CENTENNIAL PANACHÉE', HT, ob, 1939; flowers light orange with white stripes, large, dbl., moderate fragrance; Tantau, Math.; (Sangerhausen)

'TEXAS CENTENNIAL, CLIMBING', Cl HT, rb, 1936; Dixie Rose Nursery, 1942

Texas Girl(TM) *see* 'MEICIJAS'

'TEXAS GOLD', HT, my, 1935; flowers golden yellow occasionally tinged pink, large, dbl., intense fragrance; [Pres. Herbert Hoover sport]; Wolfe

'TEXAS QUEEN', HT, mp, 1965; flowers base cream, shading to light and darker pink, large, very dbl., 100 petals, intense fragrance; almost thornless; foliage light green, soft; vigorous growth; [Tip Toes sport]; Leidy; Wilson Nursery

Texas Sunrise *see* 'UMSTEX'

'TEXAS WAX', S, lp; [Appears to be a hybrid of R. chinensis X R. multiflora]

Thaïs *see* 'MEIMAJ'

Thalassa, HT, m; Dorieux, 1978

'THALIA', HMult, w, 1895; (White Rambler); flowers small blooms in clusters, dbl., early; [R. multiflora X Paquerette]; Schmitt; P. Lambert

Thalia Remontant, HMult, w; Lambert, P., 1903

Thalie La Gentille, HGal, dp; Vibert, before 1811

'THANET BALLERINA', HT, w, 1973; flowers pearl-white, blushed pink, moderate fragrance; foliage rich green; vigorous growth; [Prima Ballerina sport]; Court

Thank You *see* CHEsdeep

'THANKSGIVING', HT, ob, 1965; bud ovoid; flowers bronze shades, reverse orange-red, medium, dbl., 38 petals, 3.5 in., exhibition form; foliage dark, leathery; vigorous growth; [Fred Howard X Seedling]; Warriner, William A.; Great Western Rose Co.

Thanx Mom *see* 'JULESDEAR'

Thanx Mum *see* 'JULESDEAR'

Thari®, S, dr

'THE ADJUTANT', HT, mr, 1922; bud long, pointed; flowers bright red, dbl., intense fragrance; Pemberton

The Adjutant, Climbing, Cl HT, mr; flowers large, dbl., intense fragrance; (Sangerhausen)

'THE ALAMO', HT, mr, 1959; bud ovoid; flowers cardinal red, very dbl., 110 petals, 3.5–4.5 in., slight fragrance; foliage leathery; vigorous, upright, bushy growth; [Happiness X Independence]; Meilland, F.; Co-Operative Rose Growers

'THE ALEXANDRA', T, op, 1900; flowers light salmon-pink, large, dbl., slight fragrance; Paul, W.; (Sangerhausen)

The Alexandra Rose *see* 'AUSDAY'

The Alexandria Rose *see* 'WILALEX'

'THE ALLIES', Pol, w, 1930; flowers white suffused pale pink, open, small, borne in clusters, dbl., slight fragrance; foliage small, glossy; dwarf, bushy growth; Heers

'THE AUSTRALIAN BICENTENNIAL', HT, mr, 1987; flowers medium to deep red, well-formed, borne usually singly, dbl., 30 petals, intense fragrance; curved, green prickles; foliage medium, medium green, glossy, disease resistant; tall, branching, average growth; [Daily Sketch seedling X Red Planet]; Bell, Ronald J.; Roy H. Rumsey, Pty. Ltd., 1988

The Backpackers Rose *see* 'HORBONDARC'

'THE BAIRN', HT, dy, 1976; flowers golden yellow, very full, medium, dbl., 40–50 petals, 3.5 in.; foliage dark, leathery; dwarf growth; [Arthur Bell X Unnamed seedling]; Wood

'THE BEACON', HWich, rb, 1922; flowers bright fiery-red with white-eye, as in American Pillar, sing, semi-dbl.; vigorous, climbing (to 8 ft) growth; Paul, W.

The Beloved *see* **'PRIYATAMA'**

'THE BISHOP', C, m; flowers unusual shade of cerise-magenta, fading bluish purple, rosette, flat, early bloom; slender, upright (4-5 ft.) growth

'THE BISHOP', HT, dr, 1937; flowers bright crimson, large, dbl., intense fragrance; foliage dark; vigorous growth; GM, NRS, 1937; Dickson, A.

'THE BRIDE', T, w, 1885; flowers white tinged pink; [Catherine Mermet sport]; May

'THE BRIDE', HT, w, 1963; bud pointed, ivory-white; flowers well formed, dbl., 30 petals; foliage glossy; [White Swan X Unnamed seedling]; Herholdt, J.A.; Herholdt's Nursery

The Cambridge Rose *see* **'CANTABRIGIENSIS'**

The Care Rose *see* 'HORAPSUNMOLBABE'

The Cheshire Regiment *see* FRYyat

'THE CHIEF', HT, ob, 1940; (Chief); bud long, pointed; flowers flame, coral and copper, dbl., 35 petals, 4–6 in., intense fragrance; vigorous, bushy, spreading growth; AARS, 1940; [Charles P. Kilham X Pres. Herbert Hoover]; Lammerts, Dr. Walter; Armstrong Nursery

The Clerk, S, lp; Jerabek, Paul E., 1997

The Colwyn Rose *see* FRYnblue

The Commodore *see* **'KOMMODORE'**

The Compass Rose *see* KORwisco

The Conductor *see* **'DIRIGENT'**®

The Cottage Rose *see* 'AUSGLISTEN'

The Countryman® *see* 'AUSMAN'

The Coxswain *see* 'COCADILLY'

The Crepe Rose *see* **'PAUL PERRAS'**

The Daniel(TM) *see* 'WILDAN'

The Dark Lady *see* 'AUSBLOOM'

'THE DAZZLER', Cl HT, mr, 1955; flowers bright red, petals ruffled, single, profuse, repeated bloom; [Dainty Bess, Climbing seedling]; Marsh's Nursery

The Didgemere Rose *see* 'FERTRY'

'THE DOCTOR', HT, mp, 1936; flowers satiny pink, well-formed, dbl., 25 petals, 6 in., intense fragrance; foliage soft, light; dwarf, bushy growth; GM, NRS, 1938; [Mrs J.D. Eisele X Los Angeles]; Howard, F.H.; Dreer

'THE DOCTOR, CLIMBING', Cl HT, mp, 1950; Dyess; Reliance Rose Nursery

The Dove *see* 'TANAMOLA'

'THE DOWAGER COUNTESS OF RODEN', HT, lp, 1919; flowers bright silvery pink, dbl.; [Viscountess Enfield X George C. Waud]; Paul, W.

The Dragon's Eye *see* 'CLEDRAG'

'THE DUKE', HT, rb, 1956; bud globular; flowers carmine-red, reverse gold, blooms in clusters of 3-15, dbl., 55 petals, 5–6 in., exhibition form, slight fragrance; foliage semi-glossy; vigorous, open growth; [Applause X Peace]; Von Abrams; Peterson & Dering

The Ednaston Rose, LCl, lp

The Edwardian Lady *see* **Edith Holden**

'THE FAIRY', Pol, lp, 1932; (Fairy, Feerie); flowers pink, small blooms in clusters, dbl., recurrent bloom; foliage glossy, small; compact, spreading growth; hardy.; [Paul Crampel X Lady Gay]; Bentall, Ann; J.A. Bentall

'THE FARMERS WIFE', F, lp, 1962; bud ovoid; dbl., 38 petals, 4.5 in., cupped, blooms in clusters, intense fragrance; foliage leathery, glossy; upright, moderate growth; [Queen Elizabeth X Spartan]; Boerner; J&P

The Farquhar Rose *see* **'FARQUHAR'**

The Faun®, Pol, mp; Poulsen, 1990

The Fisherman's Cot *see* 'HARWICKLOW'

The Flower Arranger *see* 'FRYJAM'

The Fordham Rose, HT, rb; Williams, J. Benjamin, 1996

'THE FRIAR', S, lp, 1969; flowers blush, edged white, medium, 10 petals, repeat bloom, intense fragrance; foliage dark; [Ivory Fashion X Seedling]; Austin, David

The Garden Editor® *see* 'MINASCO'

'THE GARLAND', Misc OGR, w, 1835; (Wood's Garland); flowers faint yellow, pink, and white, blooms in large, long cluster, semi-dbl., mid-season bloom, moderate, musk fragrance; moderate climbing (8 ft) growth; [R. moschata X R. multiflora]; Wells

'THE GENERAL', HT, mr, 1920; flowers blood-red, flushed lighter, large, dbl., globular, moderate, damask fragrance; foliage dark, bronze, leathery; vigorous, low growth; Pemberton

'THE GENTLEMAN', HT, yb, 1958; bud long; flowers creamy yellow flushed pink, semi-dbl., intense fragrance; [Unnamed seedling X Irish Fireflame]; Hay; Marsh's Nursery

The Gift, Pol, w; Demits, 1981

The Gooseberry Rose *see* **R. STELLATA MIRIFICA**

The Herbalist™ *see* 'AUSSEM'

The Holt *see* 'MEHSHERRY'

The Hon Mrs Cat, D, lp; Scarman, 1995

The Hunter *see* **'HUNTER'**

'THE INDIAN', HT, dr, 1939; flowers very dark red, large, semi-dbl.; vigorous growth; [Sensation X Seedling]; Clark, A.; Brundrett

The J.S.E. Rose *see* KORgreyel

The Jacobite Rose *see* **'ALBA MAXIMA'**

'THE JESTER', Gr, yb, 1960; (Leerder's Harlequin); flowers yellow changing to red, compact, borne in clusters, dbl., slight fragrance; foliage dark; [Masquerade X High Noon]; Leenders, J.; British Hort. Co.

'THE KNIGHT', S, dr, 1969; flowers crimson turning purple and mauve, medium, dbl., 80 petals, flat, repeat bloom, moderate fragrance; foliage dark; [Chianti X Seedling]; Austin, David

The Lady *see* 'FRYJINGO'

The Lady Scarman, HMsk, lp; Scarman, 1995

'THE LION', HMult, dp, 1901; flowers medium, single; Paul, G.; (Sangerhausen)

'THE MARQUESS OF BRISTOL' *see* WEBlady

'THE MASTER CUTLER', HT, mr, 1954; flowers cherry-red shaded rose, well formed, very large, dbl., intense fragrance; vigorous growth; [The Doctor sport]; Liberty Hill Nurs.

The Matthew™ *see* 'JAYMATT'™

The McCartney Rose™ *see* 'MEIZELI'

'THE METEOR', HT, dr, 1887; flowers medium, semi-dbl., moderate fragrance; Evans; (Sangerhausen)

'THE MILLER', S, mp, 1981; bud globular; flowers rosette-shaped blooms borne 1-4 per cluster, dbl., 40 petals, repeat bloom, moderate fragrance; hooked, red prickles; foliage medium green; upright, bushy growth; [Baroness Rothschild X Chaucer]; Austin, David, 1970

'THE MOUNTIE', F, mr, 1949; flowers cherry-red, medium, borne in huge trusses, semi-dbl., 12–15 petals, moderate fragrance; foliage leathery, glossy, light green; bushy growth; (28); [Springtime X World's Fair]; Eddie

The Mouse *see* **'GREY PEARL'**

The Nanango Rose *see* **Kathleen Kellehan**

The New Dawn *see* **'NEW DAWN'**

The Nun *see* 'AUSNUN'

The Observer *see* FRYtango

'THE OLD FASHIONED ROSE', HT, 1930; flowers deep crimson, dbl., moderate fragrance; very vigorous growth; Archer

The Old Velvet Rose *see* **'TUSCANY'**

'THE OPTIMIST', F, yb, 1955; (Sweet Repose); bud ovoid; flowers maize-yellow tinged carmine, becoming carmine, blooms in large clusters, dbl., 27 petals, 3 in., exhibition form, moderate fragrance; foliage dense, leathery, parsley-green; vigorous, upright, bushy growth; GM, NRS, 1955; [Golden Rapture X Unnamed Floribunda seedling]; deRuiter; C-P, 1956;, Gandy, 1955

The Painter *see* 'MACTEMAIK'

The Painter, F, 1995; McGredy, Sam IV; (Cavriglia)

The People *see* **'PEOPLE'**

The People's Princess *see* 'GEEPEOP'

The Pilgrim™ *see* 'AUSWALKER'

The Polar Star *see* **'POLSTJÄRNAN'**

The Prince™ *see* 'AUSVELVET'

The Prince Imperial *see* DELmagsa

'THE PRINCESS ELIZABETH', HT, ob, 1927; flowers orange-yellow, edged deep cerise, large, dbl., moderate fragrance; [The Queen Alexandra Rose sport]; Wheatcroft Bros.

'THE PRIORESS', S, lp, 1969; flowers blush to white, yellow stamens, medium, dbl., 24 petals, cupped, repeat bloom, moderate fragrance; [La Reine Victoria X Seedling]; Austin, David

The Queen *see* **'SOUR DE S. A. PRINCE'**

'THE QUEEN', HT, ob, 1954; bud pointed; flowers orange suffused salmon, medium, dbl., 26–30 petals, moderate fragrance; very free growth; Lowe

'THE QUEEN ALEXANDRA ROSE', HT, rb, 1918; (Queen Alexandra); flowers bright red, reverse shaded old-gold, base orange, dbl., 5 in., moderate fragrance; foliage glossy, dark; bushy growth; GM, NRS, 1917; McGredy

'THE QUEEN ALEXANDRA ROSE, CLIMBING', Cl HT, rb, 1929; Lindecke; Kordes

'THE QUEEN ALEXANDRA ROSE, CLIMBING', Cl HT, rb, 1931; Harkness

The Queen Elizabeth Rose *see* **'QUEEN ELIZABETH'**®

The Queen Elizabeth Rose, Climbing *see* **'QUEEN ELIZABETH, CLIMBING'**

The Queen Mother *see* **'AUGUST SEEBAUER'**

'THE QUEEN MOTHER', F, mp, 1959; flowers soft rosy pink, borne on large trusses, dbl., 45–50 petals, 3.5 in., moderate fragrance; foliage glossy, light green; vigorous growth; [Nymph sport]; Stedman

'THE QUEEN OF PERSIA', HT, ob, 1940; flowers light orange-yellow, large, dbl., slight fragrance; (Sangerhausen)

'THE RAJAH', HT, dr, 1937; flowers deep red, dbl.; bushy growth; [Red-Letter Day X ?]; Clark, A.; NRS Victoria

The Reeve® *see* 'AUSREEVE'

The Robe *see* TANseus

The Rose Tattoo *see* **'LA ROSE TATOUÉE'**

The Roseville College Rose, HT, yb; (Roseville College, The Roseville College, **The Roseville College Rose**)

The Royal Brompton Rose *see* 'MEIVILDO'

The Seckford Rose *see* 'KORPINROB'

'THE SENATOR', HT, mr, 1980; bud ovoid, long; flowers bright medium red, petals loosely rolled outward, blooms borne usually singly, dbl., 46 petals, slight, musk fragrance; long, narrow-based, brown prickles, curved down; foliage large, leathery, wrinkled, dark; tall, upright growth; [Unnamed seedling X Suspense]; Weeks, O.L.

The Service Rose *see* **'WOMEN IN MILITARY'**

The Squire® *see* 'AUSIRE'

The St Mark's Rose *see* **'ROSE D'AMOUR'**

'THE STORK', HT, mp, 1952; bud long, pointed; flowers carmine-rose, dbl., 28 petals, 5–5.5 in., exhibition form, slight fragrance; foliage leathery; bushy growth; [Seedling X The Doctor]; Whisler; Germain's

'THE SUN', F, op, 1972; flowers salmon-orange, semi-dbl., 15 petals, 3.5 in., slight fragrance; GM, Madrid, 1973; [(Little Darling X Goldilocks) X Irish Mist]; McGredy, Sam IV; McGredy

'THE SURGEON', HT, mp, 1956; flowers darker rose-pink, more vigorous than the doctor, dbl.; [The Doctor seedling]; Verschuren; Blaby Rose Gardens

The Temptations® *see* 'WEKAQ'

The Texan *see* **'TEXAN'**

The Valois Rose *see* KORdadel

'THE VENERABLE BEDE', F, yb, 1973; flowers yellow to cherry, full, dbl., 20–30 petals, 3–3.5 in., slight fragrance; foliage small, glossy; moderate, free growth; [John Church X Bobby Shafto]; Wood; Homedale Nursery

'THE WALLFLOWER', LCl, mr, 1901; (Wallflower); flowers bright red, large, borne in large clusters, semi-dbl., late; vigorous growth; [Crimson Rambler X Beauté Inconstante]; Paul

The Wasa Star *see* **'POLSTJÄRNAN'**

The Wedding Rose *see* 'GEAAWARD'

The Wife of Bath *see* 'AUSWIFE'

The Wild One *see* 'TOMWILD'

The Work Continues *see* 'JACSHAQ'

The World *see* 'DIEKOR'

The Wyevale rose *see* 'MEIBALBIKA'®

'THE YANK', HT, mr, 1942; flowers rose-red, very large, dbl., 55–60 petals, intense fragrance; foliage leathery, dark; long, strong stems; very vigorous, upright, much branched growth; [Chieftain X Lucile Hill]; Hill, Joseph H., Co.

'THE YEOMAN', S, op, 1969; flowers salmon-pink, medium, dbl., 50 petals, flat, recurrent bloom, intense fragrance; [Ivory Fashion X (Constance Spry X Monique)]; Austin, David

'THEA HARRISON', HT, lp; Harrison, A., 1927; (Weatherly, L.)

'THEA RUSSEL', F, mr, 1954; flowers bright red, moderate fragrance; vigorous growth; [Ambassadeur Nemry X Cinnabar]; Leenders, M.

Theano, HWich, lp; Geschwind, R., 1895

'THEDA MANSHOLT', Gr, my, 1966; bud ovoid; flowers medium, dbl., slight fragrance; foliage dark; upright growth; [Peace X Garden Party]; Buisman, G. A. H.

'THELMA', HWich, op, 1927; flowers coral-pink, suffused carmine-red, blooms in clusters of 3-10, semi-dbl., 3 in., slight fragrance; few prickles; vigorous, climbing growth; [R. wichurana X Paul's Scarlet Climber]; Easlea

'THELMA AMLING', Pol, w, 1942; bud globular; flowers small, semi-dbl., slight fragrance; foliage small, leathery; short stems; vigorous, bushy growth; [Mrs R.M. Finch sport]; Amling Bros.; J.H. Hill Co.

'THELMA BADER', HT, ob, 1956; flowers orange-scarlet to salmon-pink, semi-dbl., 20 petals, 3 in., slight fragrance; foliage dark, glossy; moderately vigorous growth; [Unknown seedling X (Independence X Fashion)]; Latham

'THELMA WALTON', HT, pb, 1970; flowers deep pink, lightly edged silver, very large, very dbl., exhibition form, slight fragrance; foliage glossy, dark; vigorous, upright growth; [Red Devil sport]; Walton; Cooke

Themis, F, 1984; Gaujard; (Cavriglia)

'THEODORA', HT, dr, 1984; flowers large, semi-dbl., 16 petals; [Apricot Nectar X Dame de Coeur]; Staikov, Prof. Dr. V.; Kalaydjiev and Chorbadjiiski

'THEODORA MILCH', HWich, mp, 1906; flowers small, very dbl., moderate fragrance; Weigand, C.; (Sangerhausen)

'THÉODORE LIBERTON', HP, mr, 1887; flowers large, dbl., moderate fragrance; Soupert & Notting; (Sangerhausen)

Theodore Roosevelt *see* 'BURRED'

Theone, C, mp; Noisette

'THERESA MORLEY', HT, dp, 1928; flowers brilliant carmine-cerise, dbl., moderate fragrance; [Mme Segond Weber X Lady Battersea]; H&S

Theresa Scarman, HGal, lp; Scarman, 1996

'THERESE BAUER', HSet, mp, 1963; flowers open, large, borne in clusters, semi-dbl., profuse, repeated bloom, slight fragrance; very vigorous, upright growth; [(Hansa X R. setigera) X R. setigera]; Ludwig; Kern Rose Nursery

'THÉRÈSE BONNAVIAT', HT, lp, 1934; bud long, pointed; flowers clear pink, center coppery pink, very large, moderate fragrance; foliage purplish green; very vigorous growth; [Mrs Arthur Robert Waddell X Unnamed seedling]; Chambard, C.

'THÉRÈSE BUGNET', HRg, mp, 1950; bud conical but square-tipped; flowers rose aging pale pink, white center line on top of petals, dbl., 35 petals, 4 in., borne in small clusters, repeat bloom, moderate, spicy fragrance; foliage quilted, grey-green; vigorous, 5 - 6 ft. growth; [((R. acicularis X R. rugosa kamtchatica) X (R. amblyotis X R.rugosa plena)) X Betty Bland]; Bugnet; P.H. Wright

Therese de Lisieux® *see* ORAblan

'THÉRÈSE SCHOPPER', HT, mr, 1933; bud long, pointed, blood-red; flowers nasturtium-red, reverse yellow, open, large, semi-dbl., exhibition form, moderate fragrance; foliage leathery, dark; vigorous growth; [(Charles P. Kilham X Mev. G.A. van Rossem) X Lady Forteviot]; Kordes

'THERESE ZEIMET-LAMBERT', HT, dp, 1922; bud long, pointed; flowers deep rose, yellow ground, base orange, large, dbl., exhibition form, intense fragrance; foliage glossy, bronze; vigorous growth; [Richmond X Mrs Aaron Ward]; Lambert, P.

'THERESETTE', S, mp, 1986; flowers deep soft pink edged white, carnation-like, medium, borne in, dbl., 26–40 frilly petals, moderate fragrance; foliage small, medium green, semi-glossy, disease resistant; bushy, hardy, compact, dwarf growth; [Arctic Glow X Therese Bugnet]; James, John

'THERMIDOR', S, ly, 1909; flowers medium, dbl., moderate fragrance; Corboeuf; (Sangerhausen)

Think Pink *see* 'TINTHINK'

'THIRZA', HT, or, 1932; flowers orange-scarlet, moderate fragrance; Bentall

'THISBE', HMsk, ly, 1918; flowers chamois-yellow, rosette-shaped blooms in large clusters, dbl., recurrent bloom, intense fragrance; vigorous, bushy growth; [Daphne sport(?)]; Pemberton

'THOMAS A. EDISON', HT, pb, 1931; bud long, pointed; flowers two-toned pink, very large, dbl., cupped, moderate fragrance; very vigorous growth; Bernaix, P.; C-P

'THOMAS ANDREW ELLIOTT', HT, yb, 1995; flowers golden yellow with yellow and pink reverse, very dbl., 41 petals, 4 in., borne mostly singly, slight fragrance; some prickles; foliage medium, medium green, semi-glossy; medium (50-60 cms), upright growth; [Grandpa Dickson X (City of Gloucester X Seedling X Polestar)]; Thomas, D.

Thomas Garces, HT, 1986; Dot, Simon; (Cavriglia)

'THOMAS MILLS', HP, dp, 1873; flowers rosy crimson, large, dbl., cupped, non-recurrent, moderate fragrance; vigorous growth; Verdier, E.

Thomasville Lamarque, N, lp; (found rose)

Thomasville Old Gold, T, lp; (found rose)

'THOOM', F, op, 1994; (**Glamorgan**); flowers coral-salmon, blooms borne in large clusters, semi-dbl., 14 petals, 1.5–2.75 in., slight fragrance; some prickles; foliage medium, red when young turning to medium green, glos; low to medium (2 ft 6 in), upright, compact growth; [Old Master X Red Splendor]; Thompson, M.L.; Haynes Roses, 1995

'THOR', LCl, dr, 1940; bud ovoid; flowers crimson, dbl., 58 petals, 4–5 in., slight, damask fragrance; foliage large, leathery, dark; vigorous, climbing (8-10 ft) growth; [(Alpha X R. xanthina) X Pres. Coolidge]; Horvath; Wayside Gardens Co.

'THOR SUPREME', LCl, dr, 1951; flowers dark crimson edged deeper, 4 in., intense, damask fragrance; foliage large, leathery, dark; vigorous, climbing growth; Horvath; Wyant

Thora Hird *see* 'TONYBRAC'

'THORA MCCREA', HT, mr, 1986; dbl., 20–25 petals, exhibition form, slight fragrance; red-brown prickles; foliage dark green; well-shaped, upright, medium growth; [Pink Parfait X Crimson Glory]; Summerell, B.L.

Thoresbyana *see* **'BENNETT'S SEEDLING'**

Thornfree Wonder *see* 'NOBAL'

Thornhem®, HT, w

'THORNLESS BEAUTY', HT, dr, 1938; bud long, pointed, dark crimson-red; flowers light crimson-red, dbl., 50 petals, 4 in., moderate fragrance; thornless; foliage leathery; long, strong stems; vigorous growth; [Better Times sport]; Grillo

Thornless Beauty, F, lp; Patil, B.K., 1990

'THORNLESS BLUSH', HT, lp, 1954; bud globular; flowers blush-pink, dbl., 70 petals, 4.5 in., exhibition form, moderate fragrance; thornless stems; foliage leathery; very vigorous, upright growth; [Rosalind Russell sport]; Grillo

'THORNLESS FRINGEDALE', HT, lp, 1955; dbl., 25 petals, 4 in., moderate fragrance; thornless stems; [Thornless Beauty sport]; Grillo

'THORNLESS GLORY', LCl, lp, 1935; flowers clear pink, semi-dbl., 20 petals, 3 in., early summer bloom, slight fragrance; thornless; long stems; vigorous, climbing growth; Izzo

'THORNLESS MIRAGE', HT, mr, 1955; flowers red suffused blush-pink, dbl., 25 petals, 5 in., moderate fragrance; thornless stems; [Jewel sport]; Grillo

'THORNLESS PREMIER', HT, mp, 1955; bud globular, pointed; flowers rich pink, dbl., 75 petals, 4.5 in., moderate fragrance; thornless stems; [Victory Stripe sport X Jewel]; Grillo

'THORNLESS VICTORY STRIPE', HT, rb, 1955; bud long, pointed; flowers dark red and blush-pink, striped, dbl., 50 petals, 5 in.; thornless stems; [Victory Stripe sport X Jewel]; Grillo

Thoughtful *see* 'RESTO'

Thousand Beauties *see* **'TAUSENDSCHÖN'**

Threepenny Bit Rose *see* **R. FARRERI PERSETOSA**

'THRILLER', HT, pb, 1986; flowers white center, pink edges, reverse slight pink edging, pink h, dbl., 28–30 petals, exhibition form, intense, fruity fragrance; medium, long, light green prickles, pointed downwards; foliage medium, dark green, glossy; upright, medium, strong growth; [Lady X X Flaming Beauty]; Bridges, Dennis A., 1987

'THUMBELINA', Min, rb, 1954; flowers cherry-red, white eye, small, semi-dbl.; foliage dark, glossy; dwarf (6-8 in), bushy growth; [Eblouissant X Zee]; Moore, Ralph S.; Sequoia Nursery

Thumbelina *see* 'SUNTHUMB'

'THUNDER CLOUD', Min, or, 1979; bud ovoid; dbl., 70 petals, 1–1.5 in.; foliage glossy, leathery; dwarf, bushy, upright growth; [Little Chief X Fire Princess]; Moore, Ralph S.; Sequoia Nursery

'THUNDERBIRD', F, mr, 1958; bud urn-shaped; flowers rose-red, blooms in clusters of 3-8, dbl., 48 petals, 2.5 in., exhibition form; vigorous, spreading growth; [Skylark sport]; deVor, Paul F.; Amling Bros.

Thunderbolt *see* 'BRIBOLT'

Thungabhadra, HT, ob; Kasturi, 1985

'THUSNELDA', HRg, mp, 1886; flowers rose, semi-dbl., non-recurrent; vigorous growth; [R. rugosa alba X Gloire de Dijon]; Müller, Dr. F.

'THYRA HAMMERICH', HP, lp, 1868; flowers medium, very dbl.; Vilin; (Sangerhausen)

Thyrion, HT, 1969; McGredy, Sam IV; (Cavriglia)

Tiamo *see* SPEhewe

'TIARA', F, w, 1960; bud ovoid; dbl., 48 petals, 3 in., cupped, blooms in clusters, moderate fragrance; foliage leathery; vigorous, bushy growth; [Chic X Demure seedling]; Boerner; J&P

Tibeert, S, 1991; Delforge; (Cavriglia)

Ticino, F, mr, 1979; flowers vermilion red, large, dbl.; Huber; (Sangerhausen)

Tickle Me Pink *see* 'PIXINK'

'TICKLED PINK', HT, mp, 1963; bud long, pointed; flowers medium, dbl., exhibition form, moderate, spicy fragrance; upright, bushy growth; [Queen Elizabeth X Fashion]; Lammerts, Dr. Walter; Amling-DeVor Nursery;, Germain's

Tickled Pink *see* 'SUNTICK'

Tickles™ *see* 'DYKTICK'

'TICK-TOCK', Min, ob, 1973; flowers salmon-pink, shaded orange, open, semi-dbl., 10 petals, 2 in., slight fragrance; free growth; [New Penny X Elizabeth of Glamis]; McGredy, Sam IV; McGredy

Tidbit Rose *see* Conditorum

Tidewater *see* 'BRITIDE'

Tiecelijn, HT, 1990; Delforge; (Cavriglia)

'TIERGARTENDIREKTOR TIMM', F, dp, 1944; bud long, pointed; flowers carmine-pink, open, very large, borne in clusters, semi-dbl., slight fragrance; foliage dark, glossy; very vigorous, upright growth; [Sweetness X Hamburg]; Kordes

'TIFFANY', HT, pb, 1954; bud long, pointed; flowers pink-yellow blend, dbl., 28 petals, 4–5 in., exhibition form, intense fragrance; foliage dark; vigorous, upright growth; AARS, 1955 David Fuerstenberg Prize, ARS, 1957 GM, Portland, 1954 James Alexander Gamble Fragrance Medal, ARS, 1962; [Charlotte Armstrong X Girona]; Lindquist; Howard Rose Co.

Tiffany Lite *see* 'GELLITE'

'TIFFANY LYNN', MinFl, pb, 1985; flowers light to medium pink at edges, blending to white in center, exhibition form, 21 petals, exhibition form, slight; prickles slanted downward; foliage medium, medium green with red edges, semi-glossy; upright, bushy growth; [(Tiki X Seedling) X Party Girl]; Jolly, Nelson F.; Rosehill Farm

'TIFFANY, CLIMBING', Cl HT, pb, 1958; (Grimpant Tiffany); Lindquist; Howard Rose Co.

'TIFFIE', Min, lp, 1979; bud long, pointed; flowers soft pink, sometimes apricot, blooms borne singly, 15–20 petals, exhibition form, moderate, lilac fragrance; large, red prickles; foliage medium green; upright growth; [Little Darling X Over the Rainbow]; Bennett, Cecilia 'Dee'; Tiny Petals Nursery

'TIFTON', HT, yb, 1983; flowers brilliant yellow, petal tips red, red spreading with age, large, dbl., exhibition form, moderate fragrance; foliage large, medium, semi-glossy; bushy growth; [Fire Magic X Oregold]; Perry, Astor; Perry Roses

Tiger *see* MACgremli

'TIGER BELLE', F, ob, 1962; bud ovoid; flowers light signal-red, dbl., 25–35 petals, 2–2.5 in., cupped; moderate growth; [Orange Sweetheart X Lovelight]; Jelly; E.G. Hill Co.

Tiger Butter *see* 'PIXITER'

Tiger Cub *see* POUlcub

Tiger Paws™ *see* 'LAVTIPAWS'

Tiger Stripes *see* 'CLETIG'

Tiger Tail *see* 'JACTIGER'

'TIGLIA', S, yb, 1962; flowers creamy yellow tinged pink, dbl., recurrent bloom, moderate fragrance; [Ma Perkins X High Noon]; Leenders, J.

Tigris *see* 'HARPRIER'

'TIKI', F, pb, 1964; flowers light pink and white blend, well-formed, flora-tea, dbl., 30 petals, 3.5 in.; foliage dark; vigorous growth; [Mme Leon Cuny X Spartan]; McGredy, Sam IV; McGredy

'TILL UHLENSPIEGEL', HEg, rb, 1950; bud pointed; flowers bright red, white eye, blooms in clusters, single, 5 petals, non-recurrent; foliage large, glossy, dark reddish-green; tall (to 10 ft), arching growth; [Holstein X Magnifica]; Kordes, W. Söhne

'TILLICUM', HT, 1924; flowers deep rose-pink shaded orange, dbl., slight fragrance; [Général Jacqueminot X Old Gold]; Wilber

Tilly Aston, HT, lp; [Queen Elizabeth sport]; Nieuwesteeg, J., 1996; (Weatherly, L.)

Tilt Symphonie® *see* MEIvraivou

'TIM PAGE', HT, 1920; flowers rich daffodil yellow, moderate fragrance; Page; Easlea

Timeless™ *see* 'JACECOND'

Timeless Delight *see* 'RENTDEL'

'TIMES SQUARE', HT, ob, 1943; bud peach; flowers orange, center golden yellow, large, intense fragrance; foliage leathery, glossy; vigorous growth; [Mrs Sam McGredy X Pres. Herbert Hoover]; Lammerts, Dr. Walter; Armstrong Nursery

'TIMMIE ARKLES', F, lp, 1954; dbl., 3 in., flat, blooms in clusters of 30-40, moderate, fruity fragrance; vigorous growth; [Mrs R.M. Finch X (Improved Lafayette seedling X Rochester seedling)]; Boerner; J&P

'TIMM'S JUBILAÜMSROSE', F, or, 1975; bud globular; dbl., 27 petals, 2 in., globular, slight fragrance; foliage soft; vigorous, upright, bushy growth; [Marlena X Europeana]; Kordes; Timm

'TIMOTHY BERLEN', Min, ob, 1986; flowers orange, yellow center, reverse yellow fading dark pink, medium, semi-dbl., 20 petals, exhibition form, slight fragrance; round, green yellow-orange fruit; bayonet-shaped, light pink prickles; foliage small, medium green, semi-glossy; upright, spreading, low growth; [Anita Charles X Poker Chip]; Jolly, Marie; Rosehill Farm, 1988

'TIMOTHY EATON', HT, op, 1968; flowers salmon-pink, well-formed, slight fragrance; [Radar X Mischief]; McGredy, Sam IV

'TINA TURNER', HT, ob, 1992; dbl., 26–40 petals, 1.5–2.75 in., borne mostly singly, moderate fragrance; some prickles; foliage large, dark green, glossy; bushy (60 cms) growth; [Silver Jubilee X Doris Tysterman]; Thompson, Robert; Battersby Roses, 1990

'TINABLER', Min, mr, 1991; (**Elizabeth Abler**); flowers medium red, opening to paler red at center, micro-mini, semi-dbl., 6–14 petals, 1.5 in., exhibition form, borne mostly singly, moderate fragrance; no prickles; foliage small, medium green, semi-glossy; low (20-30 cms), bushy, compact growth; [(Christian Dior X Brian Lee) X Unknown]; Bennett, Cecilia 'Dee'; Tiny Petals Nursery, 1992

'TINAMIGO', Min, mr, 1983; (**Little Amigo**); flowers small, dbl., 35 petals, exhibition form, no fragrance; foliage small, medium green, semi-glossy; upright growth; [Futura X Orange Honey]; Bennett, Cecilia 'Dee'; Tiny Petals Nursery

'TINBAB', Min, lp, 1999; (**Big and Beautiful**); flowers deep apricot, dbl., 26–40 petals, 1–1.5 in., borne mostly singly, slight fragrance; prickles moderate; foliage medium, medium green, semi-glossy; upright, bushy, tall (5-6 ft) growth; [Futura X Jean Kenneally]; Bennett, Cecilia 'Dee'; Tiny Petals, 2000

'TINBOPPER', Min, rb, 1989; (**Teeny Bopper**); bud ovoid; flowers white with red blush on outer petals, aging red with white, semi-dbl., cupped, slight, damask fragrance; globular, yellow green orange fruit; no prickles; foliage small, medium green, semi-glossy; micro-mini, bushy growth; PP007449; [Little Squirt X Unnamed seedling]; Bennett, Cecilia 'Dee'; Tiny Petals Nursery, 1988

'TINBUFORD', Min, ly, 1993; (**Linda Buford**); flowers soft light yellow, blooms borne mostly singly, dbl., 26–40 petals, 1.5 in., slight fragrance; some prickles; foliage small, medium green, semi-glossy; medium, bushy growth; [Irish Gold X Jean Kenneally]; Bennett, Cecilia 'Dee'; Tiny Petals Nursery, 1994

'TINBUTT', Min, ob, 1985; (**Bread 'n' Butter**); flowers golden orange blended yellow, medium, dbl., 28 petals, exhibition form, moderate fragrance; foliage medium, medium green, semi-glossy; upright, bushy growth; PP006135; [Arizona X Orange Honey]; Bennett, Cecilia 'Dee'; Tiny Petals Nursery

'TINCAROL', Min, dp, 1989; (**Little Carol**); bud ovoid; flowers deep magenta-pink, reverse slightly lighter, medium, borne singly, dbl., 38 petals, exhibition form, slight, spicy fragrance; globular, green brown fruit; prickles hooked slightly downward, red, few; foliage medium, dark green, semi-glossy, disease-resistant; upright, tall growth; [Sonia X Jean Kenneally]; Bennett, Cecilia 'Dee'; Tiny Petals Nursery

'TINCHAN', Min, my, 1987; (**Dr K. C. Chan**); flowers medium yellow, fading pale yellow, urn-shaped, medium, dbl., 25–30 petals, borne usually singly, slight fragrance; globular, medium, brown fruit; slender, straight, reddish prickles; foliage medium, medium green, semi-glossy; mini-flora; upright, bushy, medium growth; [Irish Gold X Rise 'n' Shine]; Bennett, Cecilia 'Dee'; Tiny Petals Nursery, 1986

'TINCHEER', Min, ob, 1986; (**Cheer Up**); flowers deep orange, urn-shaped, dbl., 28 petals, exhibition form, borne usually singly and in small clusters, slight fragrance; globular, 1/2 in., green and brown fruit; small, red prickles; foliage medium, dark, semi-glossy; medium, upright, bushy growth; PP006456; [Futura X Bread 'n' Butter]; Bennett, Cecilia 'Dee'; Tiny Petals Nursery

'TINCHRIS', Min, mr, 1988; (**Christopher**); flowers medium red, medium, dbl., 30–35 petals, exhibition form, borne usually singly or in sprays of 3-5; globular, green brown fruit; prickles hooked slightly downward, yellow-red; foliage medium, medium green, semi-glossy; upright, medium growth; PP007191; [Futura X Big John]; Bennett, Cecilia 'Dee'; Tiny Petals Nursery

'TINCHULA', Min, dr, 1992; (**Chula Vista**); dbl., 26–40 petals, 1.5 in., exhibition form, borne mostly singly, slight fragrance; some prickles; foliage small, medium green, semi-glossy, disease resistant; long stems; medium (60-80 cms), bushy growth; [Christian Dior X Brian Lee]; Bennett, Cecilia 'Dee'; Tiny Petals Nursery, 1993

'TINCUDDLE', Min, yb, 1991; (**Cuddle Up**); bud ovoid; flowers creamy yellow with coral pink blush, medium, dbl., 25–30 petals, exhibition form, moderate, fruity fragrance; foliage medium, medium green, semi-glossy; bushy, medium growth; [Lagerfeld X My Delight]; Bennett, Cecilia 'Dee', 1987; Tiny Petals Nursery, 1990

'TINCUPIE', Min, lp, 1983; (**Cupie Doll**); flowers light pink with fine coral edging of petals, heavy substance, dbl., 25 petals, exhibition form, slight fragrance; foliage small, medium green, semi-glossy; upright, bushy growth; [Seedling X Coral Treasure]; Bennett, Cecilia 'Dee'; Tiny Petals Nursery

'TINDAD', Min, mr, 1996; (**Love 'Ya Dad**); flowers does not fade, blooms borne mostly single, long-stemmed, dbl., 26–40 petals, 1.5 in., exhibition form, intense fragrance; moderate prickles; foliage medium, dark green, semi-glossy, disease-resistant; upright, bushy, tall (70-80 cms) vigorous growth; [Carrousel X Big John]; Bennett, Cecilia 'Dee'; Tiny Petals Nursery, 1997

'TINDEB', Min, op, 1985; (**Debra Gaye**); flowers orange-pink, soft yellow reverse, medium, dbl., 38 petals, exhibition form, slight fragrance; foliage medium, medium green, semi-glossy; bushy, upright growth; PP006136; [Futura X Fairest of Fair]; Bennett, Cecilia 'Dee'; Tiny Petals Nursery

'TINDEL', F, dp, 1993; (**Del Mar Fair**); flowers deep pink, blooms borne in large clusters, dbl., 26–40 petals, 3–3.5 in., moderate fragrance; some prickles; foliage medium, medium green, semi-glossy; medium, upright, bushy growth; [Gene Boerner X Pucker Up]; Bennett, Cecilia 'Dee'; Tiny Petals Nursery, 1994

'TINDICK', Min, r, 1997; (**High Cloud**); flowers soft tan with yellow base, lighter reverse, dbl., 26–40 petals, 1.5 in., borne singly or in small clusters, moderate fragrance; foliage medium, medium green, glossy; bushy, tall (60 - 70 cm) growth; [Lagerfeld X Ernie]; Bennett, Cecilia 'Dee'

'TINDILLY', Min, MinFl, 1985; (**Dilly Dilly**); flowers lavender, mini-flora, medium, dbl., 35 petals, exhibition form, moderate fragrance; foliage medium, medium green, semi-glossy; upright growth; PP006141; [Chrysler Imperial X Plum Duffy]; Bennett, Cecilia 'Dee'; Tiny Petals Nursery

'TINDINK', Min, pb, 1983; (**Fair Dinkum**); flowers soft pink, petal margins darker, small, dbl., 25 petals, exhibition form, slight fragrance; foliage small, medium green, semi-glossy; upright, bushy growth; PP005564; [Unnamed seedling X Coral Treasure]; Bennett, Cecilia 'Dee'; Tiny Petals Nursery

'TINDONN', Min, dr, 1991; (**Brian Donn**); bud ovoid; flowers darkest ruby red to maroon, dark ruby red reverse, ages only slightly lighter, very dbl., 35–45 petals, exhibition form, moderate, damask fragrance; foliage medium, medium green, semi-glossy; bushy, medium growth; [Intrique X Big John]; Bennett, Cecilia 'Dee', 1985; Tiny Petals Nursery, 1990

'TINDREAM', Min, yb, 1987; (**Dreamy**); flowers cream, edges blushed pink, reverse cream with more intense blush, dbl., 20–25 petals, slight fragrance; globular, medium, brown fruit; slender, small, reddish prickles, slanted downward; foliage medium, medium green, semi-glossy; upright, medium growth; [Irish Gold X Party Girl]; Bennett, Cecilia 'Dee'; Tiny Petals Nursery

'TINEALLY', Min, ab, 1984; (**Jean Kenneally**®); flowers pale to medium apricot, small, dbl., 22 petals, exhibition form, borne mostly singly, slight fragrance; foliage medium, medium green, semi-glossy; long stems; upright, bushy, vigorous growth; PP005637; AOE, 1986; [Futura X Party Girl]; Bennett, Cecilia 'Dee'; Tiny Petals Nursery

'TINEKE', HT, w, 1989; bud pointed; flowers creamy white, large, borne singly, dbl., 53 petals, cupped, no fragrance; prickles yellow-green with reddish tip; foliage large, dark green, semi-glossy; upright, tall growth; [Unnamed seedling X Unnamed seedling]; Select Roses, B.V.; DeVor Nurseries, Inc., 1990

Tineke van Heule® *see* 'LENTINI'

'TINERNIE', Min, m, 1989; (**Ernie**); bud ovoid; flowers light mauve, urn-shaped, medium, dbl., 48 petals, borne occasionally singly and in clusters of 6-12, moderate, fruity fragrance; globular, yellow-brown fruit; prickles hooked slightly downward, pale yellow-brown, few; foliage medium, medium green, semi-glossy; upright, bushy, tall growth; [Blue Nile X Blue Mist]; Bennett, Cecilia 'Dee'; Tiny Petals Nursery

TINgail, F, yb, 1986; (**Gail**); flowers golden yellow with orange blush on petal tips, orange spreading with age, dbl., 38 petals, exhibition form, moderate fragrance; globular, 5/8 in., light green fruit; small reddish prickles; foliage large, medium green, semi-glossy; medium, bushy, spreading growth; [Arizona X Orange Honey]; Bennett, Cecilia 'Dee'; Tiny Petals Nursery

'TINGIRL', Min, op, 1983; (**Bloomer Girl**); flowers soft vermilion to medium pink, small, dbl., 35 petals, slight fragrance; foliage medium, medium green, semi-glossy; upright, bushy growth; PP005523; [Futura X Pink Petticoat]; Bennett, Cecilia 'Dee'; Tiny Petals Nursery

'TINGRACE', Min, w, 1991; (**Grace Seward**); bud ovoid; flowers medium, single, 5 petals, moderate, damask fragrance; foliage medium, medium green, semi-glossy; bushy, tall growth; [Watercolor X Unnamed seedling]; Bennett, Cecilia 'Dee', 1987; Tiny Petals Nursery, 1990

'TINHAT', Min, my, 1988; (G'Day, Gidday, **Madhatter**); flowers medium yellow, aging paler, medium, borne usually singly or, dbl., 25–30 petals, exhibition form, moderate, damask fragrance; globular, green brown fruit; prickles hooked slightly downward, reddish; foliage medium, medium green, semi-glossy; bushy, medium growth; [Autumn X Avandel]; Bennett, Cecilia 'Dee'; Tiny Petals Nursery

'TINHELEN', Min, mr, 1991; (**Helen**); dbl., 26–40 petals, 1.5 in., exhibition form, borne mostly singly, moderate, fruity fragrance; few prickles; foliage small, medium green, semi-glossy, disease-resistant; long cutting stems; tall (60-80 cms), upright, bushy growth; [Carrousel X Starina]; Bennett, Cecilia 'Dee'; Tiny Petals Nursery, 1992

'TINHERB', Min, m, 1987; (**Herbie**); flowers rich mauve, outer petals deep mauve at margins, urn-shaped, dbl., 25–30 petals, slight fragrance; globular, medium, brown fruit; slender, straight, average, reddish prickles; foliage medium, medium green, semi-glossy; upright, bushy, medium growth; PP006787; [Deep Purple X Dilly Dilly]; Bennett, Cecilia 'Dee'; Tiny Petals Nursery

'TINHUZZY', Min, rb, 1991; (**Little Huzzy**); flowers white with striping of pink to medium red over most upper surface, dbl., 26–40 petals, 1.5–2.75 in., exhibition form, moderate fragrance; some prickles; foliage small, medium green, semi-glossy, disease-resistant; long cutting stems; tall (60-80 cms), upright growth; [Futura X Pucker Up]; Bennett, Cecilia 'Dee'; Tiny Petals Nursery, 1992

'TINJEAN', Min, pb, 1987; (**Jean Rose**); flowers peach-pink with yellow base, reverse soft yellow, fading lighter, dbl., 20–25 petals, slight fragrance; globular, medium, brown fruit; few, slender, small, reddish prickles; foliage medium, medium green, semi-glossy; upright, medium growth; [Electron X Fairest of Fair]; Bennett, Cecilia 'Dee'; Tiny Petals Nursery

'TINJOSH', Min, lp, 1999; (**Josh**); dbl., 26–40 petals, 1–1.5 in., borne singly and in small clusters, slight fragrance; prickles moderate; foliage medium, medium green, semi-glossy; upright, bushy, medium (2-3 ft) growth; [Futura X Rise 'n' Shine]; Bennett, Cecilia 'Dee'; Bridges Roses, 2000;, Tiny Petals, 2000

'TINJUDY', MinFl, pb, 1995; (**Something for Judy**); flowers cream with deep pink blush, blooms borne mostly single, dbl., 15–25 petals, 1.5–2.75 in., moderate fragrance; foliage medium, medium green, semi-glossy; tall (2-3 ft), bushy growth; [Angel Face X Big John]; Bennett, Cecilia 'Dee'; Tiny Petals Nursery, 1995

'TINKAY', Min, pb, 1995; (**Kay Denise**); flowers soft creamy pink, pink blush margins, full, very dbl., 26–40 petals, exhibition form, borne mostly singly, moderate fragrance; some prickles; foliage small, medium green, semi-glossy; upright, bushy, tall growth; [Tiki X My Delight]; Bennett, Cecilia 'Dee'; Tiny Petals Nursery, 1995

'TINKER BELL', Min, mp, 1954; bud ovoid; flowers bright rose-pink, small, dbl., 60 petals, 1.5 in., cupped; foliage small, leathery; dwarf (8 in), bushy growth; [Ellen Poulsen X Tom Thumb]; deVink; C-P

Tinkerbell *see* 'SUNTINK'

'TINKIN', Min, ob, 1983; (**Punkin**); flowers orange with yellow eye, ages red, blooms borne singly, single, 5 petals, no fragrance; foliage small, medium green, semi-glossy; upright growth; [Orange Honey X Orange Honey]; Bennett, Cecilia 'Dee'; Tiny Petals Nursery

'TINLADY', Min, m, 1988; (**Always A Lady**); flowers pale mauve to lavender, medium, dbl., 25–30 petals, exhibition form, borne usually singly or in sprays of 3-5, moderate, damask fragrance; globular, green brown fruit; prickles hooked slightly downward, pale yellow; foliage medium, medium green, semi-glossy; bushy, medium growth; PP007187; [Deep Purple X Dilly Dilly]; Bennett, Cecilia 'Dee'; Tiny Petals Nursery, 1987

'TINLEE', Min, m, 1986; (**Brian Lee**); flowers dark red, aging mauve, urn-shaped, mini-flora, blooms borne, dbl., 28 petals, exhibition form, moderate fragrance; 5/8 in.; globular, brown-green fruit; small reddish prickles; foliage medium, dark, semi-glossy; medium, upright, bushy growth; PP006789; [Carrousel X Plum Duffy]; Bennett, Cecilia 'Dee'; Tiny Petals Nursery

'TINLESLIE', Min, mp, 1997; (**Leslie**); flowers medium pink, 3 cm, semi-dbl., 15–25 petals, 1.5 in, borne in small clusters; foliage medium, medium green, semi-glossy; bushy, 40-50 cms tall growth; [Deep Purple X Jean Kenneally]; Bennett, Cecilia 'Dee'; Tiny Petals Nursery, 1997

'TINLIPS', Min, op, 1988; (**Hot Lips**); flowers deep coral to orange, aging paler, medium, borne usually singly, dbl., 25–30 petals, exhibition form, slight, fruity fragrance; globular, green yellow-brown fruit; prickles hooked slightly downward, reddish; foliage medium, medium green, semi-glossy; upright, bushy, tall growth; [Futura X Why Not]; Bennett, Cecilia 'Dee'; Tiny Petals Nursery

'TINLOIS', Min, mp, 1985; (**Lois**); flowers medium lilac pink, medium, dbl., exhibition form, moderate fragrance; foliage medium, medium green, semi-glossy; upright, bushy growth; PP006140; [Deep Purple X Plum Duffy]; Bennett, Cecilia 'Dee'; Tiny Petals Nursery

'TINLOVE', Min, pb, 1989; (**Mother's Love**); bud ovoid; flowers pastel pink, blending to soft yellow at base, medium, dbl., 23 petals, exhibition form, slight, fruity fragrance; globular, green yellow-brown fruit; prickles straight, tapering, reddish; foliage medium, medium green, semi-glossy; upright, bushy, medium growth; PP007484; [Futura X Party Girl]; Bennett, Cecilia 'Dee'; Tiny Petals Nursery, 1988

'TINLUIS', Min, ly, 1989; (**Luis Desamero**); bud ovoid; flowers pastel yellow, medium, dbl., 28 petals, exhibition form, borne usually singly and in sprays of 3-5, slight, fruity fragrance; globular, green yellow-orange fruit; prickles straight & tapering, pale yellow, reddish base; foliage medium, medium green, semi-glossy; upright, bushy, tall growth; PP007496; [Tiki X Baby Katie]; Bennett, Cecilia 'Dee'; Tiny Petals Nursery, 1988

'TINMAD', MinFl, mp, 1985; (**Madeline Spezzano**); flowers mini-flora, medium, dbl., 45 petals, exhibition form, slight fragrance; foliage large, medium green, semi-glossy; upright, bushy growth; PP006132; [Sonia X Beauty Secret]; Bennett, Cecilia 'Dee'; Tiny Petals Nursery

'TINMARY', Min, yb, 1999; (**Ms Mary**); flowers pale yellow/apricot, reverse pale yellow, dbl., 26–40 petals, 1 in., borne mostly singly, moderate fragrance; prickles moderate; foliage medium, medium green, semi-glossy; compact, bushy, low (18-24 in) growth; [Deep Purple X Fairest of Fair]; Bennett, Cecilia 'Dee'; Tiny Petals, 1999

'TINMERE', Min, pb, 1983; (**Meredith Anne**); flowers salmon-pink, soft pink reverse, small blooms in clusters, dbl., 23 petals, exhibition form; brown to pale yellow prickles; vigorous growth; [Sonia X Tea Party]; Bennett, Cecilia 'Dee'; Tiny Petals Nursery

'TINMYDE', Min, lp, 1983; (**My Delight**); flowers small, dbl., 30 petals, exhibition form, slight fragrance; foliage medium, medium green, semi-glossy; upright, spreading growth; [Futura X Avandel]; Bennett, Cecilia 'Dee'; Tiny Petals Nursery, 1984

'TINNAUGHTY', Min, ab, 1991; (**Naughty But Nice**); bud ovoid; flowers soft apricot, medium, dbl., 20–25 petals, exhibition form, slight, damask fragrance; foliage medium, medium green, semi-glossy; bushy, tall growth; [Futura X Why Not]; Bennett, Cecilia 'Dee', 1983; Tiny Petals Nursery, 1990

Tino Rossi® *see* MEIcelna

'TINOLD', Min, w, 1992; (**Old Fashioned Girl**); flowers soft lavender to white, opening very full, similar to ogr's, dbl., 26–40 petals, 1.5–2.75 in., exhibition form, intense fragrance; some prickles; foliage small, medium green, semi-glossy; medium (60-80 cms), bushy growth; [Blue Ribbon X Unnamed Miniature seedling]; Bennett, Cecilia 'Dee'; Tiny Petals Nursery, 1993

'TINPANKY', Min, ob, 1991; (**Hanky Panky**); flowers range from orange to golden peach as they open, dbl., 26–40 petals, 1.5–2.75 in., borne mostly singly, moderate, fruity fragrance; some prickles; foliage small, medium green, semi-glossy, disease resistant; medium (40-60 cms), bushy, spreading growth; [Deep Purple X Party Girl]; Bennett, Cecilia 'Dee'; Tiny Petals Nursery, 1992

'TINPAT', Min, mp, 1985; (**Patty Sue**); flowers medium, dbl., 35 petals, exhibition form, slight fragrance; foliage medium, medium green, matt; vigorous, upright growth; [Little Darling X Little Chief]; Bennett, Cecilia 'Dee'; Tiny Petals Nursery

'TINPEACES', Min, yb, 1985; (**Little Peaces**); flowers yellow, edged pink, medium, dbl., 33 petals, exhibition form, slight fragrance; foliage medium, medium green, semi-glossy; compact, bushy growth; [Electron X Fairest of Fair]; Bennett, Cecilia 'Dee'; Tiny Petals Nursery

'TINPIN', Min, m, 1987; (**Hat Pin**); flowers pale lavender, urn-shaped, small, semi-dbl., 12–15 petals, borne usually singly, moderate, fruity fragrance; globular, small, brown fruit; straight, extremely small, pale yellow prickles; foliage small, medium green, semi-glossy; micro-mini upright, bushy, low growth; PP006790; [Angel Face X Angelglo]; Bennett, Cecilia 'Dee'; Tiny Petals Nursery, 1986

'TINPLEASURE', Min, lp, 1999; (**Ultimate Pleasure**); flowers light pink, reverse medium pink, dbl., 26–40 petals, .75–1 in., exhibition form, borne mostly singly or in small clusters, moderate fragrance; prickles moderate; foliage medium, medium green, semi-glossy; extremely vigorous, bushy, upright, tall (24-36 in) growth; [Gene Boerner X Jean Kenneally]; Bennett, Cecilia 'Dee'; Tiny Petals, 1999

'TINPOPS', Min, dp, 1983; (**Pops**); flowers deep pink, small, dbl., 20 petals, exhibition form, slight fragrance; foliage medium, medium green, semi-glossy; bushy growth; [Sheri Anne X Little Girl]; Bennett, Cecilia 'Dee'; Tiny Petals Nursery

'TINPORCE', Min, lp, 1983; (**Pink Porcelain**); flowers small blooms borne singly and in clusters, dbl., 23 petals, exhibition form, slight fragrance; foliage small, medium green, semi-glossy; upright growth; [Futura X Avandel]; Bennett, Cecilia 'Dee'; Tiny Petals Nursery

'TINPUCK', Min, or, 1984; (**Pucker Up**); bud small; flowers bright orange-red, dbl., 23 petals, exhibition form, slight fragrance; foliage medium, medium green, semi-glossy; upright, bushy growth; PP005407; [Futura X Avandel]; Bennett, Cecilia 'Dee'; Tiny Petals Nursery, 1983

'TINQUEEN', Min, pb, 1996; (**Bingo Queen**); flowers light pink with medium pink inner petals, reverse same, one-, dbl., 26–40 petals, 1.5 in., exhibition form, slight fragrance; moderate prickles; foliage medium, medium green, semi-glossy; bushy, medium (50-60 cms) growth; [Brandy X Jean Kenneally]; Bennett, Cecilia 'Dee'; Tiny Petals Nursery, 1997

'TINQUIET', MinFl, m, 1995; (**Quiet Time**); flowers lavender-pink with inner petals of soft tan, blooms borne mo, dbl., 25–40 petals, 1.5 in., slight fragrance; foliage medium, medium green, semi-glossy; bushy growth; [Lagerfeld X Ernie]; Bennett, Cecilia 'Dee'; Tiny Petals Nursery, 1995

'TINRESIST', Min, w, 1989; (**Irresistible**™); bud ovoid; flowers white with pale pink center, medium, greenish in shade, dbl., 43 petals, exhibition form, borne singly and in small clusters, moderate, spicy fragrance; globular, green yellow-brown fruit; prickles straight, yellow with red; foliage medium, medium green, semi-glossy; upright, tall growth; PP7971; [Tiki X Brian Lee]; Bennett, Cecilia 'Dee'; Tiny Petals Nursery, 1990

'TINREVENGE', Min, ob, 1996; (**Sweet Revenge**); flowers soft orange with blush of deep orange, blooms borne mostly singly, dbl., 26–40 petals, 1.5 in., slight fragrance; foliage medium, dark green, semi-glossy; tall (2-3 ft), bushy growth; [Tony Jacklin X Pucker Up]; Bennett, Cecilia 'Dee'; Tiny Petals Nursery, 1995

'TINRITA', MinFl, ly, 1996; (**Rita Applegate**); flowers light yellow to softest gold with heart of deeper yellow/gold, very dbl., 26–40 petals, 2 in., exhibition form, borne mostly singly, moderate fragrance; moderate prickles; foliage medium, medium green, semi-glossy; long-stemmed; upright, bushy, tall (80-90 cms), huge growth; [Pink Porcelain X Unknown]; Bennett, Cecilia 'Dee'; Tiny Petals Nursery, 1997

'TINRUTH', Min, w, 1988; (**Ruthie**); flowers cream-white with pink blush on outer petals, aging paler blush, dbl., 25–30 petals, exhibition form, moderate, damask fragrance; globular, green brown fruit; prickles hooked slightly downward, yellow to reddish; foliage medium, medium green, semi-glossy, aging darker; upright, tall growth; PP007189; [Sonia X Little Melody]; Bennett, Cecilia 'Dee'; Tiny Petals Nursery

'TINSASSY', Min, rb, 1986; (**Fat 'n' Sassy**); flowers white with a reddish border, reverse white, aging reddish blush, dbl., 28 petals, cupped, borne singly, slight fragrance; globular, 5/8 in., green-brown fruit; small, reddish prickles; foliage medium, medium green, semi-glossy; medium, upright, bushy growth; PP006455; [Carrousel X Sheri Anne]; Bennett, Cecilia 'Dee'; Tiny Petals Nursery

Tinseltown, HT, pb; Edwards, Eddie, 1996

'TINSHINE', Min, my, 1986; (**My Sunshine**); flowers medium yellow, aging soft orange, bright yellow stamens, single, moderate fragrance; globular, 1/2 in., green fruit; small, reddish-brown prickles; foliage medium, medium green, semi-glossy; medium, upright, bushy growth; PP006453; [Sunsprite X Fool's Gold]; Bennett, Cecilia 'Dee'; Tiny Petals Nursery

'TINSHORT', Min, dp, 1984; (**Short 'n' Sweet**); flowers deep pink, very small, dbl., 35 petals, no fragrance; foliage medium, medium green, semi-glossy; bushy, spreading growth; [Sheri Anne X Unnamed seedling]; Bennett, Cecilia 'Dee'; Tiny Petals Nursery

'TINSIT', Min, pb, 1986; (**Sitting Pretty**); bud ovoid; flowers apricot pink, yellow base, dbl., 38 petals, exhibition form, borne usually singly and in small clusters, intense, damask fragrance; globular, 1/2 in., yellowish-green fruit; small, red prickles; foliage medium, medium green, semi-glossy; medium, upright, bushy growth; PP006454; [Sonoma X Mabel Dot]; Bennett, Cecilia 'Dee'; Tiny Petals Nursery

'TINSNOOK', Min, ob, 1984; (**Snookie**); flowers deep orange, blushing red with age, very small, dbl., 33 petals, slight fragrance; foliage medium green, semi-glossy; bushy, tiny growth; [Torchy X Orange Honey]; Bennett, Cecilia 'Dee'; Tiny Petals Nursery

'TINSPICE', Min, lp, 1985; (**Sugar 'n' Spice**); flowers light peach-pink, medium, dbl., 28 petals, exhibition form, slight fragrance; foliage medium, medium green, semi-glossy; upright, bushy growth; PP006139; [Futura X Avandel]; Bennett, Cecilia 'Dee'; Tiny Petals Nursery

'TINSPRING', Min, pb, 1985; (**Spring Bouquet**); flowers crimson pink, blended with yellow, medium, exhibition form, moderate fragrance; foliage medium, medium green, semi-glossy; upright, bushy growth; [Portrait X Party Girl]; Bennett, Cecilia 'Dee'; Tiny Petals Nursery

'TINSQUIRT', Min, dy, 1984; (**Little Squirt**); flowers micro-mini, very small, semi-dbl., 14 petals, exhibition form, slight fragrance; foliage small, medium green, semi-glossy; upright, compact growth; Bennett, Cecilia 'Dee'; Tiny Petals Nursery, 1983

'TINSTALEY', Min, mp, 1989; (**Ruth Staley**); bud ovoid; flowers medium shell pink, reverse lighter, medium, borne singly and, dbl., 25 petals, exhibition form, moderate, fruity fragrance; globular, green brown fruit; prickles slender, straight, red; foliage medium, dark green, semi-glossy, disease-resistant; upright, tall growth; [Electron X Peachy Keen]; Bennett, Cecilia 'Dee'; Tiny Petals Nursery, 1991

'TINSTAR', Min, ob, 1986; (**Starstruck**); flowers deep golden orange with petals edged red, reverse medium red, dbl., 32 petals, slight fragrance; globular, 3/8 in., brownish fruit; small, thin, reddish prickles; foliage medium, medium green, semi-glossy; low, spreading growth; [October X Orange Honey]; Bennett, Cecilia 'Dee'; Tiny Petals Nursery

'TINSWEET', Min, lp, 1987; (**Sweet Pickins**); flowers pale pink, urn-shaped, medium, dbl., 35–40 petals, borne usually singly, moderate, fruity fragrance; globular, medium, brown fruit; slender, straight, small, reddish prickles; foliage medium, medium green, semi-glossy; micro-mini upright, bushy, medium growth; PP006788; [Futura X Party Girl]; Bennett, Cecilia 'Dee'; Tiny Petals Nursery

Tintagel, LCl, lp; [R. brunonii seedling? sport]; McLeod, J., 1995; (Weatherly, L.)

'TINTAI', MinFl, yb, 1989; (**Yantai**); bud ovoid; flowers pastel yellow, deeper yellow in center and highlights of pas, dbl., 28 petals, exhibition form, moderate, fruity fragrance; prickles hooked downward, yellow to red; foliage medium, medium green, semi-glossy; bushy, spreading, tall, vigorous growth; [Portrait X Party Girl]; Bennett, Cecilia 'Dee'; Tiny Petals Nursery, 1988

'TINTAIL', Min, w, 1983; (**Cottontail**); flowers small, dbl., 33 petals, blooms in large clusters, slight fragrance; foliage medium, medium green, semi-glossy; upright, bushy growth; PP005436; [Pink Petticoat X Pink Petticoat]; Strawn, Leslie E.; Pixie Treasures Min. Roses

'TINTHINK', Min, mp, 1983; (**Think Pink**); flowers golden stamens, small, dbl., 28 petals, intense, apple fragrance; foliage small, medium green, semi-glossy; bushy growth; [Electron X Little Chief]; Bennett, Cecilia 'Dee'

'TINTIN', F, or, 1964; flowers cinnabar-red, large, borne in large clusters, semi-dbl., cupped; moderate growth; Mondial Roses

Tintinara™ *see* 'DICUPTIGHT'

'TINTWEE', Min, op, 1987; (**Tweedle Dee**); flowers coral pink, reverse lighter, urn-shaped, small, borne usuall, dbl., 35–40 petals, slight fragrance; globular, small, brown fruit; no prickles; foliage small, medium, semi-glossy; micro-mini; upright, bushy, low growth; [Deep Purple X Cupcake]; Bennett, Cecilia 'Dee'; Tiny Petals Nursery, 1986

'TINVIOLET', MinEl, m, 1993; (**Violet Mist**, Lavender Mist); flowers pale soft lavender, blooms borne mostly single, dbl., 26–40 petals, 1.5–2.75 in., slight fragrance; some prickles; foliage small, medium green, semi-glossy; medium, bushy growth; [Lagerfeld X Ernie]; Bennett, Cecilia 'Dee'; Tiny Petals Nursery, 1994

Tinwell Moss, M, lp; (found rose)

'TINWILD', Min, ob, 1992; (**Wild Thing**); flowers bright orange-yellow, blooms borne mostly single, dbl., 15–25 petals, 1.5–2.75 in., moderate, fruity fragrance; some prickles; foliage small, medium green, semi-glossy, disease resistant; medium (60-80 cms), bushy growth; [Gingersnap X Baby Katie]; Bennett, Cecilia 'Dee'; Tiny Petals Nursery, 1993

'TINX', Min, pb, 1993; (**X-Rated**); flowers creamy white, edges blush a soft coral to pink, dbl., 26–40 petals, 1.5 in., borne mostly singly, intense fragrance; some prickles; foliage small, medium green, semi-glossy; medium, bushy growth; [Tiki X Baby Katie]; Bennett, Cecilia 'Dee'; Tiny Petals Nursery, 1994

Tiny Bubbles™ *see* 'JACBUB'

Tiny Dancer® *see* 'WILDANCE'

'TINY FLAME', Min, or, 1969; flowers coral-orange-red, micro-mini, very small, dbl.; foliage very small; dwarf (6 in), bushy growth; [(R. wichurana X Floradora) X New Penny]; Moore, Ralph S.; Sequoia Nursery

'TINY JACK', Min, mr, 1962; bud pointed; dbl., 28 petals, 1–1.5 in., cupped, slight fragrance; foliage leathery, dark; vigorous, bushy (12-14 in)growth; [(R. wichurana X Floradora) X (Oakington Ruby X Floradora)]; Moore, Ralph S.; Blue Ribbon Plant Co.

'TINY JILL', Min, mp, 1962; bud pointed; flowers small, dbl., 45 petals, 1.5 in., exhibition form, moderate fragrance; foliage glossy; vigorous, bushy (12-14 in) growth; [(R. wichurana X Floradora) X Little Buckaroo]; Moore, Ralph S.; Blue Ribbon Plant Co.

'TINY LOVE', Min, dr, 1980; bud ovoid; semi-dbl., 11 petals, cupped, borne 1-3 per cluster, intense fragrance; straight prickles; foliage dark; open, bushy growth; [Unnamed seedling X Unnamed seedling]; Lyon

Tiny Petals *see* 'TINYPETALS'

Tiny Stars *see* 'TRASTAR'

'TINY TEARS', Min, mp, 1979; bud ovoid; flowers pink, micro-mini, very small blooms in clusters of 10-25, single, 5 petals; foliage glossy, dark; trailing growth; [Pink Ribbon X Pink Ribbon]; Bennett, Cecilia 'Dee'; Tiny Petals Nursery

'TINY TIM', F, my, 1943; bud long, pointed; flowers clear yellow, open, small, dbl., 23 petals, exhibition form; foliage glossy; moderately vigorous, compact to open growth; [Golden Glow X Shades of Autumn]; Brownell, H.C.

'TINY TOT', Min, mp, 1955; flowers deep cerise-pink; very dwarf growth; Robinson, T.

Tiny Tot *see* BENtintot

'TINY VISIONS', Min, rb, 1981; bud long; flowers crimson with white center, slightly blooms borne singly, single, 5 petals, cupped, moderate fragrance; very tiny, hooked, brown prickles; foliage dark, apple-scented; very tiny, bushy growth; [Unnamed seedling X Unnamed seedling]; Lyon

'TINY WARRIOR', Min, rb, 1975; bud ovoid; dbl., 34 petals, 1 in., flat, slight fragrance; foliage small, glossy, very dark, embossed; upright, bushy growth; [Starburst X Little Chief]; Williams, Ernest D.; Min-Roses

'TINYPETALS', Min, rb, 1992; (**Tiny Petals**); flowers orange-red with cream reverse, blooms borne mostly single, dbl., 15–25 petals, 1.5 in., exhibition form; some prickles; foliage small, dark green, semi-glossy, disease resistant; long stems; medium (60-80 cms), bushy growth; [San Antonio X Jean Kenneally]; Bennett, Cecilia 'Dee'; Tiny Petals Nursery, 1993

'TINZEL', Min, mp, 1988; (**Zelda Lloyd**); flowers medium pink, reverse deep pink, medium, single, 5 petals, cupped, borne usually singly, slight, fruity fragrance; globular, green yellow-brown fruit; prickles hooked slightly downward, reddish; foliage medium, medium green, semi-glossy; bushy, medium growth; PP007190; [Deep Purple X Blue Mist]; Bennett, Cecilia 'Dee'; Tiny Petals Nursery

'TIP TOES', HT, op, 1948; bud pointed; flowers salmon-pink, shading to yellow at base, large, semi-dbl., exhibition form, moderate fragrance; foliage glossy; vigorous, upright growth; [(Général Jacqueminot X Dr. W. Van Fleet) X Anne Venderbilt]; Brownell, H.C.

Tip Top *see* TANopel

Tipo Ideale *see* **'MUTABILIS'**

Tipper® *see* 'JOLTIP'

'TIPPERARY', HT, my, 1917; bud long, pointed; flowers golden yellow, semi-dbl., moderate fragrance; [Mrs Aaron Ward seedling]; McGredy

'TIPSY', HT, dp, 1969; flowers soft rose, reverse geranium-red, pointed, large, dbl.; foliage dark, dull; free growth; [Basildon Belle X Seedling]; Trew, C.; Basildon Rose Gardens

Tipsy Imperial Concubine, T, pb; reintroduced, 1989

'TIP-TOP'®, Pol, w, 1909; (Baby Doll); flowers white, tipped tyrian rose, aging white and pale yellow, reve, dbl.; bushy, dwarf growth; [Trier X R. foetida bicolor seedling]; Lambert, P.

Tipu's Flame, S, rb; Viraraghavan, M.S., 1990

'TIRA-MI-SU', HT, lp, 1999; flowers pale copper and orange blend, dbl., 28 petals, 5 in., slight fragrance; 4.5 ft growth; [(Julia X seedling) X seedling]; Teranishi, K.; Itami Rose Nursery, 1999

'TITANIA', Ch, rb, 1915; flowers deep salmon-red, shaded clear yellow, small, dbl.; dwarf, bushy growth; Paul, W.

'TITANIA', Pol, lp, 1938; flowers salmon-flesh to rosy white, large, very dbl., intense fragrance; foliage leathery, dark; vigorous, bushy growth; [Mev. Nathalie Nypels X Unnamed seedling]; Leenders, M.

'TITI PARISIEN', F, dr, 1959; flowers crimson, center lighter, stamens ivory, single, 5 petals, 3 in.; bushy, low growth; [Francais seedlings X (Orange Triumph seedlings X Floradora)]; Delbard-Chabert

'TITIAN', F, dp, 1950; flowers deep pink, large blooms in clusters, dbl.; vigorous, tall growth; Riethmuller

'TITIAN, CLIMBING', Cl F, dp, 1964; Kordes

'TIVOLI', F, mp, 1955; flowers warm rose-pink, center yellow, well formed, borne in clusters, dbl., 24 petals, 3 in., moderate fragrance; foliage dark, glossy; very vigorous growth; [Poulsen's Supreme X (Souv. de Claudius Denoyel X Hvissinge-Rose)]; Poulsen, S.; McGredy

Tivoli Gardens *see* 'POULDUCE'

Tiz *see* **'TZIGANE'**

'TIZIANA', F, op, 1969; bud pointed; flowers copper-pink, large, semi-dbl., slight fragrance; foliage light green; vigorous, compact growth; [Papillon Rose X Gay Paris]; Cazzaniga, F. G.

Tobago® *see* 'DELTOGO'

Tobo® *see* 'KINBO'

Toboné *see* **'PRIMA DONNA'**®

'TOBY JO', Min, yb, 1977; flowers yellow and orange, lighter reverse, pink tips, dbl., 15–25 petals, 1 in., borne singly, no fragrance; few prickles; foliage medium, medium green, semi glossy; upright, medium (16-20 in) growth; [Old Glory sport]; Garrett, Troy O.

Toby Tristam, HMult, w

Tocade *see* 'WERINA'

'TOCCATA', F, or, 1959; flowers well-formed blooms in large clusters; low growth; [Independence X Cinnabar]; Lens

'TOCCATA', HT, mp, 1963; flowers light rose edged darker, reverse silvery, becoming redder, dbl., 27 petals, 5 in., moderate fragrance; vigorous growth; [Karl Herbst X Seedling]; Sanday, John

Today® *see* 'MACCOMPU'

'TODDLER', F, pb, 1977; flowers pink, reverse deeper, semi-dbl., 19 petals, 2 in.; foliage small, dark; very low, bushy growth; [Seedling X Seedling]; LeGrice

'TODOROKI', HT, dr, 1977; bud pointed; dbl., 33 petals, 6.5 in., exhibition form, slight fragrance; foliage dark; vigorous growth; [(Pharaon X Kagayaki) X Yu-Ai]; Keisei Rose Nurseries, Inc.

Toffee *see* 'SEATOFF'

Togo, HT, 1959; Cazzaniga, F. G.; (Cavriglia)

Toison d'Or *see* **'GOLDEN FLEECE'**

'TOISON D'OR', HT, ab, 1921; flowers apricot-yellow, shaded orange-red, dbl.; GM, Bagatelle, 1922; Pernet-Ducher

Tojo®, F, mr; McGredy, Sam IV, 1978

Tokay *see* 'ZIPTOK'

'TOKEN', HT, yb, 1933; bud ovoid; flowers glowing orange, open, large, dbl., moderate fragrance; foliage glossy; strong stems; vigorous growth; [Mme Butterfly X Premier Supreme]; Montgomery Co.

'TOKEN GLORY', HT, or, 1957; bud pointed; flowers orange, dbl., 40 petals, 4 in., moderate fragrance; foliage dark, leathery; long stems; very vigorous, upright, bushy growth; [Token Supreme sport]; Grillo

'TOKEN SUPREME', HT, ob, 1940; flowers deep orange, dbl., 35 petals, 5 in., moderate fragrance; [Token sport]; Grillo

'TOKYO', Cl F, my, 1972; flowers, single, 5 petals, 3 in., profuse may-june bloom; foliage light; vigorous growth; [Salute X Canary Bird]; Oliver, W.G.

Toledo Gold, HT, my

Toliman®, Min, m

'TOLSTOI', HSet, mp, 1938; flowers large, dbl., intense fragrance; Böhm, J.; (Sangerhausen)

'TOM BARR', HT, op, 1932; bud long, pointed; flowers salmon and scarlet, suffused yellow and orange, large, dbl., exhibition form, moderate fragrance; vigorous growth; McGredy

'TOM BRENEMAN', HT, dp, 1950; bud ovoid; flowers rose-pink, dbl., 30–40 petals, 4–4.5 in., globular, intense fragrance; foliage leathery, dark; very vigorous, upright growth; [Mauna Loa X R.M.S. Queen Mary]; Howard, F.H.; H&S

'TOM BRENEMAN, CLIMBING', Cl HT, dp, 1954; H&S

'TOM BROWN', F, r, 1964; flowers orangy-brown, reverse brownish-red, well-shaped, dbl., 32 petals, 3 in., intense fragrance; foliage leathery, dark; vigorous, bushy growth; [Unnamed seedling X Amberlight]; LeGrice

Tom Foster, HT, my, 1995; flowers medium yellow, full, large, very dbl., 26–40 petals, borne mostly singly, moderate fragrance; some prickles; foliage medium, dark green, semi-glossy; bushy, medium growth; [Gertrude Shilling X Helmut Schmidt]; Poole, Lionel; Battersby Roses, 1996

'TOM MANEY', LCl, mp, 1953; bud ovoid; flowers rose-pink, borne in clusters of 3-4, dbl., 35–40 petals, 4–5 in., cupped, non-recurrent, moderate fragrance; foliage leathery, dark; vigorous, climbing (15-20 ft.) growth; quite hardy.; [R. maximowicziana pilosa X K. of K.]; Maney; Iowa State College

'TOM PILLIBY', F, or, 1963; semi-dbl., in large clusters, slight fragrance; vigorous growth; Combe

'TOM THUMB', Min, rb, 1936; (Peon); flowers deep crimson, center white, semi-dbl., 1 in.; foliage leathery, light green; very dwarf growth; [Rouletii X Gloria Mundi]; deVink; C-P

'TOM TOM', F, dp, 1957; bud ovoid; flowers deep pink, blooms in clusters, dbl., 25 petals, 3–3.5 in., exhibition form, slight, spicy fragrance; foliage dark; vigorous, upright, bushy growth; [Improved Lafayette X Floradora]; Lindquist; Howard Rose Co.

'TOM TOM, CLIMBING', Cl F, dp, 1962; Lindquist; Howard Rose Co.

'TOM WOOD', HP, mr, 1896; flowers cherry-red, large, dbl., cupped, moderate fragrance; Dickson, A.

'TOMÁS BATÂ', HP, dr, 1932; flowers shaded crimson, passing to maroon, large, dbl., moderate fragrance; foliage leathery, glossy, dark; very vigorous growth; [Fisher Holmes X Prince Camille de Rohan]; Böhm, J.

Tomato Rose *see* **R. RUGOSA**

'TOMBOLA', F, dp, 1967; flowers deep pink, 4 in., intense fragrance; foliage dark, glossy; vigorous, upright, bushy growth; [Amor X (Ena Harkness X Peace)]; deRuiter

'TOMBOT', F, pb, 1998; (**Botanica**); flowers pink with touch of lilac, medium, very dbl., 26–41 petals, 2 in., borne in large clusters, intense fragrance; prickles moderate; foliage medium, medium green, dull; bushy, medium (3 - 3.5 ft) growth; [Avandel X Madam President]; Thomson, George L.

Tomboy *see* 'ZIPBOY'

'TOMDEL', Min, lp, 1999; (**Adelaide Tonight**); flowers pink, reverse lighter, dbl., 15–25 petals, .5–.75 in., borne in large clusters, slight fragrance; prickles moderate; foliage medium, medium green, semi-glossy; spreading, medium (18-24 in) growth; [Avandel X Little Mike]; Thomson, George L.

'TOMFED', F, lp, 1999; (**Canberra Rose**); flowers medium pink, reverse darker, dbl., 16–40 petals, 3 in., borne in large clusters, intense fragrance; prickles moderate; foliage medium, dark green, glossy; upright, medium (3-4 ft) growth; [Watercolour X Madam President]; Thomson, George L.; Ross Roses, 2000

'TOMFLO', F, ab, 1998; (**Howard Florey**); flowers apricot, lighter reverse, double, dbl., 15–25 petals, 2 in., borne in small clusters, moderate fragrance; prickles moderate; foliage medium, medium green, semi-glossy; bushy, medium (3.5 - 4.5 ft) growth; [Seduction X Apricot Nectar]; Thomson, George L.

'TOMKINS RED', F, dr, 1943; bud long, pointed; flowers deep velvety crimson turning maroon, open, borne in clusters, semi-dbl., 18 petals, moderate fragrance; foliage dark, glossy; long stems; vigorous, compact, upright growth; [(Dr. W. Van Fleet X Général Jacqueminot) X Nigrette]; Brownell, H.C.

'TOMMOP', HT, lp, 1999; (**Mary of Penola**); flowers dark pink, reverse darker, very dbl., 41 petals, 6 in., borne mostly singly, moderate fragrance; prickles moderate; foliage medium, medium green, glossy; upright, medium (3-3.5 ft) growth; [Maria Callas X Elina]; Thomson, George L.

'TOMMOUNT', HT, ab, 1996; (**Pride of Mountbarker**); flowers peach cream tinted rose edge, apricot center, lighter reverse, dbl., 26–40 petals, moderate fragrance; moderate prickles; foliage small, medium green, semi-glossy; upright, medium (5 ft x 2 ft) growth; [Gold Medal X Sylvia]; Thomson, George L.

'TOMMY BRIGHT', F, mr, 1961; bud ovoid; flowers scarlet-red, dbl., 35–40 petals, 3.5 in., cupped, moderate fragrance; foliage leathery; strong stems; very vigorous, upright growth; [Chatter seedling X Garnette Supreme]; Boerner; J&P

'TOMMY THOMPSON', HT, or, 1974; bud ovoid; dbl., 50 petals, 5 in., exhibition form, moderate, spicy fragrance; foliage glossy, very dark; moderate growth; [Queen o' the Lakes X Tropicana]; Golik; Dynarose

'TOMMY TUCKER', Min, lp, 1955; flowers silvery pink, borne in clusters, dbl., 1 in.; [Rouletti X Tom Thumb]; Robinson, T.

'TOMONE', S, pb, 1996; (**Mrs Mary Thomson**); flowers lilac pink, cream center, prominent gold stamens, reverse sl, semi-dbl., 8–14 petals, 5.5–6.5 in., intense fragrance; few prickles; foliage medium, medium green, dull; spreading, medium (5 ft x 4 ft) growth; [Dapple Dawn X Ophelia]; Thomson, George L.

'TOMRING', S, ab, 1977; (**Onkaparinga**); flowers apricot pink, reverse lighter, ages pink, dbl., 41 petals, 4–5 iin., borne in small clusters, intense, myrrh fragrance; foliage medium, medium green, semi-glossy; spreading growth; [Cymbaline X Troilus]; Thomson, George L.

Tom's Pink, Pol, mp; Horsfield, 1986

Tom's Smelly Pink, Misc OGR, lp; (found rose)

'TOMSUE', S, ob, 1997; (**Susan Irvine**); flowers rose pink, cream center, prominent gold stamens, semi-dbl., 8–14 petals, 2.5–3.5 in., borne in large clusters, moderate fragrance; foliage medium, medium green, semi glossy; bushy, medium (5 ft) growth; [Mrs. Mary Thomson X R. gigantea seedling]; Thomson, George L.

'TOMTAL', Min, pb, 1997; (**Laura Chantal**); flowers varying shades of pink, dark reverse, dbl., 26–40 petals, borne in small clusters, intense fragrance; foliage medium, medium green, semi glossy; upright, medium (30 in) growth; [Avandel X Madam President]; Thomson, George L.

'TOMTWO', F, dp, 1996; (**Linda Thomson**); flowers dark dusky pink, ruffled edge, reverse darker, blooms borne, dbl., 15–25 petals, 3.5 in., moderate fragrance; foliage medium, dark green, glossy; bushy, medium growth; [Fidelio X Showbiz]; Thomson, George L.

'TOMWILD', S, w, 1998; (**The Wild One**); flowers cream with touch of rose, reverse lighter, stamens gold, single, 4–7 petals, 4.5–5.5 in., borne in small clusters, moderate fragrance; prickles numerous; foliage large, light green, semi-glossy; upright, medium (4 -5 ft) growth; [Wild Flower X Ophelia]; Thomson, George L.

'TONEHIME', HT, w, 1983; flowers white, creamy center, large blooms borne usually singly, dbl., 30 petals, exhibition form, slight fragrance; few prickles; foliage small, dark, glossy; vigorous, slender, tall growth; [Mizuho X Sodori-Hime]; Kikuchi, Rikichi

'TONGA', HT, ob, 1955; flowers deep golden orange, outer petals veined bronze and scarlet, dbl., intense fragrance; foliage dark, leathery; vigorous, upright growth; Lowe

'TONI CORSARI', F, dr, 1961; flowers velvety deep red, medium, borne in large clusters, semi-dbl., moderate fragrance; bushy, low growth; [Red Favorite X Seedling]; Delforge

'TONI LANDER', F, ob, 1958; flowers salmon-orange, borne in large clusters, dbl., 22 petals, 3 in., slight fragrance; foliage dark; bushy growth; GM, Madrid, 1960; [Independence X Circus]; Poulsen, S.

Toni Thompson's Musk, HMsk, lp; [R. brunonii X Unknown]; Thompson, T., 1984; (Weatherly, L.)

Tonia *see* 'FRANCIBEL'

Tonic Meillandina® *see* MEItrino

'TONIGHT', HT, mr, 1972; bud ovoid; flowers bright red, large, dbl., exhibition form, slight fragrance; foliage leathery; vigorous growth; [Seedling X Forever Yours]; Warriner, William A.

Tonimbuk, HT, mp; Dawson, before 1990

'TONNERRE'®, F, dr, 1953; (Not); flowers deep velvety red, blooms in large clusters, dbl., 24 petals, 3 in.; foliage dark; vigorous, bushy growth; [Holstein X Francais]; Mallerin, C.; EFR, 1952

'TONNER'S FANCY', LCl, w, 1928; flowers white, tinted pink, large, dbl., globular, moderate fragrance; vigorous, climbing growth; [R. gigantea seedling X Seedling]; Clark, A.; Gill & Searle

Tonsina, HT, ob

'TONY JACKLIN', F, op, 1972; flowers orange-salmon, flora-tea, dbl., 30 petals, 4 in., exhibition form, slight fragrance; GM, Madrid, 1972 GM, Portland, 1986; McGredy, Sam IV; McGredy

'TONY PEACE', HT, ab, 1964; flowers buff-yellow to apricot; [Peace sport]; Brundrett

'TONY SPALDING', HT, dr, 1933; flowers brilliant crimson, semi-dbl., exhibition form, intense fragrance; foliage glossy; very vigorous growth; McGredy

'TONYBRAC', F, w, 1988; (**Thora Hird**); flowers white deepen to cream in center with pink veins, loose, medium, very dbl., 40 petals, moderate, sweet fragrance; round, amber fruit; prickles straight, brown; foliage medium, dark green, semi-glossy; bushy, medium growth; [Chinatown X Picasso]; Bracegirdle, A.J.; Rosemary Roses, 1987

Too Hot To Handle *see* 'MACLOUPRI'

Toorenburg *see* KORkojotie

Tooth of Time *see* 'RENTIME'

Tootsie *see* 'GRETOOTS'

'TOP CHOICE', Min, op, 1985; flowers medium salmon-pink, medium, dbl., 36 petals, slight fragrance; foliage small, medium green, semi-glossy; dense, spreading, bushy growth; [Gingersnap X Baby Katie]; Hardgrove, Donald L.; Rose World Originals

Top Gear *see* 'MACMANLY'

Top Gun™ *see* 'KINGUN'

'TOP HAT', F, ab, 1968; bud ovoid; flowers apricot-blush, medium, dbl., exhibition form, moderate fragrance; foliage small, dark, glossy; compact, low growth; [Ma Perkins X Garden Party]; Fankhauser; A. Ross & Son

Top Marks *see* FRYministar

'TOP SECRET', Min, mr, 1971; [Beauty Secret sport]; Moore, Ralph S.; Sequoia Nursery

Top Secret *see* MEIbolnay

Top Star®, HT, ob; Noack, Werner, 1982

Topaz *see* 'JACANT'

'TOPAZ', Pol, yb, 1937; flowers lemon-yellow, petals edged cream, reverse lemon and cream, very dbl., exhibition form, slight fragrance; foliage small, leathery; dwarf, spreading growth; [Joanna Tantau X (Prof. Gnau X Julien Potin)]; Tantau; C-P

Topaz Jewel™ *see* 'MORYELRUG'

Topaze, F, mp

'TOPAZE ORIENTALE', HT, yb, 1965; flowers maize-yellow to light pink, dbl., 5–6 in., exhibition form, moderate fragrance; tall growth; [Sultane X Queen Elizabeth]; Delbard-Chabert; Cuthbert

'TOPEKA', F, mr, 1978; bud long, pointed; dbl., 21 petals, 3 in., flat; foliage glossy; vigorous, upright growth; [Vera Dalton X Unnamed seedling]; Wisbech Plant Co.

Topis d' Orient *see* **'YESTERDAY'**®

Topkapi Palace *see* POUlthe

'TOPPER', HT, mr, 1959; bud short, pointed; flowers signal-red to crimson, to open, dbl., 25–30 petals, 3–3.5 in., exhibition form, slight fragrance; foliage dark, leathery; long, strong stems; vigorous, upright growth; [Pink Bountiful X Sister Kenny]; Hill, Joseph H., Co.

Toprose *see* 'COCGOLD'

'TOPSI'®, F, or, 1971; bud ovoid; flowers orange-scarlet, patio, medium, semi-dbl., slight fragrance; moderate, dwarf growth; GM, RNRS, 1972 PIT, RNRS, 1972; [Fragrant Cloud X Signalfeuer]; Tantau, Math.

'TOPSI'S FRIEND', F, lp, 1981; flowers rosette blooms borne 5 per cluster, semi-dbl., 15 petals, no fragrance; green prickles; foliage light green, glossy; upright growth; [Dreamland X Topsi]; Anderson's Rose Nurseries

Toque Rouge *see* 'KORRED'

'TORCH', Cl Pol, or, 1942; flowers scarlet-orange, white eye, blooms in clusters, semi-dbl., 15 petals, 1.5 in., slight fragrance; foliage glossy, dark; vigorous, climbing (6-8 ft) growth; [Unnamed orange Polyantha seedling, sport climbing sport]; deRuiter

Torch of Liberty™ *see* 'MORTORCH'

'TORCH SONG', HT, or, 1959; bud ovoid; flowers vermilion, dbl., 30–35 petals, 5 in., exhibition form, slight fragrance; foliage dark, leathery; vigorous, upright, bushy growth; [(Peace X Flor-adora) X Grand'mere Jenny]; Meilland, F.; C-P

Torche Rose® *see* 'INTERTOR'

'TORCHLIGHT', F, or, 1951; flowers brilliant orange-scarlet shaded deeper, borne in clusters to, single, 5–8 petals, 3.5 in., moderate fragrance; foliage dark; vigorous growth; [Dusky Maiden X Holstein]; LeGrice

'TORCHY', F, or, 1969; bud pointed to urn-shaped; flowers brick-orange, medium, dbl., exhibition form, moderate fragrance; foliage leathery; moderate, bushy growth; [Heat Wave X Spartan]; Armstrong, D.L.; Armstrong Nursery

'TOREADOR', HT, pb, 1919; flowers rosy red, reverse golden yellow, semi-dbl.; Paul, W.

'TORESKY', F, lp, 1931; flowers light pink with white eye, borne in large clusters, dbl., cupped, slight fragrance; foliage wrinkled, light; dwarf growth; [Perle d'Or X Antoine Rivoire]; Padrosa

'TORESKY, CLIMBING', Cl F, lp, 1956; Torre Blanca

'TORINO', F, mp, 1960; flowers spinel-rose, small to medium; vigorous growth; [Unnamed seedling X Cocorico]; Mansuino, Q.

Tornado® *see* 'KORTOR'

Toro *see* **'UNCLE JOE'**

Toro de Fuego, S, 1970; Delbard-Chabert; (Cavriglia)

'TORRERO', F, ob, 1961; flowers pink-orange-red, semi-dbl., 18 petals, 3 in., flat; [Ma Perkins X Cocorico]; Leenders, J.

Torvill & Dean *see* 'LANTOR'

'TOSCA', F, or, 1972; bud long, pointed; flowers open, medium, dbl., slight fragrance; foliage large, leathery; very vigorous, upright growth; [Seedling X Ginger]; Warriner, William A.; J&P

'TOSCANA', HT, dp, 1954; flowers deep pink, dbl., moderate fragrance; vigorous growth; Cazzaniga, F. G.; Fedi

Toscana *see* KORkunde

Toscanini *see* OLIjglu

Total Recall™ *see* 'SAVACALL'

'TOTOTE GELOS', HT, w, 1915; bud long, pointed; flowers flesh-white, center shaded chrome-yellow, dbl.; Pernet-Ducher

'TOTTIE', F, mr, 1961; flowers purplish red, borne in clusters of 6, dbl., 35–40 petals, 3 in.; vigorous, bushy growth; [Alain X Fashion]; Borgatti, G.; Sgaravatti

'TOTTY'S RED', HT, dr, 1926; flowers crimson-scarlet; [Premier sport]; Totty

Touch o' Cloves *see* 'MINITOC'

Touch o' Midas™ *see* 'MINABCO'

Touch of Class™ *see* 'KRICARLO'

Touch of Elegance *see* 'LEOTOELG'

Touch of Fire *see* 'RENFIRE'

'TOUCH OF KIWI', HT, yb, 1989; bud tapering; flowers creamy yellow, 1/4" wide orange margin on petals, dbl., slight fragrance; prickles pointed, light brown, varying size; foliage medium green; upright growth; [Kiwi Queen X Command Performance]; Cattermole, R.F.; South Pacific Rose Nursery

'TOUCH OF MAGIC', HT, or, 1977; bud globular; dbl., 40–45 petals, 5 in., exhibition form, intense fragrance; very vigorous growth; [San Francisco X Peace]; Patterson; Patterson Roses

Touch of Raspberry™ *see* 'DEVICIO'

Touch of Velvet *see* 'LEOVELV'

'TOUCH OF VENUS', HT, w, 1971; flowers near white, center shaded pink, large, dbl., exhibition form, intense fragrance; foliage large, leathery; vigorous, upright growth; [Garden Party X Sweet Afton]; Armstrong, D.L.; Armstrong Nursery

'TOUCHDOWN', Min, mr, 1988; bud ovoid; flowers reverse red with white center, aging bluish, urn-shaped, small, dbl., 40 petals, no fragrance; no fruit; no prickles; foliage small, medium green, matt; low, upright growth; [Sheri Anne X Anita Charles]; Jolly, Nelson F.; Rosehill Farm, 1989

Touché *see* 'LAVONT'

Touch-Up *see* 'FOUTOUCH'

'TOUGGOURT', F, or, 1959; bud globular; flowers orange-shrimp-pink, open, large, dbl.; foliage dark, leathery; vigorous, bushy growth; [(Gruss an Teplitz X Independence) X (Floradora X Independence)]; Arles; Roses-France

Toulouse Lautrec® *see* 'MEIREVOLT'

'TOUR DE FRANCE', F, OR, 1963; flowers deep orange-scarlet shaded blood-red, large to medium, semi-dbl.; vigorous, bushy growth; [(Alain X Orange Triumph) X unnamed seedling]; Mondial Roses

'TOUR DE MALAKOFF', C, m, 1856; flowers mauve-pink shaded purple, heavily veined, center green, moderate fragrance; sprawling (7 ft) growth; Soupert & Notting

Tour Eiffel *see* **'EIFFEL TOWER'**

Tour Eiffel 2000 *see* DELrugro

Touraine *see* 'POULANDER'

Tourbillon *see* 'DELNOLLI'

'TOURBILLON', F, rb, 1959; (French Cancan); flowers red, reverse silvery pink, well formed, borne in clusters of, dbl., 30–35 petals, 3–4 in.; very vigorous, bushy growth; RULED EXTINCT 4/85; [Michele Meilland X (Incendie X (Floradora X Orange Triumph))]; Delbard-Chabert

Tourmaline® *see* 'DELFRI'

Tournament of Roses *see* 'JACIENT'

Tournee, F, dy; flowers luminous golden yellow, medium, semi-dbl., slight fragrance; VEG; (Sangerhausen)

Tove Pedersen, F, dr

Tower Bridge *see* HARavis

'TOWN CRIER', HT, ly, 1961; bud ovoid; flowers straw-yellow, dbl., 33 petals, 5–6 in., exhibition form, moderate fragrance; foliage dark, glossy; vigorous, upright, well-branched growth; [Peace X Yellow Perfection]; Hill, Joseph H., Co.

'TOWN TALK', F, or, 1966; bud ovoid; flowers small, dbl., cupped; foliage dark, leathery; moderate growth; [(Circus X Garnette) X Spartan]; Swim & Weeks; Weeks Wholesale Rose Growers

'TOWN TALK, CLIMBING', Cl F, or, 1975; Weeks, V.E.; Weeks Wholesale Rose Growers

'TOWNSEND', HSpn, mp; dbl., non-recurrent; (Scotland), before 1900

'TOWNSWOMAN', HT, rb, 1973; flowers red-purple, reverse silver, dbl., 35 petals, 5 in., globular, intense fragrance; foliage light; free growth; [Seedling X Piccadilly]; Anderson's Rose Nurseries

'TOY BALLOON', Min, dr, 1979; bud ovoid, pointed; dbl., 48 petals, 1.5 in., exhibition form, slight fragrance; foliage dark; bushy, spreading growth; [Fairy Moss X Fire Princess]; Moore, Ralph S.; Sequoia Nursery

'TOY CLOWN', Min, rb, 1966; flowers white edged red, small, semi-dbl.; foliage small, leathery; bushy, dwarf growth; AOE, 1975; [Little Darling X Magic Wand]; Moore, Ralph S.; Sequoia Nursery

Toy Soldier *see* 'CURTOY'

'TOYLAND', Min, mr, 1977; bud ovoid; flowers indian red, semi-dbl., 10 petals, 1 in., slight fragrance; foliage tiny; compact, bushy growth; Lyon

Toynbee Hall *see* 'KORWONDIS'

'TRACAIT', Min, w, 1986; (**Caitlin**); flowers flesh pink fading white, urn-shaped, small, borne in sprays, dbl., 12–15 petals, slight, spicy fragrance; no fruit; straight, tan-brown prickles; foliage small, medium green, semi-glossy; bushy, low growth; [Fairy Moss X Fairy Moss]; Travis, Louis R., 1987

'TRACEY WICKHAM', Min, yb, 1984; flowers bright yellow, petals edged bright red, patio, small, in clusters, dbl., 30 petals, exhibition form, slight fragrance; foliage medium, medium green, semi-glossy; upright growth; [Avandel X Redgold]; Welsh, Eric; Rose Hill Roses

'TRACLA', Min, yb, 1987; (**Classy**); flowers yellow with pink overlay, small, borne singly, dbl., 13–18 petals, exhibition form, intense, spicy fragrance; round, green-orange fruit; straight, very few, tan-brown prickles; foliage small, medium green, semi-glossy; bushy, low growth; [Yellow Jewel X Yellow Jewel]; Travis, Louis R.

'TRADE WINDS', HT, rb, 1964; bud long, pointed; flowers dark red, reverse silver, dbl., 55 petals, 5 in., exhibition form, intense fragrance; foliage glossy; vigorous, tall growth; [(Multnomah X Seedling) X (Carrousel X Seedling)]; Von Abrams; Edmunds Roses

Tradescant *see* 'AUSDIR'

'TRADITION', HT, mr, 1965; flowers scarlet-crimson, dbl., 35 petals, 4.5 in.; [Detroiter X Don Juan]; Kordes, R.; McGredy

Tradition 95® *see* KORkeltin

Traditional Home® *see* 'BURWINTRADHOM'

'TRAFALGAR', Cl HT, lp; [Mons Desir X Grüss an Teplitz]; Fell, J.B., 1912; (Weatherly, L.)

'TRAFALGAR SQUARE', F, mp, 1965; flowers large, dbl.; vigorous growth; Van den Akker Bros.

'TRAILBLAZER', F, or, 1975; bud long, pointed; dbl., 22 petals, 2.5 in., exhibition form, slight fragrance; foliage glossy; [Albert X Orange Sensation]; Harvey, R.E.; Kimbrew-Walter Roses

'TRAKIIKA', HT, dr, 1974; flowers large, dbl., 25 petals, cupped, moderate fragrance; foliage dark; vigorous, upright, bushy growth; [Tallyho X Spartan]; Staikov, Prof. Dr. V.; Kalaydjiev and Chorbadjiiski

'TRAMONTO', HT, 1943; Giacomasso; (Cavriglia)

Tranquil, HT, lp; Dawson, 1977

'TRANQUILITY', Cl Pol, op; flowers coral-salmon, rosette form, small, borne in trusses; foliage light green; vigorous growth; RULED EXTINCT 11/82; [Princess van Orange sport]; Radmore

Tranquillity *see* 'BAROUT'

Transit, F, 1984; VEG; (Cavriglia)

'TRAPOM', Min, w, 1986; (**Fairy Pompons**); flowers flesh pink fading white, small, borne usually singly, very dbl., 45–60 petals, exhibition form, slight, spicy fragrance; no fruit; straight, tan-brown prickles; foliage small, medium green, matt; bushy, low growth; [Fairy Moss X Fairy Moss]; Travis, Louis R.

'TRASNO', Min, w, 1986; (**Snodoll**); dbl., 30–40 crinkled, translucent petals, borne singly, slight, fruity fragrance; no fruit; bowed, white prickles; foliage small, medium green, semi-glossy; upright growth; [Yellow Doll X Yellow Doll]; Travis, Louis R.

'TRASTAR', Min, rb, 1986; (**Tiny Stars**); flowers white with red narrow edges, small, borne usually singly and, semi-dbl., 12 petals, cupped, no fragrance; no fruit; almost thornless; foliage small, medium green, matt; micro-mini; bushy, low growth; [Magic Carrousel X Magic Carrousel]; Travis, Louis R.

'TRASUZ', Min, ob, 1986; (**Crepe Suzette**); flowers deep yellow with orange overlay, fading white with orange-red, dbl., 25–30 petals, exhibition form, moderate, fruity fragrance; no fruit; bowed, tan-brown prickles; foliage small, medium green, semi-glossy; bushy, low growth; [Orange Honey X Orange Honey]; Travis, Louis R., 1987

'TRAT', Min, w, 1988; (**Little 't'**); bud pointed; flowers white tinged pale pink, reverse white, small, borne in sprays, dbl., cupped, no fragrance; no fruit; no prickles; foliage small, medium green, matt; bushy growth; [Cinderella X Unnamed seedling]; Travis, Louis R., 1989

Träumerei® *see* 'REIKOR'

'TRAUMLAND', F, lp, 1958; (Dreamland); flowers light peach-pink, well-formed blooms in clusters, dbl., 20 petals, slight fragrance; foliage dark, leathery; upright, bushy growth; [Cinnabar Improved X Fashion]; Tantau, Math.

Traumland® *see* TANdalaum

Travemünde® *see* 'KORRANTU'

'TRAVERSER', LCl, yb, 1928; flowers yellow and cream, well-shaped blooms in clusters; vigorous, climbing growth; Clark, A.

'TRAVESTI', F, yb, 1965; flowers yellow to carmine-red, medium, in clusters, dbl., 38 petals, cupped, moderate fragrance; foliage dark; vigorous, bushy growth; [Orange Sensation X Circus]; deRuiter

'TRAVESTON', Cl HT, dr, 1953; bud long, pointed; flowers crimson, medium, semi-dbl.; vigorous growth; [Black Boy X Editor McFarland]; Ulrick, L.W.

Traviata™, *see* 'MEILAVIO'

Traviata™ *see* MEIviola

'TREASURE', HT, ab, 1929; bud pointed; flowers apricot-pink, dbl., 6 in.; foliage glossy, dark; very vigorous growth; [Golden Rapture X Fred Walker]; Fletcher; Tucker

'TREASURE CHEST', HT, my, 1968; bud ovoid; flowers large, dbl., exhibition form, slight fragrance; foliage glossy; vigorous, upright growth; [Charlotte Armstrong X Fred Howard]; Whisler, D.; Germain's

'TREASURE GOLD', HT, my, 1950; bud ovoid, pointed; flowers yellow, some petals splashed red, dbl., 38 petals, 4 in., moderate fragrance; upright, branching growth; [Pink Princess X Free Gold (HT)]; Brownell, H.C.

'TREASURE ISLAND', HT, op, 1938; bud long, pointed; flowers light salmon, reverse flaming coppery pink, base orange, large, dbl., exhibition form, moderate fragrance; foliage leathery, bronze; vigorous growth; [Comtesse Vandal X Mme Nicolas Aussel]; Raffel; Port Stockton Nursery

'TREASURE ISLAND, CLIMBING', Cl HT, op, 1941; Hennessey

'TREASURE ISLE', F, op, 1967; flowers salmon-pink, reverse coppery, small, dbl., moderate fragrance; foliage glossy; vigorous, bushy growth; [Seedling X Treasure Island]; Raffel; Port Stockton Nursery

'TREASURE TROVE', LCl, ab, 1977; flowers apricot, mauve-pink and cream, dbl., 23 petals, 1.5–2 in., cupped, intense fragrance; vigorous growth; [Kiftsgate X China (possibly Old Blush)]; Treasure

Treasure Trove *see* JAClay

Trend, HT, op; flowers large, dbl., slight fragrance; VEG; (Sangerhausen)

Tresor *see* KRInico

Tresor de Thorigny, Cl HT, w

Trevi Fountain, LCl, ob

Trevor Griffiths™ *see* 'AUSOLD'

Trianon, F, pb, 1990; Lens

Tribute *see* 'JACROSE'

'TRICENTENAIRE', HT, mp, 1949; bud long, pointed; flowers pink, center deeper, medium, dbl., 25 petals, exhibition form, moderate fragrance; foliage leathery, small; vigorous growth; [Charles P. Kilham X Neville Chamberlain]; Lens

'TRICIA', LCl, lp, 1942; bud long, pointed; flowers flesh-pink to ivory-pink, dbl., exhibition form, intense fragrance; foliage dark, glossy; ; vigorous (about 8 ft. hardy.; [New Dawn seedling]; O'Neal

'TRICIA'S JOY', LCl, op, 1973; flowers deep coral pink, large, dbl., 35 petals, moderate fragrance; foliage medium, dark, semi-glossy; vigorous growth (to 18 ft.); [Mme Caroline Testout, Climbing X Blessings]; McKirdy, J.M.; John Sanday Roses, Ltd.

Trick Or Treat *see* 'TALTRICK'

Trickster *see* 'BRITRICK'

'TRICOLORE', HGal, pb, 1827; (Reine Marguerite); flowers lilac-pink, fringed at edges, dotted and mottled white; Lahaye Pere

'TRICOLORE', HMult, pb, 1863; flowers lilac-pink, similar to tricolore (g), blooms in small clusters; more vigorous than Gallica of the same name; [R. multiflora X ?]; Robert et Moreau

'TRICOLORE DE FLANDRE', HGal, pb, 1846; flowers pale blush, striped bright pink, fading to mauve, small, dbl., moderate fragrance; Van Houtte

'TRIER'®, HMult, w, 1904; flowers rosy white, base straw-yellow, semi-dbl., moderate fragrance; height 6-8 ft; (14); [Probably Aglaia self seedling sport]; Lambert, P.

Trier 2000 *see* 'KORMETTER'

Trieste, F, 1960; Mansuino; (Cavriglia)

Trigintipetala *see* **'KAZANLIK'**

'TRIGO', HT, ab, 1931; bud long, pointed; flowers indian yellow, reverse apricot, tinted cerise, large, dbl., exhibition form, intense fragrance; foliage thick, bronze, glossy; vigorous growth; GM, NRS, 1931; Dickson, A.

'TRILBY', HT, op, 1927; flowers rich salmon, dbl.; Dobbie

Trinity *see* 'MACREDPARAP'

'TRINITY', T, w; flowers pale pink to white, gold stamens, nodding blooms borne usual, semi-dbl., 4 in.; foliage dark; vigorous, spreading growth

'TRINITY', HT, pb, 1976; flowers variegated pink and yellow, medium, dbl., globular, intense fragrance; vigorous, upright, compact growth; [Korovo X ?]; de Freitas

Trinity Rambler, LCl, dp

'TRINKET', Min, mp, 1965; flowers phlox-pink, micro-mini, small, dbl.; foliage glossy; vigorous, bushy, dwarf growth; [(R. wichurana X Floradora) X Magic Wand]; Moore, Ralph S.; Sequoia Nursery

'TRINTAGO', HT, mr, 1962; flowers cardinal-red, well formed, dbl.; [Souv. de Jacques Verschuren X Charles Mallerin]; Leenders, J.

'TRIO', F, pb, 1966; flowers gold and pink, well-formed, large, slight fragrance; [Kordes' Perfecta X Shot Silk]; Dickson, A.

Triodene *see* KORituscha

Triolet *see* ORAdon

'TRIOMPHE ANGEVIN', Pol, mr, 1934; flowers cerise-red, borne in clusters, semi-dbl., cupped, slight fragrance; foliage glossy; very vigorous growth; Délépine; Pajotin-Chédane

'TRIOMPHE BRIARD', F, op, 1958; flowers coral, borne in clusters, dbl., moderate fragrance; foliage glossy; moderate growth; [Unnamed seedling X Fashion]; Robichon

'TRIOMPHE D'ALENCON', HP, dp, 1858; flowers large, dbl.; Touvais; (Sangerhausen)

'TRIOMPHE DE BOLWYLLER', HSem, w; flowers white, center tinted yellow, large, globular, moderate fragrance; Baumann, prior to 1837

Triomphe de Caen *see* **'PRINCE ARTHUR'**

'TRIOMPHE DE FLORE', HGal, lp, 1830; flowers medium, very dbl., slight fragrance; Prévost; (Sangerhausen)

'TRIOMPHE DE FRANCE', HP, mp, 1875; flowers very large, very dbl., moderate fragrance; Garçon; (Sangerhausen)

Triomphe de Guillot Fils, T, pb; Guillot, 1861

Triomphe de la Ducher, B, lp

'TRIOMPHE DE LA GUILLOTIÈRE', S, w, 1872; flowers white tinged pink; [Probably R. roxburghii X R. odorata]; Guillot et Fils

'TRIOMPHE DE LA MALMAISON', HT, dp, 1946; flowers dark satiny pink, very large, petals reflexed, intense fragrance; foliage dark; vigorous growth; Gaujard

'TRIOMPHE DE LA TERRE DES ROSES', HP, mp, 1864; flowers violaceous pink, large, dbl.; moderately vigorous growth; Guillot Père

'TRIOMPHE DE LAFFAY', HCh, w; flowers delicate flesh, changing to white, expanded, large, very dbl.; pendulous growth habit

'TRIOMPHE DE L'EXPOSITION', HP, mr, 1855; flowers cherry-red, large, dbl., 55 petals, recurrent bloom; vigorous, bushy growth; Margottin

'TRIOMPHE DE PERNET PÉRE', HT, m, 1890; flowers reddish-violet, large, very dbl., moderate fragrance; Pernet; (Sangerhausen)

Triomphe des Noisettes, N, lp; Pere Pernet, 1887

'TRIOMPHE DU LUXEMBOURG', T, pb; flowers salmon-buff shaded rose; Hardy, 1840

'TRIOMPHE ORLÉANAIS', Pol, mr, 1912; flowers cherry-red, blooms in large clusters, semi-dbl., 1.5–2 in., slight fragrance; foliage leathery, glossy, bright; vigorous growth; Peauger

'TRIOMPHE ORLÉANAIS, CLIMBING', Cl Pol, mr, 1922; Turbat

'TRISTESSE', HT, m, 1953; (Tristeza); flowers gray-mauve, medium, dbl., 30–40 petals, flat, slight fragrance; foliage glossy; vigorous growth; [Charles P. Kilham X Betty Uprichard]; Camprubi, C.

Tristeza *see* **'TRISTESSE'**

'TRITON', HT, yb, 1978; flowers white to light yellow, petals edged pink, large, dbl., 47 petals; upright growth; [Colour Wonder X Tzigane]; Dickson, Patrick

'TRIUMPHANT', HSet, w, 1850; flowers white tinted flesh, dbl.; [R. setigera X ?]; Pierce

Trix, Pol, or; flowers medium, dbl.; VEG; (Sangerhausen)

'TROBALL', Min, mp, 1987; (**Joan Ball**); flowers small blooms in sprays of 3-6, dbl., 30 petals, exhibition form, moderate, fruity fragrance; short, thin, red-brown prickles; foliage small, dark, glossy; upright, bushy growth; [Orange Sensation X Unnamed seedling]; Robinson, Thomas, Ltd., 1989

'TROBEACH', Min, w, 1983; (**Shell Beach**); bud soft pink; flowers creamy white, blooms in clusters, dbl., 28 petals, slight fragrance; foliage small, medium green, semi-glossy; bushy growth; [Simon Robinson X Simon Robinson]; Robinson, Thomas, Ltd.

'TROBEE', Min, rb, 1982; (**Dollie B**); flowers medium red, silver reverse, small, dbl., 35 petals, no fragrance; foliage small, dark green, red edges, glossy; bushy growth; [Parkdirektor Riggers X Darling Flame]; Robinson, Thomas, Ltd., 1983

'TROBELLE', Min, my, 1982; (**Calay**); flowers lemon yellow, small, semi-dbl., moderate fragrance; foliage small, dark, glossy; upright, bushy growth; [Rumba X New Penny]; Robinson, Thomas, Ltd.; T. Robinson, Ltd.

'TROBETTE', Min, op, 1983; (**Jennie Robinson**); flowers orange flushed pink, patio, small, dbl., 35 petals, moderate fragrance; foliage small, dark, glossy; bushy growth; [Rumba X Darling Flame]; Robinson, Thomas, Ltd.; Thomas Robinson, Ltd.

'TROBGINA', Min, op, 1986; (**Gina Louise**); flowers bright orange-pink, opening to bright yellow, gold anthers, dbl., 35 petals, exhibition form, intense, damask, fruity fragrance; globular, large, orange fruit; thin, red prickles pointed down; foliage small, dark green, semi-glossy; bushy, low growth; [Orange Sensation X Unnamed seedling]; Robinson, T.

'TROBGLOW', Min, or, 1984; (**Paint-Pot**); flowers medium, dbl., 24 petals, intense fragrance; foliage small, medium green, semi-glossy; bushy growth; [Unnamed seedling X Darling Flame]; Robinson, Thomas, Ltd.; T. Robinson, Ltd.

'TROBGOLD', Min, my, 1984; (**Woodland Sunbeam**); flowers patio, medium, dbl., 23 petals, intense fragrance; few prickles; foliage small, dark, glossy; vigorous, bushy growth; [Orange Sensation X Calay]; Robinson, Thomas, Ltd.

'TROBGUERN', Min, dy, 1992; (**Guernsey Gold**); flowers golden yellow, urn-shaped, blooms borne in small clusters, dbl., 26–40 petals, 1.5–2.75 in., slight fragrance; some prickles; foliage small, medium green, semi-glossy; low (30 cms), upright, bushy growth; [Rise 'n' Shine X Seedling]; Robinson, Thomas, Ltd.; Thomas Robinson, Ltd., 1990

'TROBIC', Min, pb, 1989; (**Just Magic**); bud rounded; flowers cream to deep pink, reverse cream tinged pink, aging often strongly, dbl., 21 petals, cupped, moderate, damask fragrance; spheroid, orange fruit; prickles very thin, pointed, red aging brown; foliage small, dark green, glossy; upright, bushy, medium growth; [(Parkdirektor Riggers X New Penny) X [(Parkdirektor Riggers xNew Penny) X Unnamed seedling]]; Robinson, Thomas, Ltd., 1992

'TROBIC', Min, mr, 1982; (**Little Russel**, 'TROBRIC'); flowers small, dbl., 20 petals, no fragrance; foliage small, dark, glossy; bushy growth; [Marlena X New Penny]; Robinson, Thomas, Ltd.; T. Robinson, Ltd., 1983

'TROBINA', Min, mp, 1982; (**Pink Secret**); [Beauty Secret sport]; Robinson, Thomas, Ltd.; T. Robinson, Ltd.

'TROBINETTE', Min, ab, 1982; (**Sarah Robinson**); flowers soft apricot, small, dbl., 20 petals, intense fragrance; foliage small, dark, glossy; compact, bushy growth; [Rumba X Darling Flame]; Robinson, Thomas, Ltd.

'TROBINKA', Min, mr, 1983; (**Ginny-Lou**); flowers bright medium red, blooms in clusters, dbl., no fragrance; foliage small, dark, semi-glossy; bushy growth; [Dollie B. X Unnamed seedling]; Robinson, Thomas, Ltd.; Thomas Robinson, Ltd., 1984

'TROBLOVE', Min, dr, 1986; (**Guernsey Love**); flowers small blooms in sprays of 4-5, dbl., 35 petals, cupped, moderate, fruity fragrance; globular, medium, orange-red fruit; thin, red prickles, curving downward; foliage small, dark, glossy; upright, bushy growth; [Dollie B. X Unnamed seedling]; Robinson, Thomas, Ltd., 1990

'TROBRIC', Min, mr, 1982; (**Little Russel**, 'TROBIC'); flowers small, dbl., 20 petals, no fragrance; foliage small, dark, glossy; bushy growth; [Marlena X New Penny]; Robinson, Thomas, Ltd.; T. Robinson, Ltd., 1983

'TROBROY', S, w, 1992; (**Royal Worcester®**); flowers peachy cream, blooms borne in large clusters of 5-69, semi-dbl., 6–14 petals, 3–3.5 in., moderate fragrance; some prickles; foliage medium, dark green, very glossy; low to medium (60 cms), bushy, compact growth; [Simon Robinson X Gina Louise]; Robinson, Thomas, Ltd.; Thomas Robinson, Ltd., 1991

'TROBSA', Min, mp, 1984; (**Woodlands Lady**); flowers medium salmon-pink, small blooms in large clusters, semi-dbl., 15 petals, slight fragrance; foliage small, dark, glossy; compact, bushy growth; [Unnamed seedling X New Penny]; Robinson, Thomas, Ltd.

'TROBSPREAD', S, yb, 1992; (**Unforgettable**); flowers amber, blooms borne in large clusters, dbl., 15–25 petals, 3–3.5 in., repeat bloom, moderate fragrance; some prickles; foliage small to medium, dark green, glossy; medium (90 cms), bushy, spreading (hemisphere shape) growth; [Snow Carpet X Woodland Sunbeam]; Robinson, Thomas, Ltd.; Thomas Robinson, Ltd., 1994

TRObstar, LCl, lp; (**Starlight Express**); Robinson, 1997

'TROBSUN', Min, yb, 1982; (**Woodlands Sunrise**); flowers small, dbl., 35 petals, slight fragrance; foliage small, dark, glossy; upright, bushy growth; [Rumba X Darling Flame]; Robinson, Thomas, Ltd.

'TROBWICH', Min, mp, 1982; (**Simon Robinson**); flowers small blooms in clusters of 10-60, single, 5 petals, moderate fragrance; foliage small, dark, glossy; compact, low, bushy growth; [R. wichurana X New Penny]; Robinson, Thomas, Ltd.

'TROCADERO', HT, mr, 1964; bud pointed, bright red; foliage bronze; vigorous growth; [Karl Herbst X Seedling]; Delforge

Trocadero 'D', HT, 1989; Delforge; (Cavriglia)

Troika *see* 'POUMIDOR'

'TROIKA, CLIMBING', Cl HT, 1964; Thomas, Dr. A.S.; Rumsey

Troilus *see* 'AUSOIL'

'TROJA', HT, w, 1927; bud long, pointed, cream-yellow; flowers creamy white, dbl., moderate fragrance; vigorous, bushy growth; [Mrs Herbert Stevens sport]; Mikes Böhm, J.; Böhm

'TROJAN', Gr, yb, 1961; bud long, pointed; flowers pastel pink, reverse yellow, dbl., 40 petals, 5 in., exhibition form, moderate fragrance; foliage leathery; upright growth; [Sutter's Gold X (Mme Henri Guillot X Seedling)]; Von Abrams; Peterson x Dering

Trojan Victory *see* 'KORPERKI'

Trollhattan, HRg, lp

Trompeter von Sackingen, HMult, m; Geschwind, R., before 1890

'TROPAT', F, or, 1992; (**Temptation**); single, 5 petals, 3–3.5 in., borne in small clusters, moderate fragrance; some prickles; foliage medium, medium green, semi-glossy; low (70 cms), dense, bushy, compact growth; [Orange Sensation X Seedling]; Robinson, Thomas, Ltd.; Thomas Robinson, Ltd.

'TROPHÉE', LCl, mr, 1968; foliage glossy; vigorous, climbing growth; [Valenciennes X Etendard]; Robichon; Ilgenfritz Nursery

Tropical, F, 1966; Cazzaniga, F. G.; (Cavriglia)

Tropical Paradise *see* 'BURSRTPARA'

Tropical Passion *see* CARange

Tropical Skies *see* INTerdays

Tropical Sunrise *see* 'JOHILLGOLD'

Tropical Sunset ™ *see* 'MACTAURANG'

Tropical Twist™ *see* 'JACORCA'

Tropicana *see* 'TANORSTAR'

Tropicana, Climbing *see* 'TANGOSTAR'

Tropico Sunblaze *see* MEIglassol

Tropique *see* 'DELJIS'

'TROUBADOUR', HWich, 1911; flowers bright red, shaded maroon, large clusters, dbl., free seasonal bloom; foliage large, glossy, dark; vigorous, climbing growth; RULED EXTINCT; Walsh

'TROUBADOUR', HT, 1956; flowers rose-copper, large, dbl., free bloom, moderate fragrance; strong stems; moderate growth; RULED EXTINCT; Frères, Buyl

Troubadour of Love, HT, dr

Trpaslik, Min, lp; Strnad; (Czech Rosa Club)

'TRUBY KING', F, dp, 1965; flowers carmine, open, large, borne in large clusters, semi-dbl., intense fragrance; foliage leathery, glossy, bronze; vigorous, tall growth; [Border Queen X Helen Traubel]; Harris, L.M.; Cutler

Trudor *see* **'MICAËLA'**

Trudy Mimi™ *see* MEIkyster

True Gold ™ *see* 'LAVTRU'

'TRUE LOVE', HT, dr, 1970; flowers deep red, dbl., 30 petals, 3.5–4 in., slight fragrance; foliage glossy; moderately vigorous growth; Delbard; Laxton & Bunyard Nursery

True Love *see* 'RUTRULO'

True Vintage *see* 'SEAVIN'

'TRULY FAIR', F, w, 1953; flowers cream, center apricot, small, borne in sprays of 4-5, semi-dbl., moderate fragrance; bushy growth; Ratcliffe

'TRULY YOURS', HT, ob, 1972; flowers coral-salmon to orange, dbl., 44 petals, 5 in., globular, intense fragrance; foliage large; vigorous, upright growth; Edland Fragrance Medal, ARS, 1971; [Miss Ireland X Stella]; Robinson, H.

Trumpeter® *see* 'MACTRUM'

Truper *see* **'PERSEPOLIS'**

Trylon *see* **'SHINING STAR'**

Tschaika, HT, lp, 1959; flowers large, dbl., moderate fragrance; Klimenko, V. N.; (Sangerhausen)

TSCHaka, HT, ab, 1989; (**Charlie Chaplin**); Tschanz, E.

'TSCHERNOWA UKRAIJNA', HT, mr, 1940; flowers large, semi-dbl., slight fragrance; Kosteckij; (Sangerhausen)

Tschin-Tschin, F, mr, 1973; flowers luminous red, medium, dbl.; Meilland; (Sangerhausen)

'TSUKIAKARI', HT, w, 1984; flowers creamy white, large blooms borne 1-3 per stem, dbl., 35 petals, exhibition form, no fragrance; medium prickles; foliage medium green, semi-glossy; bushy growth; [Utage X Anne Letts]; Ota, Kaichiro

Tubantia, HT, op, 1975; flowers reddish-orange and pink, large, dbl.; Verbeek; (Sangerhausen)

'TUCKACH', MinFl, w, 1997; (**Cachet™**); flowers large, exhibition form, 26–40 petals, 2 in., exhibition form, borne mostly singly, none; foliage medium, medium green, semi-glossy; upright, tall (3 ft) growth; [Seedling X Seedling]; Tucker, Robbie; Rose Hill Nursery, 1998

'TUCKAMY', ~~Min~~ MinFl, lp, 1998; (**Amy Grant™**); AmR 3/01
flowers light pink, varies by tempera-

ture, large, dbl., 17–25 petals, 1.5–1.75 in., exhibition form, borne mostly singly, slight fragrance; prickles few, small, slightly hooked downward; foliage medium, dark green, glossy; low, upright, 18-24 in. growth; [Loving Touch X White Masterpiece]; Tucker, Robbie; Nor'East Miniature Roses, 1999

'TUCKER'S FOLLY', HT, pb, 1951; flowers glowing cerise overlaid orange, dbl., 35–40 petals, 6–7 in.; foliage bronze; very vigorous growth; Fletcher; Tucker

'TUCKER'S YELLOW', HT, dy; flowers deep golden yellow; [Max Krause sport]; Tucker

'TUCKFLIP', Min, mr, 1997; (**Miss Flippins**®); flowers medium red, deep pink reverse, dbl., 15–25 petals, 1.5 in., exhibition form, borne mostly singly, no fragrance; prickles moderate; foliage medium, dark green, glossy; bushy, medium (2 ft) growth; [Elizabeth Taylor X Kristin]; Tucker, Robbie; Bridges Roses, 1997

'TUCKROX', HT, op, 1998; (**Roxie Baby**®); flowers orange to pink, reverse lighter, very full, very dbl., 65 petals, 3–4 in., exhibition form, borne mostly singly, no fragrance; foliage dark green, glossy; upright, medium (5 ft) growth; [Elizabeth Taylor X Hoagy Carmichael]; Tucker, Robbie; Rosehill Nursery, 1999

'TUCSTAND', HT, rb, 1998; (**Standing Ovation**®); flowers white with red edge, very full, very dbl., 41 petals, 4.5 in., exhibition form, borne singly or in small clusters, slight fragrance; prickles moderate; foliage medium, dark green, semi-glossy; upright, medium (4 - 5 ft) growth; [Elizabeth Taylor X White Masterpiece]; Tucker, Robbie; Edmunds Roses, 1999

'TUCTOMMY', Min, my, 1998; (**Little Tommy Tucker**®); flowers medium yellow, reverse lighter, long-lasting, dbl., 17–25 petals, 1.5–1.75 in., exhibition form, borne in small clusters, no fragrance; prickles few, small to medium, slightly hooked downward; foliage medium, dark green, glossy; upright, low (14 - 20 in) growth; AOE, 1999; [Rise 'n' Shine X Captivation]; Tucker, Robbie; Nor'East Miniature Roses, 1999

Tudelum® *see* 'KINLUM'

'TUDOR', HT, mr, 1953; (Pechtold's Flame); flowers spectrum-red overcast scarlet-red, dbl., 20–25 petals, 5 in., exhibition form, moderate fragrance; long stems; vigorous growth; [Katharine Pechtold X Crimson Glory]; Verschuren-Pechtold; J&P

'TUDOR PRINCE', Gr, mr, 1959; (Prince Philip); bud long, pointed, chestnut; flowers bright red shaded geranium-red, large, dbl., 30–40 petals; foliage glossy, dark; vigorous, upright growth; [Independence X Buccaneer]; Leenders, J.

Tudor Rose, S, lp; [R. brunonii X Unknown]; Porter, L., 1989; (Weatherly, L.)

Tudor Sunsation *see* KORfeldwo

'TUDOR VICTORY', HT, dr, 1987; flowers dusky, dark red, reverse medium red, aging purple, medium, b, dbl., 27 petals, exhibition form, slight, damask fragrance; prickles straight, large, red; foliage large, medium green, semi-glossy; upright, tall growth; [Queen Elizabeth X John Waterer]; Bracegirdle, Derek T., 1988

'TUHUA', HMult, ab, 1979; bud small, pointed; 11 petals, borne 3-5 per cluster, early bloomer with some repeat, no fragrance; small, red-brown prickles; foliage small, glossy; vigorous, lax, tall growth; [Iceberg X Tausendschon]; Murray, Nola

Tullamore, S, lp

Tulsa, HT, lp; Perry, Astor, 1996

Tumbarumba Schoolhouse, HMult, lp

Tumbling Waters *see* POUltumb

Tumulte, HT; Kordes, R.; (Cavriglia)

Tunusblumchen, LCl, m

'TUPPERWARE', HT, pb, 1981; bud ovoid, pointed; flowers carmine pink, reverse silver, borne 1-5 per cluster, dbl., 65 petals, exhibition form, moderate fragrance; medium prickles; foliage large, dark, glossy; vigorous, upright growth; [(Pink Peace X Queen Elizabeth) X (Kordes' Perfecta X Peace)]; Williams, J. Benjamin; Tupperware Home Parties

Turbo *see* 'MEIROZRUG'

Turbo®, HRg, mp

Turbo Meidiland® *see* 'MEIROZRUG'

Turbo Rugostar® *see* 'MEIROZRUG'

Turbulance *see* 'GELANCE'

'TURENNE', HP, dr, 1861; flowers maroon; vigorous growth; [Général Jacqueminot sport]; Verdier, V.

Turenne, HGal, dp; Vibert (?), ca 1846

'TÜRKES RUGOSA SÄMLING', HRg, pb, 1923; bud long, pointed; flowers peach-pink on yellow ground, large, semi-dbl., intense fragrance; foliage dark, leathery; strong stems; vigorous (2 1/2-3 ft) growth; [Conrad Ferdinand Meyer X Mrs Aaron Ward]; Türke; Teschendorff

Turlock High *see* 'CLELOCK'

Turn of the Century *see* 'GREELIZ'

'TURNER'S CRIMSON RAMBLER', HMult, mr; (Crimson Rambler, Shi Tz-mei, Soukara-Ibara, Ten Sisters); flowers bright crimson, fading toward blue, irregular blooms in larg, dbl., heavy, non-recurrent bloom; foliage light, leathery, disposed to mildew; very vigorous, climbing (15-24 ft.) growth; (14); Turner, 1893; Turner, 1893

'TURNVATER JAHN', HP, pb, 1927; flowers white with pink, very large, dbl.; Müller, Dr. F.; (Sangerhausen)

Tuscan Beauty *see* 'HINTUSCAN'

'TUSCANY', HGal, m; (The Old Velvet Rose); flowers velvety blackish-crimson to deep purple, large blooms, semi-dbl.; vigorous, upright growth

'TUSCANY SUPERB', HGal, m; (Superb Tuscan, Superb Tuscany); [Tuscany seedling]; Rivers, prior to 1837

Tuscia® *see* BARtusc

Tut's Treasure, Min, dy

Tutta's Pink Noisette, N, lp

Tutti-Frutti *see* 'JACTUTTI'

'TUTU', Min, pb, 1978; bud ovoid; flowers pink and rose, dbl., 40 petals, 1 in., exhibition form, slight fragrance; upright, bushy growth; [Over the Rainbow X Seedling]; Rovinski & Meredith; Kingsdown Nursery

'TUTU MAUVE', F, m, 1963; flowers magenta shaded mauve and rose, well-formed, dbl., 30 petals, 4 in.; bushy, low growth; GM, Madrid, 1962; Delbard-Chabert

'TUTU PETITE', F, pb, 1967; flowers pink, reverse darker, blooms in clusters, semi-dbl., slight fragrance; foliage glossy, light; compact, bushy growth; [Rudolph Timm X ?]; Samuels

Tuxedo *see* 'AROBRISP'

Tweedle Dee *see* 'TINTWEE'

'TWEETIE', Min, lp, 1973; bud ovoid, long, pointed; flowers soft pink, small, dbl., slight fragrance; foliage small, light, leathery; moderate, dwarf, bushy growth; [Perle d'Or X Fairy Princess]; Moore, Ralph S.; Sequoia Nursery

Twenty First Century *see* 'PERTWENTY-FIRST'

Twenty-Fifth *see* 'BEATWE'

'TWICE AS NICE', HT, rb, 1984; flowers white blending to red at edges, large, dbl., 40 petals, intense fragrance; foliage large, dark, semi-glossy; [Peace X Mirandy]; Patterson, Randell E.

'TWILIGHT', HT, m, 1955; bud pointed; flowers lavender-lilac, reverse silvery, dbl., 30–35 petals, 4.5 in., exhibition form, moderate fragrance; foliage dull green; bushy, upright growth; [Grey Pearl X Lavender Pinocchio]; Boerner; J&P

Twilight *see* NOAtwi

'TWILIGHT BEAUTY', Min, m, 1977; bud long, pointed; flowers red-purple, dbl., 40 petals, 1–1.5 in., exhibition form, moderate fragrance; bushy, spreading growth; [Angel Face X Over the Rainbow]; Williams, Ernest D.; Mini-Roses

Twilight Dream *see* 'MINDREAM'

Twilight Mist, LCl, m; Robinson, 1995

Twilight Time, Min, m

Twilight Trail(TM) *see* 'MINXCO'

'TWILIGHT ZONE', Min, mr, 1985; flowers well-formed, medium, dbl., 60 petals, no fragrance; foliage small, dark, semi-glossy; bushy, dense growth; [Scarlet Knight X Big John]; Hardgrove, Donald L.; Rose World Originals

Twilight Zone, Gr, lp; Williams, J. Benjamin, 1996

Twin Peaks *see* 'WEKFRAG'

Twin Pinks *see* 'CLETWIN'

'TWINKIE', Min, lp, 1974; bud pointed; flowers light clear pink, dbl., 40 petals, 1 in.; foliage small, glossy; upright, very bushy growth; [(R. wichurana X Floradora) X Eleanor]; Moore, Ralph S.; Sequoia Nursery

Twinkle® *see* 'INTERTWIK'

Twinkle Bright, Min, lp

Twinkle Charm, Min, lp

Twinkle Eyes, Min, lp

Twinkle Pink, Min, lp

Twinkle Star, Min, lp

Twinkle Toes *see* 'LYOTO'

'TWINKLE TWINKLE', Min, ab, 1981; bud pointed; flowers white with apricot petal edges, aging dark pink, blooms born, dbl., 23 petals, slight, tea fragrance; very fine, curved prickles; foliage medium green, semi-glossy; slender, straight, upright growth; [Contempo X Sheri Anne]; Bennett, Cecilia 'Dee'; Tiny Petals Nursery

Twinkler, Min, pb; Datt, Braham, 1989

'TWINKLES', Min, w, 1954; bud flesh; flowers small, dbl., 43 petals, moderate fragrance; compact, dwarf (8 in.) growth; [Perla de Montserrat X Unnamed Polyantha seedling]; Spek; J&P

Twins®, HMsk, yb; Lens, 1994

Twirly Whirly, S, lp; Peden, R., 1999; (Weatherly, L.)

Twister *see* 'MORTWISTER'

Two Sisters(TM) *see* 'WILTWOS'

Two Thumbs Up *see* 'GELTWO'

'TWOADMIRE', HT, dr, 1991; (**American Glory**); bud ovoid, pointed; flowers cardinal red, large, cupped, 28 petals, cupped, borne usually singly, slight, damask fragrance; foliage medium, dark green, semi-glossy; upright, bushy, medium growth; PP07973; [Portland Trailblazer X Seedling]; Twomey, Jerry, 1983; DeVor Nurseries, Inc., 1991

'TWOADORE', HT, lp, 1991; (**Audrey Hepburn**); bud pointed; flowers blush pink, fading to lighter pink, large, dbl., 30–32 petals, moderate, fruity fragrance; foliage medium, dark green, glossy; upright, bushy, medium growth; PP07980; [Evening Star X Seedling]; Twomey, Jerry, 1983; DeVor Nurseries, Inc., 1992

'TWOADVANCE', S, op, 1991; (**All That Jazz**(TM)); bud pointed; flowers coral salmon blend, loose, large, semi-dbl., 10–15 petals, 4.5 in., cupped, borne in sprays of 3-5, moderate, damask fragrance; prickles moderate; foliage medium, dark green, glossy; upright, bushy, tall growth; PP7978; AARS, 1992; [Gitte X Seedling]; Twomey, Jerry, 1983; DeVor Nurseries, Inc., 1991

'TWOAEBI', HT, ob, 1999; (**Dream Orange**); dbl., 26–40 petals, 3.5 in., borne in small clusters, slight fragrance; prickles moderate; foliage medium, medium green, semi-glossy; bushy, medium (4.5 ft) growth; [Cherish X (Evening Star X Trumpeter)]; Twomey, Jerry, 1999

TWOangel, HT, pb, 1995; (**Angela Lansbury**); PP10172; [Gitte X Silver Jubilee]; Twomey, Jerry

'TWOAT', HT, lp, 1999; (**Dream Blush**); dbl., 17–25 petals, 5 in., borne in small clusters, slight fragrance; few prickles; foliage large, medium green, semi-glossy; compact, medium (4 ft) growth; [Evening Star X Marijke Koopman]; Twomey, Jerry

'TWOBE', HT, op, 1989; (**Sheer Elegance**(TM)); bud pointed; flowers soft creamy pink with dark pink edges, large, exhibition form, 30–35 petals, 4.5 in., exhibition form, borne singly, moderate, musk fragrance; prickles slightly curved, red with green; foliage large, dark green, glossy; upright, tall growth; PP7901; AARS, 1991; [Pristine X Fortuna]; Twomey, Jerry; DeVor Nurseries, Inc., 1990

TWOcherish, Gr, dp, 1995; (**Colorburst**); Twomey, Jerry

'TWODI', HT, mp, 1998; (**Charming Diana**(TM)); flowers pink, reverse medium, dbl., 15–25 petals, 6 in., borne singly, very large, moderate fragrance; prickles numerous; foliage large, dark green, glossy; bushy, medium growth; Twomey, Jerry

'TWODREAM', HT, mp, 1989; (**Endless Dream**(TM)); bud pointed; flowers medium, soft pink, large, borne singly, dbl., 32 petals, cupped, moderate, musk fragrance; prickles declining, grayish-white with black spots; foliage large, dark green, semi-glossy; upright, medium growth; PP007561; [Emily Post X Seedling]; Twomey, Jerry; DeVor Nurseries, Inc., 1990

'TWOETERN', Gr, rb, 1991; (**Eternity**); flowers red/cream bicolor, blooms borne in large clusters, dbl., moderate; some prickles; foliage medium, dark green; tall (183 cms), upright growth; PP08413; [Gitte X Seedling]; Twomey, Jerry; DeVor Nurseries, Inc., 1991

'TWOEX', HT, mp, 1998; (**Our Diana**(TM)); flowers medium pink, double (15–25 petals) drops clean; very large blooms; fragrant; foliage medium, medium green, semi-glossy, upright, tall growth, fast repeat; [Parentage not listed]; Twomey, Jerry

'TWOFAN', HT, op, 1989; (**American Fantasy**(TM)); bud ovoid; flowers salmon pink, reverse lighter, large, dbl., 32 petals, cupped, borne singly, moderate, fruity fragrance; prickles declining, yellow with red tinge; foliage medium, dark green, glossy; upright, medium growth; [Sonia X Seedling]; Twomey, Jerry; DeVor Nurseries, Inc., 1991

TWOfavor, HT, pb, 1995; (**Favorite Dream**); PP10155; [(Friendship X seedling) X (Emily Post X Royalty)]; Twomey, Jerry

'TWOFREE', S, mr, 1989; (S.H. Freedom, **Spring Hill's Freedom**); bud ovoid; flowers medium, scarlet red, medium, borne singly, dbl., 35 petals, cupped, slight, musk fragrance; prickles slightly curved, red-purple; foliage medium, medium green, semi-glossy; upright, medium growth; [Samantha X Unnamed seedling]; Twomey, Jerry; DeVor Nurseries, Inc., 1990

TWOfree, F, mr; (**Springhill Freedom**); PP07868; [Samantha X Fireburst]; Twomey, Jerry, 1998

'TWOHAVE', Gr, or, 1991; (**Proud Mary**); bud pointed; flowers scarlet red, medium, dbl., 19–21 petals, moderate, damask fragrance; foliage medium, dark green, semi-glossy; upright, bushy, medium growth; PP07981; [Unnamed seedling X Royalty]; Twomey, Jerry, 1983; DeVor Nurseries, Inc., 1991

'TWOHONOR', HT, mp, 1993; (**American Honor**); flowers full, large blooms, exhibition form, 26–40 petals, 3–3.5 in., exhibition form, borne mostly singly, slight; few prickles; foliage medium, dark green, semi-glossy; medium (110 cms), upright growth; PP09443; [Sheer Elegance X Seedling]; Twomey, Jerry, 1993; DeVor Nurseries, Inc., 1993

'TWOJOAN', HT, mp, 1999; (**Dream Pink**); flowers large, dbl., 17–25 petals, borne mostly singly, moderate fragrance; prickles moderate; foliage large, medium green, glossy; upright, medium (4 ft) growth; [White Masterpiece X Silver Jubilee]; Twomey, Jerry, 1999

TWOlad, HT, dp, 1995; (**Aladdins Dream**); PP10149; [Silver Jubilee X Evening Star]; Twomey, Jerry, 1994

'TWOLOY', Gr, pb, 1990; (**Lloyd Center Supreme**); bud pointed; flowers light pink with yellow base, dark pink blending to yellow reverse, exhibition form, 25 petals, exhibition form, moderate, fruity fragrance; foliage medium, dark green, glossy; upright, bushy, medium growth; PP07979; [Brion X Seedling]; Twomey, Jerry, 1988; DeVor Nurseries, Inc., 1990

'TWOMIN', Min, or, 1989; (**Apache Princess**); bud ovoid; flowers bright orange-red, medium, dbl., 38 petals, cupped, borne singly, slight, fruity fragrance; prickles declining, purple; foliage medium, medium green, semi-glossy; upright, medium growth; PP008064; [(Cricket X Christ 78) X Seedling]; Twomey, Jerry; DeVor Nurseries, Inc., 1990

'TWOPAUL', HT, mr, 1999; (**Dream Red**); 3.5 in., borne in small clusters, slight fragrance; prickles moderate; foliage medium, dark green, glossy; bushy, medium (4.5 ft) growth; [Esmeralda X Fireburst]; Twomey, Jerry, 1999

'TWORIGHT', F, or, 1993; (**Fireburst**); very dbl., 3–3.5 in., borne mostly singly, slight fragrance; some prickles; foliage medium, medium green, semi-glossy; medium (111 cms), bushy growth; [Seedling X Evening Star]; Twomey, Jerry; DeVor Nurseries, Inc., 1993

Two-Timer *see* 'MORSWISS'

'TWOWIN', Gr, ob, 1989; (**Winning Colors**®); bud ovoid; flowers orange-yellow blend, medium, borne singly, cupped, 60 petals, cupped, moderate, musk fragrance; prickles declining, yellow-green; foliage medium, dark green, glossy; upright, medium growth; PP07907; [Gingersnap X Marina]; Twomey, Jerry; DeVor Nurseries, Inc., 1990

'TWOYEL', HT, my, 1999; (**Dream Yellow**); dbl., 17–25 petals, 5 in., borne mostly singly, intense fragrance; prickles moderate; foliage medium, medium green, semi-glossy; upright, medium (3 ft) growth; [(Sonia X Prominent) X Whisky Mac]; Twomey, Jerry, 1999

'TWYFORD', HT, op, 1939; bud long; flowers bright salmon-pink, base gold, reverse deep salmon flushed orange, moderate fragrance; foliage dark, reddish; vigorous growth; Waterer

Tycoon *see* 'BRITY'

'TYLER'®, HT, rb, 1988; (Sebago); flowers large, dbl., 15–25 petals, slight fragrance; foliage medium, dark green, matt; upright growth; [(Tamango X Red Planet) X First Prize]; Poor, Cuyler

Tynwald *see* 'MATTWYT'

Typ Kassel, C, mp

'TYPHOO TEA', HT, rb, 1974; (Doux Parfum, Été Parfumé); flowers medium red, silver reverse, classic form, dbl., 50 petals, 5 in., moderate fragrance; foliage small, glossy; [Fragrant Cloud X Arthur Bell]; McGredy, Sam IV

'TYPHOON', HT, ob, 1972; (Taifun); flowers salmon, shaded yellow, dbl., 35 petals, 4 in., intense fragrance; [Dr. A.J. Verhage X Colour Wonder]; Kordes, R.; McGredy

'TYRIANA', HT, dp, 1963; bud pointed, ovoid; flowers rose-pink, well formed, dbl., 40 petals, 4.5–5 in., moderate fragrance; foliage leathery, dark; vigorous, upright growth; [(Happiness X Independence) X Paris-Match]; Meilland, Alain A.; URS;, Wheatcroft Bros.

'TYRIUS', HT, m, 1972; flowers tyrian purple, dbl., 20 petals, 5 in., slight fragrance; foliage glossy, bronze; [Bettina X Prima Ballerina]; Gandy, Douglas L.

'TZIGANE', HT, rb, 1951; (Tiz); bud ovoid; flowers rose-red, reverse yellow, to cactus-formed, large, dbl., cupped, moderate fragrance; foliage dark, glossy, leathery; upright, bushy growth; [Peace X J.B. Meilland]; Meilland, F.; Hennessey, 1956;, URS, 1951

'TZIGANE, CLIMBING', Cl HT, rb, 1958; Lagoona Nursery; Roseglen Nursery, 1958;, URS, 1960;, Wheatcroft Bros., 1958

U

'U.P. Hedrick', HSpn, mp, 1932; flowers pink, open, large, single, profuse, non-recurrent bloom, moderate fragrance; foliage soft, dark; vigorous (6 ft.), bushy, compact growth; [R. spinosissima altaica X Betty Bland (probably)]; Central Exp. Farm

'Uetersen'®, S, mr, 1939; (Zenith); flowers glowing red, semi-dbl., recurrent bloom; upright, bushy growth; [K. of K. X Stammler]; Tantau

Ufhofen, S, op, 1964; flowers medium, dbl.; Berger, W.; (Sangerhausen)

'Uhland', HMsk, yb, 1916; flowers reddish yellow, petals fringed, borne in clusters of 3-15; foliage pointed, like Tip-Top; [Geheimrat Dr. Mittweg X Tip-Top]; Lambert, P.

'UHLater', HRg, dp, 1989; (**Buffalo Gal**, Foxi, Foxi Pavement); flowers lavender-pink, loose form, dbl., 17–25 petals, 3 in.., borne in clusters, intense fragrance; foliage large, light green, wrinkled, shiny; upright, 3 - 4 ft. growth; Uhl, J.

UHLensch, HRg, w, 1989; (**Snow Owl**); Uhl, J.

UHLwe, HRg, dp; (**Yankee Lady**); Uhl, J., 1987

Ukrainian Dawn *see* **'Ukrainskaia Zorka'**

'Ukrainskaia Zorka', F, mr, 1955; (Ukrainian Dawn); flowers bright cinnamon-red, medium, slight fragrance; [Independence X ?]; Klimenko, V. N.

Ulla Land, Misc OGR, mr

Ulmer Münster® *see* 'KORtello'

Ulrich Brünner *see* **'Ulrich Brünner Fils'**

'Ulrich Brünner Fils', HP, dp; (Ulrich Brünner); flowers geranium-red to carmine, large, dbl., 30 petals, cupped, intense fragrance; vigorous growth; (28); Levet, A., 1882

'Ulrick's Buttercup', F, ly, 1953; bud ovoid; flowers medium, borne in clusters, dbl., cupped, intense fragrance; foliage bronze; very vigorous, bushy growth; [Yvonne Rabier X Baby Alberic]; Ulrick, L.W.

'Ulrick's Gem', F, pb; flowers deep pink and white, borne in clusters, very dbl.; foliage glossy; very vigorous, bushy growth; [Mrs Tom Henderson X Self]; Ulrick, L.W.

'Ulrick's Red', HT, lp; Ulrick, L.W., 1954; (Weatherly, L.)

'Ulrick's Smokie', F, m, 1953; bud long, pointed; flowers smoky mauve, background white, large, borne in clusters, very dbl., intense fragrance; foliage light green; bushy growth; [Mrs Tom Henderson X Tip-Top]; Ulrick, L.W.

'Ulrick's Yellow', HT, my, 1953; bud globular; flowers yellow, center darker, medium, dbl., cupped, intense fragrance; foliage light green; bushy growth; [Mrs Pierre S. duPont X Lady Hillingdon]; Ulrick, L.W.

Ulrike®, F, r

'Ulster', HP, mr; flowers salmon-red, large, dbl., slight fragrance; Dickson, A., 1899; (Sangerhausen)

'Ulster Gem', HT, my, 1917; bud long, pointed; flowers canary-yellow, large, single, slight fragrance; GM, NRS, 1916; Dickson, H.

'Ulster Monarch', HT, r, 1951; flowers apricot shaded buff, high-pointed, medium, dbl., 50 petals, slight fragrance; foliage glossy, bright green; upright growth; [Sam McGredy X Mrs Sam McGredy seedling]; McGredy, Sam IV

'Ulster Queen', F, ob, 1960; flowers salmon-orange, well formed, borne in clusters, dbl., 25 petals, 3 in., slight fragrance; vigorous growth; [Cinnabar X Independence]; McGredy, Sam IV; McGredy

'Ulster Volunteer', HT, mr, 1918; flowers brilliant cherry-red, base clear white, single, 5–6 in.; Dickson, H.

Ultimate Pink™ *see* 'JACval'

Ultimate Pleasure *see* 'TINpleasure'

Uma Rao, HT, pb; Pal, Dr. B.P., 1989

Umberglo™ *see* 'MINIumber'

'UMScharm', Min, pb, 1990; (**Single Charm**); flowers hand-painted white with pink edge, small, borne mostly singly, single, 5 petals, no fragrance; foliage small, medium green, matt; tall, upright, bushy growth; [Jennifer X Unnamed seedling]; Umsawasdi, Dr. Theera

'UMSdad', HT, dr, 1990; (**Manit**); flowers medium, borne mostly singly, semi-dbl., 6–14 petals, no fragrance; foliage medium, medium green, matt; upright, bushy, tall (150-180 cms) growth; [Unnamed seedling X Olympiad]; Umsawasdi, Dr. Theera, 1991

UMSlove, F, w, 1995; (**So In Love**); flowers near white, blooms borne mostly single, dbl., 26–40 petals, 1.5–2.75 in., moderate fragrance; few prickles; foliage medium, medium green, semi-glossy; medium, bushy, spreading growth; [Golden Wings X Unknown]; Umsawasdi, Dr. Theera

'UMSnira', Min, ab, 1991; (**Niramol**); flowers moderately blooms borne mostly singly, dbl., no fragrance; foliage medium, medium green, semi-glossy; medium, bushy growth; [Loving Touch X Unnamed seedling]; Umsawasdi, Dr. Theera

'UMSpretty', Min, dp, 1990; (**Pretty 'n' Single**); flowers deep pink, small blooms, single, 5 petals, intense fragrance; foliage medium, dark green, semi-glossy; upright, bushy, tall growth; [Nymphenburg X Libby]; Umsawasdi, Dr. Theera, 1986; Theera Umsawasdi

'UMSpriub', HT, ly, 1990; (**Priub**); flowers yellowish-cream turning to pure white, large blooms, single, 5 petals, slight fragrance; many prickles; foliage medium, medium green, semi-glossy; vigorous, upright, tall growth; [Unnamed seedling X Unnamed seedling]; Umsawasdi, Dr. Theera, 1986; Theera Umsawasdi

'UMSrum', Min, r, 1990; (**Rum Candy**); flowers brownish apricot, light apricot shaded light pink when fully, semi-dbl., 6–14 petals, slight fragrance; foliage small, light green, matt; upright, bushy, low (18 cms) growth; [Twilight Trail X Unknown seedling]; Umsawasdi, Dr. Theera, 1991

'UMStar', Min, dr, 1990; (**Ruby Star**); flowers deep red, small, borne mostly singly, single, 5 petals, no fragrance; foliage medium, dark green, semi-glossy; upright, bushy, tall (24 cms) growth; [High Spirits X Unknown seedling]; Umsawasdi, Dr. Theera, 1991

'UMSTEX', Min, yb, 1990; (**Texas Sunrise**); flowers yellow at times with pinkish edge, moderately small, dbl., 15–25 petals, no fragrance; foliage medium, medium green, semi-glossy; upright, bushy growth; [Arizona Sunset sport]; Umsawasdi, Dr. Theera, 1991

'UNA', HCan, ly, 1900; flowers buff-yellow becoming creamy white, semi-dbl., non-recurrent; vigorous growth; [Tea X R. canina]; Paul

'UNA HAWKEN', F, ly, 1972; bud ovoid; flowers butter-yellow to cream, medium, dbl., cupped, moderate fragrance; foliage glossy, leathery; vigorous, upright, bushy growth; [Arthur Bell X Arthur Bell]; Murray, Nola; Rasmussen's

'UNA WALLACE', HT, dp, 1921; flowers soft, even-toned cherry-rose, well formed, dbl., moderate fragrance; long, strong stems; vigorous growth; GM, NRS, 1920; McGredy

Uncle Bill, HT, mp; Beales, Peter, 1984

'UNCLE JOE', HT, dr, 1972; (El Toro, Toro); flowers very large, very dbl., exhibition form; foliage large, dark, leathery; very vigorous, upright, tall growth; [(Mirandy X Charles Mallerin) X Unnamed seedling]; Kern Rose Nursery, 1971

'UNCLE LOU', HT, dp, 1997; flowers deep pink, medium sized, dbl., 26–40 petals, borne singly, slight fragrance; foliage medium, medium green, semi-glossy; upright, tall (4.5 - 5 ft) growth; [Queen Elizabeth X Christian Dior]; Roth, Louis A.

'UNCLE SAM', HT, dp, 1965; flowers deep rose-pink, large, exhibition form, moderate fragrance; foliage dark, leathery, glossy; vigorous, tall growth; [Charlotte Armstrong X Heart's Desire]; Warriner, William A.; J&P

Uncle Steve *see* 'RESUNC'

Uncle Walter *see* 'MACON'

'UNDINE', HT, ob, 1901; flowers dark orange, medium, moderate fragrance; upright, bushy growth; [L'Ideal X Sunset]; Jacobs

Unermüdliche, Ch, m; Lambert, P., 1904

Unforgettable *see* 'TROBSPREAD'

Unforgettable *see* 'JACHYP'

Unforgettable *see* TANalam

UNICEF *see* COCjojo

Union Redwood Cemetary, HP, lp; (found rose)

Union-Rose St Helena® *see* 'CANLISH'

'UNIQUE', Ch, w; flowers white edged pink, compact; Laffay, M.

'UNIQUE', LCl, op, 1928; flowers bright fawn-orange-salmon, recurrent bloom; vigorous growth; [Tip-Top X Hybrid Perpetual]; Evans

'UNIQUE BLANCHE', C, w, 1778; (Blanche Unique, Vièrge de Clery, White Provence); flowers white, sometimes tinged pink, large, dbl., moderate fragrance; Grimwood

Unique de Provence *see* **'UNIQUE MOSS'**

'UNIQUE MOSS', M, w, 1844; (Unique de Provence); flowers pure white, occasionally tinted pink, well-mossed, large, dbl.; prickles shoots very spiny; [Unique Blanche sport]; Robert

'UNIQUE PANACHÉE', C, w, 1821; flowers white, faintly striped rose and lilac, large, dbl., globular; vigorous growth; [Unique Rouge sport]; Caron

'UNIQUE ROUGE', C, mp; flowers bright pink, medium

Unitarian Cemetary Hybrid China, HCh, lp; (found rose)

'UNITED NATIONS', F, op, 1949; flowers salmon-pink, open, medium, borne in clusters, dbl., 26 petals, moderate fragrance; foliage glossy; very vigorous, bushy growth; [Mev. Nathalie Nypels X Rosamunde]; Leenders, M.

United Nations Rose *see* 'HARBELLA'

'UNITY', HT, dp, 1956; flowers deep pink, reverse slightly darker, well shaped, dbl., 45 petals, 5.5 in., intense fragrance; foliage dark; vigorous, upright growth; [Red Ensign sport]; Sansum

'UNIVERSAL FAVORITE', HWich, mp, 1898; flowers soft rose, borne in large clusters, dbl., moderate fragrance; vigorous, climbing growth; [R. wichurana X Paquerette]; Horvath; W.A. Manda

'UNIVERSITÉ D'ORLÉANS', Pol, mr, 1966; bud globular; semi-dbl., cupped, borne in large clusters; vigorous growth; [? X Ronde Endiablee]; Hémeray-Aubert

'UNN', F, dp, 1972; bud ovoid; flowers deep pink, open, medium, semi-dbl., slight fragrance; foliage glossy; dwarf, moderate growth; [Rimosa X Fidélio]; Lundstad

'UNSER STOLZ', HT, dr, 1960; flowers bright crimson-scarlet, large, dbl.; foliage leathery, dark, glossy; upright growth; [Ena Harkness X Unnamed seedling]; Verschuren, A.; van Engelen

Uppingham School *see* 'FRANBURST'

Upstart *see* 'JACUP'

Uptown *see* 'TALUPTOWN'

'URANIA', HP, dr, 1906; flowers bright crimson, dbl., recurrent bloom; [American Beauty X Suzanne-Marie Rodocanachi]; Walsh

Urara *see* 'KEIFUPIRA'

'URDH', HP, mp; bud long, pointed; flowers large, dbl., 45 petals, exhibition form, intense fragrance; vigorous growth; [Victor Verdier X Papa Lambert]; Tantau, 1930; C-P, 1930

'URSEL', Pol, mr, 1938; flowers carmine red, medium, semi-dbl.; Vogel, M.; (Sangerhausen)

'URSEL TGARTH', HFt, yb, 1938; flowers yellow and red, medium, dbl.; Ketten, Gebrüder; (Sangerhausen)

Ursula® *see* 'LAPON'

'USCHI', F, dp, 1972; flowers full, medium, dbl., slight fragrance; foliage glossy, dark; vigorous, bushy growth; [Marimba sport]; Tantau, Math.

'USHA', F, dr, 1975; bud ovoid; flowers deep red, open, small, dbl., 35–40 petals, 1.5 in.; foliage soft; compact, bushy growth; [Orangeade X ?]; IARI

Usmev, HT, lp; Strnad; (Czech Rosa Club)

Usvit, F, op, 1978; flowers salmon-pink, medium, dbl., slight fragrance; Urban, J.; (Sangerhausen)

'UT TYLER ROSE', Gr, ob, 1984; flowers orange, medium, dbl., 35 petals, slight fragrance; foliage large, medium green, matt; upright growth; [Unnamed seedling X Unnamed seedling]; Weeks, O.L.; University of Texas at Tyler

Uta Maro, HT, mr; flowers red with silvery reverse, large, dbl., moderate fragrance; (Sangerhausen)

'UTAGE', HT, w, 1977; bud pointed; flowers very large, dbl., 30–35 petals, exhibition form; foliage leathery; vigorous, upright growth; [Edith Krause X Bridal Robe]; Ota, Kaichiro

Utage, HT, w

'UTRO MOSKVY', HT, mp, 1952; (Morning in Moscow, Moscow Morn); flowers soft rose tinted carmine, large, dbl., 60 petals, slight fragrance; foliage leathery, grayish; spreading growth; GM, International Exhibition, 1961; [Frau Karl Druschki X Independence]; Shtanko, E.E.

'UTTAM', HT, lp, 1969; bud pointed; flowers pastel pink, full, large, dbl.; foliage glossy; vigorous, upright growth; [Elite X Open pollination]; IARI

Uwe Seeler® *see* 'KORSEE'

V

'V FOR VICTORY', HT, ly, 1941; bud ovoid, long, pointed; flowers yellow faintly tinted orange, dbl., 45 petals, exhibition form, intense fragrance; foliage glossy; vigorous, bushy, compact to open growth; [Golden Glow X Condesa de Sástago]; Brownell, H.C.

V G Glowing HT, HT, lp

V O N Canada, F, lp

V Y F Rose, HT, lp

'V. VIVIAND MOREL', T, dr; flowers rich crimson shaded dark red and carmine, large, dbl.

Vabene *see* 'MEITROGANA'

'VAGABONDE', F, ob, 1962; bud long, pointed; flowers salmon-orange, blooms in clusters of 3 or more, dbl., 25 petals, exhibition form; foliage dark, glossy; vigorous, bushy growth; [Mannequin X Fashion]; Lens

'VAHINE', HT, or, 1964; flowers dark cardinal-red tinted orange; strong stems; vigorous, upright growth; Combe; Vilmorin-Andrieux

'VAINQUEUR', HT, m, 1937; bud ovoid, dark red; flowers velvety purple, reverse dark red, very large, very dbl., moderate fragrance; foliage bright green; long, strong stems; very vigorous growth; [Sensation seedling]; Heizmann, E.; A. Meilland

Vaire *see* **'CHÂTEAU DE VAIRE'**

Vaishnavi, HT, or; K&S, 1992

Vajont, F, 1966; Galesso, G.; (Cavriglia)

Val Boughey *see* **'VALERIE BOUGHEY'**

Val d'Authion, HT, mp; Minier, 1967

'VAL DE MOSA', HT, my, 1968; very dbl., 70 petals, moderate fragrance; foliage dark, glossy; vigorous growth; [La Jolla X Cynthia Brooke]; Ellick

Valdemar, HMult, lp; Back, 1980

Valdemossa, HT, 1991; Dot, Simon; (Cavriglia)

Vale of Clwyd *see* 'BEEVAL'

Valeccia *see* KOReklia

'VALENCE DUBOIS', HGal, mp, 1880; flowers medium, dbl.; Fontaine; (Sangerhausen)

'VALENCIA', HT, ab, 1967; flowers apricot-orange, large, dbl., exhibition form, moderate fragrance; foliage glossy, leathery; vigorous, upright growth; [Golden Sun X Chantré]; Kordes, R.; J&P

Valencia® *see* KOReklia

Valencia 89 *see* KOReklia

'VALENCIENNES', LCl, dr, 1960; bud long, pointed; flowers medium, borne in clusters, semi-dbl., slight fragrance; foliage dark, glossy; very vigorous growth; [Paul's Scarlet Climber X Seedling]; Robichon

'VALENTIN BEAULIEU', HWich, m, 1902; flowers violet-pink, medium, dbl.; Barbier; (Sangerhausen)

Valentina Borgatti, HT, 1969; Borgatti, G.; (Cavriglia)

Valentina Cortese, F, 1976; Cazzaniga, F. G.; (Cavriglia)

'VALENTINE', F, mr, 1951; flowers bright red, blooms in large clusters, semi-dbl., 18 petals, 2.5–3 in., slight fragrance; foliage dark olive-green; spreading, bushy, compact growth; [China Doll X World's Fair]; Swim, H.C.; Armstrong Nursery

Valentine Heart *see* 'DICOGLE'

Valeria Sykes *see* 'HORFLASHROB'

'VALERIE', F, w, 1932; bud pointed, yellow; flowers cream, large, borne in large clusters; foliage glossy, dark; bushy growth; GM, NRS, 1931; Chaplin Bros.

Valerie, HT, lp; Select, 1994

'VALERIE BOUGHEY', HT, op, 1960; (Val Boughey); flowers coppery salmon-pink, high pointed, dbl., 36 petals, 6 in., moderate fragrance; foliage leathery, glossy; vigorous, upright growth; [Tzigane sport]; Fryers Nursery, Ltd.

Valerie Jeanne *see* 'SAVAVAL'

Valerie June, HT, lp; Allender, Robert William, 1982

Valerie Kathleen *see* 'HORVALKATH'

'VALERIE PURVES', HT, mp, 1940; flowers pink, well formed, moderate fragrance; vigorous growth; Clark, A.

Valerie Swane *see* 'AROBIPY'

'VALETA', F, or, 1960; flowers red shaded vermilion, open, borne in clusters (up to 20), dbl.; foliage dark; vigorous growth; [Signal Red X Fashion]; deRuiter

Valfleury, F, 1971; Croix; (Cavriglia)

Valia Balkanska *see* 'MANCLASSIC'

'VALIANT', HT, mr, 1948; bud long, pointed; flowers bright red, large, dbl., 30 petals, exhibition form, moderate fragrance; foliage dark, leathery; vigorous, upright, branching growth; [Poinsettia X Satan]; Boerner; J&P

VALkita, F, dr, 1995; (**Mariska**); (VALkita)

'VALLDEMOSA', F, mr, 1956; flowers fiery red, large, dbl., 20 petals, globular; foliage glossy; vigorous, open growth; [Magrana X Radar]; Dot, M.

'VALROSE', LCl, w, 1964; flowers white suffused pink at edges, medium, borne in clusters of 3, semi-dbl., recurrent bloom; Mondial Roses

'VALSTAR', HT, dp, 1962; flowers deep pink, well formed, large, dbl., intense fragrance; foliage leathery; strong stems; Mondial Roses

'VAMP', Min, dr, 1981; single, 5 petals, cupped, borne 3 per cluster, no fragrance; pointed prickles; foliage dark; angular growth habit; [Fairy Moss X Fairy Moss]; Gatty, Joseph

'VAN ARTEVELDE', HGal, dp; flowers deep pink, petals imbricated in whorls, large, very dbl.; Parmentier, before 1847

'VAN BERGEN', S, ly, 1980; bud pointed; flowers light yellow fading to white, blooms borne 3, 5, 7 per clust, dbl., 48 petals, repeat bloom, slight, fruity fragrance; hooked, red prickles; foliage small, glossy, dark; vigorous, compact, bushy growth; [(R. wichurana X Baronne Prevost) X Fun Jwan Lo]; James, John

Van Gogh™ *see* 'WILGOGH'

Van Houtten, HP, lp

Van Huyssum, HGal, lp; Parmentier, 1800's

Van Nes *see* **'PERMANENT WAVE'**

'VAN ROSSEM'S JUBILEE', HT, op, 1937; flowers bright carmine or coral; foliage glossy, bronze; vigorous growth; Van Rossem

'VANAMALI', HT, m, 1978; bud long, pointed; flowers orchid-mauve, full, dbl., 35–40 petals, 6 in., exhibition form, moderate fragrance; foliage dark, leathery; tall, vigorous, bushy growth; [Lady X X ((Gruss an Teplitz X Unknown) X (Lake Como X Angel Face))]; Viraraghavan, M.S.

Vancouver, HT, lp; Twomey, Jerry, 1997

Vancouver Centennial *see* **'JAN WELLUM'**

'VANDA BEAUTY', HT, dy, 1971; flowers pointed, dbl., 28 petals, 3.5 in., moderate fragrance; foliage glossy, dark; very free growth; [Gertrude Gregory X Unnamed seedling]; Gregory

'VANDAEL', M, m, 1850; bud well mossed; flowers rich purple, edged lilac, large, dbl.; vigorous growth; Laffay, M.

'VANESSA', HT, pb, 1946; bud ovoid; flowers coral, reverse yellow, dbl., 25 petals, 4.5 in., intense fragrance; foliage bright green; vigorous growth; [Arch. Reventos X Lord Baden-Powell]; Leenders, M.; Longley

Vanessa Belinda *see* 'BRANDYGLOW'

Vanessa Campello® *see* FEbesa

'VANGUARD', HRg, op, 1932; flowers orange-salmon, large, dbl., moderate fragrance; foliage light, very glossy; vigorous (to 10 ft) growth; David Fuerstenberg Prize, ARS, 1934 Dr. W. Van Fleet Medal, ARS, 1933; [(R. wichurana X R. rugosa alba) X Eldorado]; Stevens, G.A.; J&P

Vanilla® *see* KORplasina

Vanilla Perfume™ *see* 'JACWOTTE'

'VANITY', HMsk, dp, 1920; flowers rose-pink, near blooms in very large sprays, single, recurrent bloom, intense fragrance; foliage rich green, leathery; very vigorous (to 8 ft.), bushy growth; (21); [Château de Clos Vougeot X Seedling]; Pemberton

Vanity, HT, lp; Halstead, 1901; Select Roses, 1999

'VANITY FAIR', HT, lp, 1942; bud long, pointed; flowers cameo-pink, large, semi-dbl., 19 petals, exhibition form, slight fragrance; foliage soft; long stems; very vigorous, upright growth; [Better Times X Golden Rapture]; Roberts; Totty, 1944

Van's Choice, F, lp; Williams, J. Benjamin, 1999

'VANTO', Pol, dr; [Dick Koster sport]; Vanto

'VARBOLE', HT, mr, 1981; (**Caesar**); bud ovoid; flowers cardinal red, blooms borne 2-3 per cluster, dbl., 33 petals, exhibition form, slight, tea fragrance; red prickles; foliage leathery, medium green, semi-glossy; upright, branched growth; [Ilona X Seedling]; van Veen, Jan; Carlton Rose Nurseries

'VARELVI', HT, ab, 1978; (**Elvira**); flowers large, dbl., 35 petals, no fragrance; foliage medium, medium green, semi-glossy; upright growth; [Zorina X Dr. A.J. Verhage]; van Veen, Jan; G. Verbeek

Variant, F, op; flowers medium, dbl.; VEG; (Sangerhausen)

'VARIEGATA DI BOLOGNA', B, rb, 1909; flowers white, striped purplish red, large blooms in clusters of 3-5, dbl., globular, moderate fragrance; vigorous (6-8 ft) growth; Bonfiglio, A.

Variegata di Bologna Rouge, B, dr; Lowe, 1984

Varieté, F, m, 1987; flowers purple/pink with lighter shading, large, dbl., slight fragrance; VEG; (Sangerhausen)

'VARIETY CLUB', F, yb, 1965; flowers yellow marked rose-red, well-formed blooms in clusters, dbl., 48 petals, slight fragrance; foliage dark; [Columbine X Circus]; McGredy, Sam IV; McGredy

'VARIORA', HT, my, 1981; (**Eloira**); flowers large, dbl., 35 petals, no fragrance; foliage medium, medium green, semi-glossy; upright growth; [Elvira X Seedling]; van Veen, Jan; G. Verbeek

'VARLON', HT, mr, 1973; (**Ilona**); dbl., 40 petals, 4–4.5 in., slight fragrance; foliage glossy, leathery; bushy growth; [Miracle X (Romantica X Edith Piaf)]; Verbeek

Varna, S, dy, 1979; flowers large, dbl.; GPG Bad Langensalza; (Sangerhausen)

Varo Iglo, HT, w, 1969; flowers medium, dbl.; Verbeek; (Sangerhausen)

Varo Rania, HT, 1969; Verbeek; (Cavriglia)

Varsha, F, w; K&S, 1992

'VASANT', HT, yb, 1980; bud pointed; flowers yellow edged pink, blooms borne 10 per cluster, exhibition form, slight, spicy fragrance; straight, brown prickles; foliage dark; upright, bushy growth; [Sweet Afton X Delhi Princess]; Division of Vegetable Crops and Floriculture

Vasavi, HT, dr; Kasturi, 1976

'VASCO DA GAMA', F, mr; flowers velvety red; [Pinocchio X Alain]; da Silva, Moreira

'VASSAR CENTENNIAL', HT, pb, 1961; flowers peach to shell-pink, dbl., 30 petals, 4.5–5 in., exhibition form, moderate fragrance; foliage dark, leathery; upright, branching growth; [Helene de Roumanie X Confidence]; Meilland, Mrs. Marie-Louise; C-P

'VATER RHEIN', HT, dr, 1922; flowers very dark red, dbl., intense fragrance; [Kynast X Unnamed seedling]; Kiese

'VATERLAND', HT, dr, 1928; flowers dark red with coppery reflexes, large, dbl., moderate fragrance; foliage bronze, leathery; vigorous growth; [National Emblem X Earl Haig]; Berger, V.; Pfitzer

'VATERTAG'®, Pol, ob, 1959; (Father's Day, Fête des Pères, Jour des Pères, Orange Muttertag); flowers salmon-orange; [Mothersday sport]; Tantau, Math.

Vatican® *see* 'DELOP'

'VEDETTE', HT, mr, 1951; flowers brilliant red, large, dbl., 28 petals, moderate fragrance; foliage leathery; [(Frau Karl Druschki X George Dickson) X Unnamed seedling]; Gaujard

'VEE BRYAN', F, w, 1998; flowers cream, edged pink maroon, full, dbl., 26–40 petals, 2 in., borne in large clusters, moderate fragrance; [Solitaire X Solitaire]; Jones, L.J.; Haynes Roses, 1997

Vee Marie *see* 'JUDVEE'

'VEILCHENBLAU', HMult, m, 1909; (Blue Rambler, Blue Rosalie, Violet Blue); flowers violet, petals streaked with white, center white, yellow sta, semi-dbl., 1.25 in., cupped, moderate fragrance; very few prickles; foliage large, pointed, glossy, light; short stems; vigorous, climbing (10-15 ft) growth; [Crimson Rambler X Erinnerung an Brod]; Schmidt, J.C.

Velay Rose, LCl, 1972; Croix; (Cavriglia)

Veldfire *see* 'KORGUST'

'VELINDRE', HT, mr, 1998; flowers red-pink with peach reverse, full, classic, dbl., 26–40 petals, 5.5 in., borne singly, moderate fragrance; some prickles; foliage large, dark green, semi-glossy; upright, bushy, medium growth; [Solitaire X seedling]; Poole, Lionel

Velingstorprosen, HGal, lp

Velizy® *see* 'DELSAMOUR'

'VELLUTO', HMoy, dr, 1934; bud long, pointed; flowers velvety dark crimson, stamens red, medium, semi-dbl., intermittent bloom; foliage dark; [R. moyesii X J.C. Thornton]; San Remo Exp. Sta.

'VELOUR', F, mr, 1967; bud ovoid; dbl., flat, moderate fragrance; foliage glossy; vigorous, upright, bushy growth; [Garnette seedling X Hawaii seedling]; Boerner; J&P

Velours Épiscopal, HCh, m; flowers violet purple/pink, large, dbl.; Roseraie de l'Hay; (Sangerhausen)

Velours Pourpre, HP, m; Verdier, 1866

'VELSHEDA', HT, lp, 1936; flowers softest rose-pink, well formed, large, dbl., moderate fragrance; foliage dark; strong, erect stems; vigorous growth; Cant, F.

'VELUTINA', HGal, m, 1810; flowers velvety purple shaded violet, golden stamens, three rows of; Van Eeden

'VELUWEZOOM', HT, dp, 1909; flowers large, dbl., moderate fragrance; Pallandt; (Sangerhausen)

Velvet Arrow, HT, dr

Velvet Beauty *see* **'BARKHATNAIA KRASAVITSA'**

'VELVET BEAUTY', HT, mr; bud high pointed; flowers currant-red, dbl., 40–55 petals, 5–5.5 in., exhibition form, moderate, clove fragrance; vigorous, upright growth; [Happiness X New Yorker]; Fisher, G.; Arnold-Fisher Co.

Velvet Cloak *see* 'SEACLOAK'

Velvet Cover *see* POUlria

Velvet Dreams *see* 'LYOET'

Velvet Flame *see* 'MEIMAUR'

Velvet Flame, Climbing, Cl HT, 1981; Orard, Joseph; (Cavriglia)

Velvet Fragrance *see* 'FRYPERDEE'

'VELVET HOUR', HT, dr, 1978; flowers oxblood-red, full, dbl., 44 petals, 3 in., moderate fragrance; foliage dark; vigorous, upright growth; LeGrice

Velvet Lady *see* 'HADPRETTIE'

Velvet Lustre *see* 'SIMPALNO'

'VELVET MIST'™, HT, m, 1990; (Fragrant Lavendar); flowers deep lavender, borne usually singly, medium, dbl., 25–35 petals, exhibition form, moderate, fruity fragrance; foliage large, medium green, matt, disease-resistant; upright, bushy, medium to tall growth; [Blue Ribbon X Shocking Blue]; Christensen, Jack E., 1984; Flowers of the Month, 1990

'VELVET QUEEN', HT, dr, 1965; flowers blood-red, pointed, 4 in., moderate fragrance; moderate growth; [Fandango X Seedling]; Herholdt, J.A.

Velvet Robe *see* **'ATOMBOMBE'**

Velvet Ruby *see* 'MACHORO'

Velvet Ruby, LCl, lp

Velvet Star *see* BARvelv

Velvet Star *see* TANalednev

'VELVET TIMES', HT, dr, 1960; bud pointed; flowers rose-red, dbl., 40–50 petals, 4.5–5 in., exhibition form, intense fragrance; foliage leathery; vigorous, upright growth; [Better Times sport]; Peters; J&P

Velvet Touch *see* 'SAVAVEL'

Velvet Treasure *see* JACvet

Velveteen *see* 'MEISOYRIS'

'VELVETIER', HT, dr, 1946; bud long, pointed; flowers deep velvety red, dbl., 28–35 petals, 4–5 in., exhibition form, moderate fragrance; foliage glossy; vigorous, upright growth; hardy for the class.; [Pink Princess X Crimson Glory]; Brownell, H.C.

Velvia *see* 'HARXAMPLE'

Vendome, LCl, lp

Vendéme *see* 'GAURA'

Vendulka, F, mp, 1974; flowers large, dbl., moderate fragrance; Vecera, L.; (Sangerhausen)

Venere®, F, mp; Barni, V.

'VENEZUELA', F, op, 1957; flowers salmon, edged dull red; very vigorous growth; [Joanna Hill X Pinocchio]; Silva

'VENISE', HT, rb, 1946; flowers red with silvery white reverse flushed salmon-carmine, large, dbl.; spreading growth; [Joanna Hill X Margaret McGredy]; Meilland, F.

Venise *see* ADAzombar

'VENLO', F, or, 1960; bud pointed; flowers medium, borne in compact clusters, semi-dbl., flat, slight fragrance; foliage dark; moderate, compact growth; [Cinnabar X Fashion]; Leenders, G.

Venrosa, HT, mr, 1973; flowers large, dbl., intense fragrance; St Zila; (Sangerhausen)

Vent des Indes® *see* MAUkeole

'VENTURE', S, mp, 1984; semi-dbl., blooms in clusters, moderate fragrance; foliage medium, light green, glossy; upright, slender growth; [(Charlotte Armstrong X (Cecilia X China Belle)) X Prevue]; James, John

'VENUS', T, m; flowers carmine purple; Schmidt, J.C., 1896

'VENUS', HT, dp, 1921; bud long, pointed; flowers carmine, edge flushed cream, very dbl., moderate fragrance; GM, NRS, 1922; [J. Barriot X Sunburst]; Bees

'VENUS', F, w, 1955; flowers pure white, well shaped, borne in large trusses; long stems; vigorous, bushy growth; [Pinocchio seedling]; Maarse, G.

Venus, M, pb; Welter, 1904

Venus *see* JACeve

Venusic® *see* 'DELDRA'

'VENUSTA PENDULA', Ayr, w, 1928; bud pink; flowers blush white, fading to creamy white, small blooms in cluster, semi-dbl., early flowering, no fragrance; vigorous (to 15 ft.) growth; Re-int. Kordes, 1928

'VENU-VAISHALI', HT, pb, 1970; bud ovoid; flowers light pink, striped white, base yellow, full, large, dbl.; foliage large, soft; very vigorous, upright growth; [Astree sport]; Deshpande

'VERA', HT, op, 1922; flowers deep salmon, shaded coral-red, dbl.; Paul, W.

'VERA ALLEN', HT, op, 1939; flowers salmon-pink, well formed, large, dbl., intense fragrance; compact growth; Dickson, A.

'VERA CRUZ', HT, pb, 1938; bud long, pointed; flowers pink shaded mauve, flushed red and yellow, large, dbl., exhibition form, slight fragrance; foliage soft; vigorous growth; [Frank Reader X Johanniszauber]; da Silva, Moreira

'VERA DALTON', F, mp, 1961; bud pointed; flowers soft pink, blooms in clusters, dbl., 24 petals, 4 in., cupped, moderate fragrance; foliage glossy, dark; vigorous, bushy growth; [(Paul's Scarlet Climber X Paul's Scarlet Climber) X (Mary X Queen Elizabeth)]; Norman; Harkness

Vera Johns® *see* 'KORVERA'

Vera Johns *see* 'KORVEJOH'

'VERASTELLA', HT, mp, 1954; flowers rose, center deeper; strong stems; vigorous growth; Giacomasso

Verbeeka, F, lp

Verbesserte Tantau's Triumph *see* **'CINNABAR IMPROVED'**

Vercors *see* 'HARVINTAGE'

'VERCORS', HT, or, 1946; flowers brilliant orient red, large, dbl., globular, moderate fragrance; vigorous growth; [(Mme Arthaud X Mme Henri Guillot) X (Comtesse Vandal X Brazier)]; Mallerin, C.; A. Meilland

Verdi *see* 'INTERPOOL'

Verdi, HMsk, lp; Lens, 1984

'VERDUN', Pol, mr, 1918; flowers vivid carmine-red, large blooms in clusters of 25-50, dbl.; vigorous, dwarf growth; Barbier

'VERDUN SUPERIOR', Pol, 1918; Barbier, A.; (Cavriglia)

'VERENA', F, dr, 1973; bud ovoid; flowers dark velvety red, center lighter, medium, dbl., intense fragrance; foliage glossy, bluish green; upright, bushy growth; [Lucy Cramphorn X Inge Horstmann]; Hetzel; GAWA

Vergenal, HP, lp

'VERJO', HT, dp, 1970; (**Forsythe**); flowers venetian pink, carmine-rose, dbl., 45–50 petals, 4.5–5 in., exhibition form, intense fragrance; foliage glossy, dark, leathery; upright growth; [Miracle X Dr. A.J. Verhage]; Verbeek

Verlaine, HT, w

'VERMILLON', HT, or, 1929; flowers scarlet tinged orange, base yellow, semi-dbl., slight fragrance; [Constance X Paul's Scarlet Climber]; Barbier; Dreer

'VERNA MACKAY', HT, ly, 1912; flowers buff to bright lemon-yellow; Dickson, A.

'VERONA', F, lp, 1963; bud long, pointed to urn-shaped; dbl., 42 petals, 2.5–3 in., exhibition form, abundant bloom; foliage leathery, dark; vigorous, bushy growth; [Spartan X Garnette]; Swim & Weeks

Veronica *see* 'KORFLÜG'

'VERONICA', HT, w, 1950; bud pale yellow; flowers snow-white, imbricated, dbl., 32 petals, 4.5 in., intense fragrance; vigorous growth; Prosser

Veronica 92®, F, mp; Noack, Werner, 1992

Veronica Kay, Min, lp; Welsh, Eric; (Weatherly, L.)

Veronika *see* 'KORFLÜG'

'VERONIQUE', F, mr, 1961; flowers bright raspberry-red, borne in clusters, single, 9 petals, 4 in.; foliage dark, glossy; vigorous, bushy growth; [Sumatra X Philippe]; Delforge

Verrystata, HT, lp; Kordes

Versailles® *see* 'DELSET'

Versailles Palace *see* POUlsail

Verschuren, HT, lp; Verschuren, 1904

'VERSCHUREN'S GLOW', Pol, mr, 1939; flowers medium, semi-dbl.; Verschuren; (Sangerhausen)

'VERSCHUREN'S PINK', HT, op, 1950; flowers salmon-pink with darker reflections, reflexed, dbl., 42 petals, 3–4 in., exhibition form, moderate fragrance; foliage glossy, dark; very vigorous growth; GM, NRS, 1949; [Mme Butterfly X Pink Pearl]; Verschuren; Gregory

Versigny *see* MASversi

Versilia *see* NIPRventyel

'VERY BUSY', Min, pb, 1973; bud long, pointed; flowers pink and yellow, small, dbl., slight fragrance; foliage small, leathery; dwarf, bushy growth; [Perle d'Or X Fairy Princess]; Moore, Ralph S.; Sequoia Nursery

Very Cherry™ *see* 'JACRENEW'

Vesely, HT, pb

'VESENII AROMAT', HT, or, 1955; (Spring Fragrance); flowers red tinted orange, base lighter, medium, dbl., 73 petals, slight fragrance; foliage dark, glossy; very vigorous, spreading growth; [Crimson Glory X Peace]; Klimenko, V. N.

Vesna, HT, lp; Vecera, L., 1970; (Czech Rosa Club)

'VESPER', F, ob, 1966; flowers orange, reverse burnt orange, blooms in clusters, dbl., slight fragrance; foliage small, blue-gray; moderate growth; LeGrice

'VESTA', F, mr, 1946; flowers currant-red, semi-dbl., moderate fragrance; [Irene X Donald Prior]; Leenders, M.

'VESTAL'S CORAL GEM', HT, op, 1939; flowers soft salmon-pink, reverse glowing carmine with coppery sheen, dbl.; foliage leathery, light; vigorous growth; [Betty Uprichard seedling]; Vestal

'VESTAL'S RED', HT, mr, 1937; flowers clear red, large, dbl., cupped, slight fragrance; foliage leathery, light; vigorous growth; Vestal

'VESTAL'S TORCHLIGHT', HT, rb, 1939; bud long, pointed; flowers red and gold, open, large, dbl., slight fragrance; foliage leathery, bronze; vigorous growth; [Pres. Herbert Hoover X Unnamed seedling]; Vestal

Vestey's Pink Tea, T, lp; (found rose)

Vestey's Yellow Tea, T, lp; (found rose)

Vesuv, F, mr, 1963; flowers medium, dbl.; Vilmorin; (Sangerhausen)

'VESUVIUS', HT, dr, 1923; bud long, pointed; flowers dark velvety crimson, large, single, 6 petals, moderate fragrance; foliage light, leathery; vigorous growth; McGredy

'VESUVIUS', F, or, 1963; flowers geranium-red, large; very vigorous growth; GM, Bagatelle, 1963; Vilmorin-Andrieux

Veterans Honor, HP; flowers purple-violet, dbl., cupped

Veteran's Honor ™ *see* 'JACOPPER'

'VEVEY', HT, my, 1953; flowers sun-yellow, large, dbl.; Heizmann & Co.

Vi Ambler, HMult, lp

Via Mala® *see* 'VIAKOR'

Via Romana *see* **'LJUBA RIZZOLI'**®

'VIAKOR', HT, w, 1977; (**Via Mala**®); bud long, pointed; dbl., 33 petals, 4 in., exhibition form, slight fragrance; foliage glossy, dark, leathery; vigorous, upright, bushy growth; [Silver Star X Peer Gynt]; Kordes, W. Söhne

'VIANDEN', HT, pb, 1932; flowers reddish old-rose and pink, reverse ochre-yellow and raw sienna, very dbl., 90–100 petals, intense fragrance; vigorous, bushy growth; [George C. Waud X Ruth]; Ketten Bros.

Vice-President Curtis *see* **'AUTUMN QUEEN'**

Viceroy *see* 'BROROY'

'VICKI KENNEDY', HT, pb, 1976; bud globular; flowers deep rose-pink, center yellow, dbl., 53 petals, 4 in., slight fragrance; foliage large, bronze, upright, bushy growth; [Queen Elizabeth X Red Lion]; Murray & Hawken; Rasmussen's

'VICKIE THORNE', HT, lp, 1972; dbl., 25 petals, 4–4.5 in., intense fragrance; foliage dark; vigorous growth; [Prima Ballerina sport]; Thorne

'VICK'S CAPRICE', HP, pb, 1891; flowers lilac-rose, striped white and carmine, large, dbl., cupped, repeat bloom, moderate fragrance; medium growth; [Archiduchesse Elisabeth d'Autriche sport]; Vick

'VICKY', HT, or, 1972; bud long, pointed; flowers orange-vermilion, moderate fragrance; [Canasta X Peace]; Gaujard

Vicky, F, lp; Noack, Werner, 1994

Vicky Brown, HT, rb; Select, 1994

'VICKY MARFÁ', HT, mp, 1958; bud ovoid; flowers begonia-pink, center yellow, large, dbl., 32 petals, exhibition form, moderate fragrance; strong stems; upright, compact growth; [(Soraya X Ellinor LeGrice) X Henri Mallerin]; Dot, Simon

'VICOMTE MAURICE DE MELLON', HT, ab, 1921; flowers apricot and yellowish salmon with coppery reflexes, washed pink, dbl., moderate fragrance; [Earl of Warwick X Sunburst]; Ketten Bros.

Vicomtesse d'Avesnes, N, mp

'VICOMTESSE DE BERNIS', T, op, 1884; flowers coppery rose to fawn and deep salmon, large, dbl., intense fragrance; Nabonnand, G.

'VICOMTESSE DE CHABANNES', LCl, rb, 1921; flowers purplish crimson, center white, forming a distinct eye, large, semi-dbl.; vigorous, climbing growth; Buatois

'VICOMTESSE DE VEZINS', HP, mp, 1867; flowers large, dbl., moderate fragrance; Gautreau; (Sangerhausen)

'VICOMTESSE PIERRE DU FOU', Cl HT, op, 1923; flowers red aging to deep coral pink, large, dbl., recurrent bloom, intense fragrance; foliage large, glossy, bronze; vigorous, climbing growth; [L'Ideal X Joseph Hill]; Sauvageot, H.

'VICTOR', HT, dp, 1918; bud long, pointed; flowers deep rose, often red, semi-dbl., intense fragrance; [Ophelia seedling X Killarney Brilliant]; Hill, E.G., Co.

Victor Borge *see* POUlvue

Victor Emmanuel, B, dr; Guillot, 1859

'VICTOR FERRANT', HT, dp, 1933; flowers carmine changing to purplish pink, base indian yellow, very, dbl., 60–70 petals, slight fragrance; vigorous growth; [C.W. Cowan X Pres. Cherioux]; Ketten Bros.

Victor Hugo® *see* 'MEIVESTAL'

'VICTOR HUGO', HP, dr; flowers carmine-red shaded purple, medium, dbl., 30 petals, globular, moderate fragrance; vigorous growth; [Charles Lefebvre(?) X ?]; Schwartz, J., 1885

'VICTOR LEMOINE', HP, dr, 1888; flowers large, dbl.; Lévêque; (Sangerhausen)

'VICTOR MAGNIN', Pol, mr, 1930; flowers bright red, borne in large clusters, dbl.; vigorous growth; Van Gelderen

'VICTOR MAYER', HT, dr, 1921; bud long, pointed; flowers blood-red, reflexes deeper; Buatois

Victor Parmentier, HGal, mp; flowers medium, dbl., moderate fragrance; Roseraie de l'Hay; (Sangerhausen)

'VICTOR TESCHENDORFF', HT, w, 1920; flowers almost pure white on pale greenish yellow ground, very large, dbl., exhibition form, moderate fragrance; foliage glossy, dark; long stems; vigorous growth; [Frau Karl Druschki X Mrs Aaron Ward]; Ebeling; Teschendorff

Victor Veladin, T, w

'Victor Verdier', HP, dp, 1859; flowers bright rose, center carmine, large, dbl., 50 petals, globular, moderate fragrance; vigorous growth; [Jules Margottin X Safrano]; Lacharme, F.

'Victor Waddilove', HT, dp, 1923; bud long, pointed; flowers bright carmine-pink, base yellow, large, very dbl.; McGredy

'Victoria', HT, dp, 1924; bud long, pointed; flowers deep rose-pink, center darker, dbl., intense fragrance; [Isobel X ?]; Prince

'Victoria', F, pb, 1946; flowers carmine, center white, large, semi-dbl., moderate fragrance; [Irene X Donald Prior]; Leenders, M.

'Victoria', HT, ly, 1947; flowers pale lemon-yellow, large, moderate fragrance; foliage dark; vigorous growth; [Golden Dawn X Phyllis Gold]; Robinson, H.; Baker's Nursery

'Victoria', Cl F, or; flowers clear geranium-red; [Unnamed seedling X Alain]; da Silva, Moreira

Victoria *see* **'Victoria de los Angeles'**

'Victoria de los Angeles', HT, or, 1952; (Victoria); bud ovoid; flowers velvety geranium-red, medium, dbl., 35 petals, moderate fragrance; vigorous, compact growth; [Cynthia X Manuelita]; Dot, Pedro

Victoria Girls, S, lp; Poulsen, 1997

Victoria Gold *see* WELgold

'Victoria Harrington', HT, rb, 1931; flowers very dark red shaded orange-brown, center lighter, large, dbl., intense, spicy fragrance; foliage leathery, dark; vigorous growth; [Diadem X Hadley]; Thomas; H&S

'Victoria Harrington, Climbing', Cl HT, rb, 1938; Mordigan Evergreen Nursery

'Victoria Hyland', HT, op, 1973; bud ovoid; flowers red-pink to coral, dbl., 34 petals, 4 in., moderate, fruity fragrance; foliage glossy; moderate, compact growth; [Unnamed seedling X Colour Wonder]; Golik; Dynarose

Victoria Park, HT, lp

'Victoria Regina', HT, yb, 1938; flowers golden yellow, reverse brownish yellow, sometimes blushed peach, dbl., 40 petals; vigorous, compact growth; [Nellie E. Hillock X Golden Dawn]; Hillock

Victorian Lace *see* 'DEVcal'

Victorian Memory, LCl, pb

Victorian Spice™ *see* HARzola

'Victoriana', F, ob, 1977; flowers orange, reverse silver, dbl., 28 petals, 5 in., slight fragrance; foliage dark; LeGrice

Victoria's Song™ *see* 'ORTvic'

'Victory', LCl, dp, 1918; flowers deep pink, center darker, large, dbl., moderate fragrance; vigorous, climbing growth; [Dr. W. Van Fleet X Mme Jules Grolez]; Undritz

'Victory', HT, dr, 1920; flowers scarlet-crimson, dbl., moderate fragrance; GM, NRS, 1919; McGredy

Victory Parade® *see* POUlvic

'Victory Red', HT, dp, 1939; flowers rose-red; [Pink Delight sport]; Elliott

'Victory Stripe', HT, rb, 1942; flowers cerise-red variegated white and light pink, dbl., 50 petals, 5 in., moderate fragrance; [Jewel sport]; Grillo

'Victory Year', S, mp, 1951; bud ovoid; flowers clear pink, open, medium, semi-dbl., profuse, non-recurrent bloom, slight fragrance; foliage leathery; very vigorous, upright growth; [Betty Bland X ?]; Wright, Percy H.

'Vida Beglan', F, ly, 1993; flowers lemon yellow with very small traces of salmon, dbl., 26–40 petals, 1.5–2.75 in., borne in small clusters, intense fragrance; many prickles; foliage medium, medium green, matt; medium (90 cms), upright growth; [Elizabeth of Glamis sport]; Beglan, M.; Beglan, 1992

Vidal Sassoon *see* 'MACjuliat'

'Vidiago', HT, mr, 1962; flowers currant-red, reverse geranium-red, large, dbl., 26 petals; very vigorous growth; [Baccará X (S'Agaro X Peace)]; Dot, Simon

'Vidyut', F, dp, 1983; flowers deep pink; [Europeana sport]; Yadava, U.N.; Tata Electric Co.

'Vie en Rose', F, lp, 1994; (La Vie en Rose); flowers bright pink, white reverse, blooms borne in large clusters, semi-dbl., 6–14 petals, 1.5–2.75 in., moderate fragrance; some prickles; foliage medium, medium green, semi-glossy; medium, spreading growth; [Charleston X Friesia]; Kameyama, Yasushi; Kame-yama, 1991

Vienna Charm *see* 'KORschaprat'

'Vienna Charm, Climbing', Cl HT, ob, 1972; Gandy, Douglas L.; Gandy Roses, Ltd.

'Vienna Maid', F, my, 1957; flowers empire-yellow, borne in large clusters, dbl., 30 petals, 2.5 in.; foliage dark, glossy; moderately bushy growth; deRuiter; Blaby Rose Gardens

Vienna Woods *see* **'Wienerwald'**

Vièrge de Clery *see* **'Unique Blanche'**

Vierge Folle *see* DELvirge

Vierlanden *see* **'Pink Delight'**

Vierländerin *see* 'KORvila'

Vieux Chateau Certan *see* DORtan

View® *see* 'LENvie'

Vif Eclat, HMsk, 1992; flowers small to medium, single; Lens, Louis; (Sangerhausen)

'Vigane', HT, rb, 1962; flowers red and light yellow bicolor, dbl.; Frères, Buyl

Vigilance *see* 'MINIvig'

'Vigilant', HT, dr, 1941; flowers very dark red; [Night seedling]; Clark, A.

Vigo, F; Moreira da Silva, A.; (Cavriglia)

'Vigoro', HT, op, 1953; flowers salmon-pink, large, dbl., 30 petals, exhibition form, moderate fragrance; foliage clear green; very vigorous, upright, compact growth; [Ophelia X Federico Casas]; Dot, Pedro

Vigorosa, S, lp; short; flowers medium sized, flat, 12 petals, moderate; medium-large, dark green, semi-glossy; [(Frankfurt am Main X Maria Callas) X Dr Faust]; St. Wagner, 1992; Res. Stn. f. Horticulture, Cluj, 1994, Romania

'Viking', HT, dr, 1968; bud ovoid; flowers crimson, dbl., 45–50 petals, 4.5–5.5 in., moderate fragrance; foliage leathery, dark; vigorous, upright growth; [Volcano X Happiness]; Moro; Ball Seed Co.

'Viking Queen', LCl, mp, 1963; flowers medium to deep pink, blooms in large clusters, dbl., 60 petals, 3–4 in., globular, recurrent bloom, intense fragrance; foliage dark, glossy, leathery; vigorous growth; [White Dawn X L.E. Longley]; Phillips; Univ. of Minn.

'Viktoria Adelheid', HT, yb, 1932; flowers golden yellow edged and shaded nasturtium-red, large, dbl., moderate fragrance; foliage leathery, glossy; dwarf growth; [Charles P. Kilham X Mev. G.A. van Rossem]; Kordes

'Vilia', F, op, 1960; flowers bright coral-pink, borne in large clusters, single, 2.5 in., moderate fragrance; foliage dark, glossy; moderate growth; GM, NRS, 1958; Robinson, H.; Gregory

'Villa de Bilbao', HT, mr, 1933; flowers cardinal-red, large, dbl., cupped; vigorous growth; [O. Junyent X Margaret McGredy]; La Florida

'Villa de Madrid', HT, or, 1965; flowers vermilion-red to poppy-red, large, dbl., 60 petals, moderate, musk fragrance; strong stems; upright growth; GM, Madrid, 1961; [Baccará X Peace]; Dot, Simon; C-P

'Villa de Sitges', HT, op, 1930; flowers pink shaded salmon; very vigorous, spreading growth; [Frau Karl Druschki X Mme Edouard Herriot]; Munné, B.

'VILLA DES TYBILLES', HRg, mr, 1899; flowers large, single; L'Hay; (Sangerhausen)

'VILLA PIA', HT, dr, 1926; bud long, pointed; flowers velvety deep red, almost black, dbl., moderate fragrance; [Pres. Vignet X Château de Clos Vougeot]; Leenders Bros.

Village Charm, HMult, lp; McLeod, J., 1990; (Weatherly, L.)

Village de Taradeau, F, lp; RVS, 1995

'VILLAGE MAID', C, pb; (Belle des Jardins, Belle Rubine, Belle Villageoise, Centifolia Variegata, Cottage Maid, La Belle Villageoise, La Rubanée, La Villiageoise, Panachee Double, R. centifolia variegata); flowers white, striped pink and purple (stripes varying in width, dbl., cupped; Vibert, 1845

'VILLANDESSA', HT, ob, 1977; bud long, pointed; flowers orange-blend, dbl., 33 petals, 4.5 in., exhibition form, slight fragrance; vigorous, upright, bushy growth; [Peer Gynt X Seedling]; Kordes, W. Söhne; Willemse

'VILLE D'ANGERS', HT, mr, 1934; bud long, pointed; flowers pure currant-red, large, semi-dbl., cupped; foliage leathery, dark; vigorous, bushy growth; [Souv. de Georges Pernet X Souv. de Claudius Denoyel]; Delaunay

Ville d'Arcis sur Aube®, HT, op; Vially

Ville de Bâle *see* 'MEIVILANIC'

Ville de Bar sur Seine *see* ORaval

'VILLE DE BORDEAUX', HT, dr, 1955; bud long; flowers dark scarlet-red, dbl., moderate fragrance; very vigorous growth; Privat

'VILLE DE BREST', HT, or, 1942; flowers fiery orange veined reddish copper, medium, semi-dbl., globular, slight fragrance; foliage bronze, glossy; vigorous growth; Gaujard

Ville de Bruxelles *see* **'LA VILLE DE BRUXELLES'**

'VILLE DE CHALONS', HT, or, 1938; flowers reddish orange shaded darker, dbl.; Champion

Ville de Chine *see* **'CHINATOWN'**®

Ville de Doué, F; (Cavriglia)

'VILLE DE GAND', HT, op, 1951; flowers deep salmon, well formed, dbl., 25 petals, 3.5 in., moderate fragrance; foliage bronze; very vigorous, upright growth; GM, Geneva, 1950; [Georges Chesnel seedling X (Mme Joseph Perraud seedling X R.foetida bicolor)]; Gaujard

Ville de Grenoble *see* **'GRENOBLE'**

Ville de Liffre, HT, mp; Adam

Ville de Londres, HGal, dp; Vibert or Robert, 1850

Ville de Lorgues, F, lp

'VILLE DE MALINES', HT, yb, 1929; bud long, pointed, yellow shaded cherry-red; flowers orange to pink shaded yellow; vigorous growth; Lens

Ville de Moulins *see* ORAlamar

'VILLE DE NANCY', HT, pb, 1940; bud long, pointed, old-rose shaded gold; flowers buff-pink, edges penciled white, reverse light pink, base yellow, dbl., 50–55 petals, cupped, slight fragrance; foliage leathery, dark; vigorous, upright growth; [Souv. de Claudius Pernet X Federico Casas]; Gillot, F.; C-P

'VILLE DE PARIS', HT, my, 1925; flowers clear bright yellow, large, dbl., globular, slight fragrance; foliage reddish green, glossy; vigorous, growth; GM, Bagatelle, 1925; [Souv. de Claudius Pernet X Unnamed seedling]; Pernet-Ducher

'VILLE DE PARIS, CLIMBING', Cl HT, my, 1935; Armstrong, J.A.

'VILLE DE PARIS, CLIMBING', Cl HT, my; Cognet

Ville de Perreux, S, 1987; Delbard; (Cavriglia)

'VILLE DE PRAGUE', HT, or, 1940; bud long, bright coral-red; flowers scarlet and copper, very large; foliage bright green; bushy growth; Chambard, C.; Orard

Ville de Roanne, HT, 1966; Dorieux; (Cavriglia)

Ville de Romilly, HT, 1995; Orard; (Cavriglia)

Ville de Romilly Sur Seine, HT, 1994; Orard; (Cavriglia)

'VILLE DE SAINT DENIS', HP, 1853; Thomas; (Cavriglia)

'VILLE DE SAVERNE', HT, or, 1937; flowers orange-scarlet, tinted brownish red, reverse tinted yellow, dbl.; Heizmann, E.

'VILLE DE TOULOUSE', HGal, mp, 1876; flowers medium, dbl.; Brassac; (Sangerhausen)

Ville de Troyes®, HT, dr

'VILLE DE VALENCIENNES', HT, or, 1954; flowers orange-copper, well formed, very large, dbl., moderate fragrance; vigorous growth; [Peace X Seedling]; Gaujard

Ville de Villeurbanne, HT, op

'VILLE DE ZURICH'®, F, or, 1967; flowers well-formed, dbl., 25 petals, moderate fragrance; foliage bright green; vigorous, bushy growth; [Miss France X Nouvelle Europe]; Gaujard

Ville d'Ettelbruck *see* 'LENIVILL'

'VILLE DU HAVRE', HT, w, 1931; flowers cream-white, washed rose-pink, base yellow, very dbl.; foliage dark; very vigorous growth; [K. of K. X Souv. de Claudius Pernet]; Cayeux, H.; Turbat

Ville du Perreux *see* 'DELRULA'

Ville du Roeuix, F, pb

Ville du Roeulx, F, dp; 19 petals, 6 in.; GM, Geneve, 1989; RvS-Melle, 1989

'VIM', HT, mp, 1963; bud pointed; flowers pink, medium, single, moderate bloom; foliage soft; moderate, bushy growth; [Charlotte Armstrong X Applause]; Wyant

'VIN ROSÉ', HT, mp, 1969; bud long, pointed; flowers light coral-pink, large, dbl., 30–35 petals, exhibition form, slight fragrance; foliage glossy; vigorous, upright growth; [Revelry X Hawaii]; Boerner; J&P

'VINCENT GODSIFF', Ch, mr; flowers deep luminous rosy-red, yellow stamens, semi-dbl., 10 petals, 2 in., cupped; foliage dark; compact, upright (to 3 ft) growth

'VINCENT VAN GOGH', Pol, or, 1969; bud ovoid; flowers medium, dbl.; foliage dark; [Allotria X Hobby]; Buisman, G. A. H.

'VINCENTE PELUFFO', HP, mr, 1902; flowers large, dbl.; Lévêque; (Sangerhausen)

'VINCENZ BERGERS WEISSE', HT, w, 1943; (Vinzens Berger's Weisse); bud long, pointed, sulphur; flowers open, very large, dbl., exhibition form, moderate fragrance; foliage glossy, bronze; very vigorous, upright growth; [Mrs Sam McGredy X Seedling]; Berger, V.; Kordes

Vindonissa®, F, ob; Huber, 1986

Vineyard Song *see* 'MORGRAPES'

'VINO DELICADO', HT, m, 1972; bud long, pointed; flowers mauve, edged purple-red, well-formed, large, dbl., slight fragrance; foliage large, leathery; upright growth; [Unnamed seedling X Mauve Melodee]; Raffel; Port Stockton Nursery

Vintage Visalia *see* 'MORLU'

Vintage Wine *see* 'POULLACK'

Vinzens Berger's Weisse *see* **'VINCENZ BERGERS WEISSE'**

'VIOLA', HT, m, 1955; flowers lilac-pink, very large, dbl., intense fragrance; foliage leathery; vigorous, upright growth; [Orange Triumph X Peace seedling]; Gaujard

Viola, HT, lp; Ghosh, 1998

Viola Lougheed *see* 'LOUVIOLA'

Violacea *see* **'LA BELLE SULTANE'**

'VIOLACÉE', M, m, 1876; flowers purple, shaded violet to grayish pink, large, dbl.; Soupert & Notting

'VIOLAINE', HT, m, 1968; bud long, pointed; flowers large, dbl., exhibition form, intense fragrance; foliage leathery; tall, vigorous growth; [Eminence X Simone]; Gaujard; Ilgenfritz Nursery

Violante, S, 1991; Michler, K. H.; (Sangerhausen)

Violet Bengal *see* **'PURPLE BENGAL'**

Violet Blue *see* **'VEILCHENBLAU'**

Violet Carson *see* 'MACIO'

Violet Dawson *see* 'RESONE'

'VIOLET FONTAINE', S, m, 1972; bud ovoid; flowers violet-purple, large, dbl., abundant, continuous bloom, slight fragrance; foliage large, soft; vigorous, upright, bushy growth; [Unknown X Unknown]; Tantau, Math.; Ahrens & Sieberz

Violet Hit *see* POUltin

'VIOLET HOOD', S, m, 1976; bud ovoid; flowers dark violet, pompon shape, semi-dbl., 18 petals, 1 in., recurrent bloom, intense fragrance; foliage ribbed, brownish; very vigorous, overhanging growth; [Robin Hood X Baby Faurax]; Lens, 1975

'VIOLET LIDDELL', HT, pb, 1904; flowers light pink to white, center and reverse coppery salmon; Schwartz, A.

'VIOLET MESSENGER', LCl, lp, 1974; flowers shell-pink, base yellow, reflexed, dbl., 30 petals, 6 in., intense fragrance; foliage large, matt green; [Spek's Yellow, Climbing X Masquerade]; Cadle's Roses

Violet Mist *see* 'TINVIOLET'

'VIOLET PARNCUTT', HT, yb, 1923; bud small, pointed; flowers brownish gold, semi-dbl.; Easlea

'VIOLET QUEEN', HP, m, 1892; flowers marbled crimson and violet; Paul, G.

'VIOLET QUEEN', HT, m, 1970; flowers deep violet-pink, pointed, dbl., 35–40 petals, 4 in., intense fragrance; foliage dark; upright, free growth; [Seedling X Violette Dot]; Northfield

Violet Ruffles, HT, lp; Davidson, 1997

'VIOLET SIMPSON', HT, op, 1930; flowers vivid prawn-pink, base yellow, dbl., moderate fragrance; foliage purple; Simpson; H&S;, Laxton Bros.

'VIOLET WILTON', HT, mp, 1930; bud very long, pointed; flowers bright rose-pink on flesh-white ground, tinted yellow, large, dbl., 35–40 petals, moderate fragrance; vigorous growth; [Gen. MacArthur X Mme Charles Lutaud]; Ketten Bros.

'VIOLETERA', HT, m, 1980; (Embruixada); bud ovoid; flowers reddish-mauve, blooms borne singly to 3 per cluster, dbl., 30 petals, cupped, intense fragrance; curved, reddish-green prickles; foliage medium, light green, matt; bushy growth; Dot, Simon; Rose Barni-Pistoia

Violett Satina *see* TANttelos

Violetta *see* 'HARQUANTUM'

'VIOLETTA', HT, m, 1957;, intense fragrance; foliage glossy; vigorous growth; [Peace X Guinee]; Croix, A.

Violetta, Pol, lp; Bruant, 1924

'VIOLETTE', HMult, m, 1921; flowers pure deep violet, blooms in large clusters, very dbl.; vigorous growth; Turbat

'VIOLETTE BOUYER', HP, w, 1881; flowers pinkish white, large, cupped, moderate fragrance; [Jules Margottin X Sombreuil]; Lacharme, F.

'VIOLETTE DOT', HT, m, 1960; flowers ageratum-blue, medium, semi-dbl., 20 petals, moderate fragrance; strong stems; spreading growth; [Rosa de Friera X Prelude]; Dot, Simon

Violette Niestlé *see* 'HARVINTAGE'

Violina® *see* TANanilov

Violine *see* 'LENDADI'

'VIOLINISTA COSTA', HT, rb, 1936; flowers red to deep purplish red, well-formed, large, moderate fragrance; vigorous growth; (28); [Sensation X Shot Silk]; Camprubi, C.

'VIOLINISTE EMILE LÉVÊQUE', HT, lp, 1897; flowers medium, dbl.; Pernet-Ducher; (Sangerhausen)

'VIOLONCELLISTE ALBERT FOURÈS', HT, ab, 1920; flowers orange-yellow, shaded buff-yellow, dbl.; [Joseph Hill X Unnamed variety]; Croibier

Viorita® *see* 'HARQUANTUM'

'VÍRA', HRg, mr, 1936; flowers bright red, intense fragrance; very vigorous growth; Böhm, J.

'VIRAGO', HSet, lp, 1887; flowers large, dbl.; Geschwind, R.; (Sangerhausen)

'VIRAMBER', HT, ab, 1997; (**Nefertiti**); flowers amber yellow blooms, semi-dbl., 15–25 petals, 5 in., borne mostly singly, slight fragrance; foliage large, medium green, semi glossy; upright, medium (3 ft) growth; [Julien Potin X First Prize]; Viraraghavan, M.S.

'VIRBROWN', F, m, 1997; (**Coffee Country**, Coorg); flowers russet and coffee brown, mauve suffusion, light yellow reverse, dbl., 15–25 petals, 4 in., borne in small clusters, slight fragrance; foliage medium, medium green, semi glossy; spreading, medium (3 ft) growth; [(Zorina X Mlle Cécile Brünner) X (The Fairy X Nordia)]; Viraraghavan, M.S.

'VIRGEN DE FARNÉS', HT, mp, 1960; flowers bright rose, reverse lighter, well formed, dbl., 26 petals, intense fragrance; strong stems; vigorous growth; [Queen Elizabeth X Virgo]; Dot, M.

Virgin, Min, w; flowers small, dbl.; Lens, Louis; (Sangerhausen)

Virginale, HP, w; Lacharme, F., 1858

Virginale, C, lp

'VIRGINIA', LCl, ob, 1934; flowers brilliant flame, suffused gold, open, large, dbl., non-recurrent, moderate fragrance; foliage large, dark; vigorous, climbing (9 ft.) growth; [Magnafrano X Eldorado]; Nicolas; C-P

Virginia *see* PEKwhina

'VIRGINIA DARE', HT, dp, 1934; bud long, pointed; flowers deep cerise-pink, large, dbl., moderate fragrance; foliage leathery, dark; long stems; very vigorous growth; [Joanna Hill X Unnamed seedling (dark red)]; Thompson's, J.H., Sons

Virginia Dare *see* CLEdare

Virginia Lee *see* 'MICLEE'

Virginia R. Coxe *see* **'GRÜSS AN TEPLITZ'**

Virginia R. Coxe, Climbing *see* **'GRÜSS AN TEPLITZ, CLIMBING'**

'VIRGINIA REEL', S, dp, 1975; bud ovoid, pointed; flowers light red, dbl., 40 petals, 4–4.5 in., cupped, moderate fragrance; foliage large, dark, leathery; erect, bushy growth; [Tickled Pink X Prairie Princess]; Buck, Dr. Griffith J.; Iowa State Univ.

Virginia Rose *see* **'R. VIRGINIANA'**

Virginian Rambler, Ayr, lp

'VIRGO', HT, w, 1947; ('VIRGO LIBERATIONEM'); flowers white, sometimes blush-pink, dbl., 30 petals, 5 in., exhibition form, slight fragrance; foliage dark, leathery; vigorous growth; GM, NRS, 1949; [Blanche Mallerin X Neige Parfum]; Mallerin, C.; Meilland-Richardier

'VIRGO LIBERATIONEM' *see* **'VIRGO'**

'VIRGO, CLIMBING', Cl HT, w, 1957; Mondial Roses

'VIROLAY', HT, ob; flowers edges orange-red, base deep yellow, large, exhibition form, intense fragrance; vigorous growth; Camprubi, C.

'VIRSPLASH', S, pb, 1998; (Bodhisatva, **Magic East**); flowers white with pink and white splashes, creamy yellow eye, dbl., 15–25 petals, 5 in., borne singly and in clusters, slight fragrance; prickles moderate, medium; foliage medium, blue green, matt; bushy, medium growth; [Honor X Priyatama]; Viraraghavan, M.S.; Hortico, 1995

Vi's Violet *see* 'MORVI'

Visa *see* 'MEIRED'

'VISCOUNT SOUTHWOOD', HT, pb, 1949; flowers china-pink shaded creamy peach to copper, dbl., 35–40 petals, 4–5 in., exhibition form; vigorous growth; [Walter Bentley X Aribau]; Cobley; Harkness

'VISCOUNTESS CHARLEMONT', HT, mp, 1937; bud salmon-rose; flowers satiny rose-pink, base deep buttercup-yellow, large, dbl., intense fragrance; foliage dark cedar green; branching growth; GM, NRS, 1936; McGredy

'VISCOUNTESS DEVONPORT', HT, dy, 1923; flowers rich indian yellow, dbl.; Hicks

'VISCOUNTESS ENFIELD', HT, pb, 1910; flowers coppery old-rose, shaded yellow; [? X Soleil d'Or seedling]; Pernet-Ducher

'VISCOUNTESS FALMOUTH', HT, pb, 1879; flowers mottled pink, very large, dbl., globular, moderate fragrance; thorny; dwarf growth; [Adam X Soupert et Notting]; Bennett

'VISCOUNTESS FOLKESTONE', HT, pb, 1886; flowers creamy silver-pink, center deep salmon-pink, dbl., moderate fragrance; Bennett

Vision *see* 'POULONI'

'VISION'®, HT, pb, 1967; flowers gold and pink, dbl., 5.5 in., slight fragrance; foliage glossy; [Kordes' Perfecta X Peace]; Dickson, A.

Vision Blanc® *see* **'ICE WHITE'**

Vista™ *see* 'SAVALAV'

Vital, HT, lp; Kordes, 1998

Vital Spark *see* 'COCACERT'

'VITTONVILLE-ROSE', HT, ob, 1945; flowers orange, base yellow, reverse lighter; vigorous growth; Mallerin, C.; A. Meilland

'VIUDA VERDAGUER', HT, ob, 1934; flowers orange, open, very large, dbl.; foliage glossy, dark; very vigorous growth; [Shot Silk X Mari Dot]; Dot, Pedro

Viva *see* 'JACIV'

Viva Romana, HT, rb; Dot

'VIVACÉ', F, or, 1974; flowers large, dbl., exhibition form, slight fragrance; foliage large, leathery; very vigorous, upright growth; [Klaus Stortebeker X ?]; Kordes

'VIVACIOUS', F, mp, 1971; flowers phlox-pink, dbl., 35 petals, 4 in., moderate fragrance; very free growth; [Tropicana X ?]; Gregory

Vivacious Dianne *see* 'CHRISJEVANS'

Vivacity *see* **'TEMPERAMENT'**

Vivaldi *see* 'LENMOBAR'

Vivaldi ™ *see* RUIdriko

Vivarose, HT, mr; Croix

'VIVASTELLA', HT, dp; flowers laque de robbie color (possibly carmine), well shaped; [Julien Potin X Sensation]; Aicardi, D.; Giacomasso

'VIVE LA FRANCE', HT, rb, 1944; bud pointed, well formed; flowers purplish red, reverse yellow, large, dbl., slight fragrance; foliage glossy; vigorous growth; GM, Bagatelle, 1943; [Shining Star X Mme Arthaud]; Mallerin, C.; A. Meilland

Vive La Suisse, F; Tschanz, E.; (Cavriglia)

'VIVIAN MORELLE', T, w; bud cream; flowers snow-white, profuse bloom

'VIVID', B, m, 1853; flowers brilliant magenta to magenta-pink, dbl., moderate fragrance; prickles robust, prickly growth; foliage glossy; height to 6 ft; Paul, A.

'VIVID', F, or, 1951; flowers brilliant orange-scarlet, borne in huge clusters, single, 5–7 petals, 3.5 in.; compact growth; LeGrice

'VIVID MASON', HT, dp, 1934; flowers vivid dark pink, base yellow, large, dbl.; foliage leathery, glossy; very vigorous growth; [Premier X Mme Alexandre Dreux]; Mason, J.A.; McLellan Co.

'VIVIEN', HT, dp, 1922; flowers deep rose-pink, dbl., intense fragrance; Paul, W.

'VIVIEN LEIGH', HT, mr, 1963; flowers crimson, dbl., 35 petals, 5 in., exhibition form, moderate fragrance; foliage dark; very free growth; [Queen Elizabeth X Detroiter]; McGredy, Sam IV; Fisons Horticulture

Vivienne Maire, D, mp

Vivre, HT, 1974; Delbard-Chabert; (Cavriglia)

Vixen *see* 'JACORMIN'

'VLAM', HT, or, 1956; flowers fiery red, dbl., moderate fragrance; vigorous growth; [Tawny Gold X Sarie Mareis]; Leenders, M.

Vlammenspel, LCl, mr

'VLASTA BURIAN', Pol, m, 1937; flowers purple/pink, small, dbl.; Böhm, J.; (Sangerhausen)

'VLTAVA', LCl, m, 1936; flowers violet, passing to purplish red, large, borne in clusters, dbl., globular; foliage glossy; very vigorous, climbing growth; [Veilchenblau X ?]; Böhm, J.

Voeux de Bonheur *see* **'BON VOYAGE'**

'VOEUX DE BONHEUR', HT, pb, 1960; (Bon Voyage); flowers creamy white, petals edged cerise-pink, reverse white, large, dbl., exhibition form, intense fragrance; foliage dark, glossy; vigorous growth; [Michele Meilland X Chic Parisien]; Delbard-Chabert

Vogelpark Walsrode *see* KORlomet

'VOGUE', F, pb, 1951; bud ovoid; flowers cherry-coral, blooms in clusters, dbl., 25 petals, 3.5–4.5 in., exhibition form, moderate fragrance; foliage glossy; vigorous, upright, bushy, compact growth; (28); AARS, 1952 GM, Geneva, 1950 GM, Portland, 1950; [Pinocchio X Crimson Glory]; Boerner; J&P

Vogue *see* PEKcourofondu

Voice of Thousands *see* 'HORSUNSMILE'

'VOIE LACTÉE', Cl HT, w, 1949; bud globular, creamy white; flowers large, dbl., intense fragrance; foliage glossy; very vigorous, climbing growth; [Frau Karl Druschki X Julien Potin]; Robichon

Voie Romaine, HT, 1977; Dot, Simon; (Cavriglia)

Voila, F, 1969; Croix; (Cavriglia)

Vol de Nuit *see* 'DELRIO'

Volare *see* 'MACVOLAR'

'VOLARE', HT, dr, 1976; bud ovoid; flowers bright red, dbl., 38–44 petals, 3.5 in., exhibition form, slight fragrance; foliage leathery; bushy, upright growth; PP004172; McDaniel, Earl; Carlton Rose Nurseries

'VOLCANO', HT, dp, 1950; bud long, pointed; flowers cherry-red, dbl., 25 petals, 6.5 in., cupped, moderate, fruity fragrance; foliage dark; vigorous, upright growth; [Charles P. Kilham X Rome Glory]; Moro; J&P

Volunteer *see* 'HARQUAKER'

'VON HÖTZENDORF', HP, pb, 1916; flowers golden rose-pink, richly tinted coppery red, dbl., moderate fragrance; vigorous growth; [Frau Karl Druschki X Beauté de Lyon]; Schmidt, J.C.

'VON LILIENCRON', HFt, pb, 1916; flowers yellowish light pink with white, reverse salmon-pink, stamen, dbl., exhibition form, profuse bloom, sometimes recurrent, moderate fragrance; foliage glossy, dark, bronze; vigorous (6 ft.) growth; [Geheimrat Dr. Mittweg X Mrs Aaron Ward]; Lambert, P.

'VON SCHARNHORST', S, ly, 1921; flowers yellow to yellowish-white, semi-dbl., recurrent bloom, slight fragrance; vigorous (6-8 ft) growth; [Frau Karl Druschki X Gottfried Keller]; Lambert, P.

Vonava Elysium, HT, lp; Strnad; (Czech Rosa Club)

Voodoo™ *see* 'AROMICLEA'

'VOORBURG', F, op, 1959; flowers salmon-pink, borne in large clusters, semi-dbl.; foliage bronze to dark; vigorous growth; [Sangerhausen X Vogue]; Buisman, G. A. H.

'VORBERGII', HFt, ly; (R. vorbergii, R. X harisonii vorbergii); flowers pale yellow, single; (28); [R. foetita X R. spinosissima]

'VOX POPULI', HT, dr, 1945; flowers velvety dark red, semi-dbl.; Mallerin, C.

Voyager *see* 'MICVOYAGE'

Vrystaat *see* TANtaatsyr

'VULCAIN', HP, dr, 1861; flowers rich dark crimson, well formed; Verdier, V.

'VULCAIN', Pol, mr, 1921; flowers deep cherry-red, dbl.; vigorous growth; Turbat

'VULCANA', LCl, dr, 1964; flowers blood-red, borne in clusters, semi-dbl.; vigorous growth; Mondial Roses

'VULCANIA', HT, mr, 1948; bud long; flowers purplish red, well formed, overlarge; foliage dark; [Matador X Principe di Piemonte]; Giacomasso

'VULGENS', Pol, mr, 1942; flowers medium, semi-dbl.; Koster; (Sangerhausen)

Vulkan, HT, dr; flowers large, dbl., slight fragrance; VEG; (Sangerhausen)

'VUURBAAK', F, or, 1946; flowers scarlet, open, borne in clusters, semi-dbl., 15 petals, 4 in.; foliage reddish green; vigorous, upright growth; [Florentina X World's Fair]; Leenders, M.; Longley

Vydehi, F, ob; Datt, Braham, 1989

'VYSLANEK KALINA', HP, mr, 1935; flowers large, dbl., intense fragrance; Böhm, J.; (Sangerhausen)

W

W. E. Lippiat, HT, dr; Dickson, Patrick, 1907

'W. Freedland Kendrick', HT, pb, 1920; flowers pink with peach tones, large, dbl., moderate fragrance; Thomas; (Sangerhausen)

'W. Freeland Kendrick', LCl, lp, 1920; (Bloomfield Endurance); flowers flesh, center peach, large, very dbl., non-recurrent, moderate fragrance; foliage dark, bronze, leathery, glossy; vigorous, semi-climbing growth; [Aviateur Bleriot X Mme Caroline Testout]; Thomas; B&A

'W.A. Bilney', HT, ab, 1927; flowers pale apricot suffused cerise, reverse yellow tinted pink, large, slight fragrance; foliage dark, leathery; vigorous growth; Easlea

W.A. Willet, HT, lp; Kernovski, V.R., 1959; (Weatherly, L.)

'W.C. Gaunt', HT, dr, 1916; bud long, pointed; flowers velvety crimson-scarlet, reflexed petals tipped scarlet, rev, dbl., moderate fragrance; Dickson, A.

'W.E. Chaplin', HT, dr, 1929; flowers crimson, deepening to maroon, large, dbl., exhibition form; vigorous growth; GM, NRS, 1930; Chaplin Bros.

'W.E. Chaplin, Climbing', Cl HT, dr, 1936; Heizmann, E.

'W.E. Wallace', HT, dy, 1922; flowers deep golden yellow, well formed, large, dbl., moderate fragrance; GM, NRS, 1922; [Gorgeous sport]; Dickson, H.

'W.G. Pountney', F, or, 1964; flowers scarlet; tall growth; [Moulin Rouge X Mrs Inge Poulsen]; Bennett, H.; Pedigree Nursery

'W.H. Cotton', HT, ab, 1946; flowers orange shaded gold, large; foliage dark; vigorous, upright growth; [Mrs Beatty X Mrs Sam McGredy]; Cobley; Leicester Roses

'W.H. Dunallan', HT, mr, 1939; flowers very rich bright red, flushed darker, semi-dbl.; vigorous growth; [Edith Clark X ?]; Clark, A.; NRS Victoria

W.J. Matthews, T, lp; Matthews, W.J., 1892

'W.R. Hawkins', HT, op, 1948; bud ovoid; flowers tango-pink, large, very dbl., exhibition form, intense fragrance; foliage glossy; vigorous, upright growth; [Crimson Glory seedling X Silver Jubilee]; Toogood

Waanrode, HT, op; Lens, 1990

'Wabash Dawn', S, op, 1978; bud tapered; flowers bright orange-pink, dbl., 34 petals, 4–4.5 in., exhibition form, repeat bloom, intense fragrance; foliage large, glossy; vigorous, upright growth; [(Queen Elizabeth X Gladiator) X (Aztec X Little Darling)]; Williams, J. Benjamin; Krider Nursery

'Wadei', HRg, mp; single; foliage rich green; weak, prostrate growth; [R. rugosa X ?]

Wagbi *see* 'BLEBAR'

'Wageningen', Pol, my, 1968; bud ovoid; flowers medium, dbl.; foliage dark; [Golden Giant X Peace]; Buisman, G. A. H.

Wagtail, S, lp; Poulsen, 1990

Waiheke™ *see* 'MACWAIKE'

Waikato *see* 'MACHORO'

Waikiki *see* 'MACWAIKE'

'Waipounamu', HT, yb, 1982; flowers creamy yellow, shaded mauve-pink on outer petals, dbl., 50 petals, exhibition form, intense fragrance; foliage bronze green, glossy; upright growth; [Peace X Blue Moon]; Cattermole, R.F.

Wait 'n' See, Min, lp; Geytenbeek, 1978; (Weatherly, L.)

Waitemata *see* 'MACWEEMAT'

Waitmata® *see* 'MACWEEMAT'

'Waitziana', S, dp; (R. X waitziana); flowers deep rose often solitary, 2.5 in.; [R. canina X R. gallica]

'Waldfee', HP, mr, 1960; flowers blood-red, camellia-shaped, blooms in small clusters, dbl., 4 in., recurrent bloom, moderate fragrance; foliage glossy; vigorous, tall (10 ft.), dense growth; Kordes, R.

'Waldtraut Nielsen', M, dp; flowers clear deep pink, large; Nielsen, 1932

Walferdange, HMsk, pb; Lens, 1992

'WALKISS', Min, mp, 1993; (**Autumn Kiss**); flowers medium salmon pink, dbl., 15–25 petals, 1.5 in., borne in small clusters, slight fragrance; few prickles; foliage small, dark green, glossy; medium, upright growth; [Ginger Meggs X Avandel]; Walsh, Richard Bruce; Fradee Nursery

Walko *see* 'DELDE'

'Walküre', HT, w, 1919; bud long, pointed; flowers cream-white, center ochre-yellow, dbl.; [Frau Karl Druschki X Mme Jenny Gillemot]; Ebeling

Walküre, S, lp; Geschwind, R., 1909

'Walkyrie', HT, dr, 1959; bud urn shaped or ovoid; flowers low-centered, large; bushy growth; [Happiness X Volcano]; Moro

Wall Street® *see* BARwast

Wallflower *see* **'The Wallflower'**

'WALLOVE', Min, w, 1994; (**Love Me**); flowers creamy white, pink edge, blooms borne in small clusters, dbl., 15–25 petals, 1.5 in., slight fragrance; few prickles; foliage small, medium green, matt; low, upright growth; [Magic Carrousel X Old Master]; Walsh, Richard Bruce; Fradee Nursery, 1992

'Walsham Gold', HT, my, 1965; flowers yellow with burnished gold and copper overtones in cool weather; LeGrice; Wayside Gardens Co.

'WALSHINE', Gr, ly, 1997; (**Wollongong Gold**); flowers light creamy yellow, flushed pink, dbl., 26–40 petals, 1.5–3 in., blooms in small clusters, slight fragrance; foliage medium, light green, semi glossy; low (1 m) bushy growth; [Violet Carson X Arizona]; Walsh, Richard Bruce

Walsh's Rambler *see* **'America'**

'Walter', HT, dp, 1949; bud long, pointed; flowers pink, reverse vivid rosy red, open, very large, semi-dbl., 20 petals, intense fragrance; foliage bronze, leathery; very vigorous, bushy growth; [Charles P. Kilham X Comtesse Vandal]; Lens

'Walter Bentley', HT, op, 1938; bud long, pointed; flowers coppery orange shaded pink, very large, dbl., exhibition form, slight fragrance; foliage leathery, glossy, dark, bronze; vigorous growth; GM, NRS, 1937; [Mrs Sam McGredy X Dame Edith Helen]; Robinson, H.; Wheatcroft Bros.

Walter Butt, S, dp; Butt, 1905

'WALTER C. CLARK', HT, dr, 1917; bud long, pointed; flowers deep maroon-crimson, shaded black, large, dbl., exhibition form, intense fragrance; foliage dark, leathery; vigorous growth; Paul, W.

'WALTER RIEGER', S, dp, 1977; bud pointed; flowers pink to reddish, medium, dbl., continuous bloom, slight fragrance; foliage glossy; vigorous growth; [Carina X Molde]; Hetzel

'WALTER ROSS', HT, my, 1970; bud ovoid; flowers large, dbl., exhibition form, moderate fragrance; foliage large, glossy, dark, bronze, leathery; vigorous, upright, bushy growth; [Mme Marie Curie X King's Ransom]; Morey, Dr. Dennison; General Bionomics

'WALTER SPEED', HT, dy, 1909; flowers deep lemon-yellow, passing to milk-white, dbl.; Dickson, A.

'WALTHAM BRIDE', LCl, w, 1905; flowers snow-white, medium, dbl., moderate fragrance; Paul, W.

'WALTHAM CLIMBER NO. 1', Cl HT, mr, 1885; flowers rosy crimson, imbricated, large, dbl., recurrent bloom, intense fragrance; vigorous, climbing growth; [Gloire de Dijon seedling]; Paul, W.

'WALTHAM CLIMBER NO. 2', Cl HT, or, 1885; flowers flame-red, tinted crimson, large, dbl., recurrent bloom; Paul, W.

'WALTHAM CLIMBER NO. 3', Cl HT, mr, 1885; flowers deep rosy crimson, imbricated, large, dbl., recurrent bloom, intense fragrance; vigorous, climbing growth; Paul, W.

'WALTHAM CRIMSON', HT, dr, 1922; flowers deep crimson; Chaplin Bros.

'WALTHAM CROSS', HT, dr, 1927; flowers glowing crimson-scarlet, semi-dbl.; Chaplin Bros.

'WALTHAM FLAME', HT, 1921; flowers deep terra-cotta, shaded bronzy orange; Chaplin Bros.

'WALTHAM RAMBLER', HMult, dp, 1903; flowers deep rosy pink, center paler, stamens bright yellow, borne i, single, early; vigorous, climbing growth; Paul, W.

'WALTHAM SCARLET', HT, dr, 1914; flowers crimson-scarlet, single; Paul, W.

'WALTRAP', Min, op, 1996; (**Bett's Little Rhapsody**); flowers medium blooms, orange-pink, very dbl., 76 petals, borne mostly singly; foliage medium, dark green with pink edge, glossy; upright, very compact, medium (3-4 ft) growth; [Bett's White Delight X Seedling]; Walters, Betty & Richard

Waltz Time *see* 'DELVOR'

'WALTZING MATILDA', HT, rb, 1965; flowers red splashed light to dark pink, well shaped, large, dbl.; vigorous growth; [Christian Dior sport]; Jack; Girraween Nursery

'WALZERTRAUM', HT, mr, 1968; dbl.; foliage dark; Tantau, Math.

'WAMBEAIRD', Gr, lp, 1999; (**Dr Charles T. Beaird**); flowers coral, medium, dbl., 17–25 petals, borne mostly singly and in small clusters, slight fragrance; prickles moderate; foliage medium, medium green, semi-glossy; upright, spreading, medium growth; Wambach, Catherine; Certified Roses, Inc., 2000

'WAMHALE', HT, pb, 1996; (**Denise Hale**); flowers light cream pink, darker edge, lighter reverse, dbl., 26–40 petals, 4.5 in., borne singly, slight fragrance; moderate prickles; foliage medium, dark green, semi-glossy; upright, medium growth; [Elsie Melton X Seedling]; Wambach, Catherine; Certified Roses, Inc., 1996

'WAMHOME', HT, pb, 1997; (**1997 Traditional Home**®); flowers light pale pink with ivory shading, dbl., 26–40 petals, slight fragrance; foliage medium, dark green, dull; medium (5 ft) growth; [Bobby Charlton X Touch of Class]; Wambach, Catherine

'WAMMARY', HT, lp, 1997; (**Mary Johnston**); flowers full, light pink, dbl., 26–40 petals, 4 in., borne mostly singly, slight fragrance; foliage medium, dark green, semi glossy; medium (4.5 ft) growth; [Elizabeth Taylor X Touch of Class]; Wambach, Alex A.

'WAMTRESS', HT, w, 1998; (**Temptress**); flowers full, white, large, very dbl., 26–40 petals, 4.5–5 in., exhibition form, borne singly, moderate fragrance; few prickles; foliage medium, medium green, semi-glossy; upright, tall (5.5 ft) growth; [Elsie Melton X Elina]; Wambach, Catherine; Certified Roses, Inc., 1999

Wanaka® *see* 'MACINCA'

'WANDA', HT, mp, 1953; bud short, pointed; flowers rose, dbl., 28–30 petals, 3.5–4.5 in., exhibition form, intense fragrance; foliage leathery; vigorous, upright growth; [Pink Bountiful X Celebrity]; Hill, Joseph H., Co.

Wanda, HT, lp

Wanderer *see* 'HORWANDER'

'WANDERIN' WIND', S, lp, 1973; flowers two-toned light pink, medium, dbl., exhibition form, intense fragrance; foliage large, glossy, dark, leathery; very vigorous, upright, bushy growth; [Dornroschen X Andante]; Buck, Dr. Griffith J.; Iowa State University

Wandering Minstrel *see* 'HARQUINCE'

'WAPEX', Pol, my, 1968; bud pointed; flowers medium, semi-dbl., borne in clusters; foliage dark; [Golden Showers X Fata Morgana]; Buisman, G. A. H.

Wapiti® *see* MEInagre

'WAR DANCE', Gr, or, 1962; bud ovoid; flowers dark orange-red, dbl., 34 petals, 4–4.5 in., exhibition form; foliage leathery; vigorous, bushy growth; [Roundelay X Crimson Glory]; Swim & Weeks; C-P

War Dance *see* 'MACFIRINLIN'

'WAR PAINT', LCl, mr, 1930; flowers large, dbl., globular, slight fragrance; vigorous, pillar or climbing (8 ft) growth; [Rhea Reid seedling X ?]; Clark, A.; Hackett

'WARANA FESTIVAL', HT, mp, 1962; flowers rich pink, base rich apricot, well formed, dbl.; vigorous growth; [Christian Dior sport]; Jack; Langbecker

Warbler, S, my, 1991; Poulsen

'WARDIDO', HT, pb, 1975; (Oratorio, **Spellbinder**); bud ovoid; flowers ivory to crimson, large, dbl., exhibition form, slight fragrance; foliage large, dark, leathery; vigorous growth; [South Seas X Unnamed seedling]; Warriner, William A.; J&P

'WARDROSA', HT, dp, 1975; (Chanterelle, **Cynthia**); bud long; flowers deep pink, dbl., 35 petals, 5 in., exhibition form, moderate fragrance; foliage large, matt, light; tall, upright growth; [Seedling X Bob Hope]; Warriner, William A.; J&P

Warley Jubilee, F, mp; Warley Rose Gardens, 1986

'WARLEYENSIS', S, mp; (R. X warleyensis); (14); [R. blanda X R. rugosa]

'WARM RAIN', Min, mr, 1985; flowers medium coral red, medium, dbl., 70 petals, exhibition form, slight fragrance; foliage small, medium green, semi-glossy; vigorous, upright, bushy growth; [Fragrant Cloud X Orange Honey]; Hardgrove, Donald L.; Rose World Originals

Warm Welcome *see* 'CHEWIZZ'

Warm Wishes *see* 'FRYXOTIC'

WARoujo, HT, mr; (**Fulgurante**®); J&P

'WARRAWEE', HT, mp, 1935; bud long, pointed; flowers flesh-pink shaded rose-pink, large, dbl., 30 petals, exhibition form, moderate fragrance; foliage glossy; vigorous growth; [Padre X Rev. F. Page-Roberts]; Fitzhardinge; C-P

'WARRIOR', F, or, 1977; flowers scarlet-red, blooms in trusses, dbl., 32 petals, 3–4 in., slight fragrance; foliage light green, semi-glossy; [City of Belfast X Ronde Endiablee]; LeGrice

'WARSZAWA', HT, ob, 1957; flowers bright orange, reverse golden yellow; foliage dark, glossy; [Carioca sport]; Grabczewski

'WARTBURG', HMult, mp, 1910; flowers pink to magenta, twisted and reflexed petals, dbl., cupped, moderate fragrance; foliage large, light, soft; strong stems; very vigorous, climbing (15-20 ft) growth; Kiese

Wartburg 77, S, my; Berger, 1977

'WARWHOOP', Min, or, 1973; bud ovoid; flowers brilliant orange-red, small, very dbl., slight fragrance; foliage small, glossy, dark; vigorous, bushy growth; [Baccará X Little Chief]; Williams, Ernest D.; Mini-Roses

Warwick Castle® *see* 'AUSLIAN'

Warwickshire *see* KORkandel

'WASAGAMING', HRg, mp, 1939; flowers clear rose, dbl., recurrent bloom, moderate fragrance; vigorous (3 ft.) growth; [(R. rugosa X R. acicularis) X Gruss an Teplitz]; Skinner

Wasastiernan *see* **'POLSTJÄRNAN'**

'WASKASOO', HRg, dr, 1963; flowers not clustered, dbl.; growth typical rugosa; (14); [Little Betty X Hansa]; Erskine

Watchfire *see* 'MINFIRE'

'WATCHUNG', HT, yb, 1932; flowers yellow tipped pink, medium, dbl., 24–26 petals, 3.5 in.; [Pres. Herbert Hoover sport]; Didato

'WATER MUSIC', LCl, dp, 1982; flowers deep pink, darker on petal edges, medium, dbl., 20 petals, slight fragrance; foliage dark, medium, glossy; spreading, climbing growth; [Handel X ?]; Bell, Ronald J.

'WATERCOLOR', Min, mp, 1975; (Watercolour); bud long, pointed; flowers bright pink, small, dbl., 26 petals, 1.5 in., exhibition form, slight fragrance; foliage small, glossy, leathery; vigorous, upright, bushy growth; [Rumba X (Little Darling X Red Germain)]; Moore, Ralph S.; Sequoia Nursery

Watercolour *see* **'WATERCOLOR'**

Waterloo, HMsk, 1996; Lens; (Cavriglia)

Watermelon Ice® *see* 'JACAIR'

Wave of Flame *see* **'HO NO-O-NO-NAMI'**

Waverland® *see* 'LENWAY'

'WAVERLEY', F, dr; flowers crimson-scarlet, borne in trusses, dbl., 22 petals, 3.5 in., slight fragrance; foliage glossy, dark; vigorous, bushy, compact growth; Norman

'WAVERLEY TRIUMPH', Pol, mp, 1951; flowers bright pink, base yellow, small, borne in clusters, semi-dbl., cupped, slight fragrance; foliage glossy, light green; vigorous, bushy growth; [Orange Triumph sport]; Poulter

Waverly Garden Club, lp; [Prima Ballerina X (Scented Bouquet X Extravaganza)]; Dawson, George

'Waves', HT, mp, 1944; (Admiral); bud ovoid; flowers rose-pink, open, dbl., 35 petals, 4.5 in., cupped, intense fragrance; foliage dark, leathery, wrinkled; vigorous growth; [Seedling (deep salmon) X Lucie Marie]; Dickson, A.; J&P

Waves of Flame *see* **'HO NO-O-NO-NAMI'**

'WAVRIA', F, or, 1973; flowers orange-crimson; [Europeana sport]; Lens

Wax Rose *see* **'DUCHESSE D'ANGOULÊME'**

Waxwing, S, mp, 1989; groundcover growth; Poulsen

'WAYSIDE GARNET', Min, dr, 1956; flowers garnet-red, small, dbl.; dwarf, compact growth; [Oakington Ruby sport]; Wayside Gardens Co.

'WE ZAIR', C, pb; flowers two-tone pink

'WEBBREAD', Min, op, 1997; (**Bob's Peach**); flowers double, medium, dbl., 15–25 petals, borne in large clusters, slight fragrance; many prickles; foliage large, medium green, glossy; bushy, tall (28in.) growth; [Robin Redbreast X Pot O'Gold]; Webster, Robert; Handley Rose Nursery, 1998

'WEBCYRENE', Min, mp, 1994; (**Sheffield Pride**); very dbl., 1.5 in., borne in small clusters, slight fragrance; some prickles; foliage small, dark green, semi-glossy; low, compact growth; [Robin Red Breast X Matangi]; Webster, Robert; Handley Rose Nurseries, 1995

'WEBEYAM', F, op, 1998; (**Jeanette Talbot**); flowers orange-pink, yellow at center, small, dbl., 15–25 petals, 2 in., borne in small clusters, slight fragrance; prickles moderate; foliage small, medium green, glossy; compact, low (18 in) growth; [(Robin Redbreast X Typhoon) X The Lady]; Webster, Robert; Handley Rose Nurseries, 1998

'WEBHEART', HT, op, 1998; (**Gedge's Glory**); flowers pale orange, salmon pink reverse, very dbl., 26–40 petals, 3–4 in., borne mostly singly, slight fragrance; prickles moderate, medium, straight; foliage medium, dark green, glossy; low, compact growth; [Dave Hesswayon X Remember Me]; Webster, Robert

'WEBLADY', HT, yb, 1998; (**The Marquess of Bristol**); flowers yellow, pink edge, yellow reverse, dbl., 17–25 petals, 4–5 in., borne mostly singly, slight fragrance; prickles moderate, medium, hooked; foliage medium, medium green, glossy; upright, medium growth; [Pristine X Remember Me]; Webster, Robert; Handley Rose Nurseries, 1999

WEBpriace, HT, pb, 1996; (**Distant Sounds**); Webb

Wedded Bliss® *see* 'SAVAWED'

'WEDDIGEN', HT, lp, 1916; flowers silver-pink, moderate fragrance; Lambert, P.

'WEDDING BELLS', HMult, w, 1906; flowers white, outer half of petals soft pink, borne in clusters, semi-dbl., cupped; foliage light; short stems; very vigorous, climbing (12-15 ft)growth; [Crimson Rambler X ?]; Walsh

'WEDDING DAY', LCl, w, 1950; flowers yellow to white flushed pink, single, blooms in huge clusters, intense fragrance; height to 20 ft(14); [R. sinowilsonii X ?]; Stern, Sir Frederich

Wedding Pink *see* 'JACBIP'

'WEDDING RING', HT, my, 1956; bud long, pointed; flowers golden yellow, dbl., 25–35 petals, 4–4.5 in., slight fragrance; foliage dark, glossy, leathery; vigorous, upright, compact growth; [Ville de Paris X Mrs Sam McGredy]; Shepherd; Bosley Nursery

'WEDDING SONG', HT, w, 1971;, slight fragrance; upright growth; [Virgo X Ivory Fashion]; Whisler, D.; Gro-Plant Industries

Wee Ack *see* 'RENACK'

Wee Barbie *see* 'JELBAR'

'WEE BETH', Min, ob, 1981; bud pointed, mossy; flowers salmon-pink, blooms borne 3-20 per cluster, semi-dbl., 11 petals, moderate, sweetbriar fragrance; red prickles; foliage small, dark, bristles on reverse; bushy growth; [Orange Silk X Fairy Moss]; Cherry, R. & Welsh; Roy H. Rumsey, Pty. Ltd.

'WEE BUTTERFLIES', Pol, pb, 1989; bud pointed; flowers medium pink with white eye, small, borne in sprays of 3-35, single, 5 petals, slight fragrance; round, small, red fruit; prickles straight, pink; foliage medium, light green, glossy; bushy, low growth; [The Fairy X Unnamed seedling]; Jerabek, Paul E., 1990

Wee Cracker *see* 'COCMARRIS'

Wee Jock® *see* 'COCABEST'

'WEE LASS', Min, mr, 1974; bud pointed; flowers blood-red, semi-dbl., 18 petals, .5–1 in.; foliage small, dark; upright, bushy growth; [Persian Princess X Persian Princess]; Moore, Ralph S.; Sequoia Nursery

'WEE MAN', Min, mr, 1974; (Tapis de Soie); flowers scarlet, medium, semi-dbl., 14 petals, 2 in., slight fragrance; foliage glossy, dark; [Little Flirt X Marlena]; McGredy, Sam IV

Wee Matt *see* 'MACWEEMAT'

Wee One, Min, lp; Tantau, 1994

Wee Topper *see* 'ANDWEE'

'WEEBLUSH', Min, pb, 1983; (**Blushing Dawn**); flowers creamy white, peach pink eye, yellow stamens, petals painted peach pink, single, 7–8 petals, moderate fragrance; foliage medium, dark, matt; low, bushy growth; [Jeanne Lajoie X Eyepaint]; Weeks, Michael W.J.

'WEEGOLD', HT, dy, 1981; (**Fine Gold**); flowers deep golden yellow, blooms borne 1-3 per stem, dbl., 26 petals, exhibition form, moderate fragrance; foliage medium, dark, glossy; branching, upright growth; [Seedling X Hawaii sport]; Weeks, Michael W.J.

Week-End, S, lp

'WEEPAINT', F, ab, 1987; (**Open Secret**); flowers peach to salmon-pink, sometimes flecked darker, shading to w, single, 5 petals, moderate fragrance; foliage medium, medium green, semi-glossy; bushy, tall growth; [Eyepaint sport]; Weeks, Michael W.J.

Weeping China Doll *see* **'CHINA DOLL, CLIMBING'**

Weepy, S, mp, 1996; flowers small; Lens, Louis; (Sangerhausen)

'WEEQUILL', Min, mp, 1992; (**Pink Quill**); flowers medium bright pink, blooms borne in large clusters, semi-dbl., 6–14 petals, 1.5 in., moderate fragrance; some prickles; foliage medium, dark green, bronze when young, semi-glossy; tall (25 cms), spreading growth; [Mr Bluebird X Unknown]; Weeks, Michael W.J.

'WEESPOT', Min, my, 1992; (**Spot 'o Gold**); flowers yellow stamens, blooms borne in small clusters, single, 5 petals, 1.5 in., slight fragrance; some prickles; foliage medium, medium green, semi-glossy; bushy (20 cms) growth; [Rise 'n' Shine X Unknown]; Weeks, Michael W.J.

Weetwood, HWich, lp; Bawden, 1983

Wehrinsel, S, dr, 1959; flowers medium, semi-dbl.; Berger, W.; (Sangerhausen)

Weigand's Crimson Rambler *see* **'NON PLUS ULTRA'**

Weight Watchers Success® *see* 'JACBITOU'

'WEIHENSTEPHAN', S, dp, 1964; flowers rich pink, blooms in large clusters, semi-dbl.; vigorous (4 1/2 ft) growth; Kordes

'WEISSE AUS SPARRIESHOOP', S, w, 1962; flowers medium, moderate fragrance; Kordes

'WEISSE BETTER TIMES', HT, w, 1940; flowers large, dbl.; Wirth; (Sangerhausen)

'WEISSE ECHO', Pol, w, 1925; flowers medium, semi-dbl.; Kiese; (Sangerhausen)

'WEISSE GRÜSS AN AACHEN', F, w, 1944; bud ovoid; flowers snow-white, large, borne in clusters, very dbl., moderate fragrance; upright, bushy growth; [Gruss an Aachen sport]; Vogel

Weisse Immensee® *see* 'KORWEIRIM'

'WEISSE MARGO KOSTER', Pol, w, 1939; flowers small to medium, dbl.; Teschendorff; (Sangerhausen)

Weisse Max Graf® *see* 'KORGRAM'

Weisse Nelkenrose, HRg, w; Münster, 1966

Weisse New Dawn, LCl, w

Weisse Repandia® *see* 'KORIANT'

Weisse Wolke® *see* KORstacha

Weisser Engel, F, w; Verschuren, 1966

'WEISSER HERUMSTREICHER', HMult, w, 1895; flowers pure white, large, borne in clusters, dbl.; vigorous, climbing growth; [Daniel Lacombe X Paquerette]; Schmidt, J.C.

Weisses Meer, F, w; flowers medium, semi-dbl., slight fragrance; VEG; (Sangerhausen)

'WEKAJAZOUL', S, lp, 1999; (**Long Tall Sally** ™); flowers buff white, reverse same, single, 5–11 petals, 3–3.5 in., borne in large clusters, moderate fragrance; prickles moderate; foliage large, dark green, semi-glossy; upright, very vertical, tall (5-7 ft) growth; [All That Jazz X R. soulieana derivative]; Carruth, Tom; Weeks Roses, 1999

'WEKAMANDA', HT, yb, 1999; (Limelight, Saint Patrick, **St Patrick**®); flowers yellow gold, shaded green in heat, dbl., 30–35 petals, 5 in., exhibition form, borne mostly singly, slight fragrance; prickles moderate; foliage medium, medium green, dull; upright, bushy, medium (3-4 ft) growth; PP9591; AARS, U.S., 1996; [Brandy X Gold Medal]; Strickland, Frank A., 1991; Weeks Roses, 1996

'WEKAQ', HT, pb, 1990; (**The Temptations**®); bud pointed; flowers pink blend, medium pink reverse, aging slightly lighter, dbl., 25–30 petals, 4 in., exhibition form, borne singly and in small clusters, moderate, fruity fragrance; oval, medium, orange fruit; prickles slightly hooked, medium, green; foliage medium, dark green, semi-glossy, disease-resistant; upright, tall growth; PP8516; Gold, ARC TG, 1989; [Paradise X Admiral Rodney]; Winchel, Joseph F., 1984; Weeks Wholesale Rose Growers, 1993

'WEKBIPUHIT', F, m, 1999; (**Purple Heart**®); flowers magenta purple, dbl., 30–35 petals, 3.5–4 in., cupped, borne in large clusters, intense, clove and spice fragrance; prickles moderate; foliage medium, medium green, matte; rounded, compact, bushy, medium (3.5 ft) growth; [Stephen's Big Purple X International Herald Tribune]; Carruth, Tom; Weeks Roses, 1999

'WEKBLUSI', HT, m, 1991; (**Stainless Steel**®); flowers clean, silvery gray lavender, well-formed, large blooms born, dbl., 26–40 petals, 5–6. in., exhibition form, borne in small clusters, intense, sweet rose fragrance; some prickles; foliage large, medium green, semi-glossy; tall to medium (130-150 cms), vigorous growth; PP10188; [Blue Nile X Silverado]; Carruth, Tom; Weeks Roses, 1995

'WEKBOROCO', S, rb, 1999; (**Rockin' Robin**®); flowers red, white and pink stripes and splashes, ruffled, very dbl., 40–45 petals, 1.5–2 in., borne in large clusters, slight, apple fragrance; prickles moderate; foliage medium, dark green, glossy; foutainous, rounded, bushy (4 ft) growth; PP10070; [Bonica X Roller Coaster]; Carruth, Tom; Weeks Roses, 1997

'WEKCALROC', F, yb, 1996; (**George Burns**™); flowers yellow striped irregularly with red, cream and pink, dbl., 26–40 petals, 3–3.5 in., borne in small clusters, moderate, fruit and citrus fragrance; numerous prickles; foliage large, clean, dark green, glossy; upright, compact, medium (3 - 3 1/2 ft) growth; PP10334; [Calico X Roller Coaster]; Carruth, Tom; Spring Hill Nurseries Co., 1997

'WEKCATLART', Min, ob, 1998; (**Gizmo**®); flowers scarlet orange with white eye, long lasting, single, 4–11 petals, 1.5 in.–3. in., borne in small clusters, slight, apple fragrance; prickles moderate, small; foliage medium, dark green, semi-glossy; rounded, compact, medium (18 in) growth; [Carrot Top X Little Artist]; Carruth, Tom; Weeks Roses, 2000

'WEKCELPEP', HT, pb, 1996; (**Sunsation**); flowers yellow striped pink, large blooms, dbl., 26–40 petals, 5 in., borne mostly singly, intense fragrance; some prickles; foliage large, light green, dull; upright, bushy, medium growth; PPAF; [Celebrity X Peppermint Twist]; Carruth, Tom; Michigan Bulb Co., 1996

'WEKCRYLAND', HT, w, 1998; (Cadillac DeVille, **Moonstone**®); flowers ivory white with a fine, delicate pink edging, very dbl., 26–40 petals, 4.5 in.–5 in., exhibition form, borne singly, slight fragrance; prickles moderate; foliage large, medium green, dull; upright, medium (120-160 cms) growth; [Crystalline X Lynn Anderson]; Carruth, Tom; Weeks Roses, 1998

'WEKCRYPLAG', F, m, 1999; (**Blueberry Hill**®); bud plump, pointed, dark lilac; flowers clear lilac, curved petals, golden stamens, single, 5–11 petals, 4 in., exhibition form, borne in small clusters, moderate, apple fragrance; prickles moderate; foliage very serrated, large, dark green, glossy; upright, medium, rounded, bushy (3-4 ft) growth; PP10072; [Crystalline X Playgirl]; Carruth, Tom; Weeks Roses, 1997

'WEKcryreg', F, pb, 1999; (**Gracie Allen**); flowers white with a pink heart, dbl., 25–30 petals, 4.5 in., borne in small clusters, slight, apple fragrance; prickles moderate; foliage medium, dark green, glossy; upright, bushy, medium (3-4 ft) growth; [Crystalline X Regensberg]; Carruth, Tom; Weeks Roses, 1998

'WEKdesc', Min, w, 1995; (**Descanso Dream**); flowers white blend, small, dbl., 15–25 petals, 1.5 in., no fragrance; few prickles; foliage small, dark green, semi-glossy; low (25-30 cms), bushy, compact growth; [Origami X Little Artist]; Carruth, Tom; Weeks Roses, 1994

'WEKdoclem', HT, dy, 1992; (**Atlanta**); flowers deep clear yellow, long-lasting color, dbl., 15–25 petals, 3–3.5 in., borne mostly singly, slight fragrance; some prickles; foliage medium, medium green, matt; medium (90-120 cms), upright, bushy, full growth; [Honor X Gingersnap]; Lemrow, Dr. Maynard W.; Weeks Roses, 1996

'WEKdykstra', Gr, yb, 1999; (**Heart O' Gold**™, Heart of Gold, Rose of Narromine); flowers deep gold surrounded by cerise pink, dbl., 35–40 petals, 4–4.5 in., exhibition form, borne in large clusters, intense, fruit and rose fragrance; prickles moderate; foliage large, medium green, semi-glossy; upright, tall (5-6 ft) growth; PP10713; [Broadway X Gold Medal]; Dykstra, Dr. A. Michael; Weeks Roses, 1997

'WEKfrag', HT, or, 1993; (**Ain't She Sweet**™); flowers large, terra-cotta orange to deep orange-red, dbl., 30–35 petals, 3–3.5 in., exhibition form, borne mostly singly, intense, spice and rose fragrance; some prickles; foliage large, medium green, matt; new growth is dark red; medium, rounded, bushy growth; PP009071; [Seedling X Dolly Parton]; Winchel, Joseph F.; Weeks Roses, 1994

'WEKgibotex', F, lp, 1999; (**John-John**™); flowers bright yellow, dbl., 17–25 petals, 3–3.5 in., borne in small clusters, slight fragrance; prickles moderate; foliage medium, light green, dull; upright, bushy, medium (3-4 ft) growth; [(seedling X Sunsprite) X (Old Master X Texas)]; Carruth, Tom; Weeks Roses, 2000

'WEKhelen', Min, ab, 1990; (**Peach Fuzz**™); bud pointed, moss; flowers apricot blend fading to pastel, medium, borne singly or in small cluster, dbl., 22 petals, exhibition form, intense, moss fragrance; prickles straight, yellow brown; foliage medium, dark green, glossy; bushy, rounded, full, medium growth; [Fairy Moss X New Year]; Carruth, Tom; Weeks Wholesale Rose Growers, 1991

'WEKibertaz', HT, lp, 1999; (**Cancun**™); flowers dark orange red, dbl., 26–40 petals, 5–6 in., borne mostly singly, slight fragrance; prickles moderate; foliage large, dark green, glossy; upright, medium (3.5-4 ft) growth; [Ingrid Bergman X All That Jazz]; Carruth, Tom; Spring Hill Nurseries, Co., 1999

'WEKjoe', HT, pb, 1993; (**Lynn Anderson**, Oh My God); flowers white edged deep pink, dbl., 26–40 petals, 5–5.5 in., exhibition form, borne mostly singly, slight fragrance; few prickles; foliage very large, medium green, matt; tall (170-180 cms), upright, bushy growth; PP9389; [Seedling X Gold Medal]; Winchel, Joseph F.; Weeks Roses, 1995

'WEKjuvoo', HT, ob, 1996; (**Chris Evert**™, Raymond Kopa); flowers clear orange-yellow blushing red, reverse with less blush, dbl., 26–40 petals, exhibition form, borne singly, moderate, fruity fragrance; moderate prickles; foliage clean, medium, dark green, semi-glossy, upright, compact, medium (3-4 ft) growth; PP10071; [Voodoo X (Katherine Loker X Gingersnap)]; Carruth, Tom; Spring Hill Nurseries Co., 1996

'WEKlips', Min, m, 1988; (**Little Paradise**™); flowers deep lavender, blushing purple, reverse deep lavender, aging lighter, dbl., 20 petals, exhibition form, slight fragrance; no fruit; prickles nearly straight, small, yellow-brown; foliage small, dark green, semi-glossy, disease resistant; upright, medium, vigorous, abundant growth; PP007580; [Shocking Blue X Helen Boehm]; Carruth, Tom; Weeks Roses, 1991

'WEKmar', HT, ab, 1992; (**Imagination**); bud large, shapely; flowers apricot orange, yellow reverse, good distinct bicoloration, dbl., 26–30 petals, 3–3.5 in., slight fragrance; some prickles; foliage large, clean, medium green, semi-glossy; medium (100-130 cms), attractive, compact, upright, bushy growth; PP8414; [Marmalade X seedling]; Winchel, Joseph F.; Weeks Roses, 1993

'WEKmeyer', LCl, mp, 1999; (**Pearly Gates**™, Pearly Gates, Climbing); flowers pastel pearl pink, very dbl., 41 petals, 5 in., borne in small clusters on old and new wood, intense, spicy fragrance; few prickles; foliage large, medium green, dull; climbing, tall (9-12 ft) growth; PP10640; [America sport]; Meyer; Weeks Roses, 1999

'WEKpearl', HT, lp, 1999; (**Pearl**); flowers pearl white, dbl., 17–25 petals, 4.5–5 in., borne mostly singly, slight fragrance; prickles moderate; foliage large, dark green, glossy; upright, bushy, medium (4-5 ft) growth; Winchel, Joseph F.; Weeks Roses, 1999

'WEKpipogop', S, w, 1999; (**Pillow Fight**™); flowers bright white, small rosettes, very dbl., 17–25 petals, 1.5–2 in., rosette, borne in large clusters, floriferous, intense, honey and rose fragrance; few prickles; foliage medium, dark green, glossy; rounded, bushy, medium (30-36 in) growth; [Pink Pollyanna X Gourmet Popcorn]; Carruth, Tom; Weeks Roses, 2000

'WEKplapep', F, rb, 1999; (**Scentimental**™); flowers striped burgundy and white, or cream, or red, opening quickly, dbl., 25–30 petals, 4–4.5 in., exhibition form, borne in large clusters, intense, spicy fragrance; prickles moderate; foliage large, quilted, medium green, matt; compact, rounded, medium (3-4 ft) growth; PP10126; AARS, U.S., 1997; [Playboy X Peppermint Twist]; Carruth, Tom; Weeks Roses, 1997

'WEKplapic', F, rb, 1999; (**Betty Boop**™); bud pointed; flowers rosy edged, yellow at base when fresh, fading to white; bright stamens, single, 6–12 petals, 4 in., borne in small cluster of 3 to 5; floriferous, moderate, fruity fragrance; prickles moderate; foliage medium, dark green, glossy, dark red new growth; rounded, bushy, medium (4 ft) growth; PP10126; AARS, U.S., 1999; [Playboy X Picasso]; Carruth, Tom; Weeks Roses, 1999

'WEKplasol', F, yb, 1999; (**Flutterbye**™); flowers multicolor yellow, coral, orange, tangerine, pink, single, 5–9 petals, 1.5–2 in., borne in large clusters, moderate, spice fragrance; prickles moderate; foliage medium, dark green, glossy; very large, rounded, can be fountainous growth; PP9715; [Playboy X (R. soulieana derivative X Sunsprite)]; Carruth, Tom; Weeks Roses, 1996

'WEKpluroco', S, m, 1997; (**Purple Reign**); flowers irregularly striped magenta and white, semi-dbl., 8–14 petals, 1.5–2.5 in., borne in large clusters, intense fragrance; foliage large, dark green, glossy; spreading, medium (60-70cms)growth; [Fragrant Plum X Roller Coaster]; Carruth, Tom; Michigan Bulb Co., 1997

'WEKquaneze', HT, lp, 1999; (**Barbra Streisand**); flowers lavender, blushing darker, slow opening, dbl., 26–40 petals, 3.5–4 in., borne in small clusters, intense fragrance; prickles moderate; foliage large, dark green, glossy; upright, bushy, medium (3-4 ft) growth; [(Blue Nile X (Ivory Tower X Angel Face)) X New Zealand]; Carruth, Tom; Weeks Roses, 2000

'WEKroalt', LCl, rb, 1999; (Crazy For You, **Fourth of July**™); flowers velvety red and white striped, ruffled, single, 10–16 petals, 4–4.5 in., borne in large clusters, moderate, apple fragrance; prickles moderate; foliage large, deep green, glossy; climbing, tall (10 ft.) growth; AARS, U.S., 1999; [Roller

Coaster X Altissimo]; Carruth, Tom; Weeks Roses, 1999

'WEKsamsou', LCl, mr, 1999; (**All Ablaze**®); bud short, ovoid; flowers medium cherry red, ruffled, full, very dbl., 35 petals, 4 in., borne in small clusters, slight, spice fragrance; foliage medium, dark green, semi-glossy; vigorous growth; climbing to 12 ft.; [Don Juan X (Rosa soulieana seedling X Trumpeter)]; Carruth, Tom

'WEKsamsou', LCl, lp, 1999; (**All Ablaze**®); dbl., 26–40 petals, 2.5–3 in., borne in large clusters, slight fragrance; prickles moderate; foliage large, medium green, dull; spreading, climbing, tall (9-12 ft) growth; hardy; [Don Juan X (R. soulieana derivitive X Trumpeter)]; Carruth, Tom; Weeks Roses, 2000

WEKslats, HT, lp, 1998; (**Pink Crystal**); Wathen; Weeks Roses, 1998

'WEKsnacare', Min, lp, 1999; (**Space Odyssey**®); flowers red with white eye, reverse white, semi-dbl., 12–16 petals, 1–1.5 in., borne in small clusters, slight fragrance; few prickles; foliage small, dark green, glossy; compact, spreading, low (16-20 in) growth; [Santa Claus X Times Square]; Carruth, Tom; Weeks Roses, 2000

'WEKsybil', Min, pb, 1989; (**Heartbreaker**); bud pointed; flowers deep pink with white base, small, borne in sprays of 3-5, dbl., exhibition form, slight fragrance; globular, small, dark orange fruit; prickles nearly straight, small, dark red-brown; foliage small, dark green, glossy; upright, bushy, medium, vigorous growth; PP007588; [Crystalline X Magic Carrousel]; Carruth, Tom; Weeks Roses, 1990

'WEKtorroc', LCl, lp, 1999; (**Rosy Outlook**); flowers deep pink striped white, dbl., 17–25 petals, 3–3.5 in., borne in large clusters, slight fragrance; prickles moderate; foliage large, dark, green, glossy; spreading, climbing, tall (10-12 ft) growth; hardy; [Tournament of Roses X Roller Coaster]; Carruth, Tom; Weeks Roses, 1999

'WEKuz', F, dp, 1990; (**Columbus**®); bud ovoid, pointed; flowers deep rose pink, very little fading, large blooms, dbl., 28–35 petals, exhibition form, borne usually singly, slight fragrance; prickles almost straight, slightly hooked, medium, pinkish-br; foliage large, medium green, dull; bushy, medium growth; PP8183; [Seedling X Bridal Pink]; Carruth, Tom; Weeks Roses, 1991

'WEKwinwin', HT, rb, 1999; (**Rosie O'Donnell**, Win Win); bud urnshaped, long; flowers velvety scarlet red with creamy yellow reverse, dbl., 30–35 petals, 5–5.5 in., exhibition form, borne mostly singly, slight fragrance; foliage large, dark green; long stems; medium, upright, slightly spreading growth; Winchel, Joseph F.; Weeks Roses, 1998

'WEKyegi', Min, lp, 1999; (**Lemon Drop**); flowers clear yellow, dbl., 26–40 petals, 1–1.5 in., borne in small clusters, slight fragrance; few prickles; foliage medium, light green, semi-glossy; upright, compact, medium (20-24 in) growth; [Gingerbread Man X seedling]; Carruth, Tom; Weeks Roses, 1999

'WELan', HT, pb, 1973; (**Angel Girl**); flowers peach-pink; [Bel Ange sport]; Wyant

'WELangel', Min, pb, 1996; (**Angelica Renae**®); flowers bright medium pink blend, dbl., 26–40 petals, exhibition form, borne mostly singly, slight fragrance; some prickles; foliage medium, dark green, matt; upright, bushy (24-30 in.) growth; AOE, 1996; [Anita Charles X Seedling]; Wells, Verlie W.; Nor'East Min. Roses, 1996

'Welch', (strain of R. multiflora), w; thornless; sometimes used as understock; Mt. Arbor Nursery

'WELco', HT, pb, 1990; (**Ashgrove Jubilee**); flowers cream, shading to deep rose at edges, 30 petals, exhibition form, borne singly, slight fragrance; foliage thick, shiny, dark green, obtuse at base; tall, upright growth; [Mascot X ((Seedling X Red Lion) X Silver Lining)]; Welsh, Eric, 1985; Rose Hill Roses, 1990

'Welcome', HT, mp, 1948; bud ovoid; flowers glistening rose-pink, very large, dbl., 35 petals, exhibition form, slight fragrance; foliage leathery; vigorous, upright, bushy growth; Dickson, A.

'Welcome Guest', HT, ab, 1984; flowers apricot with a tinge of pink around edges of petals; [Jan Guest sport]; Cox, Arthur George

Welcome Home *see* 'ANDwel'

'Welcome Stranger', F, my, 1968; flowers large, single, intense fragrance; foliage extra large, glossy, leathery; very vigorous, compact growth; [Ophelia, Climbing X Allgold]; Fankhauser; A. Ross & Son

'WELcord', Min, rb, 1997; (**Cherry Cordial**); flowers full, medium, bright red with white reverse, very dbl., 26–40 petals, borne mostly singly, slight fragrance; foliage medium, dark green, semi-glossy; upright, medium growth; [Magic Carrousel X Seedling]; Wells, Verlie W.; Kimbrew-Walter Roses, 1997

WELgold, F, lp; (**Victoria Gold**); Welsh, 1999

WELira, HT, lp; (**Woman's Day**); [Avandel X WELpa]; Welsh, Eric, 1992

'WELiz', HT, mr, 1987; (**Ida Elizabeth**); flowers large, borne singly, dbl., 30 petals, exhibition form, moderate fragrance; foliage matt; medium, bushy growth; [Red Lion X Mainauperle]; Welsh, Eric; Treloar Roses Pty. Ltd., 1990

'WELlee', Min, ab, 1994; (**Dorris Lee**); flowers apricot blend, dbl., 26–40 petals, 1.5–2.75 in., borne singly and in small clusters, moderate fragrance; some prickles; foliage dark green, semi-glossy; medium, upright growth; [Seedling X Party Girl]; Wells, Verlie W.; Wells Midsouth Roses, 1994

Weller Bourbon, B, lp

'Wellesley', HT, dp, 1905; flowers dark rosy pink, large, dbl., moderate fragrance; [Bridesmaid X Liberty]; Montgomery, A.

'Wellington', HGal, m; flowers crimson-purple, cupped; Int. prior to 1848

Well's Climber

Well's Pink and White Climber

'WELLsned', Min, pb, 1997; (**Lucille McWherter**); flowers semi double, medium, pink blend, semi-dbl., 8–14 petals, intense fragrance; foliage medium, dark green, glossy; upright, medium (18-24in.) growth; [Seedling X Seedling]; Wells, Verlie W.

'Wellworth', HT, pb, 1949; flowers peach shaded gold, pointed, dbl., 40 petals, 5 in., intense fragrance; foliage grayish green; vigorous growth; [Leontine Contenot X Golden Dawn]; LeGrice

'WELpin', HT, dp, 1992; (**Edmund Rice**); bud classic ht form; flowers light red, non-fading, blooms borne 5-7 per cluster and singly, dbl., 45–50 pointed, reflexed petals, 5 in., exhibition form, repeats quickly, moderate fragrance; foliage very red when young, dark green, shiny, leathery; bushy, upright (4') growth; [Red Lion X Pink Silk]; Welsh, Eric, 1989; Christian Bros. College, 1992

'WELqueen', Min, w, 1997; (**Memphis Queen**); flowers very full, medium, very dbl., 41 petals, borne mostly singly, slight fragrance; foliage medium, dark green, glossy; upright, medium (60cms)growth; [Miss Pearl X Pacesetter]; Wells, Verlie W.; Wells Mid-South Roses, 1997

'WELrum', Min, lp, 1994; (**Kathleen Rumble**); flowers moderately blooms borne mostly single, dbl., 15–25 petals, slight fragrance; some prickles; foliage medium, dark green, glossy; medium (6 cms), upright growth; [Seedling X Miss Pearl]; Wells, Verlie W.; Wells Midsouth Roses, 1994

'WELsam', Min, yb, 1994; (**Samantha Ruth**); very dbl., 3–3.5 in., borne mostly singly, slight fragrance; some prickles; foliage medium, dark green, semi-glossy; medium, upright growth; [(Party Girl X Magic Carrousel) X Seedling]; Wells, Verlie W.; Wells Midsouth Roses, 1994

'WELSNED', Min, pb, 1997; (**Lucille McWherter**); flowers pink blend, em-dbl. (8-14 petals), medium (1½ in.) blooms; very fragrant; foliage medium, dark green, glossy; upright, medium (18-24 in.) growth; [Seedling X Seedling]; Wells, Verlie; Wells Mid-South Roses

Welwyn Garden Glory *see* HARzumber

Wembley Stadium, F, dp

'WENCHE', HT, mr, 1997; flowers full, large, very dbl., 26–40 petals, borne mostly singly, slight fragrance; foliage medium, dark green, semi-glossy; upright, medium growth; [Silver Jubilee X Loving Memory]; Poole, Lionel

'WENDELIEN', F, mr, 1946; flowers cardinal-red, borne in clusters, dbl., 22 petals, 4 in., globular, slight fragrance; foliage bright green; vigorous, branching growth; [Donald Prior X Unnamed seedling]; Leenders, M.; Longley

Wendy *see* **'PARA TI'**

'WENDY', Pol, mp, 1949; flowers pink, open, medium, borne in clusters, dbl., moderate fragrance; foliage soft, light green; moderately vigorous, upright growth; RULED EXTINCT 11/91; [Tip-Top X Dorothy Perkins]; Heers; Pacific Nursery

Wendy *see* 'ZIPWEN'

Wendy *see* 'SUNWEND'

Wendy, MinFl, lp; Schuurman, Frank B., 1993

Wendy Ackerman Rose, HT, lp; J&P, 1998

'WENDY BARRIE', Pol, op, 1936; flowers orange-salmon, well formed, dbl.; vigorous, dwarf growth; Beckwith

'WENDY CUSSONS', HT, mr, 1963; bud long, pointed; flowers rose-red, dbl., 30 petals, 5–6 in., exhibition form, intense fragrance; foliage leathery, glossy, dark; vigorous, well-branched growth; GM, NRS, 1959 GM, Portland, 1964 Golden Rose, The Hague, 1964 PIT, NRS, 1959; [Believed to be Independence X Eden Rose]; Gregory; Ilgenfritz Nursery

'WENDY CUSSONS, CLIMBING', Cl HT, mr, 1967; Follen; Gregory

'WENDY DUCKETT', F, lp, 1999; flowers red cream, striking color, dbl., 26–40 petals, 1.2in., borne in small clusters, slight fragrance; few prickles; foliage small, medium green, glossy, impervious to rain; upright, compact, medium growth; [Sheri Anne X (Whippet X Party Girl)]; Jones, L.J.

Wendy Pease *see* 'DEBRAF'

Wendy van Wanten, F, lp; RvS-Melle, 1995

Wenlock® *see* 'AUSWEN'

'WENZEL GESCHWIND', HT, m, 1902; flowers purple/pink, medium to large, dbl., moderate fragrance; Geschwind, R.; (Sangerhausen)

'WENZEL GESCHWIND, CLIMBING', Cl HT, m, 1940; flowers purple/pink, medium to large, dbl., moderate fragrance; Vogel, M.; (Sangerhausen)

'WEOKAY', F, rb, 1986; (**Playmate**); flowers vivid scarlet, yellow center, large, borne usually singly, single, 5 petals, slight fragrance; foliage medium, dark green, glossy; upright, bushy growth; [Playboy X Unnamed seedling]; Fonda, Henry; Wee Ones Miniature Roses, 1986

'WEOPOP', Min, w, 1986; (**Gourmet Popcorn**); flowers pure white, medium, semi-dbl., 6–14 petals, slight fragrance; foliage large, dark green, glossy; upright, bushy growth; PP006809; [Popcorn sport]; Desamero, Luis; Wee Ones Miniature Roses, 1986

'WERINA', Gr, ob, 1975; (**Arizona**, Tocade); bud urn-shaped; flowers golden bronze to orangy yellow, medium, dbl., 25–30 petals, exhibition form, borne singly, intense, sweet fragrance; foliage glossy, dark, leathery; long, cutting length stems; vigorous, upright, bushy growth; AARS, 1975; [((Fred Howard X Golden Scepter) X Golden Rapture) X ((Fred Howard X Golden Scepter) X Golden Rapture)]; Weeks; C-P

'WERNER DIRKS', LCl, w, 1937; bud long, pointed; flowers ivory-white, very large, borne in clusters, dbl., exhibition form, moderate fragrance; foliage large, leathery, wrinkled; long, strong stems; very vigorous, climbing growth; [Mrs Pierre S. duPont X Daisy Hill]; Kordes

'WERNER OTTO', HT, my, 1995; flowers medium yellow, very full, very dbl., 56 petals, 4–4.5 in., borne mostly singly, slight fragrance; few prickles; foliage large, dark green, glossy; medium, upright growth; [Golden Fantasie sport]; Mungia, Larry; Montebello Rose Co., 1994

'WERNER TESCHENDORFF', F, op, 1949; bud ovoid; flowers medium, borne in clusters, dbl., cupped, slight fragrance; foliage glossy; vigorous, upright, bushy growth; [Swantje X Hamburg]; Tantau; Teschendorff

Werner von Braun *see* HELilast

'WESNIANKA', HT, w, 1941; (Spring); flowers white, base carmine-yellow; [Mme Butterfly X Mrs T. Hillas]; Costetske

Wessie Roos *see* KORbonnet

West Coast® *see* 'MACNGAURU'

'WESTBROEKPARK', F, or, 1968; flowers medium, borne in clusters, dbl.; foliage dark; [Orange Sensation X Kimono]; deRuiter

Westerland® *see* 'KORWEST'

'WESTERN GOLD', HT, my, 1932; flowers clear yellow; [Talisman sport]; Western Rose Co.

'WESTERN SUN'®, HT, dy, 1965; flowers deep yellow, dbl., 40 petals, 5 in.; foliage dark; [Golden Scepter seedling X Golden Sun]; Poulsen, Niels D.

Western Sunlight *see* 'HADSUN'

'WESTERN SUNSET', F, yb, 1958; bud pointed; flowers yellow tipped pink, becoming pink and then mahogany-red, bor, dbl., 35 petals, 1.5–2 in., cupped; foliage leathery; vigorous, upright growth; [Maxine X Masquerade]; Silva

Westfalen, S, 1969; Scholle, E.; (Sangerhausen)

Westfalengold, F, 1979; Noack, Werner; (Cavriglia)

'WESTFALENGRUSS', F, mr, 1978; bud ovoid; flowers fire-red, medium, dbl.; foliage glossy; low, bushy growth; Hubner; O. Baum

Westfalenpark® *see* 'KORPLAVI'

'WESTFIELD BEAUTY', HT, ab, 1923; bud long, pointed; flowers deep coppery apricot, tinted golden and salmon-pink, dbl., moderate fragrance; [Lady Pirrie X Mme Edouard Herriot]; Morse

'WESTFIELD FLAME', HT, or, 1925; bud long, pointed; flowers very deep flame, dbl., slight fragrance; [Mme Edouard Herriot X Diadem]; Morse

'WESTFIELD GEM', HT, dr, 1925; flowers dark maroon-crimson, dbl., slight fragrance; [Col. Oswald Fitzgerald sport]; Morse

'WESTFIELD SCARLET', HT, mr, 1931; flowers clear scarlet, open, very large, dbl., slight fragrance; foliage leathery, dark; bushy growth; GM, NRS, 1932; [Lady Inchiquin sport]; Morse

'WESTFIELD STAR', HT, w, 1922; flowers cream, fading white; [Ophelia sport]; Morse

'WESTMINSTER', HT, rb, 1960; flowers red, gold reverse, loosely-formed, large, dbl., 35 petals, intense fragrance; foliage dark; vigorous, tall growth; GM, NRS, 1961; [Gay Crusader X Peace]; Robinson, H.

Westminster Pink, HT, lp; Fryer, Gareth, 1998

'WESTMONT', Min, mr, 1958; bud pointed; flowers bright red, semi-dbl., 1.5 in.; foliage leathery, semi-glossy; small vigorous (12-18 in), bushy growth; [(R. wichurana X Floradora) X (Oakington Ruby X Floradora)]; Moore, Ralph S.; Sequoia Nursery

'WESTWARD HO!', HT, rb, 1964; flowers mahogany-red, reverse silver, dbl., 42 petals, 4.5 in., moderate fragrance; foliage very dark; vigorous, upright, compact growth; [Karl Herbst X Pink Charming]; Allen, E.M.; Sanday

Wetteriana, F, lp

'WETTRA', F, mr, 1976; flowers large blooms borne 4-15 per cluster, dbl., 20 petals, cupped, no fragrance; red prickles; foliage dark, matt; upright growth; [Pink Puff X Barcarolle]; Rijksstation Voor Sierplantenteelt

'WEZIP', HT, m, 1978; (Burning Sky, **Paradise**®, 'WEZEIP'); bud long, pointed; flowers silvery lavender shading to ruby-red at edge, well-formed, dbl., 28 petals, 3.5–4.5 in., moderate fragrance; foliage glossy, dark; upright growth; AARS, 1979 GM, Portland, 1979; [Swarthmore X Unnamed seedling]; Weeks; C-P

WHAmemo, MinFl, lp; (**Sweet Memories**)

'WHEATCROFT GIANT', HT, lp, 1962; flowers pearly pink, well-formed, large; vigorous growth; Wheatcroft Bros.

Wheatcroft's Baby Crimson *see* **'PERLA DE ALCAÑADA'**

Wheatcroft's Golden Polyantha *see* **'GOLDENE JOHANNA TANTAU'**

Wheato, F, dp; Rearsby Roses, Ltd.

Wheel Horse Classic *see* HARelite

'WHICAW', F, w, 1988; (**Mr McCawber**); bud pointed, white flushed pink; flowers medium, borne in sprays of 8-10, semi-dbl., 15–17 petals, exhibition form, slight fragrance; globular, orange fruit; prickles hooked, light brown; foliage medium, medium green, semi-glossy, disease-resistant; upright, bushy, tall, prolific growth; [(French Lace X Simplex) X (Pristine X White Angel)]; White, James J.

Whickham Highway *see* 'HORMASBRICK'

'WHIMSICAL', Min, pb, 1980; bud pointed; flowers either light pink or deeper peach-pink on same plant at same, dbl., 48 petals, exhibition form, no fragrance; no prickles; foliage medium green; upright, bushy growth; [Tiki X Baby Betsy McCall]; Strawn, Leslie E.; Pixie Treasures Min. Roses

'WHIPPED CREAM', Min, w, 1968; bud pointed, ivory; flowers small, dbl.; foliage light green, leathery; vigorous, bushy, dwarf growth; [(R. wichurana X Carolyn Dean) X White King]; Moore, Ralph S.; Sequoia Nursery

'WHIPPET', HT, pb, 1973; flowers salmon-pink, reverse lighter, dbl., 35 petals, 5 in., exhibition form, slight fragrance; moderate growth; Scott, D.H.

Whirlygig *see* 'JUSPROP'

'WHISKY', F, yb, 1964; flowers yellow shaded orange-bronze, open, large, borne in clusters, dbl., slight fragrance; foliage glossy, light green; very vigorous, upright growth; [Cognac X Arc-en-Ciel]; Delforge

Whisky *see* 'TANKY'

'WHISKY GILL', HT, ob, 1972; flowers burnt-orange to bright orange; [Whisky Mac sport]; Cobley

Whisky Mac *see* 'TANKY'

Whisky Mac, Climbing *see* 'ANDMAC'

Whisky, Climbing *see* 'ANDMAC'

'WHISPER', F, lp, 1971; flowers pale pink, semi-dbl., 10 petals, 3–3.5 in., slight fragrance; foliage light; vigorous growth; [Queen Elizabeth X Monique]; Cants of Colchester, Ltd.

Whisper Louise, Pol, lp; Robinson, 1994

Whistle Stop *see* 'MACMOSCO'

'WHITE AACHEN', F, ly, 1937; flowers buff-yellow to pure white; [Gruss an Aachen sport]; Western Rose Co.

'WHITE ALASKA', HT, w, 1960; bud long, pointed; flowers pure white, dbl., 20 petals, 4.5–5 in., exhibition form, slight fragrance; foliage leathery, dark; vigorous, upright growth; [White Ophelia X Unnamed seedling]; Hartgerink; Armacost & Royston

White America, LCl, w

White American Beauty *see* **'FRAU KARL DRUSCHKI'**

White American Beauty, Climbing *see* **Freedom**®

'WHITE ANGEL', Min, w, 1971; flowers small, dbl., exhibition form, slight fragrance; foliage small, light; vigorous, dwarf, bushy growth; AOE, 1975; [(R. wichurana X Floradora) X (Little Darling X Red Miniature seedling)]; Moore, Ralph S.; Sequoia Nursery

White Aster, Min, w; Moore, 1957

White Avalanche *see* 'WILWAVL'

White Baby Rambler *see* **'KATHARINA ZEIMET'**

'WHITE BABY STAR', Min, w, 1965; [Baby Gold Star sport]; Spring Hill Nursery; Wyant

White Banksia *see* **R. BANKSIAE BANKSIAE**

'WHITE BARONESS', HP, w; [Baroness Rothschild sport]; Paul & Son, 1882

White Bath *see* **'SHAILER'S WHITE MOSS'**

'WHITE BEAUTY', HT, w, 1965; flowers pure white; [The Doctor sport]; Brooks, M.L.; Texas Rose Research Foundation

White Belle of Portugal, LCl, w

White Bells® *see* 'POULWHITE'

White Blush, A, w; Sievers, 1988

White Bon Silène *see* **'BON SILÈNE BLANC'**

White Bonnet, F, w

'WHITE BOUGÈRE', T, w, 1898; flowers pure white; Dunlop

'WHITE BOUQUET', F, w, 1956; bud ovoid; flowers gardenia-shaped, blooms in irregular clusters, dbl., 45 petals, 4.5 in., moderate, spicy fragrance; foliage dark, glossy; bushy growth; AARS, 1957; [Glacier X Pinocchio seedling]; Boerner; J&P

'WHITE BRIARCLIFF', HT, w, 1932; (Mme Louis Lens); bud long, pointed; flowers pure white, large, dbl., exhibition form, moderate fragrance; foliage leathery; vigorous growth; [(Briarcliff X Kaiserin Auguste Viktoria) X (Briarcliff X Mrs Herbert Stevens)]; Lens; J&P

'WHITE BUTTERFLY', HT, w, 1954; bud long, pointed; flowers white, inner petals pale chartreuse, center, dbl., 24 petals, 3.5–4.5 in., cupped, moderate fragrance; foliage leathery; vigorous, compact growth; John Cook Medal, ARS, 1957; [Ophelia X Curly White]; Spanbauer

'WHITE CAP', F, w, 1954; dbl., 60 petals, 3.5–4 in., moderate fragrance; [Unnamed seedling X Break o' Day, Climbing]; Brownell, H.C.

White Captain, HRg, lp; Spek/Captain, 1998

White Cascade® *see* 'CARCADE'

'WHITE CÉCILE BRÜNNER', Pol, w, 1909; flowers white, sulfur yellow and buff; [Cécile Brunner sport]; Fraque

'WHITE CHARM', F, w, 1958; bud long, pointed; flowers white, base yellow-green, semi-dbl., 17–22 petals, 2.5–3 in., exhibition form, moderate fragrance; foliage leathery; bushy, upright growth; RULED EXTINCT 10/88; [Pinocchio X Virgo]; Swim & Weeks

White Charm® *see* 'MINAQCO'

'WHITE CHATEAU', HT, lp, 1999; bud ivory; flowers pure white, dbl., 26–40 petals, 5 in., exhibition form, slight fragrance; 4.5 ft growth; [Ophelia X seedling]; Teranishi, K.; Itami Rose Nursery, 1996

'WHITE CHATILLON', Pol, w; (Chatillon White); bud pale pink, fading to white.; [Probably a Chatillon Rose sport]

White Chipper, Min, lp

'WHITE CHRISTMAS', HT, w, 1953; bud long, pointed; flowers pure white, medium, dbl., exhibition form, moderate fragrance; foliage leathery, light green; moderate, upright growth; [Sleigh Bells X Seedling]; H&S

White Cloud *see* **'HAKUUN'**

White Cloud® *see* 'SAVACLOUD'

White Cloud *see* KORstacha

'WHITE CLOUDS', Cl HT, w, 1953; bud long, pointed; dbl., 50 petals, 4.5 in., intense fragrance; foliage leathery, dark; vigorous growth; RULED EXTINCT 8/88; [Frau Karl Druschki X Kaiserin Auguste Viktoria]; Silva

White Cochet *see* **'WHITE MAMAN COCHET'**

'WHITE COCKADE', LCl, w, 1969; flowers large, dbl., moderate fragrance; foliage glossy; vigorous, low, climbing growth; [New Dawn X Circus]; Cocker

White Colorado, F, lp; Select, 1998

'WHITE COLUMBIA, CLIMBING', Cl HT, w, 1934; Clark's Rose Nursery

White Comet *see* 'DELOBLAN'

White Cover *see* 'POULCOV'

'WHITE CROSS', LCl, w, 1950; bud ovoid; flowers white, center tinted lemon, large, very dbl., moderate fragrance; foliage leathery, glossy; very vigorous growth; [Mrs Arthur Curtiss James sport]; Hester

White Crystal®, HMsk, w; Lens, 1992

White Daily Rose *see* **'INDICA ALBA'**

'WHITE DAWN'®, LCl, w, 1949; flowers gardenia-shaped, pure white, dbl., 30–35 petals, 3–3.5 in., blooms in clusters, recurrent bloom, moderate, sweet fragrance; foliage glossy, dark green; very vigorous, climbing growth; [New Dawn X Lily Pons]; Longley, L.E.; Univ. of Minn.

White de Meaux *see* **'ROSE DE MEAUX WHITE'**

White Delight *see* 'JACGLOW'

'WHITE DEMURE', F, w, 1952; bud ovoid; dbl., 30–35 petals, 2–2.5 in., flat, slight fragrance; foliage leathery; vigorous, compact, bushy growth; [Demure sport]; Boerner

White Diamond® *see* INTeramon

'WHITE DIAN', Min, w, 1965; flowers white, sometimes light pink; [Dian sport]; Moore, Ralph S.; Sequoia Nursery

'WHITE DICK KOSTER', Pol, 1946; Grootendorst, F.J.; (Cavriglia)

White Dog, S, w

'WHITE DOROTHY', HWich, w, 1908; (White Dorothy Perkins); flowers creamy white; [Dorothy Perkins sport]; Cant, B. R.

White Dorothy Perkins *see* **'WHITE DOROTHY'**

White Dr Grill, T, lp

White Dream® *see* 'LENVIR'

'WHITE DUCHESS', HT, w, 1964; bud spiral, pointed; flowers snow-white, medium; foliage glossy; strong stems; vigorous growth; Herholdt, J.A.; Herholdt's Nursery

White Duchesse de Brabant *see* **'MME JOSEPH SCHWARTZ'**

'WHITE ELFE', F, w, 1954; bud pointed; flowers white faintly tinted pink, borne in clusters of 3-7, dbl., 40–45 petals, 2–2.5 in., exhibition form, slight fragrance; foliage dark, leathery; vigorous growth; [Rosenelfe sport]; Holmes, V.E.; Avansino;, Mortensen

'WHITE ENSIGN', HT, w, 1925; flowers large, very dbl., slight fragrance; foliage leathery, dark, glossy; vigorous growth; McGredy

'WHITE FAIRY', Min, w, 1952; flowers small blooms in clusters, very dbl.; Sequoia Nursery

White Fairy, Pol, lp

White Feather *see* 'MORFEAT'

White Festival *see* 'LAVSNOW'

'WHITE FINCH', Pol, w, 1937; flowers white, tinted pink; [Mrs R.M. Finch sport]; Stielow, F.C.

White Fleurette® *see* BRIseflu

'WHITE FLIGHT', HMult, w, 1923; flowers medium, dbl.; Koster; (Sangerhausen)

'WHITE FLIGHT', HMult, 1923; Koster, D.A.; (Cavriglia)

White Flower Carpet *see* NOAschnee

'WHITE FORCER', Pol, w, 1926; flowers larger and trusses than parent; [Jesse sport]; Spek

White Fragrant Mist, HT, w

'WHITE GARNETTE', F, w, 1952; flowers white tinged cream, borne in clusters, dbl., 30–35 petals, 2 in., slight fragrance; foliage rich green; bushy growth; [Pinocchio seedling X Garnette]; Boerner; J&P

White Gem® *see* 'MEITURUSA'

'WHITE GENE BOERNER', F, w, 1978; [Gene Boerner sport]; Takatori, Yoshiho; Japan Rose Nursery

'WHITE GOLD', LCl, w, 1943; bud long, pointed; flowers white, center yellow, petals reflexed, dbl., exhibition form; foliage dark, glossy; vigorous, climbing (20 ft) growth; [Glenn Dale X Mrs Arthur Curtiss James]; Brownell, H.C.

White Goid *see* 'COCQUIRIAM'

White Gold 98 *see* 'COCQUIRIAM'

'WHITE GROOTENDORST', HRg, w, 1962; [Pink Grootendorst sport]; Eddy

White Hedge, HRg, w

White Hermosa *see* **'MARIE LAMBERT'**

White Hero, S, lp; Rex, Dr. Robert W., 1996

'WHITE HOUSE', HT, w, 1951; bud high-centered to ovoid; flowers satiny white; foliage thick, light green; [McGredy's Yellow X Frau Karl Druschki]; Silva

White House *see* 'RESLINE'

'WHITE JEWEL', F, w, 1957; dbl., 33 petals, 4 in., cupped, blooms in clusters, moderate fragrance; vigorous, bushy growth; [Starlite seedling X Glacier seedling]; Boerner; J&P

'WHITE JOY', HT, w, 1952; bud pointed; flowers pure white, dbl., 55–65 petals, 4–5 in., flat, moderate fragrance; foliage leathery; bushy, upright growth; Spanbauer

White Junior Miss, Gr; (Cavriglia)

'WHITE KILLARNEY', HT, w, 1909;, moderate fragrance; [Killarney sport]; Waban Conservatories

'WHITE KING', Min, w, 1961; bud pointed, ovoid; flowers cream-white, small, dbl., 45 petals, 1.5 in., moderate fragrance; foliage leathery; bushy (12 in) growth; [Golden Glow (LCl) X Zee]; Moore, Ralph S.; Sequoia Nursery

White Knight *see* 'MEBAN'

'WHITE KNIGHT, CLIMBING', Cl HT, w, 1959; Komatsu; Kakujitsuen

'WHITE KNIGHT, CLIMBING', Cl HT, w, 1965; (Message, Climbing); Meilland; URS

White La France *see* **'AUGUSTINE GUINOISEAU'**

White Lady *see* **'DAME BLANCHE'**

White Lady Banks' Rose *see* **R. BANKSIAE BANKSIAE**

White Lafayette *see* **'DAGMAR SPÄTH'**

White Lightnin'™ *see* 'AROWHIF'

White Lightnin'™, Gr, w

White Look, HT, lp

'WHITE LOVE', HT, w, 1973; bud long; flowers ivory-white, large, intense fragrance; foliage dark; bushy growth; [Frau Karl Druschki X (Peace X Unknown seedling)]; Buisman, G. A. H.

'WHITE MA PERKINS', F, w, 1962; bud long, pointed; dbl., 20–40 petals, 3.5 in., borne in clusters, intense fragrance; foliage light green; very vigorous, upright growth; [Ma Perkins sport]; McDonald; Hennessey

'WHITE MADONNA', Min, w, 1973; bud long, pointed; flowers white to pale pink, dbl., 33 petals, 1 in., slight fragrance; foliage glossy, leathery; upright, bushy growth; [(R. wichurana X Floradora) X (Little Darling X Unnamed red Miniature)]; Moore, Ralph S.; Sequoia Nursery

White Magic *see* 'JACARE'

White Magic, HMsk, lp; Lens, 1989

White Magic, HT, lp

White Magic Carpet™ *see* POUlland

White Majesty *see* MEIbonver

'WHITE MAMAN COCHET', T, w, 1896; (White Cochet); flowers white, often flushed pink; [Maman Cochet sport]; Cook, J.W.

'WHITE MAMAN COCHET, CLIMBING', Cl T, w, 1907; Knight, G.; Leedle & Co., 1911

White Masterpiece *see* 'JACMAS'

White Max Graf, HKor, 1983; Kordes, R.; (Cavriglia)

White Meidiland™ *see* 'MEICOUBLAN'

White Meillandina *see* 'MEIBLAM'

'WHITE MIMOLLET', F, w, 1984; [Mimo let sport]; Ota, Kaichiro

White Mini-Wonder™ *see* 'MEIZOGREL'

White Mite *see* 'FROMITE'

White Moss *see* **'COMTESSE DE MURINAIS'**

'WHITE MOUNTAINS', LCl, w, 1958; bud globular; flowers white, becoming greenish, small, borne in clusters, dbl., profuse, intermittent bloom; foliage glossy; vigorous growth; quite hardy.; [Skinner's Rambler X Skinner's Rambler]; Risley

White Mountains, LCl, w

'WHITE MRS FLIGHT', HMult, w, 1916; flowers pure white; [Mrs F.W. Flight sport]; T. Rockford

White Mystery *see* 'DEVABE'

White New Dawn, LCl, w

White Nights *see* 'POULAPS'

White Noblesse, HT, w

'WHITE NUN', HT, w, 1968; flowers open, medium, dbl.; foliage glossy; vigorous, upright growth; [Virgo X ?]; Pal, Dr. B.P.; Indian Agric. Research Inst.

'WHITE OPHELIA', HT, w, 1920; flowers white, center faintly tinted pink, becoming white, well form, semi-dbl., moderate fragrance; [Ophelia sport]; Cleveland Cut-Flower Co.; Cleveland Cut-Flower Co.;, E.G. Hill Co.

'WHITE ORLÉANS', Pol, w, 1920; [Orléans Rose sport]; Van Eyk

White Pacific, HT, w

White Pavement, S, lp, 1991; semi-dbl., moderate fragrance; light red fruit; spreading (2.5 ft) growth; Uhl, J.

'WHITE PEARL', HT, w, 1948; bud long, pointed; flowers glistening white, open, well formed, large, dbl., 45–50 petals, slight fragrance; foliage large, soft; very vigorous, upright growth; RULED EXTINCT 4/85; Totty

White Pearl, Min, lp; [Minnie Pearl sport]; Lee, 1993

White Pearl in Red Dragon's Mouth, Ch, mr

White Penthouse, Gr, 1995; Fineschi, G.; (Cavriglia)

'WHITE PET', Pol, w, 1879; (Little White Pet); flowers small blooms in large clusters, dbl.; very dwarf growth; (14); Henderson, P.

'WHITE PET, CLIMBING', Cl Pol, w, 1894; Corboeuf

'WHITE PILLAR', LCl, w, 1958; very dbl.; pillar (8 ft) growth; [William F. Dreer X ?]; Hay; Marsh's Nursery

'WHITE PINOCCHIO', F, w, 1950; bud ovoid; dbl., 50 petals, 2.5 in., globular, moderate fragrance; vigorous growth; [Mrs R.M. Finch X Pinocchio]; Boerner; J&P

'WHITE PRINCE', HT, w, 1961; bud pointed; flowers creamy white, nearly, dbl., 50–80 petals, 5–6 in., globular, slight fragrance; foliage leathery, glossy; vigorous, upright growth; [(Blanche Mallerin X Peace) X (Peace X Frau Karl Druschki)]; Von Abrams; Peterson & Dering

White Provence *see* **'UNIQUE BLANCHE'**

'WHITE QUEEN', HT, w, 1958; flowers white, center creamy, dbl., 30 petals, 5 in., cupped, moderate fragrance; foliage leathery; vigorous, upright growth; [Starlite seedling X Glacier seedling]; Boerner; J&P

'WHITE QUEEN ELIZABETH', F, w, 1965; (Blanc Queen Elizabeth); flowers pure white, borne in clusters, 4 in., moderate fragrance; foliage light green, leathery; vigorous growth; [Queen Elizabeth sport]; Banner; North Hill Nursery

White Radiance *see* **'MARY NISH'**

'WHITE RADOX BOUQUET', S, w, 1988; bud small, rounded; flowers white with pink tinge in center, large, borne in sprays of 5, very dbl., 50 petals, moderate fragrance; prickles hooked, small, beige; foliage dark green, leathery; tall, shrub to semi-climber growth; [Radox Bouquet sport]; Melville Nurseries Pty., Ltd., 1989

White Rain *see* 'MORCAS'

White Rambler *see* **'THALIA'**

White Rock *see* BARwrock

White Rosamini, Min, w

White Rose of Finland *see* **'POLSTJÄRNAN'**

White Rose of York, A, w; (Bonnie Prince Charlie's Rose, Jacobite Rose, La Rose de York, R. procera, R. usitatissima, **R. X ALBA**); dbl., borne usually several together, summer bloom, moderate fragrance; ovoid (see endpapers), scarlet fruit; 6 ft. growth; (42); [Probably R. corymbifera X R. gallica]

'WHITE SATIN', HT, w, 1965; bud urn-shaped; flowers white, center greenish yellow, large, dbl., moderate fragrance; foliage light gray-green, leathery; vigorous, tall growth; [Mount Shasta X White Butterfly]; Swim & Weeks; Carlton Rose Nurseries

'WHITE SEDUCTION', Pol, w; [Seduction sport]

'WHITE SHEEN', F, w, 1959; bud ovoid; dbl., 25–30 petals, 2 in., borne in clusters, intense fragrance; foliage dark, glossy; upright growth; [Ma Perkins seedling X Demure seedling]; Boerner; J&P

White Simplicity® *see* JACsnow

White Sparrieshoop *see* **WEISSE AUS SPARRIESHOOP**

White Spire *see* KORechtem

White Sporting Koster, Pol, lp

'WHITE SPRAY', F, w, 1968; flowers pure white, well-formed, small blooms in clusters, moderate fragrance; [Seedling X Iceberg]; LeGrice

White Spray® *see* LENpaya

'WHITE STAR', HT, w, 1920; flowers ivory-white, base shaded lemon-yellow, intense fragrance; [Ophelia sport]; Morse

White Star of Finland *see* **'POLSTJÄRNAN'**

White Success *see* 'JELPIROFOR'

White Sunsation, S, lp; Kordes, 1996

White Sunshine *see* 'JONWHITE'

White Surprise®, S, w; Lens, 1987

'WHITE SWAN', HT, w, 1951; (Cygne Blanc); bud ovoid; flowers pure white, dbl., 30 petals, 4.5–5 in., exhibition form, moderate fragrance; foliage glossy, dark; vigorous, upright growth; [Kaiserin Auguste Viktoria seedling X White seedling]; Verschuren-Pechtold; J&P

'WHITE SWEETHEART', F, w, 1941; flowers white flushed blush-pink when half open, small, dbl., exhibition form, slight fragrance; foliage leathery; short, strong stems; vigorous, upright, bushy growth; [Rosenelfe sport]; J&P

White Symphonie, HT, w; 59 petals, 4 in.; foliage matt; strong growth; RvS-Melle, 1991

'WHITE TAUSENDSCHÖN', HMult, w, 1913; flowers white, sometimes flaked pink; [Tausendschön sport]; Paul, W.

White Treasure *see* 'HADTREASURE'

White Valerie, HT, lp

White Wedding, LCl, lp; Scarman, 1995

White Wine *see* 'RESINC'

'WHITE WINGS', HT, w, 1947; bud long, pointed; anthers chocolate-colored, blooms in large clusters, single, 5 petals, 3.5 in., moderate fragrance; foliage leathery, dark; vigorous, upright, bushy growth; [Dainty Bess X Unnamed seedling]; Krebs; H&S

White Wings, HT, lp, 1979; flowers pure white with crimson stamens, single, intense fragrance; 2.5 ft growth; Poulsen

Whitegold *see* 'COCQUIRIAM'

Whiteout *see* 'MACWHITOUT'

Whitley Bay *see* 'HORHARRYPLUS'

Whitney *see* 'BRINEY'

WHItsun, MinFl, lp, 1995; (**Pearl Anniversary**); Chessum, Paul

Whitsuntide Rose *see* **R. CINNAMOMEA PLENA**

Whoopi® *see* 'SAVAWHOOP'

Why Not *see* 'MORWHY'

'WHYTEWOLD', S, w, 1961; dbl., cupped, non-recurrent; height 3 ft; Skinner

'WICHMOSS', Cl M, lp, 1911; bud long, pointed, mossed; flowers pale blush-pink fading to white, blooms in clusters of 6-15, semi-dbl., 2 in., slight fragrance; foliage dark, leathery; vigorous, climbing growth; [R. wichurana X Salet]; Barbier

Wickwar, S, lp; Steadman, 1960

'WIENER BLUT', F, or, 1961; bud globular; flowers salmon-red, borne in clusters, dbl., 30 petals, intense fragrance; vigorous, upright growth; Horstmann

Wiener Charme® *see* 'KORSCHAPRAT'

Wiener Donaupark, F, lp

'WIENER WALZER', F, mr, 1965; (Pur Sang); bud pointed; flowers velvety bright red, large, dbl., 25–30 petals; foliage glossy; very vigorous, upright growth; Tantau, Math.

'WIENERWALD', HT, op, 1974; (Vienna Woods); bud long, pointed; flowers pink to light orange, large, dbl., moderate fragrance; foliage large, dark, leathery; vigorous, upright, bushy growth; [Colour Wonder X Seedling]; Kordes

Wieteke van Dordt *see* **'POTIFAR'**

Wife of Bath *see* 'AUSWIFE'

Wiggy *see* 'KENDANQU'

'WIJHE', F, or, 1961; dbl., borne in clusters; foliage dark; moderate growth; [Unnamed seedling X Alpine Glow]; Buisman, G. A. H.

Wil Pink Supreme, S, op

'WILACE', HT, dr, 1992; (**Peggy Rockefeller**®); flowers bright crimson to cherry red with dark smoky red on edge, dbl., 40 petals, moderate, spicy fragrance; few prickles; foliage large, dark green, semi-glossy; upright, bushy, medium growth; [Queen Elizabeth X Swarthmore]; Williams, J. Benjamin, 1991; New York Botanical Garden, 1991

'WILALEX', S, rb, 1988; (**Alexandria Rose**, City of Alexandria, The Alexandria Rose); flowers ivory to white with light red washing on petal edge, dbl., 26–40 petals; foliage large, dark green, glossy, disease resistant; upright, bushy, vigorous, strong, hardy growth; [(Queen Elizabeth X Kordes' Perfecta) X Mount Shasta]; Williams, J. Benjamin

'WILALNO', HT, dr, 1998; (**Robert Aliano**); flowers deep velvet red, dull reverse, dbl., 15–25 petals, 4.5–5 in., exhibition form, borne mostly singly, moderate fragrance; few prickles; foliage large, dark green, semi-glossy; strong, upright, tall, 4-5 ft. growth; [Chrysler Imperial X Mister Lincoln]; Williams, J. Benjamin; J. Benjamin Williams & Associates, 1999

'WILANGIL', HT, pb, 1992; (**Reynolda House**); flowers light pink with ivory and coral blend, blooms borne mostly singly, dbl., 26–40 petals, 1.5–2.75 in., intense fragrance; few prickles; foliage medium, dark green, semi-glossy, disease-resistant; upright, bushy growth; medium (3-4 ft. winterhardy.; [Royal Highness X Command Performance]; Williams, J. Benjamin

'WILBENTUR', Min, w, 1981; (**Pixie Delight**®); dbl., slight fragrance; foliage small, dark, semi-glossy; upright growth; PP005270; [Ma Perkins X Easter Morning]; Williams, J. Benjamin; C-P

'WILBERFORCE', C, dr; flowers dark crimson, large, dbl.; [Probably R. centifolia X R. gallica]; About 1840

'WILBLANK', MinFl, w, 1982; (**Stardance**®); deep yellow stamens, miniflora, dbl., 35 petals, slight fragrance; foliage small, medium green, semi-glossy; compact growth; PP005222; [Ma Perkins X (Charlie McCarthy X Easter Morning)]; Williams, J. Benjamin; C-P

WILBRID, HT, w, 2000; (**All-American Bride**); Williams, J. Benjamin

'WILCAMP', Min, or, 1990; (**Campfire**); bud pointed; flowers fiery orange neon red, reverse same, ages deeper, medium, dbl., 26 petals, exhibition form, borne singly, moderate, damask fragrance; foliage medium, dark green, semi-glossy; upright, bushy growth; [Marina X Starina]; Williams, J. Benjamin; White Rose Nurseries, Ltd., 1990

'WILCGIR', HT, ob, 1999; (Country Girl, **Emsie Girl**); bud pointed; flowers creamy white with bright orange edge, dbl., slight fragrance; [Garden Party X American Heritage]; Williams, J. Benjamin

'WILCHER', Min, mr, 1997; (**Cherry Pastel**®); flowers double, medium, bright cherry red, dbl., 15–25 petals, borne mostly singly, moderate fragrance; [Red Sunblaze X Pink Sweetheart]; Williams, J. Benjamin; Paramount Roses, 1997

'WILCREST', F, dy, 1992; (**Cedar Crest College**); flowers deep golden yellow, blooms borne in small clusters, semi-dbl., 6–14 petals, 1.5–2.75 in., slight fragrance; few prickles; foliage medium, dark green, glossy; medium (3-4 ft), upright, bushy growth; [Ivory Fashion X Sunsprite]; Williams, J. Benjamin, 1993

'WILCROWN', S, op, 1992; (**Copper Crown**); flowers orange pink, blooms, single, 5 petals, 3–3.5 in., slight fragrance; foliage medium, dark green, semi-glossy; upright (4x4), bushy growth; [Westerland X Orange Velvet]; Williams, J. Benjamin; Hortico Roses, 1992

Wild Amazone, HT, dp

Wild at Heart *see* 'KORADES'

Wild Berry Breeze® *see* 'JACRULAV'

'WILD CHERRY', HT, mr, 1984; flowers large, dbl., moderate fragrance; foliage large, dark, glossy; bushy growth; [King of Hearts X Granada]; Bridges, Dennis A.; Bridges Roses, 1985

'WILD CHILD', S, lp, 1982; bud small, pointed; single, 5 petals, borne 3-15 per cluster, repeat bloom, moderate, tea fragrance; curved, thin prickles; foliage small, light green, 9 leaflet, slightly rugose; vigorous, erect, some arching growth; [R. rugosa rubra X Dortmund seedling]; Hall, William W.

Wild Dancer® *see* 'JACINA'

'WILD FLAME', HT, or, 1973; bud ovoid; flowers medium, dbl., moderate fragrance; foliage leathery; vigorous, upright growth; [Granada X South Seas]; Meyer, C.; Ball Seed Co.

'WILD GINGER', Gr, ob, 1976; bud ovoid, pointed; dbl., 33 petals, 4–4.5 in., moderate, fruity fragrance; foliage dark, leathery; upright, bushy growth; [(Queen Elizabeth X Ruth Hewitt) X Lady Elgin]; Buck, Dr. Griffith J.; Iowa State University

'WILD HONEY', HT, ab, 1977; bud long, pointed; flowers coral and peach, dbl., 45 petals, 3–5 in., intense, spicy fragrance; foliage leathery; tall, vigorous growth; Weeks

Wild One *see* 'LYOWE'

Wild Plum® *see* 'JACWIG'

Wild Spice *see* 'JACRUWHI'

'WILD SPIRIT', S, mr, 1985; flowers bright medium red, small, single, 5 petals, non-recurrent, moderate fragrance; foliage small, dark, matt; vigorous, upright, branched growth; [Alika X R. moyesii]; James, John

Wild Thing *see* 'TINWILD'

Wild West *see* 'TANRAVENS'

'WILDAC', Gr, dr, 1986; (**Dr Adam Christman**®); flowers dark crimson red to scarlet, large blooms borne in large sprays, dbl., 26–40 petals, slight fragrance; foliage large, dark green, semi-glossy; upright growth; [Queen Elizabeth X Chrysler Imperial]; Williams, J. Benjamin

'WILDAK', MinFl, ob, 1982; (**Amber Flash**®); flowers medium, 15–25 petals, moderate fragrance; foliage medium, dark, semi-glossy; PP005271; [Zorina X Starina]; Williams, J. Benjamin; C-P

'WILDAN', HT, dr, 1989; (**The Daniel**®); bud ovoid; flowers dark velvety red, reverse burgundy, medium, borne usually singly, dbl., 36 petals, exhibition form, no fragrance; fruit not observed; few prickles; foliage large, dark green, semi-glossy; upright, medium growth; [Chrysler Imperial X Queen Elizabeth]; Williams, J. Benjamin; The Peninsula Nursery, 1990

'WILDANCE', Min, lp, 1989; (**Tiny Dancer**®); flowers bright, light coral pink, small blooms, dbl., 34 petals, exhibition form, borne usually single, slight, spicy fragrance; very few prickles; foliage small, dark green, dwarf; upright, low growth; [(Carla X Sonia) X (Circus X Ma Perkins)]; Williams, J. Benjamin; Young's American Rose Nursery, 1990

'WILDAWN', S, dp, 1995; (**Dawn Creeper**®); flowers deep pink, semi-dbl., 6–14 petals, 1.5 in., borne in small clusters, slight fragrance; some prickles; foliage small, dark green, glossy; low (10-12 in), spreading growth; [The Fairy X Sea Foam (Self) F1]; Williams, J. Benjamin; J. Benjamin Williams, 1996

Wildberry Breeze *see* 'JACRULAV'

'WILDEB', HT, pb, 1991; (**Deborah Moncrief**®); flowers pink with ivory blend, large blooms borne mostly singly, dbl., 26–40 petals, intense fragrance; few prickles; foliage large, dark green, matt; medium (12-18 in), bushy, spreading growth; [Carla X Sonia]; Williams, J. Benjamin, 1988

'WILDELS', Gr, pb, 1997; (**Delany Sisters**®); flowers ruffled petals, large, dbl., 15–25 petals, borne mostly single and in large clusters, intense fragrance; foliage large, dark green, semi-glossy; upright, bushy, tall (4.5-5ft.) growth; [Love X Handel]; Williams, J. Benjamin

Wildenfels Gelb, HFt, w; Dechan, 1929

Wildenfels Rosa, HFt, mp; Dechan, 1928

Wildest Dreams *see* 'TALWIL'

Wildfang®, S, mp; Noack, Werner, 1989

'WILDFEUER', S, mr, 1953; flowers fiery red, semi-dbl., 4.5 in.; bushy growth; Kordes

'WILDFIRE', F, mr, 1955; bud ovoid; flowers bright red, blooms in clusters, 8–10 petals, 3 in., slight fragrance; foliage leathery; vigorous, bushy, compact growth; [World's Fair X Pinocchio]; Swim, H.C.; Armstrong Nursery

Wildflower *see* 'AUSWING'

Wildlife®, S, pb; Dickson, Patrick, 1989

'WILDREM', HT, dp, 1997; (**Miss All-American Dream**®); flowers large, deep pink, full, very dbl., 26–40 petals, 5.5–6 in., borne mostly singly, intense fragrance; prickles moderate; foliage large, dark green, semi-glossy; upright, bushy, tall growth; ARS Patron Rose, 1996; [Miss All-American Beauty X Bride's Dream]; Williams, J. Benjamin; J. Benjamin Williams & Associates, 1996

'WILDWOOD', HT, op, 1937; bud long, pointed; flowers gold, salmon and bronze, semi-dbl., intense fragrance; foliage glossy, wrinkled; vigorous growth; H&S

'WILF TAYLOR', HT, ab, 1981; flowers large, dbl., 35 petals, exhibition form, slight fragrance; foliage large, dark, matt; upright growth; [Gavotte X Red Lion]; Bracegirdle, A.J.

'WILFOLK', S, ob, 1998; (**Boulie's Dream**®, Musketeer); flowers orange red inside, yellow reverse, very dbl., 26–40 petals, 4.5–5 in., borne singly and in large clusters, intense fragrance; few prickles; foliage large, dark green, semi-glossy; strong, vigorous, tall, upright, climbing, arched growth; [Folklore X self]; Williams, J. Benjamin; J. Benjamin Williams & Associates, 1999

'WILFORD', HT, dr, 1989; (**Fordham Rose**®); bud pointed; flowers deep maroon-red with deep, black, velvety tones, urn-shaped, dbl., moderate fragrance; fruit not observed; prickles few, ovoid, curved down, medium, tan; foliage large, dark green, semi-glossy, thick, disease-resis; upright, medium growth; [Chrysler Imperial X Josephine Bruce]; Williams, J. Benjamin; Fordham University, 1990

'WILFRED NORRIS', HT, mp, 1972; flowers medium salmon-pink, large, dbl., 23 petals; foliage matt; [Tropicana X Elizabeth of Glamis]; Harkness

'WILFRED PICKLES', HT, op, 1939; flowers peach shaded gold, well-formed, dbl., 26 petals, 6 in., moderate fragrance; foliage glossy, dark; stems red; very vigorous growth; [Mrs Charles Lamplough X Edith Mary Mee]; Mee; Fryer's Nursery, Ltd.

Wilfrid H. Perron *see* 'LAVSHRIMP'

'WILGMAR', S, mr, 1988; (**Maren**); flowers deep scarlet red with velvet overtones, large, dbl., 26–40 petals, slight fragrance; foliage large, dark green, semi-glossy, leathery, disease re; upright, spreading, vigorous, free-standing growth; [Red Fountain X Red Fountain]; Gimpel, W.F., 1989; J.B. Williams & Associates, 1989

'WILGOGH', S, rb, 1996; (**Van Gogh**®); flowers red and white striped, reverse pale whitish pink with stripes, dbl., 15–25 petals, 3.5–4 in., exhibition form, borne in large clusters, moderate fragrance; moderate prickles; foliage large, dark green, semi-glossy; bushy, medium (3-31/2 ft) growth; [Spinning Wheel X Handel]; Williams, J. Benjamin; J.B. Williams & Associates, 1997

'WILGOLD', Min, dy, 1997; (**Golden Earing**®); flowers double, large, mini flora, dbl., 15–25 petals, borne singly and in clusters, moderate fragrance; [Hershey Yellow X Copper Kittel]; Williams, J. Benjamin

'WILGOSH', Min, my, 1984; (**Gold Sweetheart**®); flowers deep yellow, small, dbl., 35 petals, slight fragrance; foliage small, medium green, semi-glossy; upright, bushy growth; [Sunsprite X Rise 'n' Shine]; Williams, J. Benjamin; J.B. Williams & Associates

Wilhelm *see* **'SKYROCKET'**

Wilhelm Breder *see* **'GLOWING SUNSET'**

'WILHELM HANSMANN', HKor, dr, 1955; (William Hansmann); bud ovoid; flowers deep crimson, large blooms in clusters, dbl., exhibition form, recurrent bloom, slight fragrance; foliage dark, leathery; very vigorous growth; Kordes

'WILHELM KAUTH', HT, ly, 1930; flowers large, very dbl., moderate fragrance; Kauth; (Sangerhausen)

'WILHELM KORDES', HT, op, 1922; bud long, pointed; flowers salmon, copper and golden blend, large, dbl., exhibition form, moderate fragrance; foliage bronze, leathery, glossy; bushy growth; [Gorgeous X Adolf Koschel]; Kordes, H.

'WILHELM KORDES, CLIMBING', Cl HT, op, 1927; Wood & Ingram

'WILHELM MARX', S, mr, 1939; flowers medium, dbl.; Vogel, M.; (Sangerhausen)

Wilhelm May, HT, dr, 1989; flowers velvety dark red, dbl., moderate fragrance; Michler, K. H.; (Sangerhausen)

'WILHELM TEETZMANN', F, mr, 1943; (Advance Guard); flowers intense crimson, open, borne in clusters, semi-dbl., 3.5 in.; very vigorous, upright, bushy growth; (28); [Holstein X Crimson Glory]; Kordes

Wilhelm Tell®, F, mr; Meilland, 1989

'WILHERB', HT, op, 1992; (**Lady Harriet**); flowers coral and peach blend, deep orange washings on petals, long, dbl., 26–40 petals, 3–3.5 in., intense fragrance; few prickles; foliage large, dark green, semi-glossy; tall (4-5 ft.), upright, bushy growth; [Carla X Sonia]; Williams, J. Benjamin, 1993

'WILHIRE COUNTRY', F, ob, 1981; (Willhire Country); flowers flat-topped blooms borne 6-8 per cluster, 25 petals, intense fragrance; broad, hooked, brown prickles; foliage glossy, round, dark; tall, upright growth; [Elizabeth of Glamis

X Arthur Bell]; Beales, Peter; Beales Roses, 1979

'WILHKPK', HT, ob, 1988; (**Rittenhouse**®); flowers fiery orange to copper blended, large, dbl., 26–40 petals, moderate fragrance; foliage large, dark green, glossy, disease resistant; upright, vigorous, mass-blooming, winter hardy growth; [Queen Elizabeth X Zorina]; Williams, J. Benjamin, 1989

'WILIVORY', S, w, 1995; (**Ivory Carpet**®); flowers pure white, full, small, very dbl., 26–40 petals, .5–1 in., borne in small clusters, slight fragrance; some prickles; foliage small, dark green, glossy; low, spreading, 10-12 in. growth; [(Sea Foam X The Fairy) X Star Dance]; Williams, J. Benjamin; J. Benjamin Williams & Associates, 1996

'WILJAME', HT, mr, 1998; (Burgundy Queen, **James Biddle**); flowers burgundy red, reverse medium red, light silver edges, very dbl., 26–40 petals, 4.5–5 in., exhibition form, borne mostly singly, intense fragrance; few prickles; [Pink Peace X Queen Elizabeth]; Williams, J. Benjamin; J. Benjamin Williams & Associates, 1998

'WILKBSP', S, yb, 1985; (**Brown County Splendor**®); flowers ivory to yellow blended with peach to orange-red, large, dbl., 35 petals, intense fragrance; foliage large, dark, semi-glossy; upright, spreading growth; [Paul's Lemon Pillar X (Garden Party X Command Performance)]; Williams, J. Benjamin; Krider Nursery

'WILKTWO', S, w, 1991; (**Joan Longer**®); bud pointed; flowers blush pink opening to ivory with a hint of coral pink in center, semi-dbl., cupped, moderate, damask fragrance; foliage large, dark green, semi-glossy; upright, bushy, tall growth; [Queen Elizabeth X Ivory Fashion, Climbing]; Williams, J. Benjamin, 1971; The Scott Arboretum of Swarthmore College, 1991

'WILL ALDERMAN', HRg, mp, 1954; flowers clear rose-pink, well-shaped, large, dbl., repeat bloom, intense fragrance; erect (4 ft.), bushy growth; [(R. rugosa X R. acicularis) X HT]; Skinner

Will Deb, HT, op

'WILL ROGERS', HT, dr, 1936; flowers velvety maroon-crimson, base almost black, burning in sun, dbl., 65 petals, 3 in., intense, damask fragrance; foliage leathery, light; vigorous growth; [Seedling X (Hadley X Crimson Glory)]; Howard, F.H.; H&S

'WILL ROGERS, CLIMBING', Cl HT, dr, 1940; H&S

'WILL SCARLET', HMsk, mr, 1948; flowers scarlet; [Skyrocket sport]; Hilling; Wayside Gardens Co., 1956

'WILLASH', Min, m, 1984; (**Lavender Sweetheart**®); flowers deep blue lavender, small, borne singly, dbl., 20 petals, exhibition form, intense fragrance; foliage small, dark, semi-glossy; bushy growth; [Double Feature X Unnamed Min seedling]; Williams, J. Benjamin; J.B. Williams & Associates

'WILLAVQ', Gr, m, 1995; (**William Carroll**®); flowers mauve, full, very dbl., 26–40 petals, 1.5–2.75 in., borne mostly single and in small clusters, slight fragrance; some prickles; foliage large, dark green, semi-glossy; tall (4-5 ft), upright growth; [Queen Elizabeth X Angel Face]; Williams, J. Benjamin

'WILLEA', HT, ab, 1992; (**Marylea Johnson Richards**); flowers light pink with peach tones, blooms borne mostly singly and, dbl., 26–40 petals, 1.5–2.75 in., intense fragrance; few prickles; foliage medium, dark green, semi-glossy; medium (3-4 ft), upright, bushy growth; [Royal Highness X Command Performance]; Williams, J. Benjamin

Willhire Country *see* **'WILHIRE COUNTRY'**

Willi Maass *see* **'RED RIPPLES'**

'WILLIAM A. BILNEY', HT, pb, 1927; flowers pink with light orange, large, very dbl., moderate fragrance; Easlea; (Sangerhausen)

'WILLIAM ALLEN RICHARDSON', N, yb, 1878; bud pointed; flowers orange-yellow, usually white at petal edges, medium to large, dbl., early bloom, repeating; very vigorous, climbing (12 ft.) growth; (14); [Reve d'Or seedling]; Ducher, Vve.

William and Mary, S, rb; Beales, Peter, 1988

'WILLIAM BAFFIN', HKor, dp, 1983; bud ovoid; flowers deep pink, free blooming, flat, dbl., 20 petals, 2.5 in., blooms in clusters of 30, slight fragrance; foliage small, medium green, glossy; vigorous, climbing growth; [R. kordesii seedling]; Svedja, Felicitas; Agriculture Canada

William Booth, S, lp; single, 5 petals, slight; L'Assomption, 1999

'WILLIAM BOWYER', HT, mr, 1924; bud long, pointed; flowers velvety red, intense fragrance; [Hadley X Hoosier Beauty]; Chaplin Bros.

'WILLIAM C. EGAN', LCl, lp, 1900; flowers flesh-pink, large, borne in large clusters, very dbl., non-recurrent, moderate fragrance; foliage bright, glossy; long stems; vigorous, semi-climbing growth; [R. wichurana X Général Jacqueminot]; Dawson; Hoopes, Bro. & Thomas

William Carroll® *see* 'WILLAVQ'

'WILLIAM CARTER', HT, dp, 1957; flowers deep pink, moderate fragrance; very vigorous growth; [Red Ensign sport]; Carter

'WILLIAM COBBETT', F, dp, 1968; flowers claret-rose, borne in trusses, semi-dbl., slight fragrance; vigorous growth; [Dorothy Wheatcroft sport]; Seale Rose Gardens

William David *see* 'PEAROYAL'

'WILLIAM E. NICKERSON', HT, or, 1928; flowers glowing orange-cerise, large, dbl., moderate fragrance; foliage light; vigorous growth; [Priscilla X William F. Dreer]; Easlea

'WILLIAM F. DREER', HT, lp, 1920; flowers shell-pink, base yellow, sometimes suffusing the entire flower, dbl., exhibition form, moderate fragrance; [Mme Segond Weber X Lyon Rose]; H&S

'WILLIAM F. EKAS', HT, op, 1935; bud cupped; flowers salmon-orange, suffused pink, dbl., 45–50 petals, 4 in., moderate fragrance; [Souvenir sport]; Cremer

'WILLIAM FRANCIS BENNETT', HT, mr, 1880; bud long; flowers crimson, large, dbl., moderate fragrance; Bennett

'WILLIAM GODFREY', S, lp, 1954; bud white; flowers pale pink, well shaped, large, very dbl., intermittent bloom all season; moderately vigorous, erect growth; [Altalaris X Hybrid Perpetual]; Skinner

William Grant, HGal, dp

William Griffith, HP, pb

'WILLIAM GROW', dr, 1859; flowers velvety violet/red, medium, dbl.; Laffay, M.; (Sangerhausen)

William Hansmann *see* **'WILHELM HANSMANN'**

'WILLIAM HARVEY', HT, mr, 1948; flowers scarlet, dbl., 4–6 in., exhibition form, intense fragrance; vigorous, bushy growth; [Crimson Glory X Southport]; Norman; Harkness

'WILLIAM III', HSpn, m; flowers magenta-crimson, changing to rich plum, fading to dark lilac, semi-dbl., moderate, spicy fragrance; foliage tiny, dark

'WILLIAM IV', HSpn, lp; (William the Fourth); very dbl., non-recurrent; very dwarf growth; (England), before 1906

'WILLIAM IV', HSpn, 1906; (Cavriglia)

'WILLIAM JESSE', HCh, mr; flowers red suffused violet, large, dbl.; Laffay, M., 1850

William Leech *see* 'HORTROPIC'

'WILLIAM LOBB', M, m, 1855; (Old Velvet Moss); flowers dark crimson-purple, reverse lilac-pink, fading to grayish-lilac, semi-dbl.; extremely prickly; vigorous growth; Laffay, M.

William McGown, S, lp; Peden, G.H., 1996

'WILLIAM MOORE', HT, mp, 1935; bud long, pointed; flowers soft, even-toned pink, becoming deeper, large, dbl., cupped, intense fragrance; foliage soft, light; bushy growth; McGredy

William Morris *see* 'AUSWILL'

'WILLIAM NOTTING', HT, rb, 1904; flowers salmon-red, center brighter, dbl., moderate fragrance; [Mme Abel Chatenay X Antoine Rivoire]; Soupert & Notting

'WILLIAM ORR', HT, mr, 1930; bud long, pointed; flowers crimson-scarlet, large, dbl., exhibition form, intense fragrance; foliage light, glossy, leathery; vigorous growth; McGredy

William Quarrier *see* 'COCLAGE'

'WILLIAM R. SMITH', T, pb, 1908; (Blush Maman Cochet, Charles Dingee, Jeannette Heller, Maiden's Blush, President Smith, President Wm.R. Smith); bud pointed; flowers center pale pink, outer petals creamy flesh, base citron-yellow, dbl.; foliage rich green, leathery; vigorous growth; [Maman Cochet X Mme Hoste]; Bagg; E.G. Hill Co.;, P. Henderson

William Shakespeare® *see* 'AUSROYAL'

'WILLIAM SHEAN', HT, mp, 1906; bud long, tapering; flowers clear rose-pink, veined darker, very large, dbl.; GM, NRS, 1906; Dickson, A.

'WILLIAM SILVA', HT, 1951; few prickles at base; vigorous growth; [Étoile de Hollande X Radiance]; Silva

William Stubbs *see* 'BOSNEECH'

William the Fourth *see* **'WILLIAM IV'**

'WILLIAM THOMSON', Pol, pb, 1921; flowers salmon-carmine and bright rose, small, borne in clusters; compact, bushy growth; [Maman Turbat X Ellen Poulsen]; Leenders, M.

William Walker *see* 'DEBRARO'

'WILLIAM WRIGHT WALCOTT', HT, lp, 1921; flowers outer petals darker, dbl.; [Richmond sport X Ophelia]; McGorum

William-Booth *see* AC WILLIAM-BOOTH

'WILLIAM'S DOUBLE YELLOW', HSpn, my; (Double Yellow Scots Rose); flowers pine-yellow with green carpels, small, semi-dbl., 2 in., very free blooming in spring, intense, sweet fragrance; foliage tiny, dark; growth free but spindly; Williams, John, 1828

William's Double Yellow, HSpn, my

'WILLIAMSBURG', HT, lp, 1965; bud long, pointed; flowers rose-pink, medium, dbl., cupped, moderate fragrance; foliage dark, glossy; vigorous, upright growth; [Contrast X Queen Elizabeth]; Howard, A.P.; Great Western Rose Co.

Williamsburg Days, SS, lp; Williams, J. Benjamin, 1999

'WILLIE MAE', Min, mr, 1966; flowers small, dbl., slight fragrance; foliage dark, glossy, leathery; vigorous, bushy, dwarf growth; [(R. wichurana X Carolyn Dean) X Little Buckaroo]; Moore, Ralph S.; Mini-Roses

'WILLIE WINKIE', Min, lp, 1955; flowers light rose-pink, micro-mini, small, dbl., globular; [Katharina Ziemet X Tom Thumb]; deVink; T. Robinson, Ltd.

Willits Crescendo, LCl, lp

'WILLIWAW', HT, m, 1978; bud pointed; flowers red-purple, yellow at base, dbl., 60 petals, cupped, moderate fragrance; hooked prickles; foliage glossy, dark; upright growth; [Silver Lining X Duke of Windsor]; Melle

Will-o'-the-Wisp *see* CLEmist

'WILLOWMERE', HT, op, 1913; bud long, pointed, coral-red; flowers rich shrimp-pink, center yellow, dbl., cupped; foliage light; spreading growth; [Unnamed seedling X Lyon Rose]; Pernet-Ducher

'WILLOWMERE, CLIMBING', Cl HT, op, 1924; Mermet

'WILLUCE', HT, dr, 1998; (**Luce di Todi**™); flowers dark velvet red, full, very dbl., 26–41 petals, 5 in.–5.5 in., exhibition form, borne singly, intense fragrance; few prickles; foliage large, dark green, semi-glossy; tall, upright, 4-5ft. growth; [Mister Lincoln X Miss All-American Beauty]; Williams, J. Benjamin; J. Benjamin Williams & Associates, 2000

'WILLY CHAPEL', HT, yb, 1930; flowers coppery yellow, shaded salmon and shrimp-pink; vigorous growth; Delhaye

'WILLY DEN OUDEN', Pol, or, 1938; flowers intense orange, rosette form, small, borne on trusses; dwarf, compact growth; Den Ouden

'WILMAG', Min, yb, 1989; (**Little Magician**™); bud ovoid; flowers light yellow with orange-red blend, loose, small, borne usually singly, dbl., 32 petals, exhibition form, slight, damask fragrance; fruit not observed; no prickles; foliage small, medium green, semi-glossy; low, winter-hardy growth; [Circus X Magic Carrousel]; Williams, J. Benjamin; White Rose Nurseries, Ltd., 1990

'WILMNMG', HT, dr, 1984; (**Midnight Magic**™); flowers large, dbl., 35 petals, exhibition form, slight fragrance; foliage large, dark, semi-glossy; [(Chrysler Imperial X Mister Lincoln) X (Christian Dior X Josephine Bruce)]; Williams, J. Benjamin, 1986

'WILMTHO', Gr, w, 1988; (**Mary Lyon**™); flowers pure chalk white, medium, dbl., 26–40 petals, intense fragrance; foliage large, dark green, semi-glossy, thick, disease resistant; upright, hardy growth; [(Mount Shasta X Sonia) X (White Masterpiece X Ivory Fashion)]; Williams, J. Benjamin

'WILORPAT', Min, ob, 1997; (**Orange Pastel**™); flowers double, medium, bright orange-red, reverse dull, dbl., 15–25 petals, 2.5–3 in., borne in small clusters, slight fragrance; few prickles; foliage medium, dark green, semi-glossy; compact, low (12-18in.)growth; [Red Sunblaze X Orange Sunblaze]; Williams, J. Benjamin; Paramount Roses, 1997

'WILPADA', MinFl, rb, 1984; (**Patio Dance**™); flowers medium red, white reverse, small, dbl., 20 petals, exhibition form, slight fragrance; foliage medium, dark, glossy; strong, low, bushy growth; [Winifred Coulter X White Gem]; Williams, J. Benjamin

'WILPAGO', MinFl, my, 1984; (**Patio Gold**™); flowers small, dbl., slight fragrance; foliage medium, dark, semi-glossy; low, compact, bushy growth; [Patio Patty X Rise 'n' Shine]; Williams, J. Benjamin

'WILPASN', MinFl, w, 1984; (**Patio Snow**™); flowers small, dbl., 35 petals, moderate fragrance; foliage small, medium green, semi-glossy; groundcover; very compact, spreading growth; [Sea Foam X Charlie McCarthy]; Williams, J. Benjamin

'WILPAVL', S, mp, 1988; (**Pink Avalanche**); flowers coral pink to medium red, small, borne in sprays of 20-30, dbl., 15–25 petals, slight fragrance; foliage medium, medium green, glossy; spreading, compact growth; [Sea Foam X The Fairy]; Williams, J. Benjamin, 1989

'WILPINK', F, mp, 1986; (Hot Pink Pastel, **Pink Sweetheart**™); flowers medium coral pink, well-formed, small blooms, dbl., 15–25 petals, 2–2.5 in., borne singly, moderate fragrance; prickles moderate; foliage medium, medium green, semi-glossy; upright, dwarf growth; [(Carla X Sonia) X (Circus X Ma Perkins)]; Williams, J. Benjamin

'WILPINK', Min, mp, 1997; (**Pink Sweetheart**™); flowers double, medium, dbl., 15–25 petals, borne mostly singly, moderate fragrance; foliage medium, medium green, semi-glossy; compact, low (12-18in) growth; [(Pink Sweetheart X Patio Patty) X]; Williams, J. Benjamin

'WILPLPD', Min, m, 1984; (**Plum Pudding**™); flowers blue-lavender washed red and purple, small, dbl., 35 petals, moderate fragrance; foliage small, dark, semi-glossy; upright, bushy growth; [Angel Face X Unnamed lavender seedling]; Williams, J. Benjamin

'WILPOP', S, lp, 1991; (**Pink Popcorn**™); flowers light pink to coral, blooms borne in large clusters in small, semi-dbl., 6–14 petals, 1.5 in., slight fragrance; few prickles; foliage small, medium green, matt, disease-resistant; low (12-18 in), compact, bushy growth; [R. chinensis minima X (Sea Foam X The Fairy) seedling]; Williams, J. Benjamin, 1992

'WILPURP', S, m, 1991; (**Purple Popcorn**™); flowers bluish purple, blooms borne in small, grape-like clusters, semi-dbl., 6–14 petals, 1.5 in., slight fragrance; few prickles; foliage small, medium green, matt, disease resistant; low (12-18 in), compact, bushy growth; [R. chinensis minima X (Sea Foam X The Fairy) seedling]; Williams, J. Benjamin, 1992

'WILRIB', F, mr, 1997; (Red Pastel, **Red Ribbon**®); flowers double, medium, dbl., 15–25 petals, borne in small clusters, moderate fragrance; foliage medium, medium green, semi-glossy; upright, compact, medium (14-18in.) growth; [Red Sunblaze X Pink Sweetheart]; Williams, J. Benjamin

'WILRSS', S, mr, 1987; (**Pat's Delight**); flowers crimson to light scarlet, loosely, borne in sprays of 5-7, semi-dbl., 16 petals, cupped, slight fragrance; light green prickles, curving downwards; foliage large, dark green, glossy; upright, strong, good branching growth; [Queen Elizabeth X Chrysler Imperial]; Williams, J. Benjamin

'WILSAB', Min, or, 1992; (**Gabriela Sabatini**); flowers bright fire orange-red, sweetheart, dbl., 15–25 petals, 1.5 in., blooms borne mostly singly, slight fragrance; few prickles; foliage small, dark green, semi-glossy; low (14-20"), upright, compact, dwarf type growth; [Marina X Pink Sweetheart]; Williams, J. Benjamin

'WILSCARLET', S, dr, 1995; (**Scarlet Spreader**®); flowers dark scarlet red, small, semi-dbl., 6–14 petals, 1–1.5 in., borne in small clusters, slight fragrance; some prickles; foliage small, dark green, glossy; dwarf, spreading, low (10-12 inches) growth; [Seedling X Red Fountain]; Williams, J. Benjamin; J. Benjamin Williams & Associates, 1996

'WILSCSO', Gr, yb, 1988; (**Elizabeth Scholtz**®); flowers deep yellow with orange washing, reverse yellow with orange, dbl., 26–40 petals, exhibition form, moderate, damask fragrance; foliage large, plum-red to dark green, glossy; upright, bushy growth; [(Granada X Oregold) X (Arizona X Sunblest)]; Williams, J. Benjamin

'WILSEPT', HT, pb, 1991; (**September Eighteenth**); bud ovoid; flowers shell pink, reverse deep coral pink, aging light pink, bloom, dbl., exhibition form, intense, damask fragrance; foliage large, light green, semi-glossy; medium, bushy growth; [Carla X Queen Elizabeth]; Williams, J. Benjamin, 1992

'WILSFOL', HT, mr, 1988; (**Search for Life**®); flowers large, dbl., 26–40 petals, exhibition form, slight fragrance; foliage large, dark green, semi-glossy; vigorous, hardy, abundant growth; [Miss All-American Beauty X Mister Lincoln]; Williams, J. Benjamin

'WILSHIRE', HT, mp, 1991; bud pointed; flowers medium, dbl., 25 petals, exhibition form, moderate fragrance; foliage medium, medium green, glossy; upright, tall growth; Bronze, ARC TG, 1985; [Don Juan sport]; Woodard, Joe, 1983; Kimbrew Walter Roses, 1991

'WILSLAV', HT, rb, 1987; (**Slava**®); flowers scarlet red, reverse ivory white, large, borne singly, dbl., 34 petals, exhibition form, moderate, damask fragrance; rounded, medium, pumpkin-orange fruit; medium, few, yellow-green prickles, hooked downwards; foliage large, dark green, glossy, disease resistant; upright, bushy, tall growth; [Garden Party X Love]; Williams, J. Benjamin

'WILSMA', Min, mp, 1982; (**Astra**®); flowers pink, bright yellow stamens, small, single, 5 petals, slight fragrance; foliage small, medium green, semi-glossy; upright growth; [Pinafore X (Lilibet X Fairy Queen)]; Williams, J. Benjamin; C-P

'WILSNOD', S, w, 1986; (**Snowdrift**®); flowers small blooms in large sprays, dbl., 35 petals, repeat bloom, slight fragrance; foliage medium, dark, semi-glossy; low, spreading growth; [Sea Foam X The Fairy]; Williams, J. Benjamin

'WILSPRING', Min, w, 1990; (**Spring Time**®); bud pointed; flowers ivory with light red edge, small, borne usually singly, dbl., 26 petals, exhibition form, intense, damask fragrance; foliage small, dark green, semi-glossy; bushy, low growth; [Easter Morning (F2) X Toy Clown]; Williams, J. Benjamin; White Rose Nurseries, Ltd., 1990

'WILSTAR', LCl, lp, 1999; (**Scarlet Star**); dbl., 17–25 petals, 2.5–3 in., borne in large clusters, slight fragrance; many prickles; foliage large, medium green, semi-glossy; climbing, upright, bushy, tall (12-15 ft) growth; [Red Fountain X Mister LIncoln]; Williams, J. Benjamin; Conary-Pyle, 1999

'WILSUN', F, ob, 1991; (**Tequilla Sunset**®); bud ovoid; flowers brilliant orange and yellow blend, urn-shaped, blooms borne, dbl., moderate, fruity fragrance; foliage medium, medium green, semi-glossy; upright, bushy, medium growth; [Redgold X Tropicana]; Williams, J. Benjamin, 1976; Paramount Nursery, 1992

Wiltshire *see* KORmuse

Wiltshire Pride *see* 'SANWIFE'

'WILTWOS', S, w, 1993; (**Two Sisters**®); flowers light pink to white, full, very dbl., 26–40 petals, 1.5–2.5 in., borne in large clusters, slight fragrance; some prickles; foliage medium, dark green, semi-glossy; bushy, spreading growth; [Sea Foam X The Fairy]; Williams, J. Benjamin; J. Benjamin Williams & Associates, 1993

'WILWAVL', S, w, 1988; (**White Avalanche**); flowers cream, opening to pure white, small, borne in clusters, dbl., 15–25 petals, slight fragrance; foliage medium, medium green, glossy, disease resistant; spreading, compact, uniform, winter hardy growth; [Sea Foam X The Fairy]; Williams, J. Benjamin, 1989

'WILWIND', S, rb, 1992; (**Windmill**); flowers burgundy with ivory white on reverse bicolor, blooms borne singly, single, 3–3.5 in., slight fragrance; few prickles; foliage medium, medium green, semi-glossy; medium (3-5'), bushy, spreading growth; [Handel X (Love X Double Feature)]; Williams, J. Benjamin, 1994

'WILYELP', Min, dy, 1997; (**Yellow Pastel**®); flowers double, medium, dbl., 15–25 petals, borne mostly singly, slight fragrance; foliage medium, medium green, semi-glossy; upright, compact, medium (13-18in.) growth; [Sunnydew X Rise'n'Shine]; Williams, J. Benjamin

Wimi® *see* 'TANROWISA'

Win Win *see* 'WEKWINWIN'

'WINBILEE', HT, rb, 1997; (**Better Homes & Gardens Diamond Jubilee**®); flowers full, very dbl., 26–40 petals, slight fragrance; foliage medium, medium green, semi-glossy; upright, medium (5ft.) growth; [Lynn Anderson X Seedling]; Winchel, Joseph F.

'WINBOB', HT, lp, 1993; (**Dr Bob Harvey**); dbl., 26–40 petals, 1.5–2.75 in., no fragrance; foliage medium, medium green, matt; bushy growth; Winchel, Joseph F.; Coiner Nursery

'WINBUR', HT, dr, 1987; (**American Dream**®); flowers deep red, large, dbl., 25 petals, exhibition form, borne singly, slight, damask fragrance; round, average, medium, medium orange fruit; average, dark brown, slightly recurved prickles; foliage medium, medium green, semi-glossy; upright, tall growth; Gold, ARC TG, 1986; [(My Dream X Charles Mallerin) X Seedling]; Winchel, Joseph F.; Co-Operative Rose Growers, 1990

Winchester Cathedral *see* 'AUSCAT'

'WIND CHIMES', HMsk, mp; flowers rosy pink, blooms in clusters, 1 in., recurrent bloom, intense fragrance; vigorous (15-20 ft.) growth; Lester Rose Gardens, prior to 1946; Lester Rose Gardens, 1949

'WIND SONG', HT, ab, 1968; flowers orange, large, dbl., moderate fragrance; foliage leathery; bushy growth; [Royal Sunset X Sierra Sunset]; Morey, Dr. Dennison; Country Garden Nursery

'WINDEKIND', F, mr, 1974; bud pointed; dbl., 34 petals, 3 in., cupped; foliage matt, dark; upright growth; [Colour Wonder X Geisha]; Institute of Ornamental Plant Growing

'WINDERMERE', LCl, dp, 1932; flowers carmine-rose, blooms in large clusters, semi-dbl.; vigorous growth; Chaplin Bros.

Windflower *see* 'AUSCROSS'

Windjammer *see* 'SAVSWAT'

Windlass, Pol, lp; [Simon Robinson seedling sport]; Sutherland, P, 1995; Golden Vale Nursery, 1995, Australia

Windlass, HCh, lp; Sutherland, 1996

Windmill *see* 'WILWIND'

Windmill, S, lp; Williams, J. Benjamin, 1999

'WINDONNA', HT, rb, 1993; (**Donna Darlin'**); flowers yellow blending pink to bright red, blooms, dbl., 15–25 petals, 1.5–2.75 in., slight fragrance; some prickles; foliage large, dark green, semi-glossy; medium (4 ft), bushy growth; [Seedling X Double Delight]; Winchel, Joseph F.; Coiner Nursery, 1994

Windrose®, S, lp; Noack, Werner, 1993

Windrush® *see* 'AUSRUSH'

'WINDSOR', HT, mr, 1929; flowers rich crimson-scarlet, moderate fragrance; vigorous growth; Chaplin Bros.

Windsor, T, lp

'WINDSOR CHARM', HT, lp, 1978; bud long, tulip shaped; flowers light pink to ivory, dbl., 27–30 petals, 5–5.5 in., slight fragrance; foliage leathery; vigorous, bushy growth; [First Prize sport]; LeMire, Walter

'WINDSOUNDS', HT, lp, 1976; [First Prize sport]; Scoggins; American Rose Foundation

'WINDY CITY', Min, dp, 1974; flowers deep pink, reverse lighter, small, dbl., exhibition form, slight fragrance; foliage small, bronze; upright, bushy growth; [Little Darling X (Little Darling X (R. wichurana X Unnamed seedling))]; Moore, Ralph S.; Mini-Roses

Wine and Dine™ *see* 'DICUNCLE'

Wine and Roses, S, lp; PPAF; Clements, John K., 1999

Winecup, F, m

'WINEFRED CLARKE', HT, my, 1964; flowers, dbl., 34 petals, 5.5 in., exhibition form; foliage dark, glossy; [Peace X Lydia]; Robinson, H.

Winged Fellowship *see* 'HORWINGFEL'

Winghaven's White, N, lp; (found rose)

'WINGOLD', HT, lp, 1999; (**Super Gold**™); dbl., 17–25 petals, 2.5 in., borne mostly singly, slight fragrance; few prickles; foliage large, dark green, glossy; upright, tall (36 in) growth; [Gold Medal X seedling]; Winchel, Joseph F.; Coiner Nursery, 2000

'WINI EDMUNDS', HT, rb, 1973; flowers red, reverse white, large, dbl., exhibition form, moderate fragrance; foliage dark, leathery; vigorous, upright growth; [Red Lion X Hanne]; McGredy, Sam IV; Edmunds Roses

'WINIFRED', Gr, mp, 1981; bud pointed; flowers bright medium pink, dbl., 23 petals, slight fragrance; slightly hooked, red prickles; foliage reddish when young; vigorous, tall growth; [Queen Elizabeth X Unnamed seedling]; Jerabek, Paul E.

'WINIFRED', HT, yb, 1930; flowers deep yellow, shaded peach to old-gold, large, dbl.; vigorous growth; RULED EXTINCT 4/81; Chaplin Bros.

'WINIFRED COULTER', F, rb, 1962; bud pointed; flowers vermilion-scarlet, reverse silvery, blooms in clusters, dbl., 23 petals, 2.5–3 in., flat, intense fragrance; foliage leathery, glossy, dark; vigorous, bushy growth; David Fuerstenberg Prize, ARS, 1968; [Baby Chateau X Contrast]; Kemple; Van Barneveld

Winifred Coulter, Climbing, Cl F, lp; Jarvis, after 1968

'WINKFIELD CRIMSON EMPEROR', LCl, dr, 1958; flowers dark crimson, borne in clusters, dbl., 27 petals, 3.5 in., recurrent bloom, slight fragrance; foliage leathery; vigorous, pillar growth; [Red Empress X Surprise]; Combe; Winkfield Manor Nursery

'WINNERS ALL', HT, dp, 1990; bud ovoid; flowers carmine pink with a touch of orange in center, loose, large, dbl., cupped, slight, fruity fragrance; foliage medium, medium green, semi-glossy; bushy, medium growth; [Cherry Brandy sport]; Rearsby Roses, Ltd., 1986; Rearsby Roses, Ltd.

Winner's Choice *see* JACmad

'WINNIE DAVIS', HT, mp, 1902; bud long, pointed; flowers pink with lighter reflexes, semi-dbl., moderate fragrance; [Kaiserin Auguste Viktoria X Mrs W.J. Grant]; Nanz & Neuner

Winnie Renshaw *see* 'RENWINNIE'

Winning Colors™ *see* 'TWOWIN'

'WINNIPEG PARKS', S, dp, 1990; bud pointed; flowers dark pink-red, dark pink reverse, aging dark pink, medium, semi-dbl., 22 petals, cupped, repeat bloom, slight fragrance; foliage medium, medium green, matt; medium, bushy growth; [(Prairie Princess X Cuthbert Grant) X (Unnamed seedling X Morden Cardinette)]; Collicutt, Lynn M., 1981; Agriculture Canada, 1991

'WINPTIG', F, lp, 1999; (**Perfume Tiger**™); flowers dark red and white striped, very dbl., 41 petals, 1.75 in., borne in large clusters, intense fragrance; prickles moderate; foliage medium, medium green, semi-glossy; spreading, medium (27 in) growth; [Roller Coaster X seedling]; Winchel, Joseph F.; Coiner Nursery, 2000

Winsome™ *see* 'SAVAWIN'

'WINSOME', HT, dp, 1924; flowers cherry-rose, dbl., slight fragrance; RULED EXTINCT 1/85; [Premier X Hoosier Beauty]; Hill, E.G., Co.; Vestal

'WINSOME', Cl HT, mr, 1931; bud long, pointed; flowers cherry-red, large, dbl., intense fragrance; foliage leathery; very vigorous growth; RULED EXTINCT 1/85; Dobbie

'WINSOME NATIVE', S, w, 1979; bud ovoid, pointed; dbl., 20 petals, cupped, borne singly, sparse, repeat bloom, no fragrance; no prickles; foliage finely serrated, 5 leaflet; rounded, medium growth; [Gene Boerner X R. suffulta]; Stoddard, Louis

'WINTER KING', S, dp, 1985; flowers bright deep pink, blooms borne singly, semi-dbl., 11 petals, repeat bloom, intense fragrance; foliage medium, dark, leathery, matt; compact, branching growth; hardy.; [((McGredy's Scarlet X Polly) X Frau Karl Druschki) X Northlander]; James, John

Winter Magic™ *see* 'FOUMAGIC'

Winter Princess *see* 'GELWIN'

'WINTER SUNSET', S, yb, 1997; flowers full, large, very dbl., 26–40 petals, borne in large clusters, moderate fragrance; foliage large, dark green, glossy; upright, bushy, medium growth; [Serendipity X (Country Dancer X Alexandra)]; Buck, Dr. Griffith J.

Winter Wheat™ *see* 'MINIWIN'

Wintonbury Parish *see* 'BERPAR'

'WINTONIENSIS', HMoy, mp; (R. X wintoniensis); flowers rosy pink, blooms in clusters; foliage with sweetbriar (apple) fragrance; (42); [R. moyesii X R. setipoda]; 1935

'WIRCUROB', HT, yb, 1991; (**Lizabeth's Lullabye**); flowers yellow with pink tips, moderately large, dbl., 35–40 petals, moderate fragrance; foliage large, dark green, glossy; upright, tall growth; [Queen Elizabeth X Song of Paris]; Robbins, William C., 1990

Wirrunna White Fairy, Pol, lp; (Weatherly, L.)

'WISBECH GOLD', HT, yb, 1964; flowers golden yellow edged pinkish, well-formed, dbl., 35 petals, 4 in., cupped, moderate fragrance; vigorous, compact growth; [Piccadilly X Golden Star]; McGredy, Sam IV; McGredy

Wisbech Rose Fragrant Delight® *see* **'FRAGRANT DELIGHT'**®

Wise Portia *see* 'AUSPORT'

Wishful Thinking *see* 'SEAWISH'

Wishing *see* 'DICKERFUFFLE'

Wistful *see* 'SAVAWIST'

'WITCHCRAFT', F, yb, 1961; bud ovoid; flowers yellow, reverse scarlet, borne in large clusters, dbl., 45–50 petals, 2–2.5 in., moderate fragrance; foliage dark, glossy; vigorous growth; [Masquerade X Seedling]; Verschuren; Blaby Rose Gardens

'WITCHING HOUR', F, dr, 1967; bud ovoid; flowers dark red, almost black, medium, dbl.; vigorous, bushy, compact growth; [Rapture X (F.W. Alesworth X Charles Mallerin)]; Morey, Dr. Dennison; Country Garden Nursery

With Love *see* 'ANDWIT'

With Thanks *see* 'FRANSMOOV'

Wit's End *see* 'SEAWIT'

'WIWAIT', F, w, 1991; (**Playfair**); single, borne in small clusters, slight fragrance; foliage medium, medium green, semi-glossy; medium growth; [Playgirl X Unknown seedling]; Wilke, William

'WIWANCY', HT, lp, 1992; (**Nancy Shaw**); dbl., 26–40 petals, 1.5–2.75 in., borne mostly singly, slight fragrance; some prickles; foliage medium, medium green, semi-glossy; upright, bushy, tall (4 ft) growth; [Peggy Lee X Unknown]; Wilke, William

Wizard *see* 'MACLEIDE'

Wizo® *see* 'KRIWI'

'WOBURN ABBEY', F, ob, 1962; bud ovoid; flowers orange, blooms in clusters, dbl., 25 petals, 3.5 in., cupped, moderate fragrance; foliage dark, leathery; moderate growth; [Masquerade X Fashion]; Sidey & Cobley; Harkness, 1962;, J&P, 1964

Woburn Abbey, Climbing, Cl F, lp; Brundrett, S, 1972

'WOBURN GOLD', F, dy, 1970; flowers gold; [Woburn Abbey sport]; Robinson, H.

Woburn Gold, F, lp; [Woburn Abbey sport]; Robinson, H., 1970

'WODAN', HMult, 1890; Geschwind, R.; (Cavriglia)

Wohler Road White Eglantine, HEg, lp; (found rose)

'WOLFE'S GLORIE', Pol, op, 1943; flowers salmon-pink to deep orange; vigorous growth; [Orléans Rose sport]; Wolf, Van der; Hage

'WOLFGANG VON GOETHE', HP, pb; bud very long; flowers delicate bright yellowish pink, very large, dbl., exhibition form; vigorous growth; [Frau Karl Druschki X Souv. de Claudius Pernet]; Weigand, L., 1932

Wolley-Dod's Rose *see* **'DUPLEX'**

Wollongong Gold *see* 'WALSHINE'

'WOMAN AND HOME', HT, ob, 1976; flowers orange, pointed, dbl., 33 petals, 5 in., slight fragrance; foliage glossy, dark; [Apricot Silk X Piccadilly seedling]; Gregory

Woman O'th North *see* 'KIRLON'

Woman Pioneer *see* **'PIONERKA'**

'WOMAN'S DAY', HT, or, 1966; bud ovoid; flowers scarlet-orange, medium, semi-dbl., exhibition form; foliage leathery; vigorous, upright growth; [Queen Elizabeth X Karl Herbst]; Armbrust; Langbecker

Woman's Day *see* WELira

Woman's Day, F, lp; Welsh, 1993

Woman's Hour, F, lp; Beales, Peter, 1997

Woman's Institute, HT, lp

'WOMAN'S OWN'®, Min, mp, 1973; flowers pink, small, dbl., 45 petals, 1.5 in., slight fragrance; foliage dark; [New Penny X Tip-Top]; McGredy, Sam IV; McGredy

'WOMAN'S REALM', HT, mr, 1966; flowers scarlet, 3.5–4 in., slight fragrance; foliage glossy; upright growth; [Chrysler Imperial seedling]; Gregory

Woman's Value *see* 'KORVALUE'

'WOMEN IN MILITARY', S, mr, 1993; (The Service Rose); flowers moderately full, bright red, dbl., 20–30 petals, 2.75–3 in., borne in small clusters, moderate fragrance; some prickles; foliage medium, medium green, glossy; medium (135 cms), upright growth; Jerabek, Paul E.

Women O'th North *see* 'KIRLON'

'WONDER OF WOOLIES', F, ab, 1978; flowers deep apricot, dbl., 30 petals, 4.5 in., intense fragrance; vigorous growth; [Arthur Bell X Elizabeth of Glamis]; Bees

Wonderglo™ *see* 'MINIWON'

Wonderland®, S, yb; Dickson, Patrick, 1994

'WONDERSTAR', F, my, 1960; flowers bright yellow, large, semi-dbl.; low, bushy growth; [Goldilocks X Unnamed seedling]; Mondial Roses

Wonderstripe *see* CLEwonder

Wood Lawn *see* 'FIALAWN'

Woodland Sunbeam *see* 'TROBGOLD'

Woodlands Lady *see* 'TROBSA'

Woodlands Sunrise *see* 'TROBSUN'

'WOODROW', (form of R. suffulta), mp; dbl., 60 petals, fall bloom; low growth; Hunt, Miss Alice, about 1925; P.H. Wright, 1936; (collected in Sask., Canada)

Wood's Garland *see* **'THE GARLAND'**

Woodstock *see* 'MORSTOCK'

Woody, HMsk, mr; Lens, 1978

Worjackie *see* 'KIRWORJACKIE'

Work of Art *see* 'MORART'

World Class *see* HARdimple

World Peace™ *see* 'BURWORPE'

'WORLD'S FAIR', F, dr, 1939; (Minna Kordes); flowers deep crimson fading to scarlet, blooms in clusters, semi-dbl., 19 petals, 4 in., moderate, spicy fragrance; foliage leathery; vigorous, bushy growth; AARS, 1940; [Dance of Joy X Crimson Glory]; Kordes; J&P

'WORLD'S FAIR SALUTE', HT, mr, 1964; bud ovoid; flowers crimson-red, dbl., 33 petals, 5.5 in., exhibition form, moderate fragrance; foliage leathery; vigorous, upright growth; [Mardi Gras X New Yorker]; Morey, Dr. Dennison; J&P

'WORLD'S FAIR, CLIMBING', Cl F, dr, 1941; J&P

Wörlitz, S, yb, 1962; flowers orange and yellow, large, dbl.; Berger, W.; (Sangerhausen)

'WORTHINGTON', (natural variation of R. setigera), lp; flowers clear pink; (found in Ohio)

'WORTHWHILE', HT, ob, 1973; flowers orange, reverse lighter, pointed, dbl., 33 petals, 5–6 in., slight fragrance; foliage matt, olive-green; [Gavotte X Vienna Charm]; LeGrice

Woschod, HT, w, 1973; flowers large, dbl.; Bülow; (Sangerhausen)

Wow! *see* 'PIXIWOW'

Wrams Gunnarstorpsrosen, B, dp

Wretham Rose, C, lp; Hoellering, 1997

'WRIGHT'S SALMON', HT, op, 1960; flowers salmon, dbl., 32 petals, 5 in., exhibition form; [Opera sport]; Wright & Son

WRIpic, Min, ob, 1991; (**Picaninni**); Wright

Wurzburg *see* 'KORGUST'

'WYNANDA', F, mp, 1968; bud ovoid; flowers pink, medium, dbl., slight fragrance; foliage dark; [United Nations X Seedling]; Heidemij

Wyrala, HT, lp; [Lemania X Ami Guinard]; Heers, C W, 1949

'WYRALLA', HT, dr, 1949; bud ovoid; flowers large, semi-dbl., exhibition form; vigorous, bushy growth; [Lemania X Ami Quinard]; Heers; Pacific Nursery

Xanadu, F, pb

'XAVIER OLIBO', HP, dr, 1865; flowers velvety crimson shaded purplish; [Général Jacqueminot sport]; Lacharme, F.

Xaviere *see* 'MEIVIOLA'

Xerxes *see* 'HARJAMES'

X-Rated *see* 'TINX'

Y2K *see* 'SAVYK'

Yabadabadoo® *see* 'MACYABA'

'YABUSAME', HT, lp, 1999; flowers crimson, dbl., 35 petals, 6 in., exhibition form, intense fragrance; 6 ft growth; [Big Chief X Papa Meilland]; Ohtsuki, Hironaka, 1998

'YACHIYO-NISHIKI', F, ob, 1986; flowers orange, reverse yellow, medium, borne 2-5 per stem, dbl., 28 petals, moderate fragrance; prickles curved downwards; foliage dark green, semi-glossy; upright, compact growth; [(Maxim X Myo-Joh) X Duftwolke]; Suzuki, Seizo; Keisei Rose Nursery, 1984

Yakiman, F, m

Yakimour® *see* MEIpsilon

'YAMETSU-HIME', Min, pb, 1961; flowers white edged pink, small, semi-dbl.; foliage glossy, light; dwarf growth; [Tom Thumb X Unnamed seedling]; Hebaru Itami Rose Nursery

'YAMINI KRISHNAMURTI', HT, m, 1969; flowers lilac, medium, dbl., exhibition form slight fragrance; foliage glossy; moderate growth; [Sterling Silver, Climbing X Open pollination]; Singh; Gopalsinamiengar

'YANINA', F, op, 1964; flowers salmon-pink, large, semi-dbl., 16 petals, moderate fragrance; foliage glossy; vigorous growth; [Fashion X Queen Elizabeth]; Yeoman & Sons

'YANKA', S, mp, 1927; semi-dbl., borne in clusters, non-recurrent; height 3-4 ft very hardy.; Hansen, N.E.

Yankee Doodle® *see* 'YANKOR'

Yankee Lady *see* UHLwe

'YANKOR', HT, yb, 1965; (**Yankee Doodle®**); bud urn-shaped; flowers apricot to peachy pink and butter-yellow, imbricated, dbl., 75 petals, 4 in., slight, tea fragrance; foliage glossy, dark; upright, bushy growth; AARS, 1976; [Colour Wonder X King's Ransom]; Kordes, R.; Armstrong Nursery

Yantai *see* 'TINTAI'

Yardley Baroque *see* 'BEAYAR'

Yardley English Rose *see* KORyard

'YAROSLAVNA', HT, mp, 1958; (Famous Cliff); flowers large, dbl., 60 petals, moderate fragrance; foliage dark; vigorous growth; [La Parisienne X Peace]; Sushkov & Besschetnova

'YASHWANT', HT, rb, 1986; flowers red, striped white and yellow, reverse yellow, striped red, dbl., 55–60 petals, exhibition form, slight fragrance; pale, green prickles, curving downwards; foliage glossy; vigorous, upright, bushy growth; [Suspense sport]; Patil, B.K.; K.S.G. Son's Roses, 1985

'YASNAYA POLIANA', HT, op, 1958; (Bright Meadow); flowers salmon-pink, well formed, large, dbl., 45–50 petals, moderate fragrance; foliage dark; vigorous, compact growth; [Independence X Luna]; Shtanko, I.

'YATKAN', S, mp, 1927; dbl., 2.5 in., non-recurrent; very hardy.; [Probably Gruss an Teplitz X La Mélusine]; Hansen, N.E.

'YAWA', S, op, 1940; flowers light coral-pink, borne in clusters of 4-12 on 8-12" stem, dbl., 58 petals, 2 in., free, non-recurrent bloom, moderate fragrance; tall, wide spreading (pillar to 9 ft.) growth; very hardy.; [Anci Bohmova X R. blanda]; Hansen, N.E.

Year 2000, Min, lp

'YELLER ROSE O'TEXAS', HT, dy, 1986; flowers medium, dbl., 26–40 petals, intense fragrance; foliage medium, medium green, semi-glossy; upright growth; [Arizona sport]; Wambach, Alex A.

'YELLO YO YO', F, dy, 1980; bud slender, long; dbl., 20 petals, exhibition form, borne singly and in clusters, slight, tea fragrance; foliage medium green; upright growth; [Arlene Francis X Unnamed seedling]; Weeks

YelloGlo *see* 'DELCRO'

'YELLOW ALTAI', HSpn, my, 1950; flowers bright yellow, becoming lighter, small, borne in clusters, profuse, non-recurrent bloom; foliage light green, soft; stems reddish-brown; vigorous (7 ft.), upright growth; hardy (to -60).; [R. spinosissima altaica X Persian Yellow]; Wright, Percy H.

Yellow Angel, Min, dy

Yellow Baby Rambler *see* **'GEORGE ELGER'**

'YELLOW BANTAM', Min, ly, 1960; bud pointed, yellow; flowers yellow to white, micro-mini, dbl., 25 petals, .5 in., slight fragrance; foliage glossy; bushy (10 in) growth; [(R. wichurana X Floradora) X Fairy Princess]; Moore, Ralph S.; Sequoia Nursery

'YELLOW BANTAM, CLIMBING', Cl Min, ly, 1964; Rumsey, R.H.

'YELLOW BEAUTY', HT, my, 1945; bud longer, more pointed, clear golden yellow; more uniform growth than its parent; [Golden Rapture sport]; Ruzicka's

'YELLOW BEDDER', HT, ly, 1923; flowers clear yellow to cream, dbl., slight fragrance; [Mr Joh. M. Jolles X Mme Edouard Herriot]; Van Rossem

Yellow Belinda *see* 'TANYELLDA'

'YELLOW BIRD', HT, dy, 1974; bud ovoid; dbl., 42 petals, 3.5 in., slight fragrance; upright growth; [Sunlight X Arthur Bell]; McGredy, Sam IV

Yellow Blaze *see* 'BURYELLOW'

Yellow Blush, A, ly; Sievers, 1988

Yellow Bouquet *see* 'LAVKIN'

'YELLOW BUTTERFLY', HT, ob, 1927; flowers orange-yellow; RULED EXTINCT 1/89; [Mme Butterfly sport]; Scott, R.

Yellow Butterfly *see* 'MORWINGS'

'YELLOW BUTTON'®, S, yb, 1975; bud globular; flowers medium yellow, deeper in center, rosette blooms borne 1-15 per cluster, very dbl., 90 petals, flat, repeat bloom, slight fragrance; straight, red prickles; foliage dark; bushy growth; [Wife of Bath X Chinatown]; Austin, David

Yellow Cécile Brünner *see* **'PERLE D'OR'**

Yellow Cécile Brünner, Climbing *see* **'PERLE D'OR, CLIMBING'**

'YELLOW CHARLES AUSTIN'®, S, ly, 1981; [Charles Austin sport]; Austin, David, 1979

'YELLOW CLUSTER', F, my, 1949; flowers mimosa-yellow; spreading growth; [Golden Rapture X Polyantha seedling]; deRuiter; Spek

Yellow Cochet *see* **'ALEXANDER HILL GRAY'**

Yellow Condesa de Sastago *see* **'JEAN BOSTICK'**

Yellow Condesa de Sástago *see* **'YELLOW SÁSTAGO'**

'YELLOW CONTEMPO', F, dy, 1983; [Contempo sport]; Gupta, Dr. M.N., Datta, Dr. S.K. & Banerji, Shri B.K.; National Botanical Research Institute

Yellow Cover *see* POUlgode

'YELLOW CREEPING EVERBLOOM', LCl, my, 1953;, recurrent bloom; very vigorous, pillar or groundcover growth; [((Orange Everglow X New Dawn) X Unnamed seedling) X Free Gold]; Brownell, H.C.

'YELLOW CURLS', HT, my, 1947; flowers open, dbl., moderate fragrance; foliage glossy; vigorous growth; [Golden Glow X Self]; Brownell, H.C.

'YELLOW CUSHION', F, my, 1966; bud pointed; flowers large, dbl., blooms in small clusters, moderate fragrance; foliage glossy, leathery; vigorous, bushy growth; [Fandango X Pinocchio]; Armstrong, D.L.; Armstrong Nursery, 1962

Yellow Dazzler *see* **'YELLOWHAMMER'**

Yellow Delight, HT, lp

'YELLOW DOLL', Min, ly, 1962; bud pointed; flowers yellow to cream, small, dbl., 1.5 in., exhibition form, moderate fragrance; foliage leathery, glossy; vigorous, bushy (12 in) growth; [Golden Glow (LCl) X Zee]; Moore, Ralph S.; Sequoia Nursery

'YELLOW DOLL, CLIMBING', Cl Min, ly, 1976; Kirk; Sequoia Nursery

'YELLOW DOT', HT, ob, 1938; bud short, pointed; flowers light orange-yellow, reverse coral-red, small, dbl.; foliage soft, dark; short stems; vigorous growth; [Rapture X Betsy Ross]; Hill, Joseph H., Co.

Yellow Dot *see* INTerbab

'YELLOW FAÏENCE', HT, my, 1942; flowers lemon-yellow; [Faience sport]; C-P

Yellow Fairy® *see* 'POULFAIR'

Yellow Festival *see* 'LAVLINGER'

Yellow Festival, Min, lp

Yellow Fleurette® *see* INTerwell

'YELLOW FONTAINE', S, my, 1972; bud ovoid; flowers large, dbl., free, continuous bloom, slight fragrance; foliage large, glossy; vigorous, upright growth; [Unknown X Unknown]; Tantau, Math.; Ahrens & Sieberz

Yellow Fru Dagmar Hartopp *see* 'MORYELRUG'

Yellow Gamette, Pol, lp

'YELLOW GLORIA', HT, my, 1936; bud long, pointed; flowers golden yellow, base saffron, large, dbl., exhibition form, moderate fragrance; foliage leathery; long, strong stems; vigorous growth; [Talisman sport]; Bertanzel

'YELLOW GOLD', HT, dy, 1957; bud ovoid; flowers deep golden yellow, dbl., 30–35 petals, 5 in., cupped, intense fragrance; foliage leathery, glossy; strong stems; vigorous growth; Boerner; J&P

Yellow Hammer *see* **'YELLOWHAMMER'**

Yellow Hammer, Climbing, Cl F, lp; Dorieux, 1976

Yellow Herriot *see* **'FLORENCE CHENOWETH'**

Yellow Hit *see* 'POULLOW'

Yellow Holstein *see* **'GELBE HOLSTEIN'**

'YELLOW HOOVER', HT, my, 1933; flowers pure yellow; [Pres. Herbert Hoover sport]; Western Rose Co.

Yellow Hybrid Musk, HMsk, my

Yellow Jacket *see* 'JACYEPAT'

'YELLOW JEWEL', Min, my, 1973; bud long, pointed; flowers clear yellow, red eye, medium, semi-dbl., moderate fragrance; foliage small, glossy, leathery; dwarf, bushy growth; [Golden Glow X (Little Darling X Unnamed seedling)]; Moore, Ralph S.; Sequoia Nursery

'YELLOW JOANNA HILL', HT, my, 1932; bud long, pointed; flowers well formed, dbl., 28 petals, slight fragrance; foliage light; long stems; vigorous growth; [Joanna Hill sport]; White, C.N.; White Bros.

Yellow Kaiserin Auguste Viktoria *see* **'FRANZ DEEGEN'**

Yellow Lady Banks' Rose *see* **R. BANKSIAE LUTEA**

Yellow Light, HMsk, lp; Lens, 1995

'YELLOW MAGIC', Min, my, 1970; bud long, pointed; flowers small, semi-dbl., moderate fragrance; foliage small, glossy, dark, leathery; vigorous, dwarf, upright, bushy growth; [Golden Glow X (Little Darling X Seedling)]; Moore, Ralph S.; Sequoia Nursery

Yellow Maman Cochet *see* **'ALEXANDER HILL GRAY'**

Yellow Meillandina *see* 'MEITRISICAL'

'YELLOW MELLOW', Min, my, 1984; flowers small, dbl., 35 petals, slight fragrance; foliage large, dark, glossy; bushy growth; [Rise 'n' Shine X ?]; Bridges, Dennis A.; Bridges Roses, 1985

'YELLOW MINIATURE', Min, my, 1961; flowers bright yellow, small; [Rosina seedling sport]; Wayside Gardens Co.

Yellow Mini-Wonder *see* 'LAVGLO'

'YELLOW MME ALBERT BARBIER', HP, my, 1937; flowers sunflower-yellow; [Mme Albert Barbier sport]; Lens

'YELLOW MOSS', M, dy, 1932; bud ovoid, slightly mossed, deep yellow; flowers yellow, edges tinted pink, semi-dbl., cupped, free, intermittent bloom, moderate fragrance; foliage thick; vigorous growth; [Old Moss X Hybrid Tea]; Walter, L.; J&P

Yellow Mozart, S, my

'YELLOW NECKLACE', Min, ly, 1965; bud pointed; flowers straw-yellow, small, dbl., moderate fragrance; foliage leathery; vigorous, dwarf growth; [Golden Glow (LCl) X Magic Wand]; Moore, Ralph S.; Sequoia Nursery

Yellow Ocean, LCl, lp

Yellow Ophelia *see* **'SILVIA'**

'YELLOW PAGES', HT, yb, 1971; flowers classic form, dbl., 50 petals, 4 in., moderate fragrance; foliage small, glossy; [Arthur Bell X Peer Gynt]; McGredy, Sam IV; McGredy, 1972

Yellow Parade® *see* 'POULWEE'

Yellow Pastel® *see* 'WILYELP'

Yellow Patio, Min, my

'YELLOW PERFECTION', HT, ly, 1952; bud long, pointed, lemon-yellow; flowers canary-yellow, medium, dbl., 25–30

petals, exhibition form, slight fragrance; foliage leathery, dark; vigorous, upright growth; [Pearl Harbor X Golden Rapture]; Hill, Joseph H., Co.

'YELLOW PETALS', HT, my, 1971; flowers, dbl., 30 petals, 5.5 in., exhibition form, moderate fragrance; foliage large, light; [King's Ransom X Dorothy Peach]; Robinson, H.

'YELLOW PIGMY', F, my, 1964; flowers small, dbl., moderate fragrance; foliage small, glossy; will bloom in pots; dwarf, spreading growth; [Unnamed seedling X (Eblouissant X Goldilocks)]; Moore, Ralph S.; Sequoia Nursery

'YELLOW PINOCCHIO', F, my, 1949; bud ovoid; flowers apricot-yellow, fading to cream, blooms in clusters, dbl., 45 petals, 3 in., cupped, moderate fragrance; foliage dark; vigorous, bushy growth; (28); [Goldilocks X Marionette]; Boerner; J&P

'YELLOW PINOCCHIO, CLIMBING', Cl F, my, 1959; Palmer

'YELLOW QUEEN ELIZABETH', Gr, my, 1964; flowers orange-yellow; [Queen Elizabeth sport]; Vlaeminck; Fryer's Nursery, Ltd.

Yellow Quill, MinFl, lp; Williams, J. Benjamin, 1999

Yellow Rambler *see* **'AGLAIA'**

Yellow Rambler *see* **'BROWNELL YELLOW RAMBLER'**

Yellow Ribbon *see* 'DICALOW'

'YELLOW RIVER', HT, my, 1972; flowers pointed, dbl., 30 petals, 3.5 in., slight fragrance; foliage glossy; vigorous growth; [Bettina sport]; Timmerman's Roses

Yellow River, F, lp

Yellow Robusta, HRg, 1992; Maarse, J.D.; (Cavriglia)

'YELLOW RUFFLES', HT, my, 1953; flowers, dbl., 35–50 wavy petals, 4–5 in., moderate fragrance; foliage dark, glossy; compact growth; [Orange Ruffels sport]; Brownell, H.C.

'YELLOW SÁSTAGO', HT, my, 1939; (Yellow Condesa de Sástago); [Condesa de Sástago sport]; Howard Rose Co.

'YELLOW SEA', LCl, my, 1977; bud long, pointed; flowers lemon-yellow, semi-dbl., 18–20 petals, 6 in., exhibition form, intense, honey spice fragrance; [Unnamed seedling X Ivory Fashion]; Fong; United Rose Growers

'YELLOW SHOT SILK', HT, my, 1944; [Shot Silk sport]; Beckwith

Yellow Simplicity *see* 'JACYELSH'

Yellow Splendor, F, lp; Select, 1998

'YELLOW STAR', F, my, 1965; flowers well shaped, medium; foliage glossy; moderate growth; deRuiter

Yellow Sunbeam, Min, my

Yellow Sunblaze *see* 'MEITRISICAL'

Yellow Sunsation *see* KORtisching

Yellow Sunset, HT, lp; Dawson, George, 1990

Yellow Sunset, HT, lp

Yellow Sweetheart *see* **'ROSINA'**

'YELLOW SWEETHEART', F, my, 1952; bud pointed; flowers lemon-yellow, reverse sulfur-yellow, becoming, dbl., 35 petals, 3.5–4 in., globular; foliage leathery; vigorous, branching growth; [Pinocchio X Seedling]; Boerner; J&P

'YELLOW SWEETHEART, CLIMBING', Cl F, ly, 1952; flowers apricot-yellow to creamy-yellow, small, dbl., 26 petals, exhibition form, moderate fragrance; foliage leathery, glossy; vigorous, climbing (12 ft) growth; [Étoile Luisante seedling X Goldilocks]; Moore, Ralph S.; Marsh's Nursery

'YELLOW TALISMAN', HT, ly, 1929; flowers yellow; [Talisman sport]; Amling Co.

'YELLOW TALISMAN', HT, ly, 1935; flowers pale sulfur-yellow; [Talisman sport]; Eddie

'YELLOW TALISMAN, CLIMBING', Cl HT, ly; Stocking

Yellow Tausendschön *see* **'MADELEINE SELTZER'**

'YELLOW TIFFANY', HT, yb, 1976; flowers yellow tinted pink; [Tiffany sport]; Floradale Nurs.

Yellow Triumph, HT; Herholdt; (Cavriglia)

Yellow Unique, F, lp

'YELLOW WINGS', S, my, 1986; flowers light-medium yellow, fading to cream, moderately large, born, dbl., 16 petals, moderate fragrance; straight, brown prickles; foliage medium, light green, glossy, disease resistant; bushy growth; [[(Will Scarlet X Sutter's Gold) X Paula] X Golden Wings]; James, John

Yellow Wonder *see* 'MACBERN'

'YELLOWCREST', HT, ly, 1935; bud small, pointed; flowers clear canary-yellow, dbl.; foliage light, glossy, small; vigorous growth; LeGrice

'YELLOWHAMMER', F, my, 1956; (Yellow Dazzler, Yellow Hammer); flowers golden yellow to light yellow, blooms in clusters, dbl., 48 petals, 3 in., moderate fragrance; foliage dark, glossy; vigorous, bushy growth; GM, NRS, 1954; [Poulsen's Yellow X Unnamed seedling]; McGredy, Sam IV

'YELLOWSTONE', HT, my, 1969; bud pointed; flowers soft tawny yellow to lighter yellow, full, dbl., 38–44 petals, 3.5–4 in., moderate, tea fragrance; foliage glossy, dark, leathery; tall, upright growth; [Unnamed seedling X Unnamed seedling]; Weeks

'YESTERDAY'®, Pol, mp, 1974; (Topis d' Orient); flowers lilac pink, golden stamens, blooms in trusses, semi-dbl., 13 petals, 1.5 in., slight fragrance; foliage small, glossy; bushy growth with polyantha characteristics.; ADR, 1978 GM, Baden-Baden, 1975 GM, Monza, 1974; [(Phyllis Bide X Shepherd's Delight) X Ballerina]; Harkness

'YESTERDAY REEF', HT, w, 1984; flowers white, faintly tinted pink and yellow, large, dbl., 35 petals, exhibition form, slight fragrance; foliage large, medium green, matt; narrowly upright growth; [(Tiffany X Sunblest) X Irish Gold]; Stoddard, Louis

Yesterday's Garden *see* 'MORSHAKRUG'

Yesteryear *see* HARwoey

Yeswant, HT, rb; Patil, B.K., 1985

Yeti, lp; Dawson, George

Yi-Hong Yuan, HT, rb, 1995; flowers yellow, edged salmon, reverse lighter, very dbl., 26–40 petals, 4.5 in., borne mostly singly, slight fragrance; some prickles; foliage medium, dark green, semi-glossy; upright, medium growth; [Osiria X New Bal]; Hong-Quan, Li

YKH 501, HT, w, 1997; (**Kyoto 1200**); flowers very full, large, very dbl., 41 petals, moderate fragrance; some prickles; foliage medium, medium green, glossy; medium (1-2m) growth; [Garden Party X Pristine]; Kameyama, Yasushi

'YOIMATSURI', F, ob, 1989; bud ovoid; flowers orange and yellow tinged, medium, borne in sprays of many, single, 5 petals, cupped, moderate fragrance; prickles ordinary; foliage small, light green, glossy; spreading, medium growth; [Masquerade X Subaru]; Kikuchi, Rikichi, 1990

'YOLANDE D'ARAGON', P, m, 1843; flowers bright purple-pink, large, very dbl.; Vibert

'YORIKO', Pol, lp, 1997; flowers single, small, light pink, reverse white, single, 4–7 petals, 1.5 in., slight fragrance; oval, 6 - 7 mm, deep orange fruit; few prickles; foliage small, light green, semi-glossy; spreading (45cms) growth; [Seedling X Seedling]; Ohmori, Yoriko

'YORK AND LANCASTER', D, pb; (R. damascena variegata, R. damascena versicolor, York et Lancastre); flowers petals blush white and light pink, sometimes all one, dbl.; mixed prickles; foliage downy light gray-green; tall growth; (28)

York et Lancastre *see* **'YORK AND LANCASTER'**

Yorkshire *see* KORbarkeit

Yorkshire Bank *see* 'RUTRULO'

'YORKSHIRE LADY', HT, ly, 1992; dbl., 26–40 petals, 1.5–2.75 in., exhibition form, borne mostly singly, moderate fragrance; few prickles; foliage medium, medium green, glossy; upright (95-100 cms) growth; [((Red Lion X Royal Highness) X Yellow Petals) X Piccadilly]; Thompson, Robert; Battersby Roses, 1992

Yorkshire Sunblaze *see* 'MEIBLAM'

'YOSEMITE', HT, ob, 1934; bud long, pointed; flowers scarlet-orange, reverse suffused carmine, dbl., cupped, moderate fragrance; foliage leathery; vigorous growth; [Charles P. Kilham X Mrs Pierre S. duPont]; Nicolas; J&P

'YOSHINO', HT, w, 1997; flowers double, large, dbl., 15–25 petals, 5 in., intense fragrance; some prickles; foliage medium, medium green, semi-glossy; tall (200 cms) growth; [Confidence X Garden Party]; Ohtsuki, Hironaka

Yosooi, Min, w

You 'n' Me *see* 'SEAYOU'

Youki San® *see* 'MEIDONA'

'YOUNG AMERICA', HT, dr, 1902; flowers medium, dbl., moderate fragrance; Hill, E.G., Co.; (Sangerhausen)

Young at Heart, HT, ab

Young Cale *see* 'MACINCA'

'YOUNG FRANCE', HT, lp, 1945; (Jeune France); bud long, pointed; flowers apple-blossom-pink, base flushed orange, tipped silvery pink, exhibition form, moderate fragrance; foliage soft, dark; vigorous, upright growth; [Joanna Hill X Mme Joseph Perraud]; Meilland, F.; C-P

Young Hearts *see* 'RENHEART'

'YOUNG LOVE', Min, ob, 1980; bud medium-long, pointed; flowers orange, yellow-center, borne 1-3 per cluster, semi-dbl., 18 petals, intense fragrance; tiny, curved prickles; foliage tiny, dark; compact, bushy growth; [Unnamed seedling X Unnamed seedling]; Lyon

Young Love, Min, lp; Williams, J. Benjamin, 1995

Young Malmaison, lp; Walsh, R, ca. 1994

Young Mistress *see* KORviga

Young 'n' Innocent *see* 'RENYOUNG'

Young Quinn® *see* 'MACBERN'

Young Venturer *see* 'MATTSUN'

'YOUPI', Min, my, 1966; bud ovoid; flowers lemon-chrome, semi-dbl., 12–19 petals, 1 in., cupped, moderate, fruity fragrance; foliage light green; vigorous, compact growth; [R. wichurana X Pour Toi]; Lens

'YOUR SMILE', HT, lp, 1999; dbl., 35 petals, 6 in., exhibition form, slight fragrance; foliage medium green; 3 ft growth; GM, Japan Rose Concours, 1995; [Miss Harp X Golden Sun]; Ohata, Hatsuo; Komaba Rose Garden, 1997

'YOURS TRULY', HT, mp, 1945; bud long, pointed; flowers rose-pink, base golden yellow, open, dbl., 48 petals, 4.5 in., intense, spicy fragrance; foliage leathery; very vigorous, upright growth; [Unnamed seedling X Texas Centennial]; Morris; Germain's

Youth *see* **'MOLODOST'**

'YOUTH', HT, w, 1935; bud ovoid; flowers creamy, suffused pink, very large, dbl., 60 petals, exhibition form, moderate fragrance; foliage dark; long stems; [Souv. de Claudius Pernet X My Maryland]; Cook, J.W.

Youth of the World, HT, mr; flowers large, dbl., intense fragrance; (Sangerhausen)

'YU-AI', HT, rb, 1970; bud ovoid; flowers bright red, base yellow, large, dbl., cupped, slight fragrance; foliage dark, leathery; very vigorous, bushy growth; [Sarabande seedling X Peace seedling]; Suzuki, Seizo; Keisei Rose Nursery

Yugiri, HT, pb; Suzuki, Seizo, 1987

'YUHLA', S, mr, 1927; flowers crimson, semi-dbl., non-recurrent; very hardy.; [Wild rose from Lake Oakwood (SD) X Général Jacqueminot]; Hansen, N.E.

'YUKI-MATSURI', HT, w, 1989; bud ovoid; flowers urn-shaped, large, borne usually singly, dbl., 42 petals, exhibition form, moderate fragrance; large, light, light orange fruit; prickles sparse; foliage medium, dark green, glossy; medium growth; [(Dolce Vita X Royal Highness) X Nobility]; Yokota, Kiyoshi

Yukon *see* **'SUN VALLEY'**

'YULETIDE', HT, mr, 1956; bud ovoid; flowers currant-red, blooms, dbl., 43 petals, 3.5–4.5 in., exhibition form, slight fragrance; foliage leathery, semi-glossy; vigorous, upright, bushy growth; [Hill Crest X Silver Kenny]; Hill, Joseph H., Co.

Yume, HT, ab

'YUMEOTOME', Min, lp, 1989; bud ovoid; flowers pink changing to white, dbl., 38 petals, slight fragrance; prickles lower part hollow, about 15 on a stem; foliage small, dark green; bushy growth; [Miyagino sport]; Tokumatsu, Kazuhisa

'YUSAI', HT, yb, 1989; bud ovoid; flowers bright yellow with reddish fringe, reverse changing to red w, dbl., 30 petals, exhibition form; foliage medium, dark green with bronze tinge, slightly dentate; bushy, spreading growth; [(Unnamed American Heritage seedling X Christian Dior) X Miss Ireland]; Ota, Kaichiro

'YUZEN', Gr, pb, 1982; flowers pink, blending deeper at petal edges, urn-shaped, large, bor, dbl., 50 petals, exhibition form, moderate fragrance; prickles slanted downward; foliage dark, glossy; upright growth; [Unnamed seedling X Confidence]; Suzuki, Seizo; Keisei Rose Nursery

Yvan Meneve, HT, lp; RVS, 1998

'YVAN MISSION', Pol, mp, 1922; flowers peach-blossom-pink, borne in large clusters; [Jeanny Soupert X Katharina Zeimet]; Soupert & Notting

'YVES DRUHEN', HT, dr, 1920; flowers dark velvety red, dbl., intense fragrance; [Gen. MacArthur X Château de Clos Vougeot]; Buatois

Yves Piaget® *see* 'MEIVILDO'

'YVETTE', HT, op, 1921; flowers salmon-yellow; [Mrs T. Hillas X Mme Edouard Herriot]; Buatois

Yvette, Pol, lp; Scarman, 1996

'YVONNE', HWich, pb, 1921; flowers blush-pink, base deep pink shaded yellow, dbl., exhibition form, moderate fragrance; foliage dark, glossy; very vigorous growth; GM, NRS, 1920; Cant, F.

'YVONNE CARRET', HT, ob, 1961; flowers orange-red, reverse golden yellow, dbl., 25 petals, 4.5 in.; foliage bright green; Gaujard

'YVONNE D'HUART', HT, rb, 1932; bud long, pointed; flowers coral-red to lincoln red, base chrome yellow, large, dbl., slight fragrance; very vigorous growth; Ketten Bros.

'YVONNE MILLOT', HT, ab, 1936; bud pointed; flowers apricot-yellow, large, dbl., exhibition form, slight fragrance; foliage glossy; vigorous growth; [(Pharisaer X Constance) X Feu Joseph Looymans]; Mallerin, C.; C-P

'YVONNE PLASSAT', HT, or, 1941; flowers coppery orange to glowing red, dbl.; Moulin-Epinay; Nonin

'YVONNE PRINTEMPS', HT, or, 1939; bud long, almost red; flowers orange, veined copper, reverse bright yellow; Gaujard

'YVONNE RABIER', Pol, w, 1910; flowers pure white, center slightly tinted sulfur, blooms in clusters, dbl., recurrent bloom; foliage rich green, glossy; vigorous growth; (14); [R. wichurana X Polyantha]; Turbat

'YVONNE VACHEROT', HT, w, 1905; bud long, pointed; flowers porcelain-white, strongly flushed rose-pink, imbricated, large, dbl., exhibition form; vigorous growth; [Antoine Rivoire X Souv. d'un Ami]; Soupert & Notting

'YVONNE VIRLET', HT, my, 1936; flowers golden yellow, large, dbl.; Walter, L.

Z

'Zacharly Rose', HT, pb, 1998; flowers pink, shaded peach at base, very dbl., 26–40 petals, 4.5 in., exhibition form, borne mostly singly, slight fragrance; few prickles; foliage medium, medium green, semi-glossy; long stems; very floriferous, upright, tall (4 1/2 ft) growth; [Hazel Rose X Loving Memory]; Poole, Lionel; David Lister, Ltd, 1998

Zagreb *see* 'ROGzag'

Zahov Hatabor, F, yb; Fischel, 1979

'Zaïd', S, rb, 1937; bud bright yellow and copper-red; flowers coppery flame-red, semi-dbl., non-recurrent; foliage glossy, bright green; vigorous growth; [R. foetida bicolor X R. wichurana seedling]; Chambard, C.

'Zaida', HT, op, 1922; flowers pale coral-pink, large, dbl.; Lippiatt

Zaire, M, dr; Vibert, 1849

Zambra® *see* 'MEIalfi'

Zambra® *see* MEIcurbos

Zambra '80 *see* 'MEIrilocra'

Zambra 92 *see* MEIcurbos

Zambra 93, F, 1993; Meilland, Alain A. (Cavriglia)

Zambra, Climbing® *see* 'MEIalfisar'

Zandrina, HT, 1977; Zandri, R.; (Cavriglia)

'Zani', S, mr, 1927; flowers dark crimson, with a white streak through center petals, semi-dbl., non-recurrent; height 6-8 ft; very hardy.; [(R. rugosa X Anna de Diesbach) X Tetonkaha]; Hansen, N.E.

'Zansho', HT, rb, 1975; bud ovoid; flowers scarlet and yellow, dbl., 45 petals, 6 in., exhibition form, slight fragrance; foliage dark, leathery; upright, bushy growth; [Garden Party X Christian Dior]; Takahashi, Takeshi

'Zanzibar', F, mr, 1964; flowers bright red; [Paradis X Circus]; Leenders, J.

'Zara Hore-Ruthven', HT, mp, 1932; flowers rich pink, large, dbl., cupped, slight fragrance; [Mme Abel Chatenay X Scorcher]; Clark, A.; NRS South Australia

Zärtlichkeit, S, lp, 1992; flowers small, single; Wänninger, Franz; (Sangerhausen)

'Zauberlehrling', F, op, 1963; flowers light salmon, well formed, large, dbl.; strong stems; bushy, low growth; Kordes, R.

'Zborov', Pol, dr, 1935; flowers blood-red, borne in large clusters, very dbl.; foliage glossy, dark; vigorous growth; [Corrie Koster sport]; Böhm, J.

Zburlici, F, lp; rosette, 15 petals, slight; medium large, light green, glossy; [Bonica '82 X La Sevillana]; St. Wagner, 1994; Res. Stn. f. Fruit Growing, Cluj, 1995, Romania

Zebra *see* BENraar

'Zee', Cl Min, mp, 1940; flowers pink, very small, recurrent bloom; nearly thornless; foliage very small; height 30 in; [Carolyn Dean X Tom Thumb]; Moore, Ralph S.

Zeiber House, HT, dr

Zelda Lloyd *see* 'TINzel'

'Zélia Pradel', N, w, 1860; (Estelle Pradel); flowers pure white, large, dbl.; Pradel

'Zena', HT, dr, 1986; flowers dark crimson, large blooms borne singly or several together, dbl., 25 petals, slight fragrance; foliage medium, medium green, semi-glossy; upright growth; [Unnamed seedling X New Style]; Plumpton, E.

'Zenaitta', Pol, rb, 1991; bud ovoid; flowers bright medium red with small white area at petal base, small, dbl., cupped, no fragrance; foliage small, medium green, glossy; bushy, medium growth; [Unnamed seedling X Unnamed seedling]; Jerabek, Paul E., 1985

Zenith *see* 'DELzen'

Zenith *see* **'Uetersen'**®

'Zenobia', M, mp, 1892; bud well mossed; flowers satiny pink, large, dbl., moderate fragrance; vigorous growth; Paul, W.

Zepheline *see* KORkalba

'Zéphirine Drouhin', B, mp, 1868; bud long, pointed; flowers cerise pink, base white, large, dbl., 25–30 petals, 3 in., borne mostly singly, recurrent bloom, moderate, damask fragrance; no prickles; foliage soft, light; vigorous, semi-climbing growth; Bizot

'Zephyr', F, yb, 1961; flowers yellow tipped pink, borne in clusters, dbl., 64 petals, 2.5–3 in.; foliage glossy, light; very vigorous growth; [Goldilocks X Seedling]; Verschuren; Blaby Rose Gardens

'Zeus', LCl, my, 1959; bud long, pointed; flowers medium, dbl., 25 petals, exhibition form, slight fragrance; foliage leathery, dark; vigorous (15-20 ft) growth; [Doubloons X Unnamed seedling]; Kern Rose Nursery

'Zigeunerbaron', F, dr, 1954; flowers large, dbl.; Baron, W.; (Sangerhausen)

Zigeunerblut, B, dr; Geschwind, R., 1890

Zigeunerknabe *see* **'Gipsy Boy'**

Ziggy Stardust *see* 'RENzig'

Zigri, LCl, m

Zig-Zag *see* 'PIXzig'

'Zika', S, lp, 1927; flowers shell-pink, semi-dbl., non-recurrent bloom, moderate fragrance; very hardy.; [(R. rugosa X Anna de Diesbach) X Tetonkaha]; Hansen, N.E.

'Zingaro', F, mr, 1964; flowers bright crimson, borne in clusters, semi-dbl., 12 petals, 2 in.; foliage dark; compact, low growth; [Masquerade X Independence seedling]; Sanday, John

'Zinger', Min, mr, 1978; bud long, pointed; flowers small, semi-dbl., 11 petals, 1.5 in., flat, moderate fragrance; very vigorous, spreading growth; AOE, 1979; [Zorina X Magic Carrousel]; Schwartz, Ernest W.; Nor'East Min. Roses

'ZIPalm', Min, ab, 1991; (**Almond Glory**); flowers deep apricot, dbl., 26–40 petals, 1.5 in., exhibition form, borne in small clusters, moderate fragrance; some prickles; foliage small, dark green, glossy; medium (30 cms), compact growth; [Rise 'n' Shine X High Spirits]; Zipper, Herbert; Magic Moment Miniature Roses, 1992

'ZIPamb', Min, ab, 1991; (**Amber Ribbon**); bud pointed; flowers deep apricot to deep yellow at base, reverse deep apricot to deep yellow, dbl., 22 petals, 1.75–2 in., exhibition form, slight fragrance; foliage medium, dark green, semi-glossy; upright, tall growth; [Rise 'n' Shine X Pot 'o Gold]; Zipper, Herbert; Magic Moment Miniature Roses, 1992

'ZIPBAN', Min, yb, 1987; (**Banana Split**); flowers creamy yellow blushed with pink, medium, dbl., 15–25 petals, no fragrance; foliage medium, medium green, semi-glossy; upright, bushy growth; [Little Darling X Poker Chip]; Zipper, Herbert; Magic Moment Miniature Roses

'ZIPBELL', F, or, 1988; (**Bellissima**); flowers orange-red, darker along edges, medium, dbl., exhibition form, borne singly, no fragrance; foliage medium, medium green, semi-glossy; upright growth; [Pink Petticoat X Lady Rose]; Zipper, Herbert; Magic Moment Miniature Roses, 1989

'ZIPBERRY', MinFl, rb, 1988; (**Raspberry Ice**); flowers white brushed with red, deeper red at petal edge, medium, dbl., 40 petals, no fragrance; foliage medium, dark green, semi-glossy; miniflora; upright growth; [High Spirits X Charmglo]; Zipper, Herbert; Magic Moment Miniature Roses, 1989

'ZIPBOY', Min, mp, 1988; (**Tomboy**); flowers small, single, 5 petals, no fragrance; foliage small, dark green, semi-glossy; upright, spreading growth; [Maytime X Poker Chip]; Zipper, Herbert; Magic Moment Miniature Roses, 1989

'ZIPBUB', Min, yb, 1986; (**Bubbles**); flowers medium yellow, shaded pink and coral, dbl., 35 petals, borne singly and in small clusters, moderate fragrance; foliage medium, medium green, semi-glossy; upright, spreading growth; [(Little Darling X (Roundabout X Redgold)) X (Maytime X Poker Chip)]; Zipper, H.; Magic Moment Miniature Roses

'ZIPCAN', Min, pb, 1987; (**Candice**); flowers medium, dbl., slight fragrance; foliage medium, medium green, semi-glossy; upright growth; [Libby X Queen Elizabeth]; Zipper, Herbert; Magic Moment Miniature Roses

'ZIPCLOUD', Min, w, 1989; (**Summer Cloud**); bud ovoid; flowers white with hint of pink, urn-shaped, large, borne singly and, dbl., 65 petals, exhibition form, slight fragrance; globular, medium, brown-orange fruit; prickles straight, small, few, light brown; foliage medium, medium green, semi-glossy; bushy, medium growth; [Roundabout X Erfurt]; Zipper, Herbert; Magic Moment Miniature Roses, 1990

'ZIPCOOL', Min, w, 1987; (**Cool Wave**); flowers white with hints of yellow at petal base, medium, dbl., 15–25 petals, exhibition form, no fragrance; foliage medium, medium green, matt; [Poker Chip X Pink Parfait]; Zipper, Herbert; Magic Moment Miniature Roses

'ZIPCORK', F, rb, 1991; (**Corky**); flowers white with red edges, white reverse, blooms borne in small clusters, dbl., 26–40 petals, 3–3.5 in., slight fragrance; few prickles; foliage large, dark green, semi-glossy; medium (100 cms), bushy growth; [Tamango X Seedling]; Zipper, Herbert; Magic Moment Miniature Roses, 1992

'ZIPCUSS', F, rb, 1989; (**Percussion**); bud ovoid; flowers dark red with touch of yellow at base, reverse white at base, dbl., 25 petals, exhibition form, moderate, sweet fragrance; prickles very small, tan; foliage medium, medium green, semi-glossy; mini-flora; upright, tall growth; [Sheri Anne X Deep Purple]; Zipper, Herbert; Magic Moment Miniature Roses, 1990

'ZIPDEW', Min, rb, 1991; (**Dew Drop**); flowers red with white eye, blooms borne mostly singly, single, 5 petals, 1.5 in., slight fragrance; few prickles; foliage small, dark green, semi-glossy, very disease-resista; low (30 cms), compact growth; [Sheri Anne X Priscilla Burton]; Zipper, Herbert; Magic Moment Miniature Roses, 1992

'ZIPEASY', Min, rb, 1992; (**Easy**); flowers red and white, blooms borne in small clusters, very dbl., 1.5 in., no fragrance; few prickles; foliage small, dark green, glossy; medium (35 cms), bushy growth; [Pink Petticoat X Banana Split]; Zipper, Herbert; Magic Moment Miniature Roses

'ZIPELA', Min, m, 1991; (**Elaina**); dbl., 26–40 petals, 1.5 in., exhibition form, borne in small clusters, moderate fragrance; few prickles; foliage small, dark green, semi-glossy; medium (35 cms), bushy growth; [Blue Nile X Big John]; Zipper, Herbert; Magic Moment Miniature Roses, 1992

'ZIPEND', Min, pb, 1991; (**Endora**); flowers cream, edged deep pink, blooms borne mostly singly, dbl., 26–40 petals, 1.5–2.75 in., exhibition form, slight fragrance; few prickles; foliage small, medium green, semi-glossy; medium (35 cms), upright growth; [Pristine X High Spirits]; Zipper, Herbert; Magic Moment Miniature Roses, 1992

'ZIPEST', MinFl, w, 1997; (**Esther Jasik**); flowers very full, large, very dbl., 41 petals, borne mostly singly, slight fragrance; few prickles; foliage medium, dark green, dull; upright, tall (3ft.) growth; [Shocking Blue X Pink Petticoat]; Zipper, Herbert

'ZIPFAN', S, lp, 1999; (**Summer Fantasy**); flowers light pink, reverse dark pink, dbl., 17–25 petals, 2–2.5 in., borne in small clusters, no fragrance; few prickles; foliage small, dark green, semi-glossy, resistant to mildew; bushy (3-4 ft) growth; [Maytime X Charmglo]; Zipper, Herbert; Island Roses, 1999

'ZIPGOLD', Min, dy, 1991; (**Golden Coach**); flowers small, blooms borne mostly singly, very dbl., 1.5 in., no fragrance; few prickles; foliage small, dark green, semi-glossy; medium (40 cms), upright, compact growth; [Rise 'n' Shine X Pot O'Gold]; Zipper, Herbert; Magic Moment Miniature Roses, 1992

'ZIPHIGH', Min, yb, 1991; (**Highland Lass**); flowers red and yellow with yellow reverse, blooms borne in small clusters, semi-dbl., 6–14 petals, 1.5 in., no fragrance; few prickles; foliage small, dark green, semi-glossy; low (20 cms), compact growth; [Rise 'n' Shine X High Spirits]; Zipper, Herbert; Magic Moment Miniature Roses, 1992

'ZIPHILL', Min, dp, 1989; (**Castle Hill**); bud pointed; flowers deep pink with mauve undertones, shading lighter towards base, dbl., exhibition form, no fragrance; round, small, light orange fruit; prickles curved, small, tan; foliage medium, medium green, semi-glossy, highly mildew res; bushy, low growth; [Sheri Anne X Red Devil]; Zipper, Herbert; Magic Moment Miniature Roses, 1990

'ZIPJAM', Min, op, 1986; (**Jam Session**); flowers coral pink, yellow reverse, small blooms borne singly or in, dbl., 20 petals, no fragrance; foliage small, medium green, matt; upright growth; [Poker Chip X Anytime]; Zipper, H.; Magic Moment Miniature Roses

'ZIPLIB', Min, pb, 1997; (**Liberty Miss**); flowers double, medium, dbl., 8–14 petals, borne mostly singly, moderate fragrance; few prickles; foliage medium, medium green, semi-glossy; bushy, medium (12-18in.) growth; [Seedling X Cupcake]; Zipper, Herbert

'ZIPLIME', MinFl, rb, 1991; (**Limerick**); flowers red with white reverse, blooms borne in small clusters, dbl., 15–25 petals, 1.5–2.75 in., exhibition form, slight fragrance; few prickles; foliage medium, dark green, glossy; medium (40 cms), spreading growth; [Tamango X Jennifer]; Zipper, Herbert; Magic Moment Miniature Roses

'ZIPMIA', F, mp, 1986; (**Mama Mia**); flowers mini-flora, blooms borne singly or in 3's, dbl., exhibition form, slight fragrance; foliage medium, medium green, matt; upright, bushy growth; [Sheri Anne X Sparrieshoop]; Zipper, H.; Magic Moment Miniature Roses

'ZIPMUSIC', MinFl, mp, 1988; (**Night Music**); flowers deep pink, medium, borne singly and in sprays, dbl., intense fragrance; foliage large, medium green, semi-glossy; upright growth; [Tamango X Pink Petticoat]; Zipper, Herbert; Magic Moment Miniature Roses, 1989

'ZIPNIC', Min, pb, 1991; (**Nicolina**); bud ovoid; flowers coral pink shading to white, reverse pink edge, shading to c, dbl., 25 petals, 1.25 in., exhibition form, slight fragrance; foliage small, dark green, semi-glossy; upright, bushy, tall growth; [(Dandy Lyon X Razzle Dazzle) X Pink Petticoat]; Zipper, Herbert; Magic Moment Miniature Roses, 1992

'ZIPNOTE', MinFl, pb, 1990; (**Love Note**); flowers deep pink, light red going to cream at base, reverse deep pink, dbl., 35 petals, exhibition form, slight fragrance; round, medium, orange fruit; prickles straight, small, light brown; foliage large, medium green, semi-glossy; bushy, tall growth; [Tamango X Cupcake]; Zipper, Herbert, 1986; Magic Moment Miniature Roses, 1991

'ZIPPOET', Min, rb, 1990; (**Poetry**); bud pointed; flowers red with white at base, whitereverse with some red at edges, dbl., 30 petals, exhibition form, slight fragrance; foliage small, medium green, semi-glossy; bushy, low growth; [Libby X Deep Purple]; Zipper, Herbert, 1987; Magic Moment Miniature Roses, 1991

'ZIPPINK', Min, w, 1988; (**Pink-A-ling**); flowers white edged deep pink, small, borne singly and in sprays, dbl., slight fragrance; foliage medium, dark green, glossy; upright growth; [Tamango X Avandel]; Zipper, Herbert; Magic Moment Miniature Roses, 1989

'ZIPPRO', F, pb, 1987; (**Prom Night**); flowers creamy yellow, shading to pink with deep pink flushing the edges, very dbl., slight fragrance; foliage large, medium green, matt; upright, bushy, candelabra-formed growth; [Poker Chip X Libby]; Zipper, Herbert; Magic Moment Miniature Roses

'ZIPRAM', Min, mp, 1985; (**Ramapo**); flowers small blooms borne singly, single, 5 petals, no fragrance; foliage small, medium green, matt; bushy growth; [Maytime X Libby]; Zipper, Herbert; Magic Moment Miniature Roses

'ZIPROUND', Min, mr, 1985; (**Roundabout**); flowers small blooms borne usually singly, dbl., 20 petals, no fragrance; foliage medium, medium green, matt; bushy growth; [Double Joy X Libby]; Zipper, Herbert; Magic Moment Miniature Roses

'ZIPSON', Min, mp, 1987; (Sonata, **Songfest**); flowers small, borne singly, very dbl., exhibition form, no fragrance; foliage small, medium green, matt; bushy, compact growth; [Queen Elizabeth X Baby Katie]; Zipper, Herbert; Magic Moment Miniature Roses

'ZIPTALE', Min, mp, 1988; (**Tattletale**); flowers deep pink, small, borne singly and in sprays, dbl., 33 petals, no fragrance; foliage small, dark green, semi-glossy; bushy, compact growth; [High Spirits X Charmglo]; Zipper, Herbert; Magic Moment Miniature Roses, 1989

'ZIPTEASE', Min, pb, 1988; (**Little Tease**); flowers yellow to cream at base, suffused with clear pink, darker at edges, exhibition form, no fragrance; foliage small, medium green, semi-glossy; bushy, compact growth; [High Spirits X Charmglo]; Zipper, Herbert; Magic Moment Miniature Roses, 1989

'ZIPTOK', F, lp, 1999; (**Tokay**); flowers apricot suffused with pink, reverse soft pink to apricot, dbl., 26–40 petals, 3 in., borne mostly singly, slight fragrance; prickles moderate; foliage medium, dark green, semi-glossy; upright, tall (4-5 ft) growth; [Olympic Gold X Spanish Sun]; Zipper, Herbert; Island Roses, 1999

'ZIPTOWN', Min, mp, 1986; (**Our Town**); flowers small blooms in small sprays, semi-dbl., no fragrance; foliage small, medium green, semi-glossy; spreading growth; [Maytime X Sheri Anne]; Zipper, H.; Magic Moment Miniature Roses

'ZIPVIA', Min, w, 1991; (**Bella Via**); dbl., 26–40 petals, 1.5–2.75 in., exhibition form, borne mostly singly, moderate fragrance; few prickles; foliage medium, dark green, matt; medium (40 cms), upright growth; [Rise 'n' Shine X Olympic Gold]; Zipper, Herbert; Magic Moment Miniature Roses, 1992

'ZIPVIR', F, lp, 1999; (**Irene Virag**); flowers red, white eye, reverse streaked red and white, single, 5–11 petals, 2–2.5 in., borne in small clusters, slight fragrance; prickles moderate; small, medium green, semi-glossy, new foliage mahogany red; bushy, tall (3-4 ft) growth; [seedling X Playboy]; Zipper, Herbert; Island Roses, 1999

'ZIPWEN', F, mp, 1991; (**Wendy**); dbl., 15–25 petals, 1.5–2.75 in., exhibition form, borne in small clusters, intense fragrance; few prickles; foliage medium, dark green, semi-glossy; medium (75 cms), upright growth; [Wendy Cussons X High Spirits]; Zipper, Herbert; Magic Moment Miniature Roses, 1992

'ZITA', Pol, mr, 1965; bud ovoid; flowers red shaded dark geranium, open, single; foliage dark, leathery; [Amoureuse X Seedling]; Delforge

'ZITKALA', S, mr, 1942; flowers brilliant velvety red, dbl., 25 petals, 3 in., nonrecurrent, intense fragrance; almost thornless; red wood; very hardy.; [R. blanda X Amadis]; Hansen, N.E.

'ZITRONENFALTER'®, S, my, 1956; bud ovoid; flowers golden yellow, large blooms in clusters, dbl., 20 petals, repeat bloom, moderate, fruity fragrance; globular, slightly flattened top fruit; few, curved prickles; foliage dark, leathery; upright, bushy growth; [Marchenland X Peace]; Tantau, Math.

'ZITRONENFALTER', HSpn, ly, 1940; flowers brimstone color; [(R. spinosissima altaica X Star of Persia) X Golden Ophelia]; Berger, V.; Teschendorff

Zitronenjette® *see* 'KORJONI'

'ZIZI', F, dr, 1963; flowers garnet-red, medium, borne in clusters of 3-10, semi-dbl., 15 petals, moderate fragrance; vigorous, bushy growth; [Walko X Souv. de J. Chabert]; Delbard-Chabert

Zlata, F, lp; Vecera, L., 1975; (Czech Rosa Club)

'ZLATÁ PRAHA', HT, my, 1931; ('DAS GOLDENE PRAG'); flowers golden yellow; [Kardinal Piffl sport]; Böhm, J.

'ZLATÉ JUBILEUM', HT, dy, 1938; flowers large, dbl.; Böhm, J.; (Sangerhausen)

'ZLATY DECH', HT, dy, 1936; flowers orange-yellow, large, semi-dbl., moderate fragrance; Böhm, J.; (Sangerhausen)

Zluta Zorina, HT, lp; Strnad; (Czech Rosa Club)

'ZODIAC', F, mr, 1961; flowers camellia-shape, dbl., 74 petals, 2.25–3 in., borne in clusters; foliage leathery, light green; very vigorous growth; [Red Favorite X Seedling]; Verschuren; Blaby Rose Gardens

'ZODIAC', F, pb, 1963; flowers begonia-pink, base yellow, borne in clusters of 25-30, dbl.; low, bushy growth; [Masquerade X Karl Herbst]; Kordes, R.

'ZOE', HGal, mp, 1840; flowers large, very dbl., moderate fragrance; Miellez; (Sangerhausen)

'ZOÉ', M, mp, 1861; (Moussue Partout); flowers rose-pink, globular; prickles well-mossed; Pradel

'ZOLA', S, dy, 1979; bud pointed; flowers golden yellow, semi-dbl., 18 petals, 2.5 in., intense fragrance; foliage dark; vigorous, upright growth; [Spek's Yellow X Magenta]; Sanday, John

Zola Budd *see* 'KORZOLA'

Zoltan Kodaly, HT; Mark; (Cavriglia)

Zonnekeild *see* **'SONNYCHILD'**

'ZONNEKIND', HT, my, 1941; flowers buttercup-yellow, large, dbl., moderate fragrance; [Arch. Reventos X Amalia Jung]; Leenders, M.

Zonta Rose *see* 'HARTANNA'

Zonta-Rose, F, dy, 1988; flowers medium-large, dbl.; Harkness; (Sangerhausen)

'ZORA', F, yb, 1974; flowers yellow, deep pink petal edges, aging deep pink, blooms in clusters, semi-dbl., 18 petals; foliage dark; vigorous growth; [Masquerade X Rumba]; Staikov, Prof. Dr. V.; Kalaydjiev and Chorbadjiiski

'ZORINA', F, or, 1963; (Rozorina); bud ovoid; dbl., 25 petals, 3 in., cupped, moderate fragrance; foliage glossy; vigorous, upright growth; GM, Rome, 1964; [Pinocchio seedling X Spartan]; Boerner; J&P

Zorina, HT, lp; Strnad; (Czech Rosa Club)

Zorka, F, rb, 1974; flowers carmine-red with white, ; Vecera, L.; (Sangerhausen)

'ZUBLOU', S, w, 1996; (**Louis Riel**); single, 5 petals, slight fragrance; numerous prickles; foliage medium, purplish-green, dull; bushy, tall growth; [R. rubrifolia X R. spinosissima 'Altaica']; Zubrowski, Stanley

'ZUKUNFT', F, or, 1951; flowers scarlet-red; [Lafayette sport]; Verschuren

'ZULU QUEEN', HT, dr, 1939; bud long, pointed; flowers dark maroon, large, dbl., exhibition form, moderate fragrance; foliage leathery; vigorous, bushy growth; [(Cathrine Kordes X E.G. Hill) X Fritz Hoger]; Kordes; J&P

'ZULU QUEEN, CLIMBING', Cl HT, dr, 1961; Moore

'ZULU WARRIOR', F, rb, 1980; flowers burgundy red, silver reverse, blooms borne 3-20 per cluster, semi-dbl., 9 petals, slight fragrance; large, reddish prickles; foliage dark, tough; medium growth; [Arthur Bell X Flaming Peace]; Hawken, Una

Zurella™ *see* MEIvrofix

'ZWEIBRÜCKEN', HKor, dr, 1955; bud ovoid; flowers deep crimson, large blooms in large clusters, dbl., recurrent bloom, slight fragrance; foliage dark, leathery; very vigorous, climbing growth; [R. kordesii X Independence]; Kordes

'ZWEMANIA', Gr, or, 1974; flowers light vermilion, full, dbl., 25–30 petals, exhibition form, moderate, fruity fragrance; foliage dark; very vigorous growth; PP004049; [Sweet Promise sport]; Zwemstra; Meilland

Zwergeliebe, Min, 1986; Delforge; (Cavriglia)

Zwergenfee® *see* 'KORFEE'

Zwergkönig *see* **'DWARFKING'**

Zwergkonig 78 *see* KORkonig

Zwergkönigin *see* **'QUEEN OF THE DWARFS'**

Zwergkönigin '82 *see* **'KORWERK'**